Diagnostic Electron Microscopy
A TEXT/ATLAS

Second Edition

Springer Science+Business Media, LLC

G. Richard Dickersin, M.D.

Associate Professor of Pathology
Harvard Medical School

Associate Professor of Pathology
Massachusetts Institute of Technology

Pathologist
Head of Diagnostic Electron Microscopy Unit
Massachusetts General Hospital
Boston, Massachusetts

Diagnostic Electron Microscopy

A TEXT/ATLAS

Second Edition

With 894 Illustrations

G. Richard Dickersin, M.D.
Pathologist and Head, Electron Microscopy Unit
Department of Pathology
Massachusetts General Hospital
Boston, MA 02114-2617
USA
and
Associate Professor of Pathology
Harvard Medical School
Boston, MA 02115-6092
USA
and
Associate Professor of Pathology
Massachusetts Institute of Technology
Cambridge, MA 02139
USA

Library of Congress Cataloging-in-Publication Data
Dickersin, G. Richard.
Diagnostic electron microscopy : a text/atlas / G. Richard Dickersin. — 2nd ed.
p. cm.
Includes bibliographical references and index.
ISBN 978-1-4757-7382-8 ISBN 978-0-387-21852-6 (eBook)
DOI 10.1007/978-0-387-21852-6
1. Diagnosis, Electron microscopic. 2. Tumors—Diagnosis. I. Title.
[DNLM: 1. Microscopy, Electron—methods Atlases. 2. Neoplasms—ultrastructure Atlases. 3. Cytodiagnosis—methods Atlases. 4. Kidney Glomerulus—ultrastructure Atlases. 5. Metabolic Diseases—pathology Atlases. 6. Neuromuscular Diseases—pathology Atlases. QZ 17 D549d 1999]
RB43.5.D55 1999
616.07′58—dc21
DNLM/DLC
99-32439

Printed on acid-free paper.

Production coordinated by Impressions Book and Journal Services, Inc., managed by Terry Kornak, and manufacturing supervised by Jacqui Ashri.
Typeset by Impressions Book and Journal Services, Inc., Madison, WI.

9 8 7 6 5 4 3 2 1

ISBN 978-1-4757-7382-8 SPIN 10656887

To my wife, Barbara;
my daughters,
Kay, Gail, Leslie, and Amy;
and my son, Ged.

Preface

Ten years have elapsed since the appearance of the first edition of this book, and since then a number of advances have been made in the field of diagnostic pathology. New and improved techniques have contributed to our diagnostic armamentarium and to our general understanding of various disease processes. In the field of neoplasia, immunohistochemistry has become a routine procedure in most departments of pathology. Flow cytometry has become an efficient technique for measuring ploidy, and molecular biological methods such as DNA and RNA hybridization and polymerase chain reaction (PCR) have made it possible to identify specific genetic markers for various hereditary, neoplastic, and infectious diseases. The importance of these innovations in pathology is duly recognized, but, at the same time, they have limitations, and traditional morphological studies still comprise the backbone of the pathologist's work. In this latter group of studies we include electron microscopy, which has continued to be used selectively in diagnostic workups of neoplastic, renal, neuromuscular, infectious, hereditary, and metabolic diseases. In our own experience, electron microscopy has been especially valuable in complementing immunohistochemistry or in superseding immunohistochemistry when the latter is equivocal or nonspecific.

Aside from the practical application to diagnostic work, electron microscopy has been a valuable tool for educating residents and staff. It reveals cells and tissues at very high magnification, making cell surfaces and interiors visible beyond the limits of light microscopy, a seemingly important experience in the study of normal and diseased states. The omission of this basic morphological step in the training and continuing education of pathologists would be, in our opinion, a serious deficiency.

In this second edition, we have retained the style and core components of the first edition but have updated the text and bibliography, added new topics, and replaced and supplemented photographs appropriately. The result has been a larger book and, we hope, one of broader applicability.

G. Richard Dickersin, M.D.
Boston, Massachusetts

Acknowledgments

It goes without saying that a book of this type cannot be produced by one person, at least not within a practical time frame and while simultaneously carrying out routine "service work" in a busy department of pathology. Thus, there are a number of persons who deserve special acknowledgment for their roles in making this work possible. First, Cheryl Nason and Martin Selig were at the core of the daily labor of the book. Cheryl's mastery of the computer allowed for rough drafts, countless additions and revisions, and literature searches to be accomplished in a timely manner. Her work was always outstandingly prompt and accurate, and I am most appreciative of it. Martin was a steady, consistent, right-hand man in the "rescoping," photography, and assembly of illustrations. His skills went beyond the purely technical ones, of which he is such a master, and included an academic familiarity with the particular morphology we were attempting to capture and portray. As will be noticed, Martin is listed as a contributing author of the chapter on renal glomerular disease, but he also approached a similar level of involvement on several of the other chapters. I wish to acknowledge and thank him for the high quality of his work and for his commitment to the project.

An "unsung hero" in the evolution of this book was a person whose skillful and reliable performance of routine technical chores kept our electron microscopy service viable and efficient. That person is Robert Holmes. To Rob, our emphatic expression of gratitude and appreciation.

Special appreciation also goes to the authors of Chapters 12 and 13, Shamila Mauiyyedi, Alain P. Marion, Robert B. Colvin, Umberto De Girolami, and Douglas C. Anthony. Their expertise and contributions round out this book, making it more comprehensive and authoritative than I could possibly have hoped to accomplish myself.

Finally, as for the first edition, I want to thank my wife, daughters, and son for their tolerance during weekends and vacations, when I would usually put in a few hours on "the book."

G. Richard Dickersin, M.D.
Boston, Massachusetts

Contents

Contributors

Douglas C. Anthony, M.D., Ph.D.

Chief of Neuropathology Division, Department of Pathology, Children's Hospital; Associate Professor of Pathology, Harvard Medical School, Boston, Massachusetts, USA

Robert B. Colvin, M.D.

Chief of Pathology, Department of Pathology, Massachusetts General Hospital; Benjamin Castleman Professor of Pathology, Harvard Medical School, Boston, Massachusetts, USA

Umberto De Girolami, M.D.

Chief of Neuropathology Division, Department of Pathology, Brigham & Women's Hospital; Professor of Pathology, Harvard Medical School, Boston, Massachusetts, USA

Alain P. Marion, M.D.

Department of Pathology, Hôpital Maisonneuve-Rosemont, Montreal, Quebec, Canada

Shamila Mauiyyedi, M.D.

Renal Fellow in Pathology, Department of Pathology, Massachusetts General Hospital; Clinical Fellow in Pathology, Harvard Medical School, Boston, Massachusetts, USA

Martin K. Selig, B.A.

Senior Medical Technologist, Department of Pathology, Massachusetts General Hospital, Boston, Massachusetts, USA

1

Normal Cell Ultrastructure and Function

Cells comprise various subunits, the types and arrangement of which depend on the stages and direction of differentiation. Although a morphologic and functional diversity exists among various cell types, there is also a certain consistency that all cells share (Figure 1.1).

The cell is enclosed by a membrane (plasmalemma) and has two main compartments, the nucleus and cytoplasm, which are separated by the nuclear envelope. Membranes also enclose some of the substructures of the cell, serving as mechanical barriers for metabolic units as well as having a specialized molecular composition and function of their own. The living substance composing the nucleus and cytoplasm is called the protoplasm; more specifically, the protoplasm of the nucleus is referred to as the nucleoplasm (or karyoplasm), and that of the remainder of the cell as the cytoplasm. The nucleus and cytoplasm contain substructures designated as organelles and inclusions. Organelles are living, metabolically active structures, whereas inclusions are stored or transient products of metabolism, such as lipid, glycogen, and pigment deposits. The cytoplasm also contains a cytoskeleton consisting of filaments and microtubules that contribute to the shape and contractility of the cell and to the arrangement of organelles.

Cytoplasmic Organelles

The organelles are surrounded by a selectively permeable membrane and separated by a cell sap (hyaloplasm or cytosol). Numerous chemical reactions and interchanges occur constantly at the membrane interfaces between the cell sap and the organelles. Organelles are dynamic structures, changing their size, shape, and position in the cell and, in some cases, duplicating themselves.

Ribosomes. Ribosomes are small (15–30 nm), bilobed (round or oval at low magnification) bodies that are distributed free and attached to membranes of rough endoplasmic reticulum (RER). In both locations, they occur singly (monoribosomes) and in clusters (polyribosomes or polysomes). They are composed of ribonucleic acid and have the metabolic function of synthesizing polypeptides and proteins from amino acids. The sequencing of amino acids is controlled by molecules of messenger ribonucleic acid (mRNA), transported from the nucleus. The proteins formed by the free ribosomes remain in the cell sap, and the polypeptides formed by the attached ribosomes penetrate into the channels of RER. Cells that secrete large amounts of protein have a high percentage of their ribosomes in the attached form, whereas cells that manufacture protein for rapid growth or metabolism, as in embryos and malignant neoplasms, have a high proportion of their ribosomes in the free form.

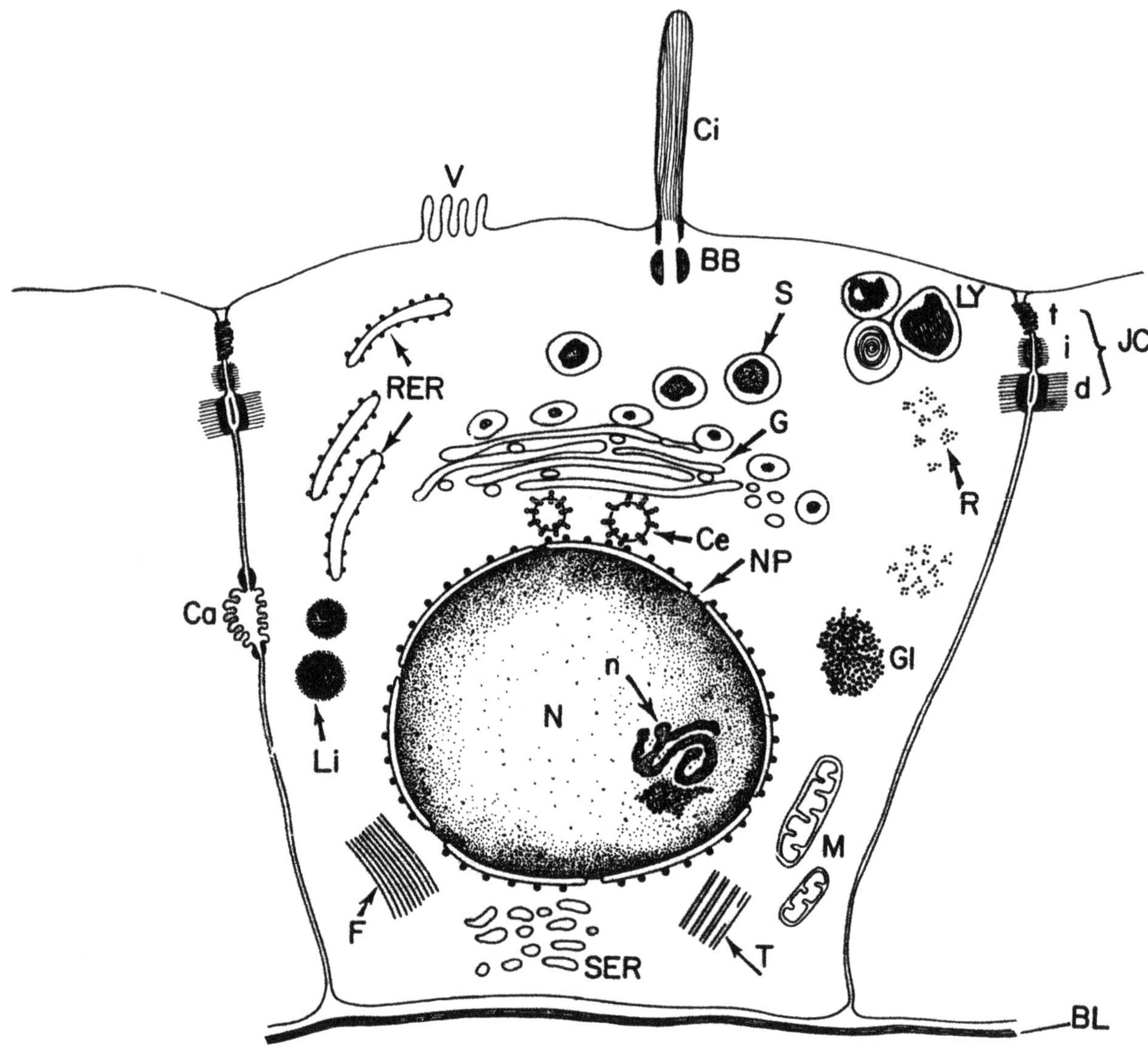

Figure 1.1. Diagram of a cell.
N = Nucleus
n = Nucleolus
NP = Nuclear pore
V = Villi
Ci = Cilium
BB = Basal body
Ca = Canaliculus
BL = Basal lamina
JC = Junctional complex
t = tight junction
i = intermediate junction
d = desmosome
Ce = Centrioles
G = Golgi apparatus
M = Mitochondria
RER = Rough endoplasmic reticulum
R = Ribosomes
SER = Smooth endoplasmic reticulum
LY = Lysosomes
S = Secretory granules
Gl = Glycogen granules
Li = Lipid vacuoles
F = Filaments
T = Microtubules

Rough endoplasmic reticulum. This organelle consists of a system of membrane-bound tubules and flattened sacs (cisternae), in which polypeptides received from the attached ribosomes are further synthesized into proteins. The proteins are then packaged into small vesicles and pass along the channels of RER to the Golgi apparatus.

Smooth (agranular) endoplasmic reticulum. This is a network of closely arranged tubules devoid of ribosomes on its limiting membranes, and it may or may not be in continuity with the RER. Its function varies from one cell type to another, but important roles include the synthesis of steroid hormones (e.g., adrenal cortex), lipid transport (intestinal epithelium), lipid and cholesterol metabolism (liver), and detoxification of lipid-soluble chemicals (liver).

Golgi apparatus. The Golgi apparatus consists of a group of side-by-side, fenestrated cisternae and many adjacent, small vesicles. The cisternae usually are curved, and the vesicles at their convex or forming face are transitional, being derived from the RER and then incorporated into the cisternae of the Golgi apparatus. The vesicles fuse into and bud off successive cisternae from the convex face to the concave or maturing face. The terminal cisternae of the maturing face are highly fenestrated and form the trans-Golgi network. While traversing the Golgi, the protein is modified (e.g., by the addition of a carbohydrate moiety) and concentrated, and vesicles are formed at the maturing face by segmentation of cisternae into secretory granules and primary lysosomes. In addition to its role in the secretory process, the Golgi apparatus is also involved in the recycling and transportation of membranes of organelles and plasmalemmas.

Mitochondria. Mitochondria usually are rod-shaped, 2–8 μm × 0.2–0.8 μm, but they can change their shape into short or long forms and from straight to curved ones. They are enclosed by a double membrane; the outer one is the usual trilaminar type, and the inner membrane has numerous infoldings called cristae. The cristae provide additional membranous surface area across which chemical transfers can take place. The morphologic and functional specialization of the inner membrane differs from that of the outer membrane. Between the cristae is matrix, which contains granules, ribosomes, strands of deoxyribonucleic acid (DNA), and enzymes. Mitochondria are mobile within the cytoplasm and may be either diffusely dispersed or concentrated in one region where their energy production is needed. They have a life cycle and self-duplicate by binary fission. The main function of mitochondria is to produce energy for the cell. They are the power source of the cell and contain the necessary enzymes for oxidative phosphorylation and cellular respiration as well as those used in the synthesis of fat and protein (for example, adenosine triphosphate and Krebs cycle enzymes).

Lysosomes. Primary lysosomes are round, oval, and irregularly shaped, membrane-bound structures, 0.25–0.5 μm in diameter, that are formed in the Golgi apparatus. They contain hydrolytic enzymes (for example, peroxidase and acid phosphatase) that are used in intra- and extracellular digestive reactions, both in pathologic and physiologic conditions. An example of lysosomal action under physiologic conditions is the engulfing and digesting of metabolic breakdown products of the cell itself, such as particles of mitochondria, endoplasmic reticulum, and other organelles. This process is known as autophagy or autophagocytosis, and it contrasts with heterophagy or heterophagocytosis, which is lysosomal digestion of solid substances taken into the cell from the extracellular environment. A lysosome engaged in autophagy or heterophagy is termed a secondary lysosome. Secondary lysosomes that contain undigested material such as certain lipid and membranes are called residual bodies. Lipofuscin (wear and tear pigment) is a common example of a substance that normally accumulates in cells in this manner. Some of these indigestible materials may be eliminated from the cell by the lysosomes transporting it to the cell surface and extruding it through the plasmalemma. The process of releasing cell products, including small vesicles and secretory granules, into the extracellular space is known as exocytosis. Pinocytosis is similar to heterophagocytosis except that the particles taken into the cells are much smaller and consist of droplets of fluid and solutes. Endocytosis encompasses both phagocytosis and pinocytosis. In both cases, the ingesting vacuoles are formed from infolding of the plasmalemma.

Centrioles. A pair (diplosome) of short, hollow rods at right angle to each other, centrioles measure about 0.3–0.5 μm long and 0.2 μm in diameter. The wall of each rod comprises nine groups of three longitudinally directed microtubules. Centrioles are usually located in the centrosome or centrosphere (cell center) region of the cytoplasm, between the nucleus and Golgi apparatus. Centrioles in this location are associated with cell division, first replicating and then migrating to opposite ends of the nucleus and becoming a part of the spindle for cell division. Centrioles also occur singly in the apical cytoplasm of some cells, where they serve as the origin or root (basal body; kinetosome; blepharoplast) of cilia. These roots are involved in the formation of the axoneme, the microtubular core of cilia, and in the metabolism of tubulins, the proteins composing the microtubules. In cells having multiple cilia, each cilium is derived from a separate centriole following proliferation of centrioles as a step in differentiation.

Peroxisomes (microbodies). These are spherical membrane-bound organelles, about 0.5 μm to 1.0 μm in diameter, which are present in various types of cells but are more numerous in metabolically active cells, such as proximal tubular epithelium of the kidney and epithelium of the liver. The matrix of peroxisomes varies from species to species, but in humans it tends to be finely granular. In certain lower animals, it may contain a paracrystalline core, or nucleoid. All the functions of peroxisomes are not known, but they contain catalases and numerous other oxidases. One function is to oxidate substrates of long-chain fatty acids, producing energy and H_2O_2. The H_2O_2 is then broken down by catalase (peroxidase).

Annulate lamellae. Annulate lamellae are parallel layers of regularly spaced cisternae with periodic, round openings with membranous diaphragms along their length. Individual cisternae with pores resemble the nuclear envelope, to which they are thought possibly to give rise. The ends of the cisternae are sometimes continuous with the cisternae of RER. The function of the annulate lamellae is not known, although it is found in germ cells and many different somatic cells, usually those that are differentiating or dividing.

Cytoskeleton (Cytoplasmic Matrix Structures)

Filaments. Most types of cells contain a framework of thin (6–7 nm), actin filaments, so-called microfilaments, as well as filaments of intermediate thickness (10 nm), such as keratin, vimentin, desmin, glial fibrillary acid protein, and neurofilaments. Those cells engaged in a contractile function (smooth and skeletal muscle) contain many more filaments, including thick (15 nm) myosin filaments. Tonofibrils are large, dense bundles of keratin filaments that occur in squamous epithelial cells. Dense bodies are sites where bundles of desmin filaments converge with actin filaments and plasmalemmas, as in smooth muscle cells. Although all classes of intermediate filaments are different biochemically, they cannot be distinguished from one another ultrastructurally.

Microtubules. Microtubules are straight structures, several micrometers long and 20–27 nm in diameter and composed of a protein called tubulin. They are present in small amounts in most types of cells and increase during mitosis and when cells undergo changes in shape. During mitosis, they form the spindle. They are also thought to serve as routes along which metabolic vesicles, organelles, and inclusions can be transported throughout the cytoplasm. Microtubules also occur in pairs or doublets, in cilia and flagella, and in triplets in centrioles and basal bodies. Microtubule formation in cells is in a dynamic state with soluble tubulin, and the number of microtubules in a cell varies with time.

Cytoplasmic Inclusions

Glycogen. Glycogen consists of irregularly shaped particles, 15–30 nm in diameter, that occur singly (beta particles) and in clusters (alpha particles). The amount of stored glycogen varies according to cell type and metabolic state. Resting skeletal muscle cells and liver parenchymal cells have a rich content of glycogen as a source of energy.

Lipid. Neutral fat occurs to some extent in most cells and is stored in various sized droplets that are not bound by a membrane. It serves as a source of energy and as a supply of carbon chain subunits in the synthesis of membranes. In cells in which lipid is synthesized, as in endocrine organs, it is present in the form of small droplets, and in cells where it is stored, as in adipocytes, it is present as a single large vacuole.

Pigment. Lipofuscin and hemosiderin are examples of pigment and are found within secondary lysosomes (phagosomes).

Nuclear Organelles

Cells cannot live without a nucleus and, therefore, will not divide, differentiate, or metabolize. The nucleus produces the RNA necessary for protein synthesis, which is required for the continuing function of the cell. The nucleus is also essential in the heredity of cells.

Chromatin and chromosomes. Chromatin is the stainable part of the nucleus and is composed of nucleic acids, especially DNA and histones. Heterochromatin is the course, clumped particles of chromatin visible during the nondividing state (interphase) of the cell. Euchromatin is finely dispersed chromatin and is more active metabolically than the heterochromatin. During mitosis, all chromatin is organized into chromosomes, which measure about 3–6 μm long and 0.5–0.8 μm in diameter.

Nucleolus. A round body, the nucleolus is usually eccentrically located in the nucleus and varies somewhat in internal structure depending on the type of cell. There may be more than one nucleolus per nucleus. Common substructures of the nucleolus include: pars amorpha (pars fibrosa and nucleolar organizer region), one or more zones of pale-staining, densely arranged 50 Å filaments; nucleolonema (pars granulosa), which surrounds the pars amorpha and consists of 15-nm granules and filaments; nucleolar-associated chromatin, a rim of chromatin immediately around and extending into the nucleolus; protein matrix; and other less consistent components, such as lipid, glycogen, and various inclusions.

The main known function of the nucleolus is to produce and process precursors of RNA. The nucleolus disperses and becomes invisible during mitosis, and it enlarges in cells that are growing or actively synthesizing protein.

Nuclear sap (karyolymph). The nuclear sap is the theoretical territory not occupied by chromatin or nucleolus; it is theoretical, because there may be euchromatin or other unknown particles, too small to be seen, that actually occupy this space.

Nuclear envelope (membrane). The nuclear envelope is the flattened sac that encloses the nucleus. It consists of inner and outer membranes, and an intervening space (perinuclear space). The two membranes meet at various foci along the circumference of the nucleus to form nuclear pores, approximately 700 Å in diameter. The pores are covered with thin, incomplete membranous diaphragms that are selectively permeable. The outer layer of the nuclear envelope has ribosomes attached to it, and the perinuclear space is continuous with the spaces of the RER. The inner aspect of the inner membrane is covered by layers of fibers that abut the peripherally located heterochromatin.

Organization of Organelles Within the Cell

The plasmalemma, or cell membrane, has three layers: outer and inner electron-dense lines, 2.5–3.0 nm thick; and a middle, lucent space, 3.5–4.0 nm thick. Similar membranes also surround organelles within the cell. The plasmalemma is composed of a bilayer of phospholipids and accompanying proteins, glycoproteins, and glycolipids. The membrane is permeable to water, gases, and small uncharged molecules but is not very permeable to charged and large uncharged molecules. The latter are transported by proteins within the membrane. Some of these proteins have a carbohydrate side chain that protrudes through the outer membrane, forming a coating or glycocalyx. The glycocalyx is especially visible on the surface of intestinal epithelial cells. It plays an important role in selective binding of external substances and other cells.

The various components of the cytoplasm are arranged in two main regions of the cell. Actin filaments form a network in the peripheral region—the ectoplasm—and intermediate filaments, microtubules, vesicles, and organelles are located in the central region—the endoplasm. In certain types of cells, the arrangement of organelles appears to correlate with function, rather than being random. An example of this relationship is in columnar epithelium lining a lumen, as in the gut, where the free luminal surface of the cells is microvillous. The microvilli increase the surface area of the cell for absorption and secretion. In the same vein, the Golgi apparatus is supranuclear in position, and secretory granules occupy the zone between the Golgi and the villous surface of the cell. Mitochondria also may be concentrated in the apical part of the cell and tend to be oriented parallel to the long axis of the cell.

Cell Attachment Sites

Zonula occludens (tight junction). These are sites of strongest attachment between cells, each junction consisting of a series of alternating points of fusion and slight separation between opposing plasmalemmas. Other than serving this purely mechanical function, it is not known if tight junctions also play a role in electrochemical communication between cells, as do some other types of junctions. The tight junction is typically, but not exclusively, found sealing the apical intercellular space between epithelial cells that line lumens.

Zonula adherens (intermediate junction). Adjacent cell membranes are about 15–20 nm apart at this junction, and there is a material of low electron density in the intercellular space between the membranes. A collection of thin actin filaments is attached to the inner surface of the junctional membrane and extends into the subjacent cytoplasm. One place where the zonula adherens is found is below and near the tight junction of epithelial cells lining lumens. Taken together, the tight junction, the intermediate junction, and the desmosome (see next paragraph) form the junctional complex (terminal bar, by light microscopy).

Macula adherens (desmosome). Maculae adherens are plaque-like, subplasmalemmal thickenings of apposing cells, with a 15–25 nm space between the plasmalemmas and a thin, dense line in the middle of the intercellular space. Extending from the thickened plasmalemma into the cytoplasm are many intermediate filaments that connect the junction with the cytoskeleton of the cell. In squamous epithelial cells, these tonofilaments course into dense bundles called tonofibrils. In addition to forming a component of the junctional complex, and to interconnecting squamous cells, desmosomes are also found between other types of epithelial cells. Hemidesmosomes may be found along the basal plasmalemma of epithelial cells that rest on a basal lamina.

Nexus (gap junction). The nexus is a plaque-like thickening of adjacent plasmalemmas, with a narrow gap of approximately 2–3 nm between them. The gap contains hexagonally packed globular subunits that contain channels that allow low resistance flow of ions and small molecules between cells. The nexus is found between epithelial type cells as well as between nerve, smooth muscle, and cardiac muscle cells.

REFERENCES

Dalton AJ, Haguenau F, eds: *Ultrastructure in Biochemical Systems.* vol 3: *The Nucleus.* Academic Press, New York, 1968.

Fawcett DW: *Bloom and Fawcett: A Textbook of Histology,* 12th ed. Chapman and Hall, New York, 1994.

Fawcett DW: *The Cell,* 2nd ed. WB Saunders, Philadelphia, 1981.

Ghadially FN: *Ultrastructural Pathology of the Cell and Matrix,* 4th ed. Butterworth-Heinemann, Boston, 1997.

Holtzman E, Novikoff AB: *Cells and Organelles.* WB Saunders, Philadelphia, 1984.

2

Selective Embryology

A knowledge of the early stages of human embryogenesis is helpful in classifying neoplasms according to cells of origin. The least differentiated tumors obviously are the most difficult to categorize, and the key to diagnosis often lies in finding cells that have sufficient cytoplasmic or surface differentiation to allow a comparison with known cytologic structures in the normal human embryo. The brief text, diagrams, and electron micrographs that follow serve to illustrate some of the basic embryonic features that are applicable in establishing the histogenesis of various neoplasms.

Embryogenesis from Fertilization Through Three Weeks

Figures 2.1A and B depict embryogenesis from fertilization to implantation. The fertilized ovum (zygote) undergoes sequential cleavages, forming a ball of 12–16 cells (morula) by the third day. Further cellular division and accumulation of extracellular fluid during the fourth through seventh days result in a blastula (blastocyst), which consists of an eccentric cavity, an outer cell mass (trophoblast), and an inner cell mass (embryoblast, Figure 2.1C). Implantation of the blastocyst into the endometrium occurs between the fifth and seventh days; by the eighth day, a bilaminar germ disc, consisting of epiblast (pre-ectoderm) and hypoblast (pre-endoderm), has formed (Figure 2.1D). During the second week of development, the amniotic cavity and yolk sac are created, and the trophoblast differentiates into two layers: cytotrophoblast and syncytiotrophoblast. By the 16th day, the primitive streak has formed in the caudal end of the embryonic disc, and cells of the epiblast in this region migrate ventrally, laterally, and cephalad to form first a loose primary mesenchyme and then a more dense third germ layer, the mesoderm (Figure 2.2A). The notochord (axial mesoderm) is formed from cephalic migration of some of the cells of the primary mesenchyme, cephalad to the primitive streak, at approximately the 18th day. By the 20th day, the mesodermal cells have aggregated into three discrete masses: paraxial, intermediate, and lateral (Figure 2.2B). The paraxial mesoderm becomes more distinct somites within a day, starting at the cranial end and progressing caudally; the intermediate mesoderm develops into nephrotomes cranially and the nephrogenic cord caudally; the lateral mesoderm divides into somatic and splanchnic layers, which become the mesothelial linings of the coelomic cavities.

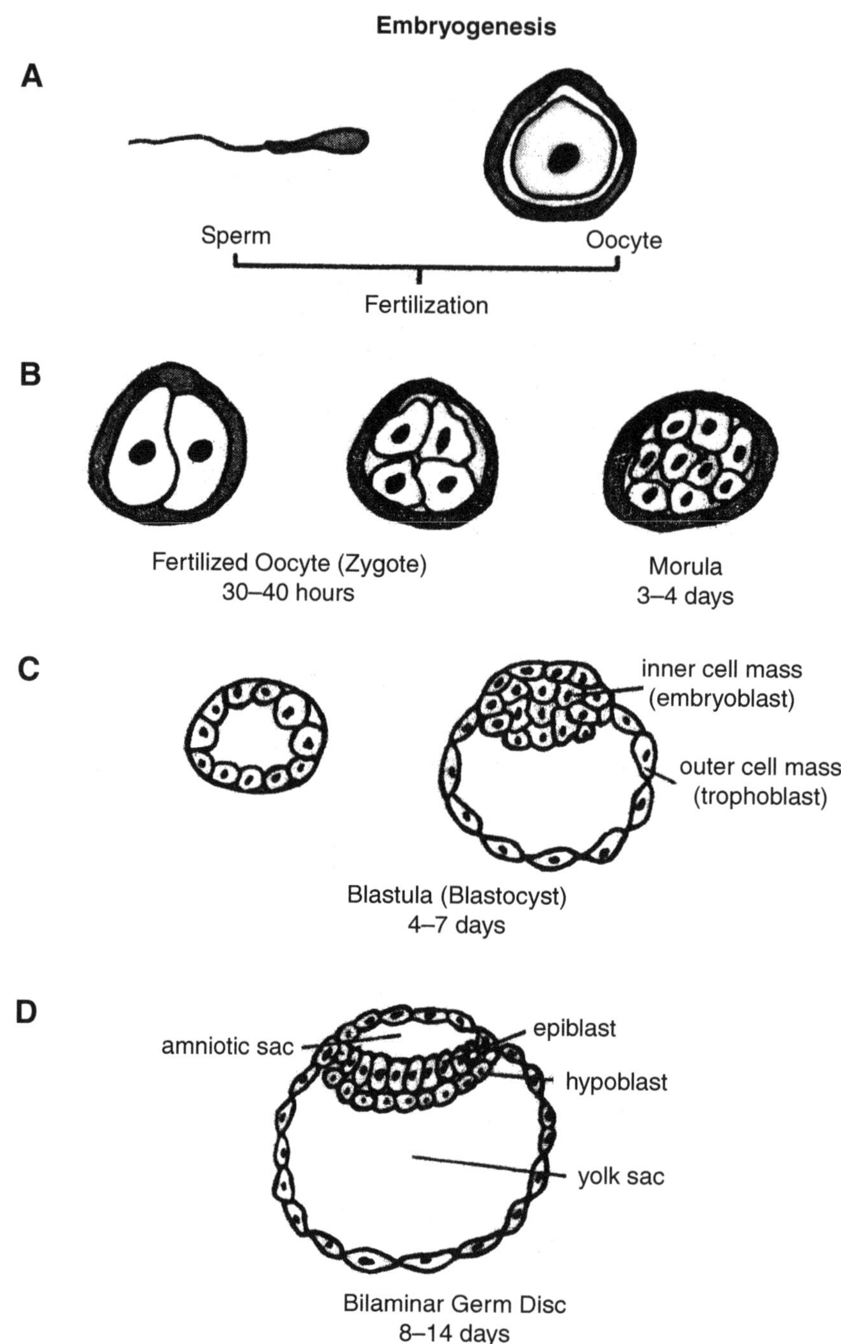

Figure 2.1. Diagram of embryogenesis, from fertilization through the 14th day.

Differentiation of the Paraxial Mesoderm

The cells of the somites have epithelial features, and in the fourth week, each somite develops a central cavity. The somatic cells continue to proliferate, become more loosely arranged, and take on a more irregular shape (secondary mesenchyme). Cells from the medial and ventral aspects of the somites migrate toward the notochord, resulting in the formation of the sclerotomes and ultimate vertebral column (Figure 2.2C). The mesenchymal cells comprising the sclerotomes have the potential to differentiate into osteoblasts, chondroblasts, and fibroblasts. The dorsal aspects of the somites become the dermatomes (the future connective tissue of the back and some of the muscles of the limbs). The remaining internal regions of the somites form the myotomes (the anlage of the muscles of the back).

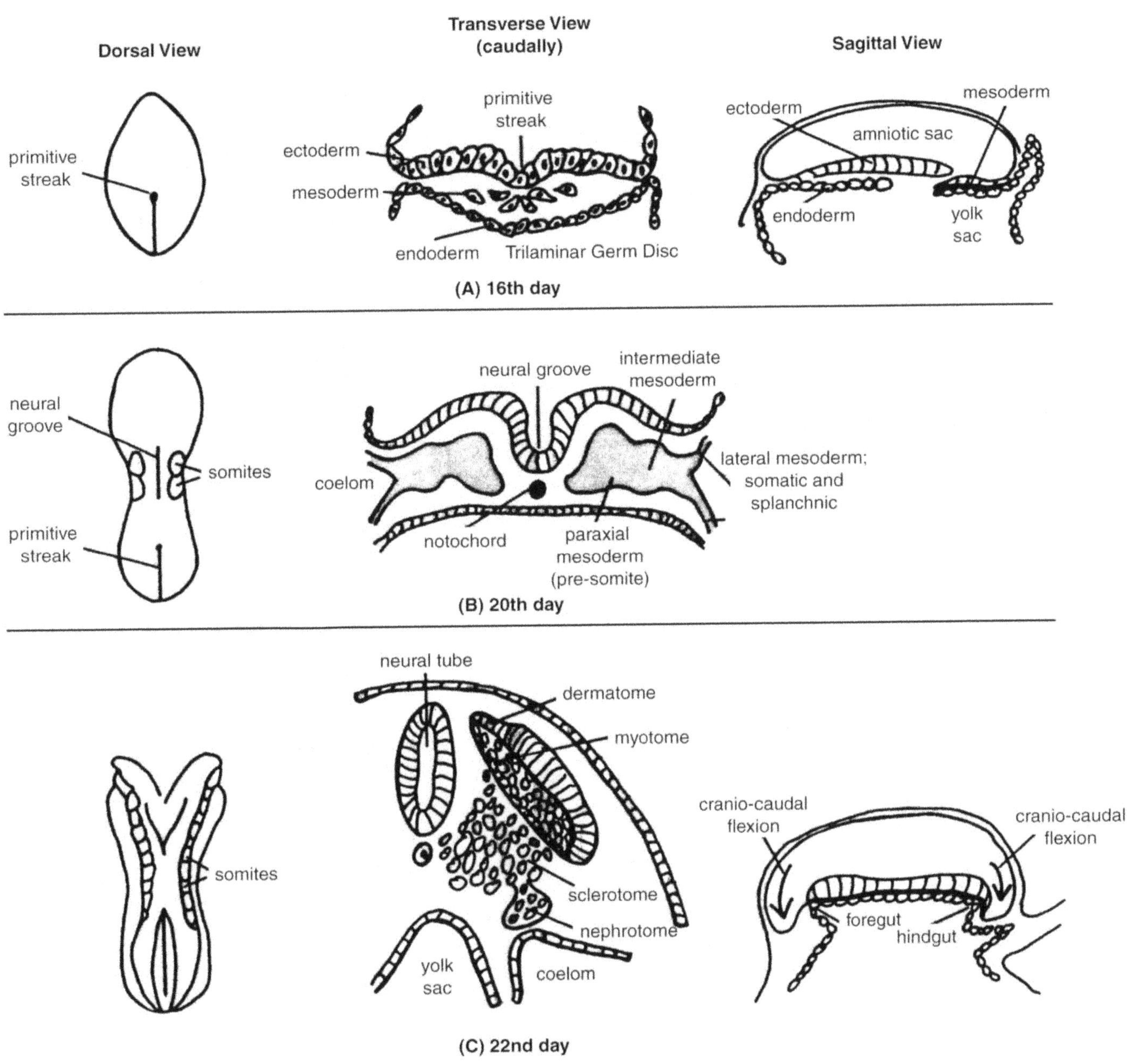

Figure 2.2. Embryogenesis during the third week of gestation. **A,** Appearance of the third germ layer, the mesoderm. **B,** Formation of discrete mesodermal masses. **C,** Migration and differentiation of cells of the mesodermal masses.

Differentiation of the Intermediate Mesoderm

The cells of the intermediate mesoderm develop into the pronephros, mesonephros, and metanephros in a cephalocaudal order and in a successive and overlapping timeframe (Figure 2.3). The pronephros is formed first, in the 7–14 somite, cranial region (occipital and cervical zones of the later fetus). This is characterized by a dorsolateral outgrowth of each intermediate cell mass (nephrotome). These buds hollow to form tubules that, on their lateral extremity, empty into the coelomic cavity, and at their medial end grow caudally and interconnect with one another, forming the pronephric

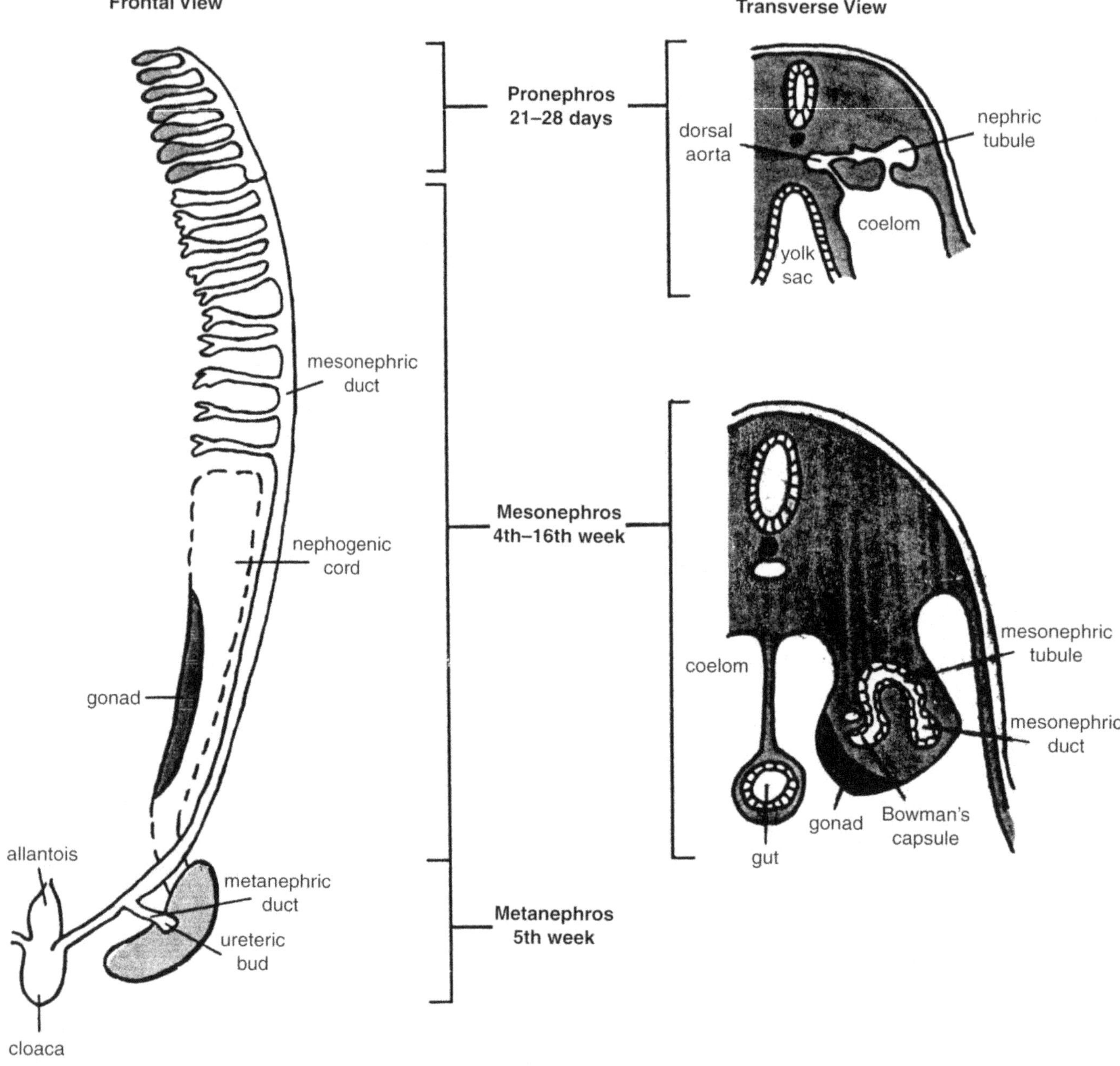

Figure 2.3. Development of the nephrogenic system from the intermediate mesoderm, from the third through fifth weeks of embryogenesis.

duct. Sprouts of the aorta grow simultaneously into the medial end of the pronephric buds to form primitive glomeruli. The pronephros reaches its peak of development in the fourth week of embryonic life and then, except for its duct, involutes. The pronephric duct remains and serves as the duct of the mesonephros, emptying into the cloaca.

The mesonephros begins to develop in the fourth and fifth weeks from the nephrogenic cord at the levels of 14–26 somites (lower cervical, thoracic, and upper lumbar regions).

Vesicles and then tubules develop in the nephrogenic cord. The dorsolateral end of each tubule joins the pronephric duct (now the mesonephric duct), and the opposite end forms a glomerulus with a branch of the aorta. Progressive cephalocaudal degeneration of the mesonephros occurs until the end of the 16th week of life, when only the mesonephric (Wolffian) duct persists in the male (vas deferens).

The metanephros begins to develop in the fifth week, in the lower lumbar region, from two primordia: an outgrowth (metanephric diverticulum, or ureteric bud) of the mesonephric duct, and the nephrogenic cord (metanephric blastema) (Figure 2.3). The nephrogenic tissue aggregates into small nodules at the tips of ingrowing collecting tubules, which are the terminal extensions of the outgrowing and dividing urogenital sinus (from the anterior part of the cloaca). The nephrogenic nodules become vesicles and then elongate into tubules, connecting with the collecting tubules on one end, and forming Bowman's capsules on the opposite end.

Differentiation of the Lateral Mesoderm

The cells of the lateral (coelomic) mesoderm separate into two layers around a central, intraembryonic coelom by the 19th day (Figures 2.2B and C). The outer somatic layer differentiates into the parietal mesothelium of the coelomic cavities and the connective tissue and skeletal muscle of the ventral body wall and portions of the limbs. The inner splanchnic layer of lateral mesoderm develops into the visceral mesothelium of the coelomic cavities, connective tissue and smooth muscle of the gastrointestinal tract, paramesonephric (Müllerian) ducts, genital ridges, adrenal cortex, and myocardium.

Comparison of Embryonic Mesodermal Differentiation with Embryonal Rhabdomyosarcoma, Wilms' Tumor, and Mesothelioma

It is during the third to eighth weeks of embryogenesis that differentiation is especially interesting to study, particularly in respect to correlating the morphology of early derivatives (Figure 2.4) of the third germ layer, the mesoderm (Figures 2.5 and 2.6), with the structure of certain mesodermally derived neoplasms. For example, myotomes (Figures 2.7 through 2.12) are morphologically recapitulated in embryonal rhabdomyosarcomas, both being composed of small round cells and early

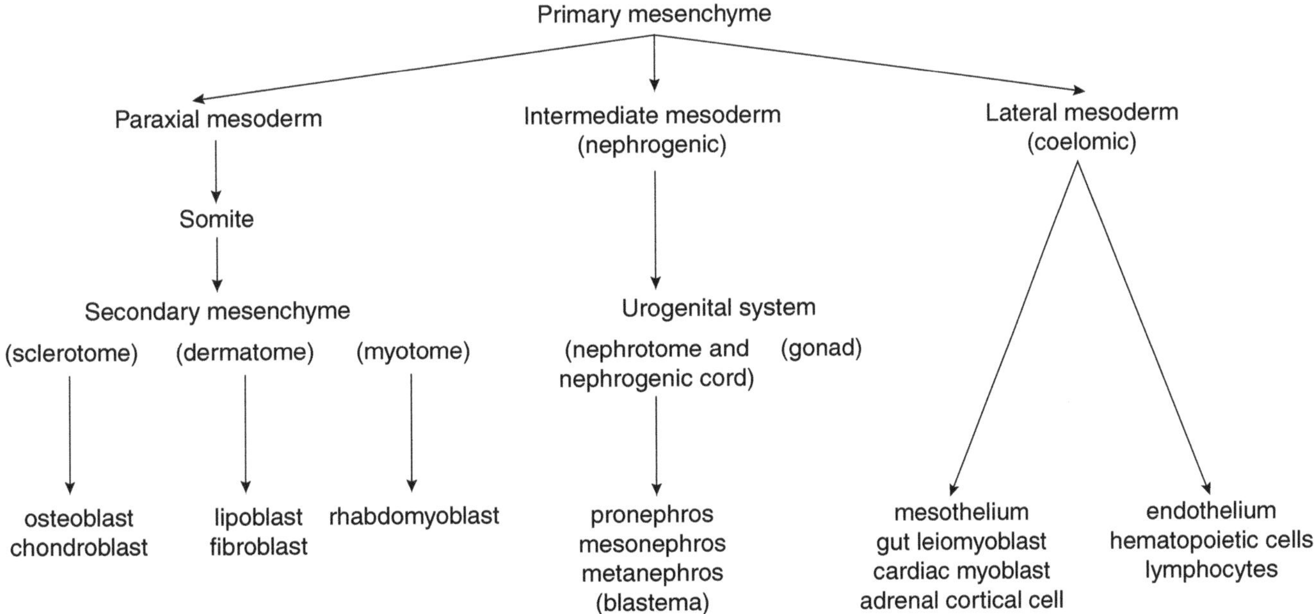

Figure 2.4. Diagram of the differentiation of primary mesenchyme.

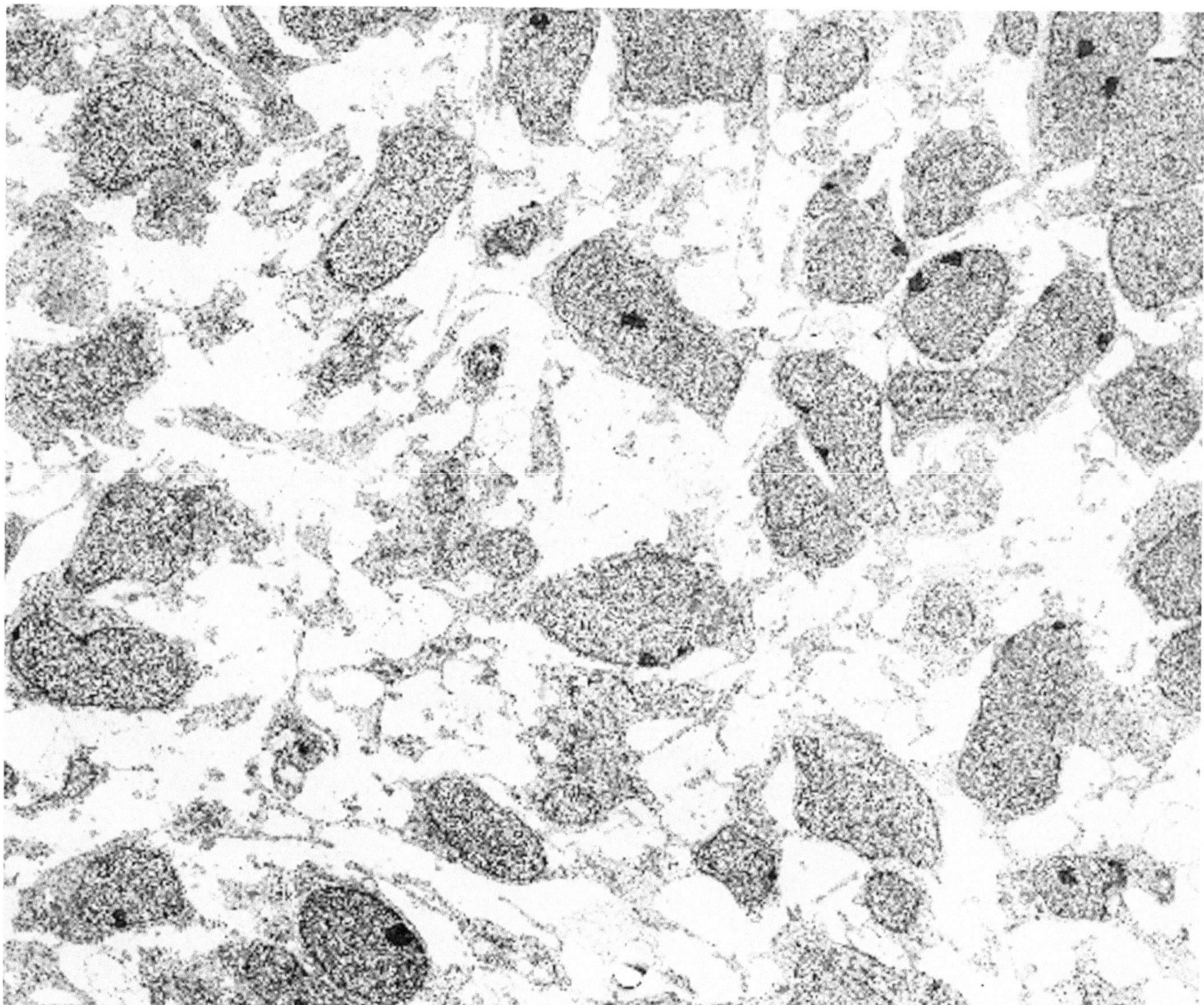

Figure 2.5. Primary mesenchyme from a 39-day-old human embryo. The cells vary in shape, are loosely arranged in an electron-lucent matrix, and have only focal contact with one another. (× 2100)

strap like cells. Likewise, a well-recognized similarity exists between metanephric blastema (Figures 2.13 through 2.16) and the undifferentiated component of Wilms' tumors. In a similar vein, intraembryonic coelimic lining cells correlate with epithelioid mesotheliomas, and, more speculatively, the subsurface cells (Figures 2.17 and 2.18) may be represented in the cells that comprise so-called fibrous mesotheliomas. (Illustrations of these and related primitive neoplasms are presented in the chapters on neoplasms).

(Text continues on page 26)

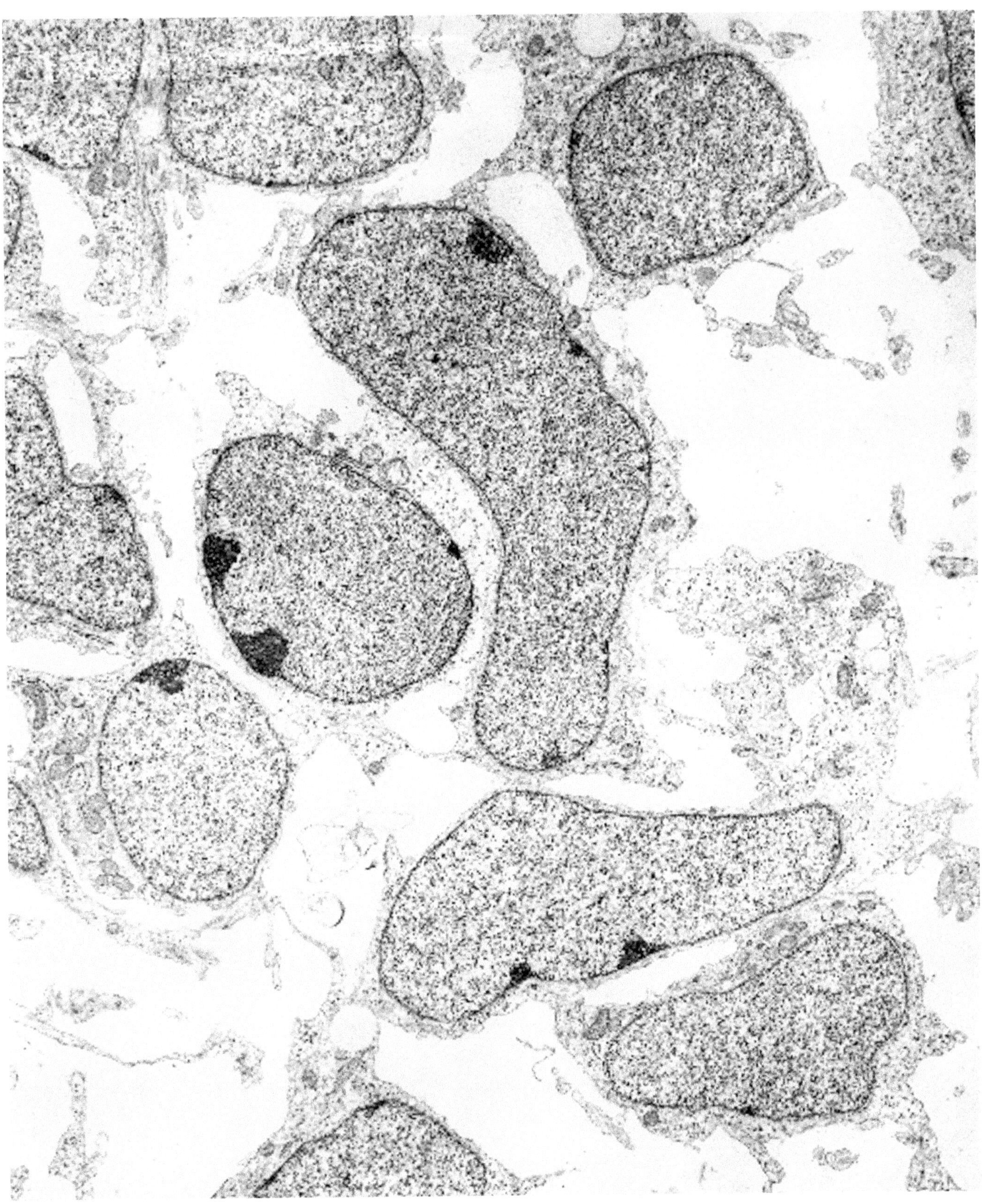

Figure 2.6. Higher magnification of primary mesenchymal cells reveals a high nuclear-cytoplasmic ratio, euchromatic nuclei, prominent nucleoli, and few cytoplasmic organelles. (× 6480)

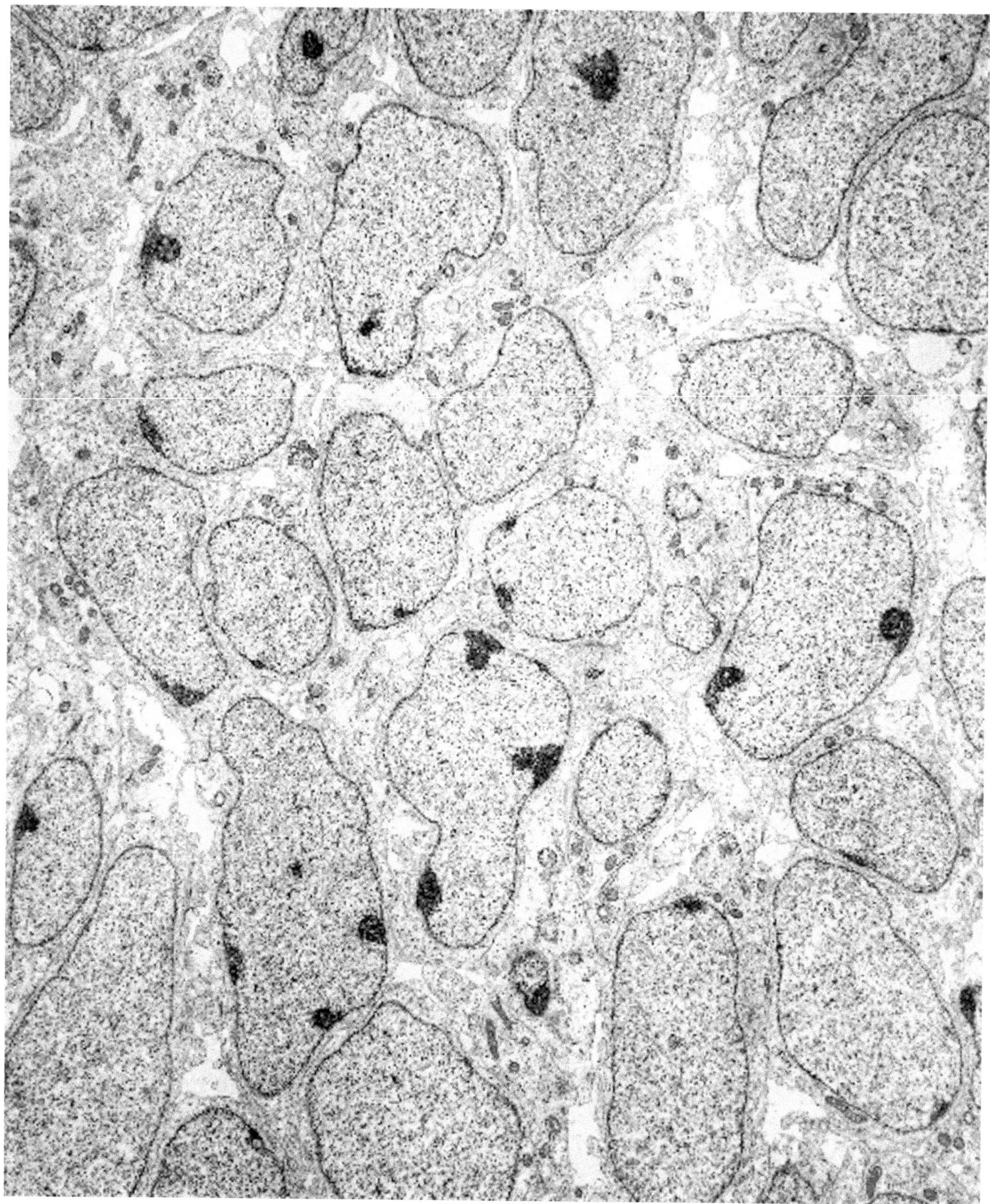

Figure 2.7. Human myotome, central dense area. The cells are polygonal and densely arranged. Their nuclei and nucleoli are similar to those of primary mesenchyme, but their cytoplasm and organelles are somewhat more abundant. (× 5510) (Permission for reprinting granted by Hemisphere Publishing Co., Dickersin GR: Embryonic ultrastructure as a guide in the diagnosis of tumors. Ultrastruct Pathol 11:609–652, 1987.)

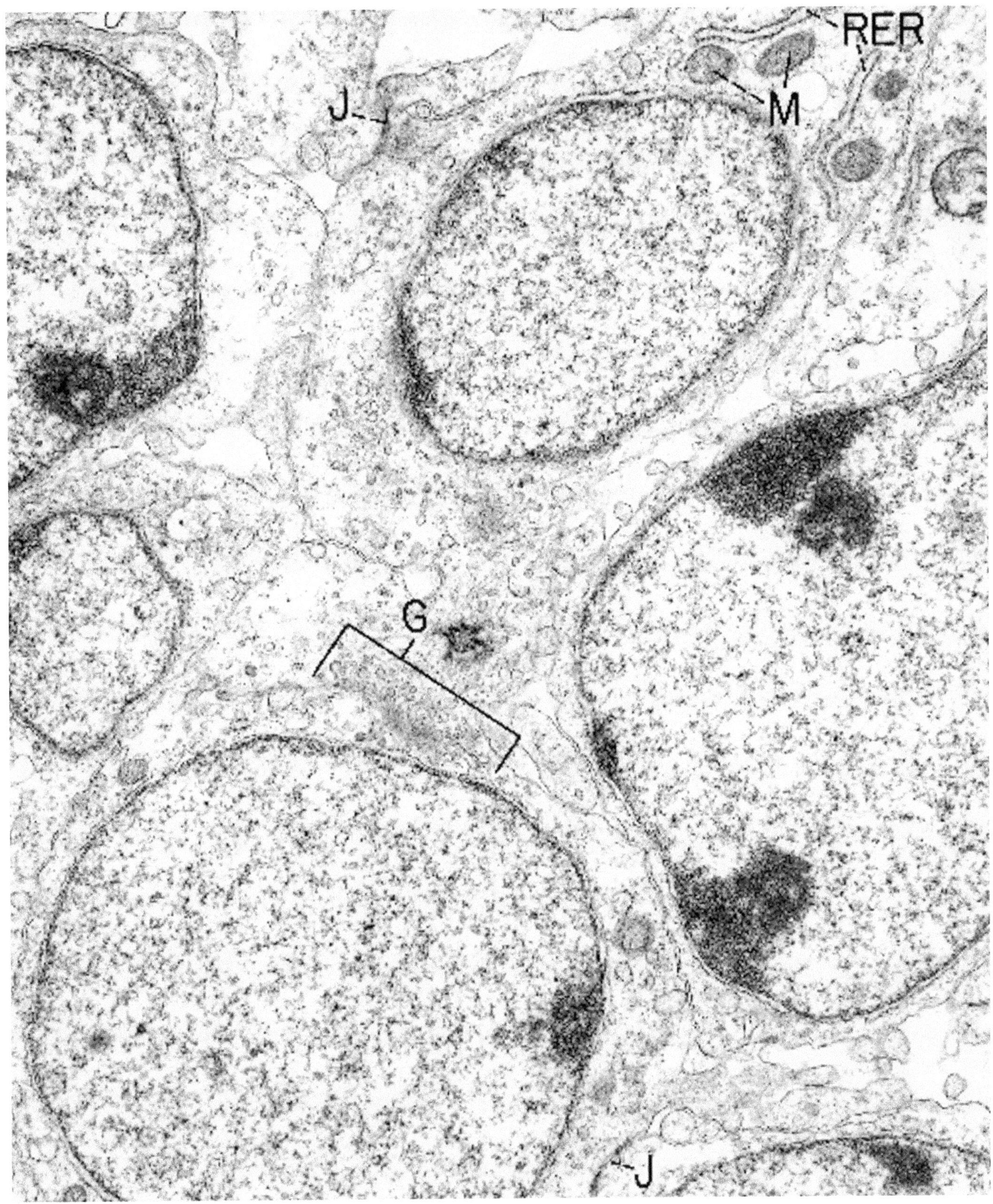

Figure 2.8. Higher power of cells of a myotome. A moderate number of organelles is visible, including Golgi apparatuses (G), mitochondria (M), and rough endoplasmic reticulum (RER). Small junctions (J) are present between the cells. (× 21,870) (Permission for reprinting granted by Hemisphere Publishing Co., Dickersin GR: Embryonic ultrastructure as a guide in the diagnosis of tumors. Ultrastruct Pathol 11:609–652, 1987.)

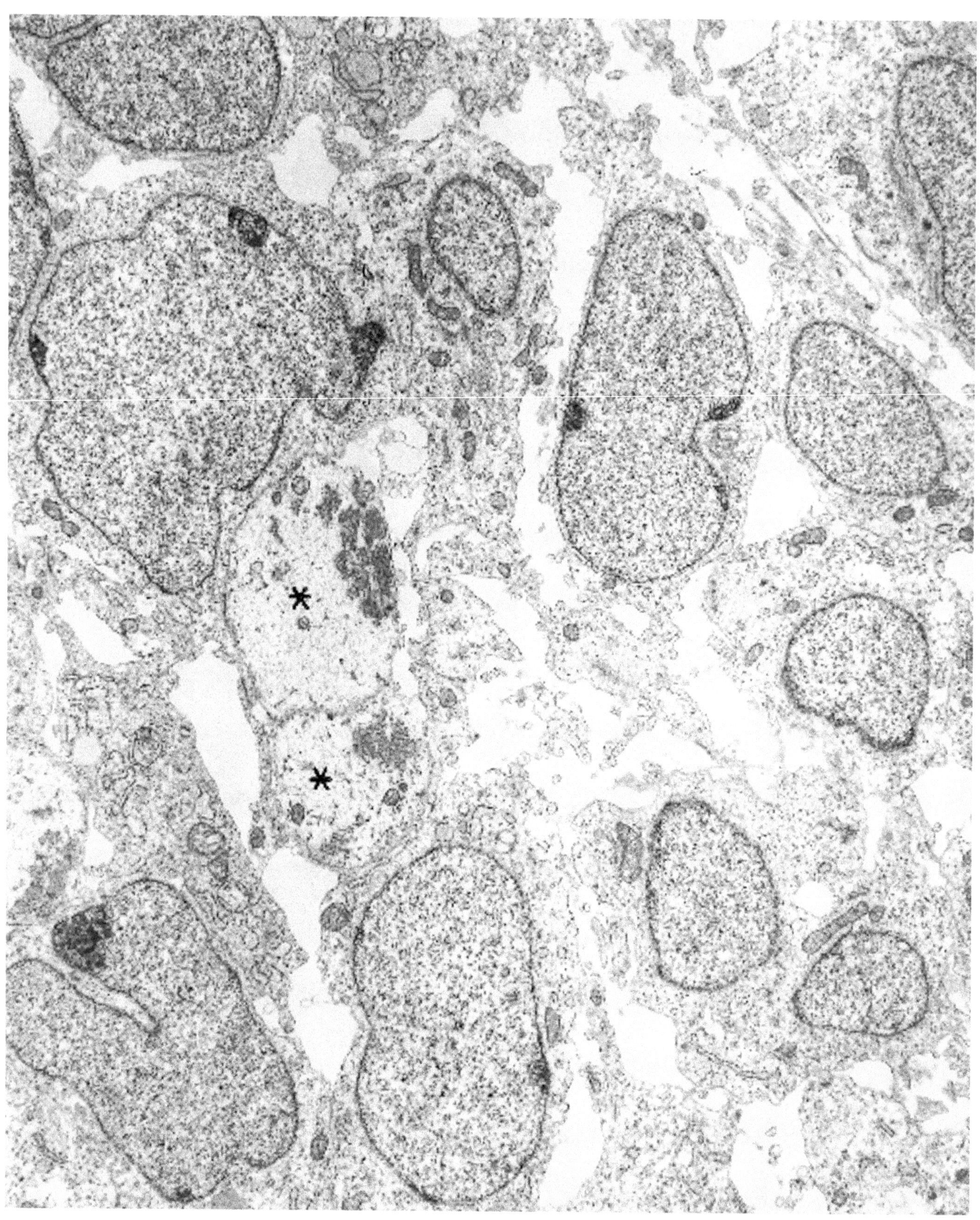

Figure 2.9. Myotome. Among the primitive-appearing cells are two cells (*) showing early myoblastic differentiation. This is characterized by a pale cytoplasm and several electron-dense areas (see Figure 2.10 for higher power). (× 7020) (Permission for reprinting granted by Hemisphere Publishing Co., Dickersin GR: Embryonic ultrastructure as a guide in the diagnosis of tumors. Ultrastruct Pathol 11:609–652, 1987.)

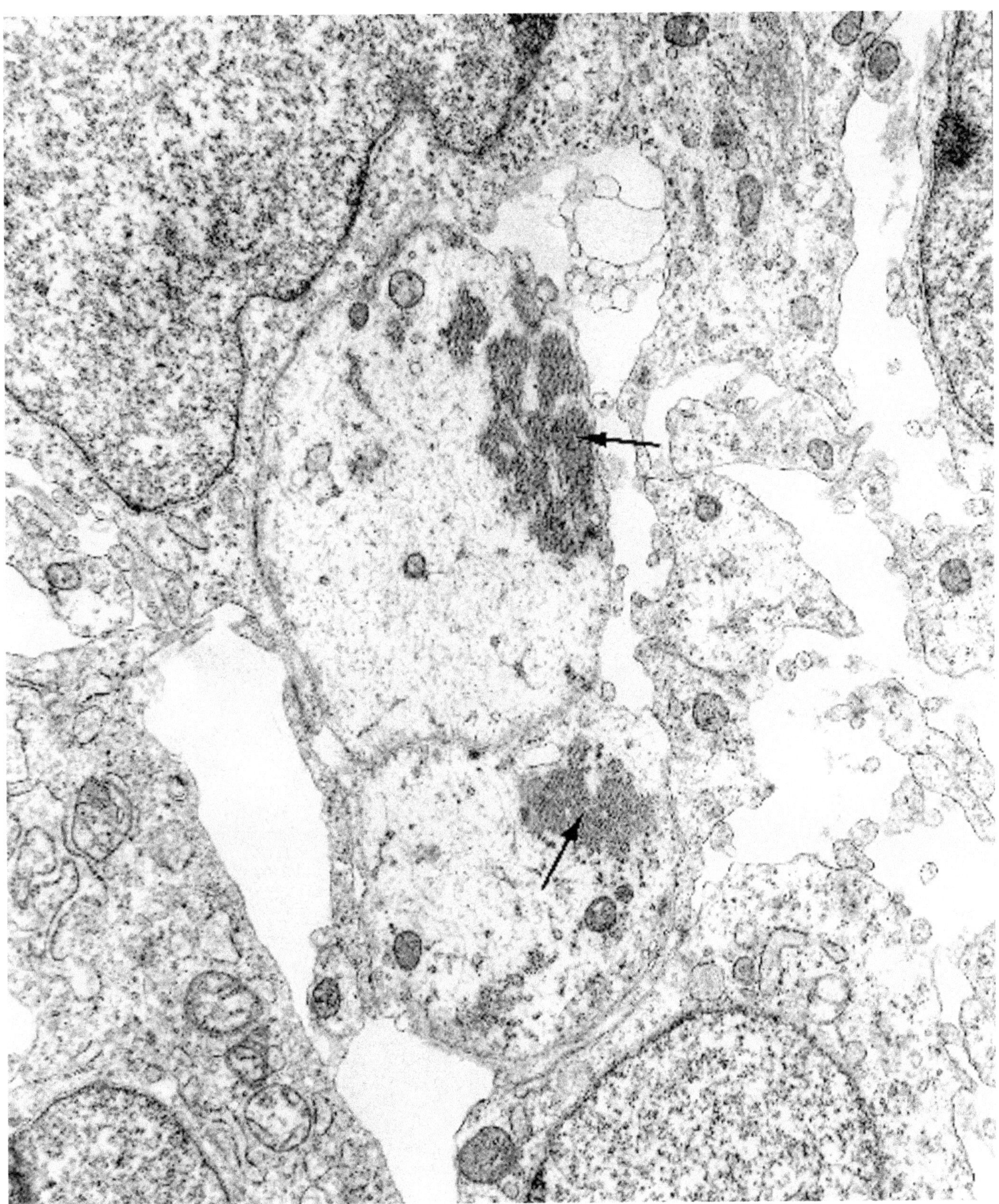

Figure 2.10. Myotome. Higher magnification of two of the pale cells in Figure 2.9 reveals most of the cytoplasm to have an open or clear background, consistent with glycogen, plus many irregularly arranged thin (actin) filaments. The dense areas (*arrows*) consist of thick (*myosin*) and thin filaments. (× 14,850)

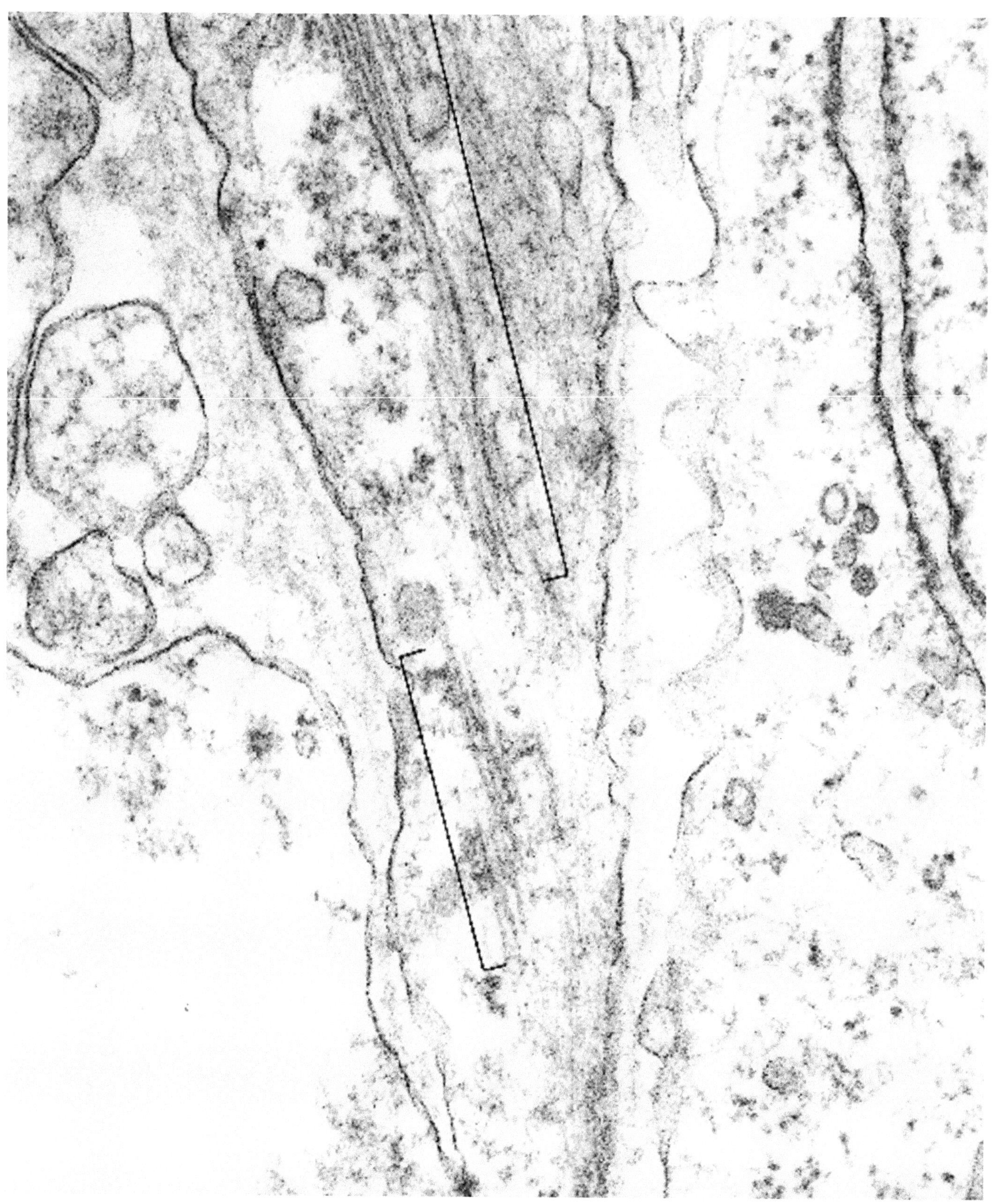

Figure 2.11. Myotome. This elongated cell is found among many primitive polygonal cells. It shows the earliest sign of skeletal muscle differentiation; that is, parallel thick filaments and closely associated rows and clusters of ribosomes (*bracketed areas*). (× 95,000) (Permission for reprinting granted by Hemisphere Publishing Co., Dickersin GR: Embryonic ultrastructure as a guide in the diagnosis of tumors. Ultrastruct Pathol 11:609–652, 1987.)

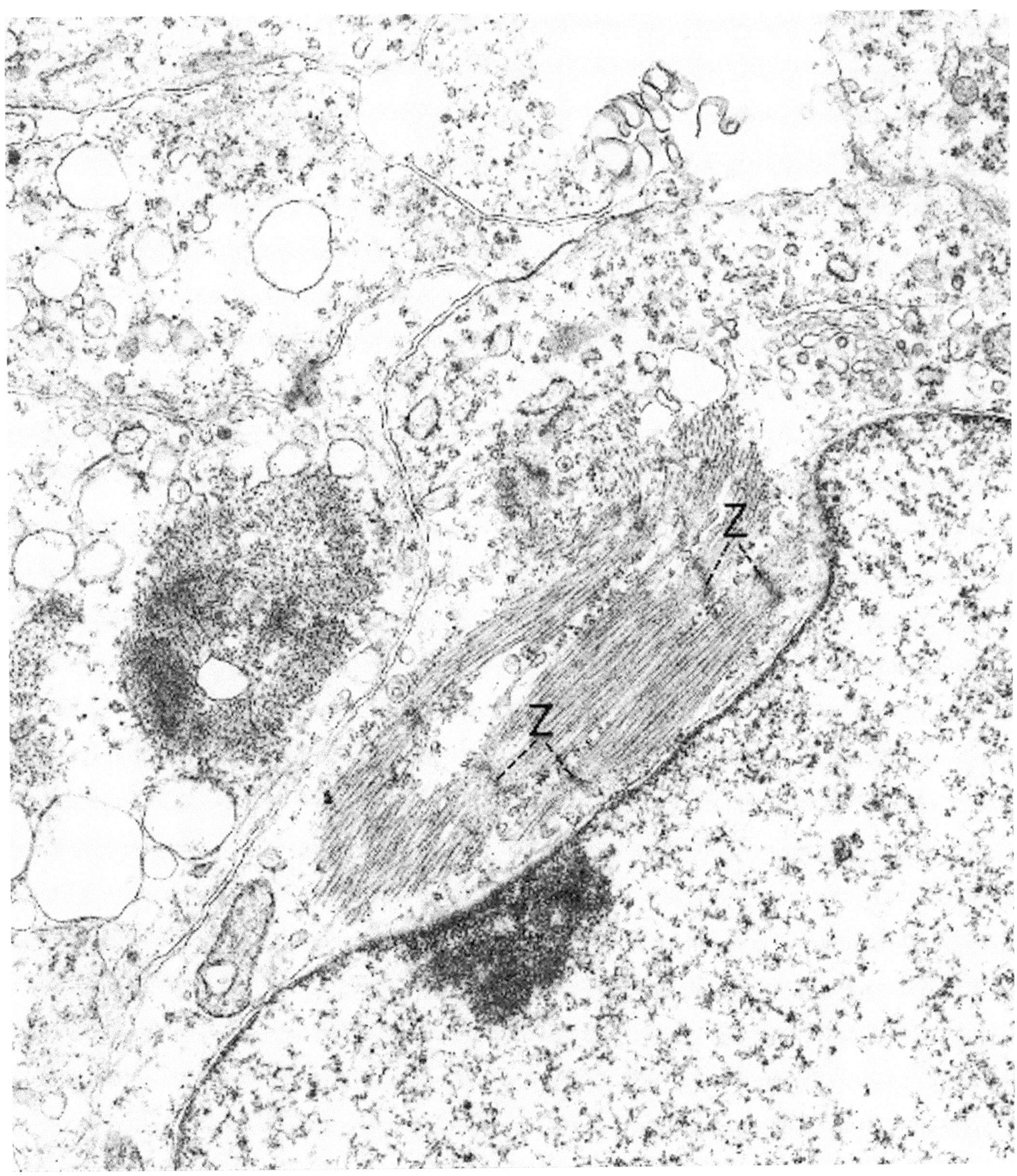

Figure 2.12. Myotome. Another recognizable myogenic cell is found among many undifferentiated ones making up the myotome. Parallel thick filaments and early Z-bands (Z) are forming sarcomeres. The electron-dense region at the left side of the field represents early sarcomeres cut transversely. (× 27,800) (Permission for reprinting granted by Hemisphere Publishing Co., Dickersin GR: Embryonic ultrastructure as a guide in the diagnosis of tumors. Ultrastruct Pathol 11:609–652, 1987.)

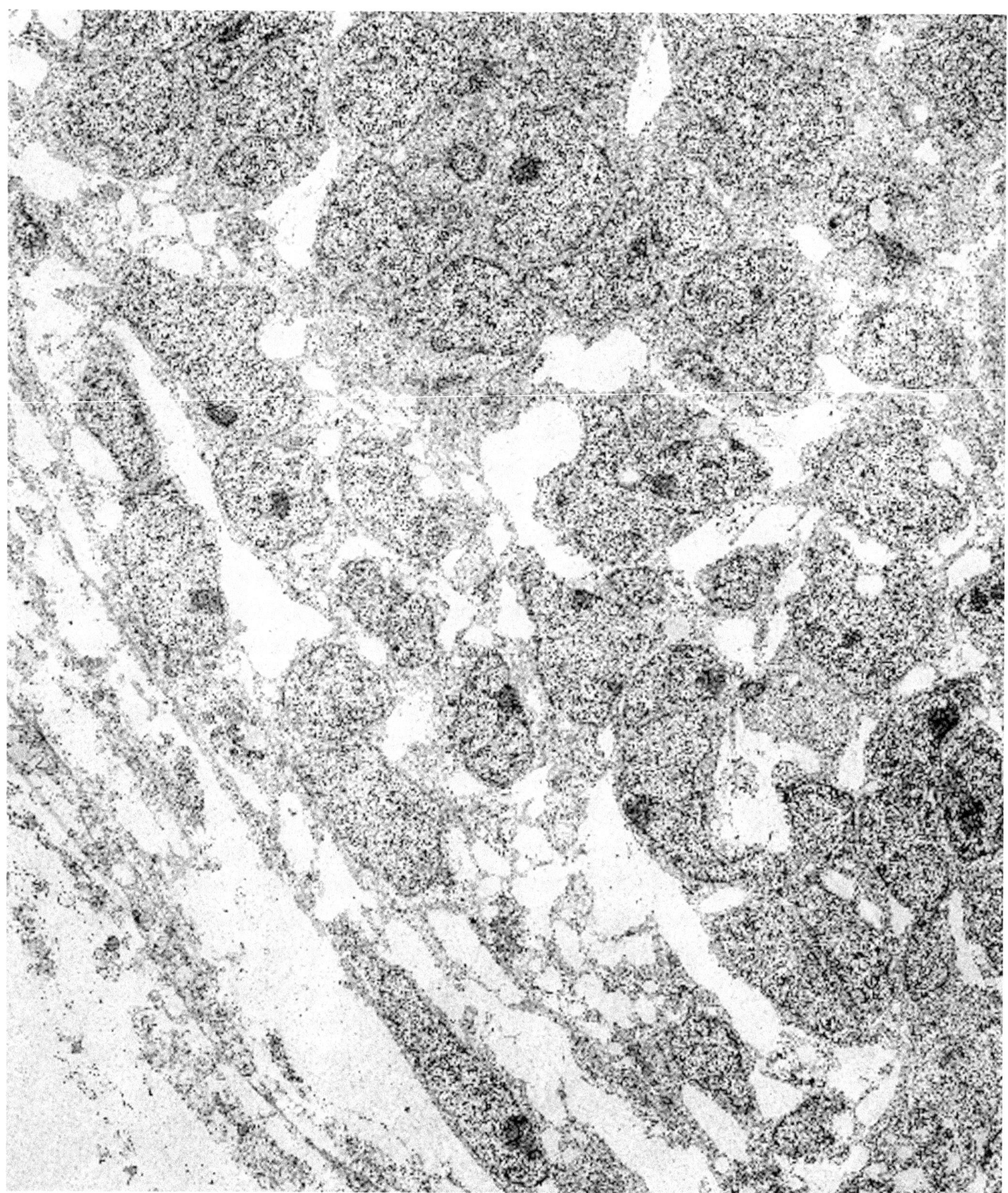

Figure 2.13. Metanephric blastema. These cells, which are derived from the intermediate mesodermal mass, generally are similar to those of the myotome (see Figures 2.7 and 2.8). They have a high nuclear-cytoplasmic ratio, euchromatic nuclei, prominent nucleoli, and scant cytoplasm. (× 3600) (Permission for reprinting granted by Hemisphere Publishing Co., Dickersin GR: Embryonic ultrastructure as a guide in the diagnosis of tumors. Ultrastruct Pathol 11:609–652, 1987.)

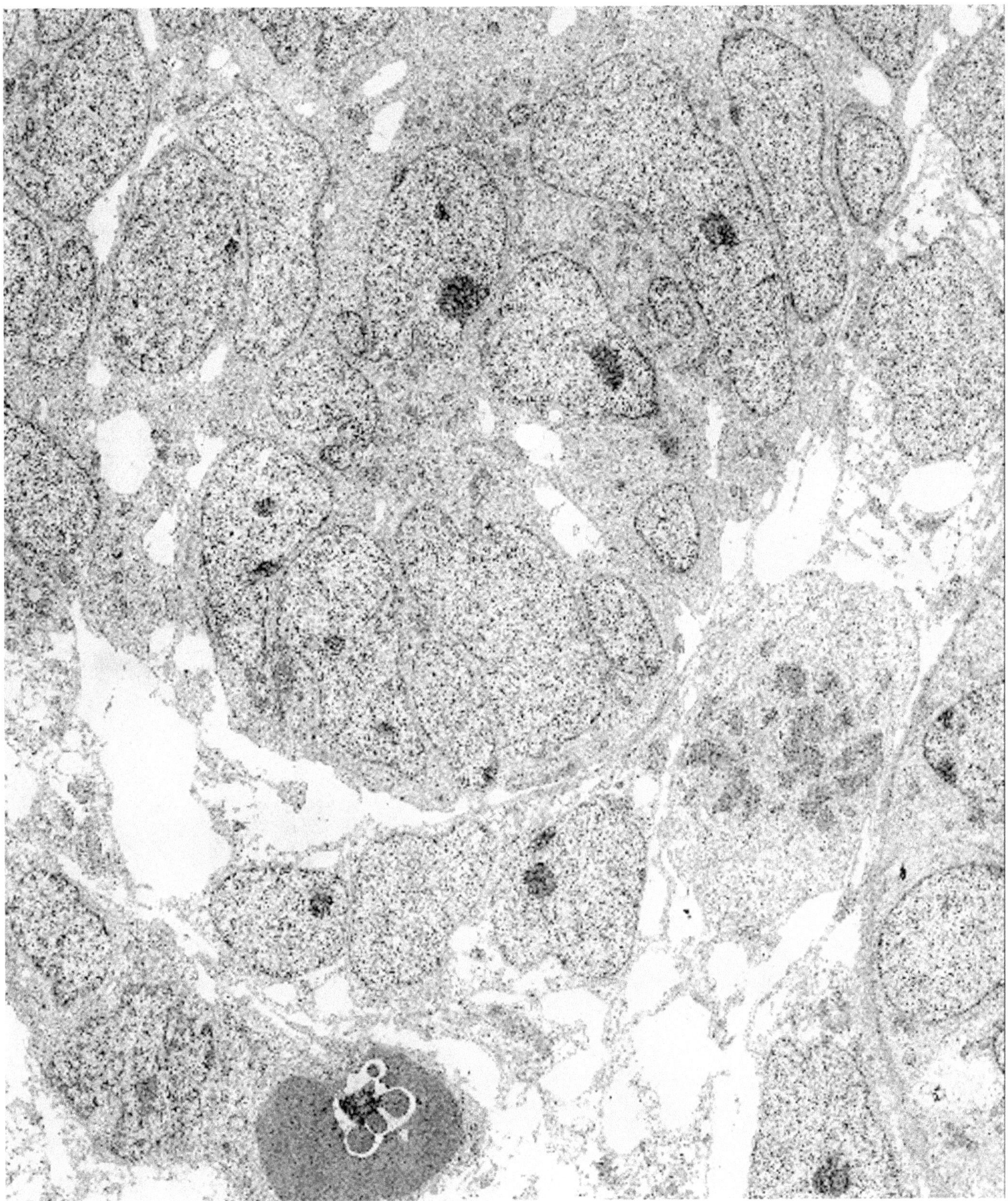

Figure 2.14. Metanephric blastema. Some of the cells in this field (*center*) have grouped together in what is consistent with a pretubule. (× 3740) (Permission for reprinting granted by Hemisphere Publishing Co., Dickersin GR: Embryonic ultrastructure as a guide in the diagnosis of tumors. Ultrastruct Pathol 11:609–652, 1987.)

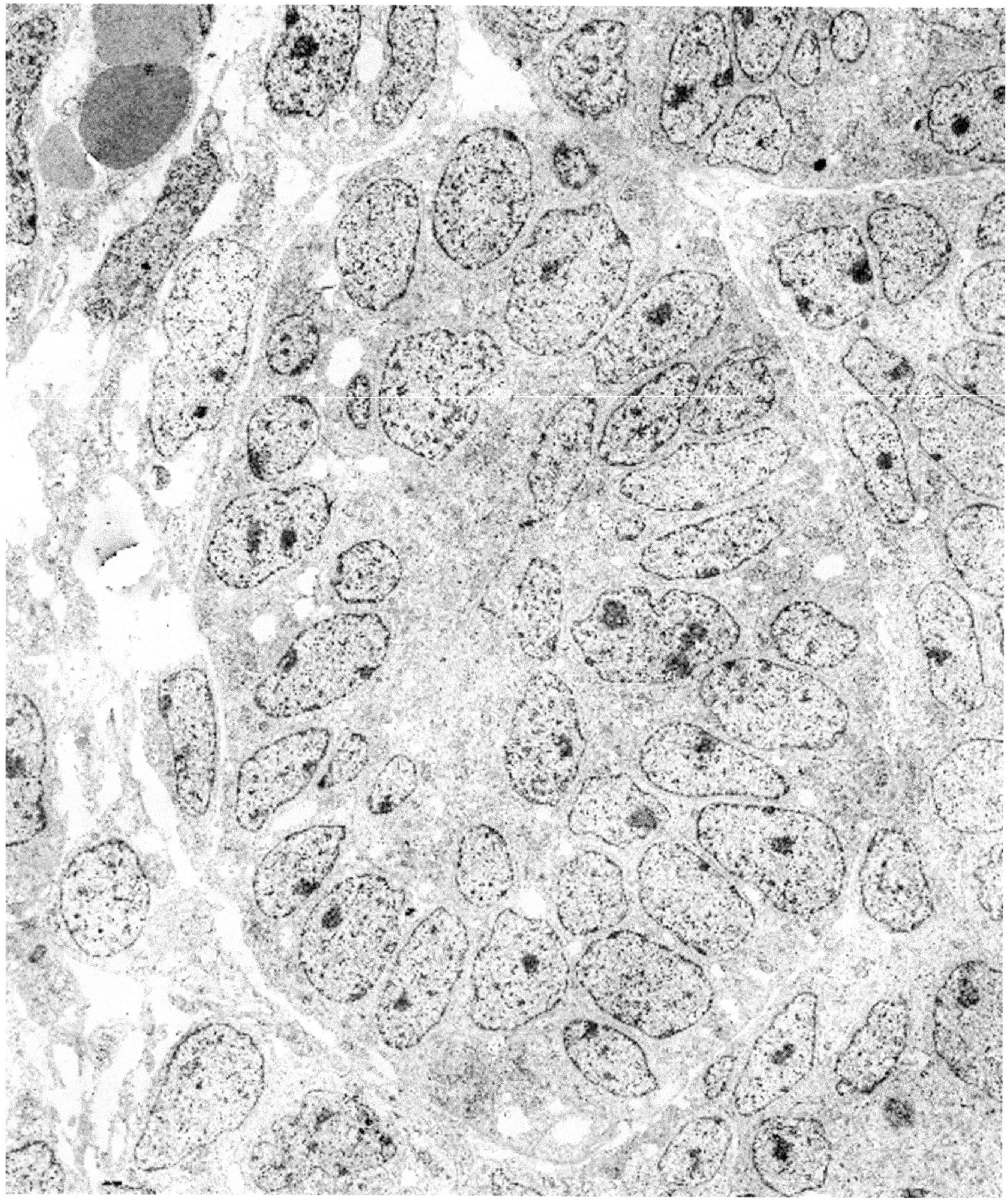

Figure 2.15. Metanephric blastema. A more discrete tubule has formed in this region, but only junctional complexes and no open lumen are demonstrable at higher power. (× 3270)

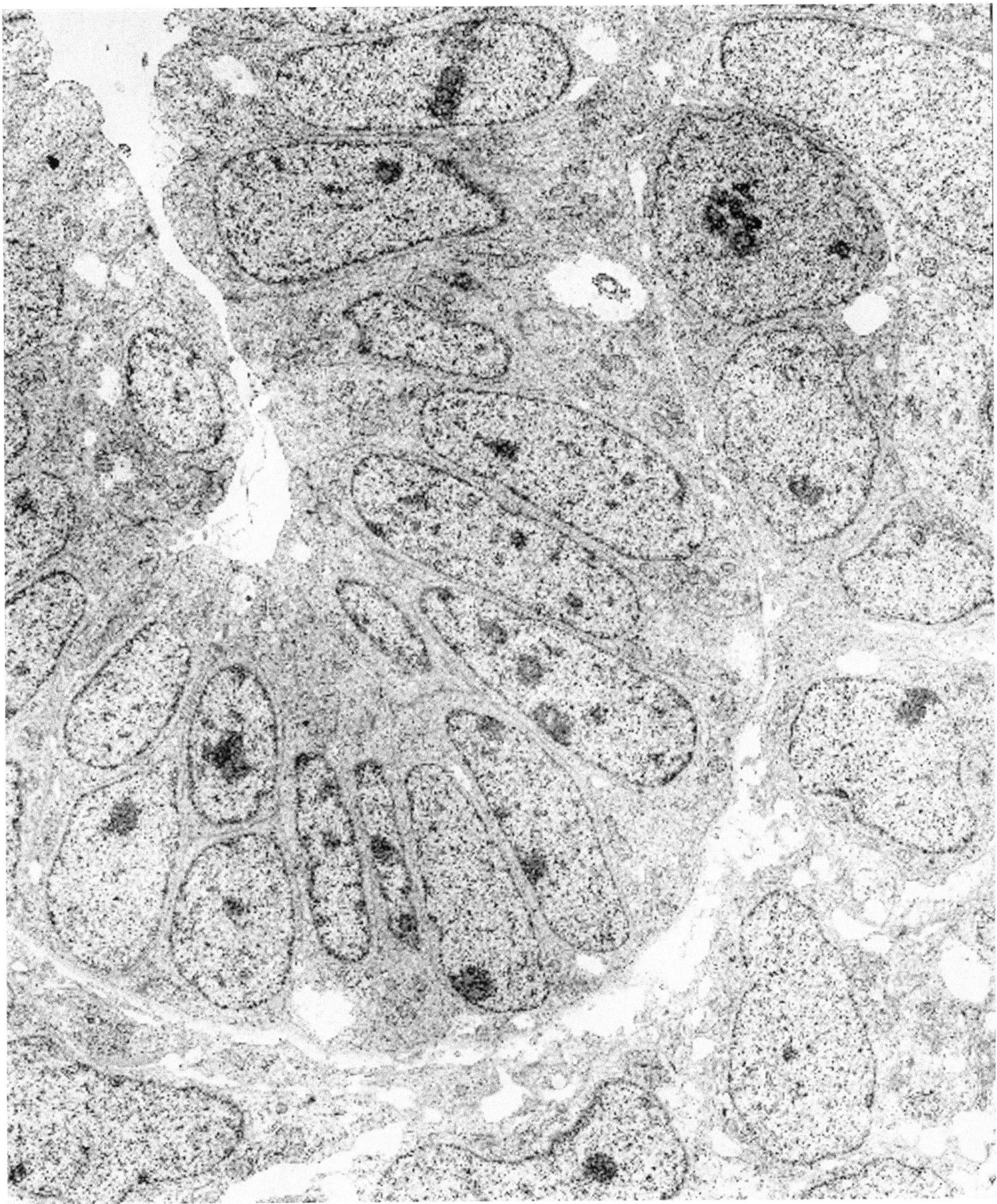

Figure 2.16. Metanephric blastema and early tubule. A definite lumen is present in this tubular form, and focal basal lamina separating the tubule from the undifferentiated blastema is visible at higher magnification. (× 4750)

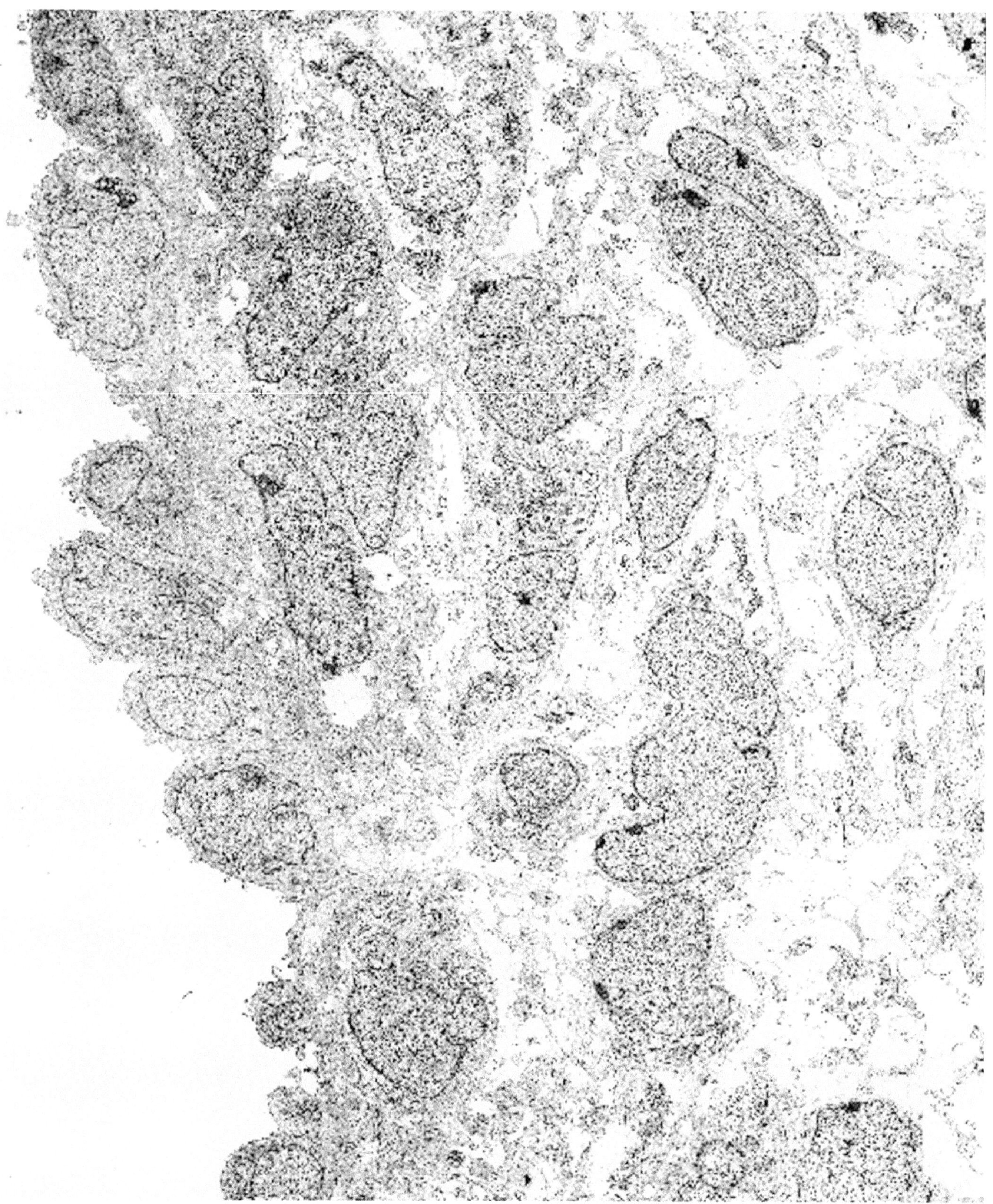

Figure 2.17. Coelomic lining. These cells derived from the lateral mesodermal mass show early mesothelial differentiation in the surface layer. No basal lamina between the surface layer and subjacent undifferentiated cells is identified at this stage of development. (× 4320) (Permission for reprinting granted by Hemisphere Publishing Co., Dickersin GR: Embryonic ultrastructure as a guide in the diagnosis of tumors. Ultrastruct Pathol 11:609–652, 1987.)

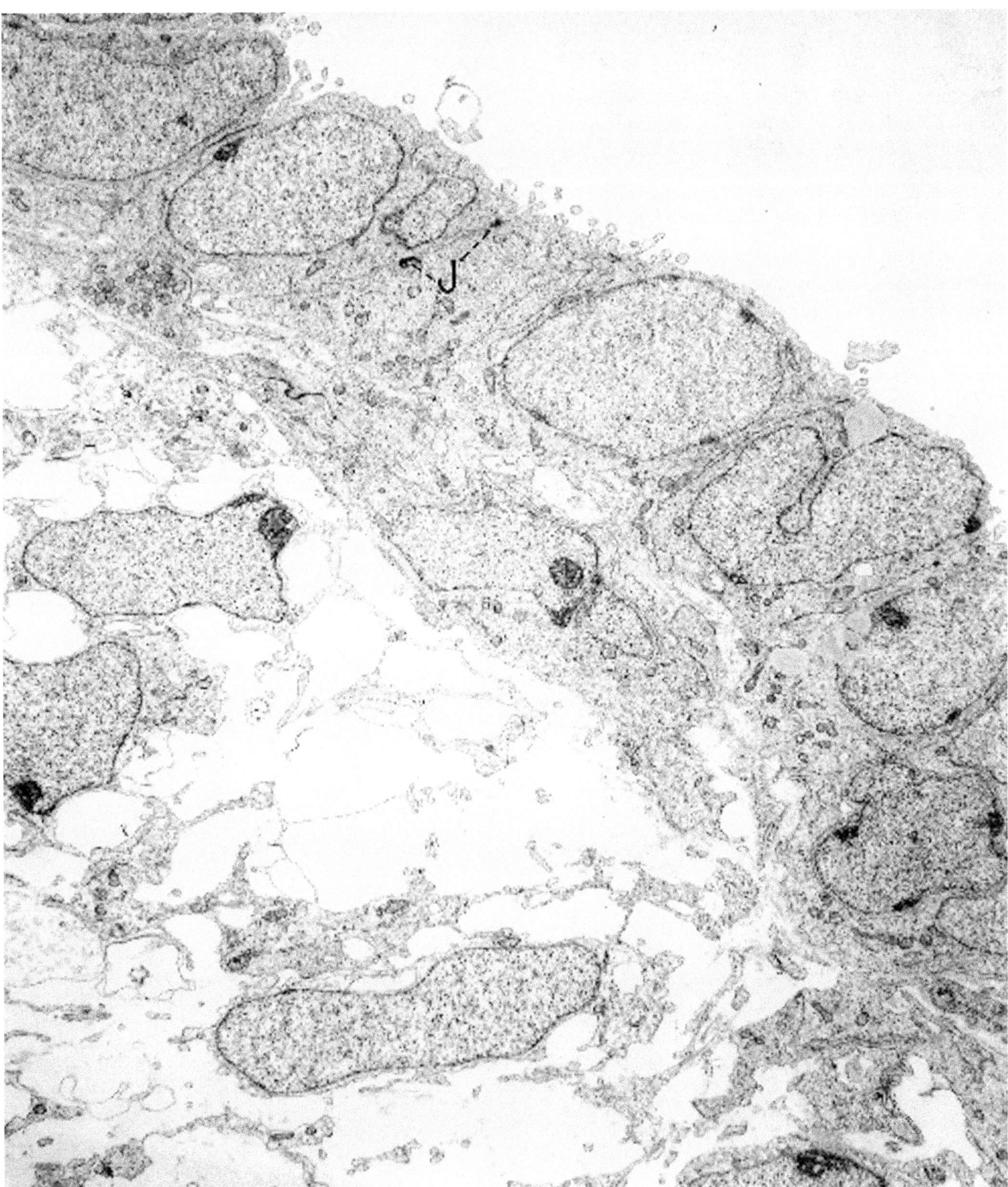

Figure 2.18. Coelomic lining. Higher magnification than Figure 2.17 shows the surface cells to have the epithelial feature of prominent junctions (J), and the deeper cells resemble primitive mesenchymal cells. (× 5940) (Permission for reprinting granted by Hemisphere Publishing Co., Dickersin GR: Embryonic ultrastructure as a guide in the diagnosis of tumors. Ultrastruct Pathol 11:609–652, 1987.)

(Text continued from page 12)

REFERENCES

Briselli M, Mark EJ, Dickersin GR: Solitary fibrous tumors of the pleura: Eight new cases and review of 360 cases in the literature. Cancer 47:2678–2689, 1981.

Corliss CE: *Patten's Human Embryology. Elements of Clinical Development.* McGraw-Hill, New York, 1976.

Dickersin GR: The contributions of electron microscopy in the diagnosis and histogenesis of controversial neoplasms. Clin Lab Med 4:123–164, 1984.

Dickersin GR: Embryonic ultrastructure as a guide in the diagnosis of tumors. Ultrastruct Pathol 11:609–652, 1987.

Fischman DA: An electron microscopic study of myofibril formation in embryonic chick skeletal muscle. J Cell Biol 32:557–575, 1967.

Friedman J, Bird ES: Electron-microscope investigation of experimental rhabdomyosarcoma. J Pathol 97:375–382, 1969.

Hay ED: The fine structure of differentiating muscle in the salamander tail. Zellforsch Mikrosk Anat 59:6–34, 1963.

Jacob F: Expression of embryonic characters by malignant cells. Ciba Found Symp 96:4–27, 1983.

Lawrence WD, Whitaker D, Sugimura H, et al: An ultrastructural study of the developing urogenital tract in early human fetuses. Am J Obstet Gynecol 167:185–193, 1992.

Nishimura H, ed: *Atlas of Human Prenatal Histology.* Igaku-Shoin, Tokyo, 1983.

Pierce GB: Neoplasms, differentiation and mutations. Am J Pathol 77:103–113, 1974.

Pierce GB: Teratocarcinoma: Model for a developmental concept of cancer. Curr Top Dev Biol 2:223–246, 1967.

Sadler TW: *Langman's Medical Embryology,* 7th ed. Williams and Wilkins, Baltimore, 1995.

Schmidt D, Dickersin GR, Vawter GF, et al: Wilms' tumor: Review of ultrastructure and histogenesis. Pathobiol Annu 12:281–300, 1982.

3

Large Cell Neoplasms

This chapter covers the salient ultrastructural features of large, round, and polygonal cell neoplasms and includes most carcinomas, melanomas and mesotheliomas and many lymphomas and some sarcomas. Many of these neoplasms appear undifferentiated by light microscopy, but most show differentiation along one cell line or another at the ultrastructural level. However, it is noteworthy that not all of the ultrastructural criteria for identifying the cell type may be present in every example of a neoplasm. As expected, the more differentiated the neoplasm, the more likely will its cells contain a broad complement of diagnostic morphologic features. Conversely, the least differentiated neoplasms may contain no cells that show any differentiated diagnostic structures, and the final electron microscopic diagnosis in these cases must rest at "undifferentiated," or "primitive," neoplasm. Usually, however, the ultrastructural findings do allow the pathologist to make a definitive diagnosis when interpreted in conjunction with the light microscopic picture and, in some cases, with the histochemical and immunohistochemical results.

Carcinoma

Various types of carcinomas have a number of distinguishing features, as described and illustrated in this chapter, but one common characteristic of all carcinomas is the presence of intercellular junctions, usually desmosomes and/or intermediate junctions. This is not to say that certain types of junctions are not present in some types of noncarcinomatous neoplasms, such as various sarcomas, but desmosomes almost always indicate epithelial differentiation.

Adenocarcinoma (and Adenoma)

(Figures 3.1 through 3.37.)

Diagnostic criteria. (1) Lumens; (2) microvilli; (3) tight junctions/junctional complexes; (4) basal lamina; (5) secretory granules; (6) prominent Golgi apparatus; (7) moderately prominent rough endoplasmic reticulum.

Additional points. Cilia may be seen in certain types of adenocarcinoma, such as those arising in Müllerian tissue, especially the *fallopian tube.* Single cilia (oligocilia) are found in many types of cells and are not diagnostic. Cilia are an important diagnostic marker for *ependymomas* and *choroid plexus* neoplasms (see Chapter 8).

Some adenocarcinomas have more-or-less specific features for their organ of origin, which may be useful in diagnosing metastatic neoplasms of unknown primary sources. How closely neoplastic cells resemble the cells of the organ of origin is determined by the level of

(Text continues on page 62)

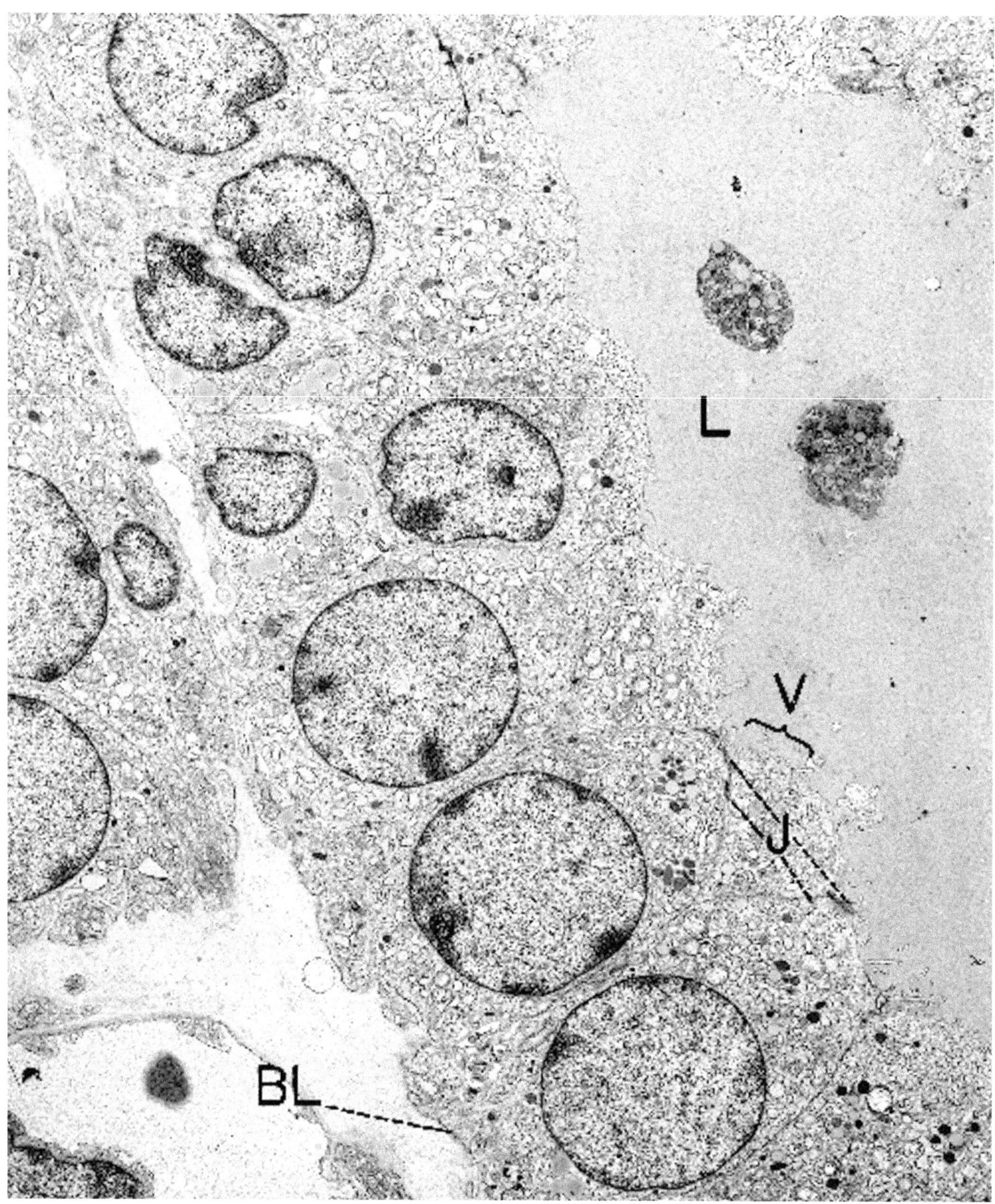

Figure 3.1. Adenoma (thyroid). An example of a well-differentiated glandular tumor. A single row of low cuboidal epithelial cells lines the lumen (L) of the gland. Microvilli (V) on the luminal surface of the cells, basal lamina (BL) along the basal aspect of the cells, and junctions (J) between cells are visible but are seen better at higher power in Figures 3.2 and 3.4. (× 5130)

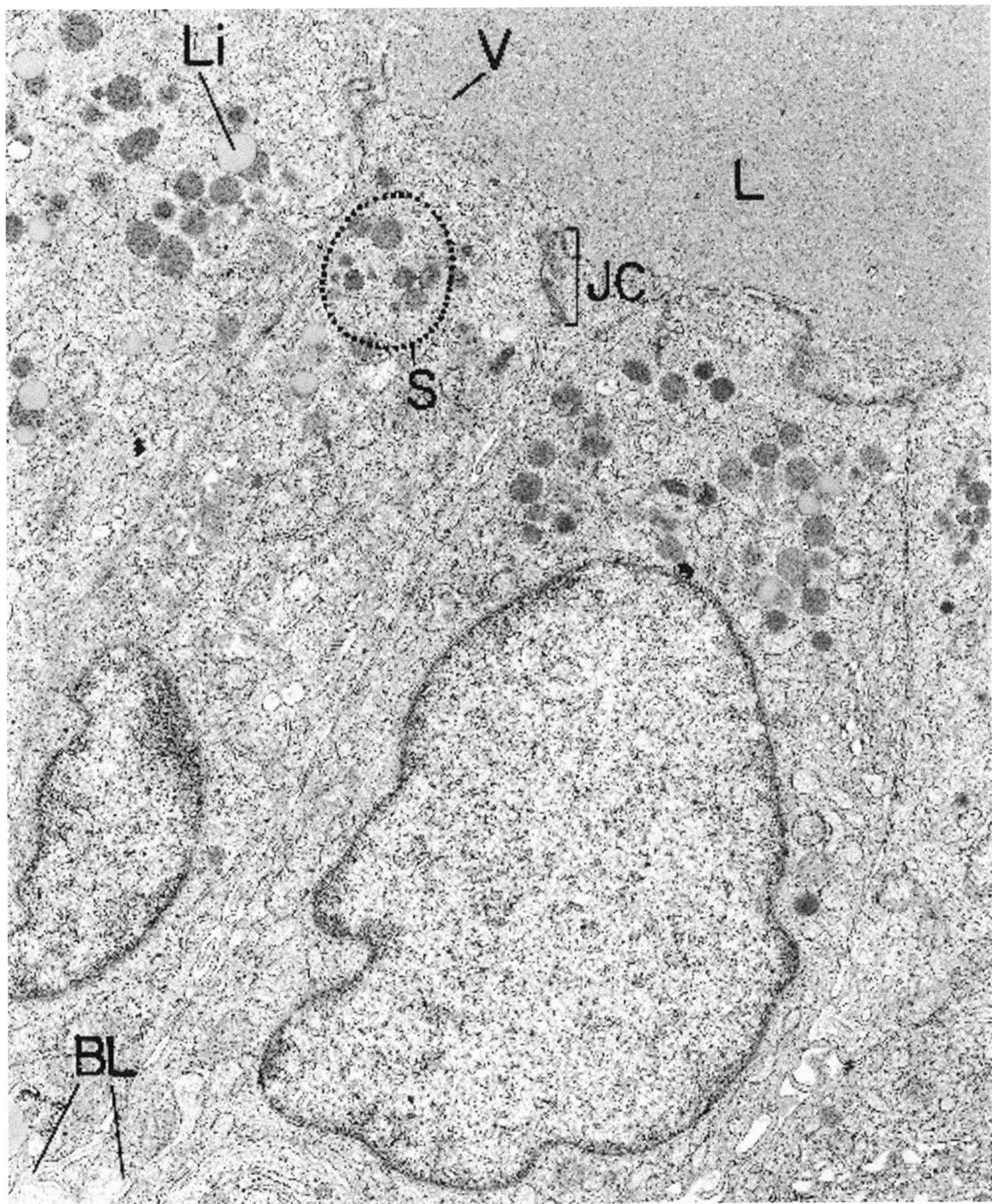

Figure 3.2. Adenoma (thyroid). Higher magnification of portions of the cells from the well-differentiated glandular tumor depicted in Figure 3.1. Microvilli (V) are sparse in this field. A small segment of basal lamina (BL) is visible along the base of cells, and junctional complexes (JC) are prominent at their luminal aspect. Numerous membrane-bound, secretory granules (S) are located in the apical cytoplasm. Non-membrane-bound lipid droplets (Li) are randomly dispersed in the cytoplasm. L = lumen. (× 14,180) (Permission for reprinting granted by WB Saunders, Dickersin GR: Electron microscopy of leukemias and lymphomas. Clin Lab Med 7:199–247, 1987.)

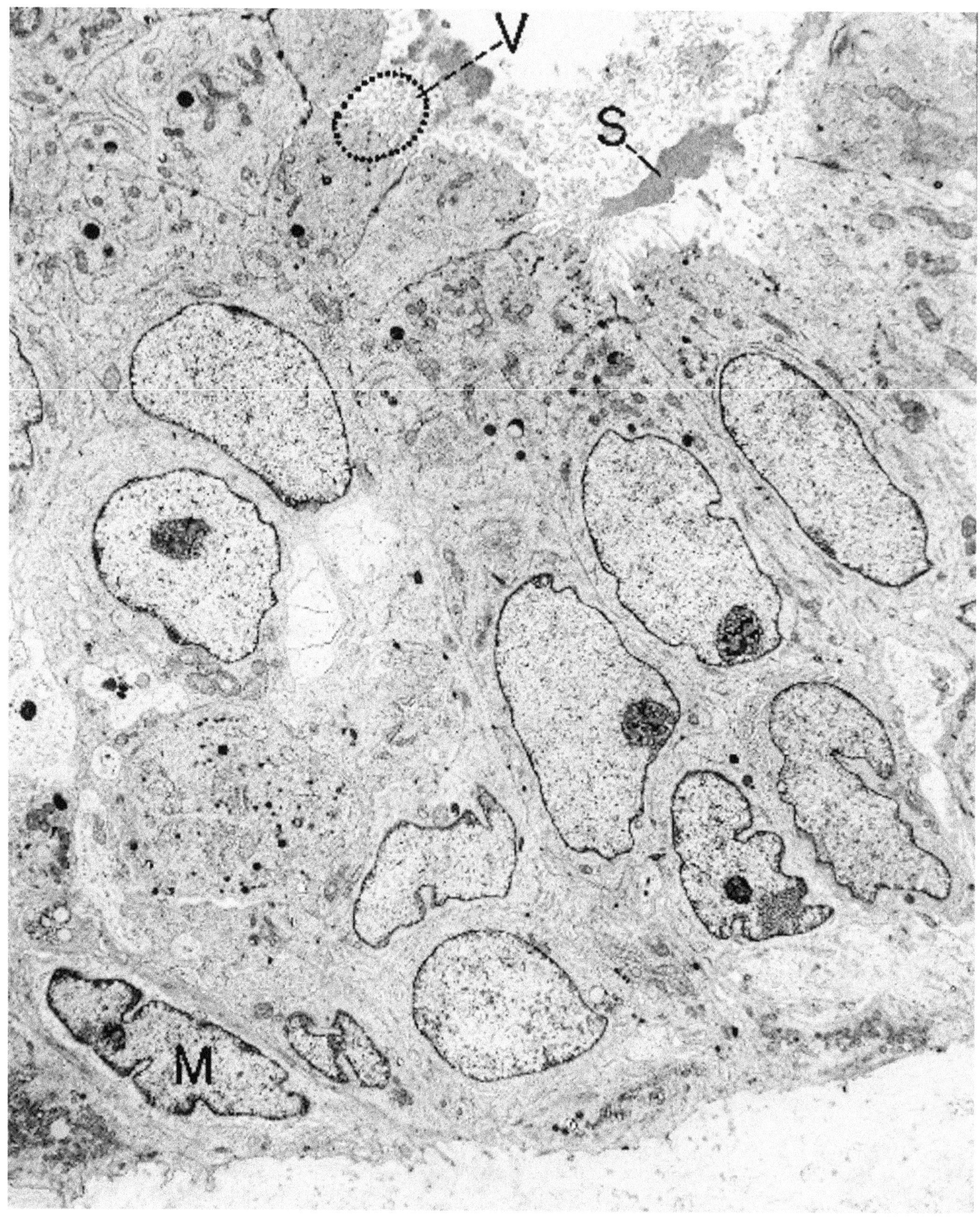

Figure 3.3. Phyllodes tumor (breast). The glands in this tumor are well differentiated but are lined by more than one layer of epithelial cells and by peripherally located myoepithelial cells (M). The lumen contains secretions (S) and is lined by cells rich in microvilli (V). (× 5320)

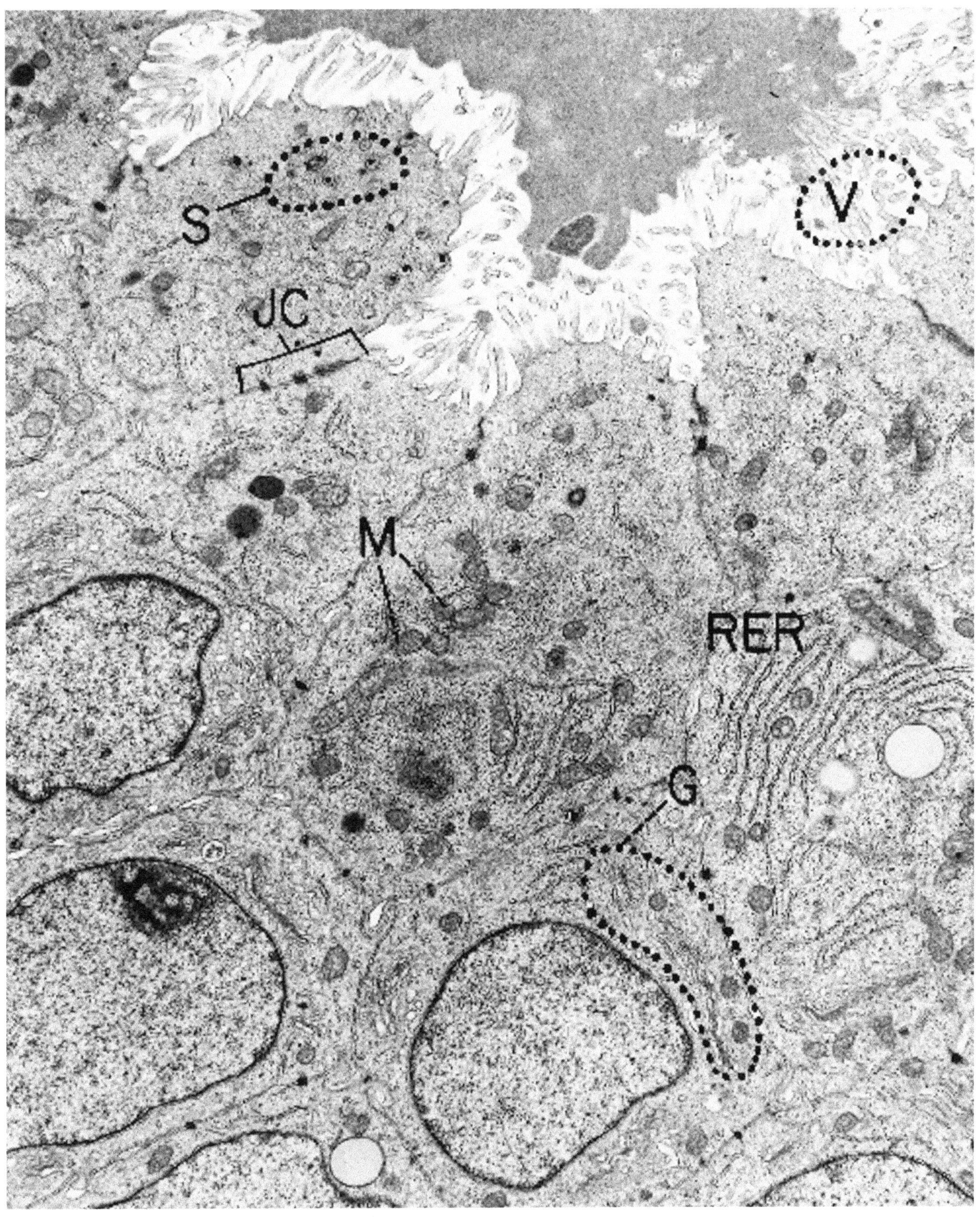

Figure 3.4. Phyllodes tumor (breast). Higher magnification of the neoplasm illustrated in Figure 3.3 shows microvilli (V) and junctional complexes (JC). Note also the secretory granules (S), Golgi apparatuses (G), rough endoplasmic reticulum (RER), and mitochondria (M). (× 11,700)

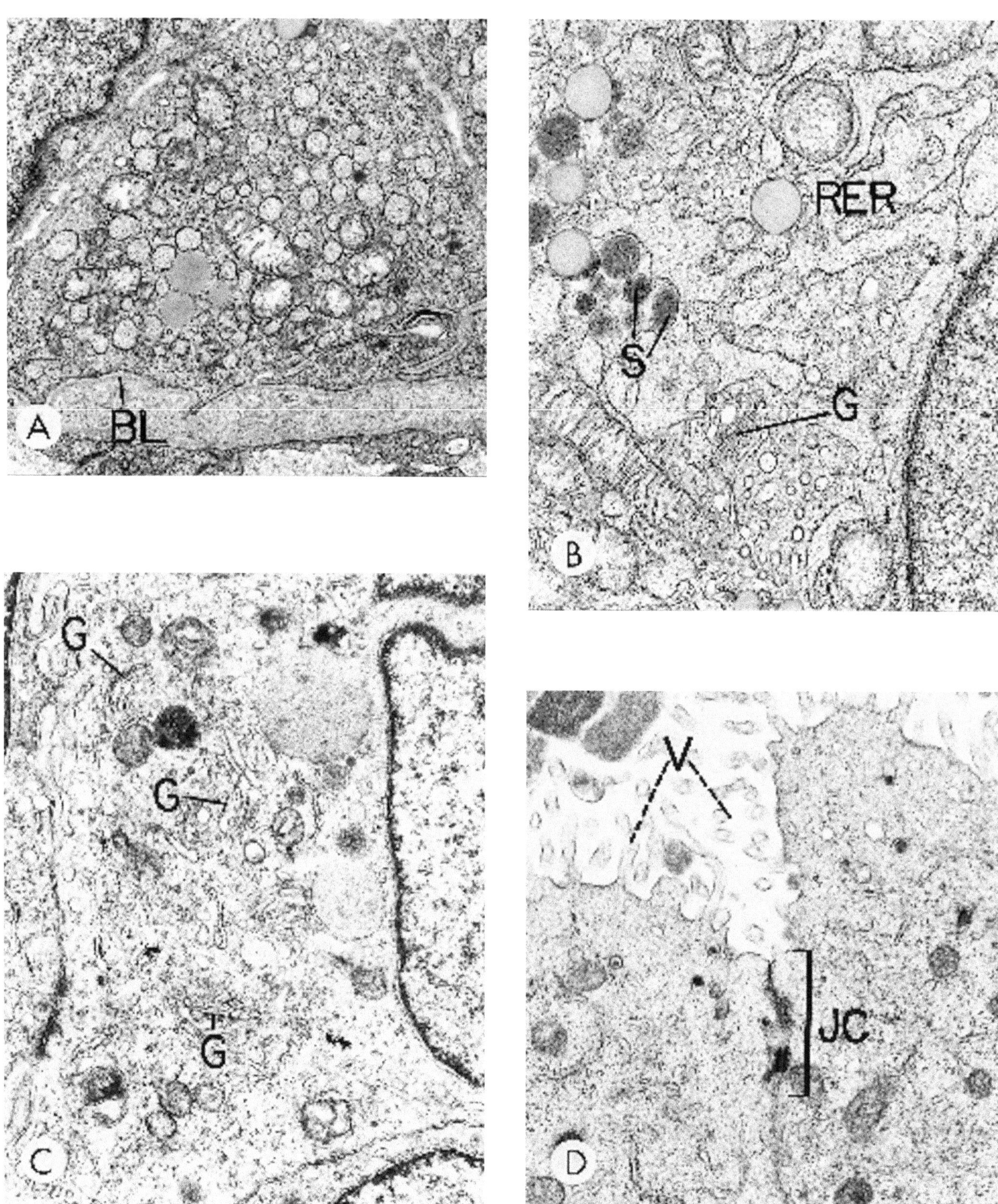

Figure 3.5. Adenomatous features at higher magnification. **A,** Basal lamina (BL). (× 12,150) **B,** Golgi apparatus (G); rough endoplasmic reticulum (RER), which is dilated and filled with medium-dense material; and secretory granules (S). (× 17,900) **C,** Golgi apparatuses (G). (× 14,280) **D,** microvilli (V) and junctional complex (JC). (× 15,250)

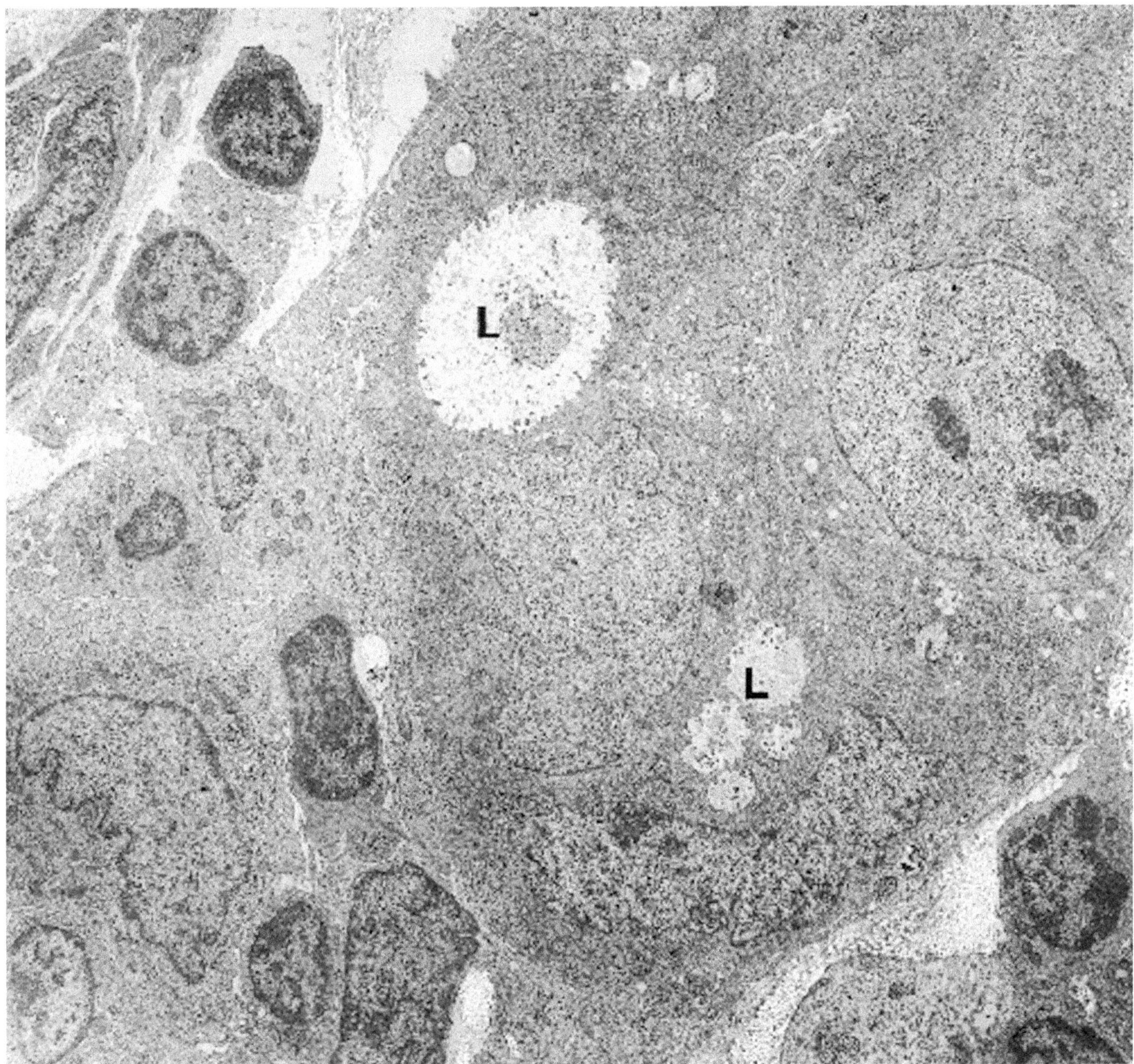

Figure 3.6. Adenocarcinoma (breast). Several cells of this infiltrating ductal carcinoma exhibit intracytoplasmic lumens (L). (× 5200)

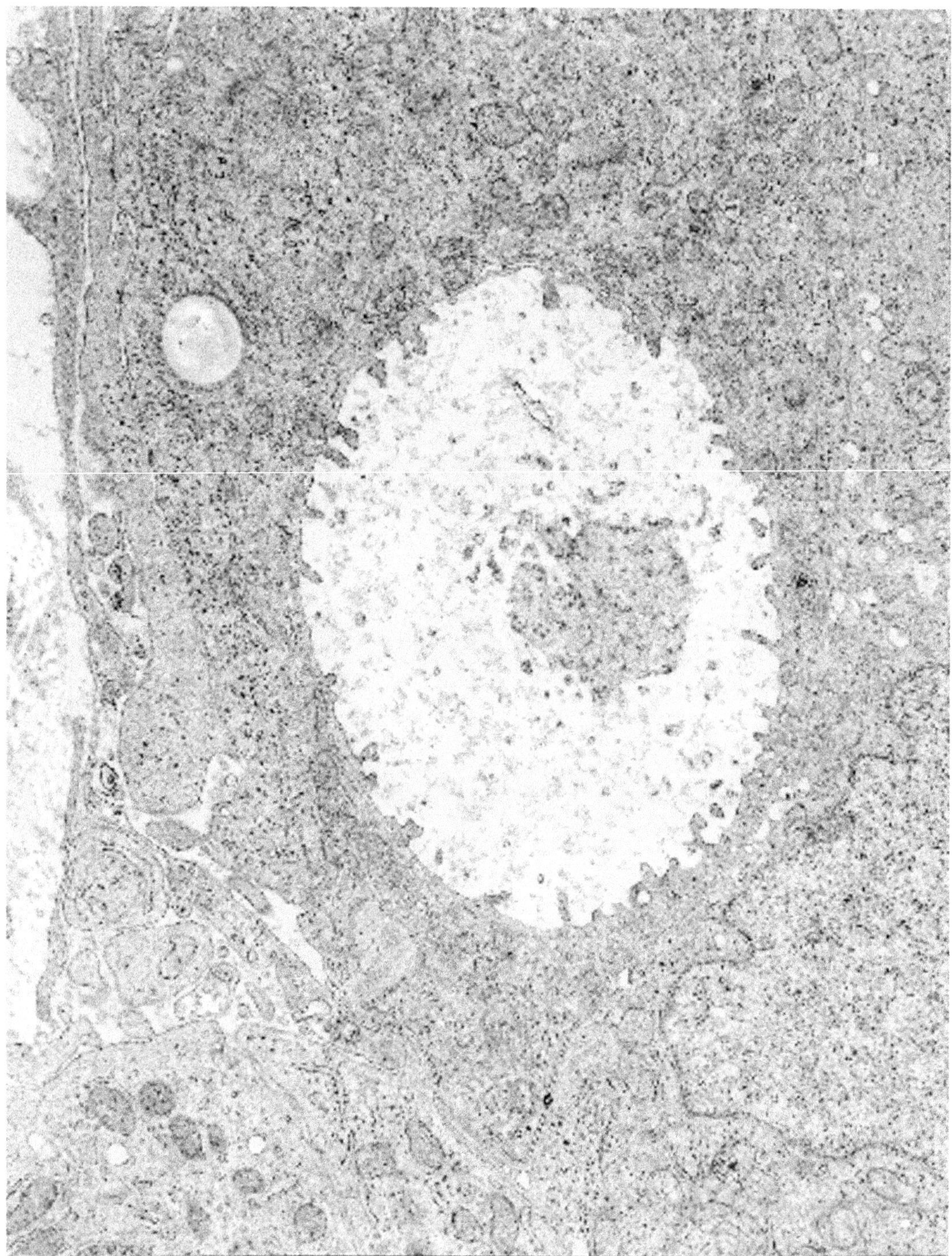

Figure 3.7. Adenocarcinoma (breast). Higher magnification of one of the cells in Figure 3.6 depicts an intracytoplasmic lumen lined by microvilli and an absence of tight junctions, which if present, would be indicative of a pseudolumen caused by invagination of the cell by extracellular matrix. (× 27,300)

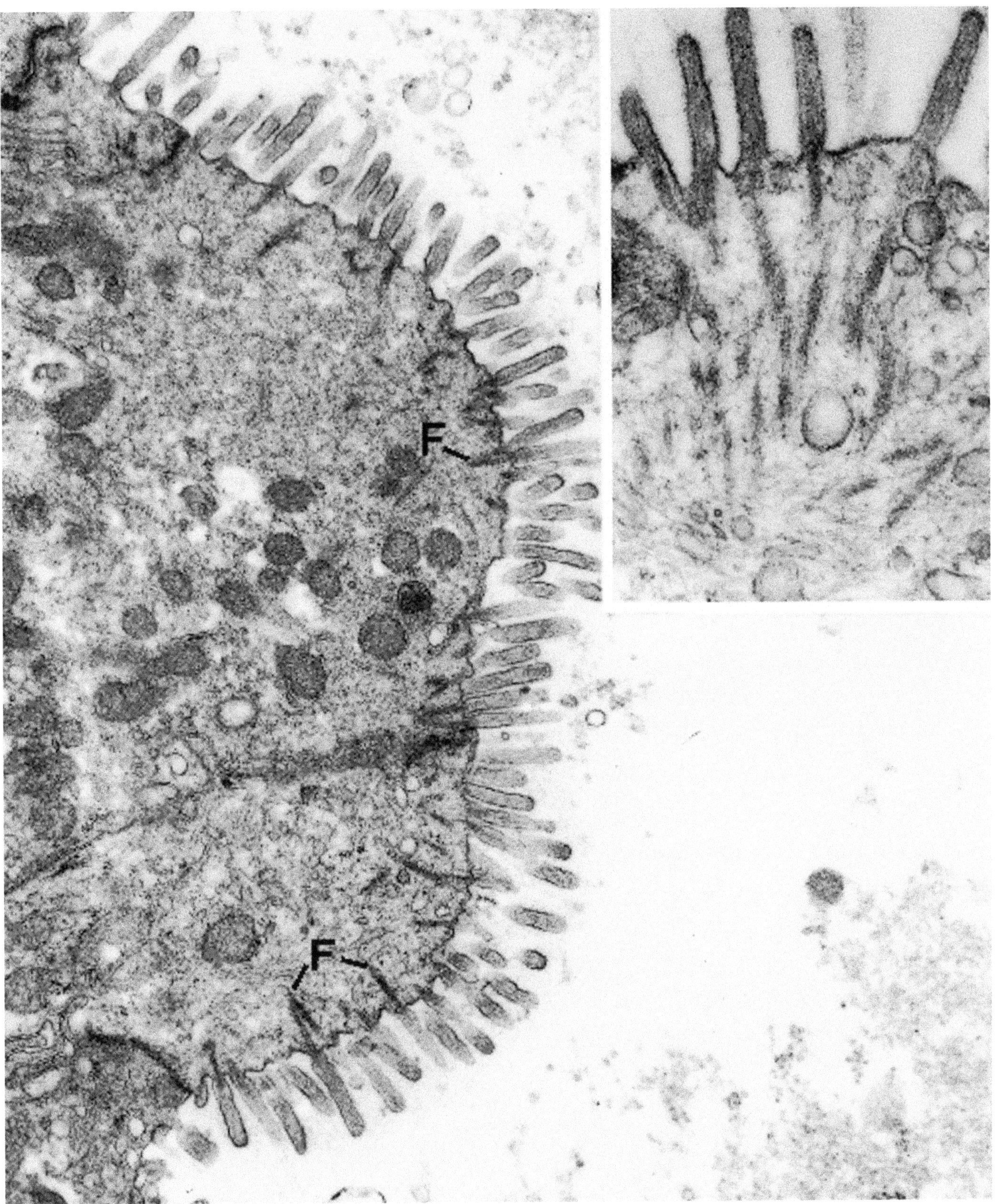

Figure 3.8. Adenocarcinoma of gastrointestinal type (ethmoid sinus). Characteristic of gastrointestinal tumors are numerous thin filaments (F) filling microvilli and extending into the subjacent cytoplasm. (× 22,300). *Inset:* higher magnification of the apical surface of a cell illustrates more clearly the filaments in the microvilli and cytoplasm. (× 47,100)

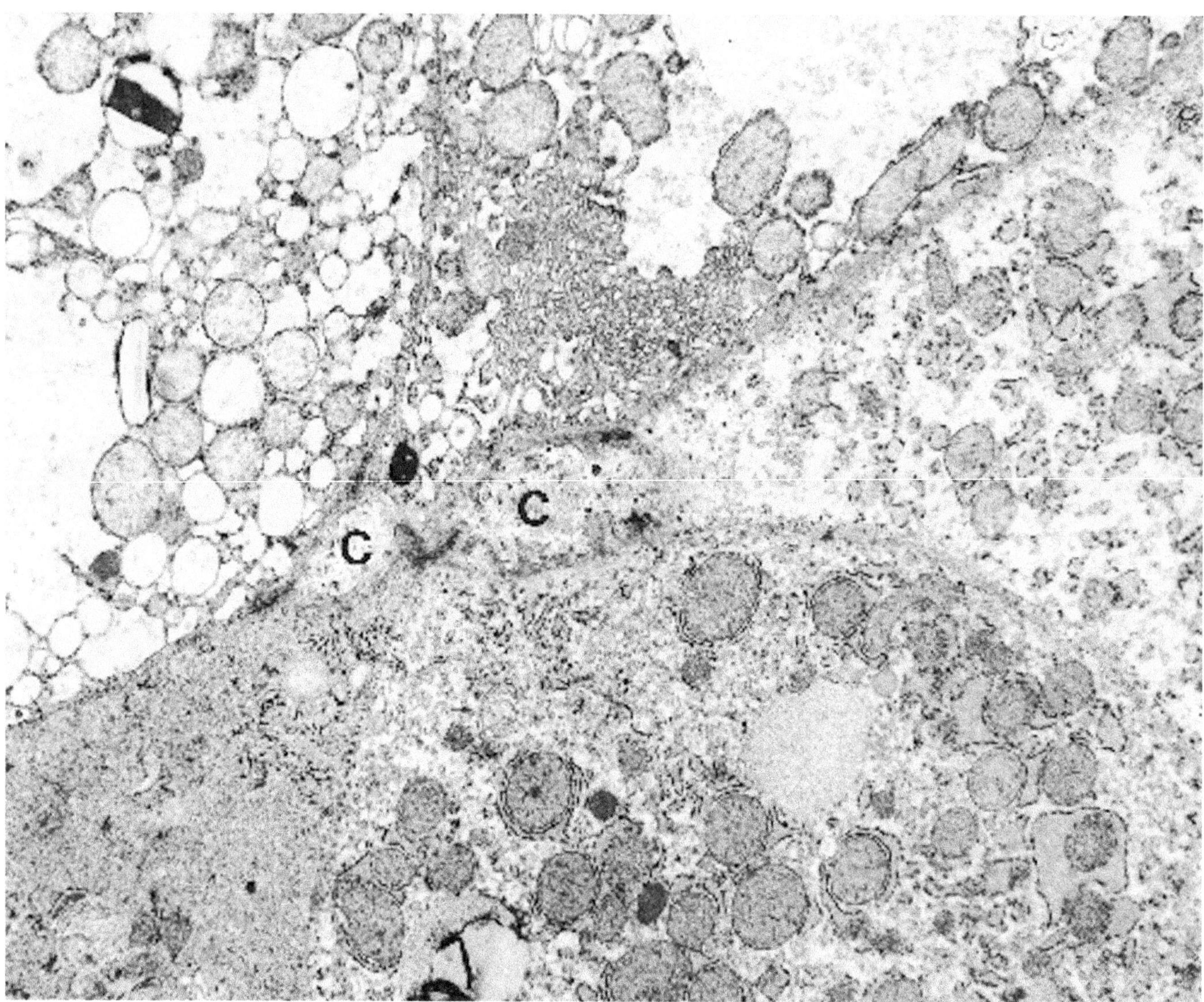

Figure 3.9. Hepatocellular carcinoma (liver). A small canaliculus (C) is identifiable among five hepatocytes. Microvilli, tight junctions and granular, membranous and vesicular luminal contents all aid in the identification of small canaliculi. (× 10,500)

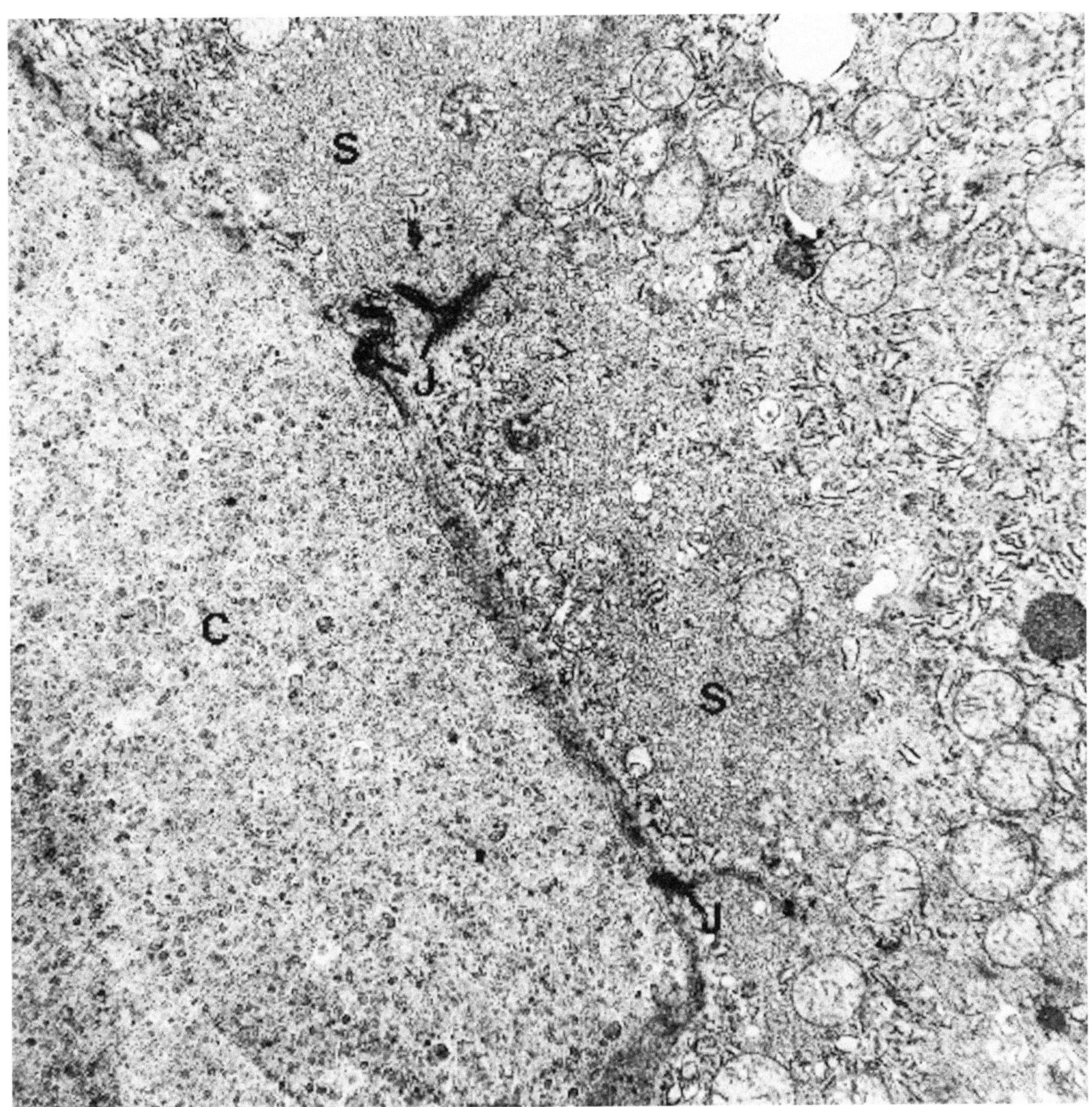

Figure 3.10. Hepatocellular carcinoma (liver). A dilated canaliculus (C) has no remaining microvilli, but tight junctions and junctional complexes (J) as well as innumerable intraluminal granules and vesicles of bile are readily identifiable. Abundant smooth endoplasmic reticulum (S), consisting of many small vesicles, occupies the apical cytoplasm of the bordering hepatocytes. (× 12,500)

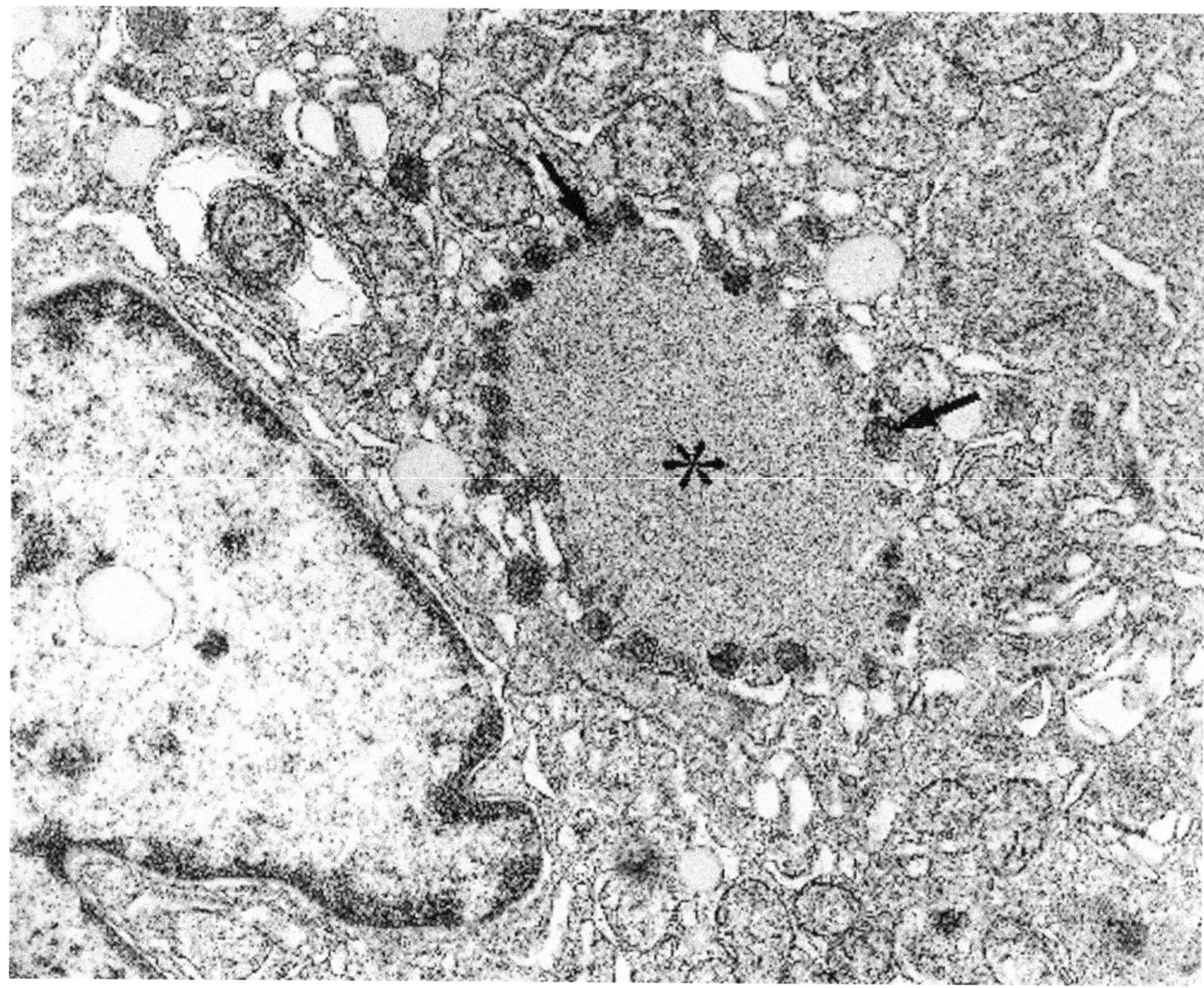

Figure 3.11. Hepatocellular carcinoma with a Mallory body (liver). This hepatocyte contains a Mallory body (*) composed of intermediate filaments and surrounded by microbodies (peroxisomes) (*arrows*). (× 20,000)

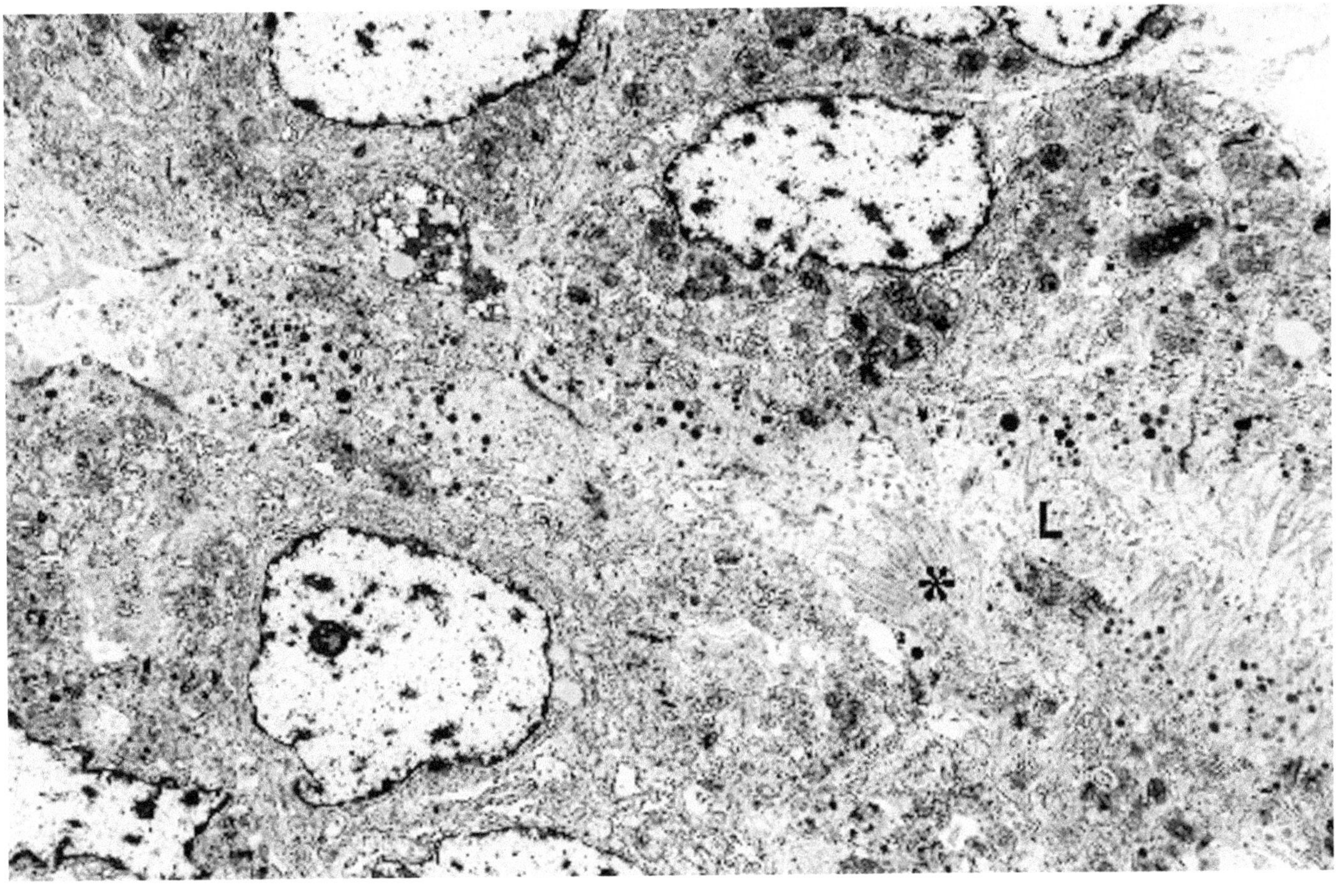

Figure 3.12. Cholangiocarcinoma (liver). Neoplastic cells surround a lumen (L), which has innumerable microvilli (*) lining it. In addition, numerous secretory granules occupy the apical cytoplasm of the luminal lining cells. (× 5700)

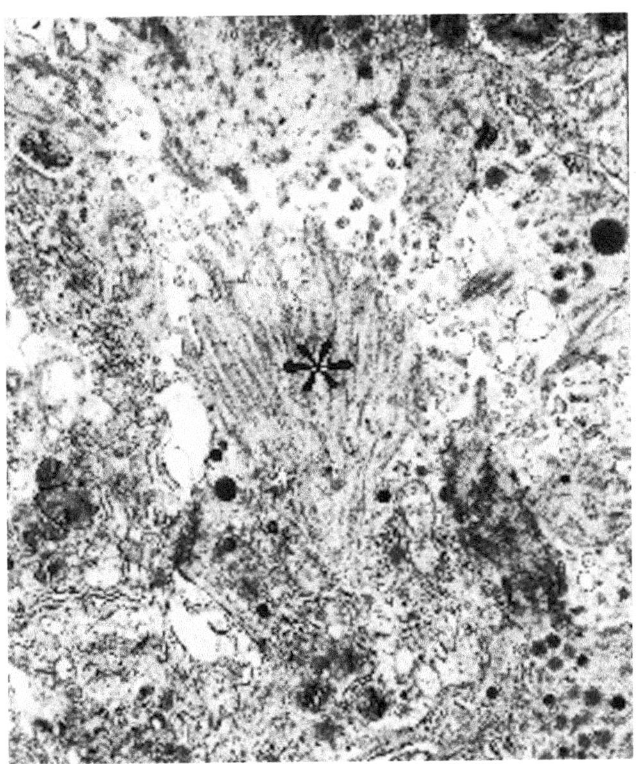

Figure 3.13. Cholangiocarcinoma (liver). Higher magnification of a portion of the lumen and lining depicted in Figure 3.12 illustrates the microvilli (*) with anchoring filaments as well as the secretory granules in the apical cytoplasm of the lining cells. (× 11,500)

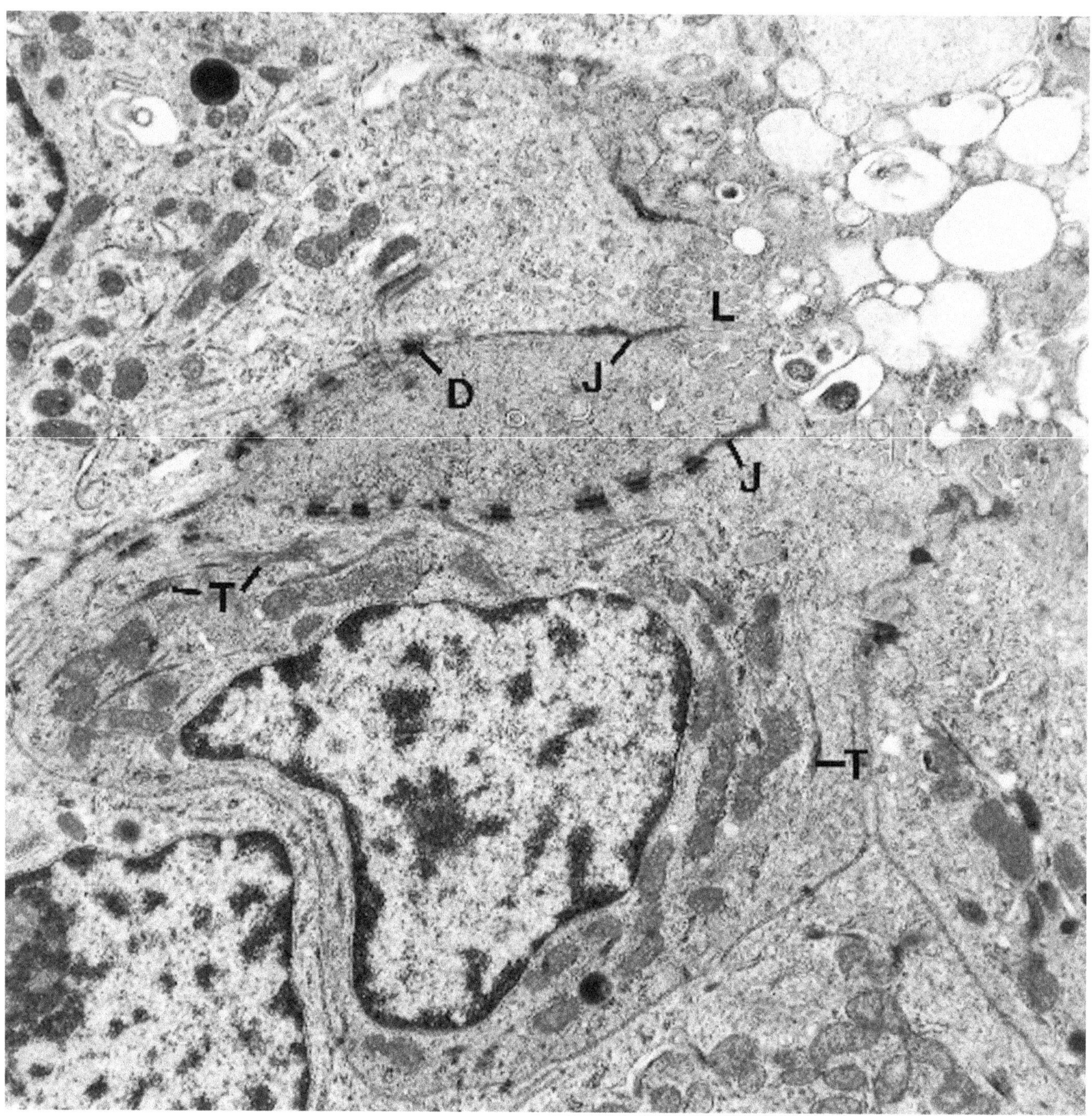

Figure 3.14. Cholangiocarcinoma (liver). In this neoplasm, lumens (L) are lined by microvilli devoid of anchoring filaments, and the cytoplasm of luminal cells contains numerous tonofibrils (T). Junctional complexes (J) and desmosomes (D) are also prominent. (× 15,400)

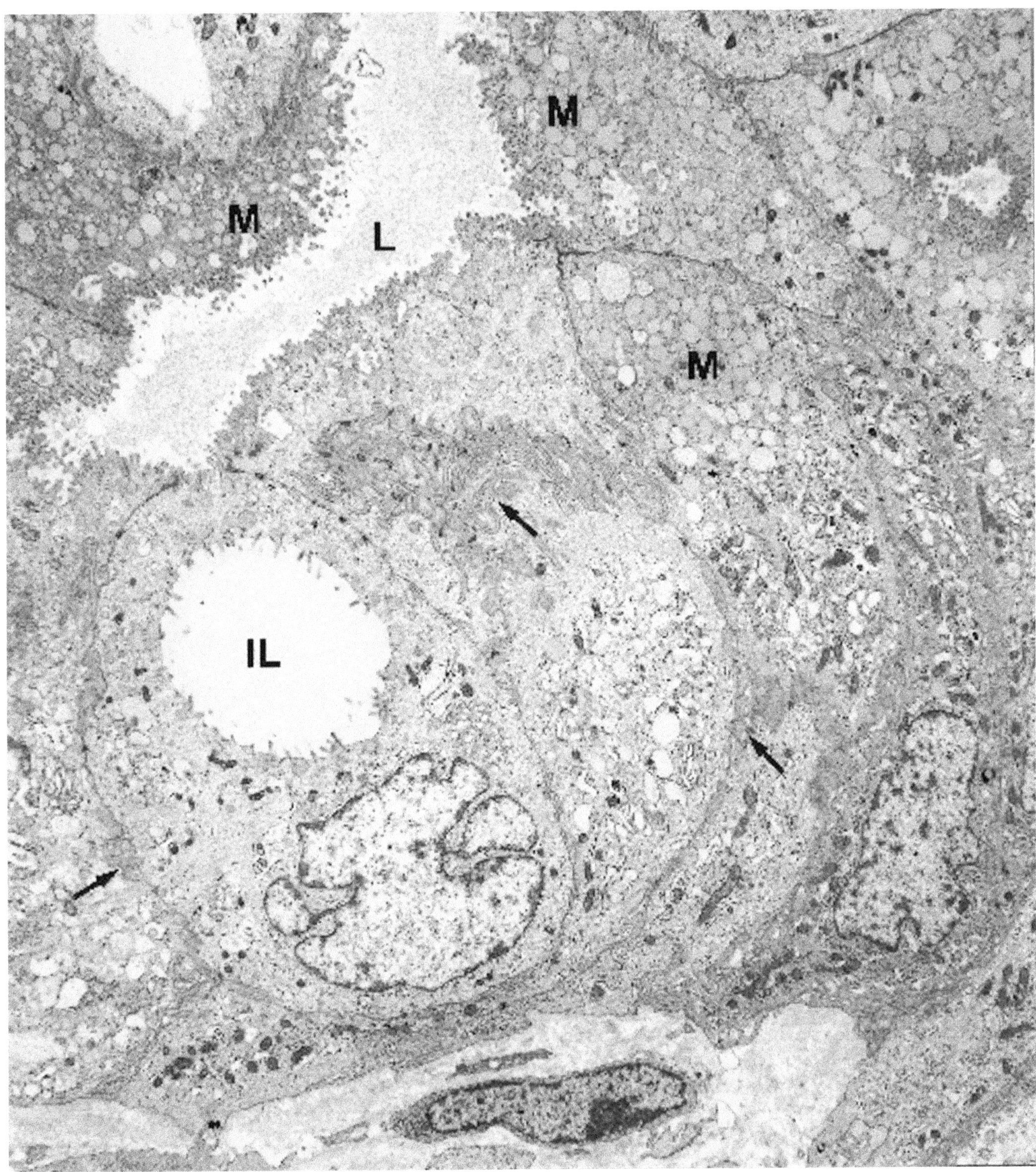

Figure 3.15. Ductal, mucinous cystadenocarcinoma (pancreas). In this field, the neoplastic cells form a cystic lumen (L) lined by innumerable microvilli. The villi have anchoring filamentous cores that are seen better in Figure 3.16. An intracytoplasmic lumen (IL), without junctional complexes, is present in one cell. Some of the cells lining the lumen have a rich collection of mucinous granules (M) in their apical cytoplasm. Lateral cell borders show a switch-backing pattern of interdigitation (*arrows*). (× 6800)

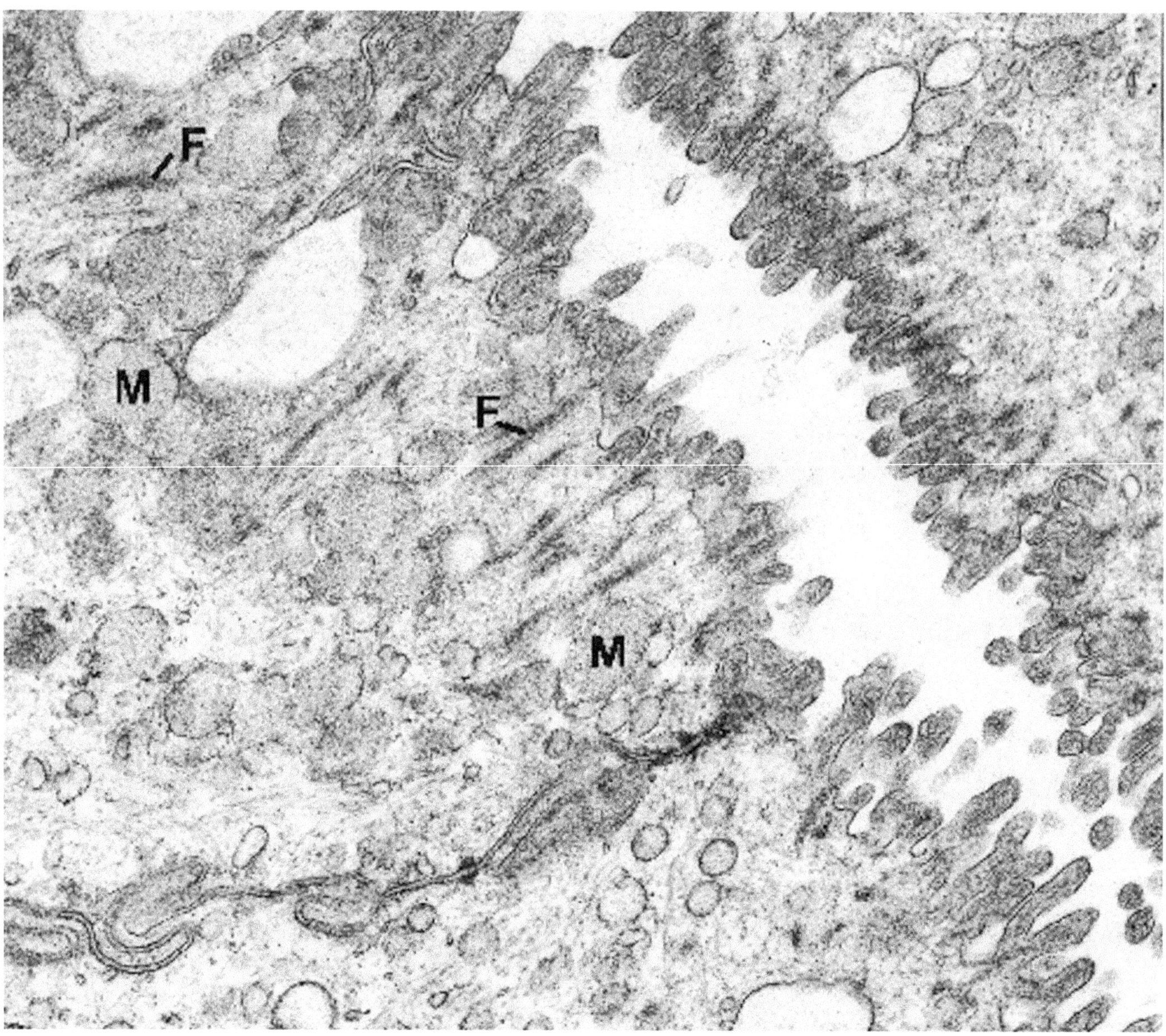

Figure 3.16. Ductal, mucinous cystadenocarcinoma (pancreas). Higher magnification of the neoplasm depicted in Figure 3.15 illustrates more clearly the microvilli with filamentous cores and rootlets (F) as well as the mucinous granules (M). (× 34,000)

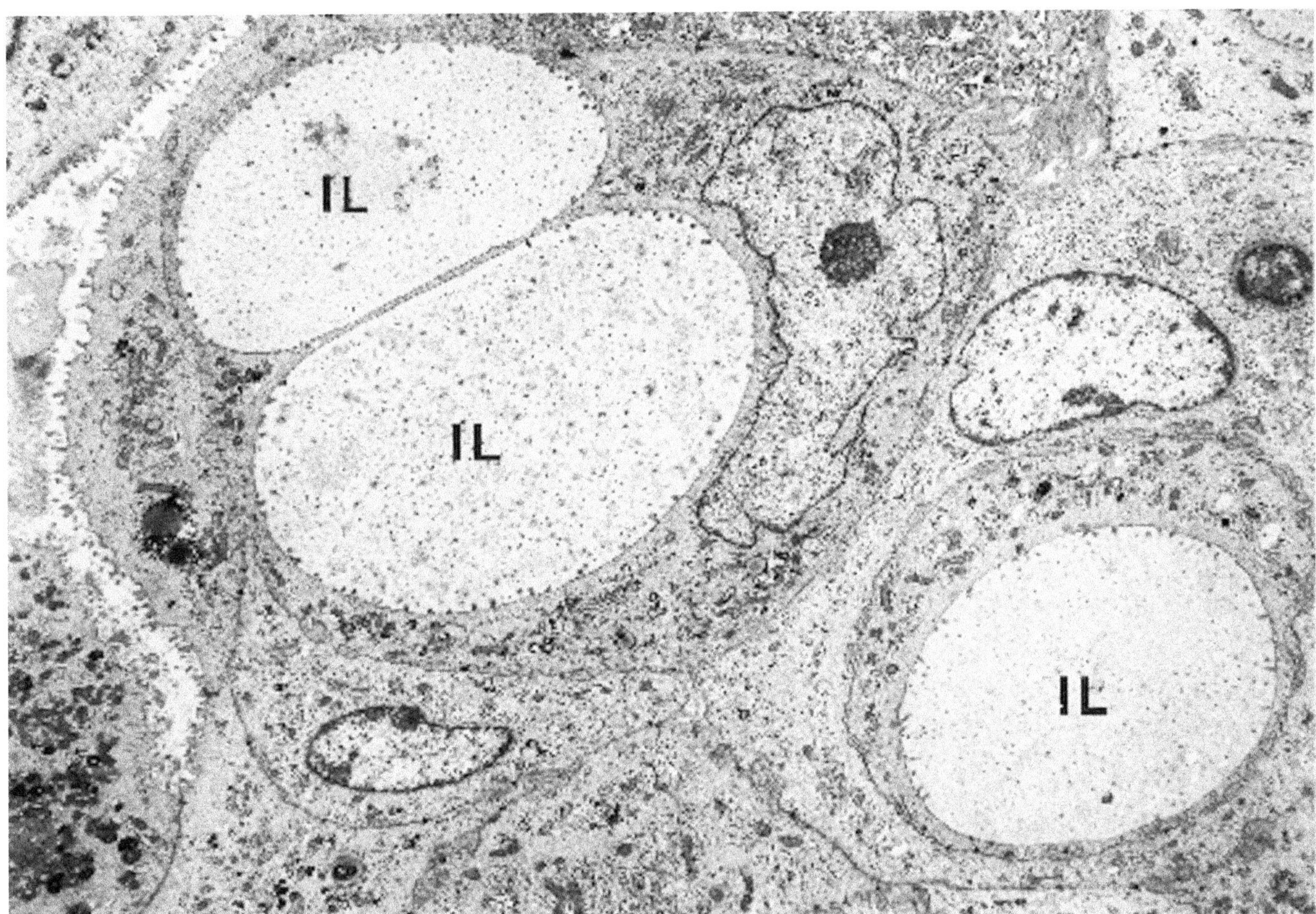

Figure 3.17. Ductal, mucinous cystadenocarcinoma (pancreas). This field is from a solid, noncystic area of the same neoplasm depicted in Figures 3.15 and 3.16. Numerous signet-ring forms that had been observed by light microscopy prove ultrastructurally to be due to true intracytoplasmic lumens (IL), devoid of junctional complexes. (× 5000)

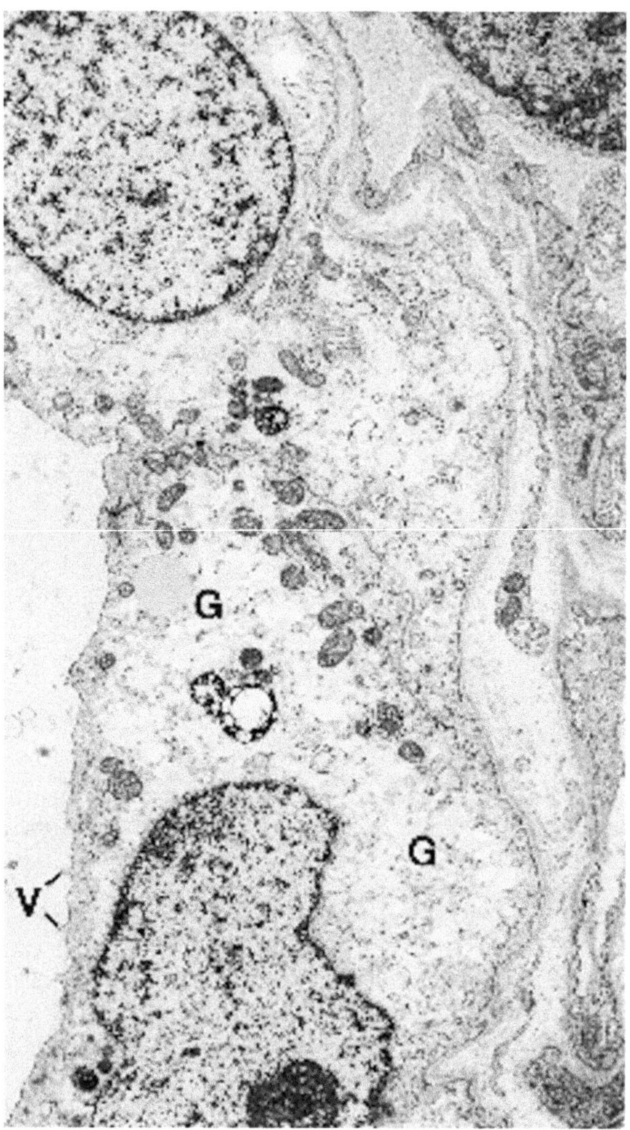

Figure 3.18. Serous microcystic cystadenoma (pancreas). Characteristic of this neoplasm are the low cuboidal, epithelial lining cells with scant, short microvilli (V), abundant cytoplasmic glycogen (G, *open spaces*) and few organelles. (× 6100)

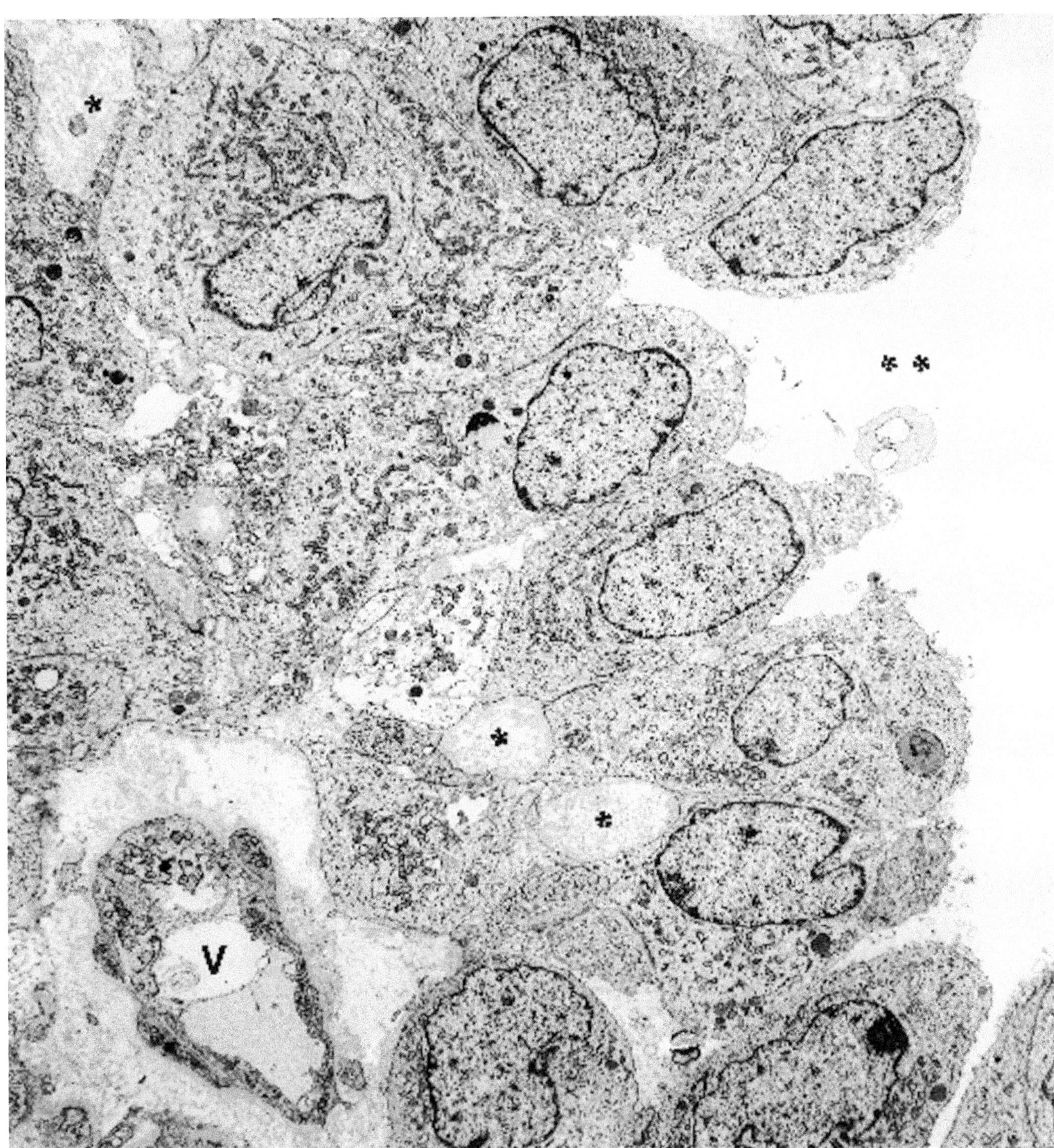

Figure 3.19. Solid and cystic (solid and papillary) tumor (pancreas). This field illustrates a pseudopapilla of neoplastic epithelial cells, with an open space on the right (**) and a blood vessel with surrounding matrix in the lower left (V). The epithelial cells are palisaded around the vessel in a pseudorosette-like fashion. The cells lining the open space do not have microvilli or tight junctions, evidence for the space being either artifact or secondary to degeneration, rather than being a true lumen. Several smaller intercellular spaces (*) are filled with a flocculent material of the same medium-density as linear basal lamina, which was identified focally in other fields and at higher magnification. (× 4400)

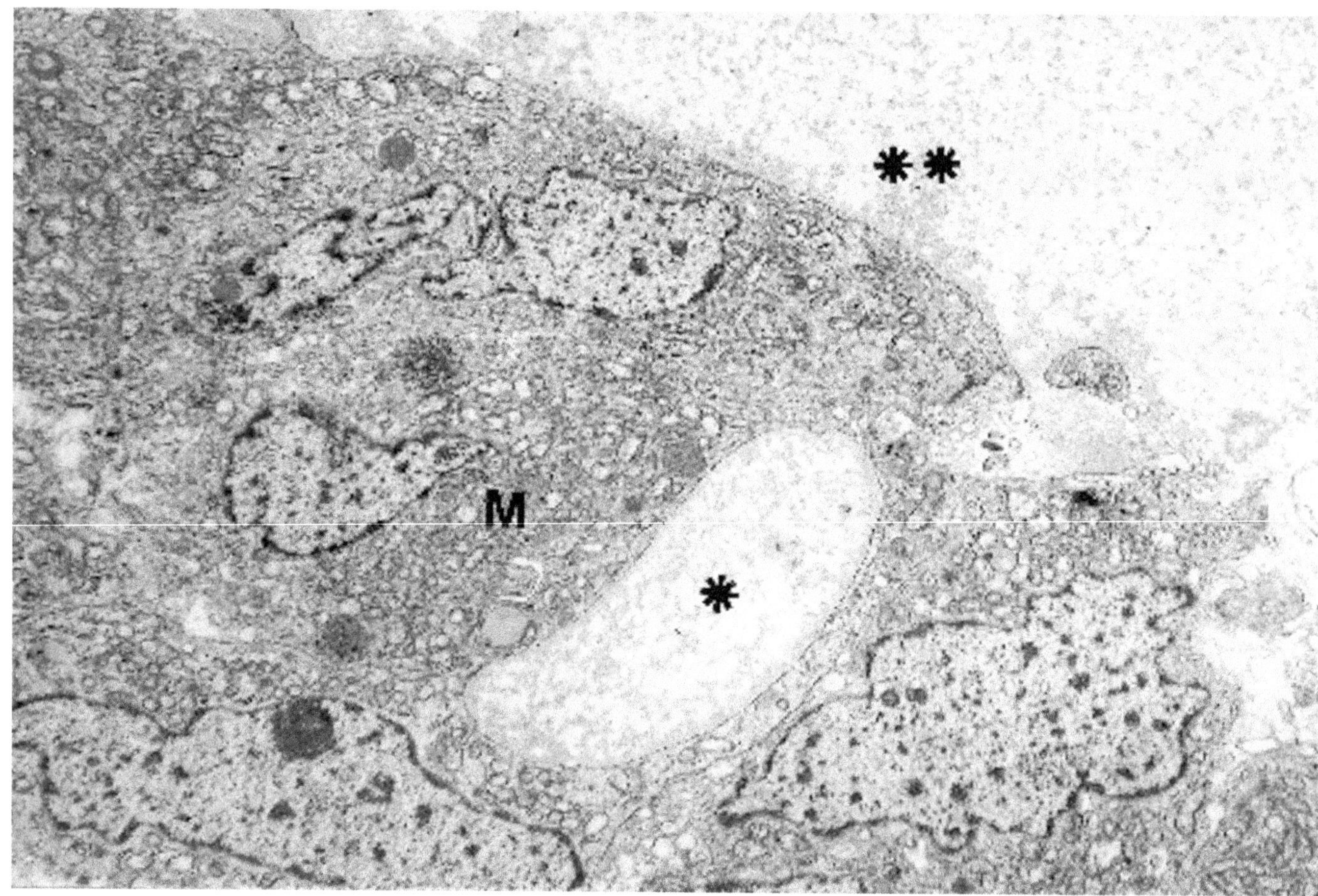

Figure 3.20. Solid and cystic (solid and papillary) carcinoma (pancreas). This region of the neoplasm is solid and microcystic, with cells being tightly apposed except for a small intercellular space (*) filled with flocculent, medium-dense material similar to the matrix (**) surrounding the solid groups. Mitochondria (M) fill the cytoplasm and represent the main organelle of the cells. (× 7500)

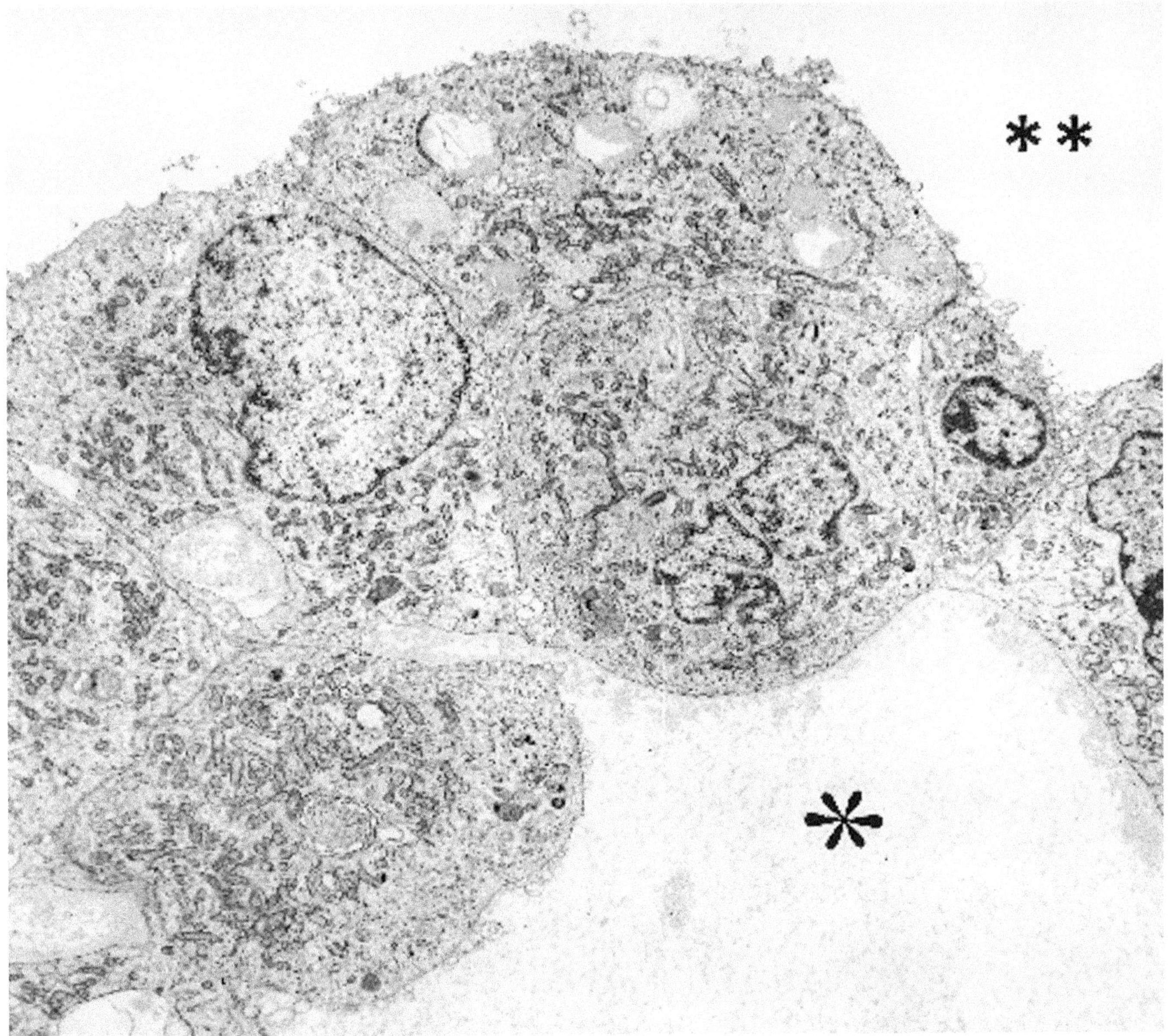

Figure 3.21. Solid and cystic (solid and papillary) carcinoma (pancreas). Higher magnification of a cluster of neoplastic cells with an intercellular cystic space filled with flocculent material (*) and an open degenerative or artefactual space (**). No well-defined junctional complexes or microvilli are present along either of the two spaces. The main organelle in the cytoplasm of the cells is mitochondria. (× 6600)

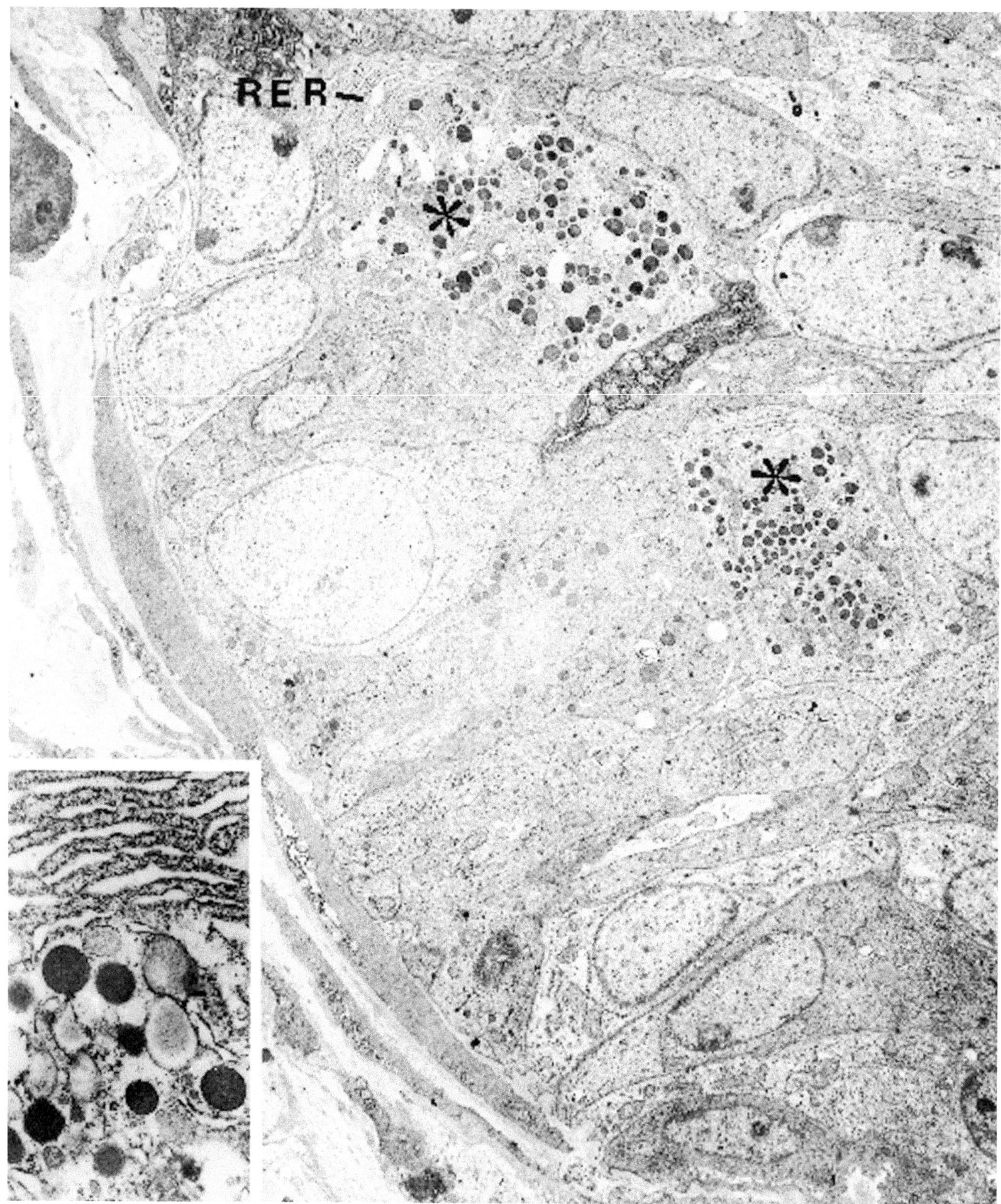

Figure 3.22. Acinar cell carcinoma (metastatic to lung from parotid gland). Among the neoplastic cells in this gland are several cells (*) having large, dense, zymogen granules. Stacked rough endoplasmic reticulum (RER), another characteristic of acinar cells, can be seen in the cell at the top of the field. The *inset* illustrates the zymogen granules and RER at higher magnification. (× 5000; *inset* × 19,000)

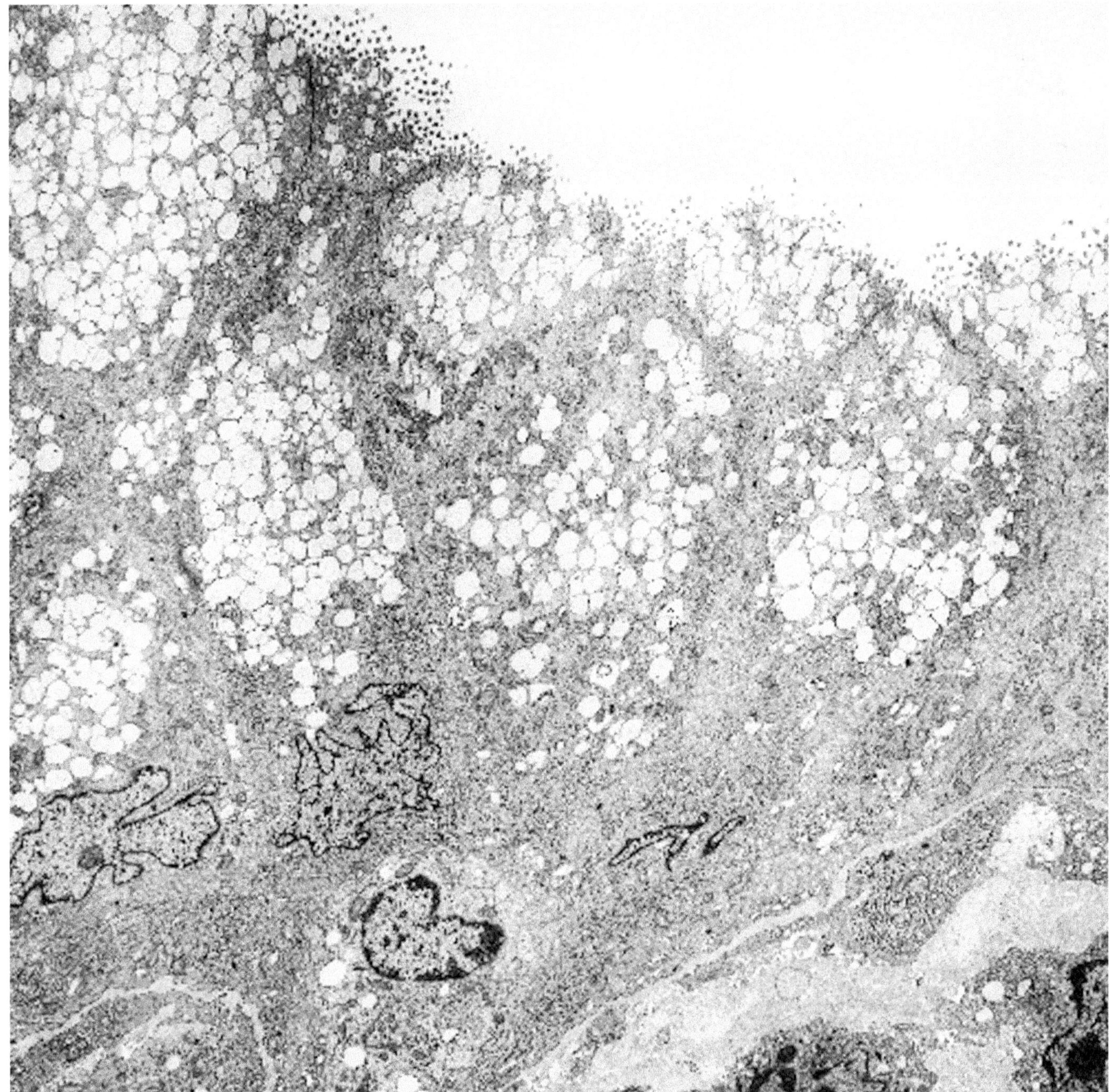

Figure 3.23. Mucinous adenocarcinoma (bronchogenic or bronchioloalveolar, lung). The malignant cells are characterized by many supranuclear, medium-dense granules of mucus. (× 5200) See higher magnification of the granules in Figure 3.24.

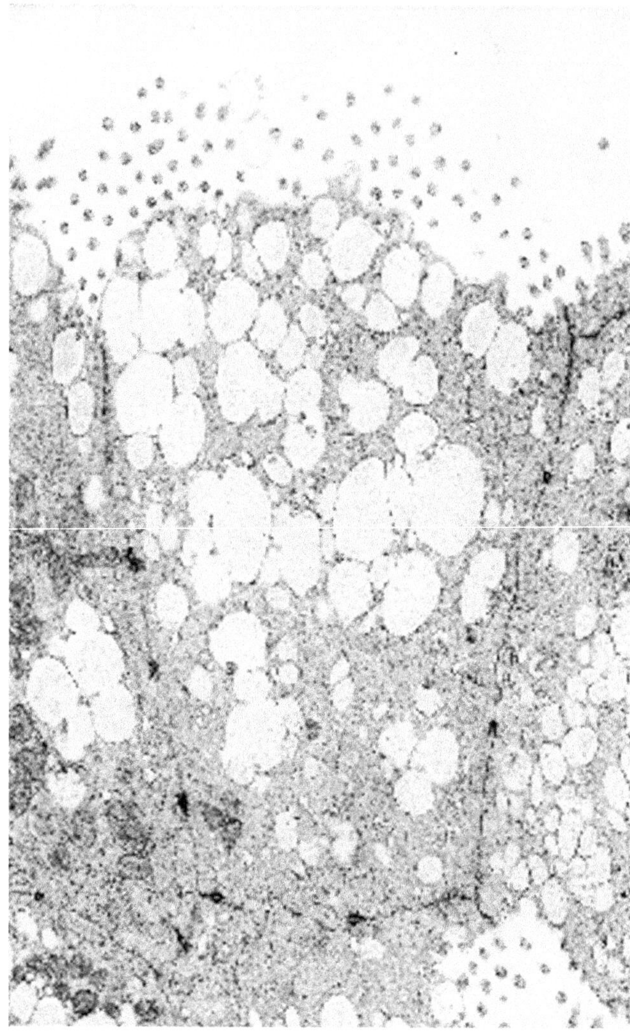

Figure 3.24. Mucinous adenocarcinoma (lung). Numerous granules of mucus fill the upper cytoplasm of the neoplastic cell. (× 12,000)

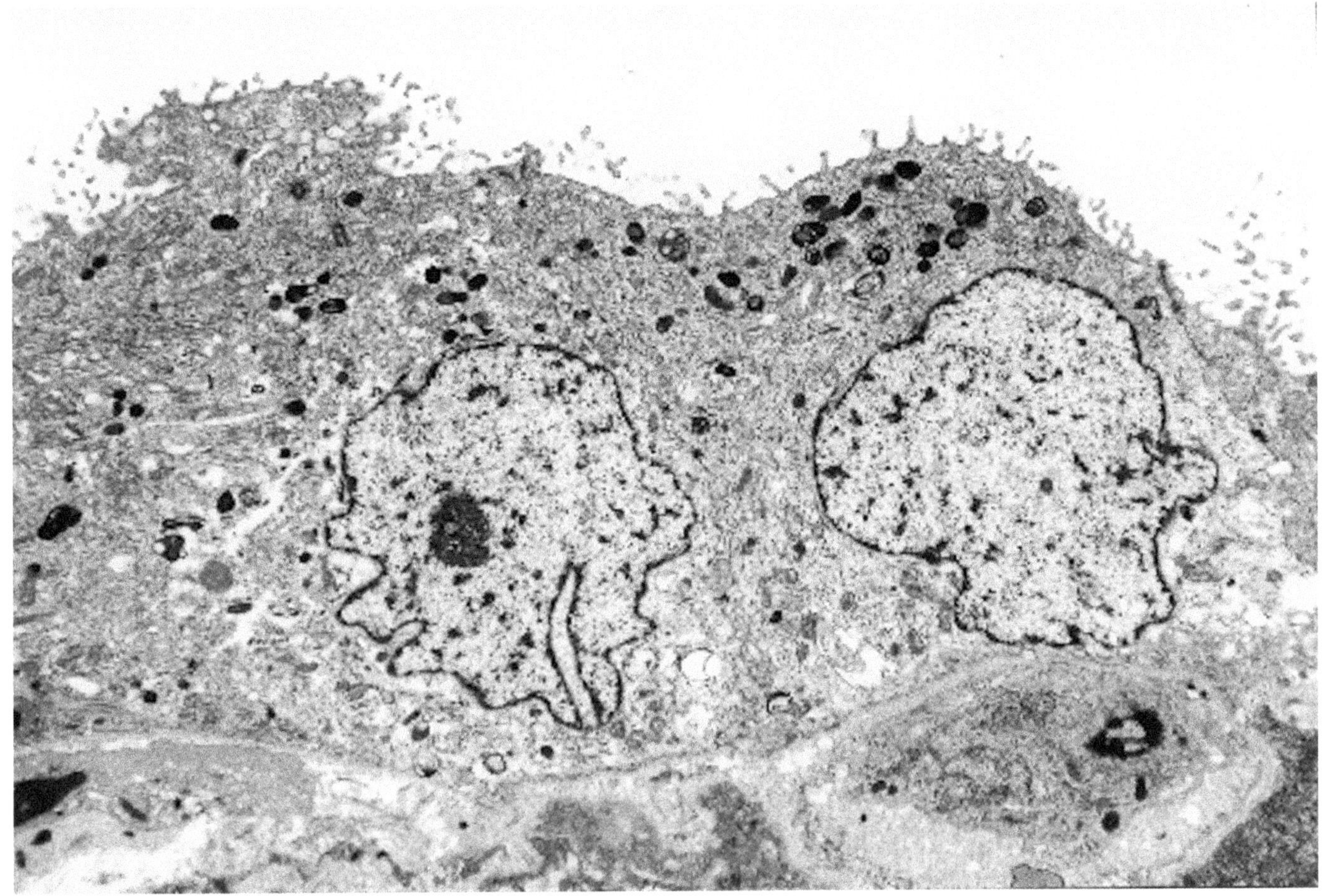

Figure 3.25. Bronchioloalveolar cell carcinoma (lung). Neoplastic Clara cells contain numerous electron-dense granules, predominantly in the apical cytoplasm. See higher magnification of another cell in Figure 3.26. (× 5800)

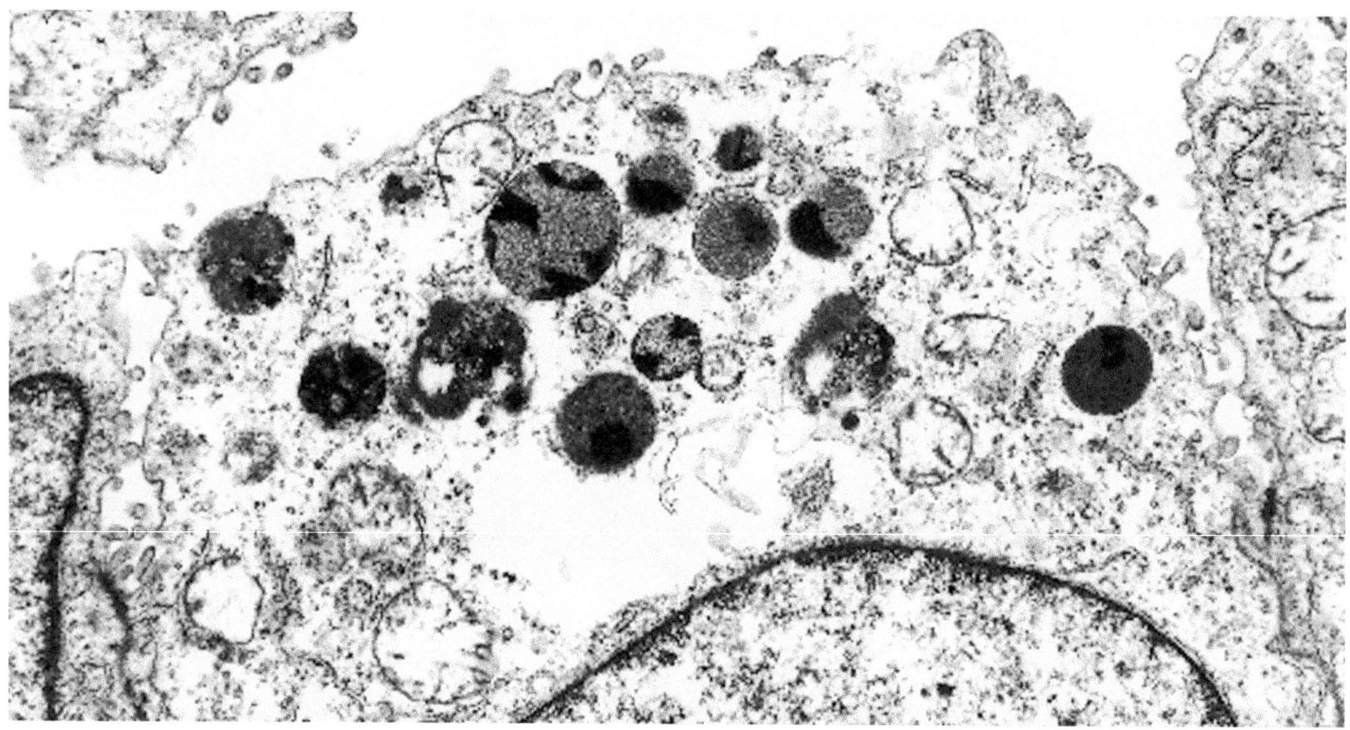

Figure 3.26. Bronchioloalveolar cell carcinoma (lung). This Clara cell exhibits characteristic membrane-bound, electron-dense granules in the apical cytoplasm. (× 14,200)

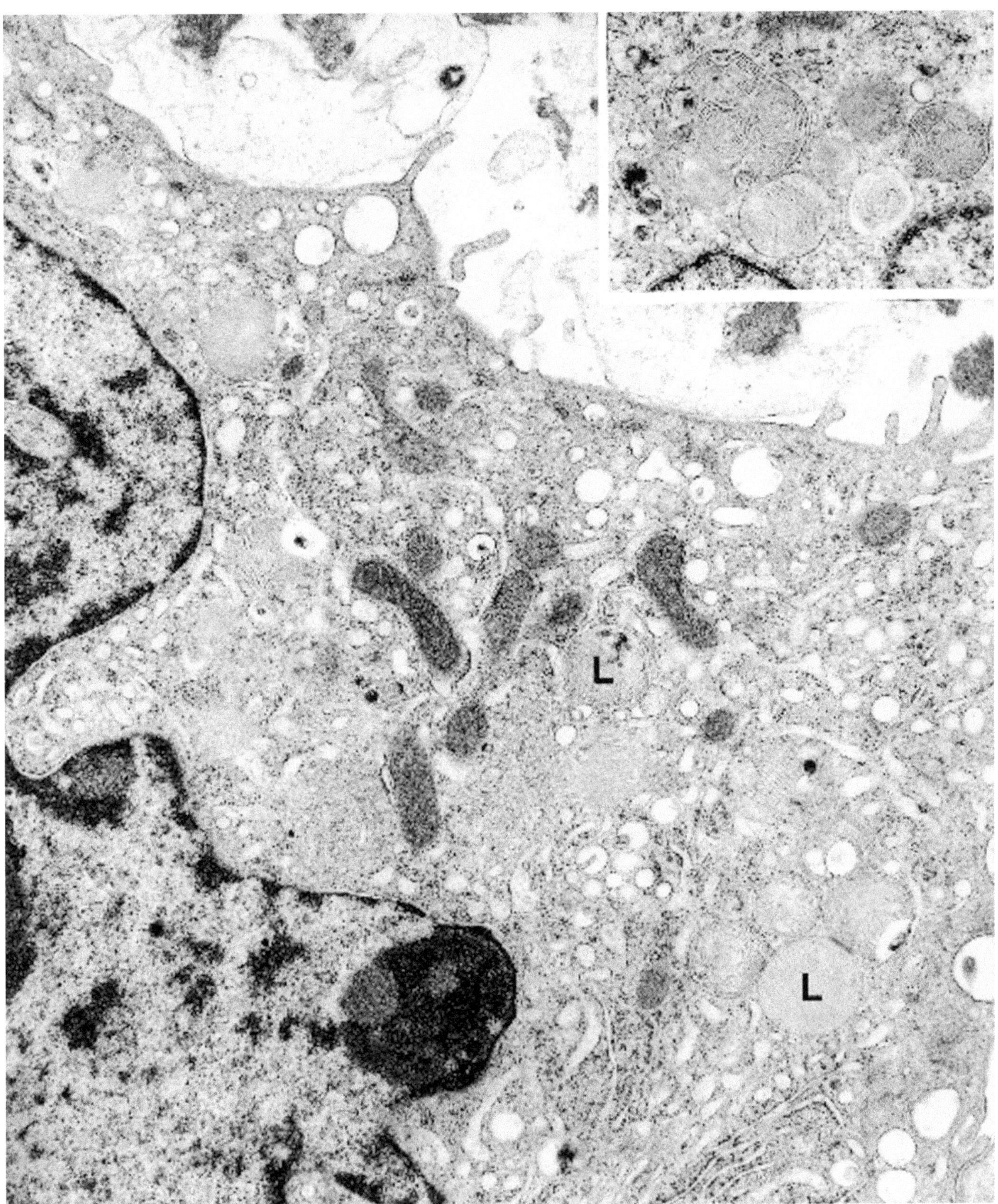

Figure 3.27. Bronchioloalveolar cell carcinoma (metastatic to a mediastinal lymph node). Characteristic of alveolar type II cells are lamellar or surfactant bodies (L) occupying the supranuclear cytoplasm of the cells. (× 22,700). *Inset:* one compound and several single lamellar bodies are prominent diagnostic features in this cell. (× 26,000)

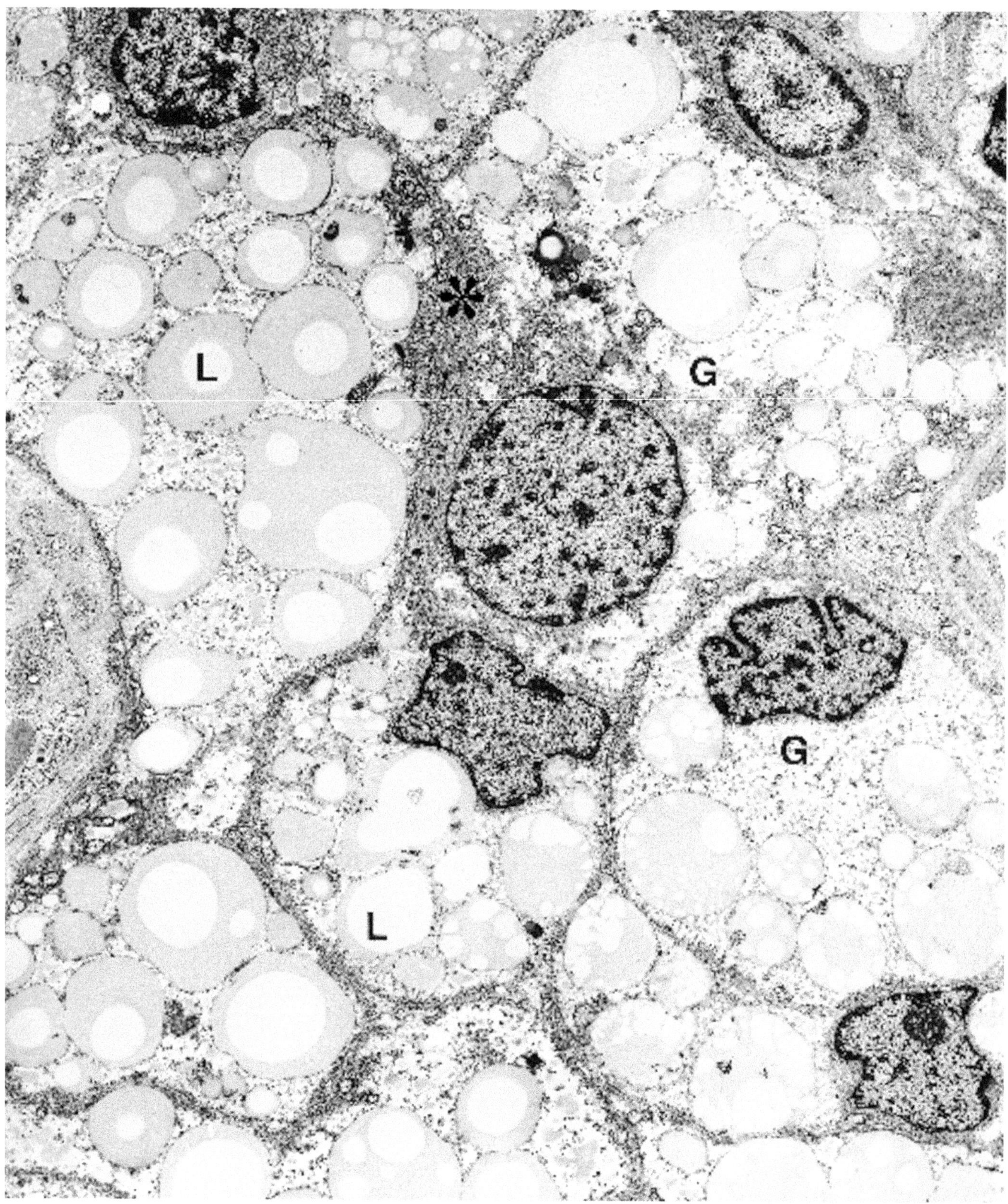

Figure 3.28. Clear cell carcinoma (kidney). The cytoplasm of the cells is rich in glycogen (G) and vacuoles of neutral lipid (L). By the method of chemical processing used, the glycogen appears as open, escalloped spaces. A small villus-lined lumen (*) is visible among several of the cells. (× 5300)

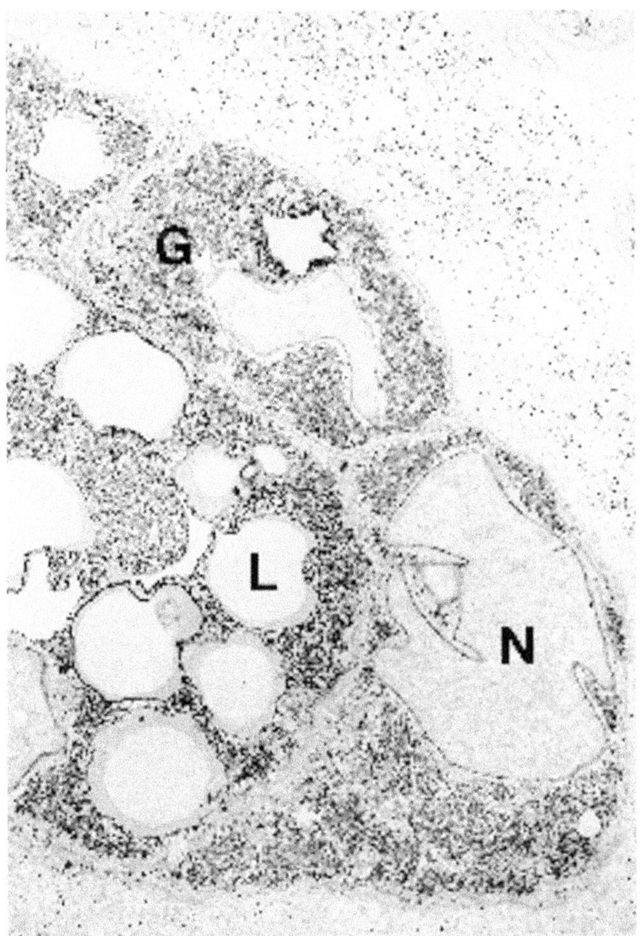

Figure 3.29. Clear cell carcinoma (kidney). By this alternative method of chemical processing, glycogen (G) has been preserved as electron-dense granules. L = lipid; N = nucleus. (× 5300)

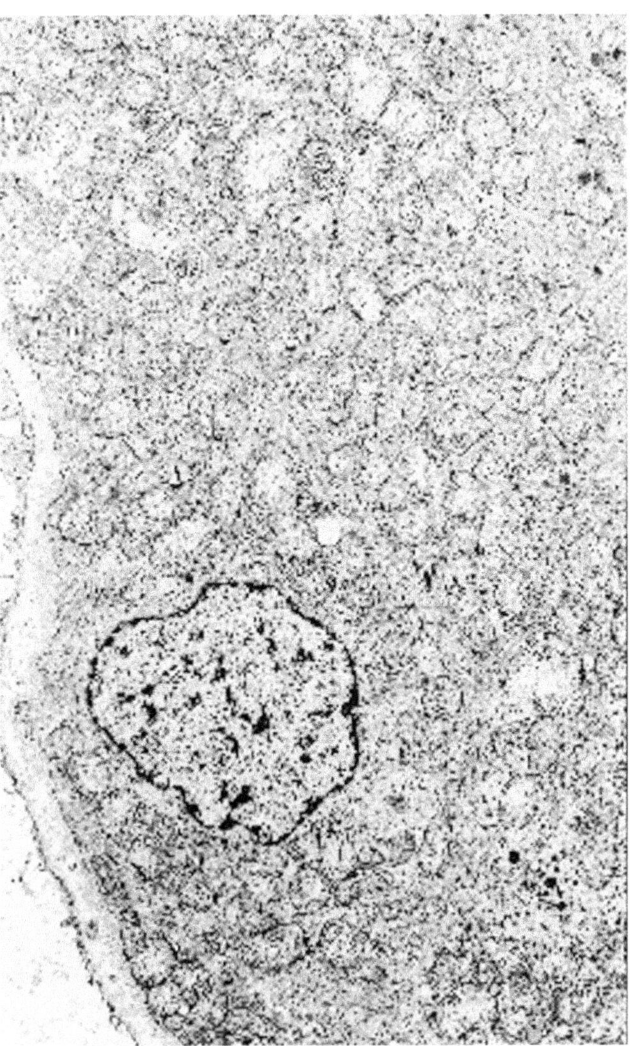

Figure 3.30. Granular cell carcinoma, reclassified as eosinophilic variant of chromophobe cell carcinoma (kidney). The cells in this neoplasm have much less glycogen and lipid than that seen in clear cell renal carcinoma, and mitochondria comprise the main constituent of the cytoplasm. (× 3600)

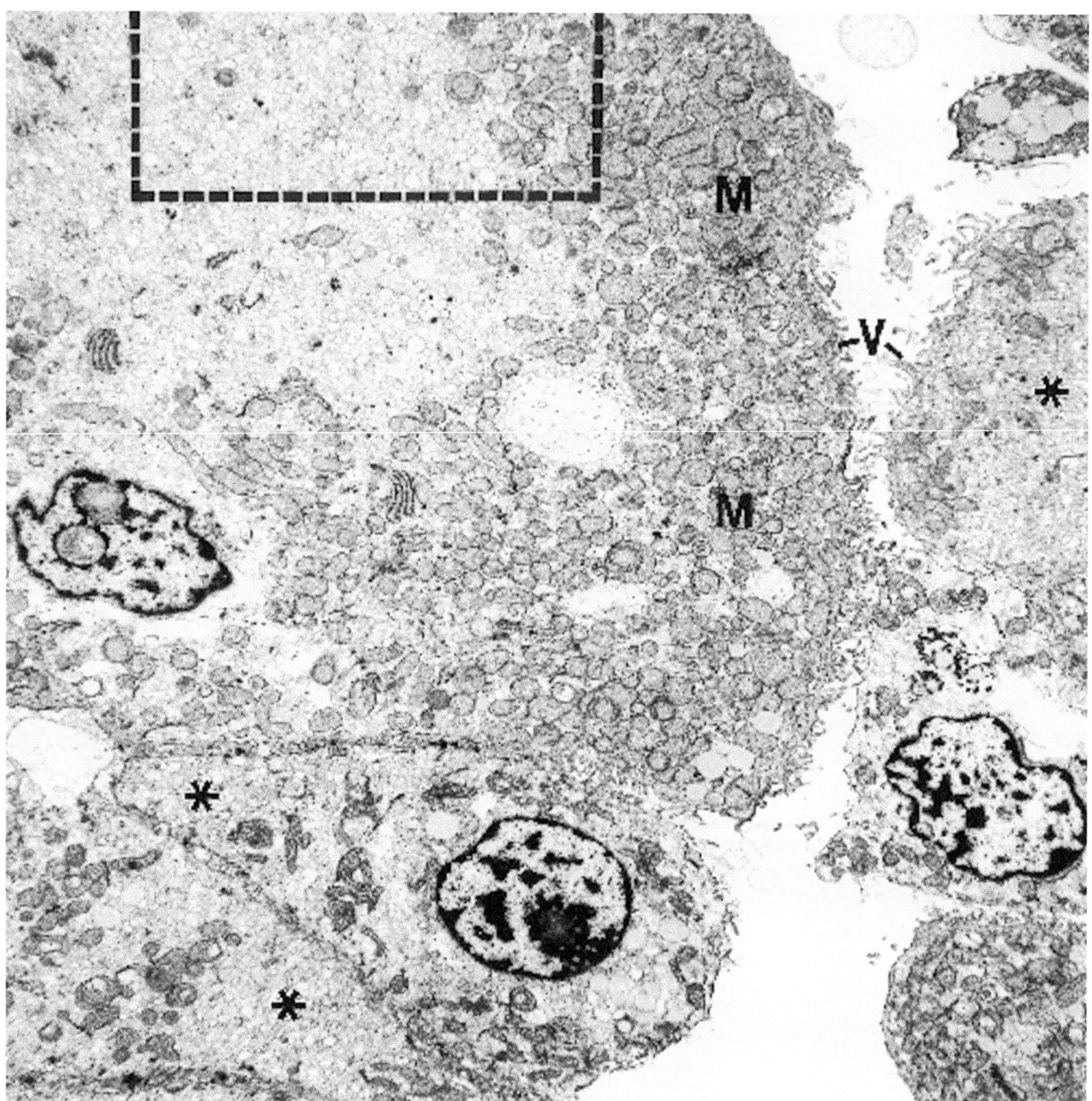

Figure 3.31. Chromophobe cell carcinoma (kidney). The main feature of the cells in this neoplasm is the innumerable small cytoplasmic vesicles (*). Numerous mitochondria (M) are also present in many of the cells. Microvilli (V) are visible along the free border of some of the cells. Higher magnification of the demarcated rectangular field is seen in Figure 3.32. (× 5600)

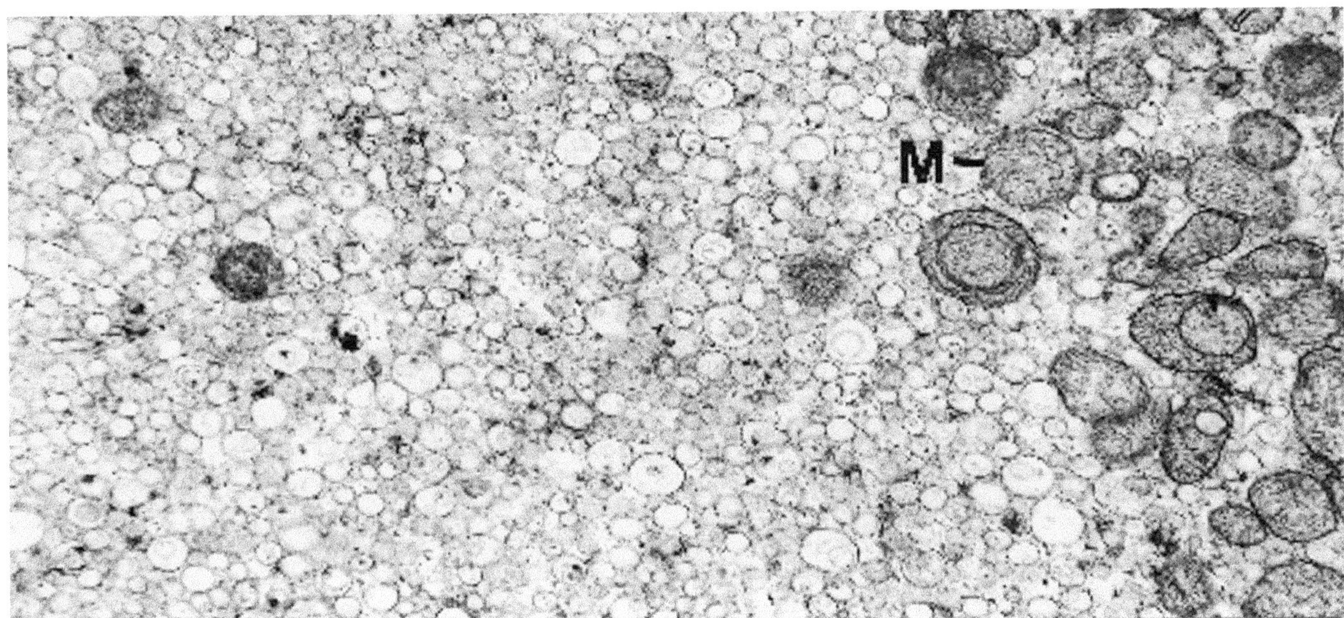

Figure 3.32. Chromophobe cell carcinoma (kidney). Higher magnification of the rectangular region in the previous photograph reveals some of the small vesicles to be clear and others to have flocculent and membranous material within them. Mitochondria (M) are moderately pleomorphic and have an abnormal number and arrangement of cristae. No definite transitional forms between small vesicles and mitochondria are apparent in this field. (× 15,900)

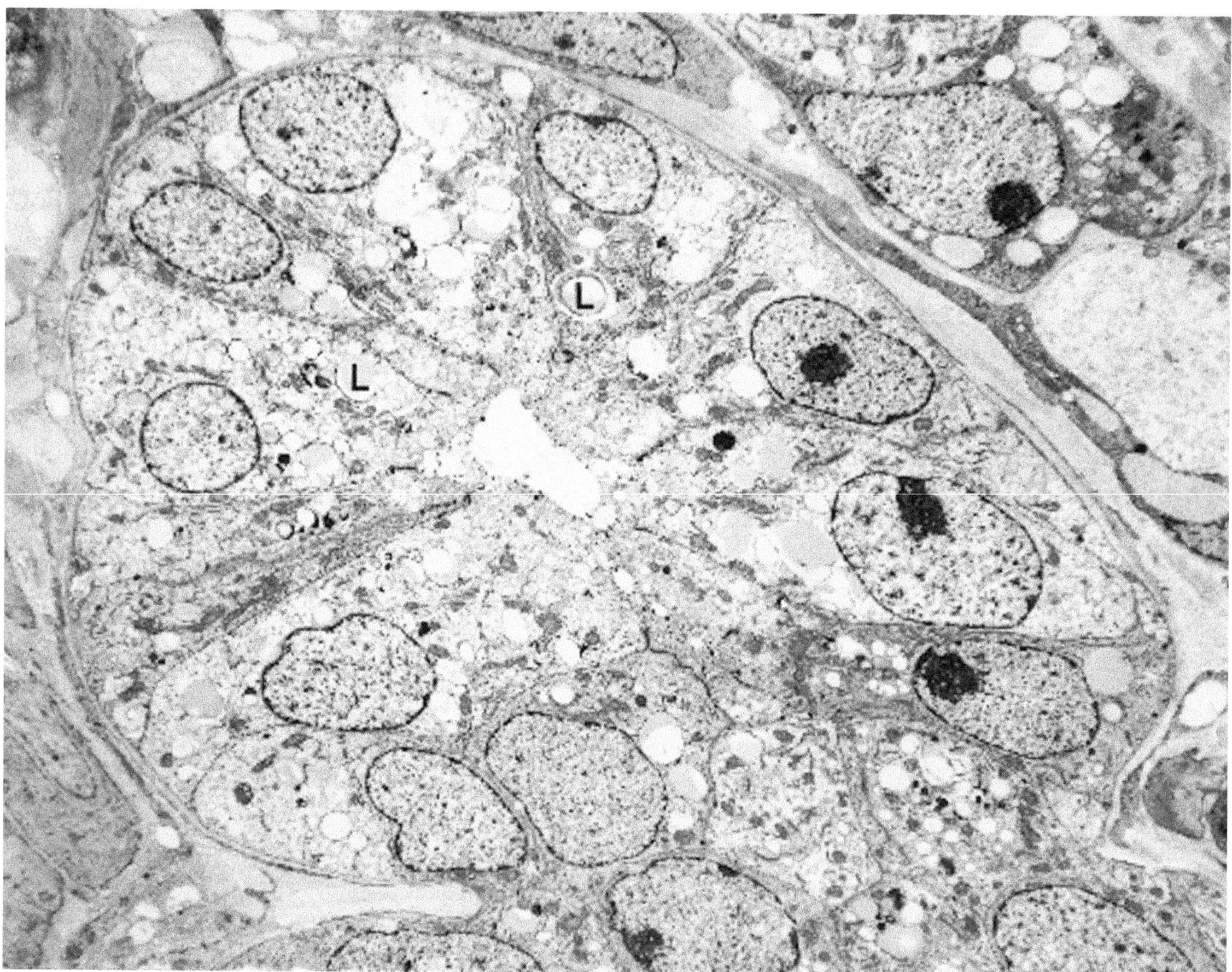

Figure 3.33. Adenocarcinoma, moderately differentiated (prostate). The epithelial lining cells in this invasive gland are single layered, without accompanying basal reserve cells. Lipid droplets (L) are visible in the cytoplasm, but secretory granules are better seen at higher magnification in Figure 3.34. (× 3500)

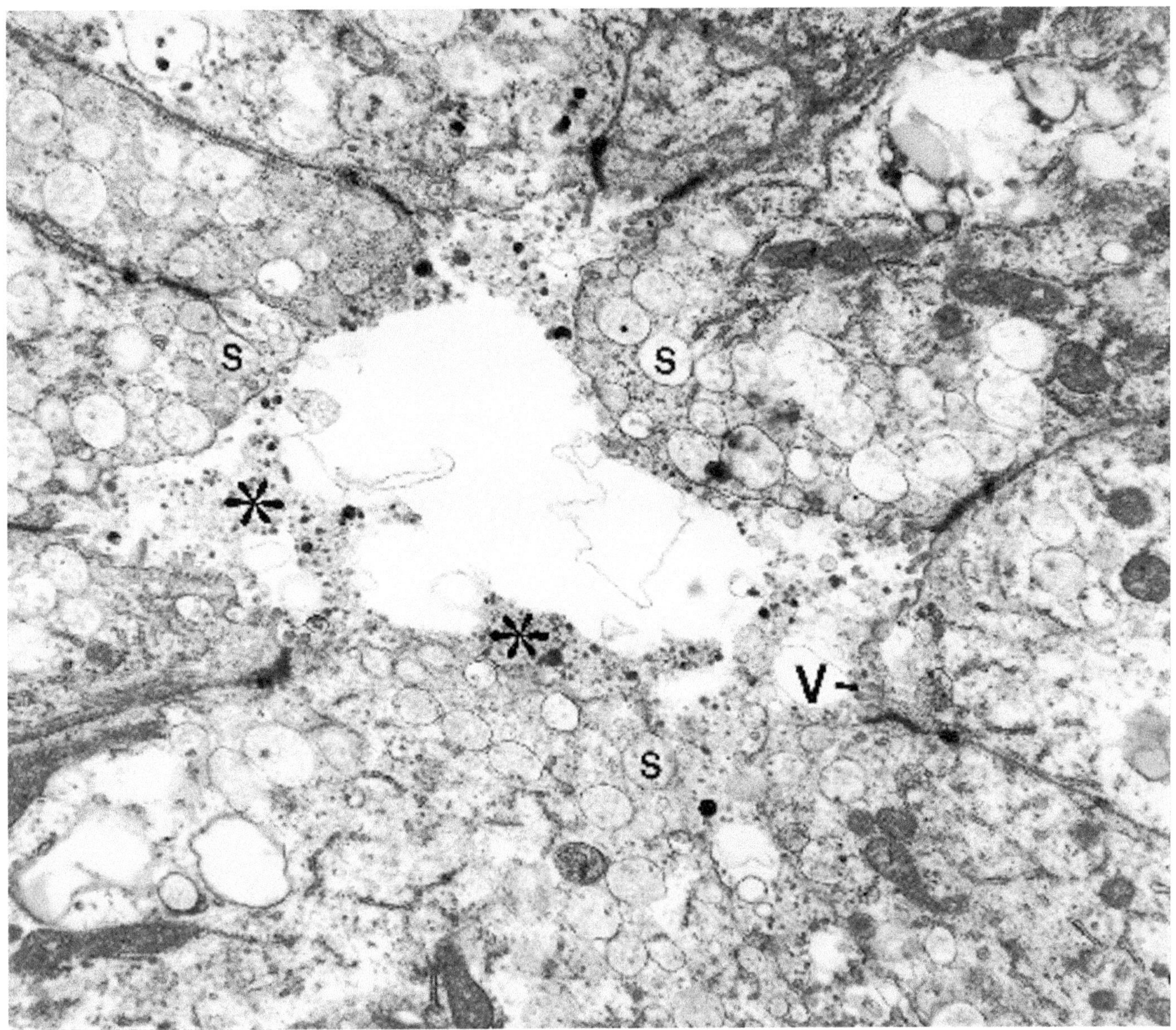

Figure 3.34. Adenocarcinoma, moderately differentiated (prostate). Innumerable secretory granules (S) occupy the apical cytoplasm of these epithelial lining cells. Particles of secretion (*) are also present in the glandular lumen. Microvilli (V) are irregularly distributed. (× 13,000)

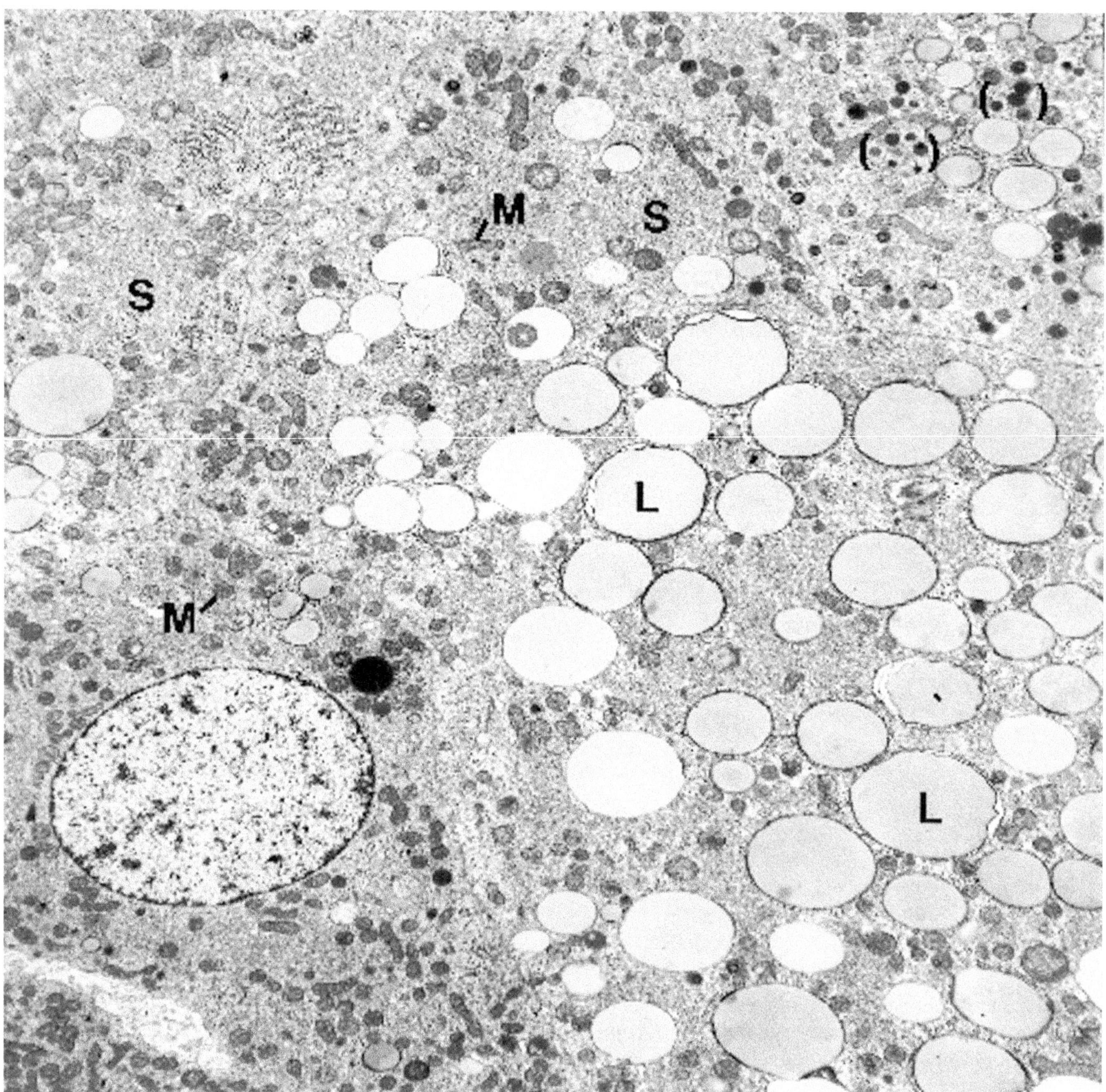

Figure 3.35. Cortical adenoma (aldosteronoma, adrenal gland). The cortical cells contain numerous organelles and lipid inclusions, some of the latter appearing to be membrane bound (L). The organelles consist of many small cisternae and vesicles of smooth endoplasmic reticulum (S), numerous small mitochondria (M), and lysosomes (*parentheses*). (× 5000)

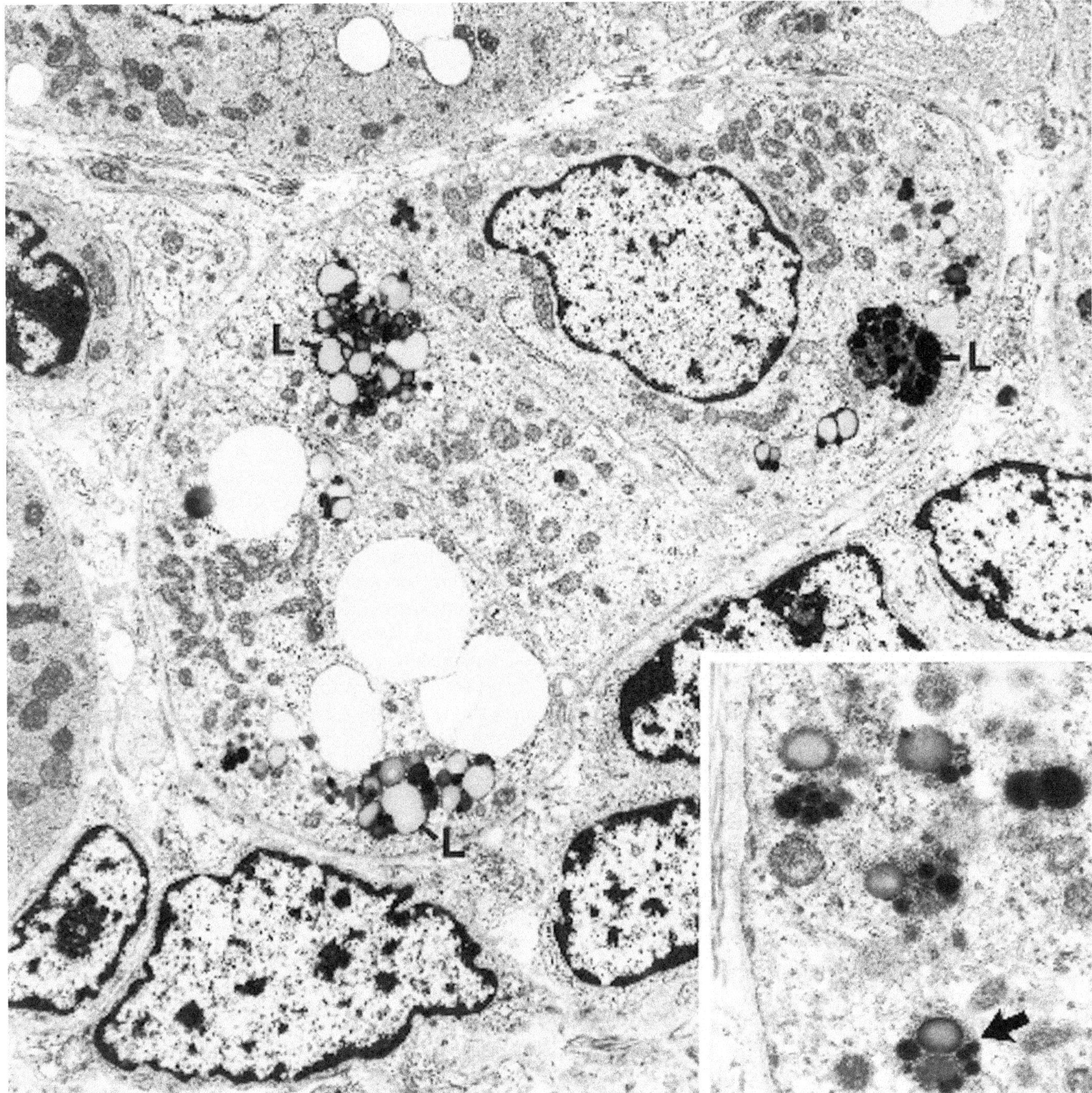

Figure 3.36. Cortical black adenoma (adrenal gland). Numerous granules of lipofuscin (L) occupy the cytoplasm of these cortical cells. (× 6800). *Inset:* higher magnification of the lipofuscin reveals a membrane focally identifiable at the periphery (*arrow*). (× 19,200)

(Text continued from page 27)

differentiation of a neoplasm. Examples of this phenomenon follow.

Thyroid follicular neoplasms have the general features of adenocarcinomas and adenomas and do not show pathognomonic structures (Figure 3.1). Thyroglobulin-containing, supranuclear granules, usually present in some of the cells lining acini (Figure 3.2), cannot be distinguished morphologically from the secretory granules of various other glandular neoplasms (Figures 3.3 through 3.5).

Breast carcinomas usually have cells with intracytoplasmic lumens, and although this feature is not specific for breast, it is probably more common there than in any other organ (Figures 3.6 and 3.7). Furthermore, intracytoplasmic lumens are found both in ductal and in lobular carcinomas. Also apropos of breast carcinomas, an outer, myoepithelial layer of cells is present in benign glands (Figure 3.3) and in *in situ* carcinoma, whereas these cells are absent in invasive and metastatic carcinomas.

Gastrointestinal carcinomas, including bile ductal and some pulmonary neoplasms (foregut derivatives), have cells with filamentous microvilli; that is, the microvilli contain cores of thin filaments, and the filaments extend as rootlets into the subjacent cytoplasm (Figure 3.8). Goblet cells and other, less frequent gastrointestinal markers consist of Paneth cells, endocrine cells, and amphicrine (combined exocrine and endocrine) cells. These gastrointestinal features may also occasionally be found in primary neoplasms arising in nongastrointestinal locations, such as *ovarian* carcinomas and *sinonasal* carcinomas (Figure 3.8). Theories of origin for primary enteric neoplasms in nonenteric sites include heterotopia, metaplasia, and pluripotential differentiation of local stem cells.

Hepatocellular carcinomas have intercellular canaliculi as an outstanding feature, and when accompanied by intraluminal and/or intracellular bile, the finding is pathognomonic (Figures 3.9 and 3.10). Bile has various forms, including homogeneous electron dense bodies and varying sized vesicles and membranous whorls (Figure 3.10). Small canaliculi are recognizable by their microvilli and tight junctions, and dilated canaliculi, by their location between cells, their contents of bile and their tight junctions; microvilli are usually attenuated or absent. Other characteristics of hepatocytes include microbodies (peroxisomes), numerous mitochondria, and abundant smooth endoplasmic reticulum (Figure 3.10). Hepatocellular carcinomas often show increased amounts of rough endoplasmic reticulum, with cisternae that are stacked or concentric and distended by electron-dense material. Mitochondria may be pleomorphic and contain crystals or large, irregularly shaped, matrical densities. Filamentous and solid cytoplasmic inclusions, corresponding respectively to Mallory hyalin (Figure 3.11) and alpha-1-antitrypsin (see Chapter 11, Figures 11.15 and 11.16), may also be present. The filaments comprising Mallory hyalin are of intermediate (10 nm) diameter and are positive immunohistochemically for keratin. The filaments are arranged in round or irregularly shaped groups but do not aggregate into tonofibrils. The groups are often surrounded by microbodies (Figure 3.11).

Bile duct carcinomas (*cholangiocarcinomas*) have a tubular, papillary, or solid pattern, and cells lining lumens have short, straight microvilli that are filled with fine filaments; the filaments anchor into the subjacent cytoplasm, similar to what is found in the stomach and intestine (Figures 3.12 and 3.13). Microvilli of this type may be absent in poorly differentiated cholangiocarcinomas. Secretory granules may occupy the apical cytoplasm of the cells, and tonofibrils may or may not be present (Figures 3.12 through 3.14).

Pancreatic exocrine carcinomas may arise from acinar cells, centroacinar cells, intercalated duct cells, intralobular duct cells, interlobular duct cells, and main pancreatic duct cells. Adenocarcinomas from main and interlobular ducts (*mucinous cystadenocarcinomas*) have cells similar to those of bile ducts and intestinal epithelium; that is, the cytoplasm contains mucin granules, and the free surface has microvilli filled with thin filaments that anchor into the subjacent cytoplasm (Figures 3.15 and 3.16). In addition, neuroendocrine cells may be present. Less differentiated large-duct carcinomas and those arising from smaller (intralobular and intercalated) ducts do not show intestinal features. Solid, noncystic areas with signet-ring cells may be present in some of these neoplasms (Figure 3.17). Glycogen-rich, *microcystic adenomas* (*serous cystadenomas*) are composed of varying sized cysts lined by cuboidal or flat epithelial cells. Most of these cells contain abundant glycogen and few other organelles (Figure 3.18), but some cells contain abundant endoplasmic reticulum. Microvilli are sparse, short, and without intestinal-type anchoring filaments. The cells resemble intercalated duct or centroacinar cells of normal pancreas. *Solid and cystic* (*solid and papillary*) *tumors* are composed of solid sheets of cells, with focal, degenerative, cystic spaces and pseudopapillae (Figures 3.19 and 3.20). There may also be smaller intercellular spaces that invaginate some of the cells, forming pseudolumens (Figures 3.19 through 3.21). True lumens lined by microvilli and tight junctions are rarely present. The cells of solid and cystic tumors are usually poorly differentiated and are probably derived from uncommitted or pluripotential cells from centroacinar and/or terminal duct epithelium. However, focal acinar cell and neuroendocrine cell differentiation, recognizable by zymogen granules and dense-core granules, respectively, may be found in some of these neoplasms. In addition, many cells are oncocytic, having numerous mitochondria (Figures 3.20 and 3.21). *Acinar (acinic) cell carcinomas* of pancreas and salivary glands are composed partially of cells having large, zymogen, or serous granules, 125–1500 nm in diameter (Figure 3.22). More pleomorphic, sometimes larger, filamentous granules, similar to those in the 12–20-week-old fetal pancreas, may also be present, and abundant stacked cisternae of rough endoplasmic reticulum are common. Other components of acinar cell carcinomas are undifferentiated stem cells and intercalated ductal cells, neither of which has exclusive ultrastructural features.

Pulmonary adenocarcinomas may be *bronchogenic* or from cells lining bronchioles and alveoli. Those arising from bronchial glands are usually mucinous, and those originating from *bronchioloalveolar* cells may be composed of any combination of three cell-types—mucus-secreting cells, Clara cells, and type II alveolar lining cells. Mucus-secreting cells are readily identifiable by characteristic granules of variable, but usually medium, density in the supranuclear cytoplasm (Figures 3.23 and 3.24). Clara cells contain varying numbers of electron-dense granules, predominantly in the supranuclear cytoplasm (Figures 3.25 and 3.26). Type II alveolar lining cells contain diagnostic lamellar (surfactant) bodies, also in the supranuclear cytoplasm (Figure 3.27).

Renal adenocarcinoma most often ultrastructurally recapitulates to varying degrees the morphology of normal proximal tubules, including long microvilli on the apical or free surface of the cell, glycocalyx in intervillous crypts, and numerous pinocytotic vesicles and vacuoles in the apical cytoplasm. Other features that are found both in normal proximal and distal tubules are infoldings of the basal plasmalemma, interdigitations of the lateral plasmalemmas, basal lamina, numerous mitochondria, varying amounts of glycogen and lipid, microbodies, and lysosomes. In renal cell carcinoma, microvilli may be present only focally and on a few cells. Extracellular or intracellular lumens may also be present. Blood vessels in renal cell carcinomas typically have fenestrated endothelium, similar to normal peritubular capillaries. In the *clear cell* form of these neoplasms, the cytoplasm is rich in neutral lipid and glycogen (Figures 3.28 and 3.29). In *granular cell* neoplasms (including *oncocytomas*), lipid and glycogen are minimal, and mitochondria are numerous and fill most of the cytoplasm (Figure 3.30). The entity previously called granular cell carcinoma is currently reclassified into oncocytoma, an eosinophilic variant of chromophobe cell carcinoma and collecting duct carcinoma. *Chromophobe cell carcinomas* represent less than 7% of all renal carcinomas and have a very distinctive ultrastructure. Cell cytoplasm contains innumerable small vesicles, 150–300 μ in diameter. Many cells also contain numerous mitochondria, and some cells may show an apparent transition between altered mitochondria and small vesicles (Figures 3.31 and 3.32).

Prostatic adenocarcinomas are composed of luminal type cells, although focal neuroendocrine cells and pure neuroendocrine neoplasms also occur. Normal glandular epithelium of the prostate includes columnar lining cells, cuboidal basal reserve cells, and scattered neuroendocrine cells. The main feature of normal and neoplastic lining cells is the presence of many secretory granules, some of which may be quite large (Figures 3.33 and 3.34). Varying numbers of lipid droplets and lysosomes are also frequently present. Secondary lysosomes often contain lipofuscin pigment that, if prominent, gives rise to the light microscopic picture referred to as melanosis and may be seen in benign and malignant prostatic epithelium. Electron microscopy substantiates the pigment as being lipofuscin and not melanin. Compared to normal prostatic epithelium, the cells comprising adenocarcinomas show a disorganized arrangement of organelles, including the secretory granules not being limited to the supranuclear cytoplasmic compartment. In addition, microvilli are not as uniform in size and distribution on the surface of the cells, and mitochondria are pleomorphic and increased in number.

Adrenal cortical adenomas and carcinomas are composed of solid groups of cells that have a wide range of ultrastructural features but usually retain some organelles and inclusions characteristic of normal cortex, albeit in abnormal amount and distribution (Figure 3.35). The normal adrenal cortex, as with other steroid producing endocrine organs, is composed of cells with abundant smooth endoplasmic reticulum, lipid droplets, mitochondria, and lysosomes. Some of the mitochondria have tubulovesicular cristae. A few small villi may be found on the free surface of the cells, but there are no microacini. Basal lamina surrounds groups of cells. Nuclei are round and have a small amount of heterochromatin. In addition, there are features characteristic of specific zones. Cells of the *zona glomerulosa* are smaller and have less cytoplasm than the cells from the other two zones. Lipid droplets are less numerous and smaller than in the zona fasciculata, and some are surrounded by a limiting membrane. Mitochondria are small, round, and elongated. The *zona fasciculata* is characterized mostly by numerous lipid droplets. Mitochondria are variable in size and shape. The *zona reticularis* has abundant lipofuscin as its main feature, and neutral lipid droplets are less numerous than in the other two zones. Mitochondria are usually elongated. There may also be stacks of rough endoplasmic reticulum and lysosomes of varying size.

Cortical adenomas and carcinomas, functional and nonfunctional, are composed of cells that usually have numerous organelles, including a moderate number of mitochondria, and in oncocytic adenomas mitochondria represent the main cytoplasmic organelle. Cristae may be tubular or lamellar, and there may be large, dense bodies in the mitochondrial matrix. Unfortunately, smooth endoplasmic reticulum may not always be present in abundance, and lipid droplets may not be numerous in some of these neoplasms. Rough endoplasmic reticulum is commonly found and is often arranged in stacks. Intercellular junctions are small, and compressed villi may be found between cells. Acini are very rare. Nuclei are usually regular in contour, contain a small amount of heterochromatin and have small nucleoli. Dense-core granules of neuroendocrine type are not expected but rarely may be present. Glycogen is not usual but may be seen in a few of these neoplasms, making the differential diagnosis between cortical carcinoma and renal cell carcinoma difficult. The combi-

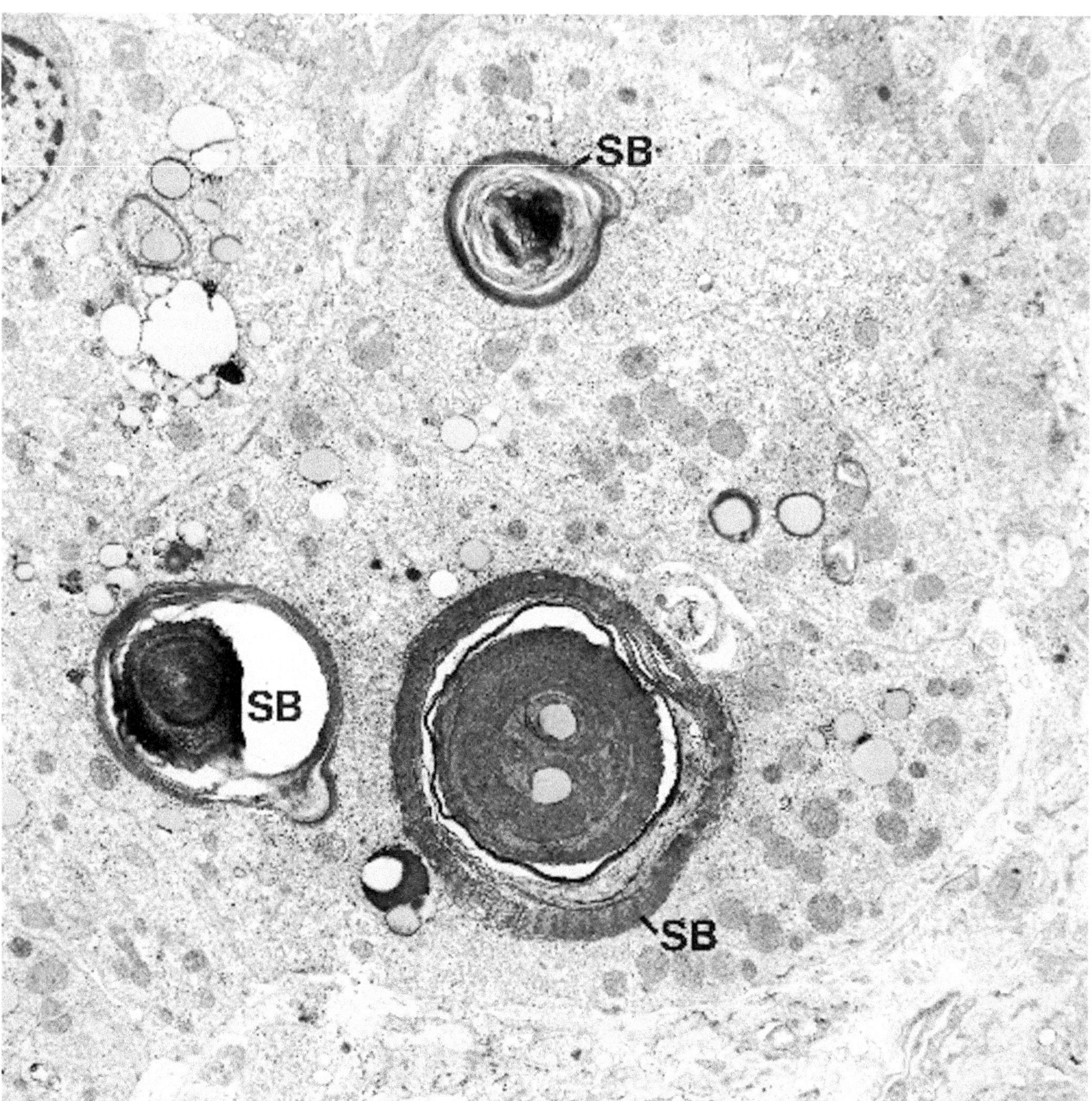

Figure 3.37. Cortical hyperplasia treated with spironolactone (adrenal gland). Several electron-dense, lamellar (spironolactone) bodies (SB) are apparent. (× 6500)

nation of glycogen and lipid is otherwise a fairly reliable finding for the diagnosis of renal cell carcinoma.

Usually, it is not possible to tell from its ultrastructure whether a cortical lesion is functioning or nonfunctioning. Furthermore, ultrastructural markers, more often than not, are unreliable in identifying specific types of functioning adenomas. On the other hand, some lesions do recapitulate the morphology of the cortical zone of origin: aldosteronomas may show the membrane-bound lipid vacuoles characteristic of the zona glomerulosa; some cortisol-secreting lesions exhibit the overabundance of lipid vacuoles typical of the zona fasciculata; and cellular proliferations producing the adrenogenital syndrome may contain many lysosomes and lipofuscin granules, characteristic of the zona reticularis. An exaggerated amount of lipofuscin is also seen in so-called black adenomas, in which the excessive pigment actually imparts a grossly visible black or brown discoloration to the involved cortex. Although a limiting membrane is often difficult to identify around the pigment, the pigment is probably in secondary lysosomes (Figure 3.36).

When hypertension in hyperaldosteronism is treated with spironolactone, an aldosterone antagonist, the adenomatous or hyperplastic adrenal cortical cells accumulate electron-dense, membranous whorls—spironolactone bodies (Figure 3.37). These structures are usually in close association with surrounding smooth and/or rough endoplasmic reticulum. Spironolactone type bodies are not morphologically specific for treated adrenal cortical hyperfunction and may be seen in other locations and circumstances, such as in hepatocytes in patients treated with phenobarbital and certain other drugs.

Squamous Cell Carcinoma

(Figures 3.38 through 3.49.)

Diagnostic criteria. (1) Desmosomes; (2) tonofibrils; (3) keratohyalin granules (in well-differentiated cells).

Additional points. Intercellular bridges may be produced if adjoining cells are partially pulled apart as a result of edema, degeneration, or autolysis (Figures 3.38 and 3.39). Basal lamina may or may not be present around groups of cells.

The epithelial cells of *thymomas* (Figures 3.40 through 3.42) and most *thymic carcinomas* are of squamous type, but the size of desmosomes and number of tonofibrils vary with the degree of differentiation and the zone of thymus from which the neoplastic cells are derived. Cortical neoplasms tend to have cells with long, slender processes; small desmosomes; and few tonofibrils. Medullary lesions (including spindle cell thymomas) have shorter cells with more abundant cytoplasm, large desmosomes, and numerous and prominent tonofibrils. The epithelial cells of some medullary thymomas may form lumens lined by microvilli and junctional complexes. Nonsquamous cell thymic carcinomas include clear cell carcinoma, mucoepidermoid carcinoma, basaloid carcinoma, lymphoepithelioma-like carcinoma, and sarcomatoid carcinoma.

So-called *undifferentiated carcinomas of the nasopharynx and paranasal sinuses* are actually poorly differentiated squamous cell carcinomas and characteristically have desmosomes but only a few tonofibrils (Figures 3.43 through 3.45).

Adamantinomas of long bones also are composed of cells exhibiting squamous differentiation (Figures 3.46 and 3.47). This is true for oval and polygonal cell components and for spindle cell portions, although a matrix containing nonneoplastic, fibroblastic spindle cells usually accompanies the islands or cords of neoplastic squamous cells. Basal lamina may surround the squamous cellular groupings.

Concomitant squamous and glandular differentiation is seen in adenocarcinomas with squamous metaplasia, adenosquamous carcinomas, and *mucoepidermoid carcinomas.* In some of these neoplasms there may be separate cell types—secretory glandular epithelium and squamous epithelium—as well as biphasic differentiation within the same cell (Figures 3.48 and 3.49).

Squamous and neuroendocrine differentiation within the same cell also may be seen in small cell carcinoma of the bronchus.

Transitional Cell (Urothelial) Carcinoma

(Figures 3.50 through 3.59.)

Diagnostic criteria. (1) Interdigitating, villus-like, lateral cell borders; (2) scalloping of luminal plasmalemmas by small invaginations that connect with elliptical apical cytoplasmic vesicles; (3) apical cytoplasmic filaments; (4) focally invaginated basal border, at points of abutment of neighboring cells; (5) desmosomes.

Additional points. A common feature of neoplastic urothelial cells is interdigitation of lateral cell membranes. Apical vesicles and plasmalemmal scalloping are also very characteristic of urothelium, but they may be scarce or inconspicuous in neoplastic tissue. Short microvilli may be present on the apical surface of cells, and varying amounts of cytoplasmic glycogen and secondary lysosomes may also be present. Tonofibrils, a sign of squamous differentiation, often are present to some degree. Most cells are polygonal, and some are elongate. Nuclei are frequently indented and irregular in shape, as is true to a lesser extent in normal, nonneoplastic urothelial cells. Nucleoli are large and multiple and consist mostly of open strands of nucleolonemas.

(Text continues on page 88)

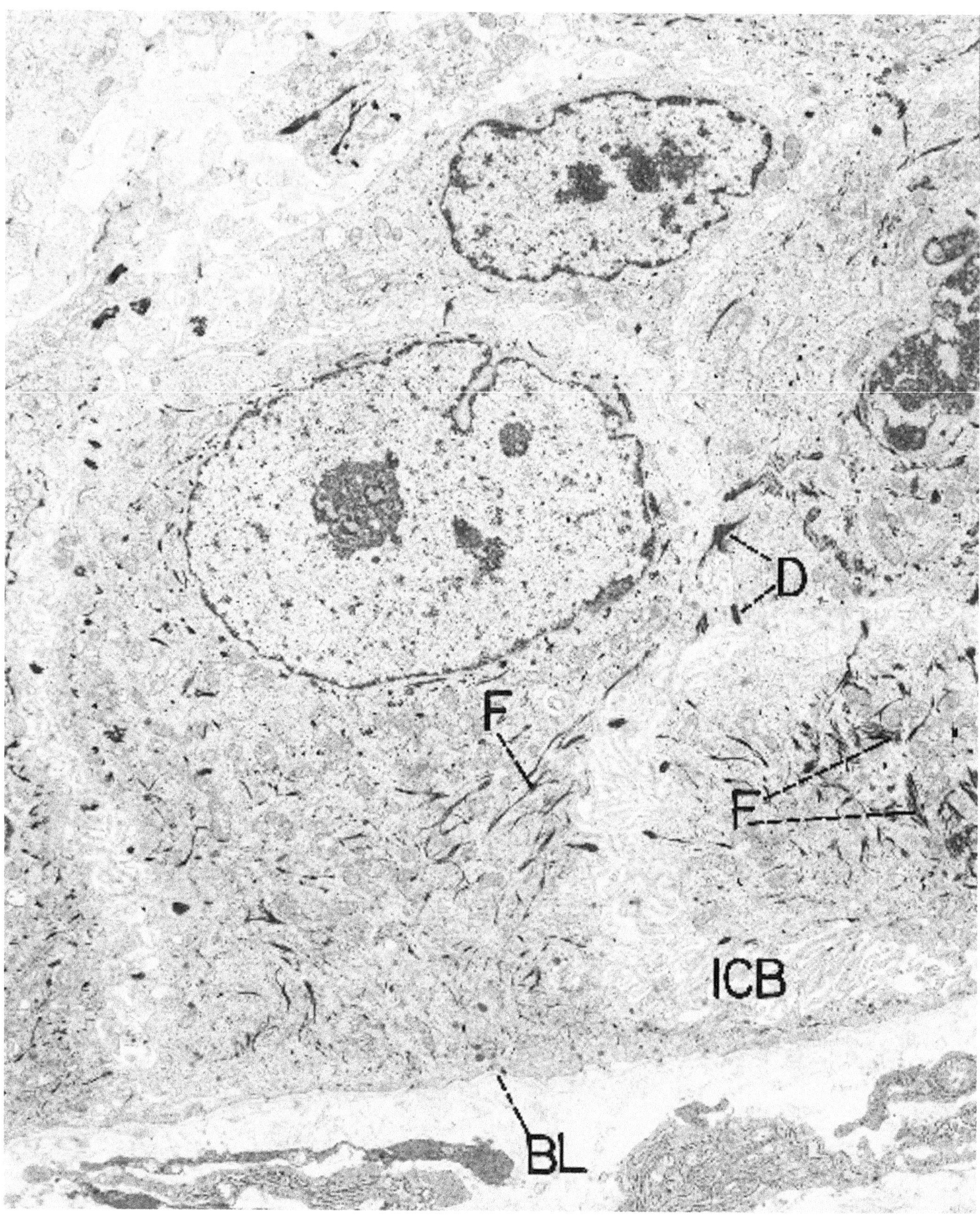

Figure 3.38. Squamous cell carcinoma (bronchus). A nest of well-differentiated squamous cells is surrounded by basal lamina (BL) and has prominent desmosomes (D) and intercellular bridges (ICB). The cytoplasm of most cells is rich in bundles of prekeratin filaments (tonofibrils) (F). (× 6700)

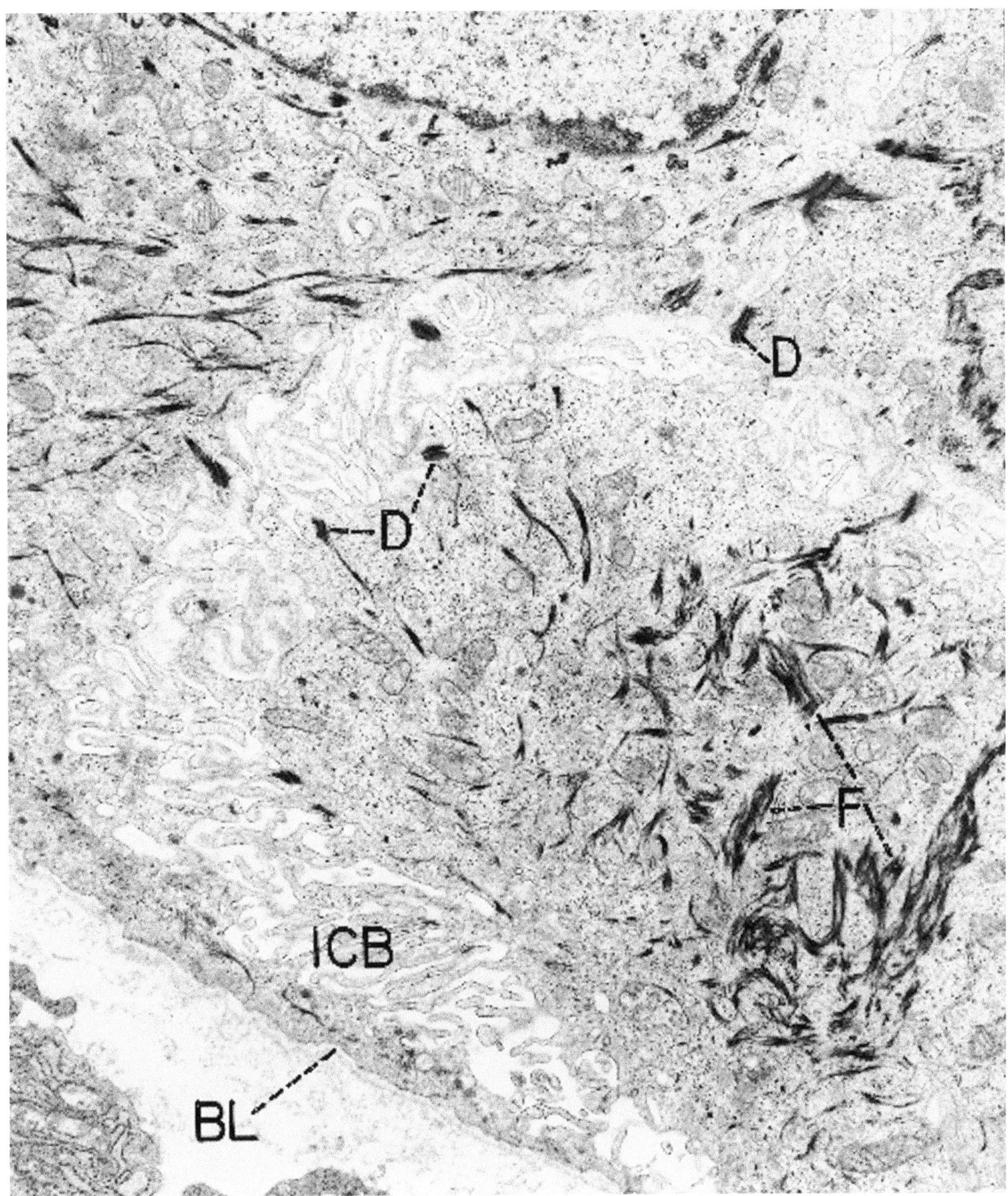

Figure 3.39. Squamous cell carcinoma (bronchus). Higher magnification of one of the cells illustrated in Figure 3.38. Desmosomes (D), intercellular bridges (ICB), and tonofibrils (F) are all prominent. Note also the basal lamina (BL) enclosing the periphery of the group of cells. (× 12,150) (Permission for reprinting granted by WB Saunders, Dickersin GR: Electron microscopy of leukemias and lymphomas. Clin Lab Med 7:199–247, 1987.)

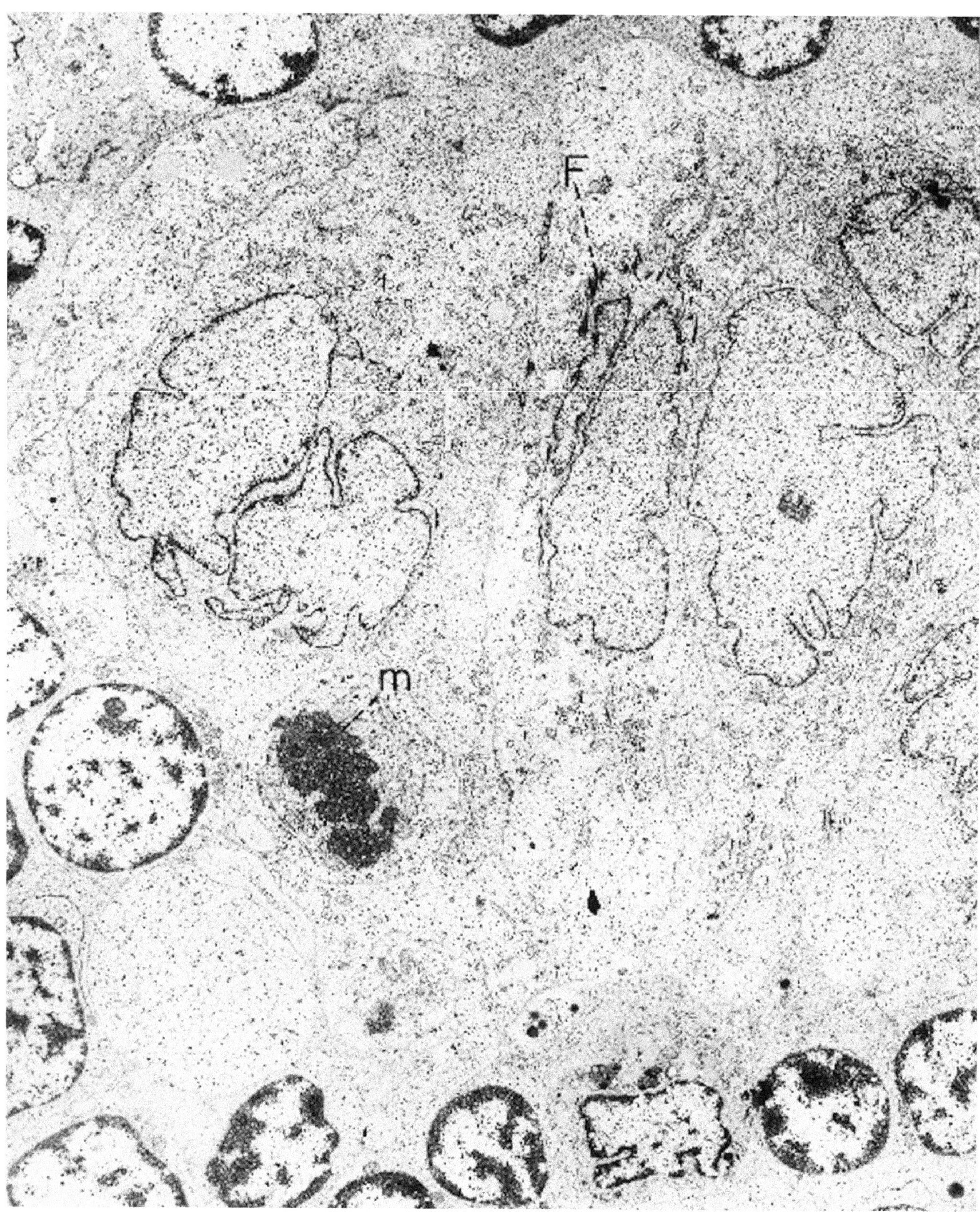

Figure 3.40. Thymoma. An island of poorly differentiated squamous cells (*center*) lies within a sea of lymphocytes (*top, bottom,* and *left*). At this magnification, tonofibrils (F) can be seen well only in one cell. Note the mitotic figure (m) in one cell. (× 4750)

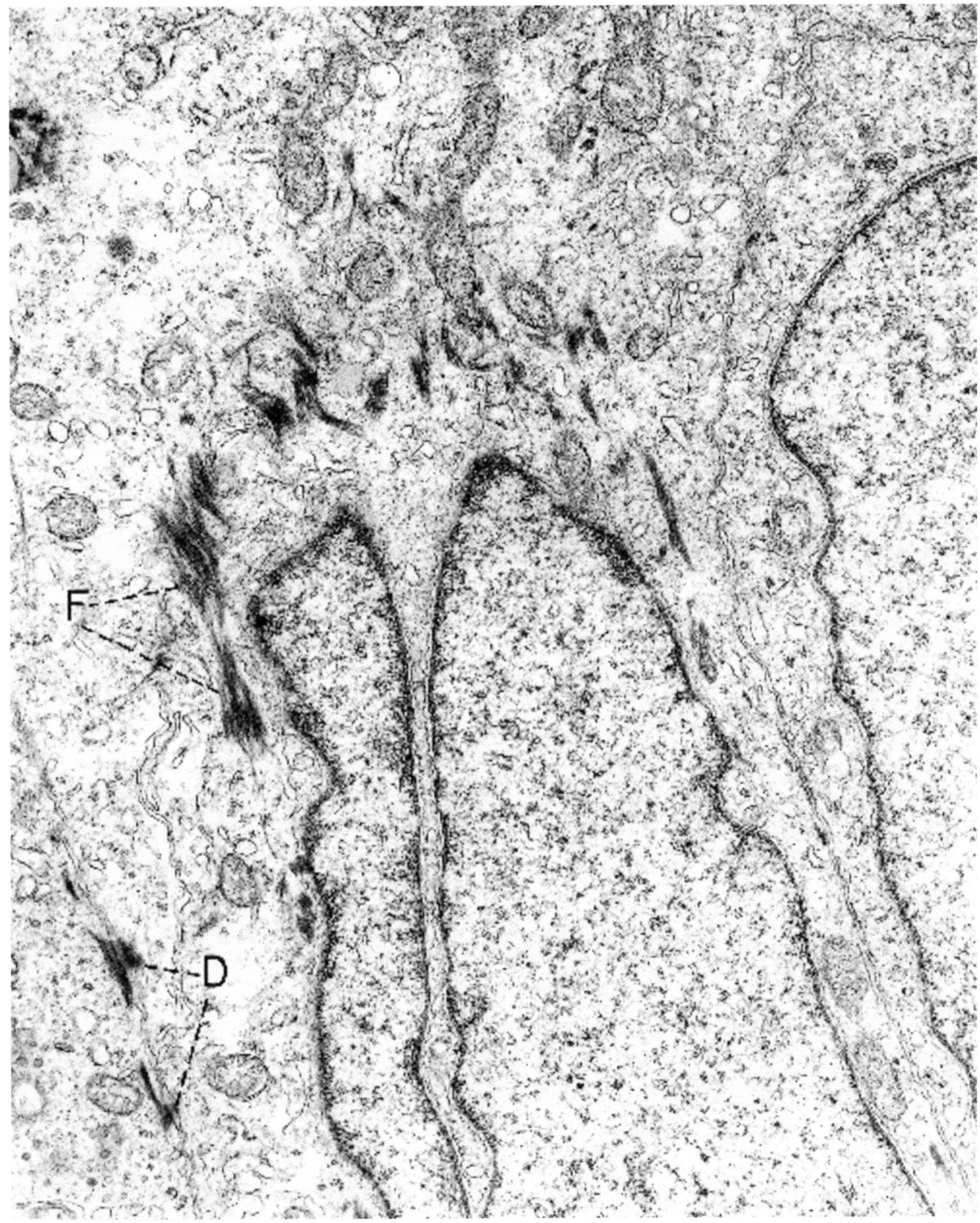

Figure 3.41. Thymoma. Higher magnification of one of the squamous cells in Figure 3.40 depicts tonofibrils (F) and desmosomes (D). (× 22,750)

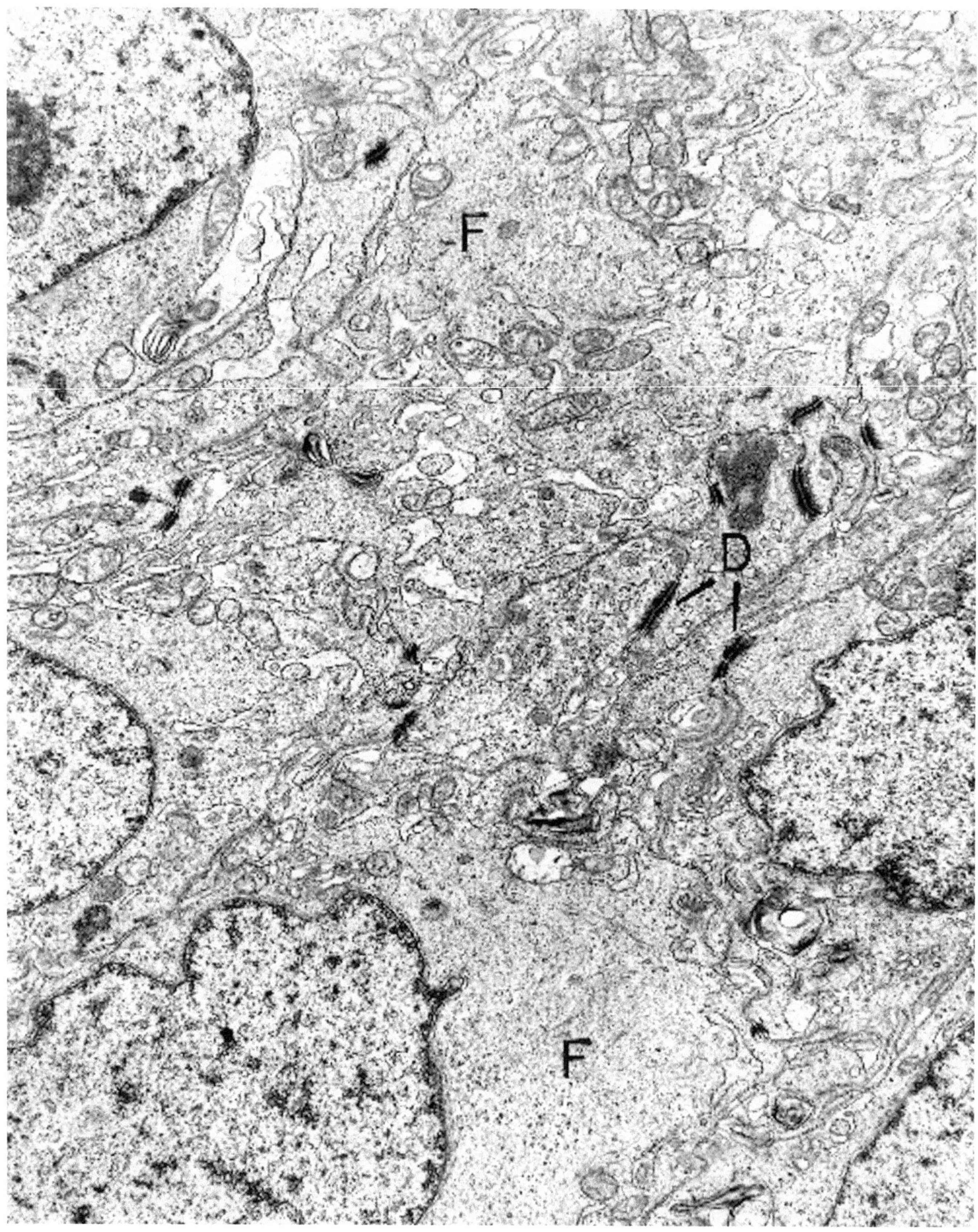

Figure 3.42. Thymoma. Poorly differentiated cells show prominent desmosomes (D) and no tonofibrils. Filaments (F) are present but in a diffuse, unbundled arrangement. (× 15,390)

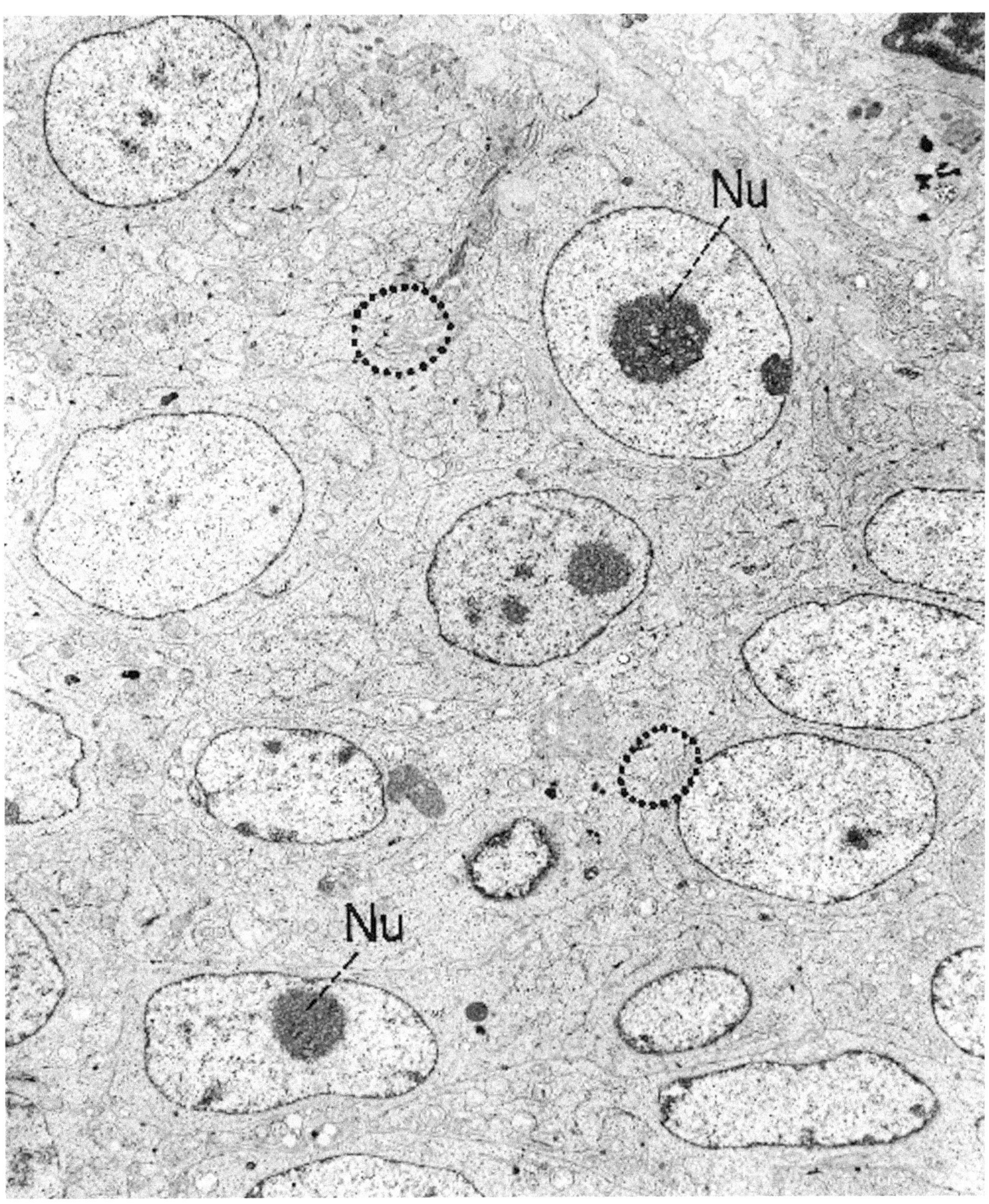

Figure 3.43. Undifferentiated (squamous) carcinoma (nasopharynx). The neoplasm is characteristically composed of large cells that have a high nuclear–cytoplasmic ratio, large nuclei with finely dispersed chromatin, and large nucleoli (Nu). The cytoplasm is primitive, with free ribosomes being the most prominent organelle. Lateral cell borders are interdigitated (*circles*) in many foci. (× 5740)

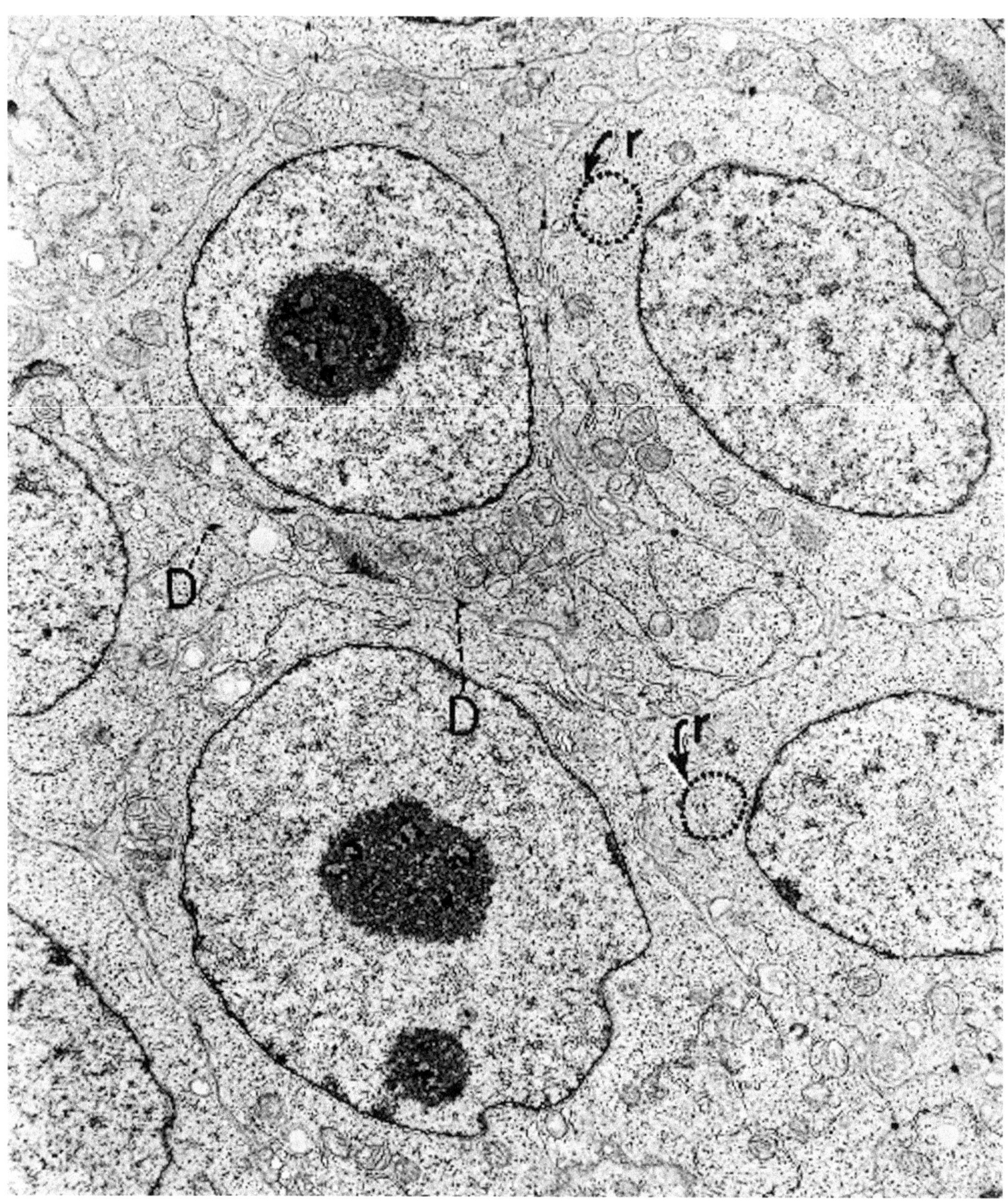

Figure 3.44. Undifferentiated (squamous) carcinoma (nasopharynx). Higher magnification of the same case as shown in Figure 3.43 illustrates the lack of differentiation of the cytoplasm. Free ribosomes (r) compose most of the cytoplasm. Desmosomes (D) are readily discernible but are small. (× 9900) (Permission for reprinting granted by WB Saunders, Dickersin GR: Electron microscopy of leukemias and lymphomas. Clin Lab Med 7:199–247, 1987.)

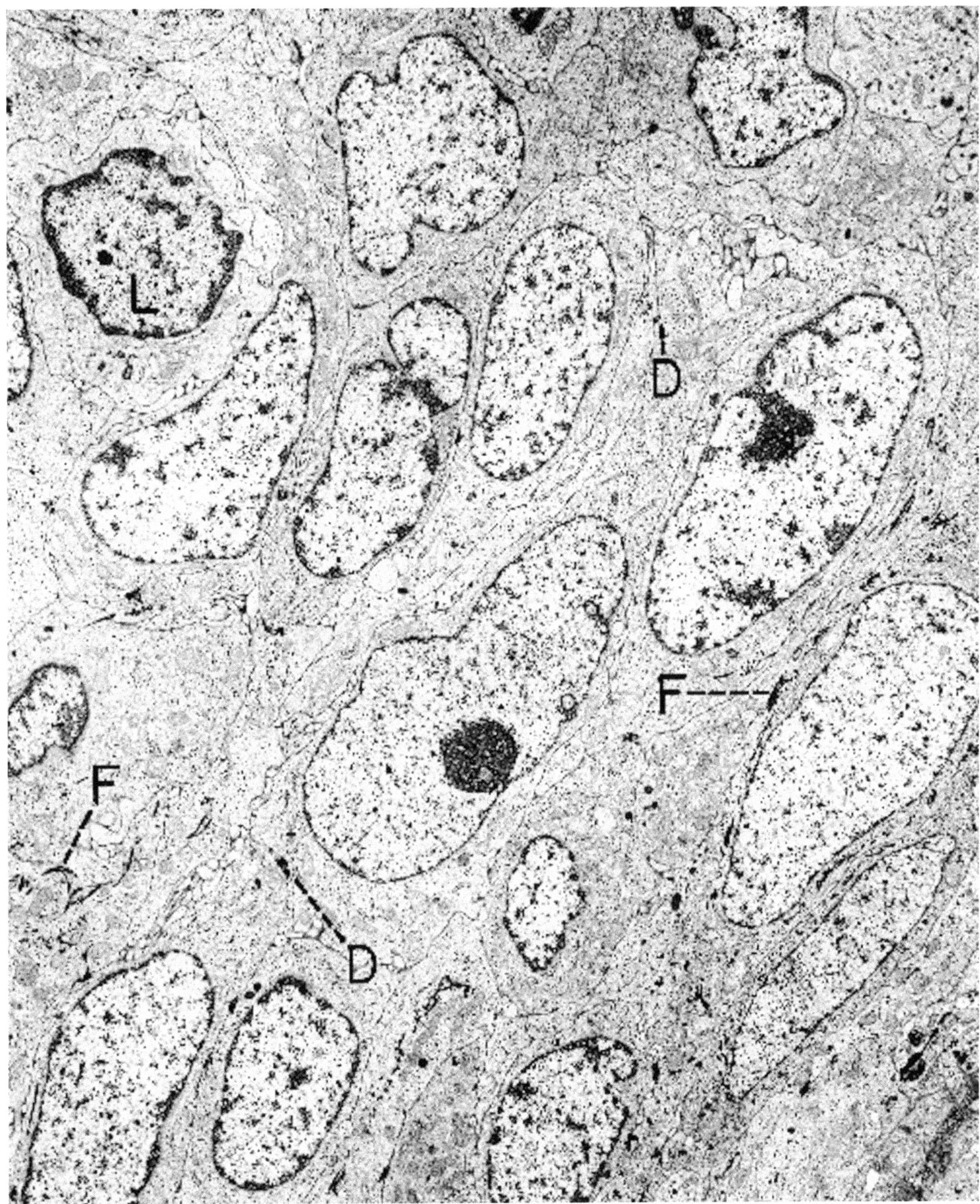

Figure 3.45. Undifferentiated (squamous) carcinoma (nasopharynx). The cells in this neoplasm are spindle shaped and show mild squamous differentiation, as indicated by the presence of a few tonofibrils (F) and desmosomes (D) of moderate size. L = lymphocyte. (× 6750)

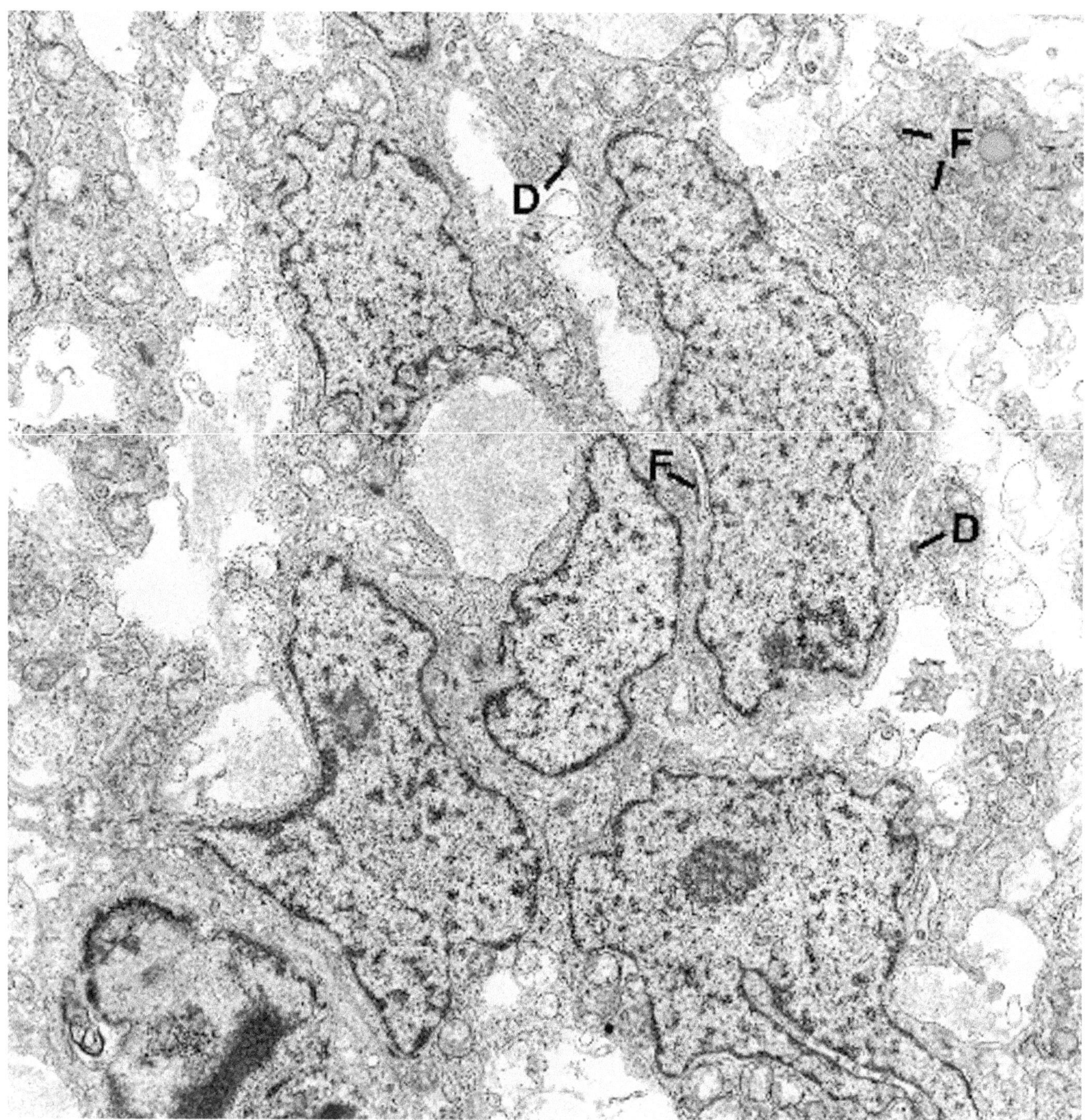

Figure 3.46. Adamantinoma (tibia). Some of the cells in this island of epithelial cells are spindle shaped but are still interconnected by desmosomes (D) and contain tonofibrils (F) in their cytoplasm. The latter are better seen at higher magnification in Figure 3.47. (× 10,400)

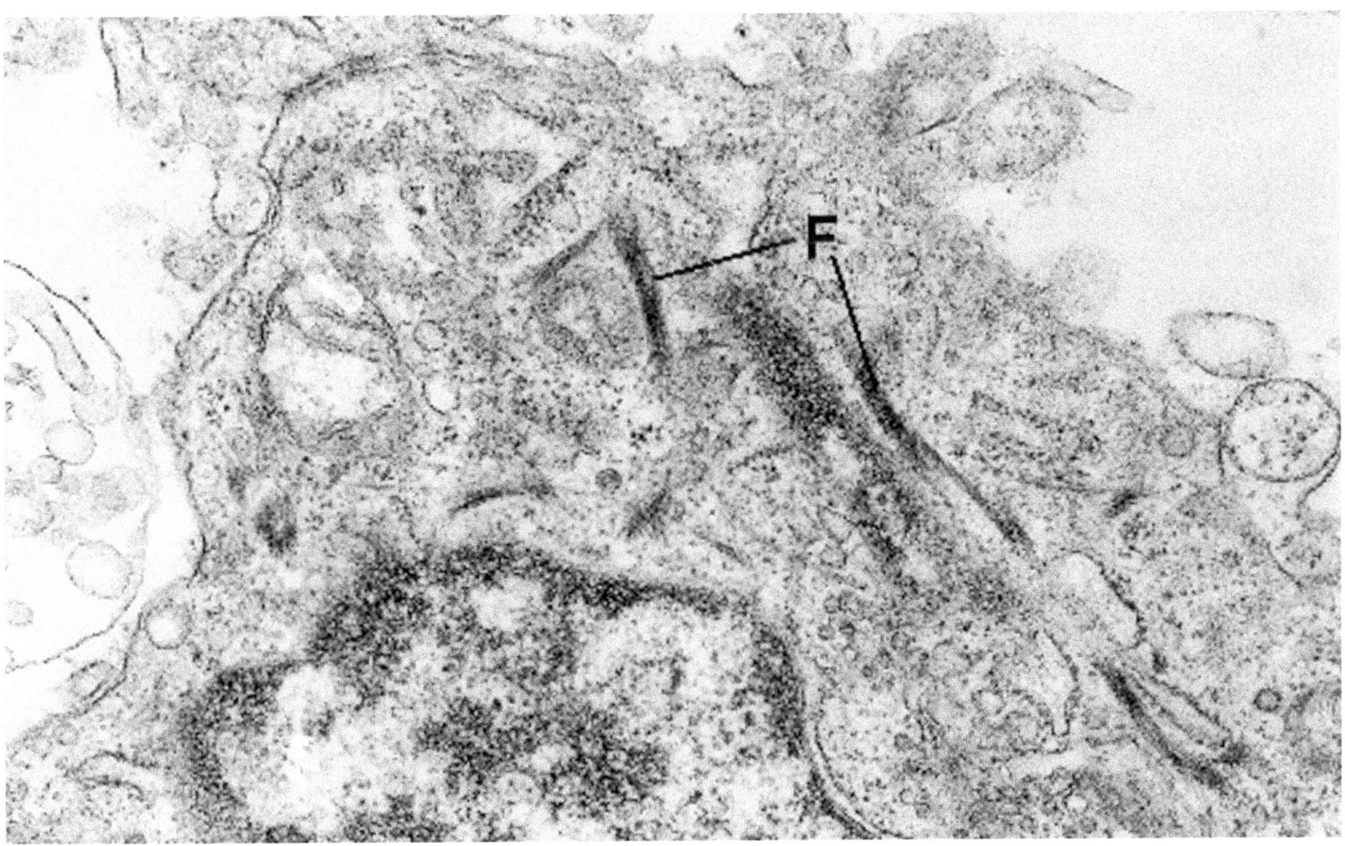

Figure 3.47. Adamantinoma (tibia). Higher magnification of a cell from the same neoplasm depicted in Figure 3.46 illustrates more clearly the cytoplasmic tonofibrils (F). (× 37,700)

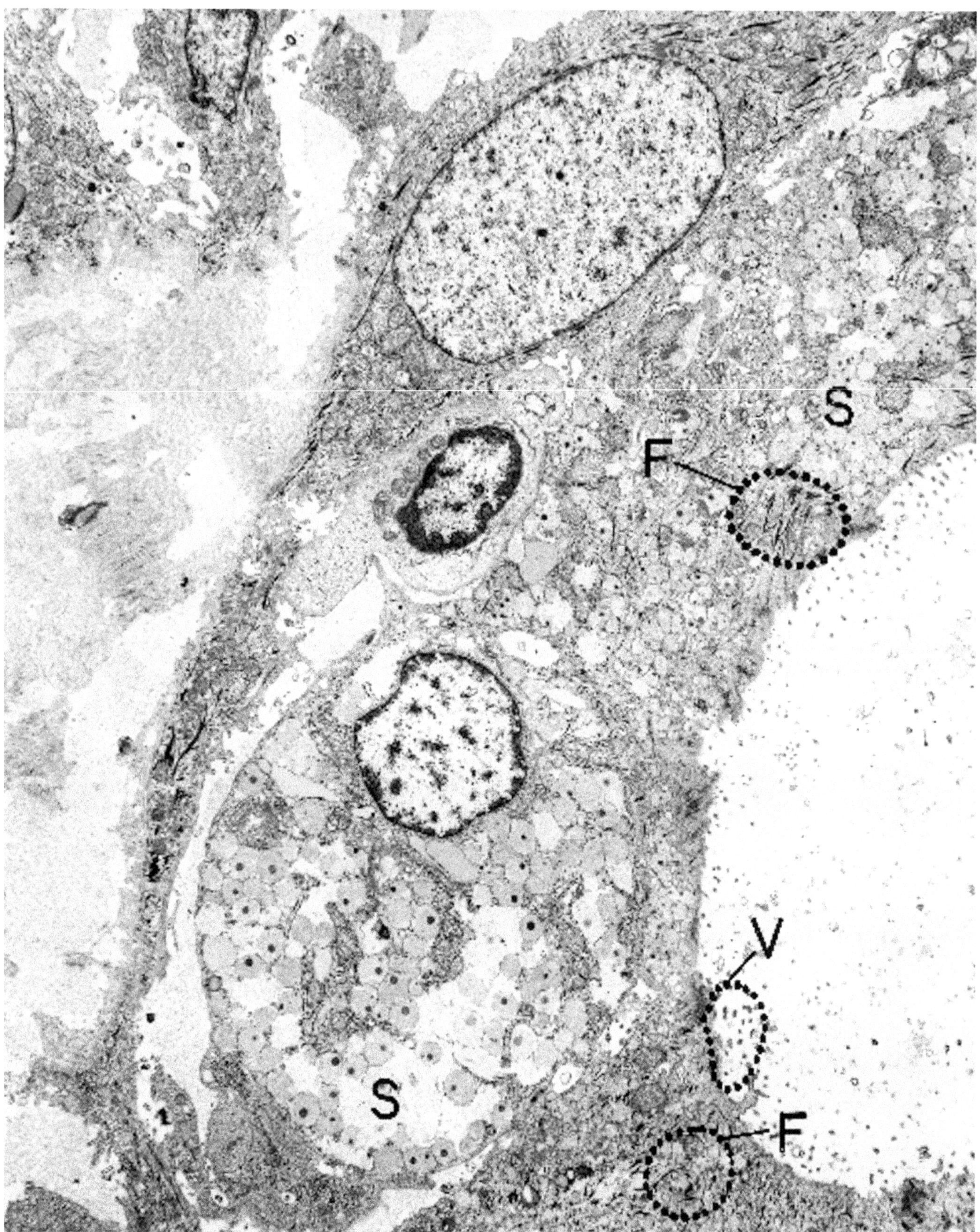

Figure 3.48. Mucoepidermoid carcinoma (bronchus). A neoplasm manifesting two lines of differentiation: glandular, with microvilli (V) and secretory (mucus) granules (S); squamous, with tonofibrils (F). (× 5320)

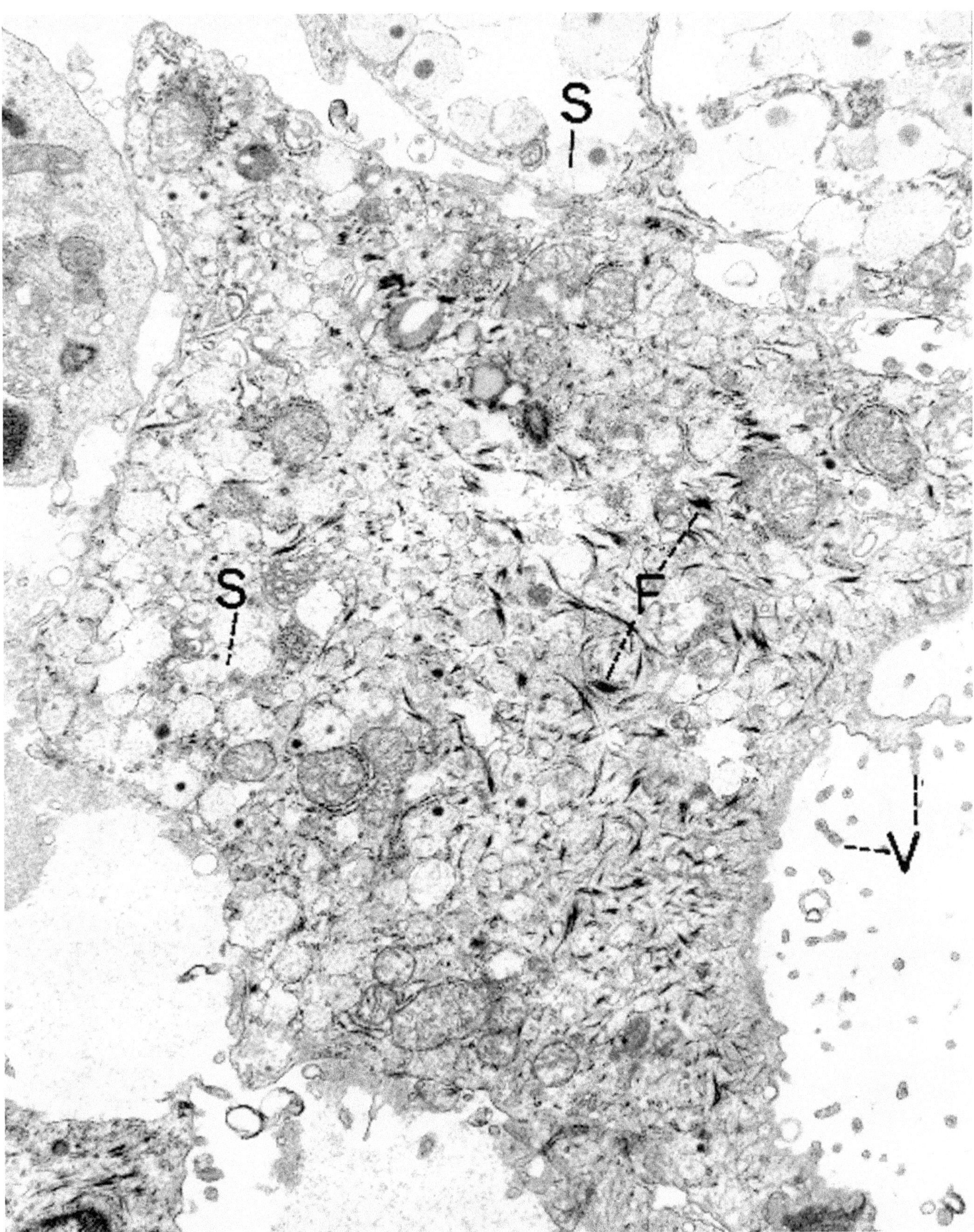

Figure 3.49. Mucoepidermoid carcinoma (bronchus). Higher magnification of one of the cells depicted in Figure 3.48, illustrates the microvilli (V), secretory granules (S), and tonofibrils (F). (× 15,960)

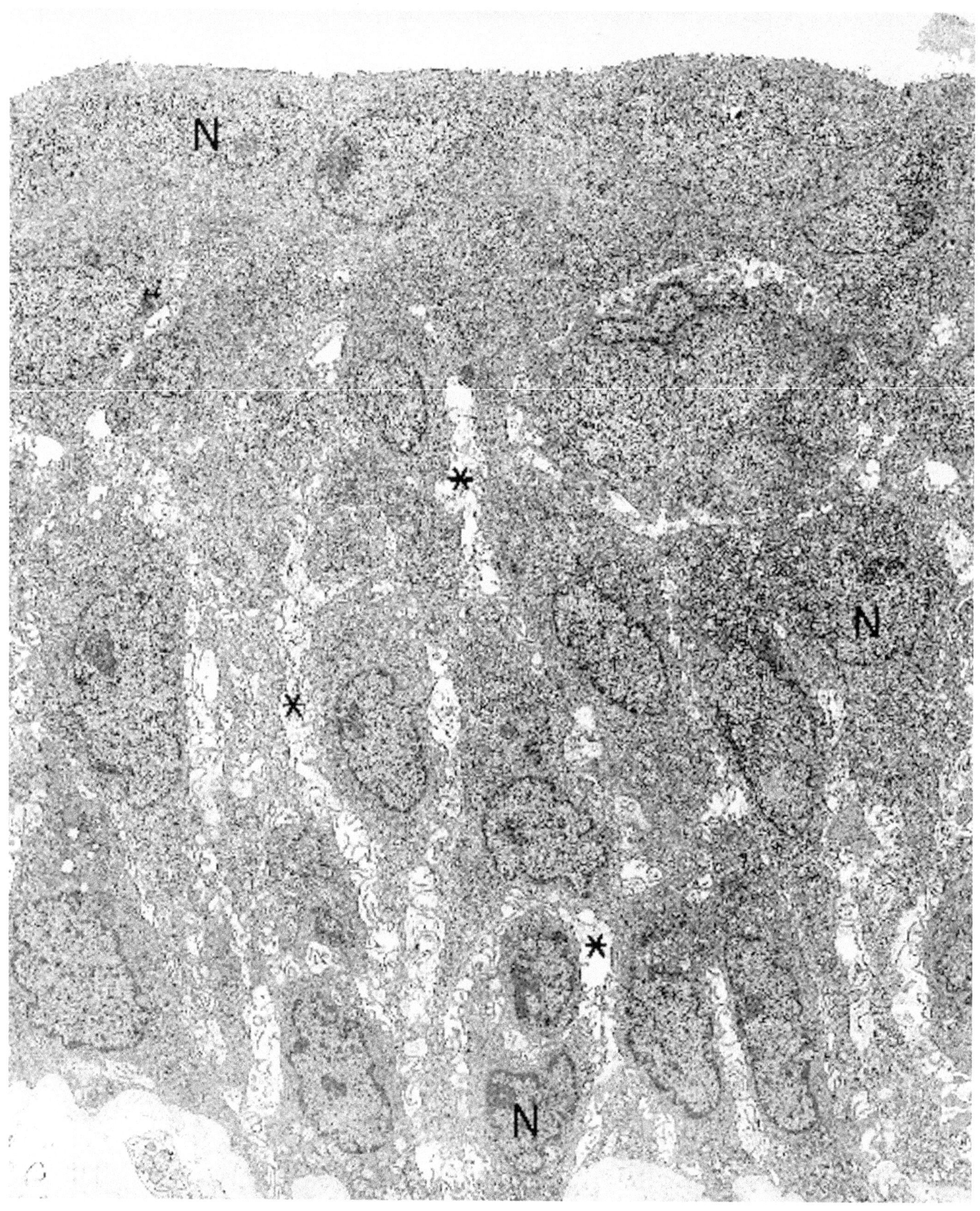

Figure 3.50. Normal urothelium (ureter). At low magnification, the characteristic transitional cell feature of interdigitating, villus-like lateral borders (*) is discernible. Also noted is the irregularly villous luminal surface (top of field) and the indented and irregularly shaped nuclei (N). (× 3700)

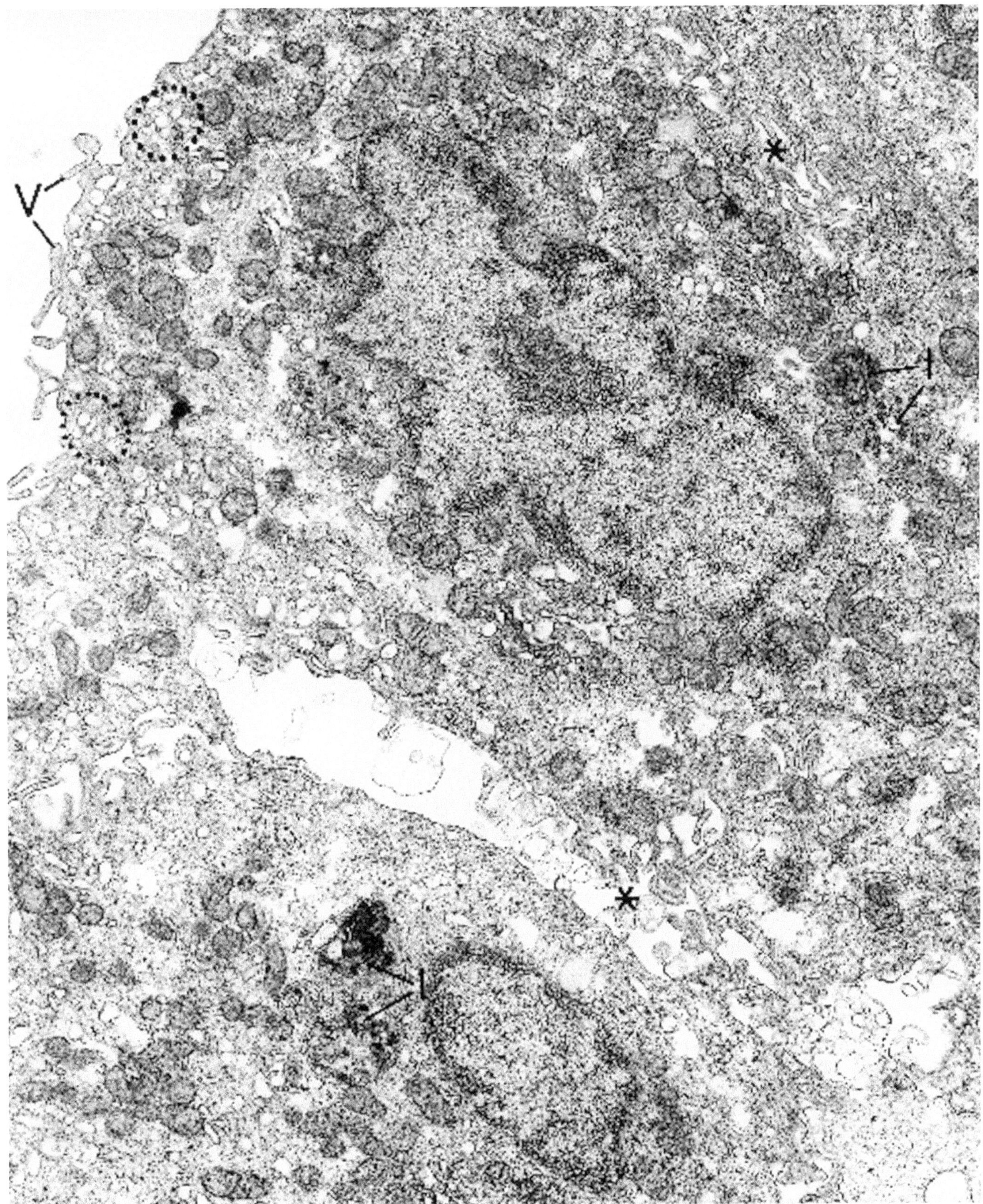

Figure 3.51. Normal urothelium (ureter). High magnification of luminal lining cells illustrates an irregularly villous surface (V), interdigitating lateral cell membranes (*), innumerable small apical vesicles (*circles*), and a few secondary lysosomes (l). (× 16,520)

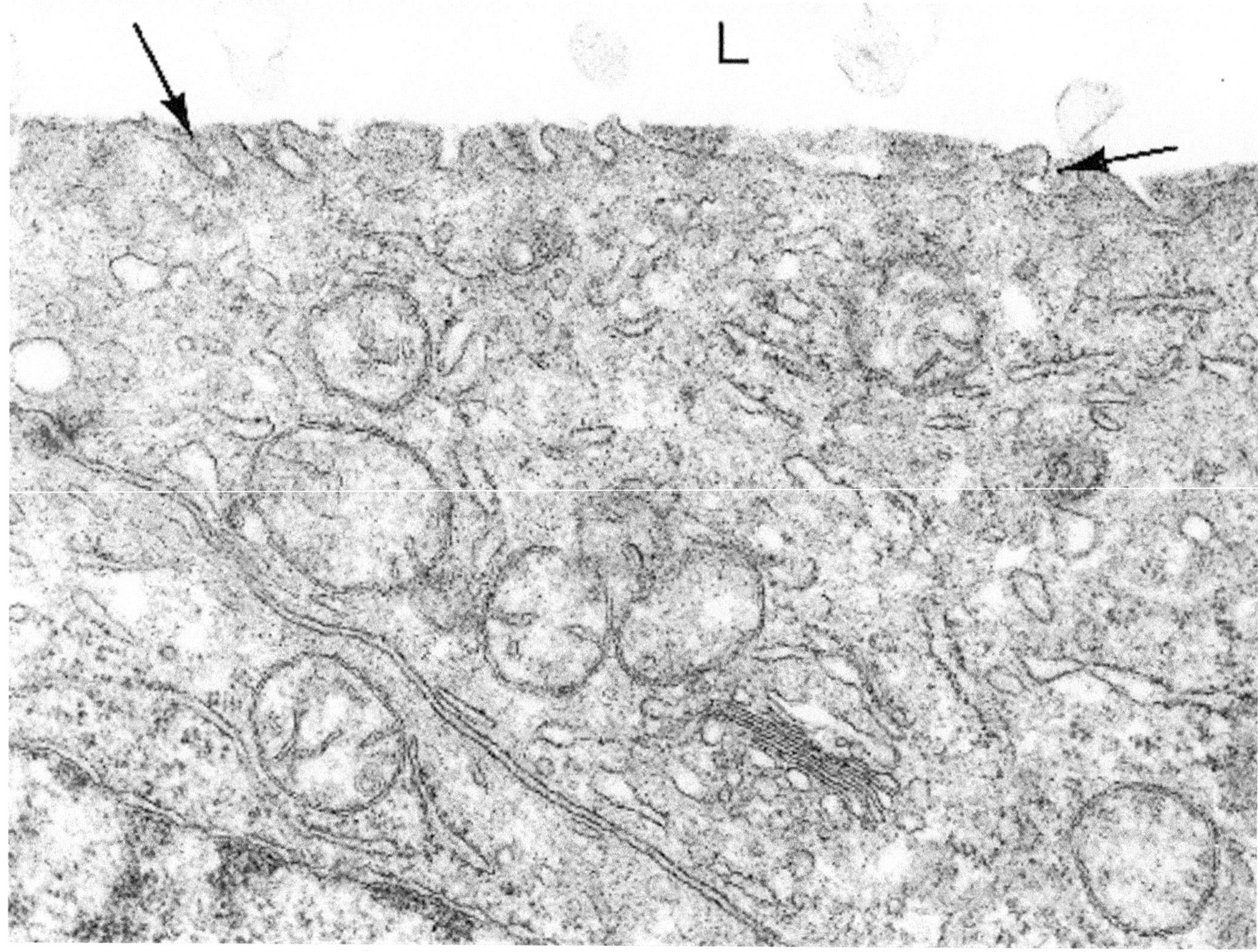

Figure 3.52. Normal urothelium (ureter). High magnification of the apical cytoplasm of a lining cell demonstrates communications between some of the small vesicles (*arrows*) and the lumen (L). (× 47,600)

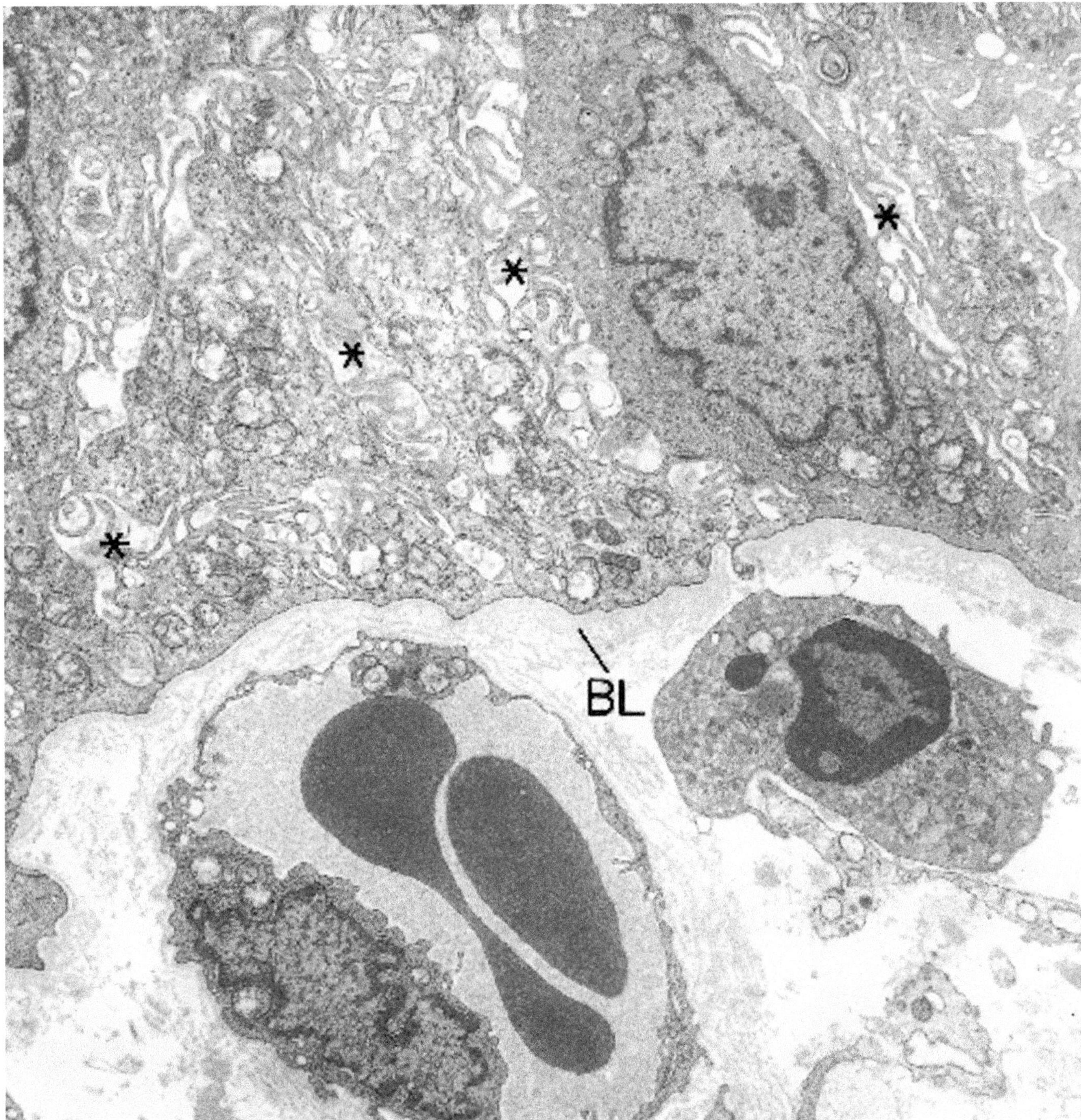

Figure 3.53. Normal urothelium (ureter). The basal side of the urothelial lining is covered by basal lamina (BL), and the lateral plasma membranes of adjacent cells show marked interdigitation of their villus-like projections (*). (× 9880)

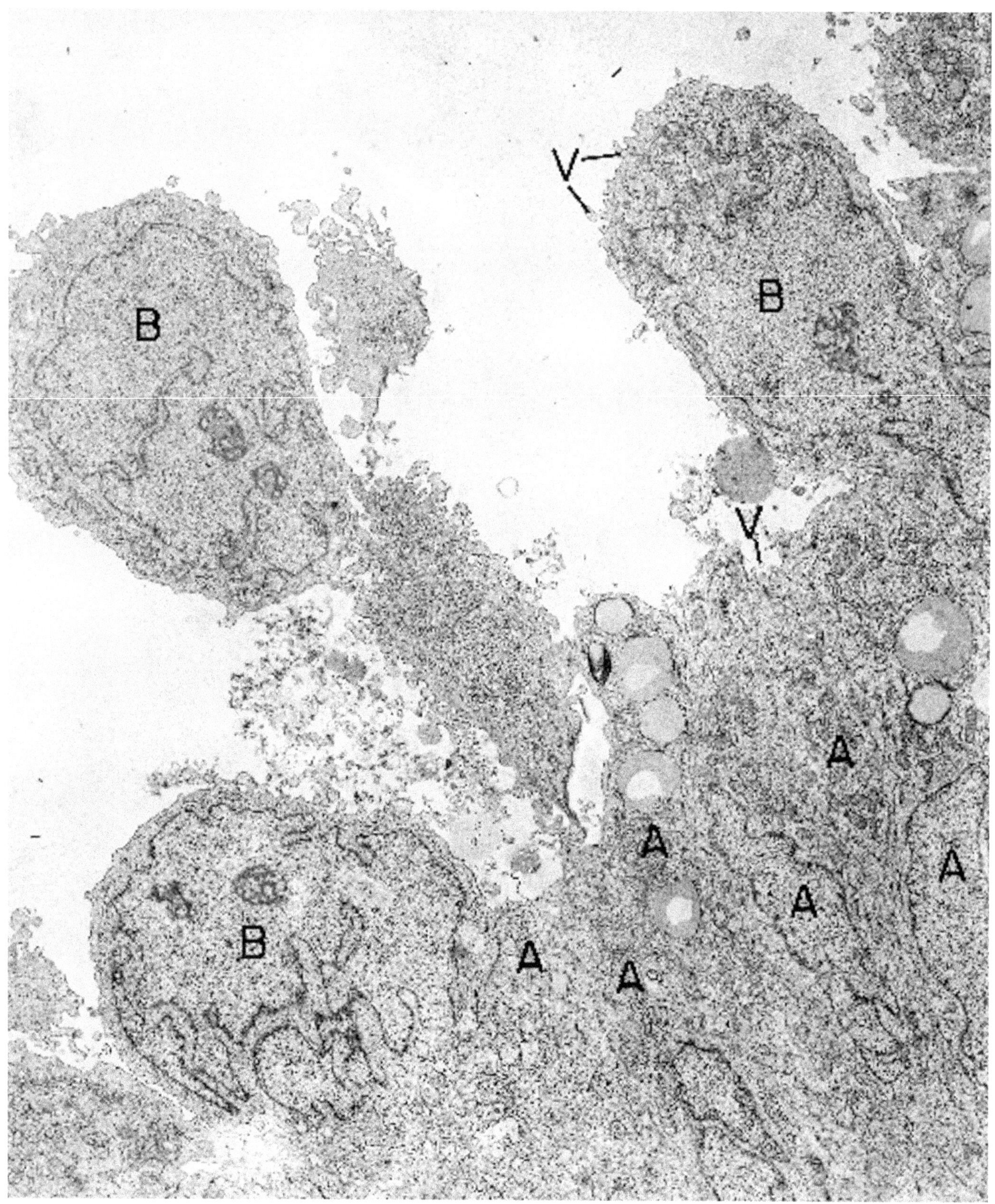

Figure 3.54. Papillary transitional cell carcinoma (renal pelvis). Some of the neoplastic cells at the surface of this papilla are narrow and long (A), and others are bulbous and pouting (B). The nuclear–cytoplasmic ratio is high, and the nuclear shape is irregular in the latter-type cells. A few microvilli (V) are present on the free surface of both types of cells. (× 6750)

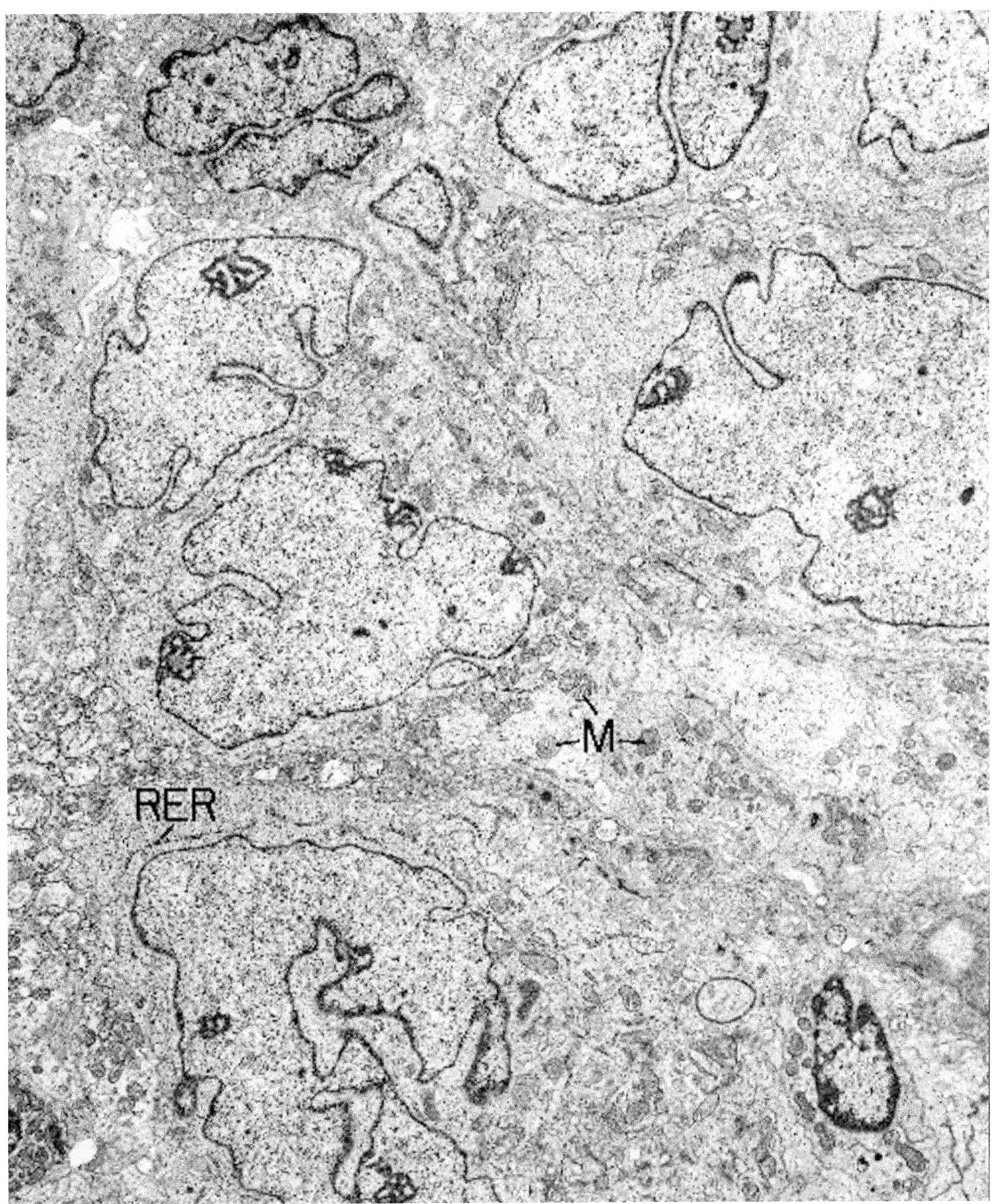

Figure 3.55. Papillary transitional cell carcinoma (renal pelvis). Neoplastic cells deep within a papilla are plump and have large and irregularly shaped nuclei with multiple open strand-like nucleoli (nucleolonemas) and a preponderance of euchromatin. The cytoplasm is composed mostly of free ribosomes plus a moderate number of mitochondria (M) and a few undilated cisternae of rough endoplasmic reticulum (RER). (× 7020)

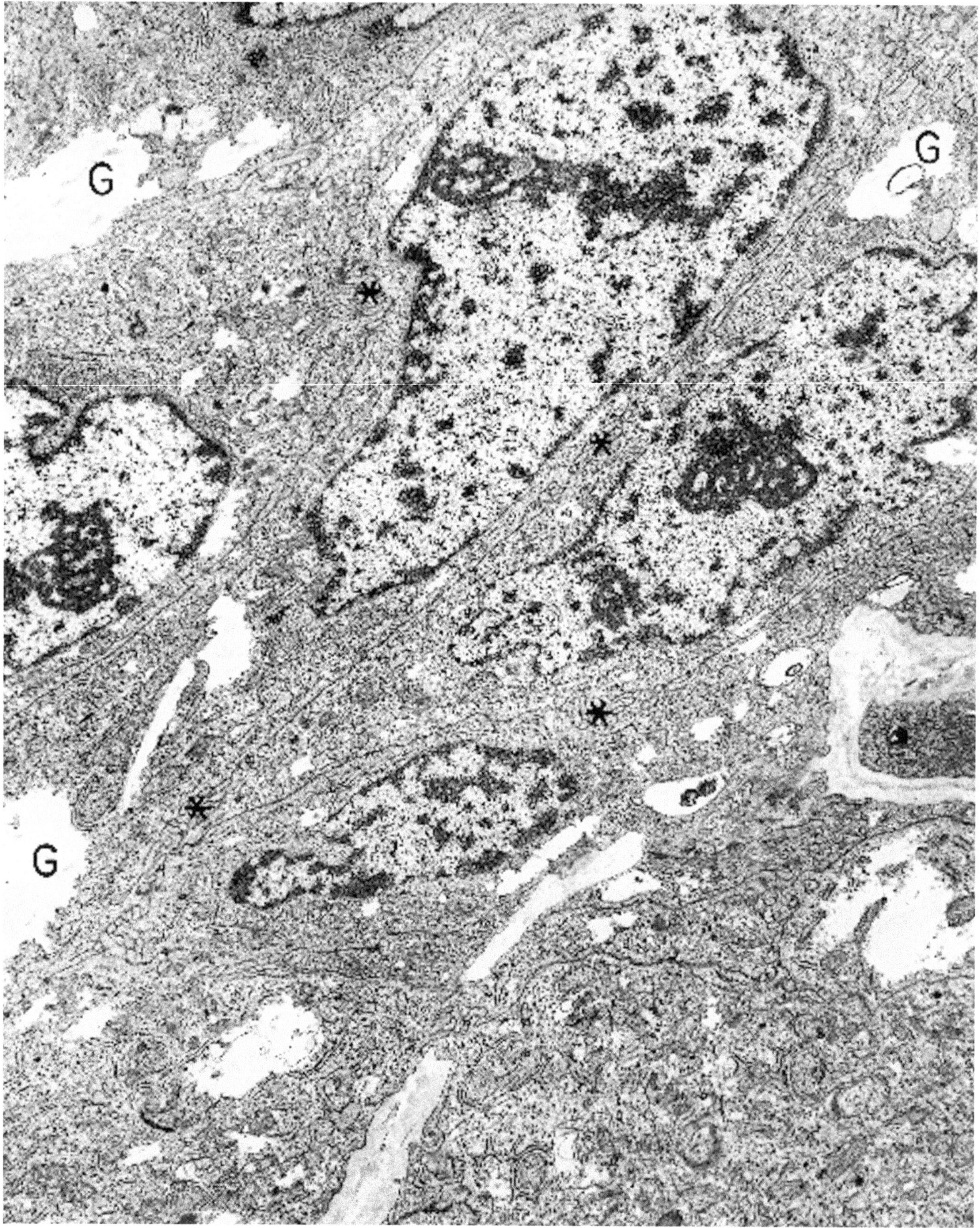

Figure 3.56. Metastatic transitional cell carcinoma from bladder (omentum). The neoplastic cells are narrow and have interdigitating lateral borders (*). The open cytoplasmic spaces represent glycogen (G). (× 10,640)

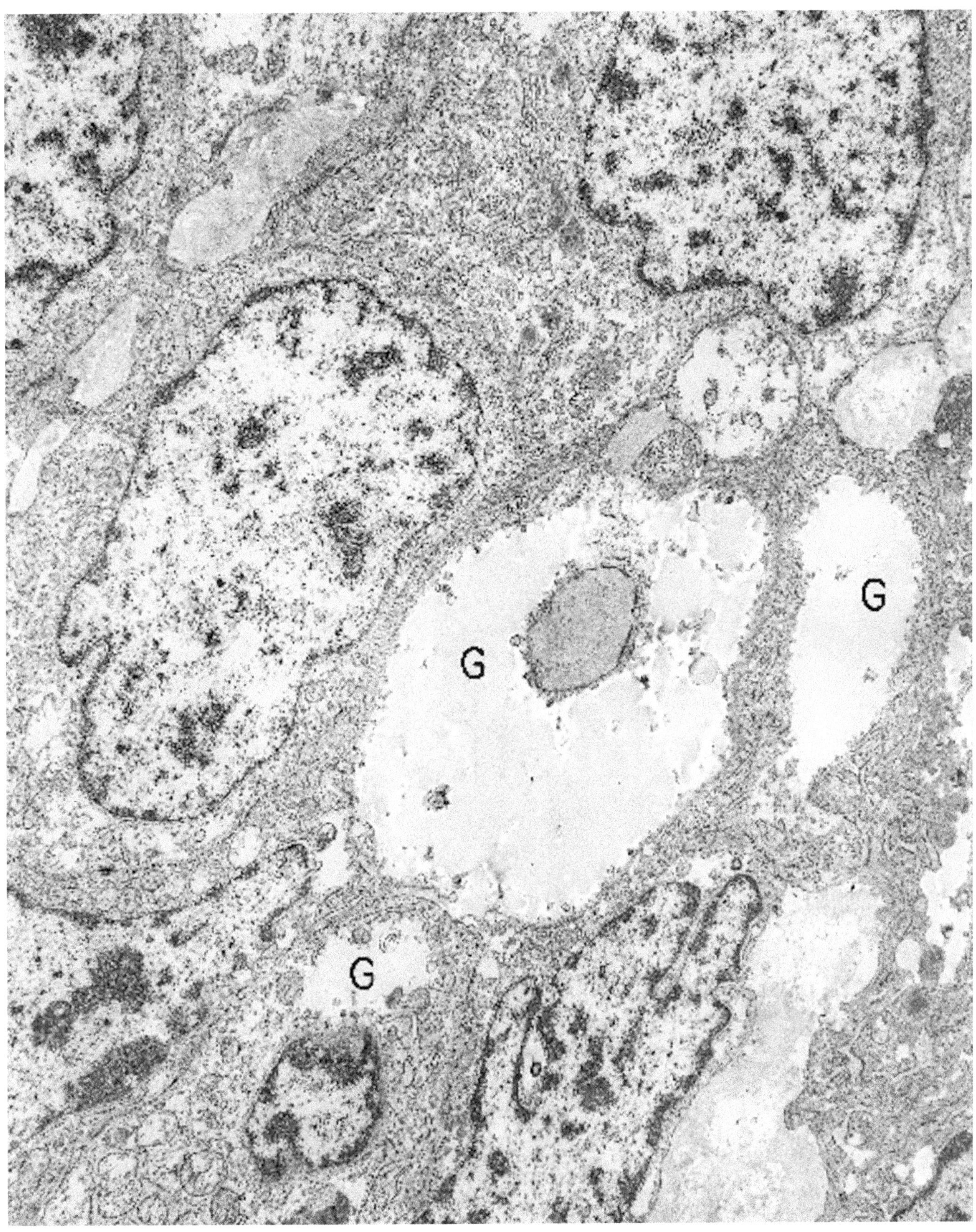

Figure 3.57. Metastatic transitional cell carcinoma from bladder (omentum). High magnification highlights the copious cytoplasmic glycogen (G) present in many of the cells of this neoplasm. (× 11,400)

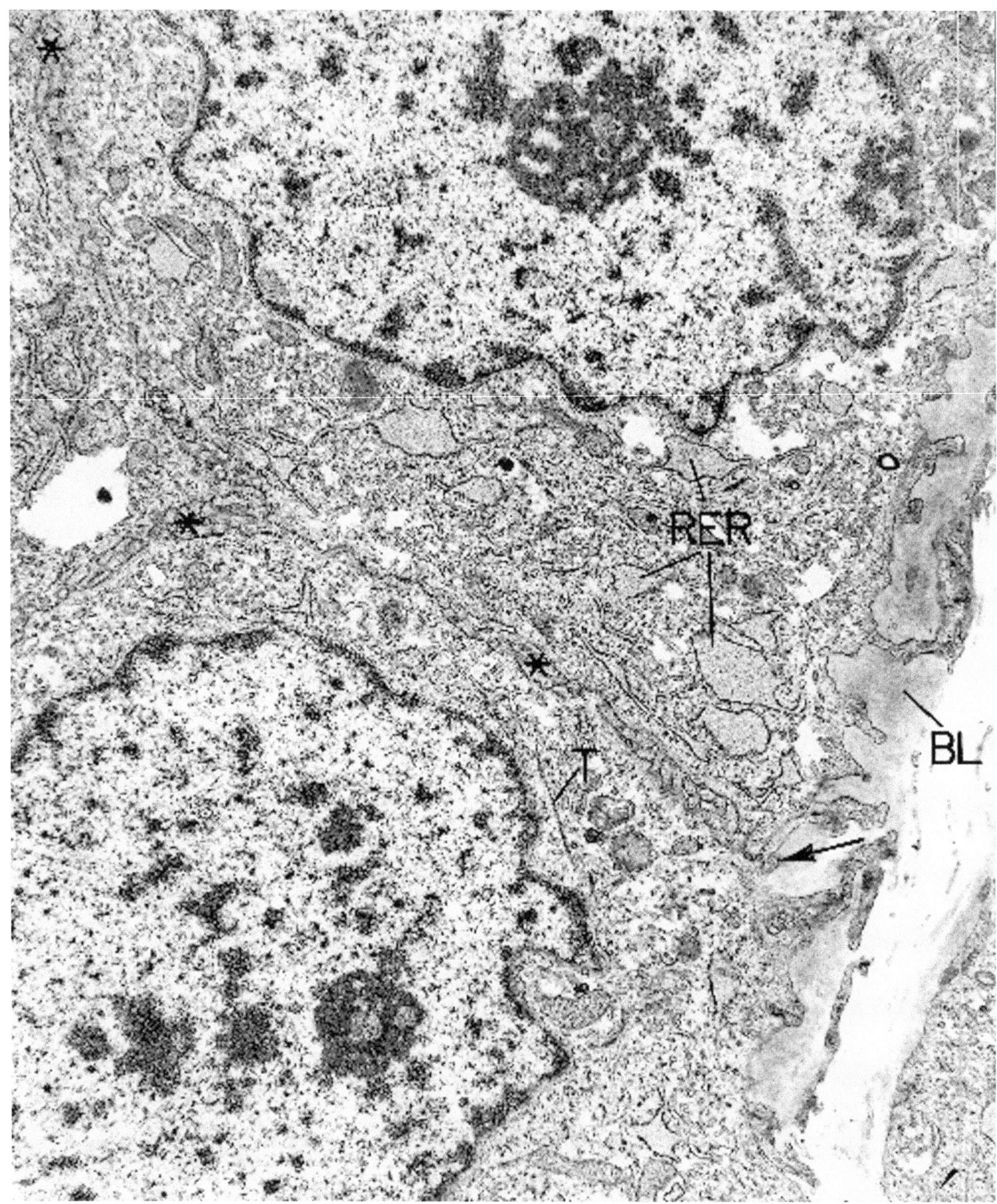

Figure 3.58. Metastatic transitional cell carcinoma from bladder (omentum). These cells at the edge of the neoplastic island illustrate interdigitating lateral borders (*), a peripheral covering by basal lamina (BL) and invagination of their basal cell membrane at the line of junction between cells (*arrow*). Additional features include tonofibrils (T) and dilated cisternae of rough endoplasmic reticulum (RER). (× 15,000)

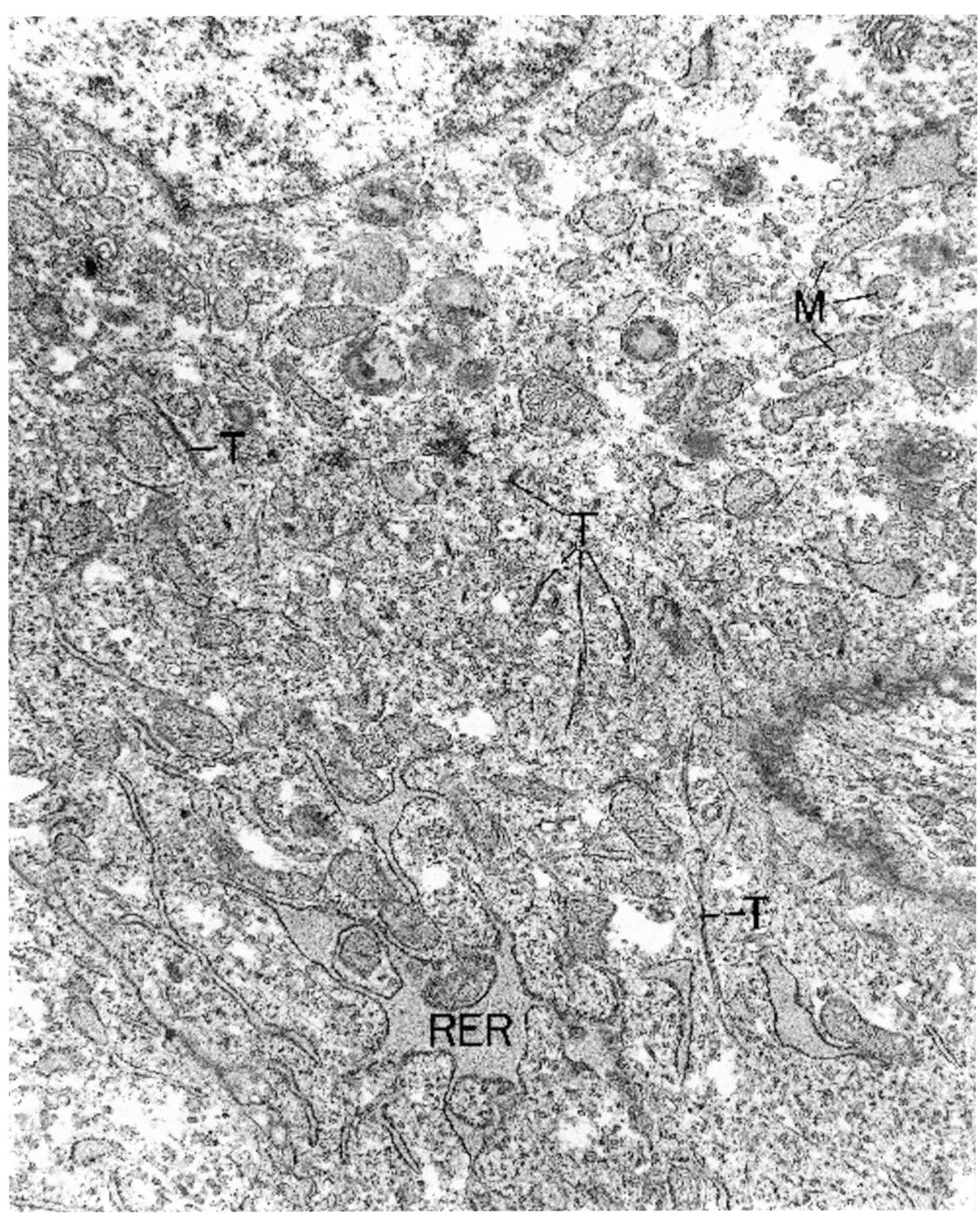

Figure 3.59. Metastatic transitional cell carcinoma from bladder (omentum). Early squamous differentiation in this cell is indicated by numerous tonofibrils (T). M = mitochondria; RER = dilated cisternae of rough endoplasmic reticulum. (× 17,700)

(Text continued from page 65)

Undifferentiated Carcinoma

(Figures 3.60 through 3.63.)

Diagnostic criteria. (1) Intercellular junctions; (2) no squamous differentiation; (3) no glandular differentiation; (4) no lymphoid or sarcomatoid differentiation (see later sections).

Additional points. Most carcinomas of the large cell, undifferentiated type at the light microscopic level can be proved to be poorly differentiated squamous cell carcinomas or adenocarcinomas when viewed through the electron microscope. This is especially true for carcinomas arising in the bronchus, nasopharynx, and paranasal sinuses, as pointed out earlier (see section on squamous cell carcinoma). However, a small percentage of neoplasms in this "undifferentiated" group appear genuinely undifferentiated at the ultrastructural level as well. Examples of these entities include germ cell tumors such as dysgerminomas and some embryonal carcinomas and a few neoplasms in which the cell line cannot be determined. Intercellular junctions are the sole feature in some of these neoplasms that allow a diagnosis of carcinoma to be made and lymphoma to be ruled out. The junctions usually are of the intermediate type and may be sparse and small. When prominent desmosomes are found, on the other hand, it is often an indication of squamous differentiation. Another epithelial feature, if the neoplasm has an insular pattern, is the presence of basal lamina surrounding groups of cells rather than individual cells.

Melanoma

(Figures 3.64 through 3.71.)

Diagnostic criteria. (1) Premelanosomes (stage I and II melanosomes); (2) melanosomes (stage III and IV); (3) atypical (or aberrant) melanosomes.

Additional points. Premelanosomes are unpigmented vesicles that include stage I and stage II melanosomes. Stage I melanosomes originate from the Golgi apparatuses as clear, round vesicles and are not diagnostic of melanoma. Stage II melanosomes consist of larger, oval or elliptical vesicles with a pathognomic internal lamellar structure. Stage III melanosomes are partially pigmented, and stage IV melanosomes are heavily pigmented and have their internal lamellar pattern completely obscured. Stage IV melanosomes and atypical melanosomes, by themselves, are often only suggestive of, or consistent with, the diagnosis, especially if they are few in number. Atypical melanosomes include a broad spectrum of morphologic types of electron-dense granules and often are difficult to distinguish from primary and secondary lysosomes. Filaments of an intermediate diameter (10 nm) are often moderately numerous in melanocytes, and arrays of microtubules within cisternae of rough endoplasmic reticulum may also be found occasionally (Figures 3.68 and 3.69). Unexpected structures such as microvilli and basal lamina may be found rarely in the cells comprising a melanoma.

Clear cell sarcoma, or *melanoma of soft parts,* has the same criteria for diagnosis as other melanomas; namely, the presence of melanosomes (Figures 3.70 and 3.71). Cytoplasmic glycogen accounts for the clear appearance of the cells at light microscopy. The cells may be polygonal and/or spindle shaped, and in some cases the spindle cells are more consistent with melanotic Schwann cells than with melanocytes. The Schwannian features include long intertwining processes, basal lamina, junctions (variable in size and number), secondary lysosomes, and long-spacing collagen (see section on Schwannoma, Chapter 6).

Balloon cell melanoma is a rare form of melanoma in which all or some of the neoplastic melanocytes are large and have copious, vacuolated cytoplasm. Some of the vacuoles disclose a melanosomal origin by the presence of striated lamellar remnants. The cause of the vacuolization is unknown but may be a degenerative change or an abnormality in melanin synthesis.

Mesothelioma

(Figures 3.72 through 3.77.)

Diagnostic criteria. (1) Numerous long, thin microvilli with a length-to-diameter ratio of 10-to-1 or higher; (2) prominent intercellular junctions, including desmosomes and junctional complexes; (3) numerous filaments, including tonofibrils; (4) glycogen; (5) intracytoplasmic lumens.

Additional points. The above criteria are characteristic of the epithelial type of mesothelioma (Figures 3.72 through 3.74). The microvilli, in addition to being on the free surface of cells lining papillae and acini, may completely surround less organoid cells and abut matrical collagen. Otherwise, a basal lamina courses along the basal plasmalemma of adjacent cells, separating them from the fibrous stroma. Cytoplasmic organelles usually include many mitochondria. Nuclei are round, and nucleoli are of moderate size. Mesotheliomas can usually be distinguished from pulmonary carcinomas by the absence of mucinous granules in the cytoplasm and by the absence of glycocalyx on the cell surface. In addition, it is rare for adenocarcinomas to have long, thin microvilli.

Less common spindle cell types of mesothelioma include sarcomatous, or mesenchymal mesothelioma, and localized fibrous tumor of the pleura. In sarcomatoid mesotheliomas, the spindle cells vary in their features, some having the prominent RER and Golgi apparatuses of fibroblasts, some having additionally the filaments and dense bodies of myofibroblasts, and others having junctions and microvilli characteristic of epithelial cells. In localized fibrous tumor of the pleura, the cells are spindle and polygonal in shape, dispersed individually and in small clusters in a matrix of dense collagen, and they may have no or only minor epithelial features such as small intercellular junctions and a few abortive microvilli (Figures 3.75 through 3.77). In addition, the cells usually have a nondescript cytoplasm, but in a small percentage of cases the cytoplasm contains prominent rough endoplasmic reticulum, consistent with fibroblasts.

Lymphoma

(Figures 3.78 through 3.86.)

Diagnostic criteria. (1) Nucleus with peripheral heterochromatin, in at least some of the cells; (2) cytoplasm composed mostly of free ribosomes and/or polyribosomes; (3) absence of intercellular junctions.

Additional points. Large cell lymphomas include high-grade follicle center lymphoma, composed of centroblasts; diffuse, large B-cell lymphoma, which has several variants including centroblastic, immunoblastic, mixed centroblastic/immunoblastic (most common), and anaplastic types; anaplastic large cell lymphoma, T- and null-cell types; and some cases of peripheral T-cell lymphoma, unspecified, subcutaneous T-cell lymphoma, and intestinal T-cell lymphoma.

Centroblasts have round nuclei with smooth or indented contours, a small amount of heterochromatin, peripheral nucleoli, and abundant cytoplasm with a predominance of ribosomes and polyribosomes and few other organelles (Figures 3.78 through 3.80). Immunoblasts have large, mostly euchromatic nuclei, large central nucleoli (frequent nucleolonemas), varying amounts (often prominent) of rough endoplasmic reticulum and numerous ribosomes and polyribosomes (Figures 3.81 and 3.82). Plasmacytoid immunoblasts have varying amounts of heterochromatin and one or two large central or peripheral nucleoli. The cytoplasm contains abundant rough endoplasmic reticulum, which is dilated and filled with a medium-dense substance (Figure 3.83). These cells have generally similar features to the plasma cells seen in plasmacytomas, multiple myeloma, and reactive plasmacytic proliferations (see Chapter 4, Figures 4.51 through 4.54).

B-cells and T-cells cannot be morphologically distinguished from one another, although T-cells tend to have more irregularly shaped nuclei.

The CD30 (Ki-1) positive anaplastic cells of T-cell, null-cell (and infrequent B-cell) lymphomas are irregularly oval and have abundant cytoplasm with a predominance of ribosomes, a moderate number of mitochondria, and a few undilated cisternae of rough endoplasmic reticulum. Nuclei are pleomorphic, indented, lobated, and euchromatic. Nucleoli are large and multiple (Figures 3.84 and 3.85). Plasmalemmas, in about 20% of cases, are raised into numerous, complicated filopodia (filiform cells, anemone cells, porcupine cells) (Figure 3.85), a feature present also in some nonanaplastic large cell (mostly B-cell) lymphomas (Figure 3.81). The ultrastructural features of the large cells of anaplastic lymphomas may be strikingly similar to those of Reed-Sternberg cells of Hodgkin's disease. Reed-Sternberg cells, depending on the plane of sectioning, may appear to have classic mirror-image, double nuclei, or they may reveal only one nucleus. Some nuclei that appear double by light microscopy are identifiable as single, bilobed nuclei with narrow isthmuses, by electron microscopy (Figure 3.86).

Although large cell lymphomas often are composed predominantly of blasts having euchromatic nuclei, they usually also contain scattered smaller cells that have the characteristic peripheral heterochromatin pattern of lymphoid nuclei (Figures 3.78 and 3.79). Other features of some large cell lymphomas are intertwining of plasmalemmas of adjacent cells and tapering, broad, polar processes (Figures 3.79 and 3.80). Intertwining of plasmalemmas is also seen with follicular dendritic cells and with paracortical interdigitating cells. Follicular dendritic cells are found in the follicular centers of normal and hyperplastic lymph nodes and have been interpreted in follicular lymphoma to indicate that the nodules derive from follicles.

Electron microscopy has elucidated the nonhistiocytic, lymphoid nature of the large neoplastic cells in large-cell lymphomas; and the term "histiocytic lymphoma", as used in the outdated Rappaport classification of lymphomas, is a misnomer. Histiocytes occur in varying numbers in large-cell lymphomas but are considered to be part of the inflammatory reaction accompanying the neoplasms. Usually, there is no difficulty encountered in distinguishing a histiocyte from a large lymphoid cell. Histiocytes have abundant cytoplasm with many different organelles. Histiocytic sarcoma, a lymphoma-like lesion composed of neoplastic histiocytes, is a very rare entity, and although it may progress to a diffuse stage, it is usually localized on first presentation (see Section on Histiocytic Disorders next).

(Text continues on page 116)

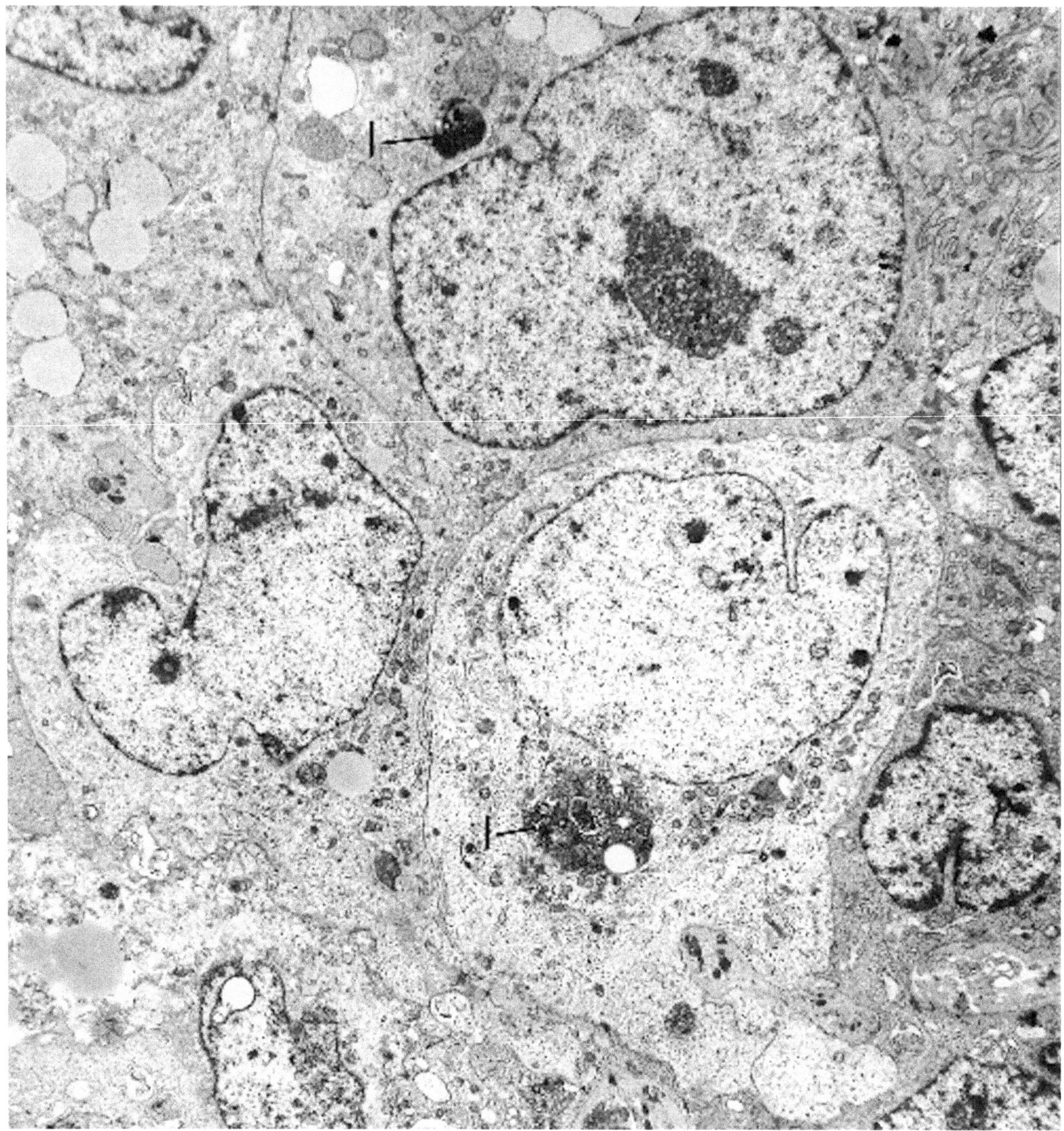

Figure 3.60. Undifferentiated large cell carcinoma (lung). The three moderately well-preserved cells in this field show an absence of squamous and glandular features and, at higher magnification (Figure 3.61), have a few small intercellular junctions. There is a high nuclear–cytoplasmic ratio, nuclei are predominantly euchromatic and have large nucleoli, and the cytoplasm is rich in free ribosomes. The large secondary lysosomes (l) could represent either a heterophagosome or an autophagosome. (× 5940)

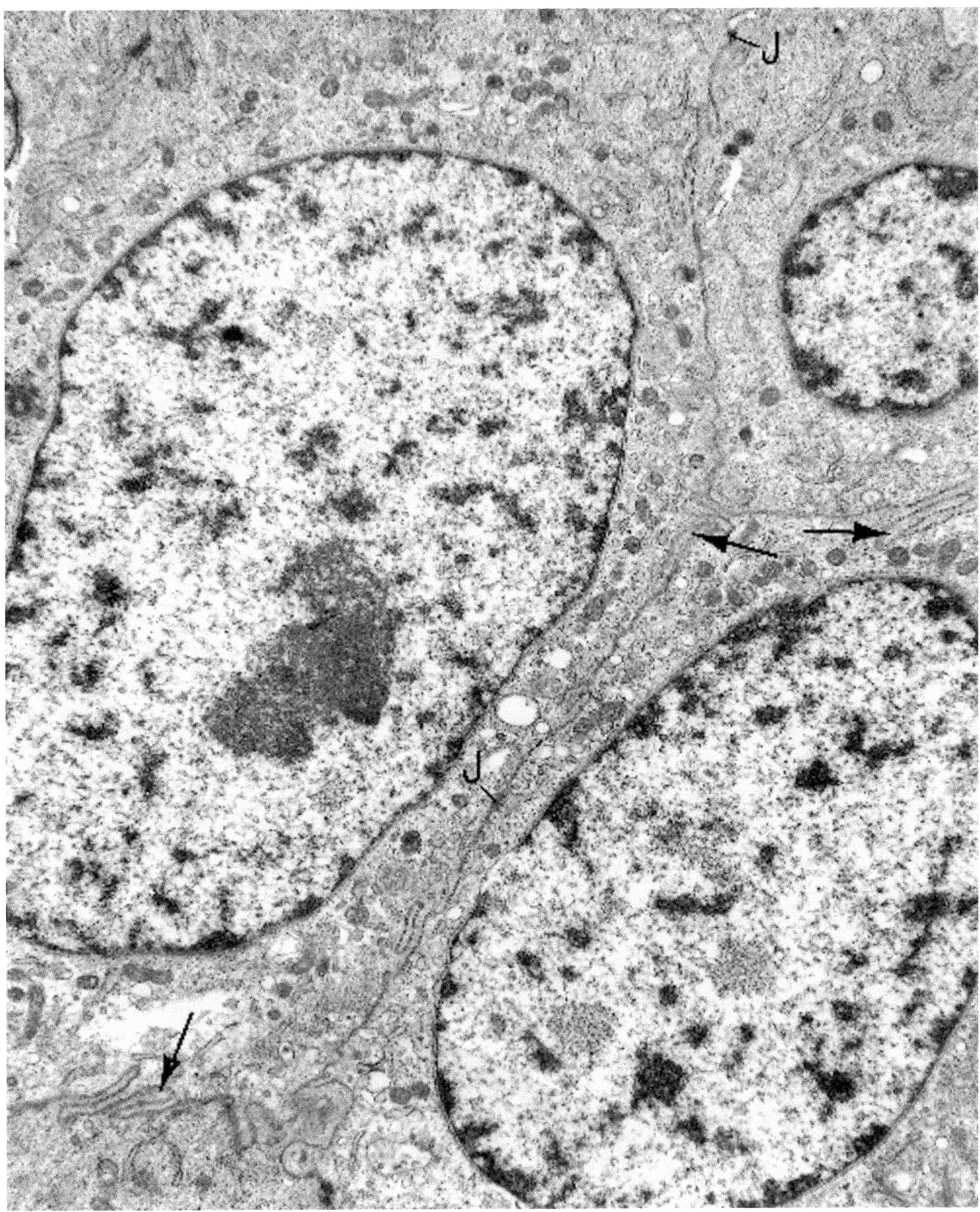

Figure 3.61. Undifferentiated large cell carcinoma (lung). Higher magnification of the same neoplasm as that depicted in Figure 3.60 shows at least two small intercellular junctions (J). In addition, cell membranes of adjacent cells show focal interdigitation (*arrows*), less diffuse than those seen in urothelial cells. (× 11,700)

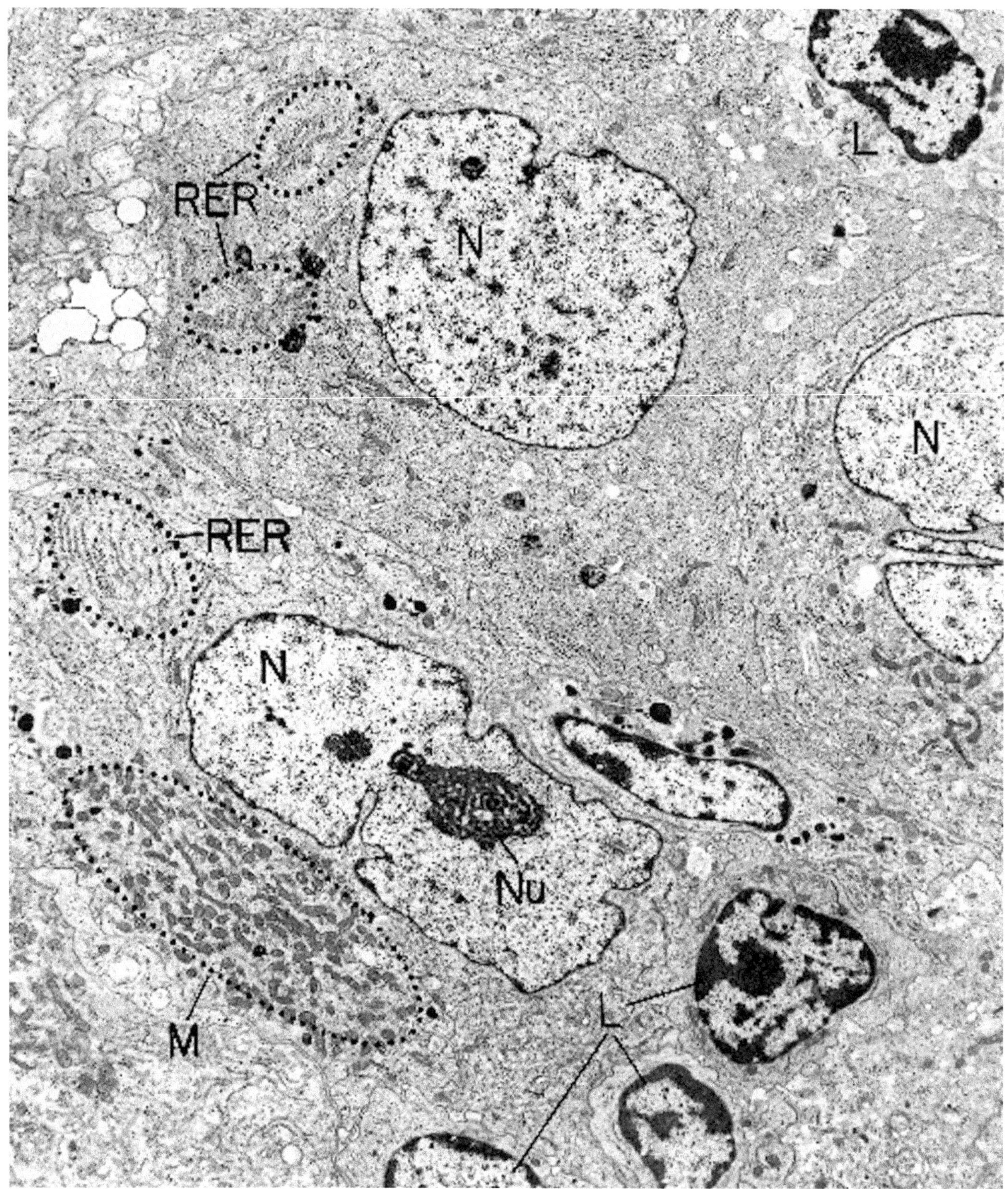

Figure 3.62. Metastatic undifferentiated large cell carcinoma (posterior mediastinum, in paratracheal lymph nodes). Several large neoplastic cells (N) are surrounded by lymphocytes (L), and the neoplastic cells have no squamous or glandular differentiation. The cytoplasm is rich in free ribosomes and also has numerous mitochondria (M) and undilated cisternae of rough endoplasmic reticulum (RER). Nuclei are large, irregularly shaped, and euchromatic, and they have large nucleoli (Nu). Junctions cannot be discerned at this low magnification. (× 5510)

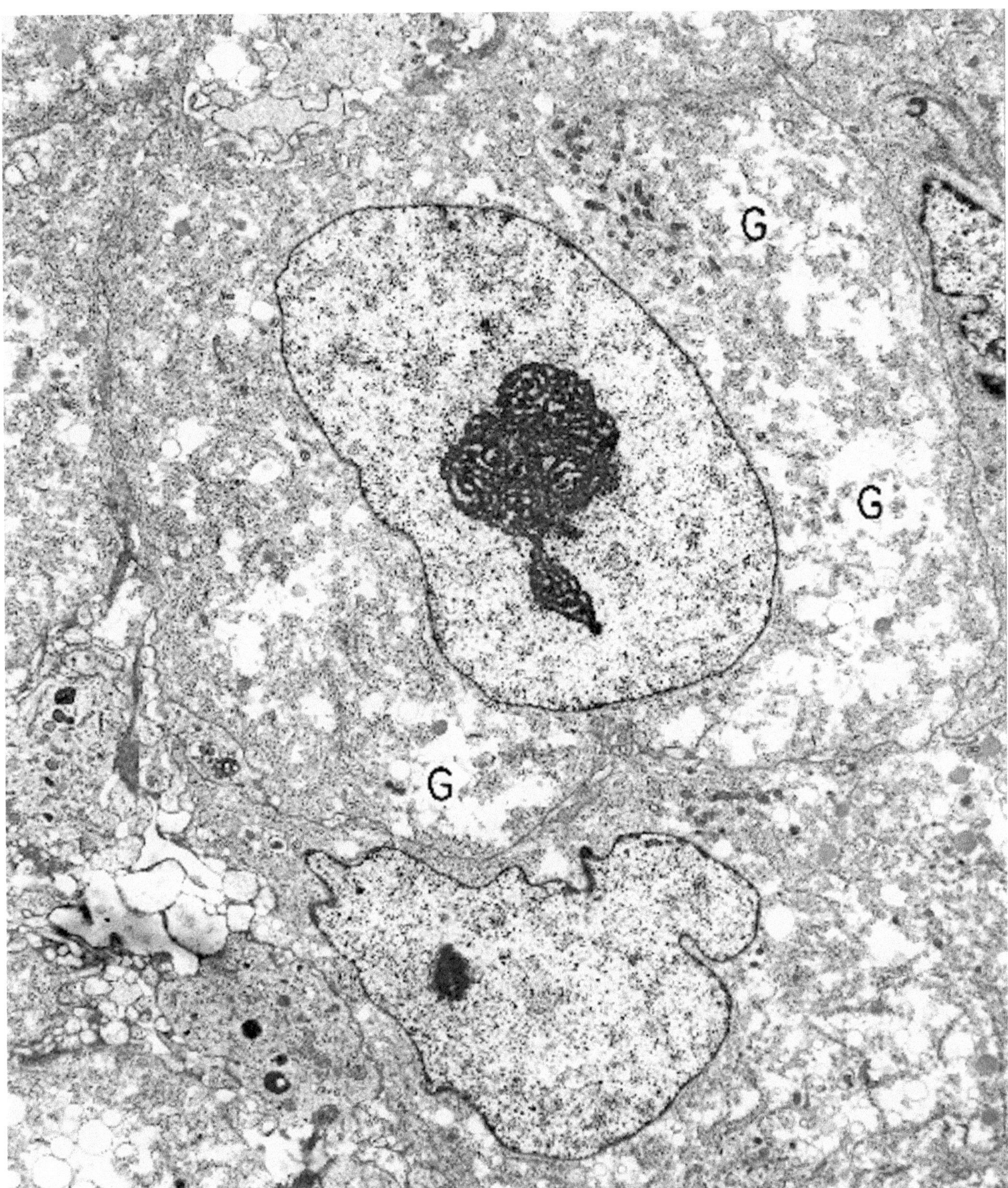

Figure 3.63. Metastatic undifferentiated large cell carcinoma (posterior mediastinum, in paratracheal lymph node). Higher magnification of the same neoplasm as that shown in Figure 3.62 highlights the lack of squamous and glandular differentiation in the cytoplasm and the inconspicuousness of intercellular junctions. The presence of copious glycogen (G, *clear spaces*) rules out lymphoma and often is present in undifferentiated large cell carcinomas. (× 5720)

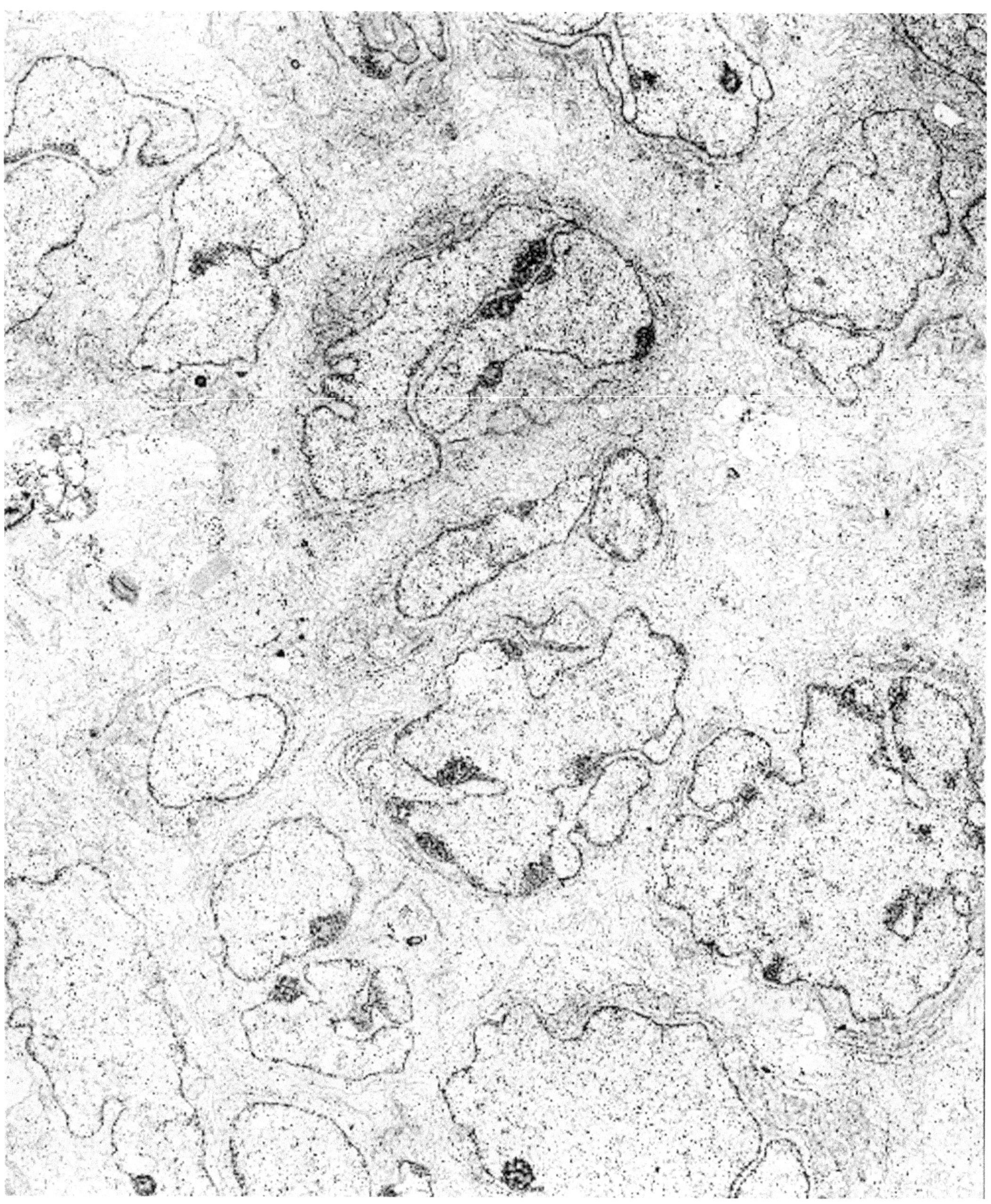

Figure 3.64. Melanoma (metastatic to soft tissue of thigh). By light microscopy and low-power electron microscopy, this poorly differentiated neoplasm was amelanotic, and only at higher power did a few cells contain diagnostic melanosomes. Note the marked nuclear pleomorphism among the cells. Also, chromatin is finely dispersed, and nucleoli are marginally located along the nuclear envelope. (× 5130)

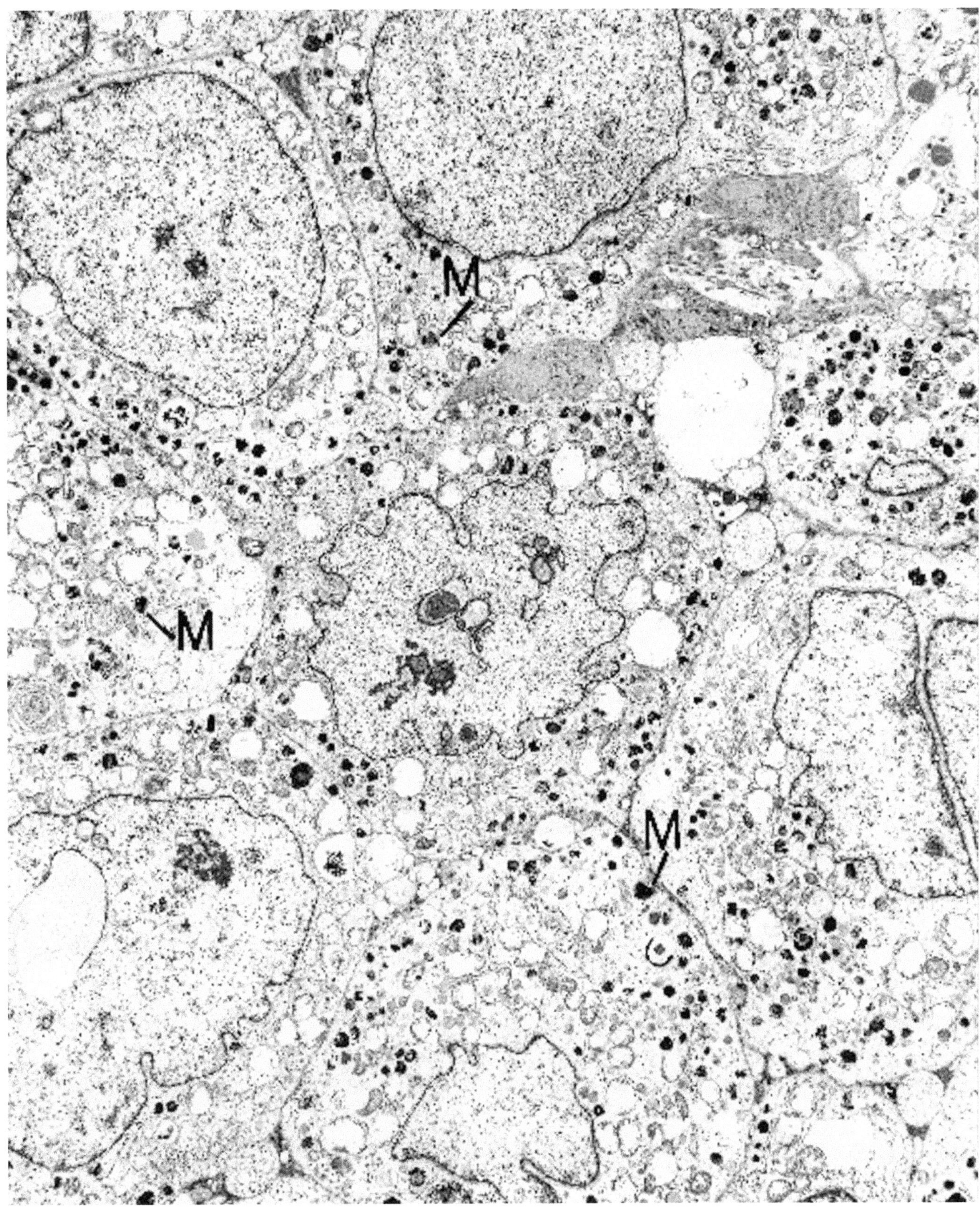

Figure 3.65. Melanoma (nasal mucosa). This neoplasm, as compared with Figure 3.64, is extremely melanotic, having many melanosomes (M) in most of the cells. Note the variation in size, shape, and density among the melanosomes. (× 5130)

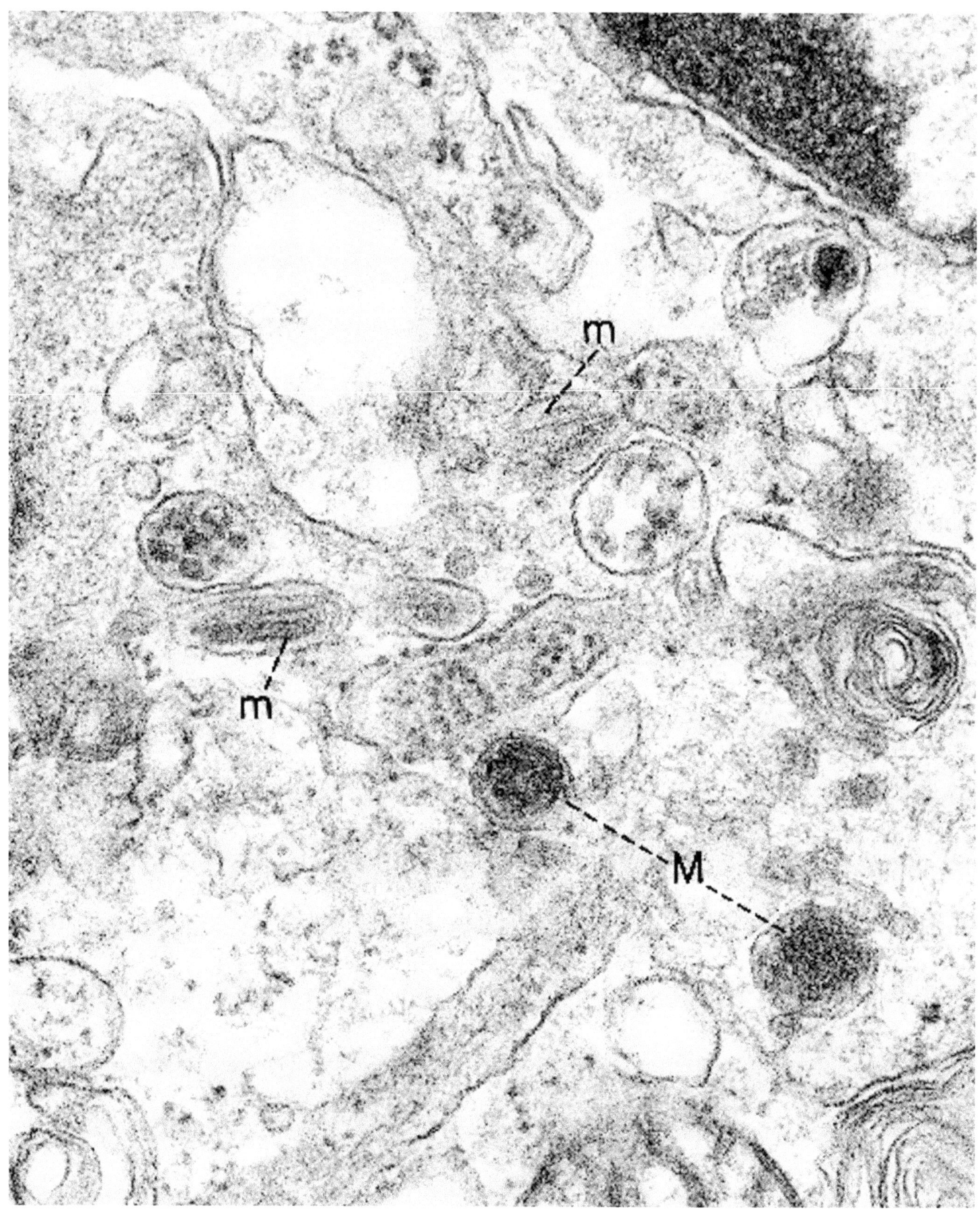

Figure 3.66. Melanoma (skin of upper eyelid). Diagnostic early-stage melanosomes (m) have a discernible internal lamellar pattern, whereas later-stage melanosomes (M) have partial or complete obliteration of that pattern by synthesized melanin pigment. (× 100,500)

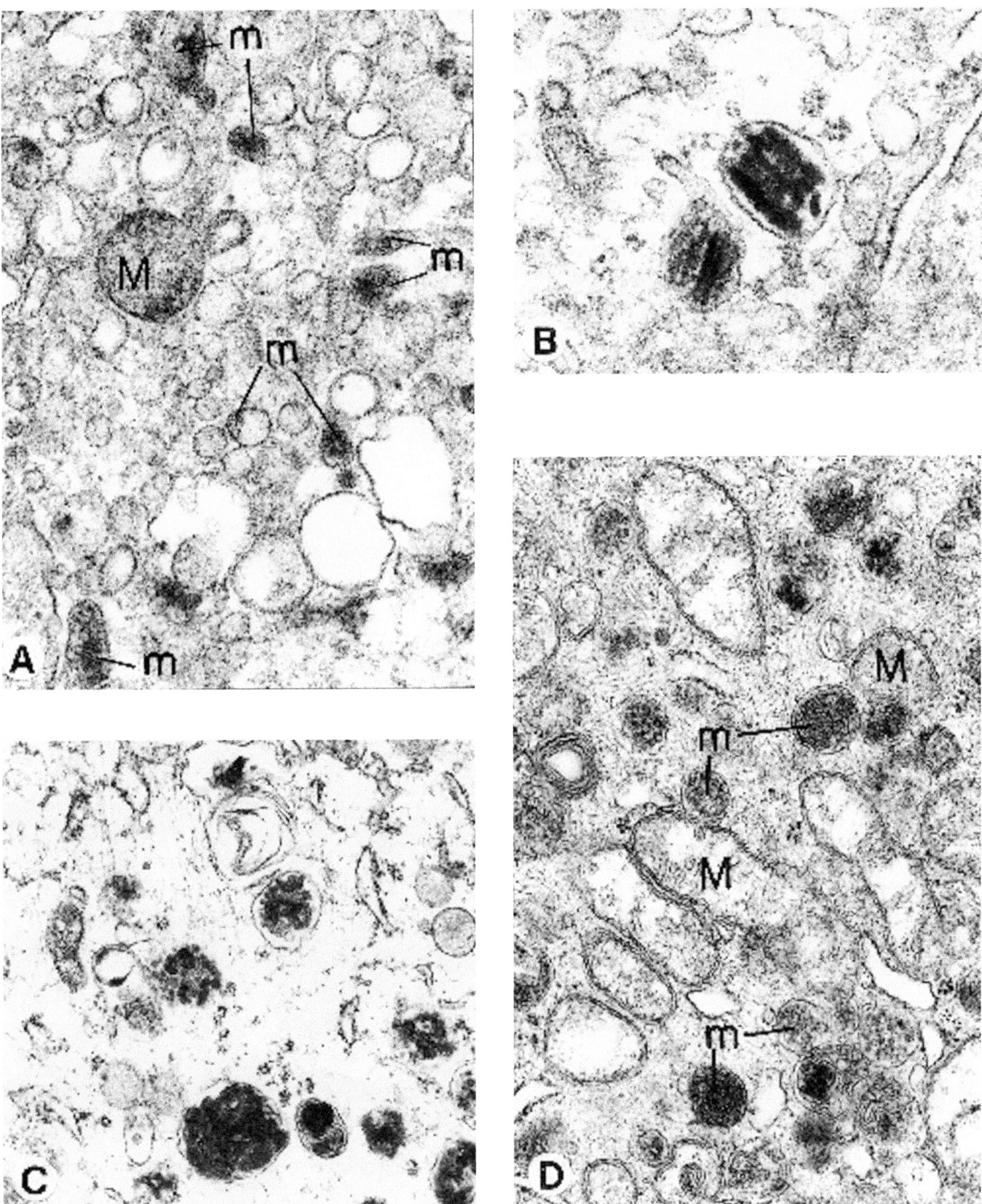

Figure 3.67. Melanomas. Examples of later-stage melanosomes from four different melanomas, illustrating the range of morphology possible in these organelles. In **A** and **D,** the size of the melanosomes (m) can be contrasted to that of mitochondria (M). **A,** × 63,000. **B,** × 64,800. **C,** × 29,700. **D,** × 40,500.

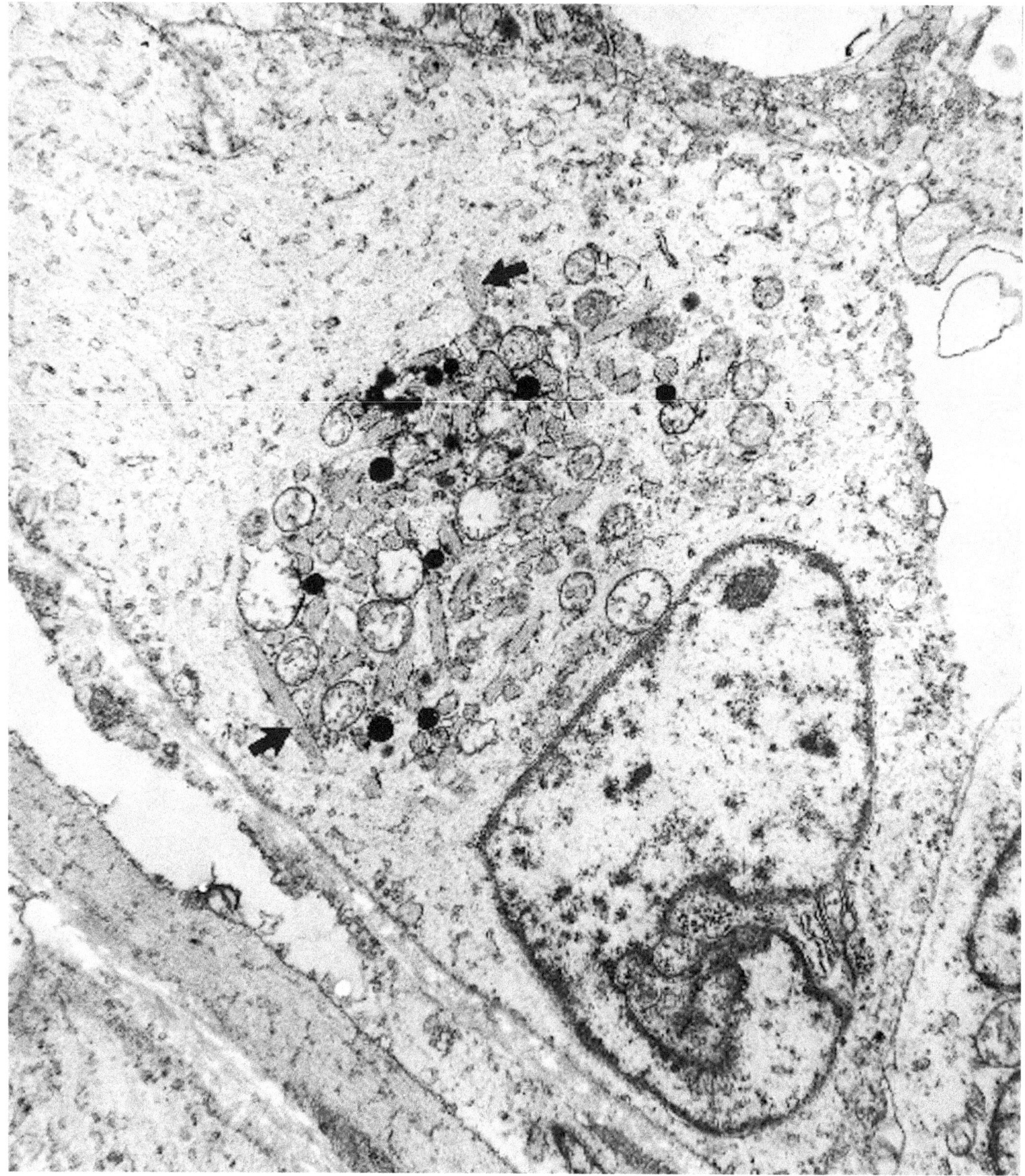

Figure 3.68. Melanoma (metastatic to inguinal lymph node). A malignant melanocyte contains a paranuclear collection of intracisternal microtubules (*arrows*) (see higher magnification in Figure 3.69). (× 14,300)

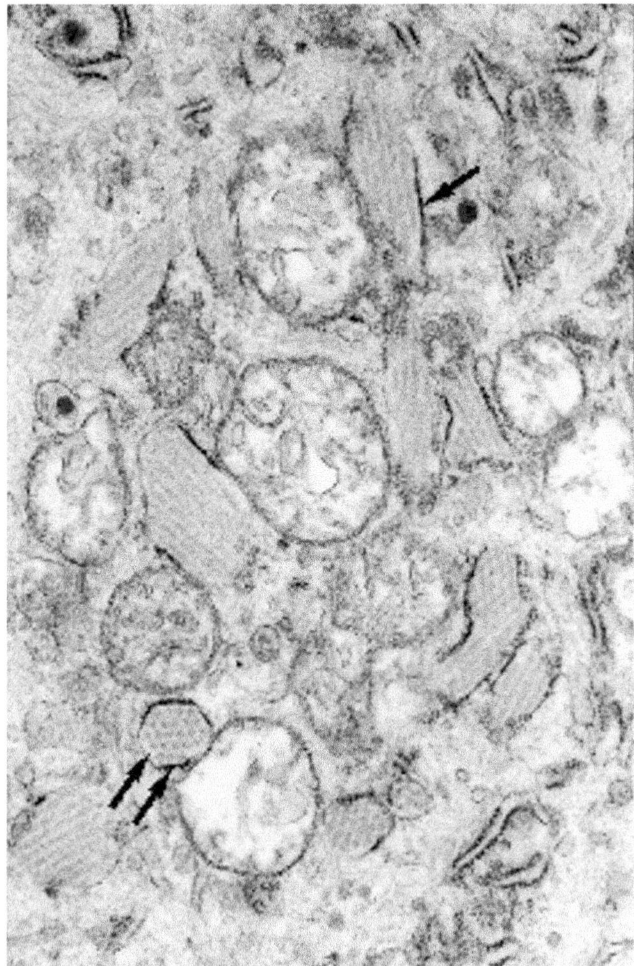

Figure 3.69. Melanoma (metastatic to inguinal lymph node). Higher magnification of a cell from the same neoplasm shown in Figure 3.68 depicts the intracisternal microtubules in longitudinal (*arrow*) and transverse (*double arrows*) directions. (× 30,000)

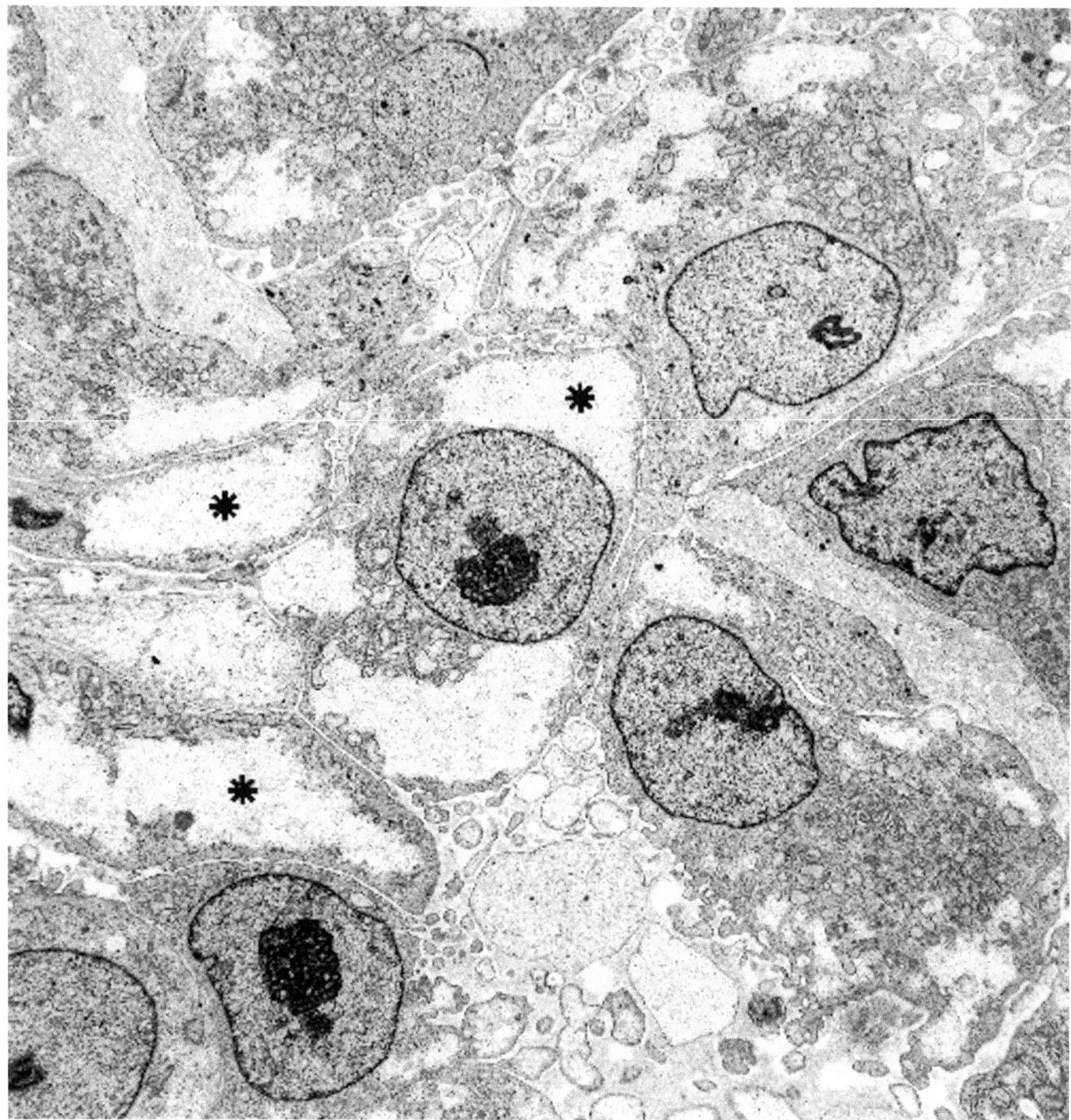

Figure 3.70. Clear cell sarcoma (lymph node, left arm). Closely arranged polygonal cells have large euchromatic nuclei with prominent nucleoli and abundant cytoplasmic glycogen (*, *open finely granular spaces*). (× 5700)

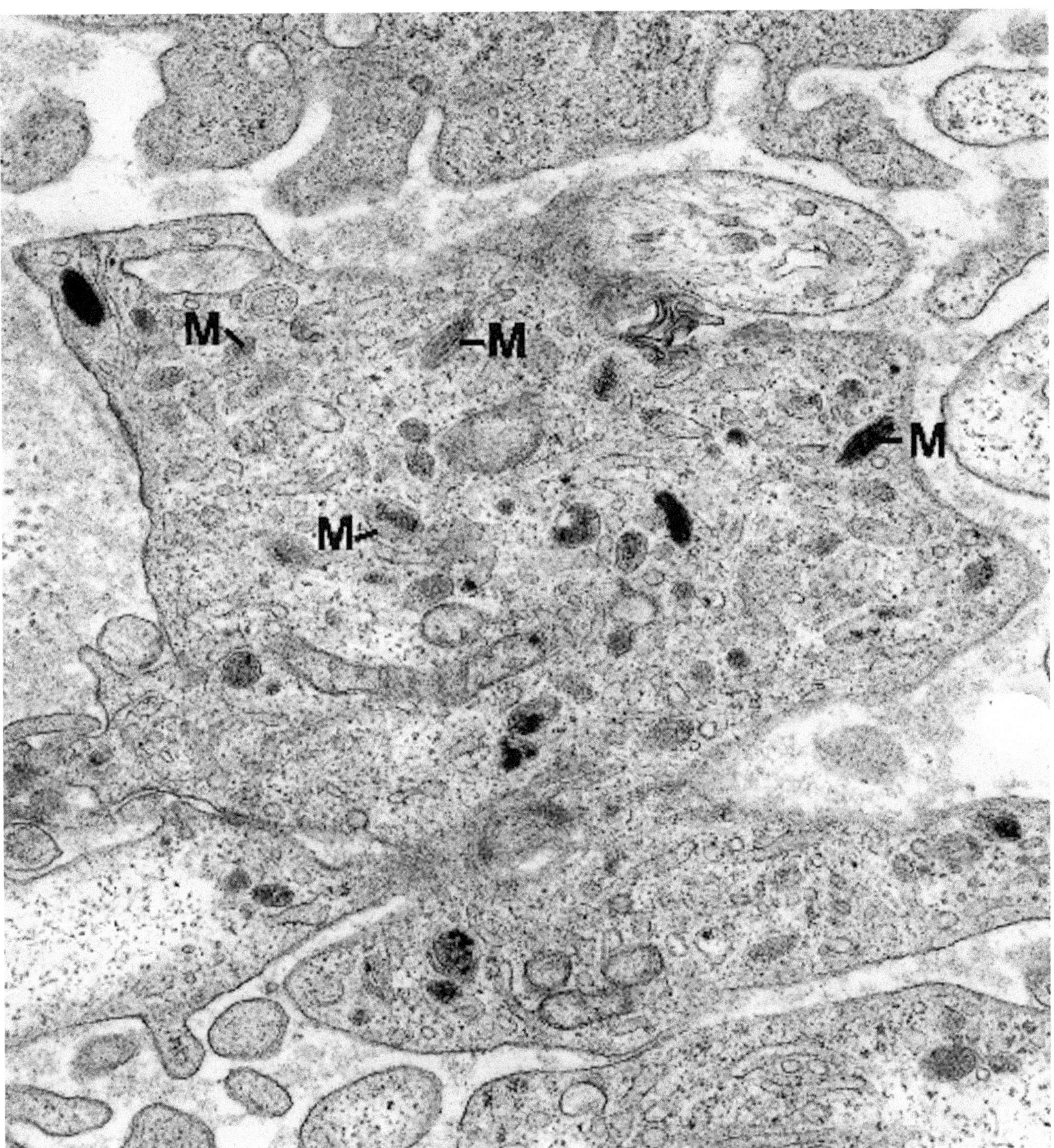

Figure 3.71. Clear cell sarcoma (lymph node, left arm). High magnification of a portion of one of the clear cells from the case illustrated in Figure 3.70 shows numerous melanosomes (M) of various stages. (× 35,000)

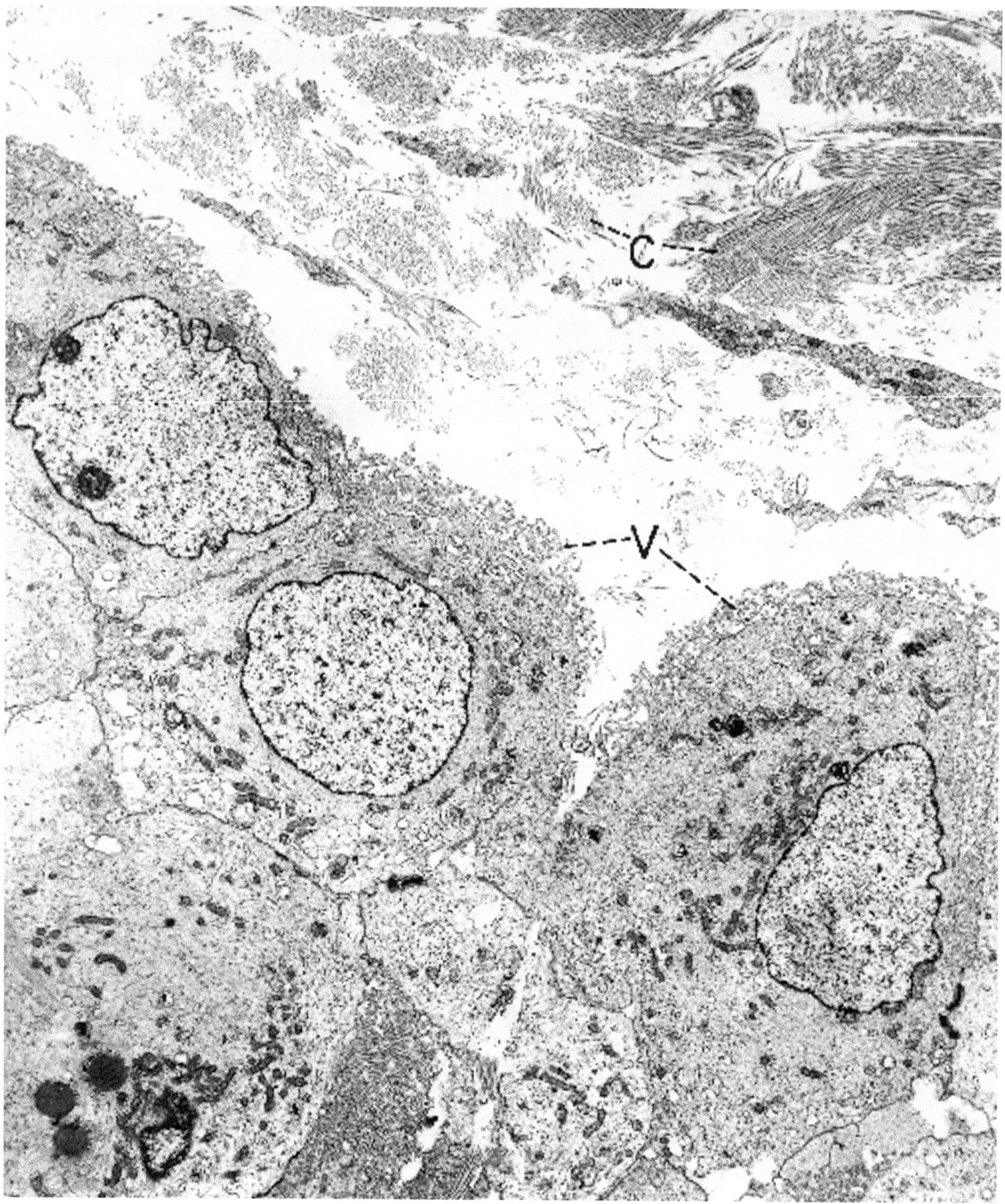

Figure 3.72. Mesothelioma (pleura and anterior thoracic wall). Epithelial type cells have a floridly villous free surface (V) and tightly apposed other surfaces. C = collagen. (× 4940)

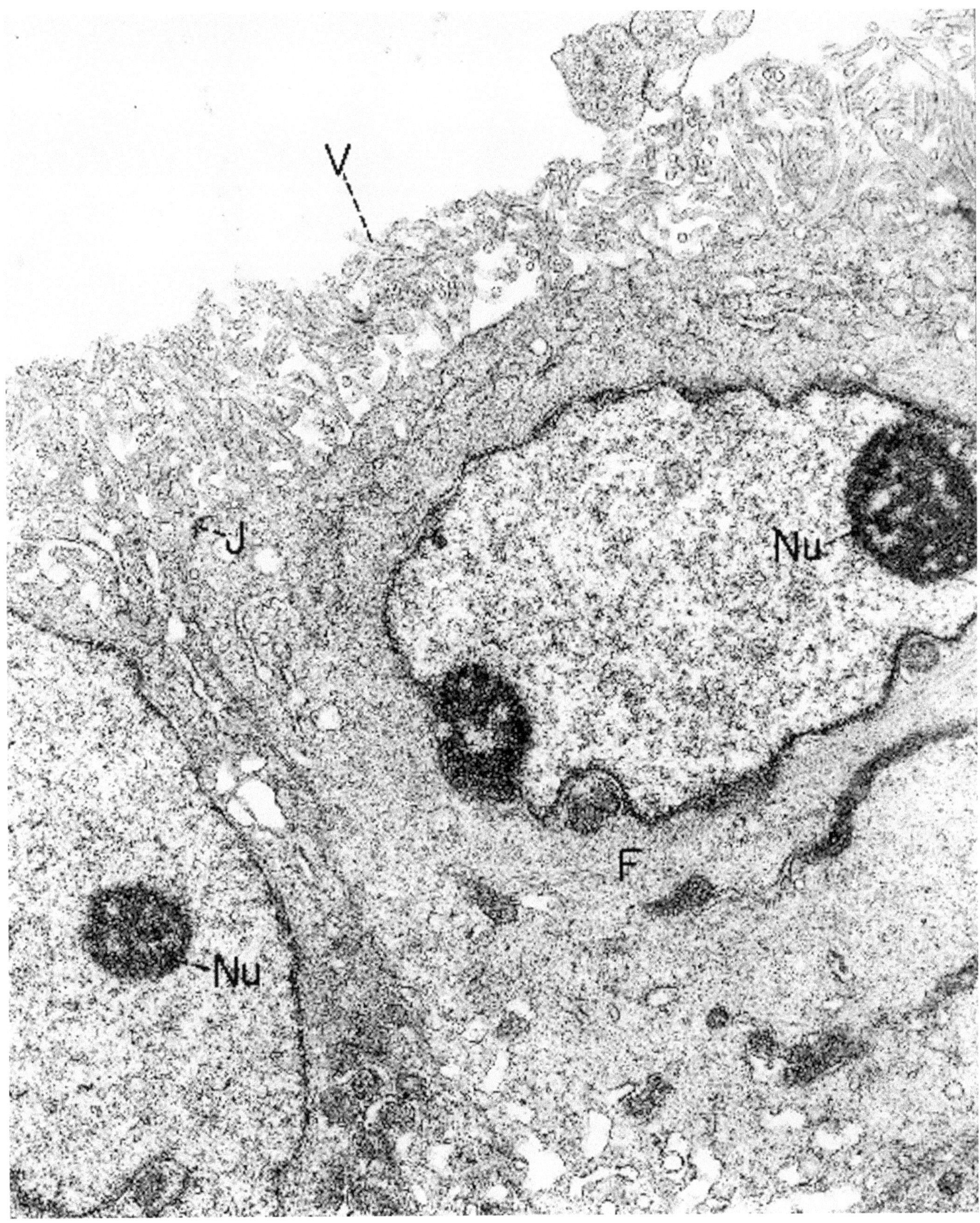

Figure 3.73. Mesothelioma (pleura and anterior thoracic wall). Higher magnification of a cell from the same neoplasm as depicted in Figure 3.72. The surface villi (V) are long and numerous. Intercellular junctions (J) and microfilaments (F) are prominent, and nucleoli (Nu) are large. (× 15,000)

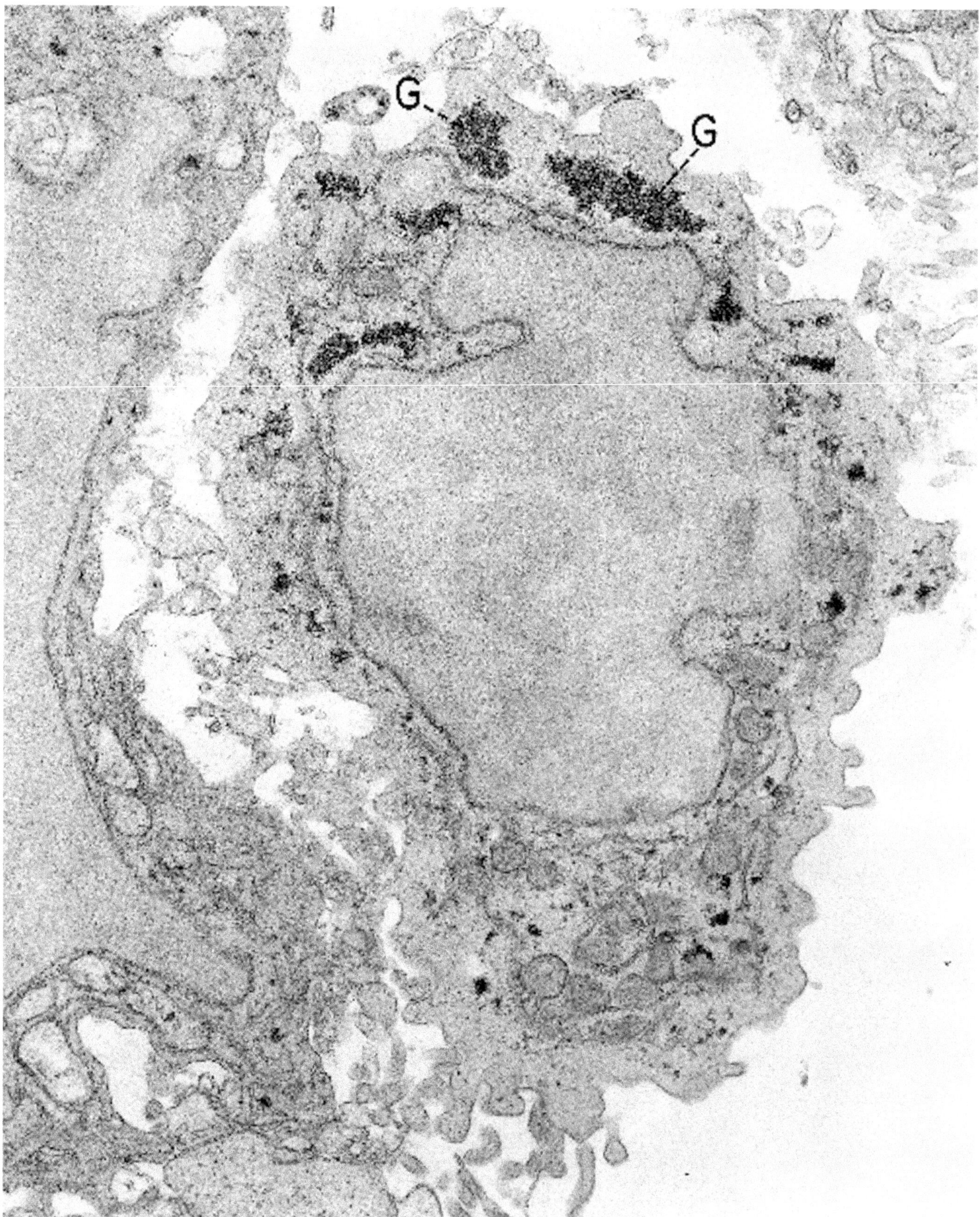

Figure 3.74. Mesothelioma (pleura). This specimen was specially processed to preserve glycogen (G) as electron-dense granules. Glycogen often is copious in the cells of mesotheliomas. (× 16,000)

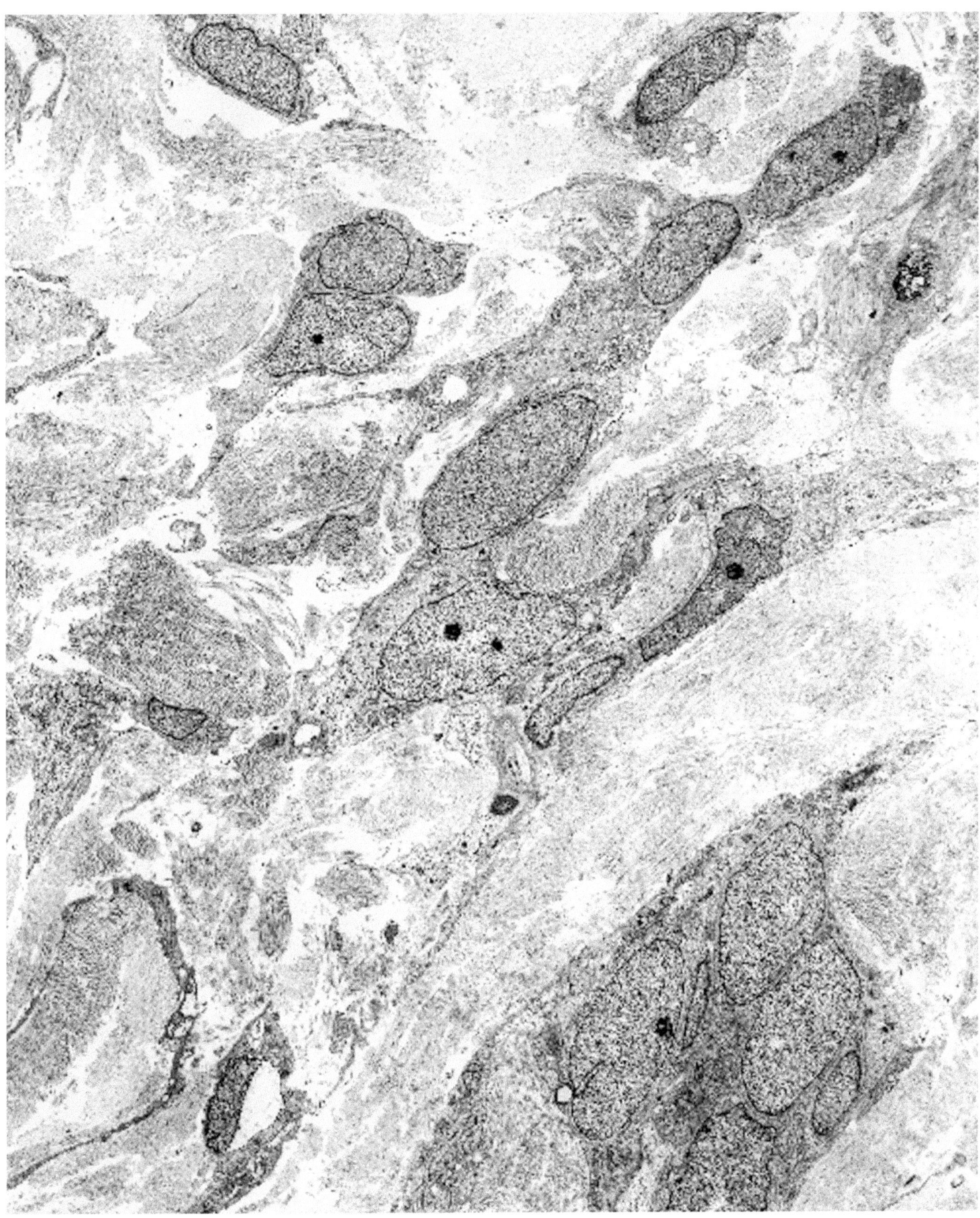

Figure 3.75. Fibrous mesothelioma (pleura). The neoplastic cells are dispersed individually and in small groups within a matrix of collagen. Within groups, some of the cells are more oval and polygonal than spindle shaped. (× 3245)

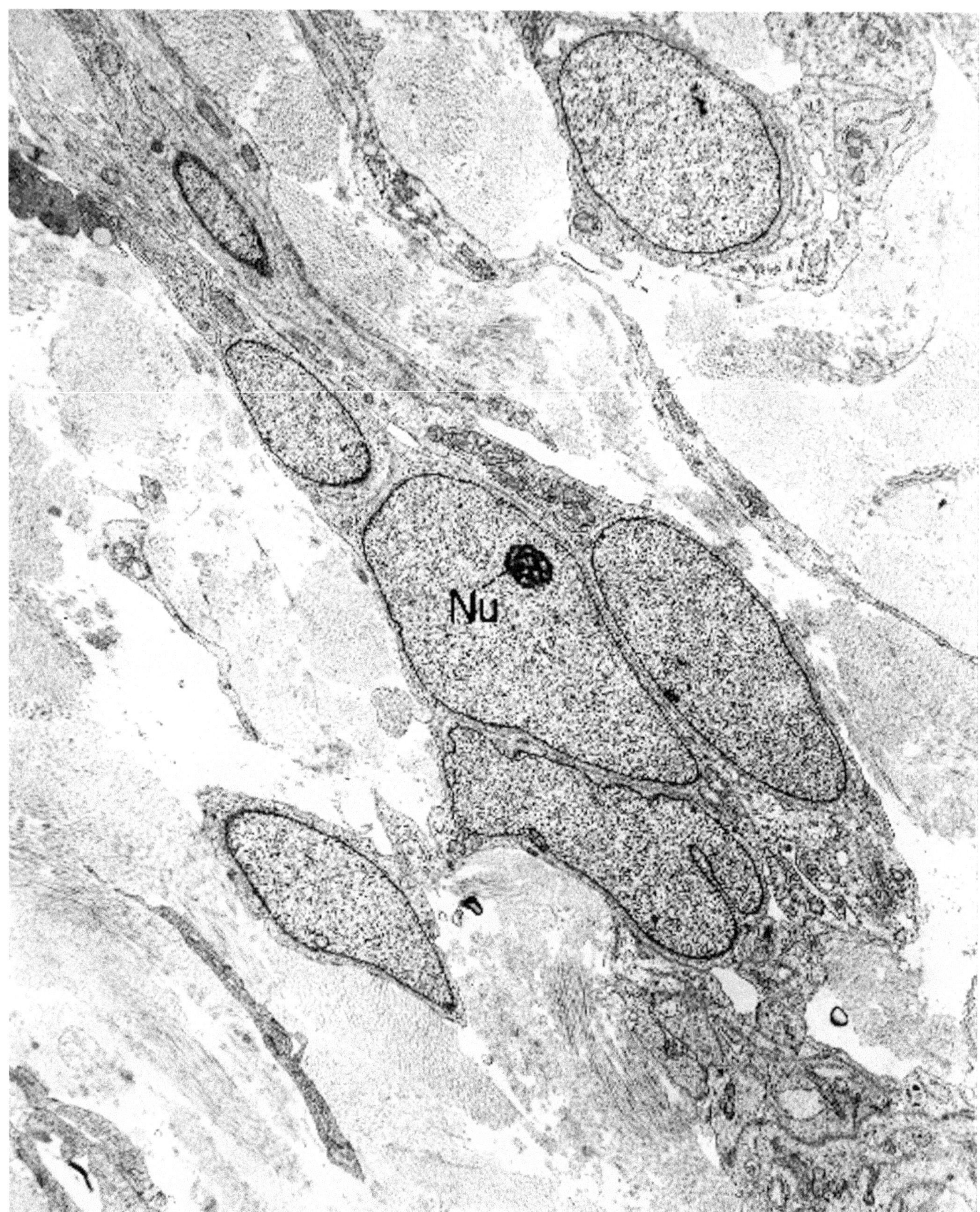

Figure 3.76. Fibrous mesothelioma (pleura). A group of neoplastic cells are tightly apposed and spindle and polygonal in shape. The high nuclear–cytoplasmic ratio, the finely dispersed chromatin, and the prominent nucleolus (Nu) all contribute to the poorly differentiated appearance of the cells. (× 5700)

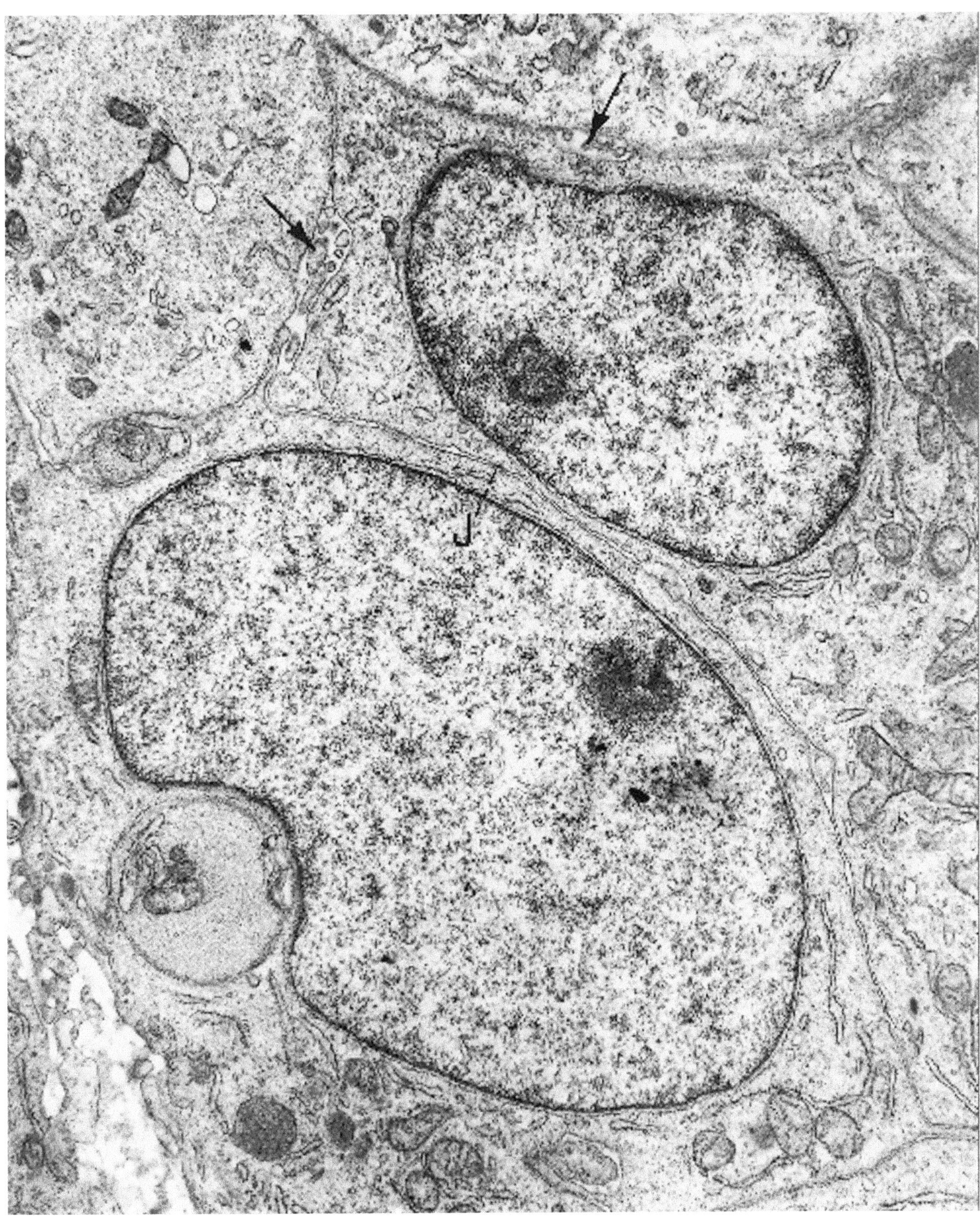

Figure 3.77. Fibrous mesothelioma (pleura). High magnification of a group of neoplastic cells depicts several epithelial-like features: polygonal shapes, small intercellular junctions (J), and small, intercellular, villus-lined spaces (*arrows*). (× 16,245)

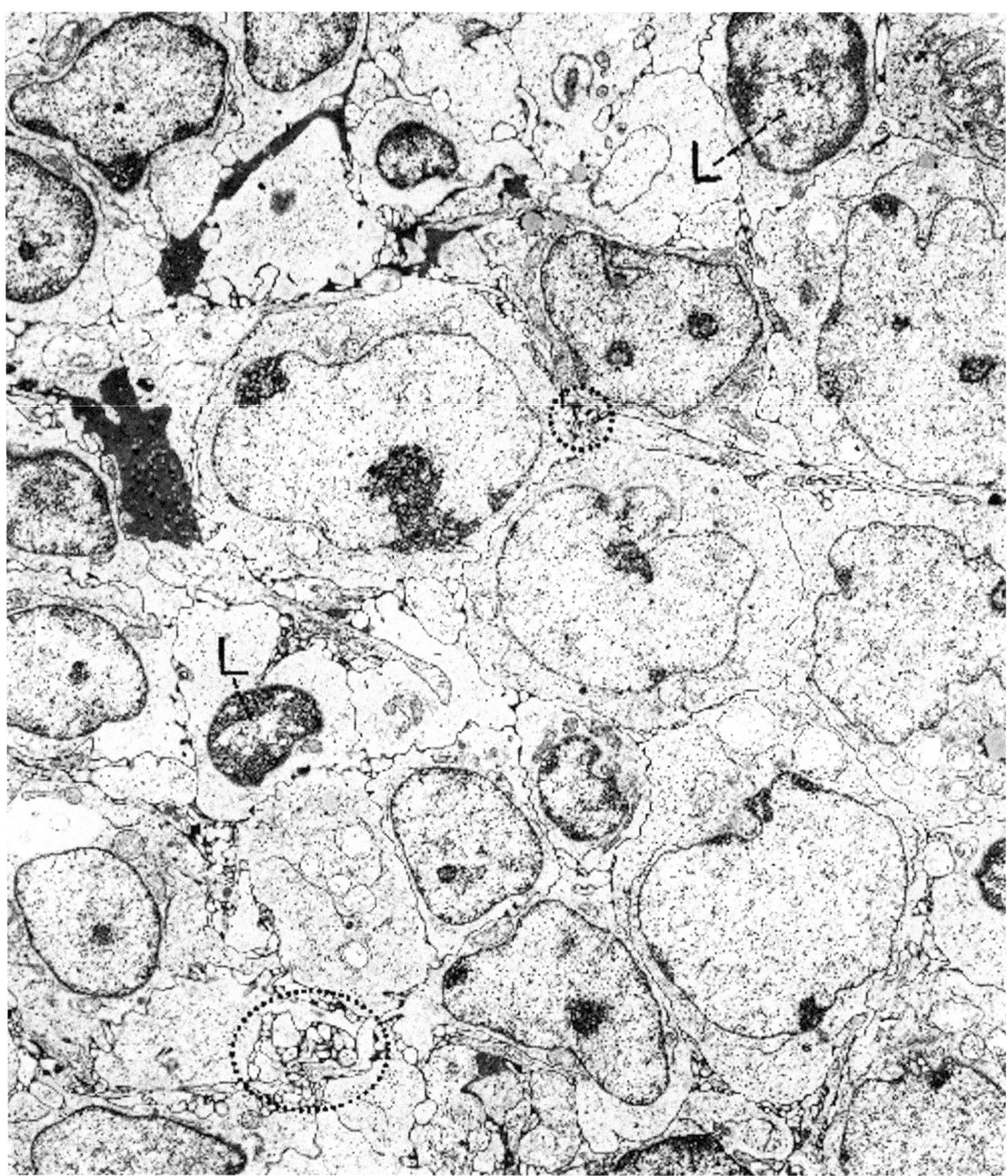

Figure 3.78. Reactive follicular center (cervical lymph node). The large cells are centroblasts and have a high nuclear–cytoplasmic ratio, finely dispersed chromatin (euchromatin), large nucleoli, and cytoplasm composed mostly of free ribosomes. Scattered among the centroblasts are smaller centrocytes having the aggregated chromatin (heterochromatin) pattern characteristic of lymphocytes (L). Although the cells are in tight apposition, there are no intercellular junctions. The abutting plasmalemmas of adjacent cells are intertwined in many foci (*circles*). (× 4990)

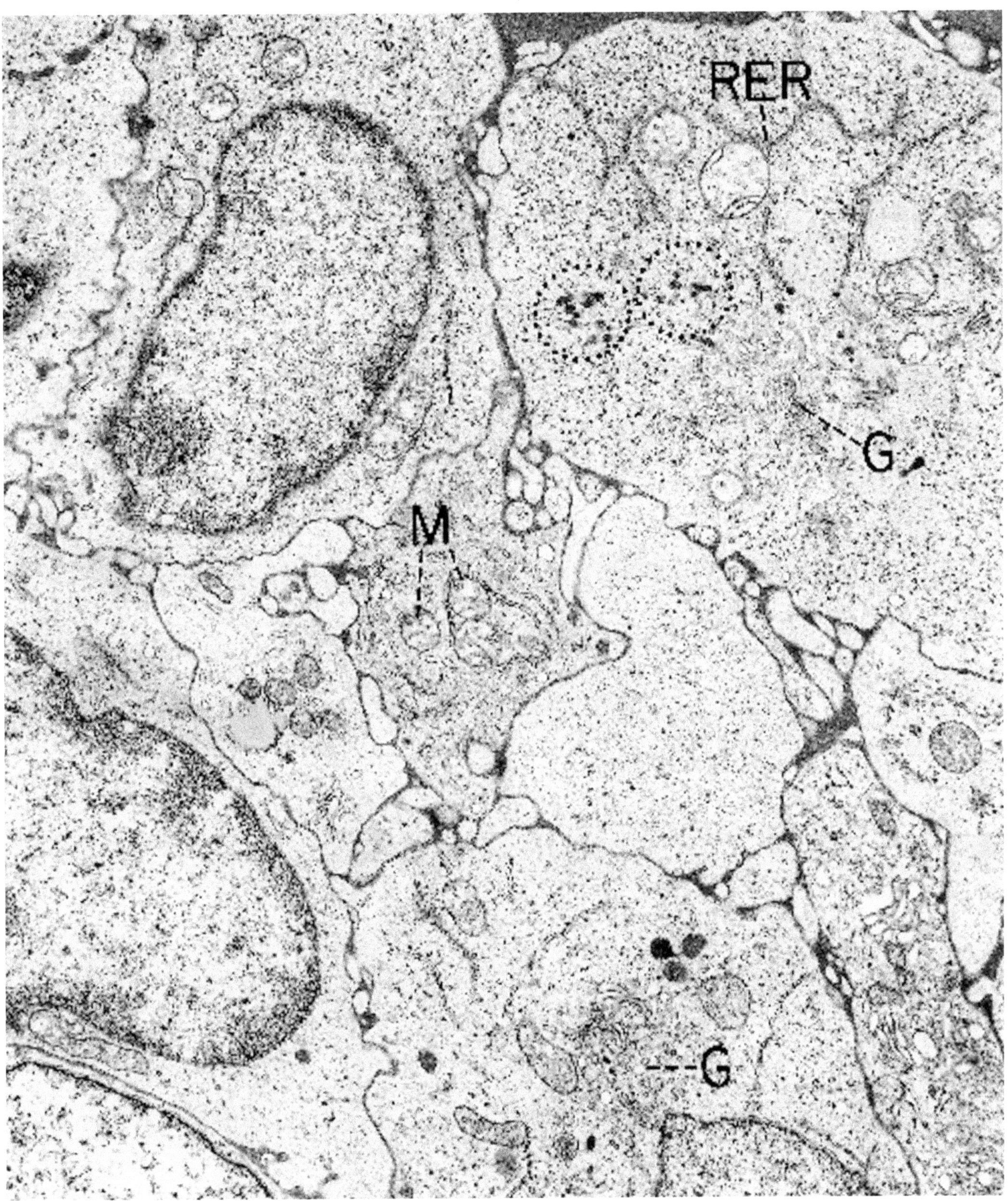

Figure 3.79. Reactive follicular center (cervical lymph node). This higher magnification of the specimen depicted in Figure 3.78 illustrates the intertwining of the plasma membranes of adjacent cells. Although free ribosomes predominate in the cytoplasm of these centroblasts, there also are a few mitochondria (M), scattered cisternae of rough endoplasmic reticulum (RER), and prominent Golgi apparatuses (G) with a few primary lysosomes (*circles*). (× 23,660)

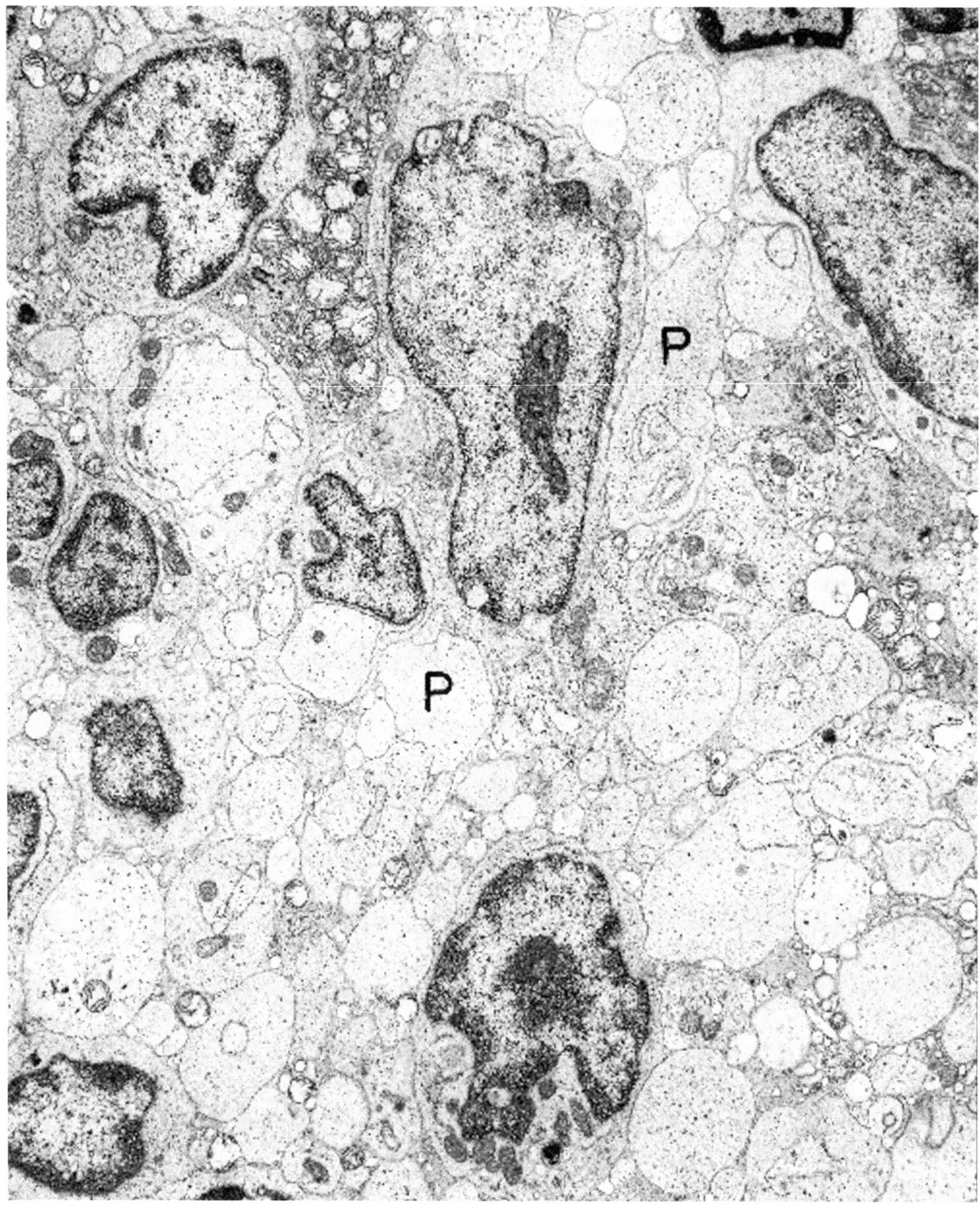

Figure 3.80. Nodular and diffuse, centroblastic large cell lymphoma (nasopharynx). This field illustrates the many broad processes (P) that lymphoid cells may exhibit. (× 7760) (Permission for reprinting granted by WB Saunders, Dickersin GR: Electron microscopy of leukemias and lymphomas. Clin Lab Med 7:199–247, 1987.)

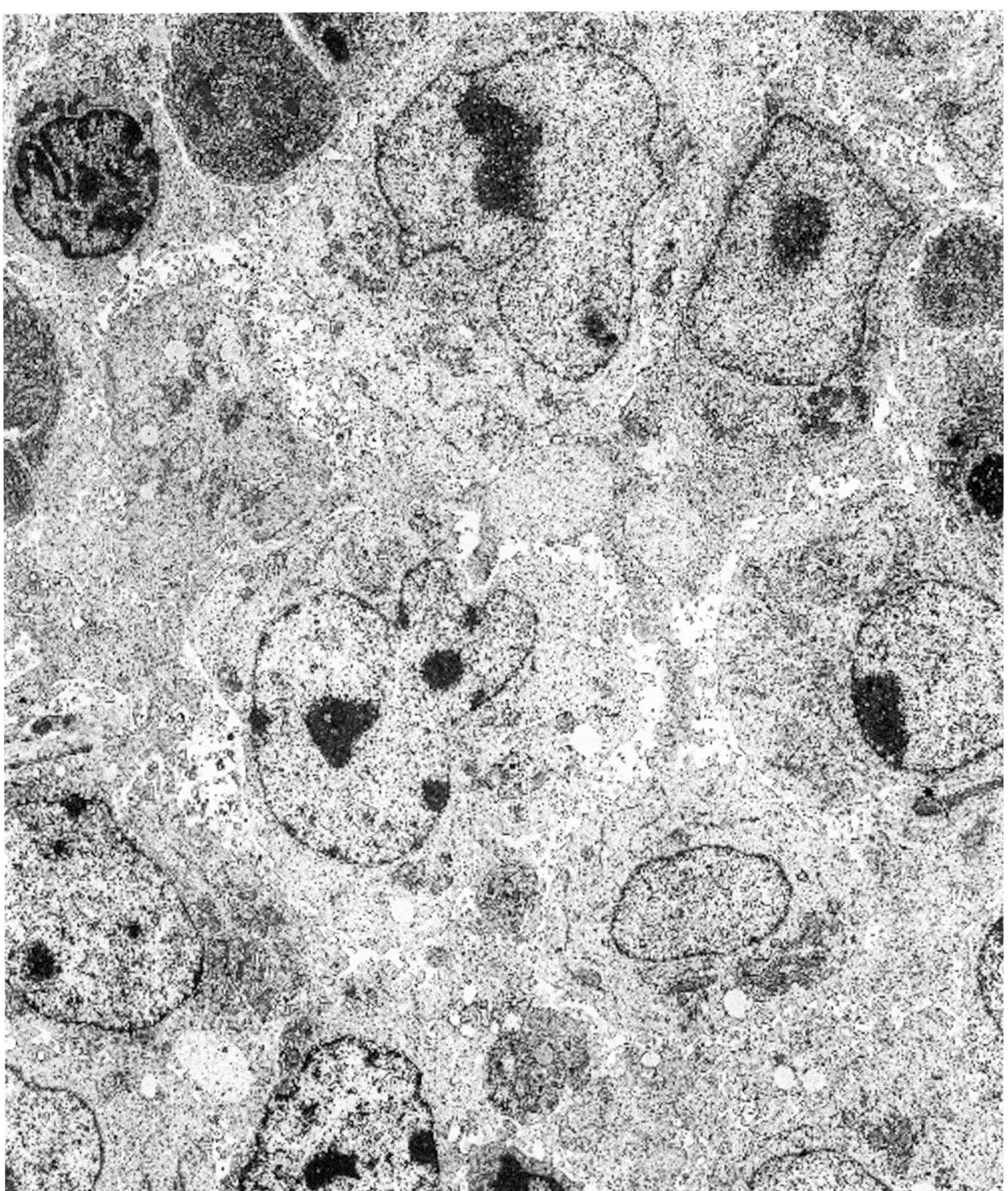

Figure 3.81. Diffuse large cell lymphoma, immunoblastic (cervical lymph node). These large lymphoid cells have a microvillous surface, similar to what would be expected in an adenocarcinoma. However, there are no junctions between the cells and no formation of microacini. (× 5250) (Permission for reprinting granted by WB Saunders, Dickersin GR: Electron microscopy of leukemias and lymphomas. Clin Lab Med 7:199–247, 1987.)

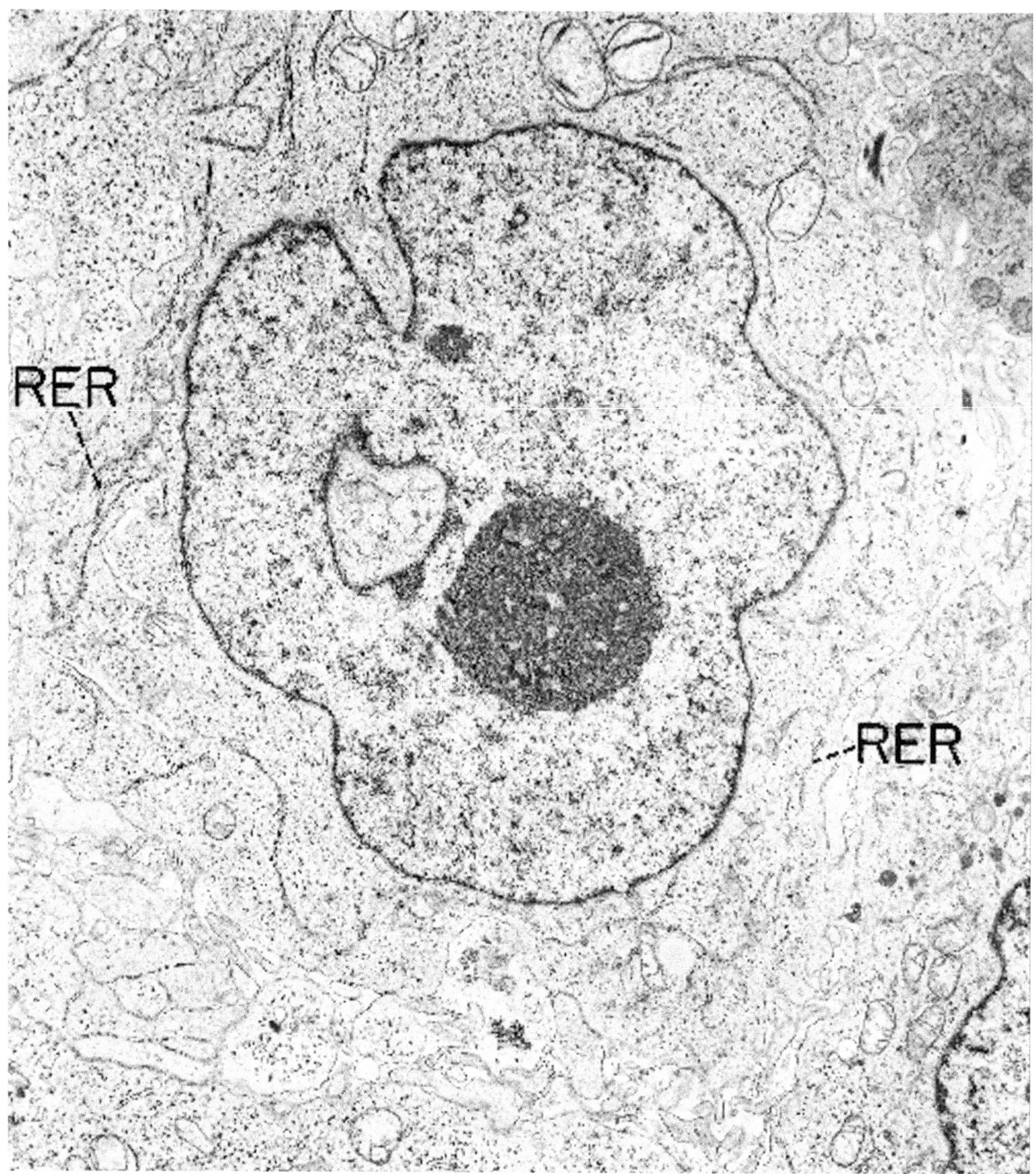

Figure 3.82. Diffuse large cell lymphoma, immunoblastic (axillary lymph node). This blast, when viewed individually and not in context with the remaining infiltrate, could be difficult to prove as a lymphoblast. The nucleus does not have the typical heterochromatin pattern of the lymphoid series, although the innumerable ribosomes and absence of intercellular junctions are clues that the cell is a lymphoblast. The moderate amount of rough endoplasmic reticulum (RER) and central, large nucleolus are consistent with an immunoblast. (× 9750) (Permission for reprinting granted by WB Saunders, Dickersin GR: Electron microscopy of leukemias and lymphomas. Clin Lab Med 7:199–247, 1987.)

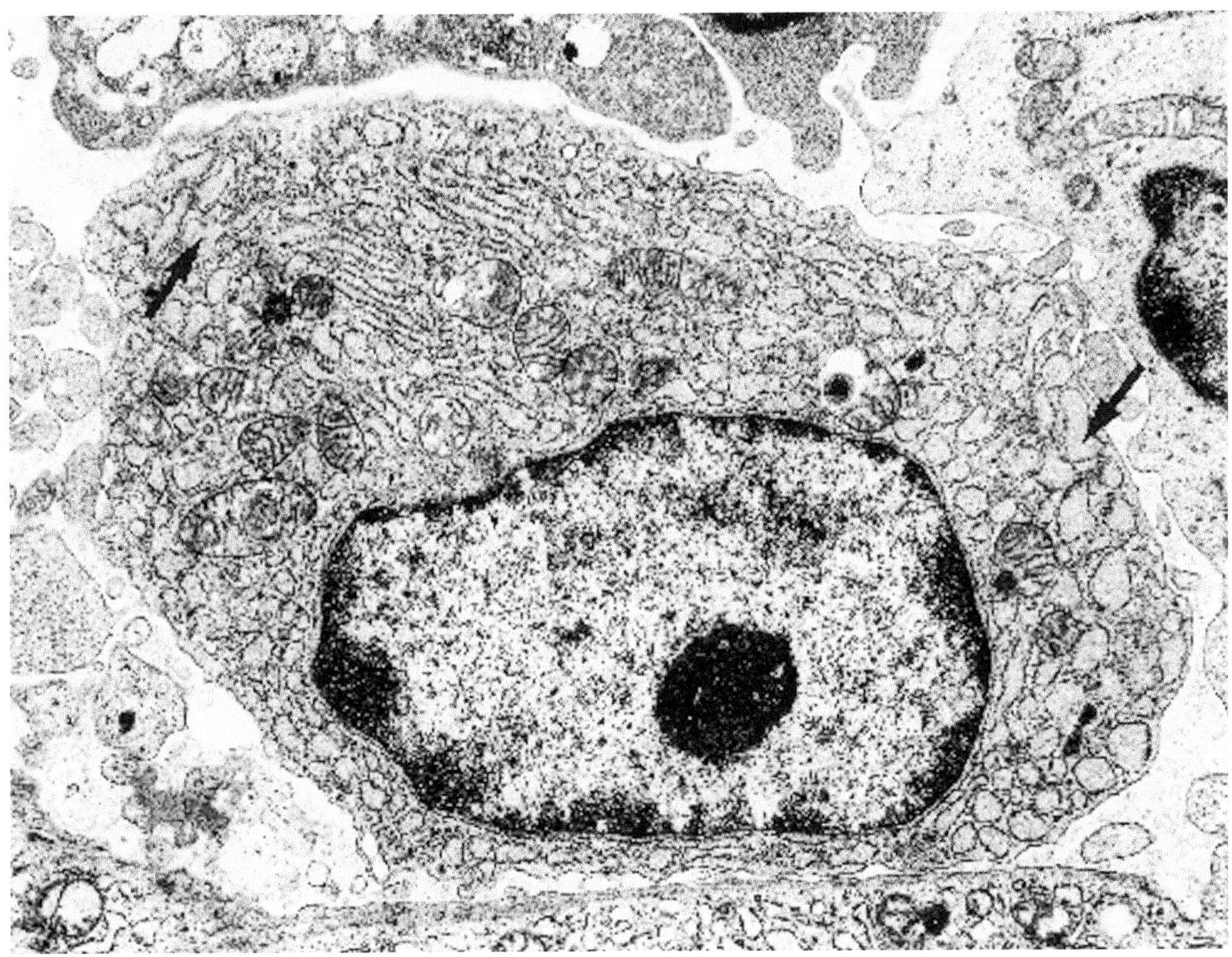

Figure 3.83. Large cell lymphoma, immunoblastic, plasmacytoid subtype (cervical lymph node). A plasmacytoid immunoblast has cytoplasm filled with dilated rough endoplasmic reticulum (*arrows*). The nucleus has prominent peripheral heterochromatin, characteristic of lymphocytes/plasma cells. A nucleolus is large and central. (× 14,200)

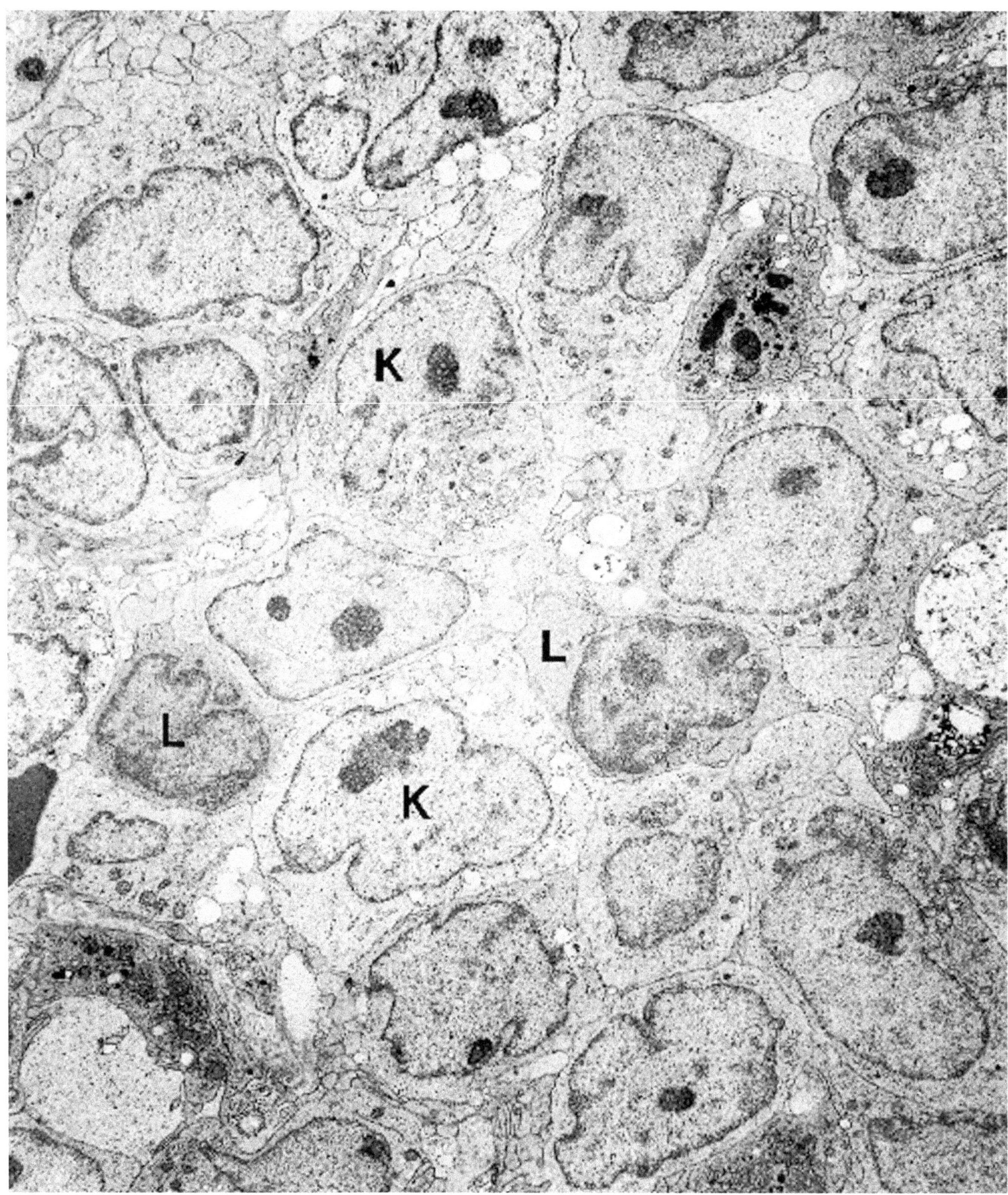

Figure 3.84. Anaplastic Ki-1 positive large cell lymphoma (retroperitoneal lymph node). Large, poorly differentiated cells (K) are interspersed with smaller cells having the chromatin pattern of lymphocytes (L). The cells are closely apposed, but there are no intercellular junctions. The poorly differentiated cells have indented and lobated nuclei. Chromatin is finely dispersed, and nucleoli are prominent and multiple. Cytoplasm contains a predominance of free ribosomes and a moderate number of mitochondria. (× 5900)

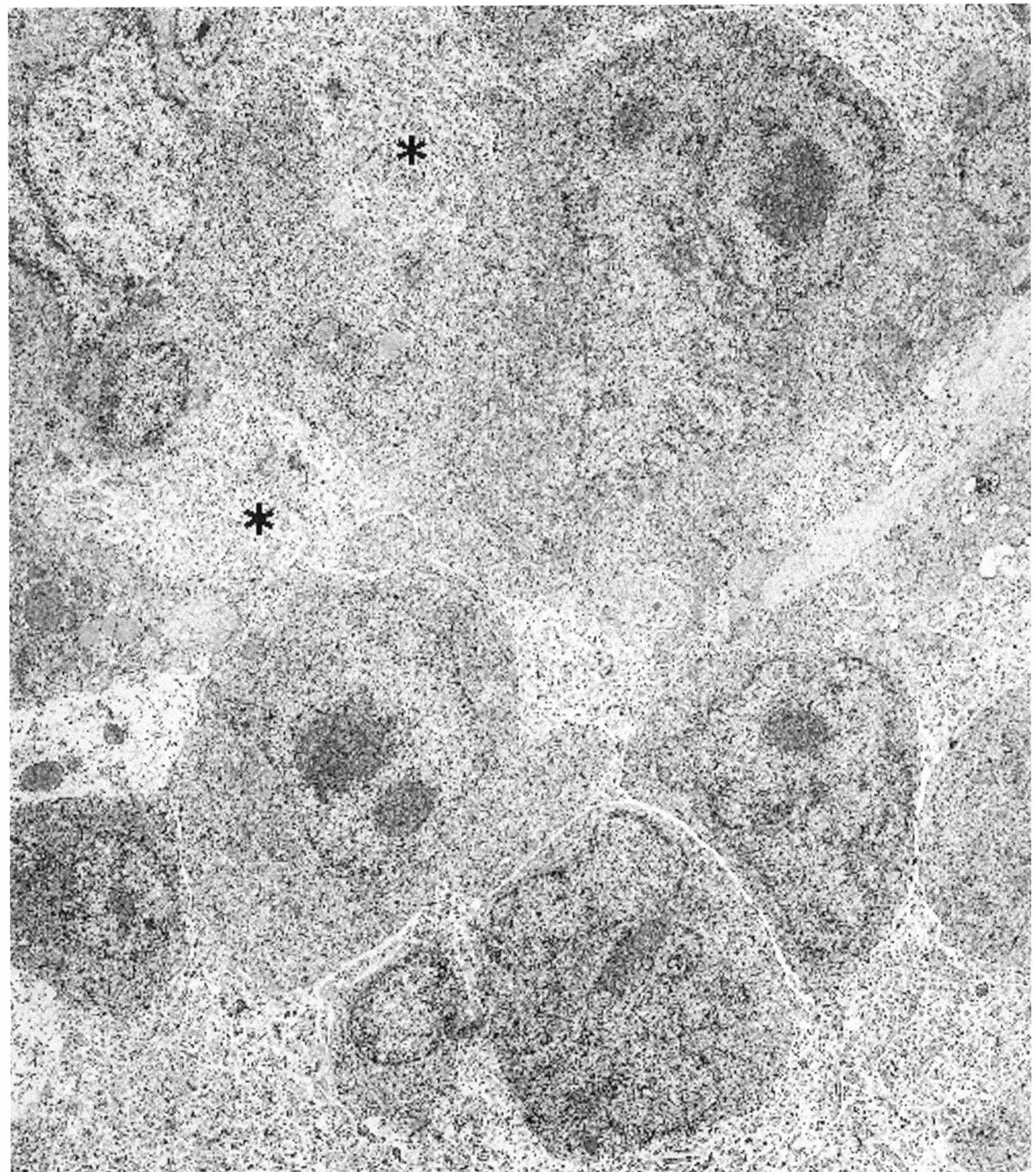

Figure 3.85. Anaplastic Ki-1 positive large cell lymphoma (retroperitoneal lymph node). Poorly differentiated large cells have pleomorphic nuclei, ribosomes, mitochondria, and a few cisternae of rough endoplasmic reticulum in the cytoplasm and innumerable filopodia (*) on their surfaces (filiform cells/anemone cells/porcupine cells). (× 5600)

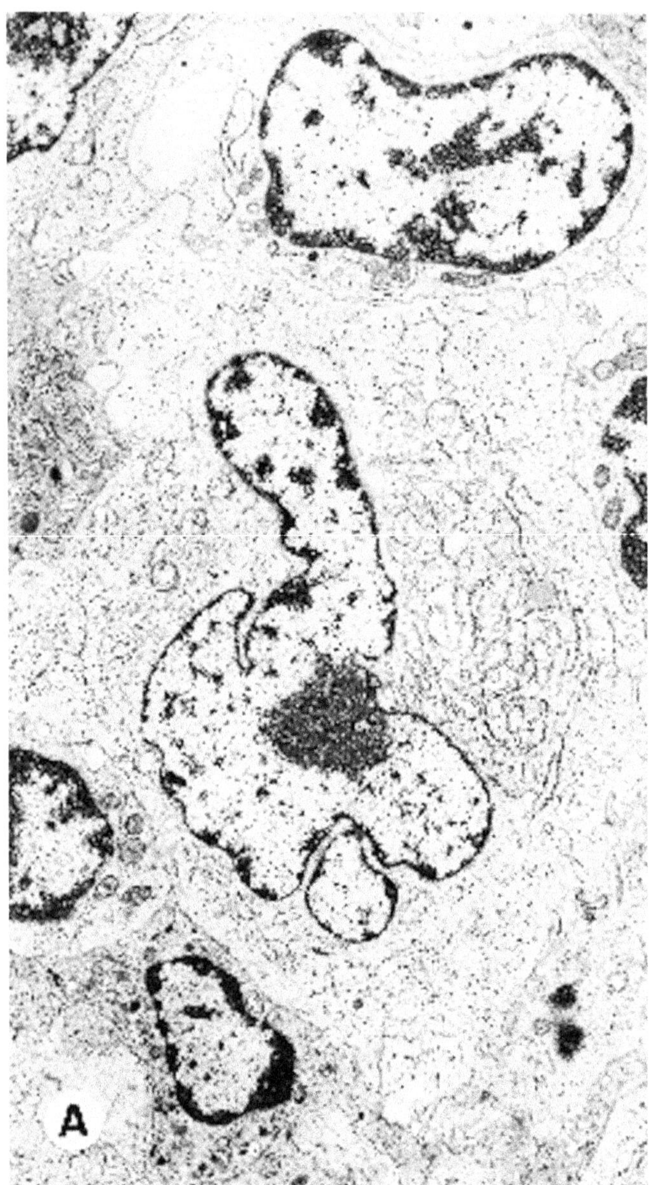

Figure 3.86. Hodgkin's disease (cervical lymph node). These four fields (**A** through **D**) illustrate the degree of lobation that may be present in the nuclei of Reed-Sternberg cells and how the plane of sectioning could result in apparent binucleation. The cytoplasmic features of all four cells are more consistent with a lymphoid line than they are with a histiocytic one (see Figure 3.87). **A,** × 6150. **B,** × 7870.

(Text continued from page 89)

Histiocytic Disorders

Macrophagic Lesions

(Figures 3.87 through 3.92.)

Diagnostic criteria. (1) Small and large cells having a copious and "busy" cytoplasm, and a surface raised into many broad pseudopods; (2) cytoplasmic organelles and inclusions include varying numbers of small (primary) and large (secondary) lysosomes, many mitochondria, a moderate amount of rough endoplasmic reticulum, prominent Golgi apparatuses, and occasional lipid droplets; (3) in hemophagocytic syndromes, phagocytosed erythrocytes, leukocytes, and/or platelets within the cytoplasm; (4) in histiocytic sarcoma, nuclei of large histiocytes are frequently large, irregular in shape, and euchromatic and have one or two prominent nucleoli.

Additional points. The question of benignancy or malignancy of histiocytes may be difficult to answer on the basis of cellular morphology alone, but usually large and irregularly shaped nuclei with euchromatin and prominent nucleoli are indicative of malignancy, rather

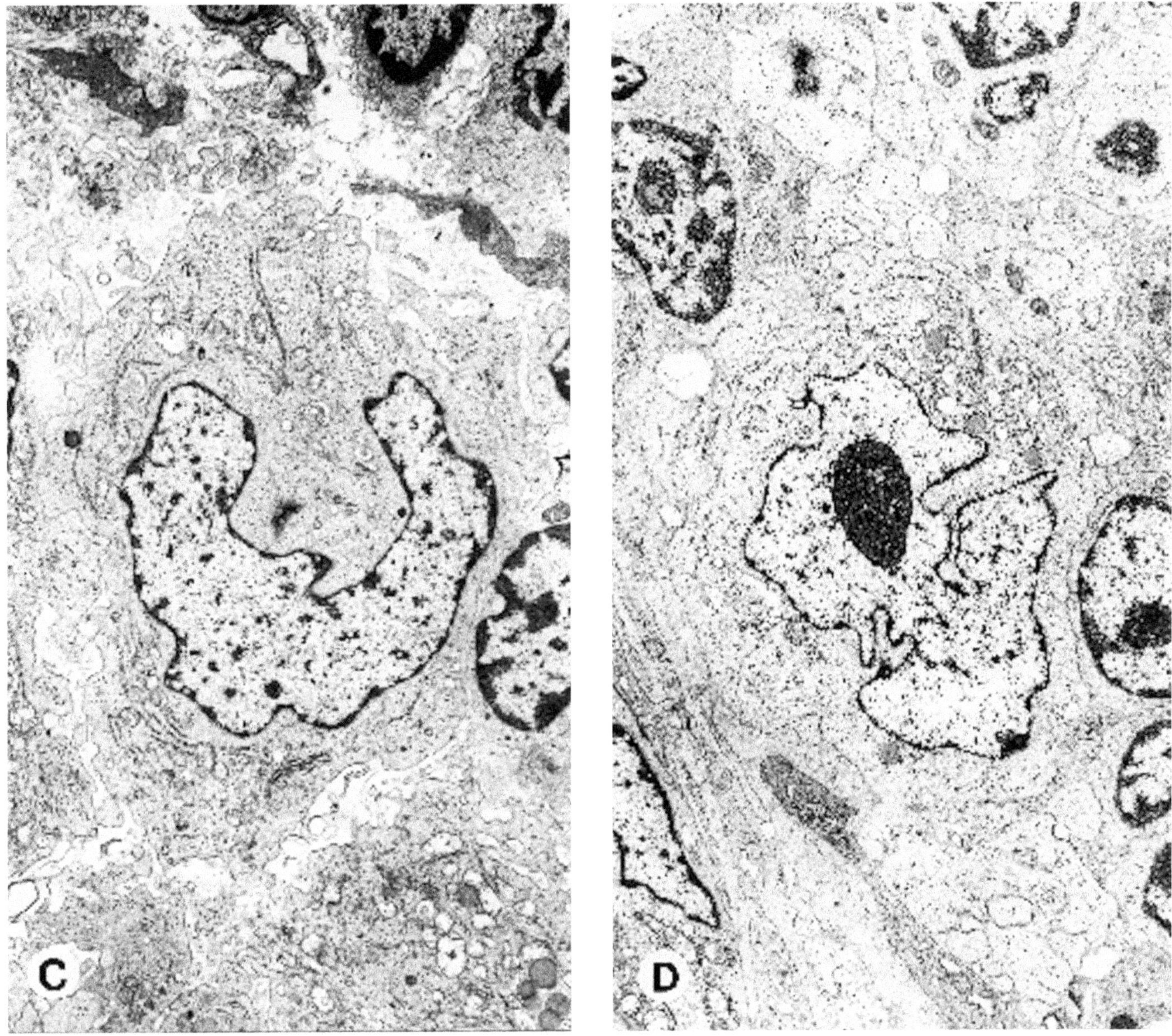

Figure 3.86. *(continued)*
C, × 6525. **D,** × 6390.

than an inflammatory reaction (Figures 3.87 through 3.92). The current classification of histiocytic disorders includes two broad categories: disorders of varied biological behavior and malignant disorders. Within each of these categories, the cell type may be a macrophage or a dendritic cell. In addition, monocytic leukemias and sarcomas are included under the malignant disorders. Macrophagic disorders encompass hemophagocytic syndromes, Rosai-Dorfman disease, multicentric reticulohistiocytosis, solitary histiocytoma, and histiocytic sarcoma (localized or disseminated).

Dendritic Cell Lesions

The cells in these lesions are not phagocytic and function to take up and deliver antigens and immune complexes to lymphoid cells. *Dendritic cell* lesions include Langerhans' cell histiocytosis and sarcoma, hyperplastic and neoplastic *follicular dendritic cell* lesions, and *interdigitating dendritic cell* proliferations.

(Text continues on page 124)

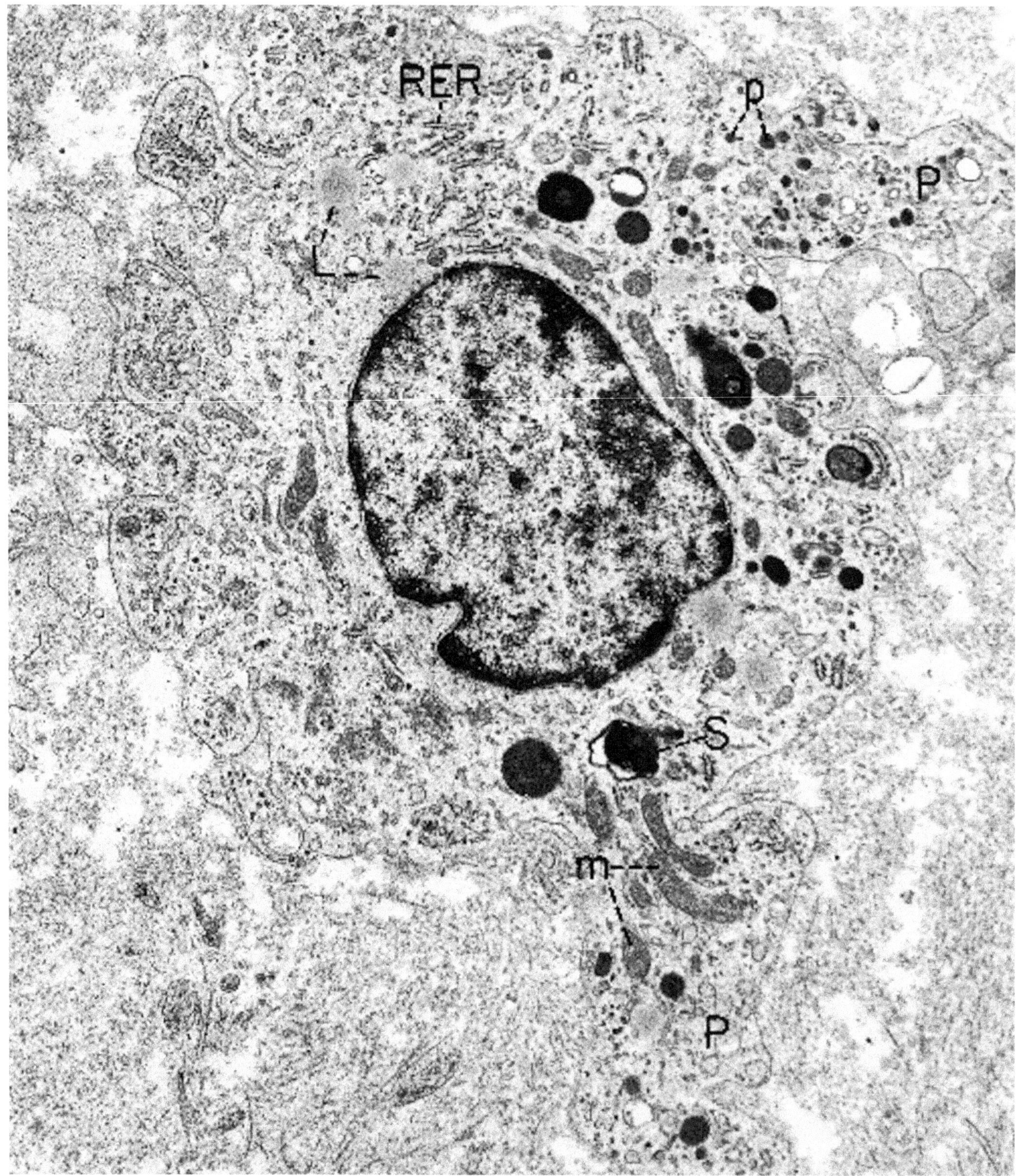

Figure 3.87. Histiocyte (axillary lymph node). The busy cytoplasm of this histiocyte is in marked contrast to the simple cytoplasm (mostly ribosomes) of the large lymphocytes seen in Figures 3.78 through 3.82. Cytoplasm containing primary (p) and secondary (S) lysosomes and cell surfaces forming pseudopods (P) are especially good markers for identifying histiocytes. Note also lipid droplets (L), mitochondria (m), and rough endoplasmic reticulum (RER). (× 12,930) (Permission for reprinting granted by WB Saunders, Dickersin GR: Electron microscopy of leukemias and lymphomas. Clin Lab Med 7:199–247, 1987.)

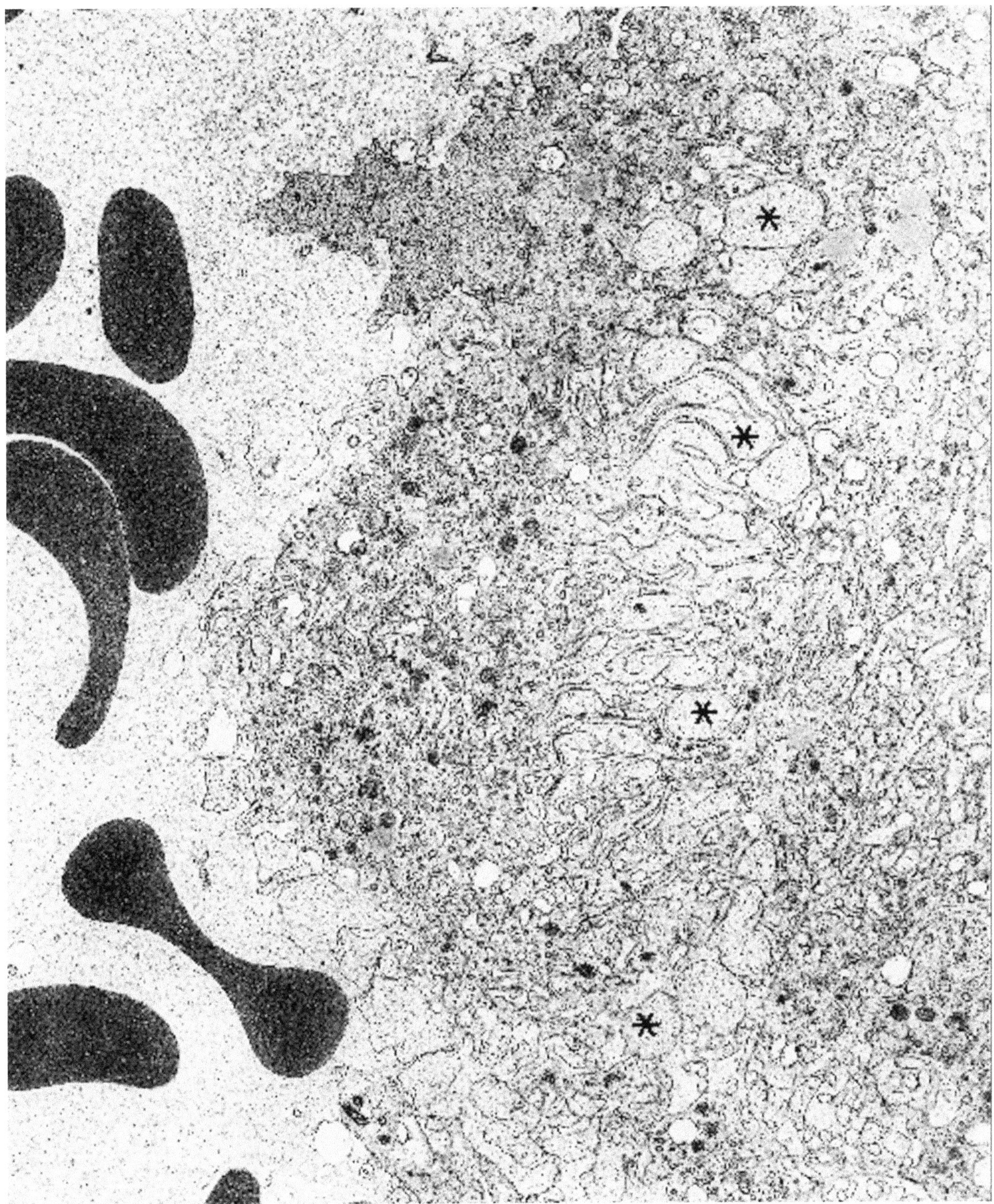

Figure 3.88. Hemophagocytic syndrome associated with acute lymphocytic leukemia (bone marrow). The filopodia (*) of two giant histiocytes, or two processes of the same histiocyte, show a plane of complicated interdigitation. The cytoplasm is rich in organelles. (× 7850) (Permission for reprinting granted by WB Saunders, Dickersin GR: Electron microscopy of leukemias and lymphomas. Clin Lab Med 7:199–247, 1987.)

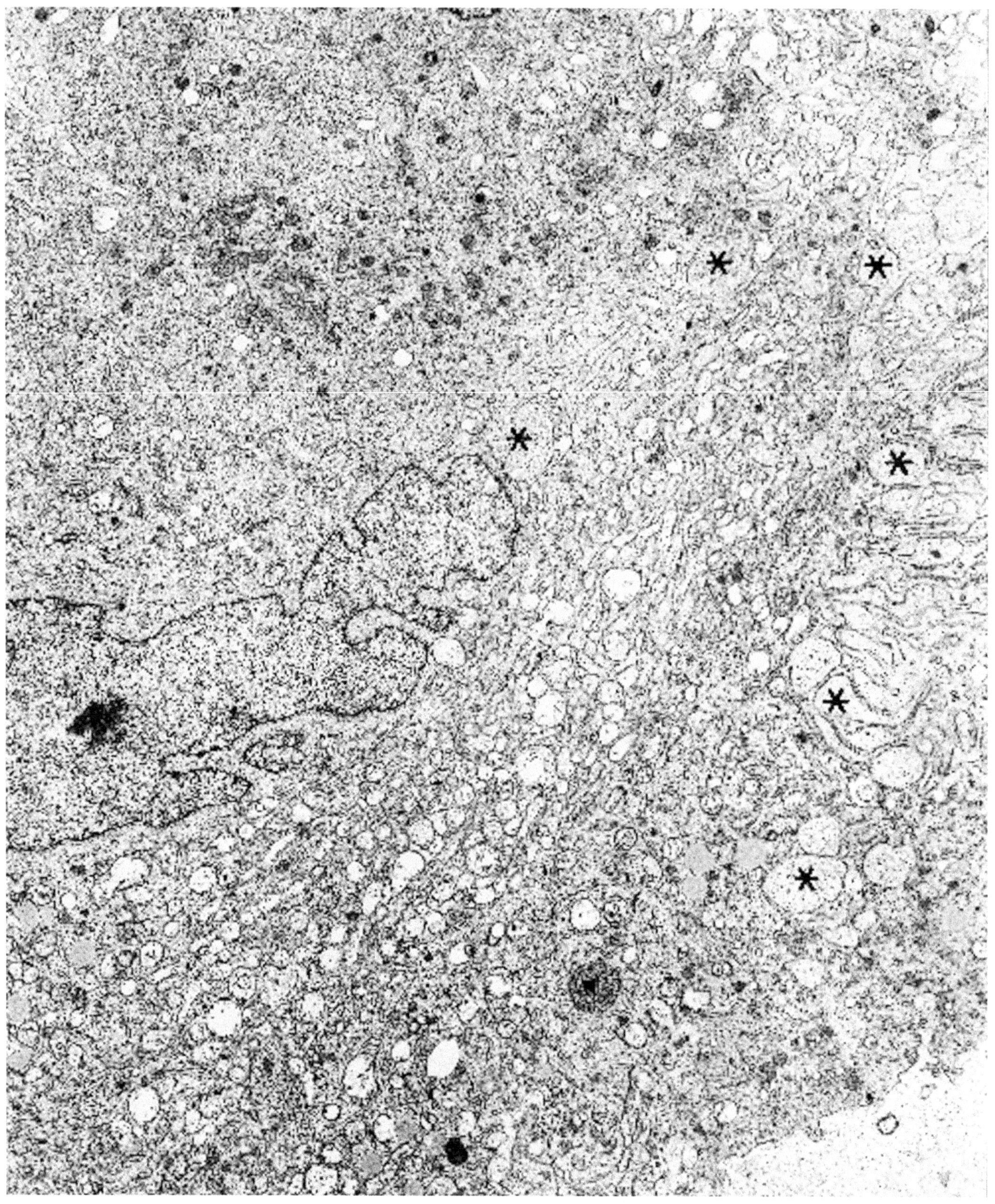

Figure 3.89. Hemophagocytic syndrome associated with acute lymphocytic leukemia (bone marrow). At least two giant histiocytes are present in this field. In addition to an intimate interdigitation of their filopodia (*) and a plethora of cytoplasmic organelles, the nucleus is large, irregularly shaped, and euchromatic and has a prominent nucleolus. (× 6750)

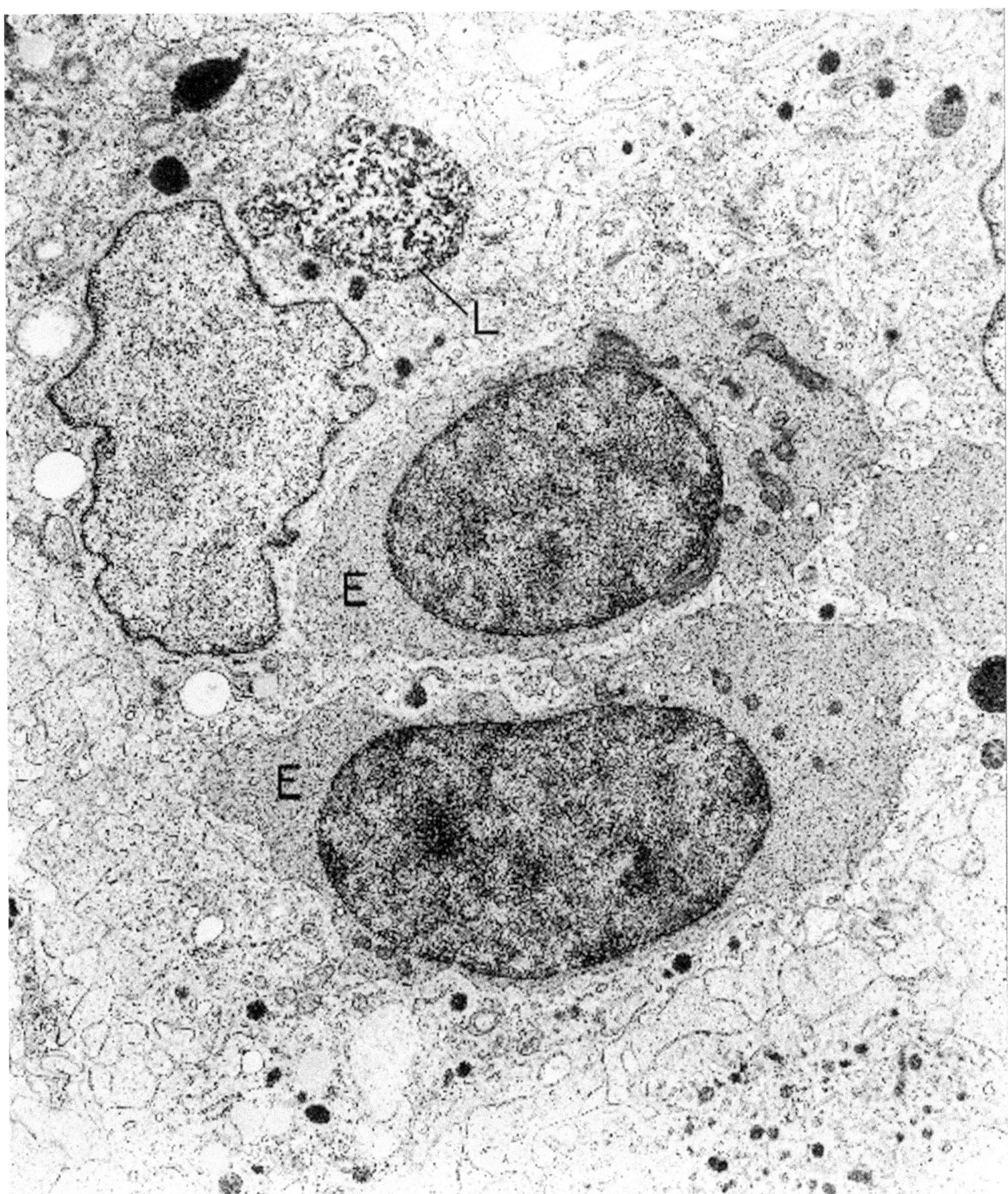

Figure 3.90. Hemophagocytic syndrome associated with acute lymphocytic leukemia (bone marrow). The cytoplasm of this histiocyte contains two phagocytosed erythrocytic precursors (E) as well as a large secondary lysosome (L) with heterogeneous contents that may be a partially digested erythrocyte or another type of phagocytosed cell. (× 8740) (Permission for reprinting granted by WB Saunders, Dickersin GR: Electron microscopy of leukemias and lymphocytes. Clin Lab Med 7:199–247, 1987.)

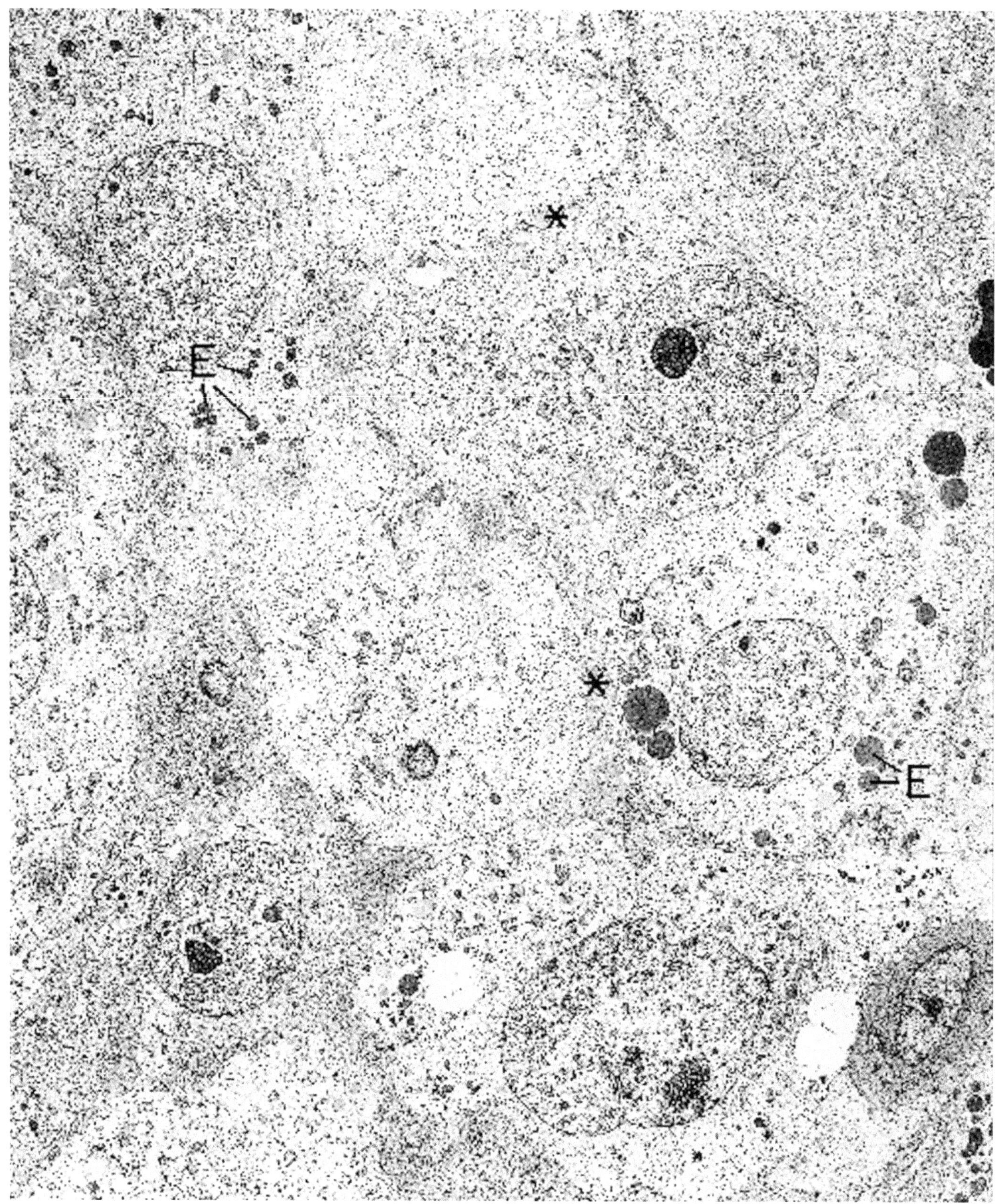

Figure 3.91. Histiocytic sarcoma (malignant histiocytosis) (femur). Characterizing these histiocytes are a solid grouping of molded oval cells with folded filopodia (*) and an absence of intercellular junctions. The cytoplasm is abundant, and organelles are plentiful. Secondary lysosomes with engulfed erythrocytic particles (E) are present in some of the cells. (× 2440)

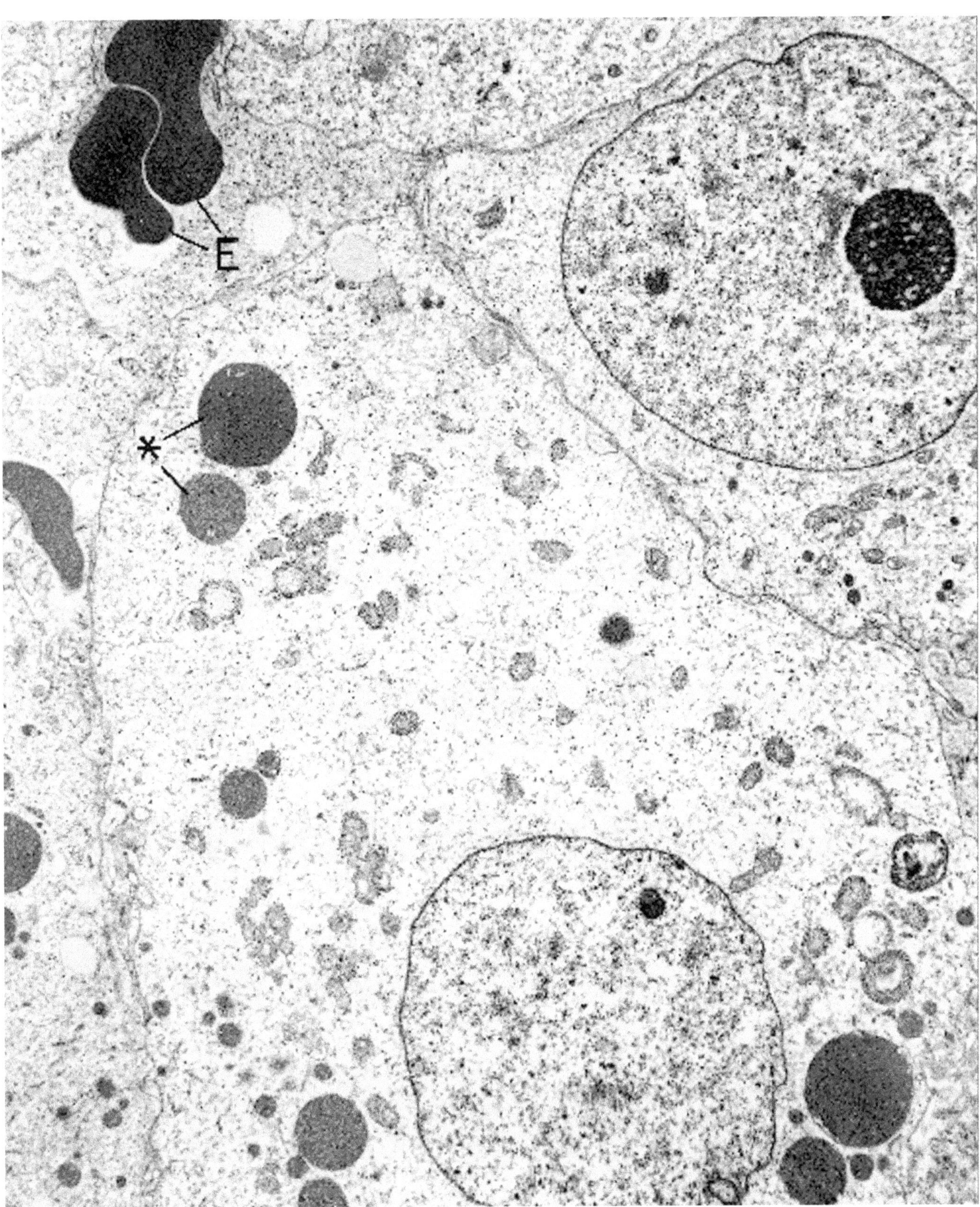

Figure 3.92. Histiocytic sarcoma (malignant histiocytosis) (femur). Higher magnification of the same neoplasm as that illustrated in Figure 3.91 highlights the phagocytosed erythrocytes (E) and erythrocytic particles (*). (× 5720) (Permission for reprinting granted by WB Saunders, Dickersin GR: Electron microscopy of leukemias and lymphomas. Clin Lab Med 7:199–247, 1987.)

(Text continued from page 117)

Langerhans' Cell Histiocytosis (Histiocytosis X)

(Figures 3.93 through 3.95.)

Diagnostic criteria. (1) Varying numbers of large (12–25 μ) mononuclear (Langerhans') cells, with low-power characteristics of a curved or oval nucleus and abundant cytoplasm; (2) at high-power, diagnostic *Birbeck* (Langerhans', X) *granules* in the cytoplasm; (3) numerous other cytoplasmic organelles including many free ribosomes, mitochondria, rough endoplasmic reticulum, and primary lysosomes; (4) filopodia at the cell surface; (5) absence of secondary lysosomes (phagolysosomes) except in rare cases.

Additional points. The most distinguishing feature of this disease is the Langerhans' or Langerhans'-like cell, a dendritic cell having the Birbeck granule as its most characteristic marker (Figure 3.95). The Birbeck granule has a three-dimensional shape of a cup or disc, or a combination of the two. The two-dimensional appearance of the granule is several-fold, including rods and tennis-racket-shaped structures. A dense, zipper-like, striated line is frequently visible running longitudinally in the middle of the granule, and may represent a cellular surface-coating similar to a glycocalyx (Figure 3.95). Some Birbeck granules can be identified as originating from invaginations of plasmalemmas (Figure 3.95), and some may possibly originate from Golgi apparatuses. Structures similar to Birbeck granules may also be found as attachment plaques between plasmalemmas of apposing Langerhans' cells. Cells with the immunophenotype and morphology of Langerhans' cells without Birbeck granules have been designated indeterminate cells.

Langerhans' cells are found normally in squamous epithelial surfaces such as skin and mucous membrane.

Follicular Dendritic Cell Lesions

(Figures 3.96 through 3.100.)

Diagnostic criteria. (1) Cells with oval or elongated nuclei, with or without indentations; (2) small amounts of heterochromatin, more concentrated at the periphery of the nucleus; (3) small nucleolus (sometimes in form of nucleolonema) is usually single and central; (4) sparse cytoplasmic organelles, mostly free ribosomes; (5) dense network of long and intertwining cytoplasmic processes; (6) desmosomes present between processes.

Additional points. Proliferations of follicular dendritic cells may be neoplastic or nonneoplastic, with the latter being very uncommon. Neoplastic lesions are usually low-grade sarcomas, but a few are high-grade sarcomas.

Interdigitating Dendritic Cell Lesions

Diagnostic criteria. (1) Cells with large, indented and pleomorphic nuclei; (2) predominance of heterochromatin subjacent to nuclear envelope; (3) sparse cytoplasmic organelles, including free ribosomes, smooth and rough endoplasmic reticulum and Golgi apparatuses; (4) interdigitating cytoplasmic processes; (5) absence of desmosomes.

Additional points. Interdigitating dendritic cells are present in T cell regions of lymph nodes and spleen, and lesions composed of them are extremely rare and usually very malignant.

Mastocytosis and Mastocytoma

(Figures 3.101 and 3.102.)

Diagnostic criteria. (1) Round, oval or spindle shaped cells with granules of varying size, internal pattern and density; (2) round, nonsegmented nucleus.

Additional points. The most characteristic morphology of mast cell granules is a lamellar or scroll-like pattern, but other nonspecific forms, including giant-size and compound granules, also may be seen. Mastocytosis may be isolated to the skin as in urticaria pigmentosa or, less commonly, systemic, in which case other organs but especially the bone marrow are involved.

Systemic mastocytosis may be associated with other myeloproliferative disorders such as acute and chronic myelocytic leukemia. Rarely, systemic mastocytosis may be a primary malignancy. Solitary mastocytomas comprise a small minority of all cases of mastocytosis. Mast cells perform some of the same biochemical and immunological functions as basophils, and both cell types derive from hematopoietic precursors in the bone marrow. However, there are some functional differences as well as distinctions in morphology and natural history. Mast cells mature and reside in connective tissue, whereas basophils mature in the bone marrow, circulate in the peripheral blood, and migrate into solid tissues in response to inflammatory stimuli.

(Text continues on page 135)

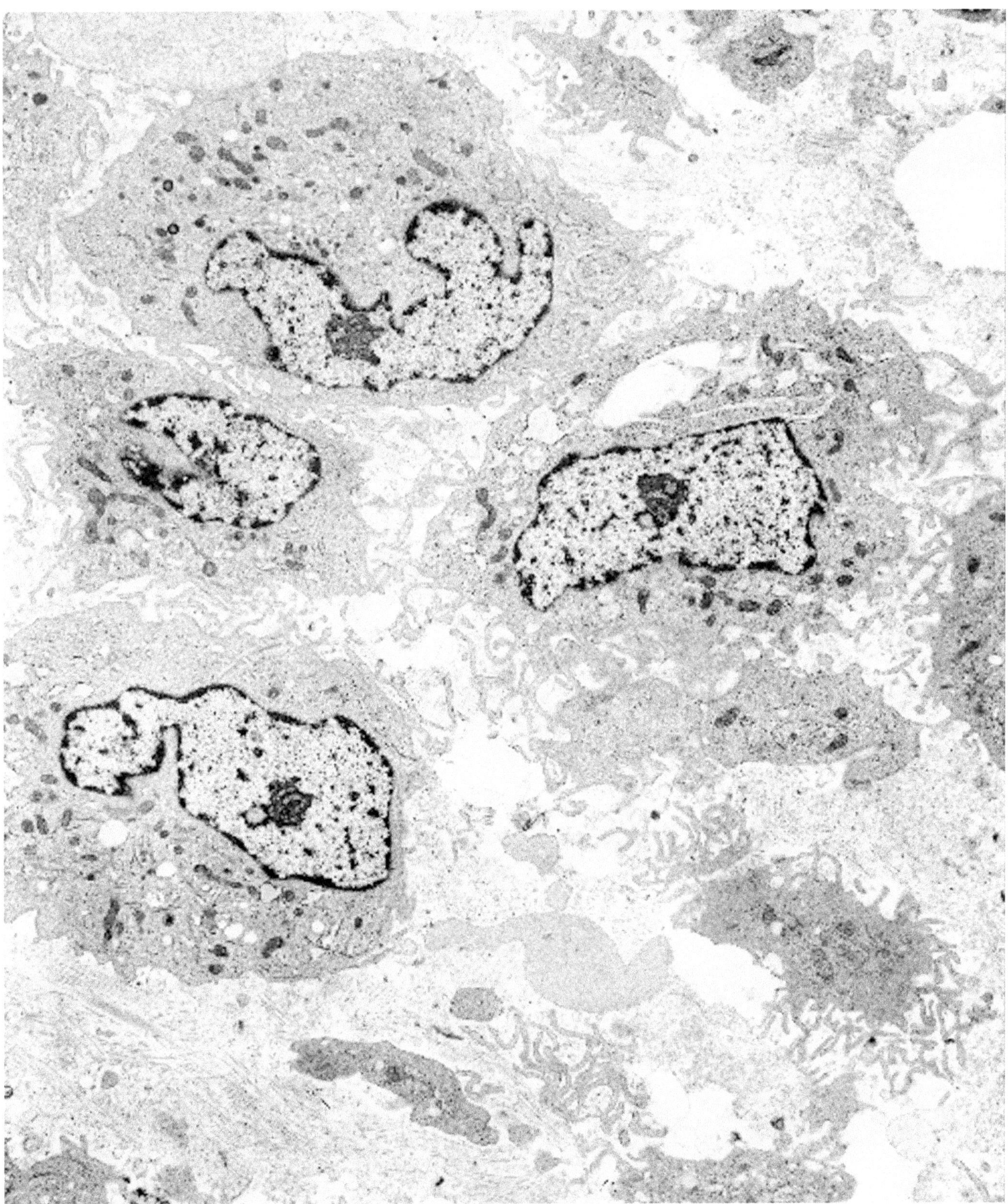

Figure 3.93. Histiocytosis X (eosinophilic granuloma, vertebral body). Several characteristic Langerhans' cells display curved and indented nuclei, copious cytoplasm, and many filopodia on their surfaces. (× 6360) (Permission for reprinting granted by WB Saunders, Dickersin GR: Electron microscopy of leukemias and lymphomas. Clin Lab Med 7:199–247, 1987.)

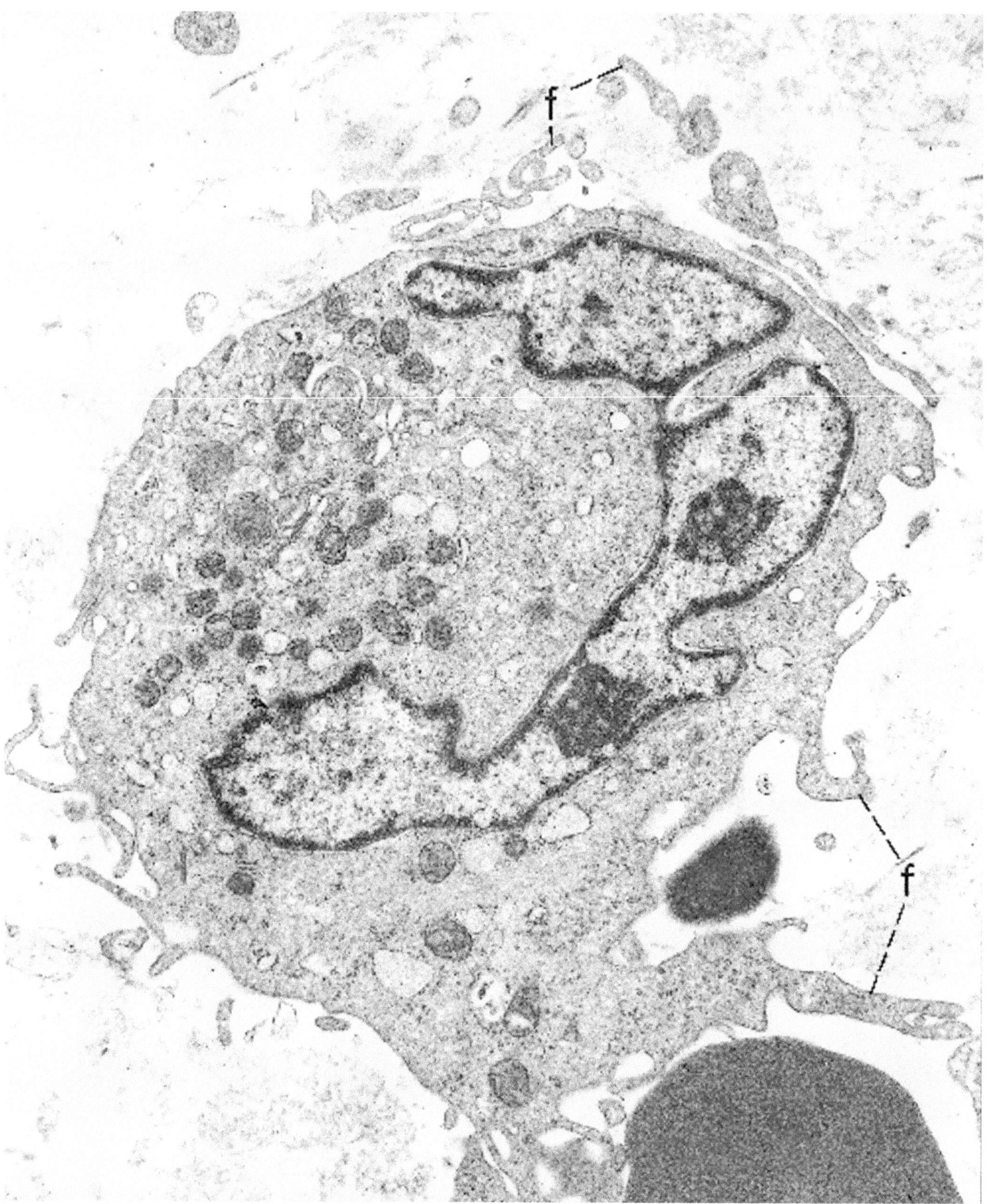

Figure 3.94. Histiocytosis X (skin). The cytoplasm of the Langerhans' cell is abundant, and many different organelles are contained. Characteristically, the nucleus is curved, and the cell surface is raised into filopodia (f). (× 13,500) (Permission for reprinting granted by WB Saunders, Dickersin GR: Electron microscopy of leukemias and lymphomas. Clin Lab Med 7:199–247, 1987.)

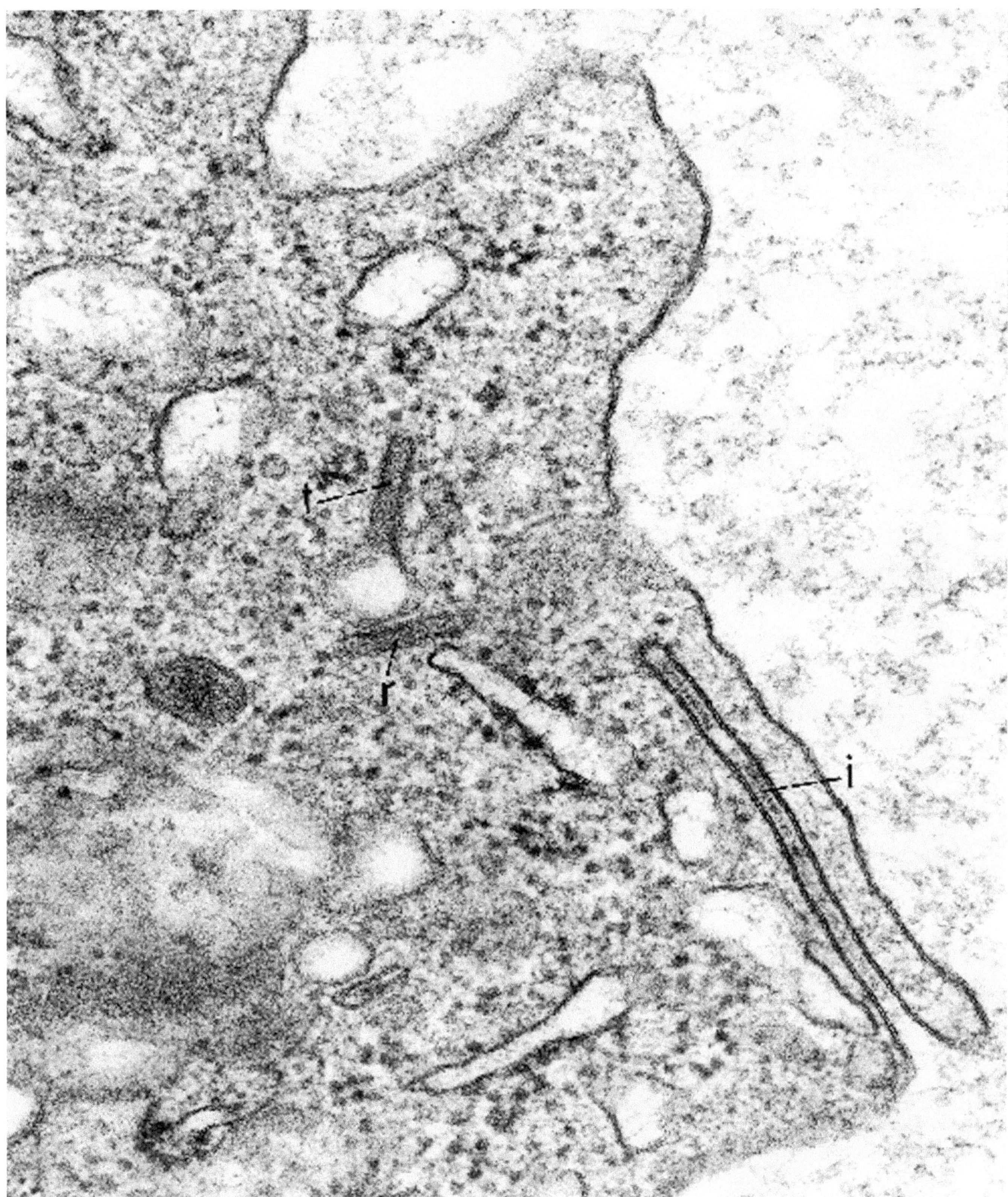

Figure 3.95. Histiocytosis X (eosinophilic granuloma, calvarium). The cytoplasm of a Langerhans' cell, seen at high magnification, illustrates several Birbeck granules, including a rod form (r), tennis-racket shape (t), and invagination of the cell membrane (i) with a glycocalyceal central linear density. (× 104,000) (Permission for reprinting granted by WB Saunders, Dickersin GR: Electron microscopy of leukemias and lymphomas. Clin Lab Med 7:199–247, 1987.)

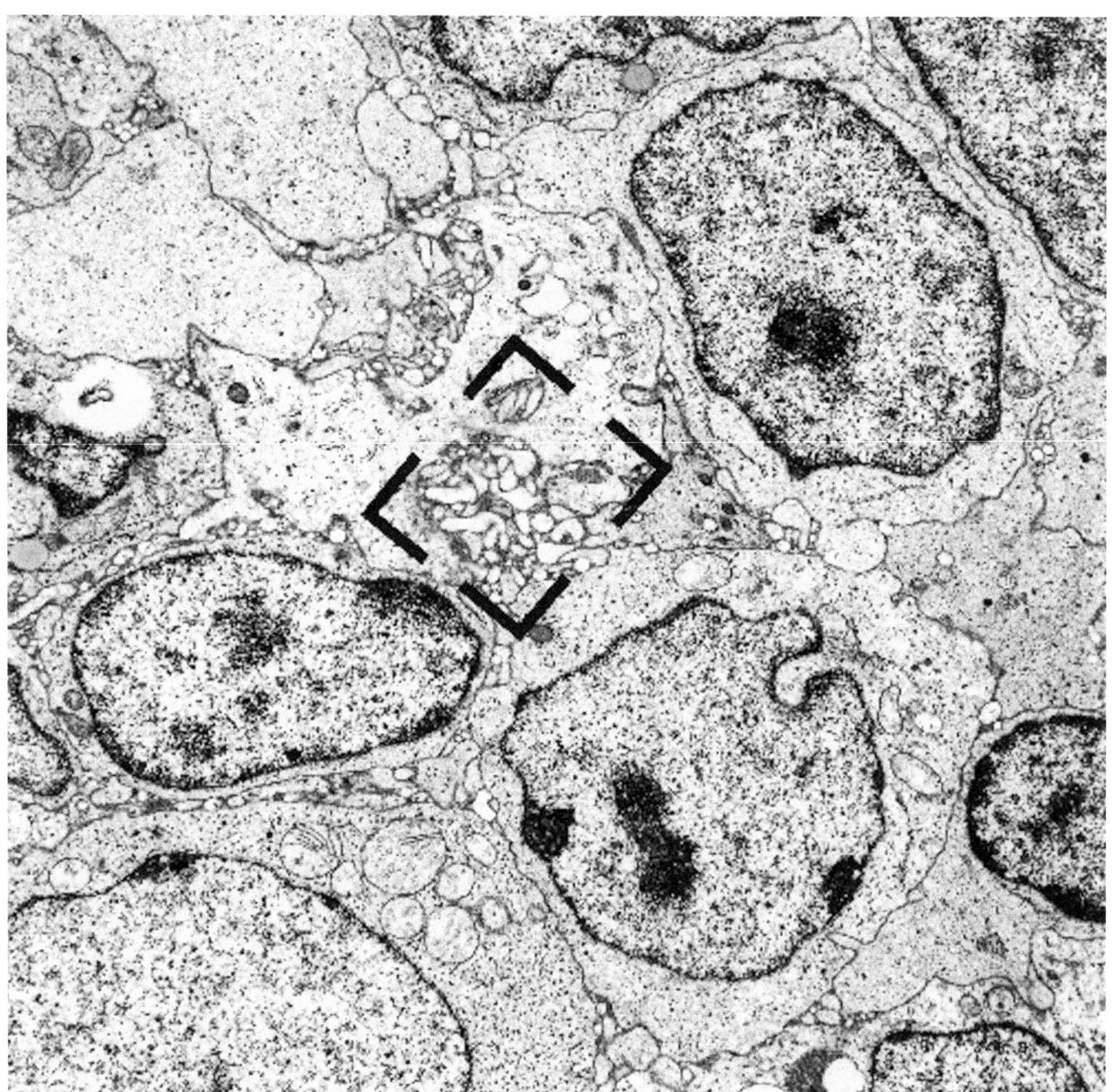

Figure 3.96. Hyperplastic lymphoid follicle (cervical lymph node). In the midst of these follicular center cells are foci of intertwining, narrow cellular processes consistent with dendritic cell processes (*brackets*). At higher magnification (see Figure 3.97), desmosomes are visible between the processes. Cytoplasm contains few organelles, mostly free ribosomes. (× 8500)

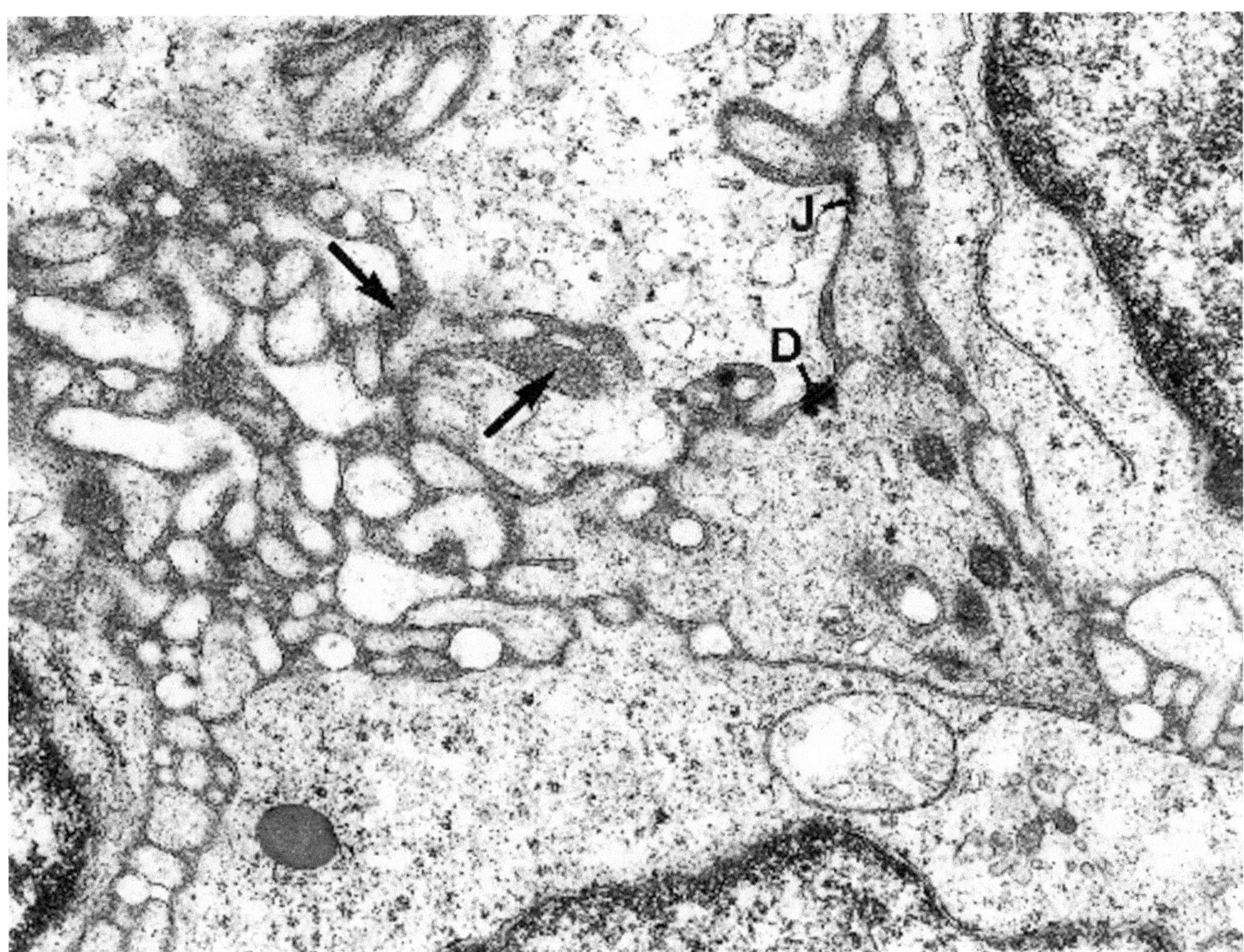

Figure 3.97. Hyperplastic lymphoid follicle (cervical lymph node). Higher magnification of the central, bracketed field in Figure 3.96 illustrates the intertwining dendritic cell processes, their fuzzy, medium-dense coating (*arrows*), a desmosome (D), and less prominent junction (J). (× 23,800)

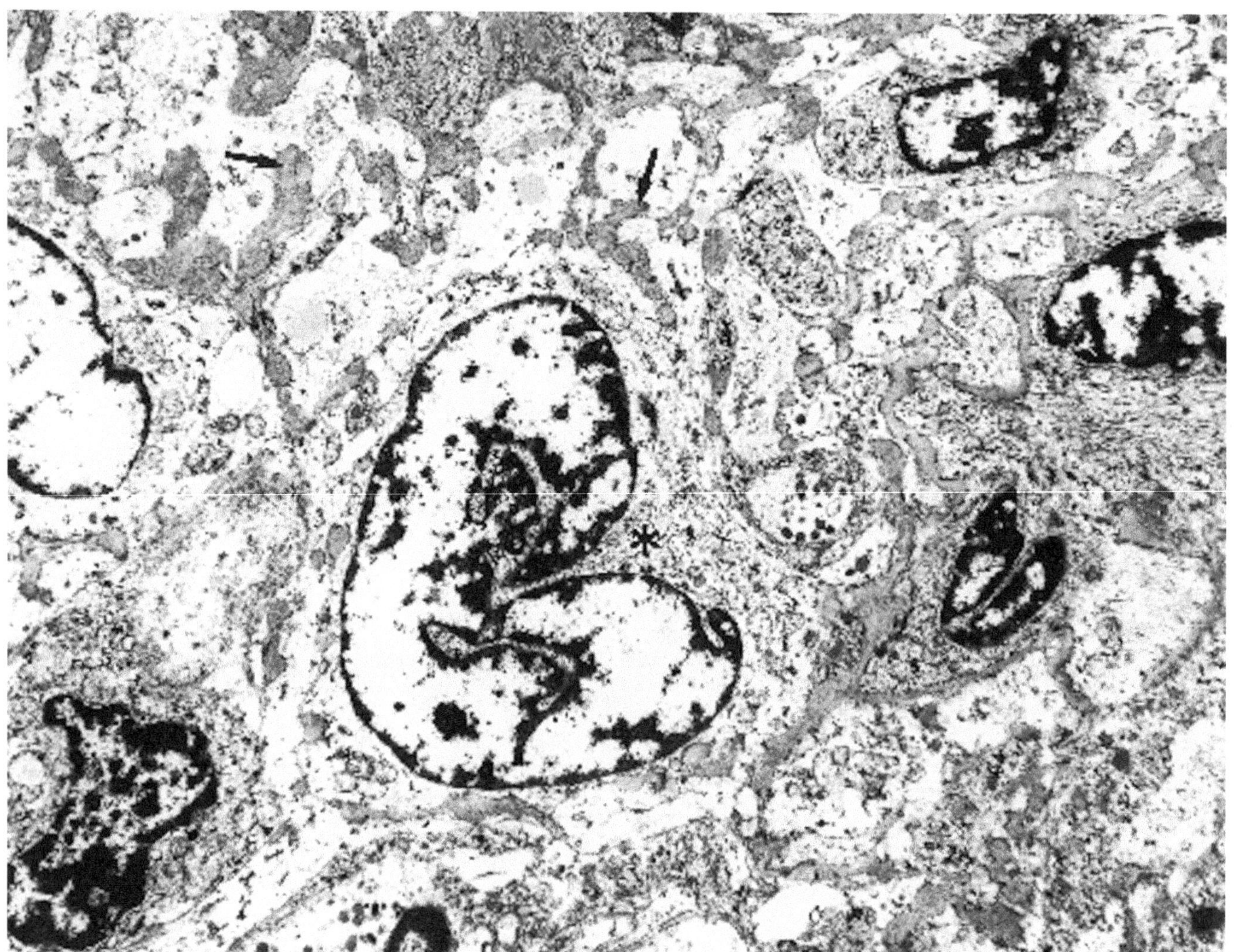

Figure 3.98. Follicular dendritic cell sarcoma (spleen). This specimen was fixed in formalin rather than glutaraldehyde, resulting in less-than-optimal cellular preservation, but still discernible are numerous cellular processes with a fuzzy, medium-dense coating (*arrows*), plus the cell body of a malignant dendritic cell (*). (× 6600)

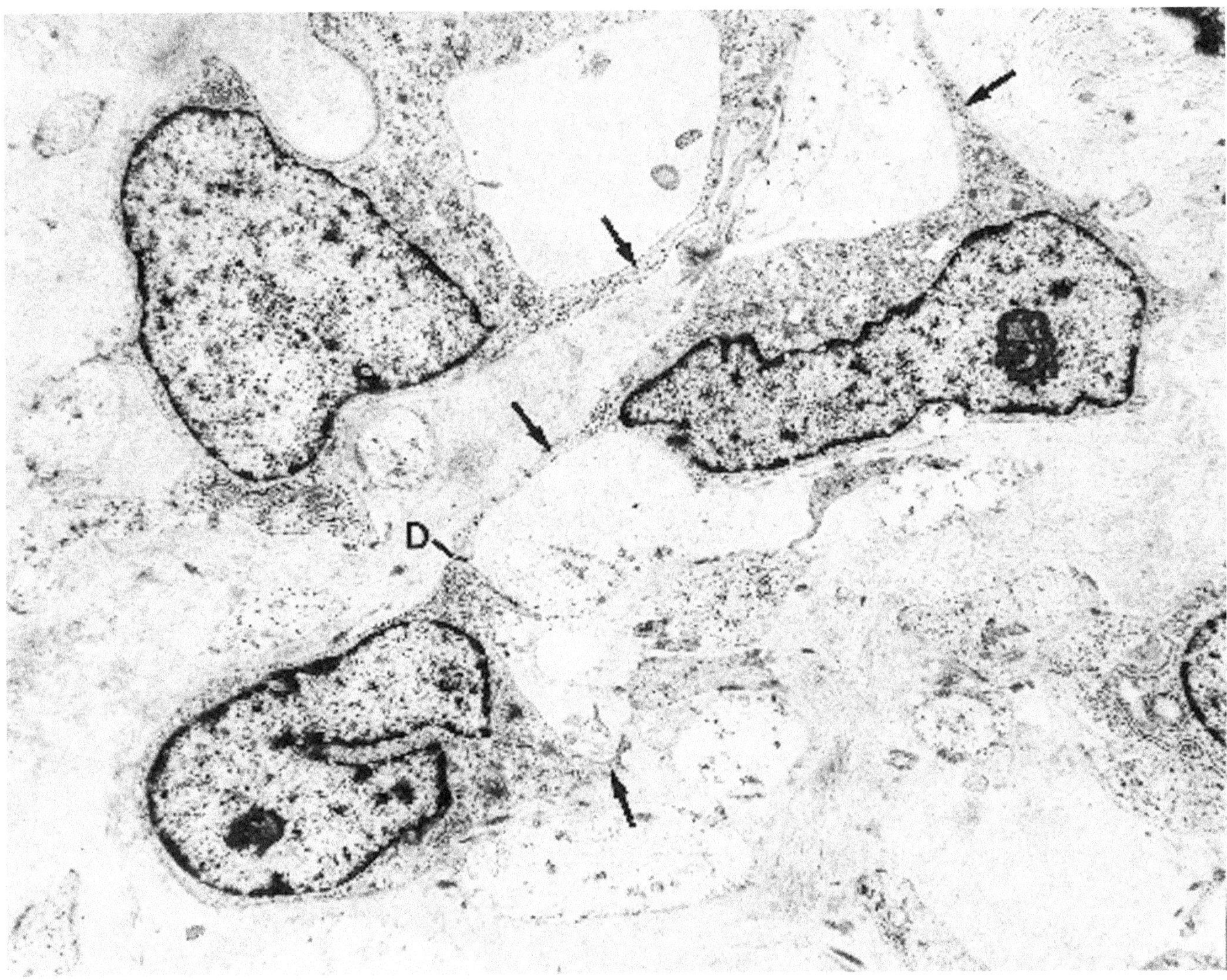

Figure 3.99. Follicular dendritic cell sarcoma (axillary lymph node). The neoplastic cells have dendritic-like processes (*arrows*), small amounts of heterochromatin, some indented nuclei, and single nucleoli. Cytoplasm is nondescript. A desmosome (D) is discernible between two processes. (× 7800)

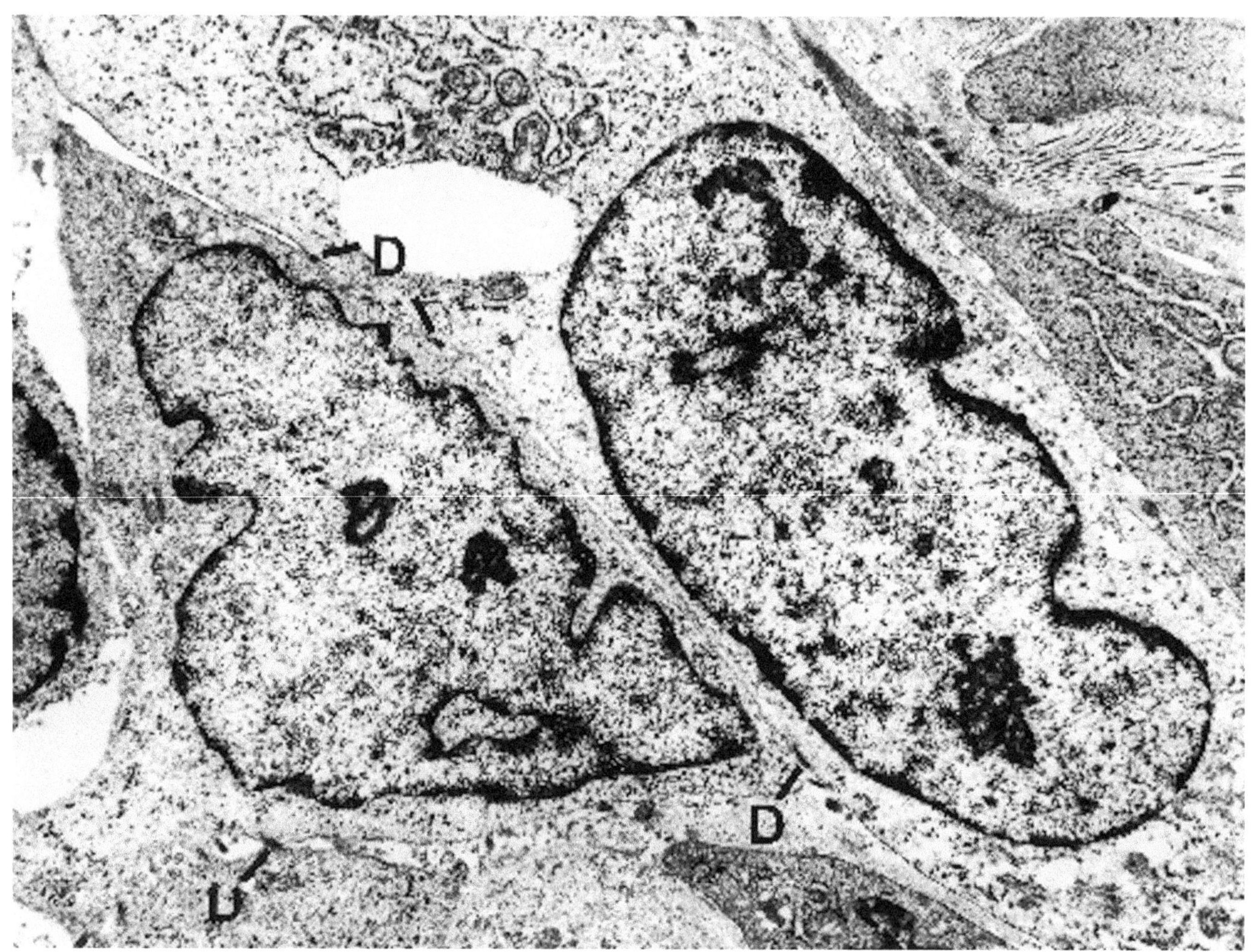

Figure 3.100. Follicular dendritic cell sarcoma (axillary lymph node). Higher magnification of two cells from the same neoplasm depicted in Figure 3.99 reveals nondescript cytoplasm with few organelles, and several desmosomes (D). (× 12,500)

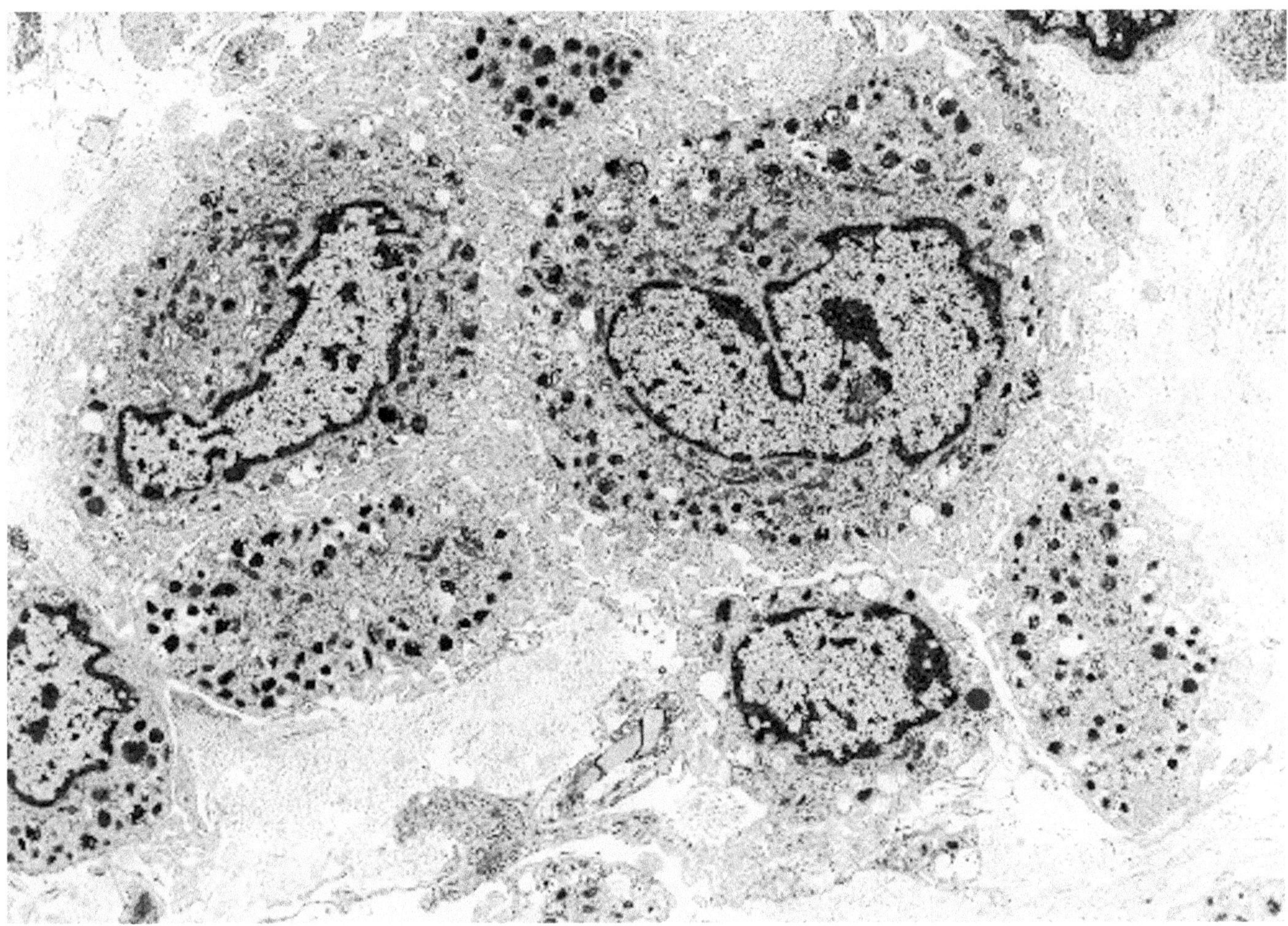

Figure 3.101. Mastocytoma (skin of leg). Several mast cells are characterized by a floridly filopodial surface and numerous cytoplasmic granules of varying density. (× 6700)

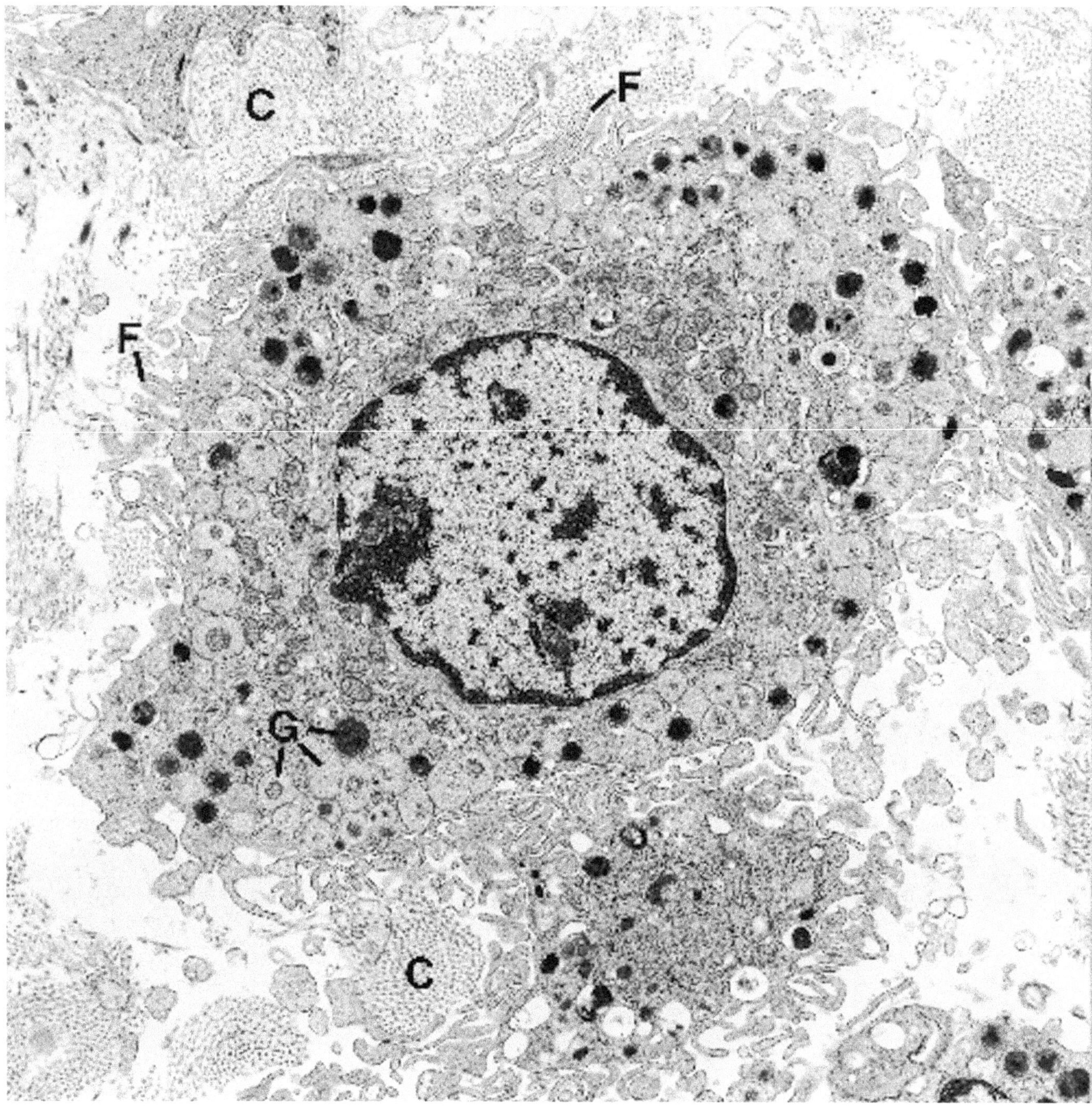

Figure 3.102. Mastocytoma (skin of leg). High magnification illustrates a mast cell with innumerable membrane-bound cytoplasmic granules (G) of varying density as well as florid filopodia (F). C = collagen. (× 11,400)

(Text continued from page 124)

REFERENCES

Adenocarcinoma

General

Hammar S, Bockus D, Remington F: Metastatic tumors of unknown origin: An ultrastructural analysis of 265 cases. Ultrastruct Pathol 11:209–250, 1987.

McDowell EM, Hess FG Jr, Trump BF: Epidermoid metaplasia, carcinoma in situ and carcinomas of the lung. In Trump BF, Jones RT, eds: *Diagnostic Electron Microscopy,* Vol 3. John Wiley and Sons, New York, 1980, pp 37–96.

Thyroid

Dickersin GR, Vickery AL, Smith SB: Papillary carcinoma of the thyroid, oxyphil cell type, "cell type" variant: A light- and electron-microscopic study. Am J Surg Pathol 4:501–509, 1980.

Johannessen JV, Sobrinho-Simoes M: The fine structure of follicular thyroid adenomas. Am J Clin Pathol 78:299–310, 1982.

Klinck GH, Oertel JE, Winship T: Ultrastructure of normal human thyroid. Lab Invest 22:2–22, 1970.

Olson JL, Penney DP, Averill KA: Fine structural studies of a human thyroid adenoma, with special reference to psammoma bodies. Hum Pathol 8:103–111, 1977.

Sobrinho-Simoes MA, Nesland JM, Holm R, et al: Transmission electron microscopy and immunocytochemistry in the diagnosis of thyroid tumors. Ultrastruct Pathol 9:255–275, 1985.

Breast

Ahmed A: *Atlas of the Ultrastructure of Human Breast Disease.* Churchill Livingstone, Edinburgh, 1978.

Battifora H: Intracytoplasmic lumina in breast carcinoma. Arch Pathol Lab Med 99:614–617, 1975.

Fisher ER: Ultrastructure of the human breast and its disorders. Am J Clin Pathol 66:291–375, 1976.

Gould VE, Jao W, Battifora H: Ultrastructural analysis in the differential diagnosis of breast tumors. Pathol Res Pract 167:45–70, 1980.

Nesland JM, Holm R: Diagnostic problems in breast pathology: The benefit of ultrastructural and immunocytochemical analysis. Ultrastruct Pathol 11:293–311, 1987.

Ozzello L: Breast. In Johannessen JV, ed: *Electron Microscopy in Human Medicine, Urogenital System and Breast,* vol 9. McGraw-Hill, London, 1979, pp 409–456.

Ozello L: Ultrastructure of the human mammary gland. In Sommers SC, ed: *Pathology Annual 1971.* Appleton-Century-Crofts, New York, 1971, pp 1–59.

Sobrinho-Simoes M, Johannessen JV, Gould VE: The diagnostic significance of intracytoplasmic lumina in metastatic neoplasms. Ultrastruct Pathol 2:327–335, 1981.

Lung

Bedrossian CWM, Weilbaecher DG, Bentinck DC, et al: Ultrastructure of human bronchioloalveolar cell carcinoma. Cancer 36:1399–1413, 1975.

Bolen JW, Thorning D: Histogenetic classification of pulmonary carcinomas. Peripheral adenocarcinomas studied by light microscopy, histochemistry, and electron microscopy. Pathol Annu 17:77–100, 1982.

Coalson JJ, Mohr JA, Pirtle JK, et al: Electron microscopy of neoplasms of the lung with special emphasis on the alveolar cell carcinoma. Am Rev Respir Dis 101:181–197, 1970.

Eimoto T, Teshima K, Shirakusa T, et al: Ultrastructure of well-differentiated adenocarcinomas of the lung with special reference to bronchioloalveolar carcinoma. Ultrastruct Pathol 8:177–190, 1985.

Greenberg SD, Smith MN, Spjut HJ: Bronchiolo-alveolar carcinoma—cell of origin. Am J Clin Pathol 63: 153–167, 1975.

Hammar SP: Adenocarcinoma and large cell carcinoma of the lung. Ultrastruct Pathol 11:263–291, 1987.

Hammar SP, Bolen JW, Bockus D, et al: Ultrastructural and immunohistochemical features of common lung tumors: An overview. Ultrastruct Pathol 9:283–318, 1985.

Kuhn C: Fine ultrastructure of bronchiolo-alveolar cell carcinoma. Cancer 30:1107–1118, 1972.

McDowell EM, Barrett LA, Glavin F, et al: The respiratory epithelium. 1. Human bronchus. J Natl Cancer Inst 61:539–549, 1978.

McDowell EM, Hess FG Jr, Trump BF: Epidermoid metaplasia, carcinoma in situ and carcinomas of the lung. In Trump BF, Jones RT, eds: *Diagnostic Electron Microscopy,* vol 3. John Wiley and Sons, New York, 1980, pp 37–96.

Sidhu GS: The ultrastructure of malignant epithelial neoplasms of the lung. Pathol Annu 17:235–266, 1982.

Gastrointestinal and Sinonasal Region

Barnes L: Intestinal-type adenocarcinoma of the nasal cavity and paranasal sinuses. Am J Surg Pathol 10:192–202, 1986.

Batsakis JG, Mackay B, Ordonez NG: Enteric-type adenocarcinoma of the nasal cavity. An electron microscopic and immunocytochemical study. Cancer 54: 855–860, 1984.

Cooper PH, Warkel RL: Ultrastructure of the goblet cell type of adenocarcinoid of the appendix. Cancer 42: 2687–2695, 1978.

Hickey WF, Seiler MW: Ultrastructural markers of colonic adenocarcinoma. Cancer 47:140–145, 1981.

Kaye GI, Fenoglio CM, Pascal RR, et al: Comparative electron microscopic features of normal, hyperplastic and adenomatous human colonic epithelium. Variations in cellular structure relative to the process of epithelial differentiation. Gastroenterology 64:926–945, 1973.

Mills SE, Fechner RE, Cantrell RW: Aggressive sinonasal lesion resembling normal intestinal mucosa. Am J Surg Pathol 6:803–809, 1982.

Nevalainen TV, Jarvi OH: Ultrastructure of intestinal and diffuse type gastric carcinoma. J Pathol 122: 129–136, 1977.

Schmid KO, Aubock L, Albegger K: Endocrine amphicrine enteric carcinoma of the nasal mucosa. Virchows Arch [A] 383:329–343, 1979.

Yamashiro K, Suzuki H, Nagayo T: Electron microscopic study of signet-ring cells in diffuse carcinoma of the stomach. Virchows Arch [A] 374:275–284, 1977.

Liver

Alpert LI, Zak FG, Werthamer S, et al: Cholangiocarcinoma: A clinicopathologic study of five cases with ultrastructural observations. Hum Pathol 5:709–728, 1974.

Becker FF: Hepatoma—Nature's model tumor. Am J Pathol 74:179–210, 1974.

Denk H, Franke WW, Dragosics B, et al: Pathology of cytoskeleton of liver cells: Demonstration of Mallory bodies (alcoholic hyalin) in murine and human hepatocytes by immunofluorescence microscopy using antibodies to cytokeratin polypeptides from hepatocytes. Hepatology 1:9–20, 1981.

Franke WW, Denk H, Schmid E, et al: Ultrastructural, biochemical, and immunologic characterization of Mallory bodies in livers of griseofulvin-treated mice. Fimbriated rods of filaments containing prekeratin-like polypeptides. Lab Invest 40:207–220, 1979.

French SW: The Mallory body: Structure, composition and pathogenesis. Hepatology 1:76–83, 1981.

Ghadially FN, Parry EW: Ultrastructure of a human hepatocellular carcinoma and surrounding non-neoplastic liver. Cancer 19:1989–2004, 1966.

Ghadially FN: Mallory's bodies. In *Ultrastructural Pathology of the Cell and Matrix. A Text and Atlas of Physiological and Pathological Alterations in the Fine Structure of Cellular and Extracellular Components,* 3rd ed, vol 2. Butterworths, London, 1988, pp 902–905.

Horie A, Kotoo Y, Hayashi I: Ultrastructural comparison of hepatoblastoma and hepatocellular carcinoma. Cancer 44:2184–2193.

Keeley AF, Iseri OA, Gottlieb LS: Ultrastructure of hyaline cytoplasmic inclusions in a human hepatoma: Relationship to Mallory's alcoholic hyalin. Gastroenterology 62:280–293, 1972.

Lapis K, Johannesson JV: Pathology of primary liver cancer. J Toxicol Environ Health 5:315–355, 1979.

Ordonez NG, Mackay B: Ultrastructure of liver cell and bile duct carcinomas. Ultrastruct Pathol 5:201–241, 1983.

Phillips MJ: Mallory bodies and the liver [editorial]. Lab Invest 47:311–313, 1982.

Phillips MJ, Poucell S, Patterson J, et al: Liver tumors and tumor-like conditions. In *The Liver. An Atlas and Text of Ultrastructural Pathology.* Raven Press, New York, 1987, pp 447–517.

Schaff Z, Lapis K, Safrany L: The ultrastructure of primary hepatocellular cancer in man. Virchows Arch [A] 352:340–358, 1971.

Smuckler EA: The ultrastructure of human alcoholic hyalin. Am J Clin Pathol 49:790–797, 1968.

Tanikawa K: *Ultrastructural Aspects of the Liver and Its Disorders,* 2nd ed. Igaku-Shoin, Tokyo, 1979, pp 338–348.

Toker C, Trevino N: Ultrastructure of human primary hepatic carcinoma. Cancer 19:1594–1606, 1966.

Pancreas

Albores-Saavedra J, Angeles-Angeles A, Nadji M, et al: Mucinous cystadenocarcinoma of the pancreas. Morphologic and immunocytochemical observations. Am J Surg Pathol 11:11–20, 1987.

Albores-Saavedra J, Gould EW, Angeles-Angeles A, et al: Cystic tumors of the pancreas. Pathol Annu 25:19–50, 1990.

Alpert LC, Truong LD, Bossart MI, et al: Microcystic adenoma (serous cystadenoma) of the pancreas. Am J Surg Pathol 12:251–263, 1988.

Bombi JA, Milla A, Badal JM, et al: Papillary-cystic neoplasms of the pancreas. Report of two cases and review of the literature. Cancer 54:780–784, 1984.

Boor PJ, Swanson MR: Papillary-cystic neoplasm of the pancreas. Am J Surg Pathol 3:69–75, 1979.

Chaudhry AP, Cutler LS, Leifer C, et al: Histogenesis of acinic cell carcinoma of the major and minor salivary glands. An ultrastructural study. J Pathol 148:307–320, 1986.

Compagno J, Oertel JE: Microcystic adenomas of the pancreas (glycogen-rich cystadenomas). A clinicopathologic study of 34 cases. Am J Clin Pathol 69: 289–298, 1978.

Echevarria RA: Ultrastructure of the acinic cell carcinoma and clear cell carcinoma of the parotid gland. Cancer 20:563–571, 1967.

Erlandson RA, Tandler B: Ultrastructure of acinic cell carcinoma of the parotid gland. Arch Pathol 93:130–140, 1972.

Gustafsson H, Carlsöö B, Henriksson R: Ultrastructural morphology and secretory behavior of acinic cell carcinoma. Cancer 55:1706–1710, 1985.

Jorgensen LJ, Hansen AB, Burcharth F, et al: Solid and papillary neoplasm of the pancreas. Ultrastruct Pathol 16:659–666, 1992.

Kay S, Schatzki PF: Ultrastructure of acinic cell carcinoma of the parotid salivary gland. Cancer 29:235–244, 1972.

Kim YI, Seo JW, Suh JS, et al: Microcystic adenomas of the pancreas. Am J Clin Pathol 94:150–156, 1990.

Klimstra DS, Heffess CS, Oertel JE, et al: Acinar cell carcinoma of the pancreas. A clinicopathologic study of 28 cases. Am J Surg Pathol 16:815–837, 1992.

Lack EE: Primary tumors of the exocrine pancreas. Classification, overview, and recent contributions by immunohistochemistry and electron microscopy [review]. Am J Surg Pathol 13:66–88, 1989.

Laitio M, Lev R, Orlic D: The developing human fetal pancreas: An ultrastructural and histochemical study with special reference to exocrine cells. J Anat 117: 619–634, 1974.

Lewandrowski K, Warshaw K, Compton C: Macrocystic serous cystadenoma of the pancreas: A morphologic variant differing from microcystic adenoma. Hum Pathol 23:871–875, 1992.

Lieber MR, Lack EE, Roberts JR, et al: Solid and papillary epithelial neoplasm of the pancreas. An ultrastructural and immunocytochemical study of six cases. Am J Surg Pathol 11:85–93, 1987.

Lo JW, Fung CHK, Yonan TN, et al: Cystadenoma of the pancreas. An ultrastructural study. Cancer 39: 2470–2474, 1977.

Miettinen M, Partanen S, Fräki O, et al: Papillary cystic tumor of the pancreas. An analysis of cellular differentiation by electron microscopy and immunohistochemistry. Am J Surg Pathol 11:855–865, 1987.

Morrison DM, Jewell LD, McCaughey WTE, et al: Papillary cystic tumor of the pancreas. Arch Pathol Lab Med 108:723–727, 1984.

Munger BL: The ultrastructure of the exocrine pancreas. In Carey LC, ed: *The Pancreas.* C.V. Mosby, St Louis, 1973, pp 17–31.

Nyongo A, Huntrakoon M: Microcystic adenoma of the pancreas with myoepithelial cells. A hitherto undescribed morphologic feature. Am J Clin Pathol 84: 114–120, 1985.

Osborne BM, Culbert SJ, Cangir A, et al: Acinar cell carcinoma of the pancreas in a 9-year-old child: Case report with electron microscopic observations. South Med J 70:370–372, 1977.

Pettinato G, Manivel JC, Ravetto C, et al: Papillary cystic tumor of the pancreas. A clinicopathologic study of 20 cases with cytologic, immunohistochemical, ultrastructural, and flow cytometric observations, and a review of the literature. Am J Clin Pathol 98:478–488, 1992.

Schlosnagle DC, Campbell WG Jr: The papillary and solid neoplasm of the pancreas. A report of two cases with electron microscopy, one containing neurosecretory granules. Cancer 47:2603–2610, 1981.

Shorten SD, Hart WR, Petras RE: Microcystic adenomas (serous cystadenomas) of pancreas. A clinicopathologic investigation of eight cases with immunohistochemical and ultrastructural studies. Am J Surg Pathol 10: 365–372, 1986.

Toner PG, Carr KE, McLay ALC: The exocrine pancreas. In Johannessen JV, ed. *Electron Microscopy in Human Medicine, Digestive System,* vol 7. McGraw-Hill, London, 1980, pp 211–245.

Webb JN: Acinar cell neoplasms of the exocrine pancreas. J Clin Pathol 30:103–112, 1977.

Kidney

Akhtar M, Kardar H, Linjawi T, et al: Chromophobe cell carcinoma of the kidney. A clinocopathologic study of 21 cases. Am J Surg Pathol 19:1245–1256, 1995.

Beckwith JB, Palmer NF: Histopathology and prognosis of Wilm's tumor: Results from the First National Wilm's Tumor Study. Cancer 41:1937–1948, 1978.

Bennington JL, Beckwith JB: Tumors of the kidney, renal pelvis, and ureter. In Firminger HI, ed: *Atlas of Tumor Pathology,* fasc 12. Armed Forces Institute of Pathology, Washington, DC, 1975, pp 156–159.

Delong WH, Sakr W, Grignon DJ: Chromophobe renal cell carcinoma. A comparative histochemical and immunohistochemical study. J Urol Pathol 4:1–8, 1996.

Dickersin GR, Colvin RB: Pathology of renal tumors. In Skinner DG, Lieskovsky, eds: *Diagnosis and Management of Genitourinary Cancer,* 2nd ed. WB Saunders, Philadelphia, 1988, pp 122–123.

Fisher ER, Horvart B: Comparative ultrastructural study of so-called renal adenoma and carcinoma. J Urol 108:382–386, 1972.

Gerharz CD, Moll R, Storkel S, et al: Establishment and characterization of two divergent cell lines derived from a human chromophobe renal cell carcinoma. Am J Pathol 146:953–962, 1995.

Hirano A, Zimmerman HM: Fenestrated blood vessels in metastatic renal carcinoma in the brain. Lab Invest 26:465–468, 1972.

Oberling C, Riviere M, Haguenau F: Ultrastructure of the clear cells in renal carcinomas and its importance for the demonstration of their origin. Nature 186:402, 1960.

Sarjanen K, Hjelt L: Ultrastructural characteristics of human renal cell carcinoma in relation to the light microscopic grading. Scand J Urol Nephrol 12:57–65, 1978.

Seljelid R, Ericsson JL: Electron microscopic observations on specialization of the cell surface in renal clear cell carcinoma. Lab Invest 14:435–447, 1965.

Storkel S, Steart PV, Drenckhahn D, et al: The human chromophobe cell renal carcinoma: Its probable relation to intercalated cells of the collecting duct. Virchows Arch [B] 56:237–245, 1989.

Sun CN, Bissada NK, White HJ, et al: Spectrum of ultrastructural patterns of renal cell adenocarcinoma. Urology 9:159–200, 1977.

Tannenbaum M: Ultrastructural pathology of human renal cell tumors. Pathol Annu 6:249–277, 1971.

Thoenes W, Storkel S, Rumpelt H-J: Human chromophobe cell renal carcinoma. Virchows Arch [Cell Pathol] 48:207–217, 1985.

Prostate

Brennick JB, O'Connell JX, Dickersin GR, et al: Lipofuscin pigmentation (so called "melanosis") of the prostate. Am J Surg Pathol 18:446–454, 1994.

Fisher ER, Jeffrey W: Ultrastructure of human normal and neoplastic prostate. Am J Clin Pathol 44:119–134, 1965.

Mao P, Nakao K, Angrist A: Human prostatic carcinoma: An electron microscopic study. Cancer Res 26: 955–973, 1966.

Mau P, Angrist A: The fine structure of the basal cell of human prostate. Lab Invest 15:1768–1782, 1966.

Sinha AA, Blackard CE: Ultrastructure of prostatic benign hyperplasia and carcinoma. Urology 2:114–120, 1973.

Srigley JR, Hartwick WJ, Edwards V, et al: Selected ultrastructural aspects of urothelial and prostatic tumors [review]. Ultrastruct Pathol 12:49–65, 1988.

Takayasu H, Yamaguchi Y: An electron microscopic study of the prostatic cancer cell. J Urol 87:935–940, 1962.

Tannenbaum M, Tannenbaum S: Ultrastructural pathology of human prostatic carcinoma. In Trump BF, Jones RT, eds: *Diagnostic Electron Microscopy,* vol 3. John Wiley and Sons, New York, 1980, pp 175–201.

Adrenal Cortex

Case Records of the Massachusetts General Hospital (Case 17-1977). N Engl J Med 296:988–994, 1977.

Case Records of the Massachusetts General Hospital (Case 23-1979). N Engl J Med 300:1322–1328, 1979.

Drachenberg CB, Lee HK, Gann DS, et al: Adrenal cortical carcinoma with adenosquamous differentiation. Report of a case with immunohistochemical and ultrastructural studies. Arch Pathol Lab Med 119:260–265, 1995.

El-Naggar AK, Evans DB, Mackay B: Oncocytic adrenal cortical carcinoma. Ultrastruct Pathol 15:549–556, 1991.

Erlandson RA, Reuter VE: Oncocytic adrenal cortical adenoma. Ultrastruct Pathol 15:539–547, 1991.

Erlandson RA, Nesland JM: Tumors of the endocrine/neuroendocrine system: An overview [review]. Ultrastruct Pathol 18:149–170, 1994.

Fawcett DW: Adrenal glands. In *Bloom and Fawcett: A Textbook of Histology,* 12th ed. Chapman and Hall, New York, 1994, pp 503–515.

Garret R, Ames RP: Black-pigmented adenoma of the adrenal gland. Report of three cases including electron microscopic study. Arch Pathol 95:349–353, 1973.

Jones EC, Pins M, Dickersin GR, et al: Metanephric adenoma of the kidney. A clinicopathological, immunohistochemical, flow cytometric, cytogenetic, and electron microscopic study of seven cases. Am J Surg Pathol 19: 615–626, 1995.

Hogan TF, Gilchrist KW, Westring DW, et al: A clinical and pathological study of adrenocortical carcinoma. Therapeutic implications. Cancer 45:2880–2883, 1980.

King DR, Lack EE: Adrenal cortical carcinoma. A clinical and pathologic study of 49 cases. Cancer 44:239–244, 1979.

Long JA, Jones AL: Observations on the fine structure of the adrenal cortex of man. Lab Invest 17:355–369, 1967.

Mackay B, El-Naggar A, Ordonez NG: Ultrastructure of adrenal cortical carcinoma. Ultrastruct Pathol 18: 181–190, 1994.

Molberg K, Vuitch F, Stewart D, et al: Adrenocortical blastoma. Hum Pathol 23:1187–1190, 1992.

Nader S, Hickey RC, Sellin RV, et al: Adrenal cortical carcinoma. A study of 77 cases. Cancer 52:707–711, 1983.

Osanai T, Konta A, Chui D, et al: Electron microscopic findings in benign deoxycorticosterone and 11-deoxycortisol-producing adrenal tumor. Arch Pathol Lab Med 114:829–831, 1990.

Page DL, Hough AJ, Gray GF: Diagnosis and prognosis of adrenocortical neoplasms. Arch Pathol Lab Med 110:993–994, 1986.

Perez-Ordonez B, Hamed G, Campbell S, et al: Renal oncocytoma: A clinicopathologic study of 70 cases. Am J Surg Pathol 21:871–883, 1997.

Rhodin JAG: Adrenal cortex. In *An Atlas of Ultrastructure.* WB Saunders, Philadelphia, 1963, pp 140–141.

Sasano H, Suzuki T, Sano T, et al: Adrenocortical oncocytoma. A true nonfunctioning adrenocortical tumor. Am J Surg Pathol 15:949–956, 1991.

Silva EG, Mackay B, Samaan NA, et al: Adrenocortical carcinomas: An ultrastructural study of 22 cases. Ultrastruct Pathol 3:1–7, 1982.

Spark RF: Low renin hypertension and the adrenal cortex. Semin Med Beth Israel Hosp 287:343–350, 1972.

Symington T: Summary of the ultrastructural features of cells found in the pathological human adrenal. In *Functioning Pathology of the Human Adrenal Gland.* E&S Livingstone Ltd, Edinburgh, 1969, pp 476–484.

Weiss LM: Comparative histologic study of 43 metastasizing and nonmetastasizing adrenocortical tumors. Am J Surg Pathol 8:163–169, 1984.

Young RH, Dunn J, Dickersin GR: An unusual oncocytic renal tumor with sarcomatoid foci and osteogenic differentiation. Arch Pathol Lab Med 112:937–940, 1988.

Squamous Cell Carcinoma

General

Chen SY, Harwick RD: Ultrastructure of oral squamous cell carcinoma. Oral Surg 44:744–753, 1977.

Inoue S, Dionne GP: Tonofilaments in normal human bronchial epithelium and in squamous cell carcinoma. Am J Pathol 88:345–354, 1977.

McDowell EM, Hess FG Jr, Trump BF: Epidermoid metaplasia, carcinoma in situ and carcinoma of the lung. In Trump BF, Jones RT, eds: *Diagnostic Electron Microscopy,* vol 3. John Wiley and Sons, New York, 1980, pp 37–96.

Thymoma

Bearman RM, Levine GD, Bensch KG: The ultrastructure of the normal human thymus: A study of 36 cases. Anat Rec 190:755–782, 1978.

Bloodworth JMB, Hiratsuka H, Hickey RC, et al: Ultrastructure of the human thymus, thymic tumors, and myasthenia gravis. Pathol Annu 10:329–391, 1975.

Eimoto T, Teshima K, Shirakusa T, et al: Heterogeneity of epithelial cells and reactive components in thymomas: An ultrastructural and immunohistochemical study. Ultrastruct Pathol 10:157–173, 1986.

Hammond EH, Flinner RL: The diagnosis of thymoma: A review. Ultrastruct Pathol 15:419–438, 1991.

Levine GD, Bensch KG: Epithelial nature of spindle-cell thymoma. An ultrastructural study. Cancer 30:500–511, 1972.

Levine GD, Med M, Rosai J, et al: The fine structure of thymoma, with emphasis on its differential diagnosis. Am J Pathol 81:49–86, 1975.

Levine GD, Rosai J: Thymic hyperplasia and neoplasia: A review of current concepts. Hum Pathol 9:495–515, 1978.

Lewis JE, Wick MR, Scheithauer BW, et al: Thymoma: A clinicopathologic review. Cancer 60:2727–2743, 1987.

Marino M, Muller-Hermelink HK: Thymoma and thymic carcinoma: Relation of thymoma epithelial cells to the cortical and medullary differentiation of the thymus. Virchows Arch [A] 407:119–149, 1985.

Moran CA, Suster S: Current status of the histologic classification of thymoma. Int J Surg Pathol 3:67–72, 1995.

Pascoe HR, Miner MS: An ultrastructural study of nine thymomas. Cancer 37:317–326, 1976.

Rosai J, Levine GD: Tumors of the thymus. In Firminger HI, ed: *Atlas of Tumor Pathology,* 2nd series, Fasc 13. Armed Forces Institute of Pathology, Washington, DC, 1976.

Snover DC, Levine GD, Rosai J: Thymic carcinoma. Five distinctive histological variants. Am J Surg Pathol 6: 451–470, 1982.

Suster S, Rosai J: Histology of the normal thymus. Am J Surg Pathol 14:284–303, 1990.

Suster S, Rosai J: Thymic carcinoma: A clinicopathologic study of 60 cases. Cancer 67:1025–1032, 1991.

Truong LD, Mody DR, Cagle PT, et al: Thymic carcinoma: A clinicopathologic study of 13 cases. Am J Surg Pathol 14:151–166, 1990.

van de Wijngaert FP, Kendall MD, Schuurman H-J, et al: Heterogeneity of epithelial cells in the human thymus. An ultrastructural study. Cell Tissue Res 237: 227–237, 1984.

Walker AN, Mills SE, Fechner RE: Thymomas and thymic carcinomas. Semin Diagn Pathol 7:250–265, 1990.

Wick MR, Weiland LH, Scheithauer B, et al: Primary thymic carcinomas. Am J Surg Pathol 6:613–630, 1982.

Adamantinoma

Knapp RH, Wick MR, Scheithauer BW, et al: Adamantinoma of bone. An electron microscopic and immunohistochemical study. Virchows Arch [A] 398:75–86, 1982.

Morai H, Yamamoto S, Hiramatsu K, et al: Adamantinoma of the tibia. Ultrastructural and immunohistochemic study with reference to histogenesis. Clin Orthop Rel Res 190:299–310, 1984.

Pieterse AS, Smith PS, McClure J: Adamantinoma of long bones: Clinical, pathological and ultrastructural features. J Clin Pathol 35:780–786, 1982.

Perez-Atayde AR, Kozakewich HPW, Vawter GF: Adamantinoma of the tibia. An ultrastructural and immunohistochemical study. Cancer 55:1015–1023, 1985.

Povysil C, Matejovsky Z: Ultrastructure of adamantinoma of long bone. Virchows Arch [A] 393:233–244, 1981.

Rosai J: Adamantinoma of the tibia. Electron microscopic evidence of its epithelial origin. Am J Clin Pathol 51:786–792, 1969.

Yoneyama T, Winter WG, Milsow L: Tibial adamantinoma: Its histogenesis from ultrastructural studies. Cancer 40:1138–1142, 1977.

Transitional Cell (Urothelial) Carcinoma

Battifore H, Eisenstein R, McDonald JH: The human urinary bladder mucosa. An electron microscopic study. Invest Urol 1:354–361, 1974.

Fulker MJ, Cooper EH, Tanaka T: Proliferation and ultrastructure of papillary transitional cell carcinoma of the human bladder. Cancer 27:71–82, 1971.

Koss LG: Some ultrastructural aspects of experimental and human carcinoma of the bladder. Cancer Res 37:20–24, 1977.

Tannenbaum M: Light and electron microscopy of urothelial cancer. Carcinoma in situ. Urology 8:498–501, 1976.

Undifferentiated Carcinoma

de-Thé G, Vuillaume M, Giovanella BC, et al: Epithelial characteristics of tumor cells in nasopharyngeal carcinoma passaged in nude mice: Ultrastructure. J Natl Cancer Inst 57:1101–1105, 1976.

Gazzolo L, de-Thé G, Vuillaume M, et al: Nasopharyngeal carcinoma. II. Ultrastructure of normal mucosa, tumor biopsies, and subsequent epithelial growth in vitro. JNCI 48:73–86, 1972.

Hammar S: Adenocarcinoma and large cell undifferentiated carcinoma of the lung. Ultrastruct Pathol 11: 263–291, 1987.

Prasad U: Cells of origin of nasopharyngeal carcinoma: An electron microscopic study. J Laryngol Otol 88: 1087–1094, 1974.

Svoboda DJ, Kirchner FR, Shanmugaratnam K: The fine structure of nasopharyngeal carcinomas. In Muir CS, Shanmugaratnam K, eds: *UICC Monograph Series,* vol 1. Munksgaard, Copenhagen, 1967, pp 163–171.

Svoboda DJ, Kirchner FR, Shanmugaratnam K: Ultrastructure of nasopharyngeal carcinomas in American and Chinese patients: An application of electron microscopy to geographic pathology. Exp Mol Pathol 4: 189–204, 1965.

Melanoma

General

Carlson JA, Dickersin GR, Sober AJ, et al. Desmoplastic neurotropic melanoma. A clinicopathologic analysis of 28 cases. Cancer 75:478–494, 1995.

Clark WH, Bretton R: A comparative fine structural study of melanogenesis in normal human epidermal melanocytes and in certain human malignant melanoma cells. In Helwig E, Mostofi FK, eds: *The Skin,* International Academy of Pathology, monograph 10. Williams and Wilkins, Baltimore, 1971, pp 192–214.

Curran RC, McCann BG: The ultrastructure of benign pigmented nevi and melanocarcinomas in man. J Pathol 119:135–146, 1976.

Erlandson RA: Ultrastructural diagnosis of amelanotic malignant melanoma: Aberrant melanosomes, myelin figures or lysosomes? Ultrastruct Pathol 11:191–208, 1987.

Hernandez FJ: Malignant blue nevus. A light and electron microscopic study. Arch Dermatol 107:741–744, 1973.

Jimbow K, Miura S, Ito Y, et al: Characterization of melanogenesis and morphogenesis of melanosomes by physio-chemical properties of melanin and melanosomes in malignant melanoma. Cancer Res 44: 1128–1134, 1984.

Klug H, Gunther JW: Ultrastructural differences in human malignant melanomata. An electron microscopical study. Br J Dermatol 86:395–407, 1972.

Mackay B, Ayala AG: Intracisternal tubules in human melanoma cells. Ultrastruct Pathol 1:1–6, 1980.

Mazur MT, Katzenstein A-LA: Metastatic melanoma: The spectrum of ultrastructural morphology. Ultrastruct Pathol 1:337–356, 1980.

Clear Cell Sarcoma

Argenyi ZB, Cain C, Bromley C, et al: S-100 protein-negative malignant melanoma: Fact or fiction? A light-microscopic and immunohistochemical study. Am J Dermatopathol 16:233–240, 1994.

Azumi N, Turner RR: Clear cell sarcoma of tendons and aponeuroses: Electron microscopic findings suggesting Schwann cell differentiation. Hum Pathol 14:1084–1089, 1983.

Benson JD, Kraemer BB, Mackay B: Malignant melanoma of soft parts: An ultrastructural study of four cases. Ultrastruct Pathol 8:57–70, 1985.

Boudreaux D, Waisman J: Clear cell sarcoma with melanogenesis. Cancer 41:1387–1394, 1978.

Chung EB, Enzinger FM: Malignant melanoma of soft parts. A reassessment of clear cell sarcoma. Am J Surg Pathol 7:405–413, 1983.

Hoffman GJ, Carter D: Clear cell sarcoma of tendons and aponeuroses with melanin. Arch Pathol 95:22–25, 1973.

Kindblom L-G, Lodding P, Angervall L: Clear-cell sarcoma of tendons and aponeuroses. An immunohistochemical and electron microscopic analysis indicating neural crest origin. Virchows Arch [A] 401:109–128, 1983.

Kubo T: Clear cell sarcoma of patellar tendon studied by electron microscopy. Cancer 24:948–953, 1969.

Mukai M, Torikata C, Iri H, et al: Histogenesis of clear cell sarcoma of tendons and aponeuroses. An electron-microscopic, biochemical, enzyme histochemical, and immunohistochemical study. Am J Pathol 114:264–272, 1984.

Ohno T, Park P, Utsunomiya Y, et al: Ultrastructural study of a clear cell sarcoma suggesting Schwannian differentiation. Ultrastruct Pathol 10:39–48, 1986.

Parker JB, Marcus PB, Martin JH: Spinal melanotic clear-cell sarcoma: A light and electron microscopic study. Cancer 46:718–724, 1980.

Ballon Cell Melanoma

Hashimoto K, Bale GF: An electron microscopic study of balloon cell nevus. Cancer 30:530–540, 1972.

Kao GF, Helwig EB, Graham JH: Balloon cell malignant melanoma of the skin. A clinicopathologic study of 34 cases with histochemical, immunohistochemical, and ultrastructural observations. Cancer 69:2942–2952, 1992.

Khalil MK: Balloon cell malignant melanoma of the choroid: Ultrastructural studies. Br J Ophthalmol 67:579–584, 1983.

Ranchod M: Metastatic melanoma with balloon cell changes. Cancer 30:1006–1013, 1972.

Sondergaard K, Henschel A, Hou-Jensen K: Metastatic melanoma with balloon cell changes: An electron microscopic study. Ultrastruct Pathol 1:357–360, 1980.

Mesothelioma (Epithelial Type)

Bedrosian CW, Bonsib S, Moran C: Differential diagnosis between mesothelioma and adenocarcinoma: A multimodal approach based on ultrastructure and immunocytochemistry. Semin Diagn Pathol 9:124–140, 1992.

Bolen JW, Hammar SP, McNutt MA: Serosal tissue: Reactive tissue as a model for understanding mesotheliomas. Ultrastruct Pathol 11:251–262, 1987.

Burns TR, Greenberg SD; Mace ML, et al: Ultrastructural diagnosis of epithelial malignant mesothelioma. Cancer 56:2036–2040, 1985.

Burns TR, Johnson EH, Cartwright J, et al: Desmosomes of epithelial malignant mesothelioma. Ultrastruct Pathol 12:385–388, 1988.

Carstens PH: Contact between abluminal microvilli and collagen fibrils in metastatic adenocarcinoma and mesothelioma. J Pathol 166:179–182, 1992.

Coleman M, Henderson DW, Mukherjee TM: The ultrastructural pathology of malignant pleural mesothelioma. Pathol Annu 24:303–353, 1989.

Dardick I, Al-Jabi M, McCaughey WTE, et al: Ultrastructure of poorly differentiated diffuse epithelial mesotheliomas. Ultrastruct Pathol 7:151–160, 1984.

Dardick I, Al-Jabi M, McCaughey WTE, et al: Diffuse epithelial mesothelioma: A review of the ultrastructural spectrum. Ultrastruct Pathol 11:503–533, 1987.

Davis JMG: Ultrastructure of human mesotheliomas. J Natl Cancer Inst 52:1715–1725, 1974.

Hammar SP: The pathology of benign and malignant pleural disease [review]. Chest Surg Clin North Am 4: 405–430, 1994.

Katsube Y, Mukai K, Silverberg SG: Cystic mesothelioma of the peritoneum. A report of five cases and review of the literature. Cancer 50:1615–1622, 1982.

Legrand M, Pariente R: Ultrastructural study of pleural fluid in mesothelioma. Thorax 29:164–171, 1974.

Leong AS, Stevens MW, Mukherjee TM: Malignant mesothelioma: Cytologic diagnosis with histologic, immunohistochemical, and ultrastructural correlation. Semin Diagn Pathol 9:141–150, 1992.

Luse SA, Spjut HJ: An electron microscopic study of a solitary pleural mesothelioma. Cancer 17:1546–1554, 1964.

McCaughey WTE, Colby TV, Battifora H, et al: Diagnosis of diffuse malignant mesothelioma: Experience of a US/Canadian mesothelioma panel. Mod Pathol 4: 342–353, 1991.

Mennemeyer R, Smith M: Multicystic, peritoneal mesothelioma. A report with electron microscopy of a case mimicking intra-abdominal cystic hygroma (lymphangioma). Cancer 44:692–698, 1979.

Suzuki Y, Chung J, Kannerstein M: Ultrastructure of human malignant diffuse mesothelioma. Am J Pathol 85:241–262, 1976.

Suzuki Y, Kannerstein M, Chung J: Ultrastructure of human mesothelioma [abstract]. Am J Pathol 70:7, 1973.

Torikata C, Kawai T, Nakayama M: Confronting cisternae and ciliated cells in malignant pleural mesothelioma: An ultrastructural study. Ultrastruct Pathol 15: 249–256, 1991.

Warhol MJ, Corson JM: An ultrastructural comparison of mesotheliomas with adenocarcinomas of the lung and breast. Hum Pathol 16:50–55, 1985.

Weidner N: Malignant mesothelioma of peritoneum. Ultrastruct Pathol 15:515–520, 1991.

Wick MR, Loy T, Mills SE, et al: Malignant epitheliod pleural mesothelioma versus peripheral pulmonary adeno-carcinoma: A histochemical, ultrastructural, and immunohistologic study of 103 cases. Hum Pathol 21: 759–766, 1990.

Mesothelioma (Spindle Cell Type)

Benisch B, Peison B, Sobel HJ, et al: Fibrous mesotheliomas (pseudofibroma) of the scrotal sac: A light and ultrastructural study. Cancer 47:731–735, 1981.

Bolen JW, Hammar SP, McNutt MA: Reactive and neoplastic serosal tissue. A light-microscopic, ultrastructural, and immunocytochemical study. Am J Surg Pathol 10:34–47, 1986.

Briselli M, Mark EJ, Dickersin GR: Solitary fibrous tumors of the pleura: Eight new cases and review of 360 cases in the literature. Cancer 47:2678–2689, 1981.

Burrig KF, Kastendieck H: Ultrastructural observations on the histogenesis of localized fibrous tumors of the pleura (benign mesothelioma). Virchows Arch [A] 403: 413–424, 1984.

El-Naggar A, Ro JY, Ayala AG, et al: Localized fibrous tumor of the serosal cavities. Immunohistochemical, electron-microscopic, and flow-cytometric DNA study. Am J Clin Pathol 92:561–565, 1989.

Hammar SP, Bolen JW: Sarcomatoid pleural mesothelioma. Ultrastruct Pathol 9:337–343, 1985.

Kawai T, Mikata A, Torikata C, et al: Solitary (localized) pleural mesothelioma. A light and electron microscopic study. Am J Surg Pathol 2:365–375, 1978.

Kay S, Silverberg SG: Ultrastructural studies of a malignant fibrous mesothelioma of the pleura. Arch Pathol Lab Med 92:449–455, 1971.

Osamura RY: Ultrastructure of localized fibrous mesothelioma of the pleura. Report of a case with histogenetic considerations. Cancer 39:139–142, 1977.

Said JW, Nash G, Banks-Shlegal S, et al: Localized fibrous mesothelioma: An immunohistochemical and electron microscopic study. Hum Pathol 15:440–443, 1984.

Steinetz C, Clarke R, Jacobs GH, et al: Localized fibrous tumors of the pleura: Correlation of histopathological, immunohistochemical and ultrastructural features. Pathol Res Pract 186:344–357, 1990.

Witkin GB, Rosai J: Solitary fibrous tumor of the mediastinum. A report of 14 cases. Am J Surg Pathol 13: 547–557, 1989.

Yousem SA, Flynn SD: Intrapulmonary localized fibrous tumor. Intraparenchymal so-called localized fibrous mesothelioma. Am J Clin Pathol 89:365–369, 1988.

Large Cell Lymphoma

Azar HA, Jaffe ES, Berard CW, et al: Diffuse large cell lymphomas (reticulum cell sarcomas, histiocytic lymphomas). Correlation of morphologic features with functional markers. Cancer 46:1428–1441, 1980.

Azar HA, Espinoza CG, Richman AV, et al: "Undifferentiated" large cell malignancies: An ultrastructural and immunocytochemical study. Hum Pathol 13:323–333, 1982.

Bacchi MM, Bacchi CE, Gown AM, et al: Diversity of morphologic, immunocytochemical, and molecular biological findings in Ki-1 (CD30-positive) large cell lymphomas. IAP Abstract #393, 1991.

Banks PM, Metter J, Allred DC: Anaplastic large cell (Ki-1) lymphoma with histiocytic phenotype simulating carcinoma. Am J Clin Pathol 94:445–452, 1990.

Bernier V, Azar HA: Filiform large-cell lymphomas: An ultrastructural and immunohistochemical study. Am J Surg Pathol 11:387–396, 1987.

Bitter MA, Franklin WA, Larson RA, et al: Morphology in Ki-(CD30)-positive non-Hodgkin's lymphoma is correlated with clinical features and the presence of a unique chromosomal abnormality, t(2;5)(p23;q35). Am J Surg Pathol 14:305–316, 1990.

Burns BF, Cripps C, Dardick I: A case of Ki-1 large cell anaplastic lymphoma with ultrastructural features. Hum Pathol 20:393–396, 1989.

Carstens HB: Anemone cell tumor revisited [editorial]. Ultrastruct Pathol 16:iii–v, 1992.

Chan JKC, Banks PM, Cleary ML, et al: A revised European-American classification of lymphoid neoplasms proposed by the International Lymphoma

Study Group. A summary version. Am J Clin Pathol 103: 543–560, 1995.

Dardick I, Srinivasan R, Al-Jabi M: Signet-ring cell variant of large cell lymphoma. Ultrastruct Pathol 5: 195–200, 1983.

Dardick I, Cavell S, Moher D, et al: Ultrastructural morphometric study of follicular center lymphocytes: I. Nuclear characteristics and the Lukes-Collins' concept. Ultrastruct Pathol 13:373–391, 1989.

Dehner LP: Here we go again: A new classification of malignant lymphomas. A viewpoint from the trenches [editorial]. Am J Clin Pathol 103:539–540, 1995.

Dickersin GR: Electron microscopy of leukemias and lymphomas. Clin Lab Med 7(1):199–247, 1987.

Erlandson RA, Filippa DA: Unusual non-Hodgkin's lymphomas and true histiocytic lymphomas. Ultrastruct Pathol 13:258–260, 1989.

Gillespie JJ: The ultrastructural diagnosis of diffuse large cell ("histiocytic") lymphoma: Fine structural study of 30 cases. Am J Surg Pathol 2:9–20, 1978.

Henry K: Electron microscopy in the non-Hodgkin's lymphomata. Br J Cancer 31:73–93, 1975.

Glick AD, Leech JH, Waldron JA, et al: Malignant lymphomas of follicular center cell origin in man. II. Ultrastructure and cytochemical studies. J Natl Cancer Inst 54:23–36, 1975.

Iossifides I, Mackay B, Butler JJ: Signet-ring cell lymphoma. Ultrastruct Pathol 1:511–517, 1980.

Kinney MC, Glick AD, Stein H, et al: Comparison of anaplastic large cell Ki-1 lymphomas and microvillous lymphomas in their immunologic and ultrastructural features. Am J Surg Pathol 14:1047–1060, 1990.

Kinney MC, Collins RD, Greer JP, et al: A small-cell-predominant variant of primary Ki-1 $(CD30)^+$ T-cell lymphoma. Am J Surg Pathol 17:859–868, 1993.

Levine GD, Dorfman RF: Nodular lymphoma: An ultrastructural of its relationship to germinal centers on a correlation of light and electron microscopic findings. Cancer 35:148–164, 1975.

Michel RP, Case BW, Moinuddin M: Immunoblastic lymphosarcoma: A light immunofluorescence, and electron microscopic study. Cancer 43:224–236, 1979.

Osborne BM, Mackay B, Butler JJ, et al. Large cell lymphoma with microvillus-like projections: An ultrastructural study. Am J Clin Pathol 79:443–450, 1983.

Rivas C, Piris MA, Gamallo C, et al: Ultrastructure of 26 cases of Ki-1 lymphomas: Morphoimmunologic correlation. Ultrastruct Pathol 14:381–397, 1990.

Said JW, Hargreaves HK, Pinkus GS: Non-Hodgkin's lymphomas: An ultrastructural study correlating morphology with immunologic cell type. Cancer 44: 504–528, 1979.

Sibley R, Rosai J, Froelich W, Battifora H: Comments by the panel. Ultrastruct Pathol 8:369–373, 1985.

Stein H, Mason DY, Gerdes J, et al: The expression of the Hodgkin's disease associated antigen Ki-1 in reactive and neoplastic lymphoid tissue: Evidence that Reed-Sternberg cells and histiocytic malignancies are derived from activated lymphoid cells. Blood 66:848–858, 1985.

Taccagani GL, Terreni MR, Rovere E, Villa E, Cantaboni A: Anaplastic large cell Ki-1 lymphoma of the stomach with villopodial projections: An immunocytochemical and ultrastructural study and review of the literature. Ultrastruct Pathol 16:291–302, 1992.

Yowell R, Hammond E: Ki-1 lymphoma: A case report. Ultrastruct Pathol 16:11–16, 1992.

Hemophagocytic Syndrome and Histiocytic Sarcoma

Chen R-L, Su I-J, Lin K-H, et al: Fulminant childhood hemophagocytic syndrome mimicking histiocytic medullary reticulosis. An atypical form of Epstein-Barr virus infection. Am J Clin Pathol 96:171–176, 1991.

Chin NW, Gangi M, Fani K, et al: Colonic histiocytic neoplasm mimicking malignant histiocytosis and presenting as intussusception. Hum Pathol 26:682–687, 1995.

Dickersin GR: Electron microscopy of leukemias and lymphomas. Clin Lab Med 7(1):199–247, 1987.

DiSant'Agnese P, Ettinger LJ, Ryan CK, et al: Histiomonocytic malignancy. A spectrum of disease in an 11-month-old infant. Cancer 52:1417–1422, 1983.

Favara BE, Feller AC, Paulli M, et al: Contemporary classification of histiocytic disorders. The WHO Committee on Histiocytic/Reticulum Cell Proliferations. Reclassification Working Group of the Histiocytic Society. Med Pediatr Oncol 29(3):157–166, 1997.

Ferster A, Corazza F, Heimann P, et al: Anaplastic large cell lymphoma of true histiocytic origin in an infant: Unusual clinical, hematological, and cytogenetic features. Med Pediatr Oncol 22:147–152, 1994.

Heustis DG, Bull BS, Hadley GG: Ultrastructure of the spleen in malignant histiocytosis. Arch Pathol Lab Med 101:239–242, 1977.

Huhn D, Meister P: Malignant histiocytosis. Morphologic and cytochemical findings. Cancer 42:1341–1349, 1978.

Isaacson P, Wright DH, Jones DB: Malignant lymphoma of true histiocytic (monocyte/macrophage) origin. Cancer 51:80–91, 1983.

Jaffe ES: Malignant histiocytosis and true histiocytic lymphoma. In *Surgical Pathology of the Lymph Nodes and Related Organs,* 2nd ed. Major Problems in Pathology, vol 16. WB Saunders, Philadelphia, 1995, pp 560–593.

Koo CH, Reifel J, Kogut N, et al: True histiocytic malignancy associated with a malignant teratoma in a patient with 46XY gonadal dysgenesis. Am J Surg Pathol 16: 175–183, 1992.

Lampert IA, Catovsky D, Bergier N: Malignant histiocytosis: A clinicopathologic study of 12 cases. Br J Haematol 40:65–77, 1978.

Lombardi L, Carbone A, Pilotti S, et al: Malignant histiocytosis: A histological and ultrastructural study of lymph nodes in six cases. Histopathology 2:315–328, 1978.

Salisbury JR, Hall PA, Williams HC, et al: Multicentric reticulohistiocytosis. Detailed immunophenotyping confirms macrophage origin. Am J Surg Pathol 14: 687–693, 1990.

Salyer J, Craven CM: Malignant histiocytosis in a patient with acquired immunodeficiency syndrome-related complex. Arch Pathol Lab Med 114:376–378, 1990.

Schouten TJ, Hustinx TWJ, Scheres JMJC, et al: Malignant histiocytosis. Clinical and cytogenetic studies in a newborn and a child. Cancer 52:1229–1236, 1983.

Tubbs RR, Sheibani K, Sebek BA, et al: Malignant histiocytosis. Ultrastructural and imunocytochemical characterization. Arch Pathol Lab Med 104:26–29, 1980.

Vilpo JA, Klemi P, Lassila O, et al: Cytological and functional characterization of three cases of malignant histiocytosis. Cancer 46:1795–1801, 1980.

Langerhans' Cell Histiocytosis (Histiocytosis X)

Bartosik J, Andersson A, Axelsson S, et al: Direct evidence for the cytomembrane derivation of Birbeck granules: The membrane-sandwich effect. Acta Derm Venereol (Stockh) 65:157–160, 1985.

Basset F, Turiaf J: Identification par la microscopie electronique de particules de nature probablement virale dans le lesions granulomateuses d'une histiocytose "X" pulmonaire. CR Acad Sci [III] 261:3701, 1965.

Birbeck MS, Breathnach AS, Everall JD: An electron microscopic study of basal melanocytes and high-level clear cells (Langerhans' cells) in vitiligo. J Invest Dermatol 37:51–64, 1961.

Bruno J, Tognetti A: Ultrastructural formation of Langerhans' cell granules in a case of histiocytosis X [letter to the editor]. Ultrastruct Pathol 13:89–90, 1989.

Callihan TR: Langerhans' cell histiocytosis (histiocytosis X). In Jaffe ES: *Surgical Pathology of the Lymph Nodes and Related Organs,* 2nd ed. Major Problems in Pathology, vol 16. WB Saunders, Philadelphia, 1995, pp 534–559.

Case Records of the Massachusetts General Hospital, Case #40-1993. N Engl J Med 329:1108–1115, 1993.

Dickersin GR: Electron microscopy of leukemias and lymphomas. Clin Lab Med 7(1):199–247, 1987.

Mierau GW, Favara BE, Brenman JM: Electron microscopy in histiocytosis X. Ultrastruct Pathol 3: 137–142, 1982.

Lieberman PH, Jones CR, Steinman RM, et al: Langerhans cell (eosinophilic) granulomatosis. A clinicopathologic study encompassing 50 years. Am J Surg Pathol 20:519–552, 1996.

Lukes RJ, Collins RD. Mononuclear phagocyte system neoplasms. In Hartman WH, ed: *Atlas of Tumor Pathology. Tumors of the hematopoietic system,* 2nd series, fasc 28. Washington, DC, Armed Forces Institute of Pathology, 1992, pp 273–304.

Lukes RJ, Collins RD: Benign hematopoietic disorders that resemble neoplasms or may develop into neoplasms. In *Atlas of Tumor Pathology. Tumors of the hematopoietic system,* 2nd series, fasc 28. Washington, DC, Armed Forces Institute of Pathology, 1992, pp 396–399.

Morales AR, Fine G, Horn RC, et al: Langerhans' cells in a localized lesion of the eosinophilic granuloma type. Lab Invest 20:412–423, 1969.

Murphy GF, Harrist TJ, Bhan AK, et al: Distribution of cell surface antigens in histiocytosis X cells. Quantitative immuno-electron microscopy using monoclonal antibodies. Lab Invest 48:90–97, 1983.

Risdall RJ, Dehner LP, Duray P, et al: Histiocytosis X (Langerhans' cell histiocytosis). Arch Pathol Lab Med 107:59–63, 1983.

Robb IA, Jimenez L, Carpenter BF: Birbeck granules or Birbeck junctions? Intercellular "zipperlike" lattice junctions in eosinophilic granuloma of bone. Ultrastruct Pathol 16:423–428, 1992.

Sagebiel RW, Reed TH: Serial reconstruction of the characteristic granule of the Langerhans' cell. J Cell Biol 36:595–602, 1968.

Schuler G, Romani N, Stingl G, et al: Coated Langerhans' cell granules in histiocytosis X cells. Ultrastruct Pathol 5:77–82, 1983.

Tornowski WM, Hashimoto K: Langerhans' cell granules in histiocytosis X. The epidermal Langerhans' cell as a macrophage. Arch Dermatol 96:298–304, 1967.

Dendritic Cell Lesions

Chan JKC, Tsang WYW, Ng CS, et al: Follicular dendritic cell tumors of the oral cavity. Am J Surg Pathol 18:148–157, 1994.

Monda L, Warnke R, Rosai J: A primary lymph node malignancy with features suggestive of dendritic reticulum cell differentiation. A report of 4 cases. Am J Pathol 122:562–572, 1986.

Nakamura S, Hara K, Suchi T, et al: Interdigitating cell sarcoma. A morphologic, immunohistologic, and enzyme-histochemical study. Cancer 61:562–568, 1988.

Perez-Ordonez B, Erlandson RA, Rosai J: Follicular cell tumor. Report of 13 additional cases of a distinctive entity. Am J Surg Pathol 20:944–955, 1996.

Vasef MA, Zaatari GS, Chan WC, et al: Dendritic cell tumors associated with low-grade B-cell malignancies. Report of three cases. Am J Clin Pathol 104:696–701, 1995.

Weiss LM, Berry GJ, Dorfman RF, et al: Spindle cell neoplasms of lymph nodes of probable reticulum cell lineage. True reticulum cell carcinoma? Am J Surg Pathol 14:405–414, 1990.

Mastocytoma and Mastocytosis

Abraham EK, Augustine J, Amma S, et al: Malignant systemic mastocytosis. Ind J Cancer 29:192–197, 1992.

Anstey A, Lowe DG, Kirby JD, et al: Familial mastocytosis: A clinical, immunophenotypic, light and electron microscopic study. Br J Dermatol 125:583–587, 1991.

Diebold J, Riviere O, Gosselin B, et al: Different patterns of spleen involvement in systemic and malignant mastocytosis. A histological and immunohistochemical study of three cases. Virchows Arch [A] 419:273–280, 1991.

Galli SJ: New concepts about the mast cell. Semin Med Beth Israel Hosp 328:257–265, 1993.

Horny HP, Rabenhorst G, Loffler H, et al: Solitary fibromastocytic tumor arising in an inguinal lymph node: The first description of a unique spindle cell tumor simulating mastocytosis. Mod Pathol 7:962–966, 1994.

Kawai S, Okamoto H: Giant mast cell granules in a solitary mastocytoma. Pediatr Dermatol 10:12–15, 1993.

Rottem M, Okada T, Goff JP, et al: Mast cells cultured from the peripheral blood of normal donors and patients with mastocytosis originate from a CD34+/Fc epsilon RI-cell population. Blood 84:2489–2496, 1994.

Soter NA: The skin in mastocytosis [review]. J Invest Dermatol 96:32S–38S; discussion 38S–39S, 1991.

Weidner N, Horan RF, Austen KF: Mast-cell phenotype in indolent forms of mastocytosis. Ultrastructural features, fluorescence detection of avidin binding, and immunofluorescent determination of chymase, tryptase, and carboxypeptidase. Am J Pathol 140:847–857, 1992.

Wood C, Sina B, Webster CG, et al: Fibrous mastocytoma in a patient with generalized cutaneous mastocytosis. J Cutan Pathol 19:128–133, 1992.

4

Small Cell Neoplasms

Neuroendocrine Carcinoma

(Figures 4.1 through 4.8.)

Diagnostic criteria. (1) Oval and/or spindle shaped cells, variably with polar processes; (2) diffuse, usually noninsular arrangement of cells; (3) high nucleocytoplasmic ratio in cell bodies; (4) intercellular junctions; (5) dense-core granules.

Additional points. Neuroendocrine neoplasms include those derived from neural crest, such as neuroblastoma, and those derived from epithelium in various parts of the body, the gastrointestinal tract and skin being exemplary sites. Two examples of small cell neuroendocrine carcinomas are oat cell carcinoma of the lung and Merkel cell carcinoma of the skin. Oat cell carcinoma originates from the bronchogenic Kulschitzki cell, an endodermal derivative, and is the malignant counterpart of the carcinoid tumor (see Chapter 9). Merkel cell carcinoma arises from cutaneous neuroendocrine cells of probable neuroectodermal derivation. These neoplasms, especially oat cell carcinoma, often have a paucity of dense-core granules, making diagnosis more difficult. Furthermore, the presence of one or two small, dense granules in a cell does not rule out the possibility of the granules being primary lysosomes, which may be seen in almost any type of cell, including lymphocytes. Difficulty in diagnosis may arise if the tissue is not well preserved and/or if there is compression or other artifact. The identification of intercellular junctions, especially desmosomes, in these situations may be very helpful in making the diagnosis of carcinoma. Also, aggregates of paranuclear intermediate filaments and/or tonofibrils may be found in some tumors, especially Merkel cell carcinomas. Another particular feature of Merkel cell tumors is that the dense-core granules are predominantly in a subplasmalemmal location.

Neuroblastoma

(Figures 4.9 through 4.13.)

Diagnostic criteria. (1) Diffuse, nonorganoid pattern of small round and oval cell-bodies with high nucleocytoplasmic ratio; (2) zones devoid of cell bodies occupied by back-to-back cellular processes (neuropil); (3) microtubules, parallel and longitudinally directed, within cellular processes; (4) intermediate filaments; (5) small, round dense-core granules (more numerous in processes than in cell bodies); (6) synaptic vesicles in processes (variable); (7) intercellular junctions.

Additional points. Neuroblastoma, a neuroectodermally derived neuroendocrine neoplasm, has a unique ultrastructural appearance. Nuclei often have an irregular contour. The neuropil is characteristic, and the bare

(Text continues on page 161)

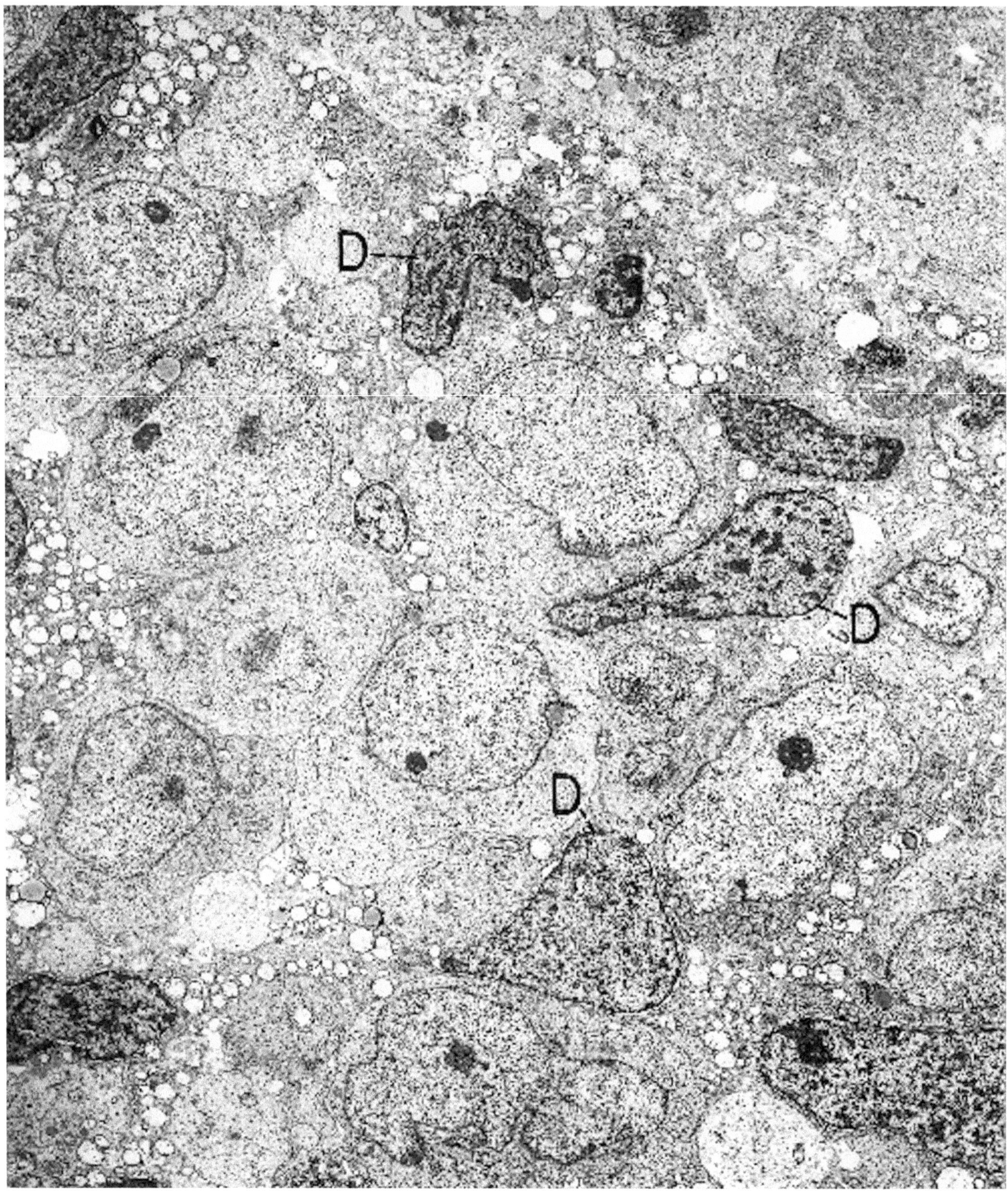

Figure 4.1. Oat-cell carcinoma (metastatic in mediastinal lymph node). The neoplastic small cells are tightly apposed and have active-appearing nuclei (euchromatin and prominent nucleoli). The dark cells (D) are examples of cell death, either *in vivo* or *in vitro*. They are characterized by their smaller volume, shrunken nuclei with aggregated chromatin, loss of plasma membrane, and swollen, membrane-bound, cytoplasmic organelles. (× 5100)

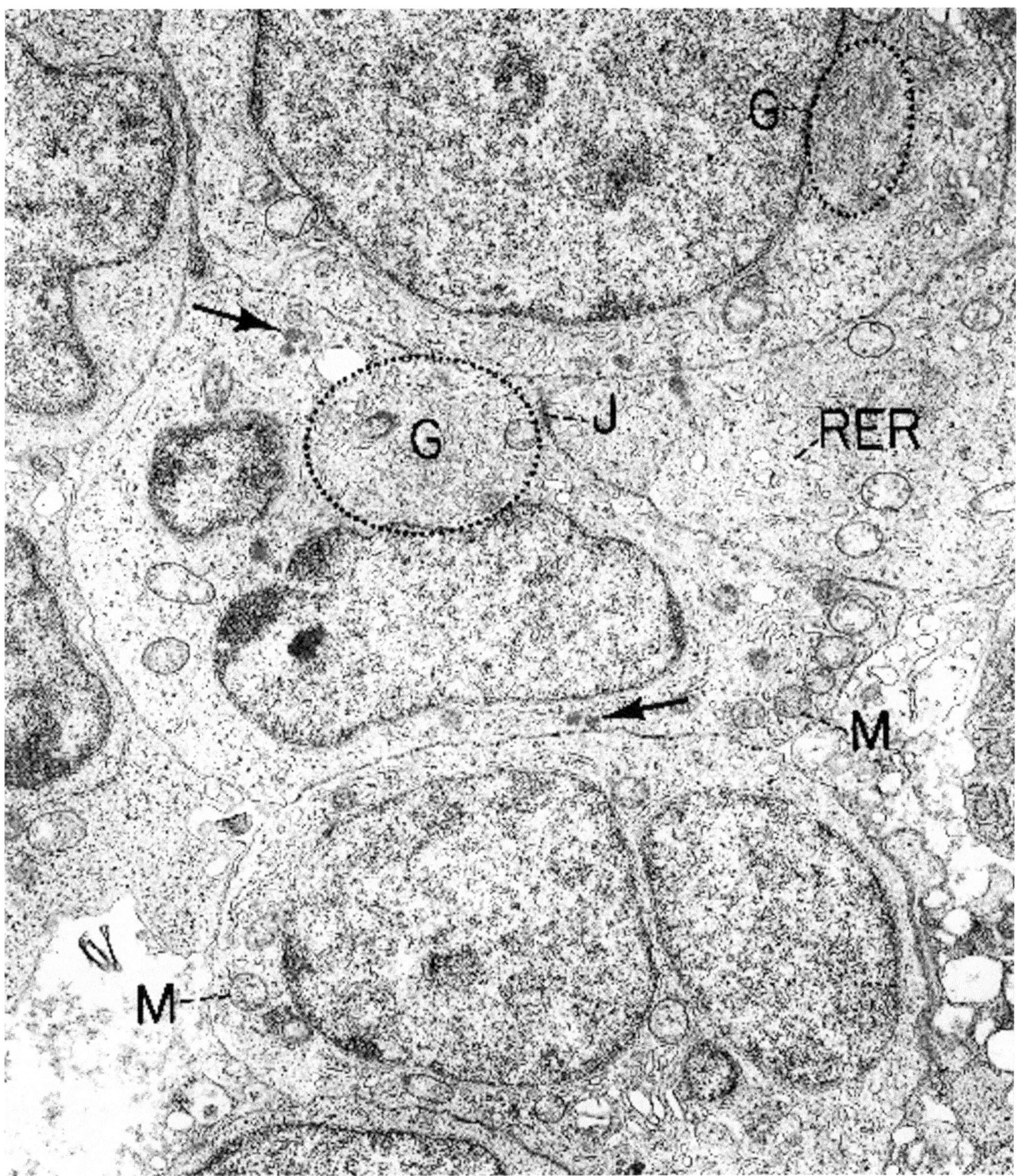

Figure 4.2. Oat-cell carcinoma (metastatic in mediastinal lymph node). The higher magnification of the neoplasm shown in Figure 4.1 illustrates intercellular junctions (J) and a few dense-core granules (*arrows*). Other nondiagnostic organelles, in addition to free ribosomes (background granules) that can be seen in small amounts, include rough endoplasmic reticulum (RER), Golgi apparatuses (G), and mitochondria (M). (× 12,100) (Permission for reprinting granted by WB Saunders, Dickersin GR: Electron microscopy of leukemias and lymphomas. Clin Lab Med 7:199–247, 1987.)

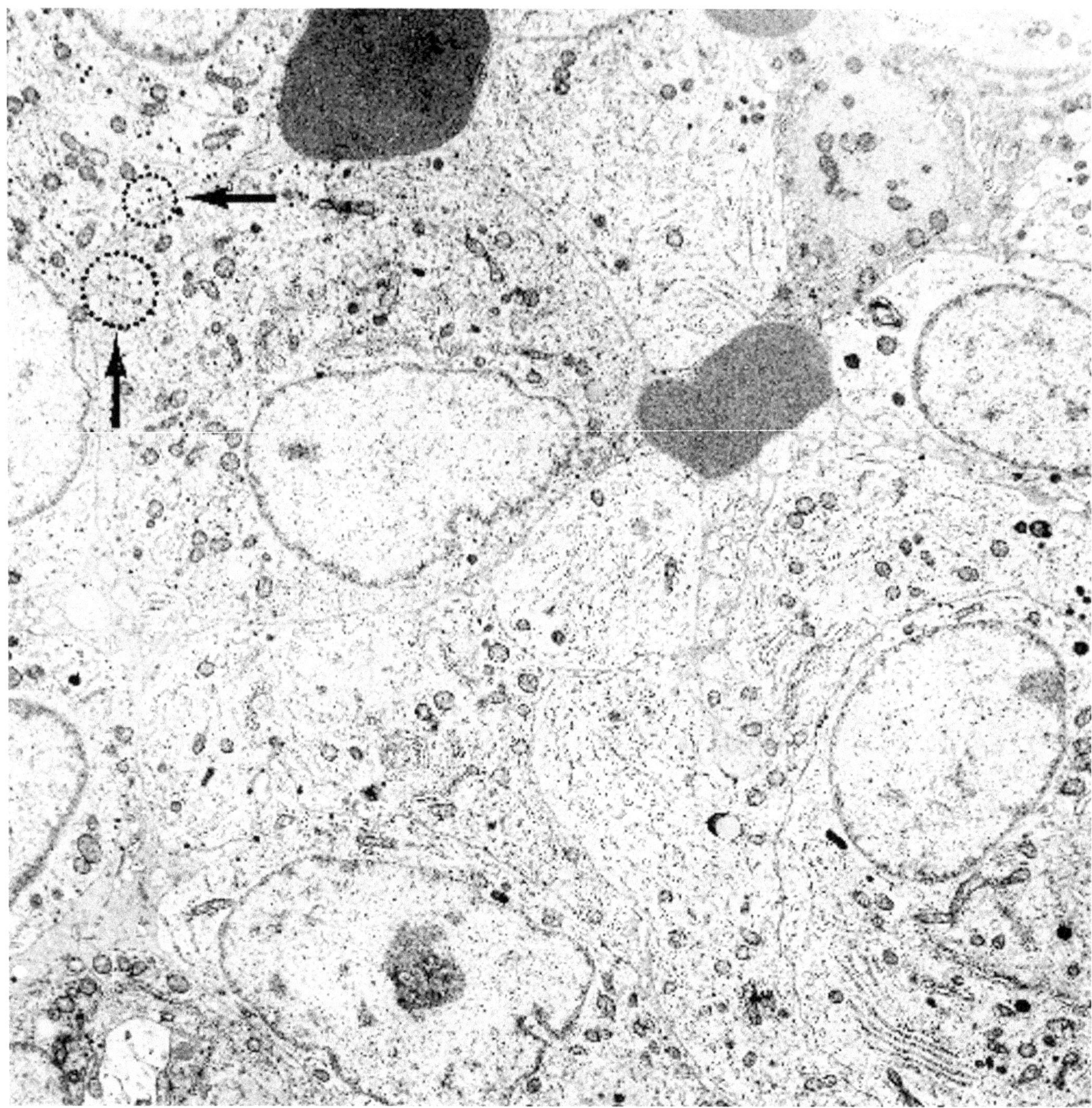

Figure 4.3. Oat-cell carcinoma (metastatic in bronchial lymph node). Contrast this better differentiated, small cell carcinoma with that depicted in Figures 4.1 and 4.2. The various cytoplasmic organelles are more numerous here, and diagnostic dense-core granules (*arrows*) are particularly easy to find (see also Figures 4.4 and 4.5). (× 5100)

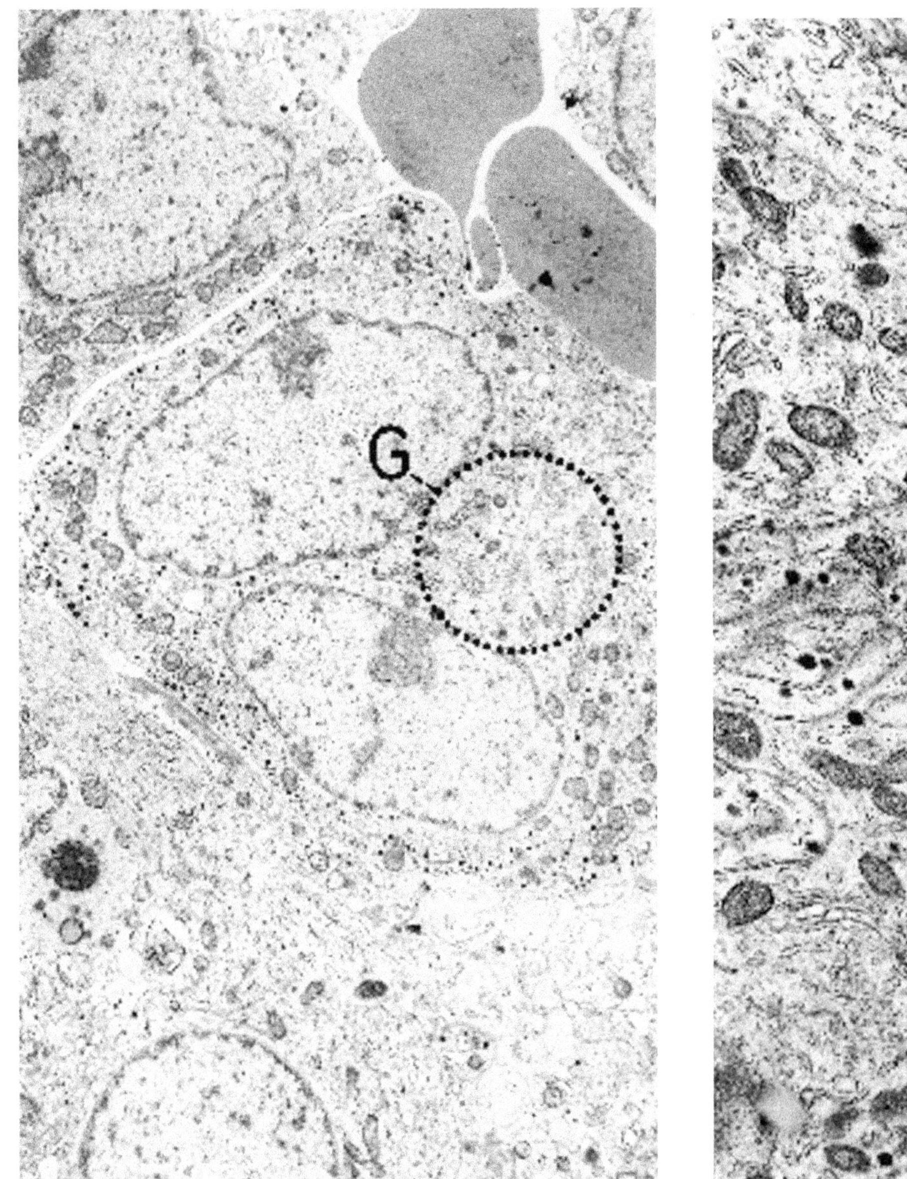

Figure 4.4. Oat-cell carcinoma (metastatic in bronchial lymph node). A binucleated neoplastic cell contains numerous dense-core granules. The large size of the Golgi apparatus (G) presumably is related to the production of granules. (× 5250)

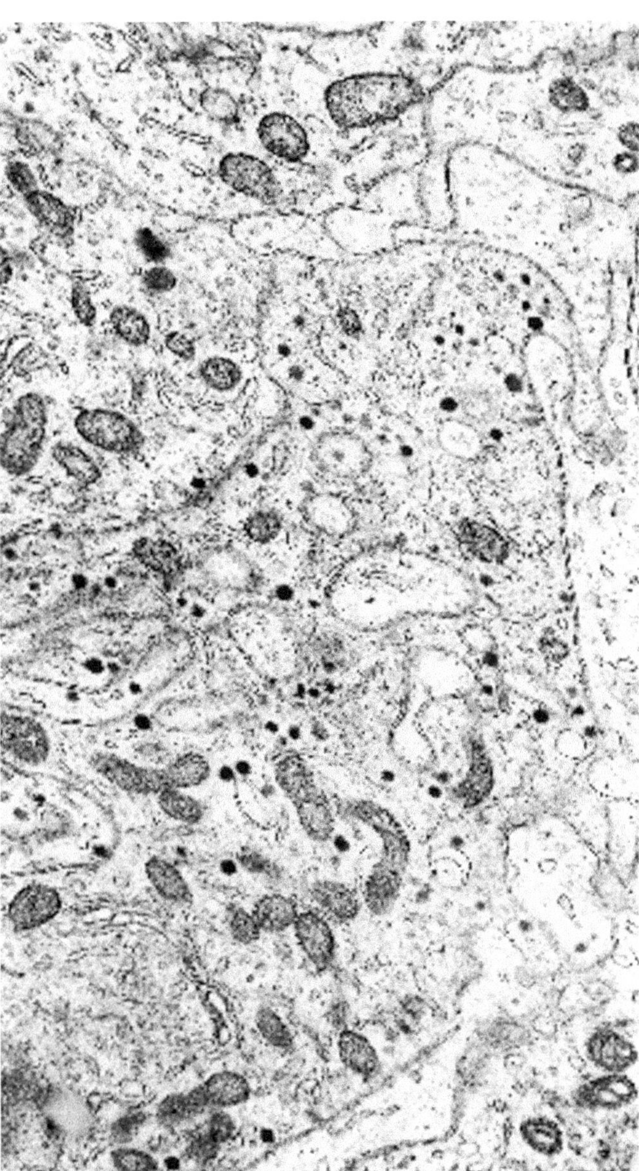

Figure 4.5. Oat-cell carcinoma (metastatic in bronchial lymph node). A higher power of the numerous dense-core granules in the same neoplasm as depicted in Figures 4.3 and 4.4. Note also the cells have long intertwining processes. (× 15,000)

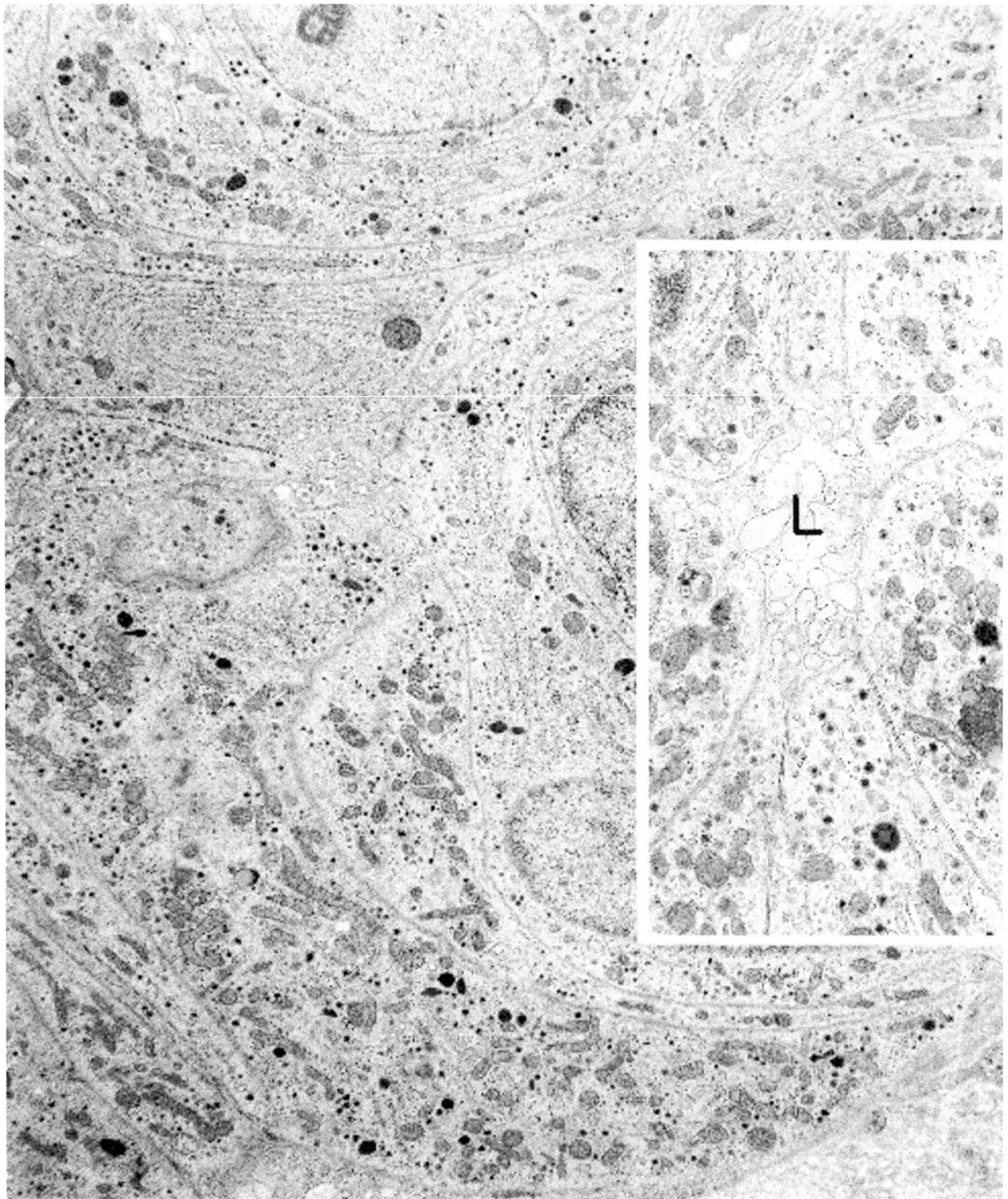

Figure 4.6. Carcinoid tumor (bronchus). The neoplastic cells contain innumerable dense-core granules, in contrast to relatively few granules in the less well-differentiated cells of oat-cell carcinoma, as depicted in Figures 4.1 through 4.5. The long cytoplasmic processes predominating in this carcinoid tumor are characteristic of the spindle-cell variant. (× 6700) *Inset:* Higher magnification of several cytoplasmic processes illustrates numerous dense-core granules and a microlumen (L). (× 12,150)

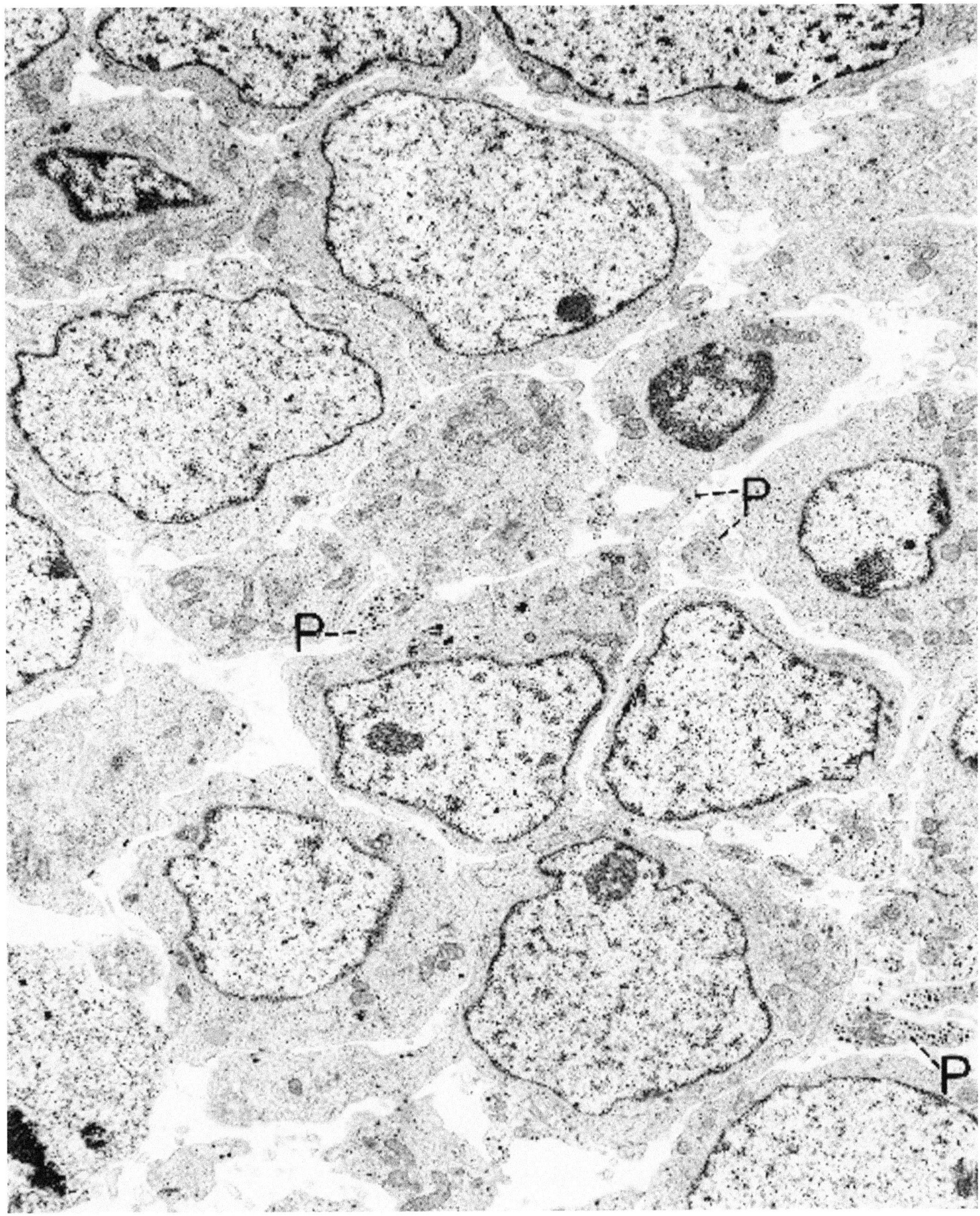

Figure 4.7. Merkel cell carcinoma (skin of eyelid). The neoplastic cells are oval and polygonal and have narrow processes (P). Dense-core granules are more numerous in the processes than in the cell bodies. (× 6800)

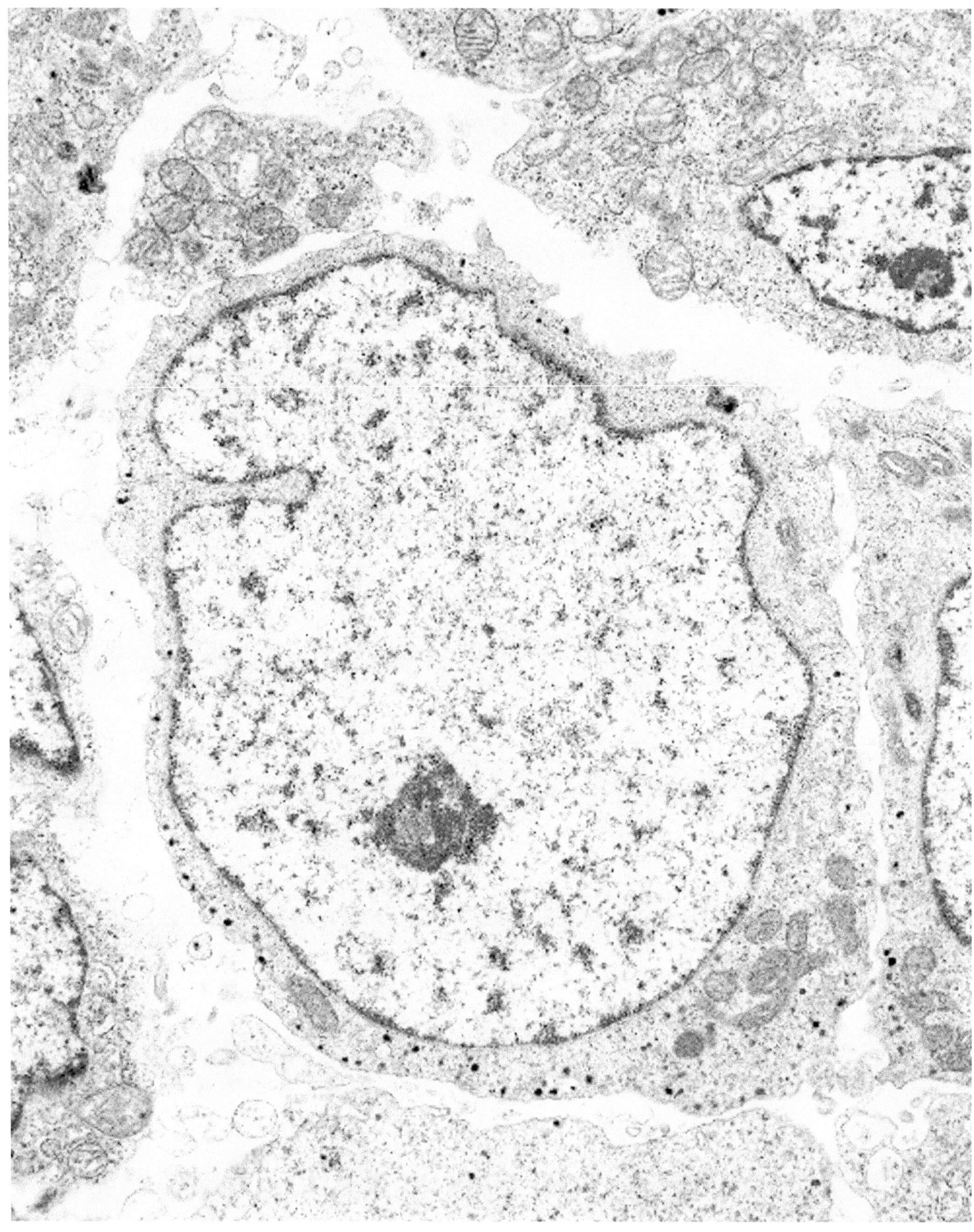

Figure 4.8. Merkel cell carcinoma (skin of eyelid). High power of a neoplastic cell illustrates the subplasmalemmal location of the dense-core granules. (× 12,000)

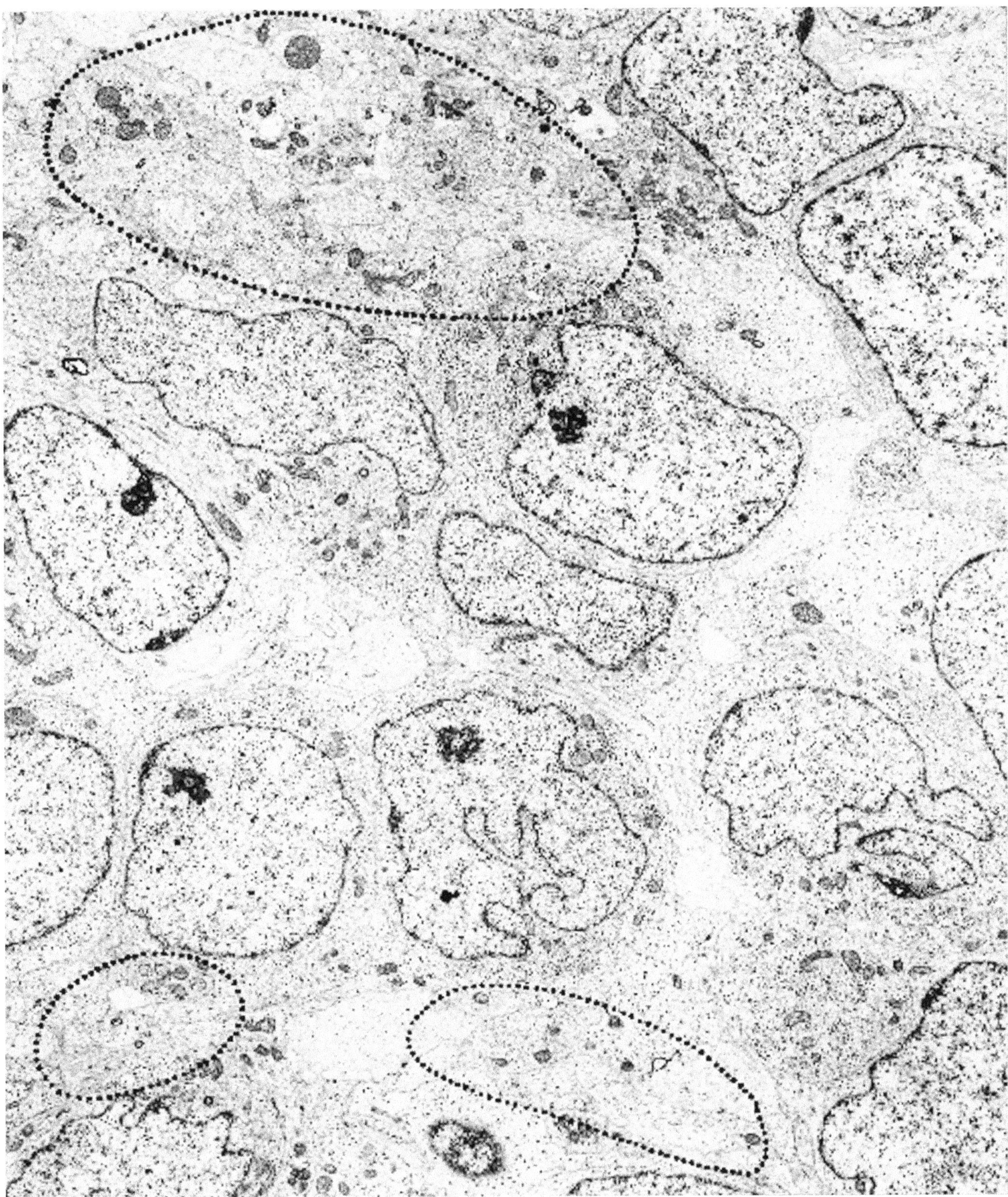

Figure 4.9. Neuroblastoma (retroperitoneum). Most of this field of the neoplasm consists of cell bodies, and there are only small zones (*encircled areas*) comprised solely of cellular processes. Within the cell bodies, note the prominent nucleoli, the finely dispersed nuclear chromatin, and the simple complement of cytoplasmic organelles (mostly ribosomes and a few mitochondria); these cellular features indicate active synthesis or division. (× 5130)

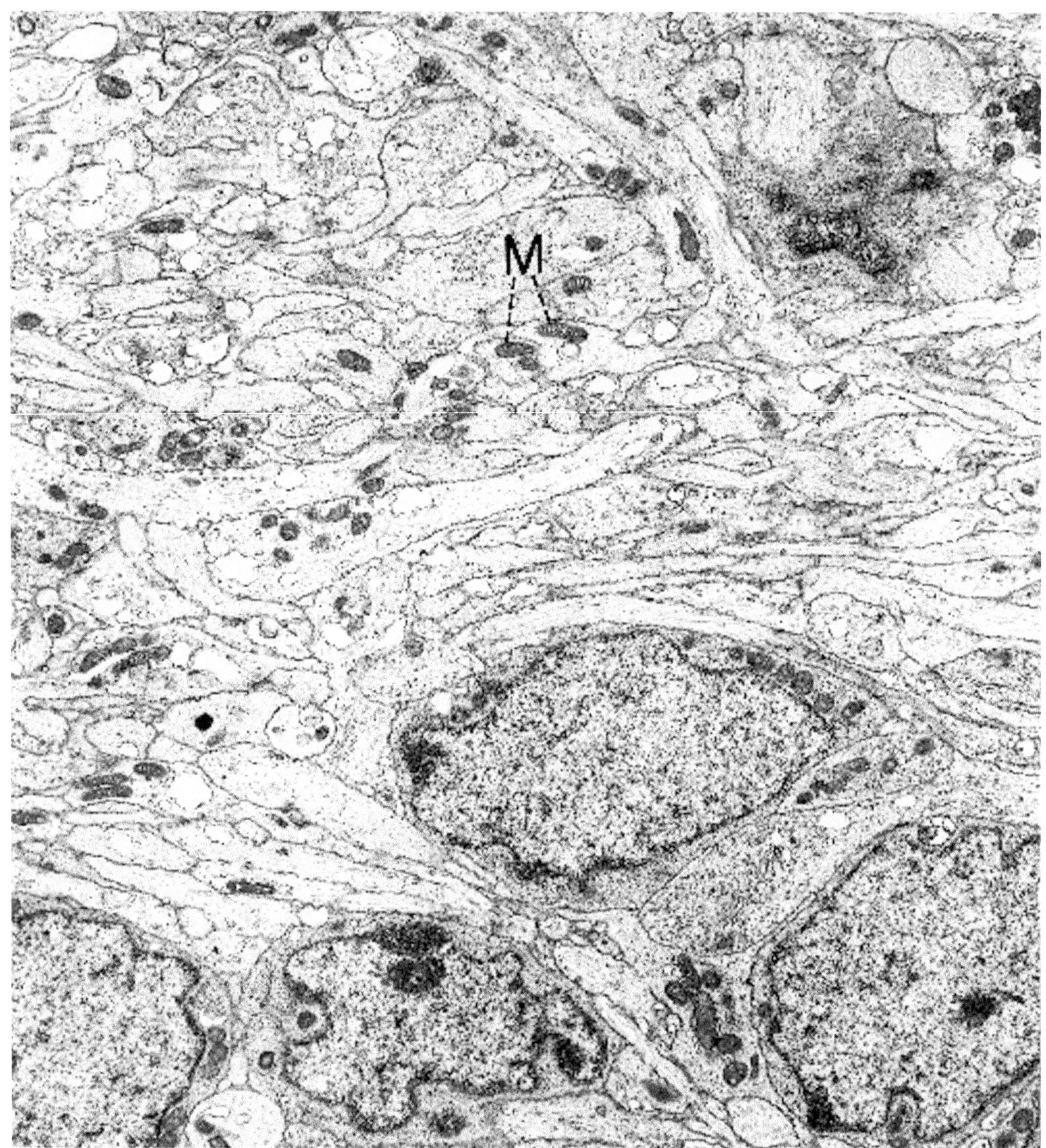

Figure 4.10. Neuroblastoma (bone marrow). The lower portion of the field is occupied mostly by cell bodies, and the upper portion, back-to-back neuritic processes. The longitudinal lines within the processes are microtubules. Mitochondria (M) are also visible in the cytoplasm of the processes. (× 9180) (Permission for reprinting granted by WB Saunders, Dickersin GR: Electron microscopy of leukemias and lymphomas. Clin Lab Med 7:199–247, 1987.)

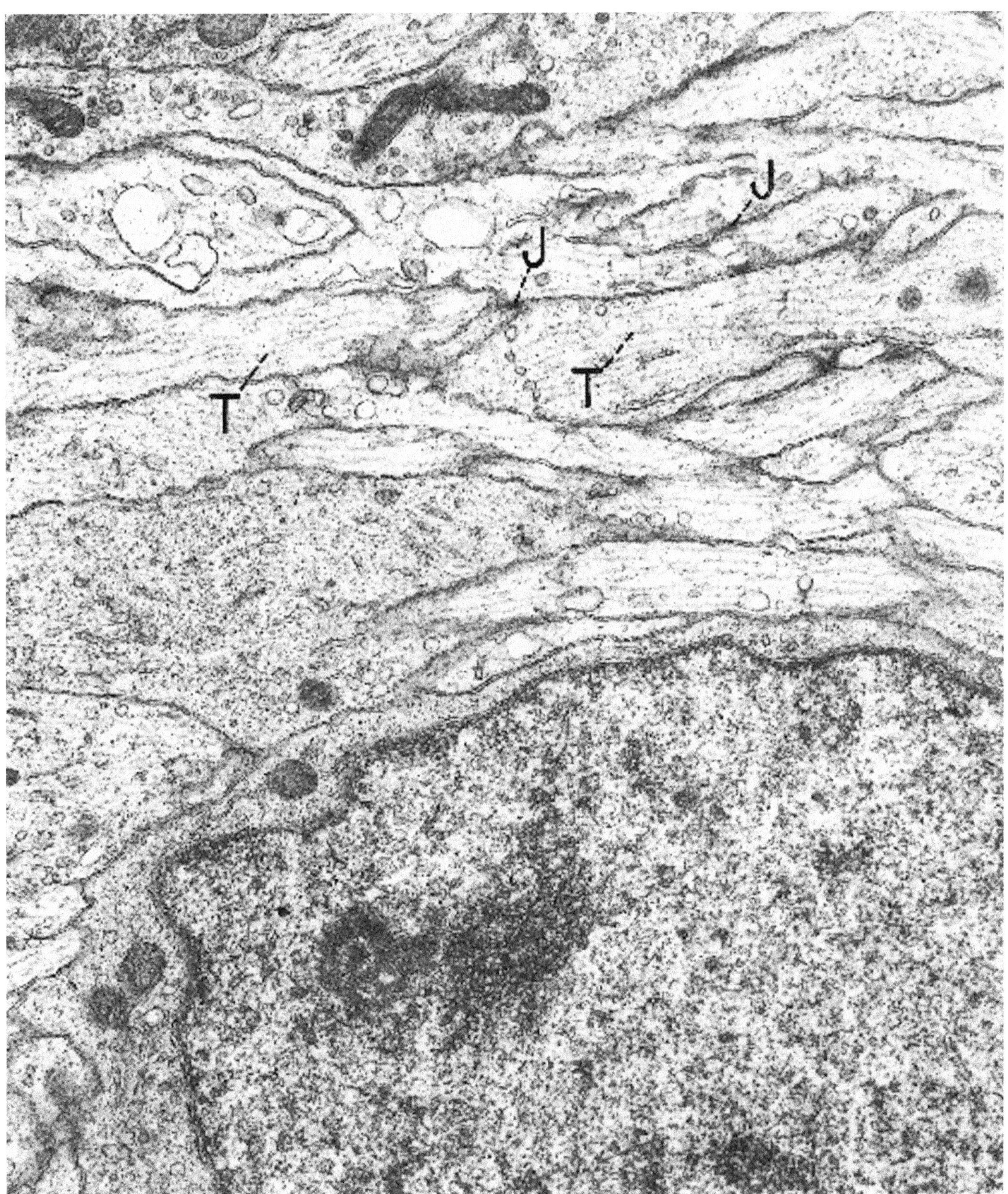

Figure 4.11. Neuroblastoma (bone marrow). Higher magnification of the same neoplasm as depicted in Figure 4.10 illustrates the neuritic processes with microtubules (T). Intercellular junctions (J) are somewhat vague in this field, and no definite dense-core granules are identified. (× 24,600) (Permission for reprinting granted by WB Saunders, Dickersin GR: Electron microscopy of leukemias and lymphomas. Clin Lab Med 7:199–247, 1987.)

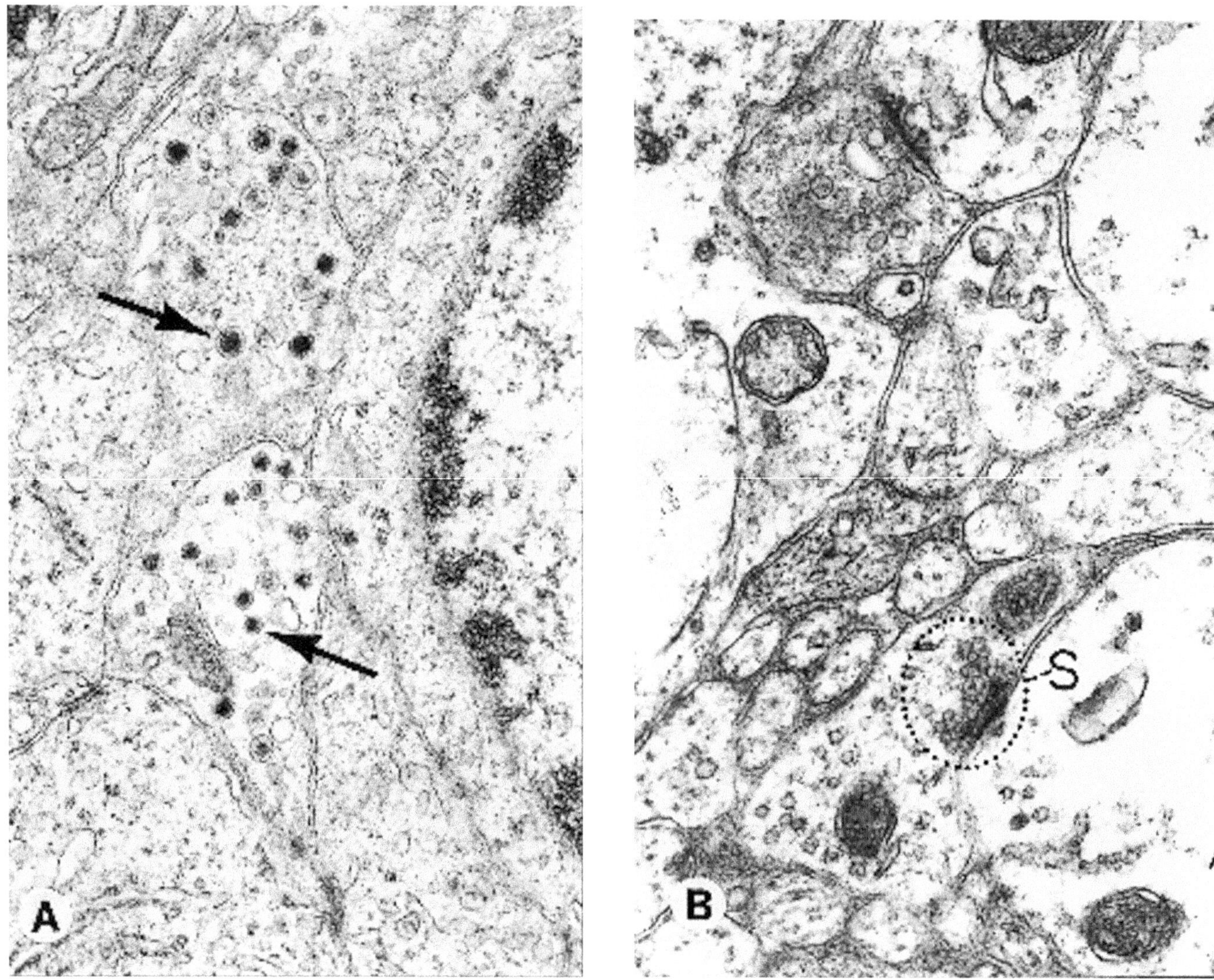

Figure 4.12. Neuroblastoma (**A,** nasal mucosa; **B** and **C,** brain). High magnification of neuritic processes. **A,** Dense-core granules (*arrows*). (× 27,700) **B,** synaptic vesicles (S). (× 46,000)

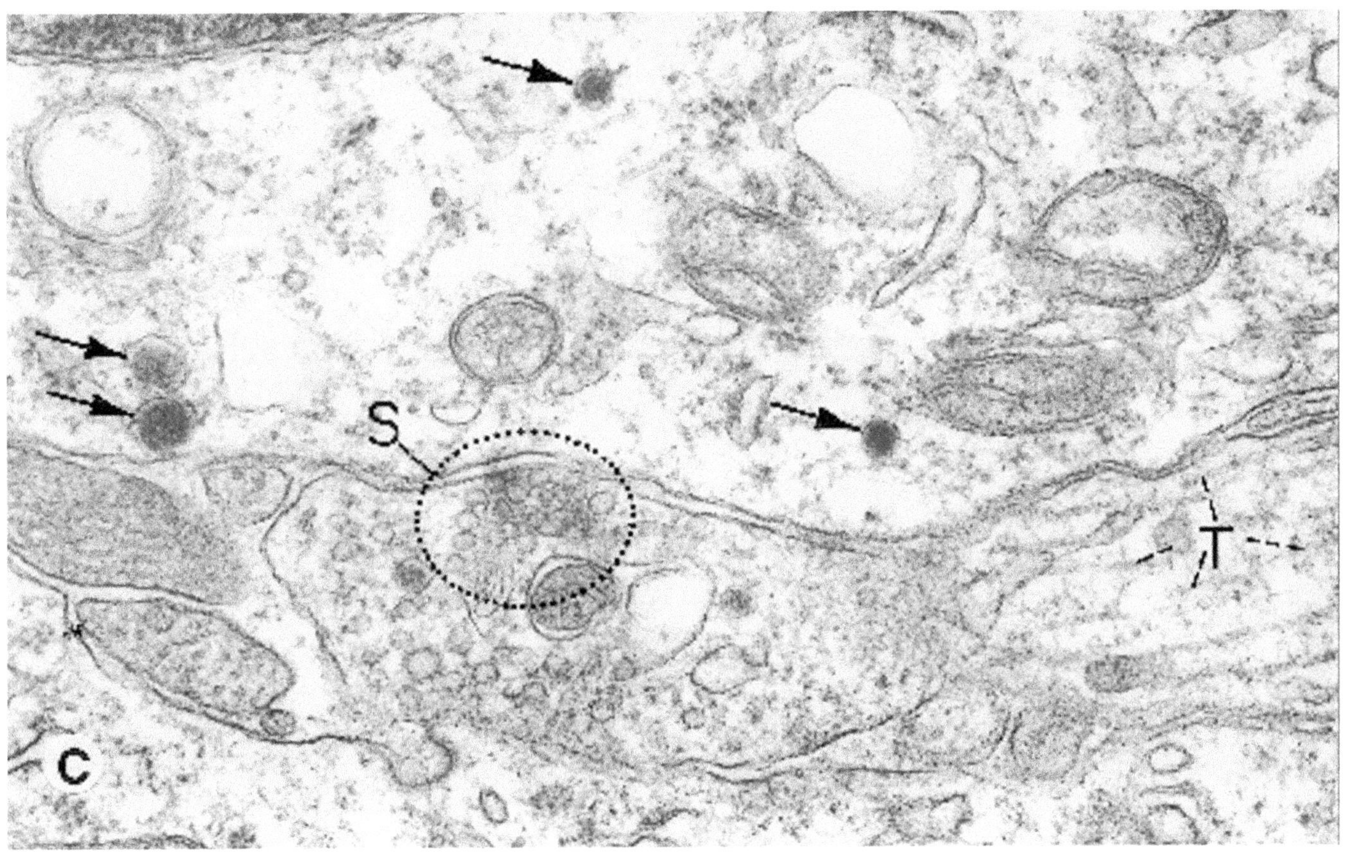

Figure 4.12. *(continued)*
C, dense-core granules (*arrows*), microtubules (T), and synaptic vesicles (S). (× 57,000) (Permission for reprinting granted by WB Saunders, Dickersin GR: Electron microscopy of leukemias and lymphomas. Clin Lab Med 7:199–247, 1987.)

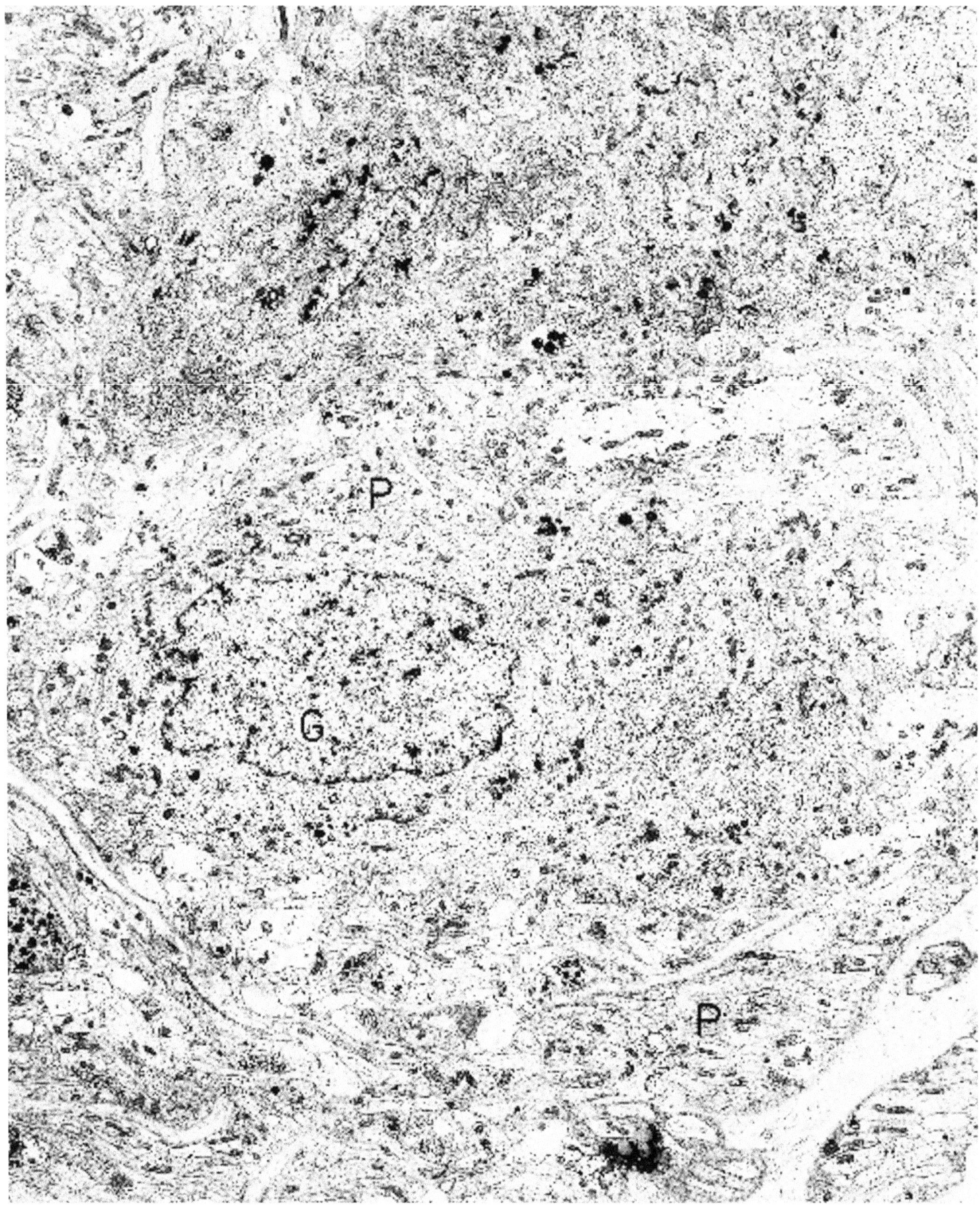

Figure 4.13. Ganglioneuroblastoma (soft tissue of thoracic wall). A mature ganglion cell (G) has copious cytoplasm with many organelles. It is surrounded by cell processes (P) of its own as well as those of less-differentiated neuroblasts composing the neoplasm. (× 4940)

(Text continued from page 147)

minimum among the criteria for diagnosis is the presence of small foci of cellular processes with microtubules (Figures 4.10 through 4.12). Dense-core granules (Figures 4.12A and C) may be scant in poorly differentiated tumors and are not necessary for diagnosis. Synaptic vesicles (Figures 4.12B and C) are the least frequent criterion encountered and are present in some better differentiated neuroblastomas and in ganglioneuroblastomas. The ganglion cells of *ganglioneuroblastomas* and *ganglioneuromas* represent mature cells that have differentiated from neuroblasts, and they are characterized ultrastructurally by their large size, round nucleus, prominent nucleolus, and voluminous cytoplasm with many organelles, including dense-core granules (Figure 4.13 and Chapter 8, Figures 8.33 through 8.35). Ganglioneuroblastomas also contain foci of Schwann cell–neurite clusters, where groups of neuritic processes are surrounded by Schwann cells and basal lamina.

The key to finding readily the diagnostic criteria in any neuroblastoma is to select an area for electron microscopic examination that is composed of neuropil (zones of apparent acellularity, by light microscopy), rather than an area occupied by cell bodies. *Olfactory neuroblastomas* (*esthesioneuroblastomas*) have generally similar ultrastructural features as those of neuroblastomas of other sites. *Neurocytomas* of the central nervous system are composed of mature unmyelinated neurons, cells further differentiated than the cells of neuroblastoma, and are described in Chapter 8.

Ewing's Sarcoma

(Figures 4.14 through 4.18.)

Diagnostic criteria. (1) Cells of uniform size and shape (oval and polygonal); (2) high nuclear–cytoplasmic ratio; (3) junctions are few, small, and inconspicuous; (4) finely dispersed chromatin (euchromatin); (5) cytoplasmic glycogen, usually copious but may be less; (6) cytoplasm composed predominantly of free ribosomes and polyribosomes.

Additional points. Ewing's sarcomas of bone and those of soft tissue have an identical ultrastructural appearance. The cells appear primitive and active; the cytoplasm is filled with ribosomes, and the nuclei are euchromatic and have small-to-large, often open nucleoli (nucleolonemas) (Figures 4.14 and 4.15). A few mitochondria and occasional areas of intermediate filaments are also present in the cytoplasm. Atypical Ewing's sarcomas show larger cells with irregularly shaped nuclei, moderate amounts of heterochromatin, and larger nucleoli. In the past, lymphoblasts were suggested as one of the possible cells of origin for Ewing's sarcomas, but subsequent immunohistochemical evidence has more or less ruled out that possibility. Furthermore, ultrastructurally it would be unusual in a lymphoblastic lymphoma to have no cells with the classic heterochromatin pattern of the lymphoid series. The vague intercellular junctions seen in Ewing's sarcoma constitute another difference between these two neoplasms, but there are examples of Ewing's sarcoma in which any semblance of a junction is difficult to find. Glycogen is the best differential criterion between Ewing's cells (Figures 4.16 through 4.18) and lymphoblasts, because it is virtually always present in the untreated former cells and absent in the latter ones.

The most popular current theory of histogenesis of Ewing's sarcoma, based on ultrastructural, immunohistochemical, and genetic evidence, is that it is a primitive neuroectodermal tumor (PNET; discussed in the next section). Although this concept may be true of some examples of Ewing's sarcoma, it is probably not true for all. The ultrastructural features seen in some Ewing's sarcomas supporting neural differentiation are occasional cells with polar processes, which may contain a few microtubules and rare dense-core type granules. Ultrastructurally, the cells resemble a stage of mesenchymal differentiation just beyond primary mesenchyme (for example, paraxial mesenchymal mass or somites, myotomes, and sclerotomes; intermediate and lateral mesenchymal masses).

Primitive Neuroectodermal Tumor

(Figures 4.19 through 4.21.)

Diagnostic criteria. (1) Oval and elongated cells; (2) polar processes (some cells); (3) high nucleocytoplasmic ratio of cell bodies; (4) irregularly shaped nuclei with varying amounts of heterochromatin; (5) small intercellular junctions; (6) varying sized nucleoli; (7) cytoplasm with mostly ribosomes and polyribosomes; (8) focal intermediate filaments (variable); (9) occasional microtubules; (10) rare or occasional dense-core granules.

Additional points. Glycogen is less frequently present and is often in lesser quantities than in typical Ewing's sarcoma. PNET is less differentiated than neuroblastoma; polar processes are less numerous and not organized in parallel arrays. Rather, irregularly oriented, intertwining processes occupy small foci and have fewer microtubules and often only rare dense-core granules. Homer Wright rosettes are rare in this poorly differentiated neoplasm. PNETs arising in bone and those in soft tissue are ultrastructurally similar. An example of PNET is the thoracic *Askin tumor* (Figures 4.19 through 4.21).

(Text continues on page 169)

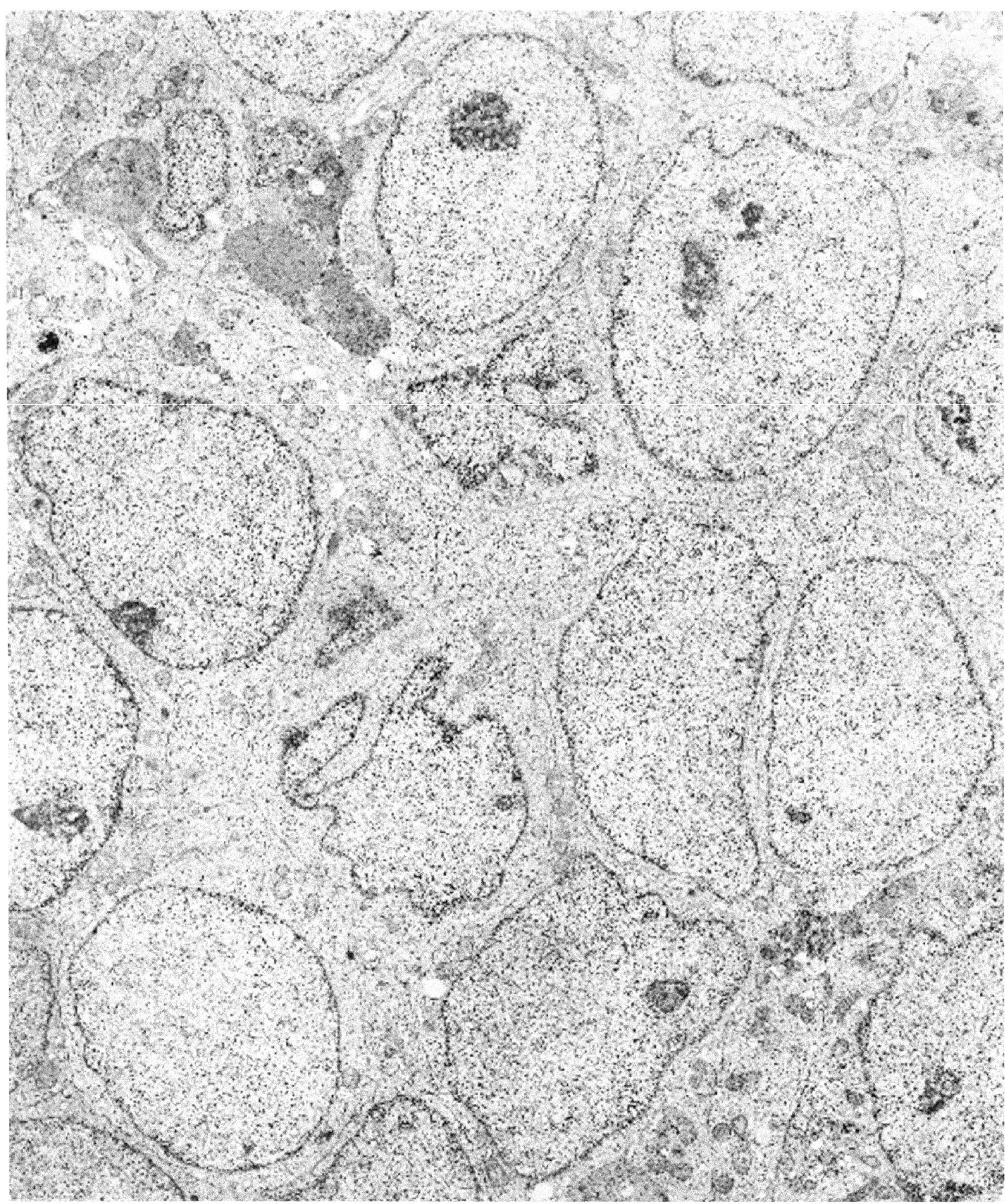

Figure 4.14. Ewing's sarcoma (soft tissue of leg). The neoplastic cells are in contiguity with one another along all borders. They are oval and polygonal cells and uniformly sized. Intercellular junctions are few, small, and inconspicuous at this magnification. The primitive nature of the cells is reflected in the high nuclear–cytoplasmic ratio, the euchromatic nuclei, the large open nucleoli, and the preponderance of free ribosomes and lack of many other organelles in the cytoplasm. (× 6250)

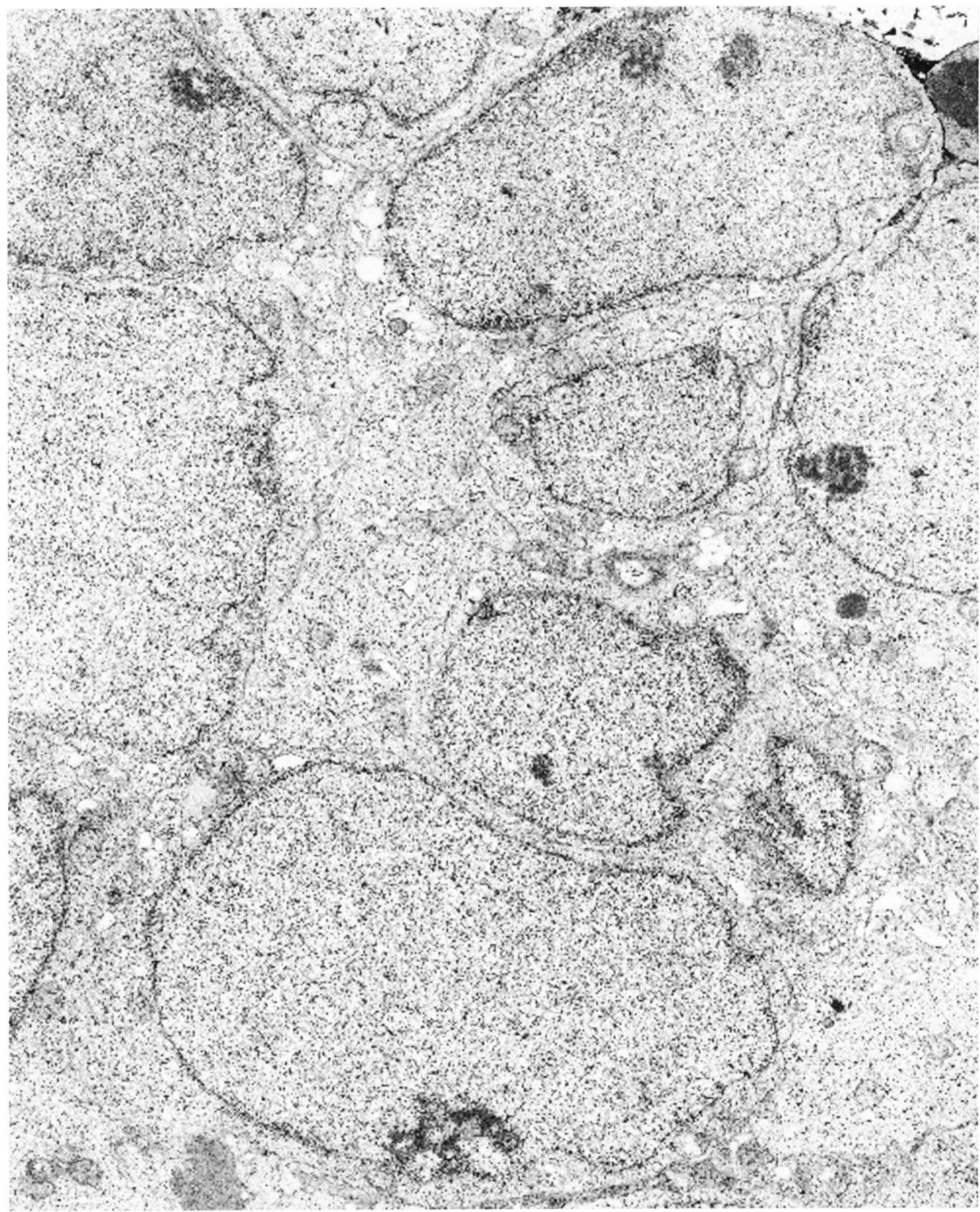

Figure 4.15. Ewing's sarcoma (soft tissue of leg). Higher magnification of the same neoplasm as depicted in Figure 4.14 shows the bland or poorly differentiated cytoplasm (mostly free ribosomes and a few mitochondria). In well-preserved cells, the chromatin in Ewing's sarcoma usually is finely dispersed (euchromatin). (× 8840)

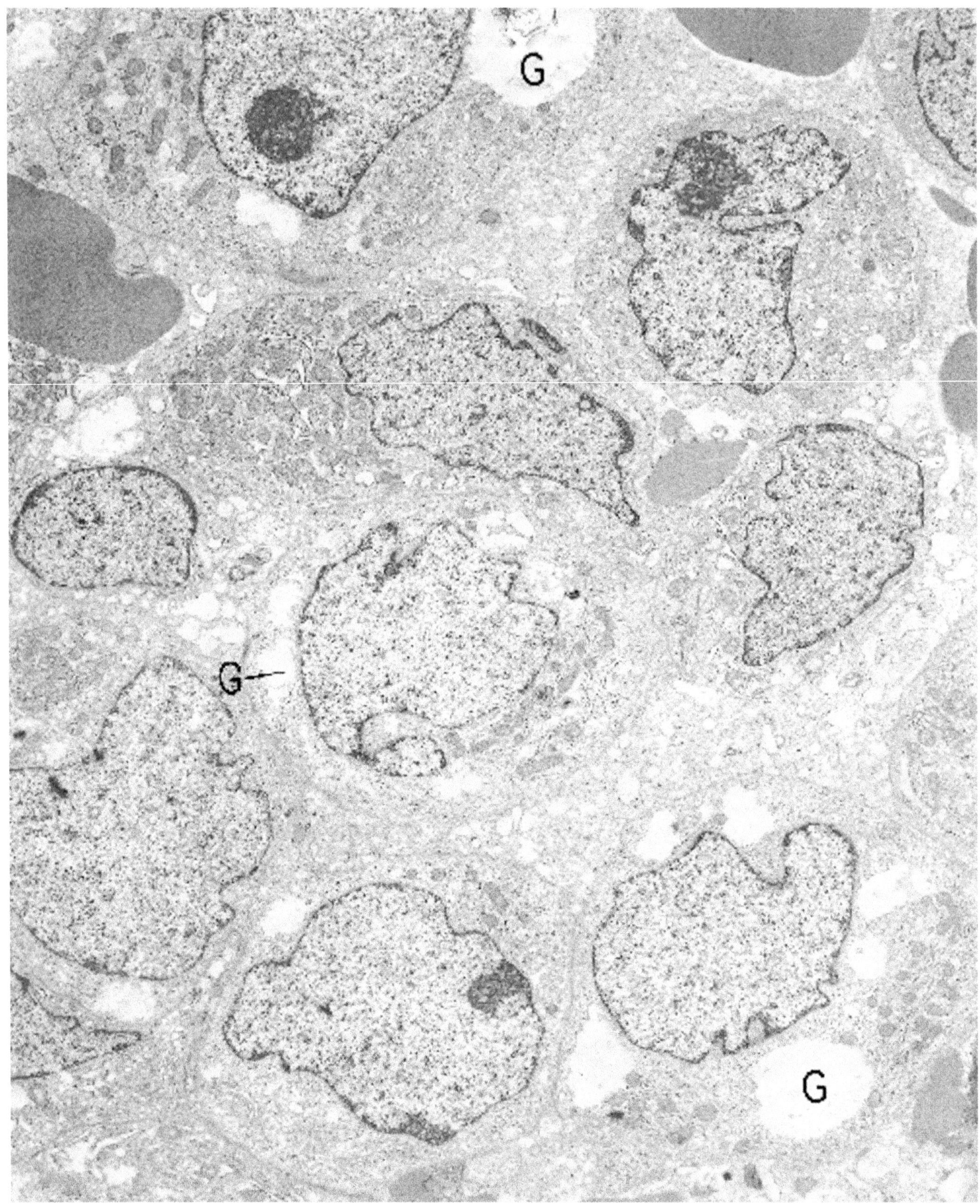

Figure 4.16. Ewing's sarcoma (ilium). This neoplasm is morphologically similar to that depicted in Figures 4.14 and 4.15, but glycogen is present in copious amounts and appears, by this method of chemical processing, as escalloped clear areas in the cytoplasm (G). (× 6250)

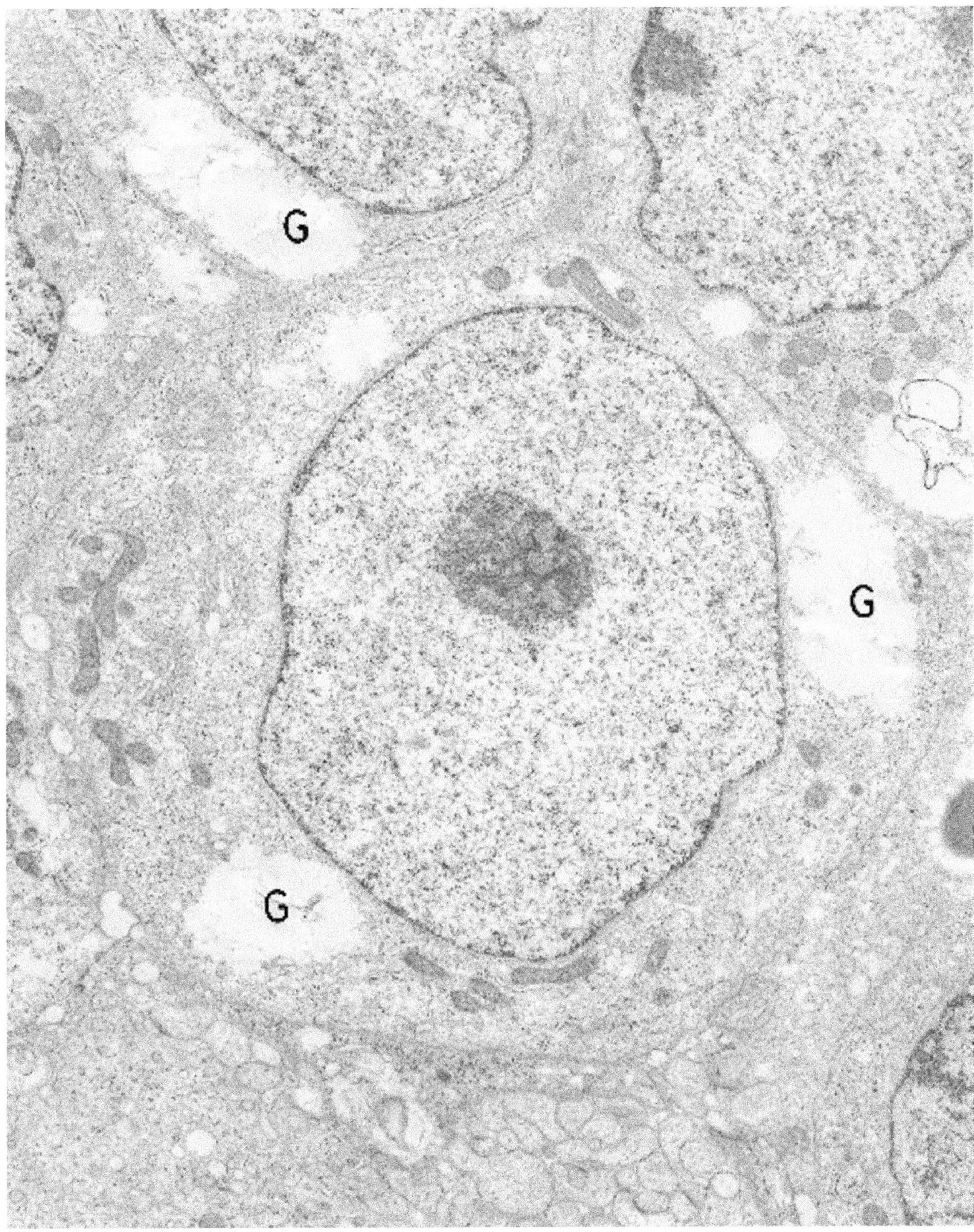

Figure 4.17. Ewing's sarcoma (ilium). This sample is from the same neoplasm as illustrated in Figure 4.16 and shows the escalloped clear spaces of glycogen (G) at a higher magnification. (× 11,250)

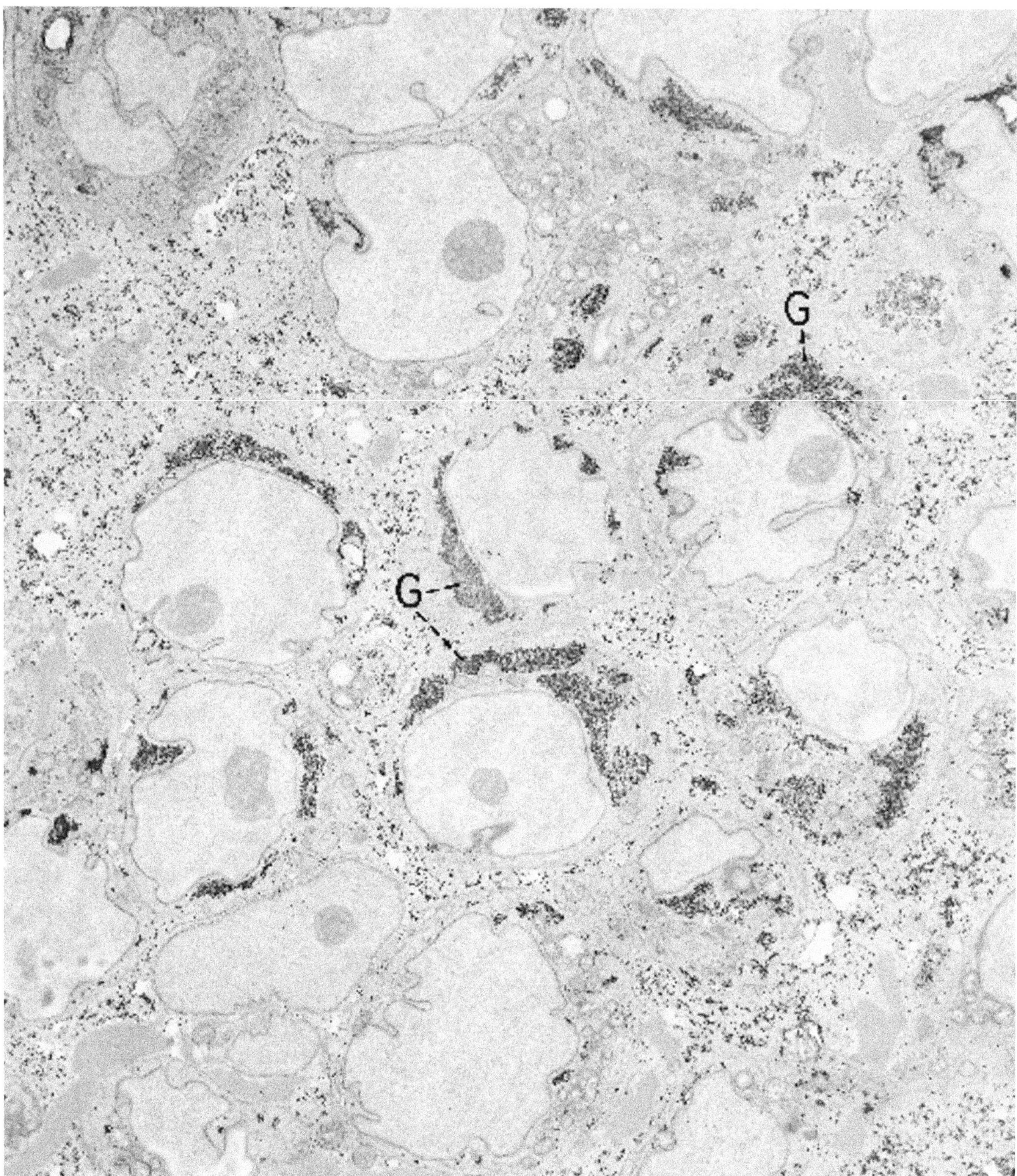

Figure 4.18. Ewing's sarcoma (ilium). This is the same neoplasm as pictured in Figures 4.16 and 4.17, but the method of chemical processing was such that glycogen (G) was preserved as electron-dense granules. This method allows for the demonstration of smaller aggregates of glycogen than would be revealed by the alternate method that results in glycogen appearing as open cytoplasmic spaces. (× 9180) (Permission for reprinting granted by WB Saunders, Dickersin GR: The contributions of electron microscopy in the diagnosis and histogenesis of controversial neoplasms. Clin Lab Med 4:123–164, 1984.)

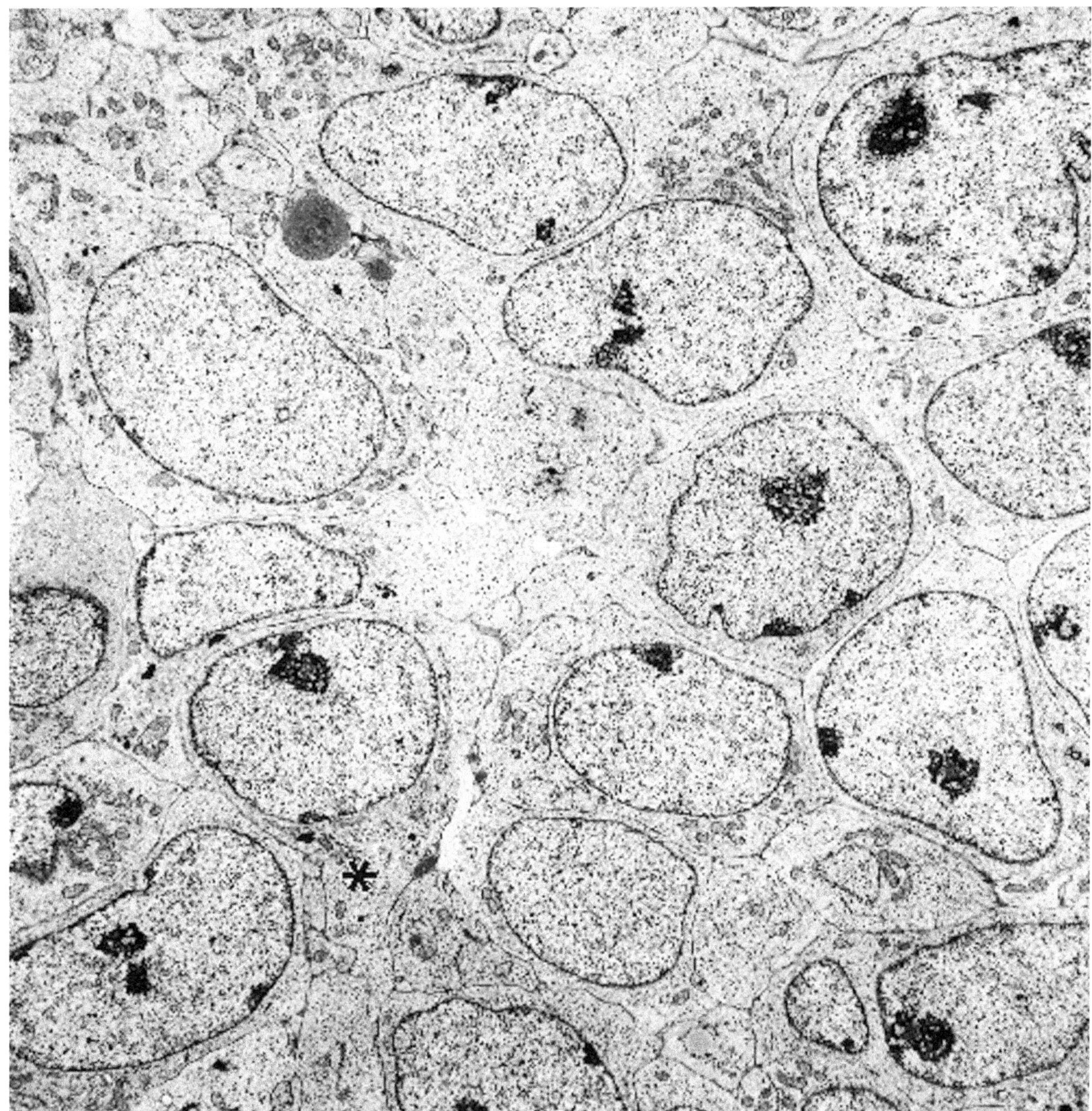

Figure 4.19. Primitive neuroectodermal tumor (Askin tumor; chest wall). The neoplastic cells are closely apposed, and diminutive junctions are inconspicuous at this magnification. The cells have a high nucleocytoplasmic ratio; nuclei are round, oval, and euchromatic, and nucleoli are prominent. Cytoplasm contains predominantly free ribosomes and is otherwise nondescript. No glycogen is apparent. One small focus of cytoplasmic processes (*) is present in this field. (× 5600)

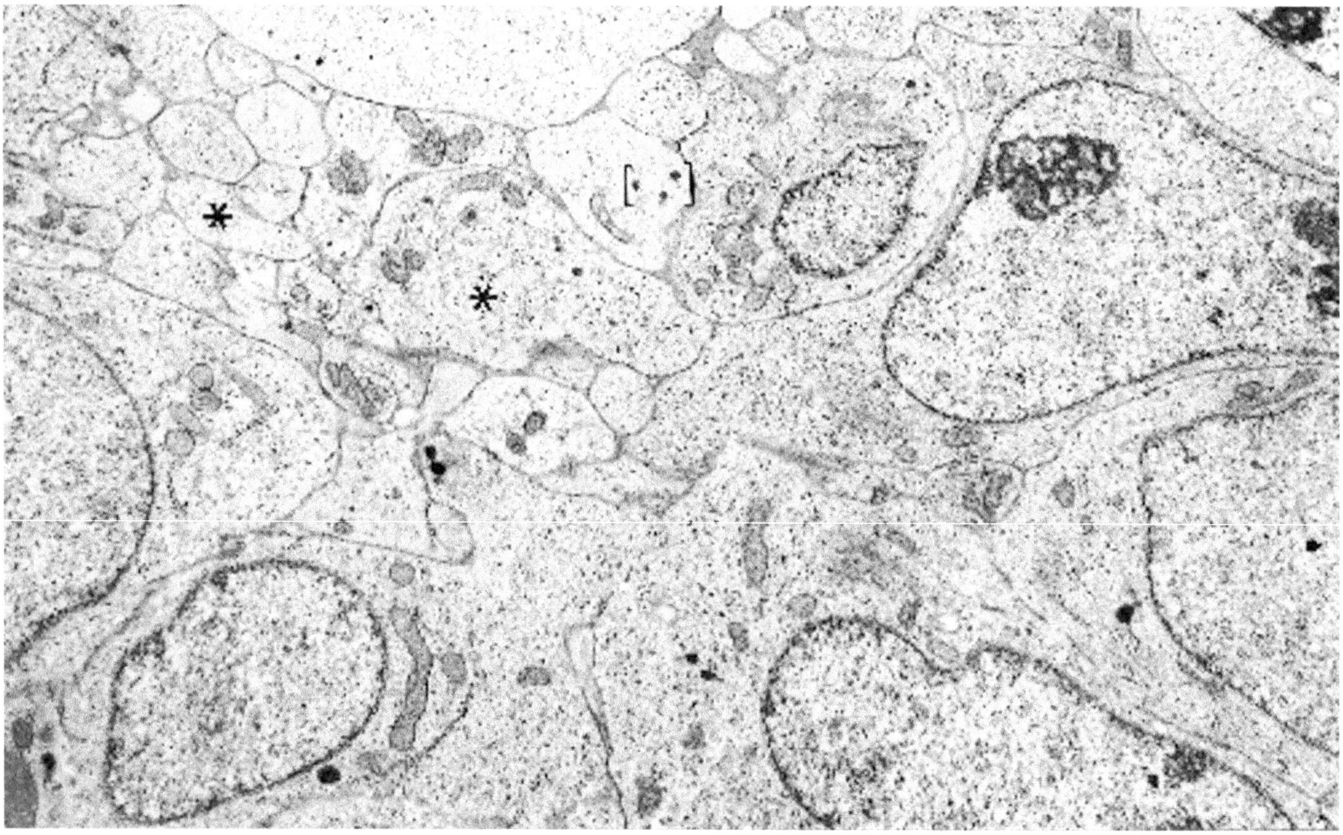

Figure 4.20. Primitive neuroectodermal tumor (Askin tumor; chest wall). Higher magnification of another field of the neoplasm depicted in Figure 4.19 illustrates a collection of closely apposed cellular processes (*), some of which contain a few small, dense granules (*bracket*). (× 9500)

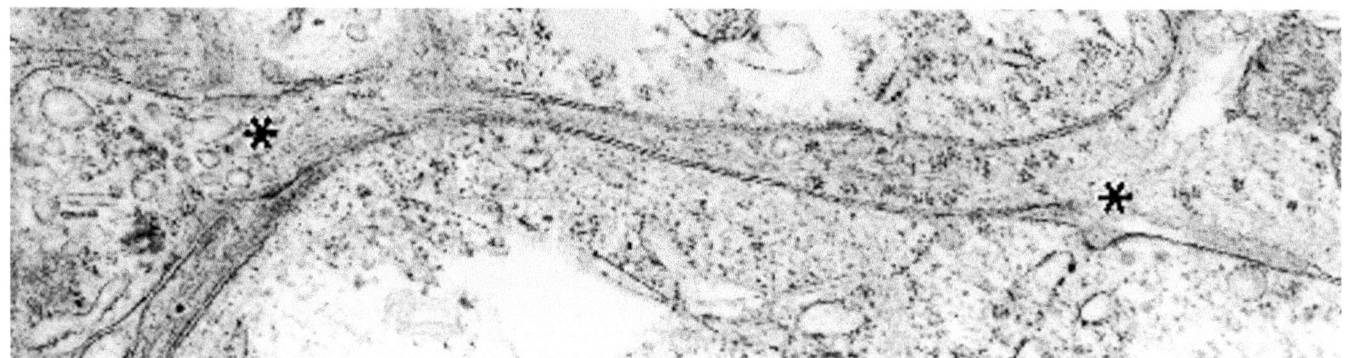

Figure 4.21. Primitive neuroectodermal tumor (Askin tumor; chest wall). This neoplasm is a different one from the one illustrated in Figures 4.19 and 4.20. Its cells contain abundant glycogen, but focally there are processes containing microtubules (between the *asterisks*), supportive evidence for neuronal differentiation. (× 30,900)

(Text continued from page 161)

Embryonal and Alveolar Rhabdomyosarcoma

(Figures 4.22 through 4.32.)

Diagnostic criteria. (1) Thick (15 nm) myosin filaments; (2) Z-band formation; (3) thick filament-ribosomal complexes; (4) basal lamina; (5) glycogen.

Additional points. Thin (6 nm) actin filaments may be seen in early rhabdomyoblasts and, by themselves, are not diagnostic of rhabdomyosarcoma. Thick, myosin filaments must accompany the thin filaments in order to establish the line of differentiation as skeletal muscle. The normal arrangement of thin and thick filaments is 12 thin ones around a central thick one, but it is usually not discernible in poorly differentiated rhabdomyoblasts.

There are examples of probable embryonal rhabdomyosarcoma in which the cells are too poorly differentiated to allow a definite diagnosis (Figure 4.22). Usually in these cases the clinical setting and the light microscopy will provide the background upon which the ultrastructure can confirm the diagnosis, but there are a few occasions when Ewing's sarcoma and rhabdomyosarcoma cannot be morphologically distinguished from one another. There are some helpful but not pathognomonic differences between these two primitive neoplasms. Rhabdomyoblasts have more pleomorphic nuclei, with more heterochromatin than have Ewing's cells. The typical Ewing's nucleus is composed completely of euchromatin. Basal lamina may form around individual rhabdomyoblasts, around groups of them (Figure 4.22), and along a row of them in the alveolar form of rhabdomyosarcoma. The cells of alveolar and embryonal rhabdomyosarcoma otherwise have similar ultrastructural features. Glycogen, as has been illustrated (Figures 4.23 and 4.24), may be just as copious in rhabdomyosarcoma as in Ewing's sarcoma. Intercellular junctions are small and infrequent in both neoplasms and, therefore, are of no diagnostic help. Like most neoplasms, if adequately sampled for electron microscopy, embryonal rhabdomyosarcoma will be composed of cells in more than one stage of differentiation, and a thorough search often will reveal an occasional cell with a few foci of thick filaments and small aggregates of Z-band material. The earliest ultrastructural clue that an undifferentiated cell is a rhabdomyoblast is the finding of thick filament-ribosomal complexes, in which a group of thick filaments are arranged in parallel, and rows of ribosomes are situated between them (Figures 4.25 through 4.27). Z-bands also are a helpful criterion for recognizing rhabdomyoblasts, and puffs of Z-band material appear soon after the appearance of thick filaments (Figures 4.25 and 4.27). Narrow and more discrete Z-bands develop as the cell differentiates further and as alignment of bands and filaments into recognizable sarcomeres occurs.

Some concept of mid- to late-stage differentiation in neoplastic rhabdomyoblasts is captured in Figures 4.28 through 4.32.

Rhabdoid Tumor

(Figures 4.33 through 4.35.)

Diagnostic criteria. (1) Diffuse, nonorganoid collections of round, oval, and polygonal cells; (2) irregularly shaped nuclei with large central nucleoli; (3) large paranuclear whorls of intermediate filaments and occasionally focal tonofibrils; (4) diminutive junctions.

Additional points. No basal lamina surrounds cells. There are no polar processes. Cytoplasm has varying amounts of dilated rough endoplasmic reticulum, primary and secondary lysosomes, and lipid vacuoles. Glycogen may rarely be present. Renal and extrarenal rhabdoid tumors have a similar ultrastructure.

Nephroblastoma (Wilms' Tumor)

(Figures 4.36 through 4.48.)

Diagnostic criteria. (1) Loose and tight groups of polygonal and elongated cells (blastema), with a high nuclear–cytoplasmic ratio and scant cytoplasm containing mostly free ribosomes; (2) blastemal cells in small groups surrounded by basal lamina (pretubules); (3) junctions and junctional complexes in the pretubules; (4) true lumens and microvilli on luminal lining cells (tubules); (5) varying numbers of secondary lysosomes in blastemal and tubular cells; (6) rare nests of epithelial cells forming glomeruloid structures (devoid of mesangial cells and capillaries); (7) flocculent, medium-dense, basal lamina-like, intercellular material; (8) areas of banded collagen and loosely arranged spindle cells having abundant, often dilated, rough endoplasmic reticulum (fibroblastic stroma).

Additional points. Because Wilms' tumors contain varying combinations of blastema, epithelium, and stroma, sampling for electron microscopy may not be completely representative. Furthermore, some neoplasms contain heterologous elements, including various types of epithelial cells, striated muscle, smooth muscle, cartilage, bone, and nerve, and the electron microscopic features of these cells are similar to those described elsewhere in this book in the sections on the primary neoplasms composed of those respective cells.

(Text continues on page 197)

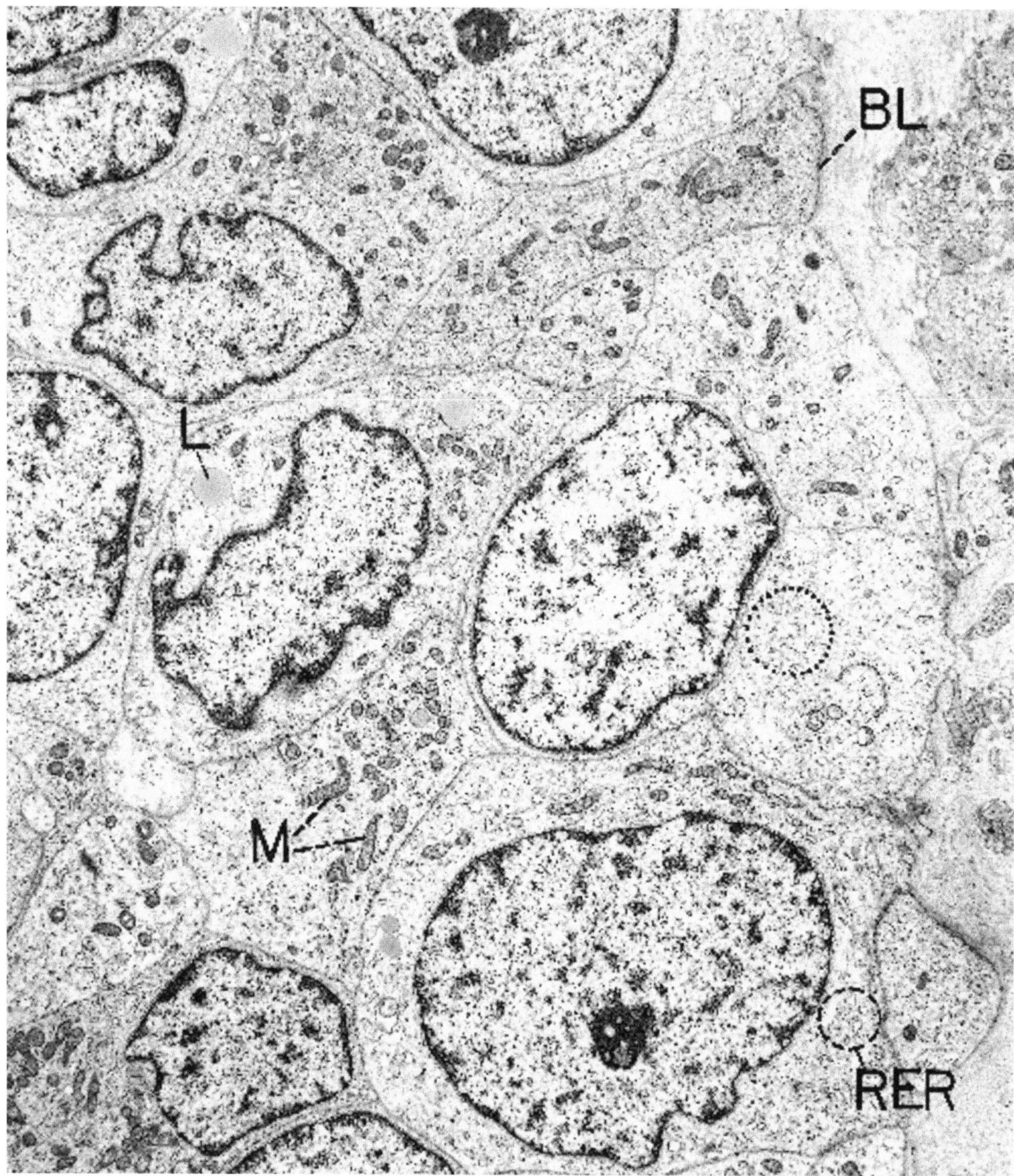

Figure 4.22. Embryonal (alveolar) rhabdomyosarcoma (metastatic to axillary lymph node). A group of undifferentiated small cells is surrounded by basal lamina (BL). The cells are oval and polygonal, are closely apposed, and have small junctions. There is a high nuclear–cytoplasmic ratio, chromatin tends to be aggregated, and nucleoli are large. Cytoplasmic organelles are sparse and include free ribosomes (*circle*), a moderate number of mitochondria (M), and a few cisternae of rough endoplasmic reticulum (RER). An occasional lipid droplet (L) also is present. (× 6750)

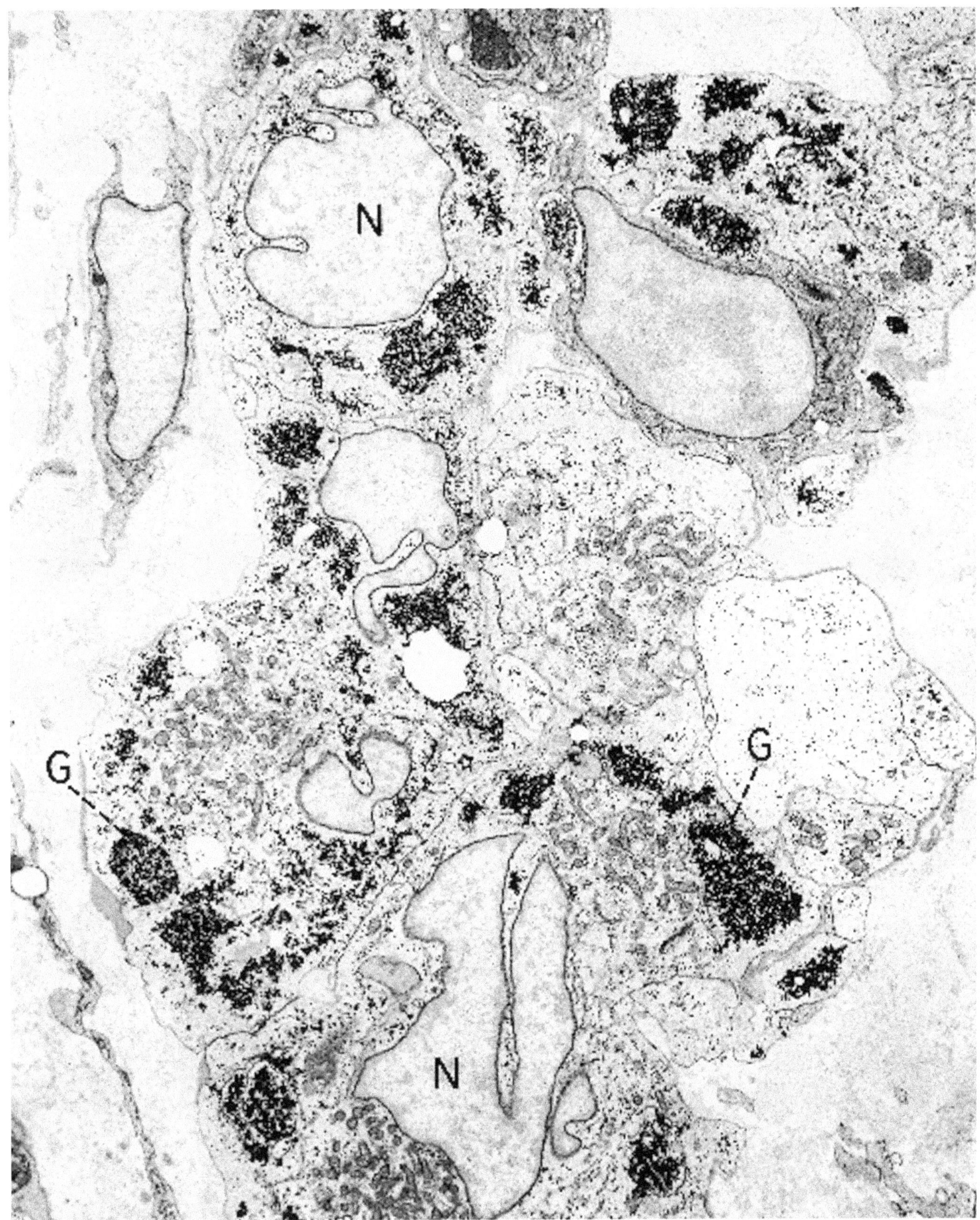

Figure 4.23. Embryonal rhabdomyosarcoma (metastatic to cervical lymph node). The tissue was processed to preserve glycogen (G) as electron-dense granules, and typically the embryonal rhabdomyoblasts are rich in glycogen. Nuclei (N) are pale or unstained by this technique. (× 6270)

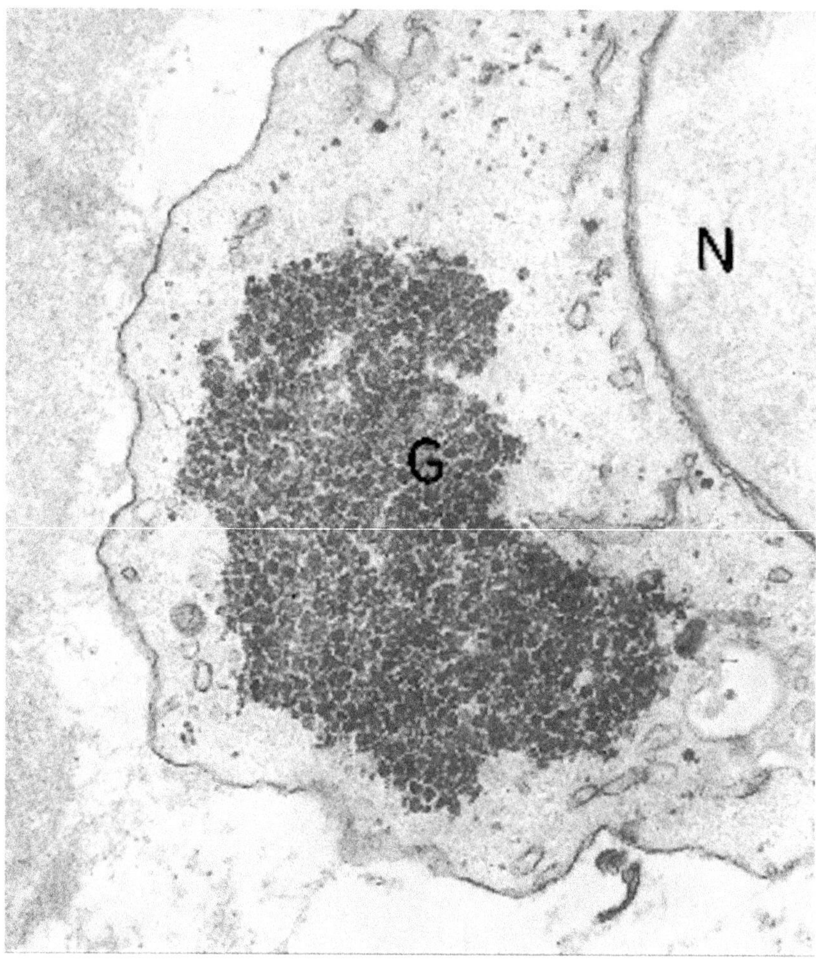

Figure 4.24. Embryonal rhabdomyosarcoma (deep fascia of forearm). High magnification of glycogen granules (G) in the cytoplasm of an embryonal rhabdomyoblast. N = nucleus. (× 33,750)

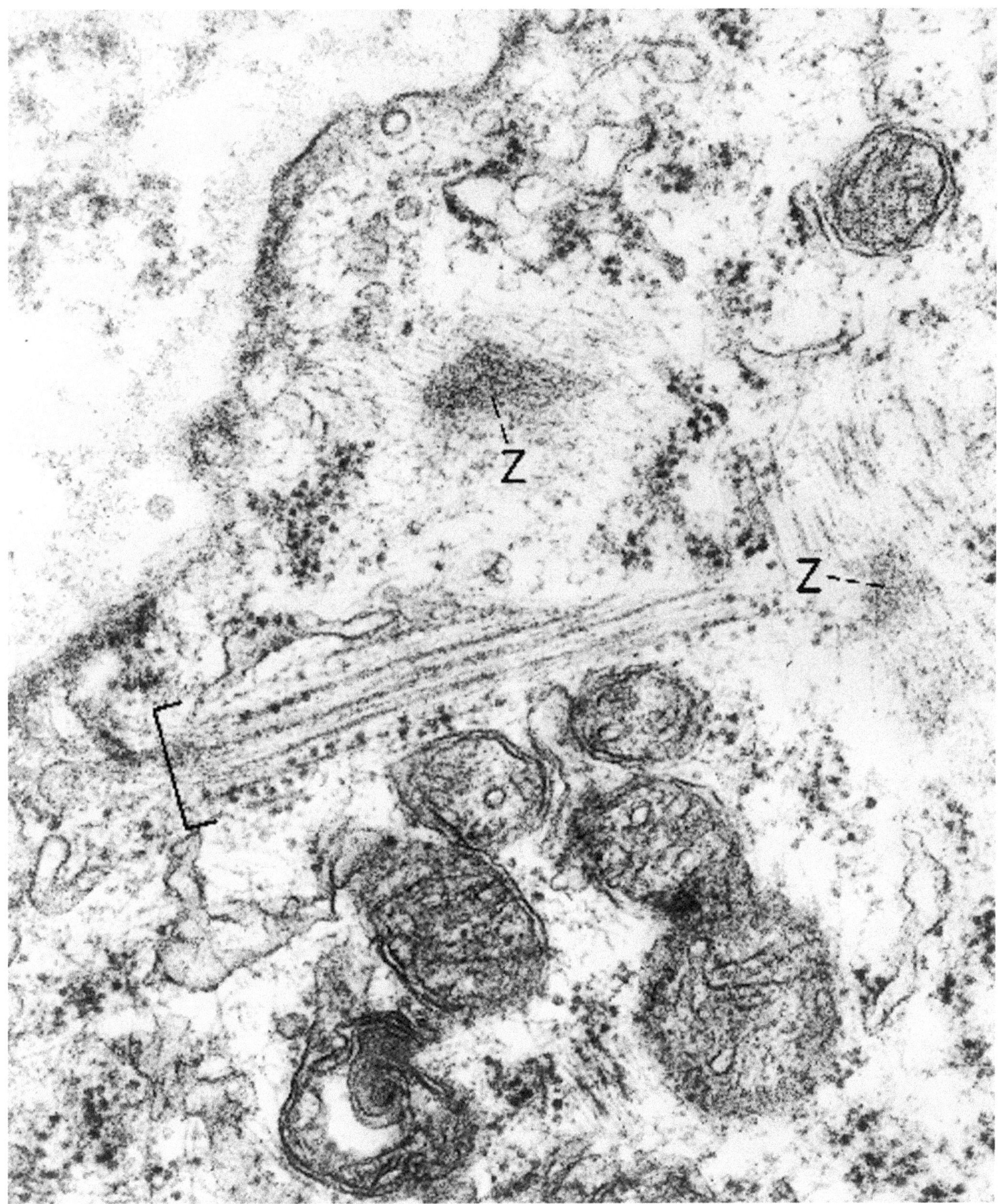

Figure 4.25. Embryonal rhabdomyosarcoma (soft tissue of forearm). High magnification of this primitive cell reveals a thick filament-ribosomal complex (to right of *bracket*), the earliest morphologic marker of skeletal muscle differentiation. Several dense puffs (Z) in the vicinity of thick filaments probably represent early Z-band formation. (× 70,000)

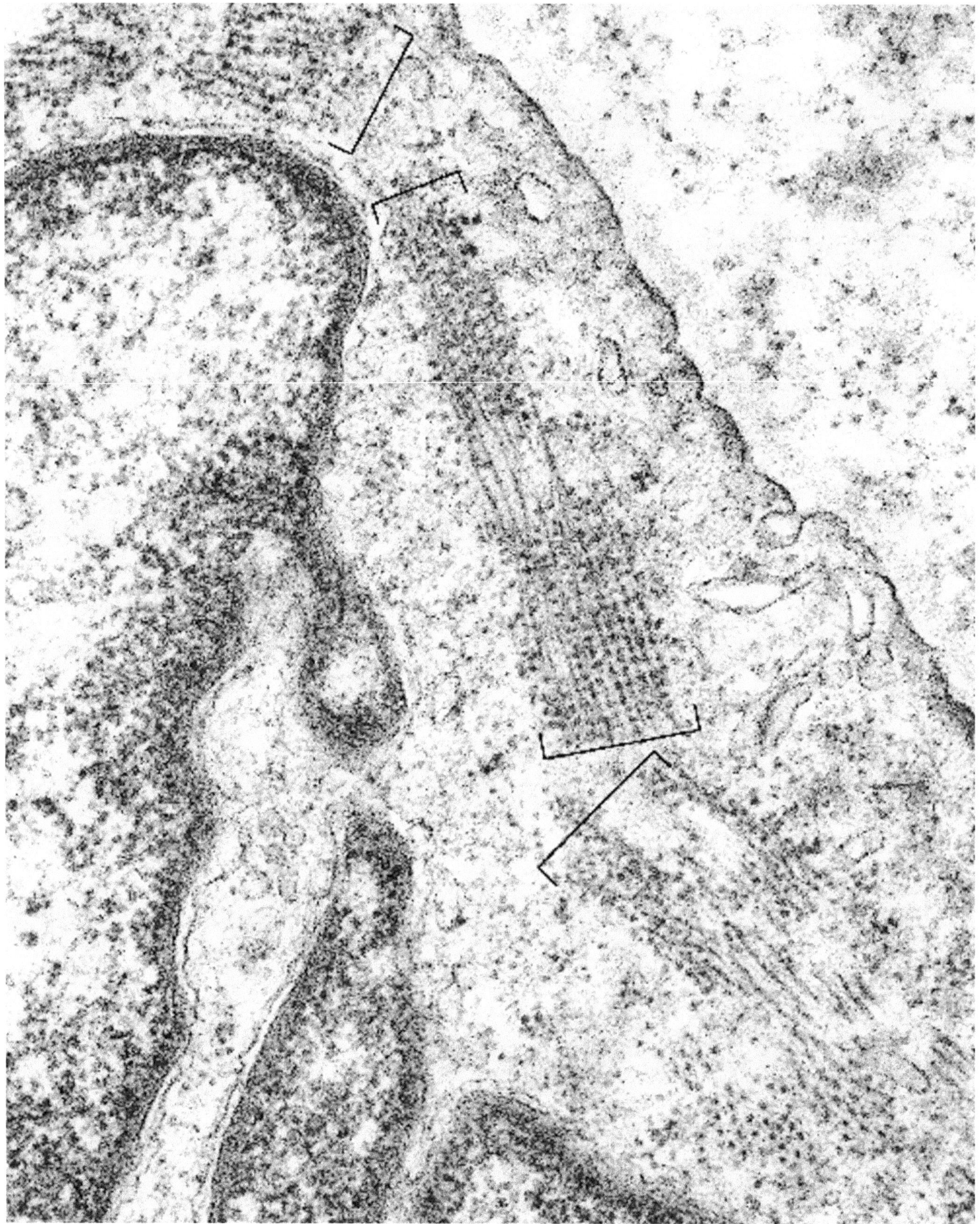

Figure 4.26. Embryonal rhabdomyosarcoma (soft tissue of forearm). Other thick filament-ribosomal complexes (*bracketed*) are somewhat more developed than the one depicted in Figure 4.25. (× 70,700)

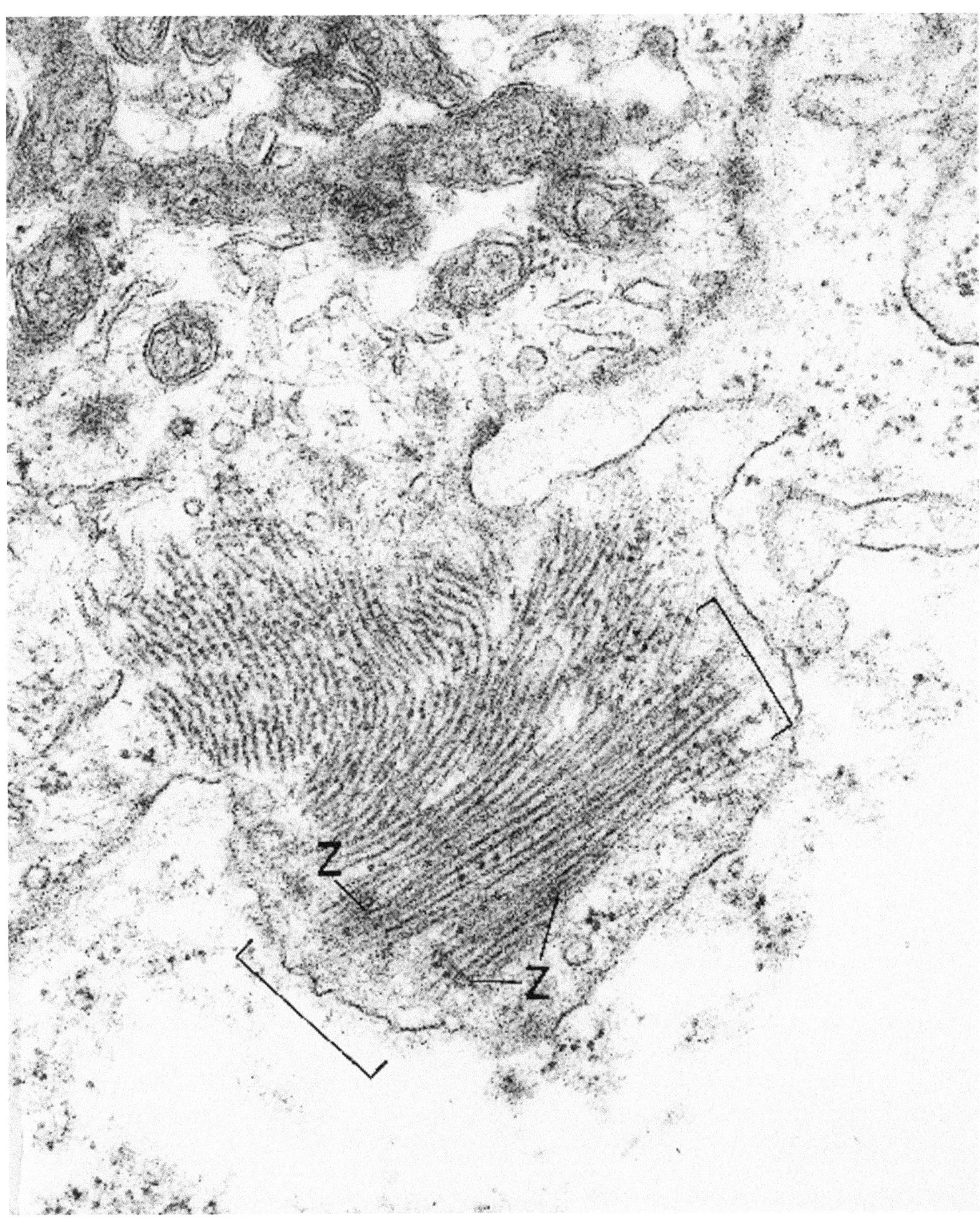

Figure 4.27. Embryonal rhabdomyosarcoma (soft tissue of forearm). Further development of thick filament-ribosomal complexes will result in early sarcomere formation; note region of parallel thick filaments with fewer ribosomes (*bracket*) and early Z-band regions (Z). (× 62,500)

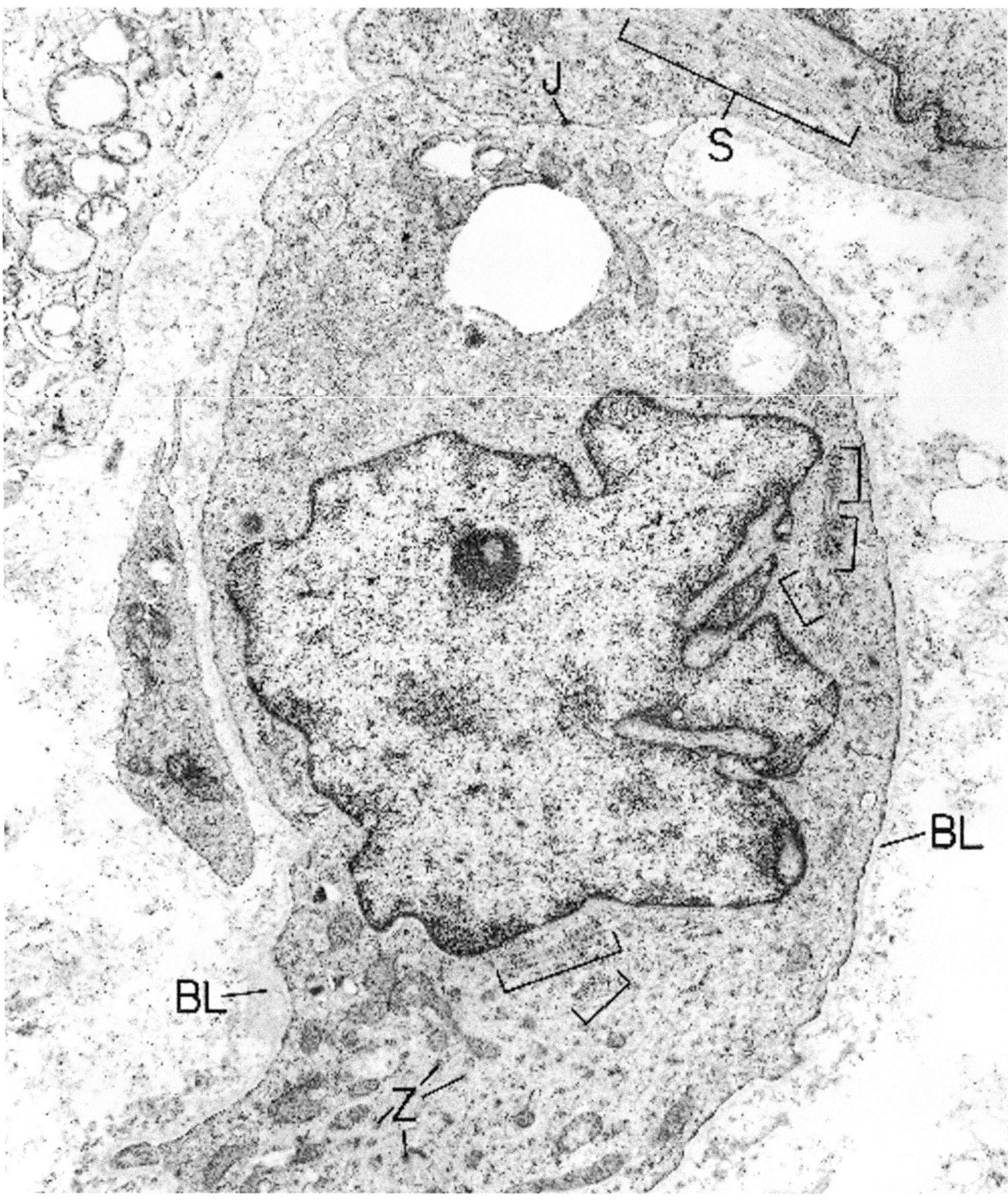

Figure 4.28. Embryonal rhabdomyosarcoma (soft tissue of forearm). Early strap cell differentiation is present in this cell and its neighbor (upper end of field). Puffs of Z-band material (Z) and thick filament-ribosomal complexes (*brackets*) can be recognized in the lower cell, and early sarcomere development (S) is already present in the upper cell. Note also the small junction (J) between the two cells, and the basal lamina (BL) around them. (× 12,900)

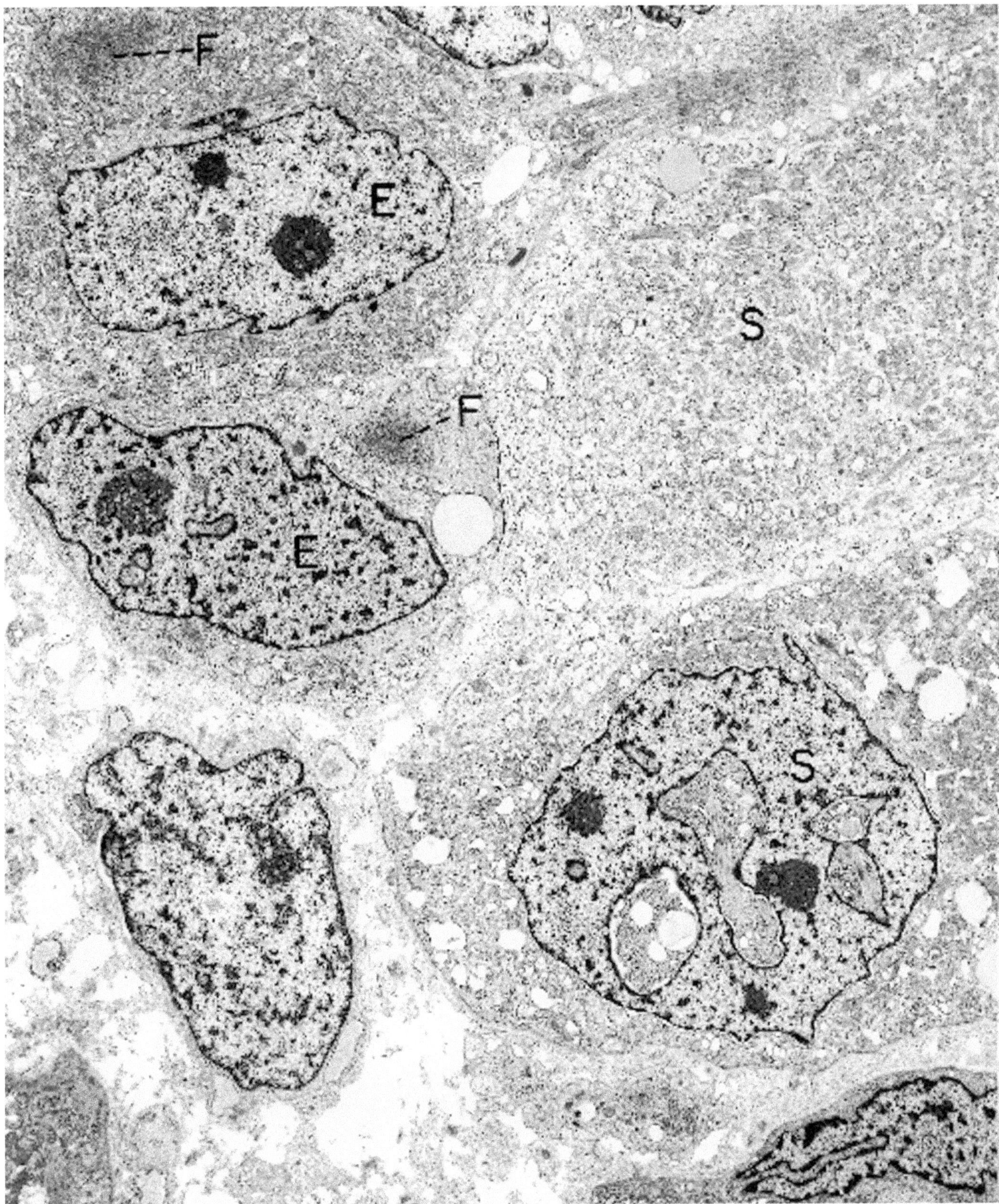

Figure 4.29. Embryonal rhabdomyosarcoma (vagina). Early skeletal muscle differentiation in embryonal type cells (E) is noted in the left part of the field. Note the moderate increase in cytoplasm in the electron-dense foci (F), which represents aggregates of developing actin and myosin filaments. In the right part of the field are portions of two early- to mid-stage strap cells (S), with numerous aggregates of filaments filling the cytoplasm. (× 5130)

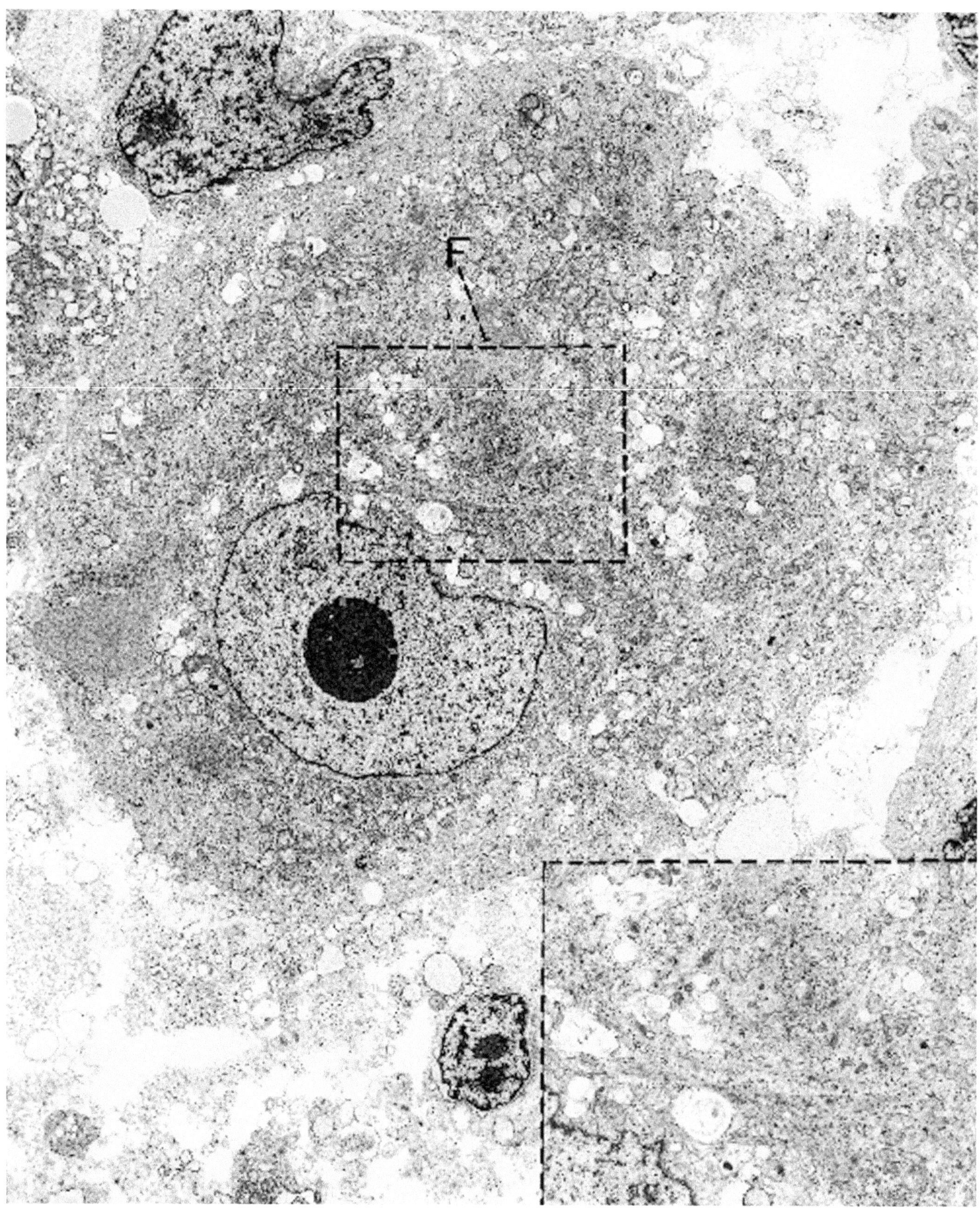

Figure 4.30. Embryonal rhabdomyosarcoma (vagina). Early strap cell differentiation is characterized by copious cytoplasm and many dense aggregates of actin and myosin filaments (F). (× 4900) *Inset* shows filaments at higher power. (× 8200)

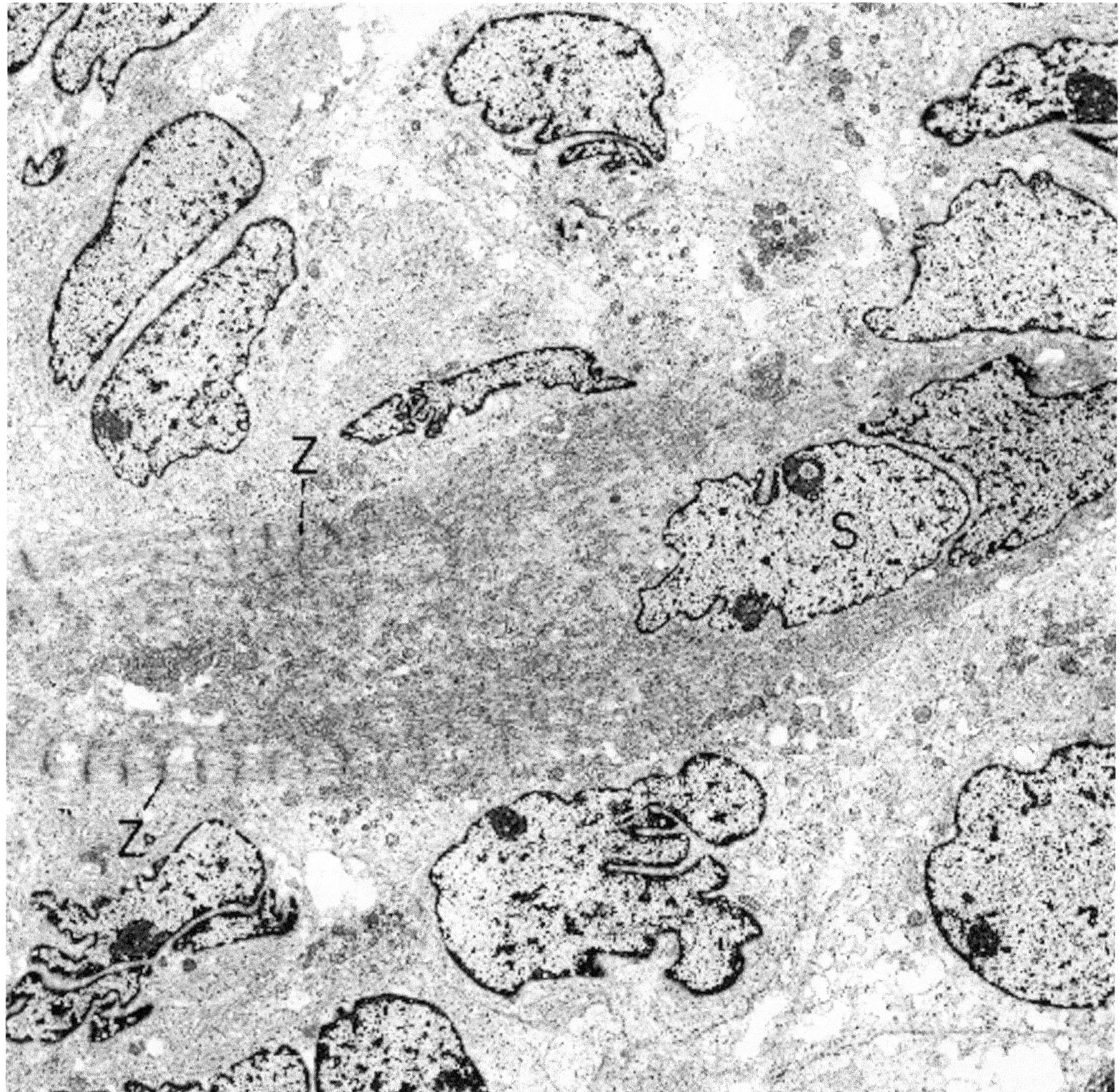

Figure 4.31. Embryonal rhabdomyosarcoma (metastatic to cervical lymph node). This field illustrates the wide range of differentiation that may exist within a single neoplasm. Most of the cells are embryonal type rhabdomyoblasts, but the elongated cell that spans the width of the field is a later-stage strap cell (S). Note that, at places, there are sarcomeres with parallel and regularly spaced Z-bands (Z). (× 7200) (Permission for reprinting granted by WB Saunders, Dickersin GR: Electron microscopy of leukemias and lymphomas. Clin Lab Med 7:199–247, 1987.)

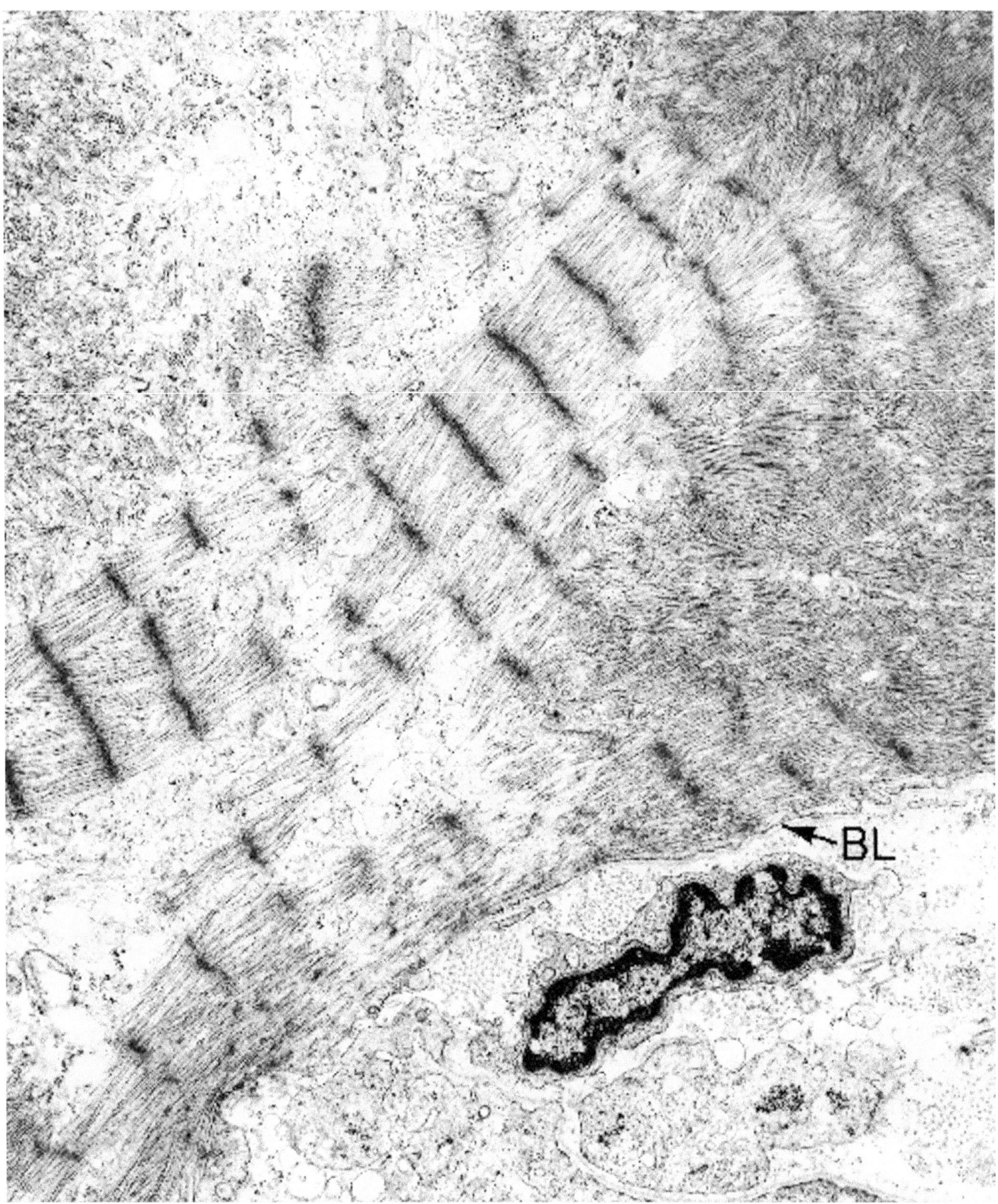

Figure 4.32. Embryonal rhabdomyosarcoma (metastatic to cervical lymph node). Higher magnification of the same neoplasm as depicted in Figure 4.31. Sarcomeres are well formed in the center and left part of the field, whereas lesser degrees of differentiation of the filaments are discernible elsewhere in the cytoplasm of this cell. Note the basal lamina (BL) along the lower edge of the cell. (× 15,670)

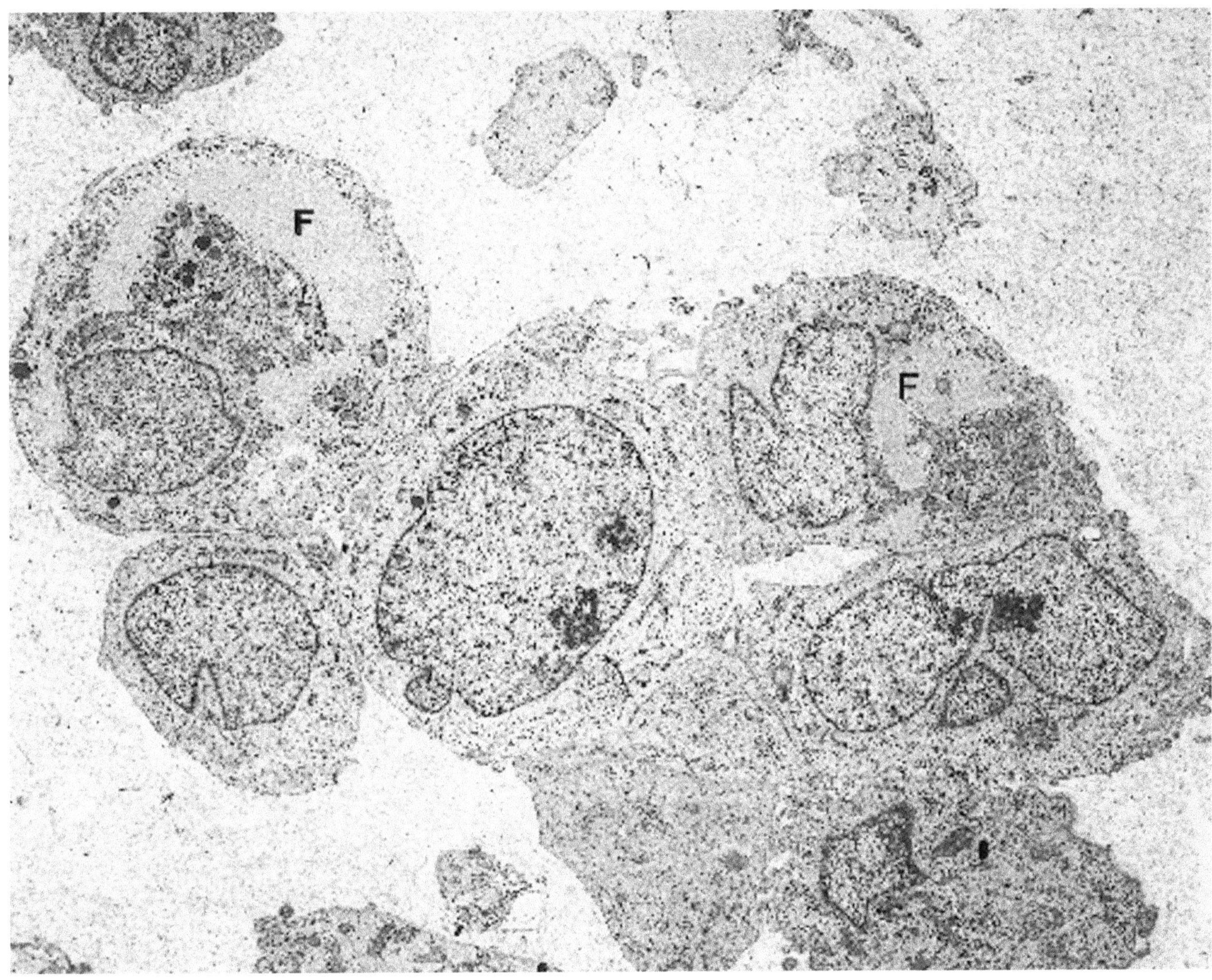

Figure 4.33. Rhabdoid tumor (orbit). A collection of round and polygonal cells have large, irregularly shaped nuclei and cytoplasm containing large areas of filaments (F). (× 5500)

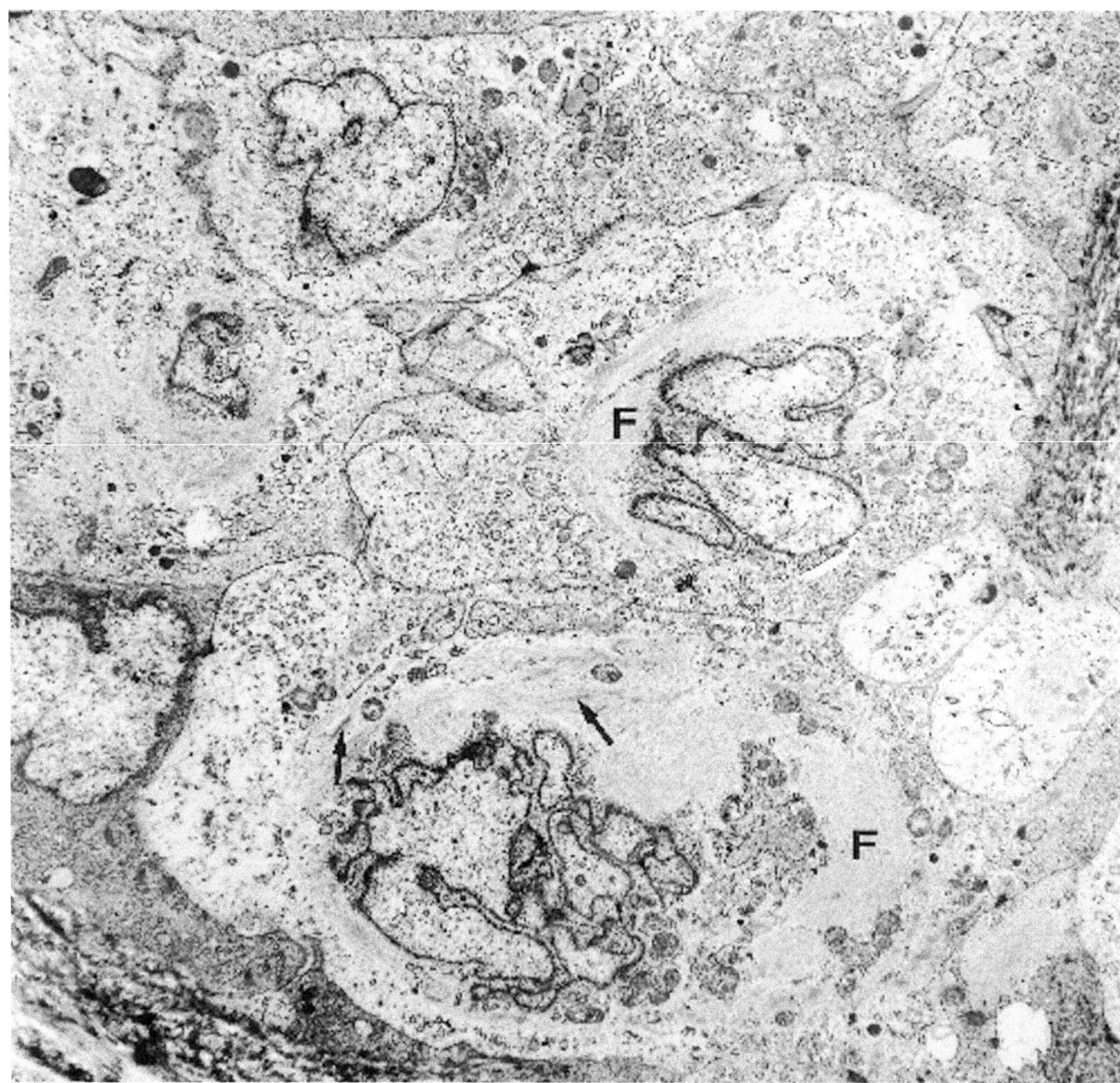

Figure 4.34. Rhabdoid tumor (abdominal wall). The neoplastic cells have markedly irregularly shaped nuclei and innumerable cytoplasmic filaments (F) with focal densities suggestive of tonofibrils (*arrow*). (× 7600)

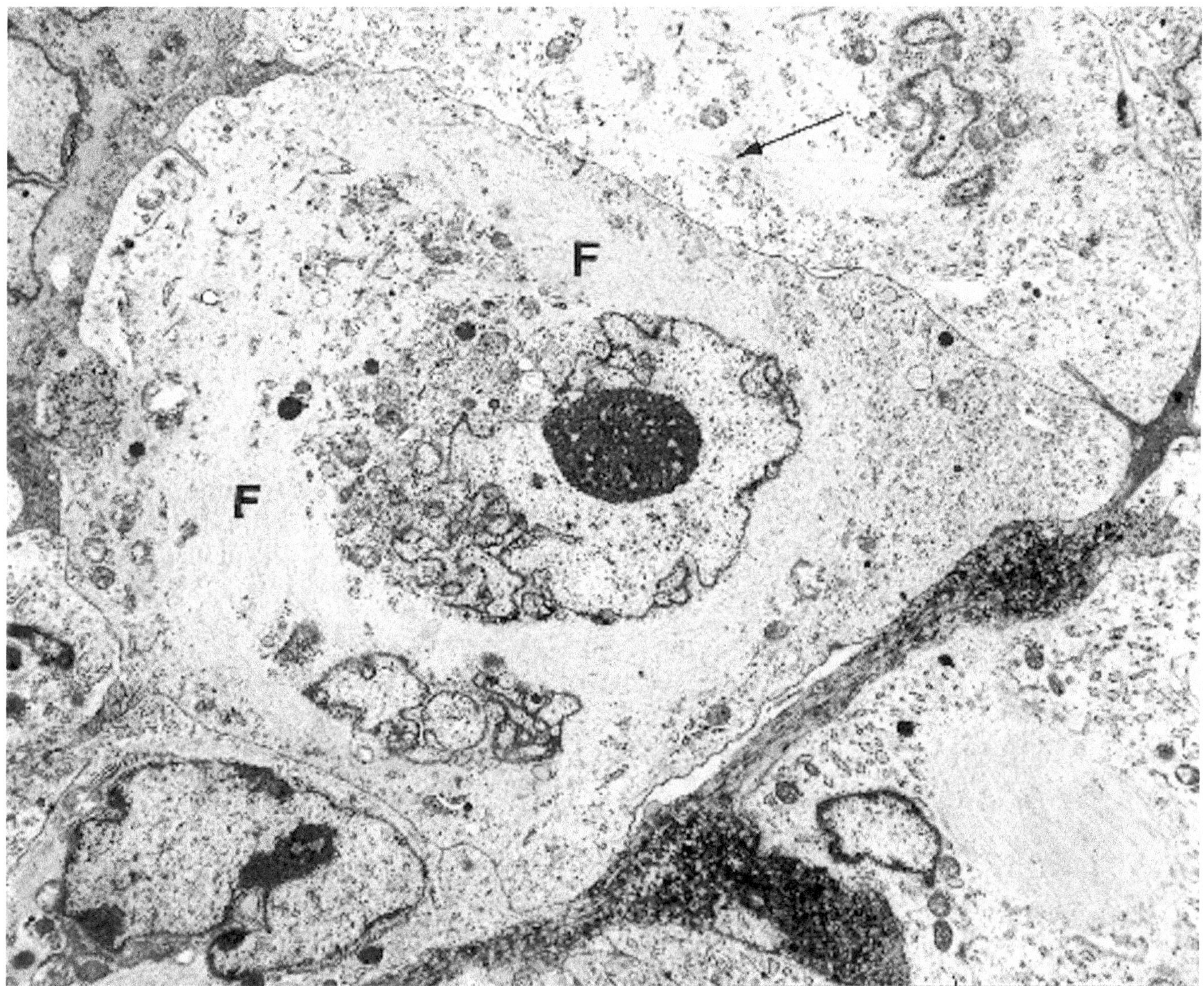

Figure 4.35. Rhabdoid tumor (abdominal wall). Higher magnification of a cell from the same neoplasm as depicted in Figure 4.34 illustrates the numerous cytoplasmic filaments (F) and an irregularly shaped nucleus with a large central nucleolus. The cell at the top of the field shows densities among the filaments suggestive of tonofibrils (*arrow*). (× 6200)

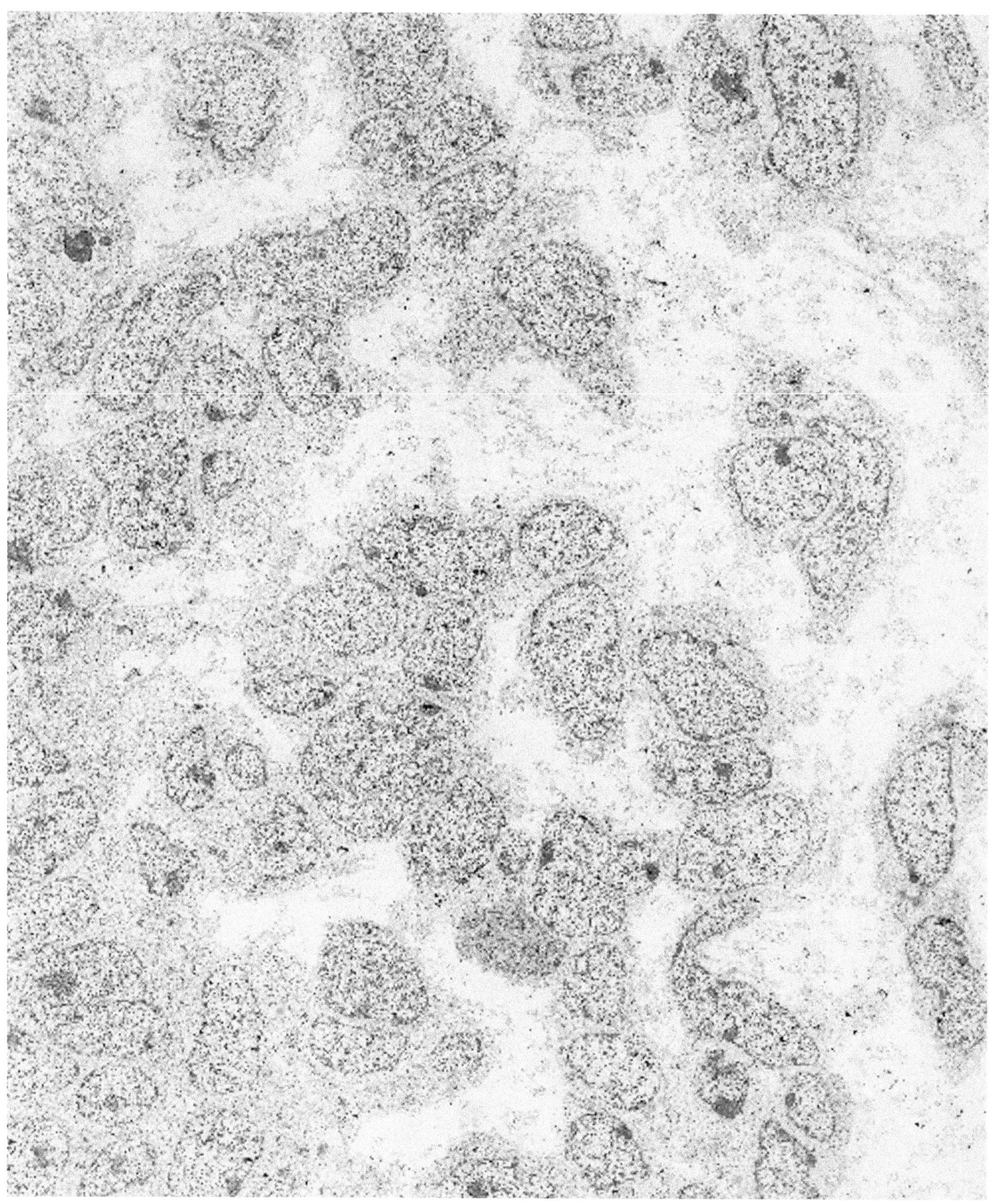

Figure 4.36. Nephroblastoma (Wilms' tumor). An area of neoplastic blastema consists of loosely arranged polygonal and elongated cells having a high nuclear–cytoplasmic ratio and a scant amount of cytoplasm. (× 2870) (Permission for reprinting granted by Hemisphere Publishing, Dickersin GR: Embryonic ultrastructure as a guide in the diagnosis of tumors. Ultrastruct Pathol 11:609–652, 1987.)

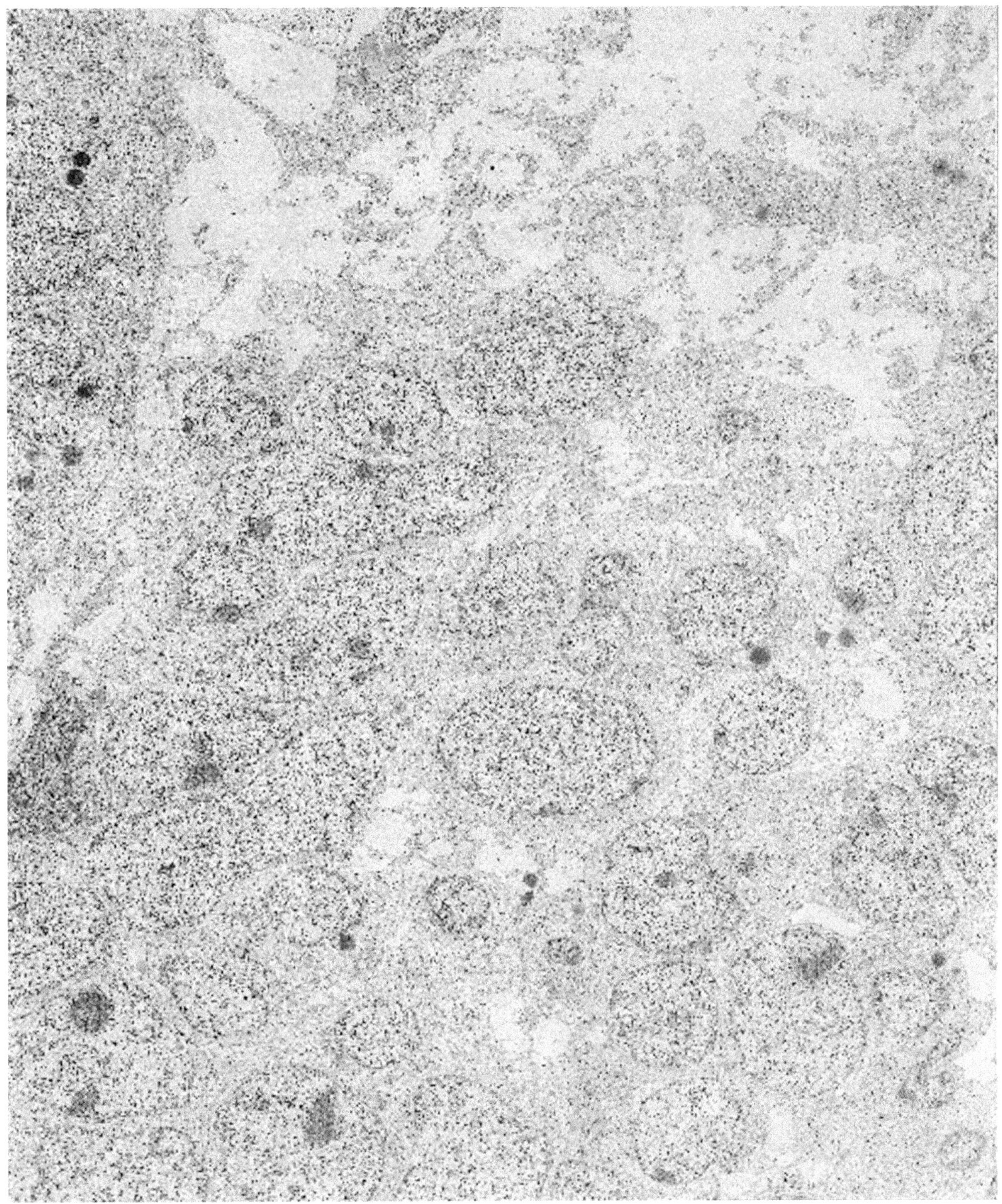

Figure 4.37. Nephroblastoma (Wilms' tumor). This region of neoplastic blastema consists of more tightly apposed cells than those illustrated in Figure 4.36. (× 3915) (Permission for reprinting granted by Hemisphere Publishing, Dickersin GR: Embryonic ultrastructure as a guide in the diagnosis of tumors. Ultrastruct Pathol 11:609–652, 1987.)

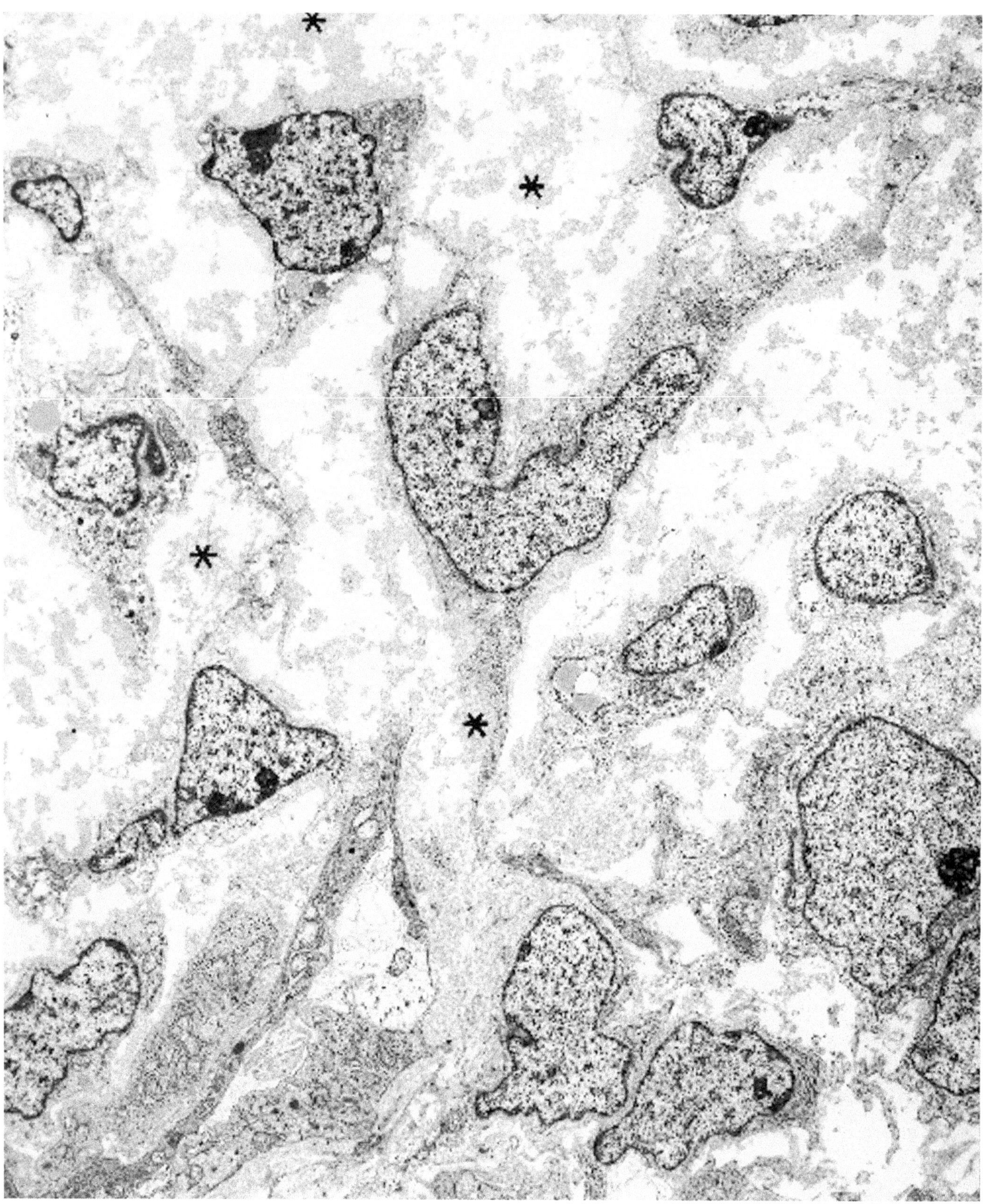

Figure 4.38. Nephroblastoma (Wilms' tumor). Higher magnification of a group of loosely arranged blastemal cells reveals their poorly differentiated character; that is, their high nuclear–cytoplasmic ratio and scant cytoplasm. The abundant flocculent, medium-dense, intercellular substance (*) is consistent with basal laminar material. (× 5620)

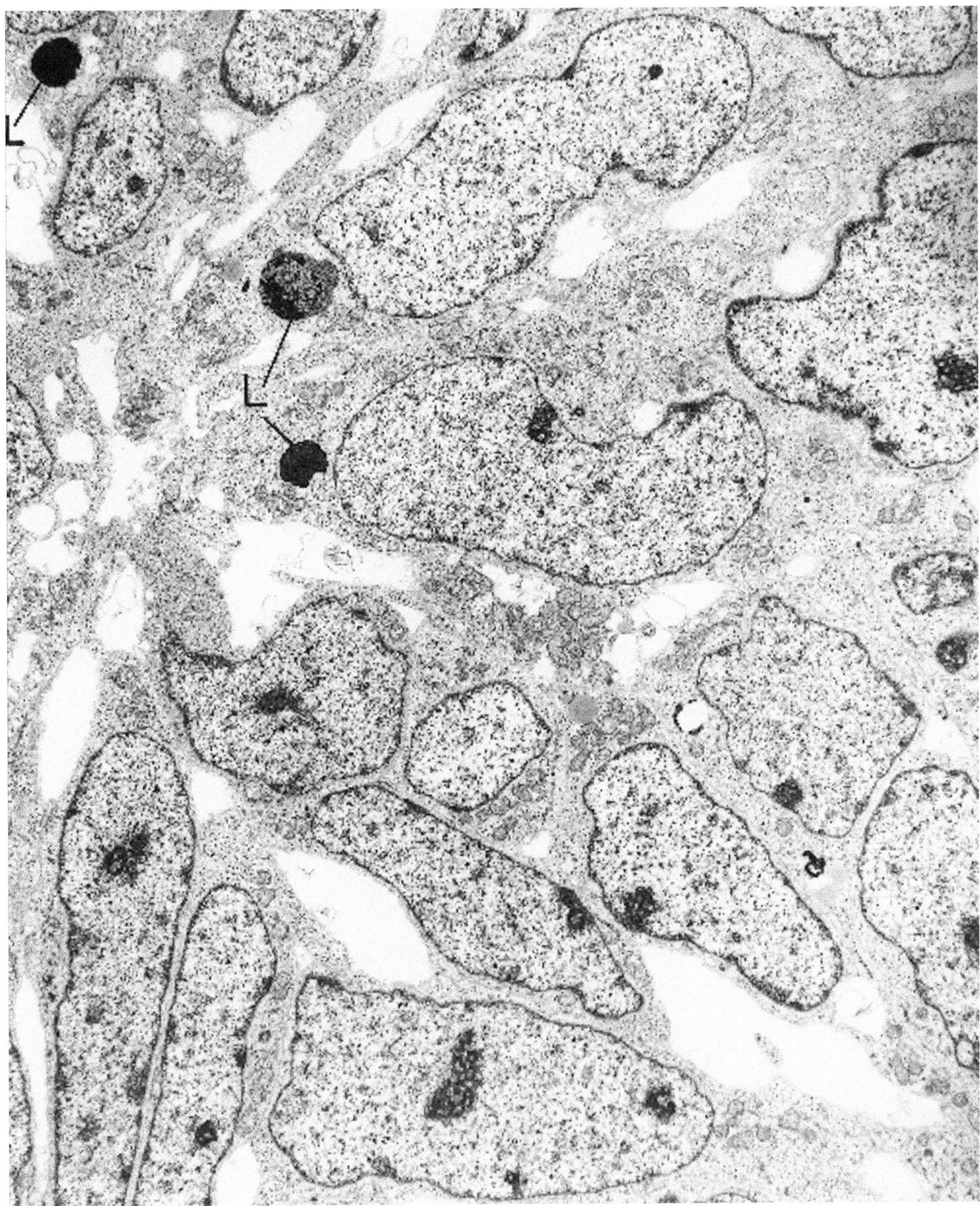

Figure 4.39. Nephroblastoma (Wilms' tumor). Higher magnification of a group of tightly apposed blastemal cells shows further differentiation of the cytoplasm than seen in the loosely arranged regions of Figure 4.38. Secondary lysosomes (L) are present in at least three of the cells. (× 5510)

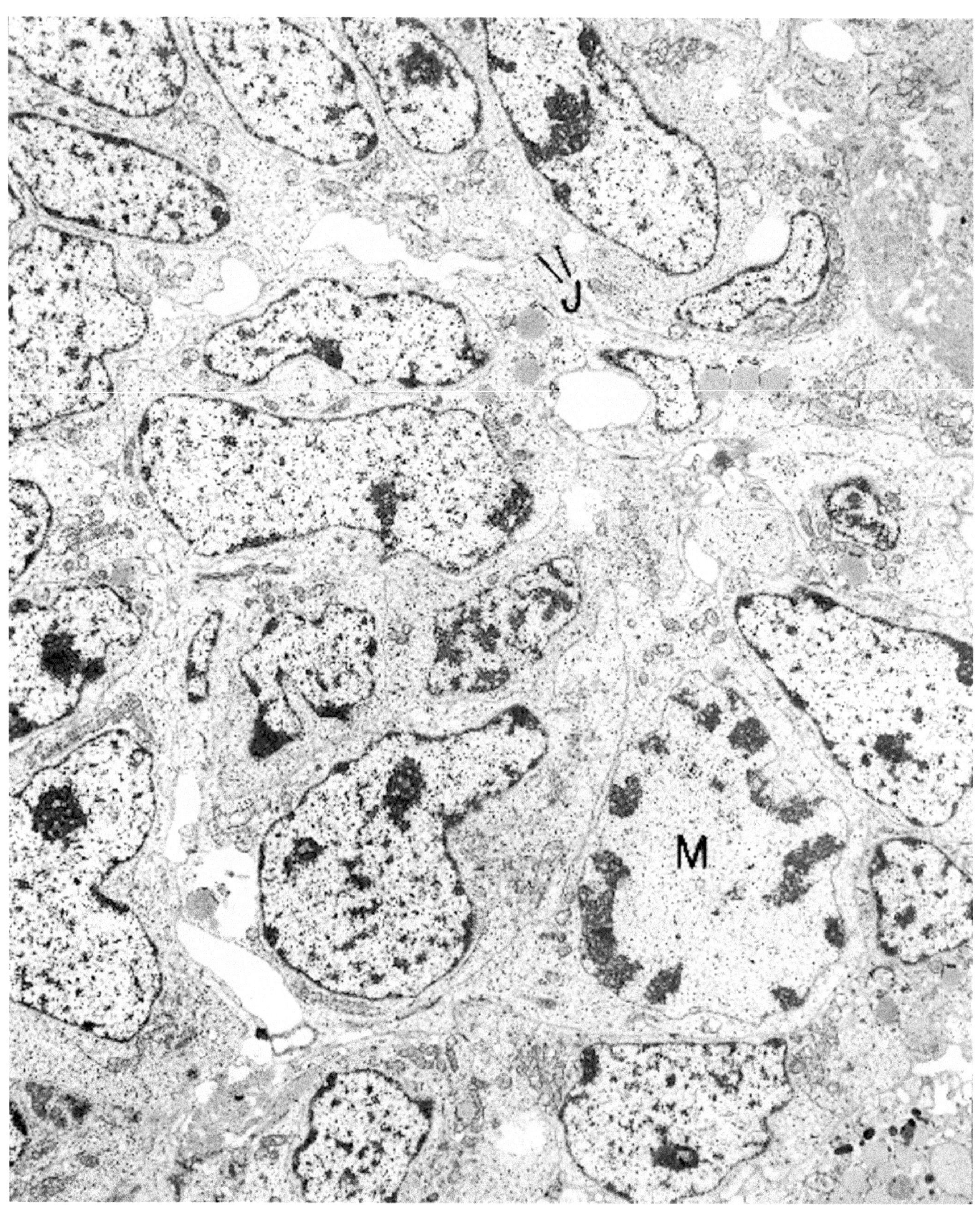

Figure 4.40. Nephroblastoma (Wilms' tumor). A group of closely apposed, polygonal cells are poorly differentiated, but focal parallel or radial alignment and a few small intercellular junctions (J) indicate early epithelial and tubular development. One cell (M) is in mitosis. (× 5930)

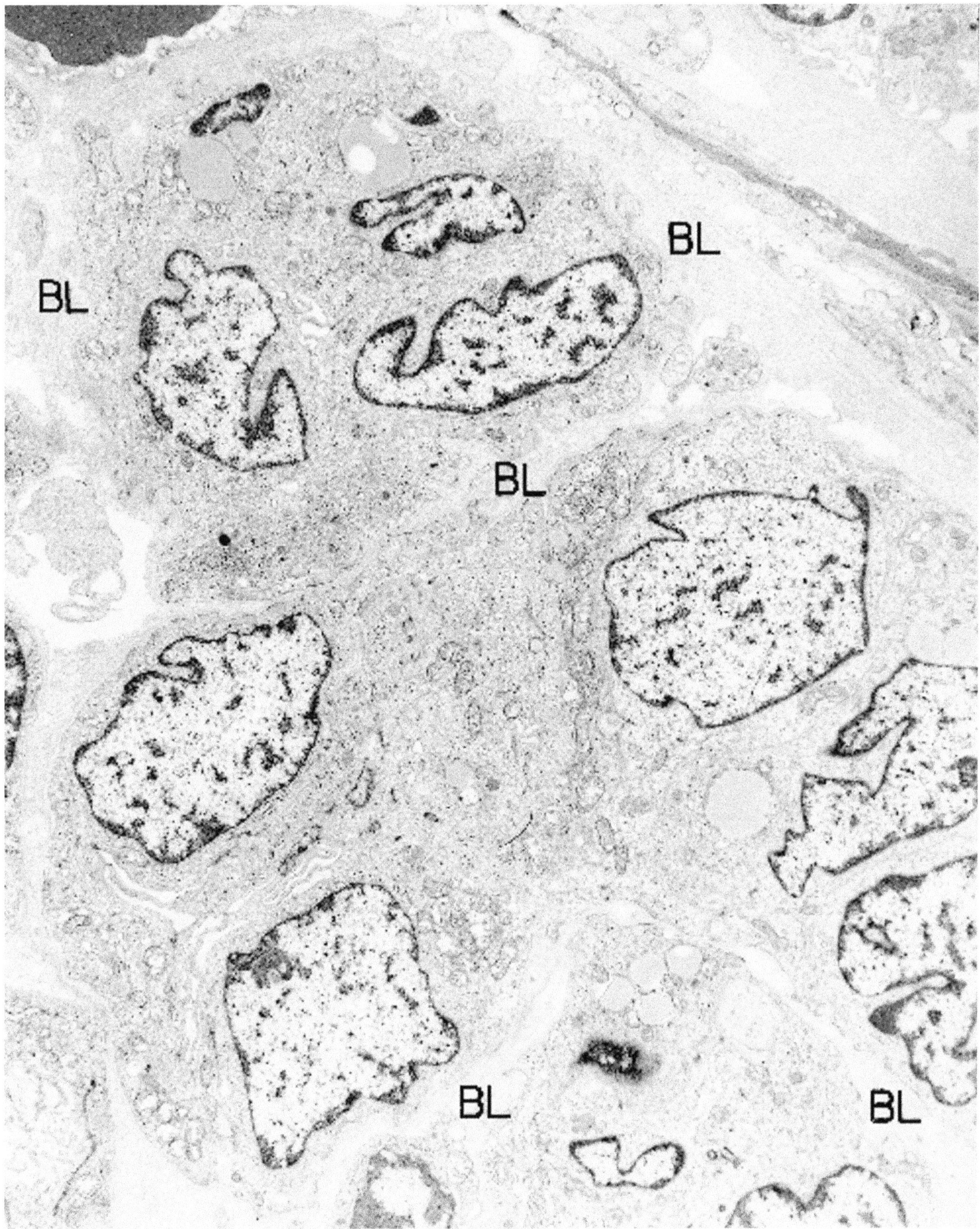

Figure 4.41. Nephroblastoma (Wilms' tumor). Early pretubules are recognizable as small groups of aligned cells surrounded by basal lamina (BL). (× 7020)

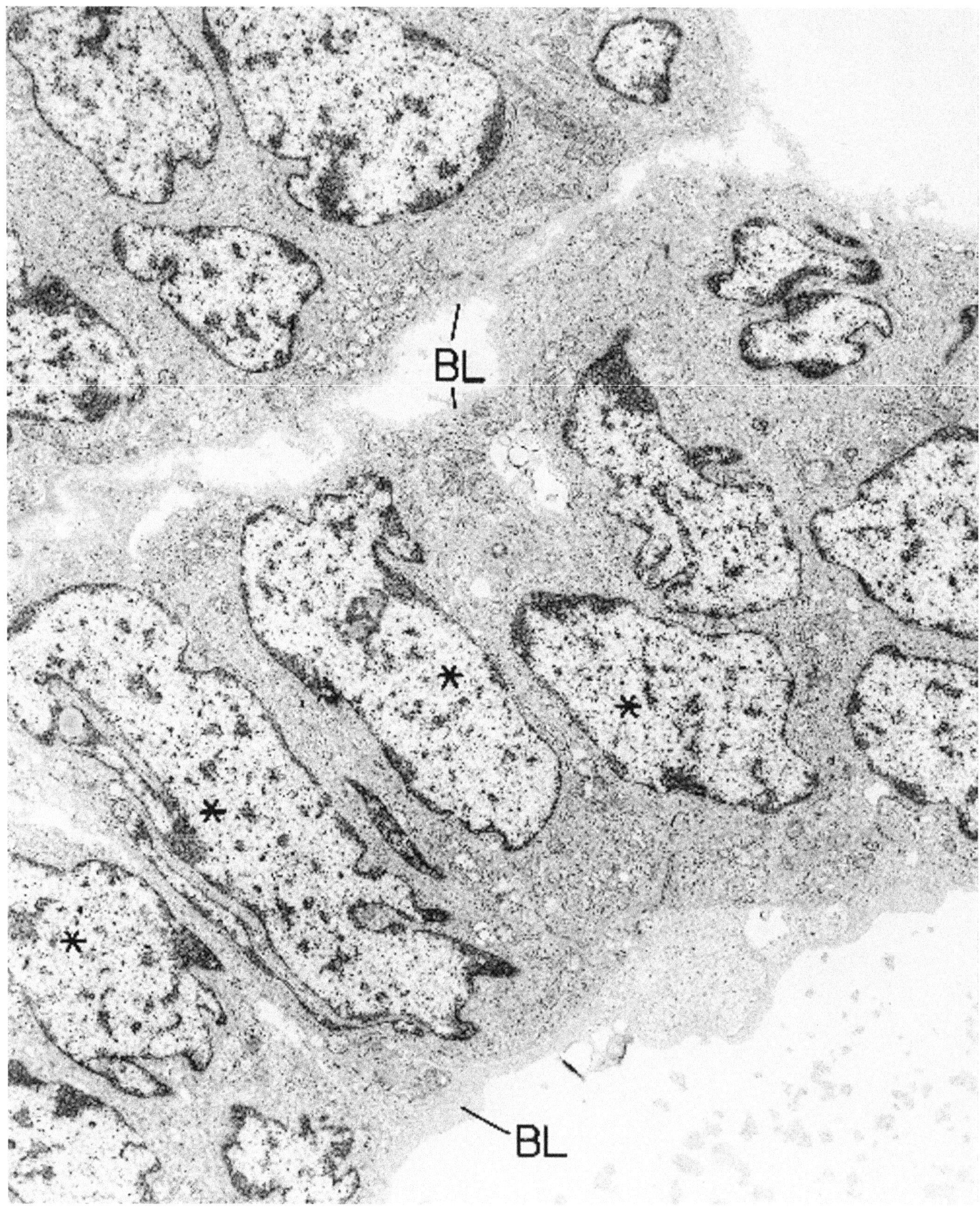

Figure 4.42. Nephroblastoma (Wilms' tumor). These two pretubule groupings illustrate focal parallel alignment of cells (*) and delimiting basal lamina (BL). Also, the cytoplasm has more organelles than that of the less-differentiated blastema of Figures 4.37 and 4.38. (× 6480)

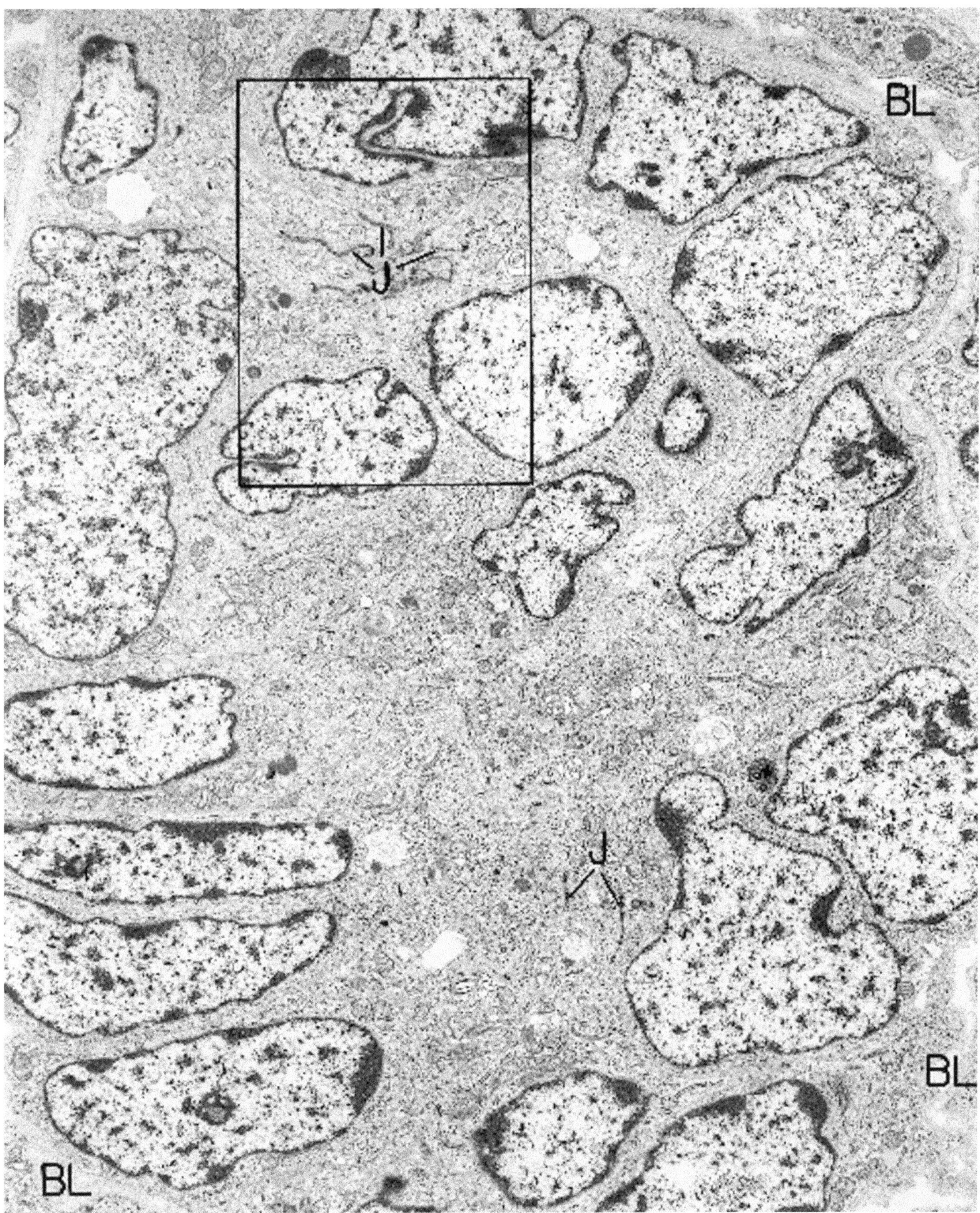

Figure 4.43. Nephroblastoma (Wilms' tumor). This pretubule contains two foci of several junctional complexes (J), the earliest indication of a lumen forming. BL = basal lamina. Delineated rectangular area is enlarged in Figure 4.44. (× 6480)

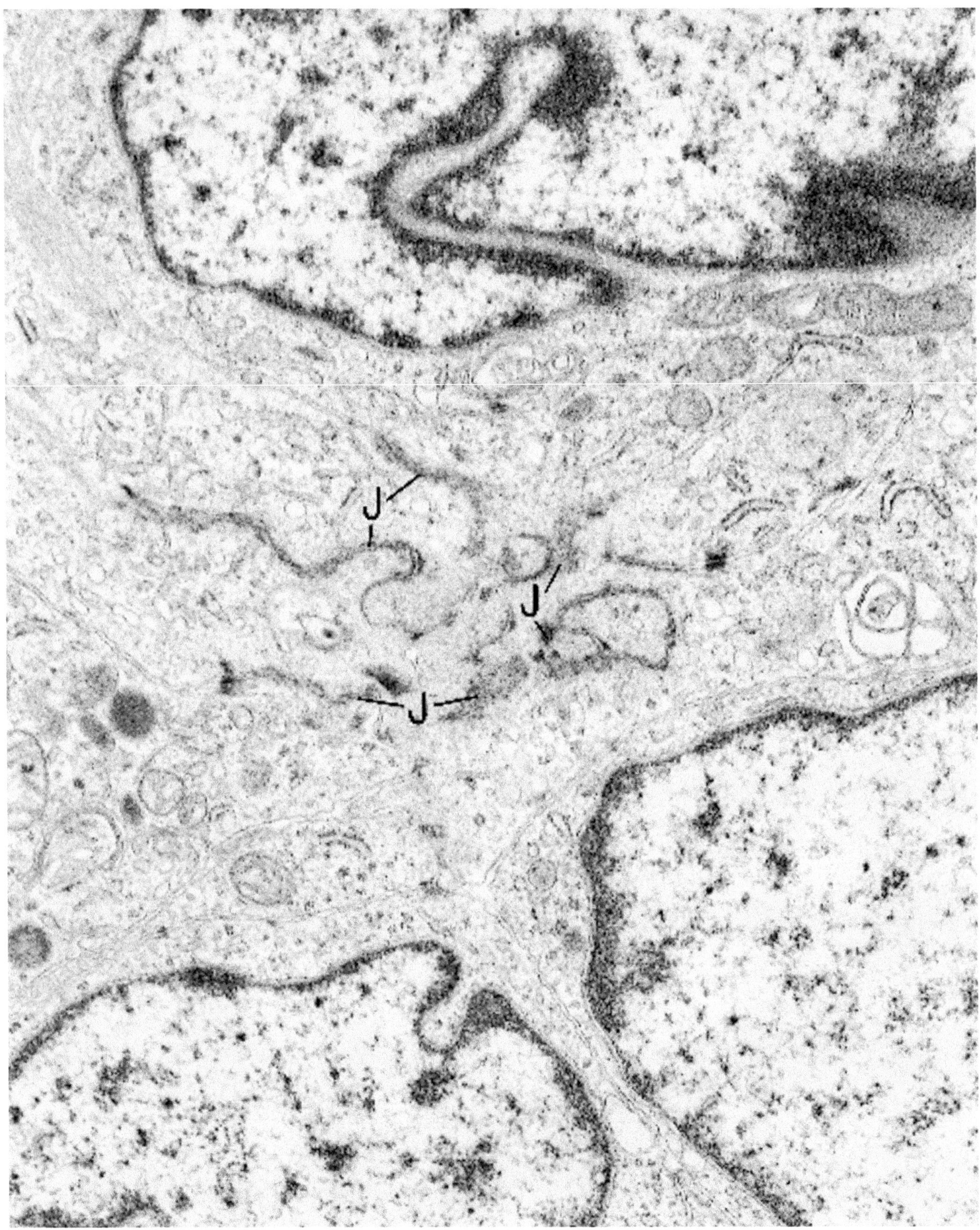

Figure 4.44. Nephroblastoma (Wilms' tumor). Enlargement of the demarcated rectangular area in Figure 4.43 illustrates a focus of early lumen formation. J = junctional complexes (× 20,640)

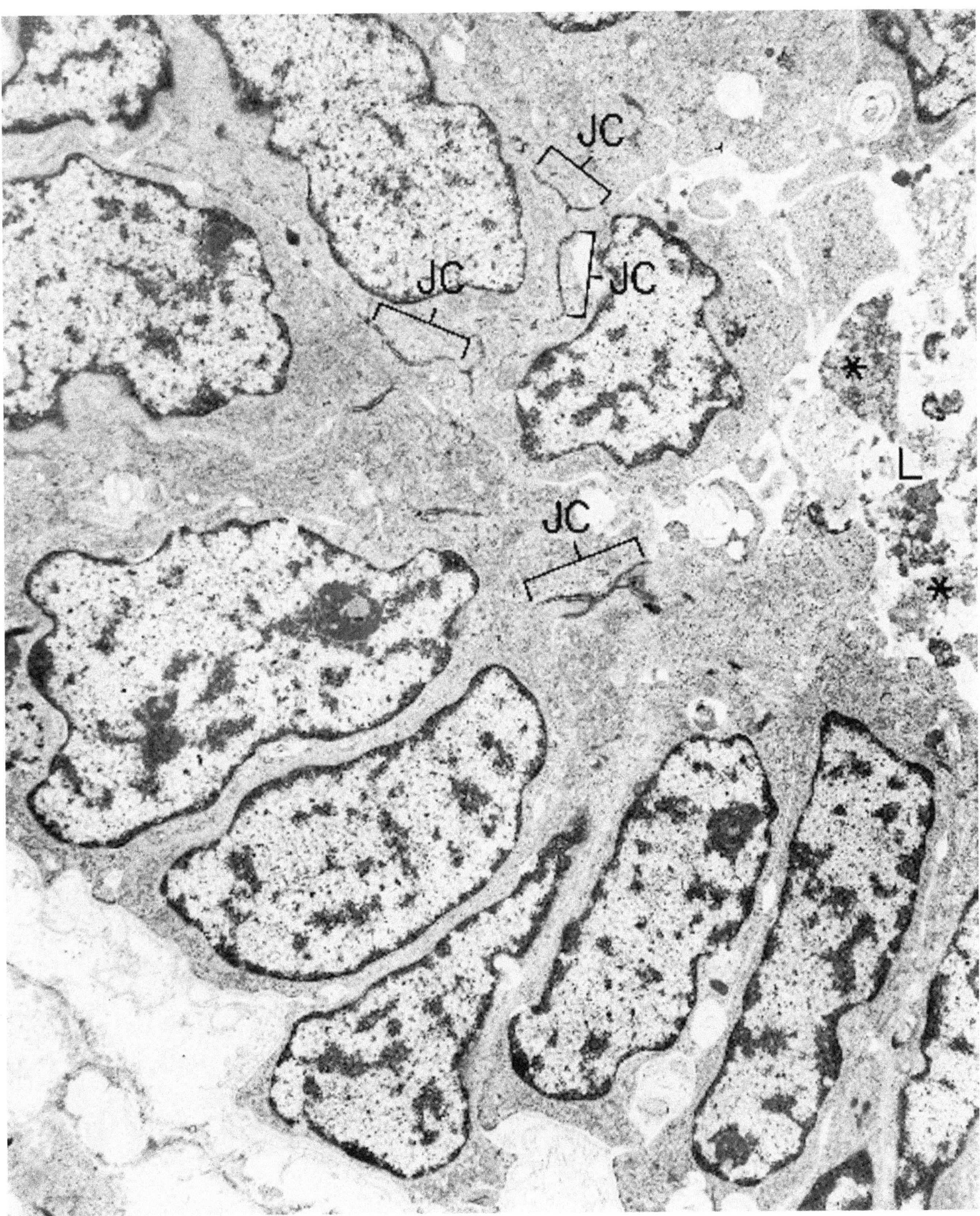

Figure 4.45. Nephroblastoma (Wilms' tumor). An early tubule has a lumen (L) with cellular debris (*) and prominent junctional complexes (JC). (× 9500) (Permission for reprinting granted by Hemisphere Publishing, Dickersin GR: Embryonic ultrastructure as a guide in the diagnosis of tumors. Ultrastruct Pathol 11:609–652, 1987.)

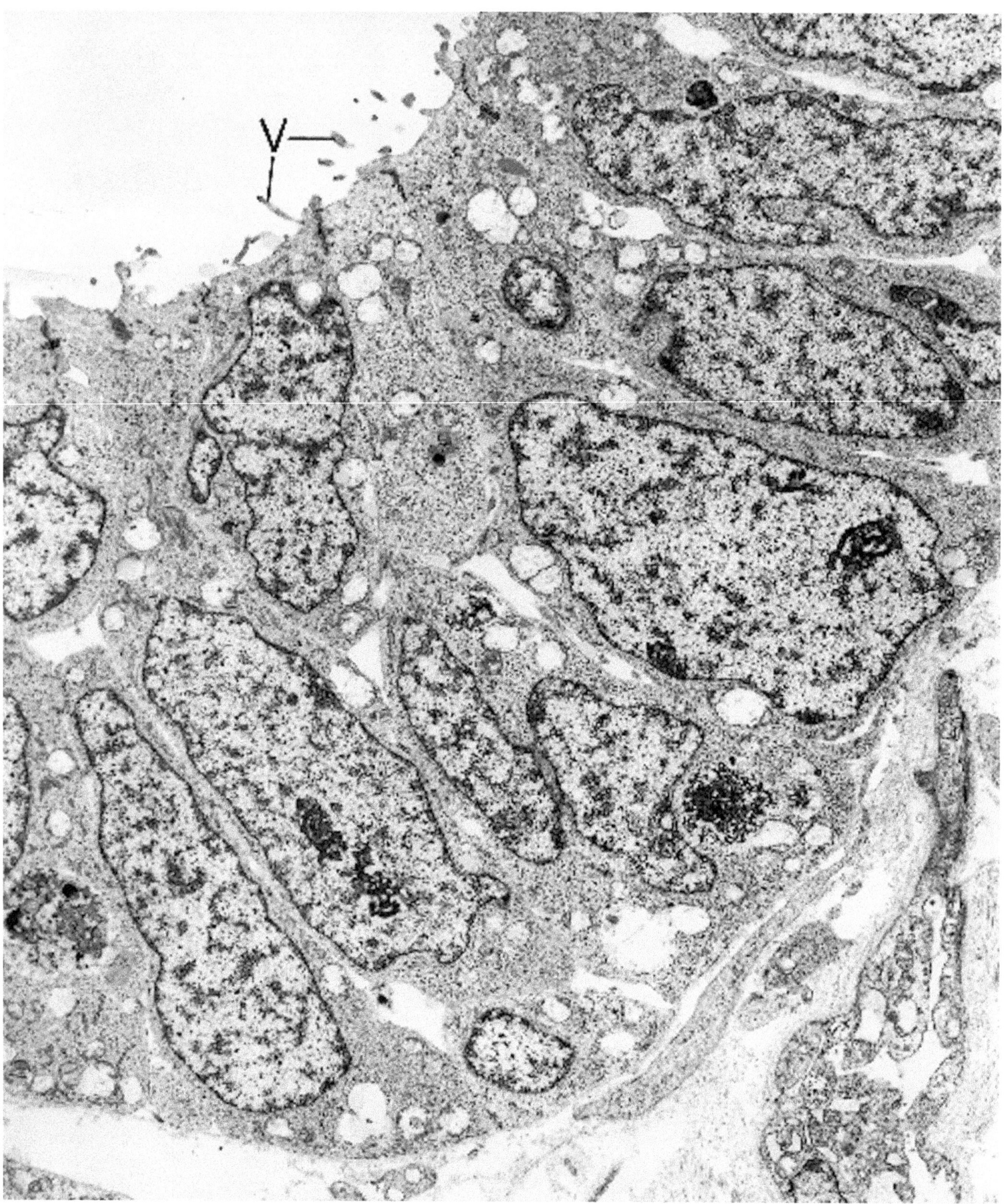

Figure 4.46. Nephroblastoma (Wilms' tumor). Moderately well-formed, neoplastic tubule has luminal lining cells with microvilli (V) on their apical surface. (× 7020) (Permission for reprinting granted by Raven Press, Schmidt D, Dickersin GR, Vawter GF, et al: Wilms' tumor: Review of ultrastructure and histogenesis. Pathobiol Annu 12:281–300, 1982.)

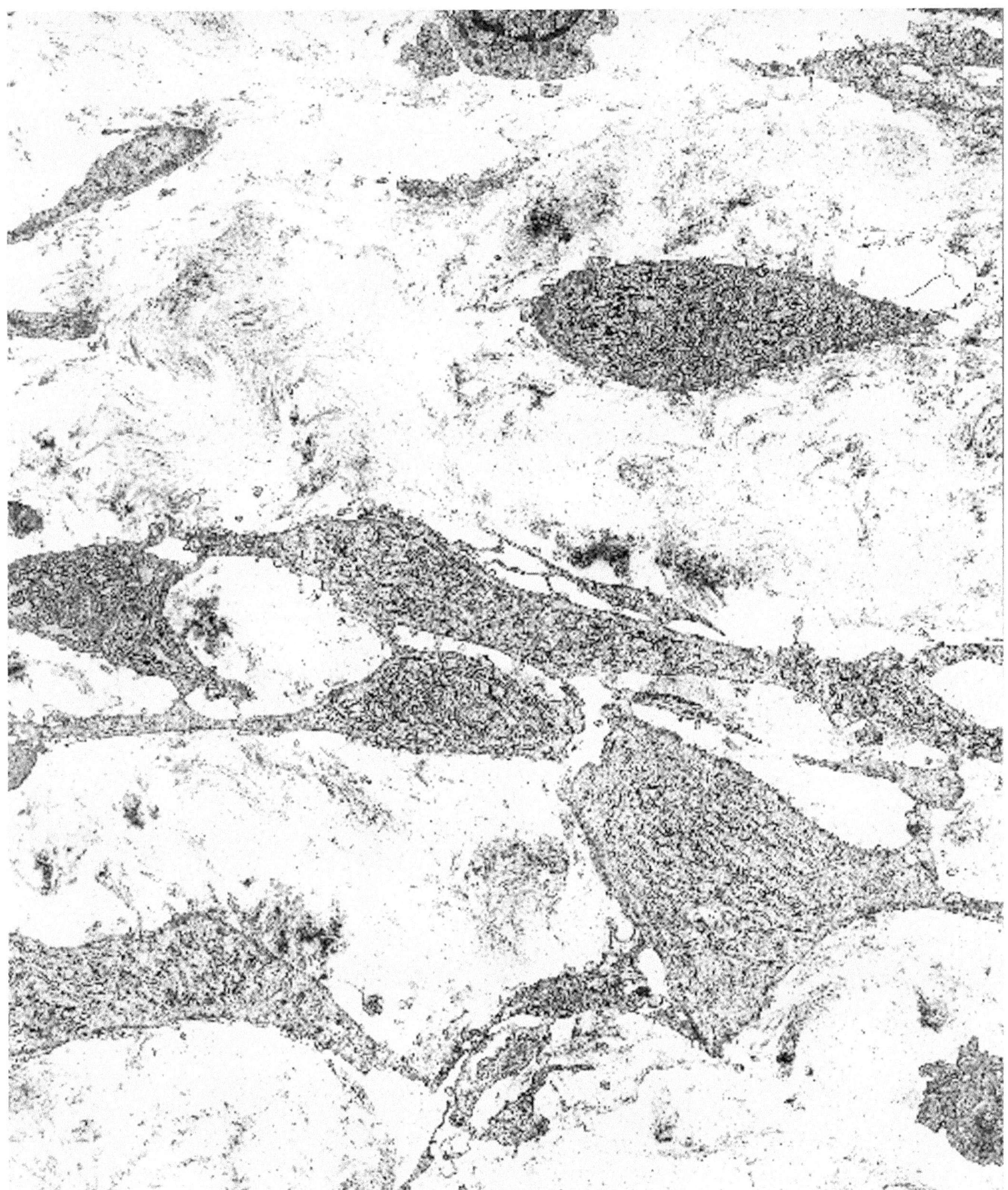

Figure 4.47. Nephroblastoma (Wilms' tumor). The stromal component of this neoplasm consists of fibroblasts with abundant rough endoplasmic reticulum and intercellular collagen. (× 4900) (Permission for reprinting granted by Hemisphere Publishing, Dickersin GR: Embryonic ultrastructure as a guide in the diagnosis of tumors. Ultrastruct Pathol 11:609–652, 1987.)

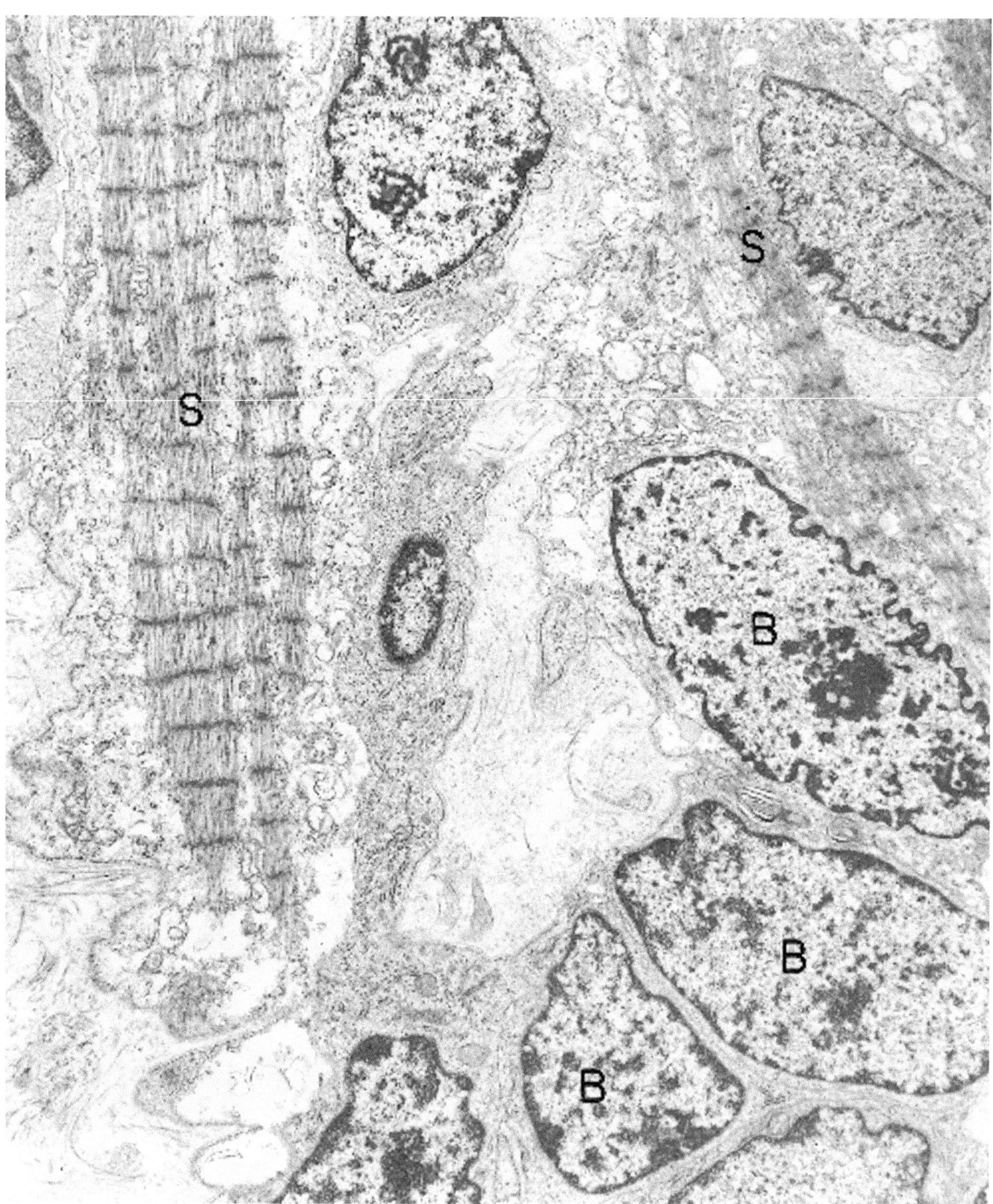

Figure 4.48. Nephroblastoma (Wilms' tumor). This neoplasm contains focal skeletal muscle differentiation, and well-differentiated strap cells (S) are found adjacent to groups of blastemal cells (B). (× 9880) (Permission for reprinting granted by WB Saunders, Dickersin GR, Colvin RB: Pathology of renal tumors. In Skinner DG, Lieskovsky G, eds: *Genitourinary Cancer,* WB Saunders, Philadelphia, 1987.)

(Text continued from page 169)

Lymphoma (Small Cell)

(Figures 4.49 and 4.50.)

Diagnostic criteria. (1) Nucleus with abundant heterochromatin, especially along the nuclear envelope; (2) cytoplasm composed predominantly of ribosomes; (3) absence of intercellular junctions.

Additional points. Small lymphocytes usually are not a problem in identification because of their typical pattern of heterochromatin (Figure 4.49). Intermediate-sized lymphocytes have less heterochromatin, but it is still recognizable as having the peripheral distribution of a lymphocyte (Figure 4.50). Lymphoblasts may show a complete absence of heterochromatin, but they almost always are mixed with other lymphoid cells having the typical heterochromatin arrangement (see Chapter 3, Figure 3.79). As a lymphocyte is transformed into a blast, the free ribosomes aggregate into clusters of polyribosomes. If one were to examine a single blast having a completely euchromatic nucleus and cytoplasm devoid of a specific line of differentiation, it would be impossible to classify that cell as lymphoid or any other cell line.

Plasmacytoma

(Figures 4.51 through 4.54.)

Diagnostic criteria. (1) Nucleus with abundant heterochromatin, similar to a small lymphocyte; (2) cytoplasm with stacks of rough endoplasmic reticulum; (3) centrosome or hof, a perinuclear area of cytoplasm devoid of rough endoplasmic reticulum and occupied by Golgi apparatus, mitochondria, and centrioles.

Additional points. Intermediate forms between lymphocytes and plasma cells may be seen in lymphomas, and these cells have less developed rough endoplasmic reticulum than do fully differentiated plasma cells (Figures 4.51 through 4.54). Intermediate lymphocytes, plasmacytoid lymphocytes, plasmablasts, and immunoblasts are terms used to represent a range of differentiation within the lymphoid line of cells. The rough endoplasmic reticulum in plasmacytoid cells often is dilated and filled with a substance of medium electron density. This substance represents active protein (immunoglobulin) synthesis and occasionally may be represented as huge spherical collections. Because the nuclear envelope is continuous with the cisternae of rough endoplasmic reticulum, these collections may be located in the envelope as well as in the cisternae (Figure 4.53). Rarely, lymphoid cells of any type may have a filopodial, villus-like surface (Figure 4.54), mimicking a truly villous surface of some epithelial cells. However, lymphoid cells never have junctional complexes or microacini. Therefore, lymphomas composed of filopodia-covered cells are readily distinguishable from epithelial neoplasms.

Desmoplastic Small Round Cell Tumor with Divergent Differentiation

(Figures 4.55 through 4.57.)

Diagnostic criteria. (1) Groups of closely apposed round and oval cells separated by bands of fibroblastic stroma; (2) discontinuous basal lamina surrounding the groups; (3) high nucleocytoplasmic ratio; (4) variable cytoplasmic composition, with ribosomes frequently predominating; (5) intermediate filaments with paranuclear whorls (variable); (6) Golgi apparatuses may be prominent in some cells; (7) dense-core granules may be present but usually are not; (8) focal glycogen present or not; (9) diminutive and intermediate junctions and occasionally desmosomes; (10) occasional microacini; (11) occasional polar processes with microtubules.

Additional points. The cellular composition of these neoplasms varies, as the term "divergent differentiation" implies. Usually there is evidence of more than one line of differentiation, with epithelial, mesenchymal, and neural features being evident ultrastructurally. The stromal component separating the islands of small round cells contains spindle cells having the morphological features of fibroblasts and myofibroblasts.

Small Cell Osteosarcoma

(Figures 4.58 and 4.59.)

Diagnostic criteria. (1) Groups and sheets of tightly apposed small, round, oval, and polygonal cells in a matrix of banded collagen and flocculent, medium-dense material; (2) small, distinct intercellular junctions; (3) high nucleocytoplasmic ratio; (4) nuclei variably with regular or irregular contours; (5) chromatin usually finely dispersed (euchromatin); (6) nucleoli of moderate or large size; (7) poorly differentiated cytoplasm with numerous ribosomes, a moderate number of small mitochondria, small-to-moderate amount of undilated or slightly dilated rough endoplasmic reticulum, occasionally prominent Golgi apparatuses, small-to-moderate number of filaments, and variable amounts of glycogen.

Additional points. Small cell osteosarcoma may be difficult or impossible to distinguish from Ewing's sarcoma if there is no evidence of hydroxyapatite in the sample studied. Malignant cartilage, present in some small cell osteosarcomas, is another feature ruling out Ewing's sarcoma. Focal spindle cell areas may be present or even predominant in some small cell osteosarcomas, and rough endoplasmic reticulum may or may not be prominent in the spindle cells.

(Text continues on page 209)

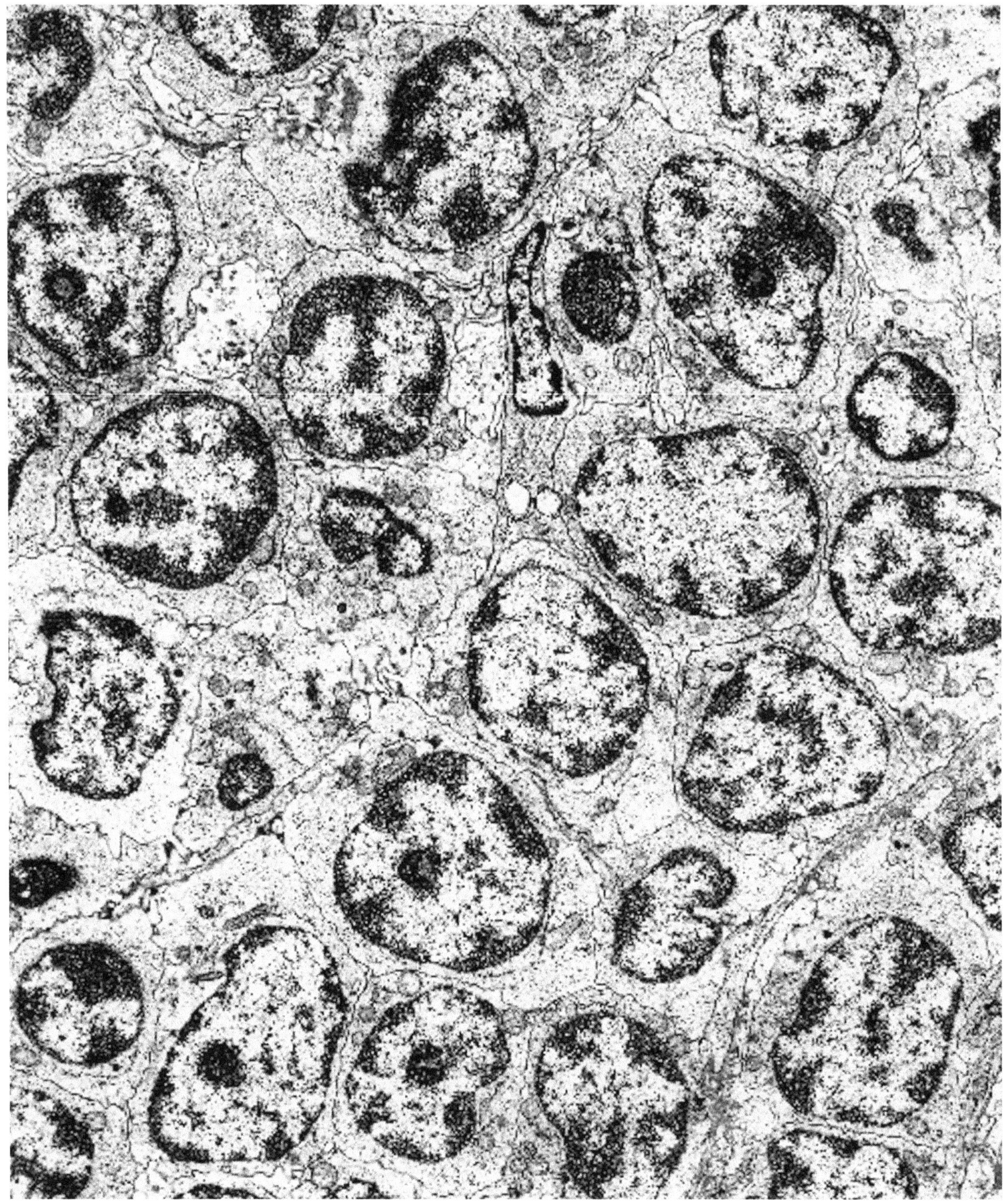

Figure 4.49. Lymphoma, well-differentiated, small cell type (cervical lymph node). This neoplasm is composed of small lymphocytes with the characteristic peripheral chromatin of the lymphoid series. (× 6250)

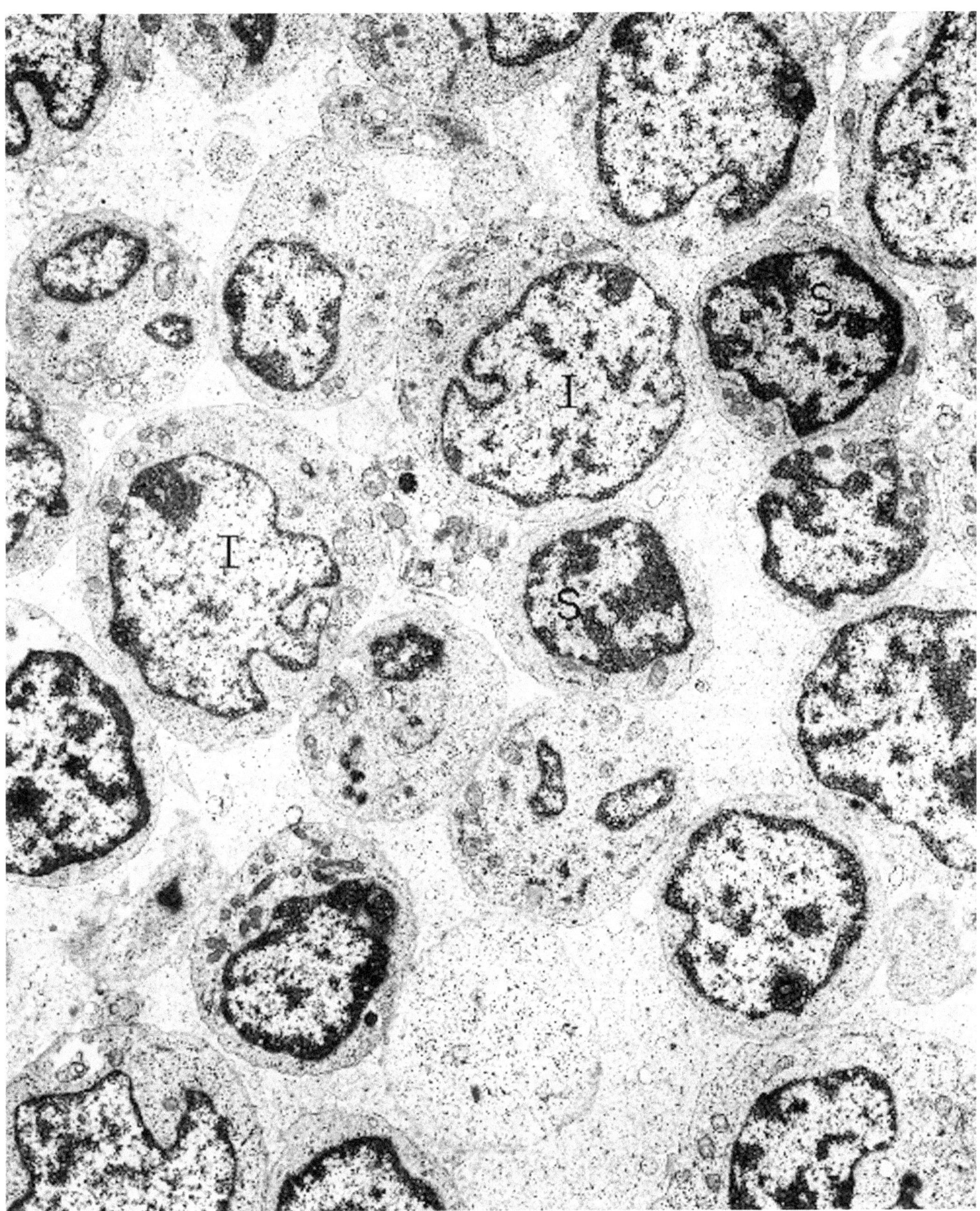

Figure 4.50. Lymphoma, poorly differentiated, small and intermediate cell type (supraclavicular lymph node). The small lymphocytes (S) have abundant peripheral chromatin, and the intermediate lymphocytes (I) have less of the same, although it is in insufficient amounts to allow the cells to be identified as lymphocytes. (× 5320)

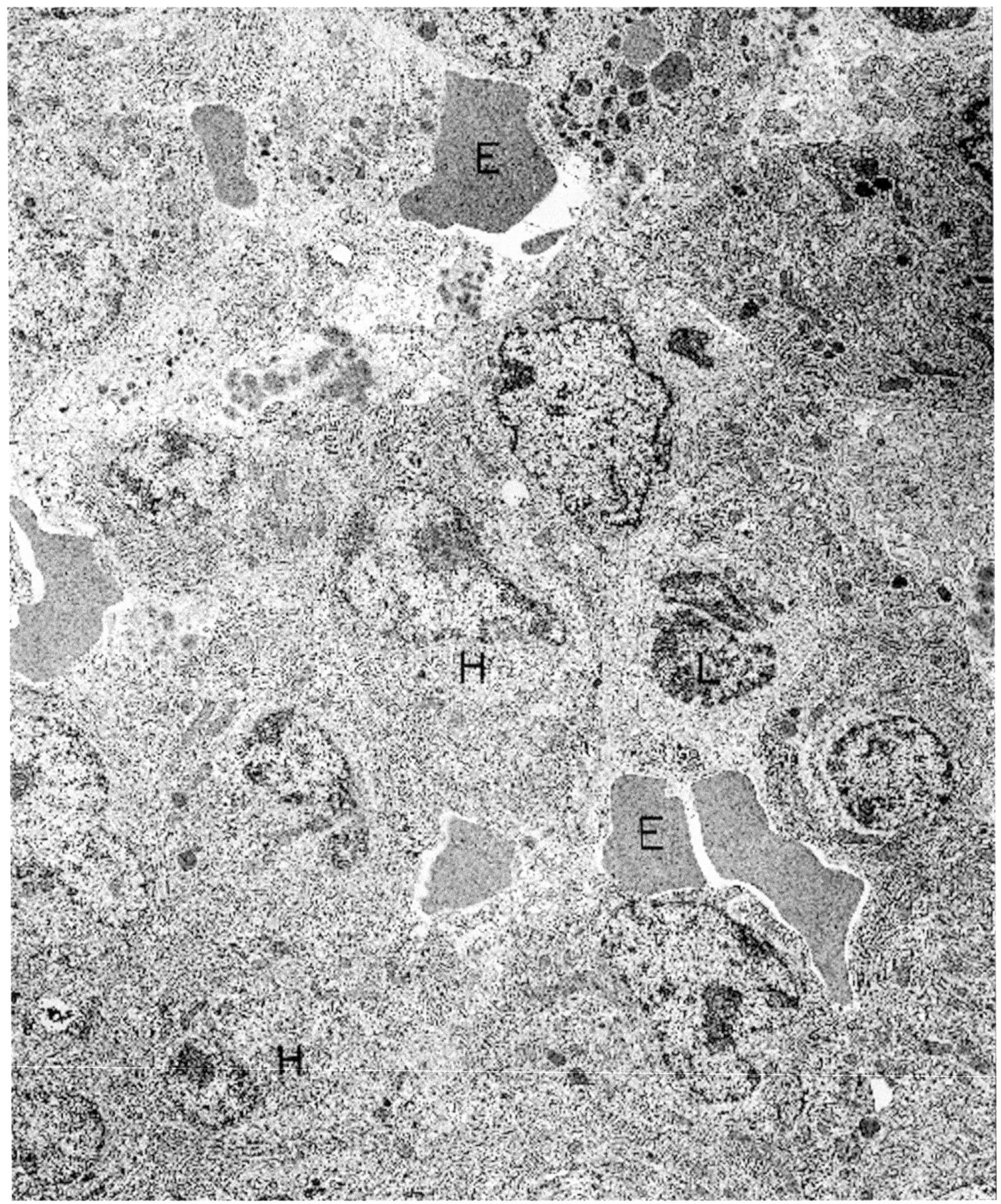

Figure 4.51. Plasmacytoma (rib). Except for several erythrocytes (E) and one lymphocyte (L), all the cells in this field are plasmacytes. Their most striking ultrastructural features include the heterochromatin pattern of the nuclei, and the abundance of stacked rough endoplasmic reticulum. Several of the cells are oriented in a direction that allows their centrosome, or hof (H), to be visible. (× 5130)

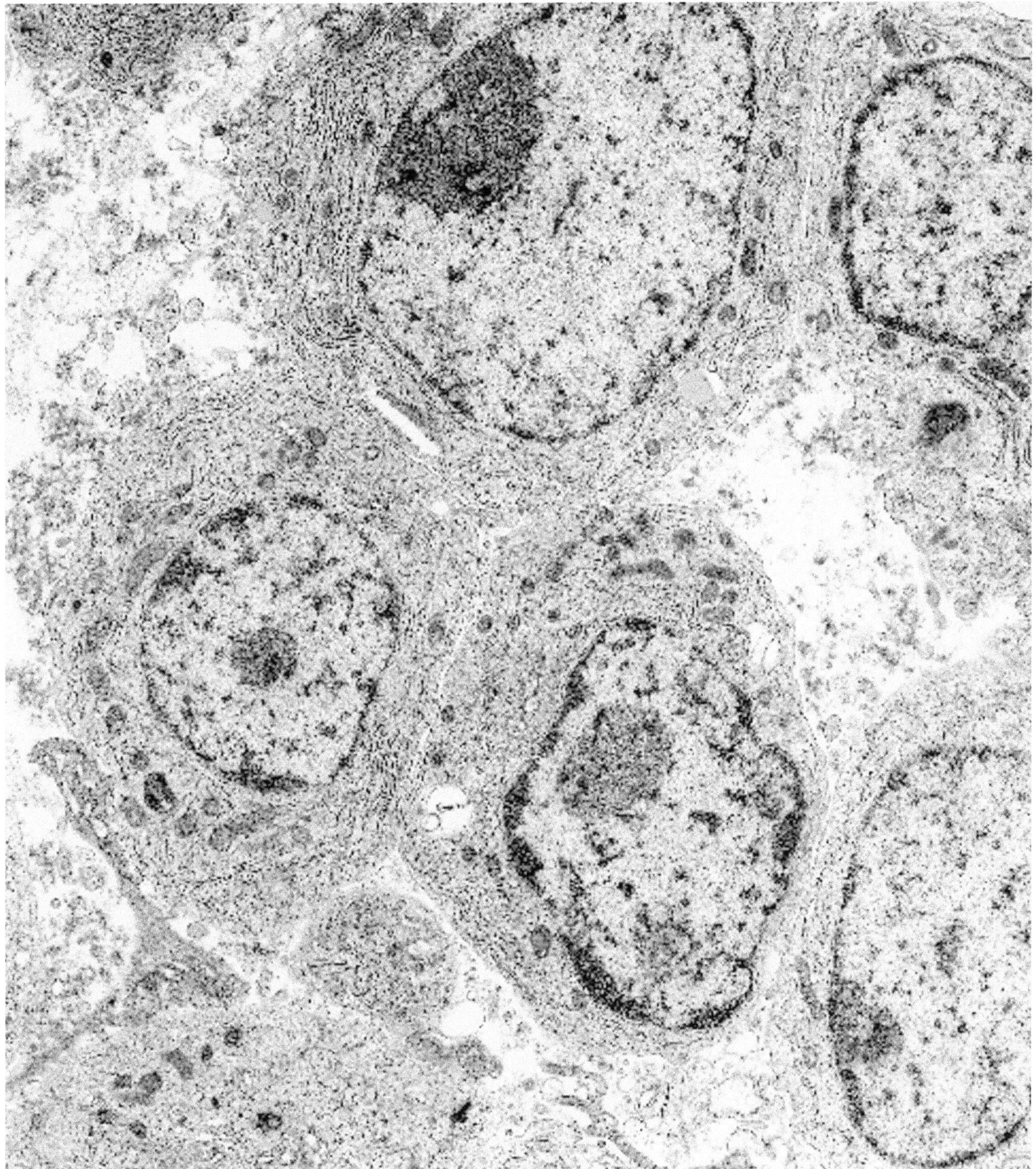

Figure 4.52. Plasmacytoma (epidural space). The plasma cells in this neoplasm are slightly less mature than those in the neoplasm depicted in Figure 4.51, manifested in this tumor by a greater nuclear–cytoplasmic ratio, more euchromatin (especially in the uppermost cell), and larger nucleoli. (× 8840)

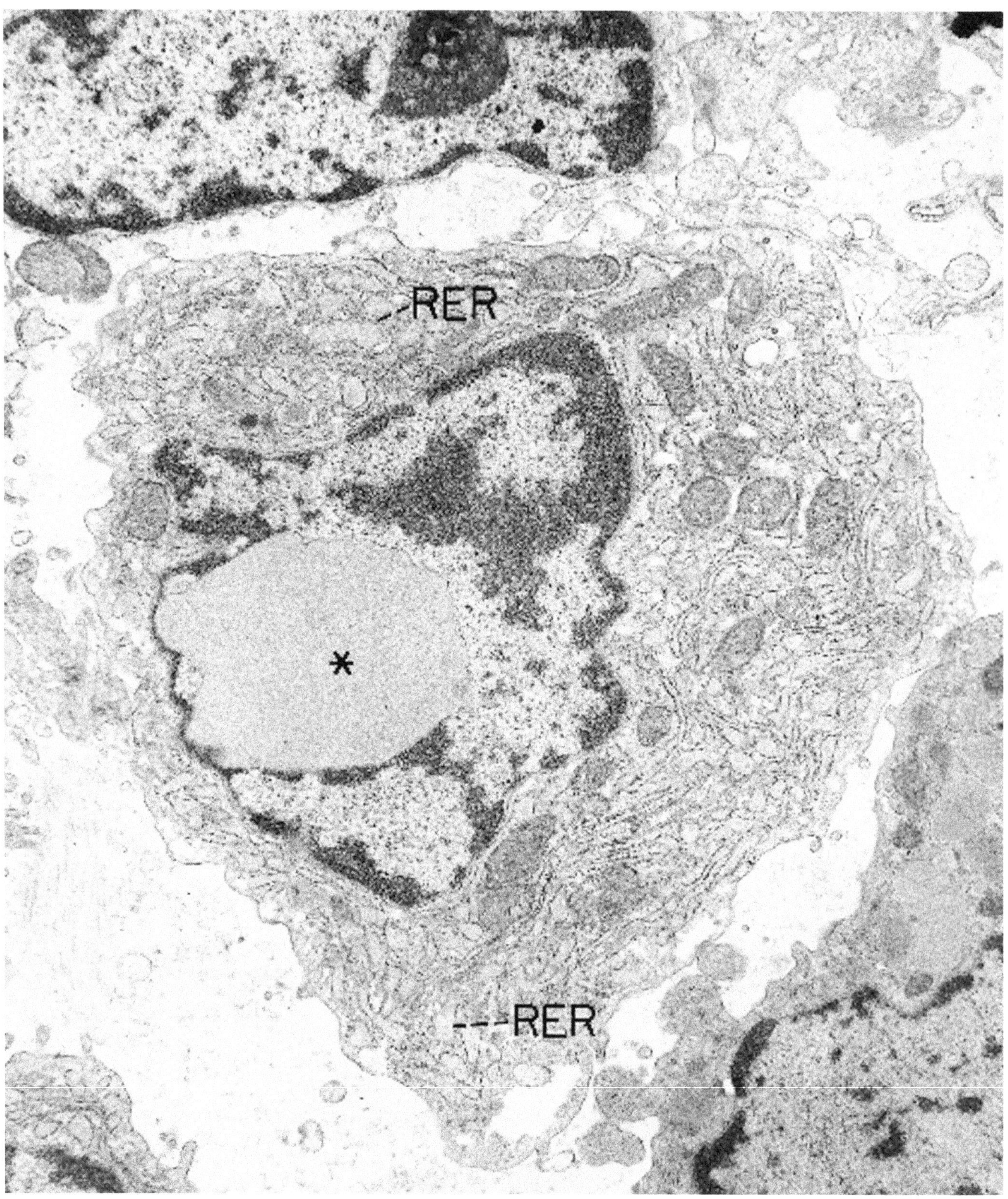

Figure 4.53. Plasmacytoma (stomach). High magnification of a plasma cell illustrates the abundant rough endoplasmic reticulum (RER), which is dilated and filled with a medium-dense material. The large pseudoinclusion (*) of the nucleus actually is a dilated and protein-rich nuclear envelope, which is continuous with the rough endoplasmic reticulum. The nuclear heterochromatin pattern is characteristic for a plasma cell. (× 14,250)

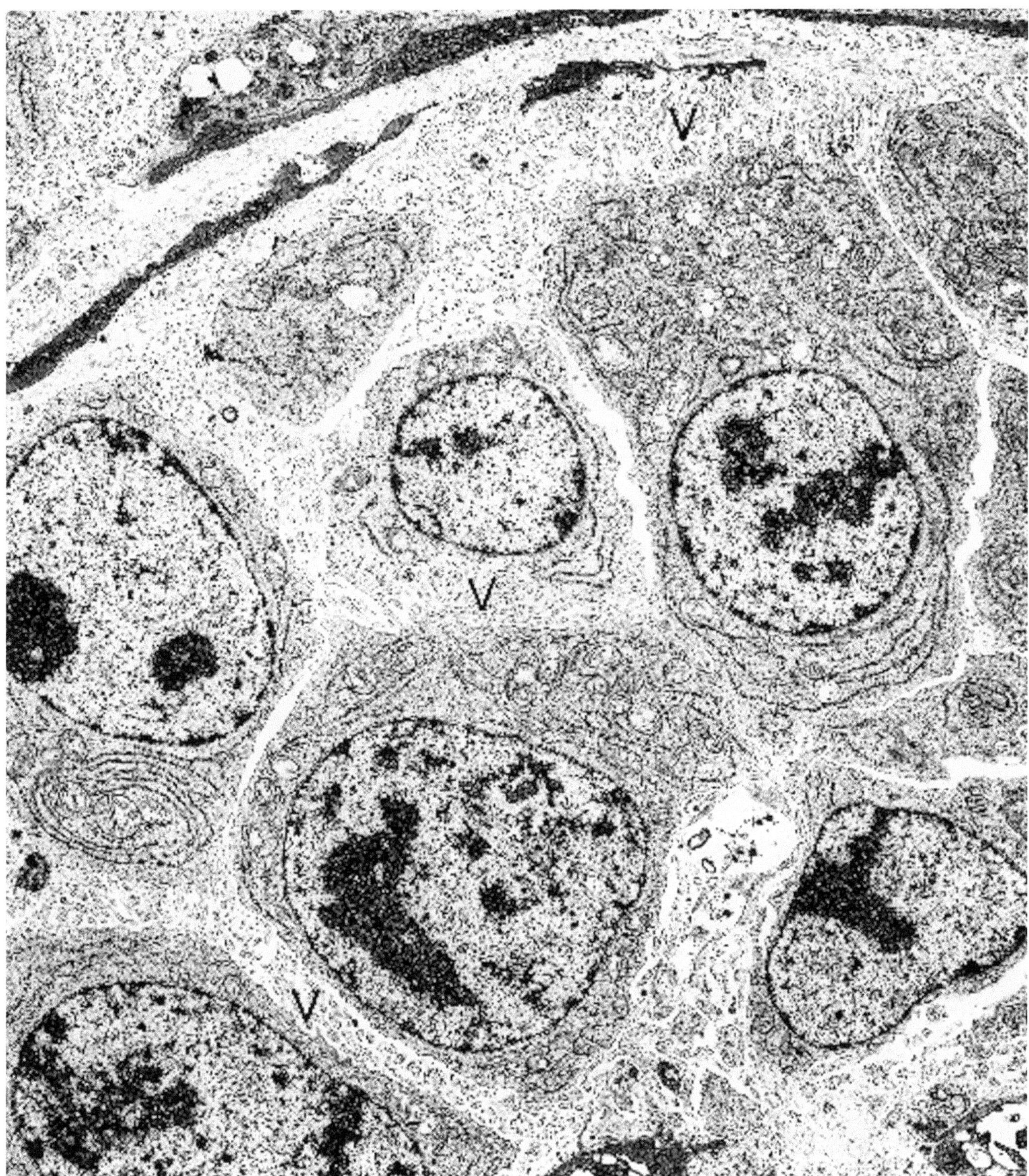

Figure 4.54. Plasmacytoma (cervical lymph node). The cells in this field qualify as plasmablasts or plasmacytoid lymphocytes, because they have more euchromatin and less rough endoplasmic reticulum than a more differentiated plasmacyte. The villous surface (V) is rare for plasma cells and may also be seen in a small percentage of nonplasmacytoid lymphocytes. (× 6750)

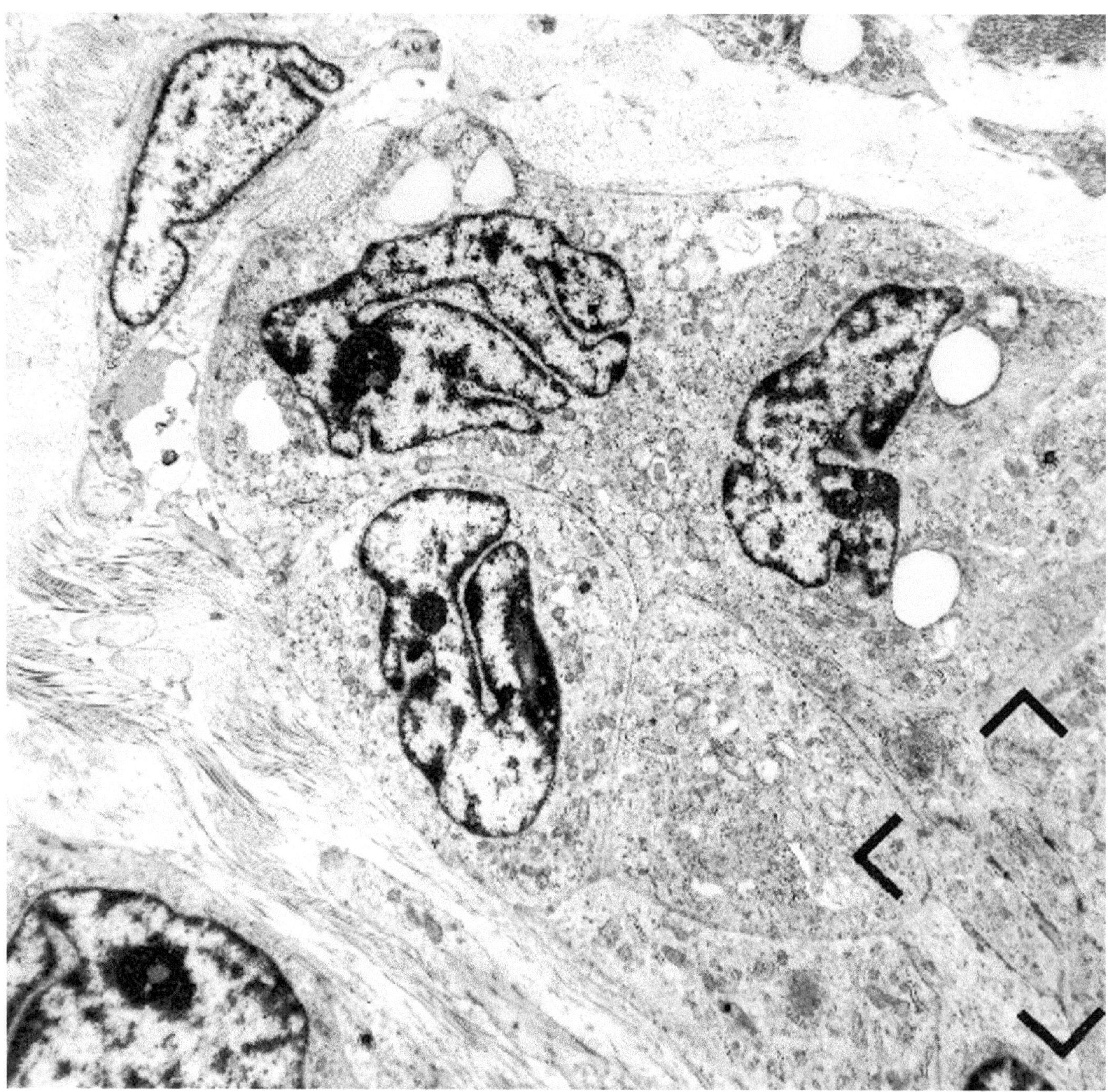

Figure 4.55. Desmoplastic small round cell tumor of divergent differentiation (ovary). An island of closely apposed, oval and polygonal cells is surrounded by abundant banded collagen. The cells have a high nuclear–cytoplasmic ratio, and nuclei are extremely pleomorphic. Even at this relatively low magnification, junctions (*brackets*) are visible; see Figure 4.56. (× 7500)

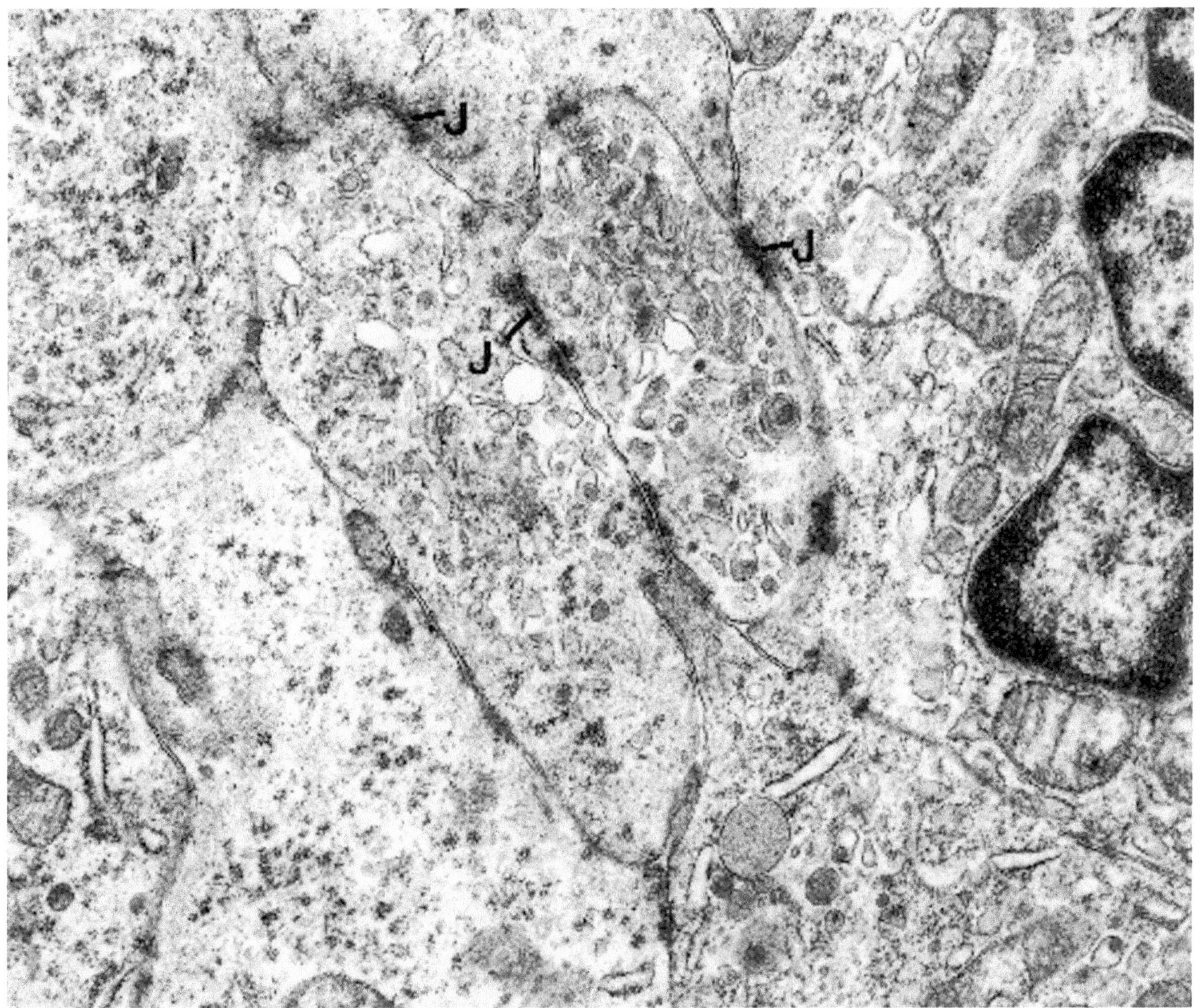

Figure 4.56. Desmoplastic small round cell tumor of divergent differentiation (ovary). Higher magnification of the bracketed area in Figure 4.55 illustrates multiple intercellular junctions (J), including desmosomes. Other organelles include free ribosomes, mitochondria, and small vesicles. (× 27,800)

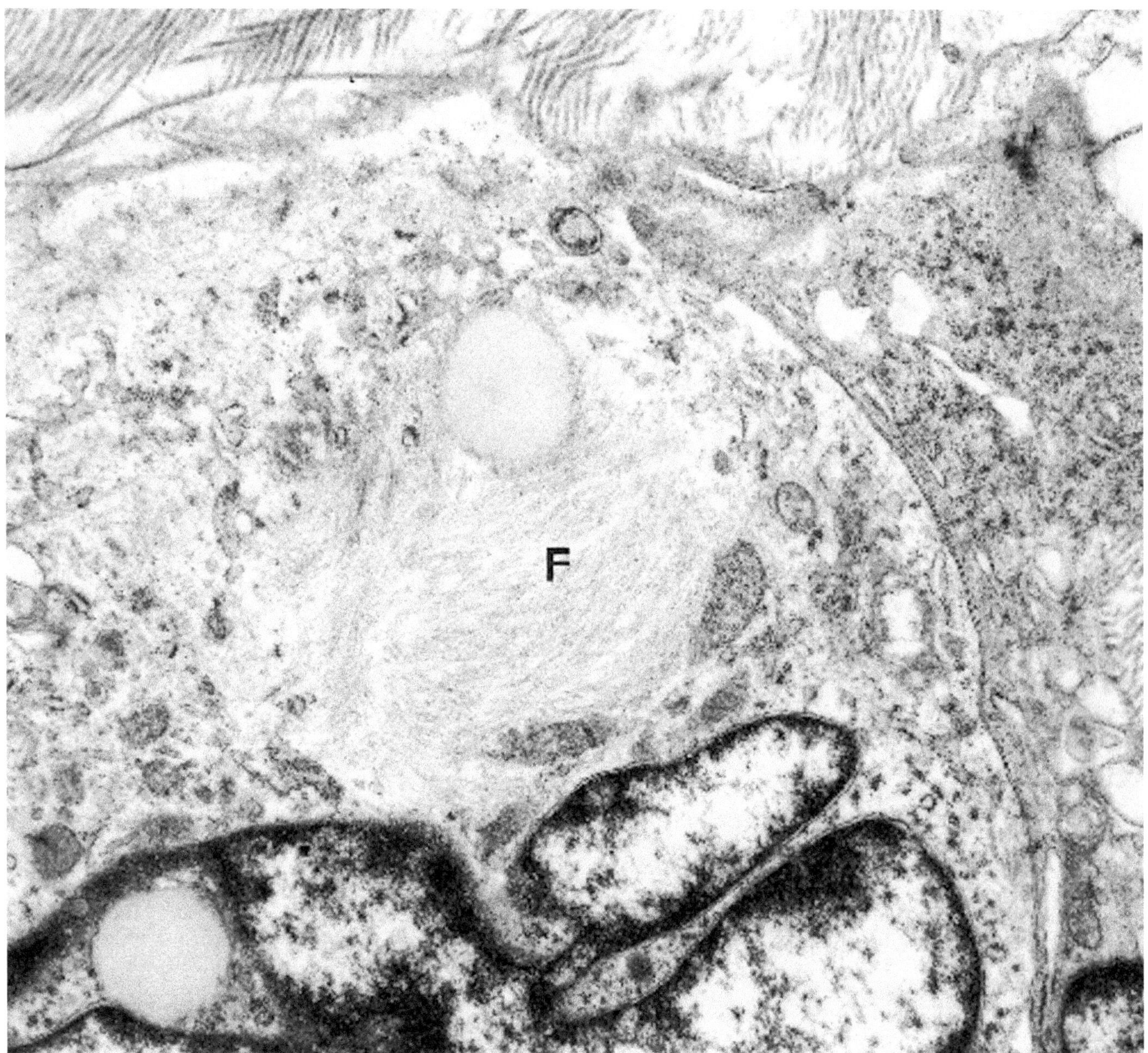

Figure 4.57. Desmoplastic small round cell tumor of divergent differentiation (ovary). High magnification of a cell from the same neoplasm as depicted in Figures 4.55 and 4.56 illustrates a whorl of paranuclear filaments (F). (× 20,400)

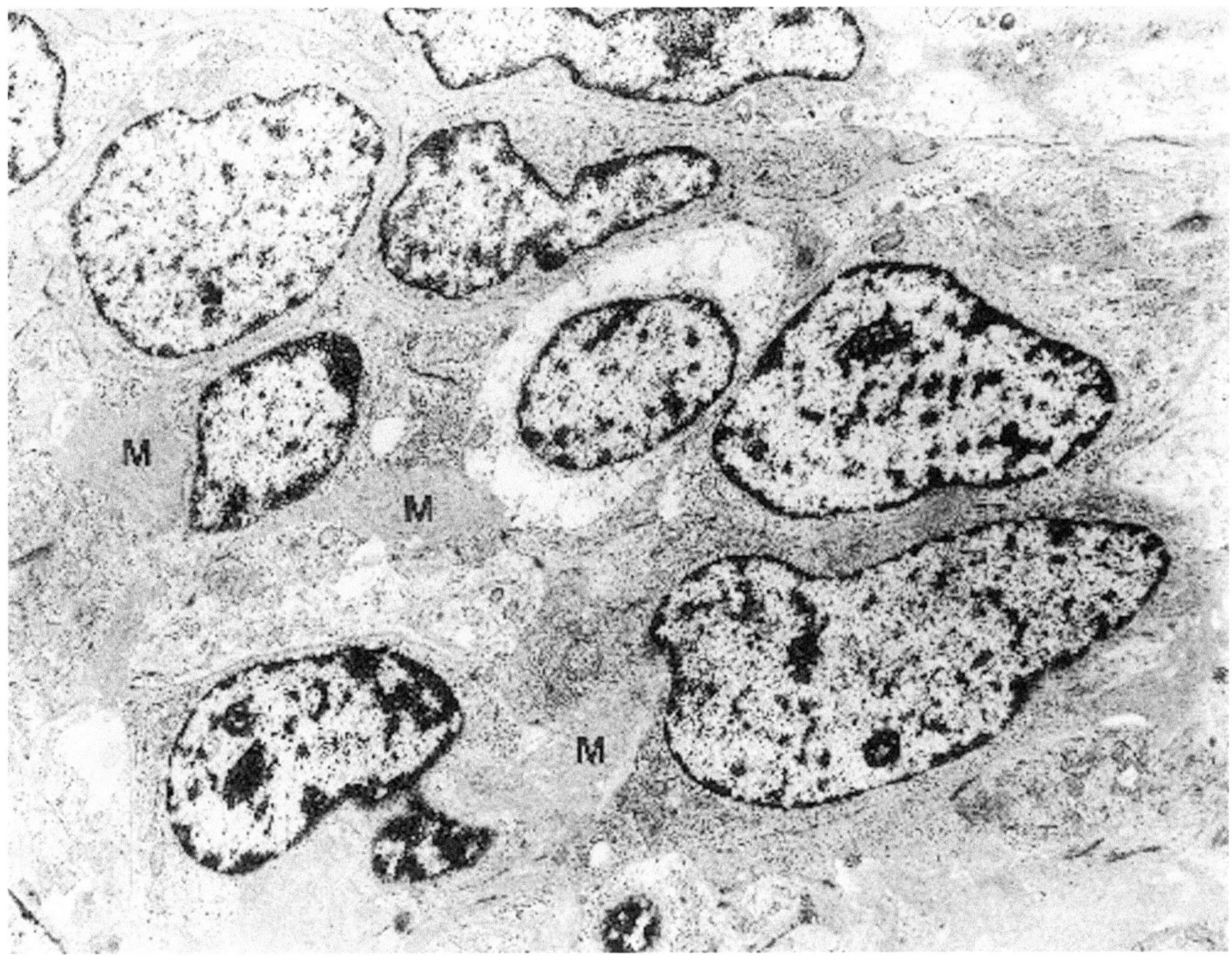

Figure 4.58. Small cell osteosarcoma (tibia). Small oval and elongated neoplastic cells have a high nuclear–cytoplasmic ratio and pleomorphic nuclei. Dense matrical material (M) completely or partially surrounds many of the cells. (× 7200)

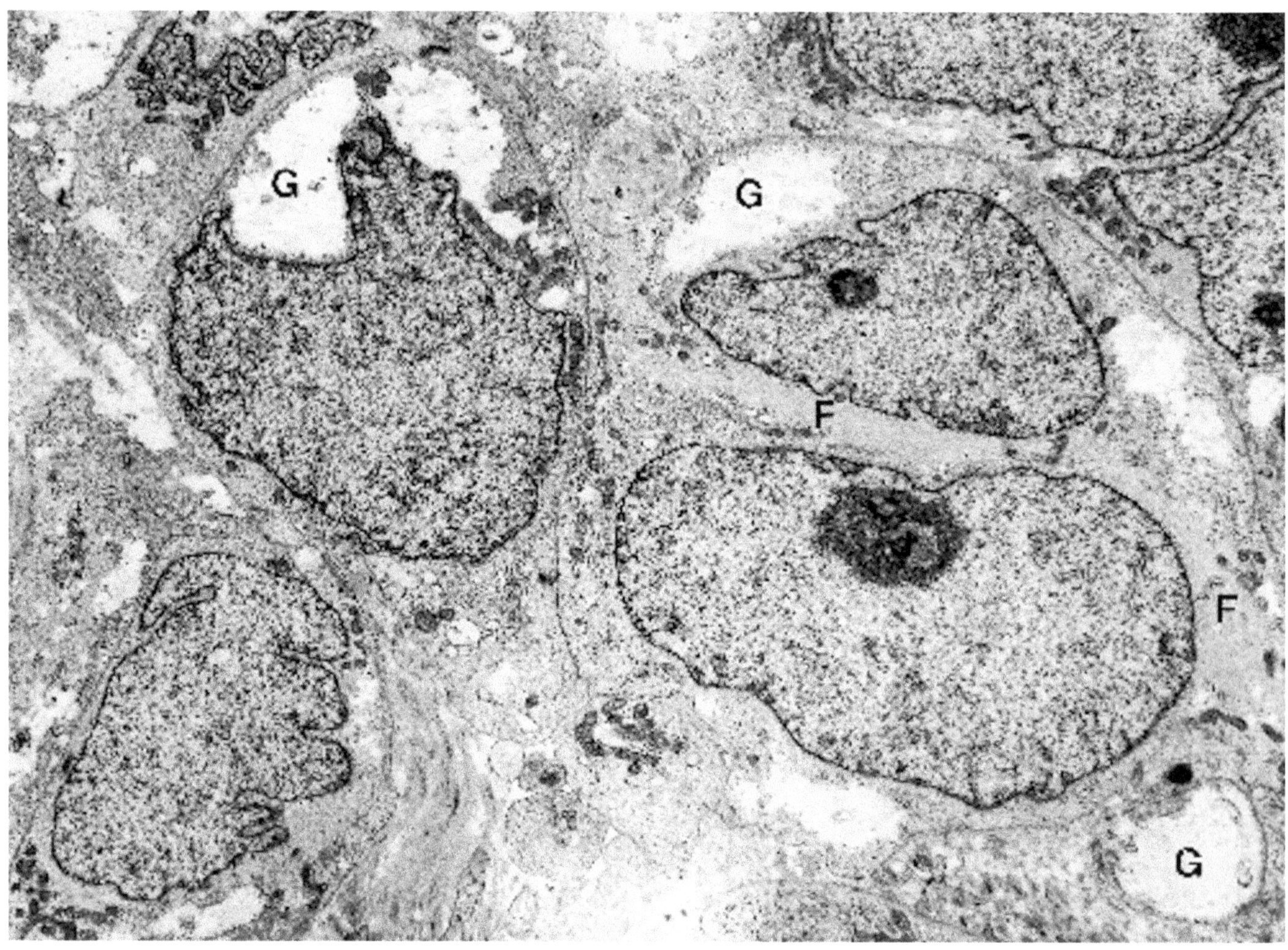

Figure 4.59. Small cell osteosarcoma (femur). A group of neoplastic oval cells illustrates a high nuclear–cytoplasmic ratio, pockets of glycogen (G), and collections of filaments (F) within the cytoplasm. Nuclei are predominantly euchromatic, and one large nucleolus is apparent. (× 7600)

(Text continued from page 197)

Mesenchymal Chondrosarcoma

(Figures 4.60 and 4.61.)

Diagnostic criteria. (1) Groups of small, poorly differentiated cells in a collagenous matrix; (2) high nucleocytoplasmic ratio; (3) irregularly shaped nuclei; (4) few cytoplasmic organelles; (5) foci of further differentiated cartilaginous cells.

Additional points. Mesenchymal chondrosarcoma is another small cell neoplasm that may be indistinguishable from small cell osteosarcoma, and hydroxyapatite, in association with prominent, banded collagen fibrils of varying diameter (osteoid), may be the only distinguishing feature.

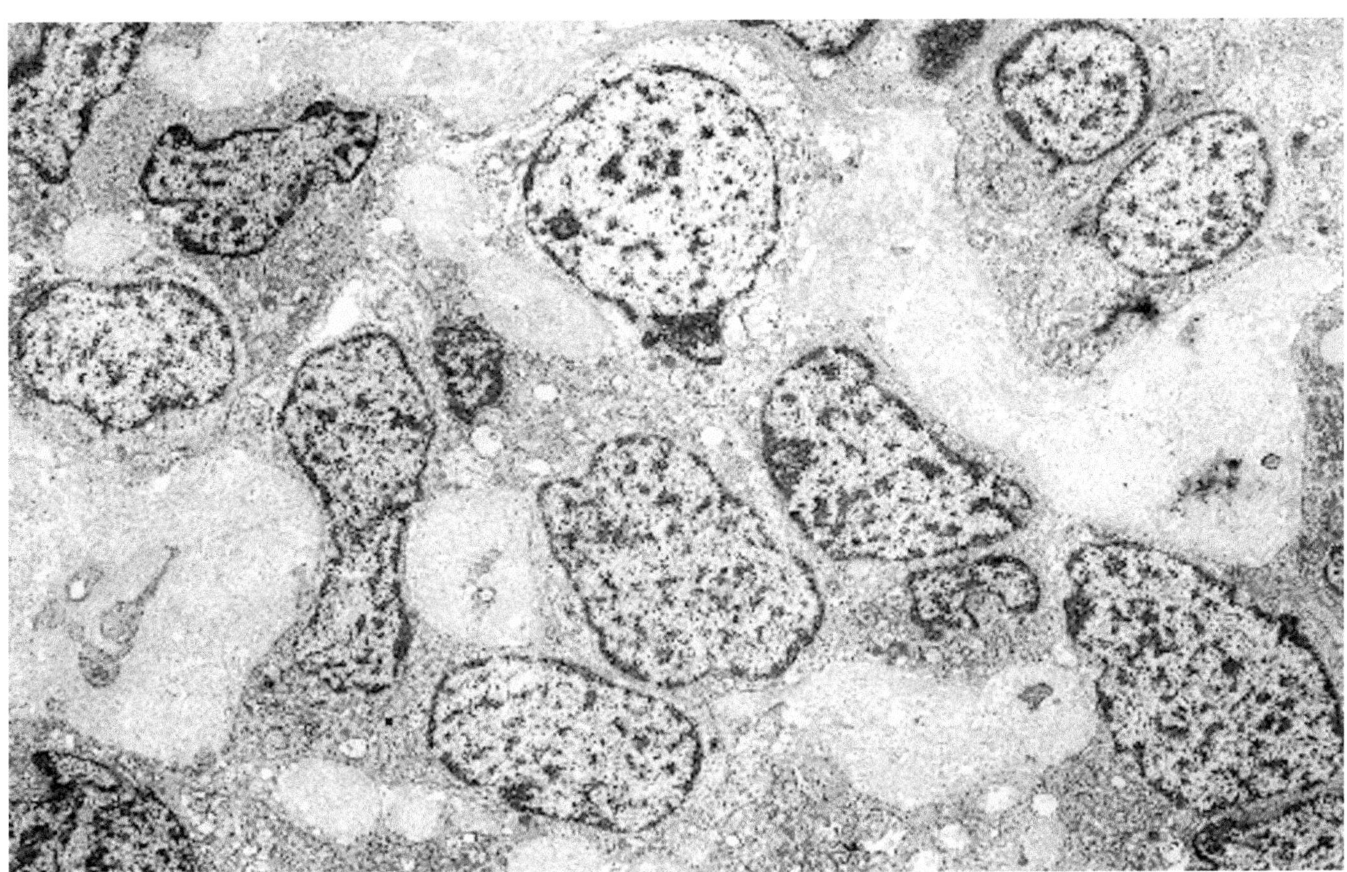

Figure 4.60. Mesenchymal chondrosarcoma (proximal tibia). Small groups of poorly differentiated cells are distributed in a medium-dense matrix. Nuclear–cytoplasmic ratios are high, nuclei vary in shape, and cytoplasm is scant. (× 5900)

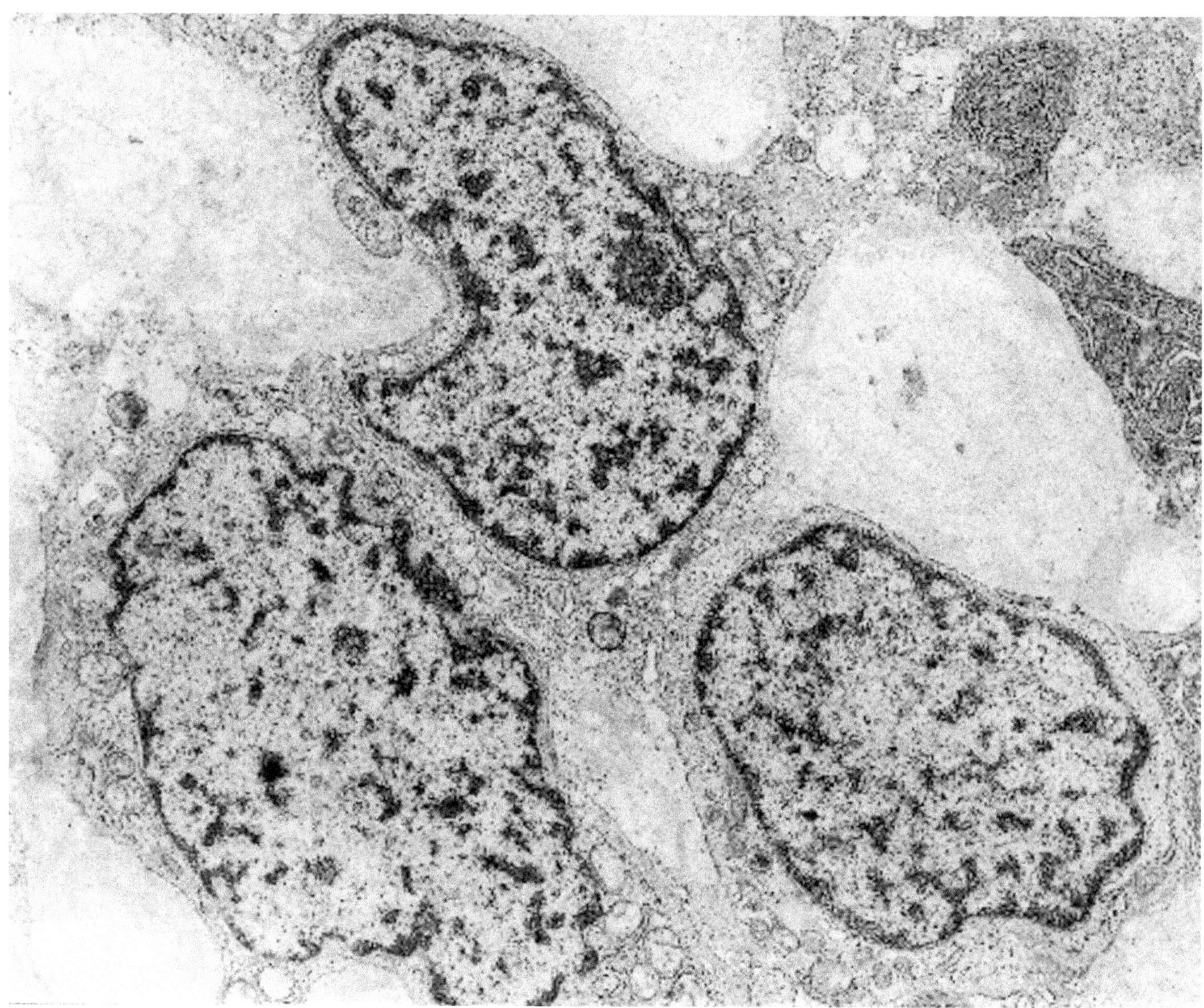

Figure 4.61. Mesenchymal chondrosarcoma (proximal tibia). Higher magnification of mesenchymal chondrocytes illustrates poorly differentiated, scant cytoplasm and intervening collagenous stroma. (× 11,900)

REFERENCES

Neuroendocrine Carcinoma: Bronchogenic Oat Cell Carcinoma

Bensch KG, Corrin B, Pariente R, et al: Oat-cell carcinoma of the lung. Its origin and relationship to bronchial carcinoid. Cancer 22:1163–1172, 1968.

Blobel GA, Gould VE, Moll R, et al: Co-expression of neuroendocrine markers and epithelial cytoskeletal proteins in bronchopulmonary neuroendocrine neoplasms. Lab Invest 52:39–51, 1985.

Elema JD, Keuning HM: The ultrastructure of small cell lung carcinoma in bronchial biopsy specimens. Hum Pathol 16:1133–1140, 1985.

Gould VE, Linnoila RI, Memoli VA, et al: Biology of disease. Neuroendocrine components of the bronchopulmonary tract: Hyperplasias, dysplasias, and neoplasms. Lab Invest 49:519–537, 1983.

Hammar SP, Bockus D, Remington F, et al: The unusual spectrum of neuroendocrine lung neoplasms. Ultrastruct Pathol 13:515–560, 1989.

Mark EJ, Quay SC, Dickersin GR: Papillary carcinoid tumor of the lung. Cancer 48:316–324, 1981.

McDowell EM, Bennett LA, Trump BF: Observations on small granule cells in adult human bronchial epithelium and in carcinoid and oat cell tumors. Lab Invest 34: 202–206, 1976.

McDowell EM, Wislon TS, Trump BF: Atypical endocrine tumors of the lung. Arch Pathol Lab Med 105: 20–28, 1981.

McDowell EM, Sorokin SP, Hoyt RF Jr, et al: An unusual bronchial carcinoid tumor: Light and electron microscopy. Hum Pathol 12:338–348, 1981.

Mills SE, Walker AN, Cooper PH, et al: Atypical carcinoid tumor of the lung. A clinicopathologic study of 17 cases. Am J Surg Pathol 6:643–654, 1982.

Nomori H, Shimosato Y, Kodama T, et al: Subtypes of small cell carcinoma of the lung: Morphometric, ultrastructural, and immunohistochemical analysis. Hum Pathol 17:604–613, 1986.

Papotti M, Gherardi G, Eusebi V, et al: Primary oat cell (neuroendocrine) carcinoma of the breast. Report of four cases. Virchows Arch [A] 420:103–108, 1992.

Pilotti S, Patriarca C, Lombardi L, et al: Well-differentiated neuroendocrine carcinoma of the lung: A clinicopathologic and ultrastructural study of 10 cases. Tumori 78:121–129, 1992.

Ranchod M, Levine GD: Spindle-cell carcinoid tumors of the lung. A clinicopathologic study of 35 cases. Am J Surg Pathol 4:315–331, 1980.

Stahlman MT, Gray ME: The ontogeny of neuroendocrine cells in human fetal lung. I. An electron microscopic study. Lab Invest 51:449–463, 1984.

Travis WD, Linnoila I, Tsokos MG, et al: Neuroendocrine tumors of the lung with proposed criteria for large-cell neuroendocrine carcinoma. An ultrastructural, immunohistochemical, and flow cytometric study of 35 cases. Am J Surg Pathol 15:529–553, 1991.

Warren WH, Memoli VA, Gould VE: Immunohistochemical and ultrastructural analysis of bronchopulmonary neuroendocrine neoplasms. II. Well differentiated neuroendocrine carcinomas. Ultrastruct Pathol 7:185–199, 1984.

Yesner R: Small cell tumors of the lung. Am J Surg Pathol 7:775–785, 1983.

Neuroendocrine Carcinoma: Merkel Cell Carcinoma

De Wolf-Peeters C, Marien K, Mebis J, et al: A cutaneous APUDoma or Merkel cell tumor? A morphologically recognizable tumor with a biological and histological malignant aspect in contrast with its clinical behavior. Cancer 46:1810–1816, 1980.

Gould VE, Dardi LE, Memoli VA, et al: Neuroendocrine carcinomas of the skin: Light microscopic, ultrastructural, and immunohistochemical analysis. Ultrastruct Pathol 1:499–509, 1980.

Hamilton J, Levine MR, Lash R, et al: Merkel cell carcinoma of the eyelid. Ophthalmic Surg 24:764–769, 1993.

Johannessen JV, Gould VE: Neuroendocrine skin carcinoma associated with calcitonin production: A Merkel cell carcinoma? Hum Pathol 11(suppl):586–590, 1980.

Moll R, Osborn M, Hartschuh W, et al: Variability of expression and arrangement of cytokeratin and neurofilaments in cutaneous neuroendocrine carcinomas (Merkel cell tumors): Immunocytochemical and biochemical analysis of twelve cases. Ultrastruct Pathol 10:473–495, 1986.

Mount SL, Taatjes DJ: Neuroendocrine carcinoma of the skin (Merkel cell carcinoma). An immunoelectron-microscopic case study. Am J Dermatopathol 16:60–65, 1994.

Rubsamen PE, Tanenbaum M, Grove AS, et al: Merkel cell carcinoma of the eyelid and periocular tissues. Am J Ophthalmol 113:674–680, 1992.

Sibley RK, Rosai J, Foucar E, et al: Neuroendocrine (Merkel cell) carcinoma of the skin. A histologic and ultrastructural study of two cases. Am J Surg Pathol 4: 211–221, 1980.

Silva E, Mackay B: Neuroendocrine (Merkel cell) carcinomas of the skin: An ultrastructural study of nine cases. Ultrastruct Pathol 2:1–9, 1981.

Tang CK, Toker C: Trabecular carcinoma of the skin. An ultrastructural study. Cancer 42:2311–2321, 1978.

Taxy JB, Ettinger DS, Wharam MD: Primary small cell carcinoma of the skin. Cancer 46:2308–2311, 1980.

Vigneswaran N, Muller S, Lense E, et al: Merkel cell carcinoma of the labial mucosa. An immunohistochemical and ultrastructural study with a review of the literature on oral Merkel cell carcinoma. Oral Surg Oral Med Oral Pathol 74:193–200, 1992.

Neuroblastoma

Askin FB, Rosai J, Sibley RK, et al: Malignant small cell tumor of the thoracopulmonary region in childhood. A distinctive clinicopathologic entity of uncertain histogenesis. Cancer 43:2438–2451, 1979.

Banerjee AK, Sharma BS, Vashista RK, et al: Intracranial olfactory neuroblastoma: Evidence for olfactory epithelial origin. J Clin Pathol 45:299–302, 1992.

Carachi R. Campbell PE, Kent M: Thoracic neural crest tumors. A clinical review. Cancer 51:949–954, 1983.

Gonzalez-Crussi F, Wolfson SL, Misugi K, et al: Peripheral neuroectodermal tumors of the chest wall in childhood. Cancer 54:2519–2527, 1984.

Hachitanda Y, Tsuneyoshi M, Enjoji M: An ultrastructural and immunohistochemical evaluation of cytodifferentiation in neuroblastic tumors. Mod Pathol 2:13–19, 1989.

Hicks MJ, Mackay B: Comparison of ultrastructural features among neuroblastic tumors: Maturation from neuroblastoma to ganglioneuroma. Ultrastruct Pathol 19: 311–322, 1995.

Linnoila RI, Tsokos M, Triche TJ, et al: Evidence for neural origin and PAS-positive variants of the malignant small cell tumor of thoracopulmonary region ("Askin tumor"). Am J Surg Pathol 10:124–133, 1986.

Louis DN, Swearingen B, Linggood RM, et al: Central nervous system neurocytoma and neuroblastoma in adults—report of eight cases. J Neuro-Oncol 9:231–238, 1990.

Romansky SG, Crocker DW, Shaw KNF: Ultrastructural studies on neuroblastoma. Evaluation of cytodifferentiation and correlation of morphology and biochemical and survival data. Cancer 42:2392–2398, 1978.

Salter JE Jr, Gibson D, Ordonez NG, et al: Neuroblastoma of the anterior mediastinum in an 80-year-old woman. Ultrastruct Pathol 19:305–310, 1995.

Schmidt D, Harms D, Burdach S: Malignant peripheral neuroectodermal tumors of childhood and adolescence. Virchows Archiv [A] 406:351–365, 1985.

Serrano-Olmo J, Tang CK, Seidmon EJ, et al: Neuroblastoma as a prominent component of a mixed germ cell tumor of testis. Cancer 72:3271–3276, 1993.

Taxy JB: Electron microscopy in the diagnosis of neuroblastoma. Arch Pathol Lab Med 104:355–360, 1980.

Taxy JB, Hidvegi DF: Olfactory neuroblastoma. An ultrastructural study. Cancer 39:131–138, 1977.

Taxy JB, Bharani NK, Mills SE, et al: The spectrum of olfactory neural tumors. A light microscopic, immunohistochemical and ultrastructural analysis. Am J Surg Pathol 10:687–695, 1986.

Ewing's Sarcoma and Primitive Neuroectodermal Tumor (PNET)

Ayala AG, Mackay B: Ewing's sarcoma: An ultrastructural study. *1977 Year Book of Pathology*. Year Book Medical, Chicago, 1977, pp 179–185.

Contesso G, Llombart-Bosch A, Terrier P, et al: Does malignant small round cell tumor of the thoracopulmonary region (Askin tumor) constitute a clinicopathologic entity? An analysis of 30 cases with immunohistochemical and electron-microscopic support treated at the Institute Gustav Roussy. Cancer 69:1012–1020, 1992.

Dehner LP: Primitive neuroectodermal tumor and Ewing's sarcoma [review]. Am J Surg Pathol 17:1–13, 1993.

Dickersin GR: The contributions of electron microscopy in the diagnosis and histogenesis of controversial neoplasms. Clin Lab Med 4:123–164, 1984.

Dickersin GR: Embryonic ultrastructure as a guide in the diagnosis of tumors. Ultrastruct Pathol 11:609–652, 1987.

Dickman PS, Liotta LA, Triche TJ: Ewing's sarcoma. Characterization in established cultures and evidence of its histogenesis. Lab Invest 47:375–382, 1982.

Dockhorn-Dworniczak B, Shafer KL, Dantcheva R, et al: Molecular genetic detection of t(11;22) (q24;12) translocation in Ewing sarcoma and malignant peripheral neuroectodermal tumors [German]. Pathologe 15: 103–112, 1994.

Friedman B, Gold H: Ultrastructure of Ewing's sarcoma of bone. Cancer 22:307–322, 1968.

Gillespie JJ, Roth LM, Wills ER, et al: Extraskeletal Ewing's sarcoma. Histologic and ultrastructural observations in three cases. Am J Surg Pathol 3:99–108, 1979.

Hachitanda Y, Tsuneyoshi M, Enjoji M, et al: Congenital primitive neuroectodermal tumor with epithelial and glial differentiation: An ultrastructural and immunohistochemical study. Arch Pathol Lab Med 114: 101–105, 1990.

Hicks MJ, Smith JD, Carter AB, et al: Recurrent intrapulmonary malignant small cell tumor of the thoracopulmonary region with metastasis to the oral cavity: Review of literature and case report. Ultrastruct Pathol 19:297–303, 1995.

Kadin ME, Bensch KG: The origin of Ewing's tumor. Cancer 27:257–273, 1971.

Kudo M: Neuroectodermal differentiation in "extraskeletal Ewing's sarcoma." Acta Pathol Jpn 39: 795–802, 1989.

Llombart-Bosch A, Blache R, Peydro-Olaya A: Ultrastructural study of 28 cases of Ewing's sarcoma: Typical and atypical forms. Cancer 41:1362–1373, 1978.

Llombart-Bosch A, Lacombe MJ, Peydro-Olaya A, et al: Malignant peripheral neuroectodermal tumours of bone other than Askin's neoplasm: Characterization of 14 new cases with immunohistochemistry and electron microscopy. Virchows Arch [A] 412:421–430, 1988.

Mahoney JP, Alexander RW: Ewing's sarcoma. A light- and electron-microscopic study of 21 cases. Am J Surg Pathol 2:283–298, 1978.

Markesbery WR, Challa VR: Electron microscopic findings in primitive neuroectodermal tumors of the cerebrum. Cancer 44:141–147, 1979.

Mawad JK, Mackay B, Raymond AK, et al: Electron microscopy in the diagnosis of small round cell tumors of bone. Ultrastruct Pathol 18:263–268, 1994.

Papierz W, Alwasiak J, Kolasa P, et al: Primitive neuroectodermal tumors: Ultrastructural and immunohistochemical studies. Ultrastruct Pathol 19:147–166, 1995.

Povysil C, Matejovsky Z: Ultrastructure of Ewing's tumor. Virchows Arch [A] 374:303–316, 1977.

Steiner GC, Lewis MM: Malignant round cell tumor of bone with neural differentiation (neuroectodermal tumor). Ultrastruct Pathol 12:505–512, 1988.

Steiner GC: Neuroectodermal tumor versus Ewing's sarcoma—immunohistochemical and electron microscopic observations. Curr Top Pathol 80:1–29, 1989.

Tsuneyoshi M, Yokoyama R, Hashimoto H, et al: Comparative study of neuroectodermal tumor and Ewing's sarcoma of the bone. Histopathologic, immunohistochemical and ultrastructural features. Acta Pathol Jpn 39:573–581, 1989.

Ushigome S, Shimoda T, Takaki K, et al: Immunocytochemical and ultrastructural studies of the histogenesis of Ewing's sarcoma and putatively related tumors. Cancer 64:52–62, 1989.

Wigger HJ, Salazar GH, Blane WA: Extraskeletal Ewing's sarcoma. An ultrastructural study. Arch Pathol Lab Med 101:446–449, 1977.

Rhabdomyosarcoma (Embryonal and Alveolar)

Churg A, Ringus J: Ultrastructural observations on the histogenesis of alveolar rhabdomyosarcoma. Cancer 41:1355–1361, 1978.

Dickersin GR: Embryonic ultrastructure as a guide in the diagnosis of tumors. Ultrastruct Pathol 11:609–652, 1987.

Erlandson RA: The ultrastructural distinction between rhabdomyosarcoma and other differentiated "sarcomas." Ultrastruct Pathol 11:83–101, 1987.

Horvat BI, Caines M, Fisher ER: The ultrastructure of rhabdomyosarcoma. Am J Clin Pathol 53:555–564, 1970.

LaValle Bundtzen J, Norback DH: The ultrastructure of poorly differentiated rhabdomyosarcoma: A case report and literature review. Hum Pathol 13:301–313, 1982.

Leuschner I, Newton WA Jr, Schmidt D, et al: Spindle cell variants of embryonal rhabdomyosarcoma in the paratesticular region. A report of the Intergroup Rhabdomyosarcoma Study. Am J Surg Pathol 17:221–230, 1993. Published erratum appears in Am J Surg Pathol 17:858, 1993.

Lombardi L, Pilotti S: Ultrastructural characterization of poorly differentiated rhabdomyosarcomas. Ultrastruct Pathol 17:669–680, 1993.

Mierau GW, Favara BE: Rhabdomyosarcoma in children: Ultrastructural study of 31 cases. Cancer 46: 2035–2040, 1980.

Morales AR, Fine G, Horn RC Jr: Rhabdomyosarcoma: An ultrastructural appraisal. Pathol Annu 7:81–106, 1972.

Prince FP: Ultrastructural aspects of myogenesis found in neoplasms. Acta Neuropathol (Berl) 43:315–320, 1981.

Seidal T, Kindblom L-G: The ultrastructure of alveolar and embryonal rhabdomyosarcoma. A correlative light and electron microscopic study of 17 cases. Acta Pathol Microbiol Immunol Scand [A] 92:231–248, 1984.

Tsokos M: The diagnosis and classification of childhood rhabdomyosarcoma [review]. Semin Diag Pathol 11: 26–38, 1994.

Rhabdoid Tumor

Kodet R, Newton WA, Sachs N, et al: Rhabdoid tumors of soft tissues: A clinicopathologic study of 26 cases enrolled on the intergroup rhabdomyosarcoma study. Hum Pathol 22:674–684, 1991.

Litman DA, Bhuta S, Barsky SH: Synchronous occurrence of malignant rhabdoid tumor two decades after Wilms' tumor irradiation. Am J Surg Pathol 17:729–737, 1993.

Ota S, Crabbe DC, Tran TN, et al: Malignant rhabdoid tumor. A study with two established cell lines. Cancer 71:2862–2872, 1993.

Parham DM, Weeks DA, Beckwith JB: The clinicopathologic spectrum of putative extrarenal rhabdoid tumors. An analysis of 42 cases studied with immunohistochemistry or electron microscopy. Am J Surg Pathol 18:1010–1029, 1994.

Perez-Atayde AR, Newbury R, Fletcher JA, et al: Congenital "neurovascular hamartoma" of the skin. A possible marker of malignant rhabdoid tumor. Am J Surg Pathol 18:1030–1038, 1994.

Tsokos M, Kouraklis G, Chandra RS, et al: Malignant rhabdoid tumor of the kidney and soft tissues. Evidence for a diverse morphological and immunocytochemical phenotype. Arch Pathol Lab Med 113:115–120, 1989.

Tsuneyoshi M, Daimaru Y, Hashimoto H, et al: The existence of rhabdoid cells in specific soft tissue sarcomas. Histopathological, ultrastructural and immunohistochemical evidence. Virchows Arch [A] 411:509–514, 1987.

Weeks DA, Beckwith JB, Mierau GW: Rhabdoid tumor. An entity or a phenotype [editorial]? Arch Pathol Lab Med 113:113–114, 1989.

Weeks DA, Beckwith JB, Mierau GW, et al: Rhabdoid tumor of kidney. A report of 111 cases from the National Wilms' Tumor Study Pathology Center. Am J Surg Pathol 13:439–458, 1989.

Weeks DA, Beckwith JB, Mierau GW, et al: Renal neoplasms mimicking rhabdoid tumor of the kidney. A report from the National Wilms' Tumor Study Pathology Center. Am J Surg Pathol 15:1042–1054, 1991.

Weeks DA, Malott RL, Zuppan CW, et al: Primitive cerebral tumor with rhabdoid features: A case of phenotypic rhabdoid tumor of the central nervous system. Ultrastruct Pathol 18:23–28, 1994. Published erratum appears in Ultrastruct Pathol 18:459, 1994.

Nephroblastoma (Wilms' Tumor)

Badini A, Buffa D, Micheletti V, et al: Morphologic studies using the optical and electron microscope and short histogenetic considerations on 8 cases of Wilms' tumor. Pathologica 68:137–151, 1976.

Balsaver A, Gibley C, Tessmer C: Ultrastructural studies in Wilms' tumor. Cancer 22:417–427, 1968.

Dickersin GR, Colvin RB: Pathology of renal tumors. In Skinner DG, Lieskovsky G, eds: *Genitourinary Cancer.* WB Saunders, Philadelphia, 1988, pp 130–142.

Litman DA, Bhuta S, Barsky SH: Synchronous occurrence of malignant rhabdoid tumor two decades after Wilms' tumor irradiation. Am J Surg Pathol 17:729–737, 1993.

Pinchuk VG, Chudokov VG, Goldschmidt B, et al: Morphology and ultrastructure of Wilms' tumor in children. Arch Pathol 36:42–48, 1974.

Rousseau MF, Nabarra B: Embryonic kidney tumors: Ultrastructure of nephroblastoma. Virchows Arch [A] 363:149–162, 1974.

Schmidt D, Dickersin GR,Vawter GF, et al: Wilms' tumor: Review of ultrastructure and histogenesis. Pathobiol Annu 12:281–300, 1982.

Tannenbaum M: Ultrastructural pathology of human renal cell tumors. Pathol Annu 6:249–277, 1971.

Tremblay M: Ultrastructure of a Wilms' tumor and myogenesis. J Pathol 105:269–277, 1971.

Williams AO, Ajayi O: Ultrastructure of Wilms' tumor (nephroblastoma). Exp Mol Pathol 24:35–47, 1976.

Lymphoma and Plasmacytoma

Bessie M: *Living Blood Cells and Their Ultrastructure.* Springer-Verlag, Berlin, 1973.

Dickersin GR: Electron microscopy of leukemias and lymphomas. Clin Lab Med 7(1): 199–247, 1987.

Henry K: Electron microscopy in the non-Hodgkin's lymphomata. Br J Cancer 31:73–93, 1975.

Kaiserling E: Ultrastructure of non-Hodgkin's lymphomas. In Lennert K, ed. *Malignant Lymphomas.* Springer-Verlag, Berlin, 1978, pp 471–528.

Mori Y, Lennert K: *Electron Microscopic Atlas of Lymph Node Cytology and Pathology.* Springer-Verlag, Berlin, 1969.

Said JW, Hargreaves HK, Pinkus GS: Non-Hodgkin's lymphomas: An ultrastructural study correlating morphology with immunologic cell type. Cancer 44: 504–528, 1979.

Zucker-Franklin D, Greaves MF, Grossi CE, et al: *Atlas of Blood Cells. Function and Pathology,* vol 2. Lea and Febiger, Philadelphia, 1981, pp 347–556.

Desmoplastic Small Round Cell Tumor with Divergent Differentiation

Dorsey BV, Benjamin LE, Rauscher F, et al: Intra-abdominal desmoplastic small round-cell tumor: Expansion of the pathologic profile. Mod Pathol 9: 703–709, 1996.

Gerald WL, Rosai J: Case 2: Desmoplastic small cell tumor with divergent differentiation. Pediatr Pathol 9: 117–183, 1989.

Gerald WL, Miller HK, Battifora H, et al: Intra-abdominal desmoplastic small round-cell tumor. Report of 19 cases of a distinctive type of high-grade polyphenotypic malignancy affecting young individuals. Am J Surg Pathol 15:499–513, 1991.

Gonzalez-Crussi F, Crawford SE, Sun C-CJ: Intra-abdominal desmoplastic small-cell tumors with divergent differentiation. Observations on three cases of childhood. Am J Surg Pathol 14:633–642, 1990.

Ordonez NG, Zirkin R, Bloom RE: Malignant small-cell epithelial tumor of the peritoneum coexpressing mesenchymal-type intermediate filaments. Am J Surg Pathol 13:413–421, 1989.

Ordonez NG, el-Naggar AK, Ro JY, et al: Intra-abdominal desmoplastic small cell tumor: A light microscopic, immunocytochemical, ultrastructural, and flow cytometric study. Hum Pathol 24:850–865, 1993. Comment in Hum Pathol 25:109–110, 1994.

Variend S, Gerrard M, Norris PD, et al: Intra-abdominal neuroectodermal tumour of childhood with divergent differentiation. Histopathology 18:45–51, 1991.

Wills EJ: Peritoneal desmoplastic small round cell tumors with divergent differentiation: A review. Ultrastruct Pathol 17:295–306, 1993.

Young RH, Eichhorn JH, Dickersin GR, et al: Ovarian involvement by the intra-abdominal desmoplastic small round cell tumor with divergent differentiation: A report of three cases. Hum Pathol 23:454–464, 1992.

Small Cell Osteosarcoma

Dickersin GR, Rosenberg AE: The ultrastructure of small-cell osteosarcoma, with a review of the light microscopy and differential diagnosis. Hum Pathol 22:267–275, 1991.

Mawad JK, Mackay B, Raymond AK: An ultrastructural study of small cell osteosarcoma [abstract]. Mod Pathol 364:61A, 1988.

Mawad JK, Mackay B, Raymond AK, et al: Electron microscopy in the diagnosis of small round cell tumors of bone. Ultrastruct Pathol 18:263–268, 1994.

Papadimitriou JC, Drachenberg CB: Ultrastructural features of the matrix of small cell osteosarcoma [letter]. Hum Pathol 25:430, 1994.

Ringus JC, Riddell RH: Small cell osteosarcoma: Ultrastructural description and differentiation from atypical Ewing's sarcoma [abstract]. Lab Invest 44:55A, 1981.

Roessner A, Grundmann E: Cytogenesis and histogenesis of malignant and semimalignant bone tumors. Stuttgart, Gustav Fischer Verlag, 1984, pp 60–63.

Roessner A, Immenkamp M, Hiddemann W, et al: Case report 331. Skeletal Radiol 14:216–225, 1985.

Mesenchymal Chondrosarcoma

Bertoni F, Picci P, Bacchini P, et al: Mesenchymal chondrosarcoma of bone and soft tissues. Cancer 52: 533–541, 1983.

Fu YS, Kay S: A comparative ultrastructural study of mesenchymal chondrosarcoma and myxoid chondrosarcoma. Cancer 33:1531–1542, 1974.

Huvos AG, Rosen G, Dabska M, et al: Mesenchymal chondrosarcoma. A clinicopathologic analysis of 35 patients with emphasis on treatment. Cancer 51: 1230–1237, 1983.

Martinez-Tello FJ, Navas-Palacios JJ: Ultrastructural study of conventional chondrosarcomas and myxoid-and-mesenchymal-chondrosarcomas. Virchows Arch 396:197–211, 1982.

Nakashima Y, Unni KK, Shives TC, et al: Mesenchymal chondrosarcoma of bone and soft tissue. A review of 111 cases. Cancer 59:2444–2453, 1986.

Salvador AH, Beabout JW, Dahlin D: Mesenchymal chondrosarcoma—observations on 30 new cases. Cancer 28:605–615, 1971.

Sato N, Minase T, Yoshida Y, et al: An ultrastructural study of extraskeletal mesenchymal chondrosarcoma. Acta Pathol Jpn 34:1355–1363, 1984.

Steiner GC, Mirra JM, Bullough PG: Mesenchymal chondrosarcoma. A study of the ultrastructure. Cancer 32:926–939, 1973.

Swanson PE, Lillemoe TJ, Manivel C, et al: Mesenchymal chondrosarcoma. An immunohistochemical study. Arch Pathol Lab Med 114:943–948, 1990.

Ushogome S, Takakuwa T, Shinagawa T, et al: Ultrastructure of cartilagenous tumors and S-100 protein in the tumors with reference to the histogenesis of chondroblastoma, chondromyxoid fibroma and mesenchymal chondrosarcoma. Acta Pathol Jpn 34:1285–1300, 1984.

5 Leukemias

Myelocytic Leukemia

(Figures 5.1 through 5.9.)

Diagnostic criteria. (1) Cytoplasmic granules; (2) demonstrable peroxidase in the cytoplasmic granules; (3) Auer rods.

Additional points. Myeloblasts are large cells with a high nuclear–cytoplasmic ratio, predominantly euchromatic nuclei, large and multiple nucleoli, and cytoplasm with free ribosomes as the main organelle (Figure 5.1). As the blasts mature, cytoplasmic volume and organelles increase, and nuclei become more heterochromatic. The presence of primary (azurophilic) granules (Figure 5.2) in leukemic cells is the single most reliable proof of a myelocytic (or monocytic, see next section) line of differentiation. Golgi apparatus, rough endoplasmic reticulum, and perinuclear cisterna, as other components of the synthesizing and secretory system in the cells, also may appear prominent, but they are not as diagnostic as the granules themselves. Granules are numerous and readily found in chronic forms of granulocytic leukemia, in which there is neutrophilia, basophilia, and eosinophilia, but they may be sparse in the acute, minimally differentiated (M0) and without differentiation (M1) blastic phases of the disease (Figures 5.1 and 5.3 through 5.5). In these phases, it may be necessary to identify myeloperoxidase in the granules, in order to exclude their being nonspecific, primary lysosomes, which are present in many types of cells, including lymphocytes. The method most used for demonstrating peroxidase is the diaminobenzadine (DAB) reaction, which is performed by incubating the specimen of marrow or blood cells with DAB *in vitro.* DAB does not react with the enzymes of primary lysosomes but does combine with peroxidase to form a reaction product that is electron dense and easily visualized (Figures 5.6 and 5.7). In this way, blasts that otherwise may be unclassifiable can be concluded to be myeloblasts and monoblasts if their granules and/or Golgi apparatuses, rough endoplasmic reticulum, and nuclear envelope show a positive DAB reaction. Any granules present in lymphoblasts would be DAB-negative. *Auer rods,* another diagnostic feature of myelocytic leukemia, form from the coalescence of azurophilic granules (Figures 5.8 and 5.9). Auer rods are present in myeloblasts in more than half of the cases of acute myeloblastic leukemia. In acute promyelocytic leukemia, cells having numerous Auer rods are frequent, and the Auer rods have a tubular rather than the more usual lamellar internal structure. Also, rough endoplasmic reticulum is dilated and focally forms parallel cisternae and stellate arrangements. Myeloblasts and promyelocytes may have aggregates of cytoplasmic filaments. Nuclei of these cells often have deep infoldings. Monocytes (see next section) are also present in most cases of chronic myelocytic leukemia.

(Text continues on page 227)

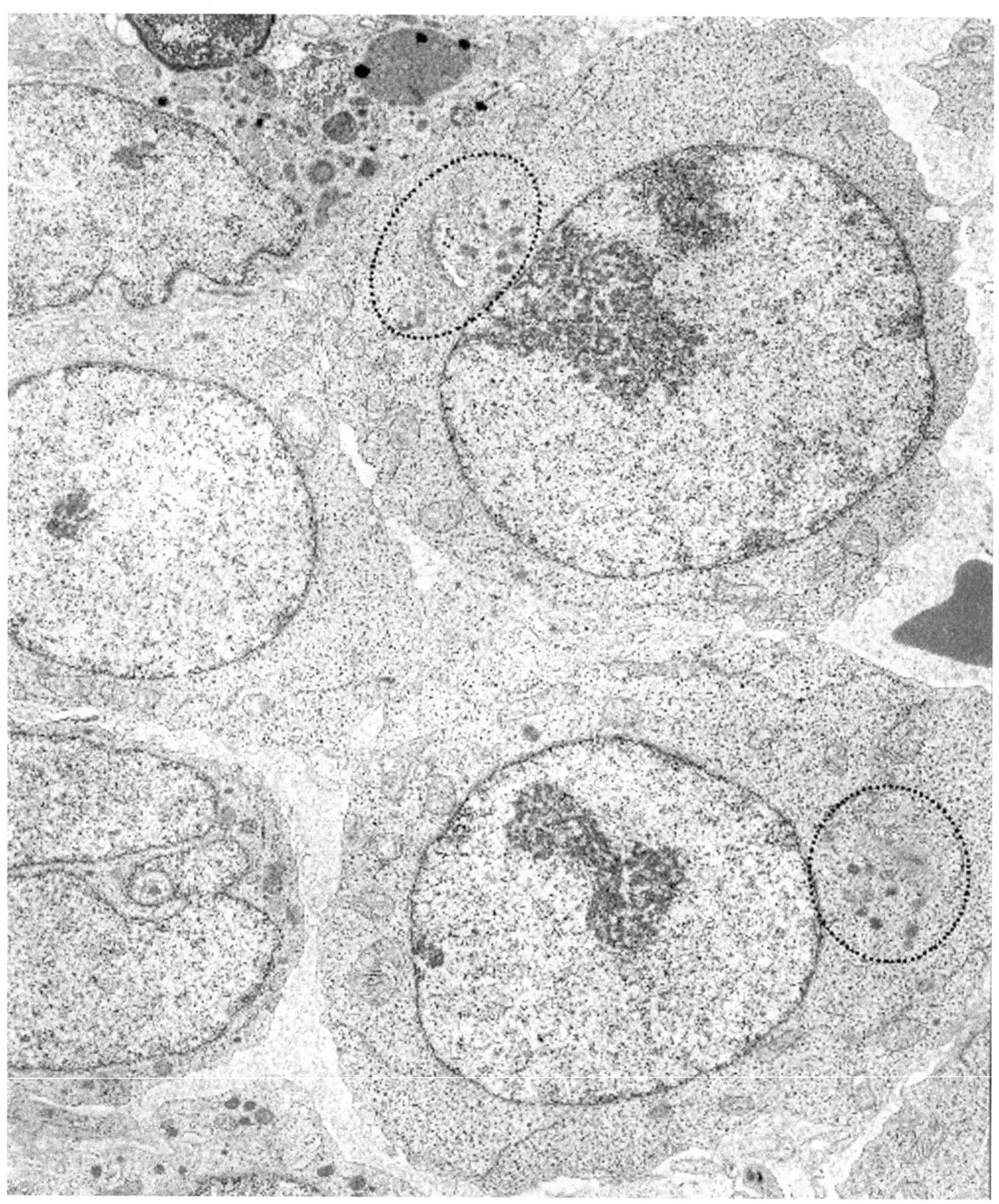

Figure 5.1. Acute myelocytic leukemia (bone marrow). Several myeloblasts contain only a few granules in association with small Golgi apparatuses (*encircled zones*). Otherwise, the cytoplasm is composed predominantly of free ribosomes. Note the euchromatic nuclei and large, open nucleoli. (× 8500)

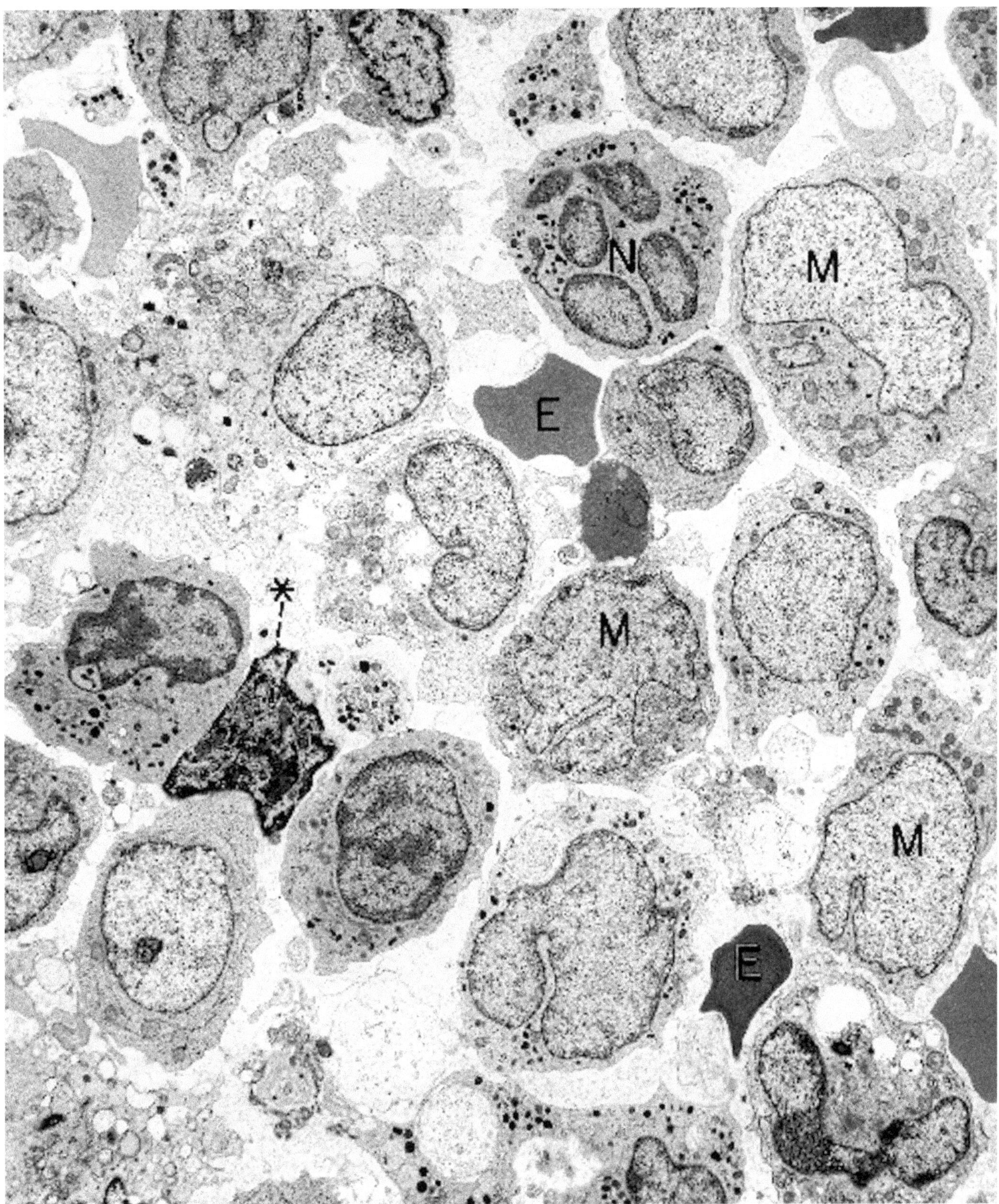

Figure 5.2. Acute myelocytic leukemia (bone marrow). This field of marrow illustrates a range of cells in the granulocytic series, from myeloblasts (M) to neutrophils (N). The blasts contain few or no granules, and the more mature cells have both large primary (azurophilic) and small secondary (specific) granules. E = erythrocytes; * = dying cell. (× 4750)

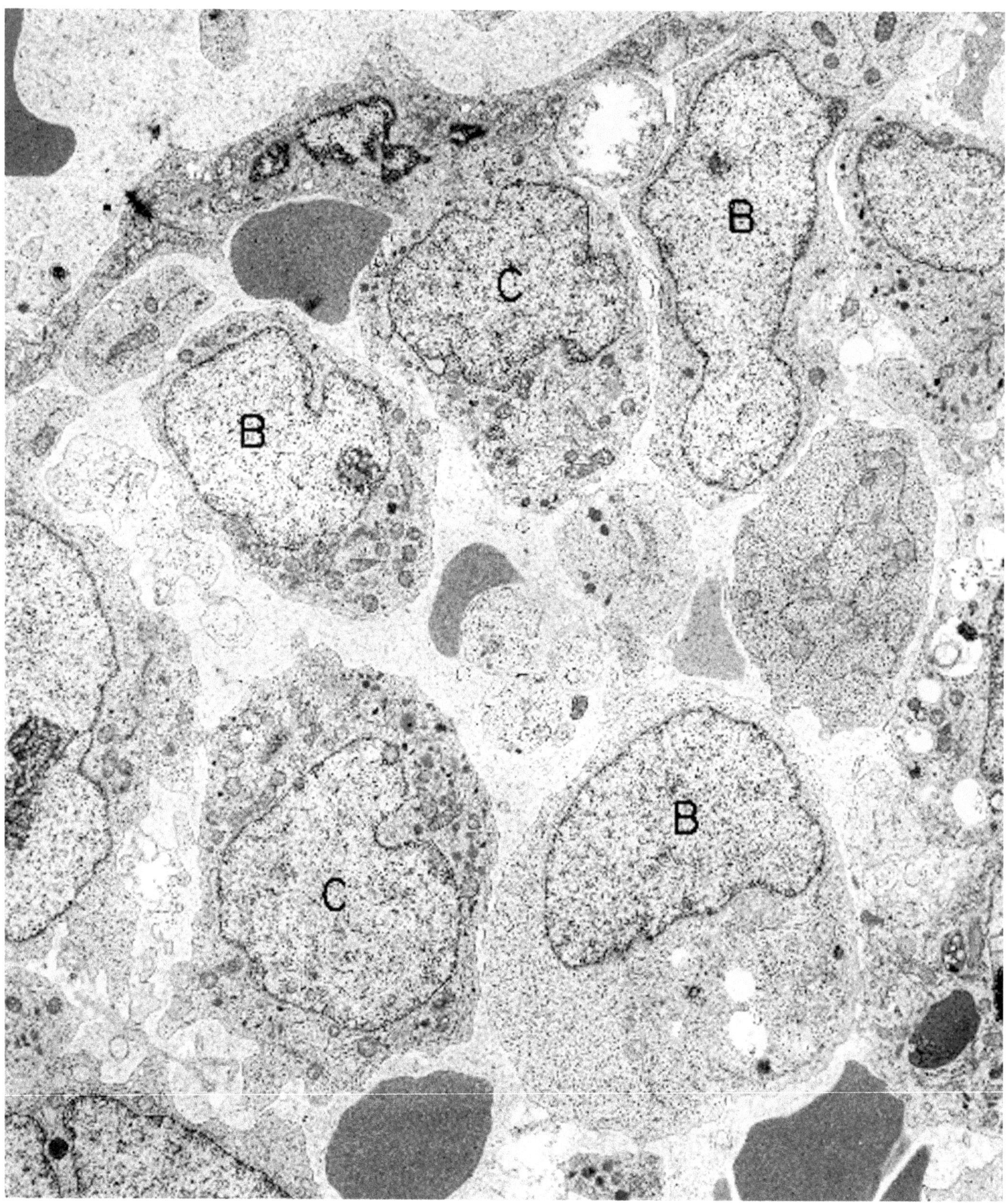

Figure 5.3. Acute myelocytic leukemia (bone marrow). This is a higher magnification of the same marrow as depicted in Figure 5.2, illustrating a field of immature granulocytic series cells. B = myeloblasts; C = myelocytes. (× 6250)

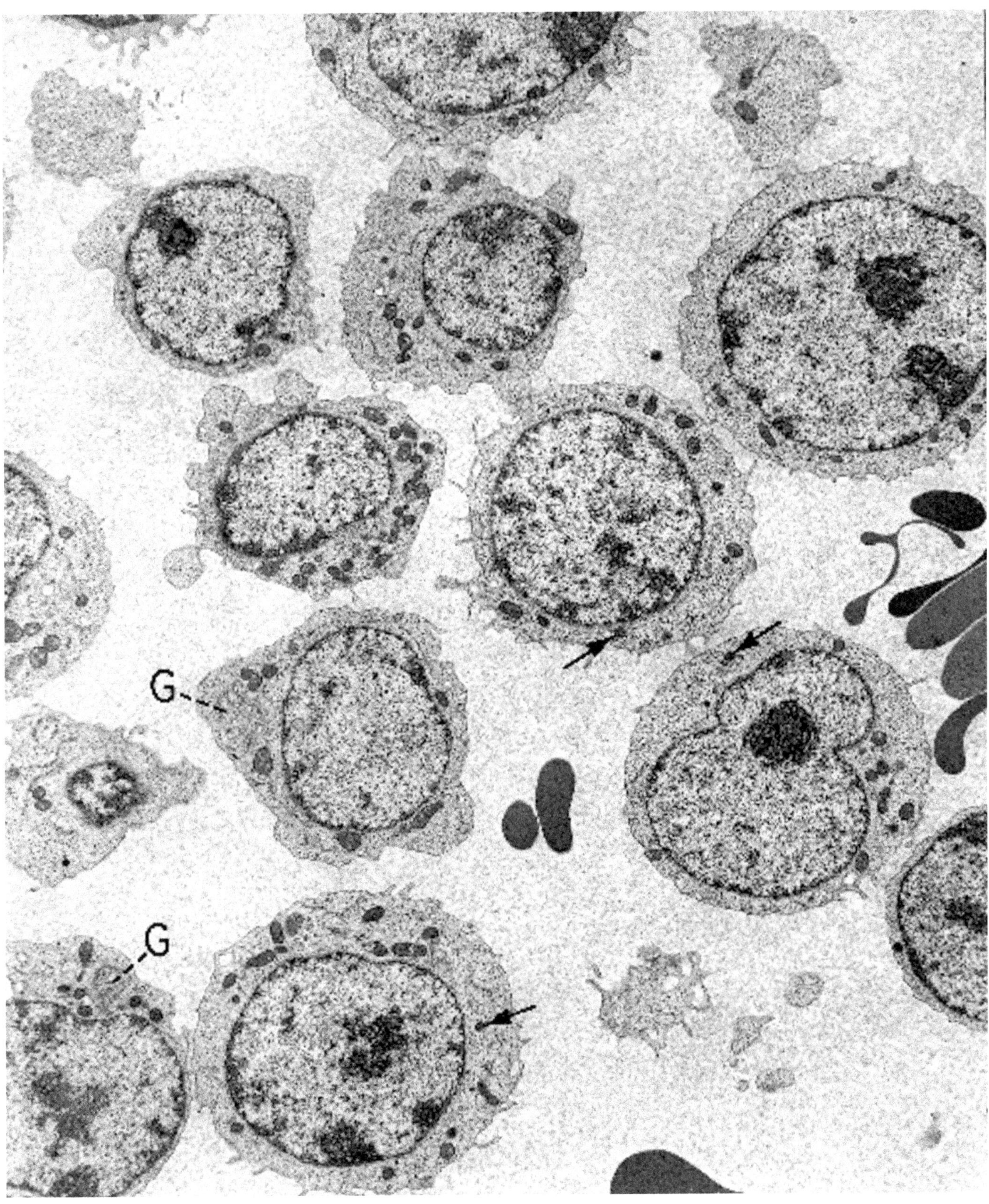

Figure 5.4. Acute myelocytic leukemia (peripheral blood). Small amounts of heterochromatin are present in the nuclei of these cells and could be responsible for an erroneous interpretation that the cells are lymphoblasts. Infrequent Golgi apparatuses (G) and only rare granules (*arrows*) can add to the diagnostic difficulties. (× 5130)

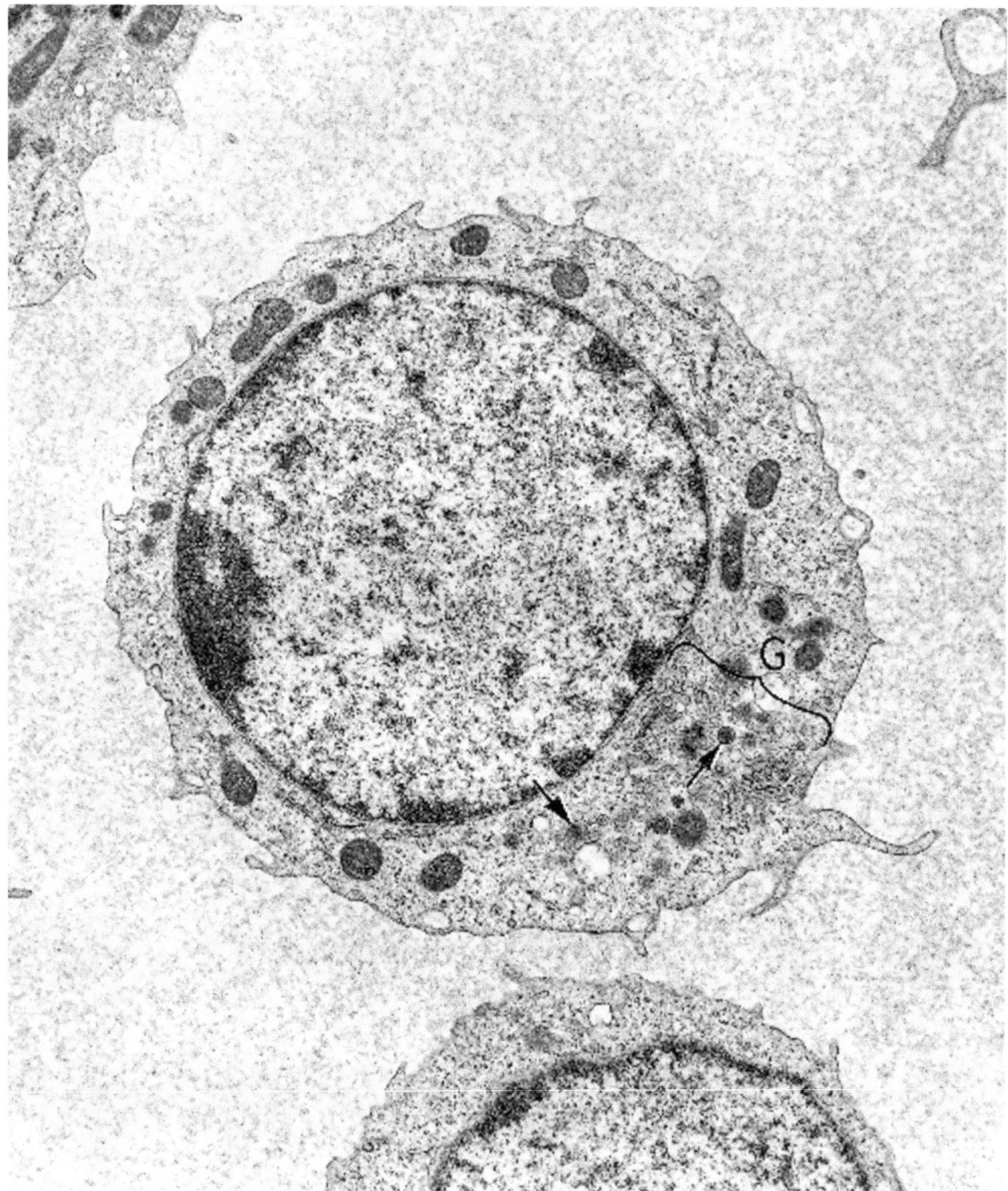

Figure 5.5. Acute myelocytic leukemia (peripheral blood). This is the same sample illustrated in Figure 5.4, but this cell has a more completely developed Golgi apparatus (G) and a few more granules (*arrows*). (× 11,500)

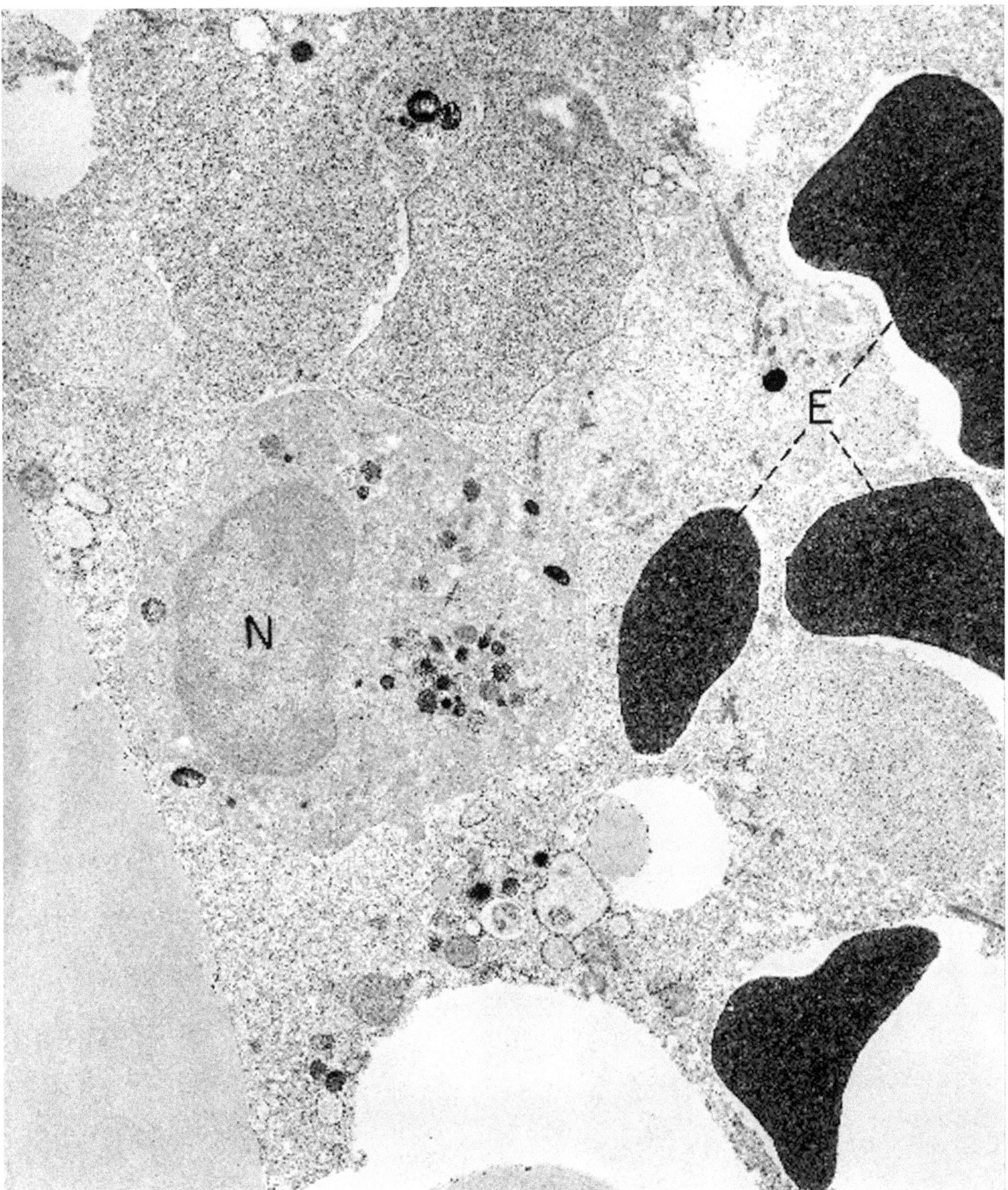

Figure 5.6. Acute myelocytic leukemia (bone marrow). A positive DAB-peroxidase reaction in the granules of this immature hematopoietic cell establishes its myelomonocytic lineage. Other cytoplasmic organelles and nucleus (N) stain lightly by this method of chemical processing, resulting in the granules being seen in more contrast. Note that neighboring erythrocytes (E) are peroxidase-positive and therein serve as a control for this method of staining. (× 11,250)

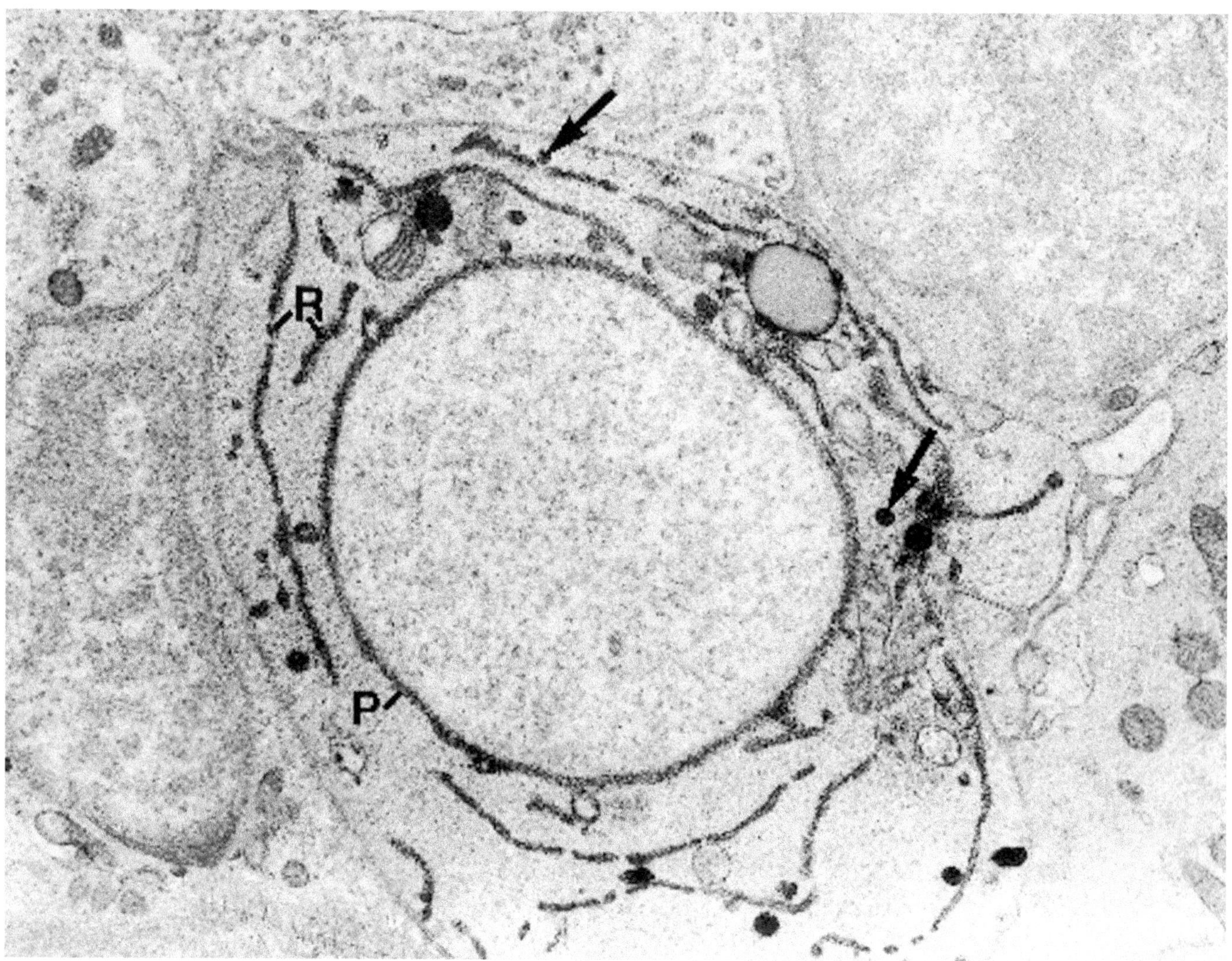

Figure 5.7. Acute myelocytic leukemia (bone marrow). This myeloblast exhibits DAB-peroxidase reaction product, not only in granules (*arrows*), but in the perinuclear envelope (P) and rough endoplasmic reticulum (R). Golgi apparatus is not present in this plane of section. (× 16,000)

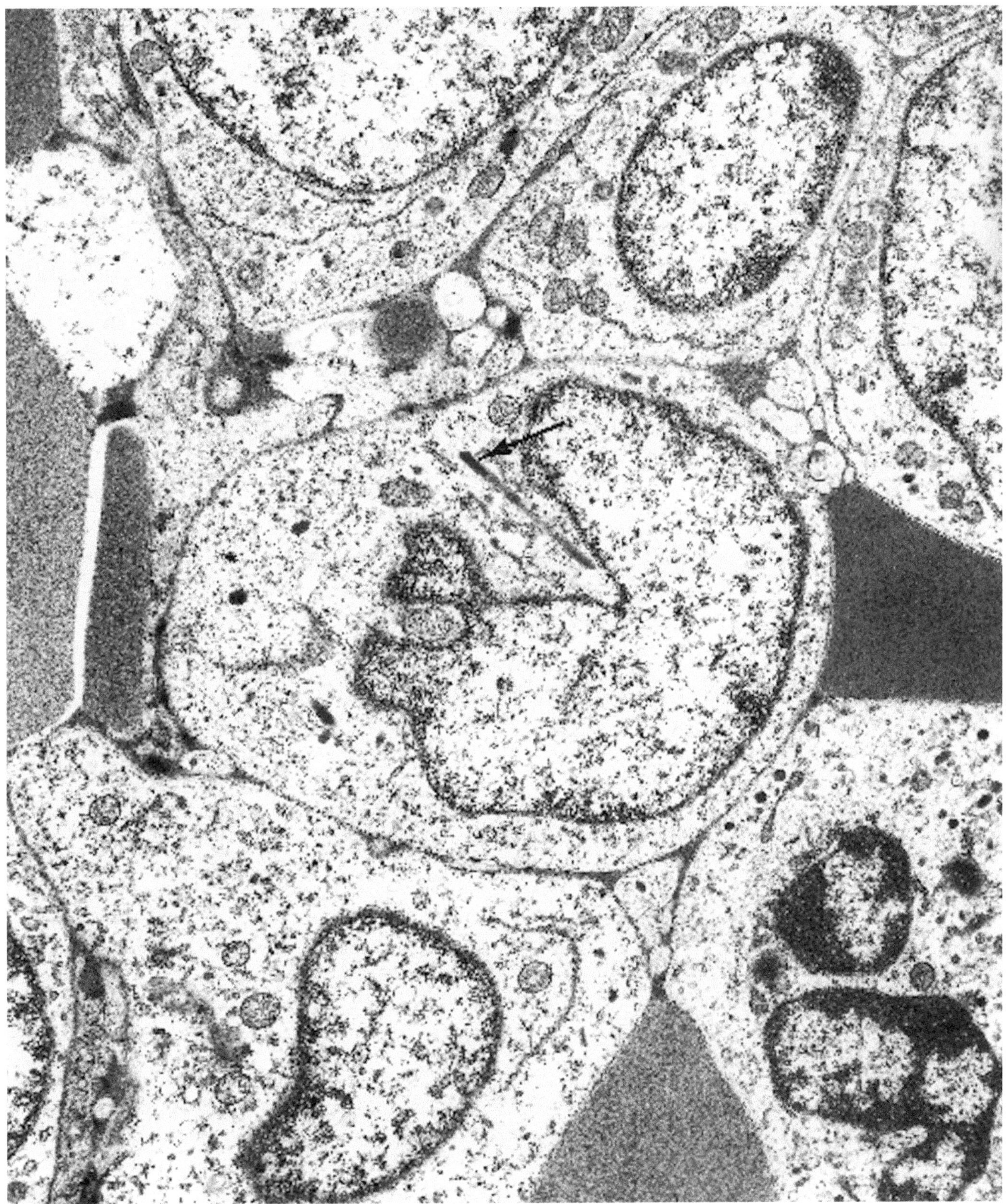

Figure 5.8. Acute myelocytic leukemia (bone marrow). The myeloblast in the center of the field contains an Auer rod (*arrow*). (× 12,500)

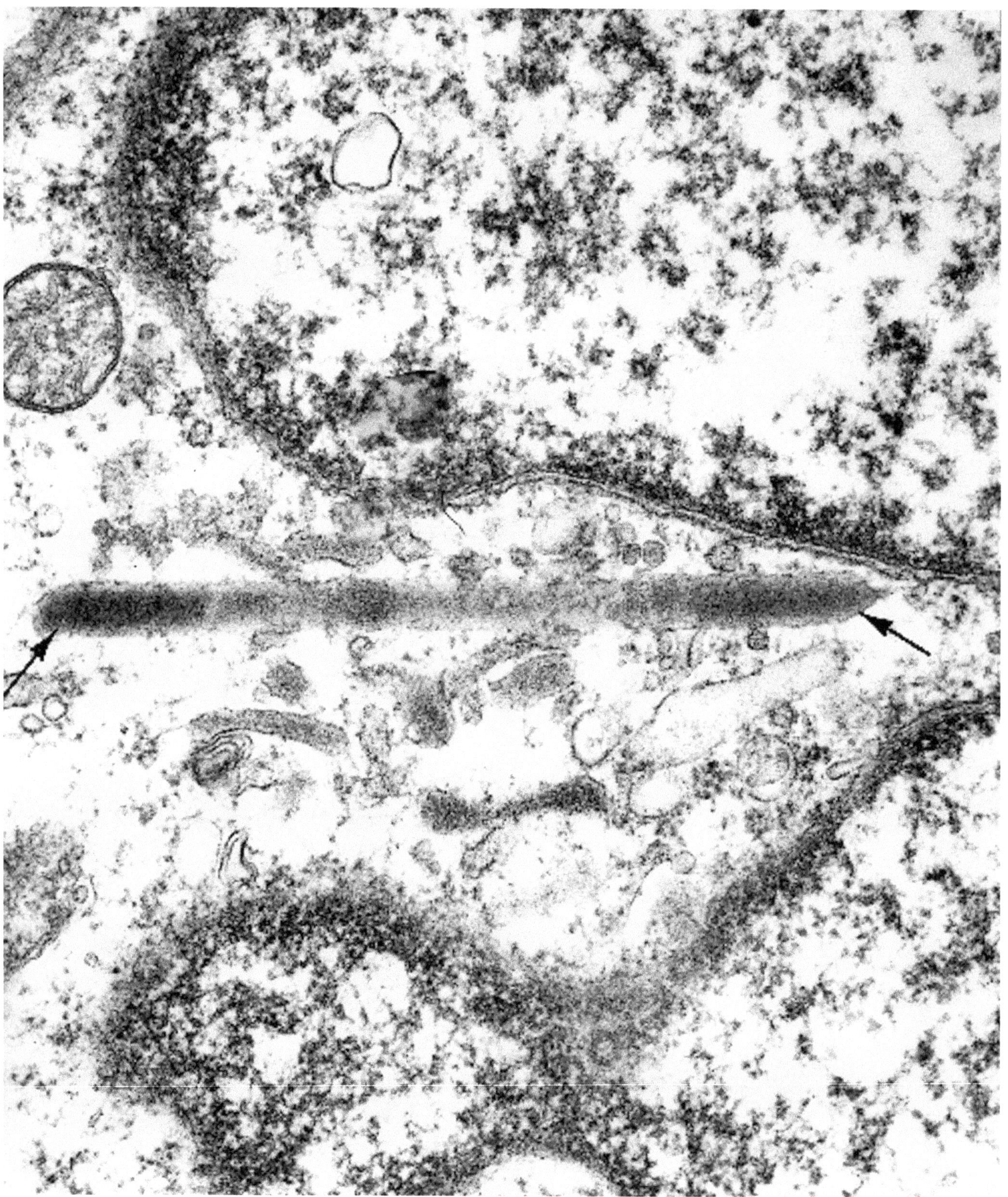

Figure 5.9. Acute myelocytic leukemia (bone marrow). Higher magnification of the Auer rod (*arrows*) in Figure 5.7 shows in better detail its limiting membrane and dense internal composition (× 57,000)

(Text continued from page 217)

Monocytic and Myelomonocytic Leukemia

(Figures 5.10 and 5.11.)

Diagnostic criteria. (1) Indented, U- or W-shaped nucleus; (2) ruffled cell surface; (3) cytoplasmic granules.

Additional points. Monoblasts are large cells with round or oval nuclei and abundant cytoplasm with free ribosomes and polysomes, a moderate number of mitochondria and cisternae of rough endoplasmic reticulum, groups of microfilaments, and varying numbers of primary granules. Auer rods are seldom present. Ribosomal lamellar type complexes (see section on hairy cell leukemia) are present in about one-fourth of cases. Nuclei are euchromatic, and nucleoli are prominent and sometimes multiple. Filopodia and pseudopods may be present at the surface of the cells. As monoblasts mature into promonocytes and monocytes, their cytoplasm develops more organelles, and nuclei become more heterochromatic, indented, and lobated. The granules in monocytes are smaller and fewer than in myelocytes. The granules are peroxidase-positive with DAB. Monocytic cells often coexist with myelocytic ones, as myelomonocytic leukemia, rather than being a pure monocytic population.

Lymphocytic Leukemia

(Figures 5.12 and 5.13.)

Diagnostic criteria. (1) Nucleus with abundant heterochromatin, especially peripherally along the nuclear envelope; (2) cytoplasm devoid of secretory granules, or a few peroxidase-negative, primary lysosomes.

Additional points. Leukemias manifesting most of the lymphoid population as small- and medium-sized lymphocytes (for example, chronic lymphocytic leukemias) usually pose no problem in diagnosis, because nuclei have a typical lymphoid pattern of heterochromatin. On the other hand, some acute lymphoblastic leukemias show a high percentage of lymphoblasts having euchromatic nuclei, and these cells may be indistinguishable from myeloblasts. As described in the section on myelocytic leukemia, the absence of peroxidase-containing granules is evidence in favor of the blasts being lymphoblasts. The chromatin pattern in most cell lines indicates whether the cell is a blast or a more mature form; euchromatin is present in blasts, and heterochromatin increases with maturation beyond blasts. However, this sequence may not always be assumed, and chromatin pattern alone may not be sufficient evidence for determining the degree of differentiation of lymphoblasts. Even the most primitive of blasts may show moderately heterochromatic nuclei. Also, cells of this type are smaller than later-stage lymphoblasts (although larger than normal, nonneoplastic, small lymphocytes) and have small or inconspicuous nucleoli. Consistent criteria for identifying the earliest blast are its high nuclear–cytoplasmic ratio and its small amount of cytoplasm. Cytoplasmic organelles consist mainly of free ribosomes and polysomes, but a Golgi apparatus and a few mitochondria and cisternae of rough endoplasmic reticulum are also present. Nuclei may be regular in contour or indented. Later-stage blasts are larger and have more cytoplasm, which often contains lipid vacuoles and, rarely, lysosomes. Nuclei are round or irregularly shaped, and nucleoli are large and multiple.

Erythrocytic Leukemia

(Figures 5.14 and 5.15.)

Diagnostic criteria. (1) Nucleus mostly euchromatic in earliest blast (proerythroblast) and heavily heterochromatic in later blast forms (polychromatophilic erythroblasts); (2) few cytoplasmic organelles, especially in the later blast forms; (3) cytoplasmic glycogen.

Additional points. The later blast forms are readily recognized because of the abundant heterochromatin resembling the nucleus of a small lymphocyte and because of the lack of many organelles in the cytoplasm. Even the ribosomes are decreased from the amount present in early erythroblasts. Invisible hemoglobin, fine particles of ferritin, and open spaces of glycogen occupy the territory between the ribosomes. Early erythroblasts contain a few mitochondria, centrioles, a Golgi apparatus, and a small number of primary lysosomes. Secondary, iron-containing lysosomes (ferritin or hemosiderin) also may be seen. In the leukemic state, the erythroblasts usually show increased ribosomes, lipid, and glycogen in the cytoplasm, and nuclei may be lobed or multilobed, variable in size, and multiple. Erythroblasts are accompanied by myeloblasts and together represent a variant of acute myelocytic leukemia in which the erythroblasts represent 30–50% of nucleated marrow cells, and the nonerythroid myeloblasts comprise another 30% or more.

(Text continues on page 234)

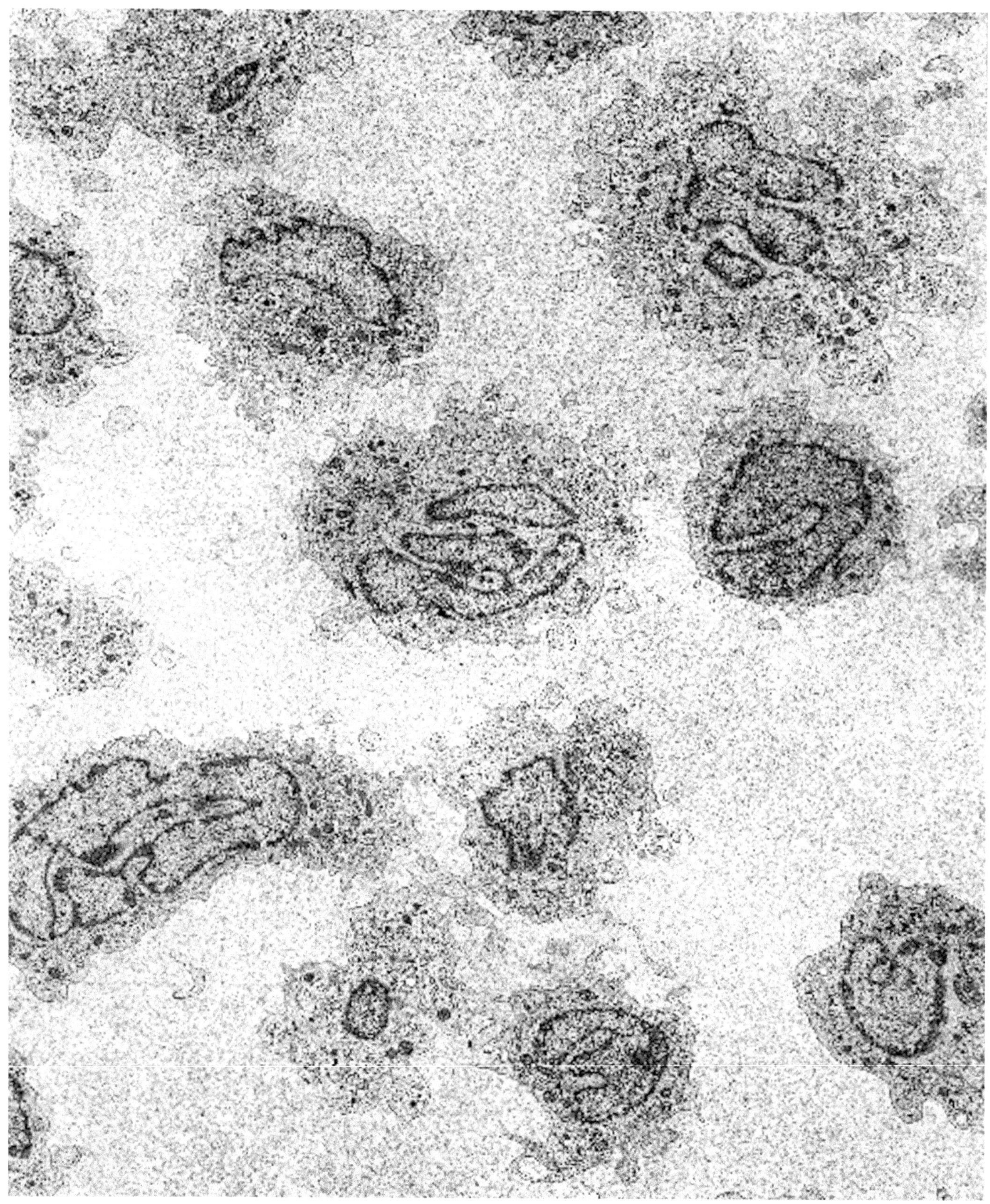

Figure 5.10. Acute monocytic leukemia (peripheral blood). These cells are promonocytes and monocytes without accompanying blasts. They exhibit characteristic ruffled (filopodia and pseudopodia) plasmalemmas, U- and W-shaped nuclei, and small cytoplasmic granules. (× 4750)

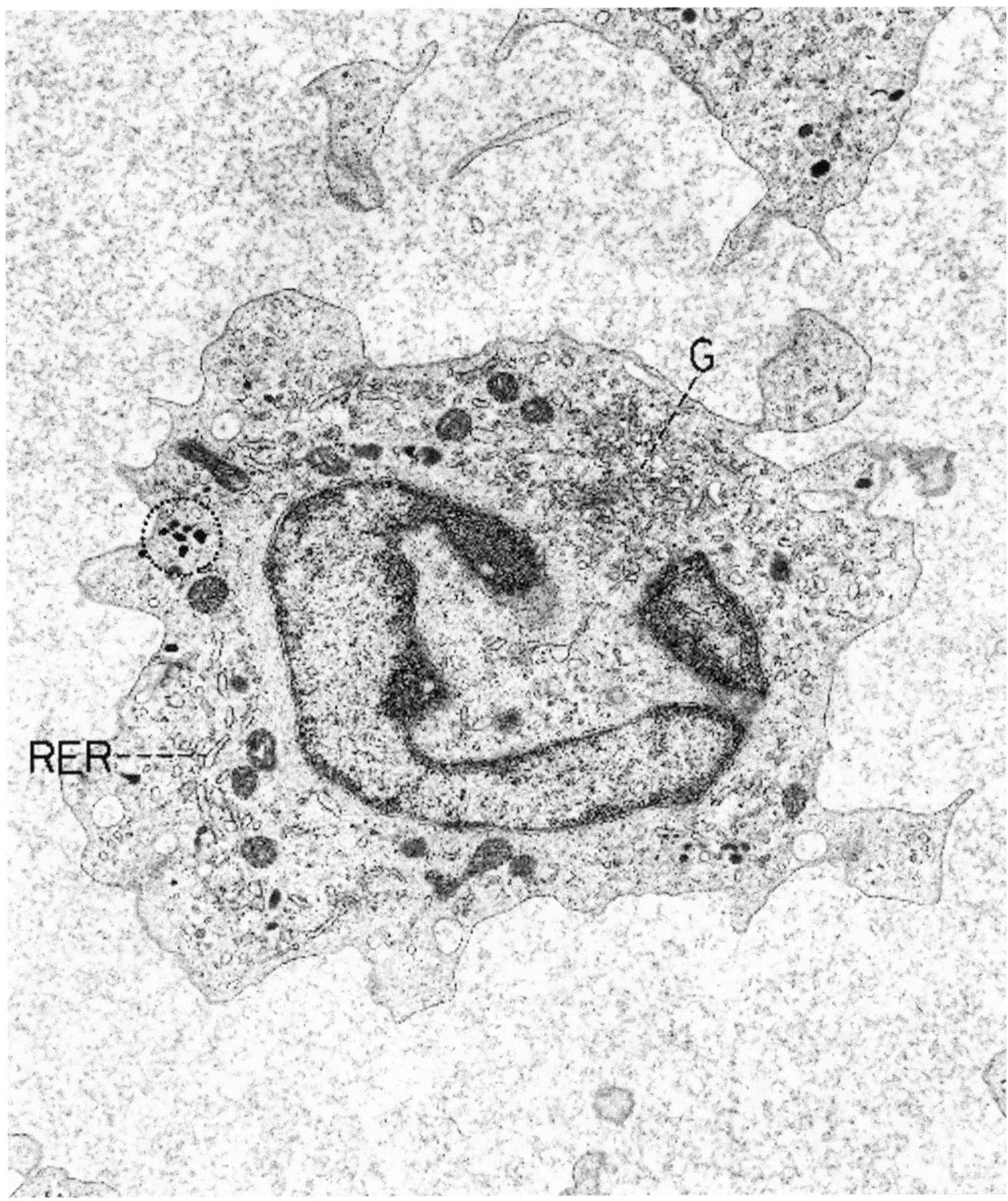

Figure 5.11. Acute monocytic leukemia (peripheral blood). Higher magnification of one of the promonocytes from the same sample as in Figure 5.10 illustrates a prominent Golgi apparatus (G) and rough endoplasmic reticulum (RER) as well as the small granules (*circle*). (× 12,375)

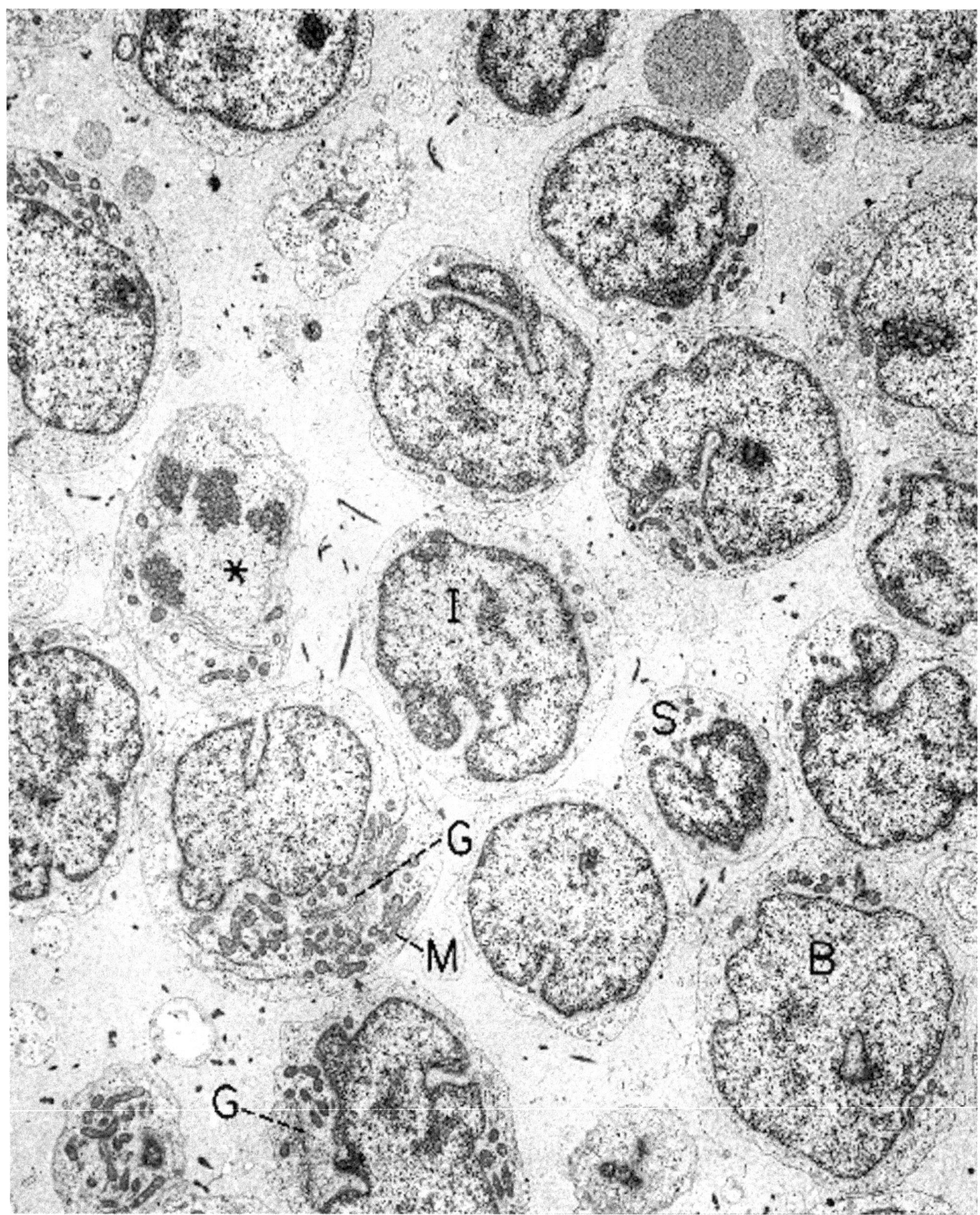

Figure 5.12. Acute lymphocytic leukemia (bone marrow). This field depicts lymphocytes in different stages of differentiation, although small lymphocytes (S) are outnumbered by intermediate lymphocytes (I) and lymphoblasts (B). All cells except the blasts have the characteristic pattern of heterochromatin subjacent to the nuclear envelope. Note that at least two of the cells have a prominent Golgi apparatus (G) and numerous mitochondria (M). One cell is undergoing mitosis (*). (× 5130)

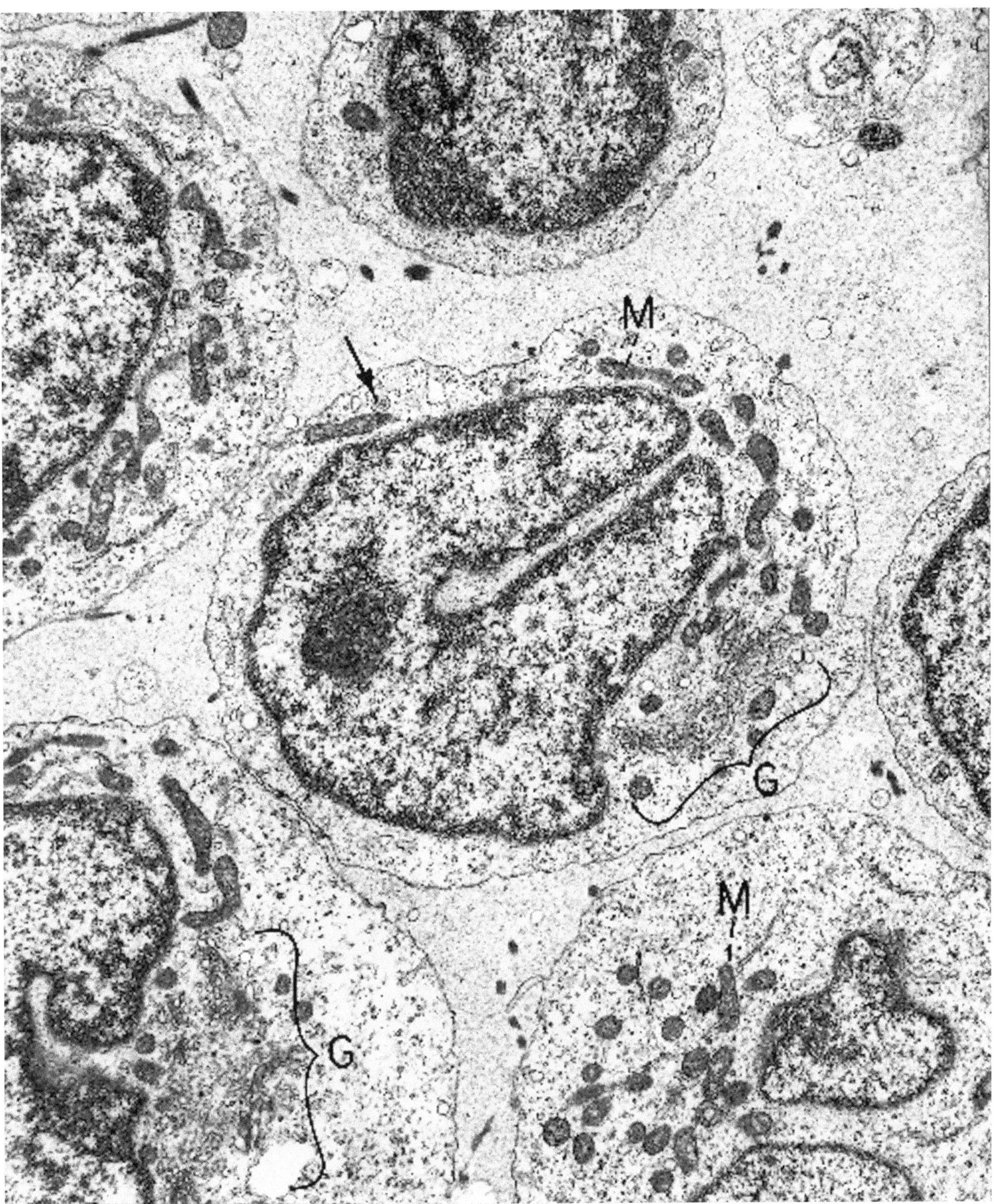

Figure 5.13. Acute lymphocytic leukemia (bone marrow). Higher magnification of the same sample illustrated in Figure 5.12 elucidates the cytoplasmic details of the lymphoid cells. In addition to a background of free ribosomes, Golgi apparatuses (G) and mitochondria (M) are also clearly discernible. A rare granule (*arrow*) is consistent with being a primary lysosome. (× 11,000)

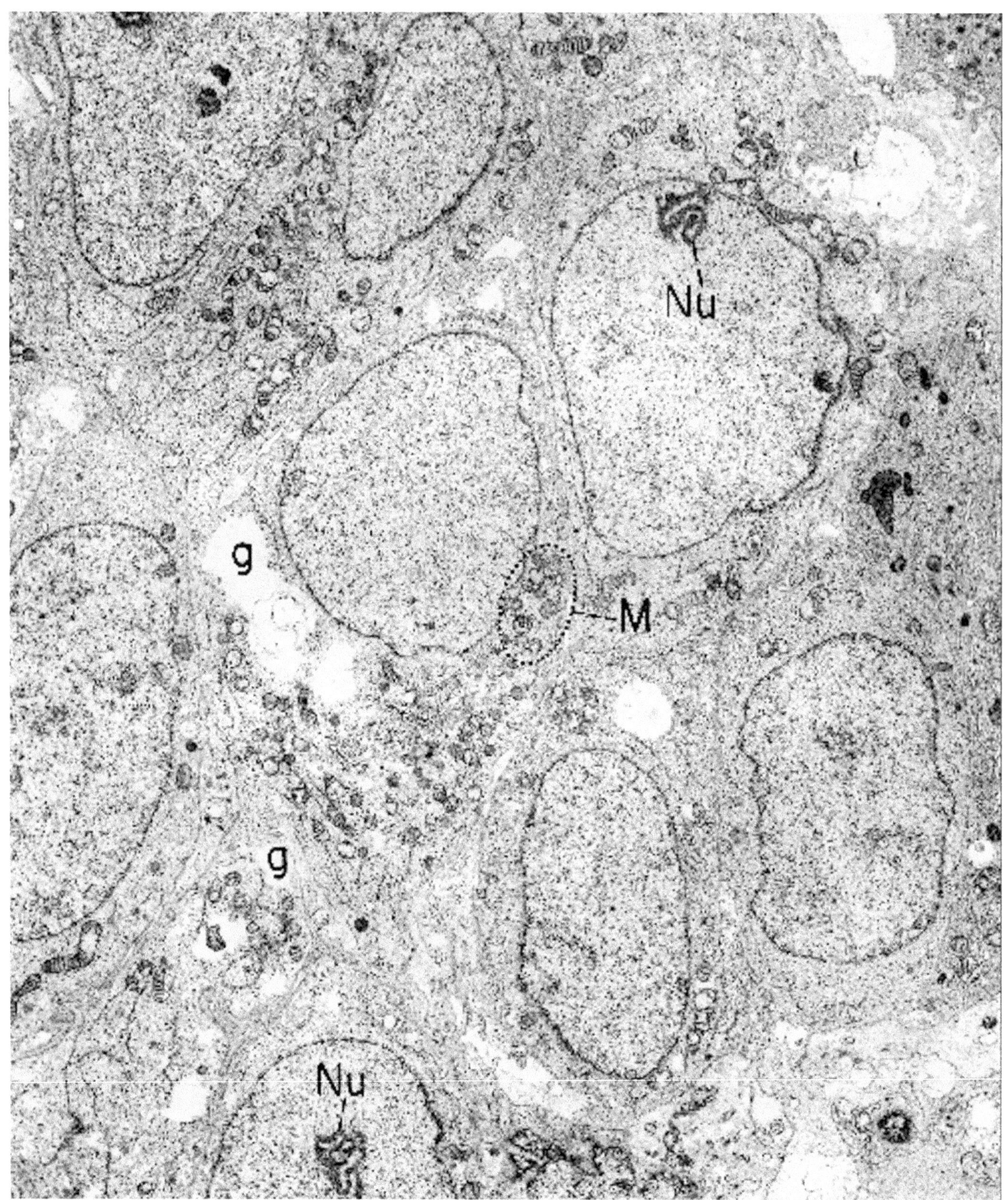

Figure 5.14. Erythroleukemia (bone marrow). A monomorphic collection of erythroblasts is seen in this field. Nuclei are euchromatic, and nucleoli consist of open nucleolonemas (Nu). Cytoplasm is rich in free ribosomes and glycogen, the latter being represented by open, clear spaces (g). Mitochondria (M) are numerous in some of the cells. (× 6100)

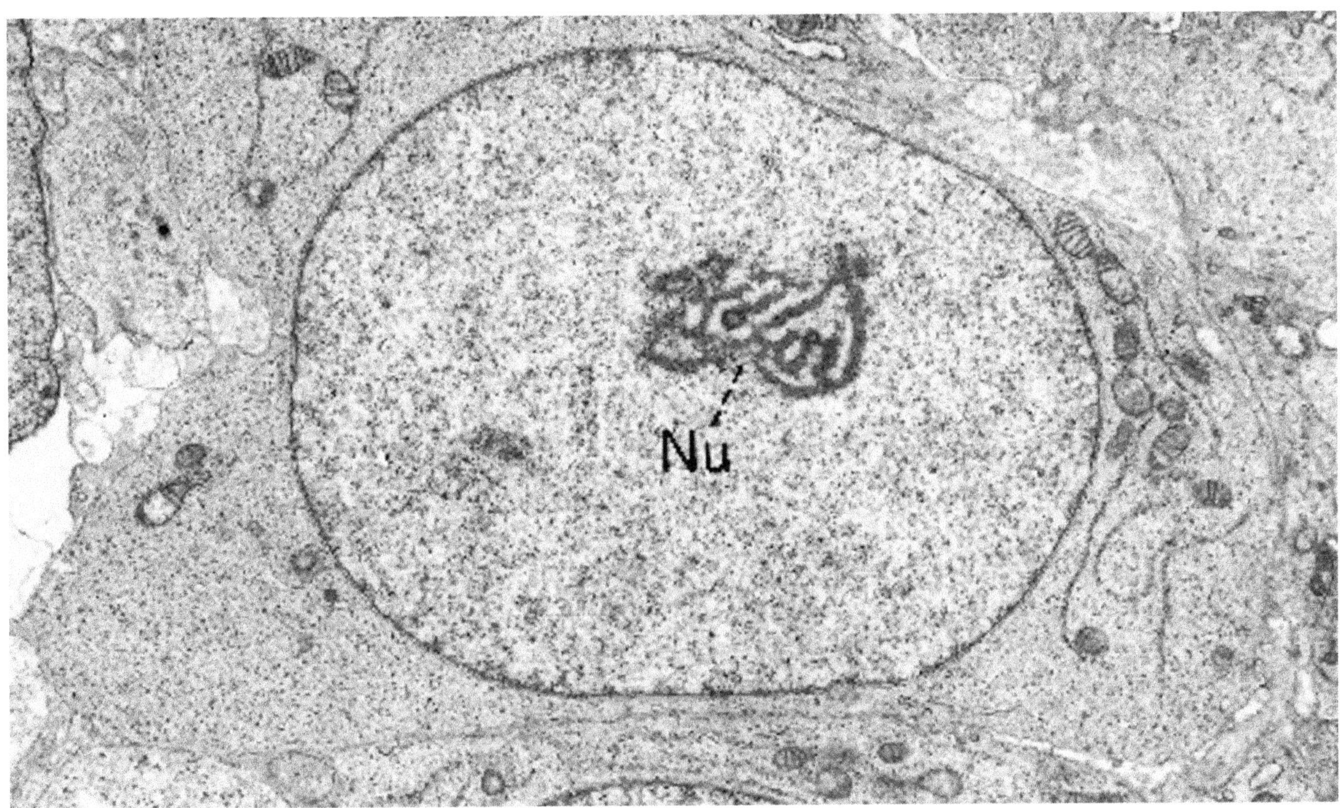

Figure 5.15. Erythroleukemia (bone marrow). Higher magnification of an erythroblast, illustrating the nucleolonema (Nu) and the predominance of free ribosomes in the cytoplasm. (× 8500)

(Text continued from page 227)

Megakaryocytic Leukemia

(Figures 5.16 through 5.20.)

Diagnostic criteria. (1) Megakaryoblasts, characterized as undifferentiated cells with a high nuclear–cytoplasmic ratio, a moderate amount of heterochromatin, and cytoplasmic protrusions; (2) megakaryocytes, usually characterized by their large size, but also may be small (micro-megakaryocytes), with (a) copious cytoplasm, (b) irregularly shaped nuclei, (c) cytoplasmic demarcation membranes and blebs (budding platelets), and (d) small dense (azurophilic) cytoplasmic granules.

Additional points. The demarcation membranes of the megakaryocyte consist of invaginations of the plasmalemma of the cell and represent platelet formation. The number of demarcation membranes increases with the maturation of the megakaryocyte. Other organelles in the cytoplasm of the megakaryoblast and megakaryocyte include many free ribosomes and polysomes, a small amount of rough endoplasmic reticulum, and a large Golgi apparatus. Platelets are normally about 2–5 μ in diameter, but in reactive and neoplastic states they may reach five or more times this size. They usually are oval and have a few pseudopods. They are devoid of a nucleus and have a busy cytoplasm that contains, in addition to small, dense (azurophilic) granules, primary lysosomes, small mitochondria, bundles of microtubules along the cell membrane, many thin and thick filaments, vesicles, glycogen, and lipid. Platelet peroxidase is demonstrable in the nuclear envelope and rough endoplasmic reticulum, using unfixed or tannic-acid-fixed specimens.

Hairy Cell Leukemia (Leukemic Reticuloendotheliosis)

(Figures 5.21 through 5.24.)

Diagnostic criteria. (1) Small-to-medium-sized lymphocytes with filopodia or villus-like (hairy) plasmalemmal projections, in samples of peripheral blood, where cells are separated by plasma and their surfaces are free (Figure 5.21), and overlapped and interdigitated projections, in samples of marrow and spleen, where cells are tightly apposed (Figure 5.22); (2) nucleus with a lymphoid chromatin pattern (abundant heterochromatin, especially along nuclear envelope); (3) *ribosome-lamellar complexes* in the cytoplasm, usually in a paranuclear location (Figure 5.22 through 5.24).

Additional points. Hairy cells are morphologically and functionally distinctive, because they are B-lymphocytes that have an ability, although weak, to phagocytize. They synthesize immunoglobulin, have a low proliferative rate, and are capable of engulfing (perhaps without digesting) erythrocytes, platelets, and latex and zymosan particles. They express monocytoid antigens, but their overall ultrastructural appearance is that of a lymphocyte; specifically, the nucleus has abundant heterochromatin, including a heavy peripheral distribution, and the cytoplasm lacks the many primary and secondary lysosomes expected in a monocyte/histiocyte. Nuclei are round, oval, or lobated, and nucleoli are small or inconspicuous. The most striking feature of the cytoplasm of hairy cells is the presence of *ribosome-lamellar complexes,* which are found in approximately half of the cases of this type of leukemia as well as in occasional cases of chronic lymphocytic leukemia (where they were first described), acute monocytic leukemia, Waldenstrom's macroglobulinemia, and Cushing's syndrome. In those cases of hairy cell leukemia having ribosome-lamellar complexes, the number of cells exhibiting the complexes ranges from less than 1% to almost 100%. The ultrastructure of these inclusions consists of cylinders of alternating and parallel rows of ribosomes and tubular lamellae, spiraling and intersecting around a central core. Other structures that are less frequently encountered in hairy cells include parallel tubular arrays in the cytoplasm, and zipper-like junctions between abutting cells.

(Text continues on page 244)

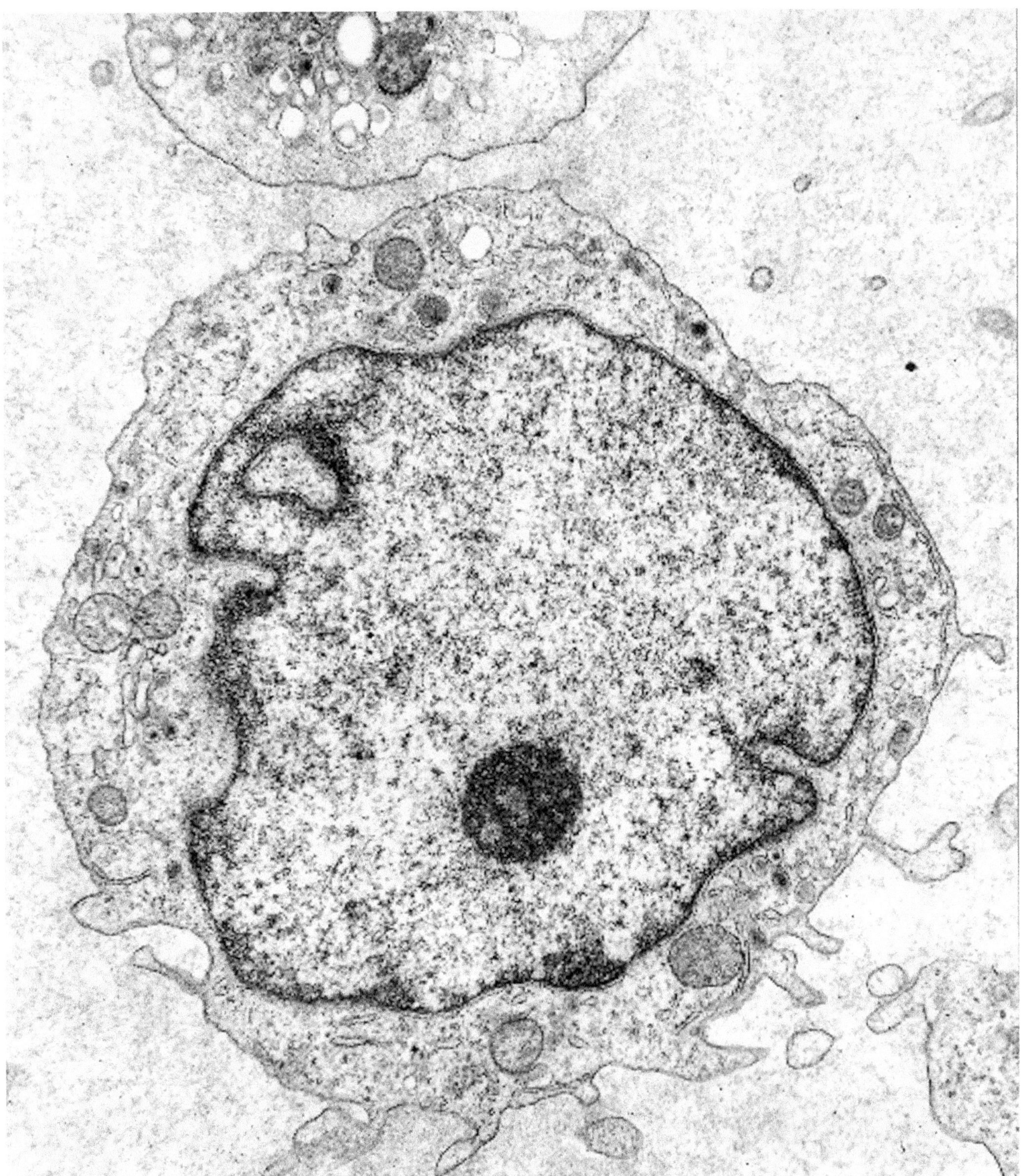

Figure 5.16. Megakaryocytic leukemia (peripheral blood). This megakaryoblast is too poorly differentiated for positive classification without the presence of neighboring megakaryocytes or special cytochemistry for platelet peroxidase. The nuclear–cytoplasmic ratio is high, the nucleus has a moderate amount of heterochromatin, there is a nonspecific complement of organelles in the cytoplasm, and the surface of the cell is focally raised into filopodia. (× 11,400) (Permission for reprinting granted by WB Saunders, Dickersin GR: Electron microscopy of leukemias and lymphomas. Clin Lab Med 7:199–247, 1987.)

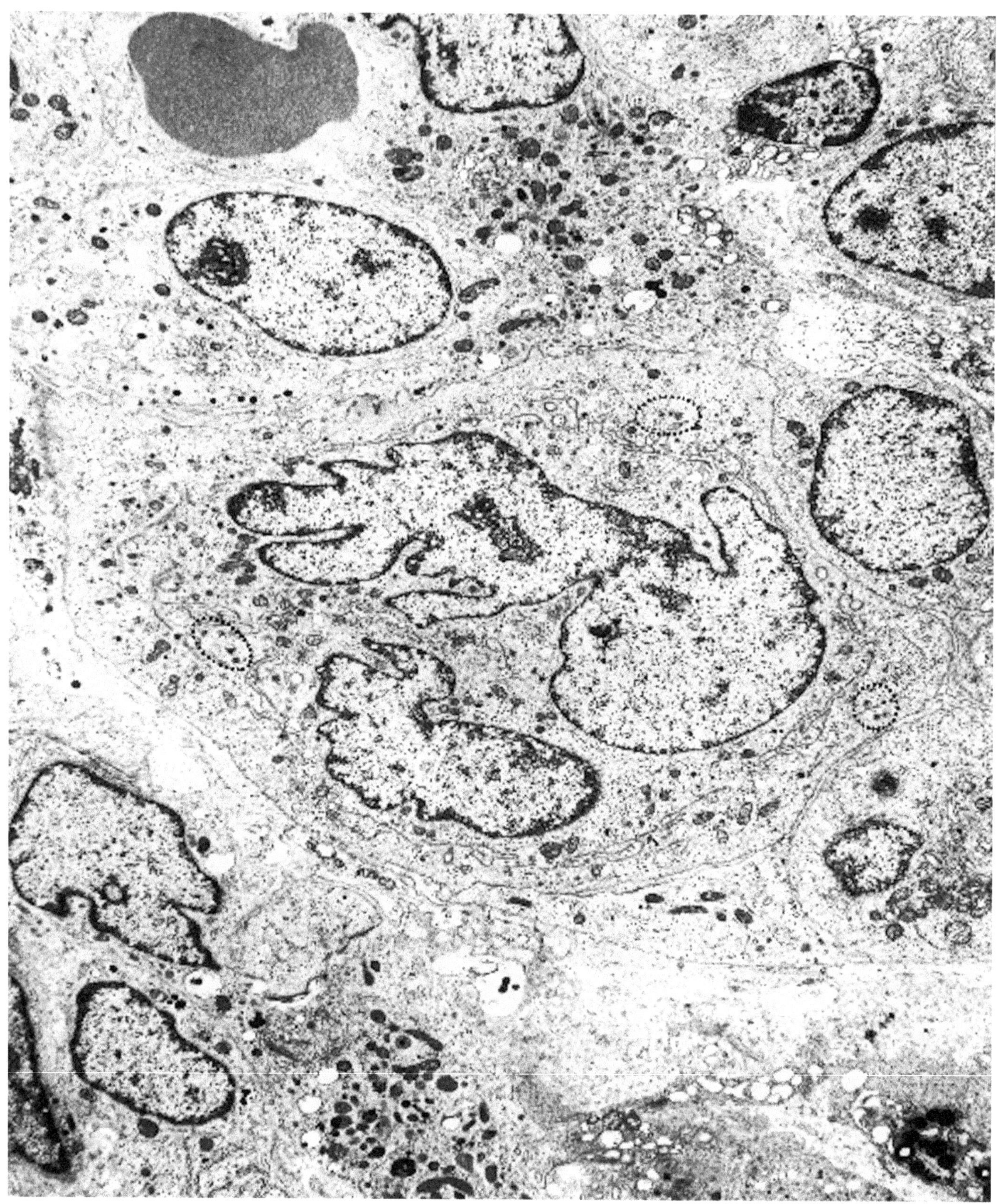

Figure 5.17. Megakaryocytic leukemia (bone marrow). A megakaryocyte is readily identifiable by its large size and irregular nucleus, contrasted with the surrounding cells of the marrow. Its cytoplasm contains many organelles, including scattered small, dense granules (*circles*), but demarcation membranes and budding platelets are not visible. (× 5320)

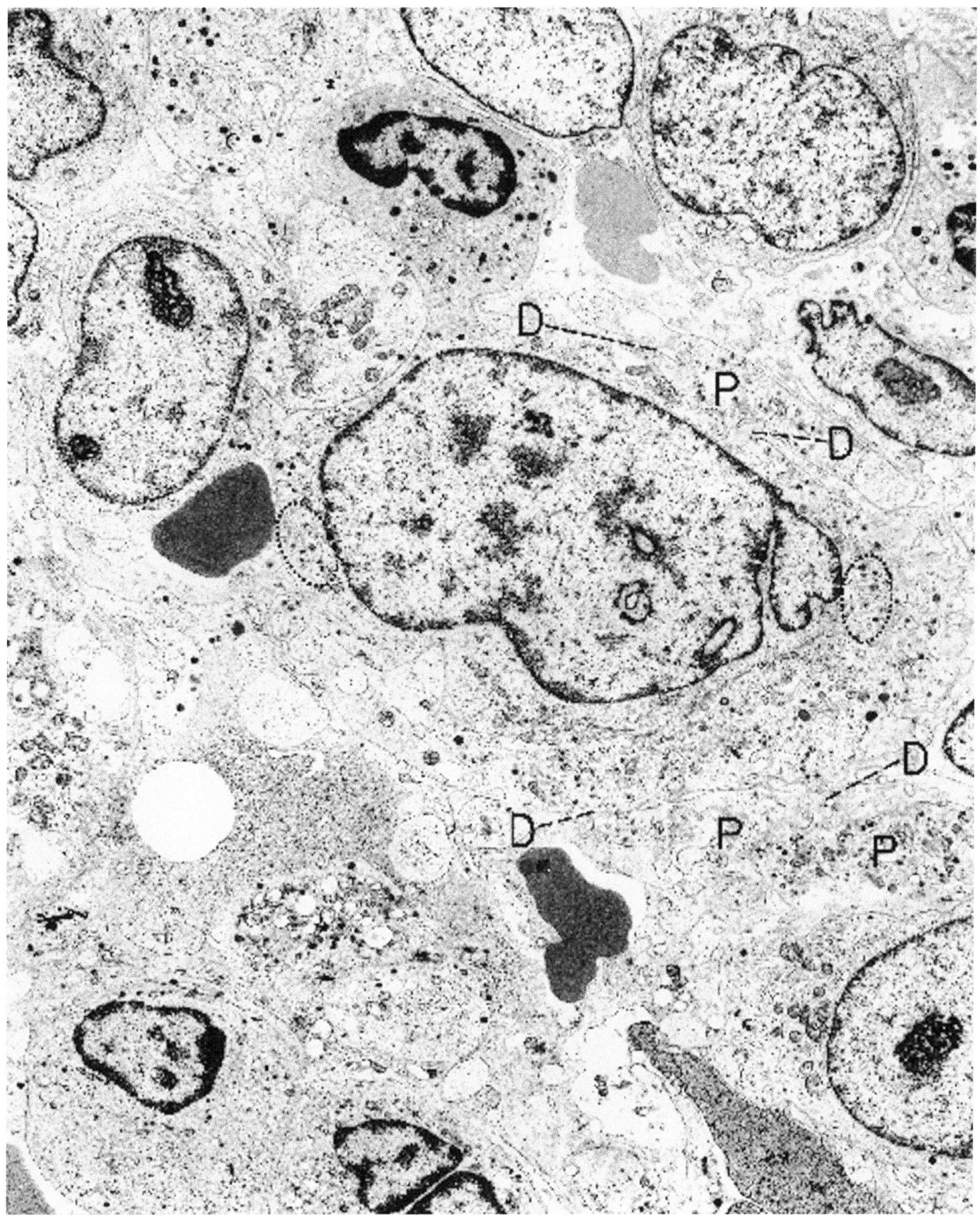

Figure 5.18. Megakaryocyte (bone marrow). A megakaryocyte exhibits numerous small, dense granules (*circles*) and demarcation membranes (D), where platelets (P) are budding from the superficial cytoplasm. (× 4940)

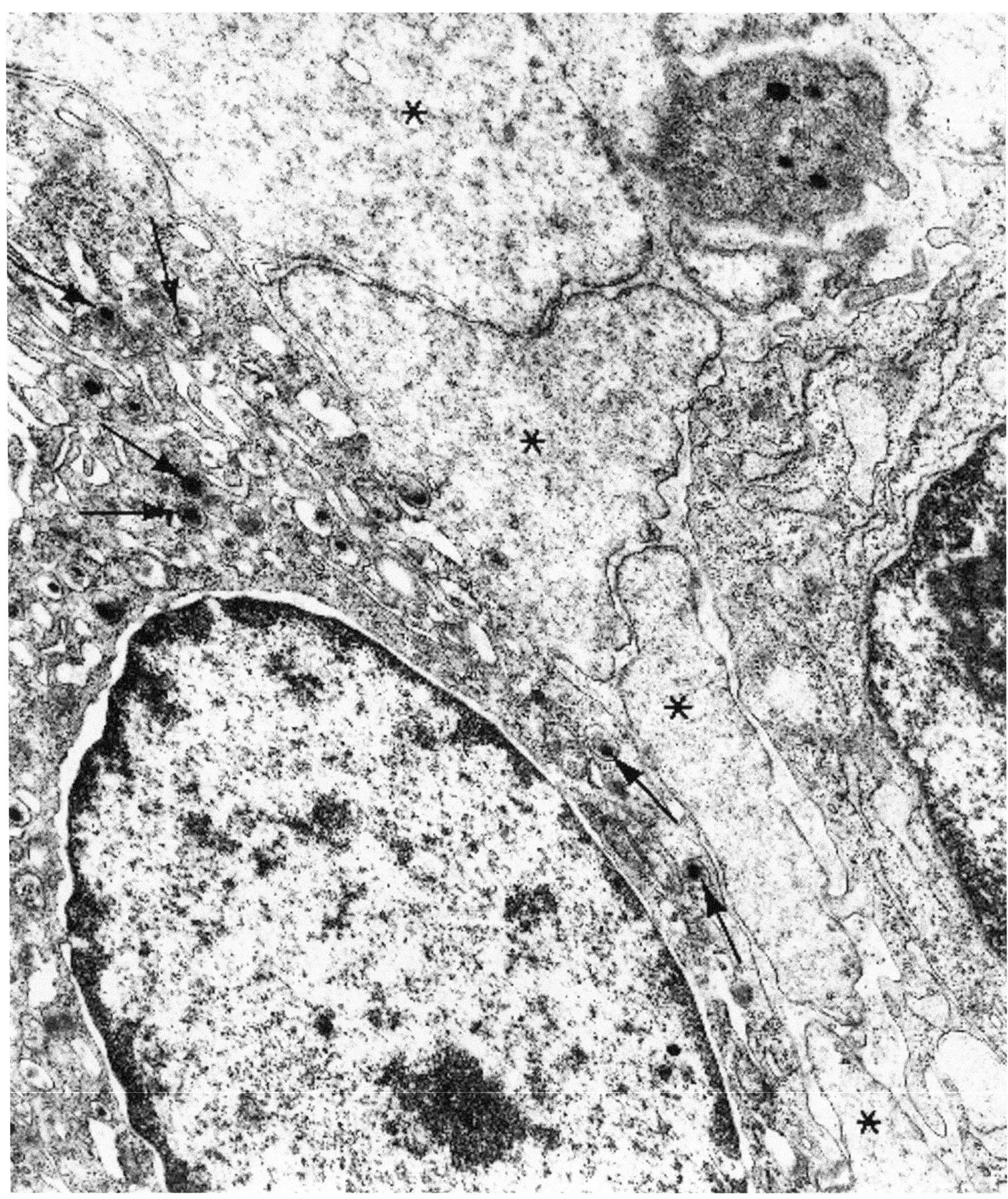

Figure 5.19. Megakaryocyte (bone marrow). High magnification of a megakaryocyte illustrates diagnostic granules (*arrows*) and cytoplasmic blebs (*). (× 22,680) (Permission for reprinting granted by WB Saunders, Dickersin GR: Electron microscopy of leukemias and lymphomas. Clin Lab Med 7(1):199–247, 1987.)

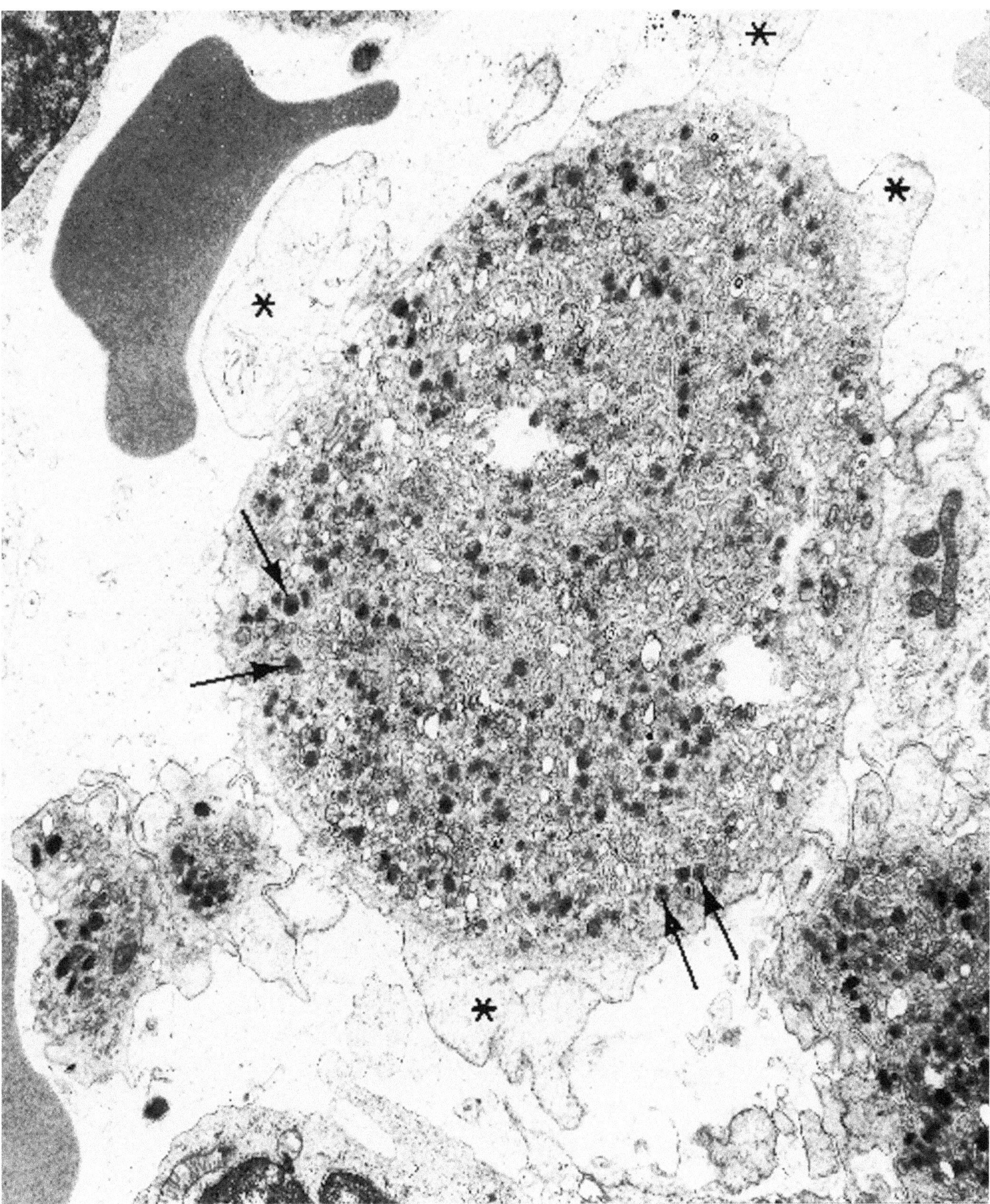

Figure 5.20. Micromegakaryocyte (bone marrow). The size of this cell is within the range expected for micromegakaryocytes, but it also could be one end of a larger cell. The cytoplasm shows blebs (*) and many dense core granules (*arrows*). (× 13,500) (Permission for reprinting granted by WB Saunders, Dickersin GR: Electron microscopy of leukemias and lymphomas. Clin Lab Med 7:199–247, 1987.)

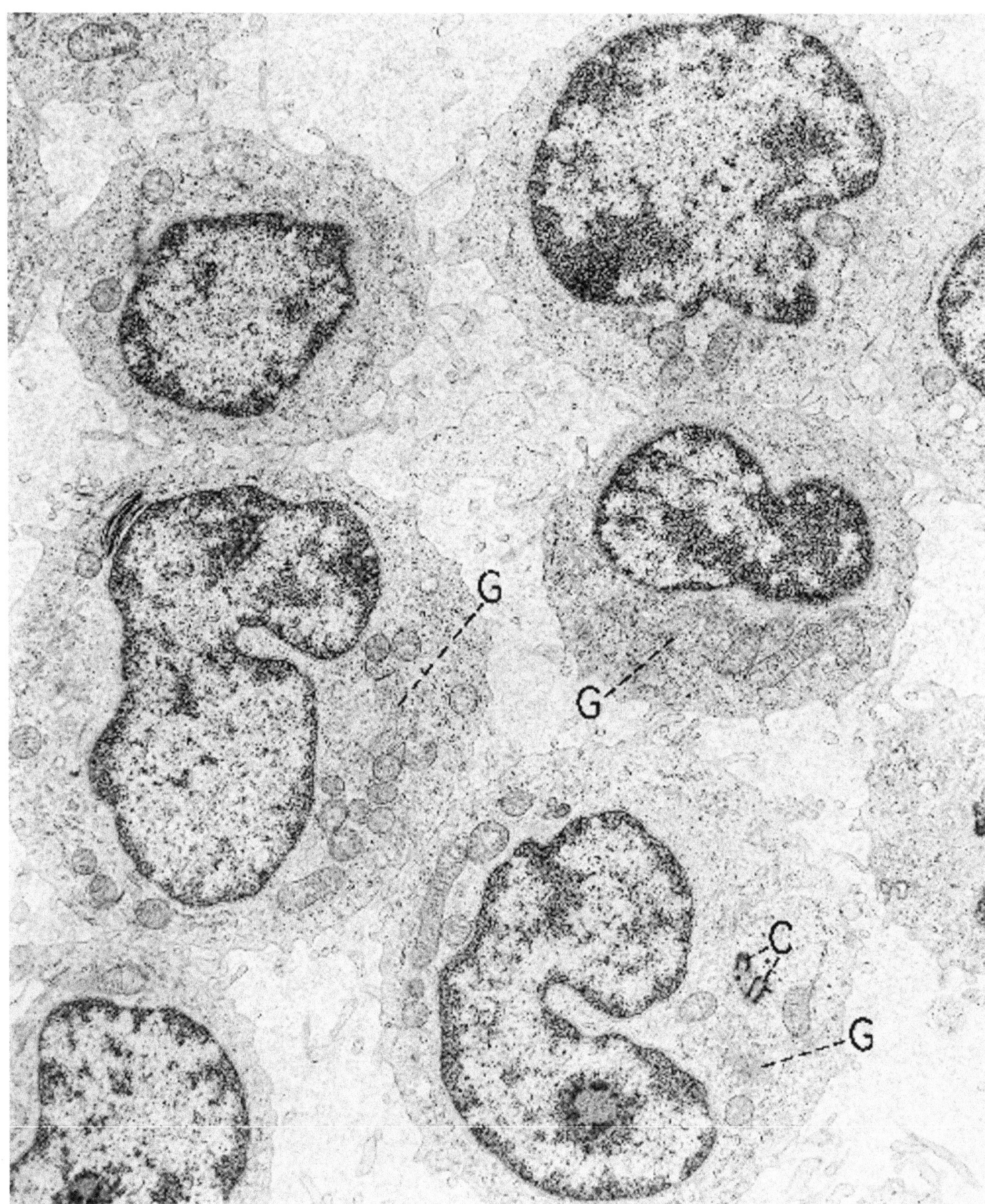

Figure 5.21. Hairy cell leukemia (peripheral blood). The cells have a villus-like, or hairy, surface, and their nuclei have abundant heterochromatin in a lymphoid type distribution. A high proportion of nuclei are indented and lobated, and Golgi apparatuses (G) and centrioles (C) are located adjacent to the indentation. (× 8500) (Permission for reprinting granted by WB Saunders, Dickersin GR: Electron microscopy of leukemias and lymphomas. Clin Lab Med 7(1):199–247, 1987.)

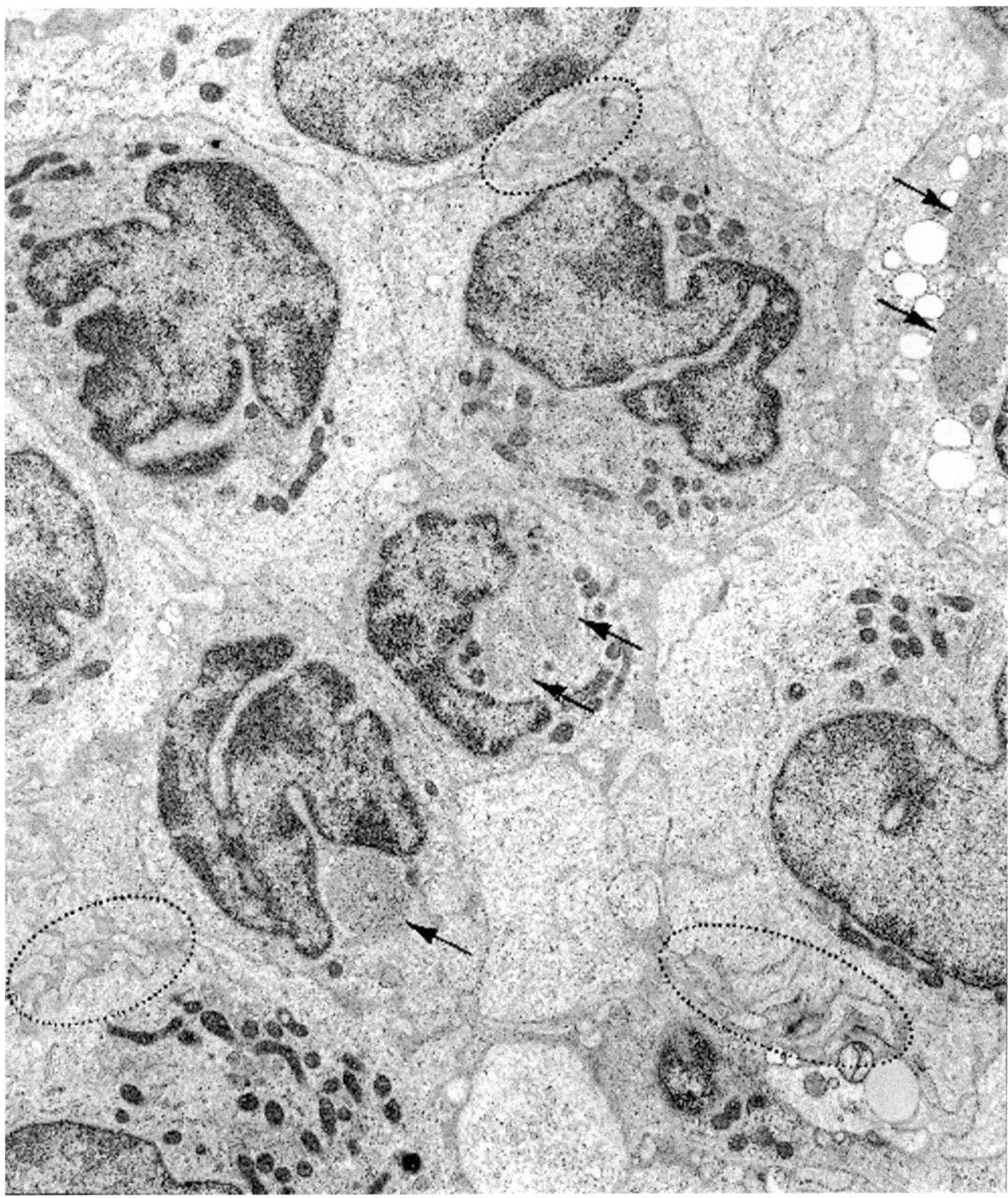

Figure 5.22. Hairy cell leukemia (bone marrow). The leukemic cells are tightly apposed in the marrow, and their hairy surfaces are overlapped and interdigitating (*circles*). Nuclei are markedly indented, and cytoplasm contains numerous juxtanuclear ribosome-lamellar complexes (*arrows*). (× 8500) (Permission for reprinting granted by WB Saunders, Dickersin GR: Electron microscopy of leukemias and lymphomas. Clin Lab Med 7(1):199–247, 1987.)

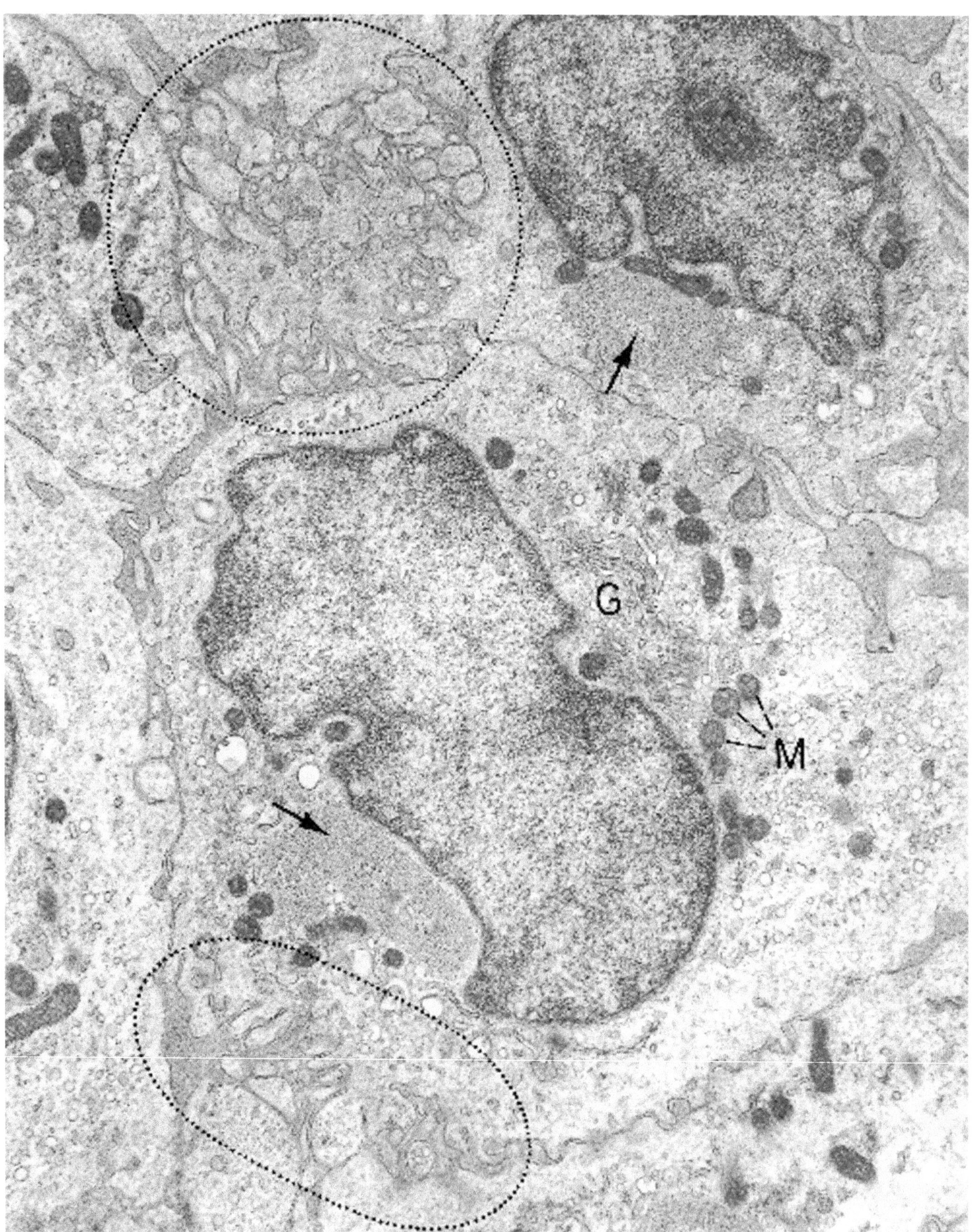

Figure 5.23. Hairy cell leukemia (bone marrow). Higher magnification of several hairy cells illustrates the overlapped surface projections (*circles*) of adjacent cells, as well as two ribosome-lamellar complexes (*arrows*). A prominent Golgi apparatus (G) and numerous mitochondria (M) also are evident in one cell. (× 14,250)

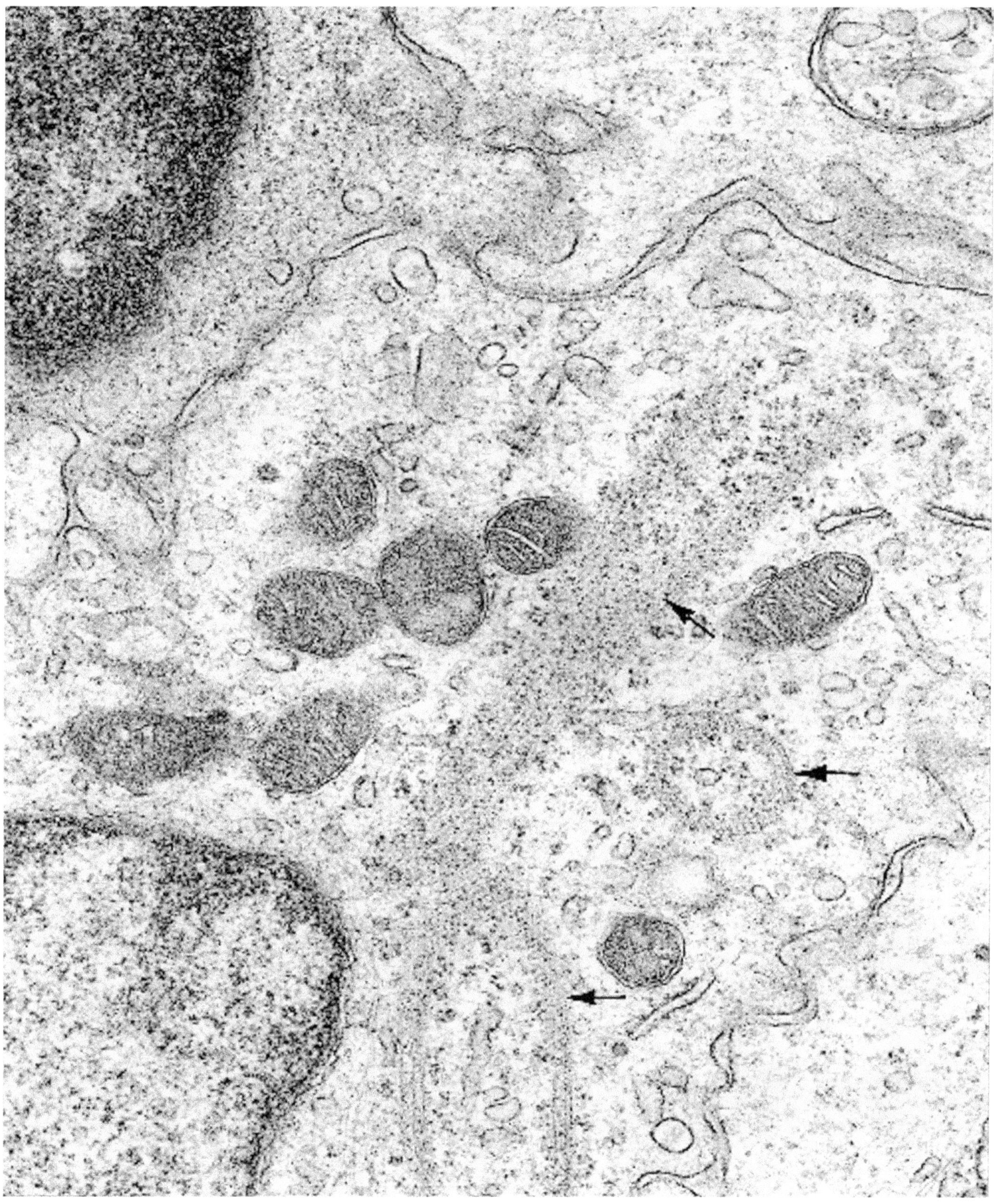

Figure 5.24. Hairy cell leukemia (bone marrow). High magnification of a juxtanuclear region of one cell depicts ribosome-lamellar complexes (*arrows*) in more than one plane of section. (× 46,500) (Permission for reprinting granted by WB Saunders, Dickersin GR: Electron microscopy of leukemias and lymphomas. Clin Lab Med 7(1):199–247, 1987.)

(Text continued from page 234)

REFERENCES

Normal Morphology and Common Leukemias

Azar HA: The hematopoietic system. In Trump BF, Jones RT, eds: *Diagnostic Electron Microscopy,* vol 2. John Wiley and Sons, New York, 1979, pp 47–161.

Behm FG: Morphologic and cytochemical characteristics of childhood lymphoblastic leukemia. Hematol Oncol Clin North Am 4:715–741, 1990.

Bennett JM, Catovsky D, Daniel MT, et al: Proposals for the classification of chronic (mature) B and T lymphoid leukaemias. French-American-British (FAB) Cooperative Group. J Clin Pathol 42:567–584, 1989.

Bessho F, Hayashi Y, Hayashi Y, et al: Ultrastructural studies of peripheral blood of neonates with Down's syndrome and transient abnormal myelopoiesis. Am J Clin Pathol 89:627–633, 1988.

Bessie M: *Living Blood Cells and Their Ultrastructure.* Springer-Verlag, Berlin, 1973.

Breton-Gorius J, Houssay D: Auer bodies in acute promyelocytic leukemia. Demonstration of their fine structure and peroxidase localization. Lab Invest 28: 135–141, 1973.

Brunning RD, McKenna RW: Tumors of the bone marrow. In Rosai J, Sobin LH, eds: *Atlas of Tumor Pathology,* 3rd series, fasc 9. Armed Forces Institute of Pathology, Washington, DC, 1994.

Brunning RD, Parkin JL, McKenna RW: Acute lymphoblastic leukemia. In Polliack A, ed. *Human Leukemias: Cytochemical and Ultrastructural Techniques in Diagnosis and Research.* Boston, Martinus Nijhoff, 1984, pp 173–218.

Buccheri V, Shetty V, Yoshida N, et al: The role of an antimyeloperoxidase antibody in the diagnosis and classification of acute leukaemia: A comparison with light and electron microscopy cytochemistry. Br J Haematol 80:62–68, 1992.

Chan WC, Link S, Mawle A, et al: Heterogeneity of large granular lymphocyte proliferations: Delineation of two major subtypes. Blood 68:1142–1153, 1986.

Cowley JC, Hayhoe FGJ: *Ultrastructure of Hemic Cells: A Cytologic Atlas of Normal and Leukemic Blood and Bone Marrow.* WB Saunders, Philadelphia, 1973.

Darbyshire PJ, Lilleyman JS: Granular acute lymphoblastic leukemia of childhood: A morphologic phenomenon. J Clin Pathol 40:251–253, 1987.

Dickersin GR: Electron microscopy of leukemias and lymphomas. Clin Lab Med 7(1):199–247, 1987.

Dvorak AM, Monahan RA, Dickersin GR: I. Hematology: Differential diagnosis of acute lymphoblastic and acute myeloblastic leukemia. Use of ultrastructural peroxidase cytochemistry and routine electron microscopy technology. Pathol Annu 16:101–137, 1981.

Eimoto T, Mitsui T, Kikuchi M: Ultrastructure of adult T-cell leukemia/lymphoma. Virchows Arch [Cell Pathol] 38:189–208, 1981.

Fradera J, Velez-Garcia E, White JG: Acute lymphoblastic leukemia with unusual cytoplasmic granulation: A morphologic, cytochemical and ultrastructural study. Blood 68:406–411, 1986.

Glick AD: Acute leukemia: Electron microscopic diagnosis. Semin Oncol 3:229–241, 1976.

Glick AD, Paniker K, Flexner JM, et al: Acute leukemia of adults. Ultrastructural, cytochemical and histologic observations in 100 cases. Am J Clin Pathol 73:459–470, 1980.

Golomb HM, Rowley JD, Vardiman JW, et al: "Microgranular" acute promyelocytic leukemia: A distinctive clinical, ultrastructural and cytogenetic entity. Blood 55:253–259, 1980.

Grogan TM, Insalaco SJ, Savage RA, et al: Acute lymphoblastic leukemia with prominent azurophilic granulation and punctate acidic nonspecific esterase and phosphatase activity. Am J Clin Pathol 75:716–722, 1981.

Hay CR, Barnett D, James V, et al: Granular common acute lymphoblastic leukemia in adults: A morphologic study. Eur J Haematol 39:299–305, 1987.

Kantarjian HM, Diesseroth A, Kurzrock R, et al: Chronic myelogenous leukemia: A concise update. Blood 82: 691–703, 1993.

Marie JP, Vernant JP, Dreyfus B, et al: Ultrastructural localization of peroxidases in "undifferentiated" blasts during the blast crisis of chronic granulocytic leukaemia. Br J Haematol 43:549–558, 1979.

McKenna RW, Parkin J, Brunning RD: Morphologic and ultrastructural characteristics of T-cell acute lymphoblastic leukemia. Cancer 44:1290–1297, 1979.

Melo JV, Hegde U, Parreira A, et al: Splenic B cell lymphoma with circulating villous lymphocytes: Differential diagnosis of B cell leukaemias with large spleens. J Clin Pathol 40:642–651, 1987.

Oshimi K: Granular lymphocyte proliferative disorders: Report of 12 cases and review of the literature. Leukemia 2:617–627, 1988.

Parkin JL, McKenna RW, Brunning RD: Philadelphia chromosome-positive blastic leukaemia: Ultrastructural and ultracytochemical evidence of basophil and mast cell differentiation. Br J Haematol 52:663–677, 1982.

Schmidt U, Mlynek ML, Leder LD: Electron-microscopic characterization of mixed granulated (hybridoid) leucocytes of chronic myeloid leukaemia. Br J Haematol 68:175–180, 1988.

Shetty V, Chitale A, Matutes E, et al: Immunological and ultrastructural studies in acute biphenotypic leukaemia. J Clin Pathol 46:903–907, 1993.

Stasi R, Del Poeta G, Venditti A, et al: Lineage identification of acute leukemias: Relevance of immunologic and ultrastructural techniques. Hematol Pathol 9: 79–94, 1995.

Stein P, Peiper S, Butler D, et al: Granular acute lymphoblastic leukemia. Am J Clin Pathol 79:426–430, 1983.

Takemori N, Hirai K, Onodera R, et al: Parallel tubular granules in human immature neutrophils—An electron microscopic study. Leuk Lymphoma 15:177–186, 1994.

Ullyot JL, Bainton DF: Azurophil and specific granules of blood neutrophils in chronic myelogenous leukemia: An ultrastructural and cytochemical analysis. Blood 44:469–482, 1974.

Yanagihara ET, Naeim F, Gale RP, et al: Acute lymphoblastic leukemia with giant intracytoplasmic inclusions. Am J Clin Pathol 74:345–349, 1980.

Zucker-Franklin D, Greaves MF, Grossi CE, et al: *Atlas of Blood Cells, Function and Pathology*, 2nd ed, vols 1 and 2. Lea and Febiger, Philadelphia, 1988.

Megakaryocytic Leukemia

Bain B, Catovsky D, O'Brien M, et al: Megakaryoblastic transformation of chronic granulocytic leukaemia: An electron microscopy and cytochemical study. J Clin Pathol 30:235–242, 1977.

Bennett JM, Catovsky D, Daniel MT, et al: Criteria for the diagnosis of acute leukemia of megakaryocytic lineage (M7): A report of the French-American-British Cooperative Group. Ann Intern Med 103:460–462, 1985.

Breton-Gorius J, Reyes F, Duhamel G, et al: Megakaryoblastic acute leukemia: Identification by the ultrastructural demonstration of platelet peroxidase. Blood 51:45–60, 1978.

Breton-Gorius J, Reyes F, Vernant JP: The blast crisis of chronic granulocytic leukaemia: Megakaryoblastic nature of cells are revealed by the presence of platelet-peroxidase-cytochemical ultrastructural study. Br J Haematol 39:295–303, 1978.

Brunning RD, McKenna RW: Tumors of the bone marrow. In Rosai J, Sobin LH, eds: *Atlas of Tumor Pathology*, 3rd series, fasc 9. Armed Forces Institute of Pathology, Washington, DC, 1994.

Innes DJ, Mills SE, Walker GK: Megakaryocytic leukemia: Identification utilizing anti-factor VIII immunoperoxidase. Am J Clin Pathol 77:107–110, 1982.

Jacobs P, LeRoux I, Jacobs L: Megakaryoblastic transformation in myeloproliferative disorders. Cancer 54: 297–302, 1984.

Koike T: Megakaryoblastic leukemia: The characterization and identification of megakaryoblasts. Blood 64:683–692, 1984.

Mirchandani I, Palutke M: Acute megakaryoblastic leukemia. Cancer 50:2866–2872, 1983.

Williams W, Weiss GB: Megakaryoblastic transformation of chronic myelogenous leukemia. Cancer 49: 921–926, 1982.

Hairy-Cell Leukemia

Bartl R, Frisch B, Hill W, et al: Bone marrow histology in hairy-cell leukemia: Identification of subtypes and their prognostic significance. Am J Clin Pathol 79: 531–545, 1983.

Braylan RC, Jaffe ES, Triche TJ, et al: Structural and functional properties of the "hairy" cells of leukemic reticuloendotheliosis. Cancer 41:210–227, 1978.

Brunning RD, McKenna RW: Tumors of the bone marrow. In Rosai J, Sobin LH, eds: *Atlas of Tumor Pathology*, 3rd series, fasc 9. Armed Forces Institute of Pathology, Washington, DC, 1994.

Burke JS, MacKay B, Rappaport H: Hairy cell leukemia (leukemic reticuloendotheliosis). II. Ultrastructure of the spleen. Cancer 37:2267–2274, 1976.

Catovsky D, Petitt JE, Galton DAG, et al: Leukaemic reticuloendotheliosis ("hairy" cell leukaemia): A distinct clinicopathologic entity. Br J Haematol 26:9–27, 1974.

Chang KL, Stroup R, Weiss LM: Hairy cell leukemia. Current status. Anat Pathol 97:719–738, 1992.

Daniel M, Flandrin G: Fine structure of abnormal cells in hairy cell (tricholeukocytic) leukemia, with special reference to their *in vitro* phagocytic capacity. Lab Invest (Paris) 30:1–8, 1974.

Zucker-Franklin D: Virus-like particles in the lymphocyte of a patient with chronic lymphocytic leukemia. Blood 21:509–512, 1963.

6

Spindle Cell Neoplasms and Their Epithelioid Variants

Fibrous Neoplasms

(Figures 6.1 through 6.17.)

Diagnostic criteria. (1) Abundant rough endoplasmic reticulum; (2) type I (banded) collagen closely surrounding the cells (Figures 6.1 through 6.3).

Additional points. Fibroblasts also have a prominent Golgi apparatus, and they may or may not have filopodia and long, tapering, polar processes. Lipid vacuoles in varying numbers may also be present in the cytoplasm. Fibroblasts often show partial differentiation toward smooth muscle cells, so-called myofibroblasts, in actively proliferating fibroblastic lesions (Figures 6.4 through 6.6), and toward histiocytes, designated facultative histiocytes, as in some examples of malignant fibrous histiocytoma (MFH) (see next section). True smooth muscle cells and true histiocytes have less rough endoplasmic reticulum than do fibroblasts, and the cisternae of the rough endoplasmic reticulum usually are not dilated. Fibroblasts also usually have more rough endoplasmic reticulum than do the spindle cells of synovial sarcoma (see section on synovial sarcoma) and solitary fibrous tumors of pleura and soft tissues (see Chapter 3, section on mesothelioma, Figures 3.75 through 3.77, and Figure 6.14). Infrequently, the cells of fibrosarcomas may be *epithelioid,* rather than spindle shaped, but the characteristic ultrastructural features of fibroblasts are still evident (Figures 6.16 and 6.17).

Fibroblasts, alone or in conjunction with other cell types, are found in a number of neoplasms other than *fibroma, fibrosarcoma,* and MFH, and these other neoplasms include *myxoma, fibromyxoid sarcoma, dermatofibrosarcoma protuberans, elastofibroma, angiofibroma, reparative granuloma, giant cell tumor of bone,* and *solitary fibrous tumors of soft tissue. Dermatofibrosarcoma protuberans,* usually a very cellular lesion with a storiform pattern, may be myxoid as well (Figure 6.7). The lesion is composed of fibroblasts and myofibroblasts (Figure 6.8), but perineurial cells possibly could be the cell type in some cases. It is still unsettled whether perineurial cells are modified fibroblasts or modified Schwann cells (see Figure 6.124). Evidence for the latter is the finding of melanosomes in the cells of a few examples of dermatofibrosarcoma protuberans, the so-called *Bednar tumor.* Fibrosarcoma may develop within dermatofibrosarcoma protuberans, as illustrated in Figures 6.9 and 6.10. In *elastofibroma,* fibroblasts are associated with matrical elastin fibers, which have a medium-dense, amorphous, and filamentous appearance (Figures 6.11 and 6.12). In *reparative granuloma* and *giant cell tumor of bone,* the giant cells have features of osteoclasts, the most notable feature being numerous mitochondria (Figure 6.13).

(Text continues on page 264)

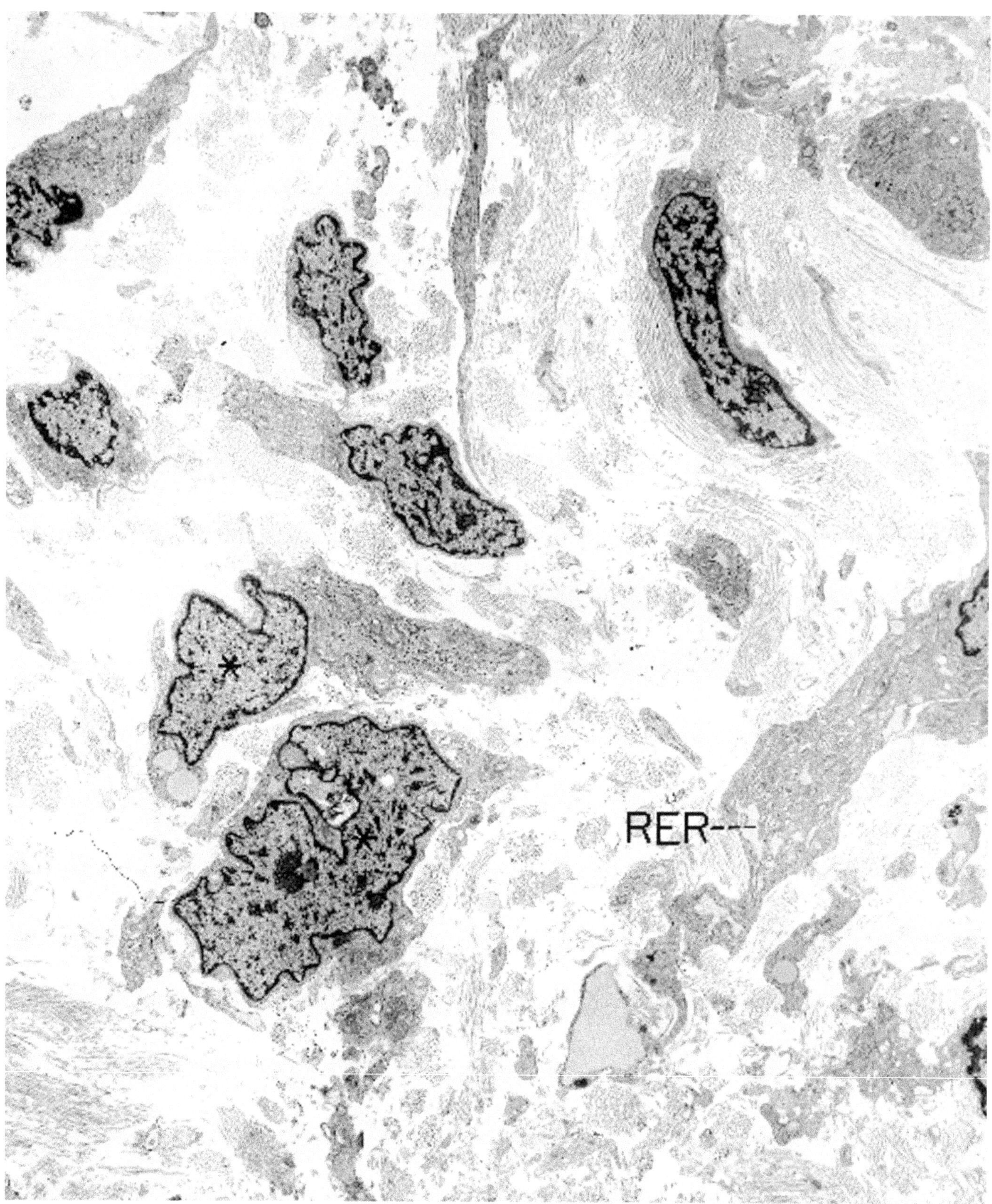

Figure 6.1. Fibrosarcoma (soft tissue of leg). This low power view illustrates an intimate admixture of spindle-shaped cells and collagenous matrix. The irregular shape of some of the nuclei (*) is a criterion for malignancy. The cytoplasm of the cells is rich in dilated rough endoplasmic reticulum (RER). (× 4940)

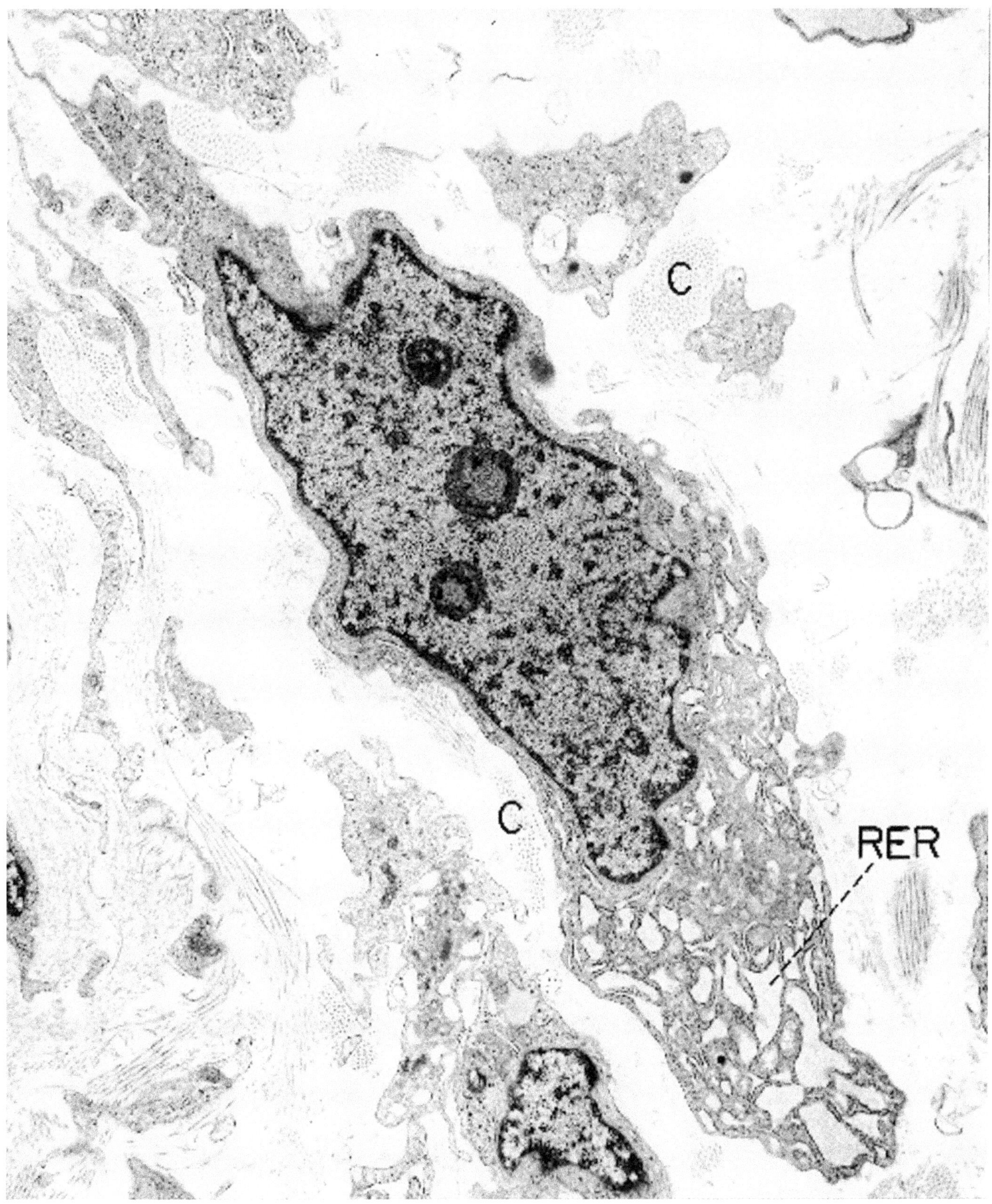

Figure 6.2. Fibrosarcoma (soft tissue of leg). Higher magnification of one of the fibroblasts from the same neoplasm as shown in Figure 6.1 accentuates the rough endoplasmic reticulum (RER) and the surrounding banded collagen (C). (× 9690)

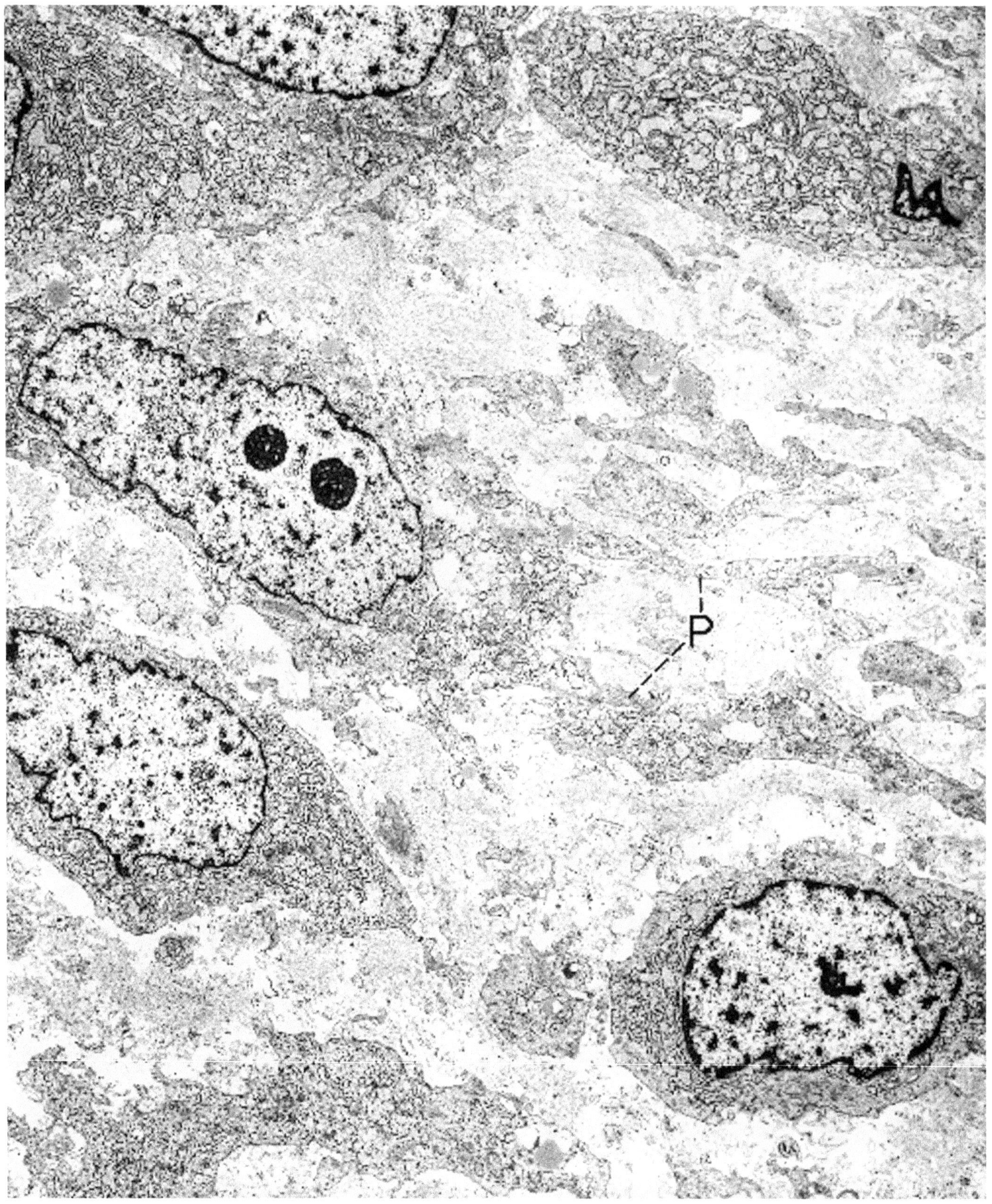

Figure 6.3. Fibrosarcoma (soft tissue of buttock). Some cells are oval, and others have long cytoplasmic processes (P). Rough endoplasmic reticulum is dilated and virtually fills the cytoplasm. (× 4940)

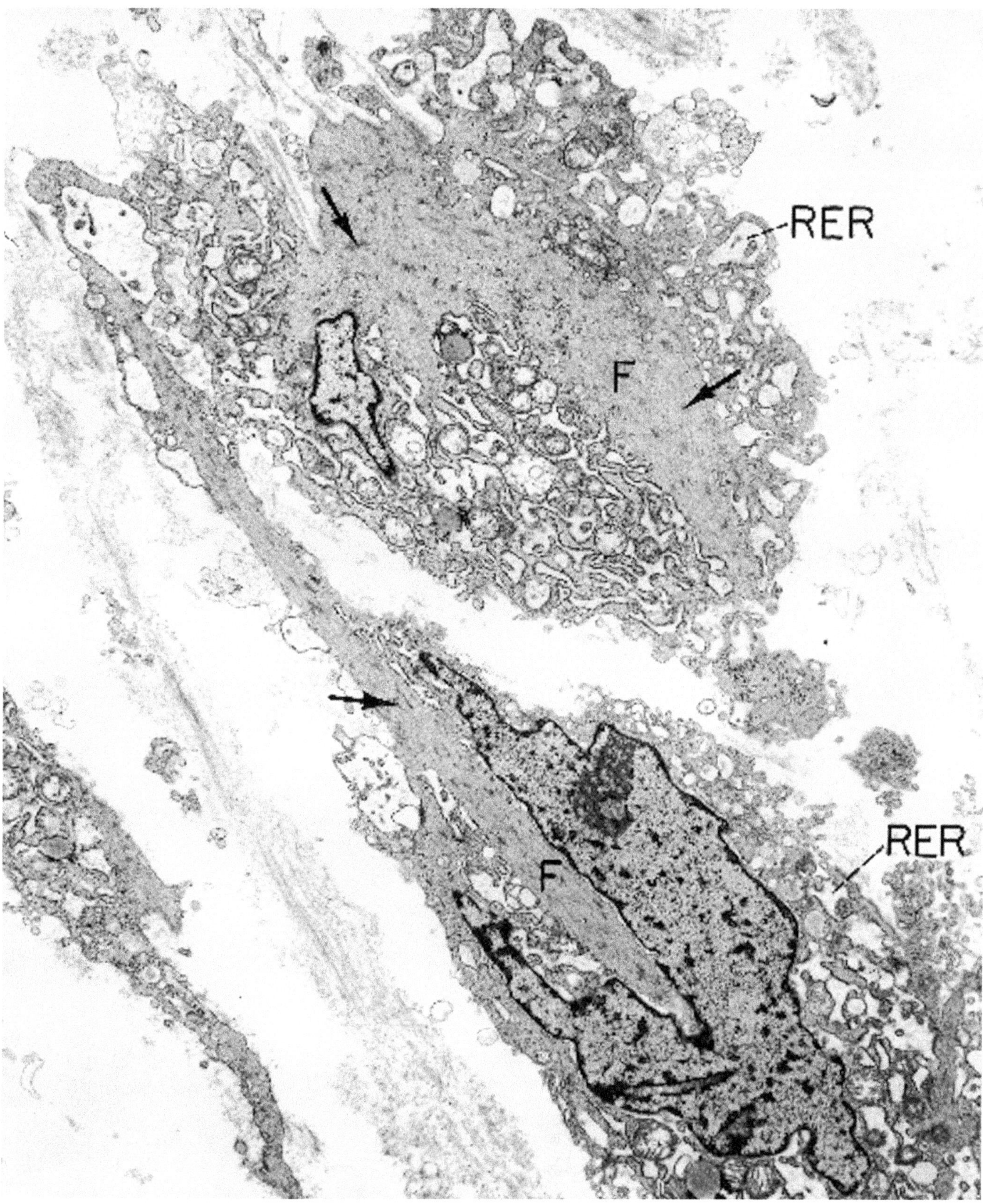

Figure 6.4. Fibrosarcoma (transverse colon). Two neoplastic fibroblasts with abundant dilated rough endoplasmic reticulum (RER) also contain large cytoplasmic areas of microfilaments (F) and associated densities, so-called dense bodies (*arrows*), qualifying as myofibroblasts. (× 4940)

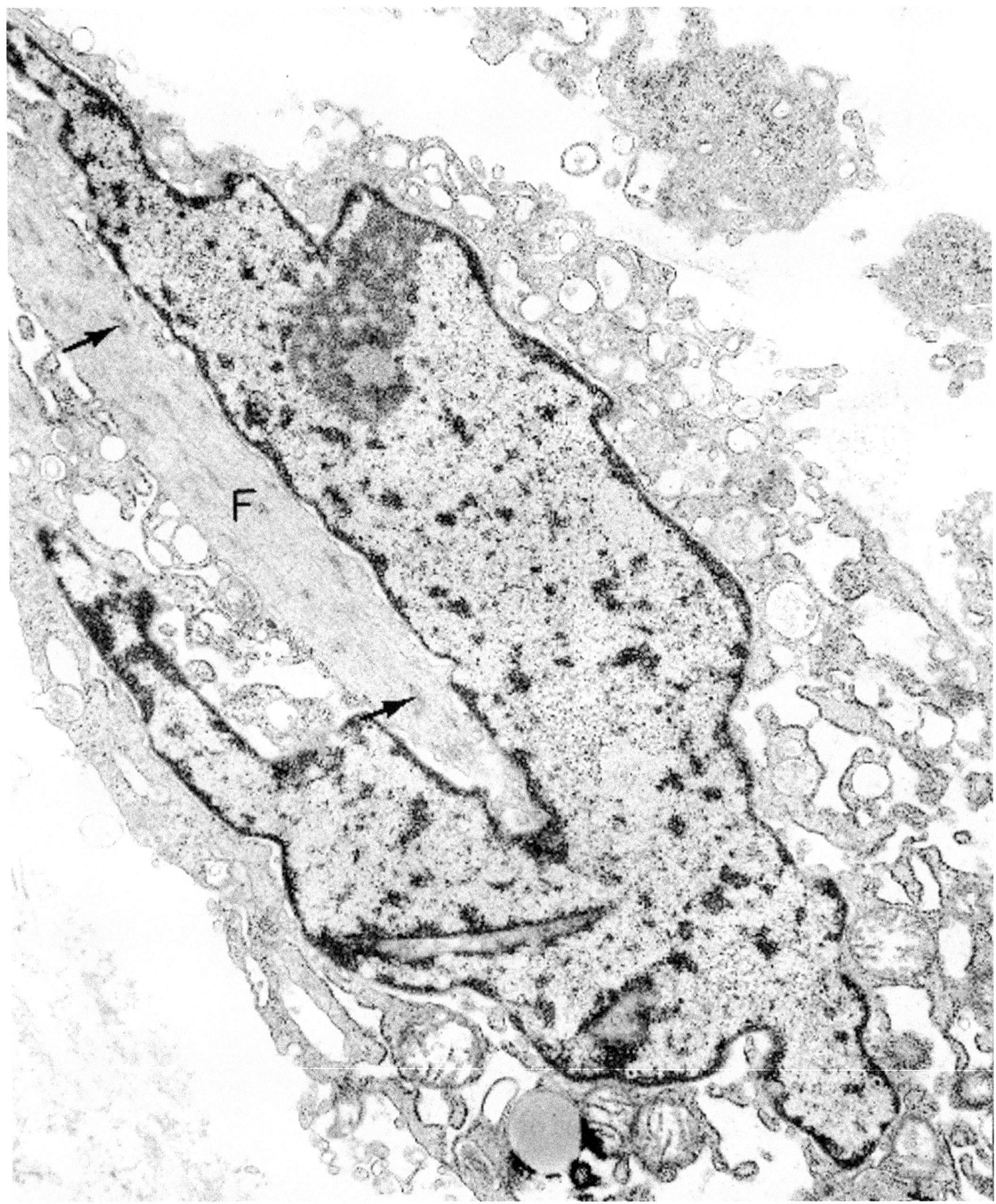

Figure 6.5. Fibrosarcoma (transverse colon). Higher magnification of one of the myofibroblasts illustrated in Figure 6.4 shows in better detail the microfilaments (F) and dense bodies (*arrows*). (× 23,600)

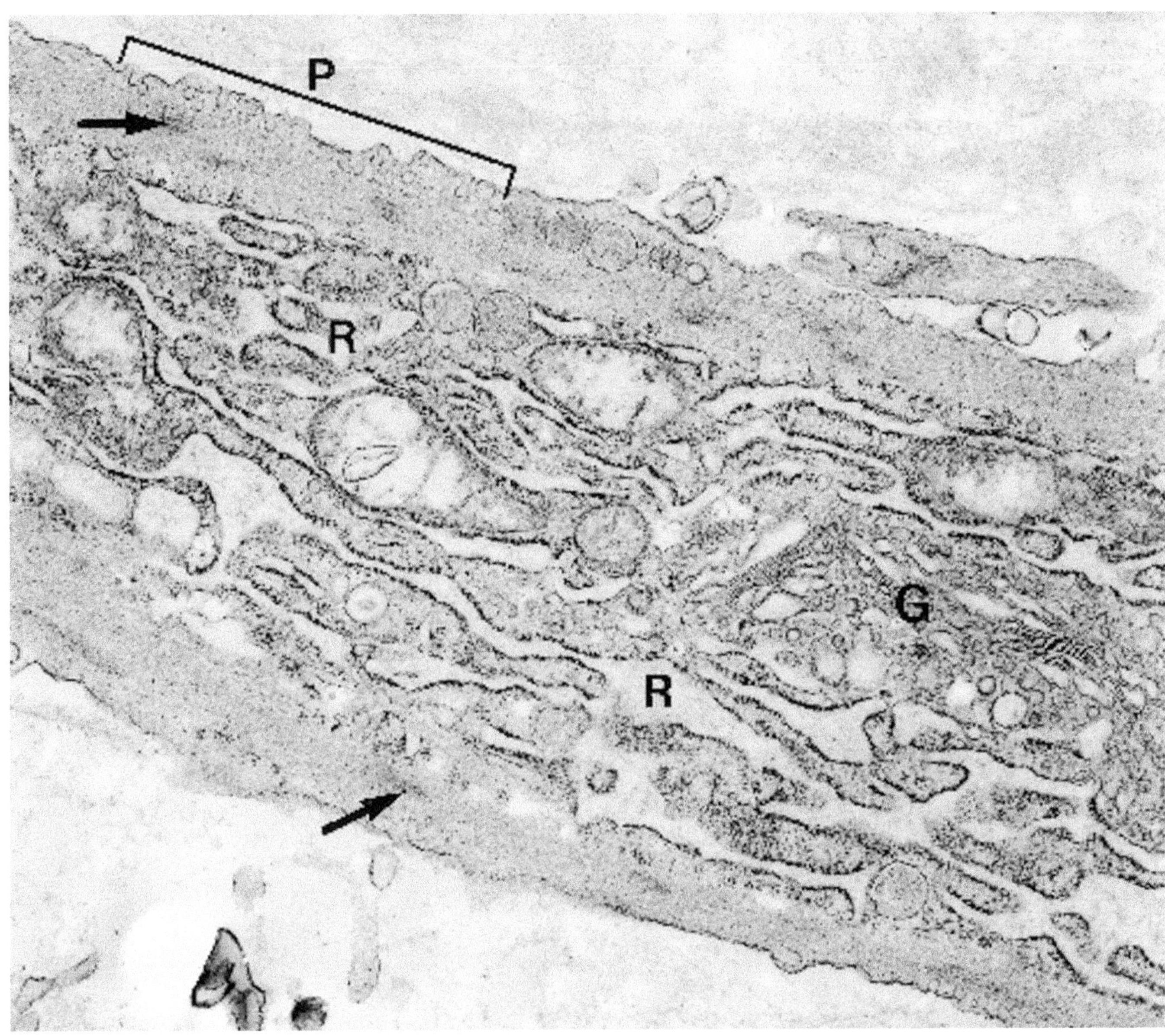

Figure 6.6. Fibrosarcoma (metastatic to liver). High magnification of a portion of a cytoplasmic process of a myofibroblast shows abundant, dilated rough endoplasmic reticulum (R) and a prominent Golgi apparatus (G), so characteristic of fibroblastic differentiation. In addition, there are numerous filaments with dense bodies (*arrows*) and numerous pinocytotic vesicles (P), typical of smooth muscle differentiation. (× 30,000)

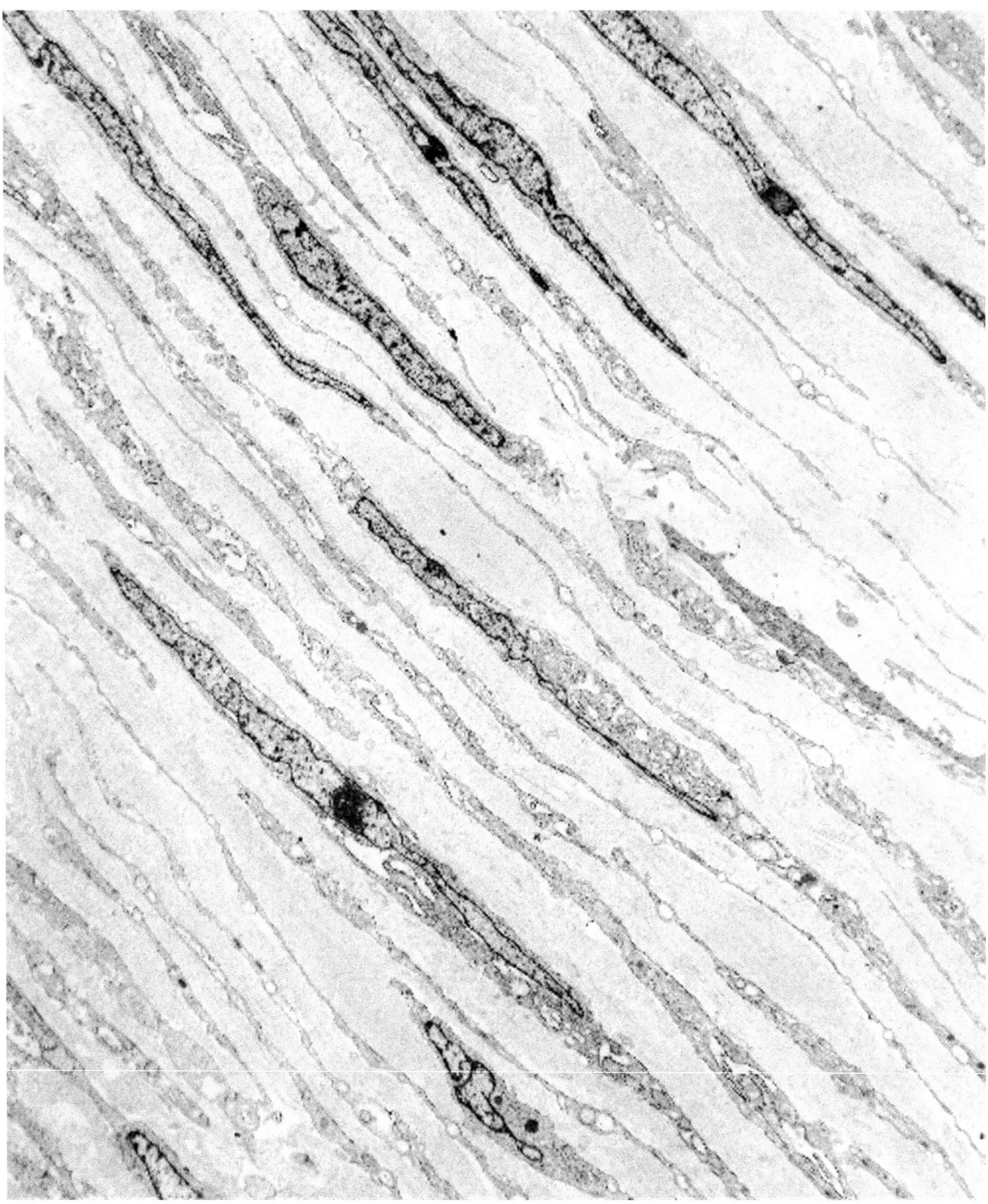

Figure 6.7. Dermatofibrosarcoma protuberans (scalp). Low magnification illustrates the spindle cells to be long and narrow and individually dispersed in a collagenous matrix. Many cells have a high nucleocytoplasmic ratio in their cell-bodies, and the amount of rough endoplasmic reticulum and number of other organelles varies from cell to cell. (× 3400)

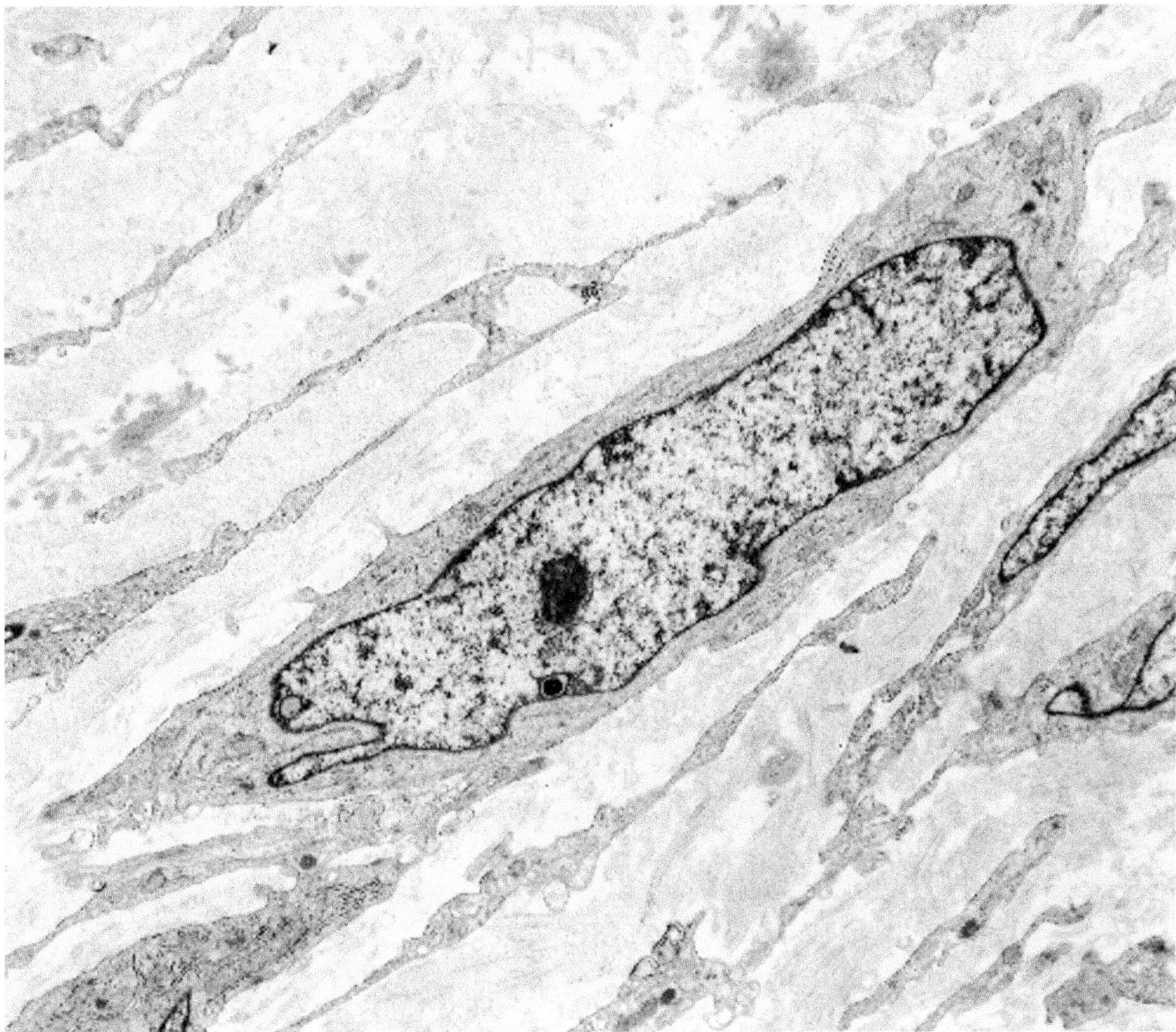

Figure 6.8. Dermatofibrosarcoma protuberans (scalp). Higher magnification of the same neoplasm as depicted in Figure 6.7 shows the cytoplasm to have a small to moderate amount of rough endoplasmic reticulum, most of which is not dilated. The cells are consistent with incompletely differentiated fibroblasts. (× 9000)

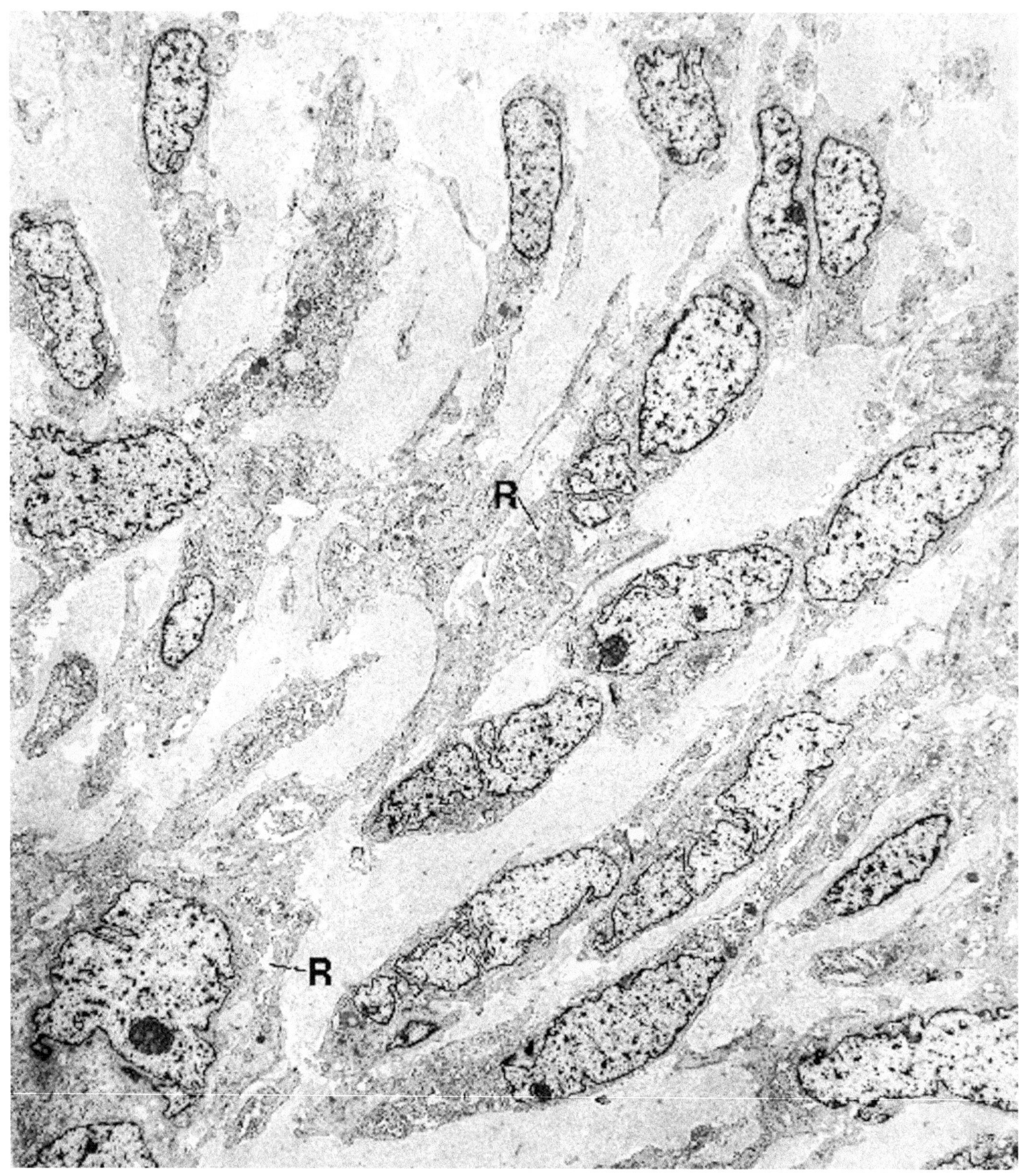

Figure 6.9. Fibrosarcoma arising in dermatofibrosarcoma protuberans (scalp). The spindle cells in this malignant area of the dermatofibrosarcoma protuberans illustrated in Figures 6.7 and 6.8 reveal a closer arrangement of cells, shorter polar processes, and more irregularly shaped nuclei. Dilated rough endoplasmic reticulum (R) is visible in some of the cells. (× 2800)

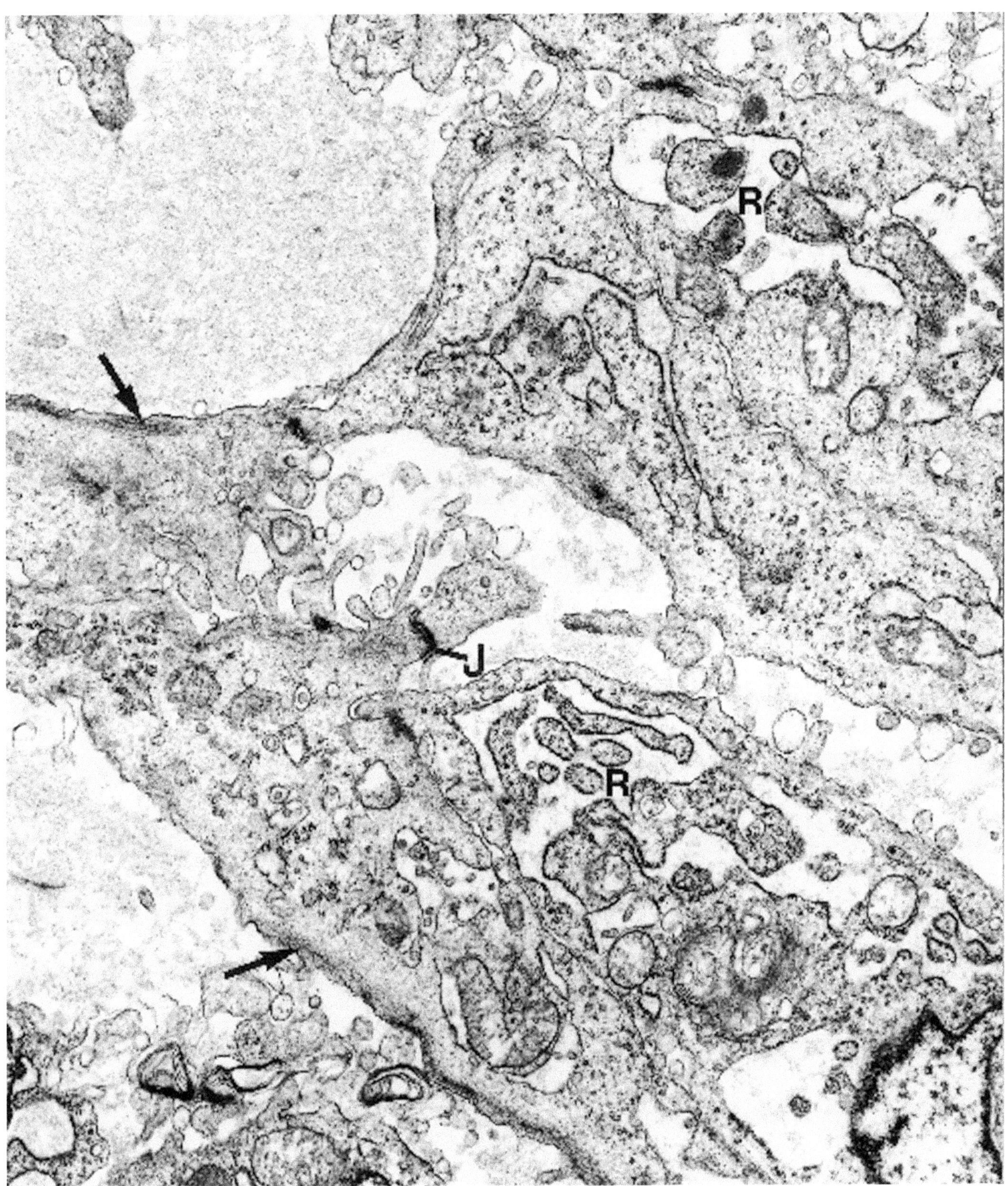

Figure 6.10. Fibrosarcoma arising in dermatofibrosarcoma protuberans (scalp). Higher magnification of one of the cells depicted in Figure 6.9 illustrates dilated rough endoplasmic reticulum (R), prominent junctions (J), and focal filaments with dense bodies (*arrows*), features consistent with smooth muscle differentiation within a fibroblast, a so-called myofibroblast. (× 18,000)

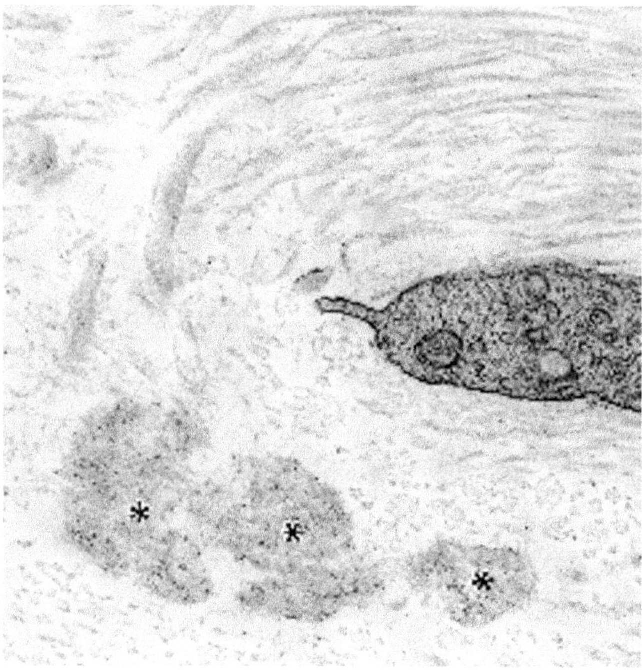

Figure 6.11. Elastofibroma (soft tissue of scapula). Low magnification of a portion of a fibroblast and surrounding collagenous matrix reveals electron-dense deposits of elastic fibers (*), seen better at higher magnification in Figure 6.12. (× 13,800)

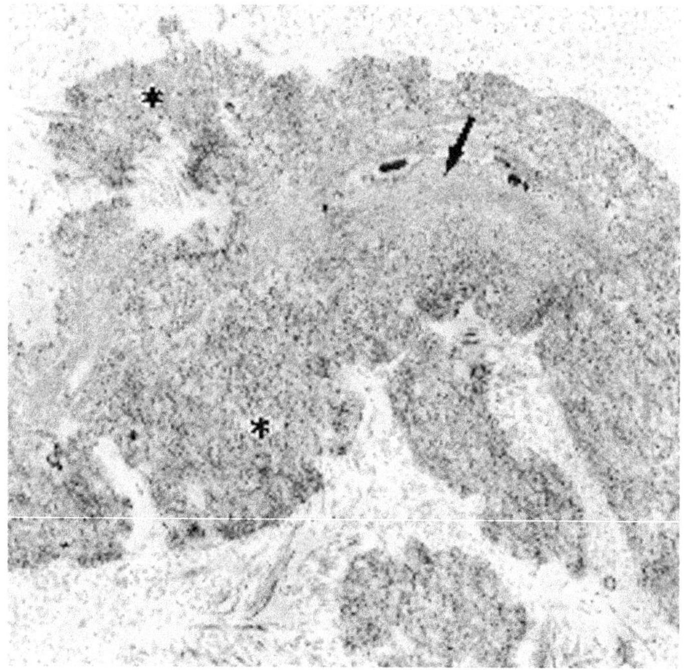

Figure 6.12. Elastofibroma (soft tissue of scapula). Higher magnification of elastic fibers reveals a central amorphous region (*arrow*) and peripheral more granular and fibrillar regions (*). (× 27,400)

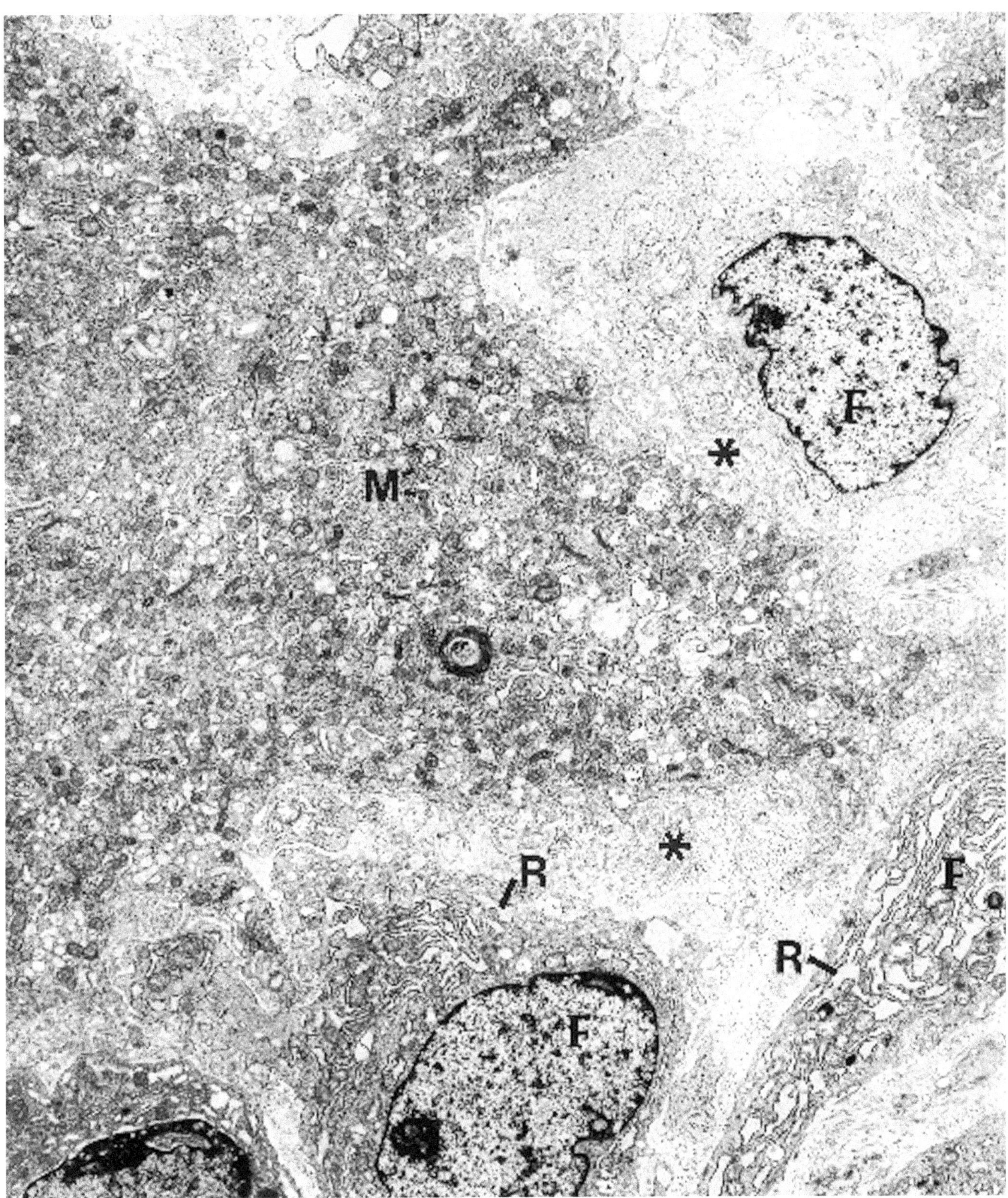

Figure 6.13. Giant cell reparative granuloma (soft tissue of buttock). Several fibroblasts (F) and an adjacent osteoclast-like giant cell exemplify this lesion. The fibroblasts are rich in dilated rough endoplasmic reticulum (R), and the giant cell has numerous filopodia on the surface (*) and many mitochondria (M) within the cytoplasm. (× 6800) (Permission for reprinting granted by *New England Journal of Medicine;* Case Records of the Massachusetts General Hospital: Weekly clinicopathologic exercises. Case 1-1986. *N Engl J Med* 314;105–113, 1986.)

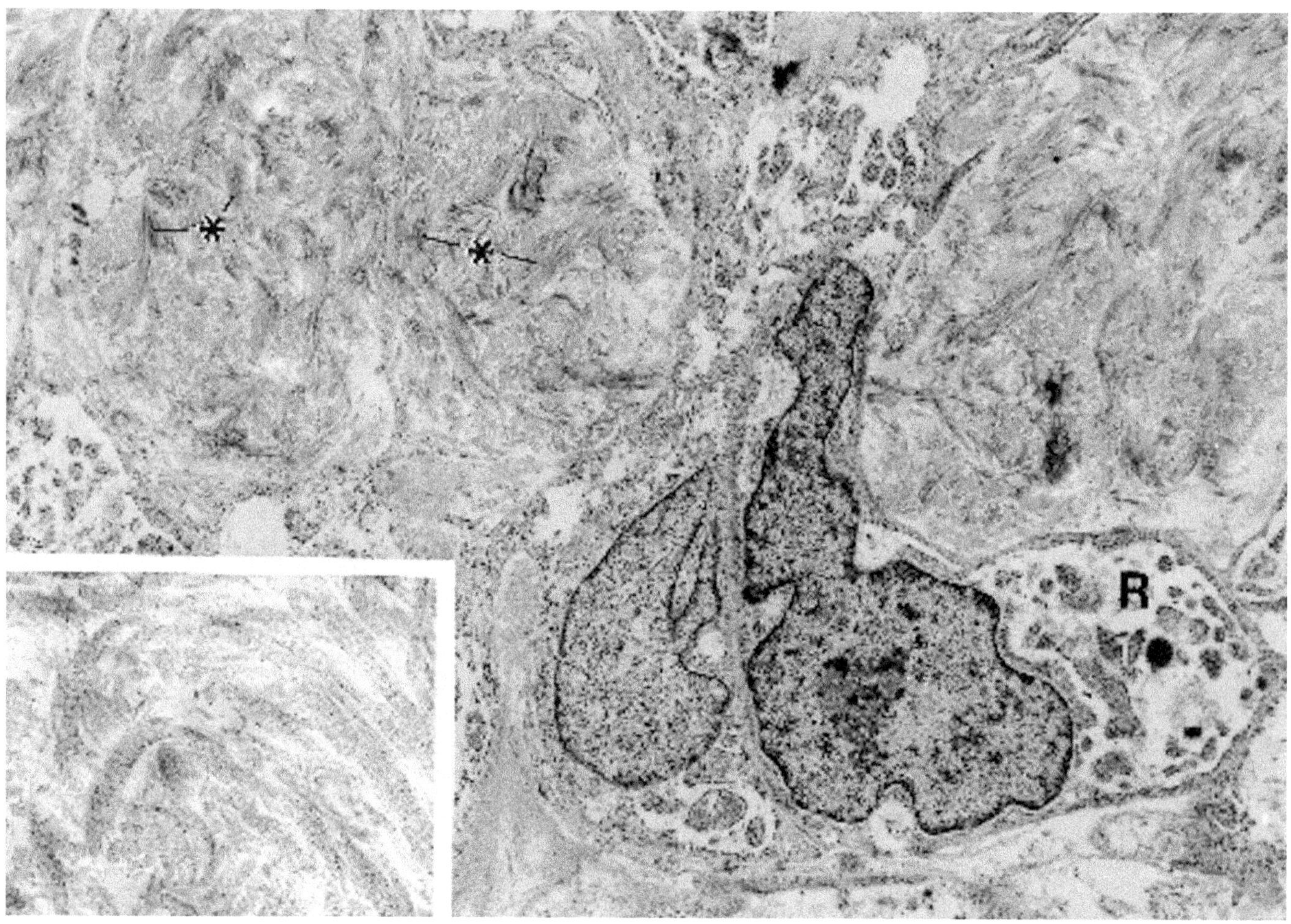

Figure 6.14. Solitary fibrous tumor of soft tissue (face). A neoplastic fibroblast characteristically contains dilated rough endoplasmic reticulum (R), and the adjacent matrix is rich in heavy, dense fibers of collagen (*), "amianthoid" fibers, seen better at higher magnification in the *inset.* (× 8900; inset × 22,000)

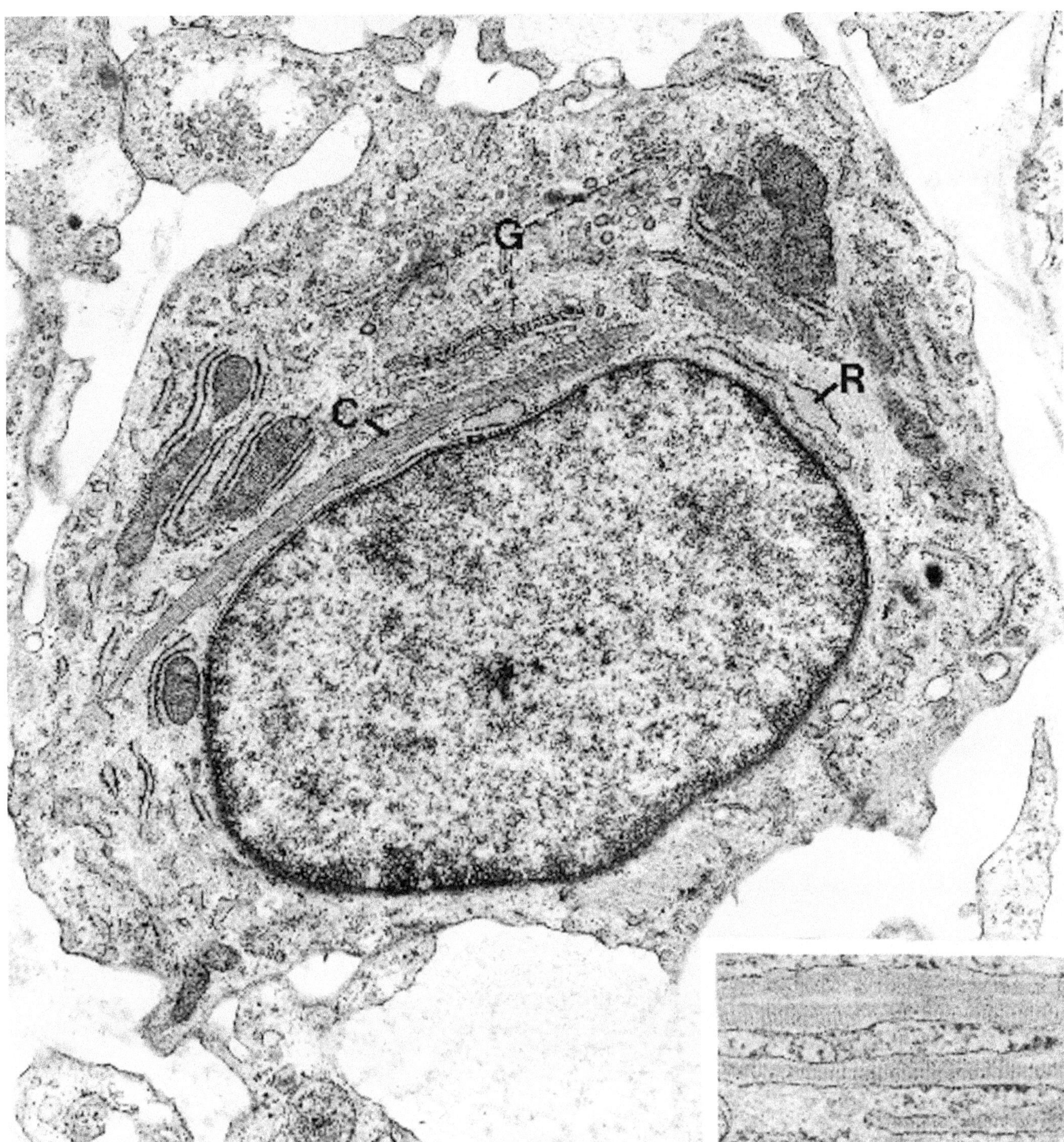

Figure 6.15. Fibrosarcoma (supraclavicular lymph node). This neoplastic fibroblast shows a moderate amount of rough endoplasmic reticulum (R), prominent Golgi apparatuses (G), and intracisternal banded collagen (C), seen at higher magnification in the *inset.* (× 19,000; inset × 90,000)

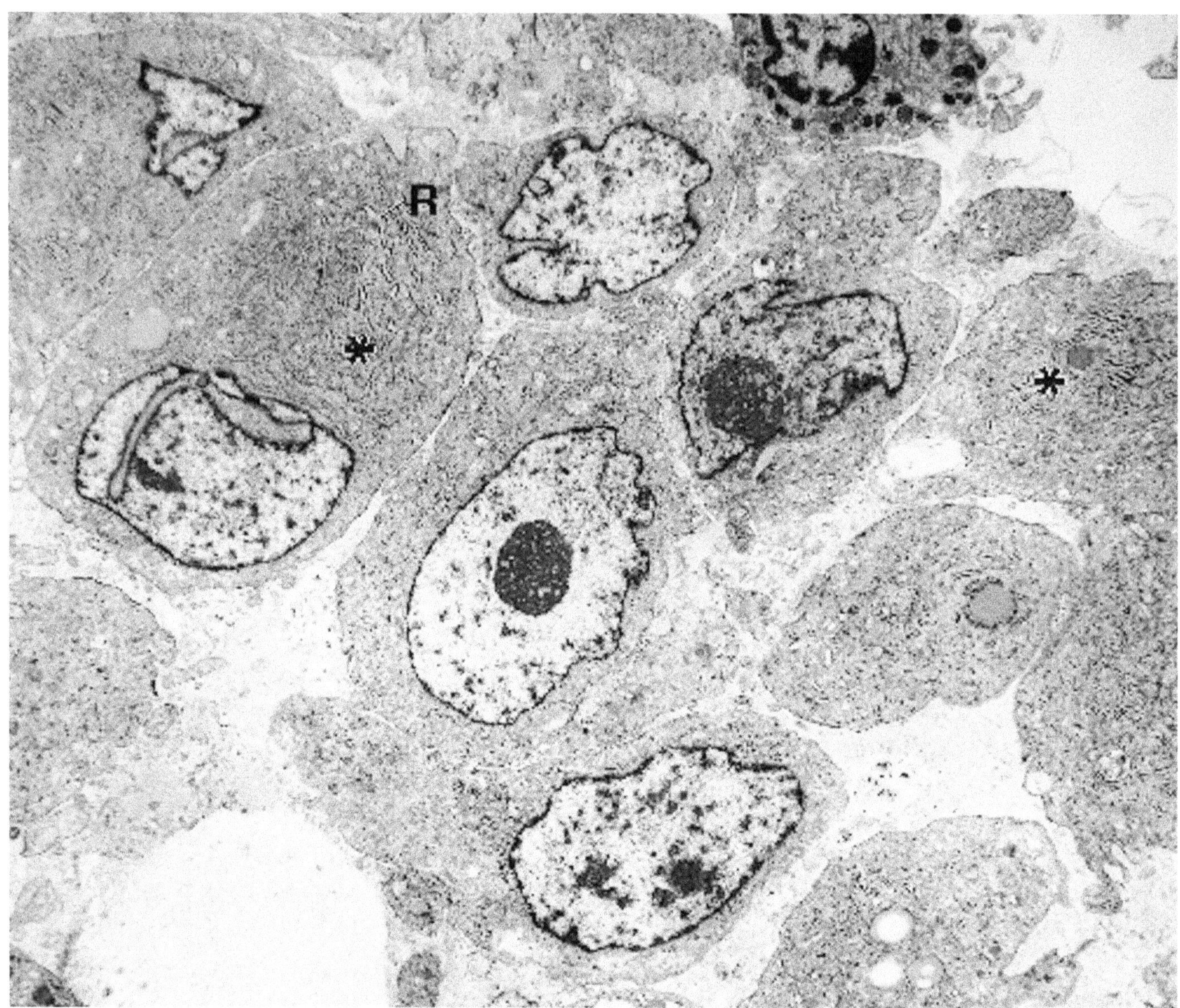

Figure 6.16. Epithelioid fibrosarcoma (pretracheal soft tissue). Several epithelioid cells have varying amounts of rough endoplasmic reticulum (R), and it is the main organelle in some of the cells (*). No line of differentiation other than a fibroblast is suggested by the ultrastructure. (× 5600)

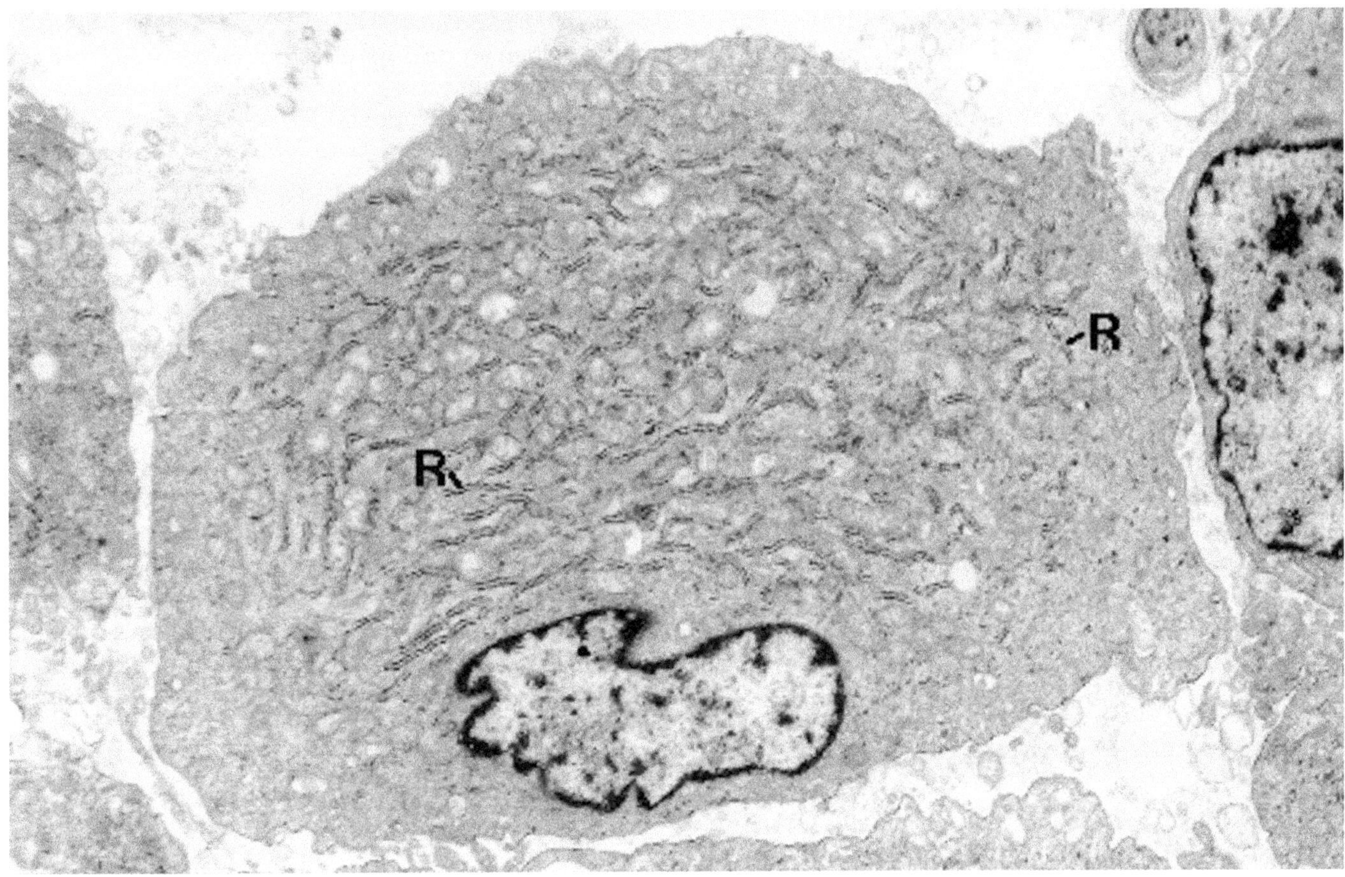

Figure 6.17. Epithelioid fibrosarcoma (pretracheal soft tissue). Higher magnification of a cell from the same neoplasm depicted in Figure 6.16 illustrates the abundant rough endoplasmic reticulum (R). (× 9300)

(Text continued from page 247)

Solitary fibrous tumors of soft tissue are similar ultrastructurally to solitary fibrous tumors of pleura and other serosal surfaces, namely that the cell type is either an identifiable fibroblast (Figure 6.14) or too poorly differentiated to be certain of cell type (See Figures 3.75 and 3.76). Giant collagen fibrils, collectively forming so-called amianthoid fibers (Figure 6.14), may be found in these and other fibroblastic neoplasms as well as in cartilagenous lesions. Intracisternal banded collagen, that is, collagen within cisternae of rough endoplasmic reticulum (Figure 6.15), may be seen occasionally in actively synthesizing fibroblasts.

Although most, by far, fibroblastic proliferations are composed of spindle cells, rare fibrosarcomas may be epithelioid, or a combination of spindle and epithelioid cells (Figures 6.16 and 6.17). Neoplasms of this type may be exceedingly difficult to diagnosis by light microscopy and immunohistochemistry, and electron microscopy is essential.

Malignant Fibrous Histiocytoma

(Figures 6.18 through 6.30.)

Diagnostic criteria. (1) Fibroblasts (spindle cells with abundant rough endoplasmic reticulum; see section on fibrosarcoma) (2) mononucleated and multinucleated giant cells of fibroblastic type (Figures 6.18 and 6.19); (3) giant cells of osteoclastic type, especially in "giant cell MFH" (Figure 6.20; also see Figure 6.13 [giant cell reparative granuloma] and Figures 6.45 and 6.46 [osteosarcoma]); (4) malignant facultative histiocytes (Figures 6.20 and 6.21) or malignant true histiocytes (Figure 6.22); (5) primitive, or poorly differentiated fibroblasts (Figure 6.23); (6) small, benign-appearing histiocytes (see Chapter 3, Figure 3.87).

Additional points. A high percentage of sarcomas diagnosed as *storiform/pleomorphic MFH* and *myxoid MFH* at the light microscopic level prove to be pure fibrosarcomas when studied by electron microscopy; both the spindle cells and the giant cells have fibroblastic features. A smaller proportion of these tumors contains giant cells that are consistent with being facultative histiocytes; that is, they have primary and secondary lysosomes as well as abundant, usually undilated, rough endoplasmic reticulum. A few tumors contain a component of malignant-appearing, true histiocytes. Small, benign-appearing histiocytes as well as other inflammatory cells may be found in varying numbers and distribution in these tumors. In *giant cell MFH,* osteoclastic type giant cells are numerous but also mixed with fibroblasts and facultative histiocytes. In *inflammatory MFH,* numerous xanthoma cells, rich in lipid (Figure 6.24), are found in combination with acute and chronic inflammatory cells and with areas of storiform/pleomorphic spindle cells. Lipid-rich fibroblasts may also be present in the usual storiform/pleomorphic form of MFH and in fibrosarcoma, and there the cells are spindle shaped rather than round, and fibroblastic rather than histiocytic (Figure 6.25). *Angiomatoid MFH* is usually described at the light microscopic level as consisting of islands of histiocytes, histiocyte-like cells, fibrohistiocytes, and/or undifferentiated mesenchymal cells, interspersed with blood-filled spaces and surrounded by a lymphoplasmacytic infiltrate. In our own experience, the neoplastic cells ultrastructurally are fibroblasts and myofibroblasts (Figures 6.26 through 6.30). Thus, although there may not be a consensus of opinion on the cell type comprising angiomatoid MFH, it appears acceptable to continue to classify neoplasms of this type in the fibroblastic group and, more specifically, with MFH.

Pleomorphic malignant fibrous histiocytomas at the light microscopic level may be confused with pleomorphic forms of rhabdomyosarcoma, leiomyosarcoma, and liposarcoma, but distinguishing among these lesions usually is readily achievable by electron microscopic examination.

(Text continues on page 278)

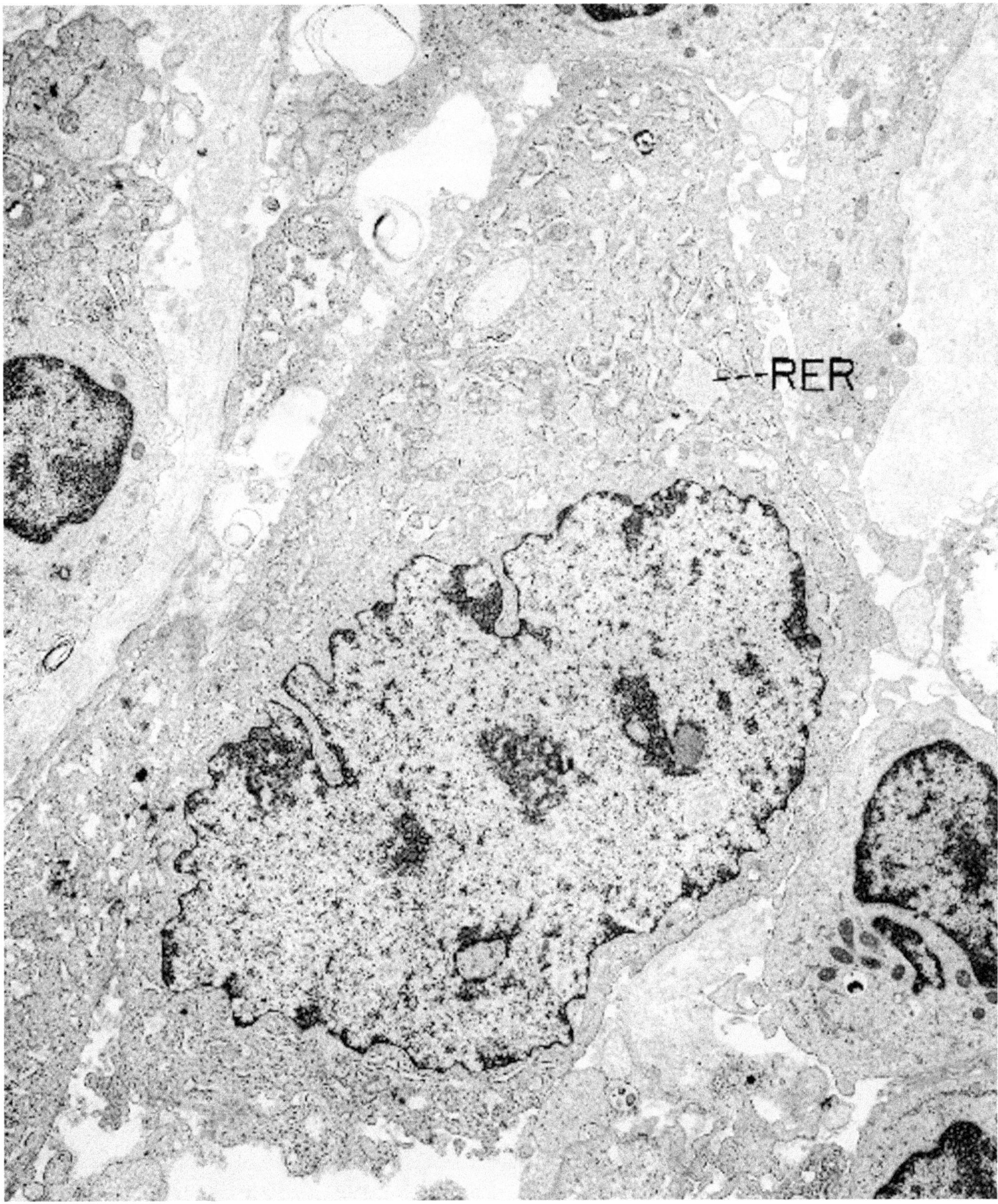

Figure 6.18. Malignant fibrous histiocytoma (soft tissue of anterior abdominal wall). This mononucleated giant cell is a fibroblast, typified by its abundant, dilated rough endoplasmic reticulum (RER). (× 6750)

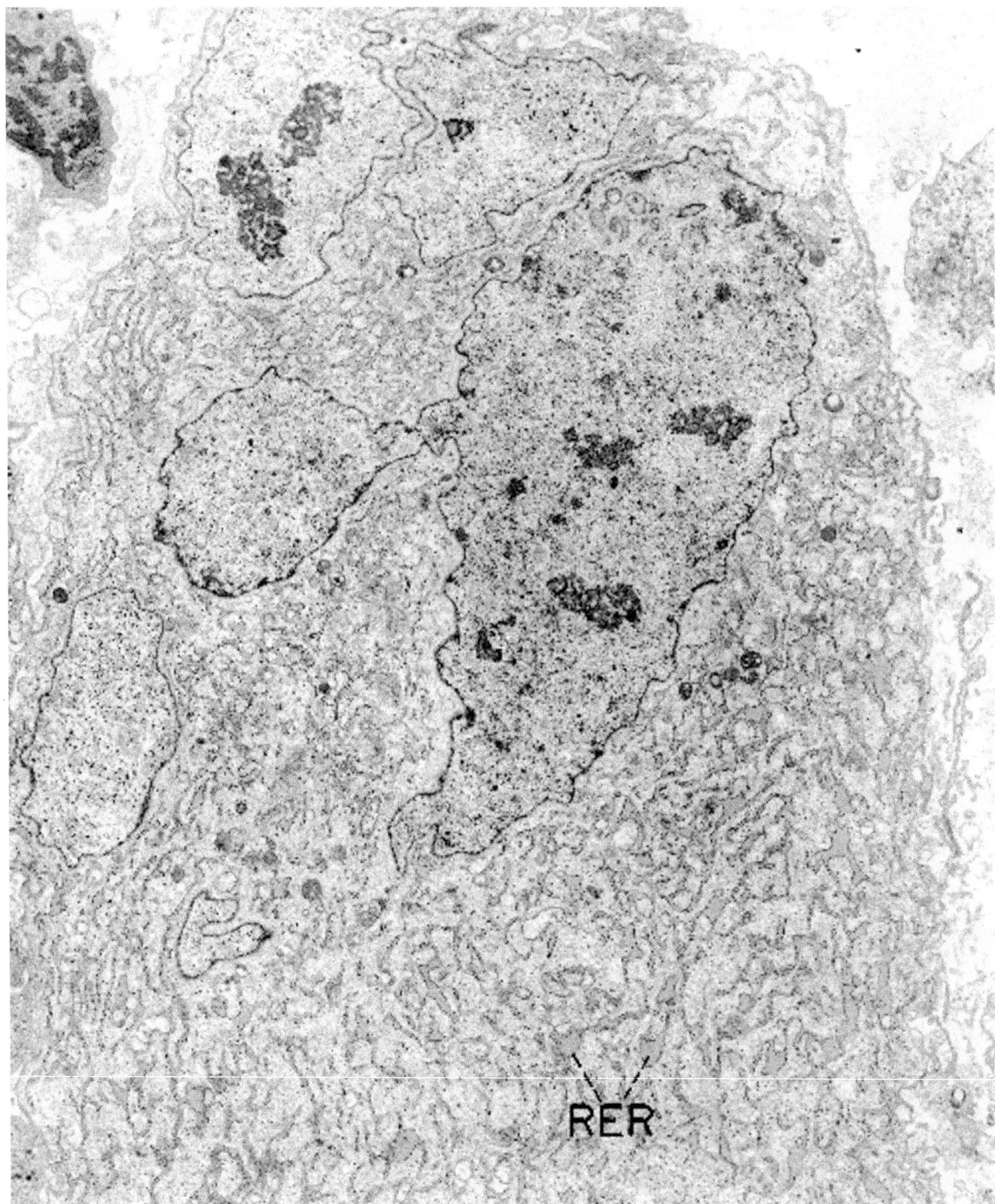

Figure 6.19. Malignant fibrous histiocytoma (soft tissue of back). This giant multinucleated fibroblast is filled with rough endoplasmic reticulum (RER) and shows no evidence of histiocytic markers. The nucleus is probably multilobed, but at the light microscopic level of magnification it probably would be interpreted as multiple nuclei. (× 5320)

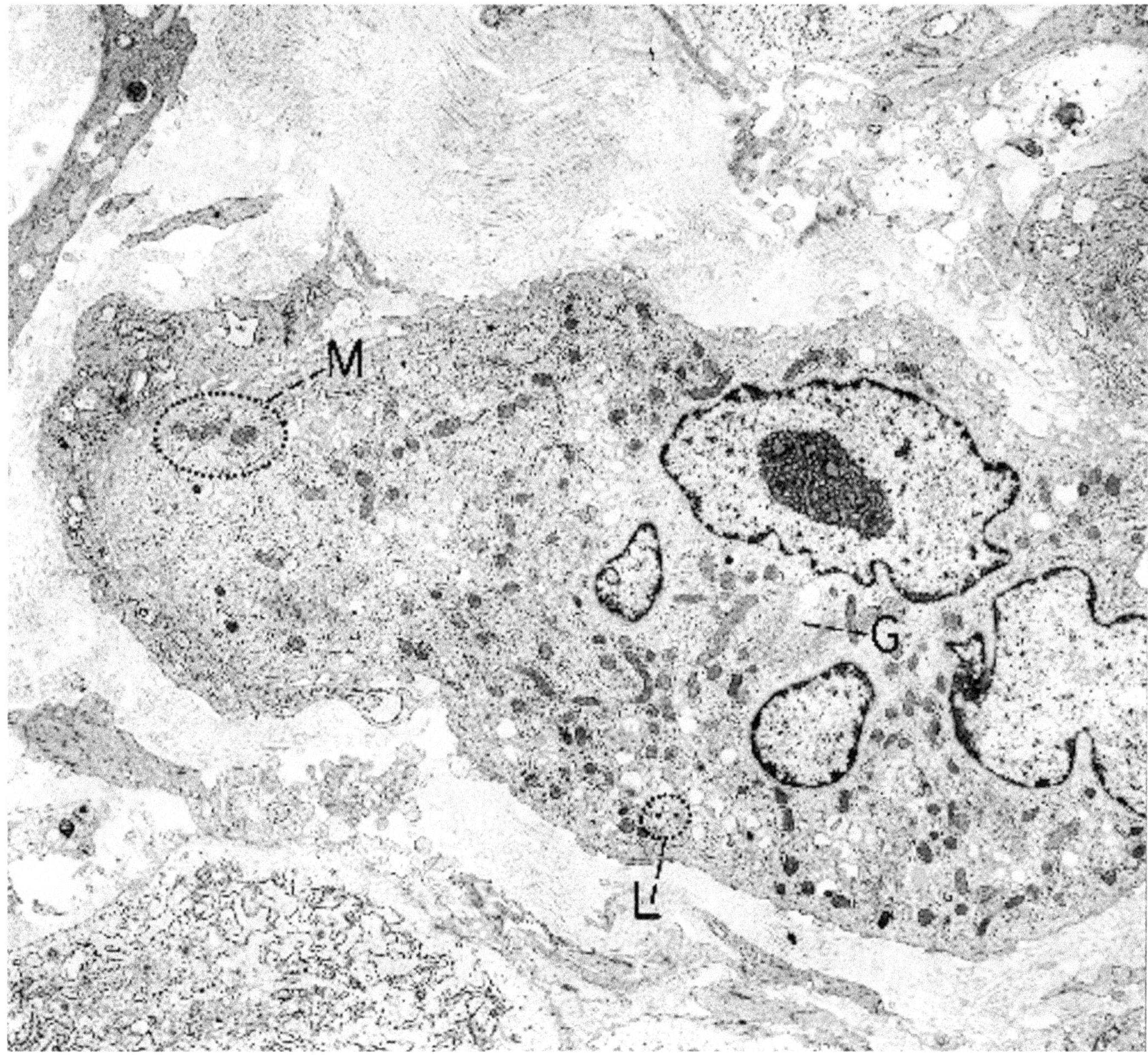

Figure 6.20. Malignant fibrous histiocytoma (rectus femoris muscle). A giant cell contains many different organelles in its cytoplasm. Rough endoplasmic reticulum is abundant but not especially dilated. The Golgi apparatus (G) is prominent, mitochondria (M) are numerous, and lysosomes (L) are moderate in number (seen at higher power in Figure 6.21. These characteristics are highly suggestive of the cell being a facultative histiocyte. (× 4940)

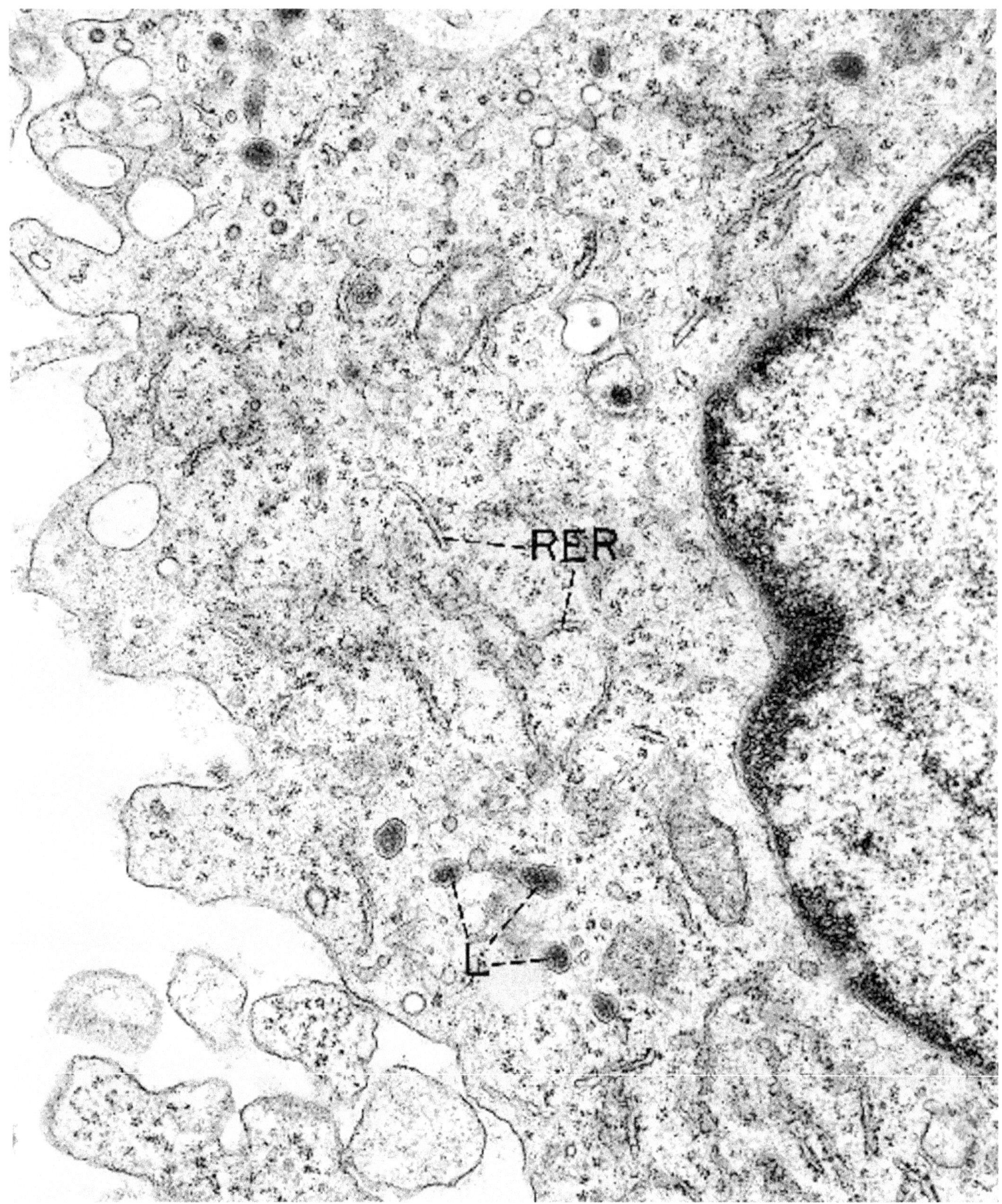

Figure 6.21. Malignant fibrous histiocytoma (rectus femoris muscle). A higher magnification of a portion of the giant cell and probable facultative histiocyte depicted in Figure 6.20 highlights the frequent primary lysosomes (L). No secondary lysosomes (phagosomes) were identified in this cell. The rough endoplasmic reticulum (RER) is plentiful but not very dilated. (× 29,000)

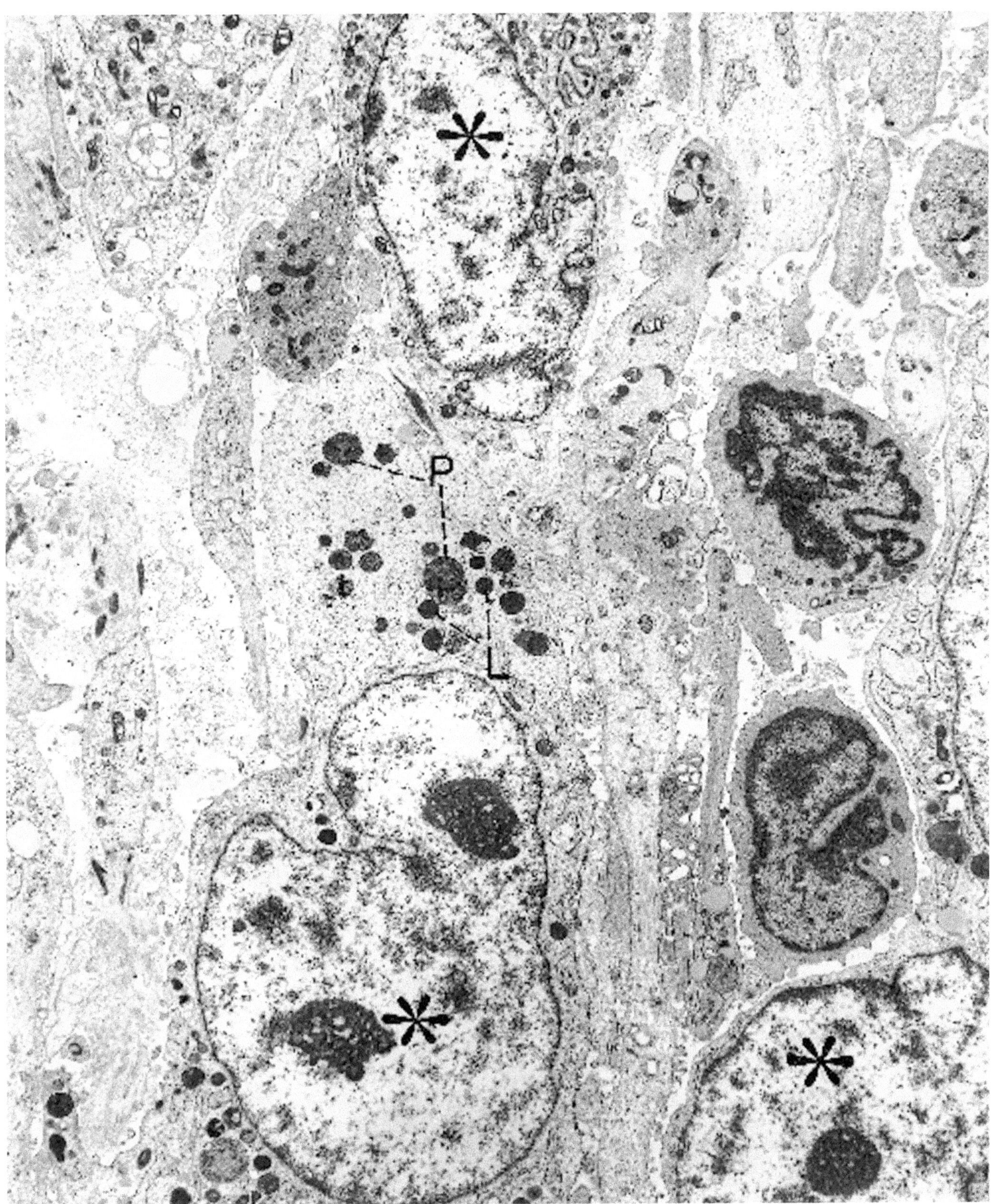

Figure 6.22. Malignant fibrous histiocytoma (parailial soft tissue). Among the fibroblasts and facultative histiocytes in this neoplasm were cells of the type depicted here (*). Their size and nuclear features are of a malignant order, and their cytoplasm contains numerous phagosomes (P) and lipid droplets (L). They are interpreted as malignant histiocytes. (× 5700)

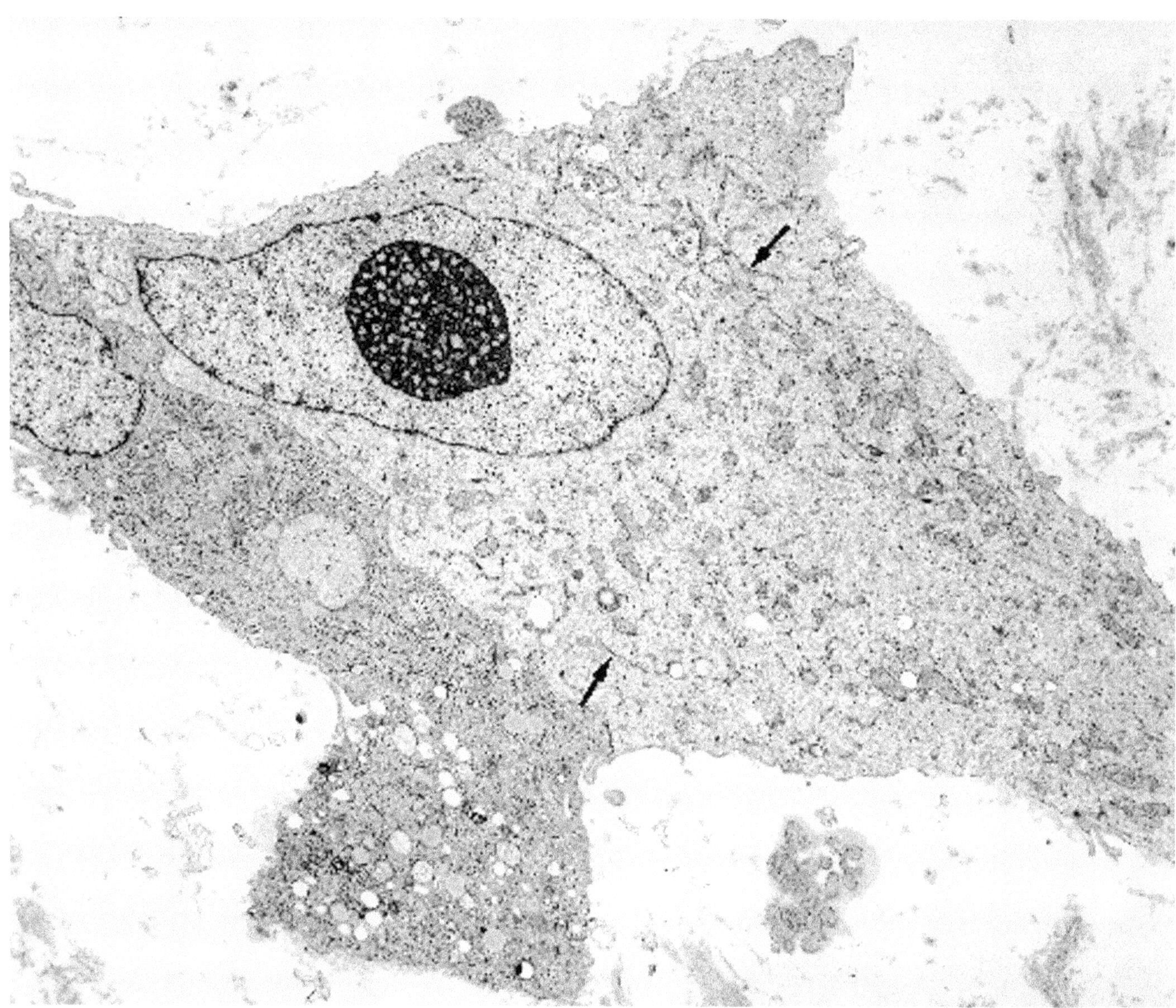

Figure 6.23. Malignant fibrous histiocytoma (liver). A primitive or poorly differentiated fibroblast has a moderate amount of undilated rough endoplasmic reticulum (*arrows*), but the predominant organelle is free ribosomes. The nucleus is euchromatic and contains a large nucleolus, features characteristic of proliferative or metabolic activity, the former being more likely in this neoplastic cell. (× 5100)

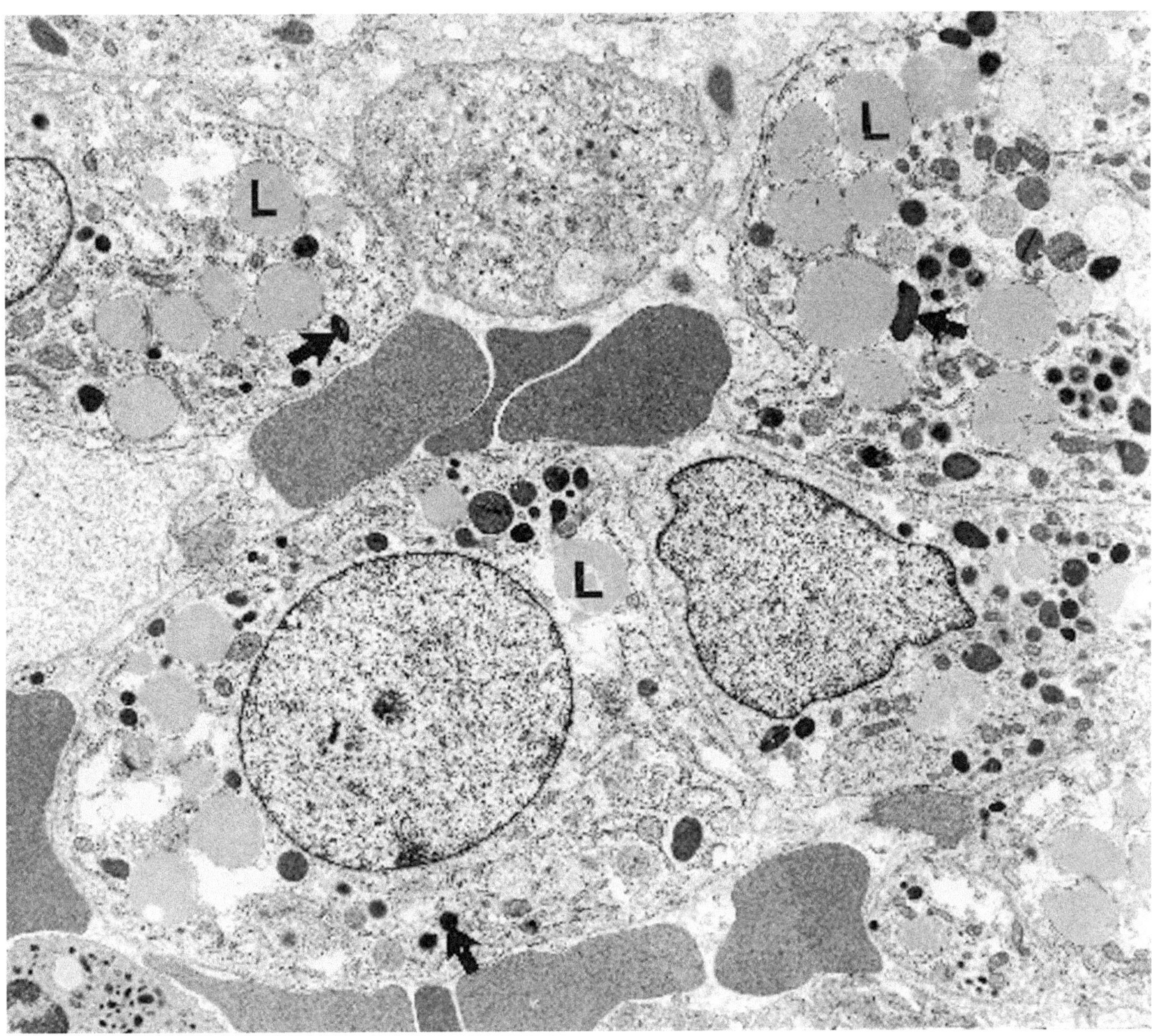

Figure 6.24. Malignant fibrous histiocytoma, inflammatory type (soft tissue of thigh). Several xanthomatous, histiocytic type cells contain numerous lipid droplets (L) and lysosomes (*arrows*) of various sizes. (× 6500)

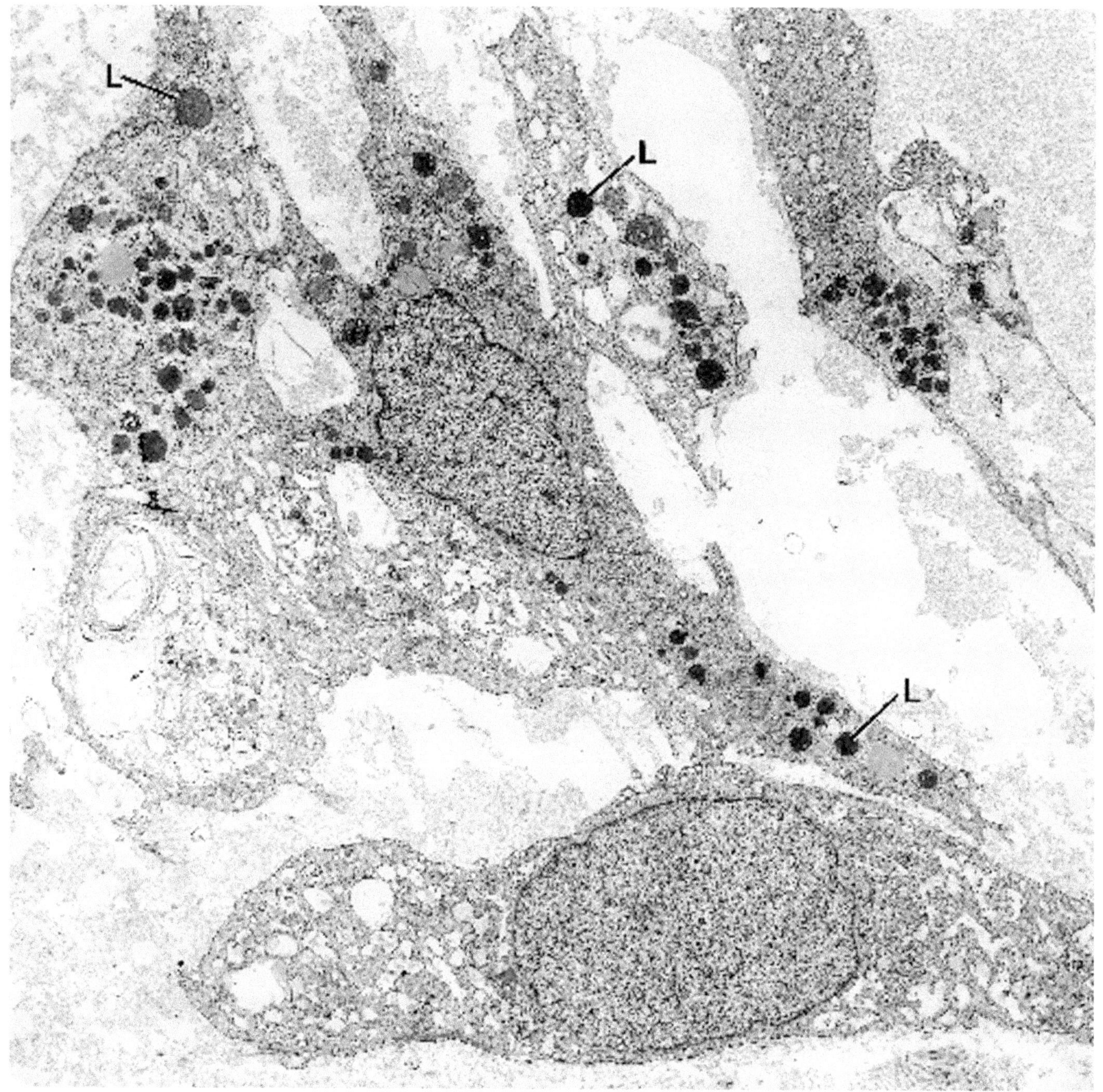

Figure 6.25. Malignant fibrous histiocytoma (metastatic to lung). Neoplastic fibroblasts have numerous lipid droplets (L) in their cytoplasm. (× 4800)

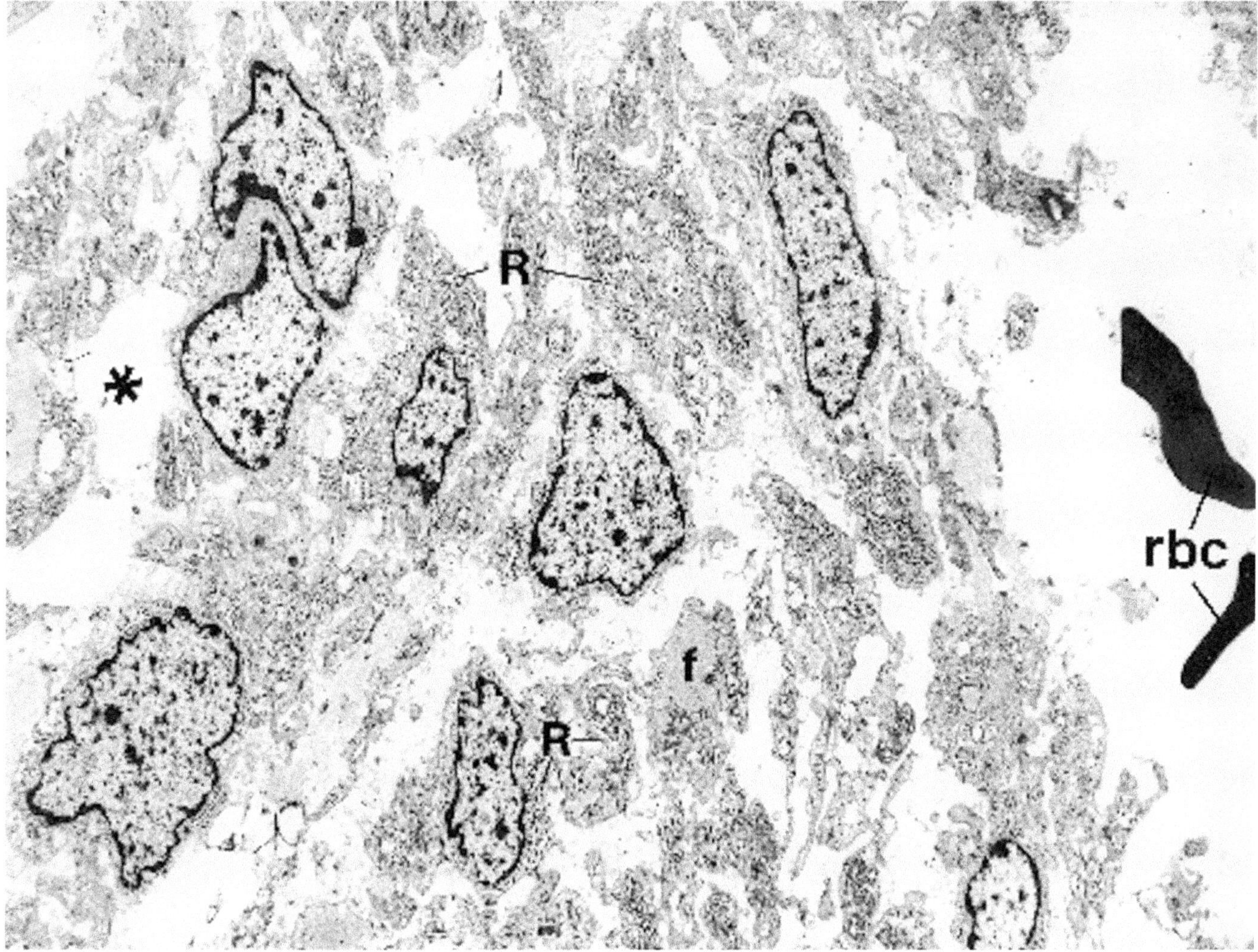

Figure 6.26. Malignant fibrous histiocytoma, angiomatoid type (soft tissue of arm). Edematous (*) and hemorrhagic (rbc) matrix separates neoplastic oval and spindle cells. Cytoplasm contains rough endoplasmic reticulum (R) (better seen at higher magnification in Figure 6.27 and focal filaments (f). Nuclei are irregularly indented and moderately heterochromatic. (× 5300)

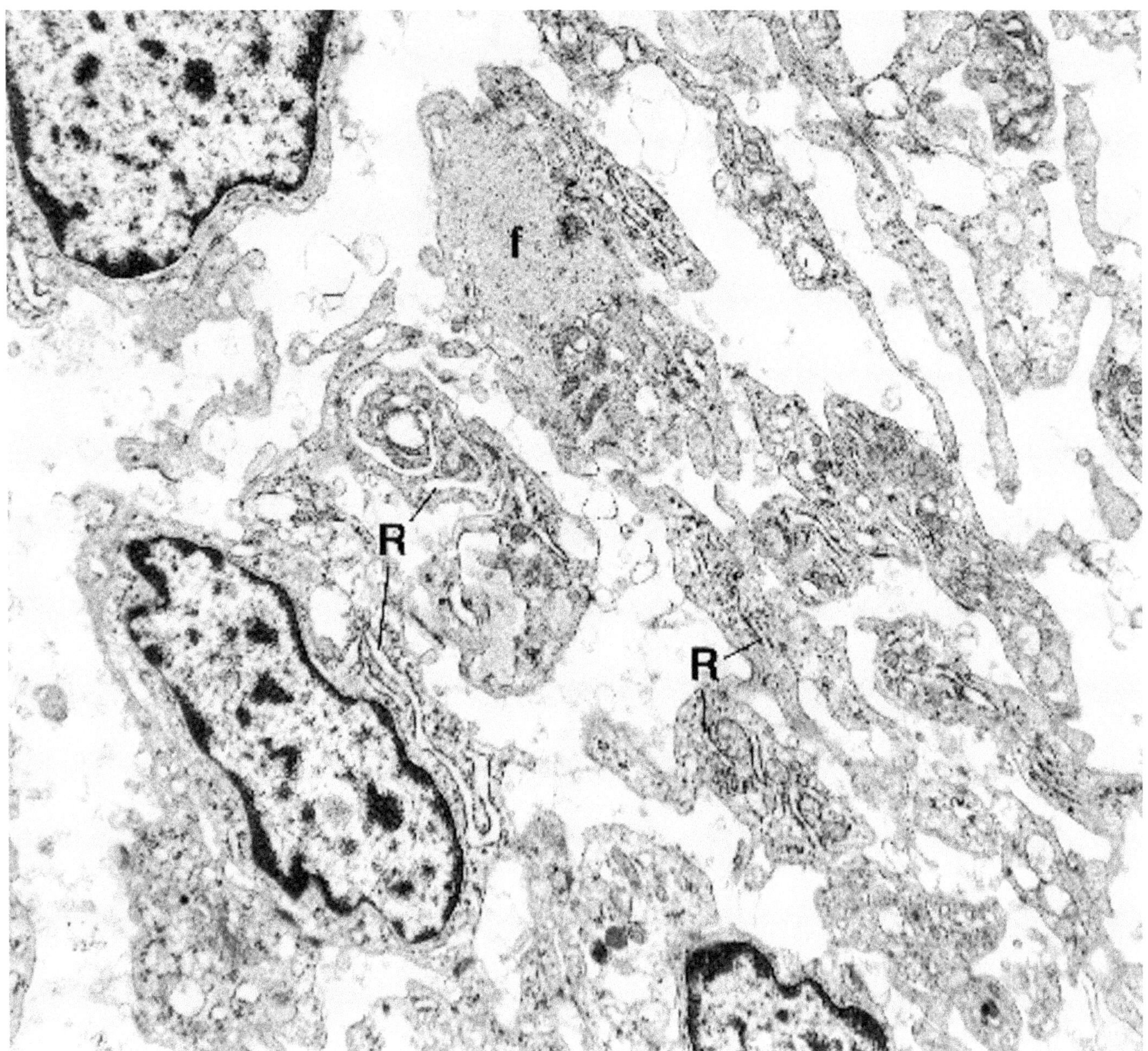

Figure 6.27. Malignant fibrous histiocytoma, angiomatoid type (soft tissue of arm). Higher magnification of the lower central field of Figure 6.26 depicts more clearly the rough endoplasmic reticulum (R) and focal filaments (f). (× 13,600)

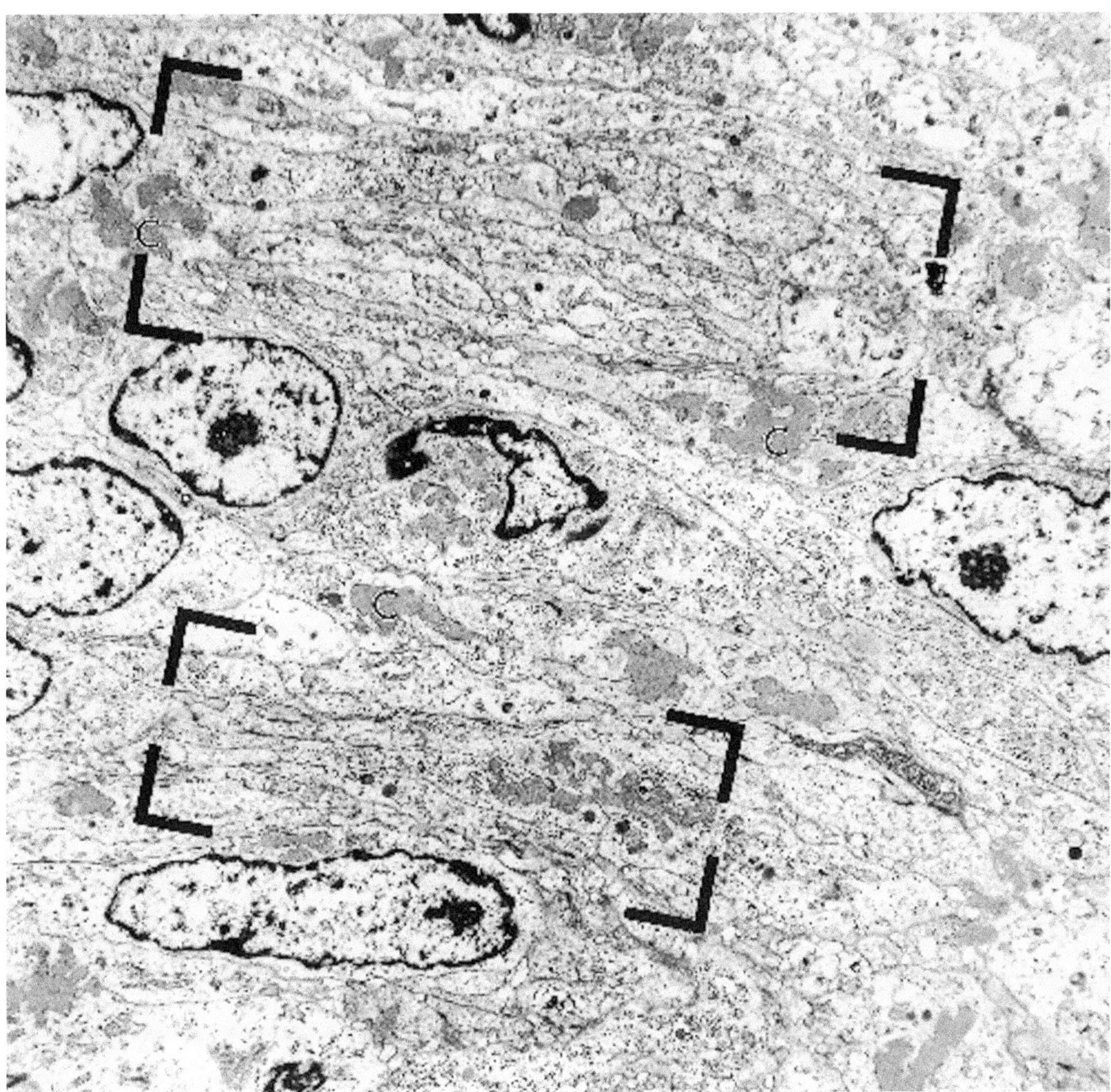

Figure 6.28. Malignant fibrous histiocytoma, angiomatoid type (soft tissue of arm). A more cellular and less edematous and hemorrhagic area of the neoplasm depicted in Figures 6.26 and 6.27 reveals groups of long, narrow, cellular processes (*brackets*) and pockets of closely apposed extracellular collagen (C). (× 5500)

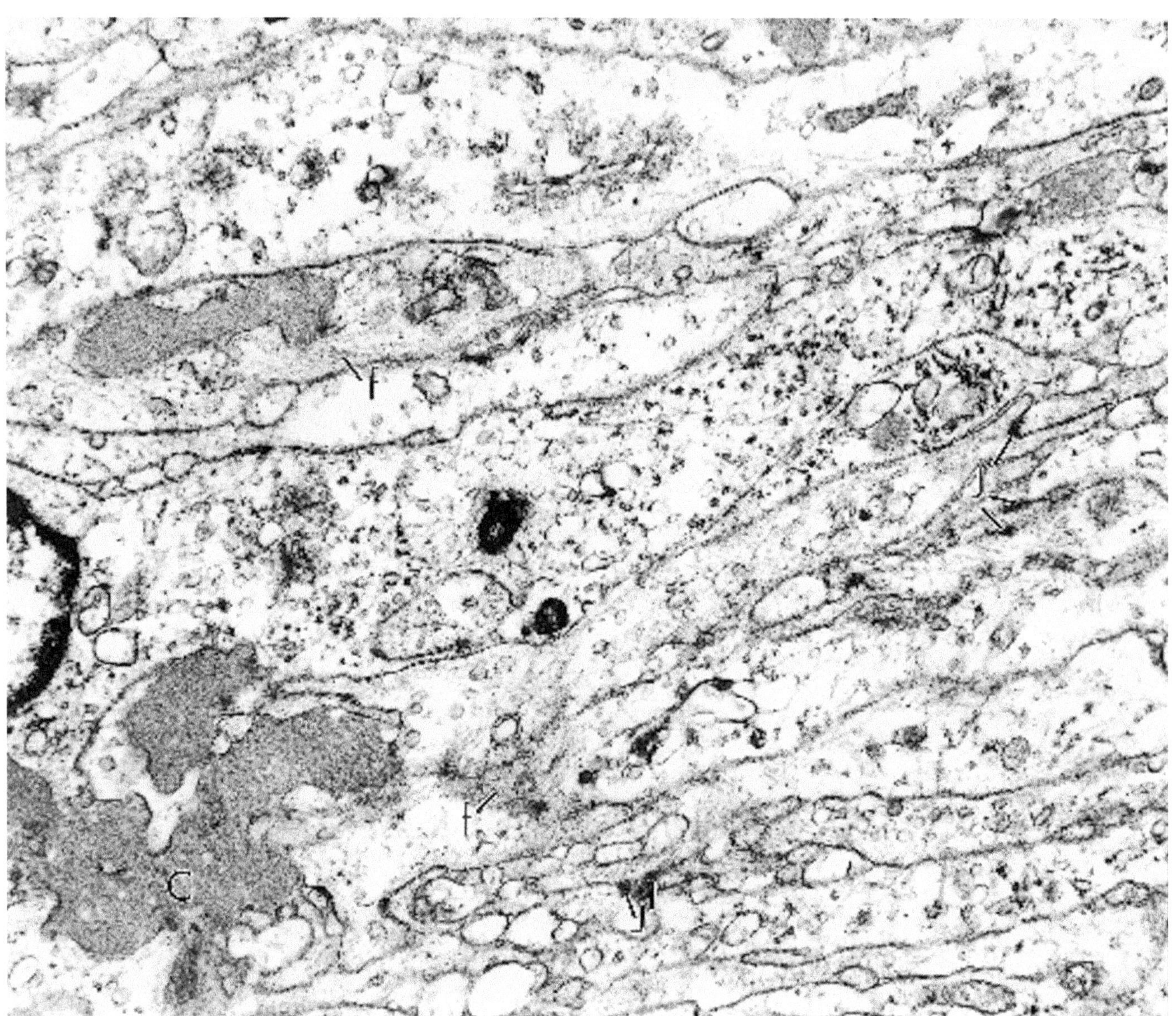

Figure 6.29. Malignant fibrous histiocytoma, angiomatoid type (soft tissue of arm). Higher magnification of the left half of the upper bracketed area in Figure 6.28 illustrates in finer detail the processes and collagen (C). In addition, focal filaments (f) and several intercellular junctions (j) are also visible. (× 19,800)

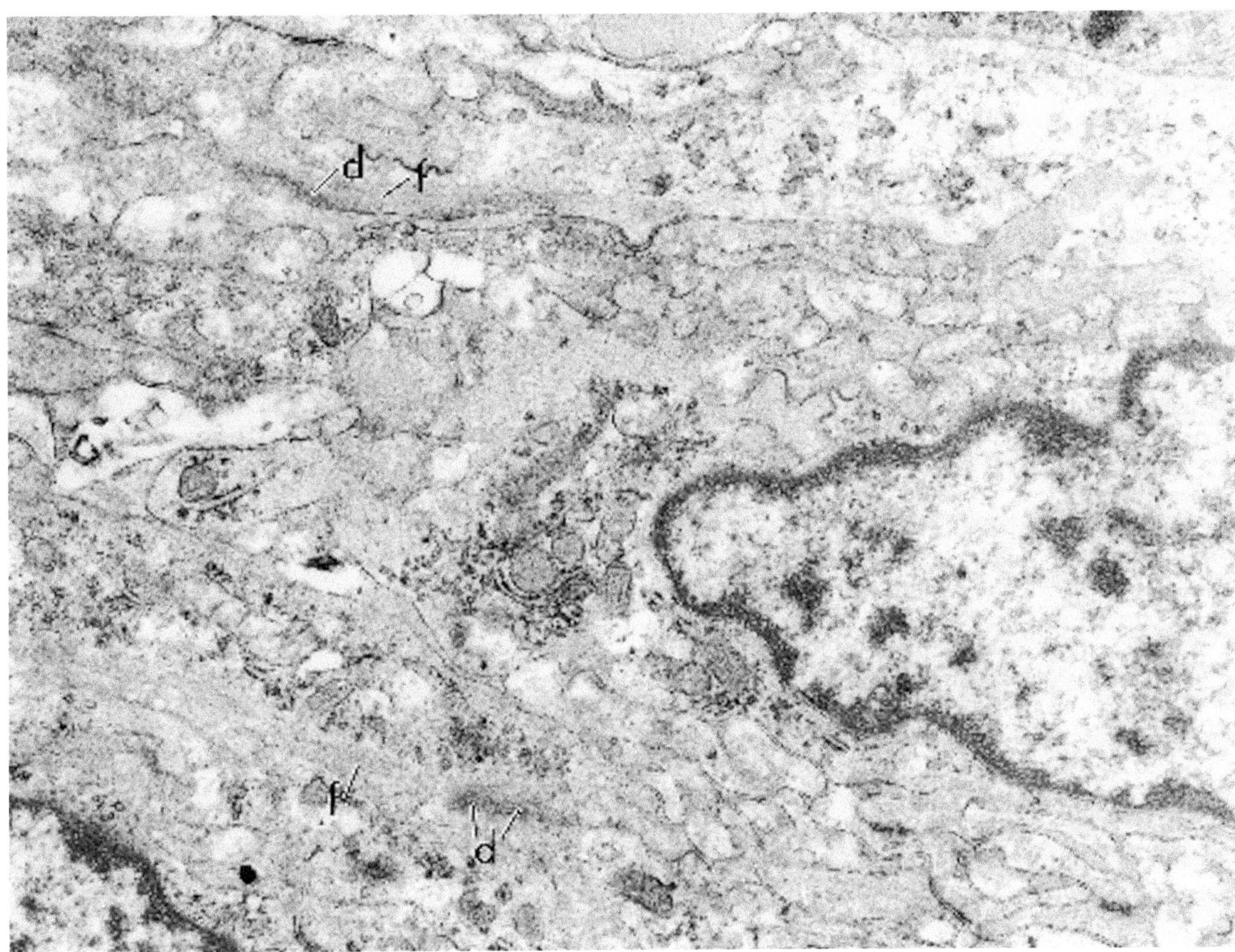

Figure 6.30. Malignant fibrous histiocytoma, angiomatoid type (soft tissue of arm). High magnification of the same neoplasm as that depicted in Figures 6.26 through 6.29 depicts cytoplasmic filaments (f) and dense bodies (d). (× 22,000)

(Text continued from page 264)

Cartilaginous Neoplasms

(Figures 6.31 through 6.36.)

Diagnostic criteria. (1) Oval and polygonal cells and, in some less differentiated neoplasms, spindle-shaped cells; (2) escalloped or villus-like cell surfaces (Figures 6.31 through 6.33); (3) clear zone between cell and visible matrix (Figure 6.33); (4) abundant dilated rough endoplasmic reticulum (Figures 6.31 through 6.33); (5) large Golgi apparatus (Figure 6.32); (6) copious cytoplasmic glycogen (Figures 6.34 and 6.35), especially in *clear cell chondrosarcoma* (Figure 6.34); (7) variable intermediate filaments.

Additional points. Chondroblasts may be difficult to distinguish from osteoblasts, although they usually contain more glycogen than osteoblasts, and they do not manufacture osteoid. Both chondroblasts and osteoblasts are closely related morphologically and functionally to fibroblasts, and this is most evident by their abundant rough endoplasmic reticulum and their active synthesis of extracellular matrix.

In *extraskeletal myxoid chondrosarcoma,* chondroblasts are widely separated by loose, flocculent matrix. Mitochondria may be numerous, and rough endoplasmic reticulum may contain arrays of microtubules (Figure 6.36). In the past, some examples of extraskeletal myxoid chrondrosarcoma were referred to as "chordoid sarcoma" and "parachordoma" because of their light microscopic resemblance to chordomas, but by electron microscopy they are seen to be composed of chondroblasts.

Mesenchymal chondrosarcoma is a small cell neoplasm that may be indistinguishable from Ewing's sarcoma and small cell osteosarcoma (see Chapter 4, Figures 4.60 and 4.61).

Chondromyxoid fibroma is a neoplasm in which chondroblasts are interspersed with fibroblasts and myofibroblasts in a myxoid stroma.

Osteoblastoma and Osteosarcoma

(Figures 6.37 through 6.46.)

Diagnostic criteria. (1) Oval and polygonal cells are more common than spindle-shaped cells (Figures 6.37 and 6.38); (2) escalloped or villus-like cell surfaces (Figures 6.38 and 6.39); (3) clear zone between cell and visible matrix (Figure 6.40); (4) abundant dilated rough endoplasmic reticulum (Figures 6.38 through 6.40); (5) large Golgi apparatus (Figure 6.40); (6) moderate to large amounts of cytoplasmic glycogen (Figure 6.41); (7) hydroxyapatite deposits on prominent fibers of collagen (osteoid) (Figure 6.42).

Additional points. Osteoblasts are morphologically similar to chondroblasts, and the presence of hydroxyapatite deposits in the extracellular matrix (Figures 6.42 through 6.44) may be the only distinguishing feature. In poorly differentiated neoplasms without much osteoid, small deposits that may not be visible or convincing by light microscopy can be identified readily at the ultrastructural level. However, the small size of the electron microscopic sample often makes the findings of the deposits a matter of chance.

Although most osteogenic and chondrogenic neoplasms are composed, at least in part, of cells having most of the morphologic features described in the diagnostic criteria section, there also are less-differentiated examples that are composed of spindle-shaped cells having some resemblance to fibroblasts (Figure 6.37). In these instances, the search for better differentiated foci and a careful reevaluation of the light microscopic picture can prove rewarding. On occasion, it may be necessary to excise from the paraffin blocks a focus of better differentiated neoplasm and reprocess it for electron microscopy. The morphologic detail of reprocessed, formalin-fixed, paraffin-embedded tissue usually will not be well preserved, but it may be adequate for establishing the exact or probable cell type of the neoplasm.

Osteoclasts sometimes are included in the sample submitted for electron microscopic study, and they are readily distinguishable from osteoblasts by their large size, multinucleation, and busy cytoplasm, especially numerous mitochondria (Figures 6.45 and 6.46).

Small cell osteosarcoma is composed of poorly differentiated, small, round, and oval cells that may be indistinguishable from the cells of mesenchymal chondrosarcoma and Ewing's sarcoma (see Chapter 4, Figures 4.58 and 4.59).

(Text continues on page 295)

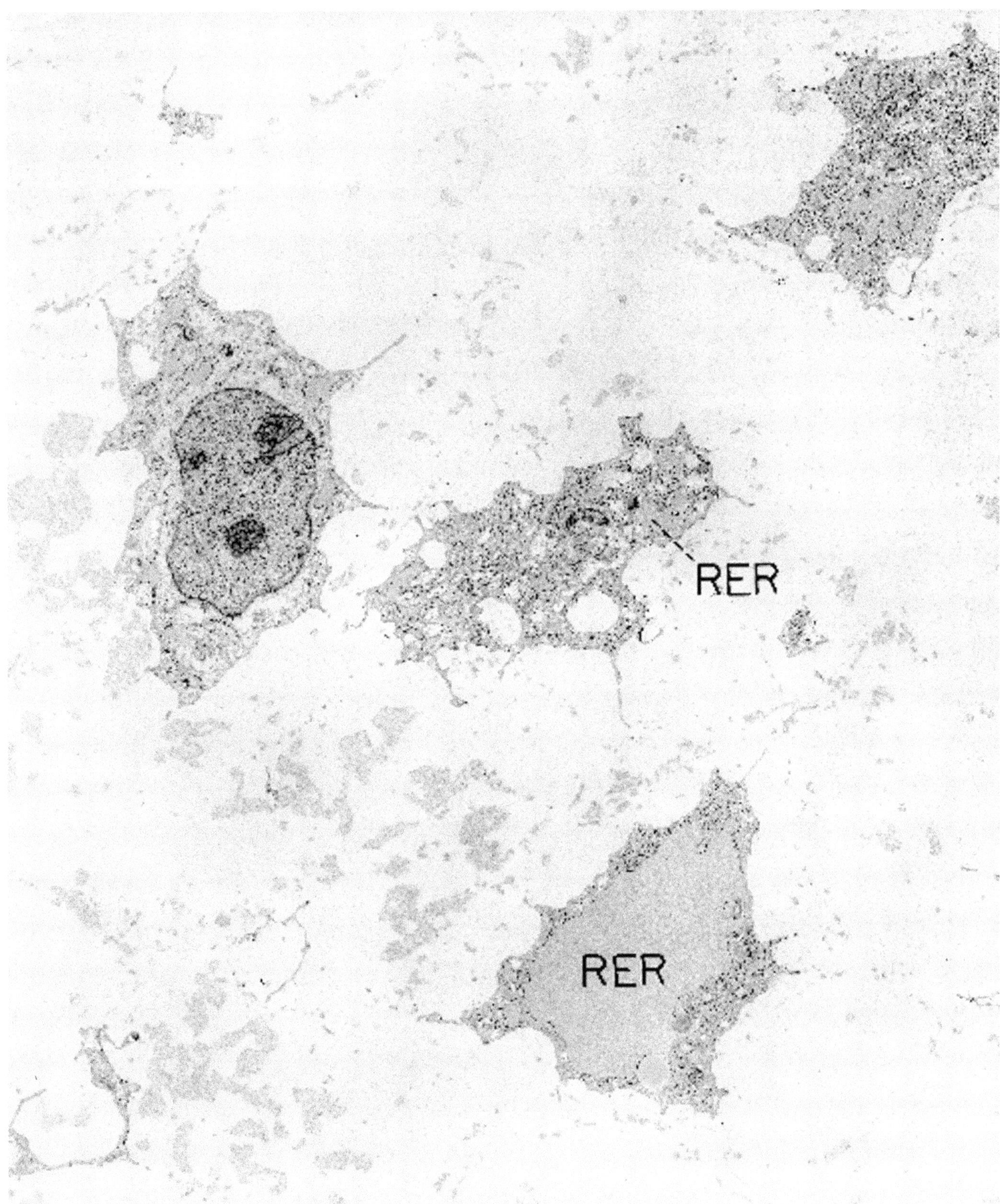

Figure 6.31. Myxoid chondrosarcoma (parasellar region of brain). The chondroblasts are widely separated in abundant matrix, and their cytoplasm is rich in dilated rough endoplasmic reticulum (RER). The medium-dense material filling the reticulum represents the proteinaceous precursor of the matrical collagen, glycoprotein, and glycosaminoglycans. (× 4275)

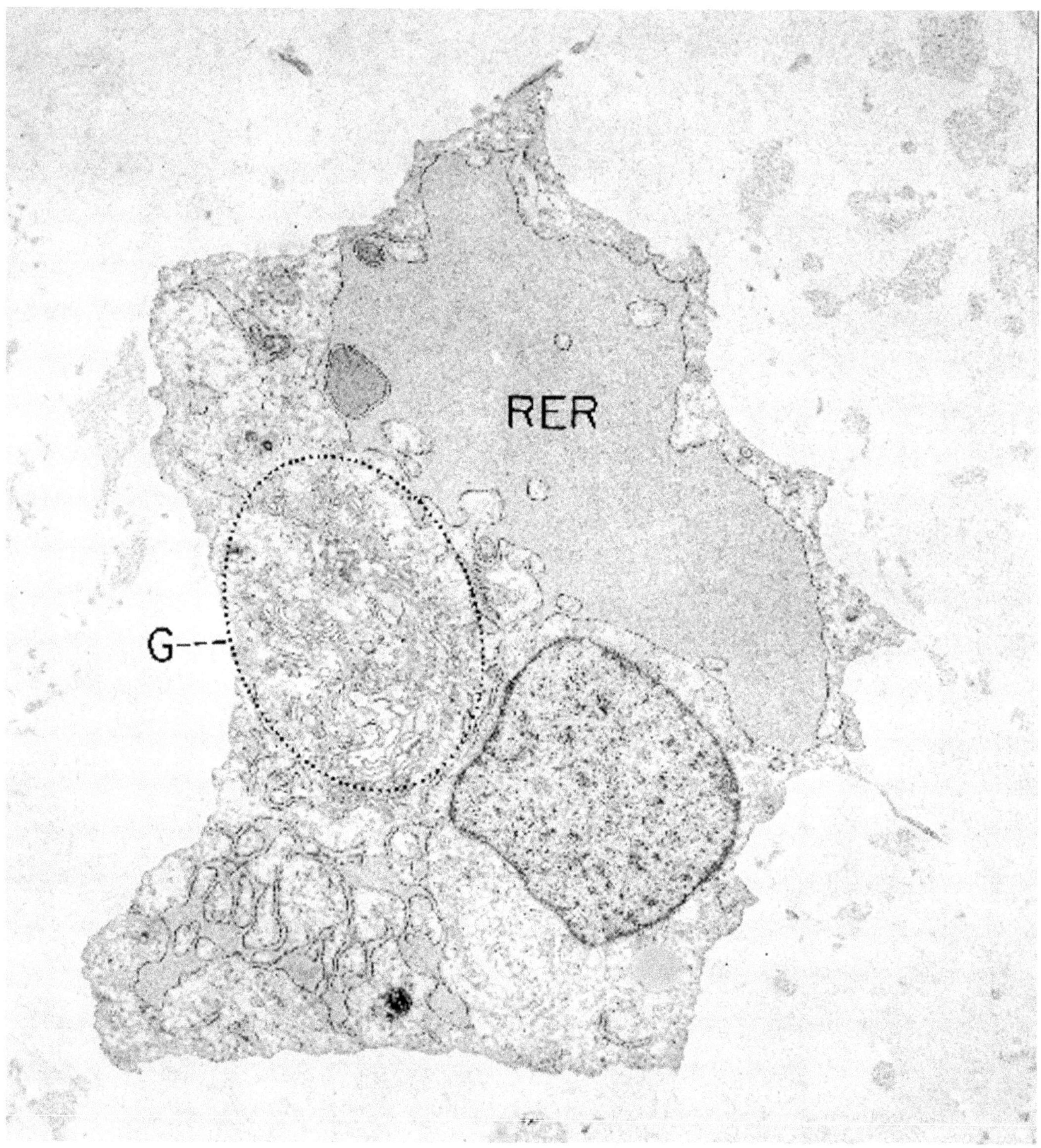

Figure 6.32. Myxoid chondrosarcoma (parasellar region of brain). This chondroblast, in addition to showing extreme dilation of the rough endoplasmic reticulum (RER), has a large Golgi apparatus (G). (× 7600)

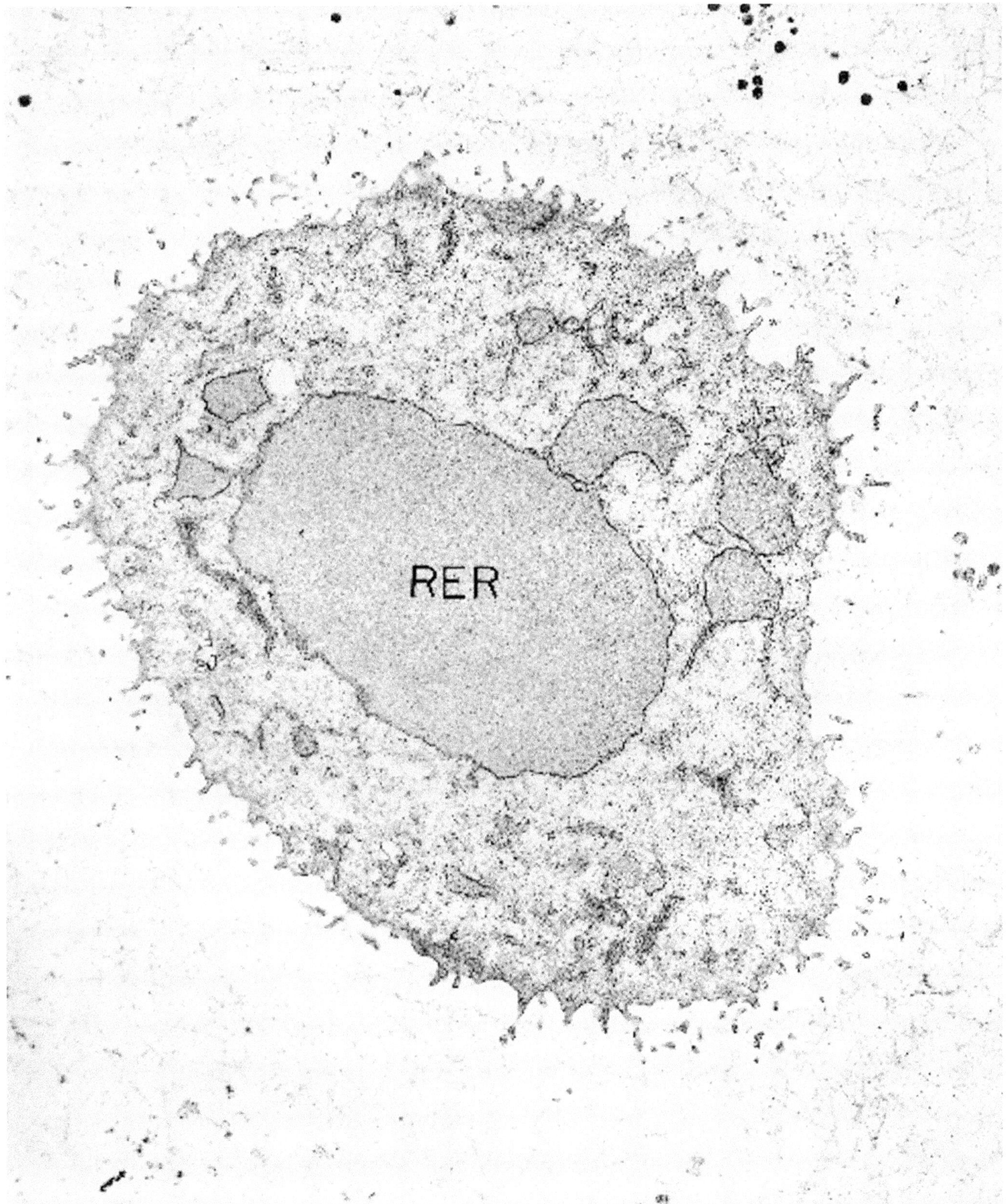

Figure 6.33. Enchondroma (tibia). A well-differentiated chondroblast has a villus-like surface and a clear zone between the cell and the collagen. RER = rough endoplasmic reticulum. (× 9200)

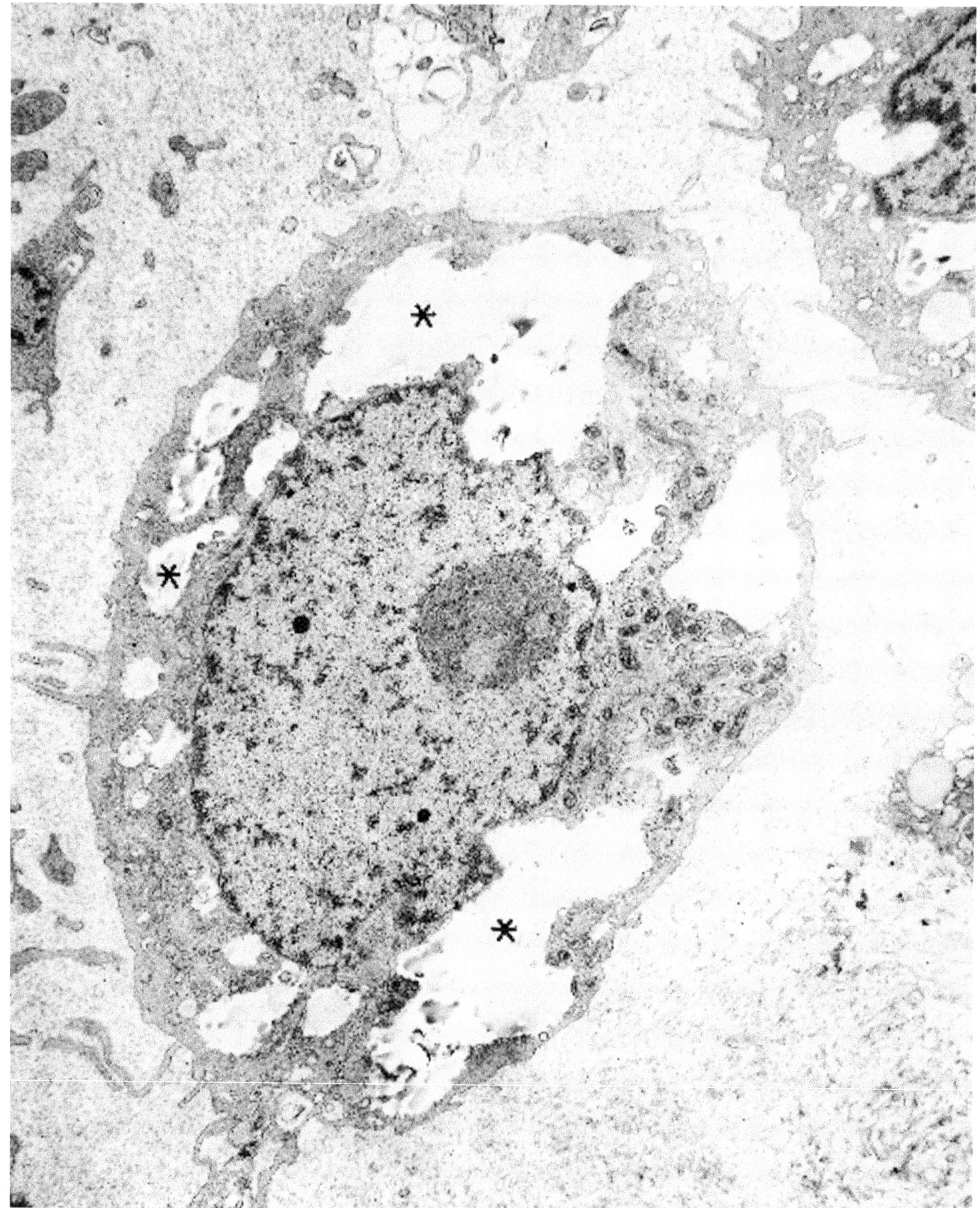

Figure 6.34. Clear cell chondrosarcoma (synovial membrane of thigh). The glassy, clear spaces (*) in the cytoplasm of this chondroblast represent glycogen that was lost during the chemical processing of the tissue. (× 9000)

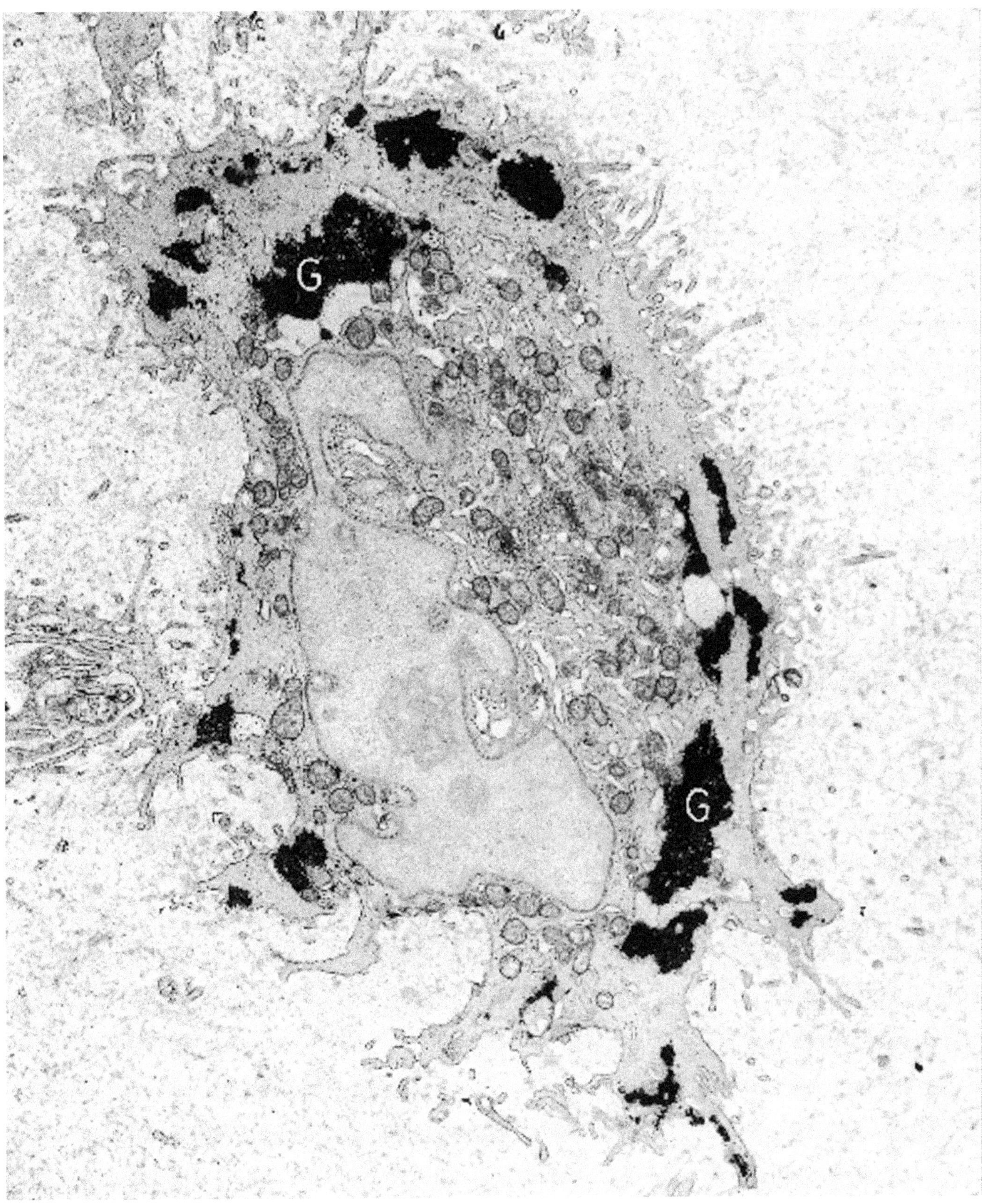

Figure 6.35. Chondroblastoma (humerus). This specimen was processed by a method that preserves glycogen as electron-dense granules, and copious amounts of glycogen (G) are evident in the cytoplasm of this neoplastic cell. (× 8200)

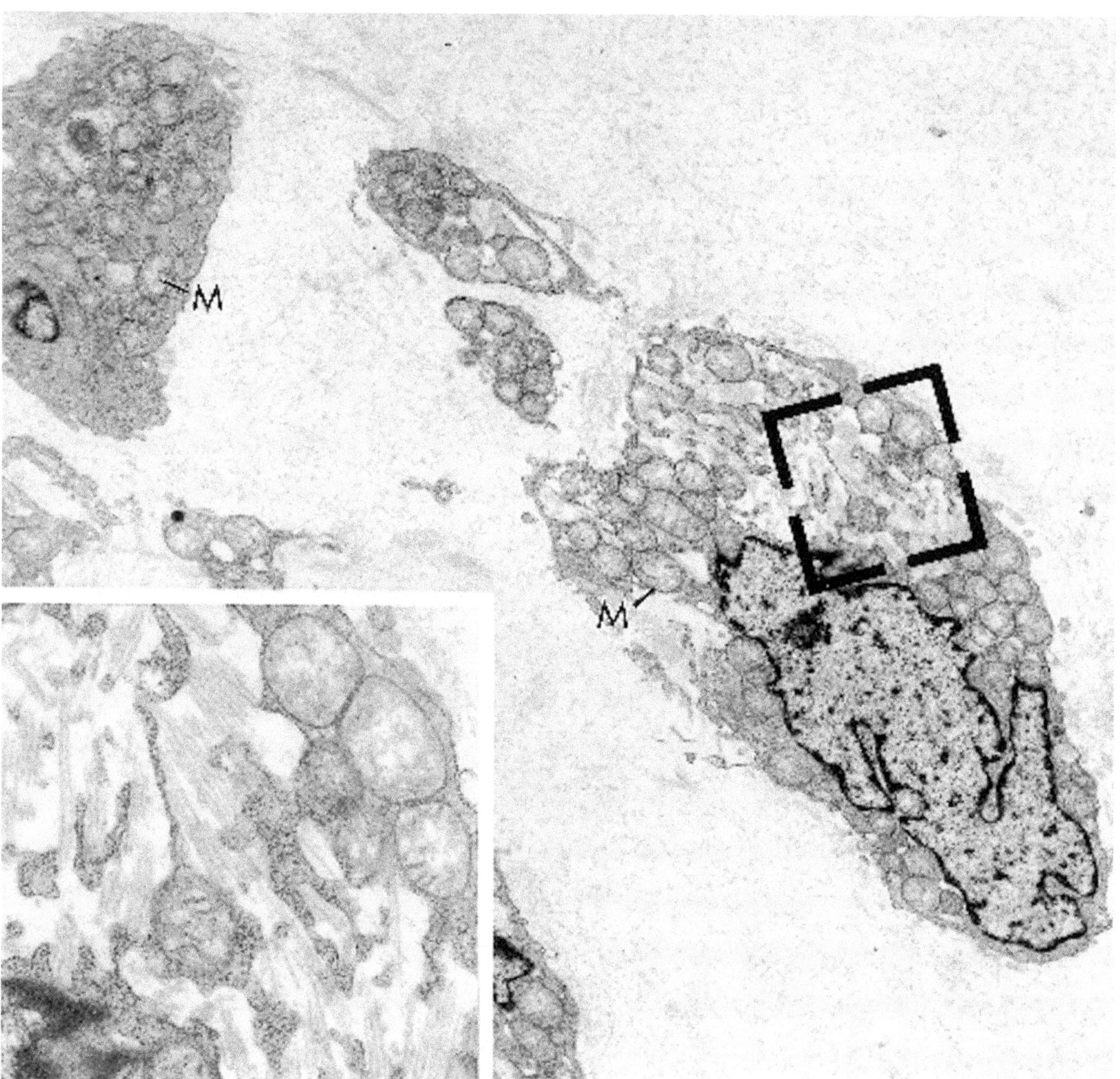

Figure 6.36. Extraskeletal myxoid chondrosarcoma (femur). Chondroblasts are widely separated by a medium-dense, flocculent matrix. Cell cytoplasm is rich in mitochondria (M), and one cell has markedly dilated rough endoplasmic reticulum (partially bracketed). (× 7800) *Inset:* high magnification of the bracketed area reveals parallel microtubules within the dilated RER. (× 23,000)

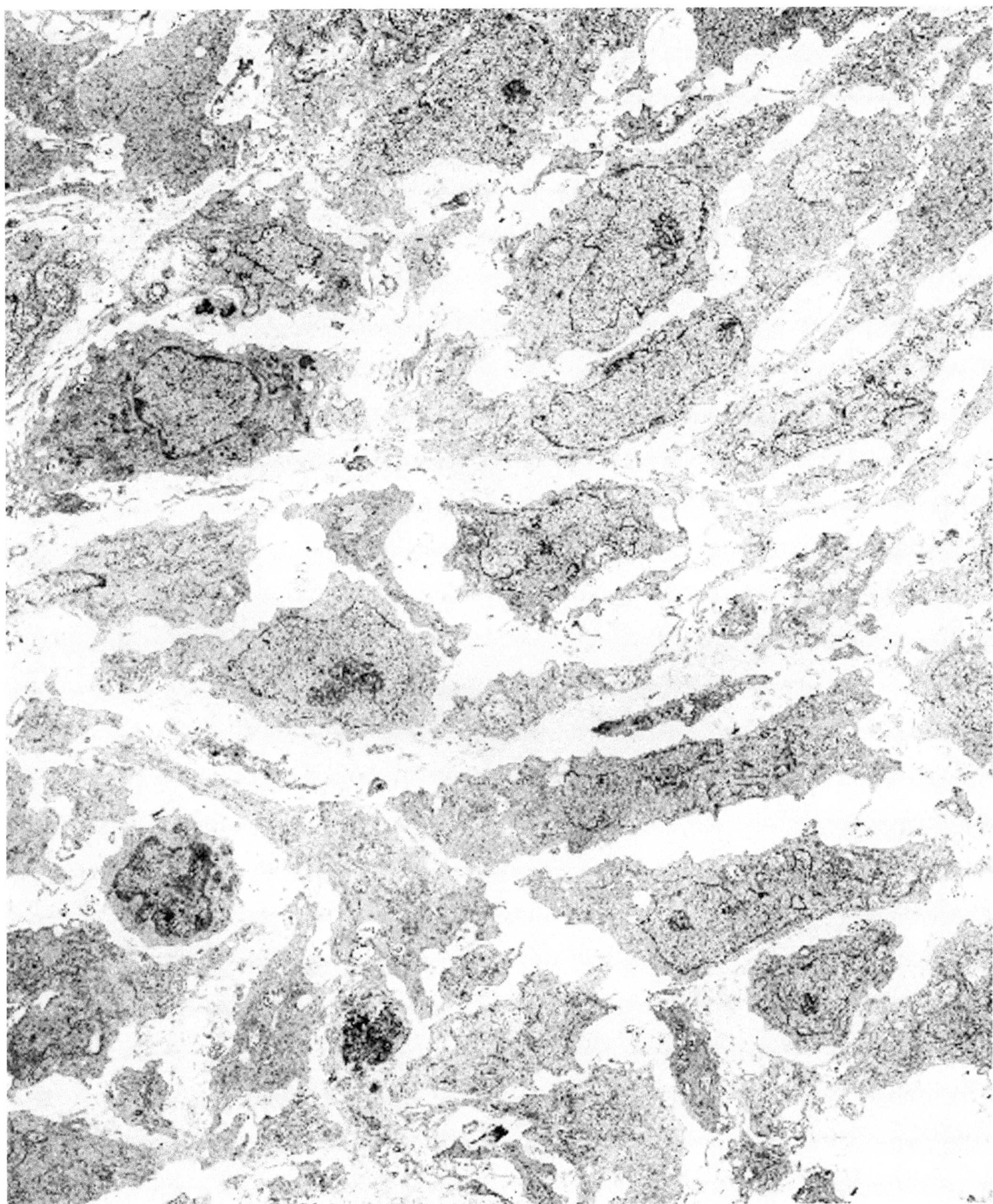

Figure 6.37. Osteosarcoma (femur). Many of the cells in this neoplasm were spindle-shaped and poorly differentiated osteoblasts, as shown, but there also were better differentiated areas composed of oval and polygonal cells (see Figure 6.38). Collagen consistent with osteoid but no hydroxyapatite is present in the extracellular matrix of this field. (× 3900)

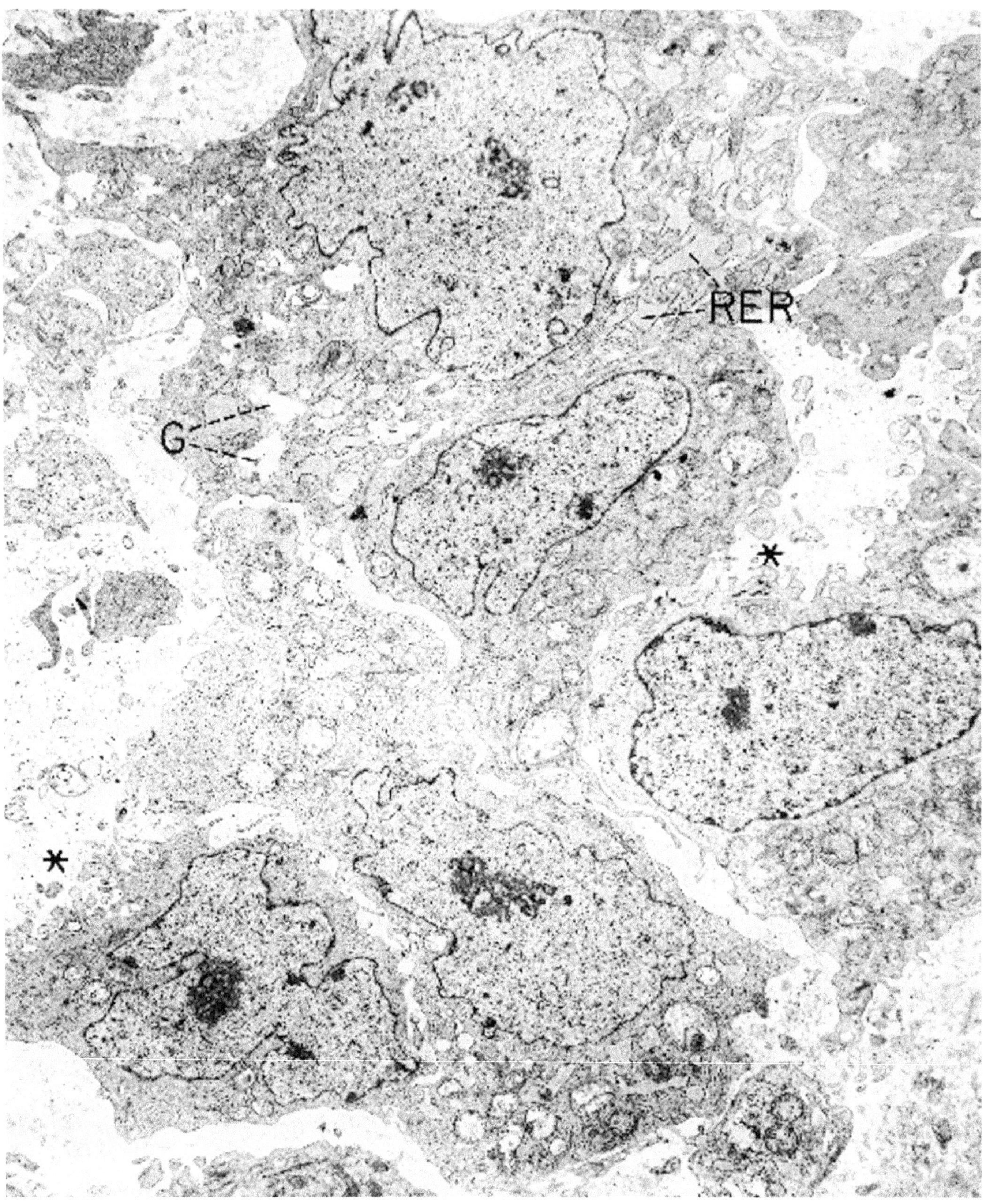

Figure 6.38. Osteosarcoma (femur). This better differentiated area of the osteosarcoma illustrated in Figure 6.37 is composed of oval and polygonal cells having abundant rough endoplasmic reticulum (RER), occasional pockets of glycogen (G), and focally villus-like cell surfaces (*). (× 5000)

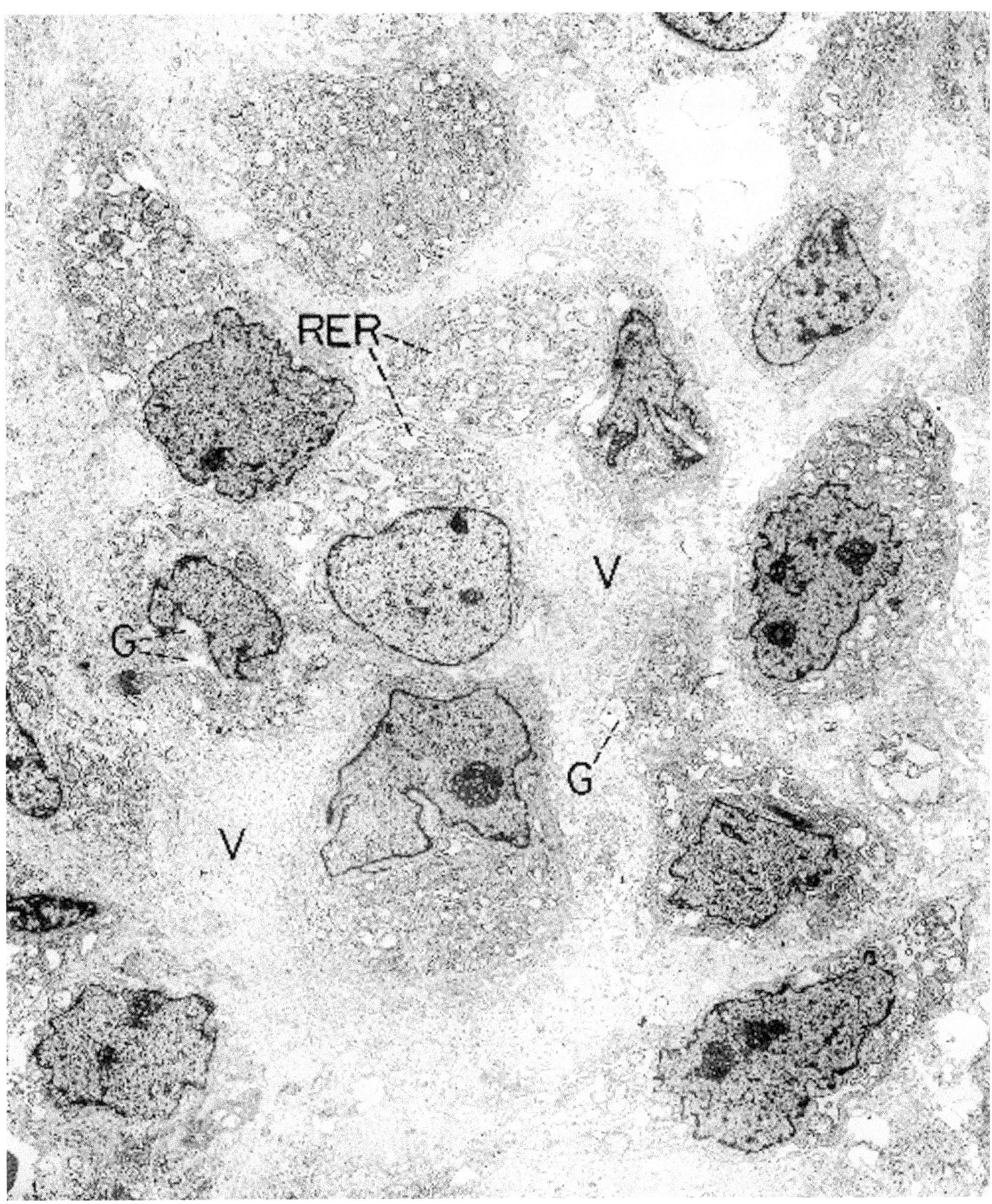

Figure 6.39. Osteosarcoma (parafemoral soft tissue). Well-differentiated osteoblasts comprise this neoplasm and are typified by their oval and polygonal shape, florid villus-like surface (V), abundant rough endoplasmic reticulum (RER), and pockets of glycogen (G). (× 5300)

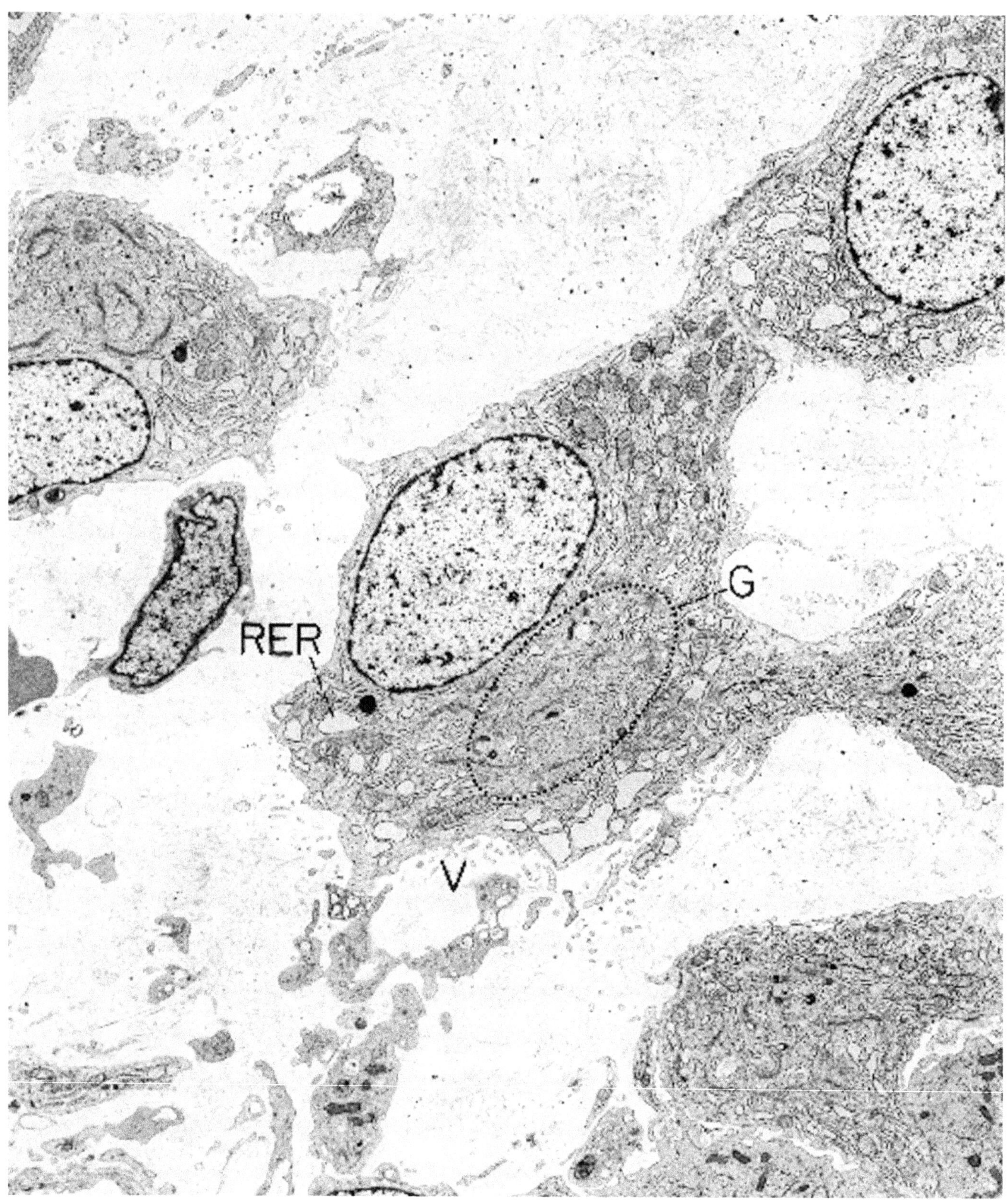

Figure 6.40. Osteoblastoma (pubic ramus). The osteoblasts in this neoplasm are even better differentiated than those in Figure 6.39, in that they have more abundant cytoplasm and more organelles within it. Golgi apparatuses (G) are large, but their detail is not clearly visible at this low magnification. Villus-like surface projections (V) and rough endoplasmic reticulum (RER) are abundant. (× 5300)

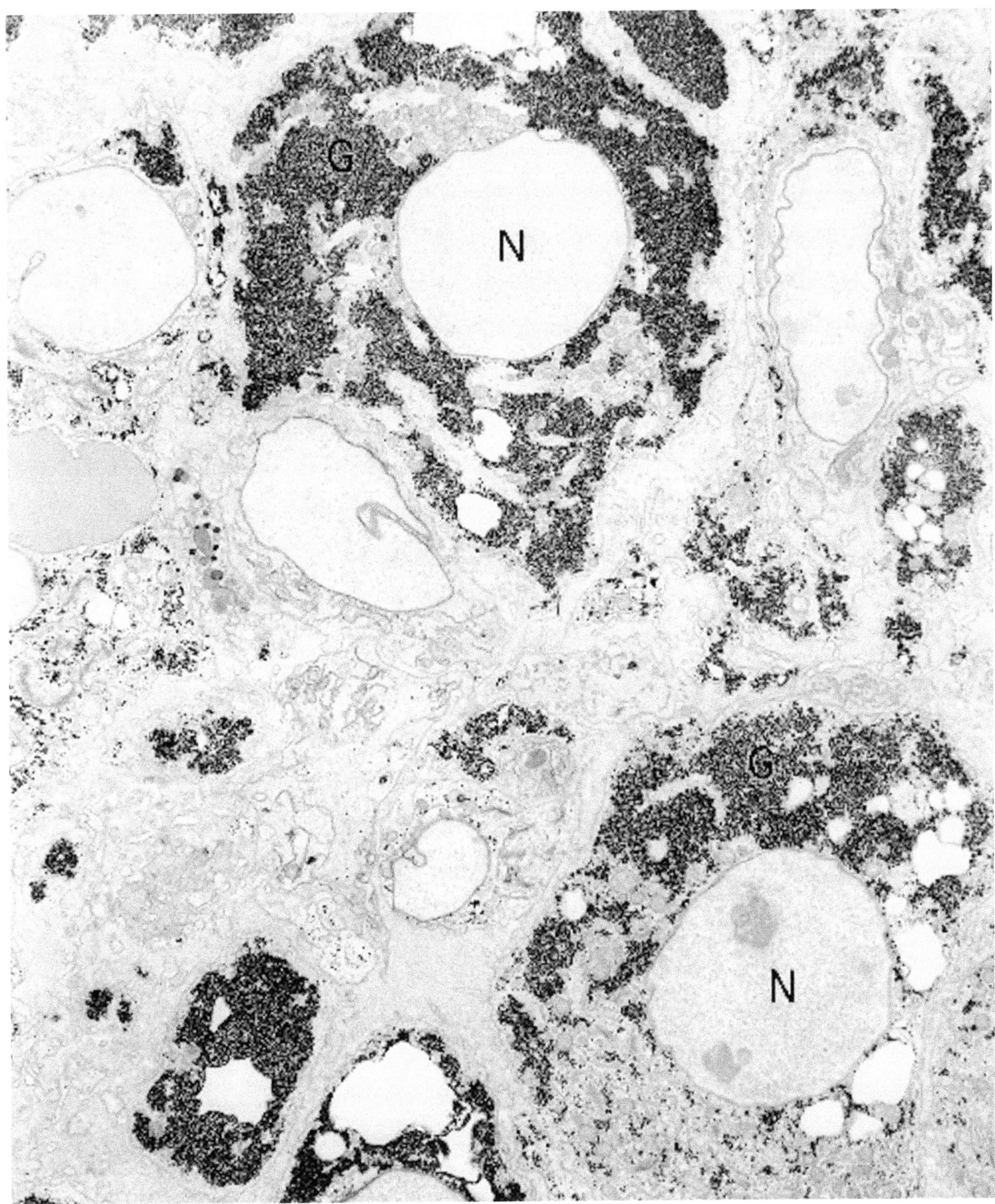

Figure 6.41. Osteoblastoma (femur). The specimen was processed by a method that preserves glycogen (G) as electron-dense granules, and these well-differentiated osteoblasts are extremely rich in this inclusion. N = nuclei. (× 6750)

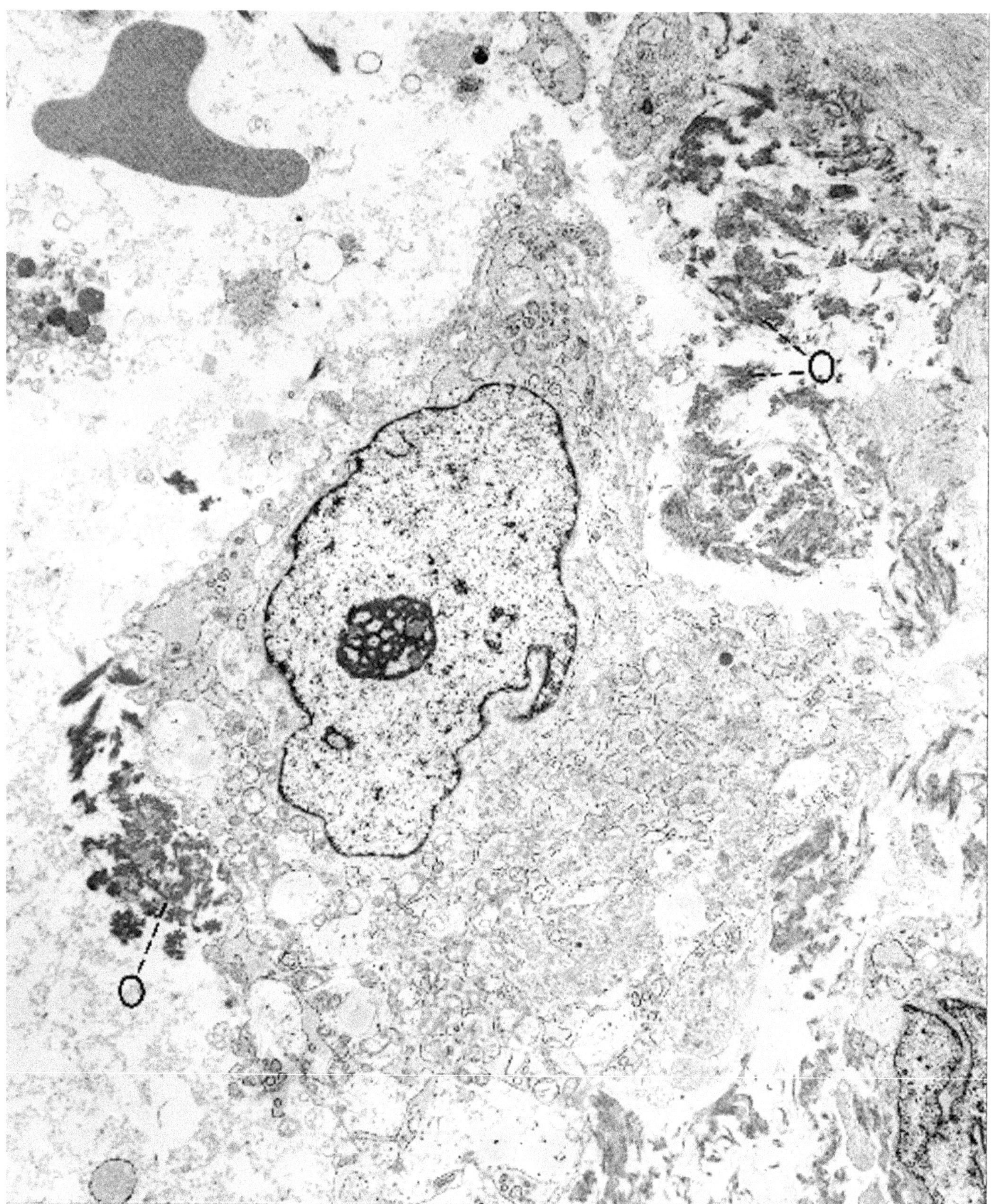

Figure 6.42. Osteosarcoma (soft tissue of thigh). Mineralizing osteoid (O), characterized by electron-dense deposits of hydroxyapatite (on prominent fibers of collagen that are not well visualized at this magnification), abuts this well-differentiated osteoblast. (× 6750)

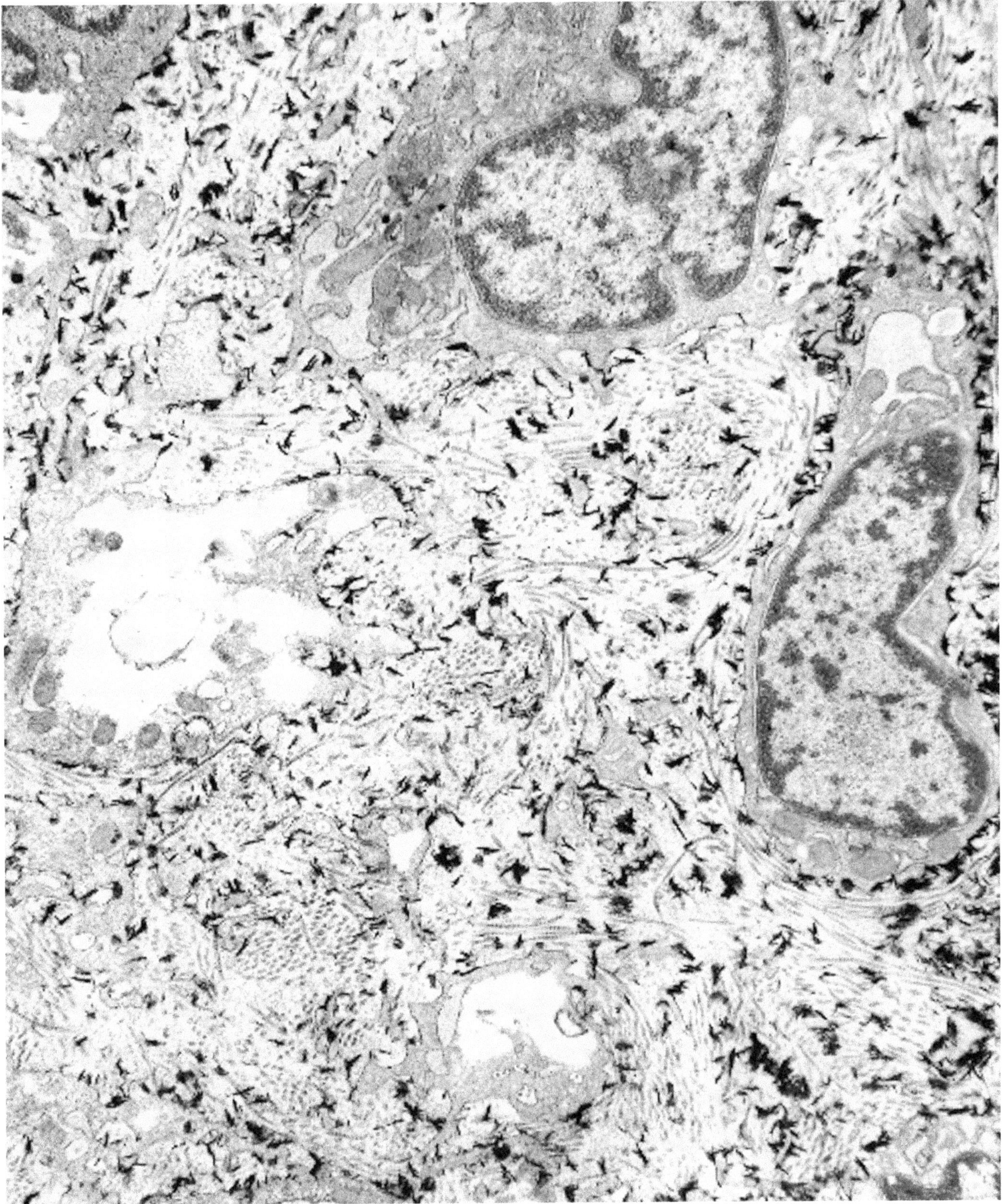

Figure 6.43. Reactive parosteal bone (femur). Innumerable deposits of hydroxyapatite are superimposed on prominent fibers of banded collagen in this zone of benign, new bone formation. (× 12,000)

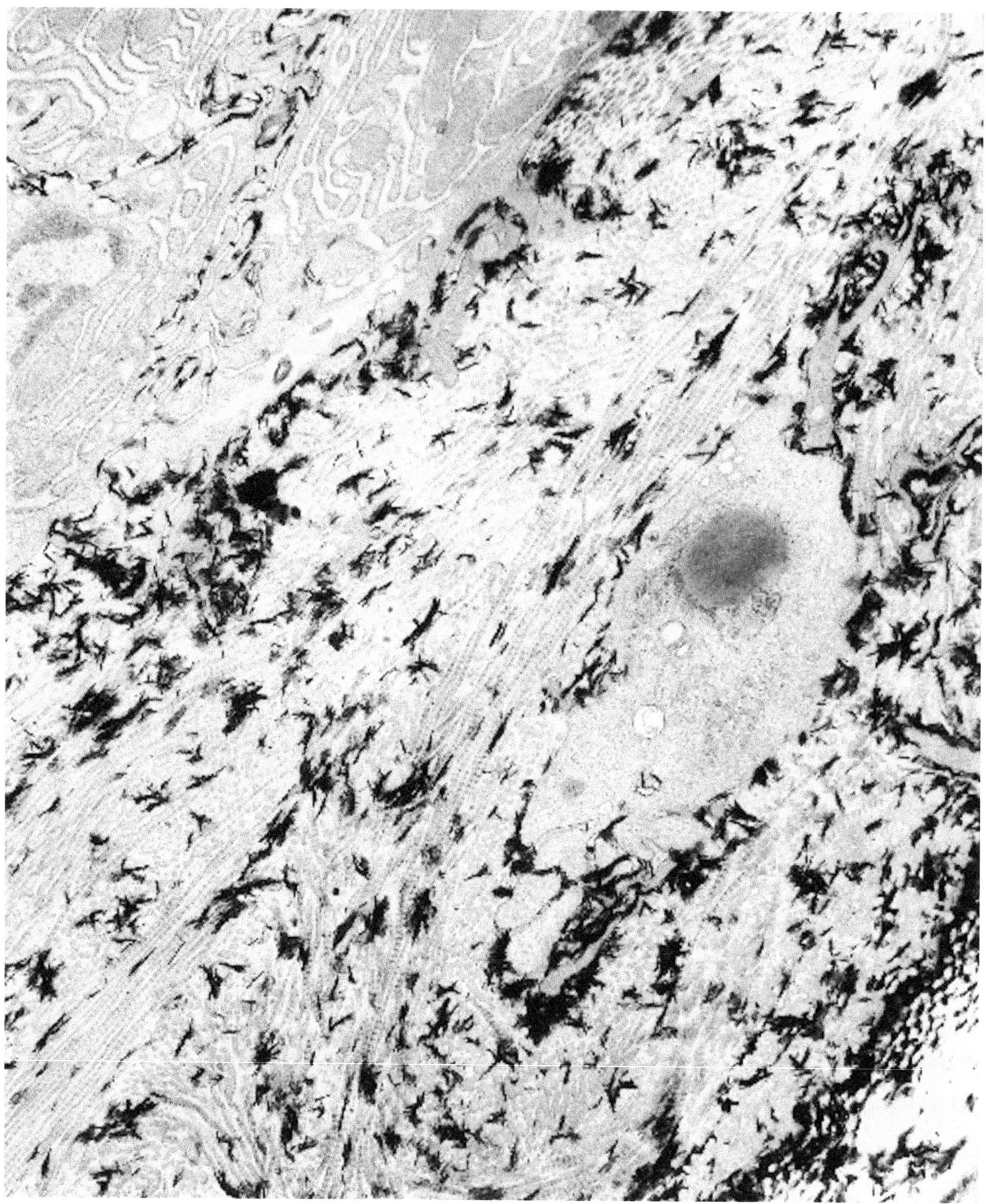

Figure 6.44. Reactive parosteal bone (femur). Higher magnification of the same specimen as depicted in Figure 6.43 shows the hydroxyapatite and prominent fibers of collagen (osteoid) in greater detail. The edge of an osteoblast, in close association with the osteoid, is seen in the upper left corner of the field. (× 27,000)

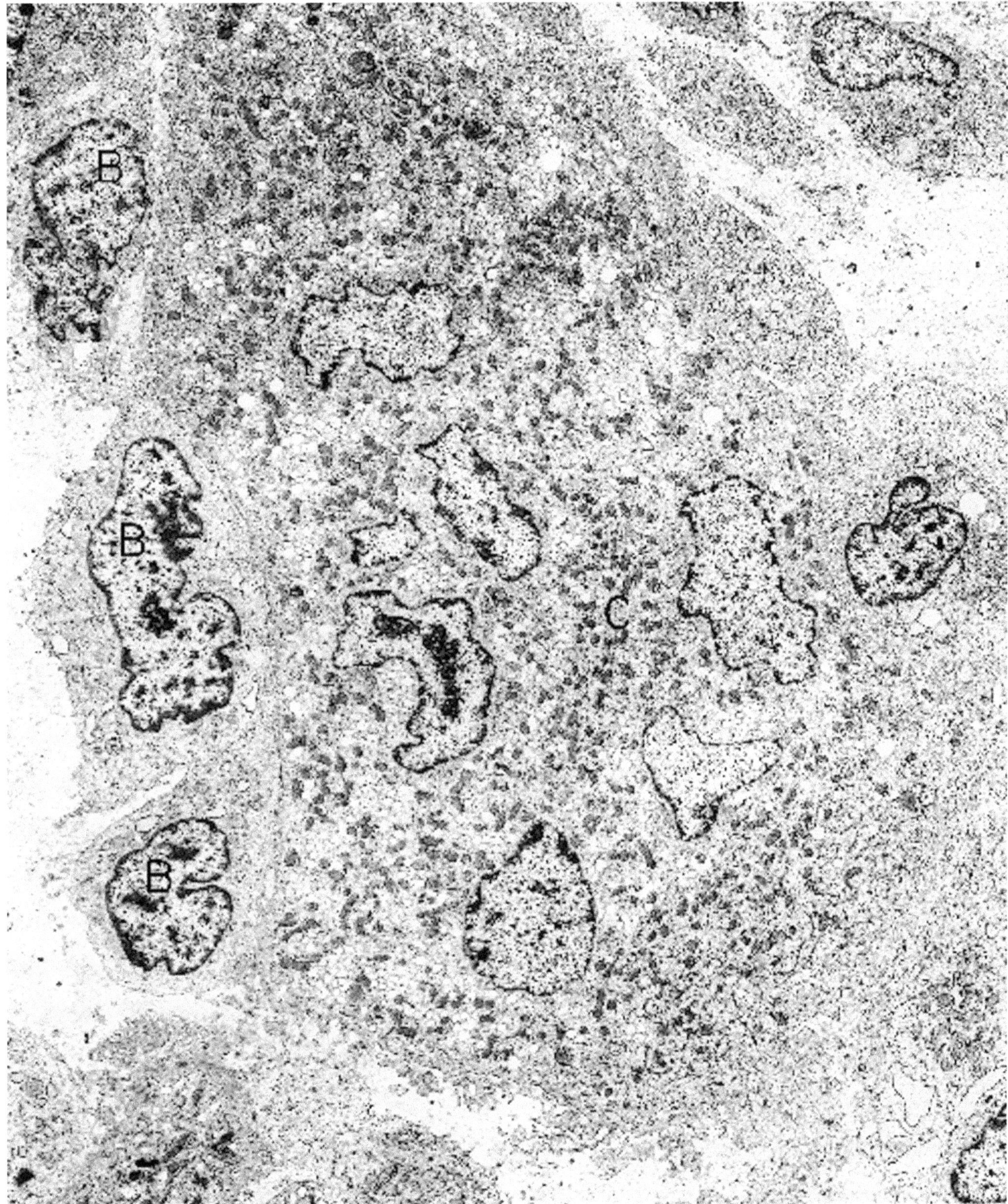

Figure 6.45. Osteogenic sarcoma (soft tissue of hip). Lying between several osteoblasts (B) is an osteoclast (C). It is of giant size and has multiple nuclei. The cytoplasm is abundant and contains many organelles. Mitochondria, seen at higher power in Figure 6.46, are especially numerous. (× 4550)

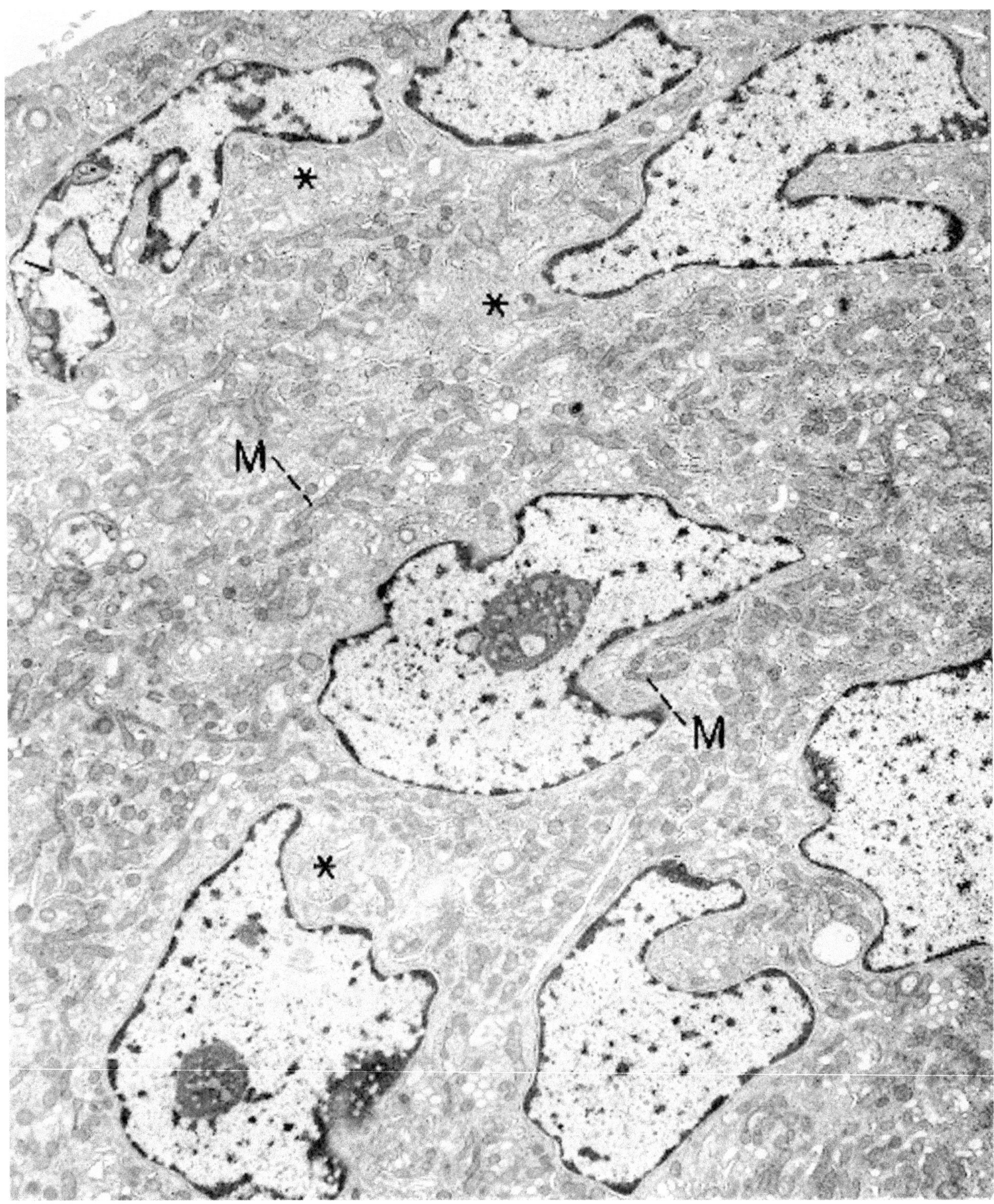

Figure 6.46. Osteogenic sarcoma (soft tissue of hip). Mitochondria (M) predominate in the busy cytoplasm of this osteoclast, and there are also many small vesicles, collapsed cisternae of rough endoplasmic reticulum, and at least three Golgi apparatuses (*), although less well seen. The villus-like surface of the cell is visible in the upper left corner of the field. (× 6750)

(Text continued from page 278)

Synovial Sarcoma

(Figures 6.47 through 6.56.)

Diagnostic criteria. (1) Biphasic components (in the classic case) of spindle-shaped stromal cells and epithelial cells arranged in solid groups or lining spaces or glands (Figures 6.47 through 6.52 and 6.55); (2) the spaces and glands are separated from the stromal cells by basal lamina (Figures 6.47, 6.48, and 6.51); (3) spindle cells are dispersed in varying amounts of matrix composed of banded collagen and amorphous, medium-dense material (Figure 6.53), but many of them also contact one another focally and have small junctions (Figures 6.54 and 6.55); (4) epithelial cells lining spaces and glands have microvilli and junctional complexes (Figures 6.47 through 6.50); (5) glands may or may not contain secretory material (Figure 6.49).

Additional points. A range of patterns may be seen among synovial sarcomas. In addition to the classic biphasic picture, there are also monophasic spindle cell tumors and, rarely, monomorphic epithelioid sarcomas. In the latter-type neoplasm and in poorly differentiated epithelial components of biphasic neoplasms, the oval and polygonal epithelioid cells often occur in solid nests. The cells have junctions but may not form microvilli or gland-like spaces (Figure 6.55). In the spindle-cell type or component of synovial sarcoma, the cells have a high nuclear–cytoplasmic ratio and do not have the ultrastructural characteristics of any other type of spindle cell (Figure 6.53). They may have small intercellular junctions. Although probably derived from primitive mesenchymal cells and possibly from fibroblasts, they possess less rough endoplasmic reticulum than do typical fibroblasts. Intermediate filaments (keratin and/or vimentin) often are demonstrable, sometimes in large amounts, both in the epithelial and spindle cells of synovial sarcomas (Figures 6.52 and 6.56). It is open to question whether the two cell types in synovial sarcoma have a common lineage. There is some evidence that in normal synovial membrane, the lining cells are macrophagic and derived from monocytes, and the underlying spindle cells are secretory and derived from mesenchymal cells.

Adipose Neoplasms

(Figures 6.57 through 6.70.)

Diagnostic criteria. (1) Lipid droplets; (2) pinocytotic vesicles; (3) cytoplasmic glycogen; (4) basal lamina; (5) intermediate filaments.

Additional points. Golgi apparatuses, varying amounts of smooth and rough endoplasmic reticulum, and mitochondria are other organelles that may be seen in lipoblasts. Most of the nonlipid cellular features listed are found in all stages of differentiation except for the very late lipoblast and mature lipocyte.

There probably are three lines of differentiation for lipoblasts: pericytes, fibroblasts, and poorly differentiated mesenchymal cells. The role of the pericyte can be studied conveniently in *myxoid liposarcoma* (Figures 6.57 through 6.59), where the vascular pattern consistently is a prominent component of the neoplasms. Here, pericytes and neighboring early lipoblasts can be seen to resemble one another closely in both size and shape and in nuclear and cytoplasmic detail. A few pericytes even contain lipid droplets (Figure 6.58). Another pertinent observation in these neoplasms is that there is a gradient of increasing differentiation in lipoblasts as their distance from capillaries increases. The main criterion for recognizing advancing differentiation is the increasing amount of lipid in the cells, and as the number of lipid droplets increases, the cytoplasmic space expands (Figures 6.60 through 6.63). Coalescence of droplets in seen in late-stage lipoblasts, and finally a single cytoplasmic vacuole is present in the mature lipocyte. Glycogen usually is present in moderate-to-heavy amounts in early- and mid-stage lipoblasts (Figures 6.64 and 6.65).

Developmental relationships similar to those noted between the lipoblast and pericyte also can be seen between the lipoblast and poorly differentiated mesenchymal cell and between the lipoblast and fibroblast. In all examples, the increasing amount of cytoplasmic lipid is the main morphologic index for maturation of the cell. The fibroblast, or fibrolipoblast, is found more frequently in *well-differentiated liposarcoma* than in *myxoid liposarcoma* and *round cell liposarcoma*. Round cells consist mostly of poorly differentiated mesenchymal cells and, to a lesser extent, early lipoblasts (Figure 6.66 and 6.67).

The cells comprising *pleomorphic liposarcomas* cover a wide spectrum that includes all stages of lipoblasts and giant cells having bizarre nuclei and varying amounts of cytoplasmic lipid (Figures 6.68 and 6.69).

Hibernomas are lipomas composed of brown fat, and the component lipocytes have copious cytoplasm that contains numerous small- and intermediate-sized lipid droplets and numerous mitochondria. Glycogen may also be present in the cytoplasm. Basal lamina can be found coating the cells (Figure 6.70).

Lipocytes and lipoblasts occur in combination with other cell types in a variety of neoplasms, including *myelolipoma*, *angiolipoma*, and *angiomyolipoma*.

In *dedifferentiated liposarcoma*, areas of well-differentiated liposarcoma are interspersed with regions of fibrosarcoma or MFH.

(Text continues on page 320)

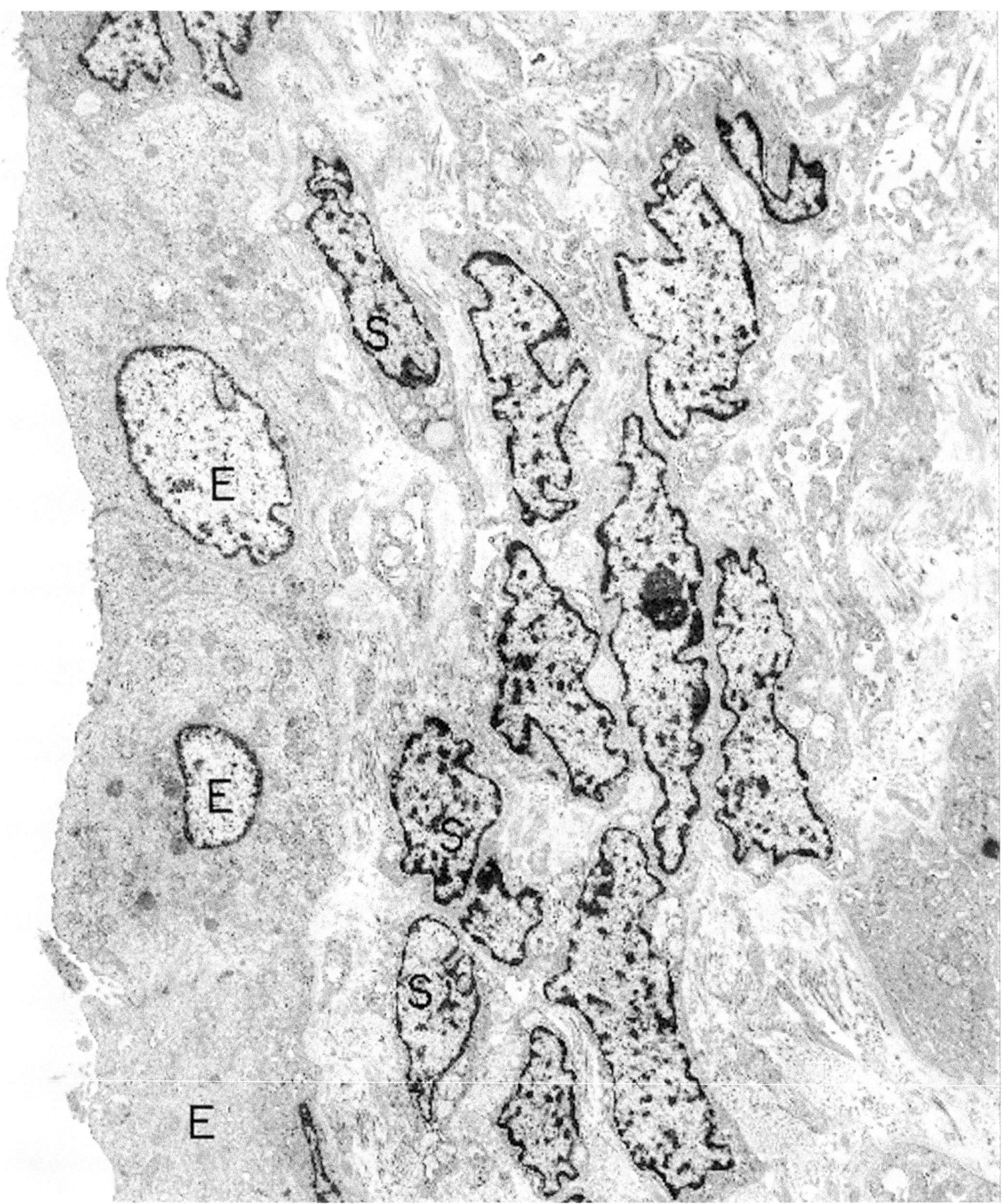

Figure 6.47. Synovial sarcoma (soft tissue of thigh). Epithelial (E) and stromal (S) cells are depicted in this biphasic neoplasm, and a sharp demarcation, including a basal lamina (barely discernible at this low magnification), separates the two cell-types. (× 5700)

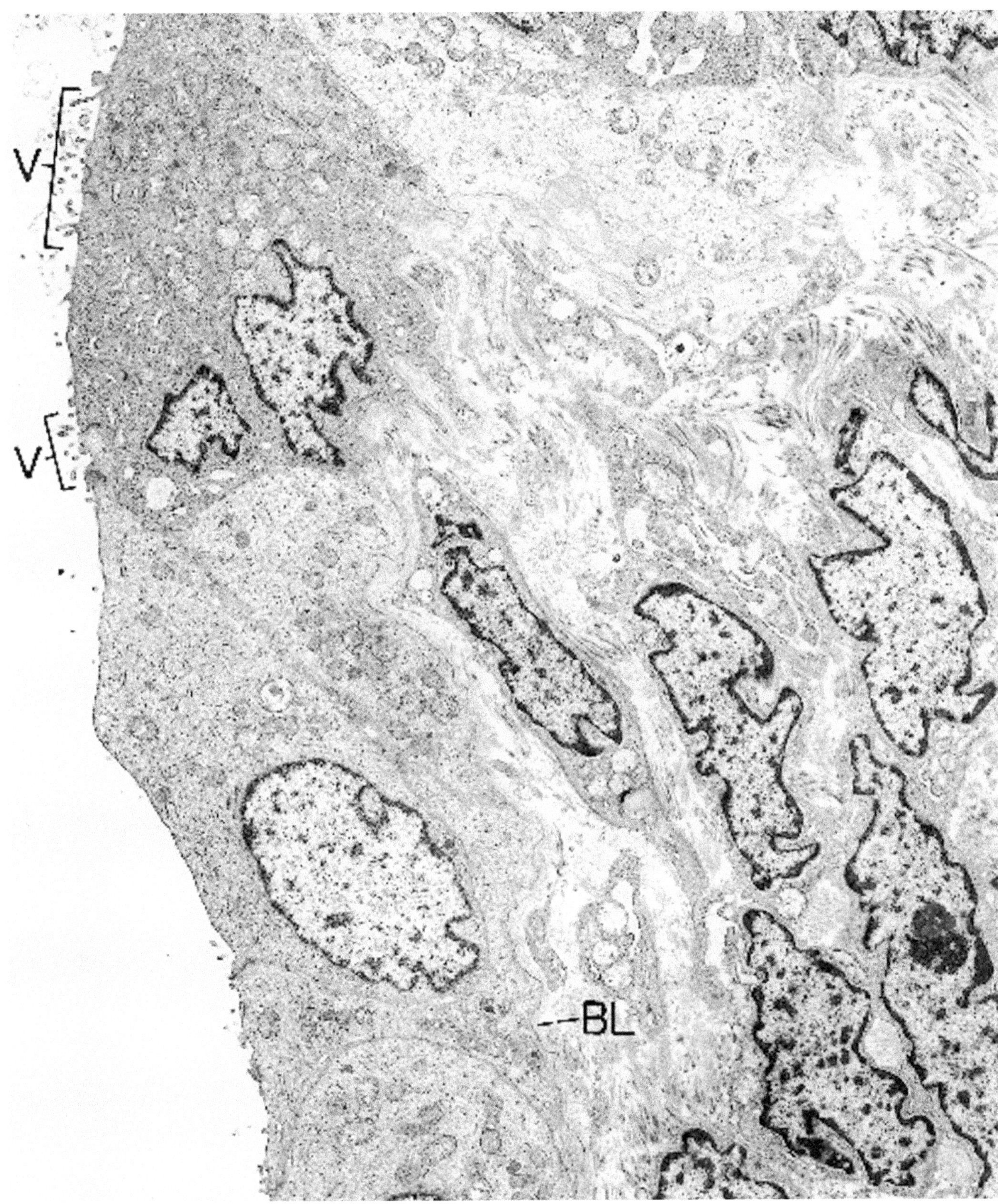

Figure 6.48. Synovial sarcoma (soft tissue of thigh). In this field, a varying population of microvilli (V) are visible, and the epithelial cells are otherwise well differentiated, including the formation of a basal lamina (BL). (× 6720)

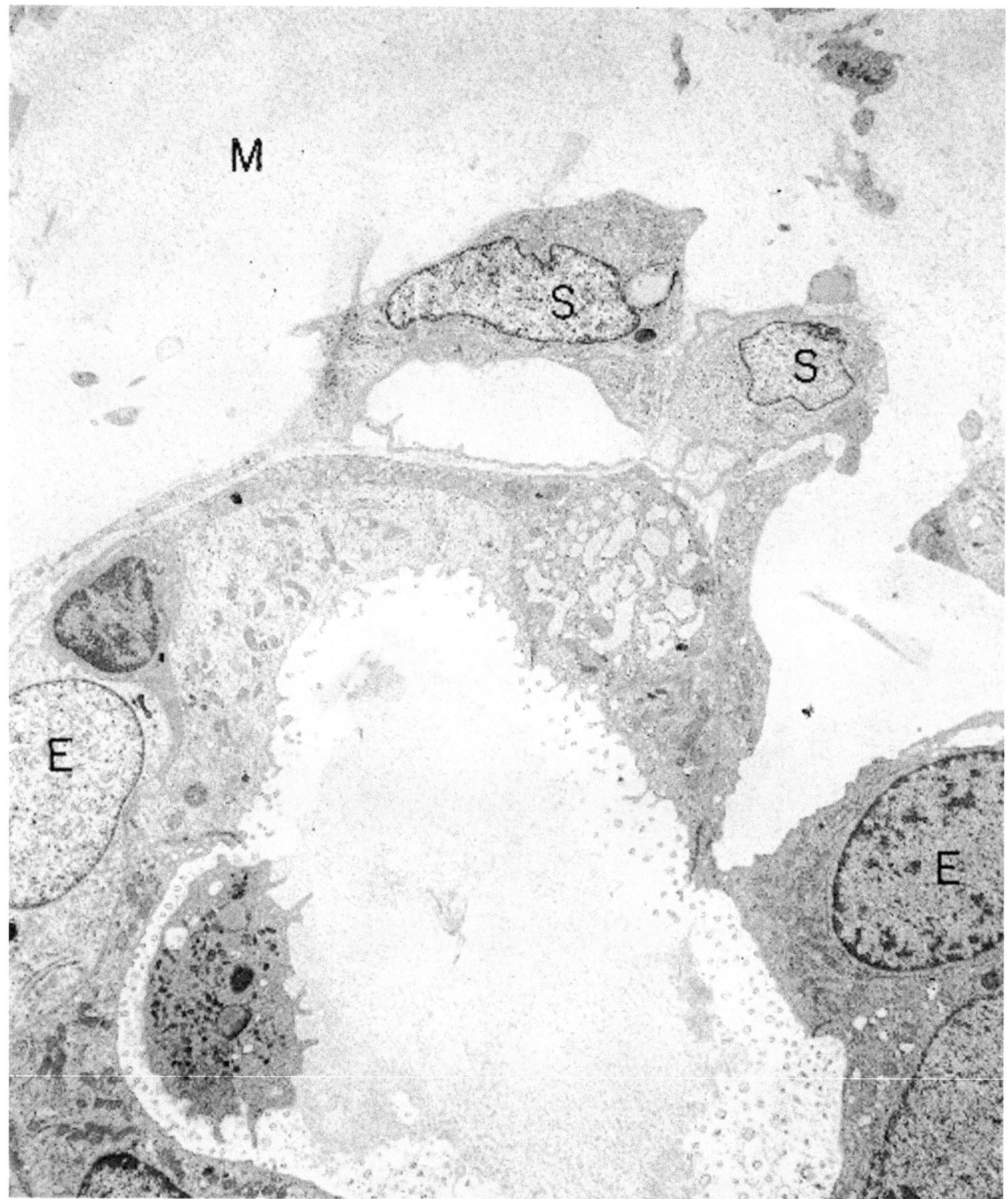

Figure 6.49. Synovial sarcoma (soft tissue of leg). A discrete acinus lined by well-differentiated epithelial cells (E) is surrounded by a stroma that is hypocellular (S) and rich in extracellular matrix (M). (× 5300)

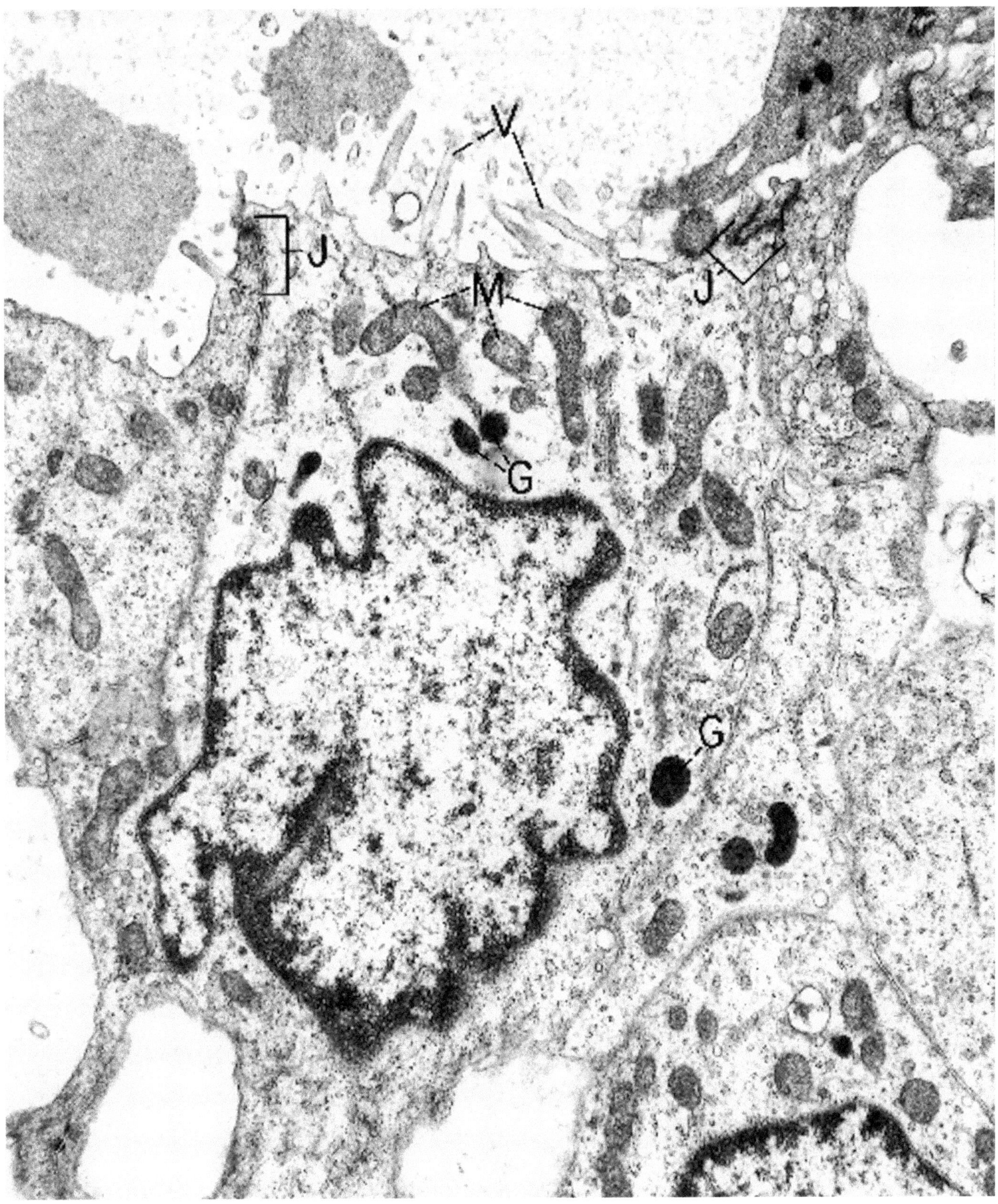

Figure 6.50. Synovial sarcoma (soft tissue of leg). High magnification of an epithelial cell of the neoplasm illustrated in Figure 6.49 shows microvilli (V) on the luminal surface and junctional complexes (J) at its apical aspect and membrane-bound granules (G) and mitochondria (M) as the outstanding organelles in the cytoplasm. (× 18,500)

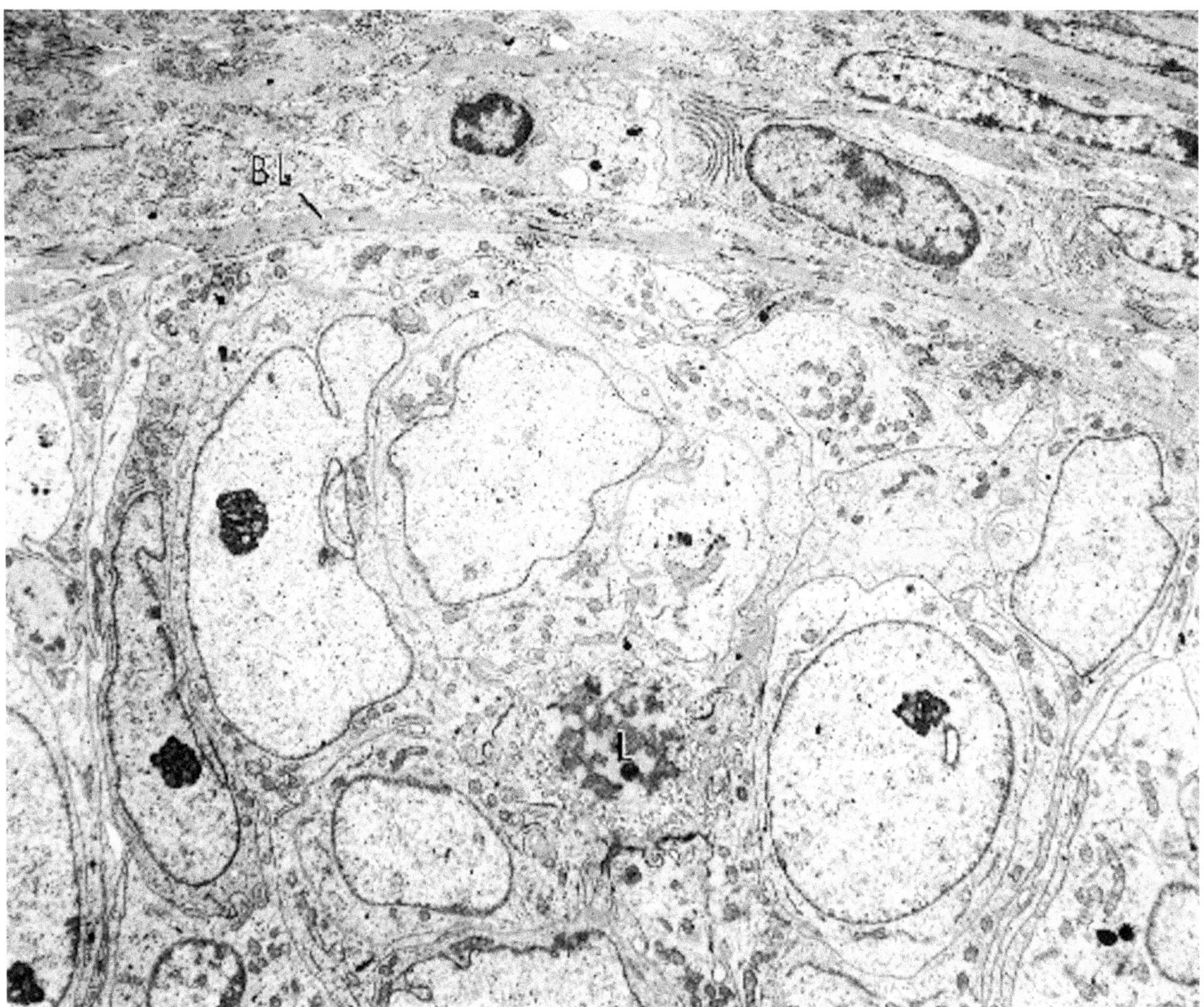

Figure 6.51. Synovial sarcoma (vulva). An island of epithelial cells in this biphasic synovial sarcoma has a central microlumen (L) and is separated from the mesenchymal component by a thick layer of basal lamina (BL). (× 4800)

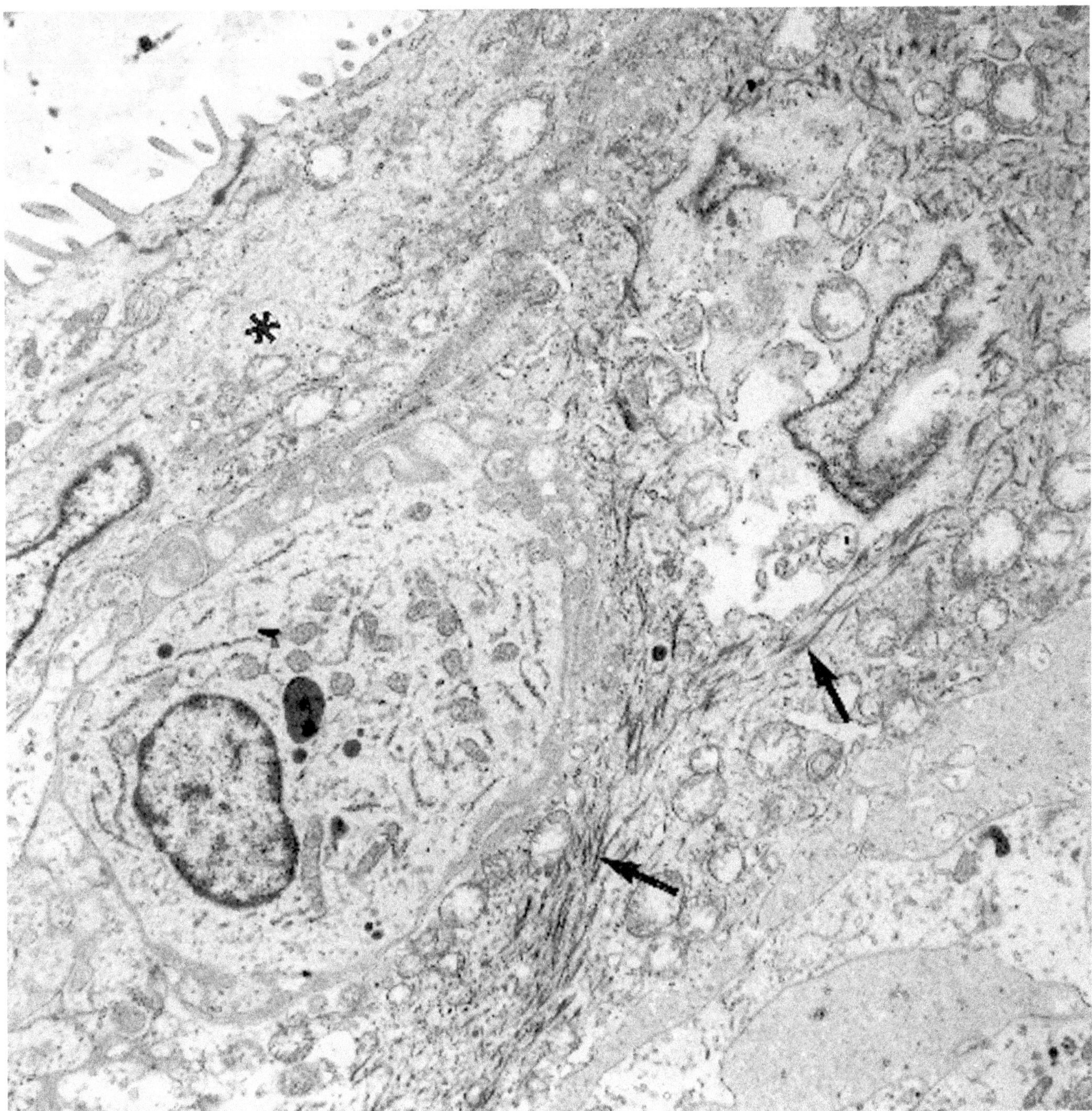

Figure 6.52. Synovial sarcoma (vulva). An epithelial cell subjacent to a luminal lining cell (*) contains numerous tonofibrils (*arrows*) in its cytoplasm. (× 8200)

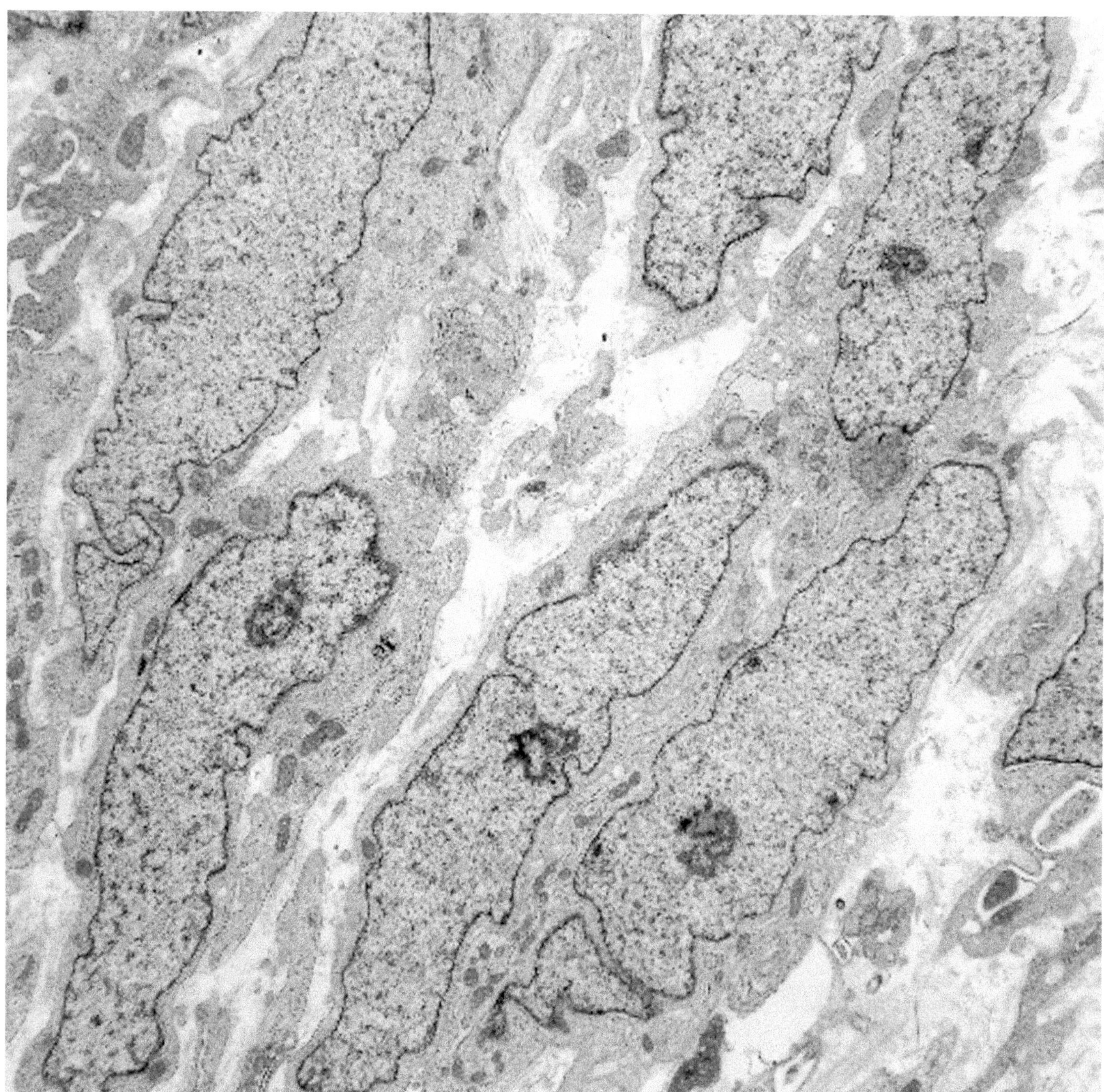

Figure 6.53. Synovial sarcoma (soft tissue of arm). A spindle cell component of a biphasic synovial sarcoma reveals the spindle cells to have a nondescript cytoplasm, devoid of fibroblastic and leiomyoblastic markers. Also, features of Schwannian differentiation such as long intertwining processes, continuous basal lamina, and long spacing collagen are absent. (× 9300)

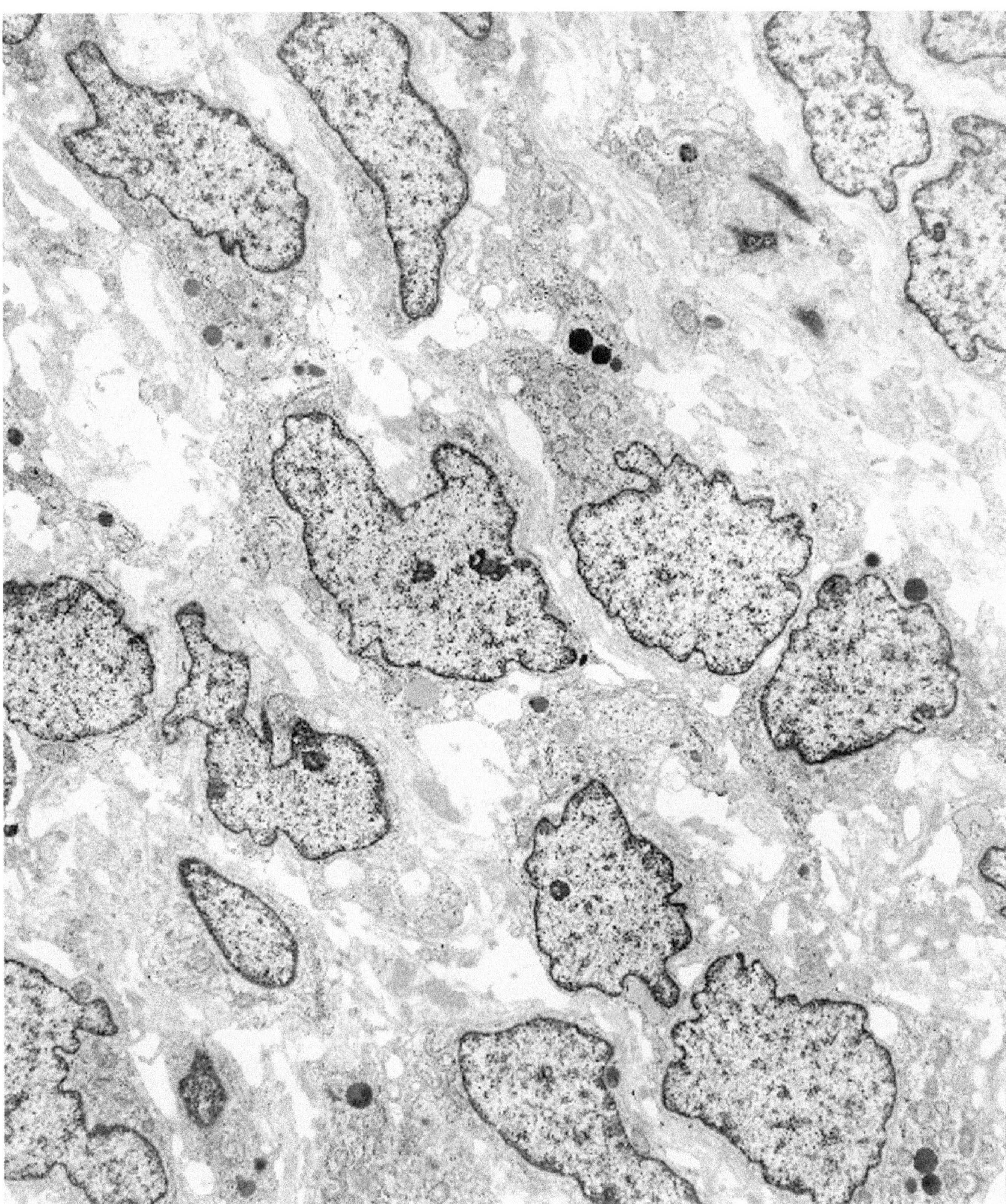

Figure 6.54. Synovial sarcoma (soft tissue of hand). The spindle cells of this monomorphic synovial sarcoma abut focally on one another and are otherwise separated by collagen and amorphous electron-dense material. (× 4940)

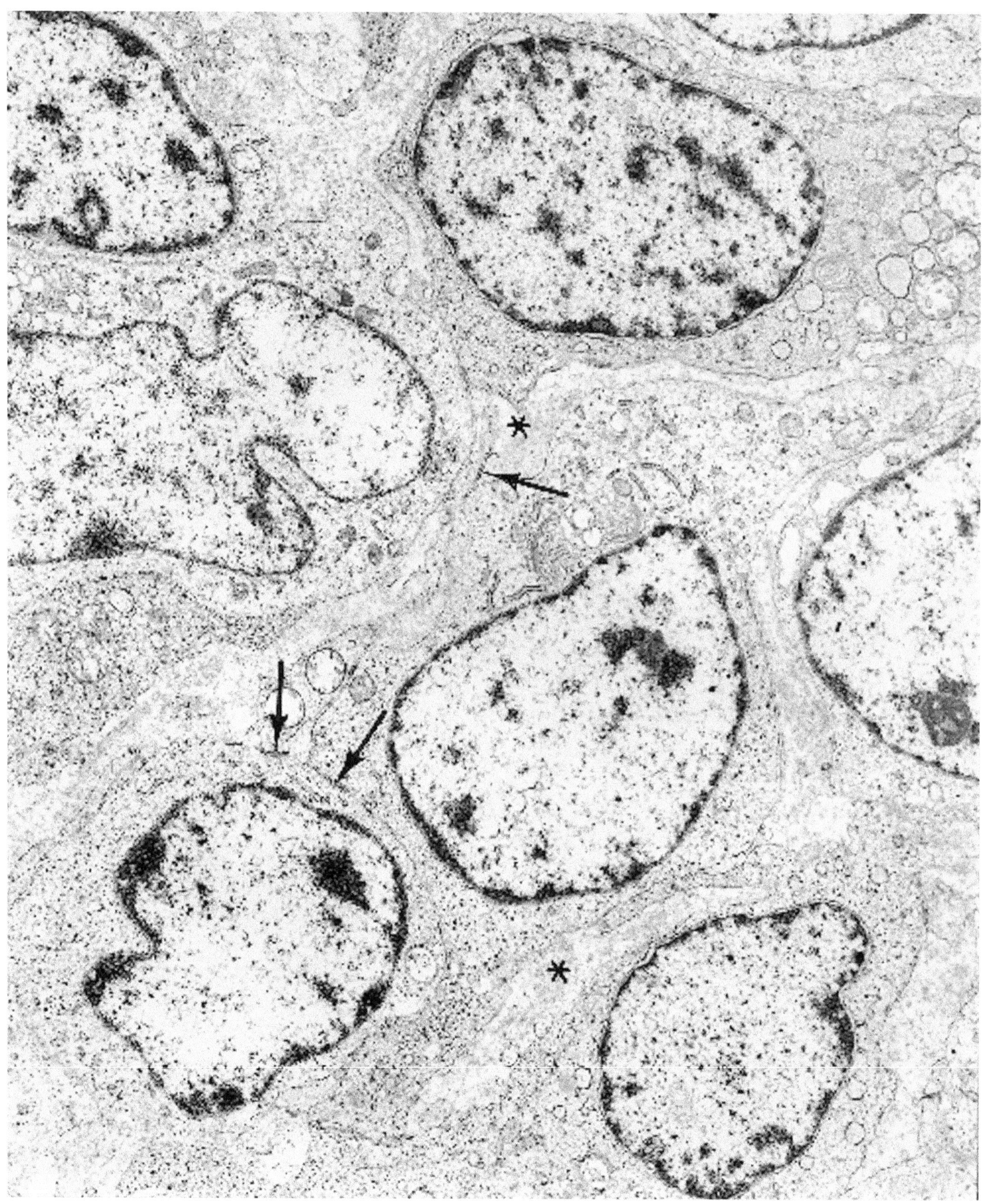

Figure 6.55. Synovial sarcoma (soft tissue of deltoid region). This predominantly monomorphic spindle cell neoplasm had focal solid areas of oval and polygonal cells with small intercellular junctions (*arrows*) and only a small amount of intercellular amorphous matrix (*). (× 9350)

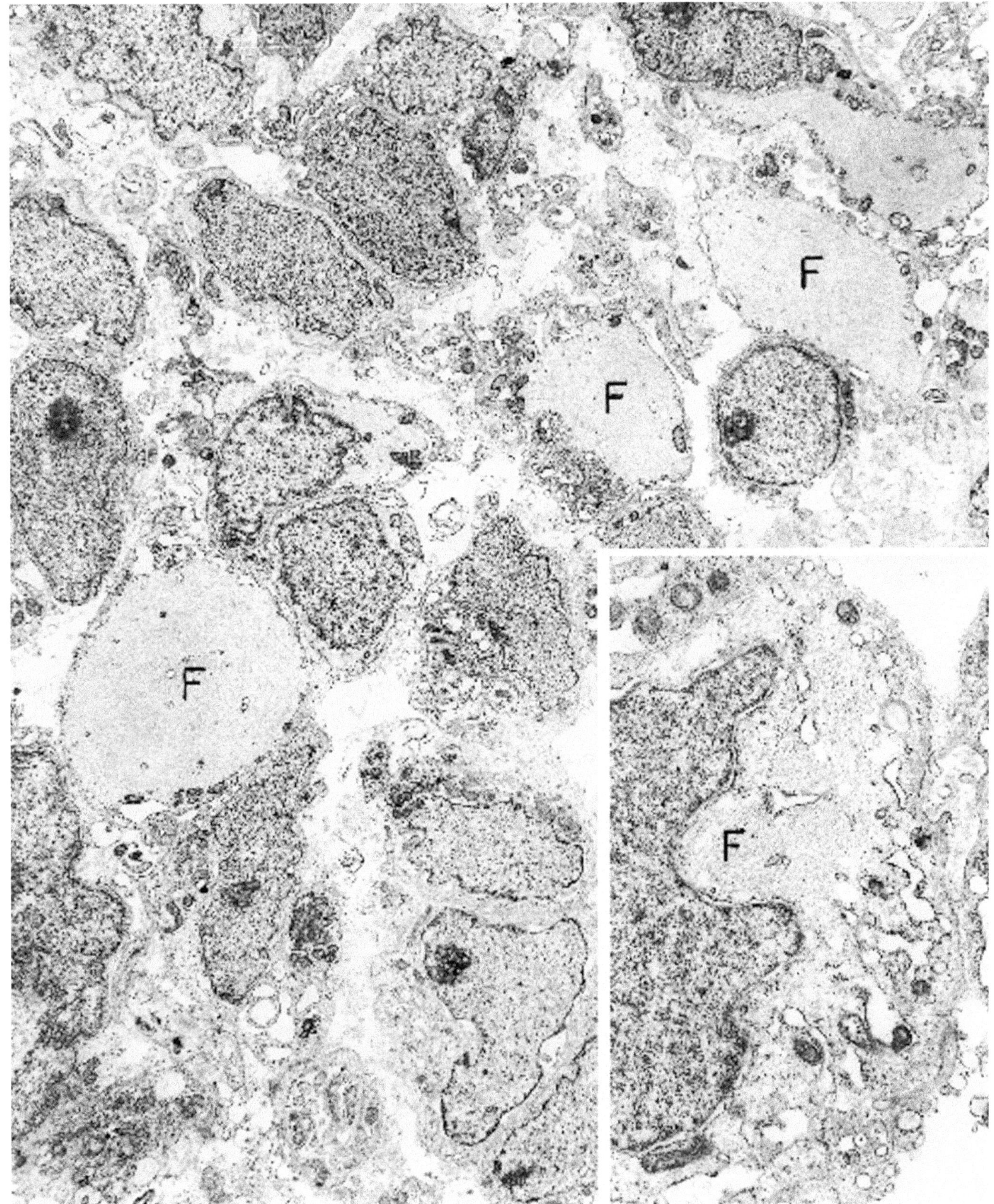

Figure 6.56. Synovial sarcoma (soft tissue of thigh). Many cells of this monophasic synovial sarcoma contain large zones of microfilaments (F and *inset*). (× 5130) (inset × 15,100)

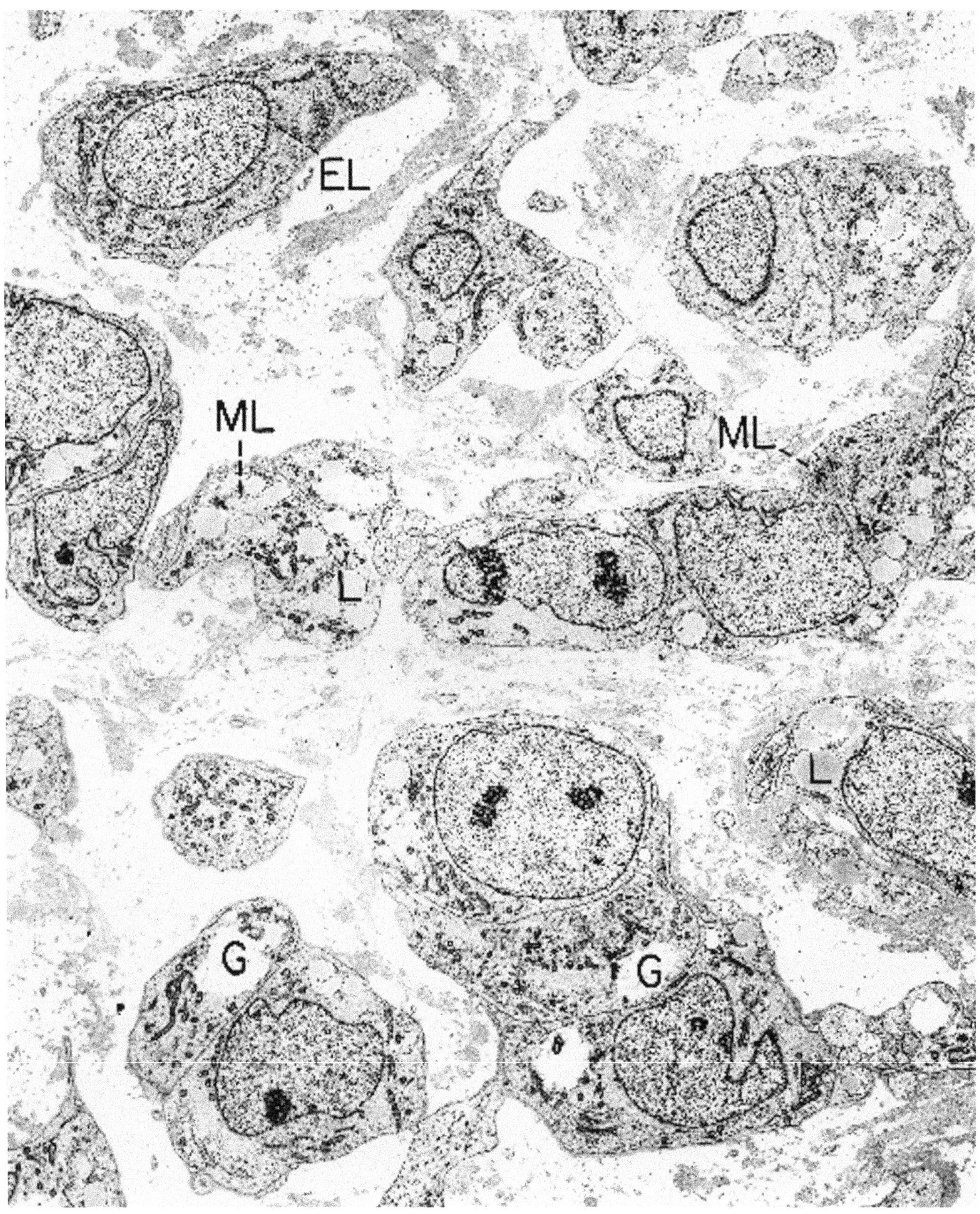

Figure 6.57. Myxoid liposarcoma (soft tissue of right knee). Lipoblasts of early (EL) and intermediate (ML) stages are dispersed individually and in columns, in a flocculent and medium-dense matrix. In addition to droplets of lipid (L), many cells also contain open spaces representing glycogen (G). (× 3960)

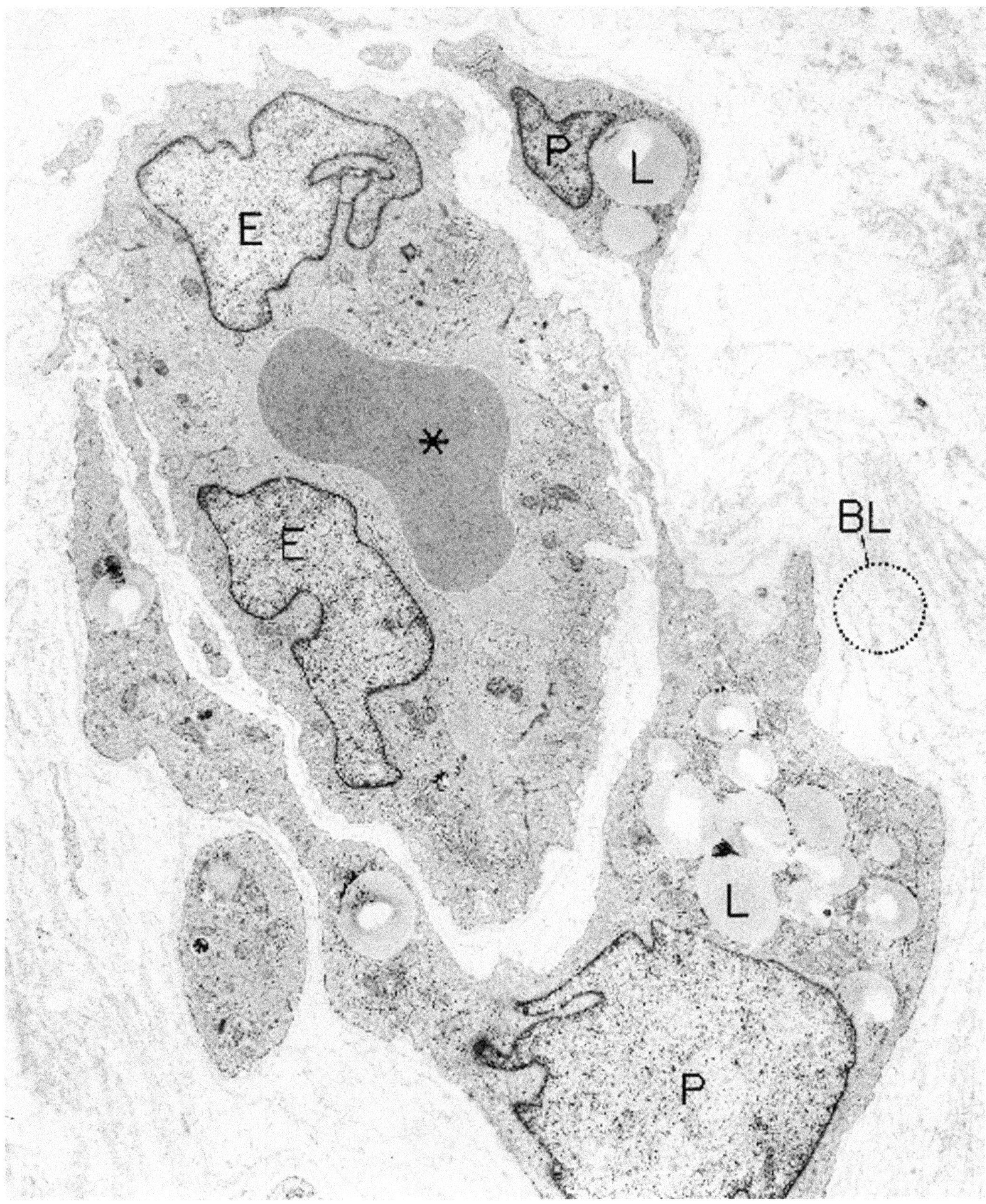

Figure 6.58. Myxoid liposarcoma (soft tissue of thigh). A capillary is partially surrounded by pericytes (P) having lipid droplets (L) and a duplicated basal lamina (BL). * = erythrocyte; E = endothelial cells. (× 7420)

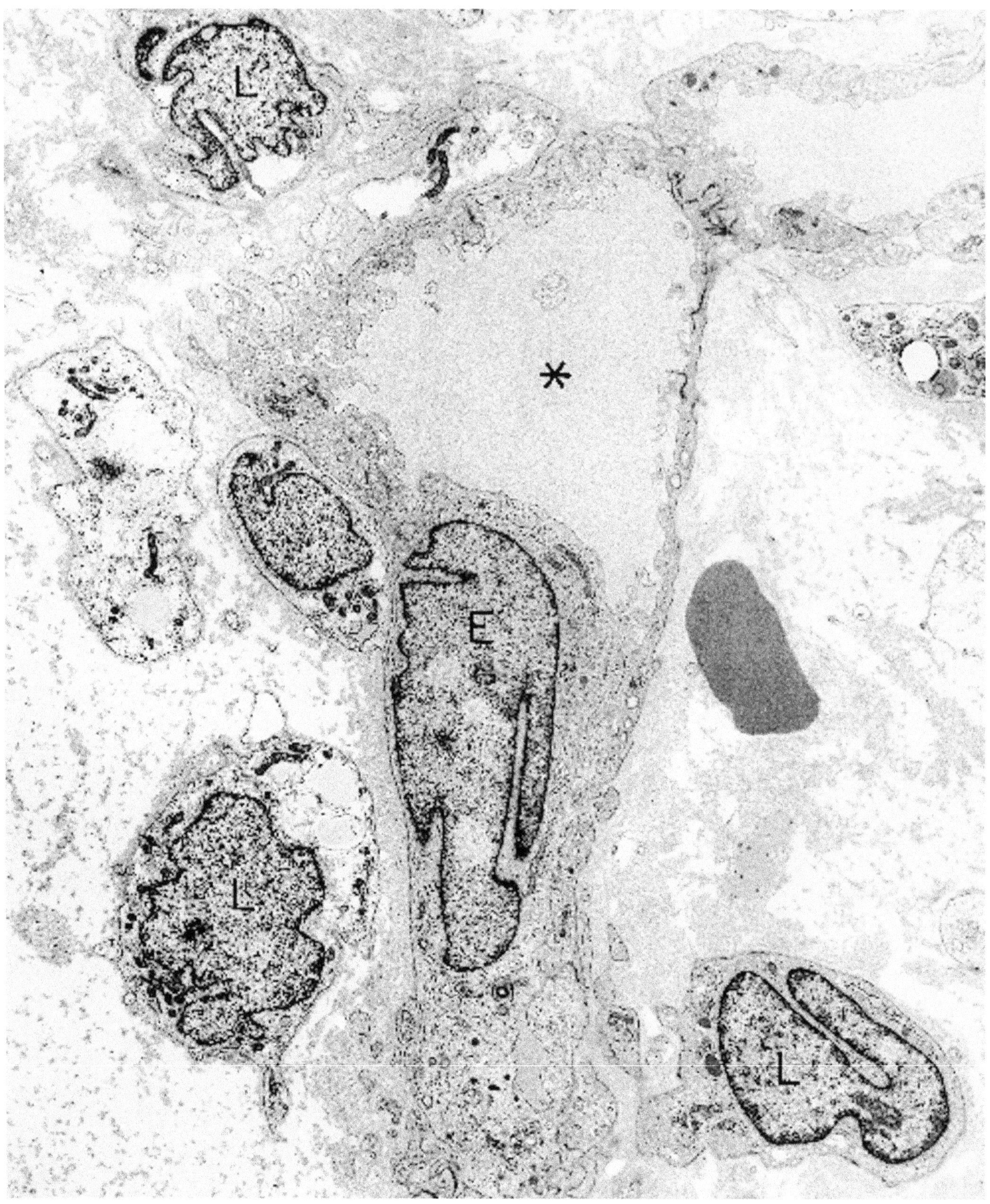

Figure 6.59. Myxoid liposarcoma (soft tissue of thigh). Several early lipoblasts (L) surround a capillary. Compare with pericytes in Figure 6.58. E = endothelial cell; * = lumen of capillary. (× 5225)

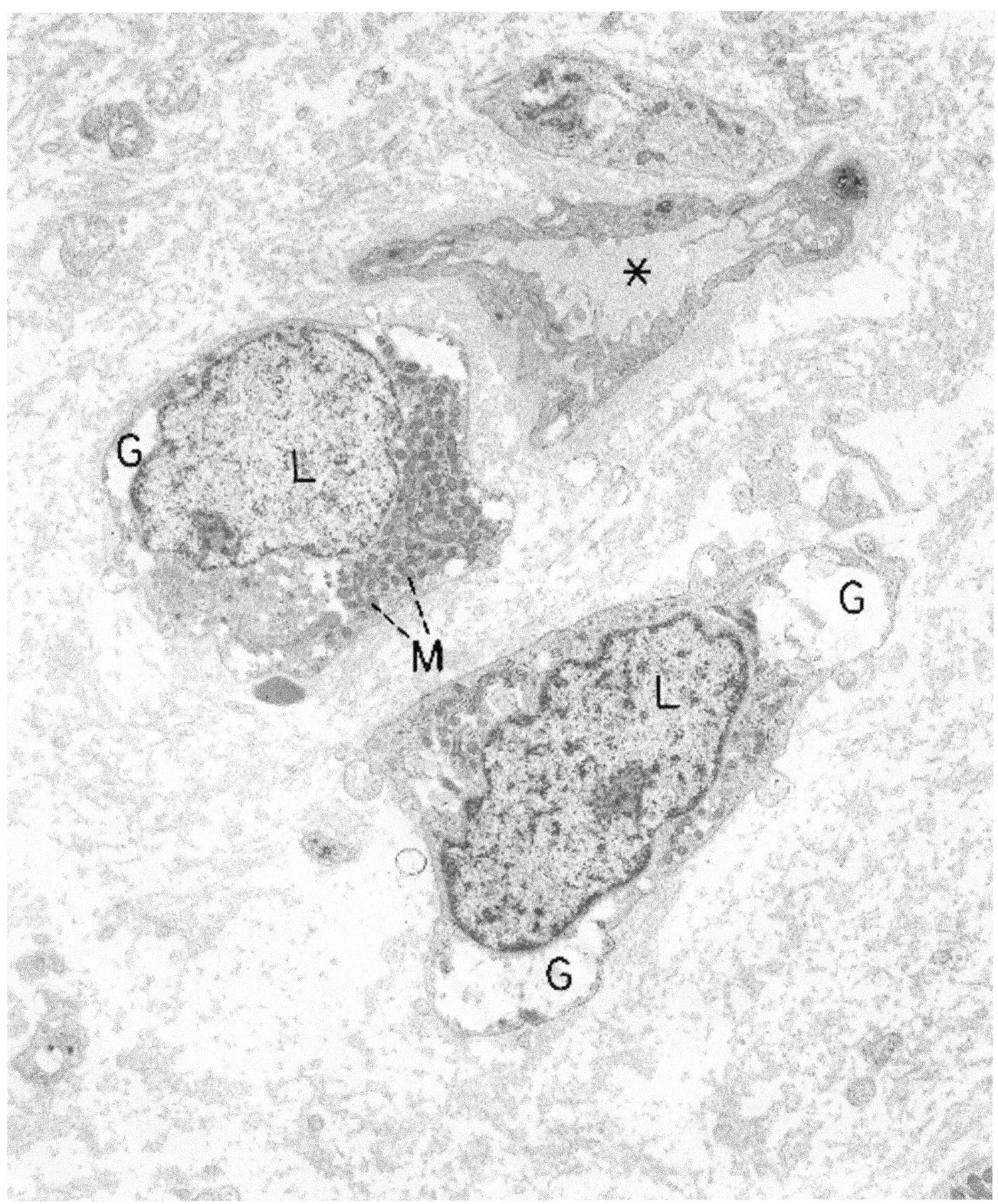

Figure 6.60. Myxoid liposarcoma (soft tissue of thigh). Early lipoblasts (L) surround a capillary (*) and show abundant glycogen clear spaces (G) and many mitochondria (M). Basal lamina around capillary and lipoblasts is diffuse and flocculent rather than discrete. (× 7000)

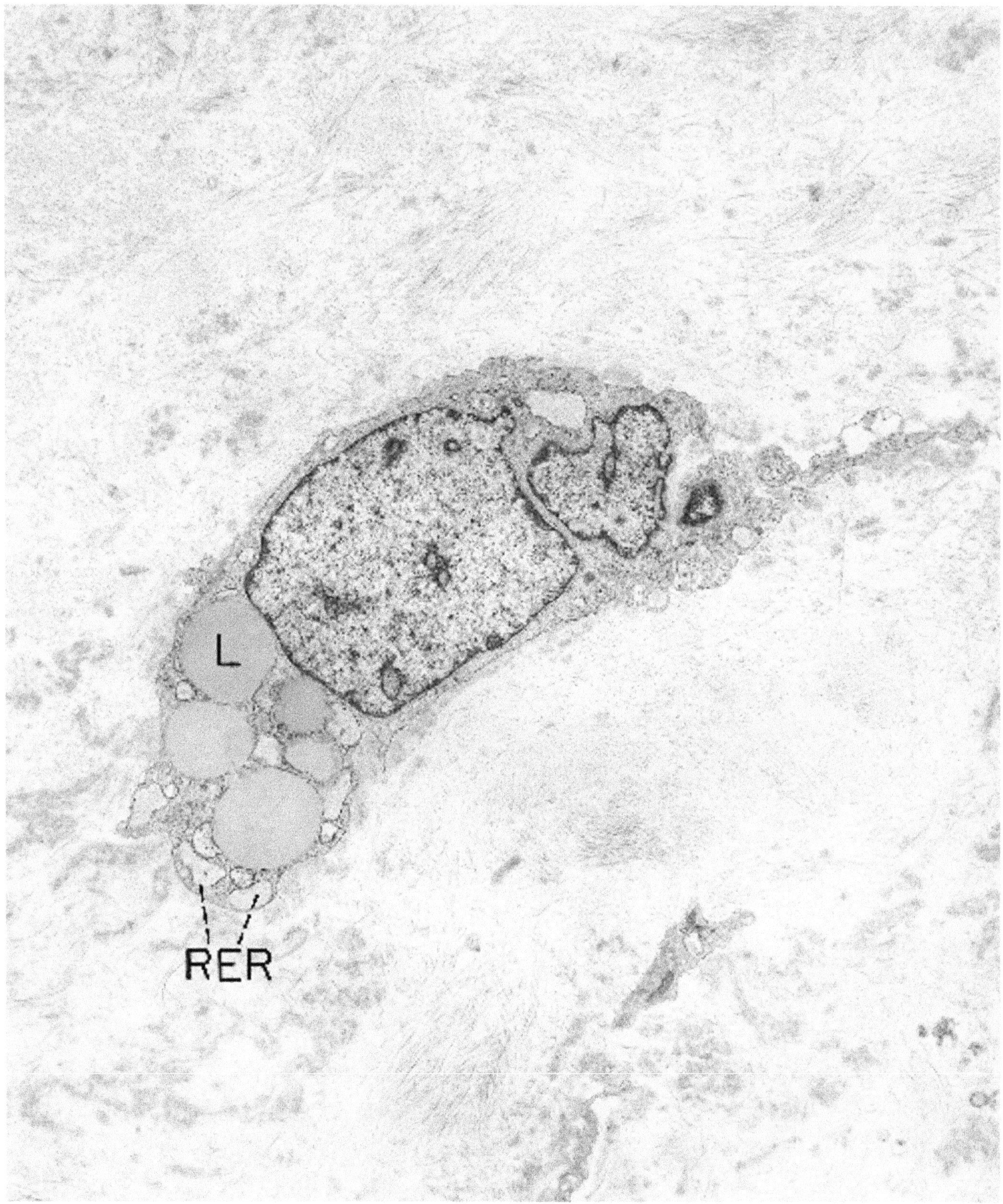

Figure 6.61. Myxoid liposarcoma (soft tissue of thigh). An early lipoblast contains prominent lipid (L) and rough endoplasmic reticulum (RER). (× 7000)

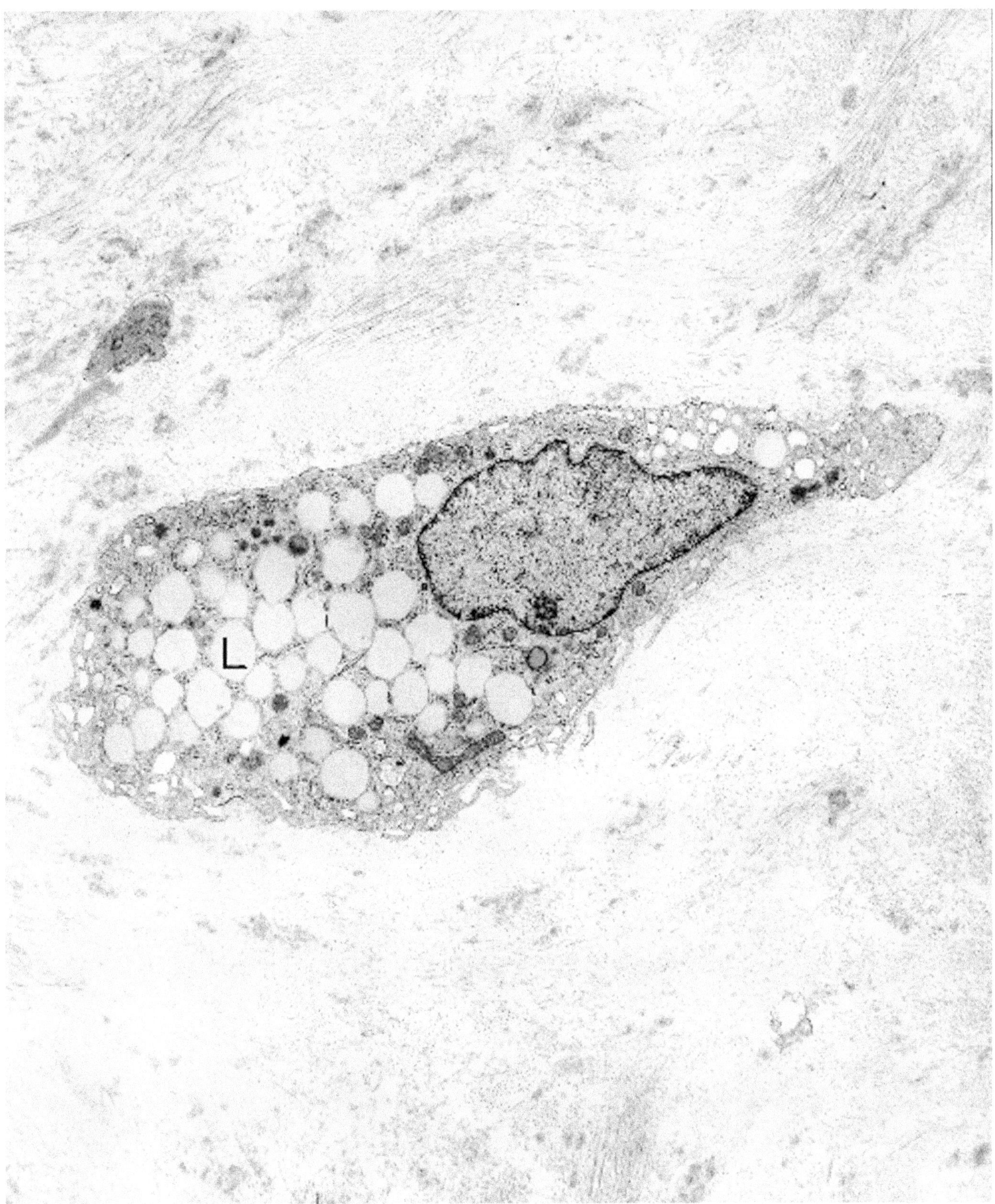

Figure 6.62. Myxoid liposarcoma (soft tissue of thigh). An intermediate-stage lipoblast contains many lipid droplets (L) as its main ultrastructural feature. Compare with early lipoblasts in Figure 6.59 through 6.61. (× 6875)

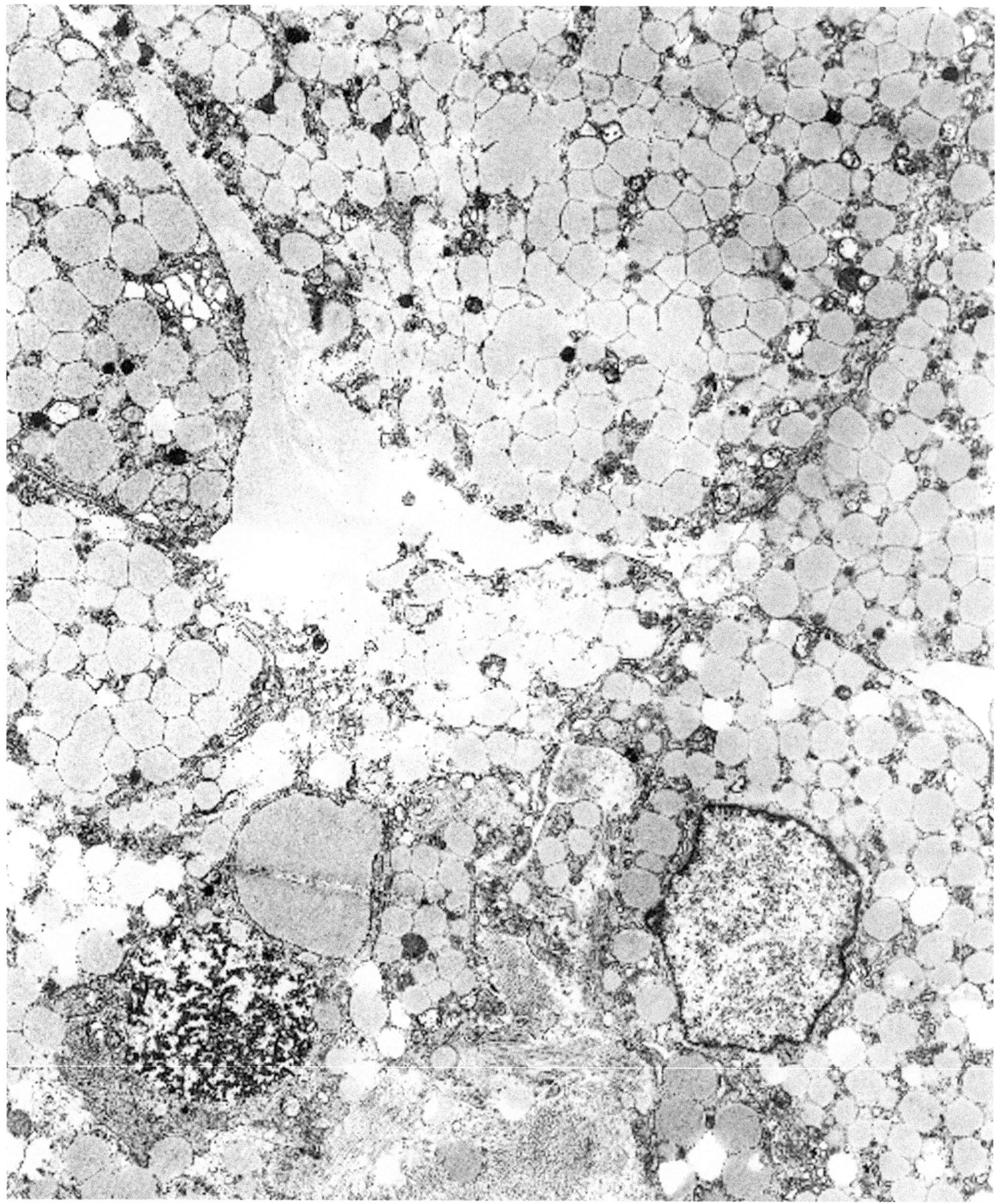

Figure 6.63. Myxoid liposarcoma (soft tissue of thigh). Several late lipoblasts have cytoplasm composed predominantly of lipid droplets. (× 5900)

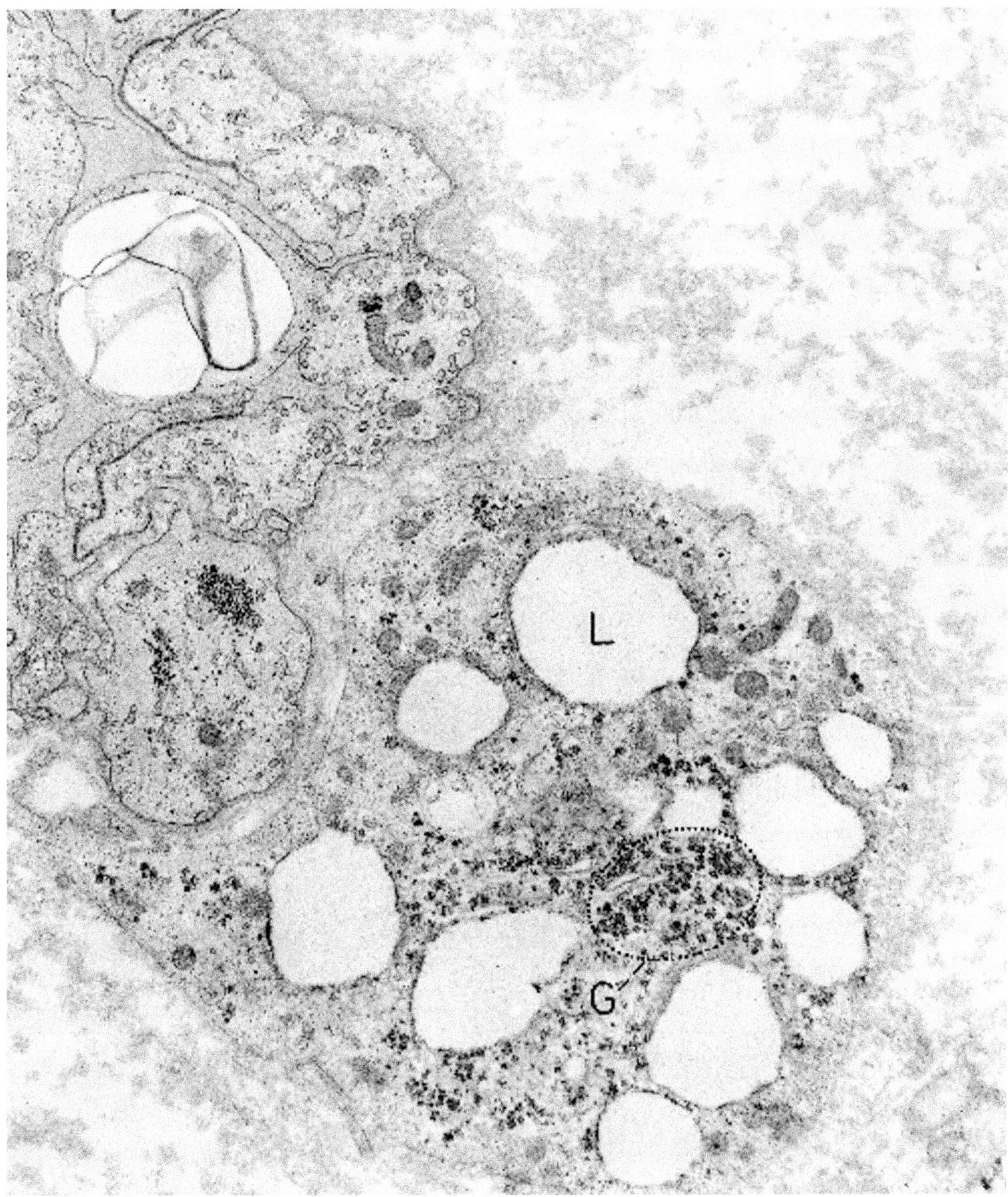

Figure 6.64. Myxoid liposarcoma (soft tissue of thigh). The specimen was processed to preserve glycogen as electron-dense granules. This early lipoblast contains a moderate amount of glycogen (G) and lipid (L). (× 18,000)

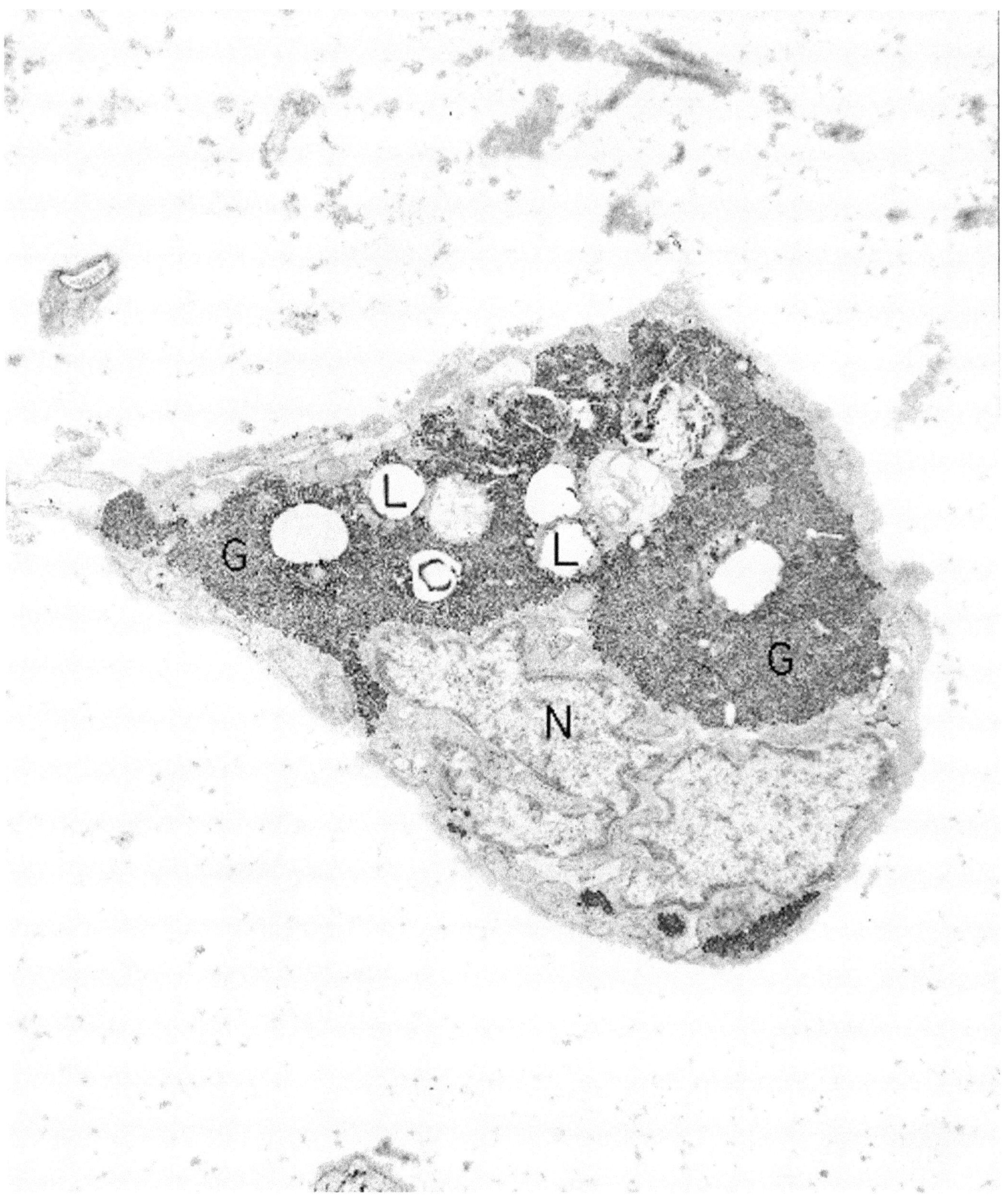

Figure 6.65. Myxoid liposarcoma (soft tissue of thigh). This early lipoblast has cytoplasm bulging with glycogen (G). L = lipid. N = nucleus. (× 13,275)

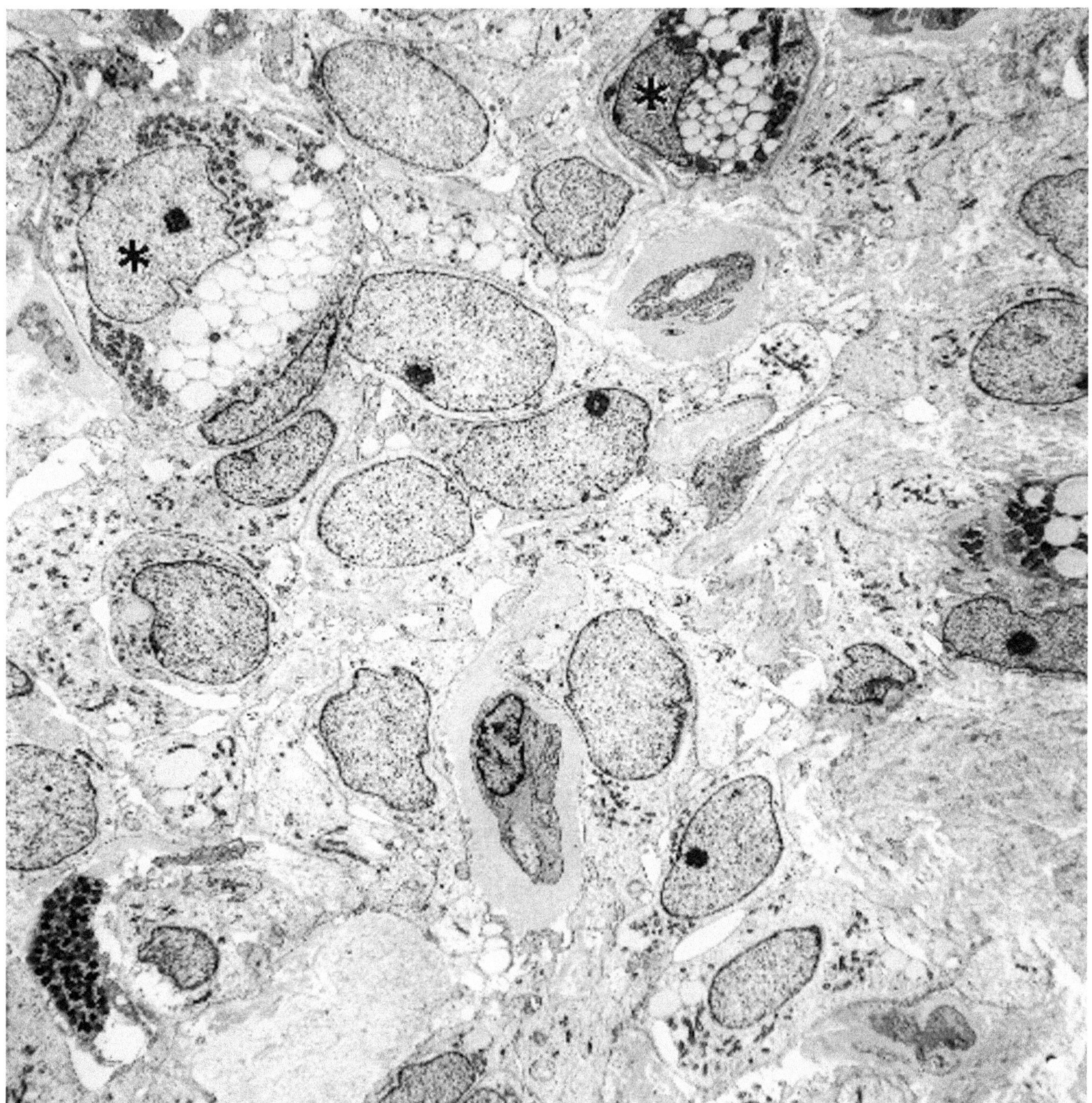

Figure 6.66. Round cell liposarcoma (soft tissue of thigh). Most of the cells in this neoplasm have no or only a few lipid droplets in their cytoplasm. The nuclear–cytoplasmic ratio is high, and nucleoli are small. A few interspersed cells (*) have more abundant cytoplasm, numerous lipid droplets, and numerous mitochondria. (× 3800)

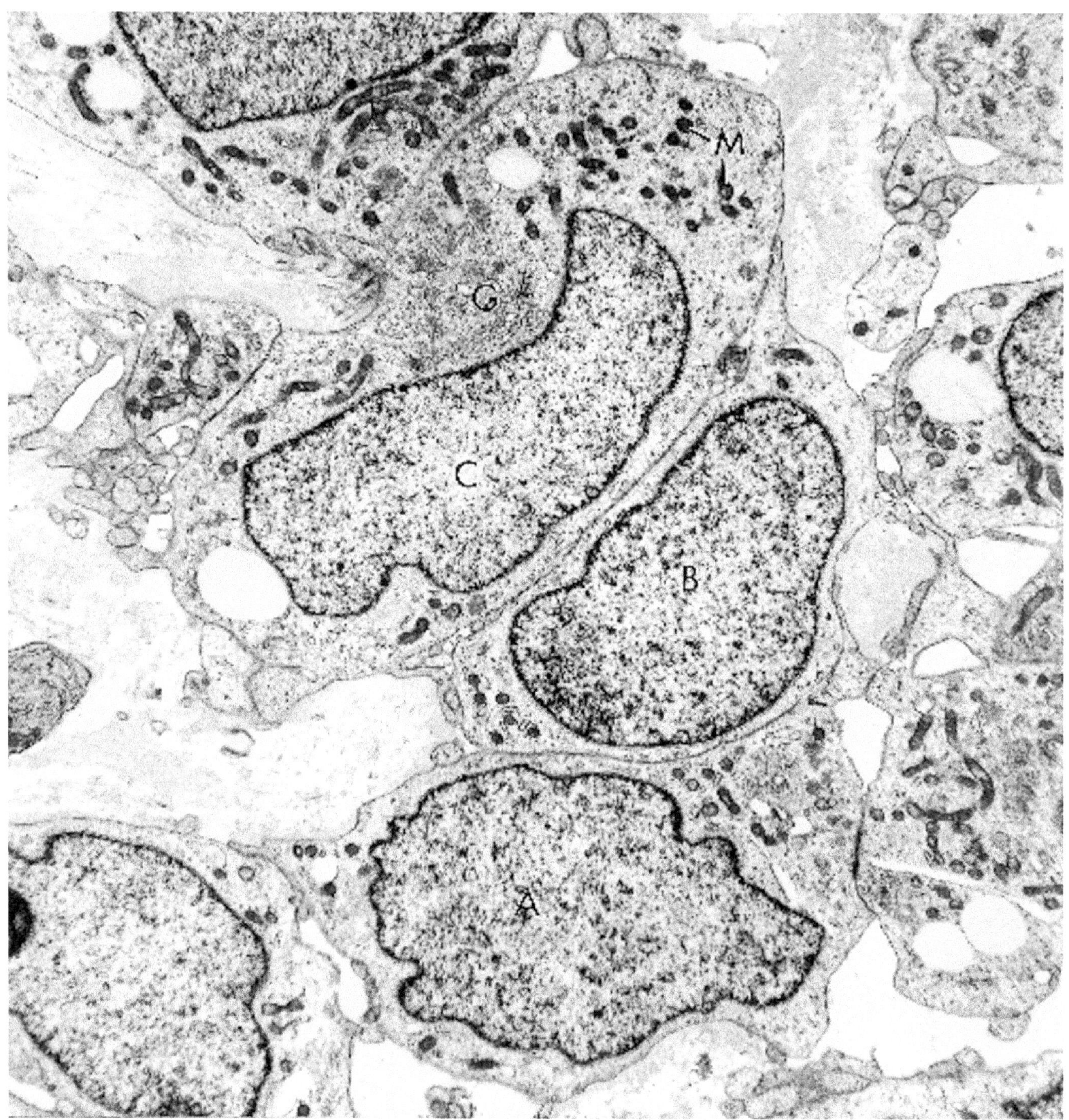

Figure 6.67. Round cell liposarcoma (soft tissue of thigh). Higher magnification of cells from the same neoplasm as depicted in Figure 6.66 illustrates two poorly differentiated lipoblasts (A and B) and one slightly more differentiated lipoblast (C) showing a rare lipid droplet, a moderate number of mitochondria (M), and a large Golgi apparatus (G). (× 9600)

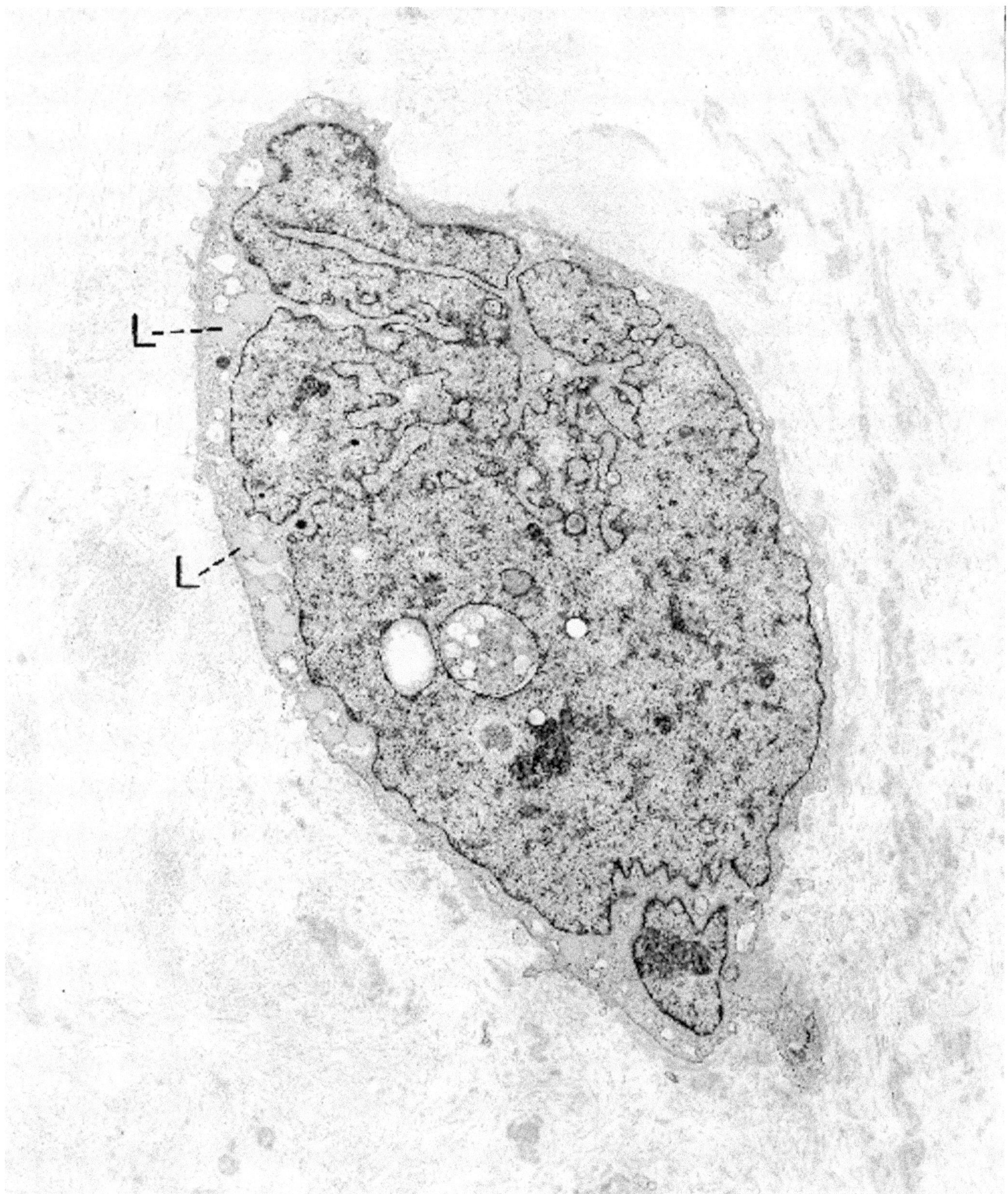

Figure 6.68. Pleomorphic liposarcoma (soft tissue of thigh). This giant lipoblast has a high nuclear–cytoplasmic ratio; a bizarre, multilobed nucleus, and a moderate number of lipid droplets (L). (× 4940)

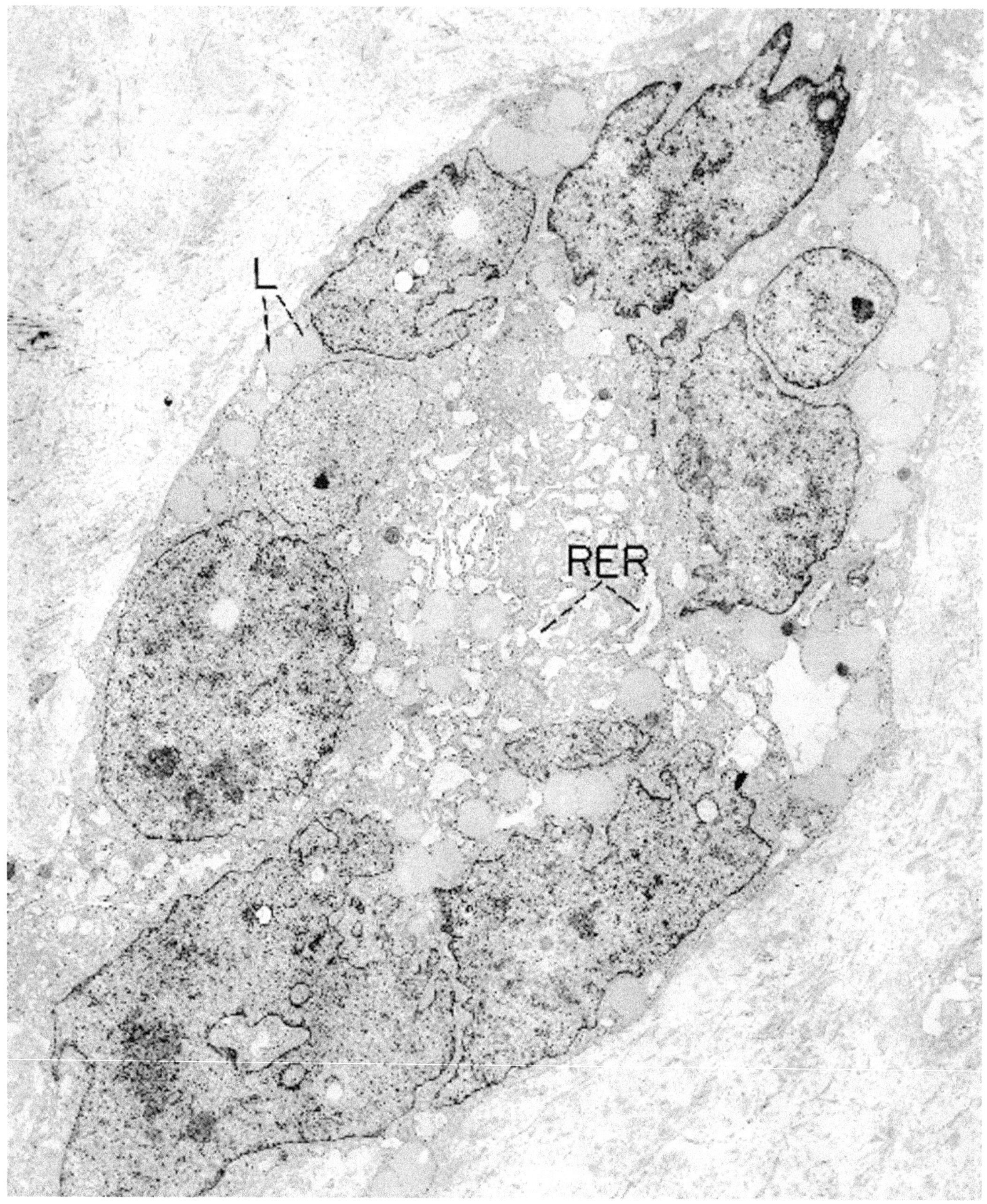

Figure 6.69. Pleomorphic liposarcoma (soft tissue of thigh). A multinucleated, giant lipoblast has copious cytoplasm with a preponderance of lipid droplets (L) and dilated rough endoplasmic reticulum (RER). (× 5130)

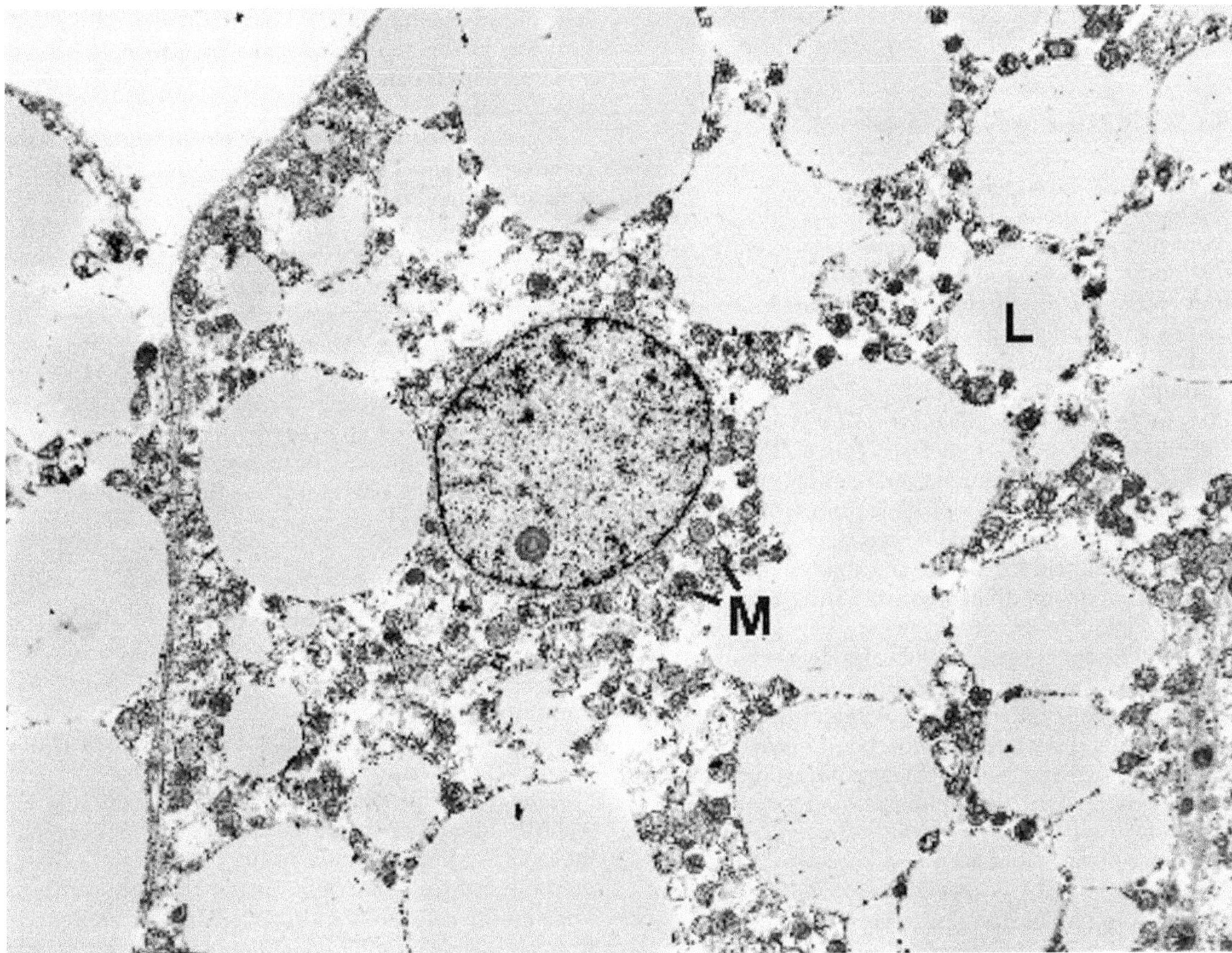

Figure 6.70. Hibernoma (soft tissue of interscapular region of back). These cells have the features of brown fat, namely, numerous mitochondria (M) and numerous small- and medium-sized lipid droplets (L). (× 5100)

(Text continued from page 295)

Smooth Muscle Neoplasms

(Figures 6.71 through 6.86.)

Diagnostic criteria. (1) Fascicular or syncytial arrangement of spindle cells, in a matrix of collagen (Figures 6.71 and 6.72); (2) basal lamina surrounding cells (Figures 6.72 and 6.73); (3) thin (6 nm) filaments and dense bodies among filaments, within the cytoplasm proper and subjacent to the plasmalemma (Figures 6.73 through 6.75); (4) pinocytotic vesicles (Figures 6.75); (5) round-ended nuclei (Figures 6.72 and 6.73); (6) contraction indentations of nuclei (Figure 6.76).

Additional points. Filaments and dense bodies tend to be more numerous in cytoplasmic processes than in cell bodies, and they are usually considered to be a minimum requirement for identifying smooth muscle cells. However, in poorly differentiated leiomyosarcomas, epithelioid leiomyosarcomas, and certain gastrointestinal stromal tumors, these filaments and dense bodies may be scant or absent, and the fulfillment of some of the other diagnostic criteria in small amounts or focally is considered supportive. Round-ended nuclei are especially valuable and, with nuclear contraction indentations, may be sufficient evidence for a "probable" or "consistent with" diagnosis. Furthermore, narrow attachment plaques may be found between cells.

An example of a poorly differentiated leiomyosarcoma having a nondescript cytoplasm, with few filaments and no definite dense bodies, is illustrated in Figure 6.77. A few poorly differentiated leiomyosarcomas have cells with few filaments and abundant rough endoplasmic reticulum, resembling fibroblasts (Figures 6.78 and 6.79). In these cases, some of the features listed other than filaments and dense bodies as well as further sampling may lead to the correct diagnosis.

Epithelioid leiomyosarcomas (leiomyoblastomas) also depend on the presence of thin filaments and dense bodies for a definite diagnosis, but the other ultrastructural characteristics of smooth muscle may be scant or absent. Basal lamina also is often present, but more often than not the cytoplasm contains numerous mitochondria as its main organelle (Figure 6.80). *Glomus tumors* also are composed of epithelioid-type, mitochondria-rich smooth muscle cells, usually in islands and small clusters separated by basal lamina and banded collagen, and often in juxtaposition to thin-walled blood vessels (Figure 6.81 and 6.82). Thin filaments, dense bodies, pinocytosis, and basal lamina are also often present and characterize the glomus cell as smooth muscle in type (Figure 6.82).

A rare form of leiomyoma and leiomyosarcoma is the *granular cell type.* These neoplasms mimic granular cell schwannomas at the light microscopic level, but by electron microscopy their granules are seen to be derived from mitochondria rather than from electron-dense secondary lysosomes (Figures 6.83 through 6.85).

A significant proportion of *gastrointestinal stromal tumors* (GIST), in our experience, are leiomyomas and leiomyosarcomas, and some are atypical and difficult to prove as being of smooth muscle cell type (Figure 6.86). In others, the cell of origin may be a fibroblast, a Schwann cell, an autonomic neuron or, questionably, an interstitial cell of Cajal (gastrointestinal pacemaker cell) (see Chapter 9).

Skeletal Muscle Neoplasms

(Figures 6.87 through 6.90) (also see Figures 4.14 through 4.24).

Diagnostic criteria. (1) Thick (15 nm), myosin filaments; (2) Z-band formation; (3) sarcomeres.

Additional points. Rhabdomyosarcomas composed of spindle-shaped cells usually are easier to diagnose than those consisting of small, round, embryonal cells or than the bizarre cells of *pleomorphic rhabdomyosarcoma.* The spindle cells are varying stages of differentiating strap cells and contain thin and thick filaments, readily identifiable Z-band material, and usually some degree of sarcomere formation. Refer again to Figures 4.28 through 4.32 for early strap cell differentiation in embryonal rhabdomyosarcomas. Later stages in the development of skeletal muscle cells are illustrated in the rhabdomyoma in Figures 6.87 and 6.88.

Spindle cell embryonal rhabdomyosarcoma is a rare neoplasm that usually occurs in children, but it may also be seen in adults. The cells are arranged in a fascicular or storiform pattern and resemble late-stage fetal myotubules. The cells are immature rhabdomyoblasts and have copious glycogen, numerous mitochondria, a moderate amount of rough endoplasmic reticulum, and focal collections of filaments with Z-band material and early sarcomere formation (Figures 6.89 and 6.90).

(Text continues on page 341)

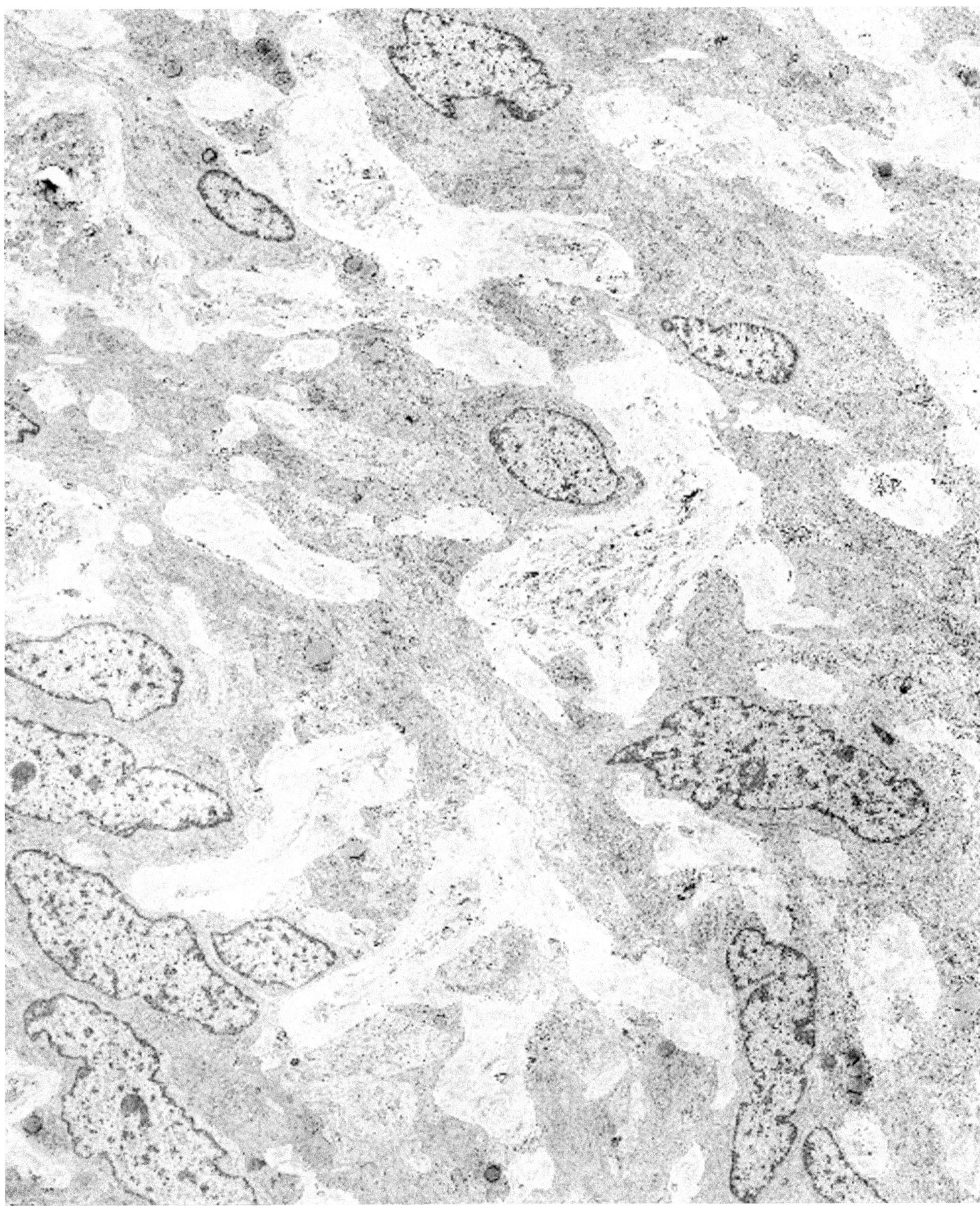

Figure 6.71. Leiomyosarcoma (pelvic soft tissue). The neoplasm is composed of a syncytium of spindle cells with focal attachments in a matrix of collagen. (× 5510)

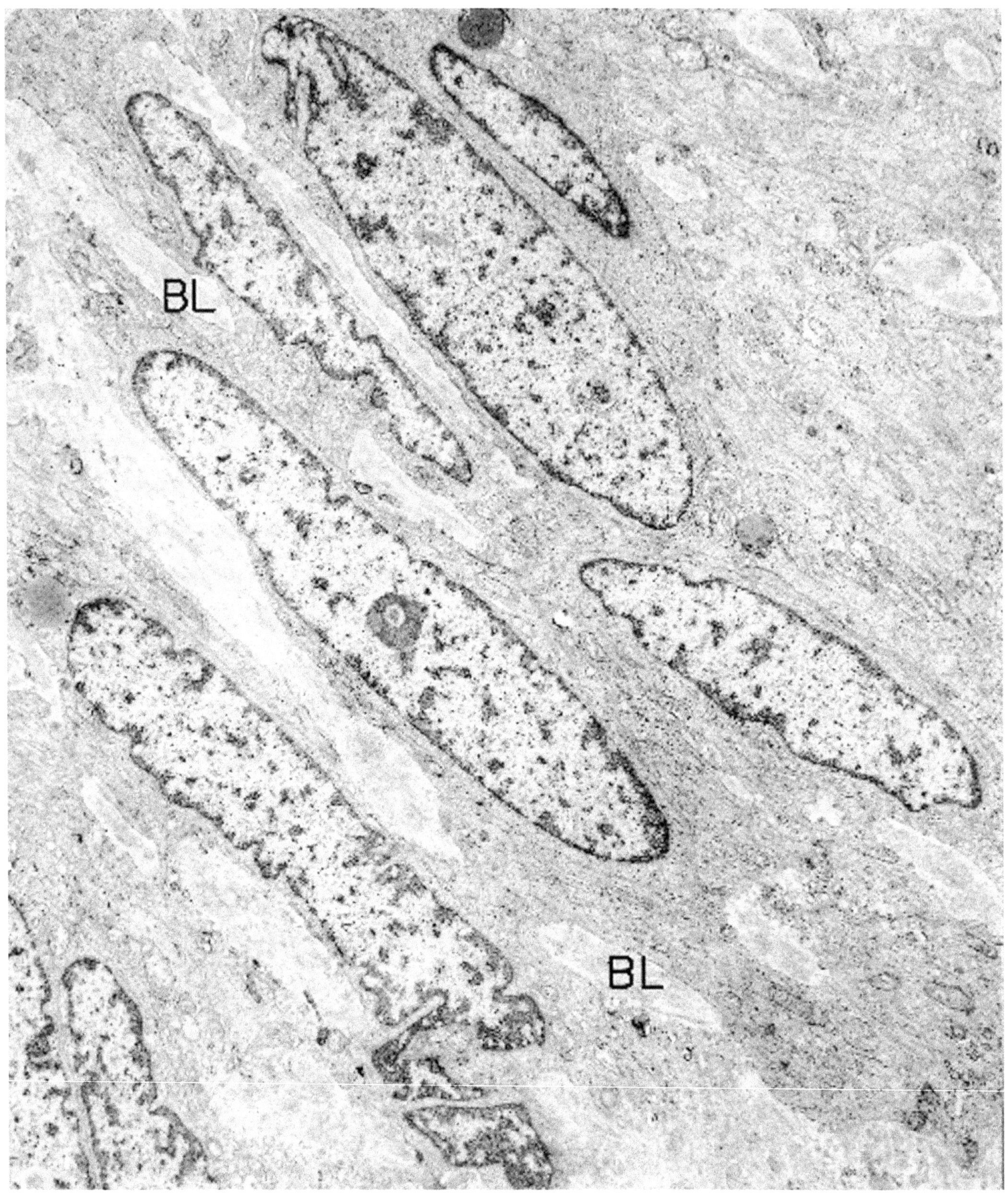

Figure 6.72. Leiomyosarcoma (pelvic soft tissue). This area of the same neoplasm as depicted in Figure 6.71 consists of a close arrangement of spindle cells rather than the loose, syncytial pattern exemplified in Figure 6.71. Basal lamina (BL) covers the free surfaces of the cells and often is diffuse rather than discrete (see Figure 6.73). Nuclei tend to have at least one rounded end, and some nuclei have shallow contraction indentations. (× 7185)

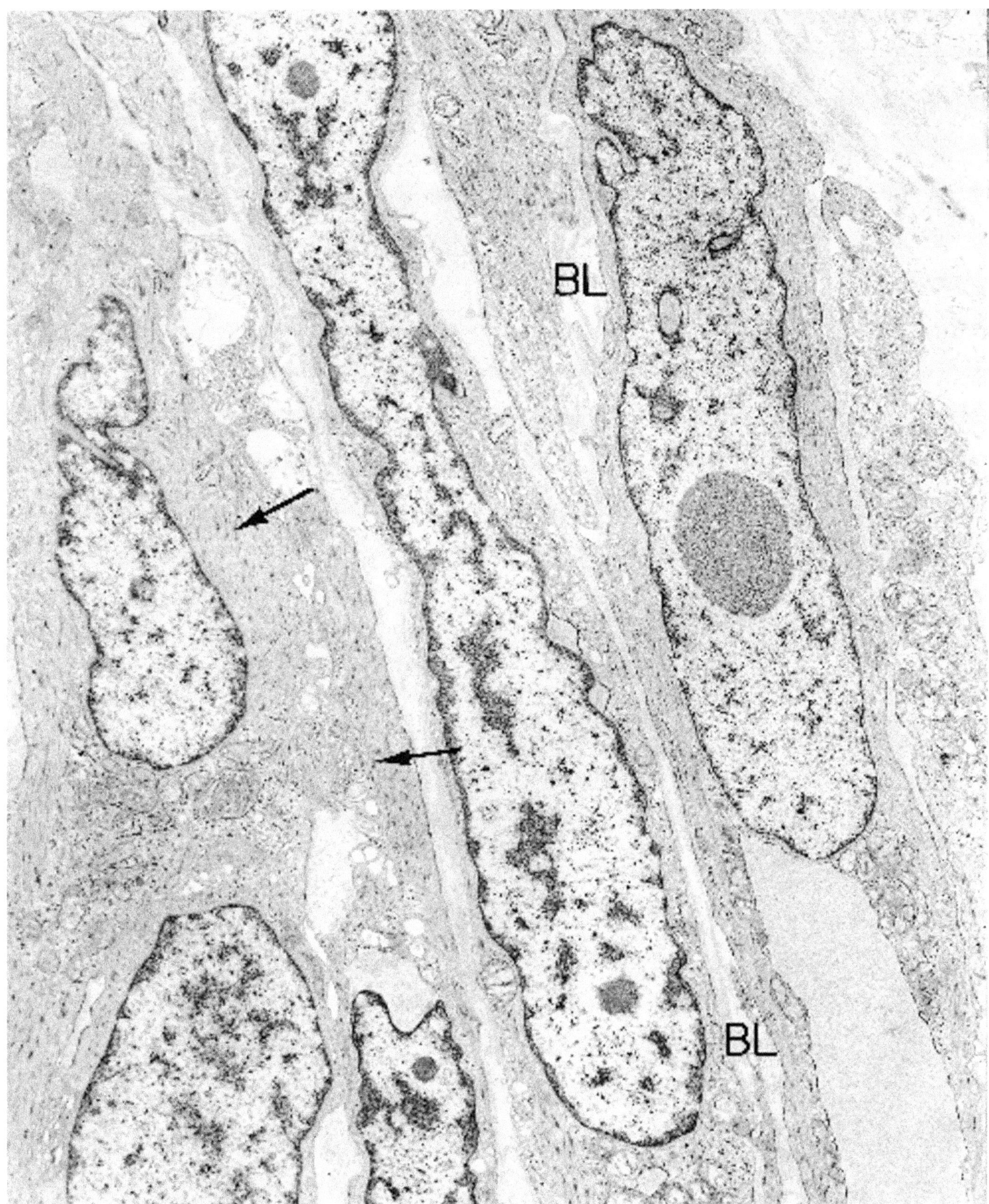

Figure 6.73. Leiomyosarcoma (metastatic to dura matter). These malignant spindle cells exhibit several markers of smooth muscle, including basal lamina (BL), innumerable thin filaments, and dense bodies of filaments (*arrows*) and round-ended nuclei. (× 9520)

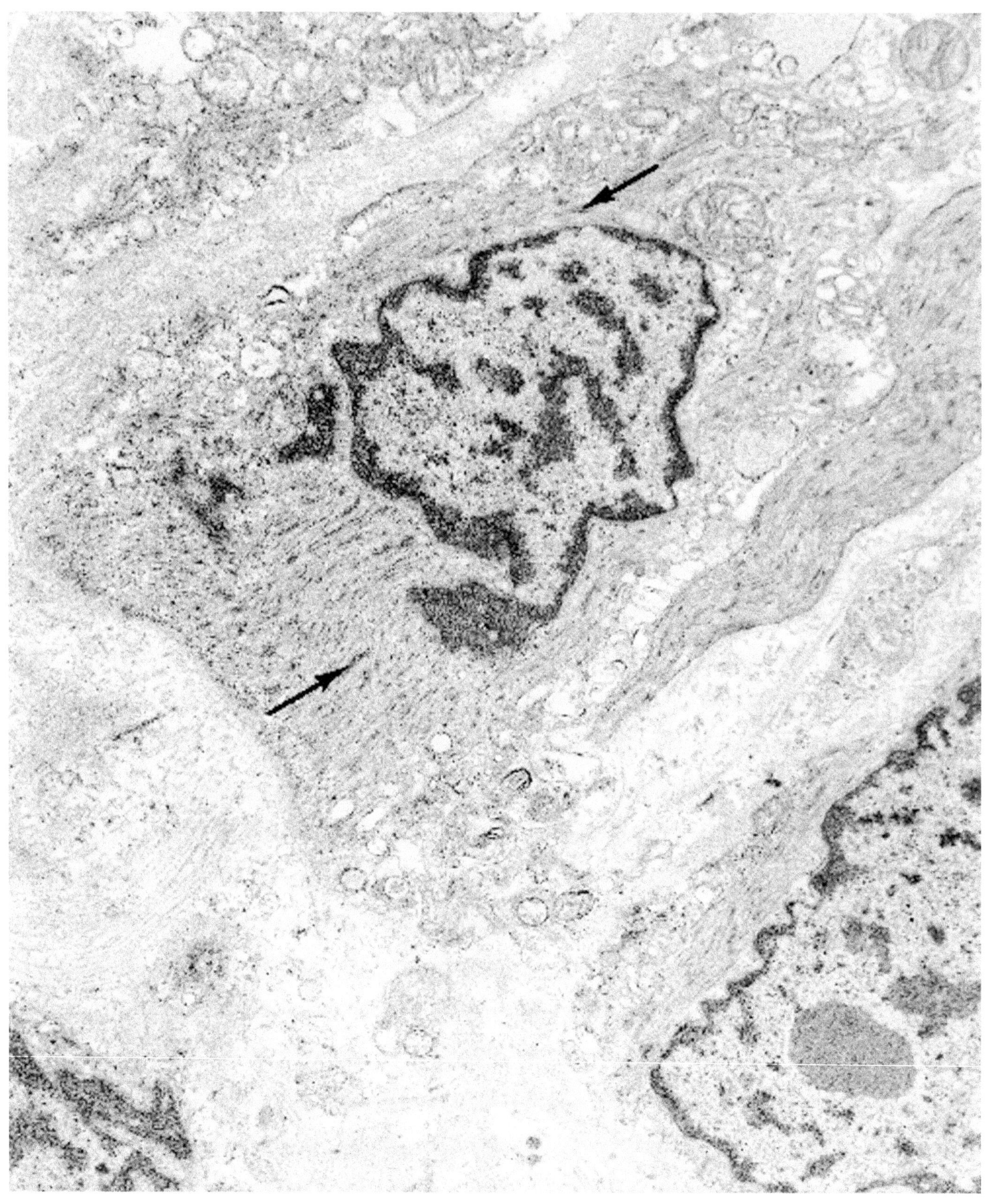

Figure 6.74. Leiomyosarcoma (metastatic to dura matter). A malignant smooth muscle cell has cytoplasm filled with thin filaments and dense bodies *(arrows)*. (× 13,500)

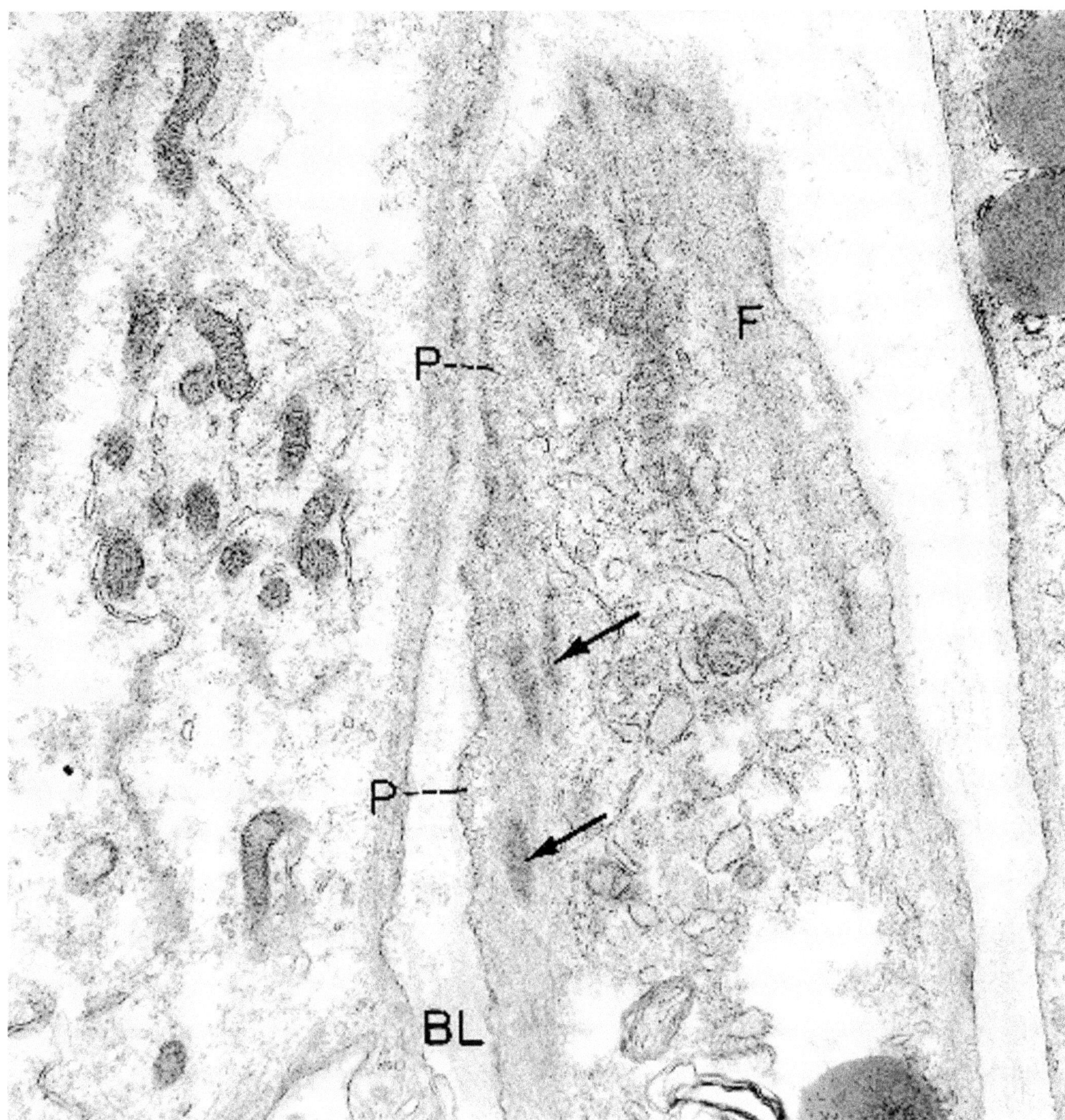

Figure 6.75. Leiomyosarcoma (metastatic to trapezius muscle). High magnification of malignant smooth muscle cells illustrates many thin filaments (F) and dense bodies (*arrows*) occupying the cytoplasm. Note also that there is peripheral compartmentalization of the filaments and that some of the dense bodies lie immediately subjacent to the plasmalemma. Pinocytotic vesicles (P) and basal lamina (BL) are easily seen at this magnification. (× 26,000)

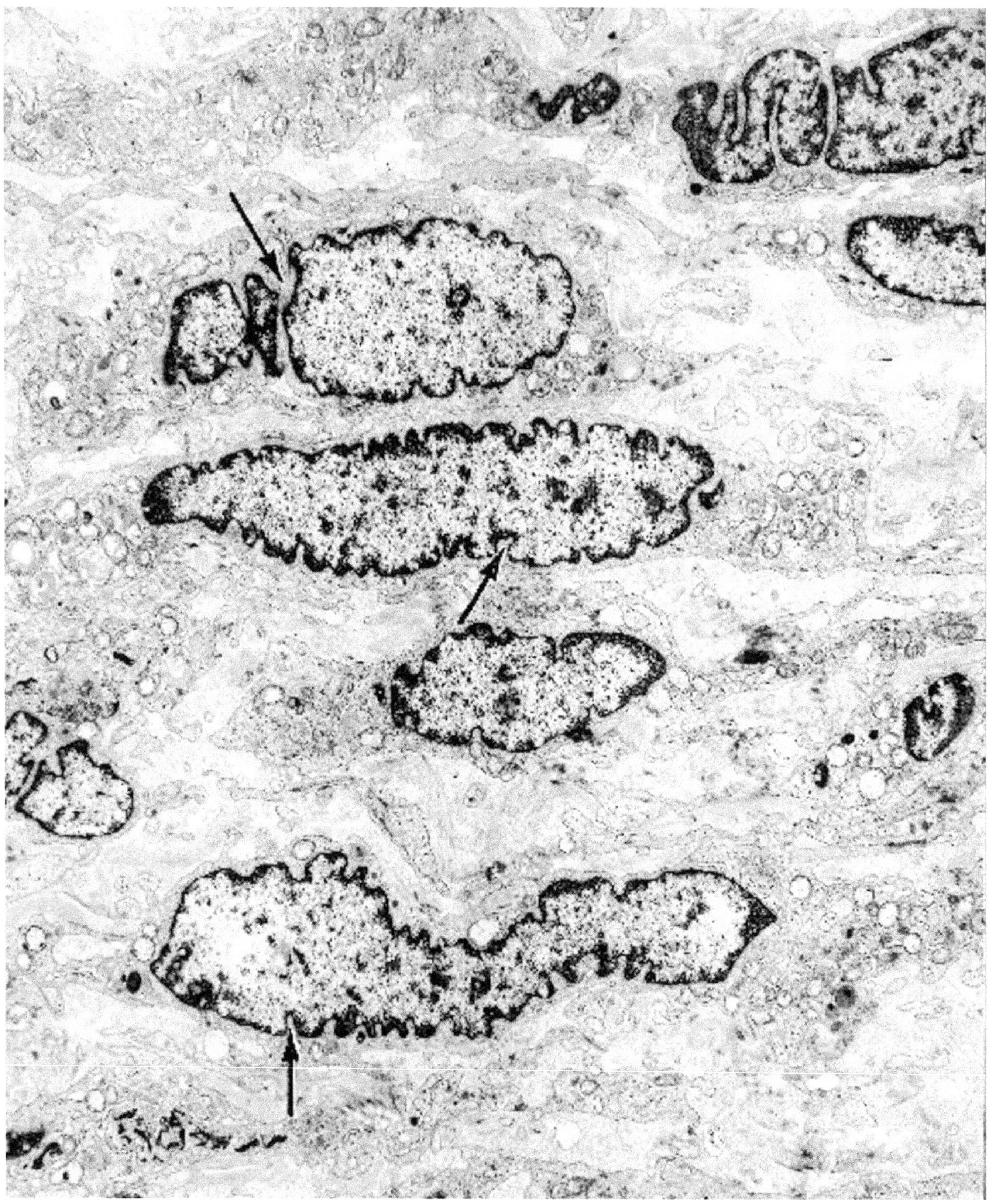

Figure 6.76. Leiomyosarcoma (tibia). This cellular spindle-cell neoplasm shows the nuclear contraction indentations (*arrows*) so often seen in smooth muscle cells. (× 6750)

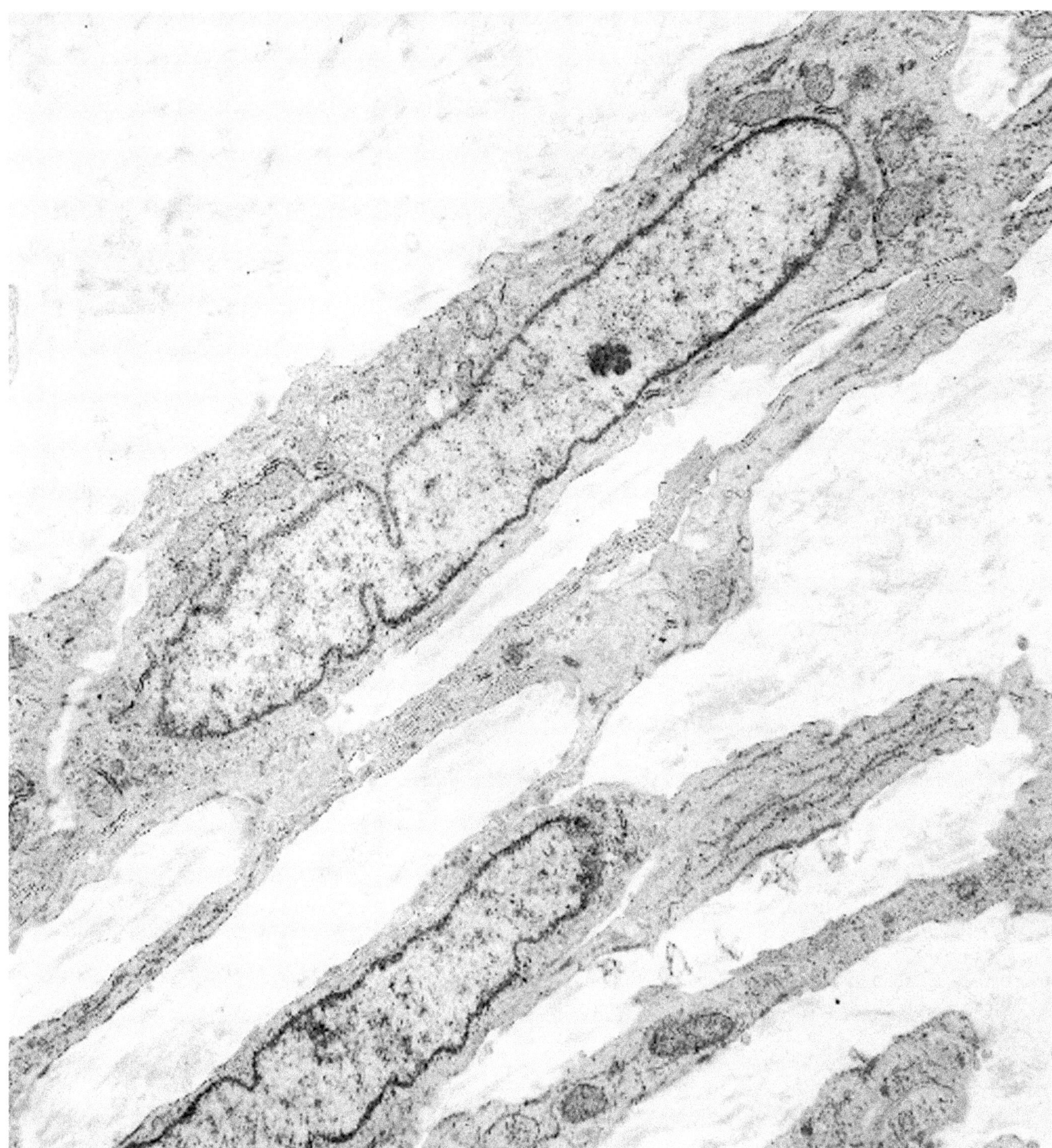

Figure 6.77. Leiomyosarcoma, nondescript type (retrovaginal tissue). Filaments are scant in these malignant spindle cells, but round-ended and contracted nuclei are salient features. (× 10,500) (Permission for reprinting granted by Taylor and Francis Publishers, Dickersin GR, Selig MK, Park YN: The many faces of smooth muscle neoplasms in a gynecological sampling: An ultrastructural study. Ultrastruct Pathol 21:109–134, 1997.)

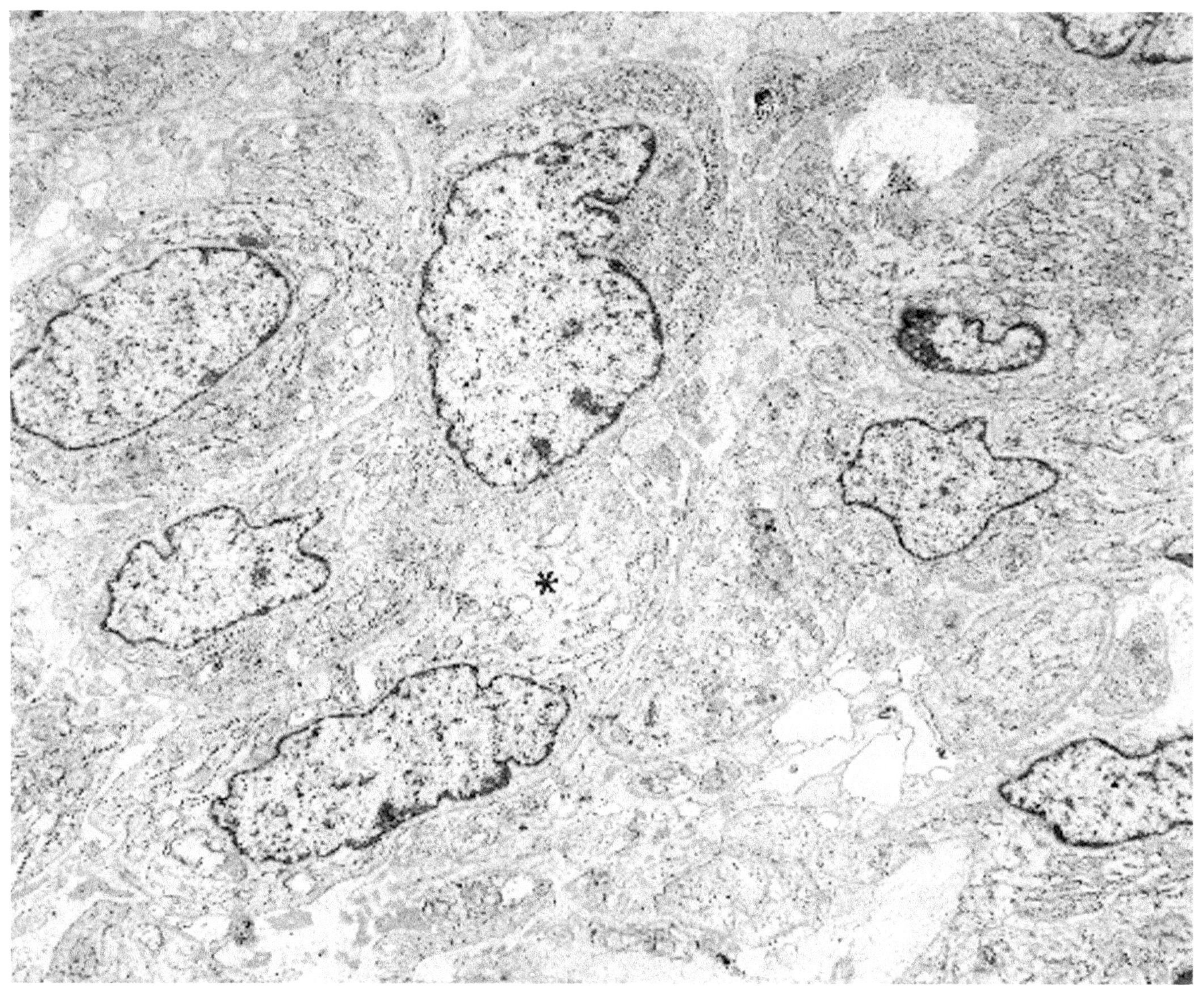

Figure 6.78. Metastatic leiomyosarcoma, fibroblast-like type (omentum). Tightly apposed, neoplastic spindle cells have abundant rough endoplasmic reticulum in their cytoplasm. A few cells contain focal collections of filaments (*) as well. (× 6800) (Permission for reprinting granted by Taylor and Francis Publishers, Dickersin GR, Selig MK, Park YN: The many faces of smooth muscle neoplasms in a gynecological sampling: An ultrastructural study. Ultrastruct Pathol 21:109–134, 1997.)

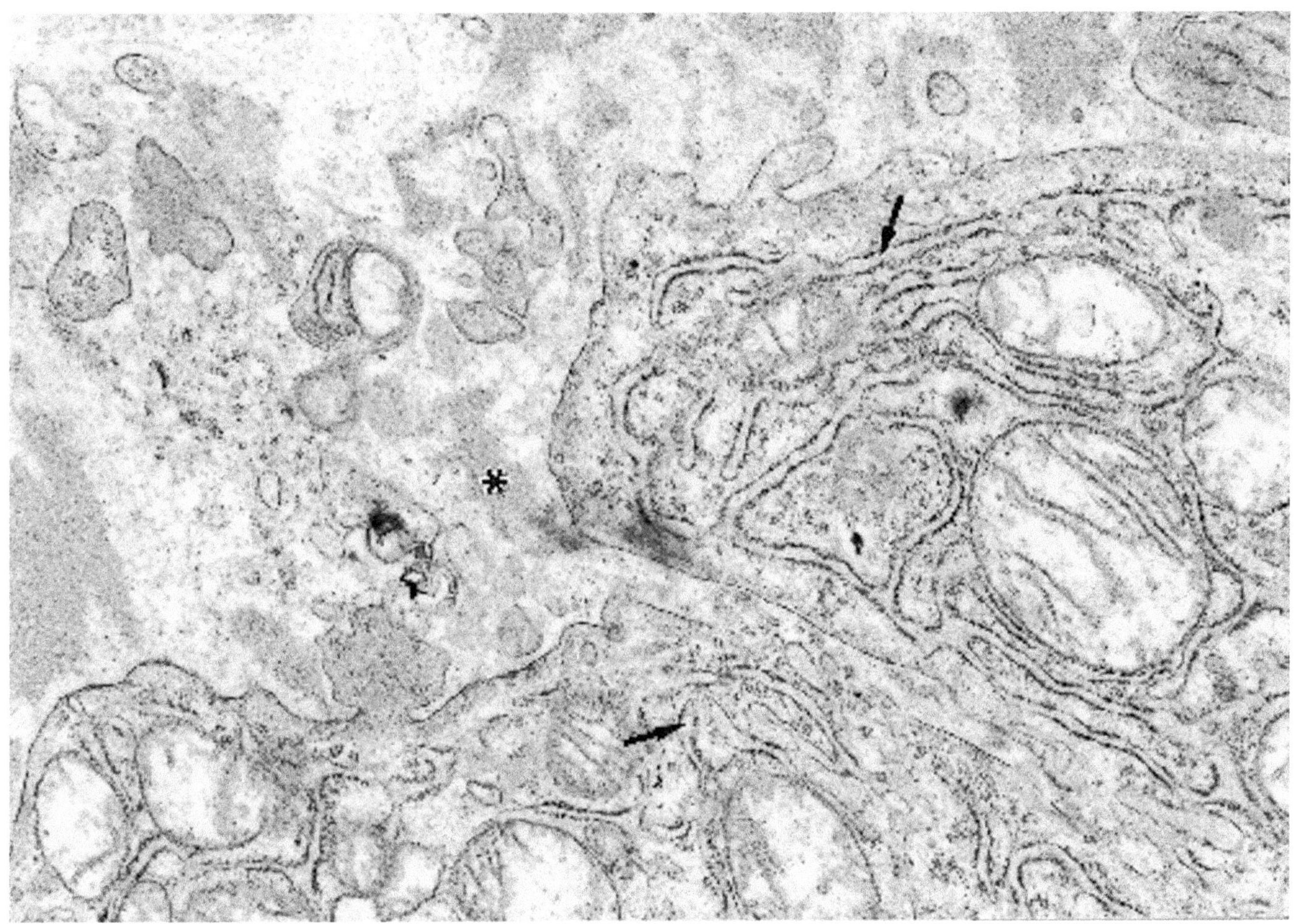

Figure 6.79. Metastatic leiomyosarcoma, fibroblast-like type (omentum). High magnification illustrates moderately dilated rough endoplasmic reticulum (*arrows*) and excessive basal lamina material (*). (× 24,900)) (Permission for reprinting granted by Taylor and Francis Publishers, Dickersin GR, Selig MK, Park YN: The many faces of smooth muscle neoplasms in a gynecological sampling: An ultrastructural study. Ultrastruct Pathol 21: 109–134, 1997.)

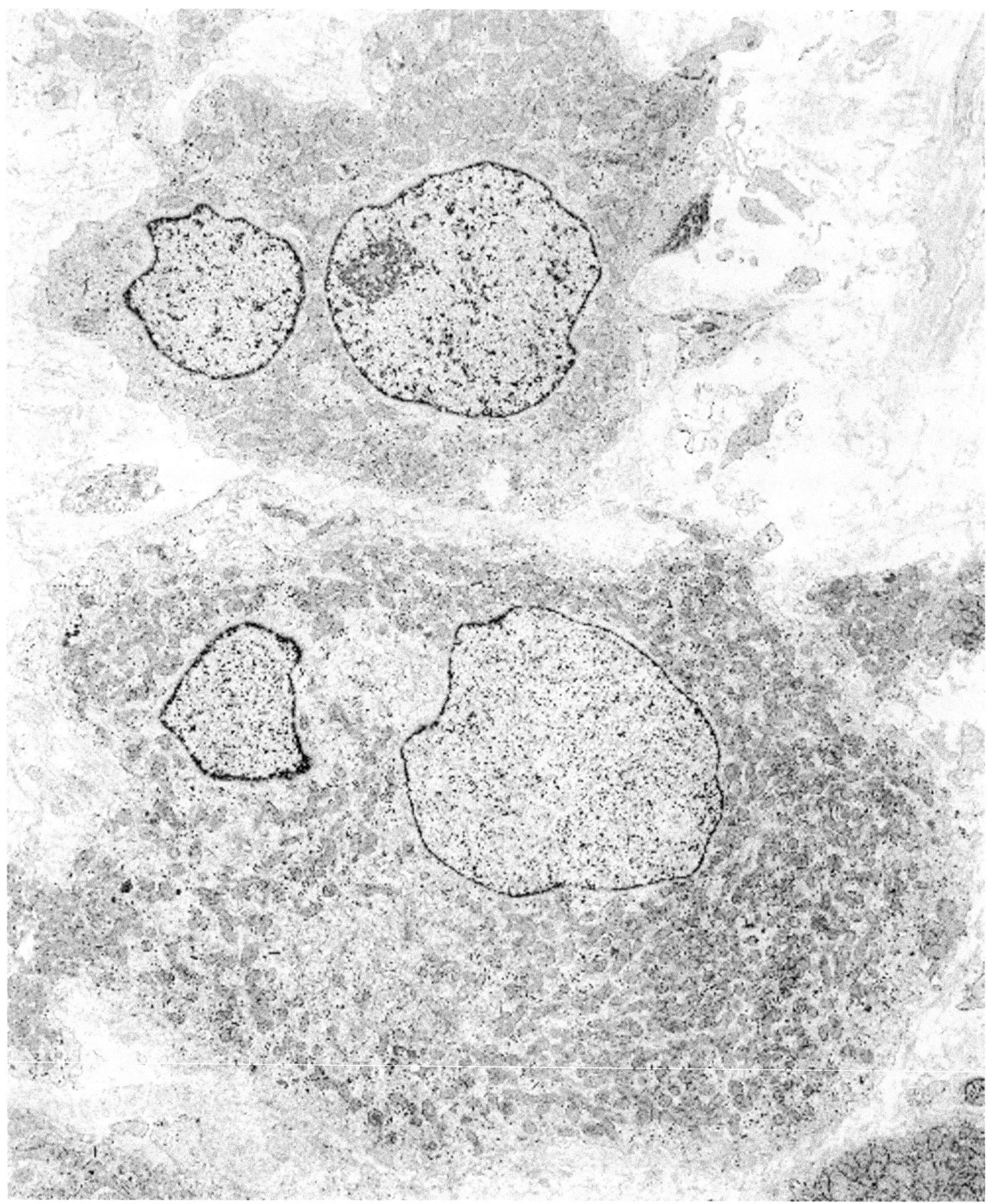

Figure 6.80. Epithelioid leiomyosarcoma (stomach). The cells of this neoplasm are polygonal rather than spindle-shaped, and their cytoplasm contains many mitochondria as the main feature. There are relatively few microfilaments. (× 5130)

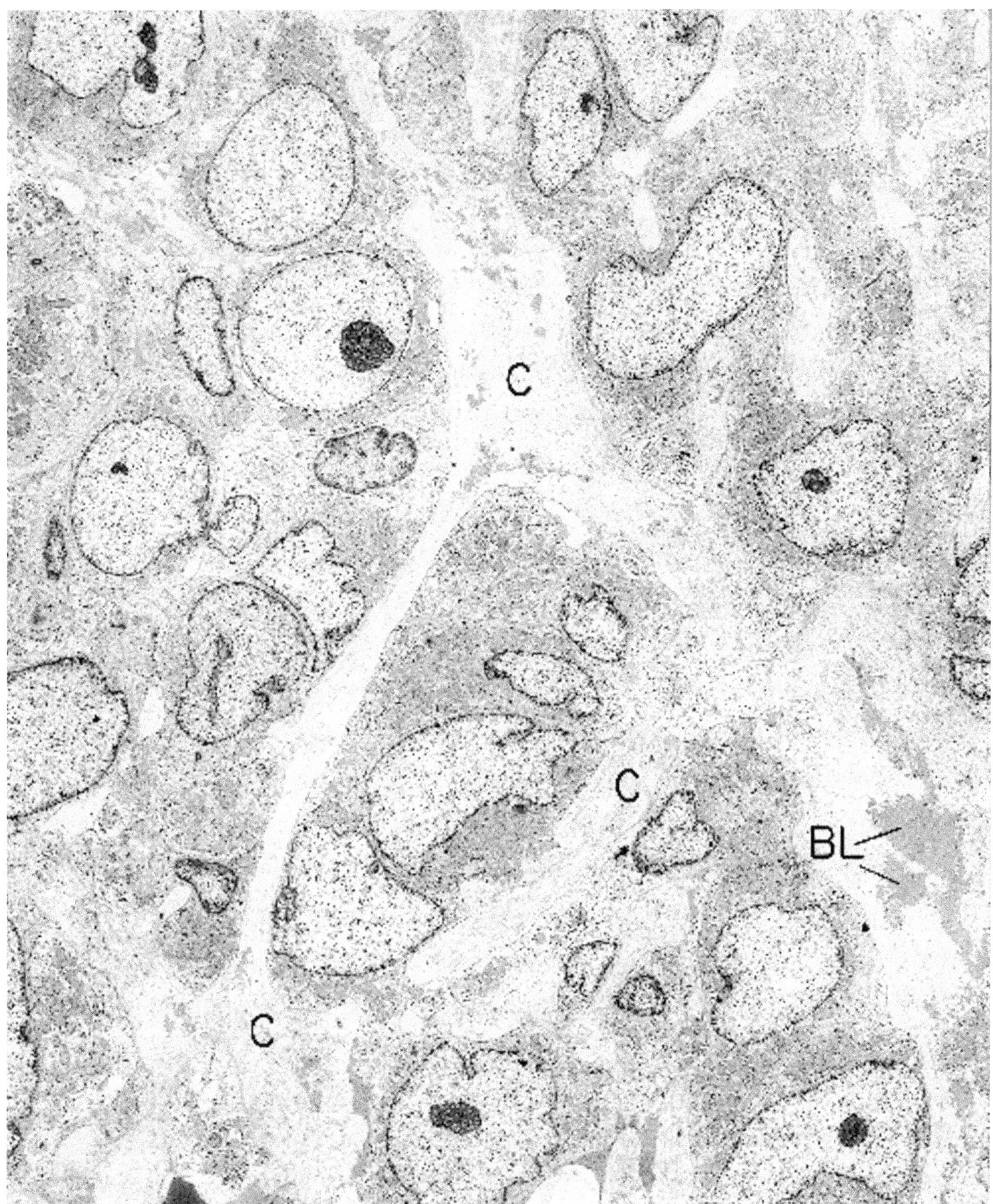

Figure 6.81. Glomus tumor (soft tissue of inguinal region). Clusters of epithelioid type cells are separated by bands of collagen (C) and excessive accumulations of basal-like material (BL). The marked density of the cytoplasm of the cells is attributable to innumerable mitochondria. (× 4250)

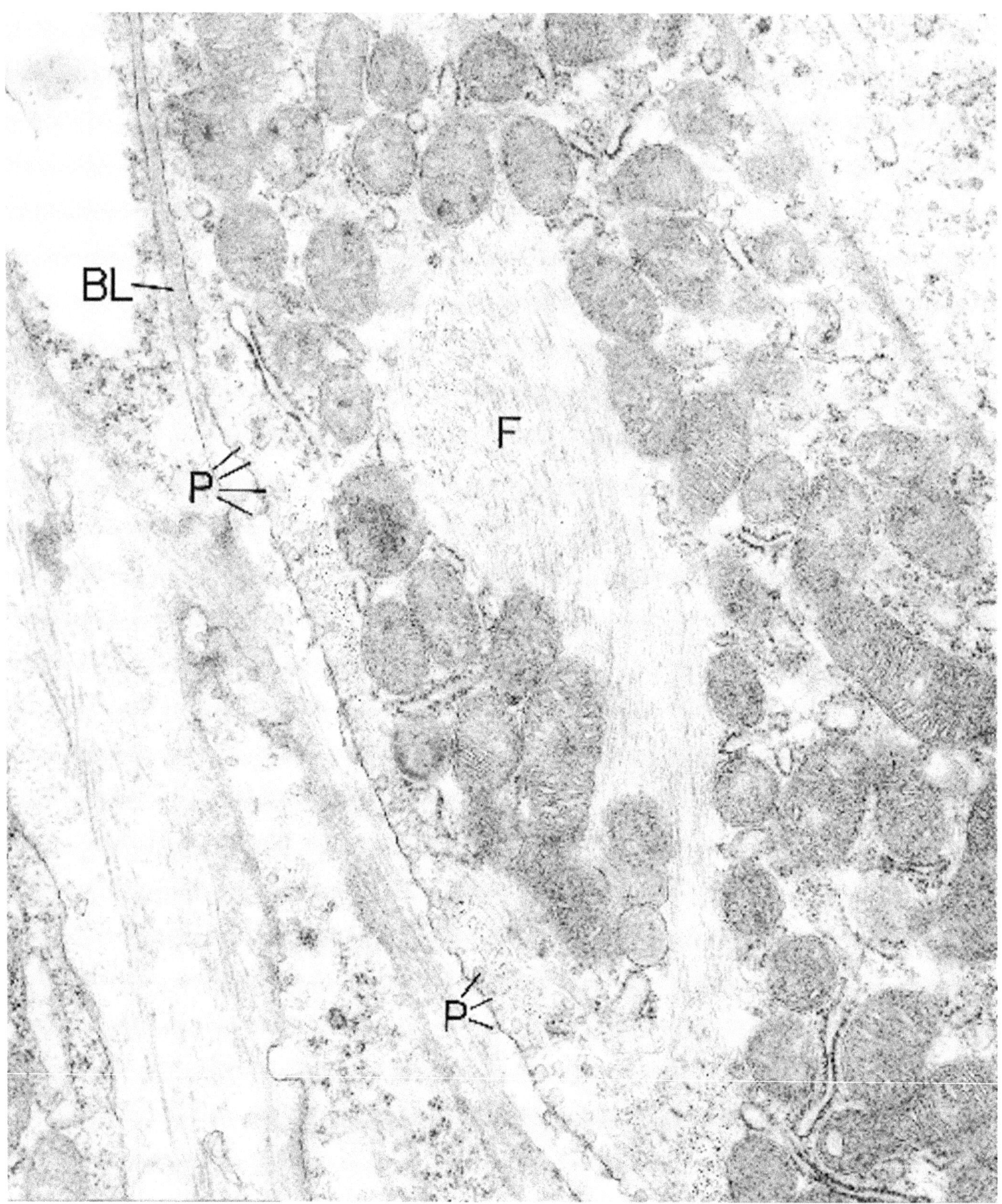

Figure 6.82. Glomus tumor (soft tissue of inguinal region). High magnification of a neoplastic glomus cell highlights the smooth muscle feature of cytoplasmic filaments (F), pinocytotic vesicles (P), and basal lamina (BL). (× 42,500)

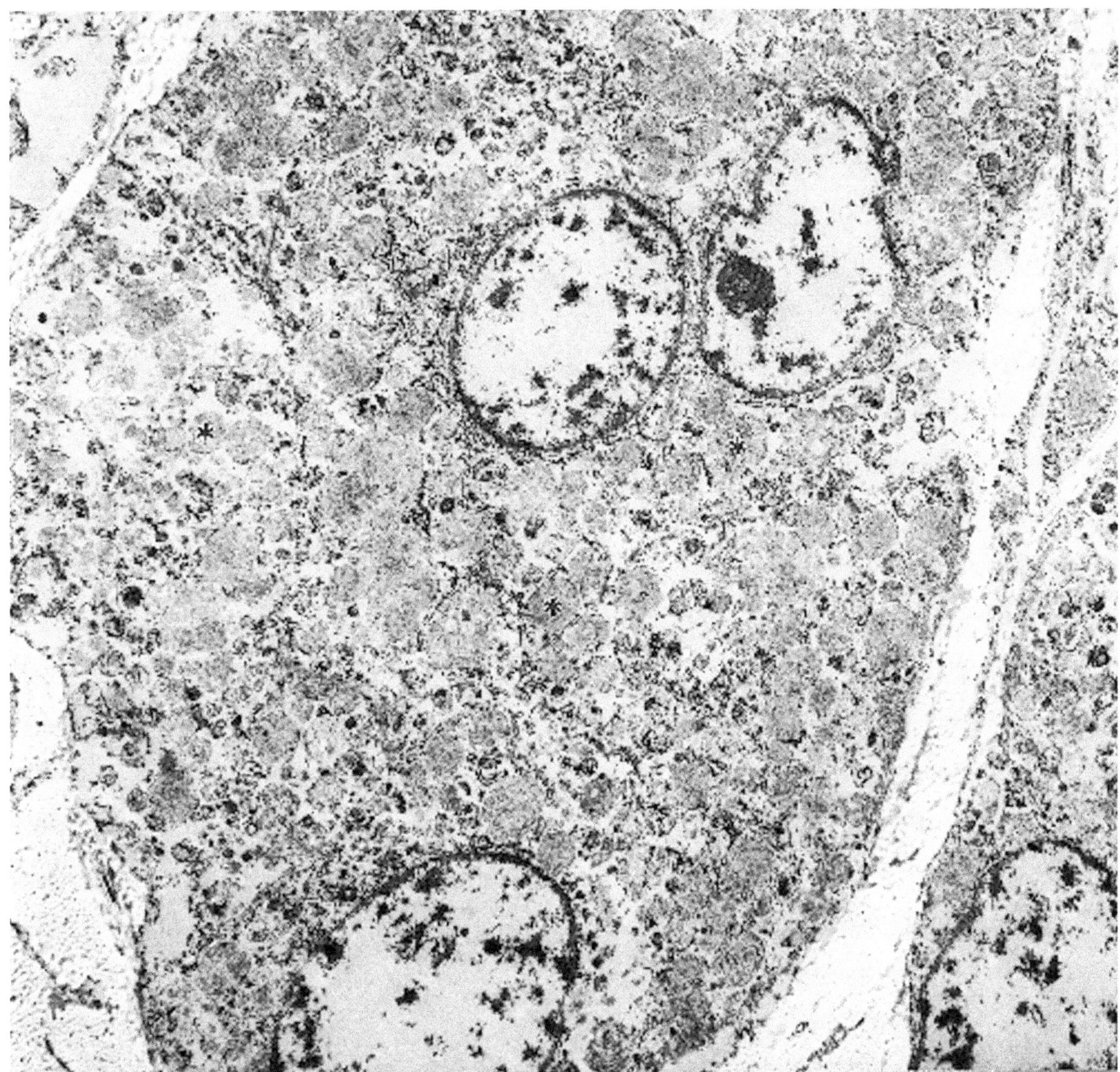

Figure 6.83. Granular cell leiomyoblastoma (uterus, formalin-fixed). Ultrastructural detail is somewhat compromised by the method of fixation, but still discernible are numerous cytoplasmic granules (*). (× 7600) (Permission for reprinting granted by Taylor and Francis Publishers, Dickersin GR, Selig MK, Park YN: The many faces of smooth muscle neoplasms in a gynecological sampling: An ultrastructural study. Ultrastruct Pathol 21: 109–134, 1997.)

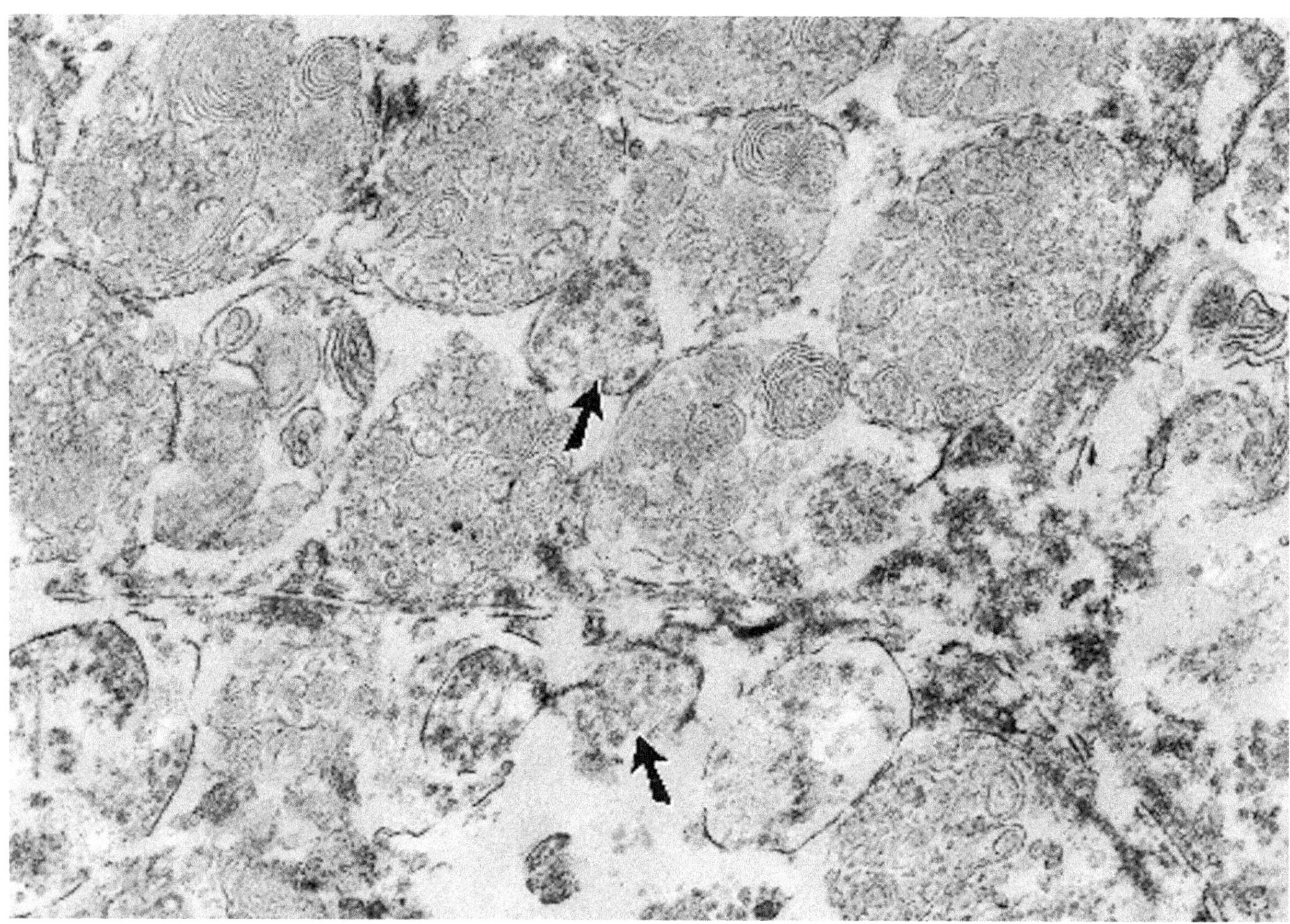

Figure 6.84. Granular cell leiomyoblastoma (uterus, formalin-fixed). High magnification of the same neoplasm as depicted in Figure 6.83 reveals the granules to contain multiple membranous whorls and, in some examples, small globular densities (*arrows*). (× 43,200) (Permission for reprinting granted by Taylor and Francis Publishers, Dickersin GR, Selig MK, Park YN: The many faces of smooth muscle neoplasms in a gynecological sampling: An ultrastructural study. Ultrastruct Pathol 21:109–134, 1997.)

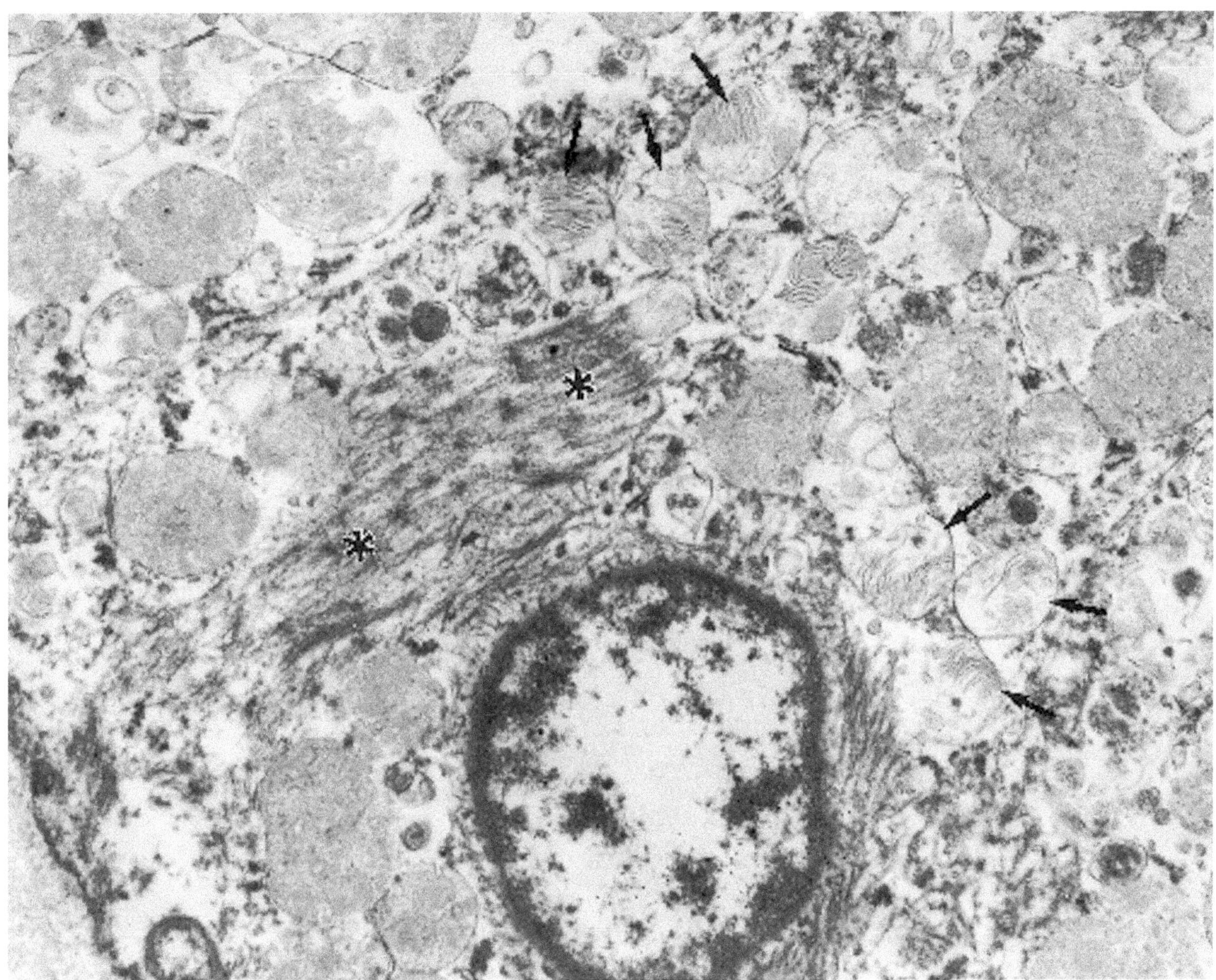

Figure 6.85. Granular cell leiomyoblastoma (uterus, formalin-fixed). This cell contains mostly simple (noncompound) granules (*arrows*), which suggest degenerating mitochondria. Filaments with densities (*) are also present. (× 24,500)) (Permission for reprinting granted by Taylor and Francis Publishers, Dickersin GR, Selig MK, Park YN: The many faces of smooth muscle neoplasms in a gynecological sampling: An ultrastructural study. Ultrastruct Pathol 21:109–134, 1997.)

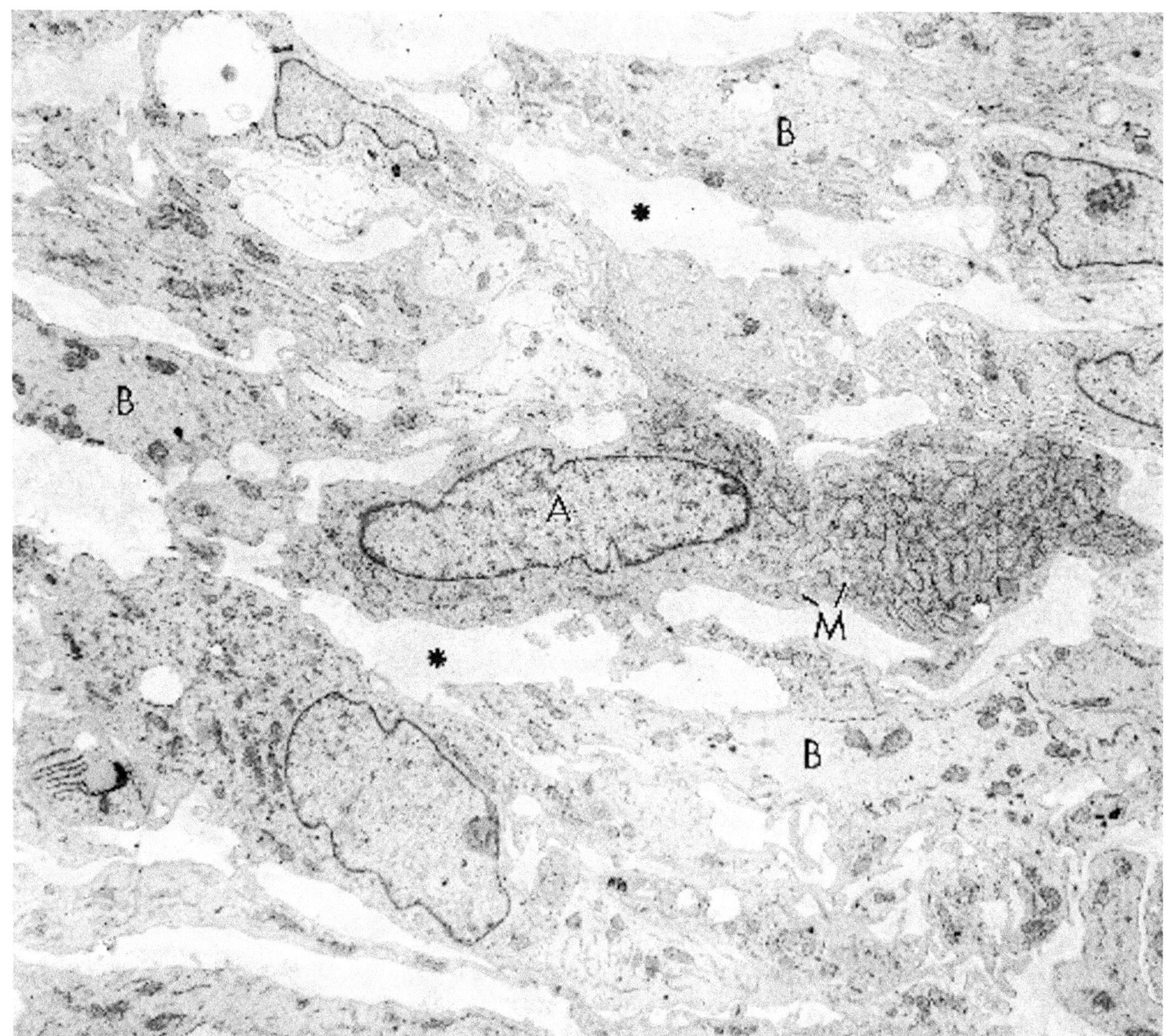

Figure 6.86. Gastrointestinal stromal tumor (colon). The neoplastic spindle cells vary in their cytoplasmic content, some (A) having numerous mitochondria (M) and others (B) having innumerable filaments (pale areas). Only a few dense bodies were found among the filaments. Nuclei tend to be round ended and mildly contracted. A flocculent matrix (*) separates the cells. These features are consistent with smooth muscle differentiation. (× 5400)

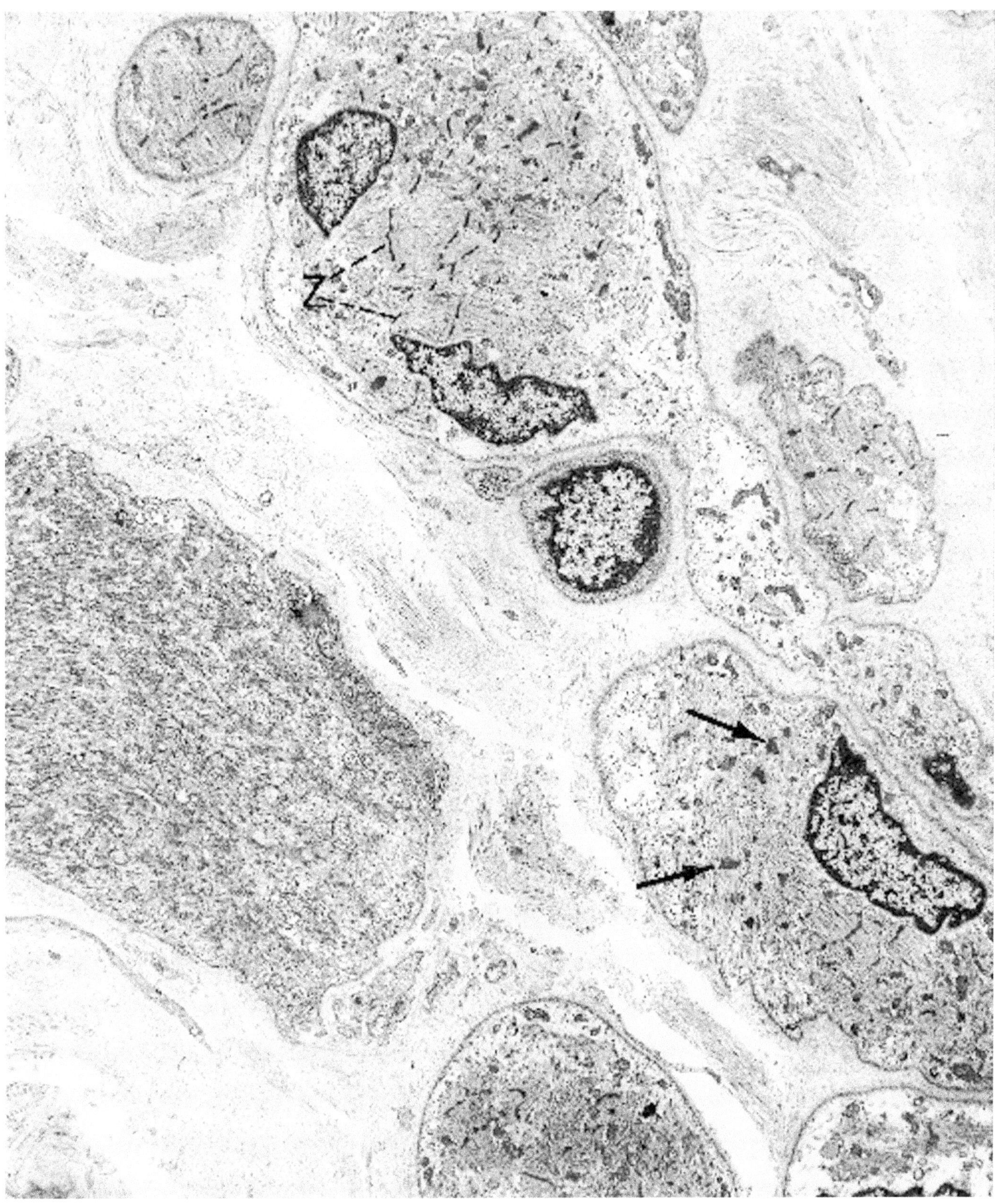

Figure 6.87. Rhabdomyoma (larynx). Moderately well-differentiated rhabdomyoblasts show early sarcomere formation, although disarray of these structural components of the cytoplasm still exists. A range of organization of Z-band material includes large, diffuse aggregates (*arrows*) as well as discrete bands (Z). (× 5500)

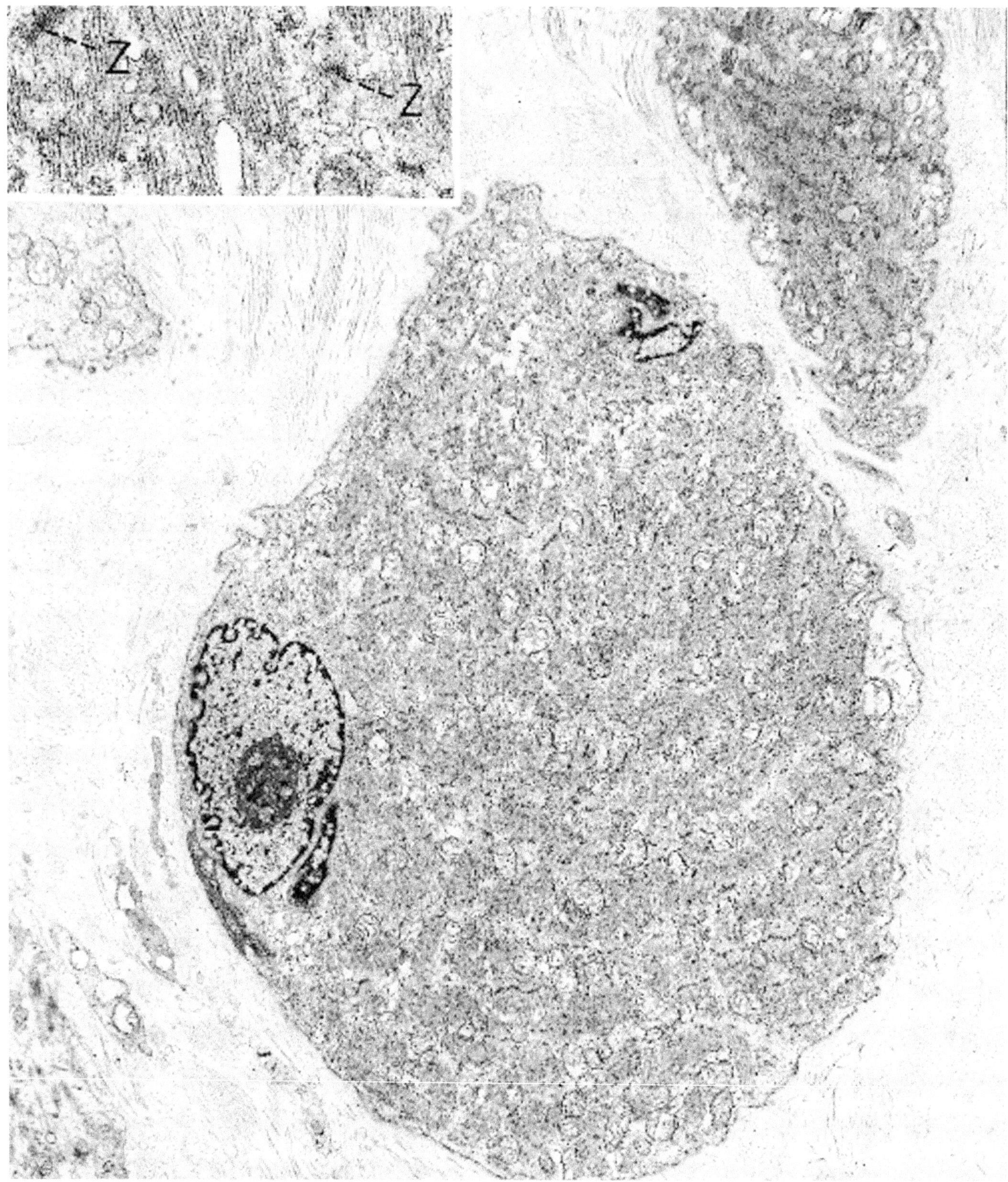

Figure 6.88. Rhabdomyoma (larynx). A differentiating rhabdomyoblast, not yet a strap cell, would be seen at the light microscopic level as a large oval cell with abundant eosinophilic cytoplasm. Note the eccentric nucleus, so characteristic of a skeletal muscle cell. Sarcomeres are numerous, although they still are in disarray. (× 7800) The *inset* is from another cell in the same neoplasm, illustrating early sarcomere formation at higher magnification. Z = Z-band material. (× 30,200)

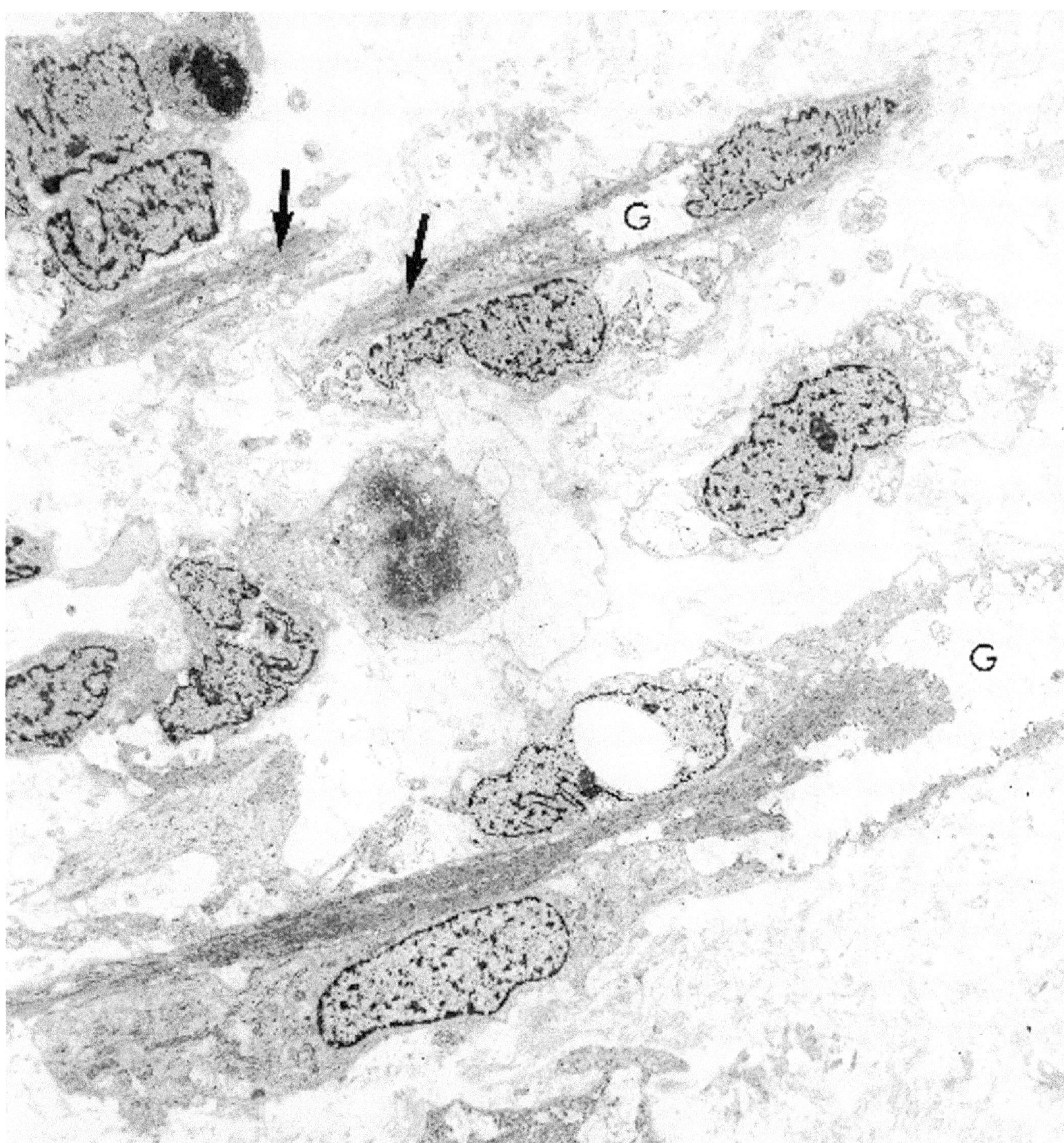

Figure 6.89. Spindle cell rhabdomyosarcoma (soft tissue of leg). Some of the neoplastic spindle cells contain numerous cytoplasmic filaments (*arrows*) and glassy escalloped spaces consistent with glycogen (G). (× 4000)

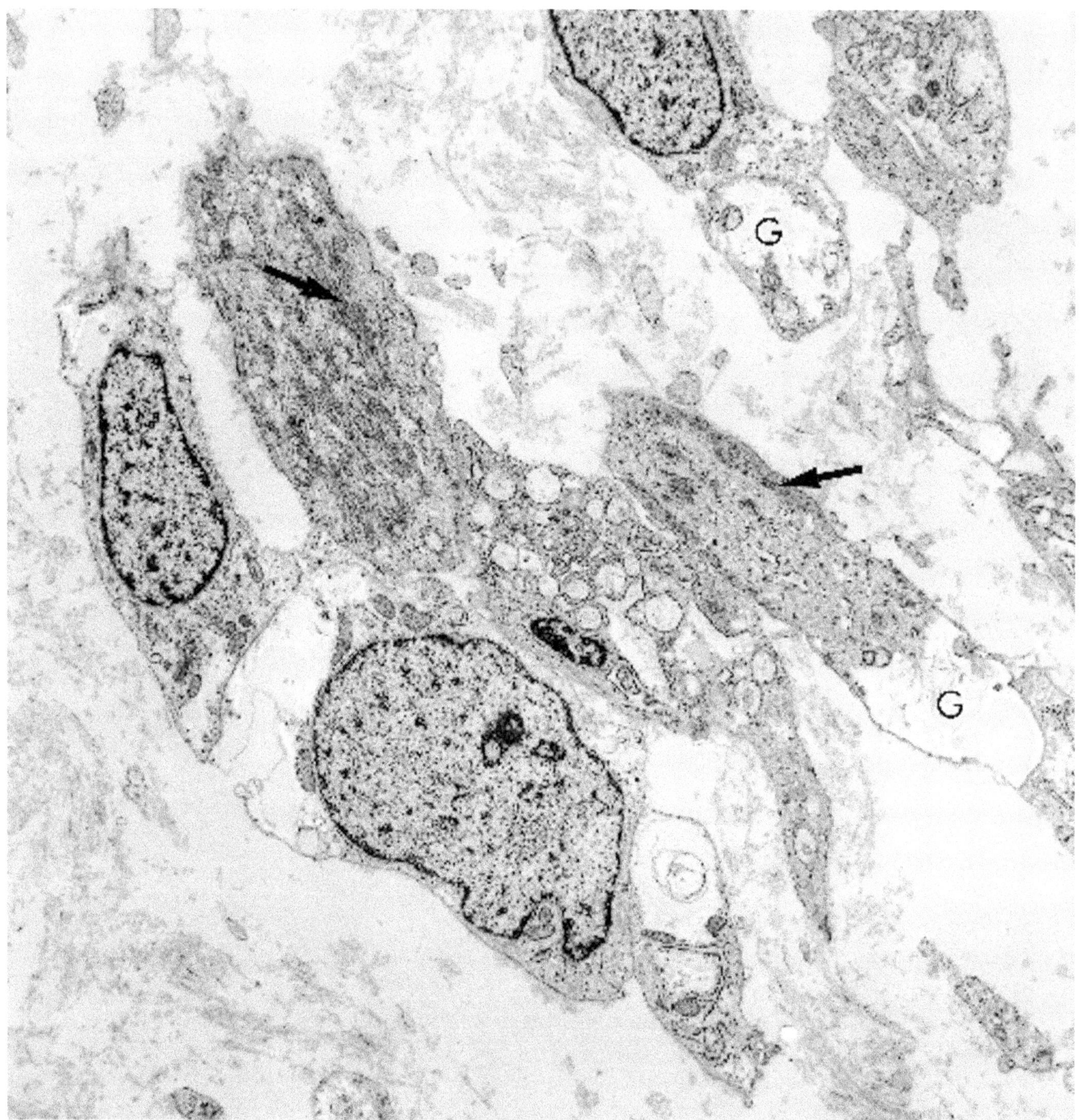

Figure 6.90. Spindle cell rhabdomyosarcoma (soft tissue of leg). Higher magnification of the same neoplasm as is depicted in Figure 6.89 shows more clearly the cytoplasmic filaments (*arrows*) and glycogen (G). (× 7000)

(Text continued from page 320)

Vascular Neoplasms

(Figures 6.91 through 6.102.)

Diagnostic criteria. (1) Endothelial cells, with varying degrees of vascular differentiation; (2) prominent junctions and junctional complexes; (3) basal lamina; (4) villus-like projections on the luminal aspect; (5) pinocytotic vesicles; (6) cytoplasmic filaments.

Additional points. Vessel formation is complete in benign vascular neoplasms, such as *angioma, angiofibroma, angiolipoma,* and *angiomyolipoma* (Figures 6.91 through 6.93), whereas it is less developed, with tortuous columns of endothelial cells and only slit-like lumens or no visible lumens in intermediate- and high-grade lesions such as *angioendothelioma* (Figures 6.94 and 6.95) and *angiosarcoma* (Figures 6.96 and 6.97). Weibel-Palade bodies, membrane-bound lysosomal-like structures, are excellent markers for normal endothelial cells, but they often are difficult to find in endothelial cell neoplasms. Otherwise, the cytoplasm of endothelial cells, in addition to pinocytotic vesicles and filaments, contains free ribosomes and a moderate number of mitochondria and cisternae of rough endoplasmic reticulum (Figure 6.95). *Epithelioid angioendothelioma* and *epithelioid angiosarcoma* have a major component of closely arranged, plump, poorly differentiated, epithelial-like cells and focal microlumens (Figures 6.96 and 6.97).

Kaposi's sarcoma in the early stages consists of a proliferation of capillaries and later, of a heavy infiltrate of capillaries, fibroblasts, and extravasated erythrocytes, in an accompanying collagenous matrix (Figures 6.98 through 6.102). An interesting feature not usually appreciated at the light microscopic level but, in our experience, identifiable by electron microscopy is that much of the erythrophagocytosis in Kaposi's sarcoma is apparently by fibroblasts acting as facultative histiocytes (Figures 6.100 and 6.101). True histiocytes may or may not be present in the infiltrates. In Kaposi's sarcoma arising in lymph nodes of patients with acquired immunodeficiency syndrome (AIDS), peculiar, small cytoplasmic inclusions—so-called tubulovesicular structures—may be found in endothelial cells and lymphocytes (Figure 6.102). These structures also occur independently of Kaposi's sarcoma in the lymph nodes of many patients with AIDS and, more generally, in persons with elevated levels of interferon.

Hemangioblastoma of the central nervous system represents a peculiar variant of angiomatous neoplasms and is covered separately in Chapter 8.

Hemangiopericytoma

(Figures 6.103 through 6.107.)

Diagnostic criteria. (1) Spindle, oval, or polygonal cells in a somewhat palisaded arrangement around capillaries (Figure 6.103); (2) basal lamina, often in abundance, covering the free surfaces of cells (Figures 6.103 through 6.106); (3) focal attachments and junctions between cells (Figure 6.107); (5) pinocytotic vesicles; (6) cytoplasmic filaments (Figures 6.106 and 6.107).

Additional points. The perithelial cells characteristically are within a narrow range of size and shape, and their cytoplasm contains a varying number of organelles in addition to filaments and pinocytotic vesicles, including rough endoplasmic reticulum, mitochondria, Golgi apparatuses, and free ribosomes. The spaces between cells and the amount of basal lamina vary within one neoplasm and from one neoplasm to another, accounting for the variable pattern of reticulin staining at the light microscopic level.

The ultrastructural criteria for identifying pericytes are the same in hemangiopericytomas from all locations and in all age groups, despite the fact that the light microscopic picture and biologic behavior may vary. Head and neck tumors tend to be better differentiated and less aggressive. Meningeal hemangiopericytomas resemble and have been misinterpreted as meningiomas. Hemangiopericytomas in children often look malignant microscopically but behave in a benign way. In all these tumors, the constituent cell is evident ultrastructurally as being a pericyte.

(Text continues on page 359)

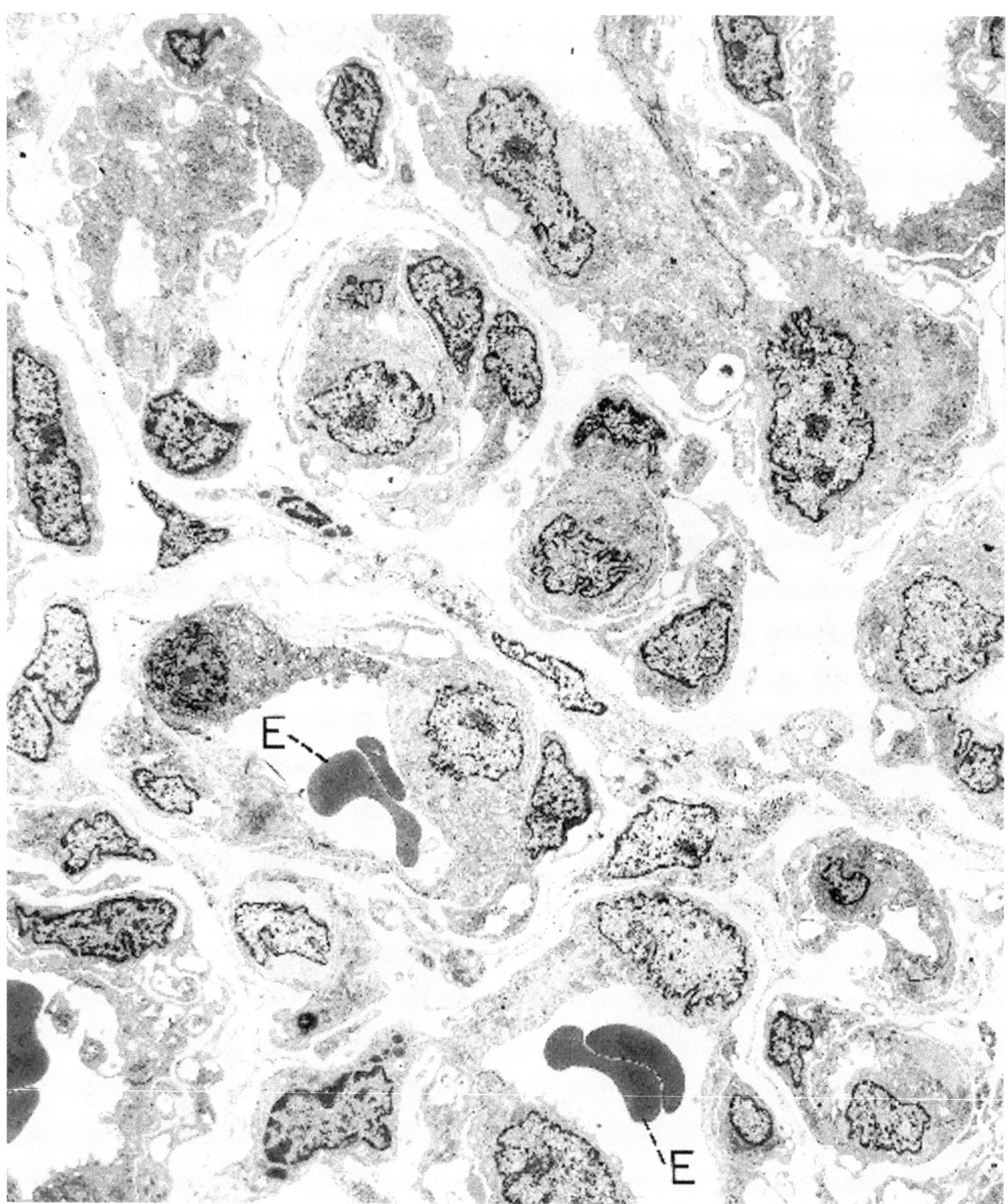

Figure 6.91. Hemangioma (scalp). This is an example of a well-differentiated endothelial cell neoplasm in which capillaries are well formed and regularly dispersed in a collagenous matrix. Erythrocytes (E) are visible in some of the lumens. (× 3750)

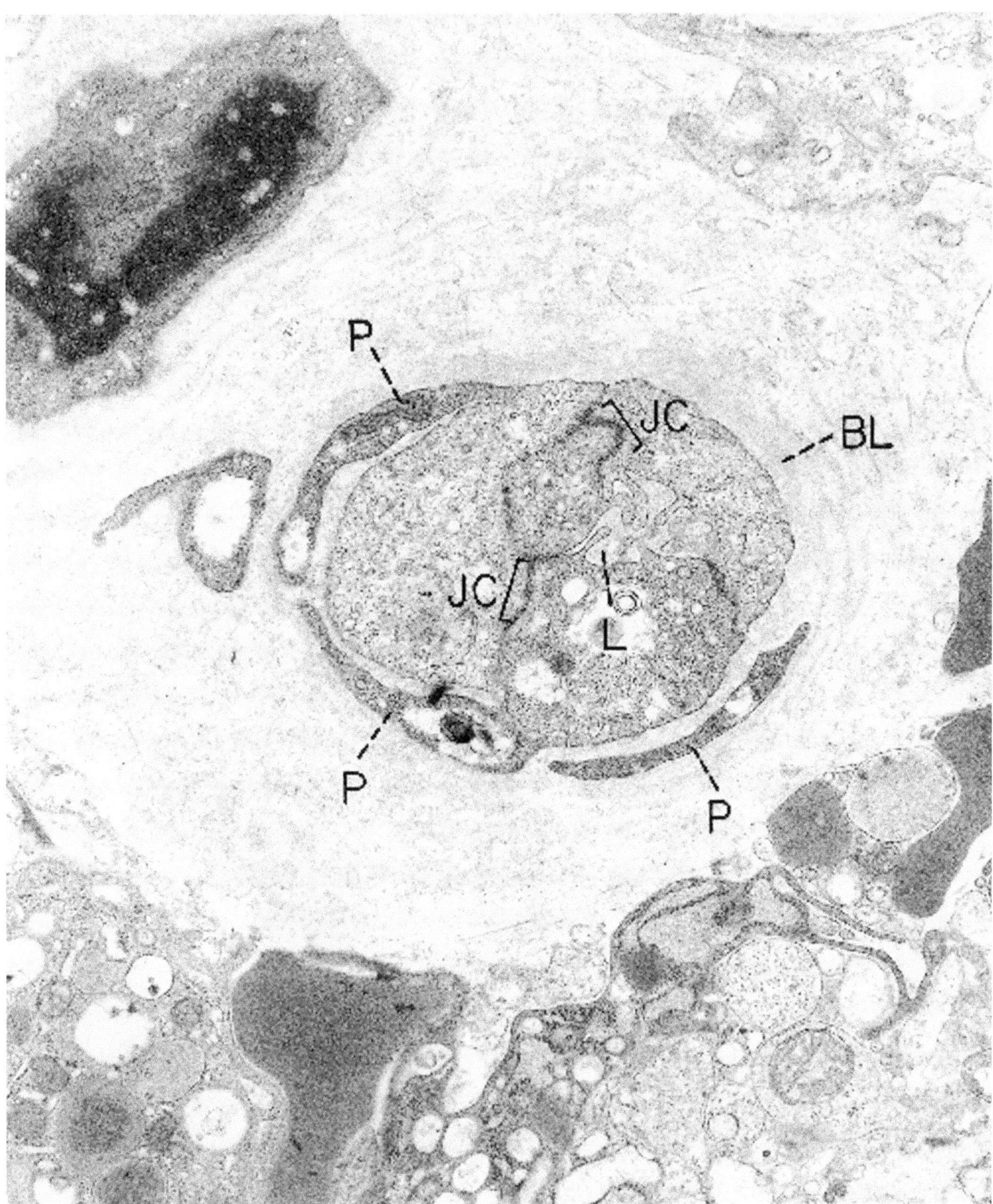

Figure 6.92. Hemangioma (scalp). Higher magnification of one of the capillaries in the hemangioma depicted in Figure 6.91 illustrates the endothelial cells to be plump and the lumen (L) to be small. The luminal surfaces of the cells have villous processes, and the lateral surfaces have prominent junctional complexes (JC). Duplicated basal lamina (BL) surrounds the endothelial cells and pericytes (P). (× 19,900)

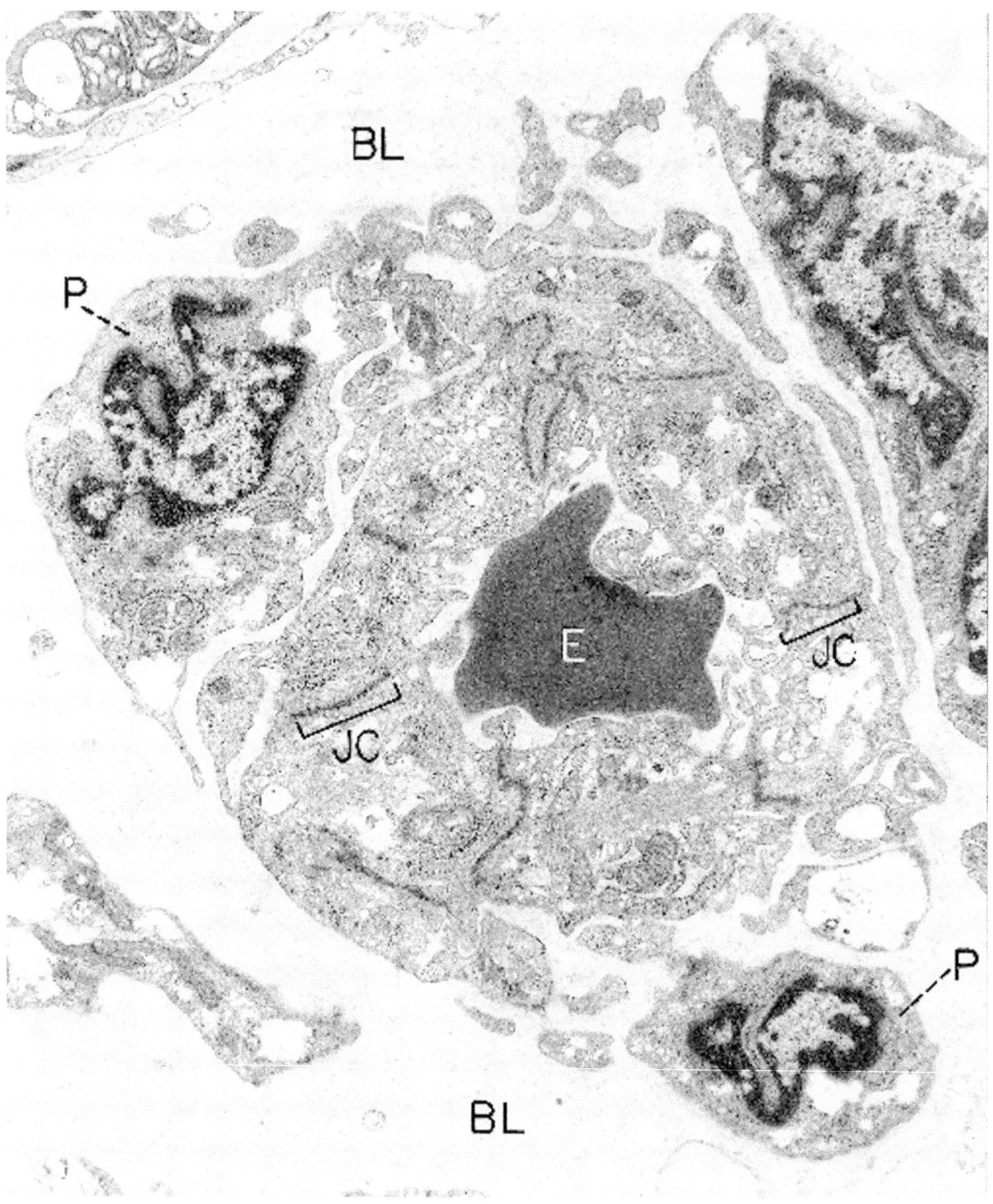

Figure 6.93. Hemangioma (scalp). This capillary contains an erythrocyte (E) in its lumen. Note the villous luminal surface and lateral junctional complexes (JC) of the endothelial cells. P = pericytes; BL = basal lamina. (× 12,600)

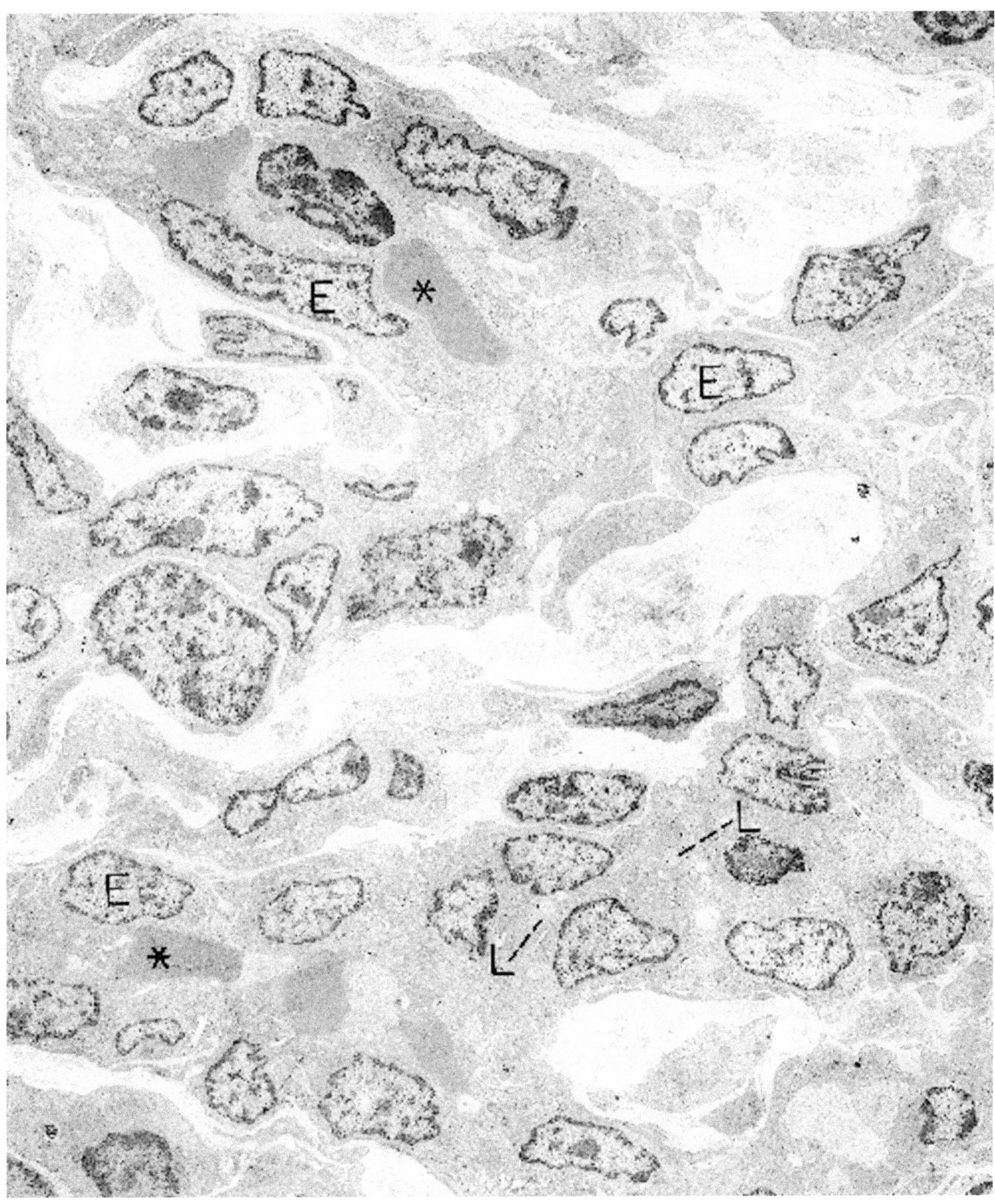

Figure 6.94. Hemangioendothelioma (trachea). Tortuous columns of endothelial cells form poorly discernible, primitive vessels with small, slit-like, or potential lumens (L). Occasional lumens are expanded by erythrocytes (*). E = endothelial cells. (× 4200)

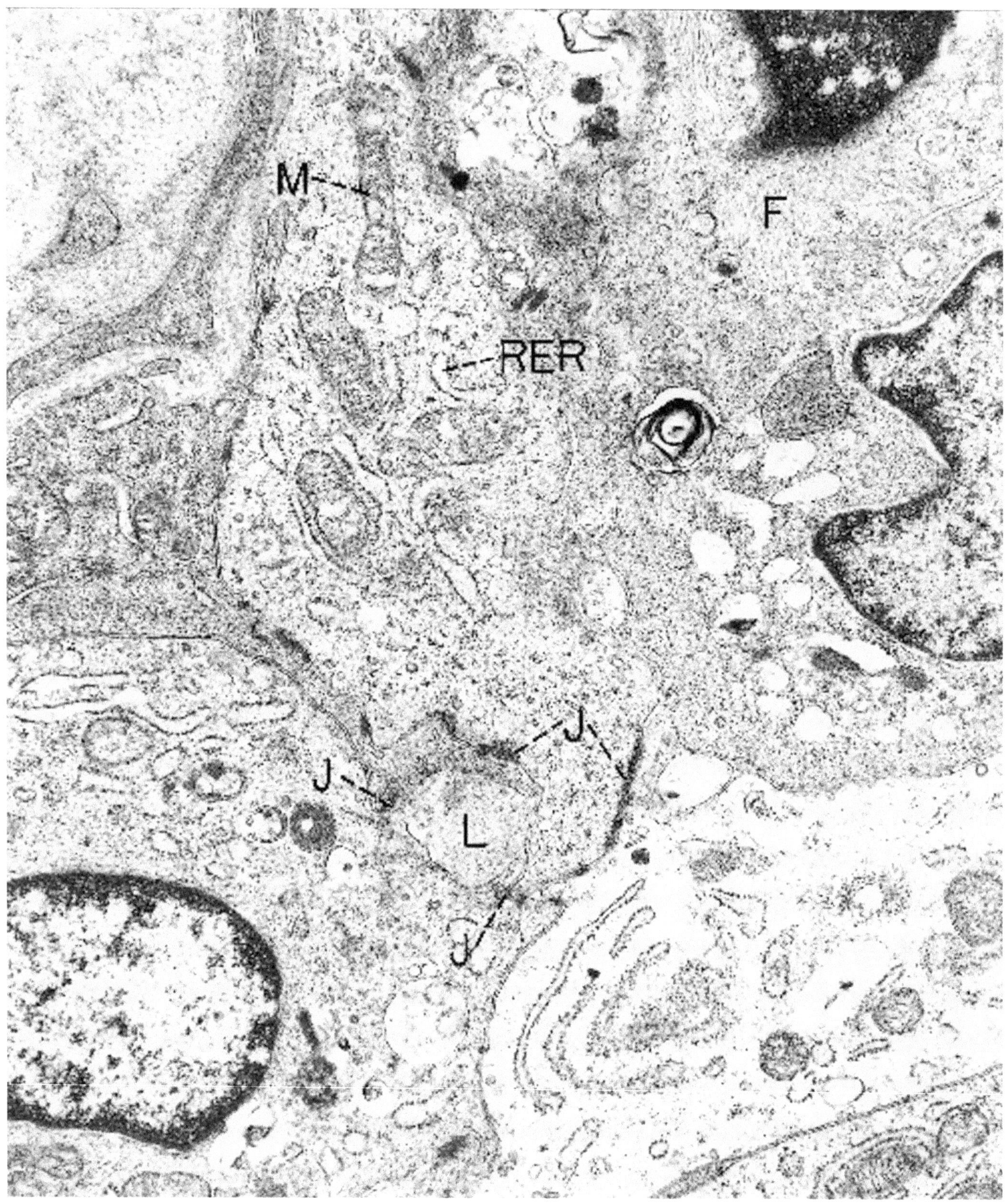

Figure 6.95. Hemangioendothelioma (trachea). High magnification allows a miniature lumen (L) in this poorly differentiated neoplasm to be visible. Tight junctions (J) between the endothelial cells help to locate the lumen. The cytoplasm contains free ribosomes and a moderate number of mitochondria (M), cisternae of rough endoplasmic reticulum (RER), and microfilaments (F). (× 22,000)

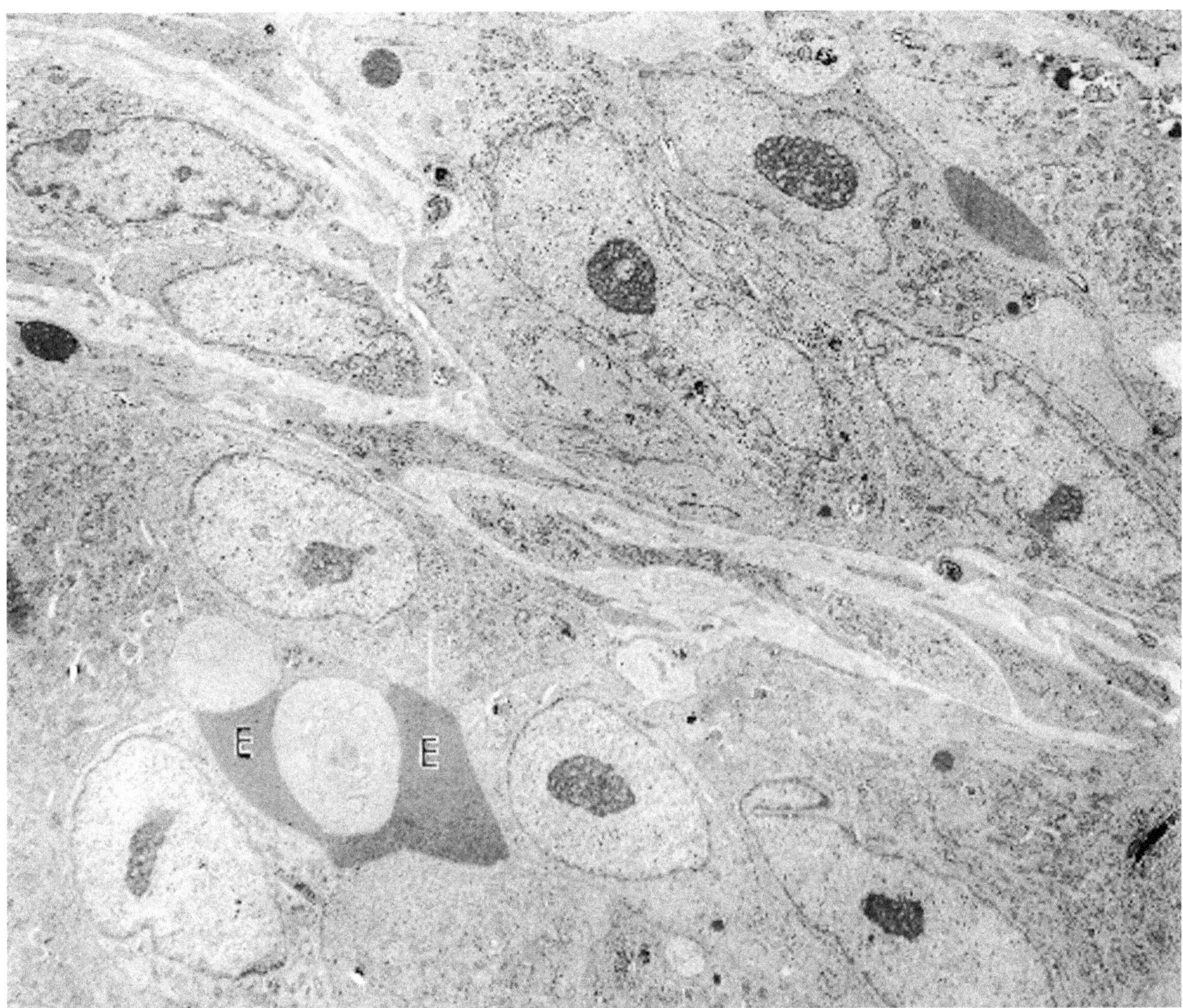

Figure 6.96. Hemangiosarcoma, epithelioid type (tibia). Groups of spindle cells and epithelioid cells have inconspicuous lumens or microlumens containing erythrocytes (E). (× 5600)

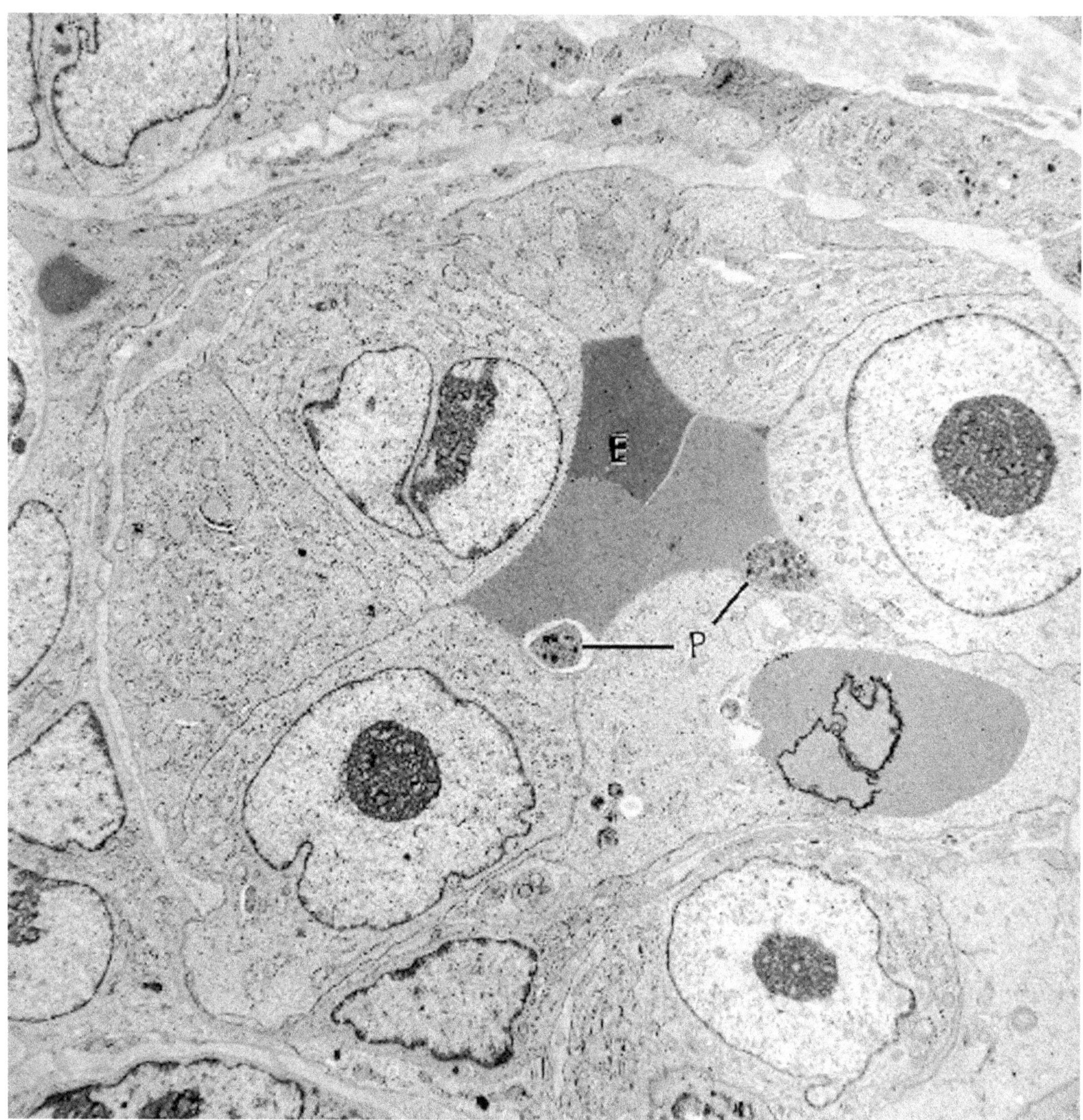

Figure 6.97. Hemangiosarcoma, epithelioid type (tibia). Higher magnification of the same neoplasm as depicted in Figure 6.96 illustrates a central lumen with erythrocytes (E) and platelets (P). The cytoplasm of the malignant cells contains a nondescript complement of organelles. (× 7400)

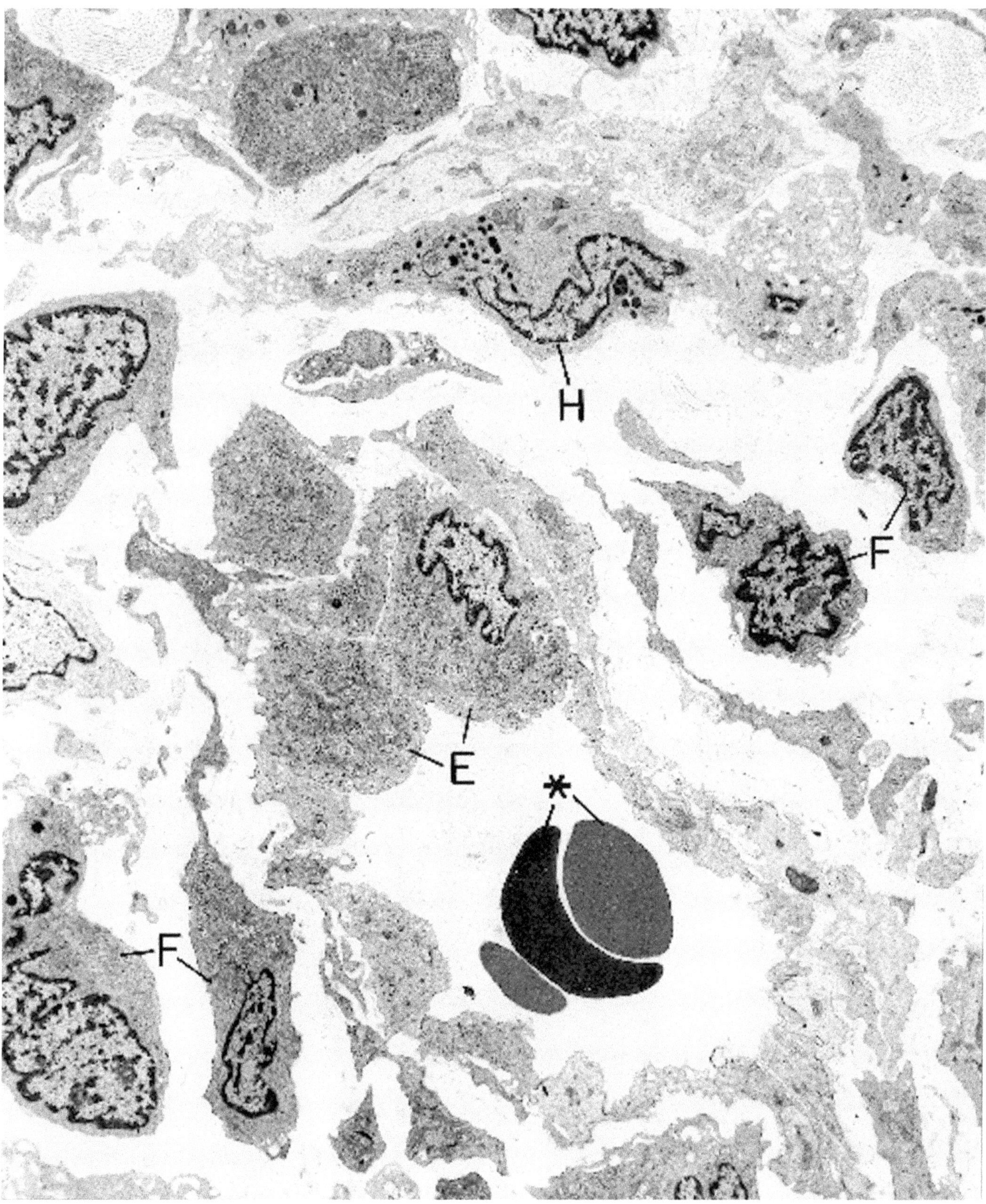

Figure 6.98. Kaposi's sarcoma (inguinal lymph node). This representative field shows a capillary filled with blood (* = erythrocytes) and lined by plump endothelial cells (E), and fibroblasts (F), histiocytes (H), and a collagenous matrix. (× 5130)

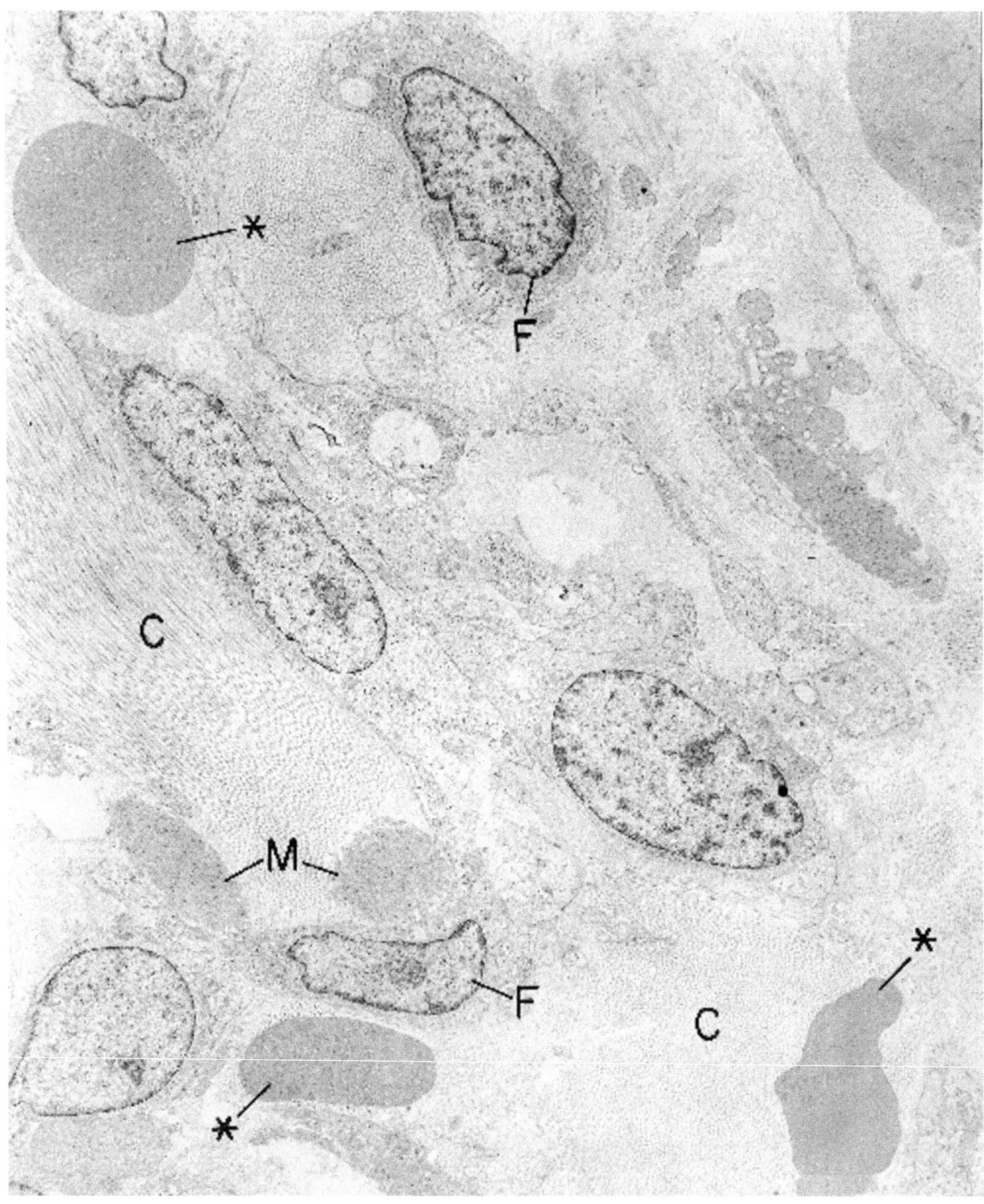

Figure 6.99. Kaposi's sarcoma (skin of shoulder). Dispersed in a matrix of dense collagen (C) are poorly differentiated fibroblasts (F) and extravasated erythrocytes (*). A few processes of smooth muscle cells (M), probably part of the arrector pili units, also are present. (× 4940)

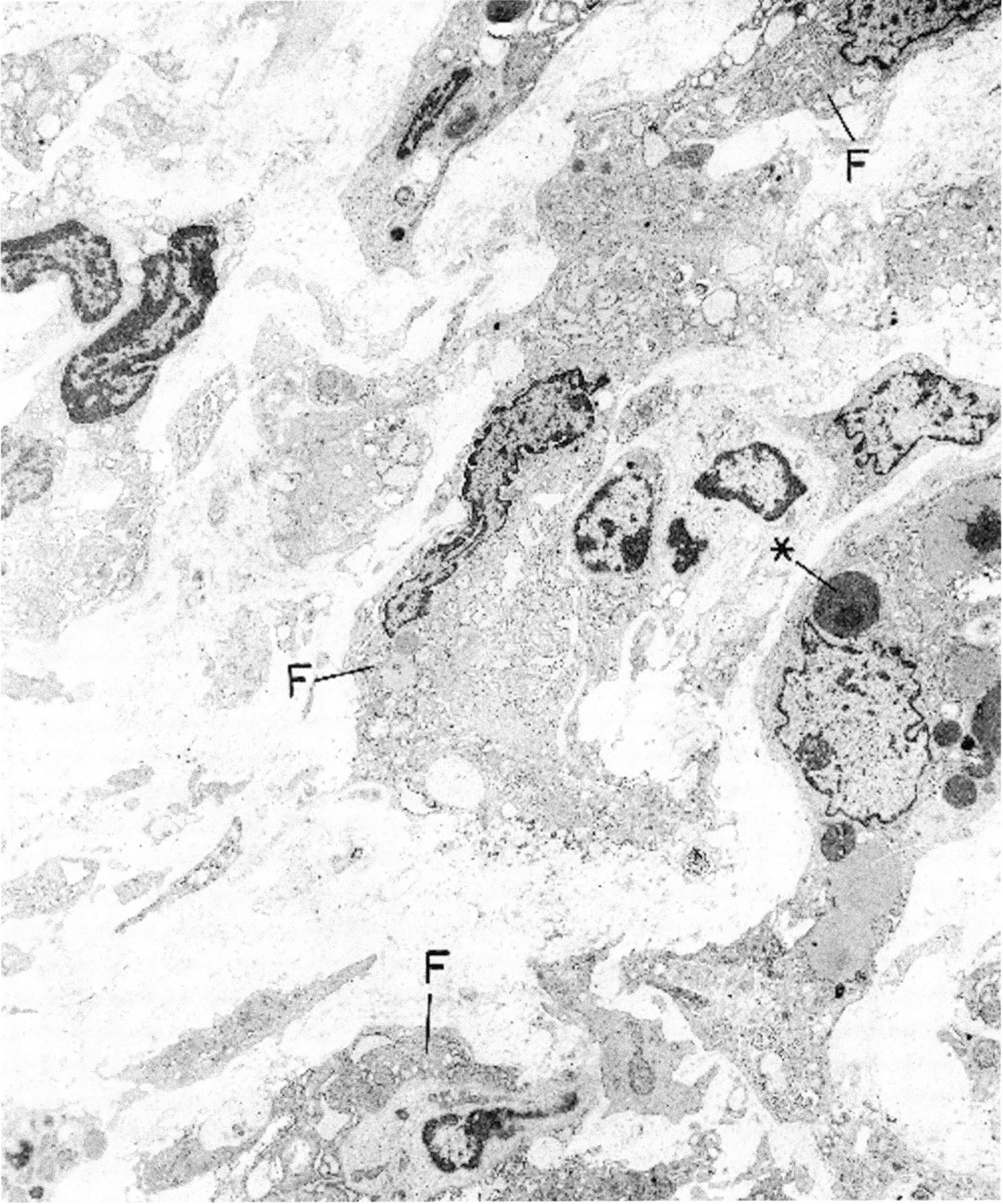

Figure 6.100. Kaposi's sarcoma (cervical lymph node). This field illustrates at low power the numerous fibroblasts (F) in the neoplastic infiltrate and the erythrophagocytic activity (*) of one of the fibroblasts. (× 4845)

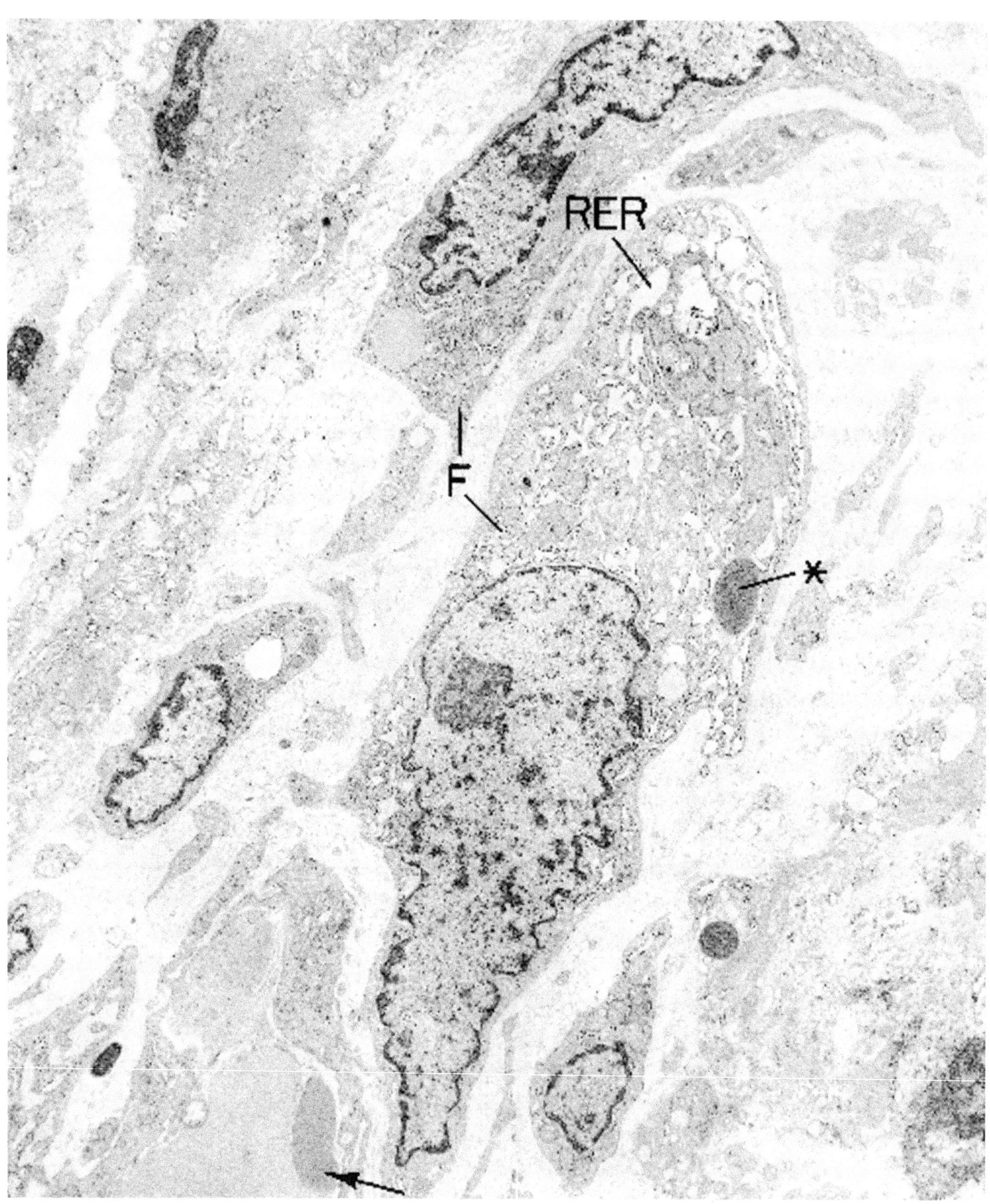

Figure 6.101. Kaposi's sarcoma (cervical lymph node). This is the same case as depicted in Figure 6.100, but at this higher magnification the active rough endoplasmic reticulum (RER) and erythrophagocytosis (*) of the fibroblasts (F) can be seen more clearly. Note the morphologic similarity between the phagocytosed erythrocytes (*) and the one in the lumen of a capillary *(arrow)*. (× 7125)

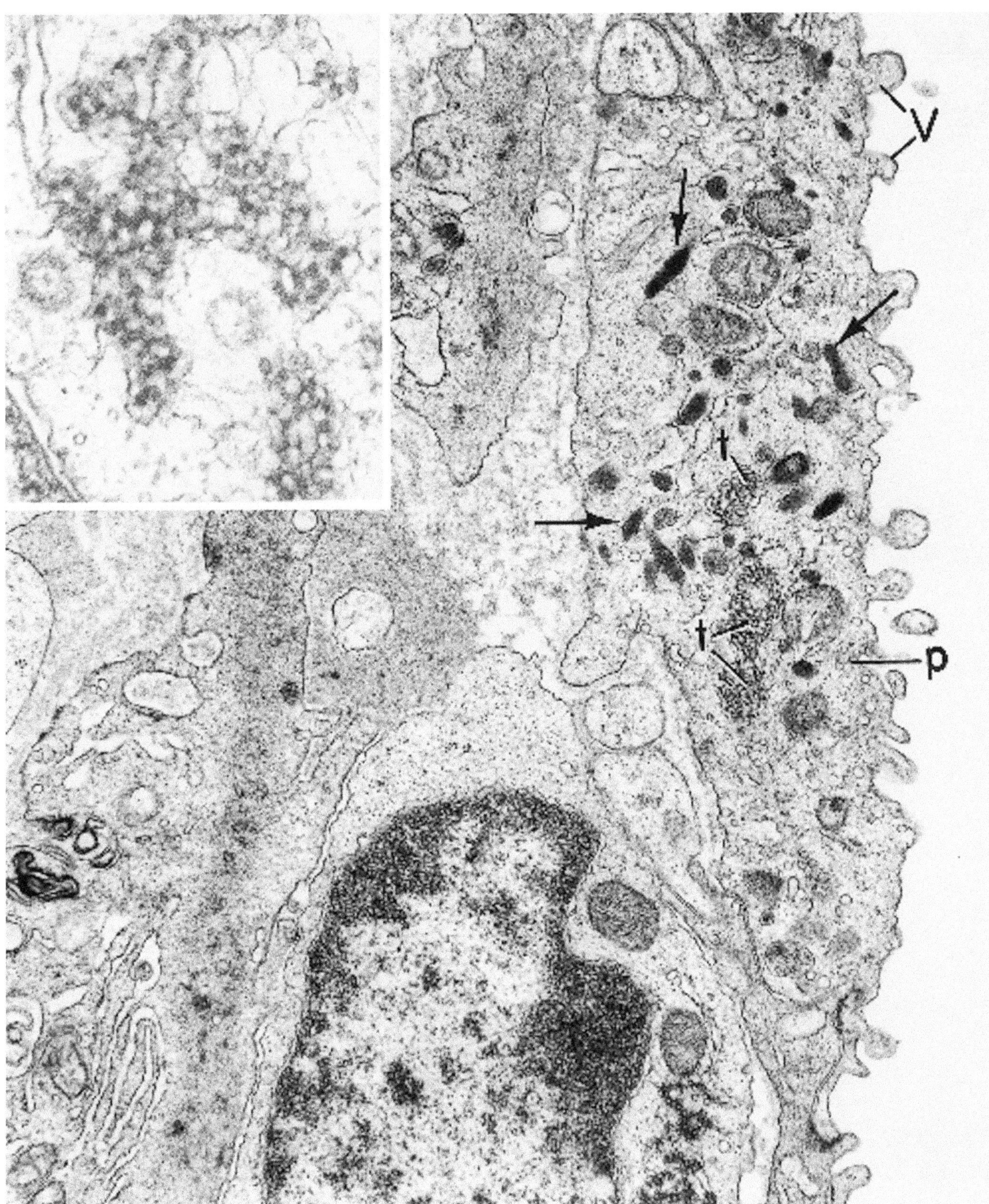

Figure 6.102. Kaposi's sarcoma (axillary lymph node). The endothelial cells in this hyperplastic and focally neoplastic lymph node contain tubulovesicular structures (t) as well as Weibel-Palade bodies (*arrows*). Note the normal villus-like projections (V) of the luminal surface of endothelial cells and the pinocytotic vesicles (p). (× 23,430) *Inset* consists of tubulovesicular structures at high magnification. (× 119,700)

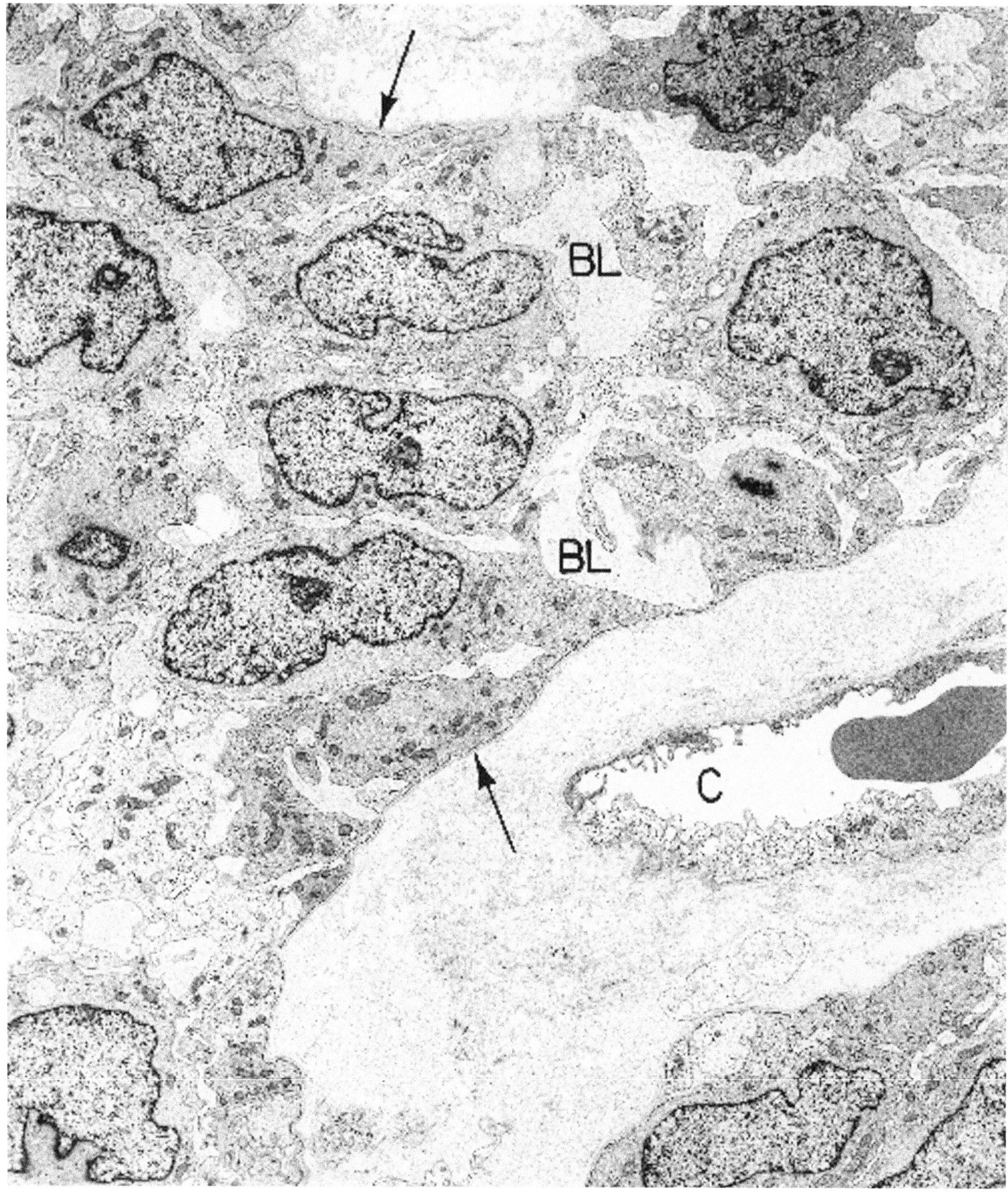

Figure 6.103. Hemangiopericytoma (sternum). Spindle and polygonal cells are arranged characteristically in a palisade around a capillary (C). The cells are focally attached to one another, and elsewhere they are separated by basal lamina that is both discrete (*arrows*) and diffuse (BL). (× 4750) (Permission for reprinting granted by WB Saunders, Dickersin GR: The contributions of electron microscopy in the diagnosis and histogenesis of controversial neoplasms. Clin Lab Med 4:123–164, 1984.)

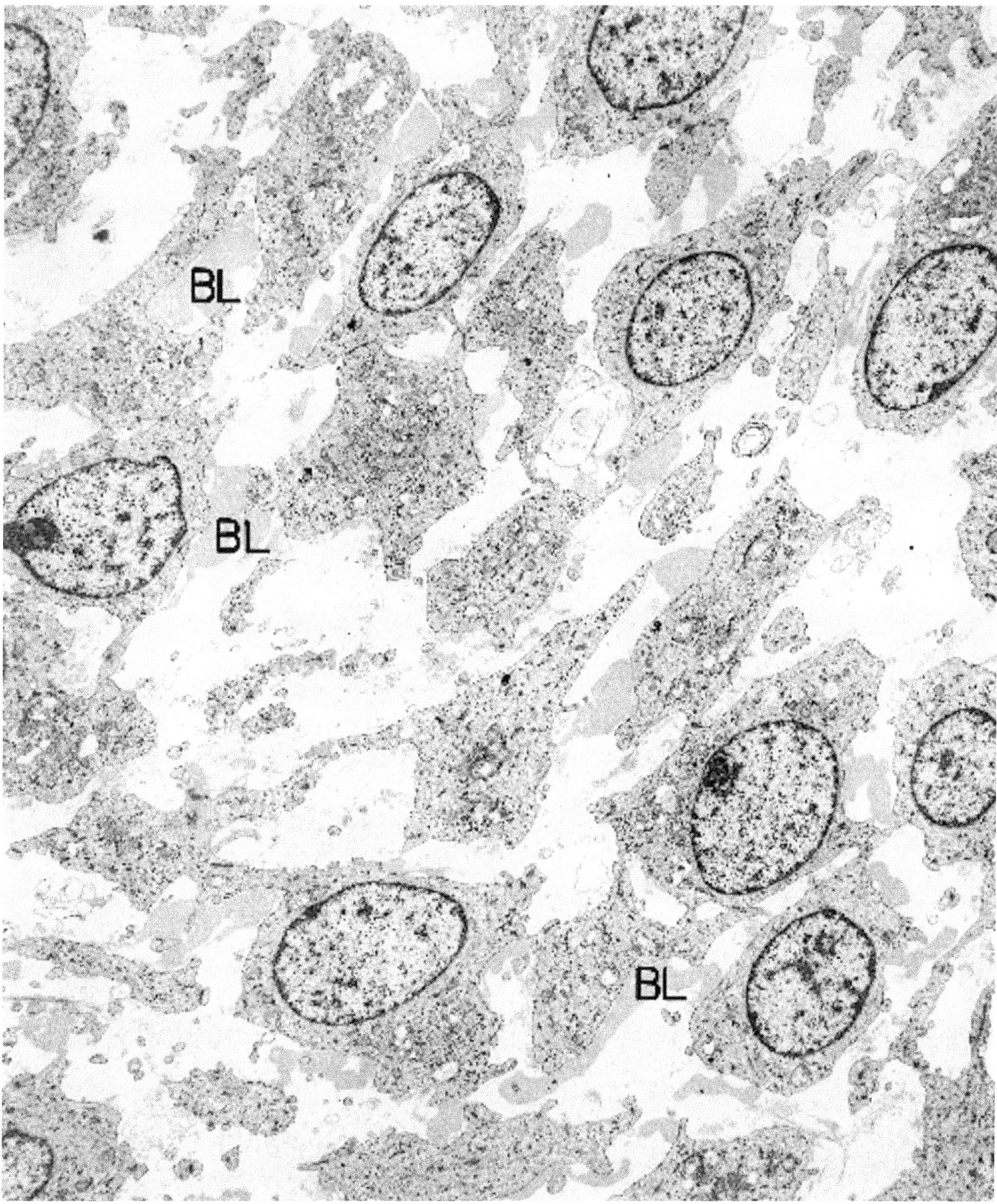

Figure 6.104. Hemangiopericytoma (nasal mucosa). This neoplasm exemplifies wide spaces of matrix and copious amounts of basal lamina (BL) between cells. (× 4845)

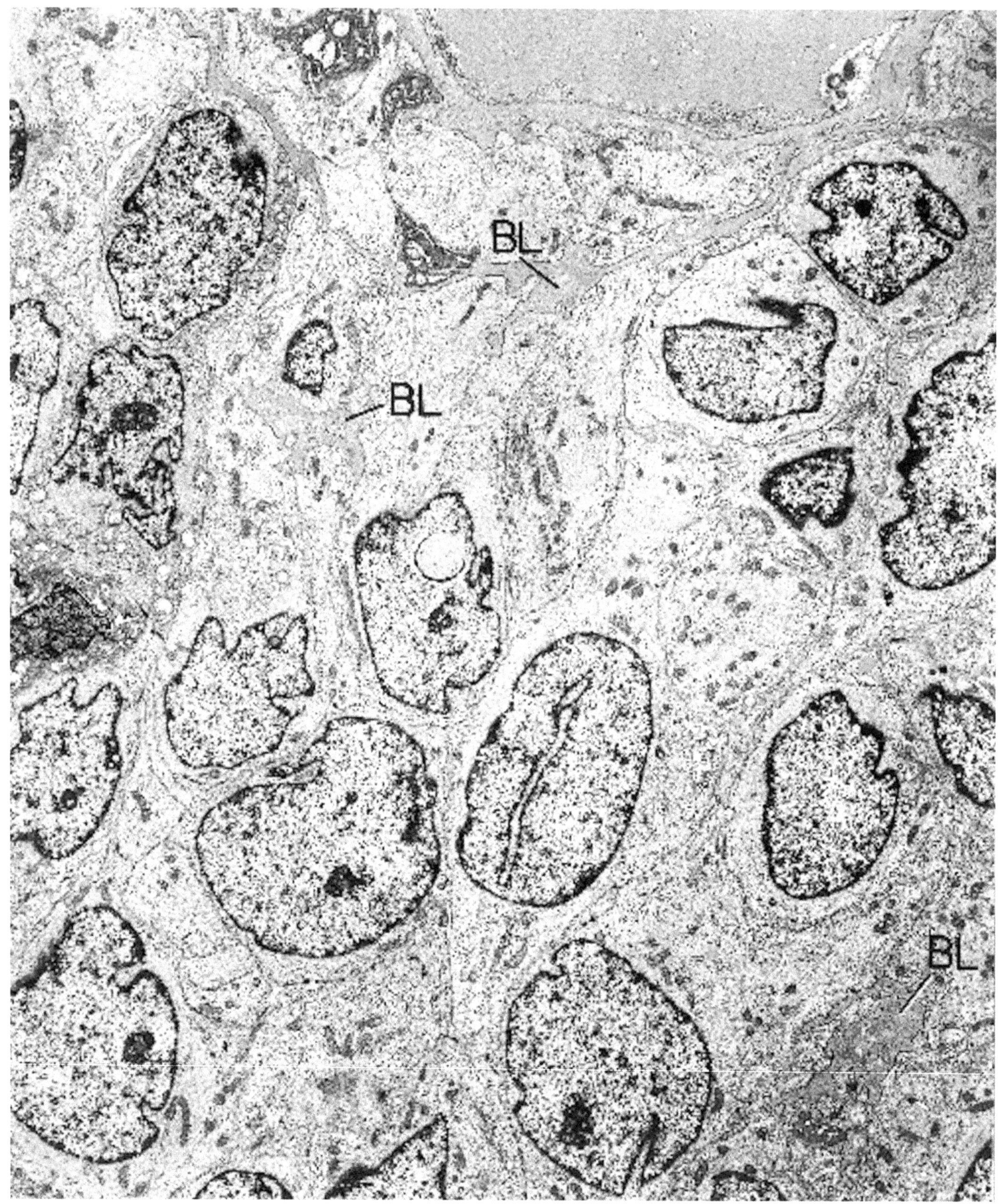

Figure 6.105. Hemangiopericytoma (sternum). This field of the same neoplasm as depicted in Figure 6.103 is composed of closely apposed cells with less extracellular space and basal lamina (BL). (× 5035)

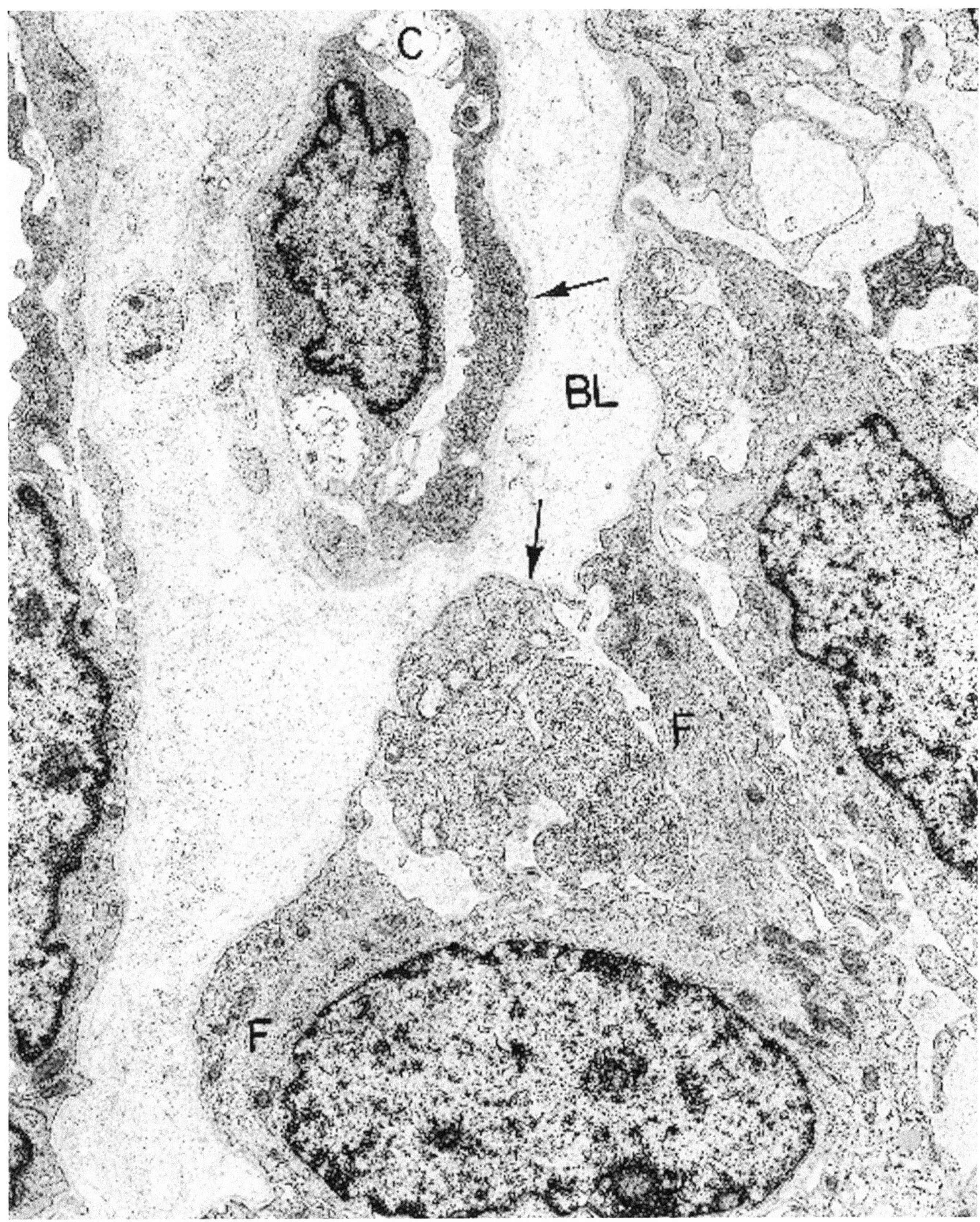

Figure 6.106. Hemangiopericytoma (sternum). Higher magnification illustrates the palisaded pericytes around a capillary (C) as well as basal lamina (diffuse = BL; discrete = *arrows*) and cytoplasmic filaments (F). (× 8670)

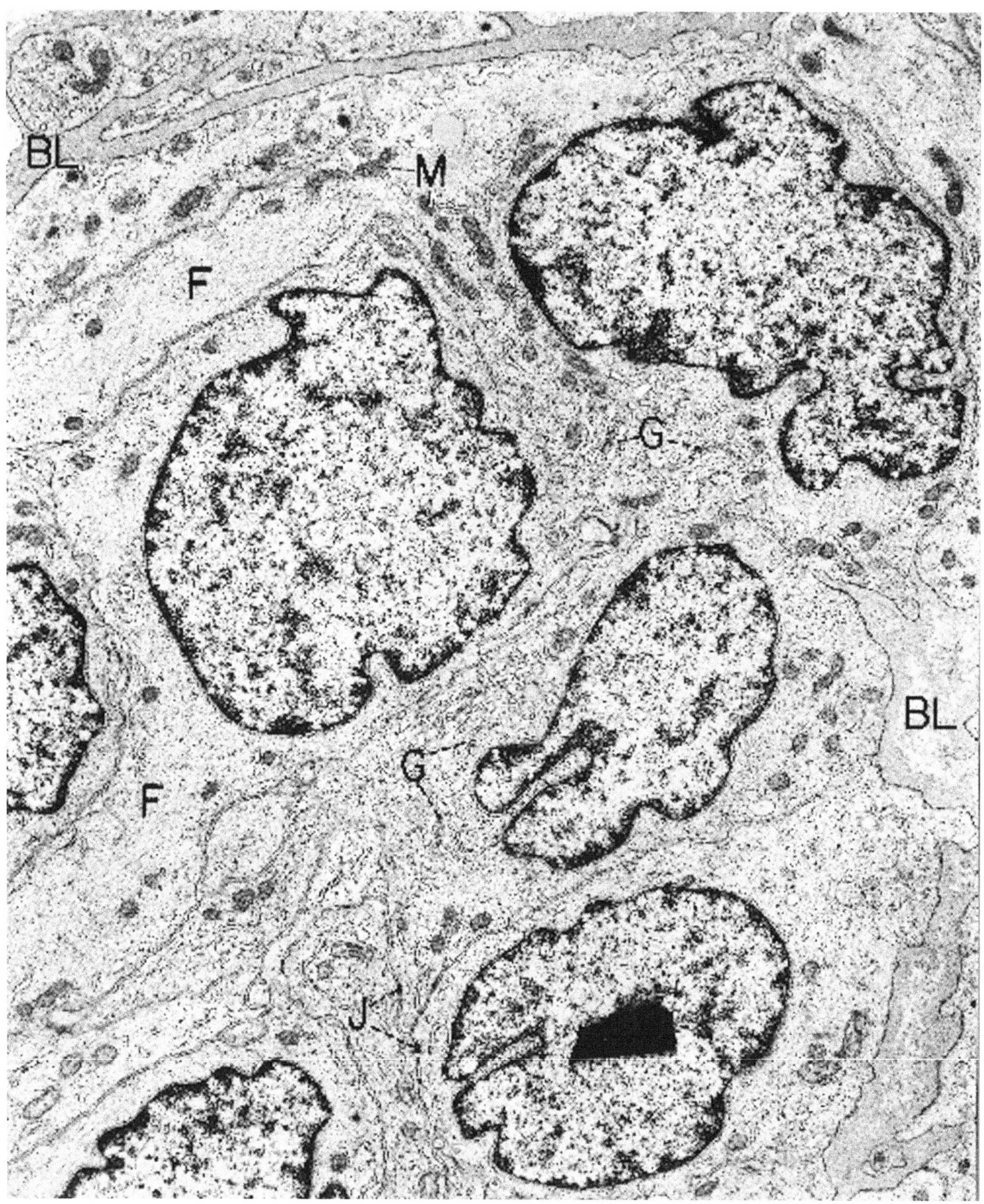

Figure 6.107. Hemangiopericytoma (sternum). Closely arranged pericytes exhibit intermediate junctions (J), many filaments (F), a moderate number of mitochondria (M), and two Golgi apparatuses (G). Cell separation and basal lamina (BL) are around groups of cells rather than individual cells. (× 11,250)

(Text continued from page 341)

Schwannoma and Malignant Schwannoma

(Figures 6.108 through 6.114.)

Diagnostic criteria. (1) Spindle cells with long, thin, intertwining processes (Figures 6.108 through 6.111); (2) basal lamina covering processes and cell bodies (Figure 6.111); (3) long-spacing collagen (Luse bodies) (Figures 6.108 and 6.112).

Additional points. It is cell processes and not cell bodies that are more diagnostic in Schwann cell neoplasms. The processes tend to wrap around collections of matrical collagen to form pseudomesaxons. The cell bodies have a high nuclear–cytoplasmic ratio, and no single organelle predominates consistently. In addition to a moderate number of mitochondria and cisternae of rough endoplasmic reticulum, intermediate filaments, occasional lysosomes, and rare dense-core granules also may be present. Secondary lysosomes usually are present in some cells of each neoplasm, and often they are quite numerous, emphasizing the Schwann cell's ability to phagocytose. Small junctions are found between cells, and long-spacing collagen, which has a periodicity significantly greater than that of the usual type I collagen, is nonspecific but often present in Schwann cell neoplasms. The ultrastructural features of Schwann cells are the same for Antoni A- and Antoni B-type tissue, except that in the more cellular Antoni A tissue the cell processes compress on one another and have less intervening matrix (Figure 6.110). Features are also the same in various types of Schwannomas, including *classical, cellular, plexiform,* and *ancient* forms. The cells of *melanotic Schwannoma* are also similar save for the presence of melanosomes.

Although benign Schwann cell neoplasms are readily identifiable by their ultrastructure, malignant ones notoriously are often more difficult to diagnose (Figures 6.113 and 6.114). Cell processes, basal lamina, and long-spacing collagen are still the most reliable criteria for identifying Schwann cells, but these features may be sparse or absent in poorly differentiated neoplasms. In these instances, it is important to keep in perspective the clinical history, physical and surgical findings, and the light microscopic picture. Spindle-cell tumors occurring in patients with von Recklinghausen's disease, or neoplasms found at operation to be arising from nerve trunks, are likely to be Schwannian in origin. If the electron microscopic study reveals a poorly differentiated spindle-cell neoplasm that does not have the markers for smooth muscle cells, fibroblasts, or other specific spindle-cell lesions considered in the light microscopic differential diagnosis, then focal and incomplete markers for Schwann cells allow a diagnosis of "consistent with malignant Schwannoma" to be made. The most useful marker in this nonspecific category, in our experience, has been the secondary lysosome.

Granular Cell Tumor

(Figures 6.115 and 6.116.)

Diagnostic criteria. (1) Groups of tightly apposed cells (Figure 6.115); (2) basal lamina surrounding groups and invaginating between cells (Figure 6.115); (3) numerous cytoplasmic accumulations of heterogeneous, osmophilic material (Figures 6.115 and 6.116).

Additional points. Most *granular cell tumors,* formerly called granular cell myoblastomas, are of Schwannian origin. They are characterized ultrastructurally by their cells containing innumerable secondary lysosomes filled with heterogeneous, osmiophilic material (Figures 6.115 and 6.116). In addition, the cells are arranged in tight groups, and the groups are covered by basal lamina. The basal lamina also invaginates and covers individual cells, giving the overall grouping a compartment-like appearance and recapitulating normal Schwann cells infolding around axons (Figure 6.115). Most of the cells consist of long processes, and some of the smaller processes can be identified as neurites. Several non-neoplastic diseases, including Wallerian degeneration and leprosy, manifest the same type of granular cells as are seen in granular cell tumors, raising the question of whether the tumors are, indeed, neoplasms. The fact that a rare one behaves in a biologically malignant manner strengthens the neoplastic view of granular cell tumors.

(Text continues on page 369)

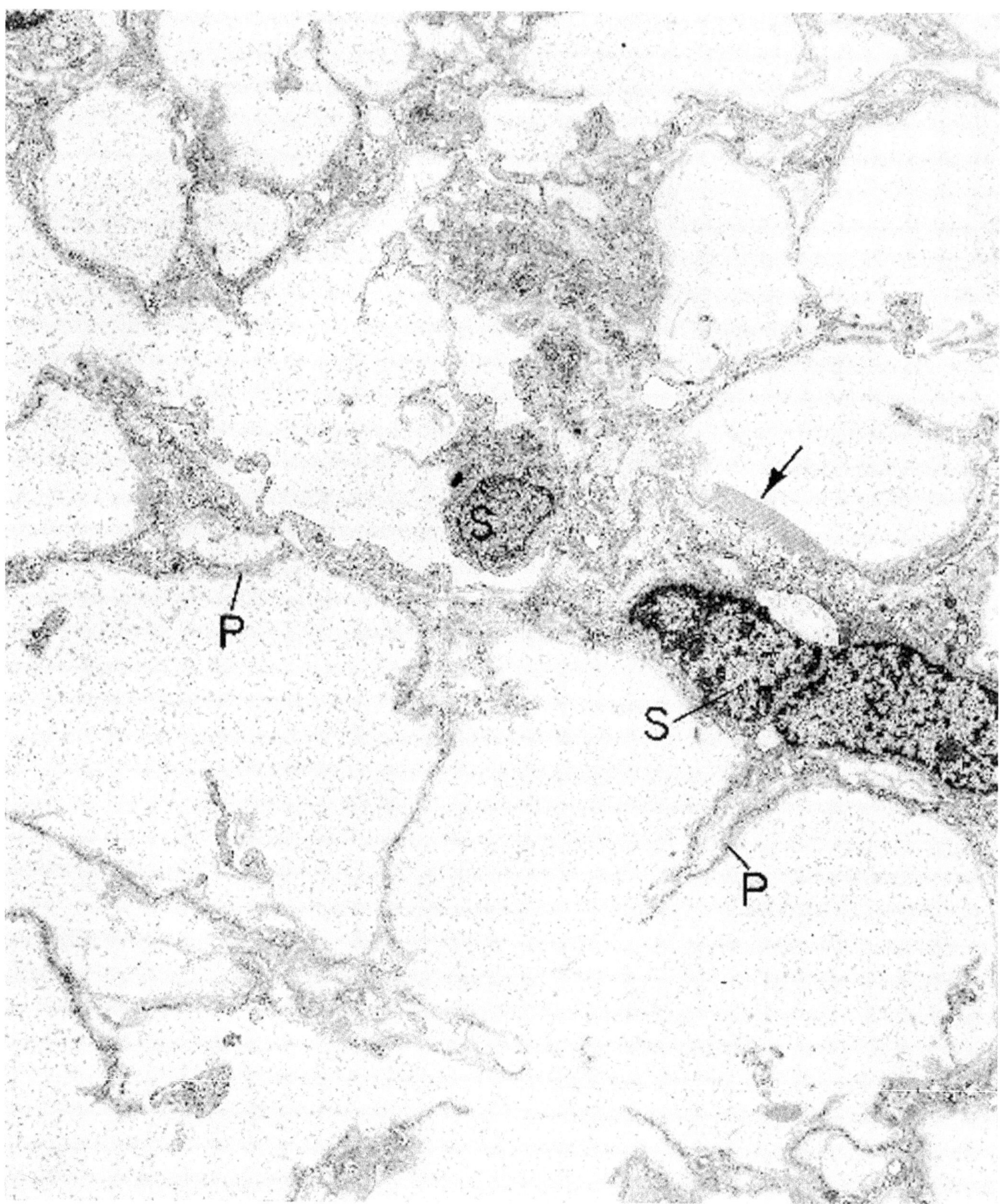

Figure 6.108. Schwannoma (cauda equina). An Antoni B region of the neoplasm is composed mostly of extracellular matrix and basal lamina-covered processes (P) of Schwann cells (S). One Luse body (*arrow*) is seen better at higher magnification in Figure 6.112. (× 4750)

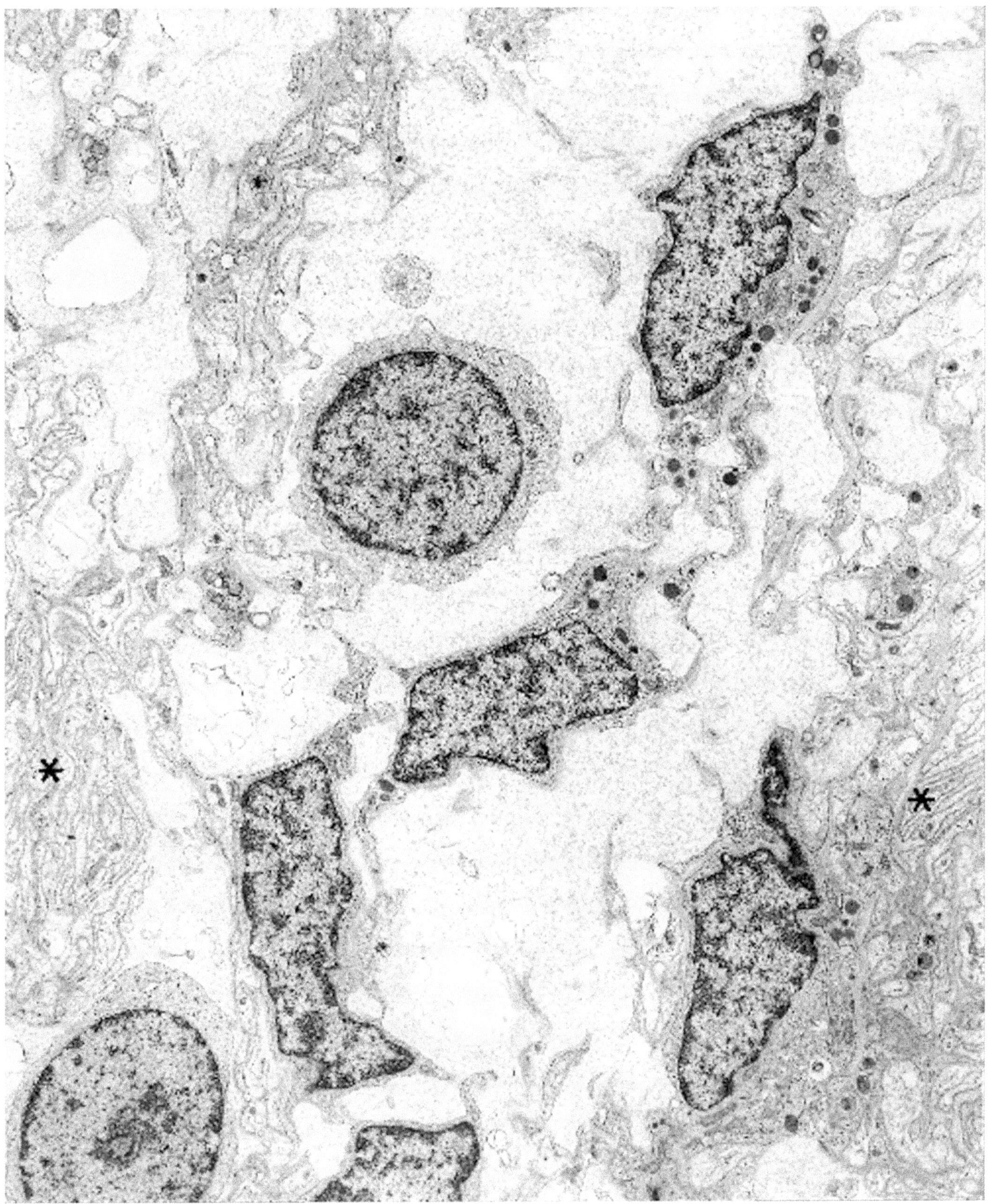

Figure 6.109. Schwannoma (retroperitoneum). Schwann cell bodies and processes are diffusely covered by basal lamina, and in some areas (*) the processes are closely packed and intertwined. (× 6250)

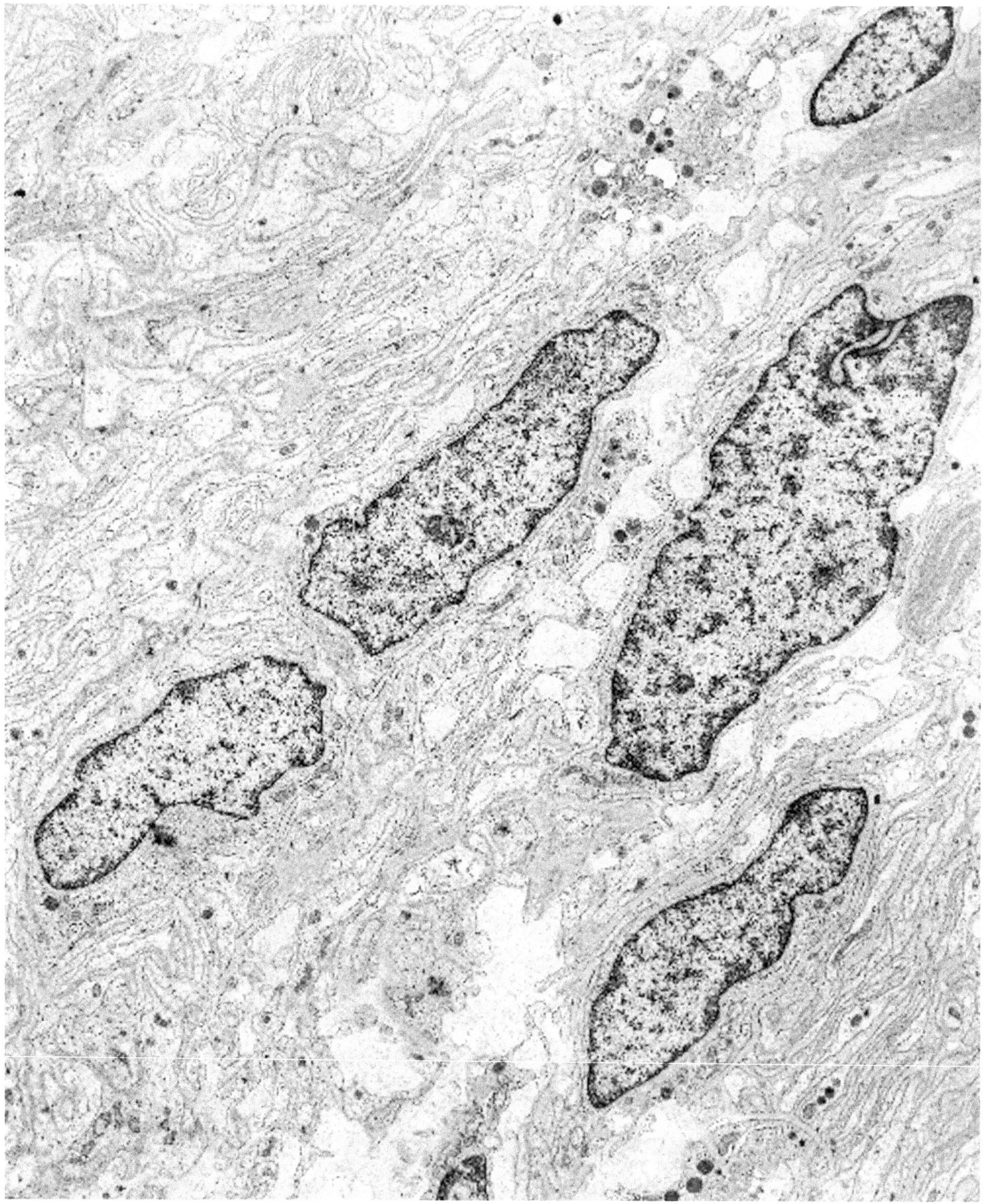

Figure 6.110. Schwannoma (retroperitoneum). Well-differentiated Schwann cells with compact, intertwined, thin processes covered by basal lamina are illustrated in this field. (× 6250)

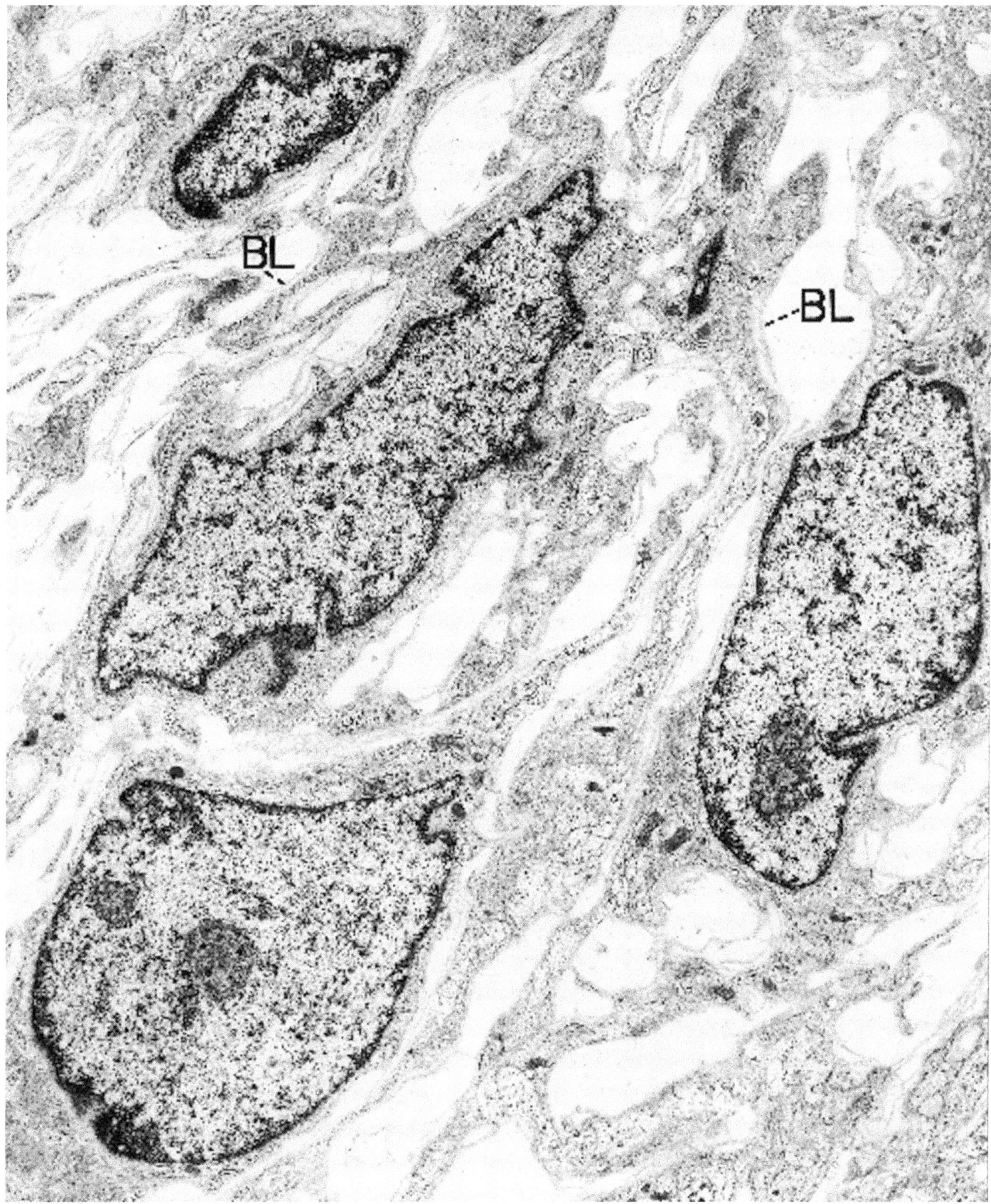

Figure 6.111. Schwannoma (cauda equina). High magnification highlights typical Schwann cell bodies and processes, with a covering of basal lamina (BL). The cell bodies have a high nuclear–cytoplasmic ratio and a nonspecific complement of cytoplasmic organelles. (× 8500)

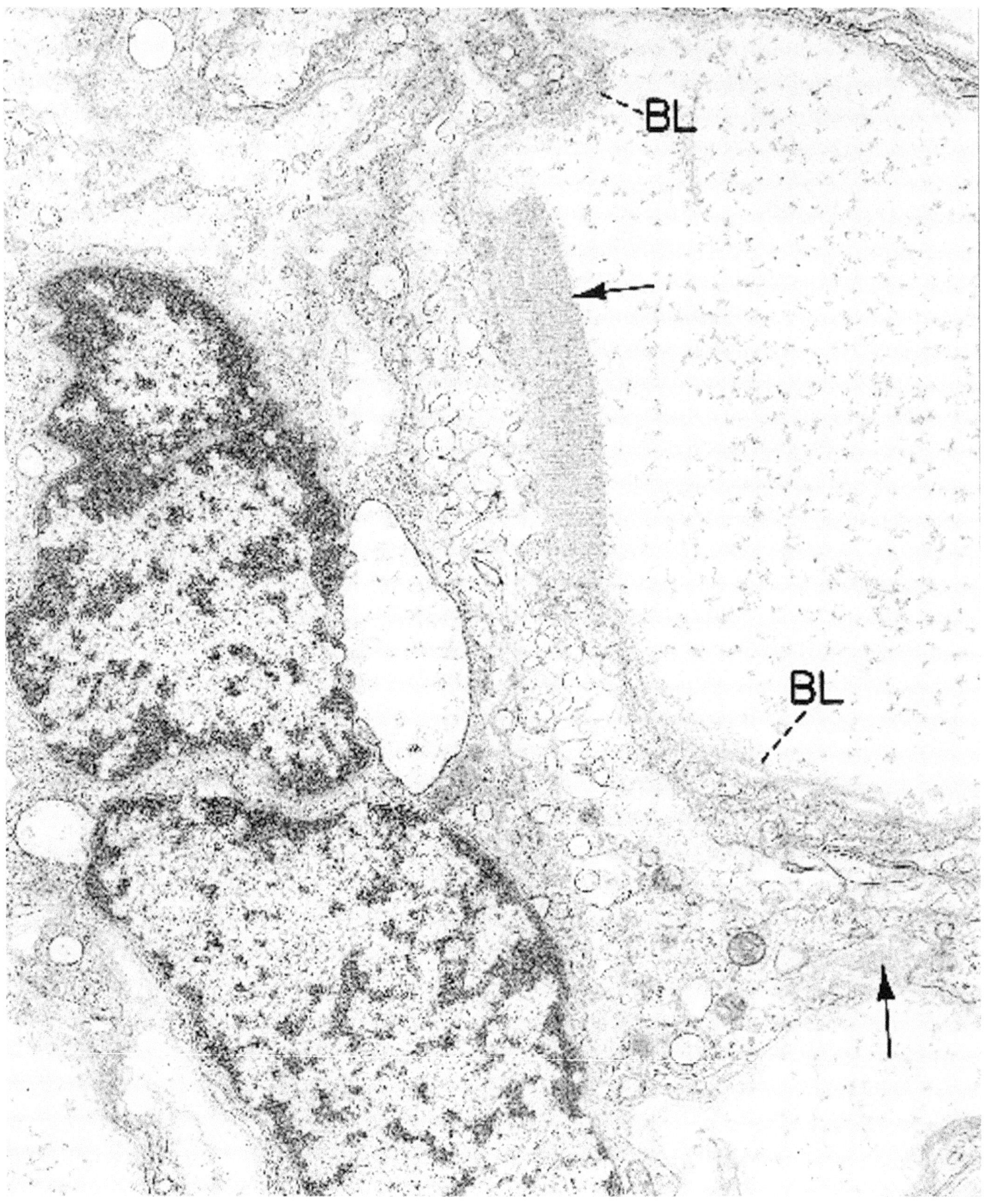

Figure 6.112. Schwannoma (cauda equina). High magnification of a Schwann cell depicts basal lamina (BL) and long-spacing collagen (Luse bodies) (*arrows*). (× 15,640)

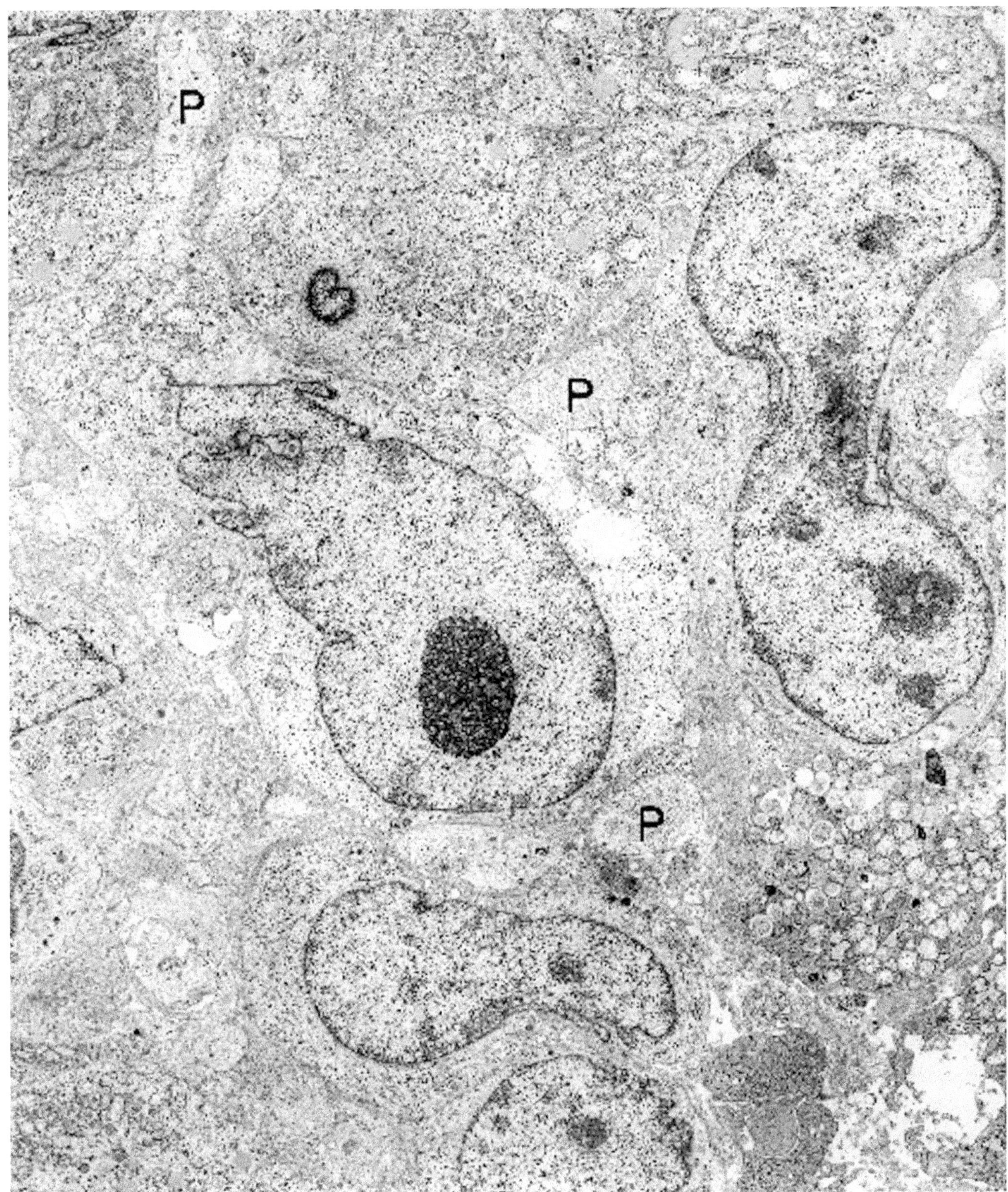

Figure 6.113. Malignant Schwannoma (third intercostal nerve). These poorly differentiated cells are difficult to prove as being Schwann cells. They are tightly apposed and are not covered by basal lamina. They have processes (P) that are broad and not intertwining. Nuclei are large and lobulated, a characteristic that is seen moderately often in malignant Schwannomas. The euchromatin and large nucleoli indicate a high degree of synthetic or mitotic activity, and in this case are consistent with the malignant state of the neoplasm. (× 4845)

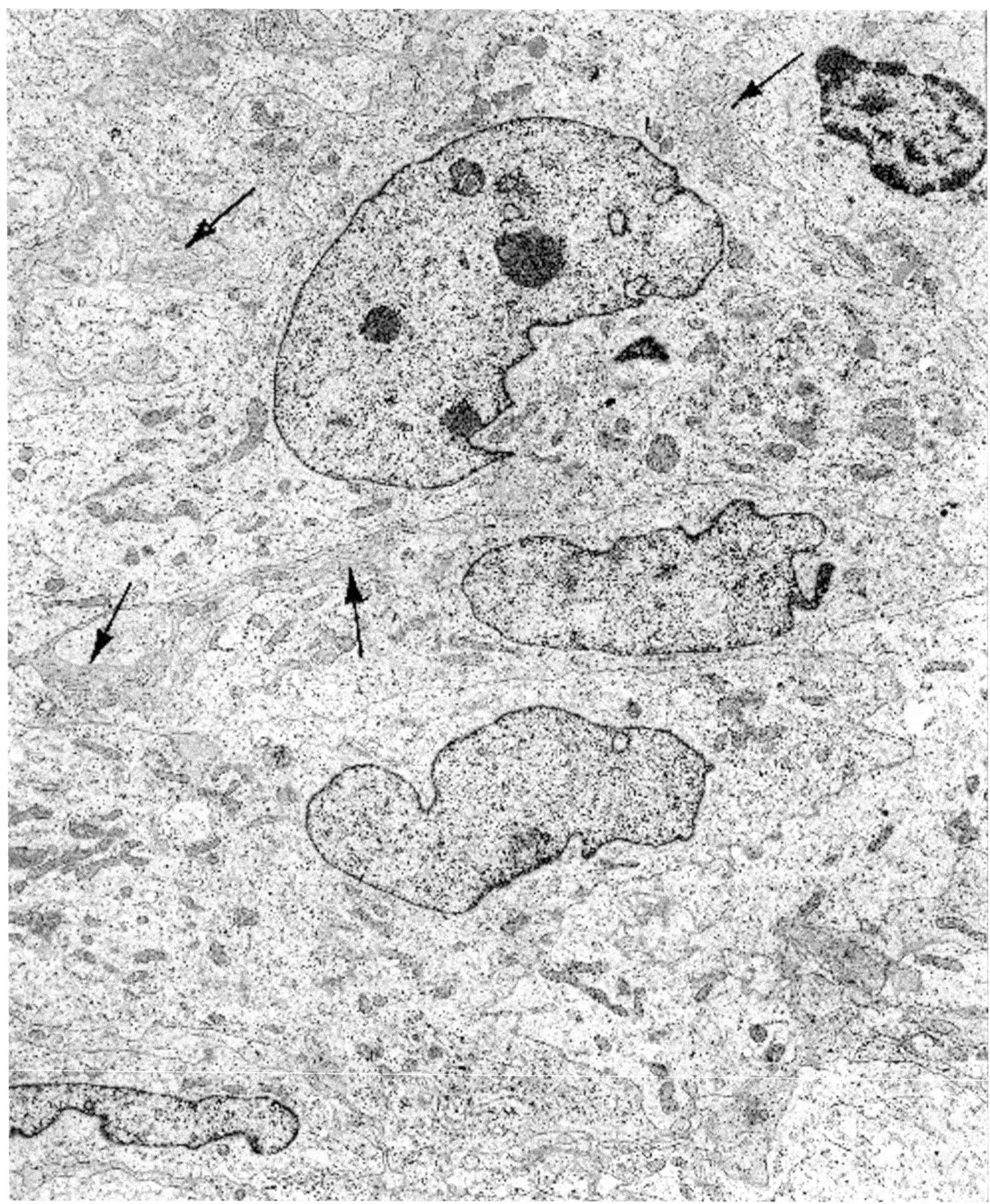

Figure 6.114. Malignant Schwannoma (abdominal wall). The cells of this tumor are similar to those of the malignant Schwannoma depicted in Figure 6.113, but here there is focal interdigitation of thin processes as well as small amounts of basal lamina covering the processes (*arrows*). (× 4750)

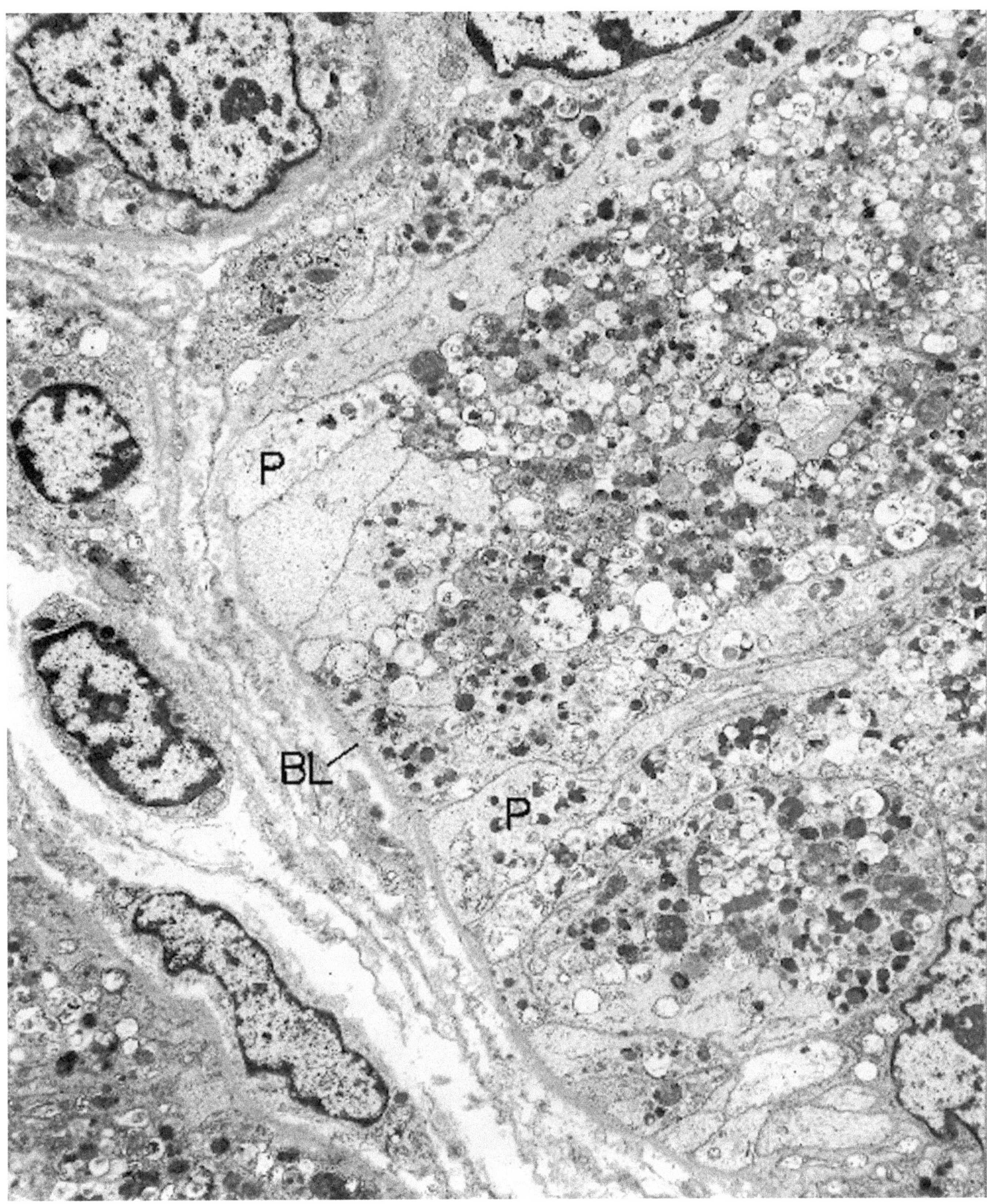

Figure 6.115. Granular cell tumor (trachea). Basal lamina (BL) surrounds a group of granular cells. They are oval and elongate, have long processes (P), and contain a myriad of osmiophilic bodies, representing phagolysosomes. (× 5724) (Permission for reprinting granted by Hemisphere Publishing, Dickersin GR: The electron microscopic spectrum of nerve sheath tumors. Ultrastruct Pathol 11:103–146, 1987.)

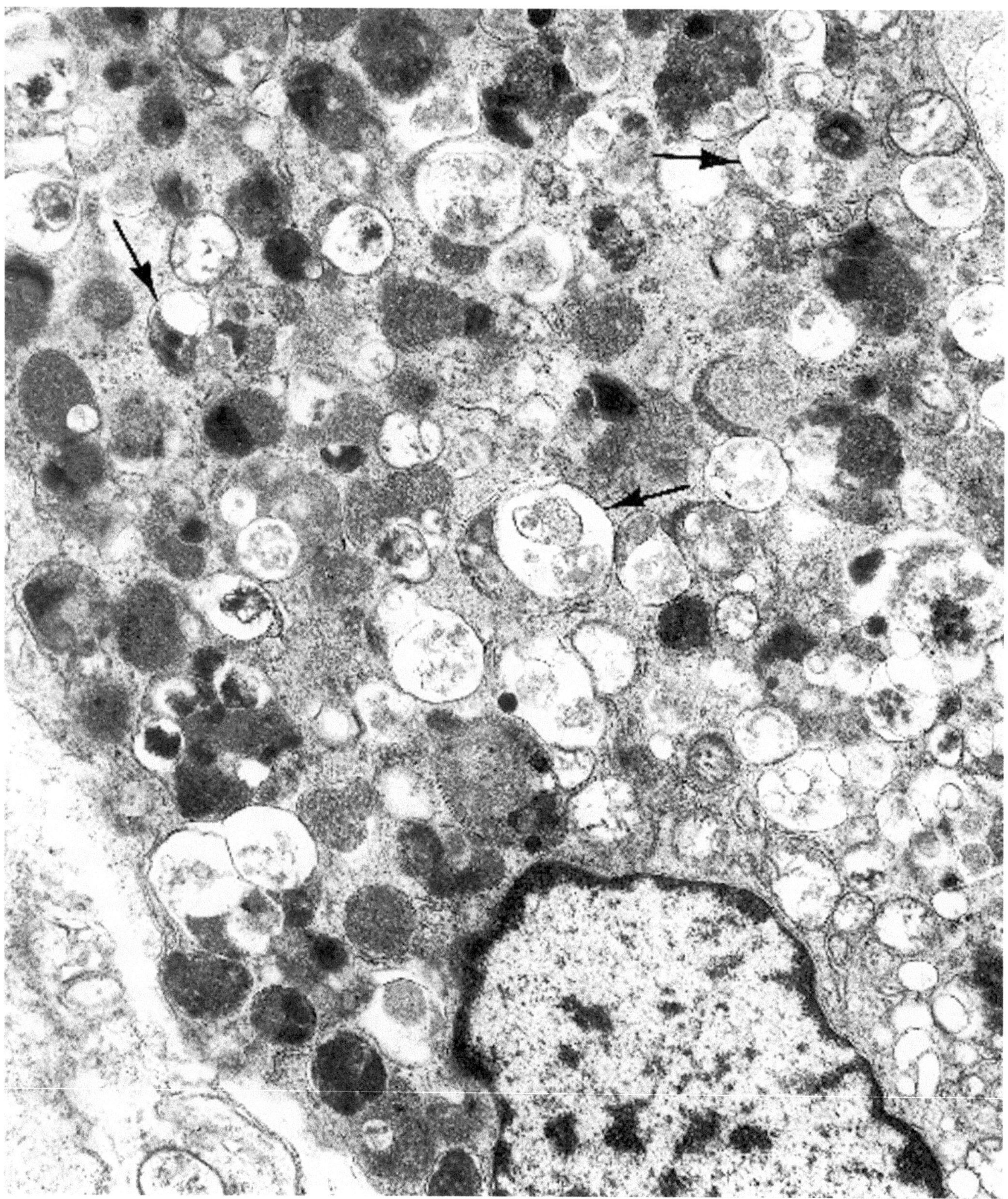

Figure 6.116. Granular cell tumor (trachea). High magnification of one of the cells from the same neoplasm as shown in Figure 6.115 illustrates the heterogeneous character of the osmiophilic material as well as the limiting membrane (*arrows*) of some of the phagolysosomes. (× 19,240) (Permission for reprinting granted by Hemisphere Publishing, Dickersin GR: The electron microscopic spectrum of nerve sheath tumors. Ultrastruct Pathol 11:103–146, 1987.)

(Text continued from page 359)

Neurofibroma

(Figures 6.117 through 6.124.)

Diagnostic criteria. A combination of nerve sheath components, including (1) Schwann cells; (2) Schwann cell-neurite complexes (myelinated and nonmyelinated); (3) fibroblasts; (4) perineurial cells; (5) a collagenous stroma of variable density (Figures 6.117 through 6.122).

Additional points. The fibroblasts in neurofibromas may have less rough endoplasmic reticulum and longer and narrower processes than have typical fibroblasts (Figures 6.117 and 6.118). The processes may even wrap around bundles of collagen fibers in a pattern similar to Schwann cells forming mesaxons around neurons, raising the question of whether these cells may actually be Schwann cells (Figures 6.119 through 6.123). However, the absence of basal lamina and the presence of classical Schwann cells in the same neoplasm are evidence against that possibility. Furthermore, these fibroblasts or fibroblast-like cells do not have pinocytotic vesicles, making a perineurial cell-type unlikely.

The *perineurial cell* in neoplasms is a somewhat controversial cell, but it is accepted as a component of neurofibroma and apparently may comprise certain other nerve sheath neoplasms; namely, *Pacinian neurofibromas* and some *dermatofibrosarcoma protuberans* and *Bednar tumors.* The normal perineurial cell has long, thin, polar processes; discontinuous basal lamina; pinocytotic vesicles; and small junctions where processes interconnect (Figure 6.124). There may also be a varying number of filaments and a moderate amount of rough endoplasmic reticulum. Some question still lingers as to whether the perineurial cell may represent a variant of the fibroblast or of the Schwann cell.

Sarcomatoid (Spindle Cell) Carcinoma

(Figures 6.125 and 6.126.)

Diagnostic criteria. (1) Spindle cells with epithelial and/or mesenchymal differentiation; (2) epithelial features include desmosomes, tonofibrils, microvilli, and lumens; (3) mesenchymal features include abundant dilated rough endoplasmic reticulum and numerous filaments with dense bodies.

Additional points. In our experience, the spindle cell component of sarcomatoid carcinomas may be one of three types: epithelial, fibroblastic/myofibroblastic or a combination of epithelial and fibroblastic/myofibroblastic. In addition to individual cells showing either epithelial or mesenchymal differentiation, bimodal differentiation may be present within the same cells (Figures 6.125 and 6.126). Neoplasms of this type often arise from epithelial surfaces but may also arise in parenchymatous organs such as the pancreas and kidney. The mesenchymal examples probably represent a metaplastic change, and a coexisting or previous carcinoma is usually demonstrable in the region of the current spindle cell neoplasm.

Sarcomatous Thymoma, Melanoma, and Mesothelioma

See Chapter 3, Large Cell Undifferentiated Neoplasms.

(Text continues on page 380)

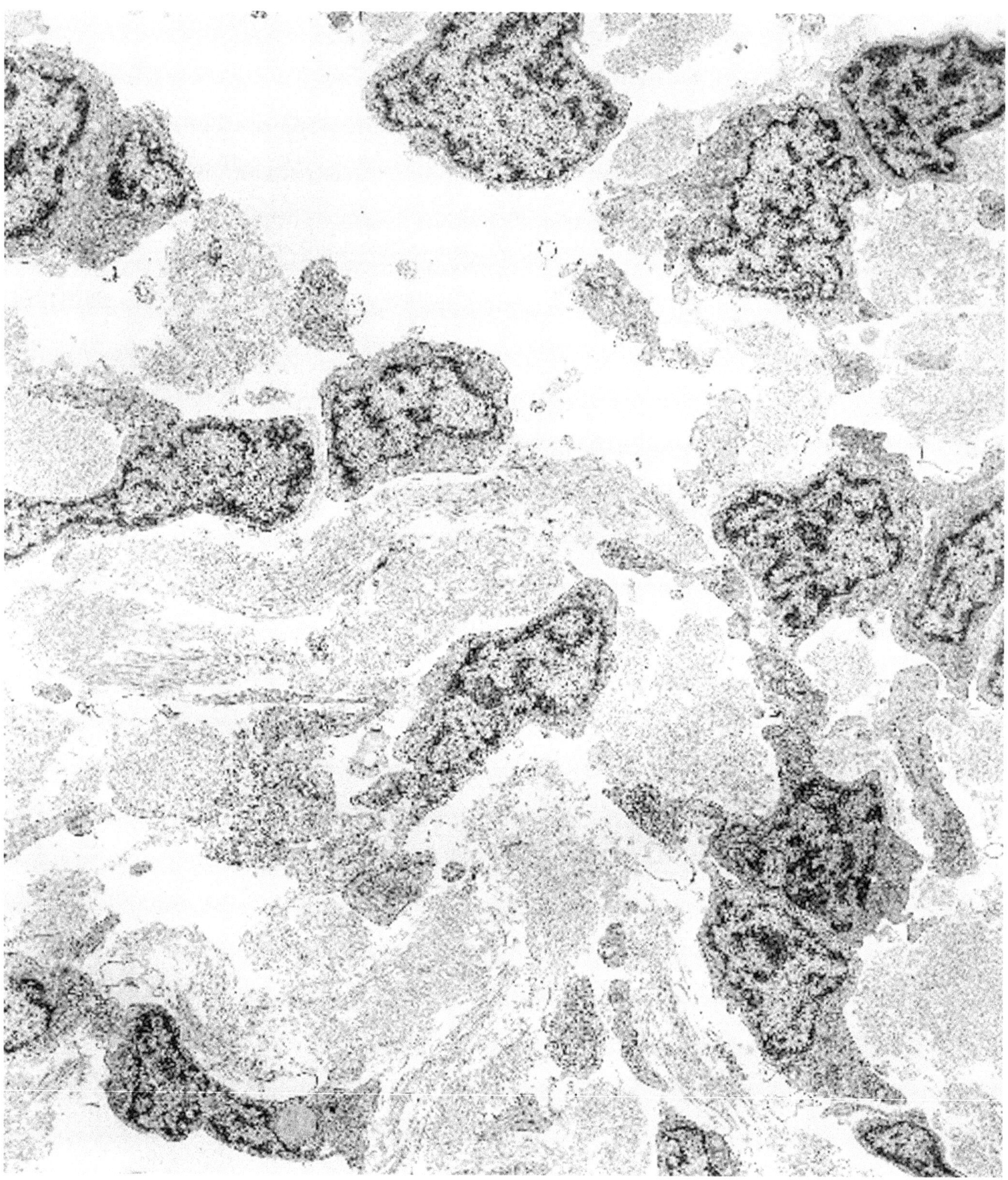

Figure 6.117. Neurofibroma (orbit). The light microscopic appearance of this neoplasm was that of a neurofibroma, and the ultrastructural characteristics of the cells are more fibroblast-like than Schwannian. The amount of rough endoplasmic reticulum is less than in a classic fibroblast. However, the cell surfaces are in close association with banded collagen, and there is no basal lamina. (× 6250) (Permission for reprinting granted by Hemisphere Publishing, Dickersin GR: The electron microscopic spectrum of nerve sheath tumors. Ultrastruct Pathol 11:103–146, 1987.)

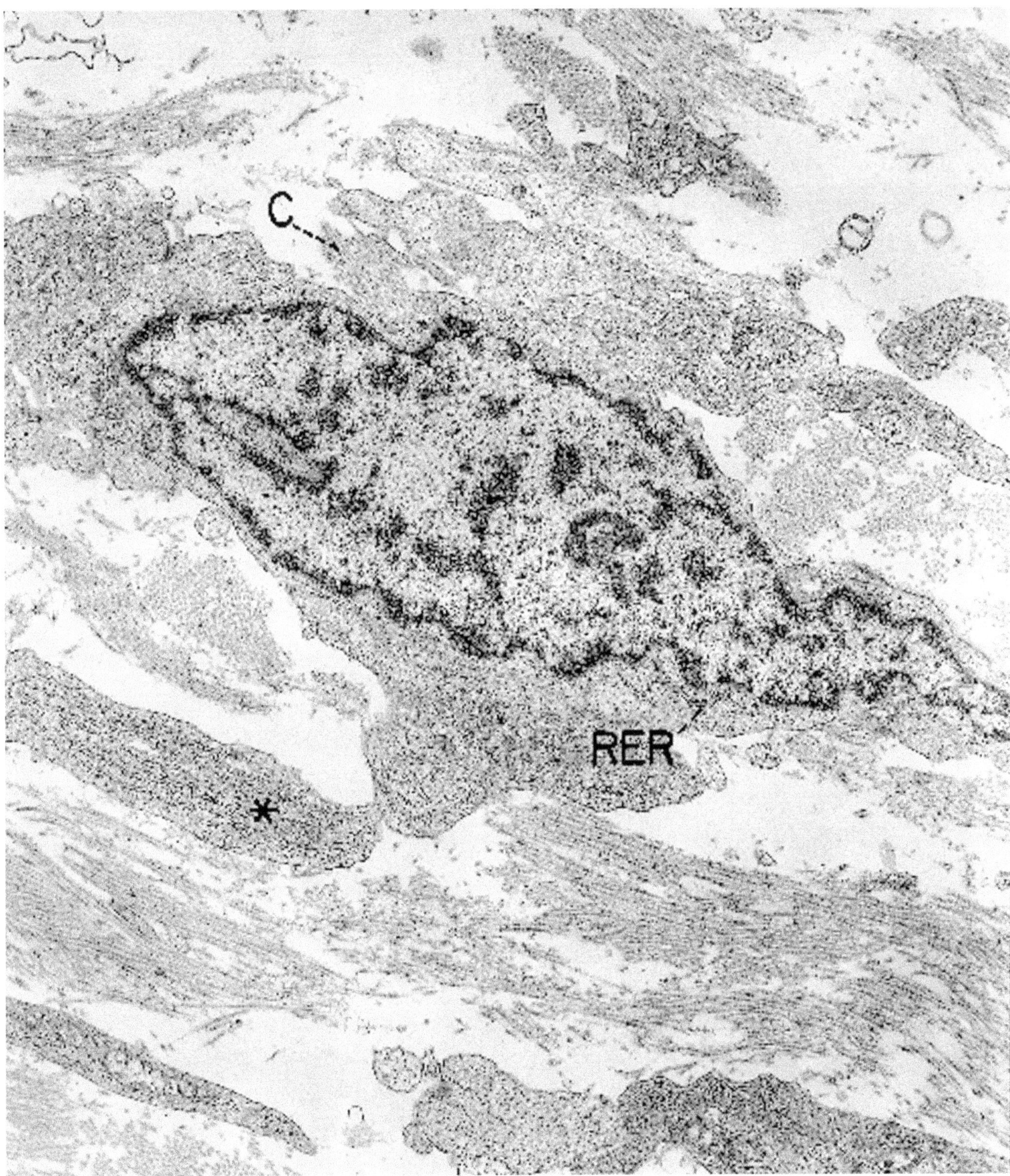

Figure 6.118. Neurofibroma (orbit). High magnification of a fibroblast-like cell illustrates only a small amount of rough endoplasmic reticulum (RER), although some of the neighboring cells (*) have more. Note the close relationship between the plasmalemma of the main cell and the banded collagen (C) of the matrix. There are no basal lamina and no pinocytotic vesicles. (× 11,250) (Permission for reprinting granted by Hemisphere Publishing, Dickersin GR: The electron microscopic spectrum of nerve sheath tumors. Ultrastruct Pathol 11:103–146, 1987.)

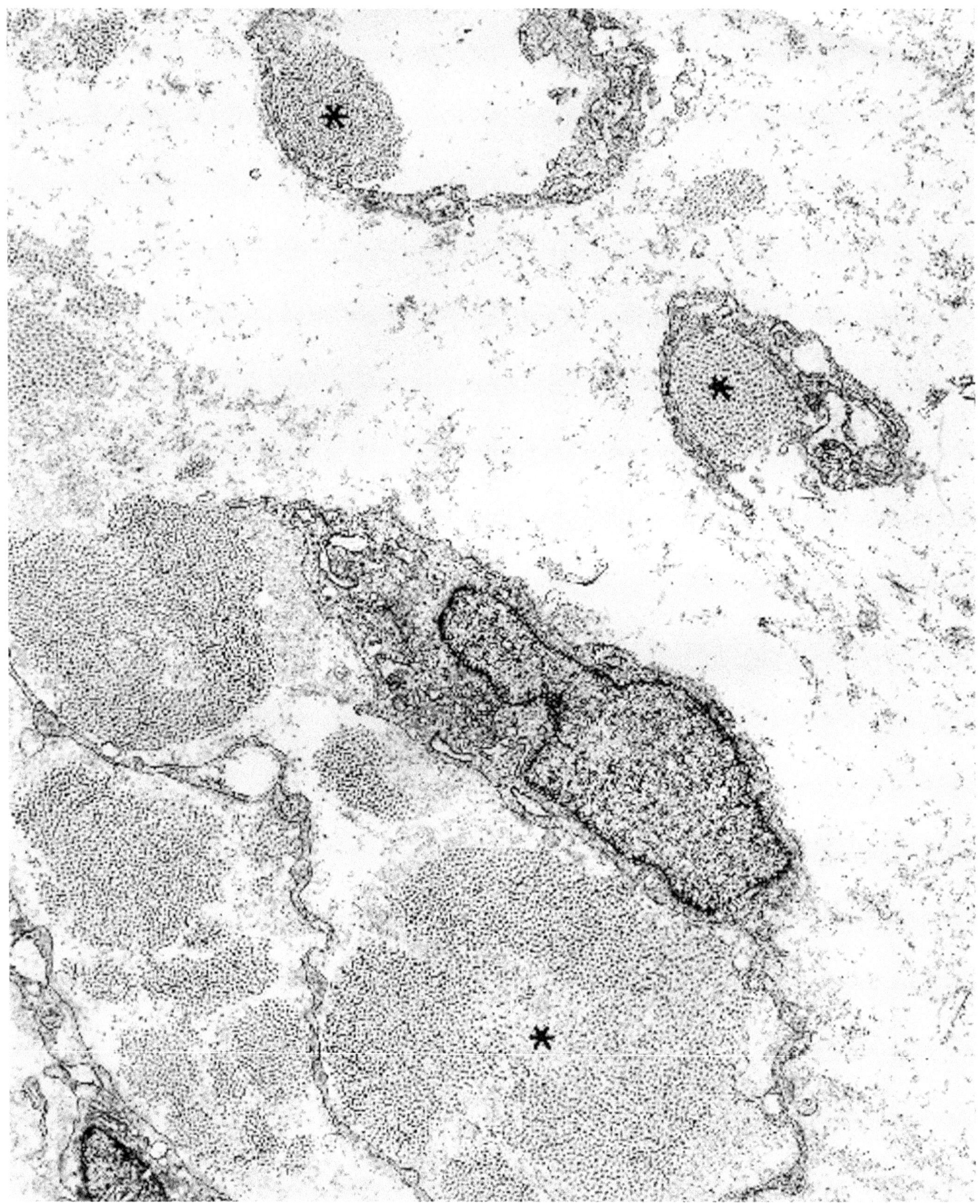

Figure 6.119. Neurofibroma (soft tissue of thigh). The neoplastic cells have long, narrow processes that wrap around bundles of collagen fibers (*), forming pseudomesaxons reminiscent of Schwann cells enclosing axons. (× 9000)

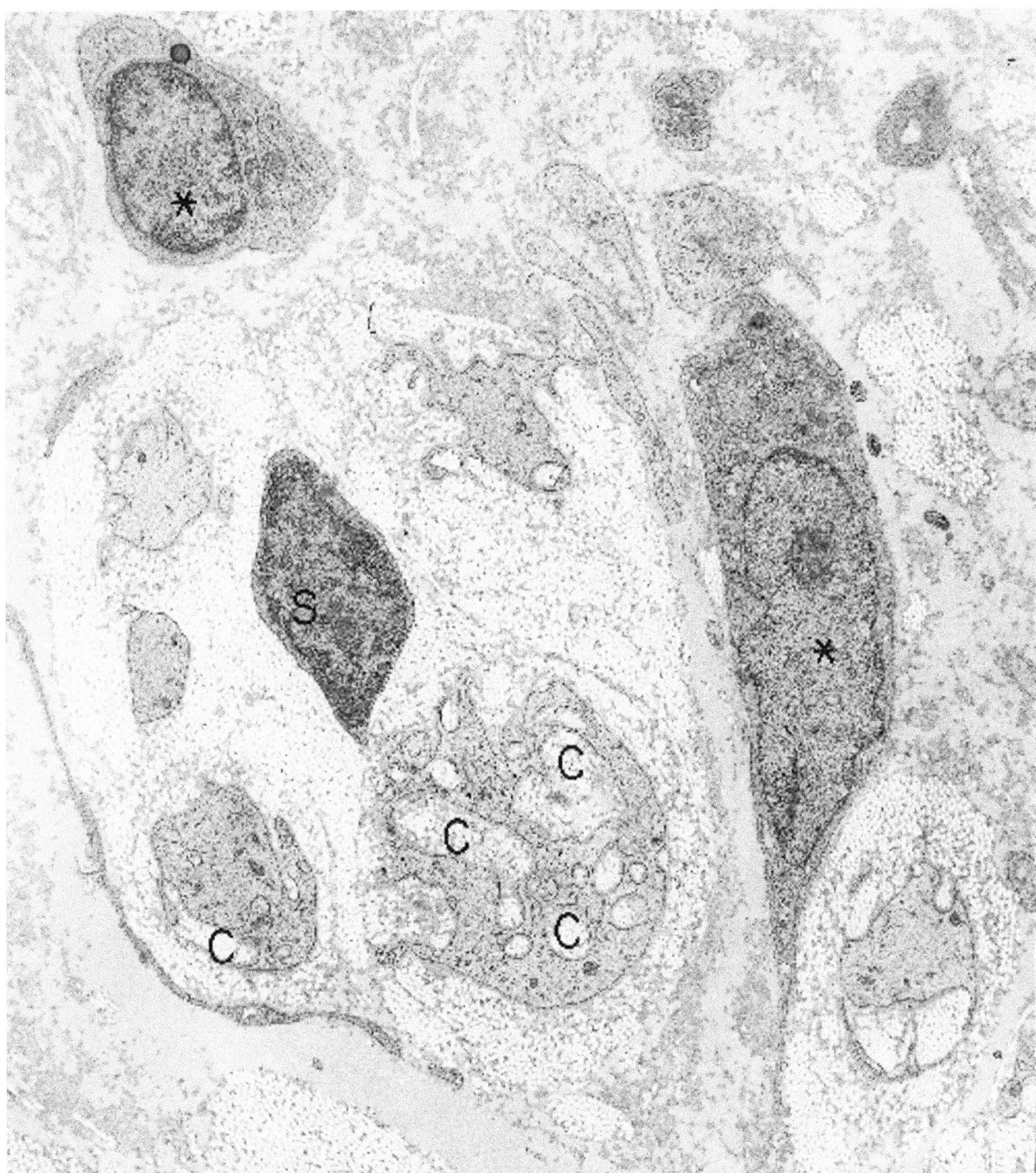

Figure 6.120. Plexiform neurofibroma (femur). The small cell (S) in the left central part of the field is consistent with being a nonneoplastic Schwann cell, but the neighboring processes are wrapped around collagen (C) rather than axons, forming pseudomesaxons. The other two cell bodies (*) are consistent with being neoplastic, but whether they are Schwannian, fibroblastic, or perineurial is difficult to answer. They have focal, fuzzy basal lamina and virtually no pinocytotic vesicles. Rough endoplasmic reticulum is moderate in amount and only mildly dilated. (× 9500) (Permission for reprinting granted by Hemisphere Publishing, Dickersin GR: The electron microscopic spectrum of nerve sheath tumors. Ultrastruct Pathol 11:103–146, 1987.)

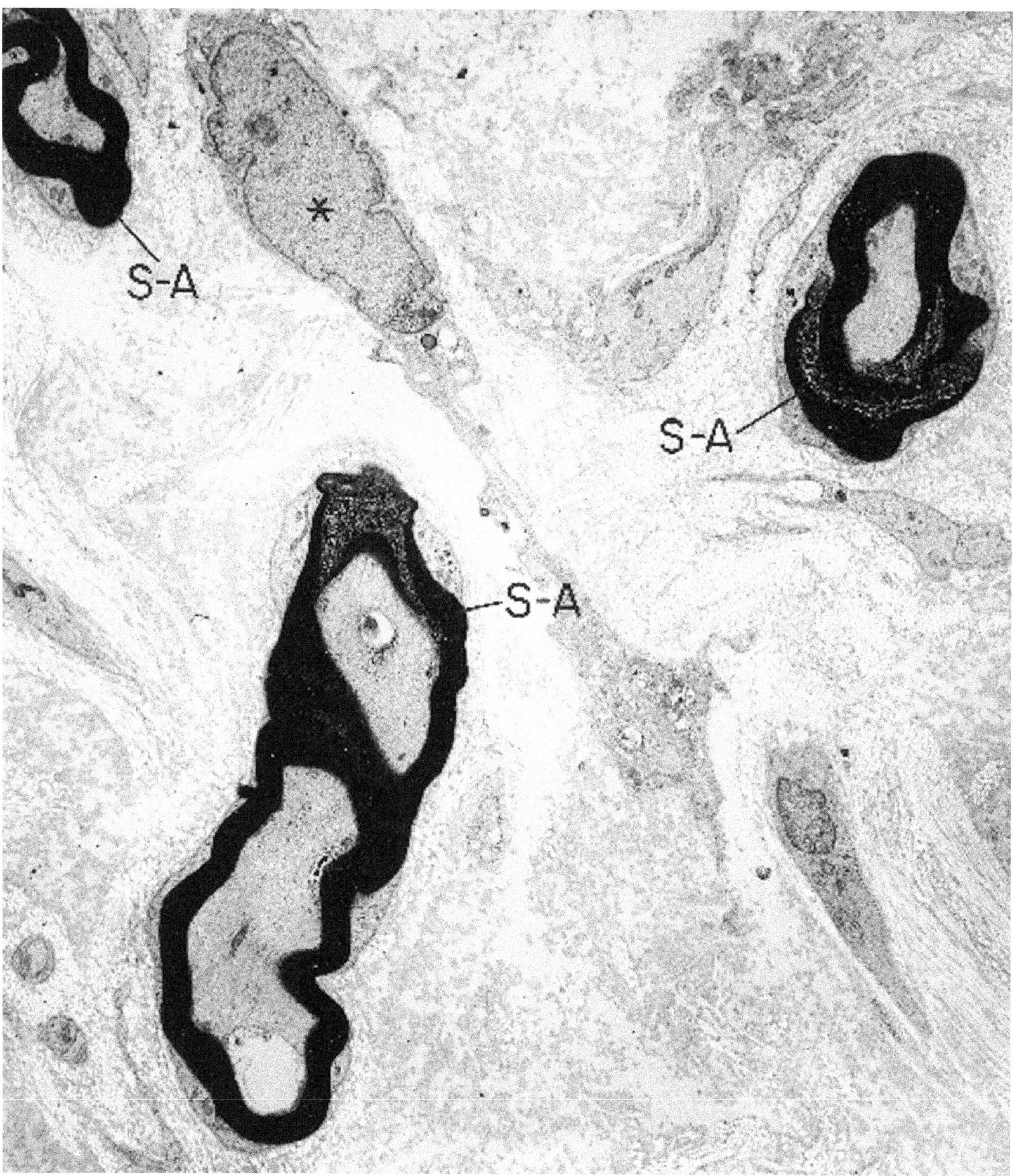

Figure 6.121. Plexiform neurofibroma (femur). This is another field of the same neoplasm as illustrated in Figure 6.120. In addition to at least one neoplastic cell of unclassified type (*), there are several residual, nonneoplastic Schwann cell-axon complexes (S-A). (× 5000) (Permission for reprinting granted by Hemisphere Publishing, Dickersin GR: The electron microscopic spectrum of nerve sheath tumors. Ultrastruct Pathol 11: 103–146, 1987.)

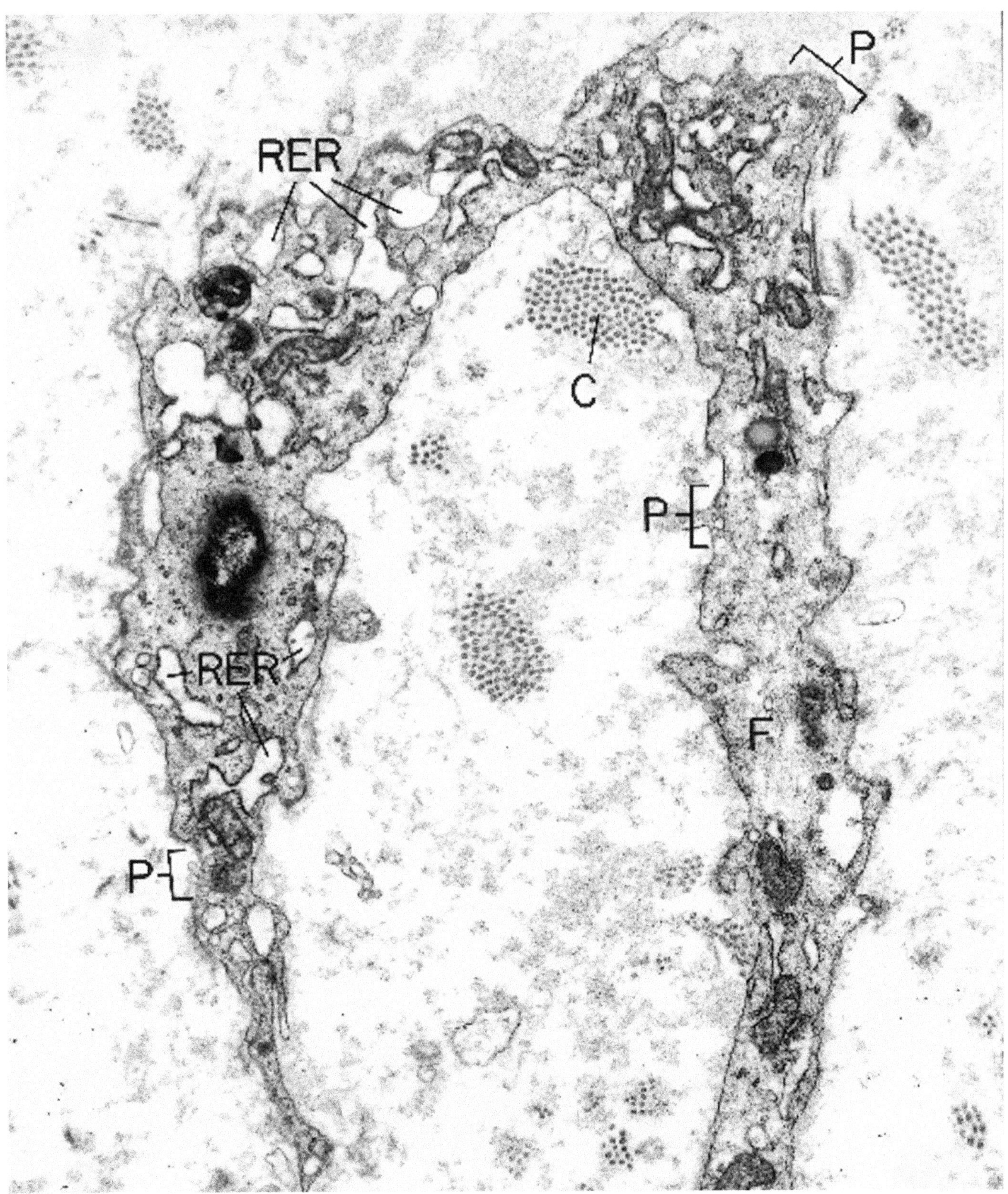

Figure 6.122. Neurofibroma (soft tissue of thigh). The cytoplasm of this tumor cell is rich in filaments (F) and contains a moderate amount of rough endoplasmic reticulum (RER). A few pinocytotic vesicles (P) also are present. Both amorphous material and banded collagen (C) border the cell and comprise the extracellular matrix. (× 17,100) (Permission for reprinting granted by Hemisphere Publishing, Dickersin GR: The electron microscopic spectrum of nerve sheath tumors. Ultrastruct Pathol 11:103–146, 1987.)

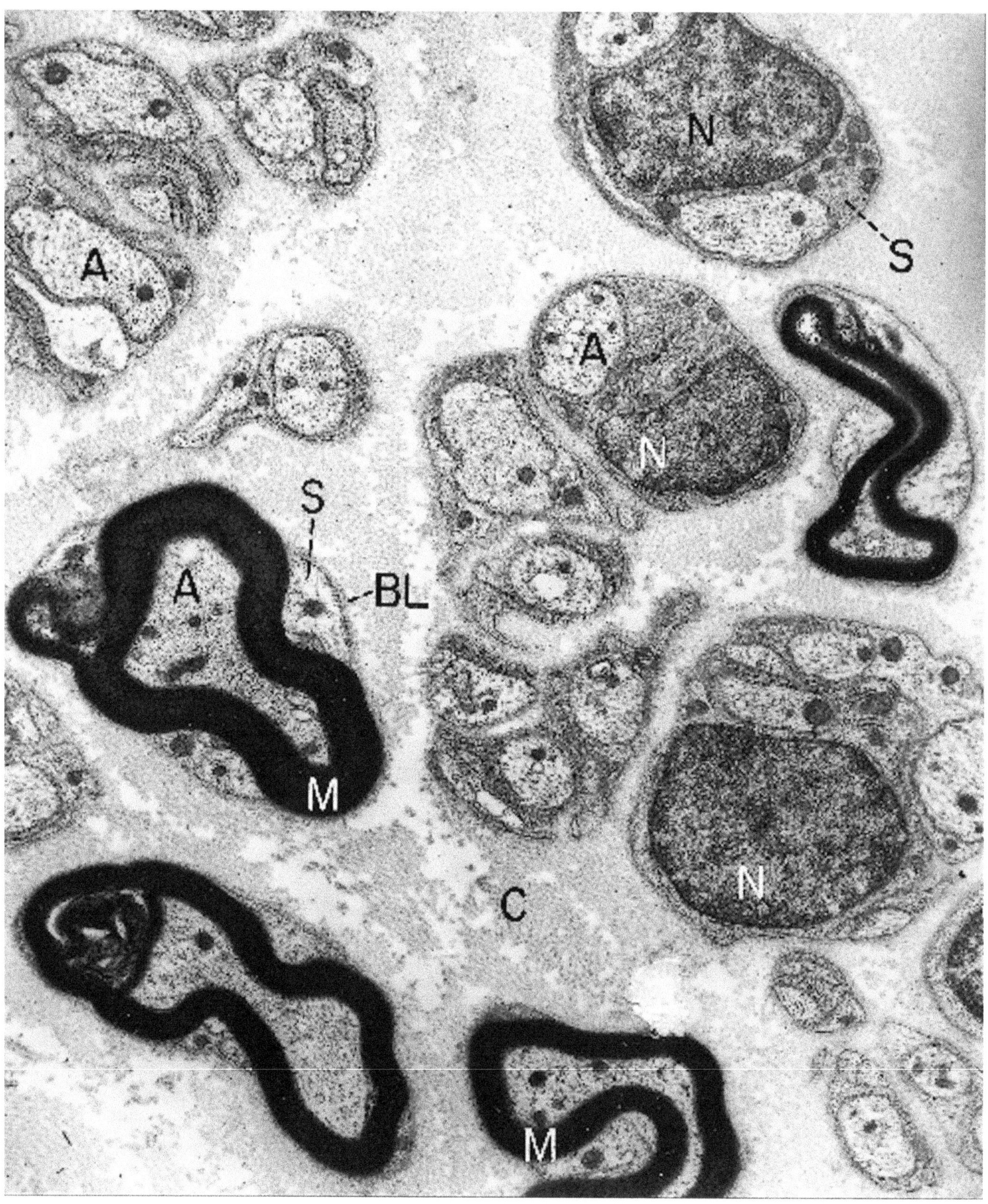

Figure 6.123. Normal sural nerve. Myelinated and unmyelinated axons (A) are ensheathed by Schwann cells (S), some of which have their nuclei (N) present at the level of this section. The myelin (M) consists of many wrappings of the Schwann cell plasmalemmas. Note the discrete basal lamina (BL) covering the Schwann cells and the abundance of mature, banded collagen (C) composing the extracellular matrix. (× 11,925)

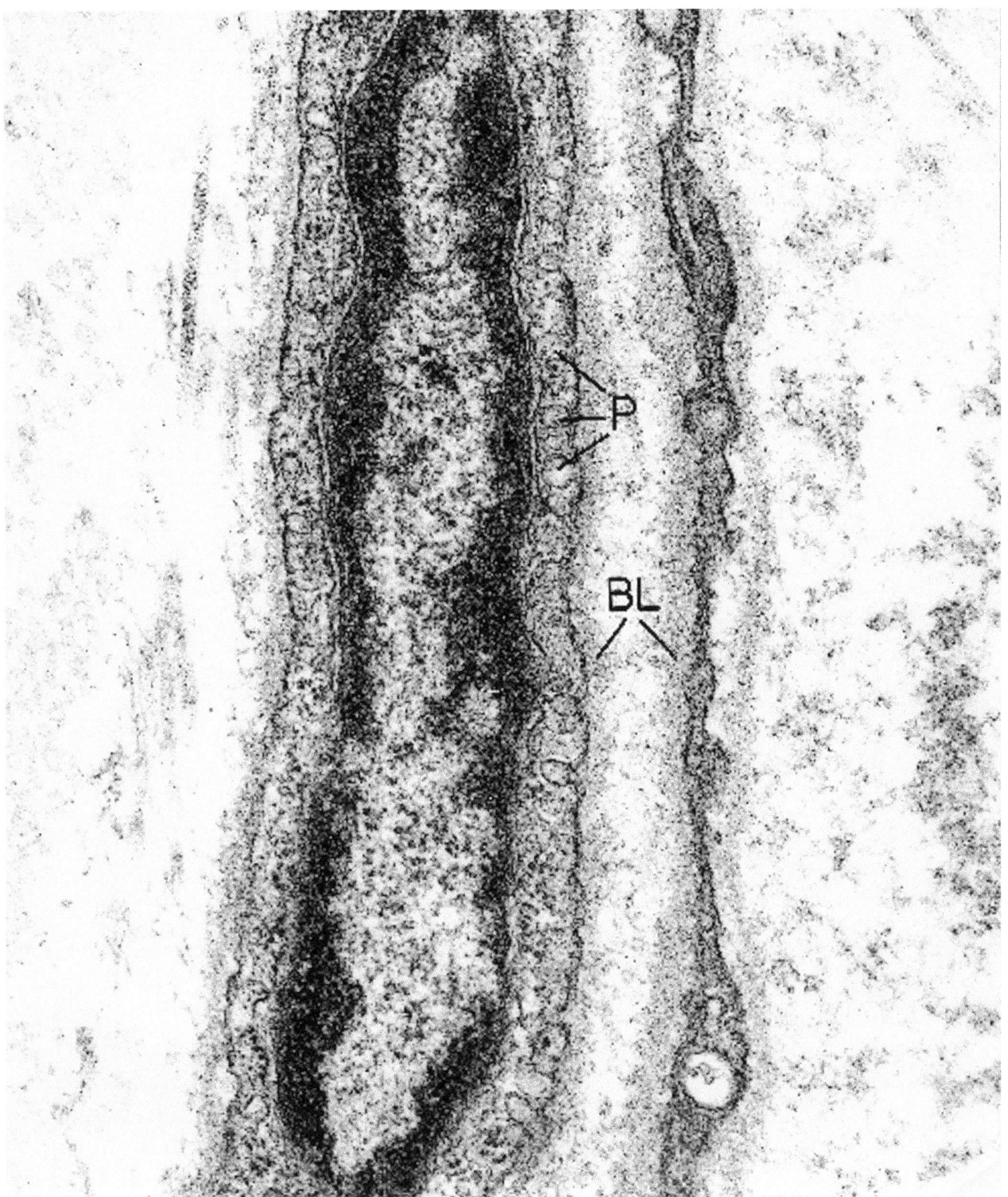

Figure 6.124. Normal perineurial cell (cutaneous nerve). These two perineurial cells, a cell body on the left and a cell process on the right, exemplify the two most characteristic features of this cell type: discontinuous and irregularly distributed basal lamina (BL) and pinocytotic vesicles (P). (× 58,700) (Permission for reprinting granted by Hemisphere Publishing, Dickersin GR: The electron microscopic spectrum of nerve sheath tumors. Ultrastruct Pathol 11:103–146, 1987.)

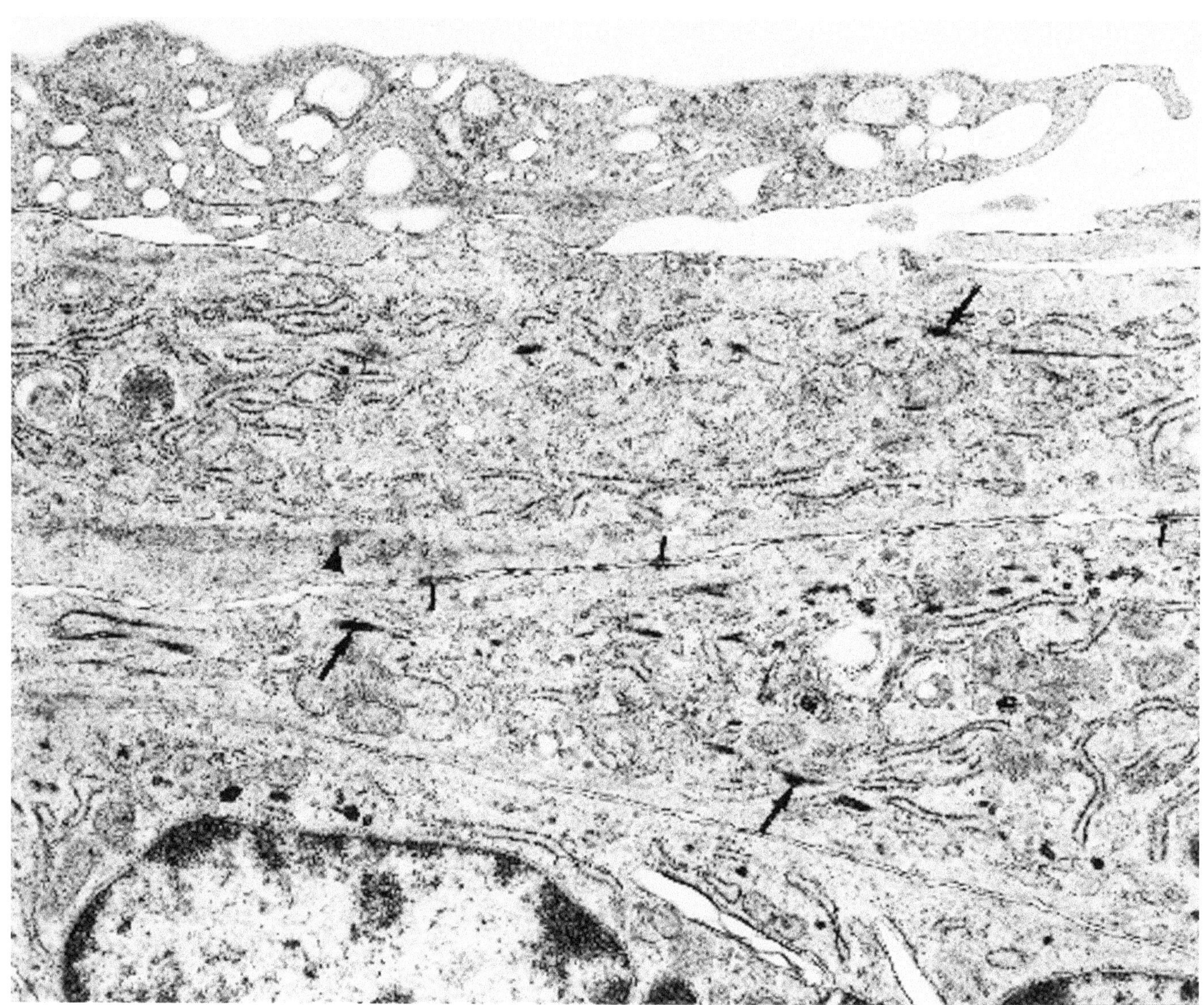

Figure 6.125. Sarcomatoid carcinoma (larynx). Malignant spindle cells have the combined epithelial and myofibroblastic features of desmosomes (*straight lines*), tonofibrils (*arrows*), and filaments with dense bodies (*arrowhead*). (× 19,000) (Permission for reprinting granted by Taylor and Francis, Balercia G, Bhan AK, Dickersin GR: Sarcomatoid carcinoma: An ultrastructural study with light microscopic and immunohistochemical correlation of 10 cases from various anatomic sites. Ultrastruct Pathol 19:249–263, 1995.)

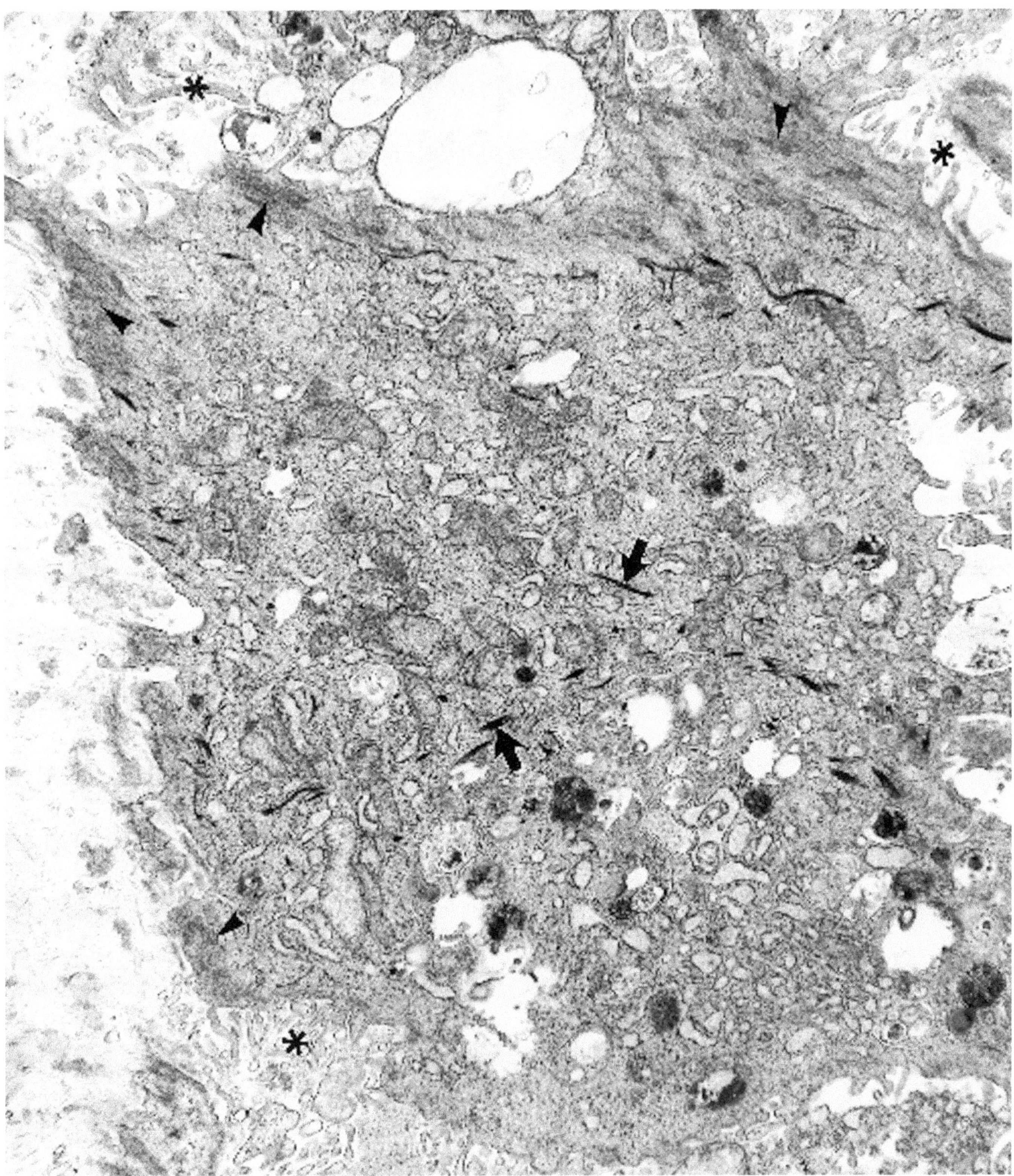

Figure 6.126. Sarcomatoid carcinoma (ureter). In addition to filaments and dense bodies (*arrowheads*) and tonofibrils (*arrows*), this malignant spindle cell has villous processes on its free surfaces (*) (× 13,000) (Permission for reprinting granted by Taylor and Francis, Balercia G, Bhan AK and Dickersin GR: Sarcomatoid carcinoma: An ultrastructural study with light microscopic and immunohistochemical correlation of 10 cases from various anatomic sites. Ultrastruct Pathol 19:249–263, 1995.)

(Text continued from page 369)

REFERENCES

Fibrosarcoma and Other Fibroblastic Neoplasms

Abdul-Karim FW, Evans HL, Silva EG: Giant cell fibroblastoma: A report of three cases. Am J Clin Pathol 83:165–170, 1985.

Akhtar M, Miller RM: Ultrastructure of elastofibroma. Cancer 40:728–735, 1977.

Alguacil-Garcia A: Giant cell fibroblastoma recurring as dermatofibrosarcoma protuberans. Am J Surg Pathol 15:798–801, 1991.

Alston SR, Francel PC, Jane JA Jr: Solitary fibrous tumor of the spinal cord. Am J Surg Pathol 21:477–483, 1997.

Angervall L, Kindblom LG, Merck C: Myxofibrosarcoma—A study of 30 cases. Acta Pathol Microbiol Scand Sect A 85:127–140, 1977.

Bangfield WG, Lee CK: Elastofibroma: An electron microscopic study. J Natl Cancer Inst 40:1067–1077, 1968.

Churg AM, Kahn LB: Myofibroblasts and related cells in malignant fibrous and fibrohistiocytic tumors. Hum Pathol 8:205–218, 1977.

Ding J, Hashimoto H, Enjoji M: Dermatofibrosarcoma protuberand with fibrosarcomatous areas. A clinicopathologic study of nine cases and a comparison with allied tumors. Cancer 64:721–729, 1989.

Dixon AY, Lee SH: An ultrastructural study of elastofibromas. Hum Pathol 11:257–262, 1980.

Dorfman DM, To K, Dickersin GR, et al: Solitary fibrous tumor of the orbit. Am J Surg Pathol 18:281–287, 1994.

Feiner H, Kaye GI: Ultrastructural evidence of myofibroblasts in circumscribed fibromatosis. Arch Pathol Lab Med 100:265–268, 1976.

Gabbiani G, Hirschel BJ, Ryan GB, et al: Granulation tissue as a contractile organ. A study of structure and function. J Exp Med 135:719–734, 1972.

Goellner JR, Soule EH: Desmoid tumors. An ultrastructural study of eight cases. Hum Pathol 11:43–50, 1980.

Govoni E, Severi B, Laschi R, et al: Elastofibroma: An in vivo model of abnormal neoelastogenesis. Ultrastruct Pathol 12:327–339, 1988.

Hasegawa T, Hirose T, Seki K, et al: Solitary fibrous tumor of the soft tissue. An immunohistochemical and ultrastructural study. Am J Clin Pathol 106:325–331, 1996.

Kindblom LG, Merck C, Angervall L: The ultrastructure of myxofibrosarcoma. A study of 11 cases. Virchows Arch [A] 381:121–139, 1979.

Kindblom LG, Spicer SS: Elastofibroma. A correlated light and electron microscopic study. Virchows Arch [A] 396:127–140, 1982.

Kumaratilake JS, Krishnan R, Lomax-Smith J, et al: Elastofibroma: Disturbed elastic fibrillogenesis by periosteal-derived cells? An immunoelectron microscopic and in situ hybridization study. Hum Pathol 22: 1017–1029, 1991.

Leak LV, Caulfield JB, Burke JF, et al: Electron microscopic studies on a human myxofibrosarcoma. Cancer Res 27:261–285, 1967.

Lopes JM, Paiva ME: Dermatofibrosarcoma protuberans. A histological and ultrastructural study of 11 cases with emphasis on the study of recurrences and histogenesis. Pathol Res Pract 187:806–813, 1991.

Mackay B: Tumors of fibroblasts and muscle cells. Ultrastruct Pathol 9:175–177, 1985.

Mentzel T, Calonje E, Wadden C, et al: Myxofibrosarcoma. Clinicopathologic analysis of 75 cases with emphasis on the low-grade variant. Am J Surg Pathol 20: 391–405, 1996.

Nielsen GP, O'Connell JX, Dickersin GR, et al: Collagenous fibroma (desmoplastic fibroblastoma): A report of seven cases. Mod Pathol 9:781–785, 1996.

Nielsen GP, O'Connell JX, Dickersin GR, et al: Solitary fibrous tumor of soft tissue. A report of 15 cases, including 5 malignant examples with light microscopic, immunohistochemical and ultrastructural data. Mod Pathol 10:1028–1037, 1997.

Papadimitriou JC, Ord RA, Drachenberg CB: Head and neck fibromyxoid sarcoma: Clinicopathological correlation with emphasis on peculiar ultrastructural features related to collagen processing. Ultrastruct Pathol 21:81–87, 1997.

Parveen T, Fleischmann J, Petrelli M: Benign fibrous tumor of the tunica vaginalis testis. Report of a case with light, electron microscopic, and immunocytochemical study, and review of the literature. Arch Pathol Lab Med 116:277–280, 1992.

Pinto A, Hwang W-S, Wong AL, et al: Giant cell fibroblastoma in childhood immunohistochemical and ultrastructural study. Mod Pathol 5:639–642, 1992.

Pulitzer DR, Martin PC, Reed RJ: Fibroma of tendon sheath. A clinicopathologic study of 32 cases. Am J Surg Pathol 13:472–479, 1989.

Ryan GB, Cliff WJ, Gabbiani G, et al: Myofibroblasts in human granulation tissue. Hum Pathol 5:55–67, 1974.

Said JW, Nash G, Banks-Shlegel S, et al: Localized fibrous mesothelioma: An immunohistyochemical and electron microscopic study. Hum Pathol 15:440–443, 1984.

Shmookler BM, Enzinger FM, Weiss SW: Giant cell fibroblastoma. A juvenile form of dermatofibrosarcoma protuberans. Cancer 64:2154–2161, 1989.

Stiller D, Katenkamp D: Cellular features in desmoid fibromatosis and well-differentiated fibrosarcomas. An electron microscopic study. Virchows Arch [A] 369: 155–164, 1975.

Suster S, Nascimento AG, Miettinen M, et al: Solitary fibrous tumors of soft tissue. A clinicopathologic and immunohistochemical study of 12 cases. Am J Surg Pathol 19:1257–1266, 1995.

van de Rijn M, Lombard CM, Rouse RV: Expression of CD34 by solitary fibrous tumors of the pleura, mediastinum, and lung. Am J Surg Pathol 18:814–820, 1994.

Waisman J, Smith DW: Fine structure of an elastofibroma. Cancer 22:671–677, 1968.

Wirman JA: Nodular fasciitis, a lesion of myofibroblasts. An ultrastructural study. Cancer 38:2378–2389, 1976.

Wrotnowski U, Cooper PH, Shmookler BM: Fibrosarcomatous change in dermatofibrosarcoma protuberans. Am J Surg Pathol 12:287–293, 1988.

Malignant Fibrous Histiocytoma

Abdelatif OM, Khankhanian NK, Crosby JH, et al: Malignant fibrous histiocytoma and malignant melanoma: The role of immunohistochemistry and electron microscopy in the differential diagnosis. Mod Pathol 2:477–485, 1989.

Alguacil-Garcia A, Unni KK, Goellner J: Malignant fibrous histiocytoma: An ultrastructural study of six cases. Am J Clin Pathol 69:121–129, 1978.

Alguacil-Garcia A: Malignant fibrous histiocytoma. An electron microscopic study of 17 cases. Virchows Arch [A] 392:135, 1981.

Argenyi ZB, Van Rybroek JJ, Kemp JD, et al: Congenital angiomatoid malignant fibrous histiocytoma. A light-microscopic, immunopathologic, and electron-microscopic study. Am J Dermatopathol 10:59–67, 1988.

Asirwatham JE, Pickren JW: Inflammatory fibrous histiocytoma: Case report. Cancer 41:1467–1471, 1978.

Barr RJ, Wuerker RB, Graham JM: Ultrastructure of atypical fibroxanthoma. Cancer 40:736–743, 1977.

Bertoni F, Capanna R, Biagini R, et al: Malignant fibrous histiocytoma of soft tissue: An analysis of 78 cases located and deeply seated in the extremities. Cancer 56:356–367, 1985.

Churg AM, Khan LB: Myofibroblasts and related cells in malignant fibrous and fibrohistiocytic tumors. Hum Pathol 8:205–218, 1977.

Costa MJ, Weiss SW: Angiomatoid malignant fibrous histiocytoma: A follow-up study of 108 cases with evaluation of possible histologic predictors of outcome. Am J Surg Pathol 14:1126–1132, 1990.

Dahlin DC, Unni KK, Matsuno T: Malignant (fibrous) histiocytoma of bone—fact or fancy? Cancer 39: 1508–1516, 1977.

Dehner LP: Malignant fibrous histiocytoma. Non-specific morphologic pattern, specific pathologic entity, or both? Arch Pathol Lab Med 112:236–237, 1988.

Dickersin GR: The contributions of electron microscopy in the diagnosis and histogenesis of controversial neoplasms. Clin Lab Med 4:123–164, 1984.

duBoulay CEH: Demonstration of alpha-1-antitrypsin and alpha-1-antichymotrypsin in fibrous histiocytoma using immunoperoxidase technique. Am J Surg Pathol 6:559, 1982.

Enjoji M, Hashimoto H, Iwasaki H: Malignant fibrous histiocytoma: A clinicopathologic study of 130 cases. Acta Pathol Jpn 30:727–741, 1980.

Enzinger FM: Angiomatoid malignant fibrous histiocytoma: A distinct fibrohistiocytic tumor of children and young adults simulating a vascular neoplasm. Cancer 44:2147–2157, 1979.

Enzinger FM: Malignant fibrous histiocytoma 20 years after Stout. Am J Surg Pathol 10:43–52, 1986.

Fisher C: The value of electron microscopy and immunohistochemistry in the diagnosis of soft tissue sarcomas: A study of 200 cases. Histopathology 16: 441–454, 1990.

Fletcher CDM: Benign fibrous histiocytoma of subcutaneous and deep soft tissue: A clinicopathologic analysis of 21 cases. Am J Surg Pathol 14:801–809, 1990.

Fletcher CDM: Angiomatoid "malignant fibrous histiocytoma": An immunohistochemical study indicative of myxoid differentiation. Hum Pathol 22:563–568, 1991.

Fletcher CDM: Pleomorphic malignant fibrous histiocytoma: Fact or fiction? A critical reappraisal based on 159 tumors diagnosed as pleomorphic sarcoma. Am J Surg Pathol 16:213–228, 1992.

Fletcher CDM: Angiomatoid fibrous histiocytoma [letters to the editor]. Am J Surg Pathol 16:426–427, 1992.

Fletcher CDM: Fact or fiction? Malignant fibrous histiocytoma [letters to the editor]. Am J Surg Pathol 16: 1023–1024, 1992.

Fu Y, Gabbiani G, Kaye G, et al: Malignant soft tissue tumors of probable histiocytic origin (malignant fibrous histiocytomas): General considerations and electron microscopic and tissue culture studies. Cancer 35:176–198, 1975.

Fukuda T, Tsuneyoshi M, Enjoji M: Malignant fibrous histiocytoma of soft parts: An ultrastructural quantitative study. Ultrastruct Pathol 12:117–129, 1988.

Hirose T, Kudo E, Hasegawa T, et al: Expression of intermediate filaments in malignant fibrous histiocytomas. Hum Pathol 20:871–877, 1989.

Hirose T, Sano T, Hizawa K: Ultrastructural study of the myxoid area of malignant fibrous histiocytomas. Ultrastruct Pathol 12:621–630, 1988.

Hoffman MA, Dickersin GR: Malignant fibrous histiocytoma. An ultrastructural study of eleven cases. Hum Pathol 14:913–922, 1983.

Huvos AG: Primary malignant fibrous histiocytoma of bone. Clinicopathologic study of 18 patients. NY State J Med 76:552–559, 1976.

Imai Y, Yamakawa M, Sato T, et al: Malignant fibrous histiocytoma: Similarities to the "fibrohistiocytoid cells" in chronic inflammation. Virchows Arch [A] 414: 285–298, 1989.

Iwasaki H, Isayama T, Johzaki H, et al: Malignant fibrous histiocytoma. Evidence of perivascular mesenchymal cell origin immunocytochemical studies with monoclonal anti-MFH antibodies. Am J Pathol 128: 528–537, 1987.

Iwasaki H, Yoshitake K, Ohjimi Y, et al: Malignant fibrous histiocytoma: Proliferative compartment and heterogeneity of "histiocytic" cells. Am J Surg Pathol 16: 735–745, 1992.

Jabi M, Jeans D, Dardick I: Ultrastructural heterogeneity in malignant fibrous histiocytoma of soft tissue. Ultrastruct Pathol 11:583–592, 1987.

Katenkamp D, Stiller D: Malignant fibrous histiocytoma of bone. Light microscopic and electron microscopic examination of four cases. Virchows Arch [A] 391: 323–335, 1981.

Kay S: Angiomatoid malignant fibrous histiocytoma: Report of two cases with ultrastructural observations of one case. Arch Pathol Lab Med 109:934–937, 1985.

Kearney MM, Soule EH, Ivins JC: Malignant fibrous histiocytoma: A retrospective study of 167 cases. Cancer 45:167–178, 1980.

Kyriakos M, Kempson RL: Inflammatory fibrous histiocytoma: An aggressive and lethal lesion. Cancer 37:1584–1606, 1976.

Lagace R, Delage C, Seemayer T: Myxoid variant of malignant fibrous histiocytoma: Ultrastructural observations. Cancer 43:526–534, 1979.

Lagace R: The ultrastructural spectrum of malignant fibrous histiocytoma. Ultrastruct Pathol 11:153–159, 1987.

Lattes R: Malignant fibrous histiocytoma. A review article. Am J Surg Pathol 6:761–771, 1982.

Leu HJ, Makek M: Angiomatoid malignant fibrous histiocytoma. Virchows Arch Pathol Anat 395:99–107, 1982.

Leu HJ, Makek M: Angiomatoid malignant fibrous histiocytoma. Arch Pathol Lab Med 110:466–467, 1986.

Limacher J, Delage C, Lagace R: Malignant fibrous histiocytoma. Clinicopathologic and ultrastructural study of 12 cases. Am J Surg Pathol 2:265–274, 1978.

Mackay B: Malignant fibrous histiocytoma [editorial]. Ultrastruct Pathol 12:3, 1988.

Meister P, Nathrath W: Immunohistochemical markers of histiocytic tumors. Hum Pathol 11:300–301, 1980.

Meister P: Malignant fibrous histiocytoma. History, histology, histogenesis. Pathol Res Pract 183:1–7, 1988.

Merkow L, Frich JC Jr, Slifkin M, et al: Ultrastructure of a fibroxanthosarcoma (malignant fibroxanthoma). Cancer 28:372–383, 1971.

Miettinen M, Soini Y: Malignant fibrous histiocytoma. Heterogeneous patterns of intermediate filament proteins by immunohistochemistry. Arch Pathol Lab Med 113:1363–1366, 1989.

O'Brien JE, Stout AP: Malignant fibrous xanthomas. Cancer 17:1445, 1964.

Patel VC, Meyer JE: Retroperitoneal malignant fibrous histiocytoma. Cancer 45:1724–1725, 1980.

Pettinato G, Manivel JC, DeRosa G, et al: Angiomatoid malignant fibrous histiocytoma: Cytologic, immunohistochemical, ultrastructural, and flow cytometric study of 20 cases. Mod Pathol 3:479–487, 1990.

Reddick RL, Michelitch H, Triche TJ: Malignant soft tissue tumors (malignant fibrous histocytoma, pleomorphic liposarcoma, and pleomorphic rhabdomyosarcoma): An electron microscopic study. Hum Pathol 10:327–343, 1979.

Roholl PJ, Kleyne J, Elbers H, et al: Characterization of tumor cells in malignant fibrous histiocytomas and other soft tissue tumors, in comparison with malignant histiocytes. Immunohistochemical study on paraffin section. J Pathol 147:87–95, 1985.

Roholl PJ, Kleyne J, Van Unnik JA: Characterization of tumor cells in malignant fibrous histiocytomas and other soft tissue tumors, in comparison with malignant histiocytes. II. Immunoperoxidase study on cryostat sections. Am J Pathol 121:269–274, 1985.

Rosenberg AE, O'Connell JX, Dickersin GR, et al: Expression of epithelial markers in malignant fibrous histiocytoma of the musculoskeletal system: An immunohistochemical and electron microscopic study. Hum Pathol 24:284–293, 1993.

Rudholm A, Syk I: Malignant fibrous histiocytoma of soft tissue. Correlation between clinical variables and histologic malignancy grade. Cancer 57:2323, 1986.

Seo IS, Frizzera G, Coates TD, et al: Angiomatoid malignant fibrous histiocytoma with extensive lymphadenopathy simulating Castleman's disease. Pediatr Pathol 6(2–3):233–247, 1986.

Silverman JS, Lomvardias S: An unusual soft tumor with features of angiomatoid malignant fibrous histiocytoma composed of bimodal CD34 and factor XIIIa positive dendritic cell subsets. Cd34 and factor XIIIa in angiomatoid MFH. Pathol Res Pract 193:59–60, 1997.

Smith MEF, Costa MJ, Weiss SW: Evaluation of CD68 and other histiocytic antigens in angiomatoid malignant fibrous histiocytoma. Am J Surg Pathol 15:757–763, 1991.

Sun CC Jr, Toker C, Breitenecker R: An ultrastructural study of angiomatoid fibrous histiocytoma. Cancer 49:2103–2111, 1982.

Taxy JB, Battifora H: Malignant fibrous histiocytoma: An electron microscopic study. Cancer 40:254–267, 1977.

Tsuneyoshi M, Enjoji M, Shinohara N: Malignant fibrous histiocytoma. An electron microscopic study of 17 cases. Virchows Arch [A] 392:135–145, 1981.

Wegmann W, Heitz PU: Angiomatoid malignant fibrous histiocytoma: Evidence for the histiocytic origin of tumor cells. Virchows Arch [A] 406:59, 1985.

Weiss SW, Enzinger FM: Myxoid variant of malignant fibrous histiocytoma. Cancer 39:1672–1685, 1977.

Weiss SW, Enzinger FM: Malignant fibrous histiocytoma: An analysis of 200 cases. Cancer 41:2250–2256, 1978.

Weiss SW: Malignant fibrous histiocytoma: A reaffirmation. Am J Surg Pathol 6:773, 1982.

Wood GS, Beckstead JH, Turner RR, et al: Malignant fibrous histiocytoma tumor cells resemble fibroblasts. Am J Surg Pathol 10:323–335, 1986.

Chondrosarcoma

Abenoza P, Neumann MP, Manivel JC, et al: Dedifferentiated chondrosarcoma: An ultrastructural study of two cases, with immunocytochemical correlations. Ultrastruct Pathol 10:529–538, 1986.

Erlandson RA, Huvos AG: Chondrosarcoma: A light and electron microscopic study. Cancer 34:1642–1652, 1974.

Fu YS, Kay S: A comparative ultrastructural study of mesenchymal chondrosarcoma and myxoid chondrosarcoma. Cancer 33:1531–1542, 1974.

Kahn LB: Chondrosarcoma with dedifferentiated foci. A comparative and ultrastructural study. Cancer 37: 1365–1375, 1976.

Levine GD, Bensch KG: Chondroblastoma—the nature of the basic cell. A study by means of histochemistry, tissue culture, electron microscopy, and autoradiography. Cancer 29:1546–1562, 1972.

Smith MT, Farinacci CJ, Carpenter HA, et al: Extraskeletal myxoid chondrosarcoma. A clinicopathologic study. Cancer 37:821–827, 1976.

Steiner GC, Mirra JM, Bullough PG: Mesenchymal chondrosarcoma. A study of the ultrastructure. Cancer 32:926–939, 1973.

Têtu B, Ordonez NG, Ayala AG, et al: Chondrosarcoma with additional mesenchymal component (dedifferentiated chondrosarcoma). II. An immunohistochemical and electron microscopic study. Cancer 58:287–298, 1986.

Ushigome S, Takakuwa T, Shinagawa T, et al: Ultrastructure of cartilaginous tumors and S-100 protein in the tumors, with reference to the histogenesis of chondroblastoma, chondromyxoid fibroma and mesenchymal chondrosarcoma. Acta Pathol Jpn 34:1285–1300, 1984.

Osteosarcoma

Aparisi T, Arborgh B, Ericsson JLE: Studies on the fine structure of osteoblastoma with notes on the localization of nonspecific acid and alkaline phosphatase. Cancer 41:1811–1822, 1978.

Ghadially FN, Mehta PN: Ultrastructure of osteogenic sarcoma. Cancer 25:1457–1467, 1970.

Rao U, Cheng A, Didolker MS: Extraosseous osteogenic sarcoma. Clinicopathological study of eight cases and review of literature. Cancer 41:1488–1496, 1978.

Reddick RL, Michelitch HJ, Levine AM, et al: Osteogenic sarcoma. A study of the ultrastructure. Cancer 45:64–71, 1980.

Reddick RL, Popovsky MA, Fantone JC, et al: Parosteal osteogenic sarcoma. Ultrastructural observations in three cases. Hum Pathol 11:373–380, 1980.

Shapiro F: Ultrastructural observations on osteosarcoma tissue: A study of ten cases. Ultrastruct Pathol 4:151–161, 1983.

Stark A, Aparisi T, Ericsson JLE: Human osteogenic sarcoma: Fine structure of hard tissue areas. Ultrastruct Pathol 8:83–102, 1985.

Stark A, Aparisi T, Ericsson JLE: Human osteogenic sarcoma: Fine structure of the fibroblastic type. Ultrastruct Pathol 7:301–319, 1984.

Stark A, Aparisi T, Ericsson JLE: Human osteogenic sarcoma: Fine structure of the osteoblastic type. Ultrastruct Pathol 4:311–329, 1983.

Steiner GC: Ultrastructure of osteoblastoma. Cancer 39:2127–2136, 1977.

Steiner GC: Ultrastructure of osteoid osteoma. Hum Pathol 7:309–325, 1976.

Williams AH, Schwinn CP, Parker JW: The ultrastructure of osteosarcoma. A review of twenty cases. Cancer 37:1293–1301, 1976.

Synovial Sarcoma

Abenoza P, Manivel JC, Swanson PE, et al: Synovial sarcoma: Ultrastructural study and immunohistochemical analysis by a combined peroxidase-anti-peroxidase/avidin-biotin-peroxidase complex procedure. Hum Pathol 17:1107–1115, 1986.

Akerman M, Willen H, Carlen B, et al: Fine needle aspiration (FNA) of synovial sarcoma—A comparative histological-cytological study of 15 cases, including immunohistochemical, electron microscopic and cytogenetic examination and DNA-ploidy analysis. Cytopathology 7:187–200, 1996.

Alvarez-Fernandez E, Escalona-Zapata J: Monophasic mesenchymal synovial sarcoma: Its identification by tissue culture. Cancer 47:628–635, 1981.

Barland P, Novikoff AB, Hamerman D: Electron microscopy of the human synovial membrane. J Cell Biol 14:207–220, 1962.

Dickersin GR: Synovial sarcoma: A review and update, with emphasis on the ultrastructural characterization of the nonglandular component. Ultrastruct Pathol 15:379–402, 1991.

Dische FE, Darby AJ, Howard ER: Malignant synovioma: Electron microscopical findings in three patients and review of the literature. J Pathol 124:149–155, 1978.

Fernandez BB, Hernandez FJ: Poorly differentiated synovial sarcoma. A light and electron microscopic study. Arch Pathol Lab Med 100:221–223, 1976.

Fisher C: Synovial sarcoma: Ultrastructural and immunohistochemical features of epithelial differentiation in monophasic and biphasic tumors. Hum Pathol 17:996–1008, 1986.

Gabbiani G, Kaye GI, Lattes R, et al: Synovial sarcoma. Electron microscopic study of a typical case. Cancer 28:1031–1039, 1971.

Ghadially FN: *Fine Structure of Synovial Joints*. Butterworths, London, 1983.

Krall RA, Kostianovsky M, Patchefsky AS: Synovial sarcoma. A clinical, pathological, and ultrastructural study of 26 cases supporting the recognition of a monophasic variant. Am J Surg Pathol 5:137–151, 1981.

Kubo T: A note on the fine structure of synovial sarcoma. Acta Pathol Jpn 24:163–168, 1974.

Lombardi L, Rilke F: Ultrastructural similarities and differences of synovial sarcoma, epithelioid sarcoma, and clear cell sarcoma of the tendons and aponeuroses. Ultrastruct Pathol 6:209–219, 1984.

Luse SA: A synovial sarcoma studied by electron microscopy. Cancer 13:312–322, 1960.

Mickelson MR, Brown GA, Maynard JA, et al: Synovial sarcoma. An electron microscopic study of monophasic and biphasic forms. Cancer 45:2109–2118, 1980.

Nielsen GP, Shaw PA, Rosenberg AE, et al: Synovial sarcoma of the vulva: A report of two cases. Mod Pathol 9:970–974, 1996.

Roth JA, Enzinger FM, Tannenbaum M: Synovial sarcoma of the neck: A follow-up study of 24 cases. Cancer 35:1243–1253, 1975.

Sajjad SM, Mackay B: Hyaline inclusions in a synovial sarcoma following intraarterial chemotherapy. Ultrastruct Pathol 3:313–318, 1982.

Schmidt D, Mackay B: Ultrastructure of human tendon sheath and synovium: Implications for tumor histogenesis. Ultrastruct Pathol 3:269–283, 1982.

Zeren H, Moran CA, Suster S, et al: Primary pulmonary sarcomas with features of monophasic synovial sarcoma: A clinicopathological, immunohistochemical, and ultrastructural study of 25 cases. Hum Pathol 26: 474–480, 1995.

Liposarcoma

Battifora H, Nunez-Alonso C: Myxoid liposarcomas: Study of ten cases. Ultrastruct Pathol 1:157–169, 1980.

Bolwen JW, Thorning D: Benign lipoblastoma and myxoid liposarcoma: A comparative light- and electron-microscopical study. Am J Surg Pathol 4:163–174, 1980.

Cinti S, Cigolini M, Bosello O, et al: A morphological study of the adipocyte precursor. J Submicrosc Cytol 16:243–251, 1984.

Desai U, Ramos CV, Taylor HB: Ultrastructural observations in pleomorphic liposarcoma. Cancer 42: 1284–1290, 1978.

Dickersin GR: The contributions of electron microscopy in the diagnosis and histogenesis of controversial neoplasms. Clin Lab Med 4:123–164, 1984.

Feldman PS: A comparative study including ultrastructure of intramuscular myxoma and myxoid liposarcoma. Cancer 43:512–525, 1979.

Fu YS, Parker FG, Kaye GI, et al: Ultrastructure of benign and malignant adipose tissue tumors. Pathol Annu 15:67–89, 1980.

Gould VE, Jao W, Gould NS, et al: Electron microscopy of adipose tissue tumors comparative features of hibernomas, myxoid and pleomorphic liposarcomas. Pathobiol Annu 9:339–357, 1979.

Iyama K, Ohzono U, Usuku G: Electron microscopical studies on the genesis of white adipocytes: Differentiation of immature pericytes into adipocytes in transplanted pre-adipose tissue. Virchows Arch [B] 31: 143–155, 1979.

Kindblom LG, Save-Soderbergh J: The ultrastructure of liposarcoma: A study of 10 cases. Acta Pathol Microbiol Scand [A] 87:109–121, 1979.

Lagace R, Jacob S, Seemayer TA: Myxoid liposarcoma: An electron microscopic study. Biological and histogenetic considerations. Virchows Arch [A] 384:159–172, 1979.

Napolitano L: The differentiation of white adipose cells: An electron microscopic study. J Cell Biol 18:663–679, 1963.

Rossouw DJ, Med M, Cinti S, et al: Liposarcoma: An ultrastructural study of 15 cases. Am J Clin Pathol 85: 649–667, 1986.

Leiomyosarcoma

Bures JC, Barnes L, Mercer D: A comparative study of smooth muscle tumors utilizing light and electron microscopy, immunocytochemical staining and enzymatic assay. Cancer 48:2420–2426, 1981.

Chen KTK, Ma CK: Intravenous leiomyoblastoma. Am J Surg Pathol 7:591–596, 1983.

Cornog JI: Gastric leiomyoblastoma. A clinical and ultrastructural study. Cancer 34:711–719, 1974.

Dickersin GR, Selig MK, Park YN: The many faces of smooth muscle neoplasms in a gynecological sampling:

An ultrastructural study. Ultrastruct Pathol 21:109–134, 1997.

Hagan C, Zukerberg LR, Bhan AK, et al: Epithelioid "smooth muscle" tumors: An immunohistochemical and ultrastructural profile. Lab Invest 19:5A, 1991.

Kay S, Still WJS: A comparative electron microscopic study of a leiomyosarcoma and bizarre leiomyoma (leiomyoblastoma) of the stomach. Am J Clin Pathol 52: 403–413, 1969.

Mazur MT, Clark HB: Gastric stromal tumors: Reappraisal of histogenesis. Am J Surg Pathol 7:507–519, 1983.

Murata Y, Tsuji M, Tani M: Ultrastructure of multiple glomus tumor. J Cutan Pathol 11:53–58, 1984.

Nevalainen TJ, Linna MI: Ultrastructure of gastric leiomyosarcoma. Virchows Archiv [A] 379:25–33, 1978.

Pulitzer DR, Martin PC, Reed RJ: Epithelioid glomus tumor. Hum Pathol 26:1022–1027, 1995.

Salazar H, Totten RS: Leiomyoblastoma of the stomach. An ultrastructural study. Cancer 25:176–185, 1970.

Shmookler BM, Lauer DH: Retroperitoneal leiomyosarcoma. A clinicopathologic analysis of 36 cases. Am J Surg Pathol 7:269–280, 1983.

Tsuneyoshi M, Enjoji M: Glomus tumor. A clinicopathologic and electron microscopic study. Cancer 50:1601–1607, 1982.

Tworek JA, Appleman HD, Singleton TP, et al: Stromal tumors of the jejunum and ileum. Mod Pathol 10: 200–209, 1997.

Weiss R, Mackay B: Malignant smooth muscle tumors of the gastrointestinal tract: Ultrastructural study of 20 cases. Ultrastruct Pathol 2:231–240, 1981.

Welsh RA, Meyer AT: Ultrastructure of gastric leiomyoma. Arch Pathol 87:71–81, 1969.

Zukerberg LR, Cinti S, Dickersin GR: Mitochondria as a feature of smooth muscle differentiation: A study of 70 smooth muscle tumors. J Submicrosc Cytol Pathol 22:335–344, 1990.

Rhabdomyosarcoma

Agamanolis DP, Dasu S, Krill CE: Tumors of skeletal muscle. Hum Pathol 17:778–795, 1986.

Dehner LP, Enzinger FM, Font RL: Fetal rhabdomyoma. An analysis of nine cases. Cancer 30:160–166, 1972.

Gold JH, Bossen EH: Benign vaginal rhabdomyoma. A light and electron microscopic study. Cancer 37: 2283–2294, 1976.

Hay ED: Electron microscopic observations of muscle dedifferentiation in regenerating amblystoma limbs. Dev Biol 1:555–585, 1959.

Hay ED: The fine structure of differentiating muscle in the salamander tail. Z Zellforsch Mikrosk Anat 59:6–34, 1963.

Konrad EA, Meister P, Hubner G: Extracardiac rhabdomyoma. Report of different types with light microscopic and ultrastructural studies. Cancer 49:898–907, 1982.

Miller R, Kurtz SM, Powers JM: Mediastinal rhabdomyoma. Cancer 42:1983–1988, 1978.

Patterson HC, Dickersin GR, Pilch BZ, et al: Hamartoma of the hypopharynx. Arch Otolaryngol 107:767–772, 1981.

Prince FP: Ultrastructural aspects of myogenesis found in neoplasms. Acta Neuropathol (Berl) 43:315–320, 1981.

Scrivner D, Meyer JS: Multifocal recurrent adult rhabdomyoma. Cancer 46:790–795, 1980.

Silverman JF, Kay S, McCue CM, et al: Rhabdomyoma of the heart. Ultrastructural study of three cases. Lab Invest 35:596–606, 1976.

Silverman JK, Kay S, Chang CH: Ultrastructural comparison between skeletal muscle and cardiac rhabdomyomas. Cancer 42:189–193, 1978.

Vye MV: The ultrastructure of striated muscle. Ann Clin Lab Sci 6:142–151, 1976.

Whitten RO, Benjamin DR: Rhabdomyoma of the retroperitoneum. A report of a tumor with both adult and fetal characteristics: A study by light and electron microscopy, histochemistry, and immunohistochemistry. Cancer 59:818–824, 1987.

Angiomatous Lesions, Kaposi's Sarcoma

Akhtar M, Bunuan H, Ali MA, et al: Kaposi's sarcoma in renal transplant recipients. Ultrastructural and immunoperoxidase study of four cases. Cancer 53: 258–266, 1984.

Angervall L, Kindblom LG, Karlsson K, et al: Atypical hemangioendothelioma of venous origin. A clinicopathologic, angiographic, immunohistochemical, and ultrastructural study of two endothelial tumors within the concept of histiocytoid hemangioma. Am J Surg Pathol 9:504–516, 1985.

Arnold G, Klein PJ, Fischer R: Epithelioid hemangioendothelioma. Report of a case with immunolectinhistochemical and ultrastructural demonstration of its vascular nature. Virchows Arch [A] 408:435–443, 1986.

Beckstead JH, Wood GS, Fletcher V: Evidence for the origin of Kaposi's sarcoma from lymphatic endothelium. Am J Pathol 119:294–300, 1985.

Capo V, Ozello L, Fenoglio CM, et al: Angiosarcomas arising in edematous extremities: Immunostaining for factor VIII-related antigen and ultrastructural features. Hum Pathol 16:144–150, 1985.

Carstens PHB: The Weibel-Palade body in the diagnosis of endothelial tumors. Ultrastruct Pathol 2:315–325, 1981.

de Araujo VC, Marcucci G, Sesso A, et al: Epithelioid hemangioendothelioma of the gingiva: Case report and ultrastructural study. Oral Surg Oral Med Oral Pathol 63:472–477, 1987.

Ehlers N, Jensen OA: Juxtapapillary retinal hemangioblastoma (angiomatosis retinae) in an infant: Light microscopical and ultrastructural examination. Ultrastruct Pathol 3:325–333, 1982.

Gibbs AR, Johnson NF, Giddings JC, et al: Primary angiosarcoma of the mediastinum: Light and electron microscopic demonstration of factor VIII-related antigen in neoplastic cells. Hum Pathol 15:687–691, 1984.

Gindhardt TD, Tucker WY, Choy SH: Cavernous hemangioma of the superior mediastinum. Report of a case with electron microscopy and computerized tomography. Am J Surg Pathol 3:353–361, 1979.

Gokel JM, Kurzl R, Hubner G: Fine structure and origin of Kaposi's sarcoma. Pathol Eur 11:45–47, 1976.

Lagace R, Leroy J-P: Comparative electron microscopic study of cutaneous and soft tissue angiosarcomas, postmastectomy angiosarcoma (Stewart-Treves syndrome) and Kaposi's sarcoma. Ultrastruct Pathol 11:161–173, 1987.

Leu HJ, Odermatt B: Multicentric angiosarcoma (Kaposi's sarcoma). Light and electron microscopic and immunohistochemical findings of idiopathic cases in Europe and Africa and of cases associated with AIDS. Virchows Arch [A] 408:29–41, 1985.

Llombart-Bosch A, Peydro-Olaya A, Pellin A: Ultrastructure of vascular neoplasms. A transmission and scanning electron microscopical study based upon 42 cases. Pathol Res Pract 174:1–41, 1982.

McNutt NS, Fletcher V, Conant MA: Early lesions of Kaposi's sarcoma in homosexual men. An ultrastructural comparison with other vascular proliferations in skin. Am J Pathol 111:62–77, 1983.

Miettinen M, Lehto VP, Virtanen I: Post-mastectomy angiosarcoma (Stewart-Treves syndrome). Light-microscopic, immunohistological, and ultrastructural characteristics of two cases. Am J Surg Pathol 7:329–339, 1983.

Pins MR, Mankin HJ, Xavier RJ, et al: Malignant epithelioid hemangioendothelioma of the tibia associated with a bone infarct in a patient with Gaucher disease. A case report. J Bone Joint Surg 77:777–781, 1995.

Rosai J, Sumner HW, Kostianovsky M, et al: Angiosarcoma of the skin. A clinicopathologic and fine structural study. Hum Pathol 7:83–109, 1976.

Scoazec JY, Degott C, Reynes M, et al: Epithelioid hemangioendothelioma of the liver: An ultrastructural study. Hum Pathol 20:673–681, 1989.

Silverberg SG, Kay S, Koss LG: Post-mastectomy lymphangiosarcoma: Ultrastructural observations. Cancer 27:100–108, 1971.

Steiner GC, Dorfman HD: Ultrastructure of hemangioendothelial sarcoma of bone. Cancer 29:122–140, 1972.

Taxy JB, Gray SR: Cellular angiomas of infancy. An ultrastructural study of two cases. Cancer 43:2322–2331, 1979.

Tomec R, Ahmad I, Fu YS, et al: Malignant hemangioendothelioma (angiosarcoma) of the salivary gland. An ultrastructural study. Cancer 43:1664–1671, 1979.

Tsuneyoshi M, Dorfman HD, Bauer TW: Epithelioid hemangioendothelioma of bone. A clinicopathologic, ultrastructural, and immunohistochemical study. Am J Surg Pathol 10:754–764, 1986.

van Haelst UJ, Pruszczynski M, ten Cate LN, et al: Ultrastructural and immunohistochemical study of epithelioid hemangioendothelioma of bone: Coexpression of epithelial and endothelial markers. Ultrastruct Pathol 14:141–149, 1990.

Waldo ED, Vuletin JC, Kaye GI: The ultrastructure of vascular tumors: Additional observations and a review of the literature. Pathol Annu 2:279–308, 1977.

Weiss SW, Enzinger FM: Epithelioid hemangioendothelioma. A vascular tumor often mistaken for a carcinoma. Cancer 50:970–981, 1982.

Weiss SW, Enzinger FM: Spindle cell hemangioendothelioma. A low-grade angiosarcoma resembling a cavernous hemangioma and Kaposi's sarcoma. Am J Surg Pathol 10:521–530, 1986

Hemangiopericytoma

Batsakis JG, Jacobs JB, Templeton AC: Hemangiopericytoma of the nasal cavity: Electron-optic study and clinical correlations. J Laryngol Otol 97:361–368, 1983.

Battifora H: Hemangiopericytoma: Ultrastructural study of five cases. Cancer 31:1418–1432, 1973.

Dickersin GR: The contributions of electron microscopy in the diagnosis and histogenesis of controversial neoplasms. Clin Lab Med 4:123–164, 1984.

Eichorn JH, Dickersin GR, Bhan AK, et al: Sinonasal hemangiopericytoma. A reassessment with electron microscopy, immunohistochemistry and long-term follow-up. Am J Surg Pathol 14:856–866, 1990.

Kuhn C, Rosai J: Tumors arising from pericytes. Ultrastructure and organ culture of a case. Arch Pathol 88: 653–663, 1969.

Lagace R, Delage C, Seemayer T: Myxoid variant of malignant fibrous histiocytoma: Ultrastructural observations. Cancer 43:526–534, 1979.

Nappi O, Ritter JH, Pettinato G, et al: Hemangiopericytoma: Histopathological pattern or clinicopathologic entity [review]? Semin Diagn Pathol 12:221–232, 1995.

Nielsen GP, Dickersin GR, Provenzal JM, et al: Lipomatous hemangiopericytoma. A histologic, ultrastructural and immunohistochemical study of a unique variant of hemangiopericytoma. Am J Surg Pathol 19: 748–756, 1995.

Ramsey HJ: Fine structure of hemangiopericytoma and hemangioendothelioma. Cancer 19:2005–2018, 1966.

Schwannoma, Neurofibroma, and Perineurioma

Abenoza P, Sibley RK: Granular cell myoma and schwannoma: Fine structural and immunohistochemical study. Ultrastruct Pathol 11:19–28, 1987.

Alguacil-Garcia A, Unni KK, Goellner JR: Histogenesis of dermatofibrosarcoma protuberans: An ultrastructural study. Am J Clin Pathol 69:427–434, 1978.

Aronson PJ, Fretzin DF, Potter BS: Neurothelioma of Gallager and Helwig (dermal nerve sheath myzoma variant): Report of a case with electron microscopic and immunohistochemical studies. J Cutan Pathol 12: 506–519, 1985.

Bedetti CD, Martinez AJ, Beckford NS, et al: Granular cell tumor arising in myelinated peripheral nerves. Light and electron microscopy and immunoperoxidase study. Virchows Arch [A] 402:175–183, 1983.

Bilboa JM, Khoury JS, Hudson AR, et al: Perineurioma (localized hypertrophic neuropathy). Arch Pathol Lab Med 108:557–560, 1984.

Chen KTK, Latorraca R, Fabich D, et al: Malignant schwannoma. A light microscopic and ultrastructural study. Cancer 45:1585–1593, 1980.

Chitale AR, Dickersin GR: Electron microscopy in the diagnosis of malignant schwannomas. A report of six cases. Cancer 51:1448–1461, 1983.

Damiani S, Koerner FC, Dickersin GR, et al: Granular cell tumour of the breast. Virchows Arch [A] 420: 219–226, 1992.

Dickersin GR: The electron microscopic spectrum of nerve sheath tumors. Ultrastruct Pathol 11:103–146, 1987.

Dupree WB, Langloss JM, Weiss SW: Pigmented dermatofibrosarcoma protuberans (Bednar tumor). A pathologic, ultrastructural, and immunohistochemical study. Am J Surg Pathol 9:630–639, 1985.

Erlandson RA: Peripheral nerve sheath tumors. Ultrastruct Pathol 9:113–122, 1985.

Erlandson RA, Woodruff JM: Peripheral nerve sheath tumors: An electron microscopic study of 43 cases. Cancer 49:273–287, 1982.

Finkel G, Lane B: Granular cell variant of neurofibromatosis: Ultrastructure of benign and malignant tumors. Hum Pathol 13:959–963, 1982.

Fisher ER, Vuzevski VD: Cytogenesis of schwannoma (neurilemoma), neurofibroma, dermatofibroma and dermatofibrosarcoma as revealed by electron microscopy. Am J Clin Pathol 49:141–153, 1968.

Fisher ER, Wecksler H: Granular cell myoblastoma—A misnomer. Electron microscopic and histochemical evidence concerning its Schwann cell derivation and nature (granular cell schwannoma). Cancer 15:936–943, 1962.

Goldstein J, Lifshitz T: Myxoma of the nerve sheath. Report of three cases, observations by light and electron microscopy and histochemical analysis. Am J Dermatopathol 7:423–429, 1985.

Hashimoto K, Brownstein MH, Jakobiec FA: Dermatofibrosarcoma protuberans: A tumor with perineural and endoneural features. Arch Dermatol 110: 874–885, 1974.

Hirose T, Sano T, Hizawa K: Ultrastructural localization of S-100 protein in neurofibroma. Acta Neuropathologica 69:103–110, 1986.

Klima M, Peters J: Malignant granular cell tumor. Arch Pathol Lab Med 111:1070–1073, 1987.

Lassmann H, Jurecka W, Lassmann G, et al: Different types of benign nerve sheath tumors. Light microscopy, electron microscopy and autoradiography. Virchows Arch [A] 375:197–210, 1977.

Lazarus SS, Trombetta LD: Ultrastructural identification of a benign perineurial cell tumor. Cancer 41:1823–1829, 1978.

Mittal KR, True LD: Origin of granular cell tumor. Intracellular myelin formation with autodigestion. Arch Pathol Lab Med 112:302–303, 1988.

Nakazato Y, Ishizeki J, Takahashi K, et al: Immunohistochemical localization of S-100 protein in granular cell myoblastoma. Cancer 49:1624–1628, 1982.

Peters A, Palay SL, Webster HdeW: *The Fine Structure of the Nervous System.* WB Saunders, Philadelphia, 1976, pp 181–230.

Stefansson K, Wollmann RL: S-100 protein in granular cell tumors (granular cell myoblastomas). Cancer 49: 1834–1838, 1982.

Taxy JB, Battifora H: Epithelioid schwannoma: Diagnosis by electron microscopy. Ultrastruct Pathol 2:19–24, 1981.

Taxy JB, Battifora H, Trujillo V, et al: Electron microscopy in the diagnosis of malignant schwannoma. Cancer 48:1381–1391, 1981.

Troncoso P, Ordonez NG, Raymond AK, et al: Malignant granular cell tumor. Immunocytochemical and ultrastructural observations. Ultrastruct Pathol 12: 137–144, 1988.

Ushigome S, Takakuwa T, Hyuga M, et al: Perineurial cell tumor and the significance of the perineurial cells in neurofibroma. Acta Pathol Jpn 36:973–987, 1986.

Uzoaru I, Firfer B, Ray V, et al: Malignant granular cell tumor. Arch Pathol Lab Med 116:206–208, 1992.

Waggener JD: Ultrastructure of benign peripheral nerve sheath tumors. Cancer 19:699–709, 1966.

Webb JN: The histogenesis of nerve sheath myxoma: Report of a case with electron microscopy. J Pathol 127:35–37, 1979.

Weiser G: An electron microscopic study of "Pacinian neurofibroma." Virchows Arch [A] 366:331–340, 1975.

Weiser G: Granularzelltumor (Granulares Neurom Feyrter) und Schwannsche Phagen. Virchows Arch [A] 380:49–57, 1978.

Weiser G: Neurofibrom and Perineuralzelle. Electronoptische Untersuchung an 9 Neurofibromen. Virchows Arch [A] 379:73–83, 1978.

Woodruff JM, Godwin TA, Erlandson RA, et al: Cellular schwannoma. A variety of schwannoma sometimes mistaken for a malignant tumor. Am J Pathol 5:733–744, 1981.

Woodruff JM, Marshall ML, Godwin TA, et al: Plexiform (multinodular) schwannoma. A tumor simulating the plexiform neurofibroma. Am J Surg Pathol 7: 691–697, 1983.

Yousem SA, Colby TV, Urich H: Malignant epithelioid schwannoma arising in a benign schwannoma. Cancer 55:2799–2803, 1985.

Sarcomatoid (Spindle Cell) Carcinoma

Balercia G, Bhan AK, Dickersin GR: Sarcomatoid carcinoma: An ultrastructural study with light microscopic and immunohistochemical correlation of 10 cases from various anatomic sites. Ultrastruct Pathol 19:249–263, 1995.

Battifora H: Spindle cell carcinoma: Ultrastructural evidence of squamous origin and collagen production by the tumor cells. Cancer 37:2275–2282, 1976.

Bird DJ, Semple JP, Seiler MW: Sarcomatoid renal cell carcinoma metastatic to the heart: Report of a case. Ultrastruct Pathol 15:361–366, 1991.

Deitchman B, Sidhu GS: Ultrastructural study of a sarcomatoid variant of renal cell carcinoma. Cancer 46: 1152–1157, 1980.

Dickersin GR, Erlandson RA: Controversial spindle cell tumors: An overview. Ultrastruct Pathol 15:315–316, 1991.

Humphrey PA, Scroggs MW, Roggli VL, et al: Pulmonary carcinomas with a sarcomatoid element: An immunocytochemical and ultrastructural analysis. Hum Pathol 19:155–165, 1988.

Lichtiger B, Mackay B, Tessmer CF: Spindle-cell variant of squamous carcinoma: A light and electron microscopic study of 13 cases. Cancer 26:1311–1320, 1970.

Morikawa Y, Tohya K, Kusuyama Y, et al: Sarcomatoid renal cell carcinoma: An immunohistochemical and ultrastructural study. Int Urol Nephrol 25:51–58, 1993.

Suster S, Huszar M, Herczeg E: Spindle cell squamous carcinoma of the lung: Immunocytochemical and ultrastructural study of a case. Histopathology 11: 871–878, 1987.

Torenbeek R, Blomjous CEM, de Bruin PC, et al: Sarcomatoid carcinoma of the urinary bladder: Clinicopathologic analysis of 18 cases with immunohistochemical and electron microscopic findings. Am J Surg Pathol 18:241–249, 1994.

Zarbo RJ, Crissman J, Venkat H, et al: Spindle-cell carcinoma of the upper aerodigestive tract mucosa: An immunohistologic and ultrastructural study of 18 biphasic tumors and comparisons with seven monophasic spindle-cell tumors. Am J Surg Pathol 10:741–753, 1986.

7

Gonadal and Related Neoplasms

Surface Epithelial–Stromal Tumors of the Ovary

Serous Tumors

(Figures 7.1 through 7.14.)

Diagnostic criteria. Epithelial cells lining (1) lumens and papillae and having (2) microvilli, (3) cilia, (4) junctional complexes, (5) basal lamina, (6) glycogen, and (7) secretory granules; stromal cells in varying numbers, with (8) fibroblastic and/or myofibroblastic differentiation, (9) lipid vacuoles (variable), (10) smooth endoplasmic reticulum (variable) and (11) tubular mitochondrial cristae (variable).

Additional points. The serous neoplasms include benign, borderline, and malignant cystadenomas; papillary cystadenomas; adenofibromas; cystadenofibromas, and surface papillary tumors. The frequency of the epithelial diagnostic criteria depends to some degree on the level of glandular or papillary differentiation of the neoplasm. For example, most benign serous tumors are lined by ciliated epithelium, although diminution in the number of ciliated cells may also be related to pressure within cystic portions of a neoplasm. Cilia are characteristic of fallopian tubal differentiation (Figure 7.1), and because they are not a component of the coelomic mesothelial lining cells from which serous tumors are ultimately derived, they are considered to be a metaplastic phenomenon in these tumors. The simplest and most common examples of this metaplastic process are the surface epithelial inclusion glands and cysts (Figures 7.2 through 7.5) from which the epithelial component of the surface epithelial–stromal tumors directly derive. In addition to glycogen and secretory granules, several nonspecific cytoplasmic structures are found in varying amounts in the epithelial cells of serous cysts and neoplasms (Figures 7.6 through 7.11). These structures include microfilaments, Golgi apparatuses, and lipid droplets. Calcific deposits, including psammoma bodies, are often present, especially in the malignant serous tumors. Poorly differentiated tumors may show irregularly shaped and slit-like glands and complex papillarity with budding of epithelium.

The stroma between the epithelial components, in polypoid excrescences of serous tumors, and in other surface epithelial–stromal tumors is composed of dense or edematous collagen, with fibroblastic cells showing varying degrees of smooth muscle differentiation (myofibroblasts; Figures 7.12 through 7.14). The stromal cells are derived from ovarian stroma and may be steroid-secreting, showing prominence of lipid and smooth endoplasmic reticulum as well as tubular cristae in mitochondria.

(Text continues on page 406)

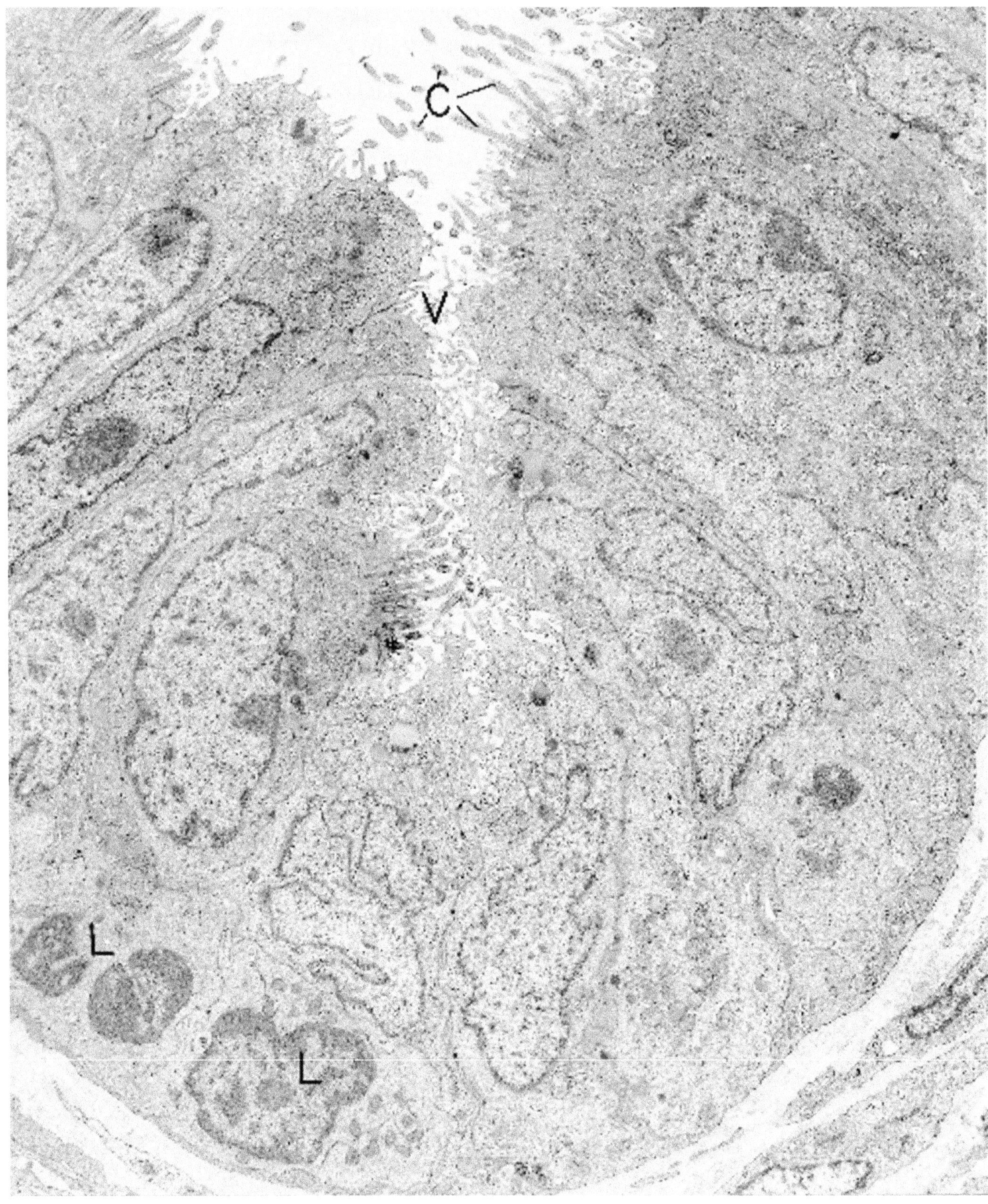

Figure 7.1. Normal fallopian tube. The epithelial lining cells are of a tall columnar type, with a florid array of microvilli (V) and cilia (C) on their luminal surface. L = lymphocytes. (× 6000)

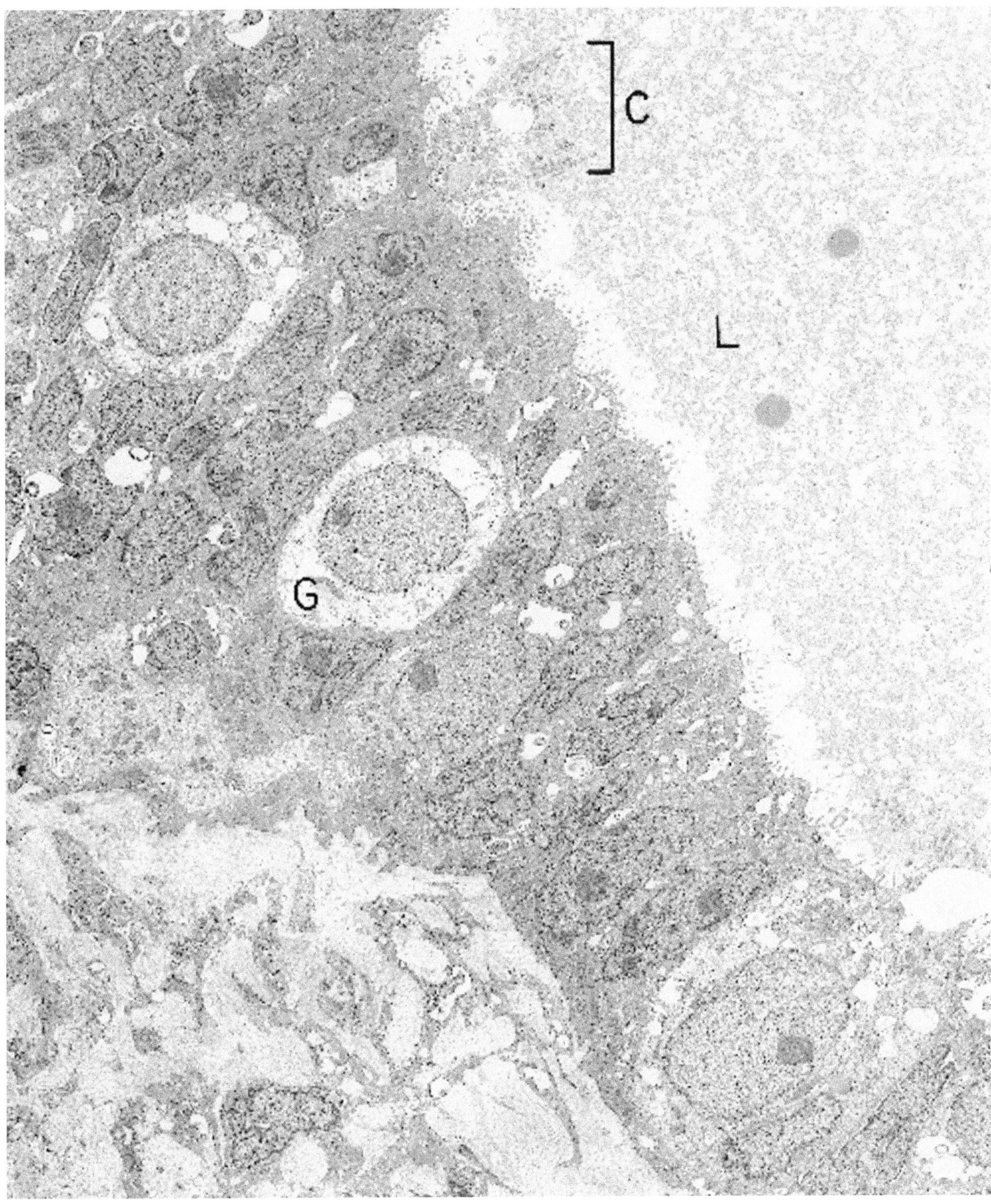

Figure 7.2. Inclusion cyst (ovary). The simplest of all the ovarian lesions derived from peritoneal surface epithelium is characterized by a central fluid-filled lumen (L) and a lining of columnar epithelial cells with microvilli and cilia (C). Some of the cells have a clear cytoplasm, indicative of glycogen (G). (× 3740)

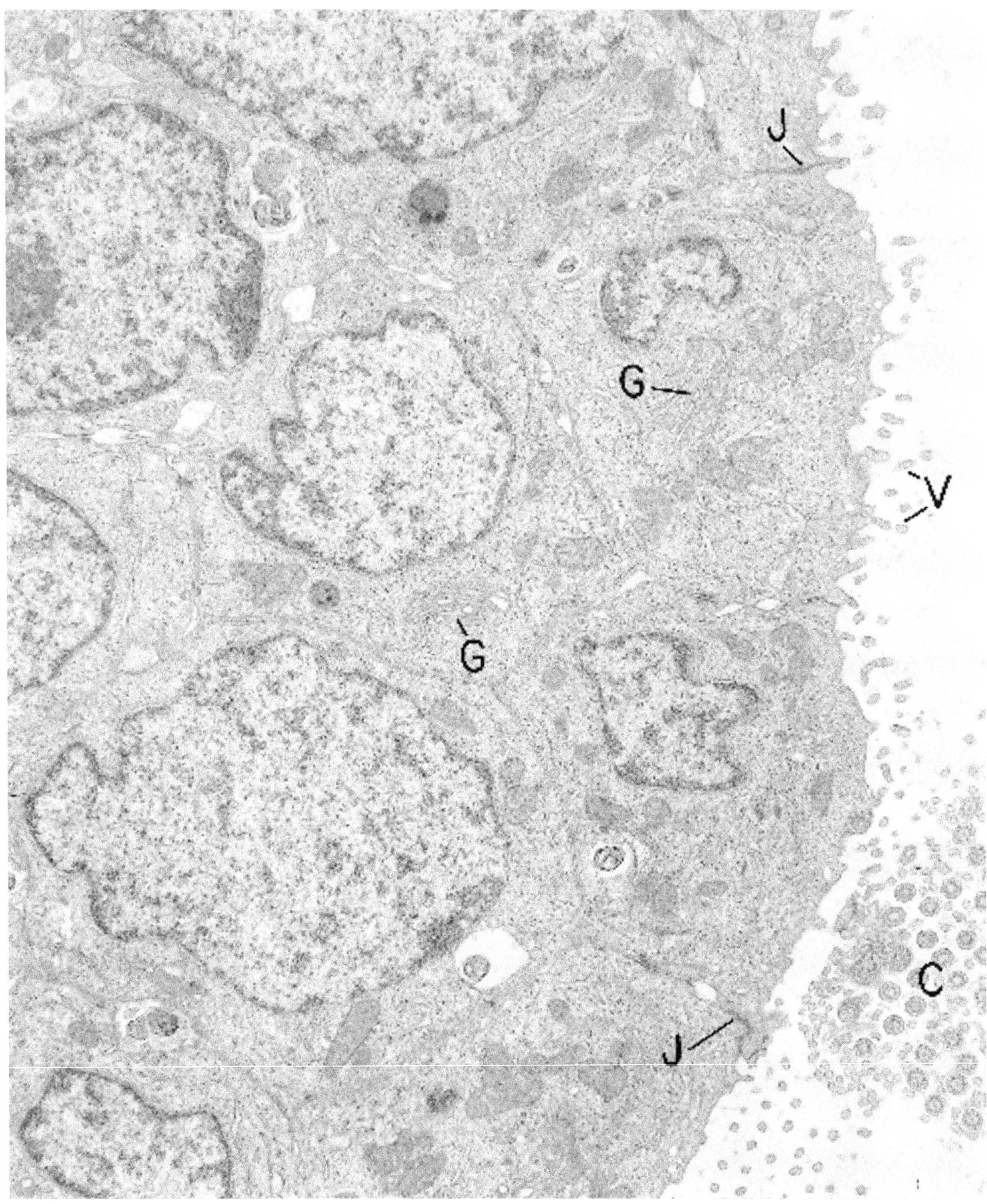

Figure 7.3. Inclusion cyst (ovary). The epithelial lining cells have many microvilli (V), and some cells have cilia (C). Intercellular junctions and junctional complexes (J) are prominent, and Golgi apparatuses (G) are large. (× 13,500)

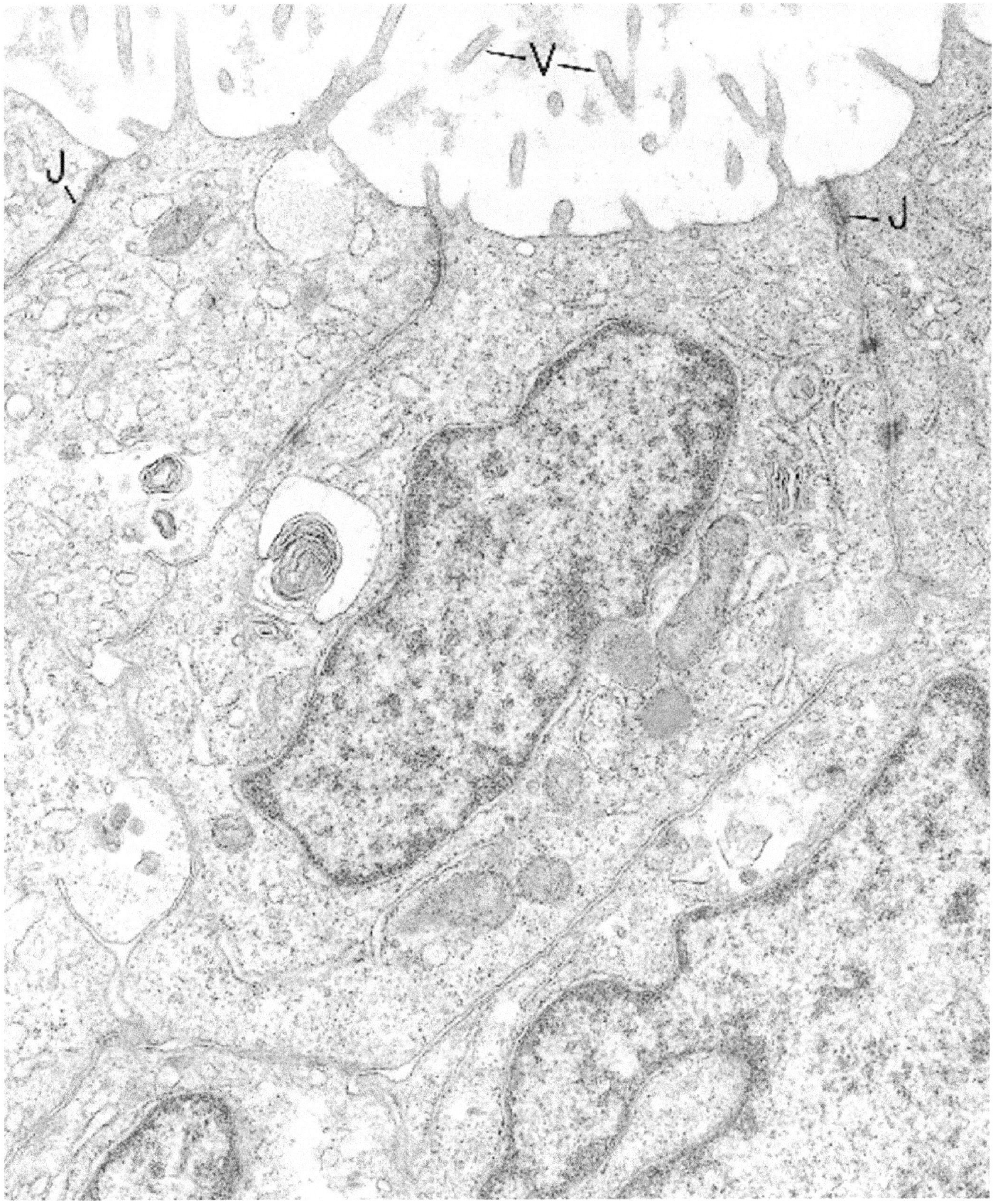

Figure 7.4. Inclusion cyst (ovary). High magnification of several lining cells illustrates long, thin microvilli (V), junctions and junctional complexes (J), and nonspecific cytoplasmic contents. (× 27,800)

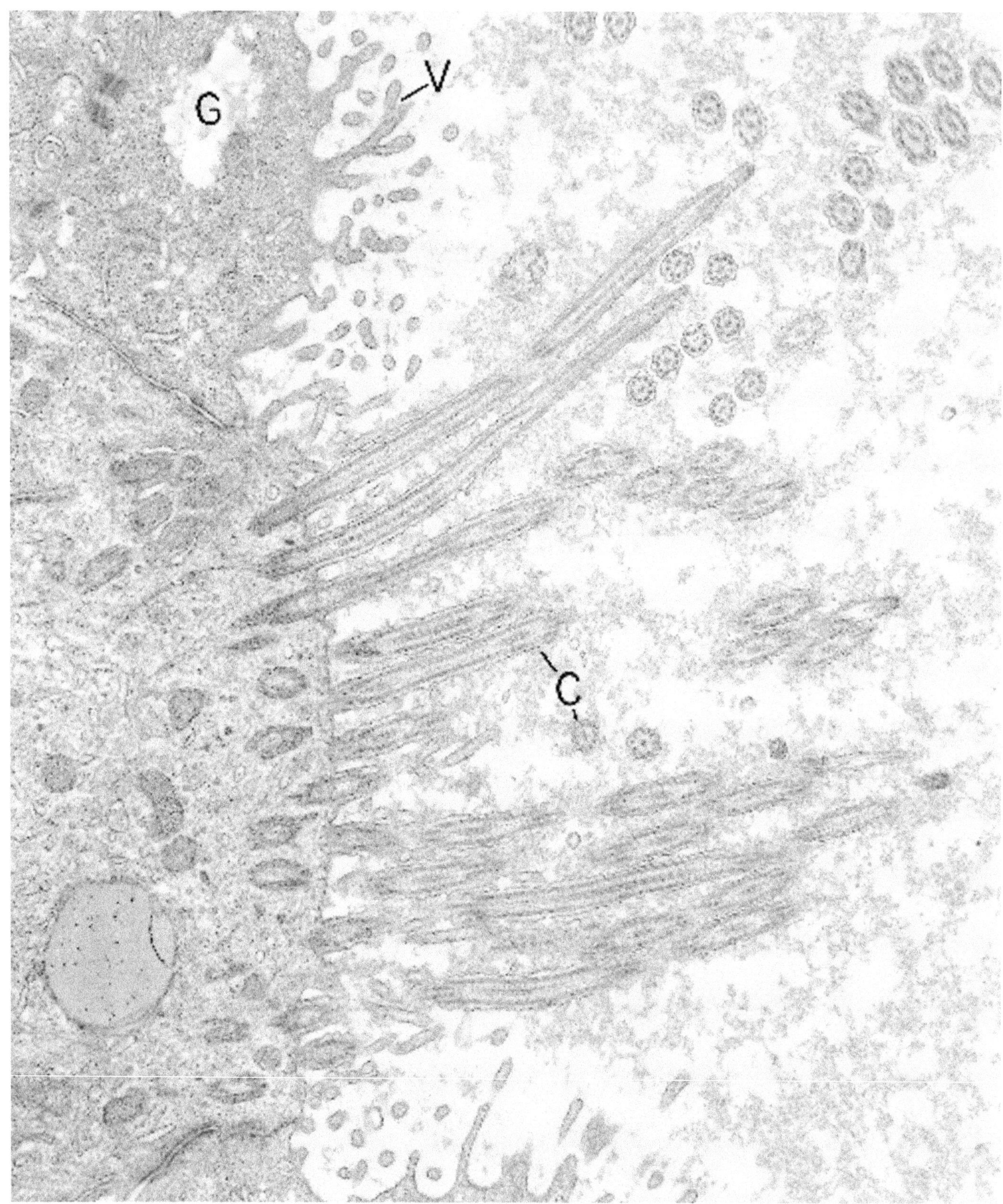

Figure 7.5. Inclusion cyst (ovary). High magnification of a ciliated lining cells highlights some of the internal structure of the cilia and basal bodies as well as the contrast in size and architecture between cilia (C) and microvilli (V). A pocket of glycogen (G) is visible in the apical cytoplasm of one cell. (× 21,900)

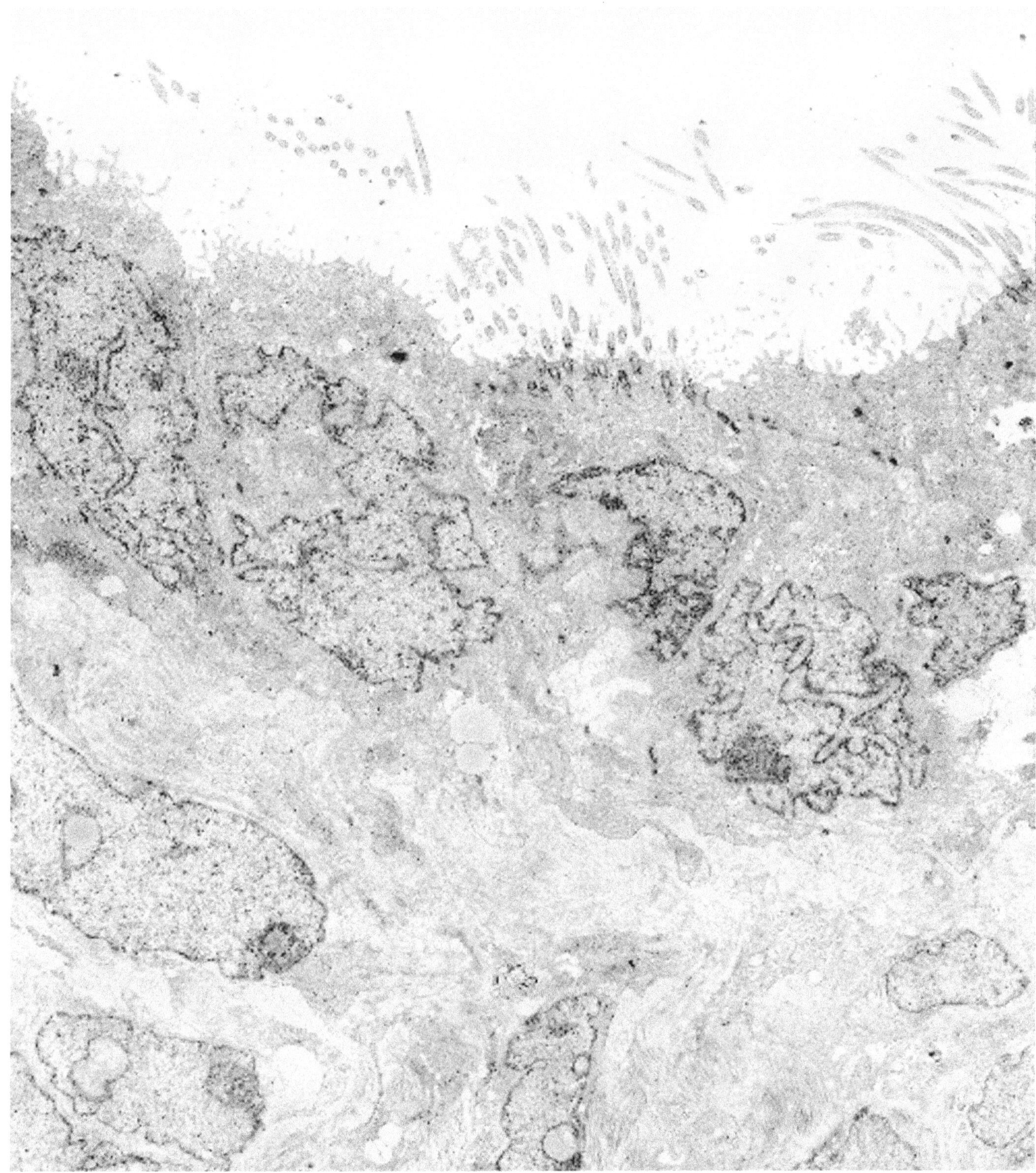

Figure 7.6. Papillary serous cystadenoma (ovary). The epithelium of this well-differentiated neoplasm is villous and ciliated and similar to the cells of the ovarian inclusion cysts (Figures 7.2 through 7.5). However, nuclei are more indented and irregular in shape. (× 6100) (Permission for reprinting granted by WB Saunders, Dickersin GR: The ultrastructure of selected gynecologic neoplasms. Clin Lab Med 7:117–156, 1987.)

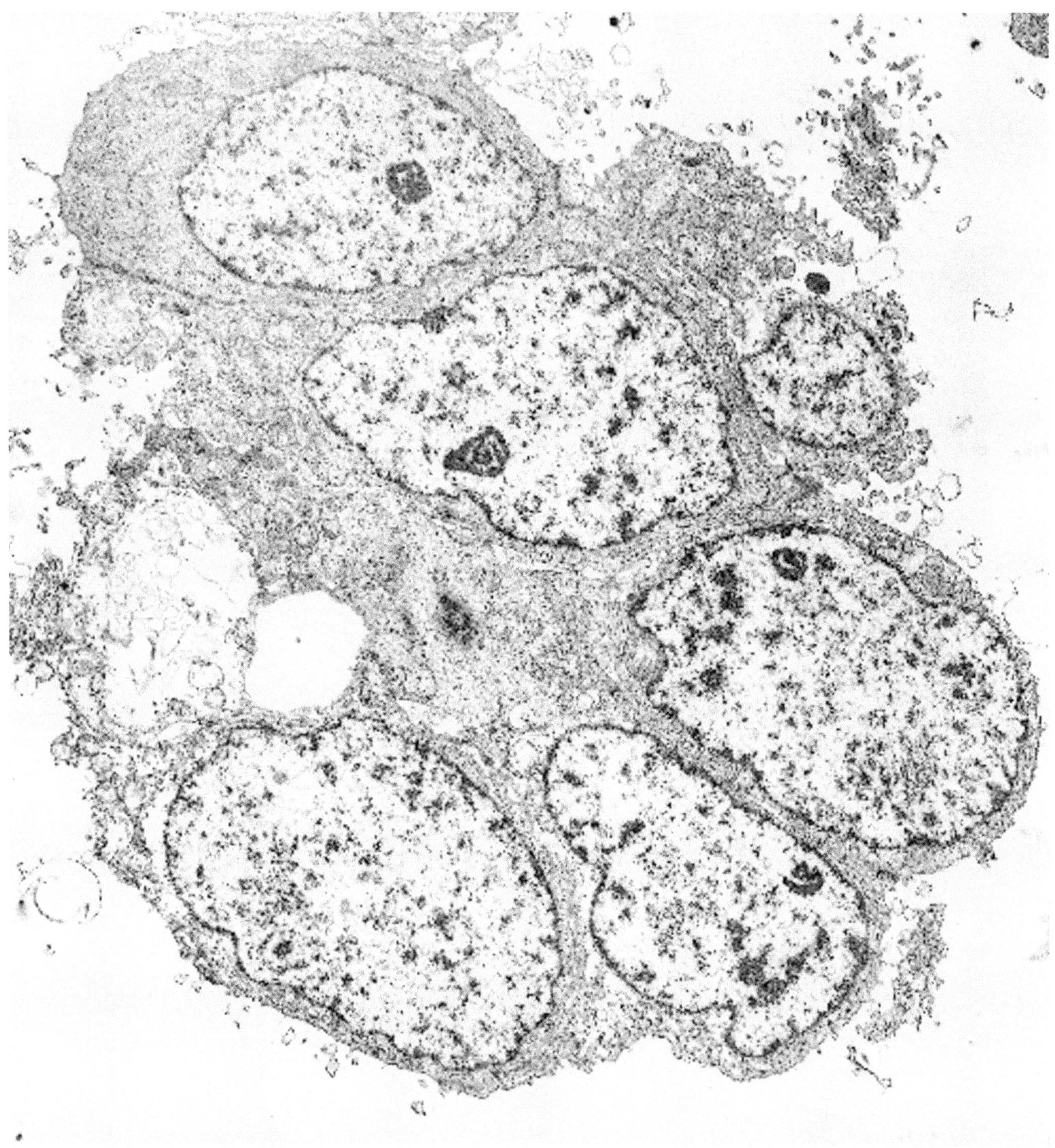

Figure 7.7. Papillary serous cystadenoma of borderline malignancy (ovary). The cells in this cluster have some of the same general features as those in the inclusion cysts and cystadenomas, but there is an absence of cilia, fewer microvilli, less differentiation of the cytoplasm, and a higher nuclear–cytoplasmic ratio. (× 6960)

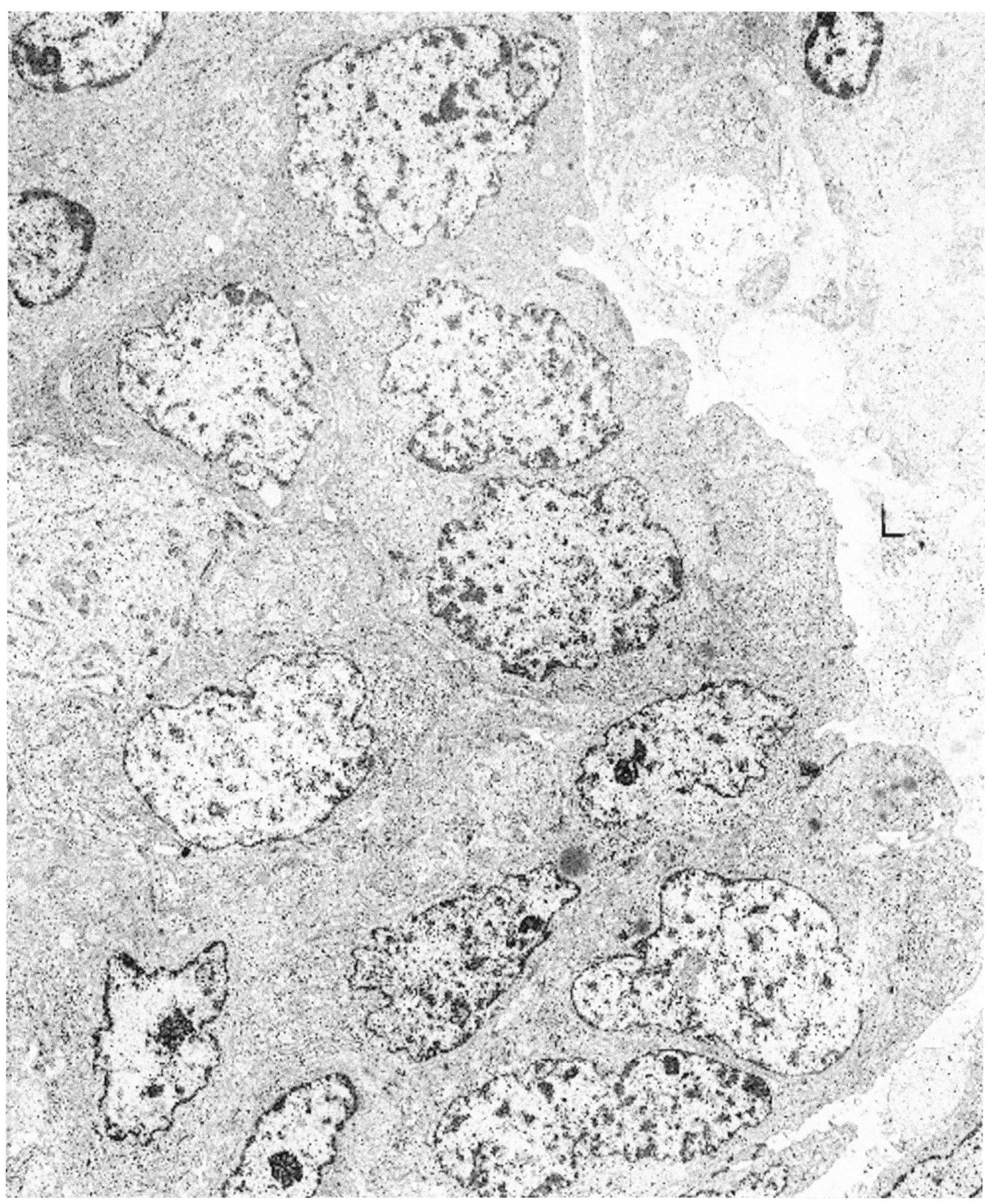

Figure 7.8. Papillary serous cystadenoma of borderline malignancy (ovary). The lining cells are multilayered and have a relatively high nuclear–cytoplasmic ratio. No cilia and only a few microvilli are present on the luminal (L) surface of the cells. (× 4320) (Permission for reprinting granted by WB Saunders, Dickersin GR: The ultrastructure of selected gynecologic neoplasms. Clin Lab Med 7:117–156, 1987.)

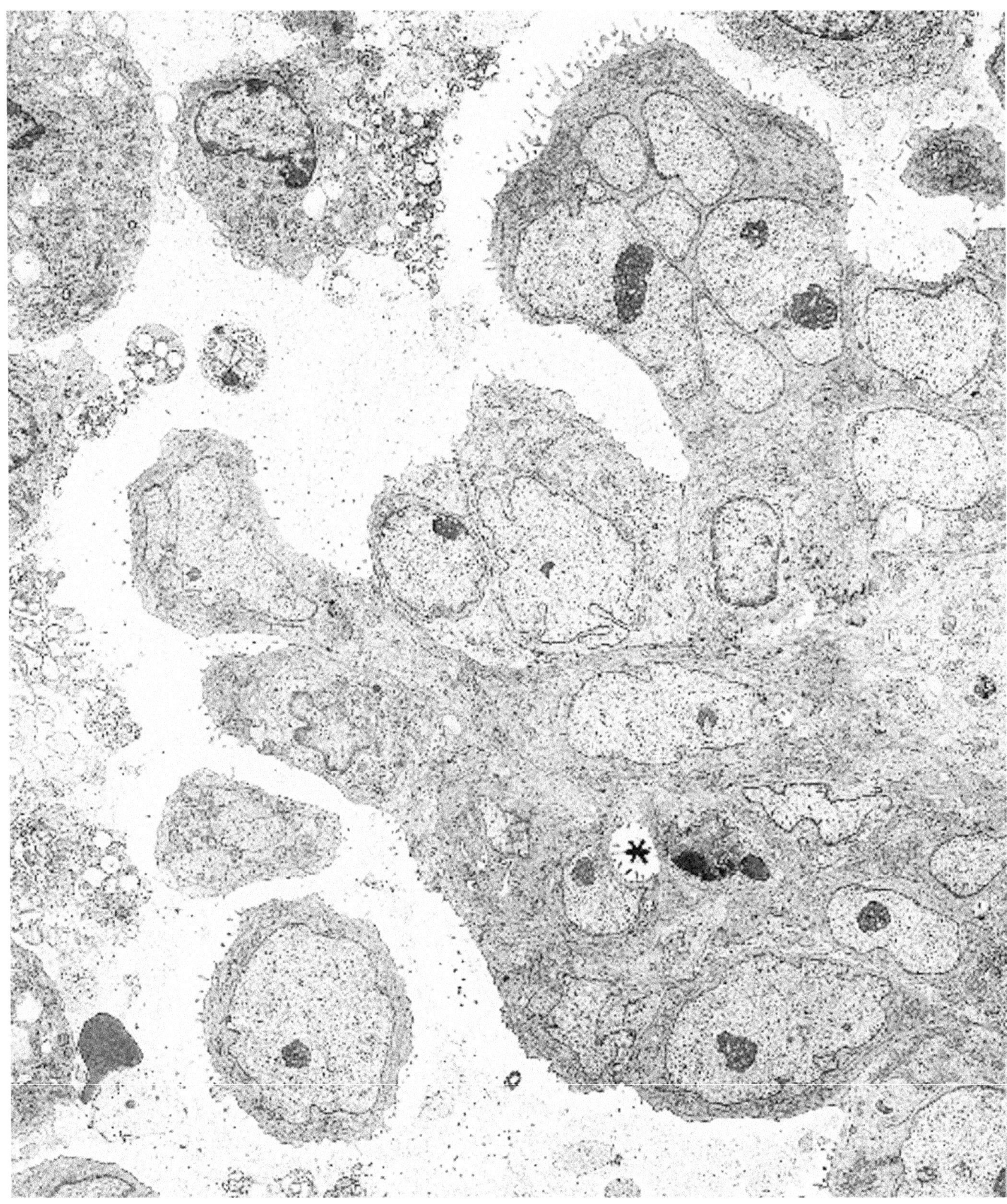

Figure 7.9. Papillary serous cystadenocarcinoma (ovary). The cells are multilayered and have an inconsistent population of microvilli and no cilia. Deep invaginations between cells produce numerous cytoplasmic pseudolumens, one (*) of which is present in this field. (× 3740)

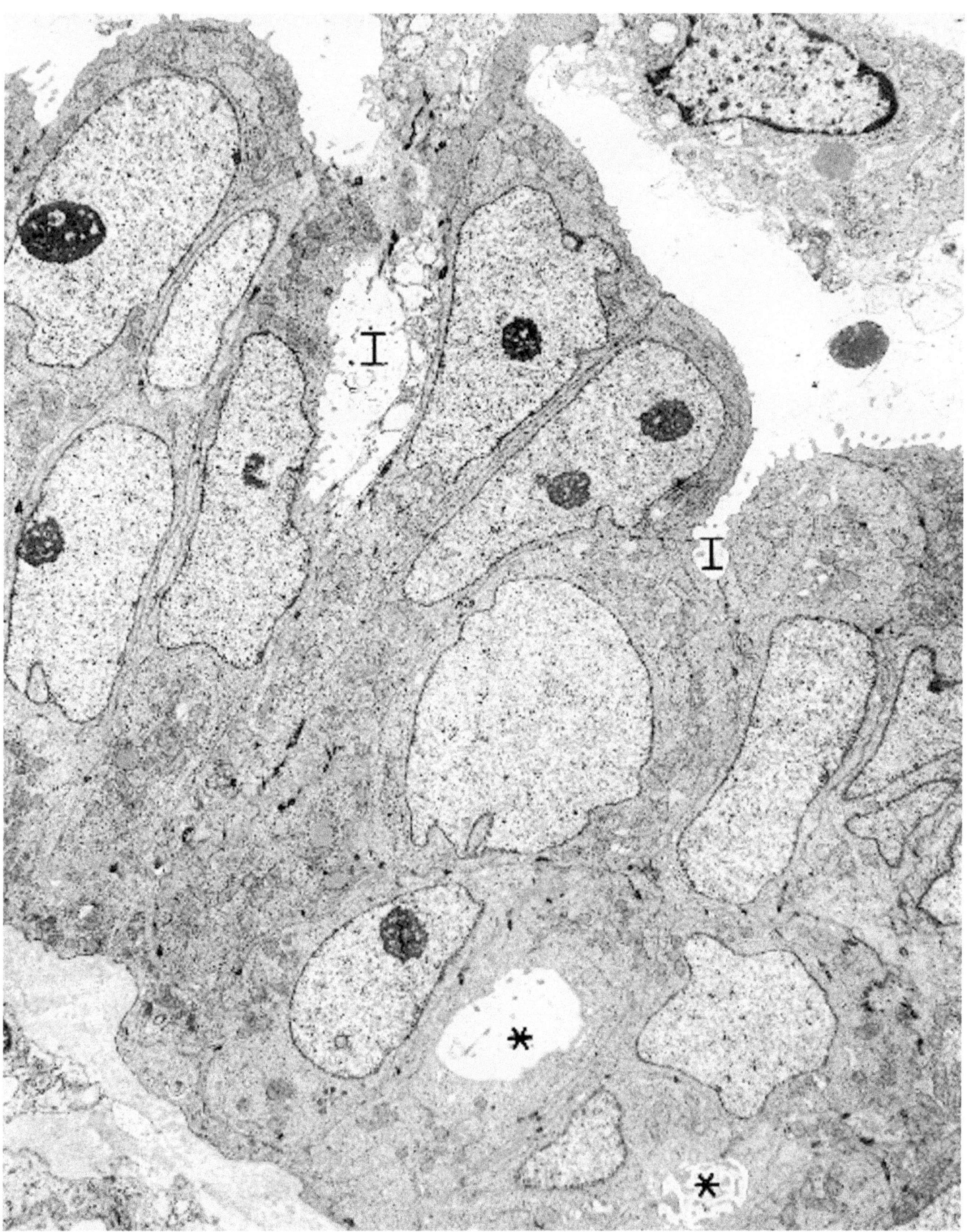

Figure 7.10. Papillary serous cystadenocarcinoma (ovary). Another field of the same neoplasm in Figure 7.9 shows two invaginations (I) of the surface epithelium and two deep cytoplasmic pseudolumens (*). (× 5940)

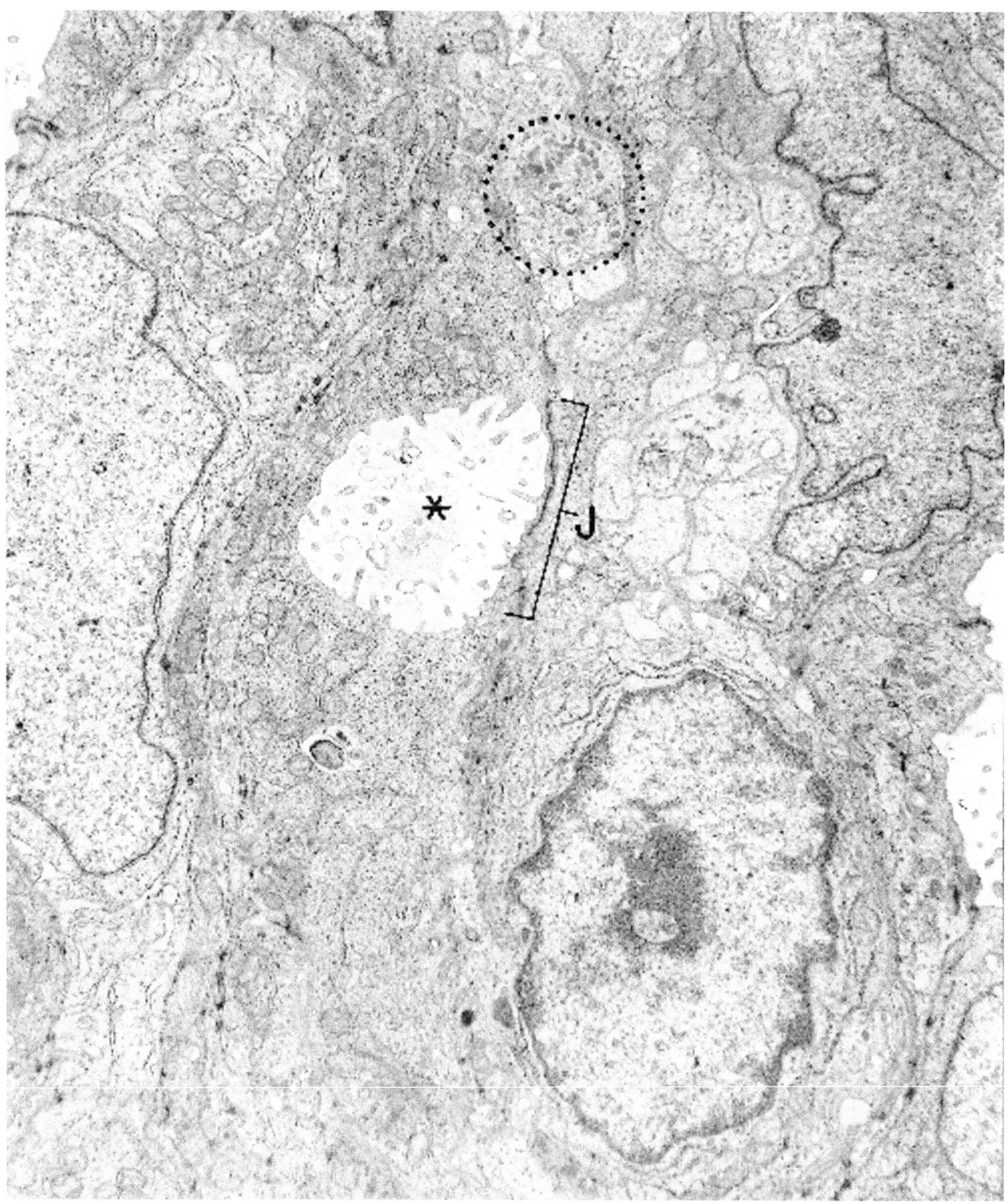

Figure 7.11. Papillary serous cystadenocarcinoma (ovary). Higher magnification of the same neoplasm in Figures 7.9 and 7.10 highlights a cytoplasmic pseudolumen (*) with long junctional complexes (J) along one of its borders. In addition, the cytoplasmic contents are discernible, and only one pocket of small secretory granules (*circle*) is present. Elsewhere also, the neoplasm had a paucity of secretory differentiation. (× 10,260)

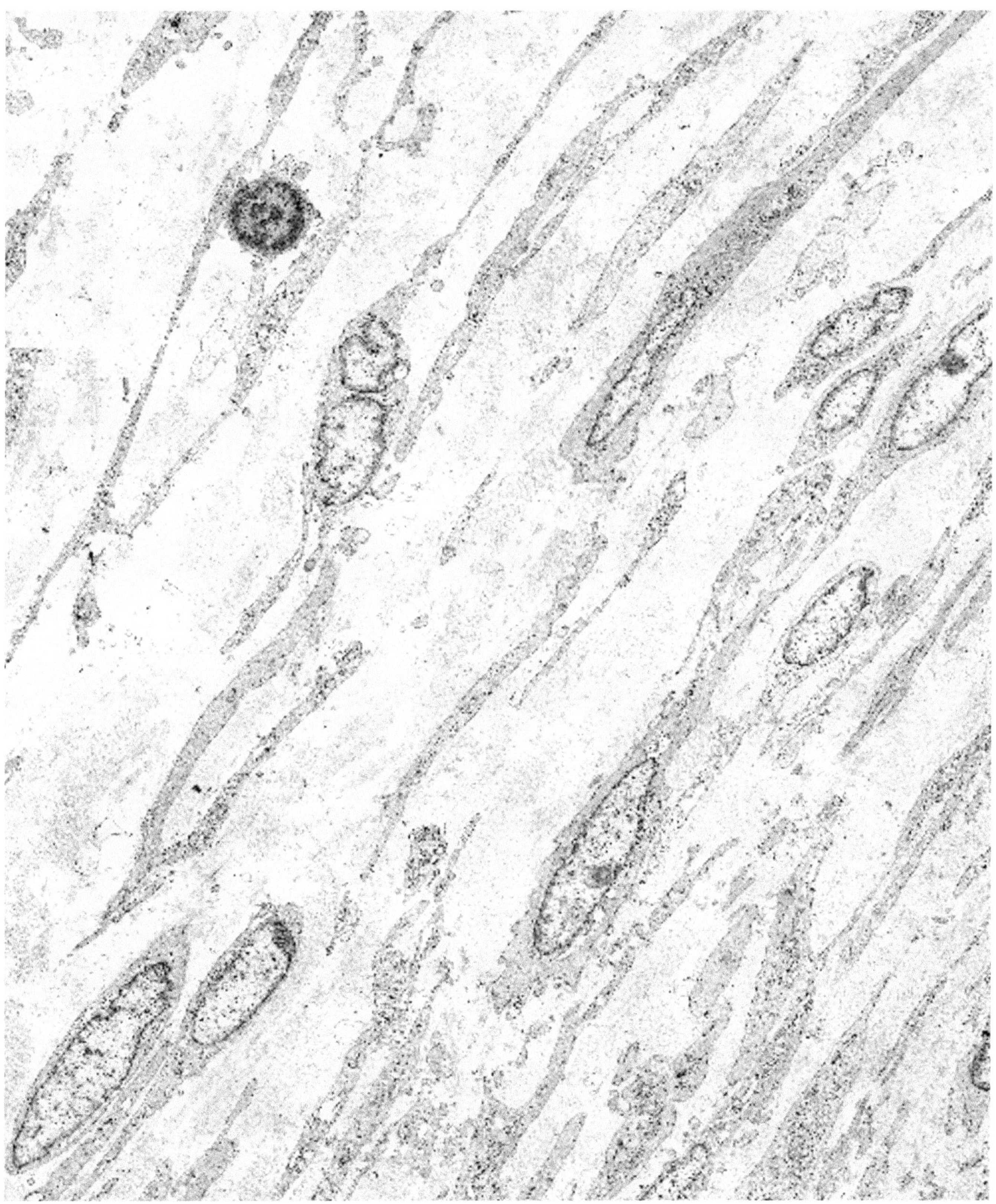

Figure 7.12. Cortical stroma (ovary). This field is representative of nonluteinized stromal cells that may be found in normal ovaries, between inclusion cysts and as a component in surface epithelial–stromal tumors. The cells have long cytoplasmic processes and are arranged regularly in a matrix of collagen. (× 2940)

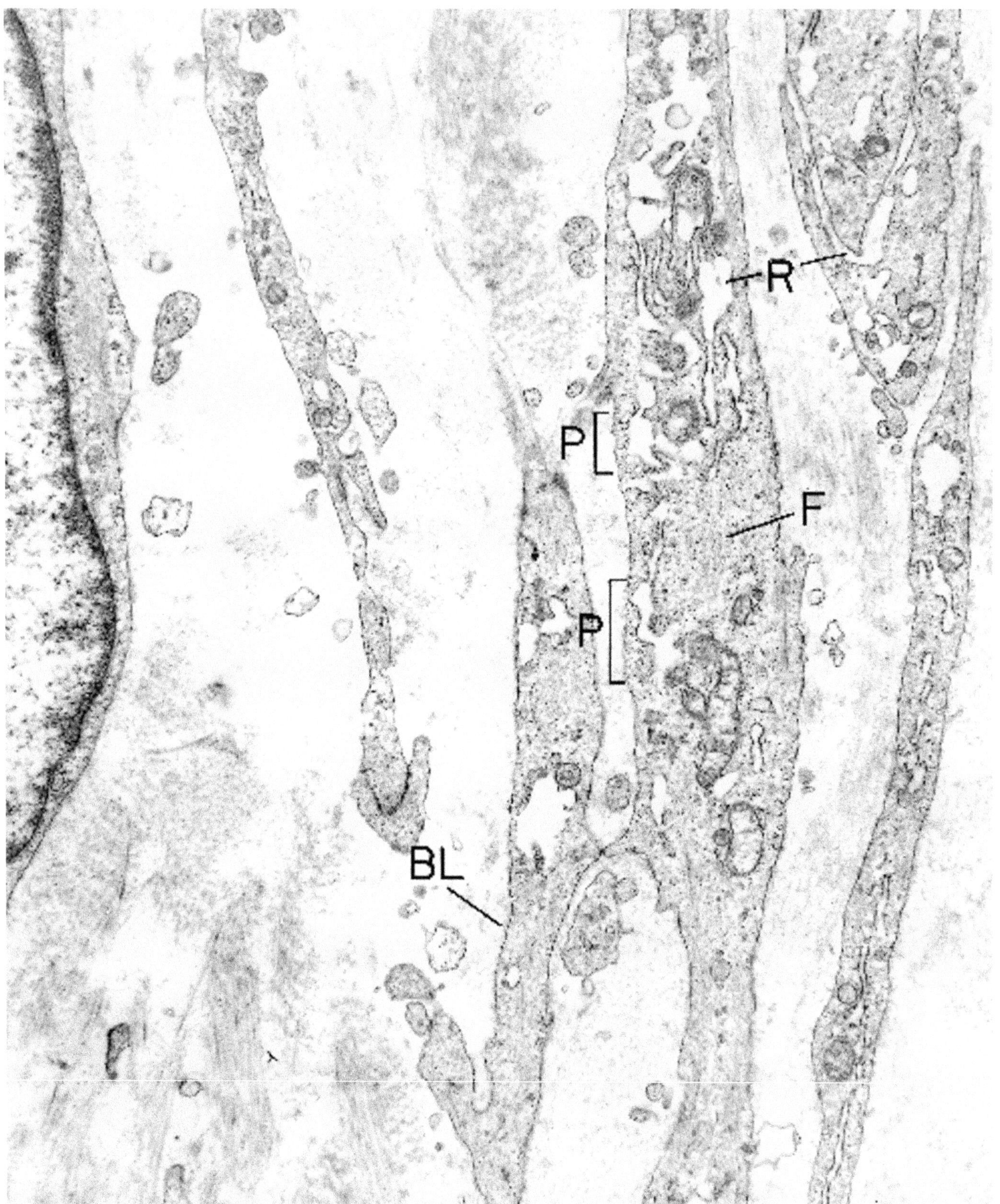

Figure 7.13. Stroma (ovary). The stromal cells are consistent with myofibroblasts—their processes having the dilated rough endoplasmic reticulum (R) of fibroblasts—and the filaments (F), pinocytosis (P), and basal lamina (BL) of smooth muscle cells. (× 42,000)

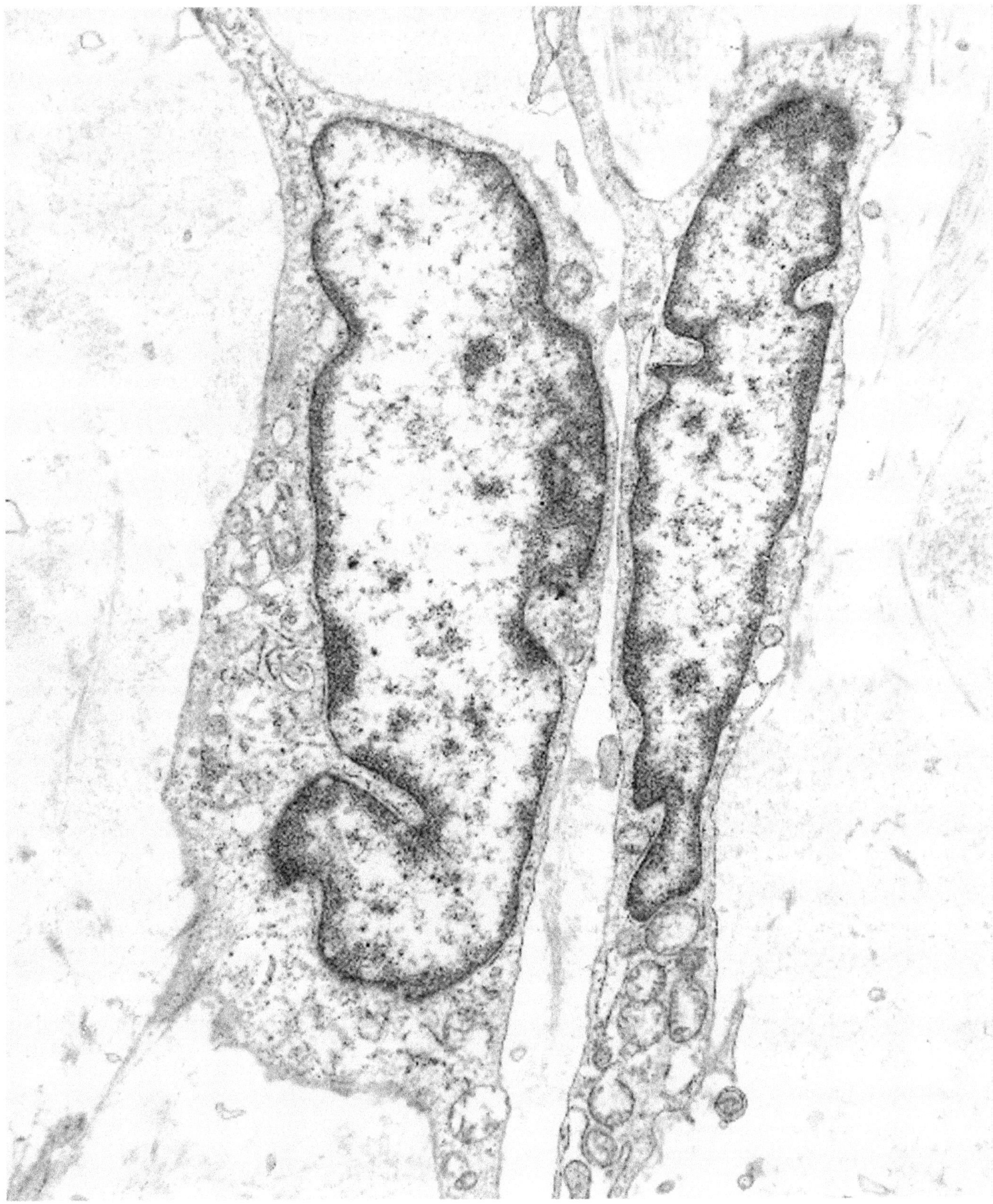

Figure 7.14. Stroma (ovary). The cell bodies of two stromal cells show fewer ultrastructural markers for fibroblasts and smooth muscle cells than do the cell processes, as depicted in Figure 7.13. (× 15,000)

(Text continued from page 391)

Mucinous Tumors

(Figures 7.15 through 7.20.)

Diagnostic criteria. Epithelial cells lining (1) lumens and occasional papillae and having (2) microvilli, (3) junctional complexes, (4) basal lamina, (5) mucinous granules.

Additional points. The epithelium of mucinous tumors is of two main types: endocervical and intestinal. Endocervical-type epithelium is characterized as tall, narrow cells with uniformly small secretory granules and basal nuclei (Figure 7.15). Membrane-bound fibrillogranular bodies, a type of secretory granule, are found near the mucinous granules and Golgi apparatus (Figure 7.16). Intestinal-type epithelium typically includes absorptive, goblet, and argyrophilic cells, but other endocrine cells (serotonin and peptide-secreting) and Paneth cells may also be present (Figures 7.17 through 7.20). The absorptive cells have microvilli that contain a core of filaments that extend into the subjacent cytoplasm (Figure 7.18). Goblet cells are characterized by a cytoplasm filled with varying-sized and coalescent mucous granules (Figures 7.17 and 7.18). Endocrine cells are located basally in the glands and are identifiable by their many dense-core granules (Figures 7.19 and 7.20). Endocervical epithelium is more common in benign mucinous tumors, and intestinal epithelium is more common in borderline tumors. Carcinomas may have either type of epithelium exclusively, a combination of the two types, or a nondescript type of mucinous epithelium. Benign tumors usually are lined by a single-cell layer of epithelium, whereas borderline and malignant tumors have a multilayered cellular lining. Papillarity and budding are generally similar to what is seen in serous tumors.

The stroma of mucinous cystadenomas usually is rich in collagen, and the cells resemble those of ovarian stroma and serous tumors, including occasionally having the luteinized features of lipid vacuoles, prominent smooth endoplasmic reticulum, and tubular mitochondrial cristae. The stroma in malignant mucinous tumors may be ovarian in type or desmoplastic, and it frequently contains inflammatory cells and pools of mucin.

Endometrioid Tumors

These tumors are composed of epithelium and/or stroma that have an appearance similar to the corresponding components of primary uterine endometrial tumors. Ovarian endometrioid tumors are probably derived mostly from surface epithelial inclusions and ovarian stroma and, to a lesser extent, from endometriosis. The epithelial tumors in this group include endometrioid carcinomas and rare endometrioid cystadenomas and endometrioid adenofibromas. Endometrioid stromal tumors, or tumors with a sarcomatous component, include endometrioid stromal sarcoma, malignant mesodermal mixed tumor, and adenosarcoma.

Endometrioid Carcinoma

(Figures 7.21 through 7.25.)

Diagnostic criteria. Epithelial cells forming glands with (1) lumens, (2) microvilli, (3) junctional complexes, (4) basal lamina, (5) bundles of paranuclear filaments, and (6) abundant glycogen; stromal cells having (7) a nondescript cytoplasm or the fibroblastic features of prominent rough endoplasmic reticulum, with or without (8) the additional myofibroblastic features of filaments and dense bodies.

Additional points. The epithelium of endometrioid carcinoma of the ovary (Figure 7.21) resembles that of adenocarcinoma of the endometrium (Figure 7.22) and normal endometrium (Figure 7.23), although the latter varies to some extent with the estrous cycle. The cells are of a tall columnar type and have a well-developed cytoplasm. Ribosomes, rough endoplasmic reticulum, and Golgi apparatuses increase in prominence during the mid and late proliferative phases of the cycle and, with mitochondria, reach a maximum in the early secretory phase. Perinuclear filaments and glycogen are abundant in the early and mid secretory phases. Cilia and tonofibrils (squamous metaplasia) may be present in well-differentiated neoplasms, and serous cells and mucinous cells of the endocervical type are seen in some cases. Oxyphilic cells with numerous mitochondria comprise some endometrioid carcinomas, and a small but significant number of endometrioid carcinomas contain neuroendocrine cells, identifiable by their cytoplasmic dense-core granules. The stroma of endometrioid carcinoma of the ovary may have the appearance of normal, luteinized, or fibrous ovarian stroma (Figures 7.24 and 7.25).

(Text continues on page 418)

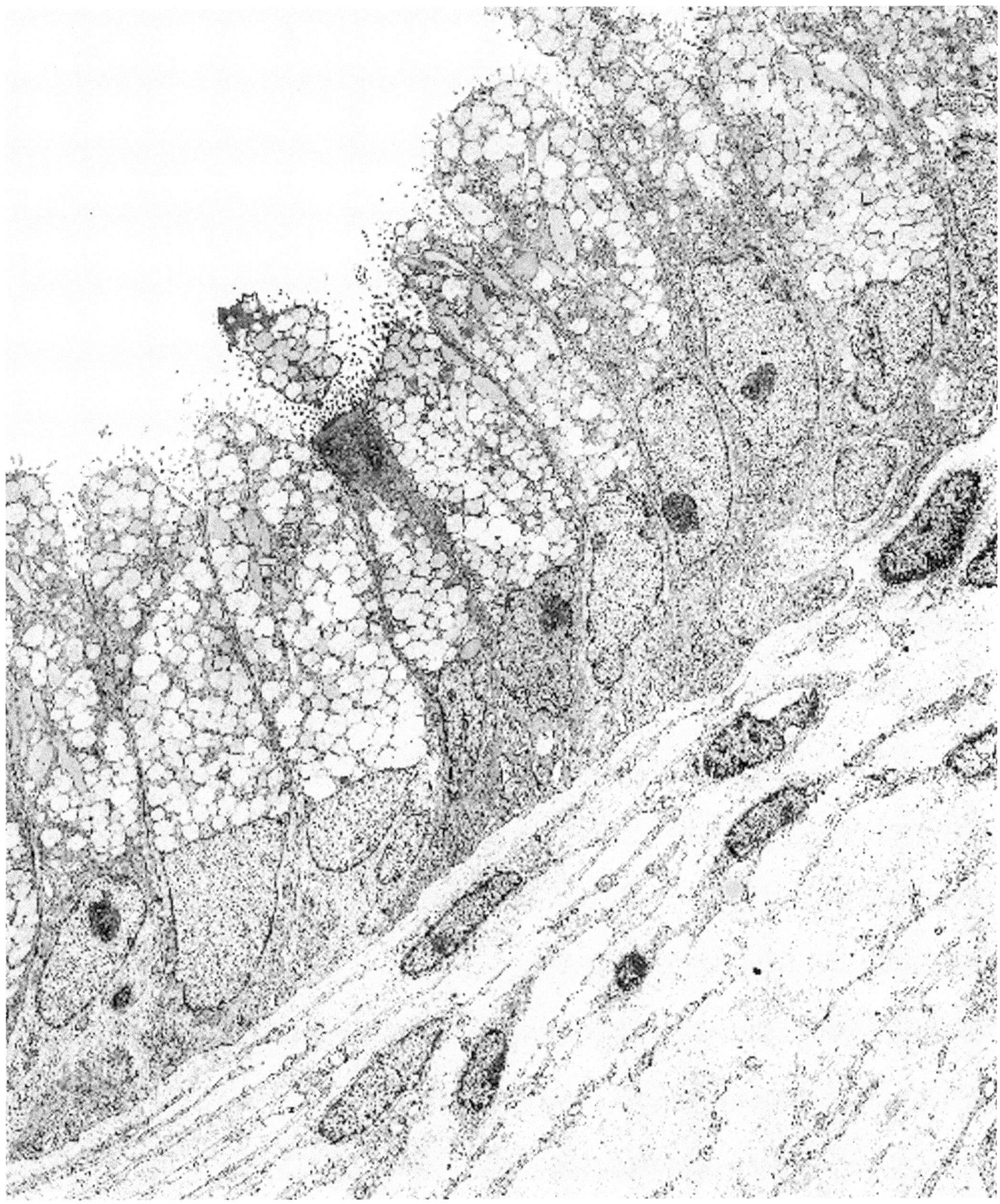

Figure 7.15. Normal endocervix. The epithelial lining cells are single layered, tall columnar, and villous. Most striking are the many secretory granules in the supranuclear cytoplasm. Nuclei are basal in the cells. (× 3600)

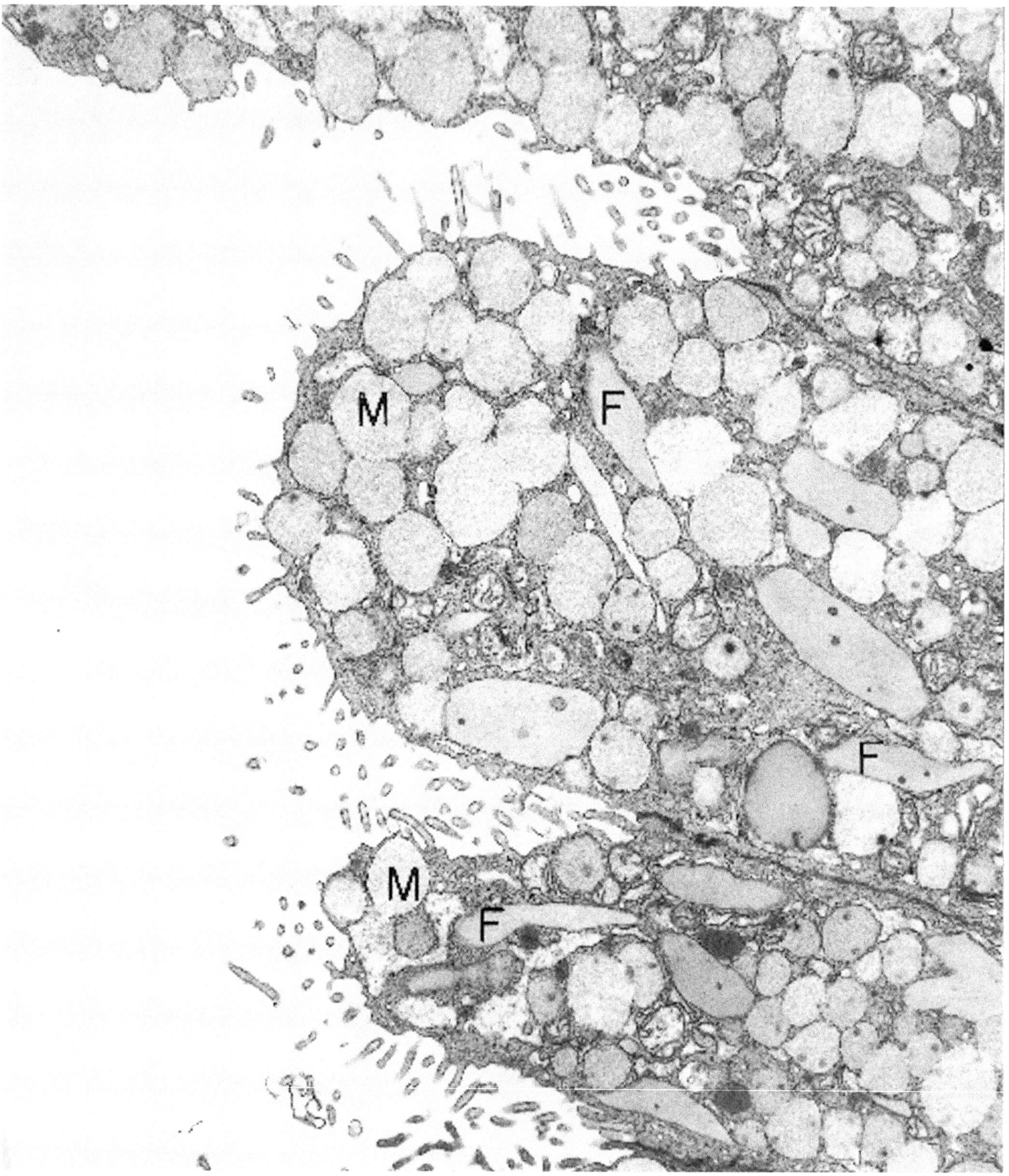

Figure 7.16. Normal endocervix. High magnification of the apical cytoplasm of several lining cells illustrates innumerable mucous granules (M) and intervening fibrillogranular bodies (F). (× 12,500)

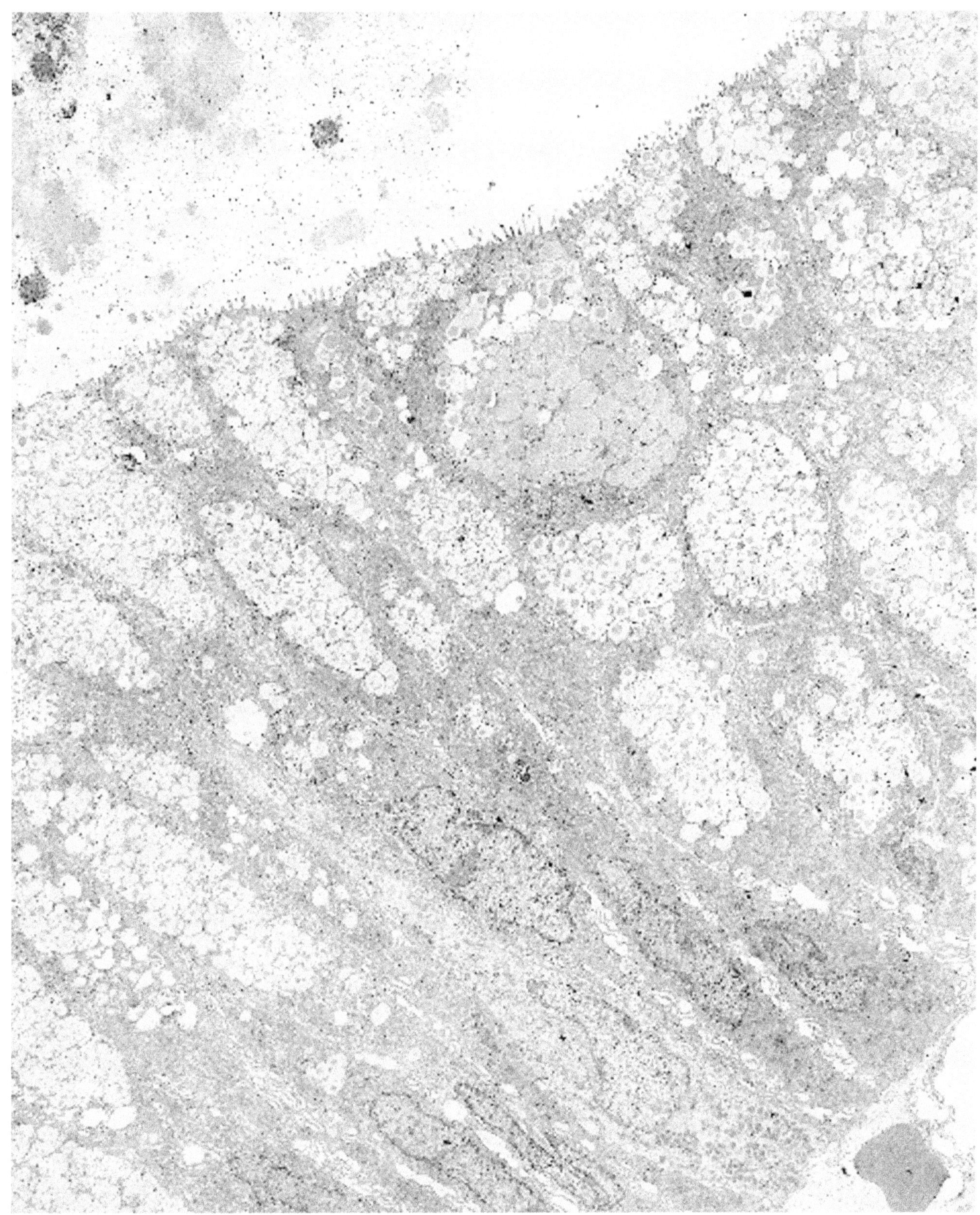

Figure 7.17. Mucinous cystadenoma of borderline malignancy (ovary). The epithelial lining cells are multilayered and are striking in regard to the copious collection of mucinous granules in their supranuclear cytoplasm. Many microvilli, but no cilia, are located on the luminal surface of the cells. (× 3600)

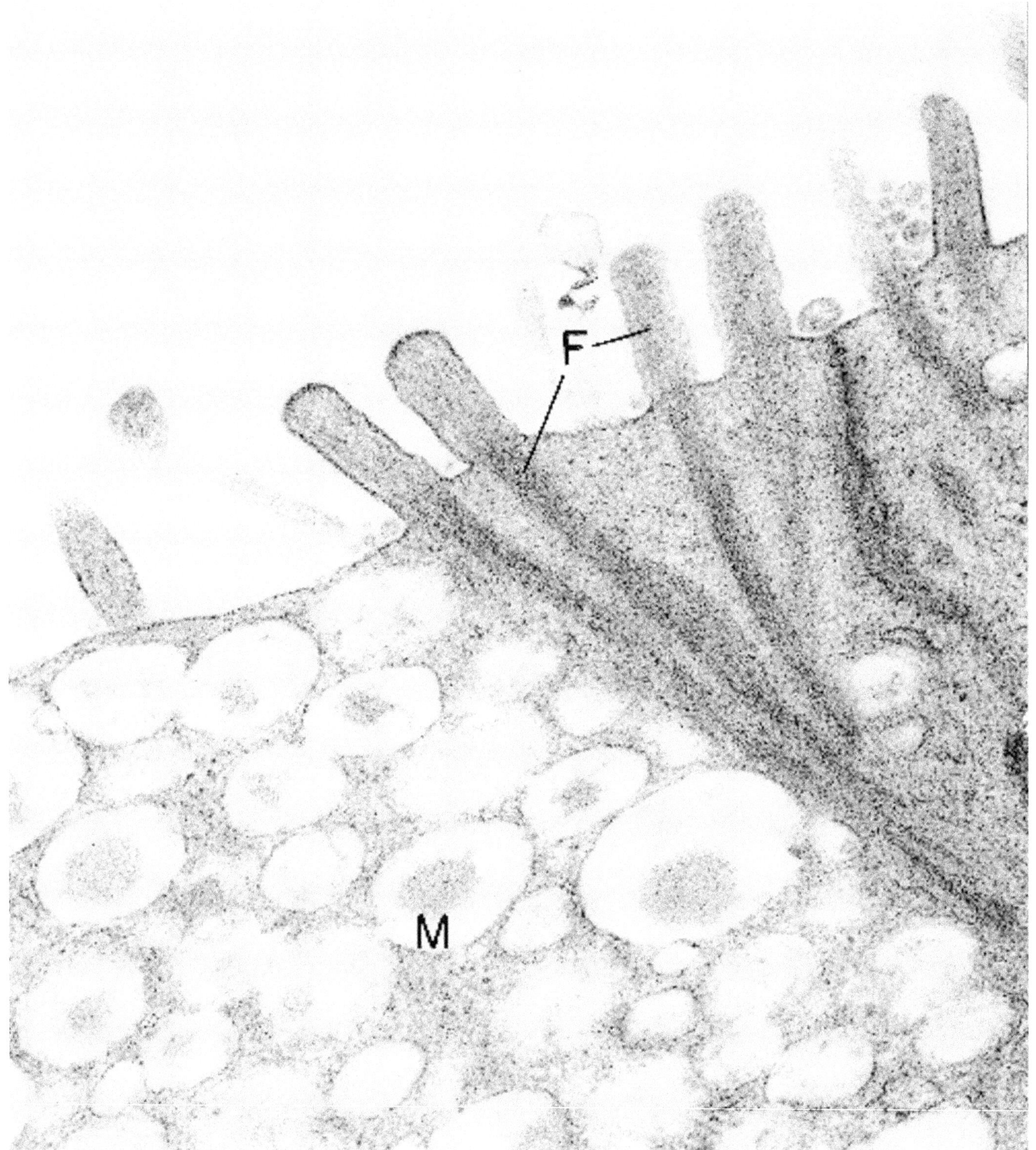

Figure 7.18. Mucinous cystadenoma of borderline malignancy (ovary). High magnification of the microvilli of the mucous epithelial cells reveals the intestinal type of anchoring filaments (F) that extend from the core of each villus into the subjacent cytoplasm. Note also one of the typical patterns of mucus in the granules (M), in which a central or eccentric condensate of secretion is surrounded by a broad halo. (× 61,500)

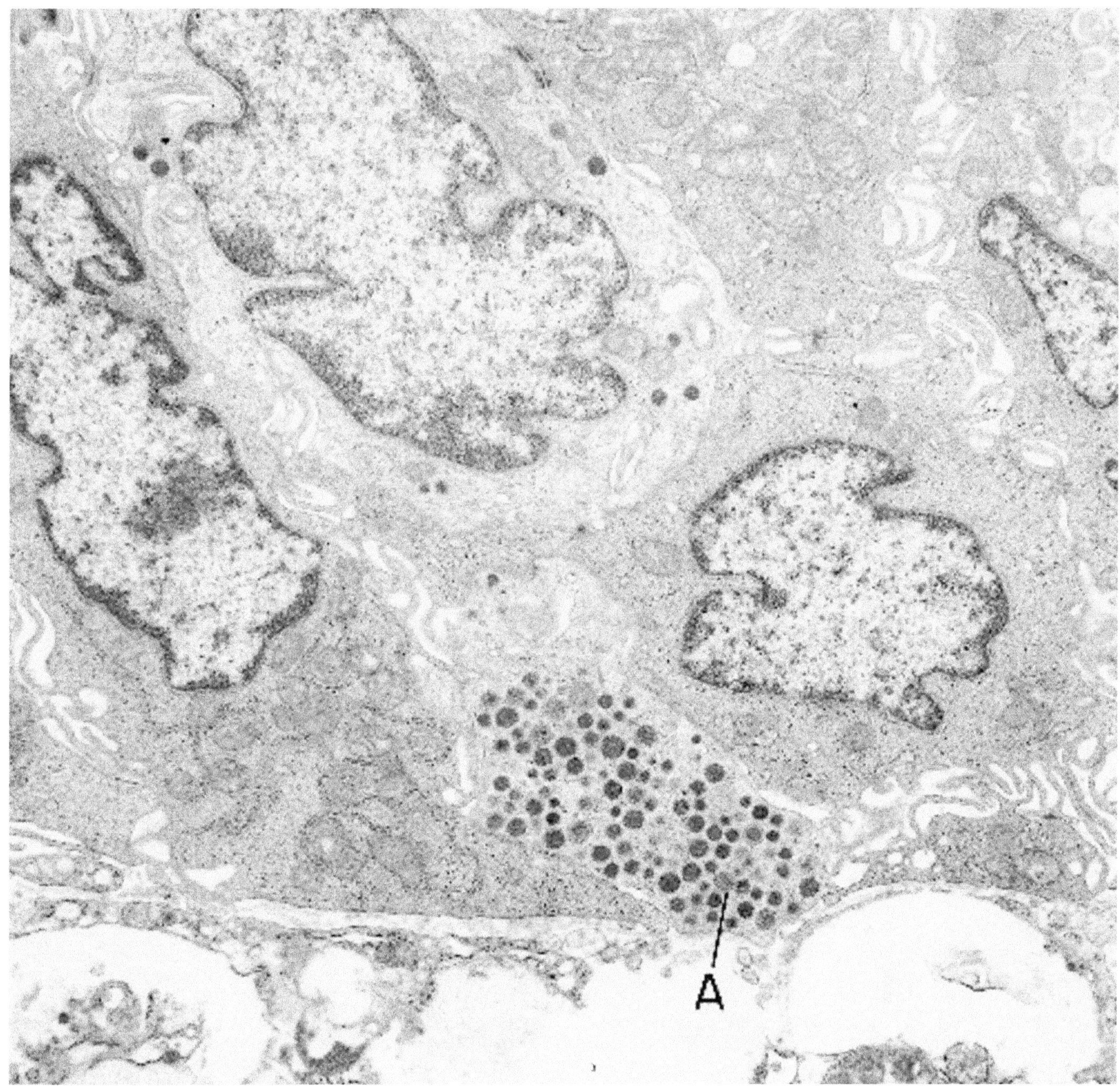

Figure 7.19. Mucinous cystadenoma of borderline malignancy (ovary). The base of this neoplastic gland contains an endocrine cell (A), readily identified by its distinctive dense-core granules. (× 13,500)

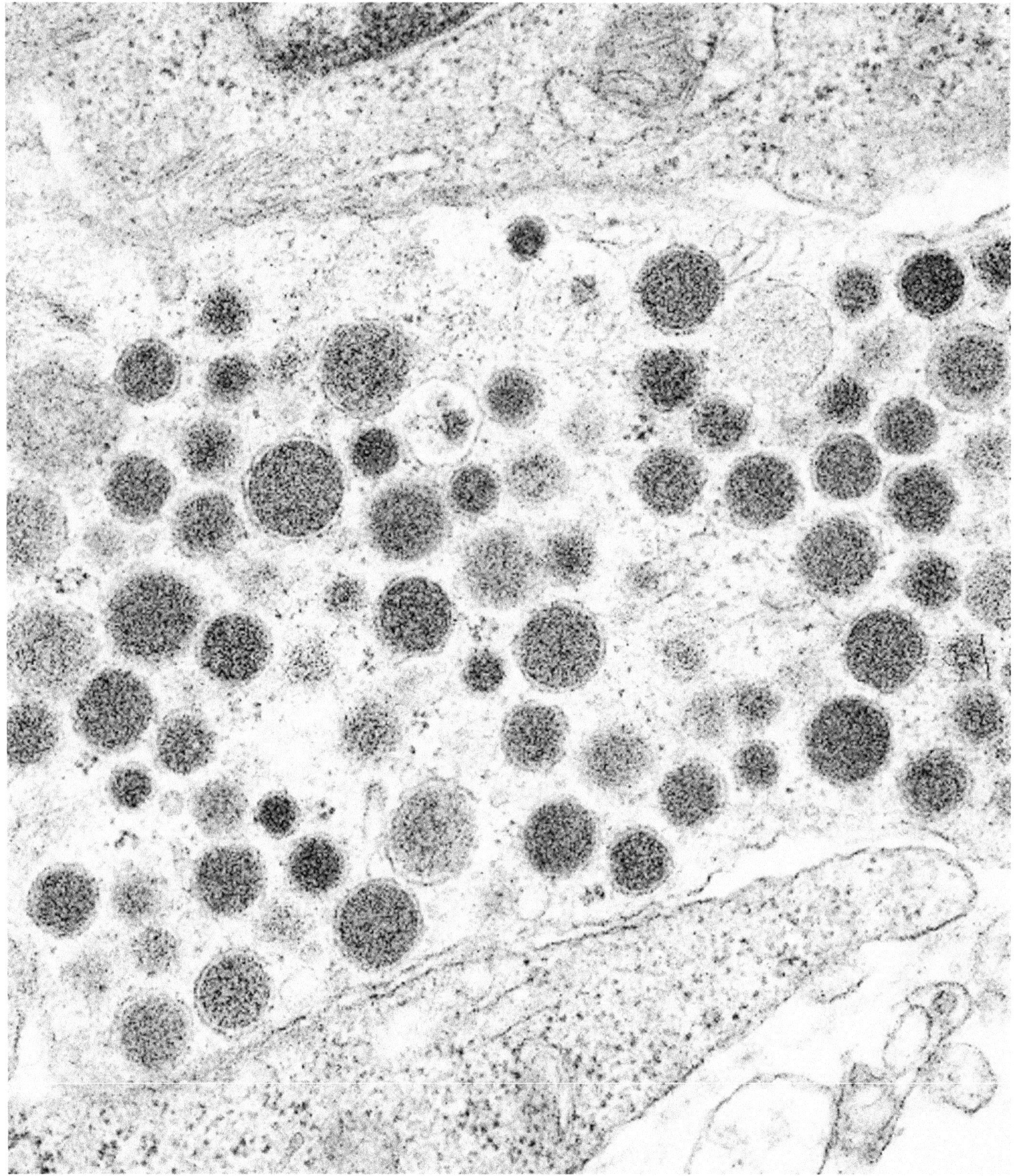

Figure 7.20. Mucinous cystadenoma of borderline malignancy (ovary). High magnification of the endocrine cell in Figure 7.19 illustrates in detail the dense-core granules. (× 61,600)

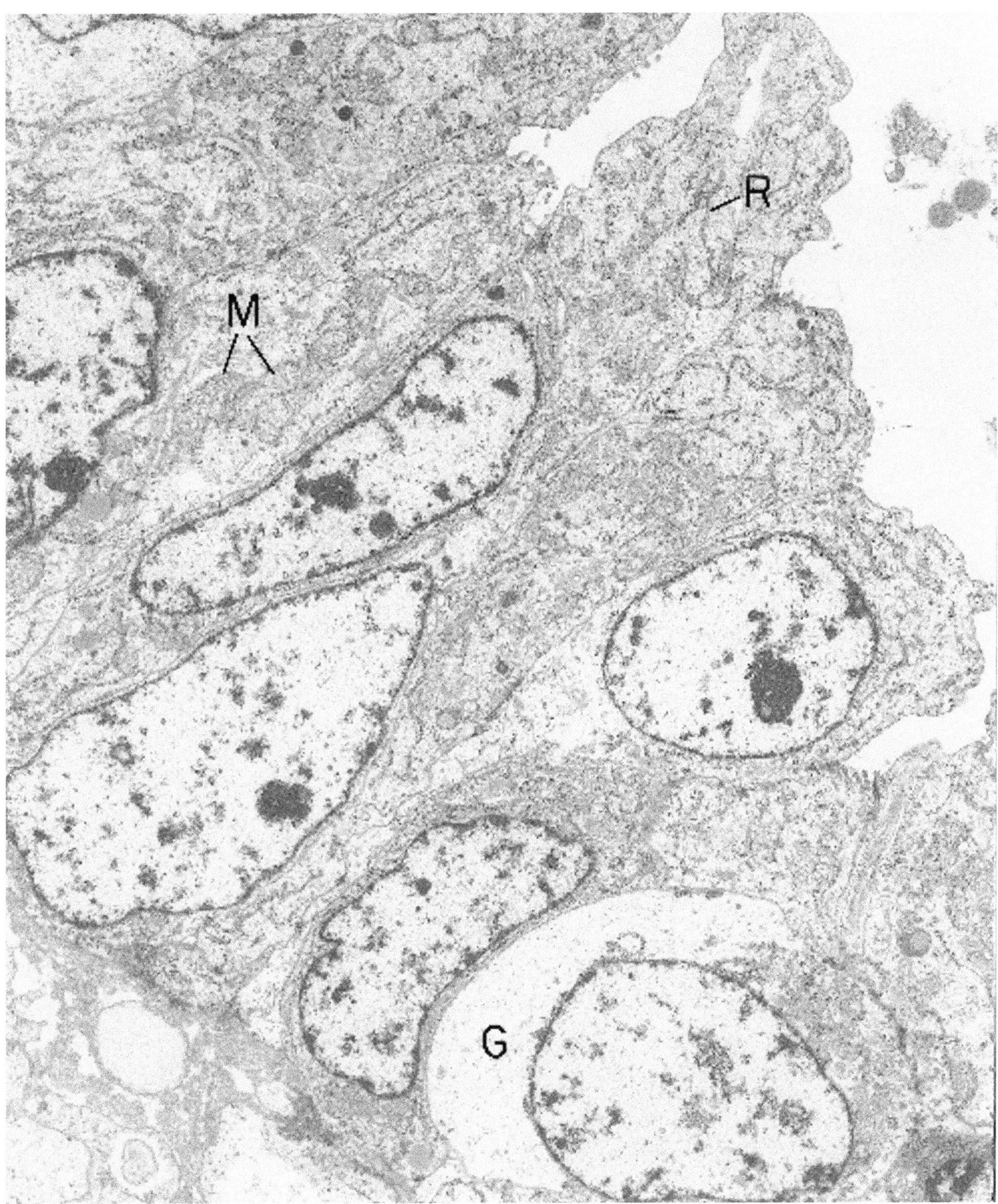

Figure 7.21. Endometrioid carcinoma (ovary). Neoplastic glands are lined by tall columnar cells that have cytoplasm with many free ribosomes, a moderate number of mitochondria (M), and undilated cisternae of rough endoplasmic reticulum (R). Glycogen (G, clear zone) is copious in some of the cells. (× 6860) (Negative for photograph courtesy of Dr. Bruce Mackay.) (Permission for reprinting granted by WB Saunders, Dickersin GR: The ultrastructure of selected gynecologic neoplasms. Clin Lab Med 7:117–156, 1987.)

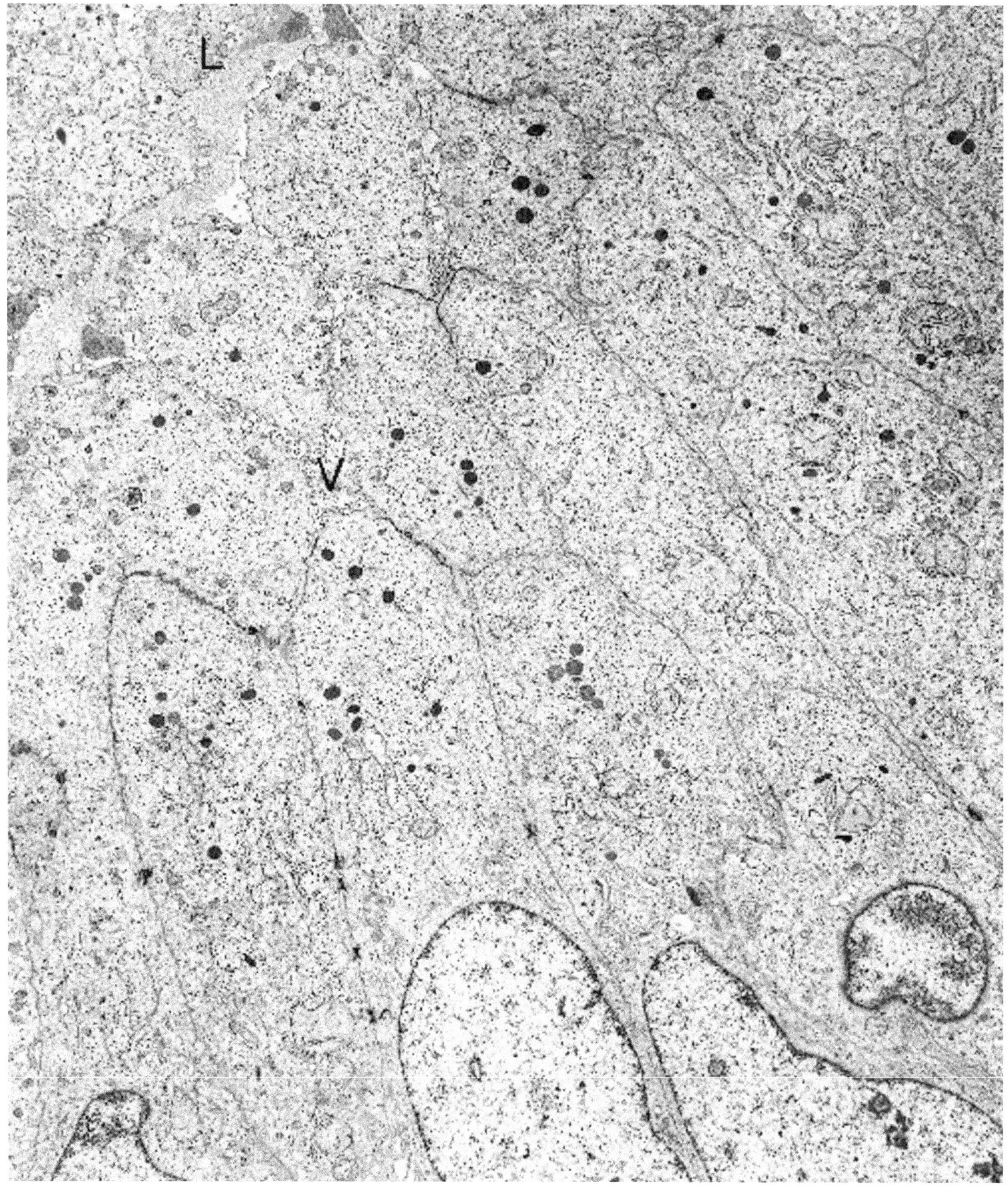

Figure 7.22. Adenocarcinoma of endometrium (uterus). Stratified columnar cells have irregularly dispersed, short, stubby microvilli (V) lining a microlumen (L). (× 5300)

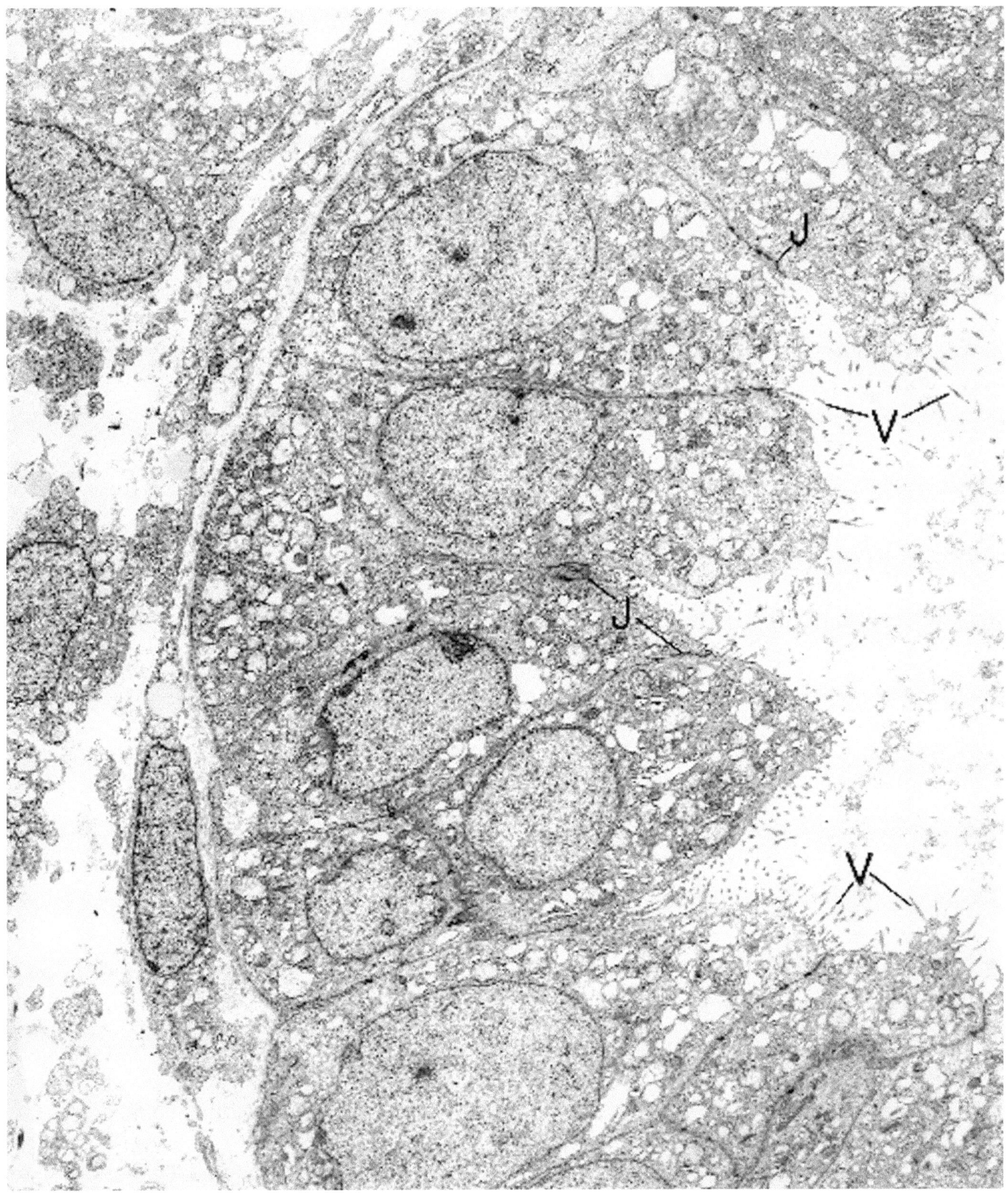

Figure 7.23. Normal endometrium (39-year-old, midsecretory endometrium). The lining cells of this gland are of the simple columnar type, with basal and midcell nuclei, many cytoplasmic organelles, junctional complexes (J), and microvilli (V). Portions of several stromal cells are present in the upper left field. (× 4176)

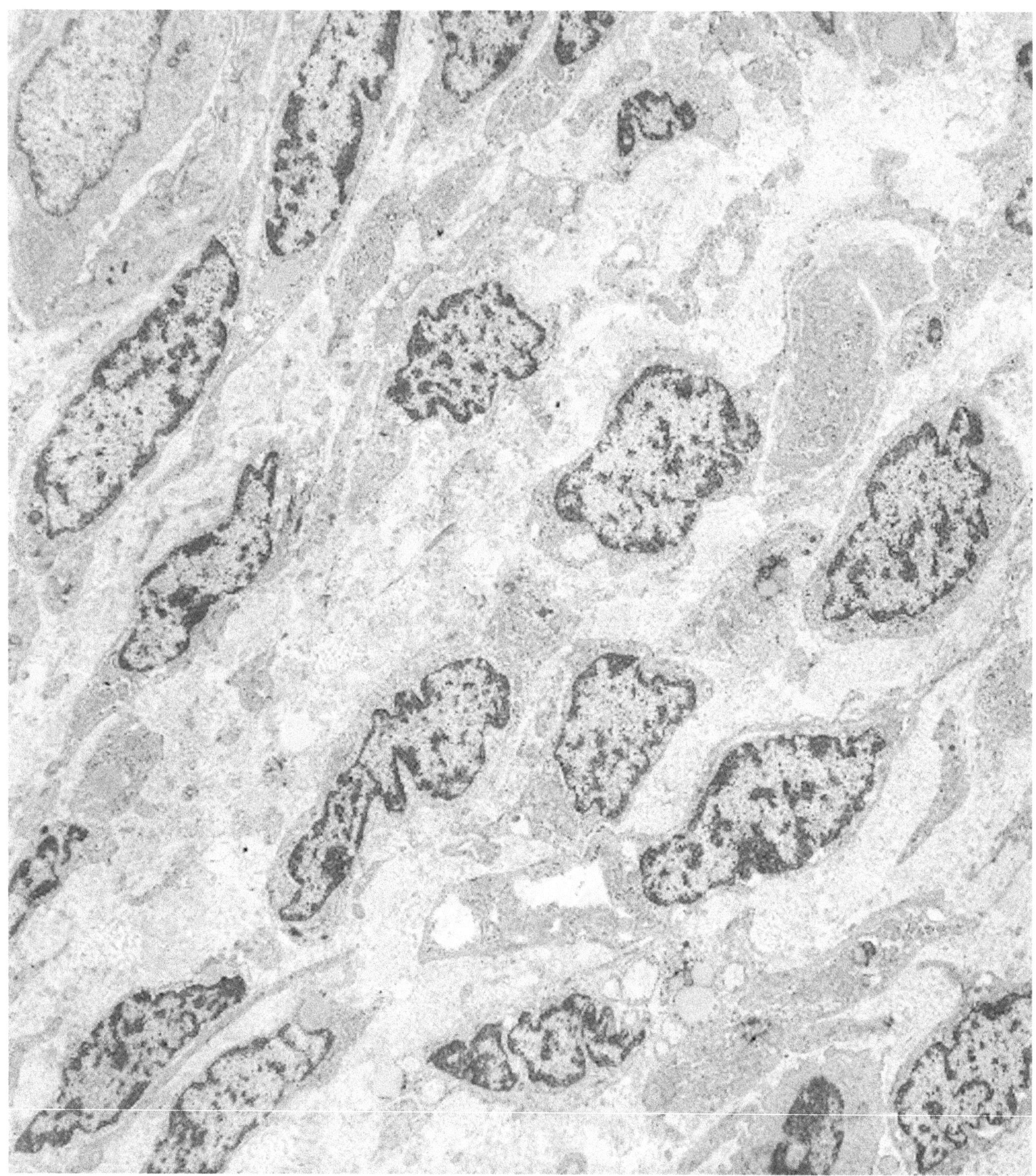

Figure 7.24. Normal endometrium (late proliferative). Oval and spindle cells are arranged individually in a matrix of banded collagen. There is a high nuclear–cytoplasmic ratio, cytoplasm is scant, and nuclei are indented. (× 5900)

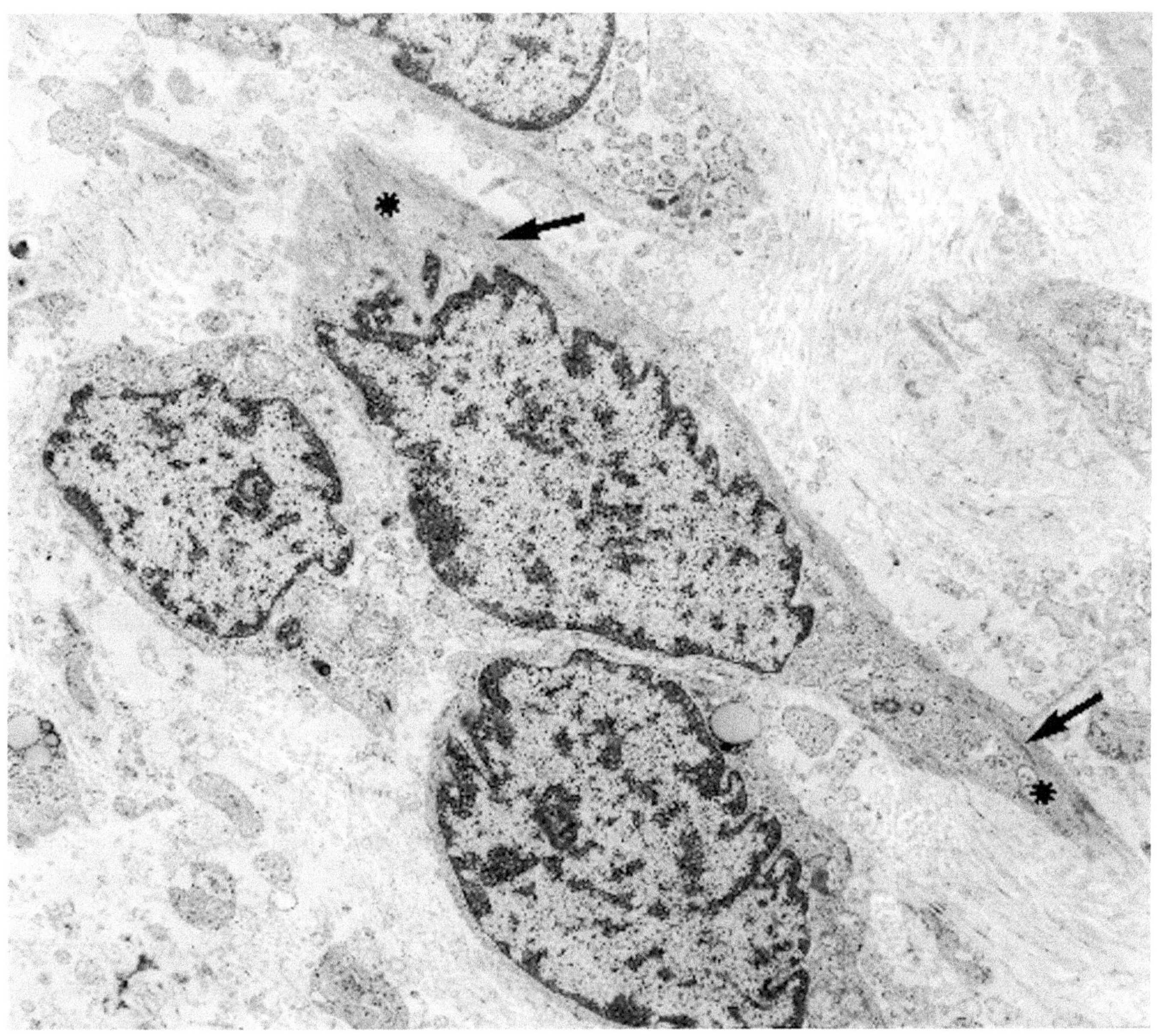

Figure 7.25. Normal endometrium (late proliferative). Higher magnification of cells from the same endometrium as depicted in Figure 7.24 reveals the cytoplasm of one cell to contain numerous peripherally located filaments (*) with dense bodies (arrows). (× 8300)

(Text continued from page 406)

Endometrioid Cystadenoma

The epithelial component of endometrioid cystadenoma is stratified, nonciliated, and devoid of mucin granules. It is generally similar to the epithelium of endometrioid carcinoma and to primary uterine endometrial tumors and normal endometrium. The stroma, however, is nondescript and, except focally in some tumors, is not of endometrioid type. Endometrioid cystadenomas are rare tumors, and their main component, endometrioid epithelium, is illustrated in Figures 7.21 through 7.23.

Endometrioid Adenofibroma

The glandular epithelium in adenofibroma is stratified or occasionally simple, and it resembles the epithelium of uterine endometrium (Figure 7.23). Squamous metaplasia with tonofibrils may be present. The stroma is fibromatous. This tumor is also very uncommon.

Endometrioid Stromal Sarcoma

(Figures 7.26 through 7.27.)

Diagnostic criteria. Diffuse infiltrate of small cells having (1) an oval or fusiform shape, (2) a high nuclear–cytoplasmic ratio and nondescript cytoplasm, with (3) a moderate number of mitochondria, (4) a moderate to large amount of rough endoplasmic reticulum, (5) infrequent small or moderate size Golgi apparatuses, (6) scant to many filaments, the latter with (7) dense bodies, (8) heterochromatic nuclei and (9) large nucleoli; (10) focal basal lamina; (11) small junctions; (12) collagenous stroma.

Additional points. The cells of ovarian endometrioid stromal sarcomas have varying degrees of similarity to uterine endometrial stromal cells in the proliferative phase (Figures 7.26 and 7.27). The similarity is more evident in low-grade tumors and may be less obvious or absent in high-grade tumors. Fibroblastic and myofibroblastic differentiation may be present in some cells. Other features that may be seen in endometrioid stromal sarcomas are lipid-rich foam cells; cords, clusters, and tubules of epithelial-like cells; and an extracellular matrix rich in banded collagen.

Malignant Mesodermal Mixed Tumor (Carcinosarcoma)

These tumors have a malignant mesenchymal component in conjunction with a Müllerian-type carcinoma. The mesenchymal component may be derived either from ovarian stromal cells or from heterologous, nonovarian type stroma. If originating from ovarian stroma, the cells have features of fibroblasts or endometrial stromal cells (described in previous paragraph). If heterologous, the stromal cells may be osteoblasts, chondroblasts, lipoblasts, or rhabdomyoblasts. The carcinoma is usually serous, endometrioid, or undifferentiated, but squamous cell, clear cell, and mucinous components may also occur. The ultrastructure of the epithelial and mesenchymal components of these tumors is illustrated earlier in this chapter and in Chapter 6.

Adenosarcoma

Adenosarcomas are composed of benign Müllerian epithelium (endometrioid, serous, mucinous, or clear-cell) and malignant stroma. The stromal component may be of endometrial stromal or fibrosarcomatous types. Heterologous mesenchymal cells may or may not be present.

Clear Cell Tumors

These tumors are of Müllerian derivation and are composed of varying combinations of clear cells, hobnail cells, cuboidal and flat cells, and signet-ring cells with mucin. The tumors may be benign, borderline, or malignant, and cellular patterns include papillary, tubular/cystic, and solid. Clear cell *cystadenocarcinomas* and *adenofibrosarcomas* are extremely rare. The benign ones have benign-appearing epithelium, and the borderline ones have malignant-appearing epithelium but without invasion.

Clear Cell Carcinoma

(Figures 7.28 through 7.30.)

Diagnostic criteria. (1) Cysts, tubules, and papillae lined by hobnail, cuboidal, or flat cells; (2) solid areas of polygonal cells; (3) short, blunt microvilli; (4) junctional complexes; (5) abundant glycogen; (6) stacks of rough endoplasmic reticulum.

Additional points. Other variable features of clear cell carcinoma include interlocking lateral borders, prominent Golgi apparatuses, numerous mitochondria, and paranuclear filaments. Nuclei are oval, chromatin is finely dispersed, and nucleoli are large and have open nucleolonemas. Clear cell carcinomas are ultrastructurally similar, whether they occur in the ovary, uterus, or vagina. The clear cells and hobnail cells in these tumors resemble the cells that line normal endometrial glands during pregnancy.

(Text continues on page 424)

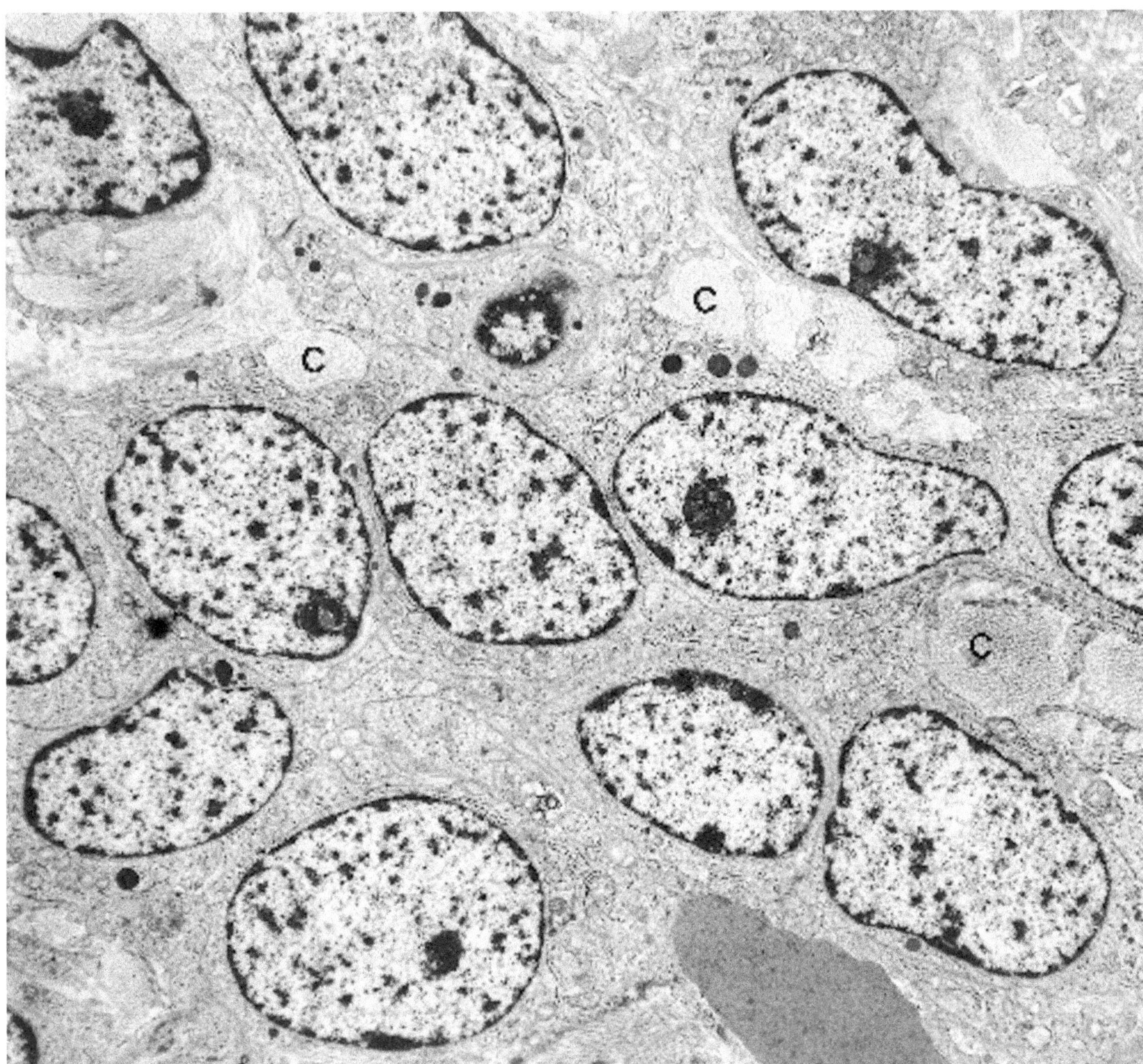

Figure 7.26. Endometrial stromal sarcoma (metastatic to lung). Oval and polygonal neoplastic cells are tightly apposed in groups, with small amounts of banded collagen (C) separating the groups. Nuclear–cytoplasmic ratio is high, and cytoplasm is scant and nondescript. Nuclei are heterochromatic and have nucleoli of moderate size. (× 6300) (Permission for publication granted by Taylor and Francis, Dickersin GR, Scully RE: Role of electron microscopy in metastatic endometrial stromal tumors. Ultrastruct Pathol 17:377–403, 1993.)

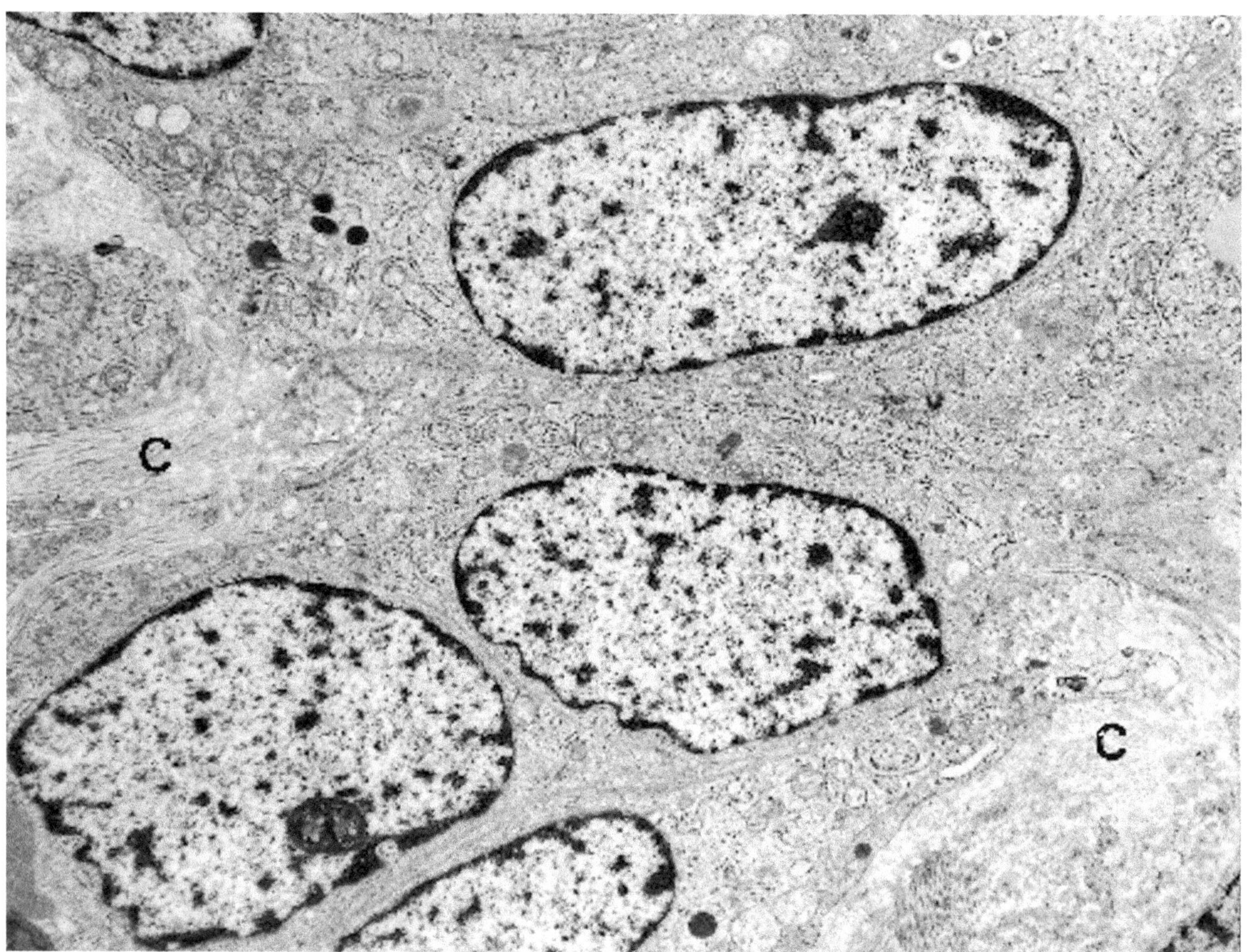

Figure 7.27. Endometrial stromal sarcoma (metastatic to lung). Higher magnification of cells from the same neoplasm as depicted in Figure 7.26 illustrates a relatively nondescript cytoplasm. Banded collagen (C) occupies the matrix between groups of cells. (× 9100)

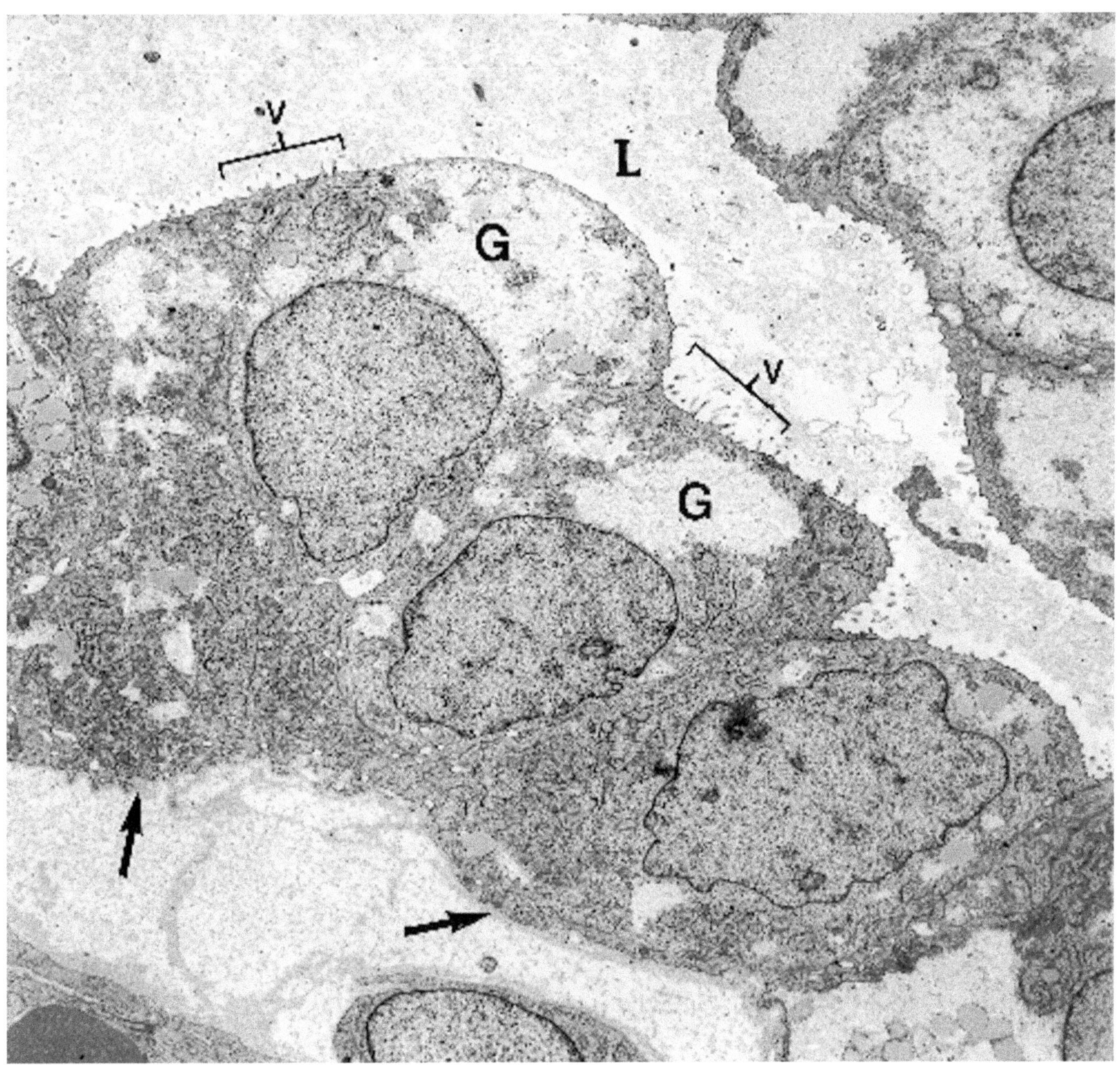

Figure 7.28. Clear cell carcinoma (ovary). Neoplastic clear cells line a lumen (L) and have irregularly distributed microvilli (V). Basal lamina (*arrows*) coats the basal aspect of the cells. The clearing in the cell cytoplasm is due to glycogen (G). Nuclei are irregularly oval and euchromatic. (× 5200)

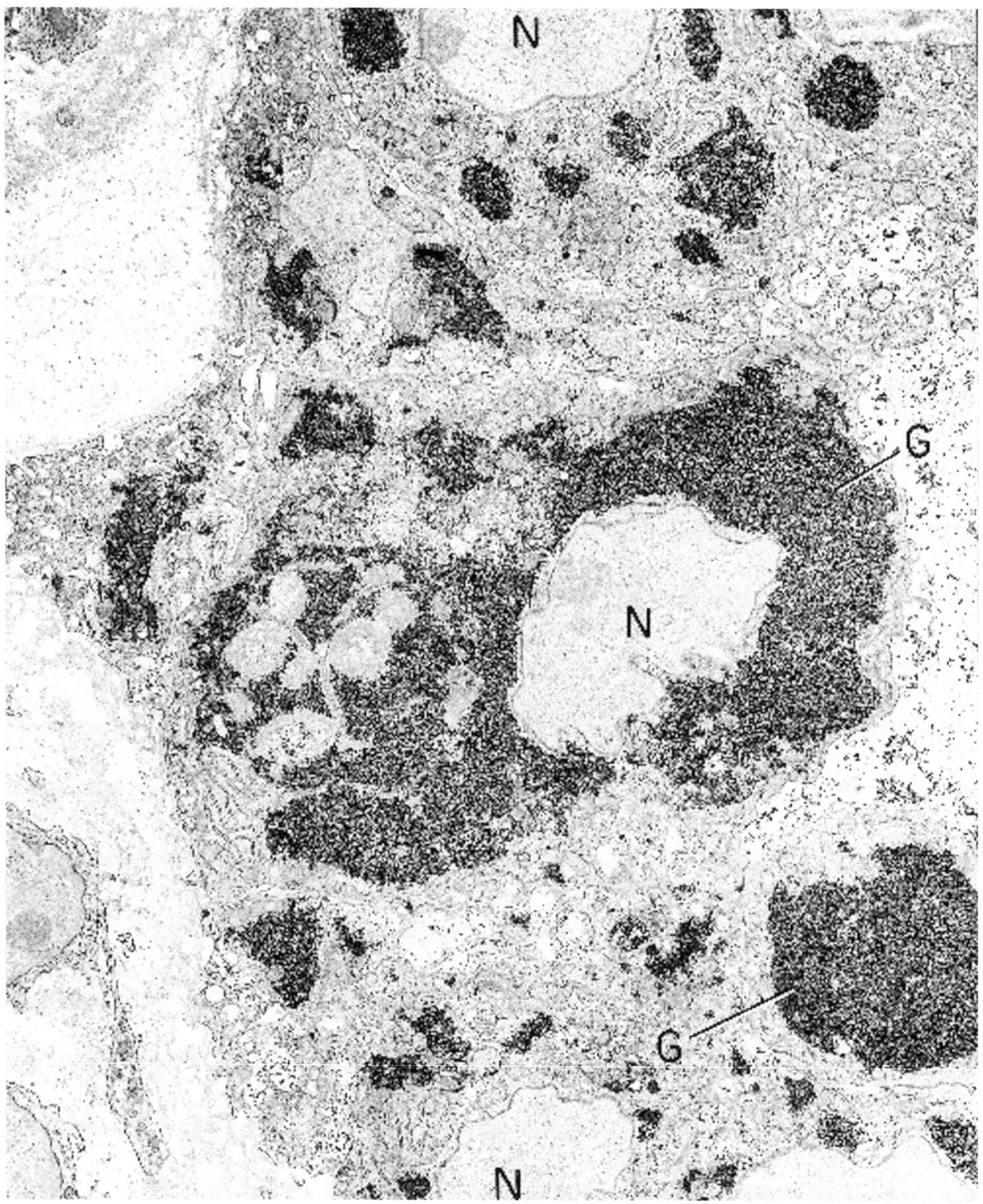

Figure 7.29. Clear cell carcinoma (cervix). The tissue in this illustration was processed by a method that preserves glycogen (G) as electron-dense granules, in contrast to the open spaces as shown in Figure 7.28. Nuclei (N) are less electron dense by this technique. (× 7200) (Permission for reprinting granted by WB Saunders, Dickersin GR: The ultrastructure of selected gynecologic neoplasms. Clin Lab Med 7:117–156, 1987.)

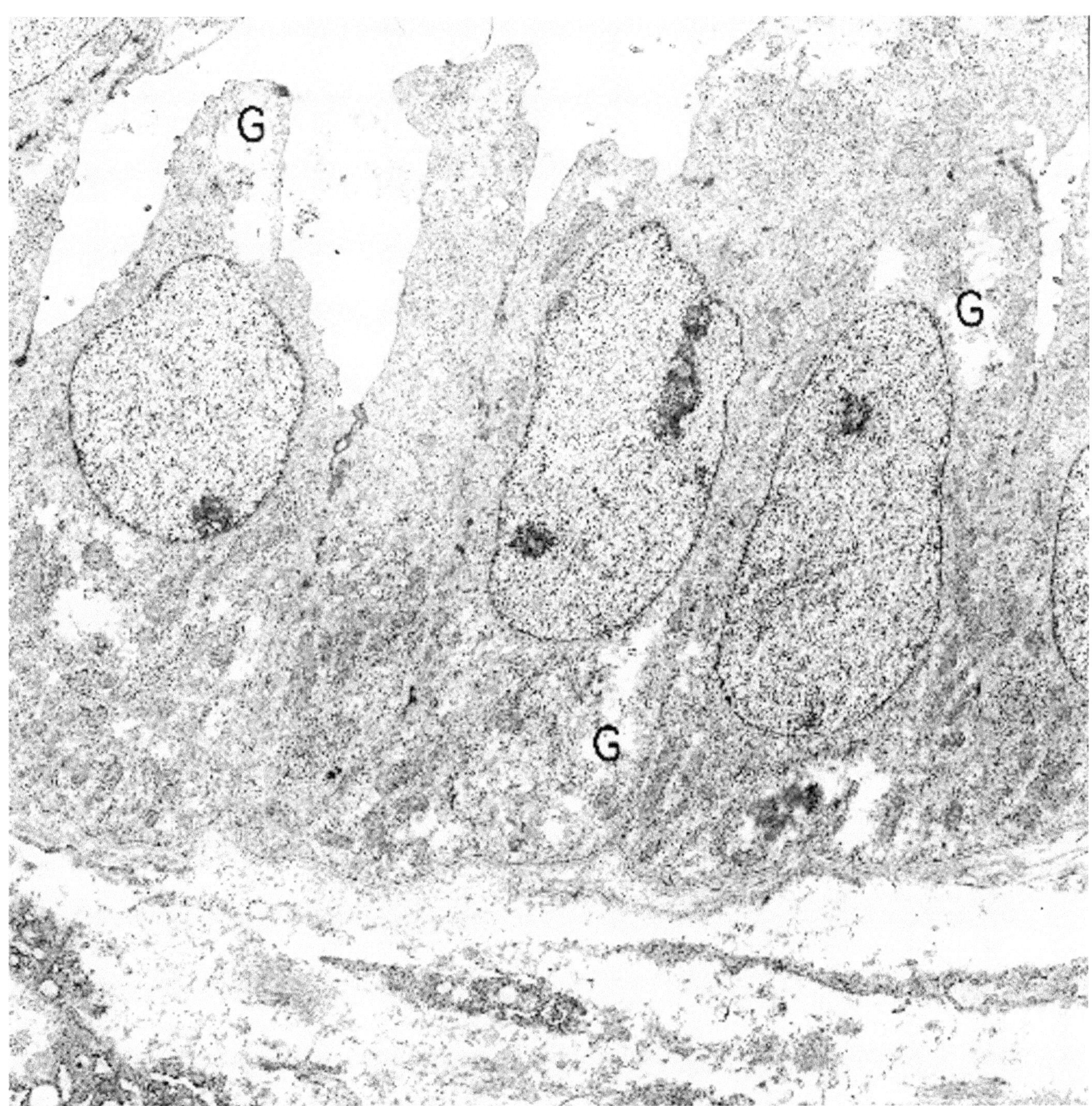

Figure 7.30. Clear cell carcinoma (cervix). The neoplastic cells lining the lumen are of the hobnail type. Glycogen (G) is abundant, nuclei are euchromatic, and nucleoli are prominent. (× 6080) (Permission for reprinting granted by WB Saunders, Dickersin GR: The ultrastructure of selected gynecologic neoplasms. Clin Lab Med 7:117–156, 1987.)

(Text continued from page 418)

Transitional Cell Tumors (Brenner and Non-Brenner Types)

(Figures 7.31 through 7.36.)

The epithelium of these tumors has urothelial features and may be benign, borderline, or malignant. In benign tumors, the urothelial-type cells are arranged in nests, and the stroma is fibromatous, resembling an ovarian fibroma. Luteinized cells with lipid droplets are occasionally present. Borderline and malignant Brenner tumors contain some benign Brenner tumor elements, whereas ovarian transitional cell carcinoma does not. Also, transitional cell carcinoma usually is mixed with other types of surface epithelial carcinomas, especially serous cystadenocarcinoma.

Diagnostic criteria. (1) Cysts and solid nests of large, oval, and polygonal cells; (2) basal lamina around cysts and nests; (3) increasing gradation in cellular size, cytoplasmic volume, and number of organelles, from basal layer to central or luminal lining cells; (4) wide intercellular spaces lined by villus-like projections; (5) amorphous, medium-dense material lining intercellular spaces; (6) numerous pinocytotic vesicles; (7) short, stubby microvilli on luminal surface of cells; (8) junctional complexes; (9) numerous small vesicles in the apical cytoplasm of luminal lining cells and in the lumens; (10) indented nuclei.

Additional points. The cytoplasm of the epithelial cells contains numerous free ribosomes, many filaments, a moderate number of mitochondria, a small amount of rough endoplasmic reticulum, occasional secondary lysosomes, and a few lipid droplets. There are no mucous or other secretory granules in these urothelial type cells, but ciliated serous and mucinous epithelial components as well as argyrophilic cells with dense-core granules are found in some Brenner tumors. Squamous metaplasia may also be seen.

Squamous Cell Tumors (Epidermoid Cyst and Squamous Cell Carcinoma)

The ultrastructural characteristics of squamous cells and the cells comprising these tumors are covered in Chapter 3.

Mixed Epithelial Tumors

These tumors are composed of combinations of various types of surface epithelial–stromal tumors in which the minority component(s) composes at least 10% of the total tumor. Examples include Brenner tumor with a mucinous cystic component, an endometrioid/clear cell mixture, and an endometrioid/serous combination, among others. The ultrastructure of the various cellular components has already been described.

Undifferentiated Carcinoma

There is very little information available about the electron microscopic appearance of large cell undifferentiated carcinomas of the ovary. The small cell types are described in Chapter 4 in the section on oat cell (neuroendocrine) carcinoma and later in this chapter in the section on small cell undifferentiated carcinoma of the hypercalcemic type.

Sex Cord–Stromal Tumors

Sex cord–stromal tumors contain a single cell type or a combination of cell types, including luteinized or nonleuteinized granulosa cells and theca cells, Sertoli cells, Leydig cells, and stromal fibroblasts. The most common of these tumors are thecomas/fibromas, and granulosa cell tumors are next in frequency.

Granulosa Cell Tumor

(Figures 7.37 through 7.44.)

Diagnostic criteria. (1) Groups of polygonal (granulosa) cells; (2) basal lamina around some of the groups; (3) intercellular junctions; (4) high nuclear–cytoplasmic ratio (relatively small amount of cytoplasm) in the adult form and lower nuclear cytoplasmic ratio (moderate to large amount of cytoplasm) in the juvenile form; (5) oval and deeply indented ("grooved") nuclei in the adult form and nonindented nuclei in the juvenile form; (6) varying numbers of spindle (theca) cells, arranged individually in a matrix of collagen; (7) the spindle cells may have abundant rough endoplasmic reticulum (fibroblasts); or (8) they may have rough endoplasmic reticulum plus filaments, dense aggregates of filaments (dense bodies), basal lamina, and pinocytotic vesicles (myofibroblasts); or (9) they may have abundant smooth endoplasmic reticulum, many lipid droplets, and mitochondria with tubular cristae (luteinized thecal cells).

Additional points. In well-differentiated granulosa cell tumors, a microcystic (Call-Exner-like) or, less commonly, a macrocystic pattern is present. Granulosa cells occasionally may be luteinized in the same way as are thecal cells (criterion no. 9). Other variable cytoplasmic features include whorled aggregates of filaments, medium-size Golgi apparatuses, a small to moderate amount of rough endoplasmic reticulum, and occasional lipid droplets. Glycogen is scant or absent. In juvenile granulosa cell tumors, the granulosa cells and theca cells both tend to be large and have abundant cytoplasm and atypical nuclei. The epithelial cells have round, nongrooved, heterochromatic nuclei and lipid-rich cytoplasm. Stromal cells are also rich in lipid. Call-Exner bodies are rare.

(Text continues on page 439)

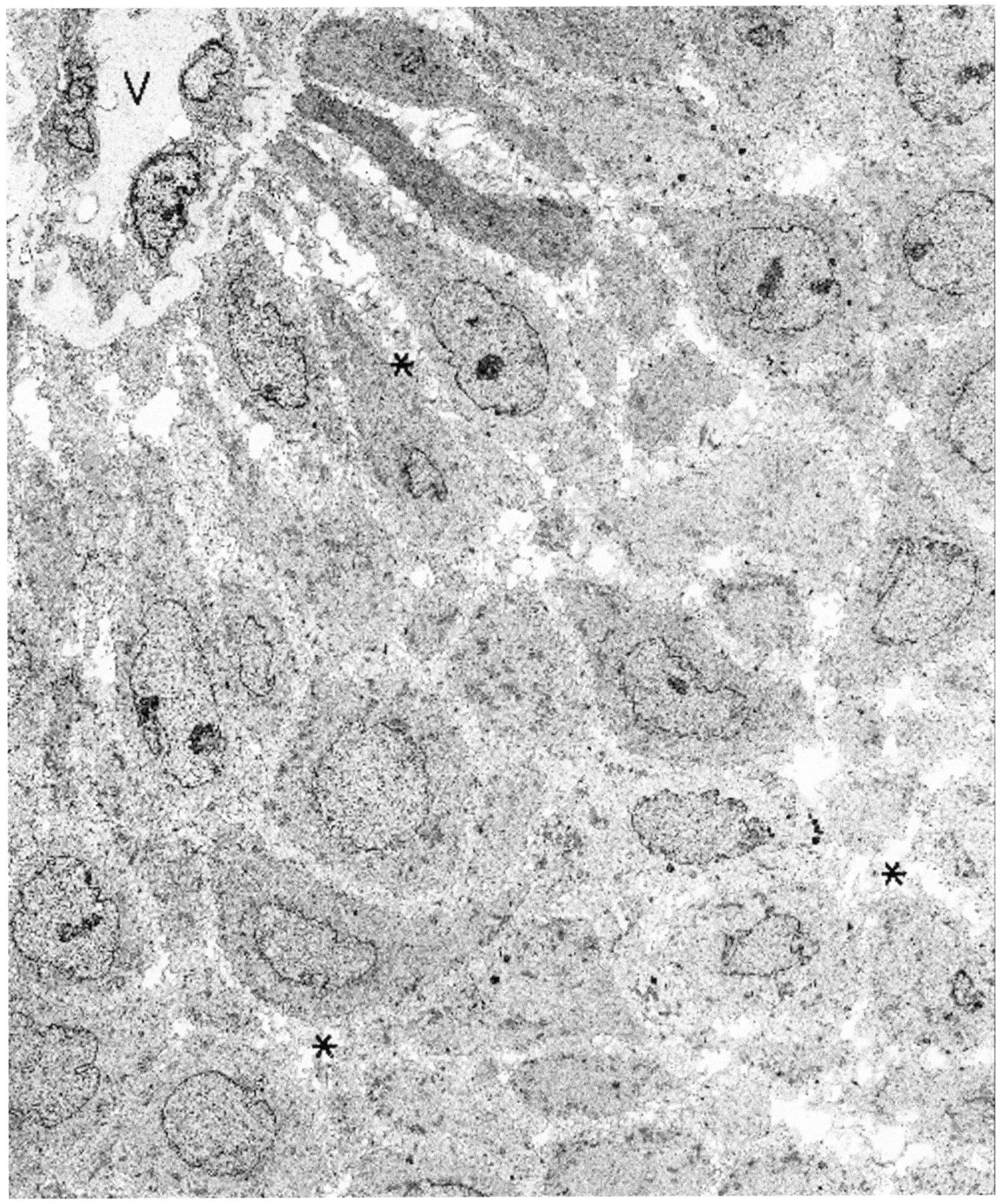

Figure 7.31. Brenner tumor (ovary). A solid nest of neoplastic cells with a gradient in shape and size, from basal zone (next to blood vessel, V) to central region. Wide intercellular spaces (*) are lined by numerous villus-like projections. (× 2725)

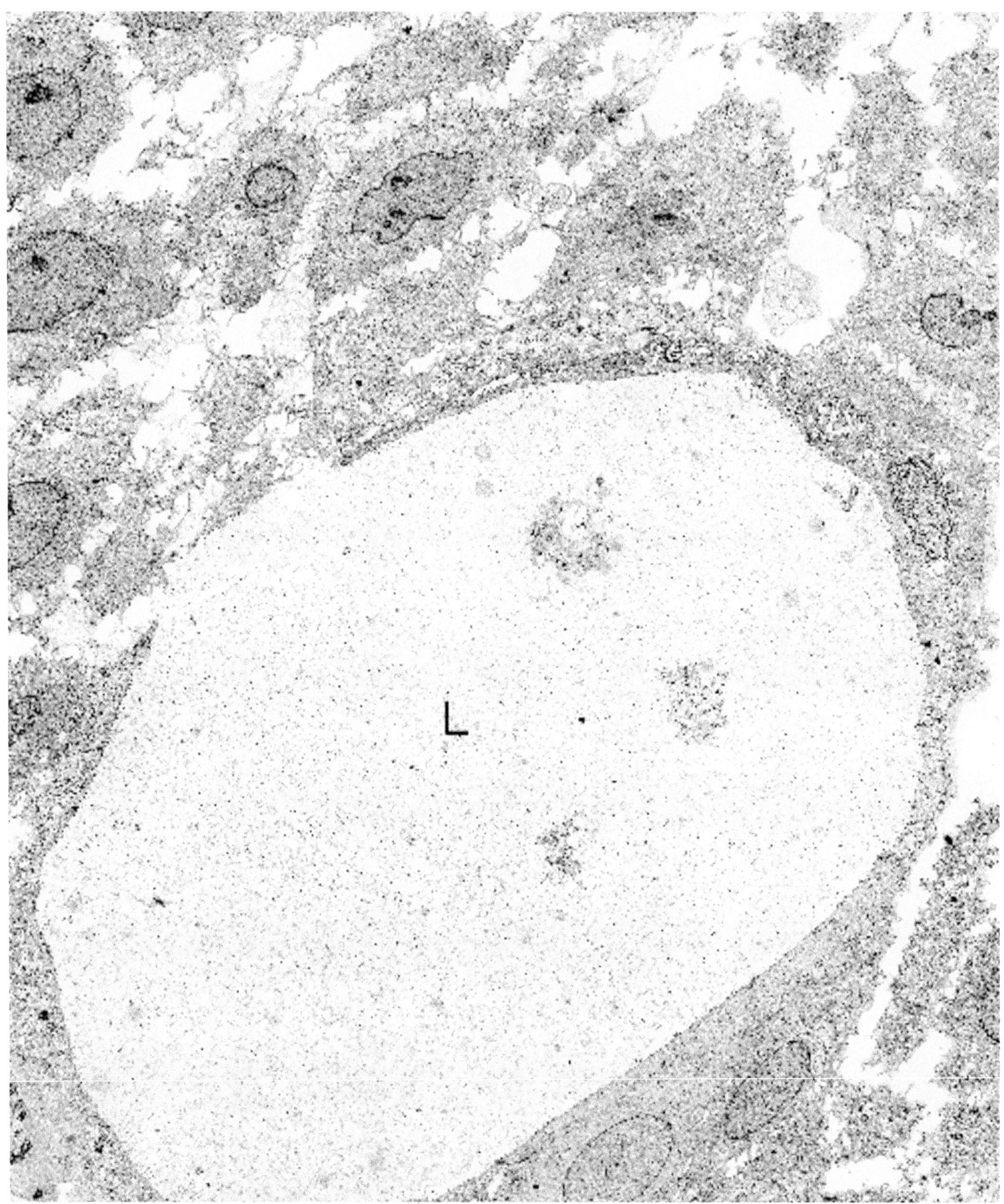

Figure 7.32. Brenner tumor (ovary). A group of neoplastic cells has a central cyst (L), which at higher power (Figure 7.35) proves to be a true lumen with microvilli and junctional complexes. Note again the wide intercellular spaces with villus-like projections. (× 2440)

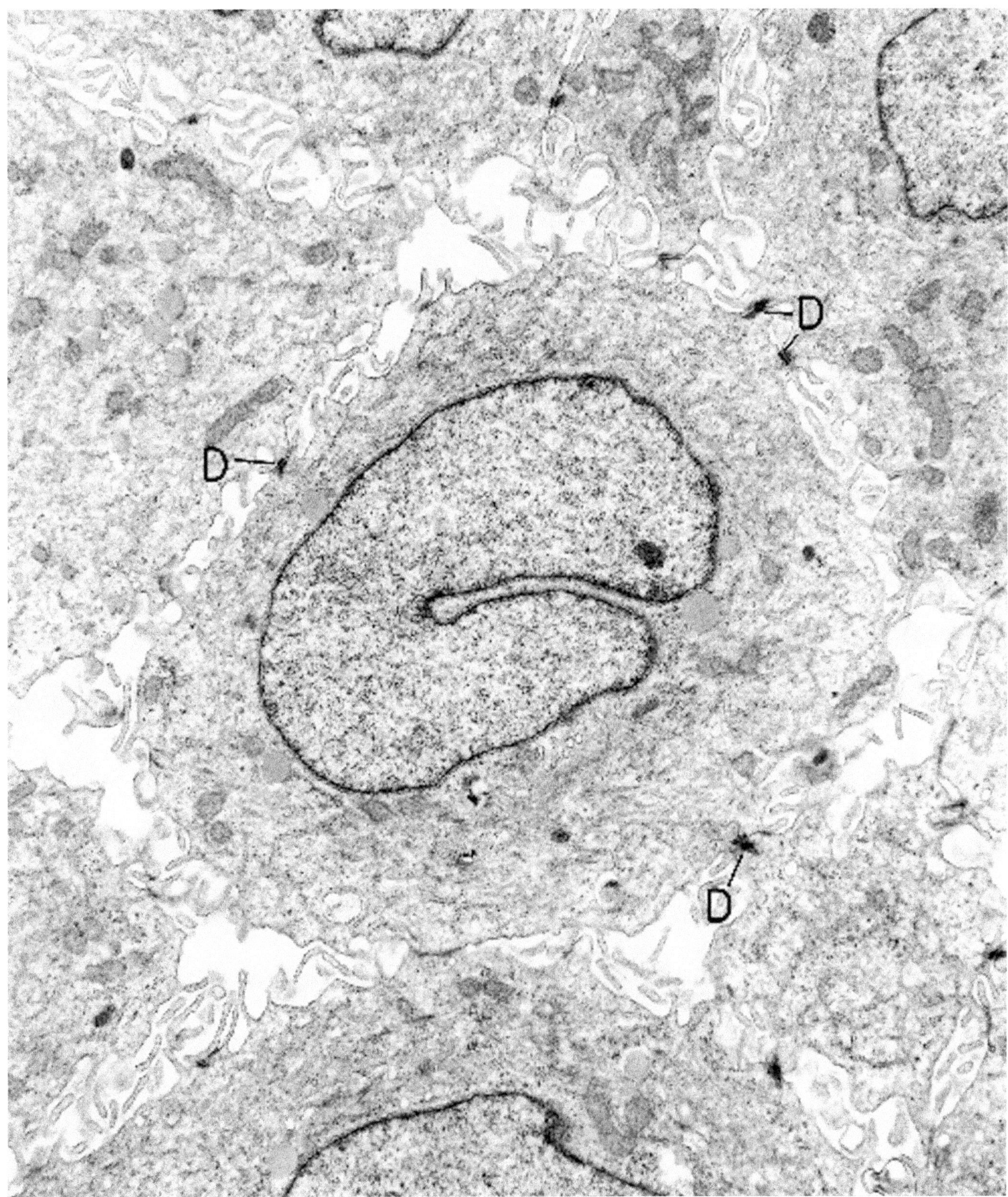

Figure 7.33. Brenner tumor (ovary). A typical neoplastic cell has a large, indented, euchromatic nucleus, and a moderate amount of cytoplasm. No particular organelle predominates, and there are no secretory granules. The plasma membrane is raised into many villus-like projections, and the intercellular space is wide. Desmosomes (D) are numerous and prominent. (× 4500)

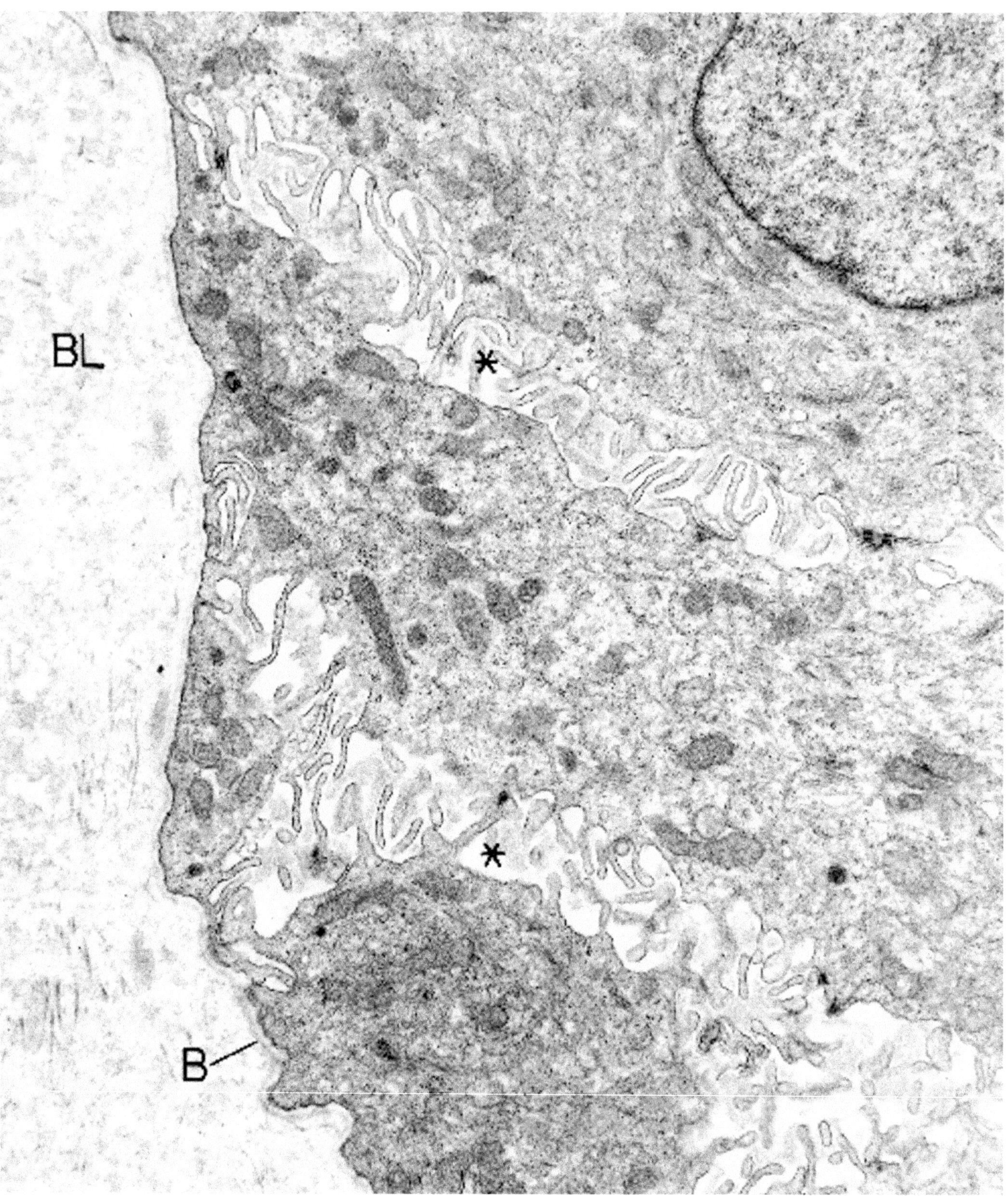

Figure 7.34. Brenner tumor (ovary). High magnification of the periphery of a nest of neoplastic cells illustrates well: discrete (B) as well as diffuse (BL) basal lamina, intercellular spaces (*) with villus-like projections, and amorphous, medium-dense coating. (× 16,500)

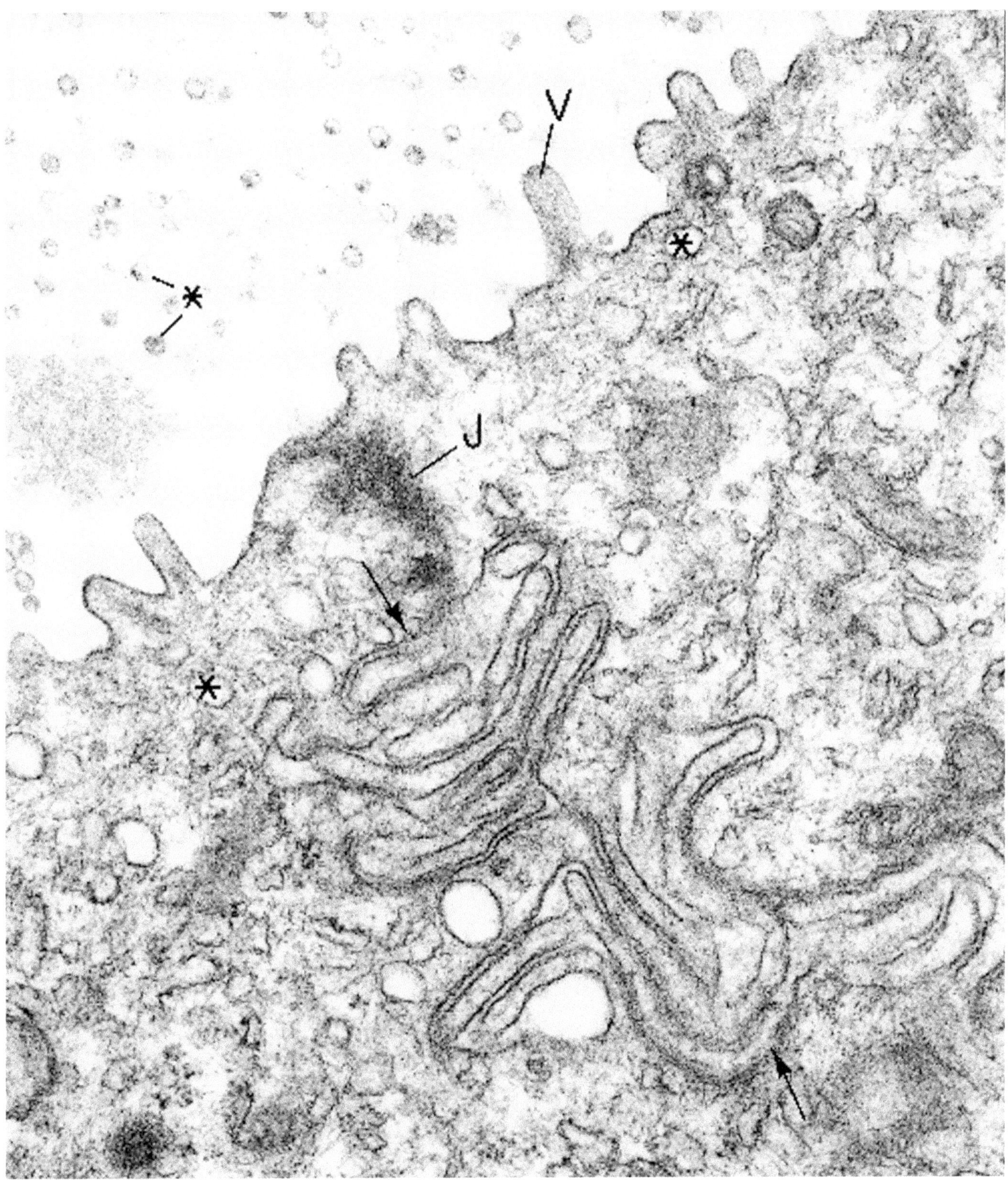

Figure 7.35. Brenner tumor (ovary). The cells lining the lumen of this cyst have short, stubby microvilli (V) and junctional complexes (J). Note also (between the *arrows*) the interdigitating lateral plasma membranes. Many small vesicles (*) are located in the apical cytoplasm of the cells, and in the lumen of the cyst. (× 58,800) (Permission for reprinting granted by WB Saunders, Dickersin GR: The ultrastructure of selected gynecologic neoplasms. Clin Lab Med 7:117–156, 1987.)

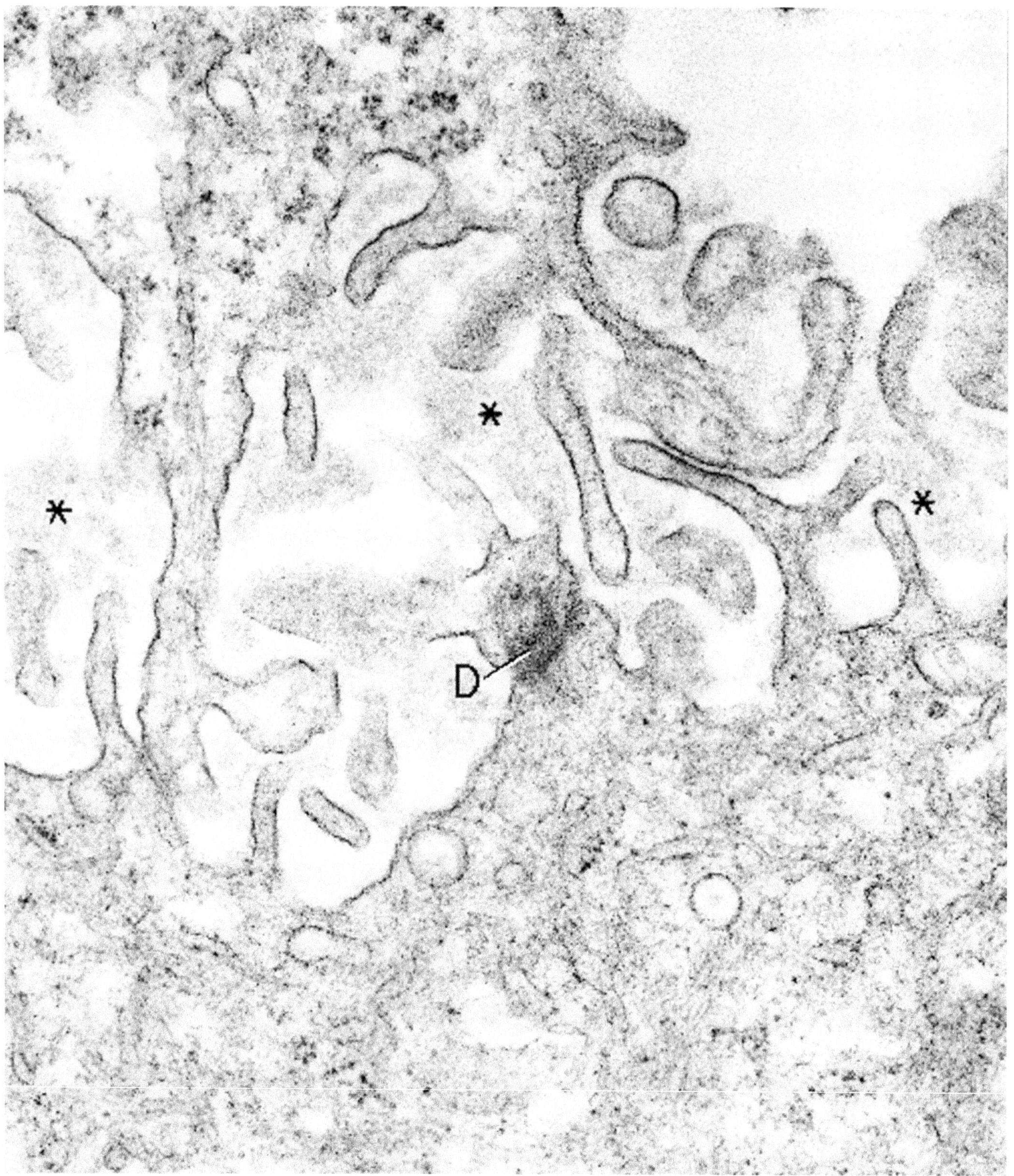

Figure 7.36. Brenner tumor (ovary). High magnification of an intercellular space depicts the villus-like projections from the cells, the amorphous coating (*) of the cellular surfaces and a desmosome (D). (× 66,000)

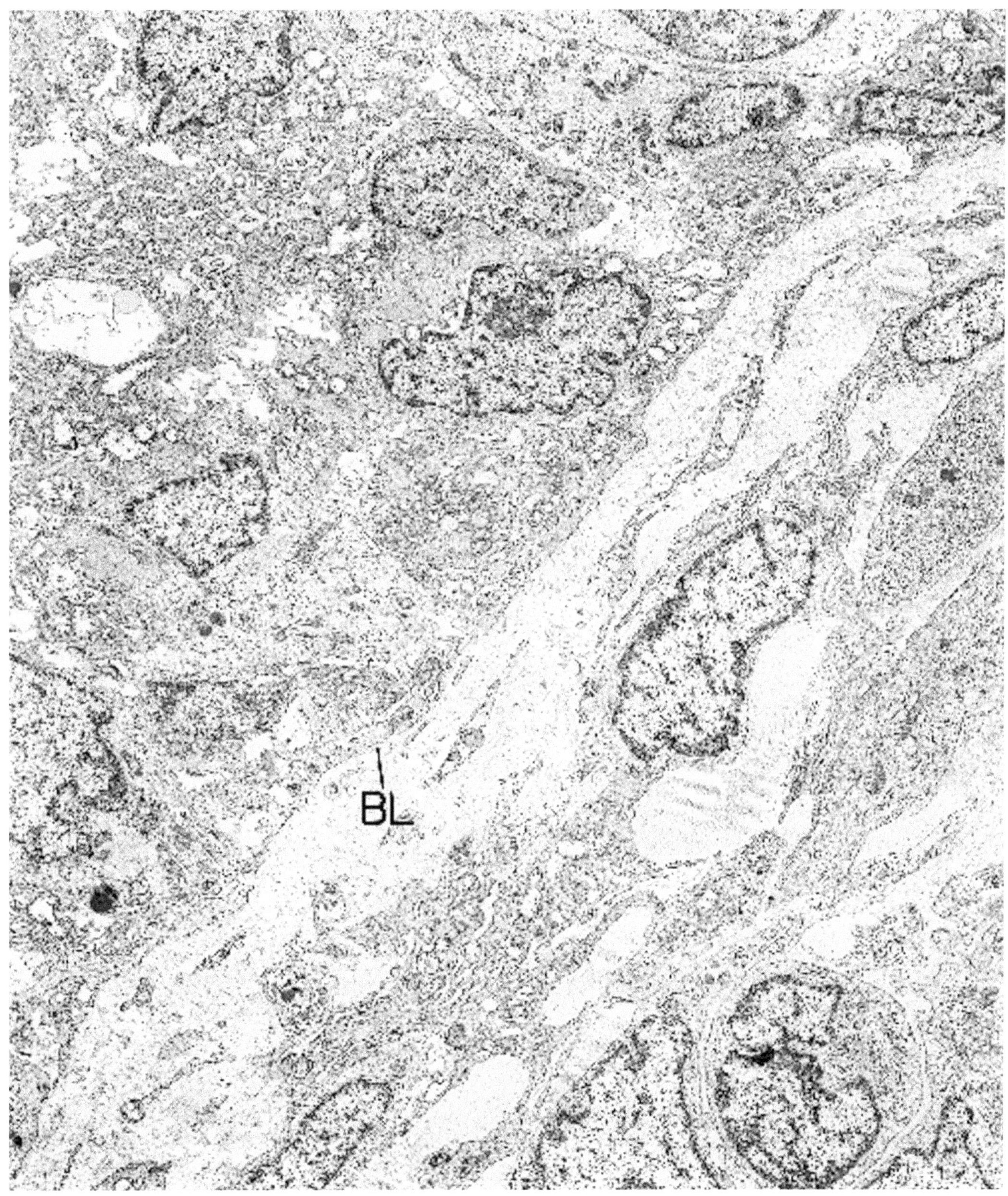

Figure 7.37. Granulosa cell tumor (ovary). A group of granulosa cells on the left is sharply demarcated by basal lamina (BL), barely discernible at this magnification, from the stroma cells on the right. (× 6570) (Permission for reprinting granted by WB Saunders, Dickersin GR: The ultrastructure of selected gynecologic neoplasms. Clin Lab Med 7:117–156, 1987.)

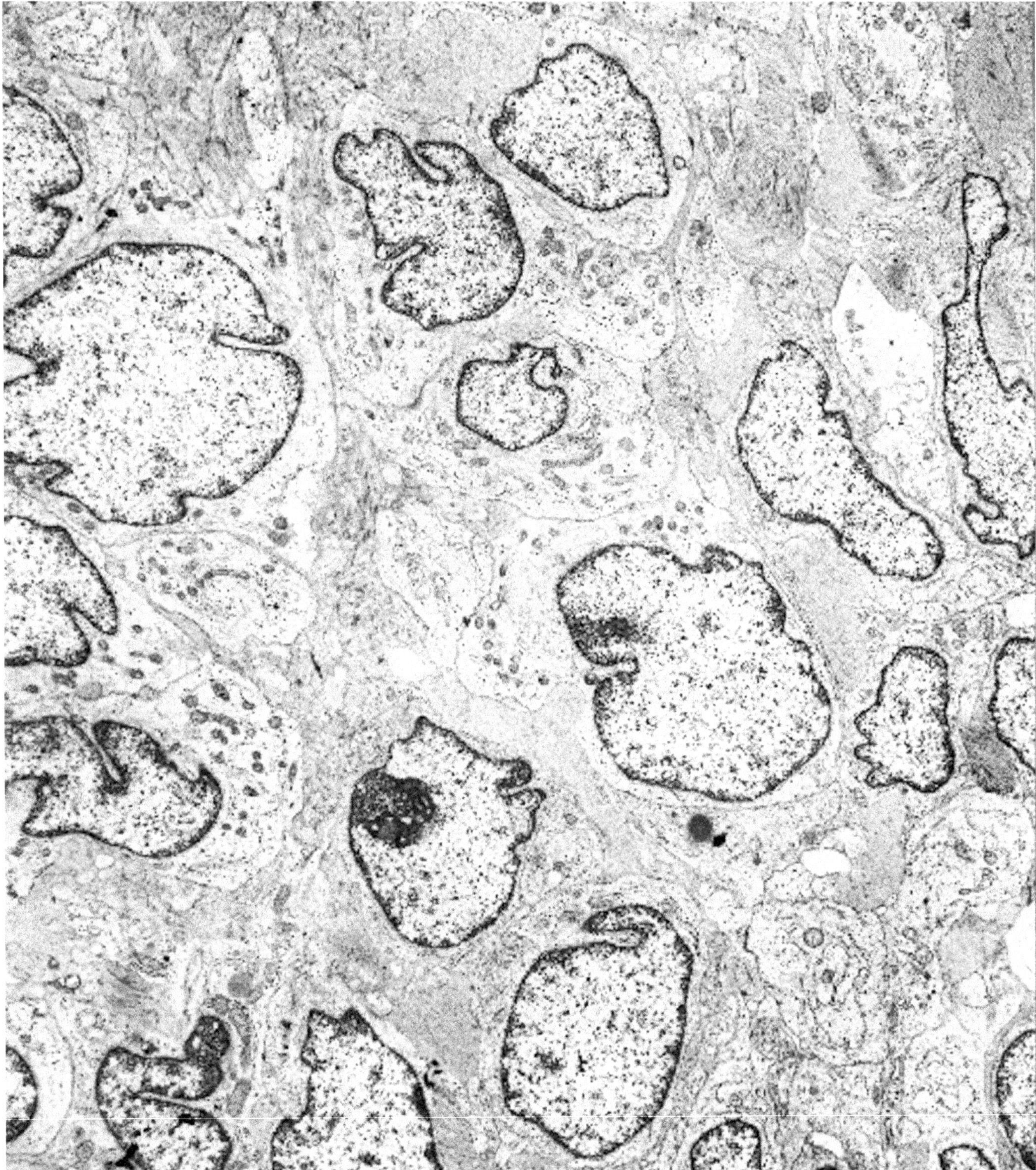

Figure 7.38. Granulosa cell tumor (ovary). The granulosa cells are polygonal and closely apposed and have a high nuclear–cytoplasmic ratio. Nuclei characteristically are indented. (× 5936)

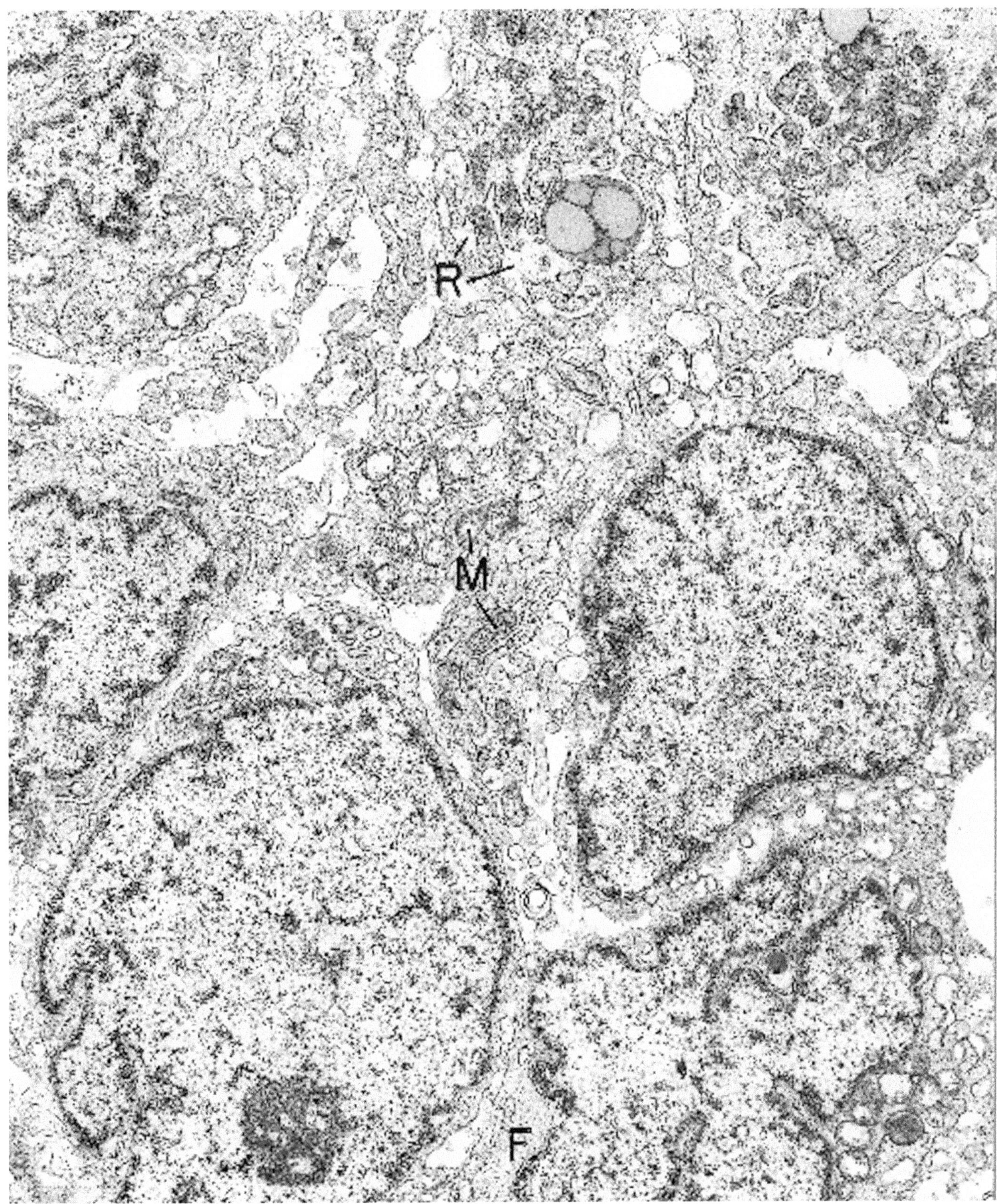

Figure 7.39. Granulosa cell tumor (ovary). The granulosa cells have a relatively small amount of cytoplasm, the main components of which are ribosomes, mitochondria (M), rough endoplasmic reticulum (R), and filaments (F). (× 12,160) (Permission for reprinting granted by WB Saunders, Dickersin GR: The ultrastructure of selected gynecologic neoplasms. Clin Lab Med 7:117–156, 1987.)

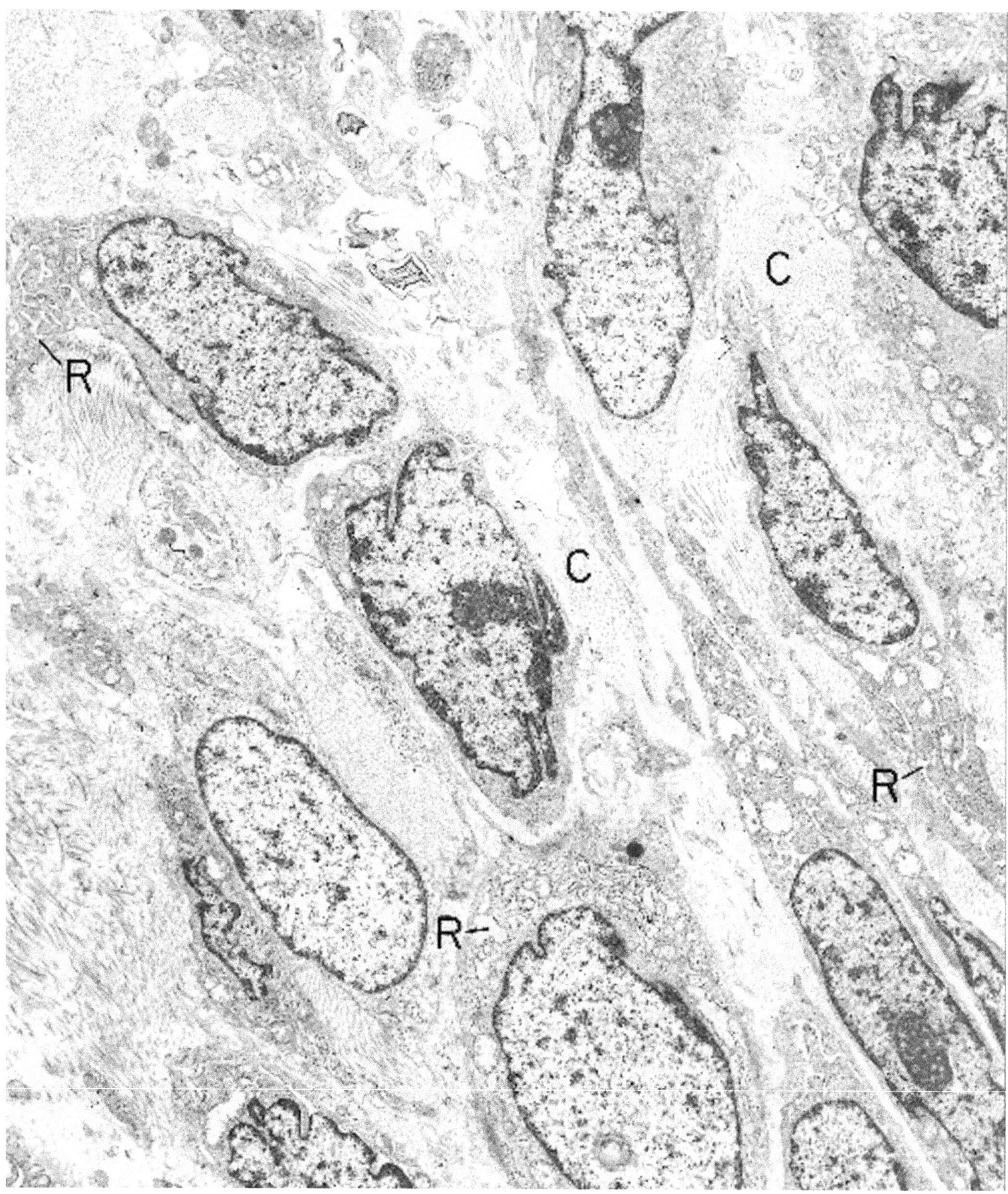

Figure 7.40. Granulosa cell tumor (ovary). The stromal (thecal) cells between the islands of granulosa cells are spindle shaped and individually dispersed in a matrix of collagen (C). At this magnification, the most prominent organelle in some of the cells is dilated rough endoplasmic reticulum (R), a fibroblastic characteristic. (× 6570) (Permission for reprinting granted by WB Saunders, Dickersin GR: The ultrastructure of selected gynecologic neoplasms. Clin Lab Med 7:117–156, 1987.)

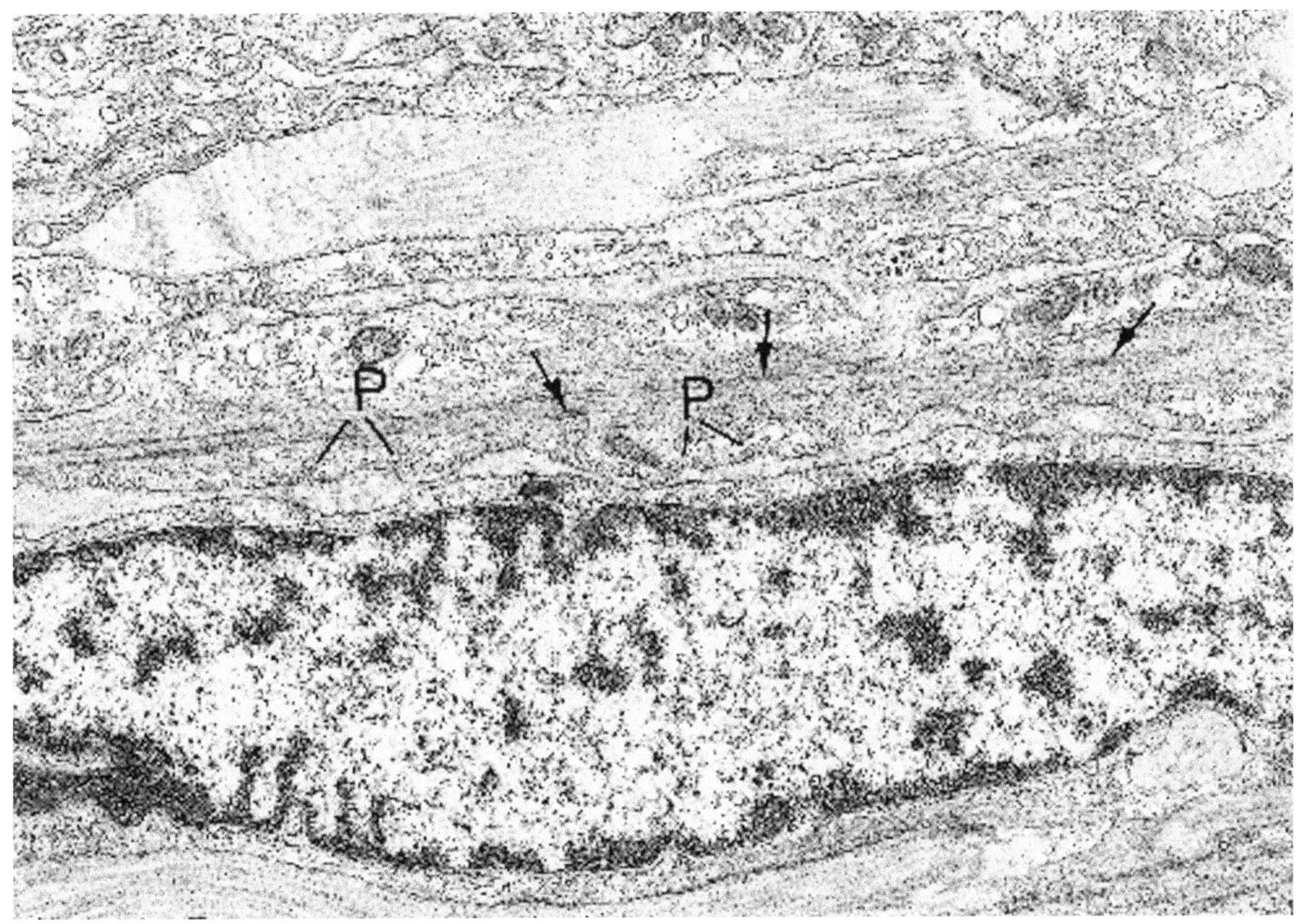

Figure 7.41. Granulosa cell tumor (ovary). Some of the stromal cells have the smooth muscle features of microfilaments, dense bodies (*arrows*), and pinocytotic vesicles (P). (× 14,000) (Permission for reprinting granted by WB Saunders, Dickersin GR: The ultrastructure of selected gynecologic neoplasms. Clin Lab Med 7:117–156, 1987.)

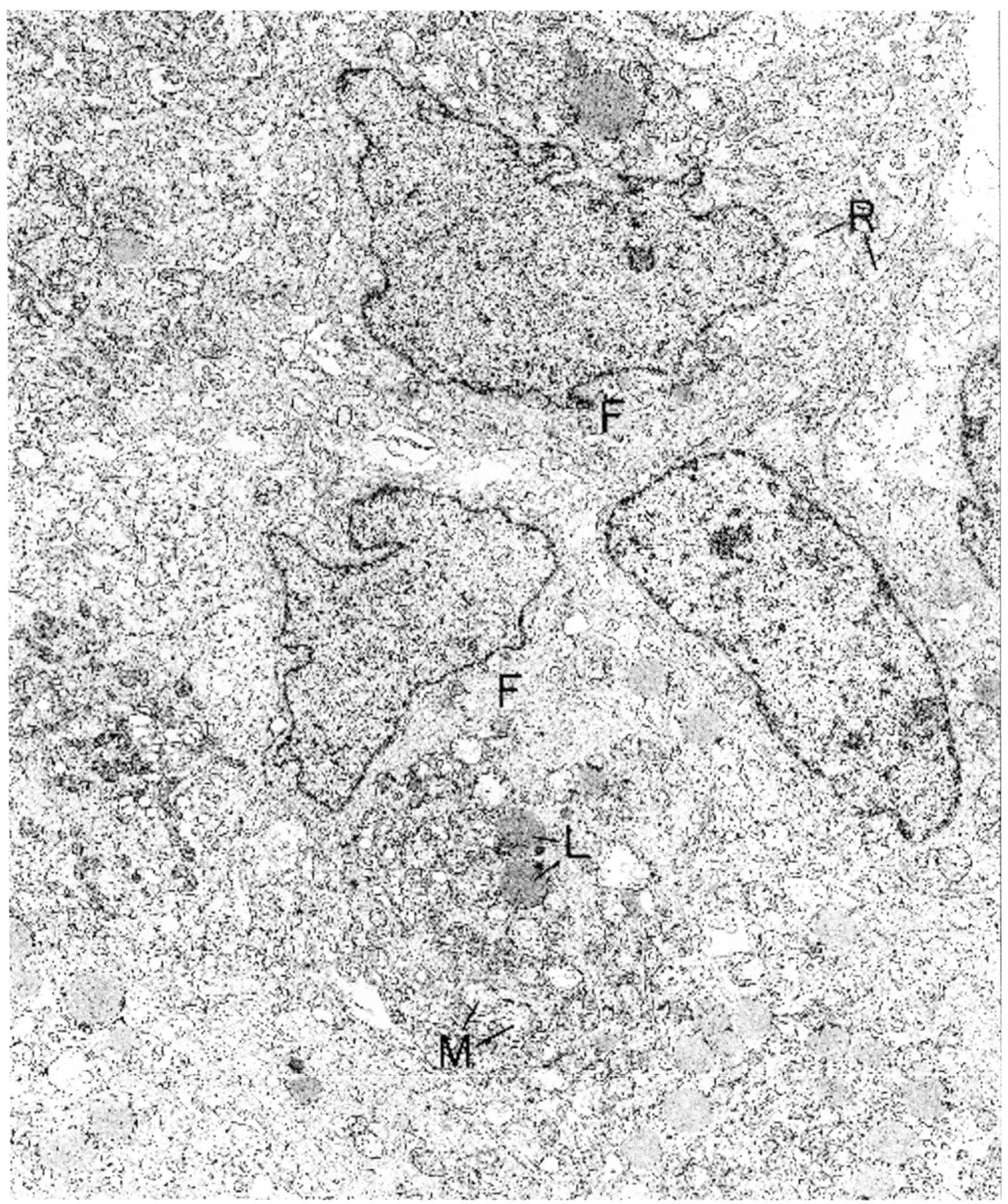

Figure 7.42. Juvenile granulosa cell tumor (ovary). The granulosa cells are large and have abundant cytoplasm and many organelles. Filaments (F), mitochondria (M), cisternae of rough endoplasmic reticulum (R), and lipid droplets (L) are all numerous. (× 7630) (Permission for reprinting granted by WB Saunders, Dickersin GR: The ultrastructure of selected gynecologic neoplasms. Clin Lab Med 7:117–156, 1987.)

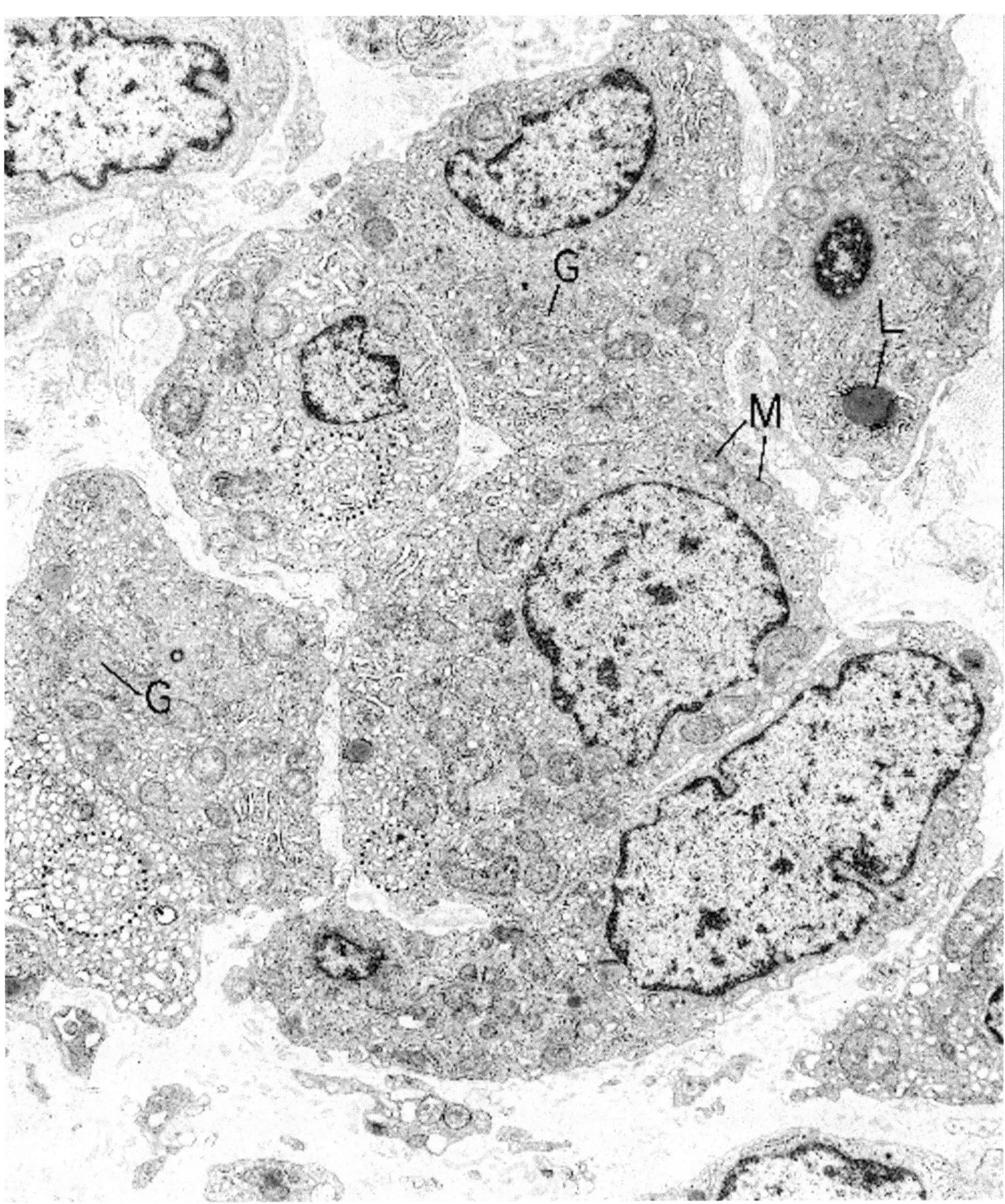

Figure 7.43. Juvenile granulosa cell tumor (ovary). These theca cells are consistent with being hormonally active, being plump and having copious cytoplasm with abundant vesicles and cisternae of smooth endoplasmic reticulum (circles), a moderate number of mitochondria (M), large Golgi apparatuses (G), and occasional droplets of lipid (L). (× 7200) (Permission for reprinting granted by WB Saunders, Dickersin GR: The ultrastructure of selected gynecologic neoplasms. Clin Lab Med 7:117–156, 1987.)

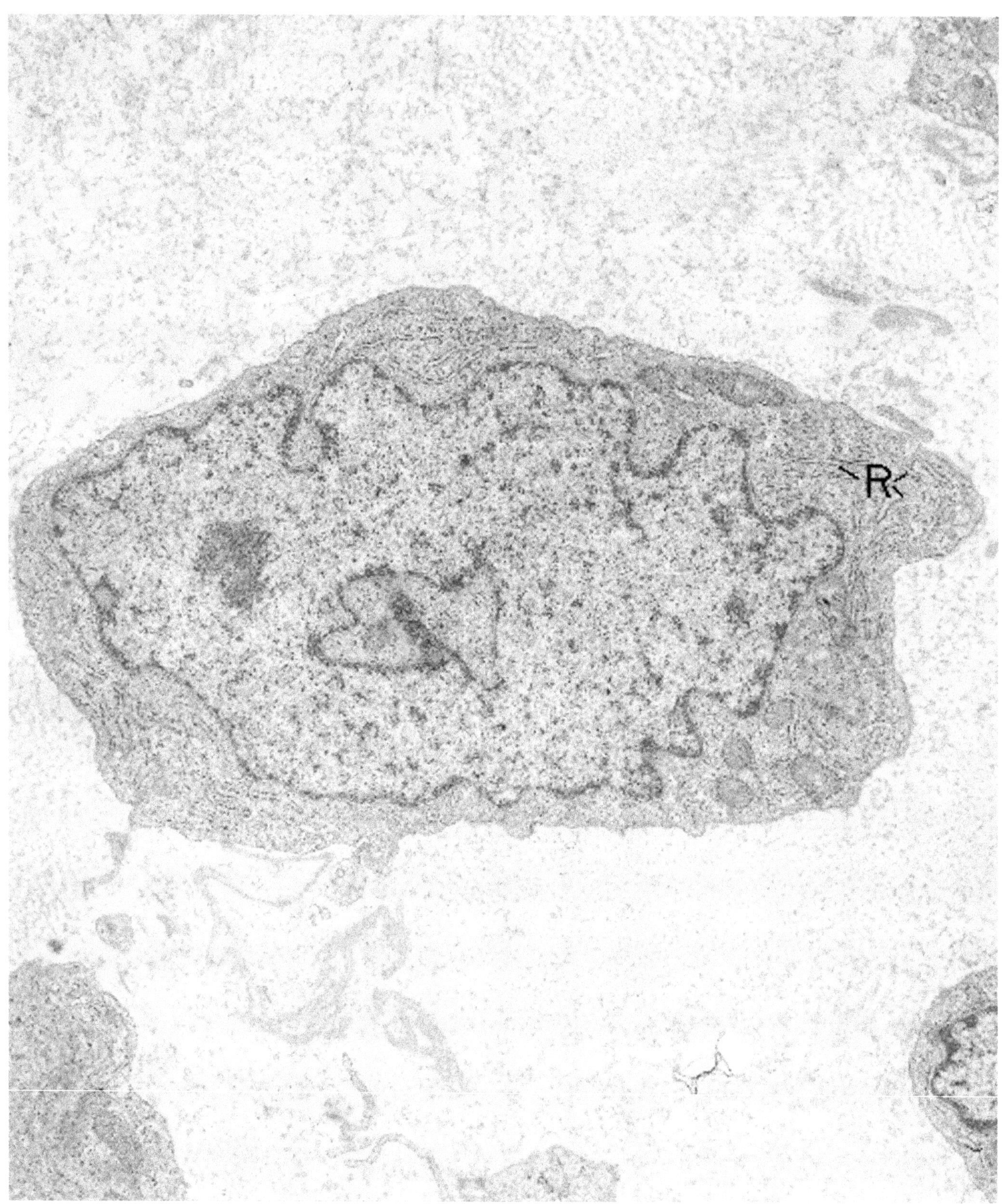

Figure 7.44. Juvenile granulosa cell tumor (ovary). This theca cell has an irregular nucleus and a less-active-appearing cytoplasm than the cells in Figure 7.43; that is, the cytoplasmic compartment is small, and undilated rough endoplasmic reticulum (R) is the main organelle. (× 11,020)

(Text continued from page 424)

Thecoma

(Figure 7.45.)

Diagnostic criteria. (1) Diffuse arrangement of oval and fusiform cells with (2) copious cytoplasm; (3) a varying number of lipid droplets; (4) smooth endoplasmic reticulum; (5) oval to elongated nuclei.

Additional points. Thecomas are composed of cells that resemble theca interna cells and may be luteinized, having numerous cytoplasmic lipid droplets, with or without prominent smooth endoplasmic reticulum. The intercellular matrix is rich in banded collagen.

Fibroma

(Figures 7.46 and 7.47.)

Diagnostic criteria. (1) Oval and spindle cells with (2) a high nucleocytoplasmic ratio; (3) scant cytoplasm; (4) scant organelles except for more conspicuous and dilated rough endoplasmic reticulum in some cells; (5) intercellular matrix of banded collagen.

Additional points. These tumors resemble fibromas in other tissues (see Chapter 6) and can be cellular and fascicular or hypocellular and edematous. A few lipid droplets may be present in the cytoplasm of the cells, but the overall amount of lipid and volume of the cytoplasm are less than in thecoma.

Signet-Ring Stromal and Related Tumors

(Figures 7.48 through 7.51.)

Diagnostic criteria. (1) Diffuse arrangement of oval and spindle nonvacuolated cells and (2) round and oval signet-ring type cells with large cytoplasmic vacuoles of variable type and eccentric nuclei; (3) absence of mucin and lipid.

Additional points. The nuclei and cytoplasm of the nonvacuolated and vacuolated cells are similar except for the absence or presence of vacuoles, respectively. The type of vacuole in any one tumor may consist of hydropic swelling (Figure 7.48), mitochondrial swelling (Figure 7.49), and pseudoinclusions from the extracellular matrix (Figures 7.50 and 7.51). Although the majority of these tumors are derived from cells consistent with an ovarian stromal origin, some tumors may be "stromal-like" by light microscopy and prove to be epithelial with junctions, ultrastructurally. The type of epithelial cell in these cases is uncertain but is most likely in the unclassified sex cord category.

Sertoli–Stromal Cell Tumors (Androblastomas)

(Figures 7.52 through 7.55.)

Diagnostic criteria. (1) Solid or hollow tubules (of Sertoli cells), surrounded by basal lamina and separated by a fibrocollagenous stroma; (2) tubular lumens lined by cuboidal or columnar cells with microvilli and junctional complexes; (3) pseudolumens filled with basal lamina; (4) interdigitated lateral cell membranes; (5) basally located nuclei; (6) rare Charcot–Böttcher filaments; (7) groups of oval or polygonal Leydig cells having copious cytoplasm; (8) abundant smooth endoplasmic reticulum; (9) lipid droplets; (10) mitochondria with tubular cristae (11) secondary lysosomes containing lipochrome pigment; (12) rare crystals of Reinke and their filamentous precursors; (13) oval nuclei with regular contours, finely dispersed chromatin, and large nucleoli.

Additional points. The ultrastructural features of these neoplasms depend on their degree of differentiation and the proportions of the two cell types present. Well-differentiated tumors form tubules and have lining cells that resemble closely the Sertoli cells of the normal testis. Tubular patterns may blend into a diffuse pattern of similar cells. In addition to the diagnostic criteria listed in the previous paragraph, well-differentiated Sertoli cells also have many free ribosomes, a small amount of rough and sometimes smooth endoplasmic reticulum, many small mitochondria, filaments, scattered and sometimes numerous lipid droplets, and a varying number of secondary lysosomes. Leydig cells in these neoplasms usually are devoid of crystals of Reinke, perhaps because of incomplete differentiation or the small size of the sample used for electron microscopy. The cytoplasmic structures that correlate with steroid hormone synthesis are smooth endoplasmic reticulum, tubular cristae in the mitochondria, and lipid droplets. Some less differentiated Sertoli–stromal cell tumors have a retiform pattern, and some contain heterologous elements such as various mesenchymal cells and gastrointestinal epithelium, including goblet cells, Paneth cells, and endocrine cells.

(Text continues on page 451)

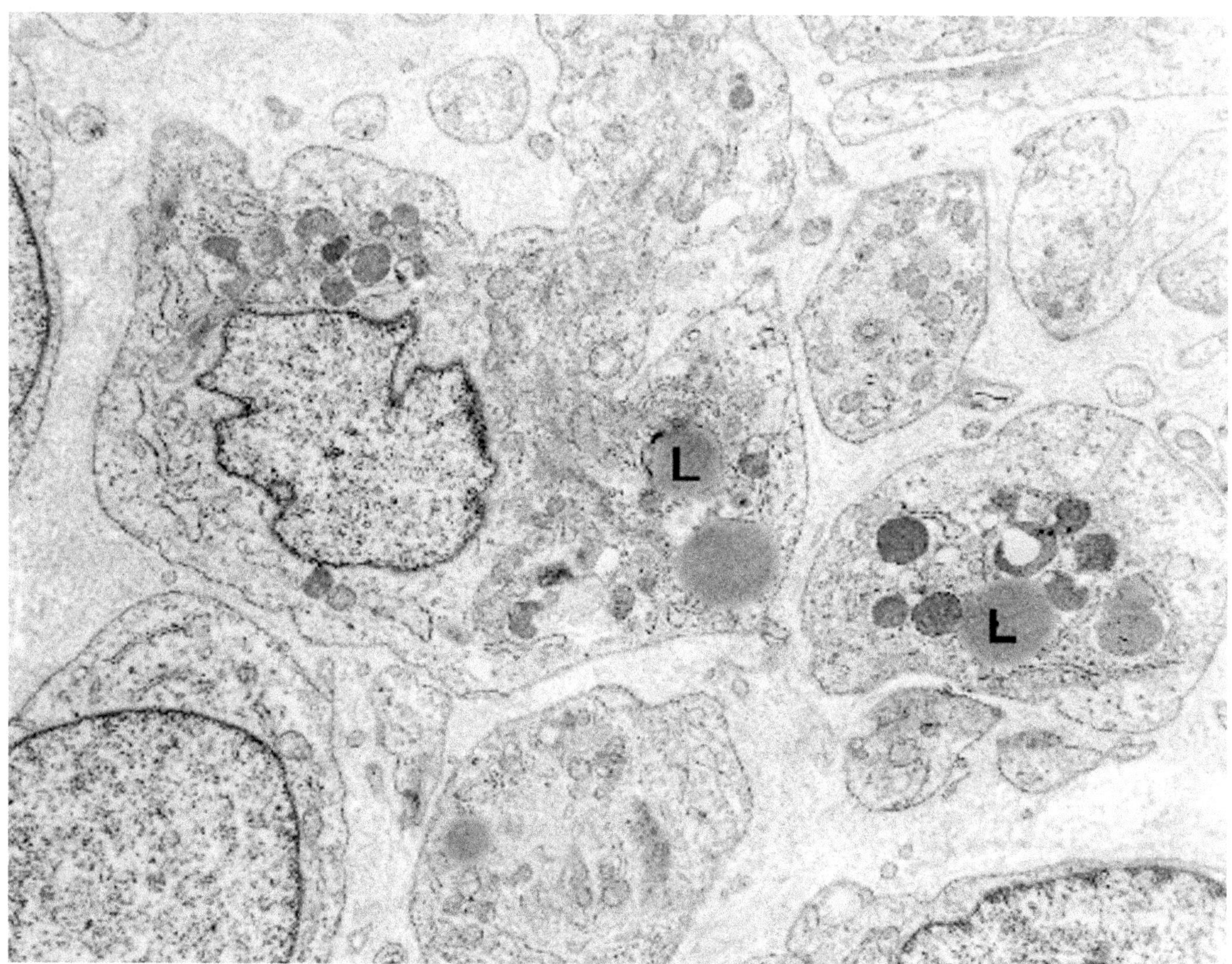

Figure 7.45. Thecoma (ovary). Neoplastic oval cells have copious cytoplasm, a moderate number of lipid droplets (L), and a mixture of other cytoplasmic organelles, including numerous small vesicles (not well seen at this magnification) and oval and irregular nuclei. (× 9800)

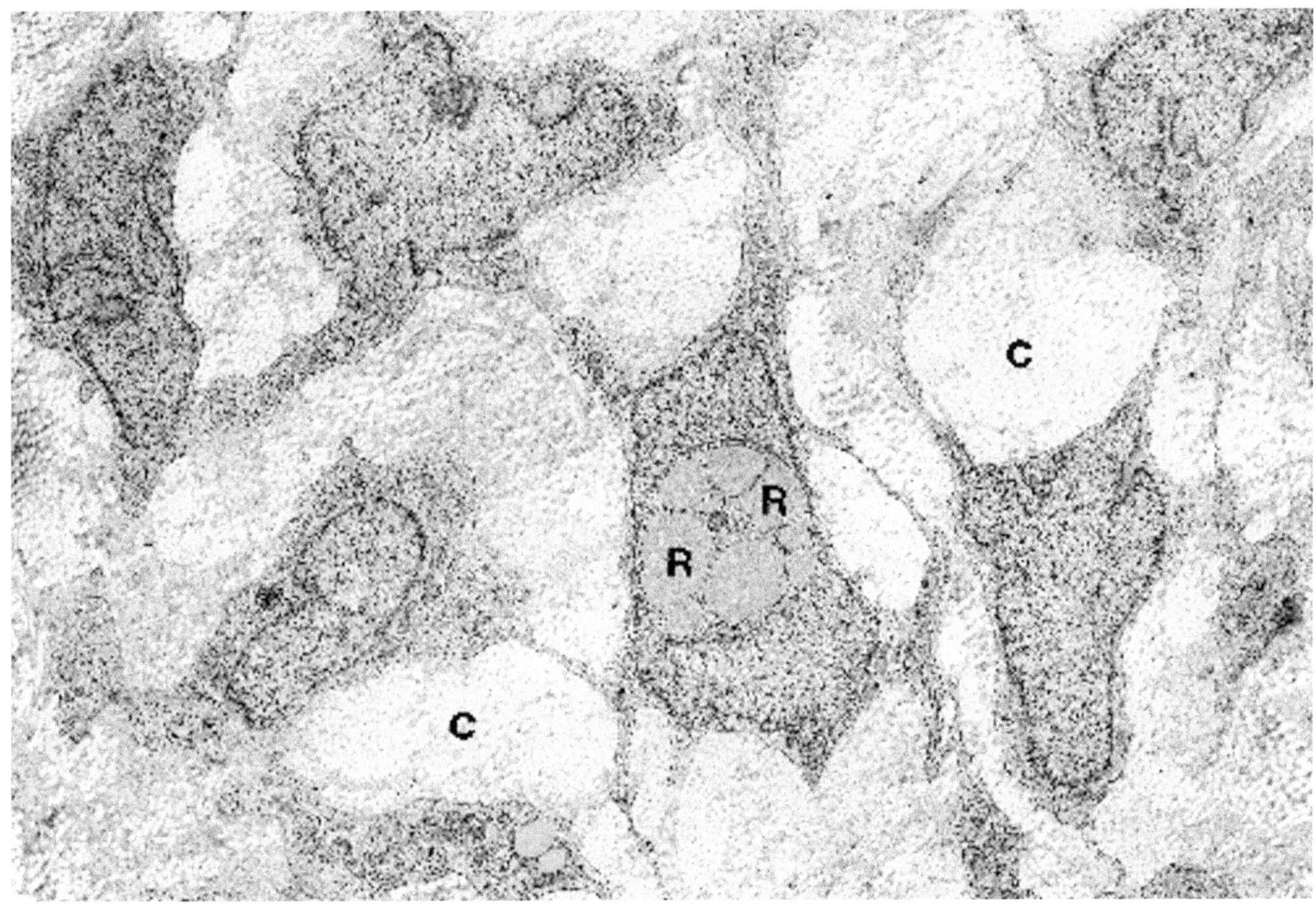

Figure 7.46. Fibroma (ovary). Oval and spindle cells have a high nuclear–cytoplasmic ratio and scant cytoplasm. One cell has a nuclear pseudoinclusion formed by invaginated cytoplasm rich in dilated rough endoplasmic reticulum (R). The extracellular matrix is rich in banded collagen (C). (× 7000)

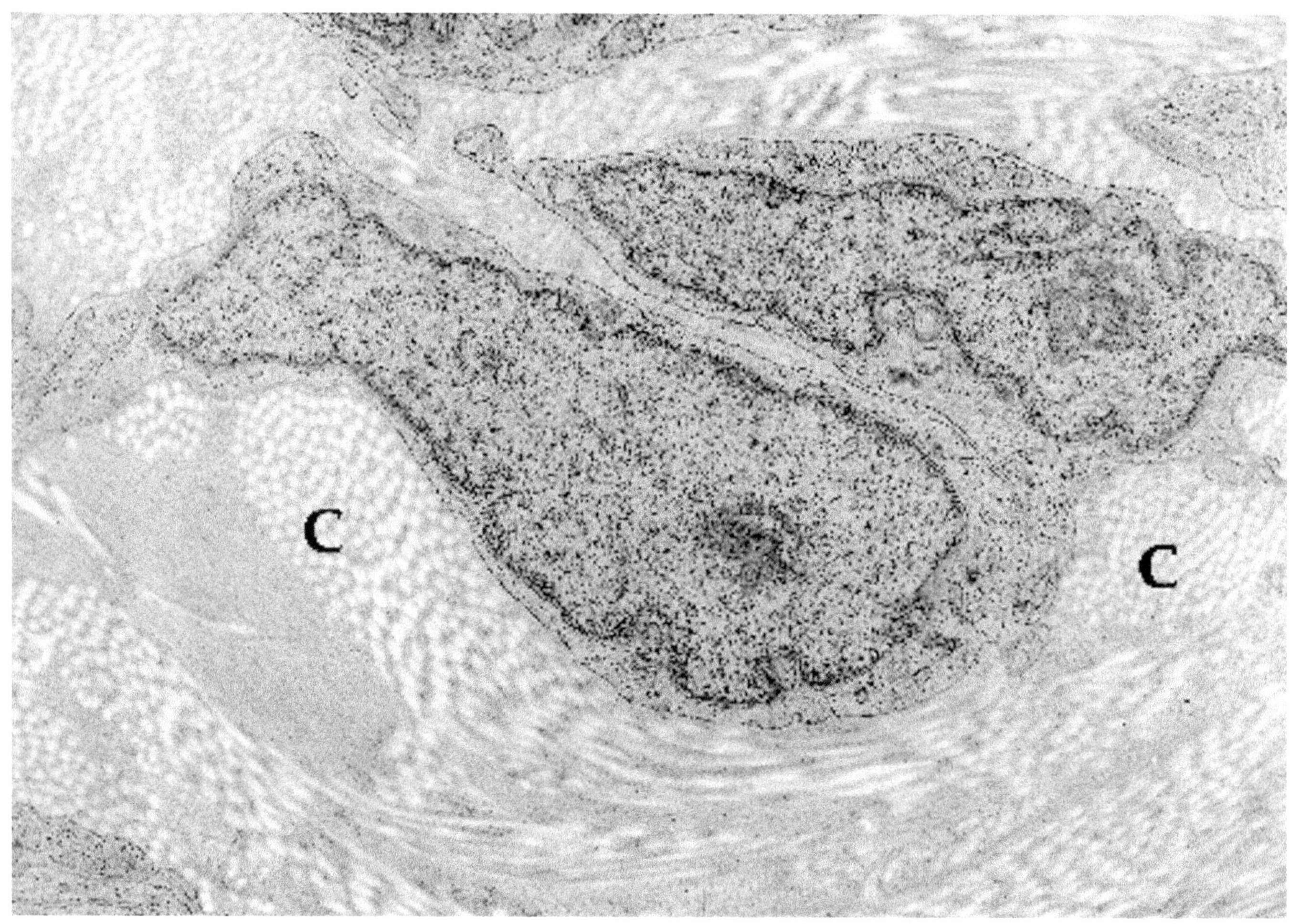

Figure 7.47. Fibroma (ovary). Higher magnification of the same neoplasm in Figure 7.46 reveals scant cytoplasm and scant organelles, notably less rough endoplasmic reticulum than in typical fibroblasts. C = banded collagen. (× 13,400)

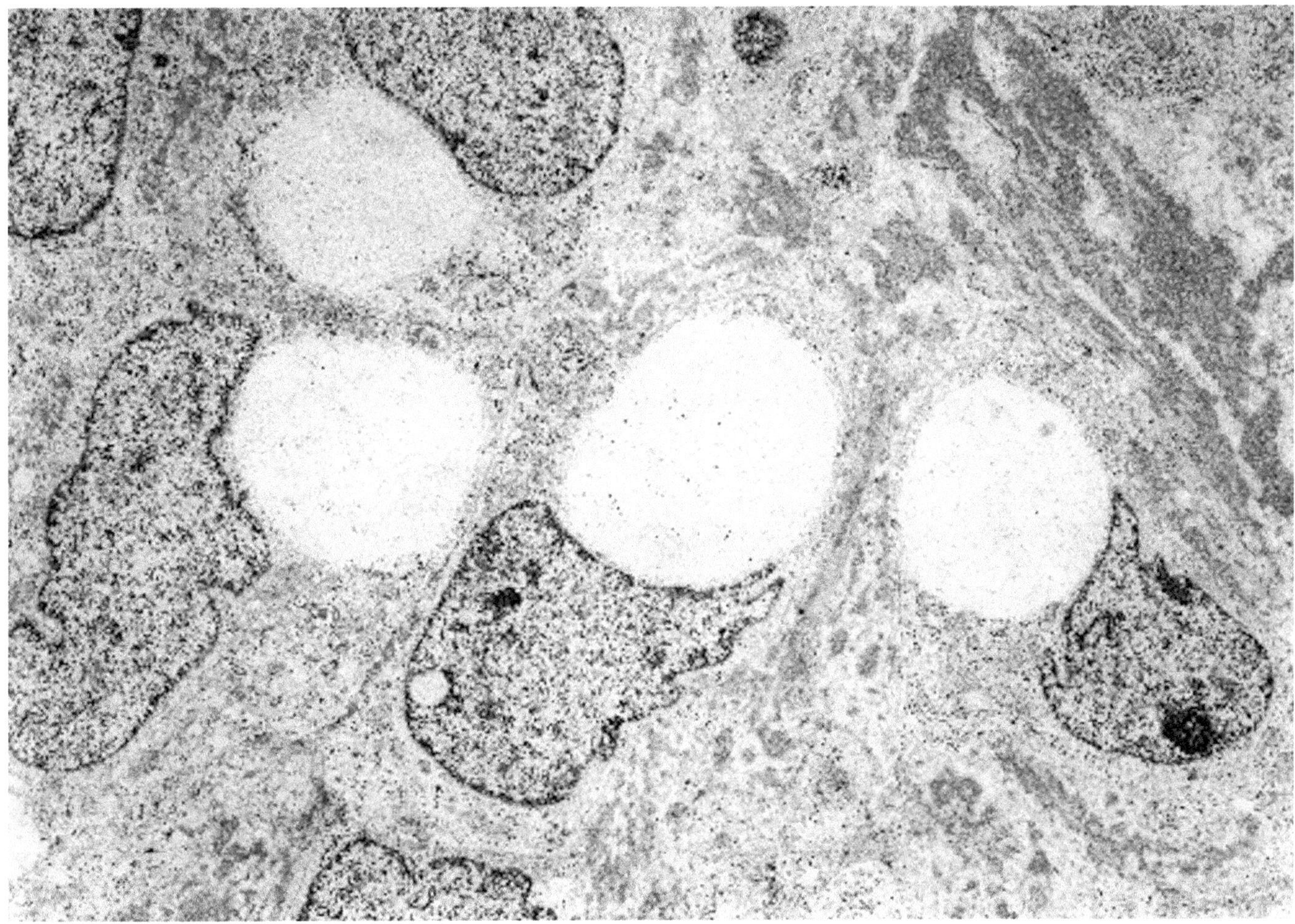

Figure 7.48. Signet-ring stromal tumor (ovary). These signet-ring stromal cells contain a single large vacuole that is not bound by a membrane. Other cells in the tumor and not depicted here had less discrete and less severe cytoplasmic clearing, supportive of a hydropic type swelling. No particular cytoplasmic organelle was dilated, and the clearing appeared to be consistent with increase in cytosol. (× 5200)

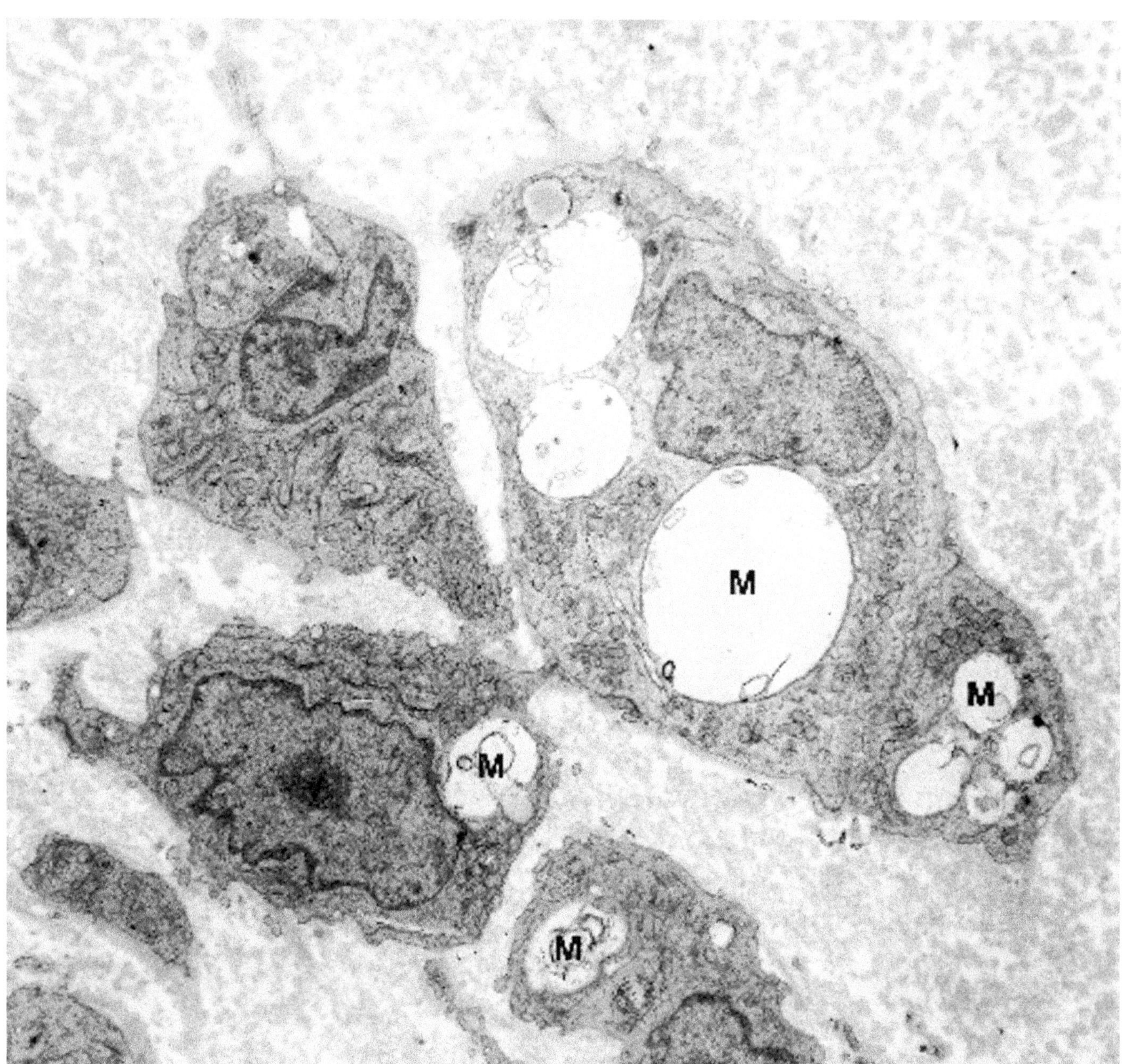

Figure 7.49. Signet-ring stromal tumor (ovary). The cells in this neoplasm show varying degrees of dilatation of mitochondria, the more severe forms resulting in signet-ring-type cells as illustrated. A gradient in size of mitochondria was evident in some cells, and most cells had a moderate to large number of mitochondria. M = swollen mitochondria, with limiting membrane and residual cristae still being evident. (× 7500) (Permission for reprinting granted by Taylor and Francis, Dickersin GR, Young RH, Scully RE: Signet-ring stromal and related tumors of the ovary. Ultrastruct Pathol 19:401–419, 1995.)

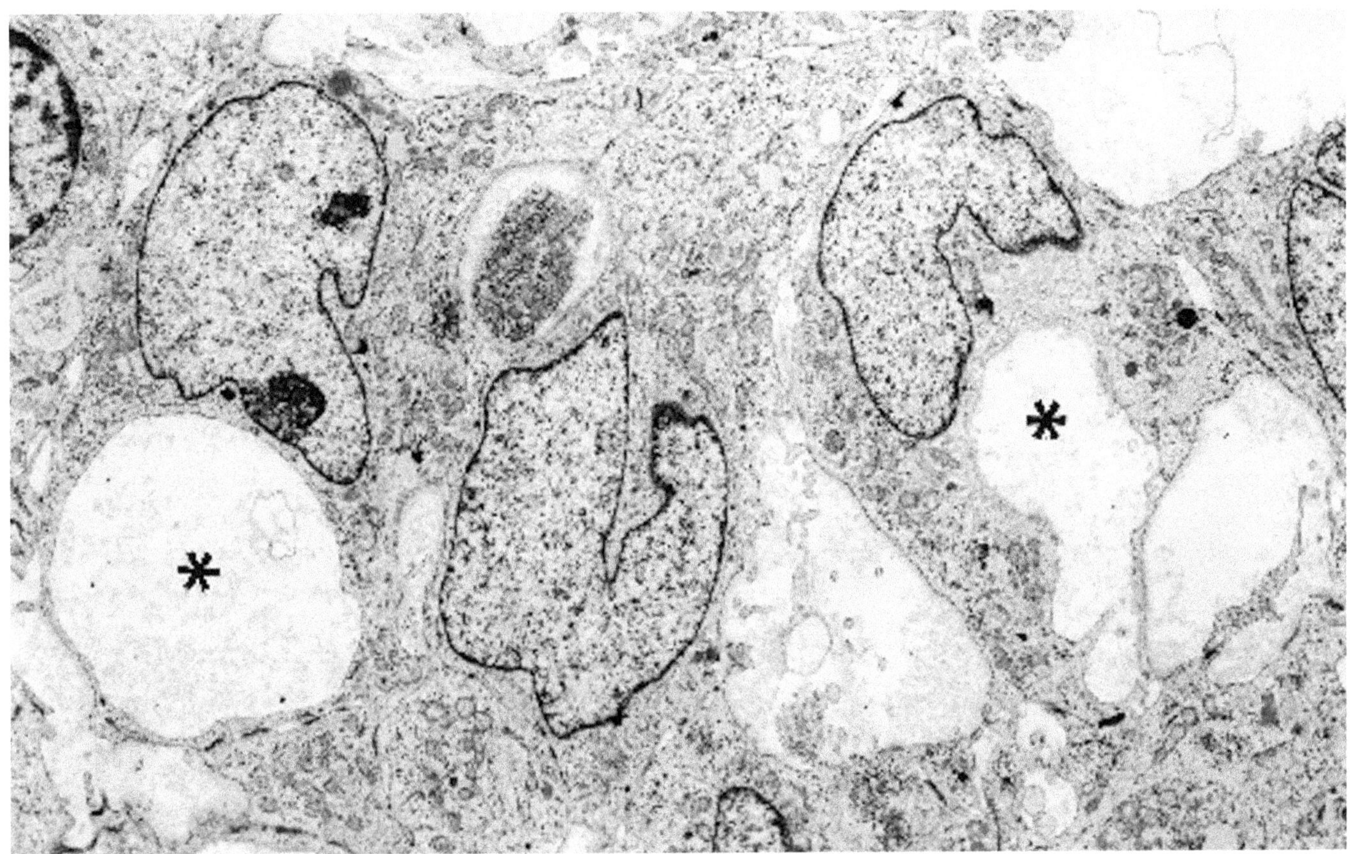

Figure 7.50. Signet-ring stromal tumor (ovary). Expansive intercellular spaces compress and indent the cytoplasm of some of the stromal cells, creating a pseudo-signet-ring form. The cytoplasm of the cell itself is not vacuolated, but this fact is not appreciable at the light microscopic level. * = intercellular spaces. (× 5600)

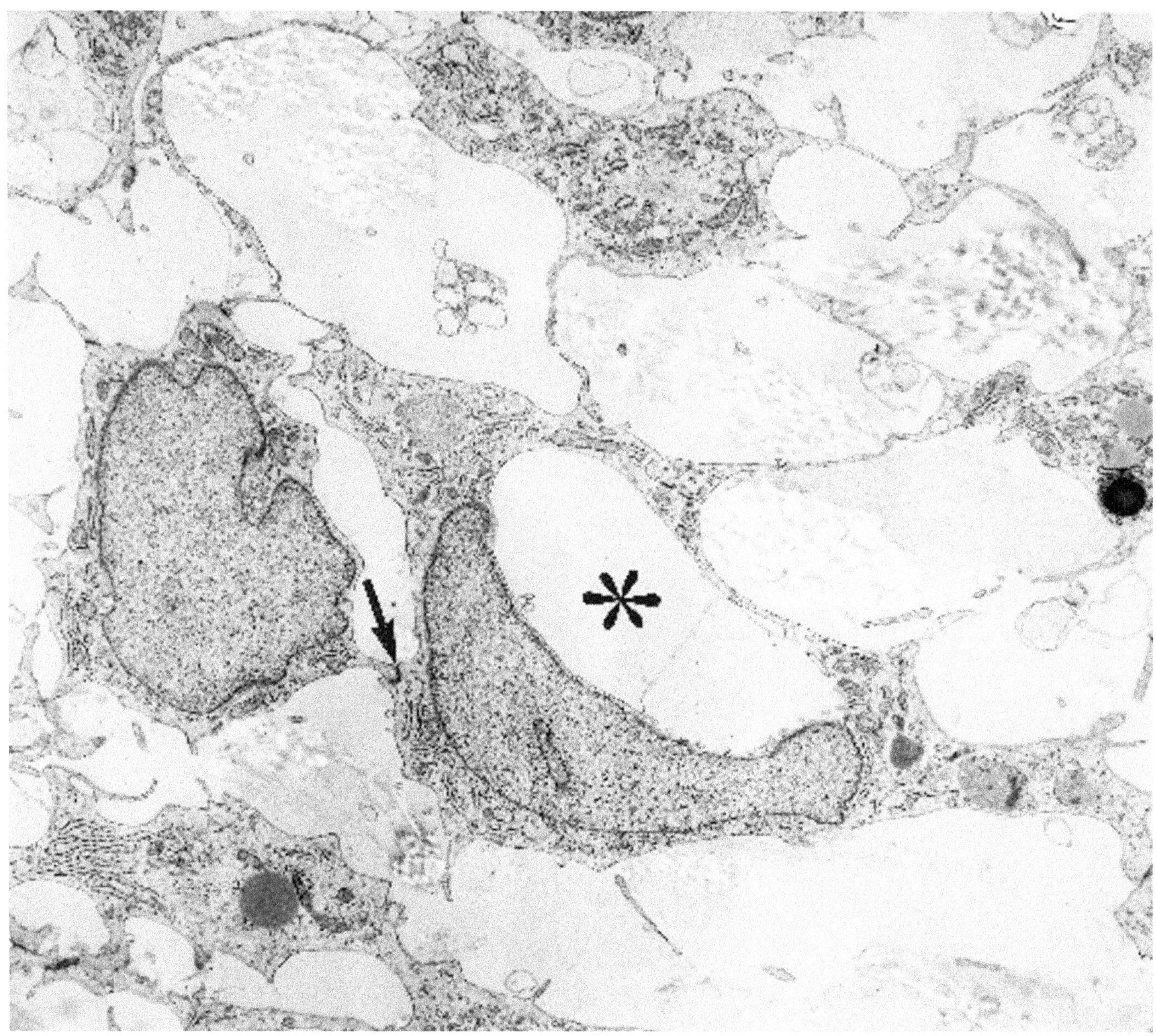

Figure 7.51. Signet-ring stromal tumor (ovary). In this neoplasm, pseudo-signet-ring forms were formed by invaginating intercellular spaces (*), similar to the same phenomenon in the tumor in Figure 7.50. Here, however, the cells proved to be epithelial rather than stromal, as evidenced by prominent intercellular junctions (*arrow*). (× 7400) (Permission for publication granted by Taylor and Francis, Dickersin GR, Young RH, Scully RE: Signet-ring stromal and related tumors of the ovary. Ultrastruct Pathol 19:401–419, 1995.)

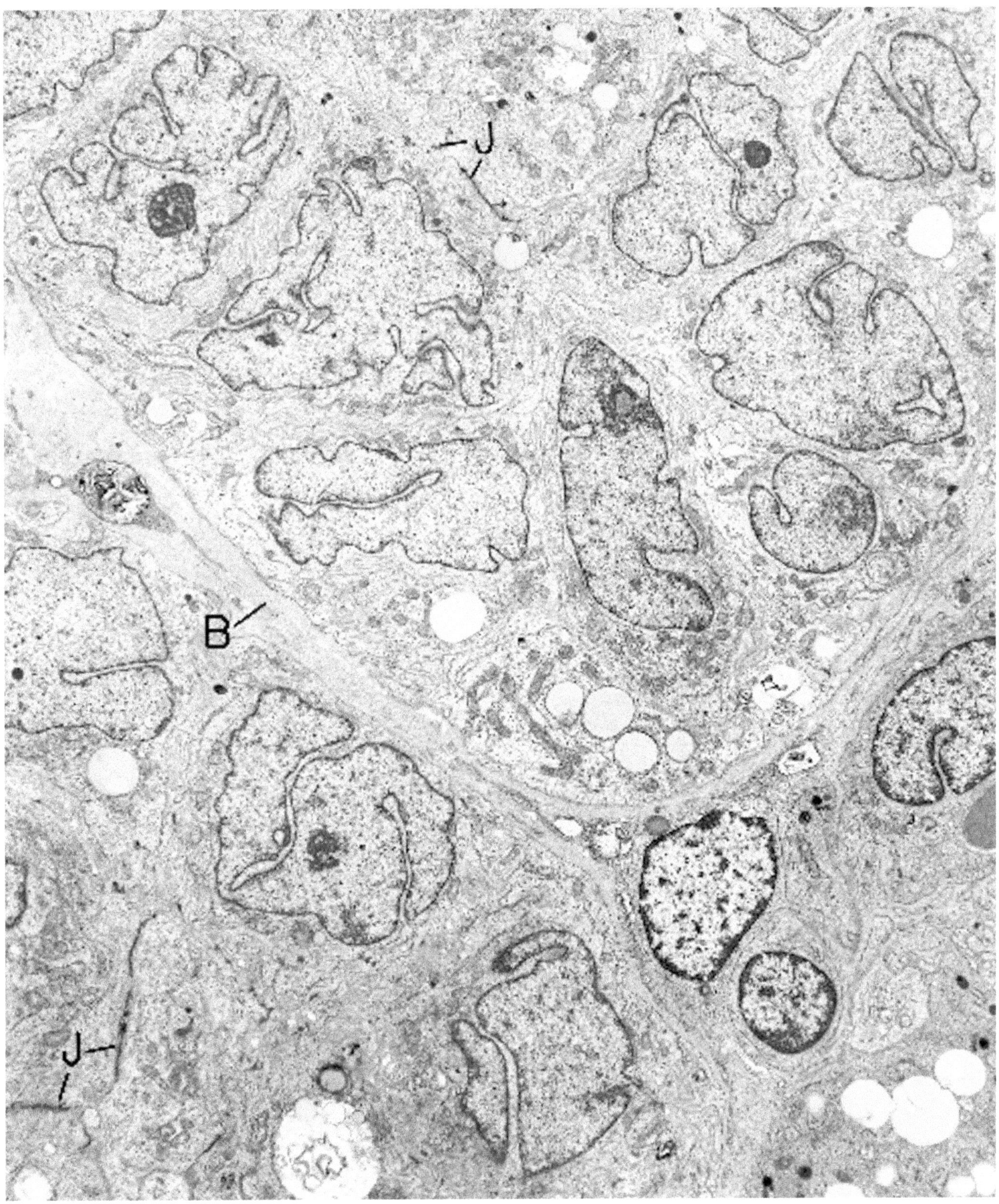

Figure 7.52. Sertoli–stromal cell tumor (ovary). Tubules of Sertoli cells are surrounded by basal lamina (B) and although appearing solid at low magnification, often have a microlumen that can be located by recognizing converging junctional complexes (J). Nuclei are markedly indented. (× 5500)

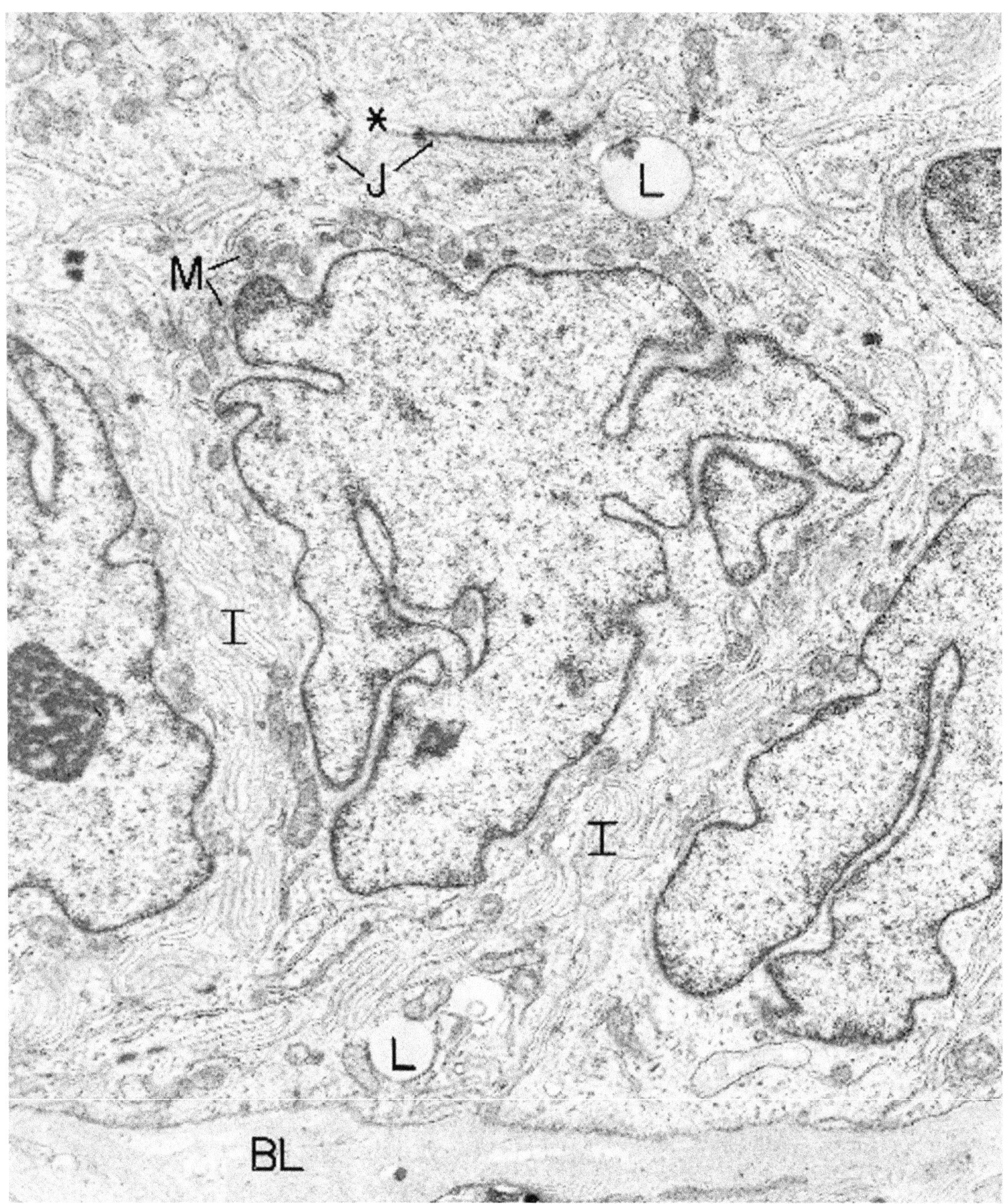

Figure 7.53. Sertoli–stromal cell tumor (ovary). Higher magnification of one of the tubules depicted in Figure 7.52 illustrates several junctional complexes (J) converging on an inconspicuous microlumen (*). Note also the marked interdigitation (I) of the lateral cell borders. Mitochondria (M) are numerous, and there are small lipid droplets (L). Basal lamina (BL) is thick. (× 13,500)

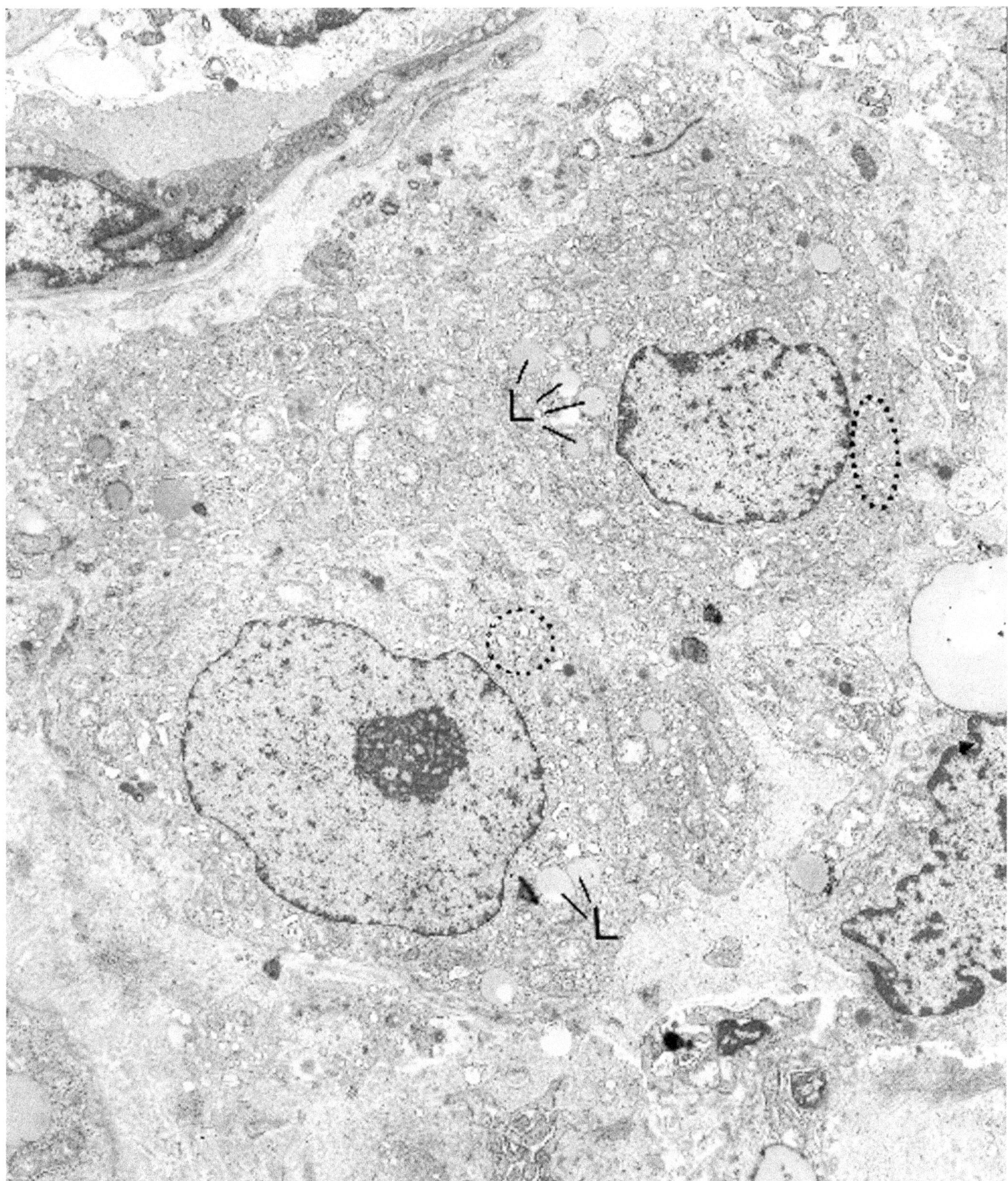

Figure 7.54. Sertoli–stromal cell tumor (ovary). A group of Leydig cells is characterized by abundant cytoplasm, innumerable vesicles of smooth endoplasmic reticulum (*circles*), many lipid droplets (L), and oval, evenly contoured nuclei with large nucleoli. (× 6240)

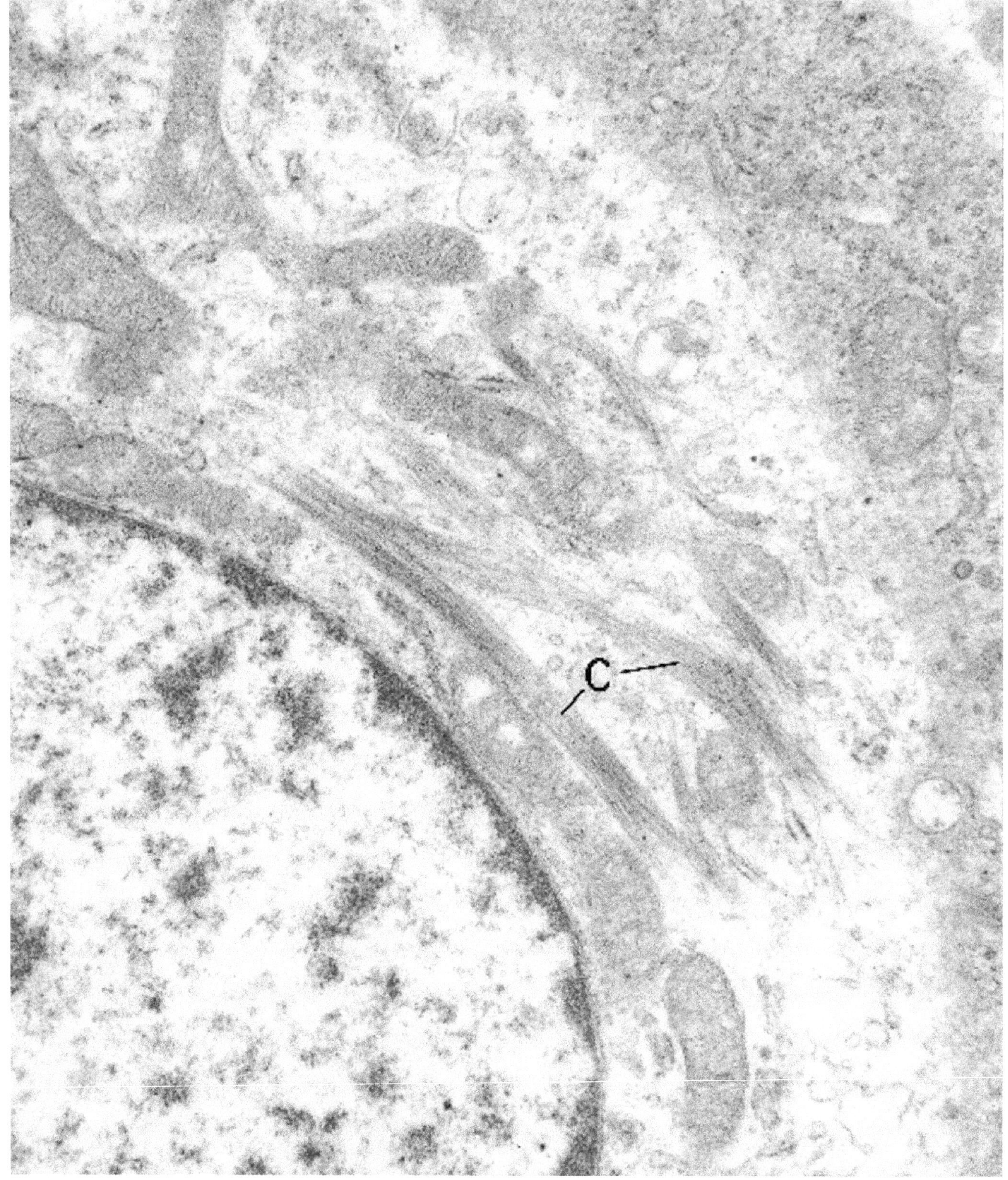

Figure 7.55. Sertoli–stromal cell tumor (ovary). Bundles of tightly arranged filaments make up diagnostic Charcot–Böttcher "crystalloids" (C) in a Sertoli cell. (× 31,100)

(Text continued from page 439)

Sex Cord Tumor with Annular Tubules

(Figures 7.56 through 7.58.)

Diagnostic criteria. (1) Tubules (simple and complex); (2) basal lamina around and in lumens of tubules; (3) irregularly shaped, indented nuclei at basal and luminal poles of epithelial lining cells; (4) nondescript cytoplasm (see next paragraph).

Additional points. Most of the tubules of this neoplasm are pseudotubules, the lumens of which contain basal lamina, but true tubules with junctional complexes and microvilli lining the lumens also may be seen. Furthermore, these structures may blend into tubules that have a closer resemblance to seminiferous tubules lined by more typical Sertoli cells; they also may blend into solid nests of cells resembling granulosa cells. The nondescript cytoplasm listed in the diagnostic criteria section includes a background of free ribosomes, a moderate amount of rough endoplasmic reticulum, many mitochondria, and many lipid droplets. Charcot–Böttcher crystalloids consisting of bundles of cytoplasmic filaments are a marker for Sertoli cells, but they are rarely found. A focally diffuse pattern of cells, including islands of lipid-rich cells, may also be present.

Gynandroblastoma

Gynandroblastomas are rare tumors in which at least 10% of a Sertoli–stromal cell tumor is composed of granulosa–stromal elements or at least 10% of a granulosa–stromal cell tumor is composed of Sertoli–stromal cells.

Sex Cord–Stromal Tumors Unclassified

Sex cord–stromal tumors that are unclassified have cells and patterns intermediate between granulosa–stromal cell tumors and Sertoli–stromal cell tumors, or they have the cells and patterns of both types of tumors.

Steroid (Lipid) Cell Tumors

(Figures 7.59 through 7.62.)

These tumors include stromal luteomas, Leydig cell tumors (nonhilar type), and hilus cell tumors.

Diagnostic criteria. (1) Cytoplasmic droplets of lipid; (2) abundant smooth endoplasmic reticulum; (3) mitochondria with tubular cristae; (4) microvilli covering most of the surface of the cells; (5) basal lamina covering the nonvillous portion of the cellular surface; (6) canalicular-like spaces between cells.

Additional points. Characteristic of steroid-type cells are lipid droplets, smooth endoplasmic reticulum, and mitochondria with tubular cristae. However, a significant proportion of these tumors contain little or no lipid. Lipochrome pigment in secondary lysosomes may or may not be present. Possible cells of origin for the steroid-cell tumor include Leydig and hilus cells, lutein (thecal or stromal) cells, and adrenal cortical cells. Leydig cells can be specifically identified if Reinke crystals are present, but most Leydig cell tumors do not contain them. A large percent of steroid cell tumors cannot be classified as any of the specific subtypes because of a lack of a typical pattern or Reinke crystals, and they are then referred to as steroid cell tumors not otherwise specified.

Germ Cell Tumors

A large majority of germ cell tumors are dermoid cysts (mature cystic teratomas). Most malignant germ cell tumors are of a single cell-type, but combinations of various cell types also occur.

Dysgerminoma (Seminoma)

(Figures 7.63 through 7.65.)

Diagnostic criteria. (1) Large round or polygonal cells in close apposition; (2) intercellular junctions; (3) large euchromatic nuclei; (4) prominent and multiple nucleoli with open nucleolonemas; (5) poorly differentiated cytoplasm (mostly free ribosomes, plus a moderate number of mitochondria, a small number of undilated cisternae of rough endoplasmic reticulum, and small Golgi apparatuses); (6) copious cytoplasmic glycogen.

Additional points. The cells resemble primordial germ cells of the embryo. They are similar, whether occurring in gonadal, mediastinal, or pineal neoplasms. Lipid droplets, secondary lysosomes, and annulate lamellae also may be present in the cytoplasm of the cells. Syncytiotrophoblastic giant cells are found among the germ cells in rare cases. No cytotrophoblast is present, however, therefore differing from a choriocarcinoma and a mixed germ cell tumor. The extracellular matrix in dysgerminomas usually contains small lymphocytes and may contain luteinized stromal cells.

(Text continues on page 462)

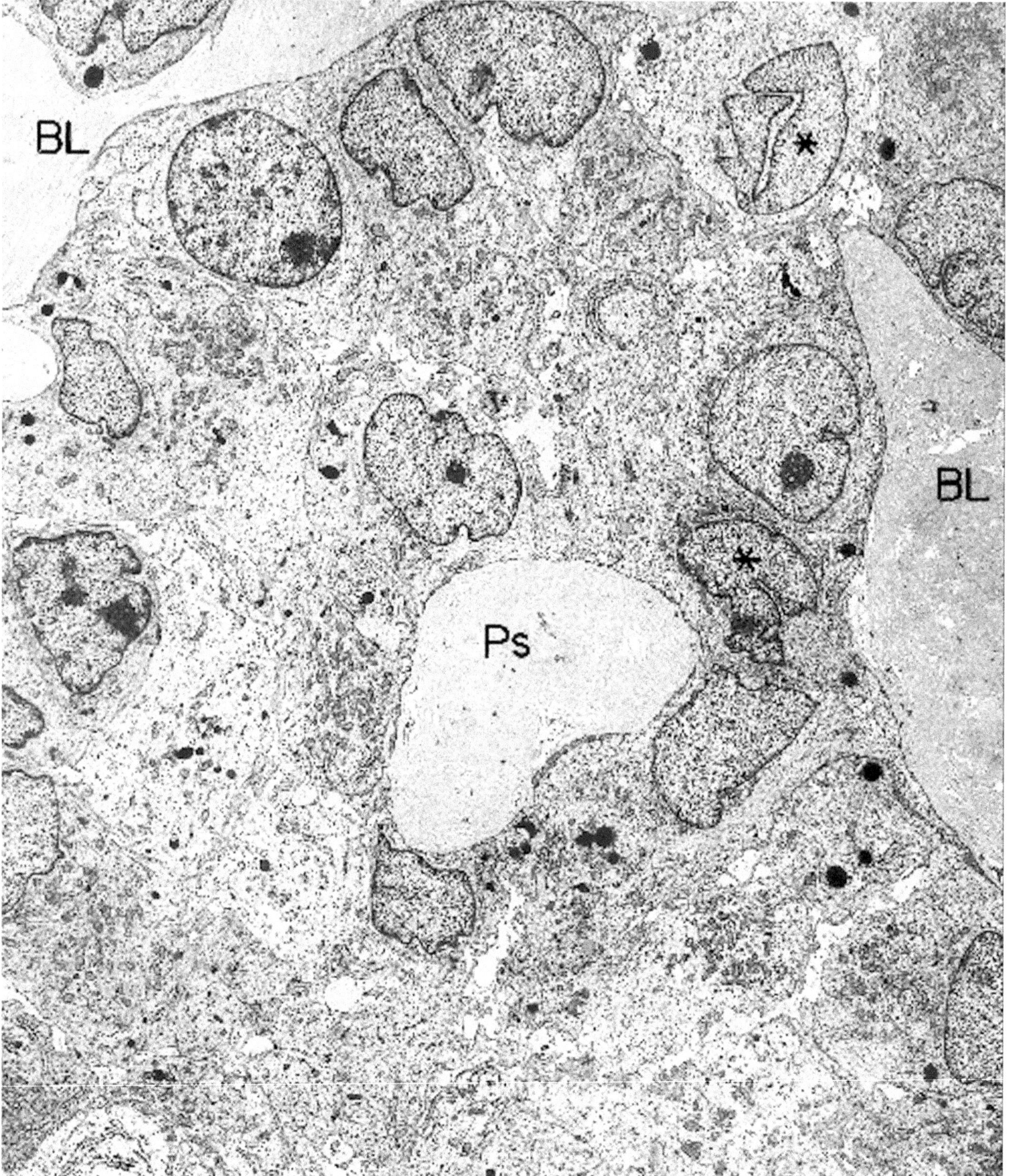

Figure 7.56. Sex cord tumor with annular tubules (ovary). Abundant basal lamina (BL) surrounds tubules and fills their pseudolumens (Ps). Nuclei are often indented and irregular in shape (*). (× 3600)

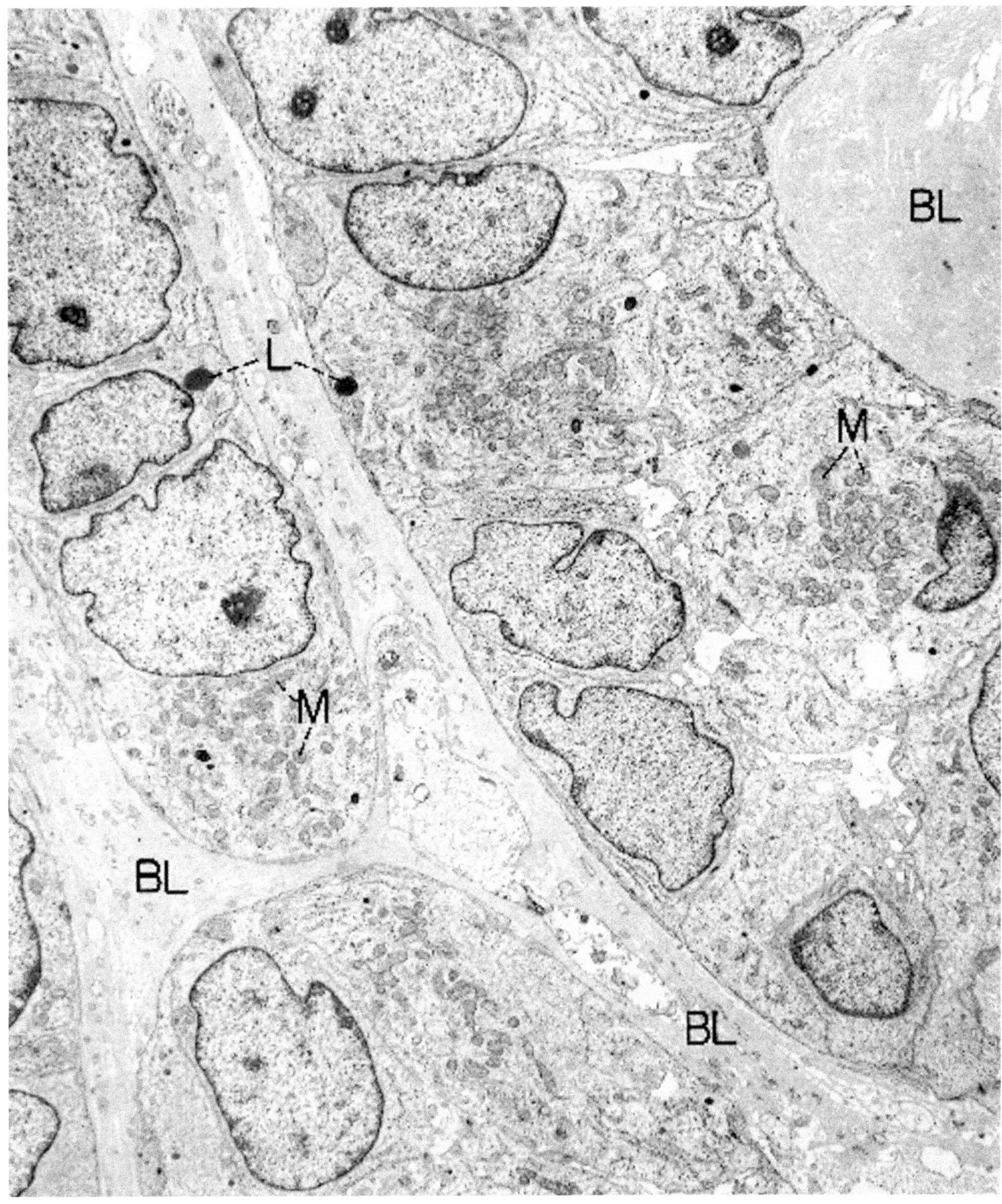

Figure 7.57. Sex cord tumor with annular tubules (ovary). Portions of several tubules are surrounded and separated by copious basal lamina (BL). Cytoplasm is moderate in amount and is identifiable at this power as containing many mitochondria (M) and scattered lipid droplets. (× 5300)

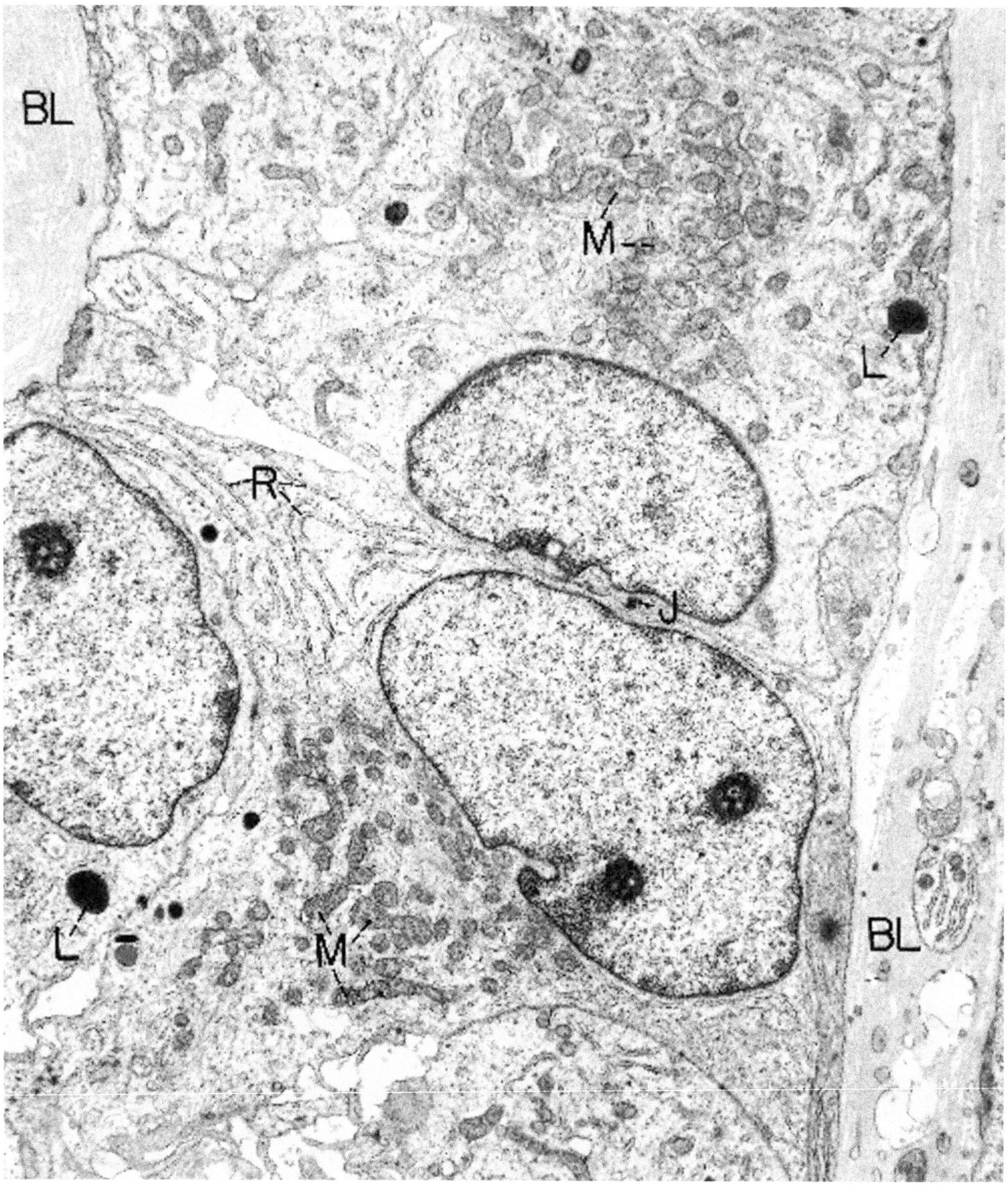

Figure 7.58. Sex cord tumor with annular tubules (ovary). Higher magnification allows the composition of the cytoplasm to be visualized. In addition to a background of free ribosomes, there are many mitochondria (M) and a moderate number of cisternae of rough endoplasmic reticulum (R). L = lipid; J = junction; BL = basal lamina. (× 9500)

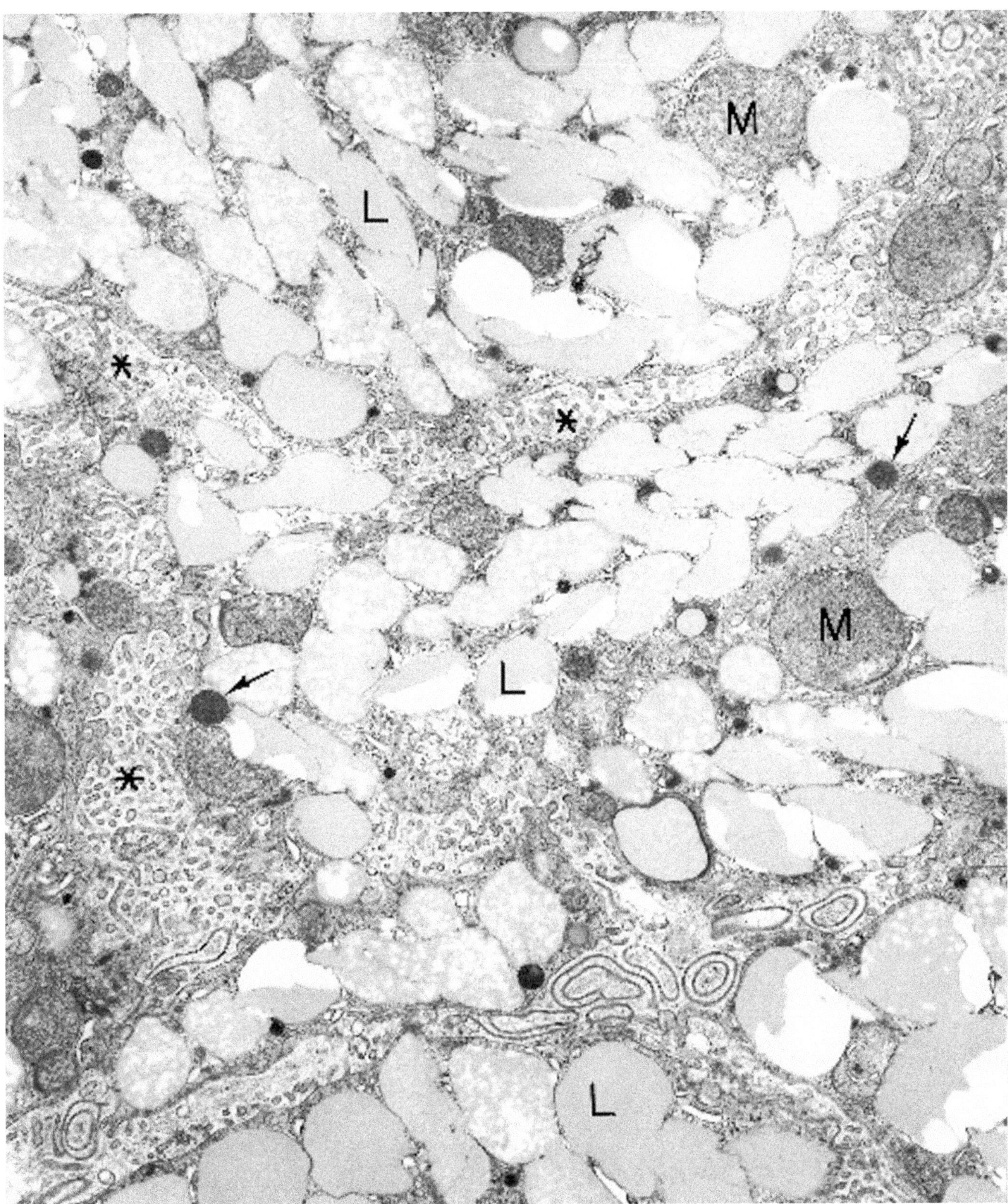

Figure 7.59. Steroid (lipid) cell tumor (luteoma) (ovary). The most striking features of the cells are the many droplets of lipid (L) in the cytoplasm and the florid collection of microvilli on the surface and filling intercellular, canalicular-like spaces (*). Mitochondria (M) are large and moderate in number. There are occasional secondary lysosomes (*arrows*). (× 10,600)

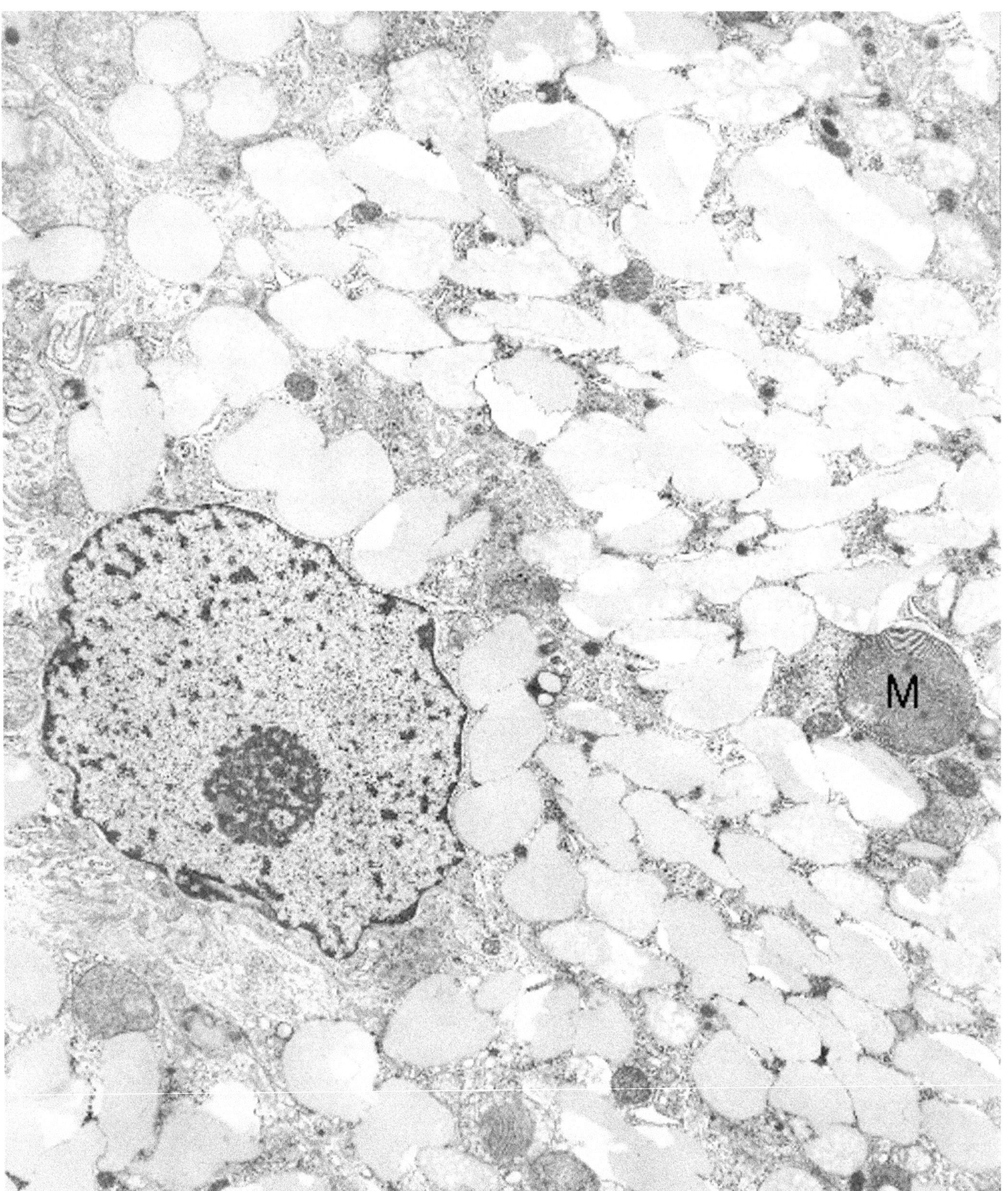

Figure 7.60. Steroid (lipid) cell tumor (luteoma) (ovary). The nuclei of lipid cells are round or oval, have a generally regular contour, and contain a prominent nucleolus. Note the large size and abnormal configuration of cristae of some of the mitochondria (M). (× 9880)

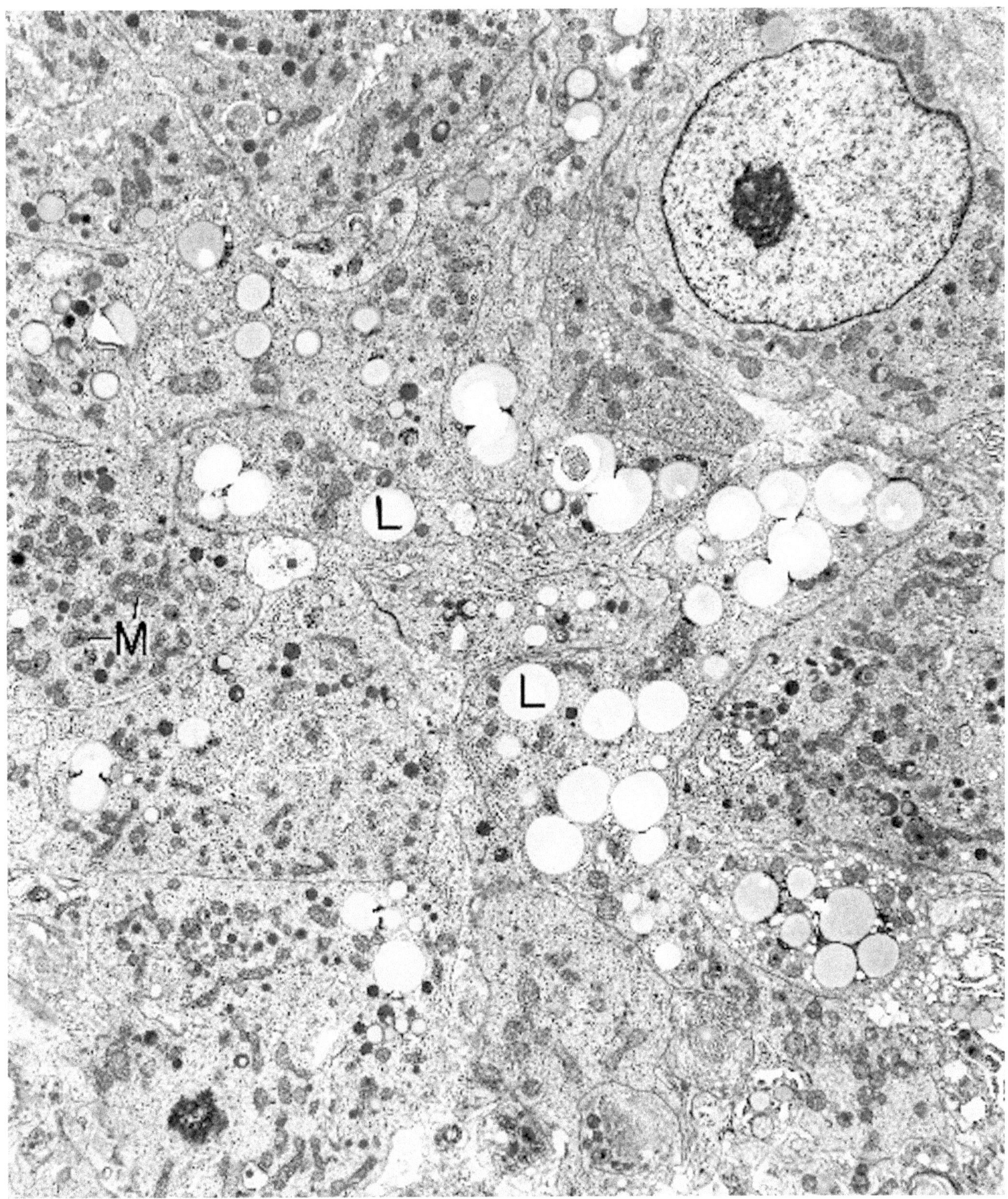

Figure 7.61. Leydig cell tumor (testis). The cells are plump and have abundant cytoplasm and round nuclei with even contours. Lipid droplets (L), vesicles of smooth endoplasmic reticulum (not well seen at this low power), and mitochondria (M) are numerous. (× 5720)

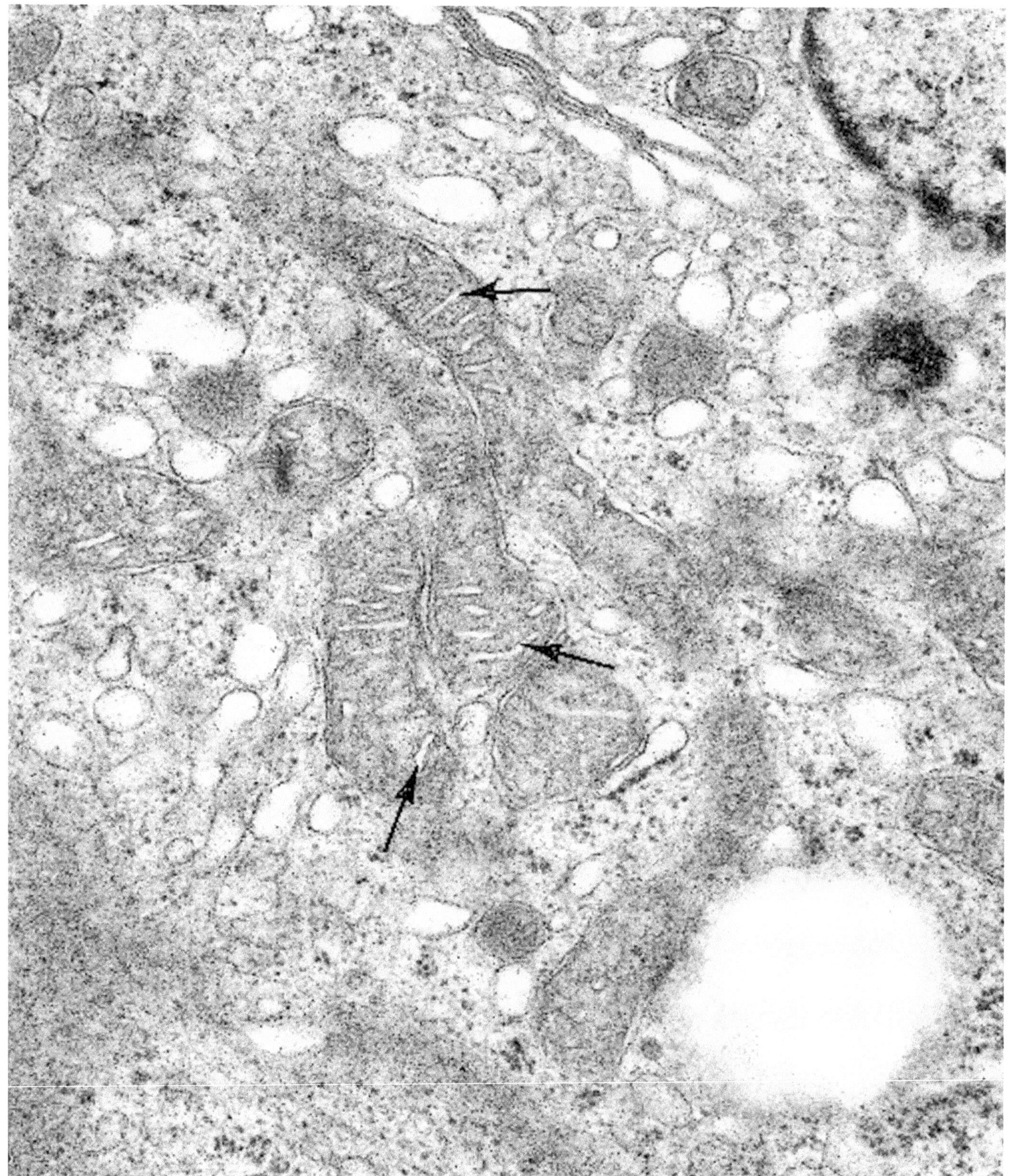

Figure 7.62. Leydig cell tumor (testis). High magnification of the cytoplasm of a neoplastic Leydig cell depicts several mitochondria with moderately tubular cristae (*arrows*). (× 48,400)

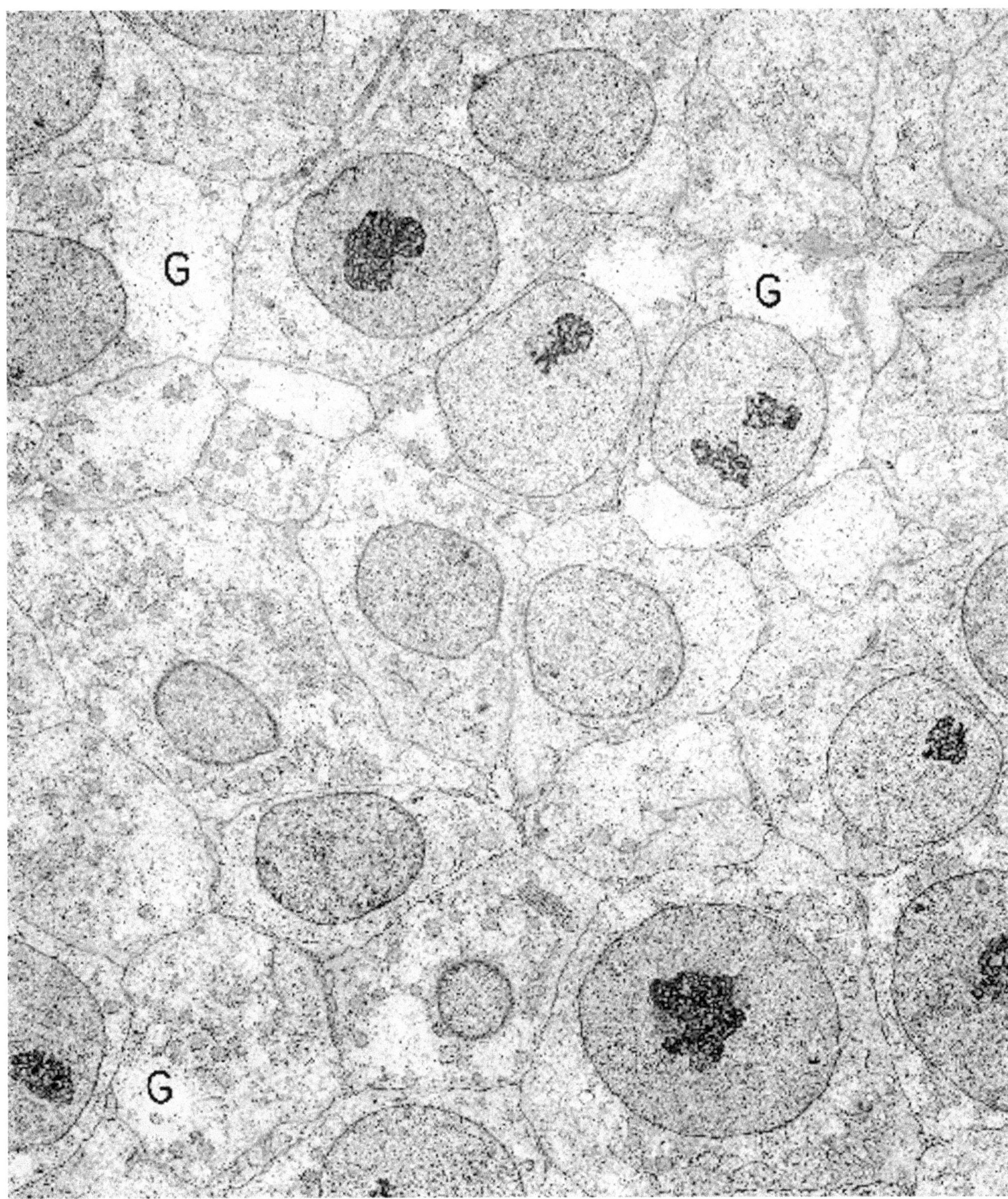

Figure 7.63. Seminoma (testis). The cells are large, polygonal, and in close apposition. Nuclei are euchromatic, and nucleoli are large and have an open nucleolonema. The cytoplasm contains a background of free ribosomes, copious glycogen (G, clear open spaces), and few other organelles. (× 3600)

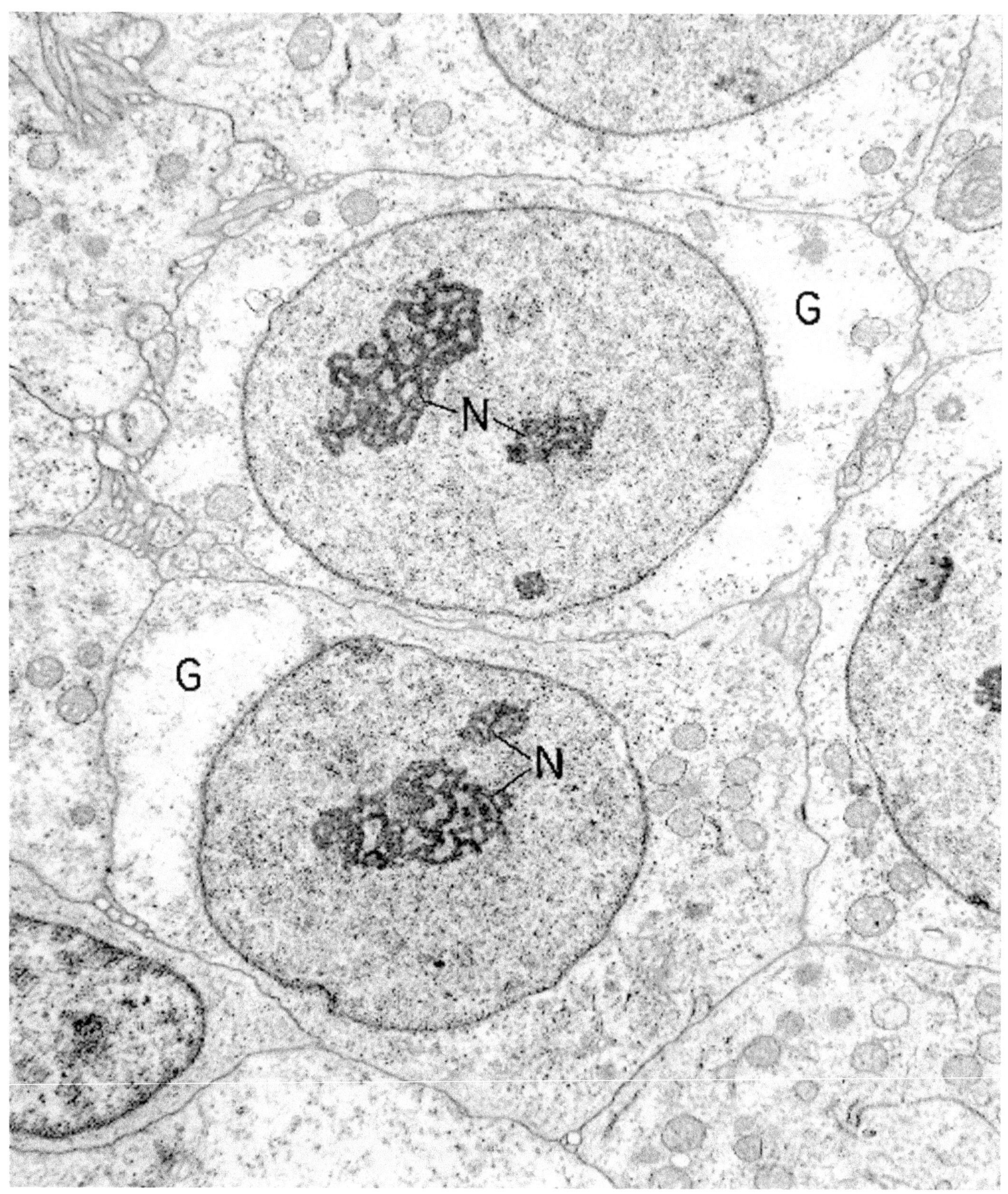

Figure 7.64. Seminoma (testis). High magnification illustrates the primitive nature of the cells, with a high nuclear–cytoplasmic ratio, finely dispersed chromatin, open nucleolonemas (N), abundant glycogen (G), and few cytoplasmic organelles. (× 9500)

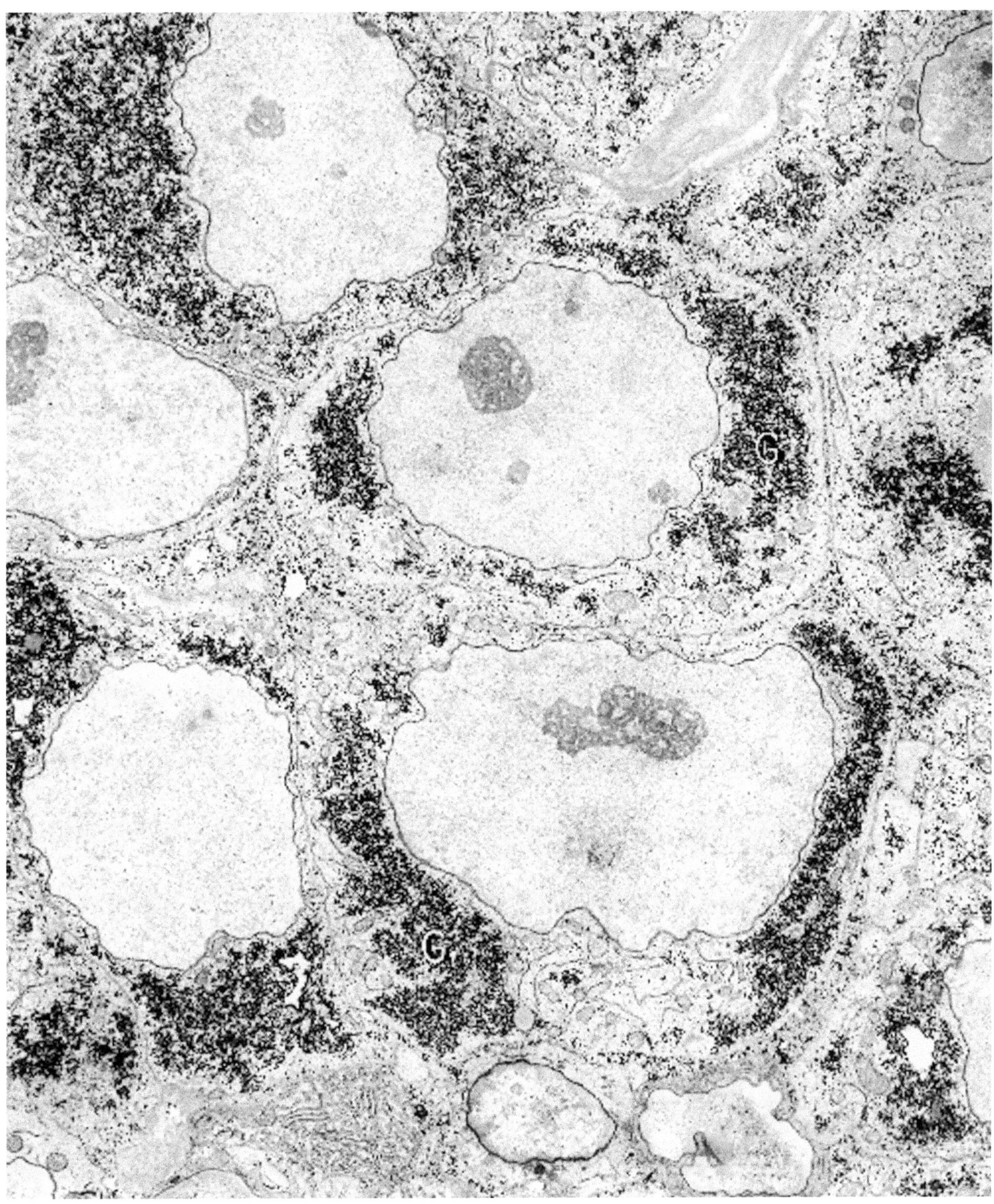

Figure 7.65. Seminoma (testis). This specimen was processed by a method that preserved glycogen (G) as electron-dense granules, making possible a more accurate assessment of the large quantity of this inclusion that is present in the cells. (× 6150)

(Text continued from page 451)

Yolk Sac Tumor (Endodermal Sinus Tumor)

(Figures 7.66 through 7.72.)

Diagnostic criteria. (1) Epithelial-type cells arranged along a network of interconnecting spaces (reticular pattern); (2) profuse basal lamina material coating the cells and sometimes forming globules and pseudoinclusions in the cells; (3) rough endoplasmic reticulum, sometimes dilated and filled with material of similar density to that of basal lamina; (4) microvilli on cells lining some of the spaces (true lumens); (5) narrow, villus-lined canalicular spaces between the lateral surfaces of cells; (6) simple papillae with a central blood vessel and a covering of tall primitive cells, projecting into spaces lined by flat, cuboidal, or hobnail cells (Schiller-Duval bodies); (7) abundant cytoplasmic glycogen; (8) occasional lipid droplets; (9) large globular densities; and (10) large, irregular nuclei with prominent, multiple nucleoli and open nucleolonemas.

Additional points. In addition to the most common reticular pattern, solid and glandular patterns representative of embryonic rather than extraembryonic differentiation, also may be present, in part or exclusively, in yolk sac tumors. Ultrastructurally, these elements are consistent with *hepatic* and *enteric* differentiation. The former have the polygonal shape, central nucleus, collection of organelles, and even the globular densities of alpha-1-antitrypsin, characteristic of hepatocytes. The intestinal type cells are characterized by microvilli having a core of anchoring filaments and a covering of glycocalyx. Goblet cells, endocrine cells, Paneth cells, and gastric parietal cells also may be present. A rare pattern that may be seen in yolk sac tumors is the *endometrioid variant,* in which the glandular pattern is suggestive of endometrioid carcinoma. Adenofibromatous, parietal (Reichert's membrane-like), mesenchymal, solid, and papillary-carcinoma-like patterns may also be found rarely.

Embryonal Carcinoma

(Figures 7.73 through 7.74.)

Diagnostic criteria. (1) Large cells in solid, papillary, or tubuloglandular arrangement; (2) high nuclear–cytoplasmic ratio; (3) irregularly shaped nuclei with large nucleoli; (4) in solid areas, cytoplasm has mostly free ribosomes and mitochondria; (5) in tubular or glandular areas, cytoplasm has additional organelles, including vesicles of smooth endoplasmic reticulum and cisternae of rough endoplasmic reticulum; (6) in tubular and glandular cells, microvilli on luminal surface, and junctional complexes are well developed; (7) basal lamina surround, glands.

Additional points. The cells in the solid arrangements have a poorly differentiated cytoplasm and a simple surface structure, whereas those of the glands and tubules have a cytoplasm with more diverse organelles and specialized plasmalemmas with microvilli, junctional complexes, and basal lamina. Globular densities, similar to those seen in yolk sac tumors, may also be present in the cells of embryonal carcinoma. Syncytial trophoblastic giant cells may be dispersed among the other germ cell components in these neoplasms.

Choriocarcinoma and Placental Site Tumor (and Normal Placenta)

(Figures 7.75 through 7.80.)

Choriocarcinoma may be gestational or of germ cell origin, and it is composed of cytotrophoblast and/or intermediate trophoblast, interspersed with syncytiotrophoblast. Although choriocarcinoma may be the sole component of a tumor, more frequently it is present with other germ cell types in a mixed tumor.

Diagnostic criteria. (1) Cytotrophoblastic cells are round, oval, and polygonal; (2) nuclear–cytoplasmic ratio is high; (3) cytoplasm is scant, with few organelles; (4) nuclei are euchromatic; (5) desmosomes interconnect cells; (6) intermediate trophoblastic cells have a single nucleus that often is indented and segmented and is predominantly euchromatic; (7) nucleoli are prominent; (8) cytoplasm is abundant and has numerous organelles; (9) the free surface of the cells is raised into microvilli; (10) adjacent cells have intermediate junctions and desmosomes; (11) syncytial trophoblastic cells have multiple heterochromatic nuclei and (12) abundant cytoplasm with (13) numerous free ribosomes; (14) prominent rough endoplasmic reticulum; (15) large and multiple Golgi apparatuses; (16) a moderate number of mitochondria with tubular cristae and (17) varying numbers of lipid droplets.

Teratoma (Immature and Mature; Monodermal)

Most teratomas are composed of more than one embryonic cell line, but a small percentage is monodermal, that is, purely ectodermal (e.g, neuroectodermal tumors, sebaceous tumors) or purely endodermal (e.g., struma ovarii, carcinoid tumor). If any of the components of a teratoma is embryonal, the tumor is classified as "immature." "Mature" teratomas are almost all dermoid cysts. Examples of various mature tissues include cutaneous, respiratory, and gastrointestinal epithelium, and examples of embryonal tissues encompass neuroectodermal, rhabdomyoblastic, and chondroblastic elements. All these cell types are illustrated in other chapters of this book.

(Text continues on page 478)

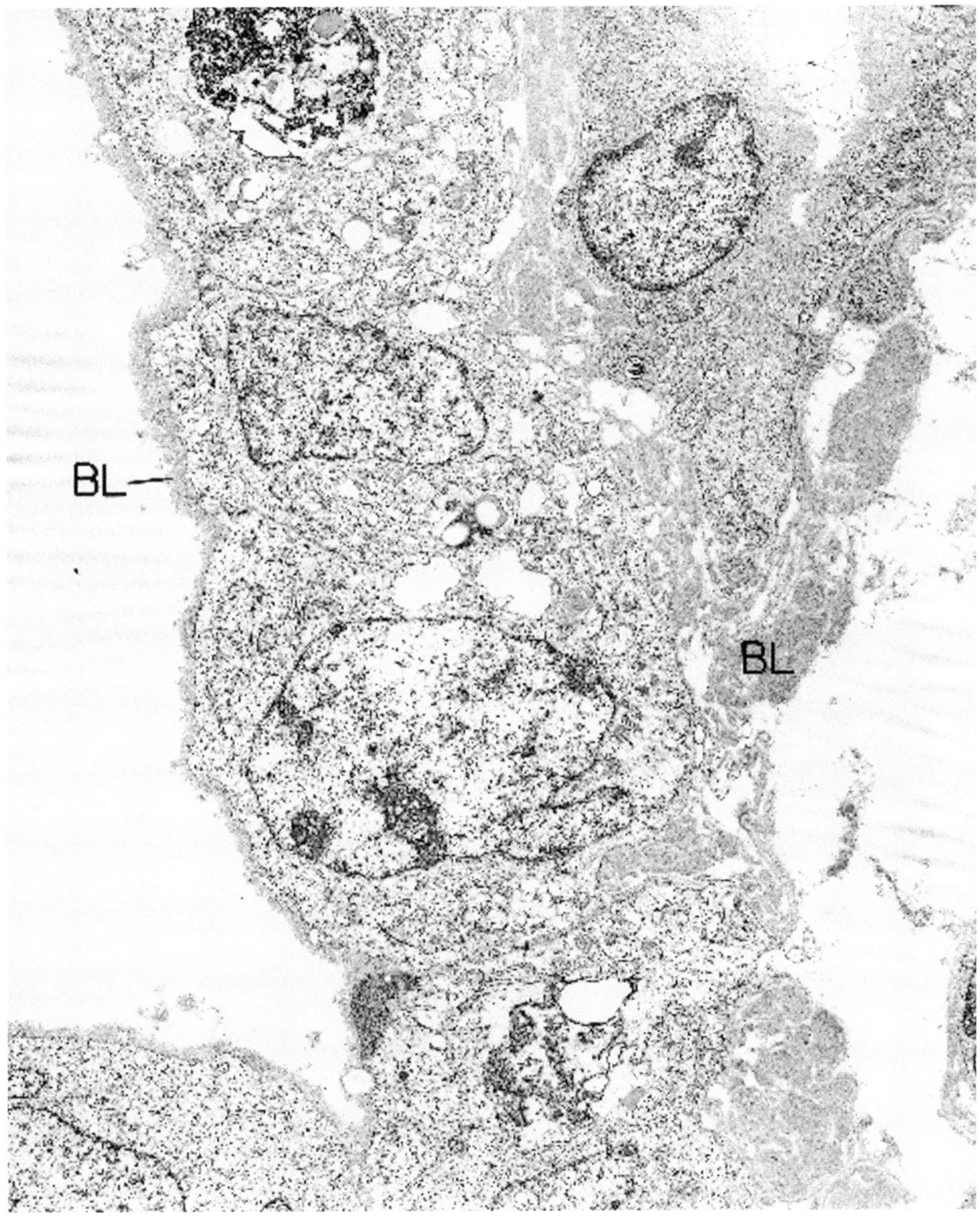

Figure 7.66. Yolk sac tumor (vagina). The cells line spaces and are covered on both sides with thick basal lamina (BL). (× 5300) (Permission for reprinting granted by WB Saunders, Dickersin GR: The ultrastructure of selected gynecologic neoplasms. Clin Lab Med 7:117–156, 1987.)

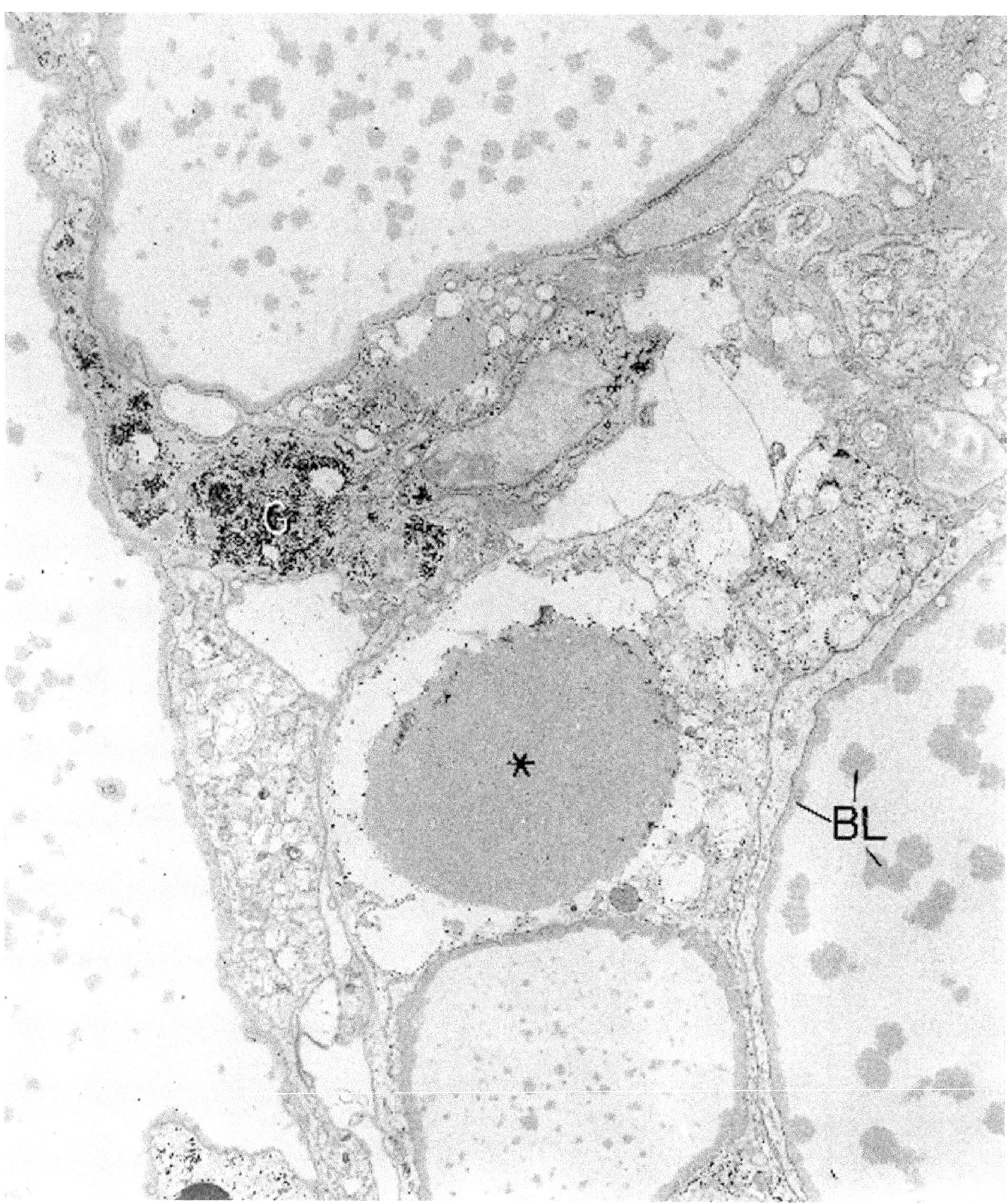

Figure 7.67. Yolk sac tumor (suprarenal tissue). The specimen was processed to preserve glycogen (G) as electron-dense granules. Abundant basal lamina (BL), including a large intracytoplasm pseudoinclusion of it (*), are readily discernible. (× 6480)

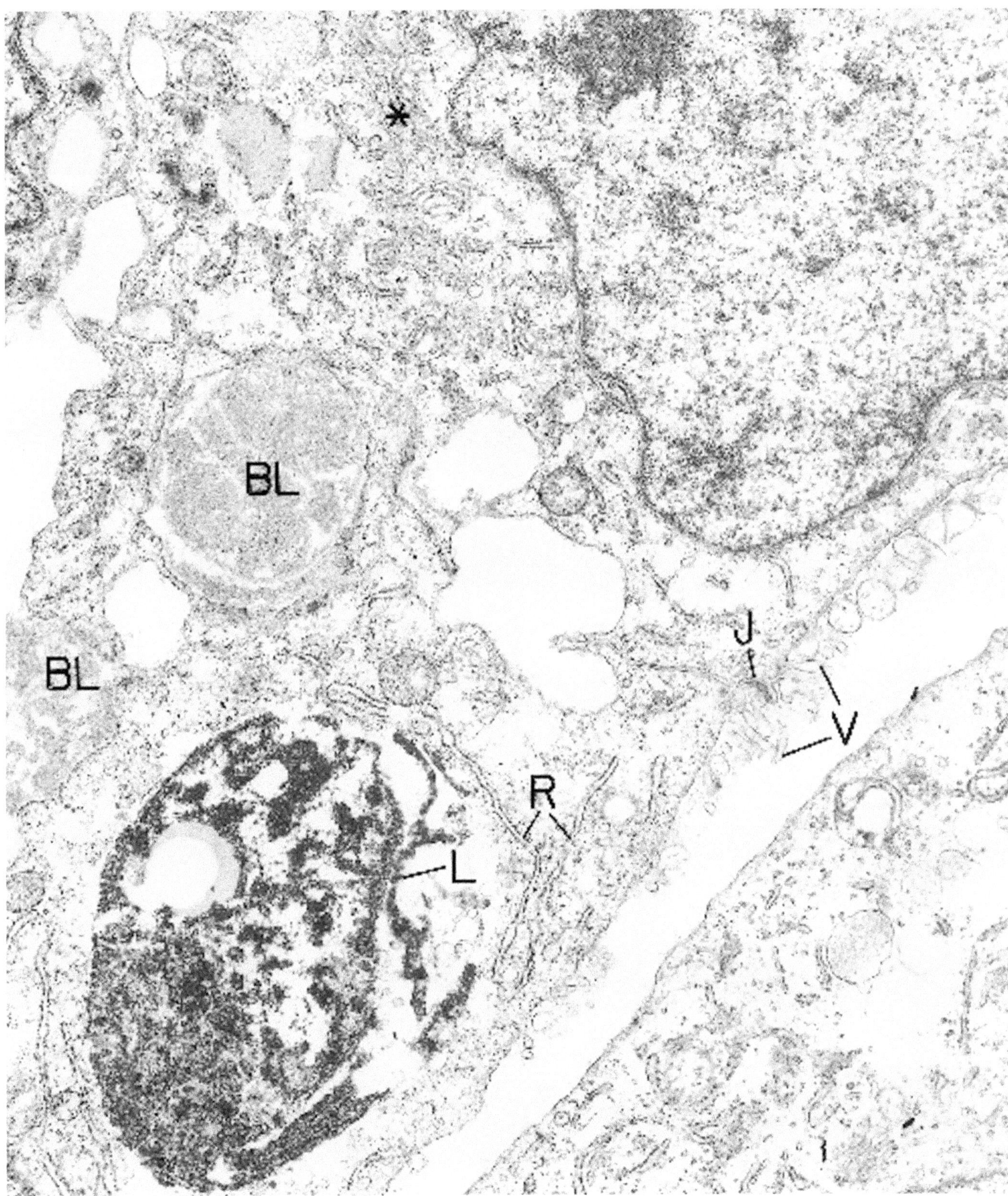

Figure 7.68. Yolk sac tumor (vagina). The large, dense, cytoplasmic globule (L) in this cell is a secondary lysosome. A pseudoinclusion filled with basal lamina (BL) is located just above it. Notice the Golgi apparatus (*) and scattered cisternae of rough endoplasmic reticulum (R). A few microvilli (V) partially cover one surface of the cell, and a diminutive junctional complex (J) is present. (× 13,750) (Permission for reprinting granted by WB Saunders, Dickersin GR: The ultrastructure of selected gynecologic neoplasms. Clin Lab Med 7:117–156, 1987.)

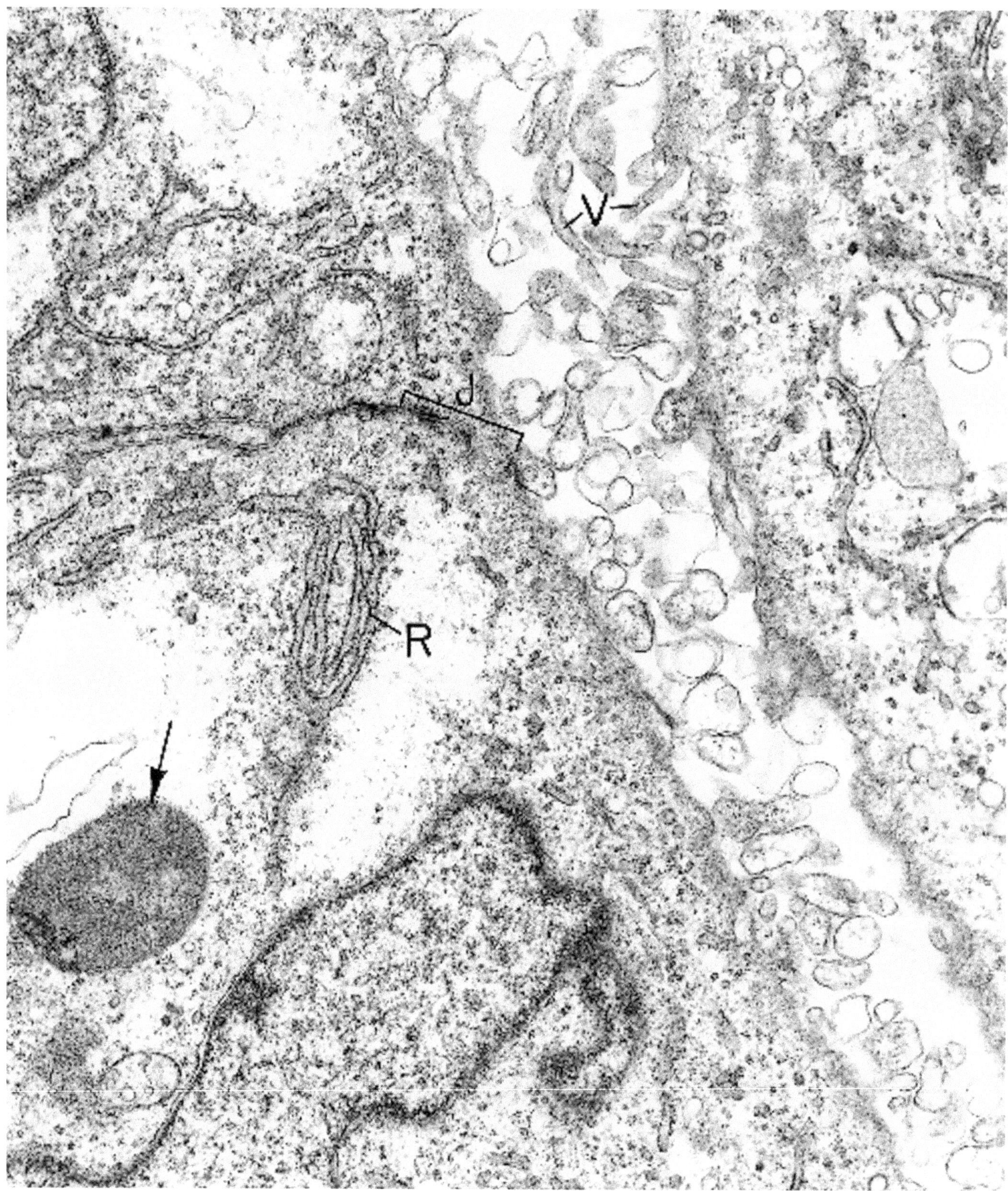

Figure 7.69. Yolk sac tumor (vagina). Another area from the same neoplasm in Figure 7.68 shows at high power a cytoplasmic pseudoinclusion (*arrow*), dilated rough endoplasmic reticulum with medium-dense contents (R), a junctional complex (J), and microvilli (V). (× 20,350) (Permission for reprinting granted by WB Saunders, Dickersin GR: The ultrastructure of selected gynecologic neoplasms. Clin Lab Med 7:117–156, 1987.)

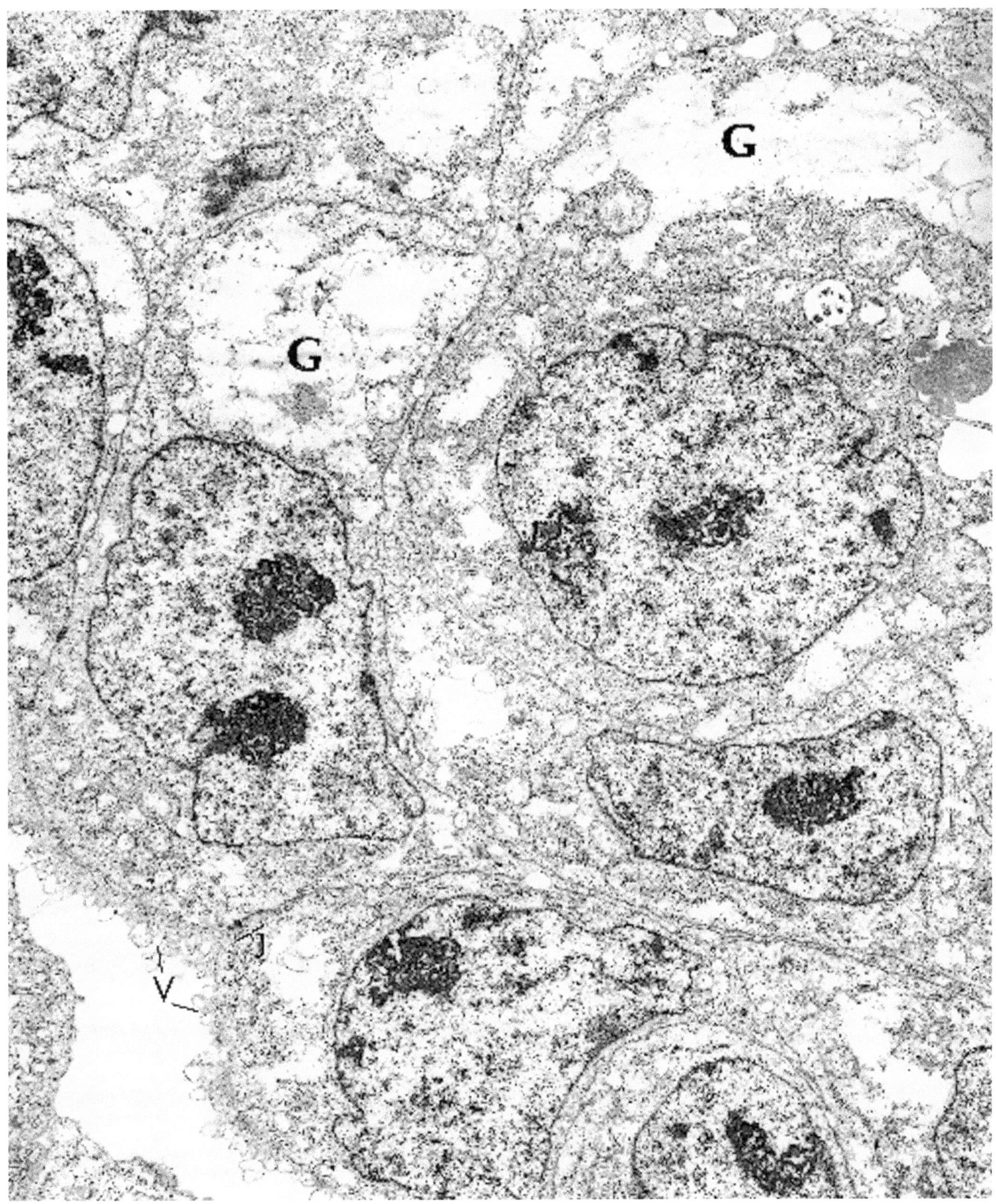

Figure 7.70. Yolk sac tumor (vagina). In addition to microvilli (V) and junctional complexes (J), abundant glycogen (G, open spaces) is present in the neoplastic cells. Nuclei are large and irregularly shaped, and nucleoli are prominent and multiple and have open nucleolonemas. (× 5300) (Permission for reprinting granted by WB Saunders, Dickersin GR: The ultrastructure of selected gynecologic neoplasms. Clin Lab Med 7:117–156, 1987.)

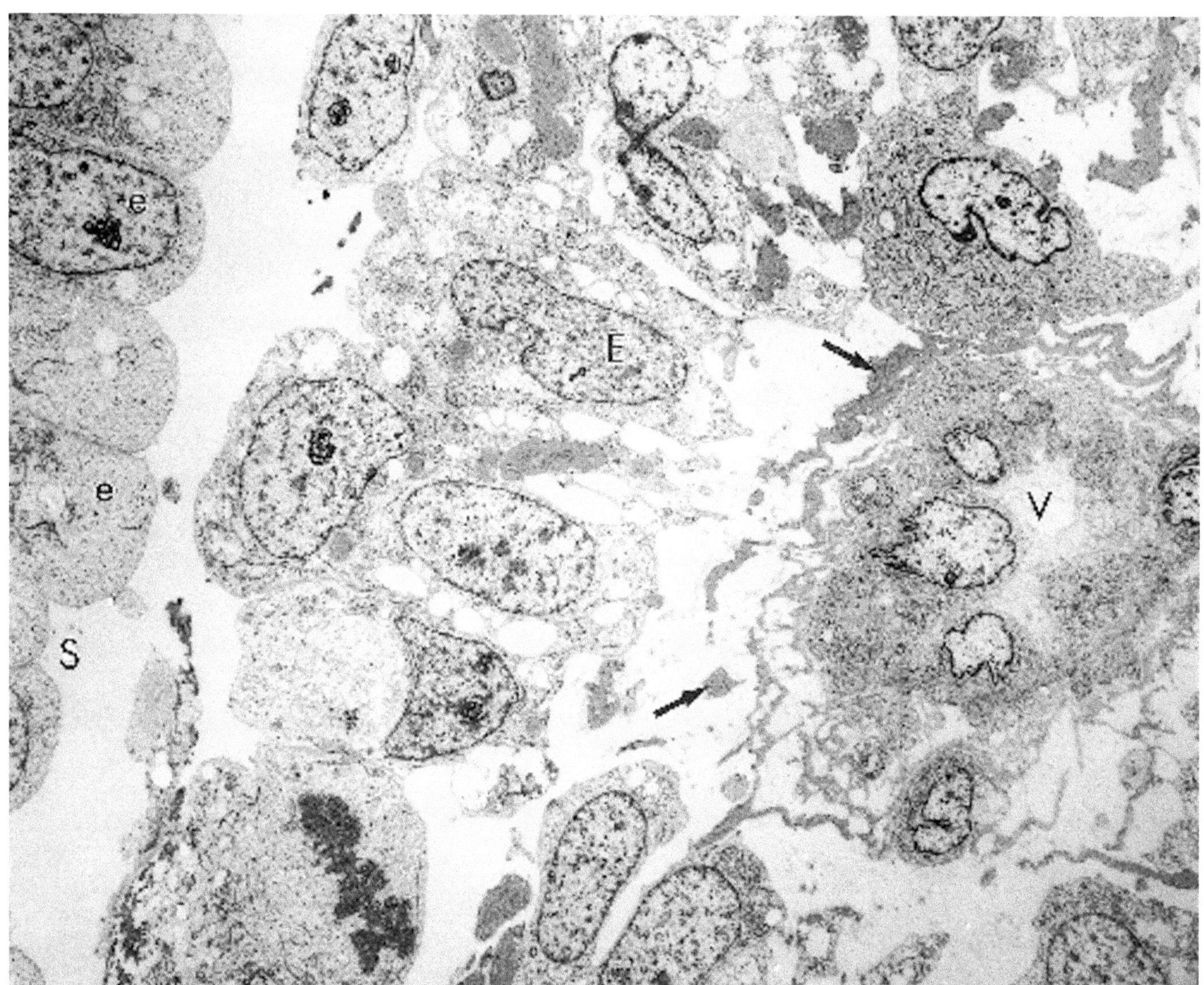

Figure 7.71. Yolk sac tumor (mesentery). A Schiller-Duval body is characterized by a papilla composed of a central blood vessel (V), a perivascular space with basal lamina (*arrows*), a surface layer of columnar epithelial cells (E) that project into an open peripheral space (S). An external layer of cuboidal epithelial cells (e) is just visible at the edge of the field. (× 3400)

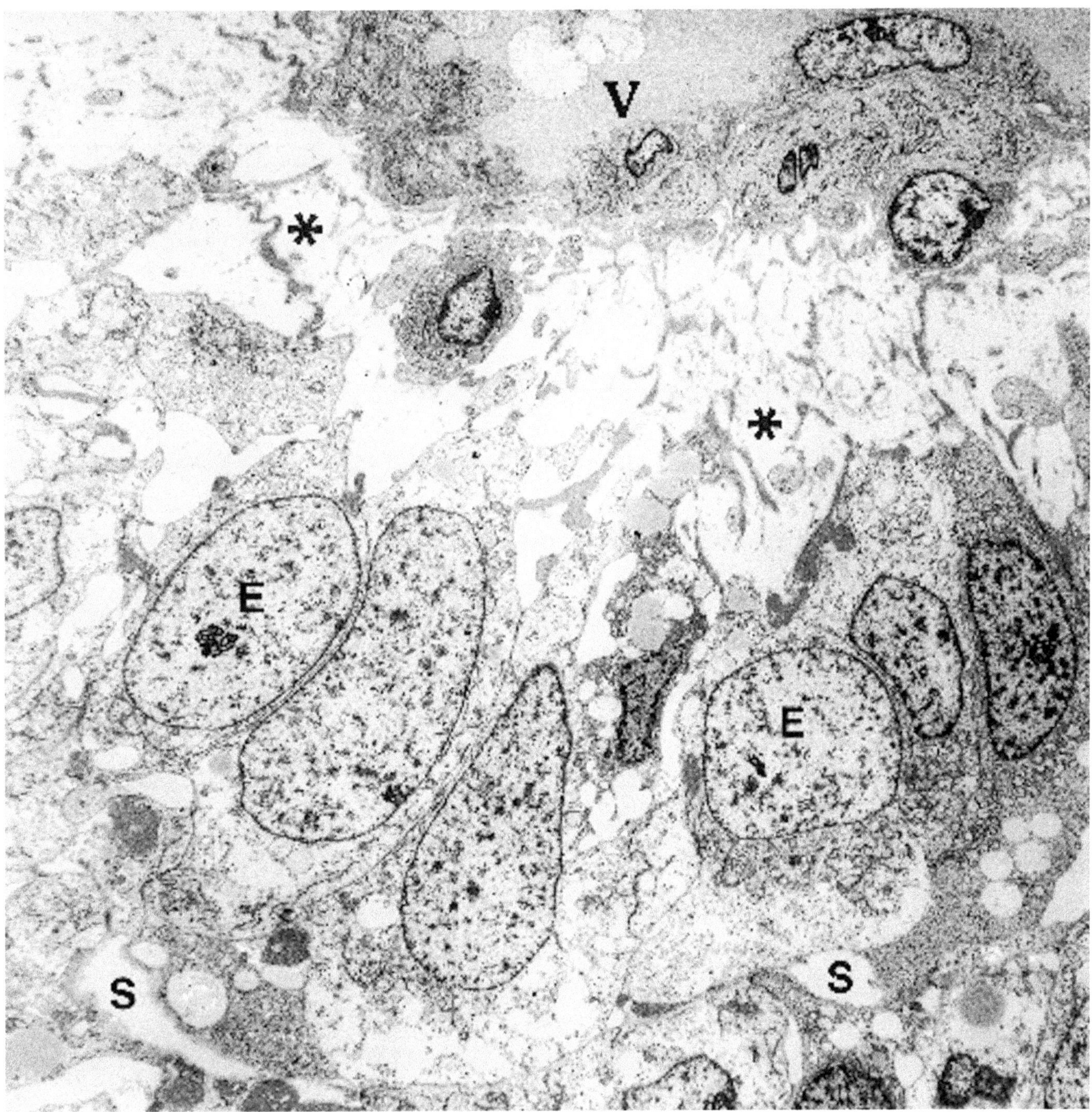

Figure 7.72. Yolk sac tumor (mesentery). Another Schiller-Duval body is shown at higher magnification than that depicted in Figure 7.71. V = vascular lumen; * = perivascular basal lamina; E = surface epithelium of papilla; S = peripapillary space. (× 4500)

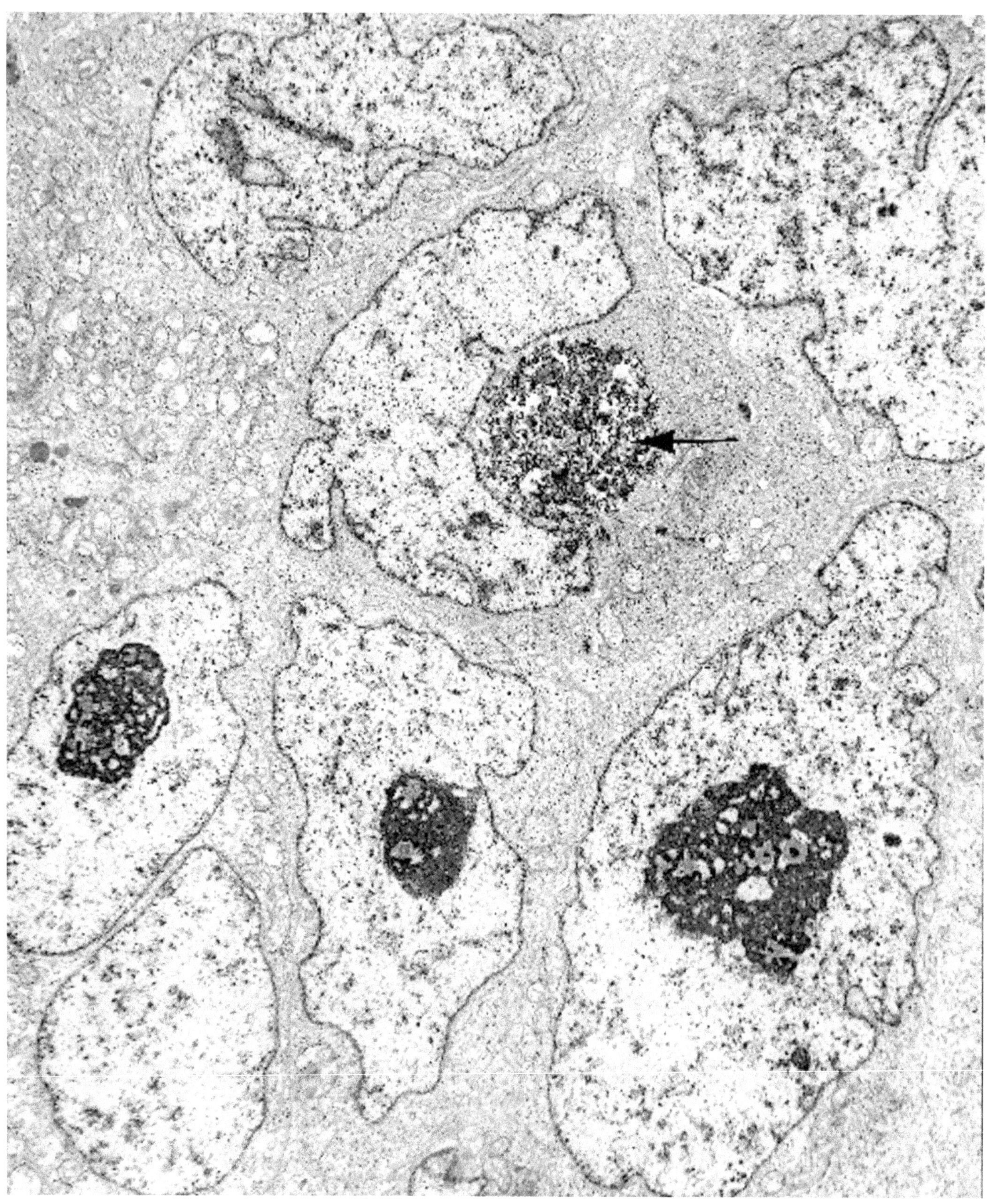

Figure 7.73. Embryonal carcinoma (testis). Large cells with large, irregular nuclei, large nucleoli, and poorly differentiated cytoplasm are arranged in a solid sheet. The large, dense, paranuclear body (*arrow*) is a secondary lysosome. (× 5500) (Permission for reprinting granted by WB Saunders, Dickersin GR: The ultrastructure of selected gynecologic neoplasms. Clin Lab Med 7:117–156, 1987.)

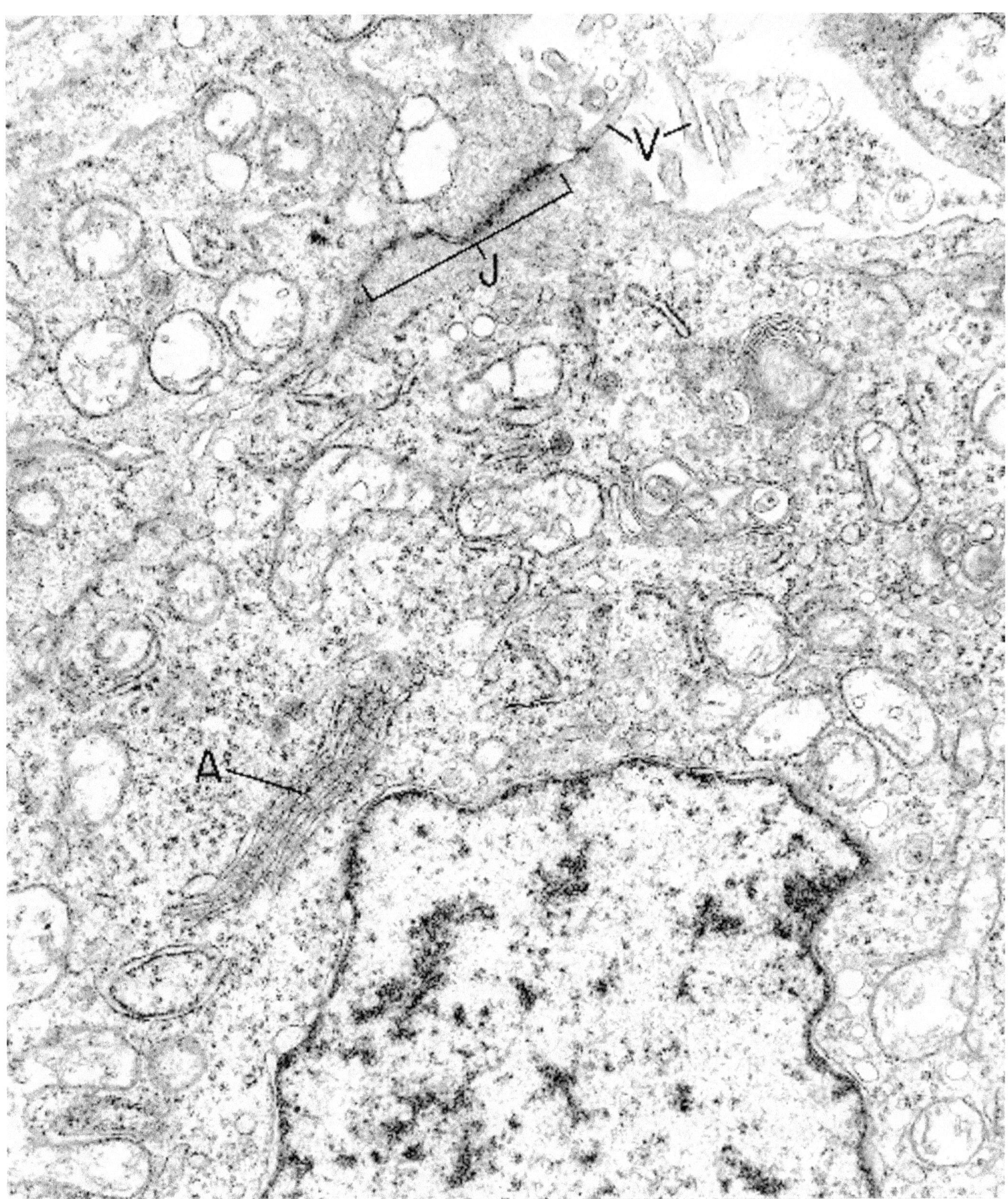

Figure 7.74. Embryonal carcinoma (testis). The same neoplasm as illustrated in Figure 7.73 shows focal tubule formation, with the lining cells having microvilli (V), junctional complexes (J), and copious cytoplasm with many organelles, including annulate lamellae (A). (× 21,460) (Permission for reprinting granted by WB Saunders, Dickersin GR: The ultrastructure of selected gynecologic neoplasms. Clin Lab Med 7:117–156, 1987.)

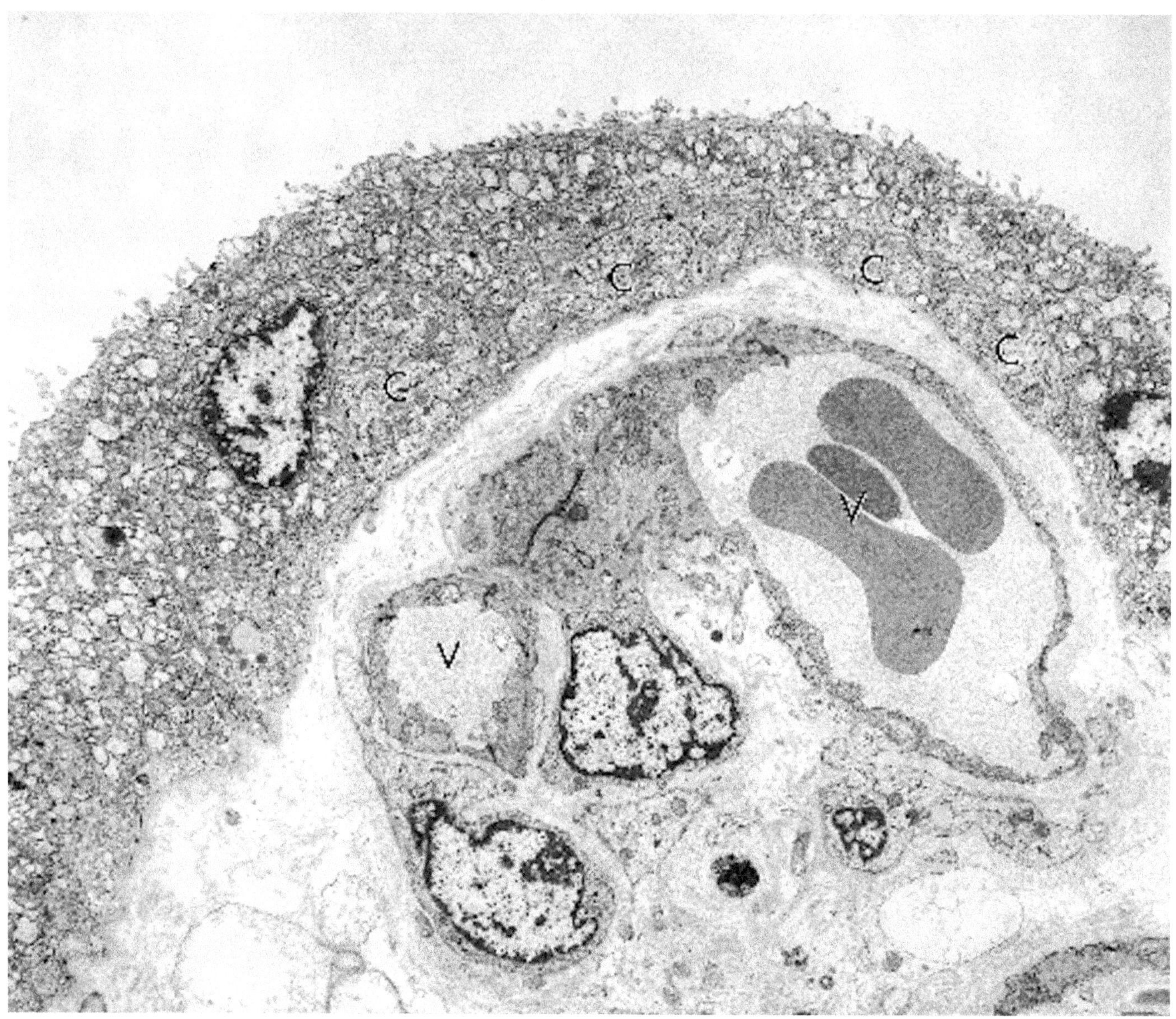

Figure 7.75. Normal full-term placenta. A chorionic villus is covered by syncytial trophoblast and four poorly discernible underlying cytotrophoblastic cells (C). The cytotrophoblast shows fewer organelles in the cytoplasm, mainly free ribosomes, a few cisternae of undilated rough endoplasmic reticulum, and a few mitochondria. By contrast, the overlying syncytial trophoblast has a rich collection of dilated rough endoplasmic reticulum. Numerous microvilli project from the luminal surface of the cell. V = blood vessels. (× 4600)

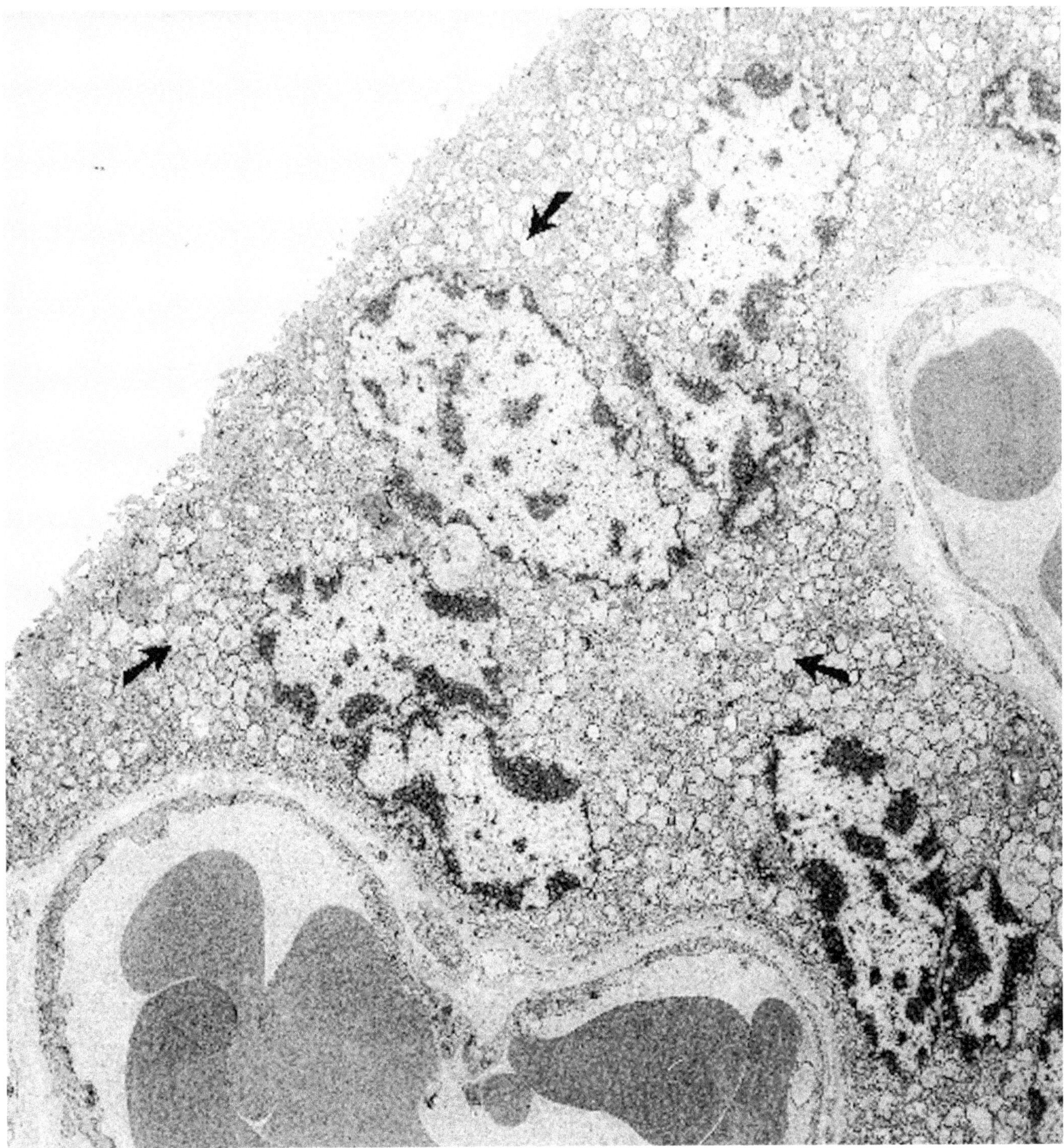

Figure 7.76. Normal full-term placenta. Higher magnification of the surface of a chorionic villus illustrates syncytial trophoblast filled with dilated rough endoplasmic reticulum (*arrows*). (× 7400)

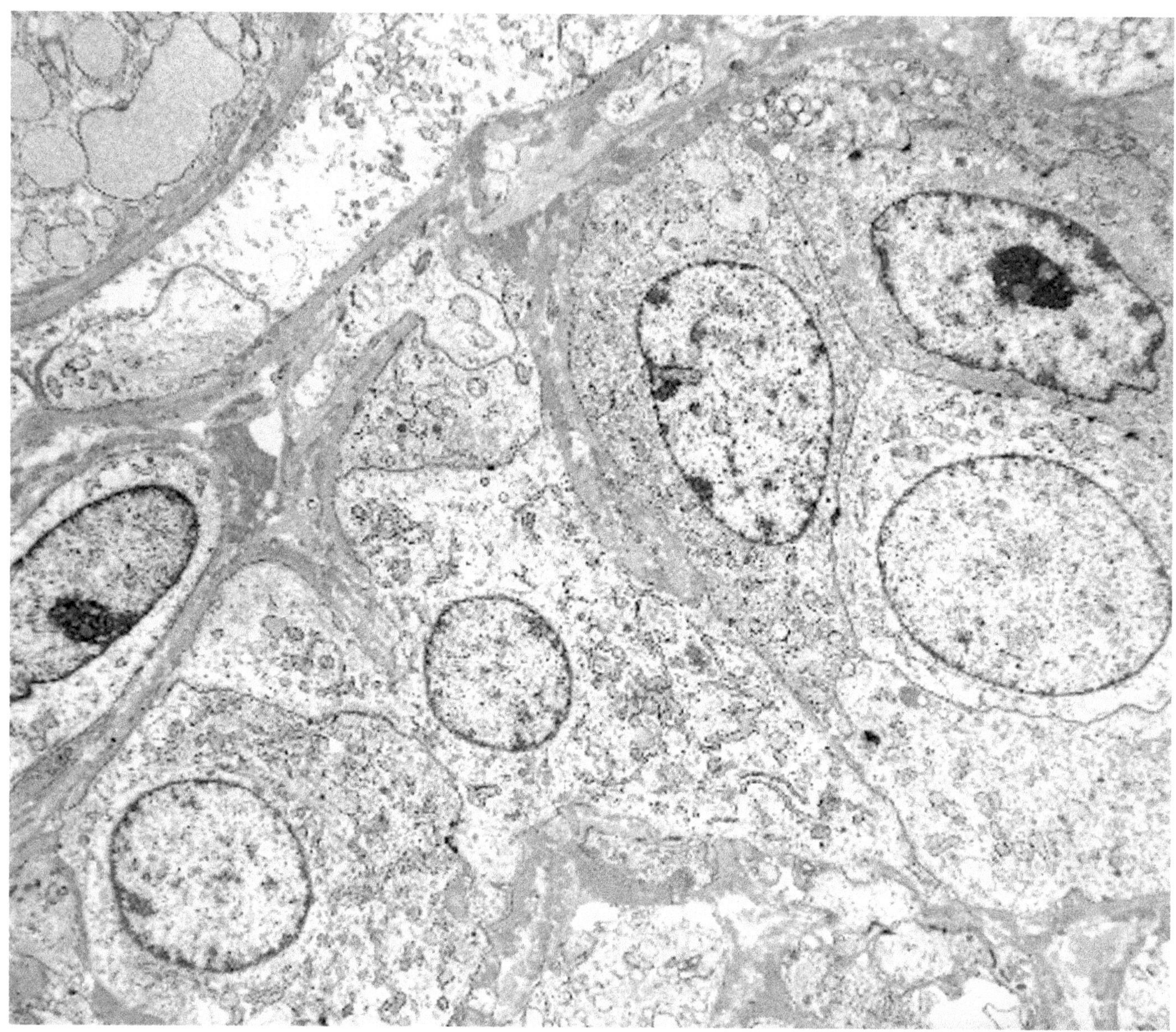

Figure 7.77. Placental site trophoblastic tumor (trophoblastic pseudotumor) (uterus). The high nuclear–cytoplasmic ratio, predominantly euchromatic nuclei and bland cytoplasm characterize these cells as cytotrophoblasts. (× 3600)

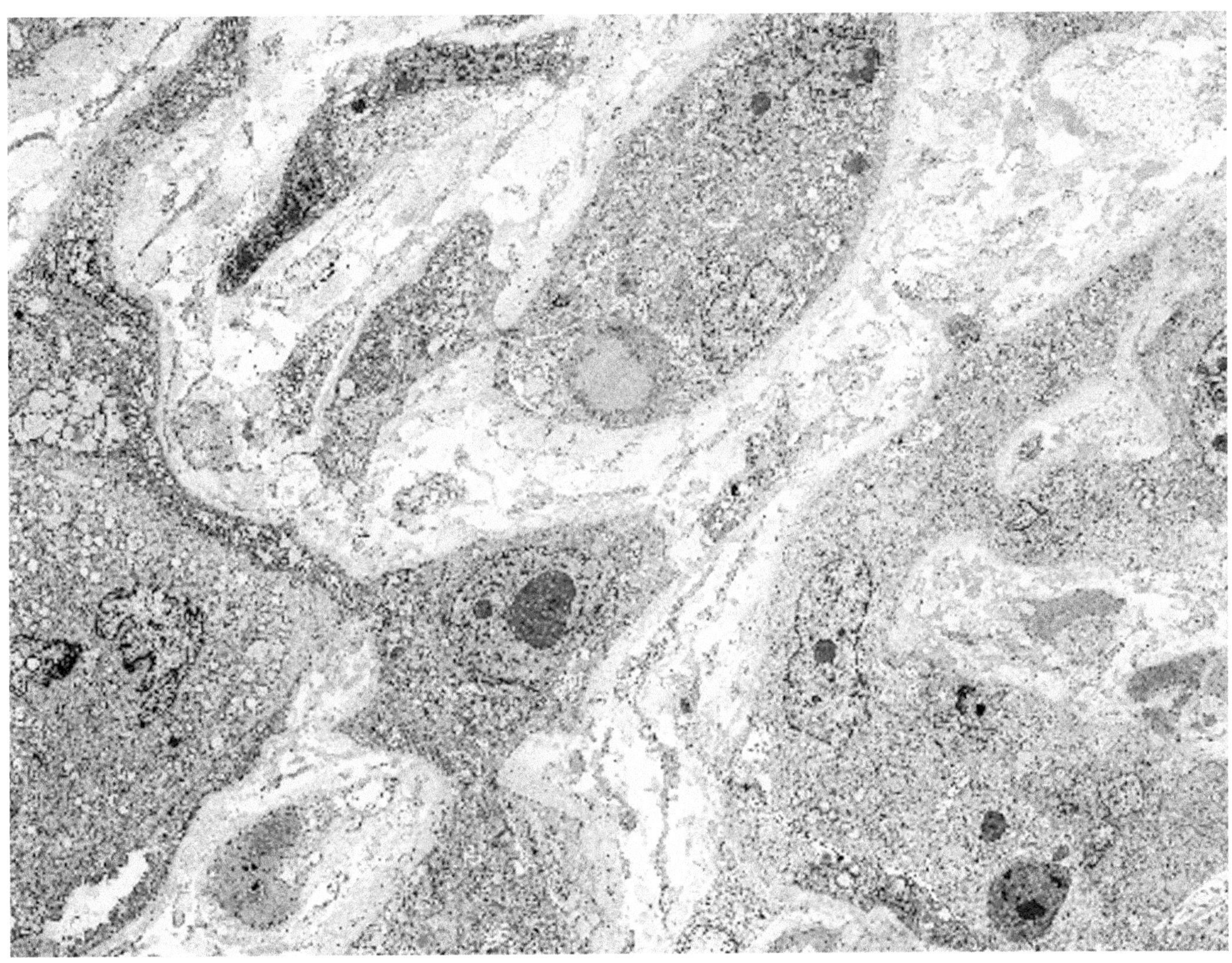

Figure 7.78. Placental site trophoblastic tumor (trophoblastic pseudotumor) (uterus). Intermediate trophoblastic cells are dispersed individually and in small groups in a loose matrix. The cells have abundant cytoplasm and numerous organelles. (× 2300)

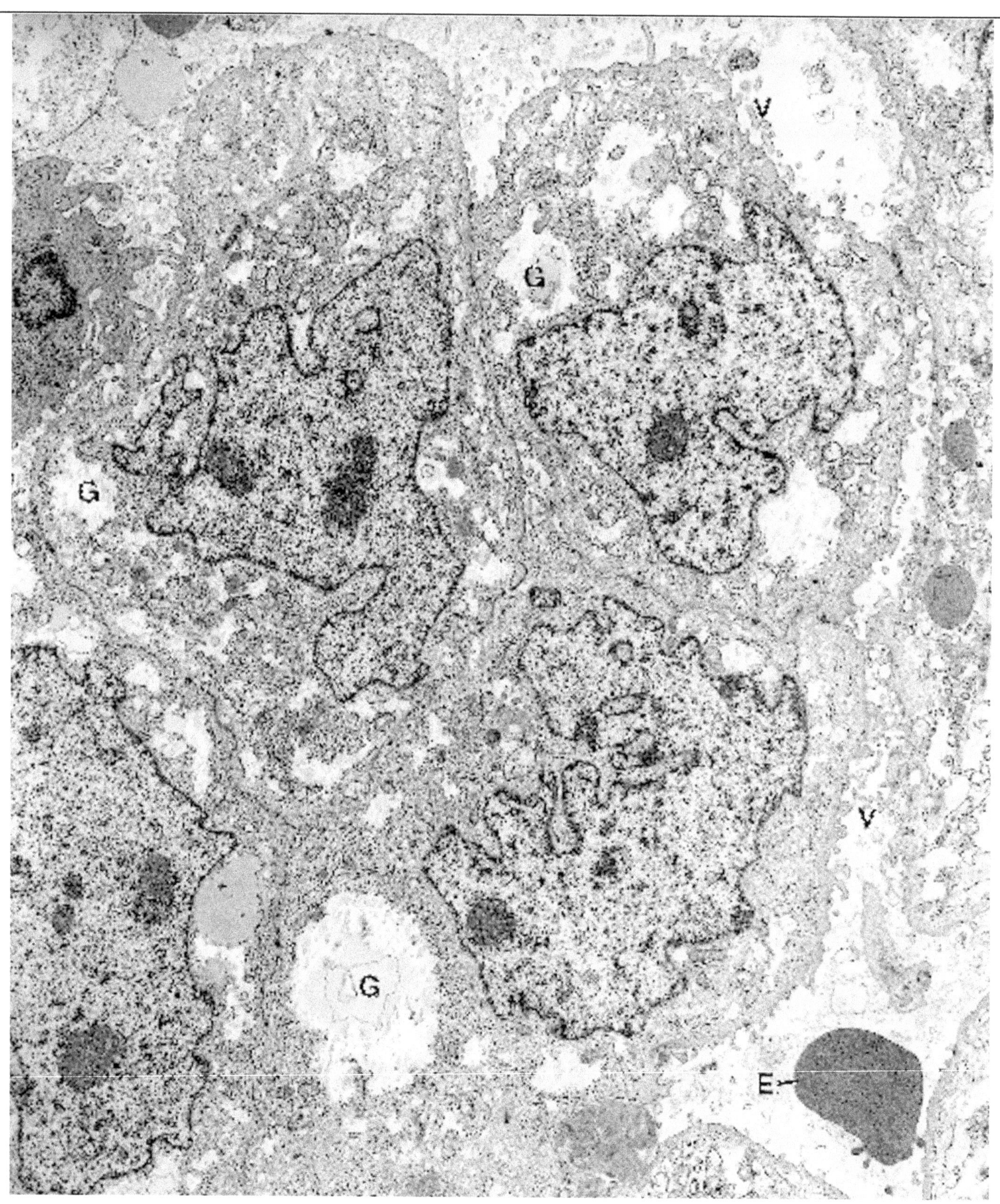

Figure 7.79. Placental site trophoblastic tumor (trophoblastic pseudotumor) (metastatic to lung). Intermediate trophoblastic cells are tightly apposed and have villi on their free surface (V), irregularly indented and segmented nuclei, prominent nucleoli, and moderately copious cytoplasm. The open cytoplasmic spaces are consistent with glycogen (G). An erythrocyte (E) marks a vascular sinus. (× 6300)

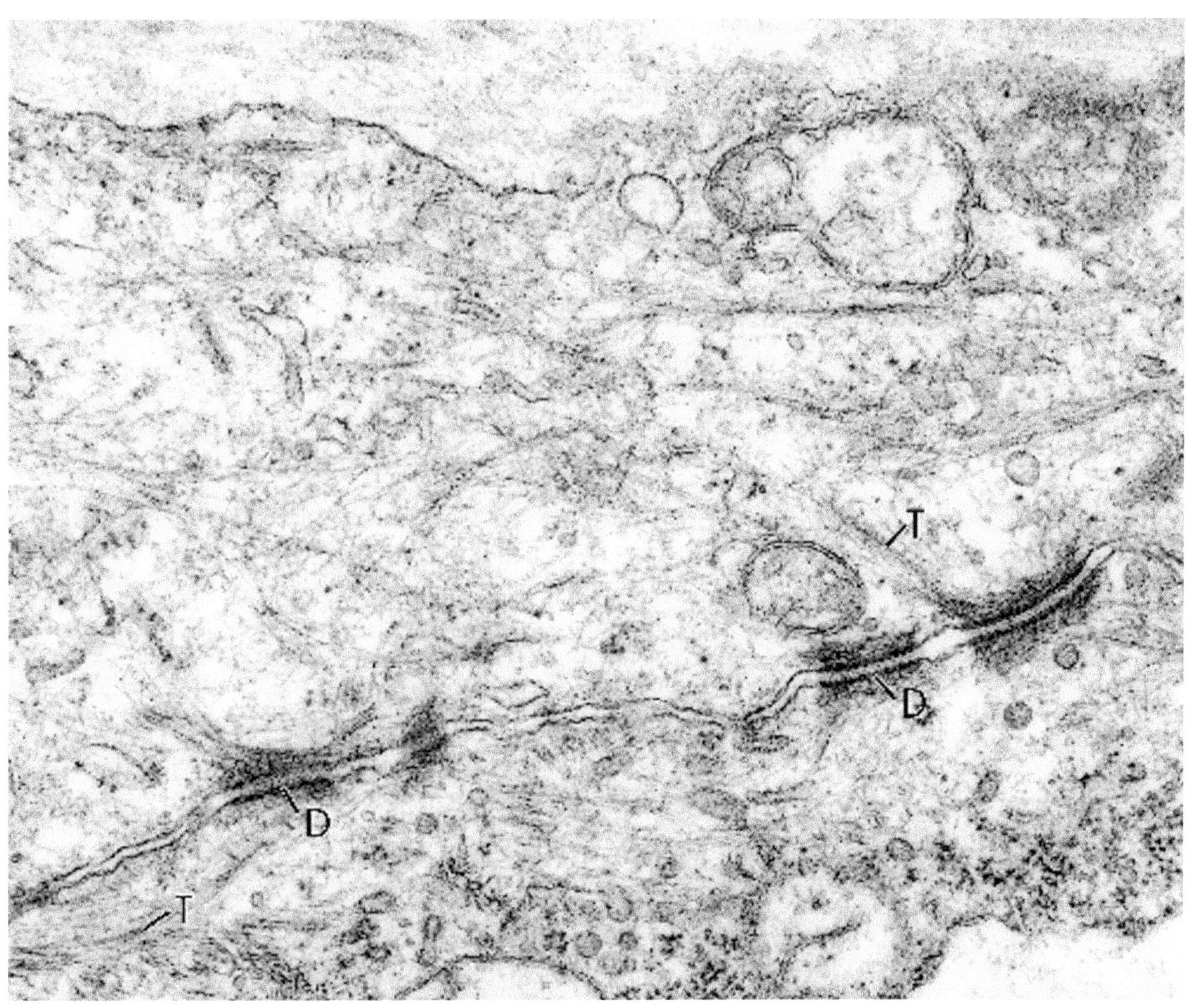

Figure 7.80. Placental site trophoblastic tumor (trophoblastic pseudotumor) (uterus). High magnification of the same tumor as that depicted in Figure 7.78 illustrates frequent desmosomes (D) and tonofibrils (T). (× 56,000)

(Text continued from page 462)

Gonadoblastoma

Gonadoblastoma is a neoplasm composed of a mixture of germ cells and smaller sex cord stromal cells. The two cell types are mixed in various patterns within nests, and the nests are separated by a fibrous stroma that usually contains lutein and Leydig-type cells. The germ cells are most commonly of the dysgerminoma type, and the sex cord cells are immature granulosa or Sertoli cells. The ultrastructure of the various cell types comprising this tumor are illustrated in earlier sections of this chapter.

Adenomatoid Tumor

(See Figures 3.72 through 3.74.)

Diagnostic criteria. (1) Gland-like and slit-like spaces lined by low cuboidal and columnar cells, or solid cords of large, oval, or polygonal cells in a scanty matrix of fibroblasts and collagen; (2) long, thin (sometimes branching) microvilli; (3) copious cytoplasm; (4) glycogen; (5) filaments, including bundles of filaments (tonofibrils); (6) large nuclei with prominent nucleoli; (7) prominent desmosomes and junctional complexes; (8) basal lamina; (9) lateral intercellular spaces.

Additional points. These benign neoplasms represent peritoneal mesotheliomas of the genital region, and their ultrastructure is the same as the epithelial (nonfibrous) type of mesothelioma found elsewhere in the peritoneum and pleura (see Chapter 3). The long, thin microvilli and the absence of secretory granules are the most characteristic features in distinguishing these neoplasms from adenocarcinomas of various organs, such as the ovary and lung.

Tumors of Uncertain Origin and Miscellaneous Tumors

Ovarian Small Cell Carcinoma, Hypercalcemia Type

(Figures 7.81 through 7.82.)

Diagnostic criteria. (1) Diffuse sheets (most common pattern), or nests, cords, and follicle-like groups of small cells in a sparse collagenous matrix; (2) basal lamina focally surrounding groups of cells; (3) closely apposed oval and polygonal cells with intermediate junctions and desmosomes; (4) high nuclear–cytoplasmic ratio; (5) slightly elongated, angular, and indented nuclei with small amounts of heterochromatin and occasional large nucleoli; (6) broad cytoplasmic processes with most of the organelles; (7) numerous free ribosomes; (8) moderate number of mitochondria; (9) consistent prominence of dilated rough endoplasmic reticulum; (10) paranuclear whorls of filaments (variable).

Additional points. No glycogen has been identified in the cells of these neoplasms, nor have dense-core cytoplasmic granules been found, as would be expected in neuroendocrine small cell carcinomas. Microacini, lined by cells with short, blunt microvilli, are found rarely.

Small Cell Carcinoma, Pulmonary Type

(See Figures 4.1 through 4.5.)

Diagnostic criteria. (1) Islands of closely apposed small oval and spindle shaped cells; (2) high nuclear–cytoplasmic ratio with scanty cytoplasm; (3) intermediate junctions; (4) euchromatic nuclei; (5) predominance of free ribosomes in cytoplasm; (6) few or no dense core granules in most cases; (7) filaments and tonofibrils in a few cells.

Additional points. Golgi apparatuses may be large in the cells of some tumors, but rough endoplasmic reticulum is not prominent (as distinguished from small cell carcinoma of the hypercalcemic type). Rarely, these small cell tumors may contain other elements such as endometrioid carcinoma, mucinous cysts, squamous cell carcinoma, and Brenner tumor.

Tumor of Probable Wolffian Origin

(Figures 7.83 through 7.84.)

Diagnostic criteria. (1) Any of several patterns—tubular, cystic, trabecular, and vacuolar; (2) tubular pattern—tubules surrounded by basal lamina and lined by small epithelial-type cells with (a) microvilli, (b) junctional complexes, (c) rough endoplasmic reticulum, (d) many microfilaments, (e) a few lysosomes, and (f) indented nuclei.

Additional points. The ultrastructure of the tubules resembles that of human fetal and adult mesonephric tissues and human mesonephric rests. Characteristic of Wolffian (as opposed to Müllerian) differentiation are the formation of tubules, the small size of the cells, interdigitation of lateral cell membranes in association with infolded basal plasma membranes, absence of cilia, and paucity of glycogen. The tubules may be closely arranged or separated by fibrous stroma.

(Text continues on page 483)

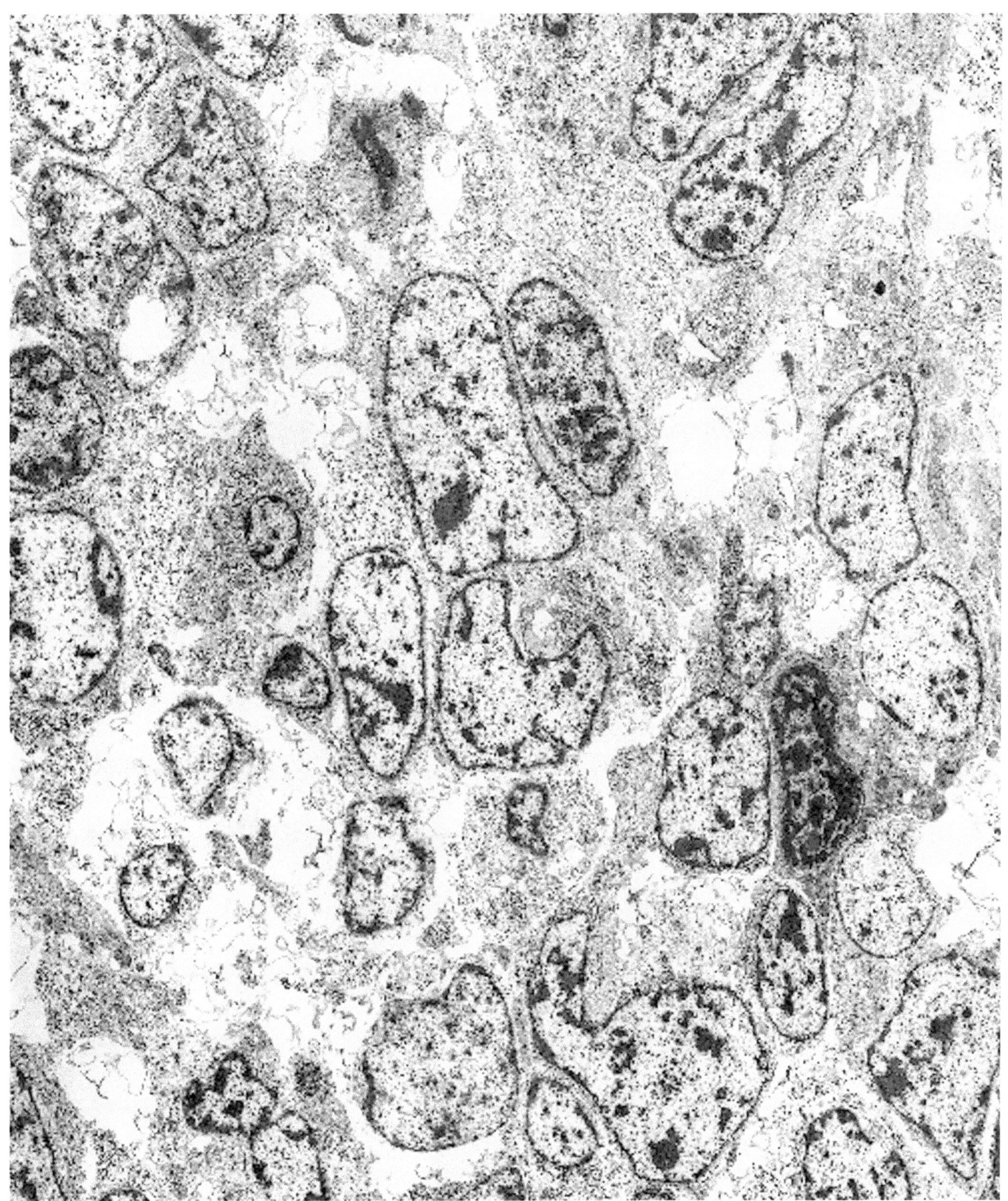

Figure 7.81. Small cell carcinoma (with hypercalcemia; ovary). The cells are small, oval and polygonal, and closely arranged. The nuclear–cytoplasmic ratio is high, the cytoplasm being relatively scant. (× 4465). (Permission for reprinting granted by WB Saunders, Dickersin GR: The ultrastructure of selected gynecologic neoplasms. Clin Lab Med 7:117–156, 1987.)

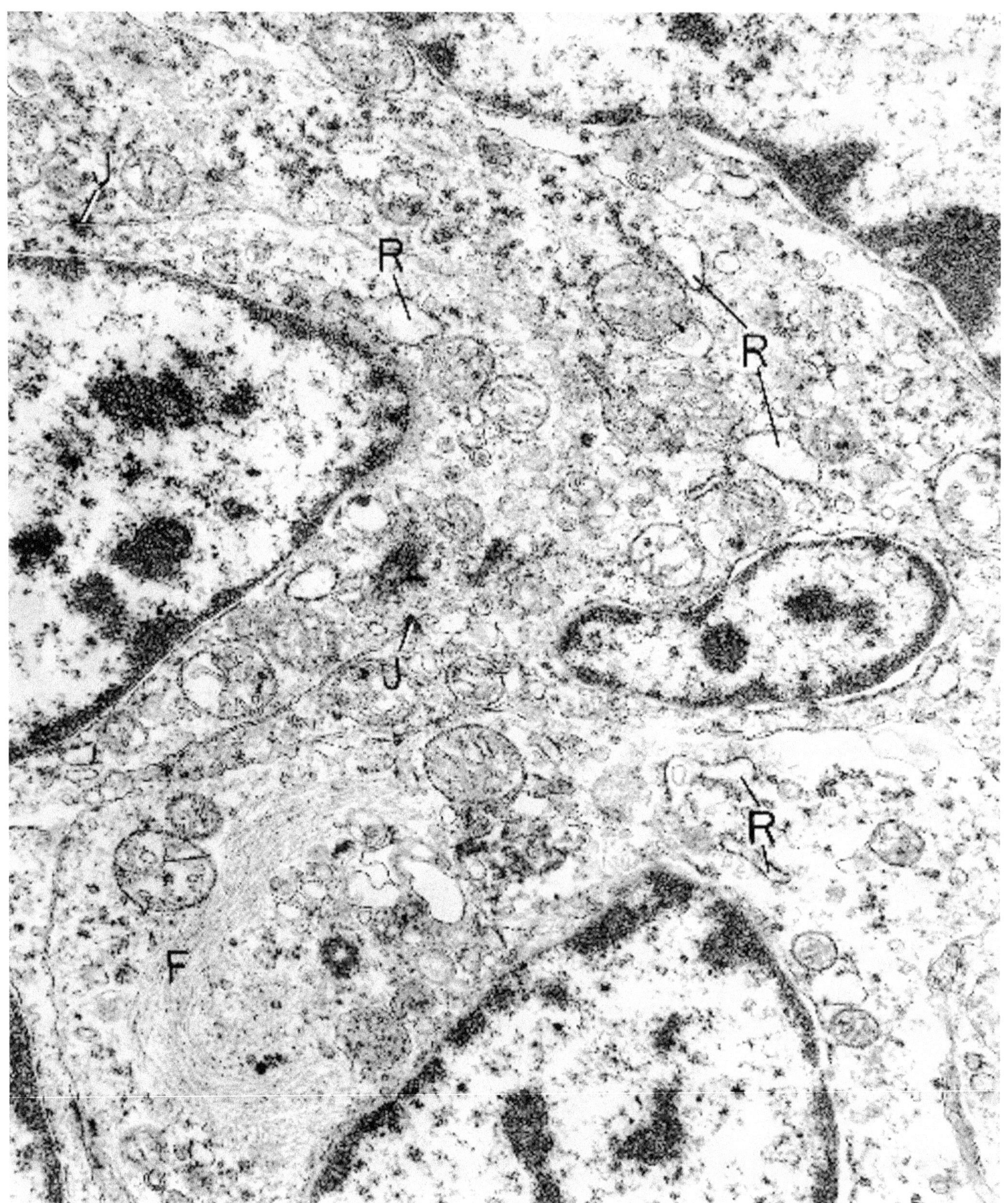

Figure 7.82. Small cell carcinoma (with hypercalcemia; ovary). This higher magnification of several neoplastic cells depicts dilated cisternae of rough endoplasmic reticulum (R), a whorl of paranuclear filaments (F), other nonspecific organelles and several junctions (J). (× 21,900) (Permission for reprinting granted by WB Saunders, Dickersin GR: The ultrastructure of selected gynecologic neoplasms. Clin Lab Med 7:117–156, 1987.)

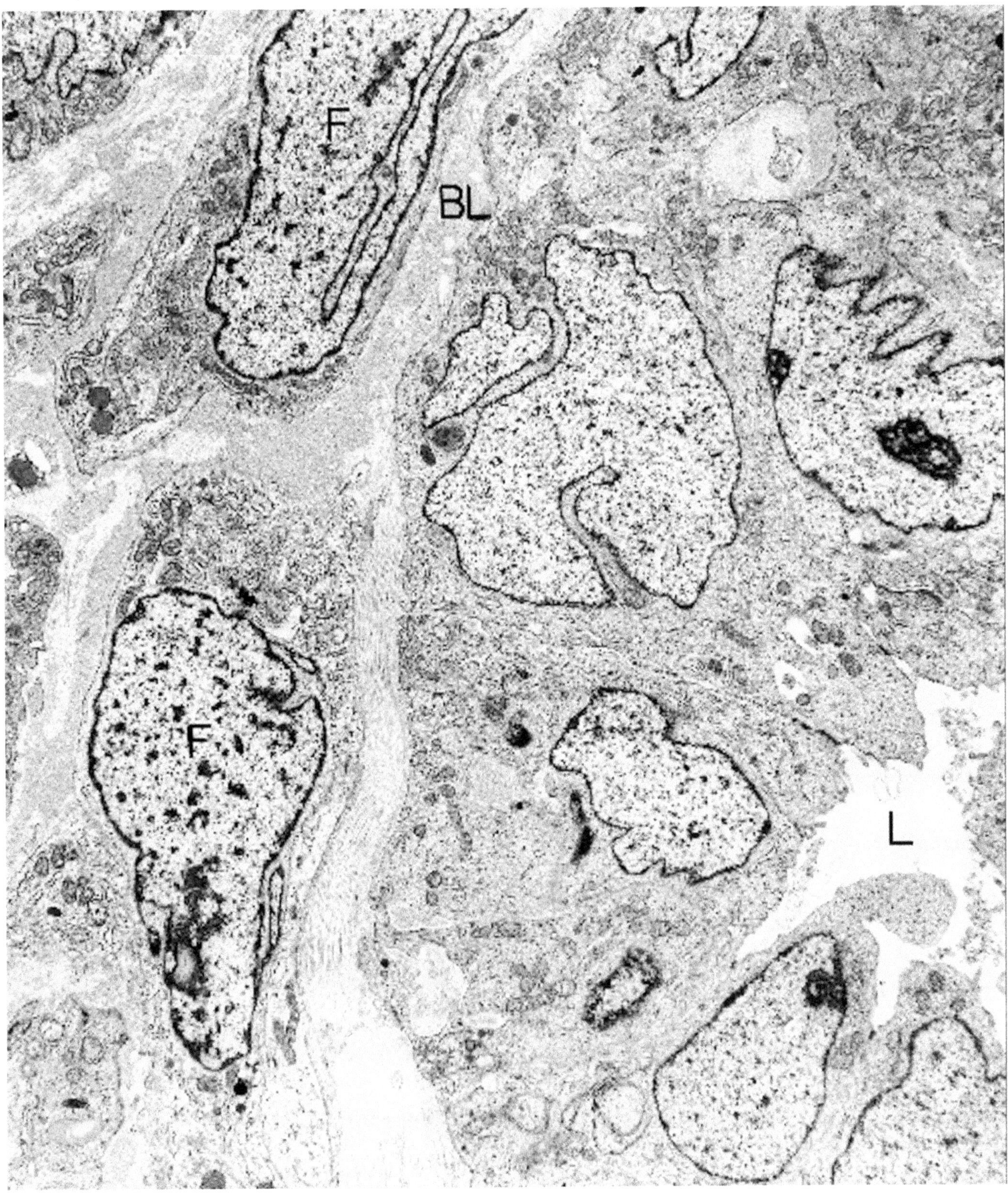

Figure 7.83. Tumor of probable Wolffian origin (ovary). The neoplastic cells form a tubule with a true lumen (L) and a surrounding basal lamina (BL). Nuclei are markedly indented. The cells are smaller than the surrounding fibroblasts (F). (× 7000) (Permission for reprinting granted by WB Saunders, Dickersin GR: The ultrastructure of selected gynecologic neoplasms. Clin Lab Med 7:117–156, 1987.)

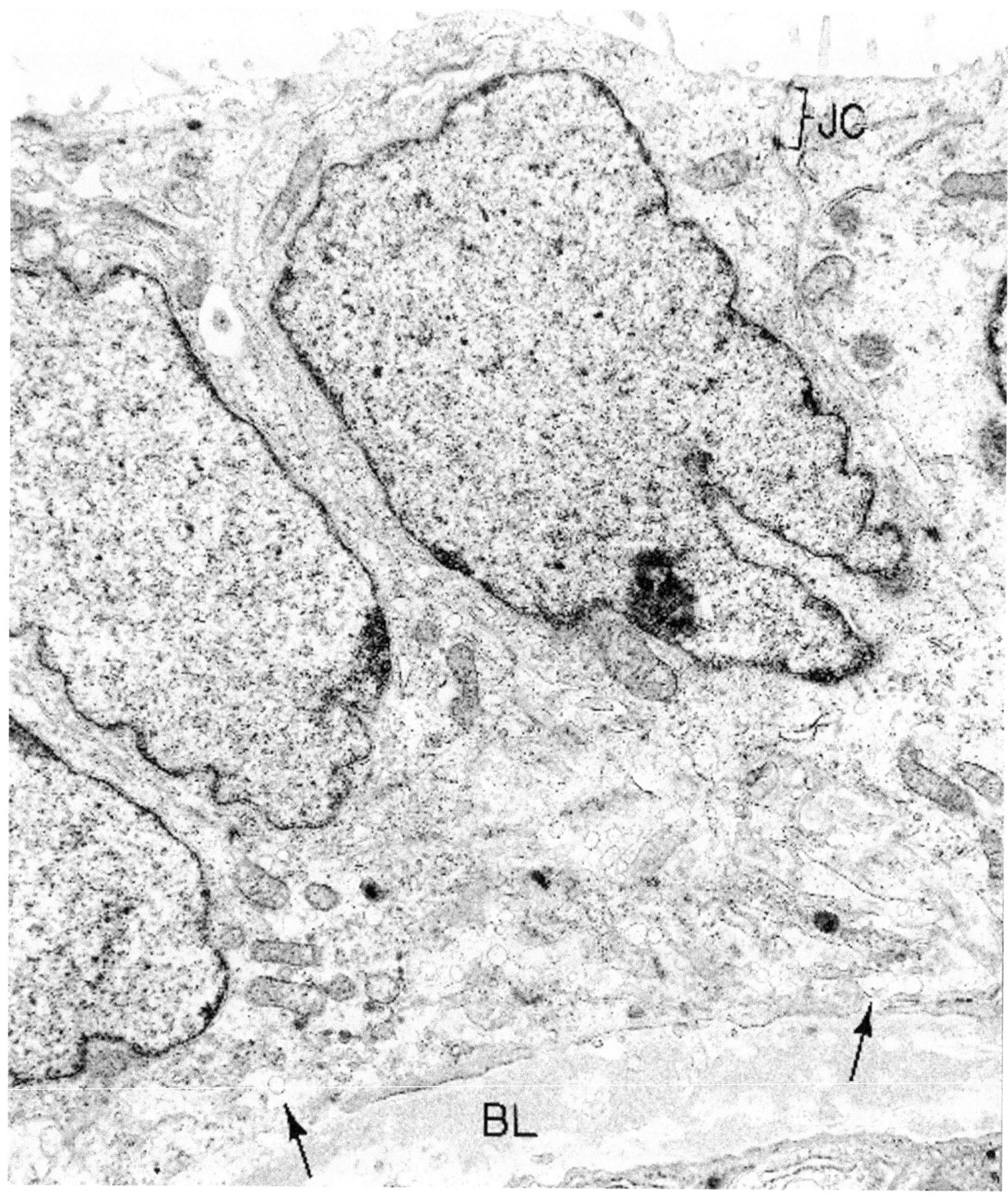

Figure 7.84. Tumor of probable Wolffian origin (ovary). Details of a tubular lining cell include irregularly distributed microvilli, junctional complexes (JC), a thick basal lamina (BL), and an infolded basal plasmalemma (*arrows*) in continuity with interdigitating lateral plasma membranes. (× 13,500) (Permission for reprinting granted by WB Saunders, Dickersin GR: The ultrastructure of selected gynecologic neoplasms. Clin Lab Med 7:117–156, 1987.)

(*Text continued from page 478*)

REFERENCES

Serous Tumors

Blaustein A: Papillary serous tumors of the ovary: An electron microscopic study. Gynecol Oncol 4:314–323, 1976.

Dickersin GR: The ultrastructure of selected gynecologic neoplasms. Clin Lab Med 7:117–156, 1987.

Fenoglio CM: Overview article: Ultrastructural features of the common epithelial tumors of the ovary. Ultrastruct Pathol 1:419–444, 1980.

Fenoglio CM, Castadot M, Ferenczy A, et al: Serous tumors of the ovary. I. Ultrastructural and histochemical studies of the epithelium of the benign serous neoplasms, serous cystadenoma and serous cystadenofibroma. Gynecol Oncol 5:203–218, 1977.

Gondos B: Electron microscopic study of papillary serous tumors of the ovary. Cancer 27:1455–1464, 1971.

Scully RE, Young RH, Clement PB: Tumors of the ovary, maldeveloped gonads, fallopian tube, and broad ligament. In Rosai J, ed: *Atlas of Tumor Pathology,* 3rd series. Armed Forces Institute of Pathology, Washington, DC, 1998.

White PF, Merino MJ, Barwick KW: Serous surface papillary carcinoma of the ovary: A clinical, pathologic, ultrastructural, and immunohistochemical study of 11 cases. Pathol Annu 20:403–418, 1985.

Mucinous Tumors

Dickersin GR: The ultrastructure of selected gynecologic neoplasms. Clin Lab Med 7:117–156, 1987.

Fenoglio CM, Ferenczy A, Richart RM: Mucinous tumors of the ovary. Ultrastructural studies of mucinous cystadenomas with histogenetic considerations. Cancer 36:1709–1722, 1975.

Fenoglio CM, Ferenczy A, Richart RM: Mucinous tumors of the ovary. II. Ultrastructural features of mucinous cystadenocarcinomas. Am J Obstet Gynecol 125: 990–999, 1976.

Langley FA, Cummins PA, Fox H: An ultrastructural study of mucin secreting epithelia in ovarian neoplasms. Acta Pathol Microbiol Scand [A] 80(suppl 233):76–78, 1972.

Scully RE, Young RH, Clement PB: Tumors of the ovary, maldeveloped gonads, fallopian tube, and broad ligament. In Rosai J, ed: *Atlas of Tumor Pathology,* 3rd series. Armed Forces Institute of Pathology, Washington, DC, 1998.

Stenback K: Morphology of ovarian mucinous cystadenoma: Surface ultrastructure, development and biologic behavior. Int J Gynaecol Obstet 18:157–167, 1980.

Szymanska K, Szamborski J, Miechowiecka N, et al: Malignant transformation of mucinous ovarian cystadenomas of intestinal epithelial type. Histopathology 7:497–509, 1983.

Endometrioid Tumors

Cummins PA, Fox H, Langley FA: An electron microscopic study of the endometrioid adenocarcinoma of the ovary and a comparison of its fine structure with that of normal endometrium and of adenocarcinoma of the endometrium. J Pathol 13:165–173, 1974.

Dickersin GR: The ultrastructure of selected gynecologic neoplasms. Clin Lab Med 7:117–156, 1987.

Fenoglio CM, Puri S, Richart RM: The ultrastructure of endometrioid carcinoma of the ovary. Gynecol Oncol 6:152–164, 1978.

Scully RE, Young RH, Clement PB: Tumors of the ovary, maldeveloped gonads, fallopian tube, and broad ligament. In Rosai J, ed: *Atlas of Tumor Pathology,* 3rd series. Armed Forces Institute of Pathology, Washington, DC, 1998.

Clear-Cell Carcinoma

Dickersin GR, Welch WR, Erlandson R, et al: Ultrastructure of 16 cases of clear cell adenocarcinoma of the vagina and cervix in young women. Cancer 45: 1615–1624, 1980.

Dickersin GR: The ultrastructure of selected gynecologic neoplasms. Clin Lab Med 7:117–156, 1987.

Ohkana K, Amasaki H, Terashima Y, et al: Clear cell carcinoma of the ovary. Light and electron microscopic studies. Cancer 40:3019–3029, 1977.

Okagaki T, Richart RM: Mesonephroma ovarii (hypernephroid carcinoma): Light and ultrastructural study of one case. Cancer 26:453–461, 1970.

Roth LM: Clear-cell adenocarcinoma of the female genital tract. A light and electron microscopic study. Cancer 33:990–1001, 1974.

Scully RE, Young RH, Clement PB: Tumors of the ovary, maldeveloped gonads, fallopian tube, and broad ligament. In Rosai J, ed: *Atlas of Tumor Pathology,* 3rd series. Armed Forces Institute of Pathology, Washington, DC, 1998.

Scully RE, Barlow JF: Mesonephroma of the ovary: Tumor of Müllerian nature related to the endometrioid carcinoma. Cancer 20:1405–1417, 1967.

Silverberg SG: Ultrastructure and histogenesis of clear cell carcinoma of the ovary. Am J Obstet Gynecol 115: 384–400, 1973.

Brenner Tumor

Aguirre P, Scully RE, Wolfe HJ, et al: Argyrophil cells in Brenner tumors: Histochemical and immunohistochemical analysis. Int J Gynaecol Pathol 5:223–234, 1986.

Arey LB: The origin and form of Brenner tumor. Am J Obstet Gynecol 81:743–751, 1961.

Dickersin GR: The ultrastructure of selected gynecologic neoplasms. Clin Lab Med 7:117–156, 1987.

Fetissof F, Berger G, Dubois MP, et al: Endocrine cells in the female genital tract. Histopathology 9:133–145, 1985.

Kennedy M, Holch ST, Bock J: Bilateral malignant Brenner tumor. A light and electron microscopic study. Acta Pathol Microbiol Immunol Scand [A] 92:161–166, 1984.

Klemi PJ: Epithelial mucosubstances and argyrophil cells in Brenner tumors. Acta Pathol Microbiol Immunol Scand [A] 85:819–825, 1977.

Langley FA, Cummins PA, Fox H: An ultrastructural study of mucin secreting epithelia in ovarian neoplasms. Acta Pathol Microbiol Immunol Scand [A] 80: 76–86, 1972.

Roth LM: Fine structure of the Brenner tumor. Cancer 27:1482–1488, 1971.

Scully RE, Young RH, Clement PB: Tumors of the ovary, maldeveloped gonads, fallopian tube, and broad ligament. In Rosai J, ed: *Atlas of Tumor Pathology,* 3rd series. Armed Forces Institute of Pathology, Washington, DC, 1998.

Shevchuk MM, Fenoglio CM, Richart RM: Histogenesis of Brenner tumors. I. Histology and ultrastructure. Cancer 46:2607–2616, 1980.

Shikary A, Petrelli M, Hamilton P, et al: The Brenner tumor. Arch Pathol Lab Med 105:207–213, 1981.

Silverberg SG, Wilson MA: Ultrastructure of the Brenner tumor. Am J Obstet Gynecol 112:91–100, 1972.

Granulosa–Theca Cell Tumor

Bjersing L, Frankendal B, Anstrom T: Studies on a feminizing ovarian mesenchymoma (granulosa cell tumor). I. Aspiration biopsy cytology, histology, and ultrastructure. Cancer 32:1360–1369, 1973.

Dickersin GR: The ultrastructure of selected gynecologic neoplasms. Clin Lab Med 7:117–156, 1987.

Gaffney EF, Majmudar B, Hertzler GL, et al: Ovarian granulosa cell tumors—immunohistochemical localization of estradiol and ultrastructure, with functional correlations. Obstet Gynecol 61:311–319, 1983.

Gondos B: Ultrastructure of a metastatic granulosa–theca cell tumor. Cancer 24:954–959, 1969.

Gondos B, Monroe SA: Cystic granulosa cell tumor with massive hemoperitoneum. Light and electron microscopic study. Obstet Gynecol 38:683–689, 1971.

MacAulay MA, Weliky I, Schulz RA: Ultrastructure of a biosynthetically active granulosa cell tumor. Lab Invest 17:562–570, 1967.

Pederson PH, Larsen JF: Ultrastructure of a granulosa cell tumour. Acta Obstet Gynecol Scand 49:270–272, 1970.

Roth LM, Nicholas TR, Ehrlich CE: Juvenile granulosa cell tumor. A clinicopathologic study of three cases with ultrastructural observations. Cancer 44:2194–2205, 1979.

Salazar H, Gonzales-Angulo A: Ultrastructural diagnosis in gynecologic pathology. Clin Obstet Gynecol 11: 25–77, 1984.

Scully RE, Young RH, Clement PB: Tumors of the ovary, maldeveloped gonads, fallopian tube, and broad ligament. In Rosai J, ed: *Atlas of Tumor Pathology,* 3rd series. Armed Forces Institute of Pathology, Washington, DC, 1998.

Takeuchi H, Sodemoto Y, Ushigome S: Juvenile granulosa cell tumor associated with rapid distant metastases. Acta Pathol Jpn 33:537–545, 1983.

Toker C: Ultrastructure of granulosa cell tumor. Am J Obstet Gynecol 100:388–392, 1968.

Young RH, Dickersin GR, Scully RE: Juvenile granulosa cell tumor. A clinicopathological analysis of 125 cases. Am J Surg Pathol 8:575–596, 1984.

Stromal Tumors

Dickersin GR, Scully RE: Role of electron microscopy in metastatic endometrial stromal tumors. Ultrastruct Pathol 17:377–403, 1993.

Dickersin GR, Young RH, Scully RE: Signet-ring stromal and related tumors of the ovary. Ultrastruct Pathol 19:401–419, 1995.

Scully RE, Young RH, Clement PB: Tumors of the ovary, maldeveloped gonads, fallopian tube, and broad ligament. In Rosai J, ed: *Atlas of Tumor Pathology,* 3rd series, fasc 22. Armed Forces Institute of Pathology, Washington, DC, 1998.

Scully RE: Ovarian tumors with functioning stroma. In H Fox, M Wells, eds: *Haines & Taylor Obstetrical and Gynaecological Pathology,* 4th ed, vol 2. Churchill Livingstone, New York, 1995.

Suarez A, Palacios J, Burgos E, et al: Signet-ring stromal tumor of the ovary: A histochemical, immunohistochemical and ultrastructural study. Virchows Arch [A] 422:333–336, 1993.

Sertoli-Leydig Cell Tumor

Able ME, Lee JC: Ultrastructure of a Sertoli-cell adenoma of the testis. Cancer 23:481–486, 1969.

Dickersin GR: The ultrastructure of selected gynecologic neoplasms. Clin Lab Med 7:117–156, 1987.

Jenson AB, Fechner RE: Ultrastructure of an intermediate Sertoli-Leydig cell tumor. A histogenetic misnomer. Lab Invest 21:527–535, 1969.

Kalderon EE, Tucci JR: Ultrastructure of a human chorionic gonadotropin- and adrenocorticotropin-responsive functioning Sertoli-Leydig cell tumor (Type I). Lab Invest 29:81–89, 1973.

Ramzy I, Bos C: Sertoli cell tumors of the ovary. Cancer 38:2447–2456, 1976.

Roth LM, Cleary RE, Rosenfield RL: Sertoli-Leydig cell tumor of the ovary, with an associated mucinous cystadenoma. An ultrastructural and endocrine study. Lab Invest 31:648–657, 1975.

Schnoy H: Ultrastructure of a verilizing ovarian Leydig cell tumor. Virchows Arch [A] 397:17–27, 1982.

Scully RE, Young RH, Clement PB: Tumors of the ovary, maldeveloped gonads, fallopian tube, and broad ligament. In Rosai J, ed: *Atlas of Tumor Pathology,* 3rd series. Armed Forces Institute of Pathology, Washington, DC, 1998.

Tavassoli FA: A combined germ cell–gonadal stromal-epithelial tumor of the ovary. Am J Surg Pathol 7:73–84, 1983.

Sex-Cord Tumor with Annular Tubules

Ahn GH, Chic JG, Lee SK: Ovarian sex cord tumor with annular tubules. Cancer 57:1066–1073, 1986.

Astenso-Osuna C: Ovarian sex cord tumor with annular tubules. Case report with ultrastructural findings. Cancer 54:1070–1075, 1984.

Crissman JD, Hart WR: Ovarian sex cord tumors with annular tubules. An ultrastructural study of three cases. Am J Clin Pathol 75:11–17, 1981.

Dickersin GR: The ultrastructure of selected gynecologic neoplasms. Clin Lab Med 7:117–156, 1987.

Hertel BF, Kempson RL: Ovarian sex cord tumors with annular tubules: An ultrastructural study. Am J Surg Pathol 1:145–153, 1977.

Scully RE, Young RH, Clement PB: Tumors of the ovary, maldeveloped gonads, fallopian tube, and broad ligament. In Rosai J, ed: *Atlas of Tumor Pathology,* 3rd series. Armed Forces Institute of Pathology, Washington, DC, 1998.

Tavassoli FA, Norris HJ: Sertoli tumors of the ovary: A clinicopathological study of 28 cases with ultrastructural observations. Cancer 46:2281–2297, 1980.

Young RH, Welch WR, Dickersin GR, et al. Ovarian sex cord tumor with annular tubules. Cancer 50:1384–1402, 1982.

Lipid-Cell Tumor

Dickersin GR: The ultrastructure of selected gynecologic neoplasms. Clin Lab Med 7:117–156, 1987.

Green JA, Maqueo M: Histopathology and ultrastructure of an ovarian hilar cell tumor. Am J Obstet Gynecol 96:478–485, 1966.

Ishida L, Okagaki T, Tagatz GE, et al: Lipid cell tumor of the ovary: An ultrastructural study. Cancer 40: 234–243, 1977.

Koss LG, Rothschild EO, Fleisher M, et al: Masculinizing tumor of the ovary, apparently with adrenocortical activity. Cancer 23:1245–1258, 1969.

Merkow LP, Slifkin M, Acevedo HF, et al: Ultrastructure of an interstitial (hilar) cell tumor of the ovary. Obstet Gynecol 37:845–859, 1971.

Muechler EK, Gillette MB, Cary D, et al: Ovarian lipoid cell tumor. Steroid hormones and ultrastructure. Diagn Gynecol Obstet 4:309–315, 1982.

Scully RE, Young RH, Clement PB: Tumors of the ovary, maldeveloped gonads, fallopian tube, and broad ligament. In Rosai J, ed: *Atlas of Tumor Pathology,* 3rd series. Armed Forces Institute of Pathology, Washington, DC, 1997.

Dysgerminoma (Seminoma) and Embryonal Carcinoma

Andrews PW, Damjanov I, Simon D, et al: Pluripotent embryonal carcinoma clones derived from the human teratocarcinoma cell line Tera-2. Differentiation *in vivo* and *in vitro.* Lab Invest 50:147–162, 1984.

Dickersin GR: The ultrastructure of selected gynecologic neoplasms. Clin Lab Med 7:117–156, 1987.

Ferenczy A, Richart RM: *Female Reproduction System. Dynamics of Scan and Transmission Electron Microscopy.* John Wiley and Sons, New York, 1974, pp 336–338.

Gondos B, Bhiraleus P, Hobel CJ: Ultrastructural observations on germ cells in human fetal ovaries. Am J Obstet Gynecol 110:644–652, 1971.

Koide O, Iwai S: An ultrastructural study on seminoma cells. Acta Pathol Jpn 31:755–766, 1981.

Markesberg WR, Brooks WH, Milsow L, et al: Ultrastructural study of the pineal germinoma *in vivo* and *in vitro.* Cancer 37:327–337, 1976.

Scully RE, Young RH, Clement PB: Tumors of the ovary, maldeveloped gonads, fallopian tube, and broad ligament. In Rosai J, ed: *Atlas of Tumor Pathology,* 3rd series. Armed Forces Institute of Pathology, Washington, DC, 1998.

Endodermal Sinus Tumor (Yolk Sac Tumor)

Cohen MD, Mulchahey KM, Molnar JJ: Ovarian endodermal sinus tumor with intestinal differentiation. Cancer 57:1580–1583, 1986.

Dickersin GR: The ultrastructure of selected gynecologic neoplasms. Clin Lab Med 7:117–156, 1987.

Dickersin GR, Oliva E, Young RH: Endometrioid-like variant of ovarian yolk sac tumor with foci of carcinoid: An ultrastructural study. Ultrastruct Pathol 19:421–429, 1995.

Klemi PJ, Meurman L, Gronroos M, et al: Clear cell (mesonephroid) tumors of the ovary with characteristics resembling endodermal sinus tumor. Int J Gynaecol Pathol 1:95–100, 1982.

Nakanishi N, Kawahara E, Kijikawa K, et al: Hyaline globules in yolk sac tumor. Histochemical, immunohistochemical and electron microscopic studies. Acta Pathol Jpn 32:733–739, 1982.

Prat J, Bhan AK, Dickersin GR, et al: Hepatoid yolk sac tumor of the ovary (endodermal sinus tumor with hepatoid differentiation): A light microscopic, ultrastructural and immunohistochemical study of seven cases. Cancer 50:2355–2368, 1982.

Nogales-Fernandez F, Silverberg SG, Blaustein PA, et al: Yolk sac carcinoma (endodermal sinus tumor). Cancer 39:1462–1474, 1977.

Salazar H, Kanbour A, Tobon H, et al: Endoderm cell derivatives in embryonal carcinoma of ovary: An electron microscopic study of two cases. Am J Pathol 74: 108a, 1973.

Scully RE, Young RH, Clement PB: Tumors of the ovary, maldeveloped gonads, fallopian tube, and broad ligament. In Rosai J, ed: *Atlas of Tumor Pathology,* 3rd series. Armed Forces Institute of Pathology, Washington, DC, 1998.

Sekine T, Ito H, Tanaka T, et al: Studies on the histogenesis of experimentally induced yolk sac tumor in rats. Nippon Sanka Fujinka Gakkai Zasshi 35: 1999–2006, 1983.

Takashina T: An ultrastructural study of human ovarian yolk sac tumor—an investigation of its histogenesis compared to that of human yolk sacs. Acta Obstet Gynaecol Jpn 33:916–924, 1981.

Teilum G: Classification of endodermal sinus tumor (mesoblastoma vitellinum) and so-called "embryonal carcinoma" of the ovary. Acta Pathol Microbiol Immunol Scand [A] 64:407–429, 1965.

Teilum G, Albrechtsen R, Norgaard-Pedersen B: Immunofluorescent localization of alpha fetoprotein synthesis in endodermal sinus tumor (yolk sac tumor). Acta Pathol Microbiol Scand [A] 82:586–588, 1974.

Ulbright TM, Roth LM, Brodhecker CA: Yolk sac differentiation in germ cell tumors. A morphological study of 50 cases with emphasis in hepatic, enteric and parietal yolk sac features. Am J Surg Pathol 10:151–164, 1986.

Female Adnexal Tumor of Probable Wolffian Origin

Brescia RJ, Cardoso de Almeida P, Fuller AF Jr, et al: Female adnexal tumor of probable Wolffian origin with multiple recurrences over 16 years. Cancer 56: 1456–1461, 1985.

Demopoulos RI, Sitelman A, Flotte T, et al: Ultrastructural study of a female adnexal tumor of probable Wolffian origin. Cancer 46:2273–2280, 1980.

Dickersin GR: The ultrastructure of selected gynecologic neoplasms. Clin Lab Med 7:117–156, 1987.

Kariminejad MH, Scully RE: Female adnexal tumor of probable Wolffian origin: A distinctive pathologic entity. Cancer 31:671–677, 1973.

Scully RE, Young RH, Clement PB: Tumors of the ovary, maldeveloped gonads, fallopian tube, and broad ligament. In Rosai J, ed: *Atlas of Tumor Pathology,* 3rd series. Armed Forces Institute of Pathology, Washington, DC, 1998.

Taxy JB, Battifora H: Female adnexal tumor of probable Wolffian origin: Evidence for low grade malignancy. Cancer 37:2349–2354, 1976.

Young RH, Scully RE: Ovarian tumors of probable Wolffian origin: A report of 11 cases. Am J Surg Pathol 7: 125–135, 1983.

Small-Cell Carcinoma with Hypercalcemia

Dickersin GR, Kline IW, Scully RE: Small cell carcinoma of the ovary with hypercalcemia. A report of eleven cases. Cancer 49:188–197, 1982.

Dickersin GR: The ultrastructure of selected gynecologic neoplasms. Clin Lab Med 7:117–156, 1987.

Dickersin GR, Scully RE: An update on the electron microscopy of small cell carcinoma of the ovary with hypercalcemia. Ultrastruct Pathol 17:411–422, 1993.

Dickersin GR, Scully RE. Ovarian small cell tumors. An electron microscopic review. Ultrastruct Pathol 22: 199–226, 1998.

Fortune MA, Ireland K: Small cell carcinoma of the ovary: A review of eight cases with clinicopathologic correlation and ultrastructural findings [abstract]. Am J Clin Pathol 86:399, 1986.

McMahon JT, Hart WR: An ultrastructural study of ovarian small cell carcinomas [abstract]. Lab Invest 56: 49, 1987.

Scully RE, Young RH, Clement PB: Tumors of the ovary, maldeveloped gonads, fallopian tube, and broad ligament. In Rosai J, ed: *Atlas of Tumor Pathology,* 3rd series. Armed Forces Institute of Pathology, Washington, DC, 1998.

Adenomatoid Tumor

Carlier MT, Dardick I, Lagace AF, et al: Adenomatoid tumor of uterus: Presentation in endometrial curettings. Int J Gynecol Pathol 5:69–74, 1986.

Ferenczy A, Fenoglio J, Richart RM: Observations on benign mesothelioma of the genital tract (adenomatoid tumor). A comparative ultrastructural study. Cancer 30: 244–260, 1972.

Mackay B, Bennington JL, Skaglund RW: The adenomatoid tumor: Fine structural evidence for a mesothelial origin. Cancer 25:109–115, 1971.

Marcus JB, Lynn JA: Ultrastructural comparison of an adenomatoid tumor, lymphangioma, hemangioma and mesothelioma. Cancer 25:171–175, 1970.

Salazar H, Kanbour A, Burgess F: Ultrastructure and observations on the histogenesis of mesotheliomas "adenomatoid tumors" of the female genital tract. Cancer 29:141–152, 1972.

Scully RE, Young RH, Clement PB: Tumors of the ovary, maldeveloped gonads, fallopian tube, and broad ligament. In Rosai J, ed: *Atlas of Tumor Pathology,* 3rd series. Armed Forces Institute of Pathology, Washington, DC, 1998.

Taxy JB, Battifora H, Oyasu R: Adenomatoid tumors: A light microscopic, histochemical, and ultrastructural study. Cancer 34:306–316, 1974.

8

Central Nervous System Neoplasms

Meningioma

(Figures 8.1 through 8.10.)

Diagnostic criteria. (1) Long, interdigitating cellular processes (Figures 8.1 through 8.5); (2) numerous cytoplasmic intermediate filaments (Figures 8.3, 8.4, and 8.6); (3) numerous intercellular junctions, including prominent desmosomes (Figures 8.3, 8.5, and 8.6).

Additional points. The diagnostic ultrastructural criteria apply regardless of whether the light microscopic type of meningioma is *meningothelial, fibrous,* or *mixed.* The cellular processes often run in parallel, with interdigitations being focal (Figures 8.2 and 8.3). The nuclei of the cells may be elongated (Figure 8.2) or oval (Figure 8.1). The derivation of the cells comprising most meningiomas is thought to be from the arachnoid layer of the meninges, which normally is characterized by elongated cells with interweaving processes and junctions. The cellular component of the dura mater is the fibroblast, and those of the pia are fibroblast-like cells and macrophages. The *psammoma bodies* that occur so commonly in meningiomas and normal arachnoid villi sometimes appear to form on a nidus of degenerative cellular debris, and other times, as a deposition of hydroxyapatite crystals on collagen, resembling the formation of osteoid (Figures 8.7 through 8.9). A number of rare forms of meningioma exist. *Microcystic meningiomas* are characterized by long cellular processes being arranged around extracellular spaces, which may be empty or may contain an amorphous matrix (Figure 8.10). *Secretory meningiomas* have cells arranged in gland-like groups, intracellular and extracellular lumens, and lining cells with microvilli and tonofibrils. The lumens, including the intracellular lumens, may contain granular material (hyaline inclusions or "pseudopsammoma bodies," by light microscopy). *Clear cell meningiomas* are composed of the same type of meningiothelial cell already described, but the cytoplasm contains copious glycogen. A group of tumors formerly designated as *angioblastic meningiomas* have been shown by electron microscopy to be hemangiopericytomas and hemangiomas. The ultrastructure of these lesions is similar to that of hemangiopericytomas and hemangiomas outside the nervous system (see Chapter 6, Spindle Cell Neoplasms).

(Text continues on page 499)

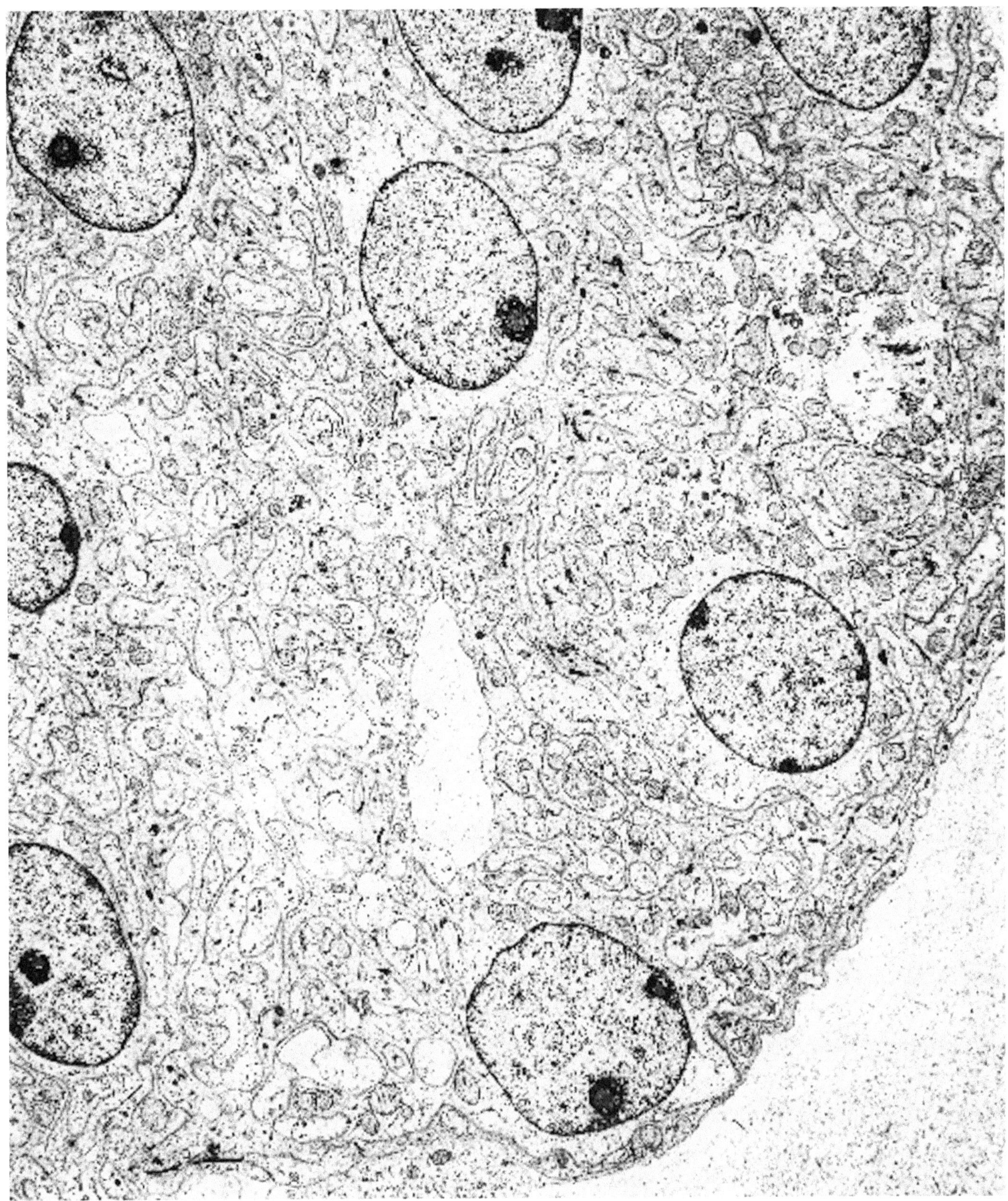

Figure 8.1. Meningioma, meningotheliomatous type (cerebrum). Cell bodies contain oval nuclei with predominantly euchromatin and nucleoli of small to moderate size. Cell processes are innumerable and are characteristically long, thin, and intertwining. (× 5130)

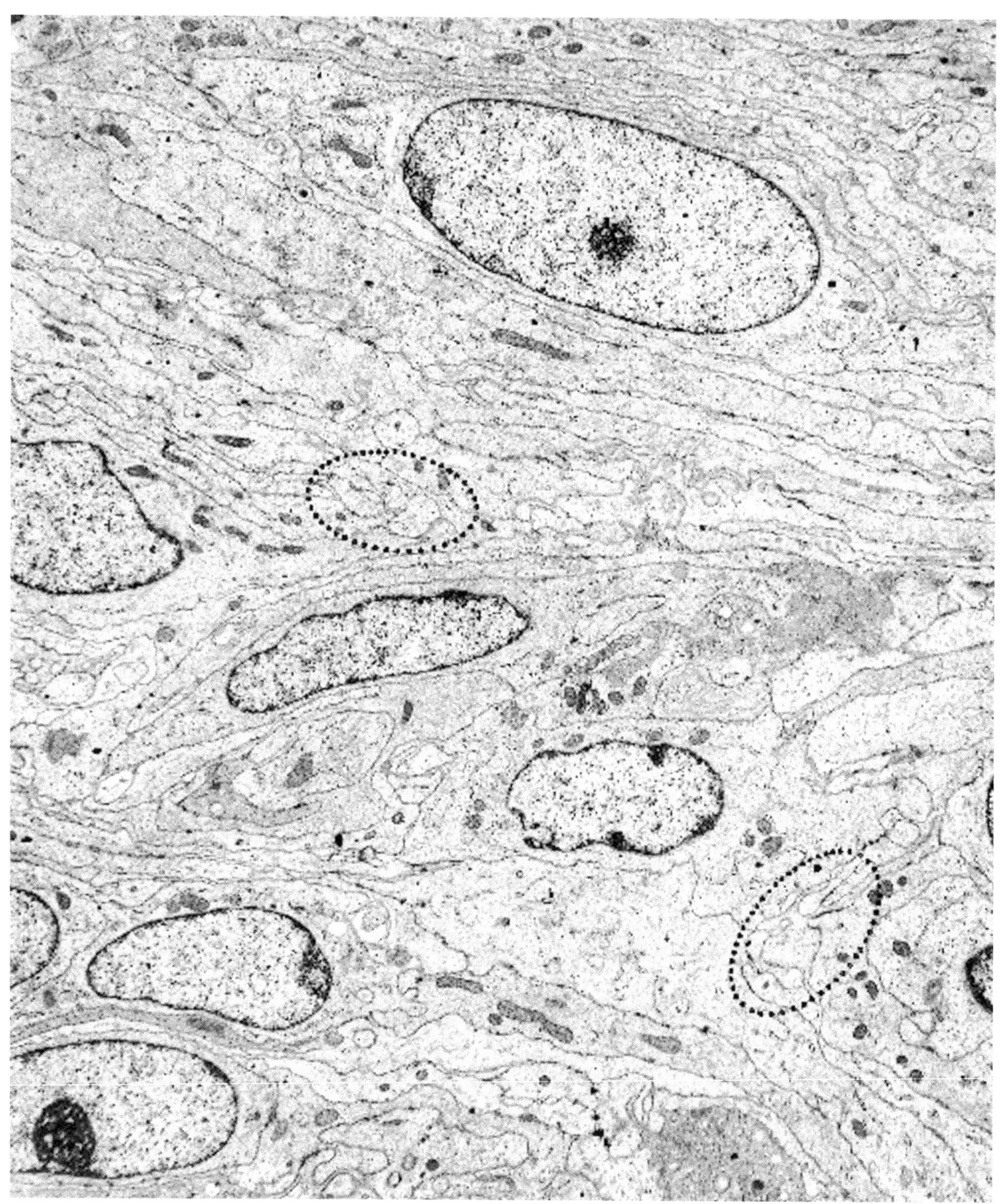

Figure 8.2. Meningioma, mixed type (cerebrum). Cell bodies contain elongated nuclei, in contrast to the oval nuclei seen in Figure 8.1. Cell processes are in parallel arrangement in many areas and with interdigitations (*dotted enclosures*) focally. (× 4940)

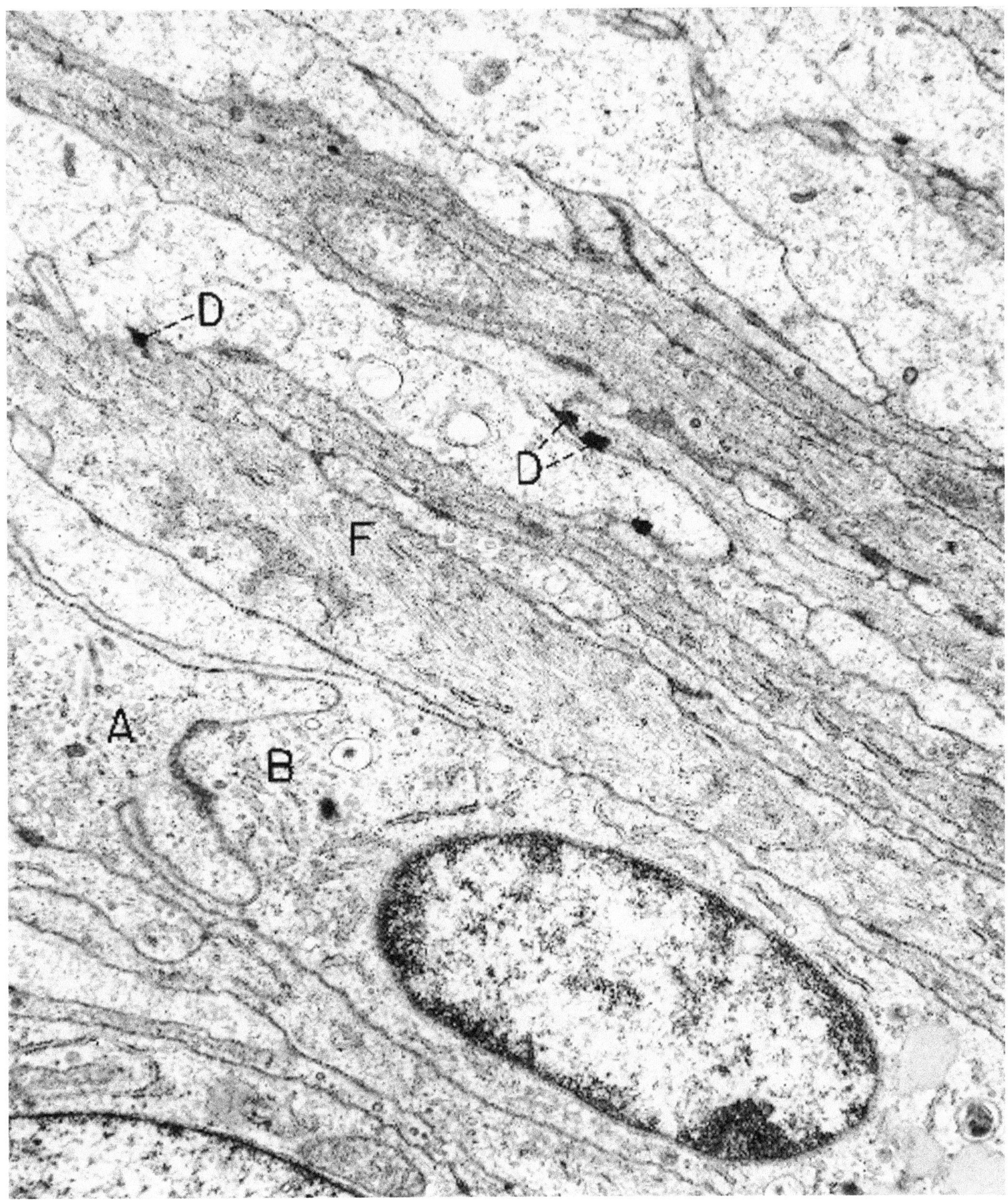

Figure 8.3. Meningioma, mixed type (cerebrum). Higher power of cell processes illustrates an intimate interdigitation between two cells (A and B), numerous filaments (F), and several desmosomes (D). (× 15,675)

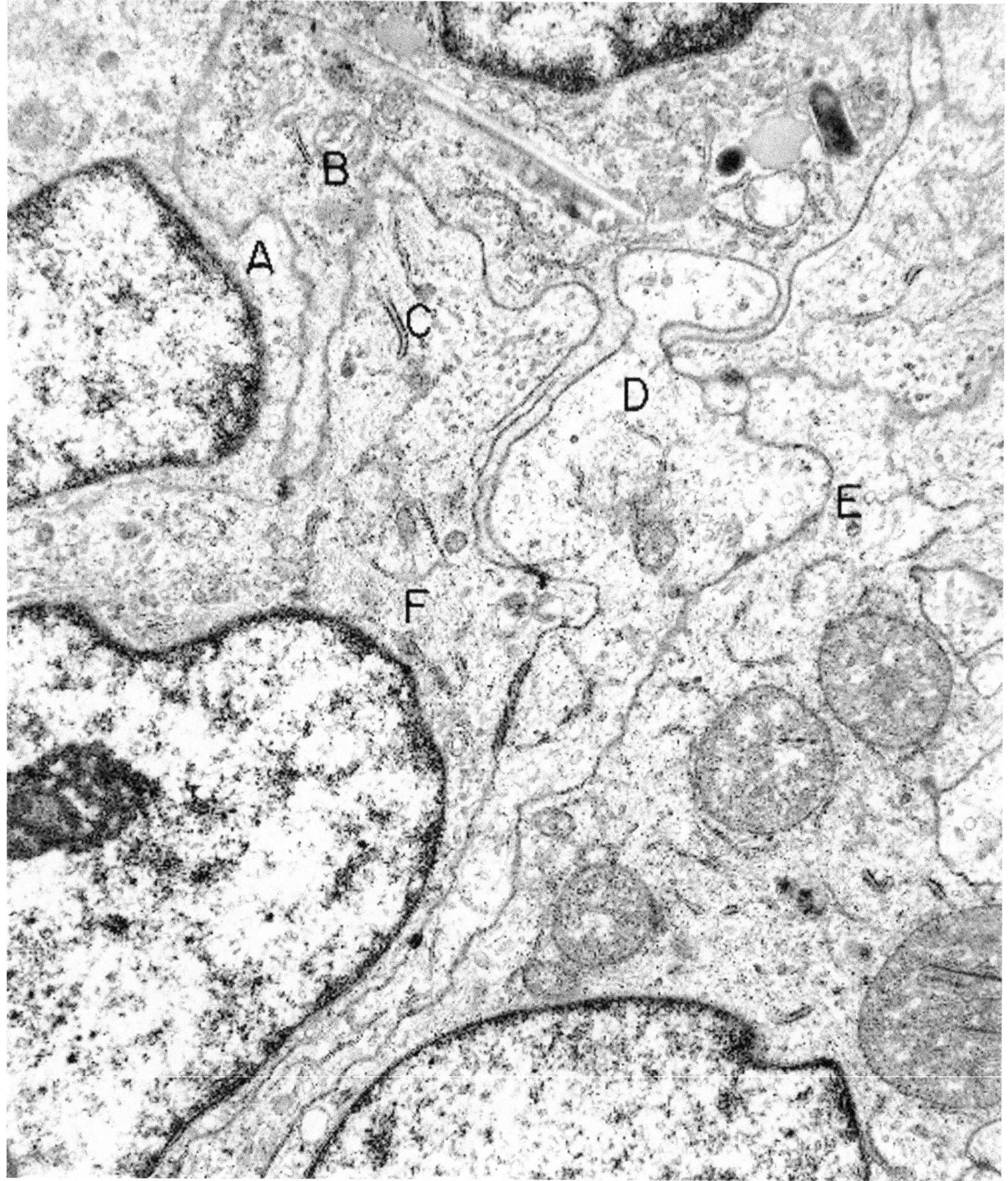

Figure 8.4. Meningioma, mixed type (cerebrum). Several neoplastic cells (A through E) are interdigitated and contain varying amounts of cytoplasmic filaments (F). (× 15,960)

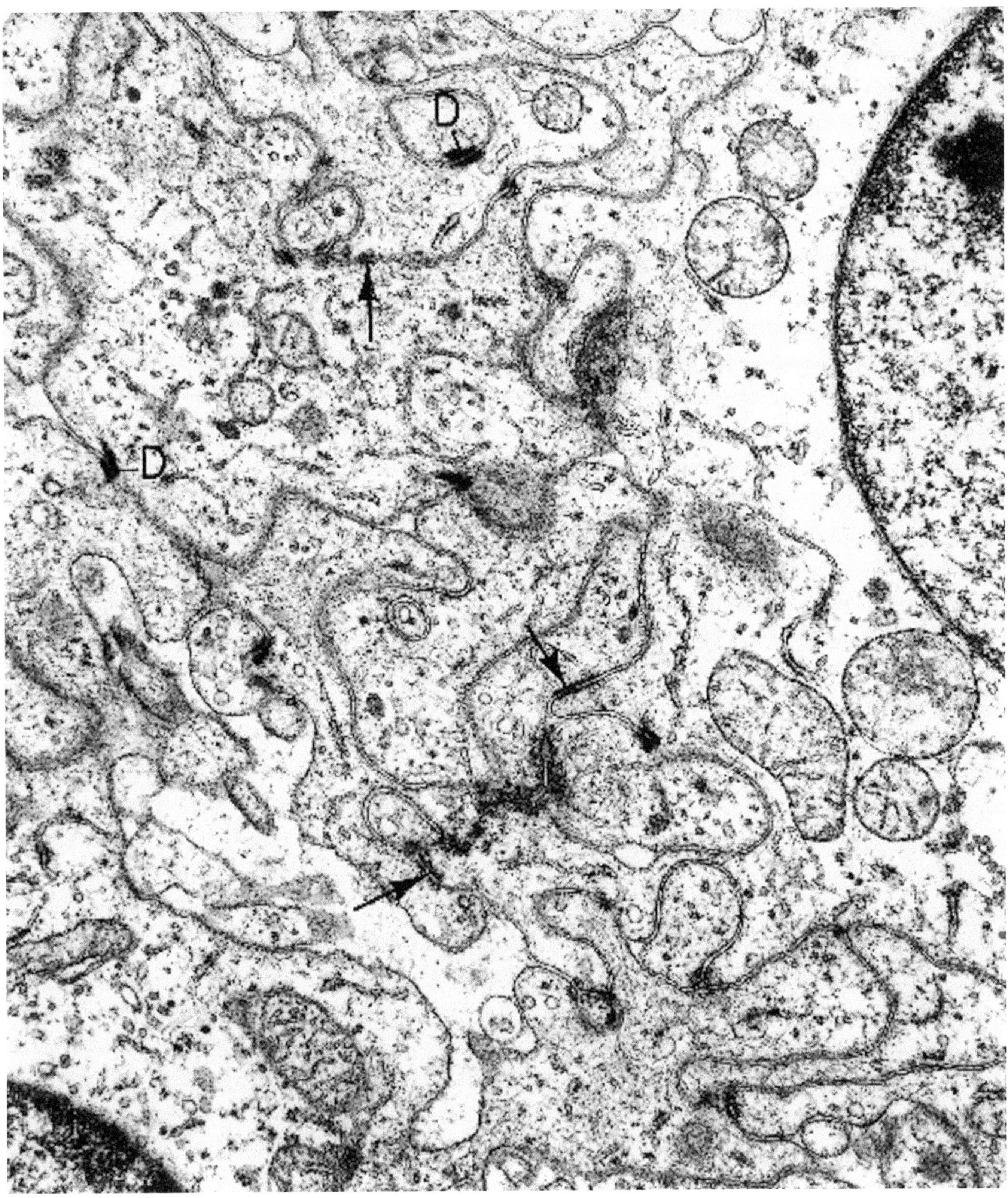

Figure 8.5. Meningioma, meningotheliomatous type (cerebrum). High power illustrates the numerous interdigitations of cell processes and the frequent intercellular junctions of several types (*arrows*), including desmosomes (D). (× 25,480)

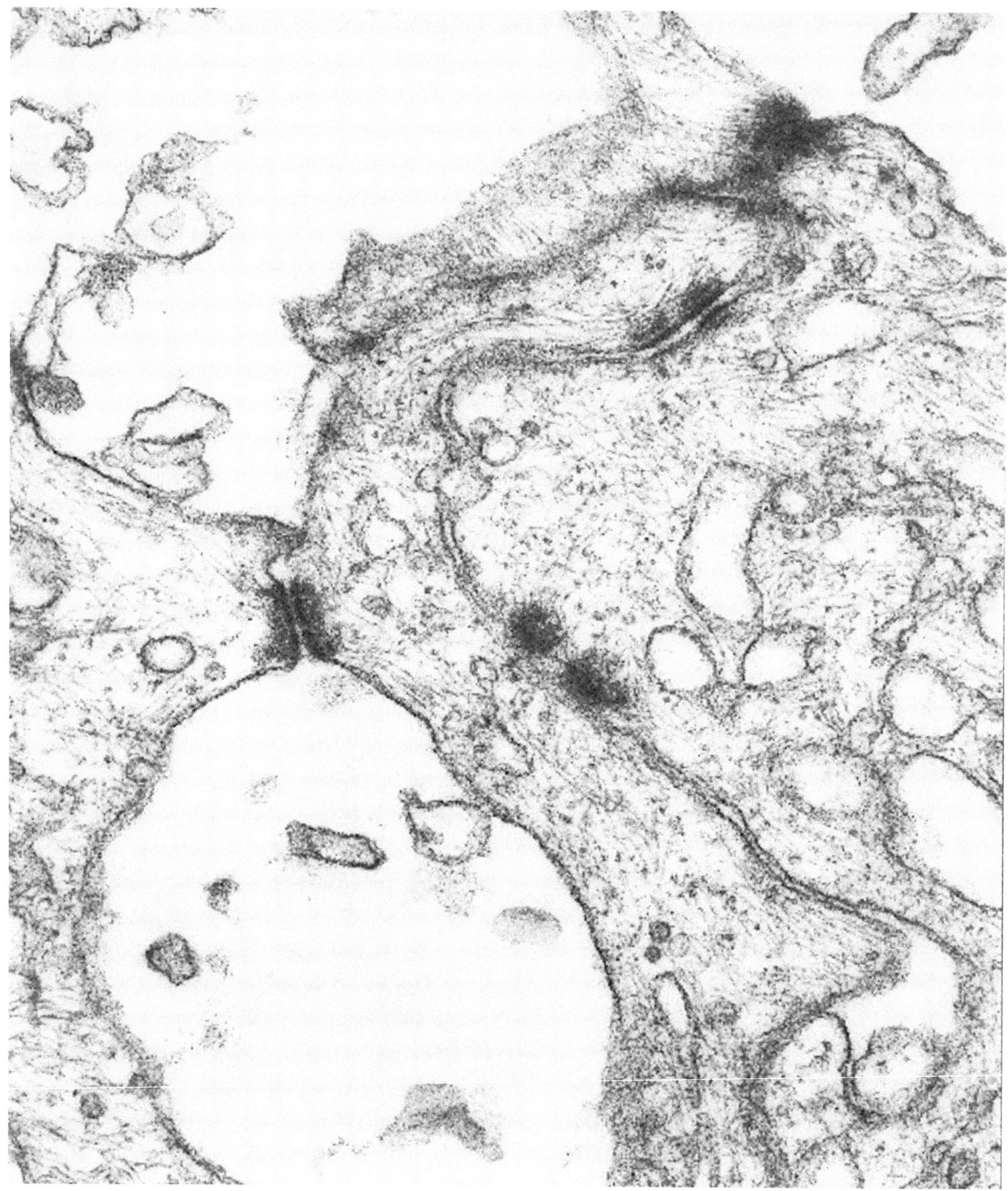

Figure 8.6. Meningioma, angioblastic type (cerebrum). Higher power of the intercellular junctions and filaments characteristic of these neoplasms. (× 67,500)

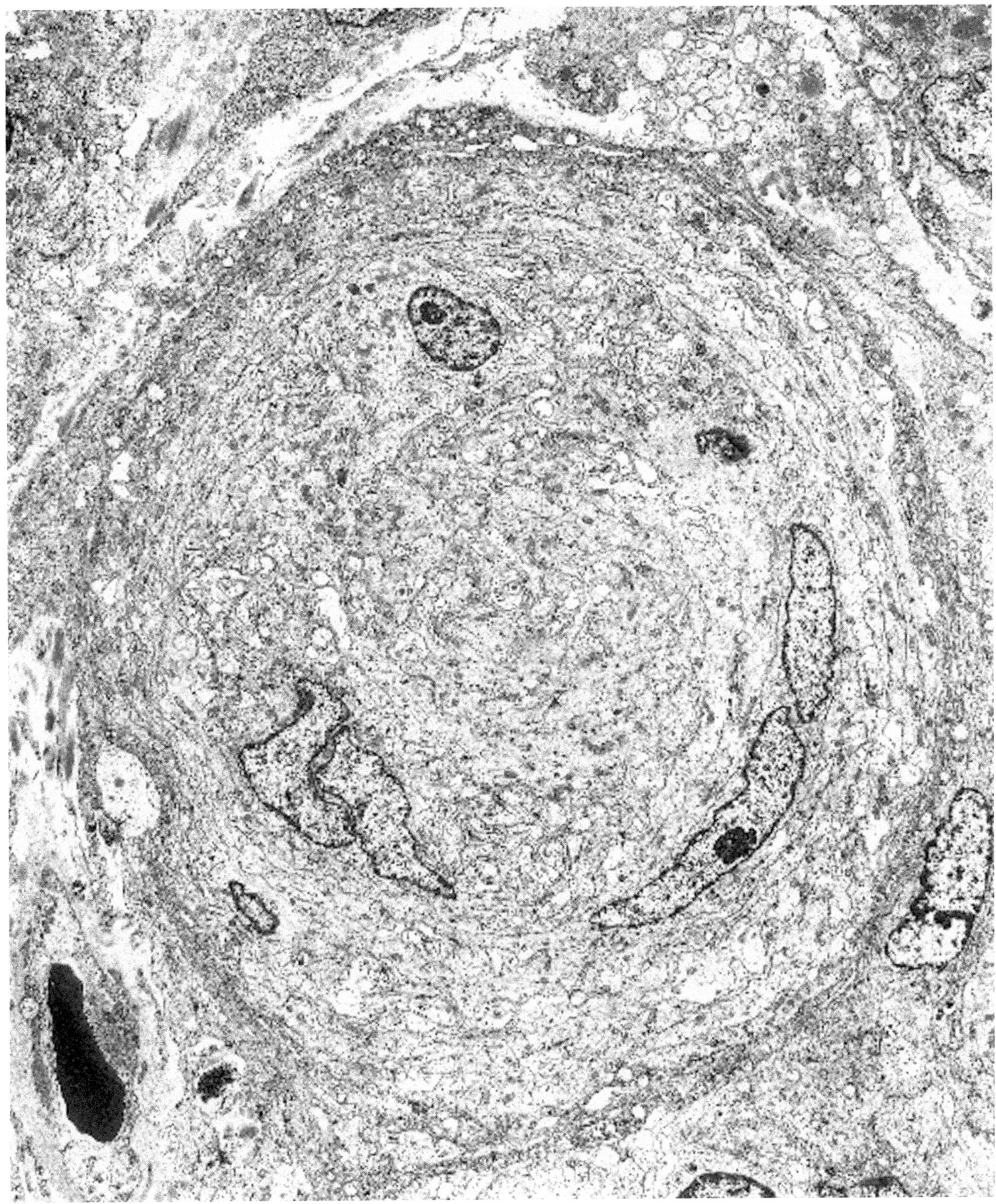

Figure 8.7. Meningioma, meningotheliomatous type (cerebellum). A concentric whorl of viable meningeal-type cells may be the future site of formation of a psammoma body. (× 3900)

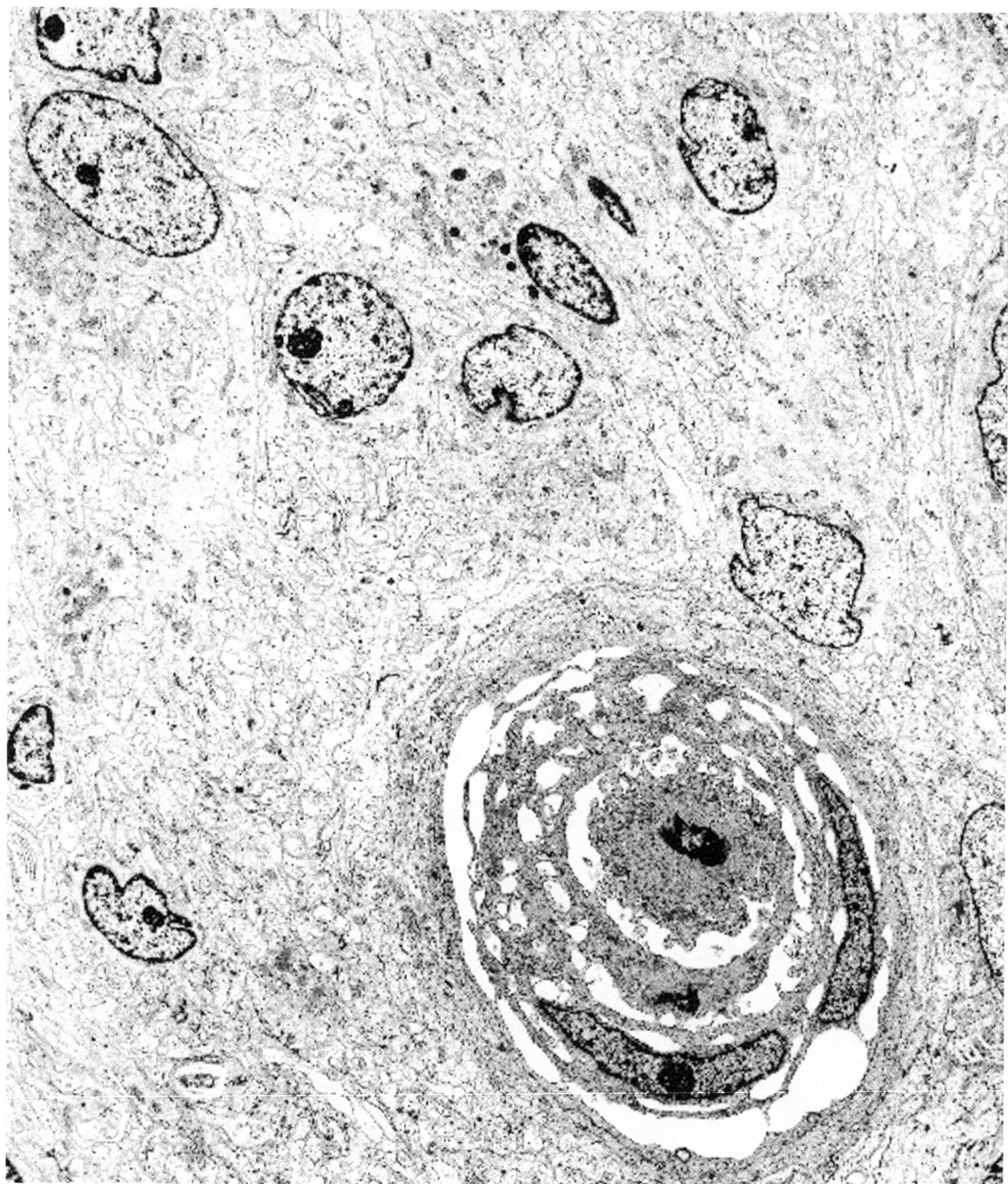

Figure 8.8. Meningioma, meningotheliomatous type (cerebellum). In this concentric whorl (lower right part of field), the cells are partially degenerated, marked by their shrunken size and darker nucleoplasm and cytoplasm (compare with the well-preserved cells in the other parts of the field). There is no evidence of calcification in the whorl at this stage. (× 3900)

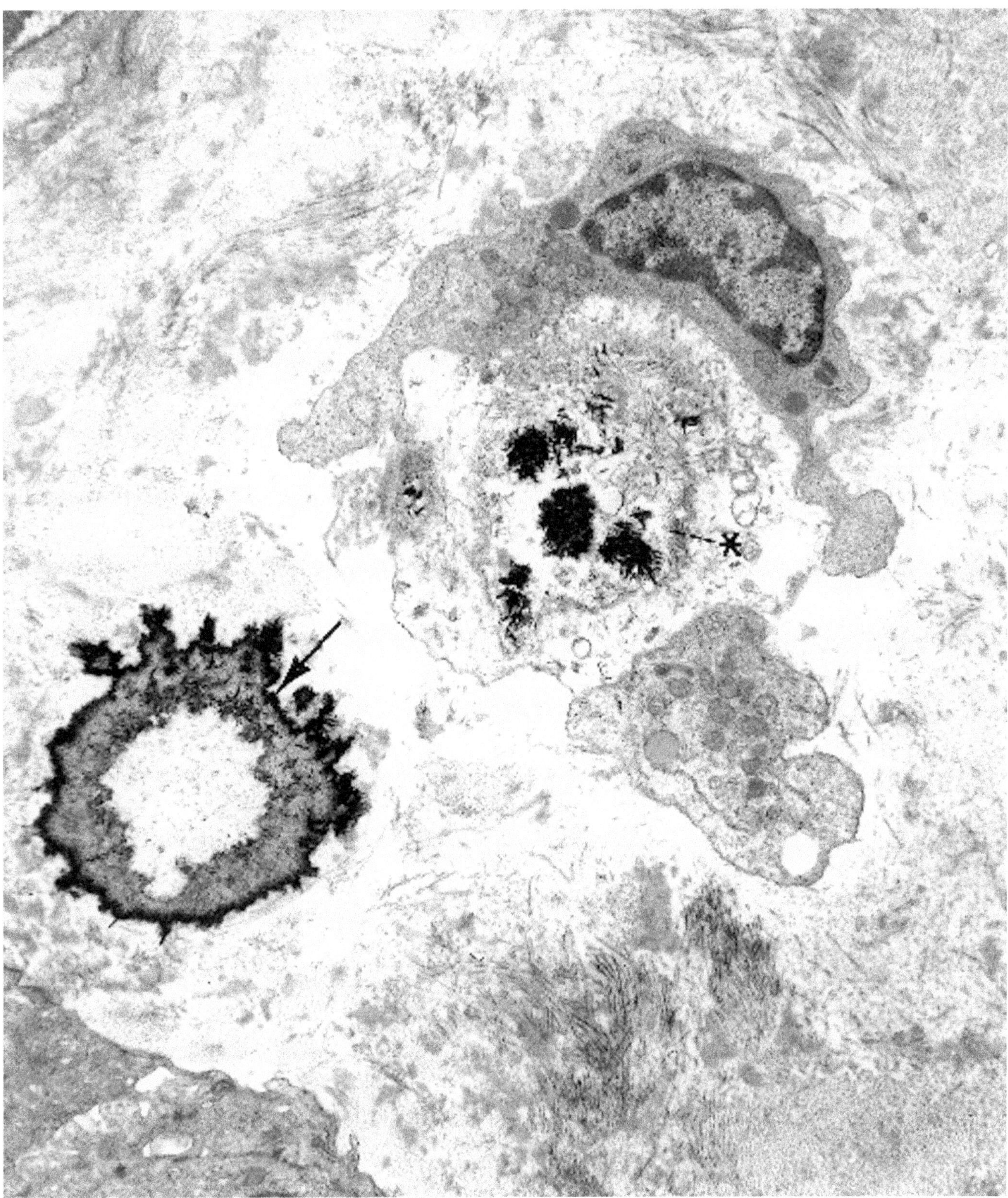

Figure 8.9. Meningioma, meningotheliomatous type (cerebellum). Two psammoma bodies are present in this field. The one in the lower left (*arrow*) is well formed and consists of a ring of electron-dense, crystalline material and a core of collagen and amorphous substance. The one in the right center (*) is immature and is composed of collagen and hydroxyapatite crystals partially surrounded by cells. (× 8840)

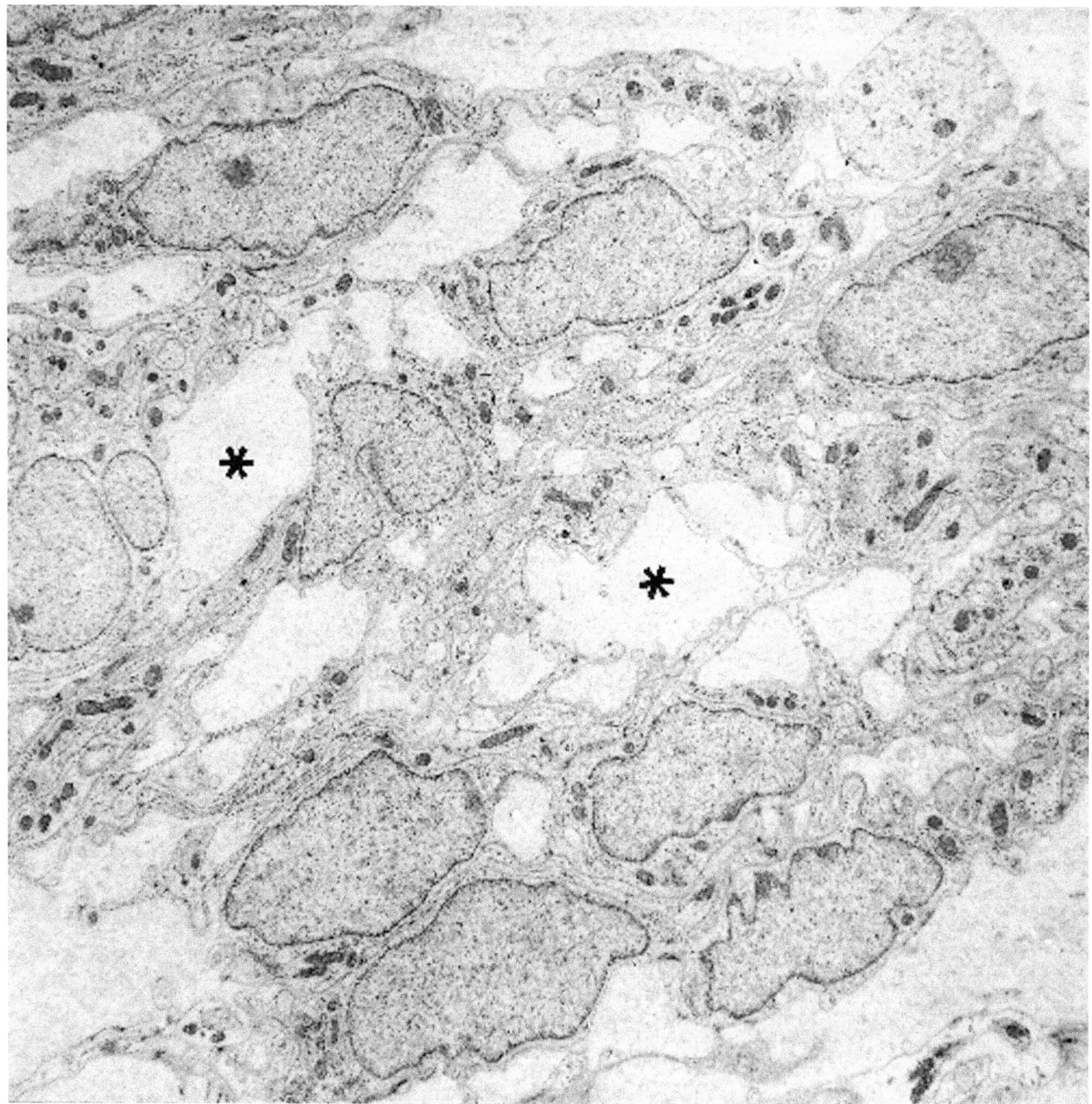

Figure 8.10. Meningioma, microcystic (right frontal meninges). Long cellular processes encircle extracellular spaces (*) that contain flocculent matrix. (× 5800)

(Text continued from page 488)

Astrocytoma

(Figures 8.11 through 8.17.)

Diagnostic criteria. (1) Cells usually widely separated in an amorphous matrix; (2) small cell bodies with oval or elongated nuclei; (3) innumerable cell processes filled with intermediate filaments (Figures 8.11 and 8.15).

Additional points. Fibrillary, protoplasmic, and *gemistocytic astrocytes* are morphologic variants of a single cell type, with the fibrillary cell having the classic filament-filled processes, the protoplasmic form being more irregular in shape and having less filaments, and the gemistocytic variant having many filaments and secondary lysosomes, in a plump cell body (Figure 8.16). *Piloid astrocytes* have broader processes than those of fibrillary astrocytes, and the processes interdigitate focally with one another. The number of filaments in the processes of piloid astrocytes varies from few to many. Intercellular junctions are found in astrocytomas, but they are usually infrequent and small. *Rosenthal fibers* are markers for nonneoplastic and neoplastic astrocytes and are composed of aggregates of electron-dense, granular material in the midst of cytoplasmic filaments (Figures 8.13 and 8.14). Whereas most astrocytomas consist of cells loosely arranged in an amorphous matrix, some tumors are more cellular, with the astrocytes having a back-to-back arrangement (Figures 8.13 and 8.15). *Astroblastomas* have a characteristic perivascular arrangement of cells that mimics the pseudorosette pattern of ependymomas. In addition, the cells may have microvilli and basal lamina, but cilia and junctions are less frequent than in ependymomas. *Pleomorphic xanthoastrocytomas* are composed of spindle-shaped cells and pleomorphic giant cells with numerous intermediate filaments and a varying number of lysosomes and lipid droplets (Figure 8.17). Large secondary lysosomes correlate with the granular bodies seen by light microscopy. Basal lamina surrounds cells and groups of cells. *Subependymal giant cell astrocytoma,* frequently associated with tuberous sclerosis, is composed of giant cells that contain numerous intermediate filaments, numerous lysosomes, prominent smooth and rough endoplasmic reticulum, occasional membrane-bound crystals, and nuclear inclusions and pseudoinclusions. A few dense core granules and microtubules may also be found in some cells. *Infantile desmoplastic astrocytoma* is a rare, collagen-rich, and predominantly paucicellular neoplasm in which the cells mimic fibroblasts at the light microscopic level but prove to be astrocytic ultrastructurally.

Oligodendroglioma

(Figures 8.18 through 8.21.)

Diagnostic criteria. (1) Closely arranged, round or oval cells (Figure 8.18) with (2) round or oval, predominantly euchromatic central nuclei; (3) cytoplasm containing free ribosomes, a few to moderate number of mitochondria, and microtubules and a few filaments (Figures 8.19 and 8.20); (3) occasional cells with short, broad processes.

Additional points. Oligodendrogliomas may be pure or mixed with various neoplastic astrocytes. Even when pure, groups of neoplastic oligodendrocytes often are interspersed with nonneoplastic astrocytes and preexisting neuritic processes. Oligodendrocytes are readily distinguishable from astrocytes by the former's round or oval shape, pale nucleus and cytoplasm, absence of long, thin processes, and absence of many filaments. The cytoplasm and nuclei of oligodendrocytes frequently have a watery or washed-out appearance. Nucleoli usually are of moderate size. Microtubules often are identifiable but in small numbers. Some cells, "granular cells," may contain primary and secondary lysosomes. There are no intercellular junctions or surface microvilli. Oligodendrogliomas may be difficult to distinguish from central neurocytomas by light microscopy alone, but electron microscopy and immunohistochemistry are usually definitive. Although the cells of both tumors have microtubules, dense-core granules and synaptic vesicles are present only in neurocytoma. In addition, immunohistochemistry for synaptophysin is positive only in neurocytoma and not in oligodendroglioma. The extracellular regions of oligodendrogliomas often contain concentric microcalcifications, but these may also be present in neurocytomas (Figure 8.21).

(Text continues on page 510)

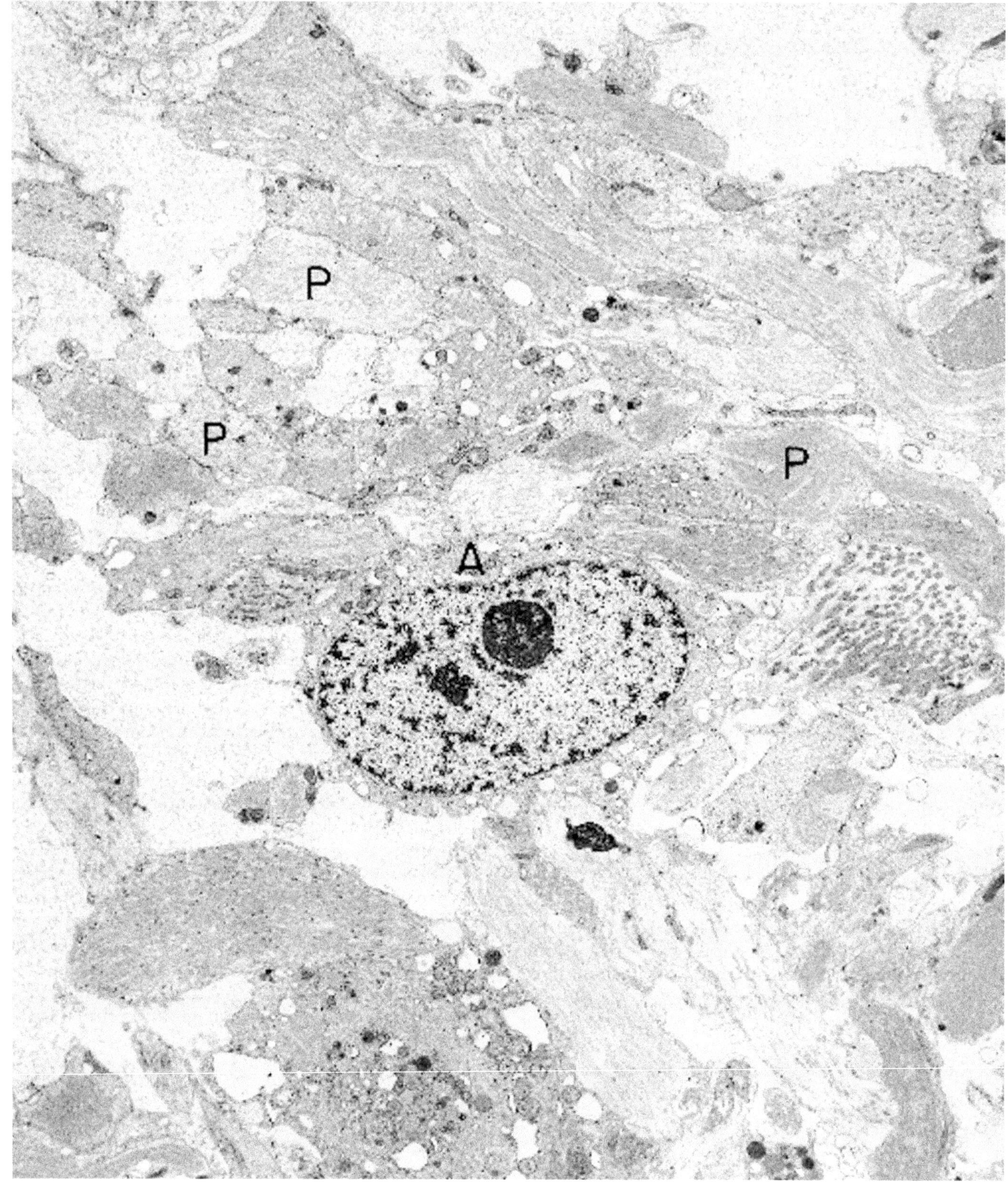

Figure 8.11. Astrocytoma, well differentiated (cerebellum). This field shows close packing of astrocytic processes (P), both to each other and to a cell body (A). (× 4940)

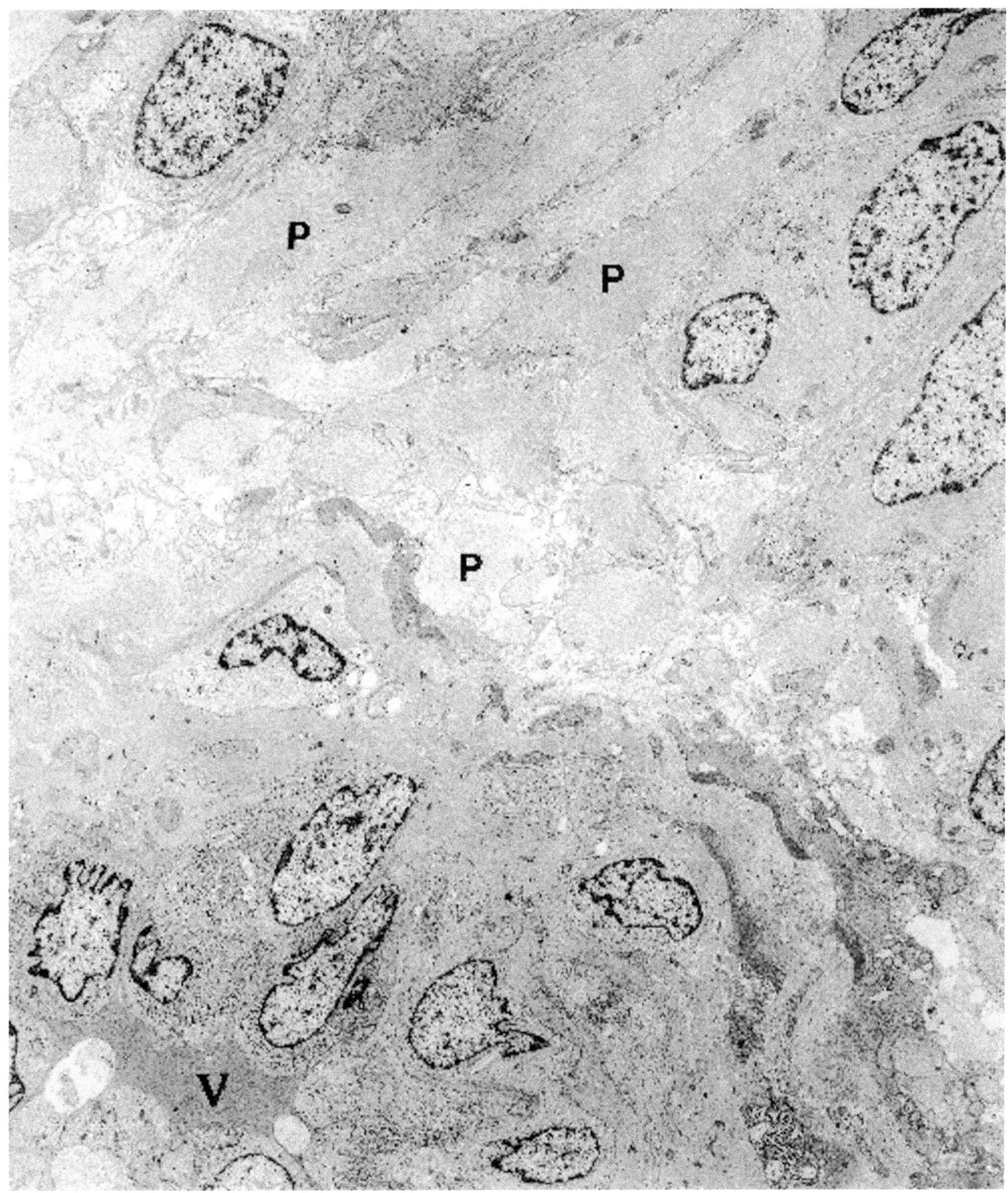

Figure 8.12. Astroblastoma (cerebrum). This field depicts a portion of a pseudorosette, with astroblastic processes (P) palisading toward a central blood vessel (V). The processes contain mostly filaments, not individually identifiable at this low magnification. (× 3700)

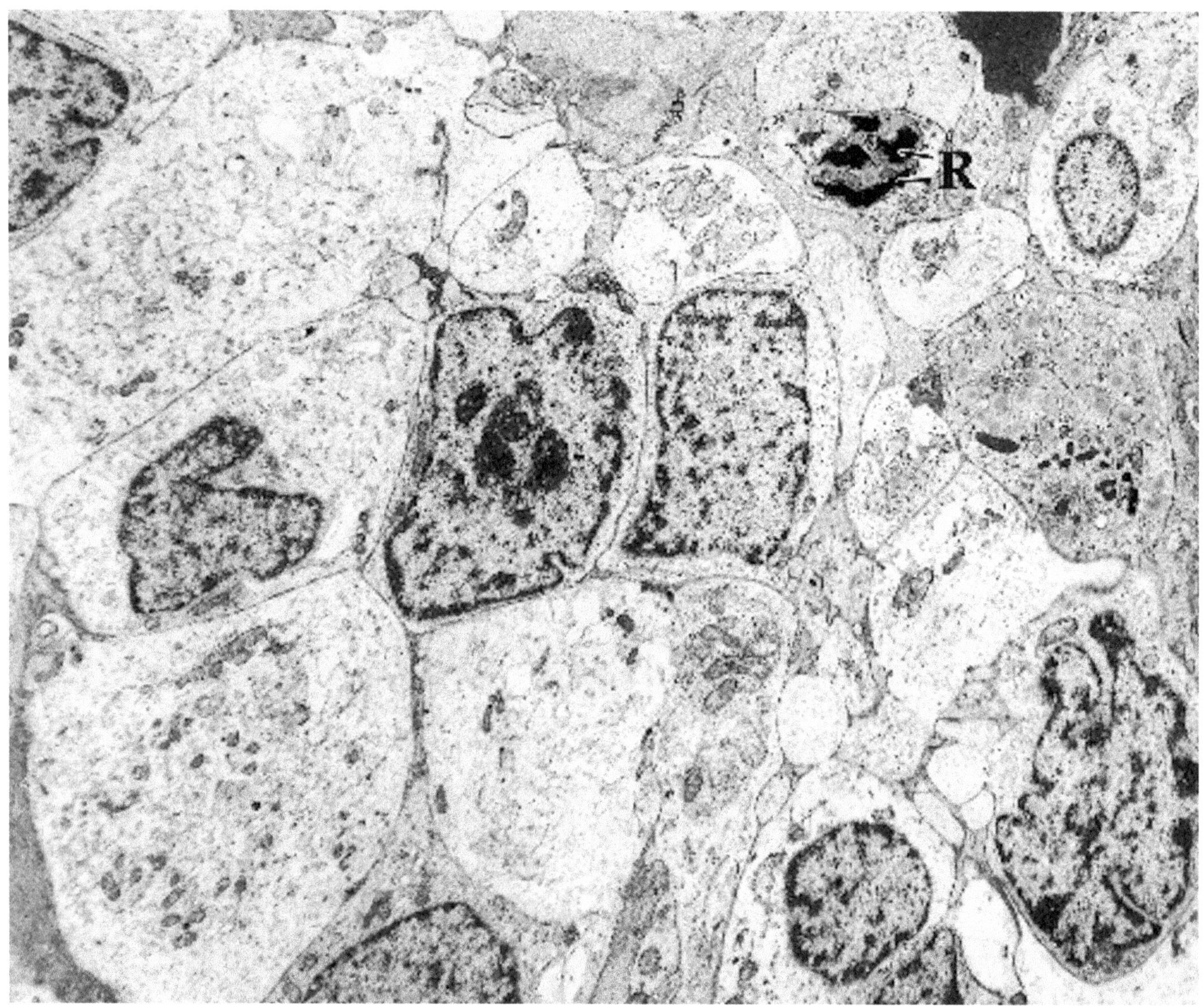

Figure 8.13. Astrocytoma, grade IV (cerebrum). Closely apposed cells have a high nuclear–cytoplasmic ratio in the cell bodies, plus numerous polar processes. Rosenthal fibers (R) are present in one of the cells. (× 5900)

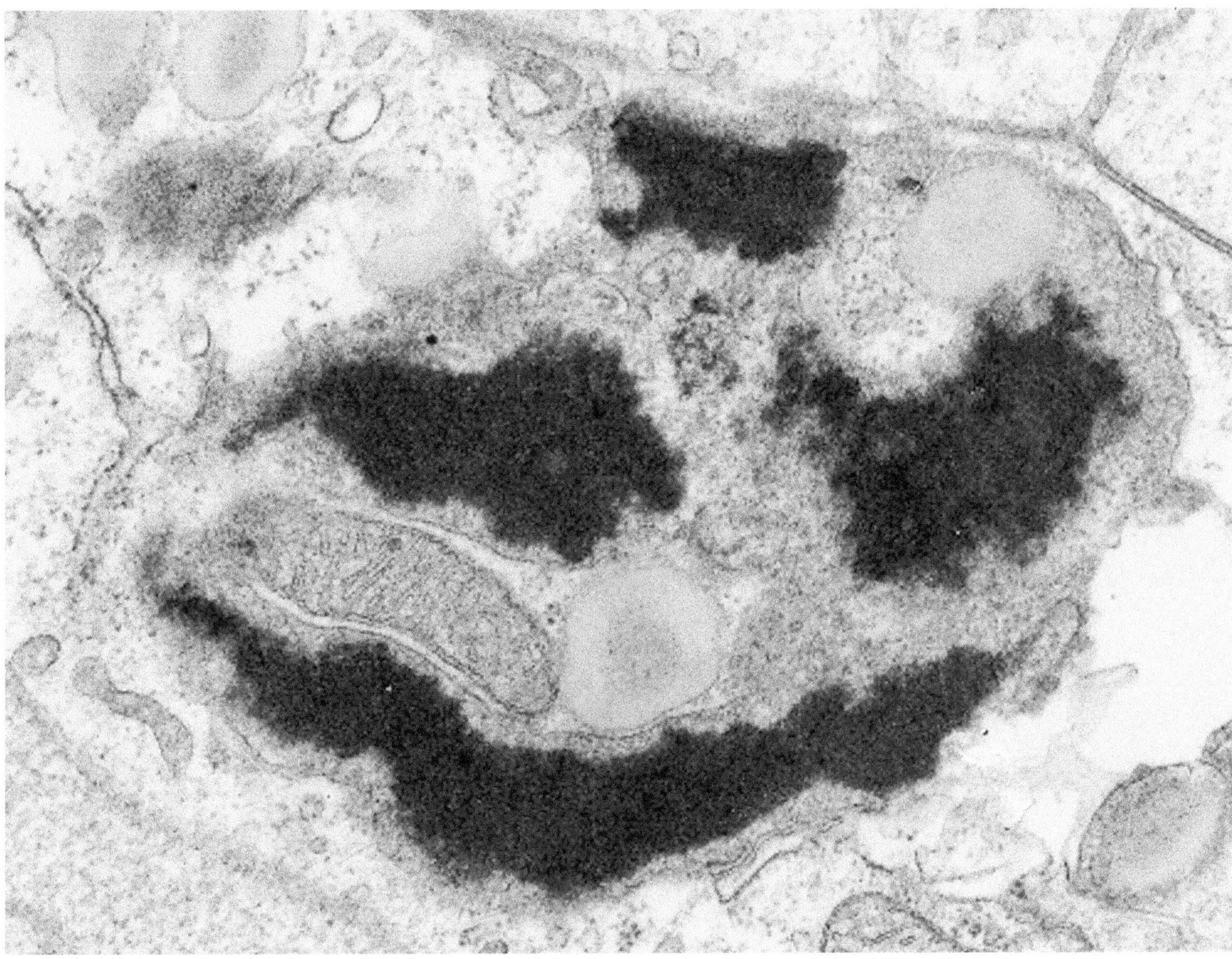

Figure 8.14. Astrocytoma, grade IV (cerebrum). High magnification of the Rosenthal fibers depicted in (Figure 8.13) reveals their composition to be of a granular, electron-dense material. (× 50,000)

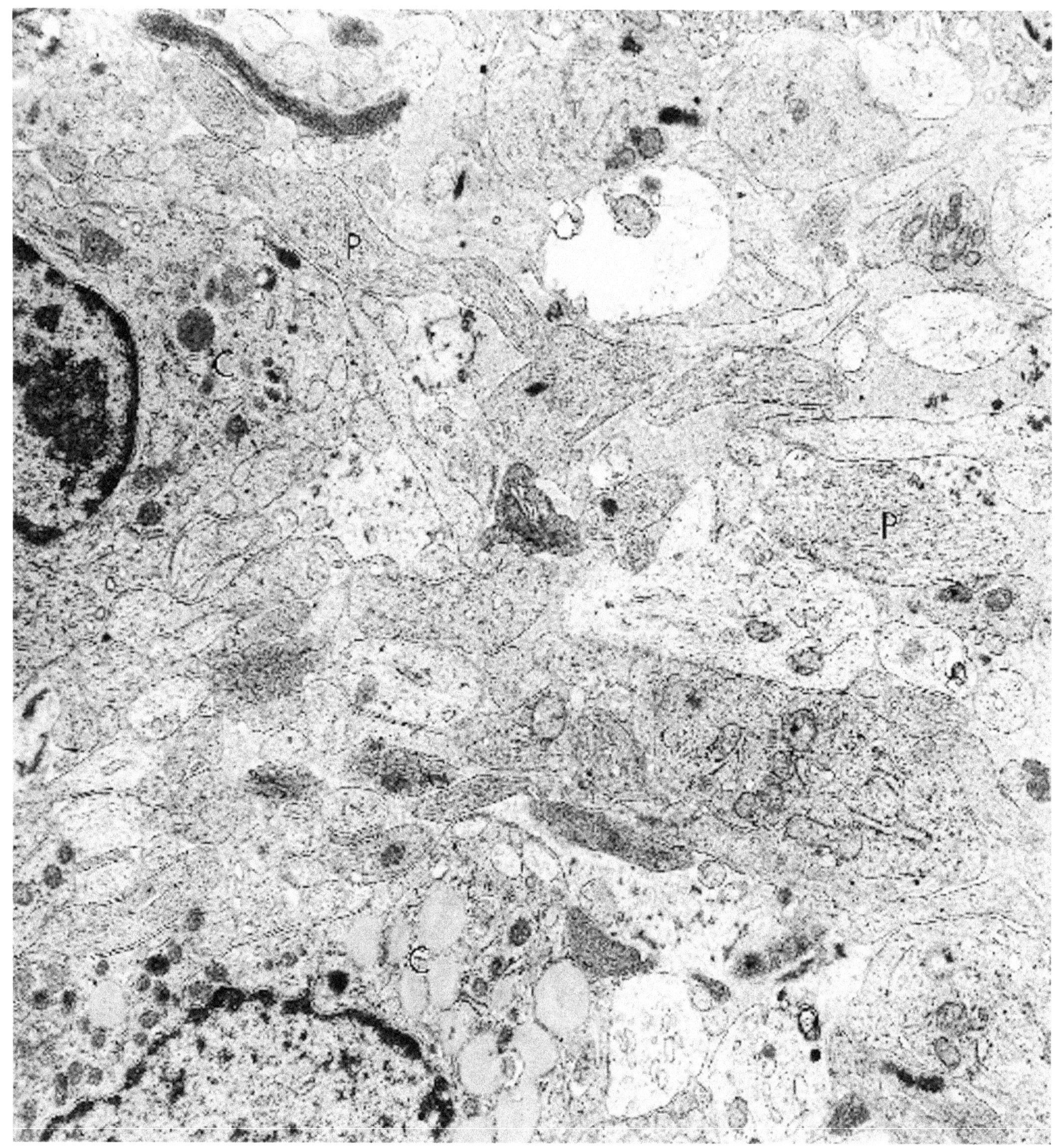

Figure 8.15. Astrocytoma, grade IV (cerebrum). Cell bodies (C) and processes (P) are closely apposed, and the processes show characteristic intermediate filaments. (× 13,100)

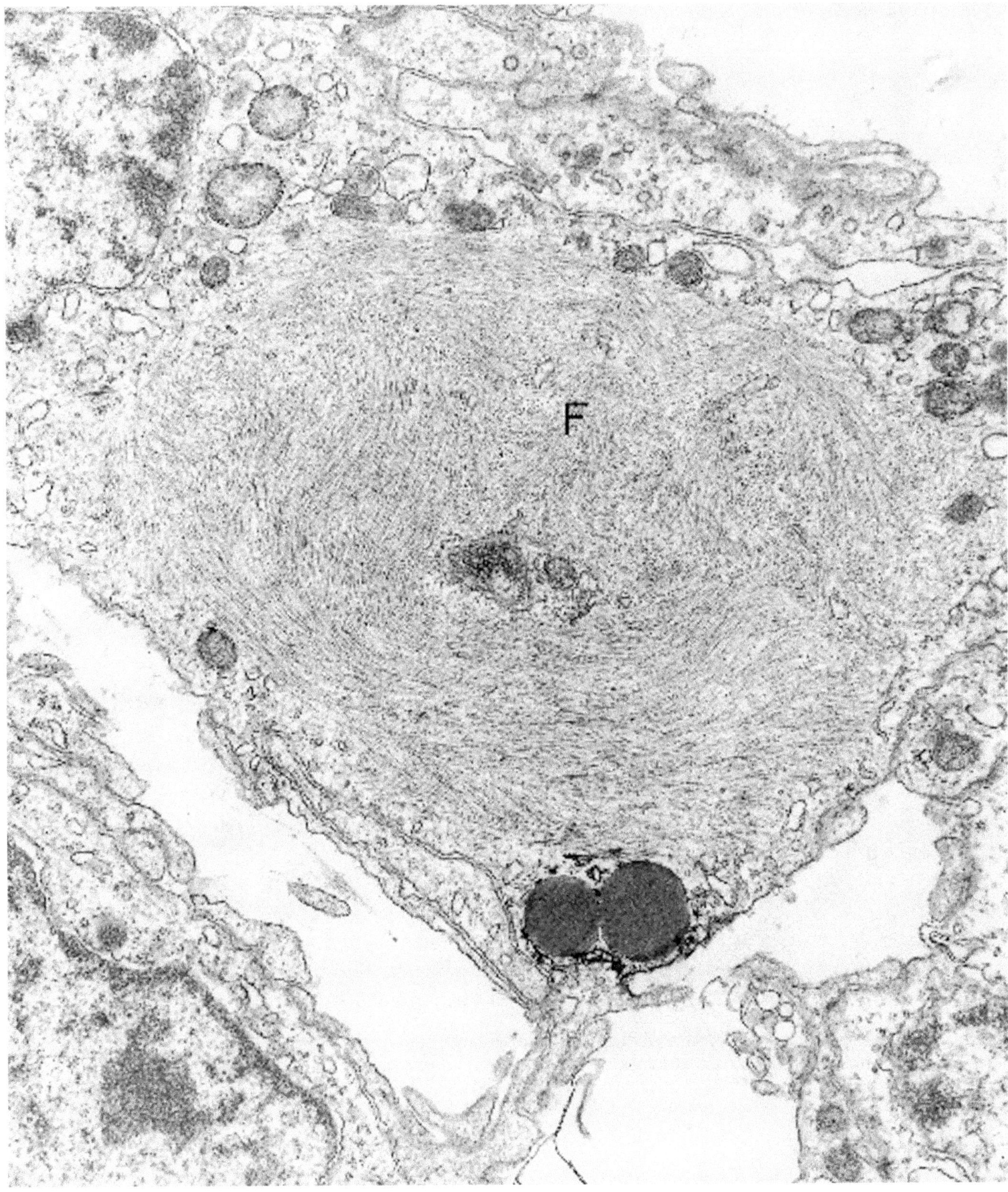

Figure 8.16. Astrocytoma, poorly differentiated (cerebrum). High magnification of a gemistocyte illustrates the large volume of cytoplasm that is occupied by filaments (F). (× 39,000)

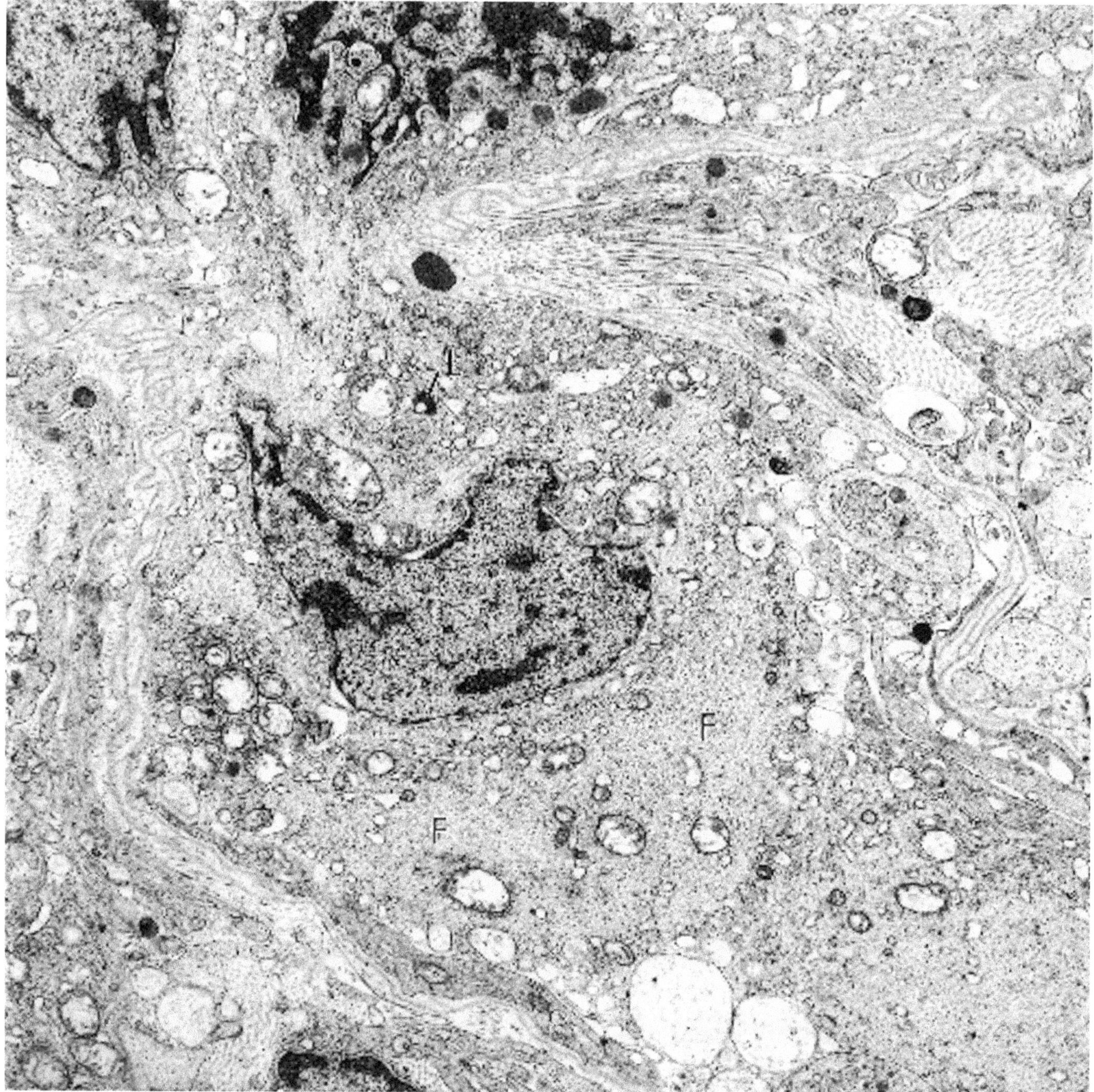

Figure 8.17. Pleomorphic xanthoastrocytoma (cerebrum). A large neoplastic astrocyte has an irregularly shaped, heterochromatic nucleus and copious cytoplasm with filaments (F) and a few lipid droplets (L). (× 12,700)

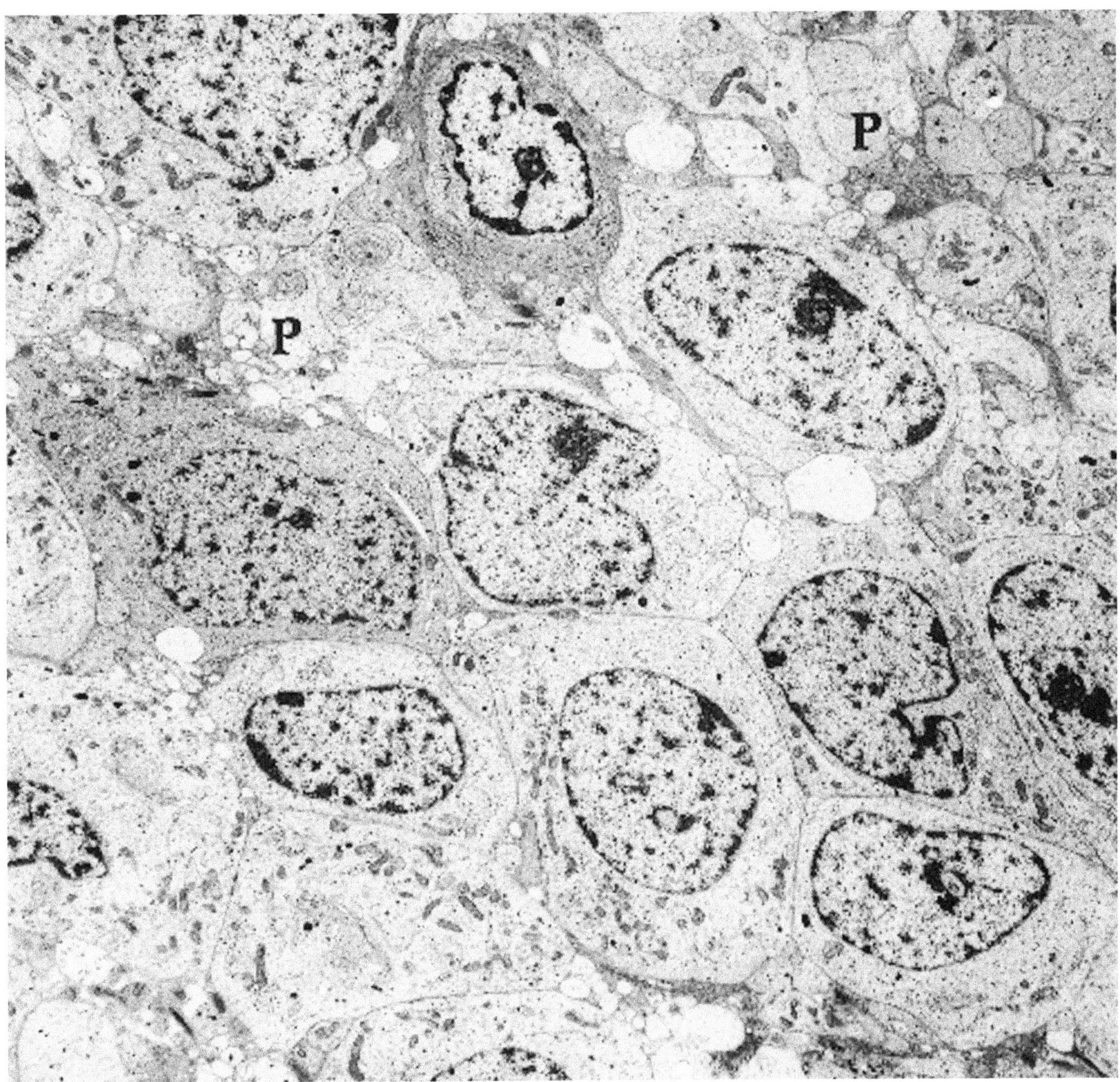

Figure 8.18. Oligodendroglioma (septum pallucidum and lateral ventricle). Closely apposed round and oval cells have irregularly oval nuclei with nucleoli of moderate size. Cytoplasm is moderate in amount, and a few clusters of processes (P), some of which are neuritic, are included in this field. (× 6200)

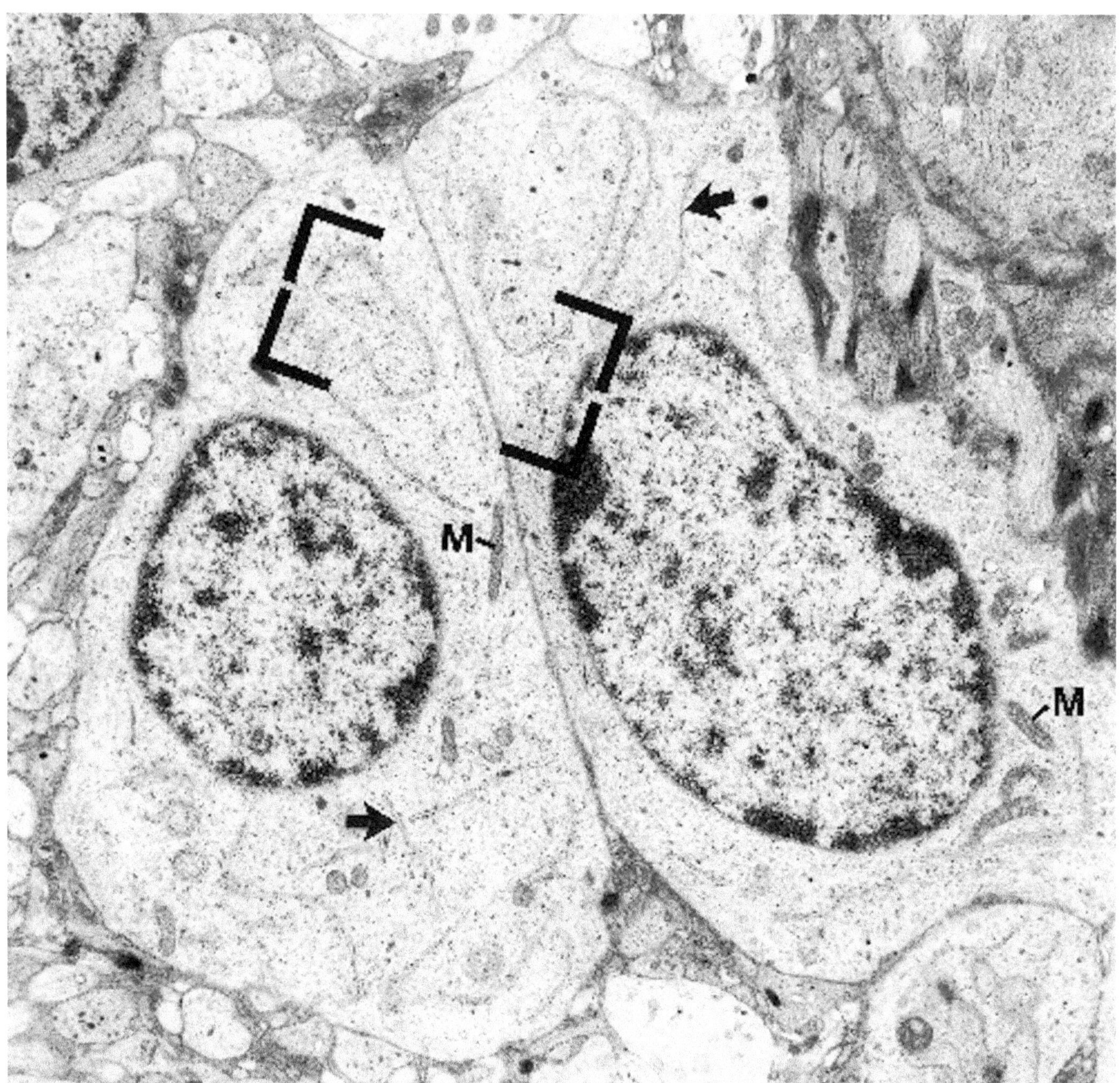

Figure 8.19. Oligodendroglioma (septum pallucidum and lateral ventricle). Two oligodendrocytes from the same neoplasm in Figure 8.18 have cytoplasm containing predominantly free ribosomes and a few mitochondria (M) and cisternae of rough endoplasmic reticulum (*arrows*). The bracketed area is enlarged in Figure 8–20 to show microtubules. (× 13,600)

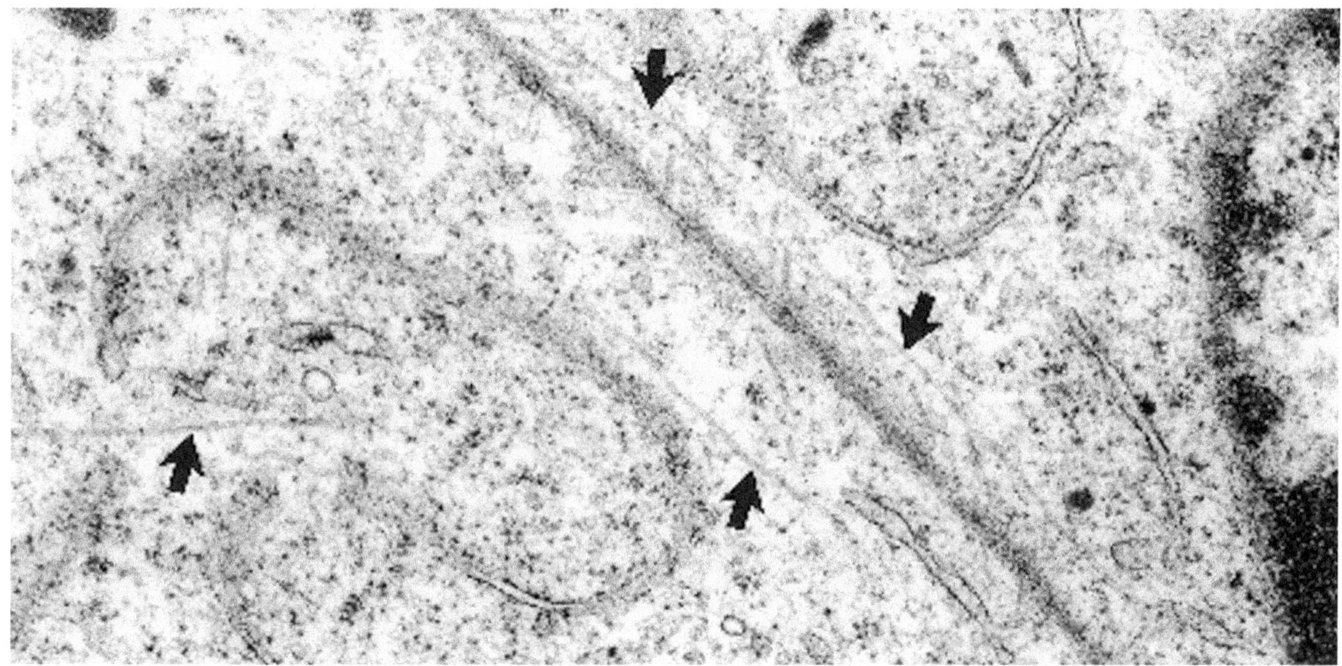

Figure 8.20. Oligodendroglioma (septum pallucidum and lateral ventricle). High magnification of the bracketed area in Figure 8.19 illustrates a few microtubules (*arrows*). (× 45,000)

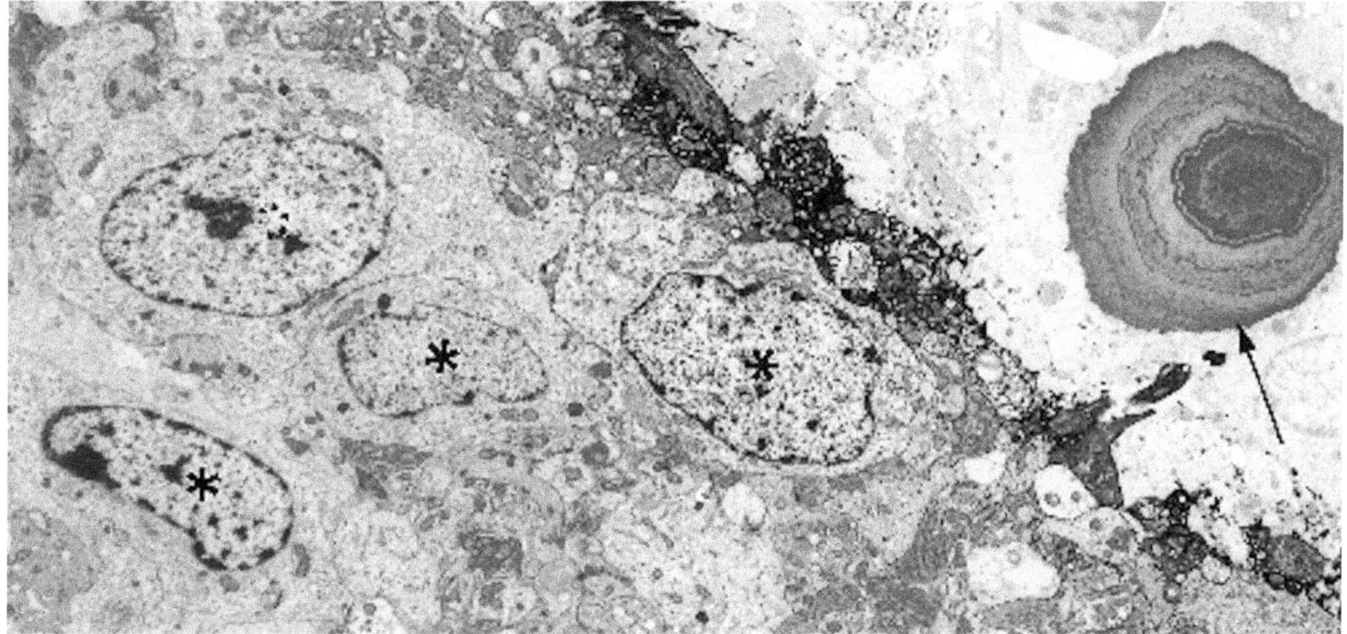

Figure 8.21. Oligodendroglioma-like neurocytoma (cerebrum). A laminated calcific deposit (*arrow*) is present extracellularly, adjacent to a collection of neoplastic neuronal cells, which at this low magnification resemble oligodendrocytes (*). Although this neoplasm was initially interpreted as an oligodendroglioma, electron microscopy revealed long polar processes and synaptic vesicles (but no dense-core vesicles), and immunohistochemistry disclosed strong reactivity with synaptophysin. (× 3600)

(Text continued from page 499)

Ependymoma (and Subependymoma)

(Figures 8.22 through 8.29.)

Diagnostic criteria. (1) Small, oval, or elongated cells forming solid sheets (Figure 8.22), rosettes (Figures 8.23 through 8.25), and pseudorosettes (Figures 8.26 and 8.27); (2) long, intermediate intercellular junctions, and apical tight junctions; (3) microvilli lining lumens (rosettes); (4) intracytoplasmic lumens; (5) cilia and basal bodies (blepharoplasts) in subluminal cytoplasm of cells; (6) broad cellular processes extending toward blood vessels (pseudorosettes); (7) numerous filaments filling the processes.

Additional points. Ependymal cells have combined epithelial (junctions, microvilli, and cilia) and astrocytic (processes with filaments) features. Long processes with filaments are especially prominent in *subependymomas,* which also contain astrocytes. *Myxopapillary ependymomas* are composed of elongated cells with interdigitating processes and a covering of basal lamina (Figures 8.28 and 8.29). They resemble choroid plexus papillomas (a neoplasm of a related cell), except for having less well-formed papillae and abundant cytoplasmic filaments. A flocculent extracellular matrix separates groups of ependymal cell processes and surrounds blood vessels. *Clear cell (vacuolated) ependymomas* are very rare and at the light microscopic level mimic clear cell oligodendrogliomas, but ultrastructurally they have the same identifying features described for classic ependymomas. *Tanycytic ependymomas* are paucicellular, with the cells being elongated and having long, fibrillar processes in a fascicular arrangement.

Choroid Plexus Neoplasms

(Figures 8.30 through 8.33.)

Diagnostic criteria. (1) Papillae with fibrovascular cores and a covering of single-layered columnar or cuboidal cells; (2) microvilli, blepharoplasts, and cilia at apical surface of epithelial cells; (3) basal lamina separating epithelium from subjacent collagenous stroma.

Additional points. The papillomas closely resemble normal choroid plexus (Figures 8.30 and 8.31), and the carcinomas range from well-differentiated adenocarcinomas to completely undifferentiated adenocarcinomas. The choroid plexus forms embryogenetically from infolded ependymal epithelium, and in fully differentiated brain, the cells from the ependyma and those of the choroid still have certain features in common. These features are epithelial in type and include intercellular junctions, junctional complexes, microvilli, and cilia. Therefore, neoplasms from these two origins may resemble one another; for example, choroid plexus papilloma and myxopapillary ependymoma. However, ependymal cells have the additional trait of sometimes differentiating along a glial line, resulting in the formation of many intermediate filaments in their cytoplasm. These dual astrocytic and epithelial features, when present, usually allow ependymomas to be distinguished from purely epithelial, choroid plexus neoplasms (Figures 8.32 and 8.33). Rarely, choroid plexus neoplasms appear by light microscopy to be mucin-producing, but we are unaware of any electron microscopic studies that have demonstrated actual mucous granules in the cytoplasm of the cells. Another rare occurrence is for choroid plexus tumors to be pigmented, either with melanin or with lipofuscin.

Neuronal and Mixed Neuronal Glial Neoplasms, Including Embryonal Forms

Gangliocytoma (Central Ganglioneuroma)

Gangliocytoma is a neoplasm composed completely or mostly of mature ganglion cells (see Figures 8.35 and 8.36 in the next section). Cell bodies may be diffusely scattered within neuropil, or they may occur in clusters or irregular arrangements in a collagenous matrix. Varying numbers of glial cells (predominantly astrocytes) are present as part of the stroma, and a spectrum in the number of glial components exists between pure gangliocytoma and ganglioglioma (see section on ganglioglioma). Calcospherites may be present in the stroma and/or walls of blood vessels.

Ganglioneuroma (Peripheral)

(Figures 8.34 through 8.36.)

Diagnostic criteria. (1) Groups of unmyelinated and myelinated axons with surrounding Schwann cells; (2) scattered ganglion cells with neuritic processes, large euchromatic nuclei, large nucleoli, and copious cytoplasm; (3) numerous cytoplasmic organelles, including free ribosomes, rough endoplasmic reticulum, prominent Golgi apparatuses, mitochondria, lysosomes, and dense-core granules; (4) numerous microtubules and intermediate filaments; (5) satellite (Schwann-like) cells surrounding some of the ganglion cells.

Additional points. Peripheral ganglioneuromas involve mostly autonomic nerves and ganglia. They and central ganglioneuromas (gangliocytomas, see previous section) are similar neoplasms in regard to the presence of mature ganglion cells as the main component, but the surrounding stroma differs with respect to being Schwannian or glial, respectively.

(Text continues on page 526)

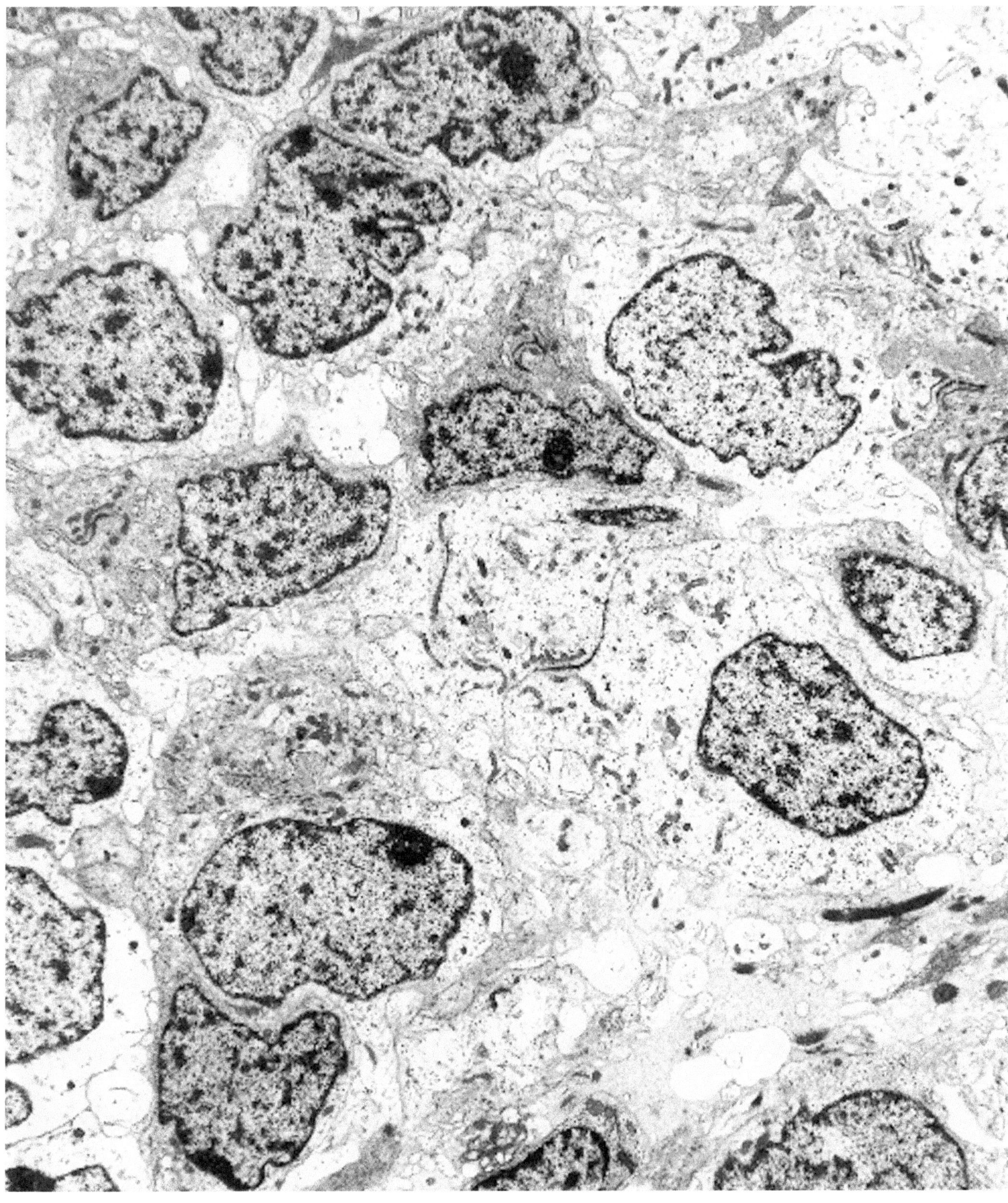

Figure 8.22. Ependymoma (cerebellum). Closely packed, oval, and molded cells have a high nuclear–cytoplasmic ratio and form no rosettes or pseudorosettes in this area. (× 5130)

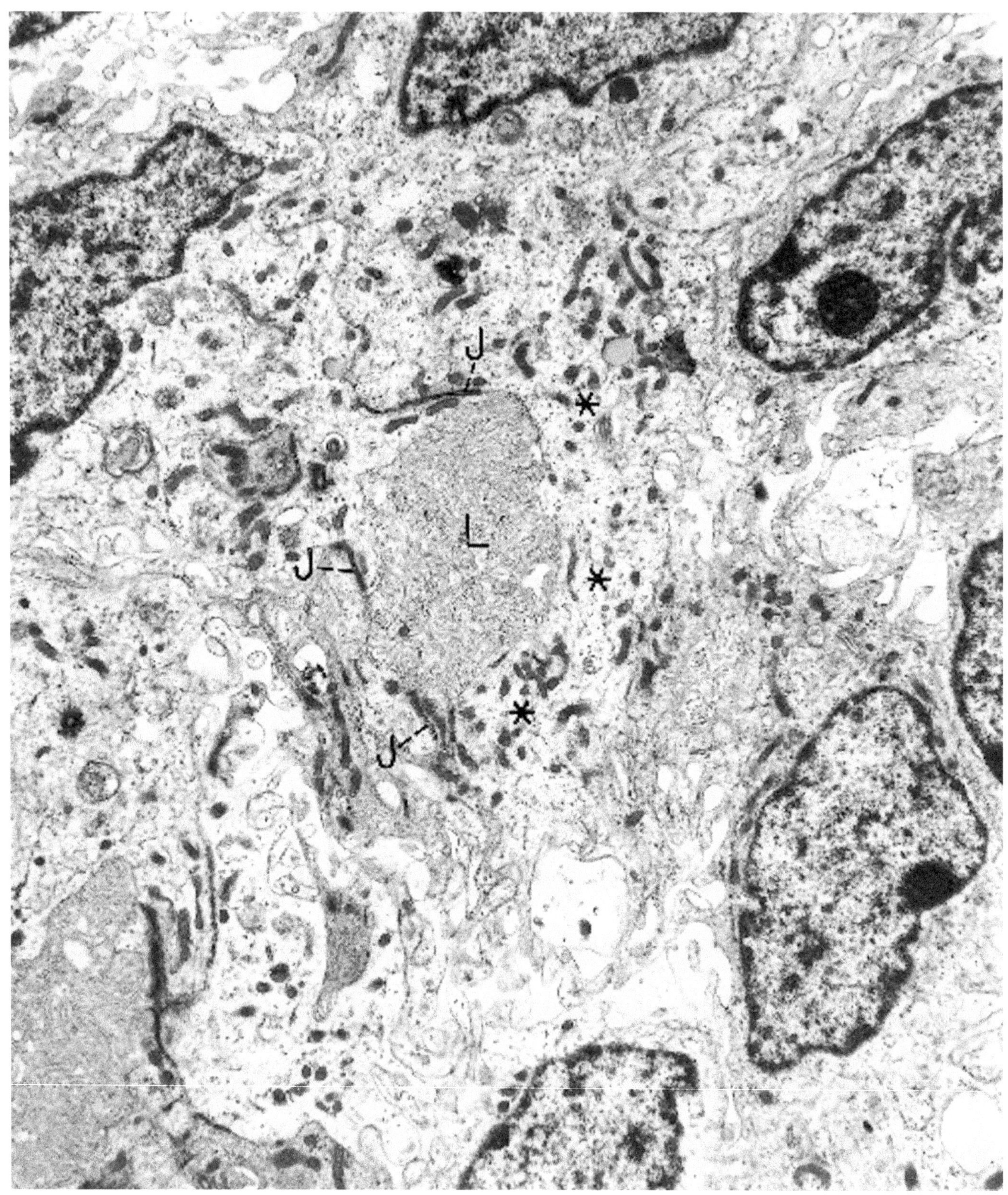

Figure 8.23. Ependymoma (cerebellum). A group of epithelial type ependymal cells form a true rosette, with microvilli from the apical surface of the cells packing a small lumen (L). Tight junctions (J) around the lumen are prominent and often are useful in locating a very small lumen. Only three cells border the lumen depicted, and one cell (*) has a broad process that wraps around one-half of the circumference of the lumen. (× 8840)

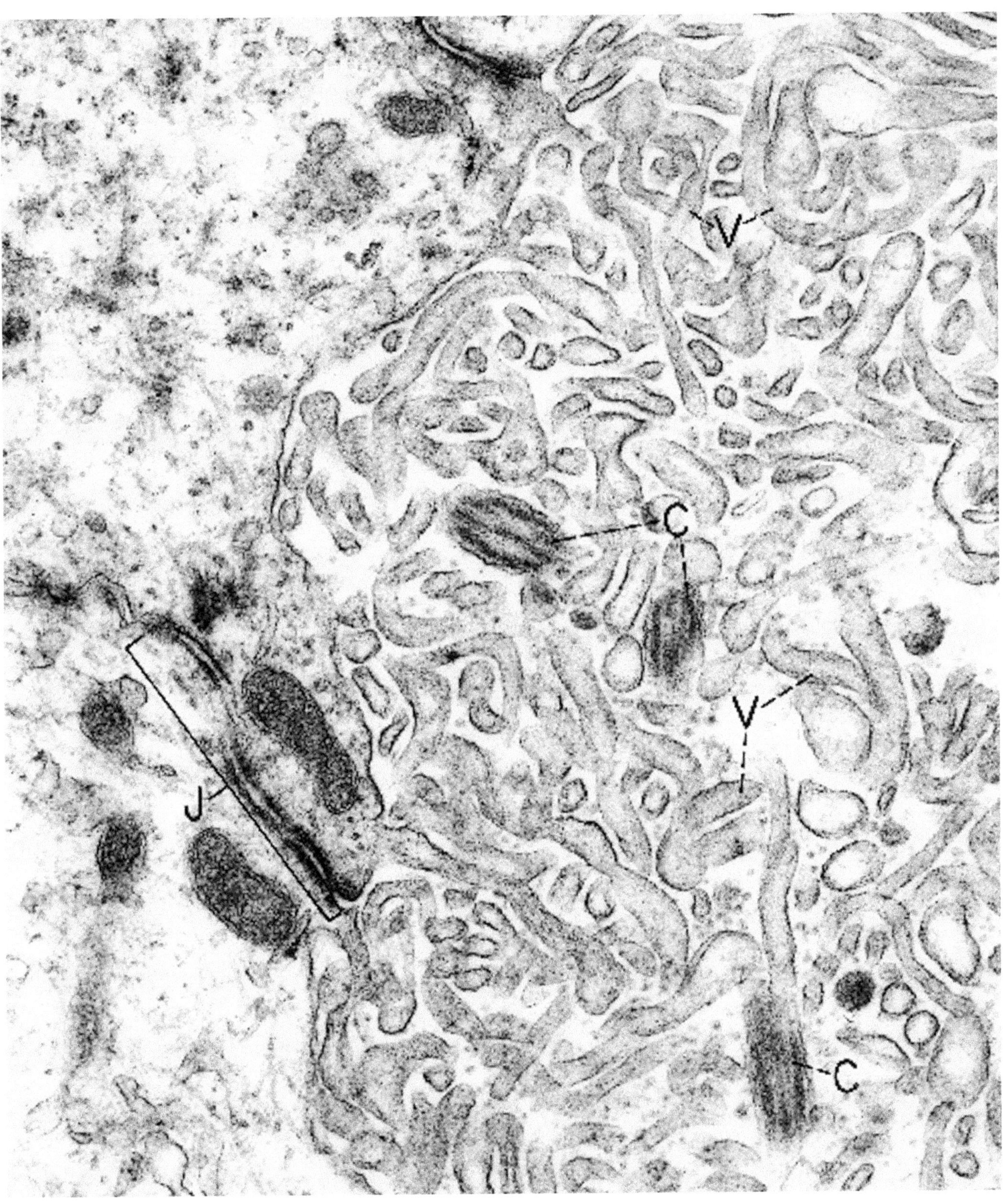

Figure 8.24. Ependymoma (cerebellum). High magnification of a rosette shows an extensive collection of microvilli (V), cilia (C), and junctional complexes (J) that line the lumen. (× 52,250)

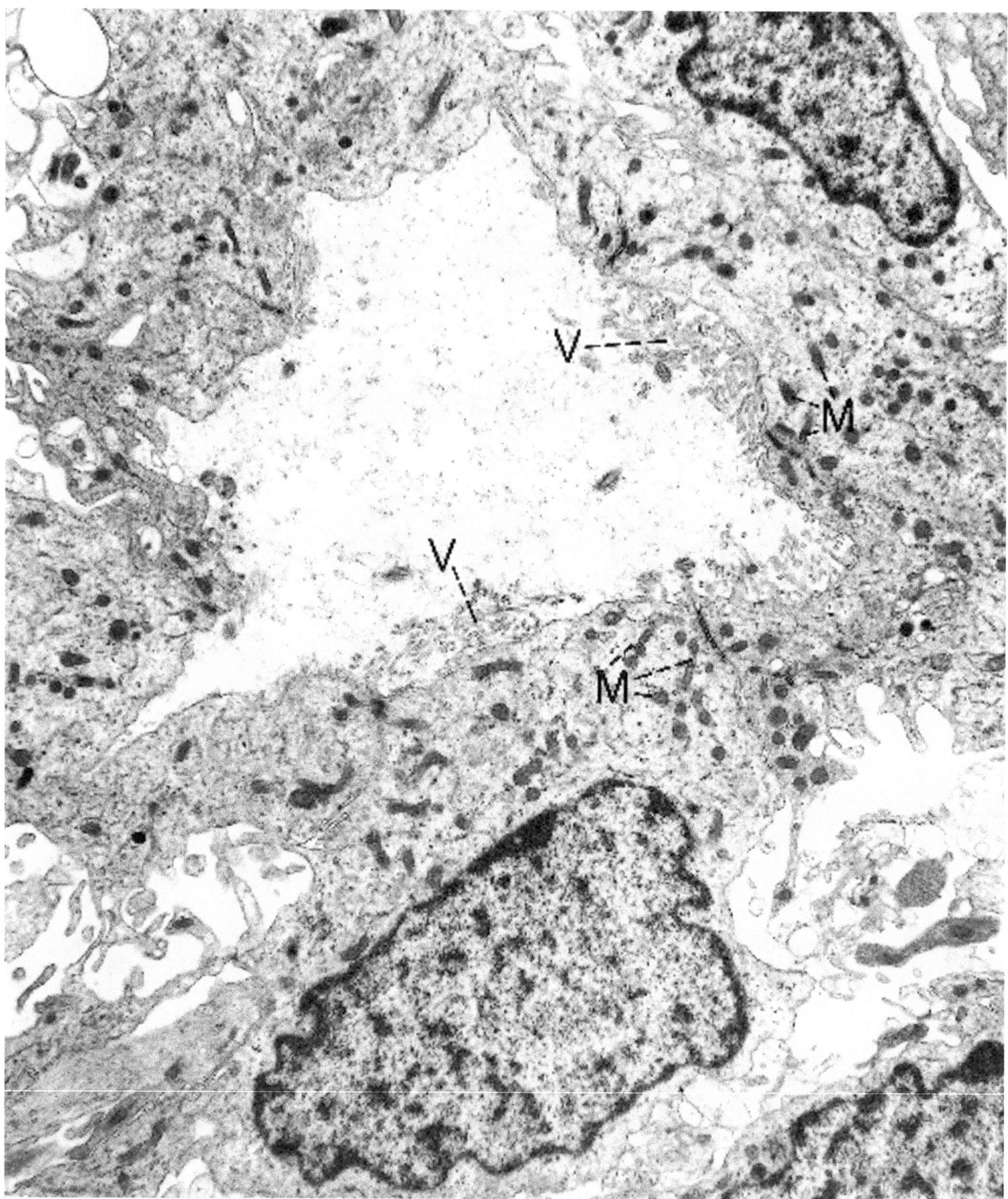

Figure 8.25. Ependymoma (cerebellum). This rosette has a larger lumen and fewer microvilli (V) than those depicted in Figures 8.23 and 8.24. The lining cells have a high nuclear–cytoplasmic ratio and cytoplasm that contains mostly ribosomes and mitochondria (M). (× 9350)

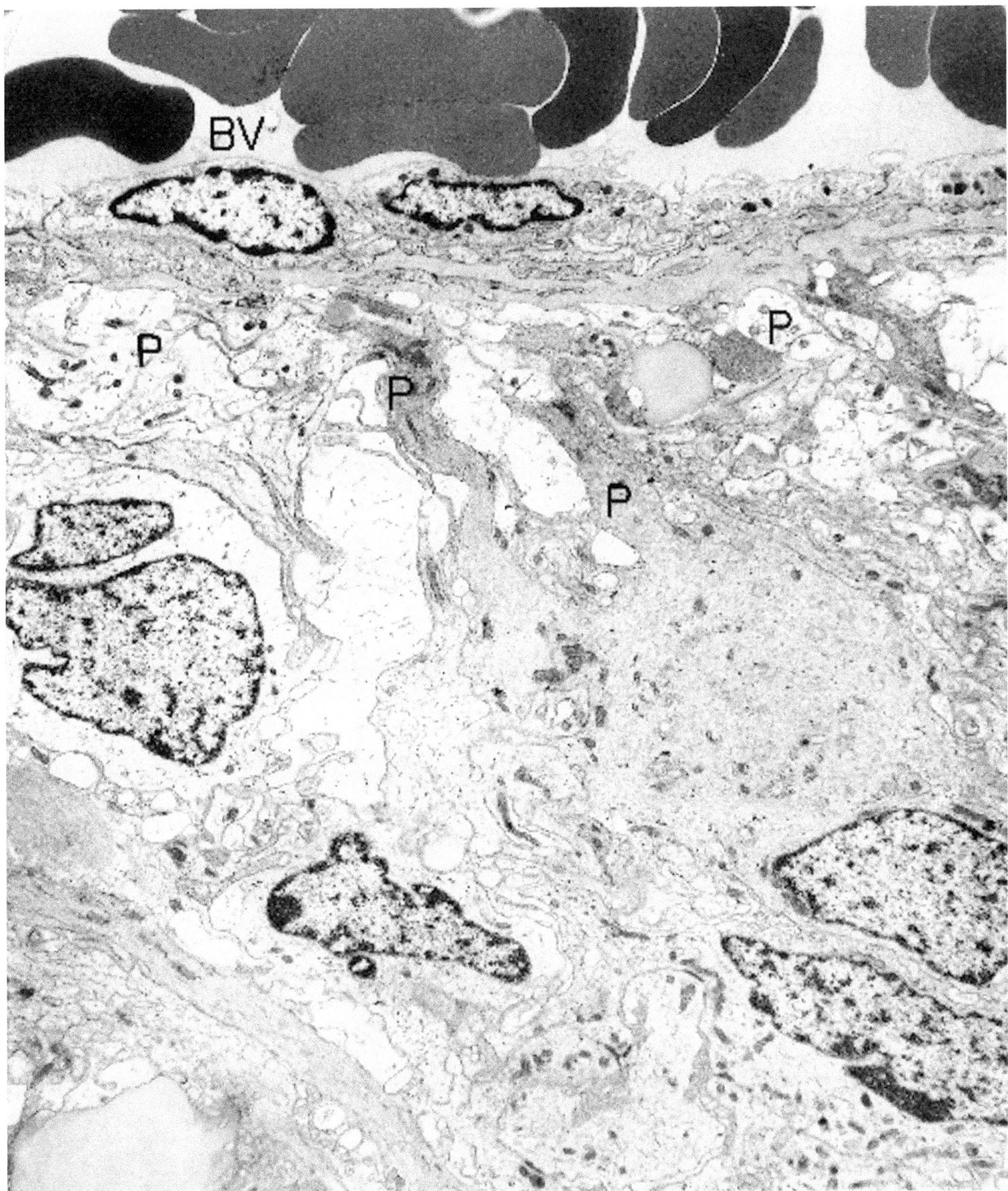

Figure 8.26. Ependymoma (cerebellum). A pseudorosette is formed by ependymal cell processes (P) arranged around a blood vessel (BV). The processes are filled with glial-type filaments (pale gray component of cytoplasm). (× 11,700)

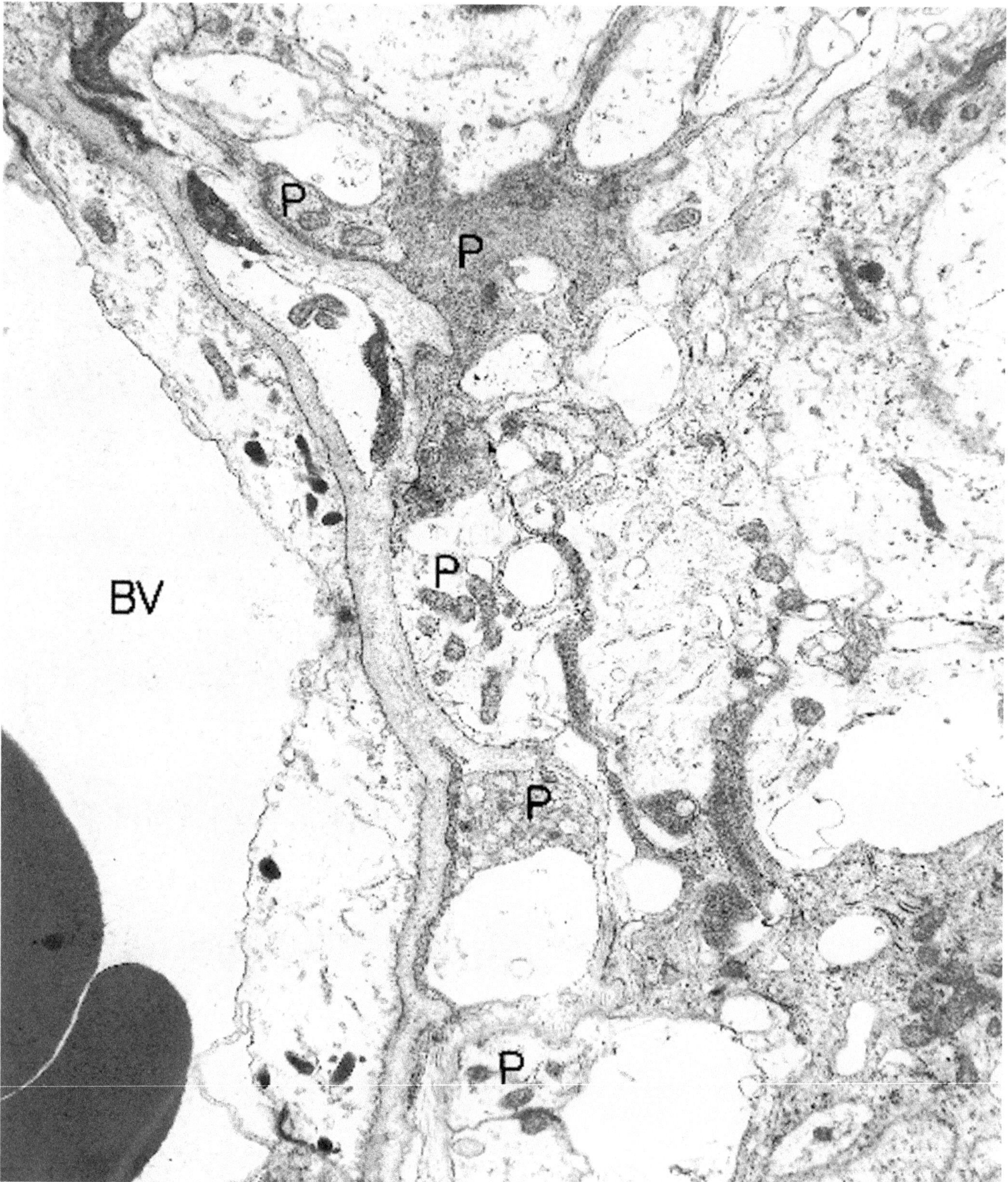

Figure 8.27. Ependymoma (cerebellum). High magnification of a pseudorosette highlights a blood vessel (BV) and surrounding ependymal cell processes (P). (× 12,600)

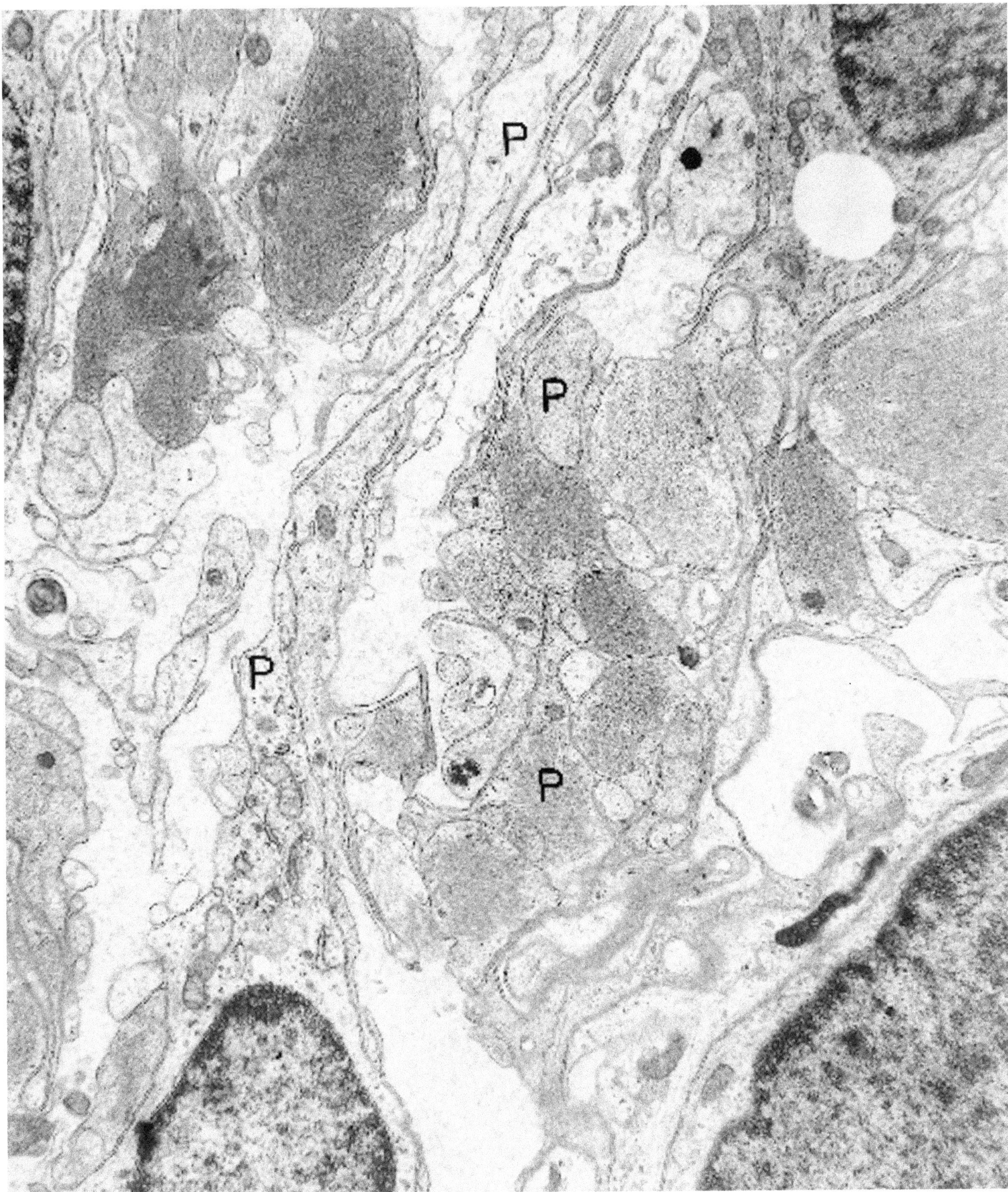

Figure 8.28. Ependymoma, myxopapillary type (cauda equina). Innumerable interdigitating processes (P) filled with filaments and covered by basal lamina (visualized better in Figure 8.29) characterize this ependymoma. (× 12,600)

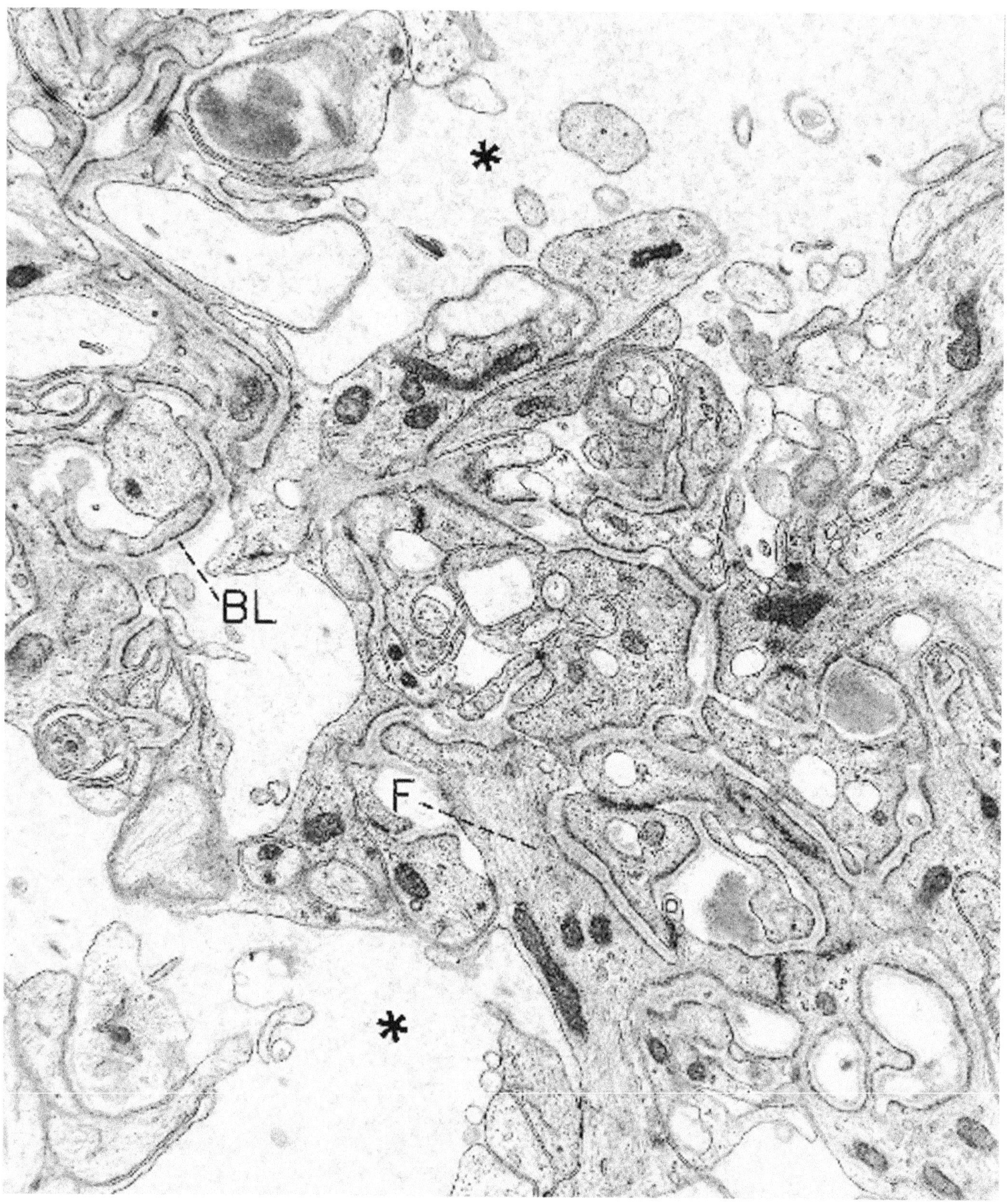

Figure 8.29. Ependymoma, myxopapillary type (cauda equina). Higher magnification of the same neoplasm as depicted in Figure 8.28 shows the intertwining cellular processes with the cytoplasmic filaments (F), a covering of basal lamina (BL), and a flocculent extracellular matrix (*). (× 15,400)

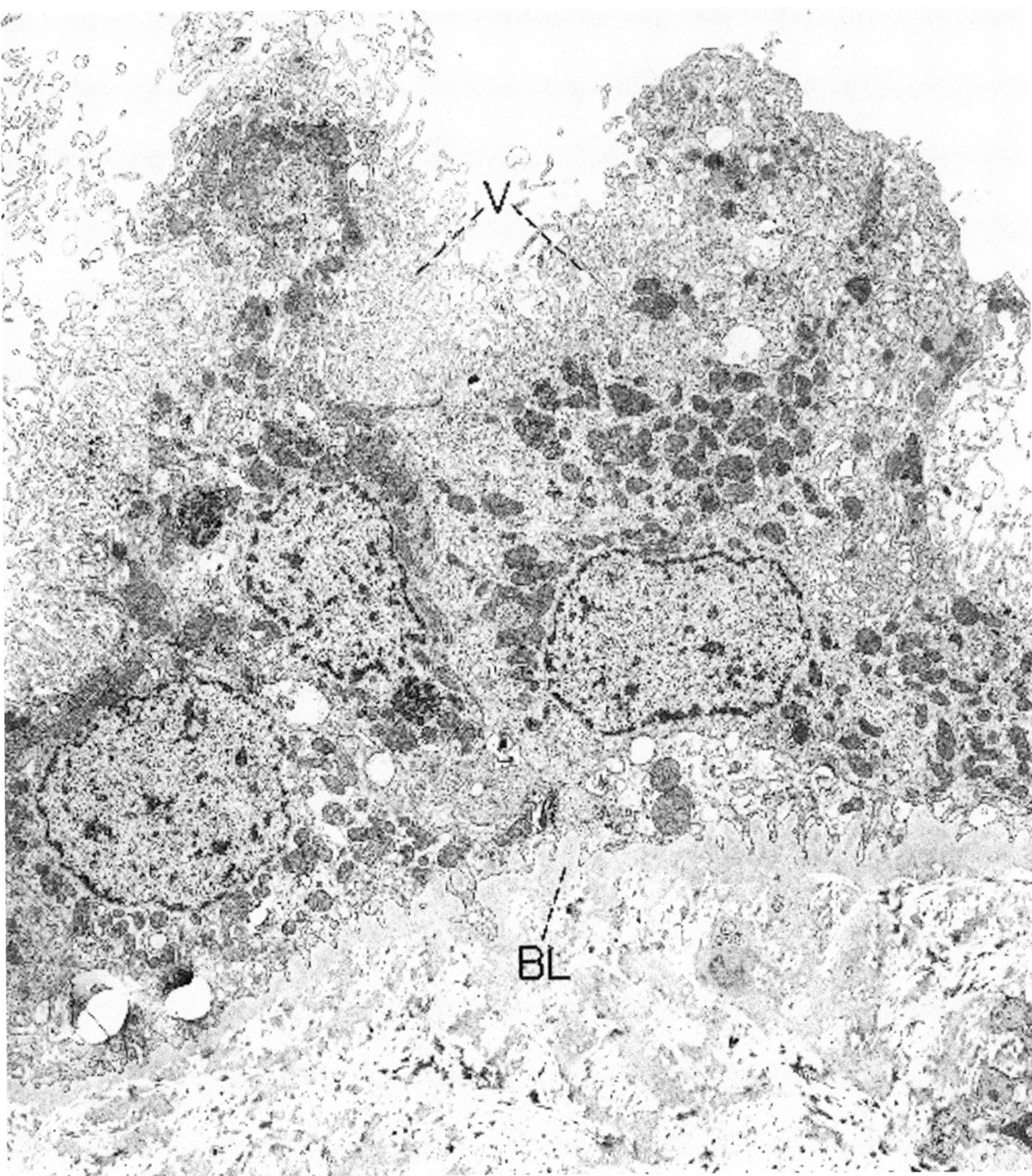

Figure 8.30. Normal choroid plexus (fourth ventricle). The cellular architecture and pattern of papillomas of the choroid plexus are similar to those of the nonneoplastic, normal counterpart. The cells are of epithelial type, having microvilli (V) on their free surfaces, junctions, and junctional complexes at their lateral and abutting borders (not discernible at this magnification), and basal lamina (BL) along their interface with a collagenous stroma. (× 5130)

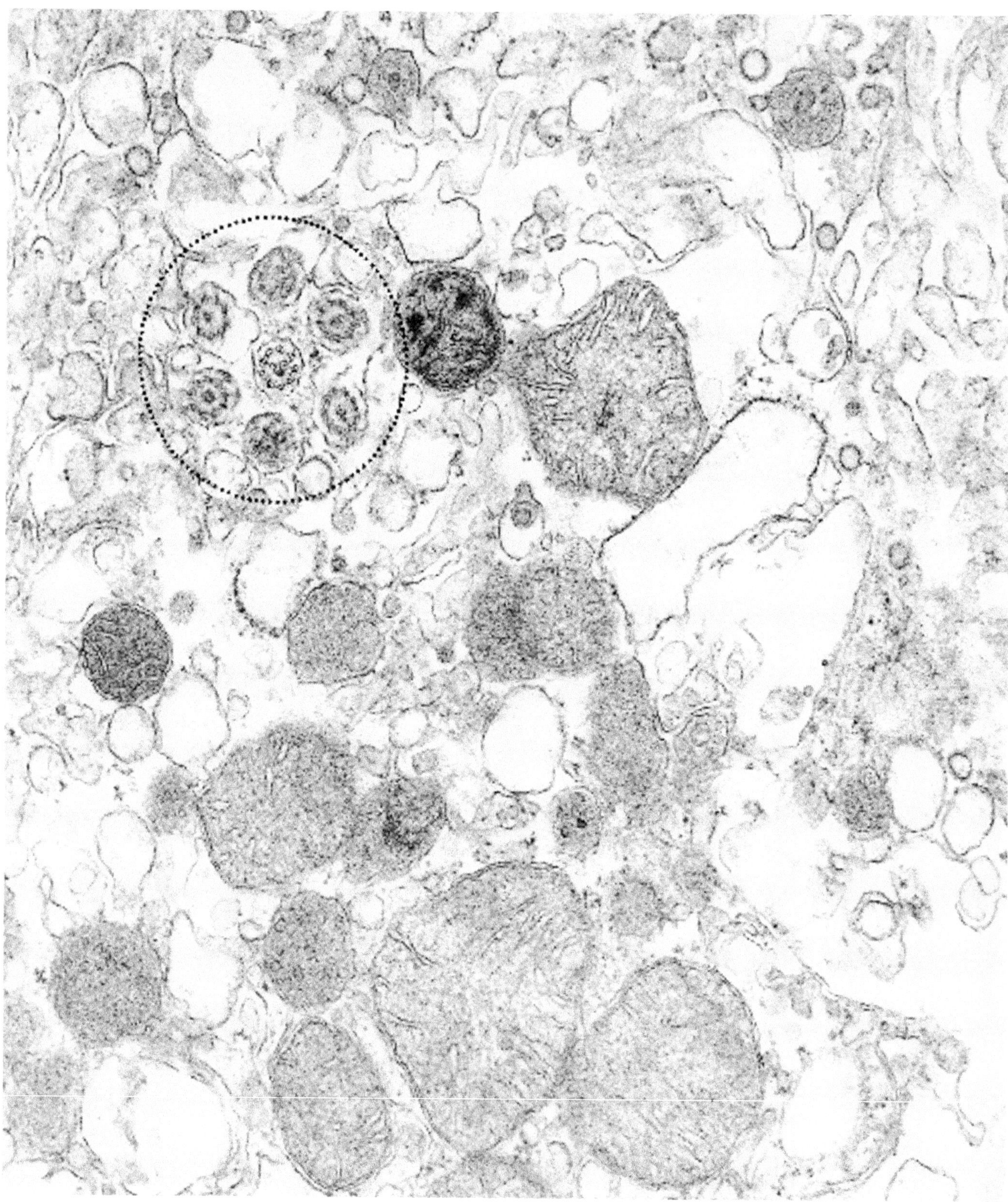

Figure 8.31. Normal choroid plexus (fourth ventricle). Higher magnification of the same tissue as illustrated in Figure 8.30 shows several cilia (*circle*) arising from the villous surface of a cell. (× 47,000)

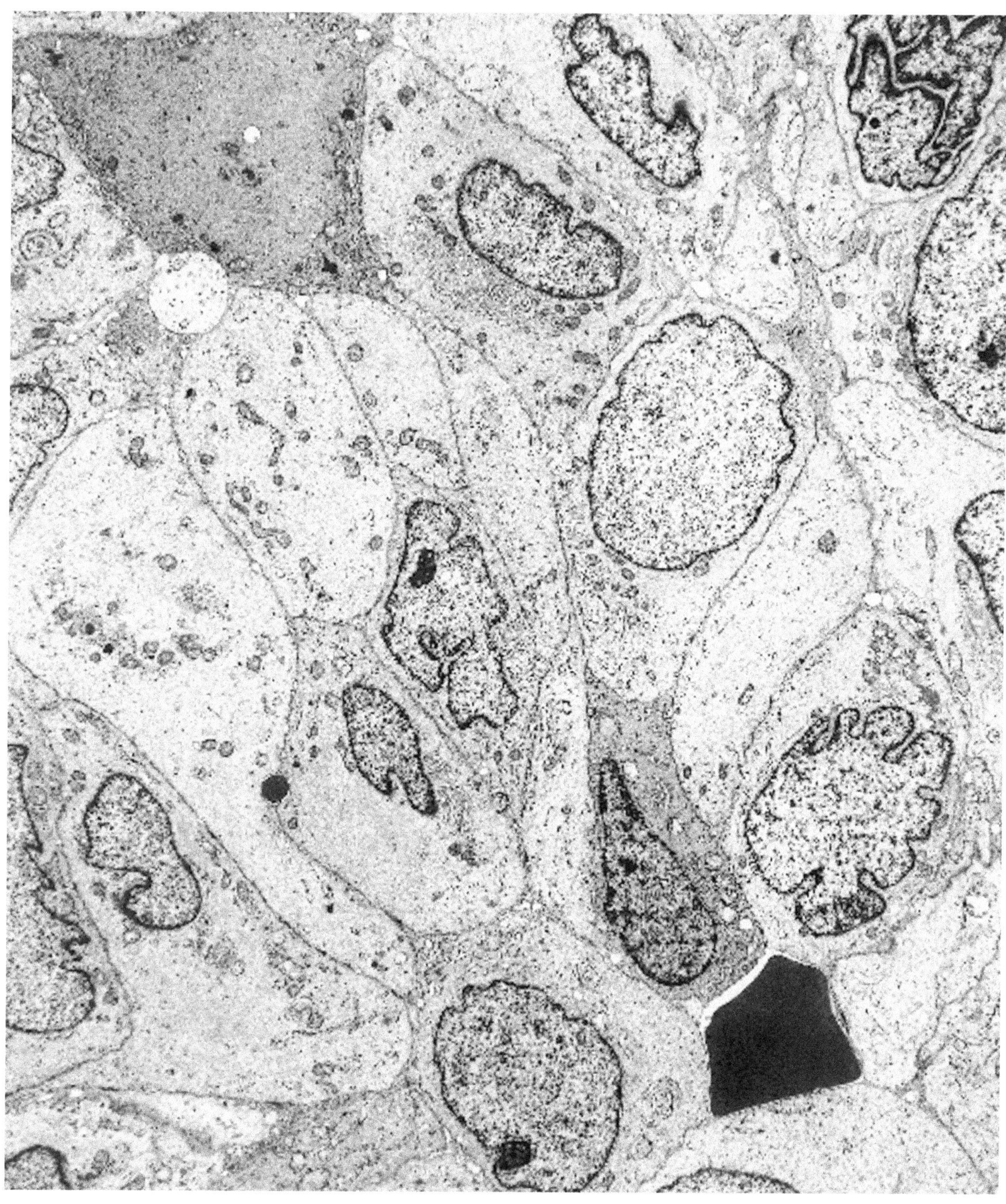

Figure 8.32. Choroid plexus carcinoma (cerebellopontine angle). This neoplasm had a mixed papillary and solid pattern, and the photograph is from a solid area. The cells are closely apposed and have numerous junctions and intermediate filaments (both of which are seen better at higher magnification in Figure 8.33). The filaments suggest the alternative diagnosis of ependymoma, but the location and light microscopy of the neoplasm were in favor of choroid plexus carcinoma. (× 5225)

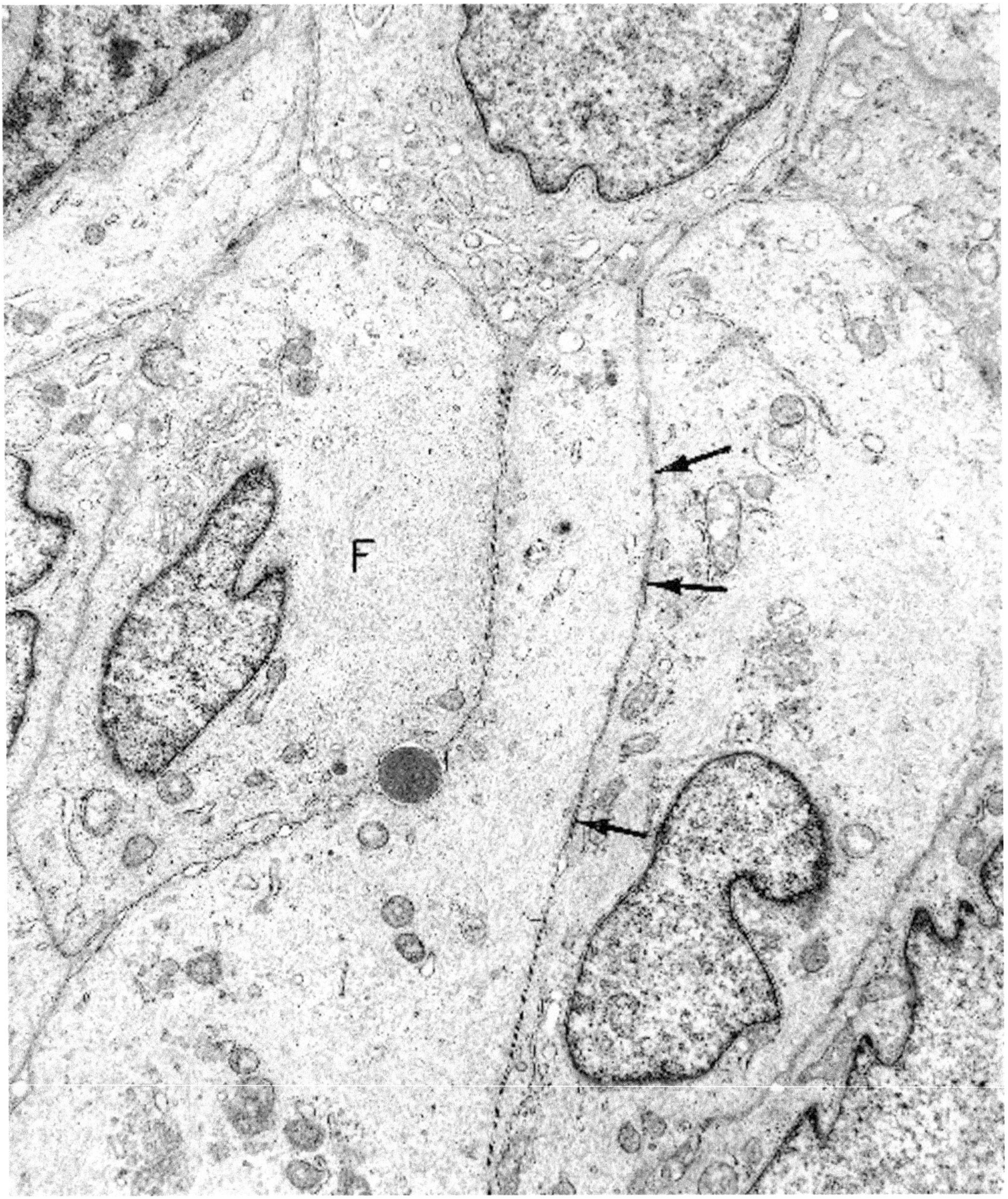

Figure 8.33. Choroid plexus carcinoma (right cerebellopontine angle). This higher magnification of several of the cells in Figure 8.32 shows many junctions (*arrows*) and innumerable filaments (F). (× 12,150)

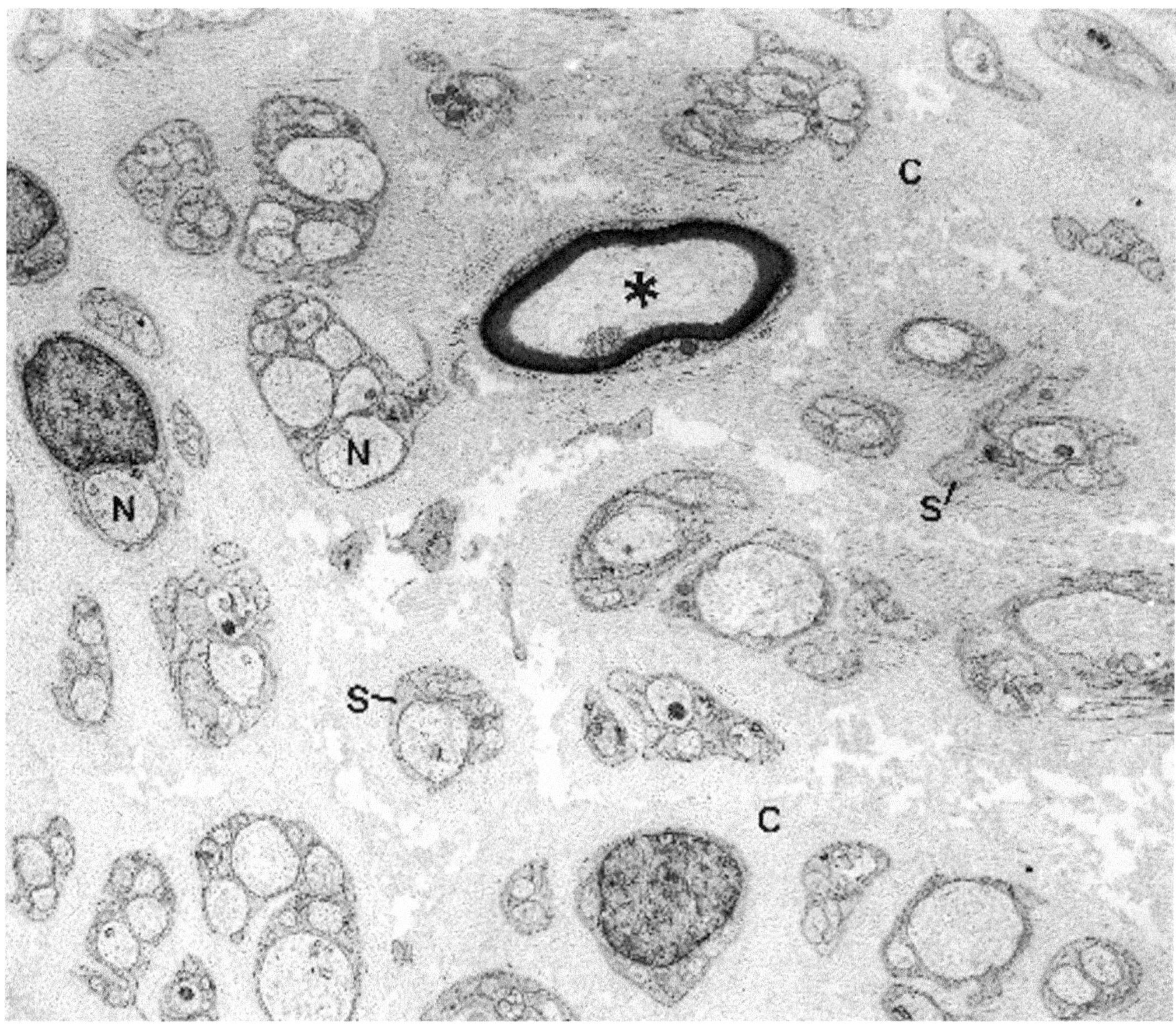

Figure 8.34. Ganglioneuroma (adrenal gland). Low magnification reveals groups of neuritic processes (N) surrounded by Schwann cells (S). One myelinated neurite (*) is included in the field. The nuclei pictured are those of Schwann cells. No ganglion cells are present in this particular field. The extracellular matrix is rich in banded collagen (C). (× 7000)

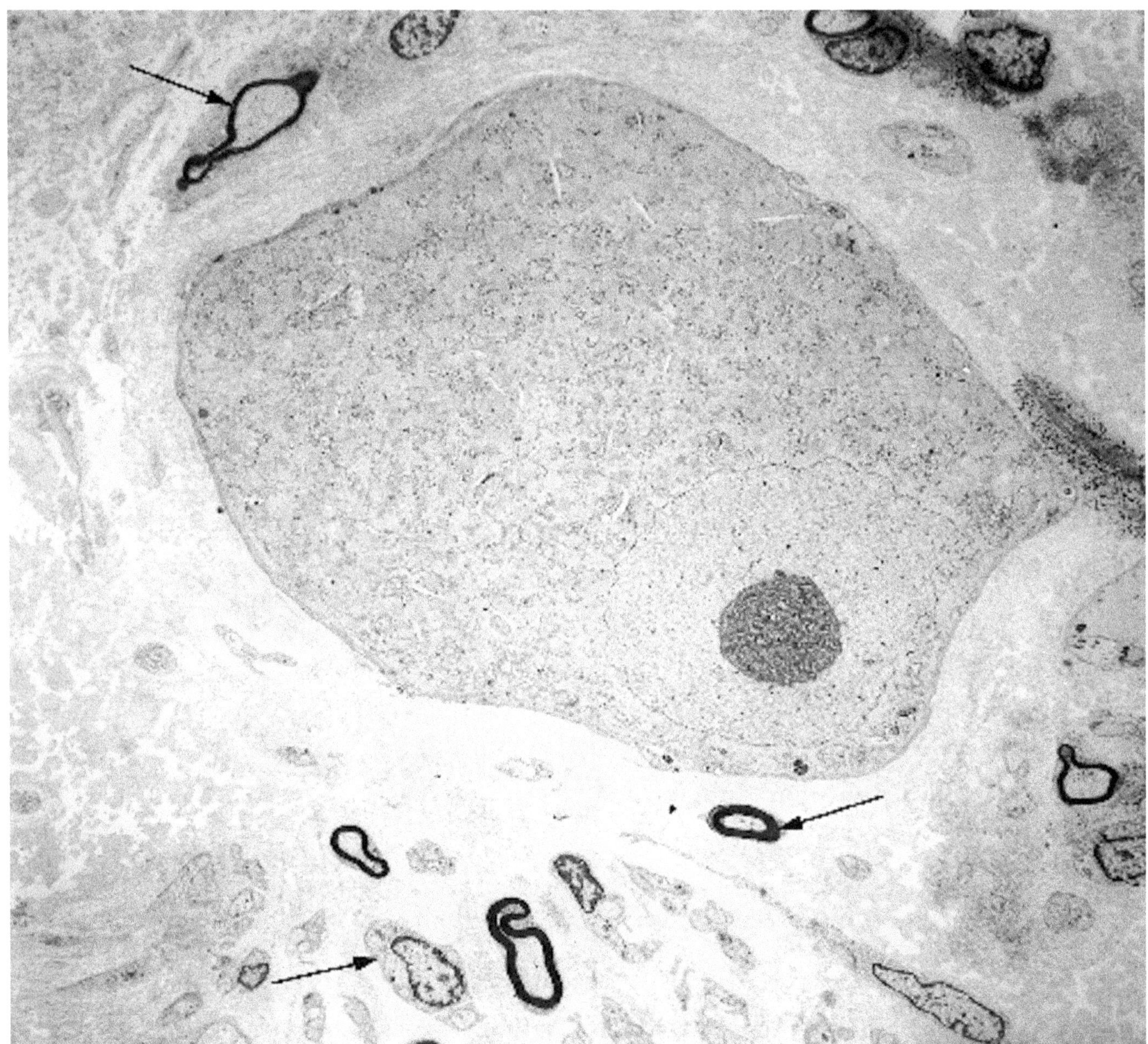

Figure 8.35. Ganglioneuroma (adrenal gland). The huge size of this ganglion cell is realized when contrasted to the size of surrounding Schwann cell/neurite complexes (*arrows*). The cytoplasm of the ganglion cells is copious and filled with innumerable organelles (seen better at higher magnification in Figure 8.36). (× 3300)

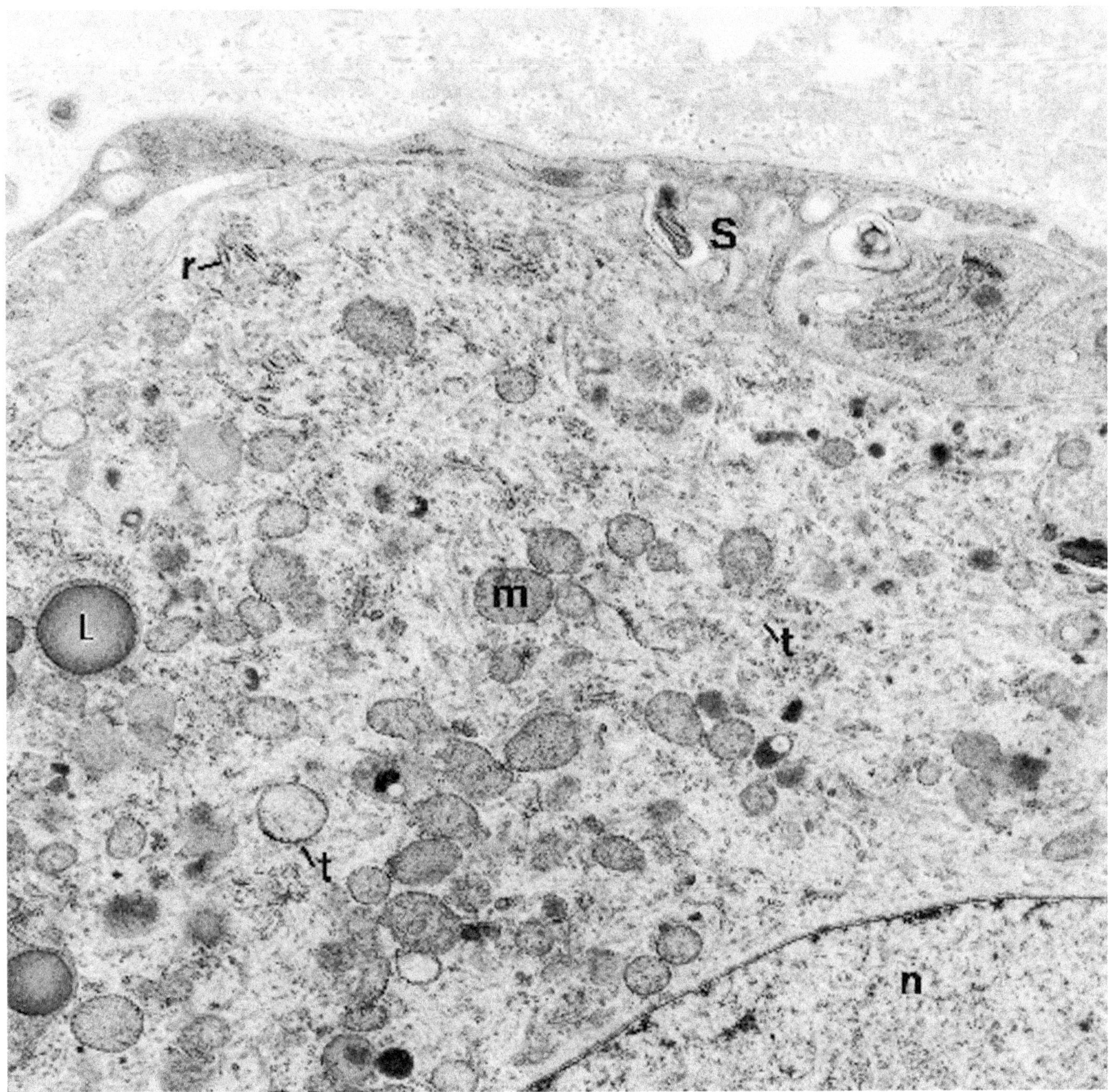

Figure 8.36. Ganglioneuroma (adrenal gland). High magnification of this ganglion cell reveals a cytoplasm filled with numerous organelles and cytoskeletal elements. m = mitochondria; t = microtubules; r = rough endoplasmic reticulum; L = lipid droplet; n = nucleus; S = Schwann cell cytoplasm. (× 19,000)

(Text continued from page 510)

Ganglioglioma

Ganglioglioma is a neoplasm composed of a combination of ganglion cells and glial cells, the latter more commonly being astrocytes (Figures 8.11 through 8.16) but also, to a lesser extent, oligodendrocytes (Figures 8.18 through 8.21).

Central Neurocytoma

(Figures 8.37 and 8.38.)

Diagnostic criteria. (1) Zones of contiguous neuritic processes (neuropil); (2) microtubules; (3) intermediate filaments; (4) dense-core granules; (5) clear, synaptic vesicles; (6) junctions.

Additional points. The cells of neurocytomas tend to be somewhat more differentiated than those of neuroblastomas. Nuclei are more uniform in size and shape, and cytoplasm often contains synaptic vesicles in neurocytomas. Homer Wright rosettes, with central cores of neuritic processes, may be present both in neuroblastomas and in neurocytomas. Mitoses and necrosis of tumor cells are less common in neurocytomas.

Neuroblastoma

Neuroblastoma, ependymoblastoma, primitive neuroectodermal tumor, and medulloblastoma are all classified as "embryonal" tumors by the World Health Organization. Neuroblastoma is described and illustrated in the section on small cell undifferentiated neoplasms

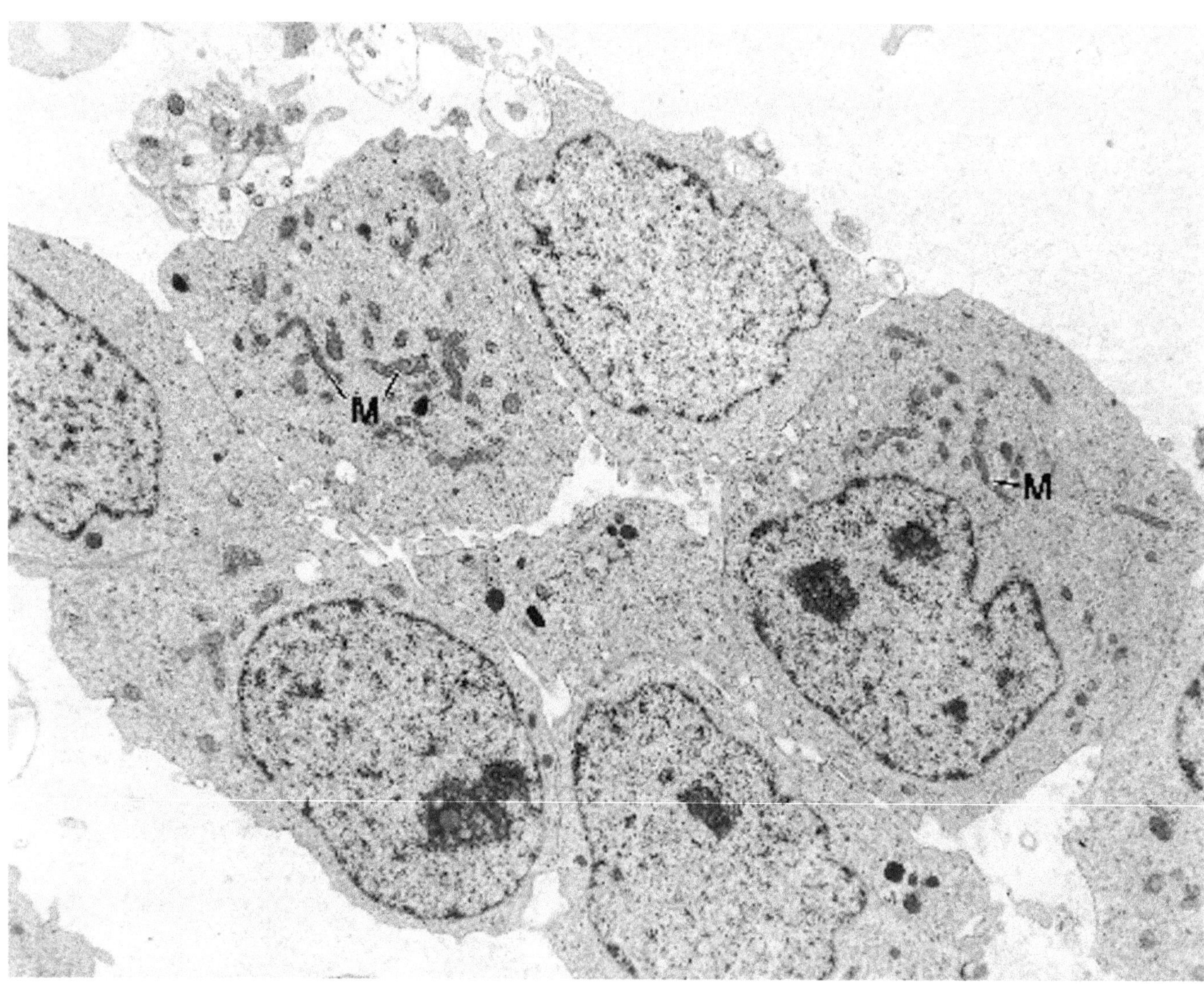

Figure 8.37. Central neurocytoma (occipital cerebrum). These cell bodies have a moderate amount of cytoplasm and irregularly oval nuclei. At this magnification, the cytoplasm is composed of a background of free ribosomes and a moderate number of mitochondria (M). (× 7700)

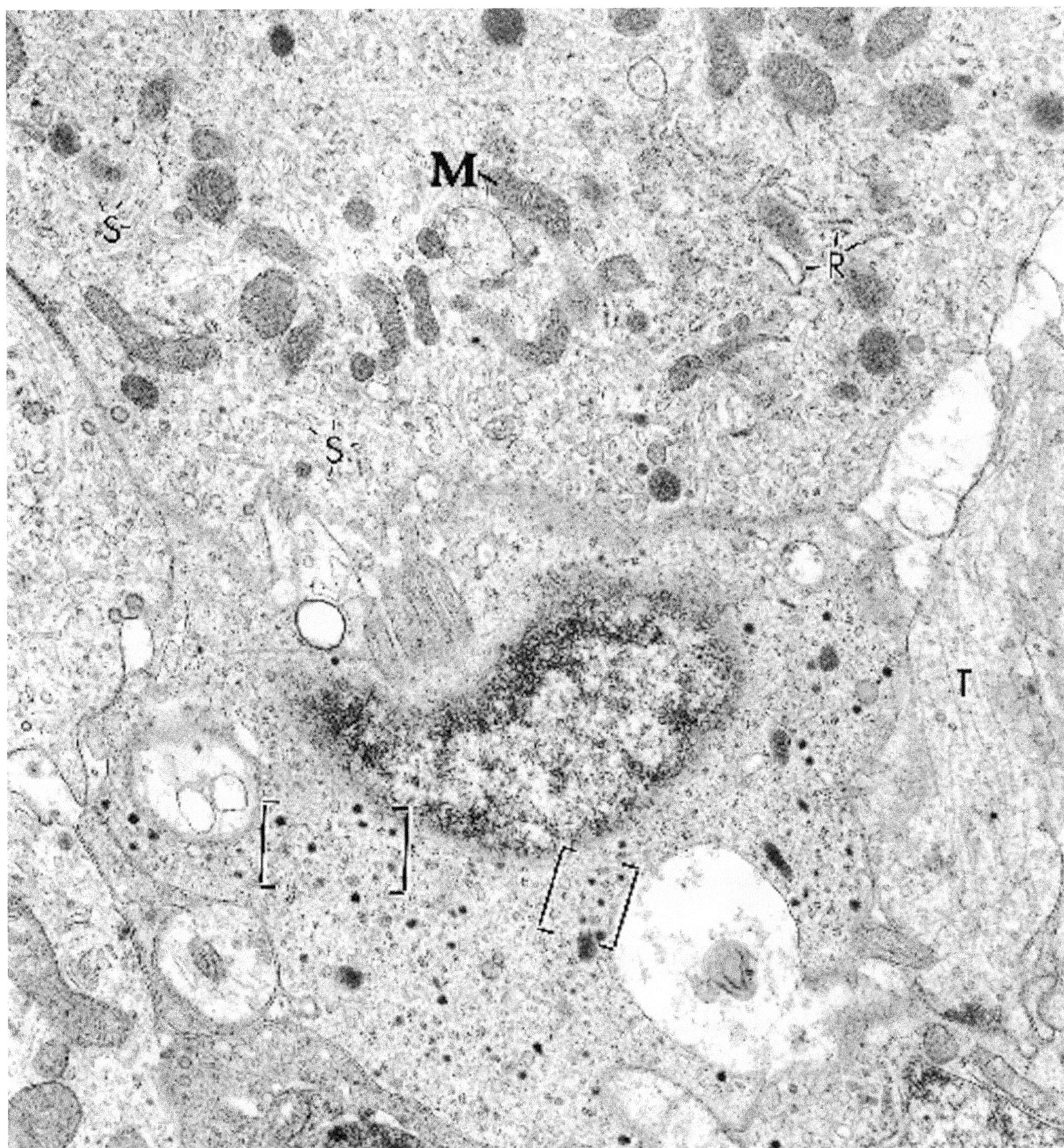

Figure 8.38. Central neurocytoma (occipital cerebrum). Higher magnification of the same neoplasm as that depicted in Figure 8.37 reveals the cytoplasm to contain more than just ribosomes and mitochondria; in addition, there is an abundance of smooth endoplasmic reticulum (S), a small amount of rough endoplasmic reticulum (R), numerous dense core granules (*brackets*), and focal collections of microtubules (T). (× 21,200)

(see Chapter 4, Figures 4.9 through 4.12). Neuroblastomas are generally the same, whether central or peripheral, although the centrally located lesions are rare, less likely to show maturation to ganglion cells, and often mixed with varying numbers of nonneoplastic or neoplastic astrocytes and oligodendrocytes. Olfactory neuroblastoma (esthesioneuroblastoma) is composed of neuroblasts either in a diffuse or a nesting arrangement. Sustentacular (Schwann-like) cells may be found surrounding groups of cells. The neuroblasts have the same ultrastructural features already described, but rosettes, ganglion cells, nuclear pleomorphism, and necrosis are rare.

Ganglioneuroblastoma

Diagnostic criteria. (1) Neuroblasts with rosettes and abundant neuropil (see section on neuroblastoma, Chapter 4, Figures 4.9 through 4.12); (2) scattered ganglion cells (see sections on ganglioneuroblastoma, Figure 4.13, and ganglioneuroma, Figures 8.35 and 8.36); (3) cells transitional between neuroblasts and ganglion cells; (4) Schwann cells surrounding groups of unmyelinated neurites.

Additional points. These are rare neuroblastic tumors in which neuronal differentiation to intermediate cells and mature ganglion cells is extensive, as opposed to neuroblastomas with only focal ganglion cells.

Ependymoblastoma

This tumor is characterized by numerous rosettes and intervening zones of small undifferentiated cells. The ependymal cells lining the rosettes are similar to the cells in classic ependymoma, except that in ependymoblastoma the cells are in more than one layer, and mitoses are frequent (see section on ependymoma, Figures 8.22 through 8.29).

Primitive Neuroectodermal Tumor

(See Figures 4.19 through 4.21.)

Central primitive neuroectodermal tumors (PNETs) include a broader group of neoplasms than peripheral PNETs and, therefore, have a more diverse histology and ultrastructure. Central PNETs are small cell tumors of childhood that arise predominantly in the cerebellum and have the potential for divergent differentiation into neuronal (neuroblastic), astrocytic, ependymal, muscular, and melanotic lines. Rare rhabdoid tumors, so-called atypical teratoid-rhabdoid tumors, may also be included in the PNET category. In addition to having small cells with large aggregates of whorled intermediate filaments as well as neuroepithelial, epithelial, and mesenchymal elements, these tumors also contain components of more typical PNET/medulloblastoma.

Medulloblastoma

(Figures 8.39 through 8.41.)

Diagnostic criteria. (1) Small primitive cell bodies, with relatively few cytoplasmic organelles and a high nuclear–cytoplasmic ratio; (2) cells with focal back-to-back narrow cellular processes (Figures 8.39 and 8.40); (3) microtubules within cellular processes (Figure 8.41); (4) small intercellular junctions; (5) elongated, angular, and indented, heterochromatic nuclei.

Additional points. This cerebellar neoplasm is classified as a central PNET, which arises in the roof of the fourth ventricle or in the cerebellum and may be completely undifferentiated or have neuronal and/or glial differentiation. Neuronal differentiation consists of long cellular processes and microtubules, similar to those seen in neuroblastoma. Homer Wright rosettes and even ganglion cells may also be present. However, dense-core granules are usually absent or rare in medulloblastomas. Glial differentiation is into astrocytes and oligodendrocytes. Rarely, medulloblastomas may show focal striated or nonstriated muscle differentiation (*medullomyoblastomas*), or they may be pigmented (*melanotic medulloblastomas*).

Germinoma (and Embryonal Carcinoma, Choriocarcinoma, and Teratoma)

See the section on germ cell tumors in Chapter 7.

Except for very rare forms, this group of neoplasms is described in Chapter 7, and the ultrastructure is the same wherever they occur, in gonads, retroperitoneum, mediastinum, or pineal region.

(Text continues on page 532)

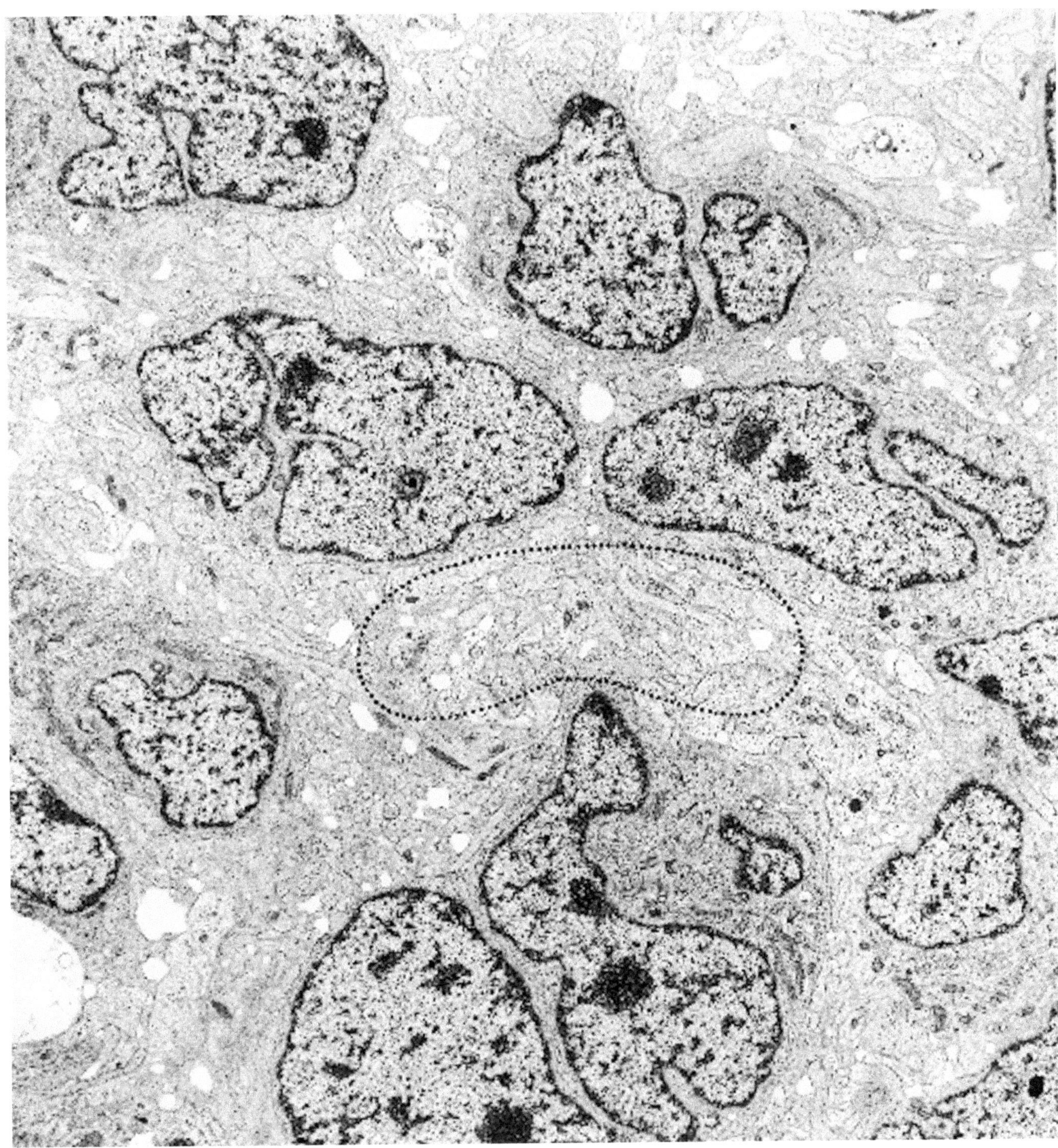

Figure 8.39. Medulloblastoma (cerebellum). The cells have the high nuclear–cytoplasmic ratio of primitive cells, but the innumerable narrow processes (within *dotted line*) between cell bodies are one of the diagnostic features of neuronal differentiation. Nuclei are elongated, angular and indented, and heterochromatic. (× 6500)

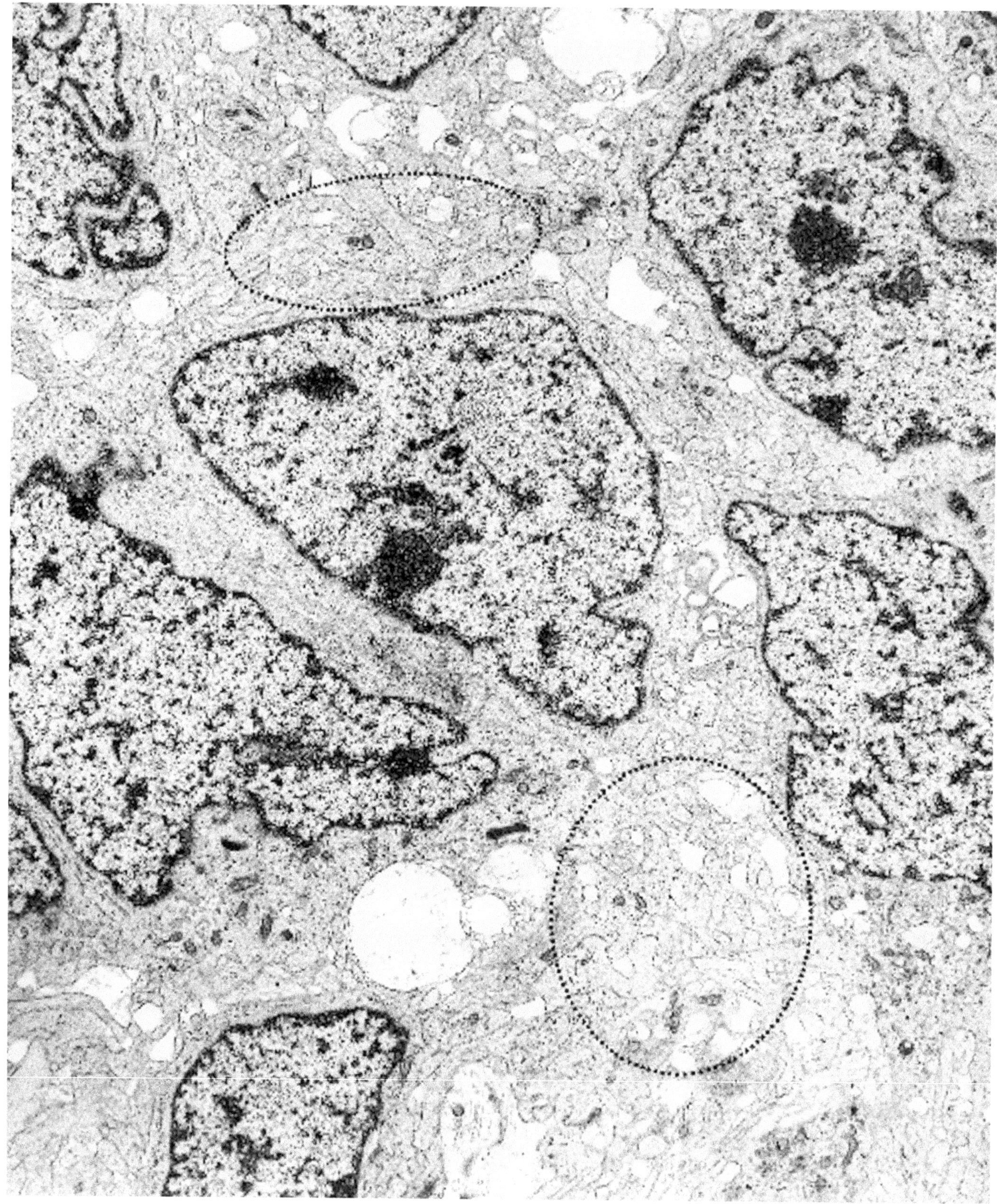

Figure 8.40. Medulloblastoma (cerebellum). Innumerable narrow processes create small zones of neuropil (within *dotted lines*) between primitive cell bodies. (× 8500)

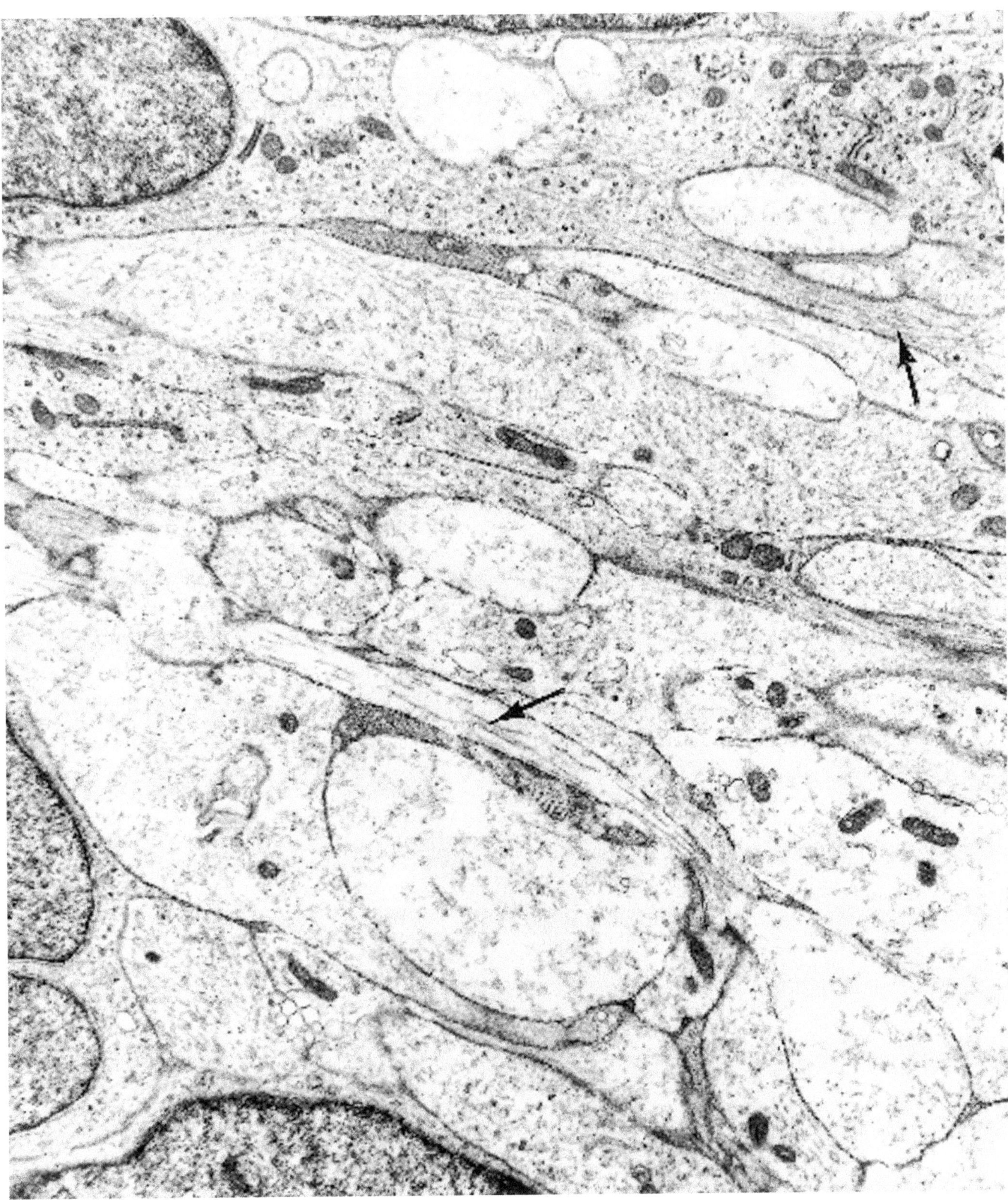

Figure 8.41. Medulloblastoma (cerebellum). High magnification of neoplastic cell processes depicts diagnostic microtubules (*arrows*). (× 13,500)

(Text continued from page 528)

Pineocytoma and Pineoblastoma

(Figures 8.42 through 8.49.)

Diagnostic criteria. (1) Groups of oval and polygonal cell bodies, with the groups being surrounded by basal lamina (Figures 8.42 and 8.43); (2) cell processes with varying numbers of intermediate filaments, microtubules, and dense-core granules (Figures 8.47 through 8.49).

Additional points. Rosettes (radially arranged cells having their processes converge on a central space) and pseudorosettes (radially directed cells having their processes converge on a blood vessel) may be present in more differentiated examples (pineocytomas) of these neoplasms. Other features to be noted include prominent intercellular junctions (Figures 8.47 and 8.49), microlumens (Figure 8.44), intracytoplasmic pseudolumens (Figure 8.42 *inset,* 8.45), moderate amounts of cytoplasmic glycogen (Figure 8.45), and a moderate number of mitochondria (Figures 8.44 and 8.47). In addition to neuronal differentiation, pineocytomas and the less-differentiated pineoblastomas may also show glial differentiation.

Hemangioma

See the section on angiosarcoma in Chapter 6.

The ultrastructure of the capillary hemangioma is similar to the description of this benign lesion in Chapter 6.

Hemangioblastoma

(Figures 8.50 and 8.51.)

Diagnostic criteria. (1) A field of thin-walled blood vessels separated by a cellular stroma; (2) the endothelial cells lining the vessels have junctions, basal lamina, pinocytotic vesicles, villus-like projections on their luminal surface, and sometimes Weibel-Palade (lysosome-like) bodies; (3) the stromal cells are plump and usually possess many lipid droplets and microfilaments in the cytoplasm (Figures 8.50 and 8.51).

Additional points. The capillaries of hemangioblastomas of the central nervous system (usually the cerebellum) are well formed, and the stromal component comprising a solid sea of large, polygonal cells with copious cytoplasm is unique (Figures 8.50 and 8.51). The exact line of differentiation of these cells is uncertain. Some of the stromal cells with the most lipid show shrinkage and hyperchromatism of their nuclei and swelling of their membrane-bound, cytoplasmic organelles, suggesting that the lipid is a degenerative change. However, intact cells also usually contain some lipid, albeit apparently less than in the degenerating cells. Given that the lipid could be a feature of cell death, a satisfactory classification of the stromal cells of hemangioblastomas is still lacking. The cells are not typical of histiocytes, because there is an insufficient variety of cytoplasmic organelles, including lysosomes and mitochondria. Moreover, they do not take up metallic stains, as would be expected of histiocytes. Pericytes are a possible related cell, because they and the stromal cells in question have filaments and small intercellular junction. On the other hand, basal lamina and a wider matrix between cells would be expected to accompany more typical pericytes. Endothelial and glial cells are other theoretical possibilities, but they do not have a close enough morphologic resemblance to the stromal cells, and their immunocytochemistry has been mixed in regard to exhibiting glial fibrillary acid protein, S-100 protein, and factor VIII, markers for glial, neural, and endothelial cells, respectively. Finally, the idea that the stromal cells could represent a stem cell for the endothelial cells has insufficient evidence to be convincing, and the classification of the stromal cell in hemangioblastoma currently goes unanswered.

(Text continues on page 543)

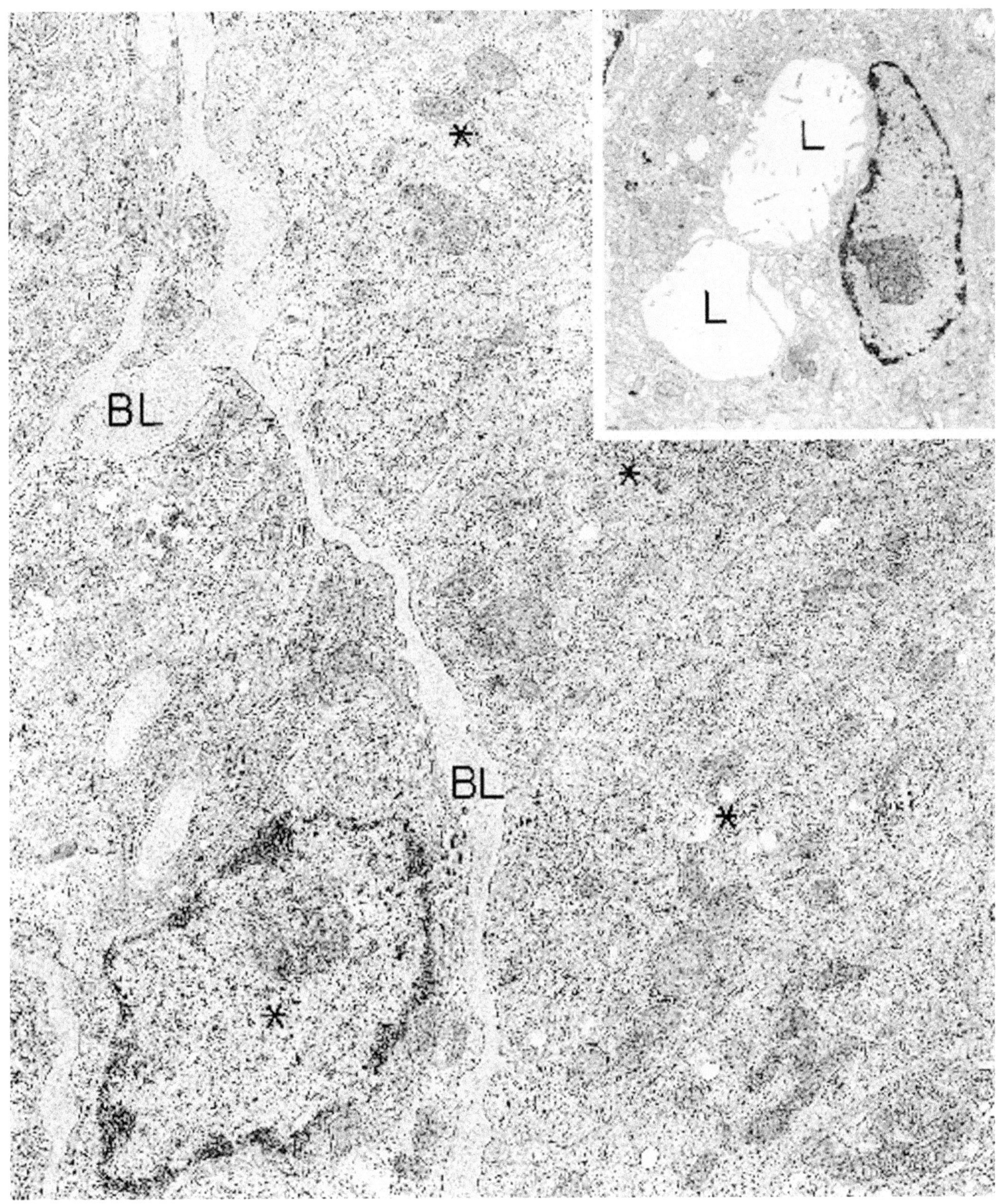

Figure 8.42. Pineal gland. This specimen was interpreted as probably a normal pineal gland, adjacent to a neoplasm, the two being similar to one another ultrastructurally. Groups of polygonal and elongated parenchymal cells (*) are surrounded by basal lamina (BL). (× 9500) The *inset* shows one or two intracytoplasmic pseudolumens (L) in a parenchymal cell. (× 5400)

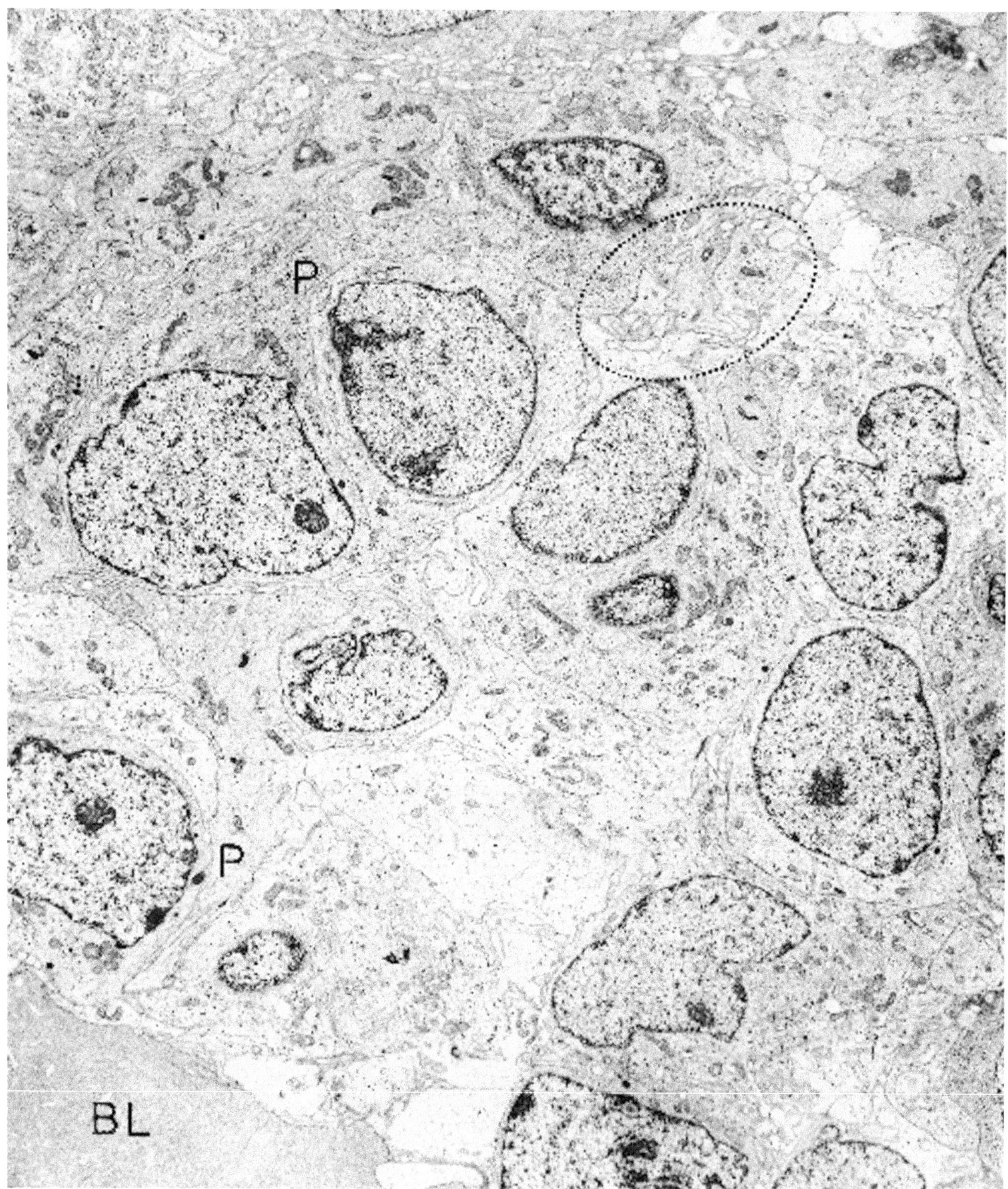

Figure 8.43. Pineocytoma (region of pineal gland). The cell bodies are oval and polygonal, and some of them are cut in a plane that allows the origin of their processes (P) to be seen. The processes extend outward and intertwine with one another (*dotted circle*). Basal lamina (BL) surrounds groups of cells. (× 4960)

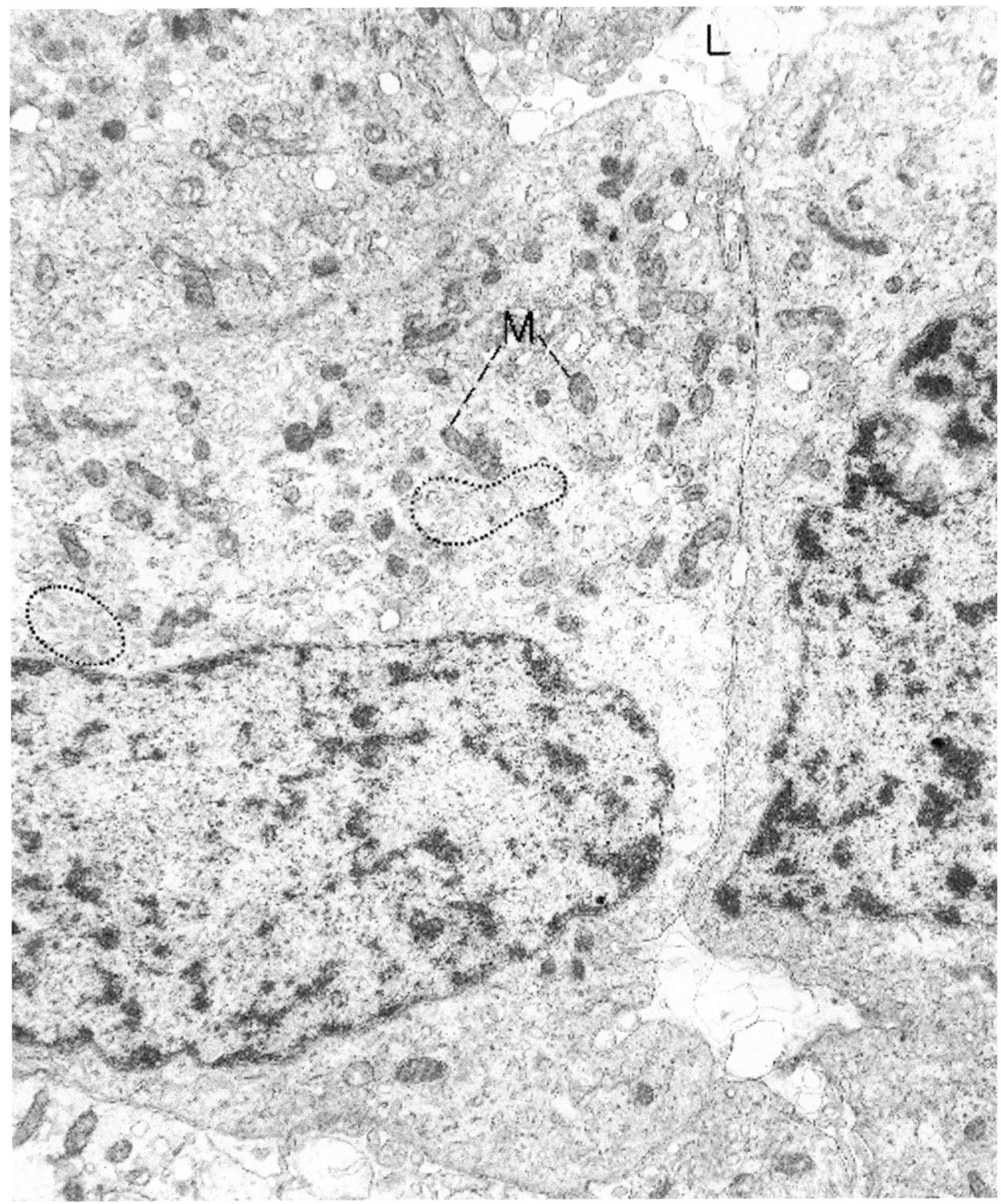

Figure 8.44. Pineocytoma (region of pineal gland). This neoplasm showed more differentiation than the one illustrated in Figure 8.43, characterized by more copious cytoplasm, more organelles, and formation of microlumens (L). There also is compartmentalization of mitochondria (M) and smooth endoplasmic reticulum (*circles*) in a supranuclear position. (× 12,600)

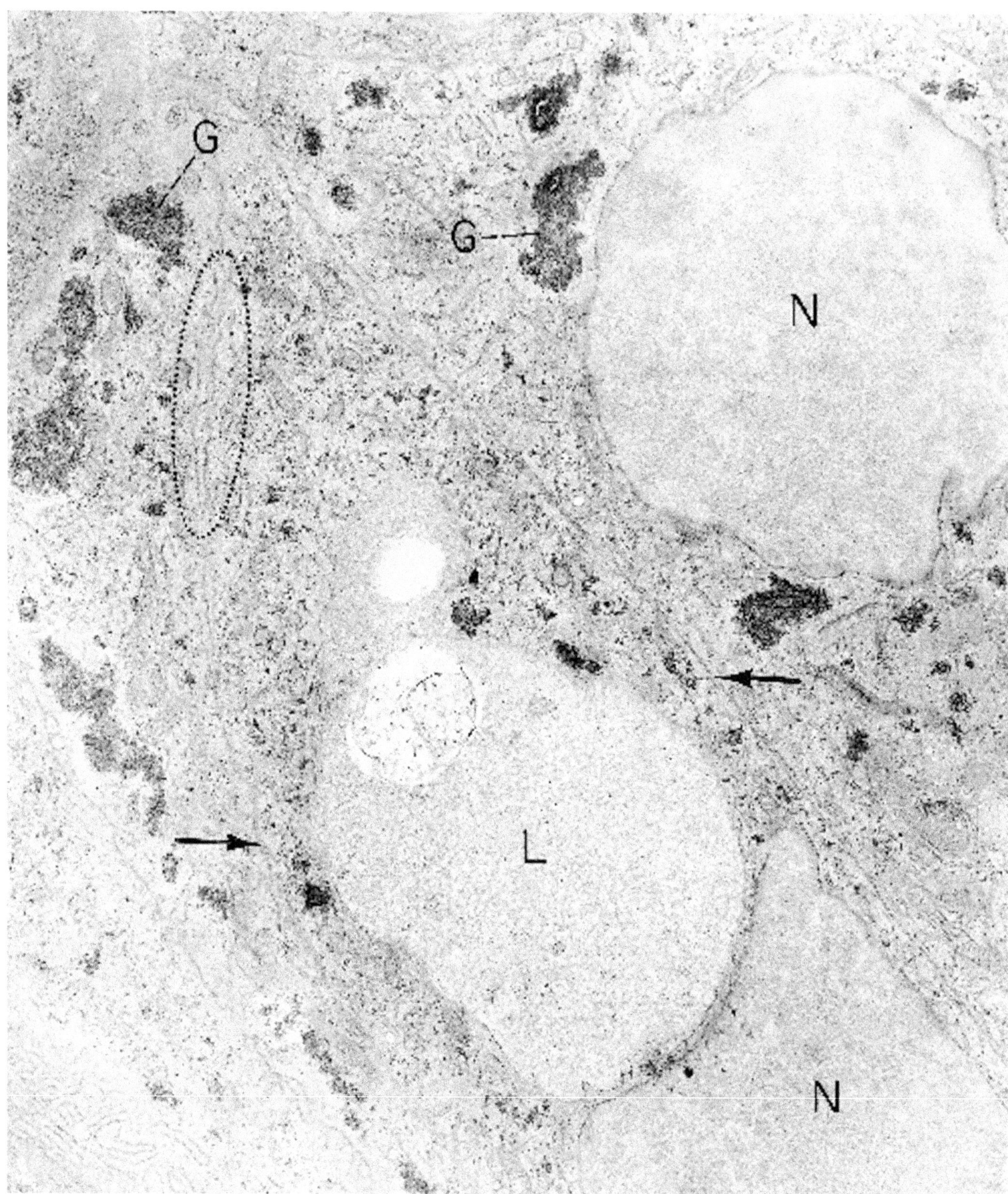

Figure 8.45. Pineoblastoma (pineal body). This specimen was processed by a method that preserved glycogen (G) as electron-dense granules. Other structures are less dense by this technique but are still visible, for example, nuclei (N), intracytoplasmic pseudolumen (L), and lateral cell membranes (*arrows*) with focal interdigitations (*circle*). (× 11,250)

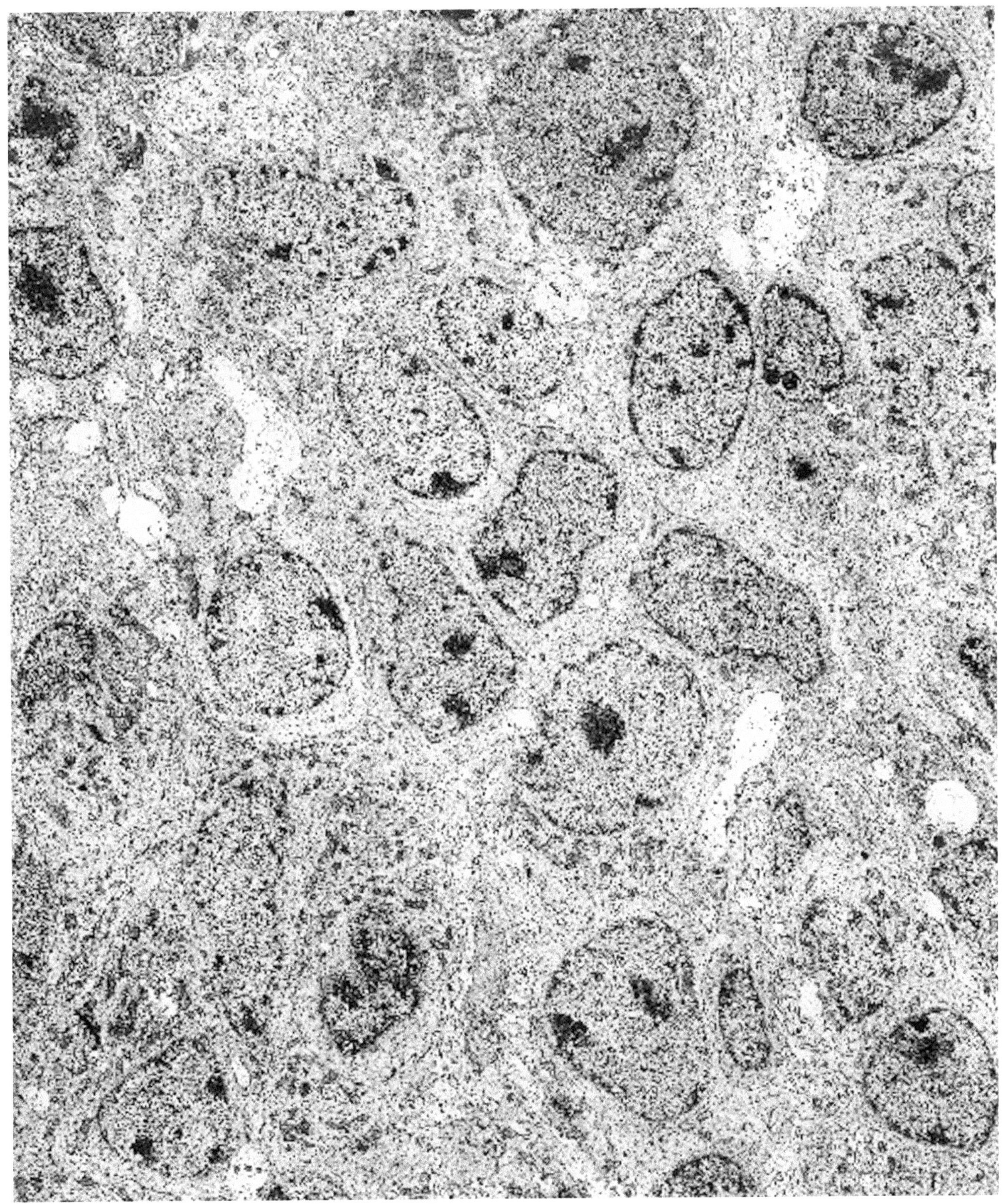

Figure 8.46. Pineoblastoma (pineal gland). Although less well preserved, this neoplasm was identifiable as being similar to, but less differentiated than, the pineocytomas depicted in Figures 8.43 and 8.44. There is a high nuclear–cytoplasmic ratio, an absence of a grouping arrangement with surrounding basal lamina, and an absence of lumens, rosettes, and pseudorosettes. Narrow, interdigitating cell processes are present, however, and are seen better at the higher magnification of Figure 8.47. (× 4275)

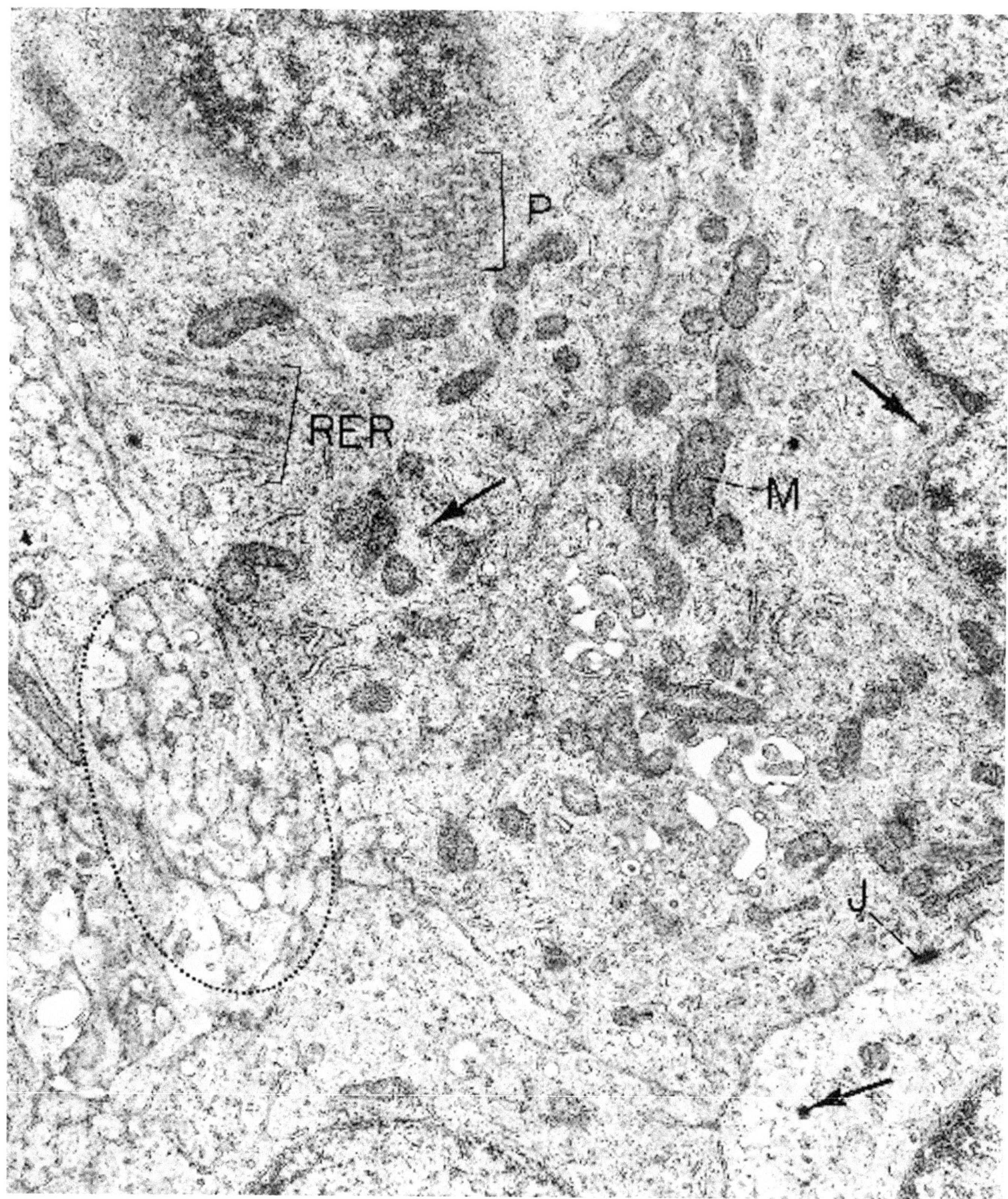

Figure 8.47. Pineoblastoma (pineal gland). Higher magnification of the neoplasm depicted in Figure 8.46 illustrates the various cytoplasmic organelles, including mitochondria (M), rough endoplasmic reticulum (RER), and dense-core granules (*arrows*); nuclear pores (P) cut tangentially; intercellular junctions (J); intertwined cell processes (*broken line* enclosure). (× 18,460)

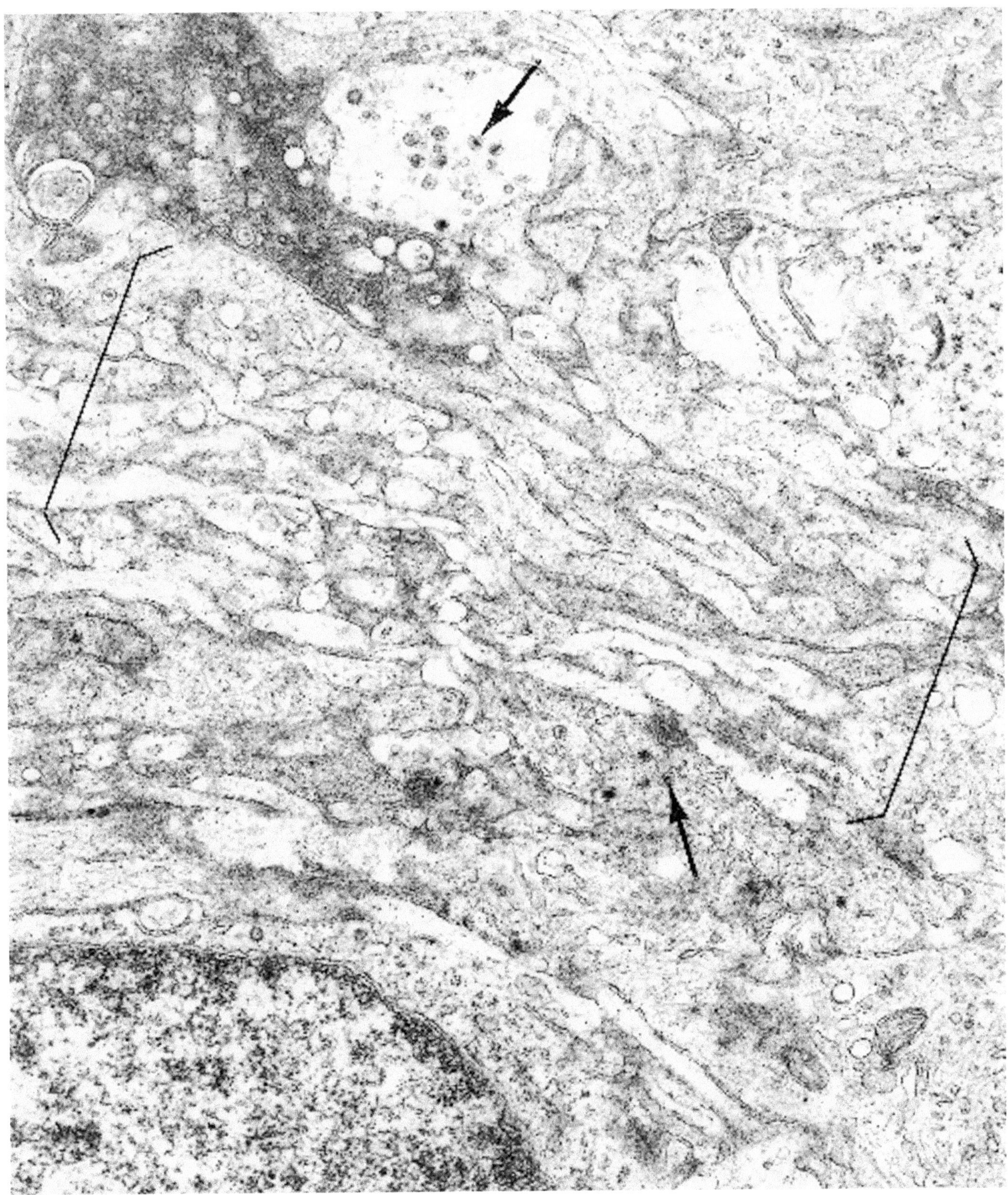

Figure 8.48. Pineoblastoma (pineal gland). High magnification of the neoplasm shown in Figures 8.46 and 8.47 depicts numerous intertwined cell processes (*brackets*) and scattered dense-core granules (*arrows*). (× 28,600)

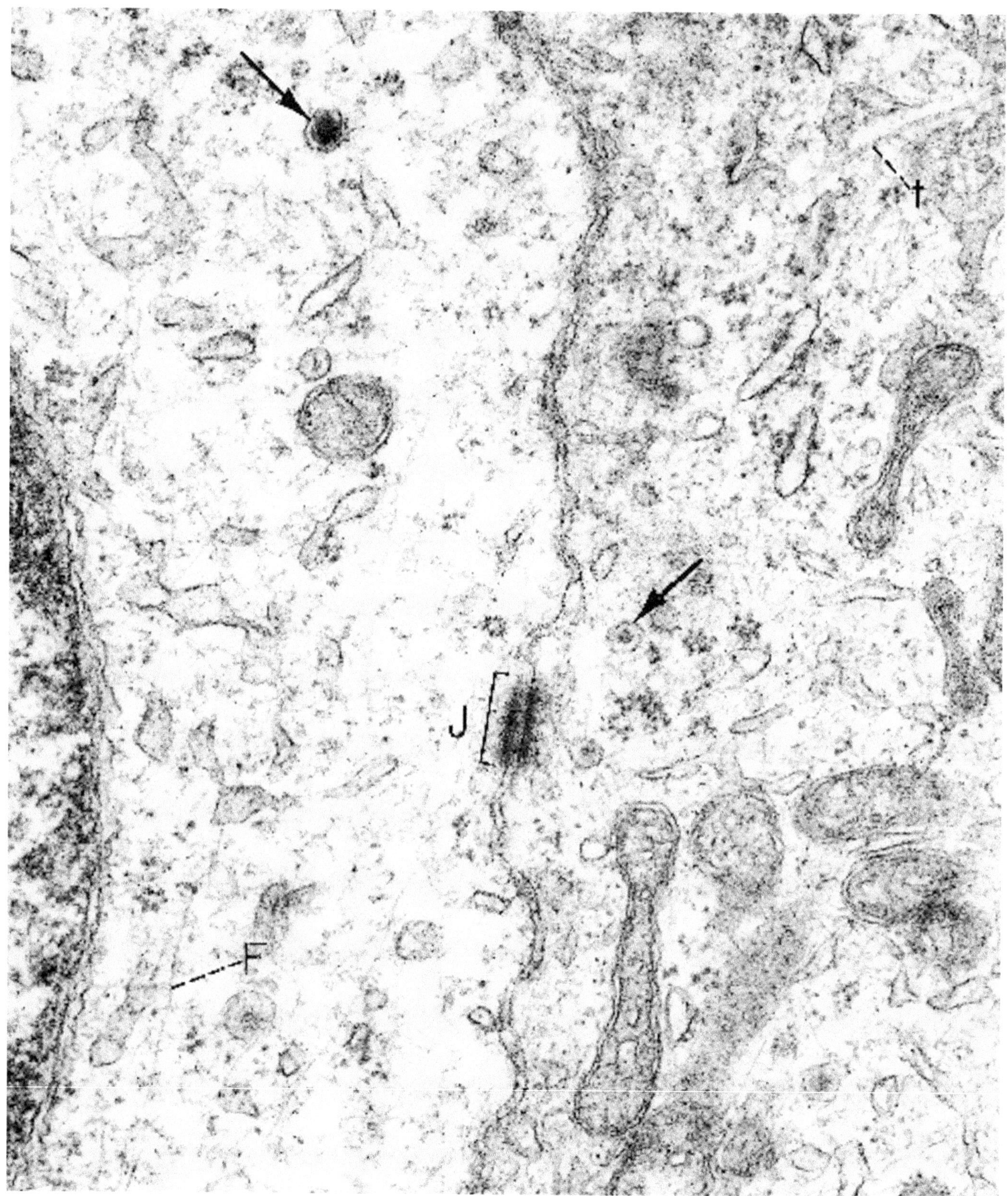

Figure 8.49. Pineoblastoma (pineal gland). Still higher magnification of the neoplasm illustrated in Figures 8.46 through 8.48 shows well an intercellular junction (J), several dense-core granules (arrows), microfilaments (F), and sparse microtubules (t). (× 62,250)

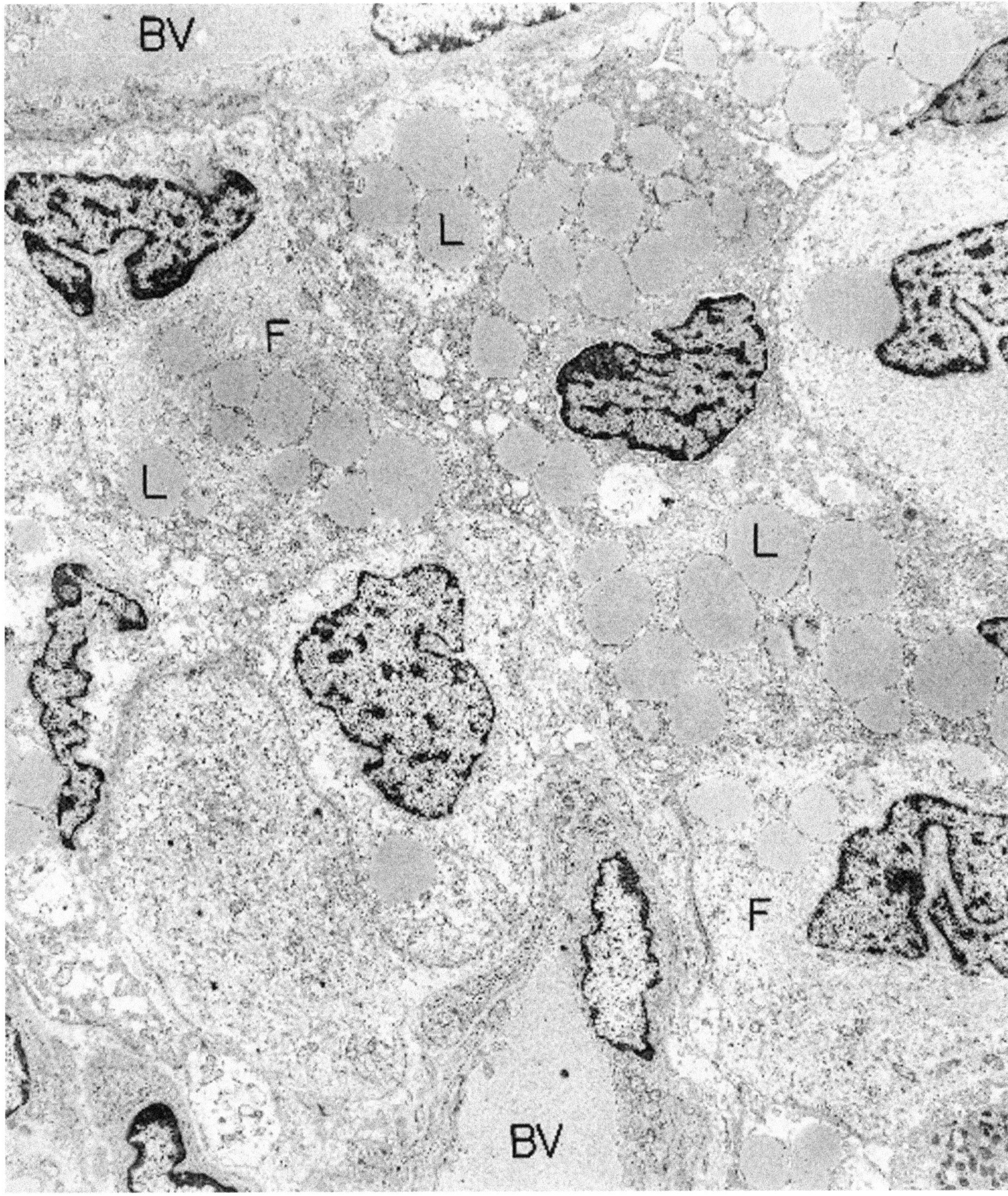

Figure 8.50. Hemangioblastoma (cerebellum). Between the well-formed blood vessels (BV) are back-to-back stromal cells, most of which contain numerous lipid droplets (L) and microfilaments (F). (× 4940)

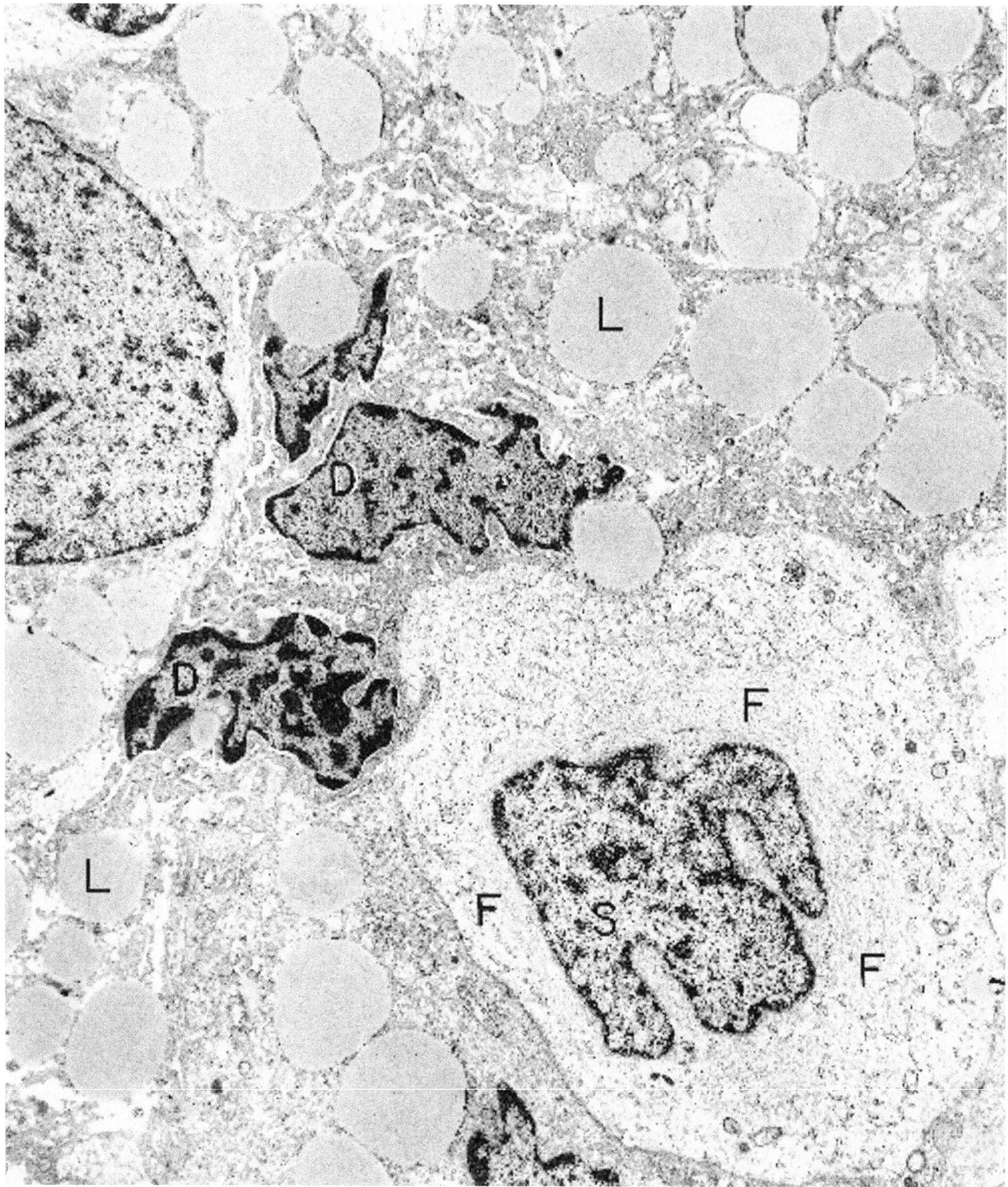

Figure 8.51. Hemangioblastoma (cerebellum). The lipid droplets (L) are more numerous in degenerating cells (D), characterized by shrunken nuclei with aggregated chromatin and dilated cytoplasmic organelles. A relatively well-preserved stromal cell (S) contains no apparent lipid, and microfilaments (F) are abundant. (× 7250)

(Text continued from page 532)

Pituitary Adenoma

(Figures 8.52 through 8.61.)

Diagnostic criteria. (1) Diffuse sheets of closely arranged, round, and polygonal cells with intercellular junctions; (2) varying numbers and sizes of secretory granules; (3) prominent Golgi apparatuses; (4) rough endoplasmic reticulum often well developed, including parallel stacks and whorls (nebenkern); (5) paranuclear fibrous bodies (globoid aggregates of intermediate filaments).

Additional points. Somatotrophic (growth-hormone producing) adenomas usually are acidophilic by light microscopy, correlating by electron microscopy with at least a moderate number of secretory granules. When the granules are numerous, they also tend to be large (300–600 nm), and when they are sparse, they tend to be smaller (100–250 nm). When the granules are numerous, the Golgi apparatus and rough endoplasmic reticulum also are prominent. *Prolactinomas* usually have only a few small granules (chromophobic or weakly acidophilic), but in some tumors the granules may be very large (range, 500 to 800 nm). *Adrenocorticotrophic* adenomas usually have some granules that are tear shaped. The cells of these tumors may also contain large, dense, secondary lysosomes ("enigmatic bodies") as well as numerous intermediate keratin filaments, including tonofibrils (Crooke's hyaline, by light microscopy). Misplaced exocytosis of granules is a characteristic of pituitary adenomas, especially prolactinomas, and it consists of extrusion of granules into lateral intercellular spaces rather than via the normal route through the cell membranes facing adjacent capillaries. Granules may be located predominantly subjacent to the plasmalemma; this is often true of the *gonadotrophic, thyrotrophic,* and *null cell* adenomas, in which granules are small and sparse. Some adenomas are of a mixed cell type, as in the combined somatotrophic and prolactin type, and any combination of sparsely and heavily granulated cells may occur.

(Text continues on page 554)

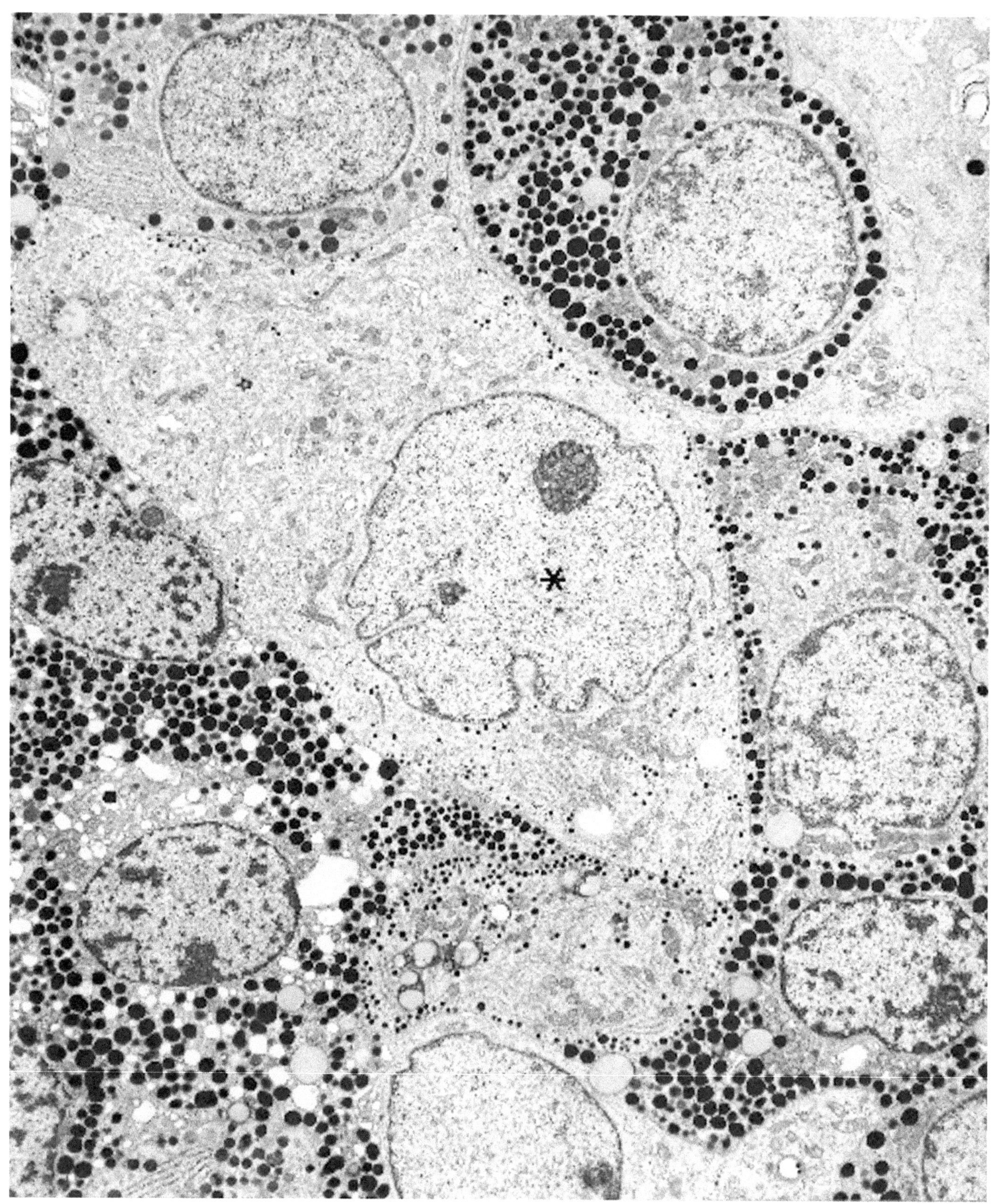

Figure 8.52. Pituitary gland, normal. The normal parenchymal cells are polygonal and in close apposition. Granules are present in all the cells, but the number and size of the granules vary widely. The large cell (*) in the center of the field has only a few small granules and would be interpreted by light microscopy to be chromophobic. (× 5940)

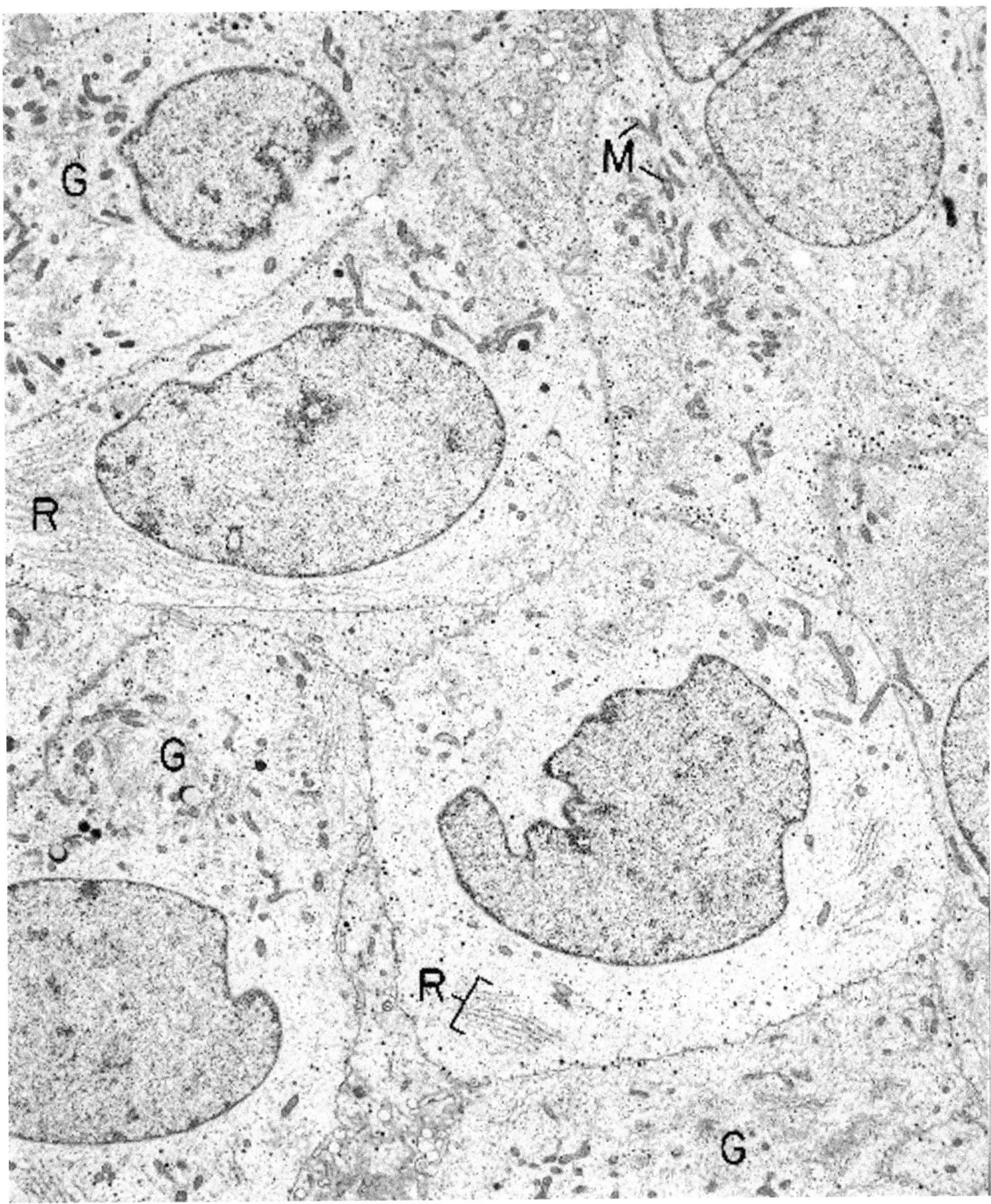

Figure 8.53. Pituitary adenoma (nonfunctioning). The neoplastic cells have a moderate number of small granules, many of which are located close to the plasmalemmas. Rough endoplasmic reticulum (R) is prominent and arranged in stacks. Golgi apparatuses (G) are large, and mitochondria (M) are numerous. (× 5720)

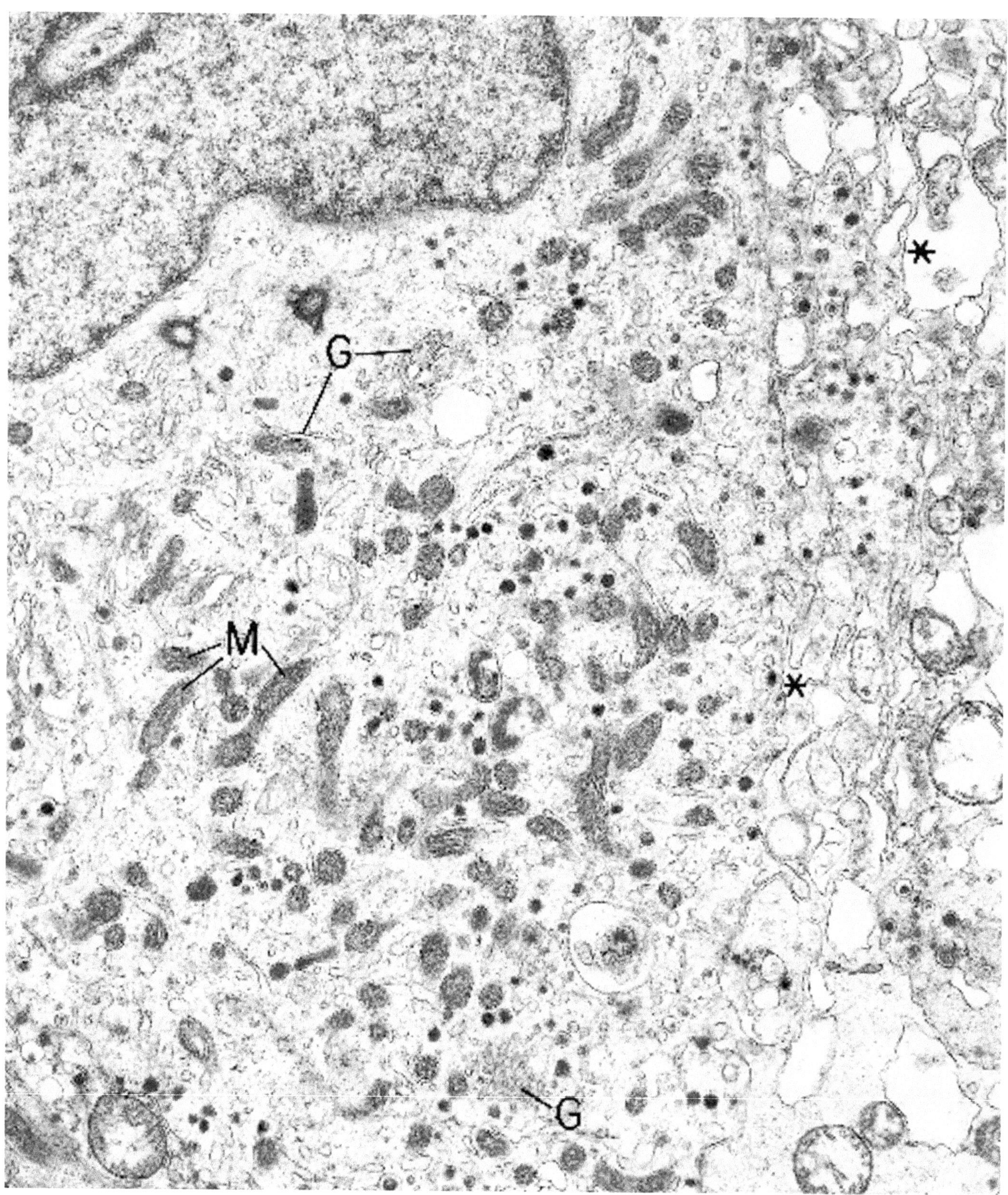

Figure 8.54. Pituitary adenoma (nonfunctioning). High magnification of a cell from the same neoplasm in Figure 8.53 shows the complexity of the cytoplasm. The secretory granules are of the small, dense-core type and appear to be extruding into the space (*) between two cells. Other prominent organelles include Golgi apparatuses (G) and mitochondria (M). (× 20,000)

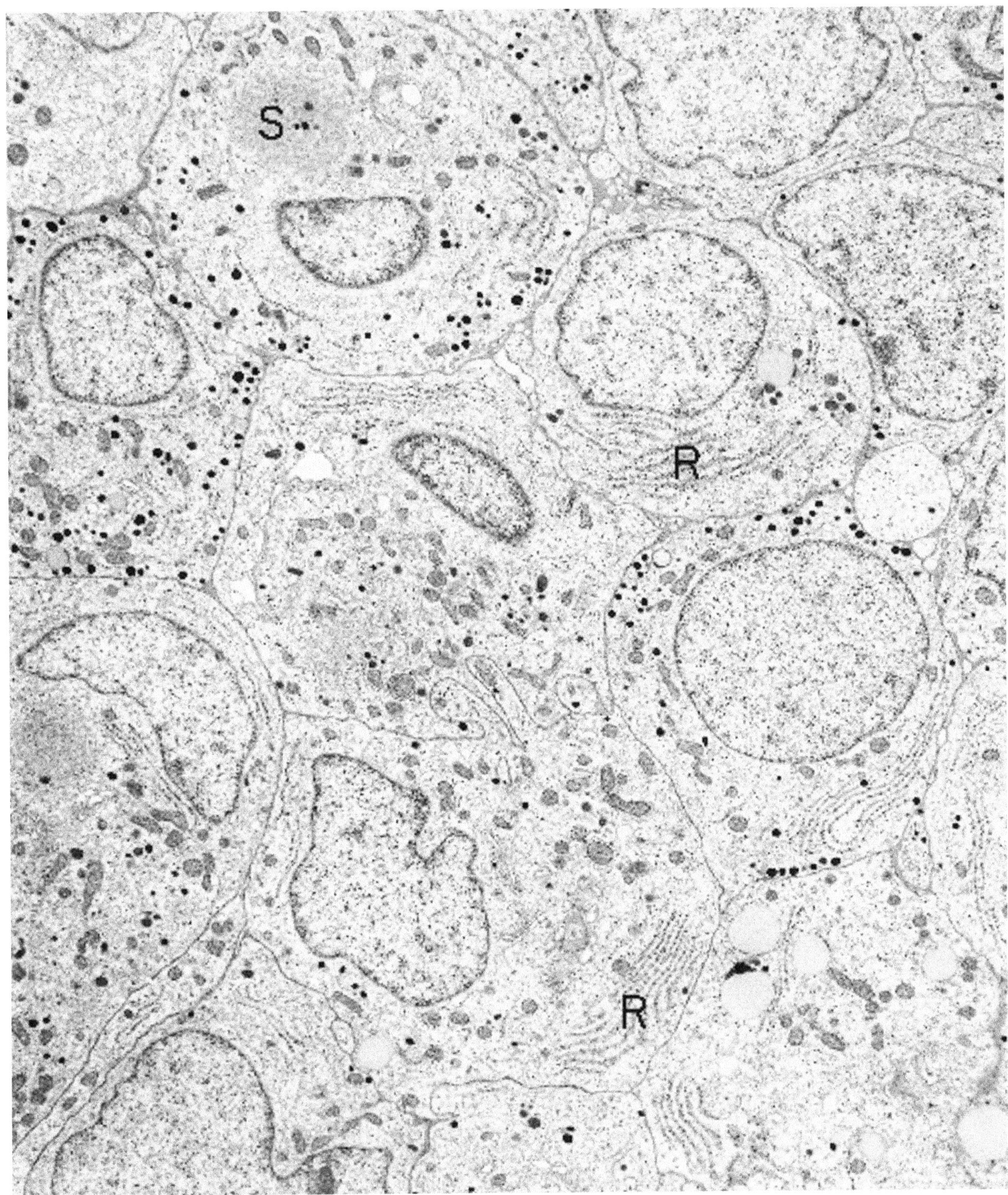

Figure 8.55. Pituitary adenoma (prolactin-secreting). At this magnification, the cells are of uniform size and shape and are characterized by their small granules, stacks of rough endoplasmic reticulum (R), and aggregates of smooth endoplasmic reticulum (S; see Figure 8.56). (× 5940)

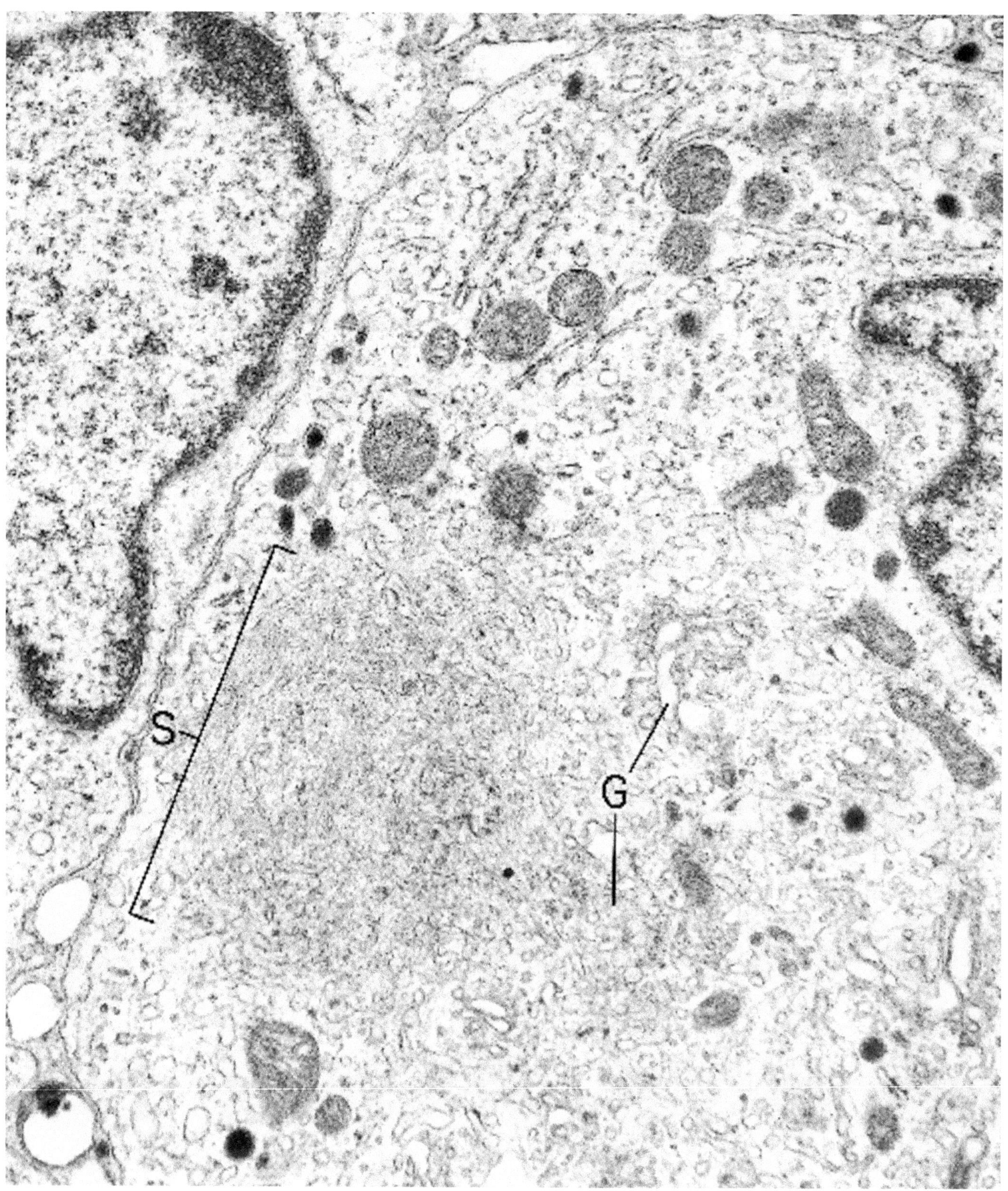

Figure 8.56. Pituitary adenoma (prolactin-secreting). Higher magnification of a cell of the same neoplasm in Figure 8.55 shows a large aggregate of smooth endoplasmic reticulum (S) as well as a prominent Golgi apparatus (G). (× 20,200)

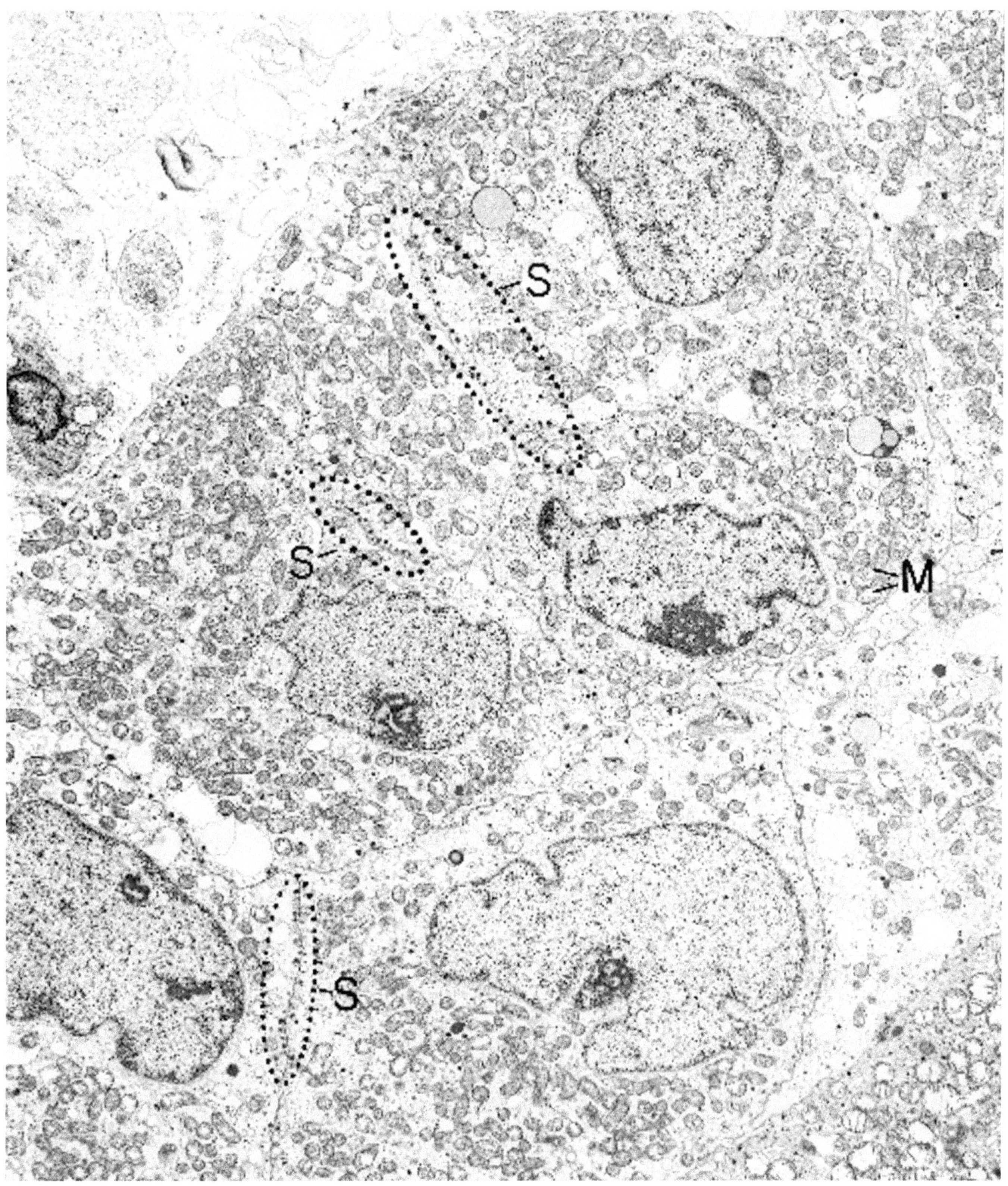

Figure 8.57. Pituitary adenoma (nonfunctioning). The cells are of the oncocytic type, having abundant and eosinophilic cytoplasm by light microscopy, and many mitochondria (M) by electron microscopy. Secretory granules also are present (S), but they are not numerous and are small and localized to the subplasmalemmal region of the cytoplasm. (× 6570)

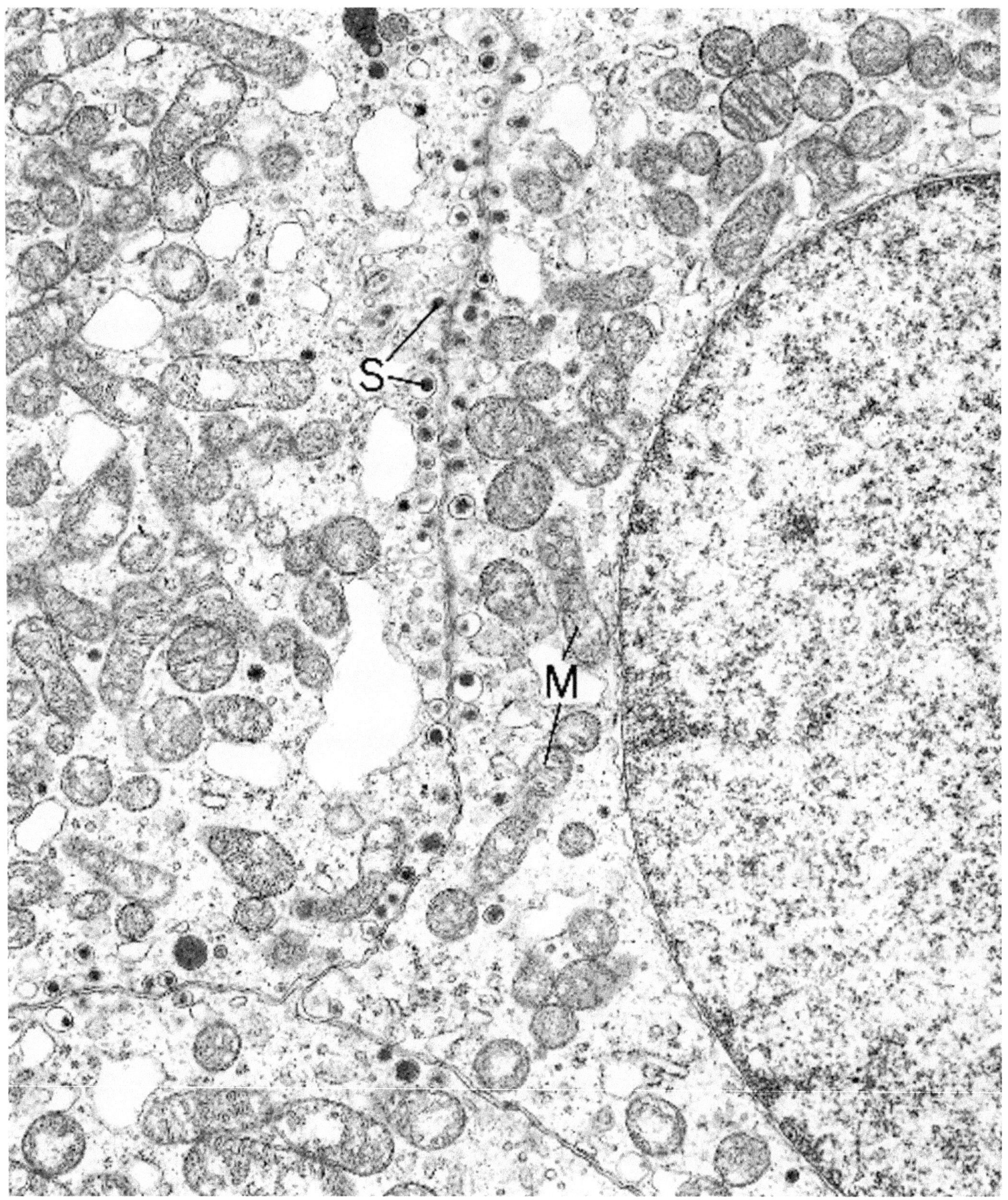

Figure 8.58. Pituitary adenoma (nonfunctioning). Higher magnification of the neoplasm shown in Figure 8.57 highlights the many mitochondria (M) and the subplasmalemmal location of the dense-core granules (S). (× 23,000)

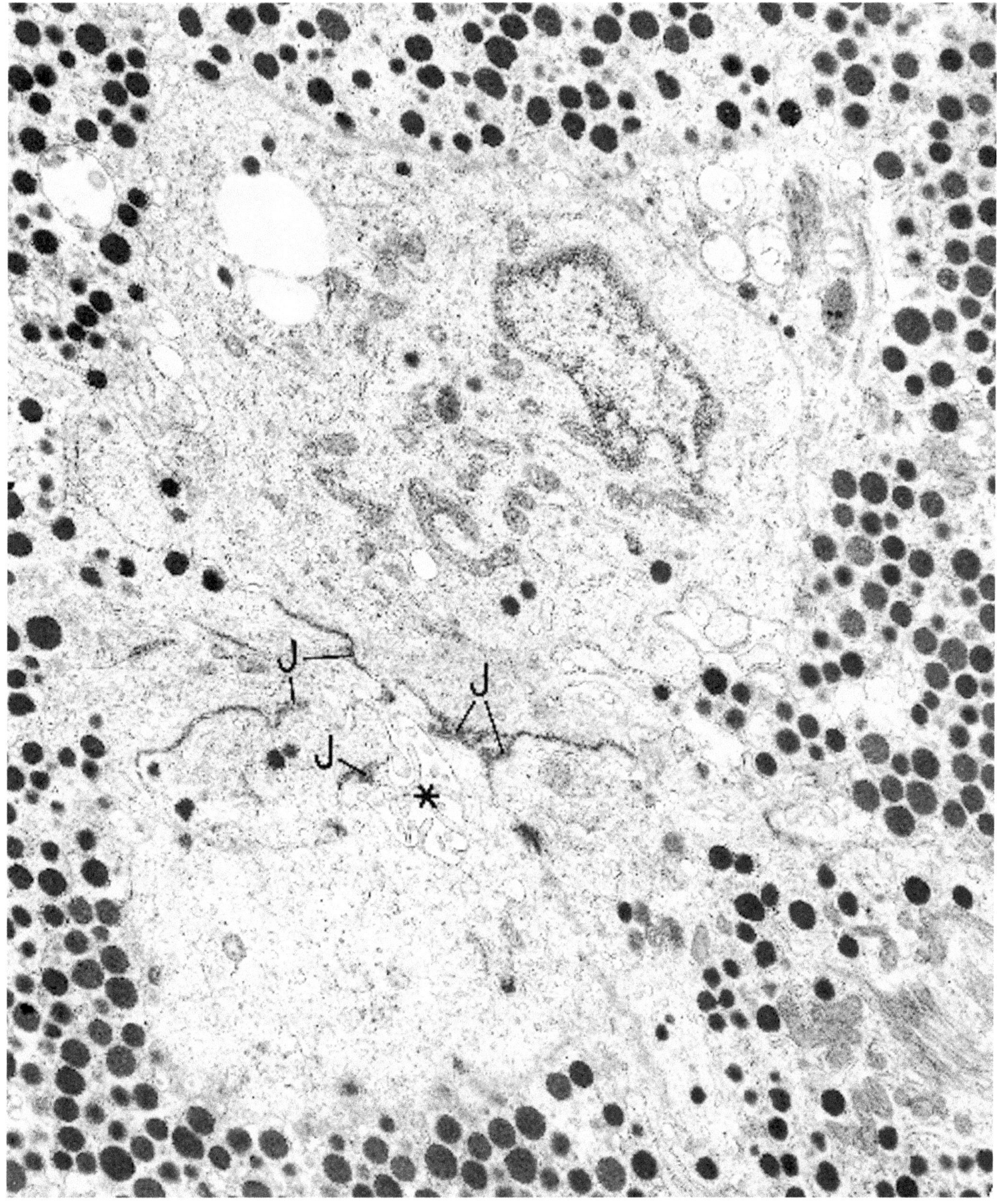

Figure 8.59. Pituitary adenoma (corticotrophic). The cells of this neoplasm focally form a villus-lined microacinus (*), readily identifiable even at low magnification by converging junctional complexes (J). (× 15,340)

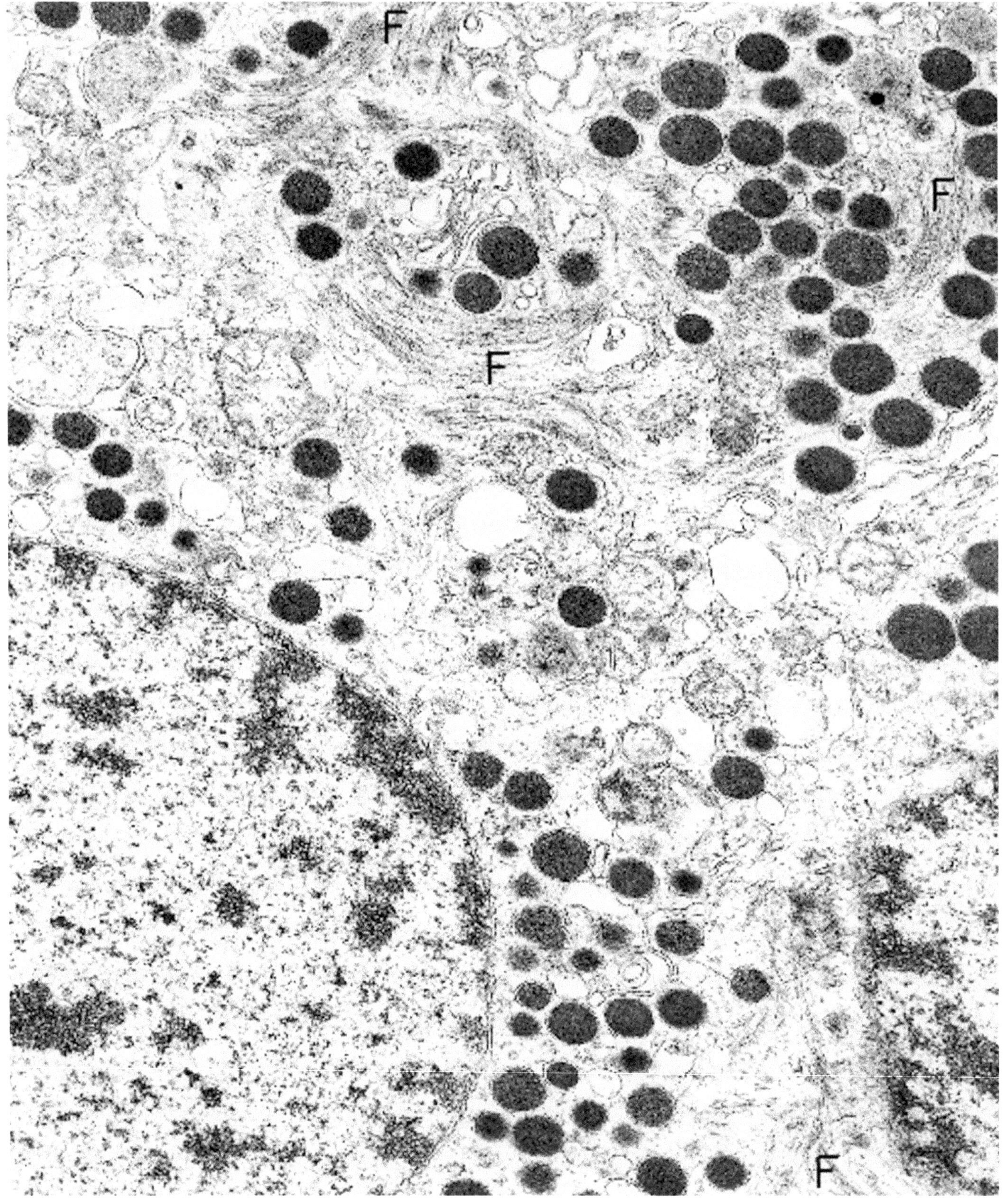

Figure 8.60. Pituitary adenoma (corticotrophic). Many of the cells of this neoplasm are similar to the one depicted, with a cytoplasm rich in filaments (F). (× 25,900)

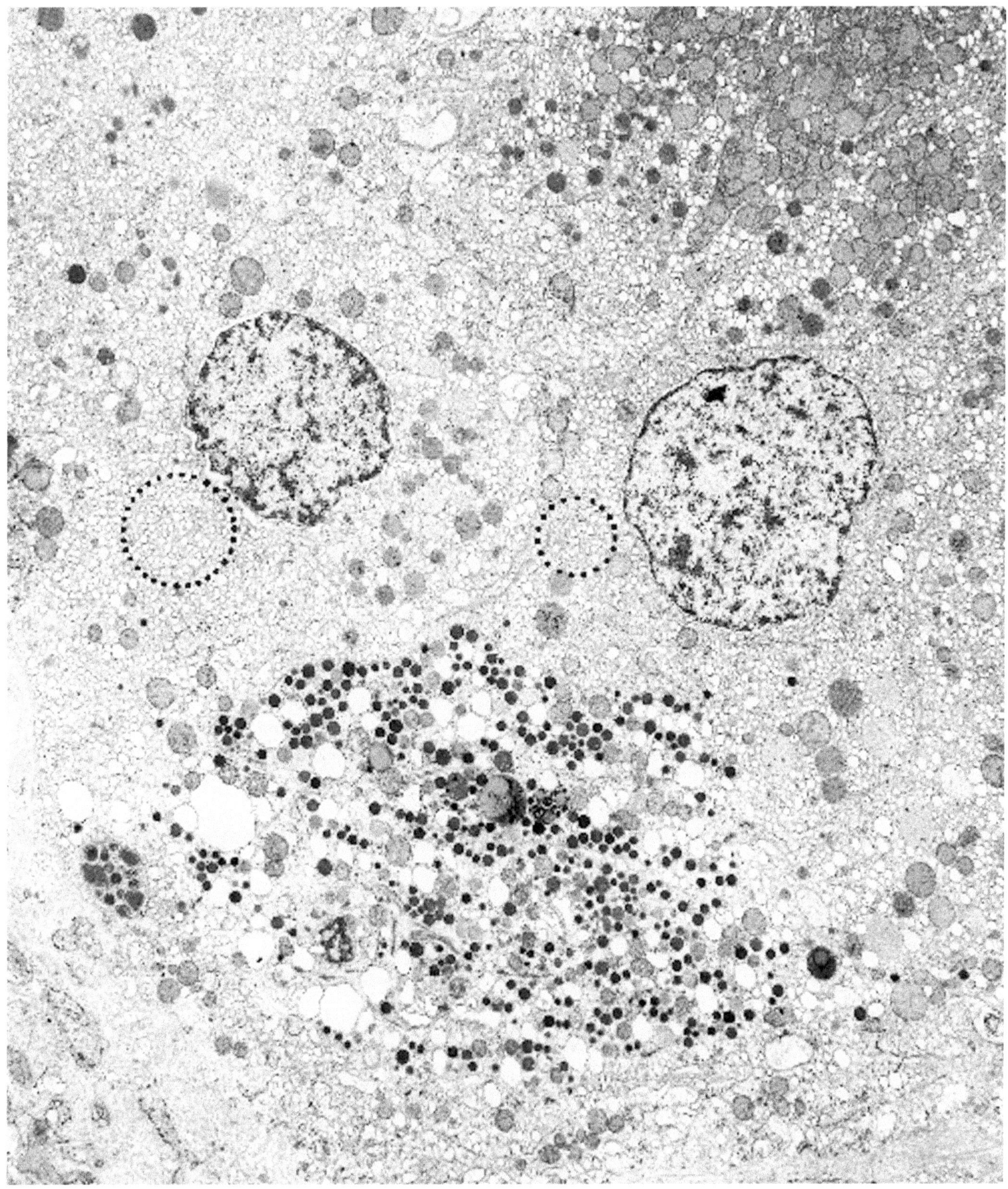

Figure 8.61. Pituitary adenoma (corticotrophic). This neoplasm appeared oncocytic at light microscopy, but ultrastructurally the cytoplasm was rich in vesicles of smooth endoplasmic reticulum (*dotted circles*) rather than mitochondria. (× 6150)

(Text continued from page 543)

REFERENCES

General

Burger PC, Scheithauer BW: *Atlas of Tumor Pathology. Tumors of the Central Nervous System,* 3rd series, fasc 10. Armed Forces Institute of Pathology, Washington, DC, 1994, p 452.

Dolman CL: *Ultrastructure of Brain Tumors and Biopsies: A Diagnostic Atlas.* Praeger, New York, 1984.

Hirano A: Some contributions of electron microscopy to the diagnosis of brain tumors. Acta Neuropathol 43: 119–128, 1978.

Kleihues P, Burger PC, Scheithauer BW: *Histological Typing of Tumors of the Central Nervous System,* 2nd ed. In Collaboration with Sobin LH and Pathologists in 14 Countries. Springer-Verlag, New York, 1993, p 112.

Peters A, Palay SL, Webster HDeF: *The Fine Structure of the Nervous System: The Neurons and Supporting Cells.* WB Saunders, Philadelphia, 1976.

Rubinstein LJ: Tumors of the central nervous system. HL Firminger, ed: *Atlas of Tumor Pathology,* 2nd series, fasc 6. Armed Forces Institute of Pathology, Washington, DC, 1972, pp 1–400.

Meningioma

Alguacil-Garvia A, Pettigrew NM, Sima AA: Secretory meningioma. A distinct subtype of meningioma. Am J Surg Pathol 10:102–111, 1986.

Cervós-Navarro J, Vasquez J: An electron microscopic study of meningiomas. Acta Neuropathol (Berl) 13: 301–323, 1969.

Humeau C, Vic P, Sentein P, et al: The fine structure of meningiomas: An attempted classification. Virchows Arch [A] 382:201–216, 1979.

Huszar M, Fanburg JC, Dickersin GR, et al: Retroperitoneal malignant meningioma. A light microscopic, immunohistochemical, and ultrastructural study. Am J Surg Pathol 20:492–499, 1996.

Jay V, Edwards V: Filamentous aggregates in a meningioma. Ultrastruct Pathol 20:577–583, 1996.

Kallem Z, Fitzpatrick MM, Ritter JH: Primary pulmonary meningioma. Report of a case and review of the literature. Arch Pathol Lab Med 121:631–636, 1997.

Kepes JJ: Electron microscopic studies of meningiomas. Am J Pathol 39:499–510, 1961.

Kubota T, Sato K, Yamamato S, et al: Ultrastructural study of the formation of psammoma bodies in fibroblastic meningioma. J Neurosurg 60:512–517, 1984.

Kubota T, Sato K, Kabuto M, et al: Clear cell (glycogen-rich) meningioma with special reference to spherical collagen deposits. Noshuyo Byori 12:53–60, 1995.

Kubota Y, Ueda T, Kagawa Y, et al: Microcystic meningioma without enhancement or neuroimaging—case report. Neurol Med Chir (Tokyo) 37:407–410, 1997.

Prinz M, Patt S, Mitrovics T, et al: Clear cell meningioma: Report of a spinal case. Gen Diagn Pathol 141: 261–267, 1996.

Saito A, Nakazato Y, Hirato J, et al: Intracytoplasmic chromophobe inclusion bodies in an anaplastic meningioma. Acta Neuropathol 93:421–425, 1997.

Tsuchida T, Matsumoto M, Shirayama Y, et al: Observation of psammoma bodies in cultured meningiomas: Analysis of three-dimensional structure using scanning and transmission electron microscopy. Ultrastruct Pathol 20:241–247, 1996.

Zorludemir S, Scheithauer BW, Hirose T, et al: Clear cell meningioma. A clinicopathologic study of a potentially aggressive variant of meningioma. Am J Surg Pathol 19:493–505, 1995.

Angioblastic Meningioma (Hemangiopericytoma)

Cappabianca P, Maiuri F, Pettinato G, et al: Hemangiopericytoma of the spinal canal. Surg Neurol 15: 298–302, 1981.

Goellner JR, Laws ER, Soule EH, et al: Hemangiopericytoma of the meninges: Mayo Clinic experience. Am J Clin Pathol 70:375–380, 1978.

Horton BC, Urich H, Rubinstein LJ, et al: The angioblastic meningioma. A reappraisal of nosological problems. J Neurol Sci 31:387–410, 1977.

Pena CE: Meningioma and intracranial hemangiopericytoma. A comparative electron microscopic study. Acta Neuropathol 39:69–74, 1977.

Pena CS: Intracranial hemangiopericytoma. Acta Neuropathol (Berl) 33:279–284, 1975.

Popoff NA, Malinin TI, Rosomoff HL: Fine structure of intracranial hemangiopericytoma and angiomatous meningioma. Cancer 34:1187–1197, 1974.

Rubinstein LJ: Relationship of angioblastic meningioma to hemangiopericytoma. In Hartmann WH, Cowan WR, eds: *Tumors of the Central Nervous System,* Supplement to *Atlas of Tumor Pathology,* 2nd series, fasc 6. Armed Forces Institute of Pathology, Washington, DC, 1982, pp S9–S13.

Astrocytoma

Duffell D, Farber L, Chou S, et al: Electron microscopic observations on astrocytomas. Am J Pathol 43:539–554, 1963.

Ebhart G, Cervos-Navarro J: The fine structure of cells of astrocytomas of various grades of malignancy. Acta Neuropathol 7:88–90, 1981.

Hirose T, Scheithauer BW, Lopes MB, et al: Tuber and subependymal giant cell astrocytoma associated with tuberous sclerosis: An immunohistochemical, ultrastructural, and immunoelectron and microscopic study. Acta Neuropathol 90:387–399, 1995.

Hosokowa Y, Tsuchihashi Y, Okabe H, et al: Pleomorphic xanthoastrocytoma. Ultrastructural, immunohistochemical, and DNA cytofluorometric study of a case. Cancer 68:853–859, 1991.

Hossmann KA, Wechsler W: Ultrastructural cytopathology of human cerebral gliomas. Oncology 25: 455–480, 1971.

Iwaki T, Fukui M, Kondo A, et al: Epithelial properties of pleomorphic xanthoastrocytomas determined in ultrastructural and immunohistochemical studies. Acta Neuropathol 74:142–150, 1987.

Kubota T, Hirano A, Sato K, et al: The fine structure of astroblastoma. Cancer 55:745–750, 1985.

Kuhajda FP, Mendelsohn G, Taxy JB, et al: Pleomorphic xanthoastrocytoma: Report of a case with light and electron microscopy. Ultrastruct Pathol 2:25–32, 1981.

Rubinstein LJ, Herman MM: The astroblastoma and its possible cytogenic relationship to the tanycyte. An electron microscopic, immunohistochemical, tissue- and organculture study. Acta Neuropathol 78:472–483, 1989.

Scheithauer BW, Bruner JM: The ultrastructural spectrum of astrocytic neoplasms. Ultrastruct Pathol 11: 535–581, 1987.

Weldon-Linne CM, Victor TA, Groothuis DR, et al: Pleomorphic xanthoastrocytoma. Ultrastructural and immunohistochemical study of a case with a rapidly fatal outcome following surgery. Cancer 52:2055–2063, 1983.

Oligodendroglioma

Cenacchi G, Giangaspero F, Cerasoli S, et al: Ultrastructural characterization of oligodendroglial-like cells in central nervous system tumors. Ultrastruct Pathol 20:537–547, 1996.

Cervos-Navarro J, Pehlivan N: Ultrastructure of oligodendrogliomas. Acta Neuropathol 7:91–93, 1981.

Chorneyko K, Maguire, J, Simon GT: Oligodendroglioma with granular cells: A case report. Ultrastruct Pathol 22:79–82, 1998.

Garcia JH, Lemmi H: Ultrastructure of oligodendroglioma of the spinal cord. Am J Clin Pathol 54: 757–765, 1970.

Hossmann KA, Wechsler W: Ultrastructural cytopathology of human cerebral gliomas. Oncology 25: 455–480, 1971.

Kawano N, Yada K, Aihara M, et al: Oligodendroglioma-like cells (clear cells) in ependymoma. Acta Neuropathol 62:141–144, 1983.

Luse SA: Electron microscopic studies of brain tumors. Neurology 10:881–905, 1960.

Min KW, Scheithauer BW: Oligodendroglioma: the ultrastructural spectrum. Ultrastruct Pathol 18:47–60, 1994.

Raimondi AJ: Ultrastructure and the biology of human brain tumors. In Krayenbuhl H, Maspes PE, Sweet WH, eds: *Progress in Neurological Surgery,* vol. 1, Year Book, Chicago, 1966, pp 1–63.

Sarkar C, Roy S, Tandon PN: Oligodendroglial tumors. An immunohistochemical and electron microscopic study. Cancer 61:1862–1866, 1988.

Slowik F, Jellinger K, Gaszo L, et al: Gliosarcomas: Histological, immunohistochemical, ultrastructural, and tissue culture studies. Acta Neuropathol 67:201–210, 1985.

Takei Y, Mirra SS, Miles ML: Eosinophilic granular cells in oligodendrogliomas: An ultrastructural study. Cancer 38:1968–1976, 1976.

Ependymoma (and Subependymoma)

Goebel HH, Carvioto H: Ultrastructure of human and experimental ependymomas. J Neuropathol Exp Neurol 31:54–71, 1972.

Guccion JG, Saini N: Ependymoma: Ultrastructural studies of two cases. Ultrastruct Pathol 15:159–166, 1991.

Hirano A, Ghatak NR, Zimmerman HM: The fine structure of ependymoblastoma. J Neuropathol Exp Neurol 32:144–152, 1973.

Kawano N, Yada K, Yagashita S: Clear cell ependymoma. A histological variant with diagnostic implications. Virchows Arch [A] 415:467–472, 1989.

Kindblom LG, Lodding P, Hagmar B, et al: Metastasizing myxopapillary ependymoma of the sacrococcygeal region. A clinico-pathologic, light- and electron microscopic, immunohistochemical, tissue culture, and cytogenetic analysis of a case. Acta Pathol Microbiol Immunol Scand [A] 94:79–90, 1986.

Liu HM, McLone DG, Clark S: Ependymomas of childhood (II). Electron-microscopic study. Childs Brain 3:281–296, 1977.

Rawlinson DG, Herman M, Rubinstein LJ: The fine structure of a myxopapillary ependymoma of the filum terminale. Acta Neuropathol (Berl) 25:1–13, 1973.

Rubinstein LJ: The definition of the ependymoblastoma. Arch Pathol 90:35–45, 1970.

Specht CS, Smith TW, DeGirolami U, et al: Myxopapillary ependymoma of the filum terminale. A light and electron microscopic study. Cancer 58:310–317, 1986.

Stern JB, Helwig EB: Ultrastructure of subcutaneous sacrococcygeal myxopapillary ependymoma. Arch Pathol Lab Med 105:524–526, 1981.

Takei Y, Manuelidis EE: Electron microscopic study of human ependymomas [abstract]. J Neuropathol Exp Neurol 33:179, 1975.

Choroid Plexus Neoplasms

Aguilar D, Martin JM, Aneiros J, et al: The fine structure of choroid plexus carcinoma. Histopathology 7:939–946, 1983.

Carter LP, Beggs J, Waggener JD: Ultrastructure of three choroid plexus papillomas. Cancer 30:1130–1136, 1972.

Diengdoh JV, Shaw MD: Oncocytic variant of choroid plexus papilloma. Evolution from benign to malignant "oncocytoma." Cancer 71:855–858, 1993.

Dohrmann GJ, Bucy PC: Human choroid plexus. A light and electron microscopic study. J Neurosurg 33: 506–516, 1970.

Hamilton RL: Case of the month. May 1996—hydrocephalus in a 9 month old infant. Brain Pathol 6: 533–534, 1996.

Masuzawa T, Shimabukuru H, Yoshimizi N, et al: Ultrastructure of a disseminated choroid plexus papilloma. Acta Neuropathol 54:352–344, 1981.

Matsushima T: Choroid plexus papillomas and human choroid plexus: A light and electron microscopic study. J Neurosurg 59:1054–1062, 1983.

McComb RD, Burger PC: Choroid plexus carcinoma. A report of a case with immunohistochemical and ultrastructural observations. Cancer 51:470–475, 1983.

Nakashima N, Goto K, Tsukidate K, et al: Choroid plexus papilloma. Light and electron microscopic study. Virchows Arch [A] 400:201–211, 1983.

Navas JJ, Battifora H: Choroid plexus papilloma: Light and electron microscopic study of three cases. Acta Neuropathol (Berl) 44:235–239, 1978.

Reimund EL, Sitton JE, Harkin JC: Pigmented choroid plexus papilloma. Arch Pathol Lab Med 114:902–905, 1990.

Vajtai I, Varga Z, Bodosi M, et al: Melanotic papilloma of the choroid plexus: Report of a case with implications for pathogenesis. Noshuyo Byori 12:151–154, 1995.

Neurocytoma

Diraz A, Tada M, Sawamura Y, et al: Dystrophic axonal formation (spheroid body) in central neurocytoma. Case report. Neurol Med Chir (Tokyo) 33:568–571, 1993.

Hassoun J, Gambarelli D, Grisoli F, et al: Central neurocytoma. Acta Neuropathol (Berl) 56:151–156, 1982.

Ishiuchi S, Tamura M: Central neurocytoma: An immunohistochemical, ultrastructural and cell culture study. Acta Neuropathol 94:425–435, 1997.

Louis DN, Swearingen B, Linggood RM, et al: Central nervous system neurocytoma and neuroblastoma in adults. Report of eight cases. J Neurooncol 9:231–238, 1990.

Maiuri F, Spaziante R, DeCaro ML, et al: Central neurocytoma: Clinico-pathological study of 5 cases and review of the literature. Clin Neurol Neurosurg 97: 219–228, 1995.

Miyagami M, Nakamura S: Ultrastructure of microvessels and tumor cell processes on central neurocytoma. Brain Tumor Pathol 14:79–83, 1997.

Nishio S, Tashima T, Takeshita I, et al: Intraventricular neurocytoma: Clinicopathological features of six cases. J Neurosurg 68:665–670, 1988.

Wilson AJ, Leaffer DH, Kohout ND: Differentiated cerebral neuroblastoma: A tumor in need of discovery. Hum Pathol 16:647–649, 1985.

Medulloblastoma

Alwasiak J, Papierz W, Kolasa P, et al: Ultrastructure of the primitive neuroectodermal tumors (PNET). Polish J Pathol 45:129–138, 1994.

Bhattacharjee M, Hicks J, Langford L, et al: Central nervous system atypical teratoid/rhabdoid tumors of infancy and childhood. Ultrastruct Pathol 21:369–378, 1997.

Boesel CP, Suhan JP, Sayers MP: Melanotic medulloblastoma. Report of a case with ultrastructural finding. J Neuropathol Exp Pathol 37:531–543, 1978.

Camins MB, Cravioto HM, Epstein F, et al: Evidence for astrocytic and neuronal differentiation. Neurosurgery 6:398–411, 1980.

Davis DG, Wilson D, Schmitz M, et al: Lipidized medulloblastoma in adults. Hum Pathol 24:990–995, 1993.

Giangaspero F, Cenacchi G, Roncaroli F, et al: Medullocytoma (lipidized medulloblastoma). A cerebellar neoplasm of adults with favorable prognosis. Am J Surg Pathol 20:656–664, 1996.

Gulotta F, Kersting G: The ultrastructure of medulloblastoma in tissue culture. Virchows Arch [A] 356: 111–118, 1972.

Holl T, Kleihues P, Yasargil MG, et al: Cerebellar medullomyoblastoma with advanced neuronal differentiation and hamartomatous component. Acta Neuropathol 82:408–413, 1991.

Jay V, Zielenska M, Lorenzana A, et al: An unusual cerebellar primitive neuroectodermal tumor with t(11;22) translocation: Pathological and molecular analysis. Pediatr Pathol Lab Med 16:119–128, 1996.

Jay V, Edwards V, Halliday W, et al: "Polyphenotypic" tumors in the central nervous system: Problems in nosology and classification. Pediatr Pathol Lab Med 17:369–389, 1997.

Katsetos CD, Mei-Liu H, Zacks SI: Immunohistochemical and ultrastructural observations on Homer Wright (neuroblastic) rosettes and the "pale islands" of human cerebellar medulloblastomas. Hum Pathol 19: 1212–1227, 1988.

Matakas F, Cervós-Navarro J, Gulotta F: The ultrastructure of medulloblastomas. Acta Neuropathol (Berl) 16:271–284, 1970.

Moss TH: Evidence for differentiation in medulloblastomas appearing primitive on light microscopy: An ultrastructural study. Histopathology 7:919–930, 1983.

Orlandi A, Marino B, Brunori M, et al: Lipomatous medulloblastoma. Clin Neuropathol 16:175–179, 1997.

Papierz W, Alwasiak J, Kolasa P, et al: Primitive neuroectodermal tumors: Ultrastructural and immunohistochemical studies. Ultrastruct Pathol 19:147–166, 1995.

Rorke LB: The cerebellar medulloblastoma and its relationship to primitive neuroectodermal tumors. J Neuropathol Exp Neurol 42:1–15, 1983.

Roy S: An ultrastructural study of medulloblastoma. Neurol Ind 25:226–229, 1977.

Smith TW, Davidson RI: Medullomyoblastoma. A histologic, immunohistochemical and ultrastructural study. Cancer 54:323–332, 1984.

Walter GF, Brucher JM: Ultrastructural study of medulloblastoma. Acta Neuropathol (Berl) 48:211–214, 1979.

Pineocytoma and Pineoblastoma

Hassoun J, Gambarelli D, Pellissier JF, et al: Germinomas of the brain. Light and electron microscopic study. A report of seven cases. Acta Neuropathol 7:105–108, 1981.

Hassoun J, Gambarelli D, Peragut JC, et al: Specific ultrastructural markers of human pinealomas: A study of four cases. Acta Neuropathol 62:31–40, 1983.

Hassoun J, Devictor B, Gambarelli D, et al: Paired twisted filaments: A new ultrastructural marker of human pinealomas? Acta Neuropathol 65:163–165, 1984.

Kees UR, Spagnolo D, Hallam LA, et al: A new pineoblastoma cell line, PER-480 with der(10)t(10;17), der(16)t(1;16), and enhanced MYC expression in the absence of gene amplification. Cancer Genet Cytogenet 100:159–164, 1998.

Kline KT, Damjanov I, Katz SM, et al: Pineoblastoma: An electron microscopic study. Cancer 44:1692–1699, 1979.

Markesbery WR, Brooks WH, Milson L, et al: Ultrastructural study of the pineal germinoma in vivo and in vitro. Cancer 37:327–337, 1976.

Markesbery WR, Haugh RM, Young AB: Ultrastructure of pineal parenchymal neoplasms. Acta Neuropathol 55:143–149, 1981.

Nielson SL, Wilson SB: Ultrastructure of a "pineocytoma." J Neuropathol Exp Neurol 34:148–158, 1975.

Ramsey HJ: Ultrastructure of a pineal tumor. Cancer 18:1014–1025, 1965.

Germ Cell Tumors

Ejeckam G, Norman MG, Ivan LP: Case report and ultrastructural study of intracranial embryonal carcinoma. Can J Neurol Sci 5:447–450, 1978.

Hassoun J, Gambarelli D, Choux M, et al: Macrophagic activity in intracerebral germinoma. Human Pathol 11:207–210, 1980.

Markesbery WR, Brooks WH, Milsow L, et al: Ultrastructural study of pineal germinoma *in vivo* and *in vitro*. Cancer 37:327–337, 1976.

Matsutani M: Ultrastructural study of pinealoma in view of its histogenesis. Brain Nerv 28:41–56, 1976.

Ramsey HJ: Ultrastructure of a pineal tumor. Cancer 18:1014–1025, 1965.

Takei Y, Pearl GS: Ultrastructural study of intracranial yolk sac tumor: With special reference to the oncologic phylogeny of germ cell tumors. Cancer 48:2038–2046, 1981.

Tani E, Ikeda K, Kudo S, et al: Specialized intercellular junctions in human intracranial germinomas. Acta Neuropathol (Berl) 27:139–151, 1974.

(See additional references under Chapter VII, "Gonadal and Related Neoplasms")

Hemangioblastoma

Brodkey JA, Buchignani JA, O'Brien TF: Hemangioblastoma of the radial nerve: Case report. Neurosurgery 36:198–200, 1995.

Cancilla PA, Zimmerman HM: The fine structure of a cerebellar hemangioblastoma. J Neuropathol Exp Neurol 24:621–628, 1965.

Chaudhry AP, Montes M, Cohn GA: Ultrastructure of a cerebral hemangioblastoma. Cancer 42:1834–1850, 1978.

Gokden M, Roth KA, Carroll SL, et al: Clear cell neoplasms and pseudoneoplastic lesions of the central nervous system. Semin Diagn Pathol 14:253–269, 1997.

Ho KL: Ultrastructure of cerebellar capillary hemangioblastoma. I. Weibel-Palade bodies and stromal cell histogenesis. J Neuropathol Exp Neurol 43:592–608, 1984.

Ishwar S, Taniguchi RM, Vogel FS: Multiple supratentorial hemangioblastomas. Case study and ultrastructural characteristics. J Neurosurg 25:396–405, 1971.

Kawamura J, Garcia JH, Kanijyo Y: Cerebellar hemangioblastoma: Histogenesis of stromal cells. Cancer 21: 1528–1540, 1973.

Omulecka A, Lach B, Alwasiak J, et al: Immunohistochemical and ultrastructural studies of stromal cells in hemangioblastoma. Folia Neuropathol 33:41–50, 1995.

Pituitary Adenomas

Hampton TA, Scheithauer BW, Rojiani AM, et al: Salivary gland-like tumors of the sellar region. Am J Surg Pathol 21:424–434, 1997.

Horvath E, Kovacs K: Ultrastructural classification of pituitary adenomas. Can J Neurol Sci 3:9–21, 1976.

Horvath E, Kovacs K, Killinger DW, et al: Silent corticotroph adenomas of the human pituitary gland: A histologic, immunocytologic and ultrastructural study. Am J Pathol 98:617–638, 1980.

Ikeda H, Yoshimoto T, Ogawa Y, et al: Clinico-pathological study of Cushing's disease with large pituitary adenoma. Clin Endocrinol 46:669–679, 1997.

Iwase T, Nishizawa S, Baba S, et al: Intrasellar neuronal choristoma associated with growth hormone-producing pituitary adenoma containing amyloid deposits. Hum Pathol 26:925–928, 1995.

Kojima Y, Arita J, Kuwana N: Mammosomatotroph adenoma cells secrete both growth hormone and prolactin. Neurol Med Chir 35:791–896, 1995.

Kovacs K, Horvath E, Corenblum B, et al: Pituitary chromophobe adenomas consisting of prolactin cells: A histologic, immunocytological and electron microscopic study. Virchows Arch [A] 366:113–123, 1975.

Lack B, Rippstein P, Benott BG, et al: Differentiating neuroblastoma of pituitary gland: Neuroblastic transformation of epithelial adenoma cells. Case report. J Neurosurg 85:953–960, 1996.

Robert F: Electron microscopy of human pituitary tumors. In Tindall GT, Collins WF, eds: *Clinical Management of Pituitary Disorders*. Raven Press, New York, 1979, pp 113–131.

Robert F, Hardy J: Prolactin-secreting adenomas: A light and electron microscopical study. Arch Pathol Lab Med 99:625–633, 1975.

Robert F, Pelletier G, Hardy J: Pituitary adenomas in Cushing's disease: Histologic, ultrastructural and immunocytochemical study. Arch Pathol Lab Med 102: 448–455, 1978.

Sakaguchi H, Koshyama H, Sano T, et al: A case of nonfunctioning pituitary adenoma resembling so-called silent corticotroph adenoma. Endocr J 44:329–333, 1997.

Scheithauer BW: Surgical pathology in the pituitary: The adenomas (part I). In Sommers SC, Rosen PP, eds: *Pathology Annual,* Part 1, 1984, pp 317–374.

Scheithauer BW: Surgical pathology in the pituitary: The adenomas (part II). In Sommers SC, Rosen PP, eds: *Pathology Annual,* part 2, 1984, pp 269–329.

Wong K, Raisanen J, Taylor SL, et al: Pituitary adenoma as an unsuspected clival tumor. Am J Surg Pathol 19:900–903, 1995.

9

Miscellaneous Neoplasms

Neuroendocrine Neoplasms

See also sections on neuroblastoma, Merkel cell carcinoma, and oat cell carcinoma in Chapter 4.

The ultrastructural common denominator of neuroendocrine neoplasms is the presence of cytoplasmic dense-core (neurosecretory) granules. Some of the tumors, such as neuroblastoma, medullary carcinoma of the thyroid, pheochromocytoma, extra-adrenal paraganglioma, and Merkel cell tumor are derived from neural crest cells. Other neuroendocrine neoplasms, such as pulmonary and gastrointestinal carcinoids and pancreatic islet cell tumors, are derived from endodermal cells.

Carcinoid/Islet Cell Neoplasms

(Figures 9.1 through 9.8.)

Diagnostic criteria. (1) Oval and/or spindle-shaped cells, variably with long processes; (2) usually an insular arrangement of cells with basal lamina surrounding the islands; (3) islands usually solid but may have lumens with tight junctions and microvilli; (4) intermediate intercellular junctions and desmosomes; (5) numerous cytoplasmic dense-core granules; (6) variable intermediate filaments.

Additional points. Usually, carcinoids and islet cell tumors pose no problem in identification because of their many dense-core granules. However, some tumors, such as atypical carcinoids, are less well differentiated and may contain few granules, making diagnosis more difficult. Carcinoid tumors derived from foregut have small, round dense-core granules; those from midgut have predominantly larger, pleomorphic granules; and tumors from the hindgut have round small and large granules. Pancreatic islet cells of the alpha-1 (delta) and alpha-2 types have round dense-core granules of overlapping size and cannot be reliably distinguished from one another. Beta cells, on the other hand, have both nonspecific type granules and very distinctive ones with angular crystalline cores (Figures 9.7 and 9.8).

Some neoplasms in this group are composed of a combination of endocrine and exocrine cells, as in adenocarcinoids of the gastrointestinal tract. Furthermore, both endocrine and exocrine differentiation may be found within individual (amphicrine) cells, exemplified in neoplasms of various organs, such as gastrointestinal (GI) and respiratory tracts.

(Text continues on page 569)

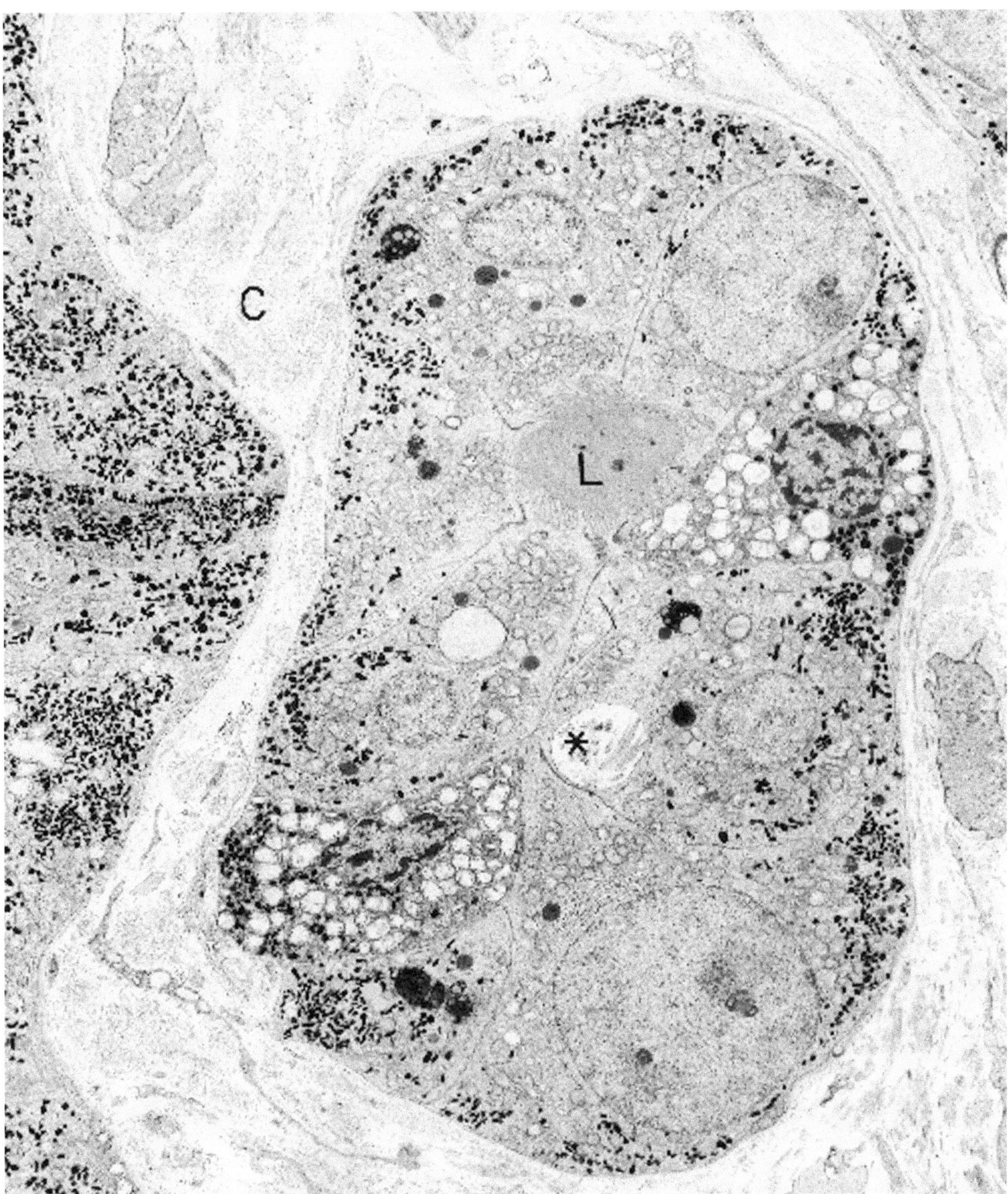

Figure 9.1. Carcinoid tumor (ileum). Islands of polygonal cells are dispersed in a matrix of collagen (C). A microlumen (L) is present in the island in the center of the field, and one cell contains a cytoplasmic pseudolumen (*). Most conspicuous are the many electron-dense granules that occupy the basal cytoplasm of the cells. (× 5500)

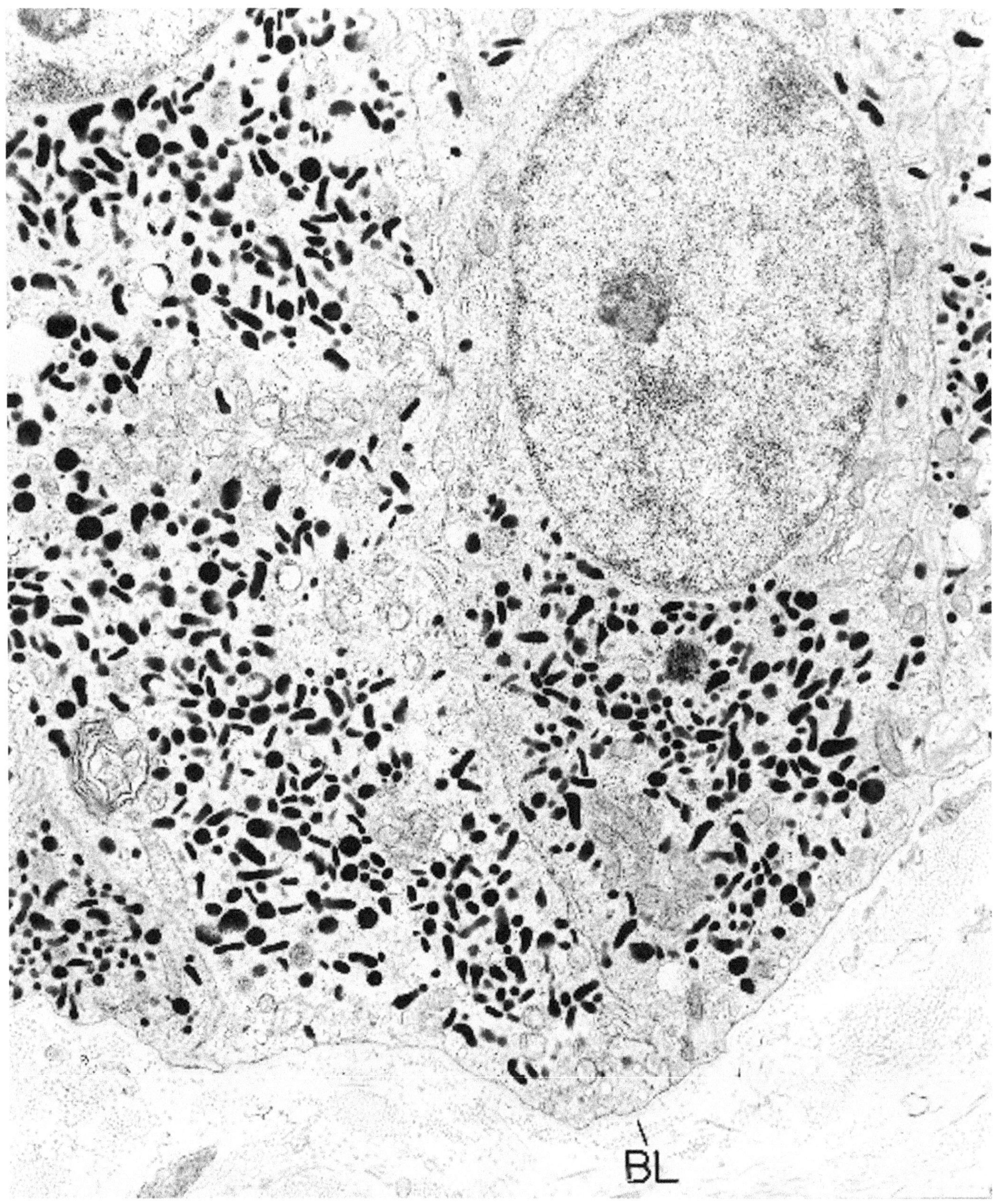

Figure 9.2. Carcinoid tumor (ileum). Higher magnification of the neoplasm illustrated in Figure 9.1 reveals the large and pleomorphic nature of the dense-core granules. A discrete basal lamina (BL) around the island of cells also is visible. (× 14,750)

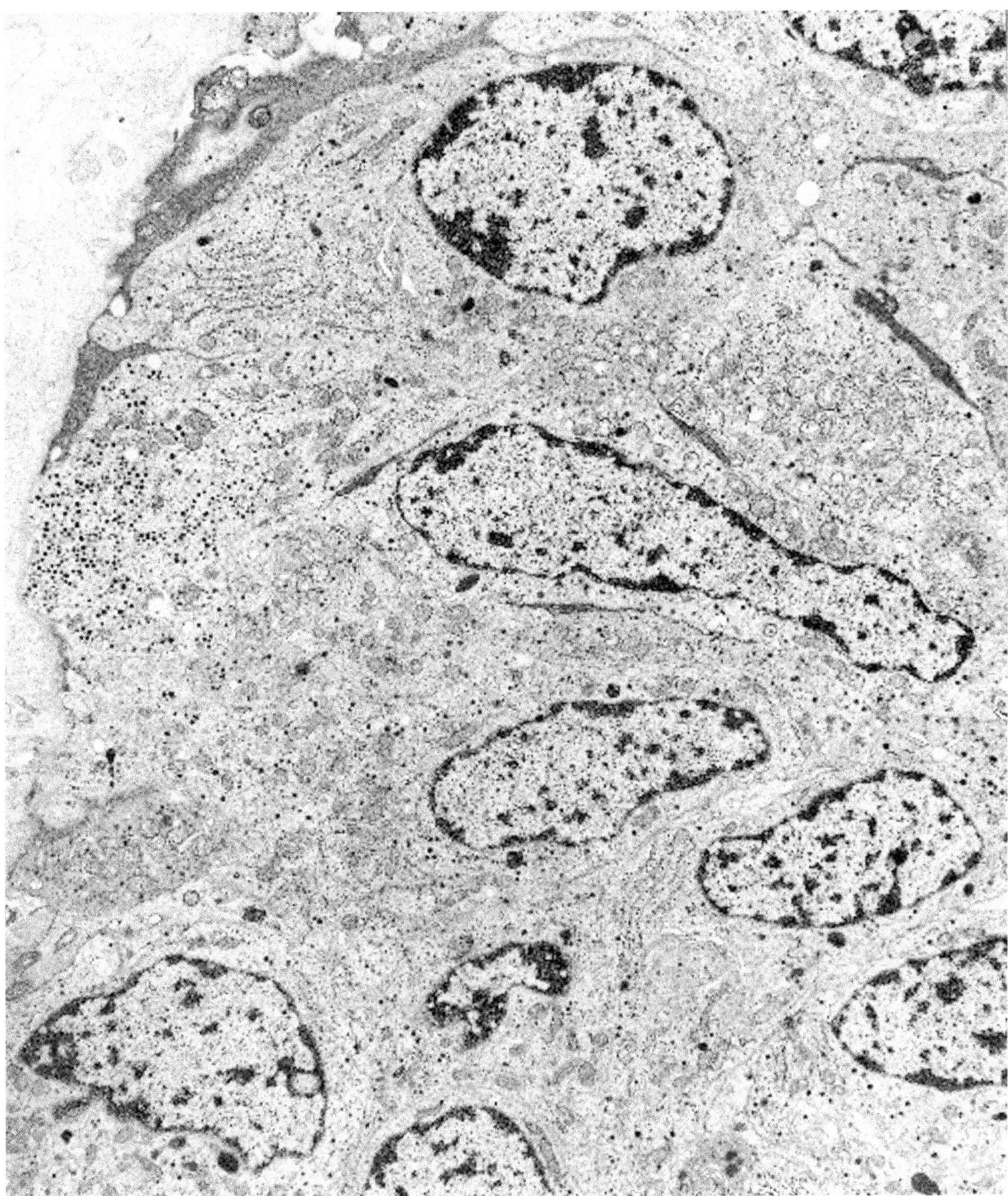

Figure 9.3. Carcinoid tumor (bronchus). An island of polygonal and elongated cells depicts many dense-core granules, and the granules in this foregut-derived neoplasm are smaller, more frequently round, and less pleomorphic than those of the midgut carcinoid illustrated in Figure 9.2. (× 7300)

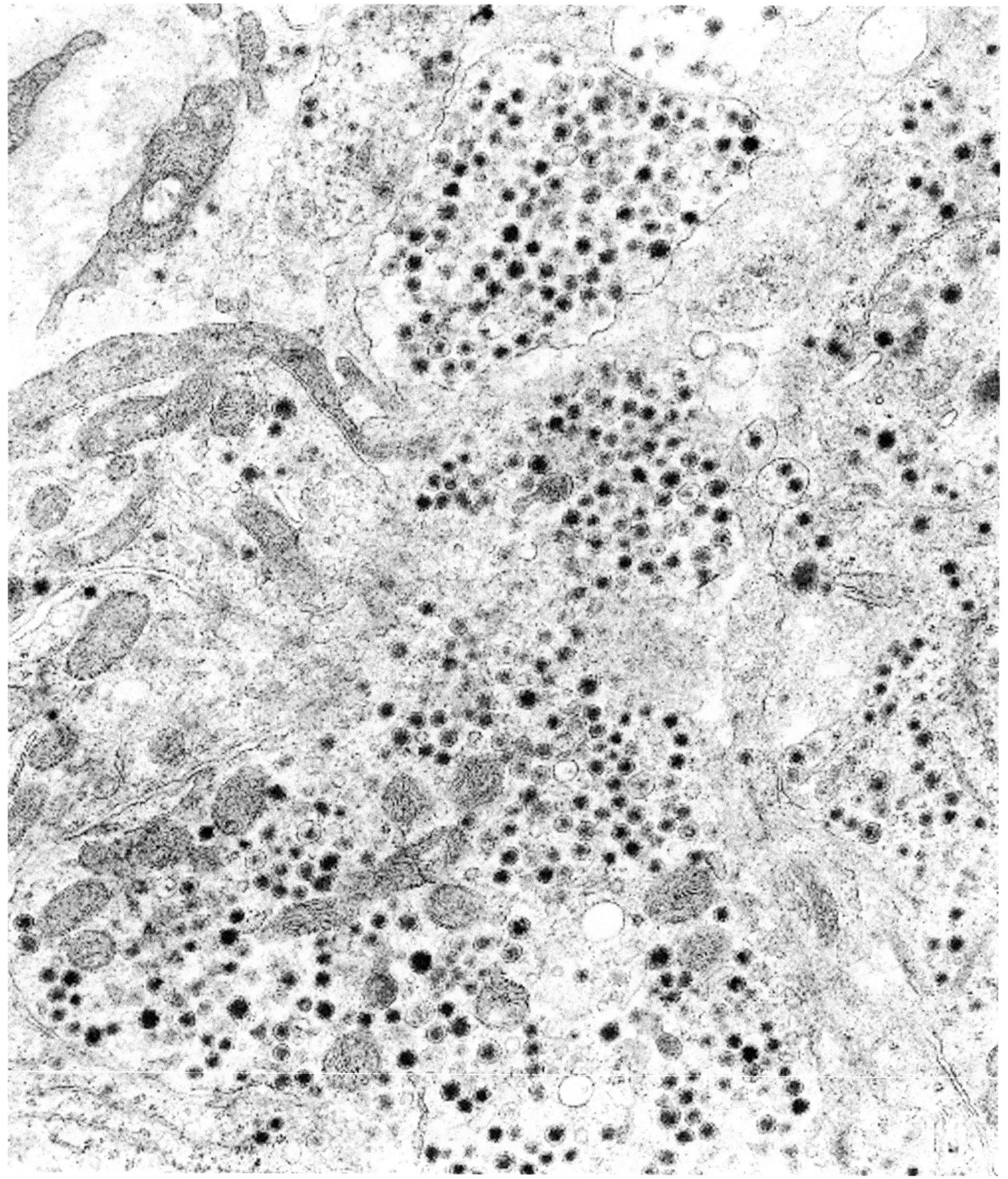

Figure 9.4. Carcinoid tumor (bronchus). Higher magnification of the neoplasm shown in Figure 9.3 highlights the round, uniformly sized, dense-core granules that typify foregut-derived carcinoid tumors. (× 21,870)

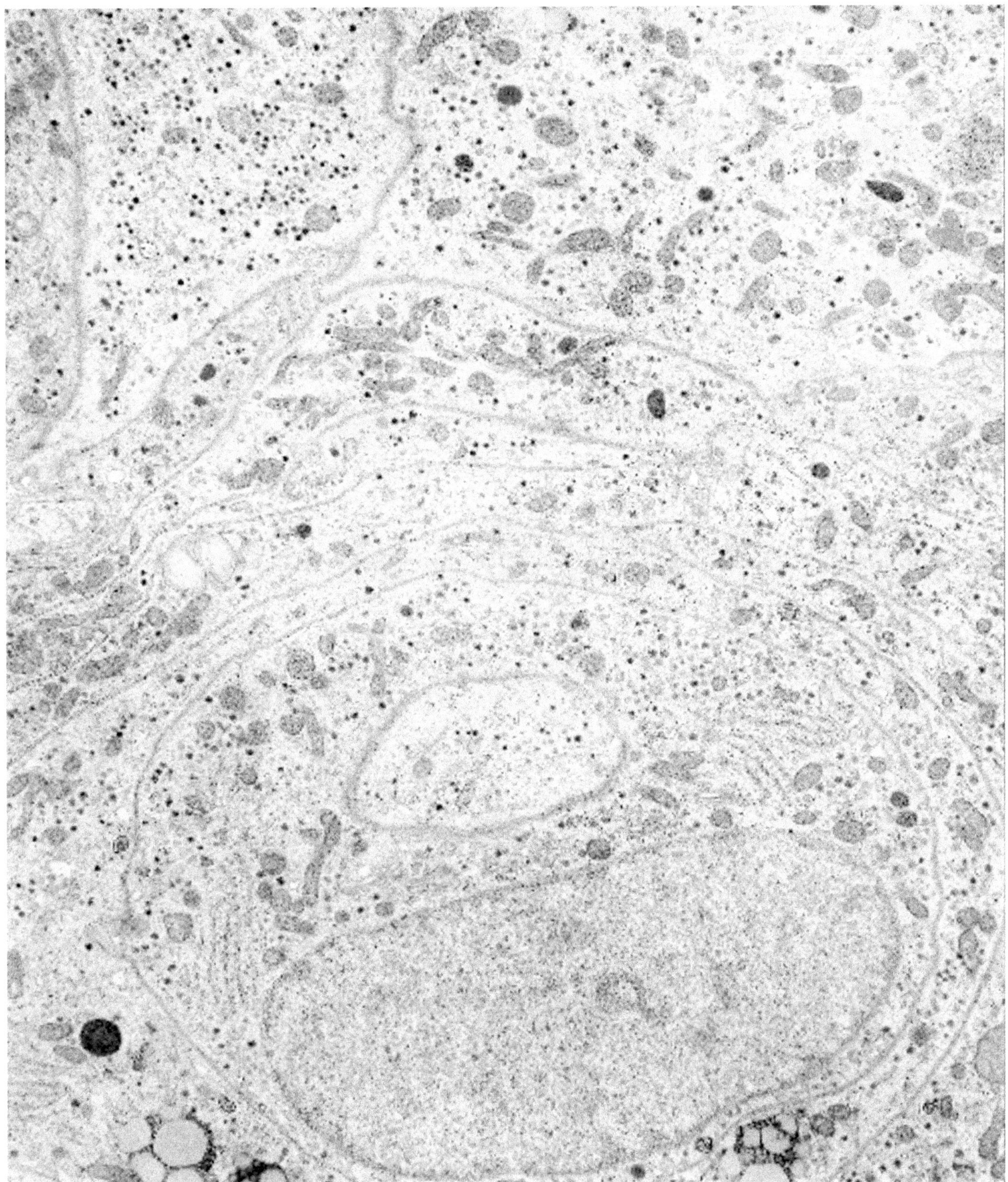

Figure 9.5. Carcinoid tumor, spindle-cell type (bronchus). The neoplastic cells have long, narrow processes that run parallel and in whorls. The diagnostic dense-core granules abound in all the cells. (× 10,260)

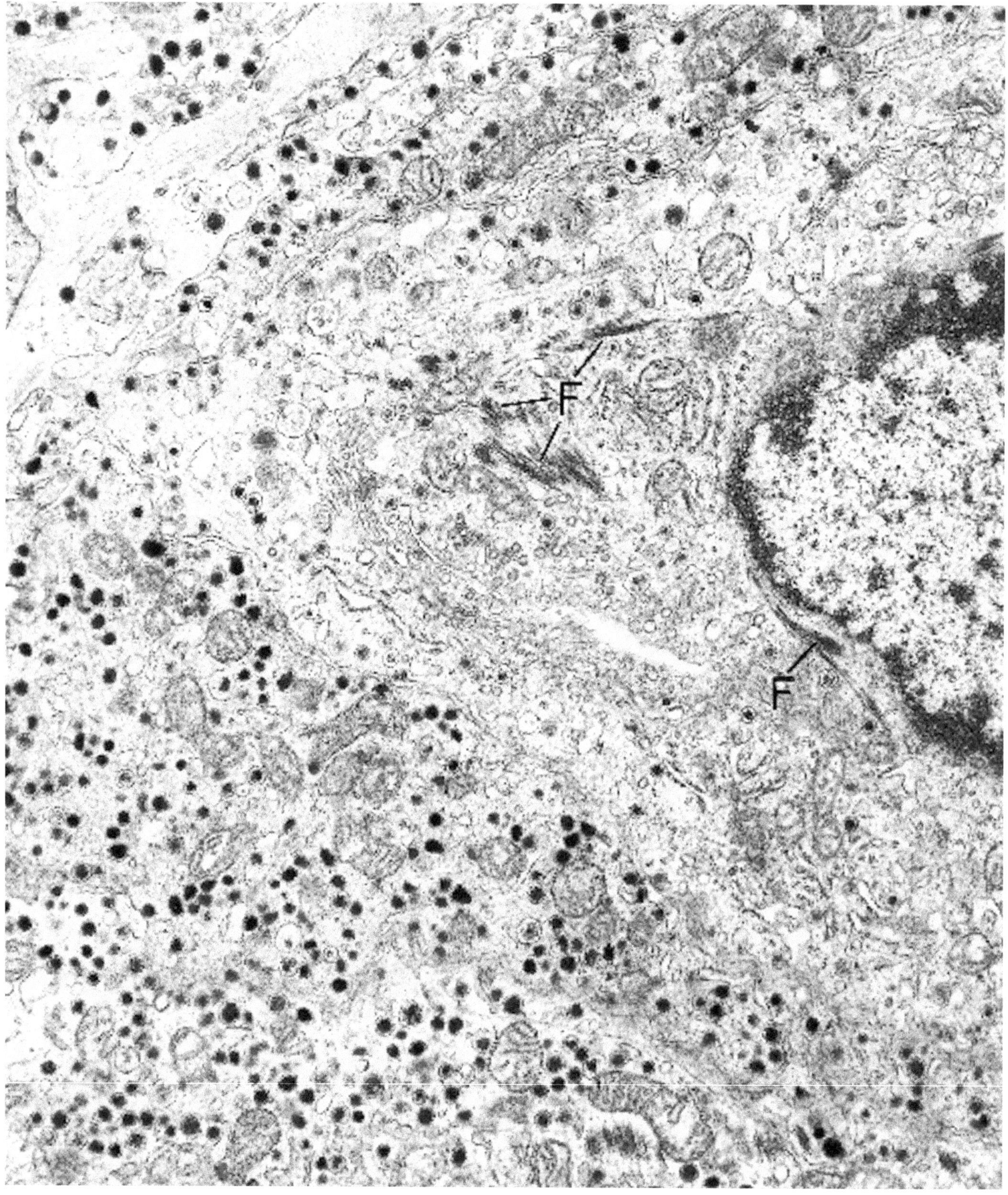

Figure 9.6. Carcinoid tumor (bronchus). In addition to having innumerable dense-core granules, some of the cells may contain bundles of filaments (tonofibrils) (F). (× 20,000)

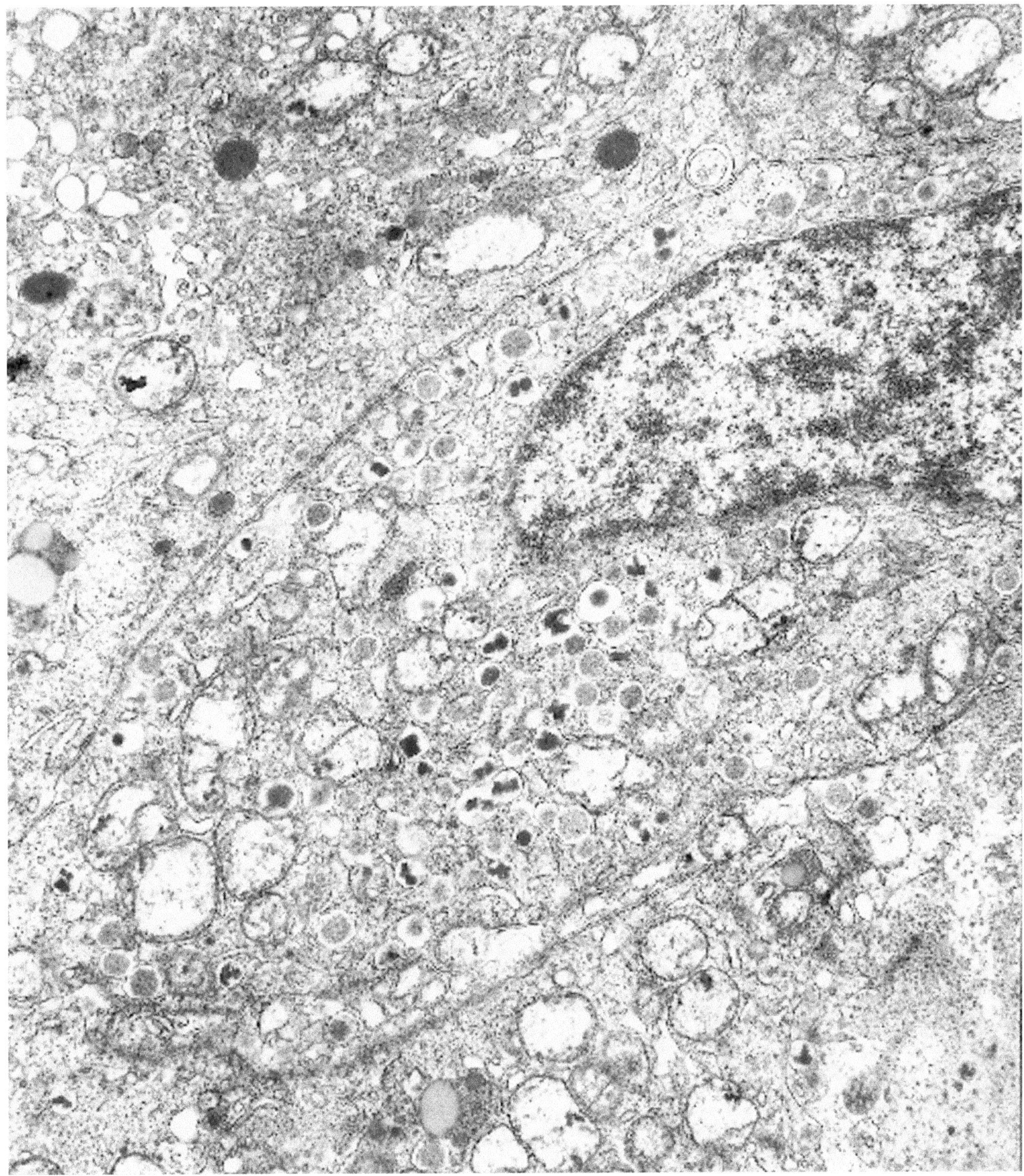

Figure 9.7. Islet (beta) cell tumor (pancreas). The neoplastic cell in the center of the field contains numerous dense-core type granules, some of which have a round, medium-dense secretory product, and others a round or angular, crystalline material. (× 19,240)

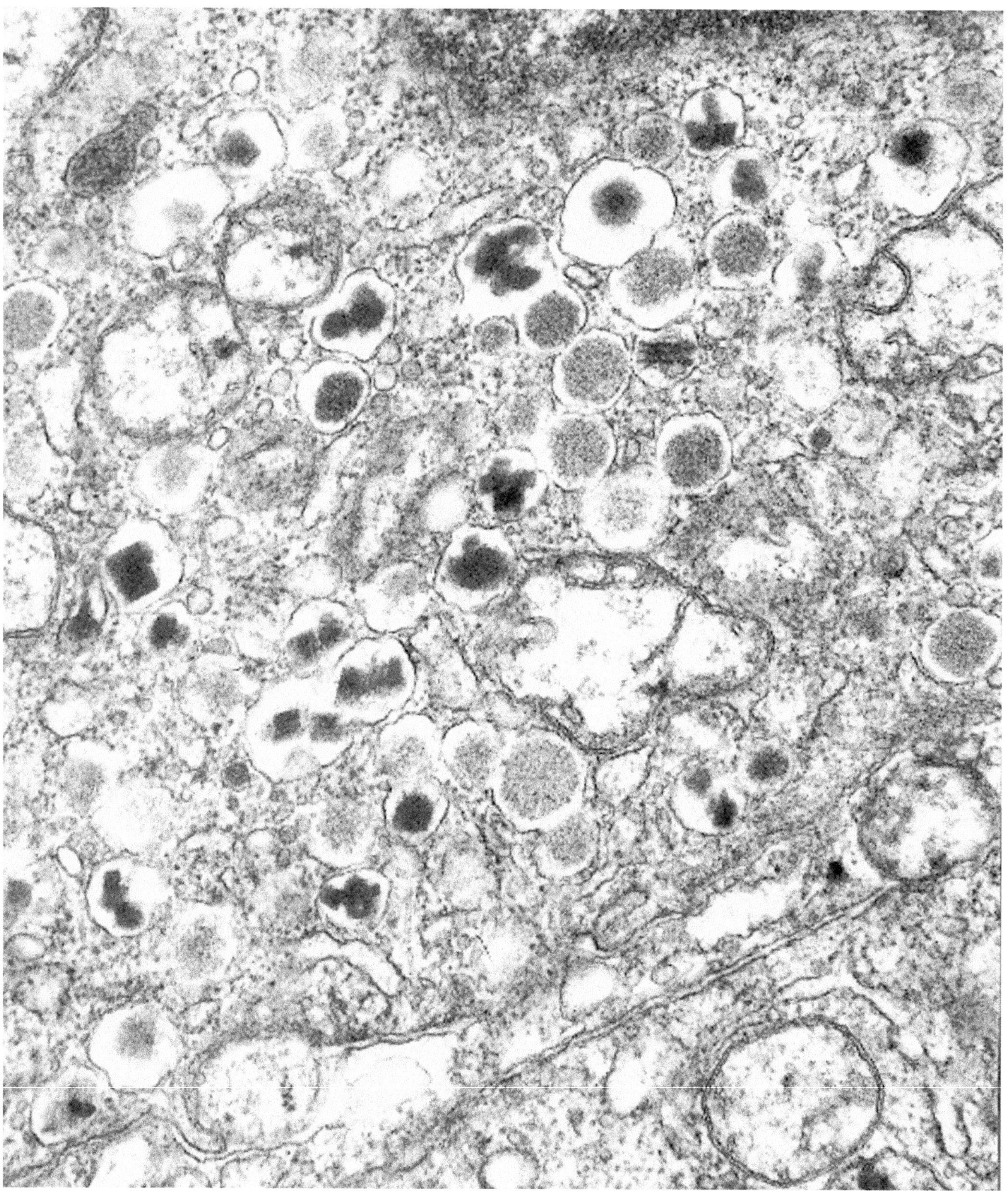

Figure 9.8. Islet (beta) cell tumor (pancreas). High magnification of a portion of the cell shown in Figure 9.7 illustrates the characteristic beta-type granules. (× 50,300)

(Text continued from page 560)

Medullary (C-Cell) Carcinoma of the Thyroid

(Figures 9.9 through 9.12.)

Diagnostic criteria. (1) Islands of polygonal cells surrounded by basal lamina; (2) dense-core granules of two types: type I—large, medium-dense, and no halo, and type II—small, very dense, and with a halo between the secretory material and the limiting membrane of the granule; (3) intercellular collections of nonbranching fibrils of amyloid may or may not be present.

Additional points. Type I and type II granules may be present within the same cell, although one or the other usually predominates. The two types of granules represent different stages of secretory activity in the cells, type I being associated with a storage phase and type II being related to synthesizing and secretory phases. In cells of the latter type, other signs of secretory activity are also present, such as prominent Golgi apparatuses and many cisternae of rough endoplasmic reticulum.

Parathyroid Carcinoma and Adenoma

(Figures 9.13 through 9.16.)

Diagnostic criteria. (1) Trabeculae and islands of polygonal (chief) cells surrounded by basal lamina; (2) intercellular junctions; (3) interdigitation of lateral cell membranes; (4) dense secretory granules; (5) annulate lamellae; (6) varying numbers of oncocytes, characterized by an overabundance of mitochondria, occurring in groups among the chief cells.

Additional points. Prominent Golgi apparatuses and smooth and rough endoplasmic reticulum vary with the secretory state of the cells. Glycogen is usually present and, in "clear" cells, is copious. A few droplets of lipid, less than in normal parathyroid cells, single cilia, lumens, and intracytoplasmic pseudolumens also may be present. Distinguishing between parathyroid adenomas and carcinomas is not reliably interpretable from their ultrastructural features.

Paraganglioma (Chemodectoma), Extra-adrenal

(Figures 9.17 through 9.20.)

Diagnostic criteria. (1) Groups of oval or polyhedral chief cells (Zellballen) surrounded by basal lamina; (2) round dense-core granules; (3) prominent Golgi apparatuses; (4) rarely, sustentacular (Schwann-like) cells, with filaments, arranged concentrically at the periphery of the groups.

Additional points. Chief cells have interweaving cytoplasmic processes and also may have paranuclear filaments and many mitochondria (oncocytes). Mitochondria may be swollen and irregular in shape. Sustentacular cells are found only in highly differentiated paragangliomas, and they are polygonal, triangular, or elongate and have long polar processes with intermediate filaments. They may contain secondary lysosomes and lipofuscin. Axons are distributed between, but not within the groups of cells. Ganglion cells are present infrequently.

Pheochromocytoma (Adrenal Paraganglioma)

(Figures 9.21 through 9.23.)

Diagnostic criteria. (1) Groups of polygonal cells surrounded by basal lamina and thin bands of collagen with thin-walled blood vessels; (2) abundant cytoplasm with many large, pleomorphic, dense-core granules; (3) prominent Golgi apparatuses.

Additional points. The dense-core granules are often clear or only partly filled by the electron-dense secretory product (Figure 9.22). This feature and the pleomorphism (Figure 9.23) of the granules make pheochromocytoma a specifically identifiable member of the neuroendocrine group of neoplasms. Both epinephrine- and norepinephrine-containing granules are present in the neoplastic cells but overlap in their morphologic features, thus making specific identification of granules unreliable. Sustentacular cells are not readily discernible in pheochromocytomas.

Monomorphic Adenoma

These epithelial and myoepithelial neoplasms occur in salivary, eccrine, apocrine, and mammary glands (see sections on glandular and ductal epithelial ultrastructure in Chapter 3, and myoepithelial ultrastructure in pleomorphic adenoma and adenoid cystic carcinoma in this section). In the salivary glands specifically, monomorphic adenomas may be composed purely of luminal epithelial cells or dually of epithelial and myoepithelial cells. Subtypes of monomorphic adenoma include *papillary cystadenoma lymphomatosum* (*Warthin's tumor, adenolymphoma*), *oncocytoma, basal cell adenoma, clear cell adenoma, myoepithelioma,* and others. The cells in Warthin's tumor are epithelial and oncocytic, containing numerous mitochondria, similar to oncocytic cells described elsewhere in this book (Chapter 3, Figure 3.30). Basal cell adenomas may be composed purely of basal-type reserve cells without differentiation into myoepithelial cells or glandular epithelial cells, or they may consist of basal cells mixed with myoepithelial and/or epithelial cells. Clear cell adenomas and carcinomas are composed of epithelial cells with glycogen-rich cytoplasm.

(Text continues on page 585)

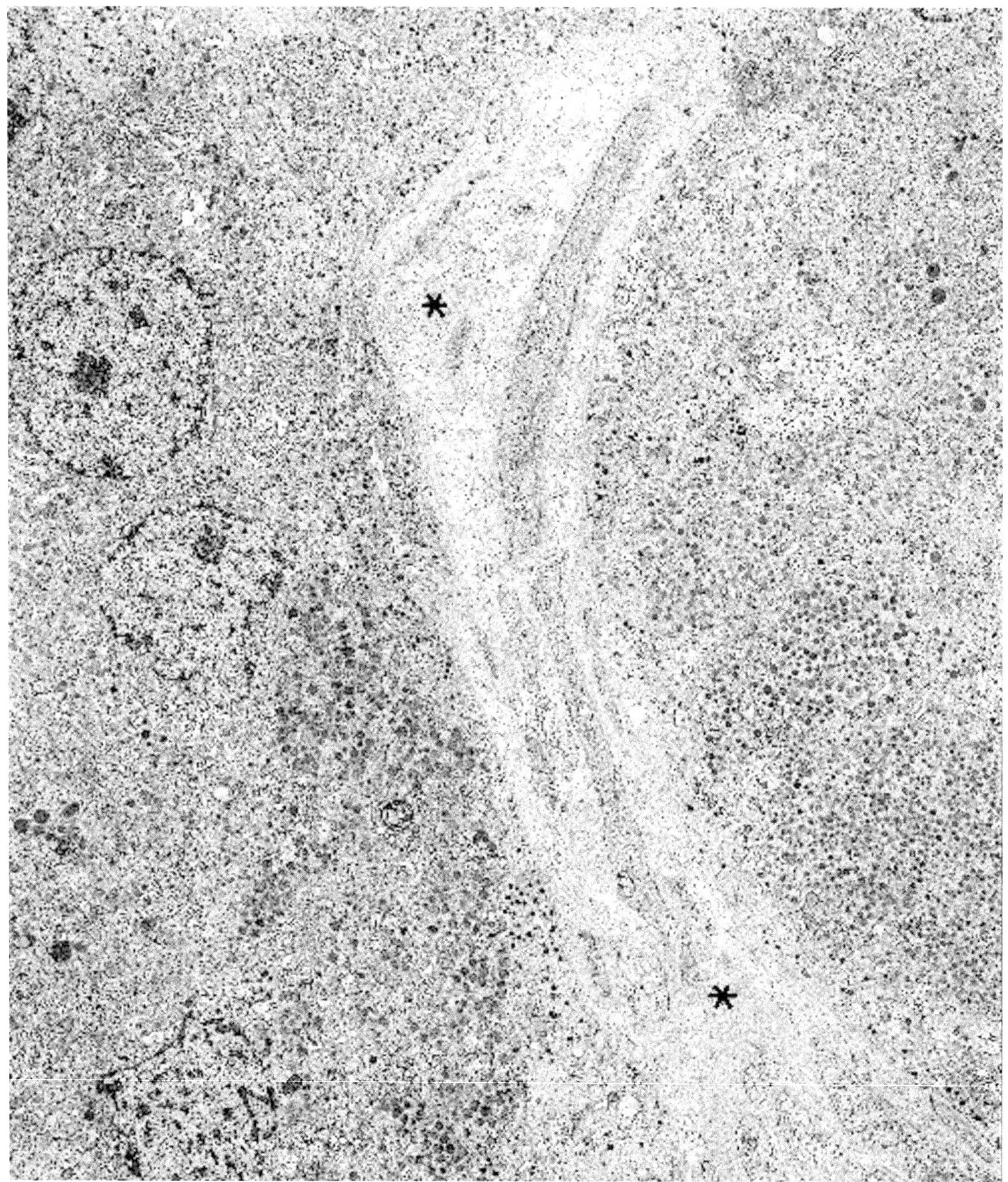

Figure 9.9. Medullary (C-cell) carcinoma (thyroid gland). Portions of two islands of polygonal cells are separated by a collagenous stroma (*). Most striking are the innumerable round granules in the cytoplasm of the cells. Even at this low magnification, the granules can be discerned as being of two types: large and medium-dense and small and very dense. (× 5720)

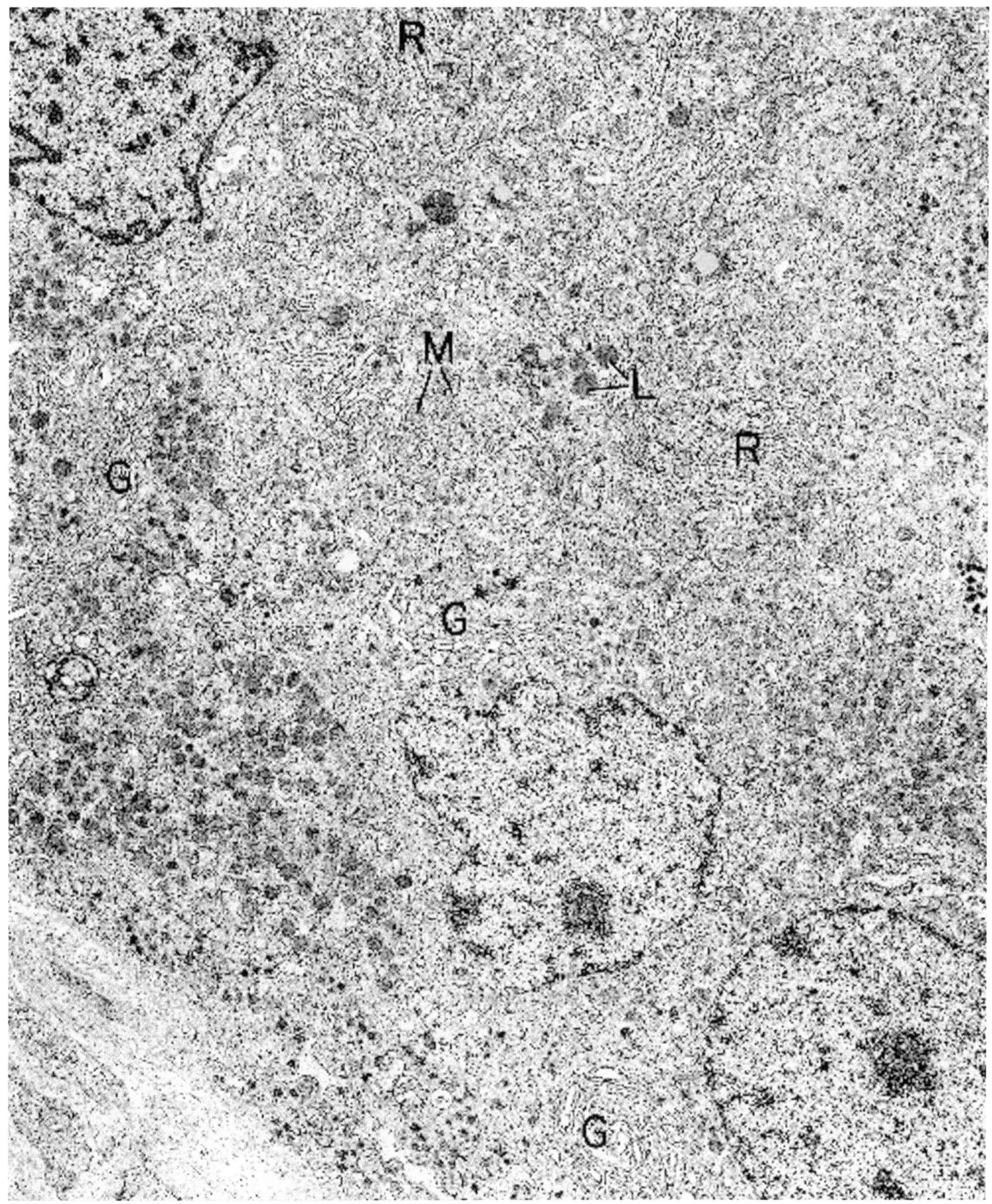

Figure 9.10. Medullary (C-cell) carcinoma (thyroid gland). In addition to many granules of both types (large and medium-dense; small and very dense), numerous other organelles are found in the cytoplasm of the cells: Golgi apparatuses (G), rough endoplasmic reticulum (R), mitochondria (M), and lipid droplets (L). (× 10,260)

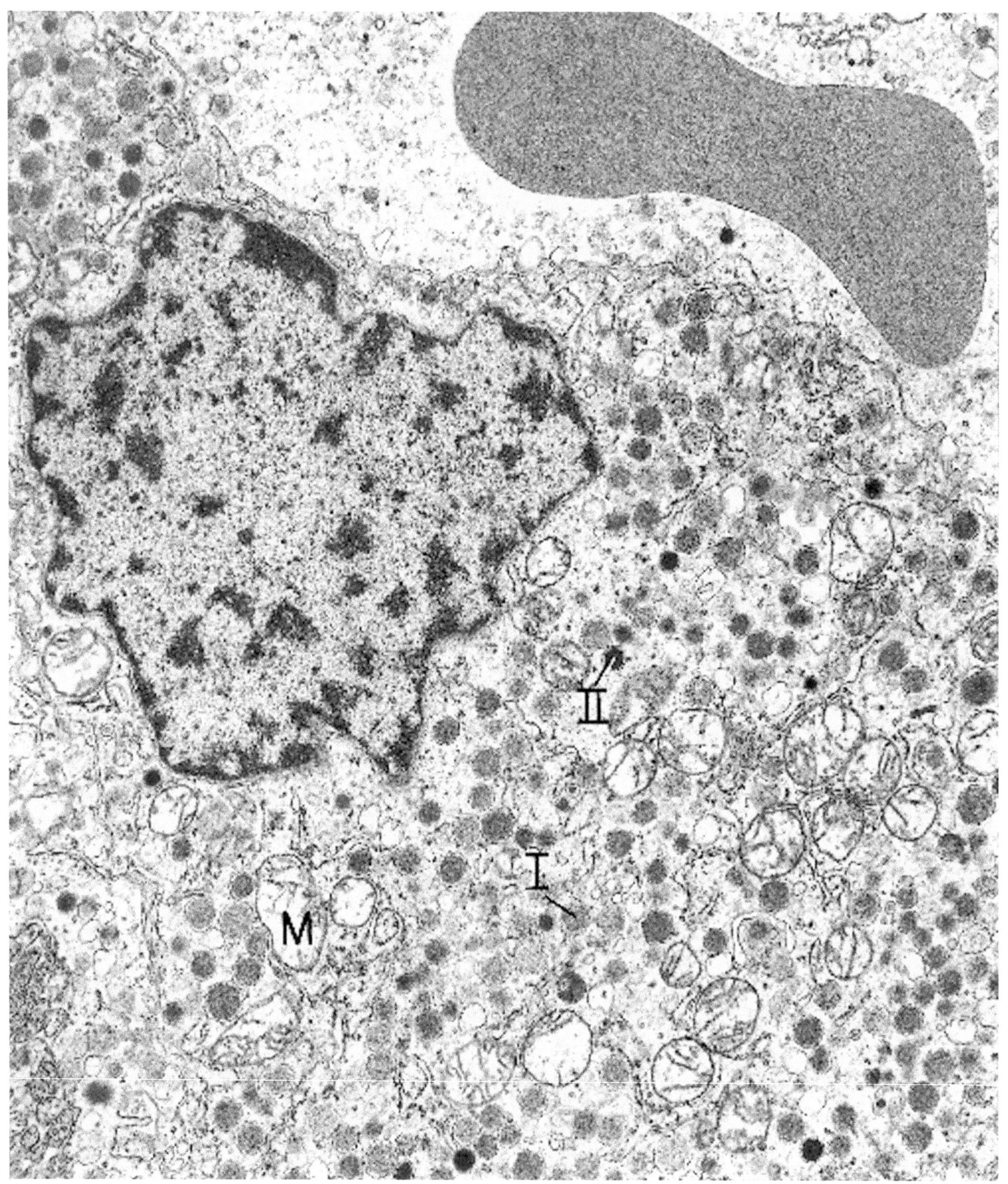

Figure 9.11. Medullary (C-cell) carcinoma (thyroid gland). The small, dark, type II granules more consistently have halos around their secretory product than do the larger, medium-dense, type I granules. M = mitochondria. (× 15,930)

Figure 9.12. Medullary (C-cell) carcinoma (thyroid gland). An example of amyloid, frequently found in the extracellular matrix of medullary carcinomas, consists of a dense network of nonbranching, 7- to 10-nm diameter filaments. (× 81,000)

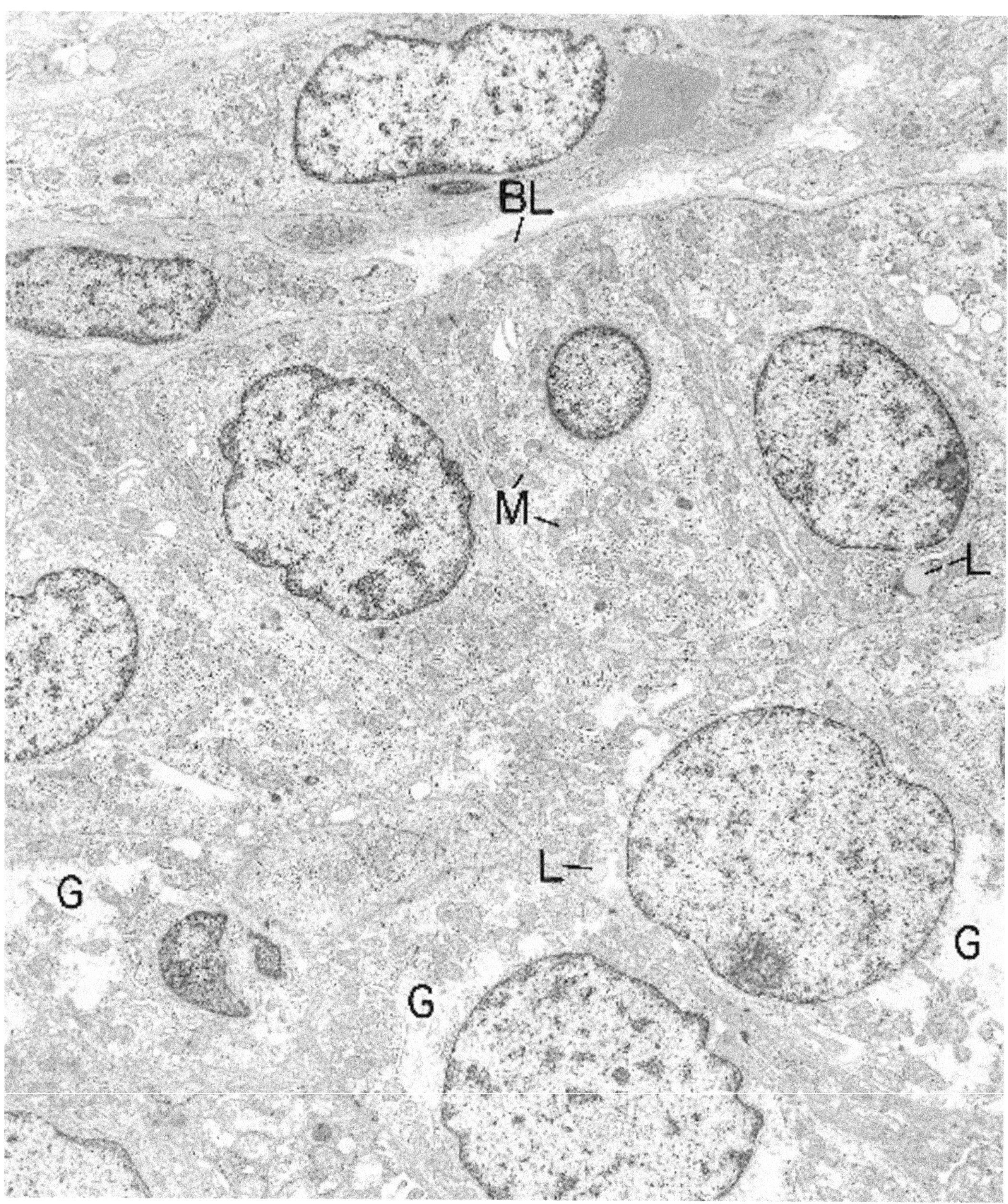

Figure 9.13. Parathyroid carcinoma (metastatic to lung). This carcinoma is well-differentiated and consists of islands of polygonal chief cells surrounded by basal lamina (BL). The cytoplasm contains many organelles, including many mitochondria (M), and focal inclusions of glycogen (large clear spaces, G), and a few droplets of lipid (L). (× 7020)

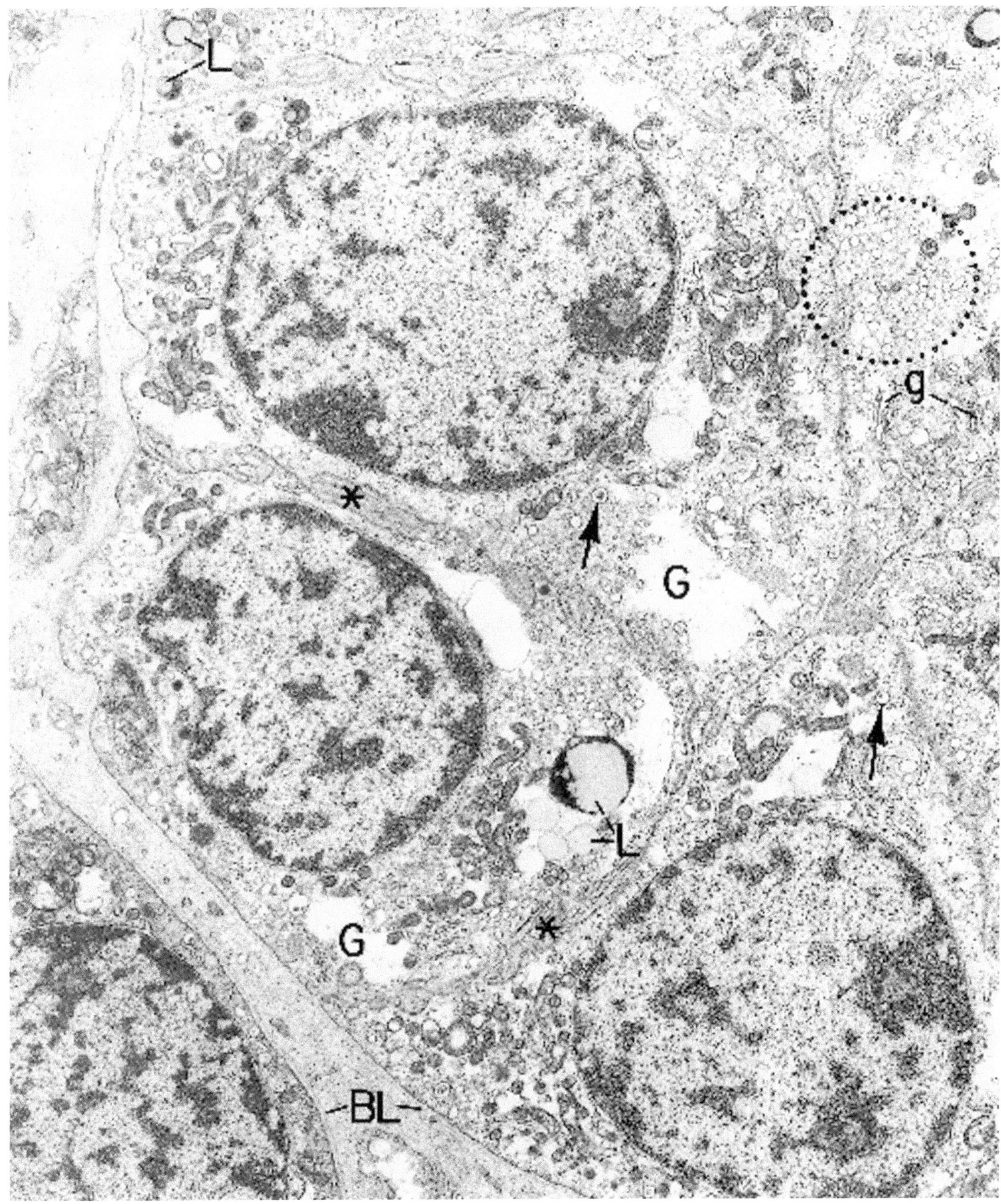

Figure 9.14. Parathyroid carcinoma (metastatic to lung). Higher magnification of several neoplastic cells illustrates many cytoplasmic organelles, including many small and dense or empty granules (*arrows* and *circle*) and Golgi apparatuses (g). Lipid (lipofuscin) (L) and glycogen inclusions (G), folded lateral plasmalemmas (*), and a discrete basal lamina (BL) also are visible. (× 13,650)

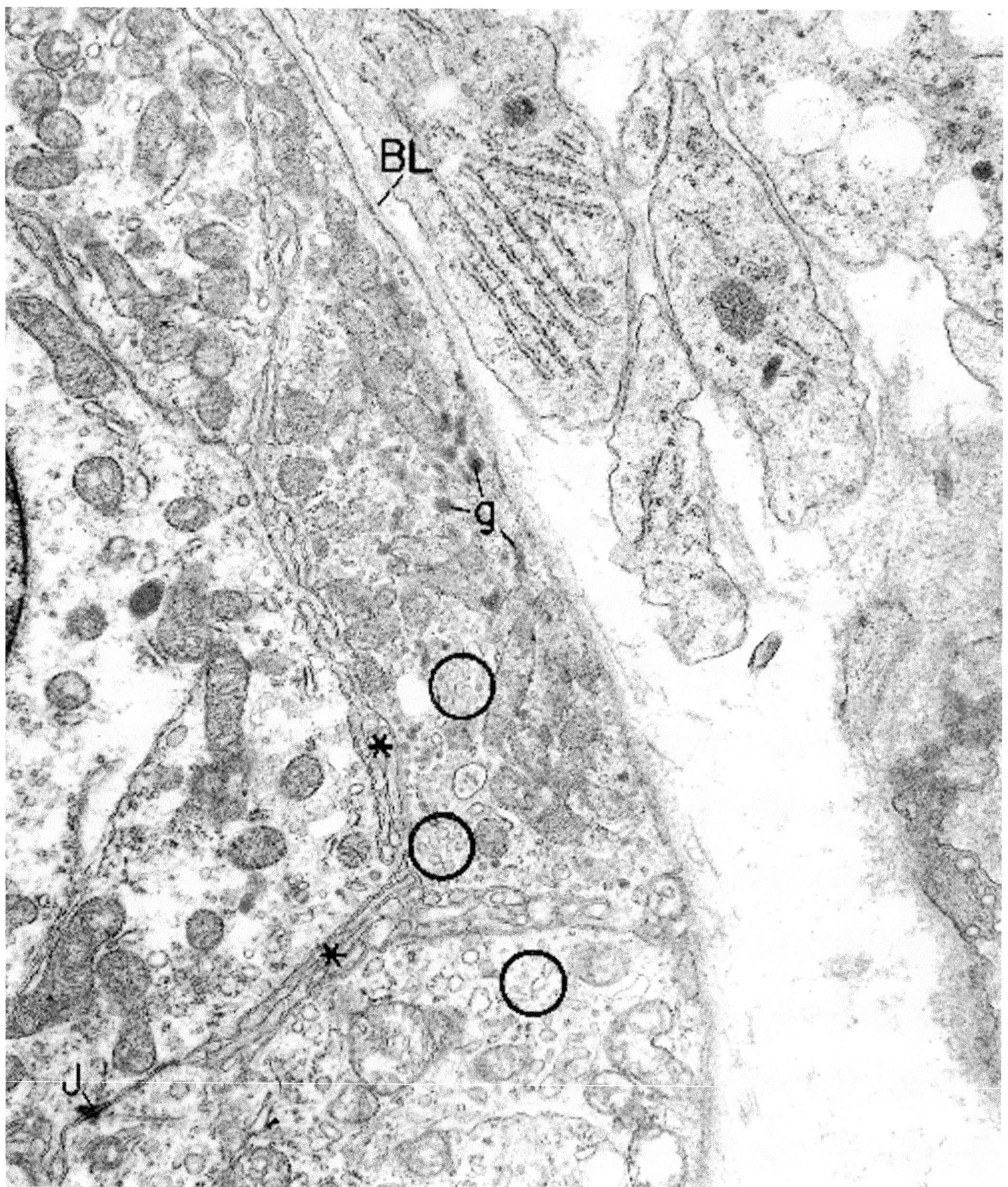

Figure 9.15. Parathyroid carcinoma (metastatic to lung). High magnification of the periphery of an island of neoplastic chief cells depicts a limiting basal lamina (BL), infolded lateral plasmalemmas (*), a prominent junction (J), many vesicles and cisternae of smooth endoplasmic reticulum (*circles*), and small membrane-bound granules (g). (× 7100)

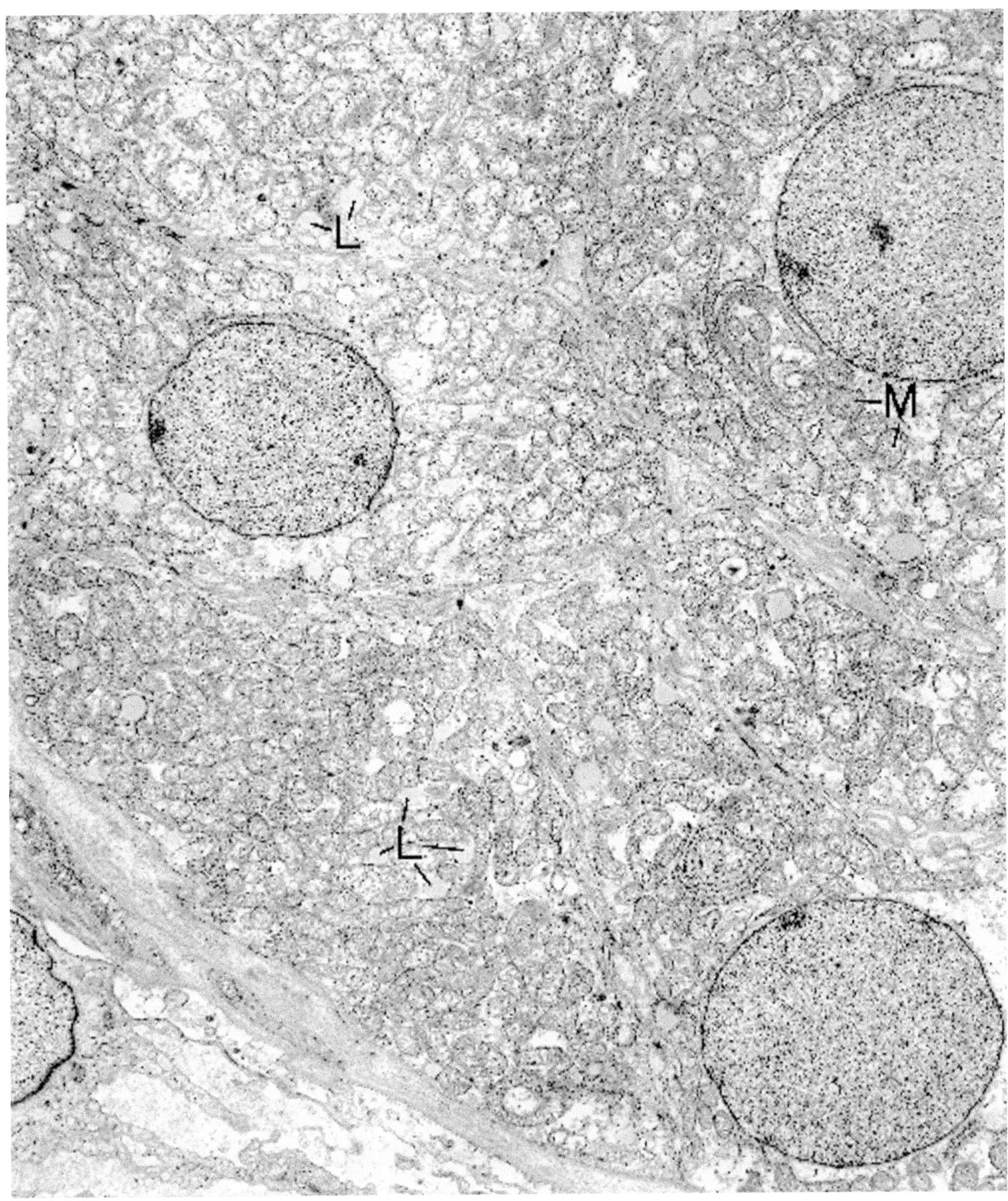

Figure 9.16. Parathyroid adenoma, oncocytic type. The oncocytes are characterized by innumerable mitochondria (M) filling their cytoplasm. Scattered lipid droplets (L) also are present. (× 7020)

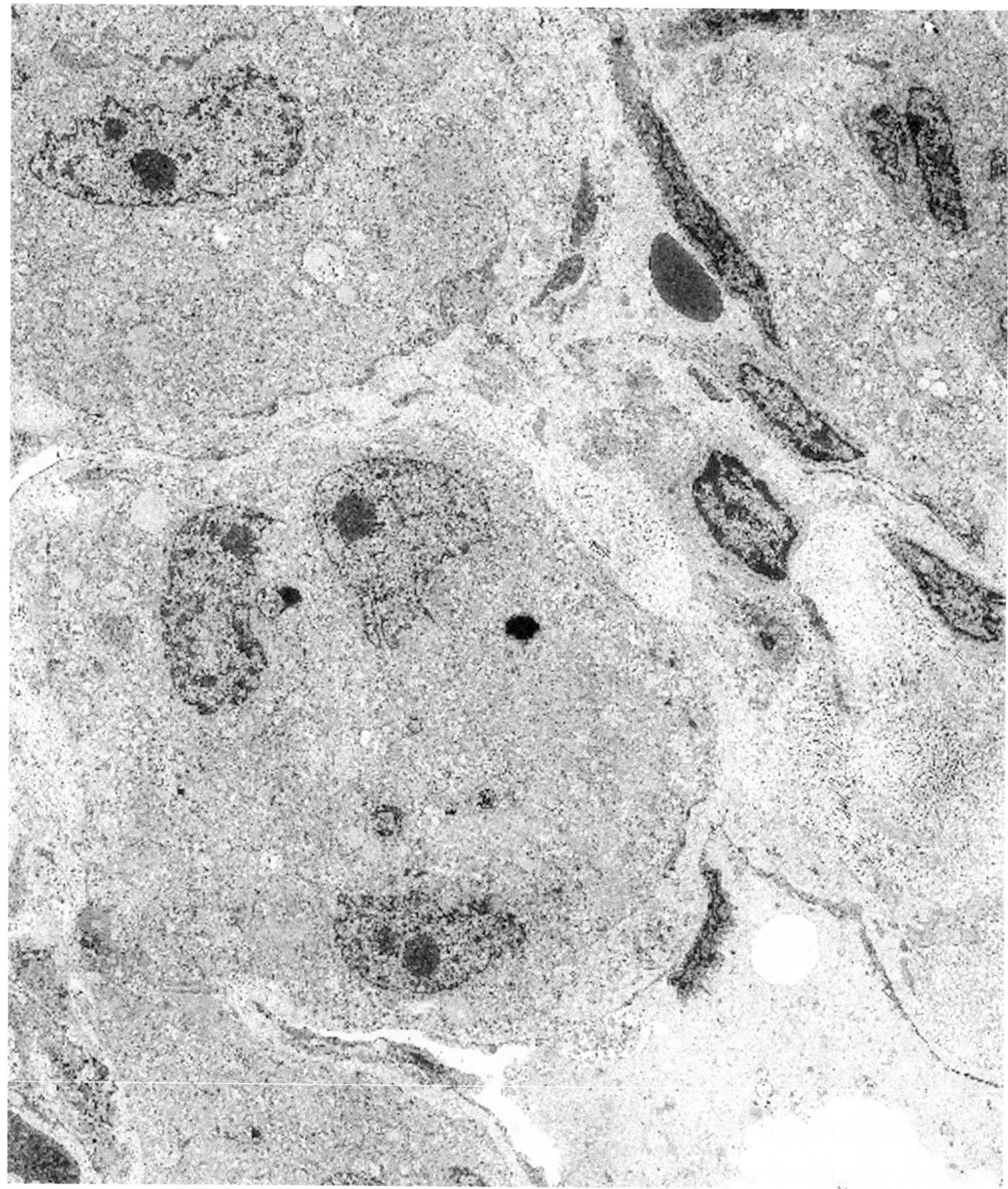

Figure 9.17. Paraganglioma (carotid body). Discrete balls of oval and polygonal (chief) cells are distributed in a matrix of collagen. (× 2440)

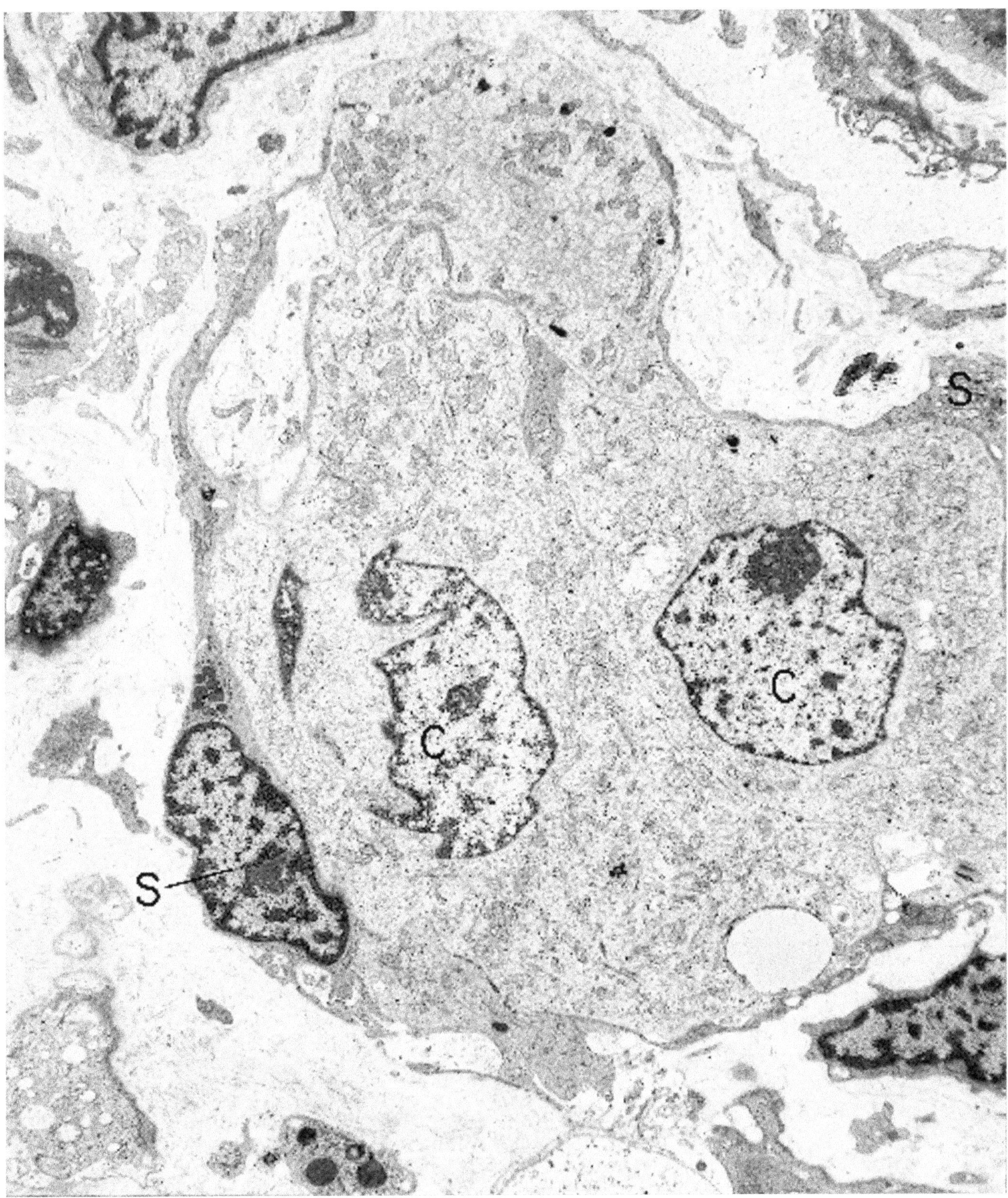

Figure 9.18. Paraganglioma (carotid body). Within this Zellball a clear distinction is discernible between chief cells (C) and sustentacular cells (S), the later being spindle shaped and located peripherally and concentrically. (× 5940)

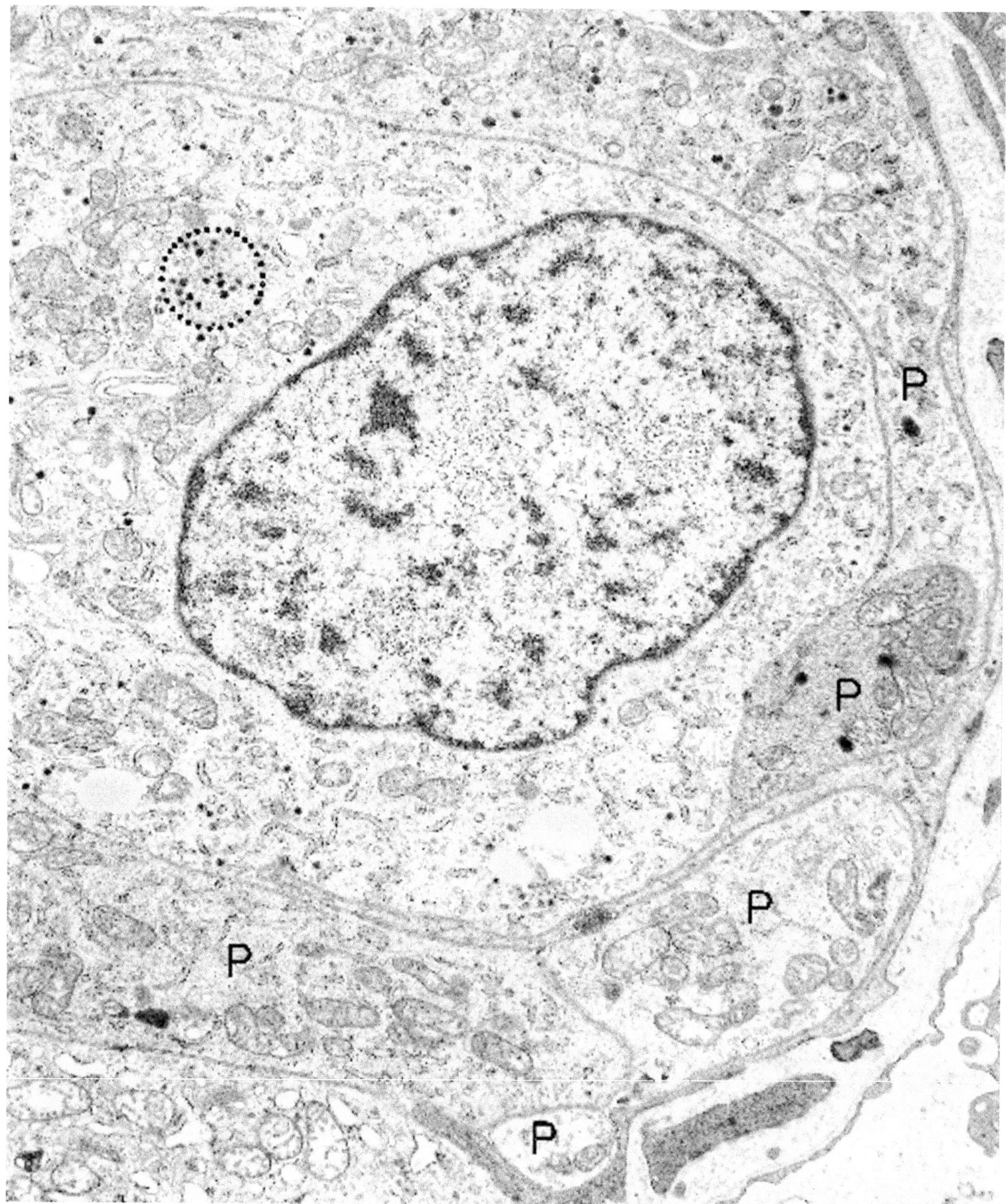

Figure 9.19. Paraganglioma (carotid body). This region of a Zellball illustrates the characteristics of the interweaving cytoplasmic processes (P) of the chief cells. A moderate number of dense-core granules (*circle*) is visible. (× 14,000)

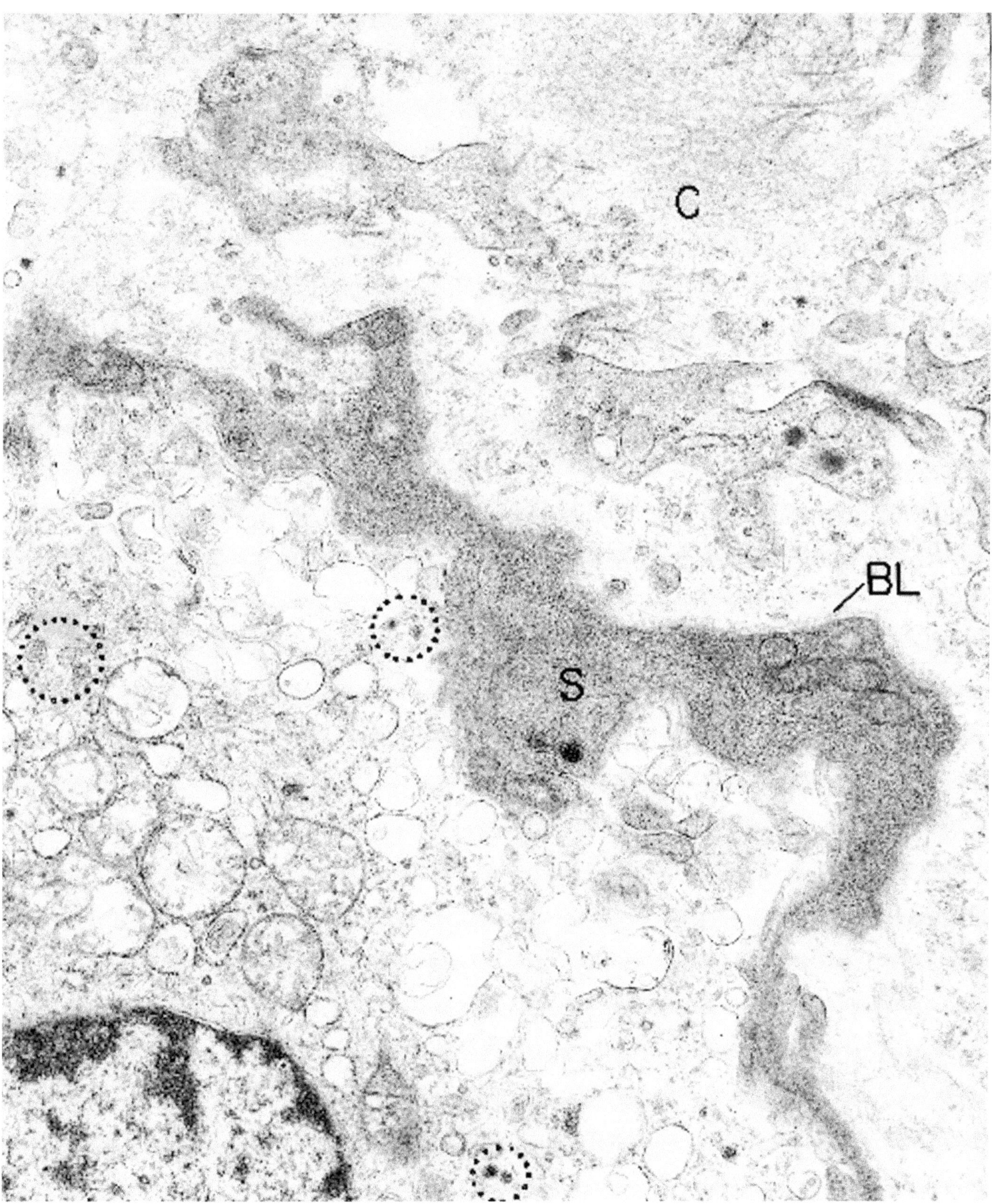

Figure 9.20. Paraganglioma (carotid body). High magnification of the periphery of a Zellball depicts an enwrapping sustentacular cell (S). The dark cytoplasm is related to innumerable microfilaments. A basal lamina (BL) is visible focally between the sustentacular cell and the collagenous matrix (C). A few dense-core granules (*circles*) are present in the cytoplasm of the chief cell. (× 3380)

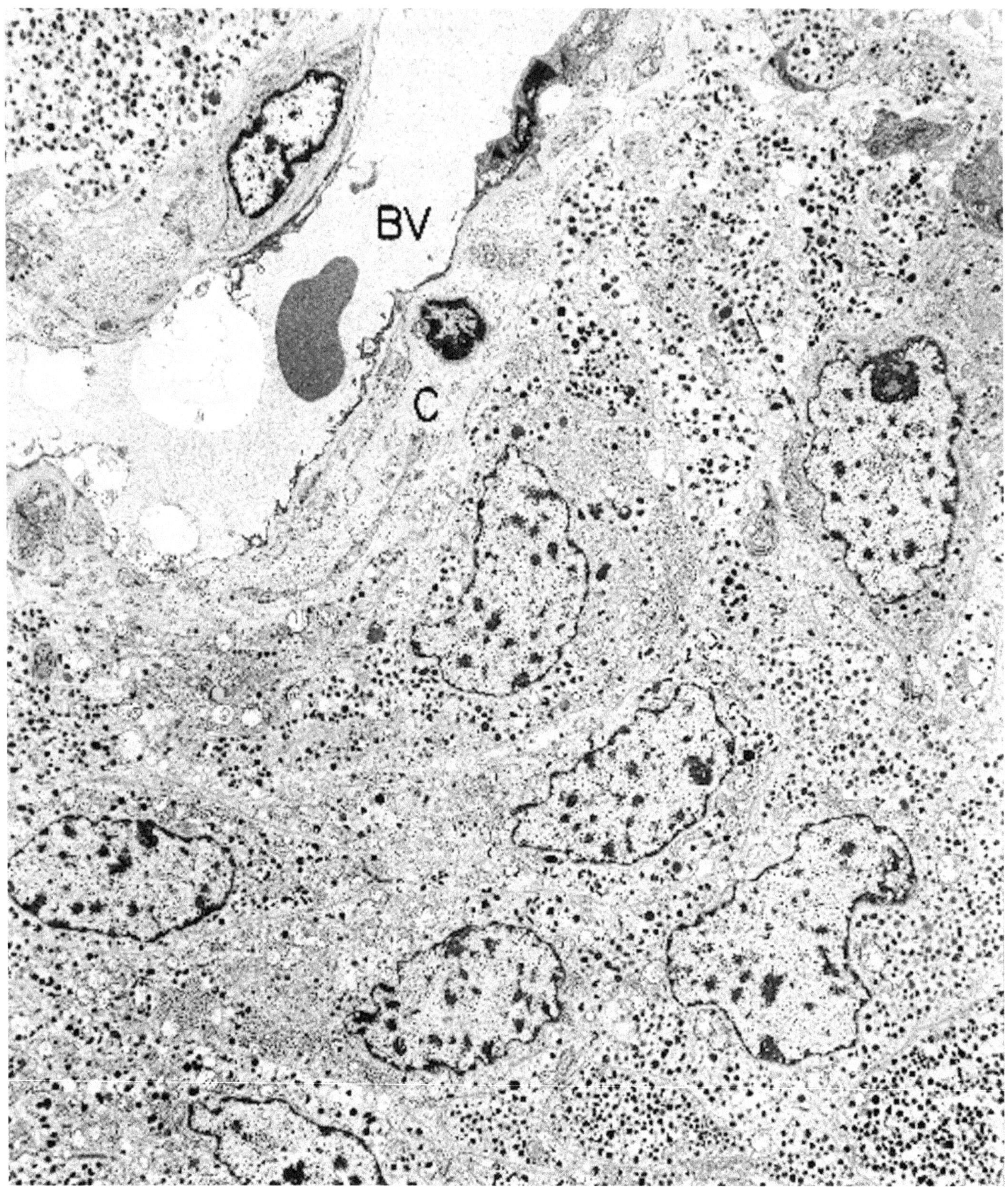

Figure 9.21. Pheochromocytoma (adrenal gland). A group of polygonal cells is sharply demarcated from the surrounding collagen (C) and blood vessel (BV). Most striking are the many electron-dense granules in the cytoplasm of the cells. (× 5500)

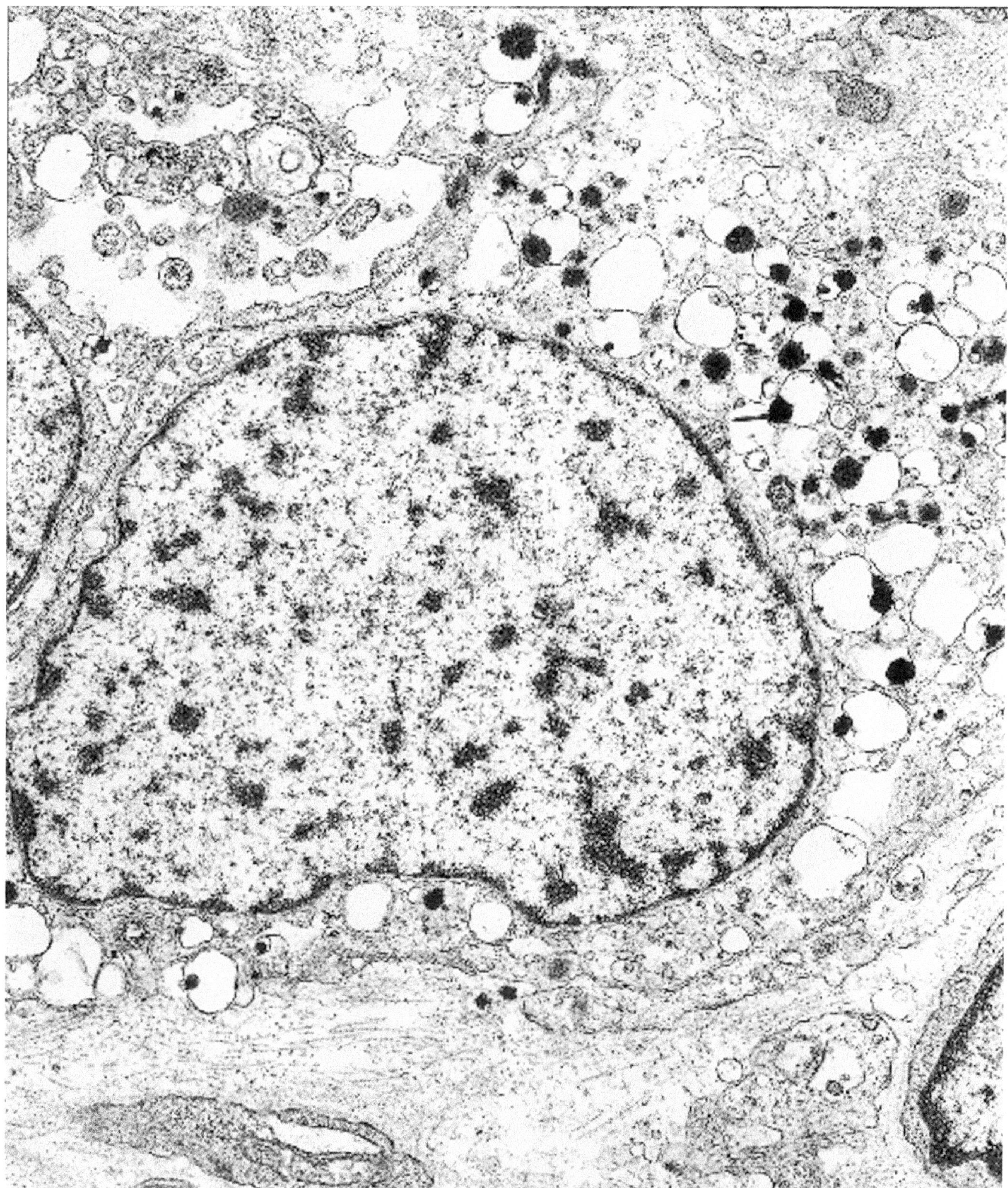

Figure 9.22. Pheochromocytoma (adrenal gland). Characteristic of the granules in this type of neuroendocrine tumor are their large size, pleomorphism, and partial emptiness. (× 20,000)

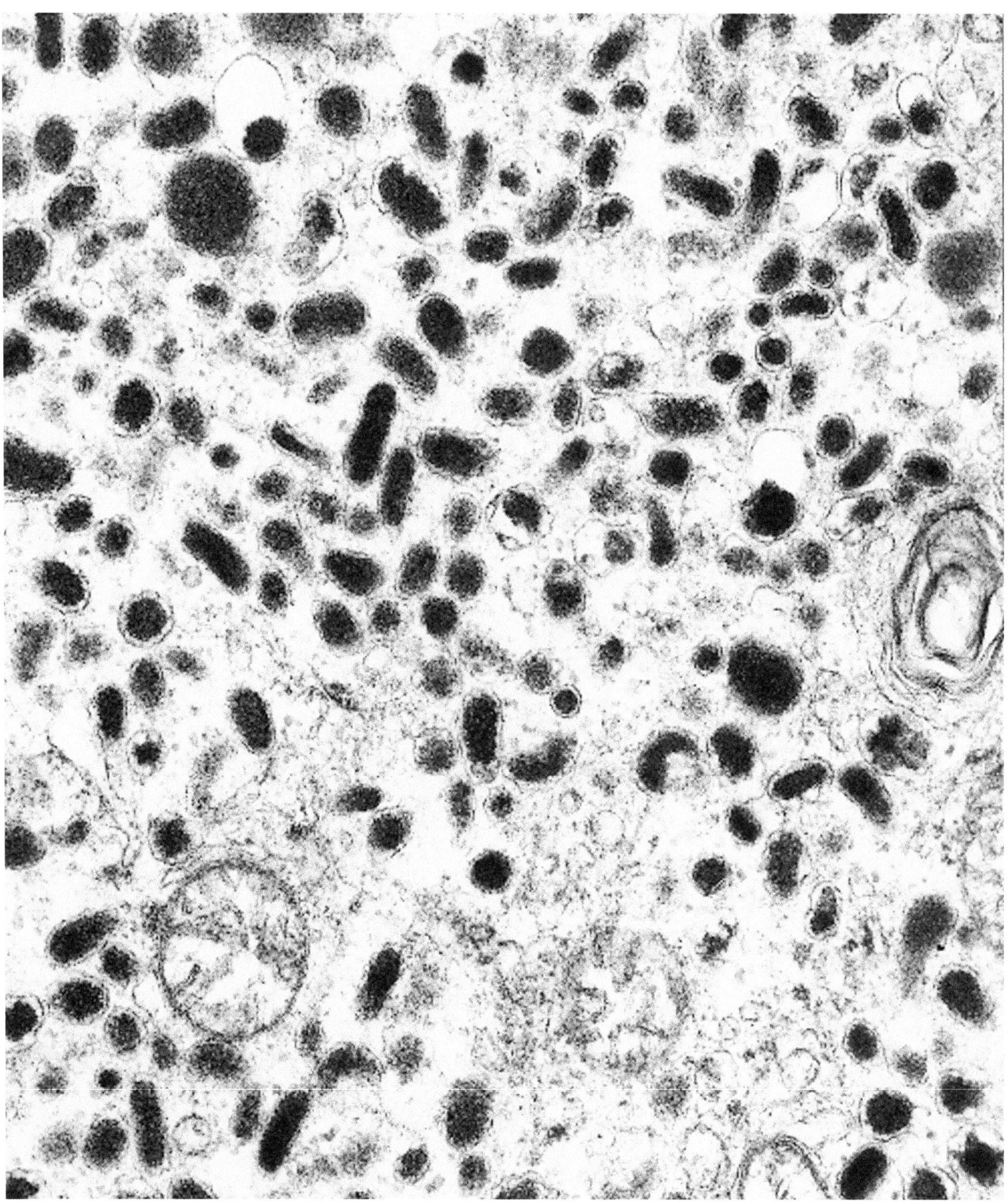

Figure 9.23. Pheochromocytoma (adrenal gland). In contrast to Figure 9.22, most of the dense-core granules in this cell are filled with secretory product. However, their large size and pleomorphism are again well exemplified. (× 43,500)

(Text continued from page 569)

Pleomorphic Adenoma

(Figures 9.24 through 9.28.)

Diagnostic criteria. (1) Nests or sheets of two types of cells (epithelial and myoepithelial), in a prominent matrix of banded collagen, amorphous and medium-dense basal lamina, and clear glassy glycosaminoglycans (Figure 9.24); (2) flat, cuboidal, or columnar (epithelial) cells with microvilli and junctional complexes, lining central acinus-like and duct-like spaces (Figure 9.25); (3) spindle, stellate, or irregularly shaped (myoepithelial) cells with desmosomes, thin and intermediate filaments, and dense bodies (dense aggregates of filaments) surrounding the central lining cells (Figure 9.26); (4) secretory granules (serous and/or mucous) in the epithelial lining cells and secretory products of varying density in the lumens (Figure 9.25).

Additional points. The epithelial cells lining lumens usually are single layered, and the myoepithelial cells surrounding them are multilayered. The myoepithelial cells may be sharply demarcated from the adjacent matrix by basal lamina, or they may extend irregularly into the matrix where they become separated, sometimes widely, and have thin lateral cytoplasmic processes and focal basal lamina (Figure 9.27). Myoepithelial cells may be well differentiated, with the desmosomes, filaments, and dense bodies described in the previous paragraph, or they may be poorly differentiated (Figure 9.28), or even metaplastic to squamous epithelial cells with tonofibrils and keratin cysts and to chondrocytes with abundant glycogen and numerous dilated cisternae of rough endoplasmic reticulum.

Adenoid Cystic Carcinoma

(Figures 9.29 through 9.35.)

Diagnostic criteria. (1) Islands of polygonal cells, with numerous sieve-like spaces scattered through the islands (Figures 9.29, 9.31, and 9.32); (2) basal lamina surrounding the islands and filling some of the spaces (pseudolumens) (Figures 9.29, 9.31, and 9.32); (3) true lumens comprising some of the spaces, with epithelial-type lining cells that have microvilli and junctional complexes (Figure 9.30); (4) myoepithelial cells, circumferentially arranged at the periphery of the epithelial cells and characterized by many cytoplasmic filaments, dense bodies, and hemidesmosomes along the subjacent basal lamina; (5) in some tumors, luminal lining cells that have short, stubby microvilli and many cytoplasmic intermediate filaments and tonofibrils (Figures 9.34 and 9.35).

Additional points. The most specific criteria for recognizing this neoplasm are the profuse basal laminar material filling some of the spaces and the myoepithelial cells. The myoepithelial cells show a range of differentiation, with the less differentiated ones not exhibiting the characteristic filaments and dense bodies. Others may show squamous metaplasia, with tonofibrils (Figure 9.33). The ultrastructure of adenoid cystic carcinomas is similar wherever they occur—salivary glands, trachea and bronchi, or breast. The short, blunt type of microvilli described under the fifth diagnostic criterion is more characteristic of ductal- than glandular-type epithelium. Glandular villi are longer and more slender, and are also found in these and other salivary gland-type neoplasms.

Mucoepidermoid Carcinoma

(See also Chapter 3, Figures 3.48 and 3.49)

(Figures 9.36 through 9.41.)

Diagnostic criteria. (1) Solid and cystic arrangements of two types of cells, epithelial and myoepithelial (Figure 9.36); (2) some epithelial cells contain many mucous granules (goblet cells) and (3) others are of the luminal lining type described in pleomorphic adenoma and adenoid cystic carcinoma (Figure 9.37); (4) purely squamous cells with tonofibrils and no mucous granules; (5) tonofibrils in some of the mucous cells (squamous metaplasia); (6) myoepithelial cells with desmosomes, filaments, and dense bodies; (7) tonofibrils in myoepithelial cells (squamous metaplasia) (Figures 9.36 through 9.38).

Additional points. The proportion of mucous cells, squamous cells, myoepithelial cells, and squamous metaplastic cells varies among neoplasms in the mucoepidermoid category. Also, myoepithelial cells may be absent or difficult to identify with certainty when there is squamous metaplasia.

(Text continues on page 604)

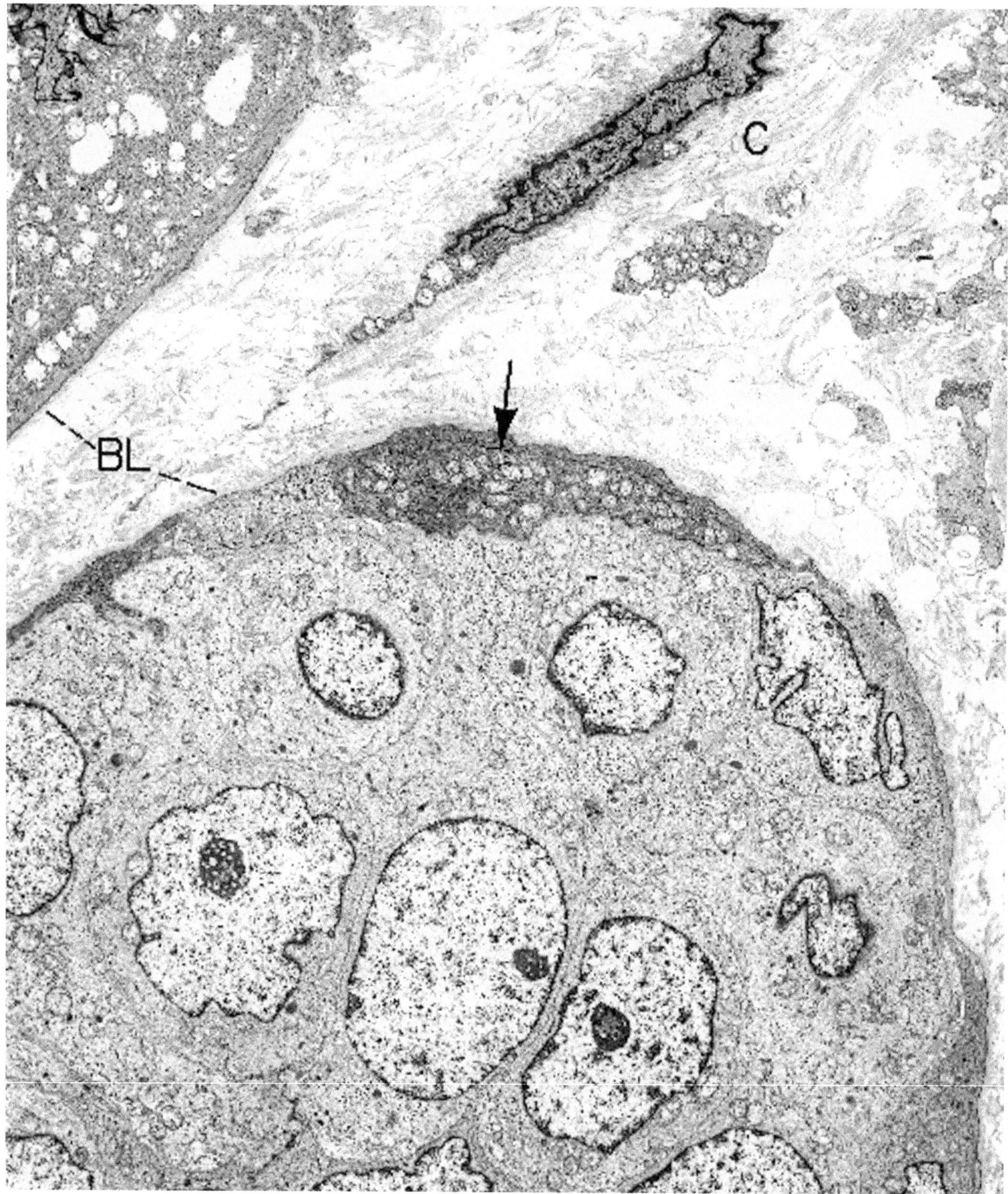

Figure 9.24. Pleomorphic adenoma (lung). Islands of polygonal cells are surrounded by basal lamina (BL) and separated by a matrix rich in collagen (C). The dark cell (*arrow*) at the periphery of one island is oriented circumferentially and represents a myoepithelial cell. (× 5720)

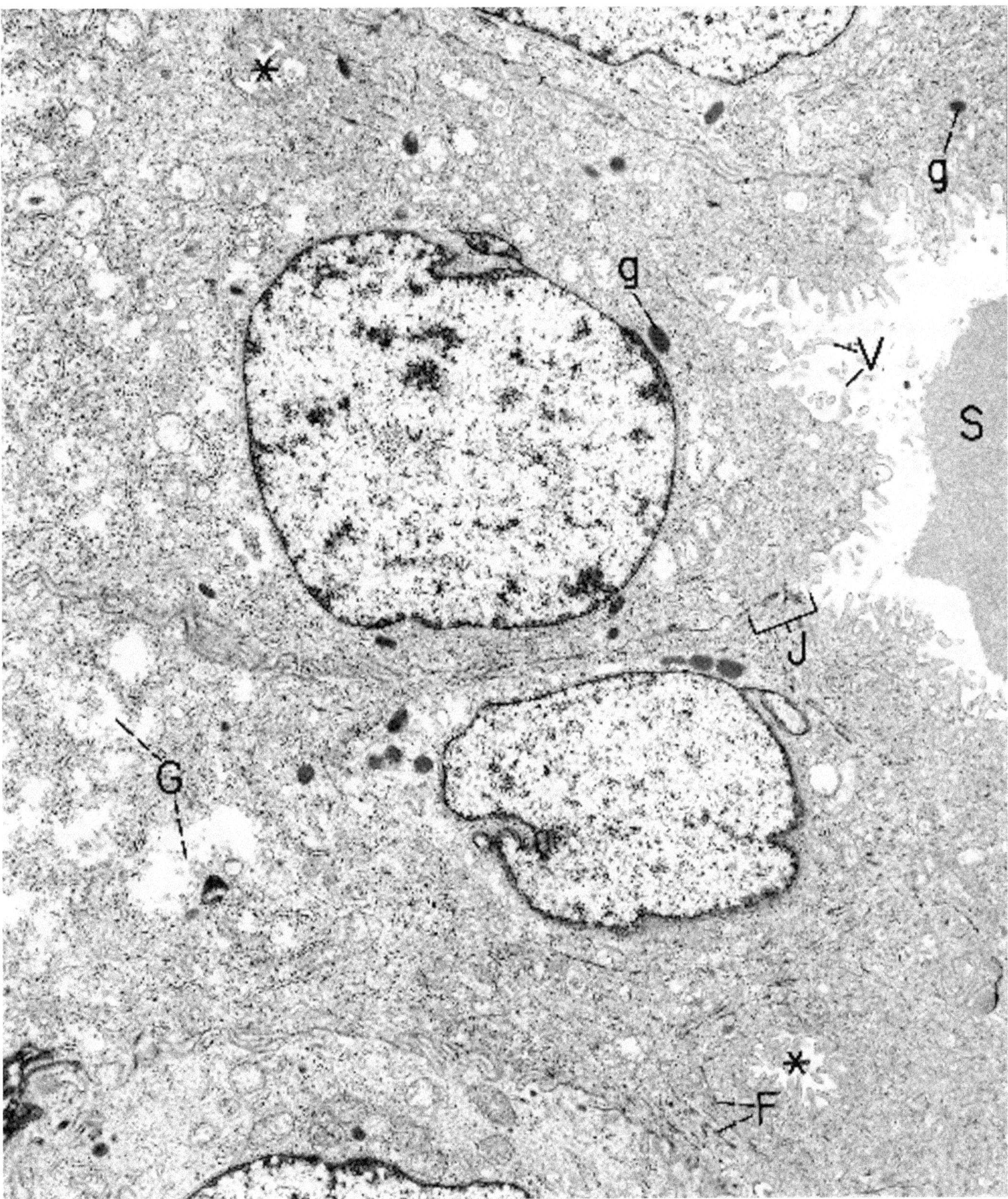

Figure 9.25. Pleomorphic adenoma (lung). High magnification of the center of an island of neoplastic cells reveals a true lumen with junctional complexes (J), microvilli (V), and intraluminal secretory products (S). Also discernible in the cytoplasm of the cells are secretory granules (g), open spaces of glycogen (G), bundles of filaments (F), and pseudolumens (*). (× 10,260)

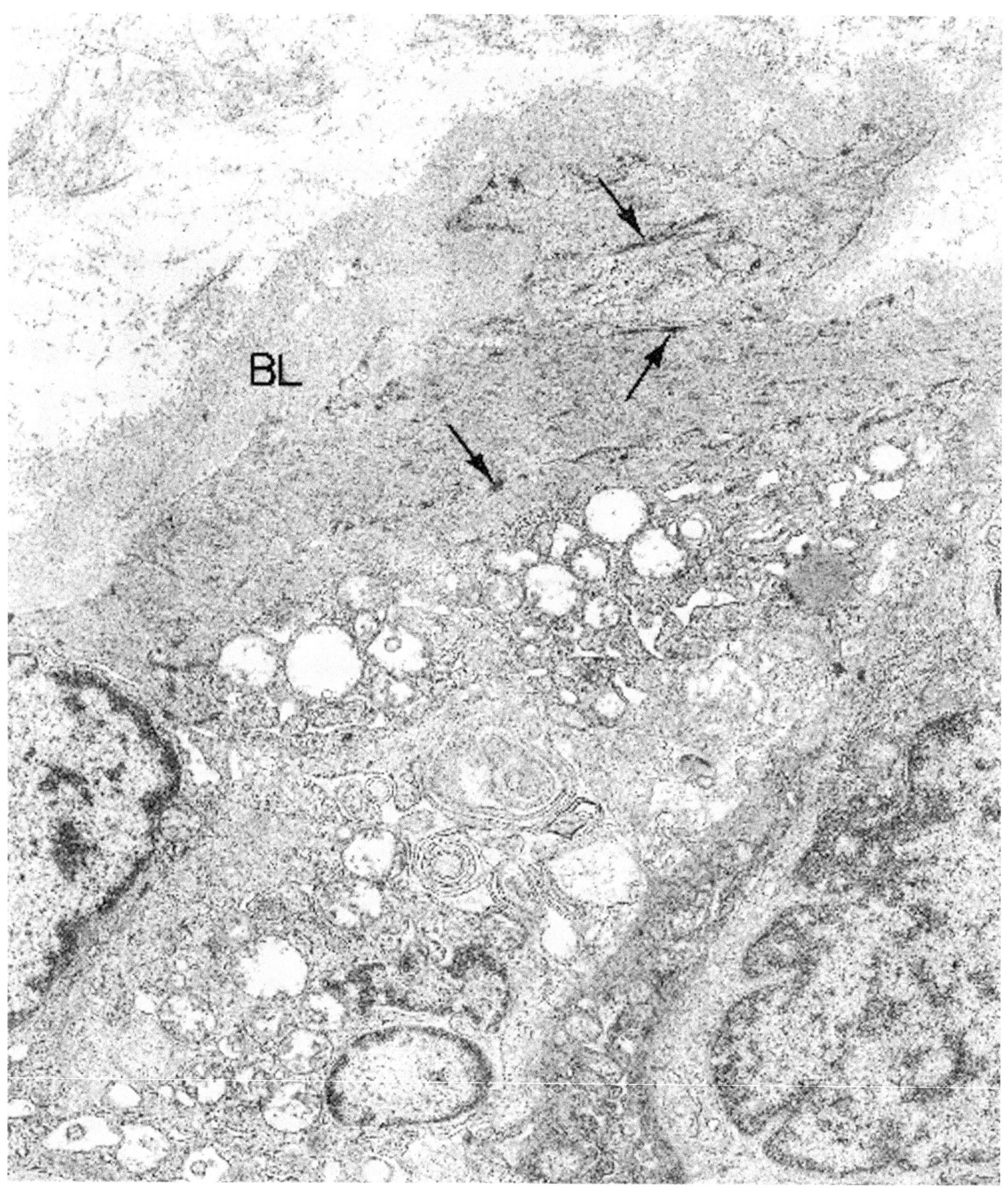

Figure 9.26. Pleomorphic adenoma (lacrimal gland). A myoepithelial cell at the periphery of an island of neoplastic polygonal cells is filled with filaments and dense aggregates (dense bodies) (*arrows*) of filaments. A thick layer of basal lamina (BL) borders the myoepithelial cells. (× 14,600)

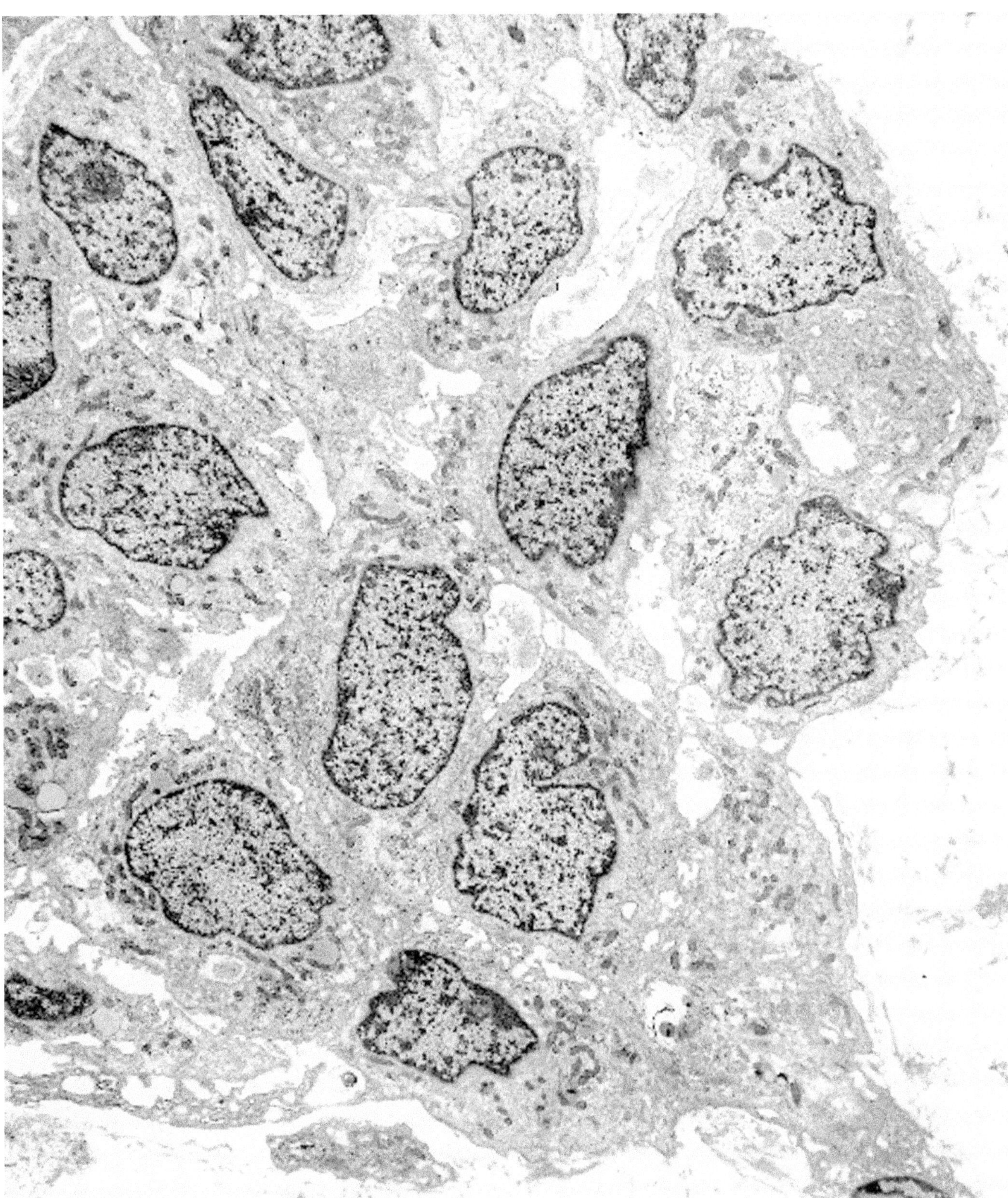

Figure 9.27. Pleomorphic adenoma (parotid gland). An island of irregularly shaped neoplastic cells is incompletely demarcated from the surrounding edematous stroma by basal lamina. Most of the cells are separated from one another, and their lateral borders are raised into narrow processes. (× 5300)

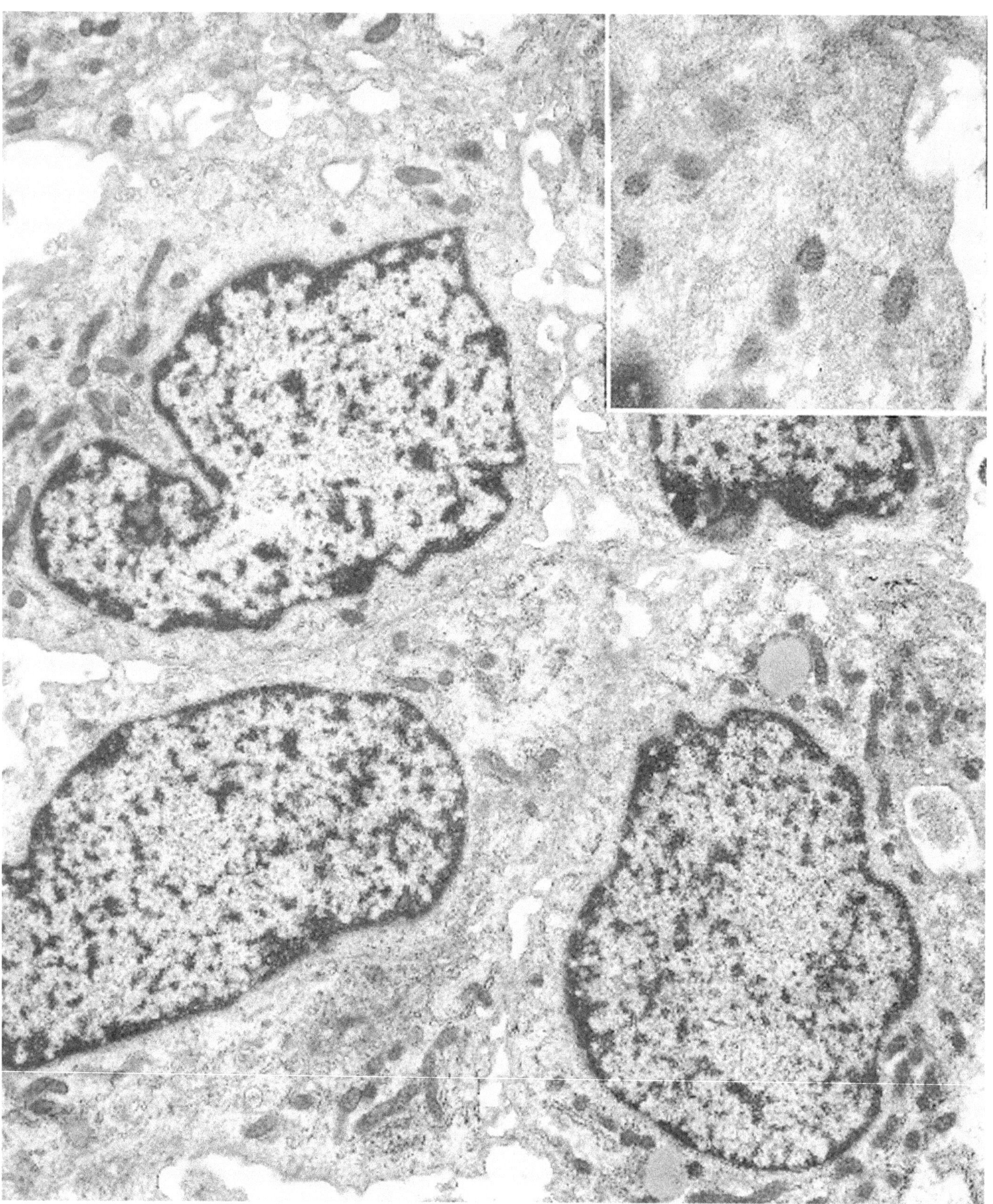

Figure 9.28. Pleomorphic adenoma (parotid gland). Higher magnification of the island of cells illustrated in Figure 9.27 reveals a moderate number of cytoplasmic filaments (see *Inset*) but no dense aggregates of them. The cells are myoepithelial in type but less differentiated than their nonneoplastic counterpart. (× 12,500) (*inset* × 24,500)

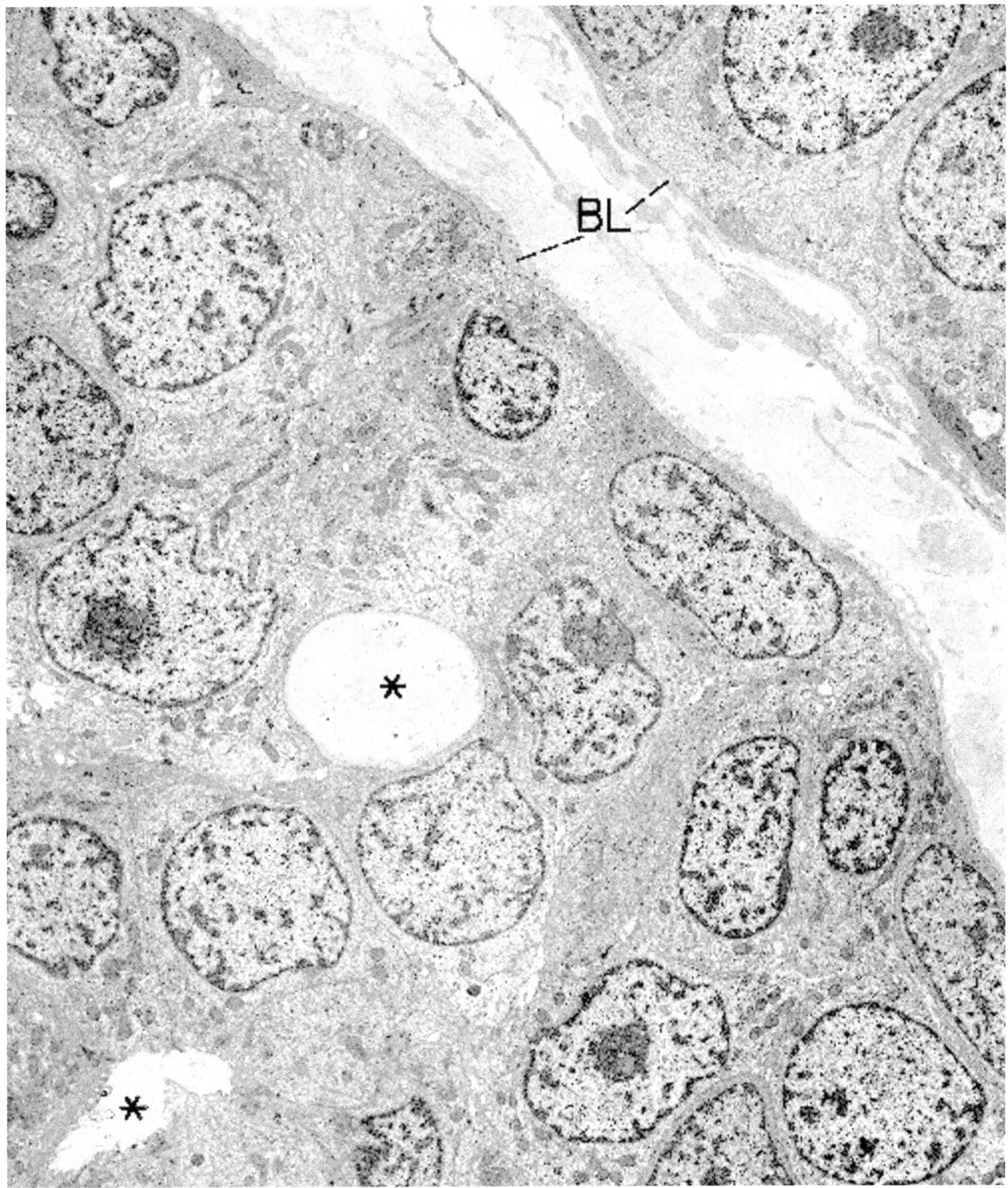

Figure 9.29. Adenoid cystic carcinoma (bronchus). Islands of polygonal cells are surrounded by basal lamina (BL) and contain scattered spaces (*). Even at this low power the upper space is discernible as being a pseudolumen filled with basal lamina. (× 5720)

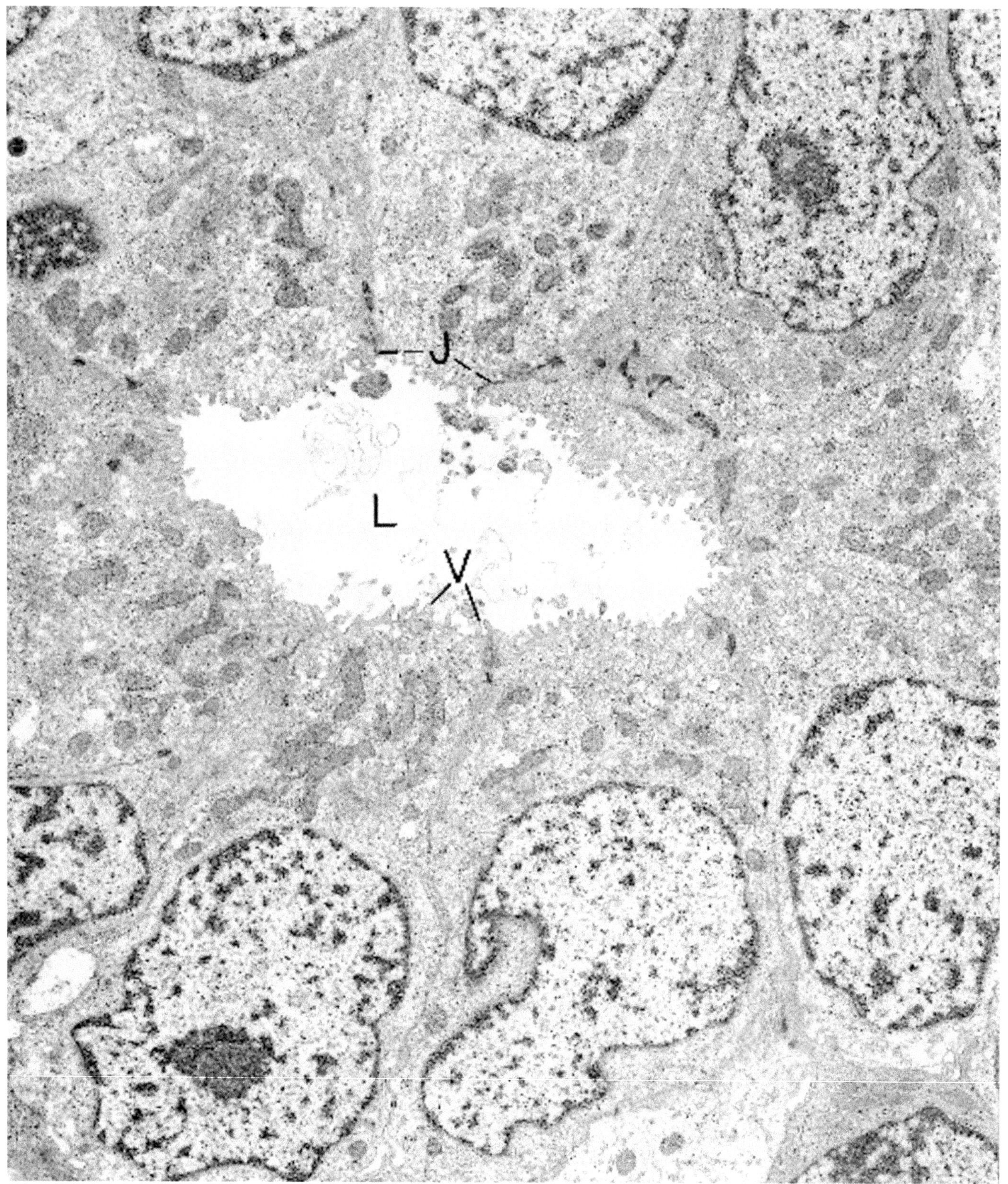

Figure 9.30. Adenoid cystic carcinoma (bronchus). This space represents a true lumen (L), characterized by microvilli (V) and junctional complexes (J). (× 5500)

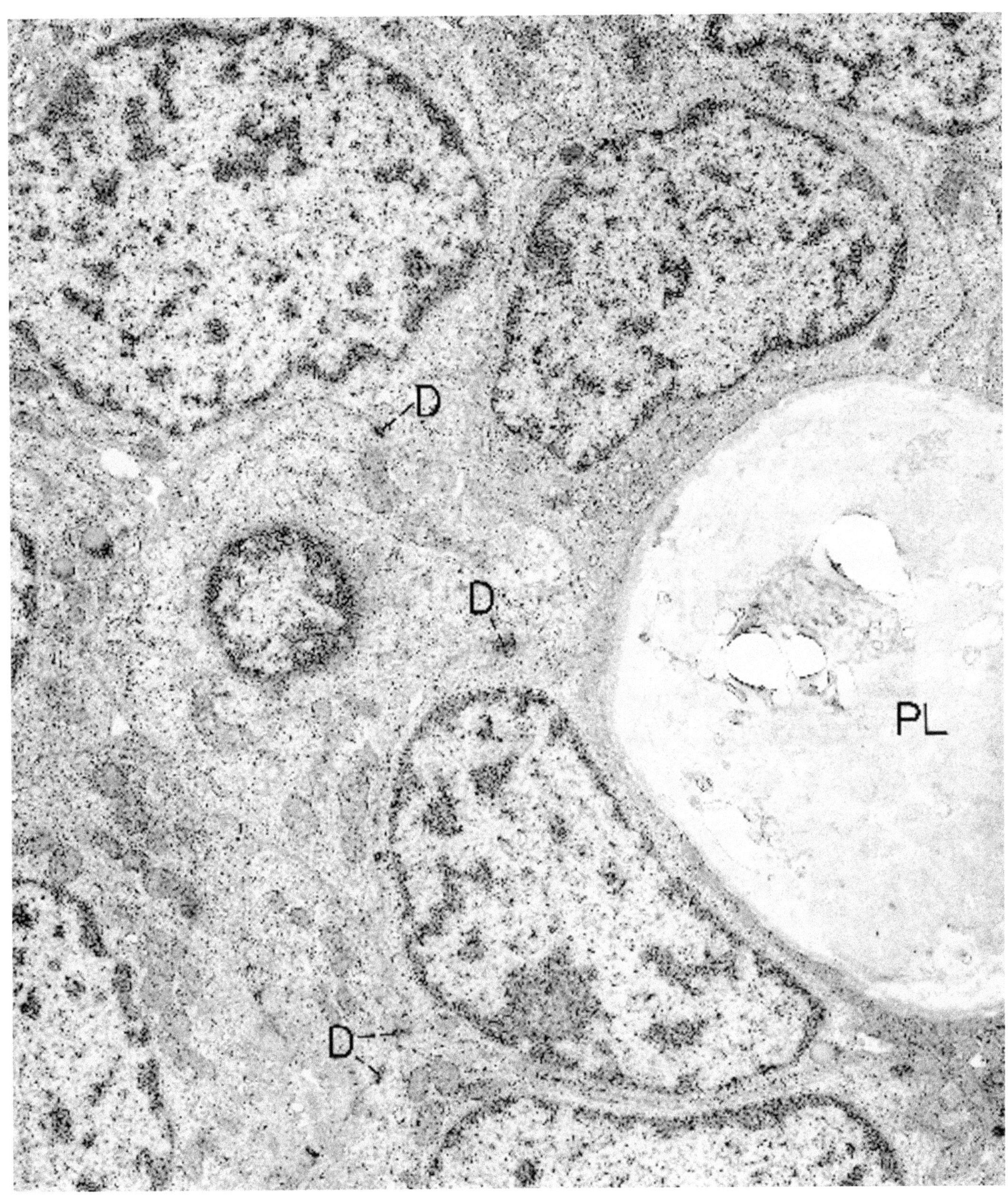

Figure 9.31. Adenoid cystic carcinoma (bronchus). Contrasted to the true lumen illustrated in Figure 9.30, this pseudolumen (PL) is filled with basal laminar material, and the adjacent cells have no microvilli or junctional complexes. The cells are poorly differentiated myoepithelial cells, and although desmosomes (D) are conspicuous, cytoplasmic filaments are not prominent. (× 9100)

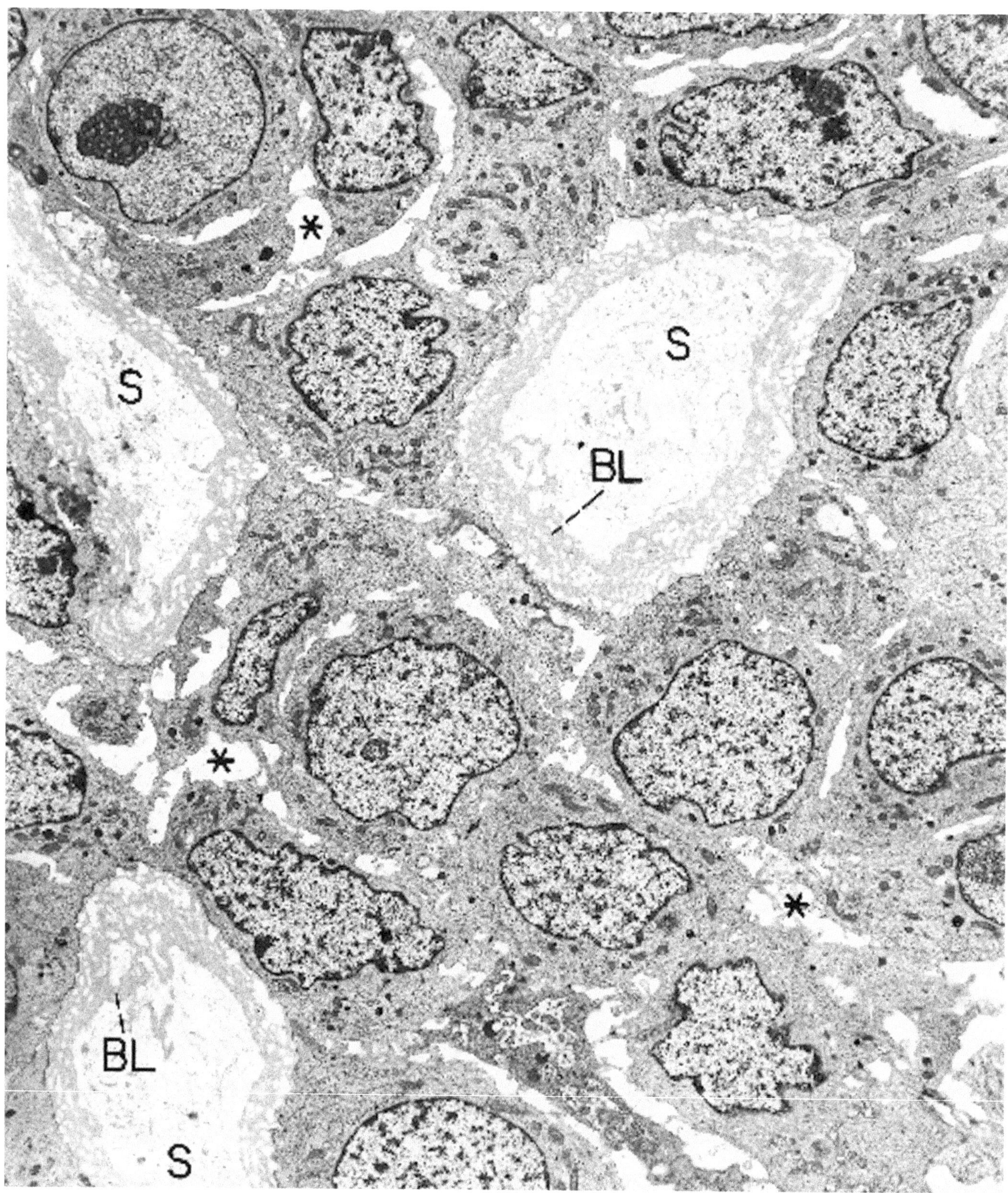

Figure 9.32. Adenoid cystic carcinoma (breast). Islands of poorly differentiated myoepithelial cells are separated by irregular cyst-like spaces (S) filled with basal laminar material (BL). The cells tend to be separated from one another and to have lateral cytoplasmic processes projecting into the intercellular spaces (*). (× 5090)

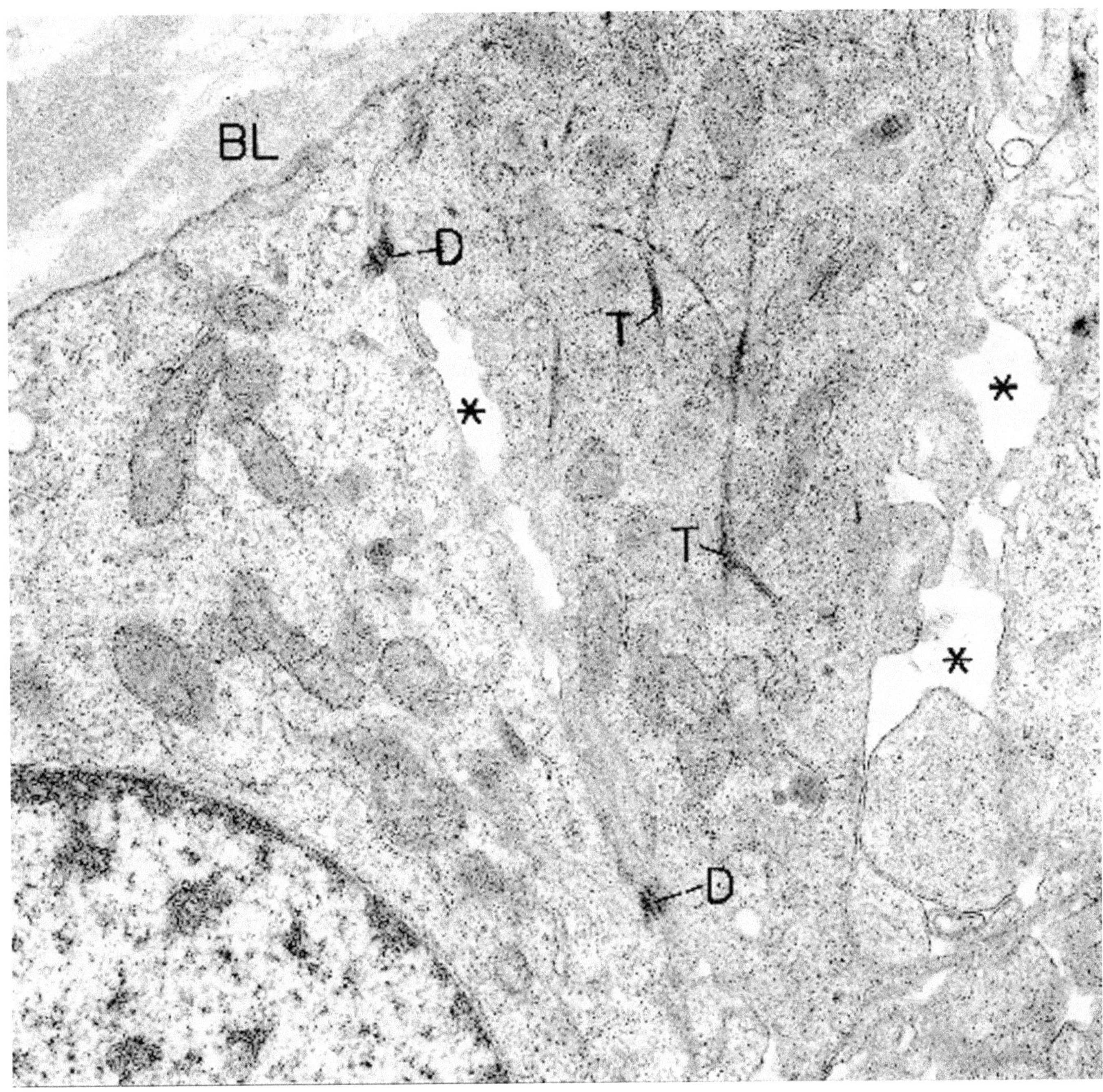

Figure 9.33. Adenoid cystic carcinoma (bronchus). High magnification of myoepithelial cells shows a few tonofibrils (T), consistent with mild squamous metaplasia. Desmosomes (D), basal lamina (BL), and intercellular spaces (*) also are evident in this field. (× 21,200)

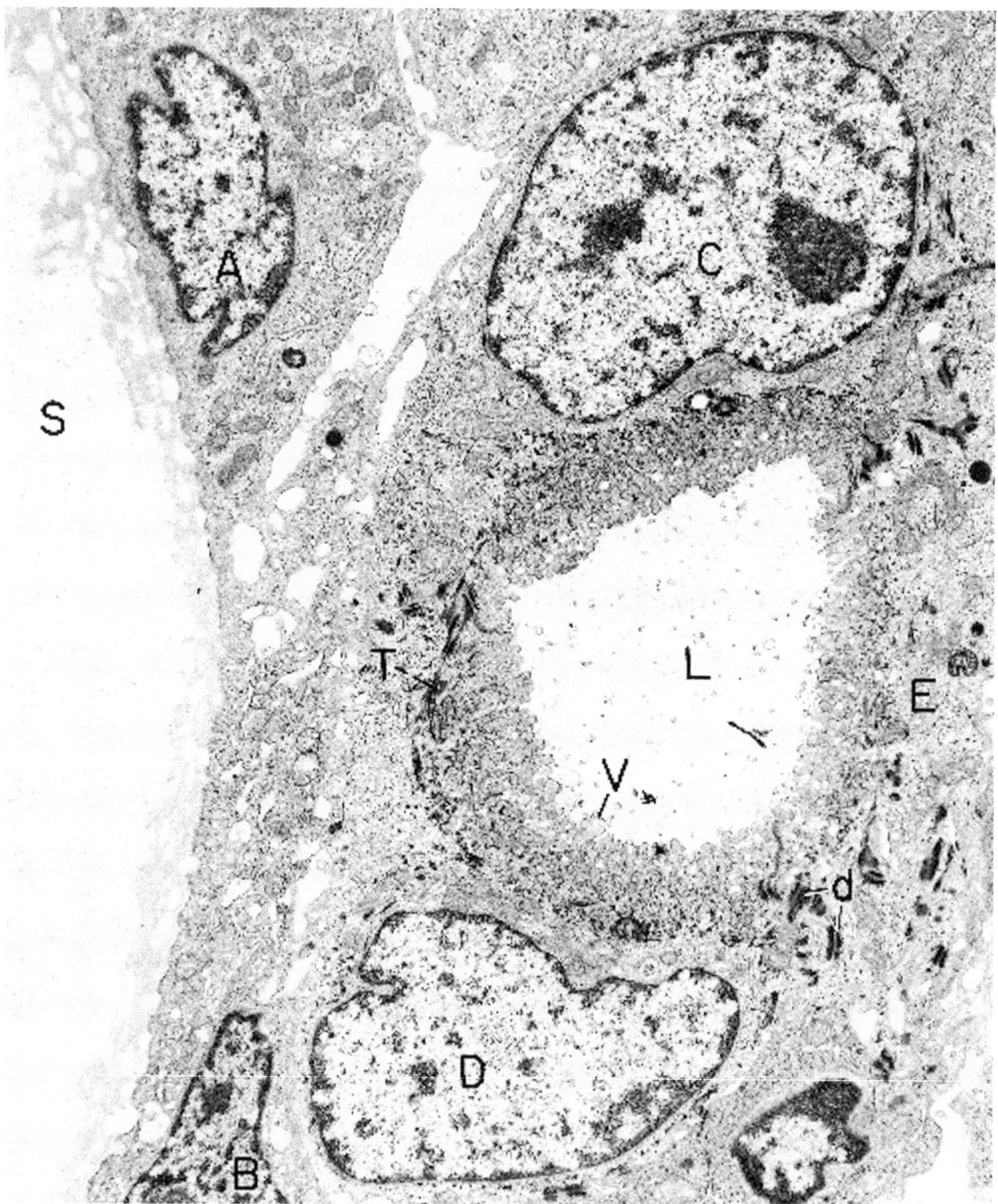

Figure 9.34. Adenoid cystic carcinoma (breast). The two cells (A and B) lining the basal-lamina-filled spaces (S) on the left are consistent with myoepithelial cells, whereas the ones (C, D, and E) lining the true lumen (L) on the right have tonofibrils (T), extra-large desmosomes (d), and many short, blunt microvilli (V). (× 11,000)

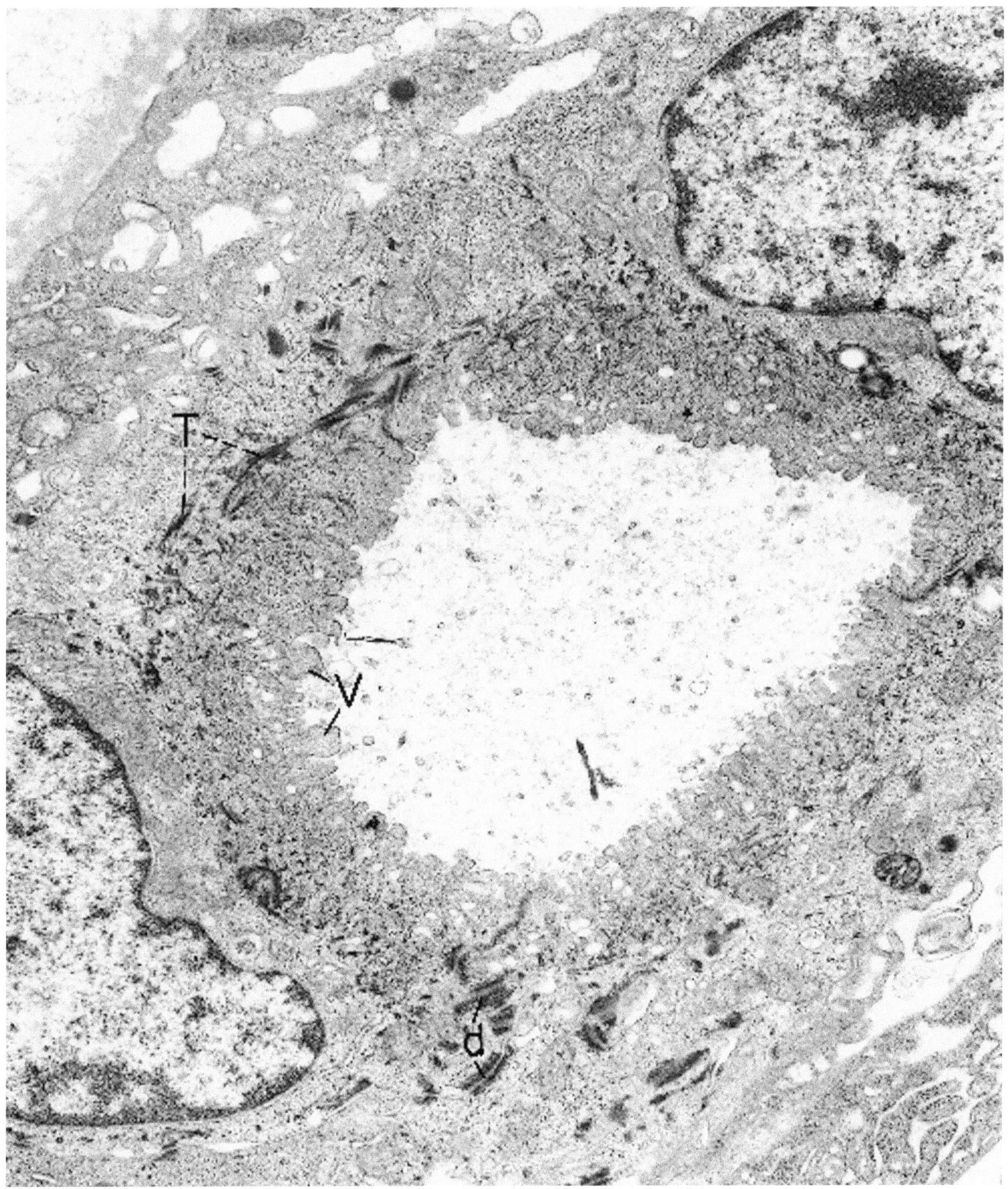

Figure 9.35. Adenoid cystic carcinoma (breast). Higher magnification of the luminal lining cells of Figure 9.34 provides more detail of the extra-large desmosomes (d) and the short, blunt microvilli (V). T = tonofibrils. (× 15,900)

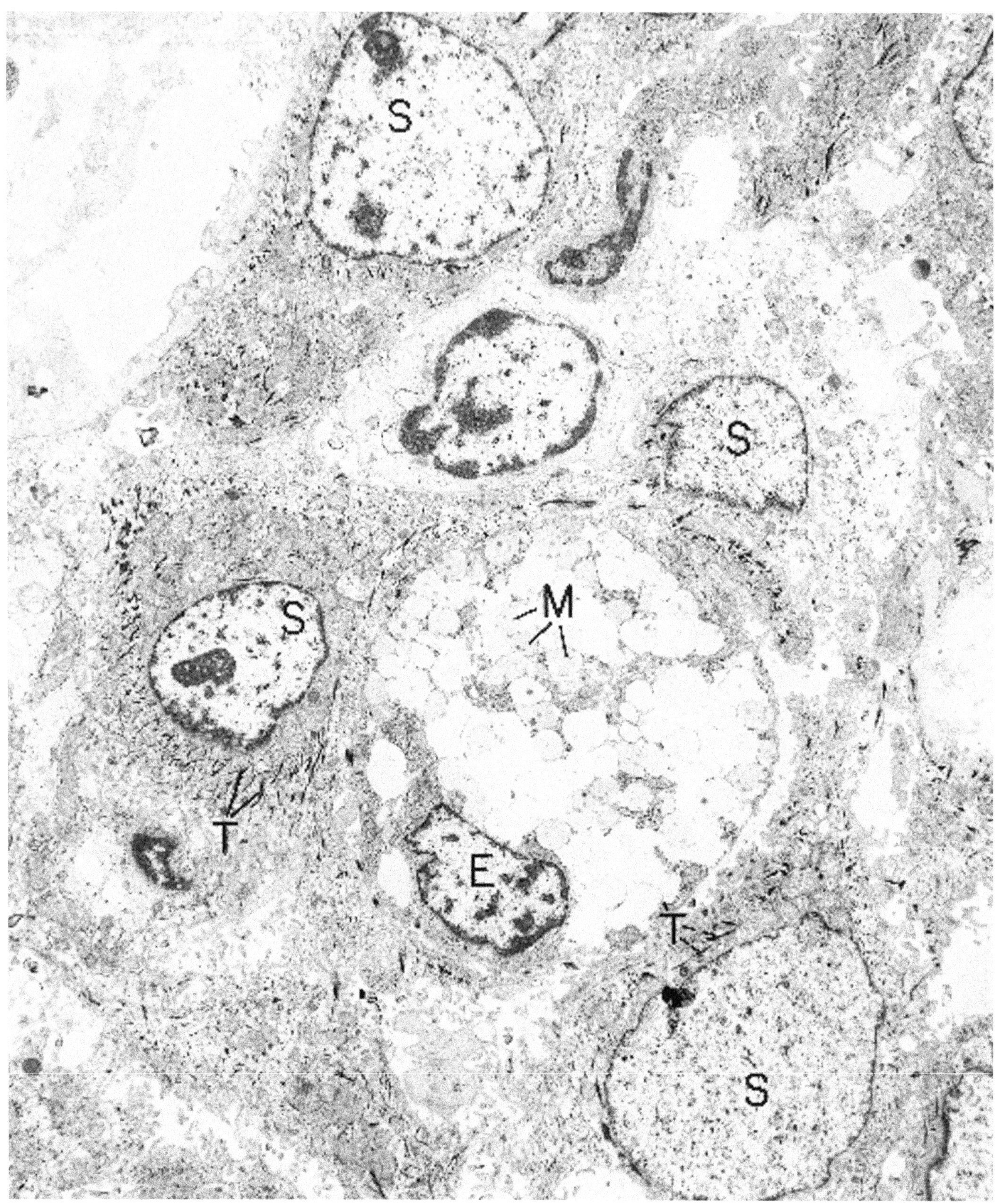

Figure 9.36. Mucoepidermoid carcinoma (bronchus). A solid island of metaplastic squamous cells (S) and a goblet-type epithelial cell (E). The tonofibrils (T) mark the squamous cells, and the mucous granules (M) characterize the goblet cell. (× 4880)

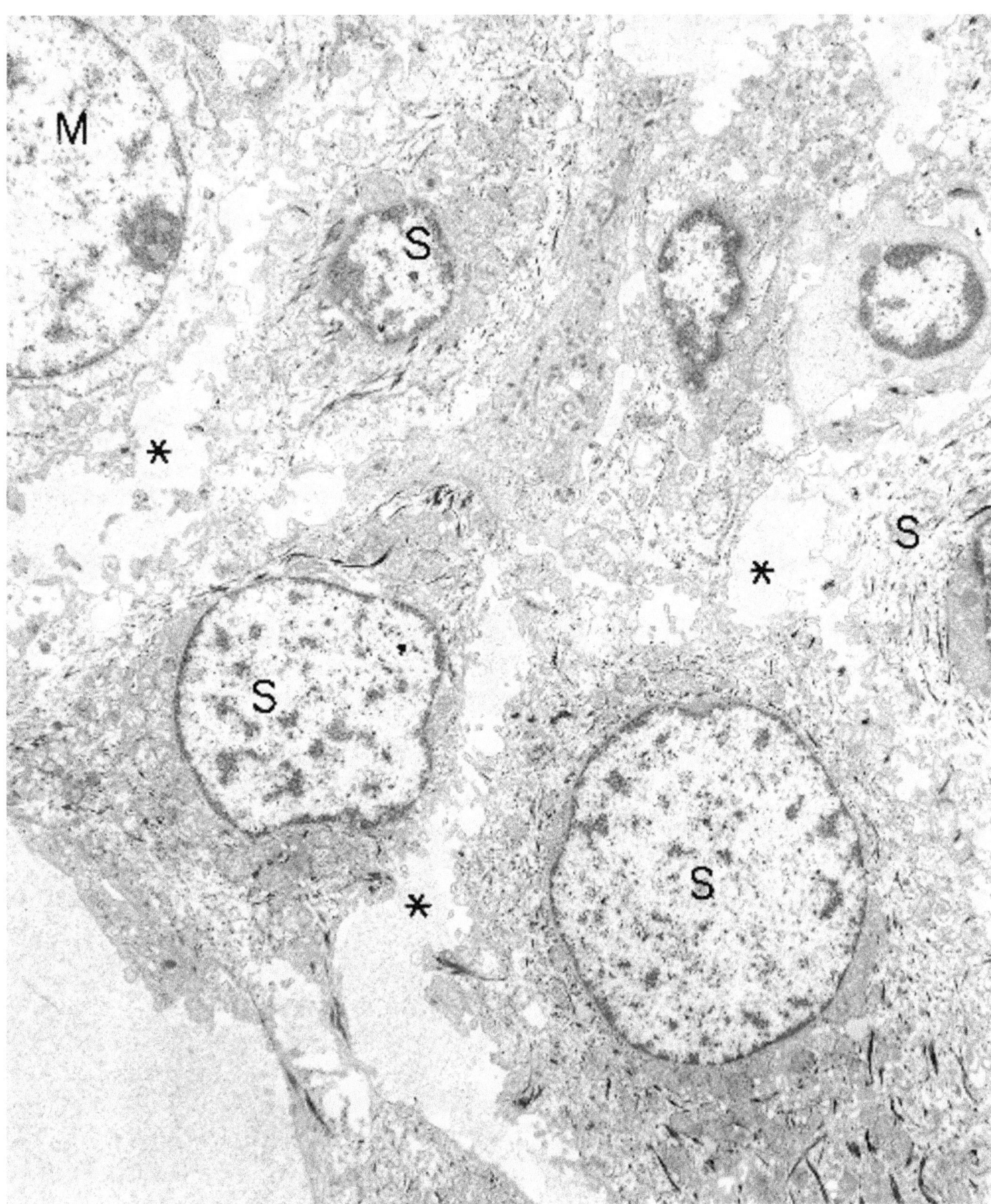

Figure 9.37. Mucoepidermoid carcinoma (bronchus). This field of myoepithelial (M) and metaplastic squamous (S) cells illustrates the tendency of the cells to separate and for the intercellular spaces (*) to be filled by basal laminar material and lined by villus-like cellular projections. (× 6960)

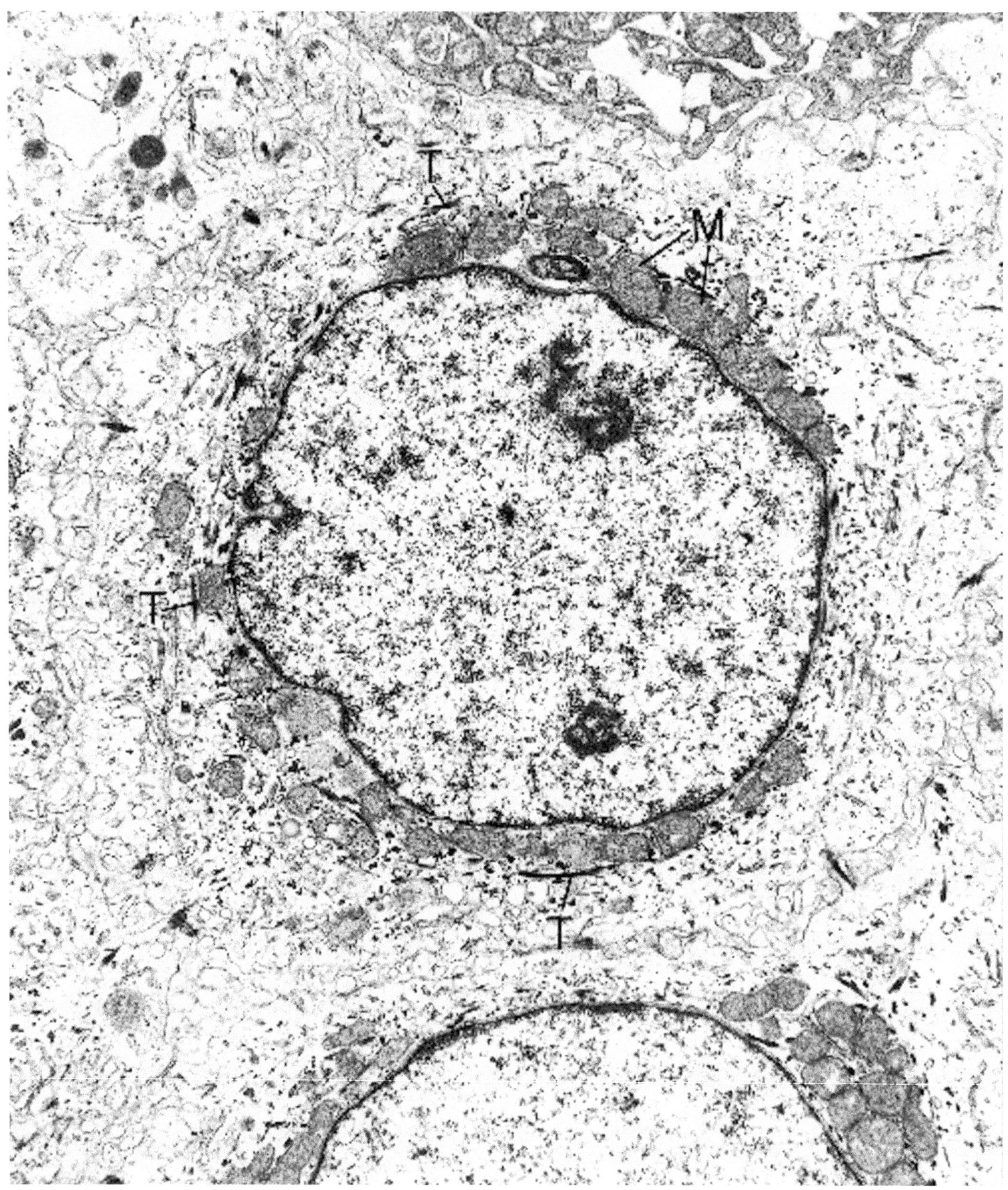

Figure 9.38. Mucoepidermoid carcinoma (bronchus). Higher magnification of two squamous-type cells reveals the tonofibrils (T) to have a predilection for a perinuclear, concentric arrangement. The mitochondria (M) also are located predominantly in the perinuclear region. (× 14,500)

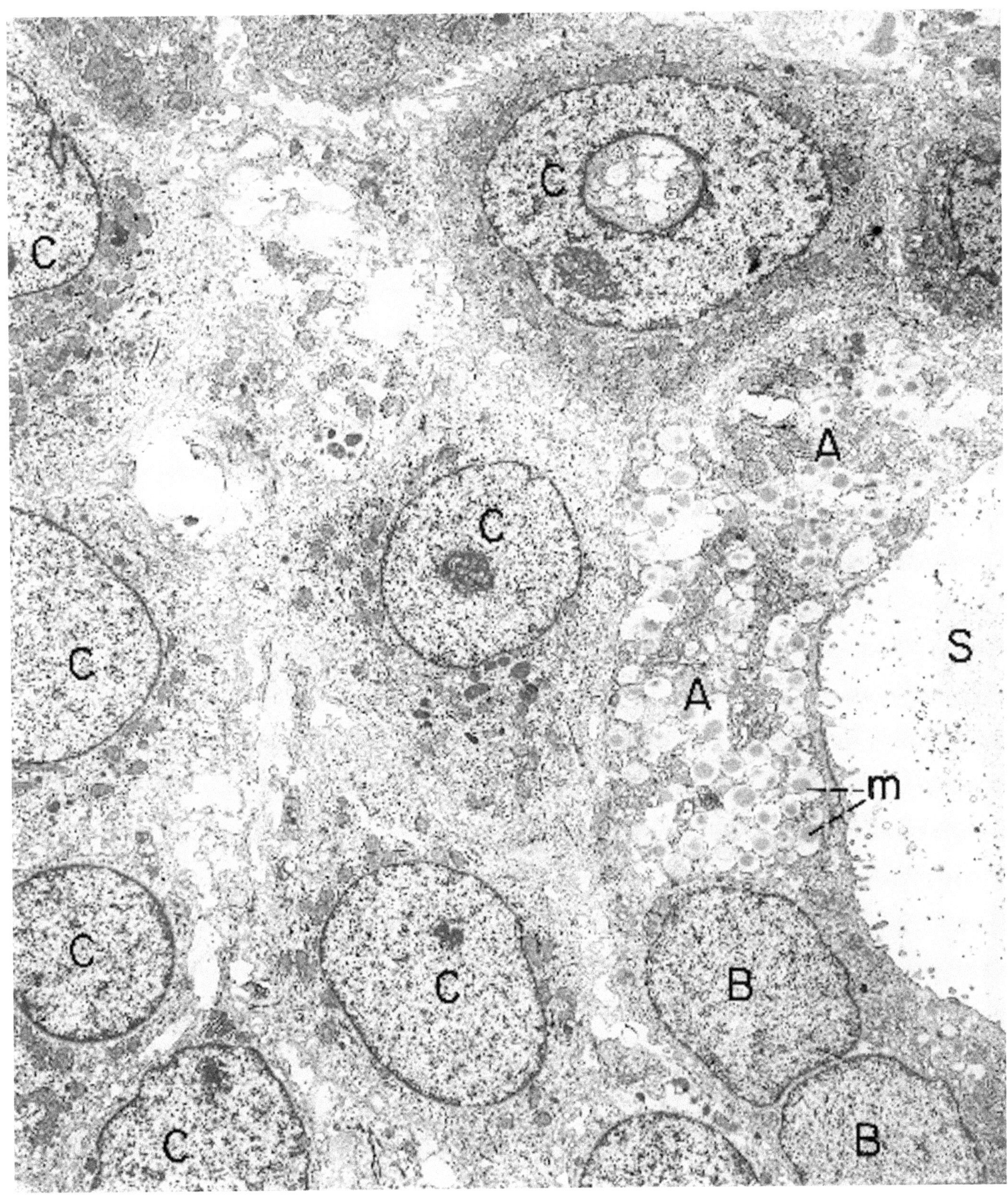

Figure 9.39. Mucoepidermoid carcinoma (bronchus). A small gland-like space (S) is lined by two types of epithelial cells (A and B), both of which have microvilli, but only one (A) of which contains mucinous granules (m). The cells (C) adjacent to the lining cells show mild squamous differentiation (seen better at higher magnification in Figure 9.40). (× 4880)

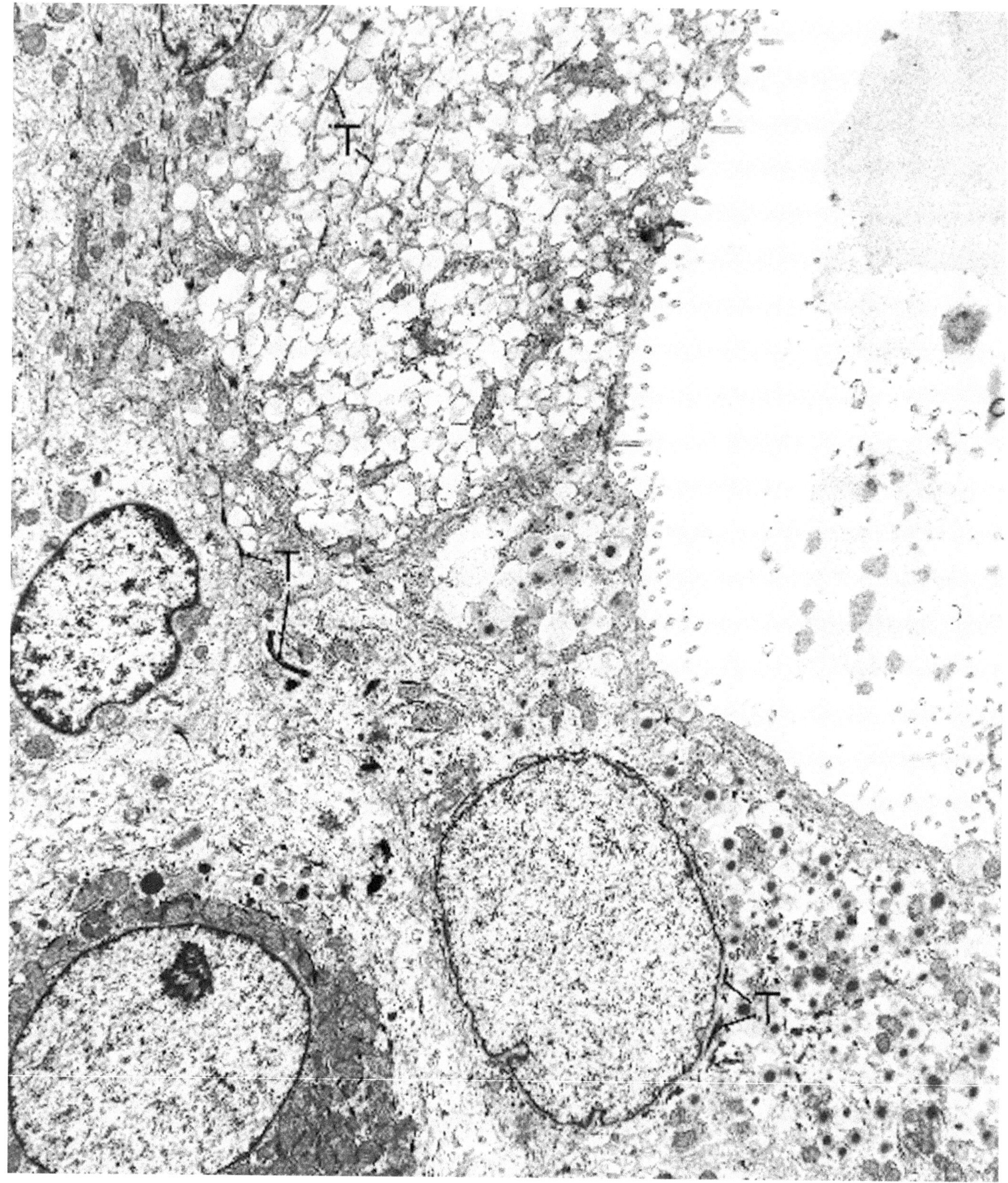

Figure 9.40. Mucoepidermoid carcinoma (bronchus). Another field of the neoplasm depicted in Figure 9.39 reveals that some of the mucous lining cells also contain tonofibrils (T). (× 6960)

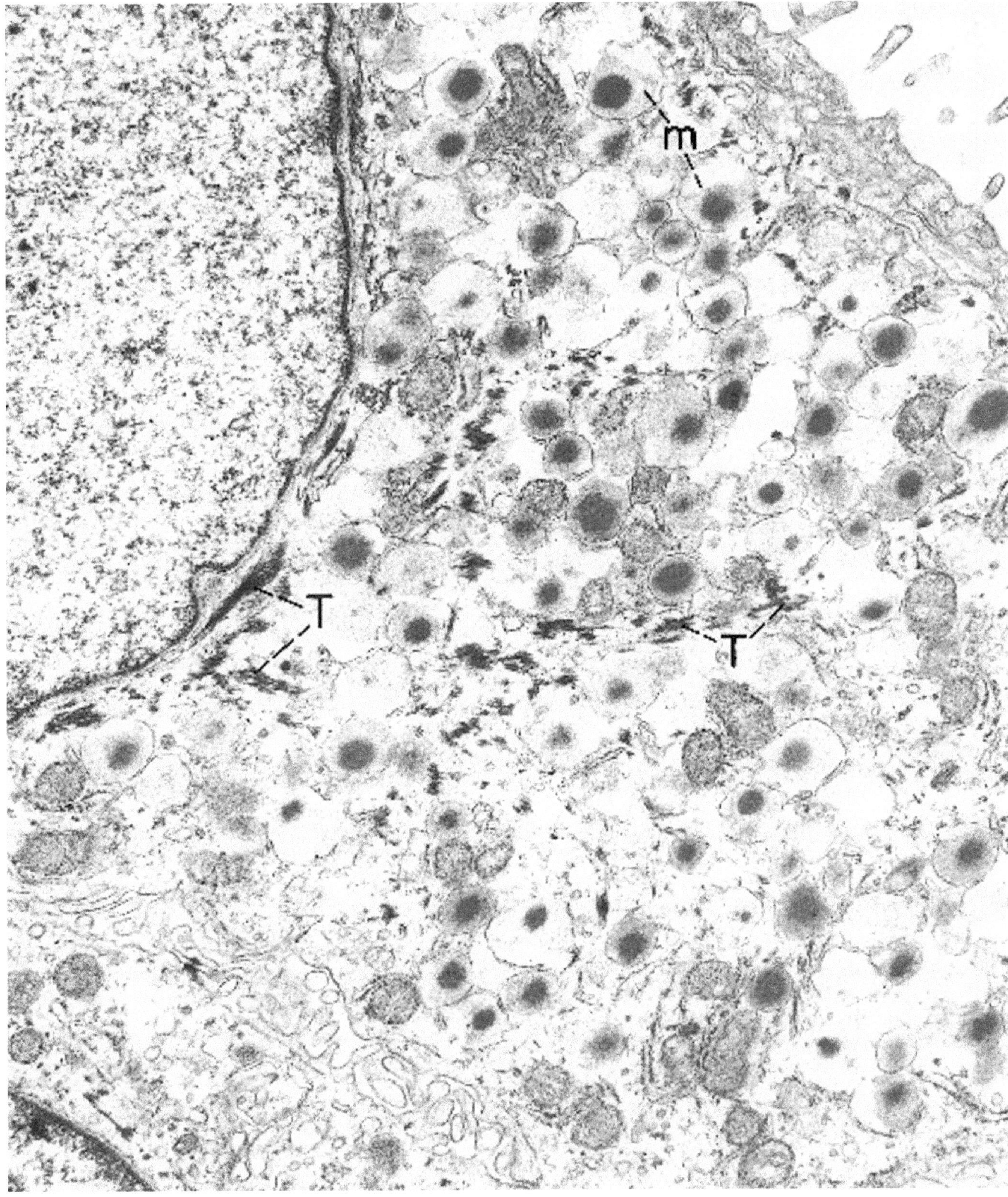

Figure 9.41. Mucoepidermoid carcinoma (bronchus). High magnification reveals granules of mucus (m) and tonofibrils (T) within the same glandular lining cell. (× 23,500)

(Text continued from page 585)

Alveolar Soft-Part Sarcoma

(Figures 9.42 through 9.45.)

Diagnostic criteria. (1) Small groups of polyhedral cells with abundant cytoplasm, surrounded by basal lamina and thin-walled blood vessels (Figures 9.42); (2) small junctions; (3) small, villus-like projections focally between cells, but no true lumens; (4) numerous mitochondria (Figure 9.42); (5) stacked rough endoplasmic reticulum (Figure 9.42); (6) well-developed Golgi apparatuses (Figure 9.43); (7) small-to-large round granules, usually in the region of the Golgi (Figure 9.43); (8) large polygonal, often rhomboid crystals (Figures 9.43 and 9.44); (9) varying amounts of glycogen (Figure 9.42).

Additional points. The crystals are the most distinctive feature of the neoplastic cells. They have an internal structure of parallel filaments with a constant periodicity, and a limiting membrane is often visible. They develop progressively from the round granules, and transitional forms are usually identifiable (Figure 9.45). Rough endoplasmic reticulum and glycogen vary in amount. Some of the groups of cells have central, degenerating cells that account for the pseudolumens seen by light microscopy. Morphologic evidence is still incomplete to prove any of the theories of histogenesis of alveolar soft-part sarcoma, including smooth muscle, nonchromaffin paraganglia, and renal juxtaglomerular-like cells. Theories of origin that probably can be ruled out are those that suggest skeletal muscle and Schwann cells.

Chordoma

(Figures 9.46 through 9.50.)

Diagnostic criteria. (1) Round and polyhedral cells arranged singly and in cords and clusters, in a flocculent (mucopolysaccharide) matrix (Figure 9.46); (2) cytoplasmic processes of one cell wrapping around an adjacent cell, leaving an intervening space (Figures 9.47 and 9.48); (3) villus-like projections from portions of the cell surface (Figures 9.46 to 9.49); (4) basal lamina on perivascular side of cell; (5) prominent desmosomes (Figures 9.48 and 9.49); (6) varying number and size of cytoplasmic vacuoles produced by dilated rough endoplasmic reticulum; (7) many mitochondria, often in one compartment of the cytoplasm; (8) mitochondria–rough endoplasmic reticulum complexes, in which a cisterna of rough endoplasmic reticulum arches over and follows the contour of a mitochondrion (Figure 9.49); (9) large accumulations of glycogen in some cells (Figures 9.46, 9.47, 9.49, and 9.50); (10) intermediate filaments, sometimes in bundles; (11) indentations (pseudoinclusions) of the nucleus by the cytoplasm; (12) invaginations of the cytoplasm by the extracellular matrix (Figure 9.46).

Additional points. Other cellular features include varying amounts of smooth endoplasmic reticulum, prominent Golgi apparatuses (Figure 9.49), and pinocytotic vesicles. The stellate and physaliferous cells comprising chordomas are the same cell type when viewed ultrastructurally. Their difference is in the extent of vacuolization, and the vacuoles seen at the light microscopic level prove ultrastructurally to be the result of several factors: dilatation of rough endoplasmic reticulum, cytoplasmic glycogen, and invaginations of the cytoplasm by the extracellular matrix. Furthermore, the intercellular spaces created by one cell wrapping around another may be discerned as intracellular vacuoles at the light microscopic level. Although dilated rough endoplasmic reticulum filled with medium-dense material represents active production of proteinaceous substances such as collagen, glycosaminoglycan (mucopolysaccharide), and glycoproteins, dilated empty cisternae of rough endoplasmic reticulum represent a degenerative or artifactual change. The dilated reticulum in physaliferous cells may be of both types, but the empty type is more usual. *Chondroid chordoma* is somewhat controversial in regard to cell type(s), but a variant of chordoma rather than a form of chondrosarcoma is more probable. *Parachordoma,* a rare nonaxial and extraskeletal tumor, has the same ultrastructural features as axial chordoma.

(Text continues on page 613)

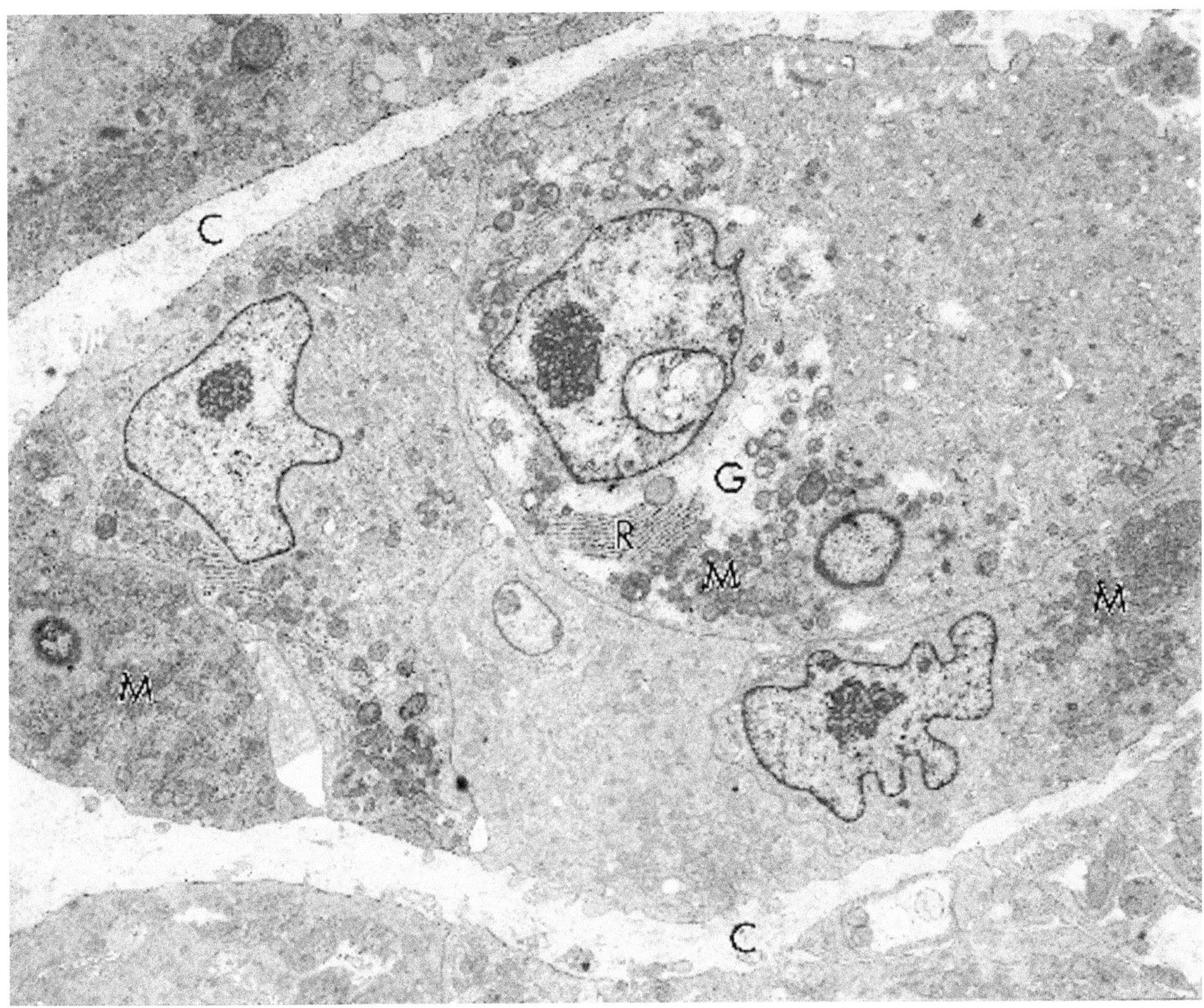

Figure 9.42. Alveolar soft-part sarcoma (soft tissue of thigh). Groups of closely apposed oval and polygonal cells are separated by bands of collagenous matrix (C). The cells have abundant cytoplasm with numerous mitochondria (M), open spaces of glycogen (G), and collections of stacked rough endoplasmic reticulum (R). (× 5300)

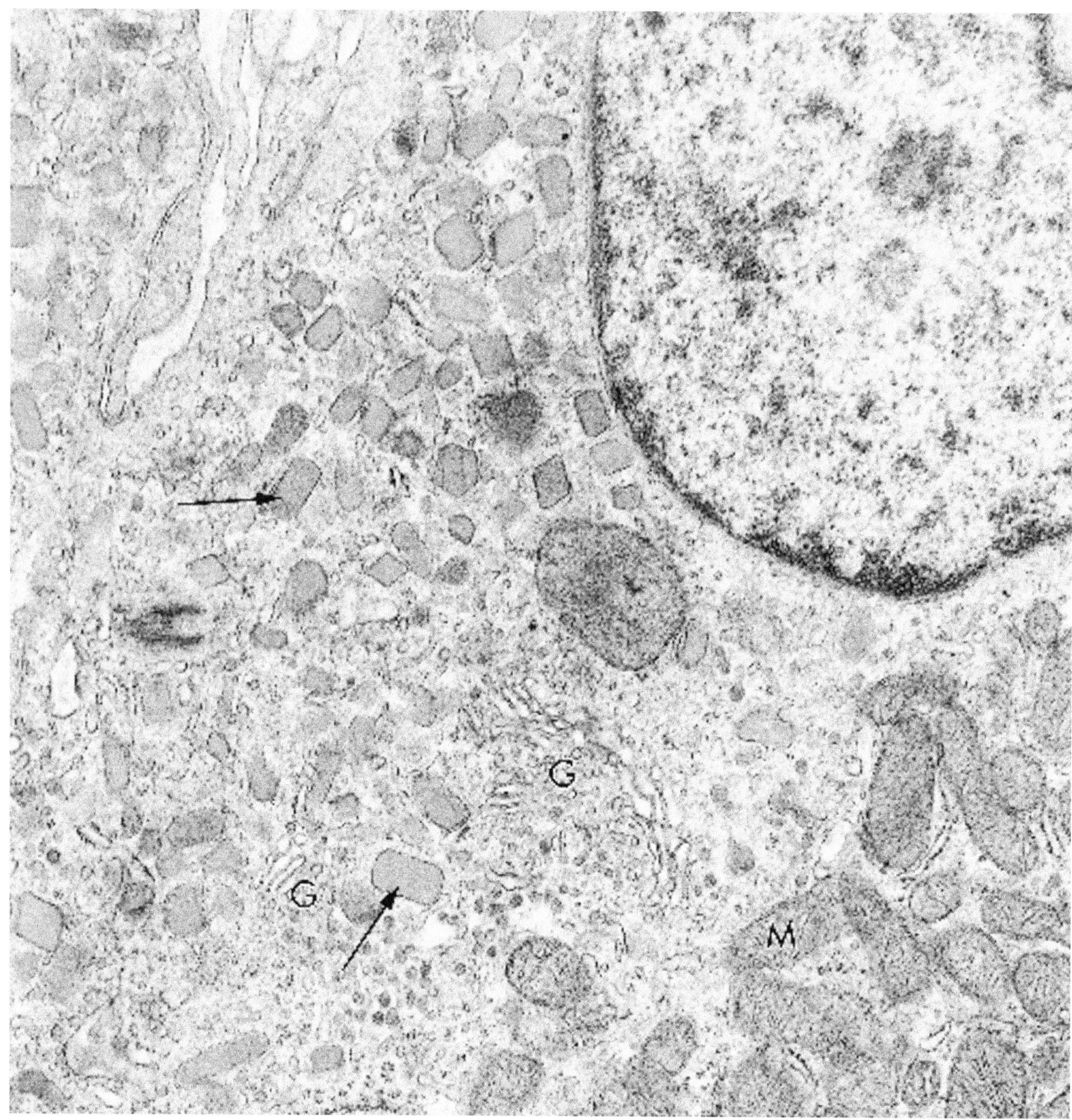

Figure 9.43. Alveolar soft-part sarcoma (soft tissue of thigh). High magnification of one of the neoplastic cells depicts a large Golgi apparatus (G) with numerous small clear and dense granules, numerous mitochondria (M), and numerous angular, crystalline granules (*arrows*). (× 38,000)

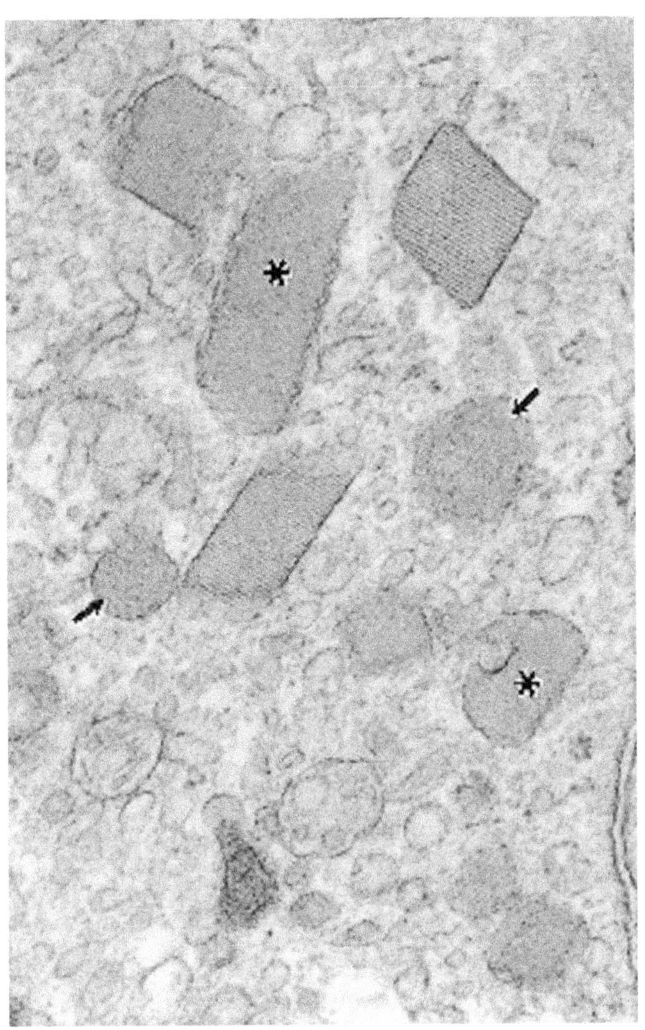

Figure 9.44. Alveolar soft-part sarcoma (soft tissue of thigh). High magnification of several round granules (*arrows*) and rhomboid crystalline granules (*) reveals the latter to have an internal linear pattern. (× 56,000)

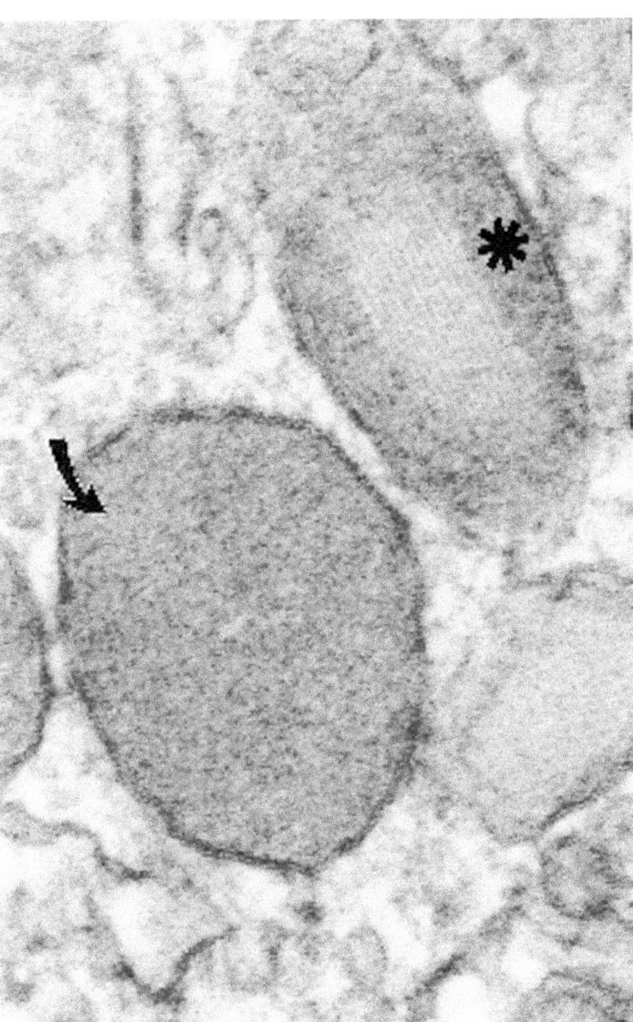

Figure 9.45. Alveolar soft-part sarcoma (soft tissue of thigh). Ultra-high magnification of two crystals shows one to be of the round or oval type (*arrow*) and the other to be transitional from round to angular type (*). (× 85,000)

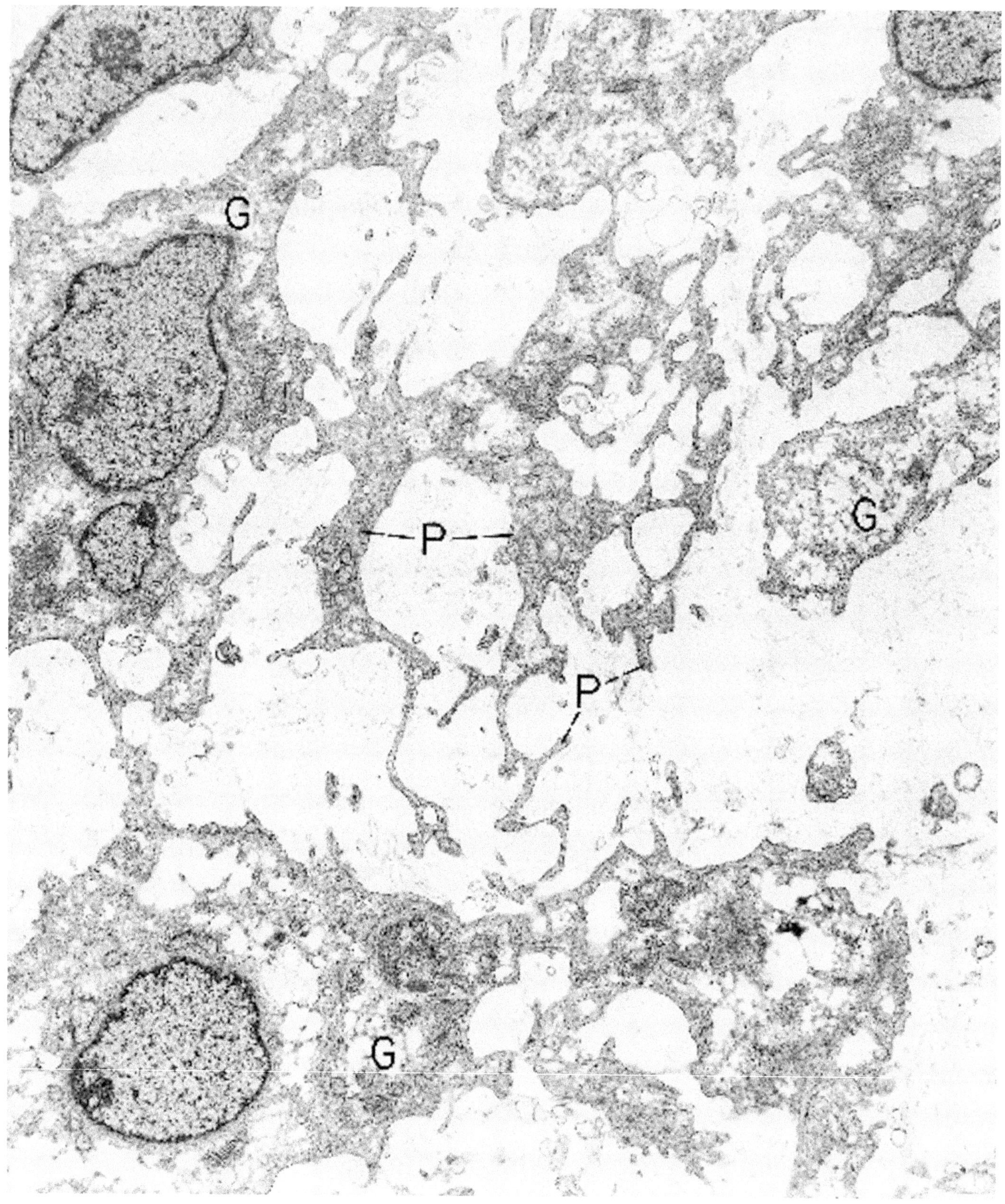

Figure 9.46. Chordoma (cervical vertebra). Clusters of neoplastic cells are distributed in a clear and flocculent matrix. The cells have numerous complicated processes (P) and escalloped open spaces of glycogen (G) in the cytoplasm. (× 7290)

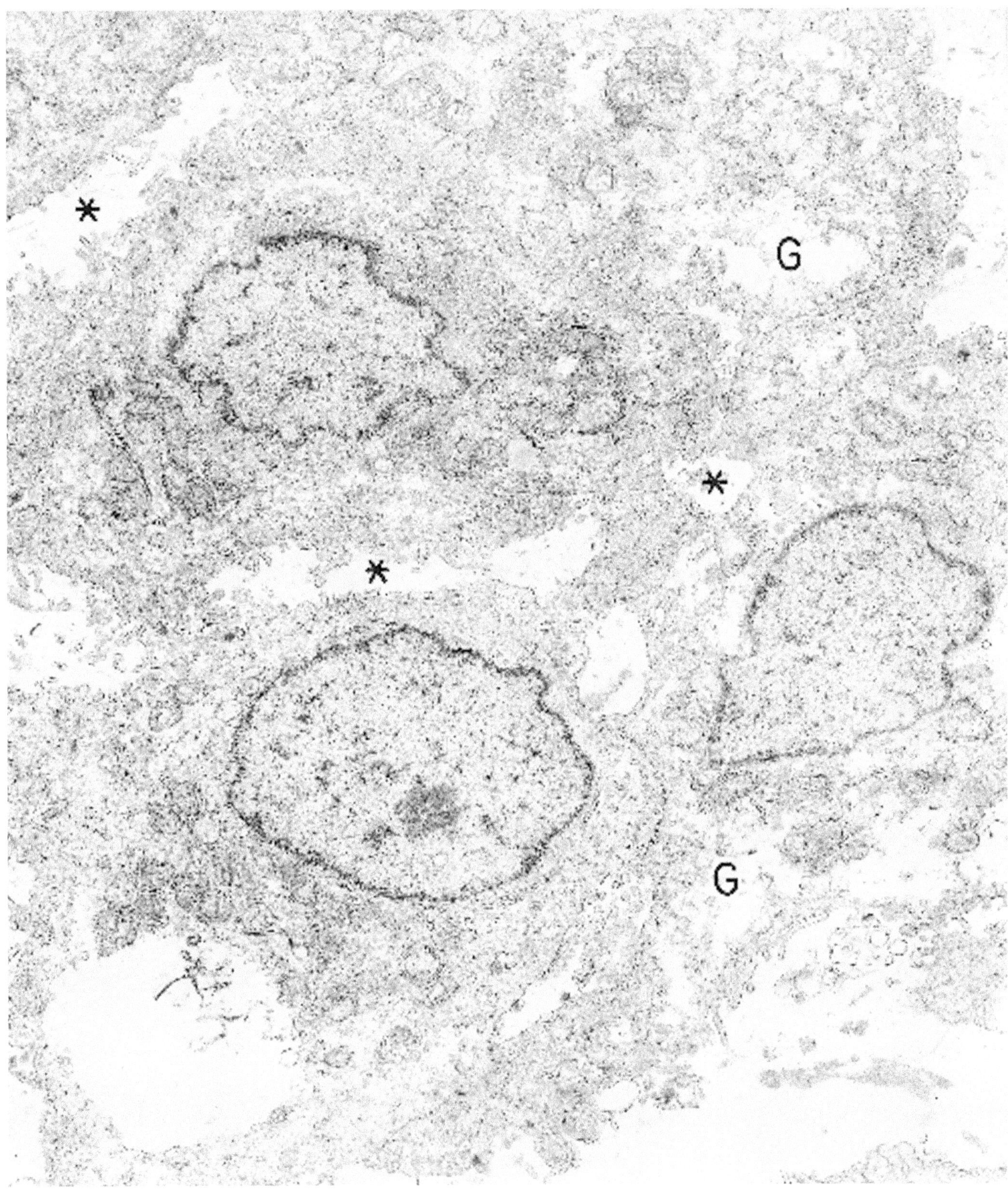

Figure 9.47. Chordoma (lumbar vertebra). Round and polygonal neoplastic cells focally abut one another but elsewhere are separated by channel-like spaces (*). The florid villus-like processes are recognized more easily in the spaces than where adjacent cell borders abut one another. The cytoplasm contains many organelles, and escalloped clear spaces of glycogen (G) are extensive. (× 9880)

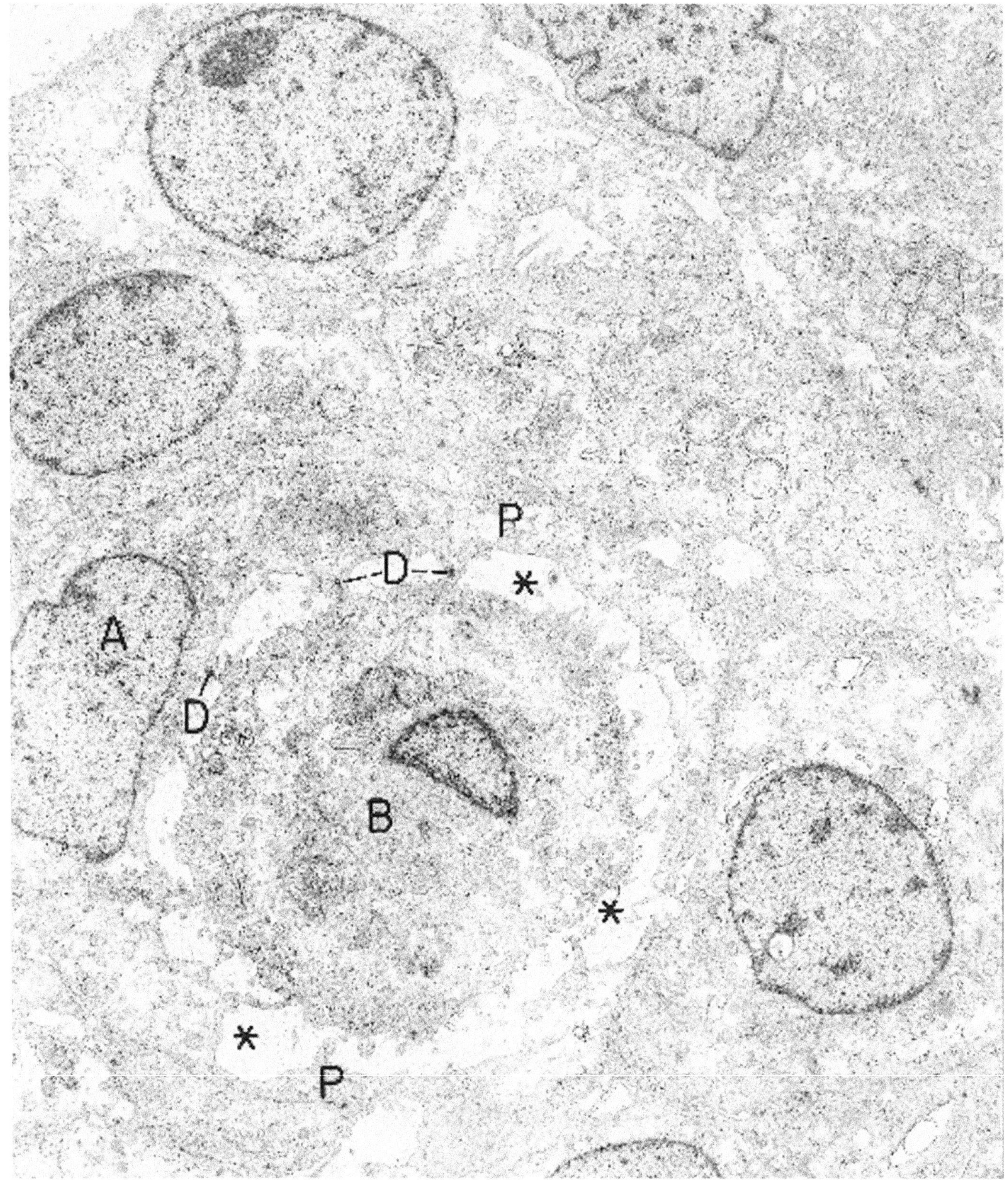

Figure 9.48. Chordoma (lumbar vertebra). Processes (P) of one neoplastic cell (A) tend to wrap around another cell (B), creating a channel-like intercellular space (*). Villus-like projections off the main processes (P) are connected by desmosomes (D) to similar projections from the enwrapped cell. Nuclei usually are oval and have a small amount of heterochromatin and nucleoli of moderate size. (× 7560)

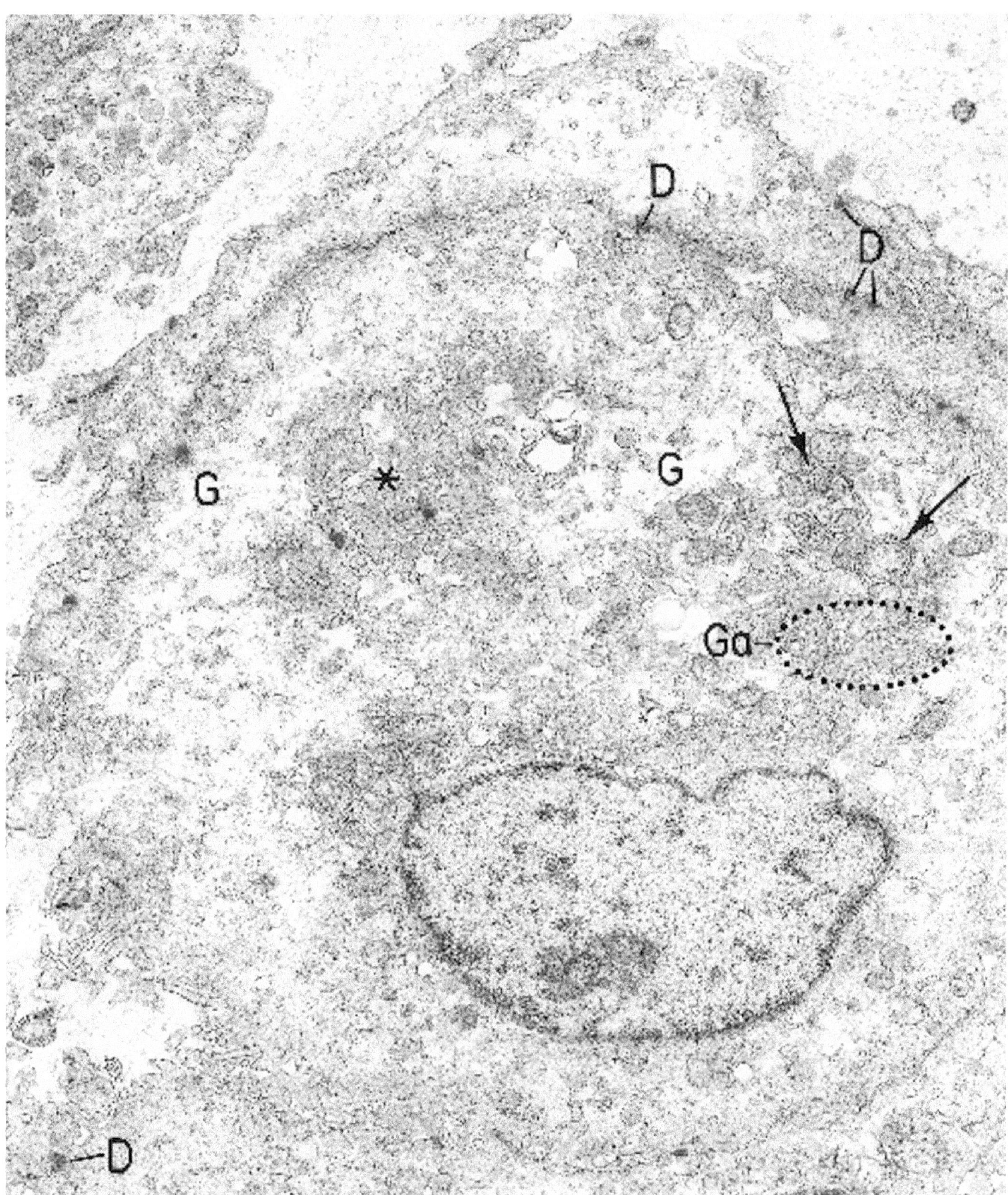

Figure 9.49. Chordoma (lumbar vertebra). High magnification of a neoplastic cell illustrates several characteristic features: copious cytoplasmic glycogen (G, the clear, glassy spaces); mitochondria–rough endoplasmic reticulum complexes (*arrows*); a prominent Golgi apparatus (Ga); a pseudolumen (*) produced by an invagination of the cell by the extracellular matrix; and large desmosomes (D). (× 13,000)

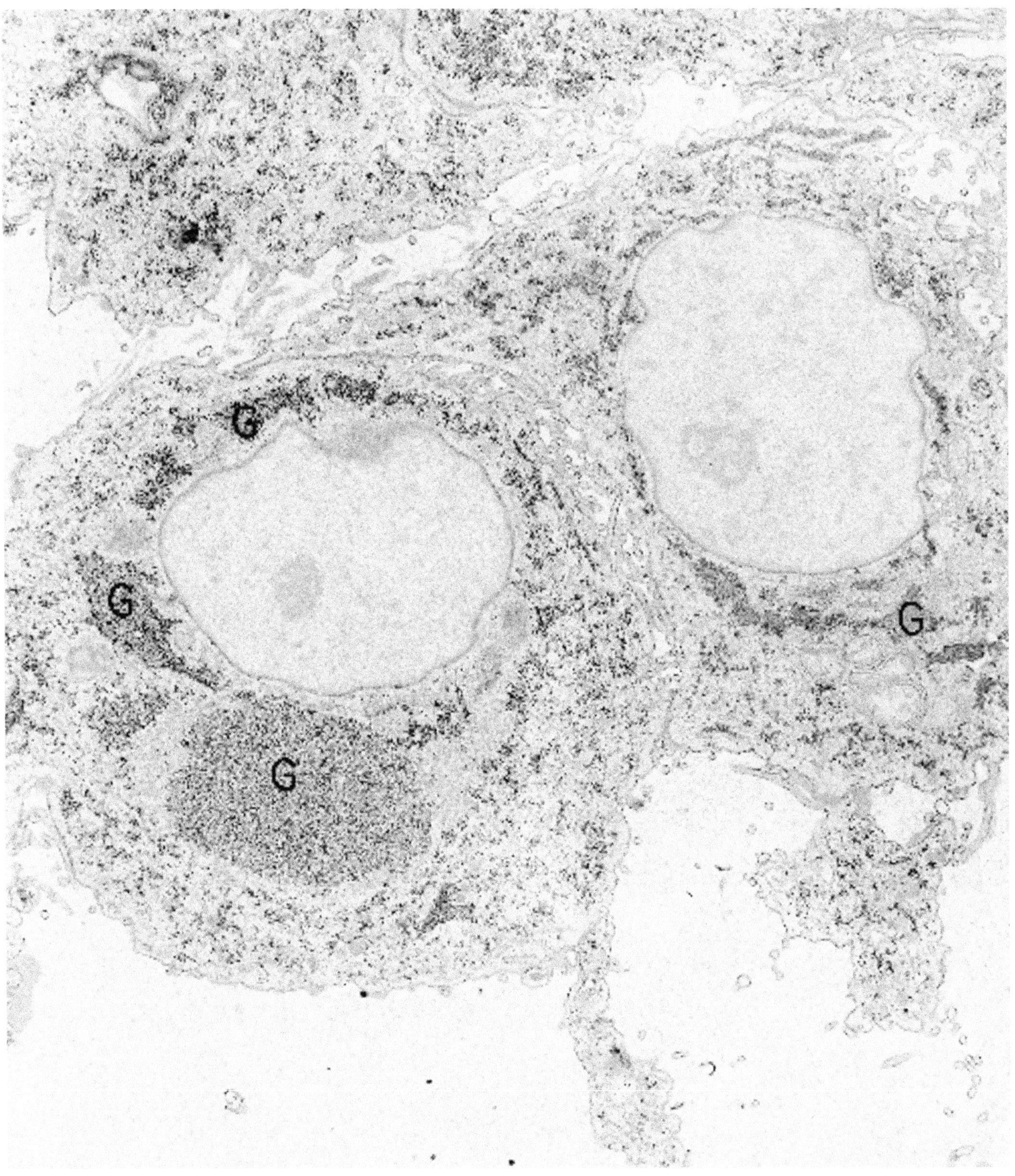

Figure 9.50. Chordoma (lumbar vertebra). This specimen was processed to preserve glycogen as electron-dense granules (G), and the cytoplasm of the neoplastic cells is demonstrated to be rich in this inclusion. (× 6480)

(Text continued from page 604)

Epithelioid Sarcoma

(Figures 9.51 through 9.53.)

Diagnostic criteria. (1) Groups of tightly apposed, large, and oval and polygonal cells with (2) round, oval, or indented nuclei; (3) abundant cytoplasm; with (4) numerous intermediate filaments and tonofibrils; (5) a moderate number of organelles including free ribosomes, mitochondria, and rough endoplasmic reticulum; (6) intercellular junctions, including desmosomes; (7) inconsistent, focal basal lamina around cellular groups; (8) flocculent and collagenous matrix separating epithelioid cellular groups and containing pleomorphic fibroblasts, myofibroblasts, and facultative histiocytes.

Additional points. Variable other features of the epithelioid cells are primary and secondary lysosomes, inclusions of lipid and glycogen, and filopodia at free surfaces of cells. Paranuclear whorls of intermediate filaments, creating a rhabdoid appearance both at electron microscopic and light microscopic levels, may also be found.

The mesenchymal components of epithelioid sarcomas, that is, the fibroblastic family of cells between the epithelioid groupings, are controversial in regard to being malignant versus reactive.

Hepatoblastoma

(Figures 9.54 through 9.58.)

Diagnostic criteria. (1) Small, fetal-type hepatocytes in cords and/or sheets; (2) canaliculi between some cells; (3) vascular sinuses adjacent to hepatoblasts; (4) a moderate amount of cytoplasm with glycogen, variable lipid, and few organelles, including free ribosomes and a moderate number of mitochondria and cisternae of rough endoplasmic reticulum; (5) round and oval nuclei with multiple nucleoli; (6) oval and spindle embryonal cells with scanty cytoplasm and few organelles; and (7) pleomorphic nuclei with large nucleoli.

Additional points. Hepatoblastomas most commonly are composed of epithelial-type cells, but a mixture of epithelial and mesenchymal cells also occurs in some tumors. The epithelial hepatoblastomas are subdivided into *fetal, embryonal,* and *small cell anaplastic* subtypes. Combinations of cellular subtypes may be found in any one neoplasm. The fetal component resembles fetal liver, with cords of two hepatocyte thickness, occasional canaliculi between cells, and sinuses adjacent to the cords. Sheets of cells may also be present. No bile ducts or portal triads accompany the other components. The embryonic subtype or component of hepatoblastoma is composed of less differentiated cells, in varying noncord arrangements. The nuclear–cytoplasmic ratio is high, and glycogen is less abundant than in the fetal type cells. The anaplastic subtype of hepatoblastoma is composed of small, poorly differentiated cells in varying arrangements and with heterochromatic nuclei, scant cytoplasm, and few organelles. The mesenchymal component of mixed hepatoblastomas consists of oval and spindle cells that may show fibroblastic, osteoblastic, chrondroblastic, or rhabdomyoblastic differentiation. Focal neuroendocrine differentiation, characterized by cells with dense-core granules, may be present both in mixed and in pure epithelial hepatoblastomas.

Embryonal Sarcoma of Liver

(Figures 9.59 and 9.60.)

Diagnostic criteria. (1) Undifferentiated, primitive mesenchymal type cells; (2) fibroblasts, myofibroblasts, and facultative histocytes; (3) variably, leiomyoblastic, rhabdomyoblastic, lipoblastic, and epithelial (squamous) cells.

Additional points. Other terms for this neoplasm have been *undifferentiated sarcoma,* and *malignant mesenchymoma.* The neoplastic cells may contain round, membrane-bound, electron-dense bodies that correspond to hyaline bodies visible by light microscopy and that stain for serum proteins such as alpha-1-antitrypsin.

Gastrointestinal Stromal Tumor

(Figures 9.61 through 9.64.)

Diagnostic criteria. (1) Poorly differentiated spindle and/or epithelioid cells; (2) dispersed individually or in tight apposition; with (3) nondescript cytoplasm and nucleus; or with (4) features of leiomyoblasts, fibroblasts, Schwann cells, or autonomic nerve cells.

Additional points. Although many of these tumors are being diagnosed by light microscopy alone, simply as "GI stromal tumor," many of them can be subclassified by electron microscopy. In our experience, many GI stromal tumors have either obvious or subtle ultrastructural features of smooth muscle (see Chapter 6 for descriptions and illustrations of leiomyoblasts, fibroblasts, and Schwann cells).

We have seen one example of a smooth muscle type of GI stromal tumor with so-called skeinoid fibers, peculiar collagen fibers in a tangled arrangement between tumor cells (Figure 9.64). These fibers are more often present in gastrointestinal autonomic nerve (GAN) tumors (see next section) but may also be seen in Schwann cell tumors and other GI stromal tumors.

(Text continues on page 628)

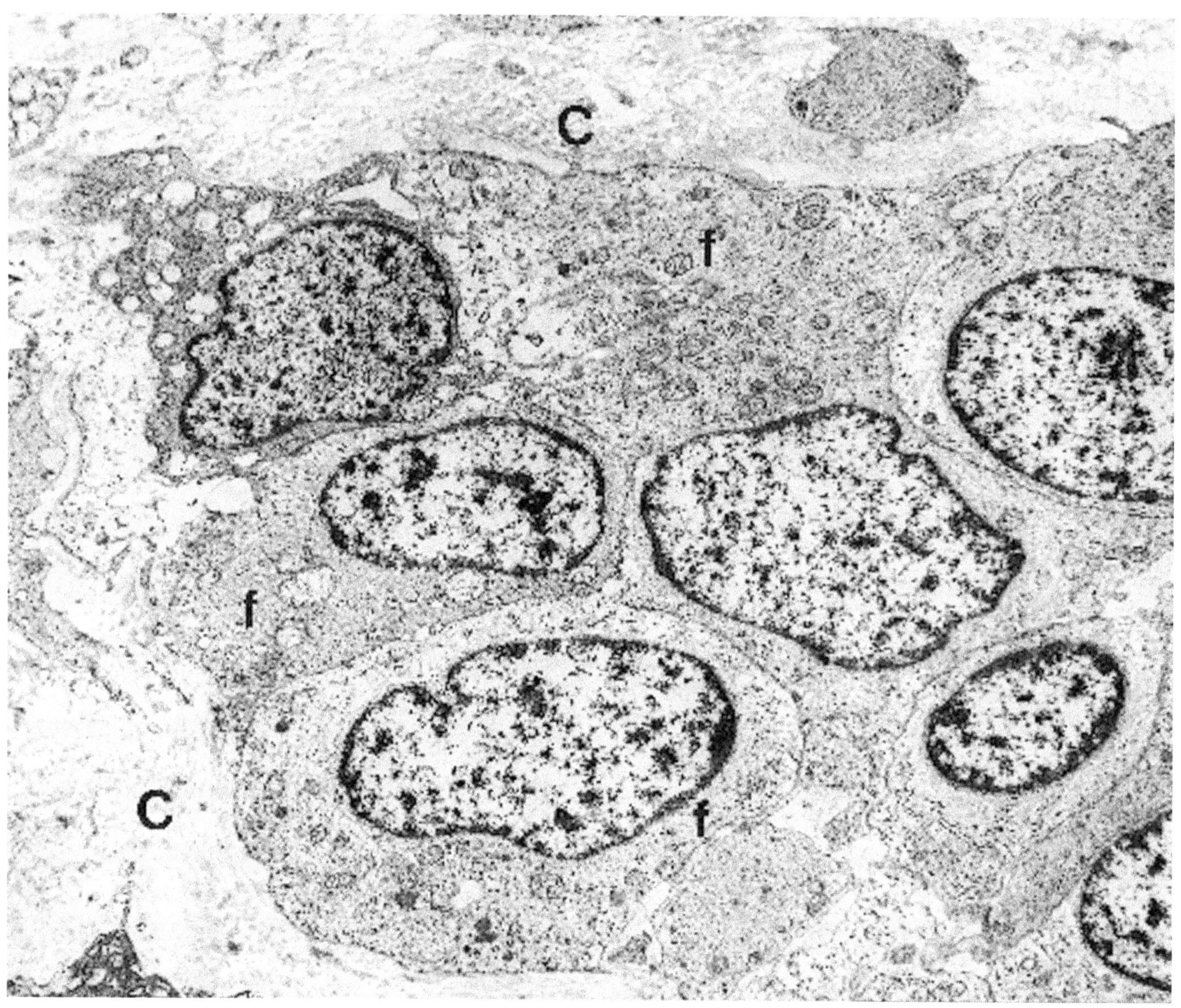

Figure 9.51. Epithelioid sarcoma (hypothenar soft tissue). A group of large, oval and polygonal, tightly apposed cells is surrounded by a collagenous matrix (C). The cells have oval and indented nuclei without conspicuous nucleoli. Cytoplasm is abundant and contains a moderate number of organelles and focal collections of filaments (f). (× 7000)

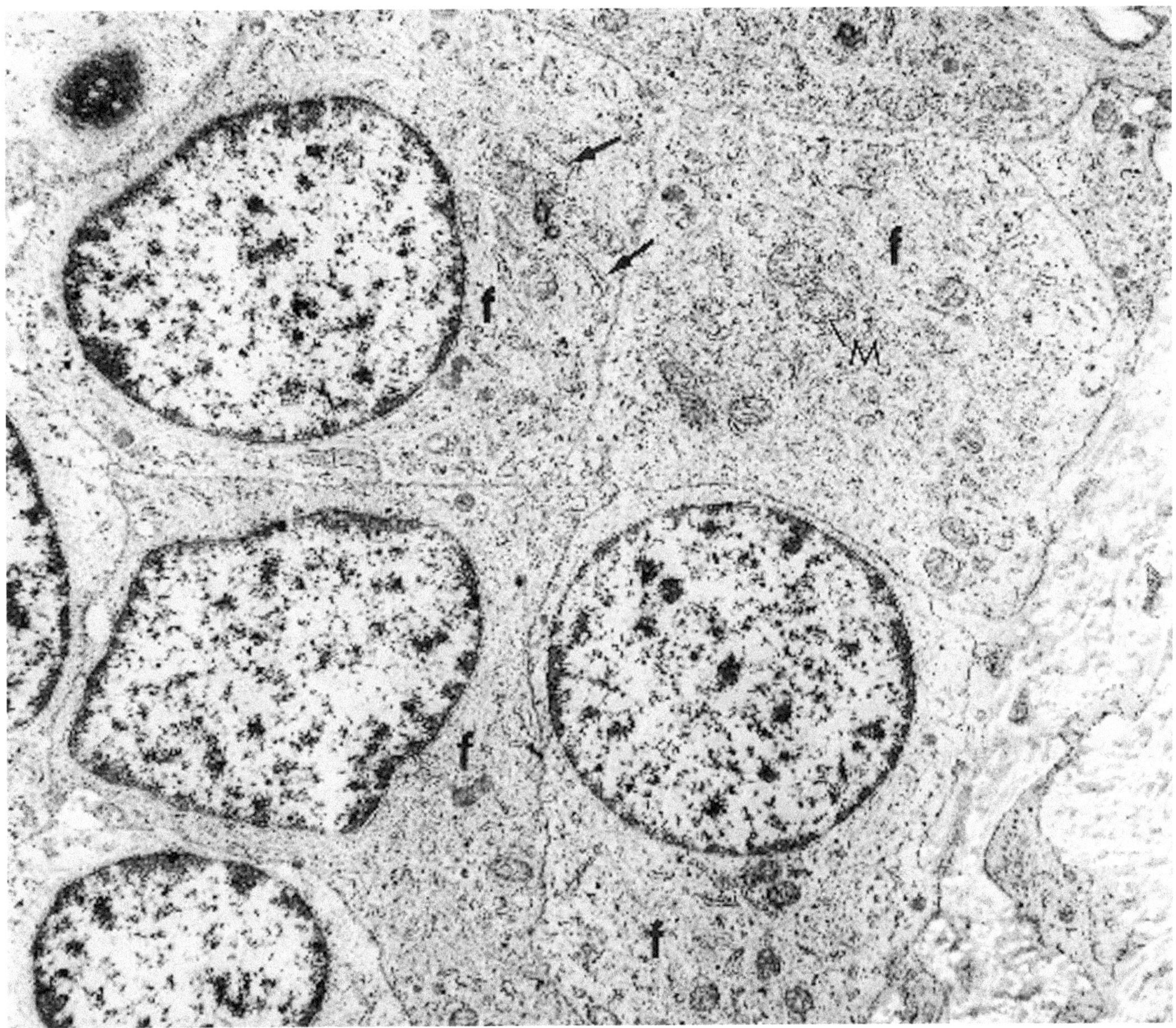

Figure 9.52. Epithelioid sarcoma (hypothenar soft tissue). Four large polygonal cells have numerous filaments (f) filling their cytoplasm. Other organelles include a few mitochondria (M) and a few cisternae of rough endoplasmic reticulum (*arrows*). (× 10,500)

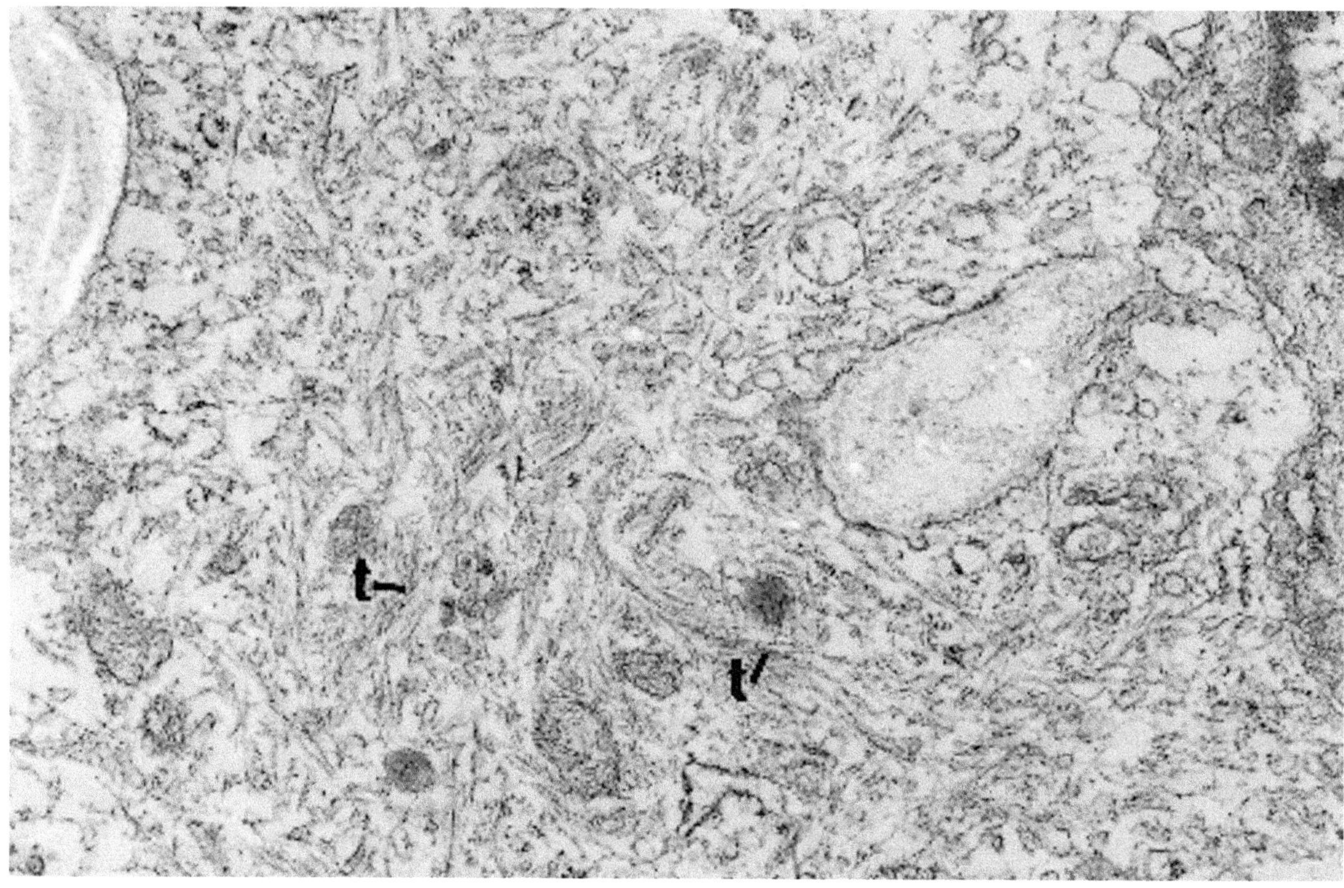

Figure 9.53. Epithelioid sarcoma (hypothenar soft tissue). High magnification of an area of cytoplasm with many filaments reveals the tendency of aggregation of the filaments into bundles or tonofibrils (t). (× 27,000)

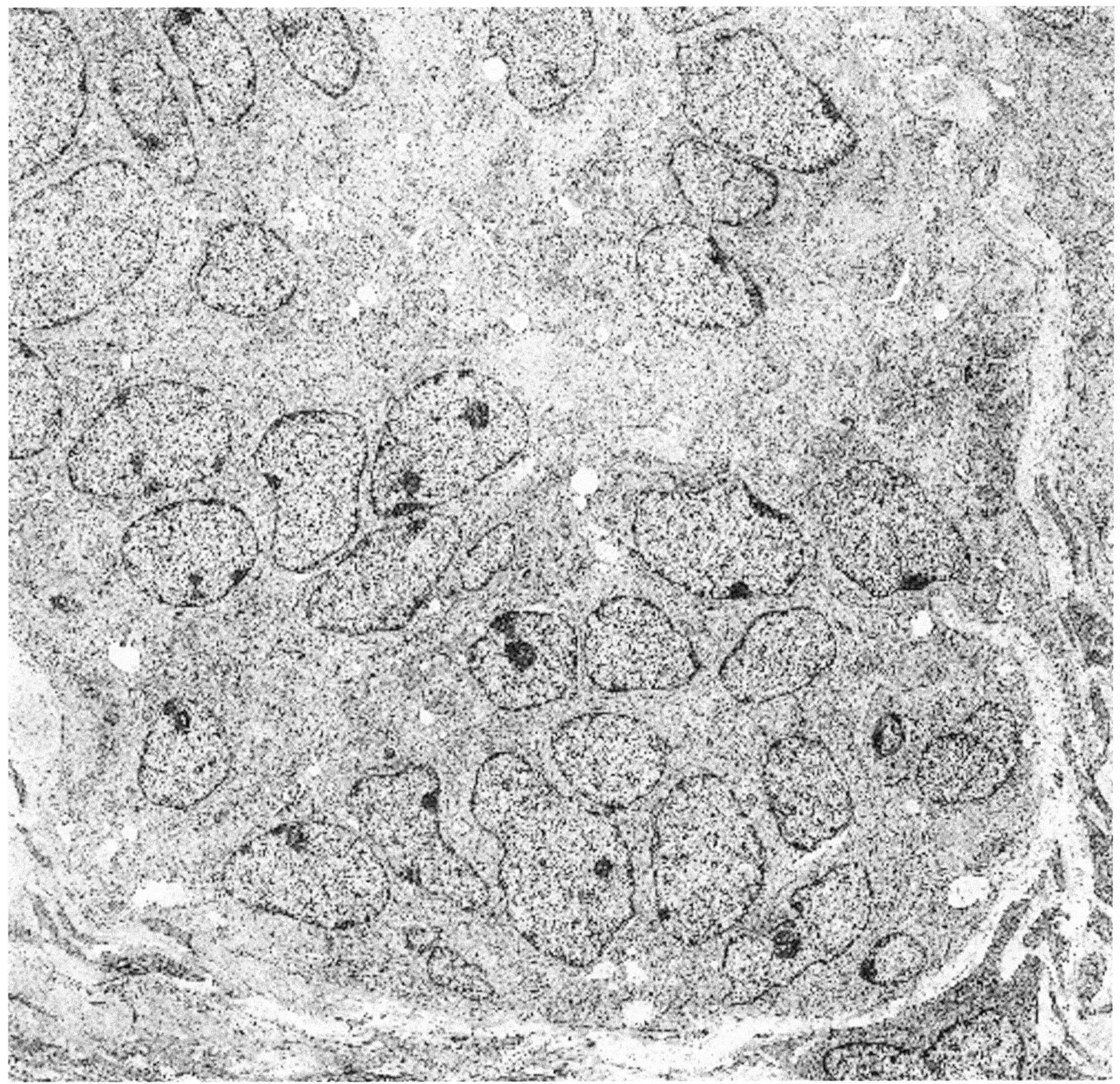

Figure 9.54. Hepatoblastoma, embryonal (liver). A diffuse collection of small oval and elongated cells are tightly apposed and have a high nuclear–cytoplasmic ratio. Nuclei are pleomorphic and predominantly euchromatic, and nucleoli are of moderate size and, in some cells, multiple. (× 3600)

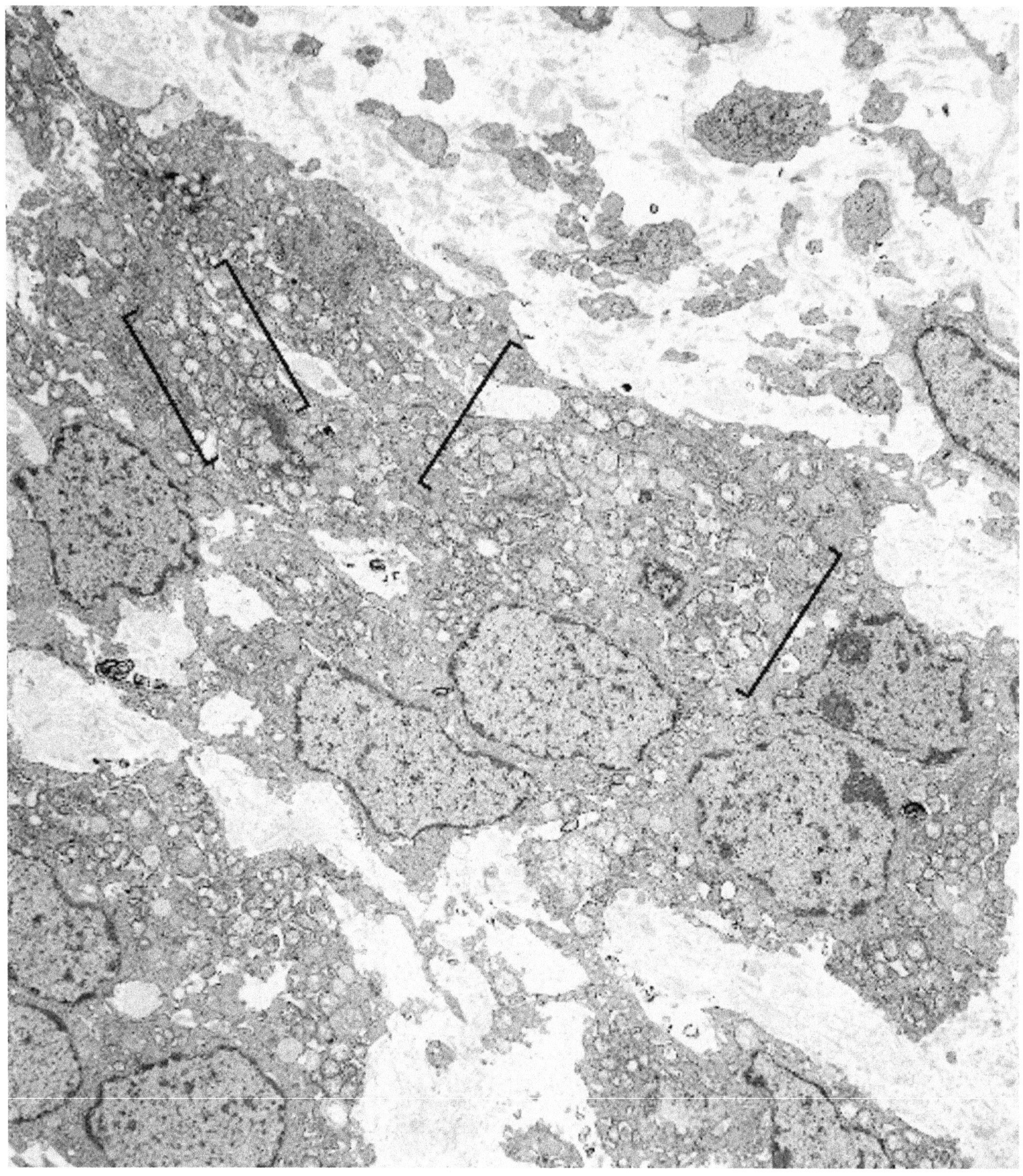

Figure 9.55. Hepatoblastoma, embryonal (liver). An island of neoplastic small cells reveals some of the cells to be elongated and for the cytoplasm to contain numerous mitochondria (*brackets*). (× 6000)

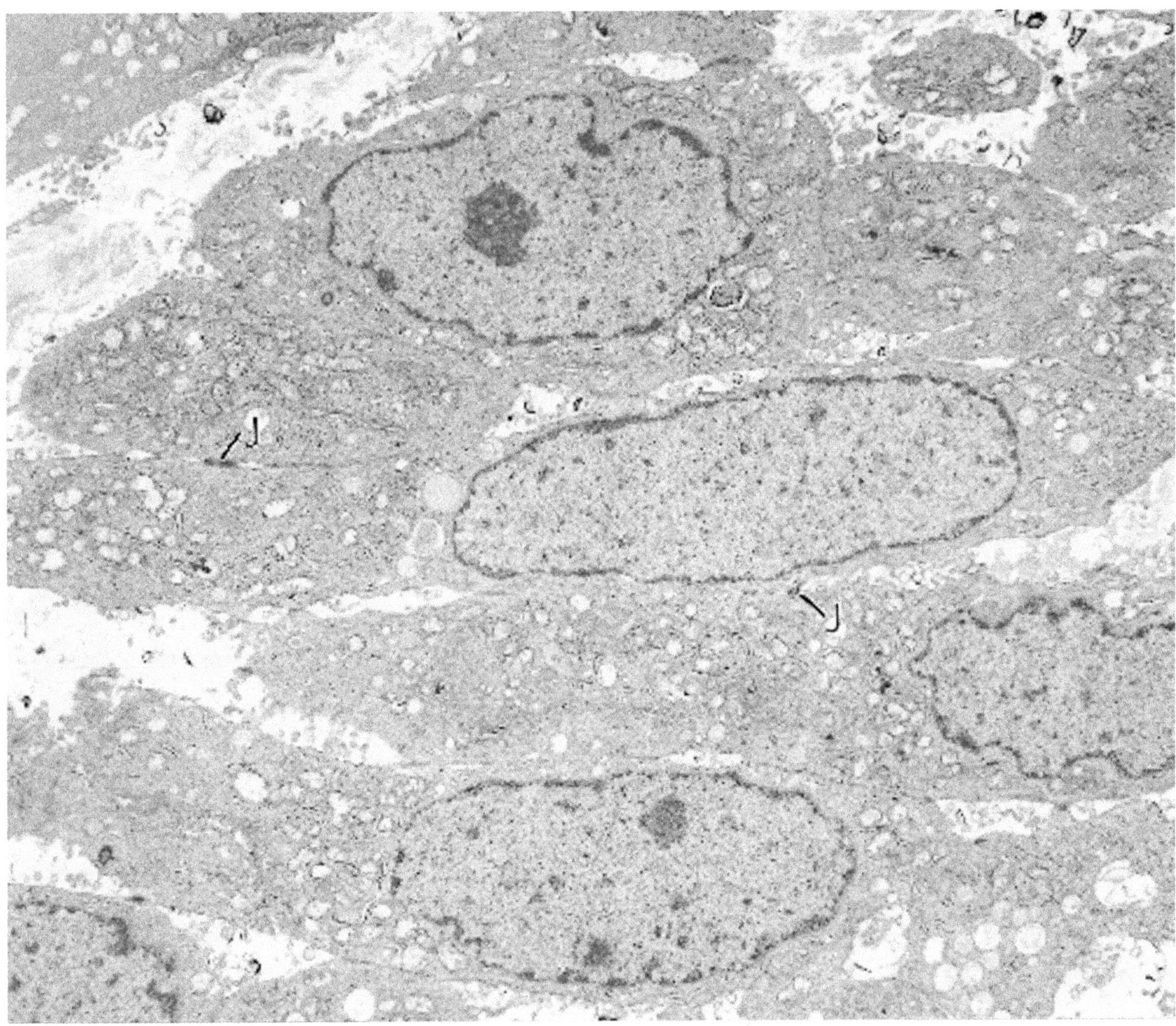

Figure 9.56. Hepatoblastoma, embryonal (liver). Slightly higher magnification of elongated cells from the same neoplasm in Figure 9.55 again illustrates the numerous mitochondria as well as intercellular junctions (J). (× 6800)

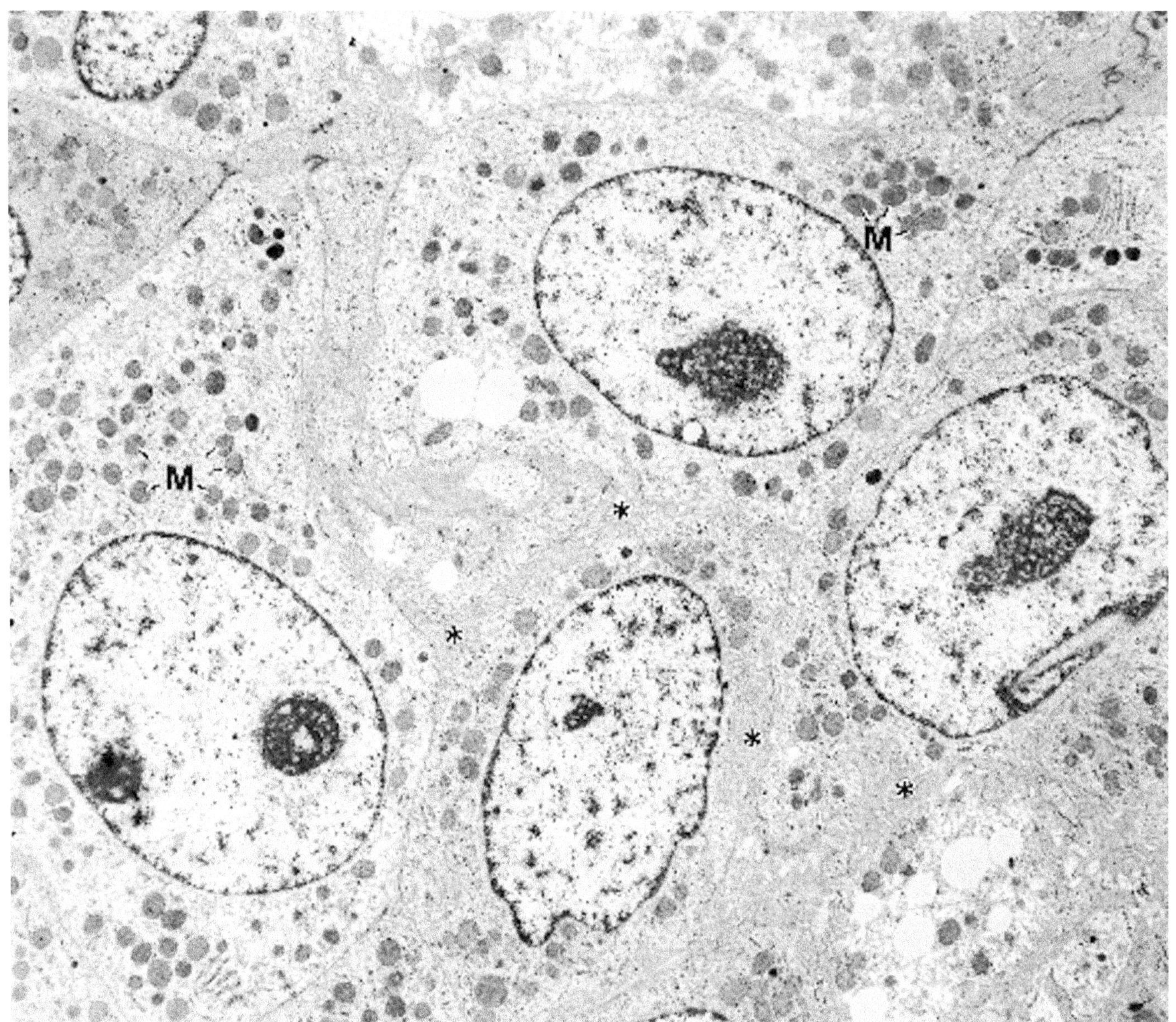

Figure 9.57. Hepatoblastoma, fetal (liver). The cells in this fetal type of hepatoblastoma are larger and have more cytoplasm than those depicted in the embryonal type in Figures 9.54 through 9.56. Nuclei are oval and usually regular in contour, and cytoplasm has a mixture of organelles, including numerous mitochondria (M). Intercellular canaliculi are identifiable focally by the villous character of the plasma membranes (*). (× 6000)

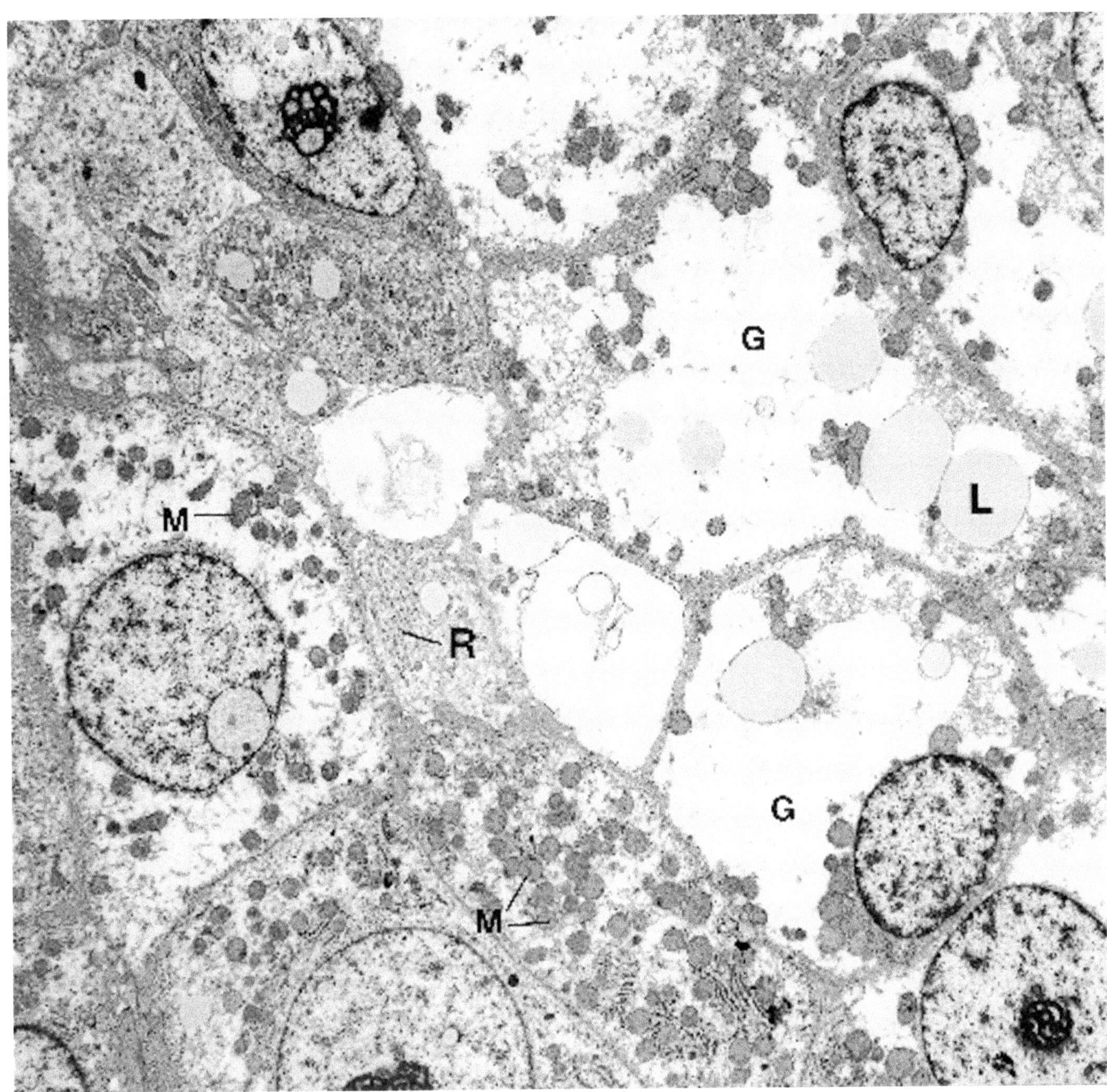

Figure 9.58. Hepatoblastoma, fetal (liver). These fetal-type hepatocytes have a distinct resemblance to normal hepatocytes and contain numerous mitochondria (M), open spaces of glycogen (G), scattered lipid droplets (L), and collections of endoplasmic reticulum (R). (× 5100)

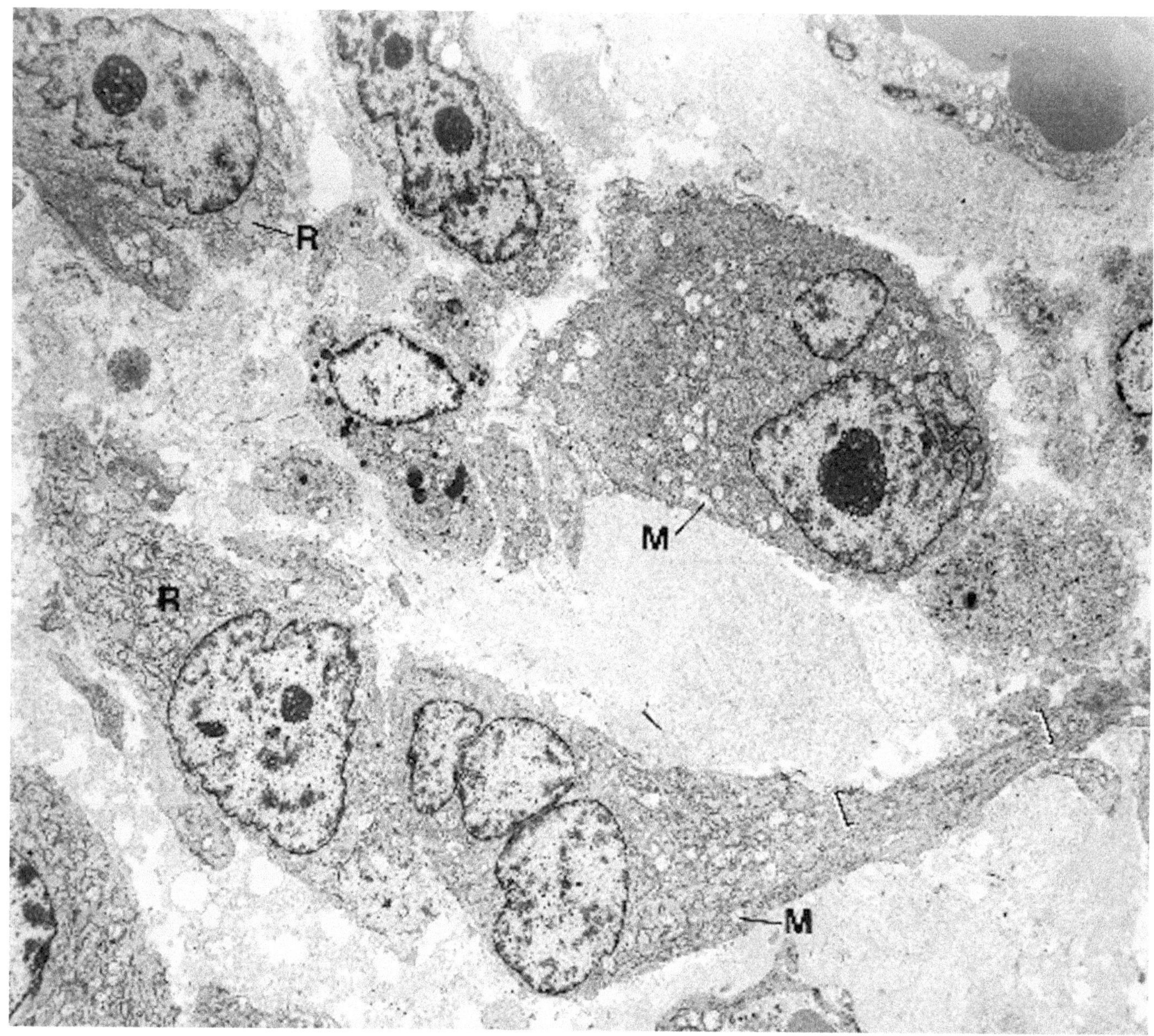

Figure 9.59. Embryonal sarcoma of liver. Several poorly differentiated neoplastic cells have pleomorphic nuclei, large nucleoli, and a mixture of cytoplasmic organelles, including mitochondria (M), rough endoplasmic reticulum (R), and focal filaments (*bracket*). (× 3500)

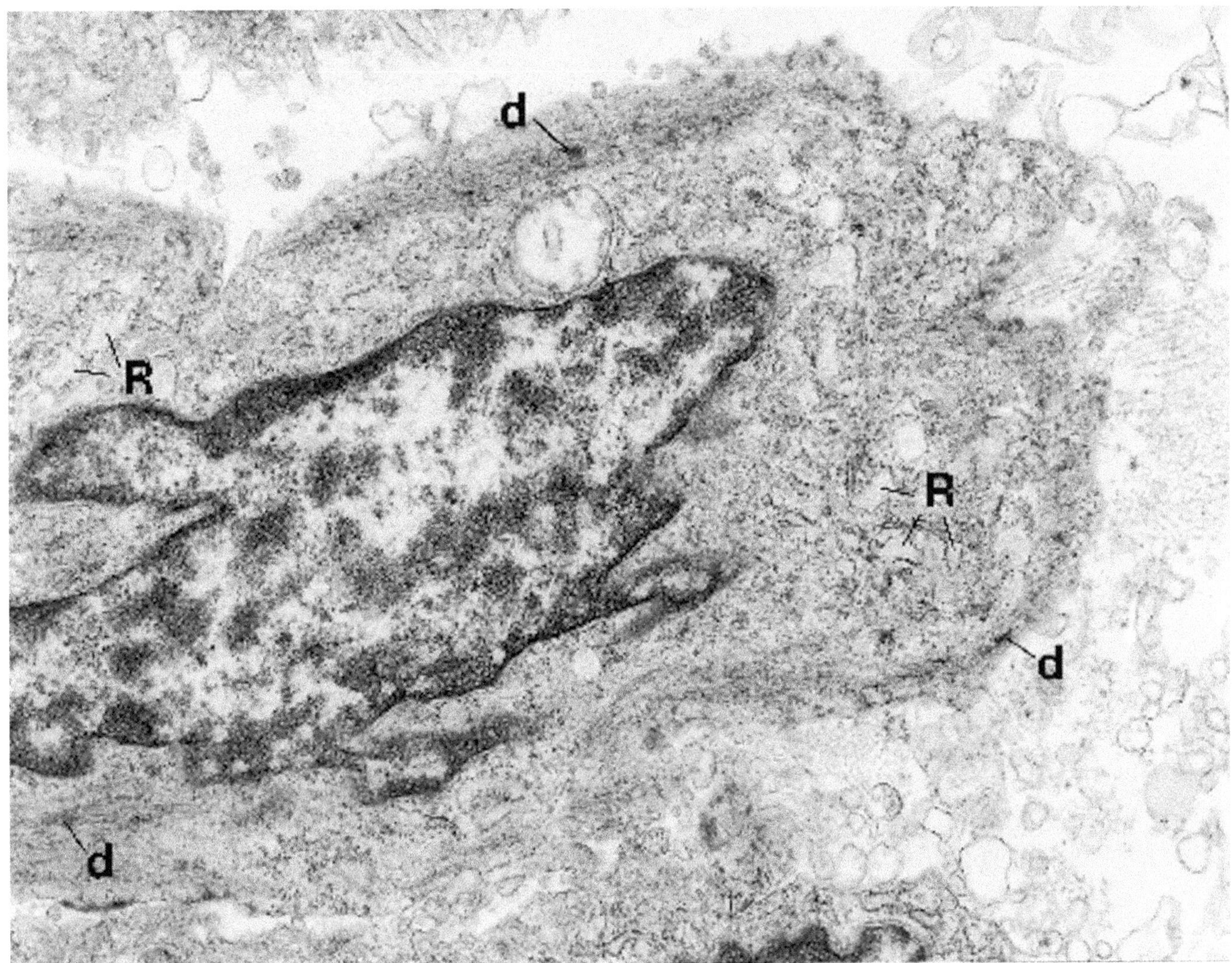

Figure 9.60. Embryonal sarcoma of liver. High magnification of a neoplastic cell depicts dilated rough endoplasmic reticulum (R) and peripherally located intermediate filaments with associated dense bodies (d), characteristic of a myofibroblast. (× 21,000)

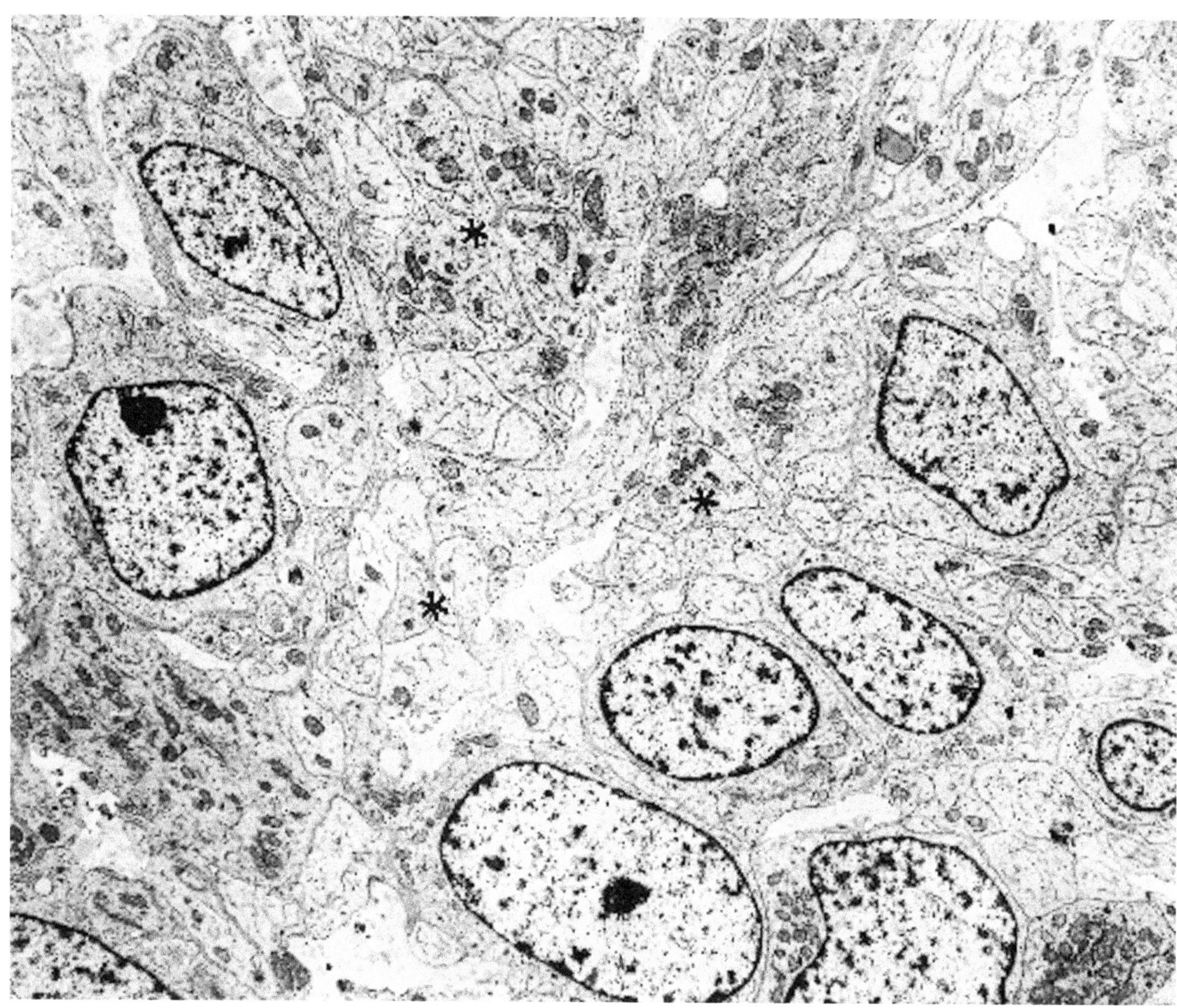

Figure 9.61. Gastrointestinal stromal tumor, low grade and nondescript type (jejunum). Neoplastic oval and elongated cells have long processes that focally aggregate in areas (*) devoid of cell bodies. Although suggestive of neural differentiation, no microtubules, dense-core granules, or synaptic vesicles were found. (× 5200)

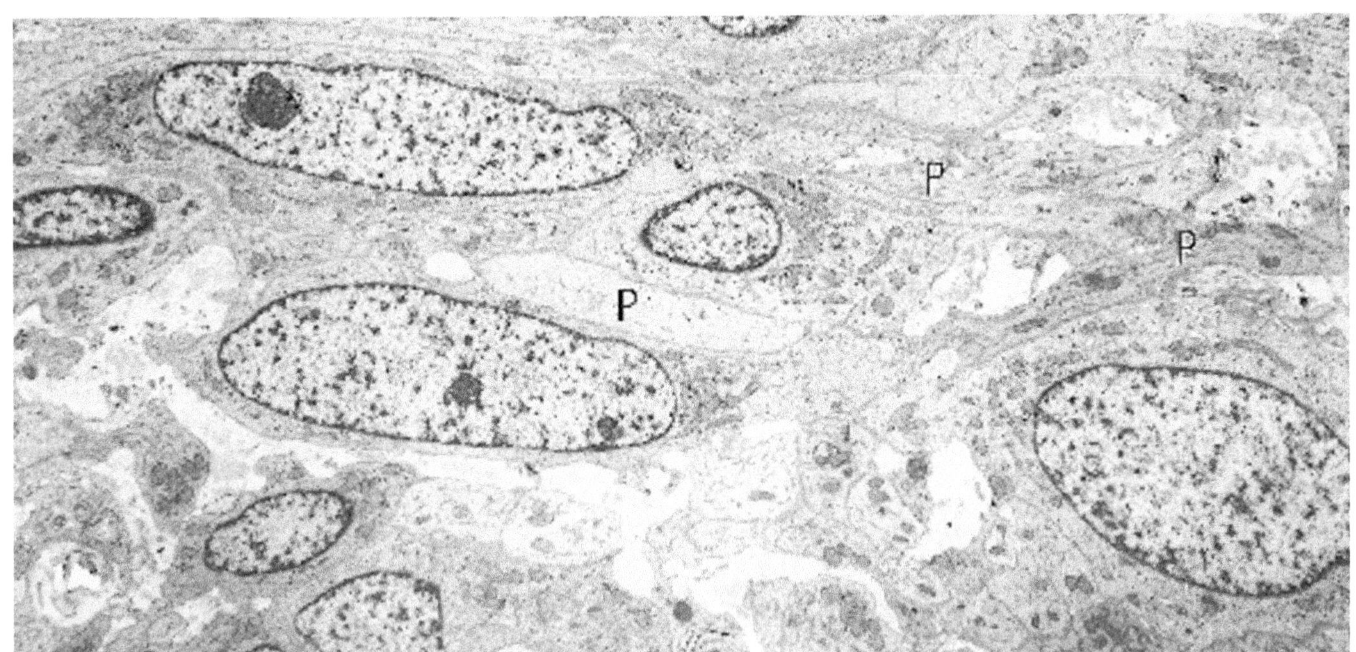

Figure 9.62. Gastrointestinal stromal tumor, low grade and nondescript type (jejunum). The neoplastic cells have a high nuclear–cytoplasmic ratio and long polar processes (P). Round-ended nuclei suggest smooth muscle differentiation, but cytoplasmic features (see Figure 9.63) are not conclusive for smooth muscle. (× 5200)

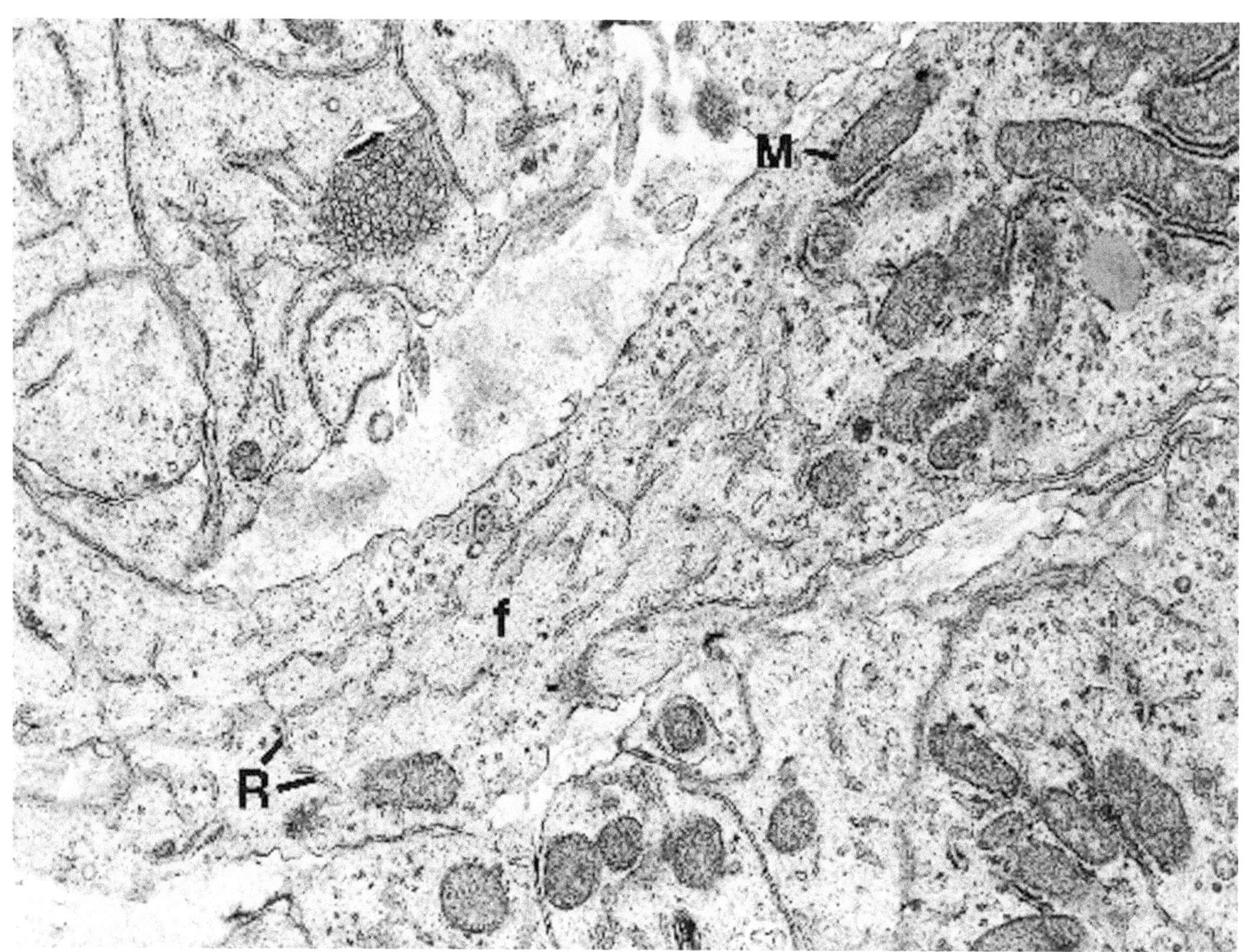

Figure 9.63. Gastrointestinal stromal tumor, low grade and nondescript type (jejunum). High magnification of one of the polar processes of a cell reveals numerous filaments (f) without dense bodies. Rough endoplasmic reticulum (R) is also moderately prominent, and mitochondria (M) are numerous in some compartments of the cells. (× 22,000)

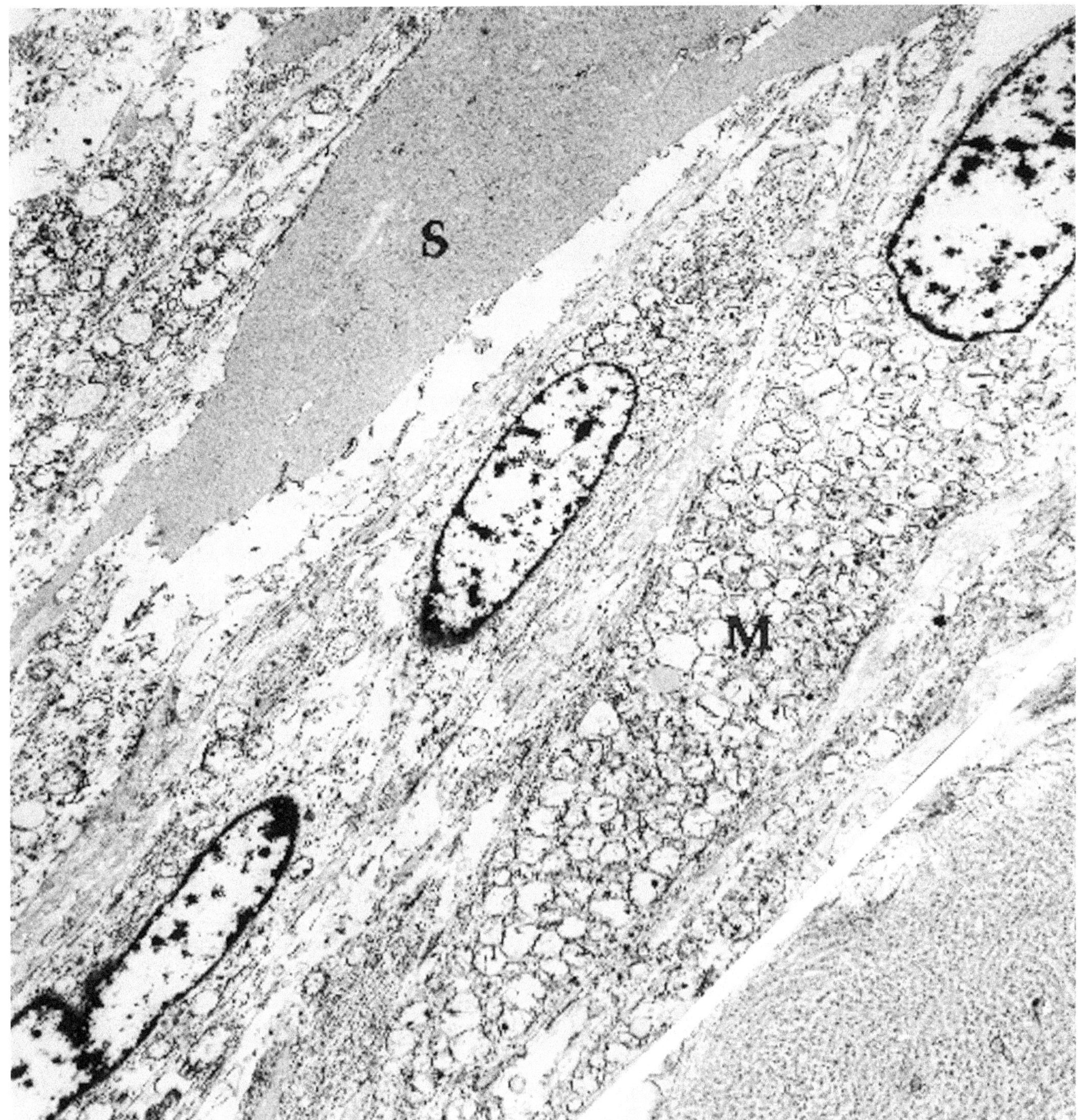

Figure 9.64. Gastrointestinal stromal tumor, smooth muscle type, with skeinoid fibers (ileum). This tissue was removed from paraffin, and ultrastructural detail is not optimally present. However, round-ended nuclei and innumerable mitochondria (M) are supportive of smooth muscle differentiation. Between the cells are large aggregates of skeinoid fibers (S). (× 5300) *Inset:* Higher magnification of skeinoid fibers (× 25,000)

(Text continued from page 613)

Gastrointestinal Autonomic Nerve Tumor (Plexosarcoma)

(Figure 9.65.)

Diagnostic criteria. (1) Spindle and/or epithelioid cells, loosely or tightly arranged; with (2) oval or elongated nuclei and small nucleoli; (3) long interdigitating, cytoplasmic processes (axons); with (4) small intercellular junctions; (5) intermediate filaments; (6) microtubules; (7) synaptic boutons and vesicles; and (8) dense-core granules.

Additional points. GAN tumors probably arise from submucosal and myenteric plexuses. They may be difficult to diagnose with certainty, but microtubules and convincing synaptic vesicles are important criteria. Schwann cells, with characteristic basal lamina, accompany the neurones in a minority of cases. Skeinoid fibers, defined in the section on GI stromal tumors, are more frequently present in GAN tumors (Figure 9.65).

A tumor that may be related to the GAN tumor is one composed of cells suggesting derivation from the interstitial cells of Cajal, also referred to as GI pacemaker cells. Ultrastructurally, these cells have some overlapping features of smooth muscle cells and autonomic nerve cells. They have interdigitating filopodia, basal lamina, junctions, numerous mitochondria, abundant smooth endoplasmic reticulum, large Golgi apparatuses, varying numbers of thin filaments and dense bodies, microtubules, and variable dense-core granules and synaptic vesicles.

Pulmonary Blastoma

(Figures 9.66 through 9.70.)

Diagnostic criteria. (1) Undifferentiated blastema with a high nuclear–cytoplasmic ratio and sparse cytoplasmic organelles; and (2) biphasic epithelial and mesenchymal differentiation; with (3) some epithelial cells being cuboidal and columnar and occurring in solid and/or glandular and tubular arrangements; and having (4) copious glycogen; (5) few cytoplasmic organelles; and (6) other epithelial cells showing differentiation to Clara cells, type II alveolar cells and neuroendocrine cells (see Chapter 3); (7) embryonal mesenchymal cells, elongated and spindle shaped; with (8) a few diminutive junctions; (9) varying lines of cytoplasmic differentiation, including fibroblastic (prominent rough endoplasmic reticulum), rhabdomyoblastic (thin and thick filaments and Z-bands, with sarcomere formations), leiomyoblastic (thin and intermediate filaments with dense bodies, etc.), chondroblastic (abundant rough endoplasmic reticulum, cytoplasmic glycogen, and escalloped plasmalemma), and osteoblastic (abundant rough endoplasmic reticulum, glycogen, and hydroxyapatite (see Chapter 6).

Additional points. Epithelial differentiation into Clara cells is characterized primarily by the cells having apical large, dense, secretory granules. Type II alveolar cells are identifiable by cytoplasmic lamellar (surfactant) bodies. Other markers for pulmonary epithelium include cilia and microvilli with GI-type anchoring filaments (see Chapter 3, Figure 3.8).

Juxtaglomerular Cell Tumor

(Figures 9.71 and 9.72.)

Diagnostic criteria. (1) Round and elongated cells; with (2) surrounding basal lamina; (3) infrequent, diminutive junctions; (4) prominent Golgi apparatuses; (5) moderately prominent rough endoplasmic reticulum; (6) round, angular, and rhomboid secretory (renin-containing) granules; (7) crystalline secretory product in rhomboid granules; (8) varying numbers of thin filaments; (9) round, mildly indented, euchromatic nuclei, and small- to medium-sized nucleoli.

Additional points. Mast cells and groups of unmyelinated neurites may be present among the juxtaglomerular cells.

(Text continues on page 637)

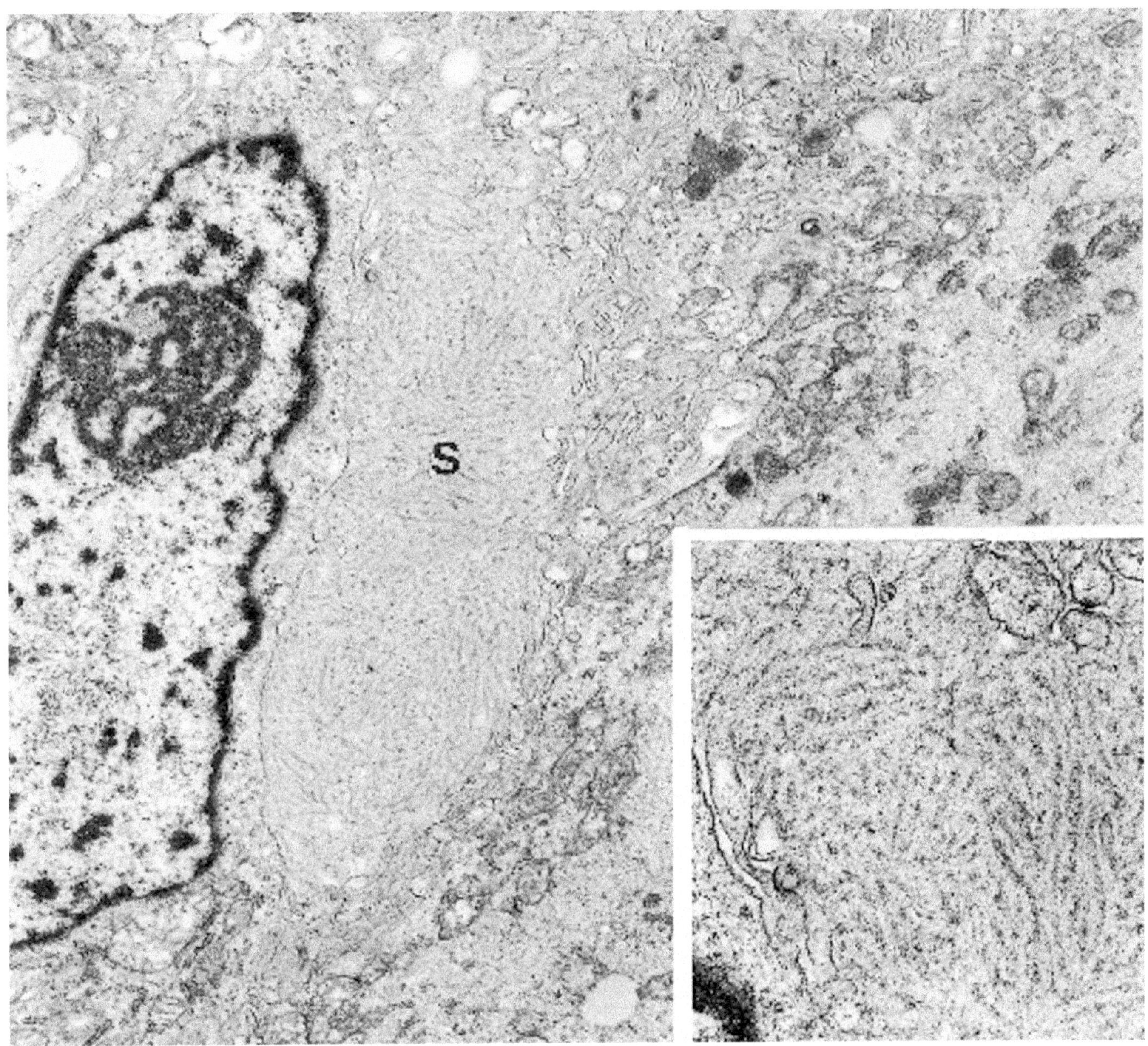

Figure 9.65. Gastrointestinal autonomic nerve tumor (peritoneum, metastatic from stomach). This controversial neoplasm has a number of features, such as polar processes with microtubules, suggestive of autonomic nerve origin, but smooth muscle differentiation could not be ruled out 100%. Shown here is an intercellular group of skeinoid fibers (S). (× 15,000) *Inset:* Higher magnification of the skeinoid fibers. (× 36,000)

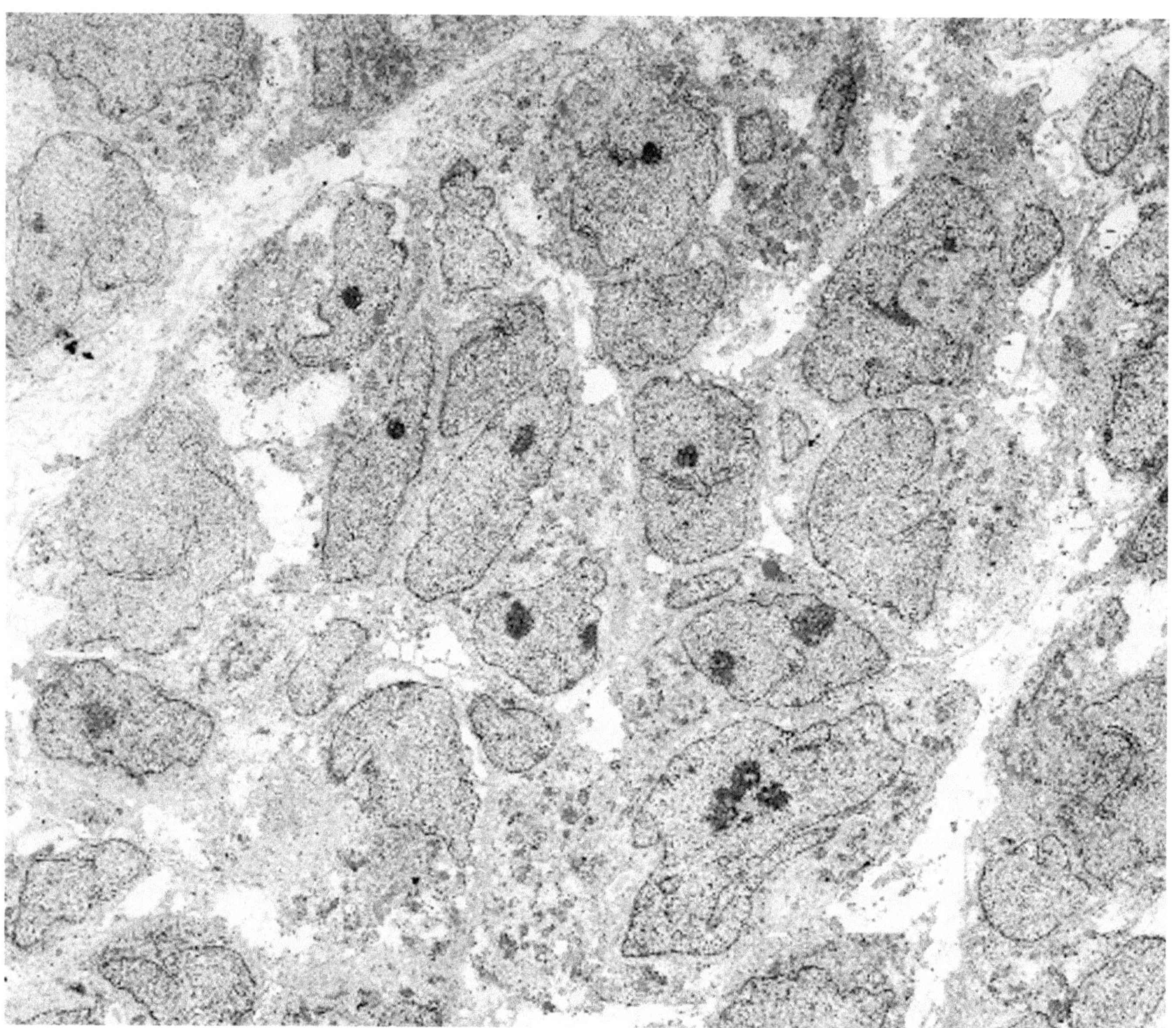

Figure 9.66. Pulmonary blastoma, blastema (seventh cervical vertebra, metastatic from lung). Undifferentiated blastema cells have a high nuclear–cytoplasmic ratio, euchromatic and irregularly shaped nuclei, prominent and multiple nucleoli, and scanty cytoplasm with few organelles. (× 3200)

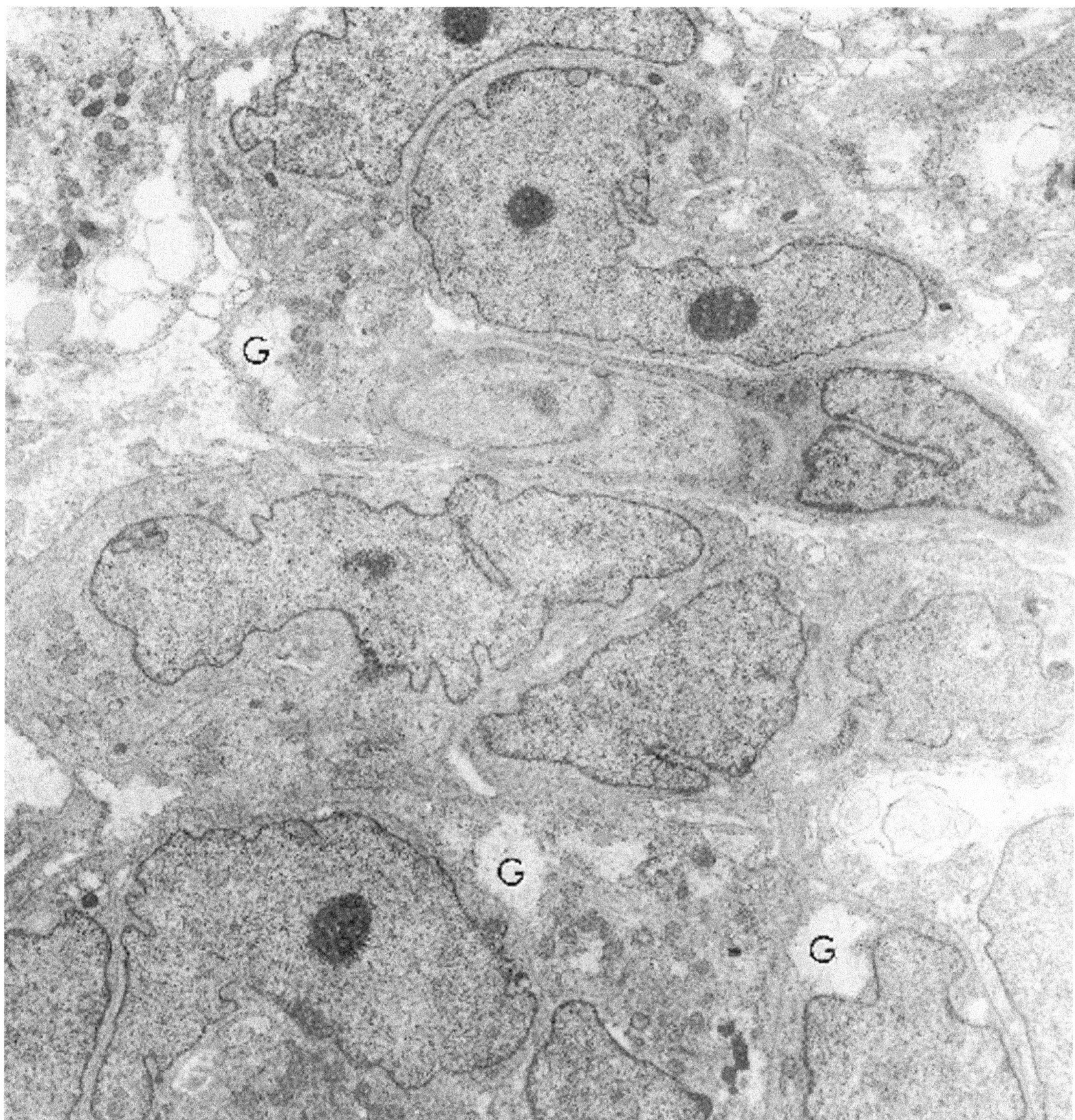

Figure 9.67. Pulmonary blastoma, blastema (seventh cervical vertebra, metastatic from lung). Higher magnification than pictured in Figure 9.66 shows several blastema cells demonstrating scanty cytoplasm and few organelles. Pockets of glycogen (G) are visible in some of the cells. (× 7000)

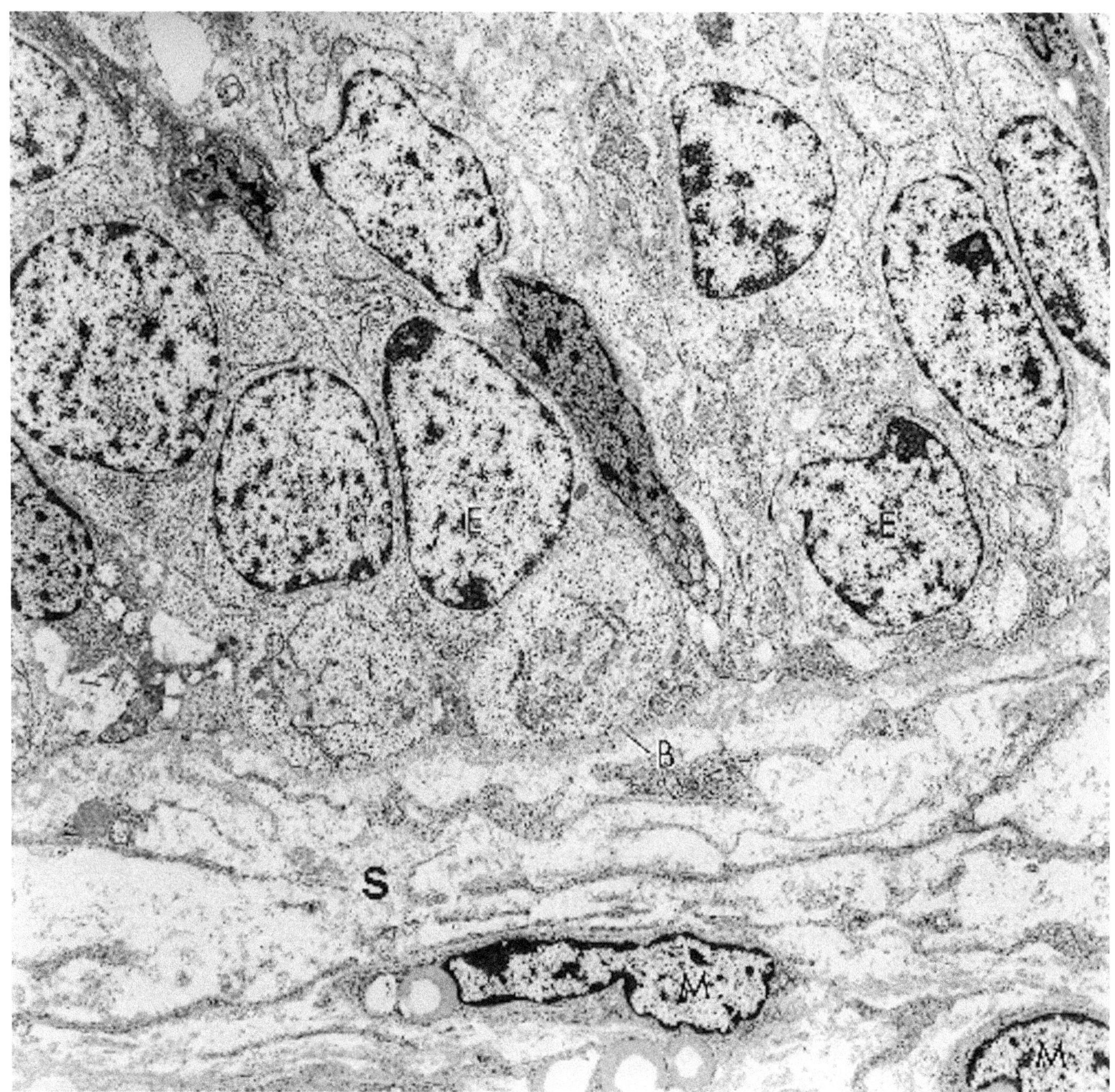

Figure 9.68. Pulmonary blastoma, epithelium and stroma (lung). A group of epithelial cells (E) is separated by basal lamina (B) from a mesenchymal type stroma with widely separated poorly differentiated mesenchymal cells (M). The epithelial cells are cuboidal and columnar and contain ribosomes and few other cytoplasmic organelles. (× 5000)

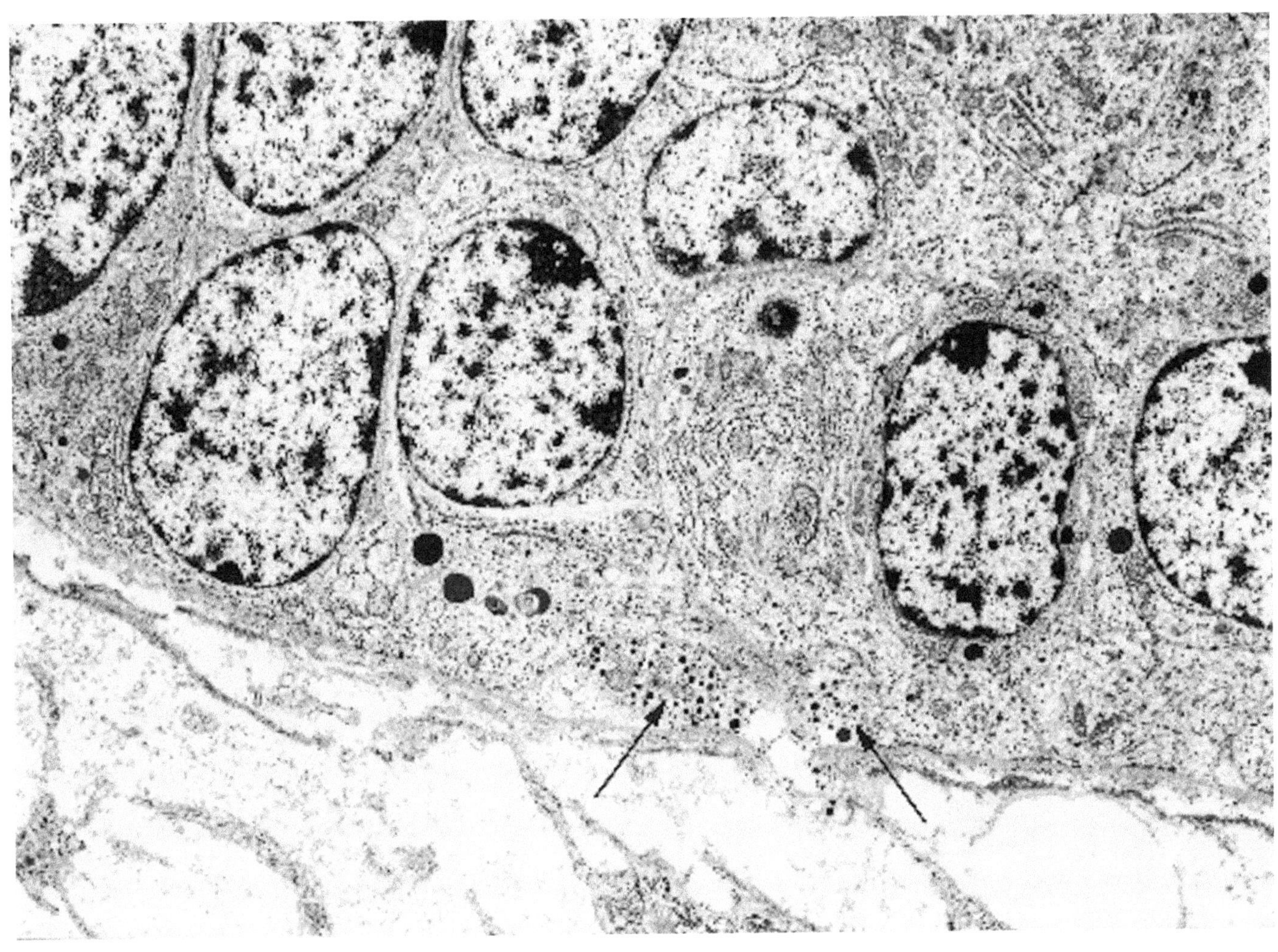

Figure 9.69. Pulmonary blastoma, epithelium and stroma (lung). Peripheral in this group of epithelial cells are cells containing neuroendocrine type granules (*arrows*). The stroma is very loose and mesenchymal in appearance. (× 5800)

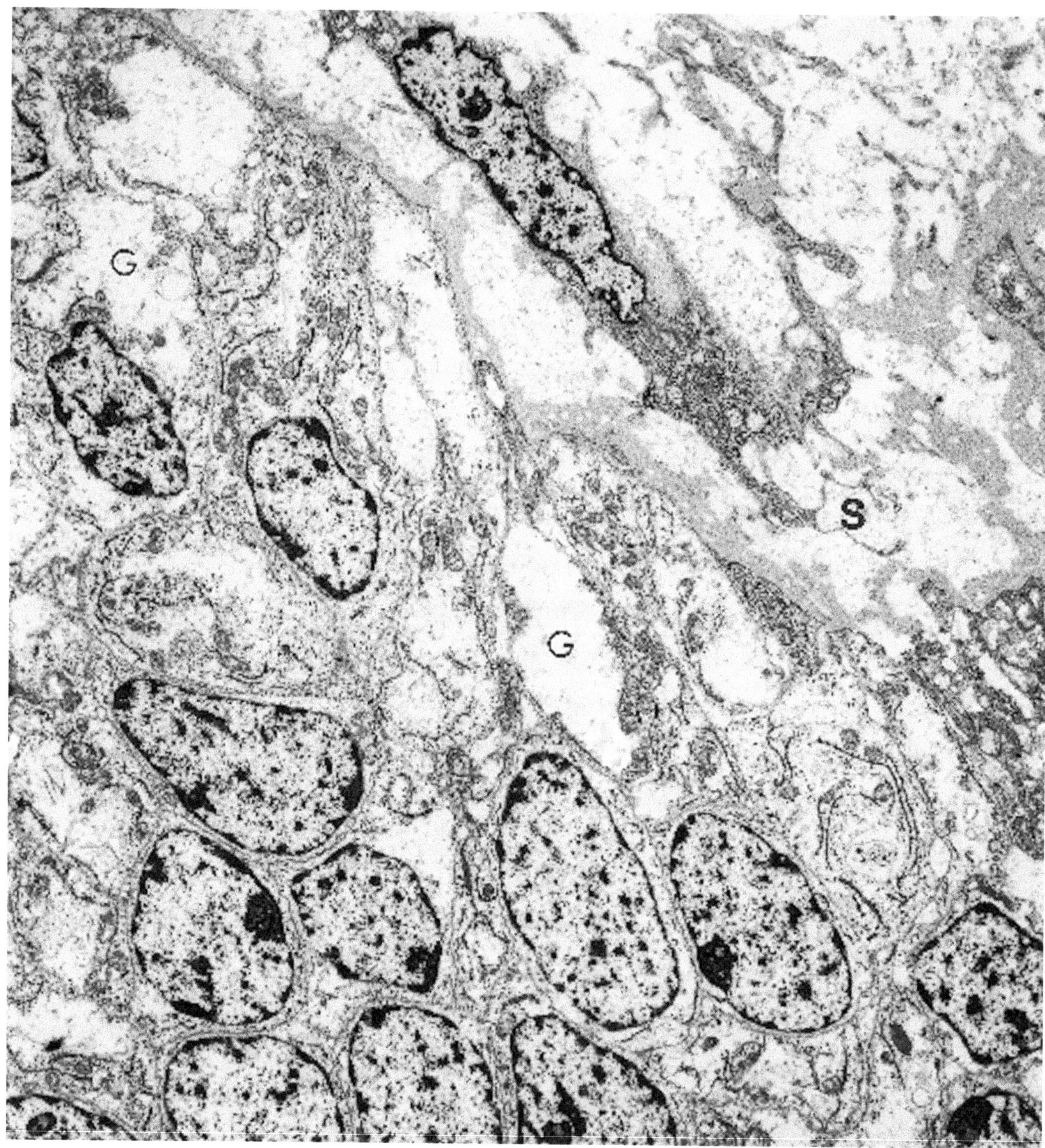

Figure 9.70. Pulmonary blastoma, epithelium and stroma (lung). In this field, the neoplastic cells are more blastematous than epithelial appearing and blend subtly into the adjacent mesenchymal type stroma (S). Glycogen (G) is represented by the open clear spaces in some of the cells. (× 5800)

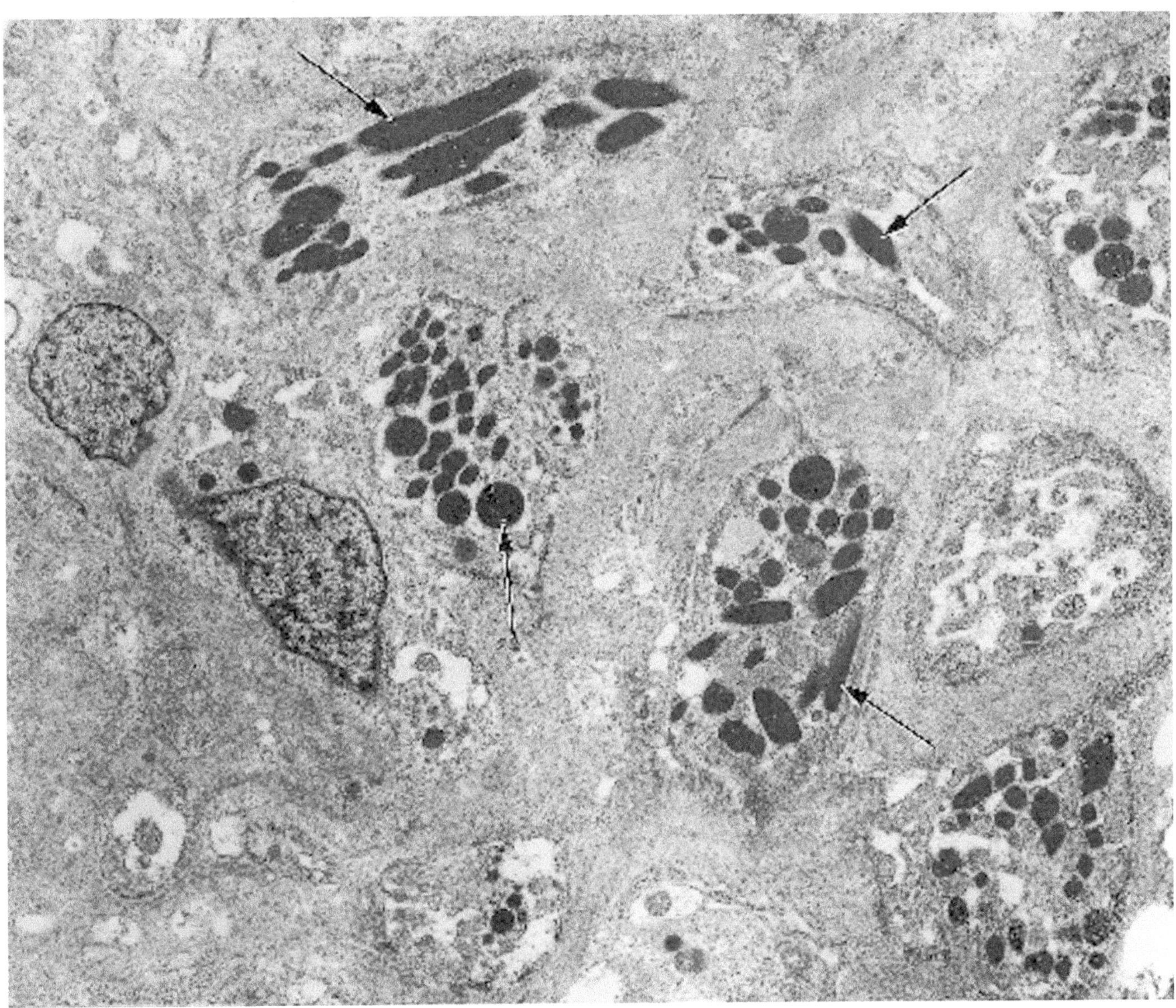

Figure 9.71. Juxtaglomerular cell tumor (kidney). Tightly clustered round and oval cells contain numerous round and elongated, electron-dense granules (*arrows*). (× 5500)

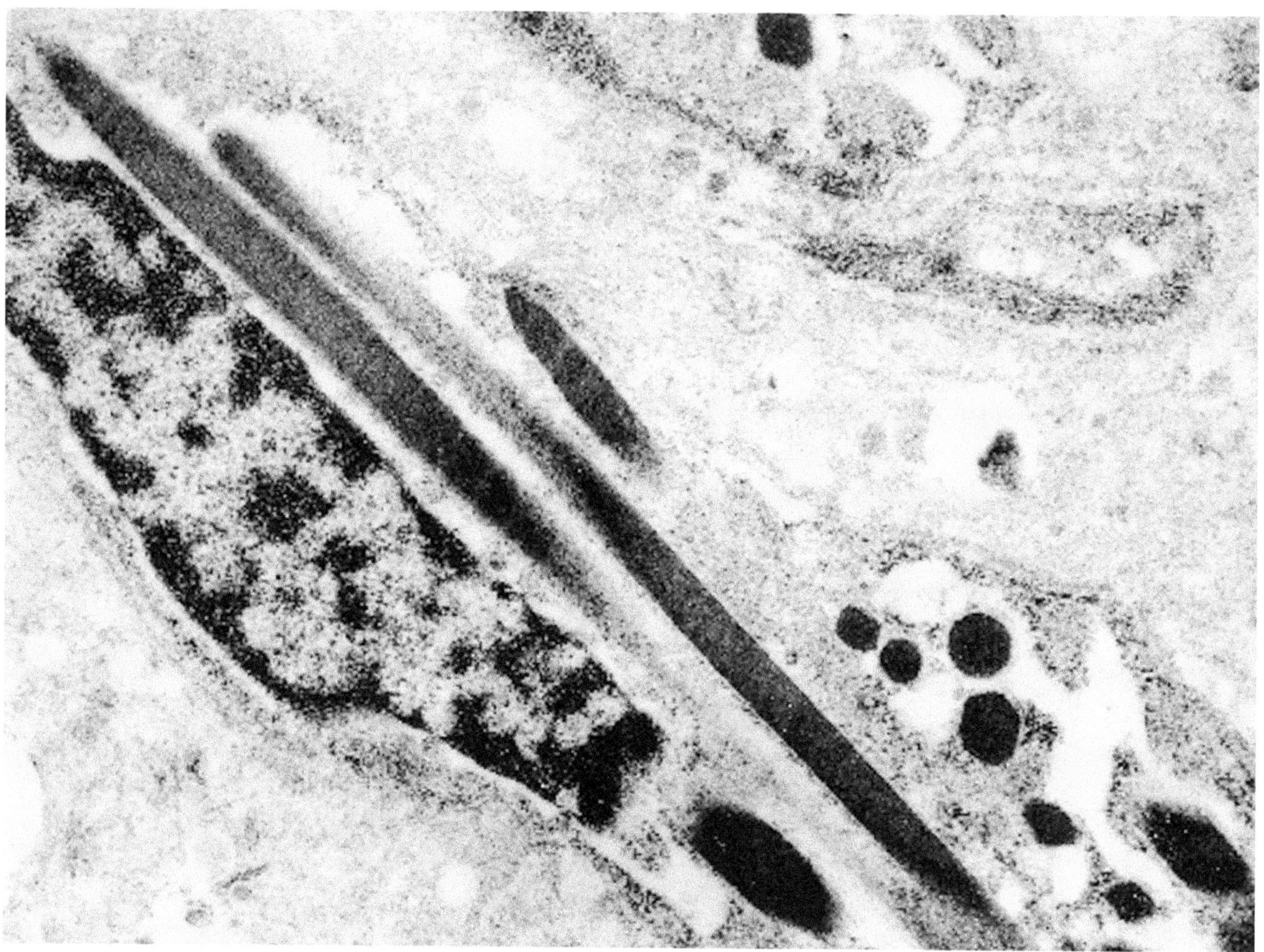

Figure 9.72. Juxtaglomerular cell tumor (kidney). High magnification of one of the neoplastic juxtaglomerular type cells illustrates round and crystalline, rhomboid and needle-shaped granules in the cytoplasm. (× 13,800)

(Text continued from page 628)

REFERENCES

Neuroendocrine Neoplasms (APUD-omas)

General

Nagle RB, Payne CM, Clark VA: Comparison of the usefulness of histochemistry and ultrastructural cytochemistry in the identification of neuroendocrine neoplasms. Am J Clin Pathol 85:289–296, 1986.

Pearse AGE: Common cytochemical and ultrastructural characteristics of cells producing polypeptide hormones (the APUD series) and their relevance to thyroid and ultimobranchial C cells and calcitonin. Proc R Soc B 170:71–80, 1968.

Pearse AGE: The cytochemistry and ultrastructure of polypeptide hormone-producing cells of the APUD series and the embryologic, physiologic and pathologic implications of the concept. J Histochem Cytochem 17:303–313, 1969.

Carcinoid/Islet Cell Tumor

Abt AB, Carter SL: Goblet cell carcinoid of the appendix. An ultrastructural and histochemical study. Arch Pathol Lab Med 100:301–306, 1976.

Akagi T, Fujii Y: Histology, ultrastructure, and tissue culture of human insulinomas. Cancer 47:417–424, 1981.

Black WC: Enterochromaffin cell types and corresponding carcinoid tumors. Lab Invest 19:473–486, 1968.

Black WC, Haffner HE: Diffuse hyperplasia of gastric argyrophil cells and multiple carcinoid tumors. An historical and ultrastructural study. Cancer 21:1080–1099, 1968.

Campbell IL, Harrison LC, Ley CJ, et al: Nesidioblastosis and multifocal pancreatic islet cell hyperplasia in an adult. Clinicopathologic features and *in vitro* pancreatic studies. Am J Clin Pathol 84:534–541, 1985.

Carvalheira AF, Welsch U, Pearse AGE: Cytochemical and ultrastructural observations on the argentaffin and argyrophil cells of the gastro-intestinal tract in mammals, and their place in the APUD series of polypeptide-secreting cells. Histochemie 14:33–46, 1968.

Creutzfeldt W, Arnold R, Creutzfeldt C, et al: Pathomorphologic, biochemical, and diagnostic aspects of gastrinomas (Zollinger-Ellison syndrome). Hum Pathol 6:47–76, 1975.

DeLellis RA, Dayal Y, Wolfe JH: Carcinoid tumors: Changing concepts and new perspectives. Am J Surg Pathol 8:295–300, 1984.

Forssmann WG, Orci L, Pictet R, et al: The endocrine cells in the epithelium of the gastrointestinal mucosa of the rat. An electron microscope study. J Cell Biol 40: 692–715, 1969.

Goldenberg VE, Goldenberg NS, Benditt EP: Ultrastructural features of functioning alpha- and beta-cell tumors. Cancer 24:236–247, 1969.

Gould VE, Benditt EP: The pancreatic islet and its tumors. Pathol Annu 8:207–216, 1973.

Gould VE, Chejfec G: Neuroendocrine carcinomas of the colon. Ultrastructural and biochemical evidence of their secretory functions. Am J Surg Pathol 2:31–38, 1978.

Gould VE, Valaitis J, Trujillo Y, et al: Neuroendocrinoma of the jejunum: Electron microscopic and biochemical analysis. Cancer 46:713–717, 1980.

Guarda LA, Silva EG, Ordonez NG, et al: Clear cell islet. Am J Clin Pathol 79:512–517, 1983.

Hammar S, Sale G: Multiple hormone producing islet cell carcinomas of the pancreas. A morphological and biochemical investigation. Hum Pathol 6:349–362, 1975.

Hernandez FJ, Fernandez BB: Mucus-secreting colonic carcinoid tumors: Light- and electron-microscopic study of three cases. Dis Colon Rectum 17:387–396, 1974.

Kaneko H, Yanaihara N, Ito S, et al: Somatostatinoma of the duodenum. Cancer 44:2273–2279, 1979.

Lacy PE: The pancreatic beta cell. Structure and function. N Engl J Med 276:187–195, 1967.

Larsson LI: Endocrine pancreatic tumors. Hum Pathol 9:401–416, 1978.

Like AA: The ultrastructure of the secretory cells of the islets of Langerhans in man. Lab Invest 16:937–951, 1967.

Liu TH, Tseng HC, Zhu Y, et al: Insulinoma. An immunocytochemical and morphologic analysis of 95 cases. Cancer 56:1420–1429, 1985.

Lloyd RV, Caceres V, Warner TF, et al: Islet cell adenomatosis. A report of two cases and review of the literature. Arch Pathol Lab Med 105:198–202, 1981.

Millikin PD: Extraepithelial enterochromaffin cells and Schwann cells in the human appendix. Arch Pathol Lab Med 107:189–194, 1983.

Mills SE, Allen MS, Cohen AR: Small-cell undifferentiated carcinoma of the colon. A clinicopathological study of five cases and their association with colonic adenomas. Am J Surg Pathol 7:643–651, 1983.

O'Brian DS, Dayal Y, DeLellis RA, et al: Rectal carcinoids as tumors of the hindgut endocrine cells. A morphological and immunohistochemical analysis. Am J Surg Pathol 6:131–142, 1982.

Radi MJ, Fenoglio-Preiser CM, Chiffelle T: Functioning oncocytic islet-cell carcinoma. Report of a case with electron-microscopic and immunohistochemical confirmation. Am J Surg Pathol 9:517–524, 1985.

Reid JD, Yuh SL, Petrelli M, et al: Ductuloinsular tumors of the pancreas: A light, electron microscopic and immunohistochemical study. Cancer 49:908–915, 1982.

Rodriguez FH, Sarma DP, Lunseth JH: Goblet cell carcinoid of the appendix. Hum Pathol 13:286–288, 1982.

Solcia E, Capella C, Buffa R, et al: Endocrine cells of the gastrointestinal tract and related tumors. Pathobiol Annu 9:163–204, 1979.

Ulich T, Cheng L, Lewin KJ: Acinar-endocrine cell tumor of the pancreas. Report of a pancreatic tumor containing both zymogen and neuroendocrine granules. Cancer 50:2099–2015, 1982.

Warner TF, Block M, Hafez R, et al: Glucagonomas. Ultrastructure and immunocytochemistry. Cancer 51:1091–1096, 1983.

Warner TF, Seo IS: Goblet cell carcinoid of appendix. Ultrastructural features and histogenetic aspects. Cancer 44:1700–1706, 1979.

Weichert RF, Roth LM, Harkin JC: Carcinoid-islet cell tumor of the duodenum and associated multiple carcinoid tumors of the ileum. An electron microscopic study. Cancer 27:910–918, 1971.

Wilander E, Westermark P, Grimelus L: Intracellular and extracellular fibrillary structures in gastroduodenal endocrine tumors. Ultrastruct Pathol 1:49–54, 1980.

Medullary (C-Cell) Carcinoma of the Thyroid

Bussolati G, Monga G: Medullary carcinoma of the thyroid with atypical patterns. Cancer 44:1769–1777, 1979.

Chan AS, Conen PE: Ultrastructural observations on cytodifferentiation of parafollicular cells in the human fetal thyroid. Lab Invest 25:249–259, 1971.

DeLellis RA, Nunnemacher G, Wolfe HJ: C-cell hyperplasia. An ultrastructural analysis. Lab Invest 36: 237–248, 1977.

DeLellis RA, May L, Tashjian AH, et al: C-cell granule heterogeneity in man. An ultrastructural immunocytochemical study. Lab Invest 38:263–269, 1978.

DeLellis RA, Nunnemacher G, Bitman WR, et al: C-cell hyperplasia and medullary thyroid carcinoma in the rat. An immunohistochemical and ultrastructural analysis. Lab Invest 40:140–154, 1979.

Dominguez-Malagon H, Macias-Martinez V, Molina-Cardenas H, et al: Amphicrine medullary carcinoma of the thyroid with luminal differentiation: Report of an immunohistochemical and ultrastructural study. Ultrastruct Pathol 21:569–574, 1997.

Fernandes BJ, Bedard YC, Rosen I: Mucus-producing medullary cell carcinoma of the thyroid gland. Am J Clin Pathol 78:536–540, 1982.

Gould VE, Benditt EP: Ultrastructural and functional relationships of some human endocrine tumors. Pathol Annu 8:205–230, 1973.

Hales M, Rosenau W, Okerlund MD, et al: Carcinoma of the thyroid with a mixed medullary and follicular pattern. Morphologic, immunohistochemical, and clinical laboratory studies. Cancer 50:1352–1359, 1982.

Harach HR, Williams ED: Glandular (tubular and follicular) variants of medullary carcinoma of the thyroid. Histopathology 7:83–97, 1983.

Holm R, Sobrinho-Simoes M, Nesland JM, et al: Medullary carcinoma of the thyroid gland: An immunocytochemical study. Ultrastruct Pathol 8:25–41, 1985.

Holm R, Farrants GW, Nesland JM, et al: Ultrastructural and electron immunohistochemical features of medullary thyroid carcinoma. Virchows Arch [A] 414: 375–384, 1989.

Huang SN, Goltzman D: Electron and immunoelectron microscopic study of thyroidal medullary carcinoma. Cancer 41:2226–2235, 1978.

Landon G, Ordonez NG: Clear cell variant of medullary carcinoma of the thyroid. Hum Pathol 16:844–847, 1985.

Meyer JS, Hutton WE, Kenny AD: Medullary carcinoma of the thyroid gland. Subcellular distribution of calcitonin and relationship between granules and amyloid. Cancer 31:433–441, 1973.

Nesland JM, Holm R, Sund BS, et al: A thyroid tumor in a 57-year-old man. Ultrastruct Pathol 10:197–202, 1986.

Sobrinho-Simoes M, Sambade C, Nesland JM, et al: Lectin histochemistry and ultrastructure of medullary carcinoma of the thyroid gland. Arch Pathol Lab Med 114:369–375, 1990.

Tateishi R, Takahashi Y, Noguchi A: Histologic and ultracytochemical studies on thyroid medullary carcinoma. Diagnostic significance of argyrophil secretory granules. Cancer 30:755–763, 1972.

Weiss LM, Weinberg DS, Warhol MJ: Medullary carcinoma arising in a thyroid with Hashimoto's disease. Am J Clin Pathol 80:534–538, 1983.

Parathyroid Tumors

Aguilar-Parada E, Gonzalez-Angulo A, Del Peon L, et al: Functioning microvillous adenoma of the parathyroid gland containing nuclear pores and annulate lamellae. Hum Pathol 16:511–516, 1985.

Altenahr E: Functional cytology and ultrastructural pathology of parathyroid glands in animal experiments. Curr Topics Pathol 56:2–51, 1972.

Altenahr E, Saeger W: Light and electron microscopy of parathyroid carcinoma. Report of three cases [abstract]. Virchows Arch [A] 360:107–122, 1973.

Bedetti CD, Dekker A, Watson CG: Functioning oxyphil cell adenoma of the parathyroid gland: A clinicopathologic study of ten patients with hyperparathyroidism. Hum Pathol 15:1121–1126, 1984.

Bichel P, Thomsen OF, Askyaer SAA, et al: Light and electron microscopic investigation of parathyroid carcinoma during dedifferentiation. Virchows Arch [A] 386:363–370, 1980.

Boquist L: Annulate lamellae in human parathyroid adenoma. Virchows Arch [B] 6:234–246, 1970.

Cinti S, Balercia G, Zingaretti MC, et al: The normal human parathyroid gland. A histochemical and ultrastructural study with particular reference to follicular structures. J Submicrosc Cytol 15:661–679, 1983.

Cinti S, Colussi G, Minola E, et al: Parathyroid glands in primary hyperparathyroidism: An ultrastructural study of 50 cases. Hum Pathol 17:1036–1046, 1986.

Cinti S, Osculati F: Ribosome–lamellae complex in the adenoma cells of the human parathyroid gland. J Submicrosc Cytol 14:521–524, 1982.

Cinti S, Osculati F, LoCascio V: Submicroscopic aspects of the chief cells in a case of parathyroid adenoma. J Submicrosc Cytol 12:293–300, 1980.

Elliott RL, Arthelger RB: Fine structure of parathyroid adenomas: With special reference to annulate lamellae and septate desmosomes. Arch Pathol Lab Med 81: 200–212, 1966.

Faccini JM: The ultrastructure of parathyroid glands removed from patients with primary hyperparathyroidism: A report of 40 cases, including four carcinomata. J Pathol 102:189–199, 1970.

Holck S, Pedersen NT: Carcinoma of the parathyroid gland. A light and electron microscopic study. Acta Pathol Microbiol Immunol Scand [A] 89:297–302, 1981.

Kovacs K, Horvath E, Murray TM: Large clear cell adenoma of the parathyroid gland associated with primary hyperparathyroidism. A light and electron microscopic study. J Submicrosc Cytol 9:323–328, 1977.

Marshall RB, Roberts DK, Turner RA: Adenomas of the human parathyroid. Light and electron microscopic study following selenium-75 methionine scan. Cancer 20:512–524, 1967.

Nilsson O: Studies on the ultrastructure of the human parathyroid glands in various pathological conditions. Acta Pathol Microbiol Immunol Scand [A] 263(suppl): 5–88, 1977.

Obara T, Fujimoto Y, Yamaguchi K, et al: Parathyroid carcinoma of the oxyphil cell type. A report of two cases, light and electron microscopic study. Cancer 55: 1482–1489, 1985.

Rodriguez FH, Sarma DP, Lunseth JH, et al: Primary hyperparathyroidism due to an oxyphil cell adenoma. Am J Clin Pathol 80:878–880, 1983.

Roth SI: Recent advances in parathyroid gland pathology. Am J Med 50:612–622, 1971.

Roth SI, Capen CC: Ultrastructural and functional correlation of the parathyroid gland. Int Rev Exp Pathol 13:161–221, 1974.

Sweet JM, Gardiner GW, Kovacs K, et al: Parathyroid carcinoma with conspicuous nuclear charges. J Submicrosc Cytol 11:495–502, 1979.

Paraganglioma

Chaudhry AP, Haar JG, Koul A, et al: A nonfunctioning paraganglioma of vagus nerve. An ultrastructural study. Cancer 43:1689–1701, 1979.

Gallivan MVE, Chun B, Rowden G, et al: Intrathoracic paravertebral malignant paraganglioma. Arch Pathol Lab Med 104:46–51, 1980.

Glenner GG, Grimley PM: Tumors of the extra-adrenal paraganglion system (including chemoreceptors). In Firminger HI, ed: *Atlas of Tumor Pathology,* 2nd series, fasc 9. Armed Forces Institute of Pathology, Washington, DC, 1974, pp 43–51.

Hull MT, Roth LM, Glover JL, et al: Metastatic carotid body paraganglioma in von Hippel-Lindau disease. Arch Pathol Lab Med 106:235–239, 1982.

Jago R, Smith P, Heath D: Electron microscopy of carotid body hyperplasia. Arch Pathol Lab Med 108:717–722, 1984.

Kahn LB: Vagal body tumor (nonchromaffin paraganglioma, chemodectoma, and carotid body-like tumor) with cervical node metastasis and familial association. Ultrastructural study and review. Cancer 38:2367–2377, 1976.

Lack EE, Cubilla AL, Woodruff JM: Paragangliomas of the head and neck region. A pathologic study of tumors from 71 patients. Hum Pathol 10:191–218, 1979.

Lack EE, Stillinger RA, Colvin DB, et al: Aortico-pulmonary paraganglioma. Report of a case with ultrastructural study and review of the literature. Cancer 43:269–278, 1979.

Lange E, Johannessen JV: Histochemical and ultrastructural studies of chemodectoma-like tumors in the cod (*Gadus morrhua* L). Lab Invest 37:96–104, 1977.

Macadam RF: The fine structure of a human carotid body tumour. J Pathol 99:101–104, 1969.

Merino MJ, LiVolsi VA: Malignant carotid body tumors: Report of two cases and review of the literature. Cancer 47:1403–1414, 1981.

Olson JL, Salyer WR: Mediastinal paragangliomas (aortic body tumor). A report of four cases and a review of the literature. Cancer 41:2405–2412, 1978.

Robertson DI, Cooney TP: Malignant carotid body paraganglioma: Light and electron microscopic study of the tumor and its metastases. Cancer 46:2623–2633, 1980.

Sonneland PRL, Scheithauer BW, LeChago J, et al: Paraganglioma of the cauda equina region. Clinicopathologic study of 31 cases with special reference to immunocytology and ultrastructure. Cancer 58:1720–1735, 1986.

Toker C: Ultrastructure of a chemodectoma. Cancer 20:271–280, 1967.

Warren WH, Lee I, Gould VE, et al: Paragangliomas of the head and neck: Ultrastructural and immunohistochemical analysis. Ultrastruct Pathol 8:333–343, 1985.

Pheochromocytoma

Brown WJ, Barajas L, Waisman J, et al: Ultrastructural and biochemical correlates of adrenal and extra-adrenal pheochromocytoma. Cancer 29:744–759, 1972.

Lamovec J, Memoli VA, Terzakis JA, et al: Pheochromocytoma producing immunoreactive ACTH with Cushing's syndrome. Ultrastruct Pathol 7:41–48, 1984.

Misugi K, Misugi N, Newton A Jr: Fine structural study of neuroblastoma, ganglioneuroblastoma, and pheochromocytoma. Arch Pathol 86:160–170, 1968.

Page DL, DeLellis RA, Hough AJ: Tumors of the adrenal. In Hartmann WH, ed: *Atlas of Tumor Pathology,* fasc 23. Armed Forces Institute of Pathology, Washington, DC, 1986, pp 183–206.

Pollock WJ, Lim JY, McClellan SL: An adrenal gland tumor in a 24-year-old woman. Ultrastruct Pathol 10: 369–376, 1986.

Shin WY, Groman GS, Berkman JI: Pheochromocytoma with angiomatous features. A case report and ultrastructural study. Cancer 40:275–283, 1977.

Sisson JC, Kaliff V, Thompson NW, et al: Pheochromocytomas and intraabdominal paragangliomas [abstract]. Lab Invest 49:52, 1983.

Tannenbaum M: Ultrastructural pathology of adrenal medullary tumors. Pathol Annu 5:145–171, 1970.

Watanabe H, Burnstock G, Jarrott B, et al: Mitochondrial abnormalities in human pheochromocytoma. Cell Tissue Res 172:281–288, 1976.

Yokoyama M, Takayasu H: An electron microscopic study of the human adrenal medulla and pheochromocytoma. Urol Int 24:79–95, 1969.

Monomorphic Adenoma

Alos L, Cardesa A, Bombi JA, et al: Myoepithelial tumors of salivary glands: A clinicopathologic, immunohistochemical, ultrastructural, and flow cytometric study. Semin Diagn Pathol 13:138–147, 1996.

Chaudhry AP, Cutler LS, Satchidanand S, et al: Monomorphic adenomas of the parotid glands. Their ultrastructure and histogenesis. Cancer 52:112–120, 1983.

Chaudhry AP, Cutler LS, Satchidanand S, et al: Ultrastructure of monomorphic adenoma (ductal type) of the minor salivary glands. Arch Otolaryngol 109:118–122, 1983.

Dardick I, Kahn HJ, van Nostrand AW, et al: Salivary gland monomorphic adenoma. Ultrastructural, immunoperoxidase, and histogenetic aspects. Am J Pathol 115:334–348, 1984.

Dardick I, van Nostrand AW: Myoepithelial cells in salivary gland tumors—revisited. Head Neck Surg 7: 395–408, 1985.

Dardick I, Cavell S, Boivin M, et al: Salivary gland myoepithelioma variants. Histological, ultrastructural, and immunocytological features. Virchows Arch [A] 416: 25–42, 1989.

Dardick I, Thomas MJ, van Nostrand AW: Myoepithelioma—New concepts of histology and classification: A light and electron microscopic study. Ultrastruct Pathol 13:187–224, 1989.

Dardick I: Myoepithelioma: Definitions and diagnostic criteria. Ultrastruct Pathol 19:335–345, 1995.

Franquemont DW, Mills SW: Plasmacytoid monomorphic adenoma of salivary glands. Absence of myogenous differentiation and comparison to spindle cell myoepithelioma. Am J Surg Pathol 17:146–153, 1993.

McCluggage G, Sloan J, Cameron S, et al: Basal cell adenocarcinoma of the submandibular gland. Oral Surg Oral Med Oral Pathol Oral Radiol Endod 79:342–359, 1995.

Milchgrub S, Gnepp DR, Vuitch F, et al: Hyalinizing clear cell carcinoma of salivary gland. Am J Surg Pathol 18:74–82, 1994.

Yaku Y, Mori Y, Kanda T, et al: Ultrastructural study of glycogen-rich oxyphilic adenoma of the nasopharyngeal minor salivary gland. Virchows Arch [A] 407: 151–158, 1985.

Youngberg G, Rao MS: Ultrastructural features of monomorphic adenoma of the parotid gland. Oral Surg Oral Med Oral Pathol 47:458–461, 1979.

Pleomorphic Adenoma

Ballance WA, Ro JY, El-Naggar AK, et al: Pleomorphic adenoma (benign mixed tumor) of the breast. An immunohistochemical, flow cytometric, and ultrastructural study and review of the literature. Am J Clin Pathol 93:795–801, 1990.

Chisholm DA, Waterhouse JP, Kraucunas E, et al: A quantitative ultrastructural study of the pleomorphic adenoma (mixed tumor) of human minor salivary glands. Cancer 34:1631–1641, 1974.

Cuadros CL, Ryan SS, Miller RE: Benign mixed tumor (pleomorphic adenoma) of the breast: Ultrastructural study and review of the literature. J Surg Oncol 36: 58–63, 1987.

Dardick I, van Nostrand AW, Jeans MT, et al: Pleomorphic adenoma. I. Ultrastructural organization of "epithelial" regions. Hum Pathol 14:780–797, 1983.

Dardick I, van Nostrand AW, Jeans MT, et al: Pleomorphic adenoma. II. Ultrastructural organization of 'stromal' regions. Hum Pathol 14:798–809, 1983.

Dardick I, van Nostrand AW: Myoepithelial cells in salivary gland tumors—revisited. Head Neck Surg 7: 395–408, 1985.

Dardick I, Daley TD, van Nostrand AW: Basal cell adenoma with myoepithelial cell-derived "stroma": A new major salivary gland tumor entity. Head Neck Surg 8:257–267, 1986.

Dardick I, Hardie J, Thomas MJ, et al: Ultrastructural contributions to the study of morphological differentiation in malignant mixed pleomorphic tumors of salivary gland. Head Neck 11:5–21, 1989.

Dardick I, Gliniecki MR, Heathcote JG, et al: Comparative histogenesis and morphogenesis of mucoepidermoid carcinoma and pleomorphic adenoma. An ultrastructural study. Virchows Arch [A] 417:405–417, 1990.

Erlandson RA, Cardon-Cardo C, Higgins PJ: Histogenesis of benign pleomorphic adenoma (mixed tumor) of the major salivary glands: An ultrastructural and immunohistochemical study. Am J Surg Pathol 8:803–820, 1984.

Hampton TA, Scheithauer BW, Rojiani AM, et al: Salivary gland-like tumors of the sellar region. Am J Surg Pathol 21:424–434, 1997.

Lam RMY: An electron microscopic histochemical study of the histogenesis of major salivary gland pleomorphic adenoma. Ultrastruct Pathol 8:207–223, 1985.

Mackay B, Ordonez NG, Batsakis JG, et al: Pleomorphic adenoma of parotid with myoepithelial cell predominance. Ultrastruct Pathol 12:461–468, 1988.

Orenstein JM, Dardick I, van Nostrand AW: Ultrastructural similarities of adenoid cystic carcinoma and pleomorphic adenoma. Histopathology 9:623–638, 1985.

Palmer RM, Lucas RB, Langdon JD: Ultrastructural analysis of salivary gland pleomorphic adenoma, with particular reference to myoepithelial cells. Histopathology 9:1061–1076, 1985.

Adenoid Cystic Carcinoma

Chaudhry AP, Leifer C, Cutler LS, et al: Histogenesis of adenoid cystic carcinoma of the salivary glands. Light and electron microscopic study. Cancer 58:72–82, 1986.

Fonseca I, Soares J: Basal cell adenocarcinoma of minor salivary and seromucous glands of the head and neck region. Semin Diagn Pathol 13:128–137, 1996.

Gould VE, Miller J, Jao W: Ultrastructure of medullary, intraductal, tubular and adenocystic breast carcinomas: Comparative patterns of the myoepithelial differentiation and basal lamina deposition. Am J Pathol 78: 401–416, 1975.

Hewan-Lowe K, Dardick I: Ultrastructural distinction of basaloid-squamous carcinoma and adenoid cystic carcinoma. Ultrastruct Pathol 19:371–381, 1995.

Hoshino M, Yamamoto I: Ultrastructure of adenoid cystic carcinoma. Cancer 25:186–198, 1970.

Koss LG, Brannan CD, Ashikari R: Histologic and ultrastructural features of adenoid cystic carcinoma of the breast. Cancer 26:1271–1279, 1970.

Lawrence JB, Mazur MT: Adenoid cystic carcinoma: A comparative pathologic study of tumors in salivary gland, breast, lung, and cervix. Hum Pathol 13:916–924, 1982.

Orenstein JM, Dardick I, van Nostrand AW: Ultrastructural similarities of adenoid cystic carcinoma and pleomorphic adenoma. Histopathology 9:623–638, 1985.

Ormos J, Halasz A: Electron microscopic study of adenoid cystic carcinoma. Ultrastruct Pathol 15:149–157, 1991.

Quizilbash AH, Patterson MC, Oliveira KF: Adenoid cystic carcinoma of the breast. Light and electron microscopy and a brief review of the literature. Arch Pathol Lab Med 101:302–306, 1977.

Seifert G: Classification and differential diagnosis of clear and basal cell tumors of the salivary glands. Semin Diagn Pathol 13:95–103, 1996.

Takeuchi J, Sobue M, Katoh Y, et al: Morphologic and biologic characteristics of adenoid cystic carcinoma cells of the salivary gland. Cancer 38:2349–2356, 1976.

Tandler B: Ultrastructure of adenoid cystic carcinoma of salivary gland origin. Lab Invest 24:504–512, 1971.

Van der Kwast TH, Vuzevski VD, Ramaekers F, et al: Primary cutaneous adenoid cystic carcinoma: Case report, immunohistochemistry, and review of the literature. Br J Dermatol 118:567–577, 1988.

Mucoepidermoid Carcinoma

Chen SY: Ultrastructure of mucoepidermoid carcinoma in minor salivary glands. Oral Surg 47:247–255, 1979.

Dardick I, Daya D, Hardie J, et al: Mucoepidermoid carcinoma: Ultrastructural and histogenetic aspects. Oral Pathol 13:342–358, 1984.

Klacsman PG, Olson JL, Eggleston JC: Mucoepidermoid carcinoma of the bronchus. An electron microscopic study of the low grade and the high grade variants. Cancer 43:1720–1733, 1979.

Mullins JD, Barnes RP: Childhood bronchial mucoepidermoid tumors. A case report and review of the literature. Cancer 44:315–322, 1979.

Nicolatou O, Hartwick RD, Putong P, et al: Ultrastructural characterization of intermediate cells of mucoepidermoid carcinoma of the parotid. Oral Surg 48: 324–336, 1979.

Stafford JR, Pollock WJ, Wenzel BC: Oncocytic mucoepidermoid tumor of the bronchus. Cancer 54:94–99, 1984.

Tomita T, Lotuaco L, Talbott L, et al: Mucoepidermoid carcinoma of the subglottis. An ultrastructural study. Arch Pathol Lab Med 101:145–148, 1977.

Alveolar Soft Part Sarcoma

Auerbach HE, Brooks JJ: Alveolar soft part sarcoma. A clinicopathologic and immunohistochemical study. Cancer 60:66–73, 1987.

DeSchryver-Kecskemeti K, Kraus FT, Engelman W, et al: Alveolar soft part sarcoma—A malignant angioreninoma. Histochemical immunocytochemical, and electron microscopic study of four cases. Am J Surg Pathol 6:5–18, 1982.

Ekfors TO, Kalimo H, Rantakokko V, et al: Alveolar soft part sarcoma: A report of two cases with some histochemical and ultrastructural observations. 43: 1672–1677, 1979.

Fisher ER, Reidbord H: Electron microscopic evidence suggesting the myogenous derivation of the so-called alveolar soft part sarcoma. Cancer 27:150–159, 1971.

Font RL, Jurco S, Zimmerman LE: Alveolar soft part sarcoma of the orbit: A clinicopathologic analysis of seventeen cases and a review of the literature. Hum Pathol 13:569–579, 1982.

Mathew T: Evidence supporting neutral crest origin of an alveolar soft part sarcoma. An ultrastructural study. Cancer 50:507–514, 1982.

Mukai M, Iri H, Hakajima T, et al: Alveolar soft part sarcoma. A review on its histogenesis and further studies based on electron microscopy, immunohistochemistry, and biochemistry. Am J Surg Pathol 7:679–689, 1983.

Schmidt D, Mackay B, Sinkovics JG: Retroperitoneal tumor with vertebral metastasis in a 25-year-old female. Ultrastruct Pathol 2:383–388, 1981.

Shipkey FH, Lieberman PH, Foote FW, et al: Ultrastructure of alveolar soft part sarcoma. Cancer 17: 821–830, 1964.

Unni KK, Soule EH: Alveolar soft part sarcoma: An electron microscopic study. Mayo Clin Proc 50:591–598, 1975.

Welsh RA, Bray DM, Shipkey FH, et al: Histogenesis of alveolar soft part sarcoma. Cancer 29:191–204, 1972.

Chordoma

Carstens PH: Chordoid tumor: A light, electron microscopic, and immunohistochemical study. Ultrastruct Pathol 19:291–295, 1995.

Cancilla P, Morecki R, Hurwitt ES: Fine structure of a recurrent chordoma. Arch Neurol 11:289–295, 1964.

Erlandson RA, Tandler B, Lieberman PH, et al: Ultrastructure of human chordoma. Cancer Res 28: 2115–2125, 1968.

Friedmann I, Harrison DFN, Bird ES: The fine structure of chordoma with particular reference to the physaliphorous cell. J Cell Biol 15:116–125, 1962.

Ishida T, Oda H, Oka T, et al: Parachordoma: An ultrastructural and immunohistochemical study. Virchows Arch [A] 422:239–245, 1993.

Jeffrey PB, Biava CG, Davis RL: Chondroid chordoma: A hyalinized chordoma without cartilaginous differentiation. Am J Clin Pathol 103:271–279, 1995.

Kay S, Schatzki PF: Ultrastructural observation of a chordoma arising in the clivus. Hum Pathol 3:403–413, 1972.

Mierau GW, Weeks, DA: Chondroid chordoma. Ultrastruct Pathol 11:731–737, 1987.

Miettinen M, Gannon FH, Lackman R: Chordoma-like soft tissue sarcoma in the leg: A light and electron microscopic and immunohistochemical study. Ultrastruct Pathol 16:577–586, 1992.

Murad TM, Murthy MSN: Ultrastructure of a chordoma. Cancer 25:1204–1215, 1970.

Niwa J, Hashi K, Minase T: Immunohistochemical and electron microscopic studies on intracranial chordomas: Difference between typical chordomas and chondroid chordomas. Noshuyo Byori 11:15–21, 1994.

Pena CE, Horvat BL, Fisher ER: The ultrastructure of chordoma. Am J Clin Pathol 53:544–551, 1970.

Persson S, Kindblom LG, Angervall L: Classical and chondroid chordoma. A light-microscopic, histochemical, ultrastructural and immunohistochemical analysis of the various cell types. Pathol Res Pract 187:828–838, 1991.

Shin HJ, Mackay B, Ichinose H, et al: Parachordoma. Ultrastruct Pathol 18:249–256, 1994.

Spjut HJ, Luse SA: Chordoma: An electron microscopic study. Cancer 17:643–656, 1964.

Suster S, Moran CA: Chordomas of the mediastinum: Clinicopathologic, immunohistochemical, and ultrastructural study of six cases presenting as posterior mediastinal masses. Hum Pathol 26:1354–1362, 1995.

Valderrama E, Kahn LB: Chondroid chordoma. Electron microscopic study of two cases. Am J Surg Pathol 7: 625–632, 1983.

Epithelioid Sarcoma

Enzinger FM: Epithelioid sarcoma. A sarcoma simulating a granuloma or a carcinoma. Cancer 26:1029–1041, 2970.

Fisher C. Epithelioid sarcoma: The spectrum of ultrastructural differentiation in seven immunohistochemically defined cases. Hum Pathol 19:265–275, 1988.

Frable WJ, Kay S, Lawrence W, et al: Epithelioid sarcoma. An electron microscopic study. Arch Pathol 95: 8–12, 1973.

Guillou L, Wadden C, Coindre JM, et al: "Proximal-type" epithelioid sarcoma, a distinctive aggressive neoplasm showing rhabdoid features. Clinicopathologic, immunohistochemical, and ultrastructural study of a series. Am J Surg Pathol 21:130–146, 1997.

Ishida T Oka T, Matsushita H, et al: Epithelioid sarcoma: An electron-microscopic, immunohistochemical and DNA flow cytometric analysis. Virchows Arch [A] 421:401–408, 1992.

Machinami R, Kikuchi F, Matsushita H: Epithelioid sarcoma. Enzyme histochemical and ultrastructural study. Virchows Arch [A] 397:109–120, 1982.

Meis JM, Mackay B, Ordonez NG: Epithelioid sarcoma: An immunohistochemical and ultrastructural study. Surg Pathol 1:13–31, 1988.

Mills SE, Fechner RE, Bruns DE, et al: Intermediate filaments in eosinophilic cells of epithelioid sarcoma. A light-microscopic, ultrastructural, and electrophoretic study. Am J Surg Pathol 5:195–202, 1981.

Mukai M, Torikata C, Iri H, et al: Cellular differentiation of epithelioid sarcoma. An electron-microscopic, enzyme-histochemical, and immunohistochemical study. Am J Pathol 119:44–56, 1985.

Hepatoblastoma

Abenoza P, Manivel C, Wick MR, et al: Hepatoblastoma: An immunohistochemical and ultrastructural study. Hum Pathol 18:1025–1035, 1987.

Gonzalez-Crussi F, Manz HJ: Structure of a hepatoblastoma of pure epithelial type. Cancer 29:1272–1280, 1972.

Horie A, Kotoo Y, Hayashi I: Ultrastructural comparison of hepatoblastoma and hepatocellular carcinoma. Cancer 44:2184–2193, 1979.

Ito J, Johnson WW: Hepatoblastoma and hepatoma in infancy and childhood. Arch Pathol 87:259–266, 1969.

Silverman JF, Fu YS, McWilliams NB, et al: An ultrastructural study of mixed hepatoblastoma with osteoid elements. Cancer 36:1436–1443, 1975.

Stiller D, Holzhausen HJ: Congenital hepatoblastoma of mixed type. Light and electron microscopy studies. Zentralbl Allegmeine Pathol Patholog Anat 128: 317–326, 1983.

Warfel KA, Hull MT: Hepatoblastomas: An ultrastructural and immunohistochemical study. Ultrastruct Pathol 16:451–461, 1992.

Embryonal Sarcoma of Liver

Abramowsky CR, Cebelin M, Choudhury A, et al: Undifferentiated (embryonal) sarcoma of the liver with alpha-1-antitrypsin deposits: Immunohistochemical and ultrastructural studies. Cancer 45:3108–3113, 1980.

Aoyama C, Hachitanda Y, Sato JK, et al: Undifferentiated (embryonal) sarcoma of the liver. A tumor of uncertain histogenesis showing divergent differentiation. Am J Surg Pathol 15:615–624, 1991.

Chou P, Mangkornkanok M, Gonzalez-Crussi F: Undifferentiated (embryonal) sarcoma of the liver: Ultrastructure, immunohistochemistry, and DNA ploidy analysis of two cases. Pediatr Pathol 10:549–562, 1990.

Cozzutto C, De Bernardi B, Comelli A, et al: Malignant mesenchymoma of the liver in children: A clinicopathologic and ultrastructural study. Hum Pathol 12: 481–485, 1981.

Keating S, Taylor GP: Undifferentiated (embryonal) sarcoma of the liver. Ultrastructural and immunohistochemical similarities with malignant fibrous histiocytoma. Hum Pathol 16:693–699, 1985.

Lack EE, Schloo BL, Azumi N, et al: Undifferentiated (embryonal) sarcoma of the liver. Clinical and pathologic study of 16 cases with emphasis on immunohistochemical features. Am J Surg Pathol 15:1–16, 1991.

Miettinen M, Kahlos T: Undifferentiated (embryonal) sarcoma of the liver. Epithelial features as shown by immunohistochemical analysis and electron microscopic examination. Cancer 64:2096–2103, 1989.

Parham DM, Kelly DR, Donnelly WH, et al: Immunohistochemical and ultrastructural spectrum of hepatic sarcomas of childhood: Evidence for a common histogenesis. Mod Pathol 4:648–653, 1991.

Stanley RJ, Dehver LP, Hesker AE: Primary malignant mesenchymal tumors (mesenchymoma) of the liver in childhood. Cancer 32:973–984, 1973.

Stocker JT, Ishak KG: Undifferentiated (embryonal) sarcoma of the liver. Report of 31 cases. Cancer 42:336–348, 1978.

Gastrointestinal Stromal Tumor

Erlandson RA, Klimstra DS, Woodruff JM: Subclassification of gastrointestinal stromal tumors based on evaluation by electron microscopy and immunohistochemistry. Ultrastruct Pathol 20:373–393, 1996.

Flinner RL, Hammond EA: Gastrointestinal stromal tumor of the duodenum: A case report. Ultrastruct Pathol 15:503–507, 1991.

Mazur MT, Clark HB: Gastric stromal tumors. Reappraisal of histogenesis. Am J Surg Pathol 7:507–519, 1983.

Gastrointestinal Autonomic Nerve (GAN) Tumor

Dhimes P, Lopez-Carreira M, Ortega-Serrano MP, et al: Gastrointestinal autonomic nerve tumor and their separation from other gastrointestinal stromal tumors: An ultrastructural and immunohistochemical study of seven cases. Virchows Arch [A] 426:27–35, 1995.

Donner LR: Gastrointestinal autonomic nerve sheath tumors: A common type of gastrointestinal stromal neoplasm. Ultrastruct Pathol 21:419–424, 1997.

Faussone-Pellegrini MS, Pantalone D, Cortesini C: Smooth muscle cells, interstitial cells of Cajal and myenteric plexus interrelationships in the human colon. Acta Anat 139:3–44, 1990.

Garcia-Rostan y Perez GM, Montes Diaz M, Garcia Bragado F: Jejunal stromal tumor with skeinoid fibers or myenteric plexoma: A case report. Pathol 47:794–800, 1997.

Herrera GA, Cerezo L, Jones JE, et al: Gastrointestinal autonomic nerve tumors: "Plexosarcomas." Arch Pathol Lab Med 113:846–853, 1989.

Herrera GA, Pinto de Moraes H, Grizzle WE, et al: Malignant small bowel neoplasm of enteric plexus derivation (plexosarcoma). Light and electron microscopic study confirming the origin of the neoplasm. Dig Dis Sci 27:275–284, 1984.

Kindblom LG, Remotti HE, Aldenborg F, et al: Gastrointestinal pacemaker cell tumor (GIPACT). Gastrointestinal stromal tumors show phenotypic characteristics of the interstitial cells of Cajal. Am J Pathol 152:1259–1269, 1998.

Lam KY, Law SY, Chu KM, et al: Gastrointestinal autonomic nerve tumor of the esophagus. A clinicopathologic, immunohistochemical, ultrastructural study of a case and review of the literature. Cancer 78:1651–1659, 1996.

Lauwers GY, Erlandson RA, Casper ES, et al: Gastrointestinal autonomic nerve tumors. A clinicopathological, immunohistochemical, and ultrastructural study of 12 cases. Am J Surg Pathol 17:887–897, 1993.

MacLeod CB, Tsokos M: Gastrointestinal autonomic nerve tumor. Ultrastruct Pathol 15:49–55, 1991.

Matsumoto K, Min W, Yamada N, et al: Gastrointestinal autonomic nerve tumors: Immunohistochemical and ultrastructural studies in cases of gastrointestinal stromal tumor. Pathol Int 47:308–314, 1997.

Min K-W: Small intestinal stromal tumors with skeinoid fibers. Clinicopathological, immunohistochemical, and ultrastructural investigations. Am J Surg Pathol 16: 145–155, 1992.

Ojanguren I, Ariza A, Navas-Palacios JJ: Gastrointestinal autonomic nerve tumor: Further observations regarding an ultrastructural and immunohistochemical analysis of six cases. Hum Pathol 27:1311–1318, 1996.

Rumessen JJ, Mikkelsen HB, Qvortup K, et al: Ultrastructure of interstitial cells of Cajal in circular muscle of human small intestine. Gastroenterology 104: 343–350, 1993.

Rumessen JJ, Mikkelsen HB, Thuneberg L: Ultrastructure of interstitial cells of Cajal associated with deep muscular plexus of human small intestine. Gastroenterology 102:56–68, 1992.

Rumessen JJ, Peters S, Thuneberg L: Light and electron microscopical studies of interstitial cells of Cajal and muscle cells at the submucosal border of human colon. Lab Invest 68:481–495, 1993.

Shek TW, Luk IS, Loong F, et al: Inflammatory cell-rich gastrointestinal autonomic nerve tumor. An expansion of its histologic spectrum. Am J Surg Pathol 20:325–331, 1996.

Thuneberg L: Interstitial cells of Cajal: Intestinal pacemaker cells? In Beck F, Hild W, van Limborgh, et al. eds: *Advances in Anatomy: Embryology and Cell Biology.* Berlin, Springer-Verlag, 1982, pp 1–130.

Walker P, Dvorak M: Gastrointestinal autonomic nerve (GAN) tumor. Ultrastructural evidence of a newly recognized entity. Arch Pathol Lab Med 110:309–316, 1986.

Pulmonary Blastoma

Fung CH, Lo JW, Yonan TN, et al: Pulmonary blastoma. An ultrastructural study with a brief review of the literature and a discussion of pathogenesis. Cancer 39: 153–163, 1977.

Heckman CJ, Truong LD, Cagle PT, et al: Pulmonary blastoma with rhabdomyosarcomatous differentiation:

An electron microscopic and immunohistochemical study. Am J Surg Pathol 12:35–40, 1988.

Inoue H, Kasai K, Shinada J, et al: Pulmonary blastoma. Comparison between its epithelial components and fetal bronchial epithelium. Acta Pathol Jpn 42:884–892, 1992.

Jackson MD, Albrecht R, Roggli VL, et al: Pulmonary blastoma: An ultrastructural and histochemical study. Ultrastruct Pathol 7:259–268, 1984.

Korbi S, Boyo AM, Dusmet M, et al: Pulmonary blastoma. Immunohistochemical and ultrastructural studies of a case. Histopathology 11:753–760, 1987.

Kradin RL, Young RH, Dickersin GR, et al: Pulmonary blastoma with argyrophil cells and lacking sarcomatous features (pulmonary endodermal tumor resembling fetal lung). Am J Surg Pathol 6:165–172, 1982.

Marcus PB, Dieb TM, Martin JH: Pulmonary blastoma: An ultrastructural study emphasizing intestinal differentiation in lung tumors. Cancer 49:1829–1833, 1982.

McCann MP, Fu Y-S, Kay S: Pulmonary blastoma. A light and electron microscopic study. Cancer 38: 789–797, 1976.

Nakatani Y, Dickersin GR, Mark EJ: Pulmonary endodermal tumor resembling fetal lung: A clinicopathologic study of five cases with immunohistochemical and ultrastructural characterization. Hum Pathol 21: 1097–1107, 1990.

Nakatani Y, Kitamura H, Inayama Y, et al: Pulmonary endodermal tumor resembling fetal lung. The optically clear nucleus is rich in biotin. Am J Surg Pathol 18: 637–642, 1994.

Sciot R, Dal Cin P, Brock P, et al: Pleuropulmonary blastoma (pulmonary blastoma of childhood): Genetic link with other embryonal malignancies? Histopathology 24:559–563, 1994.

Spencer H: Pulmonary blastomas. J Pathol Bacteriol 82:161–165, 1961.

Wick MR, Ritter JH, Humphrey PA: Sarcomatoid carcinomas of the lung: A clinicopathologic review. Am J Clin Pathol 108:40–53, 1997.

Yang P, Morizumi H, Sano T, et al: Pulmonary blastoma: An ultrastructural and immunohistochemical study with special reference to nuclear filament aggregation. Ultrastruct Pathol 19:501–509, 1995.

Yazawa T, Ogata T, Kamma H, et al: Pulmonary blastoma with a topographic transition from blastic to more differentiated areas. An immunohistochemical assessment of its embryonic nature using stage-specific embryonic antigens. Virchows Arch [A] 419:513–518, 1991.

Juxtaglomerular Cell Tumor

Barajas L: The development and ultrastructure of the juxtaglomerular apparatus. J Ultrastruct Res 15: 400–413, 1966.

Barajas L, Bennett CM, Connor G, et al: Structure of a juxtaglomerular cell tumor: The presence of a neural component. A light and electron microscopic study. Lab Invest 37:357–368, 1977.

Barajas L: Anatomy of the juxtaglomerular apparatus. Am J Physiol 237:F333–F343, 1979.

Biava CG, West M: Fine structure of normal human juxtaglomerular cells. I. General structure and intercellular relationships. Am J Pathol 49:679–721, 1966.

Biava CG, West M: Fine structure of normal human juxtaglomerular cells. II. Specific and nonspecific cytoplasmic granules. Am J Pathol 49:955–979, 1966.

Camilleri JP, Hinglais N, Bruneval P, et al: Renin storage and cell differentiation in juxtaglomerular cell tumors: An immunohistochemical and ultrastructural study of three cases. Hum Pathol 15:1069–1079, 1984.

DeSchryver-Kecskemeti K, Kraus FT, Engleman W, et al: Alveolar soft-part sarcoma—A malignant angioreninoma: Histochemical, immunocytochemical, and electron microscopic study of four cases. Am J Surg Pathol 6:5–18, 1982.

Hasegawa A: Juxtaglomerular cells tumor of the kidney: A case report with electron microscopic and flow cytometric investigation. Ultrastruct Pathol 21:201–208, 1997.

Kodet R, Taylor M, Vachalova H, et al: Juxtaglomerular cell tumor. An immunohistochemical, electron-microscopic, and in situ hybridization study. Am J Surg Pathol 18:837–842, 1994.

Lindop GBM, Stewart JA, Downie TT: The immunocytochemical demonstration of renin in a juxtaglomerular cell tumour by light and electron microscopy. Histopathology 7:421–431, 1983.

MacCallum DK, Conn JW, Baker BL: Ultrastructure of renin-secreting juxtaglomerular cell tumor of the kidney. Invest Urol 11:65–74, 1973.

Phillips G, Mukherjee TM: A juxtaglomerular cell tumour: Light and electron microscopic studies of a renin-secreting kidney tumour containing both juxtaglomerular cells and mast cells. Pathology 4:193–204, 1972.

Squires JP, Ulbright TM, DeSchryver-Kecskemeti K, et al: Juxtaglomerular cell tumor of the kidney. Cancer 53: 516–523, 1984.

Tetu B, Vaillancourt L, Camilleri JP, et al: Juxtaglomerular cell tumor of the kidney: Report of two cases with a papillary pattern. Hum Pathol 24:1168–1174, 1993.

10

Infectious Agents

Bacteria

(Figures 10.1 through 10.12.)

Diagnostic criteria (of bacterial rods and cocci, in general). (1) Outer cell wall (in most genera [Figure 10.5], but absent in *Mycoplasma*); (2) presence or absence of loose, fuzzy, extracellular capsule; (3) presence or absence of flagella or pili (fimbria) on the outer surface of the cell; (4) inner cell membrane; (5) central nuclear region (nucleoid), without a limiting membrane; (6) dense cytoplasm composed mostly of ribosomes, but also a varying number of vesicles formed from the inner cell membrane (mesosomes), storage vacuoles, and endospores; (7) in *Whipple's disease,* the rods are free and within macrophages (Figure 10.4); (8) also in Whipple's disease the macrophages contain secondary lysosomes filled with degenerating bacteria and serpiginous membranes (Figures 10.4, 10.6, and 10.8).

Additional points. Bacteria are prokaryotic cells and, therefore, have a simple internal structure, including a nonmembrane-bound nucleus and a lack of membrane-bound cytoplasmic organelles, such as mitochondria and endoplasmic reticulum. Gram-positive bacteria have a thicker outer cell wall than do gram-negative bacteria, and the latter usually have an additional outer, wrinkled cell membrane. Although most bacteria are readily demonstrable by light microscopy, some are small and, therefore, more visible by electron microscopy. *Legionella pneumophila* (2.5–5 μm long and 0.3–1.0 μm in diameter) is an example of one of these (Figures 10.1 through 10.3). The bacilli found in Whipple's disease also are small (about 1–2 μm long and 0.25 μm in diameter) and are barely visible by light microscopy. They resist usual methods of culturing and, therefore, were not classifiable before the availability of newer techniques, such as nucleotide sequencing and polymerase chain reaction on rDNA and bacterial antisera profiles. Current evidence indicates that the Whipple bacillum represents a single genus and species (*Tropheryma whippelii*) rather than multiple ones, as was thought previously. The bacilli have the thick (about 20 nm) outer cell wall of gram-positive bacilli but, in addition, have a trilaminar membrane on the external surface of the wall. It is the glycoprotein- or polysaccharide-rich (PAS-positive) cell walls of degenerated bacteria that comprise the serpiginous membranes in the classic Whipple's macrophages. Bacteria also may be found less frequently in cells other than macrophages, including epithelial, endothelial, smooth muscle, lymphocytic, plasmocytic, and mast cells. When organs other than the intestine are involved in Whipple's disease, the same diagnostic ultrastructural findings of

(Text continues on page 659)

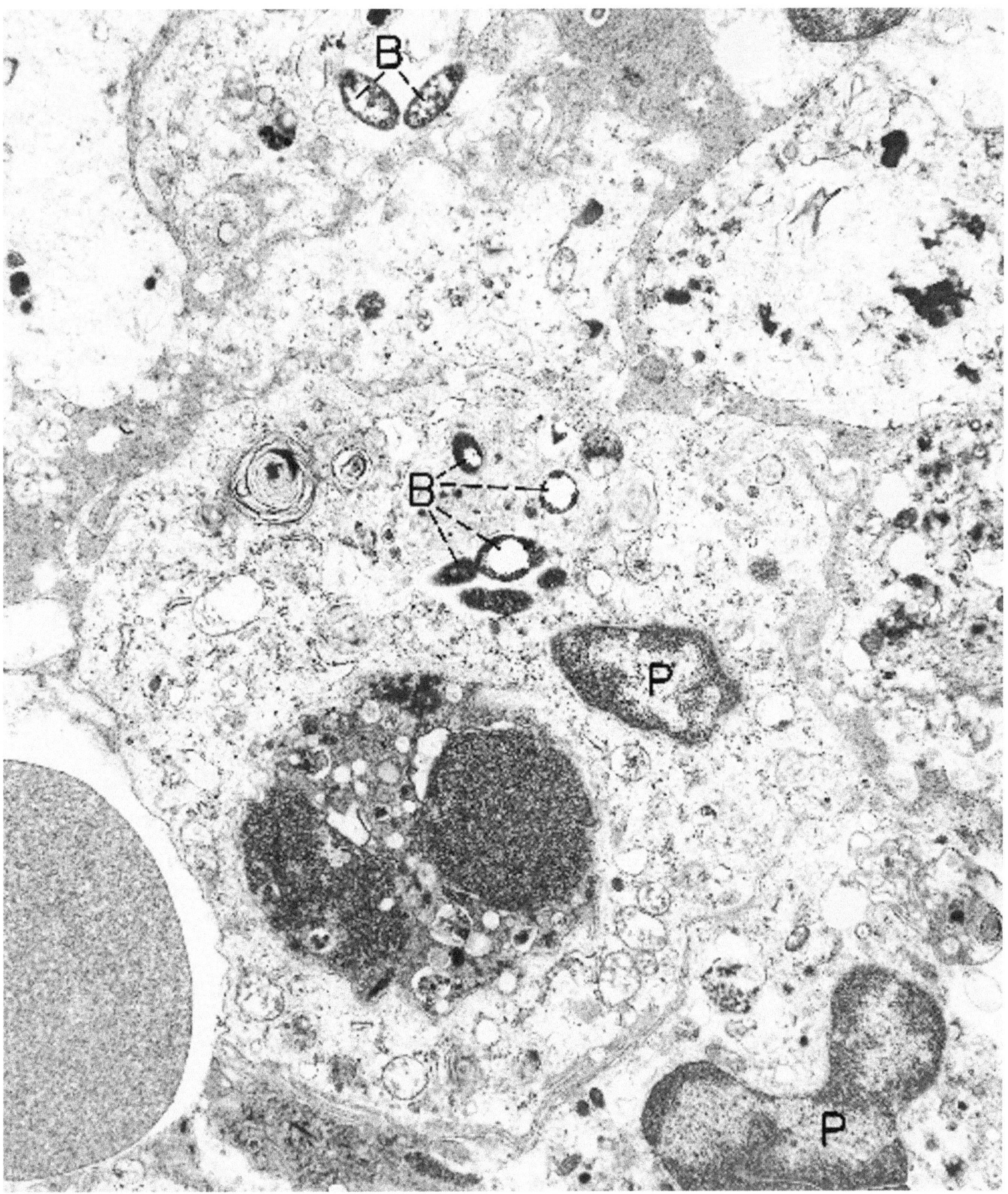

Figure 10.1. Bacterial pneumonia, *Legionella pneumophilia* (lung). Low power of this alveolar exudate shows intact and degenerating bacterial rods (B), predominantly within phagocytic cells. P = nuclei of phagocytic cells. (× 10,640)

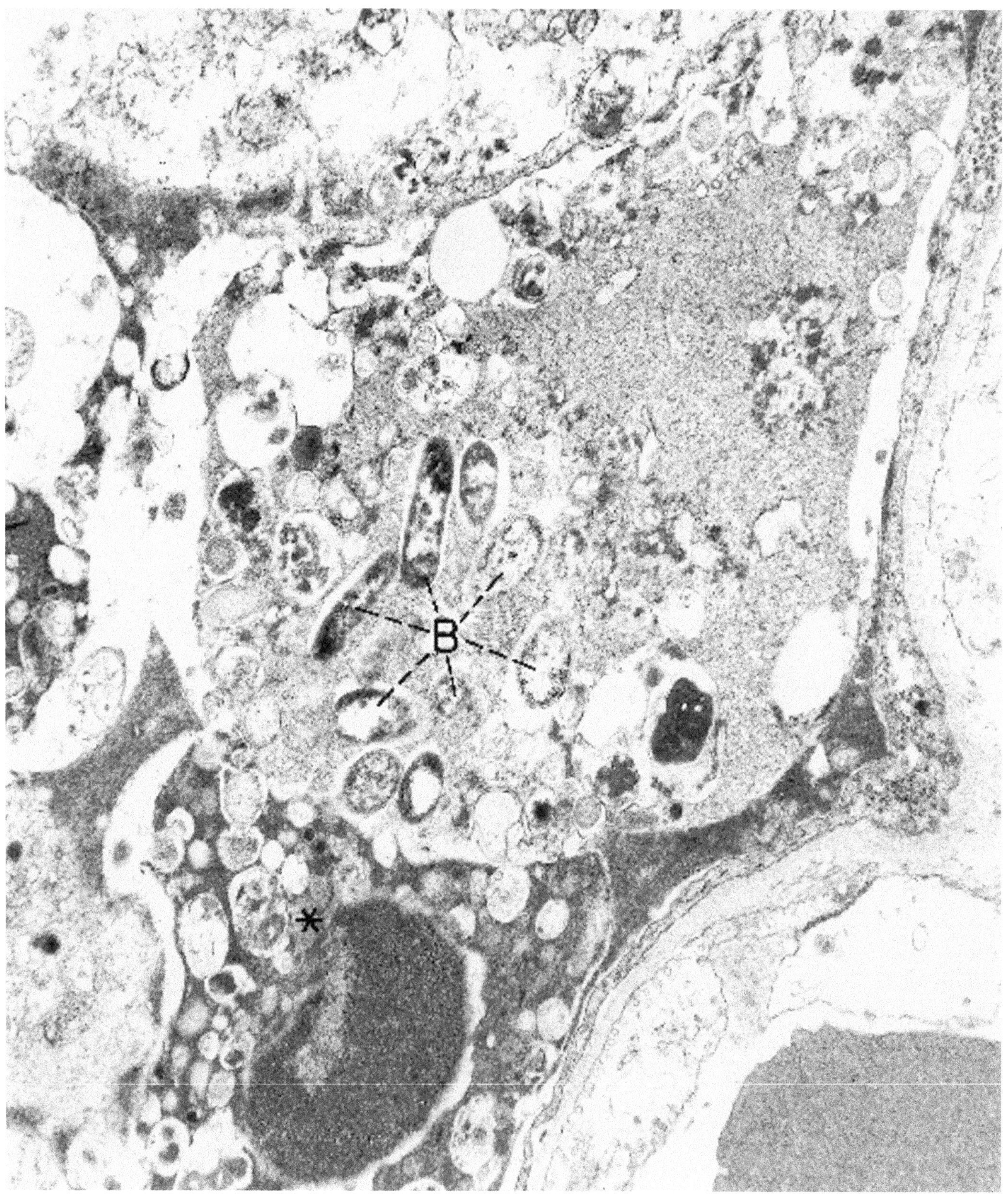

Figure 10.2. Bacterial pneumonia, *Legionella pneumophilia* (lung). At this higher magnification of the same lung in Figure 10.1, it is apparent that most of the bacteria (B) are degenerating. Some of the osmiophilic debris, as is present in the lower part of the field (*), is presumed to represent completely necrotic organisms. (× 19,240)

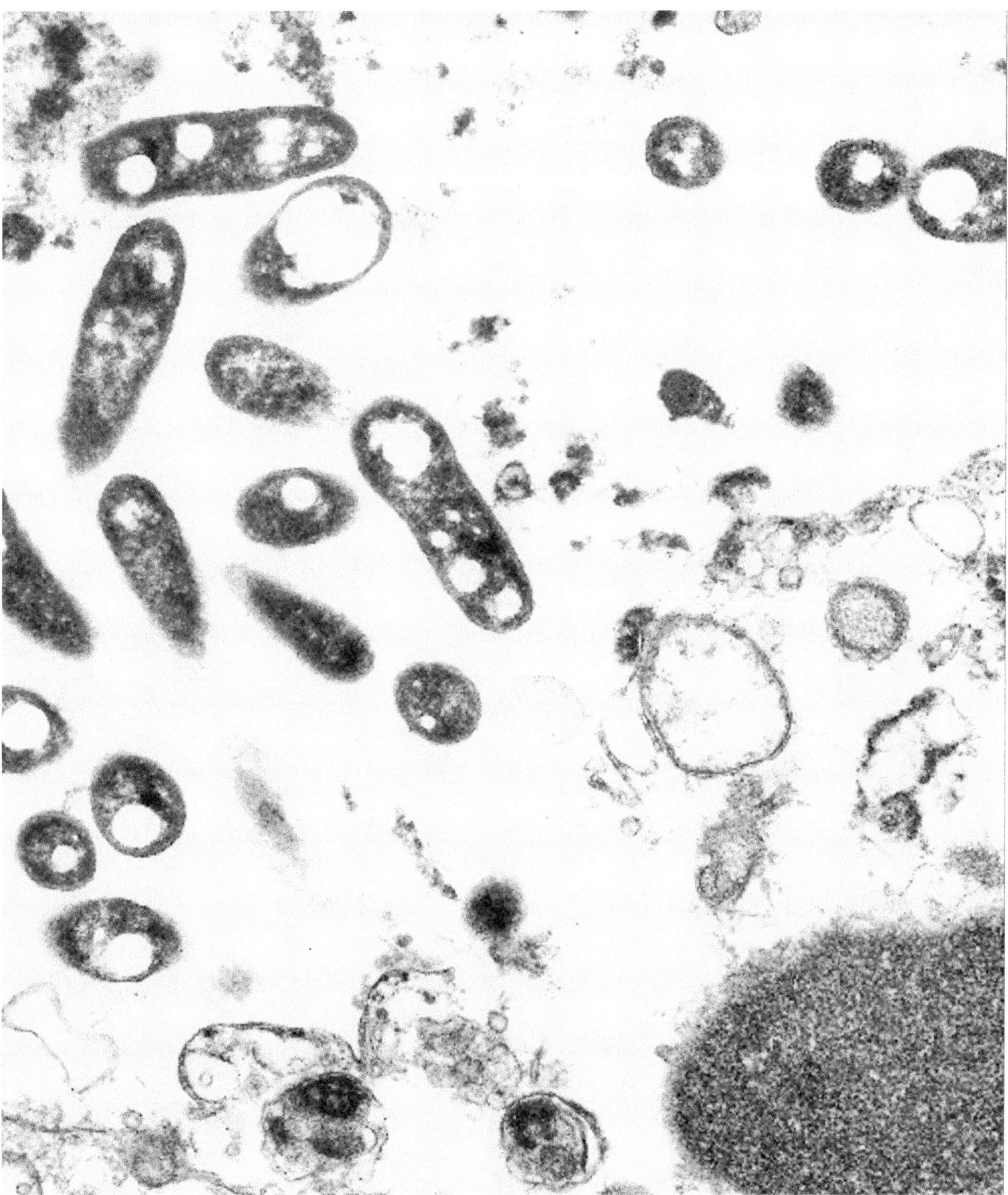

Figure 10.3. Bacterial pneumonia, *Legionella pneumophilia* (lung). The only substructures of these poorly preserved bacteria that are discernible are the outer cell membrane and the cytoplasmic vacuoles. Also, there are no apparent flagella or pili and no extracellular capsule. (× 36,700)

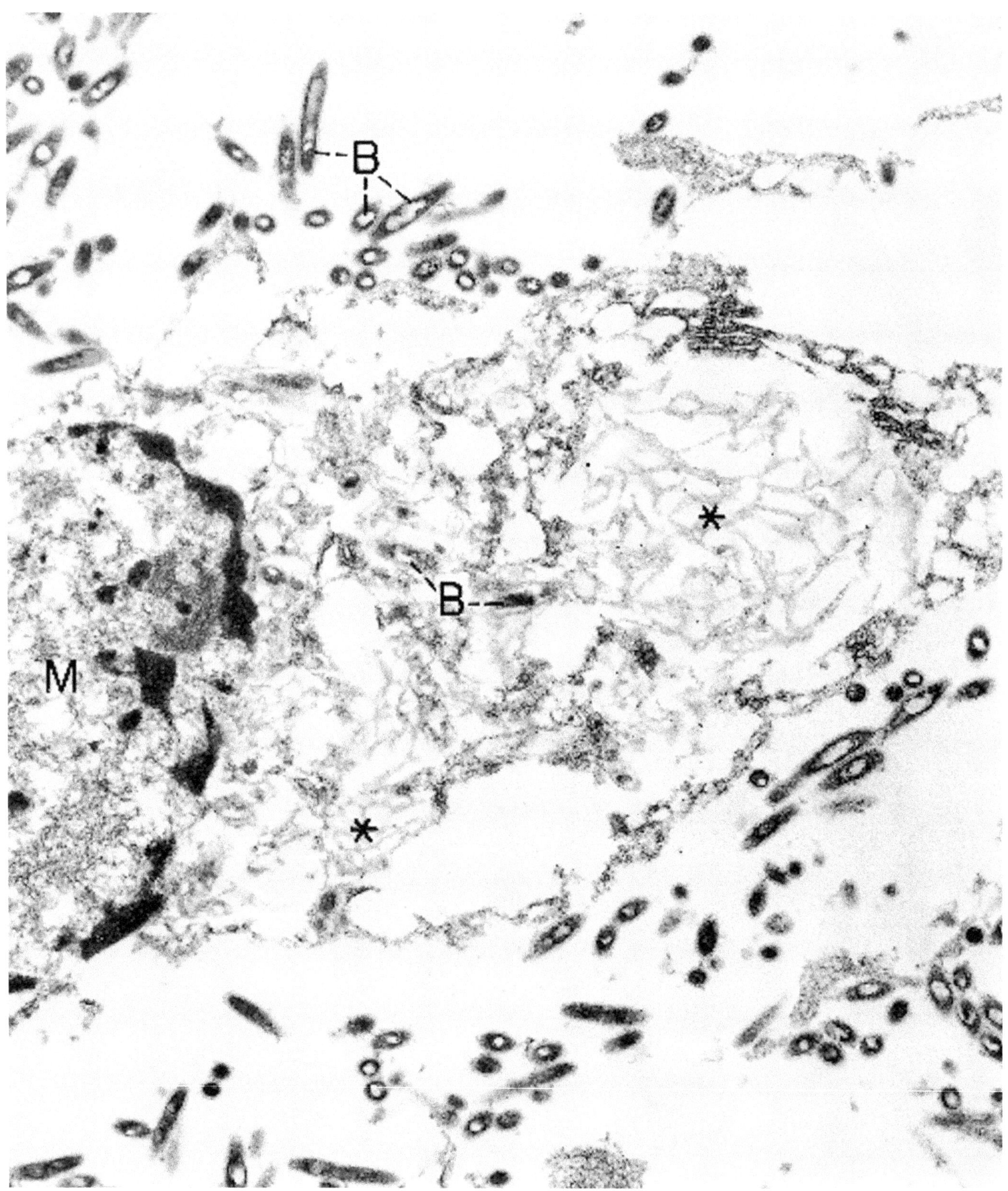

Figure 10.4. Whipple's disease (jejunum, tissue retrieved from paraffin). The lamina propria of the jejunal mucosa contained numerous Whipple's type macrophages (M = macrophage nucleus) and bacterial rods (B). Most of the intact bacilli are extracellular, whereas those in the macrophage are in various stages of degeneration, including the end-stage of serpiginous membranes (*). (× 16,500)

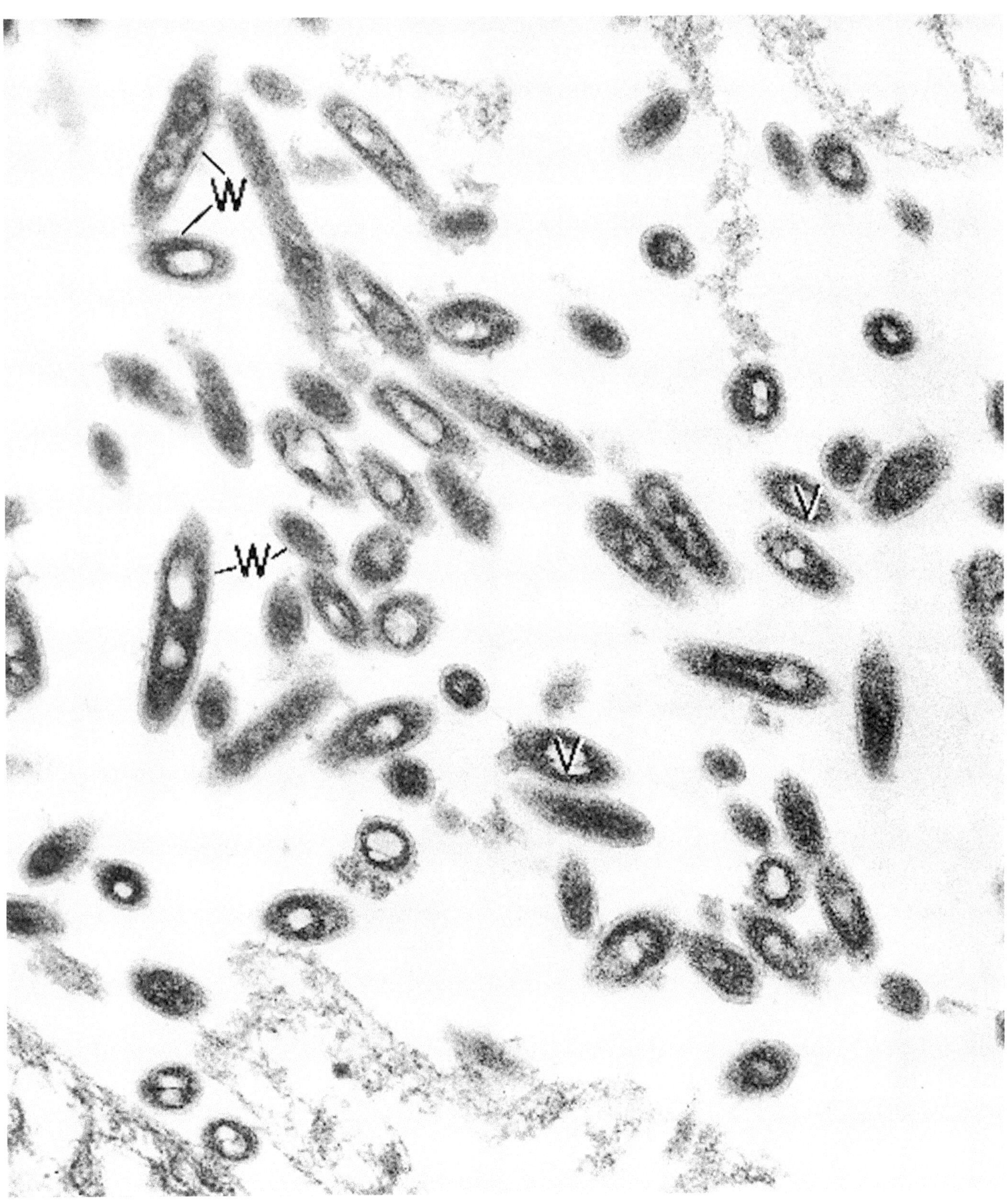

Figure 10.5. Whipple's disease (jejunum, tissue retrieved from paraffin). High magnification of extracellular bacilli allows the cell wall (W) and cytoplasmic vacuoles (V) to be seen clearly. (× 45,000)

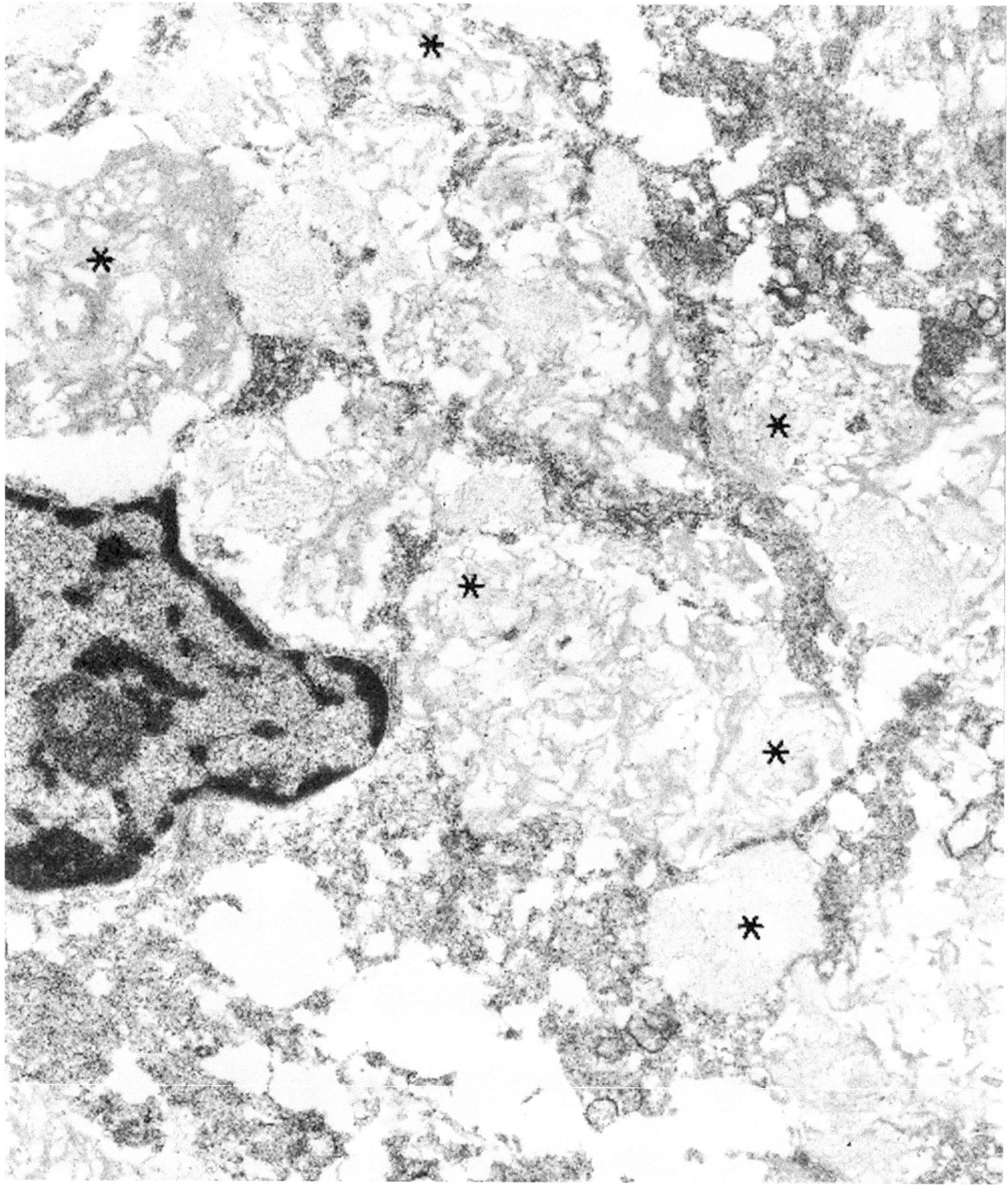

Figure 10.6. Whipple's disease (jejunum, tissue retrieved from paraffin). The cytoplasm is filled with confluent, serpiginous membranous bodies (*). (× 15,930)

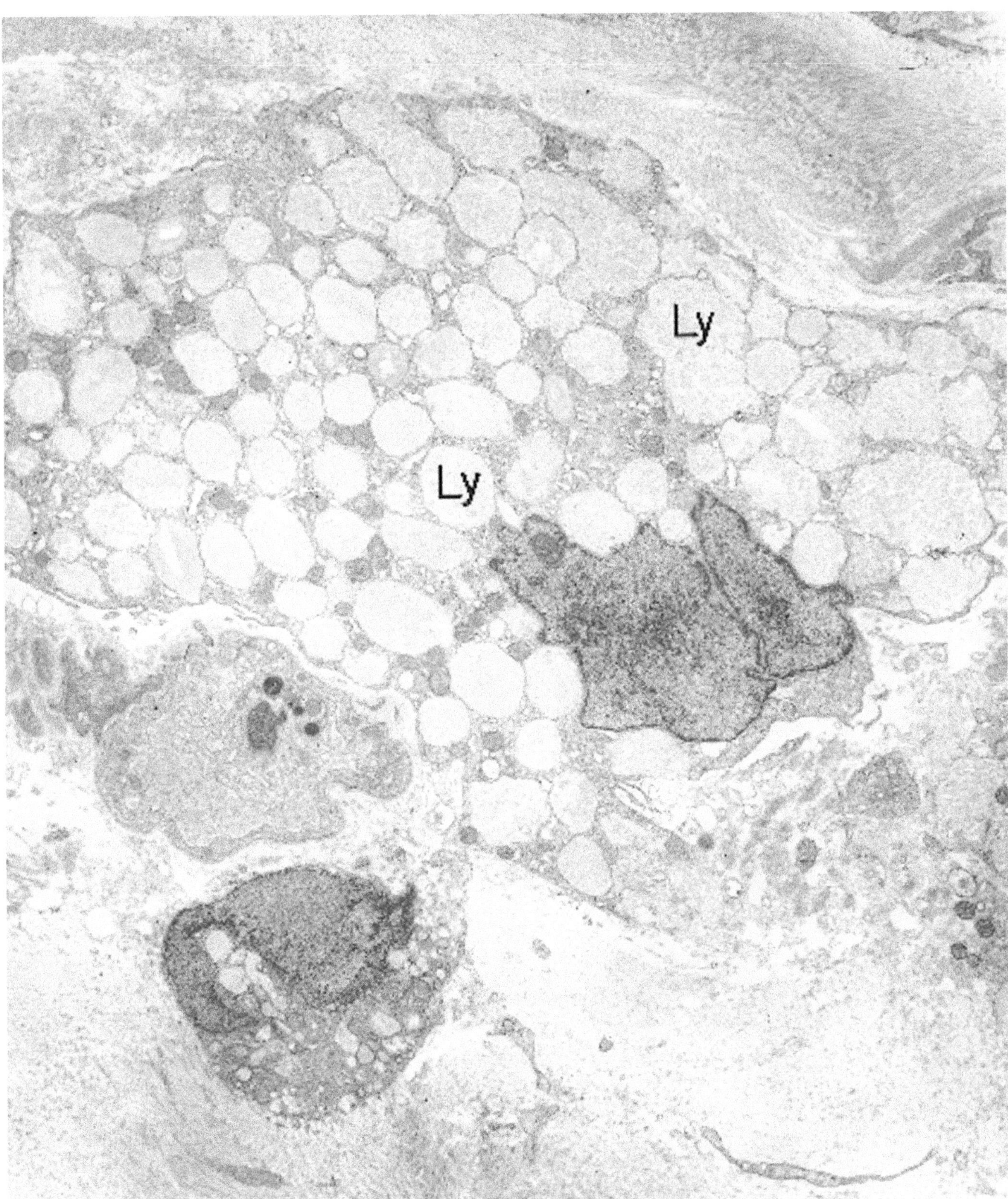

Figure 10.7. Whipple's disease (heart). Endomyocardial biopsies in this case showed scattered histiocytes filled with distended, membrane-bound, medium-dense bodies (Ly) that on higher magnification (see Figure 10.8) proved to be secondary lysosomes with serpiginous membranes and rare bacteria. (× 6750)

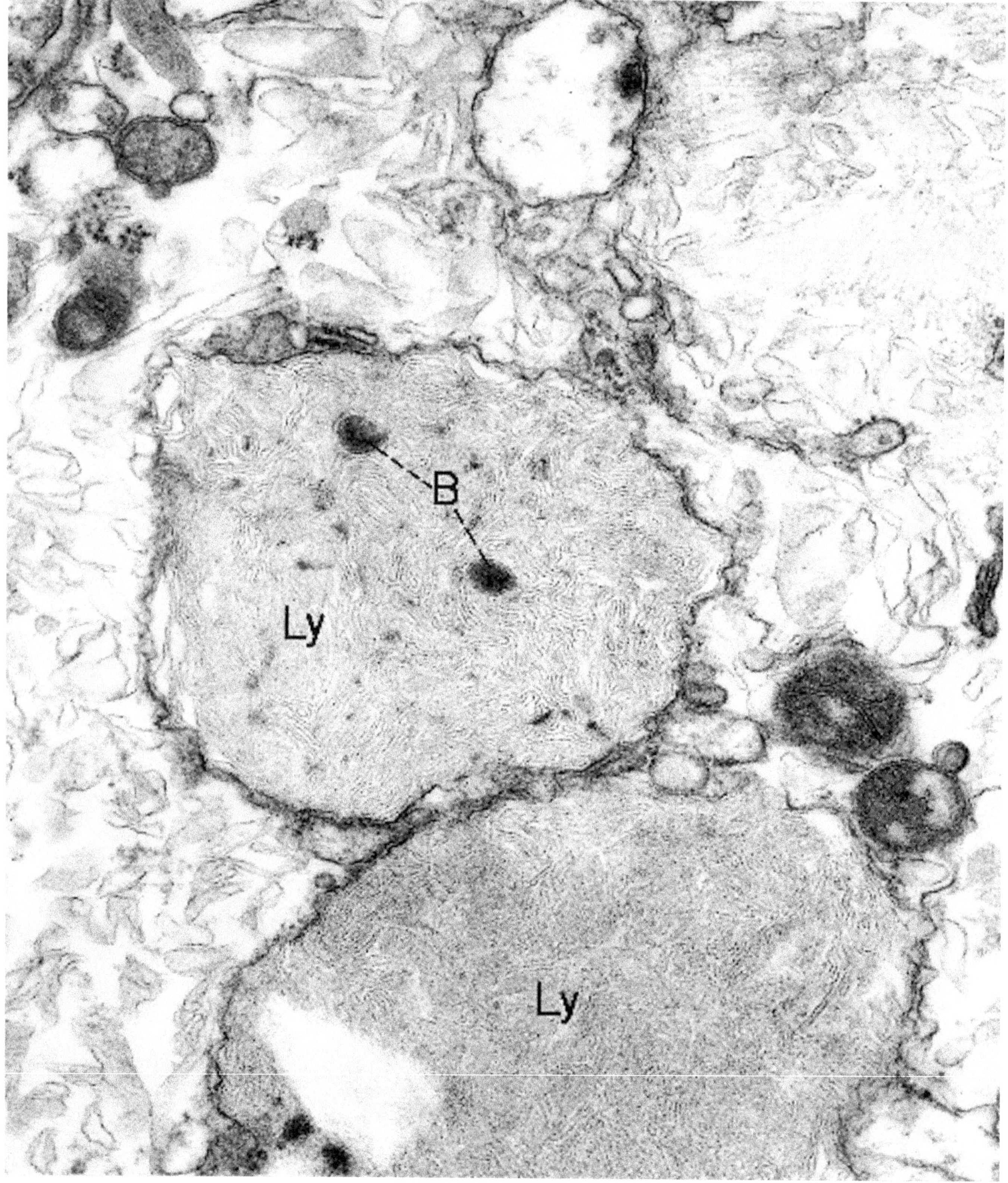

Figure 10.8. Whipple's disease (heart). High magnification of a myocardial macrophage reveals two secondary lysosomes (Ly) containing serpiginous membranes and two bacteria (B). (× 51,900)

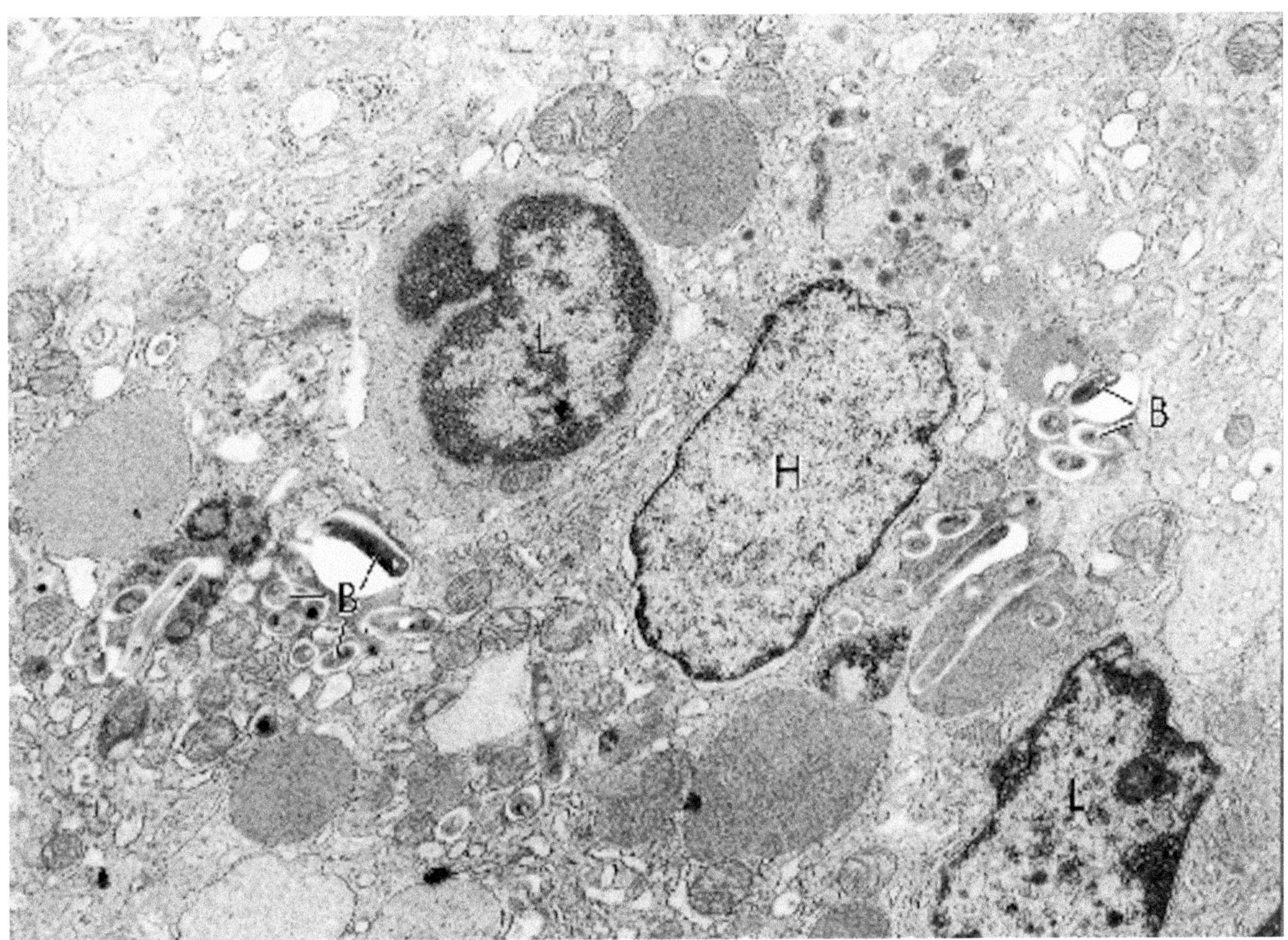

Figure 10.9. *Mycobacterium avium-intracellulare* (duodenum). The lamina propria of the mucosa contains lymphocytes (L) and histiocytes (H), the latter of which have copious cytoplasm, and within the cytoplasm are numerous bacteria (B) in secondary lysosomes. (× 9300)

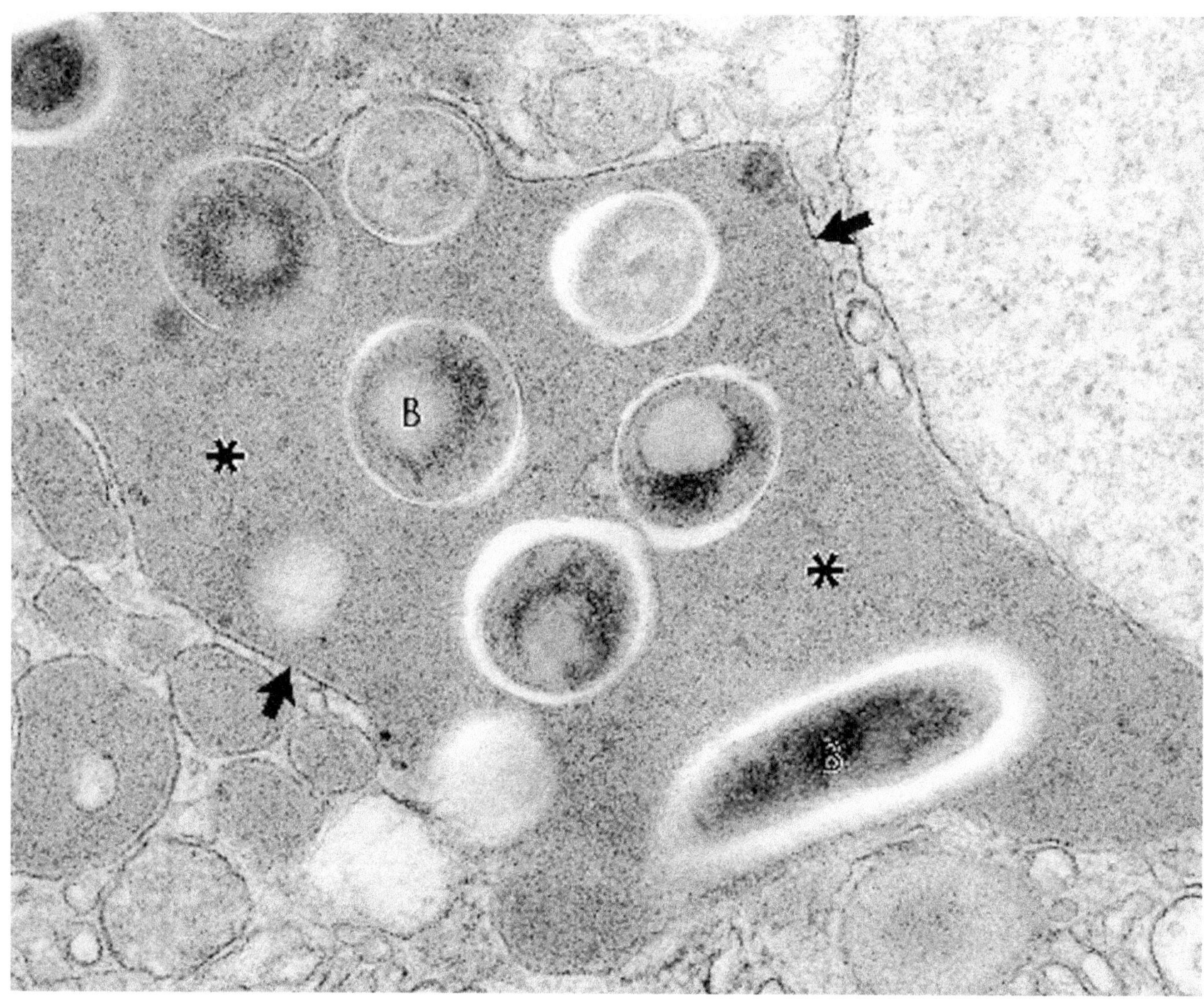

Figure 10.10. *Mycobacterium avium-intracellulare* (duodenum). High magnification of bacteria from the same specimen as depicted in Figure 10.9 shows bacilli (B) within histiocytic lysosomes (with medium-dense surrounding contents [*] and limiting membranes [*arrows*]). The bacilli have a characteristic adjacent clear halo. (× 57,000)

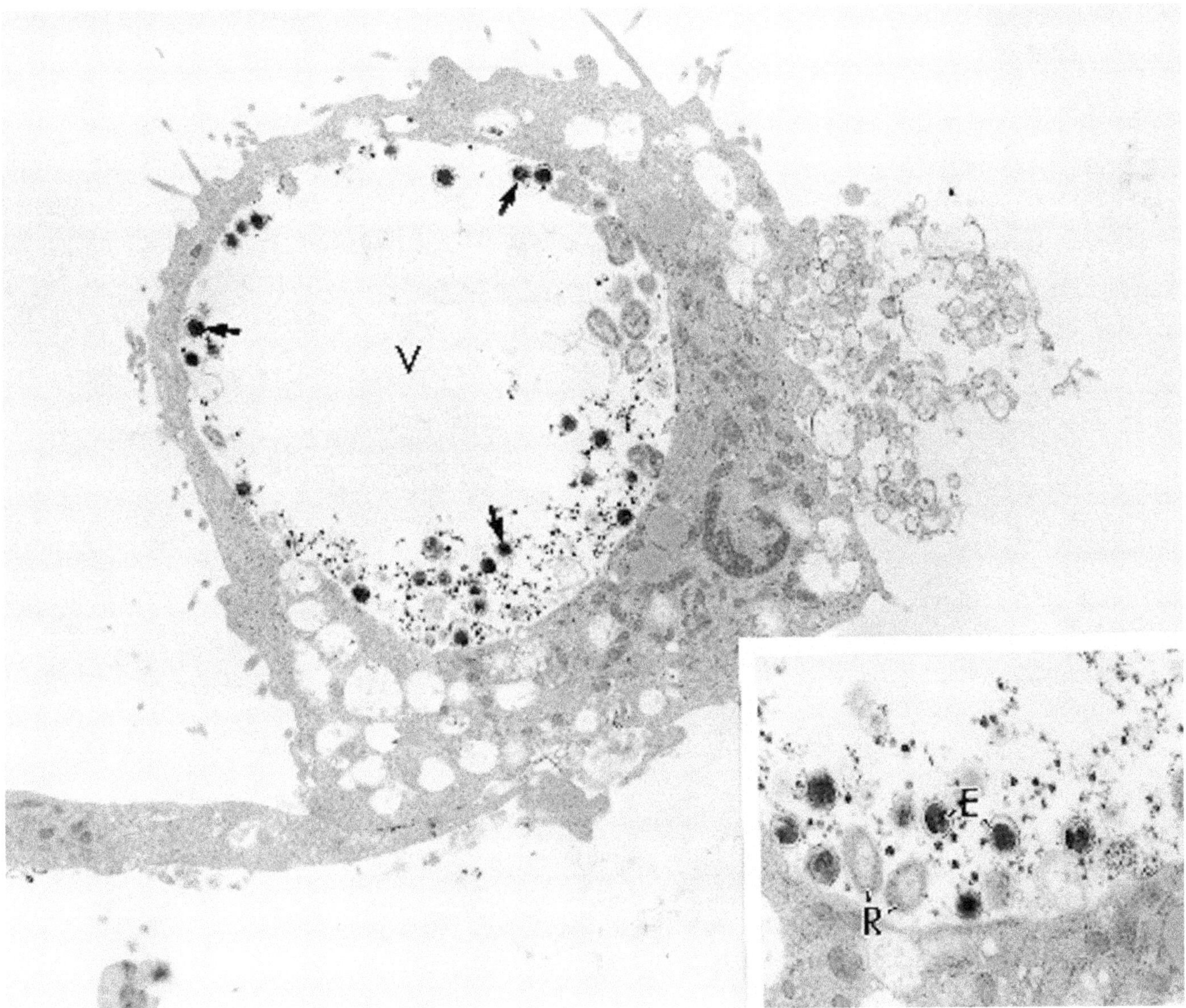

Figure 10.11. *Chlamydia trachomatis* (culture). An infected host cell contains a large cytoplasmic inclusion vacuole (V), which harbors numerous chlamydia (*arrows*). At higher magnification, the *inset* reveals two stages in the growth cycle of chlamydia: elementary bodies (E) and reticulate bodies (R). (× 9800; *inset* × 21,000) (Epon blocks generously contributed by William Taylor, M.D., and Lynne A. Farr, B.A., Rhode Island Hospital, Providence, Rhode Island.)

(Text continued from page 648)

intracellular and extracellular bacilli and macrophages with membrane-filled phagosomes are found.

The morphology of bacteria may vary among species and within a given genus, for example, *Mycobacterium tuberculosis, M. bovis, M. leprae, M. avium-intracellulare,* and others. The organisms may be straight, curved, or branched rods, or they may be coccobacilli. They range from 1–6 μm long and are less than 1 μm in diameter. *M. avium-intracellulare* is coccobacillary, or almost coccoid (Figures 10.9 and 10.10), and it is characteristically surrounded by a clear halo. In addition, artifactual open spaces adjacent to the organisms, secondary to incomplete penetration by epoxy embedding medium, are frequently present.

Chlamydia trachomatis is a bacterium that, before 1960, was thought to be a virus. It has an outer membrane and resembles gram-negative bacteria. Cytoplasmic inclusion bodies in infected host cells are identifiable by

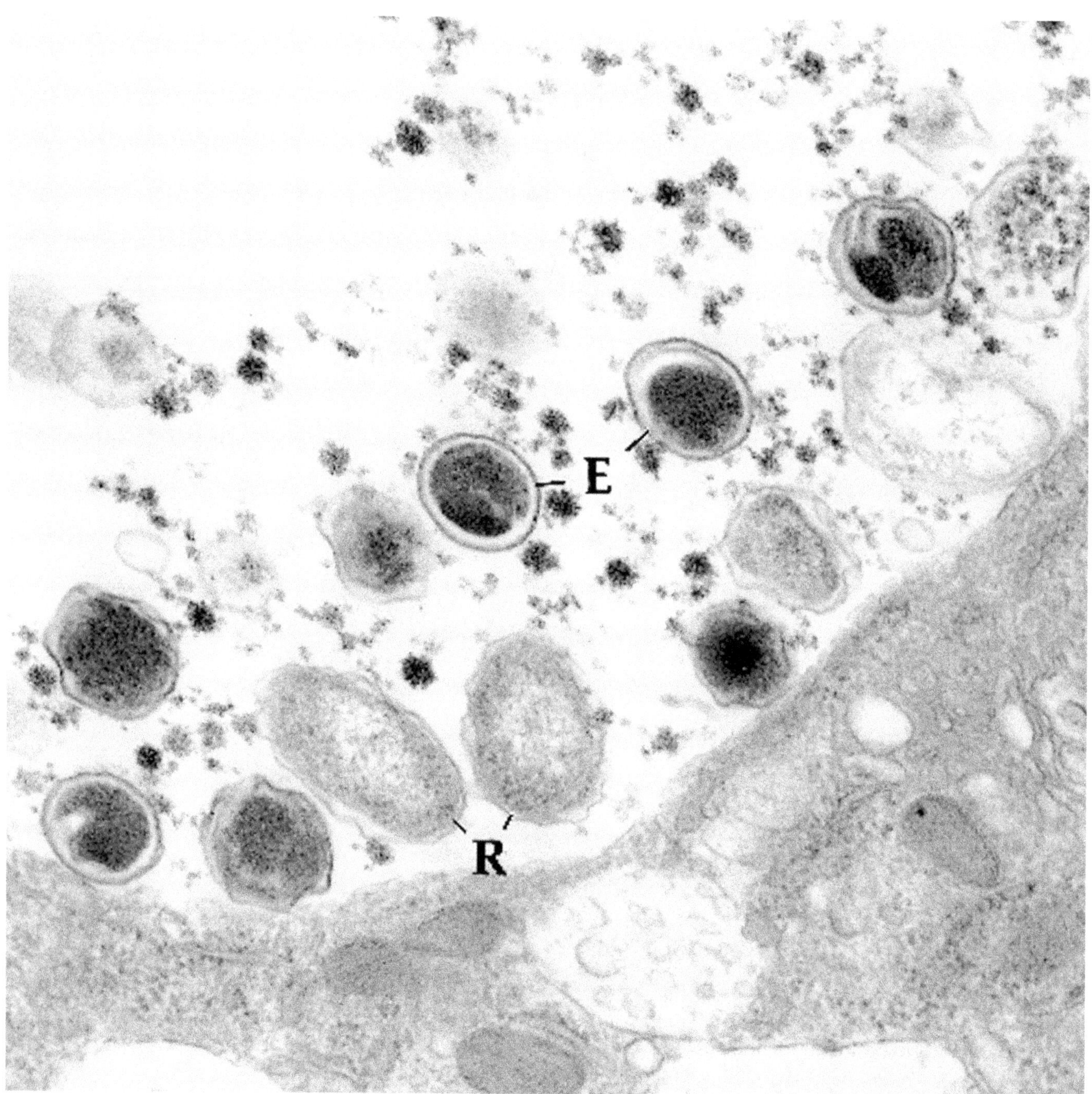

Figure 10.12. *Chlamydia trachomatis* (culture). Higher magnification of an inclusion vacuole illustrates in more detail the internal structure of elementary bodies (E) with dense cores, and more diffusely granular reticulate bodies (R). (× 71,000) (Epon blocks generously contributed by William Taylor, M.D., and Lynne A. Farr, B.A., Rhode Island Hospital, Providence, Rhode Island.)

electron microscopy as membrane-bound vacuoles (secondary lysosomes) that contain elementary bodies, reticulate bodies, and intermediate forms (Figures 10.11 and 10.12). Elementary bodies represent the infectious phase of the life cycle of *Chlamydia,* and they enter the host cell by endocytosis and phagocytosis. They are 200–350 nm in diameter. Reticulate bodies are the metabolic and replicating phase of the cycle. They replicate by binary fission and then condense into elementary bodies. They are 800–1000 nm in diameter. Inclusion bodies undergo lysis and/or rupture at the end of the cycle and are then exocytosed from the host cell.

Viruses

(Figures 10.13 through 10.30.)

Diagnostic criteria (of viruses in general). (1) Intracellular and/or extracellular elliptical, strand-like, round, or polygonal structures (viruses, viral bodies, virions, nucleocapsids) measuring 20–300 nm in diameter; (2) basic morphology of the virion consists of a central, electron-dense core (DNA- or RNA-containing nucleoid) and an outer shell (capsid), which may have more than one layer (Figures 10.14 and 10.15); (3) viruses distributed randomly or in an organized, crystalline, or lattice-like arrangement (Figures 10.16 through 10.20).

Additional points. Some viruses have a capsule or complex coat surrounding them. Examples of these are the *herpes* (Figures 10.13 through 10.15) and *poxvirus* groups, whereas *papovavirus* (e.g., *JC virus* that causes progressive multifocal leukoencephalopathy [Figs. 10.18 and 10.19] and *papilloma virus*) and *adenovirus* (Figure 10.20) exist as naked, unenveloped nucleocapsids. Papovaviruses tend to have a random, rather than straight-line arrangement of individual virions, whereas the converse is true for adenovirus, which is usually in paracrystalline arrays. Poxviruses are large (220–450 nm long and 140–260 nm in diameter), herpes and adenoviruses are intermediate in size (100–150 nm and 70–90 nm in diameter, respectively), and papovaviruses are small (45–53 nm). The herpes viruses (simplex, varicella-zoster, cytomegalovirus, and Epstein-Barr) are 120–200 nm in diameter. They are examples of viruses that replicate in the nucleus of the host cell as naked nucleocapsids and then pass through the nuclear envelope, from which they gain their own outer envelope before passing into the cytoplasm of the cell. The host cells that are infected by viruses undergo nuclear, cytoplasmic, and plasmalemmal degeneration. The nuclear changes consist of central clearing of the nucleoplasm and peripheral aggregation of chromatin. In the cytoplasm there is swelling of organelles and destruction of their membranes as well as generalized hydropic swelling of the cell. The plasmalemma disintegrates focally but progressively, and the cell is lysed.

Morphology is useful in identifying viruses, but overlapping size and shape among them make other, more specific diagnostic methods necessary. *In situ* hybridization, Southern blot analysis, and polymerase chain reaction for DNA characterization are examples of these.

Hepatitis B virus (Figures 10.16 and 10.17) is 42 nm in diameter and consists of a central core (composed of circular and partially double-stranded DNA and nucleocapsid core antigen), plus an outer shell of surface antigen. *Influenza A and B viruses* (Figure 21) are about 80 to 120 nm in diameter, and *paramyxoviruses* (Figures 10.22 and 10.23) measure about 150–250 nm; they are round and elongated. Paramyxoviruses include *parainfluenza, measles, mumps,* and *respiratory syncytial viruses. Enteroviruses* (Figures 10.24 and 10.25) are in the family of Picornaviridae and include *polio, coxsackie,* and *echo* viruses. They measure 24–30 nm in diameter and are unencapsulated and spherical or polyhedral. The virions are assembled in the host cell cytoplasm, and clusters of virions may be found within a membrane-bound vesicle. *Eastern equine encephalitis* virus (Figure 10.26) is an alpha virus in the family Togaviridae. It is an enveloped virus measuring 40–70 nm in diameter. It multiplies in the cytoplasm and buds from cellular membranes.

Human immunodeficiency virus (HIV) (Figures 10.27 and 10.28) is a member of the lentivirus family of retroviruses. It is spherical and 100–130 nm in diameter, and it has a characteristic conical dense core (nucleoid) in its mature form. The virus develops on the cell membrane and cytoplasmic vacuolar membranes of monocytes/macrophages and on vacuolar membranes of lymphocytes. The outer unit membrane of the virion is derived from host cell membrane. Surface spikes develop during budding and insert into the viral coat. The virus leaves the cell or vacuolar membrane as an immature ring form and then develops its core.

Rabies virus, a member of the RNA Rhabdoviridae family and *Lyssavirus* genus, is bullet-shaped, 130–220 nm long, and 60–80 nm in diameter (Figures 10.29 and 10.30). The virions tend to aggregate around or in a granular, electron-dense material in the cytoplasm of neurones, the total configuration constituting a Negri body. Individual virions can be seen budding off membrane-bound organelles.

(Text continues on page 680)

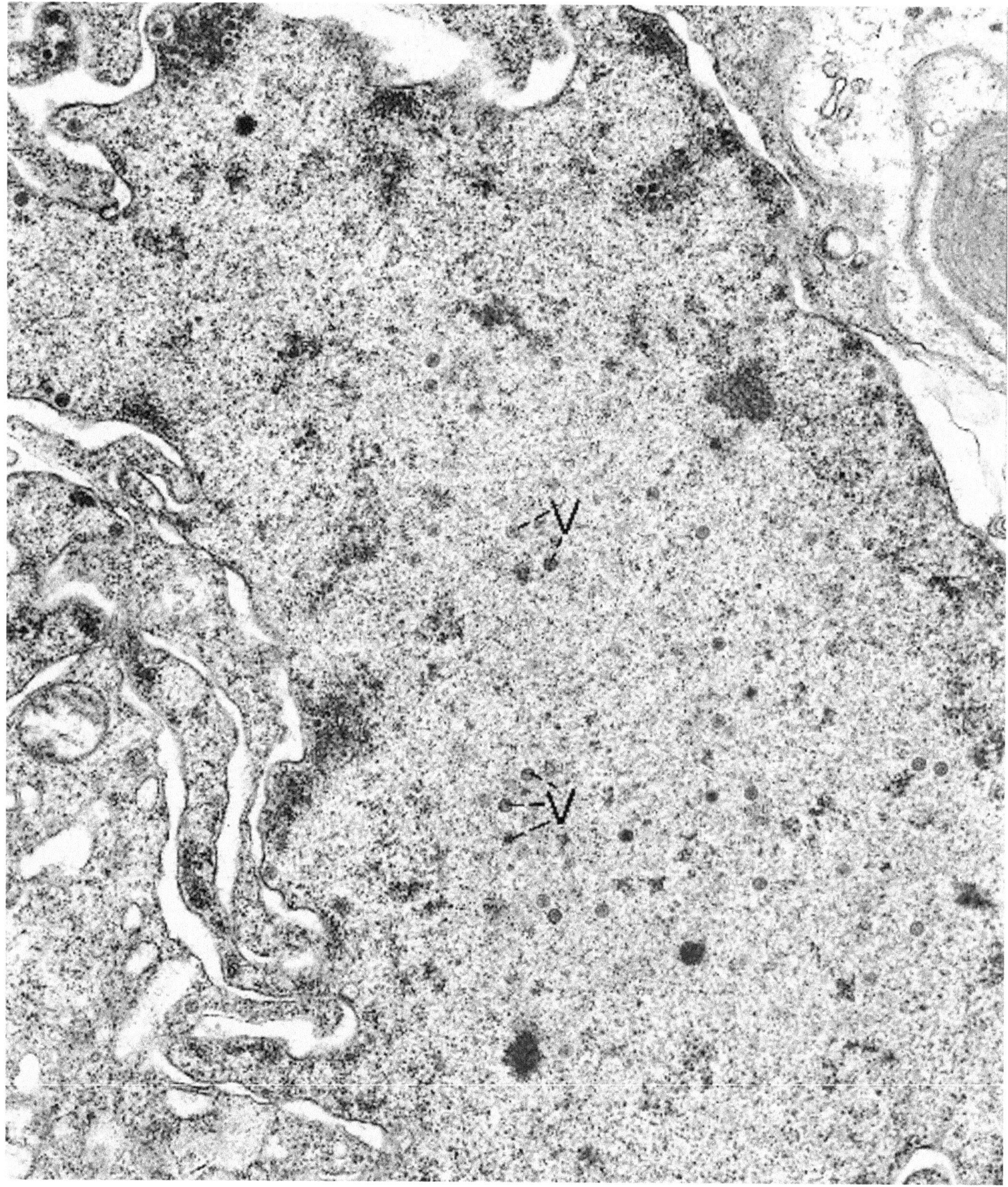

Figure 10.13. Herpes simplex encephalitis (cerebrum). A nucleus of a neuron contains numerous randomly dispersed virions (V). (× 28,800)

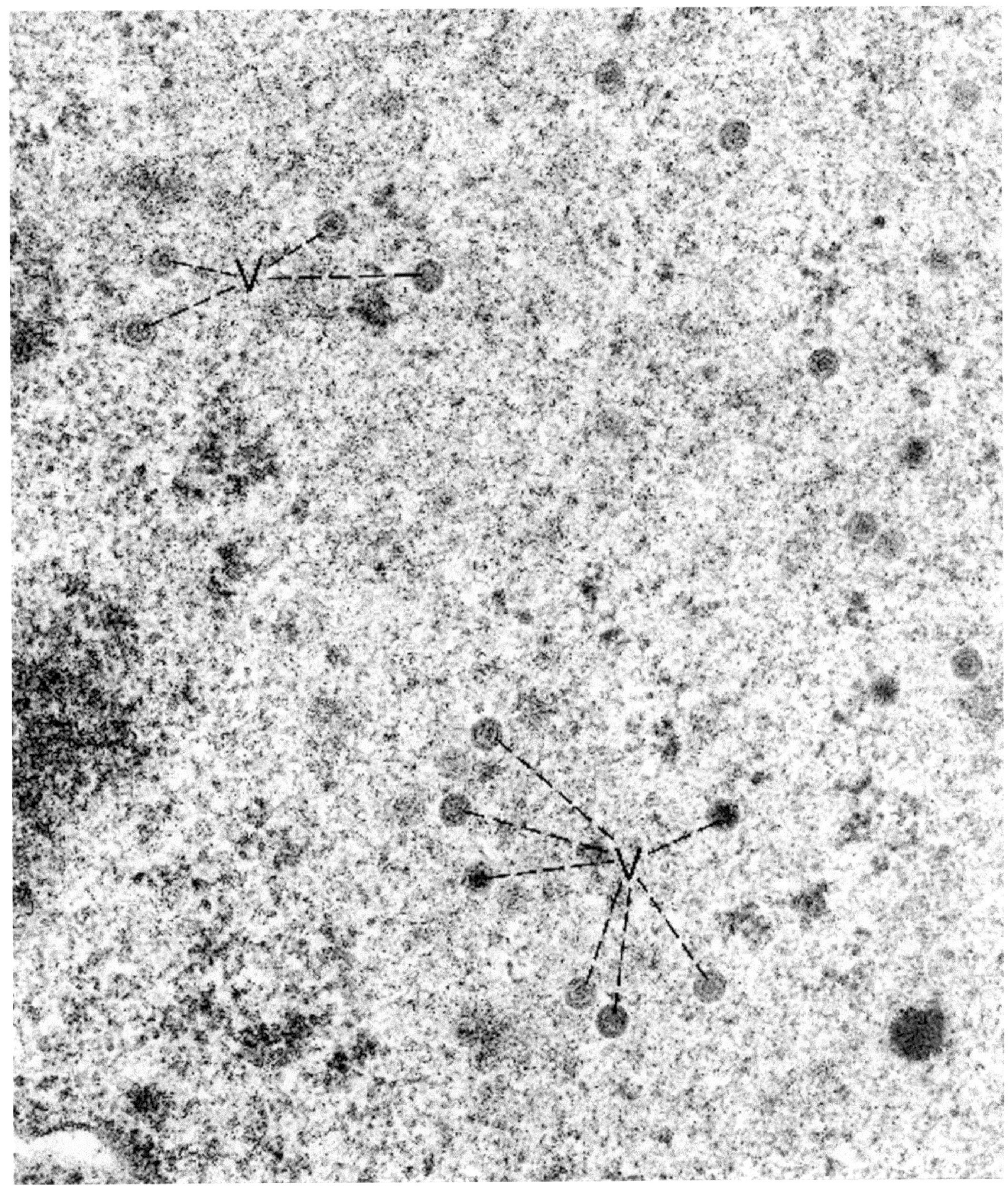

Figure 10.14. Herpes simplex encephalitis (cerebrum). Higher magnification of the virions (V) in Figure 10.13 shows their basic structure of a central dense nucleoid and an outer three-layered capsid. (× 63,800)

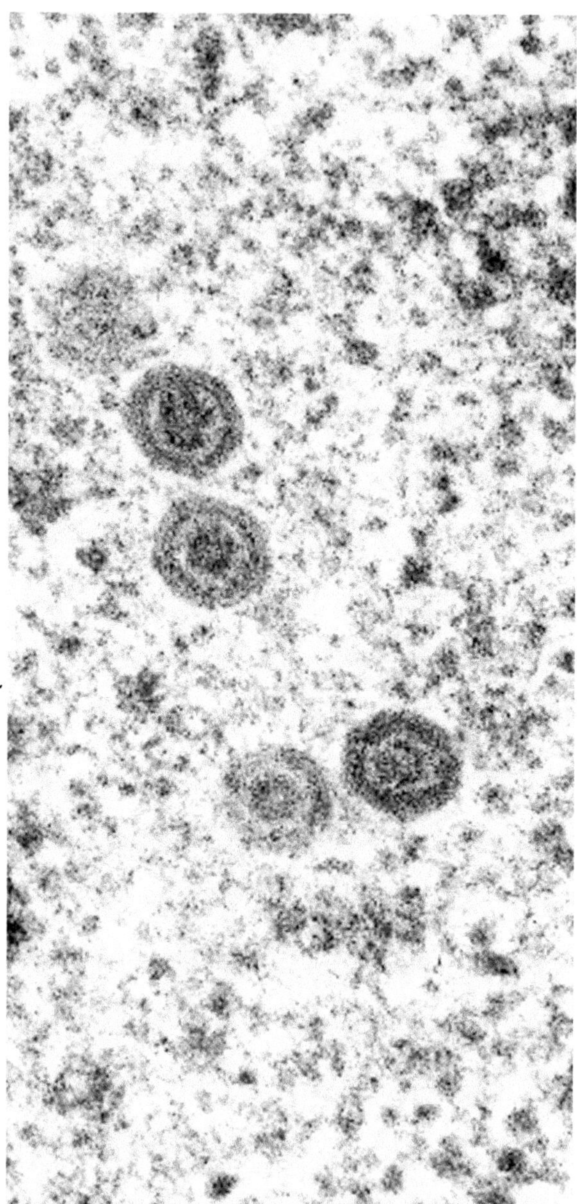

Figure 10.15. Herpes simplex encephalitis (cerebrum). High magnification of several virions accentuates their basic nucleocapsid structure and hexagonal shape. (× 155,000)

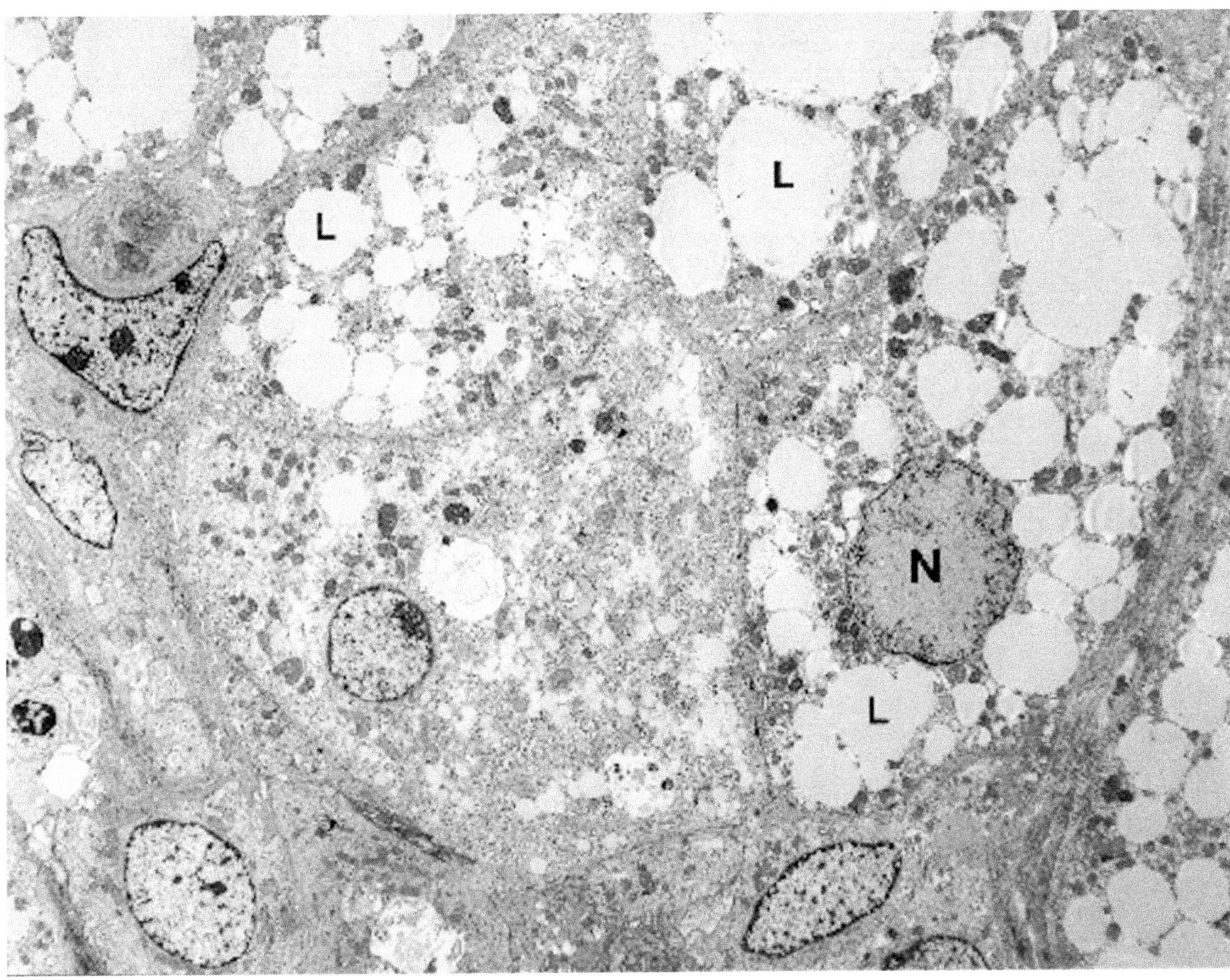

Figure 10.16. Hepatitis B virus (liver). Hepatocytes show numerous cytoplasmic droplets of lipid (L), and one nucleus (N) has a ground-glass appearance, seen at higher magnification in Figure 10.17. (× 3800)

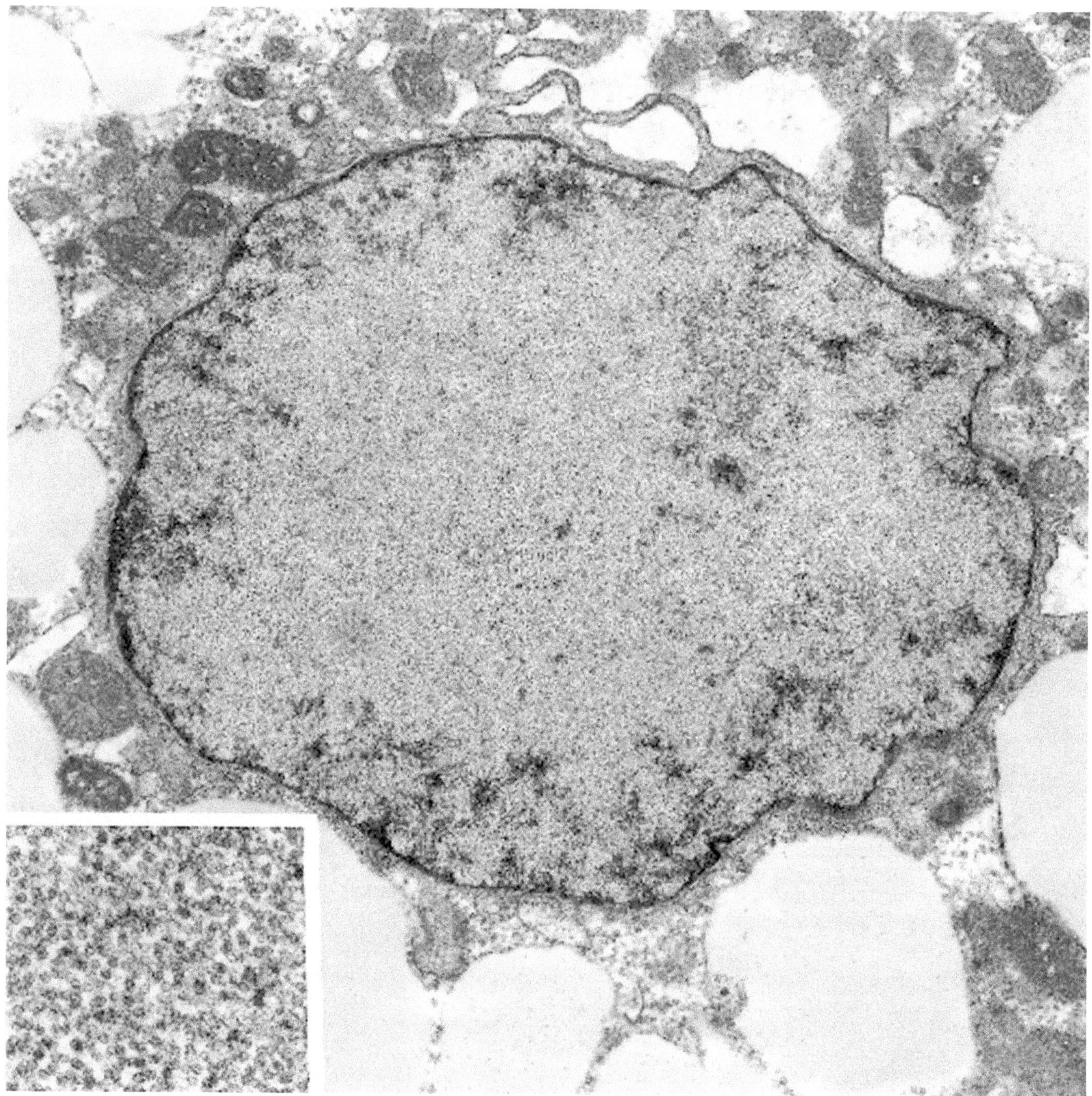

Figure 10.17. Hepatitis B virus (liver). High magnification of a hepatocyte nucleus reveals a diffusely granular texture, which in the *inset* is discernible as innumerable virions. (× 19,000; *inset:* × 88,000)

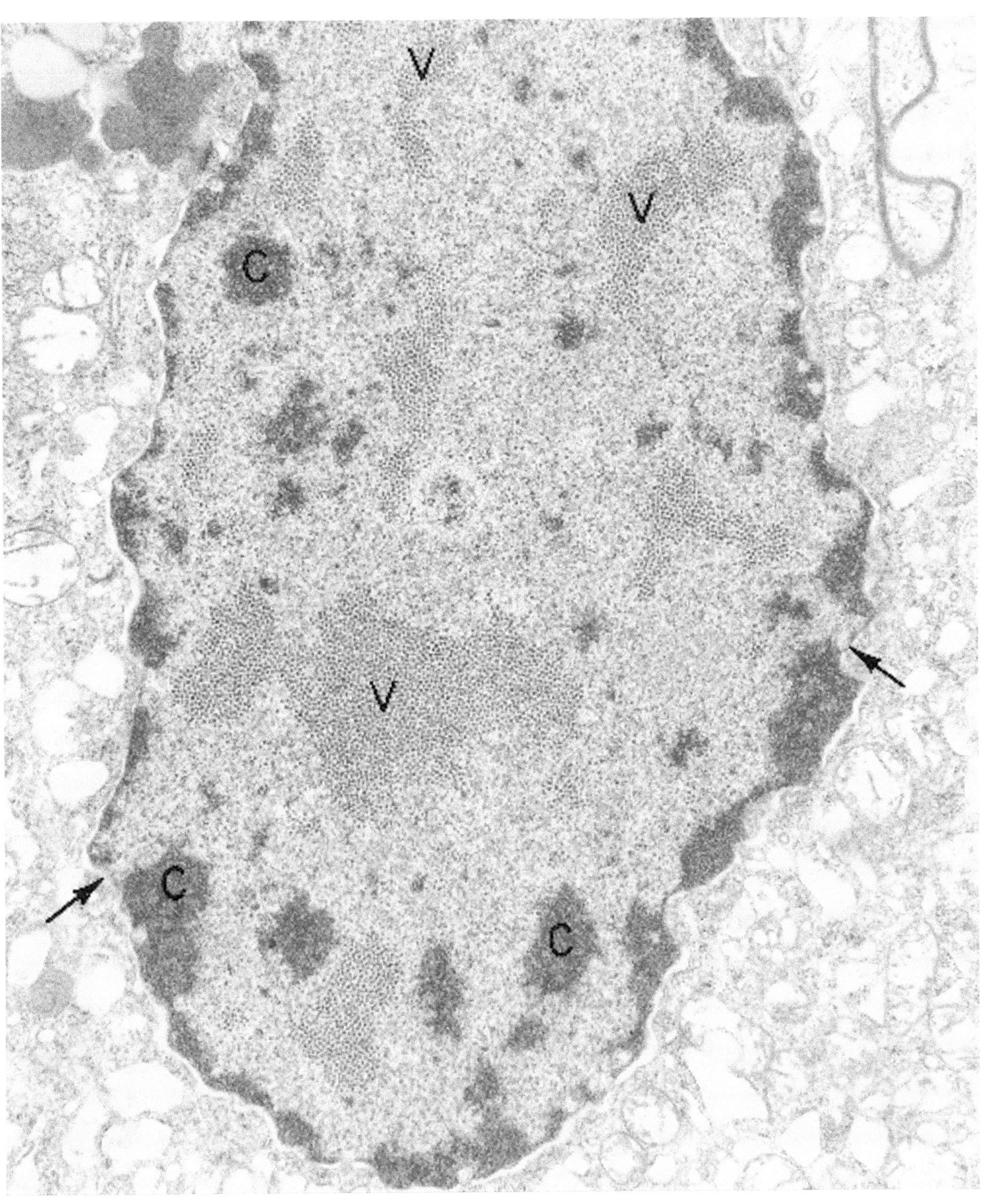

Figure 10.18. Papovavirus, in progressive multifocal leukoencephalopathy (cerebrum). A degenerating nucleus of an oligodendrocyte shows focal loss of the nuclear envelope (*arrows*), aggregation of chromatin (C), and packets of virions (V). (× 21,870)

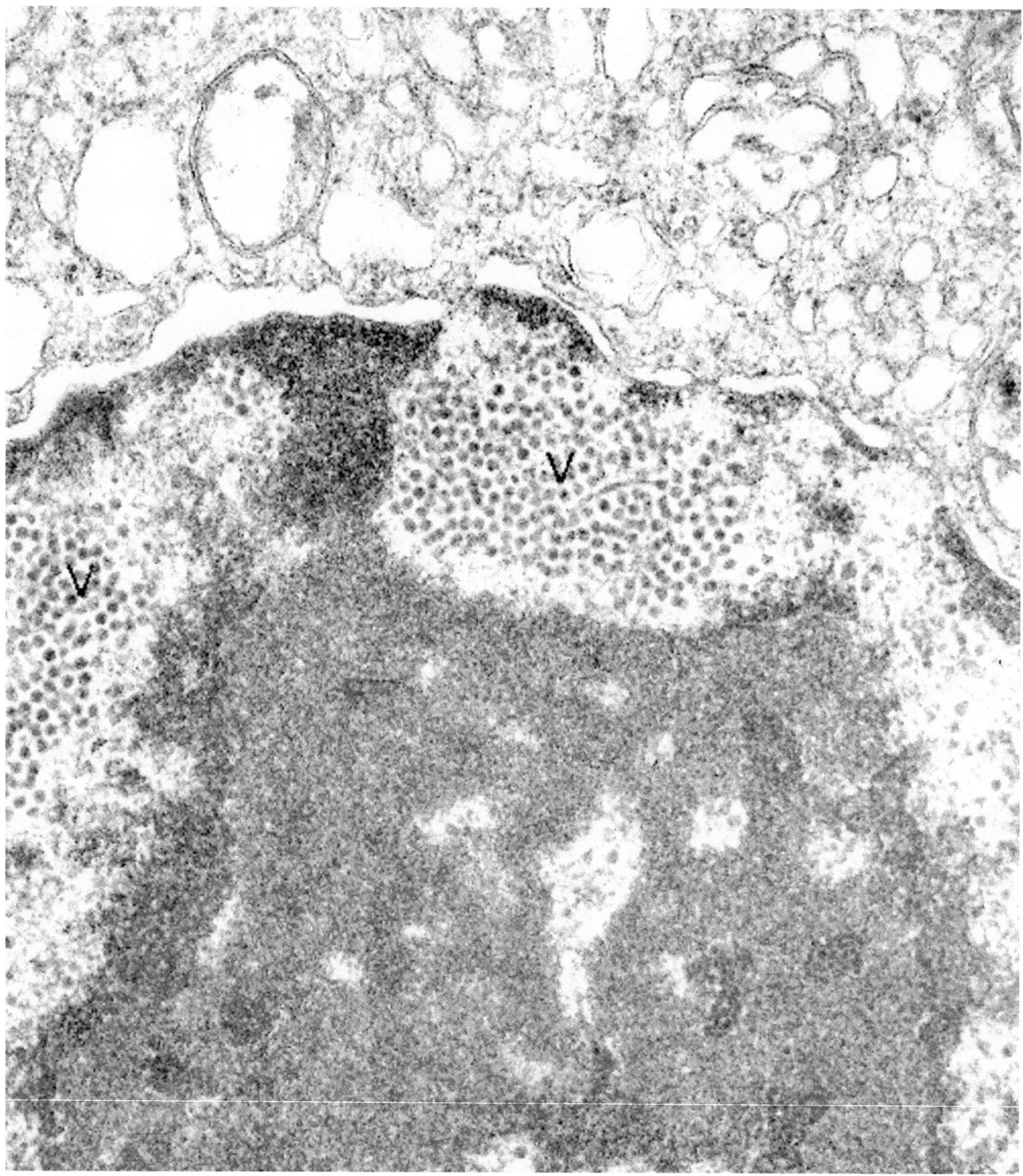

Figure 10.19. Papovavirus, in progressive multifocal leukoencephalopathy (cerebrum). High magnification of groups of virions (V), from the same cerebral specimen as in Figure 10.18. (× 66,120)

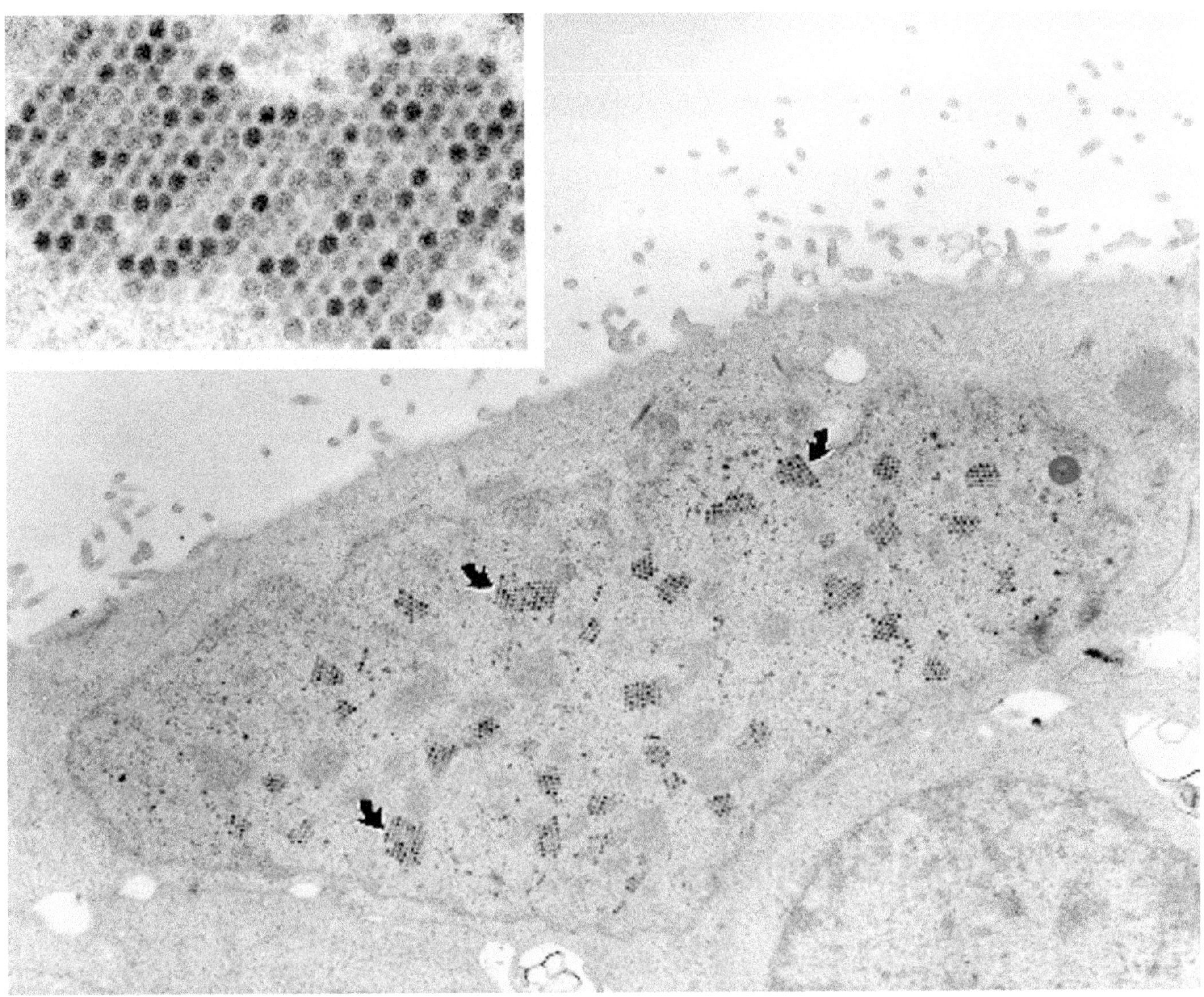

Figure 10.20. Adenovirus (tissue culture). The nucleus of this cultured cell contains numerous packets of virions in a regular, lattice-like arrangement (*arrows*). (× 14,000). The *inset* shows the virions at higher magnification. (× 57,000) (Epon blocks generously contributed by William Taylor, M.D., and Lynne A. Farr, B.A., Rhode Island Hospital, Providence, Rhode Island.)

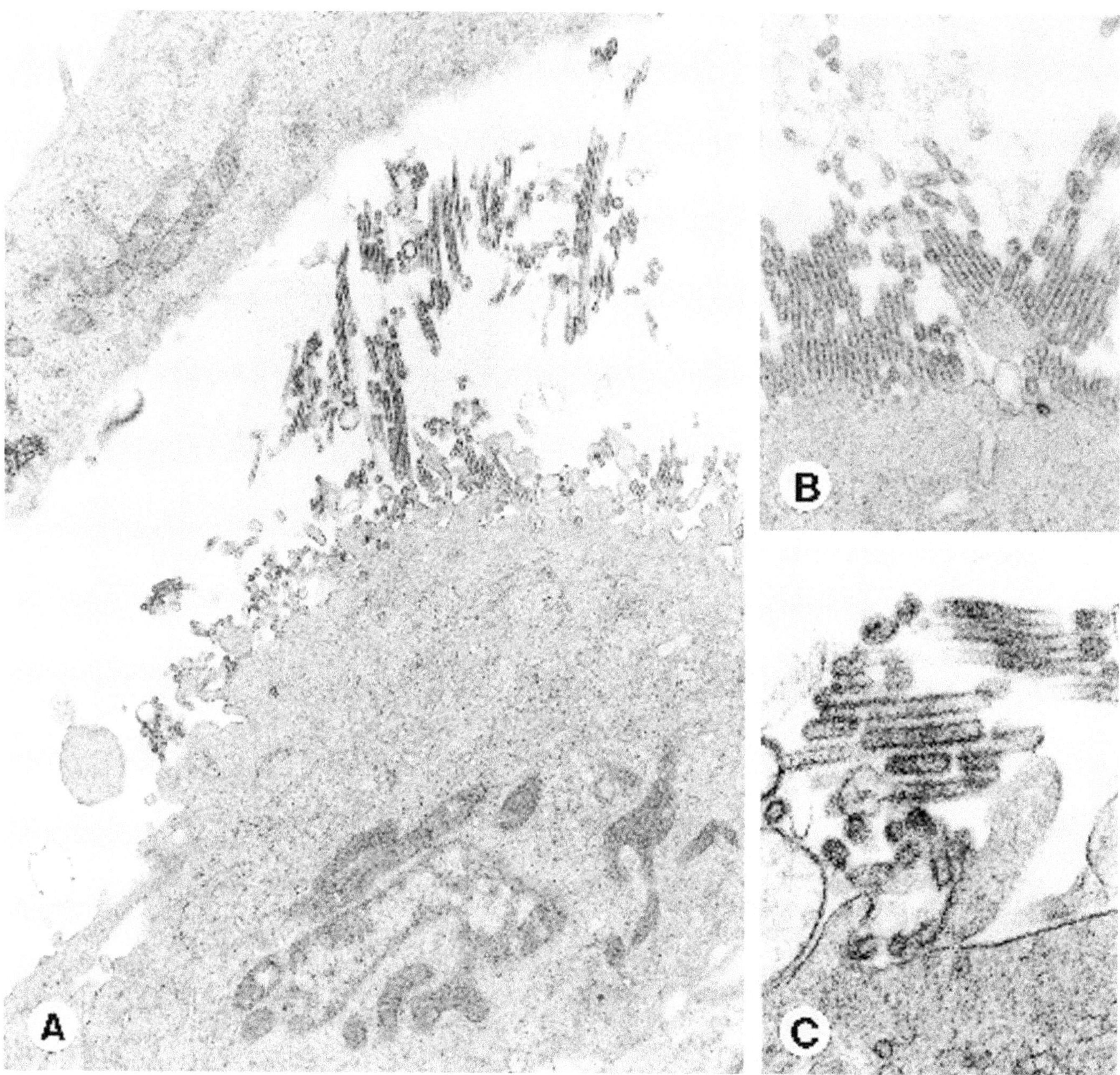

Figure 10.21. Influenza A virus (tissue culture). Three magnifications (**A, B,** and **C**) illustrate round and elongated virions, at and beneath the cell surface. (**A,** × 16,000. **B,** × 34,000. **C,** × 86,000) (Epon blocks generously contributed by William Taylor, M.D., and Lynne A. Farr, B.A., Rhode Island Hospital, Providence, Rhode Island.)

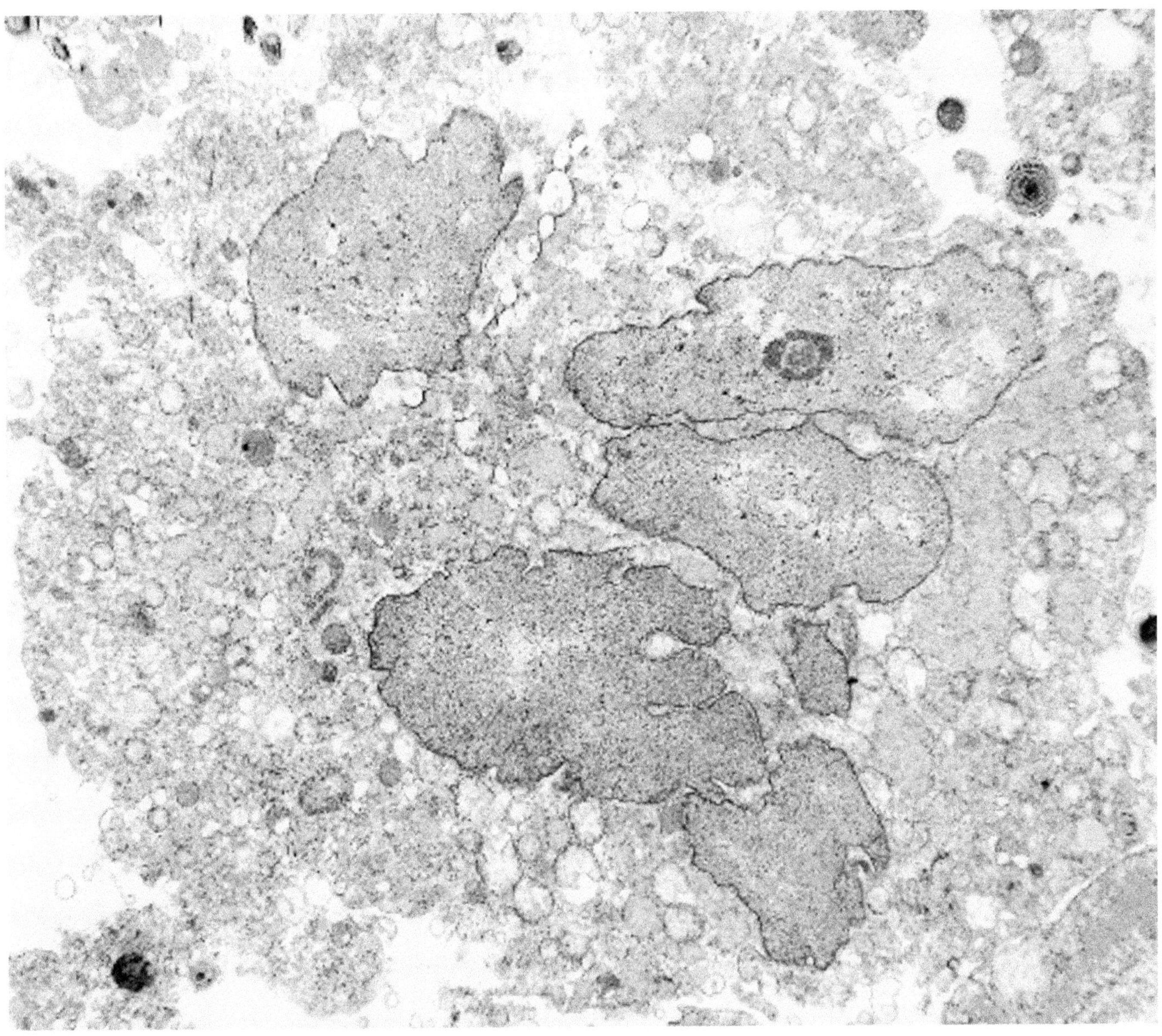

Figure 10.22. Paramyxovirus (parainfluenza) (lung). An infected type II pneumocyte is of giant size and has multiple nuclei with a ground-glass texture. (× 8900)

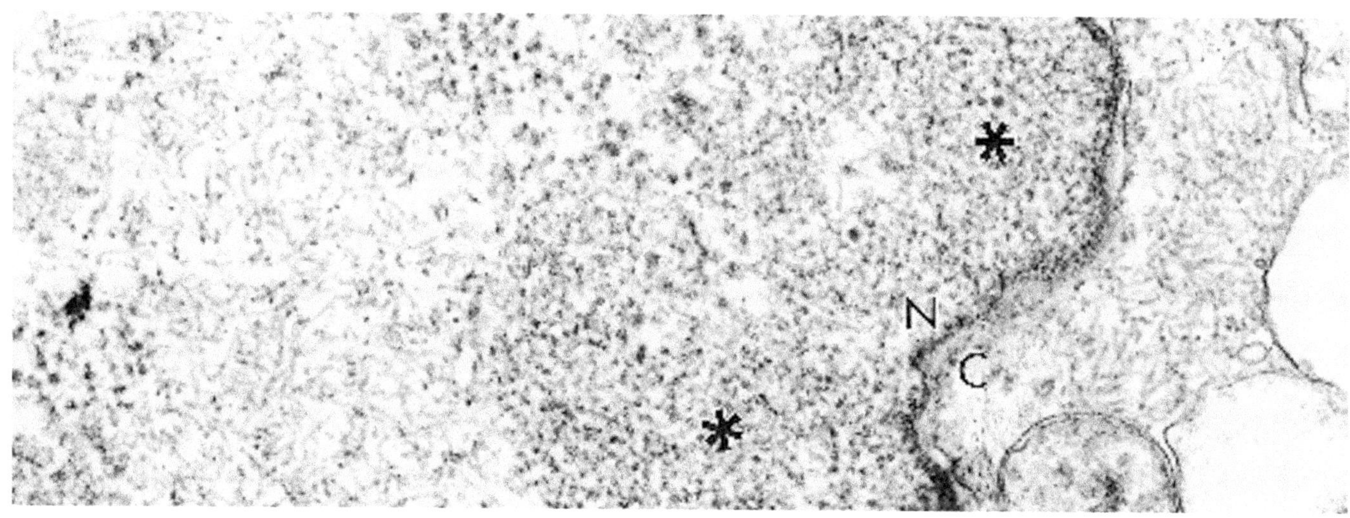

Figure 10.23. Paramyxovirus (parainfluenza) (lung). High magnification of one of the nuclei (N) and adjacent cytoplasm (C) of the giant cell shown in Figure 10.22 reveals diffusely dispersed virions. * = chromatin. (× 52,000)

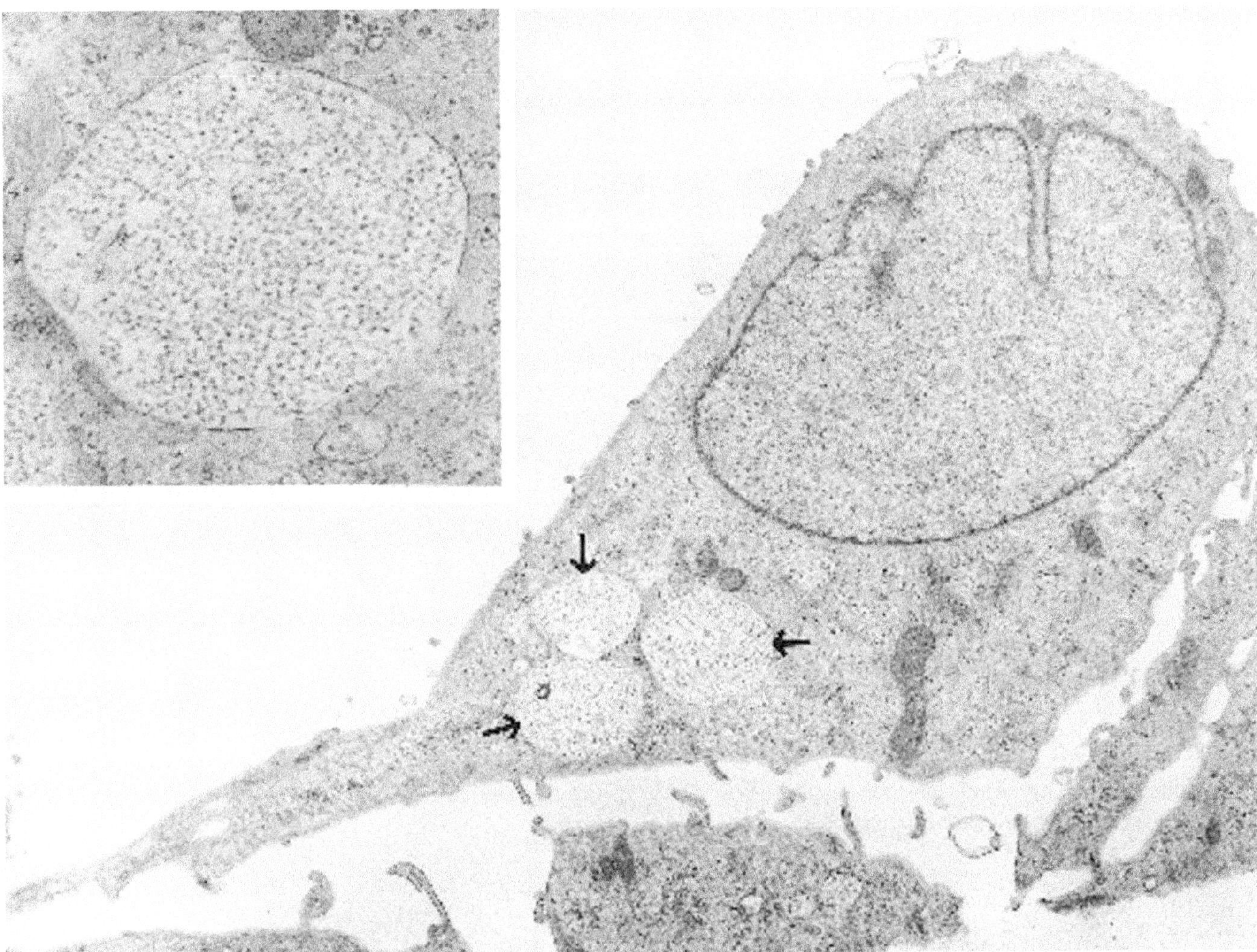

Figure 10.24. Enterovirus (coxsackie A) (tissue culture). A cultured cell contains three membrane-bound vesicles (*arrows*), which at higher magnification (*inset*) are seen to contain innumerable spherical virions. (× 10,800; *inset,* × 37,000) (Epon blocks generously contributed by William Taylor, M.D., and Lynne A. Farr, B.A., Rhode Island Hospital, Providence, Rhode Island.)

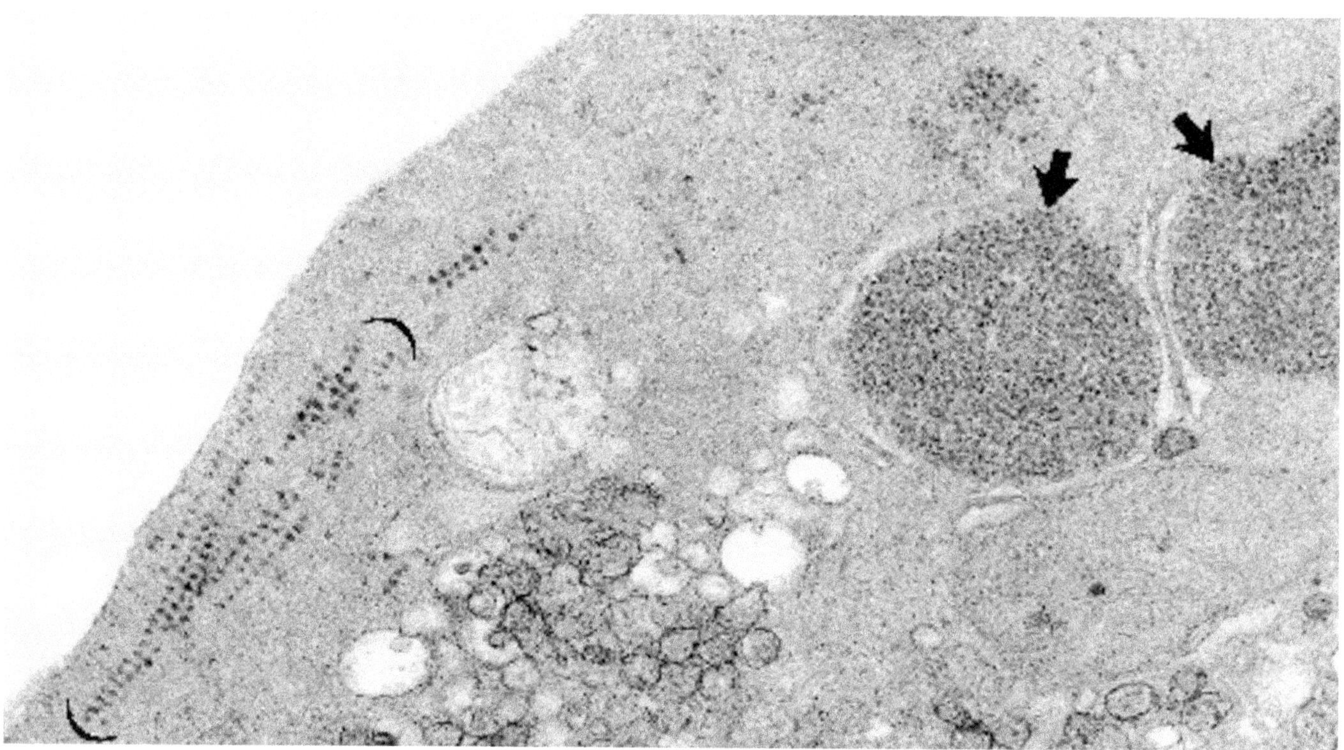

Figure 10.25. Enterovirus (echovirus) (tissue culture). Numerous virions occupy the cytoplasm of this cell, some being enclosed in membrane-bound vesicles (*arrows*) and others lying free in the cytosol (*parentheses*). (× 65,000) (Epon blocks generously contributed by William Taylor, M.D., and Lynne A. Farr, B.A., Rhode Island Hospital, Providence, Rhode Island.)

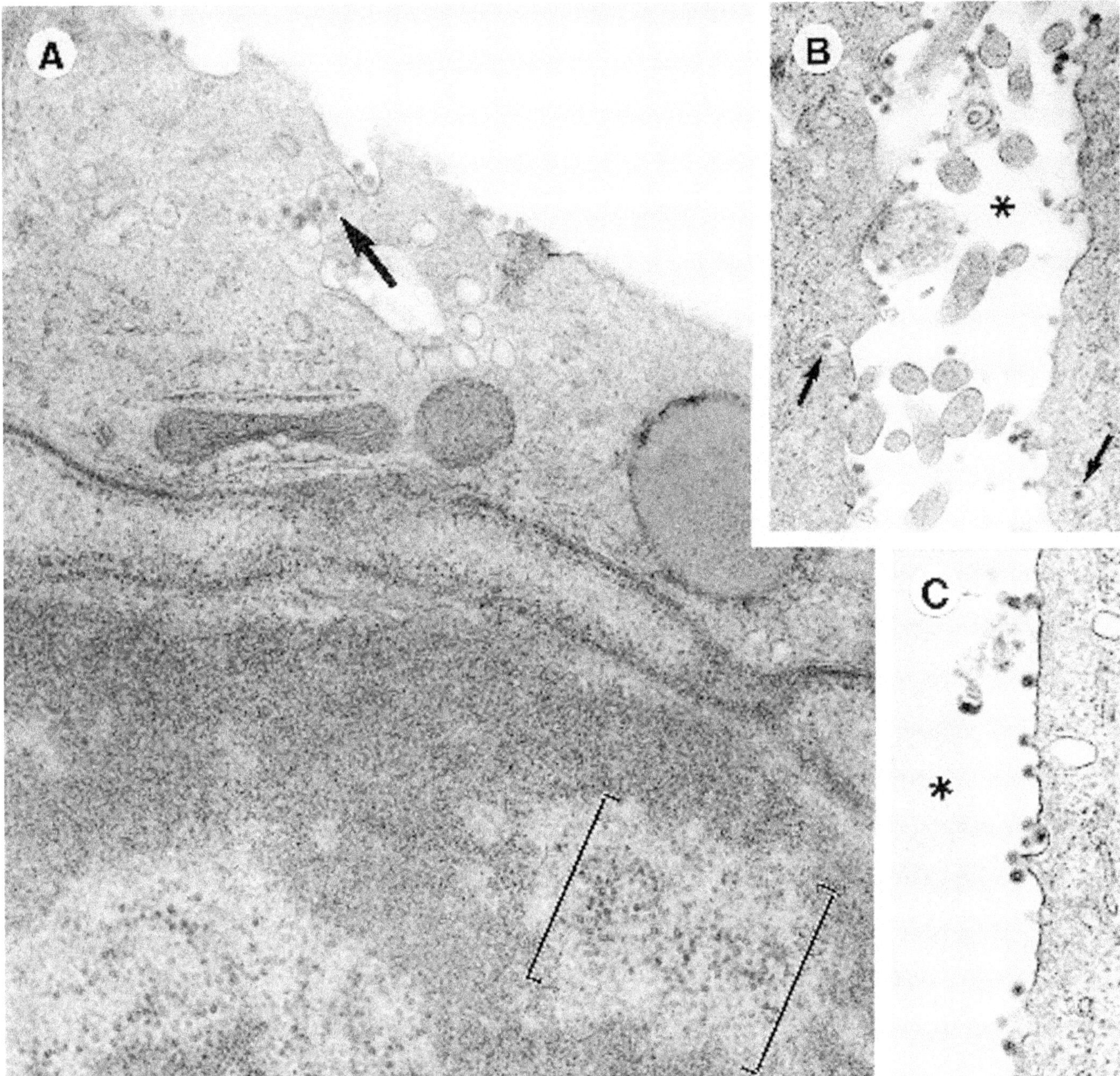

Figure 10.26. Eastern equine encephalitis virus (tissue culture). Most of these virions are at the cell surface, where they acquire an outer envelope from the plasmalemma (arrows, **A** and **B**). Outer envelopes are also acquired from the endoplasmic reticulum when virions are still in the cytoplasm. Precursor nucleocapsids, not seen in this photograph, lie free in the cytosol. Possible precursors may also be present in the nucleus (brackets, **A**) late in infection. Some of the virions have been expressed into the extracellular space (*, **B** and **C**). (**A,** × 53,000. **B,** × 46,000. **C,** × 61,000) (Epon blocks generously contributed by William Taylor, M.D., and Lynne A. Farr, B.A., Rhode Island Hospital, Providence, Rhode Island.)

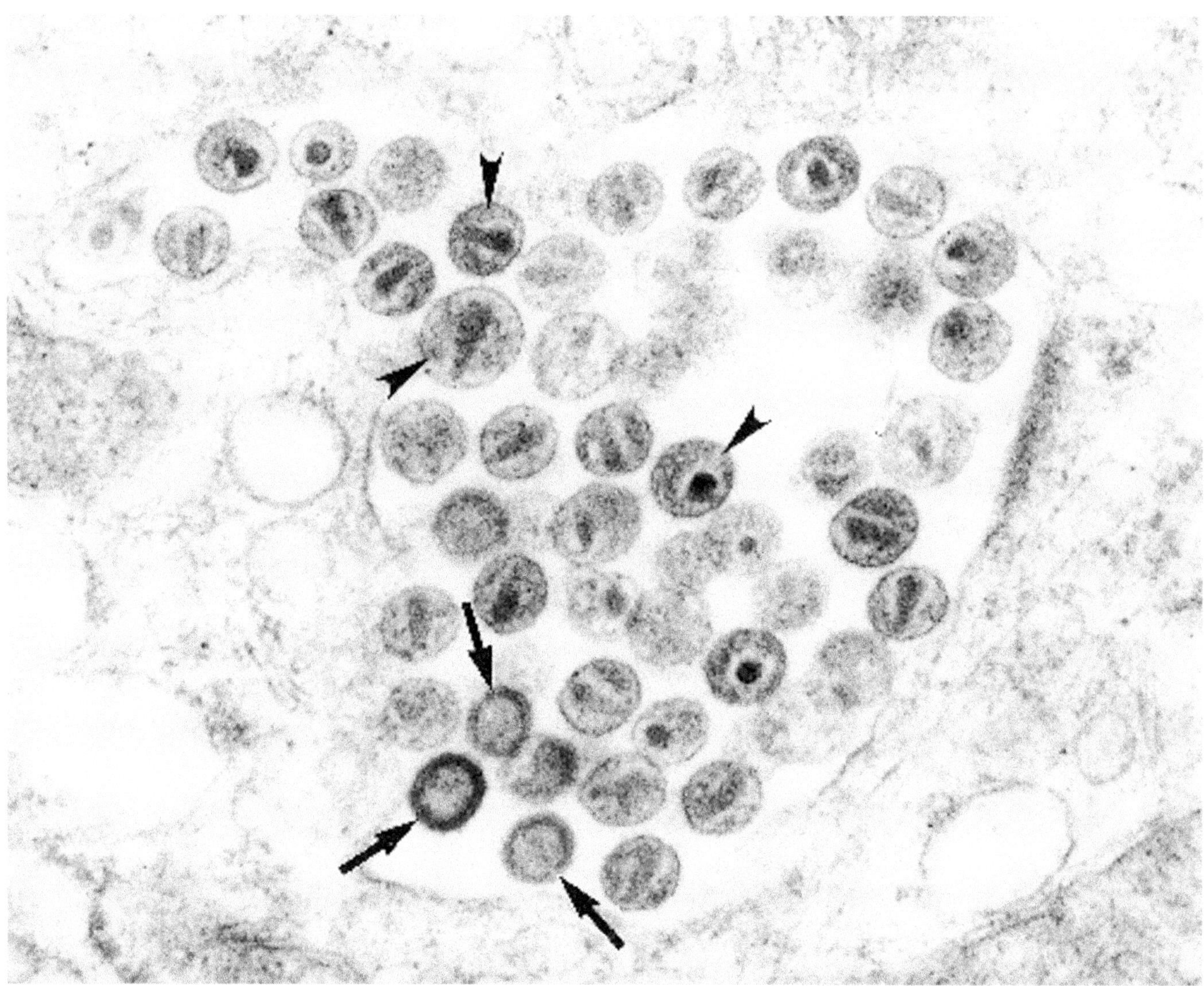

Figure 10.27. HIV-1 (cell culture). A vacuole in the cytoplasm of a macrophage contains several ring-shaped, immature HIV-1 particles (*arrows*) and numerous mature particles with conical nucleoids (*arrowheads*). Note that the appearance of the asymmetric conical nucleoids varies with their orientation and plane of sectioning. (× 104,000) (Photograph generously contributed by Jan M. Orenstein, M.D., Department of Pathology, George Washington University Medical Center, Washington, DC.)

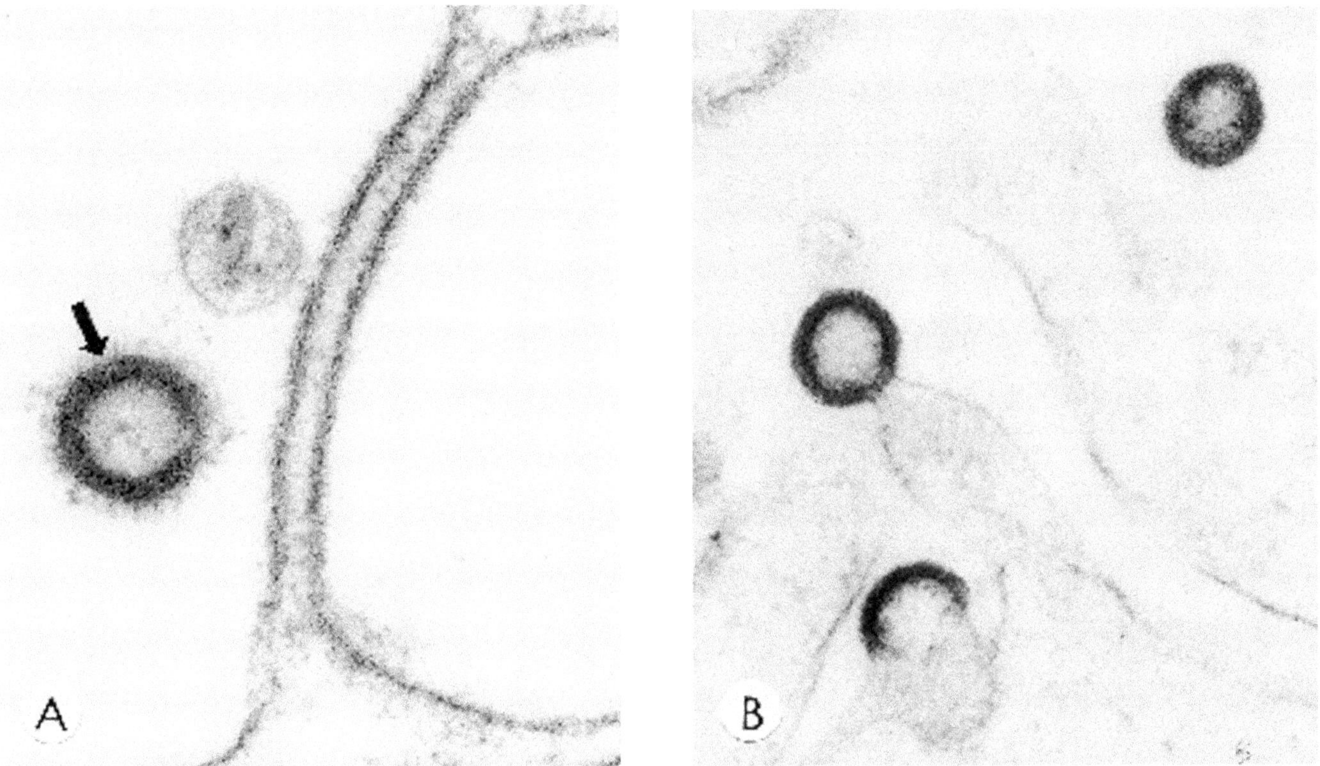

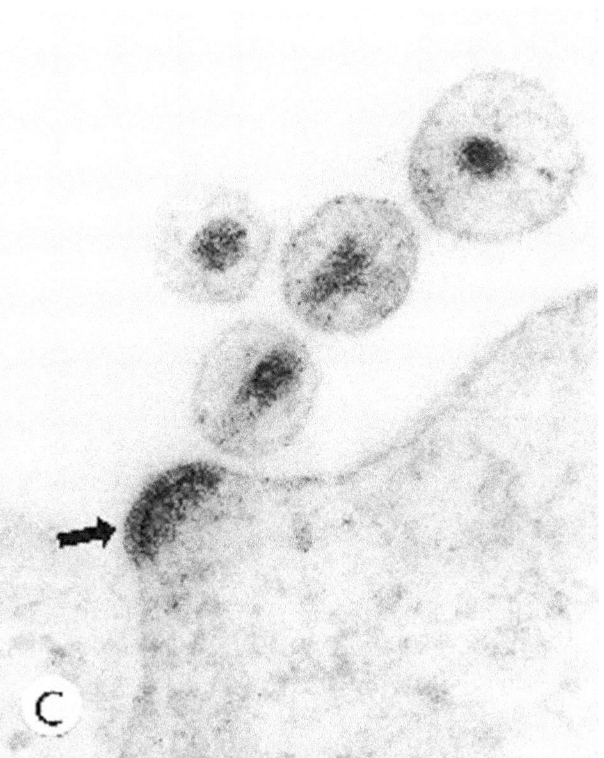

Figure 10.28. HIV-1 (cell culture). **A,** Spikes are especially prominent on the surface of this immature HIV-2 particle (*arrow*) budding into a macrophage cytoplasmic vacuole. (× 190,000) **B,** Three HIV-1 particles are at different stages of budding. The virion contains a protease that processes the gag protein of the immature particle into the mature conical nucleoid. (× 104,000) **C,** Several mature virions are associated with a budding particle on the plasma membrane (*arrow*) of a cultured lymphocyte. Note the variability in size of the virions and their conical nucleoid with an electron-dense, broad end. (× 170,000) (Photographs generously contributed by Jan M. Orenstein, M.D., Department of Pathology, George Washington University Medical Center, Washington, DC.)

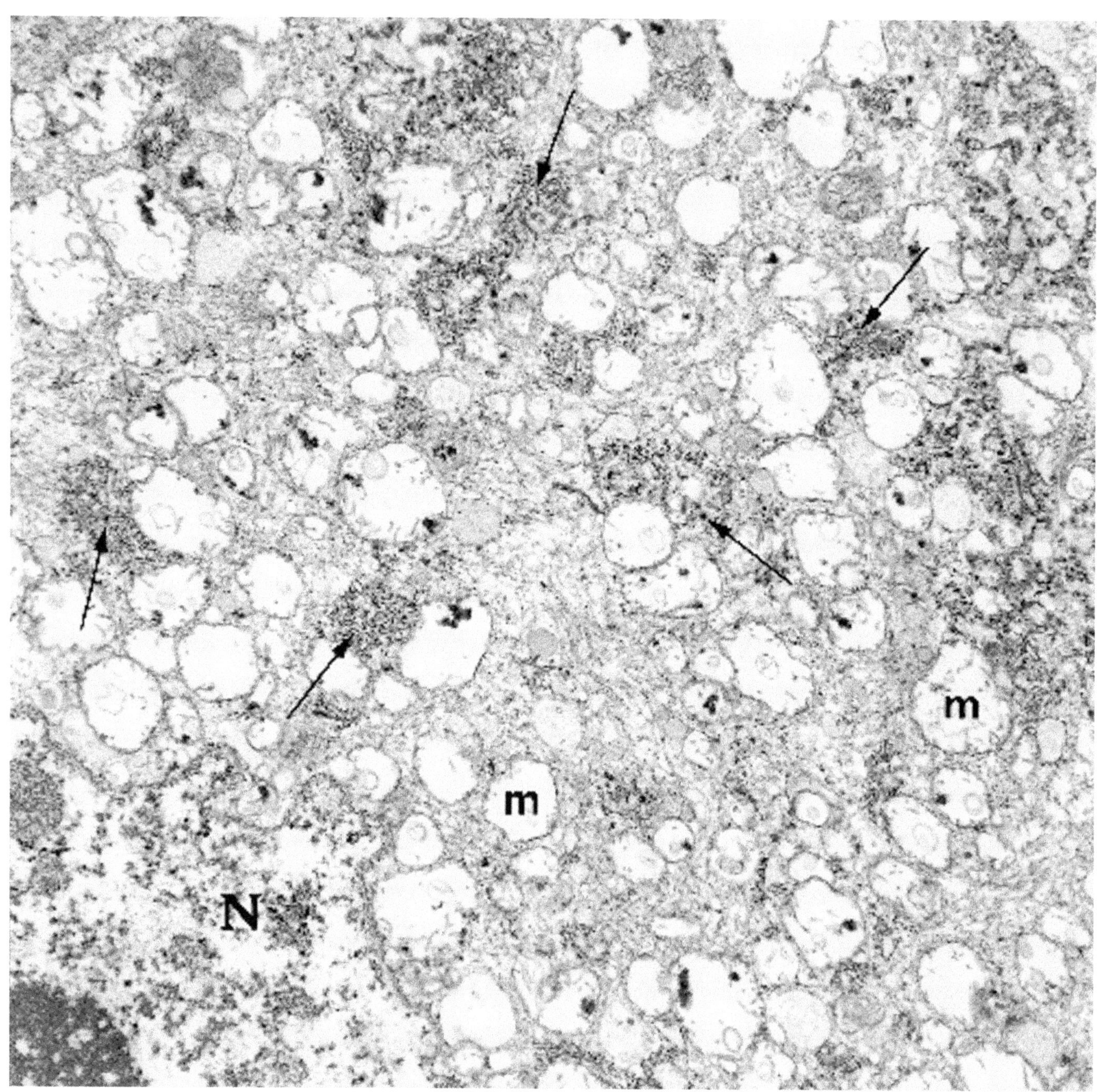

Figure 10.29. Rabies virus (cerebellum): The cytoplasm of this Purkinje cell contains packets of electron-dense, granule-like structures (*arrows*), which represent the rabies viral particles, seen in more detail in Figure 10.30. m = mitochondria; N = nucleus. (× 13,400)

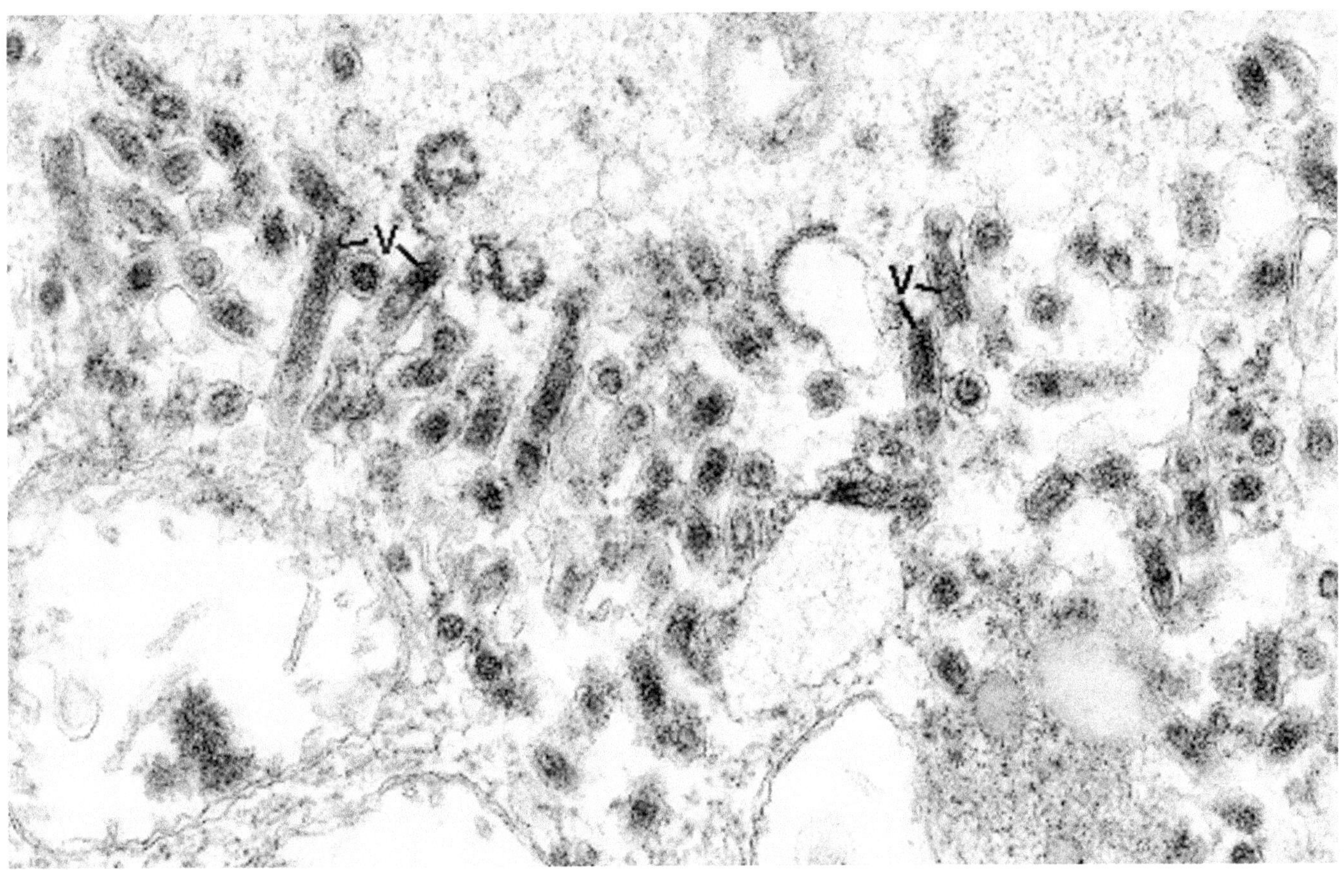

Figure 10.30. Rabies virus (cerebellum). High magnification of rabies virions (V) illustrates their bullet shape when captured in longitudinal axis. (× 57,000)

(Text continued from page 661)

Protozoa

Pneumocystis carinii

(Figures 10.31 through 10.35.)

Diagnostic criteria. (1) Alveolar exudate containing cysts, trophozoites, and a lesser number of macrophages (Figure 10.31); (2) cysts are spherical (intact) and crescentic (collapsed), are about 4 μm in diameter, have a wall composed of an outer dense layer and an inner less-dense layer, and may be empty or enclosing sporozoites (Figures 10.31 through, 10.33); (3) trophozoites may be intra- or extracystic, are pleomorphic, measure 1.5–12 μm long, have a nucleus and cytoplasm, and may have elaborate foldings of their plasmalemmas (Figures 10.33 and 10.34).

Additional points. The life cycle of *Pneumocystis carinii* includes the trophozoite (a vegetative form attached to type I alveolar lining cells), cysts (from enlarging trophozoites), and sporozoites (developing within enlarging cysts). The cysts rupture and release the sporozoites, which become trophozoites. The elaborate foldings of the trophozoite and cyst walls is not completely understood but may represent metabolic membrane activity and/or, possibly, the site of viral replication (Figures 10.34 and 10.35).

Toxoplasma gondii

(Figures 10.36 and 10.37.)

Diagnostic criteria. (1) Ovoid parasites (about 3 × 2 μm), distributed freely and within macrophagic parasitophorous vacuoles (12 μm cysts) (Figure 10.36); (2) parasites (sporozoites) have a triple-layered cell membrane and a single posterior nucleus and nonspecific and specific cytoplasmic organelles; (3) characteristic organelles include a conoid (conical opening) at the anterior end of the organism, micronemes (convoluted groups of secretory tubules) also at the anterior end, and rhoptries (club-shaped storage sacs in the anterior third), which empty by ducts into the conoid (Figure 10.37).

Additional points. Routine histologic techniques do not allow a specific identification of *Toxoplasma* organisms, and electron microscopy is useful in making the diagnosis. This is especially true when immunologic methods are not applicable, as in immunosuppressed patients and in the postmortem state when serologic testing is not available. Immunohistochemistry and DNA determination by polymerase chain reaction are also useful for specific identification of the organisms. The fine structural features of *Toxoplasma,* such as the method of multiplication, its size at various stages, the number of micronemes (30–50) and the number (5–9), morphology, and site of rhoptries, allow it to be distinguished from other sporozoites in the *Apicomplexa* genera. By light microscopy, the cysts and organisms of *Toxoplasma gondii* are similar to those of *Trypanosoma cruzi,* but by electron microscopy the two parasites are distinctly different. Specifically, conoids and endodyogenous division (internal budding into two daughter cells) are not seen in trypanosomes. In addition, trypanosomes have a kinetoplast, and *Toxoplasma* does not.

Cryptosporidium parvum

(Figures 10.38 through 10.43.)

Diagnostic criteria. (1) At low power, 3–4 μm spherical structures (cryptosporidia) along gastrointestinal (especially small intestinal) epithelium (Figure 10.38); (2) at higher power, some spherical structures attached to the epithelial cells, and some unattached near epithelial cell surface; (3) attachment site of epithelial cell usually raised, electron-dense, and devoid of microvilli (Figures 10.39 and 10.40); (4) epithelial cell surface raised at edges of attachment site to enclose partially or completely (parasitophorous vacuole) the spherical structures (Figure 10.39); (4) spherical parasites of varying morphology, depending on stage in life cycle, including macrogametes (with clear polysaccharide granules and dark dense granules) (Figure 10.43), trophozoites (with nucleus and cytoplasmic organelles) (Figures 10.41 and 10.42), schizont with daughter organisms (merozoites) (Figures 10.41 and 10.42); (5) round and elliptical merozoites, extracellular and unattached (following exit from schizont); (6) secondary lysosomes (phagosomes) frequently in apical cytoplasm of epithelial cell of attachment site.

Additional points. The life cycle of cryptosporidia includes trophozoites undergoing three nuclear divisions to form eight daughter cells. This process is known as schizogony, the overall organism is the schizont, and the contained daughter cells are called merozoites. The merozoites mature, the schizont ruptures, and the merozoites escape and invade other epithelial cells. Once in other cells, the merozoites develop into trophozoites, which produce second-generation merozoites. The merozoites escape the new schizont, invade new epithelial cells, and this time mature into macrogametes or microgametes. The gametes form an infectious form of oocysts, which pass from the body in the feces and become available to other hosts per orum.

(Text continues on page 694)

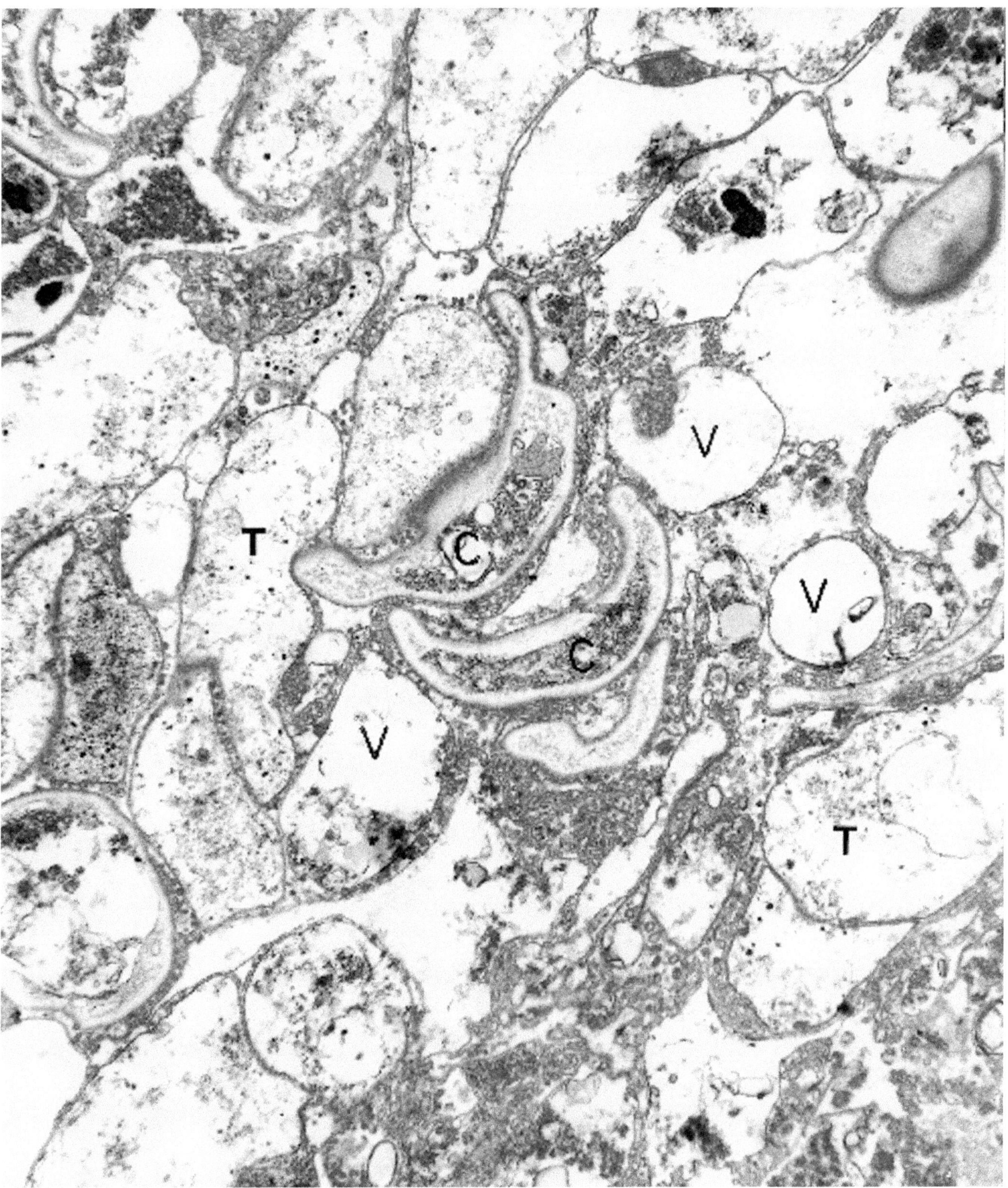

Figure 10.31. *Pneumocystis carinii* (pulmonary alveolar exudate). Two thick-walled, crescentic cysts (C) are surrounded by trophozoite processes (T) and vacuoles (V), the latter of which may represent macrophagic vacuoles. (× 13,000)

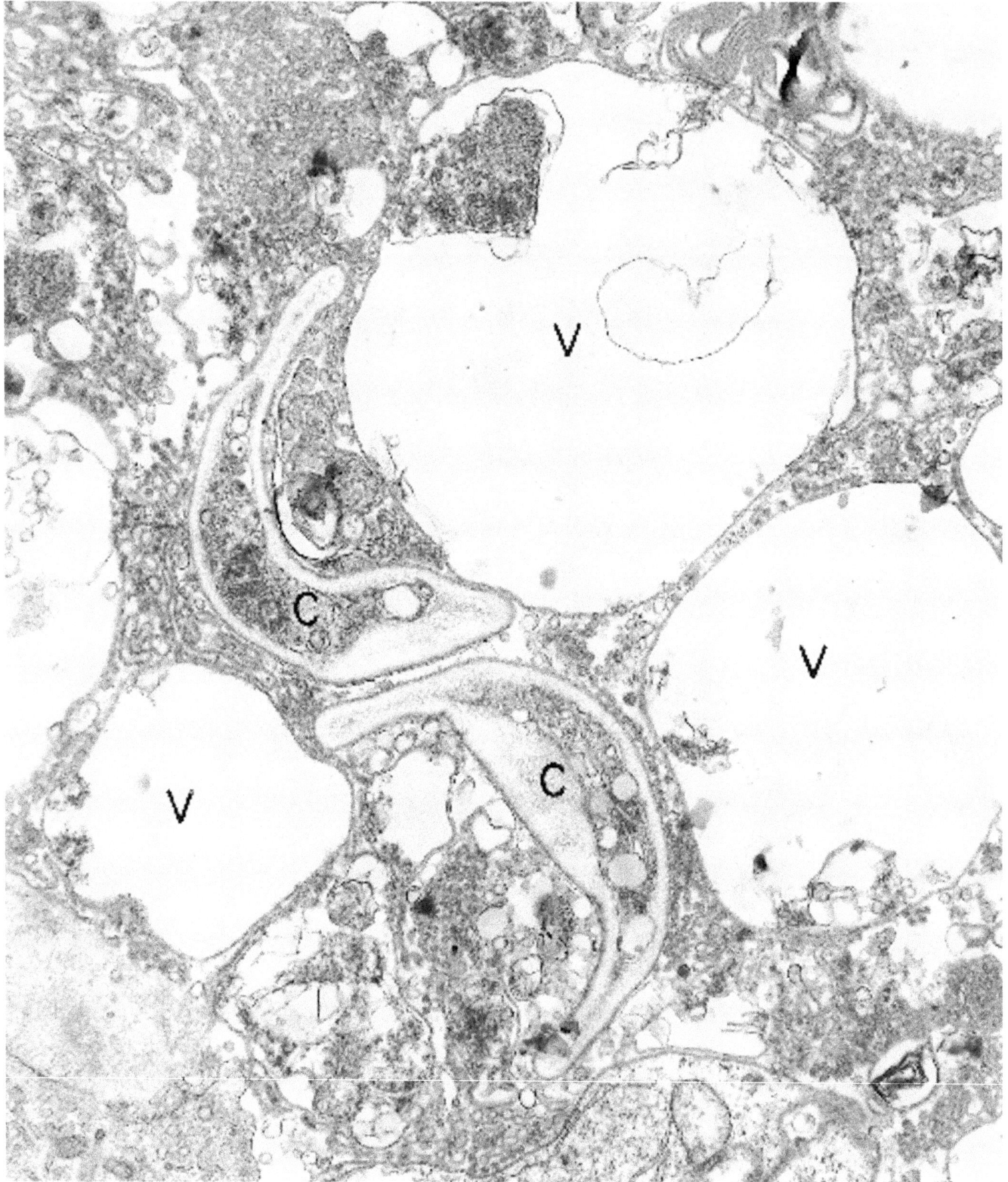

Figure 10.32. *Pneumocystis carinii* (pulmonary alveolar exudate). Several empty vacuoles (V) and two crescentic cysts (C) with internal structure are present in this field. (× 28,840)

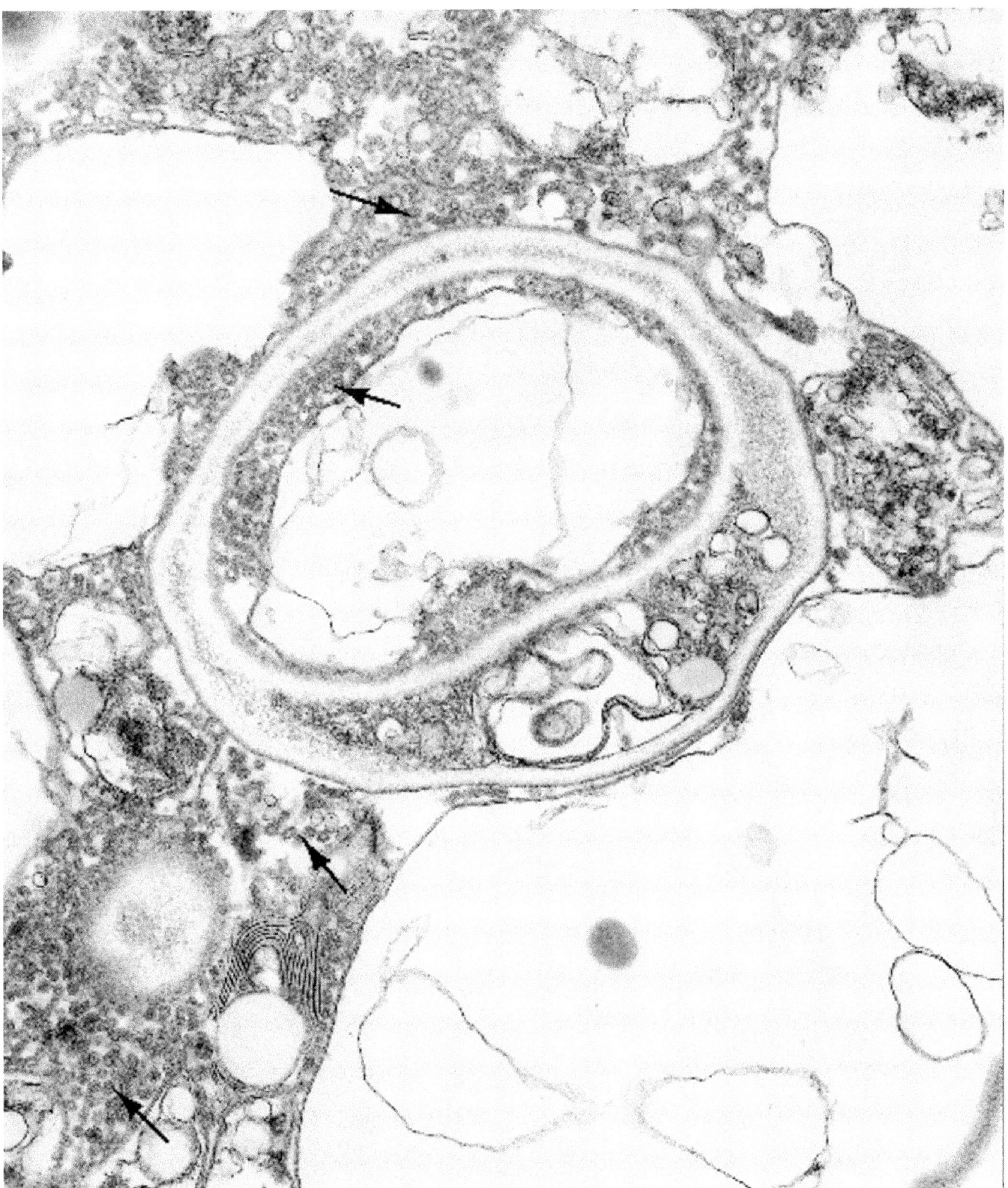

Figure 10.33. *Pneumocystis carinii* (pulmonary alveolar exudate). Cyst wall shows a florid array of membrane folding (small circlets at *arrows*); also seen at higher magnification in Figure 10.34. (× 39,440)

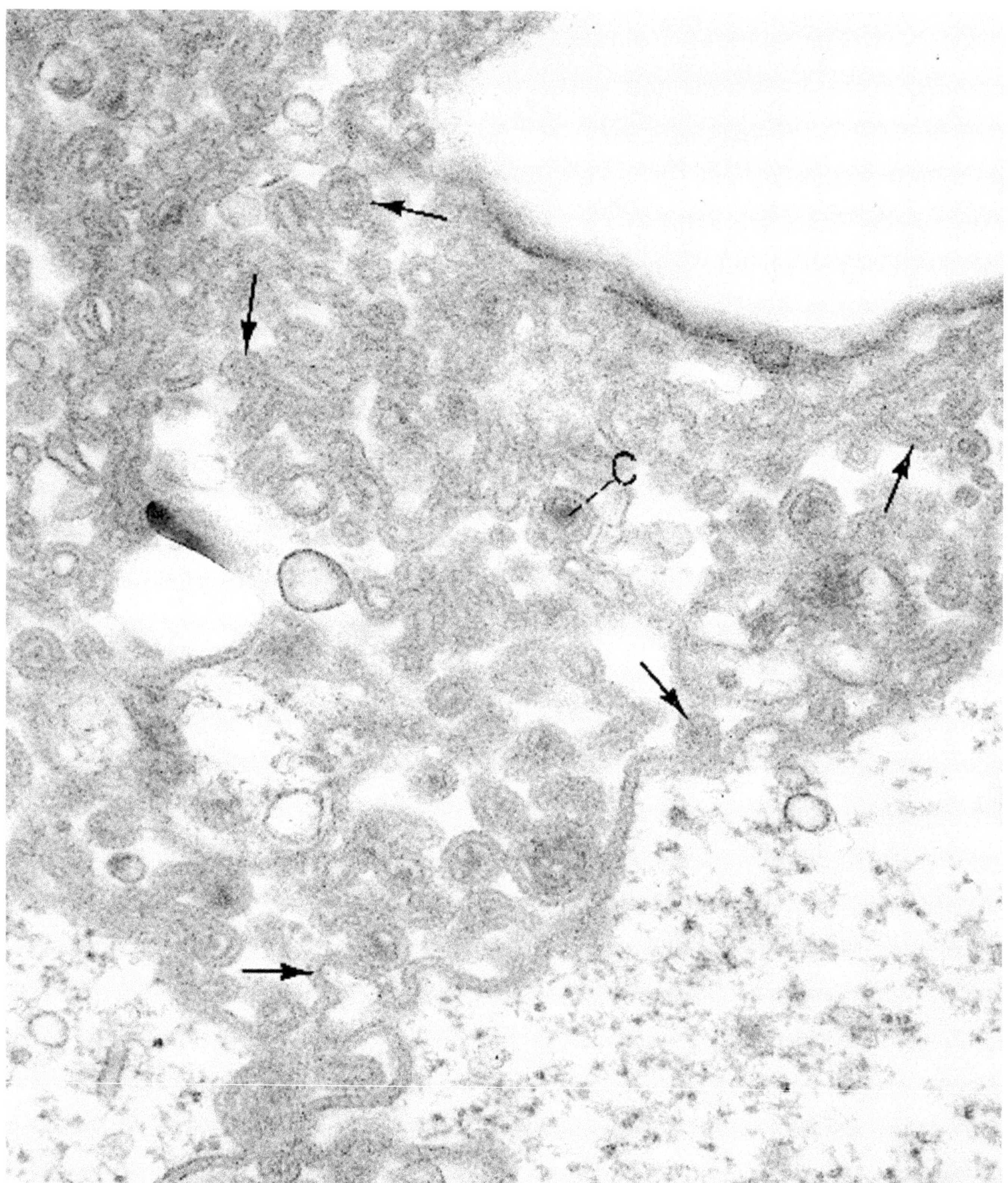

Figure 10.34. *Pneumocystis carinii* (pulmonary alveolar exudate). High magnification of a portion of a cyst highlights the elaborate foldings (*arrows*) of its limiting membrane. The electron-dense cores (C) in some of the folds suggest virions, but they may represent parts of the limiting membrane that are tangentially cut. (× 112,300)

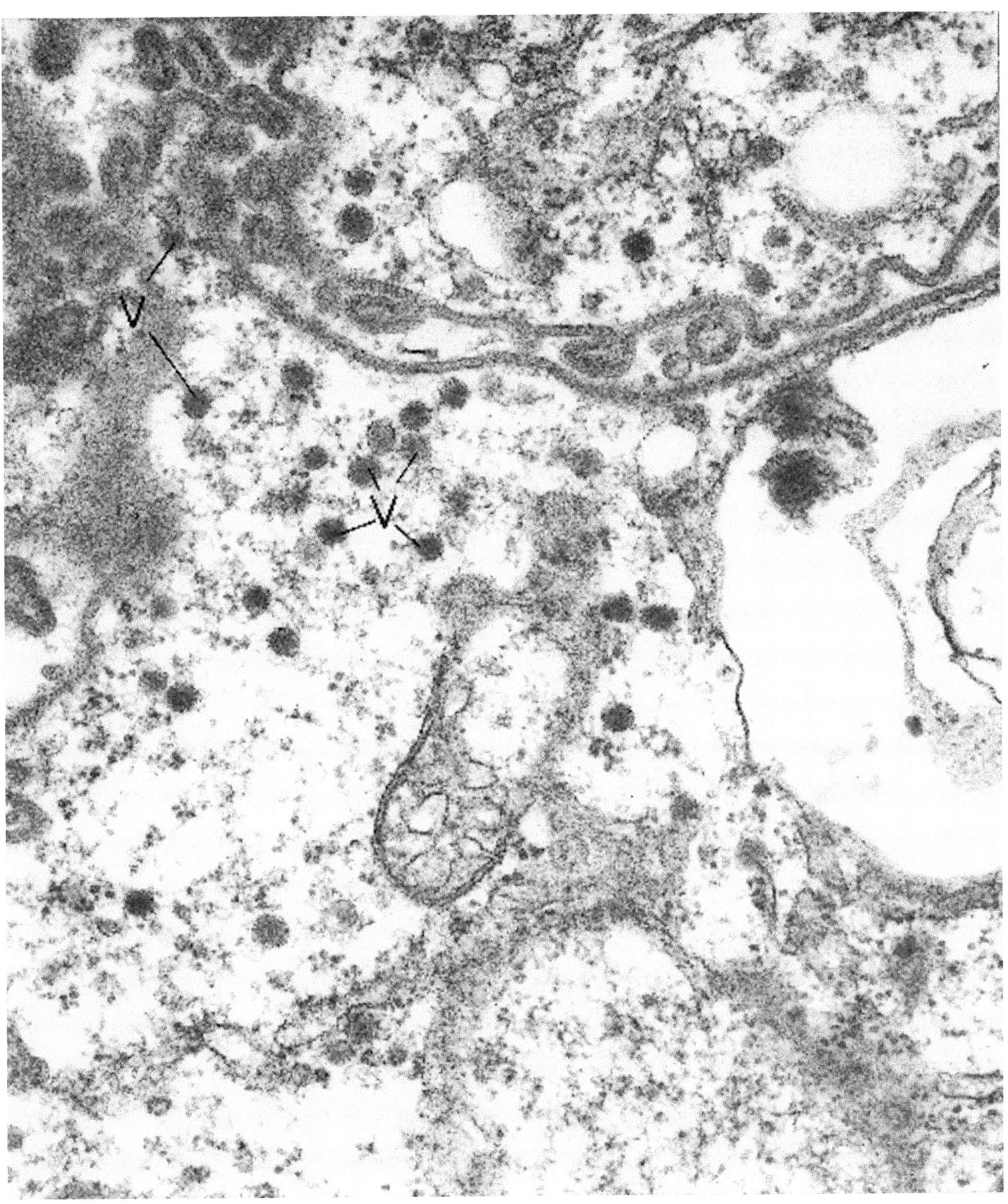

Figure 10.35. *Pneumocystis carinii* (pulmonary alveolar exudate). This exudate contains cytomegalovirus accompanying pneumocystis organisms, and the virions (V) may be replicating in the folded membranes of the protozoan. (× 62,120)

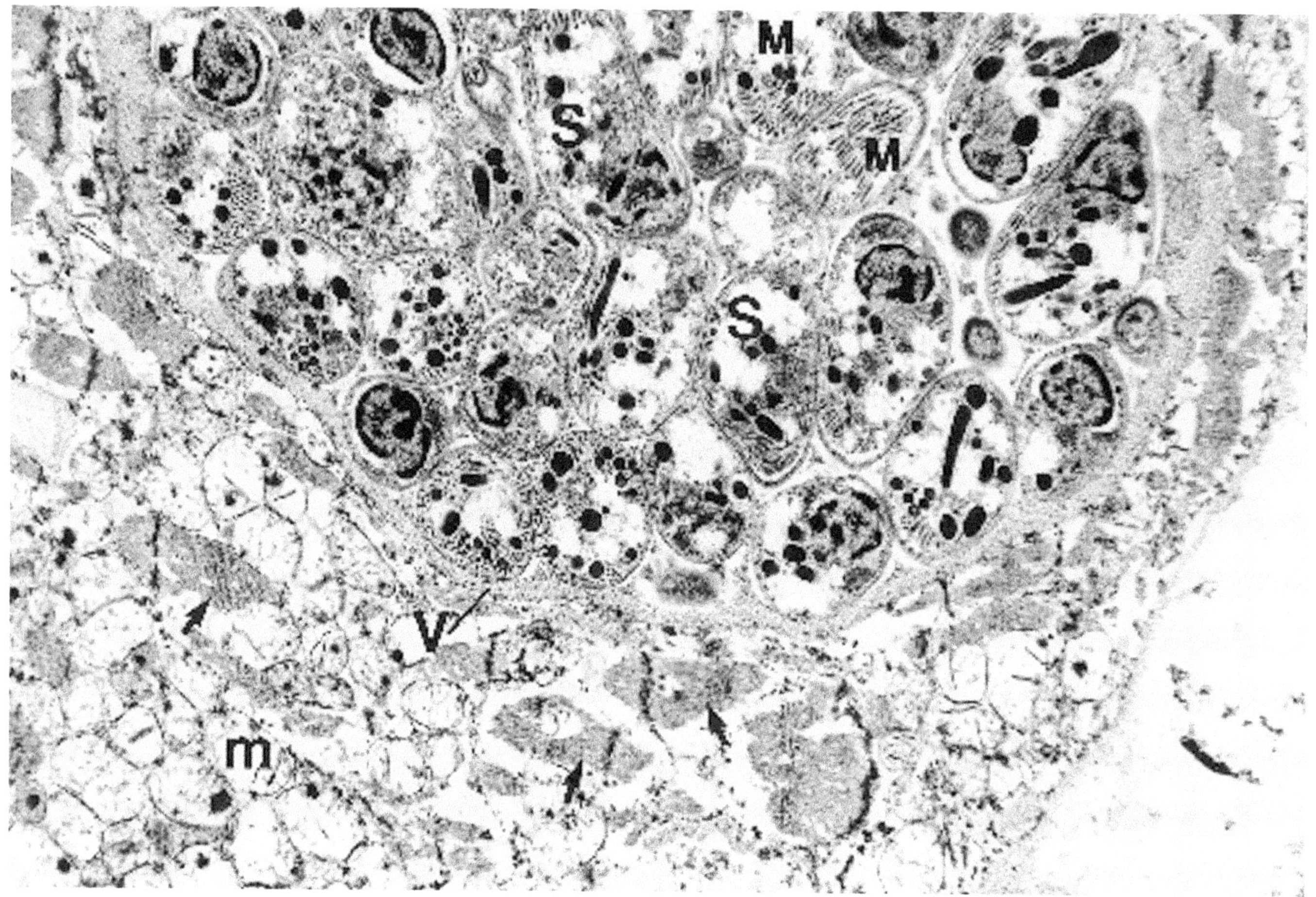

Figure 10.36. *Toxoplasma gondii* (skeletal muscle). An intramuscular cyst (parasitophorous vacuole) contains numerous sporozoites (S). Daughter cells (merozoites, M) are present in some of the parasites. *Arrows* = sarcomeres in muscle cell; m = mitochondria; V = wall of parasitophorous vacuole. (× 11,000)

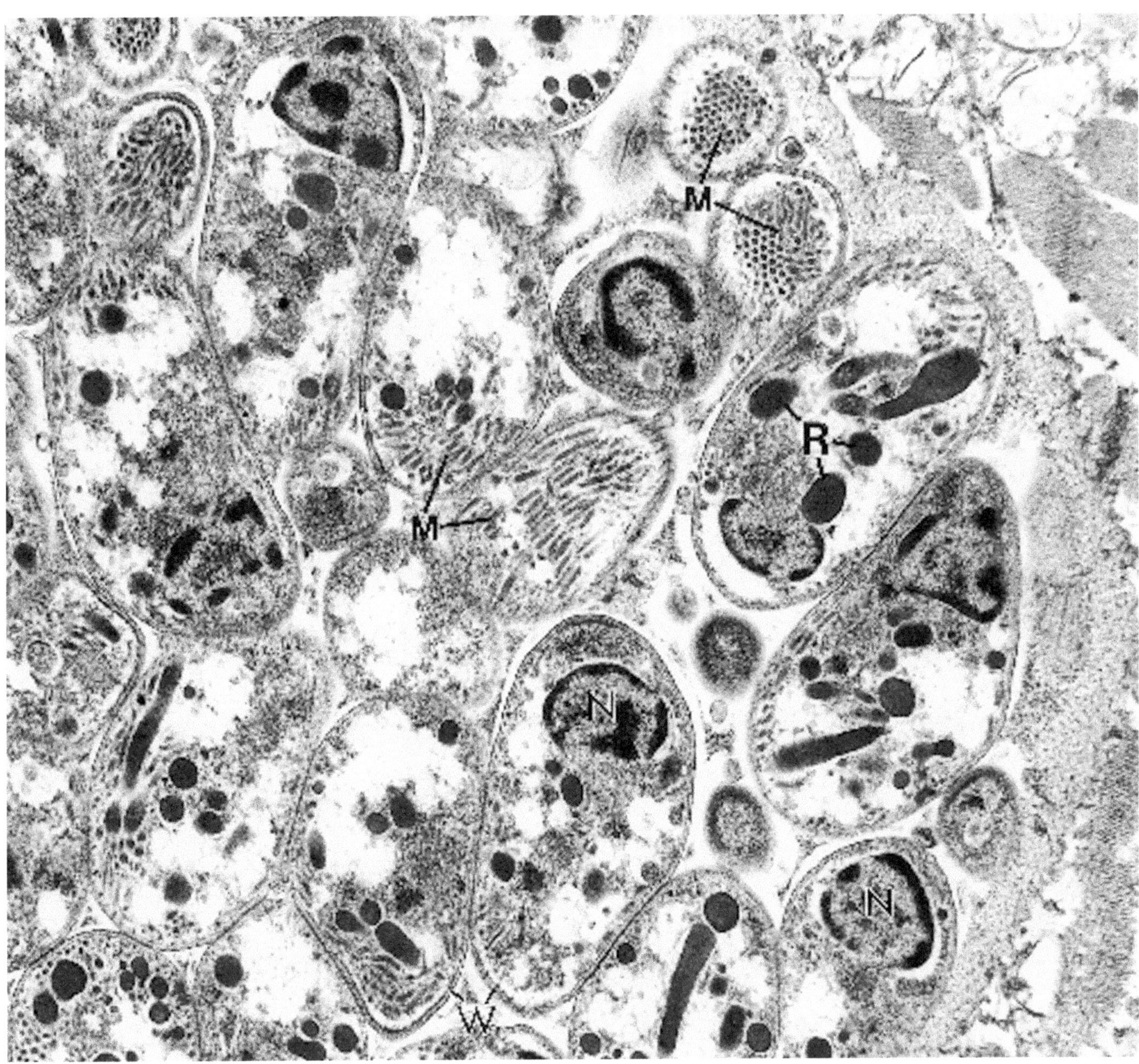

Figure 10.37. *Toxoplasma gondii* (skeletal muscle). Higher magnification of several of the sporozoites shown in Figure 10.36 illustrates the cell wall (W), rhoptries (R), micronemes (M), and nuclei (N). (× 20,000)

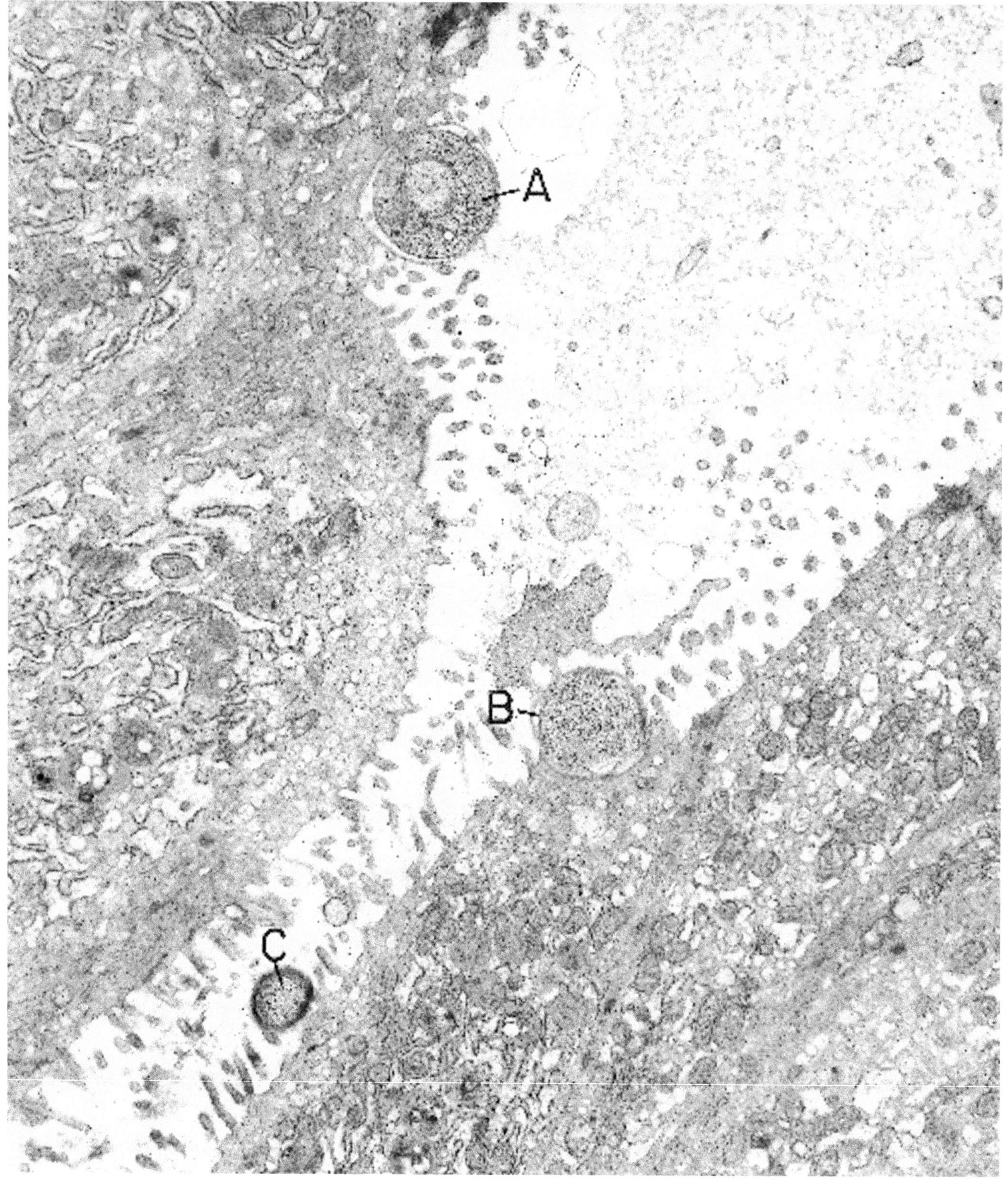

Figure 10.38. *Cryptosporidium* (stomach). Spherical microorganisms (A, B, and C) are dispersed along the mucosal surface, and some (A and B) appear to be intimately attached to epithelial cells. (× 11,780)

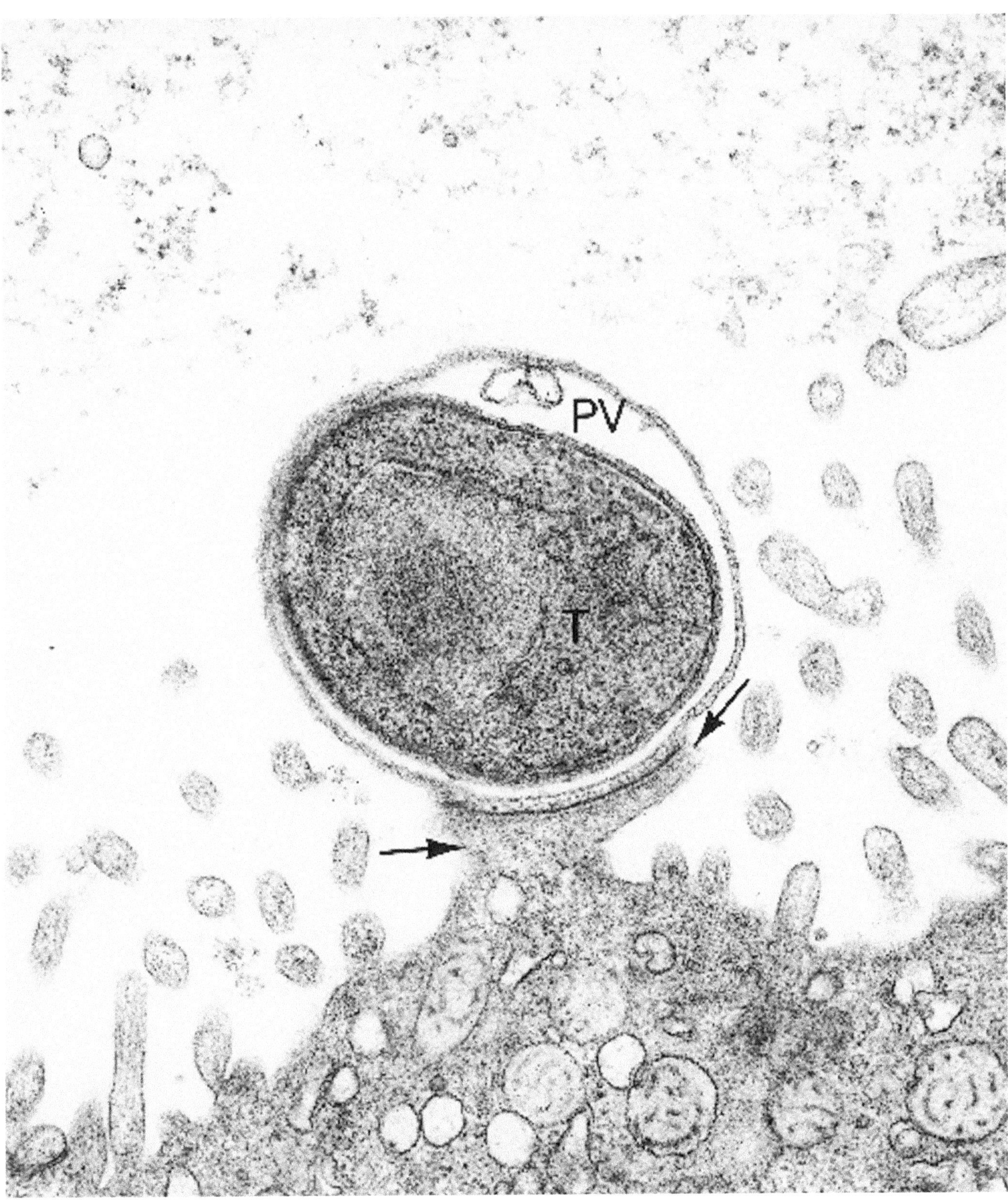

Figure 10.39. *Cryptosporidium* (jejunum). A trophozoite (T) is attached to a raised, villous portion (between *arrows*) of the epithelial cell. The edges of the attachment site seemingly extend around the trophozoite, forming a parasitophorous vacuole (PV). (× 48,400)

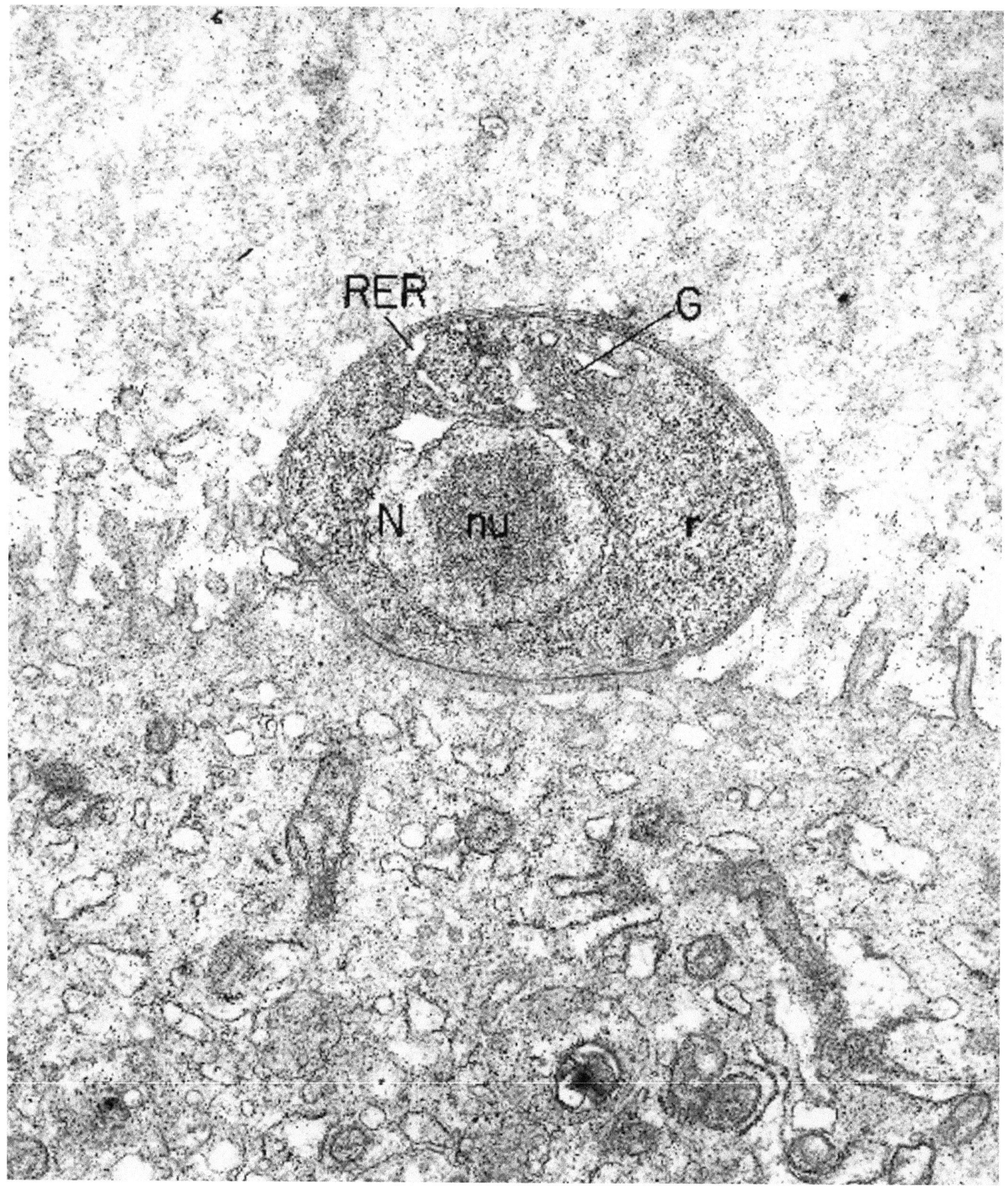

Figure 10.40. *Cryptosporidium* (stomach). High magnification of a trophozoite allows its cellular substructure to be readily visualized. N =nucleus; nu = nucleolus; r = ribosomes; RER = rough endoplasmic reticulum; G = Golgi apparatus. (× 29,870)

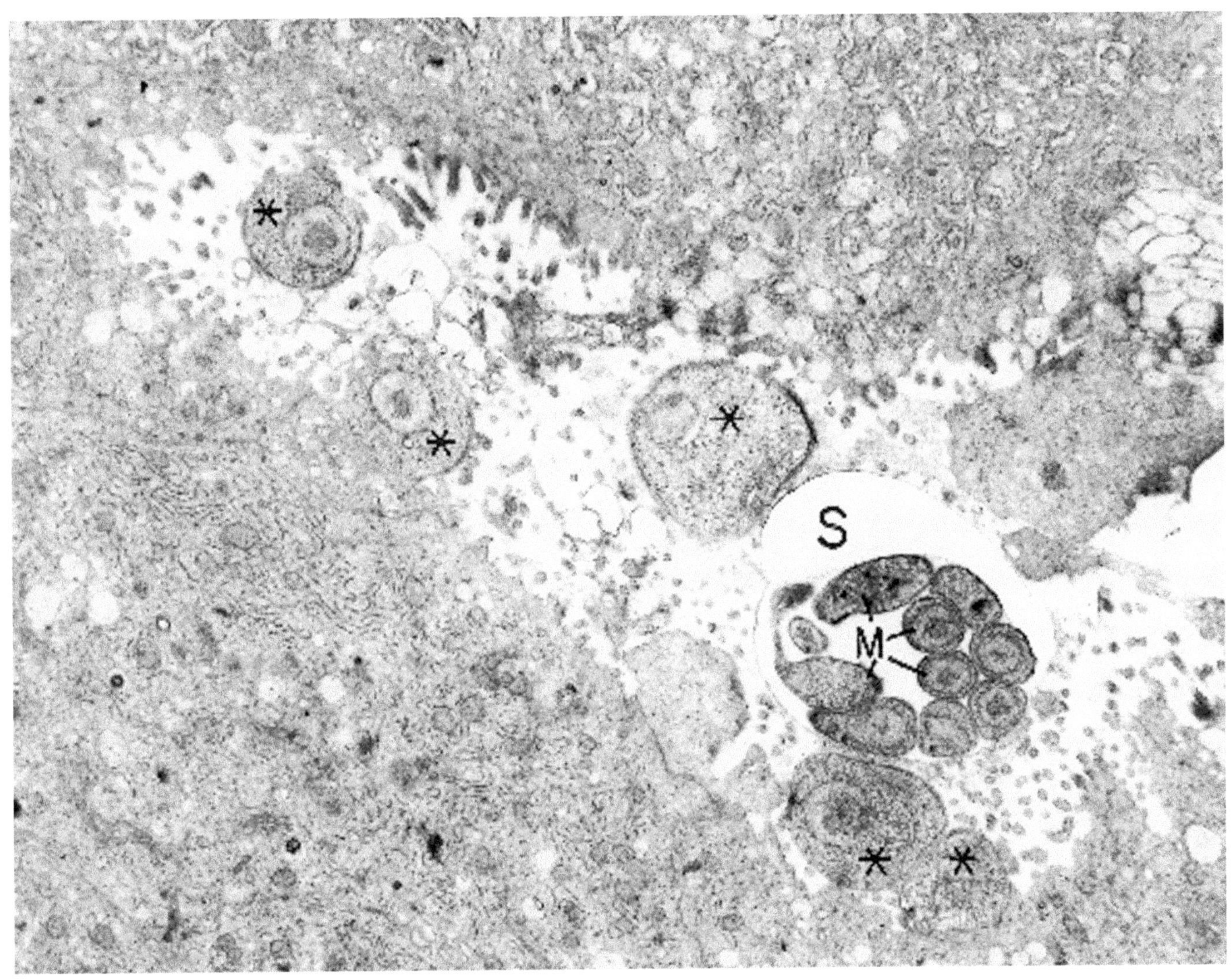

Figure 10.41. *Cryptosporidium* (stomach). Several spherical microorganisms (*), including a schizont (S) with merozoites (M) are evident along the gastric epithelial surface. (× 9450)

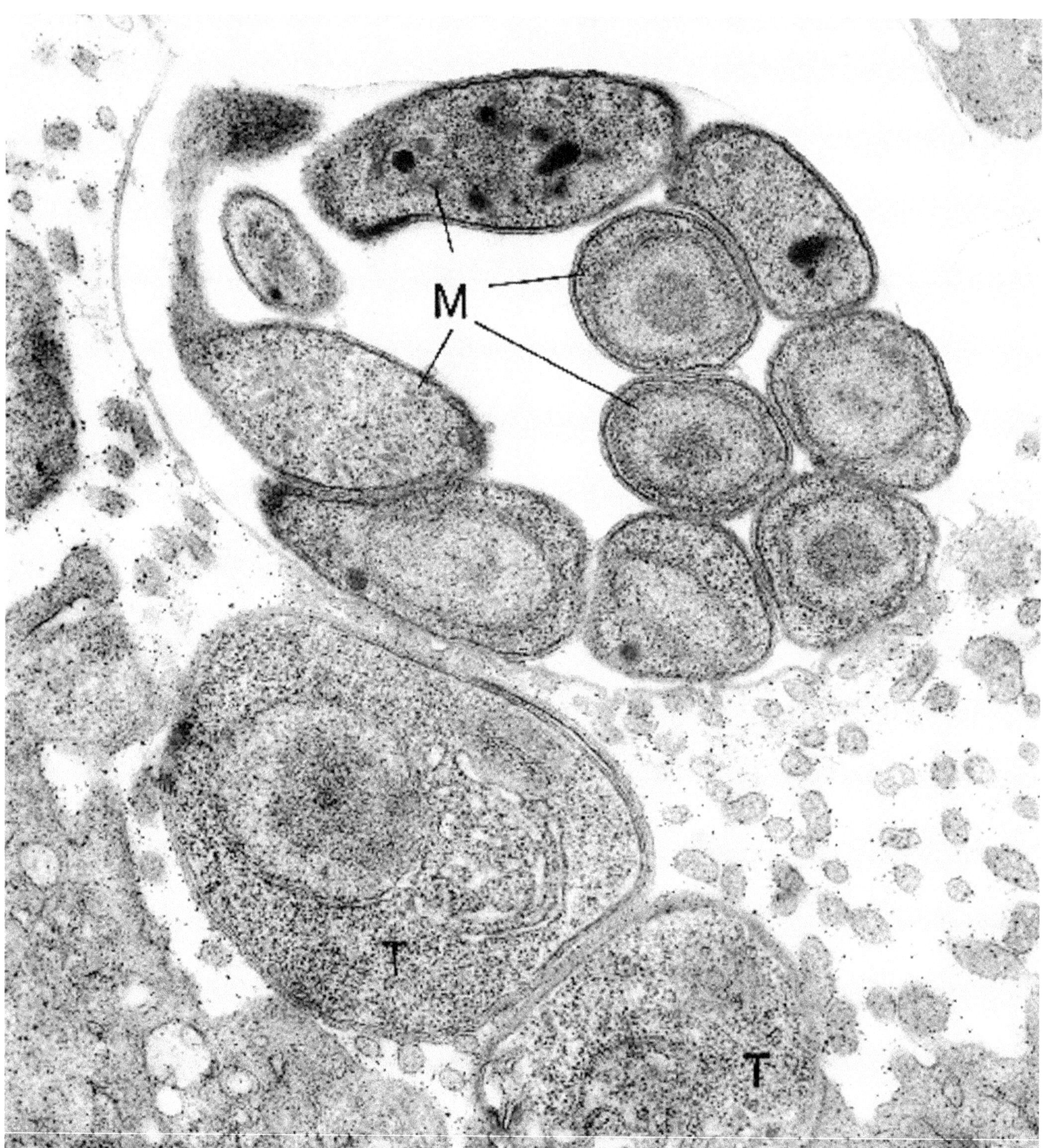

Figure 10.42. *Cryptosporidium* (stomach). This high-power field of an epithelial surface contains two trophozoites (T) and a schizont with merozoites (M). (× 35,360) (Permission for reprinting granted by *American Journal of Medicine,* Blumberg RS, Kelsey P, Perrone T, et al: Cytomegalovirus- and cryptosporidium-associated acalculous gangrenous cholecystitis. Am J Med 76:1118–1123, 1984.)

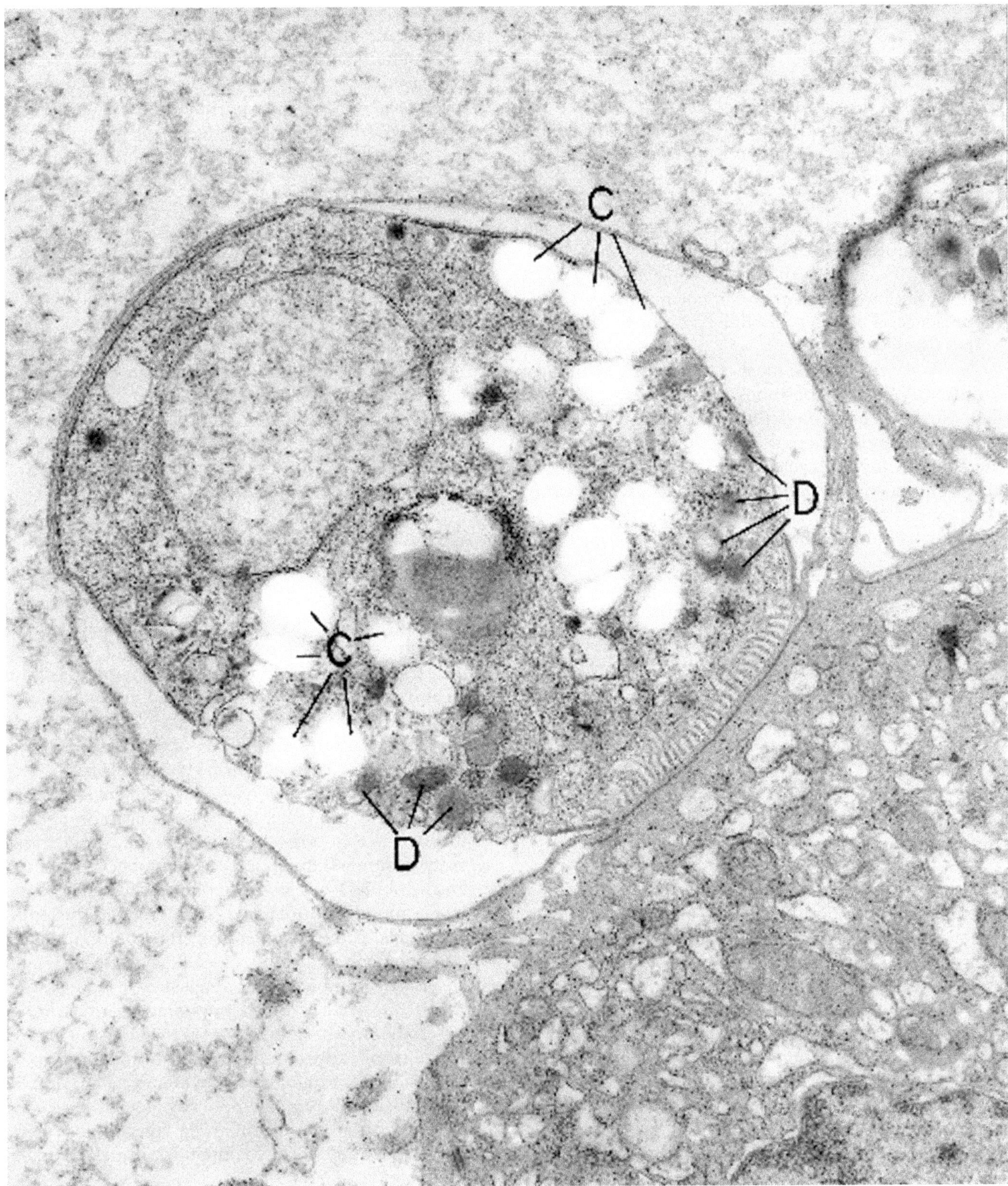

Figure 10.43. *Cryptosporidium* (stomach). A macrogamete is characterized predominantly by its clear, polysaccharide granules (C) and its dark, dense granules (D). (× 25,100)

(Text continued from page 680)

Trypanosoma cruzi

(Figures 10.44 through 10.45.)

Diagnostic criteria. (1) Oval and elliptical, 10–30 μm organisms with a nucleus, cytoplasmic organelles, and characteristic kinetoplast and flagellum (Figures 10.44 and 10.45); (2) kinetoplast is curved, electron-dense (DNA-containing) structure, at base of basal body of flagellum (Figure 10.45); (3) single long mitochondrion, coursing subjacent to cell membrane, from kinetoplast to posterior compartment of cytoplasm (Figure 10.45); (4) presence or absence of a free (exterior to cell membrane) flagellum.

Additional points. The most characteristic structures for identifying trypanosomes are the kinetoplast and flagellum. The position of the kinetoplast in relation to the nucleus and the length of the flagellum vary with the stage in the life cycle of the organism. Epimastigotes, promastigotes, and amastigotes have a kinetoplast and flagellar origin anterior to the nucleus, whereas trypomastigotes have theirs posterior to the nucleus. The flagellum exits the cell through a pocket in the plasmalemma, and, in the amastigote stage, the flagellum is so short that it does not protrude above the pocket. In the promastigote stage, the flagellum extends above the pocket but is short. In the epimastigote and trypomastigote stages, it is long and courses tightly against the outer edge of the cell wall before becoming free at the anterior end of the cell. Intermediate forms of these various stages in the life cycle of *Trypanosoma* also are seen, both in culture and *in vivo*.

Microsporida (M. *enterocytozoon bieneusi* and M. *septata intestinale*)

(Figures 10.46 through 10.49.)

Diagnostic criteria. (1) Round and oval spores, 0.5–5 μm in diameter, in apical cytoplasm of host cell (e.g., enterocyte); (2) specialized extrusion apparatus in the form of a single coiled polar tube, in spores; (3) in *M. enterocytozoon bieneusi,* the polar tube coil is seen in cross-sections as a double row in the posterior half of the spore; and (4) in *M. septata intestinalis,* the polar tube coil is a single row; (5) the polar tube straightens in the anterior half of the spore and connects to an anchoring disc; (6) ribosomes and lamellar and tubular polarplast in spore cytoplasm; (7) stacked polarplast surrounds the straight section of the polar tube; (8) no mitochondria; (9) single or double nucleus in spore; (10) triple-layered spore wall; (11) spores in direct contact with cytoplasm of enterocyte in *M. enterocytozoon bieneusi;* (12) spores in a septated parasitophorous vacuole in *M. septata intestinalis.*

Additional points. Microsporidia spores enter the host cell (e.g., absorptive enterocytes, biliary duct epithelium, corneal epithelium, etc.) and proliferate (merogony) by binary fission. Meronts are small, round, and difficult to distinguish from the surrounding cytoplasm of the host cell because of their thin cell membranes and relative electron-lucent cytoplasm. Their cytoplasm contains empty clefts, and there are 1 to 6 nuclei. The spore-forming phase (sporogony) of the life cycle is identifiable by the presence of electron-dense discs, which later aggregate longitudinally to form the polar tube. Nuclei divide, and the organism enlarges and breaks into sporoblasts. These mature into spores. Spores frequently are surrounded by host cell mitochondria. Spores infect new cells by extruding their polar tube, which penetrates adjacent cell membranes and injects infected sporoplasm.

Giardia lamblia

(Figure 10.50.)

Diagnostic criteria. (1) Pear-shaped, dorsally convex and ventrically concave thophozoites; (2) measuring 9.5–21 μm long and 5–15 μm wide; (3) lying free in lumen and crypts of small intestine and in cysts in large intestine; (4) having rigid cytoskeleton; (5) evenly spaced microtubules (50–60 nm apart), linked by microribbons; (6) adhesion disc on ventral surface; (7) two symmetrically placed nuclei, with prominent karysome; (8) four pairs of flagella, three posterior and one ventral, posterior to adhesion disc; (9) mid-line, transversely positioned median bodies (compact collections of microtubules); (10) absence of mitochondria, peroxisomes, smooth endoplasmic reticulum, and nucleoli.

Additional points. Giardia is a protozoan with three different morphological types—*G. agilus, G. muris,* and *G. duodenalis* (lamblia). Only the last type is illustrated here. The life cycle consists of two stages—free trophozoites in the small gut, and cysts in the large gut and feces. The trophozoites multiply in the small gut by binary fission. Mature cysts are oval and round and 8–12 μm long and 7–10 μm wide. They have a dense wall, a finely granular cytoplasm and a layer of peripheral vacuoles or tubules. They contain four nuclei, usually at one pole.

(Text continues on page 702)

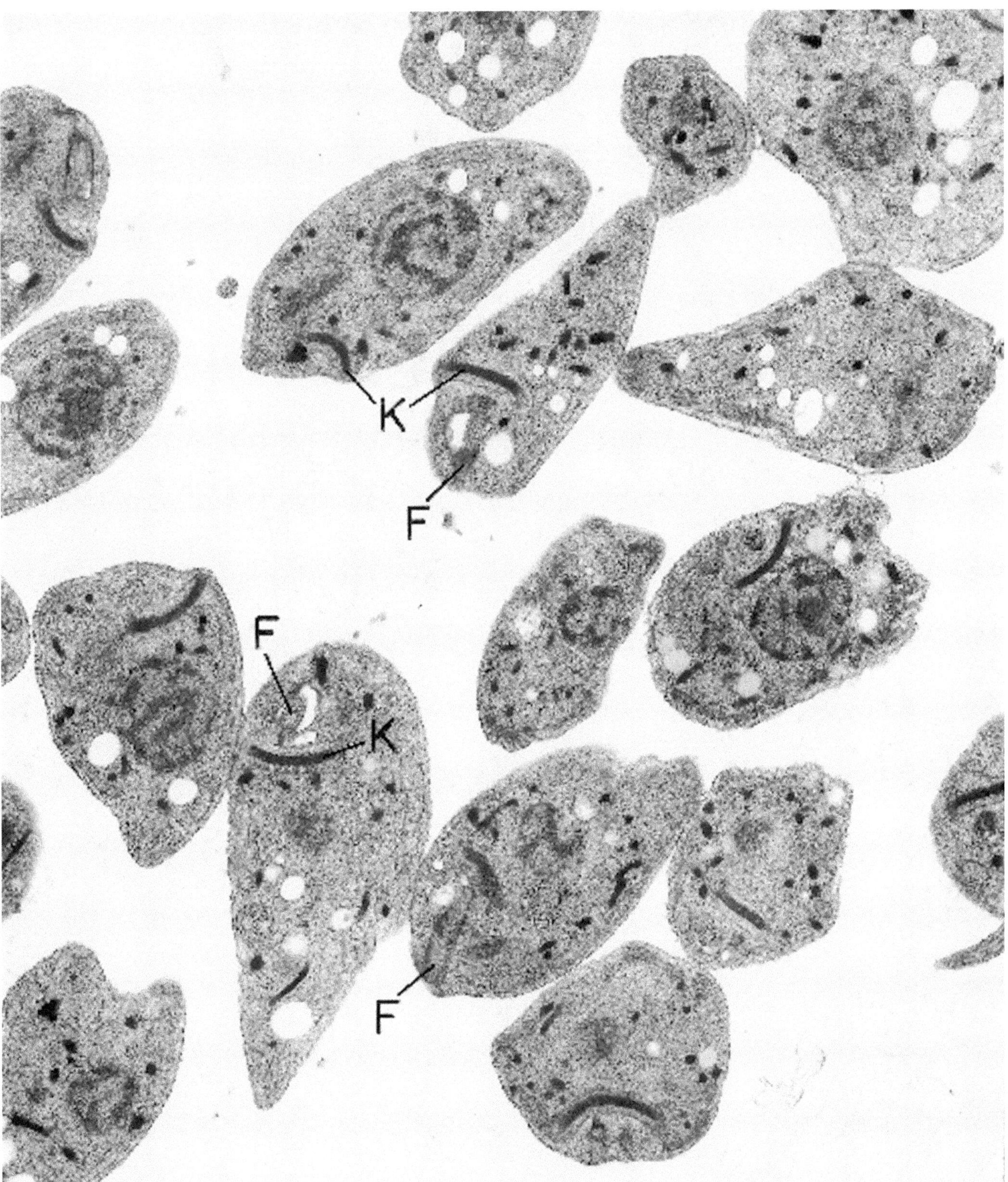

Figure 10.44. *Trypanosoma cruzi* (*in vitro* culture). A low-power field of trypanosomes illustrates their oval and elliptical (and teardrop) shapes and the presence of a nucleus and various cytoplasmic organelles. Most characteristic of the organisms are the kinetoplasts (K) and flagella (F). (× 12,500)

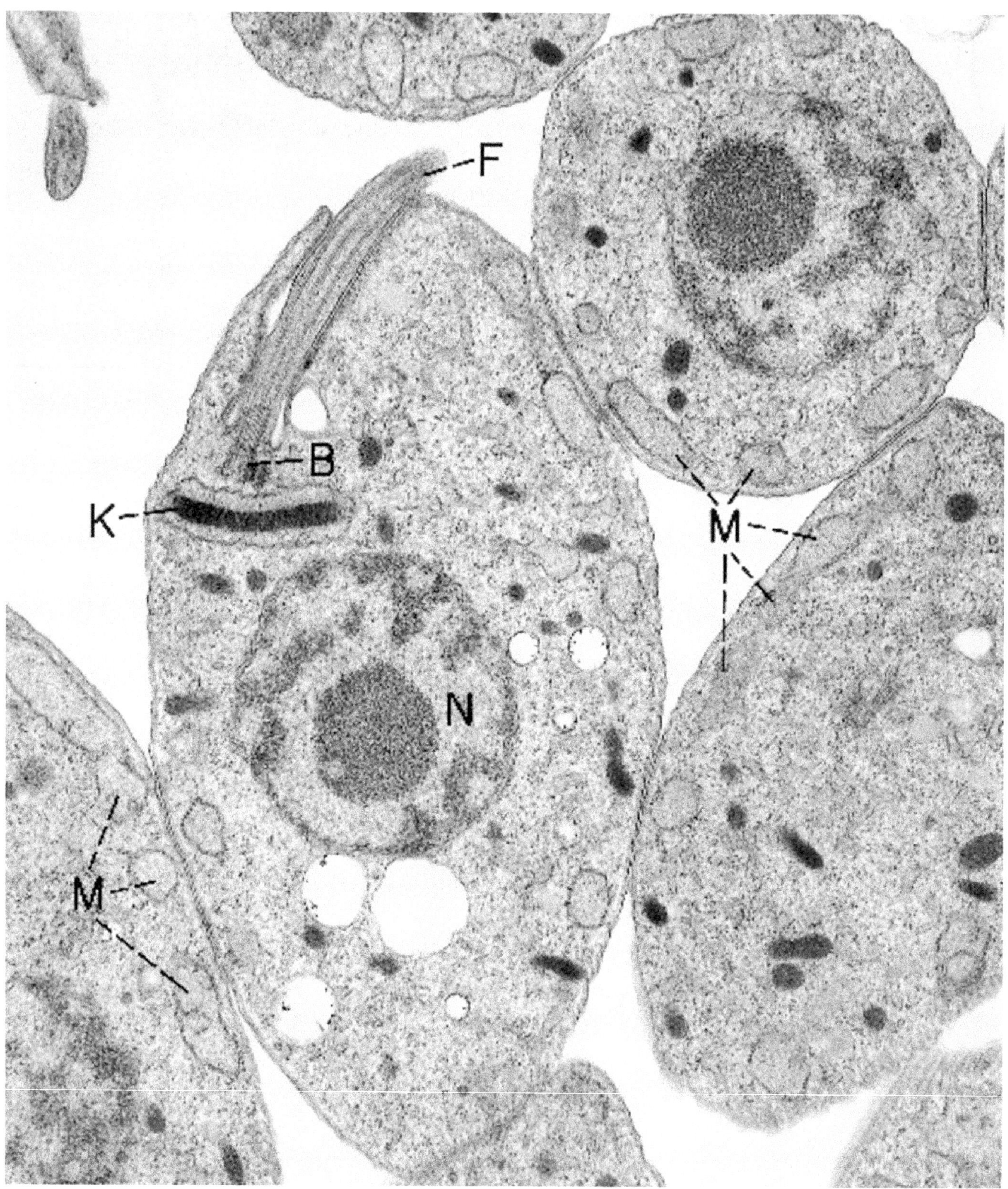

Figure 10.45. *Trypanosoma cruzi* (*in vitro* culture). High magnification allows a clear view of the parasites' internal structure. K = kinetoplast; F = flagellum; B = basal body; M = undulating, single mitochondrion; N = nucleus. (× 24,000)

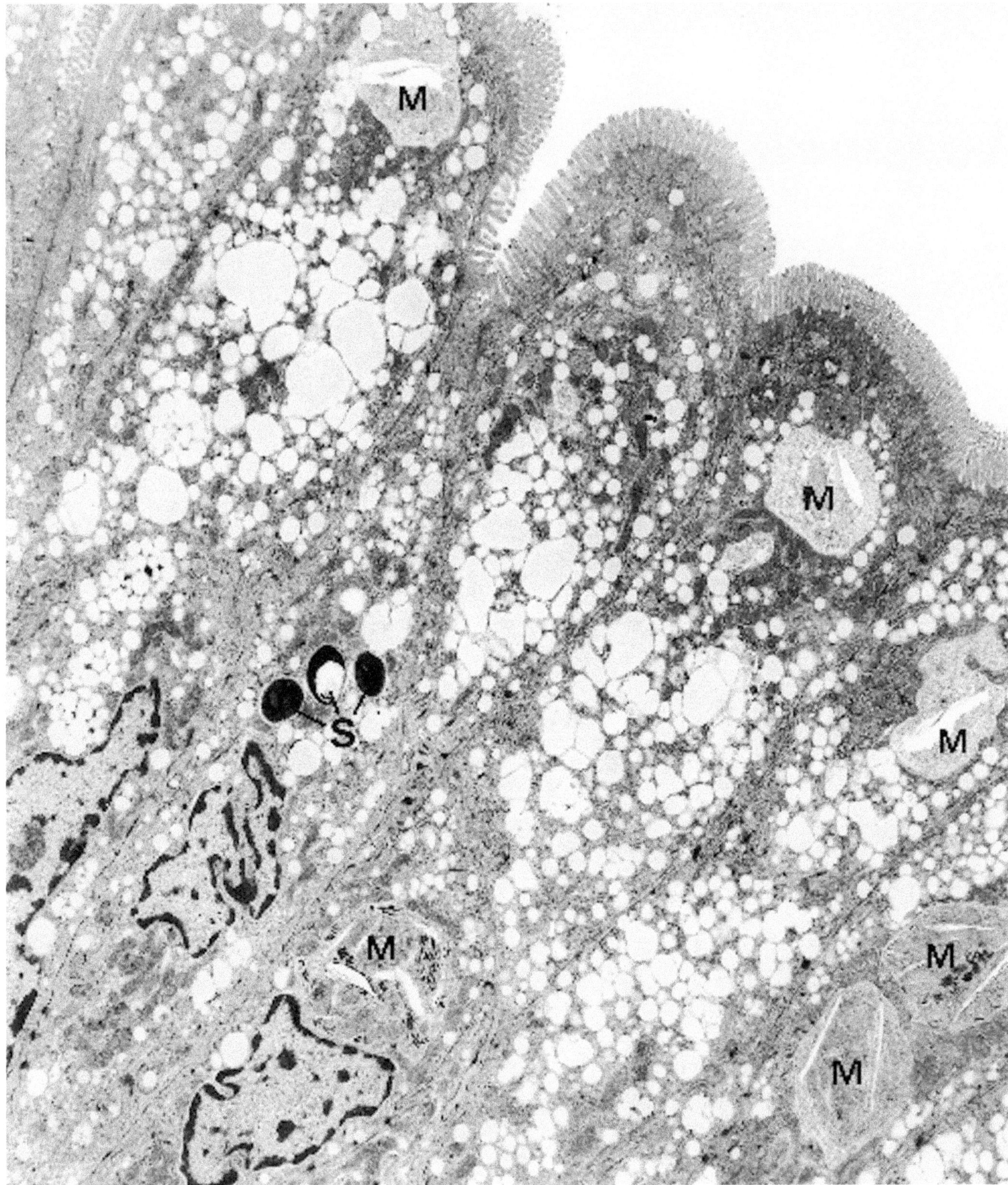

Figure 10.46. Microspora, enterocytozoon bieneusi (duodenum). These enterocytes are infected with microsporidia spores (S) and meronts (M). (× 7400)

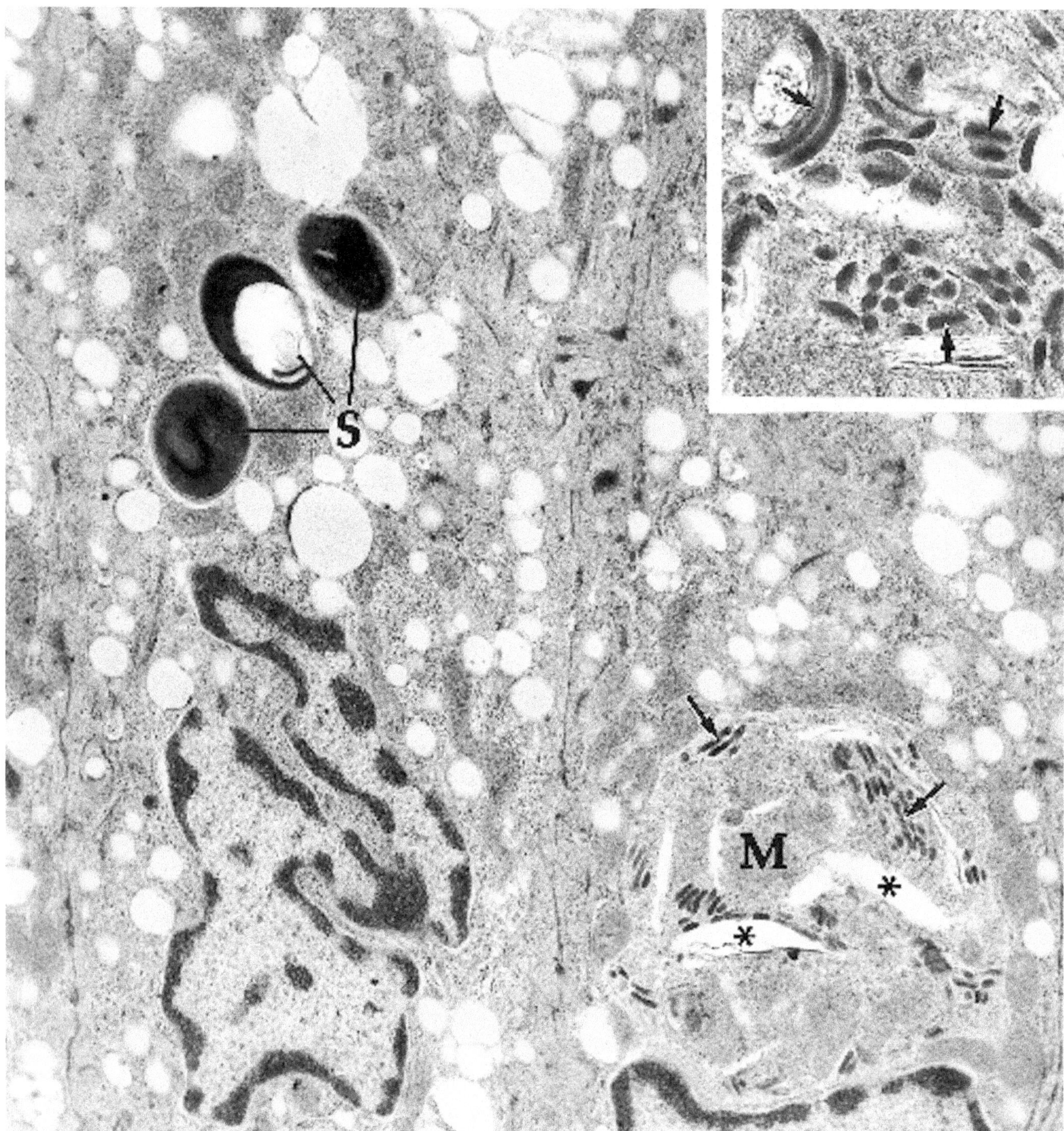

Figure 10.47. Microspora, enterocytozoon bieneusi (duodenum). Higher magnification of portions of two enterocytes shown in Figure 10.46 illustrates three spores (S) and a zone of sporogony where meronts (M) are transforming into spores. The electron-dense discs (*arrows*), seen at higher magnification in the *inset,* will aggregate to form the future polar tube. Clear clefts (*) are characteristic of the cytoplasm of meronts. (× 20,000; *inset:* × 36,000)

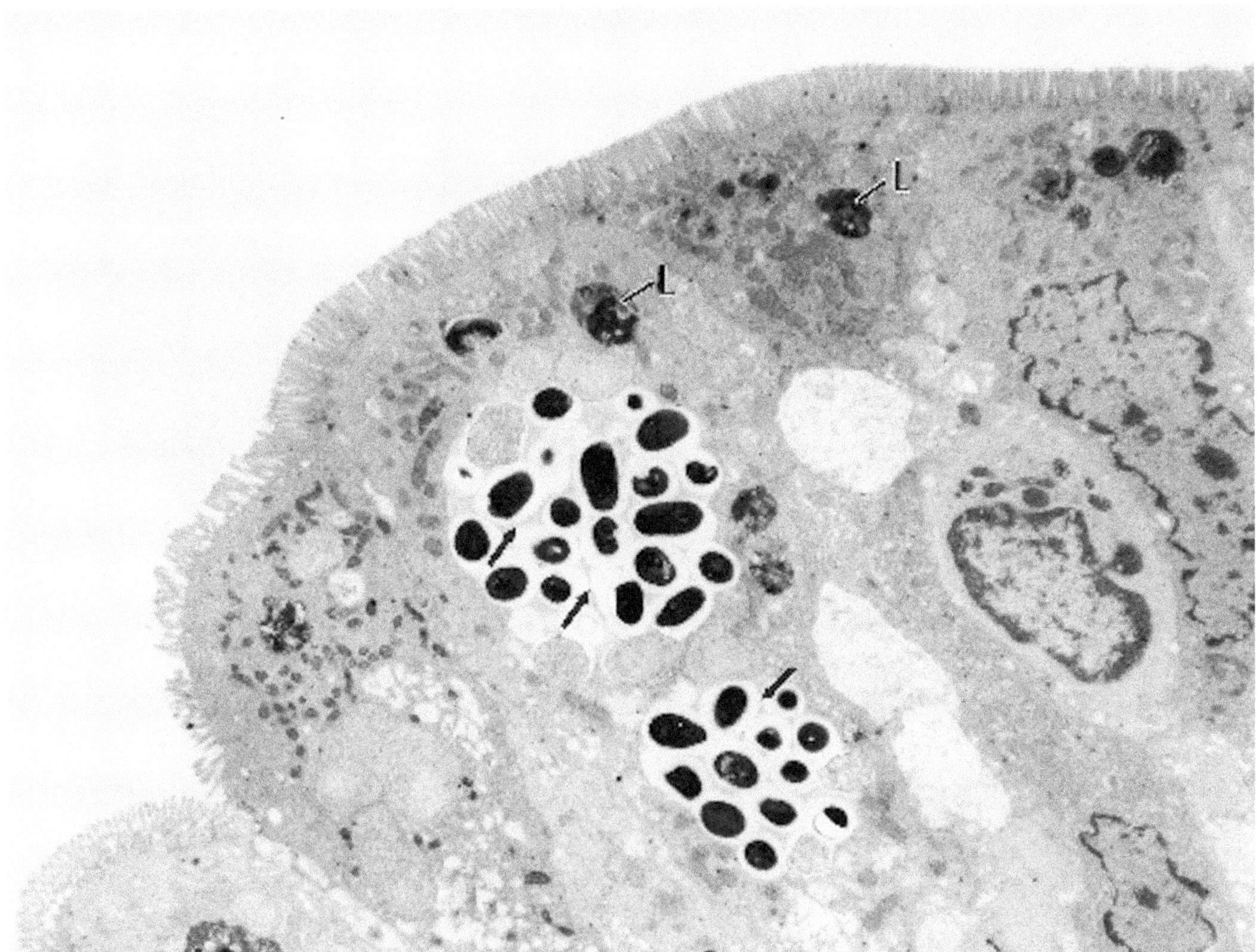

Figure 10.48. Microspora, septata intestinale (duodenum). Spores are grouped in parasitophorous vacuoles in the supranuclear cytoplasm of an enterocyte. Septae (*arrows*) are faintly visible surrounding individual spores. Secondary lysosomes (L) occupy the subluminal cytoplasm. (× 5800)

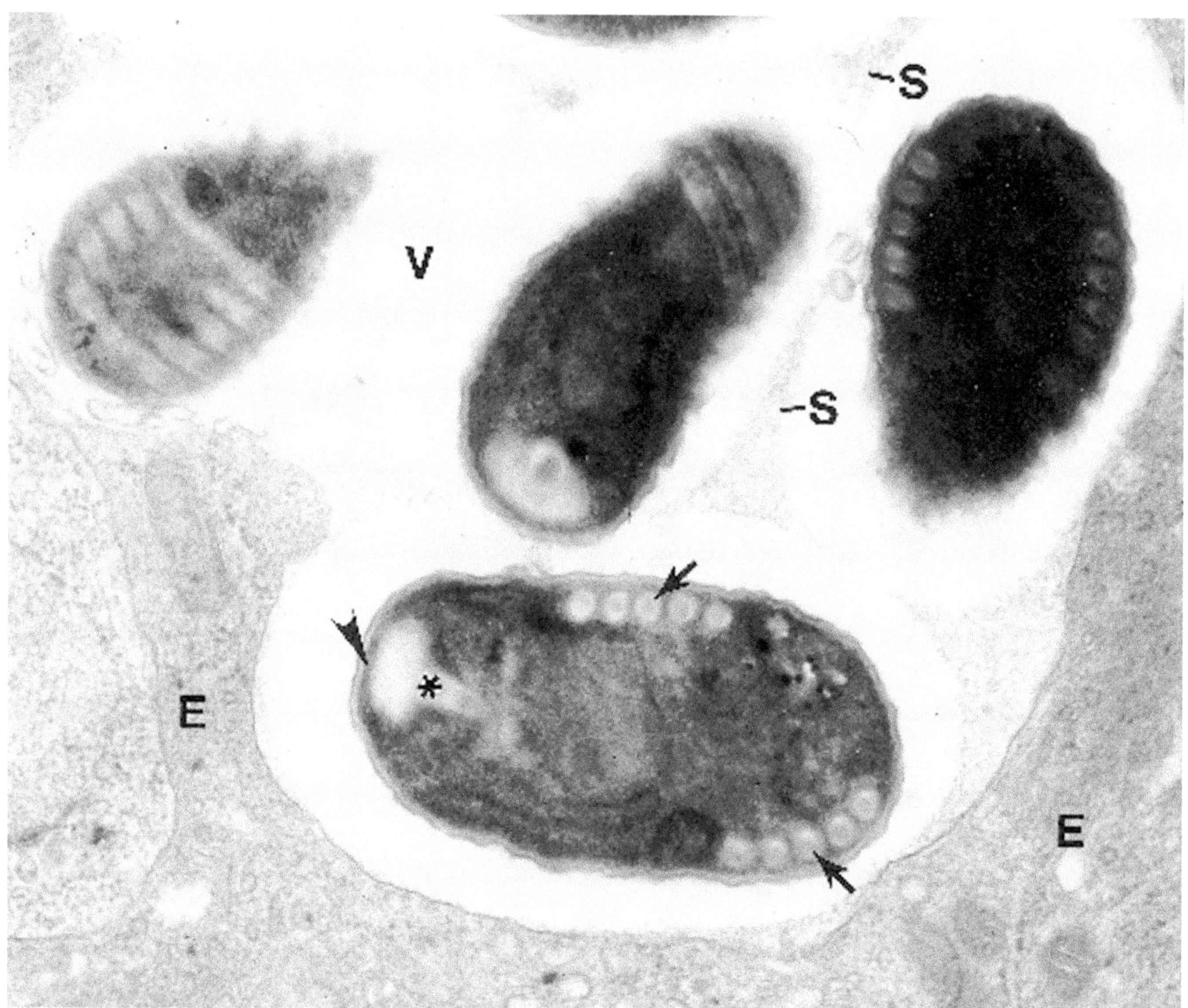

Figure 10.49. Microspora, septata intestinale (duodenum). High magnification of four spores depicts the double row of cross-sections of coiled polar tubule (*arrows*) in the posterior half of the organism. Note also the straight part of the polar tubule (*) and the anchoring disc (*arrowhead*), in the anterior aspect of the spores. S = septae, in parasitophorous vacuole (V); E = enterocyte cytoplasm. (× 50,000)

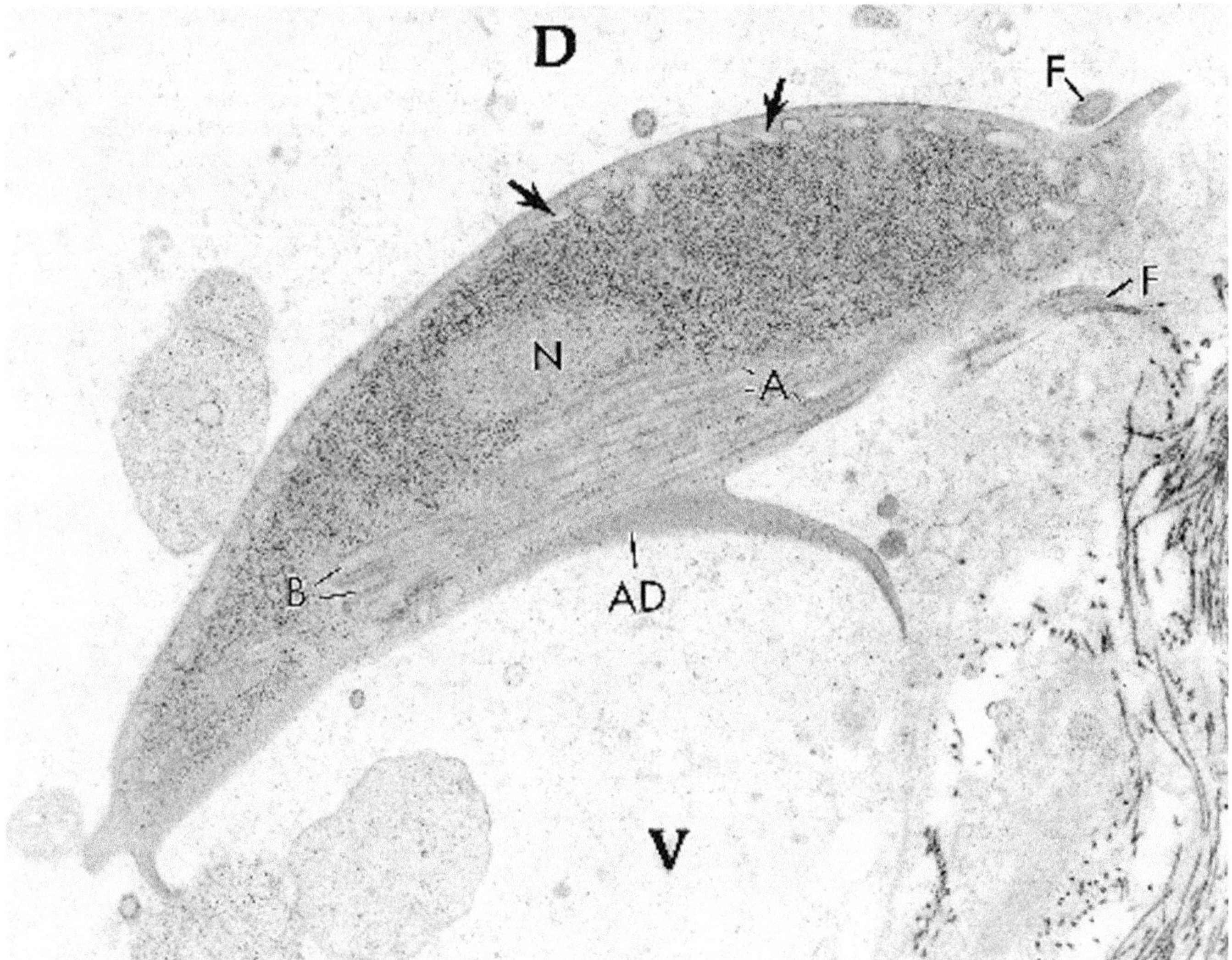

Figure 10.50. Giardia lamblia (tissue culture). An exemplary trophozoite is convex dorsally (D) and concave ventrally (V). Visible substructures in this plane of section include cytoplasmic vacuoles (*arrows*), basal bodies (B), axonemes (A) of basal bodies/flagella, flagella (F), ventral adhesion disc (AD), and nucleus (N). Most of the electron-dense cytoplasm is composed of ribosomes and granules of glycogen. (× 19,600) (Epon blocks generously contributed by William Taylor, M.D., and Lynne A. Farr, B.A., Miriam Hospital, Providence, Rhode Island.)

(Text continued from page 694)

Fungi

Histoplasma capsulatum

(Figures 10.51 through 10.53.)

Diagnostic criteria. (1) Oval yeast forms, 2–4 μm in diameter; (2) thin cell wall and no true capsule (see additional points in next paragraph); (3) a single nucleus; (4) extracellular and intracellular (within histiocytes) location of organisms.

Additional points. The capsule seen by light microscopy is an artefact due to shrinkage. When examined at the ultrastructural level, the organisms have a clear halo between their visible cytoplasm and their thin cell wall (Figures 10.51 through 10.53).

(Text continues on page 705)

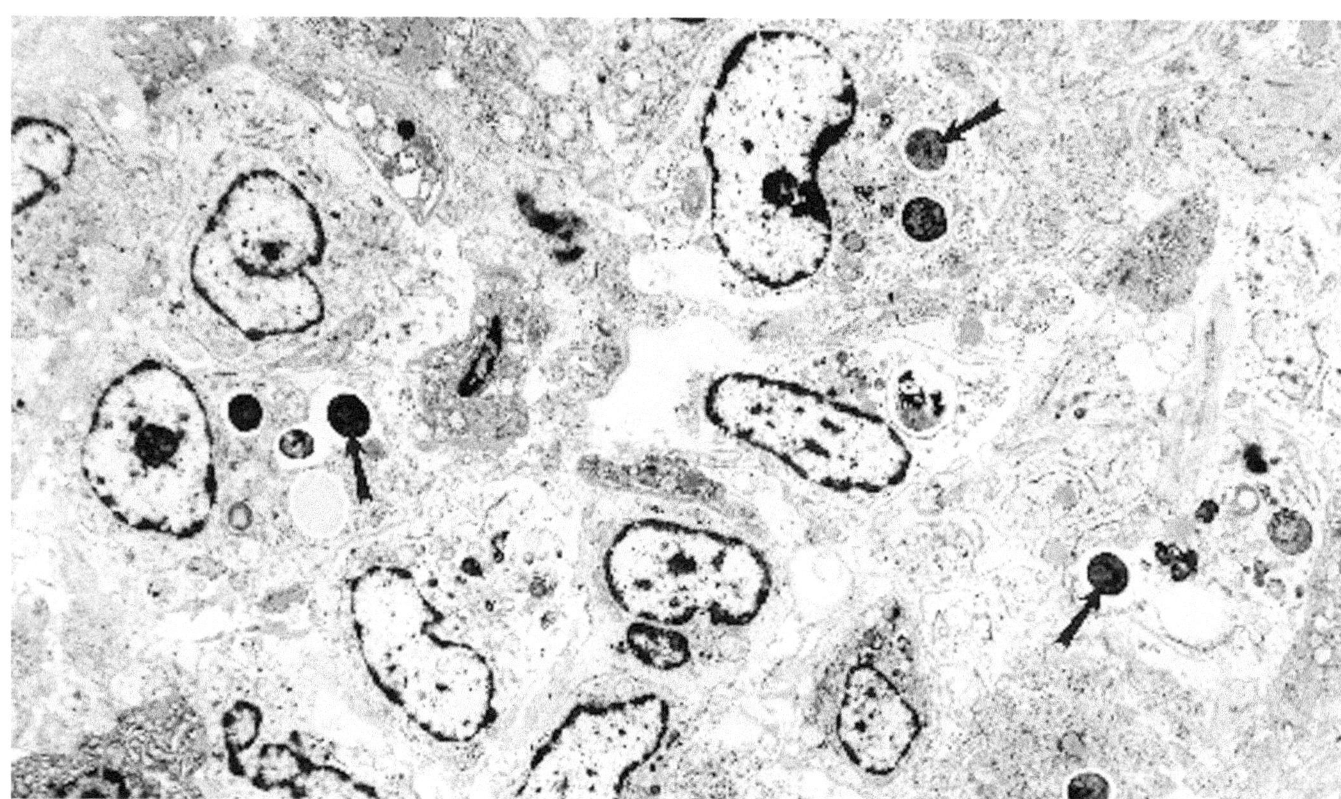

Figure 10.51. *Histoplasma capsulatum* (supraclavicular lymph node). Electron-dense, oval, and round yeast forms (*arrows*) are visible in the cytoplasm of histiocytes. Clear halos appear to surround the organisms at this magnification, but Figures 10.52 and 10.53 prove that the halos are actually cleared cytoplasm of the parasite, surrounded by its cell membrane. (× 3600)

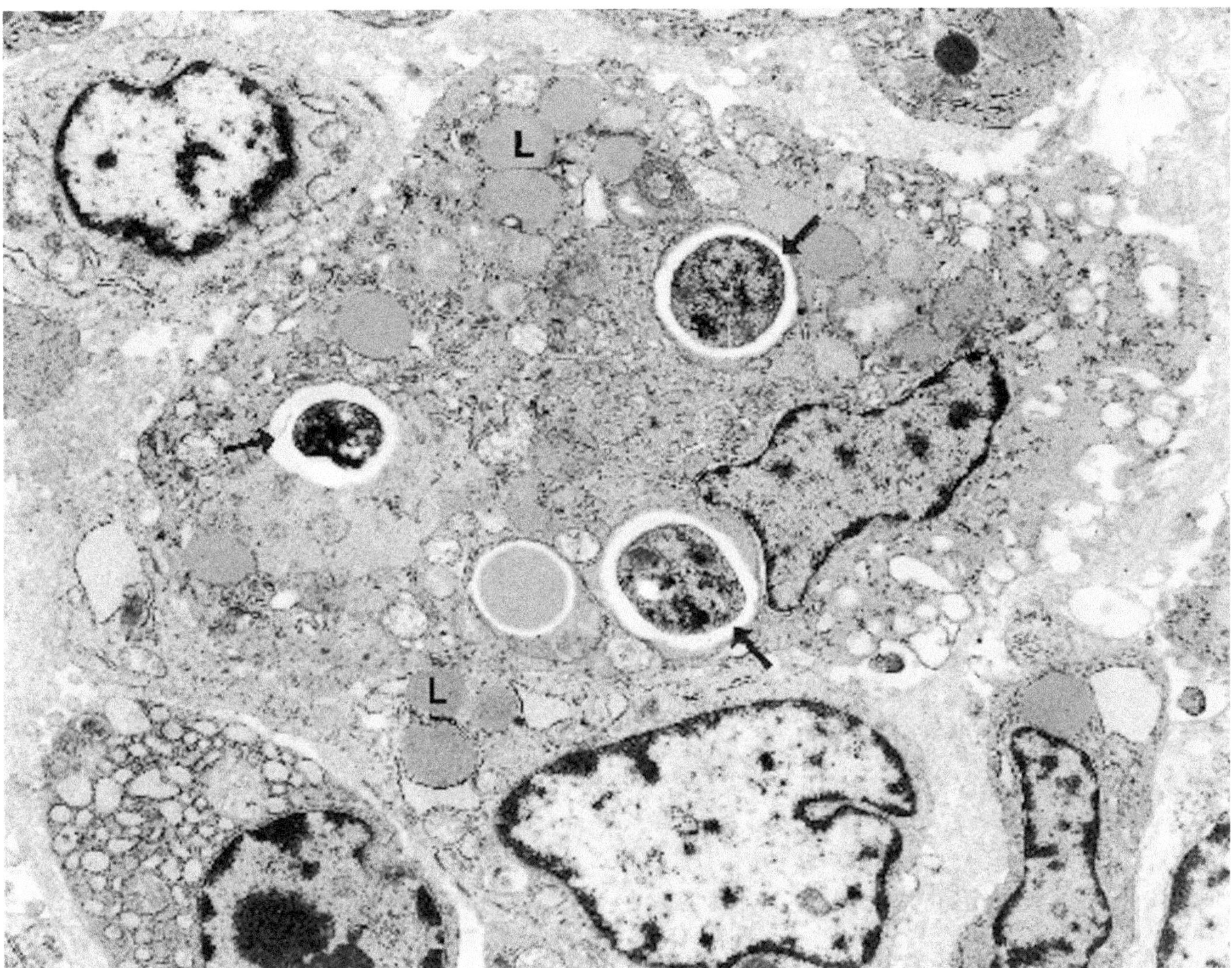

Figure 10.52. *Histoplasma capsulatum* (supraclavicular lymph node). This histiocyte has copious cytoplasm, which contains numerous organelles, a moderate number of lipid droplets (L), and three yeast forms of *H. capsulatum* (*arrows*) with clear peripheral cytoplasm. (× 8800)

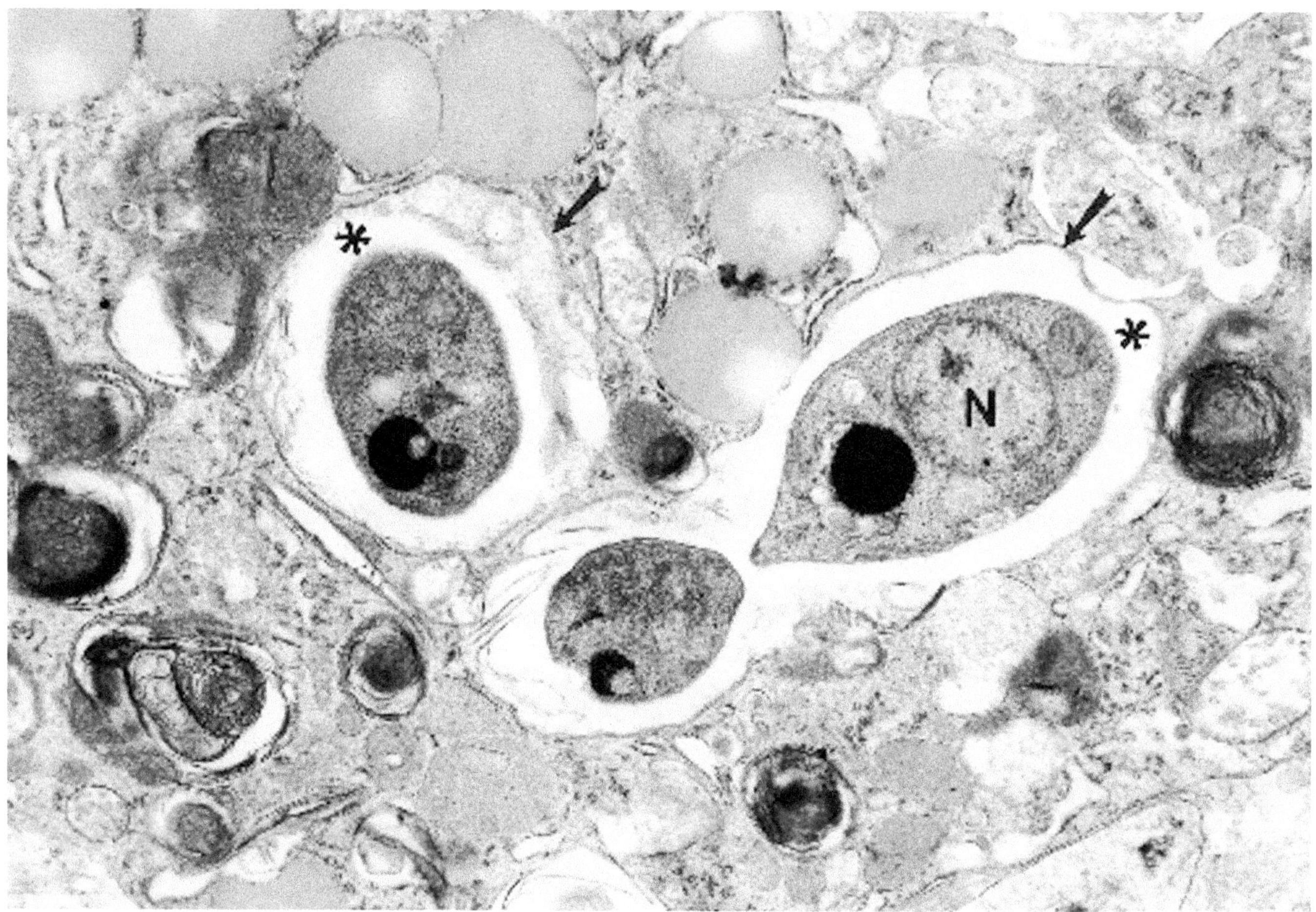

Figure 10.53. *Histoplasma capsulatum* (supraclavicular lymph node). High magnification of parasitic yeast forms illustrates details of their internal structure. N = nucleus; * = clear, peripheral, cytoplasmic halo; *arrows* = parasitic cell membrane. (× 20,000)

(Text continued from page 702)

REFERENCES

Bacteria: General

Cali A: General microsporidian features and recent findings on AIDS isolates. J Protozool 38:625–630, 1991.

Canning EU, Hollister WS: Microsporidia of mammals—widespread pathogens or opportunistic curiosities? Parasitol Today 3:267–276, 1987.

Payne CM: Electron microscopy in the diagnosis of infectious diseases. In Connor DH, Chandler FW, Schwartz DA, et al., eds: *Pathology of Infectious Diseases,* vol 1. Appleton and Lange, Stamford, CT, 1997, pp 9–34.

Weber R, Bryan RT, Schwartz DA, et al: Human microsporidial infections. Clin Microbiol Rev 7:426–461, 1994.

Bacteria: *Legionella pneumophila*

Katz SM, Bodsky I, Kahn B: Legionnaire's disease. Ultrastructural appearance of the agent in a lung biopsy specimen. Arch Pathol Lab Med 103:261–264, 1979.

Katz SM, Hashemi S, Brown KR, et al: Pleomorphism of Legionella pneumophila. Ultrastruct Pathol 6: 117–129, 1984.

White HJ, Sun CN, Hui AN: An ultrastructural demonstration of the agent of Legionnaire's disease in human lung. Hum Pathol 10:96–99, 1979.

Bacteria: Whipple's Disease

Case Records of the Massachusetts General Hospital (Case 37-1997): Whipple's disease involving the small intestine and mesenteric and mediastinal lymph nodes. N Engl J Med 337:1612–1619, 1997.

Cohen AS, Schimel EM, Hold PR, et al: Ultrastructural abnormalities in Whipple's disease. Proc Soc Exp Biol Med 105:411–414, 1960.

Chears WC, Ashworth CT: Electron microscopic study of the intestinal mucosa in Whipple's disease. Demonstration of encapsulated bacilliform bodies in the lesion. Gastroenterology 41:129–138, 1961.

Dobbins WO, Kawanishi H: Bacillary characteristics in Whipple's disease: An electron microscopic study. Gastroenterology 80:1468–1475, 1981.

Johnson L, Diamond I: Cerebral Whipple's disease. Diagnosis by brain biopsy. Am Soc Clin Pathol 74: 486–490, 1980.

Haubrich WS, Watson JHL, Sieracki J: Unique morphologic features of Whipple's disease. A study by light and electron microscopy. Gastroenterology 39:454–468, 1960.

Lie JT, Davis JS: Pancarditis in Whipple's disease. Electron microscopic demonstration of intracardiac bacillary bodies. Am J Clin Pathol 66:22–30, 1976.

McAllister HA, Fenoglio JJ: Cardiac involvement in Whipple's disease. Circulation 52:152–156, 1975.

Relman DA, Schmidt TM, MacDermott RP, et al: Identification of the uncultured bacillus of Whipple's disease. N Engl J Med 327:293–301, 1992.

Roberts DM, Themann H, Knust FJ, et al: An electron microscopic study of bacteria in two cases of Whipple's disease. J Pathol 100:249–255, 1970.

Silbert SW, Parker E, Horenstein S: Whipple's disease of the central nervous system. Acta Neuropathol (Berl) 36:31–38, 1976.

Silva MT, Macedo PM, Nunes JFM: Ultrastructure of bacilli and the bacillary origin of the macrophagic inclusions in Whipple's disease. J Gen Microbiol 131: 1001–1013, 1985.

Southern JF, Moscicki RA, Magro C, et al: Lymphedema, lymphocytic myocarditis, and sarcoid-like granulomatosis. Manifestations of Whipple's disease. JAMA 261:1467–1470, 1989.

Watson JHL, Haubrich WS: Bacilli bodies in the lumen and epithelium of the jejunum in Whipple's disease. Lab Invest 21:347–357, 1969.

Wilcox GM, Tronic BS, Schecter DJ, et al: Periodic acid–Schiff-negative granulomatous lymphadenopathy in patients with Whipple's disease. Localization of the Whipple bacillus to noncaseating granulomas by electron microscopy. Am J Med 83:165–170, 1987.

Yardley JH, Hendrix TR: Combined electron and light microscopy in Whipple's disease. Demonstration of "bacillary bodies" in the intestine. Bull John Hopkins Hosp 109:80–98, 1961.

Viruses: General

Almeida JD, Waterson AP: Viruses. In Johannessen JV, ed: *Electron Microscopy in Human Medicine,* vol 3, Infectious Agents. McGraw-Hill, New York, 1980, pp 3–45.

Cheville NF: Cytopathology in viral diseases. In Melnick JL, ed: Monographs in Virology, vol 10. S Karger, Basel, 1975, pp 1–235.

Couvreur Y, Leeman M, Ketelbant P: "Viral" intranuclear inclusions. Ultrastruct Pathol 7:67–71, 1984.

Dalton AH, Haguenau F: *Ultrastructure of Animal Viruses and Bacteriophages*. An Atlas. Academic Press, New York, 1973.

Doane FW, Anderson N: *Electron Microscopy in Diagnostic Virology.* A Practical Guide and Atlas. Cambridge University Press, Cambridge, MA, 1987, pp 1–178.

Finkelstein SD: Polyomaviruses and PML. In Connor DH, Chandler FW, Schwartz DA, et al, eds: *Pathology of Infectious Diseases,* vol 1. Appleton and Lange, Stamford, CT, 1997, pp 265–272.

Greenlee JE: Polyomavirus. In Richman DD, Whitley RJ, Hayden FG, eds: *Clinical Virology.* Churchill Livingstone, New York, 1997, pp 549–567.

Greenlee JE: Progressive multifocal leukoencephalopathy—progress made and lessons relearned [editorial]. N Engl J Med 338:1378–1380, 1998.

Grimley PM, Henson DE: Electron microscopy in virus infections. In Trump BF, Jones RT, eds: *Diagnostic Electron Microscopy,* vol 4. John Wiley, New York, 1983, pp 1–73.

Hsiung GD: *Diagnostic Virology,* 3rd ed. Yale University Press, New Haven, CT, 1982.

Madeley CR: *Virus Morphology.* Williams and Wilkins, Baltimore, 1972.

Mirra SS, Takei Y: Ultrastructural identification of virus in human central nervous system disease. In Zimmerman HM, ed: *Progress in Neuropathology.* Grune and Stratton, New York, 1976, pp 69–88.

Nesland JM, Kjellevoid K, Finseth I, et al: Spherical bodies. Ultrastruct Pathol 7:73–78, 1984.

Sobel HJ, Marquet E, Schwarz R: Extracellular virus-like particles. Ultrastruct Pathol 5:93–98, 1983.

Sobel HJ, Marquet E, Seely JC: Virus-like particles in hamster liver. Ultrastruct Pathol 5:257–259, 1983.

Williams RC, Fisher HW: *An Electron Micrographic Atlas of Viruses.* Charles C Thomas, Springfield, IL, 1974.

Wolinskey JS: Viral disease. In Johannessen JV, ed: *Electron Microscopy in Human Medicine,* vol 6. McGraw-Hill, New York, 1979, pp 54–84.

Viruses: Herpes Group

Baringer JR: Tubular aggregates in endoplasmic reticulum in herpes simplex encephalitis. N Engl J Med 285:943–945, 1971.

Baringer JR, Griffith JF: Experimental herpes encephalitis: Crystalline arrays in endoplasmic reticulum. Science 165:1381, 1969.

Chou SM, Cherry JD: Ultrastructure of Chowdry type A inclusions. I. In human herpes simplex encephalitis. Neurology 17:575–586, 1967.

Davis GL, Hawrisiak MM: Experimental cytomegalovirus infection and the developing mouse inner ear. *In vivo* and *in vitro* studies. Lab Invest 37:20–29, 1977.

Fong CKY, Lucia H, Bia FJ, et al: Histopathologic and ultrastructural studies of disseminated cytomegalovirus infection in strain 2 guinea pigs. Lab Invest 49: 183–194, 1983.

Hashida Y, Yunis EJ: Re-examination of encephalitic brains known to contain intranuclear inclusion bodies: Electron microscopic observations following prolonged fixation in formalin. Am J Clin Pathol 53:537–543, 1970.

Itabashi HH, Bass DM, McCulloch JR: Inclusion body of acute inclusion encephalitis. Arch Neurol 14:493–505, 1966.

Kasnic G, Sayeed A, Azar HA: Nuclear and cytoplasmic inclusions in disseminated human cytomegalovirus infection. Ultrastruct Pathol 3:229–235, 1982.

Kelley GR, Ashizawa T, Gyorkey F: Herpes simplex virus encephalitis. A case with dysplastic plasma cell infiltration. Arch Pathol Lab Med 110:82–85, 1986.

Kristensson K, Enerback L, Sourander P: Histochemical and ultrastructural properties of the classical inclusion body in neurones infected with herpes simplex virus. Acta Pathol Microbiol Scand [A] 78:595–604, 1970.

Lee FK, Nahmias AJ, Stagno S: Rapid diagnosis of cytomegalovirus infection in infants by electron microscopy. N Engl J Med 299:1266–1270, 1978.

Lee FK, Takei Y, Dannenbarger J, et al: Rapid detection of viruses in biopsy or autopsy specimens by the pseudoreplica method of electron microscopy. Am J Surg Pathol 5:565–572, 1981.

McGinley DM, Posalaky Z: Microtubulorecticular structures in the cerebral vasculature of a patient with herpes simplex encephalitis. Ultrastruct Pathol 7:177–183, 1984.

Nakamura Y, Yamamoto S, Tanaka S, et al: Herpes simplex viral infection in human neonates: An immunohistochemical and electron microscopic study. Hum Pathol 16:1091–1097, 1985.

Norris FH: An ultrastructural study of herpes simplex virus encephalitis. J Neuropathol Exp Neurol 31: 611–623, 1972.

Watson DH: Morphology. In Kaplan AS, ed: *The Herpes Viruses*. Academic Press, New York, 1973, pp 27–43.

White CL, Taxy JB: Early morphologic diagnosis of herpes simplex virus encephalitis: Advantages of electron microscopy and immunoperoxidase staining. Hum Pathol 14:135–139, 1983.

Viruses: HIV

Gelderblom HR, Hausmann EHS, Ozel M, et al: Fine structure of human immunodeficiency virus (HIV) and immunolocalization of structural proteins. Virology 156:171–176, 1987.

Gelderblom HR, Ozel M, Hausmann EHS, et al: Fine structure of human immunodeficiency virus (HIV), immunolocalization of structural proteins and virus-cell relation. Micron Microsc 19:41–60, 1988.

Gelderblom HR, Ozel M, Pauli G: Morphogenesis and morphology of HIV structure-function relations. Brief review. Arch Virol 106:1–13, 1989.

Gonda M, Wong-Staal F, Gallo RC, et al: Sequence homology and morphologic similarity of HTLV-III and visna virus, a pathogenic lentivirus. Science 227: 173–177, 1985.

Marx PA, Munn RJ, Joy KI: Methods in laboratory investigation. Computer emulation of thin section electron microscopy predicts an envelope-associated icosadeltahedral capsid for human immunodeficiency virus. Lab Invest 58:112–118, 1988.

Munn RJ, Marx PA, Yamamoto JK, et al: Ultrastructural comparison of the retroviruses associated with human and simian acquired immunodeficiency syndrome. Lab Invest 53:194–199, 1985.

Nakai M, Goto T: Ultrastructure and morphogenesis of human immunodeficiency virus. J Electron Microsc 45: 247–257, 1996.

Orenstein JM: Ultrastructural pathology of human immunodeficiency virus infection. Ultrastruct Pathol 16:179–210, 1992.

Perotti ME, Tan X, Phillips DM: Directional budding of human immunodeficiency virus from monocytes. J Virol 70:5916–5921, 1996.

Tacchetti C, Favre A, Moresco L, et al: HIV is trapped and masked in the cytoplasm of lymph node follicular dendritic cells. Am J Pathol 150:533–542, 1997.

Takasaki T, Kurane I, Aihara H, et al: Electron microscopic study of human immunodeficiency virus type 1 (HIV-1) core structure: Two RNA strands in the core of mature and budding particles. Arch Virol 142:375–382, 1997.

Viruses: Papovirus

Hadfield MG, Martinez AJ, Gilmartin RC: Progressive multifocal leukoencephalopathy with Paramyxovirus-like structures, hirano bodies and neurofibrillary tangles. Acta Neuropathol (Berl) 27:277–288, 1974.

Ho K, Garancis JC, Paegle RD, et al: Progressive multifocal leukoencephalopathy and malignant lymphoma of the brain in a patient with immunosuppressive therapy. Acta Neuropathol (Berl) 52:81–83, 1980.

Jenson AB, Sommer S, Payling-Wright C, et al: Human papillomavirus. Frequency and distribution in plantar and common warts. Lab Invest 47:491–497, 1982.

Miller JR, Barrett RE, Britton CB, et al: Progressive multifocal leukoencephalopathy in a male homosexual with T-cell immune deficiency. N Engl J Med 307:1436–1438, 1982.

Rowson KEK, Mahy BWJ: Human papova (wart) virus. Bacteriol Rev 31:110–131, 1967.

Silverman L, Rubinstein LJ: Electron microscopic observations on a case of progressive multifocal leukoencephalopathy. Acta Neuropathol (Berl) 5:215–224, 1965.

Zu Rhein GM: Polyoma-like virions in a human demyelinating disease. Acta Neuropathol (Berl) 8:57–68, 1967.

Zu Rhein GM, Chou SM: Particles resembling papova viruses in human cerebral demyelinating disease. Science 148:1477–1479, 1965.

Protozoa: *Pneumocystis carinii*

Barton EG Jr, Campbell WG Jr: Further observations on the ultrastructure of *Pneumocystis*. Arch Pathol 83: 527–534, 1967.

Barton EG Jr, Campbell WG Jr: *Pneumocystis carinii* in lungs of rats treated with cortisone acetate: Ultrastructural observations relating to the life cycle. Am J Pathol 54:209–236, 1969.

Bedrossian CWM: Ultrastructure of *Pneumocystis carinii:* A review of internal and surface characteristics. Semin Diagn Pathol 6:212–237, 1989.

Campbell WG Jr: Ultrastructure of *Pneumocystis* in human lung: Life cycle in human pneumocystosis. Arch Pathol 93:312–324, 1972.

Case Records of the Massachusetts General Hospital (Case 46-1984): *Pneumocystis carinii* pneumonia. N Engl J Med 311:1303–1310, 1984.

Gutierrez Y: The biology of *Pneumocystis carinii.* Semin Diagn Pathol 6:203–211, 1989.

Ham EK, Greenberg DG, Reynolds RC, et al: Ultrastructure of *Pneumocystis carinii.* Exp Mol Pathol 14: 362–372, 1971.

Haque A, Pattner SB, Cook RT, et al: *Pneumocystis carinii.* Taxonomy as viewed by electron microscopy. Am J Clin Pathol 87:504–510, 1987.

Hasleton PS, Curry A, Rankin EM: *Pneumocystis carinii* pneumonia: A light microscopical and ultrastructural study. J Clin Pathol 34:1138–1146, 1981.

Huang SN, Marshall KG: *Pneumocystis carinii* infection: A cytologic, histologic and electron microscopic study of the organism. Am Rev Dis 102:623–635, 1970.

Proce RA, Hughes WT: Histopathology of *Pneumocystis carinii* infestation and infection in malignant disease in childhood. Hum Pathol 5:737–752, 1974.

Watts JC, Chandler FW: *Pneumocystis carinii* pneumonitis. The nature and diagnostic significance of the methenamine silver-positive "intracystic bodies." Am J Surg Pathol 9:744–751, 1985.

Watts JC, Chandler FW: Pneumocystosis. In Connor DH, Chandler FW, Schwartz DA, et al, eds: *Pathology of Infectious Diseases,* vol 2. Appleton and Lange, Stamford, CT, 1997, pp 1241–1252.

Yoneda K, Walzer PD: Interaction of *Pneumocystis carinii* with host lungs: An ultrastructural study. Infect Immun 29:692–703, 1980.

Protozoa: *Toxoplasma gondii*

Sheffield HG, Melton ML: The fine structure and reproduction of *Toxoplasma gondii.* J Parasitol 54:209–226, 1968.

Cerezo L, Alvarez M, Proce G: Electron microscopic diagnosis of cerebral toxoplasmosis: Case report. J Neurosurg 63:470–472, 1985.

Chobotar B, Scholtyseck E: Ultrastructure. In Long PL, ed. *The Biology of the Coccidia.* Baltimore, MD, University Park Press, 1982, pp 101–165.

Frenkel JK: Toxoplasmosis. In Connor DH, Chandler FW, Schwartz DA, et al, eds: *Pathology of Infectious Diseases,* vol 2. Appleton and Lange, Stamford, CT, 1997, pp 1261–1278.

Gavin MA, Wanko T, Jacobs L: Electron microscope studies of reproducing and interkinetic toxoplasma. J Protozool 9:222–234, 1962.

Ghatak NR, Zimmerman HM: Fine structure of toxoplasma in human brain. Arch Pathol Lab Med 95: 276–283, 1973.

Powell HC, Gibbs CJ, Lorenzo AM, et al: Toxoplasmosis of the central nervous system in the adult. Electron microscopic observation. Acta Neuropathol (Berl) 4: 211–216, 1978.

Scholtyseck E. Ultrastructure. In Hammond DM, Long PD, eds: *The Coccidia. Eineria, Isopora, Toxoplasma and Related Genera.* Baltimore, MD, University Park Press, 1973, pp 81–144.

Protozoa: *Cryptosporidium parvum*

Bird RG, Smith MD: Cryptosporidiosis in man: Parasite life cycle and fine structural pathology. J Pathol 132: 217–233, 1980.

Blumberg RS, Kelsey P, Perrone T, et al: Cytomegalovirus- and cryptosporidium-associated acalculous gangrenous cholecystitis. Am J Med 76:1118–1123, 1984.

Orenstein JM: Cryptosporidiosis. In Connor DH, Chandler FW, Schwartz DA, et al, eds: *Pathology of Infectious Diseases,* vol 2. Appleton and Lange, Stamford, CT, 1997, pp 1147–1158.

Perrone TL, Dickersin GR: The intracellular location of cryptosporidia. Hum Pathol 14:1092–1093, 1983.

Ribeiro HC, Teichberg S, Sun T, et al: Ultrastructure of human cryptosporidial infection: A review. In Sun T, ed: *Prog Clin Parasitol.* New York, Field & Wood, 1:143–158, 1989.

Vetterling JM, Takeuchi A, Madden PA: Ultrastructure of *Cryptosporidium wrairi* from the guinea pig. J Protozool 18:248–260, 1971.

Protozoa: *Trypanosoma cruzi*

Jadin JM, Jadin JB: Protozoa. In Johannessen JV, ed: *Electron Microscopy in Human Medicine,* vol 3, Infectious Agents. McGraw-Hill, New York, 1980, pp 292–308.

Lack EE, Filie A: American trypanosomiasis. In Connor DH, Chandler FW, Schwartz DA, et al, eds: *Pathology of Infectious Diseases,* vol 2. Appleton and Lange, Stamford, CT, 1997, pp 1297–1304.

Steiger RF: On the ultrastructure of *Trypanosoma (Trypanozoon) brucei* in the course of its life cycle and some related aspects. Acta Trop 30:64–168, 1973.

Vickerman K: The fine structure of *Trypanosoma congolense* in its bloodstream phase. J Protozool 16:54–69, 1969.

Wright KA, Lumsden WHR: The fine structure of *Trypanosoma (Trypanozoon) brucei*. Trans R Soc Trop Med Hyg 63:13–14, 1964.

Microsporidia

Bryan RT, Cali A, Owen RL, et al: Microsporidia: Opportunistic pathogens in patients with AIDS. In Sun T, ed: *Progress in Clinical Parasitology*, 2. Philadelphia, PA, Field & Wood, 1990, pp 1–26.

Cali A, Kotler DP, Orenstein JM: *Septata intestinalis* N.G., N. Sp., and intestinal microsporidian associated with chronic diarrhea and dissemination and AIDS patients. J Euk Microbiol 40:101–112, 1993.

Canning EU, Hollister WS: Royal Society Tropical Medicine and Hygiene Meeting at Manson House, London, 16 March 1989. New intestinal protozoa—coccidia and microsporidia. *Enterocytozoon bieneusi* (microspora): Prevalence and pathogenicity in AIDS patients. Trans R Soc Trop Med Hyg 84:181–186, 1990.

Case Records of the Massachusetts General Hospital (Case 51-1993): Intestinal and renal microsporidiosis due to *Septata intestinalis*. N Engl J Med 329:1946–154, 1993.

Gan R: Case no. 5-6695-699, Microsporidia. Microbiol Ref Lab, Investigator, Spring 1992, April.

Orenstein JM, Chiang J, Steinberg W, et al: Intestinal microsporidiosis as a cause of diarrhea in human immunodeficiency virus-infected patients. A report of 20 cases. Hum Pathol 21:475–481, 1990.

Orenstein JM, Tenner M, Cali A, et al: A microsporidian previously undescribed in humans, infecting enterocytes and macrophages, and associated with diarrhea in an acquired immunodeficiency syndrome patient. Hum Pathol 23:722–728, 1992.

Peacock CS, Blanshard C, Tovey DG, et al: Histological diagnosis of intestinal microsporidiosis in patients with AIDS. J Clin Pathol 44:558–563, 1991.

Pol S, Romana CA, Richard S, et al: Microsporidia infections in patients with the human immunodeficiency virus and unexplained cholangitis. N Engl J Med 328: 95–99, 1993.

Schwartz DA, Visvesvara GS, Leitch GJ, et al: Pathology of symptomatic microsporidial (*Encephalitozooan hellem*) bronchiolitis in the acquired immunodeficiency syndrome. A new respiratory pathogen diagnosed from lung biopsy, bronchoalveolar lavage, sputum, and tissue culture. Hum Pathol 24:937–943, 1993.

Histoplasma Capsulatum

Chandler FW, Watts JC: Histoplasmosis capsulati. In Connor DH, Chandler FW, Schwartz DA, et al, eds: *Pathology of Infectious Diseases*, vol 2. Appleton and Lange, Stamford, CT, 1997, pp 1007–1016.

Garrison RG, Lane JW, Field MF: Ultrastructural changes during the yeastlike to mycelial-phase conversion of *Blastomyces dermatitidis* and *Histoplasma capsulatum*. J Bacteriol 101:628–635, 1970.

Garrison RG, Lane JW, Johnson DR: Electron microscopy of the transitional conversion cell of *Histoplasma capsulatum*. Mycopathologia Mycologia Applicata 44:121–129, 1971.

11

Genetic and Metabolic Diseases

Storage Diseases

(Figures 11.1 through 11.27.)

Diagnostic criteria. (1) Cytoplasmic accumulations of electron-dense material in parenchymal and/or phagocytic cells; (2) usually a single membrane surrounds the electron-dense material (secondary lysosomes); (3) varying types and configurations of the electron-dense material, depending on subtype of disease.

Additional points. This group of genetically transmitted, autosomal recessive diseases includes several dozen disorders. Most are lysosomal and result from a particular lysosomal enzyme being absent or deficient. Without the enzyme, normally engulfed endogenous and exogenous material cannot be hydrolyzed, and the specific substrate accumulates in secondary lysosomes. The lysosomal overload may lead to cellular injury and malfunction and ultimately to cell death. Parenchymal cells so affected may spill their contents into the extracellular space, where phagocytic cells ingest them. To a lesser extent, other cells, such as endothelial cells, fibroblasts, and Schwann cells, also may be involved in the disease process. Various organs may be affected in this group of diseases, depending on their particular metabolic function and need for the missing enzyme. The ultrastructural common denominator in all the subtypes of lysosomal storage diseases is an increase in the quantity and size of secondary lysosomes, and the lysosomes contain accumulated substrate. The appearance of the substrate in the lysosomes often is specific for one disease or for several diseases of one subtype. Subtypes include *sphingolipidosis* (*gangliosidosis,* as in *Tay-Sachs disease*) (Figures 11.1 through 11.3); *sulfatidosis* (as in *metachromatic leukodystrophy*), *galactosidosis* (as in *Krabbe's disease* and *Fabry's disease*) (Figures 11.4 through 11.6); *glucosidosis* (as in *Gaucher's disease*) (Figures 11.7 and 11.8); *sphingomyelinosis* (as in *Niemann-Pick disease*) (Figures 11.9 and 11.10); *mucopolysaccharidosis* (as in *Hurler's syndrome, Hunter's syndrome,* and *Maroteaux-Lamy syndrome*) (Figures 11.11 and 11.12); *Morquio's syndrome* (Figures 11.13 and 11.14); *Sanfilippo's syndrome* (Figures 11.15 and 11.16); *hyalouronidase deficiency* (Figures 11.17 and 11.18); *mucolipidosis; neuronal ceroid lipofuscinosis* (Figures 11.19 and 11.20); *cholesterol ester storage disease* (Figures 11.21 and 11.22); *cerebrotendinous xanthomatosis* (a *lipid storage disease*) (Figure 11.23); *mannosidosis* (Figure 11.24); *adrenoleukodystrophy* (long-chain fatty acid accumulation secondary to impaired peroxisomal beta oxidation) (Figure 11.25); *glycogenosis, type II,* or *Pompe's disease* (the other glycogenoses have the glycogen accumulate in the cytoplasm proper, rather than in secondary lysosomes) (Figures 11.26 and 11.27).

(Text continues on page 738)

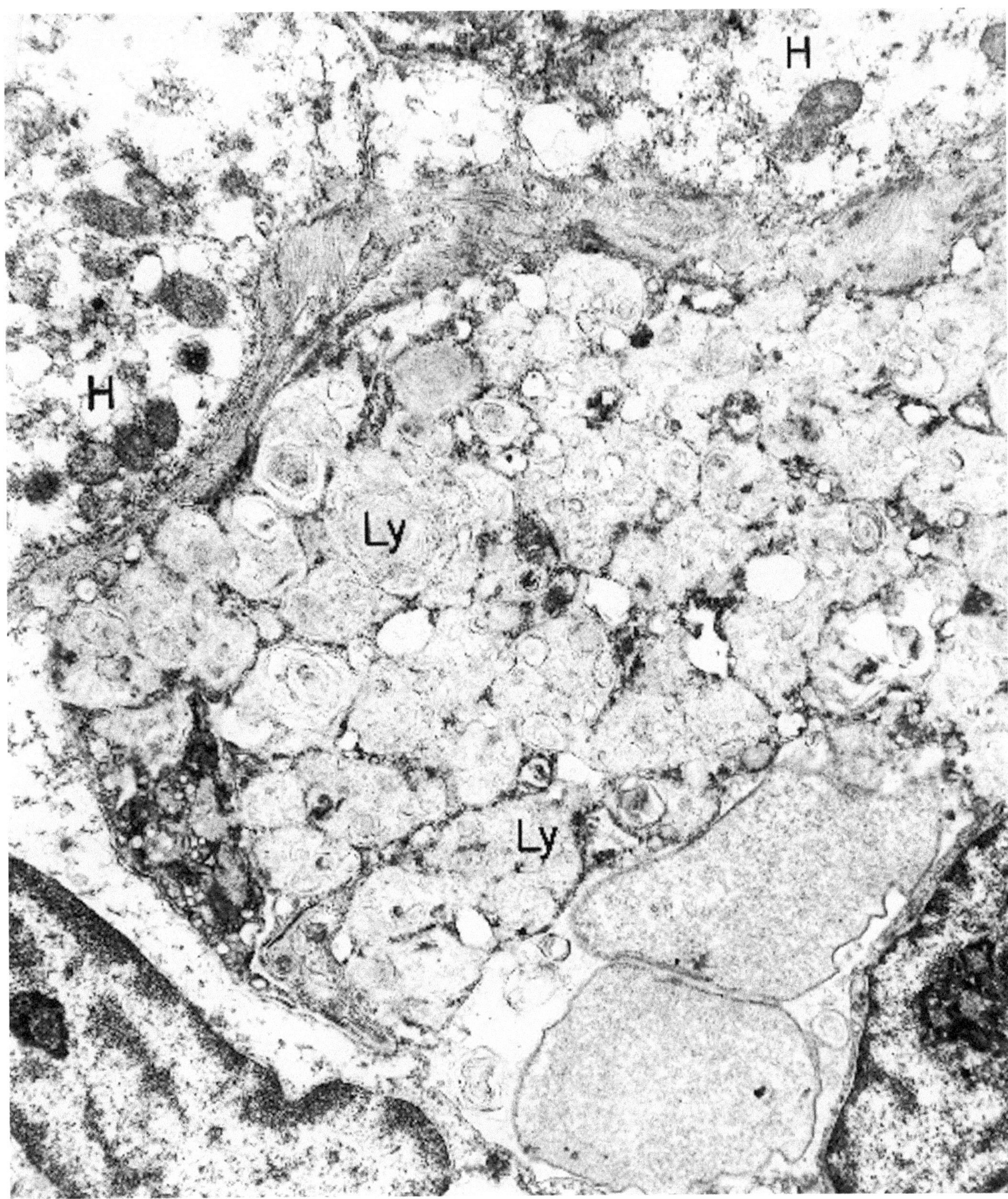

Figure 11.1. Tay-Sachs disease (liver). A von Kupffer cell has cytoplasm filled with distended and confluent secondary lysosomes (Ly) that contain concentric (ganglioside-rich) membranes and other dense osmiophilic material. Portions of two adjacent hepatocytes (H) show less involvement. (× 13,000)

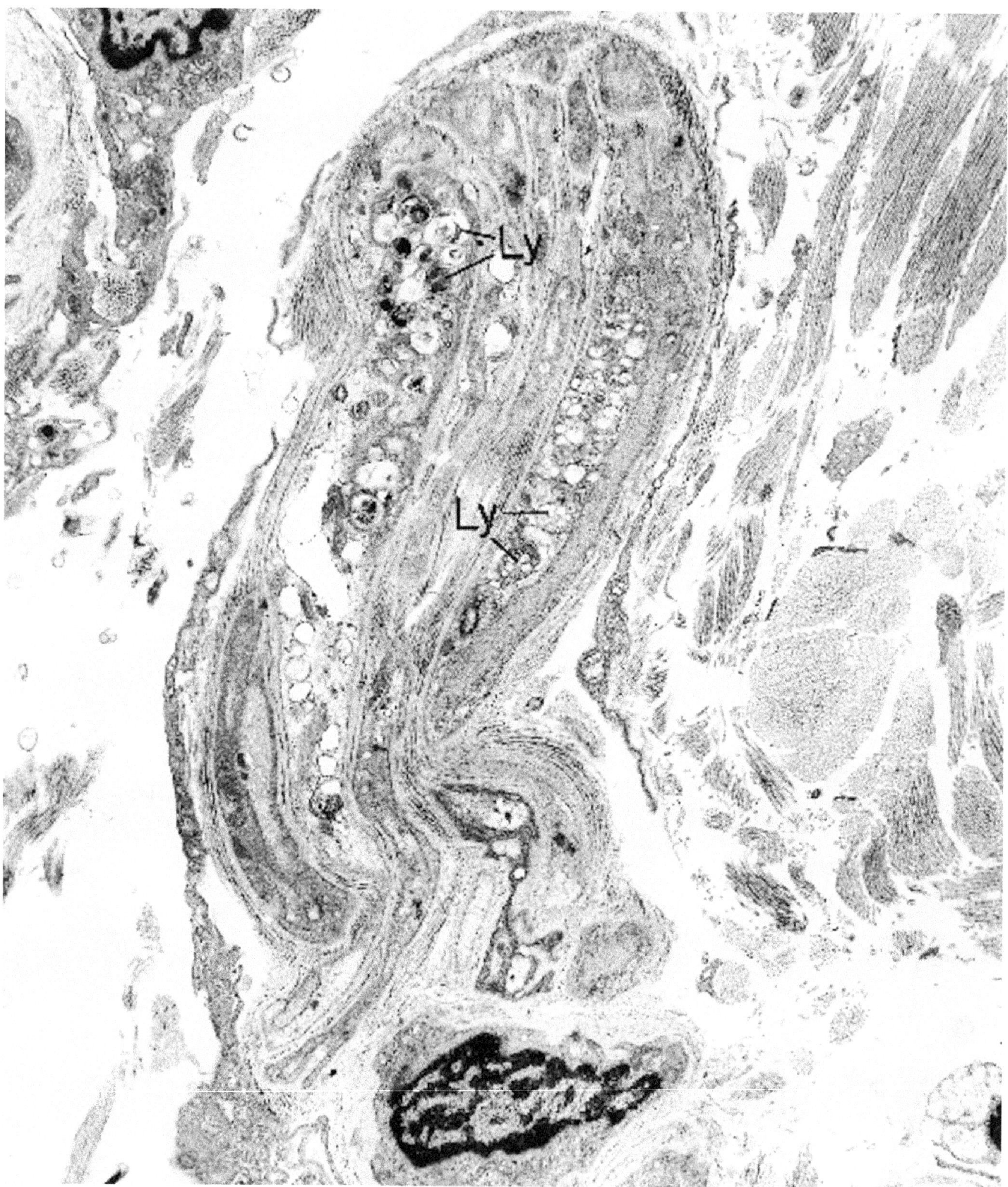

Figure 11.2. Tay-Sachs disease (cutaneous nerve). This specimen is from the same patient whose liver was illustrated in Figure 11.1, and it shows a dermal nerve having Schwann cells packed with electron-dense membranous bodies (Ly). (× 10,260)

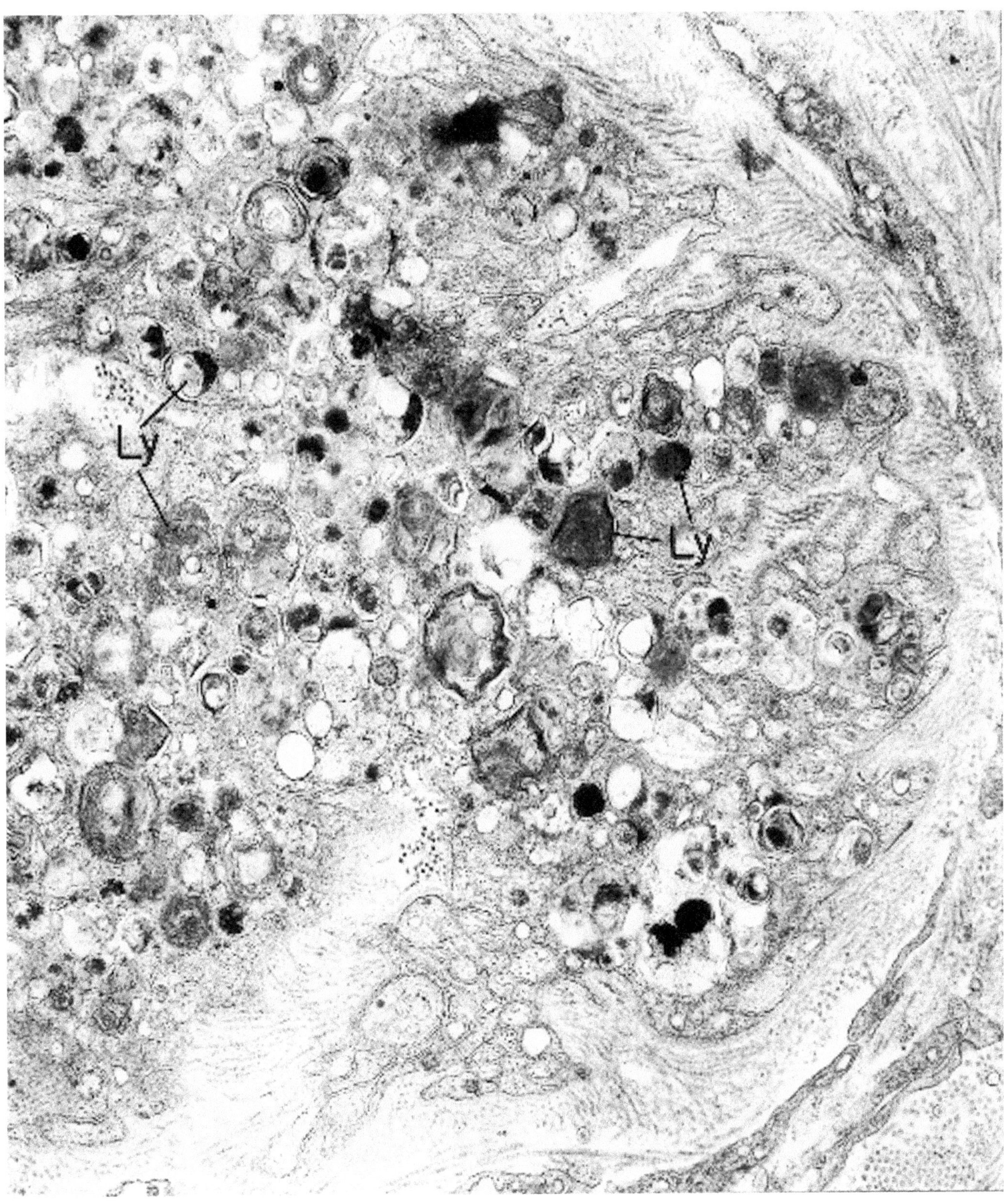

Figure 11.3. Tay-Sachs disease (appendix). This specimen is from the same patient whose tissues were depicted in Figures 11.1 and 11.2. Autonomic nerve and ganglia have Schwann cells filled with intralysosomal osmiophilic material (Ly). (× 20,000)

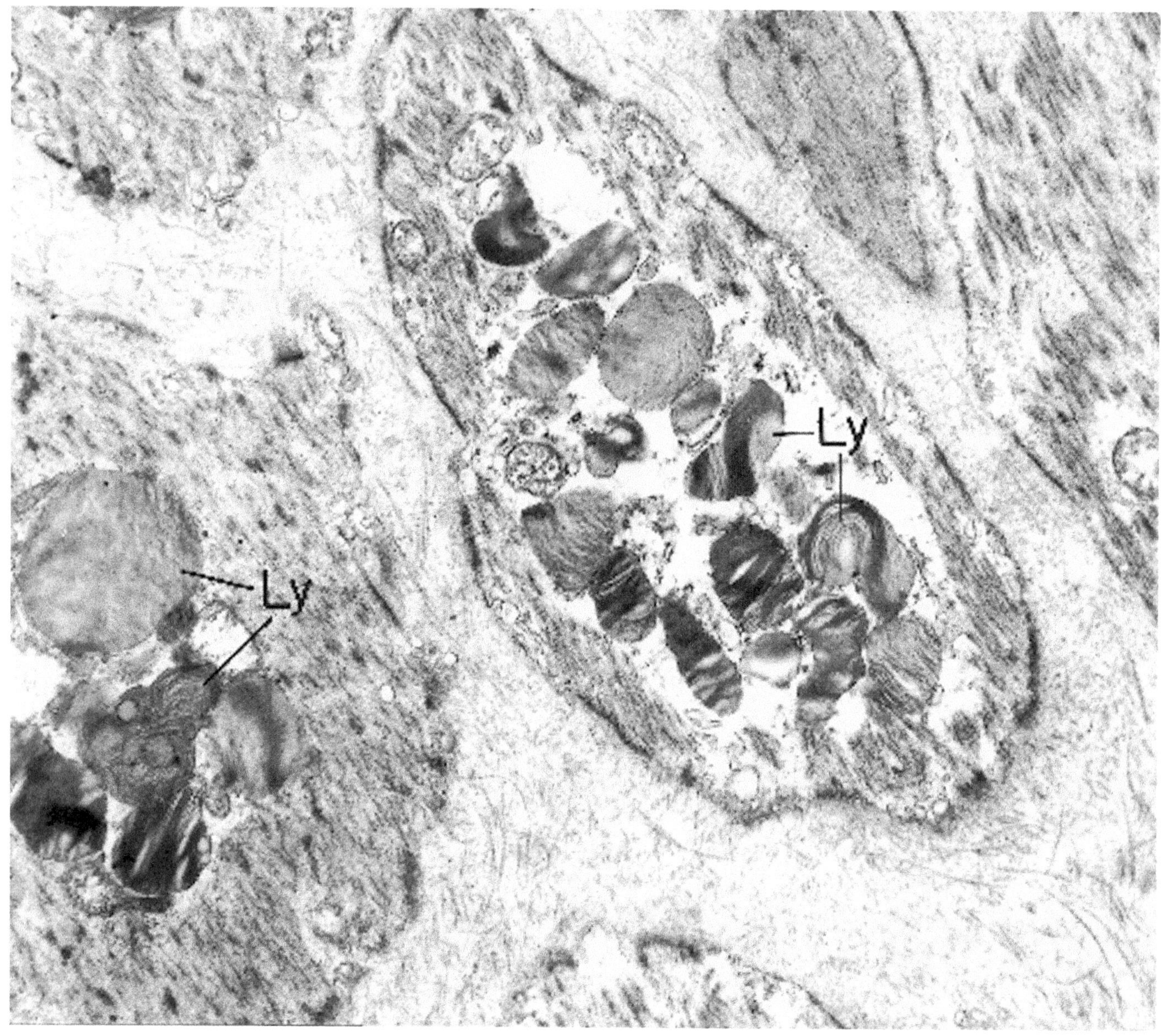

Figure 11.4. Fabry's disease (jejunum). Smooth muscle cells of this bowel wall contain many electron-dense, lamellar and solid bodies (Ly). (× 5720)

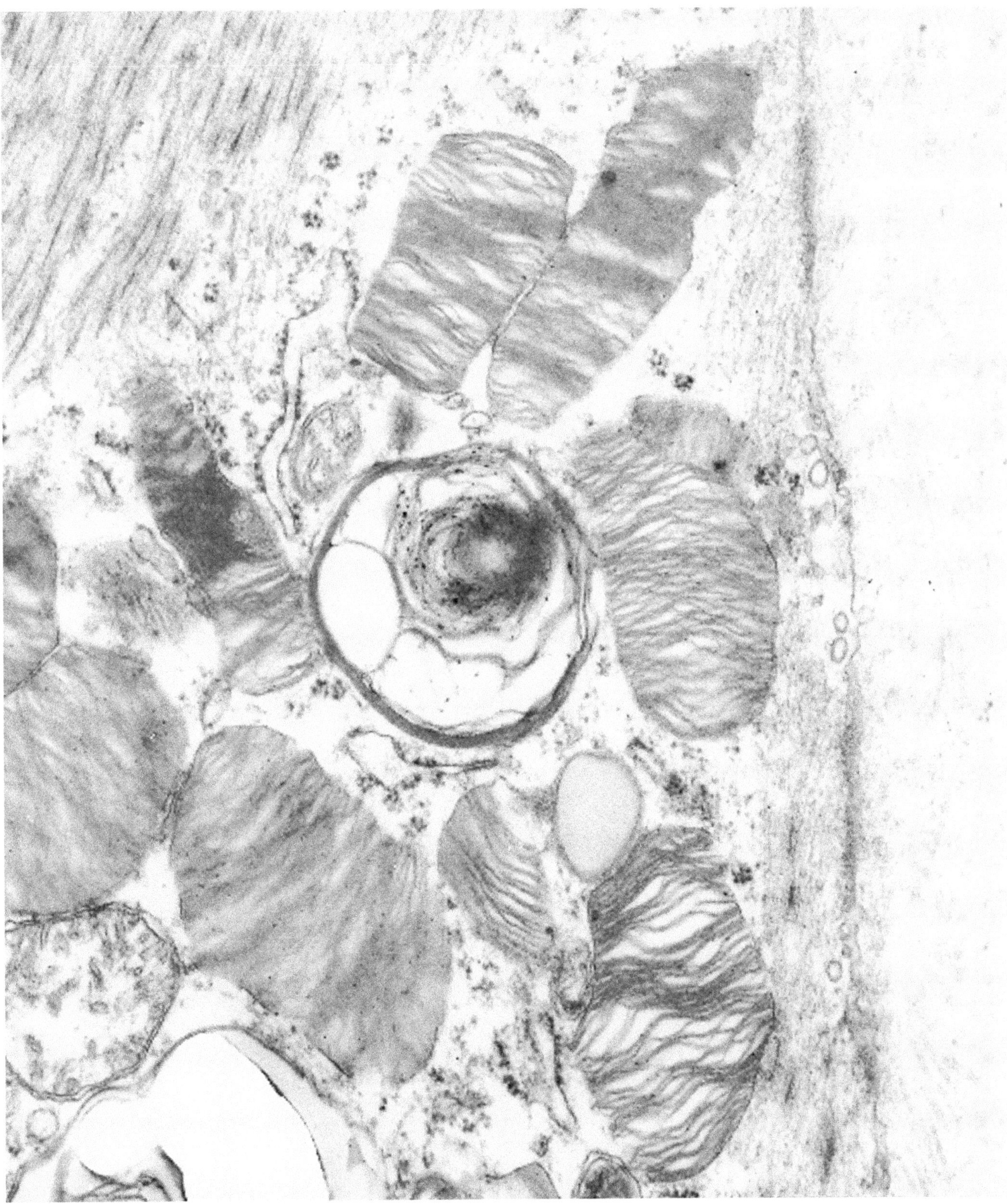

Figure 11.5. Fabry's disease (jejunum). Higher magnification of the lysosomal bodies depicted in Figure 11.4 illustrates their lamellar character. (× 48,400)

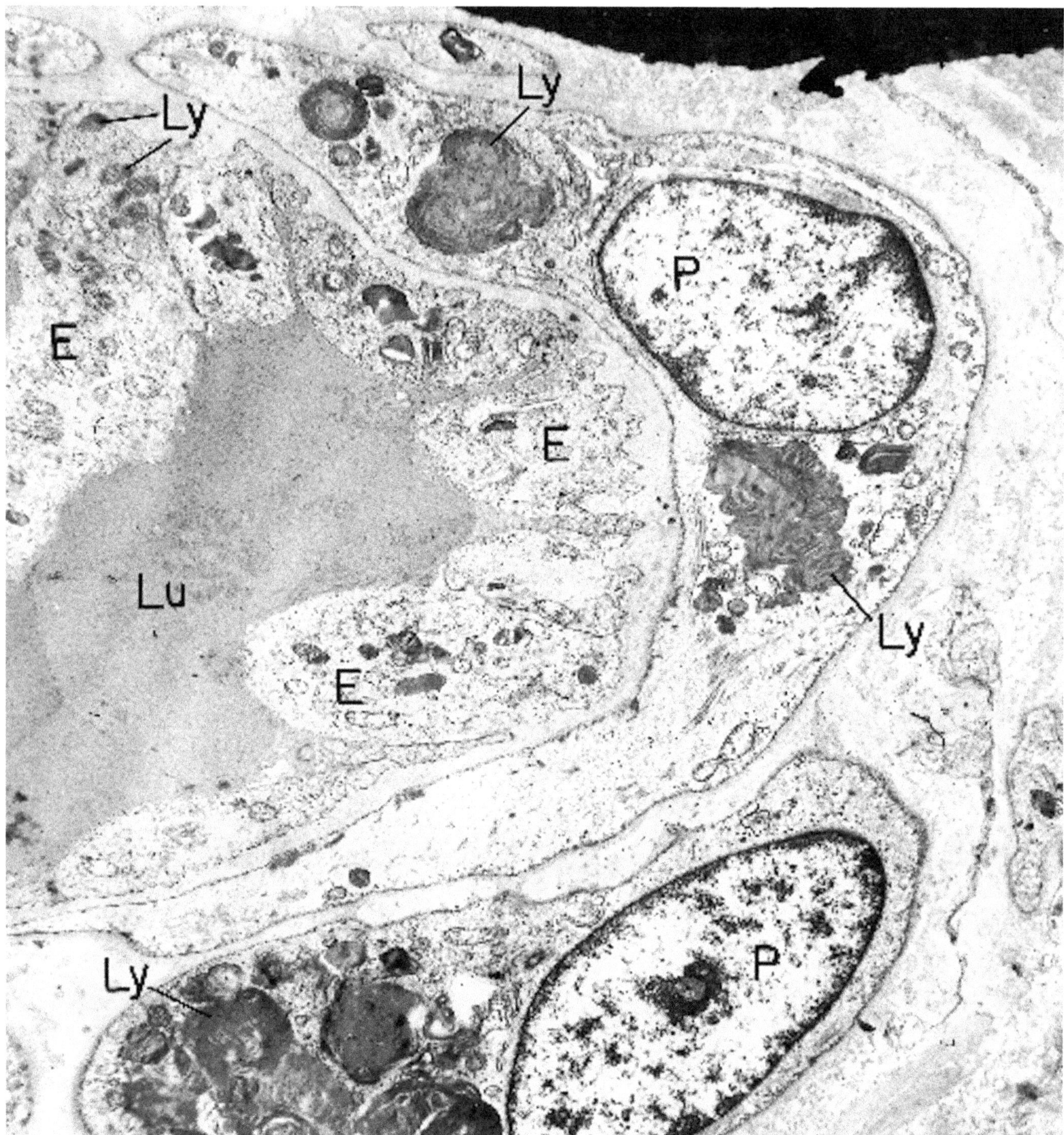

Figure 11.6. Fabry's disease (jejunum). This capillary has endothelial cells (E) and pericytes (P) that contain an increased number and size of lysosomes and excessive lysosomal substrate (Ly). L = lumen of capillary. (× 7020)

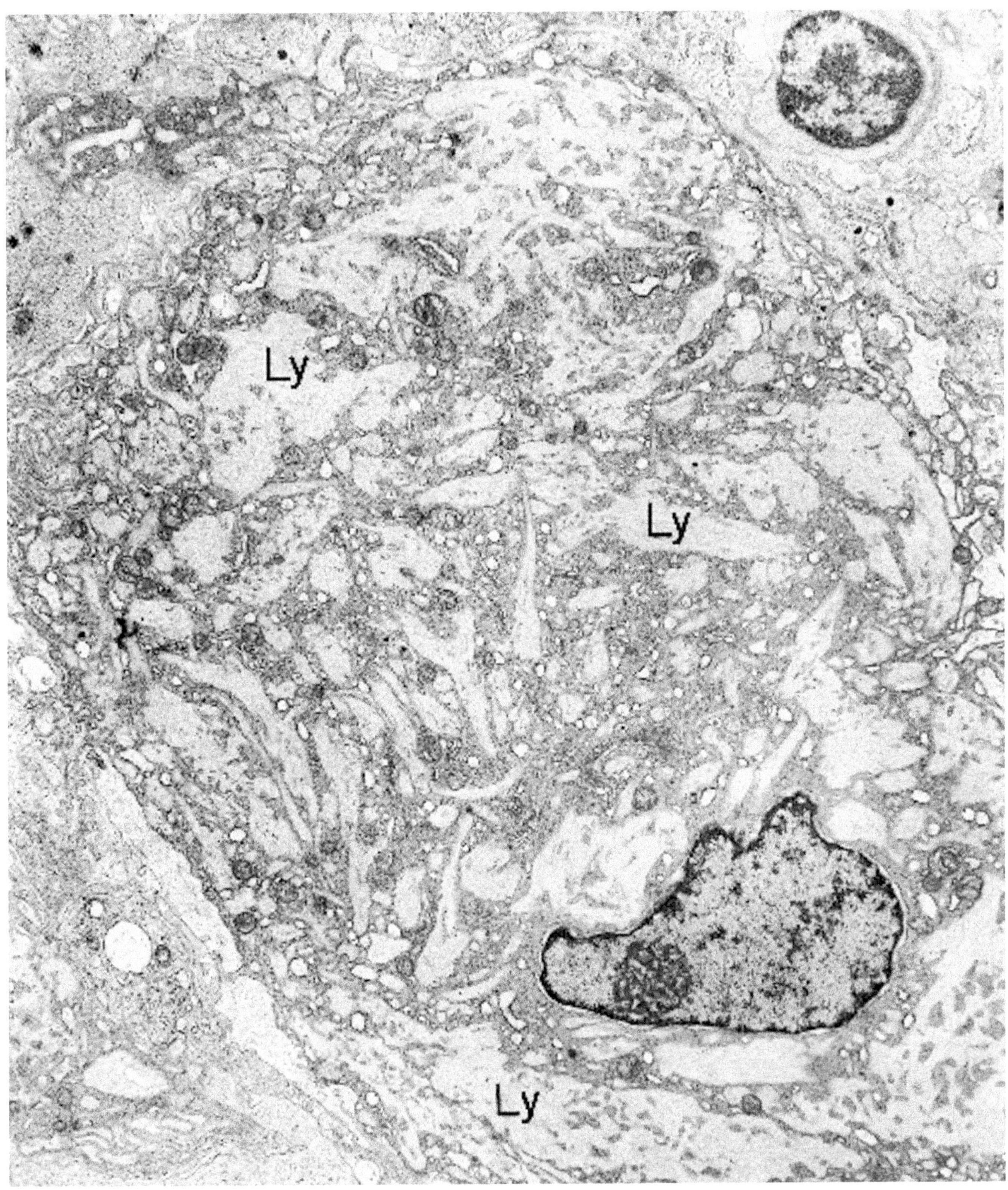

Figure 11.7. Gaucher's disease (bone marrow). This typical Gaucher cell has copious cytoplasm with innumerable, elongated lysosomes with substrate (Ly). The substrate consists of sheaves of tubules (seen better at higher magnification in Figure 11.8). (× 6720)

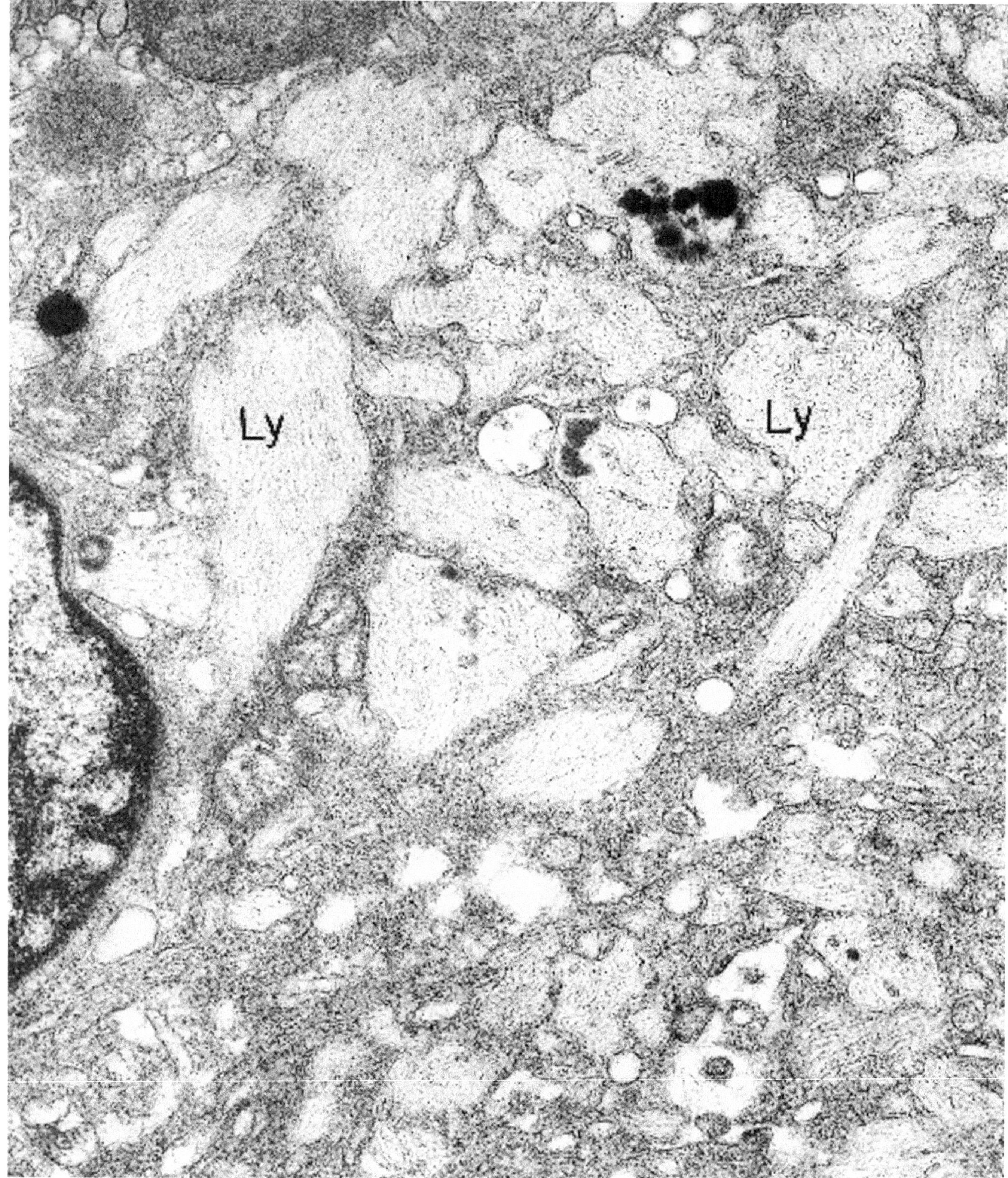

Figure 11.8. Gaucher's disease (liver). At this magnification, the tubular character of the lysosomal contents (Ly) is distinctly visible. (× 38,000)

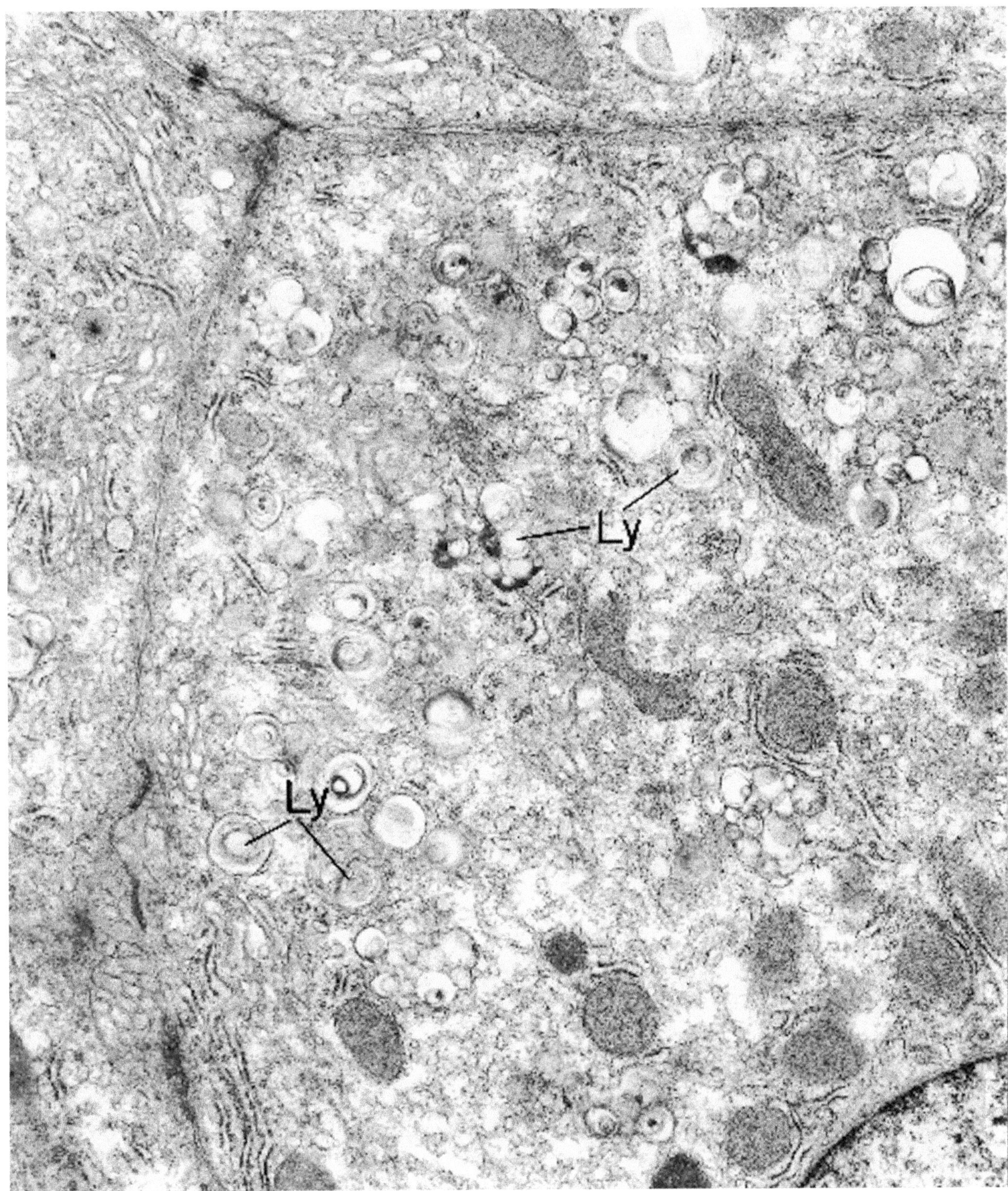

Figure 11.9. Niemann-Pick disease (liver). Mixed membranous and more solid osmiophilic inclusions (Ly) occupy much of the cytoplasm of this hepatocyte. (× 20,770)

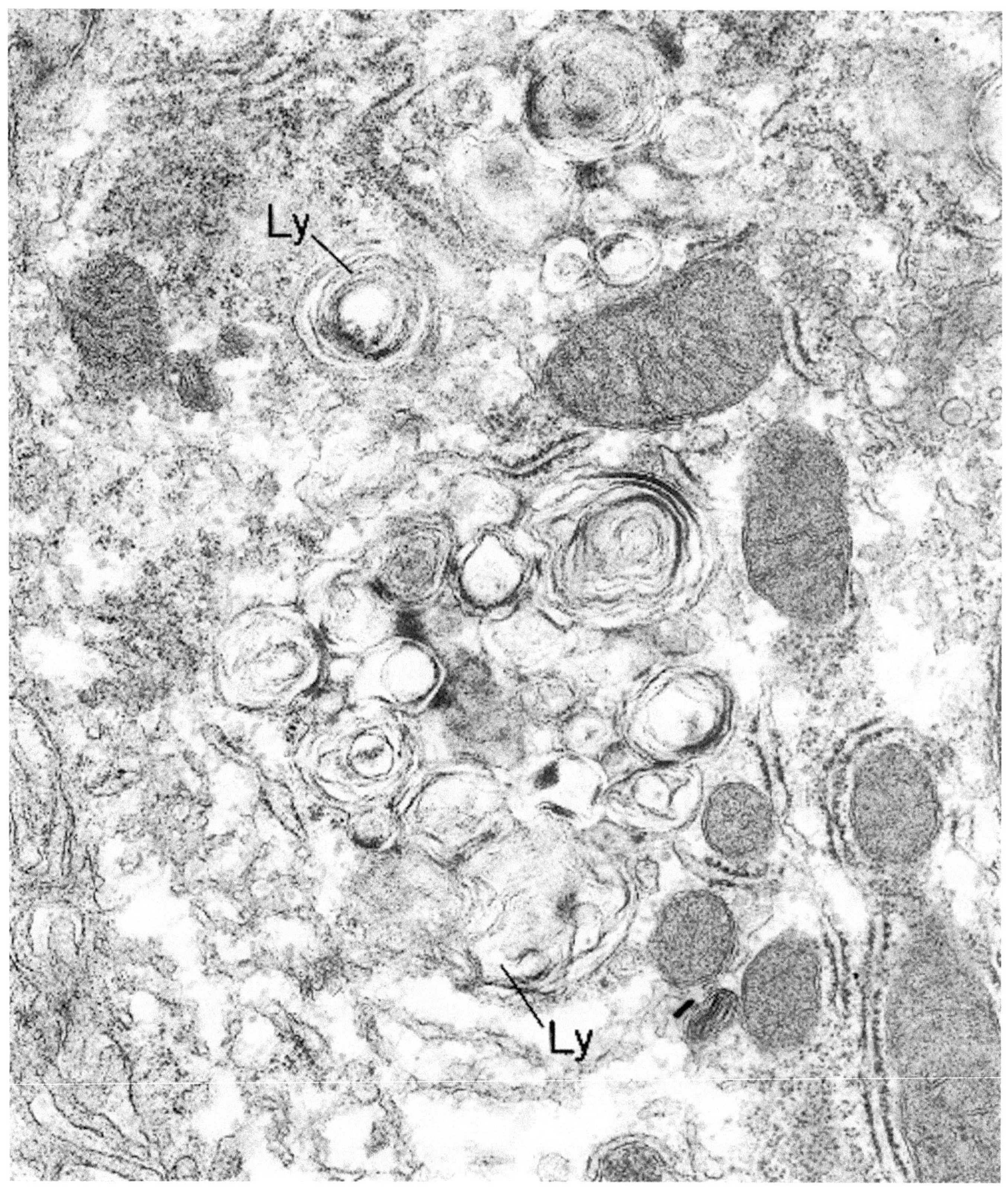

Figure 11.10. Niemann-Pick disease (liver). The concentric membranous and solid (sphingomyelin-rich) inclusions (Ly) are sharply demarcated, but the limiting membrane of the lysosomes frequently is obscured. (× 36,700)

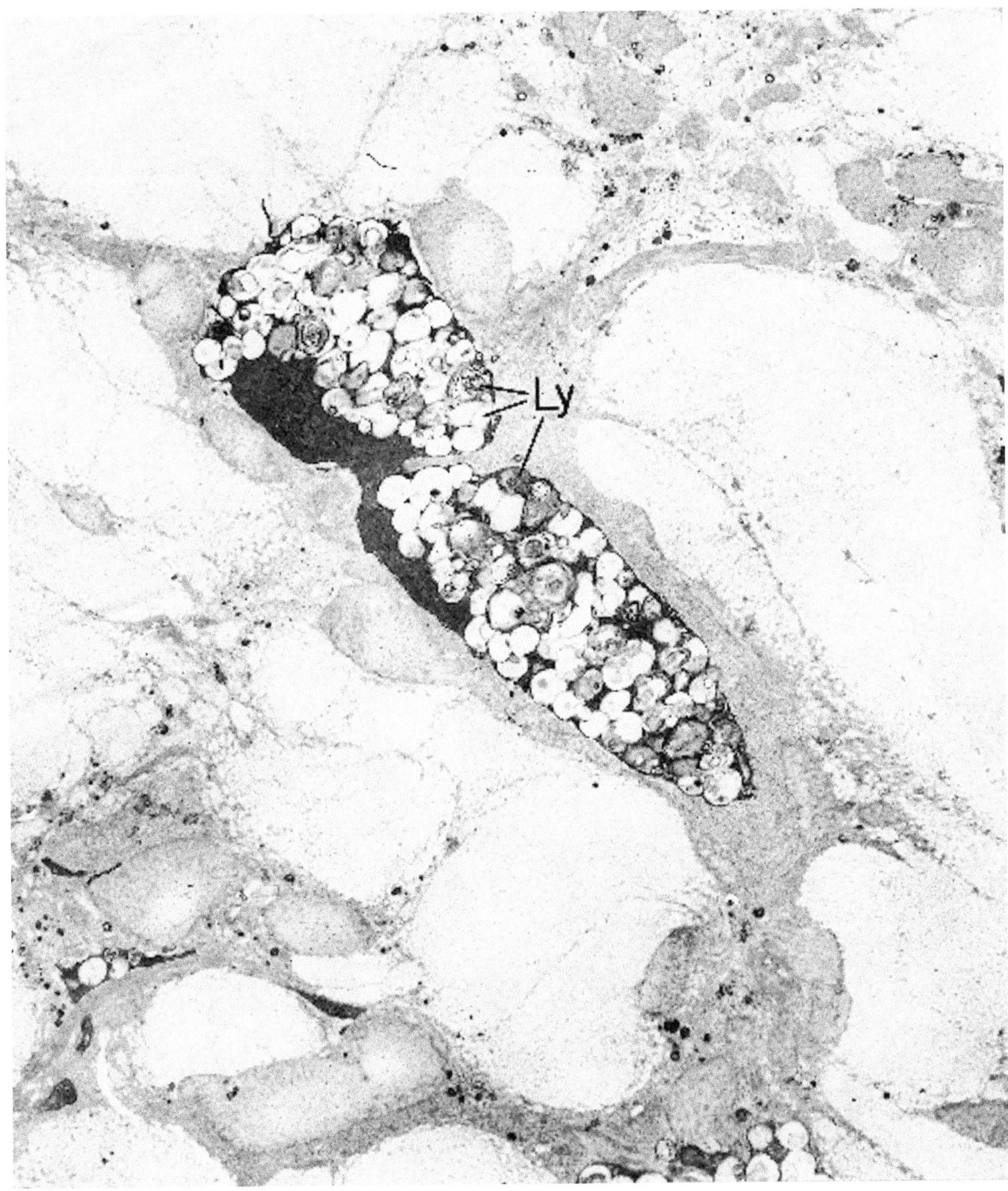

Figure 11.11. Maroteaux-Lamy syndrome; mucopolysaccharidosis type VI (ligamentum flavum). The fibroblasts of this ligament and of the spinal dura mater were packed with inclusions (Ly). The inclusions consist mostly of open spaces, but whorled membranes and solid osmiophilic material also are contained. (× 7020)

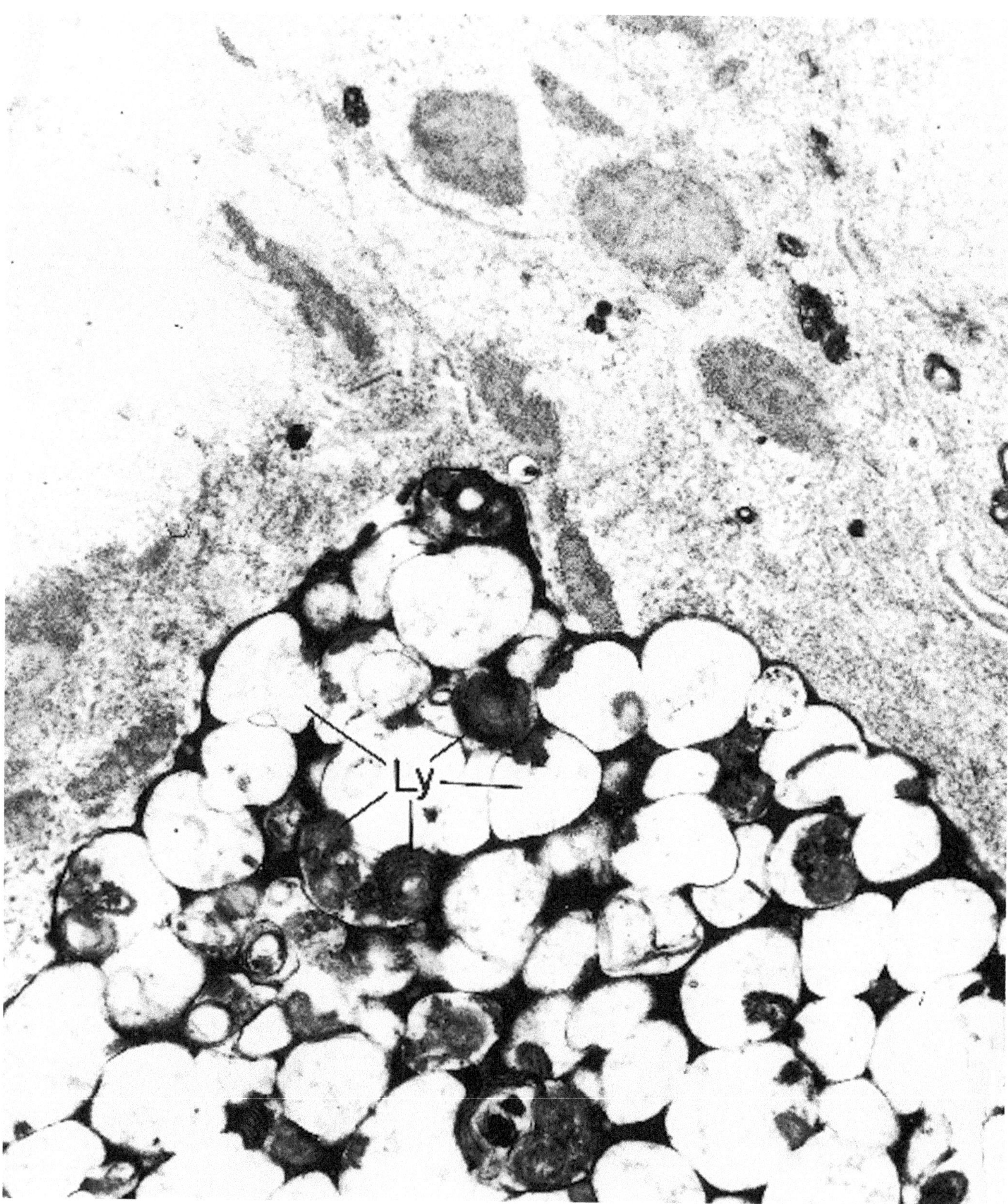

Figure 11.12. Maroteaux-Lamy syndrome; mucopolysaccharidosis type VI (ligamentum flavum). High magnification of a fibroblast illustrates the mixed open and electron-dense composition of the lysosomal inclusions (Ly). (× 22,200)

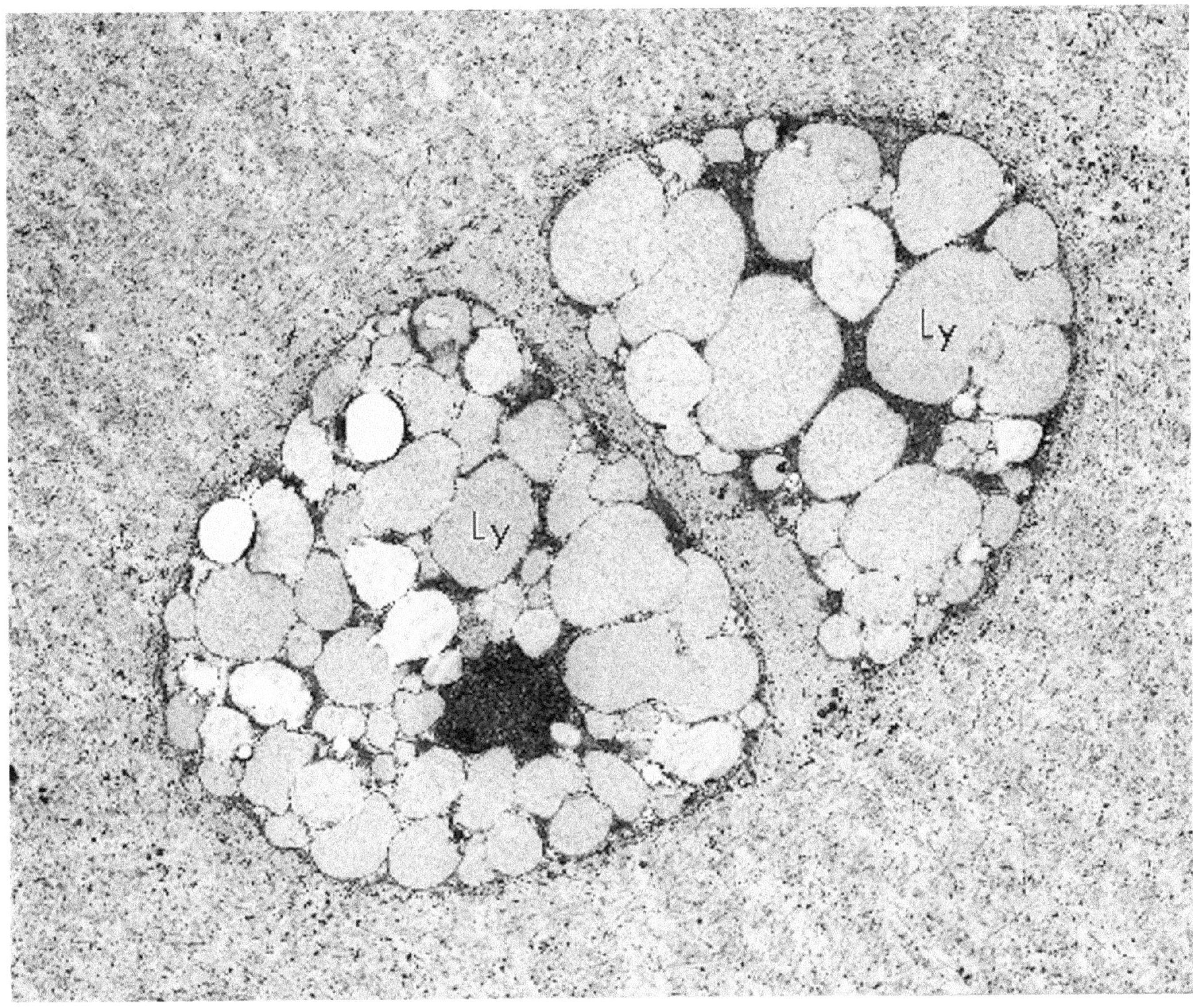

Figure 11.13. Morquio's syndrome (femoral head). Two chondrocytes have cytoplasm filled with lysosomes (Ly), which contain a flocculent medium-dense material. (× 4500)

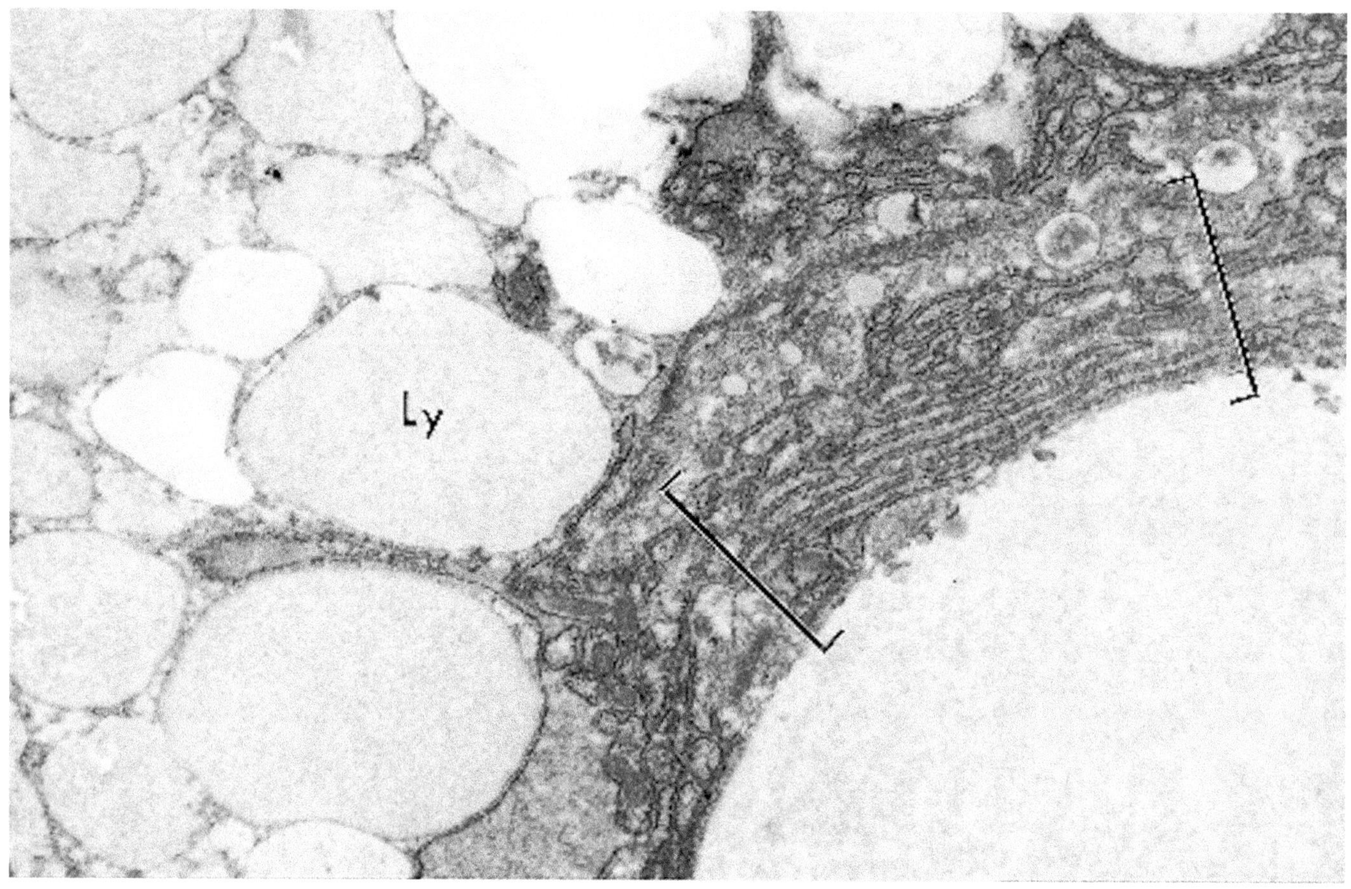

Figure 11.14. Morquio's syndrome (femoral head). High magnification of the lysosomes (Ly) shown in Figure 11.13 illustrates that they do not represent dilated rough endoplasmic reticulum (*brackets*). (× 14,800)

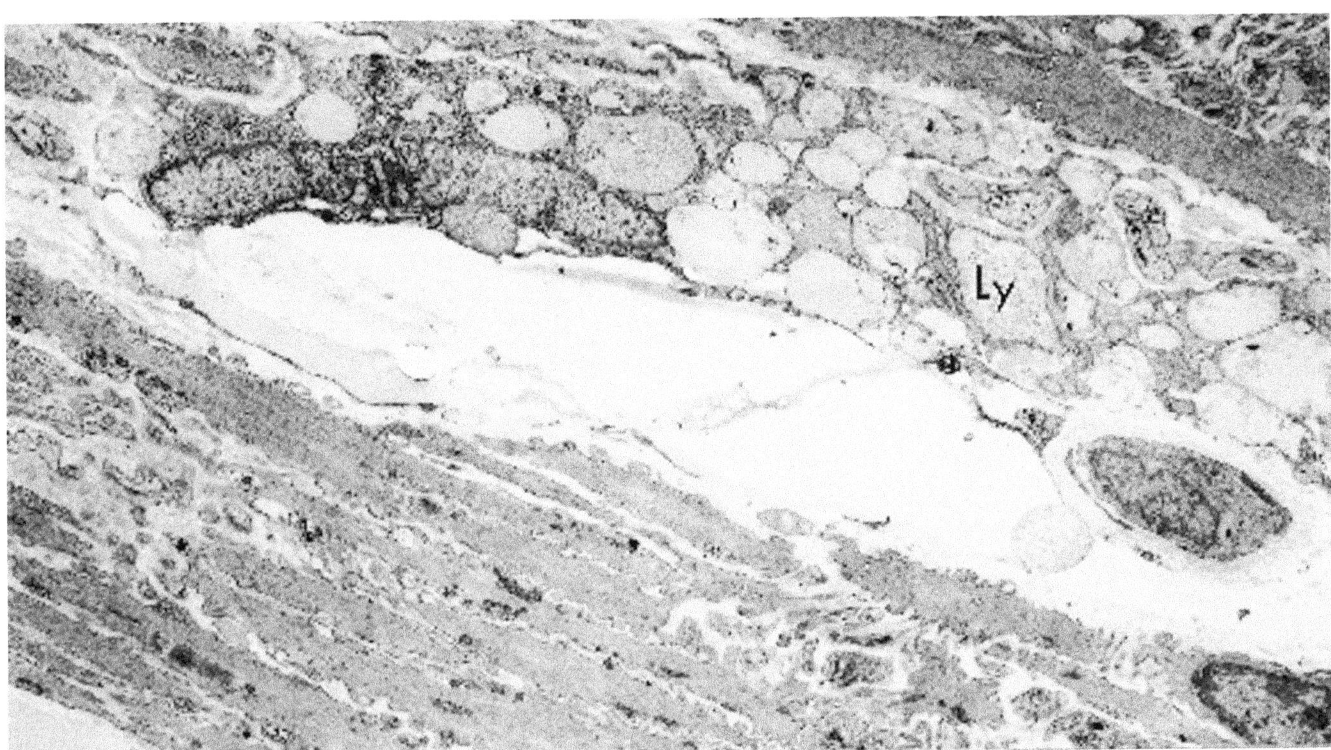

Figure 11.15. Sanfilippo's disease (skin, with smooth muscle). The cytoplasm of a smooth muscle cell is filled with membrane-bound lysosomal (Ly) vesicles—most of which contain a flocculent material. Fibroblasts were similarly involved. (× 4500)

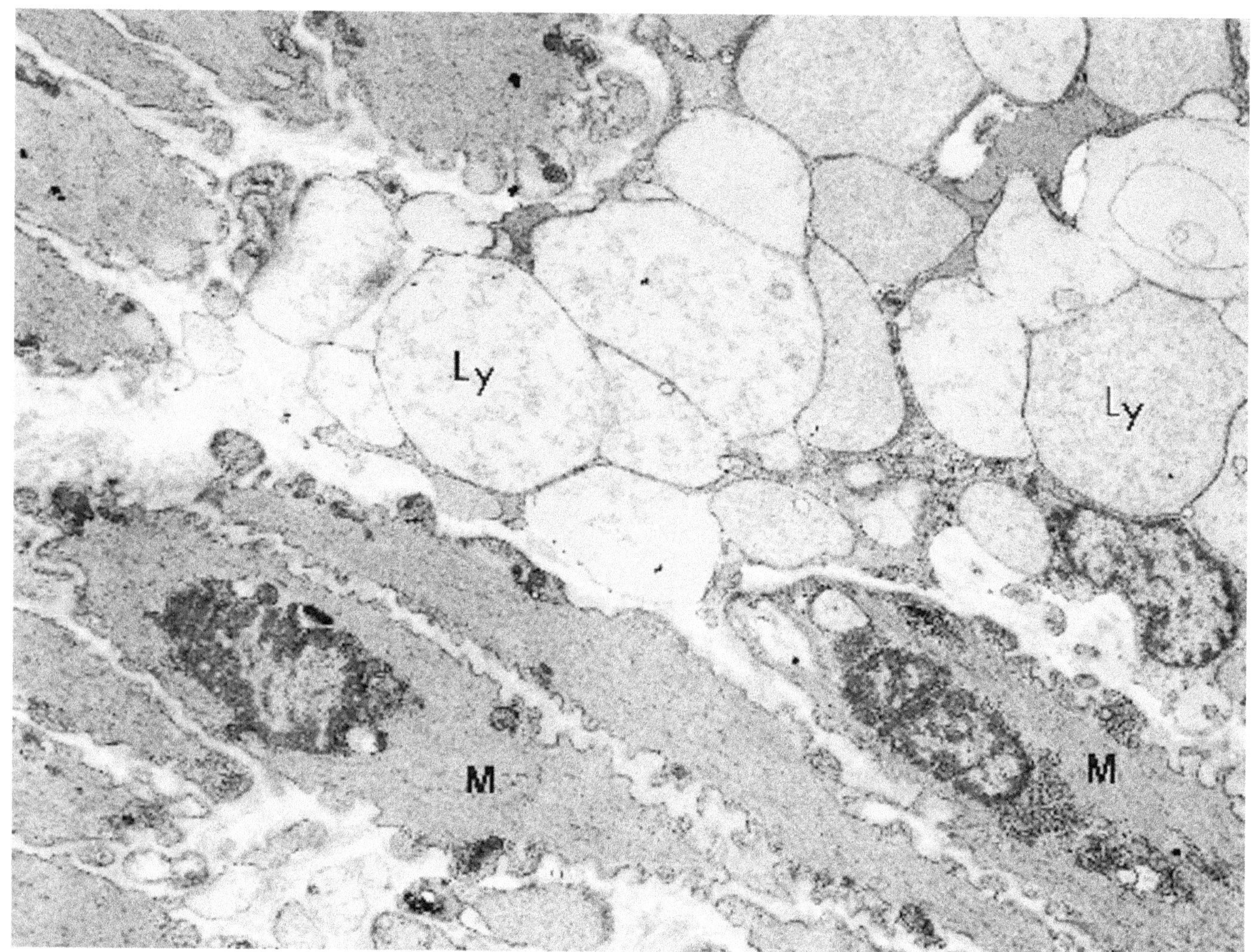

Figure 11.16. Sanfilippo's disease (skin, with smooth muscle). Higher magnification than in Figure 11.15 illustrates normal smooth muscle cells (M) as well as ones having most of their cytoplasm occupied by lysosomes (Ly). Most of the lysosomes are distended by a flocculent substrate consistent with mucopolysaccharide. (× 8000)

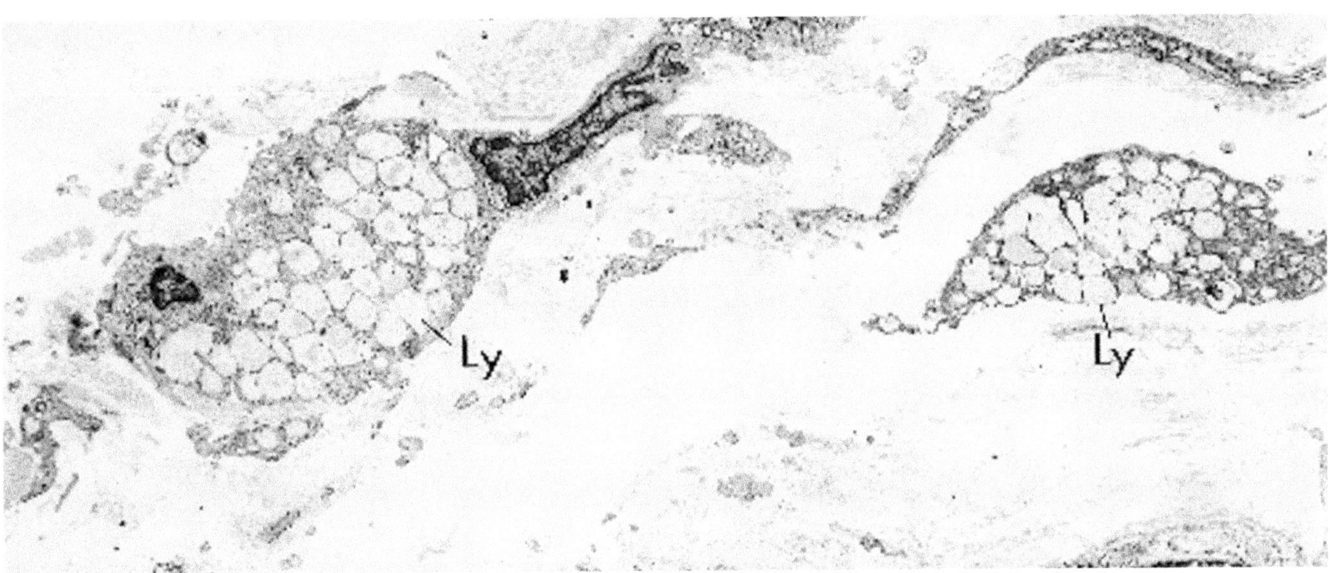

Figure 11.17. Hyaluronidase deficiency (soft tissue of finger). Two connective tissue fibroblasts contain numerous membrane-bound vesicles (Ly), which have a medium-dense material within them. (× 5000)

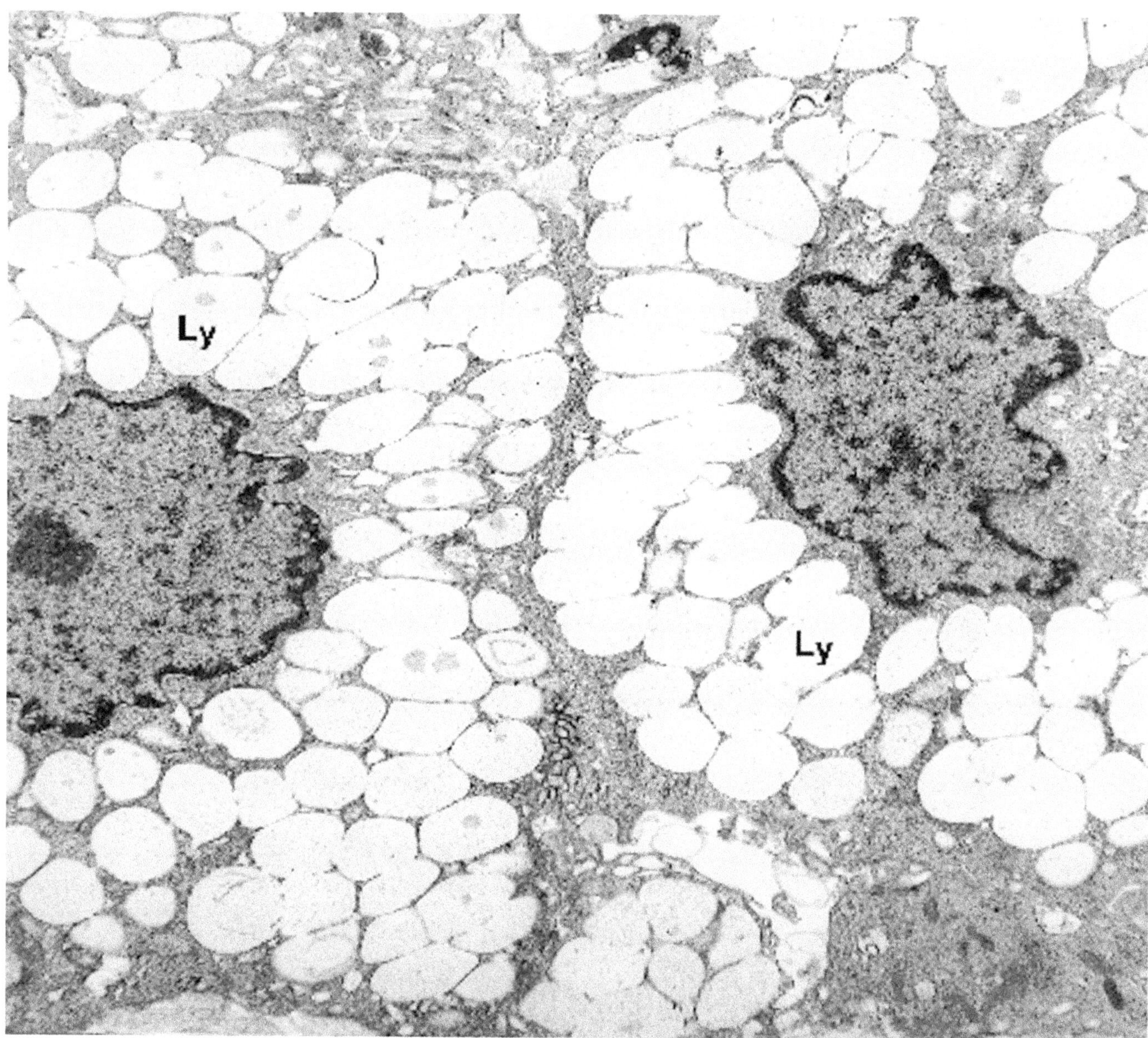

Figure 11.18. Hyaluronidase deficiency (soft tissue of finger). Two connective tissue histiocytes are filled with vesicles (Ly), which contain a medium-dense, flocculent material. (× 9500)

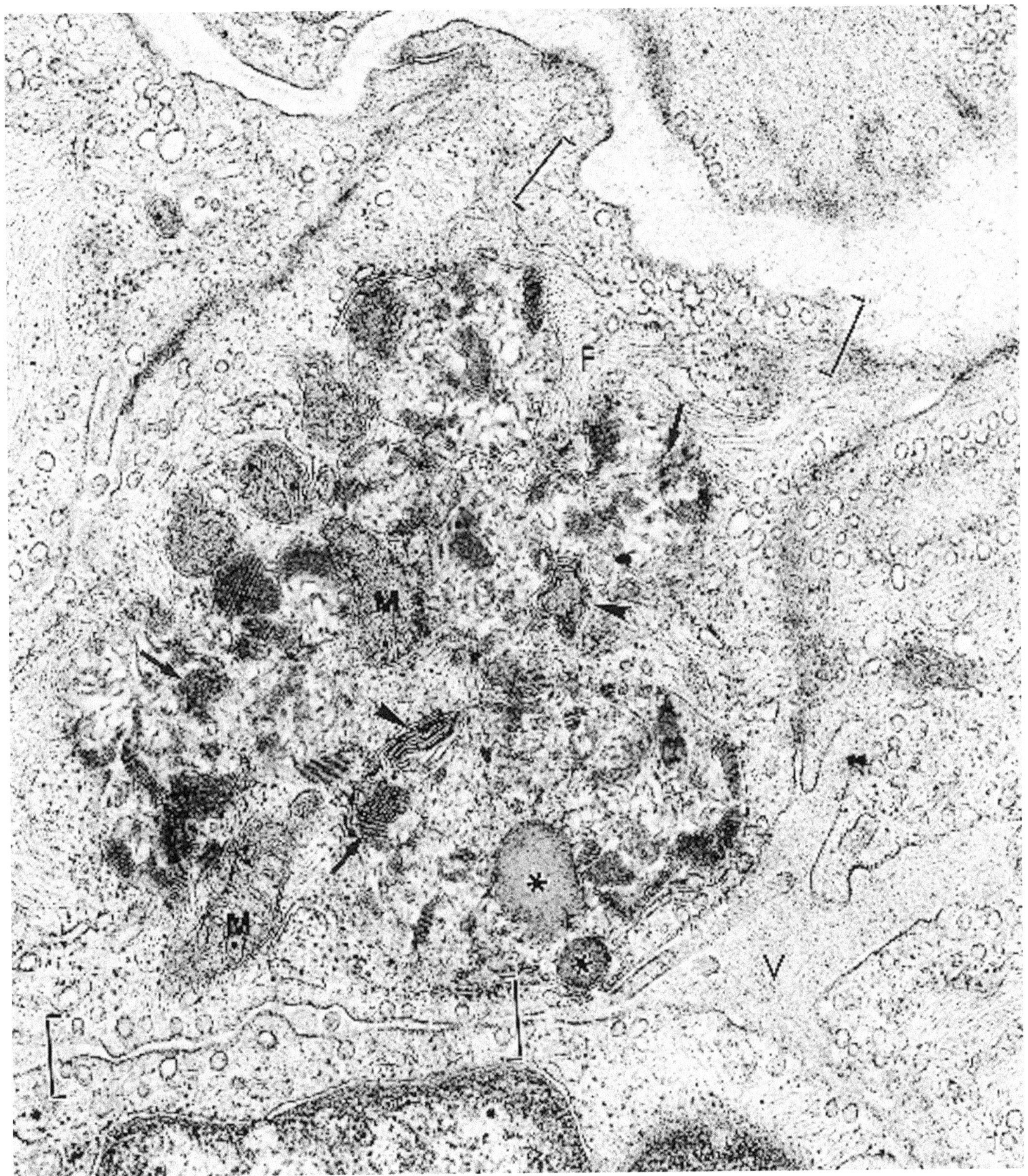

Figure 11.19. Neuronal ceroid lipofuscinosis (skin). A dermal endothelial cell contains membrane-bound inclusions of varying composition, including medium-dense amorphous and/or granular material (*), fingerprint-like lamellar structures (*arrows*), and myelin-like figures (*arrowheads*). M = mitochondria; F = filaments; *brackets* = pinocytotic vesicles; V = vascular lumen. (× 47,000)

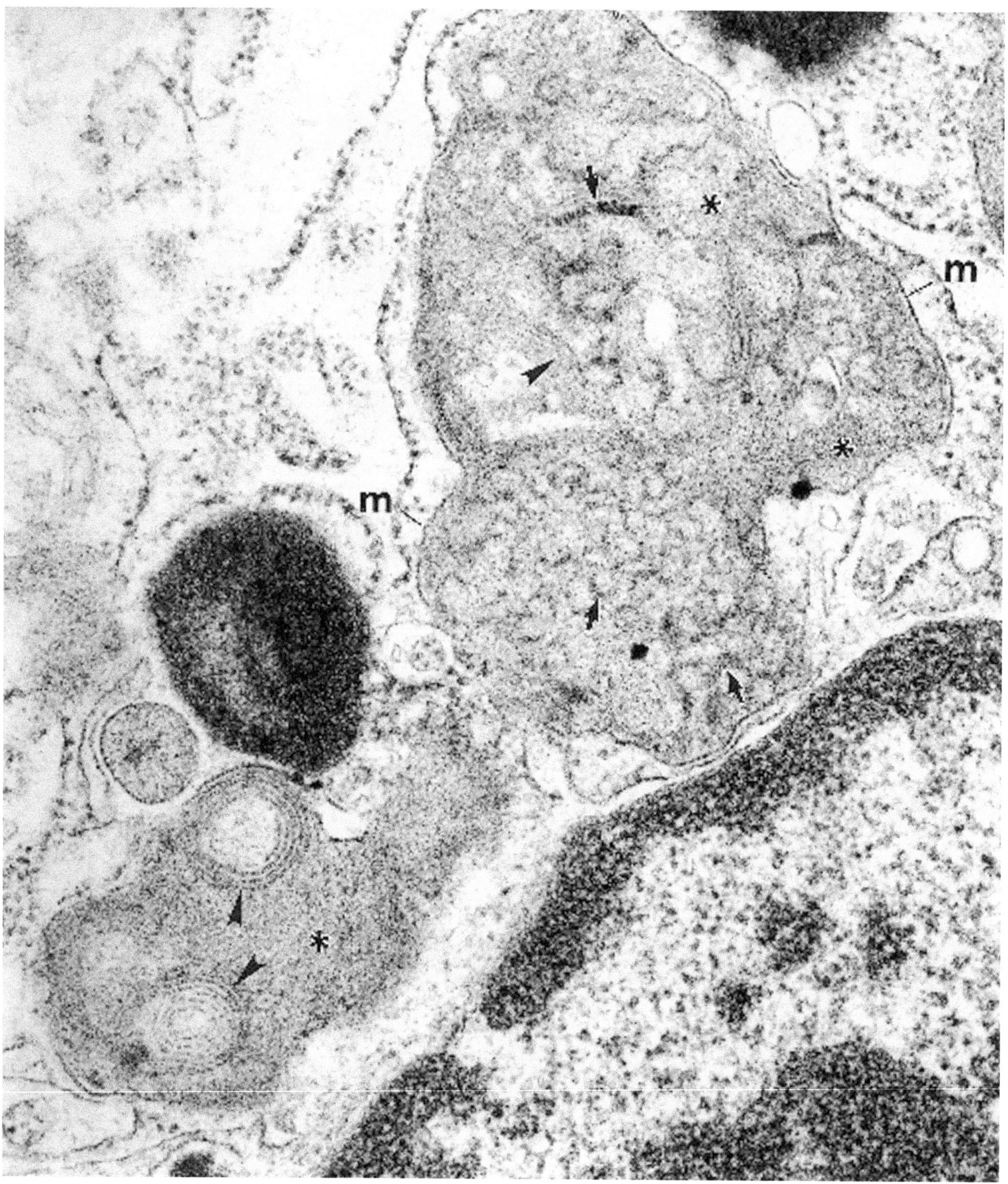

Figure 11.20. Neuronal ceroid lipofuscinosis (skin). Endothelial cell inclusions are membrane-bound (m) and of granular (*), curvilinear (*arrows*), and whorled tubular structures (*arrowheads*). (× 65,000)

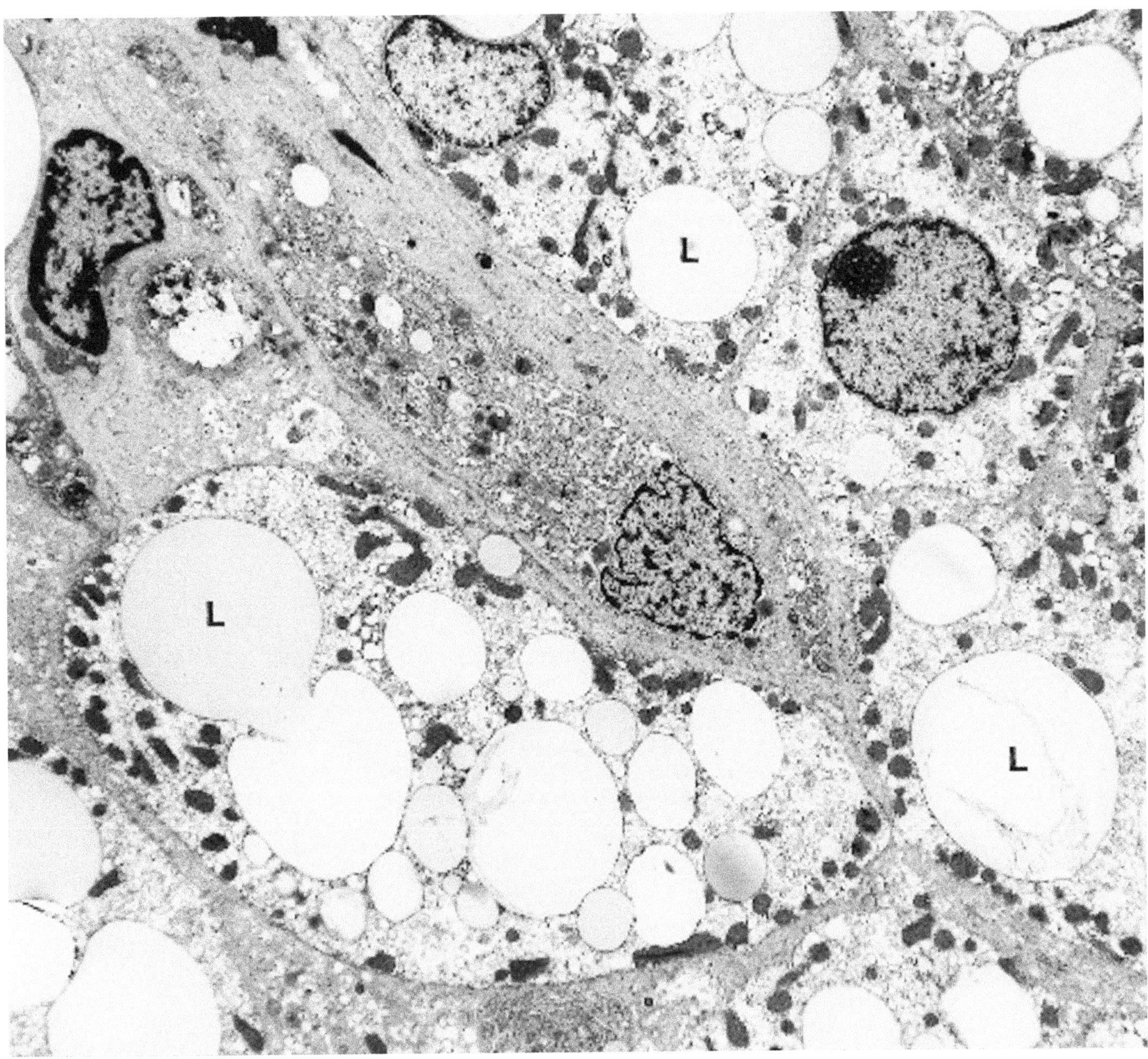

Figure 11.21. Cholesterol ester storage disease (liver). Membrane-bound lipid (L) occupies much of the cytoplasm of hepatocytes. (× 5900)

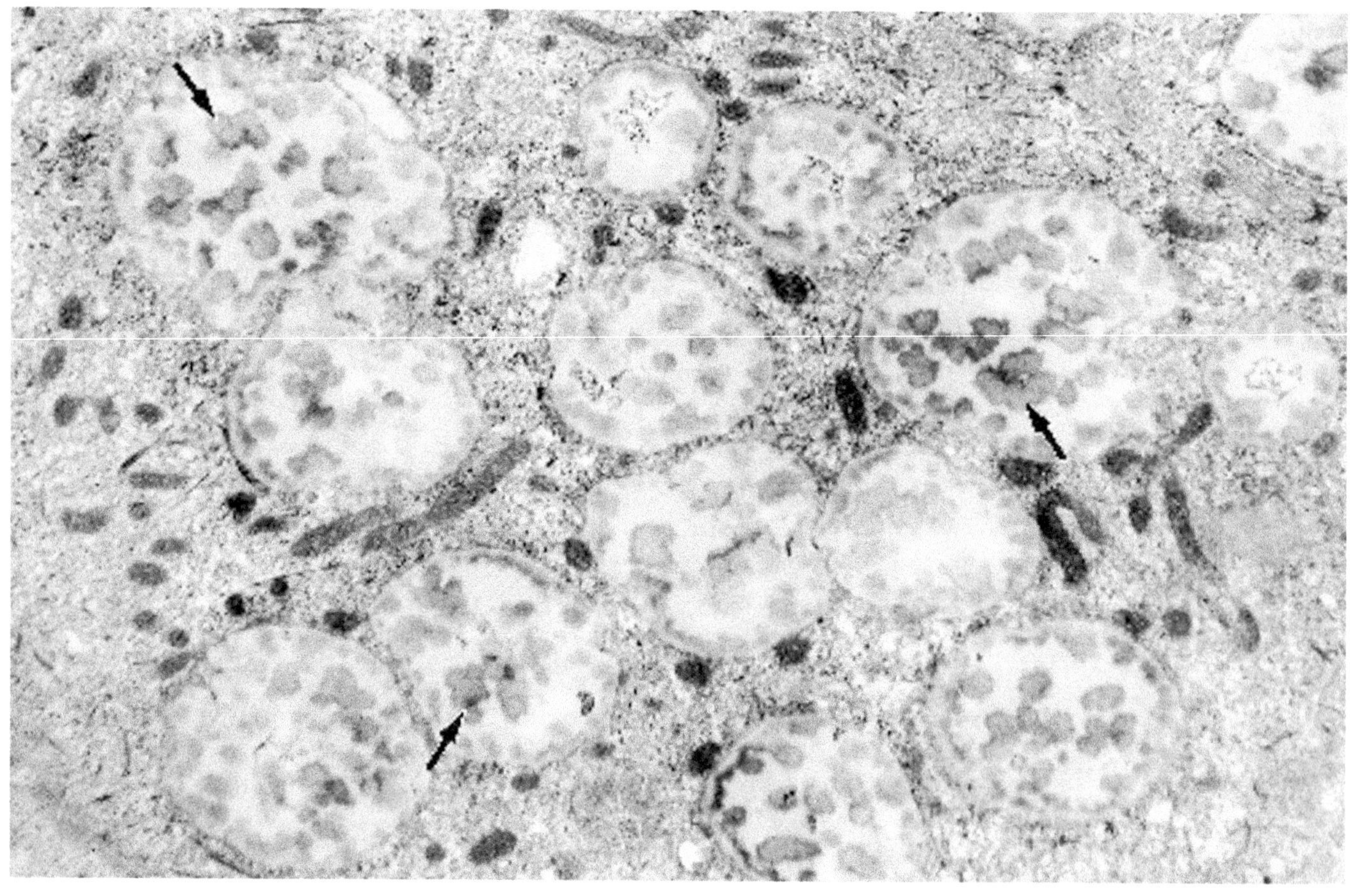

Figure 11.22. Cholesterol ester storage disease (liver). Hepatocytic cytoplasm contains numerous membrane-bound vesicles that have a medium-dense, crystal-like material (*arrows*) within them. (× 19,000)

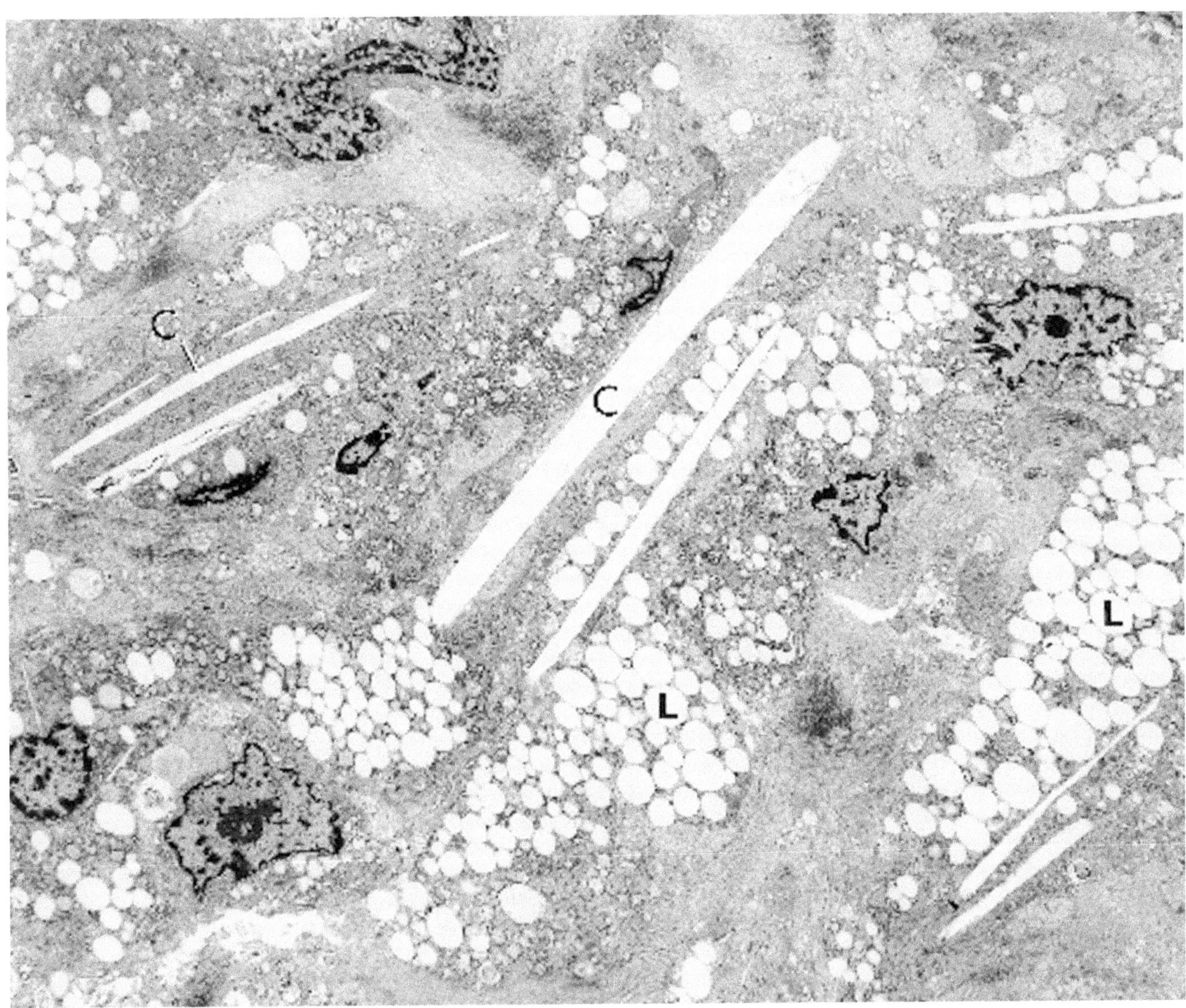

Figure 11.23. Cerebrotendinous xanthomatosis (Achilles tendon). Numerous droplets of neutral lipid (L) as well as acicular cholesterol (C) occupy the cytoplasm of histiocytes and fibroblasts in this tendon. The postmortem brain in this case also showed an infiltration of histiocytes with similar cytoplasmic inclusions. (× 4100)

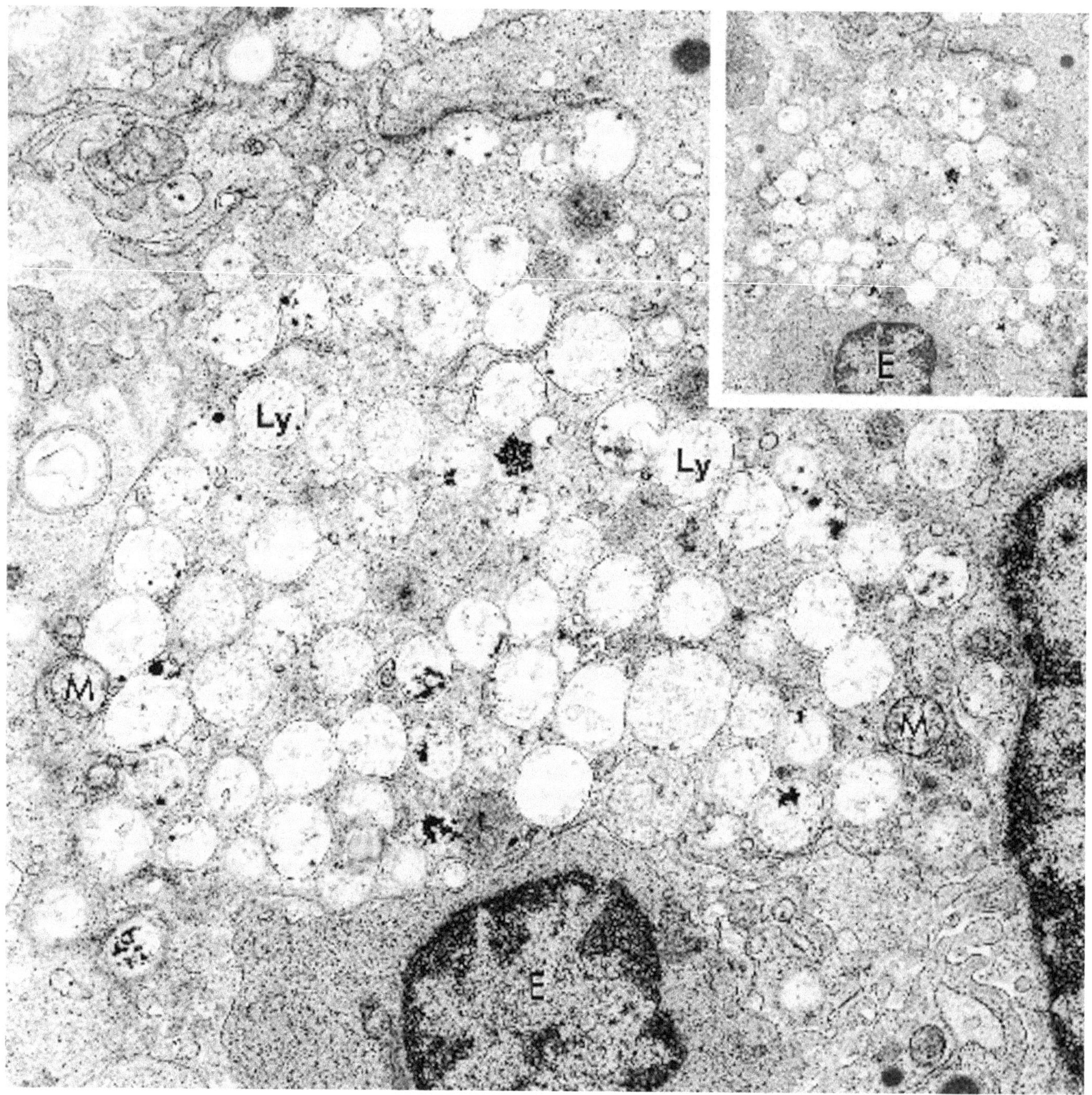

Figure 11.24. Mannosidosis (bone marrow). A histiocyte contains numerous single-membrane-bound vesicles (Ly) with flocculent interiors. Mitochondria (M), by contrast, are bound by a double-membrane. E = normoblast. (× 47,000) (*inset* × 5000)

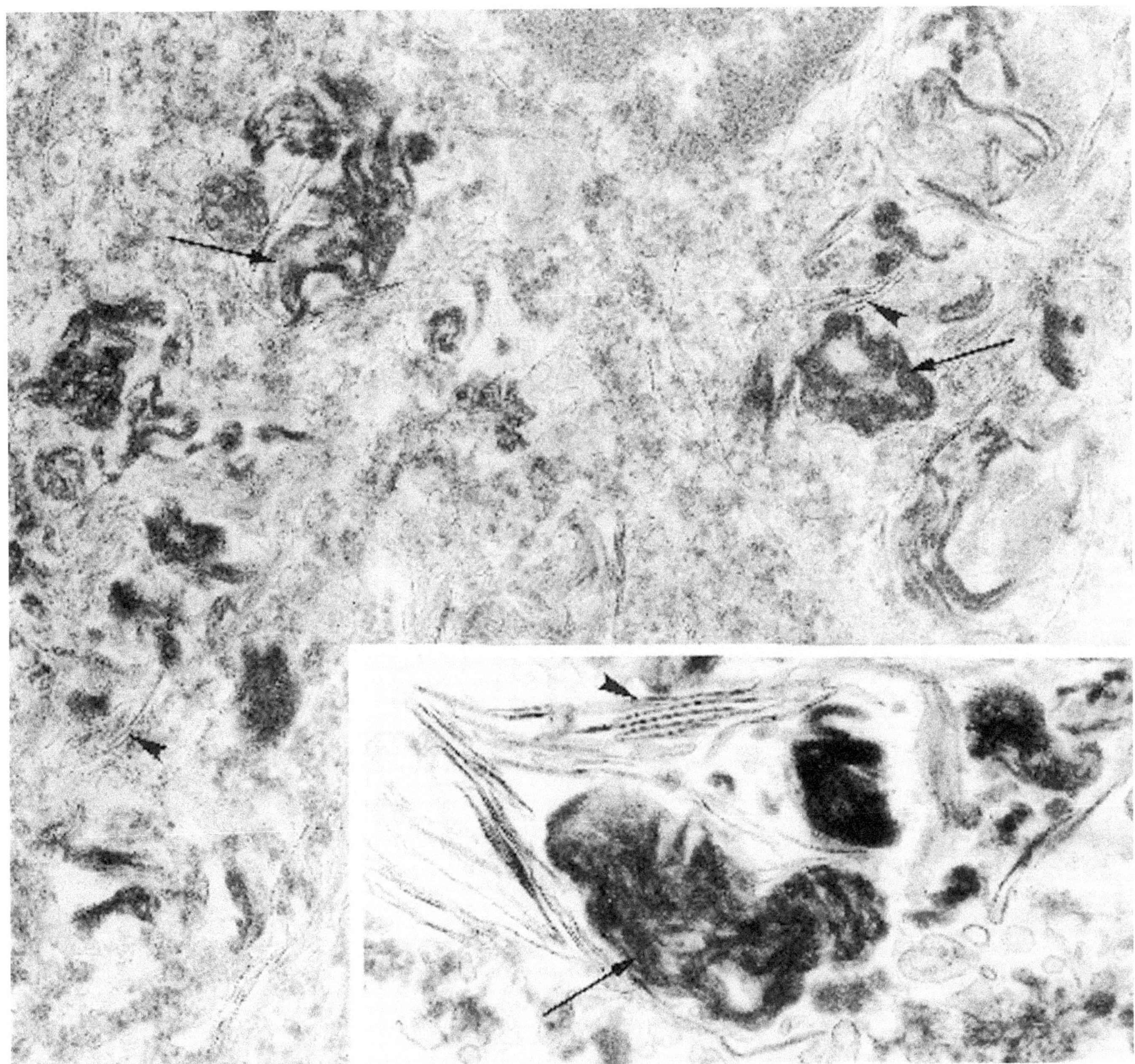

Figure 11.25. Adrenoleukodystrophy (cerebral white matter). This brain showed demyelination and inflammation, and macrophages contained numerous cytoplasmic inclusions composed of rope-like densities (*arrows*), consistent with myelin debris, and arrays of linear and curved lamellae (*arrowheads*). Lamellae were usually paired and had a clear space between them. (× 28,000) (*inset* × 45,000)

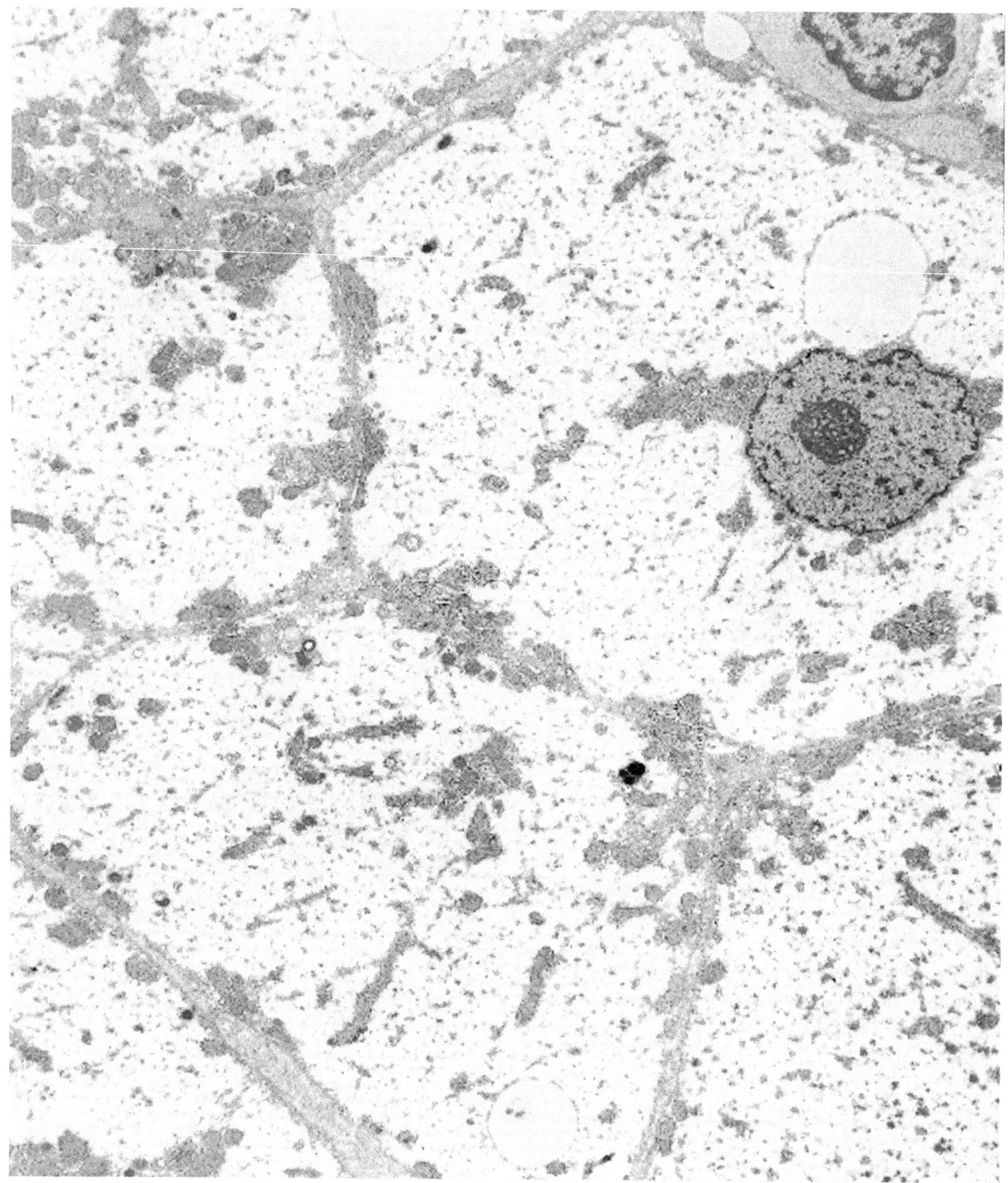

Figure 11.26. Glycogen storage disease, nonlysosomal (liver). These hepatocytes have such an excessive amount of glycogen (*open spaces*) in the cytoplasm that other organelles are compressed and inconspicuous. The glycogen is free rather than being membrane bound, as would be true of type II glycogenosis (Pompe's disease). (× 5090)

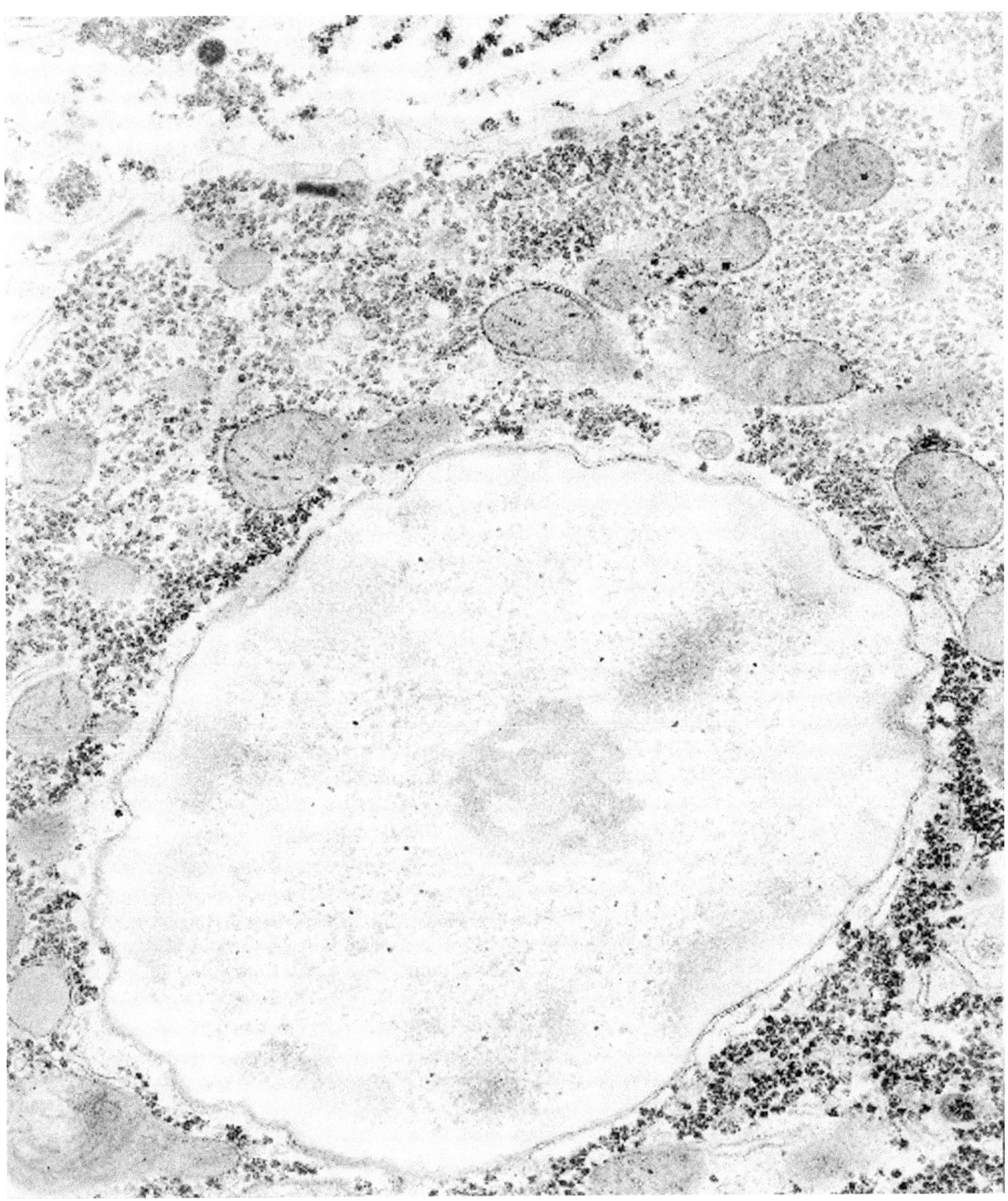

Figure 11.27. Glycogen storage disease, nonlysosomal (liver). This specimen was processed by a method that preserves glycogen as electron-dense granules, and it substantiates that the open spaces in the hepatocytes of Figure 11.26 represent that substance. (× 21,060)

(Text continued from page 710)

Erdheim-Chester Disease (Fibroxanthomatosis)

(Figures 11.28 and 11.29.)

Diagnostic criteria. (1) Bony and/or soft tissue infiltrates of lipid-laden histiocytes, fibroblasts, and multinucleated giant cells.

Additional points. This is a rare, focal or systemic disease characterized by xanthomatous infiltrates and fibrosis. Long bones (metaphyseal marrow), soft tissue, and parenchymatous organs (e.g., lung) may be involved. In some cases, infiltrates of macrophages devoid of lipid suggest a primary disease of macrophages rather than of lipid. Although Langerhans-type histiocytes have been intermixed with phagocytic type histiocytes in a few of these cases, other morphological, immunohistochemical, and clinical features favor the concept of Langerhans histiocytes and Erdheim-Chester disease being two separate entities.

Porphyria

(Figures 11.30 and 11.31.)

Diagnostic criteria. (1) Crystalline inclusions in the cytoplasm of hepatocytes, ductular epithelium, and Kupffer cells and in bile ductular and canalicular lumens; (2) stasis may or may not be present.

Additional points. Porphyria may be an inborn error in metabolism or acquired. Porphyrins, pigments present in hemoglobin, myoglobin, and cytochromes, are abnormally metabolized and accumulated in various tissues, such as skin and liver.

Crystals of protoporphyria may be present in the cytosol, in the endoplasmic reticulum, and in secondary lysosomes. Although crystalline deposits are characteristic of the disease, they may not be present in early stages. Other less-specific changes include dilated smooth and rough endoplasmic reticulum, a variety of mitochondrial changes, and bile canalicular distention with blunting of microvilli and bile pigment in lumens. Only one of five subtypes of porphyria is illustrated here, congenital *erythropoietic protoporphyria.*

Alpha-1-Antitrypsin Deficiency

(Figures 11.32 and 11.33.)

Diagnostic criteria. (1) Hepatocytes (periportal) have smooth and rough endoplasmic reticulum markedly distended by a homogeneous, finely granular material of medium electron density (Figures 11.32 and 11.33).

Additional points. Alpha-1-antitrypsin is a protease inhibitor that neutralizes elastases, collagenases, trypsin, chymotrypsin, and other proteolytic enzymes, thus preventing excessive cellular autodigestion. In an alpha-1-antitrypsin deficiency state, the enzymes accumulate and ultimately destroy the cells, resulting in cirrhosis. It may be difficult to discern whether the accumulated enzyme is in the smooth or rough endoplasmic reticulum because of marked distention and distortion of cisternae. In the lung, alveolar septae undergo elastolysis, enhancing septal rupture and emphysema.

Pulmonary Alveolar Proteinosis

(Figures 11.34 and 11.35.)

Diagnostic criteria. (1) Intra-alveolar exudate composed of whorled membranous bodies, other vesicular and amorphous osmiophilic particles, and droplets of neutral lipid (Figures 11.34 and 11.35); (2) varying number of histiocytes with engulfed components of the exudate (Figure 11.34).

Additional points. The diagnostic characteristic of the exudate in pulmonary alveolar proteinosis is the whorled membranous body, also referred to as a myelin-like figure, multilamellated structure, and surfactant body. It resembles tubular myelin and the surfactant body of type II pneumocytes and is composed of concentric phospholipid membranes and intermembranous amorphous protein. Most of the exudate is acellular, and histiocytes with phagocytosed membranous bodies are not necessary for making the diagnosis. Bronchoalveolar lavage fluid is rich in exudate as well as accessible, and its removal is therapeutic as well as diagnostic.

(Text continues on page 747)

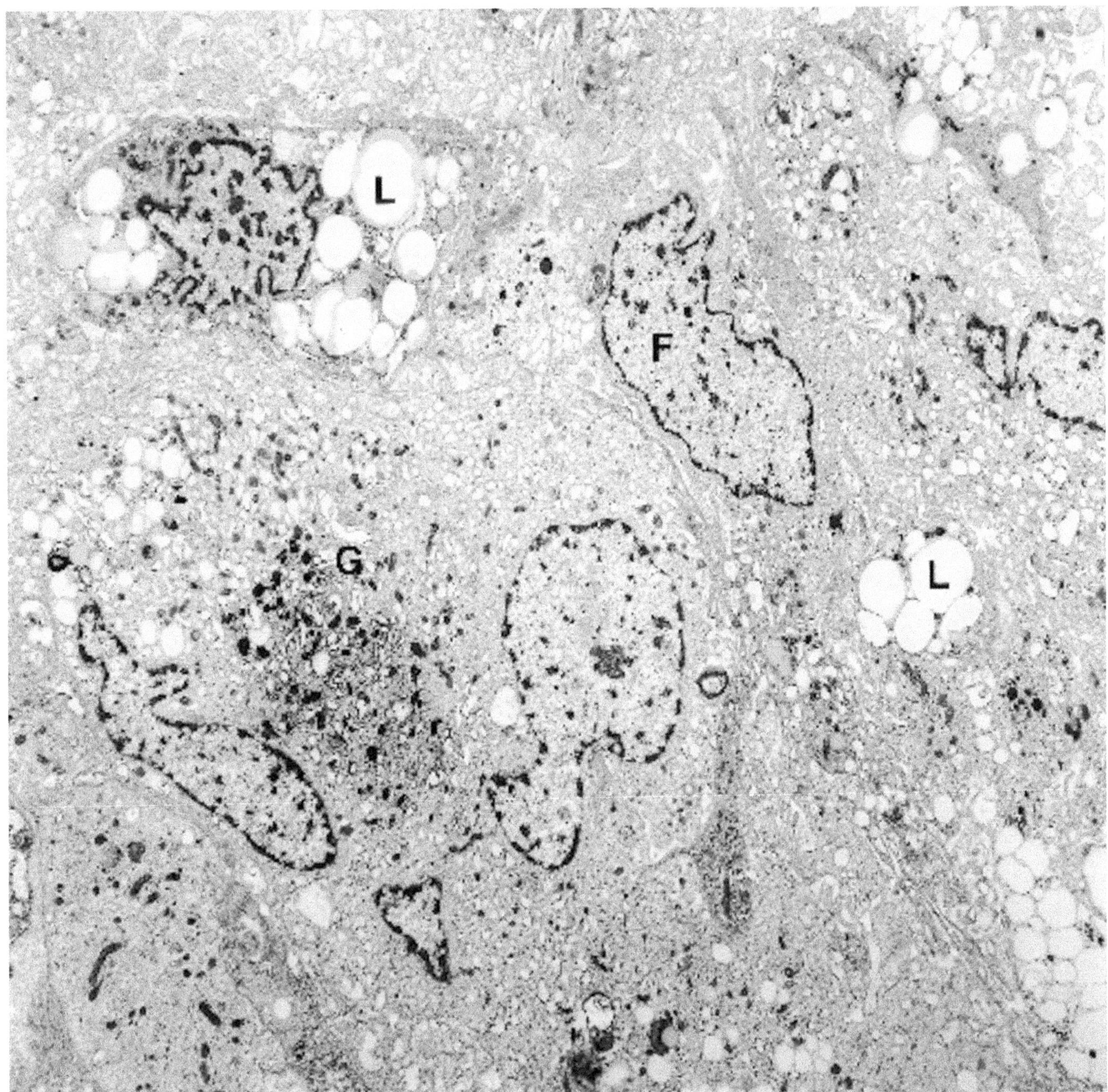

Figure 11.28. Erdheim-Chester disease (fibroxanthomatosis) (femur). Fibroblasts (F) and one multinucleated giant cell (G) contain numerous neutral lipid droplets (L). (× 5600)

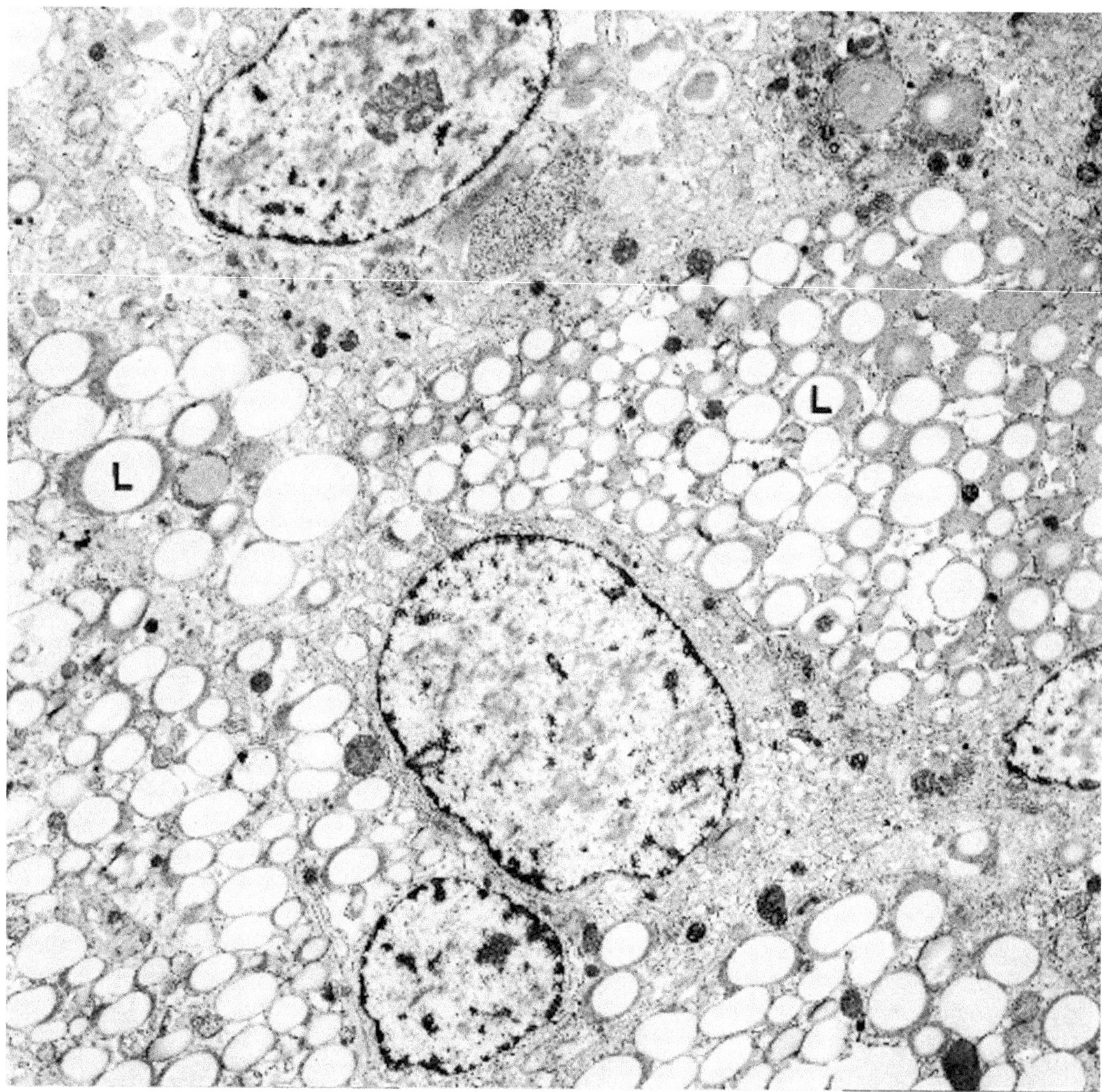

Figure 11.29. Erdheim-Chester disease (fibroxanthomatosis) (femur). High magnification of several fibroblasts shows distention of their cytoplasm by numerous non-membrane-bound neutral lipid droplets (L). (× 7200)

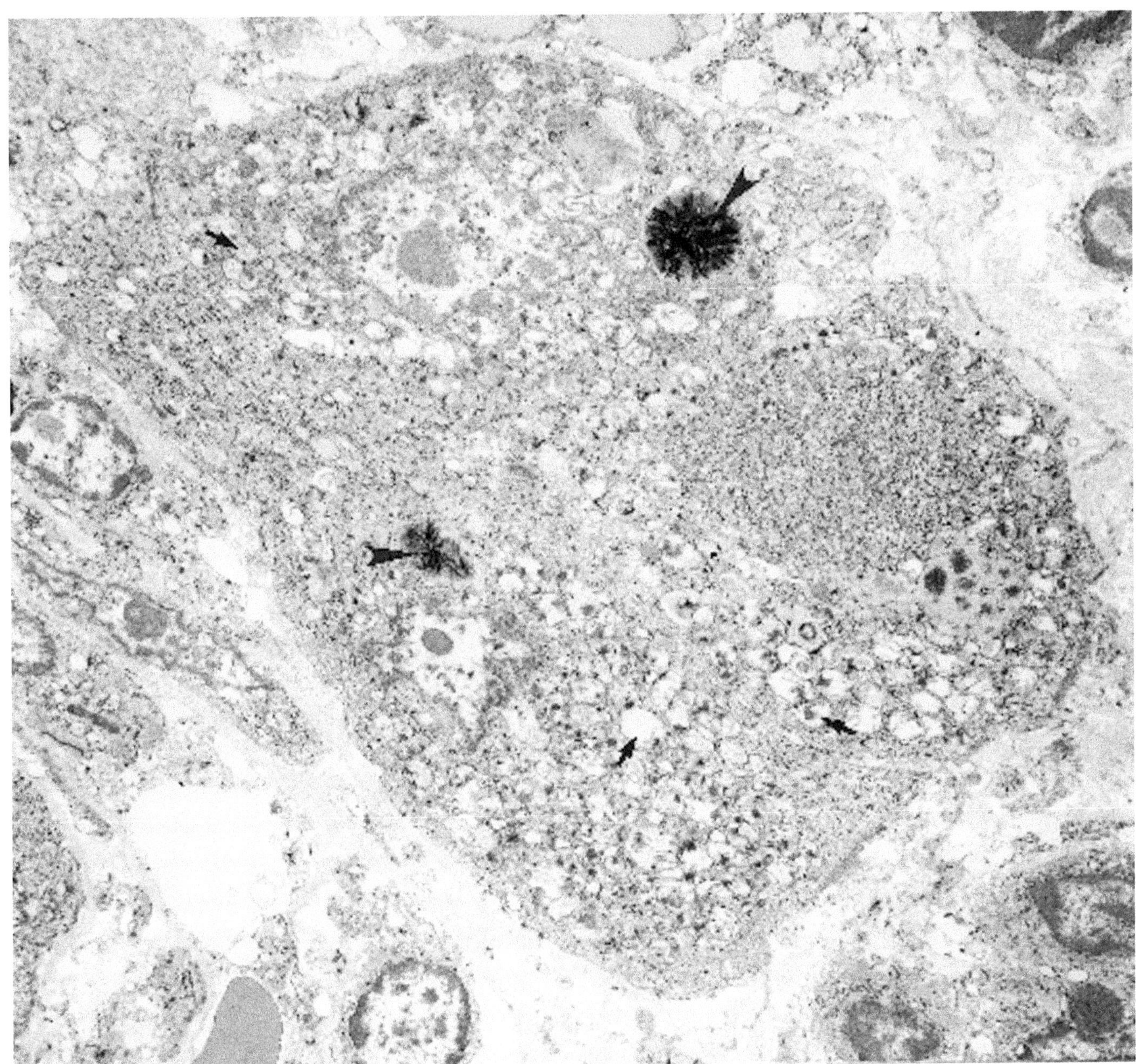

Figure 11.30. Erythropoietic protoporphyria (liver). Poorly preserved hepatocytes in a cirrhotic liver, which had delayed fixation, contain numerous dilated mitochondria (*arrows*) and several star-burst, crystalline deposits (*arrowheads*). Similar deposits were also present in von Kupffer cells, bile ductal epithelial cells and lumens, and bile canalicular lumens. These deposits were autofluorescent, and red with polarized light. (× 4500)

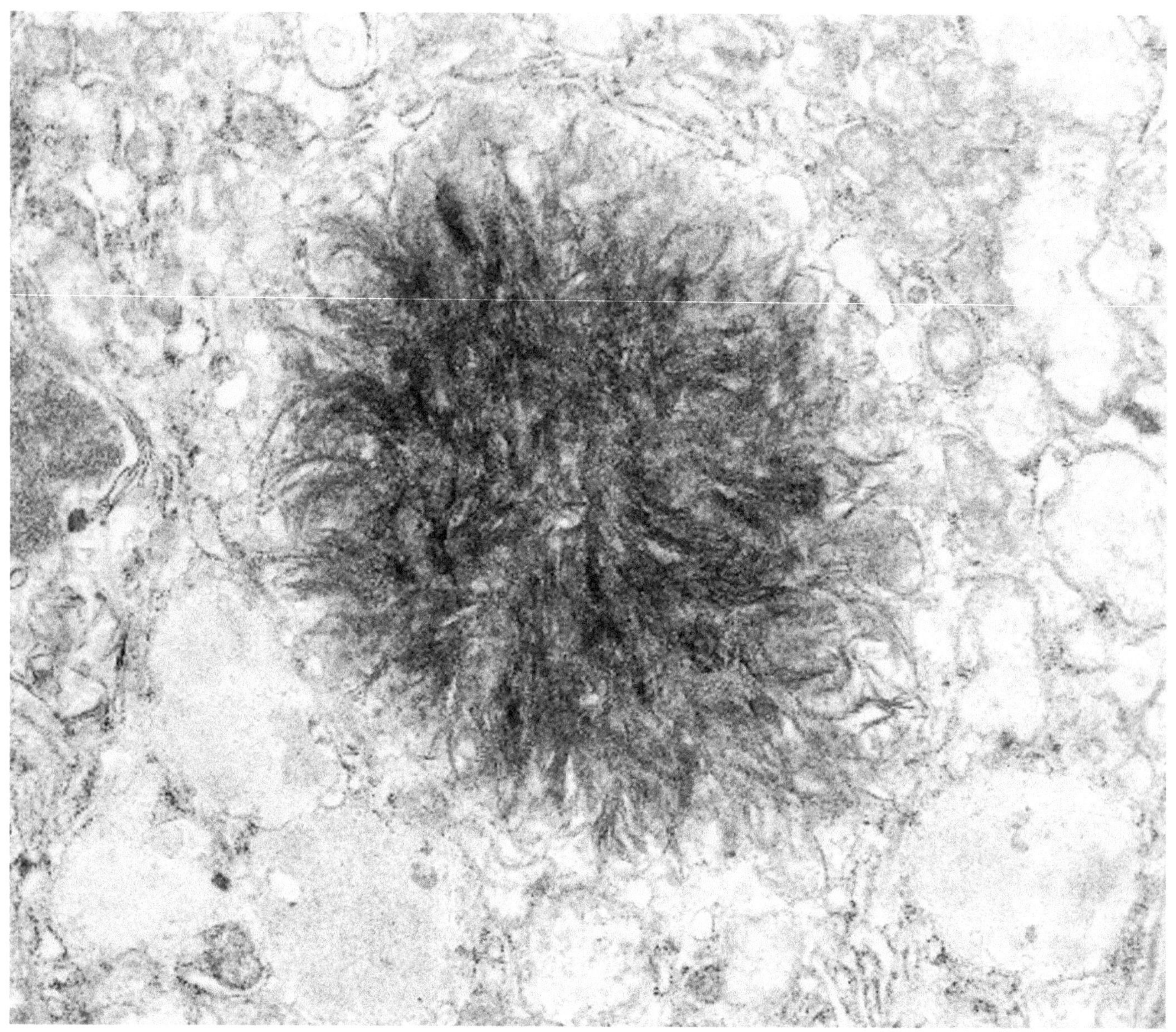

Figure 11.31. Erythropoietic protoporphyria (liver). High magnification depicts in detail a filamentous crystalline deposit. (× 25,000)

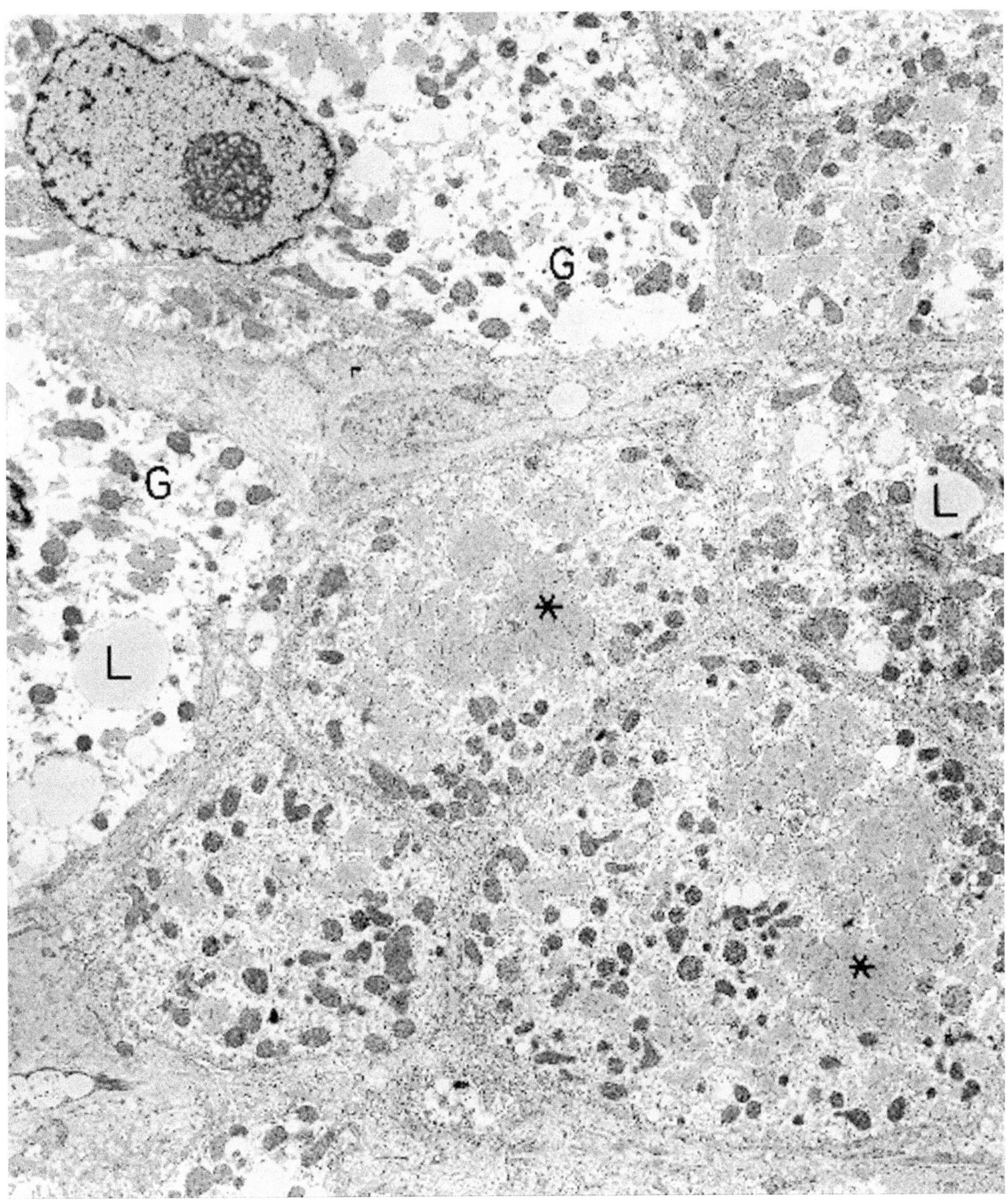

Figure 11.32. Alpha-1-antitrypsin deficiency (liver). The hepatocytes have markedly dilated cisternae of endoplasmic reticulum, which are filled with a medium-dense material (*). G = open spaces of glycogen; L = droplets of neutral lipid. (× 5940)

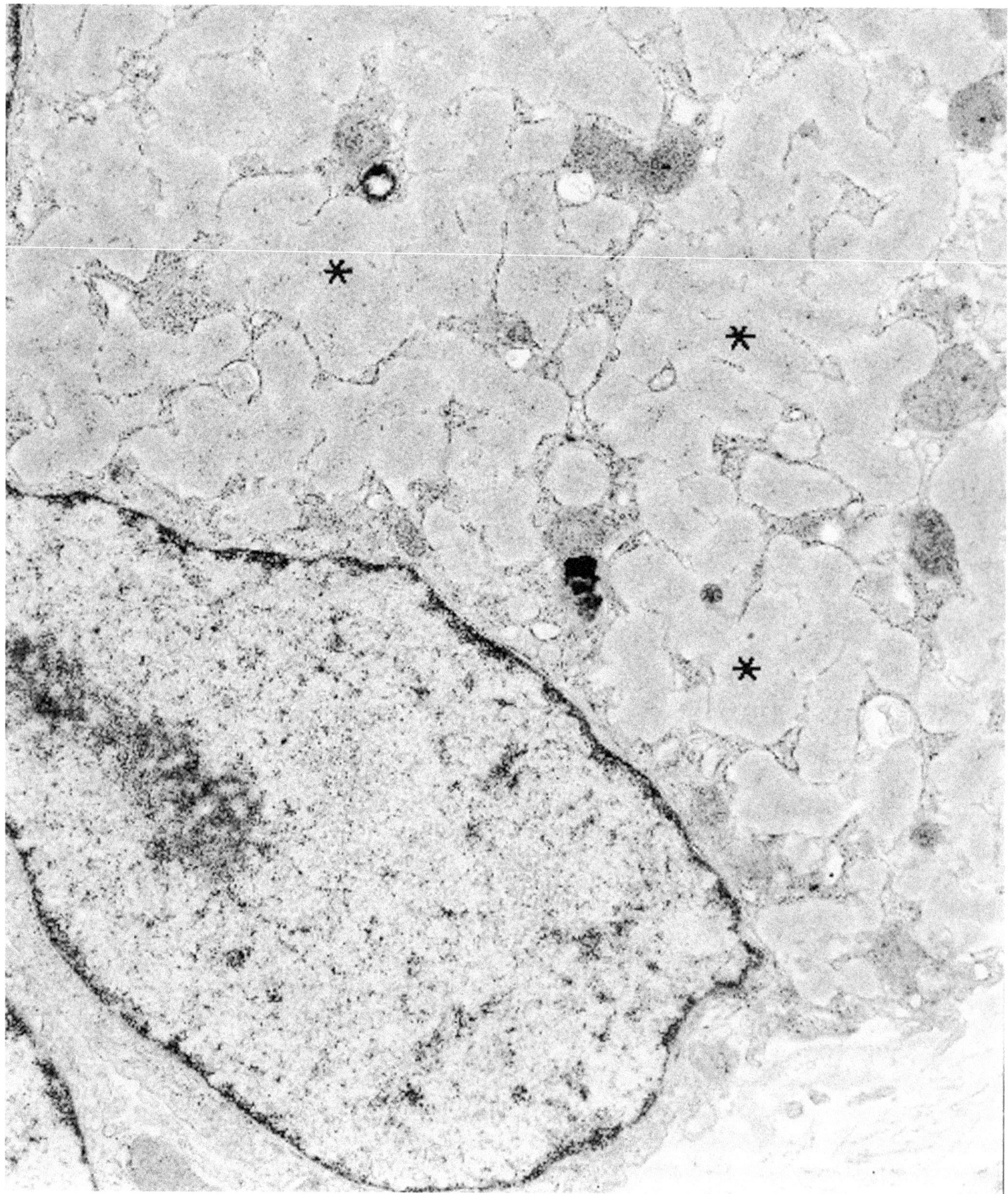

Figure 11.33. Alpha-1-antitrypsin deficiency (liver). High magnification of a hepatocyte highlights the dilated and filled endoplasmic reticulum (*). (× 18,500)

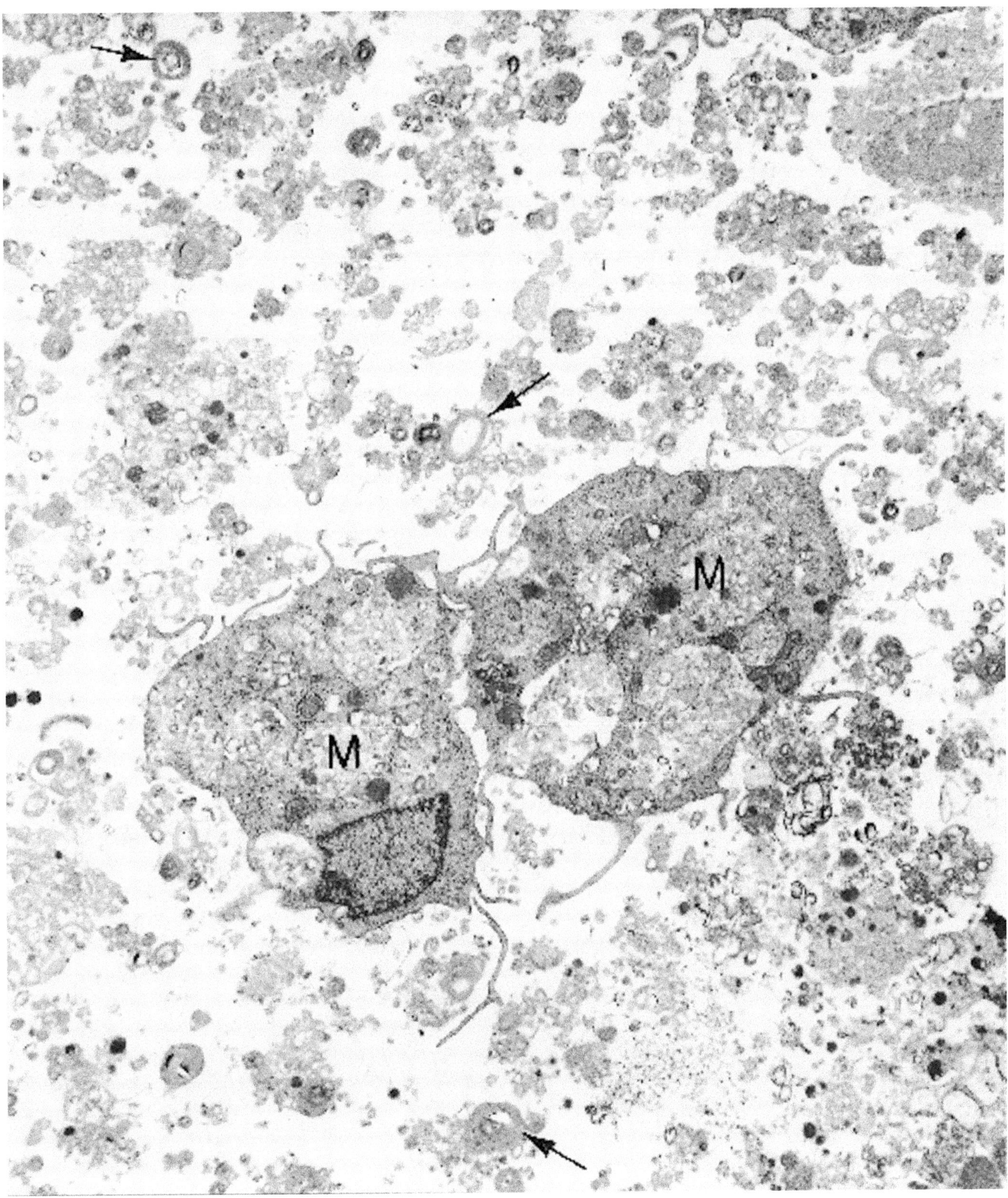

Figure 11.34. Pulmonary alveolar proteinosis (bronchoalveolar lavage fluid). A mixed alveolar exudate is free of microorganisms and rich in whorled membranous structures (*arrows*). The material is found free and within macrophages (M). (× 7560)

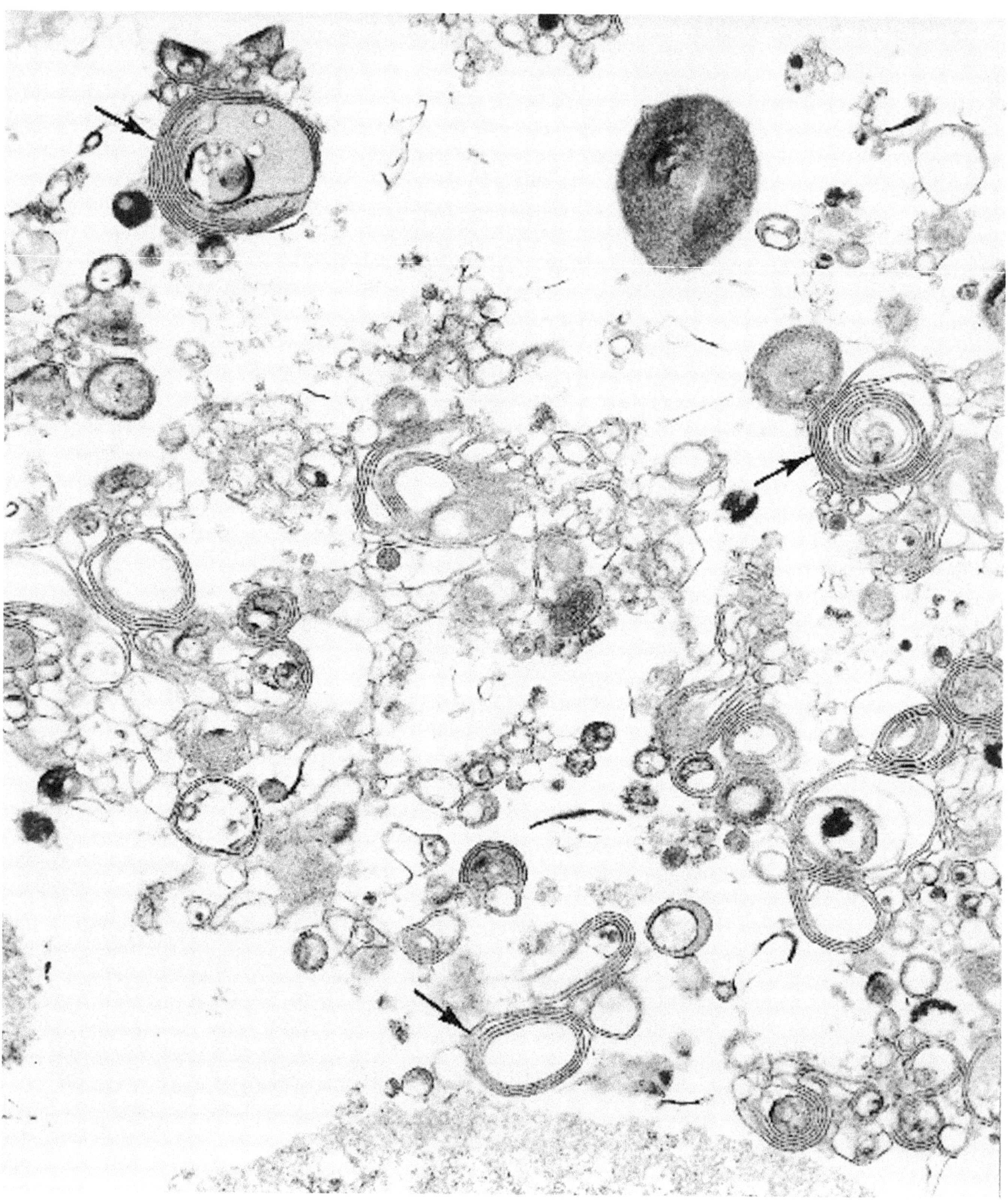

Figure 11.35. Pulmonary alveolar proteinosis (bronchoalveolar lavage fluid). Higher magnification of the alveolar exudate depicted in Figure 11.34 shows the preponderance of membranous (surfactant) bodies (*arrows*). (× 22,680)

(Text continued from page 738)

Mitochondrial Abnormalities

(Figures 11.36 through 11.41.)

Mitochondrial abnormalities may involve various organs but are often indexed under mitochondrial "myopathies." Specifically, the skeletal myopathies are covered in this book in Chapter 13. In this chapter, we exemplify some of the mitochondrial changes that may be found in systems other than skeletal muscle. Most of these disorders are genetic, and the morphological changes are usually attributable to biochemical abnormalities in substrate usage, oxidative phosphorylation, and/or respiratory chain reactions.

Diagnostic criteria. Various alterations in normal mitochondrial morphology, including (1) clustering; (2) enlargement; (3) unusual shapes; (4) increase or decrease in number of cristae; (5) unusual directions and patterns of cristae; (6) increased matrical substance; (7) abnormal inclusions (granular matrical globules and intermembranous or intracristal crystalline structures); (8) various nonmitochondrial changes, such as cytoplasmic inclusions of glycogen and lipid.

Additional points. Leigh's disease is a rare, progressive mitochondriopathy that involves the central nervous system and skeletal muscle as well as a number of other organs such as heart (myocardium), liver, and kidney. Mitochondria may be so increased as to produce an oncocytic appearance to the cells.

Intestinal pseudo-obstruction, which is often accompanied by ophthalmoplegia, is characterized by lipid vacuolar degeneration of myocytes and replacement fibrosis in the muscularis externa and muscularis mucosae of the gut. In some cases, myocytes may also contain cytoplasmic inclusions, which by light microscopy are translucent and PAS-positive, and by electron microscopy consist of aggregates of myofilaments, which with time degenerate further into a granular and homogeneous substance. These changes are in addition to some of the various mitochondrial changes of the types enumerated in the section on diagnostic criteria.

Drug toxicity is another cause of mitochondrial abnormalities even though most of the mitochondrial disorders are genetic. Some of the drugs in this category are alcohol, hydrazine, and certain antiretoviral drugs. The specific mitochondrial changes are among the types already described in the section on diagnostic criteria.

Wilson's disease

(Figures 11.42 through 11.44.)

Diagnostic criteria. Early (asymptomatic stage) changes in hepatocytes include (1) enlarged, pleomorphic mitochondria with increased matrical density, increased matrical granules, dilated cristae, dilated intracristal space, and various crystalline and/or noncrystalline inclusions; (2) increased number, size, and pleomorphism of peroxisomes; (3) variable increase in vesiculation of smooth endoplasmic reticulum; (4) increase in neutral lipid vacuoles. Additional, later (symptomatic stage) hepatocytic changes include (5) increased lipofuscin, especially in pericanalicular region; (6) copper deposits within lysosomes, characterized as multivesiculated densities.

Additional points. In the early, asymptomatic stage of Wilson's disease, copper is diffusely dispersed in the cytosol of hepatocytes and is not discernible by electron microscopy. In the later, symptomatic stage, copper is visible in lysosomes and has the characteristic morphology described in criterion 6 and illustrated in Figure 11.43. Other intermediate changes in the liver include focal necrosis of hepatocytes, cirrhosis, hemosiderin deposits, and filamentous deposits of Mallory's hyaline.

Amyloidosis

See the section on medullary (C-cell) carcinoma of the thyroid in Chapter 9 and sections on renal and muscular diseases in Chapters 12 and 13.

Diagnostic criteria. (1) Extracellular collections of randomly arranged, nonbranching, 7.0–10 nm in diameter filaments.

Additional points. Amyloid has the same ultrastructural appearance in whichever organs and sites it is deposited, and electron microscopy is useful in demonstrating small amounts that may be difficult to see by light microscopy.

(Text continues on page 757)

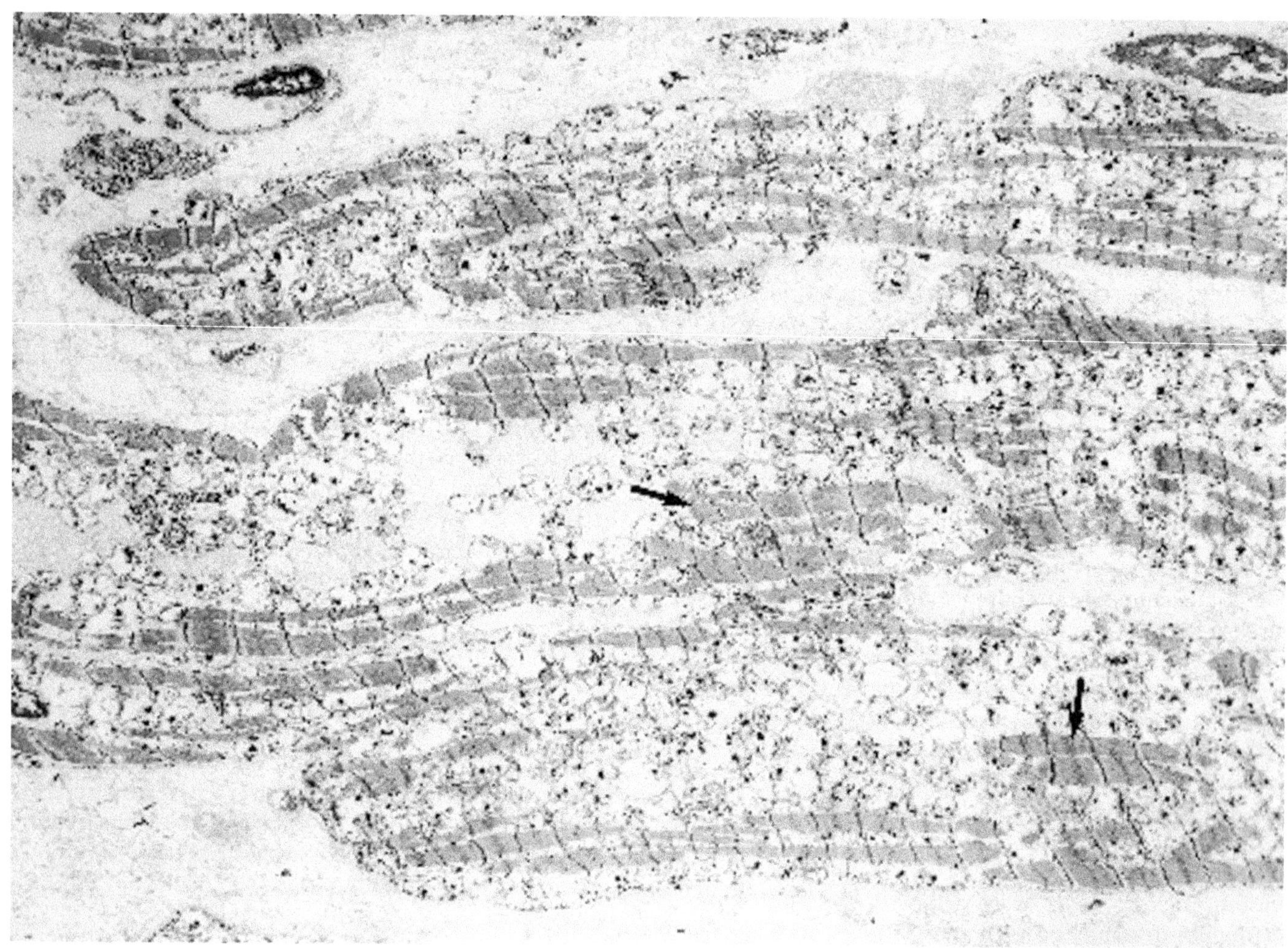

Figure 11.36. Mitochondrial abnormality in Leigh's disease (heart). This postmortem sample of myocardium shows myocytes with drop-out of myofibrils (*arrows*) and replacement by innumerable, dilated mitochondria (seen better at high magnification in Figure 11.37). (× 4900)

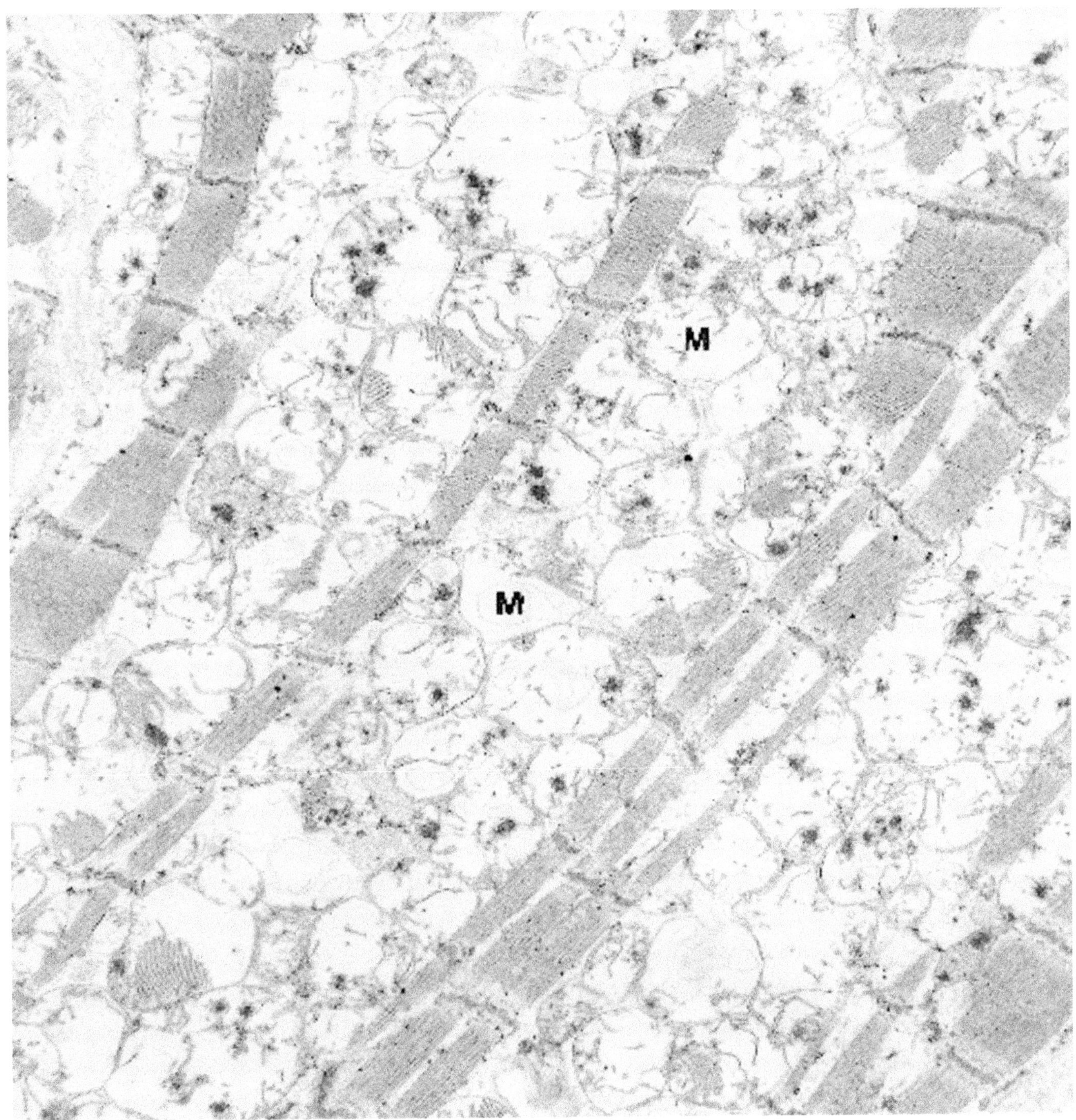

Figure 11.37. Mitochondrial abnormality in Leigh's disease (heart). Remaining myofibrils vary in size and are separated by an increased number of mitochondria (M). The mitochondria are dilated and have altered patterns of cristae. (× 13,600)

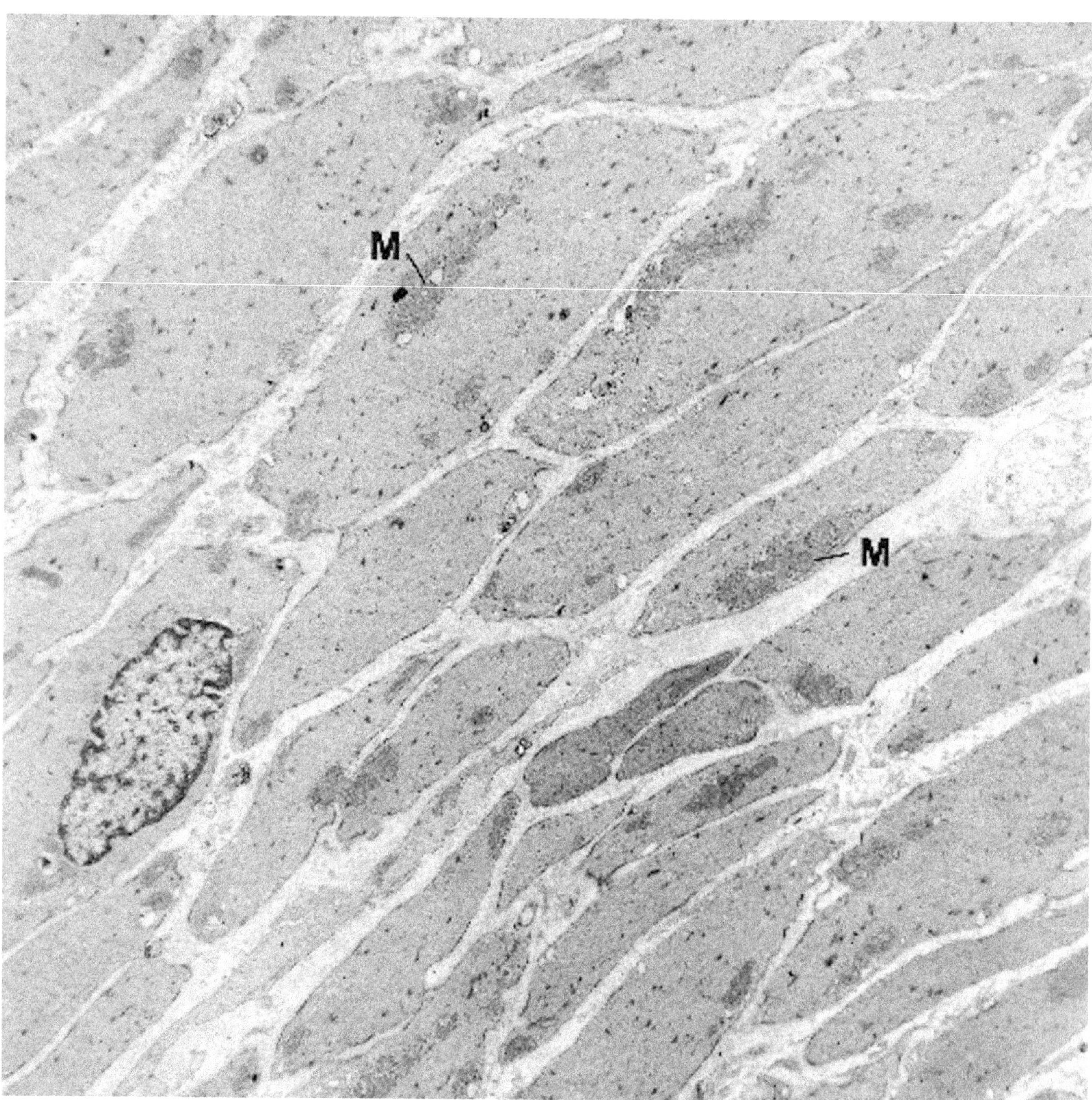

Figure 11.38. Mitochondrial abnormality in chronic intestinal pseudo-obstruction (jejunum). The leiomyocytes of the muscularis propria have an irregular distribution and arrangement of mitochondria, which occur in groups (M) but usually not in the normal paranuclear polar position. (× 7300)

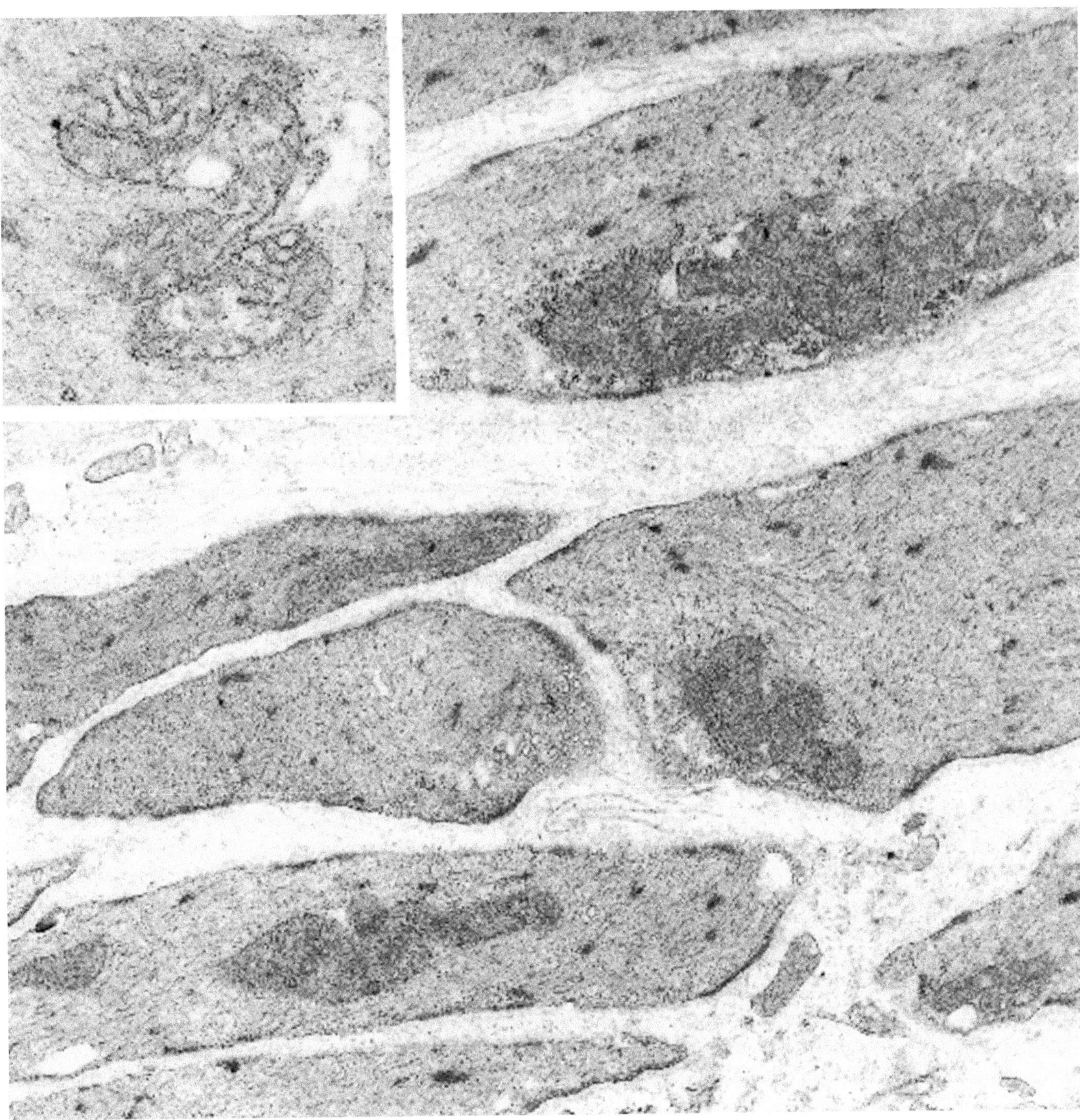

Figure 11.39. Mitochondrial abnormality in chronic intestinal pseudo-obstruction (jejunum). Higher magnification of some of the leiomyocytes shown in Figure 11.38 illustrates mitochondrial aggregation and variation in size and shape. The *inset* depicts irregularly shaped mitochondria with an abnormal arrangement of cristae. Similar changes were present in some of the Schwann cells of the myenteric plexus. (× 21,000) (*inset* × 50,000)

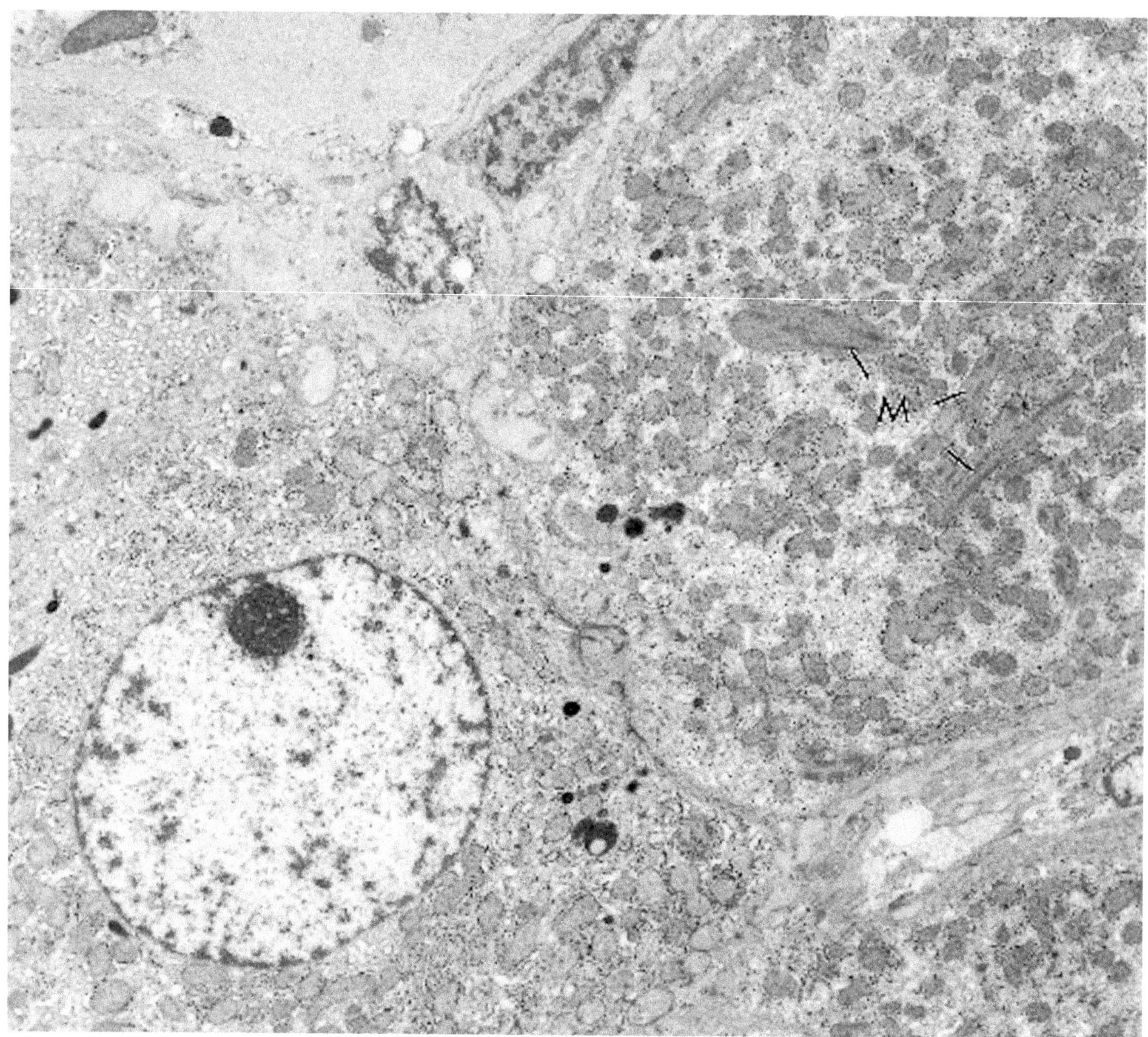

Figure 11.40. Mitochondrial abnormality in (antiretroviral) drug toxicity (liver). Some of the mitochondria (M) of the hepatocytes are of abnormal size and shape, best exemplified in the cell in the upper right field. (× 6600)

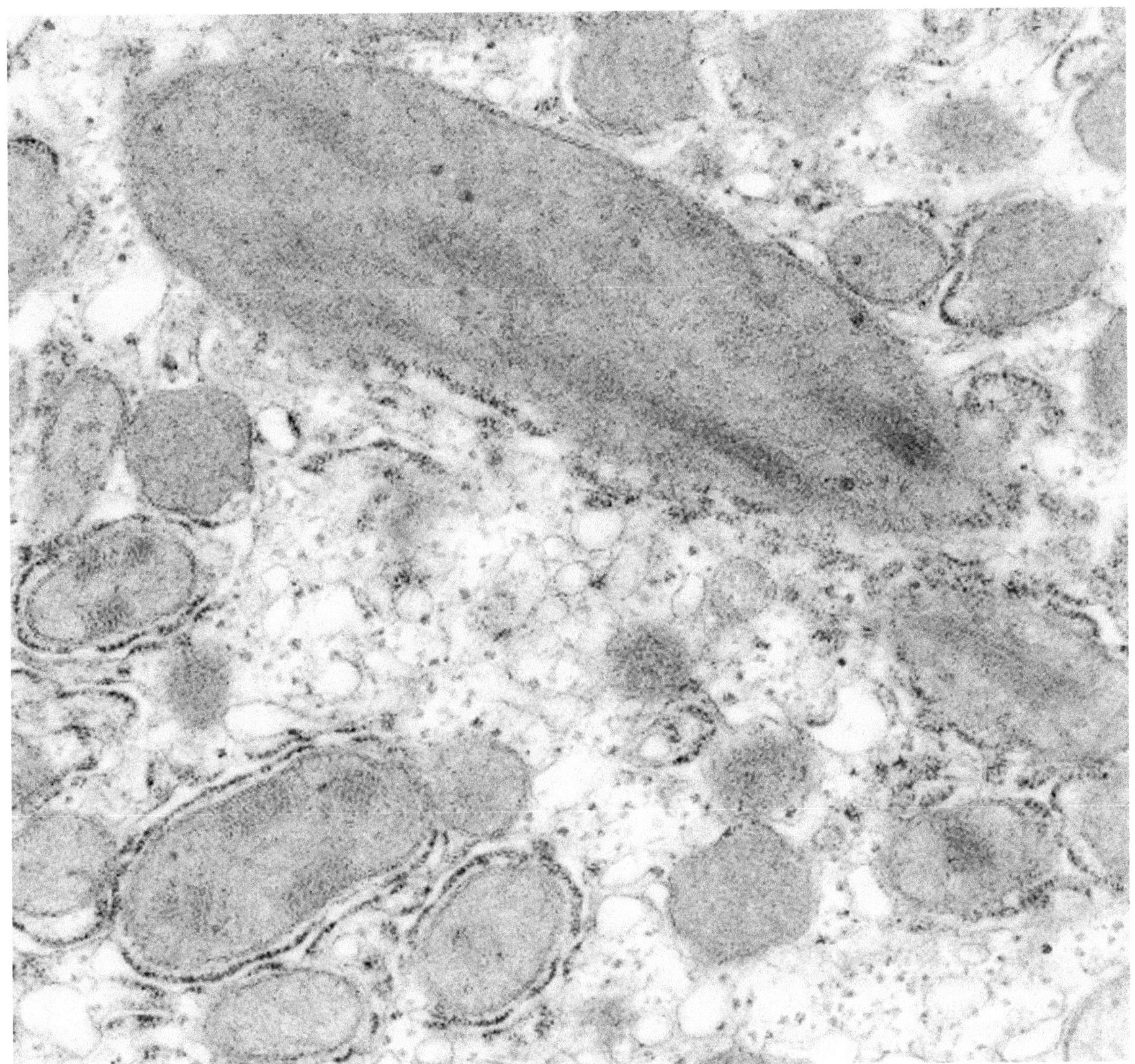

Figure 11.41. Mitochondrial abnormality in (antiretroviral) drug toxicity (liver). Higher magnification of several of the mitochondria seen in Figure 11.40 illustrates their large size as well as abnormally arranged cristae, often in tubular arrays. (× 45,000)

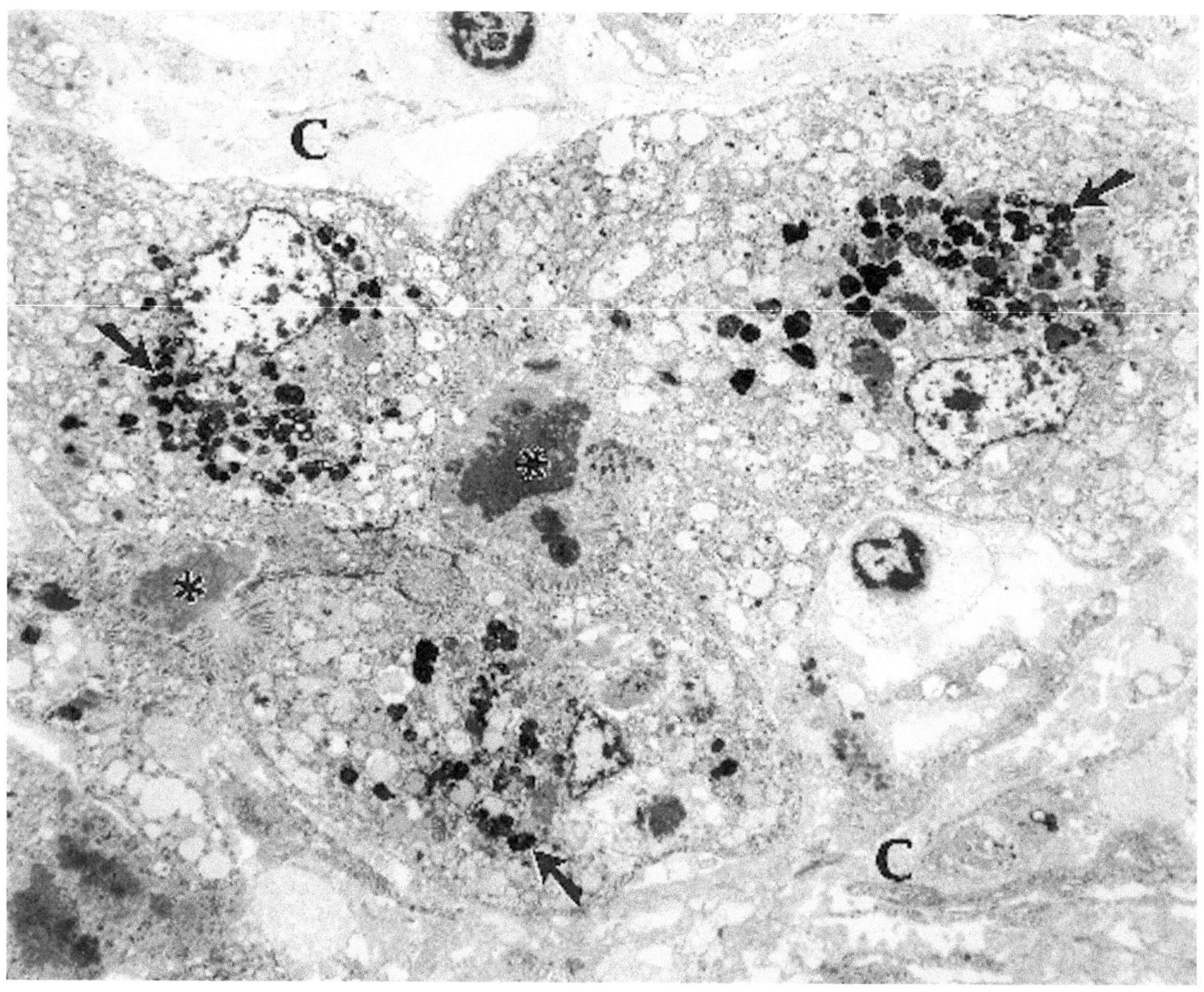

Figure 11.42. Wilson's disease (liver). A cirrhotic liver in late-stage Wilson's disease shows a nodule of hepatocytes surrounded by fibrocollagenous tissue (C). The hepatocytes contain numerous cytoplasmic electron-densities (*arrows*), and canaliculi (*) are distended by bile. (× 3500)

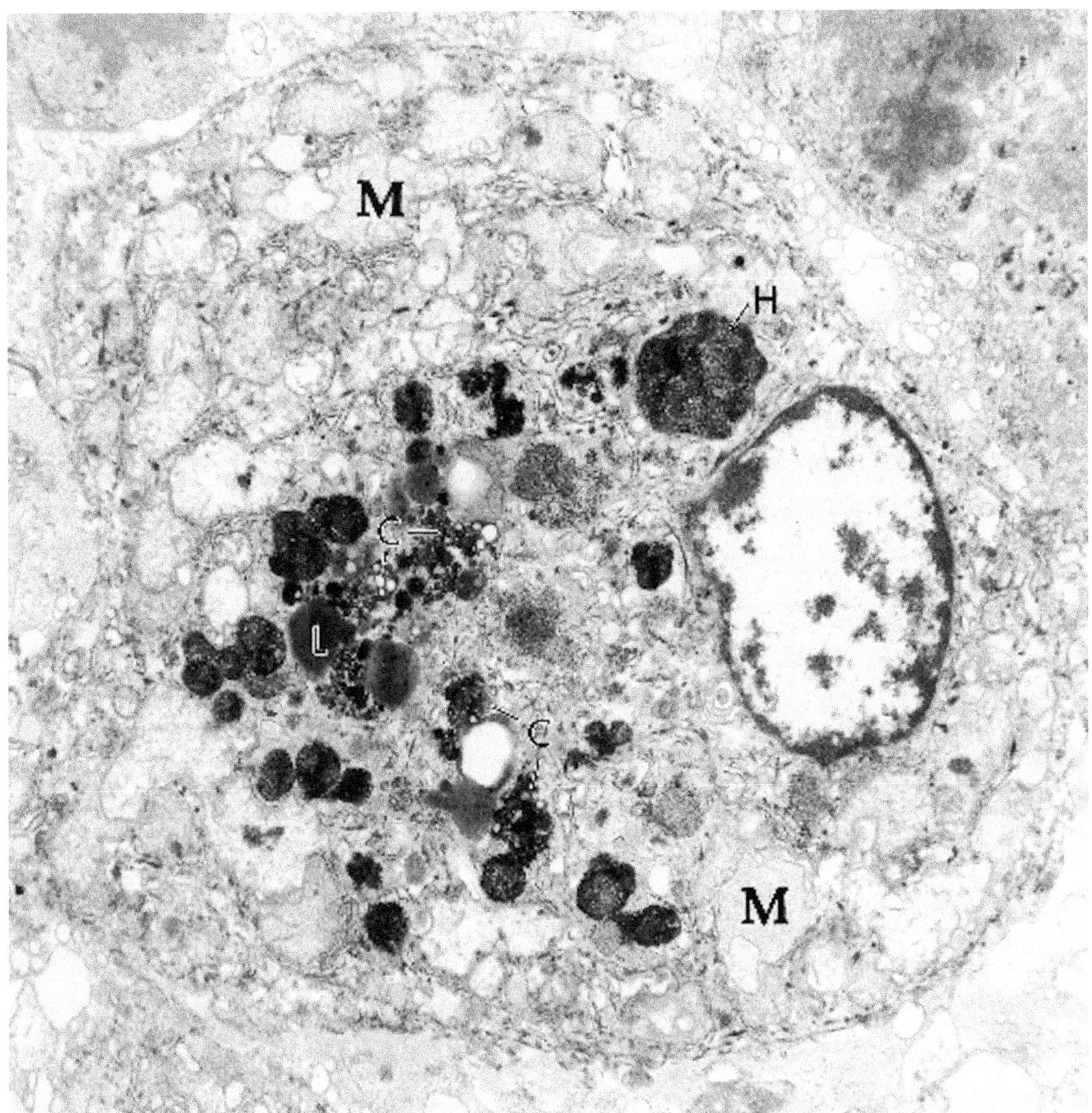

Figure 11.43. Wilson's disease (liver). Higher magnification of a hepatocyte from the same liver as depicted in Figure 11.42 illustrates the composition of the various cytoplasmic electron densities: some are smooth and represent neutral lipid (L); some are finely granular and represent hemosiderin (H); some are vesiculated and represent copper (C). Lipofuscin, shown in Figure 11.44, was also present in some cells. Note also the enlarged and pleomorphic mitochondria (M) with decreased and irregularly directed cristae. (× 12,400)

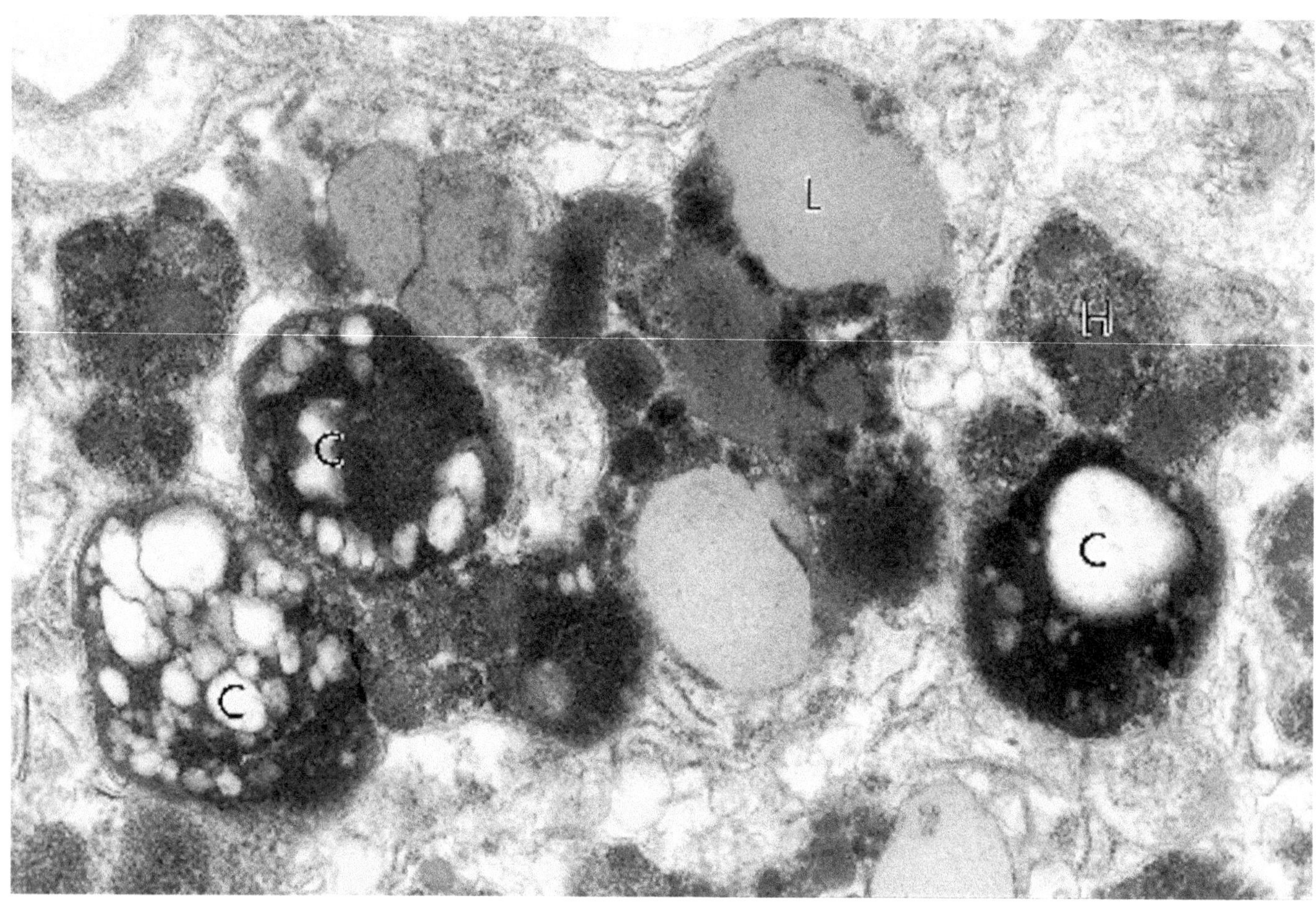

Figure 11.44. Wilson's disease (liver). High magnification of hepatocytic cytoplasm illustrates well the internal structure of the various electron densities shown in (Figures 11.42 and 11.43). L = lipofuscin; H = hemosiderin; C = copper. (× 39,000)

(Text continued from page 747)

Amiodarone Toxicity

(Figures 11.45 through 11.47.)

Diagnostic criteria. (1) Histiocytic infiltrates in lung, liver, and other organs; (2) lysosomal lamellar bodies in histiocytes and parenchymatous cells.

Additional points. Amiodarone has been used in the treatment of ventricular arrhythmias, notwithstanding its possible toxicity to various tissues, especially lung and liver. Pulmonary changes include interstitial pneumonia, foamy histiocytes in alveoli, and lamellar bodies in the histiocytes. In the liver, the drug produces progressive changes: first, cholestasis and perivenular infiltration by granular histiocytes; and later, hepatocellular necrosis and cirrhosis, with granular histiocytes. The histiocytes contain lysosomes with lamellar bodies. The mechanism of toxicity is by way of inhibiting phospholipase activity, which leads to phospholipid accumulation in the form of lamellar bodies in lysosomes. These toxic effects are more cumulative than dose related, and lysosomal lamellar bodies are identifiable before any clinical signs become manifest.

Adriamycin Toxicity

(Figures 11.48 and 11.49.)

Diagnostic criteria. (1) Dilatation of myocytic sarcoplasmic reticulum and transverse tubular system; (2) numerous lipid droplets in myocytic cytoplasm; (3) loss of myofibrils and sarcomeres; (4) nucleolar fragmentation and segregation.

Additional points. Adriamycin, effectively used in the treatment of numerous neoplasms, is cumulatively toxic to the heart. The mechanism of cardiomyopathy may be through peroxidation of cardiac lipid. The severity of myocardial damage is increased by previous or concomitant mediastinal irradiation. In addition, irradiation induces changes in small blood vessels. Capillary endothelial cells show swelling and cytoplasmic blebs, and arterioles and venules show duplication of basal lamina.

Hemosiderosis

(Figures 11.50 and 11.51.)

Diagnostic criteria. (1) Extracellular or intracellular, irregularly shaped deposits of loosely dispersed granules of electron-dense material.

Additional points. Within cells, usually histiocytes, the hemosiderin deposits are within secondary lysosomes, but the limiting membrane of the lysosomes may be difficult to discern because of its close apposition to the electron-dense hemosiderin. The hemosiderin-laden lysosomes are referred to as "siderosomes," and they may be single or compound. Hemosiderin consists of multiple aggregates of ferritin, which is composed of an outer protein shell (apoferritin) and an inner region of inorganic iron.

Cholestasis

(Figures 11.52 through 11.55.)

Diagnostic criteria. Intrahepatic, intracanalicular, and/or intraductal deposits of electron-dense substance (bile), which may have several patterns, including (1) finely granular; (2) fibrillar; and/or (3) whorled lamellar.

Additional points. The granular and fibrillar substance represents bilirubin, and the lamellar material is consistent with being phospholipid, cholesterol, and conjugated bile salts. Secondary hepatocytic changes in cholestasis include replacement of pericanalicular microfilaments by an amorphous material, distention and hypertrophy of Golgi apparatuses, and hypertrophy of smooth endoplasmic reticulum.

"Melanosis" (Lipofuscinosis) Coli and Prostaticus

(Figures 11.56 and 11.57.)

Diagnostic criteria. (1) Apoptosis of surface epithelial cells (colon); (2) phagocytosed apoptotic structures by intraepithelial histiocytes (colon); (3) granular, electron-dense material (lipofuscin), plus droplets of pale or medium-dense, homogeneous material (neutral lipid) in round and irregularly shaped vesicles (lysosomes), in epithelial cells and histiocytes (colon and prostate).

Additional points. The pigment in these conditions was earlier thought to be melanin or melanin-like, but more recently electron microscopy and electron-probe energy dispersive radiography have demonstrated lipofuscin. Histochemical evidence supports the concept that lipofuscin and ceroid are the same substance but in different stages of oxidation. Other substances such as silicates and titoneum, or hemosiderin may also be found in "melanosis" of the ileum, and iron sulfide in "melanosis" of the duodenum. In the gastrointestinal tract and prostate, lipofuscin-laden histiocytes are present in the subepithelial stroma as well as in the epithelium. In the prostatic epithelium, the pigment is located predominantly in a subnuclear position. Both benign and malignant epithelium may be involved.

(Text continues on page 771)

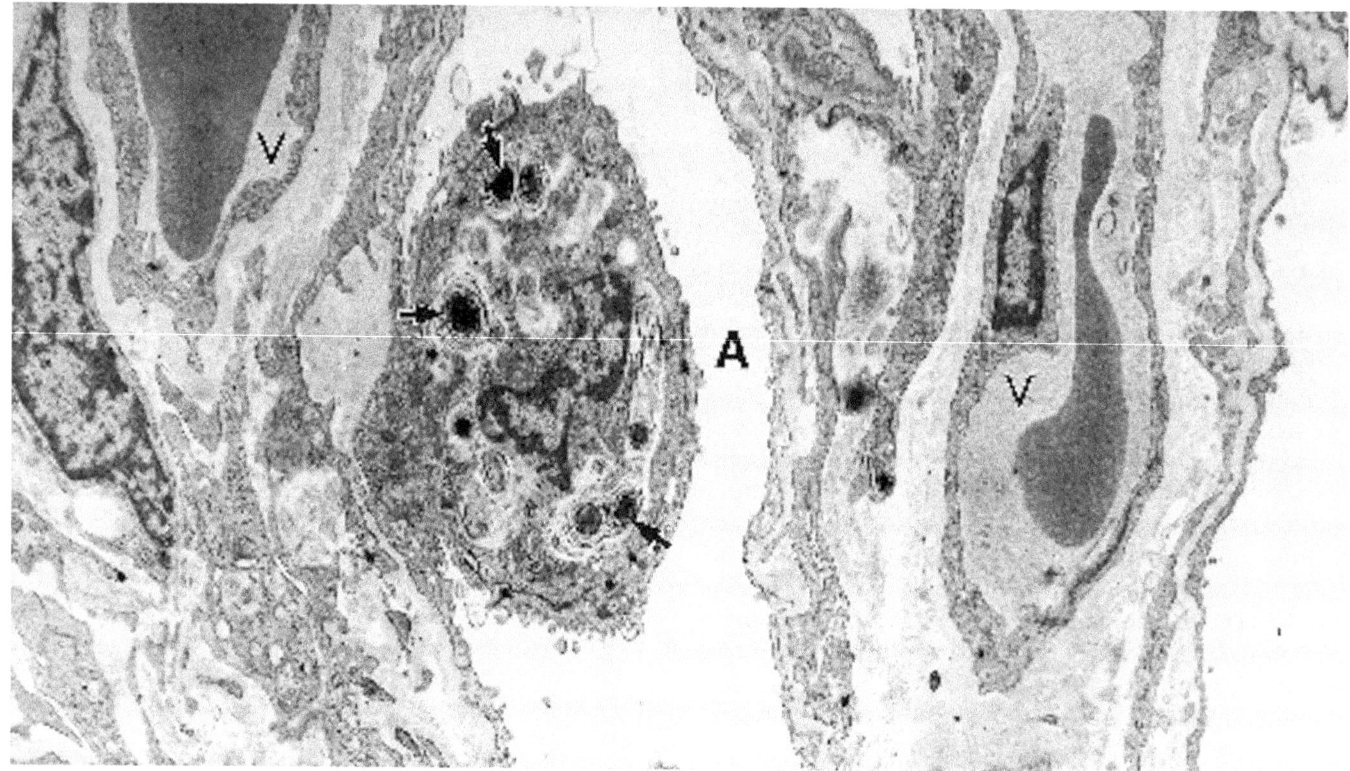

Figure 11.45. Amiodarone toxicity (lung). A type II pneumocyte contains numerous cytoplasmic lamellar (surfactant) bodies (*arrows*). A = alveolar space; V = septal blood vessels. (× 7200)

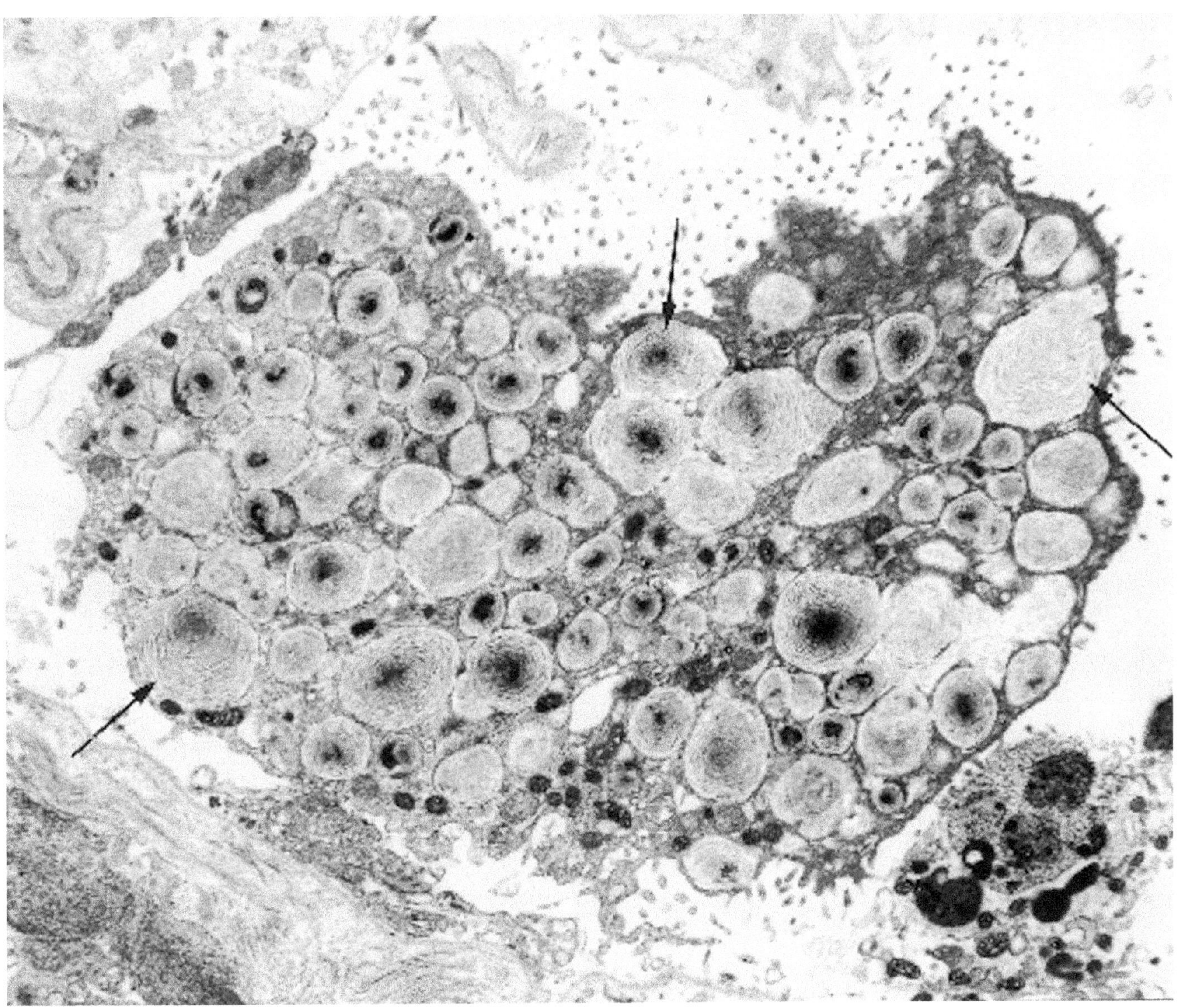

Figure 11.46. Amiodarone toxicity (lung). High magnification of a desquamated type II pneumocyte illustrates an excessive number of lamellar bodies (*arrows*) filling the cytoplasm. (× 9700)

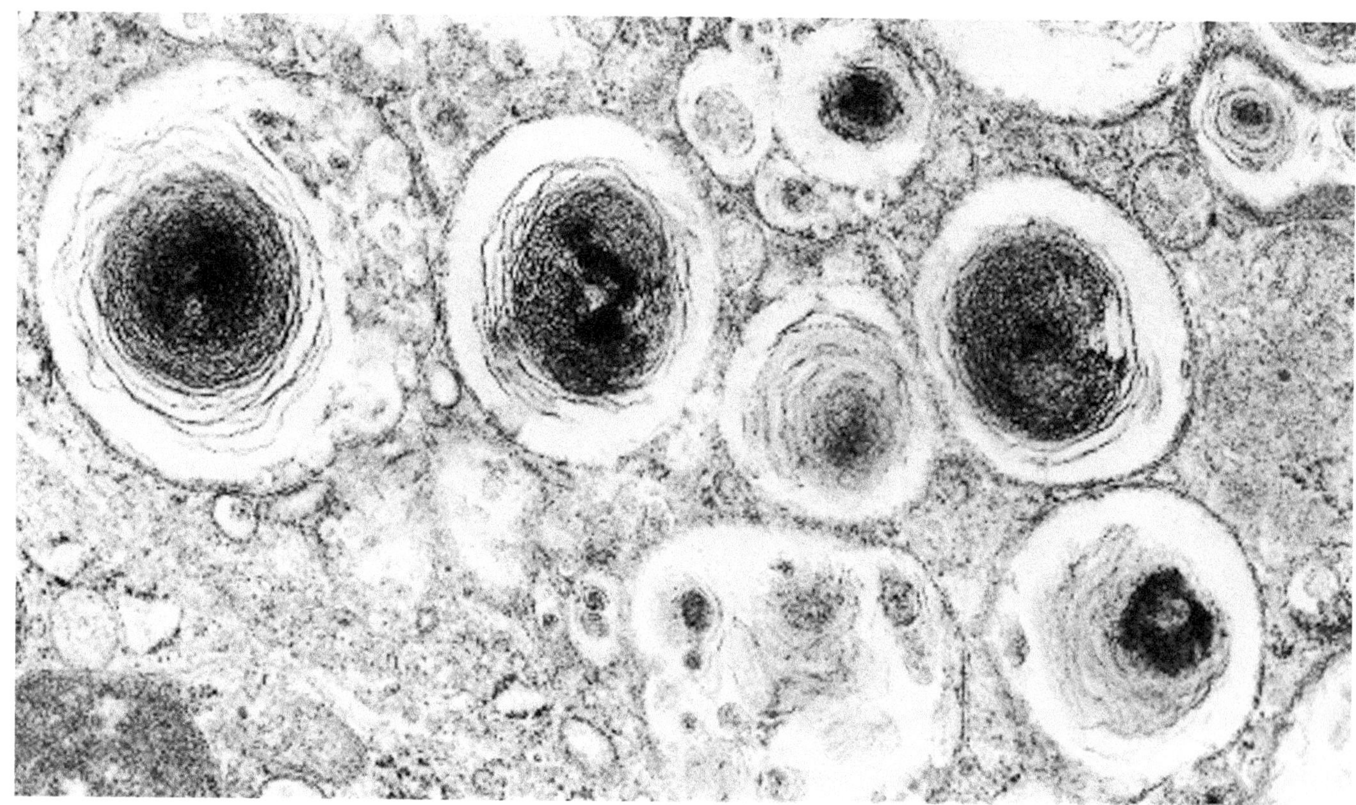

Figure 11.47. Amiodarone toxicity (lung). High magnification of type II pneumocytic cytoplasm depicts clearly the fine structural detail of several lamellar bodies. (× 45,000)

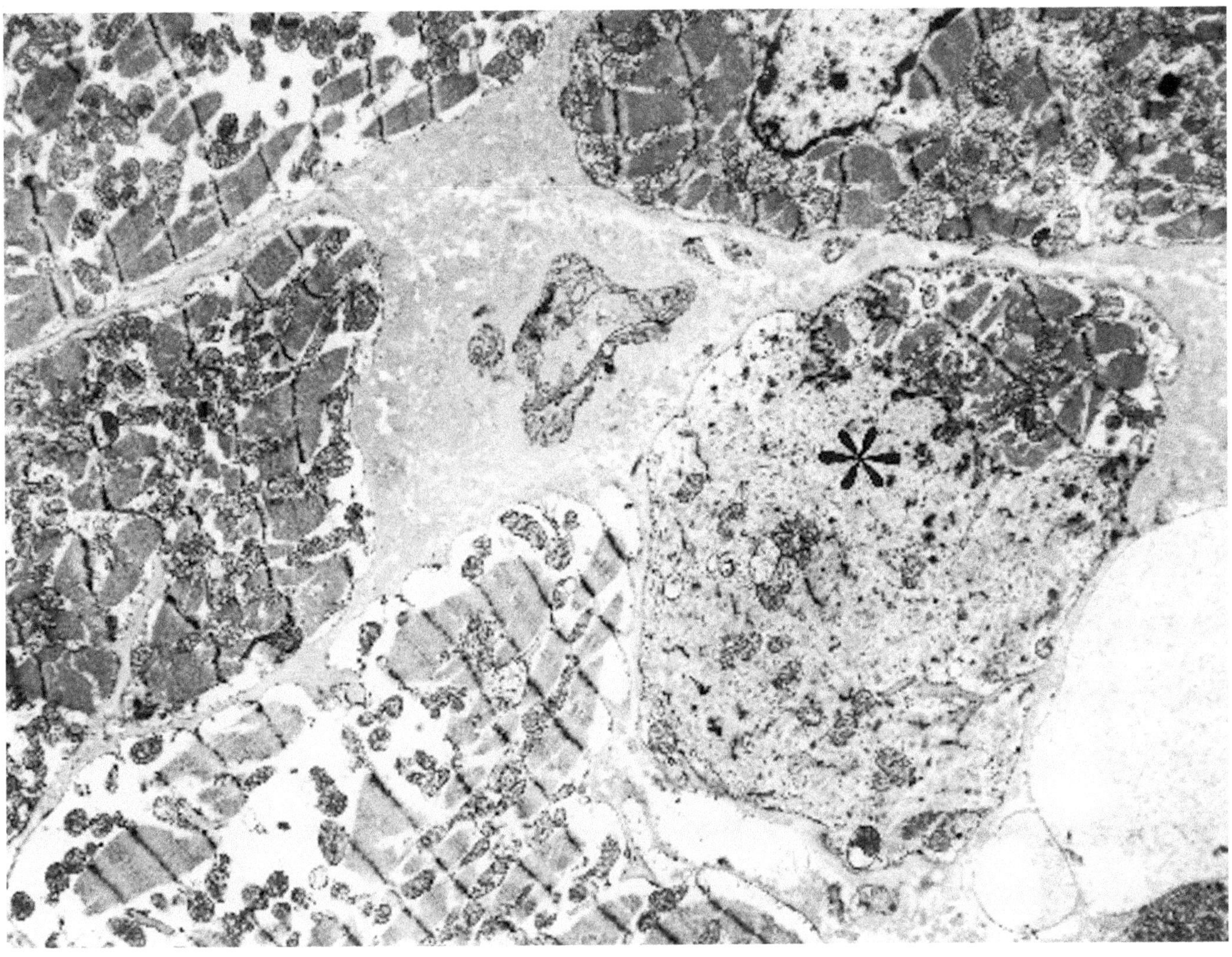

Figure 11.48. Adriamycin toxicity (heart). Focal myocytes (*) show a loss of myofibrils and sarcomeres. These changes are consistent with, but not specific for Adriamycin and/or x-ray effect. (× 5000)

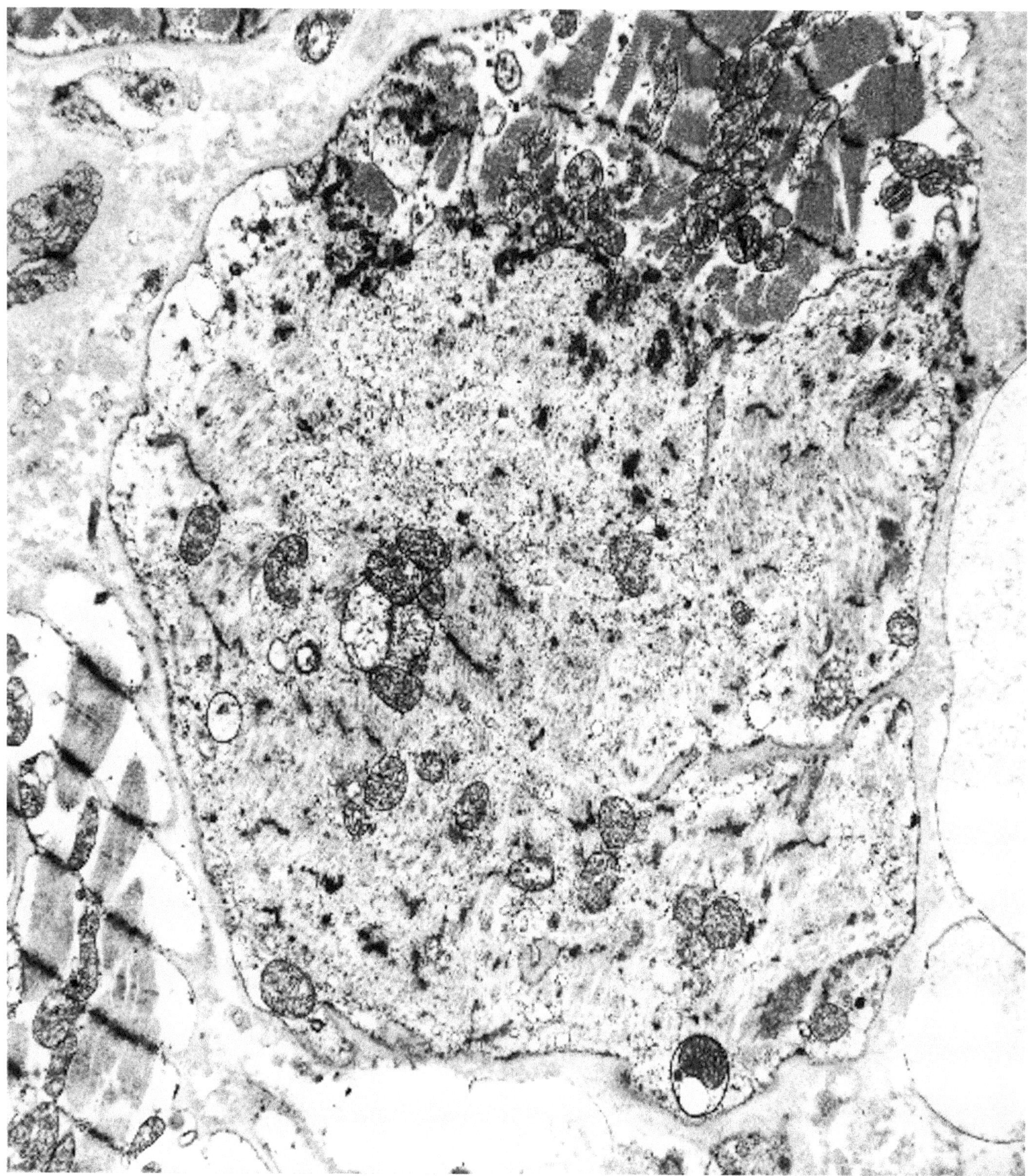

Figure 11.49. Adriamycin toxicity (heart). High magnification of the abnormal myocyte shown in Figure 11.48 illustrates again the myofibril loss. (× 10,500)

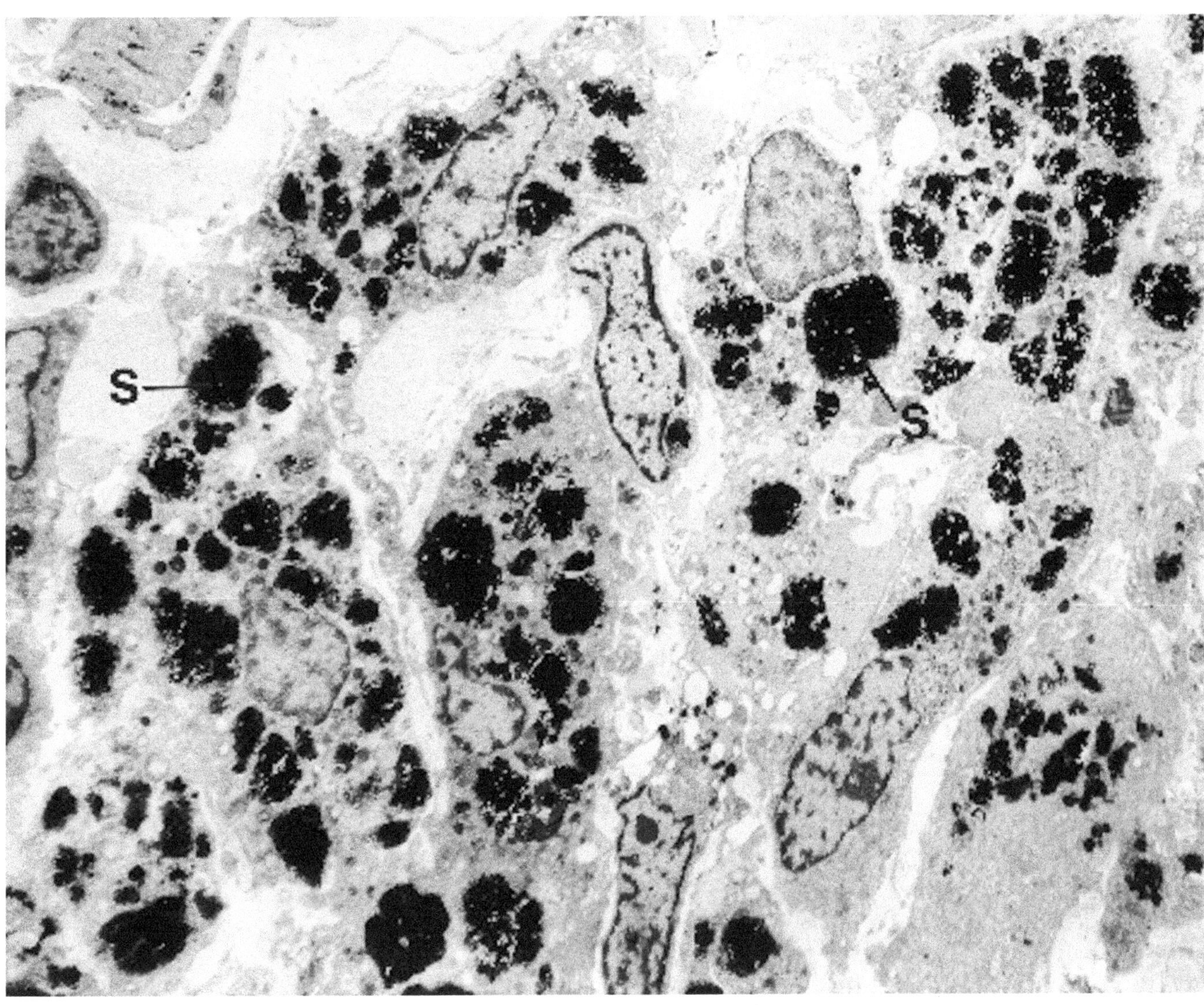

Figure 11.50. Hemosiderosis (lung). Alveolar and septal macrophages contain innumerable electron-dense deposits, siderosomes (S). (× 7400)

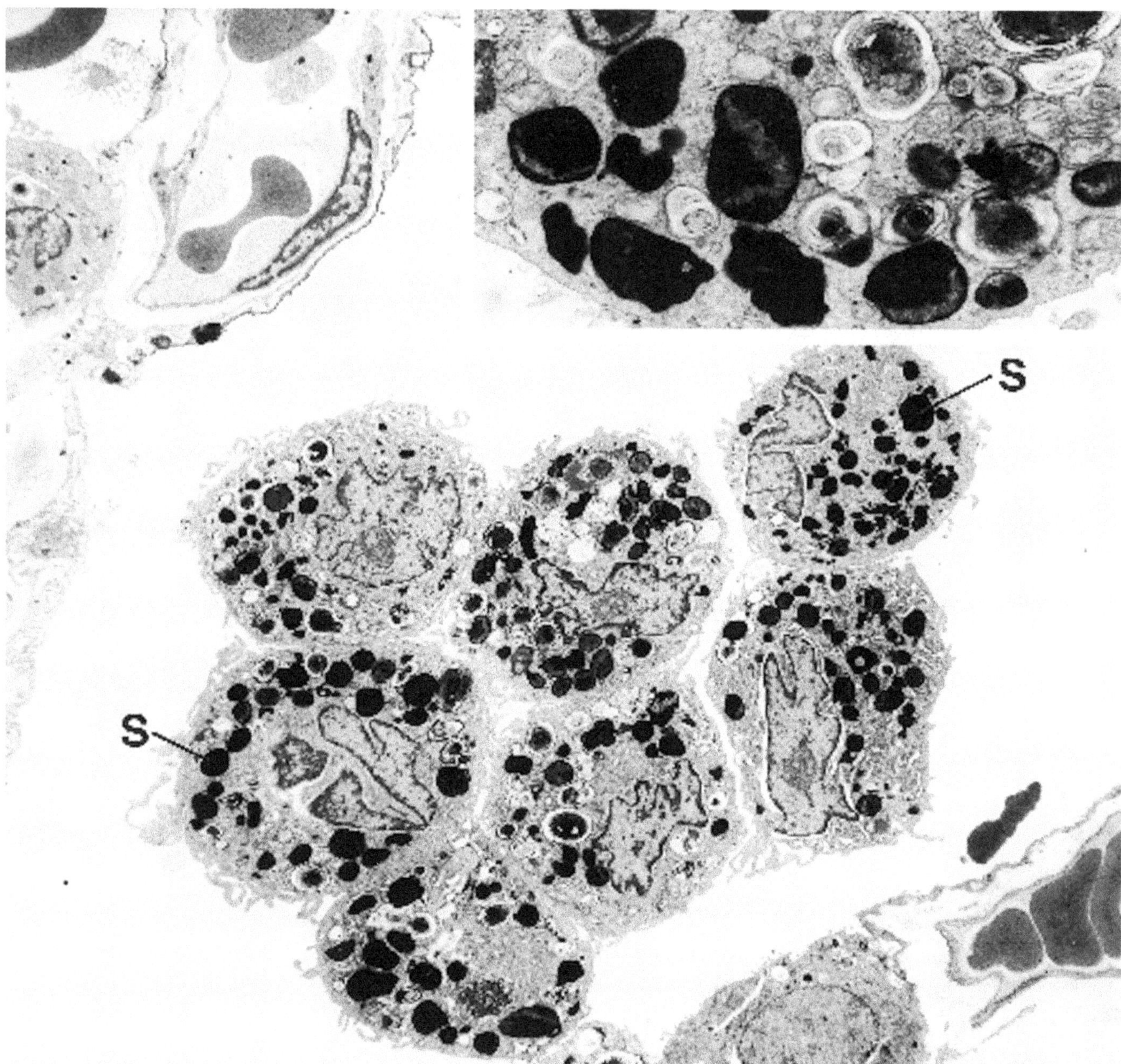

Figure 11.51. Hemosiderosis (lung). An alveolar space contains a cluster of hemosiderin-laden macrophages. S = siderosomes. (× 4200) *Inset:* High magnification of siderosomes in the cytoplasm of an alveolar macrophage reveals both a smooth and membranous internal pattern. (× 15,800)

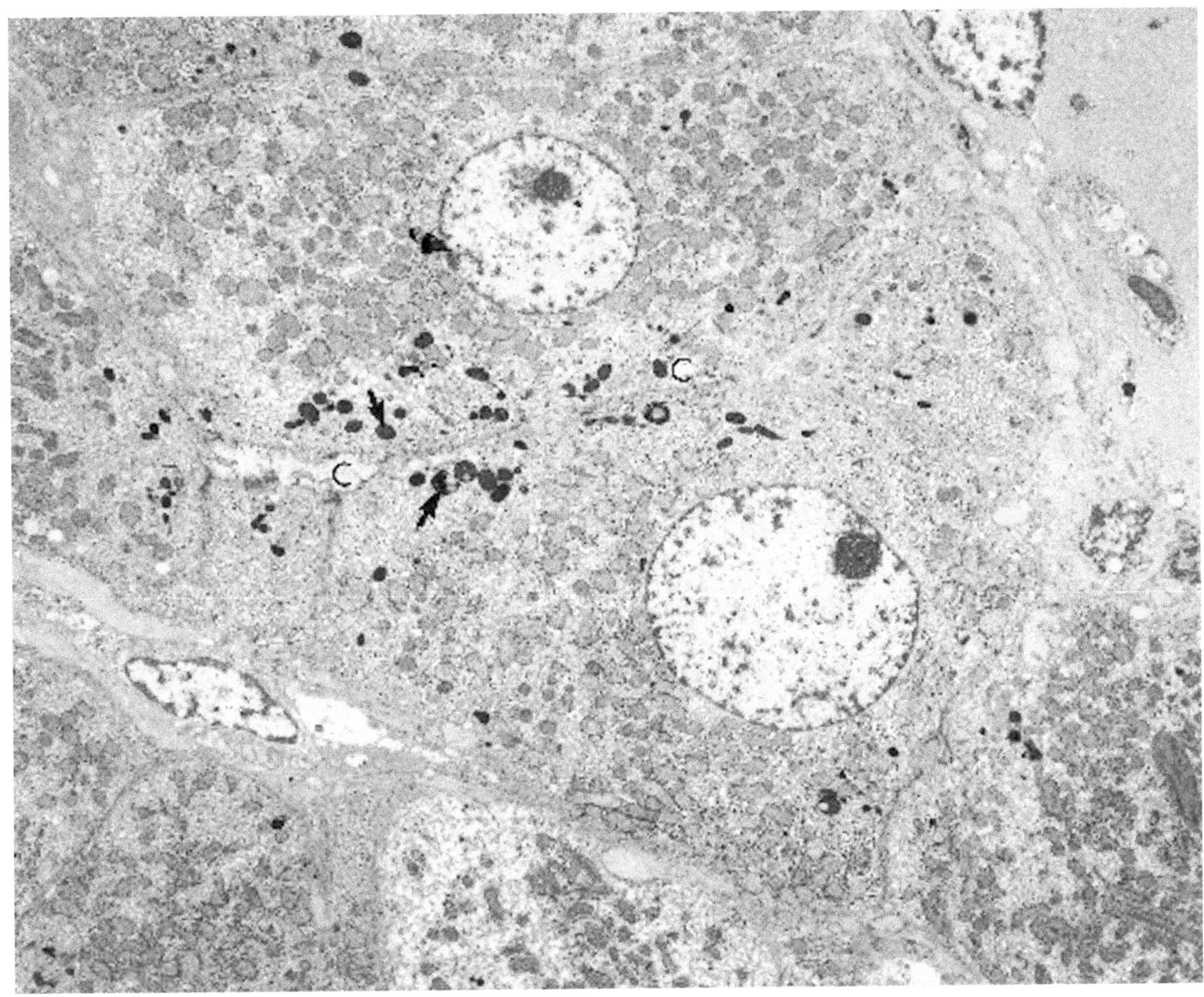

Figure 11.52. Cholestasis (liver). Hepatocytes contain a moderate number of electron-dense globules of bile pigment (*arrows*) in their apical cytoplasm, adjacent to bile canaliculi (C). (× 4000)

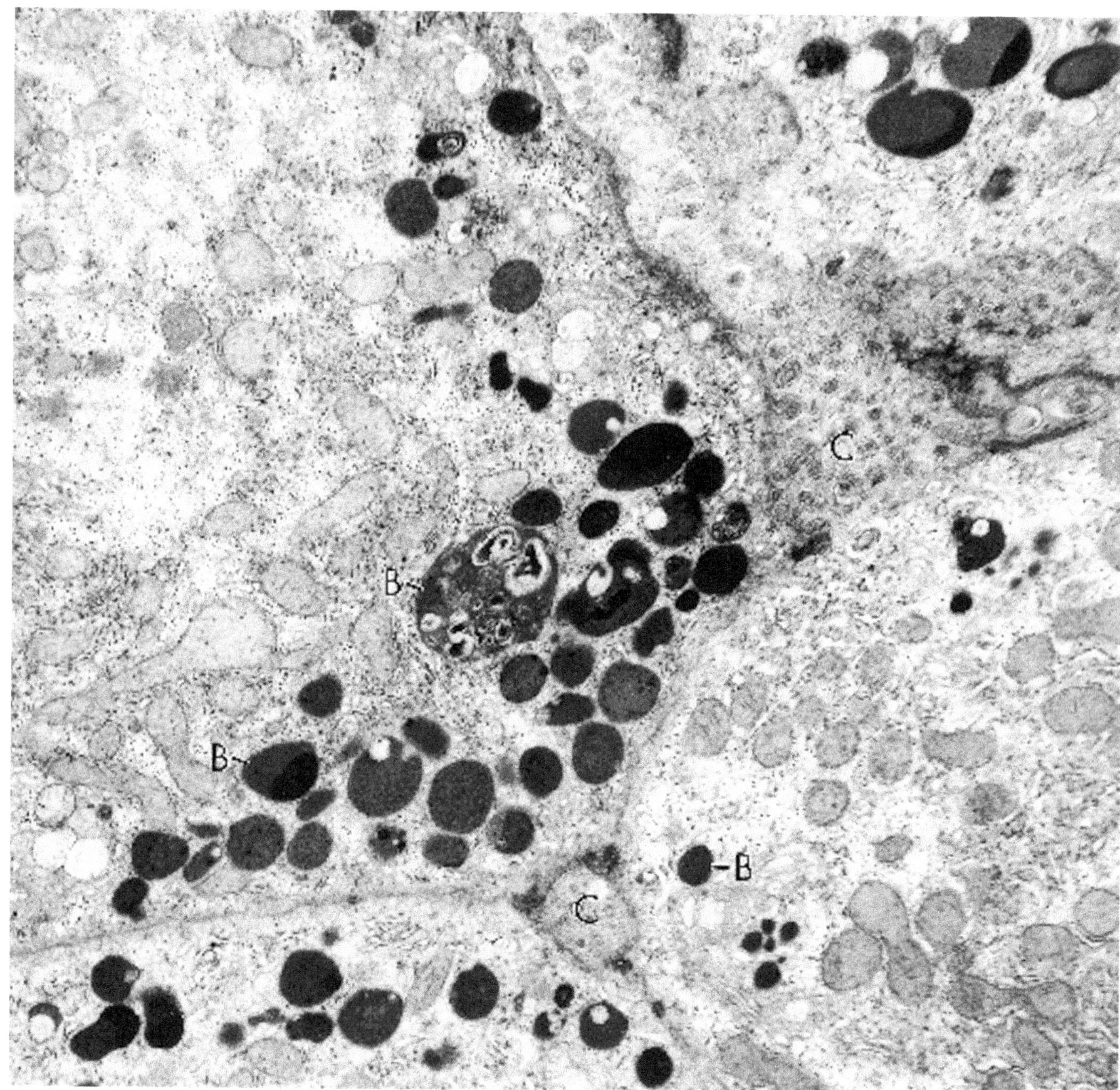

Figure 11.53. Cholestasis (liver). Higher magnification of the same liver illustrated in Figure 11.52 reveals the bile pigment (B) to be smooth, finely granular or heterogeneous, electron-dense material. C = canaliculi. (× 11,000)

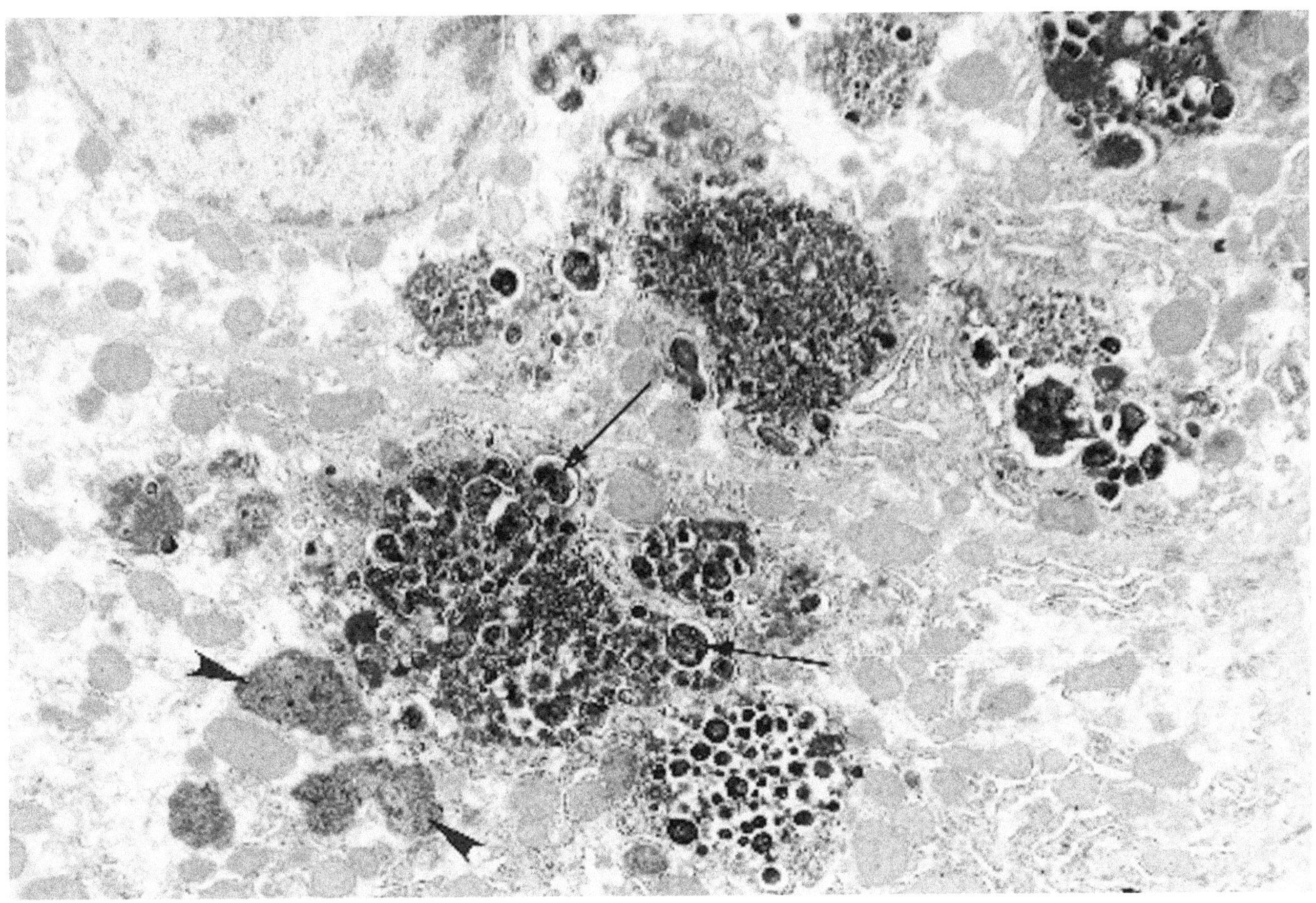

Figure 11.54. Cholestasis (liver). Hepatocytes contain numerous cytoplasmic inclusions of bile pigment, some of which is membranous (*arrows*) and others of which are finely granular (*arrowheads*). (× 10,300)

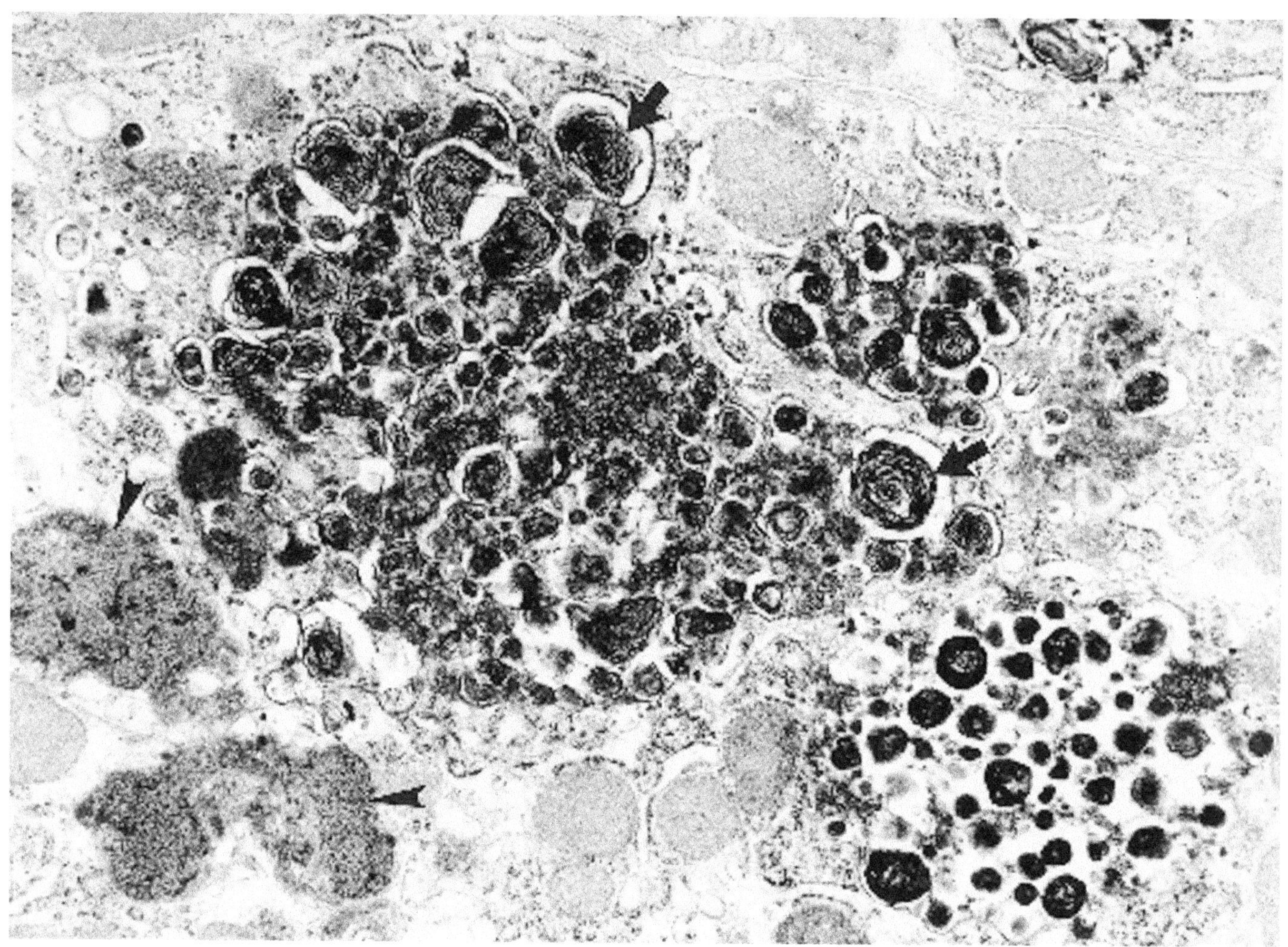

Figure 11.55. Cholestasis (liver). Higher magnification of the bile pigment shown in Figure 11.54 illustrates in detail its membranous (*arrows*) and granular (*arrowheads*) textures. (× 22,000)

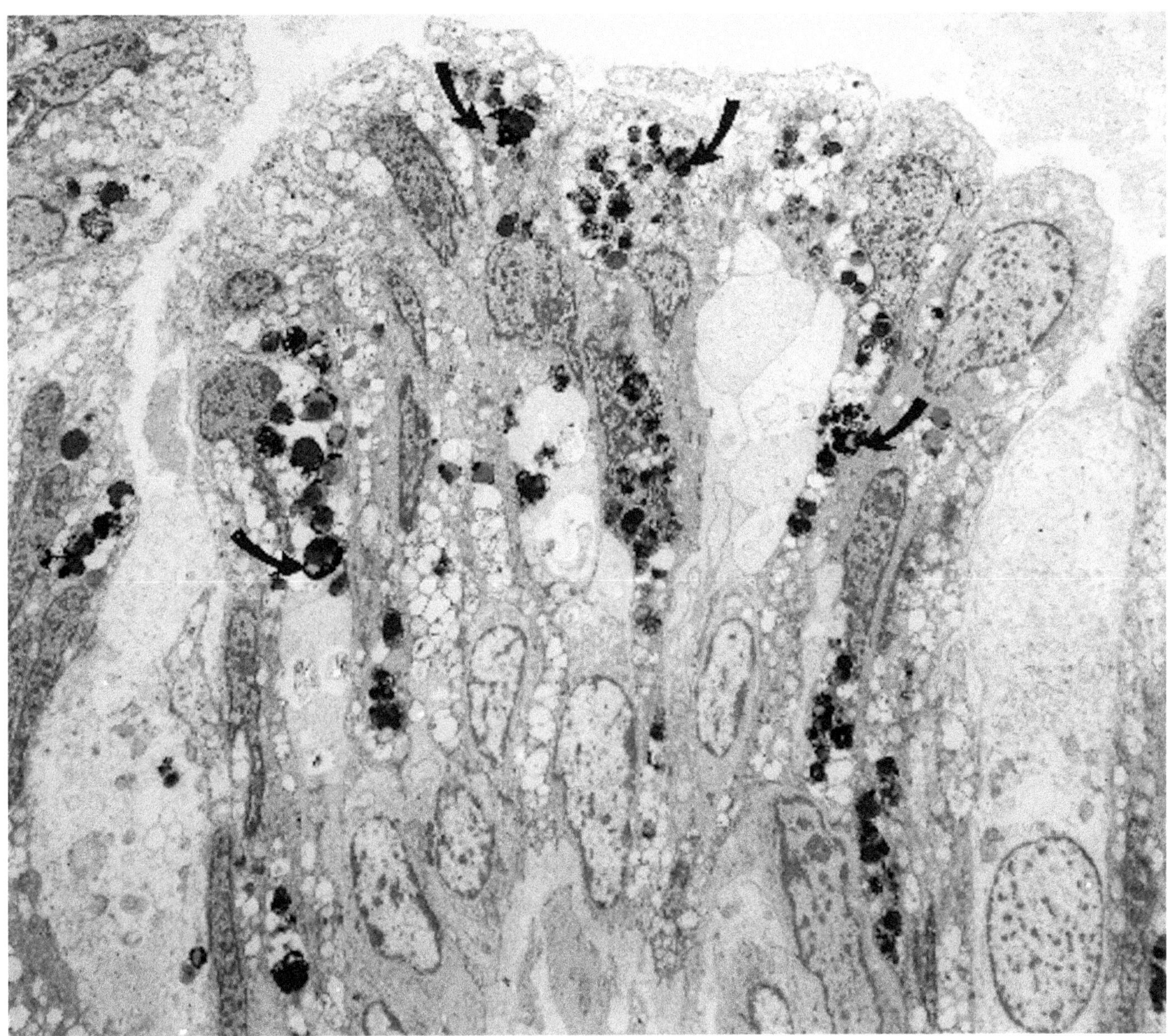

Figure 11.56. Melanosis (prostate gland). Prostatic epithelial cells contain numerous electron-dense granules (*arrows*) in their cytoplasm. (× 4200)

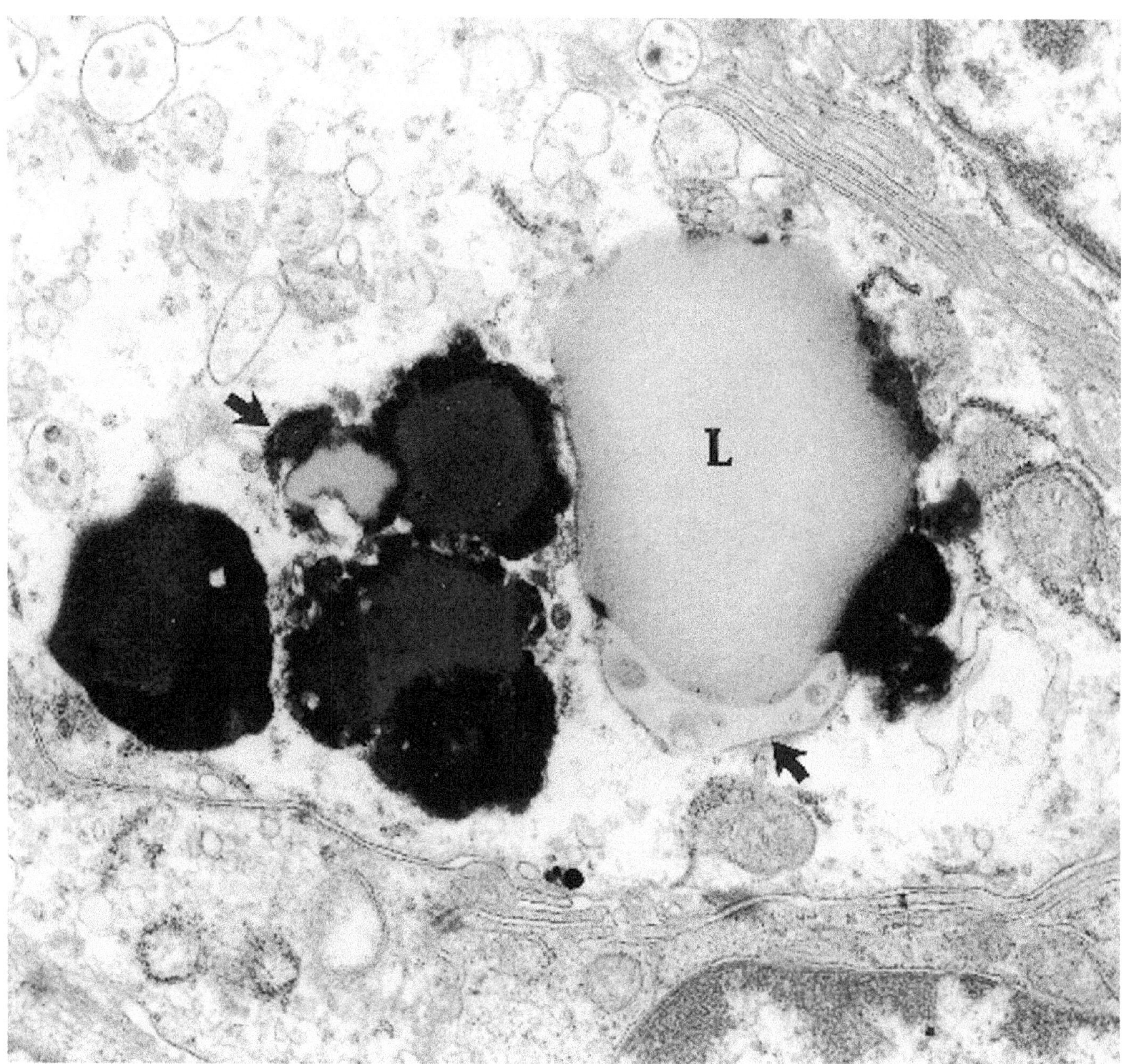

Figure 11.57. Melanosis (prostate gland). High magnification of several electron-dense granules of the type depicted in Figure 11.56 reveals them to be lipofuscin, often with a pale smooth component of neutral lipid (L) and a darker globular component. Lysosomal limiting membranes (*arrows*) are discernible in some of the granules. (× 33,000)

(Text continued from page 757)

Primary Ciliary Dyskinesia

(Figures 11.58 through 11.60.)

Diagnostic criteria. Respiratory epithelium with ciliary axonemes having (1) an absence or significant decrease of outer and/or inner dynein arms (normal average = two or more per cilium).

Additional points. Primary ciliary dyskinesia, or the immotile cilia syndrome, includes a heterogeneous group of hereditary disorders having certain clinical, functional, and ultrastructural features in common. Repeated respiratory tract infections, including combinations of chronic rhinitis, sinusitis, otitis media, bronchitis, and bronchiectasis are characteristic. Dextrocardia or situs inversus is present in many of these patients. *Kartagener's syndrome* specifically consists of situs inversus, chronic sinusitis, and bronchiectasis. Male infertility caused by defective flagella of sperm and ectopic pregnancy secondary to faulty cilia of fallopian tubular epithelium may also be evident in primary ciliary dyskinesia. Respiratory tract infections may be accompanied by various nongenetic ciliary defects, including abnormal numbers and arrangements of peripheral and central tubules, megacilia, compound cilia, and defective or absent radial spokes. Decreased ciliary beat frequency is also a useful measure of a genetic etiology (normal minimum = 10 Hz).

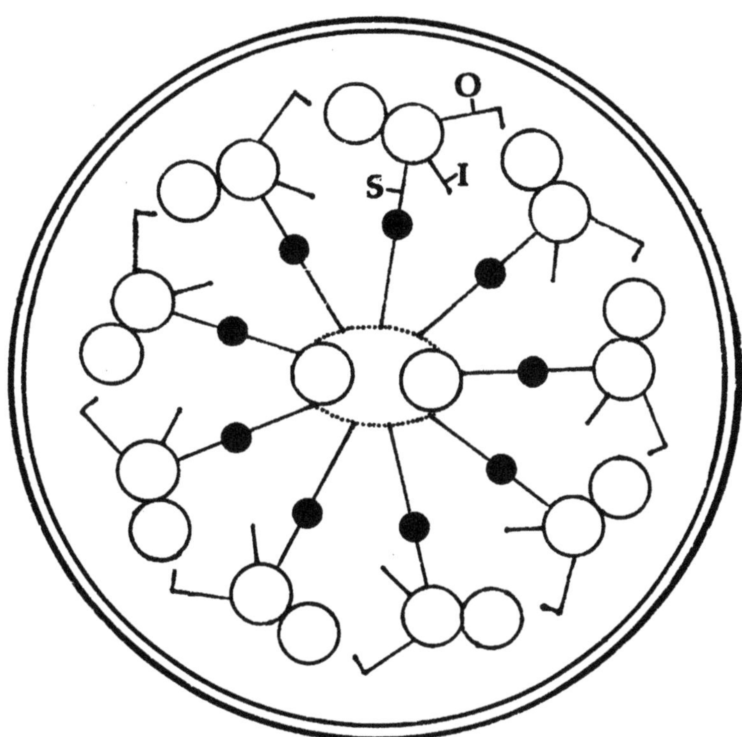

Figure 11.58. Normal human cilium (diagram). This simplified diagram of a normal cilium depicts nine peripheral and one central pair of tubules, with outer dynein arms (O), inner dynein arms (I), and spokes (S).

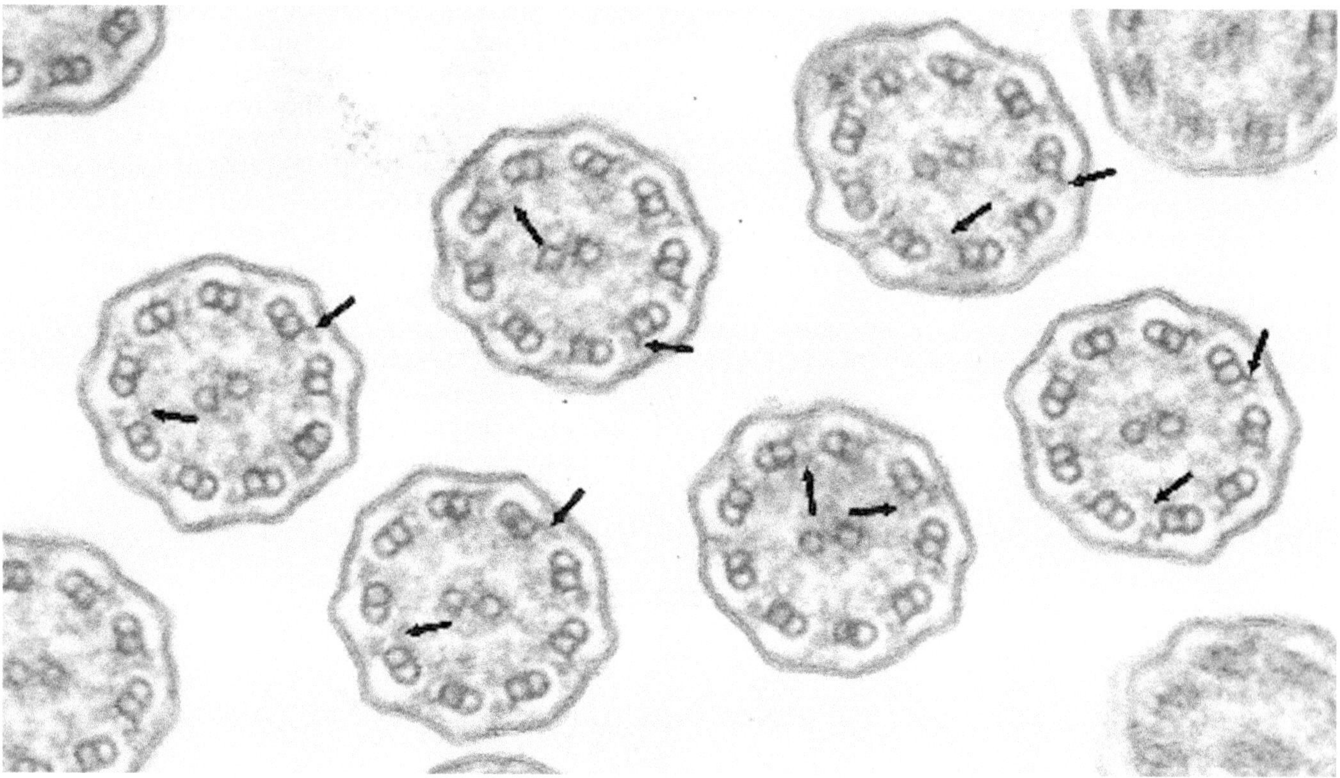

Figure 11.59. Normal cilia (nasal mucosa). Outer and inner dynein arms (*arrows*) are apparent in these cilia. Inner arms are usually blurred and less distinct than outer arms. (× 150,000)

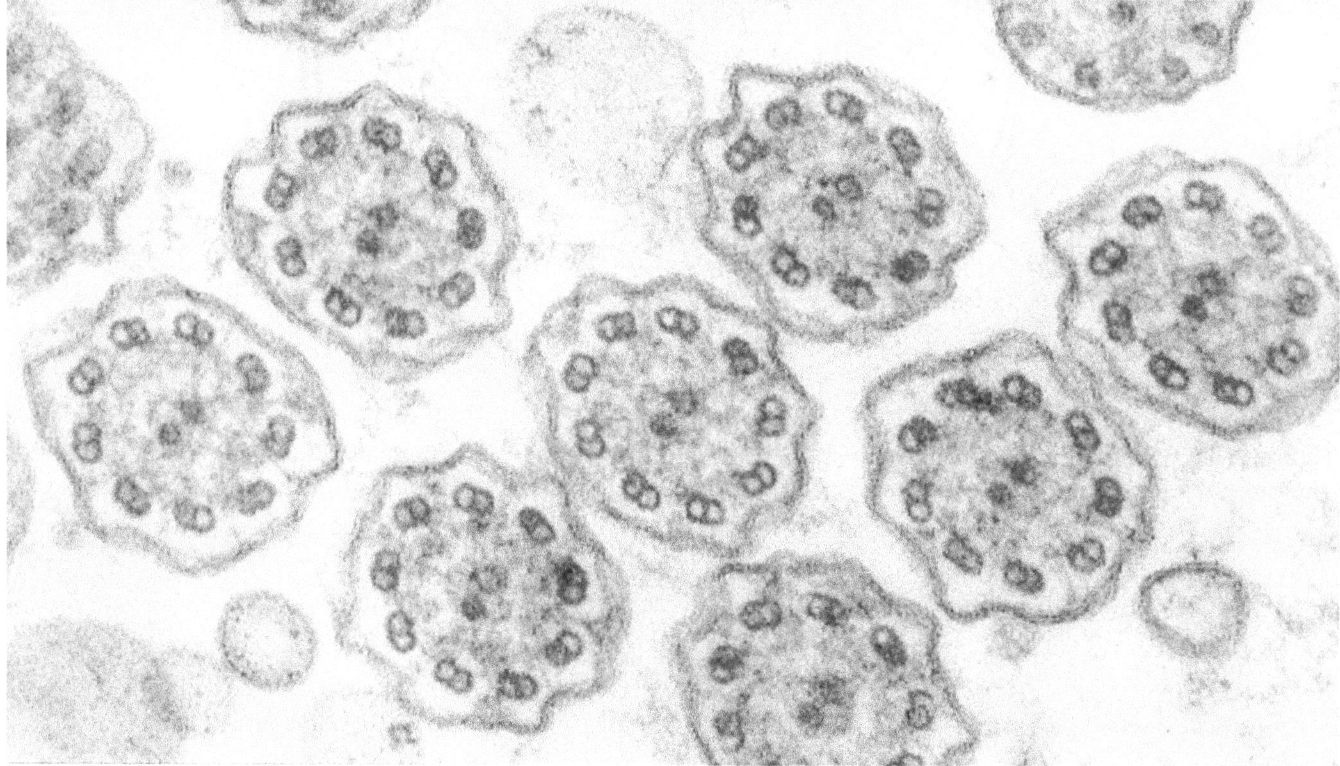

Figure 11.60. Primary ciliary dyskinesia (nasal mucosa). Definite dynein arms are absent in the cilia of this patient, who also had dextrocardia. (× 150,000)

REFERENCES

Storage Diseases: General

Bioulac P, Coquet M, Fontan D, et al: Interest of ultrastructural study of skin and muscle biopsies in inborn storage diseases. A report of 18 cases. Ann Dermatol Venereol 107:137–148, 1980.

Endo H, Miyazakai T, Asano S, et al. Ultrastructural studies of the skin and cultured fibroblasts in I-cell disease. J Cutan Pathol 14:309–317, 1987.

Glew RH, Basu A, Prence EM, et al: Biology of disease. Lysosomal storage diseases. Lab Invest 53:250–269, 1985.

Hammel I, Alroy J: The effect of lysosomal storage diseases on secretory cells: An ultrastructural study of pancreas as an example. J Submicrosc Cytol Pathol 27: 143–160, 1995.

Hers HG: Inborn lysosomal diseases. Gastroenterology 48:625–633, 1965.

Hers HG, Van Hoof F: *Lysosomes and Storage Diseases.* Academic Press, New York, 1973.

Kamensky E, Philippart M, Cancilla P, et al: Cultured skin fibroblasts in storage disorders. An analysis of ultrastructural features. Am J Pathol 73:59–80, 1973.

Kolodny EH, Boustany RM: Storage diseases of the reticuloendothelial system. In Nathan DG, Oski FA, eds, *Nathan & Oski's Hematology of Infancy and Childhood.* WB Saunders, New York, 1987, pp 1212–1247.

Mierau GW, Weeks DA: Role of electron microscopy in the diagnosis of metabolic storage diseases affecting the nervous system of children. Ultrastruct Pathol 21: 345–354, 1997.

O'Brien JS, Bernett J, Veath ML, et al: Lysosomal storage disorders. Diagnosis by ultrastructural examination of skin biopsy specimens. Arch Neurol 32:592–599, 1975.

Prasad A, Kaye EM, Alroy J: Electron microscopic examination of skin biopsy as a cost-effective tool in the diagnosis of lysosomal storage diseases. J Child Neurol 11:301–308, 1996.

Tay-Sachs and Related Diseases

Adachi M, Schneck L, Volk BW: Ultrastructural studies of eight cases of fetal Tay-Sachs disease. Lab Invest 30: 102–112, 1974.

Baker HJ, Mole JA, Lindsey JR, et al: Animal models of human ganglioside storage diseases. Fed Proc 35: 1193–1201, 1976.

Krivit W, Desnick RJ, Lee J, et al: Generalized accumulation of neutral glycosphingolipids with G_{M2} ganglioside accumulation in the brain. Sandoff's disease (variant of Tay-Sachs disease). Am J Med 52:763–770, 1972.

Prasad A, Kaye EM, Alroy J: Electron microscopic examination of skin biopsy as a cost-effective tool in the diagnosis of lysosomal storage diseases. J Child Neurol 11:301–308, 1996.

Schneck L, Volk BW, Saifer A: The gangliosidosis. Am J Med 46:245–262, 1969.

Schneck L, Wallace BJ, Saifer A, et al: A clinical, biochemical and electron microscopic study of late infantile amaurotic family idiocy. Am J Med 39:285–295, 1965.

Stern J: The induction of ganglioside storage in nervous system cultures. Lab Invest 26:509–514, 1972.

Volk BW, Adachi M, Schneck L: The gangliosidosis. Hum Pathol 6:555–569, 1975.

Yuasa T, Fukuma M, Takashima S, et al: Cultured skin fibroblasts in lipidoses. Enzymatic, histochemical, and ultrastructural relationship in Fabry's, Tay-Sachs, and Sandhoff's diseases. Arch Pathol Lab Med 104:321–327, 1980.

Fabry's Disease

deVeber GA, Schwarting GA, Kolodny EH, et al: Fabry disease: Immunocytochemical charcterization of neuronal involvement. Ann Neurol 31:409–415, 1992.

Frost P, Tanaka Y, Spaeth GL: Fabry's disease—glycolipid lipidosis. Histochemical and electron microscopic studies of two cases. Am J Med 40:618–627, 1966.

Savi M, Olivetti G, Neri TM, et al: Clinical, histopathological, and biochemical findings in Fabry's disease. A case report and family study. Arch Pathol Lab Med 101: 536–539, 1977.

Schatzki PF, Kipreos B, Payne J: Fabry's disease. Primary diagnosis by electron microscopy. Am J Surg Pathol 3:211–219, 1979.

Wandall A, Hashold L, Sorensen SA: Electron microscopic observations on cultured fibroblasts from Fabry heterozygotes and hemizygotes. Ultrastruct Pathol 3:51–58, 1982.

Gaucher's Disease

Adachi M, Volk BW: Gaucher disease in mice induced by conduritol-B-epoxide. Morphologic features. Arch Pathol Lab Med 101:255–259, 1977.

Brady RO, King FM: Gaucher's disease. In Hers HG, Van Hoof F, eds: *Lysosomes and Storage Diseases*. Academic Press, New York, 1973, pp 381–394.

Jordan SW: Electron microscopy of Gaucher cells. Exp Mol Pathol 3:76–85, 1964.

Lee RE, Ellis LD: The storage cells of chronic myelogenous leukemia. Lab Invest 24:261–264, 1971.

Niemann-Pick Disease

Boustany RN, Kaye E, Alroy J: Ultrastructural findings in skin from patients with Niemann-Pick disease, type C. Pediatr Neurol 6:177–183, 1990.

Long RG, Lake BD, Pettit JE, et al: Adult Niemann-Pick disease. Its relationship to the syndrome of the sea-blue histiocyte. Am J Med 62:627–635, 1977.

Lynn R, Terry RD: Lipid histochemistry and electron microscopy in adult Niemann-Pick disease. Am J Med 37:987–994, 1964.

Vethamany VG, Welch JP, Vethamany SK: Type D Niemann-Pick disease (Nova Scotia variant). Ultrastructure of blood, skin fibroblasts, and bone marrow. Arch Pathol Lab Med 93:537–543, 1972.

Mucopolysaccharidosis

Aleu FP, Terry RD, Zellweger H: Electron microscopy of two cerebral biopsies in gargoylism. J Neuropathol Exp Neurol 24:304–317, 1965.

Alroy JM, Jones MZ, Rutledge JC, et al: The ultrastructure of skin from a patient with mucopolysaccharidosis IIID. Acta Neuropathol 93:210–213, 1997.

Belcher RW: Ultrastructure of the skin in the genetic mucopolysaccharidosis. Arch Pathol Lab Med 94: 511–518, 1972.

Callahan WP, Lorincz AE: Hepatic ultrastructure in the Hurler syndrome. Am J Pathol 48:277–298, 1966.

DeCloux RJ, Friederici HHR: Ultrastructural studies of the skin in Hurler's syndrome. Arch Pathol Lab Med 88:350–358, 1969.

Haust MD, Gordon BA: Ultrastructural and biochemical aspects of the Sanfilippo syndrome—type III genetic mucopolysaccharidosis. Connect Tissue Res 15:57–64, 1986.

Iwamoto M, Nawa Y, Maumenee IH, et al: Ocular histopathology and ultrastructure of Morquio syndrome (systemic mucopolysaccharidosis IV A). Graefes Arch Clin Exp Ophthalmol 228:342–349, 1990.

Lagunoff D, Gritzka TL: The site of mucopolysaccharide accumulation in Hurler's syndrome. An electron microscopic and histochemical study. Lab Invest 15: 1578–1588, 1966.

Lavery MA, Green WR, Jabs EW, et al: Ocular histopathology and ultrastructure of Sanfilippo's syndrome, type III-B. Arch Ophthalmol 101:1263–1274, 1983.

Loeb H, Jonniaux G, Resibois A, et al: Biochemical and ultrastructural studies in Hurler's syndrome. J Pediatr 73:860–874, 1968.

Loeb H, Tondeur M, Toppet M, et al: Clinical, biochemical and ultrastructural studies of an atypical form of mucopolysaccharidosis. Acta Pediatr Scand 58:220–228, 1969.

Maynard JA, Cooper RR, Ponsetti IV: Morquio's disease (mucopolysaccharidosis type IV). Ultrastructure of epiphyseal plates. Lab Invest 28:194–205, 1973.

Natowicz MR, Short MP, Wang Y, et al: Clinical and biochemical manifestations of hyaluronidase deficiency. N Engl J Med 335:1029–1033, 1996.

Resnick JM, Whitley CB, Leonard AS, et al: Light and electron microscopic features of the liver in mucopolysaccharidosis. Hum Pathol 25:276–286, 1994.

Turki I, Kresse H, Scotto J, et al: Sanfilippo disease, type C: Three cases in the same family. Neuropediatrics 20: 90–92, 1989.

Van Hoof F, Hers HG: L'ultrastructure des cellules hépatiques dans la maladie de Hürler (Gargoylisme). C R Acad Sci (Paris) 259:1281–1283, 1964.

Wallace BJ, Kaplan D, Adachi M, et al: Mucopolysaccharidosis type III. Morphologic and biochemical studies of two siblings with Sanfilippo syndrome. Arch Pathol 82:462–473, 1966.

Lysosomal Storage Disease: Other

Rutsaert J, Menu R, Resibois A: Ultrastructure of sulfatide storage in normal and sulfatase-deficient fibroblasts *in vitro*. Lab Invest 29:527–535, 1973.

Neuronal Ceroid-Lipofuscinoses

Dolman CL, McLeod PM, Chang EC: Lymphocytes and urine in ceroid lipofuscinosis. Arch Pathol Lab Med 104:487–490, 1980.

Zeman W: The Neuronal Ceroid-Lipofuscinoses. In Zimmerman HM, ed: *Progress in Neuropathology,* vol 3. Grune and Stratton, New York, 1976, pp 203–223.

Cholesterol Ester Storage Disease

Elleder M, Ledvinova J, Cieslar P, et al: Subclinical course of cholesterol ester storage disease (CESD) diagnosed in adulthood. Report on two cases with remarks on the nature of the liver storage process. Virchows Arch[A] 41:357–365, 1990.

Gasche C, Aslanids C, Kain R, et al: A novel variant of lysosomal acid lipase in cholesterol ester storage disease associated with mild phenotype and improvement on lovastatin. J Hepatol 27:744–750, 1997.

Cerebrotendinous Xanthomatosis

Arpa J, Sanchez C, Vega A, et al: Cerebrotendinous xanthomatosis diagnosed after traumatic subdural haematoma. Rev Neurol 23:675–678, 1995.

Berginer VM, Salen G, Shefer S: Long-term treatment of cerebrotendinous xanthomatosis with chenodeoxycholic acid. N Engl J Med 26:1649–1652, 1984.

Bjorkham I, Skrede S: Familial diseases with storage of sterols other than cholesterol. In Scriver CR, Beaudet AL, Sly WS, et al, eds, *The Metabolic Basis of Inherited Disease,* 6th ed. McGraw-Hill, 1989, pp 1283–1293.

Chang WN, Cheng YF: Nephrolithiasis and nephrocalcinosis in cerebrotendinous xanthomatosis: Report of three siblings. Eur Neurol 35:55–57, 1995.

Cruysberg JR, Wevers RA, van Engelen BG, et al: Ocular and systemic manifestations of cerebrotendinous xanthomatosis. Am J Ophthalmol 120:597–604, 1995.

Dotti MT, Manneschi L, Federico A: Mitochondrial enzyme deficiency in cerebrotendinous xanthomatosis. J Neurol Sci 129:106–108, 1995.

Grundy SM: Cerebrotendinous xanthomatosis. N Engl J Med 26:1694–1695, 1984.

Kuramoto T, Suemura Y, Kihira K, et al: Determination of the glucurono-conjugated position in bile alcohol glucuronides present in a patient with cerebrotendinous xanthomatosis. Steroids 60:709–712, 1995.

Nagai Y, Hirano M, Mori T, et al: Japanese triplets with cerebrotendinous xanthomatosis are homozygous for a mutant gene coding for the sterol 27-hydroxylase (Arg441Trp). Neurology 46:571–574, 1996.

Okada J, Oonishi H, Tamada H, et al: Gallium uptake in cerebrotendinous xanthomatosis. Eur J Nucl Med 22: 1069–1072, 1995.

Segev H, Reshef A, Clavey V, et al: Premature termination codon at the sterol 27-hydroxylase gene causes cerebrotendinous xanthomatosis in a French family. Hum Genet 95:238–240, 1995.

Siebner HR, Berndt S, Conrad B: Cerebrotendinous xanthomatosis without tendon xanthomas mimicking Marinesco-Sjoegren syndrome: A case report. J Neurol Neurosurg Psychiatry 60:582–585, 1996.

Soffer D, Benharroch D, Berginer V: The neuropathology of cerebrotendinous xanthomatosis revisited: A case report and review of the literature. Acta Neuropathol 90:213–220, 1995.

Mannosidosis

Dickersin GR, Lott IT, Kolodny EH, et al: A light and electron microscopic study of mannosidosis. Hum Pathol 11:245–256, 1980.

Ishigami T, Schmidt-Westhausen A, Philipsen HP, et al: Oral manifestations of alpha-mannosidosis: Report of a case with ultrastructural findings. J Oral Pathol Med 24:85–88, 1995.

Adrenoleukodystrophy

Black VH, Cornacchia L: Stereological analysis of peroxisomes and mitochondria in intestinal epithelium of patients with peroxisomal deficiency disorders: Zellweger's syndrome and neonatal-onset adrenoleukodystrophy. Am J Anat 177:107–118, 1986.

Gullotta F, Hughes JL, Wittkowski W, et al: Differentiation of rare leukodystrophies by post-mortem morphological and biochemical studies: Female adrenoleukodystrophy-like disease and late-onset Krabbe disease. Neuropediatrics 27:37–41, 1996.

Hughes JL, Poulos A, Robertson E, et al: Pathology of hepatic peroxisomes and mitochondria in patients with peroxisomal disorders. Virchows Arch [A] 416:255–264, 1990.

Hughes JL, Crane DI, Robertson E, et al: Morphometry of peroxisomes and immunolocalization of peroxisomal proteins in the liver of patients with generalized peroxisomal disorders. Virchows Archiv [A] 423:459–468, 1993.

Lu JF, Lawlor AM, Watkins PA, et al: A mouse model for X-linked adrenoleukodystrophy. Proc Natl Acad Sci USA 94:9366–9371, 1997.

Martin JJ, Ceuterick C, Martin L, et al: Skin and conjunctival biopsies in adrenoleukodystrophy. Acta Neuropathol 38:247–250, 1977.

Mito T, Takada K, Akaboshi S, et al: A pathological study of a peripheral nerve in a case of neonatal adrenoleukodystrophy. Acta Neuropathol 77:437–440, 1989.

Molzer B, Gullotta F, Harzer K, et al: Unusual orthochromatic leukodystrophy with epithelial cells (Norman-Gullotta): An increase of very long chain fatty acids in brain discloses a peroxisomal disorder. Acta Neuropathol 86:187–189, 1993.

Powell H, Tindall R, Schultz P, et al: Adrenoleukodystrophy. Electron microscopic findings. Arch Neurol 32:250–260, 1975.

Powers JM, Schaumburg HH: Adrenoleukodystrophy. Similar ultrastructural changes in adrenal cortical and Schwann cells. Arch Neurol 30:406–408, 1974.

Powers JM, Schaumburg HH: Adreno-leukodystrophy (sex-linked Schilder's disease). A pathogenetic hypothesis based on ultrastructural lesions in adrenal cortex, peripheral nerve and testis. Am J Pathol 76:481–491, 1974.

Powers JM, Schaumburg HH: The adrenal cortex in adreno-leukodystrophy. Arch Pathol 96:305–310, 1973.

Roels F, Pauwels M, Poll-The BT, et al: Hepatic peroxisomes in adrenoleukodystrophy and related syndrome: Cytochemical and morphometric data. Virchows Arch [A] 413:275–285, 1988.

Schaumburg HH, Powers JM, Raine CS, et al: Adrenoleukodystrophy. A clinical and pathological study of 17 cases. Arch Neurol 32:577–591, 1975.

Takeda S, Ohama E, Ikuta F: Adrenoleukodystrophy—Early ultrastructural changes in the brain. Acta Neuropathol 78:124–130, 1989.

Tanaka K, Yamano T, Shimada M, et al: Electron microscopic study on biopsied rectal mucosa in adrenoleukodystrophy. Neurology 37:1012–1015, 1987.

Thomas PK, Landon DN, King RHM: Diseases of the peripheral nerves. In Graham DI, Lantos PL, eds, *Greenfield's Neuropathology,* 6th ed, vol 2. Arnold, London, 1997, pp 367–487.

Wray SH, Cogan DG, Kuwabara T, et al: Adrenoleukodystrophy with disease of the eye and optic nerve. Am J Ophthalmol 82:480–485, 1976.

Glycogenosis Type II
(and Nonlysosomal Glycogenoses)

Baudhuin P, Hers HG, Loeb H: An electron microscopic and biochemical study of type II glycogenosis. Lab Invest 13:1139–1152, 1964.

Griffin JL: Infantile acid maltase deficiency. II. Muscle fiber hypertrophy and the ultrastructure of end-stage fibers. Virchows Arch [B] 45:37–50, 1984.

Hug G, Schubert WK: Glycogenosis type II. Glycogen distribution in tissues. Arch Pathol Lab Med 84:141–152, 1967.

Hug G, Soukup S, Ryan M, et al: Rapid prenatal diagnosis of glycogen-storage disease type II by electron microscopy of uncultured amniotic-fluid cells. N Engl J Med 310:1018–1022, 1984.

Kuzuya T, Matsuda A, Yoshida S, et al: An adult case of type lb glycogen-storage disease. Enzymatic and histochemical studies. N Engl J Med 308:566–569, 1983.

McAdams AJ, Hug G, Bove KE: Glycogen storage disease, types I to X. Criteria for morphologic diagnosis. Hum Pathol 5:463–487, 1974.

Ullrich K, Grobe H, Korinthenberg R, et al: Severe course of glycogen storage disease type II (Pompe disease) without development of cardiomegalia. Pathol Res Pract 181:627–632, 1986.

Erdheim-Chester Disease (Fibroxanthomatosis)

Devouassoux G, Lantuejoul S, Chatelain P, et al: Erdheim-Chester disease: A primary macrophage cell disorder. Am J Respir Crit Care Med 157:650–653, 1998.

Fink MG, Levinson DJ, Brown NL, et al: Erdheim-Chester disease. Case report with autopsy findings. Arch Pathol Lab Med 115:619–623, 1991.

Ono K, Oshiro M, Uemura K, et al: Erdheim-Chester disease: A case report with immunohistochemical and biochemical examination. Hum Pathol 27:91–95, 1996.

Veyssier-Belot C, Cacoub P, Caparros-Lefebvre D, et al: Erdheim-Chester disease. Clinical and radiologic characteristics of 59 cases. Medicine 75:157–169, 1996.

Porphyria

Bloomer JR, Enriquez R: Evidence that hepatic crystalline deposits in a patient with protoporphyria are composed of protoporphyrin. Gastroenterology 82: 569–573, 1982.

Bruguera M, Esquerda JE, Mascaro JM, et al: Erythropoietic protoporphyria. A light, electron, and polarization microscopical study of the liver in three patients. Arch Pathol Lab Med 100:587–589, 1976.

Fonia O, Weizman R, Coleman R, et al: PK 11195 aggravates 3,5-diethoxycarbonyl-1,4-digydrocollidine-induced hepatic porphyria in rats. Hepatology 24: 697–701, 1996.

Gschnait F, Konrad K, Honigsmann H, et al: Mouse model for protoporphyria. I. The liver and hepatic protoporphyrin crystals. J Invest Dermatol 65:290–299, 1975.

Kemmer C, Riedel H, Kostler E, et al: Paracrystalline needle-shaped cytoplasmic liver cell inclusions in chronic hepatic porphyria—Light and electron microscopic examination of liver biopsy specimens. Zentralbl Allgemeine Pathol Patholog Anat 127:253–264, 1983.

MacDonald DM, Germain D, Perrot H: The histopathology and ultrastructure of liver disease in erythropoietic protoporphyria. Br J Dermatol 104:7–17, 1981.

Mooyaart BR, de Jong GM, van der Veen S, et al: Hepatic disease in erythropoietic protoporphyria. Dermatologica 173:120–130, 1986.

Nakanuma Y, Wada M, Kono N, et al: An autopsy case of erythropoietic protoporphyria with cholestatic jaundice and hepatic failure, and a review of literature. Virchows Arch [A] 393:123–132, 1982.

Rademakers LH, Cleton MI, Kooijman C, et al: Early involvement of hepatic parenchymal cells in erythrohepatic protoporphyria? An ultrastructural study of patients with and without overt liver disease and the effects of chenodeoxycholic acid treatment. Hepatology 11:449–457, 1990.

Rademakers LH, Cleton MI, Kooijman C, et al: Ultrastructural aspects of the liver in erythrohepatic protoporphyria. Curr Probl Dermatol 20:154–159, 1991.

Shapiro SH, Wessely Z, Klavins JV: Hepatocyte mitochondrial alterations in griseofulvin fed mice. Ann Clin Lab Sci 10:33–45, 1980.

Alpha-1-Antitrypsin Deficiency

Bhan AK, Grand RJ, Colten HR, et al: Liver in alpha-1-antitrypsin deficiency: Morphologic observations and in vitro synthesis of alpha-1-antitrypsin. Pediatr Res 10:35–40, 1976.

Campra JL, Craig JR, Peters RL, et al: Cirrhosis associated with partial deficiency of alpha-1-antitrypsin in an adult. Ann Intern Med 78:233–238, 1973.

Case Records of the Massachusetts General Hospital. Weekly clinicopathological exercises. Case 24-1980. N Engl J Med 302:1405–1413, 1980.

Case Records of the Massachusetts General Hospital. Weekly clinicopathological exercises. Case 41-1972. N Engl J Med 287:763–768, 1972.

Feldman G, Bignon J, Chahinian P, et al: Hepatocyte ultrastructural changes in alpha-1-antitrypsin deficiency. Gastroenterology 67:1214–1224, 1974.

Fukuda Y, Masuda Y, Ishizaki M, et al: Morphogenesis of abnormal elastic fibers in lungs of patients with panacinar and centriacinar emphysema. Hum Pathol 20:652–659, 1989.

Gordon HW, Dixon J, Rogers JC, et al: Alpha-1-antitrypsin (A1-AT) accumulation in livers of emphysematous patients with A1-AT deficiency. Hum Pathol 3: 361–370, 1972.

Gourley MF, Gourley GR, Gilbert EF, et al: Alpha 1-antitrypsin deficiency and the PiMS phenotype: Case report and literature review. J Pediatr Gastroenterol Nutr 8:116–121, 1989.

Hultcrantz R, Mengarelli S: Ultrastructural liver pathology in patients with minimal liver disease and alpha 1-antitrypsin deficiency: A comparison between heterozygous and homozygous patients. Hepatology 4: 937–945, 1984.

Hultcrantz R, Jelf E, Nilsson LH: Minimal liver disease in young persons with homozygous and heterozygous alpha-1-antitrypsin deficiency. Scand J Gastroenterol 19:389–393, 1984.

Ishak KG, Jenis EH, Marshall ML, et al: Cirrhosis of the liver associated with alpha-1-antitrypsin deficiency. Arch Pathol Lab Med 94:445–455, 1972.

Lieberman J, Mittman C, Gordon HW: Alpha-1-antitrypsin in the liver of patients with emphysema. Science 175:63–65, 1972.

Lomas DA, Evans DL, Finch JT, et al: The mechanism of Z alpha-1-antitrypsin accumulation in the liver. Nature 357:605–607, 1992.

Martorana PA, Brand T, Gardi C, et al: The pallid mouse. A model of genetic alpha-1-antitrypsin deficiency. Lab Invest 68:233–241, 1993.

Nemeth A, Strandvik B, Glaumann H: Alpha-1-antitrypsin deficiency and juvenile liver disease. Ultrastructural observations compared with light microscopy and routine liver tests. Virchows Arch [B] 44: 15–33, 1983.

Scotto JM, Stralin HG, Alagille D: Alpha-1-antitrypsin deficiency in children: Liver ultrastructure and speculations. Virchows Arch [A] 369:19–27, 1975.

Sharp HL: Alpha-1-antitrypsin deficiency. Hosp Pract 6:83–96, 1971.

Yunis EJ, Agostini RM, Glew RH: Fine structural observations of the liver in alpha-1-antitrypsin deficiency. Am J Pathol 82:265–285, 1976.

Pulmonary Alveolar Proteinosis

Bedrossian CWM, Luna MA, Conklin RH, et al: Alveolar proteinosis as a consequence of immunosuppression. Hum Pathol 11:527–535, 1980.

Burkhalter A, Silverman JF, Hopkins MB III, et al: Bronchoalveolar lavage cytology in pulmonary alveolar proteinosis. Am J Clin Pathol 106:504–510, 1996.

Case Records of the Massachusetts General Hospital. Weekly clinicopathological exercises. Case 19-1983. Pulmonary reticulonodular disease with consolidation and abscess formation. N Engl J Med 308:1147–1156, 1983.

Costello JF, Moriarty DC, Branthwaite MA, et al: Diagnosis and management of alveolar proteinosis: The role of electron microscopy. Thorax 30:121–132, 1975.

Gilmore LB, Talley FA, Hook GE: Classification and morphometric quantitation of insoluble materials from the lungs of patients with alveolar proteinosis. Am J Pathol 133:252–264, 1988.

Hattori A, Kuroki Y, Katoh T, et al: Surfactant protein A accumulating in the alveoli of patients with pulmonary alveolar proteinosis: Oligomeric structure and interaction with lipids. Am J Respir Cell Molec Biol 14: 608–619, 1996.

Hook GER, Bell DY, Gilmore LB, et al: Composition of bronchoalveolar lavage effluents from patients with pulmonary alveolar proteinosis. Lab Invest 39:342–357, 1978.

Hook GER, Gilmore LB, Talley FA: Dissolution and reassembly of tubular myelin-like multilamellated structures from the lungs of patients with pulmonary alveolar proteinosis. Lab Invest 55:194–208, 1986.

Hook GER, Gilmore LB, Talley FA: Multilamellated structures from the lungs of patients with pulmonary alveolar proteinosis. Lab Invest 50:711–725, 1984.

Knight DP, Knight JA: Pulmonary alveolar proteinosis in the newborn. Arch Pathol Lab Med 109:529–531, 1985.

Kuhn C, Gyorkey F, Levine BE, et al: Pulmonary alveolar proteinosis: A study using enzyme histochemistry, electron microscopy and surface-tension measurement. Lab Invest 15:492–509, 1966.

Mikami T, Yamamoto Y, Yokoyama M, et al: Pulmonary alveolar proteinosis: Diagnosis using routinely processed smears of bronchoalveolar lavage fluid. J Clin Pathol 50:981–984, 1997.

Sosolik RC, Gammon RR, Julius CJ, et al: Pulmonary alveolar proteinosis. A report of two cases with diagnostic features in bronchoalveolar lavage specimens. Acta Cytol 42:377–383, 1998.

Takemura T, Fukuda Y, Harrison M, et al: Ultrastructural, histochemical, and freeze-fracture evaluation of multilamellated structures in human pulmonary alveolar proteinosis. Am J Anat 179:258–268, 1987.

Mitochondrial Abnormalities

Adachi K, Matsuhashi T, Nishizawa Y, et al: Suppression of the hydrazine-induced formation of megamitochondria in the rat liver by coenzyme Q10. Toxicol Pathol 23:667–676, 1995.

Bioulac-Sage P, Parrot-Roulaud F, Mazat J, et al: Fatal neonatal liver failure and mitochondrial cytopathy (oxidative phosphorylation deficiency): A light and electron microscopic study of the liver. Hepatology 18: 839–846, 1993.

Hayasaka K, Takahashi I, Kobayashi Y, et al: Effects of valproate on biogenesis and function of liver mitochondria. Neurology 36:351–356, 1986.

Kimura A, Yoshida I, Yamashita F, et al: The occurrence of intramitochondrial Ca^{2+} granules in valproate-induced liver injury. J Pediatr Gastroenterol Nutr 8:13–18, 1989.

Koch OR, Bedetti CD, Gamboni M, et al: Functional alterations of liver mitochondria in chronic experimental alcoholism. Exp Mol Pathol 25:253–262, 1976.

Amyloidosis

Bladen HA, Nylen MU, Glenner GG: The ultrastructure of human amyloid as revealed by the negative staining technique. J Ultrastruct Res 14:449–459, 1966.

Bourgeois N, Buyssens N, Goovaerts G: Ultrastructural appearance of amyloid. Ultrastruct Pathol 11:67–76, 1987.

Cohen AS: Amyloidosis. N Engl J Med 277:522–530; 574–583; 628–638, 1967.

Durie BGM, Persky B, Soehnlen BJ, et al: Amyloid production in human myeloma stem-cell culture, with morphologic evidence of amyloid secretion by associated macrophages. N Engl J Med 307:1689–1692, 1982.

Glenner GG: Amyloid deposits and amyloidosis. The β-fibrilloses, part I. N Engl J Med 302:1283–1292, 1980.

Glenner GG: Amyloid deposits and amyloidosis. The β-fibrilloses, part II. N Engl J Med 302:1333–1343, 1980.

Kisilevsky R, Axelrad M, Corbett W, et al: The role of inflammatory cells in the pathogenesis of amyloidosis. Lab Invest 37:544–553, 1977.

Linke RP, Nathrath WBJ, Wilson PD: Immuno-electron microscopic identification and classification of amyloid in tissue sections by the post-embedding protein-A gold method. Ultrastruct Pathol 4:1–7, 1983.

Case Records of the Massachusetts General Hospital. Weekly clinicopathological exercises. Case 5-1980. N Engl J Med 302:336–344, 1980.

Case Records of the Massachusetts General Hospital. Weekly clinicopathological exercises. Case 43-1985. N Engl J Med 313:1070–1079, 1985.

Shirahama T, Cohen AS: An analysis of the close relationship of lysosomes to early deposits of amyloid. Ultrastructural evidence in experimental mouse amyloidosis. Am J Pathol 73:97–114, 1973.

Shirahama T, Cohen AS: High-resolution electron microscopic analysis of the amyloid fibril. J Cell Biol 33:679–708, 1967.

Tischler AS, Compagno J: Crystal-like deposits of amyloid in pancreatic islet cell tumors. Arch Pathol Lab Med 103:247–251, 1979.

Yano BL, Hayden DW, Johnson KH: Feline insular amyloid. Ultrastructural evidence for intracellular formation by non-endocrine cells. Lab Invest 45:149–156, 1981.

Zucker-Franklin D, Franklin EC: Intracellular localization of human amyloid by fluorescence and electron microscopy. Am J Pathol 59:23–41, 1970.

Mitochondrial Abnormalities: General

Cervera R, Bruix J, Bayes A, et al: Chronic intestinal pseudoobstruction and ophthalmoplegia in a patient with mitochondrial myopathy. Gut 29:544–547, 1988.

DiMauro S, Bonilla E, Zeviani M, et al: Mitochondrial myopathies. Ann Neurol 17:521–538, 1985.

Fogel SP, DeTar MW, Shimada H, et al: Sporadic visceral myopathy with inclusion bodies. A light-microscopic and ultrastructural study. Am J Surg Pathol 17:473–481, 1993.

Li V, Hostein J, Romero NB, et al: Chronic intestinal pseudoobstruction with myopathy and ophthalmoplegia. A muscular biochemical study of a mitochondrial disorder. Dig Dis Sci 37:456–463, 1992.

Lindal S, Torbergsen T, Borud O, et al: Mitochondrial diseases and myopathies: A series of muscle biopsy specimens with ultrastructural changes in the mitochondria. Ultrastruct Pathol 16:263–275, 1992.

Sengers RC, Stadhouders AM, Trijbels JM: Mitochondrial myopathies. Clinical, morphological and biochemical aspects. Eur J Pediatr 141:192–207, 1984.

Stadhouders AM, Sengers RC: Morphological observations in skeletal muscle from patients with a mitochondrial myopathy. J Inherit Metab Dis 10(suppl 1):62–80, 1987.

Leigh's Disease

Jung KC, Myong NH, Chi JG, et al: Leigh's disease involving multiple organs. J Korean Med Sci 8:214–220, 1993.

Kremer I, Lerman-Sagie T, Mukamel M, et al: Light and electron microscopic retinal findings in Leigh's disease. Ophthalmologica 199:106–110, 1989.

Langes K, Frenzel H, Seitz R, et al: Cardiomyopathy associated with Leigh's disease. Virchows Arch [A] 407: 97–105, 1985.

Seitz RJ, Langes K, Frenzel H, et al: Congenital Leigh's disease: Panencephalomyelopathy and peripheral neuropathy. Acta Neuropathol 64:167–171, 1984.

Tsai ML, Hung KL, Chen TY: Subacute necrotizing encephalomyelopathy (Leigh's disease): Report of a case. J Formos Med Assoc 89:799–802, 1990.

Chronic Intestinal Pseudo-obstruction

Alstead EM, Murphy MN, Flanagan AM, et al: Familial auronomic visceral myopathy with degeneration of muscularis mucosae. J Clin Pathol 41:424–429, 1988.

Cervera R, Bruix J, Blesa R, et al: Chronic intestinal pseudoobstruction and ophthalmoplegia in a patient with mitochondrial myopathy. Gut 29:544–547, 1988.

Fogel SP, DeTar MW, Shimada H, et al: Sporadic visceral myopathy with inclusion bodies. A light microscopic and ultrastructural study. Am J Surg Pathol 17:473–481, 1993.

Li V, Hostein J, Romero NB, et al: Chronic intestinal pseudoobstruction with myopathy and ophthalmoplegia. A muscular biochemical study of a mitochondrial disorder. Dig Dis Sci 37:456–463, 1992.

Martin JE, Swash M, Kamm MA, et al: Myopathy of internal anal sphincter with polyglucosan inclusions. J Pathol 161:221–226, 1990.

Venizelos ID, Shousha S, Bull TB, et al: Chronic intestinal pseudoobstruction in two patients. Overlap of features of systemic sclerosis and visceral myopathy. Histopathology 12:533–540, 1988.

Wilson's Disease

Epstein O, Arborgh B, Sagiv M, et al: Is copper toxic in primary biliary cirrhosis? J Clin Pathol 34:1071–1075, 1981.

Faa G: The role of the pathologist in the diagnosis and monitoring of Wilson's disease. Pathologica 88:102–110, 1996.

Hayashi H, Sternlieb I: Lipopolysosomes in human hepatocytes. Ultrastructural and cytochemical studies of patients with Wilson's disease. Lab Invest 33:1–7, 1975.

Lapis K: Metabolic disorders, In Johannessen JV, ed: *Electron Microscopy in Human Medicine,* vol 8, The Liver. McGraw-Hill, New York, 1979, pp 40–45.

Myers BM, Predergast FG, Holman R, et al: Alterations in hepatocyte lysosomes in experimental hepatic overload in rats. Gastroenterology 105:1814–1823, 1993.

Sternlieb I: Mitochondrial and fatty changes in hepatocytes of patients with Wilson's disease. Gastroenterology 55:354–367, 1968.

Sternlieb I, van den Hamer CJL, Morell AG: Lysosomal defect of hepatic copper excretion in Wilson's disease (hepatolenticular degeneration). Gastroenterology 64:99–105, 1973.

Sternlieb I, Quintana N, Volenberg I, et al: An array of mitochondrial alterations in the hepatocytes of Long-Evans Cinnamon rats. Hepatology 22:1782–1787, 1995.

Tanikawa K: Fine structure of the liver in Wilson's disease [abstract]. J Electron Microsc (Tokyo) 17:92, 1968.

Tanikawa K, Abe H, Miyakoda U: Electron microscopic observation and X-ray analysis of deposit granules in the hepatocytes. J Clin Electron Microsc (Tokyo) 6:332, 1973.

Tanikawa K: *Ultrastructural Aspects of the Liver and Its Disorders,* 2nd ed. Igaku-Shoin, Tokyo, 1978, pp 281–285.

Tanikawa K: The liver. In: Papadimitriou JM, Henderson DW, Spagnolo DV, eds: *Diagnostic Ultrastructure of Non-Neoplastic Diseases.* Churchill Livingstone, New York, 1992, pp 335–351.

Amiodarone Toxicity

Dake MD, Madison JM, Montgomery CK, et al: Electron microscopic demonstration of lysosomal inclusion bodies in lung, liver, lymph nodes, and blood leukocytes of patients with amiodarone pulmonary toxicity. Am J Med 78:506–512, 1985.

Hostetler KY, Vande Berg J, Aldern KA, et al: Effect of amiodarone on the phospholipid and lamellar body content of lymphoblasts in vitro and peripheral blood lymphocytes in vivo. Biochemic Pharmacol 1: 1007–1013, 1991.

Kennedy JI, Myers JL, Plumb VJ, et al: Amiodarone pulmonary toxicity. Clinical, radiologic, and pathologic correlations. Arch Int Med 147:50–55, 1987.

Macarri G, Felciangeli G, Berdini V, et al: Canalicular cholestasis due to amiodarone toxicity. A definite diagnosis obtained by electron microscopy. Ital J Gastroenterol 27:436–438, 1995.

Rigas B, Rosenfeld LE, Barwick KW, et al: Amiodarone hepatotoxicity. A clinicopathologic study of five patients. Ann Intern Med 104:348–351, 1986.

Shepherd NA, Dawson AM, Crocker PR, et al: Granular cells as a marker of early amiodarone hepatoxicity: A pathological and analytical study. J Clin Pathol 40: 418–423, 1987.

Adriamycin Toxicity

Billingham ME, Bristow MR, Glatstein E, et al: Adriamycin cardiotoxicity: Endomyocardial biopsy evidence of enhancement by irradiation. Am J Surg Pathol 1:17–23, 1977.

Billingham ME, Mason JW, Bristow MR, et al: Anthracycline cardiomyopathy monitored by morphologic changes. Cancer Treat Rep 62:865–872, 1978.

Ewer MS, Ali MK, Mackay B, et al: A comparison of cardiac biopsy grades and ejection fraction estimations in patients receiving Adriamycin. J Clin Oncol 2:112–117, 1984.

Fujita K, Shinpo K, Yamada K, et al: Reduction of Adriamycin toxicity by ascorbate in mice and guinea pigs. Cancer Res 42:309–316, 1982.

Merski JA, Daskal I, Busch H: Effects of Adriamycin on ultrastructure of nucleoli in the heart and liver cells of the rat. Cancer Res 36:1580–1584, 1976.

Hemosiderosis

Cheville NF: *Ultrastructural Pathology.* Iowa State University Press, Ames, IA, 1994, pp 317–318.

Ghadially FN: *Ultrastructural Pathology of the Cell and Matrix,* 3rd ed. Butterworth-Heinemann, Boston, 1997, pp 668–675.

Cholestasis

Schaff Z, Lapis K: Cholestasis. In Johannessen JV, ed, *Electron Microscopy in Human Medicine: The Liver, The Gallbladder, and Biliary Ducts,* vol 8. McGraw-Hill, New York, 1979, pp 80–88.

Melanosis

Balazs M: Melanosis coli. Ultrastructural study of 45 patients. Dis Colon Rectum 29:839–844, 1986.

Brennick JB, O'Connell JX, Dickersin GR, et al: Lipofuscin pigmentation (so-called "melanosis") of the prostate. Am J Surg Pathol 18:446–454, 1994.

Gallagher RL: Intestinal ceroid deposition—"brown bowel syndrome." A light and electron microscopic study. Virchows Arch [A] 389:143–151, 1980.

Ghadially FN, Walley VM: Pigments of the gastrointestinal tract: A comparison of light microscopic and electron microscopic findings. Ultrastruct Pathol 19: 213–219, 1995.

Hall M, Eusebi V: Yellow-brown spindle bodies in mesenteric lymph nodes: A possible relationship with melanosis coli. Histopathology 2:47–52, 1978.

Ro JY, Grignon DJ, Ayala AG, et al: Blue nevus and melanosis of the prostate. Electron-microscopic and immunohistochemical studies. Am J Clin Pathol 90: 530–535, 1988.

Walker NI, Smith MM, Smithers BM: Ultrastructure of human melanosis coli with reference to its pathogenesis. Pathology 25:120–123, 1993.

Ciliary Dysgenesis

Afzelius BA: Cilia and flagella that do not conform to the 9 + 2 pattern. I. Aberrant members within normal populations. J Ultrastruct Res 9:381–392, 1963.

Afzelius BA: A human syndrome caused by immotile cilia. Science 193:317–319, 1976.

Afzelius BA: Genetical and ultrastructural aspects of the immotile-cilia syndrome. Am J Hum Genet 33:852–864, 1981.

Camner P, Mossberg B, Afzelius B: Evidence for congenitally non-functioning cilia in the tracheobronchial tract in two subjects. Am Rev Respir Dis 112:807–809, 1975.

Carlen B, Stenram U: Ultrastructural diagnosis in the immotile cilia syndrome. Ultrastruct Pathol 11:653–658, 1987.

Costle A, Millepied M-C, Chapelin C, et al: Incidence of primary ciliary dyskinesia in children with recurrent respiratory diseases. Ann Otol Rhinol Laryngol 106: 854–858, 1997.

Dombi VH, Walt H: Primary ciliary dyskinesia, immotile cilia syndrome, and Kartagener syndrome: Diagnostic criteria. Schweiz Med Wochensch 126:421–433, 1996.

Eliasson R, Mossberg B, Camner P, et al: The immotile-cilia syndrome. N Engl J Med 297:1–6, 1977.

Halbert SA, Patton DL, Zarutskie PW, et al: Function and structure of cilia in the fallopian tube of an infertile woman with Kartagener's syndrome. Hum Reprod 12:55–58, 1997.

Kartagener M: Zur pathogenese der bronchiektasien. I. Bronchiektasien bei situs viscerum inversus. Beitr Klin Tuberk 83:489–501, 1933.

Lurie M, Rennert G, Goldenberg S, et al: Ciliary ultrastructure in primary ciliary dyskinesia and other chronic respiratory conditions: The relevance of microtubular abnormalities. Ultrastruct Pathol 16:547–553, 1992.

Mierau GW, Agostini R, Beals TF, et al: The role of electron microscopy in evaluating ciliary dysfunction: Report of a workshop. Ultrastruct Pathol 16:245–254, 1992.

Pedersen H, Mygind N: Absence of axonemal arms in nasal mucosa cilia in Kartagener's syndrome. Nature 262:494–495, 1976.

Rautiainen M, Collan Y, Nuutinen J, et al: Ciliary orientation in the "immotile cilia" syndrome. Eur Arch Otorhinolaryngol 247:100–103, 1990.

Smallman LA, Gregory J: Ultrastructural abnormalities of cilia in the human respiratory tract. Hum Pathol 17: 848–855, 1986.

Teknos TN, Metson R, Chasse T, et al: New developments in the diagnosis of Kartagener's syndrome. Otolaryngol Head Neck Surg 116:68–74, 1997.

12

Renal Glomerular Disease

Shamila Mauiyyedi
Martin K. Selig
Alain P. Marion
Robert B. Colvin

The contribution of electron microscopy to the study of normal and abnormal glomerular morphology has been medically significant. In the early 1960s and 1970s, before immunofluorescence microscopy was routine in renal biopsy evaluation, many studies provided justification for the role of electron microscopy. In the early 1970s, Siegel et al. (1973) conducted a study on 213 native renal biopsies to show how electron microscopy provided the diagnosis or additional information in 48% of their cases. In early 1996, Haas (1997) reevaluated the use of routine electron microscopy on 233 native renal biopsies and showed that 45% of their cases required EM for diagnosis, confirmation, or additional diagnostic information. It is not surprising to see that this figure has not changed, as new features invisible by light and immunofluorescence microscopy are being defined and new diseases are being discovered at the ultrastructural level. All native renal biopsies should be processed for electron microscopy, although ultrastructural examination is required chiefly for the glomerular diseases. In some cases electron microscopy is required for diagnosis in entities such as minimal change disease, early diabetic nephropathy, membranous lupus nephritis versus nonlupus membranous glomerulonephritis, membranoproliferative glomerulonephritis, postinfectious glomerulonephritis, Alport's syndrome, thin basement membrane disease, fibrillary glomerulopathies, chronic allograft glomerulopathy, and nephropathy associated with human immunodeficiency virus (HIV) versus idiopathic collapsing glomerulopathy. With the routine use of immunofluorescence, diseases less likely to require electron microscopy for diagnosis include immunoglobulin (Ig)A nephropathy, pauci-immune crescentic glomerulonephritis, diffuse proliferative glomerulonephritis, amyloidosis, and acute interstitial nephritis. Usually a combination is optimal to arrive at a confident diagnosis.

In this chapter, the glomerular diseases are grouped by key ultrastructural appearances, with the intent that this organization will promote use of the text as a diagnostic guide.

The Normal Glomerulus

(Figures 12.1 and 12.2.)

The renal glomeruli are spheroidal structures, 150–200 μm in diameter, that consist of a central tuft of anastomosing capillaries separated by supporting cells and matrix (the mesangium) and surrounded by Bowman's capsule (Figure 12.1). Four cell types are present: the endothelial cells, the visceral epithelium (podocytes), the parietal epithelium (lining Bowman's capsule), and mesangial cells.

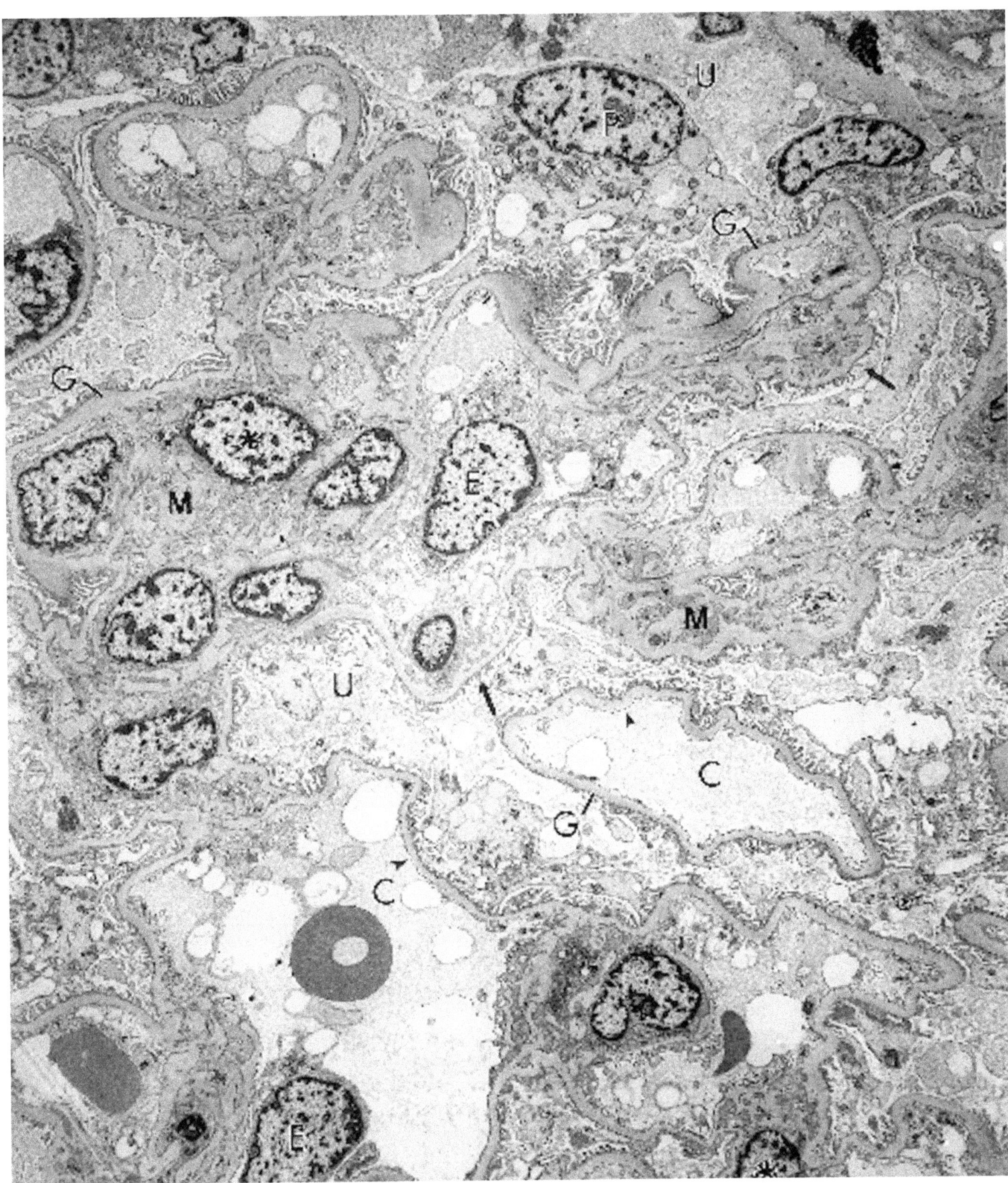

Figure 12.1. Normal glomerulus. The capillary loops (C) surround the mesangium (M), comprising mesangial cells (*) and matrix. The endothelial cells (E) have small amount of cytoplasm except in the region of the nucleus, which tends to be on the mesangial side of the capillaries. The fenestrated endothelial cytoplasm appears as a series of "dashes" (*arrowheads*) along the inner side of the glomerular basement membrane (G). The podocytes (P) are in the urinary space (U) and rest on the outer side of the glomerular basement membrane by means of foot processes (*arrows*). (× 4600)

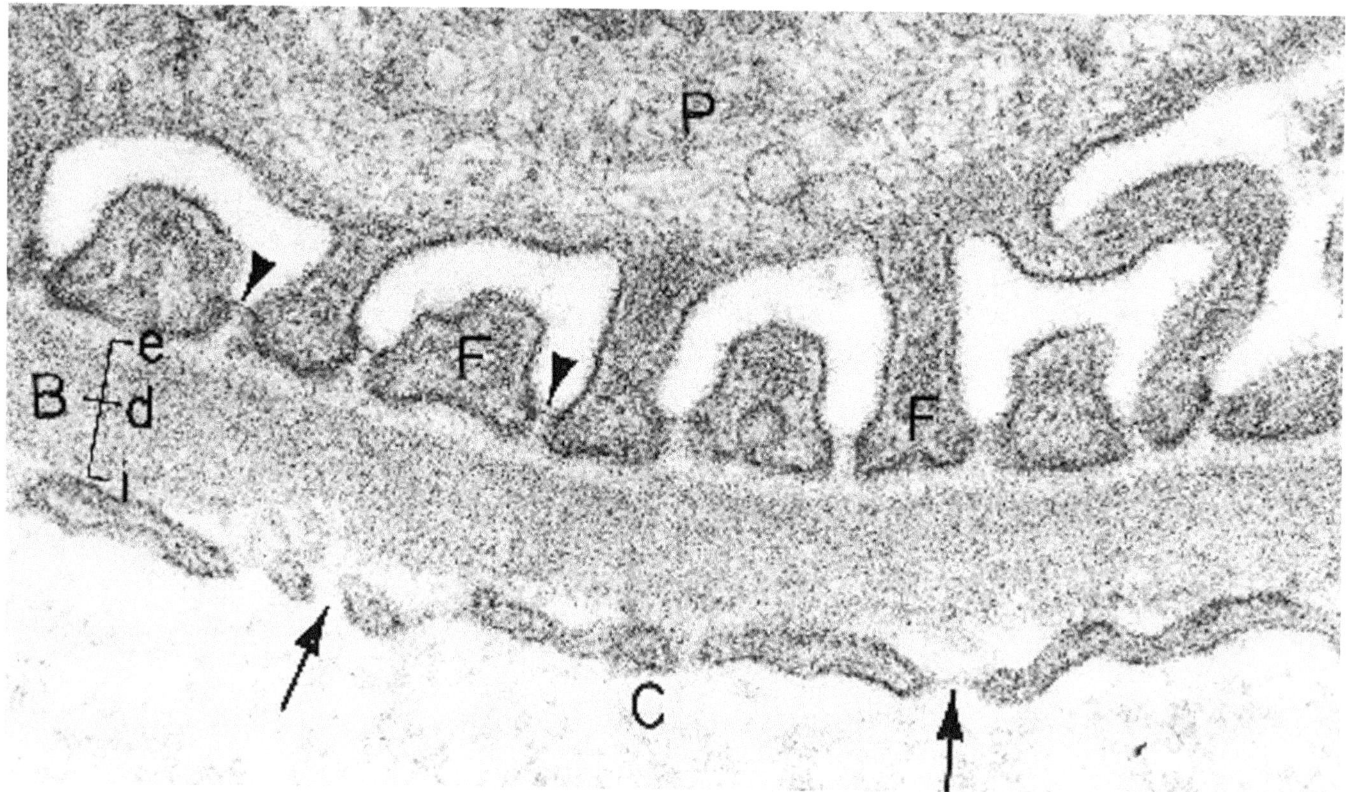

Figure 12.2. The renal capillary filtration barrier. A podocyte (P) extends foot processes (F) along the outer, urinary space side of the GBM (B). On the opposite, inner side of the GBM, the fenestrated cytoplasm of an endothelial cell lines the capillary lumen (C). The GBM is uniform in thickness and substructure. The capillary filtration barrier consists of three elements: (1) the endothelial layer with multiple fenestrae 50–100 nm in diameter (*arrows*); (2) the GBM, which has an inner layer, the lamina rara interna (i); an outer layer, the lamina rara externa (e); and a denser central layer, the lamina densa (d); (3) the filtration slits that are bridged by a thin diaphragm (*arrowheads*), in between the podocyte foot processes. Cytoplasmic filaments extend from the podocyte cell body into the feet. (× 68,400)

The glomerular filtration barrier (Figure 12.2) is composed of fenestrated endothelial cells, the glomerular basement membrane, and the filtration slit pores between podocyte foot processes, which function as three progressively finer filters. The mean area per glomerulus of the filtration surface has been reported to be 0.136 mm^2 in humans (Osterby et al. 1980). The glomerular capillary surface is formed by attenuated, thin endothelium that appears honeycombed on scanning electron microscopy as a result of the fenestrations or pores (50–100 nm in diameter). These fenestrae exclude the formed elements in blood; thin diaphragms have been demonstrated across the pores but are not significant barriers to macromolecules.

The glomerular basement membrane (GBM) is a continuous sheet, comprising three layers: the central dense homogeneous band, the lamina densa, and a zone of lesser density on either side; the lamina rara externa; adjacent to the podocytes; and the lamina rara interna, adjacent to the endothelial cells (Figure 12.2). The GBM excludes macromolecules 8 nm in diameter (albumin), particularly those that are negatively charged. The glomerular extracellular matrix (including GBM and mesangial matrix) are composed of collagen, noncollagenous structural glycoproteins, and proteoglycans. Type IV collagen is the major collagenous component of GBM and mesangial matrix. Six genetically distinct alpha(IV) collagen chains have been described in basement membranes, with variable distribution in the glomerulus. The alpha-1(IV) and alpha-2(IV) collagen chains are diffusely expressed in the mesangium but are faint in the GBM. The alpha-3(IV), alpha-4(IV), and alpha-5(IV) collagen chains are codistributed in the GBM. Using electron microscopic immunohistochemistry in human kidney, Zhu et al. (1994) showed that the alpha-1(IV) collagen chain was distributed mainly along the endothelial side of the GBM and the mesangial matrix. In contrast, the alpha-3(IV) collagen chain was detected throughout the GBM thickness but was absent in the mesangial matrix. The alpha-6(IV) collagen chain is absent in the GBM. Heparan sulfate and sialoglycoproteins in the GBM and on the surface of podocytes and endothelial cells contribute a negative charge carrier, demonstrable with colloidal iron.

The thickness of the GBM increases up to about age 30 years and then declines somewhat. The adult GBM mean thickness is 326 +/− 45 nm in women (N = 59) and 373 +/− 42 nm in men (N = 59; adult range 240–460 nm), measured from the cell membrane of the endothelial cell to the cell membrane of the epithelial cell (Steffes et al. 1983). These values are harmonic means plus or minus standard deviations, calculated by morphometric techniques (arithmetic means give somewhat higher values). Vogler et al. (1987) found that both GBM and lamina densa increase rapidly in width during the first two years of life; from 169 +/− 32 nm and 98 +/− 23 nm, respectively, at birth; to 245 +/− 49 nm and 189 +/− 42 nm, respectively, at age two years; after which the increase in thickness is more gradual, reaching 285 +/− 39 nm for GBM and 219 +/− 42 nm for lamina densa by age 11. The GBM may have occasional focal changes that have no known pathologic significance, such as scattered subendothelial fibrils, particulate vesicular debris, membranous ribbons, and focal irregular thickening. These probably represent in part the wear and tear of normal function.

The podocytes (Mundel and Kriz 1995) cover the outer, urinary surface of the GBM and join the parietal epithelium covering Bowman's capsule at the glomerular hilus. The podocytes form secondary and tertiary processes that branch terminally to form thin, club-shaped terminal processes called pedicles (or foot processes) and anchor to the GBM (Figure 12.2). The foot processes interdigitate in a complicated manner with similar processes from the same and adjacent cells, best appreciated by scanning electron microscopy. Because of three-dimensional orientation and sectioning, a continuation of each foot process to the podocyte cell body may not be seen. Foot processes are spread throughout the capillary loops. Microfibrils, 2–10 nm wide in the foot processes (Figure 12.2), may have contractile functions. The spaces between adjacent foot processes are called filtration slits, and at the junction of foot processes and GBM, the slits are bridged by a thin diaphragm (slit diaphragm) that has a zipper-like appearance *en face*, the final barrier for bulk water flow (Figure 12.2; [Schneeberger et al. 1975]). The podocyte luminal membrane and slit diaphragms are covered by a thick coat rich in sialoglycoproteins such as podocalyxin and podoendin that impart a high surface negative charge. This surface charge contributes to maintenance of the interdigitating pattern of foot processes. The podocyte function includes synthesis and maintenance of the GBM and endocytosis of filtered proteins.

The mesangium forms the trunk and branches that support the glomerular capillaries in continuity with the juxtaglomerular apparatus at the hilus. The mesangial cells have a small, densely staining nucleus and long processes that interdigitate between capillary loops. The contractile apparatus of mesangial cells consist of microfilament bundles located predominately in mesangial cell processes (Kriz et al. 1990). The thickest bundles of microfilaments occurs in the juxtacapillary cell process. No GBM separates the mesangium from the endothelial cells (Figure 12.1), and thus the mesangium is in continuity with the subendothelial space, forming a path of least resistance for particulate material trapped between the endothelium and the GBM. Judging from animal studies, some cells in the mesangium are bone-marrow derived mononuclear phagocytes. The mesangial cells are surrounded by a matrix material similar to but not identical to the peripheral GBM. The mesangial matrix is filled by a faintly fibrillar material with less electron density than the lamina densa of the GBM, but both have similar staining characteristics. It is not uncommon to find occasional amorphous electron-dense deposits in the mesangium without apparent pathologic significance.

Location of Electron-Dense Deposits

(Figure 12.3.)

Before discussing the various glomerular diseases, a brief introduction to the four basic types of deposits that may be encountered in renal glomerular pathology are illustrated in Figure 12.3. *Subepithelial deposits* are located along the outer urinary aspect of the GBM and underneath the visceral epithelial cells (podocytes). These are typical of membranous glomerulonephritis (idiopathic, secondary forms, or class V lupus nephritis) and postinfectious glomerulonephritis. *Subendothelial deposits* are located between the inner, capillary side of the GBM and the endothelium. These deposits are characteristically seen in type I membranoproliferative glomerulonephritis and class IV lupus nephritis. *Mesangial deposits* are present in the mesangial matrix, usually adjacent to mesangial cells and are classically seen in IgA nephropathy. *Intramembranous deposits* are located within the GBM as exemplified by dense deposit disease.

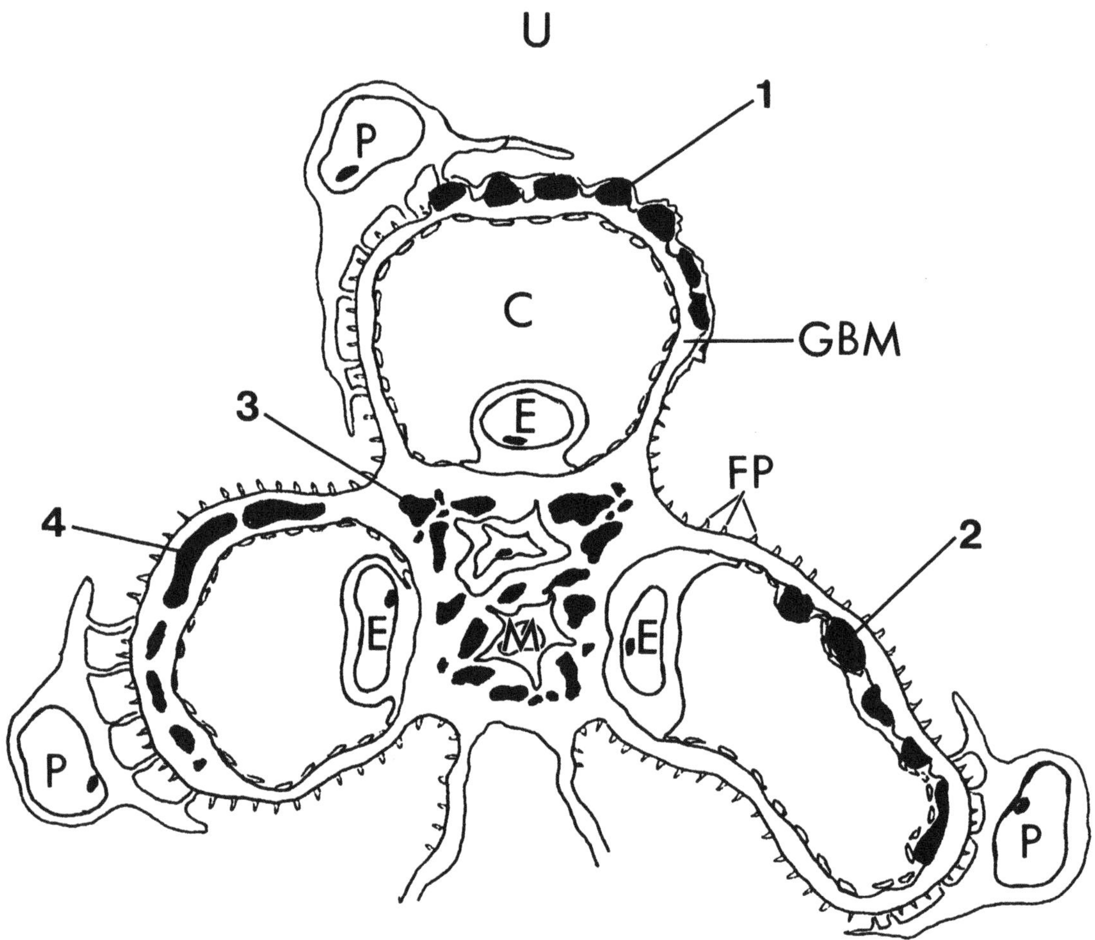

Figure 12.3. Location of electron-dense deposits. 1 = Subepithelial deposits; 2 = subendothelial deposits; 3 = mesangial deposits; 4 = intramembranous deposits; P = podocyte; FP = foot processes; GBM = glomerular basement membrane; E = endothelial cell; M = mesangial cell; C = capillary lumen; U = urinary space.

Diseases with Scant or no Glomerular Deposits

Minimal Change Disease (Lipoid Nephrosis)

(Figures 12.4 and 12.5.)

Diagnostic criteria. (1) Widespread effacement (broadening) of foot processes of podocytes over most of the GBM (Figures 12.4 and 12.5) (Bohman et al. 1984; Yoshikawa et al. 1982); (2) normal GBM thickness and appearance; (3) enlarged podocytes with vacuolated cytoplasm containing lipid droplets, numerous mitochondria, well-developed Golgi apparatuses, rough endoplasmic reticulum, and microvilli (villous hypertrophy) on the urinary surface (Figure 12.5); (4) no focal, segmental glomerular scars with tubular atrophy.

Additional points. Microfilaments are condensed at the base of swollen podocytes. Podocytes may be lifted from the GBM. Rarely, focal thinning and loosening of the GBM are present. Occasionally, the lamina rara interna is increased secondary to subendothelial lucency.

Biopsies taken in remission show return of the foot processes; in fact, the amount of proteinuria is directly correlated with the extent of foot process loss (Powell 1976). The foot process loss is sometimes called fusion; however, because the foot processes actually are retracted as the cell body settles down on the GBM, the term effacement, better defines this foot process loss. Rapid foot process effacement can be produced experimentally by polycations (protamine) or neuraminidase, owing to loss or neutralization of the negative charge on the podocyte surface (Seiler et al. 1977).

By light microscopic examination, the podocytes are typically somewhat enlarged with a basophilic cytoplasm. Proximal tubules contain PAS-positive reabsorption droplets and neutral fat (hence the original

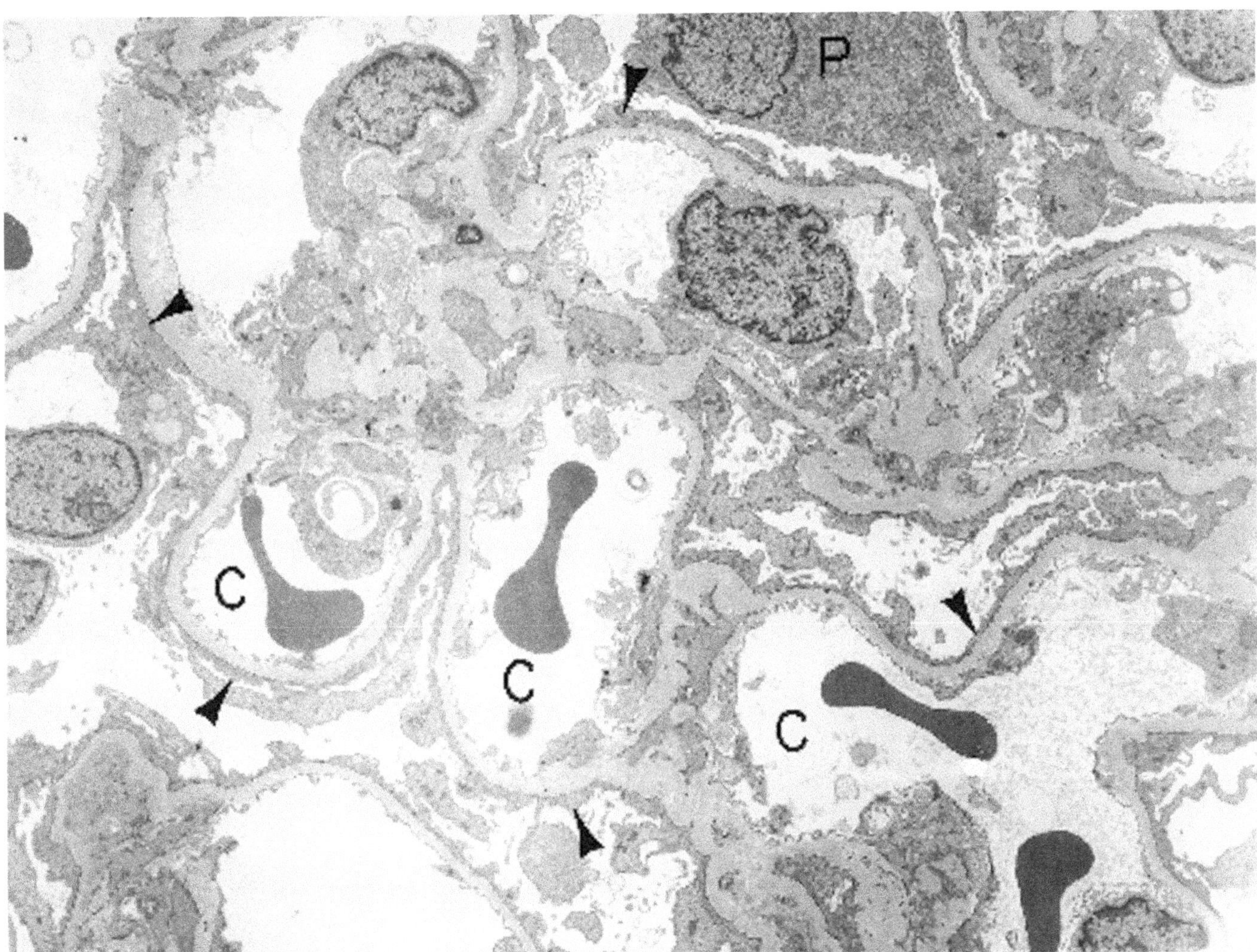

Figure 12.4. Minimal change disease (32-year-old man with nephrotic syndrome). Low power view of a glomerulus with widespread podocyte foot processes effacement (*arrowheads*). The podocytes (P) show a mild degree of villous hypertrophy. The capillary loops (C) contain red blood cells. The mesangium, endothelium, and GBM are normal. (× 4028)

name, "lipoid nephrosis"). Focal glomerular sclerosis must be excluded, insofar as possible, by step sections. Isolated globally sclerotic glomeruli without tubular atrophy can be found in normal children and have no significance. However, tubular atrophy and interstitial fibrosis (even without glomerular scars) suggest underlying focal glomerular sclerosis. In adults, global sclerosis and tubular atrophy are found as part of "benign nephrosclerosis." These tend to be grouped, characteristic of their vascular pathogenesis. The immunofluorescence studies are negative or have small amounts of mesangial IgM and C3 (a normal finding). Some classify the latter as IgM nephropathy (see next section).

IgM Nephropathy

(Figure 12.6.)

Diagnostic criteria. (1) Electron-dense mesangial deposits; (2) variable degree of mesangial hypercellularity and increased matrix; (3) effacement of foot processes in patients with nephrotic range proteinuria.

Additional points. The deposits are restricted to the mesangium, and may be ill-defined, merging with the mesangial matrix. In one report, no deposits were identifiable by electron microscopy in 36%, despite positive immunofluorescence (Cohen et al. 1978). Light microscopy shows mild glomerular changes, with mild to severe mesangial hypercellularity. These lesions differ

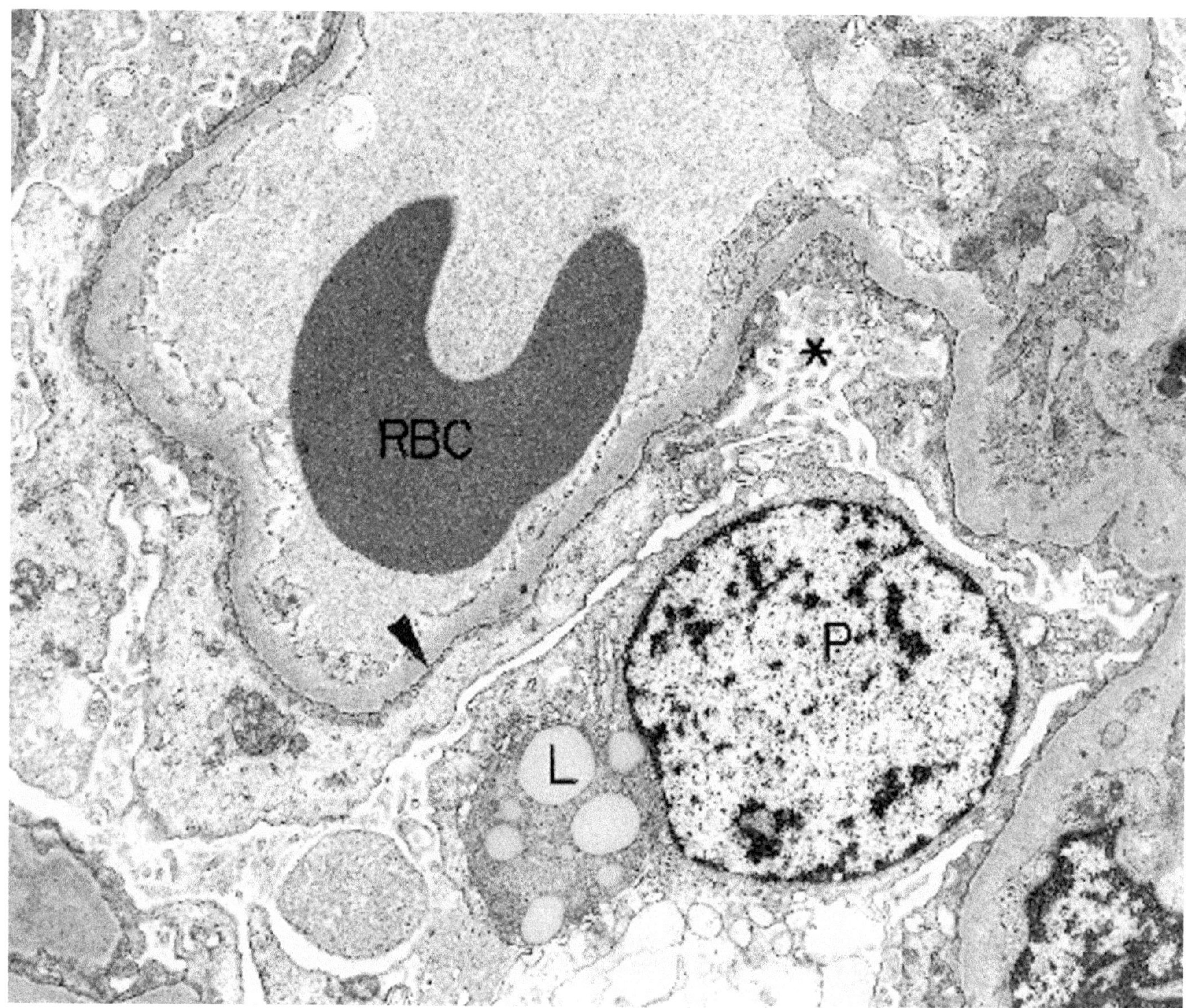

Figure 12.5. Minimal change disease (57-year-old man with a 2-week history of nephrotic syndrome [6 g/day proteinuria] and normal renal function). The podocyte (P) shows effacement of foot processes (*arrowhead*), villous hypertrophy (*), and cytoplasmic lipid accumulation (L). A red blood cell (RBC) is in the capillary lumen. (× 9180)

from focal sclerosis only in the absence of segmental scars.

The defining feature is mesangial IgM staining by immunofluorescence microscopy, which, as noted, can be a normal finding. Patients who present with proteinuria sometimes also present with microscopic hematuria. Therefore, many (including the authors) have not accepted this as a distinct clinicopathologic entity. Most cases probably represent focal glomerular sclerosis or minimal change disease. About 50% (7/16) of children with frequent relapsing nephrotic syndrome had mesangial IgM deposits (Trachtman et al. 1987). Prominent IgM has been noted in children with minimal change disease who had been resistant to prednisone and responded to prednisolone (Herrin and Colvin, unpublished).

Focal and Segmental Glomerulosclerosis (Primary and Secondary Types)

(Figures 12.7 and 12.8.)

Diagnostic criteria. (1) Effacement of podocyte foot processes; (2) swollen and vacuolated (lipid and proteinaceous fluid) podocytes with formation of microvilli on their free surface (Figure 12.7); (3) segmental areas of glomerular sclerosis with collapse, wrinkling, and

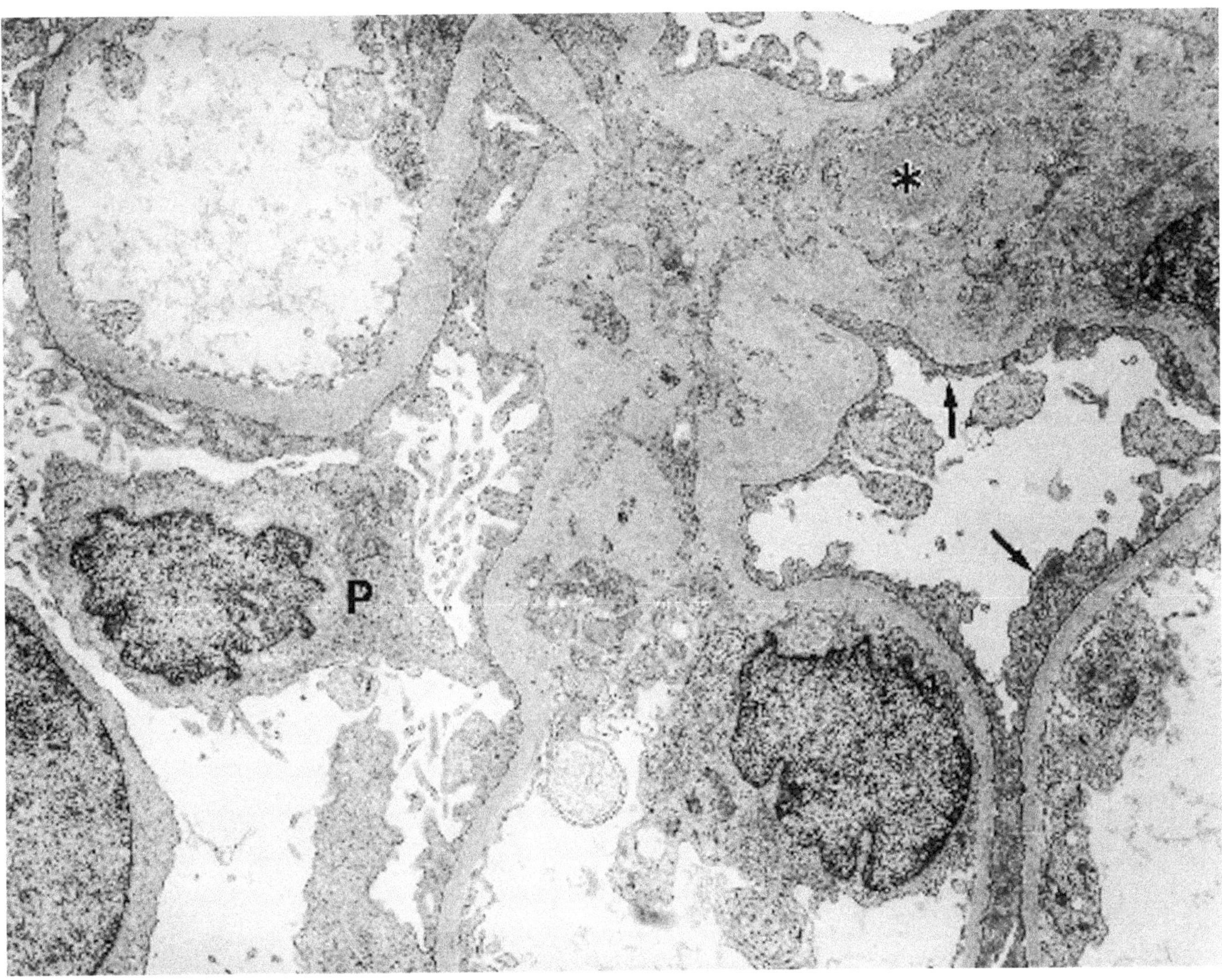

Figure 12.6. IgM nephropathy (12-year-old girl with an 8-year history of steroid-dependent nephrotic syndrome and normal renal function). Extensive effacement of podocyte foot processes (*arrows*) and villous hypertrophy of podocytes (P) are present, indistinguishable from minimal change disease or focal glomerular sclerosis. In addition, amorphous mesangial deposits are present (*). Mesangial deposits stained for IgM by immunofluorescence. The GBM is normal. (× 5400)

thickening of the GBM, typically beginning at the perihilar region (Figure 12.8A); (4) amorphous hyaline deposits, often with lipid droplets, in capillary spaces of scarred segments (Figure 12.8B), corresponding to the hyaline seen by light microscopy; (5) increased mesangial cells and matrix may be present.

Additional points. The glomerular basement membrane in nonsclerotic areas may have focal thickening of the lamina rara interna and occasional splitting with mesangial interposition. Subepithelial new GBM lamination may be present in segments where the podocyte has lifted off of the GBM (Figure 12.8A). Scattered electron-dense deposits may be present in subendothelial, intramembranous, and mesangial location. One of the distinguishing characteristics of primary, idiopathic focal segmental glomerulosclerosis (FSGS) is the presence of widespread foot process effacement over nonsclerotic otherwise normal-appearing lobules, in addition to over the sclerotic lobules (D'Agati 1994; Schwartz and Korbet 1993). Secondary forms of FSGS are commonly associated with glomerular hypertrophy related to structural and functional adaptations, including intrarenal vasodilation and increases in glomerular capillary pressure and plasma flow, leading to compensatory hyperfiltration (Rennke and Klein 1989). Secondary focal segmental glomerulosclerosis typically has a diffusely thickened GBM and frequent capillary loops with intact foot processes (Fig 12.8C).

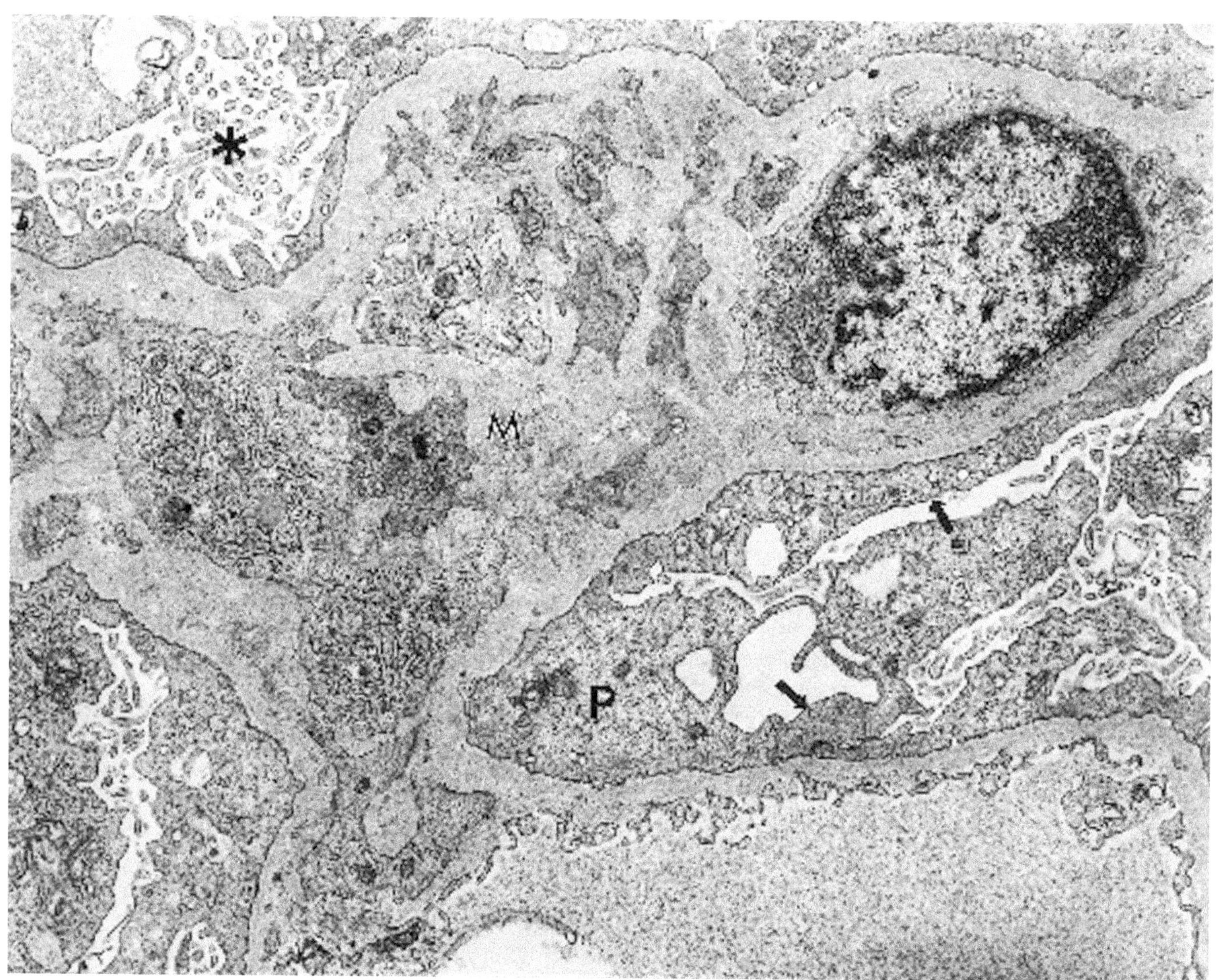

Figure 12.7. Focal and segmental glomerulosclerosis, classic type (16-month-old boy with recent-onset nephrotic syndrome not responsive to a 4-week treatment with steroids). Segmental collapse of the capillary lumina with mesangial sclerosis along with podocyte (P) foot process effacement (*arrows*) and villous hypertrophy (*) are present. The mesangium (M) is expanded and contains more cell processes, matrix, and focal electron-dense deposits. (× 10,500)

Light microscopy shows collapse and sclerosis of a portion of affected glomeruli, particularly adjacent to the hilum, with deposition of amorphous eosinophilic material on the inner aspect of capillary loops (hyalinosis), adhesions to Bowman's capsule, foam cells, and rarely crescents. The glomerular sclerosis is characteristically more frequent among juxtamedullary glomeruli. Immunofluorescence shows the presence of IgM and C3 in sclerotic areas and occasionally in the mesangium of nonscarred glomeruli accompanied by other immunoglobulin classes.

The primary form of this disease is caused by a blood-borne factor, as proved by the rapid and dramatic recurrence in about 40% of patients who received renal transplants. Such cases show foot process loss in the earliest biopsies (1 week) and adhesions and segmental sclerosis in 4–6 weeks. Cases with prominent mesangial hypercellularity tend to have a worse prognosis and recur more frequently after transplantation.

Secondary focal segmental sclerosis can be seen in association with reflux nephropathy, unilateral renal agenesis (Kiprov et al. 1982), diabetes mellitus, membranous glomerulonephritis, Alport's syndrome, IgA nephropathy, and obesity (Verani 1992). These lesions are believed to arise from hyperfiltration injury, a common final pathway of end-stage renal disease. Secondary focal sclerosis rarely, if ever, recurs in transplants. Segmental adhesions between the tuft and

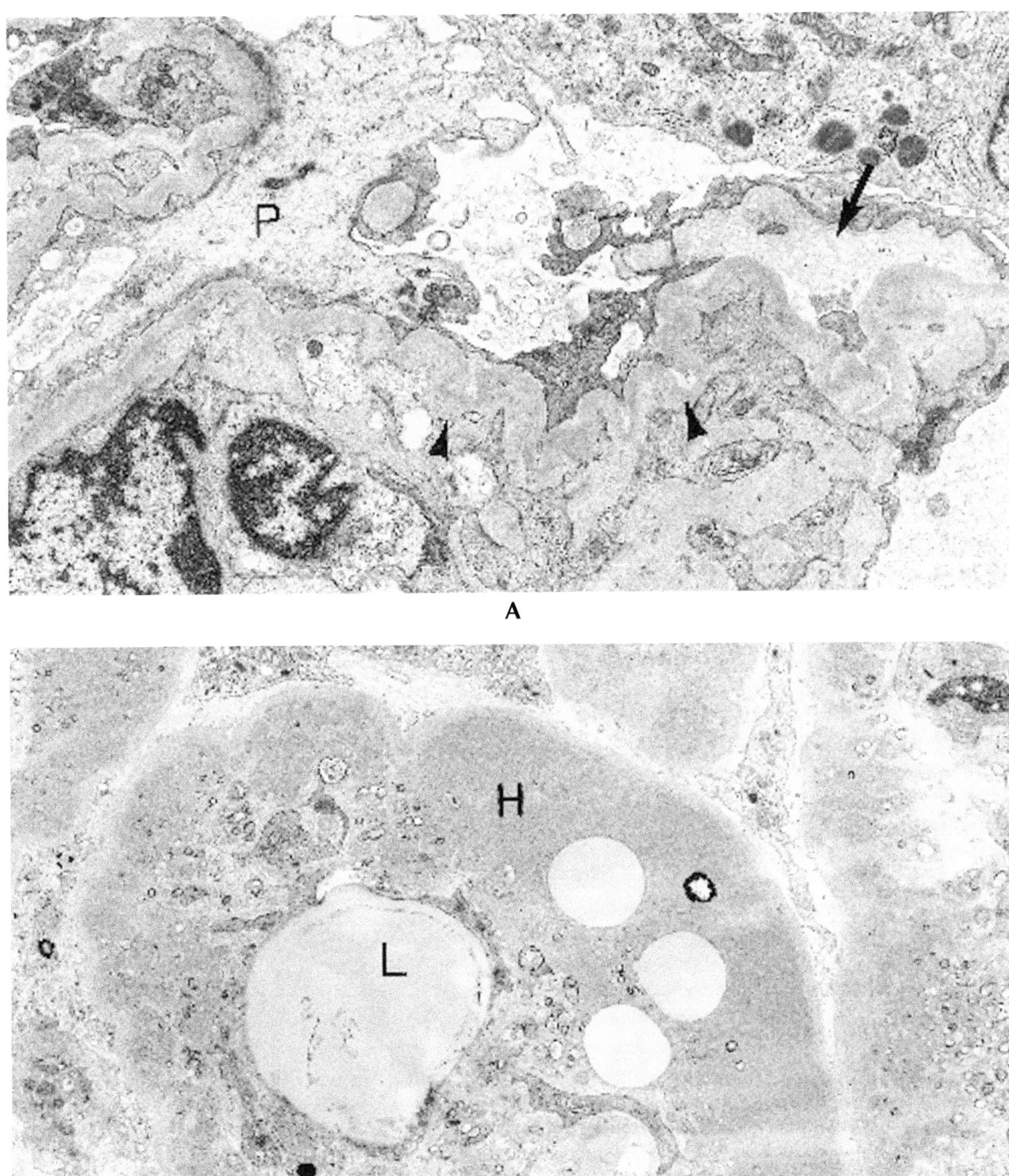

Figure 12.8. Focal and segmental glomerulosclerosis: **A** (69-year-old man with 2.5 g/day proteinuria and microscopic hematuria), the GBM is wrinkled and collapsed (*arrowheads*). The podocyte (P) is separated from the GBM with underlying laminated basement membrane material and lucency (*arrow*); foot processes are effaced. (× 10,260). **B** (recurrence of FSGS in an allograft), Massive subendothelial amorphous deposits (H) admixed with lipid vacuoles (L) are present in the segmental glomerular scar that correspond to hyaline. Note the smooth, rounded contours of hyaline deposits (× 9000).

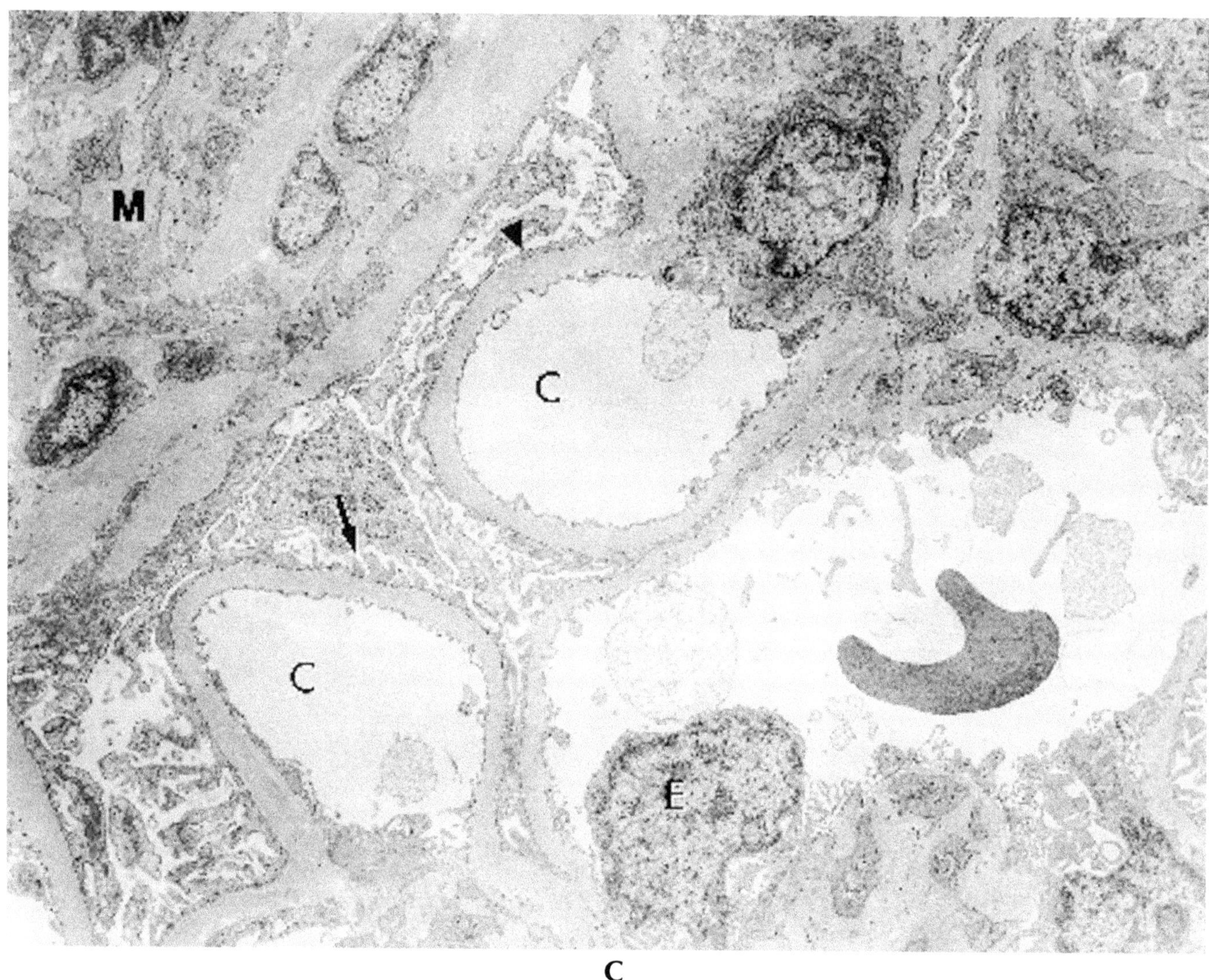

Figure 12.8. *(continued)*
C, Focal segmental glomerulosclerosis, secondary type (18-year-old male with unilateral renal agenesis, biopsy of solitary kidney for proteinuria, 1.5 g/day), widespread GBM thickening. Segmental podocyte foot process effacement (arrowhead) with areas of preserved foot processes (*arrow*). C = capillary lumen; E = endothelial cell; M = mesangium. (× 7400)

Bowman's capsule at the orifice of the proximal tubule have been termed glomerular tip lesions (Howie and Brewer 1984). These lesions may have a better prognosis than the usual form of focal glomerular sclerosis, although this point is controversial.

Focal Segmental Glomerulosclerosis, Collapsing Variant, Including HIV-Associated Nephropathy and Heroin Abuse Nephropathy

(Figures 12.9 through 12.13.)

Diagnostic criteria. (1) Segmental collapse of capillary loops and often global collapse of the glomerular tuft (implosive retraction) (Figure 12.9) without expansion, unlike classic FSGS; (2) wrinkling and folding of the GBM; (3) widening of Bowman's space; (4) podocyte proliferation and marked hypertrophy over the collapsed loops, filling the Bowman's space, usually with numerous reabsorption droplets.

Additional points. This distinct, malignant form of FSGS, now called collapsing FSGS, was originally described in patients with HIV infection and acquired immunodeficiency syndrome (AIDS) (Chander et al. 1987; D'Agati et al. 1989; Cohen and Nast 1988) as HIV nephropathy (Figures 12.11 through 12.13) and in patients who abuse heroin (Grishman and Churg 1975) as heroin nephropathy. Cases without HIV infection or

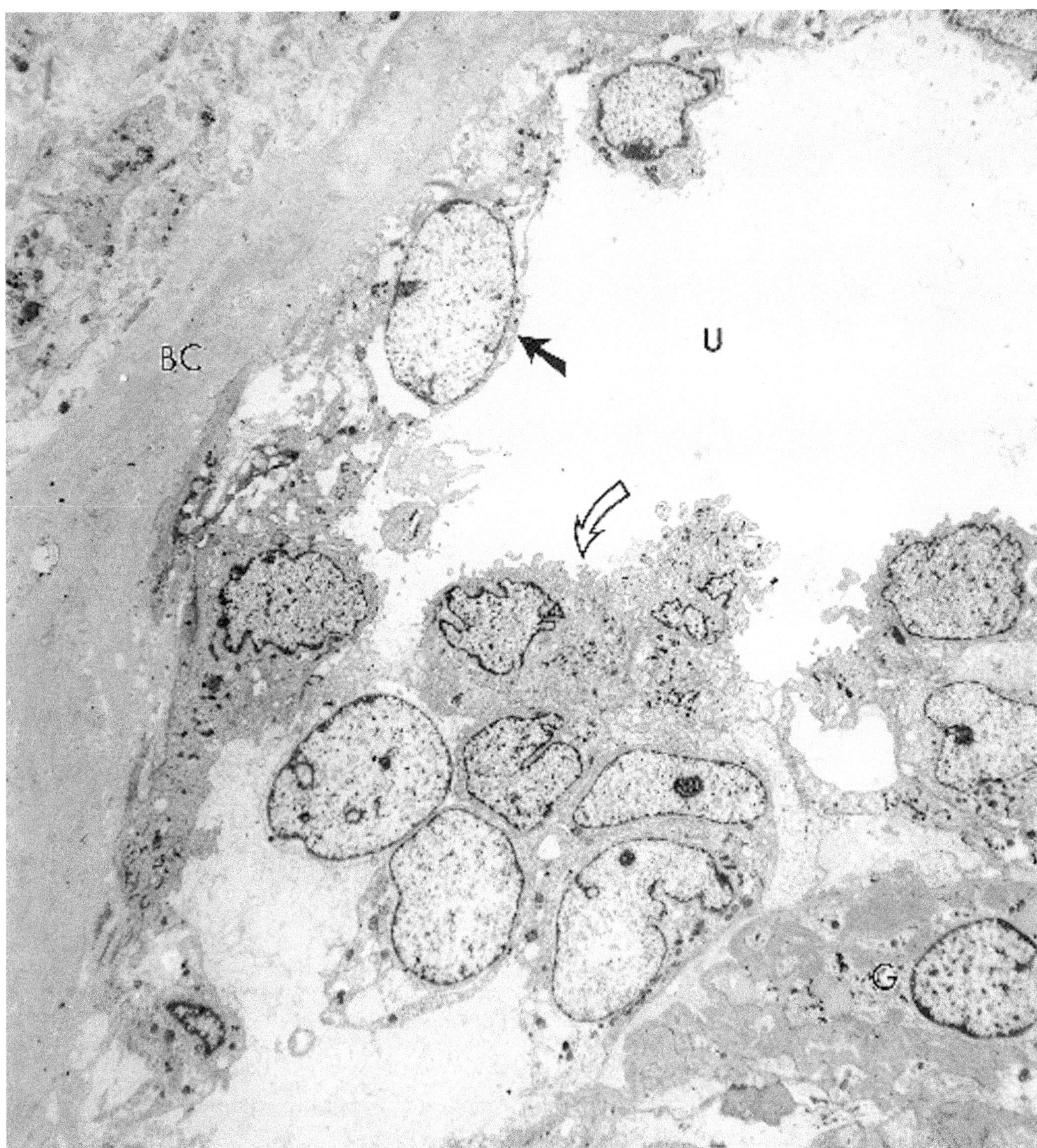

Figure 12.9. Focal and segmental glomerulosclerosis, collapsing variant (recurrence, 3.5 months post renal transplant in a 59-year-old man). Collapsed glomerular tuft (G) with reactive podocyte hyperplasia (*open arrow*) and reactive glomerular parietal epithelial cells (*arrow*). BC = Bowman's capsule; U = urinary space. (× 4600).

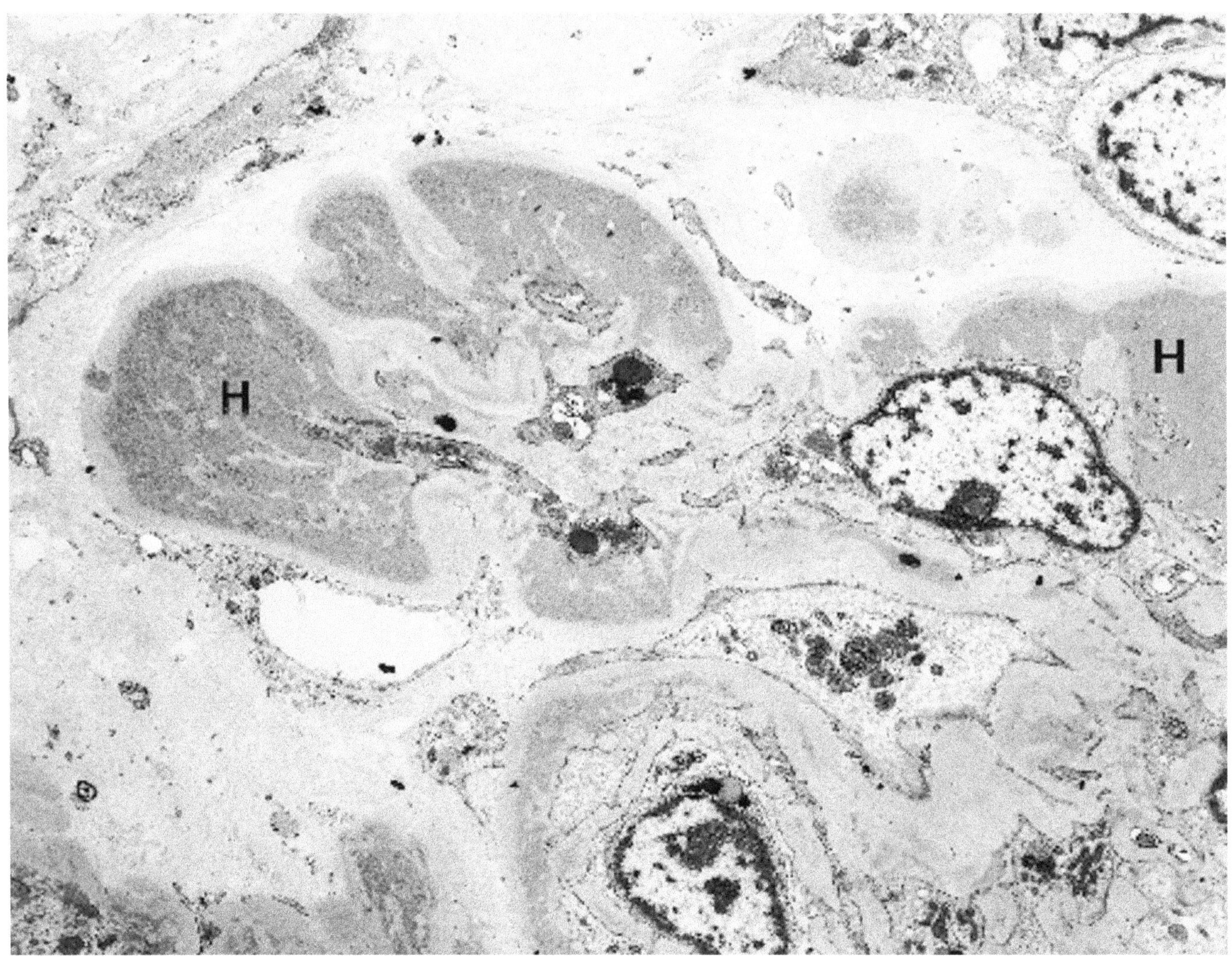

Figure 12.10. Focal and segmental glomerulosclerosis, collapsing variant (56-year-old female who received cyclosporine A after heart-lung transplant). Collapsed capillary loop with hyalinosis (H). (× 5500)

heroin abuse have been recognized (Detwiler et al. 1994) and represent a distinct entity (idiopathic type); African American racial predominance has been observed in this idiopathic type, as well as in HIV and heroin nephropathy. Endothelial tubuloreticular inclusions help distinguish HIV from non-HIV cases (Figure 12.12). Subendothelial deposits may be found (Figures 12.12 and 12.13A). Intracapillary foam cells as in classic FSGS (Figure 12.13A) are seen. Nuclear bodies or various intranuclear inclusions have been described in HIV nephropathy as well as heroin nephropathy (Figure 12.13B). In HIV nephropathy, the nuclei may also show granulofibrillary changes.

Given the recent rise in the number of cases being diagnosed with this rapidly progressive form of FSGS, possible environmental or infectious factors in genetically predisposed individuals may play a role. Collapsing glomerulopathy may also occur *de novo* after transplant (Meehan et al. 1998). We have also observed collapsing glomerulopathy in association with cyclosporine toxicity (Figure 12.10) (Mauiyyedi et al. 1999).

Congenital Nephrotic Syndrome

(Figure 12.14.)

Diagnostic criteria. (1) Extensive effacement of podocyte foot processes; (2) lamina rara interna of the GBM may be irregularly widened by fine flocculent electron-dense material with focal duplication of the lamina densa (Figure 12.14B); (3) thin lamina densa, compared with age-matched controls, although the overall thickness of the GBM is within normal limits (Autio-Harmainen and Rapola 1983); (4) mesangial matrix slight to markedly increased; (5) mesangial cells

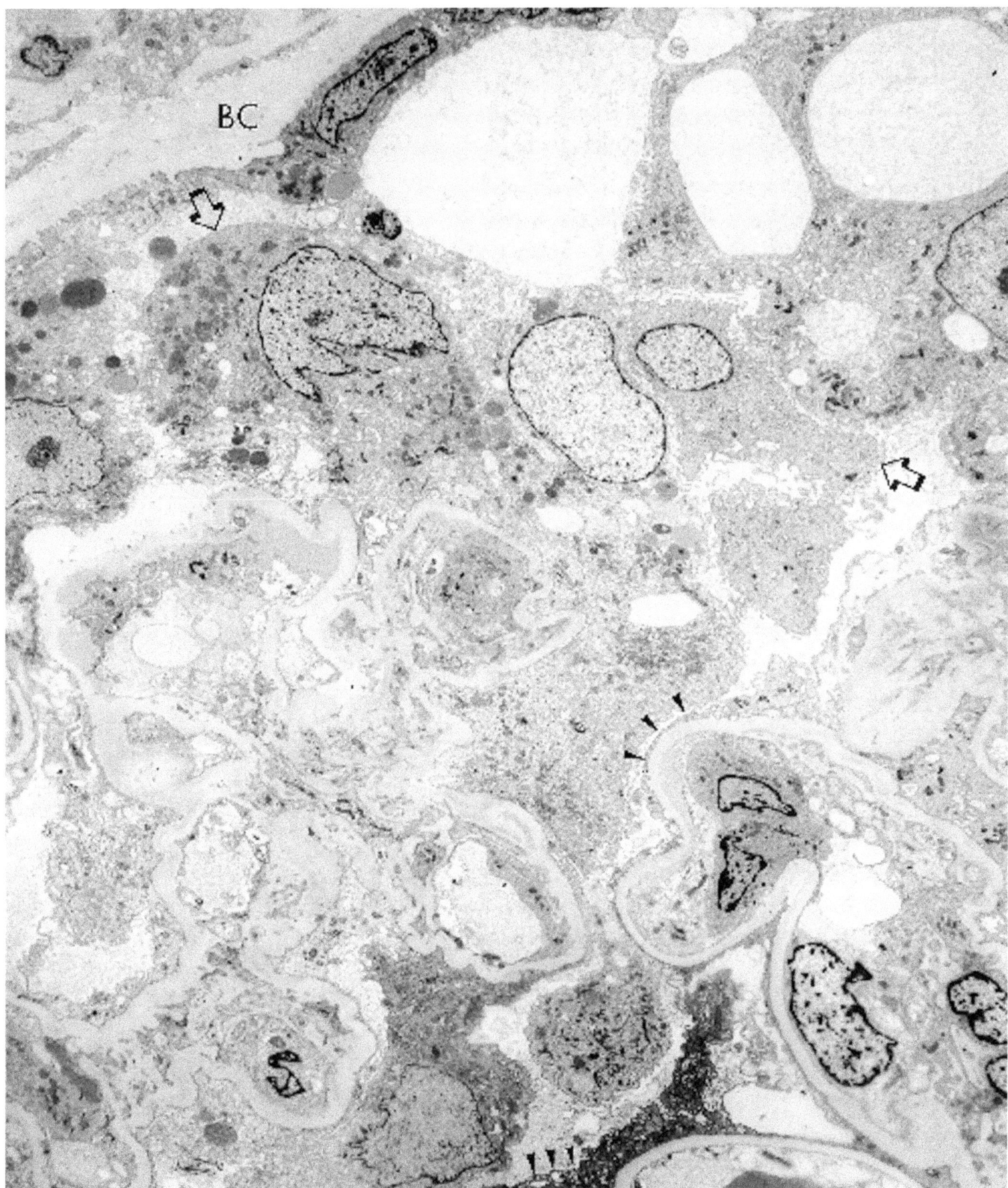

Figure 12.11. HIV-associated nephropathy (focal and segmental glomerulosclerosis, collapsing variant) (30-year-old male HIV-positive, with proteinuria of 11.6 g/day). Reactive and swollen podocytes that fill the Bowman's space (*open arrows*) with diffuse foot process effacement (*arrowheads*) and segmental collapse of the capillary loops with wrinkling, duplication, and thickening of the GBM and scattered subendothelial deposits. BC = Bowman's capsule. (× 3900).

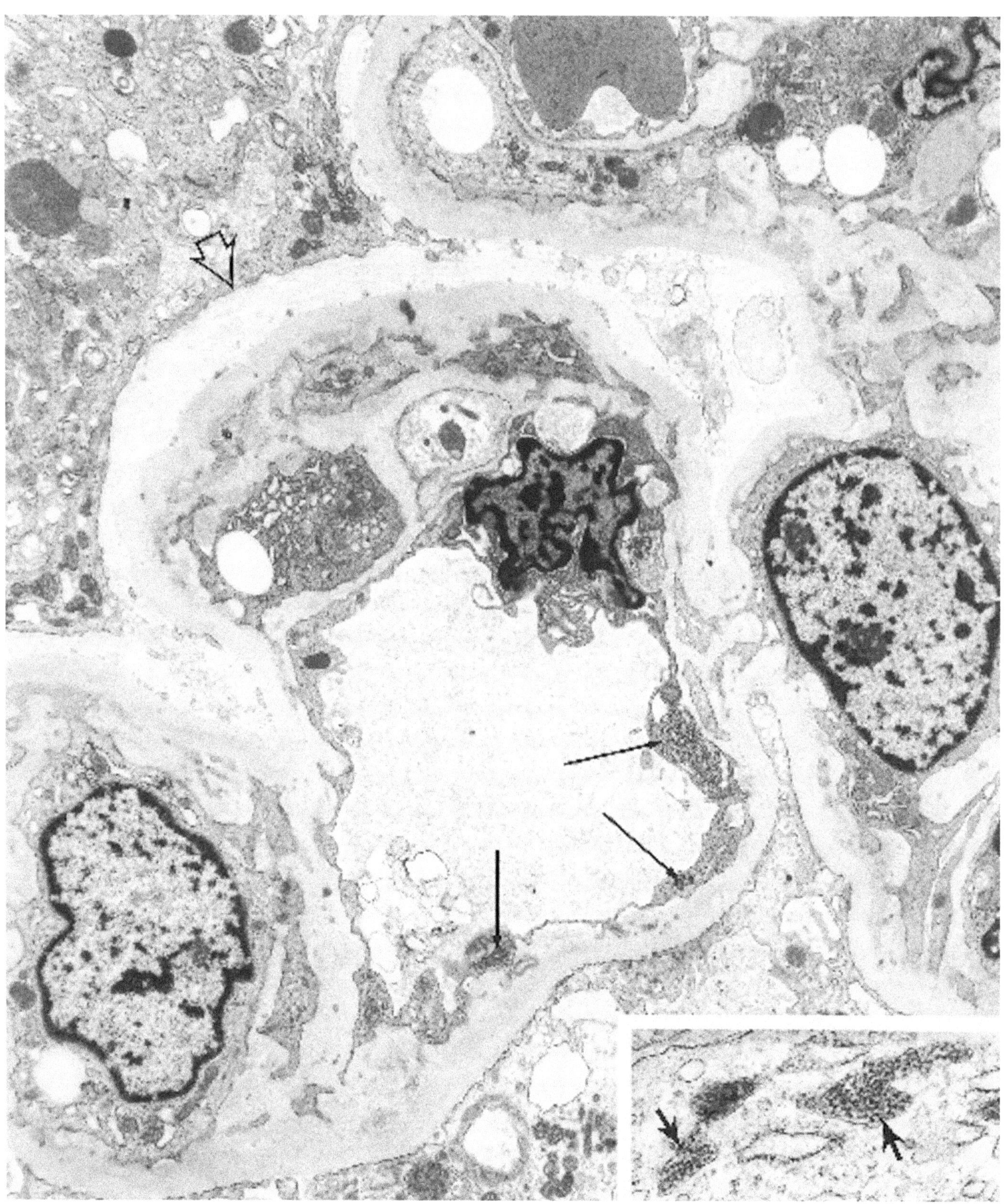

Figure 12.12. HIV-associated nephropathy (same patient as in Figure 12.11). The podocytes are reactive with foot process effacement. The GBM beneath the podocyte is prominently laminated and lucent (*open arrow*). GBM duplication with cellular interposition and few subendothelial deposits are present. The endothelial cells are reactive and contain many tubuloreticular structures (*arrows*), a higher magnification of which is shown in the *inset*. (× 9300) (*inset* × 30,000).

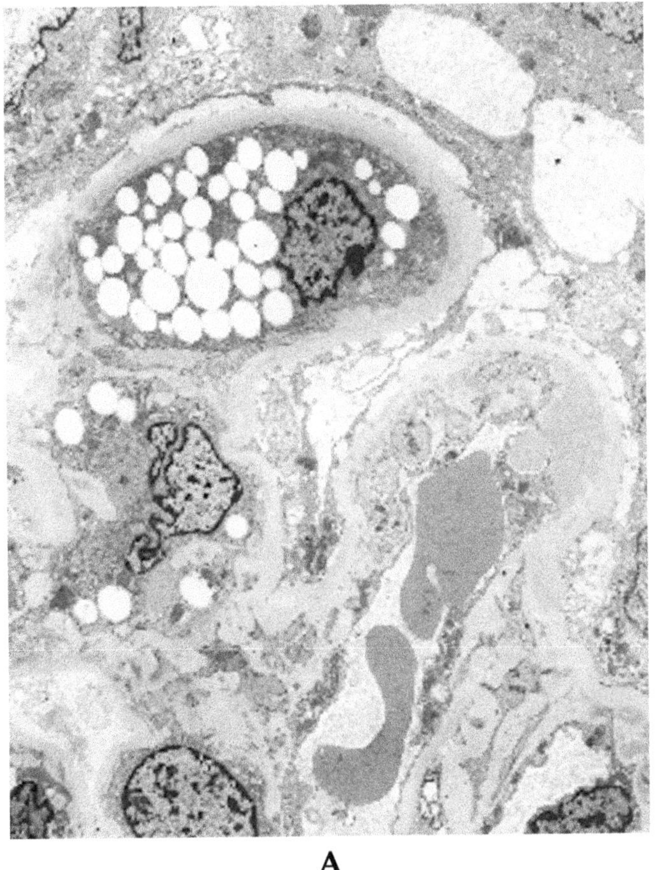

A

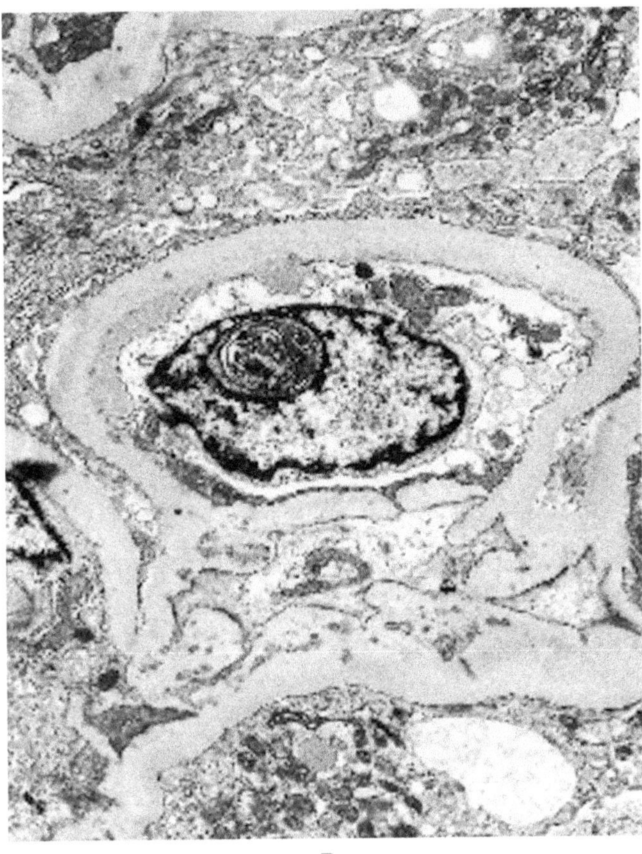

B

Figure 12.13. HIV-associated nephropathy (same patient as in Figure 12.11). **A,** Subendothelial foam cell with many lipid vacuoles. Adjacent capillary loop with subendothelial pale deposits (× 3700). **B,** Intranuclear inclusion in nucleus of an endothelial cell (× 7500).

with electron-lucent vacuoles; (6) endothelial cells enlarged with superficial processes projecting in capillary lumens.

Additional points. The thickness of the GBM varies from one glomerulus to another. The GBM is formed during embryogenesis by the fusion of a lamina densa formed by the endothelial cells and another formed by the epithelial cells (in some species the layers remain separate), therefore, the GBM of the fetus may appear split (Autio-Harmainen 1981). Duplication of the lamina densa in congenital nephrotic syndrome may therefore be a sign of immaturity.

This diagnosis can be made on fetal kidneys; the amniotic fluid contains high levels of alpha-fetoprotein owing to fetal proteinuria. The thin lamina densa and podocyte changes are best seen in the more mature glomeruli. The normal fetal lamina densa is 49 +/− 4 nm.

The other forms of congenital nephrotic syndrome have been less well documented. Nonhereditary forms are characterized by diffuse mesangial sclerosis and by mesangial hypercellularity. Other types that occur in the first year of life, but rarely in the first three weeks, are minimal change disease and FSGS (Habib and Bois 1973).

Diabetic Nephropathy

(Figures 12.15 and 12.16.)

Diagnostic criteria. (1) Diffuse homogeneous thickening of the GBM; (2) mesangial hypercellularity in the early stage of lesion; (3) increased intercapillary mesangial matrix (volume and density), which in the later stage form rounded expansions called Kimmelstiel Wilson nodules (Kimmelstiel and Wilson 1936) containing mesangial fibrils, lipids, and cell debris.

Additional points. No GBM thickness threshold has been established; 150% of that predicted for age or 450 nm (99th percentile) will have a few false-positives but not identify mild cases. Isolated segments (3%) of the GBM may be thin and laminated (Osterby et al. 1987). The podocyte foot processes are wider on average in diabetes (352 versus 224 nm; [Osterby et al. 1987]), but

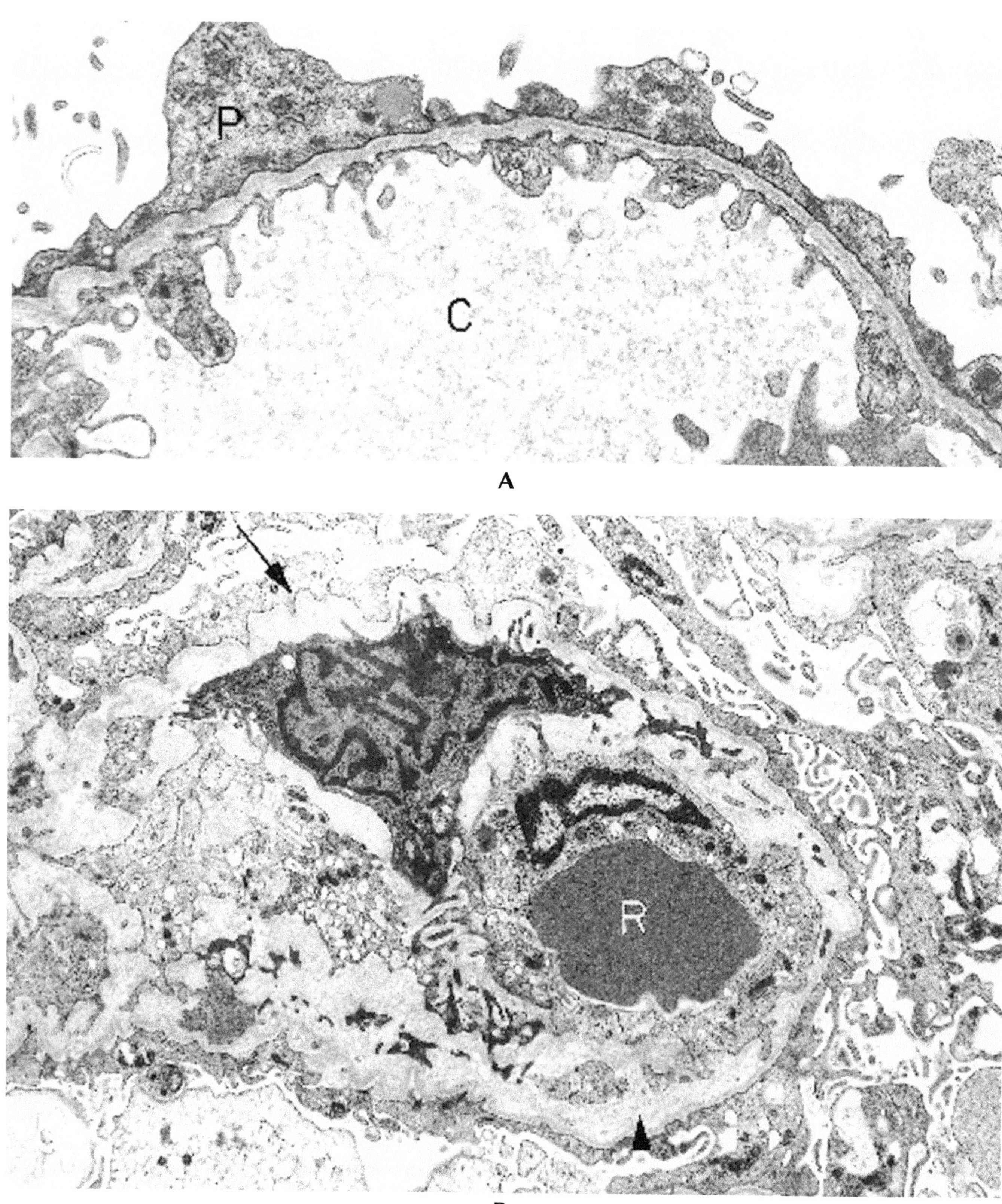

Figure 12.14. Congenital nephrotic syndrome. **A,** (10-day-old infant with proteinuria, hypoalbuminemia, and anasarca), Podocyte (P) foot processes are effaced. The lamina densa is thin. C = capillary lumen. (× 16,520). **B** (same patient as in **A**). The lamina rara interna of the GBM is irregularly widened by fine flocculent laminated material (*arrowhead*). The GBM is wrinkled. Podocytes show loss of foot processes (*arrow*) and villous hypertrophy. R = erythrocyte. (× 7980)

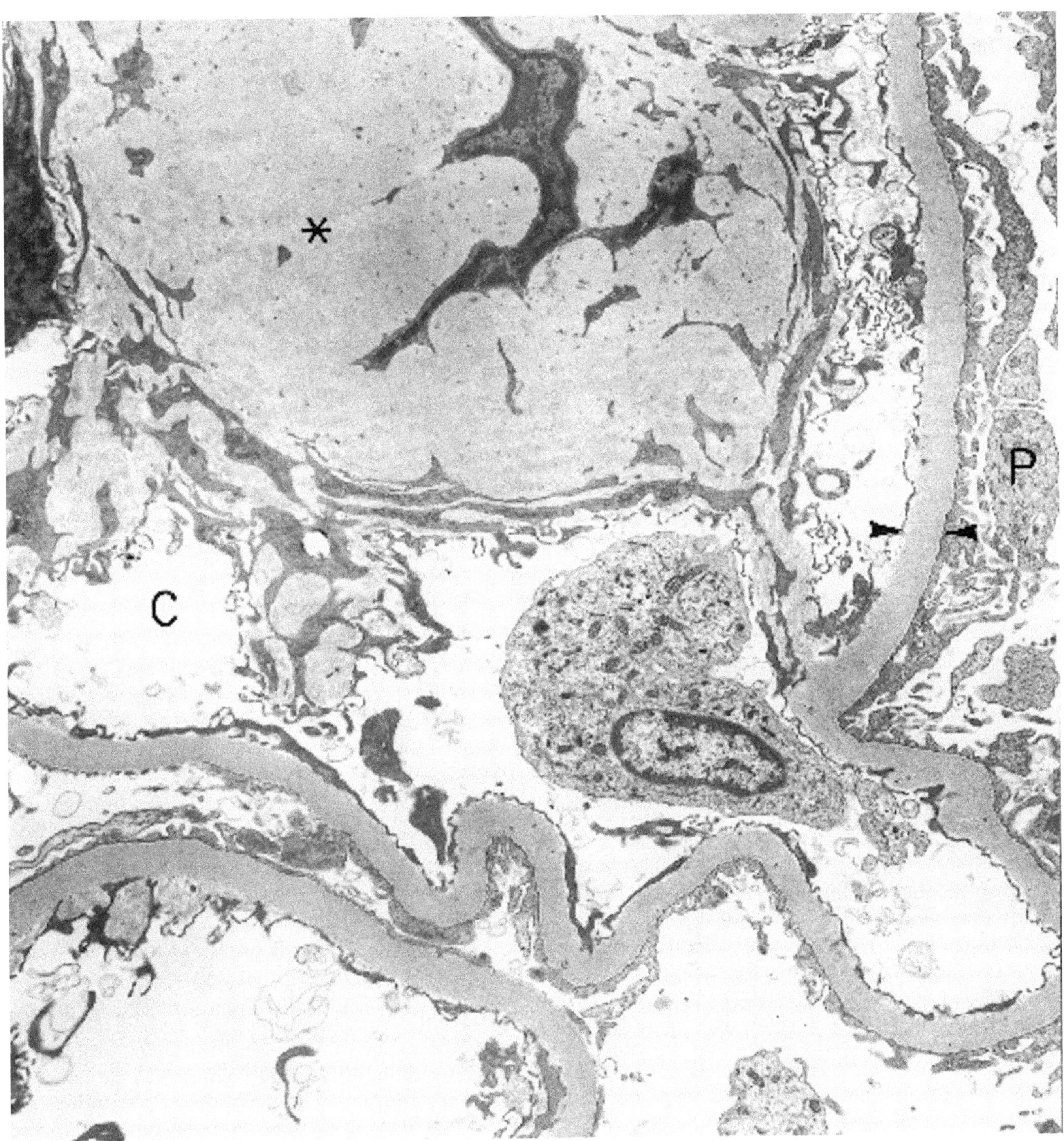

Figure 12.15. Diabetic glomerular sclerosis (51-year-old man with insulin-dependent diabetes mellitus). The GBM is uniformly thickened, about 950 nm (*arrowheads*), but otherwise of normal appearance, with nodular mesangial expansion (*). Podocytes (P) have segmental foot process loss. C = capillary lumen. (× 6840)

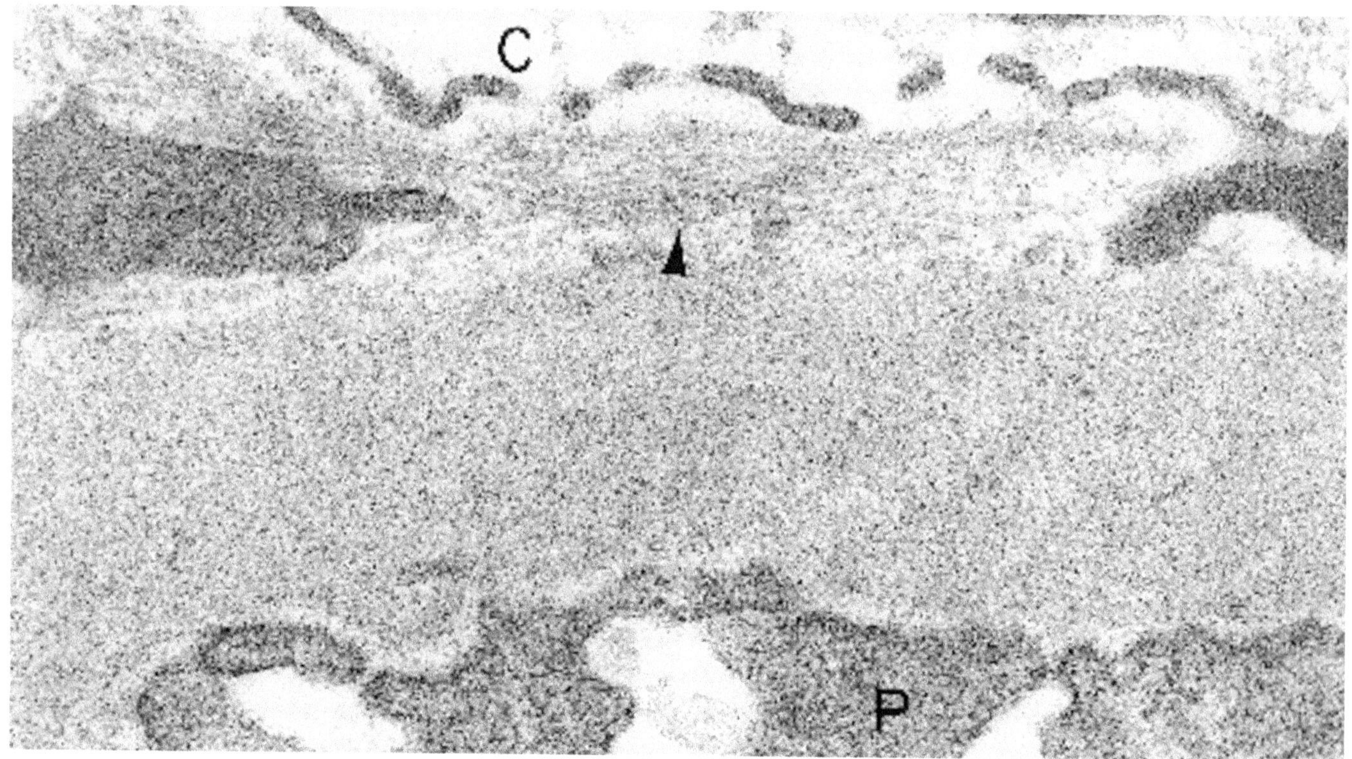

Figure 12.16. Diabetic glomerular sclerosis. The GBM is uniformly thickened but otherwise of normal appearance. Subendothelial fibrillar material is present, which measures 8–10 nm in thickness (*arrowhead*). This is a nonspecific change seen also in focal sclerosis. C = capillary lumen; P = podocyte foot process. (× 68,400)

there appears to be no correlation between foot process width and proteinuria. Extensive foot process effacement is not characteristic of diabetic glomerular sclerosis. Mesangial cell interposition may be present. Occasional fibrillar subendothelial material (Figure 12.16) or subepithelial and intramembranous granules may be present. Immune complexes are rare. Accumulation of fine granular hyaline material admixed with lipids may occur on the endothelial side of the GBM (fibrin caps) and between Bowman's capsule and the parietal epithelium (capsular drops). Hyaline deposits in both the afferent (more extensive) and efferent arterioles are present in the subendothelial space and media. Tubular basement membranes also are thickened with a layered structure (nonspecific).

The increase in mesangial matrix (nodular expansion) and GBM thickening are considered two different expressions of a fundamental basement membrane abnormality (Osterby 1986), that ultimately leads to solidification and loss of capillary surface. Many factors contribute to the pathogenesis of diabetic glomerulosclerosis, including hyperfiltration, hypertension, and poor metabolic control of glycemia. Three theories considered in the pathophysiology include (1) changes in the polyol-inositol metabolism pathway, (2) nonenzymatic glycation of proteins, and (3) direct influence of hyperglycemia on the synthesis of matrix components (Schleicher and Olgemoller 1992) via cytokines such as transforming growth factor beta. Immunofluorescence studies show bright linear deposition of albumin and IgG along the GBM and tubular basement membrane. Scattered small deposits of IgM and C3 may be seen in the mesangium.

Electron microscopy is necessary to exclude other causes of nodular glomerular lesions such as systemic light chain deposition disease, amyloidosis, other fibrillary glomerulopathies and membranoproliferative glomerulonephritis. The diffuse thickening of the GBM was once thought to be pathognomonic for diabetes, but it has been noted in conditions of physiologic hypertrophy (single kidneys, cyanotic congenital heart disease [Figure 12.17]).

Thin Glomerular Basement Membrane Disease (Benign Familial Hematuria)

(Figures 12.18, 12.19, and 12.21B.)

Diagnostic criteria. (1) Diffuse, uniformly thin GBM, compared with age-matched controls (Basta-Jovanovic et al. 1990; Hill et al. 1974) due mainly to the decreased width of the lamina densa (Figures 12.18 and 12.19);

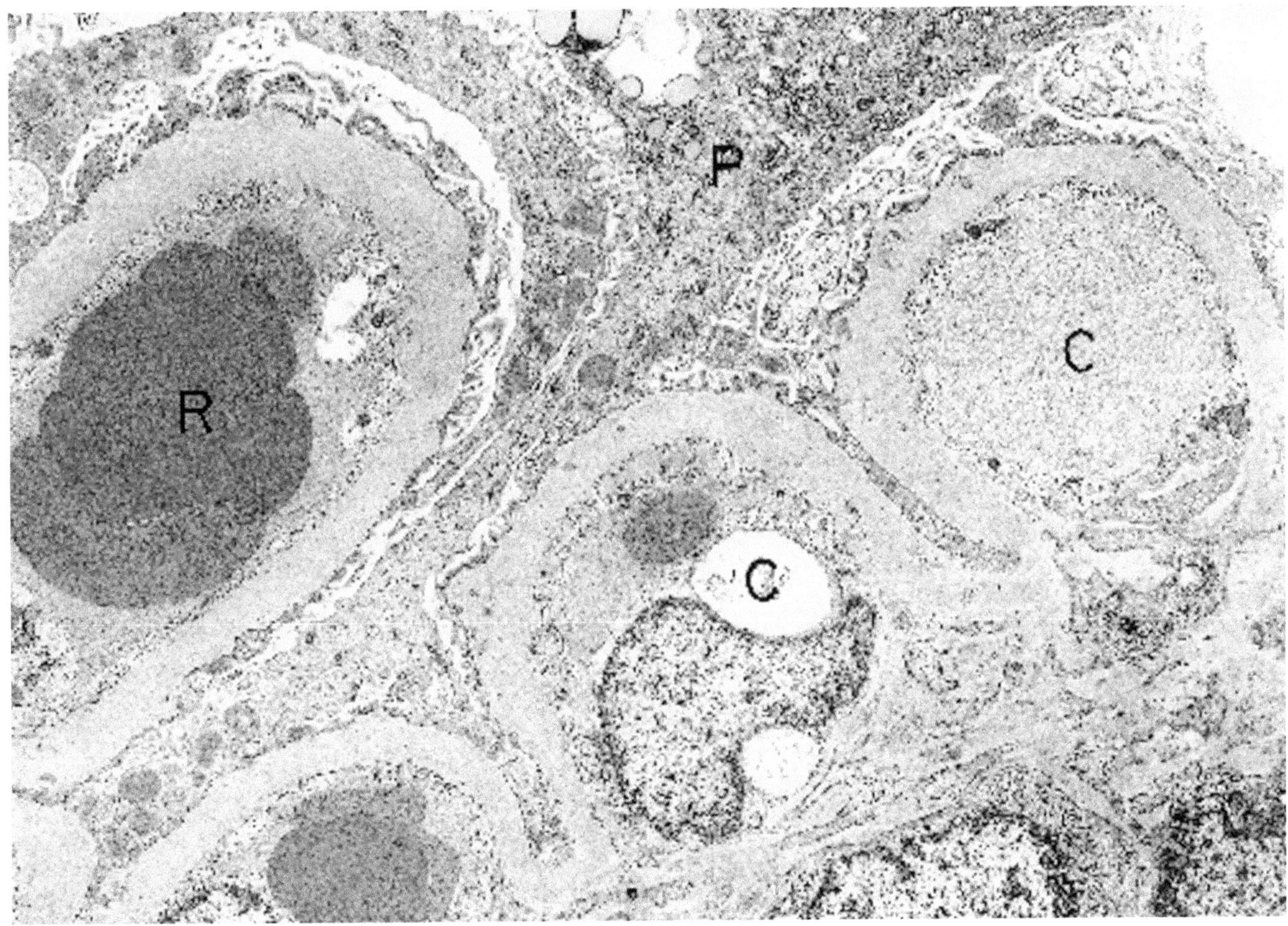

Figure 12.17. Cyanotic congenital heart disease (37-year-old man with tetralogy of Fallot, still cyanotic after Blalock procedure). There is marked, diffuse, uniform thickening of the GBM (600–950 nm) with little or no lamination, which mimics diabetic nephropathy. Scattered deposits/debris are embedded in the GBM. The podocytes (P) have little foot process loss. Red cells (R) are in capillaries (C). (× 6360)

(2) smooth, nonserrated outer and inner contours of the GBM (Figure 12.21B); (3) absence of diffuse lamination of the GBM.

Additional points. The width of the GBM can be determined by using a grid overlay and calculating the harmonic mean of the GBM (from cell surface to cell surface) where it intersects the grid (Steffes et al. 1983). Normal values using this methodology were given in the section on the normal glomerulus. There is no "official" threshold for thin basement membrane disease, although values <240 nm and <280 nm in adults, women and men respectively, and <150 nm in children ≥ two years old may be used (two standard deviations below normal). Another approach is that of Milanese et al. (1984) who found overlap of the mean GBM width between normal children and those with benign familial hematuria; discrimination was better using the frequency of GBM measurements <200 nm. A "rule of toes" is that the normal thickness of the GBM should not be less than the width of a foot process. Mild, focal thickening and splitting of the GBM may be present in some cases, with effacement of podocyte foot processes. Paramesangial thickening and wrinkling have been noted. These features are probably related to previous rupture of the peripheral capillary wall and repair.

Figure 12.21 compares thin GBM disease with Alport's syndrome. Thin GBMs may be a feature of both, especially in young children. Lamination and basket weaving of the GBM with thickening, serrated outer and inner GBM contours are features of Alport's syndrome (see next section) and are absent in thin GBM disease. In early cases these diseases may have a similar appearance on electron microscopy. Immunofluorescence microscopy study of the GBM for the alpha-3 chain of type IV collagen, using Goodpasture's serum, is helpful in the differential diagnosis; alpha-3 (IV) col-

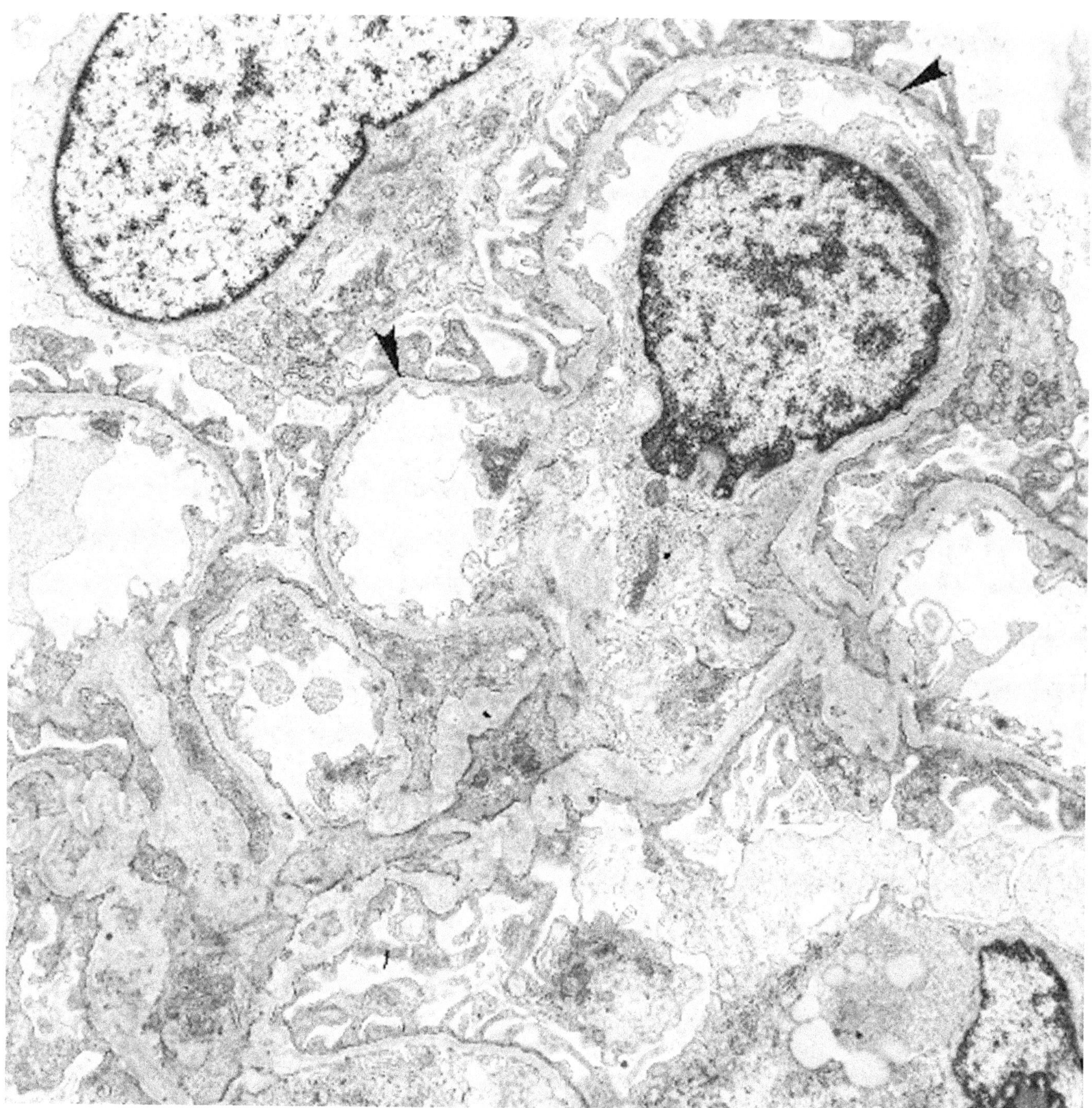

Figure 12.18. Thin GBM disease (16-year-old boy with a 6-year history of microscopic hematuria). The GBM is markedly thin, about 171–183 nm (*arrowheads*), but otherwise normal with relatively smooth contours. Paramesangial thickening and wrinkling are also seen. The podocytes and mesangium are unremarkable. (× 8175)

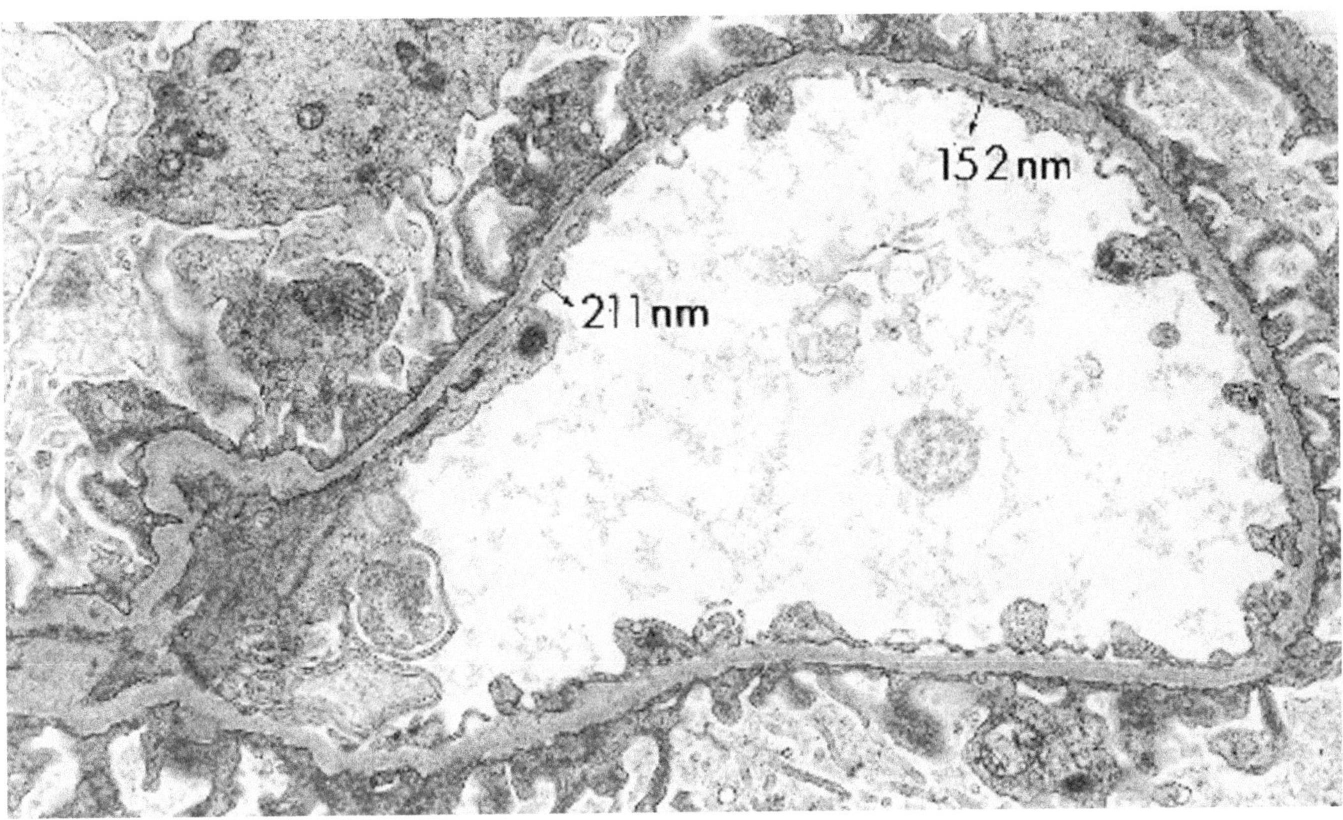

Figure 12.19. Thin GBM disease. The GBM is diffusely thin, about 152–211 nm. (× 11,800)

lagen staining is present in thin GBM disease and absent in Alport's syndrome. Clinical presentation, family history, and evolution become the definitive indicators of diagnosis and prognosis. A thin GBM can be seen in various nonhereditary glomerulopathies, but only focally. Light microscopic examination is normal, aside from red cell casts, and immunofluorescence studies are generally unremarkable.

Thin basement membrane disease accounts for about 30% of the cases presenting as persistent, asymptomatic hematuria (Bodziak et al. 1994). In most families, an autosomal dominant inheritance is seen. In some, an autosomal recessive pattern is seen, and sporadic cases have been noted. Genetic analysis may help more definitive classification. Some cases due to a defect in COL4A4 presenting with benign familial hematuria have been described (Lemmink et al. 1996) and may be heterozygote carriers of autosomal recessive Alport's syndrome.

Typically, patients do not develop renal insufficiency.

Alport's Syndrome (Hereditary Nephritis)

(Figures 12.20 and 12.21A)

Diagnostic criteria. (1) Extensive thickening of the GBM (up to 800–1200 nm) with distortion of the lamina densa due to lamination, splitting, fragmentation, and formation of a net-like pattern (basket weaving), enclosing electron lucent areas that focally contain small granules or microparticles (Yoshikawa et al. 1981; Spear 1974), approximately 5–10 nm in diameter; (2) segments of peripheral capillary loops with extreme thinning of the GBM (Figure 12.15A); (3) serrated outer and inner contours of the GBM.

Additional points. Podocytes are enlarged and contain numerous organelles and clear vacuoles; there may be effacement of foot processes and electron-dense condensation of the cytoplasm. The mesangial matrix may be increased with focal extension of mesangial cells and matrix along peripheral capillary loops (interposition). In some cases, fine granular deposits are present in the mesangium. Electron-dense deposits corresponding with immune complexes are rare. Rupture of the GBM with repair by endothelial and epithelial cells may be seen. Bowman's capsule and tubular basement membranes show similar focal thickening and splitting, but lamination at these sites may be seen in other diseases.

Each of these changes, taken as an isolated finding, is not specific for Alport's syndrome and may be seen in several glomerulopathies in the reparative stage, although rarely to the extent in hereditary nephritis (Hill et al. 1974). When all the GBM changes are present and

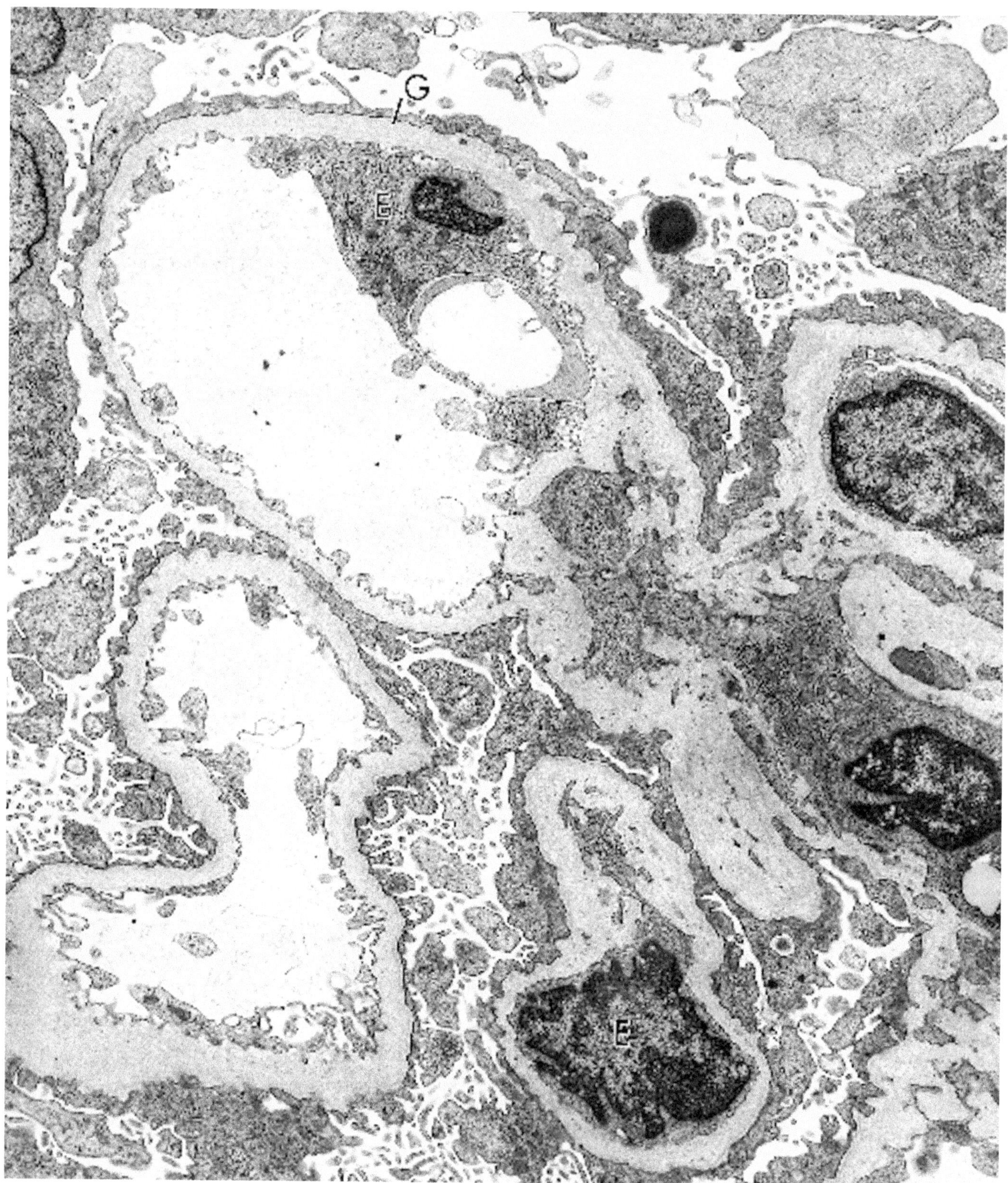

Figure 12.20. Hereditary nephritis (Alport's syndrome) (14-year-old boy with hematuria, proteinuria, and family history of chronic renal failure). The GBM (G) has variable thickening and thinning, with lamina densa laminations and lucency. Scattered granular microparticles are present in the GBM. E = endothelial cell. (× 9500). This case lacked reactivity with sera from patients with anti-GBM nephritis.

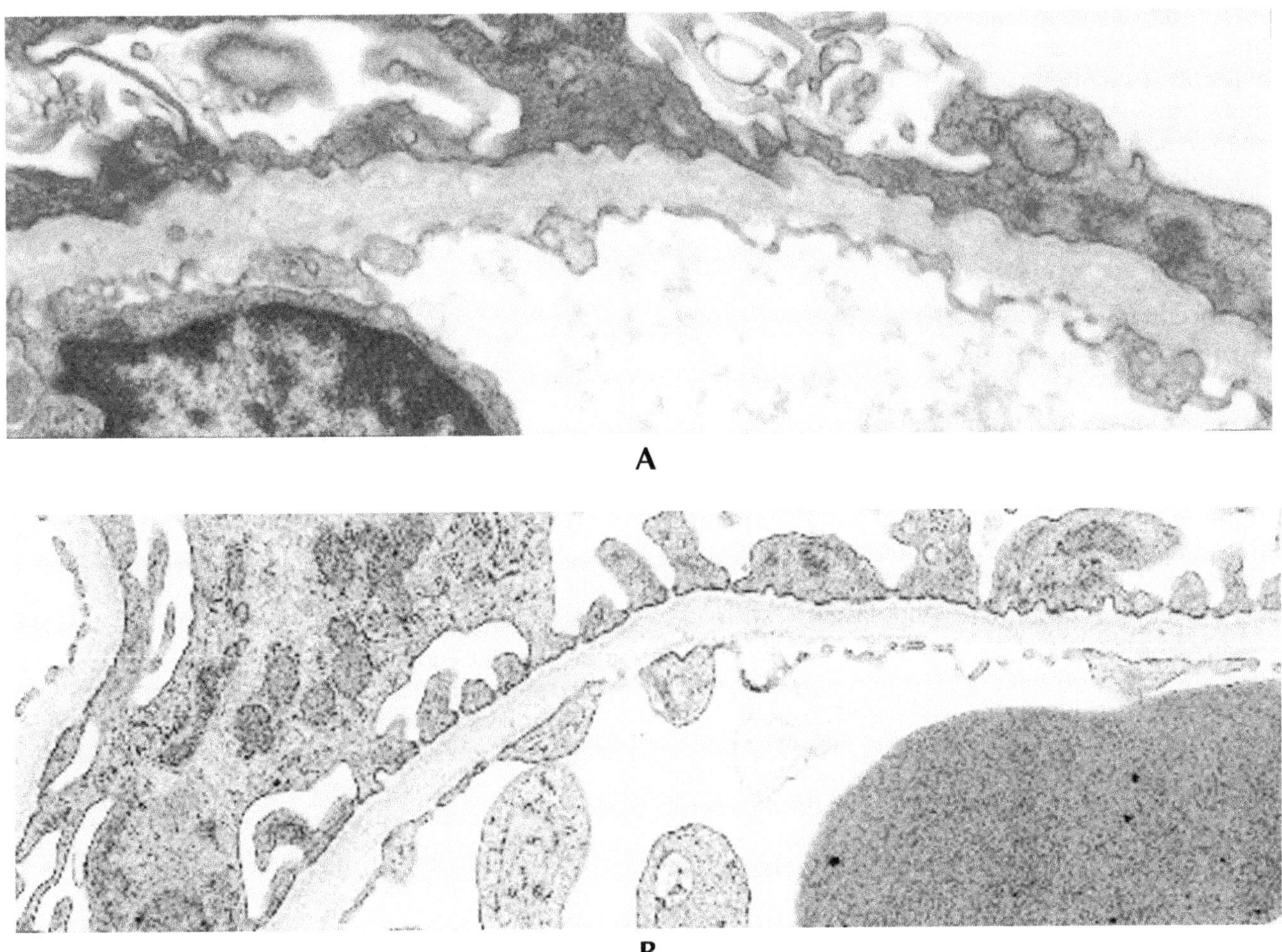

Figure 12.21. Comparison of hereditary nephritis and thin GBM disease. **A,** Hereditary nephritis (29-year-old male). The GBM is thin with prominent lamination of the lamina densa, serrated outer and inner contours of the GBM, and scattered microparticles. (× 20,500) **B,** Thin GBM disease (12-year-old male with hematuria and red blood cell casts in urine). The GBM is thin with relatively uniform lamina densa, without irregularities of the outer and inner GBM borders. (× 20,500) *Note:* Also compare **A** and **B** with Figure 12.76 where subendothelial (lamina rara interna) laminations can be seen in severe hypertension.

are widespread, the electron microscopy picture strongly suggests Alport's syndrome (Grünfeld 1985; Gubler et al. 1981; Gregory et al. 1996). Immunofluorescence microscopy for the alpha-3 chain of type IV collagen shows absence of staining along the GBM in Alport's syndrome and is used as a diagnostic test.

Extensive GBM thickening and splitting are present in 80% of patients with Alport's syndrome. Other cases show extreme thinning of the GBM. Splitting of the GBM is seen in older children and adults, and the degree of splitting and proteinuria increase with age, especially in men. Classical Alport's syndrome is inherited in an X-linked manner (85% of the cases), with clinical features of hematuria, sensorineural deafness, ocular defects, and a progression to renal failure (Kashtan and Michael 1996). Sensorineural deafness, however, is no longer a prerequisite for diagnosis of Alport's syndrome. In 1990, Hostikka et al. (1990) identified an "Alport gene" located at Xq22 designated COL4A5, encoding the alpha-5 chain of type IV collagen, mutations of which are implicated in pathophysiology. The alpha-5 chain is necessary for expression of the alpha-3 chain of type IV collagen. An autosomal dominant form accounts for about 15% of the cases due to mutations in the COL4A3 and COL4A4 genes on chromosome 2 (Jefferson et al. 1997), encoding the alpha-3 and alpha-4 chains of type IV collagen, respectively. An autosomal recessive variety has also been described with defects in COL4A3 and COL4A4 genes (Mochizuki et al. 1994).

By light microscopy, glomeruli usually appear normal or have mild segmental cell proliferation. Rumpelt et al. (1992) have described fetal-like glomeruli, smaller capillary loops, and less intense basement membrane staining in their cases of Alport-type glomerulopathy. Tubular atrophy often is seen in advanced cases. Immunofluorescence studies are negative. Renal transplantation has induced in 20% of reported cases formation of anti-GBM antibodies sometimes with crescentic glomerulonephritis, due to the foreign alpha-3 or alpha-5 antigens in the donor kidney.

Glomerular Diseases with Prominent Crescents

Crescentic glomerulonephritis can be categorized into three types based on pathophysiologic mechanisms: (1) immune-complex-mediated, such as postinfectious glomerulonephritis, Henoch-Schönlein purpura, and membranoproliferative glomerulonephritis; (2) antibody to GBM antigens; (3) pauci-immune glomerulonephritis, antineutrophil cytoplasmic antibody (ANCA) associated (Davies et al. 1982; Hall 1984; Jennette and Falk 1990), such as Wegener's granulomatosis, microscopic polyarteritis, and similar vasculitides. The crescent itself is similar in these various rapidly progressive diseases, although the primary injury to the GBM (or Bowman's capsule, perhaps) is different and categorized into the above by immunofluorescence studies. The clotting system is involved in the pathogenesis of crescents. Various cell types accumulate in Bowman's space to form crescents, including parietal epithelial cells (the major component), monocytes/macrophages, neutrophils, and occasionally multinucleated giant cells. The appearance of crescents is influenced by their duration, but not by the specific type of glomerular disease. Distortion and fragmentation of the GBM by the crescentic process or the necrotizing disease are seen (Bonsib 1988) as well as fibrin deposition in the early stage.

Wegener's Granulomatosis

(Figure 12.22.)

Diagnostic criteria. (1) Crescent formation with associated fibrin in the urinary space; (2) segmental fibrinoid necrosis; (3) fragmented GBM; (4) rarely electron-dense amorphous subendothelial, mesangial, and subepithelial deposits; (5) breaks in Bowman's capsule.

Additional points. Fibrin can be present in capillary lumens, blood vessel walls, and interstitium. Granulomas and arteritis are occasionally found. Other vascular changes include swelling and detachment of the endothelium with subendothelial fibrin and platelets (not shown). ANCA is almost always present.

Light microscopy demonstrates a focally necrotizing glomerulonephritis (Weiss and Crissman 1984) with some crescents. Granulomatous-type crescents (crescents containing numerous epitheloid histiocytes and multinucleated giant cells) are not specific for Wegner's granulomatosis but are more common than in cases of microscopic polyarteritis. Yoshikawa and Watanabe (1984) reported granulomatous crescents in 13 of their 24 cases. Jennette et al. (1989) reported such crescents more often with C-ANCA-mediated processes rather than P-ANCA. Vasculitis often is present (Serra et al. 1984). Immunofluorescence studies usually are negative (hence the "pauci-immune" term).

The diagnosis of Wegener's granulomatosis relies on the histologic triad of necrotizing granulomatous inflammation of upper or lower respiratory tracts, arteritis, glomerulonephritis, and the presence of ANCA in the serum. C-ANCA (antiproteinase 3) is present in 90% of cases, and P-ANCA (antimyeloperoxidase) is present in most of the others. These antibodies may initiate vasculitis by inducing primed neutrophils and monocytes to release free radicals and lysosomal granules that injure vascular endothelium (van der Woude et al. 1990). Other diseases with a similar electron microscopy and immunofluorescence appearance are microscopic polyarteritis (Jennette et al. 1990; D'Agati et al. 1986), which is usually associated with P-ANCA, and idiopathic crescentic glomerulonephritis. These cannot be differentiated on histologic grounds alone, unless granulomas or vasculitis are present.

Anti-Glomerular Basement Membrane Nephritis (Goodpasture's Syndrome)

(Figure 12.23.)

Diagnostic criteria. (1) Crescent formation; (2) fractures of the GBM; (3) collapse of glomerular tuft; (4) subendothelial lucency of lamina rara interna and widening of the GBM; (5) reactive endothelium; (6) absent or rare small subendothelial deposits.

Additional points. The GBM may be thickened, wrinkled, and occasionally interrupted. The GBM discontinuities are oval or circular defects of various sizes in three dimensions (Bonsib 1985). GBM splitting with interposition of mesangial cells can be present. With progressive glomerular injury, occlusion of capillary lumens by fibrin and inflammatory cells occurs with endothelial detachment. This phenomenon is implicated in the pathophysiology of anti-GBM nephritis. By using immunoelectron microscopic techniques, Sisson and colleagues have shown the distribution of IgG antibody along the inner aspect of the GBM in the region of lamina rara interna (Sisson et al. 1982). Focal effacement of podocyte foot processes may be present.

Anti-GBM antibody disease (Bolton 1996; Senekjian et al. 1980) is a focally necrotizing crescentic glomeru-

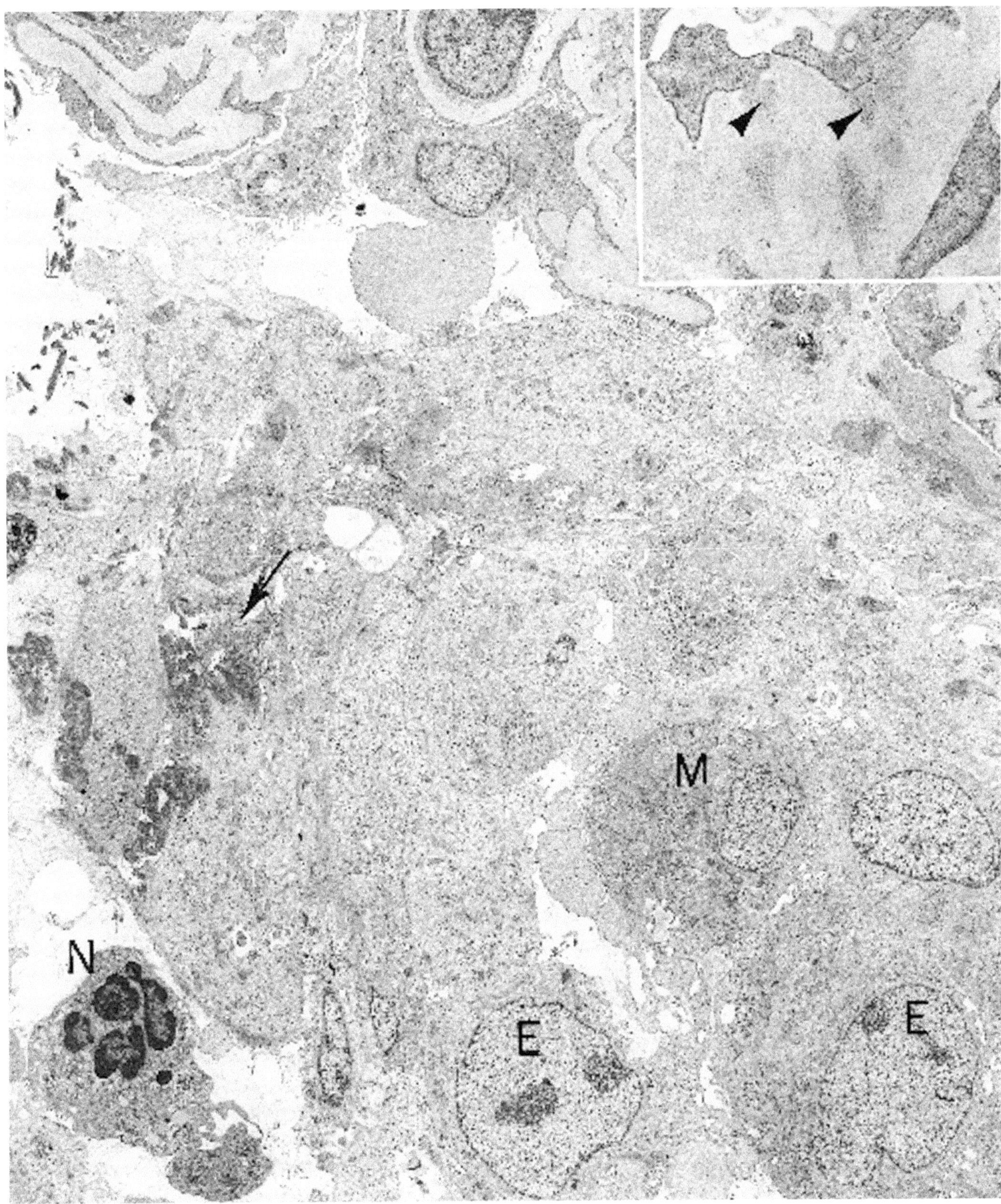

Figure 12.22. Wegener's granulomatosis (76-year-old woman with a 2-week history of acute renal failure, active urinary sediment, and bilateral pulmonary infiltrate). Adjacent to the glomerular tuft is a portion of cellular crescent with reactive epithelial cells (E) with cell junctions, phagocytic macrophages (M), and occasional polymorphonuclear neutrophilic leukocytes (N) in close contact in a matrix of fibrin. Strands of basement membrane may surround the cell periphery. *Arrow* = fibrin. (× 4176). *Inset:* Subepithelial deposits (*arrowheads*). (× 22,680).

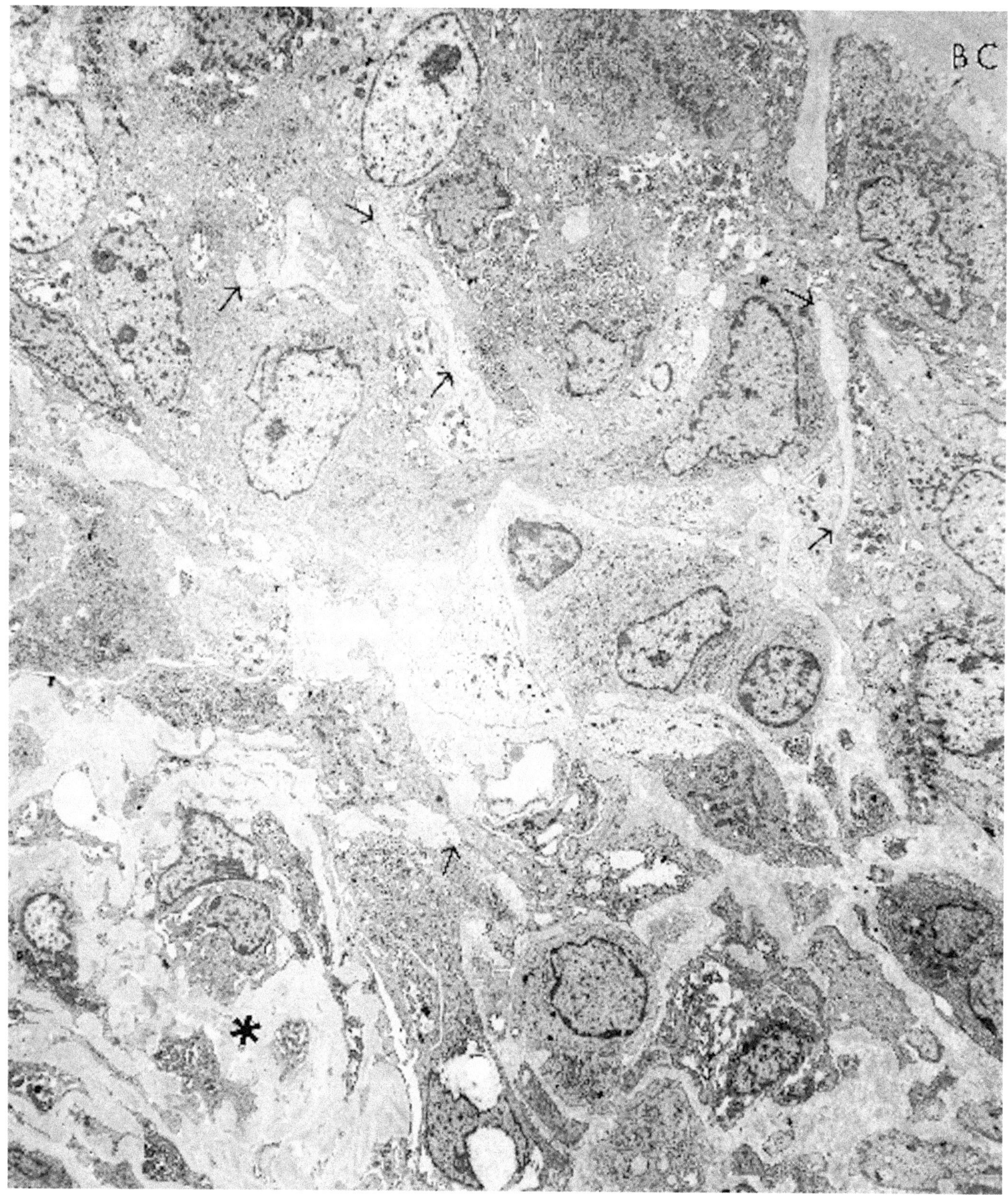

Figure 12.23. Anti-GBM nephritis (36-year-old female with necrotizing crescentic glomerulonephritis and circulating anti-GBM antibodies). The glomerular tuft is collapsed (*), and many areas of GBM fragmentation and discontinuity (*arrows*) are amid a cellular crescent, giving a gnarled appearance to the damaged capillary loops. BC = Bowman's capsule. (× 4600)

lonephritis characterized by autoantibodies against an epitope (Goodpasture antigen) in the GBM. This autoantigen (Kalluri et al. 1995) has been identified as the noncollagenous domain of the alpha-3 chain of type IV collagen (NC1α3(IV)), encoded by the gene COL4A3 located at chromosome 2q35–37 (Hudson et al. 1993). Goodpasture's syndrome (Rosenblum and Colvin 1993; Bazari and Mauiyyedi in press) is the presence of pulmonary hemorrhage and glomerulonephritis in a patient with circulating anti-GBM antibodies.

The ultrastructural and light microscopy findings are nonspecific. The diagnosis requires demonstration of linear staining for IgG along the GBM by direct immunofluorescence and the detection of antibodies directed against the GBM in the serum, or in the renal eluate by Western blot analysis and enzyme-linked immunosorbent assay (McCluskey et al. 1995).

Diseases with Prominent Amorphous Dense Deposits

Postinfectious Glomerulonephritis

(Figures 12.24 and 12.25.)

Diagnostic criteria. (1) Dome-shaped large subepithelial electron-dense deposits (humps); (2) granulocytes

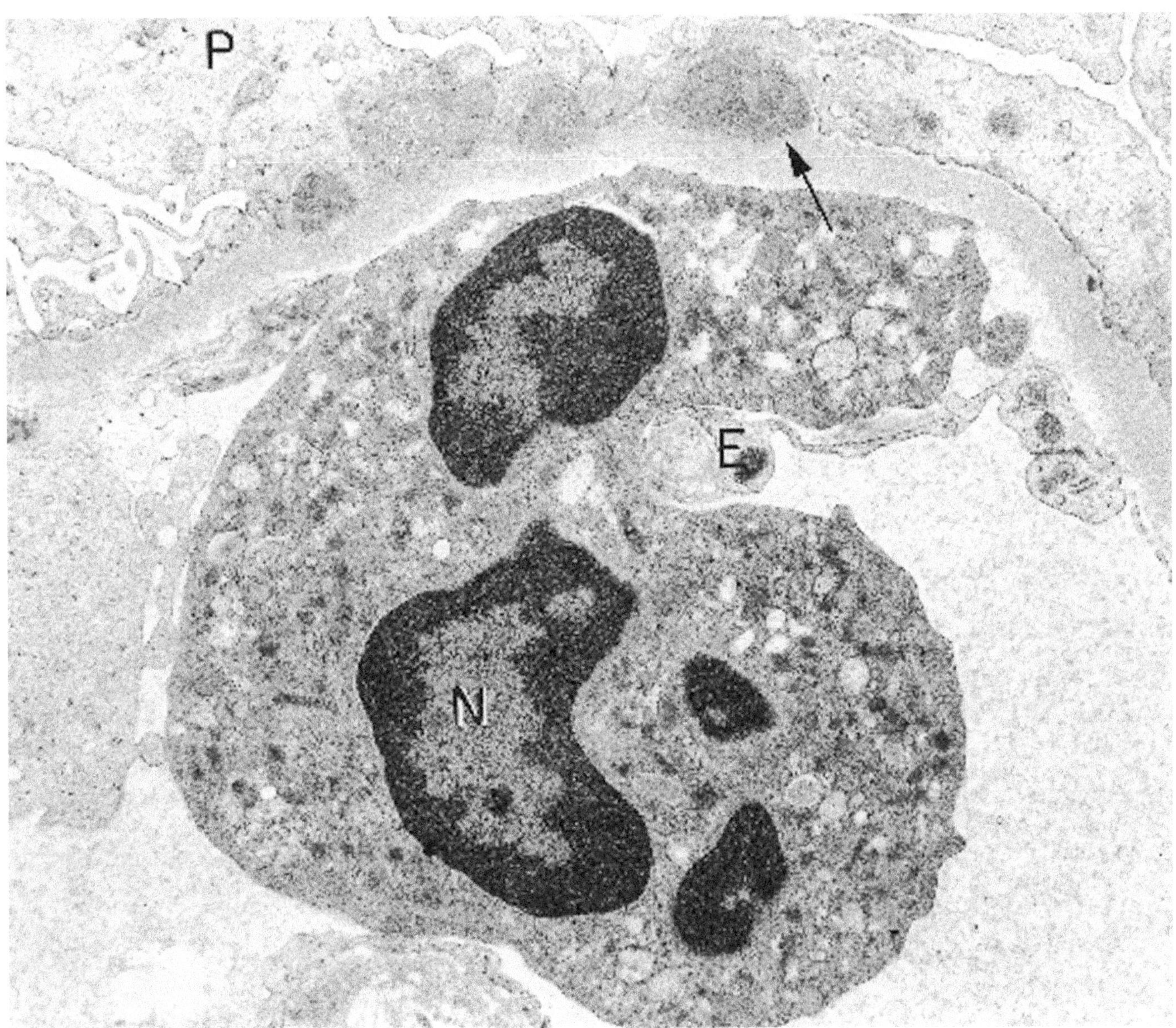

Figure 12.24. Acute postinfectious glomerulonephritis. (13-year-old boy with poststreptococcal glomerulonephritis). A polymorphonuclear neutrophilic leukocyte (N) invaginates under the endothelium (E). Hump-like electron-dense deposits are present in subepithelial location (*arrow*). P = podocyte. (× 16,000)

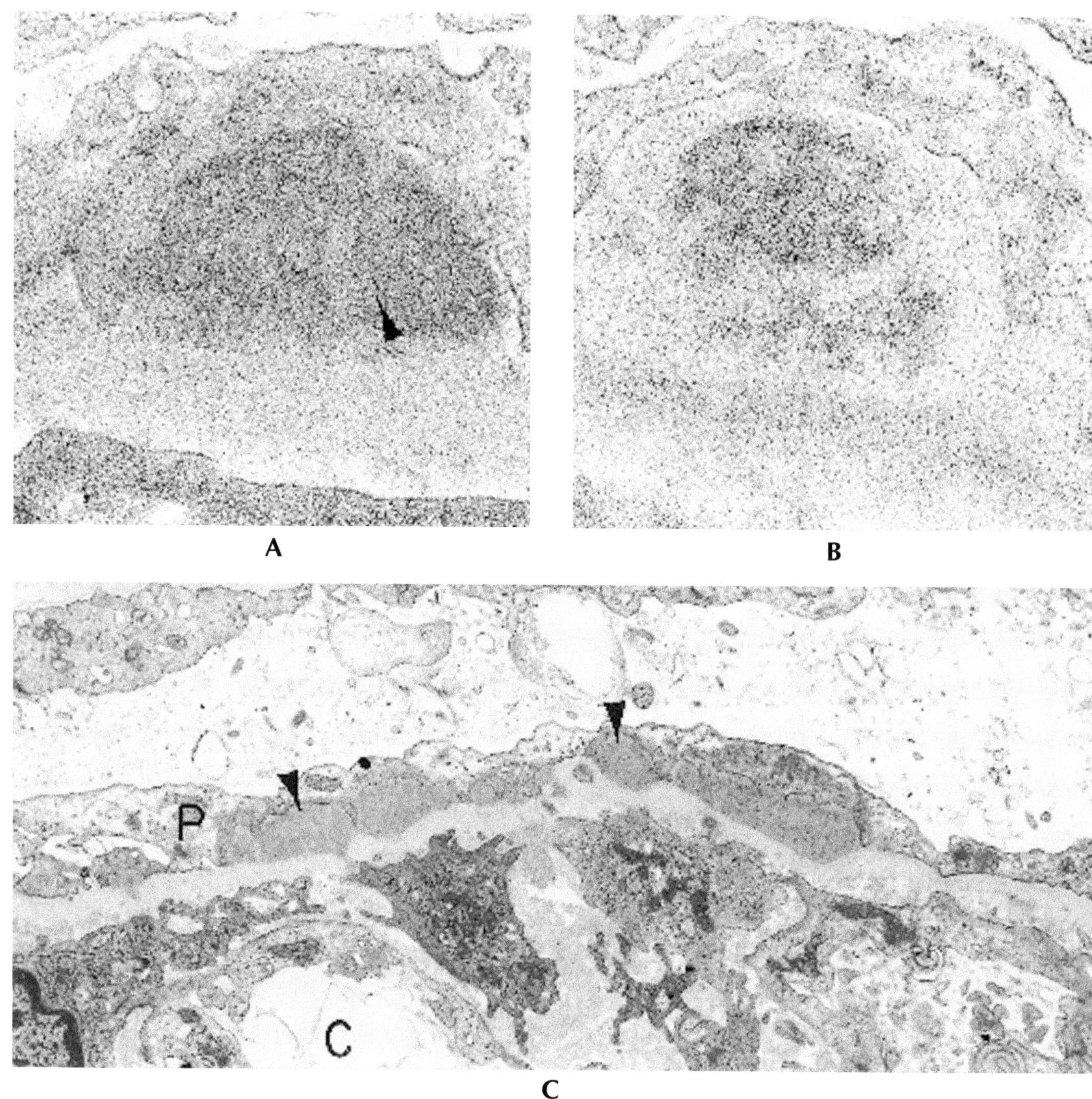

Figure 12.25. Acute postinfectious glomerulonephritis. **A** (same case as Figure 12.24), Subepithelial hump (*arrowhead*). (× 42,230) **B** (same case as Figure 12.24), Subepithelial hump with partial dissolution. (× 42,230) **C** (8-year old boy with acute hypertension and oliguria), Extensive, confluent subepithelial electron-dense deposits and atypical humps (*arrowheads*) on the GBM. C = capillary lumen; P = podocyte. (× 11,240)

and monocytes in the capillary lumina and in the mesangium.

Additional points. Less-extensive subendothelial, mesangial, or intramembranous deposits can be found, depending on the stage of disease (Sorger et al. 1982). The subepithelial deposits are in close contact with visceral epithelial cells (podocytes). The podocyte cytoplasm, immediately adjacent to the deposits, usually is of increased electron density (due to fibrils), with foot process effacement (Figure 12.24). This reaction is a general response of podocytes in contact with immune complexes (seen also in membranous glomerulonephritis, lupus nephritis, and membranoproliferative glomerulonephritis). The electron density of deposits can be variable; the granularity may be coarse to fine (Churg and Grishman 1972). Areas of lucency are probably caused by dissolution of deposits (Figure 12.25B). Occasional cases have particularly large and confluent subepithelial deposits (atypical humps; Figure 12.25C), which are associated with a poorer prognosis. The mesangium is expanded, but more by increased cells and loose edema than by matrix or sclerosis.

The GBM generally shows no marked abnormalities, but it may have segmentally one or more of the following: simple thickening, affecting the entire profile or relatively lengthy segments; thickening with areas of rarefaction; separation of the endothelium from the luminal side of the GBM by a clear space or, more frequently, by amorphous material of variable density; foci of complete disarrangement and even rupture of the GBM. Duplication and mesangial interposition, typical of membranoproliferative glomerulonephritis and the glomerulonephritis of chronic infections, are seldom found.

Cellular crescents and fibrin may be present in Bowman's space. In the healing phase (≥ six weeks), endothelial cell swelling is decreased, the infiltrate is absent, and the humps usually disappear (Tornroth 1976). The peripheral GBM may be left with irregular segmental thickening.

Light microscopic examination shows hypercellularity of the glomerular tuft owing to an increased quantity and swelling of endothelial and mesangial cells and infiltration by polymorphonuclear neutrophils and mononuclear cells. An acute interstitial nephritis typically is present, and red cell casts often are numerous.

Electron microscopic findings correlate with the immunofluorescence pattern and stage of disease and have been described into three types (Sorger 1982). The *starry sky* immunofluorescence pattern has a sprinkling of fine granular deposits in relation to capillary walls and mesangium. By electron microscopy, subendothelial and mesangial deposits predominate, with rare intramembranous or subepithelial deposits; therefore, humps are rare. This pattern occurs most frequently in early phases of the disease (first to third week). The *garland* immunofluorescence pattern has densely packed, sometimes confluent deposits located in the region of the capillary walls. Electron microscopy shows numerous humps. This pattern occurs at various times during the course of the disease. The *mesangial* immunofluorescence pattern is associated with EM deposits in that location and usually is seen in later stages of evolution of the disease.

These findings are observed in a variety of postinfectious glomerulonephritis, including that associated with group A beta-streptococcus (Tejani and Ingulli 1990), *Staphylococcus aureus,* and *Pneumococcus.* The glomerulonephritis of continued infection (as in shunt nephritis, subphrenic abscess, or endocarditis) may show this pattern or, more typically, that of membranoproliferative glomerulonephritis, type I.

Membranous Glomerulonephritis

(Figures 12.26 through 12.29.)

Diagnostic criteria. (1) Subepithelial amorphous electron-dense deposits along virtually all the glomerular capillaries (Rosen 1971); (2) scant or absent mesangial or subendothelial deposits (Graham and Nagle 1983).

Additional points. The appearance varies with the stage of the disease. The sequence of deposition and reabsorption of deposits has been divided into four stages (Ehrenreich and Churg 1968). In *stage I* (Figure 12.26) isolated, scattered electron-dense deposits are present on the epithelial side of the GBM. These resemble humps, except for the early cupping of the deposit with newly formed basement membrane. Overlying podocytes have focal effacement of foot processes and microvillous transformation. Dense fibrils are present in the cytoplasm of foot processes in contact with the deposits. In *stage II* (Figure 12.27), subepithelial deposits are numerous, evenly distributed over the capillary loops, and separated by radial extensions of the lamina densa, forming spikes. In *stage III,* spikes fuse over the subepithelial deposits (Figure 12.28A), embedding them into a thickened basement membrane. The amorphous deposits become coarsely granular and somewhat lucent; patches of GBM lucency are seen (Figure 12.28B), representing areas of dissolution. Occasional mesangial deposits may be present. In *stage IV* (Figure 12.29), deposits become rarified, and most disappear, leaving an irregularly thickened lamina densa, which may return toward normal.

Membranous glomerulonephritis can be seen as a primary renal disease or secondary to other diseases (lupus, certain infections, and drug allergy). The secondary forms of membranous glomerulonephritis generally have sparser deposits (Graham and Nagle 1983).

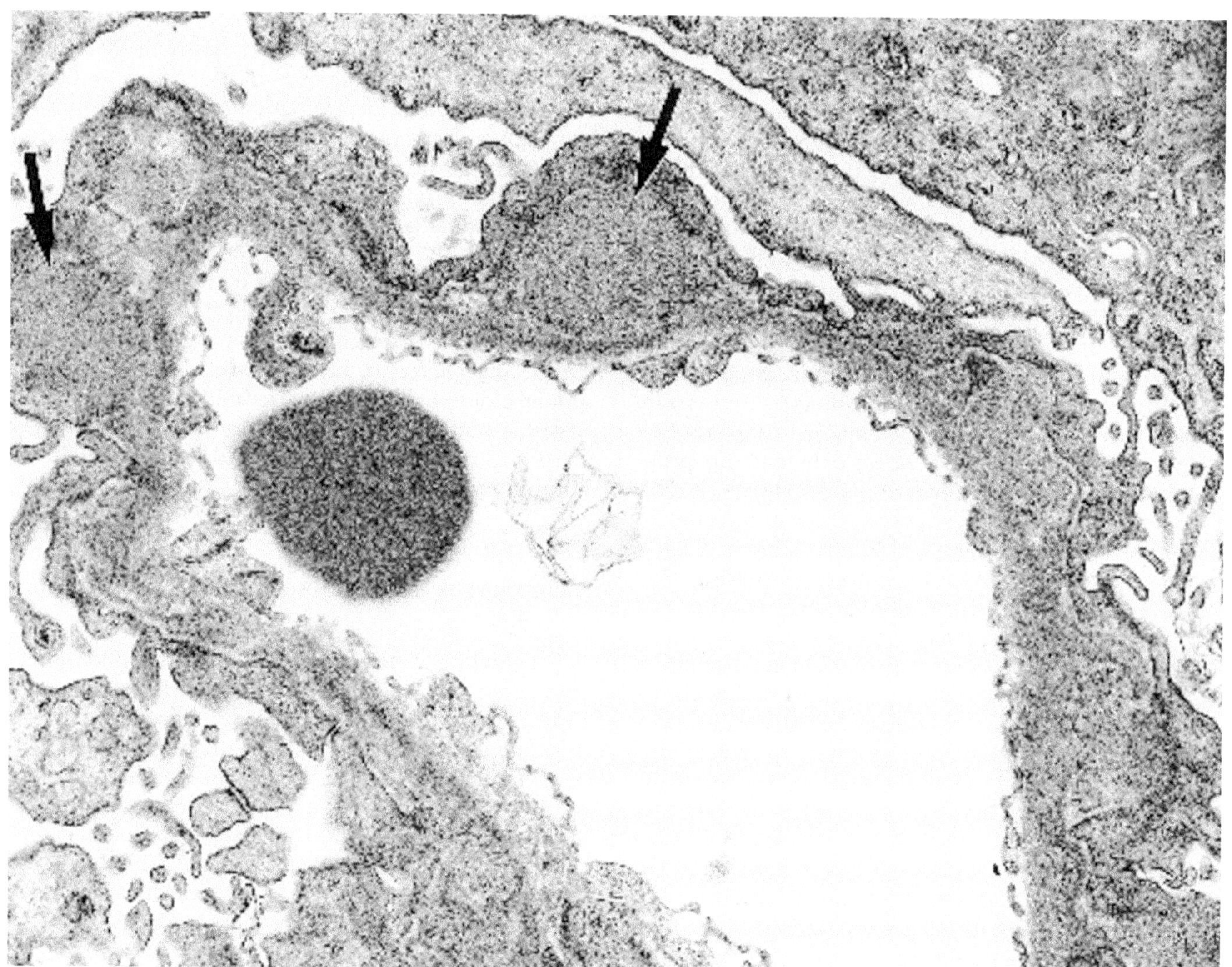

Figure 12.26. Membranous glomerulonephritis stage I (15-year-old girl with nephrotic syndrome). Large subepithelial deposits (*arrows*) are present under the podocyte. Although no fully formed spikes of basement membrane are seen, a few fine basement membrane laminae adjacent to the deposits form a collar. This reaction helps distinguish stage I membranous deposits from the humps found in acute glomerulonephritis. (× 18,500)

If mesangial and subendothelial deposits are also present, the possibility of lupus nephritis should be considered. Features that are more characteristic of lupus membranous glomerulonephritis are a fingerprint pattern of deposits and endothelial tubuloreticular structures. Electron microscopy and immunofluorescence are useful in early stages of the disease (especially in stage I), when the light microscopy is similar to minimal-change disease. The presence of fine granular IgG staining along peripheral capillary loops by immunofluorescence corresponds to the small electron-dense deposits by electron microscopy and will indicate the diagnosis. The presence of intracapillary leukocytes (neutrophils and monocytes) or subendothelial lucent material suggests superimposed renal vein thrombosis; the authors have also noted loss of endothelial fenestrae in this clinical setting.

By light microscopy, diffuse thickening of the GBM may be seen, although this may be impossible to detect in early stages, partly because the deposits themselves are not PAS-positive. Trichrome stains (for collagen) in 2-μ sections are useful in the absence of electron microscopy (deposits stain red). Immunofluorescence studies show intense, diffuse granular staining of the glomerular basement membrane for IgG and C3 and the other immunoglobulins (IgM, IgA, IgE) to a variable degree.

The epimembranous (or subepithelial) pattern of immune complex deposition is not specific for a particular disease, but rather is a common glomerular pattern

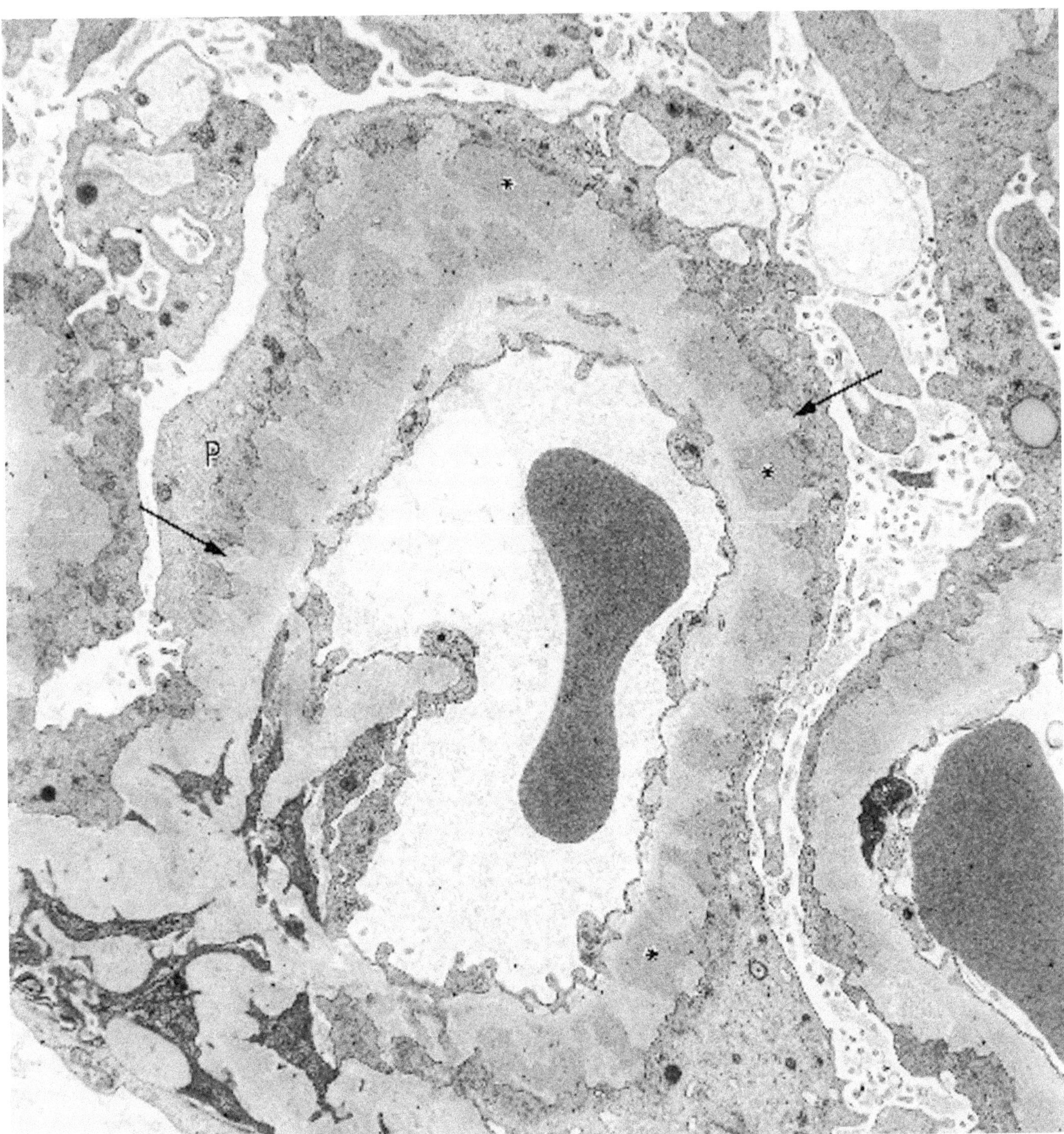

Figure 12.27. Membranous glomerulonephritis stage II (61-year-old female with proteinuria and edema). Numerous subepithelial deposits (*) are regularly scattered along the GBM. These deposits are separated by spikes or columns of basement membrane (*arrows*). The podocytes (P) show marked local reaction to the deposits and have foot process effacement. The subendothelial space is widened and lucent. (× 12,000)

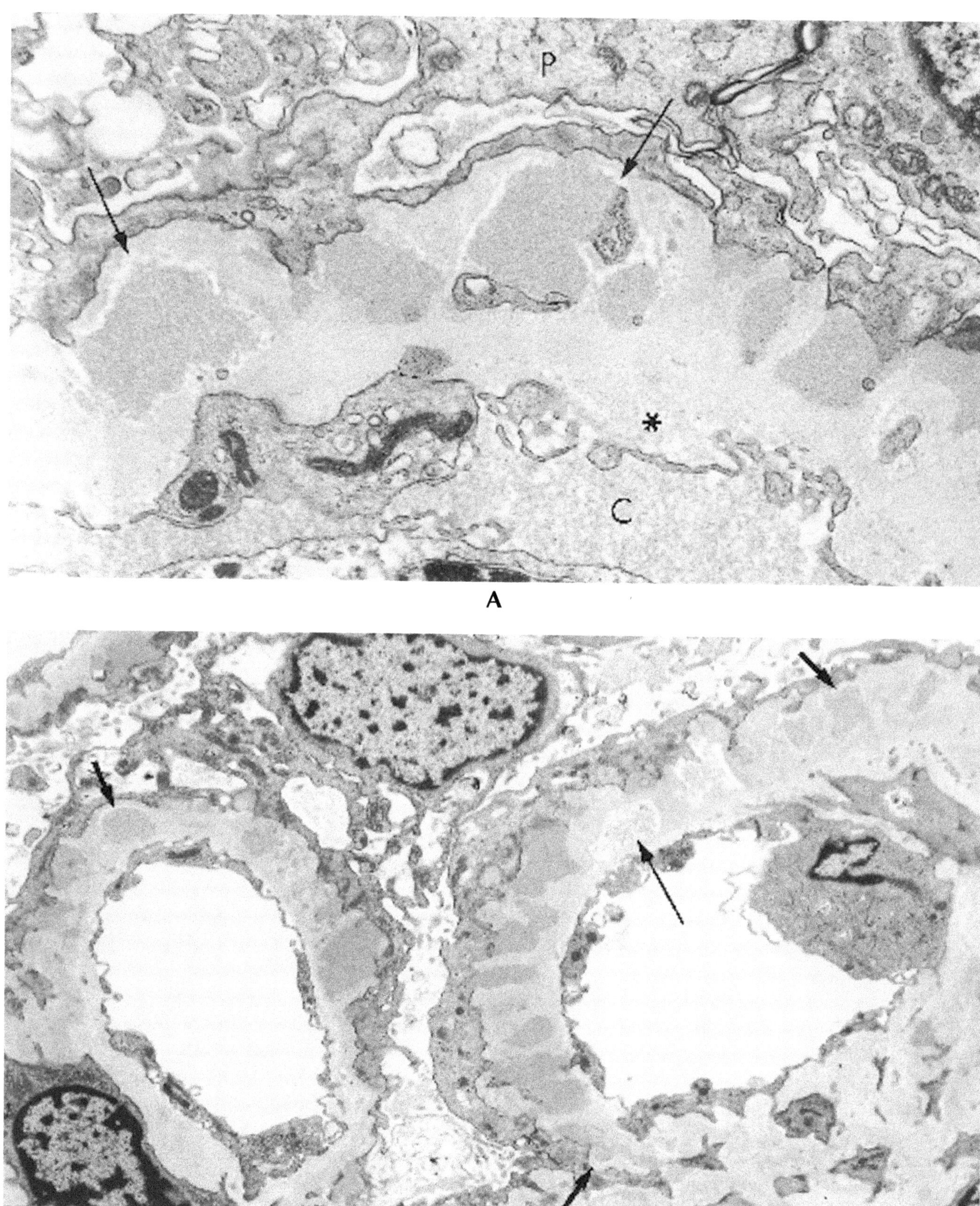

Figure 12.28. Membranous glomerulonephritis stage III. **A** (51-year-old male, proteinuria for 6 years, developed nephrotic syndrome recently, with proteinuria up to 11 g/day and creatinine of 1.7 mg/dL), Arch of new basement membrane over the deposits (*arrows*). * = subendothelial lucent expansion; C = capillary lumen; P = podocyte. (× 25,000). **B** (52-year-old female with proteinuria and hypertension), The subepithelial deposits are covered by a bridge of newly formed GBM (*short arrows*) and appear to be undergoing dissolution (*long arrow*). (× 8500)

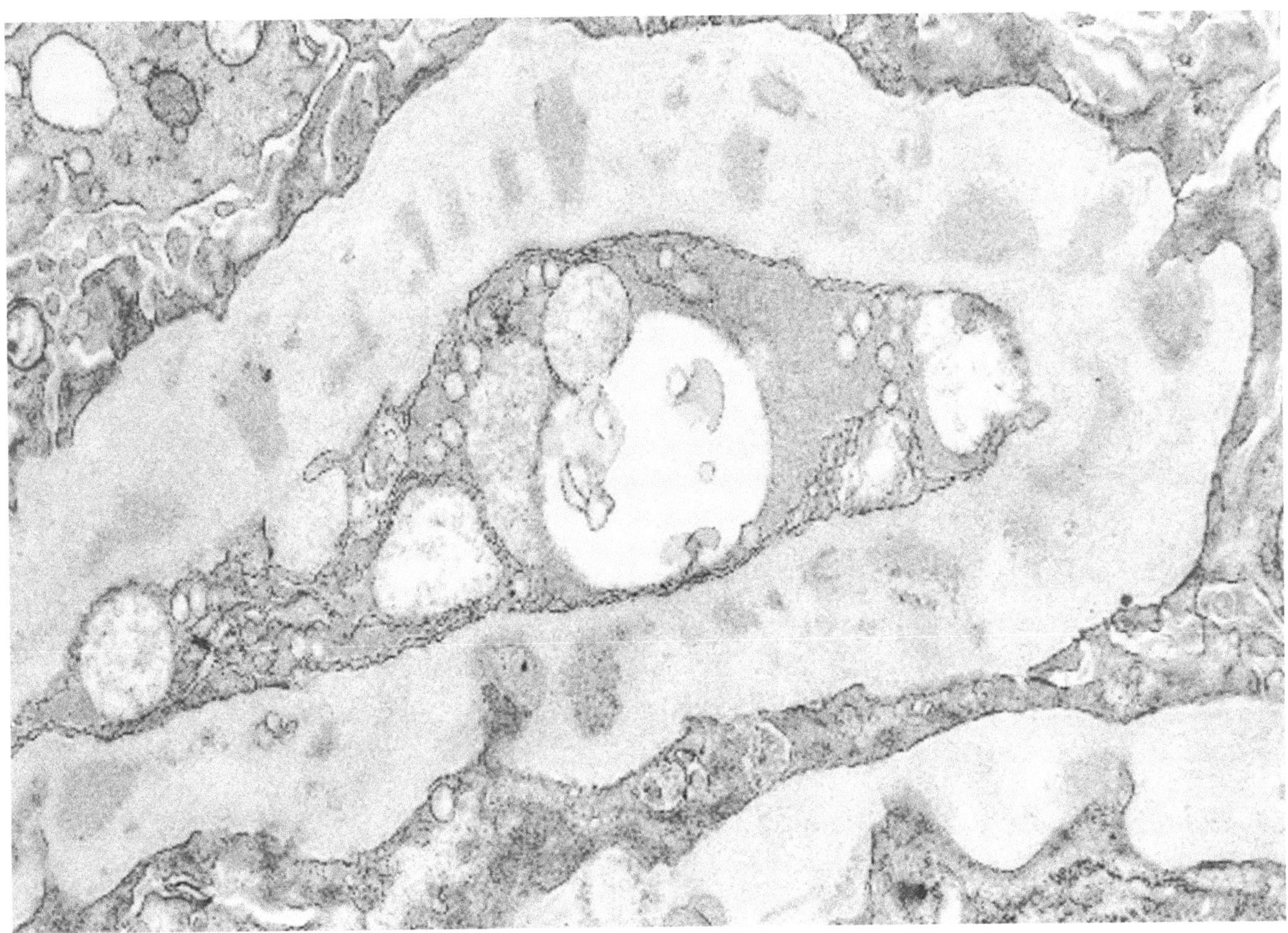

Figure 12.29. Membranous glomerulonephritis late stage III or early stage IV (42-year-old female, history of nephrotic syndrome since 18 years of age, biopsy-proved membranous glomerulonephritis 5 years before this nephrectomy for transplant). The deposits are further embedded in the thickened GBM, with areas of dissolution, cellular debris, and particles. (× 19,000)

of response to a variety of inciting conditions: lupus, renal allograft rejection, infection (hepatitis B virus [Collins et al. 1983], hepatitis C virus, syphilis), chemotherapy (gold salts, captopril, or penicillamine), possibly carcinomas and mercury compounds. The proposed pathogenesis, supported by experimental studies in rats with Heymann nephritis, is an autoimmune response to an antigen, as yet unknown, on the surface of the foot processes or planted in the GBM from the blood (Cavallo 1994).

Membranoproliferative Glomerulonephritis, Type I

(Figures 12.30 through 12.32)

Diagnostic criteria. (1) Subendothelial and mesangial electron-dense deposits; (2) extension of cells (mesangial or mononuclear phagocytes) or their processes along capillary walls in the subendothelial space ("interposition") (Figure 12.30); (3) duplication of the GBM with a new basement membrane layer formed underneath the endothelium (Figure 12.30) enclosing the deposits; (4) podocyte foot process effacement.

Additional points. In membranoproliferative glomerulonephritis (MPGN) the GBM duplication is due to extension of mesangial/mononuclear cells along the subendothelial space, splitting the basement membrane (double contours of capillary walls by light microscopy), hence the term mesangiocapillary glomerulonephritis (Churg et al. 1970; Kincaid-Smith 1973). The space occupied by the interpositioned cells is delimited by the old, normal-appearing basement membrane on the epithelial side and a newly formed basement membrane on the endothelial side. These cells are usually between the electron dense deposits and the endothelial cells. Scattered subepithelial deposits are often present.

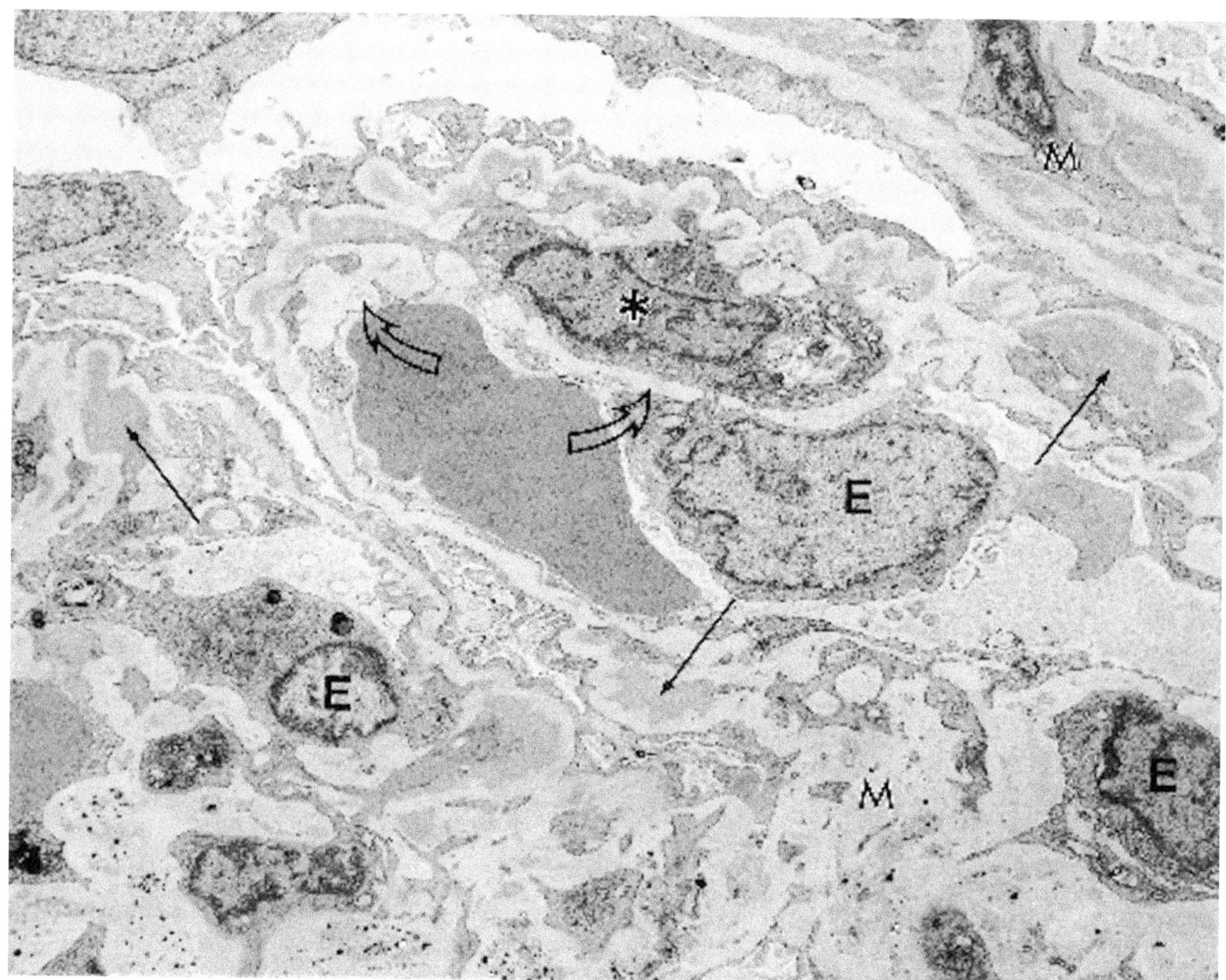

Figure 12.30. Membranoproliferative glomerulonephritis type I (42-year-old female diagnosed with MPGN I three years ago; hepatitis C and B virus negative; initially treated with steroids, cytoxan, and then trial of alpha-interferon therapy; rebiopsied to stage disease). Numerous subendothelial electron-dense deposits are present (*solid arrows*). The GBM is duplicated with new basement membrane formation (*open arrows*) toward the endothelial side and cellular interposition (*). The mesangium (M) is expanded. E = endothelial cell. (× 8300).

The mesangial cells usually are hyperplastic and hypertrophied, and mesangial matrix is increased. Extensive increase in mesangial matrix may give lobulated appearance to the glomerular tuft (so-called lobular glomerulonephritis). Endothelial cells may be increased in early proliferative lesions. The later biopsies, after clinical improvement, show less cellularity, fewer deposits, and an irregularly thickened GBM. If seen only at this late, nondescript stage, classification is difficult or impossible.

Immunofluorescence studies show coarse staining of the peripheral capillary walls with subendothelial accentuation, predominantly for C3, but also for IgG and IgM. C3 also may be found in the mesangium. Serum C3 levels fluctuate and often are low.

MPGN type I is immune-complex mediated in that the classical complement pathway is activated. The antigen is unknown (idiopathic/primary MPGN type I). Many morphologically similar cases of chronic immunologic injury such as those associated with chronic infections (hepatitis C virus, hepatitis B virus, shunt nephritis, schistosomiasis, candidiasis), systemic disease (lupus, sickle cell anemia), malignancy, alpha-1-antitrypsin deficiency, and mixed cryoglobulinemia are examples of secondary MPGN type I. Johnson et al. (1993) described eight patients with MPGN type I with

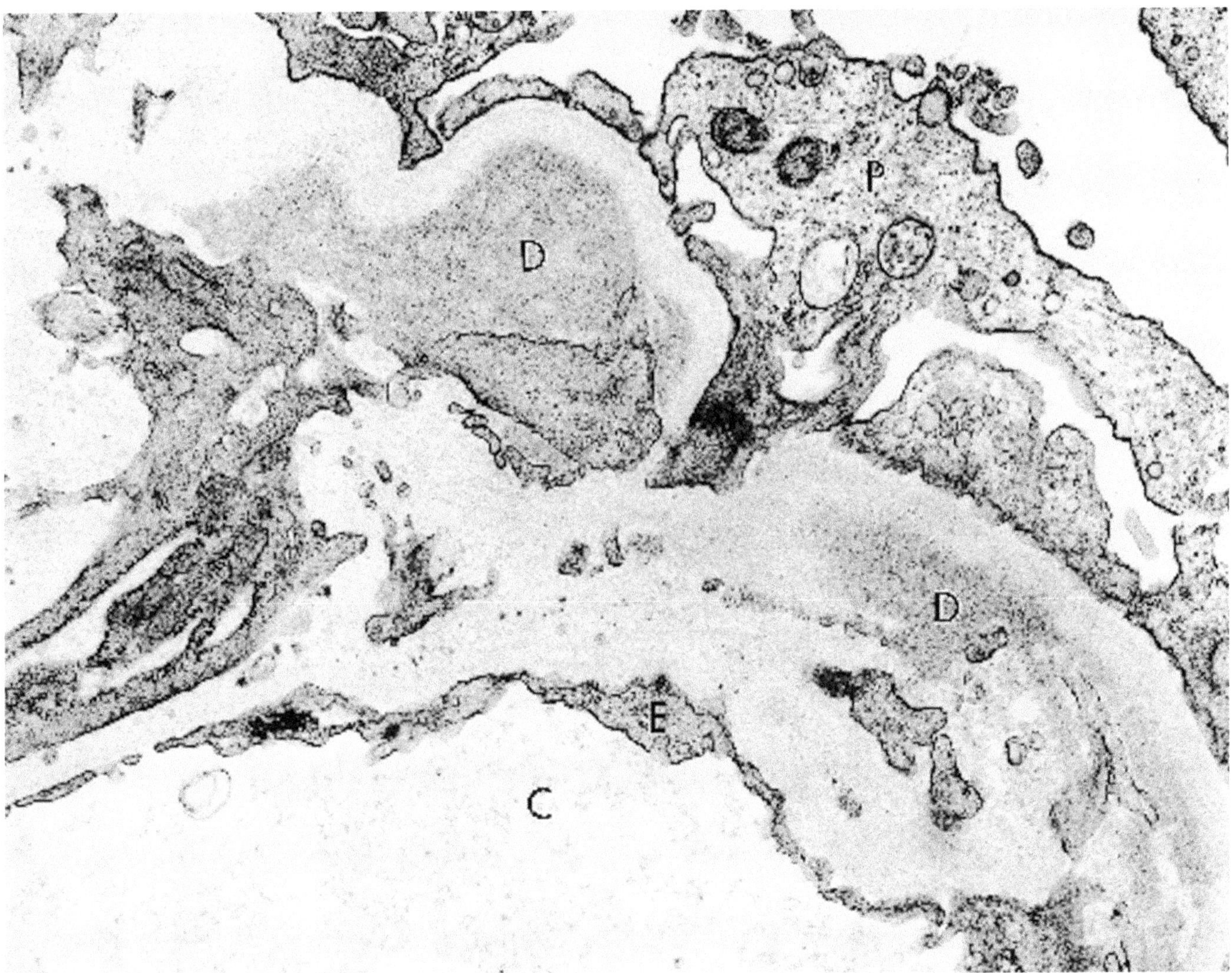

Figure 12.31. Membranoproliferative glomerulonephritis type I (same patient as in Figure 12.30; higher magnification). Note the flat edge of the deposits (D) against the original basement membrane and irregular edge toward the endothelial side. C = capillary lumen; E = endothelium; P = podocyte. (× 26,000)

hepatitis C virus ribonucleic acid (RNA) detected in serum. Cryoprecipitate was detected in five patients; three patients also had cryoglobulin substructure to the deposits detected by electron microscopy.

Variant patterns have been described in which subepithelial and intramembranous deposits are present (Figure 12.32). Burkholder et al. (1973) noted that some cases otherwise similar to type I have, in addition, numerous large subepithelial deposits. Strife et al. (1977) described a pattern characterized by contiguous subepithelial and subendothelial deposits, associated with duplication and layering of lamina-densa-like material that gives a disrupted, moth-eaten appearance to the basement membrane in silver-impregnated sections; their patients were older and had a more favorable prognosis. They called this pattern type III MPGN.

The significance of these other patterns and their relationship to MPGN type I is unclear. Many (including the authors) regard the moth-eaten appearance as part of the spectrum of type I. The best evidence is that individual cases may have both patterns in adjacent capillary loops (Figure 12.32). Type III may be asymptomatic more often with less classical pathway complement activation. Until distinct pathophysiologic or clinicopathologic features are identified, the separation seems artificial and the two patterns appear to be vari-

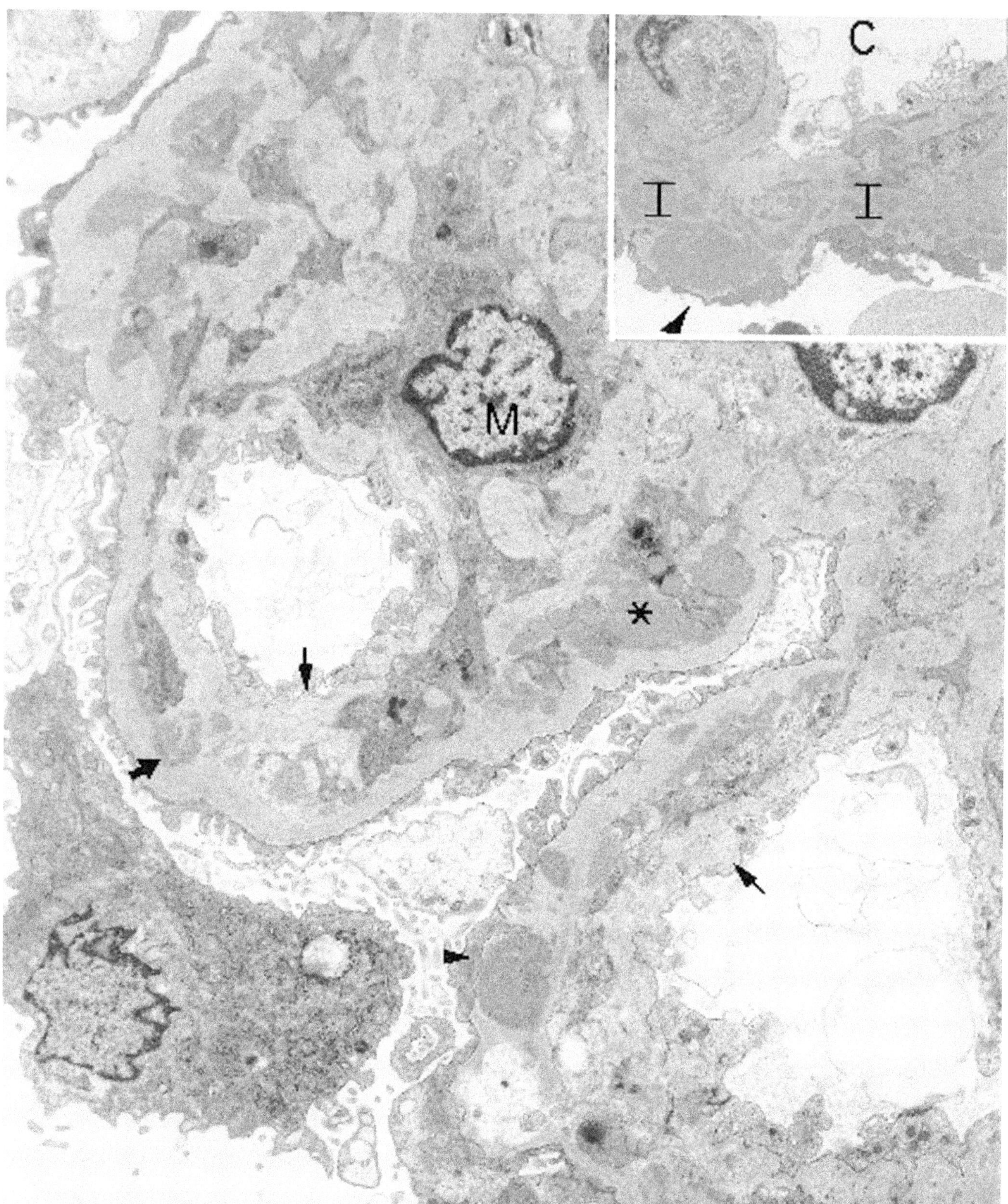

Figure 12.32. Membranoproliferative glomerulonephritis types I/III (9-year-old girl with 3 g/day proteinuria, hematuria, and normal complement studies). At low power, the numerous electron-dense mesangial (*), subendothelial (*heavy arrow*), and subepithelial (*arrowhead*) deposits can be appreciated. The capillary wall is markedly widened, and newly formed GBM can be seen under the endothelium (*thin arrows*), also called duplication of GBM. A mesangial cell (M) interposes its process in the widened capillary wall. (× 8640) *Inset* (same patient): The GBM is permeated by numerous irregular deposits, both subepithelial (*arrowhead*) and intramembranous (I). C = capillary lumen. (× 5940)

ants of the same disease (Jackson et al. 1987). Type II, dense deposit disease, is a distinct disease discussed later in this chapter.

IgA Nephropathy (Berger's Disease)

(Figures 12.33 and 12.34.)

Diagnostic criteria. (1) Amorphous electron-dense deposits in the mesangial matrix, usually surrounding the mesangial cell and frequently contiguous to the GBM; (2) extension of deposits into the paramesangial subendothelial space with occasional cellular interposition; (3) focal segmental mesangial cell hyperplasia and increase in matrix.

Additional points. Three types of mesangial cells have been observed in IgA nephropathy (Ng 1981) *Type I*—proliferative mesangial cells with numerous free ribosomes, smooth endoplasmic reticulum, mature Golgi apparatus, and centrioles, found only in the proliferative forms of IgA nephropathy. *Type II*—phagocytic mesangial cells with spinous cytoplasmic processes, rough endoplasmic reticulum, many dense bodies, and lipid inclusions, found in all types of IgA nephropathy. *Type III*—resting cells with minimal cytoplasm and poorly developed organelles, found only occasionally in minimal or focal sclerosing lesions. As the amount of mesangial deposits increases, the mesangial cells appear actively phagocytic. Why the deposits preferentially locate in the mesangium and how this leads to glomerular injury is not clearly understood. The deposits contain immunoglobulins (predominately IgA) and complement. The quantity of deposits is variable and does not correlate directly with histologic severity.

Deposits seen occasionally in the peripheral capillary loops may be subendothelial, subepithelial, or, rarely, intramembranous. Their presence is prognostically unfavorable and is associated with more severe glomerular lesions (Lee et al. 1989), proteinuria, and impairment of renal function when compared with cases without such deposits (Lee et al., 1982). Deposits in Bowman's capsule, and rarely in the tubular basement membrane have also been described (Hara et al. 1980; Hulette and Carstens 1985). The deposits reported rarely in arterioles, peritubular capillaries, and venules have not been proved to be related. Various GBM alterations such as attenuation, lytic attenuation, disruption garland-shaped widening, and dome-shaped widening have been described (Morita and Sakaguchi 1988). GBM splitting and thinning are also associated with a poorer prognosis. Podocyte foot process flattening and effacement with villous transformation and endothelial cell swelling may be observed.

Light microscopy shows a spectrum of possible lesions, from essentially normal glomeruli, to the usual focal or diffuse mesangioproliferative glomerulonephritis associated with segmental glomerular scarring, and to focal crescents.

Diagnosis requires demonstration of copious mesangial deposits by immunofluorescence that contain predominately IgA (Berger and Hinglais 1968; Berger 1969), usually with C3 and fibrin and often with IgG and IgM. Henoch-Schönlein purpura has lesions that are similar to IgA nephropathy although they usually involve the GBM to a greater degree. Patients with liver disease may have IgA deposits in the mesangium in the absence of any marked glomerular disease. Patients with AIDS also may have a predisposition to IgA nephropathy (Beaufils et al. 1995).

The IgA deposits usually recur in allografts but rarely cause loss of the graft. The IgA deposits can be reabsorbed within a few weeks, as shown in kidneys transplanted with preexisting IgA deposits.

The specificity of the IgA antibodies is unknown but is presumed to be either an exogenous antigen (bacterial and viral antigens, food and airborne antigens), a self-antigen (autoimmune diseases), or an *in situ* mesangial antigen. The exact pathogenesis of IgA nephropathy is not known. Circulating immune complexes, defects in immune regulation, mesangial or bacterial antigen, abnormal IgA molecule, mesangial cell dysfunction, mucosal barrier defect, and genetic factors have all been implicated (Galla 1995).

Henoch–Schönlein Purpura

(Figures 12.35 and 12.36.)

Diagnostic criteria. (1) Electron-dense deposits, mostly in the mesangium but also in subendothelial and, occasionally, subepithelial locations (Figure 12.35C) (Heaton et al. 1977); (2) focal thickening and splitting of the GBM with effacement of podocyte foot processes.

Additional points. Mesangial hypercellularity and increased matrix are usual features. Occasionally, cellular crescents are numerous.

The light microscopic appearance may vary from normal to a diffuse proliferative glomerulonephritis with crescents, although a focal segmental glomerulonephritis is most common. Prognosis is a function of the extent of crescents, although recovery can occur even with 75% crescents (Yoshikawa et al. 1981). Immunofluorescence demonstrates the presence of IgA in the mesangium; in late stages of the disease (after a few months), the immunofluorescence may become negative (in contrast to IgA nephropathy). The deposits along the GBM favor Henoch-Schönlein purpura over Berger's disease, though not decisively. It is usually the entire clinicopathological picture of a polysymptomatic syndrome (vasculitis, purpura, arthralgia, gastroin-

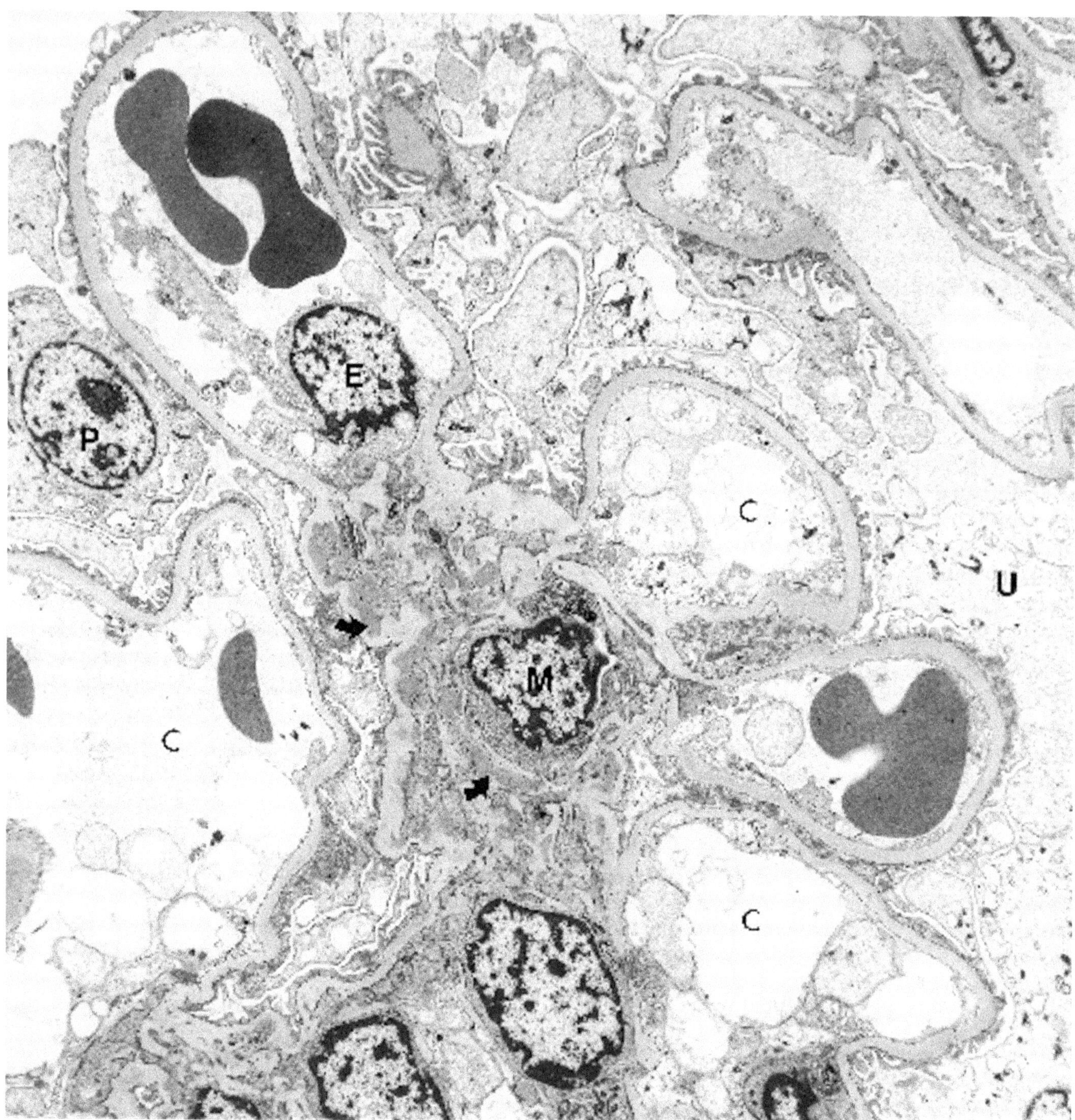

Figure 12.33. IgA nephropathy (36-year-old male with microscopic hematuria). Numerous electron-dense deposits (*arrows*) are seen in the mesangium. M = mesangial cell. Endothelial (E) and epithelial (P) cells are normal. C = capillary lumen; U = urinary space. (× 5300)

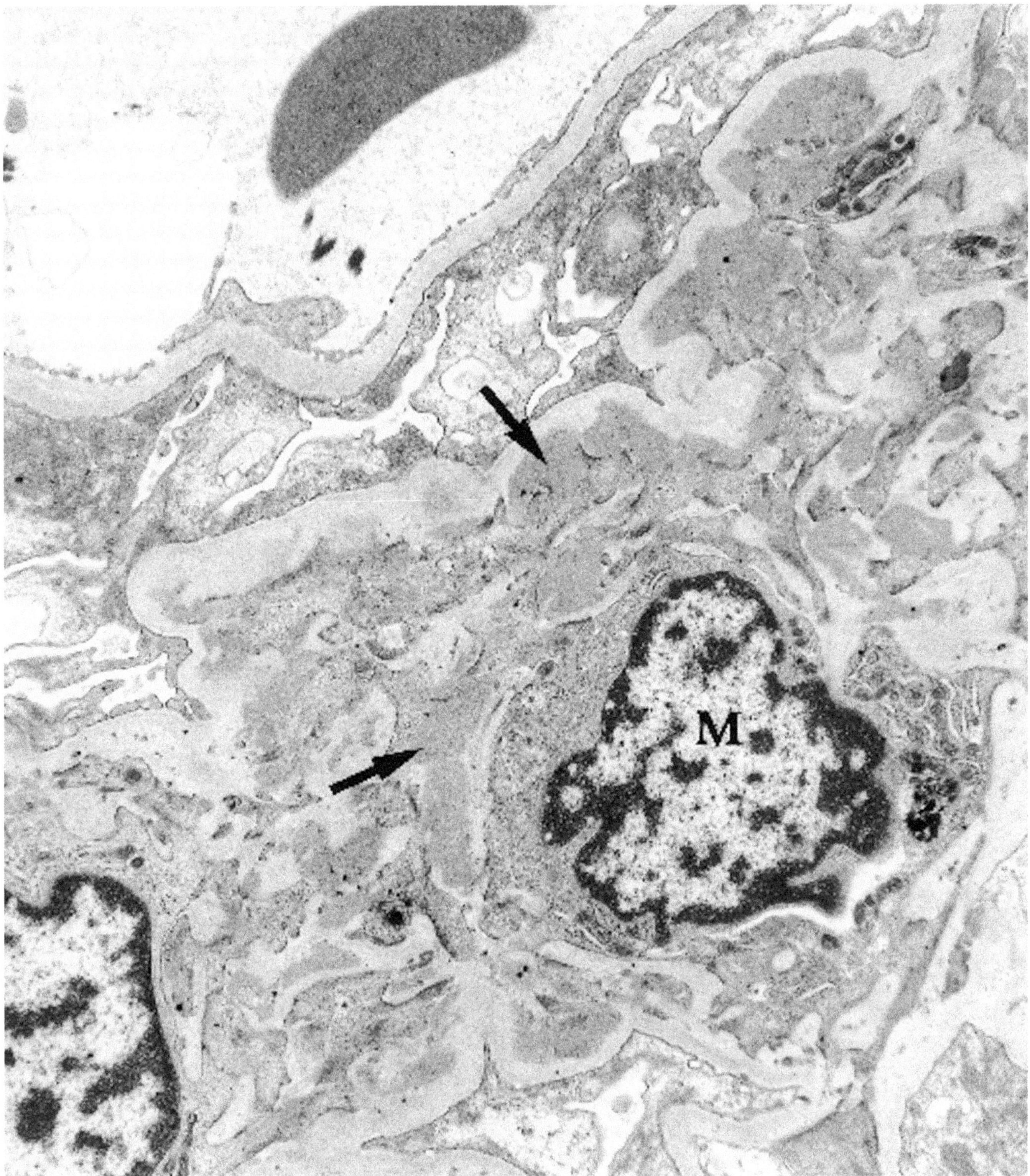

Figure 12.34. IgA nephropathy (same case as in Figure 12.33). Higher magnification of mesangium; electron-dense deposits (*arrows*) surround the reactive mesangial cell (M). (× 16,800)

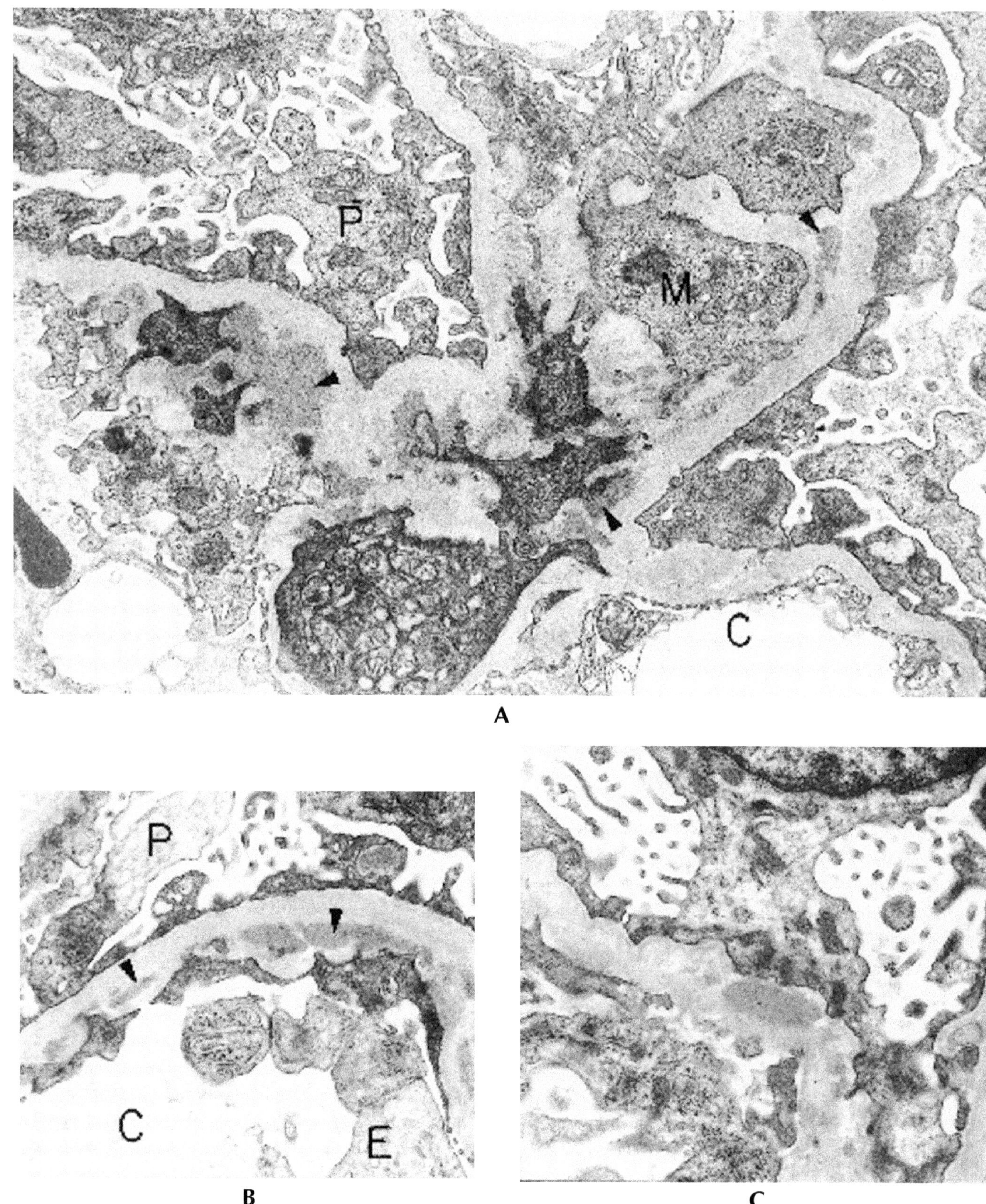

Figure 12.35. Henoch-Schönlein purpura. **A** (6-year-old girl with purpura, active urinary sediment, and nephrotic range proteinuria), The mesangium (M) contains numerous irregular, electron-dense deposits (*arrowheads*). This pattern is indistinguishable from IgA nephropathy. P = podocyte; C = capillary lumen. (× 14,800) **B** (5-year-old boy with purpura, arthritis, and nephritis for 14 weeks); Subendothelial deposits (*arrowheads*). P = podocyte; C = capillary lumen; E = endothelial cell. (× 13,570) **C** (same patient as in Figure 12.35B); A subepithelial electron-dense deposit is present. The overlying podocyte has microvillous transformation. (× 14,420)

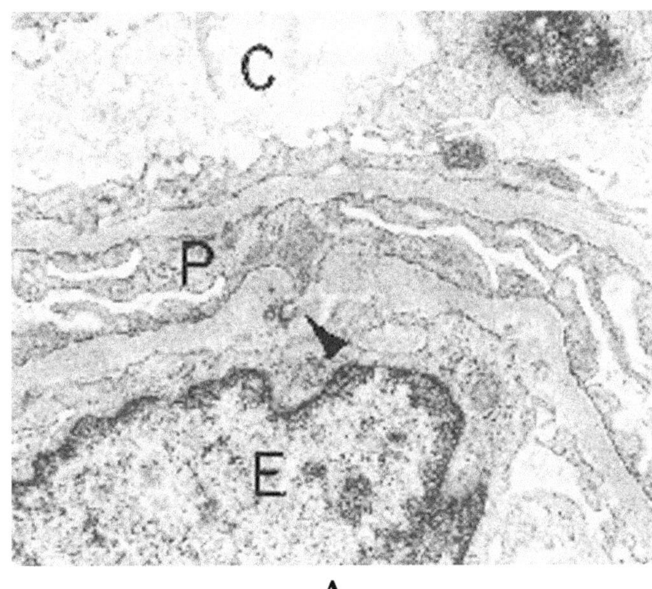

A

B

Figure 12.36. Henoch-Schönlein purpura. **A** (7-year-old boy), Break in the GBM (*arrowhead*). P = podocyte; C = capillary lumen; E = endothelial cell. (× 10,100). **B** (same patient as in **A**), The GBM is focally extremely thin and laminated, resembling Alport's syndrome. P = Podocyte; C = capillary lumen. (× 13,770).

testinal, and neurologic symptoms) that helps to separate Henoch-Schönlein purpura from Berger's disease, which otherwise may be indistinguishable morphologically (Mihatsch et al. 1984). Deposits are sometimes sparse in Henoch-Schönlein purpura (versus Berger's disease). Henoch-Schönlein purpura rarely recurs, in contrast to IgA nephropathy.

Systemic Lupus Erythematosus

(Figures 12.37 through 12.48.)

Diagnostic criteria. The World Health Organization (WHO) morphologic classification of lupus nephritis (LN), with recent modification (Churg et al. 1995), is divided into six classes and is based primarily on light microscopic appearance. Each class has a distinctive ultrastructural appearance by the pattern of deposition of electron-dense deposits in the glomeruli (Pirani 1985; Bruijn 1994).

Class I. Normal or minimal disease: (1) Morphologically normal glomeruli; (2) occasional mesangial widening with or without electron-dense deposits; (3) focal GBM thickening; (4) focal podocyte foot process effacement (Hill 1992).

Class II. Mesangial lupus nephritis: (1) Predominately mesangial electron-dense deposits, usually located in peripheral or the paramesangial waist regions of the mesangium; (2) there may be small subendothelial deposits immediately adjacent to the mesangium; (3) the GBM is usually not thickened; (4) this class is subdivided into those without significantly increased mesangial cellularity (mild, *class IIa*) and those with a prominent increase in mesangial cellularity and matrix (*class IIb*).

Class III (Figure 12.37). Focal segmental glomerulonephritis: (1) Segmentally proliferative, necrotizing, or sclerotic lesions involving less than 50% of the glomeruli, which may go through stages of development (Grishman and Churg 1982); (2) mesangial electron dense deposits with a focal or diffuse increase in mesangial cells and matrix; (3) scattered subendothelial and subepithelial deposits; (4) occasionally crescents with GBM disruption.

Class IV (Figures 12.38 through 12.40). Diffuse proliferative glomerulonephritis: (1) Large subendothelial deposits that bulge into the capillary lumens (Figures 12.38 and 12.39), corresponding to the "wireloop" appearance of the thickened capillary walls; (2) widespread mesangial deposits; (3) diffuse mesangial, endocapillary, or mesangiocapillary hypercellularity, involving more than 50% of the glomeruli, with an increase in mesangial cells and matrix and proliferation of endothelial cells (Figure 12.40); (4) cellular interposition (mesangial cell or monocyte) in the subendothelial space contributing to double contour of GBM with duplication and new basement membrane formation, toward the endothelial side; (5) scattered subepithelial wedge-shaped (Figure 12.48B) or typically penetrating deposits and intramembranous deposits; (6) podocyte foot process effacement; (7) inflammatory cell infiltration (neutrophils and monocytes); (8) tubular basement membrane deposits (Figure 12.48A) often present (50%).

Class V (Figures 12.41 through 12.43). Membranous glomerulonephritis: (1) Abundant subepithelial electron dense deposits that follow the same stages of development as idiopathic membranous glomerulonephritis, becoming incorporated into the basement membrane in the later stages; (2) mesangial deposits (usual); (3) few, if any, subendothelial deposits; (4) diffuse podocyte foot process effacement. Membranous lupus nephritis is subclassified into class Va (purely membranous) and class Vb (associated with class II lesions). It was previously subclassified into minimally proliferative (classes Va and Vb) type and those with focal or diffuse proliferative changes (classes Vc and Vd); the latter being associated with poor prognosis (Adler et al. 1990). The modified version of WHO lupus nephritis classification now categorizes classes Vc and Vd in class IV.

Class VI (Figure 12.44). Advanced sclerosing glomerulonephritis in lupus nephritis may have diffusely thickened GBM, with areas of collapse, wrinkling, sclerosis, and scattered, scant immune deposits.

Tubulointerstitial nephritis. Rare cases of lupus nephritis present with predominant acute and chronic tubulointerstitial inflammation with only minor glomerular lesions. Electron dense deposits are found along the tubular basement membrane, which is thickened, in the interstitium, and along the GBM (Singh et al. 1996). Rare cases with acute tubular necrosis have been noted (McCluskey in press). Although tubulointerstitial nephritis associated with proliferative or active glomerular lesions is well known, its predominance in the absence of significant glomerular lesions is rare and noteworthy.

Distinctive features of lupus nephritis. Several distinctive features of lupus nephritis may be seen in any of the classes. (1) *Fingerprints* (Figure 12.45): This substructural detail of the electron dense deposits is pathognomonic of lupus nephritis. It was first demonstrated by Grishman et al. (1967). These "fingerprints" appear as closely packed parallel, curvilinear bands with a thickness ranging from about 10–20 nm. The distance between center of one band to the next one tends to be about 19–33 nm. The exact nature of fingerprints and the pathogenesis of its pattern are unknown. It may be related to presence of cryoglobulins (Kim et al. 1981) or oligoclonal immunoglobulin. They are found in about 5% (Alpers et al. 1984) to 19% (Tojo et al. 1993) cases of lupus nephritis. The authors have also noted unusual microtubular and paracrystalline substructures in the deposits (Figures 12.46 and 12.47) that resemble microtubules of immunotactoid glomerulopathy rather than the usual fingerprints. These may also be related to (variants of) cryoglobulins, monoclonal immunoglobulins, or immune complexes (2) *Tubuloreticular structures/inclusions* (Venkataseshan 1991) are found in the kidney in the endothelial cytoplasm of glomerular capillaries and the peritubular capillaries (Figures 12.39, 12.40, 12.42 and 12.51), lymphocytes, peripheral blood mononuclear cells, and rarely interstitial cells and podocytes. Tubuloreticular inclusions are frequently found in lupus nephritis cases, especially during active disease or exacerbations (see next section) (3) *Extra-glomerular electron dense deposits* may be seen along the tubular basement membrane (Figure 12.48A) in any segment of the tubule (Schwartz et al. 1982a). Lamination, duplication, and irregularity of the tubular basement membrane (TBM) with incorporation of deposits and cellular debris suggest immune-mediated tubular damage (anti-TBM antibodies to *in situ* antigen or immune complex deposition), followed by repair and thickening of the TBM. Deposits may also be seen around peritubular capillaries and in the interstitium with associated interstitial nephritis. (4) *Hematoxylin bodies,* which are pathognomonic of lupus erythematosus are the tissue equivalents of lupus erythematosus cells. They occur in the mesangium or capillary loops and comprise of remnants of nuclei (clumped chromatin), cytoplasmic components (vesicles, vacuoles, degenerating granules, and glycogen) (Cohen and Zamboni 1977), and probably electron dense deposits (Morita and Sakaguchi 1984). They are found in fewer than 5% of biopsies (5) *Microspherical and microtubular curvilinear structures* are seen in visceral epithelial and tubular cells or along the GBM. They are present in several renal diseases and may be cell-membrane fragments rather than viral particles (see also section on spherical microparticles).

Additional points. Membranous lupus nephritis (WHO class V) cannot always be differentiated by electron microscopy from idiopathic membranous glomerulonephritis; certain features help to favor one over the other. Membranous lupus nephritis more often has mesangial hypercellularity; deposition of immune complexes in mesangium (96% in lupus membranous glomerulonephritis versus 11% in non–lupus membranous glomerulonephritis) (Jennette 1983), subendothelium, and tubular basement membrane (Schwartz et al. 1982b); IgA (in addition to IgG) in glomerular immune deposits; and tubuloreticular structures in glomerular endothelial cells. The subepithelial deposits in lupus nephritis tend to be wedge shaped and penetrate the GBM (Figures 12.41 and 12.48B). In membranous lupus nephritis, the subepithelial deposits tend to be homogenous and uniform in size and are distributed diffusely (type I subepithelial deposits). In contrast, type II subepithelial deposits, associated with proliferative lupus nephritis, are irregular in size and distribution with variable density (Schwartz et al. 1982; Hill 1992) and tend to be smaller.

The glomerular lesions of lupus nephritis are dynamic. Transition from one class to another either spontaneously or in response to therapy, have been observed (Hayslett and Carey 1995). Though overlap between classes may exist, the worst prognostic features should be used in the final class designation. Resolution of active lupus nephritis may lead to basement membrane changes and focal segmental glomerulosclerosis. We have seen a case of membranous lupus nephritis (Figure 12.43A and B) that after seven years of therapy showed resolution of deposits, extensive lucency, lamination, and scalloping of the GBM (Figure 12.43C and D), resembling hereditary nephritis (compare with Figures 12.20 and 12.21A); focal segmental glomerulosclerosis with mesangial deposits was also seen.

The light microscopic appearance of lupus nephritis is variable and is the basis for the WHO classification. Immunofluorescence findings also vary according to class of involvement and correspond to the location of deposits seen by electron microscopy. The distinctive feature of lupus nephritis is a "full house" pattern of immunofluorescence staining in the glomeruli for IgG, IgA, IgM, Clq, and C3. Tubular basement membranes, interstitium, and blood vessels also are sites of localization of immunoglobulins and complement, as extraglomerular deposits are detected by immunofluorescence in about 50% of cases (Schwartz et al. 1982a).

Pathogenesis and nature of electron-dense deposits in lupus nephritis. The exact pathogenesis of lupus nephritis is complex, multifactorial, and not entirely explained. Many investigators have suggested deposition of preformed circulating immune complexes in the glomeruli; most notable candidates for this hypothesis were deoxyribonucleic acid (DNA)/anti-DNA complexes. Another hypothesis is the presence of intrinsic glomerular antigens to which the antinuclear antibodies bind to directly *in situ* to form immune deposits. The possibility that these antinuclear antibodies cross-react with native antigens (heparan sulfate, laminin, collagen type IV, and fibronectin) has been proposed (Vlahakos et al. 1992; Budhai et al. 1996). A more recent hypothesis suggests that (1) binding of anti-DNA antibodies to the GBM is indirect and mediated by nucleosomes, and (2) heparan sulfate and collagen type IV act as ligands for such nucleosome-complexed autoantibodies (Berden 1997). Nucleosomes are short lengths of DNA wrapped around a histone octamer, anchored by H1 histone, and linked together with other nucleosomes by histone-free DNA, giving a "beads-on-a-string" appearance to the chromatin by electron microscopy at high magnification. Nucleosomes may be released into circulation by apoptosis; an increase of which may suggest defects in phagocytosis. Using nucleosome-specific monoclonal antibodies as a probe, van Bruggen et al. (1997) identified nucleosomes in glomerular deposits in 45% of cases of class IV lupus nephritis, whereas histones were present in 100%. Malide et al. (1993), using specific colloidal gold and electron microscopic immunocytochemistry, demonstrated DNA molecules in deposits along the capillary wall and mesangium and in cell nuclei of lupus nephritis patients, supporting the role of DNA as an antigenic component of the immune complexes.

It is conceivable that different mechanisms, related to the nature and amount of the antigens and the autoantibodies, influence the location of immune deposits, type of glomerular pattern of disease, specific clinical syndrome, and prognosis. Subendothelial and mesangial deposits may result from immune complex deposition. Their contact with circulating inflammatory cells in glomerular capillaries would be more likely to result in inflammation or necrosis and an active urinary sediment (class III and IV). Subepithelial deposits may form as a result of antibody reaction with an *in situ* GBM antigen, thereby disrupting filtration properties of the GBM and resulting in massive proteinuria (class V).

(Text continues on page 836)

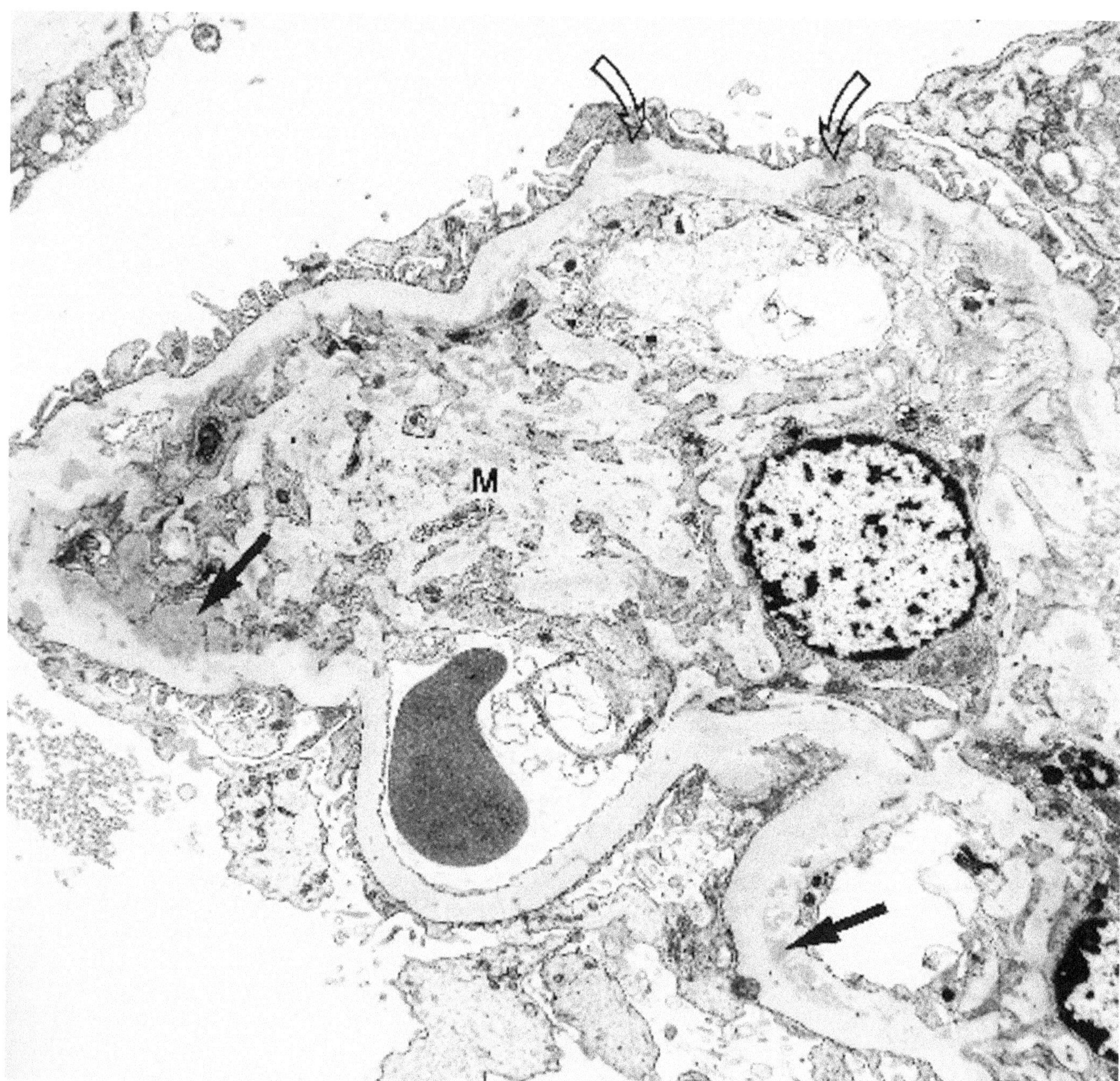

Figure 12.37. Lupus nephritis, WHO class III, focal proliferative type (30-year-old male with systemic lupus erythematosus and active urinary sediment with positive lupus anticoagulant and anticardiolipin antibody). Segmental sclerosis of capillary loops with mesangial prominence and electron dense deposits in mesangial (M), subendothelial (*solid arrows*), and subepithelial (*open arrows*) locations. (× 7800)

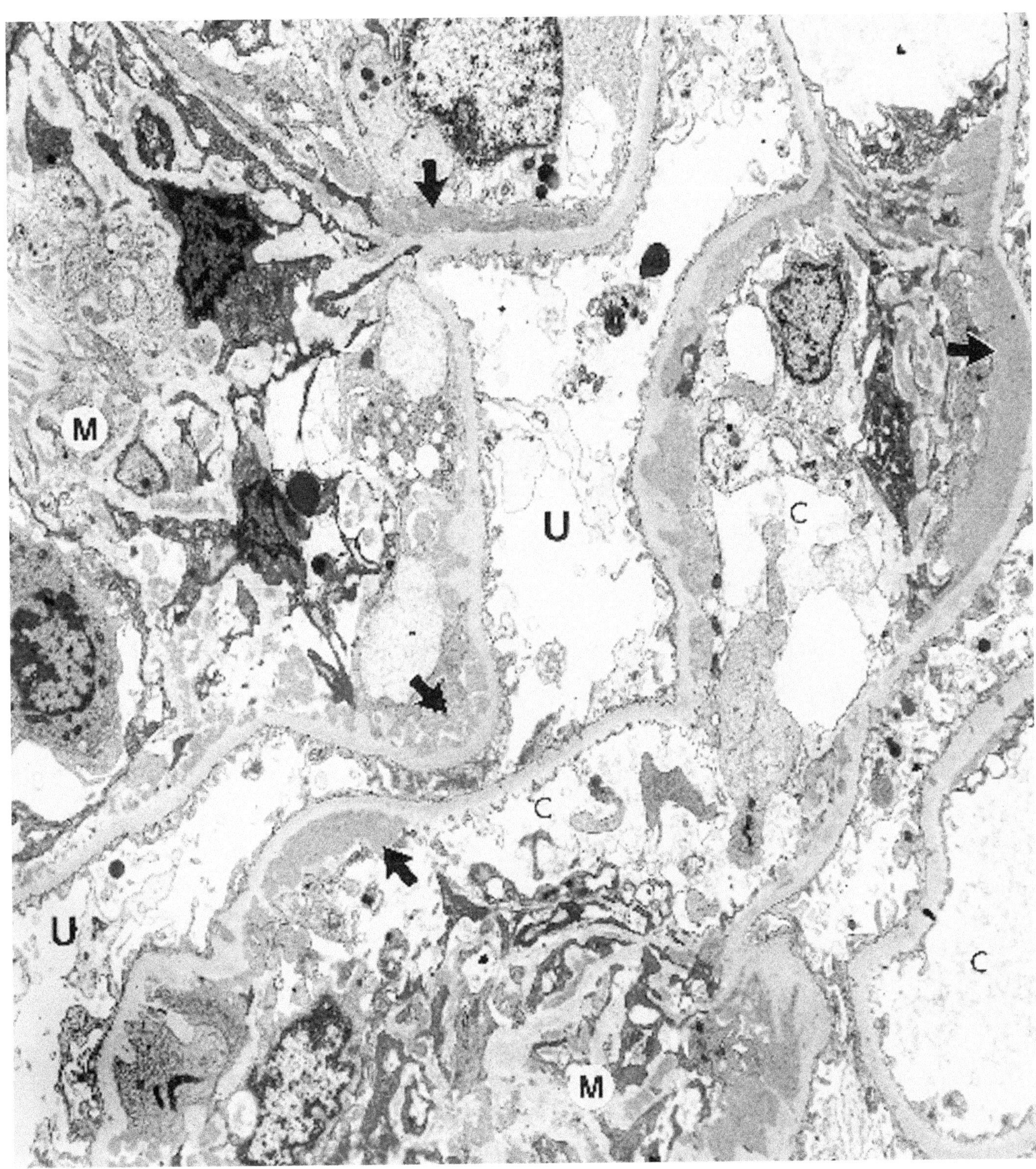

Figure 12.38. Lupus nephritis, WHO class IV, diffuse proliferative type (23-year-old female with systemic lupus erythematosus since childhood; now has proteinuria, edema, and hypertension). Numerous large electron dense deposits are present in subendothelial locations (*arrows*) with duplication of the GBM; deposits are also present in the expanded mesangium (M). U = urinary space; C = capillary lumen. (× 5500)

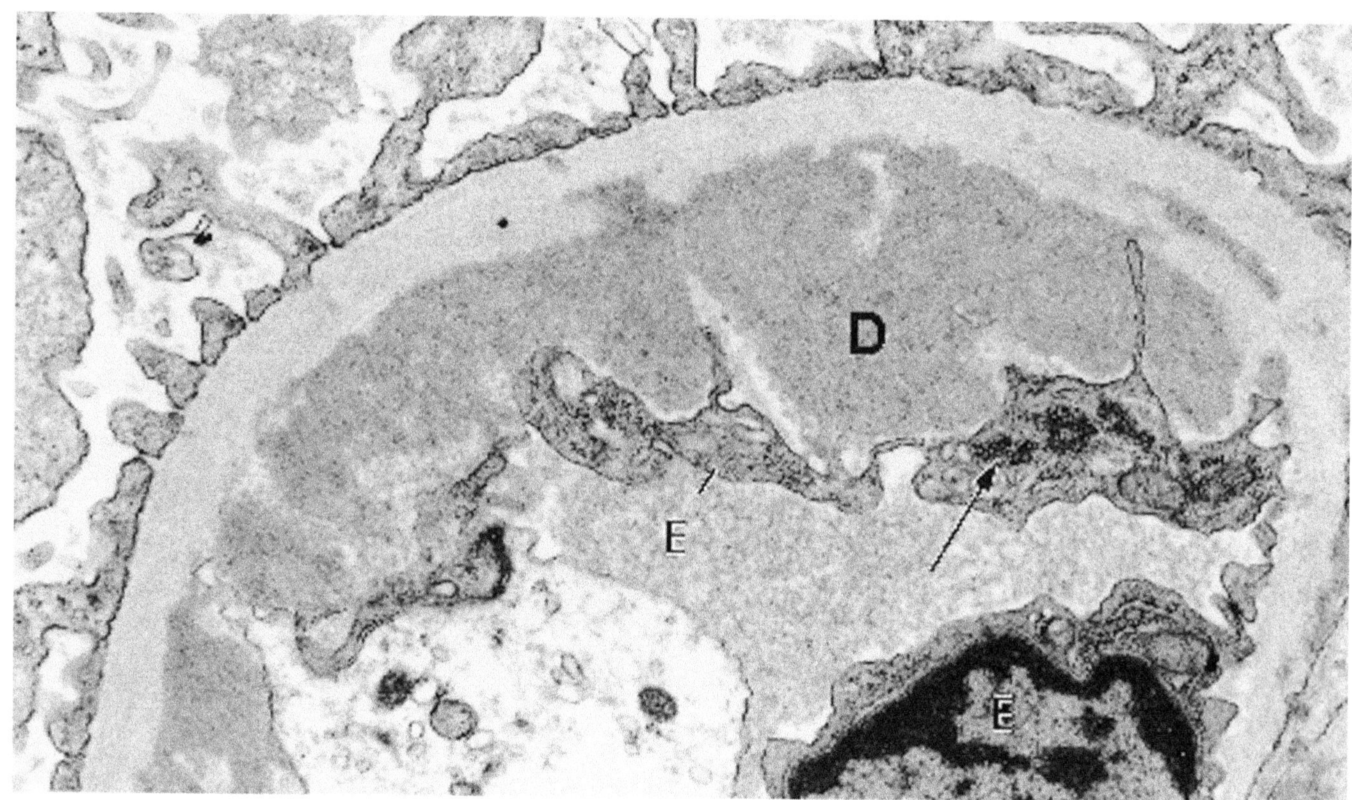

Figure 12.39. Lupus nephritis, WHO class IV (same patient as in Figure 12.38). Higher magnification of the subendothelial deposits (D) lying in between the GBM and endothelium (E). Note the presence of tubuloreticular structures (*arrow*) in the endothelial cytoplasm. (× 27,000)

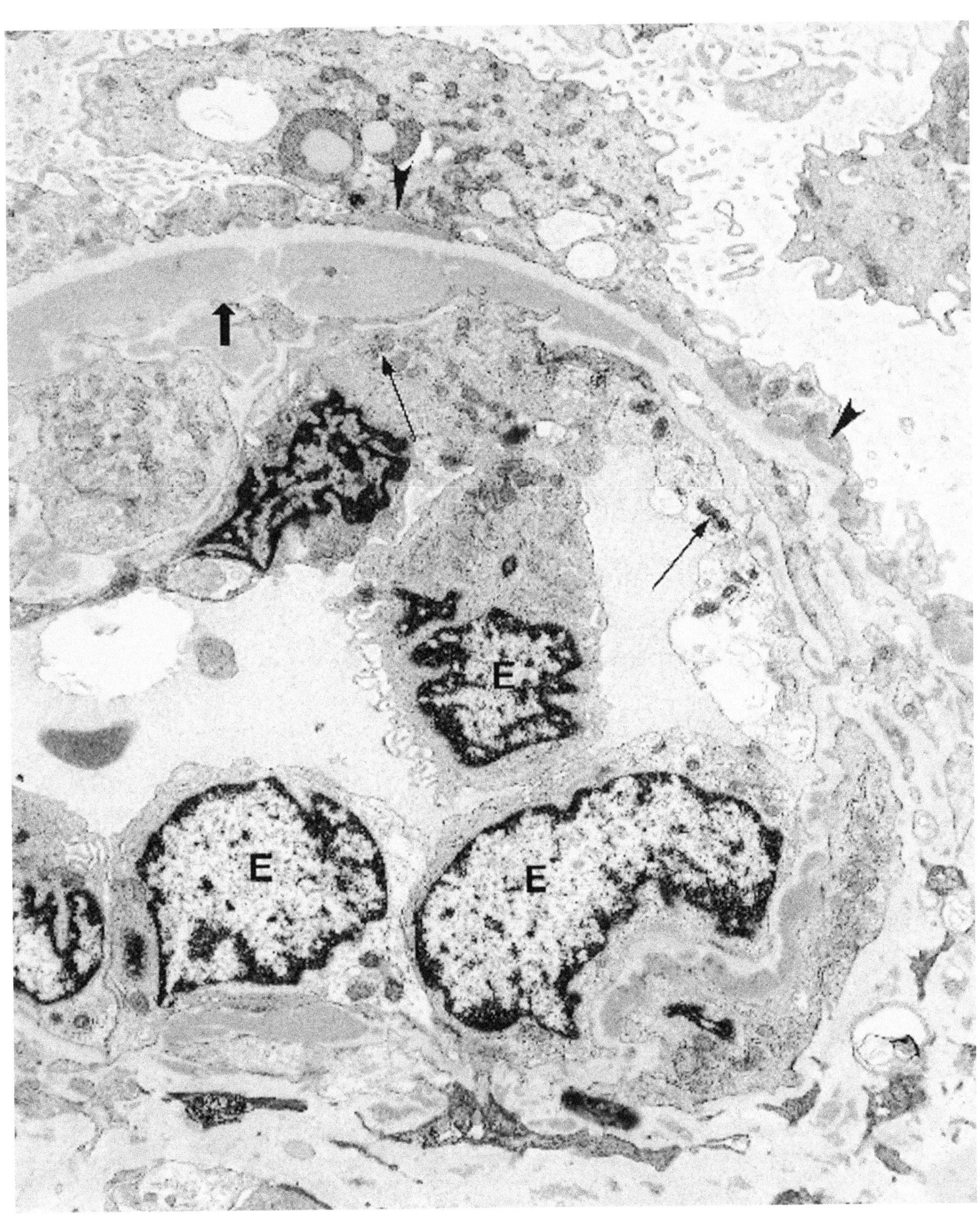

Figure 12.40. Lupus nephritis, WHO class IV (37-year-old female). Prominent glomerular endothelial cell (E) proliferation in a capillary loop; up to four endothelial cells are seen in this capillary. Tubuloreticular inclusions are present in endothelial cytoplasm (*thin arrows*); numerous deposits are in the subendothelial (*thick arrow*) region; and there are a few subepithelial deposits (*arrowheads*) as well. (× 9800)

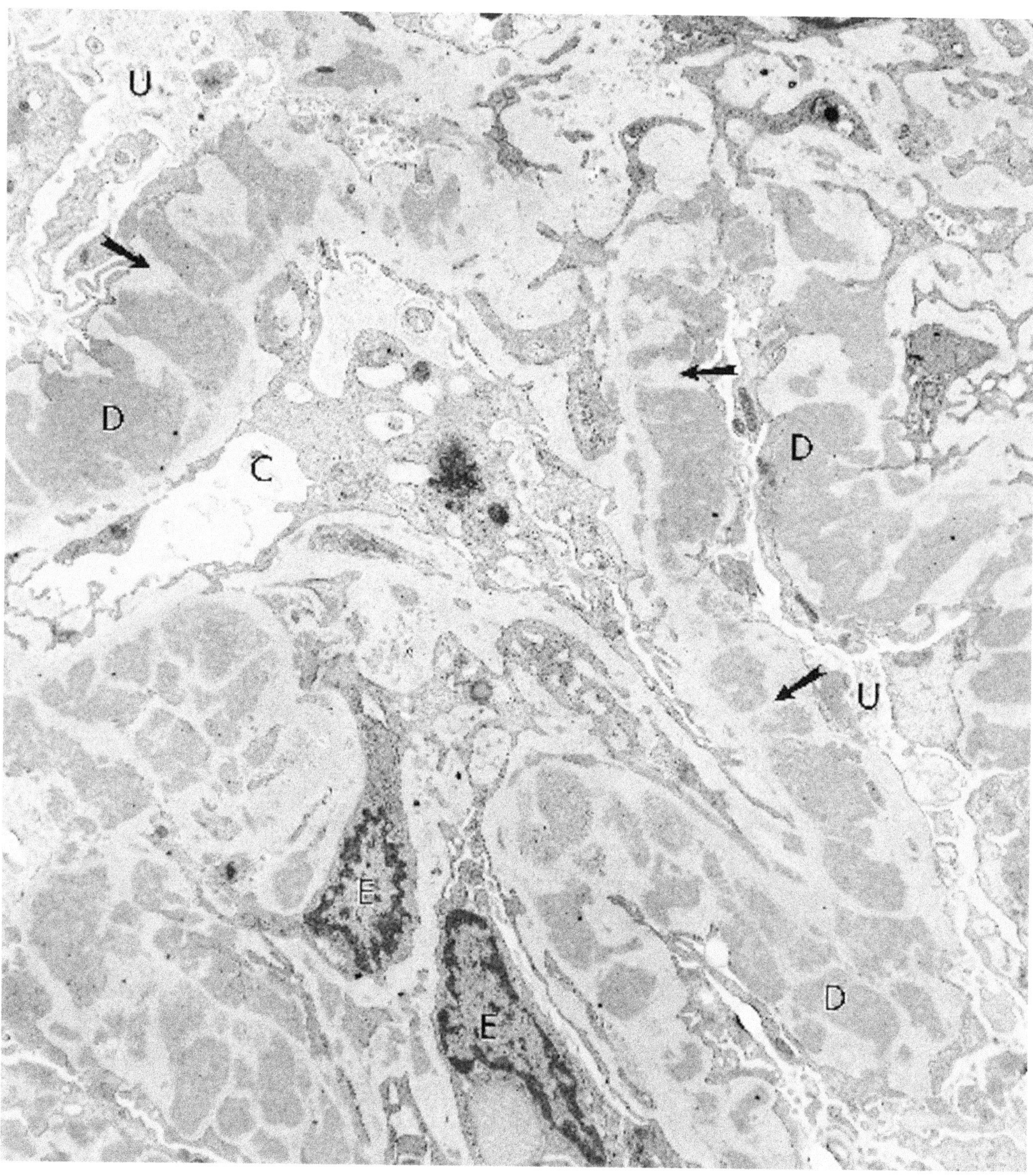

Figure 12.41. Lupus nephritis, WHO class V, membranous type (17-year-old female with autoimmune hepatitis and nephrotic syndrome since age 9 years; has hypocomplementemia and proteinuria of 6–10 g/day). Numerous irregular subepithelial electron-dense deposits (D) are present, some embedded in the GBM, with spikes of newly formed GBM in between (*arrows*) them. C = capillary lumen; E = endothelium; U = urinary space. Diffuse foot process effacement is present. (× 9700)

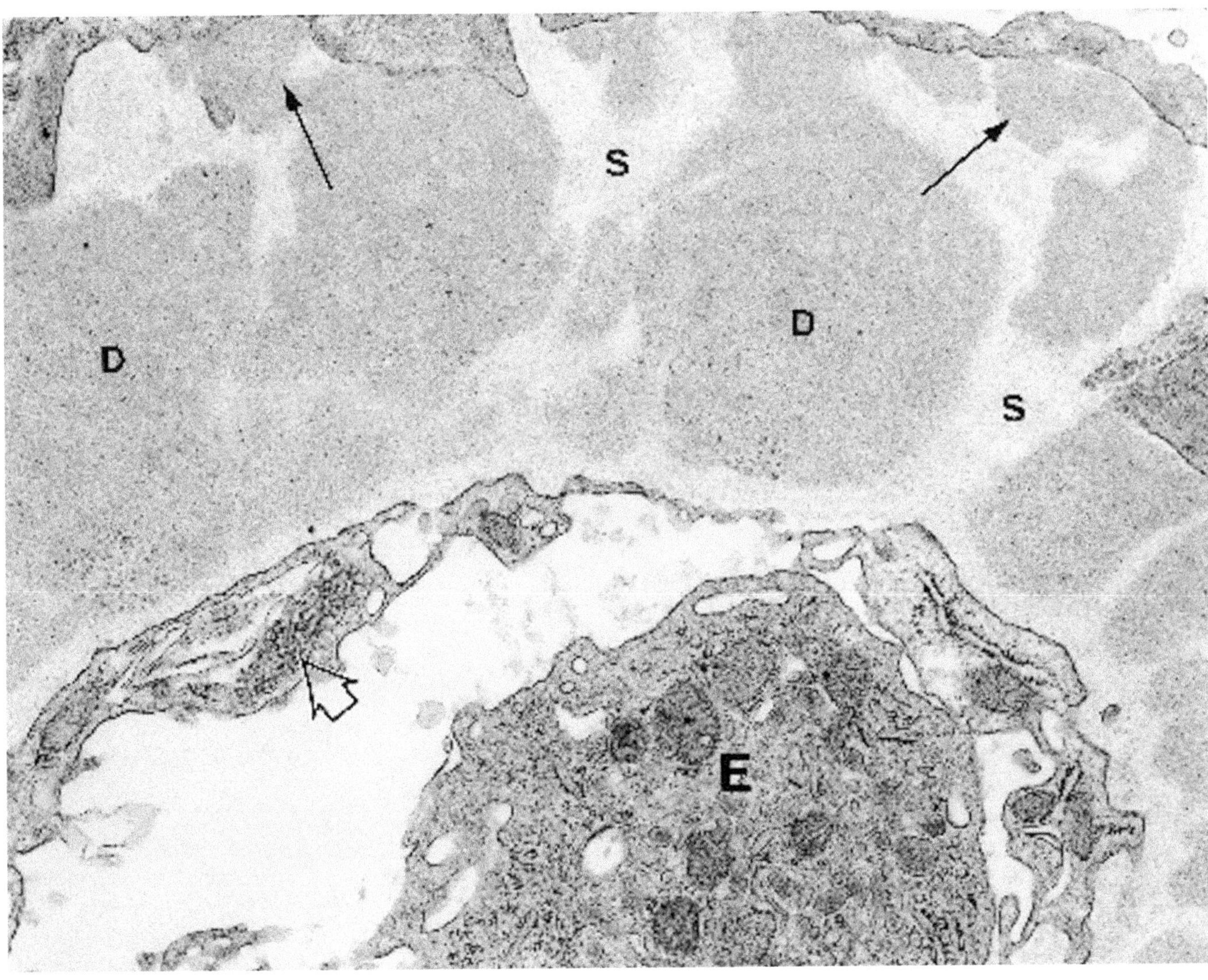

Figure 12.42. Lupus nephritis, WHO class V (same case as in Figure 12.41). Higher magnification of subepithelial deposits (D) with intervening spikes (S) and focal fingerprinting (*thin arrow*). Tubuloreticular structures (*open arrow*) in endothelial cell (E). (× 26,000)

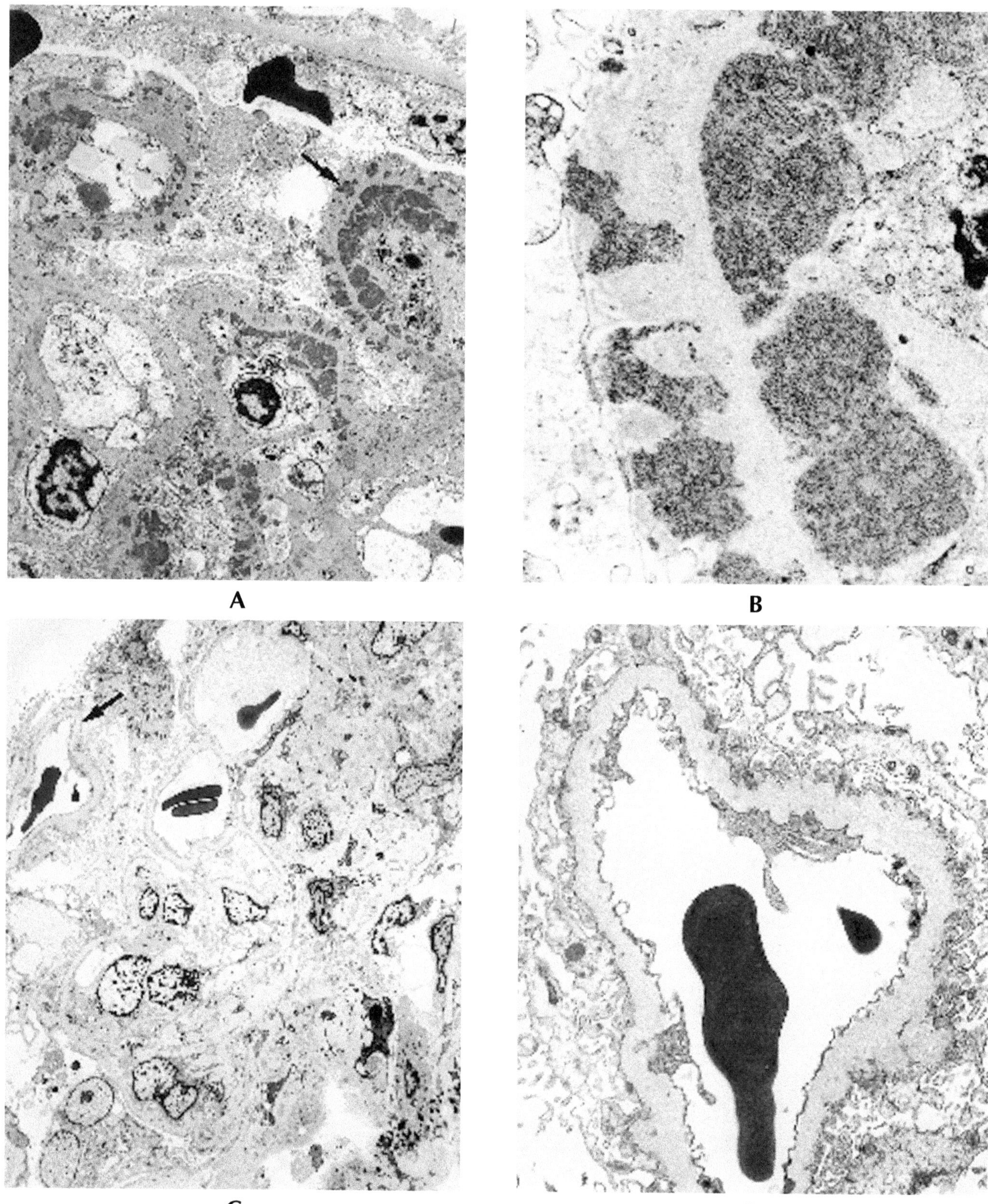

Figure 12.43. Lupus nephritis, transformation from one class to another. **A** (30-year-male with WHO class V, membranous-type lupus nephritis), Widespread subepithelial and subendothelial deposits. (× 2500) **B** Higher magnification of area marked by *arrow* in **A,** showing the deposits on both sides of the thickened and irregular GBM, with fingerprint substructure. (× 23,000) **C** (same patient as in **A** and **B,** 7 years after chemotherapy), Focal segmental glomerulosclerosis. (× 1900) **D,** Higher magnification of the capillary loop marked with an *arrow* in **C,** showing prominent lamination and irregularity of the GBM, reminiscent of Alport's syndrome, without any subepithelial or subendothelial deposits. (× 7700)

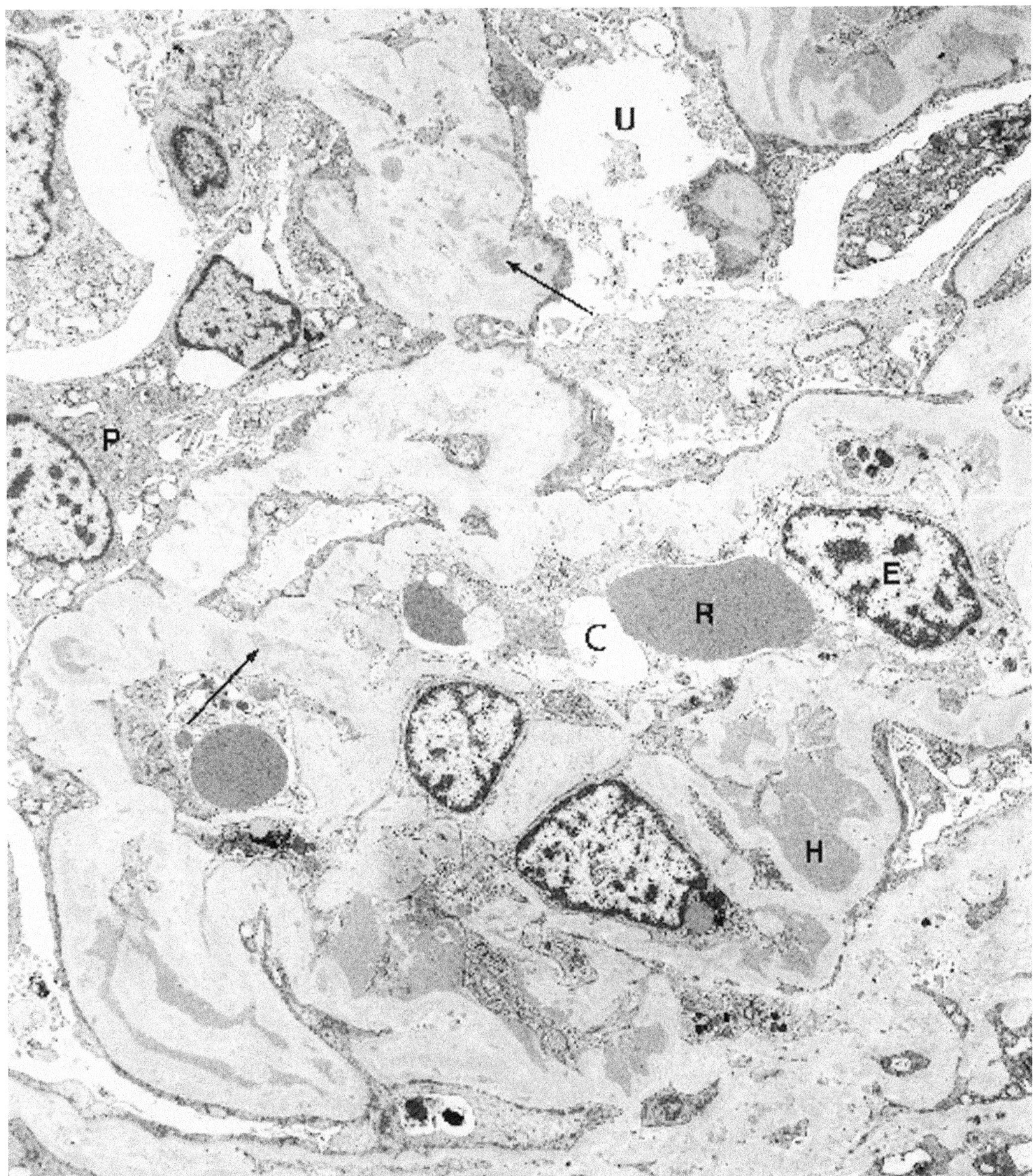

Figure 12.44. Lupus nephritis, WHO class VI, advanced sclerosing type (23-year-old female with history of class IV, diffuse proliferative lupus nephritis 1 year ago). Widespread irregular thickening of the GBM with hyalinosis (H) and sclerosis, scattered residual deposits (*arrows*), and diffuse podocyte foot process effacement. R = red blood cell; U = urinary space; P = podocyte; C = capillary lumen; E = endothelial cell. (× 5300)

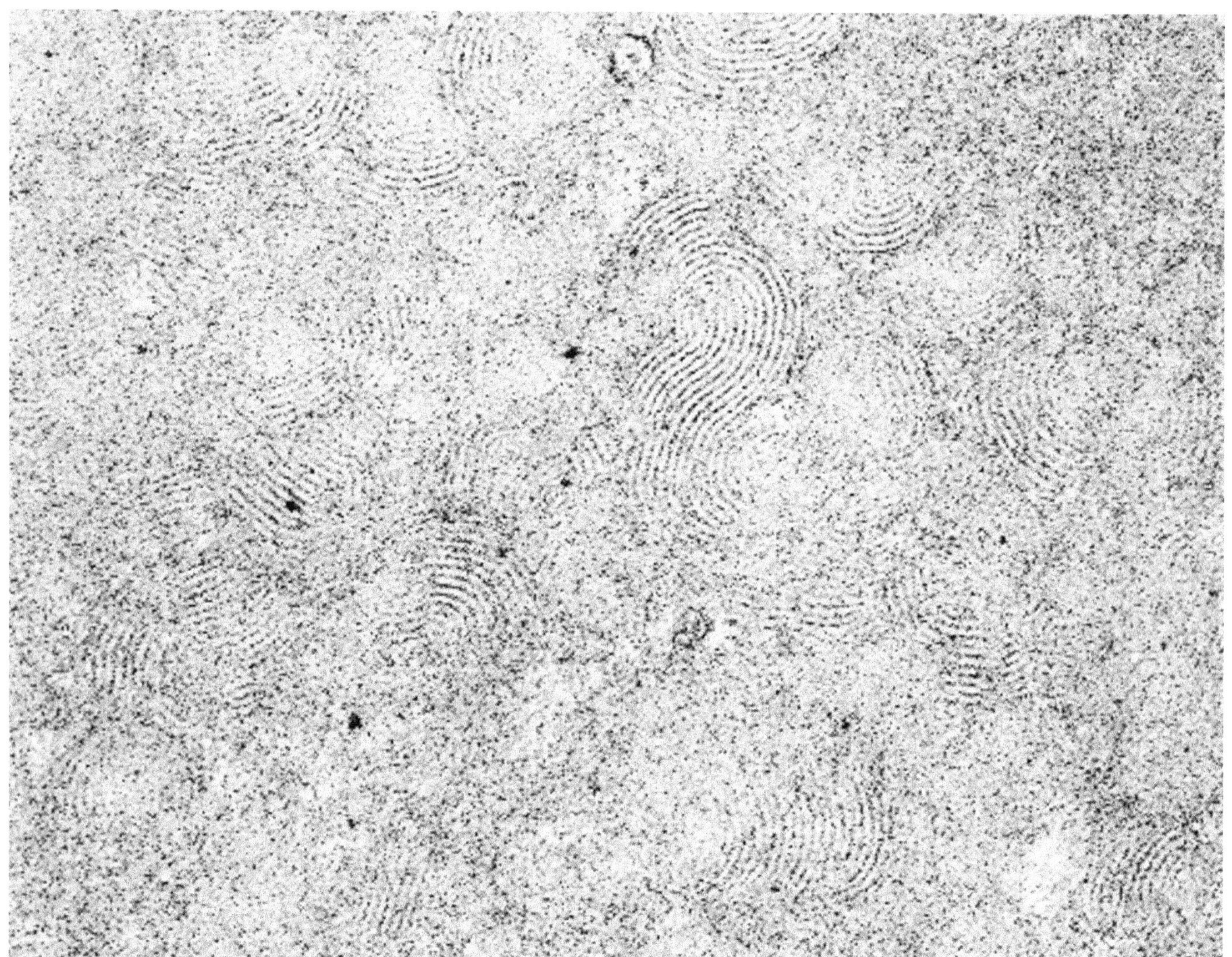

Figure 12.45. Lupus nephritis, deposit substructure (same patient as in Figure 12.41). Fingerprinting or curvilinear pattern of arrangement of microtubules (diameter 18.1 nm) in the deposits, seen at high magnification, are characteristic of lupus deposits. (× 73,000)

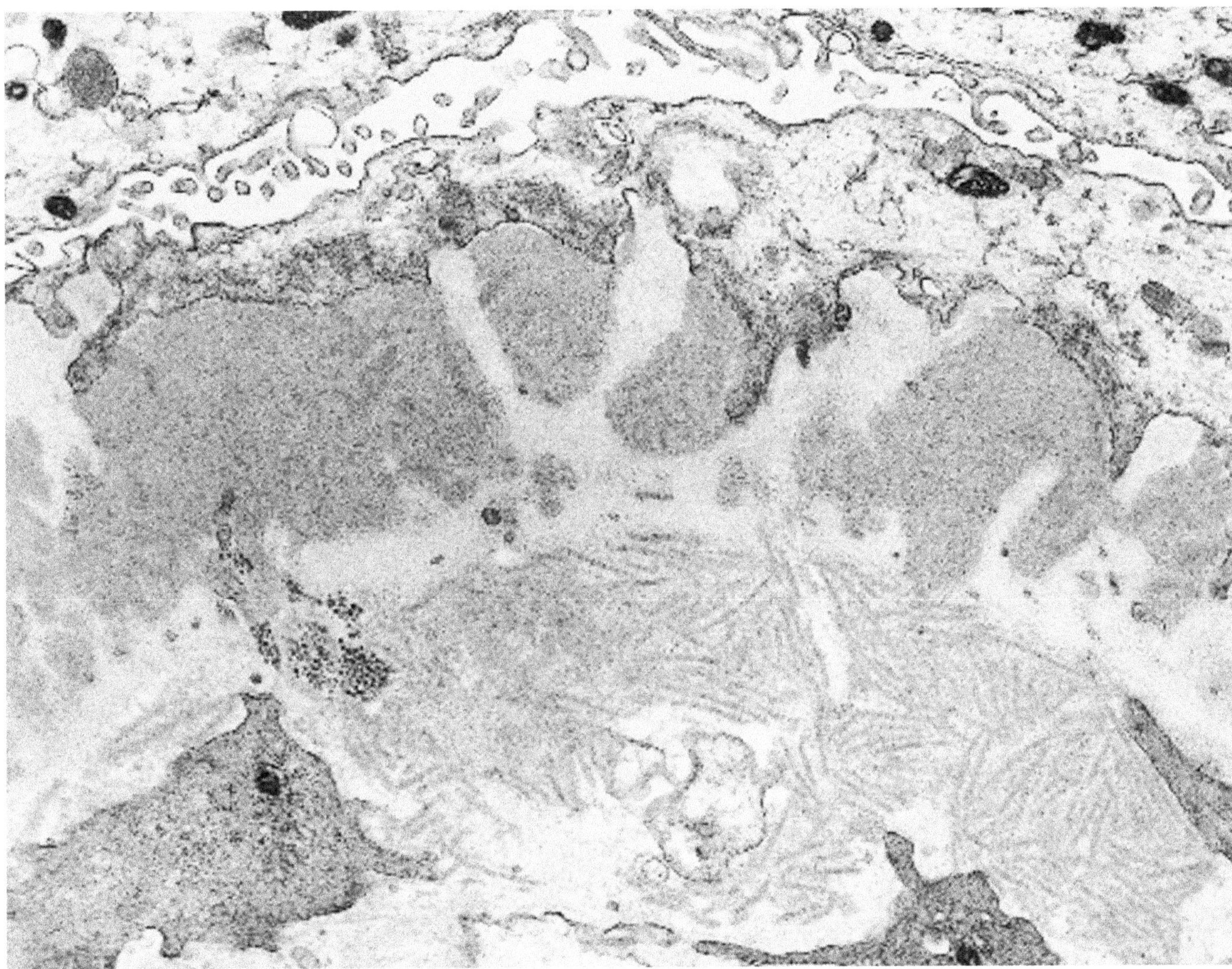

Figure 12.46. Lupus nephritis, deposit substructure (49-year-old female with systemic lupus erythematosus and 8 g/day proteinuria over 1 month). Unusual paracrystalline microtubular substructural detail of these deposits (average outer diameter of each microtubule, 23 nm). Although they may represent cryoglobulins, the patient did not have detectable cryoglobulins in the serum. (× 21,000)

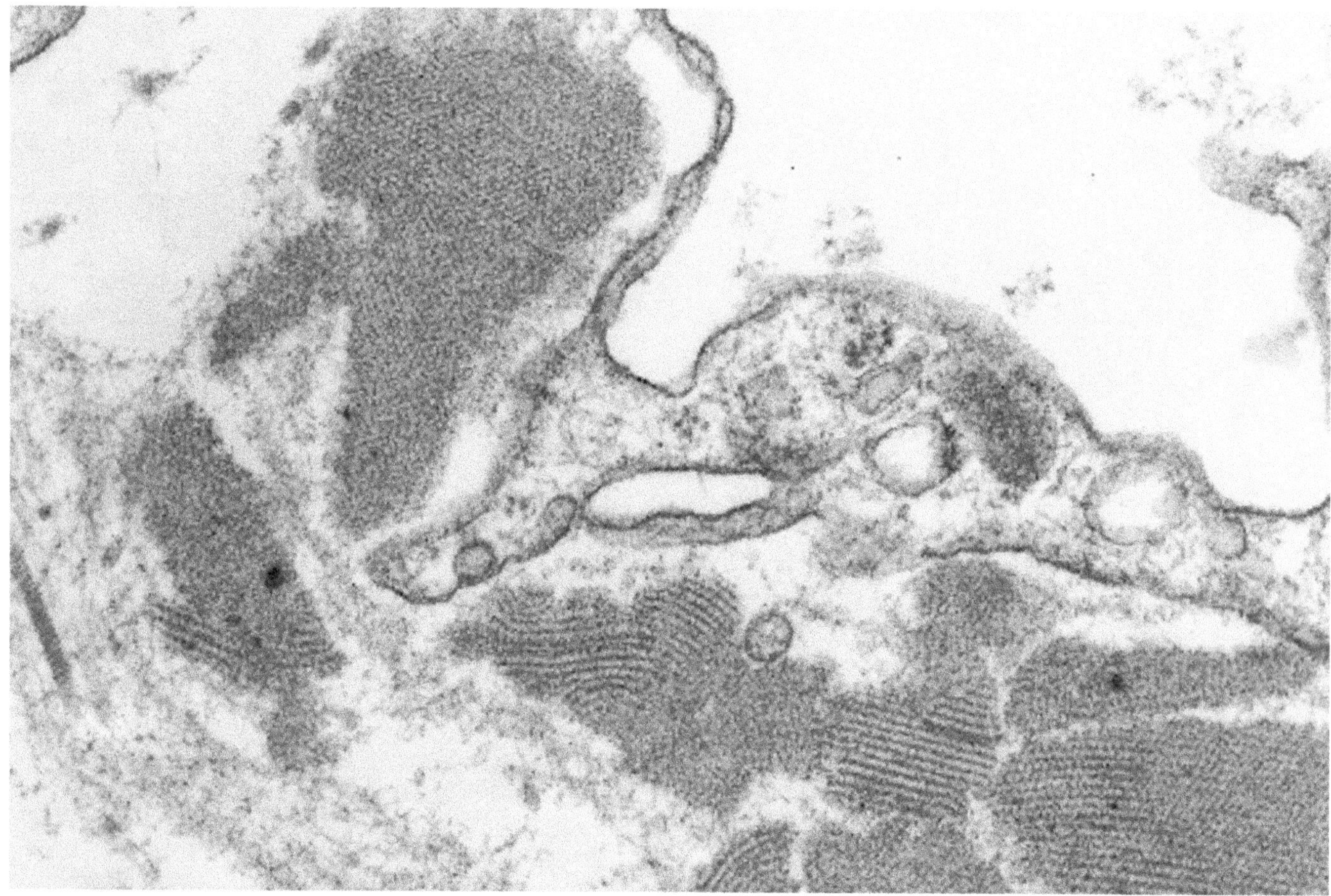

Figure 12.47. Lupus nephritis, deposit substructure (36-year-old female with systemic lupus erythematosus, hematuria, proteinuria, and recurrent leucocytoclastic vasculitis). Another case with unusual paracrystalline microtubular detail of the deposits at high magnification; each microtubule has an average diameter of 20.6 nm. These appear to form an organized sheet-like structure, unlike the scattered separated microtubules of case in Figure 12.46. (× 63,000)

(Text continued from page 825)

Tubuloreticular Inclusions and Tubular Confronting Cisternae

(Figures 12.49 through 12.51; see also 12.39, 12.40, 12.42, and 12.12.)

Diagnostic features. (1) *Tubuloreticular inclusions (TRIs)* (Figure 12.49) are membrane-bound, interlacing fine microtubular structures (20–25 nm in diameter) found in the kidney in the cytoplasm of endothelial cells of glomerular capillaries (Figures 12.39, 12.40, and 12.42) and peritubular capillaries (Figure 12.51). They have also been described in lymphocytes (Klippel et al. 1985; Rich 1981), peripheral blood mononuclear cells (Rich 1981), interstitial cells, and rarely mesangial cells and epithelial cells. They have been noted in pulmonary vascular endothelium in patients with systemic lupus pneumonitis (Lyon et al. 1984). (2) *Tubular confronting cisternae* (TCC) (Ghadially 1988) or cylindric confronting cisternae (not shown) are tubular or test-tube-shaped structures up to 2000 nm in length and 180–300 nm in diameter (ringed on cross-section) that are located in the cell cytoplasm. They appear as two dense lamellae forming a tubular structure, with cytoplasmic material in the hollow of the tube. Each dense lamina (about 20 nm) is sandwiched by two apposing (confronting) rough endoplasmic reticulum cisternae. They are found in conditions and locations similar to the TRI.

Additional points. Once thought to represent viral particles (they resemble paramyxovirus), TRIs are now regarded as a reaction to alpha-interferon (Rich 1981; Rich et al. 1986) and may represent aggregates of interferon molecules. They can be induced *in vitro* in cells by alpha-interferon, but not gamma-interferon (Rich et al. 1983). The interferon (endogenous) in systemic lupus erythematosus and HIV is an unusual alpha-interferon in that it is acid labile like gamma-interferon, but is neutralized by antibodies to alpha-interferon. The authors have seen a case with tubuloreticular inclusions devel-

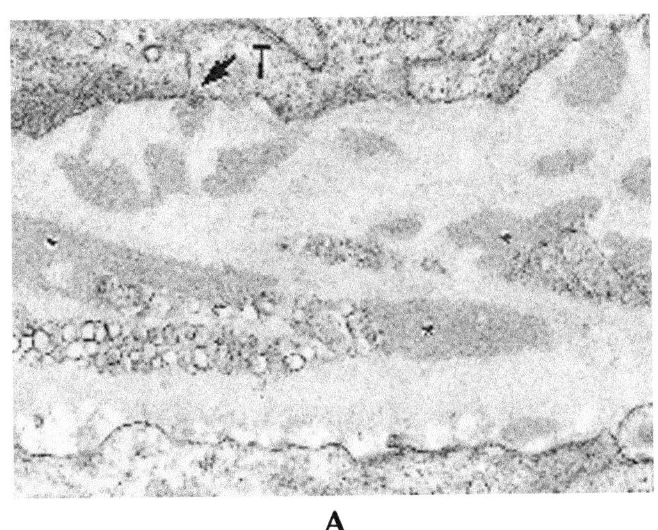

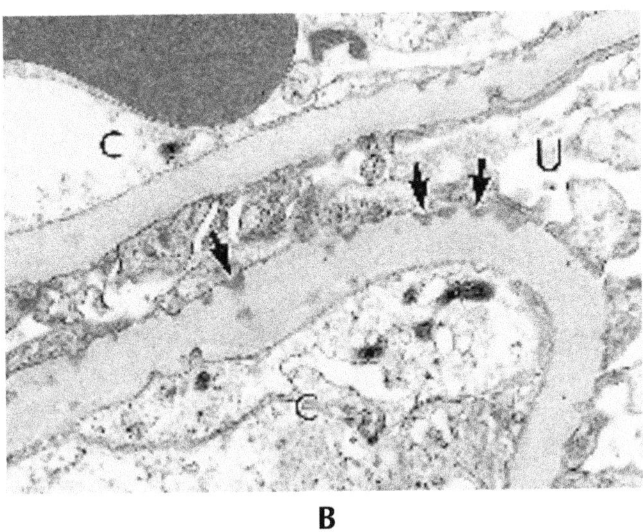

Figure 12.48. Lupus nephritis. **A,** Electron-dense deposits in tubular basement membrane. Deposits (*arrow*) are between the tubular epithelial cell (T) and the tubular basement membrane. Others are in the tubular basement membrane or the interstitium (*). (× 21,630) **B** (33-year-old female with lupus nephritis WHO class IIB; prominent mesangial deposits and increased mesangial cellularity [not shown], Wedge-shaped subepithelial deposits (*arrows*); such triangular, wedge-shaped subepithelial deposits, usually more elongated and penetrating (not shown), are frequently seen in association with lupus nephritis. U = urinary space; C = capillary space. (× 15,800)

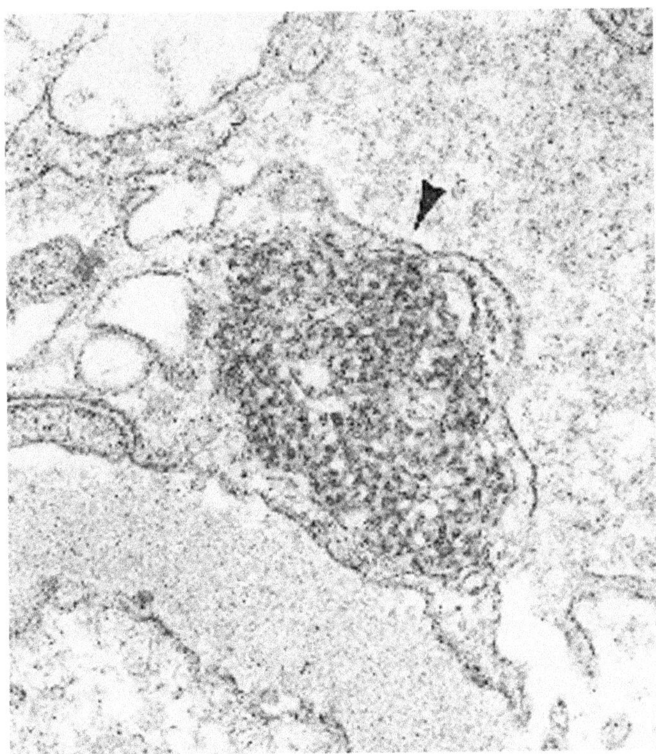

Figure 12.49. Tubuloreticular inclusions or structures (*arrowhead*). These inclusions are present in the endothelial cytoplasm and consist of anastomosing microtubules 20–25 nm in diameter with a limiting membrane that is continuous with endoplasmic reticulum. These are seen in association with high levels of circulating alpha-interferon. (× 41,200)

oping after exogenous alpha-interferon use (Figures 12.50 and 12.51). Schaff et al. (1973) demonstrated the presence of primarily phospholipid and acidic glycoprotein in the inclusions that were not digested by ribonuclease or deoxyribonuclease, arguing against a viral origin. Many have also considered TRIs as cell degenerative products. The exact nature of TRIs is not known as they have not yet been purified or characterized at the molecular level. More recently, Rich (1995) demonstrated in Raji and Daudi cell lines that alpha-interferon induces tubuloreticular inclusions in association with a 36-kD protein (p36), a candidate protein for further characterization. Although not specific, TRIs are most frequent in systemic lupus erythematosus and HIV infection/AIDS (Figure 12.12); they can also be seen with exogenous alpha-interferon therapy, as mentioned earlier.

TCC (intracytoplasmic) have been found in lymphocytes, monocytes/macrophages, and endothelium in various diseases such as adult T-cell leukemia, multiple sclerosis, hepatitis C virus (HCV) infection in animal models and in human lupus nephritis (Venkataseshan et al. 1991) and viral infections (HCV and HIV). TRIs and TCC may be created by the same stimulus, possibly viral induction, and the latter may be derived from TRIs (Luu et al. 1989).

In lupus nephritis, TRIs and TCC are associated with active glomerular and tubulointerstitial lesions containing abundant deposits, in clinically active recent-onset disease or during exacerbation (Venkataseshan et al. 1991).

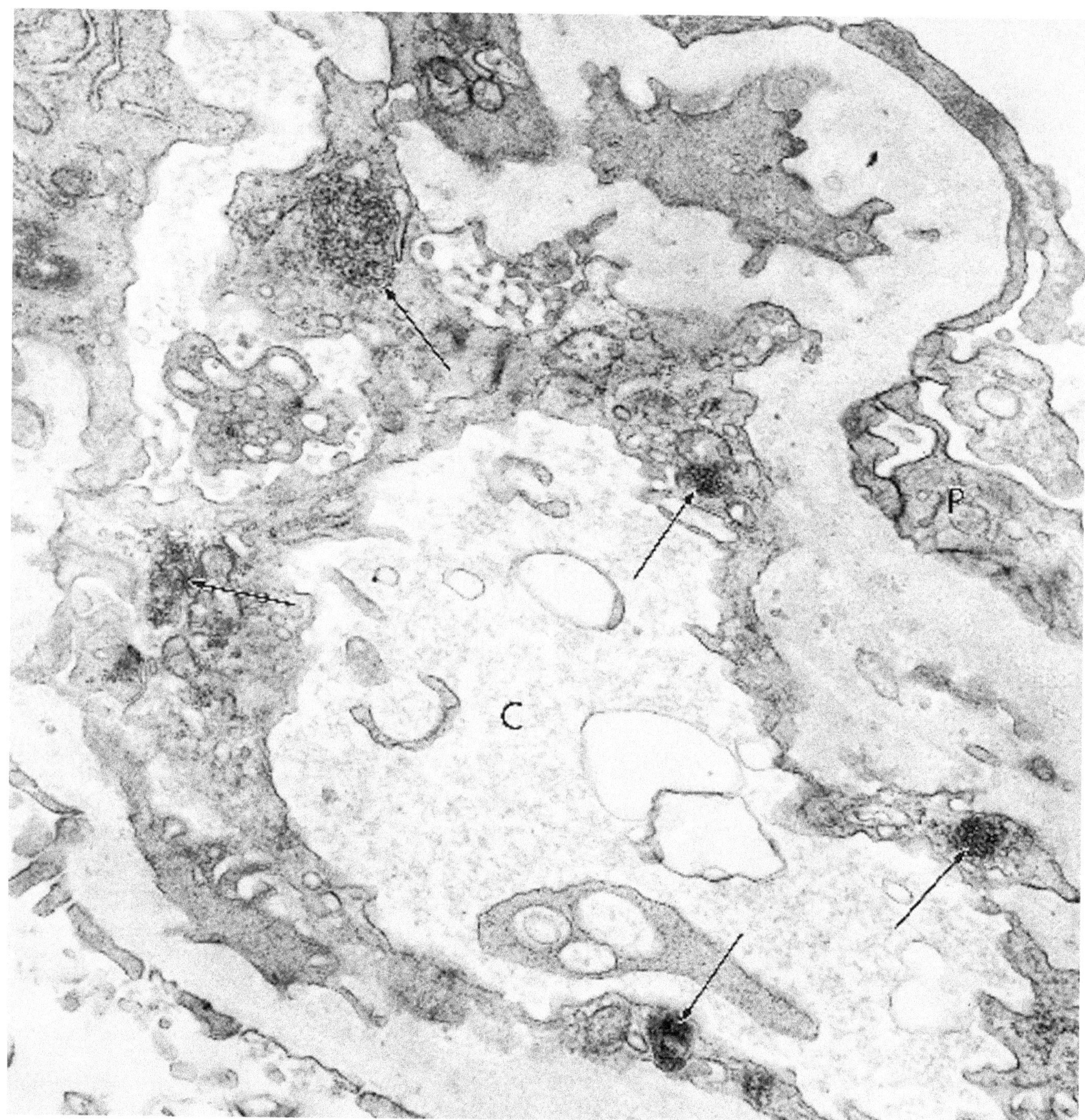

Figure 12.50. Tubuloreticular inclusions (same case as in Figure 12.30; patient with MPGN type I, who was given exogenous alpha-interferon therapy). Capillary loop (C) with numerous tubuloreticular inclusions in endothelial cytoplasm (*arrows*). P = podocyte. (× 26,000)

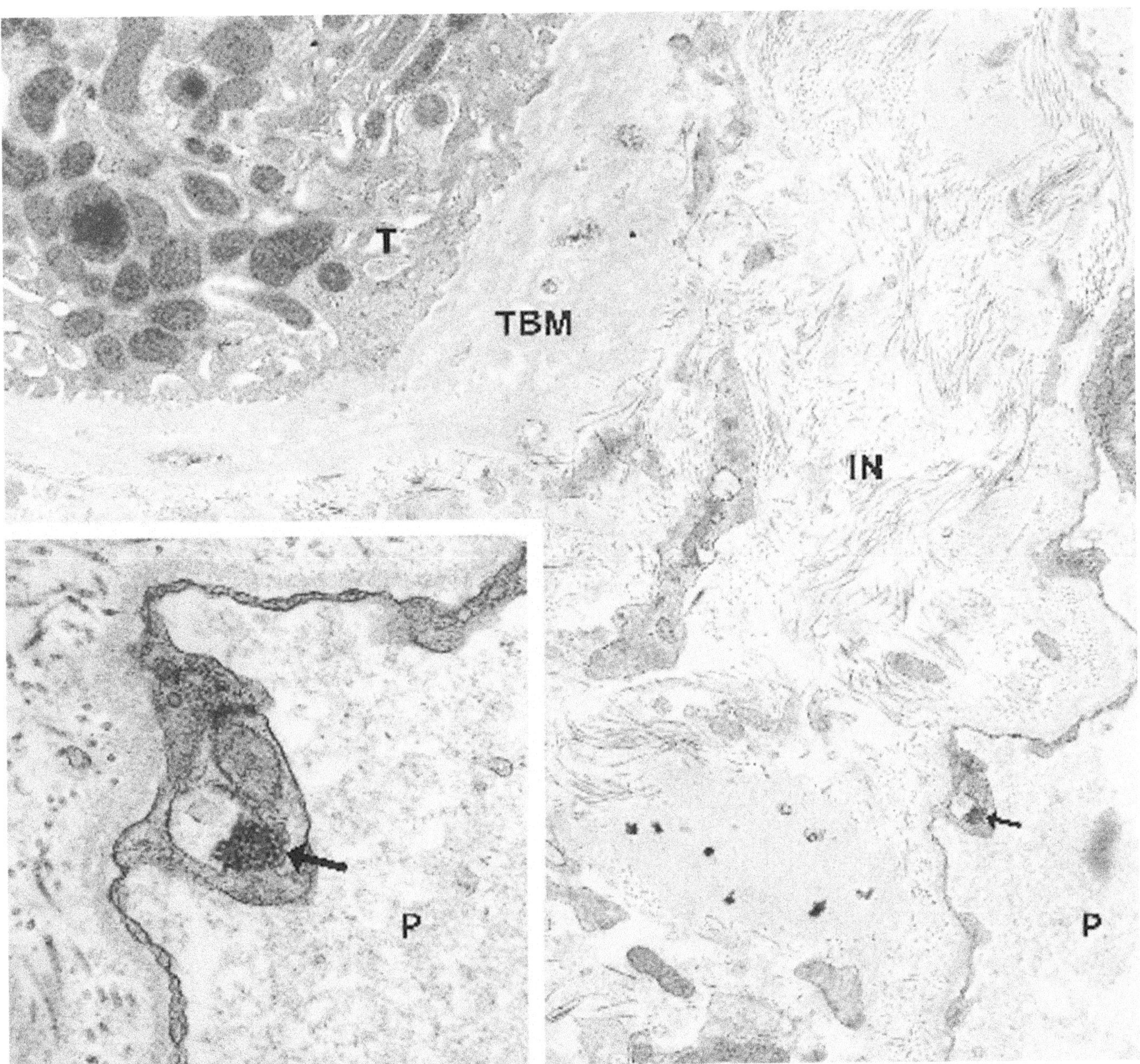

Figure 12.51. Tubuloreticular inclusions (same case as in Figure 12.30). Tubuloreticular inclusion bodies (*arrows*) in a peritubular capillary (P) endothelial cell cytoplasm. IN = interstitium; TBM = tubular basement membrane; T = tubule cell. (× 9800) *Inset* shows higher magnification of the inclusions (× 31,000)

Diseases with Distinctive Ultrastructural Deposits

Electron microscopy is necessary to detect and define deposits with unusual ultrastructure.

Dense-Deposit Disease (Membranoproliferative Glomerulonephritis, Type II)

(Figures 12.52 through 12.54.)

Diagnostic criteria. (1) Dense-deposit disease (Galle 1962; Berger and Galle 1963; Churg et al. 1979) is characterized by accumulation in the GBM of uniquely electron-dense material in a continuous, elongated, ribbon-like pattern or in small nodular aggregates within the irregularly thickened lamina densa (Figures 12.52 and 12.53). These dense deposits usually involve long segments of basement membrane, but occasionally only a few loops are involved; (2) the same type of deposit is characteristically also present in the mesangium, Bowman's capsule (Figure 12.52), tubular base-

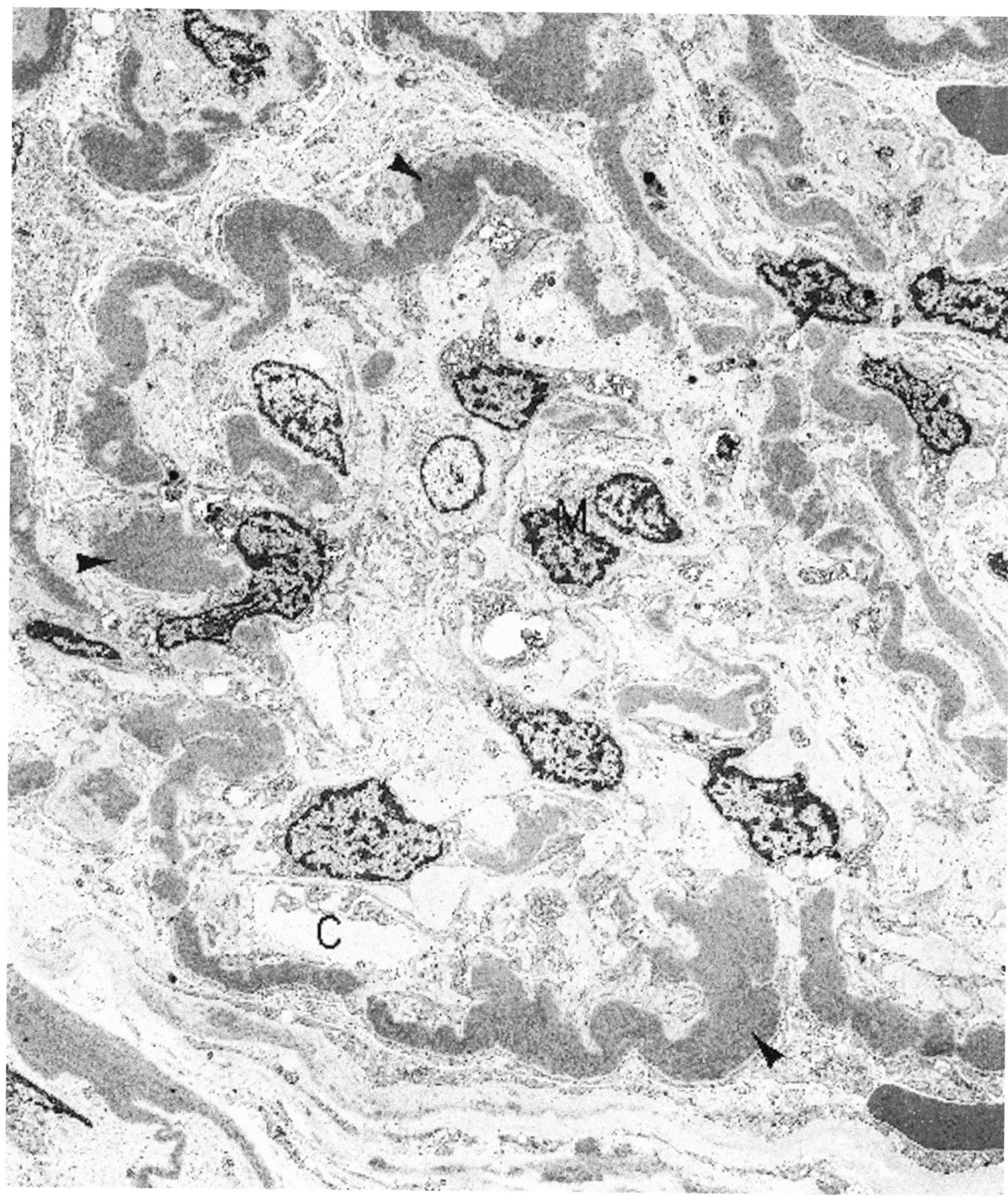

Figure 12.52. Dense deposit disease (MPGN type II) (17-year-old girl with nephrotic syndrome and hypocomplementemia). Extensive elongated, broad, and continuous deposits of highly electron-dense material are present in the GBM in an intramembranous location (*arrowheads*). The mesangial cells (M) are hyperplastic, and the mesangium is expanded with scattered deposits. C = capillary lumen. Deposits along Bowman's capsule are seen in the lower left corner. (× 3600)

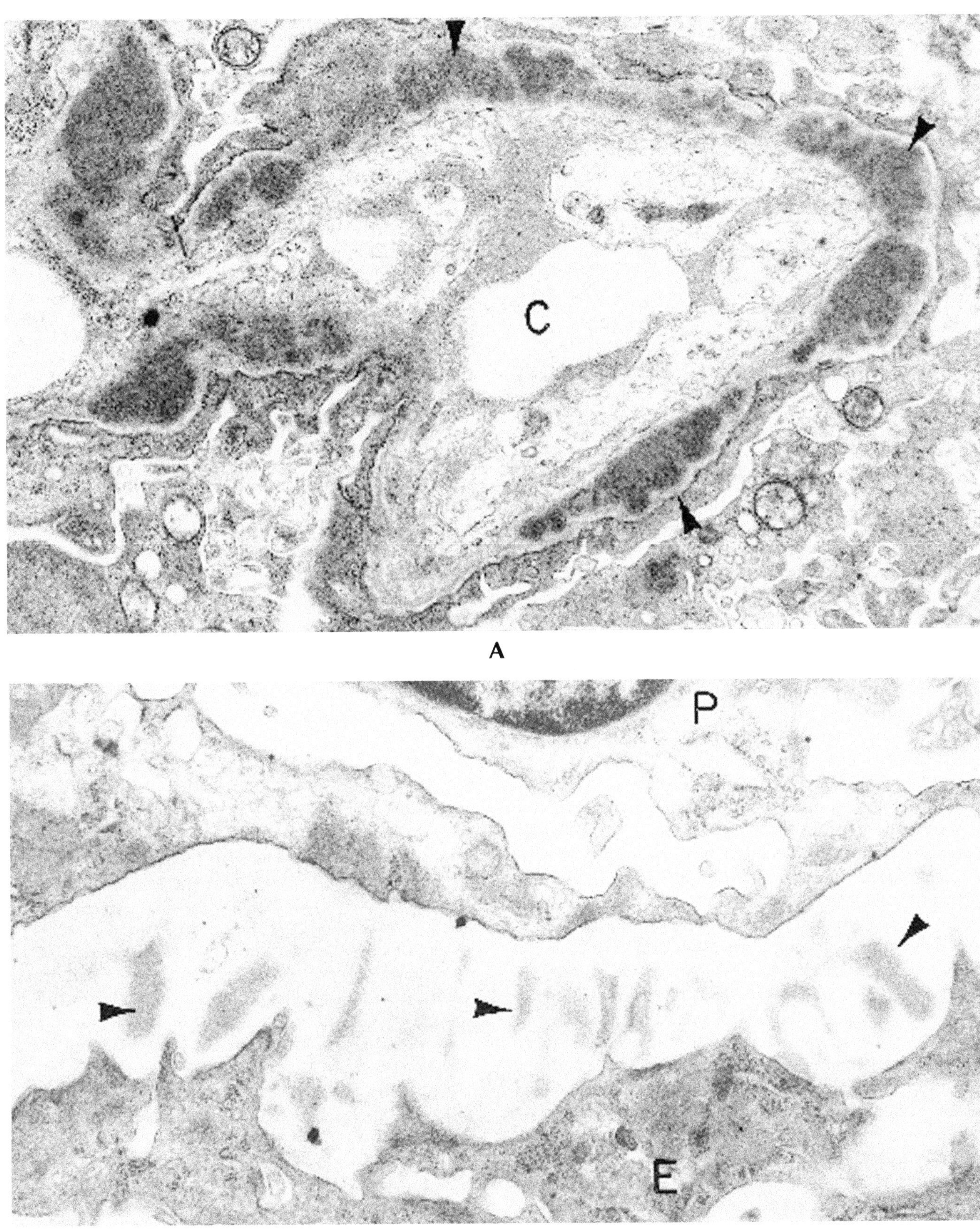

Figure 12.53. Dense-deposit disease (MPGN type II). **A** (8-year-old boy), Irregular electron-dense deposits (*arrowheads*) permeate the GBM. C = capillary lumen. (× 17,000) **B** (6-year-old girl with early dense deposit disease; father had classical dense-deposit disease), Small, irregular electron-dense deposits (*arrowheads*) occupy the GBM. P = podocyte, E = endothelium. (× 24,400)

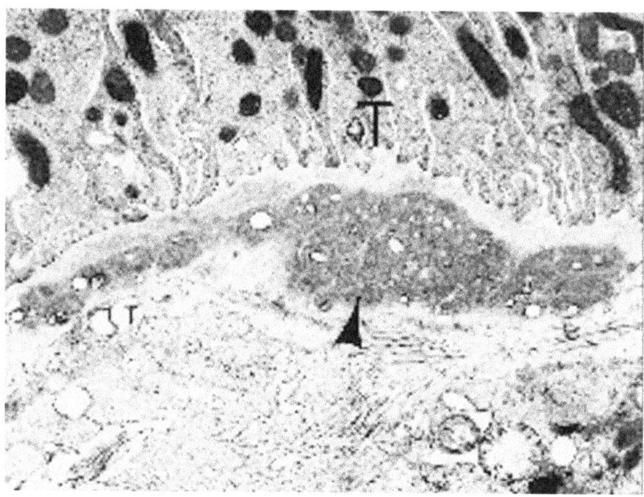

Figure 12.54. Dense-deposit disease (MPGN type II) (14-year-old girl with low C3, hematuria, and proteinuria). Electron-dense material in tubular basement membrane (*arrowhead*). T = proximal tubular cell. (× 8000)

ment membranes (Figure 12.54), and occasionally in the walls of arterioles and peritubular capillaries; (3) similar dense deposits have been noted in splenic sinusoidal basement membranes (Thorner and Baumal 1982) and in the eye involving the choriocapillaris basement membranes and Bruch's membrane (Duvall-Young et al. 1989). Retinal pigment alterations have also been described (Michielsen et al. 1990–1991).

Additional points. On scanning electron microscopy of acellular GBMs from patients with dense-deposit disease, Weidner and Lorentz (1986) found the GBM to be uniquely rigid and thickened, corresponding to areas containing the dense intramembranous deposits. They also found scattered crater-like deformities along the epithelial side of the GBM, corresponding to subepithelial deposits. Variable proliferation of mesangial cells with increased mesangial matrix may be present. Foot processes may be effaced. Occasionally widespread crescents may be seen, simulating rapidly progressive glomerulonephritis. In the early stages of the disease, there may be neutrophils in capillaries and fine granular subepithelial deposits similar in shape and electron density to humps, resembling postinfectious glomerulonephritis. Spikes rarely are formed by the glomerular basement membrane. These amorphous subepithelial deposits are less dense than the "dense deposits." The disease recurs in 90% of transplants; characteristically, the dense deposits precede detectable C3 accumulation, suggesting that C3 deposition is a secondary event.

The exact nature of these dense deposits is not known. Controversy exists whether these represent true deposits or lamina densa alterations. Various factors have been implicated in their composition, including glycoproteins, properdin, and *N*-acetylglucosamine. They do not, however, contain immune complexes, complement components, collagen, or the C3 nephritic factor. This material does not react with Goodpasture antibodies. Muda et al. (1988) suggest the presence of a highly osmiophilic lipid component in these dense deposits without any glycoprotein alterations. They showed that the electron density of dense deposits was higher than normal lamina densa, due to the stronger osmium affinity of the former. Support for a systemic factor in the etiology of this disease comes from presence of extrarenal deposits and very high recurrence rate after transplant as well as its association with partial lipodystrophy (disease of lipid metabolism).

By light microscopy, there usually is refractile thickening of the GBM, with a variable increase in mesangial cellularity and matrix. The membranoproliferative pattern is not found in all the cases, and the histology is variable; Sibley and Kim (1984) described focal segmental, necrotizing glomerulonephritis in 5 of 16 patients with dense-deposit disease. Ultrastructurally, the dense deposits in this group were in paramesangial GBM and mesangium. Davis et al. (1977) described two cases of a form of mesangial proliferation associated with hypocomplementemia. Irregular intramembranous electron-dense deposits, separated by varying lengths of normal-appearing basement membrane, were present primarily in the lamina densa (Figure 12.53B). Subendothelial deposits, mesangial interposition, and GBM splitting were not observed, and it was thought that when present, these features probably were variants of dense-deposit disease. This disease is thus better classified as dense intramembranous deposit disease (distinct from both MPGN type I and III) rather than MPGN type II.

Immunofluorescence shows the presence of C3 in peripheral capillary loops, usually as double linear or railroad-track-like staining of the borders, but not the deposit itself. C3 staining may be seen in the mesangium (mesangial rings), along the tubular basement membrane or Bowman's capsule. Occasionally IgM, IgG, or IgA may also be noted. Occasional cases are familial, and some may be related to infection (e.g., meningococcus) or to acquired partial lipodystrophy (Swainson et al. 1983). In terms of natural history, McEnery and McAdams (1988) showed, in four of six pediatric patients followed for an average 14 years, movement of dense deposits from lamina densa toward lamina interna ("dropping off") and then toward mesangial matrix where it was cleared by mesangial cells. The appearance of new lamina densa further suggested normal slow turnover of GBM.

Amyloidosis

(Figures 12.55 through 12.58A.)

Diagnostic criteria. Numerous haphazardly arranged, nonbranching fibrils about 7.5–10 nm in diameter and up to 1 μ in length (Figure 12.58A) (Dikman et al. 1981). Because other fibrils may resemble amyloid, a positive Congo red stain by light microscopy is required for diagnosis.

Additional points. The fibrils may have a beaded appearance with a periodicity of 5.5 nm; however, this feature is usually not seen. The fibrils may be loosely arrayed or form haystack-like organization perpendicular to basement membranes. Amyloid deposition is predominately extracellular. Amyloid deposits may be found in glomeruli, blood vessels, interstitium, and peritubular basement membrane (Jones et al. 1986). In the glomerulus, amyloid fibrils are deposited in the mesangium and later extend from the mesangium along and into the GBM (Figure 12.55). Fibrils in the subepithelial space often are arranged in parallel arrays or bundles, perpendicular to the GBM (haystacks) (Figure 12.56A) and have silver affinity by light microscopy appearing as long, irregular "spikes." These spikes or spicules are different from those in membranous glomerulonephritis, where the GBM in between the deposits form the shorter and more regular spikes that stain with silver. The argyrophylic subepithelial spicular amyloid deposits may be a result of amyloid interaction with renal epithelial basement membrane glycoprotein (Notling and Campbell 1981). Such spicules may also be present in Bowman's capsule or inside the tubular basement-membrane (Notling and Campbell 1981), sites where spikes are not seen in membranous glomerulonephritis. Podocyte foot process effacement and loss of endothelial cell fenestrae, especially adjacent to deposits, are readily identified. Amyloid deposits in glomeruli distort the normal architecture, leading to expansion of mesangium, widening of basement membranes, and subsequent obliteration of capillary lumens and cells. In later stages, sclerosis results with residual deposits.

Tubular amyloid deposits are seen along the tubular basement membrane (Figure 12.57), most prominently along distal tubules and loops of Henle. Associated interstitial and vasa recta deposits contribute to eventual tubular atrophy. Certain rare hereditary forms have longer and more irregular, predominantly tubulointerstitial deposits. Vascular amyloid deposits involve the media and adventitia, relatively sparing the intima and lumens initially, but later obliterating the lumens. Arteries and veins of any caliber may be affected. For a comparison between the different types of fibrils and microtubules seen in amyloid and nonamyloid glomerulopathies, see Figures 12.58 and 12.59.

The light microscopy for renal amyloidosis parallels the location of deposits, forming nodular masses and/or capillary wall thickening in the glomeruli with or without associated tubular basement membrane and vascular wall thickening. The characteristic homogeneous, eosinophilic amyloid deposits by hematoxylin and eosin (H&E) stain appear orange-red by Congo red stain and show apple-green birefringence on polarization. X-ray crystallography demonstrates a cross-beta-pleated sheet structure for the fibrils. Amyloid deposits are weakly PAS-positive but silver-negative. Immunofluorescence studies using antisera against amyloid-associated protein (AA) and light chains help categorize the subtype of amyloid.

Recent classifications define amyloidosis by the chemical nature of the precursor protein [WHO-IUIS Subcommittee 1993; Falk, et al. 1997; Gilmore et al. 1997) and are divided into systemic, localized, and hereditary groups. To date, about 18 different amyloid fibril proteins have been described in humans. Though diverse in protein composition, the unifying four features of all types of amyloid are (1) fibrils, 7.5–10 nm in diameter, (2) cross-beta-pleated sheet structure; (3) Congo-red-induced birefringence; and (4) presence of amyloid P-component. Amyloid P-component, derived from serum amyloid P-component, consists of donut-shaped pentameric units stacked together at right angles to the amyloid fibrils, probably conferring stability. Biochemically, amyloid is composed of about 95% fibril proteins, about 5% P-component and probably lesser amounts of other substances such as heparan sulfate (Moss et al., 1994) and apolipoprotein E. The differing fibril proteins and therefore type of amyloid are categorized into AL (light chain amyloid, primary, which may be localized or systemic), AA (amyloid-associated protein, in secondary amyloidosis and familial Mediterranean fever), ATTR or A-transthyretin (prealbumin, in familial amyloid polyneuropathy, or senile cardiac amyloidosis), A-beta-2-microglobulin (hemodialysis associated), beta-2-amyloid protein (Alzheimer's disease), A-IAPP (islet amyloid polypeptide), and A-calcitonin (medullary thyroid carcinoma). In general, AL and AA types are most frequent.

In renal amyloidosis, AA type accounts for about half the cases, and the AL type accounts for most of the remaining cases (Noel et al. 1987); rare cases are attributed to ATTR (familial amyloid polyneuropathy of Finnish type). AL amyloid is associated with plasma cell dyscrasias, multiple myeloma, and Waldenström's macroglobulinemia. AL amyloid contains monoclonal light chains or light chain fragments and is more frequently associated with lambda (λ) light chains, especially of the lambda variable region (V) subgroup VI type (VλVI), found exclusively in AL(λ) amyloidosis

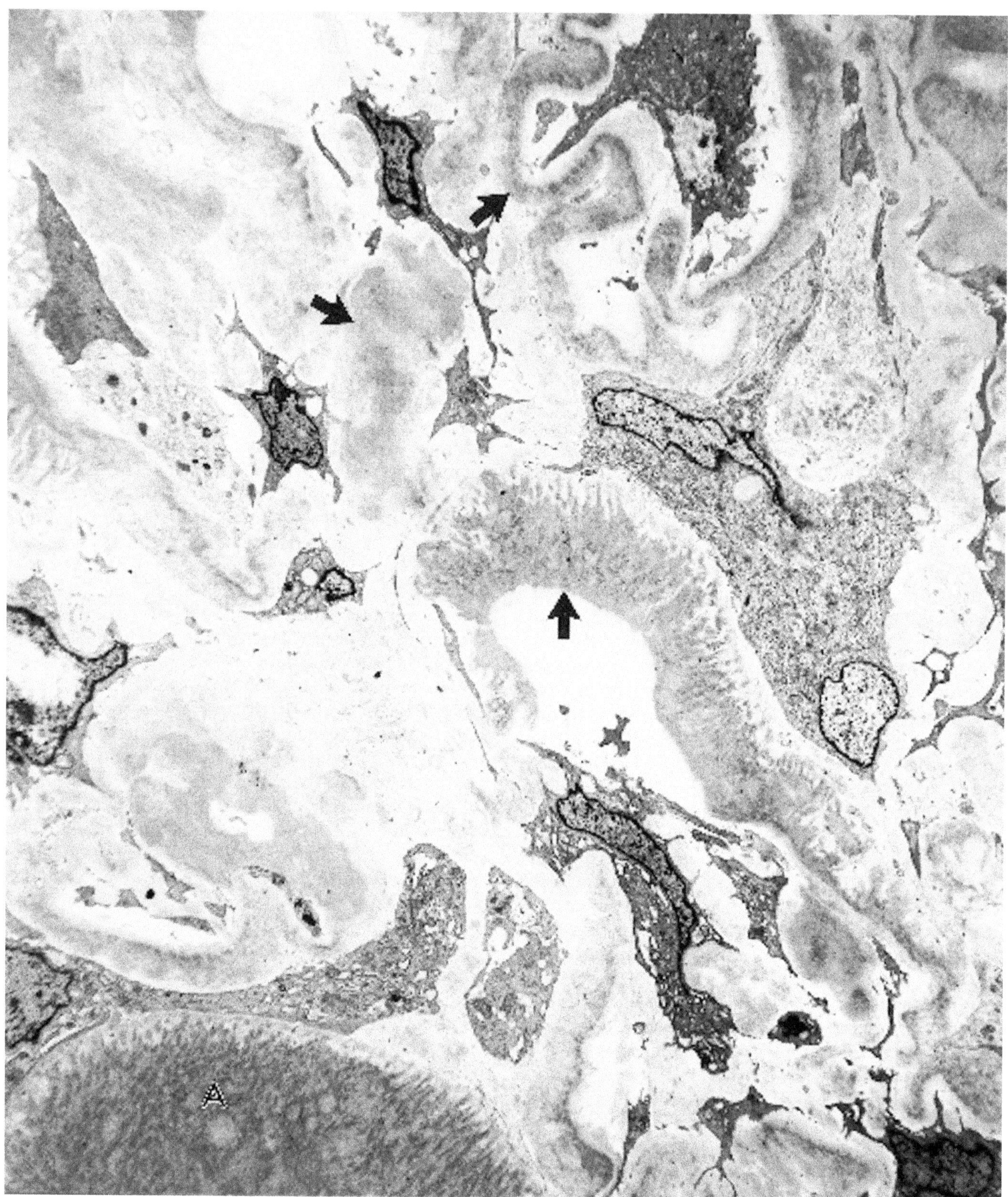

Figure 12.55. Amyloidosis (84-year-old female, 5-year history of hypertension, new-onset nephrotic range proteinuria of 6.7 g/day with associated loss of filtration). Amyloid deposits (*arrows,* A) in the mesangium and along the capillary loops in this glomerulus are obliterating the normal architecture. (× 4500)

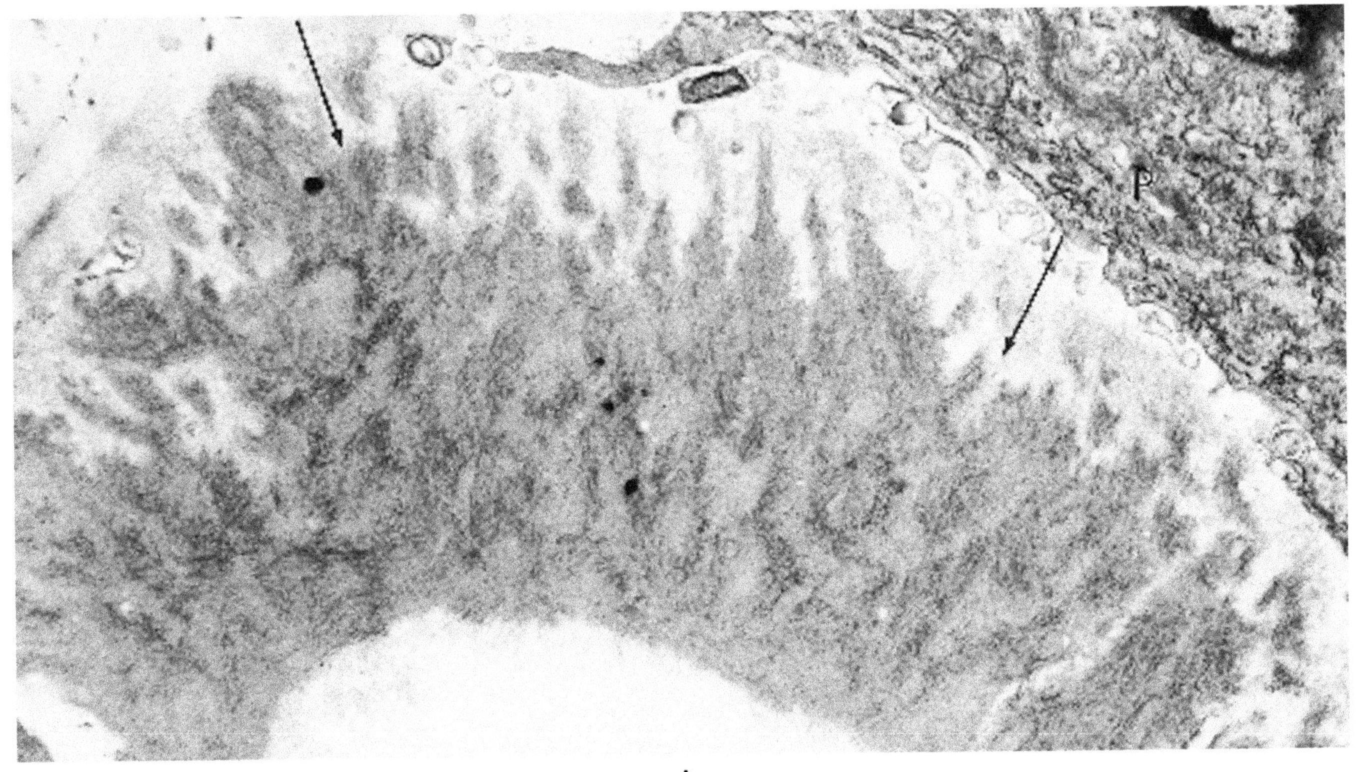

A

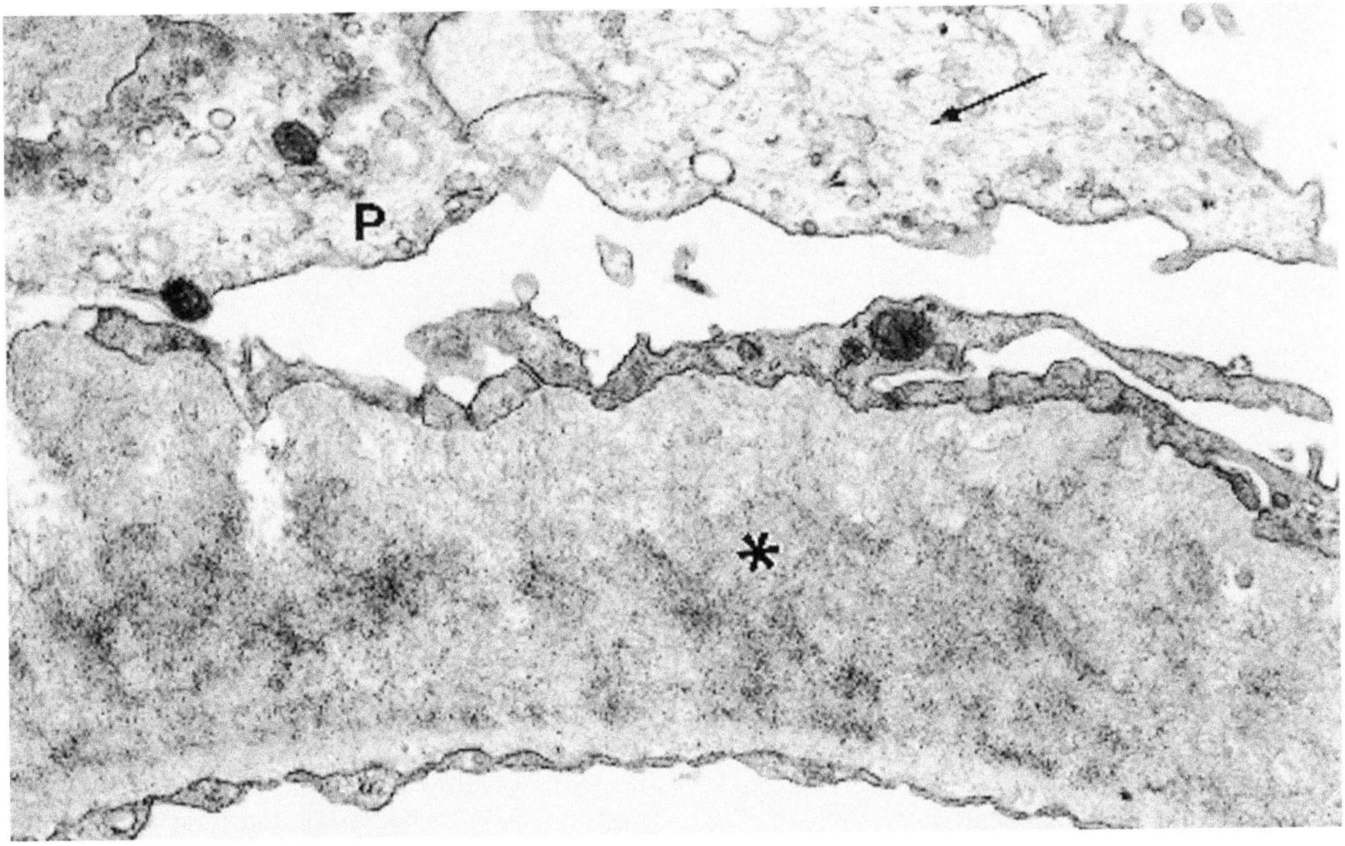

P

*

B

Figure 12.56. Amyloidosis (same case as in Figure 12.55). **A,** Bundles of amyloid fibrils arranged haphazardly, tend to form "haystacks" perpendicular to the GBM (*arrows*). P = podocyte. (× 15,000) **B,** Haphazardly arranged amyloid fibrils along the GBM (*). Note that the fibril diameter or thickness (8 nm) is very close to that of the actin filaments in podocytes (*arrow*), a useful internal measure. P = podocyte. (× 21,000)

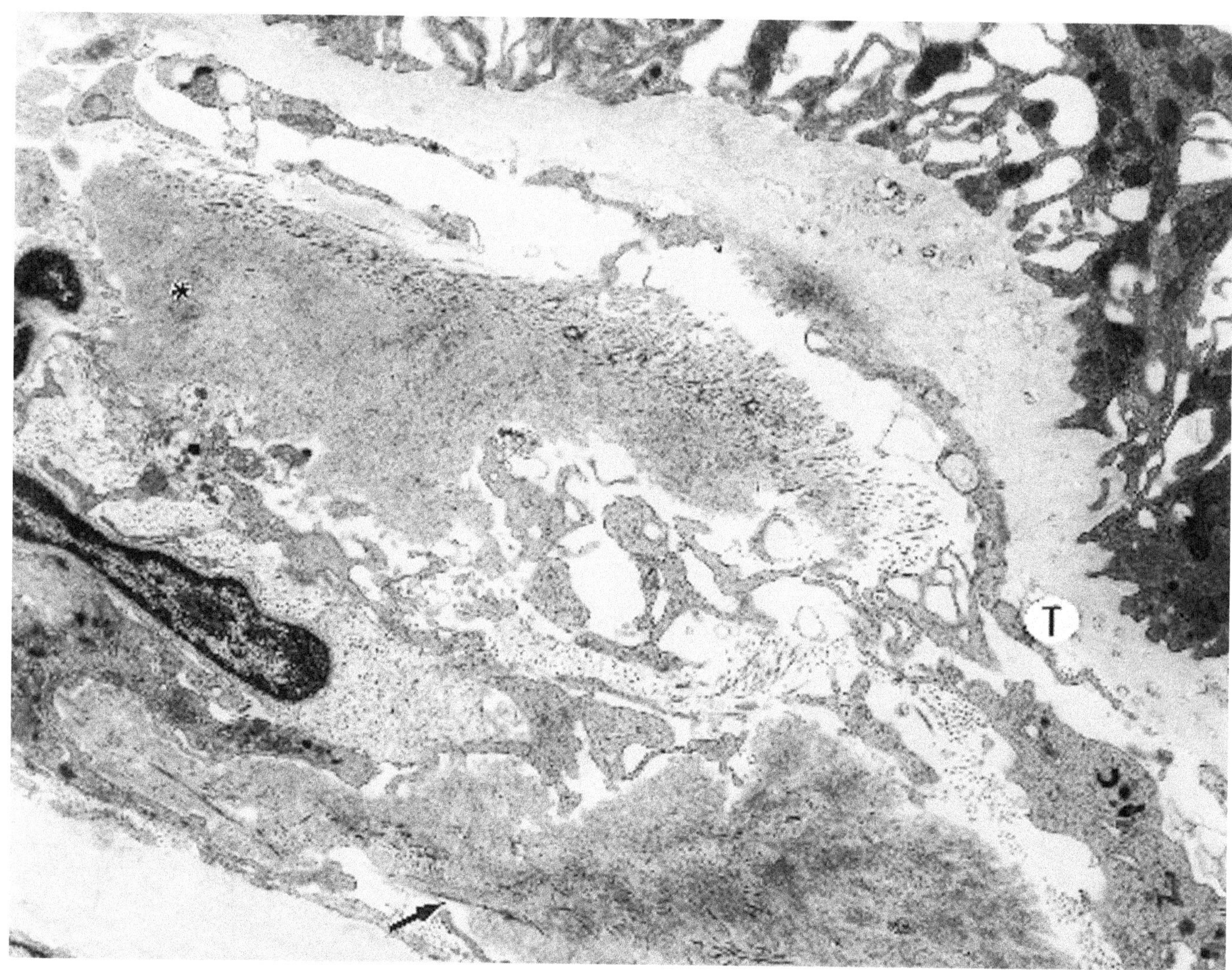

Figure 12.57. Amyloidosis (same case as in Figure 12.55). Interstitial deposits of amyloid fibrils (*) with interspersed collagen fibers (*arrow*). Amyloid fibrils also permeate the tubular basement membrane (T). (× 9800)

(Ozaki et al. 1994). The reason for lambda predominance is not exactly known, but it has been suggested that certain amyloidogenic Bence Jones proteins may have a unique amino acid sequence that allows cleavage of the variable segment forming amyloid fibrils (Glenner 1980). Similarly, certain serum-amyloid-associated proteins would be amyloidogenic. AA is derived from the precursor serum-amyloid-associated protein.

Fibrillary Glomerulonephritis

(Figures 12.60 and 12.58B.)

Diagnostic criteria. (1) Widespread fibrillary deposits in glomeruli that are in the GBM, subepithelial or subendothelial regions, mesangium, and Bowman's capsule; less frequently around tubules (Churg and Venkataseshan 1993); (2) fibril diameter about 13–20 nm, without periodicity or branching, are larger (Duffy et al. 1983) and slightly straighter than amyloid fibrils; (3) fibrils are randomly scattered against a lucent background expanding the mesangium or glomerular capillary wall.

Additional points. Amorphous electron-dense deposits are usually absent, and diffuse podocyte foot process effacement is usually present. At first glance, the fibrils resemble amyloid fibrils; however, at higher magnifications the thicker diameter and straighter fibrils help to separate the two. The distinguishing factor is complete absence of staining by Congo red and thioflavin T stains at light microscopy in these "non-amyloid" fibrillary deposits. They differ from immunotactoid deposits by the lack of thicker, obviously microtubular structures. At extremely high magnifica-

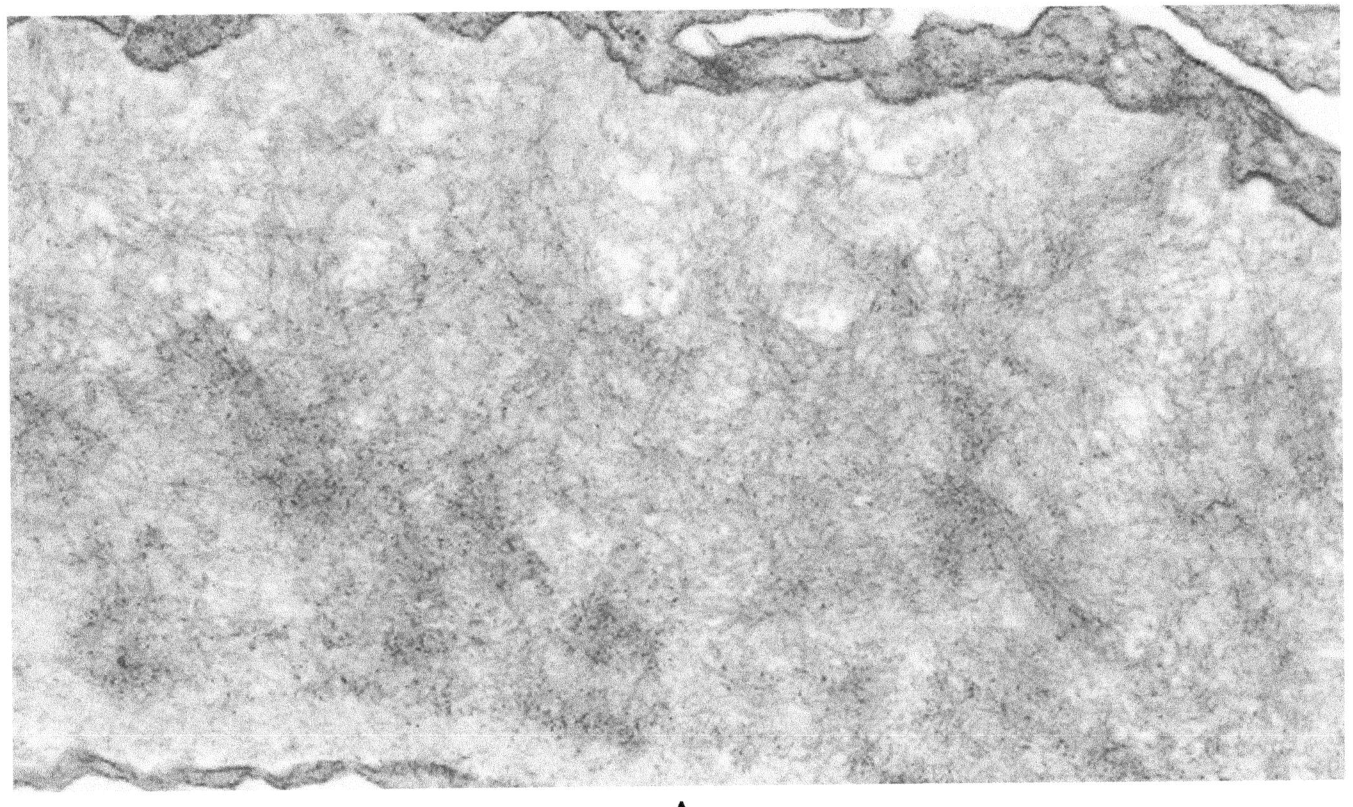

A

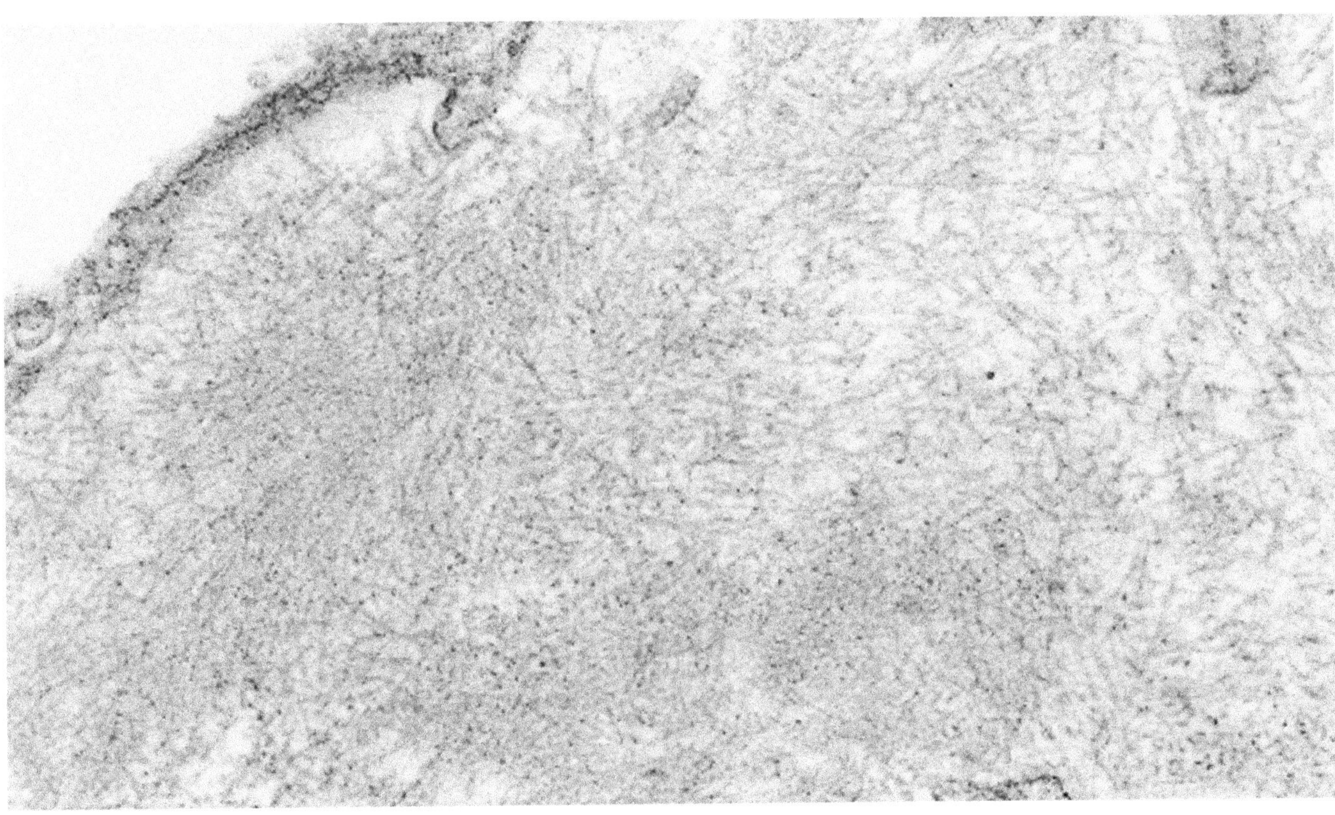

B

Figure 12.58. Comparison of amyloidosis and fibrillary glomerulonephritis. **A,** Amyloidosis (same case as in Figure 12.55). Haphazardly arranged amyloid fibrils that are about 6–8 nm in diameter. Congo red stain was positive. (× 45,000) **B,** Fibrillary glomerulonephritis (same case as in Figure 12.60). Straighter and slightly thicker fibrils, about 12–16 nm in diameter in a lucent background. Congo red stain was negative. (45,000)

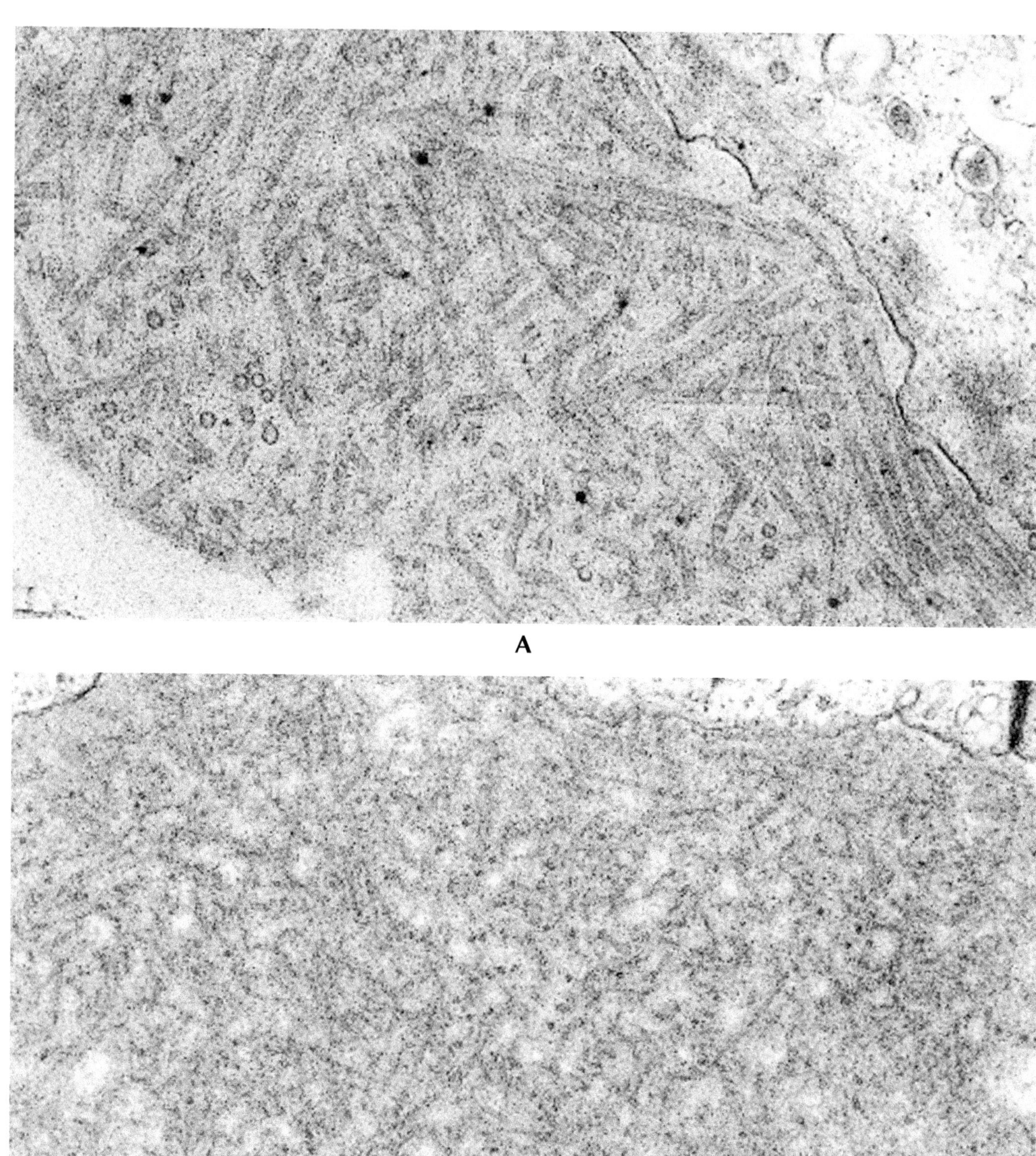

Figure 12.59. Comparison of immunotactoid glomerulopathy and mixed cryoglobulinemia. **A,** Immunotactoid glomerulopathy. These microtubules have a diameter of 48–55 nm, with an electron-lucent center. They are much thicker than the amyloid fibrils. (× 45,000) **B,** Mixed cryoglobulinemia (same case as in Figure 12.62). Straight or curved microtubules, about 20–24 nm in diameter, also have a lucent center. On cross-section, they have 8–12 spokes emanating from the perimeter, making the cross-sectional outer diameter about 33 nm. (× 45,000) Note that unlike amyloidosis and fibrillary glomerulonephritis, the background of these deposits appears amorphous and electron dense.

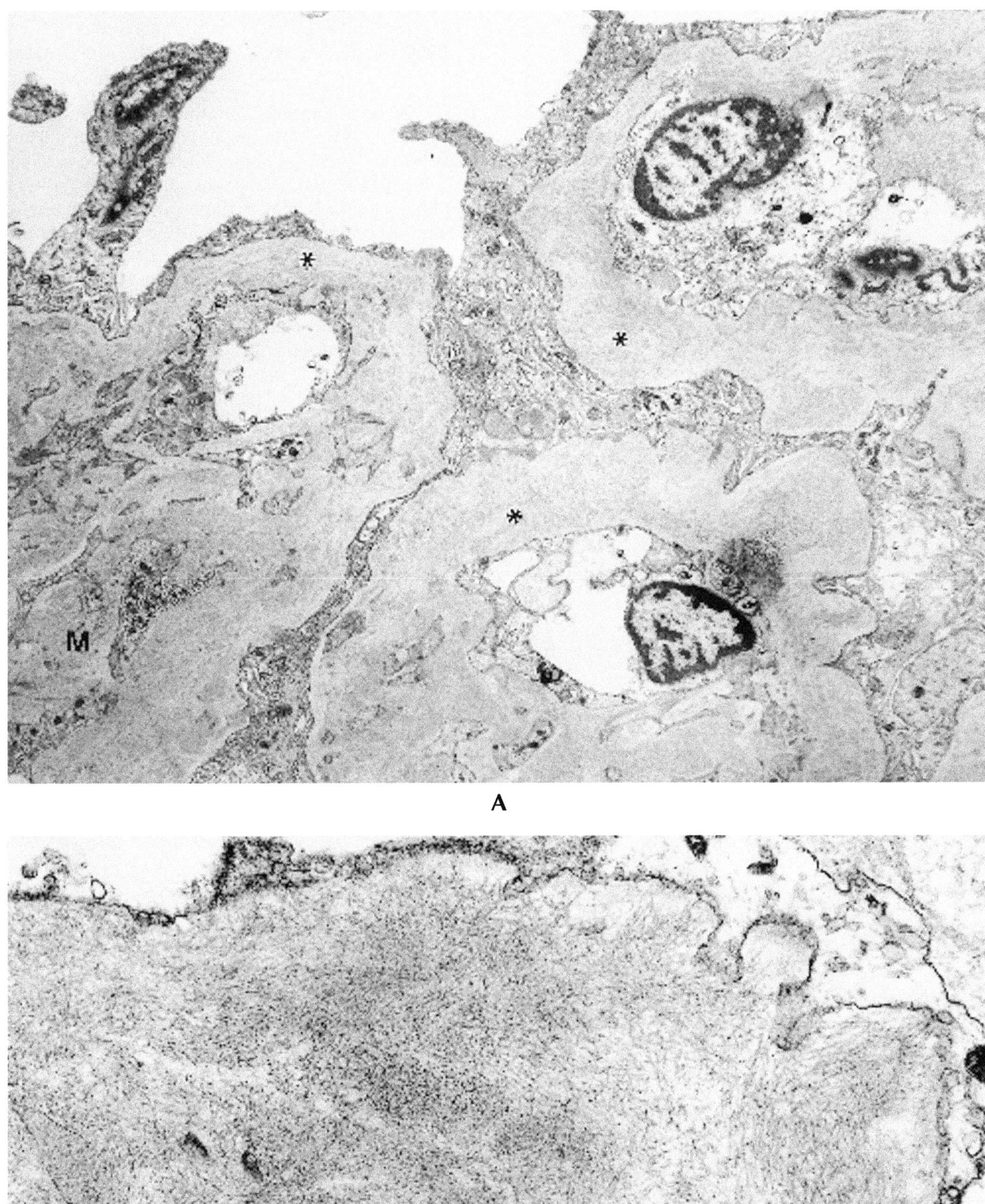

Figure 12.60. Fibrillary glomerulonephritis (54-year-old female with nephrotic syndrome and hyperlipidemia who is negative for hepatitis B or C viruses). **A,** shown are three capillary loops with expansion of the capillary wall by fibrillary deposits (*) that are also present in the mesangium (M). (× 5400) **B,** higher magnification shows irregularly arranged fibrils, about 12–16 nm, against a lucent background. A Congo red stain was negative. (× 24,500)

tions, though, even these thinner fibrils may in fact have a microtubular appearance. A negative serum cryoglobulin test is required for the diagnosis because cryoglobulin deposits can be morphologically indistinguishable from fibrils of fibrillary glomerulonephritis.

Light microscopy is variable and shows mesangioproliferative pattern, crescentic glomerulonephritis or membranous thickening of the glomerular capillary walls. By immunofluorescence, in the majority of cases IgG and C3 staining are found; IgG4 is the dominant subclass (Iskandar et al. 1992). The majority have polyclonal IgG; the few monoclonal deposits have been IgG kappa. Yang et al. (1992) have shown the fibrils themselves contain immunoglobulins (heavy and light chains) and C3 by immunoelectron microscopy. They also observed presence of amyloid P component along or in the fibrils, but not heparan sulfate, collagen type IV, fibronectin, or fibrillin in the deposits. Some cases of fibrillary glomerulonephritis without detectable immunoglobulin content have also been reported (Churg 1993), suggesting variety of precursors.

In most of these cases, a specific underlying systemic disease is not found. Occasional cases have elevated antinuclear antibodies, multiple myeloma (Fogo et al. 1993), or hepatitis C infection (Coroneos et al. 1997; Markovitz et al. 1998). The fibrillary deposits are lim-

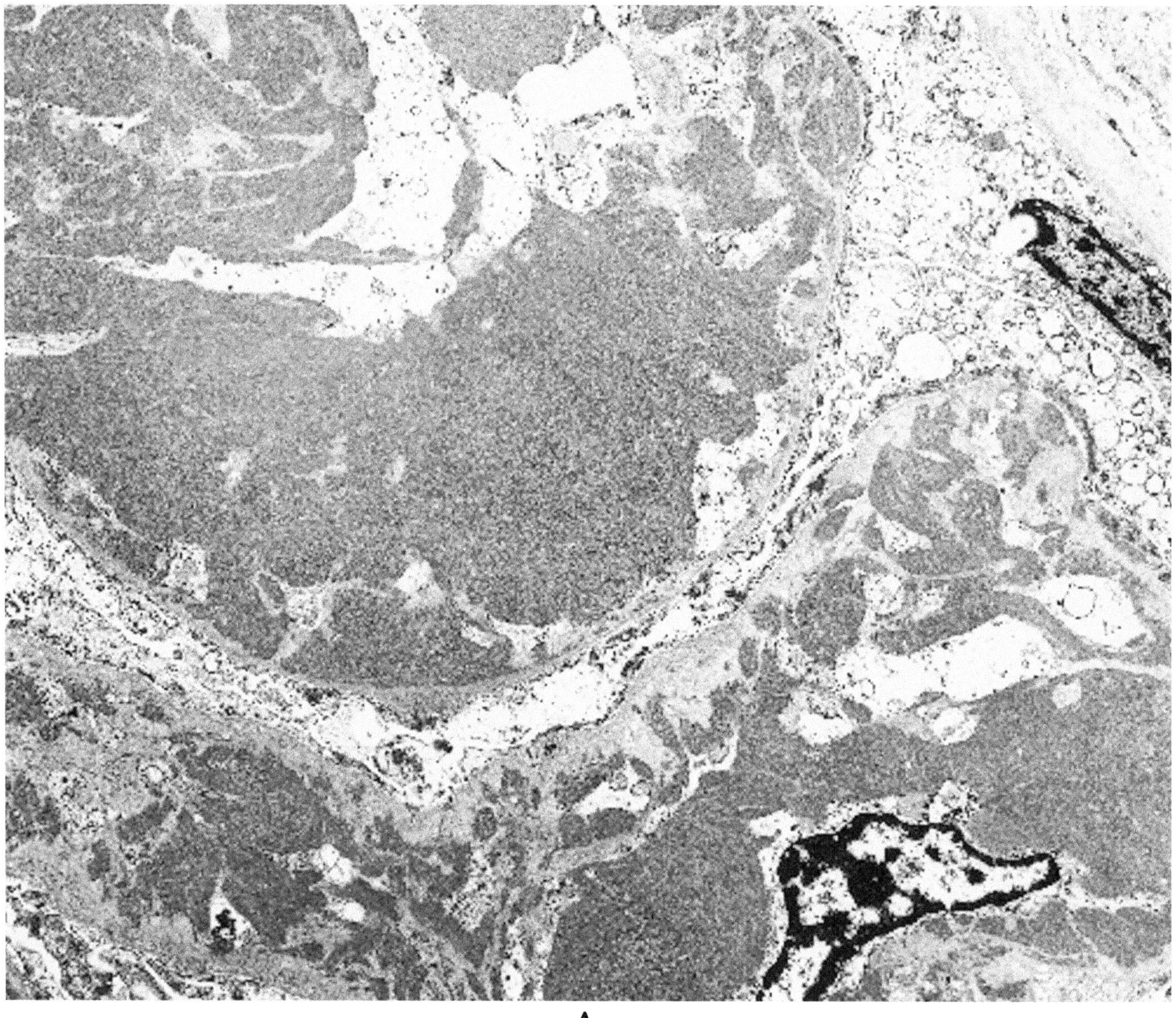

A

Figure 12.61. Immunotactoid glomerulopathy. **A,** At low power, the glomerulus is filled with dense deposits that are principally subendothelial and mesangial. (× 4000).

ited to the kidney in most cases (in contrast to amyloidosis); rare cases of extrarenal involvement (lung and liver) have been reported.

Immunotactoid Glomerulopathy

(Figures 12.61 and 12.59A.)

Diagnostic criteria. Microtubular deposits of 16–90 nm diameter (Fogo et al. 1993) with hollow, electron lucent centers and thick walls, typically arranged in organized parallel arrays, sometimes with a lattice-like pattern along the GBM in subepithelial, subendothelial, or intramembranous locations and in the mesangium.

Additional points. In comparison to fibrillary glomerulonephritis, immunotactoid fibrils are thicker, with an outer diameter usually around 30 nm, and have a prominent microtubular or cylindrical structure. Podocyte foot processes are usually effaced.

By light microscopy, proliferative, lobular, or membranous patterns of glomerular involvement have been described, usually without crescents. Immunofluorescence shows granular staining for IgG and C3 along the capillary loops and the mesangium in most cases, and rarely IgA and IgM. Congo red stains are negative. In the majority, glomerular immunoglobulin deposits have a single light chain.

The exact pathogenesis of immunotactoid glomerulopathy is not entirely known. None have cryoglobulinemia. Many cases are associated with serologic gammopathy or a lymphoproliferative disorder (Alpers 1992), in contrast to fibrillary glomerulonephritis. Fogo et al. (1993) observed the association of lymphoplasmacytic disorders, for example, chronic lymphocytic leukemia and Sjögren's syndrome with immunotactoid glomerulopathy. Lai et al. (1989) described a case of mixed connective tissue disease with immunotactoid glomerulopathy and fingerprint deposits. We have observed a case of systemic lupus erythematosus with immunotactoid deposits in the subepithelial regions (Figure 12.59A). Few cases have elevated antinuclear antibodies.

Immunotactoid glomerulopathy was originally introduced by Schwartz and Lewis (1980) and applied to crystalline rod-like particles (tactoids) with an immunoglobulin content (immuno-) deposited in an ordered, nonrandom orientation in the glomeruli. The term was later expanded (Korbet et al. 1985) to include cases with fibrillary glomerulonephritis (FGN) as defined by Duffy et al. (1983). As the number of cases of both type accumulate, comparison of clinicopathologic features has led many observers to consider these as two distinct entities (Fogo et al. 1993; Alpers 1993); others believe them to be variations of one disease (Korbet et al. 1994; Schwartz 1993). This controversy will remain until the pathogenesis is better understood.

Cryoglobulinemic Glomerulopathy

(Figures 12.62 through 12.64; 12.59B.)

Diagnostic criteria. (1) Glomerular endocapillary proliferation with leukocytes, predominately mononuclear phagocytic cells (monocytes) and fewer neutrophils; (2) cellular interposition (monocytes) in between endothelium and GBM or the subendothelial deposits; (3) duplication of the GBM; (4) electron-dense deposits, predominately in the subendothelial region, often massive and sometimes forming intraluminal pseudothrombi; (5) characteristic substructure of deposits, which differs with the type of cryoglobulin involved.

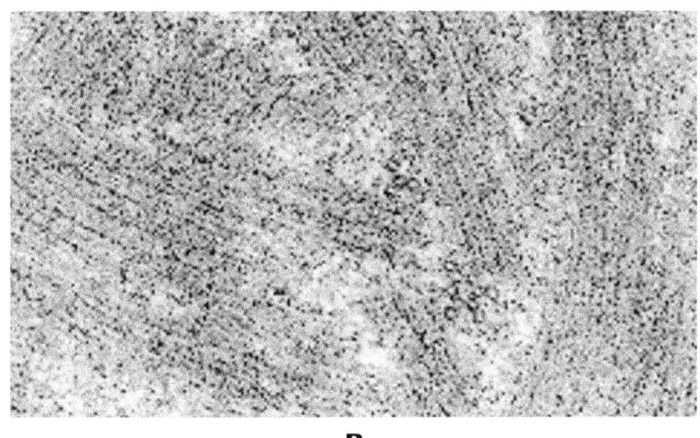

B

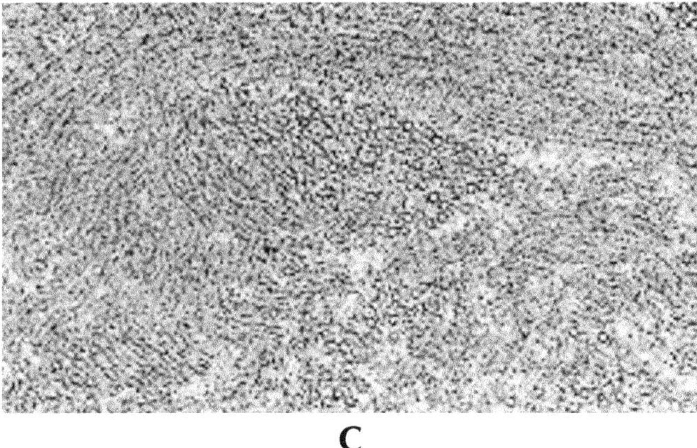

C

Figure 12.61. *(continued)*
B and **C,** The substructure is revealed at higher power, with characteristic lattice and tubular arrangements. Outer diameter of the microtubules about 23–25 nm. (× 60,000) (Photomicrographs kindly provided by Dr. Charles Alpers and Dr. Eleonora Galvanek, Brigham and Women's Hospital, Boston, Mass.)

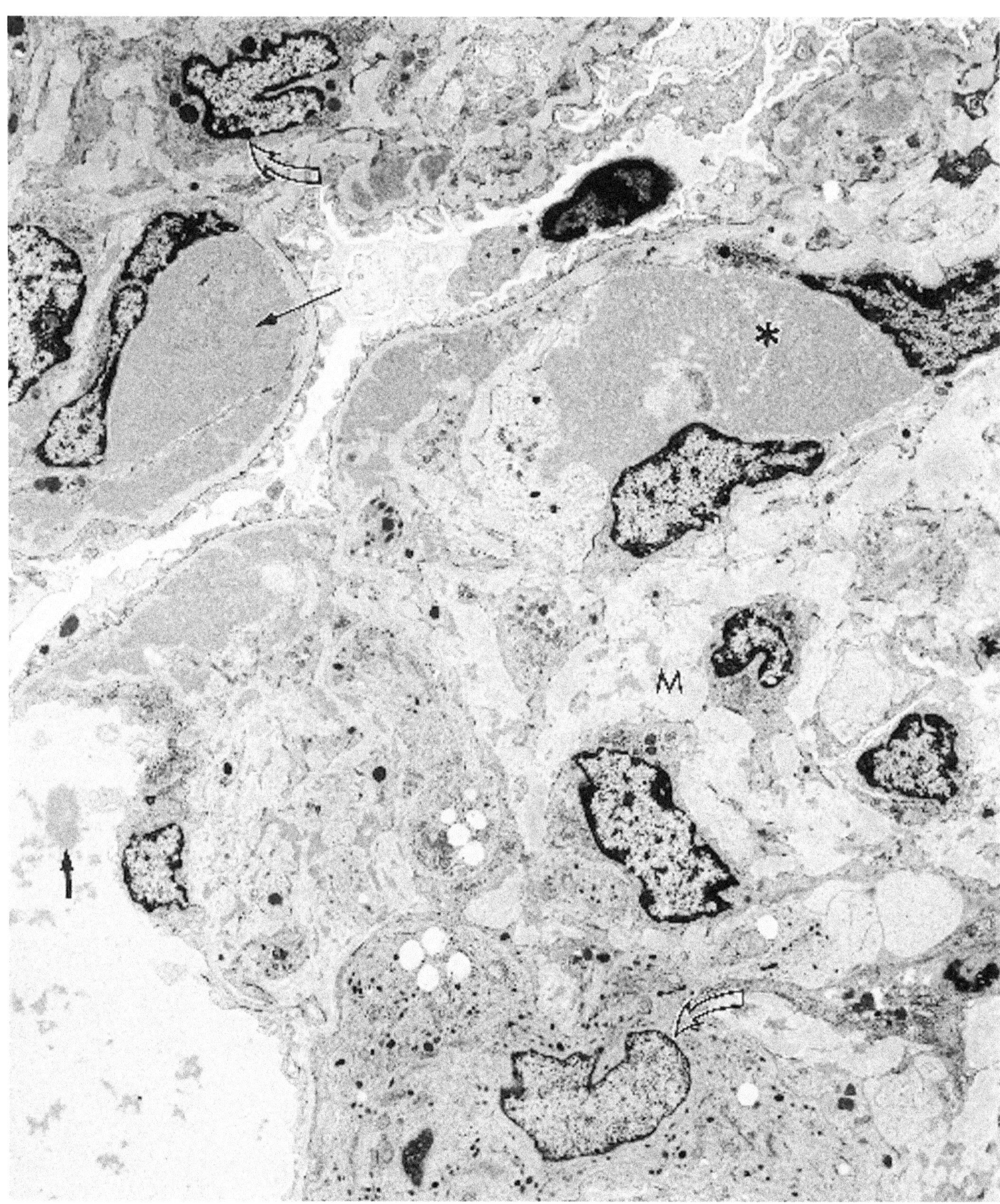

Figure 12.62. Mixed cryoglobulinemia (49-year-old woman with mixed IgM-IgG, type II, cryoglobulinemia). Large mesangial (*) and subendothelial (*long arrow*) electron-dense deposits are present. Podocytes have segmental foot process effacement. The mesangium (M) is expanded and contains many active-appearing mononuclear inflammatory cells (*open arrows*). Similar mononuclear inflammatory cells are also seen in subendothelial regions (see Figure 12.64). Flocculent aggregated material (*short arrow*) in capillary lumens have microtubular structures similar to the cryoglobulins. (× 5500)

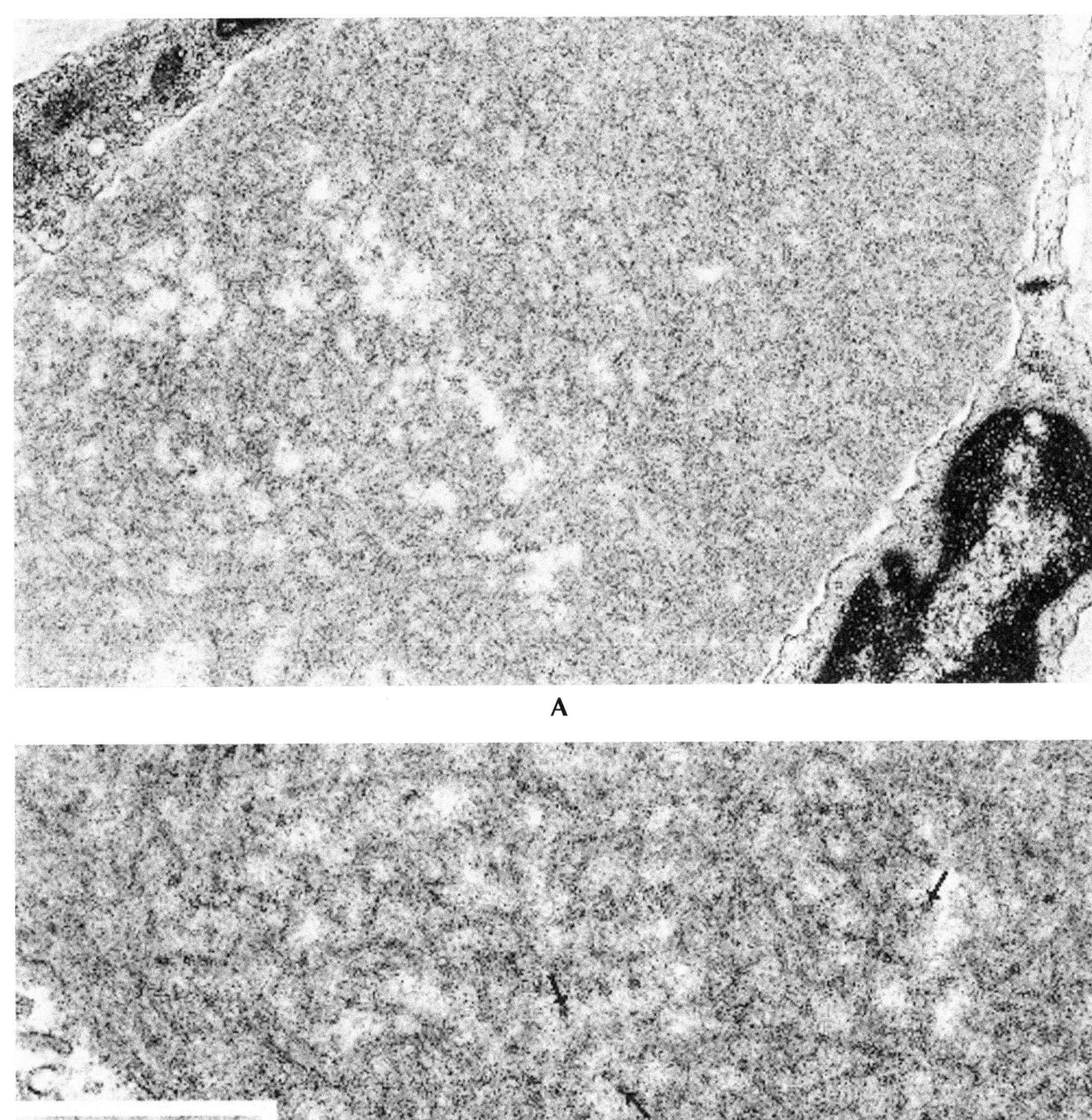

Figure 12.63. Mixed cryoglobulinemia (same case as in Figure 12.62). **A,** Deposits are composed of randomly arranged microtubules with an amorphous granular background (in comparison, amyloid and fibrillary glomerulonephritis tend to have a lucent background; see Figures 12.58 and 12.59). (× 26,000). **B,** Higher magnification of deposits to show microtubular profiles (22–24.6 nm) and annular, centrally hollow cross-sections that are surrounded by a wall from which 8–12 spokes emanate (about 33.8 nm) (*arrows*). (× 65,000). Higher magnification is shown in *inset.* (× 120,000)

Mixed cryoglobulinemia, type II, monoclonal IgM (rheumatoid factor), and polyclonal IgG (Figures 12.62 through 12.64). (a) The glomerular deposits usually have a microtubular substructure with slightly curved cylinders in pairs, each about 25 nm in diameter (Figure 12.63); (b) in cross-section, the cylinders appear like annular structures with an electron-lucent (hollow) center around which 8–12 spokes project (Figure 12.63B) (Cordonnier et al. 1983; Feiner and Gallo 1977; Stoebner et al. 1979).

IgG Cryoglobulinemia, type I, monoclonal immunoglobulin subclass (not shown). (a) two types of deposits have been described, the first being straight fibrils (each 9–12 nm) forming bundles, 80 nm wide; in cross-section, the bundles appear cross-hatched; (b) the second type of deposits are tubular structures in a fingerprint-like array.

Additional points. Deposits can also be in the mesangium, but rarely in the subepithelial space. The background of the microtubular deposits may contain amorphous material, probably representing degenerating or degrading crystalline structures. Mononuclear cells (monocytes) with ingested electron-dense material are commonly found in glomerular capillaries (Figure 12.64A); phagolysosomes with engulfed cryoglobulin structures (Figure 12.64B) have been observed, supporting the role of monocytes in clearing the deposits. Aggregates of these microtubular structures similar to the deposits, have also been observed in capillary lumens (Monga, et al. 1987). Similar or identical ultrastructural features can be observed in the serum cryoprecipitates and corresponding glomerular deposits (Szymanski et al. 1994). Rhomboid crystals may be seen in the cytoplasm of endothelial cells, mesangial cells, and occasionally podocytes. These crystals can be intracellular or extracellular, with a suggestion of a periodicity at the edges. Podocyte foot processes are usually effaced. The endothelium is reactive.

Light microscopy shows endocapillary proliferation with monocytic infiltration and, often, thickening and splitting of the GBM. This morphology has been termed exudative MPGN or cryoglobulinemic glomerulonephritis in cases of mixed IgG-IgM cryoglobulinemia. The massive monocytic infiltration is characteristic and not seen to this degree in the other glomerulonephritides. Gesualdo et al. (1997) have shown increased glomerular synthesis of monocyte chemotactic peptide-1. Hyaline pseudothrombi, formed by the large subendothelial deposits containing the cryoglobulins, may occlude the capillary lumens. In some cases, mild segmental mesangial proliferation or lobular glomerulonephritis with less monocytic infiltration may be seen. Tubulointerstitial nephritis may be present.

Glomerular involvement is most frequent and distinctive in type II cryoglobulinemia. Glomerular involvement is rare in type I cryoglobulinemia, and only few cases have been reported. Review of literature shows cases of type I cryoglobulinemia (usually IgG) with morphology and deposit substructures as described earlier. A rare case of monoclonal IgM kappa cryoglobulinemia (type I) has been reported (Tomiyoshi et al. 1998), with fibrils (forming bundles of 7–20 fibrils) in glomerular macrophages but not in the amorphous subendothelial deposits. In another case of IgG kappa type I cryoglobulinemia (Ishimura et al. 1995), the glomerular capillary lumens were occluded by electron-dense material without fibrils or crystals with few subendothelial deposits. Type III cryoblobulinemia (polyclonal IgG and polyclonal IgM) is found in heterogeneous immunologic and infectious diseases and rarely, if ever, has distinct glomerular pathology.

Cryoglobulins are immunoglobulins that precipitate in cold environment. They may have a single monoclonal immunoglobulin component (type I; usually IgG or rarely IgM) or may have two or more immunoglobulins (mixed). The mixed cryoglobulins are divided into type II where a monoclonal immunoglobulin (usually IgM kappa) acts as an antiglobulin against a polyclonal IgG. In type III cryoglobulinemia, both the IgG and the antiglobulin components are polyclonal. Type I cryoglobulinemia is associated with multiple myeloma, Waldenström's macroglobulinemia, and other lymphoproliferative disorders. Types II and III cryoglobulinemia are associated with viral (especially hepatitis C virus, as discussed next), bacterial, parasitic infections; autoimmune diseases such as systemic lupus erythematosus, Sjögren's syndrome, rheumatoid arthritis, scleroderma; lymphoproliferative disorders such as chronic lymphocytic leukemia; glomerulonephritis and chronic liver disease.

Most cases of essential (idiopathic) mixed cryoglobulinemia, types II and III, (about 30% cases of mixed cryoglobulinemia overall) are hepatitis C virus related (Angello et al. 1992; Bloch 1992; Fornasieri and D'Amico 1996; Miescher et al. 1995; Szymanski et al. 1994); up to 91% cases in one report (Ferri et al. 1995). Angello et al. (1992) found HCV RNA in the cryoprecipitate of four patients with anti-HCV antibodies. HCV has been recently considered oncogenic and may induce an abnormal proliferation of B-lymphocytes (Ferri et al. 1995), leading to cryoglobulinemia. Bloch (1992) has suggested that HCV may be involved in the pathogenesis of Sjögren's syndrome associated with mixed cryoglobulinemia (type II). Cryoglobulinemic HCV-related glomerulonephritis may recur in renal allografts, in a MPGN pattern or with more advanced changes.

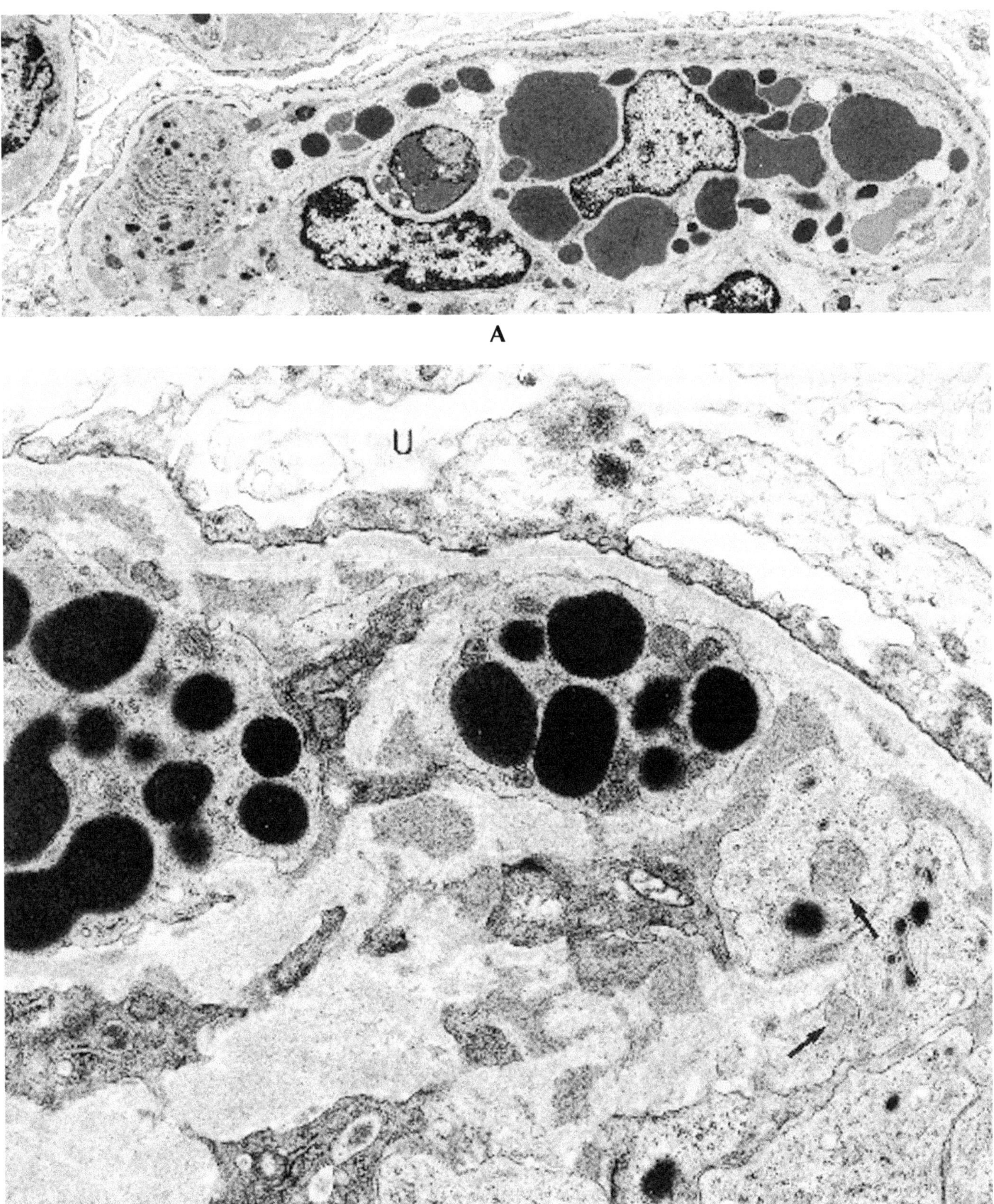

Figure 12.64. Mixed cryoglobulinemia (same case as in Figure 12.62). **A,** Mononuclear phagocytic cells are closely associated with subendothelial deposits and contain abundant phagoytosed electron-dense material and reactive cytoplasm. (× 5500) **B,** Higher magnification of another similar area to show cryoglobulins within phagolysosomes (*arrow*) of monocyte cytoplasm, in close association with the subendothelial cryoglobulin deposits. U = urinary space. (× 20,100)

Systemic Light Chain Deposition Disease

(Figures 12.65 and 12.66.)

Diagnostic criteria. (1) Fine or coarse granular, osmiophilic, electron-dense deposits along the tubular basement membrane (TBM) and GBM, which aggregate to form a line running along the outer aspect of the TBM (Figure 12.66B) and inner aspect of the GBM (Figures 12.65 and 12.66A); (2) finely granular electron-dense mesangial deposits, similar to those in criterion 1; (3) increase in mesangial matrix components in cases with prominent mesangial nodules (Bruneval et al. 1985) with scattered microfibrils and GBM-like material, different from the deposits; (4) granular deposits in the basement membranes around smooth muscle cells of arteries and arterioles (Tubbs et al. 1981; Kirkpatrick et al. 1986).

Additional points. Increase in mesangial cellularity also may be noted. Podocyte foot process effacement may be present. The tubular deposits are predominantly found in the distal tubules and collecting tubules, but in later stages may also involve the proximal tubules. The TBM may become extremely thickened with tubular atrophy. Older deposits tend to give a laminated appearance, sometimes with basement membrane duplication. The tubular deposits may be more prominent than the glomerular deposits, which may be absent altogether in some cases. Myeloma casts rarely have been noted. Deposits around peritubular capillaries, interstitium, and Bowman's capsule may be noted. Extrarenal deposits are found in almost any organ, more frequently involving heart, liver, and spleen, similar to the distribution of secondary amyloidoses. The disease recurs in renal allografts (Figure 12.66A).

The first description of systemic light chain deposition disease was by Randall et al. in 1976, who confirmed the monoclonal light chain content (kappa light chain in both patients) of these lesions in the kidney and many other organs (Randall et al. 1976). Numerous reports have followed since then (Strom et al. 1994; Bangerter and Murphy 1987; and others). In the majority of cases, the granular deposits are attributed to monotypic light chains only (light chain deposition disease, LCDD). Few cases also show associated heavy chain deposition (light and heavy chain deposition disease, LHCDD), and more recently, very rare cases of only heavy chain deposition (HCDD) without a light chain component have been described. Many authors have proposed using the term monoclonal immunoglobulin deposition disease (MIDD), Randall's type, to describe LCDD, LHCDD, and HCDD (Preud'homme et al. 1994; Ronco et al. 1997). The unifying factors in LCDD, LHCDD, and HCDD are finely granular, nonfibrillar, electron-dense deposits that do not contain the amyloid P component and are Congo-red-negative. These features are in contrast to AL amyloid, as discussed earlier.

By light microscopy, the glomeruli show mesangial nodules in about 60% of the cases, resembling nodular diabetic glomerulosclerosis. The glomeruli may be normal or have slight mesangial increase or lobularity. The TBM appears thick in spite of relatively normal-appearing tubular epithelium, or tubular atrophy may be seen. Congo red stain is negative. Direct immunofluorescence shows linear monotypic staining for the light chain. The mesangial nodules may or may not show staining, while GBM staining is variable.

In the majority of MIDD cases, the underlying process is multiple myeloma (53–93%, various series). In 15–30% cases of MIDD, an "M" component is not detected, which may be due to low levels of production, rapid tissue deposition, or increased degradation. In those cases, especially LCDD without demonstrable myeloma, a predominance of one light chain type may be found in the bone marrow plasma cells. In about 80% of cases of LCDD (or rather MIDD), monotypic kappa light chains are found in the deposits, whereas lambda light chains are seen in most of the remaining cases. These kappa light chains are more frequently of the kappa variable region subgroup IV (VKIV type, which are not exclusive to LCDD and can be found in multiple myeloma without renal disease. In HCDD, lambda light chains are associated with the heavy chain in serum. The probable pathophysiology of these light chains (and/or heavy chains) forming such granular deposits (rather than fibrils) includes structural anomalies, glycosylation, or affinity for extracellular constituents. The absence of amyloid P component from these granular deposits (in contrast to fibrillar amyloid deposits) has been consistently seen by immunohistochemical methods and may be an important factor in the mechanism of deposition and nonfibril nature (Gallo et al. 1988). Interestingly, multiple myeloma patients without LCDD initially, develop it after therapy with melphalan, which is known to induce mutations in the light chains. Occasional cases of combined LCDD in the kidney and amyloidosis (focal) elsewhere (Buxbaum 1992) have been observed and raise questions about the same light chain present in two forms under different physicochemical conditions versus the presence of two separate proliferating B-cell clones.

Monoclonal Gammopathy

Monoclonal immunoglobulins or their fragments (light chains or heavy chains) when present in serum and/or urine or CSF are referred to as "M" component or paraproteins. These paraproteins are usually elaborated by B-cell lymphoproliferative disorders, such as plasma

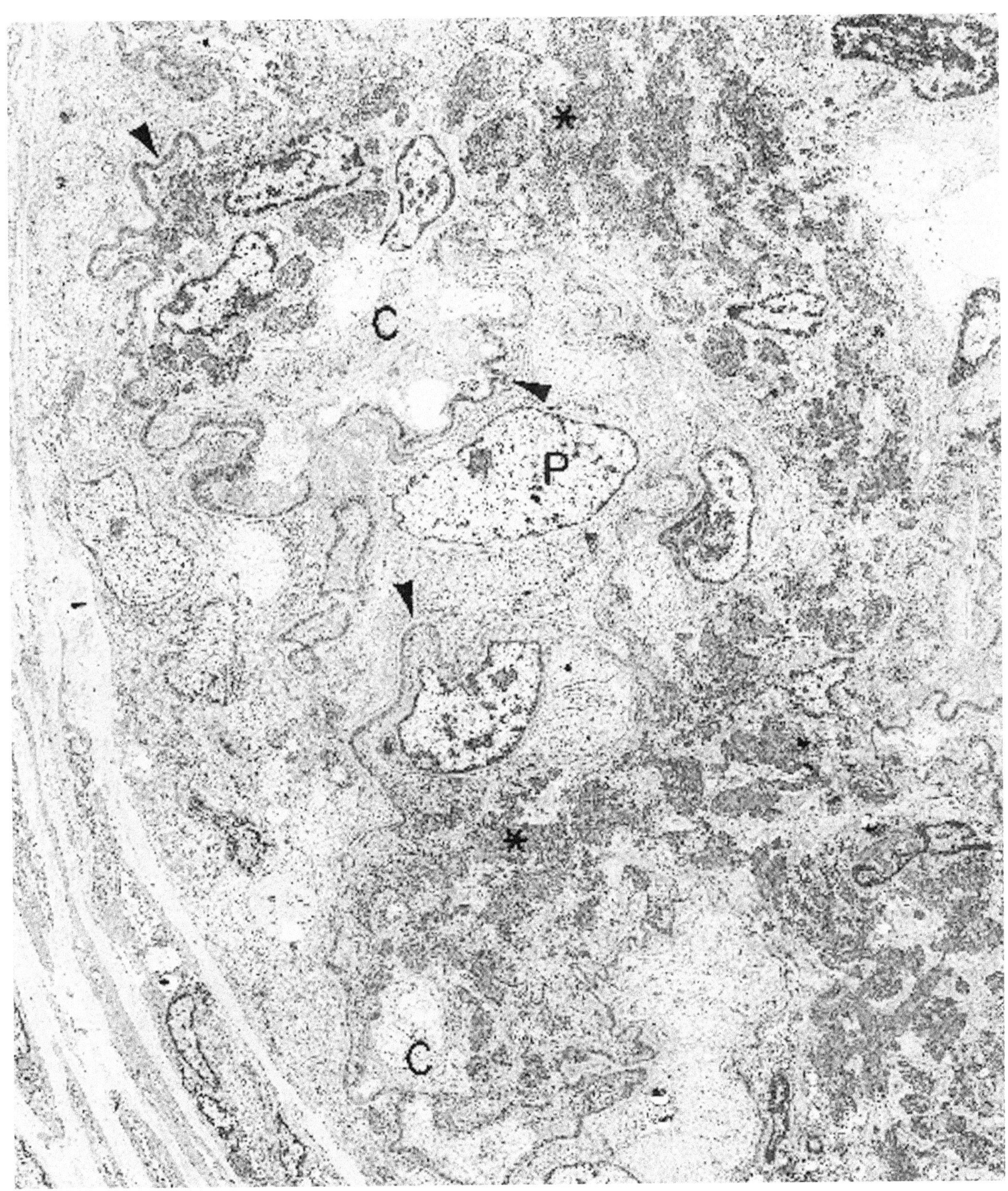

Figure 12.65. Systemic LCDD (54-year-old woman with kappa LCDD). Extensive deposition of extremely electron-dense, finely granular material in a band-like fashion is seen along the GBM (*arrowheads*) and in the mesangium (*). P = podocyte; C = capillary lumen. (× 3480)

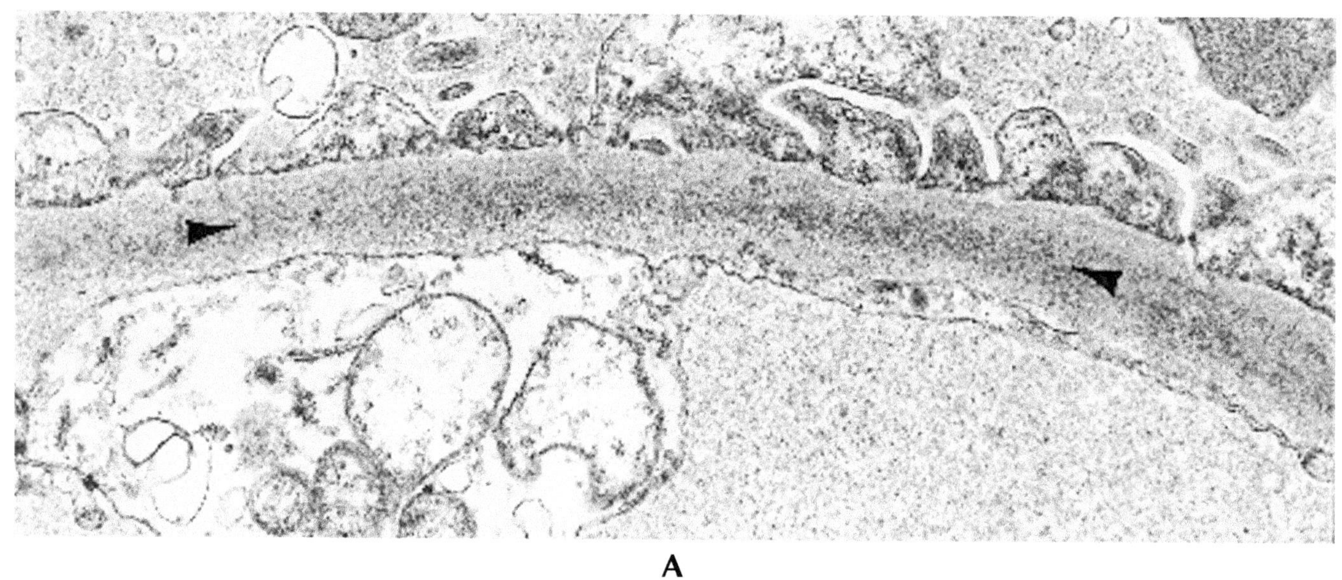

A

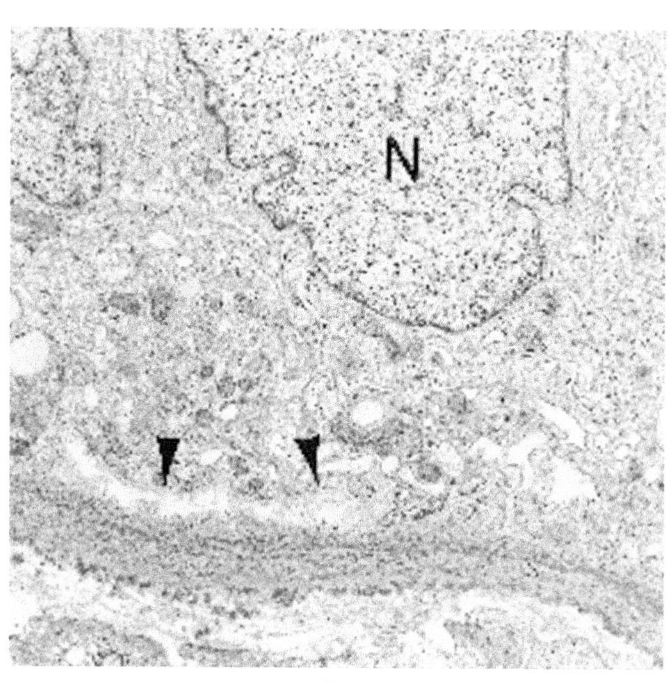

B

Figure 12.66. Systemic LCDD. **A** (recurrent in the allograft of patient illustrated in Figure 12.65), High magnification of the glomerular capillary wall shows fine granular electron-dense material (*arrowheads*) permeating the GBM. (× 19,000) **B** (81-year-old woman with kappa LCDD), High magnification of base of tubule shows fine granular electron-dense deposits in tubular basement membrane. N = nucleus of a tubular cell. Newly formed tubular basement membrane is between the deposit and the epithelium (*arrowheads*). (× 10,608)

cell myeloma, Waldenström's macroglobulinemia, or chronic lymphocytic leukemia as well as certain autoimmune diseases such as lupus and rheumatoid arthritis, and may be deposited in tissues, especially the kidney. Depending upon their structure, chemical properties, and interaction with tissues, these paraproteins may form fibrils (light chain amyloidosis, fibrillary glomerulonephritis), microtubules (cryoglobulinemic glomerulonephritis, immunotactoid glomerulopathy), or granular deposits (systemic light chain deposition disease), as discussed earlier. These deposits may sometimes occur in the absence of an identifiable "M" component, probably due to rapid tissue deposition. The above deposition diseases can occur in the absence of typical multiple myeloma and may be seen in indolent myeloma, monoclonal gammopathy of undetermined significance (MGUS), or in benign monoclonal B-cell proliferations.

In multiple myeloma, in addition to the above described deposits, the paraproteins may form *casts* (myeloma cast nephropathy) or *crystals* (myeloma-associated Fanconi's syndrome), leading to tubulopathies. The term "myeloma kidney" therefore is inclusive of many morphologically distinct processes affecting the kidneys in multiple myeloma and should not be restricted to myeloma cast nephropathy.

Nail Patella Syndrome (Hereditary Osteo-onychodysplasia)

(Figure 12.67.)

Diagnostic criteria. (1) Bundles of cross-striated, type I collagen fibers (Morita et al. 1973), with a periodicity of about 40–60 nm and varying lengths (Taguchi et al. 1988) deposited in the lamina densa and rarae interna and externa; (2) irregular thickening of the GBM, usually without cellular interposition; (3) electron-lucent spaces in the lamina densa, referred to as a "moth-eaten appearance" of the GBM; (4) presence of collagen fibers within the lucent areas.

Additional points. Collagen fibers or bundles may also be found in the subendothelial space, mesangium, and rarely peritubular interstitium; the mesangial areas are less affected than the glomerular capillary loops (this is in contrast to collagen type III collagenofibrotic glomerulopathy, discussed in the next section). Amorphous, subendothelial, and subepithelial deposits have been described in rare cases. Podocyte foot process effacement may be seen. The presence of type I collagen fibers in the basement membrane is not specific for this disease and can be found in sclerotic glomeruli in other diseases; however, type I collagen fibers are not found in otherwise normal areas of GBM as a nonspecific finding. It is not known whether the deposited collagen is synthesized locally or deposits from plasma.

Early stages of the disease may be manifested only by proteinuria with normal-appearing glomeruli by light microscopy, but with ultrastructural changes. Capillary wall thickening may be present. Later, sclerosis occurs and in some cases leads to extensive glomerular obsolescence. Immunofluorescence results are nonspecific, with focal deposits of IgM and C3 along peripheral capillary walls.

Nail patella syndrome is a rare, autosomal dominant, pleiotropic disorder characterized by nail hypoplasia, hypoplastic or absent patellae, dysplasia of radial head, iliac horns, and nephropathy in about 30% of the cases. Ultrastructural glomerular lesions of nail patella syndrome in the absence of a clinical renal syndrome have been described (Taguchi et al. 1988). The exact identity and location of the nail patella syndrome gene (NPS gene) is unknown. Linkage of the NPS gene locus to chromosome 9q34 has been shown (Campeau et al. 1995; McIntosh et al. 1997).

Collagen Type III Collagenofibrotic Glomerulopathy

(Figures 12.68 and 12.69.)

Diagnostic criteria. Abundant collagen fibers (type III) (1) are deposited in the expanded mesangium and subendothelial space, without involving the lamina densa; (2) are spiral shaped, curved, or frayed (but not straight) on longitudinal section, and comma shaped or worm-like (but not circular) on cross-section; (3) have a banding periodicity (distance between the transverse striations) of about 40–60 nm; (4) are arranged in irregular bundles.

Additional points. The lamina densa is intact, and the GBM is of normal thickness, without the moth-eaten appearance seen in nail patella syndrome. GBM duplication and cellular interposition may be seen. Such collagen deposits have not been found in extraglomerular locations such as tubules, blood vessels, or interstitium. These type III collagen fibers are different from the normal type III collagen fibers found in the renal interstitium and blood vessels, which are straight. In addition, the normal GBM and mesangium lack type III collagen. These fibers are distinctly different from the microtubular structures of immunotactoid glomerulopathy or cryoglobulinemia and microfibrils of amyloid or fibrillary glomerulopathy. Amorphous electron-dense deposits are absent. Podocyte foot process effacement is usually present.

By light microscopy, a lobular or MPGN-like architecture is noted but without significant hypercellular-

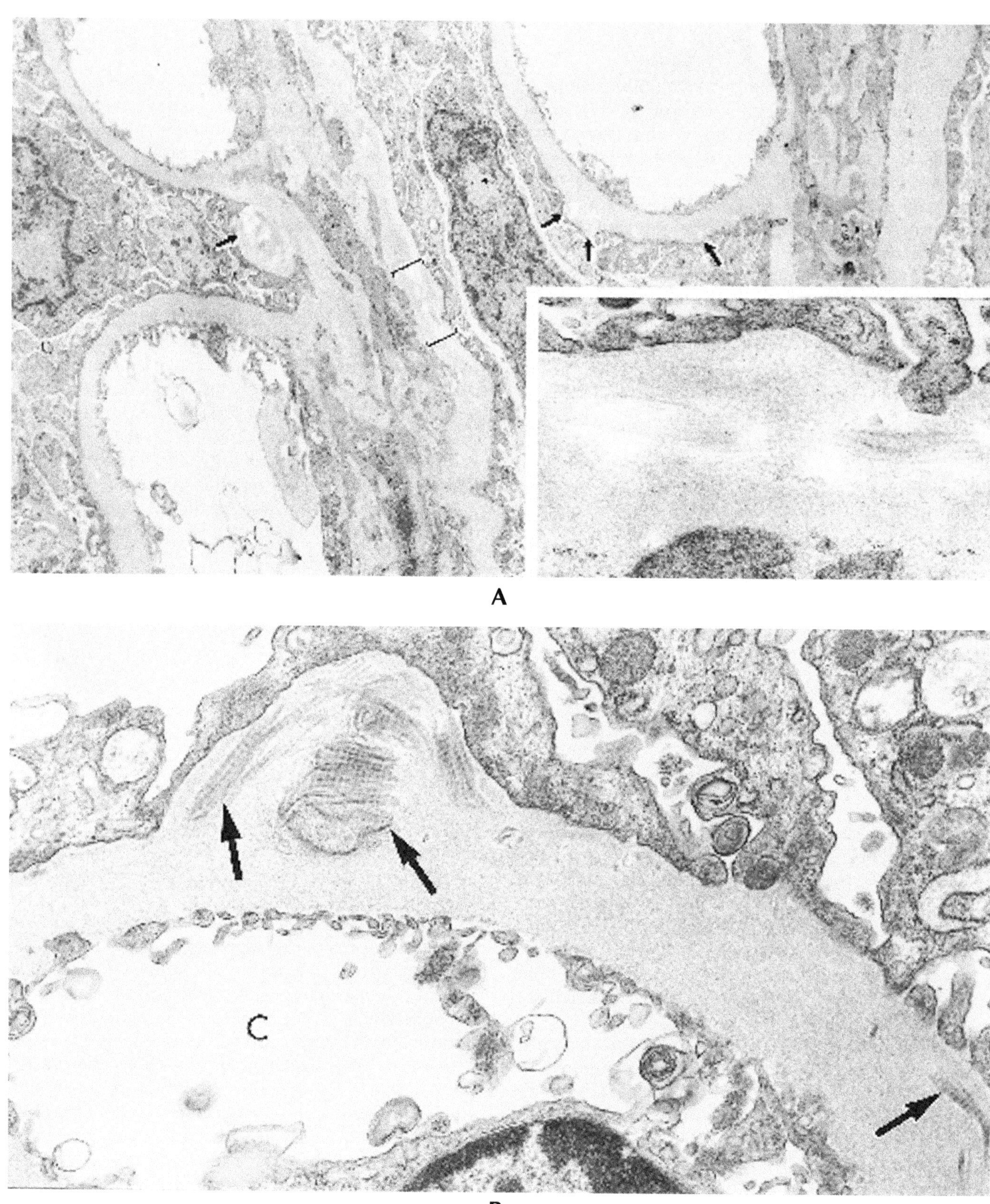

Figure 12.67. Nail patella syndrome (16-year-old female with episodic low-grade proteinuria and hematuria). **A,** GBM thickening with areas of lucency (*arrows*) appearing "moth-eaten." (× 6100) *Inset* shows higher magnification of GBM marked by *brackets* in **A** showing lucent foci and collagen fibers. (× 22,000) **B,** Higher magnification of GBM in another region showing subepithelial deposits of collagen fibers (*arrows*) with individual fiber width of 37 nm and periodicity of 42 nm. Associated podocyte foot process effacement is seen. C = capillary lumen. (× 29,000)

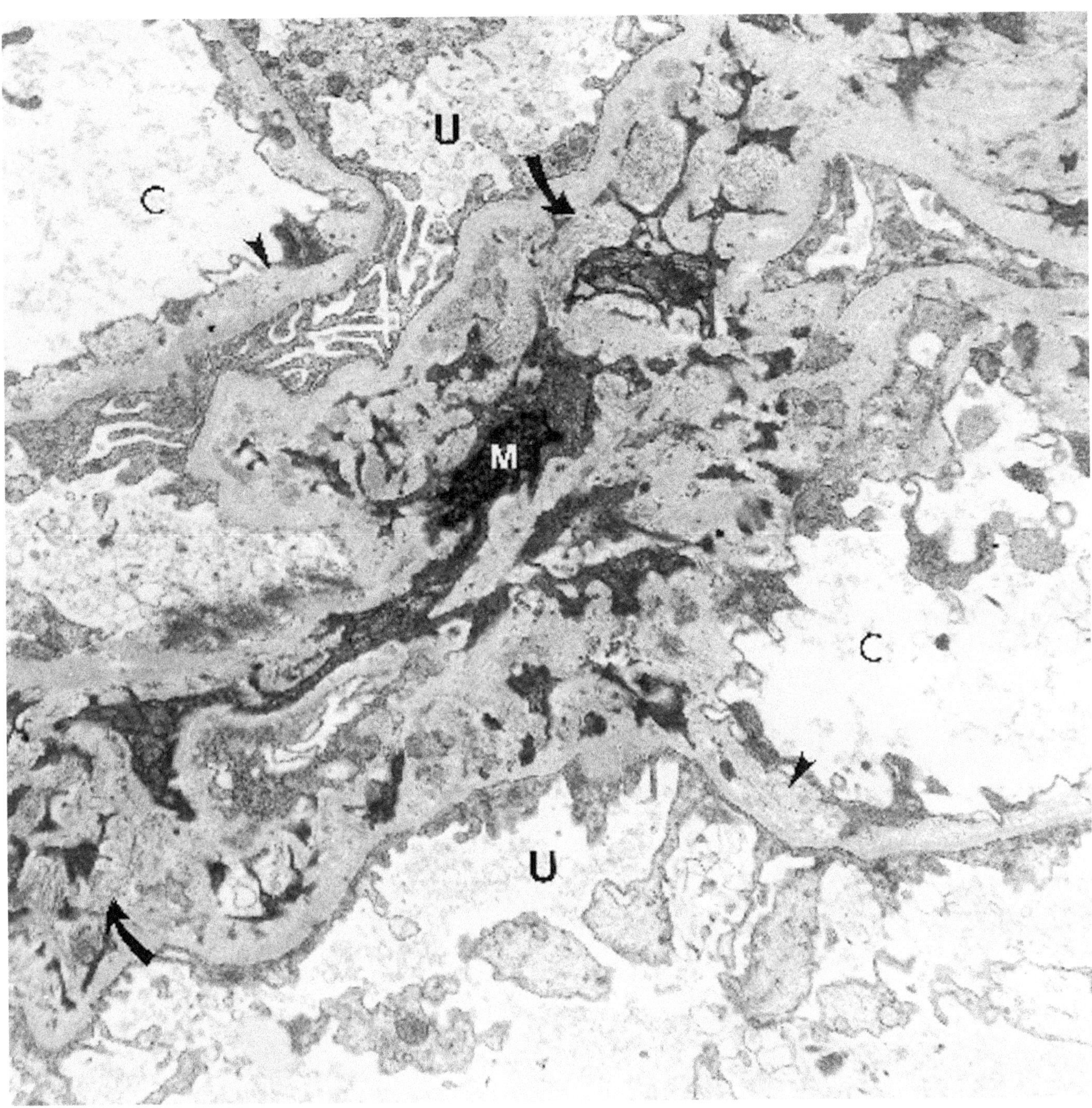

Figure 12.68. Collagen type III collagenofibrotic glomerulopathy (24-year-old male with microscopic hematuria, low-grade proteinuria of 800 mg/day, and a normal creatinine clearance). Mesangial deposits of abundant collagen fibers (*arrows*). Subendothelial regions (*arrowheads*) also contain some collagen fibers. Note the lack of moth-eaten appearance of the original GBM that is seen in nail patella syndrome. C = capillary lumens; U = urinary space; M = mesangial cell. (× 10,500).

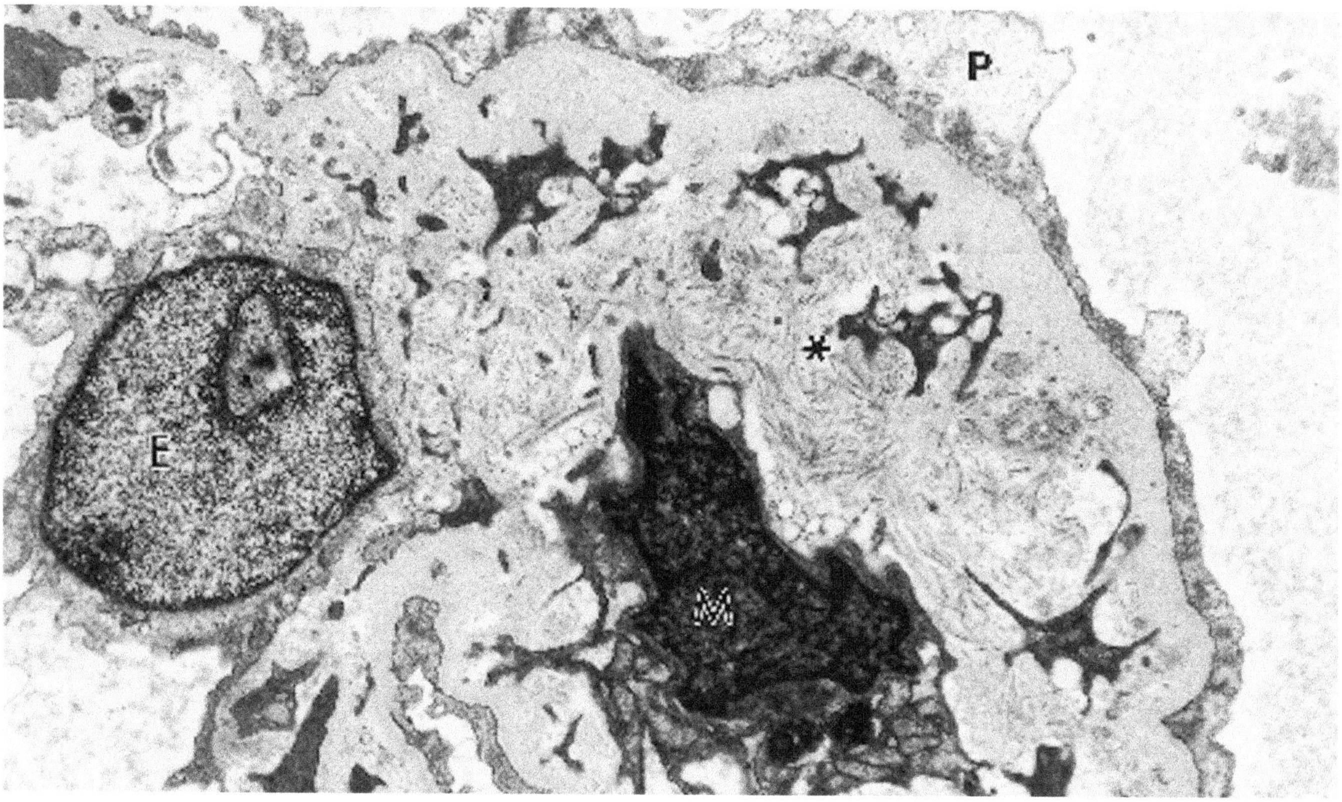

A

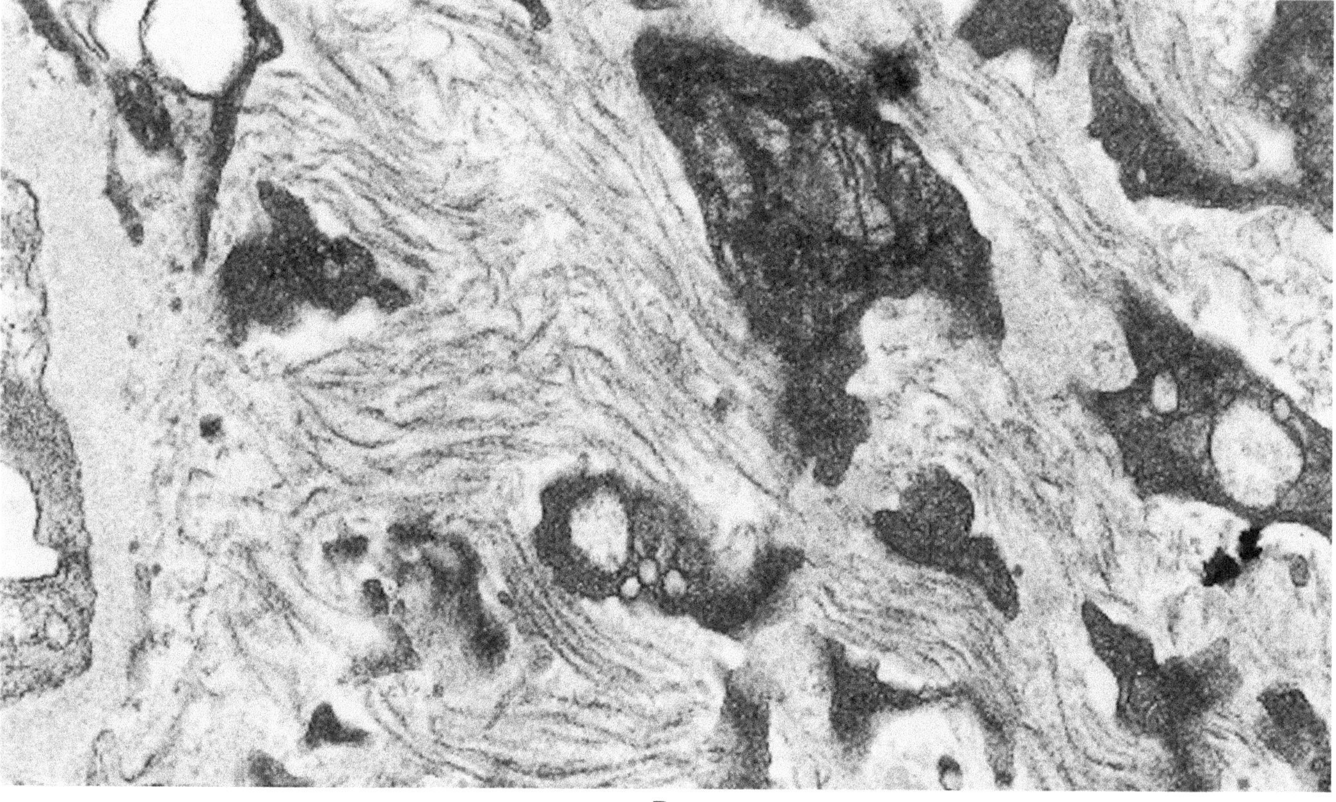

B

Figure 12.69. Collagen type III collagenofibrotic glomerulopathy (same case as in Figure 12.68). **A,** Deposition of collagen in the mesangium (*), mesangial cell (M), endothelial cell (E), and podocyte (P). (× 11,900) **B** High magnification of mesangial collagen fibers that are thick (21–29 nm) and wavy and have cross striations or periodicity (42 nm) and that are variably bundled. (× 38,000)

ity. The capillary lumens are reduced in size. The expanded mesangium may show very pale PAS and silver staining material, strong aniline blue staining has also been described. Immunofluorescence may show minimal IgG and C3. Specific immunofluorescence for collagen type III is strongly positive in areas corresponding to deposition of the abnormal fibers.

Glomerulopathy with deposition of type III collagen was first described by Ikeda et al. in 1990 and subsequently was named collagenofibrotic glomerulopathy (Arawaka and Yamanaka 1991). It is a rare, relatively new glomerular disease (Abt and Cohen 1996; Imbasciati et al. 1991). Patients have isolated renal involvement, without the extrarenal abnormalities of nail, patella, and the skeletal system that are usual in the nail patella syndrome. Cases with similar clinicopathological features were described in Japanese and English literature but without immunohistological documentation of type III collagen. This disease appears to be familial, with an autosomal recessive mode of inheritance (Tamura et al. 1996); sporadic cases have been described. The source of type III collagen is unknown but may be extraglomerular, because serum procollagen type III peptide (pIIIp) is markedly elevated. Serum pIIIp is a useful noninvasive indicator of type III collagenofibrotic glomerulopathy (Gubler et al. 1993). Slight nonspecific elevations of pIIIp may be seen in a variety of diseases from stimulated collagen synthesis.

Fabry's Disease

(Figures 12.70 and 12.71.)

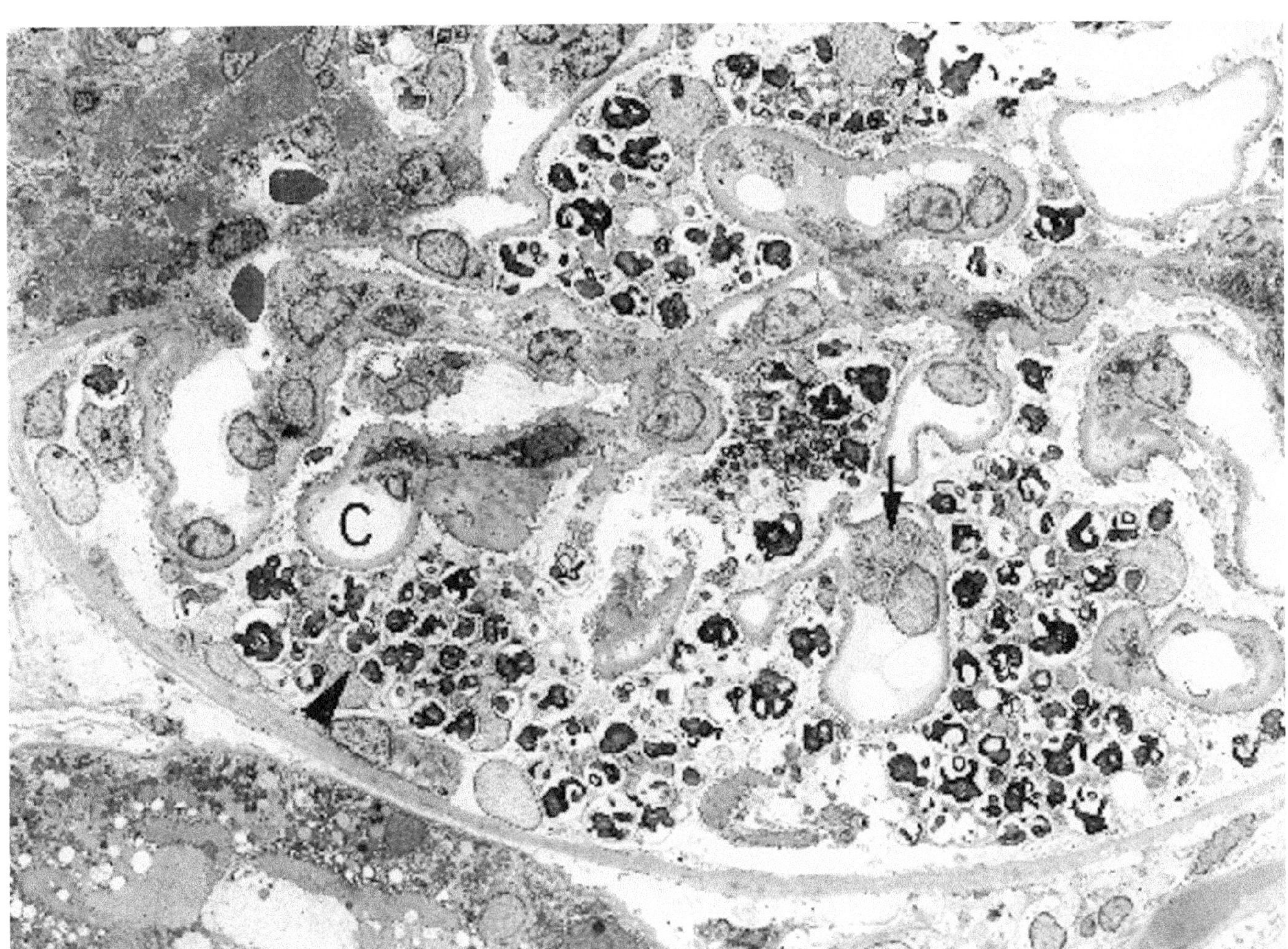

Figure 12.70. Fabry's disease. Numerous densely stained, large, laminated inclusion bodies, each surrounded by a unit membrane, are present in the podocytes (*arrowhead*) and endothelial cells (*arrow*). These inclusions are almost always in lysosomes and surrounded by a unit membrane. C = capillary lumen. (× 1800) (Permission for reprinting granted by the Association des Medecins de Langue Francaise du Canada, Paquin JG, Camirand P, Mandalenakis N, et al: Fabry's disease: Histologic and ultrastructural study. Union Med Can 104:1377–1382, 1975.)

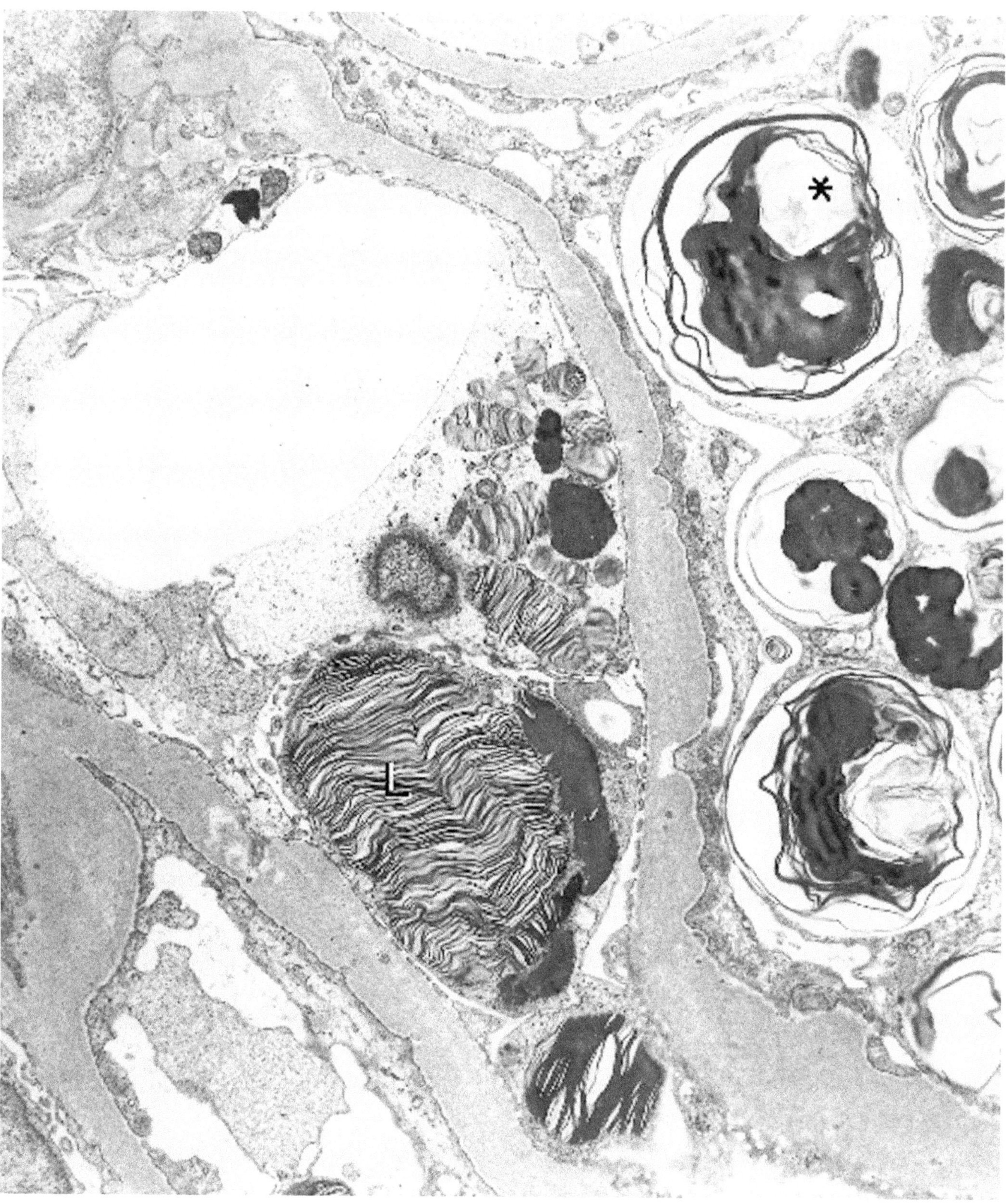

Figure 12.71. Fabry's disease. Higher magnification of a glomerular segment shows large, rounded, concentrically laminated inclusions (myelin bodies) in podocyte cytoplasm (*). The endothelium also shows these inclusions; here they appear more ovoid with parallel laminations (L) (zebra bodies). The difference in lamination may reflect orientation of these inclusions. Their fine structure appears to be glycolipid. (× 15,000) (Courtesy of Dr. Nicolas Mandalenakis, Hospital du Sacre-Couer, Montreal, Quebec, Canada.)

Diagnostic criteria. Electron-dense inclusions (1) in all types of glomerular cells, most abundant in podocytes; (2) laminated, usually 300–1000 nm in diameter (even up to 10 μ); (3) usually surrounded by a single-unit membrane, although some appear to be free in the cytoplasm (Faraggiana et al. 1981; Gubler et al. 1978).

Additional points. Two types of inclusions are seen. The most common ones are coarsely laminated, usually round with a central onion-skin structure called "myelin bodies," or ovoid with a parallel arrangement of dense layers (Gubler et al. 1978; Pacquin et al. 1975) called "zebra bodies." The alternating light and dark layers have a periodicity of 4–5 nm. These inclusions are particularly abundant in podocytes (Figure 12.71). The second type of inclusions are denser, more compact electron-dense deposits, some of which contain paracrystalline arrays (Burkholder et al. 1980). This type is more frequent in glomerular capillary endothelial and mesangial cells as well as in parietal epithelial cells of Bowman's capsule.

Inclusions are also present in tubular epithelial cells, especially in distal tubules, endothelial and smooth muscle cells of arteries, and interstitial cells, and can be detected in the urine (Tubbs et al. 1981).

The GBM often shows segmental wrinkling and thickening, with widening of the lamina rara interna by electron-lucent, fluffy, slightly fibrillary material. Adjacent podocytes have extensive effacement of foot processes. In advanced cases of the disease, there is collapse and sclerosis of glomerular tuft, but vacuolated cells may still be present.

By light microscopy, the most characteristic change is fine vacuolization of the cells of the glomerular tuft, most noticeable in podocytes but also in parietal epithelium of Bowman's capsule, endothelial cells, and mesangial cells. Immunofluorescence studies show focal deposits of IgM and C3 in arterial walls. Heterozygotes may have normal renal morphology or lesions similar to homozygotes, although milder (Gubler et al. 1978).

Fabry's disease is an X-linked disorder caused by a deficiency in the lysosomal enzyme alpha-galactosidase A, which leads to the accumulation of sphingolipids (ceramide) in cells. Recurrence in allografts is detectable by electron microscopy but is usually not clinically significant.

Cystinosis

(Figure 12.72.)

Diagnostic criteria. Crystalline or irregular cytoplasmic inclusions, corresponding to cystine; (1) predominately in interstitial cells (probably macrophages), also in epithelial cells of glomeruli and tubules, endothelial cells of capillaries, arterioles and glomeruli, and in smooth muscle cells of arterioles (Spear 1974); (2) sometimes surrounded by a unit membrane or seen in lysosomes; (3) rectangular, hexagonal, or triangular in form (Scotto and Stralin 1977); (4) dark cells, laden with cystine, predominately affecting interstitial cells (probably macrophages) and visceral epithelial cells; (5) multinucleated podocytes (Spear 1974).

Additional points. Dark cells are rarely also seen in epithelial cells of the loop of Henle and collecting ducts (Spear et al. 1971). The darkening of the "dark cells" is due to fine granular material in the cytoplasm, and nucleus, and the dark cytoplasmic inclusions, where the granules probably represent the reaction product of osmium tetroxide with cystine (a sulfur-containing amino acid), as postulated by Spear et al. (1971). By electron probe study they found high sulfur content, in sulfide form, in dark cells and therefore postulated that the material was most likely cystine, which reacted intensely with osmium tetroxide (used in fixation). Light microscopic examination shows progressive glomerular sclerosis, chronic interstitial nephritis with tubular degeneration, swan-neck atrophy of proximal tubules with vacuolization, multinucleated podocytes, and crystals or spaces. Cystinosis usually does not recur in the renal allograft; however, cystine crystals and dark cells predominately in interstitial cells and the mesangium have been found, probably in host-derived leukocytes (Spear et al. 1989).

Cystinosis is the most common of a group of lysosomal transport disorders, with a defect in cystine transport out of the lysosomal membranes. The cystinosis gene (CTNS) has been mapped to chromosome 17p13 (Cystinosis Collaborative Research Group 1995; Town et al. 1998). CTNS encodes an integral membrane protein, cystinosin, with features of a lysosomal membrane protein. Eleven different mutations, most commonly deletion, are described and predict loss of protein function (Town et al. 1998).

Glomerulopathy of Sickle Cell Disease/Trait

(Figure 12.73.)

Diagnostic criteria. (1) Few scattered, mesangial and subendothelial amorphous electron dense deposits; (2) increase in mesangial cellularity and matrix; (3) double contour of GBM and mesangial cell interposition; (4) abnormal morphology of red blood cells forming sickle shapes with characteristic microfibrillary hemoglobin S crystals.

Additional points. The diagnostic criteria describe the rare membranoproliferative variant (McCoy 1969; Pardo et al. 1975) of sickle cell disease/trait involving the glomerulus. More commonly, patients show con-

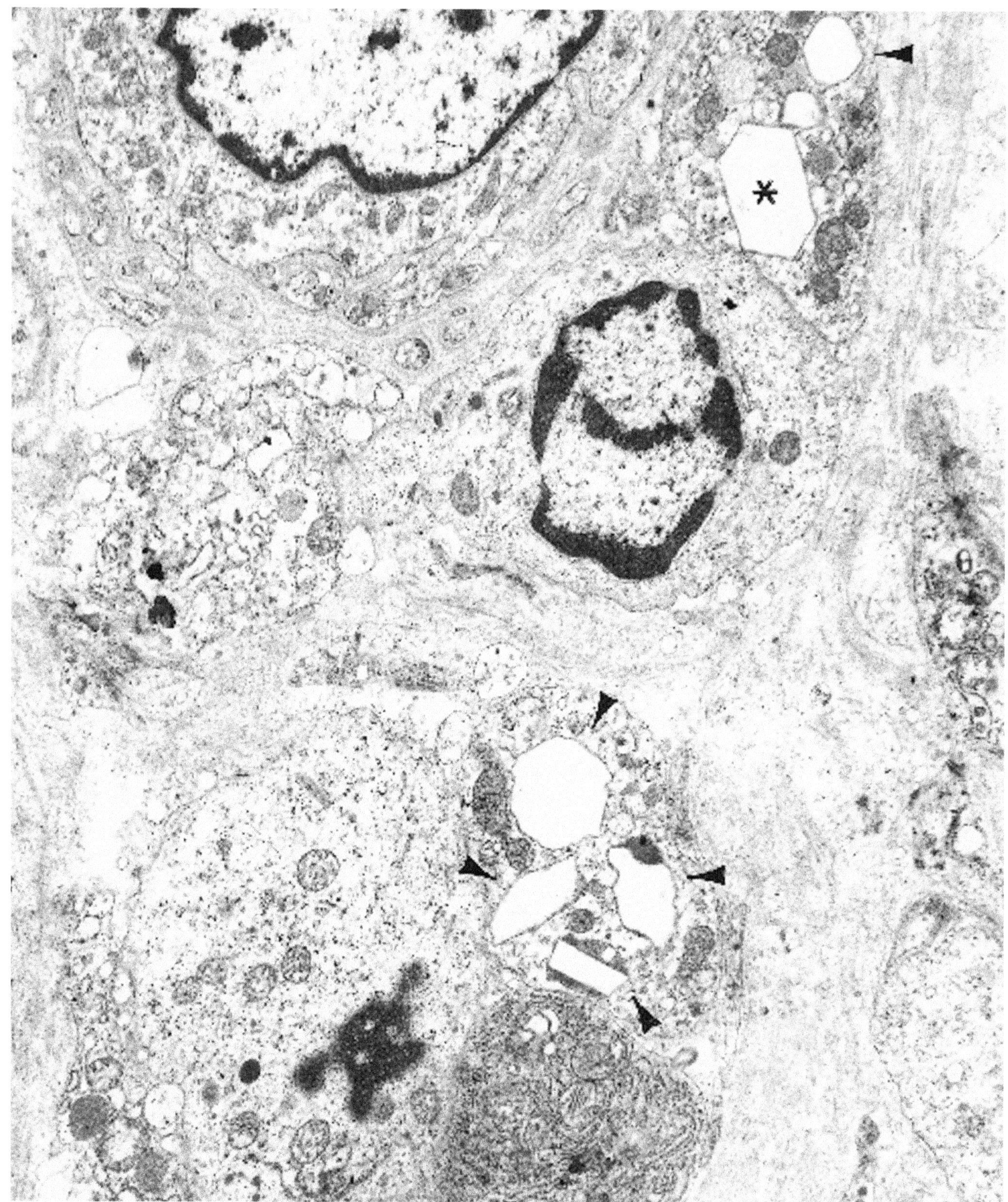

Figure 12.72. Cystinosis (7-year-old boy with cystinosis and end-stage renal failure). Scattered cells in the interstitium contain rectangular (*arrowheads*) and hexagonal (*) profiles of crystals. These spaces appear empty on routine electron microscopic processing. (× 9990)

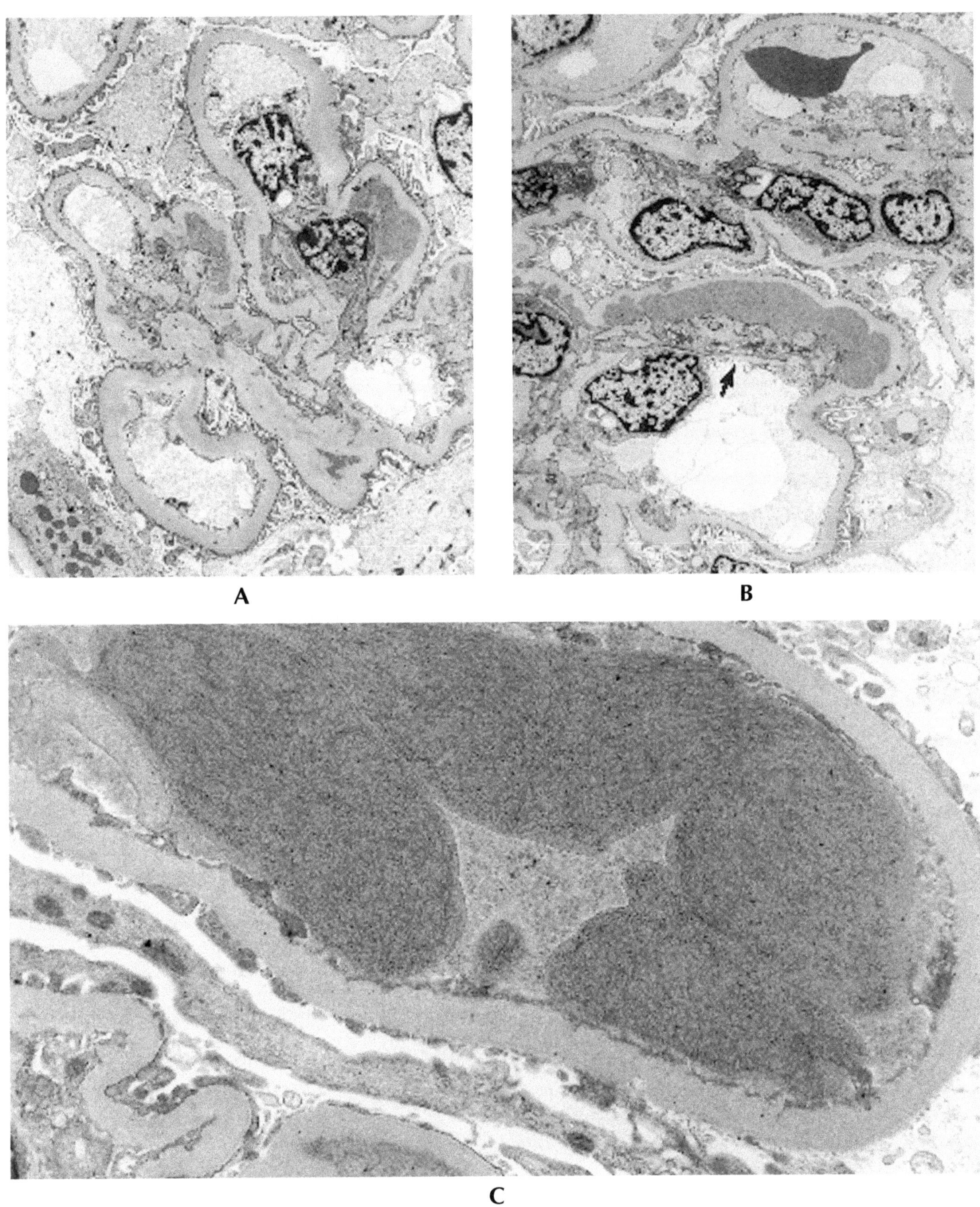

Figure 12.73. Sickle cell disease/trait associated MPGN (19-year-old male with long-standing history of proteinuria). **A,** Mesangial deposits without any substructure. (× 3900) **B,** Subendothelial deposit and duplication of the GBM (*arrow*). Sickle-shaped red blood cell in a capillary loop at top of figure. (× 3900). **C,** Higher magnification of a capillary loop containing the characteristic sickle red blood cells containing fibrillar crystals (about 12–16 nm) of hemoglobin S. (× 12,200)

gestion of glomerular capillaries and peritubular capillaries, often containing sickle-shaped red blood cells. Podocyte foot processes may be focally effaced. By light microscopy, many glomeruli appear hypertrophied (Bernstein and Whitten 1960). Renal cortical infarctions and papillary necrosis are seen and may be associated with tubular necrosis, especially in acute cases. Focal segmental or global glomerulosclerosis or mesangial hypercellularity have been described in sickle cell disease (homozygous SS genotype), and therefore hyperfiltration injury has been invoked in the pathophysiology. Iron deposits may also be found in the mesangium and have been considered in the pathogenesis of this lesion by some (McCoy 1969) but not others (Pardo et al. 1975). Some investigators have proposed intracapillary erythrocyte fragmentation (Antonovych 1971; Elfenbien et al. 1974) in the pathophysiology leading to glomerular injury.

Diseases with Endothelial Reaction

Thrombotic Microangiopathy (in Hemolytic Uremic Syndrome, Thrombotic Thrombocytopenic Purpura, Scleroderma, Malignant Hypertension, Rejection, and Cyclosporine Toxicity)

(Figures 12.74 through 12.79.)

Thrombotic microangiopathy (TMA) is the morphological description of a lesion characterized by microvascular (capillaries and arterioles) thrombosis with accompanying basement membrane changes and endothelial injury. It is usually a systemic condition with prominent glomerular involvement. Thrombotic microangiopathy thus encompasses both hemolytic uremic syndrome and thrombotic thrombocytopenic purpura. TMA is also described in association with acute phase of scleroderma, systemic lupus erythematosus, malignant hypertension, preeclampsia/eclampsia (discussed separately), allograft rejection, and cyclosporine toxicity. These diseases have similar light and electron microscopic features and their differential diagnosis is based primarily on clinical findings. Distinction of etiologies on pathologic grounds alone is not yet feasible. The ultrastructural features (Hsu and Churg 1980; Sinclair et al. 1976) are as follows.

Diagnostic criteria. (1) Swollen, vacuolated, and focally detached glomerular endothelial cells with loss of fenestrations (Figure 12.75) so that capillaries are occluded by the endothelial cell; (2) creation of a subendothelial space that accumulates fine granular or fibrillar electron-dense and lucent material composed of fibrin, fragments of erythrocytes and platelets, and the cytoplasmic processes of endothelial and mesangial cells (Figures 12.74 and 12.79); (3) thickened GBM with double contours and/or wrinkling; (4) fibrin, red blood cells, and platelets in glomerular, arteriolar, and arterial lumina or in their walls (Figures 12.77 and 12.78).

Additional points. The subendothelial region of the GBM may show lamination (Figure 12.76), especially in a case of severe hypertension. The mesangium may be expanded and permeated by proteinaceous material similar to that seen in capillary loops (Figures 12.74 and 12.79). Podocyte foot process effacement may be seen. The arteriolar and arterial endothelium appears reactive and swollen. Arterial intimal thickening results from the accumulation of myoepithelial cells and matrix. Myoepithelial cells migrate from the media; they are spindle shaped with smooth muscle differentiation manifested by the presence of a basement membrane, pinocytotic vesicles, cytoplasmic myofilaments, dense bodies, and subplasmalemmal dense plaques. The matrix may be mature collagen, fibrillar material without periodicity, electron-dense granules, or electron-lucent areas. Late changes in the glomeruli include prominent duplication of the GBM and mesangial hypercellularity. Severe cases may have renal cortical necrosis (hemolytic uremic syndrome) or focal infarction (scleroderma and malignant hypertension).

By light microscopy, interlobular arteries and arterioles have moderate to severe thickening of their wall secondary to marked expansion of the intima or to intraluminal thrombi. Fibrinoid necrosis predominates in the smallest interlobular arteries and arterioles. Glomeruli have swollen endothelial cells and thickened peripheral capillary walls. The glomerular lesion progresses to give a fibrillar appearance to the tuft and, eventually, sclerosis. Immunofluorescence shows the presence of fibrin in vessel walls and glomeruli, with inconsistent admixture of immunoglobulins, especially IgM and C3.

(Text continues on page 875)

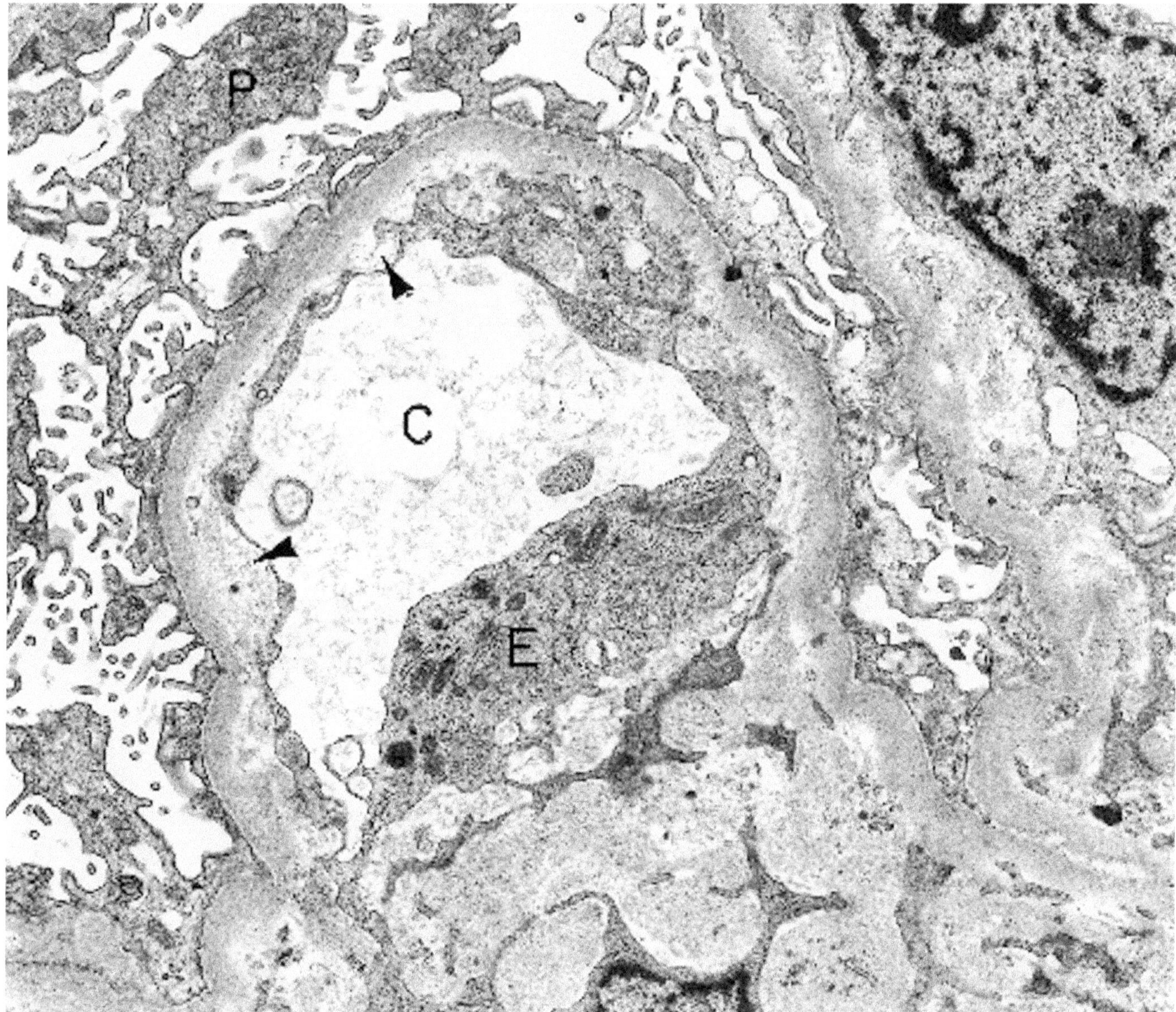

Figure 12.74. Thrombotic microangiopathy, hemolytic uremic syndrome. The subendothelial space is diffusely widened and filled with fine granular material (*arrowheads*). The endothelial cell (E) is reactive with abundant cytoplasmic organelles and loss of fenestrations. Mesangium appears edematous. C = capillary lumen; P = podocyte. (× 9000) (Courtesy of Dr. Walter Schurch, Hotel-Dieu, Montreal, Quebec, Canada.)

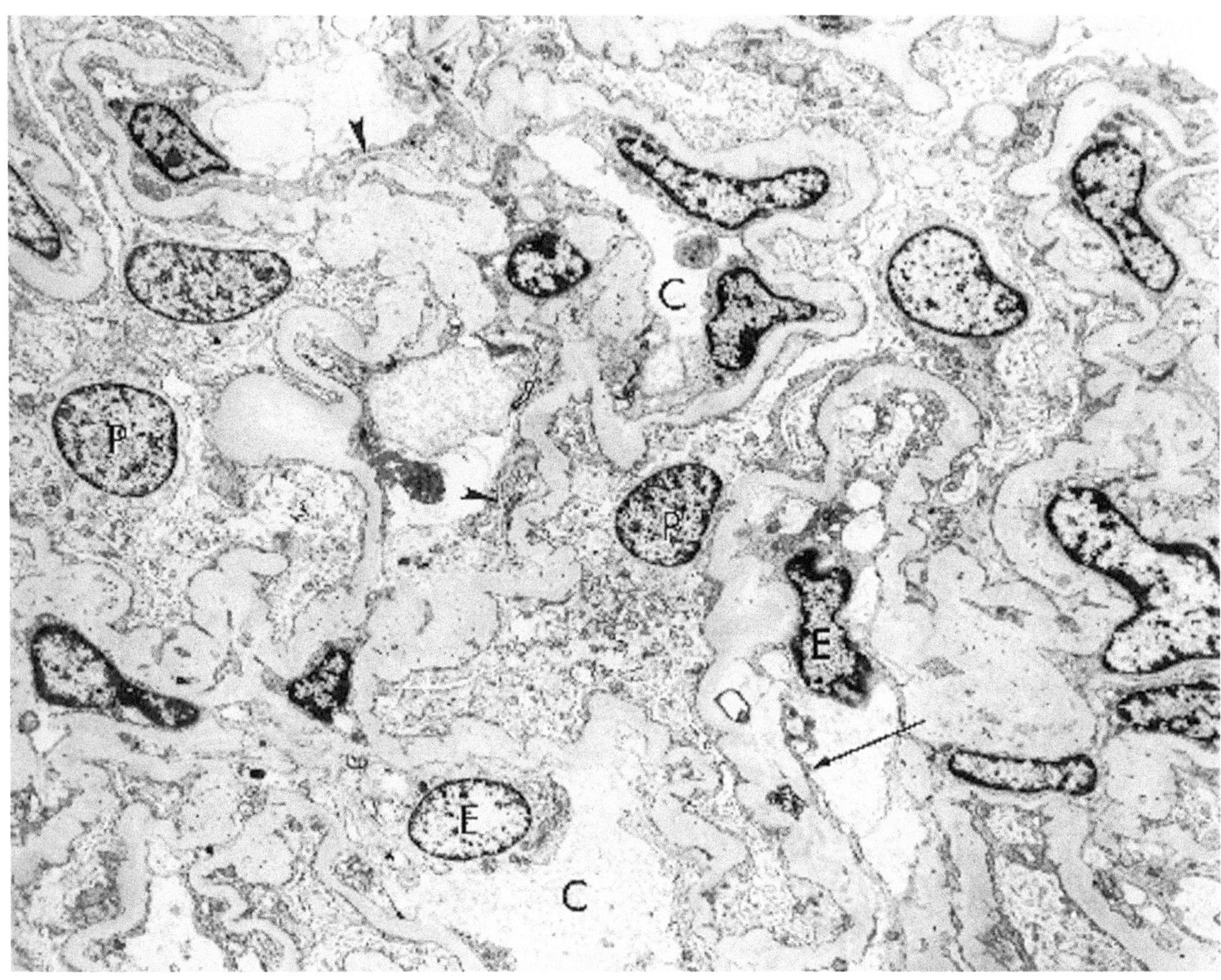

Figure 12.75. Thrombotic microangiopathy associated with malignant hypertension (33-year-old male with malignant hypertension, proteinuria, and dysmorphic red blood cells on a peripheral blood smear). Glomerulus with wrinkling of the GBM and focal duplication (*arrowhead*). The endothelium (E) is reactive with loss of fenestrae (*arrow*). P = podocyte; C = capillary loop. (× 3500)

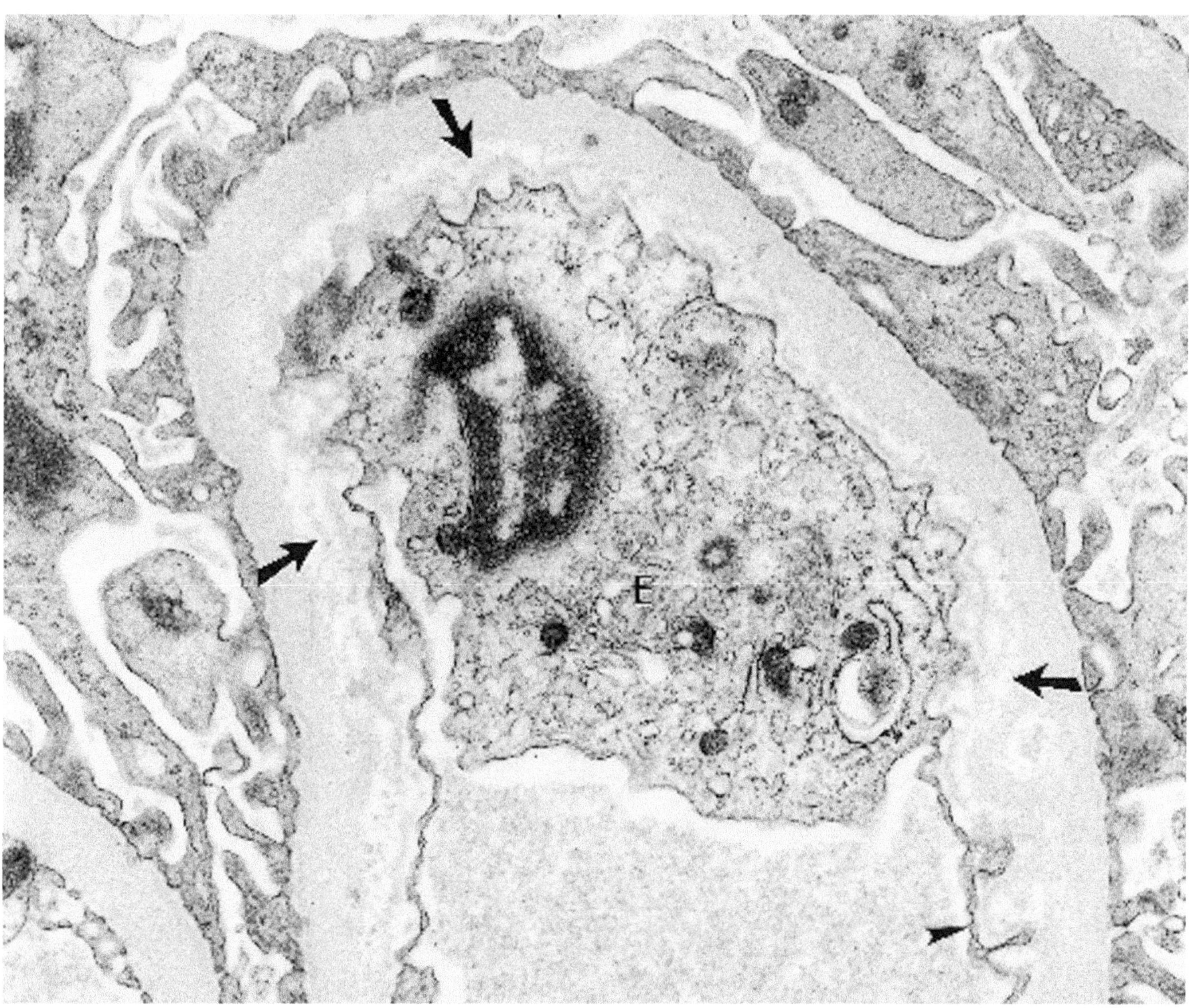

Figure 12.76. Thrombotic microangiopathy of malignant hypertension (50-year-old female with microscopic hematuria, proteinuria, and hypertension). High magnification of a glomerular capillary loop shows reactive endothelium (E) with increase in the number of cytoplasmic organelles and loss of fenestrae (*arrowhead*). The GBM in the subendothelial region is laminated and fragmented (*arrows*). (× 21,400) Note how this differs from Alport's syndrome (Figures 12.20 and 12.21).

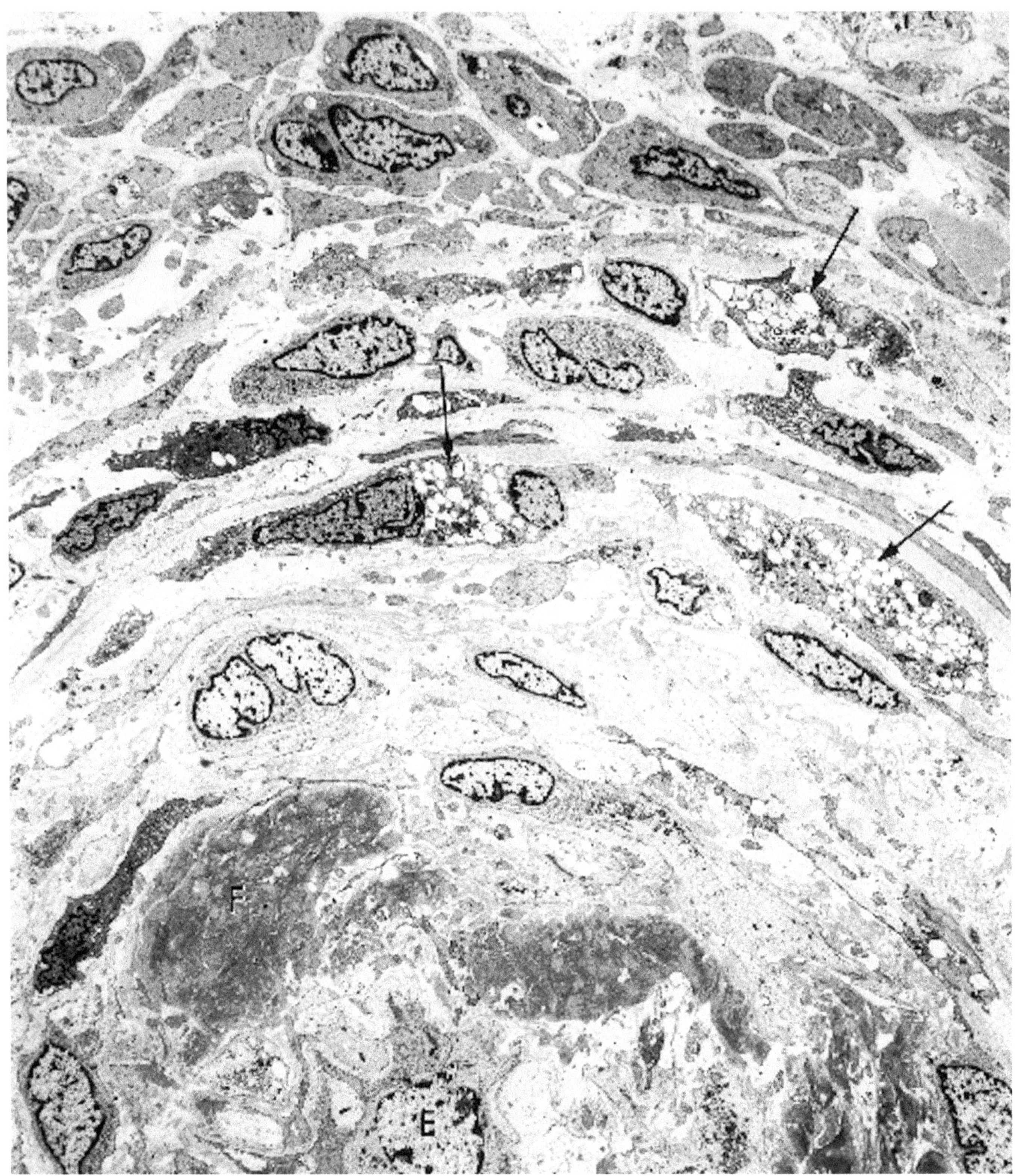

Figure 12.77. Thrombotic microangiopathy of malignant hypertension (same case as in Figure 12.75). Section of an artery that shows medial hypertrophy and intimal proliferation, with lipid accumulation in the myointimal cells (*arrows*). Collagen fibers traverse between the myocytes. Fibrin (F) is in the intima, and the endothelium (E) is swollen. (× 3900)

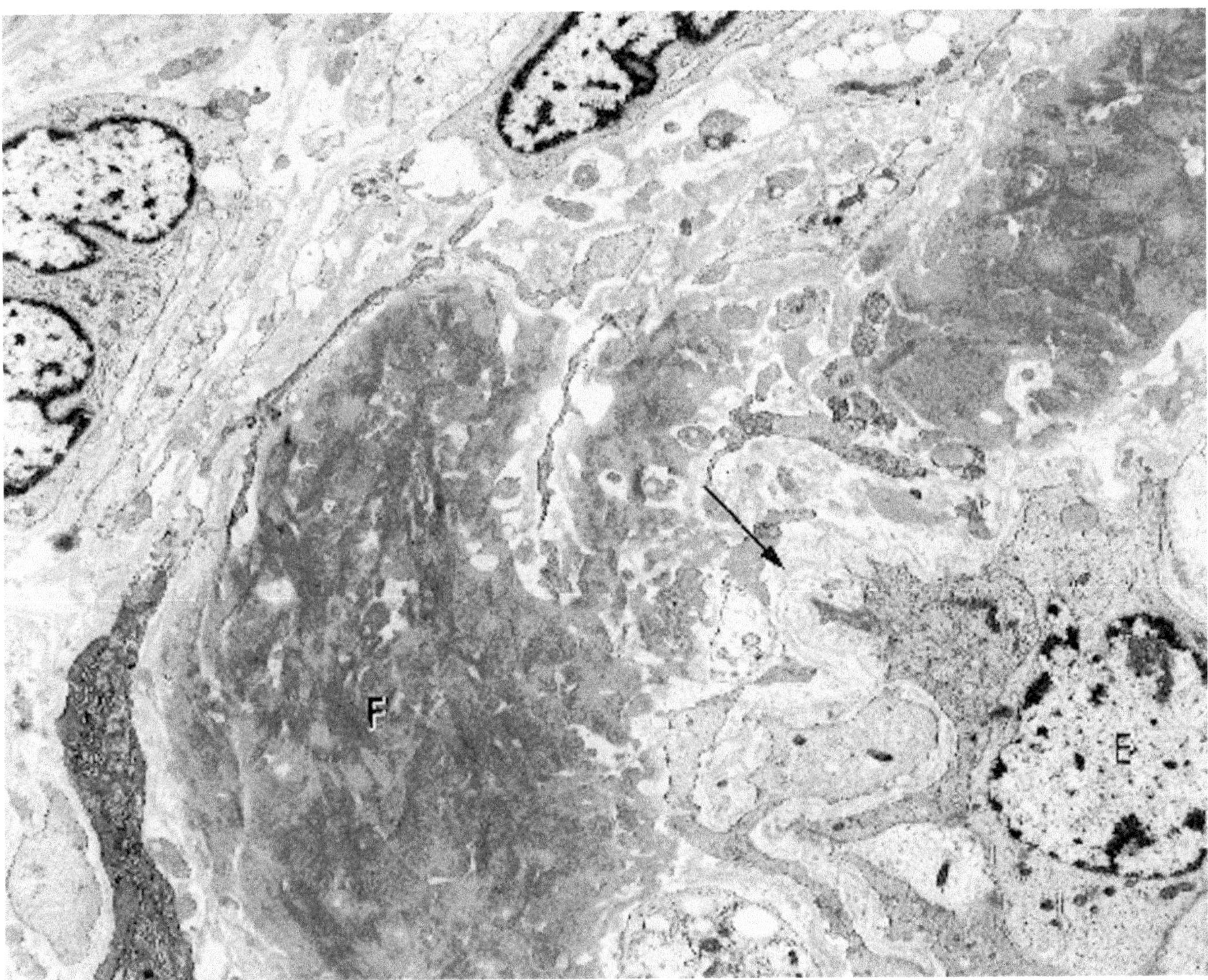

Figure 12.78. Thrombotic microangiopathy of malignant hypertension (higher magnification of Figure 12.77). Endothelial swelling (E) with subendothelial lamination and lucency (*arrow*) and fibrin deposition (F). (× 7500)

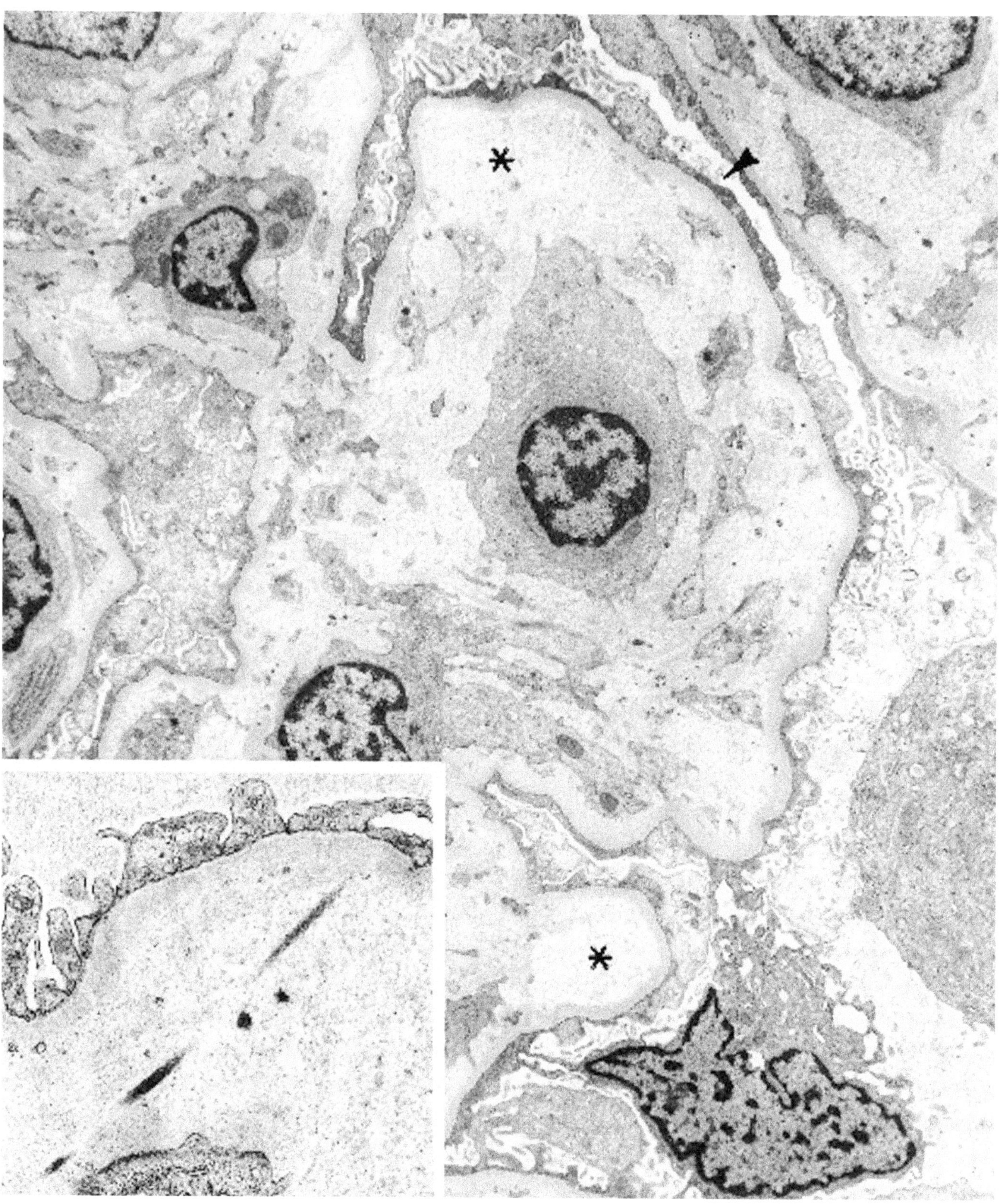

Figure 12.79. Scleroderma nephropathy, acute (27-year-old woman with acute renal failure, hypertension, and 4+ proteinuria; subsequent skin biopsy showed scleroderma). Capillary loops are collapsed and their lumina filled with lucent, amorphous granular-fibrillar matrix-like material (*) and cell processes. Epithelial cells are enlarged and have effacement of foot processes (*arrowhead*). (× 6750) *Inset:* Fibrin strands are present in the subendothelial space, admixed with fine granular background material. (× 10,640)

(Text continued from page 868)

Eclampsia/Preeclampsia

(Figure 12.80 through 12.82.)

Diagnostic criteria. (1) Markedly swollen endothelial cells, with narrowing of capillary lumens known as glomerular capillary endotheliosis (Spargo et al. 1959); (2) loss of endothelial fenestrae; (3) amorphous and fibrillar subendothelial deposits, with expansion of the lamina rara interna (Figure 12.81) (Tribe et al. 1979; Gaber et al. 1994); (4) swollen mesangial cells and increased mesangial matrix; mesangial cell interposition contributes to duplication of GBM (Tribe et al. 1979); (5) focal podocyte foot process effacement.

Additional points. The capillary loops have a cigar-shaped appearance with ballooning of the tips (Gaber et al. 1994) (Figure 12.81) secondary to the endothelial swelling and subendothelial accumulation of material. In addition, endothelial cytoplasmic vacuolization and hypertrophy of organelles, especially lysosomes with accumulation of neutral lipids, has been described. The mesangium may have electron-dense deposits. Crescents have been noted in rare instances. Tubulointerstitial lesions are usually not seen. Arterial lesions are variable and related to the severe hypertension.

Light microscopy shows diffusely enlarged, bloodless glomeruli, with endothelial swelling and narrowed capillary lumens, without hypercellularity (Sheehan 1980). The swollen glomerular compartments leads to

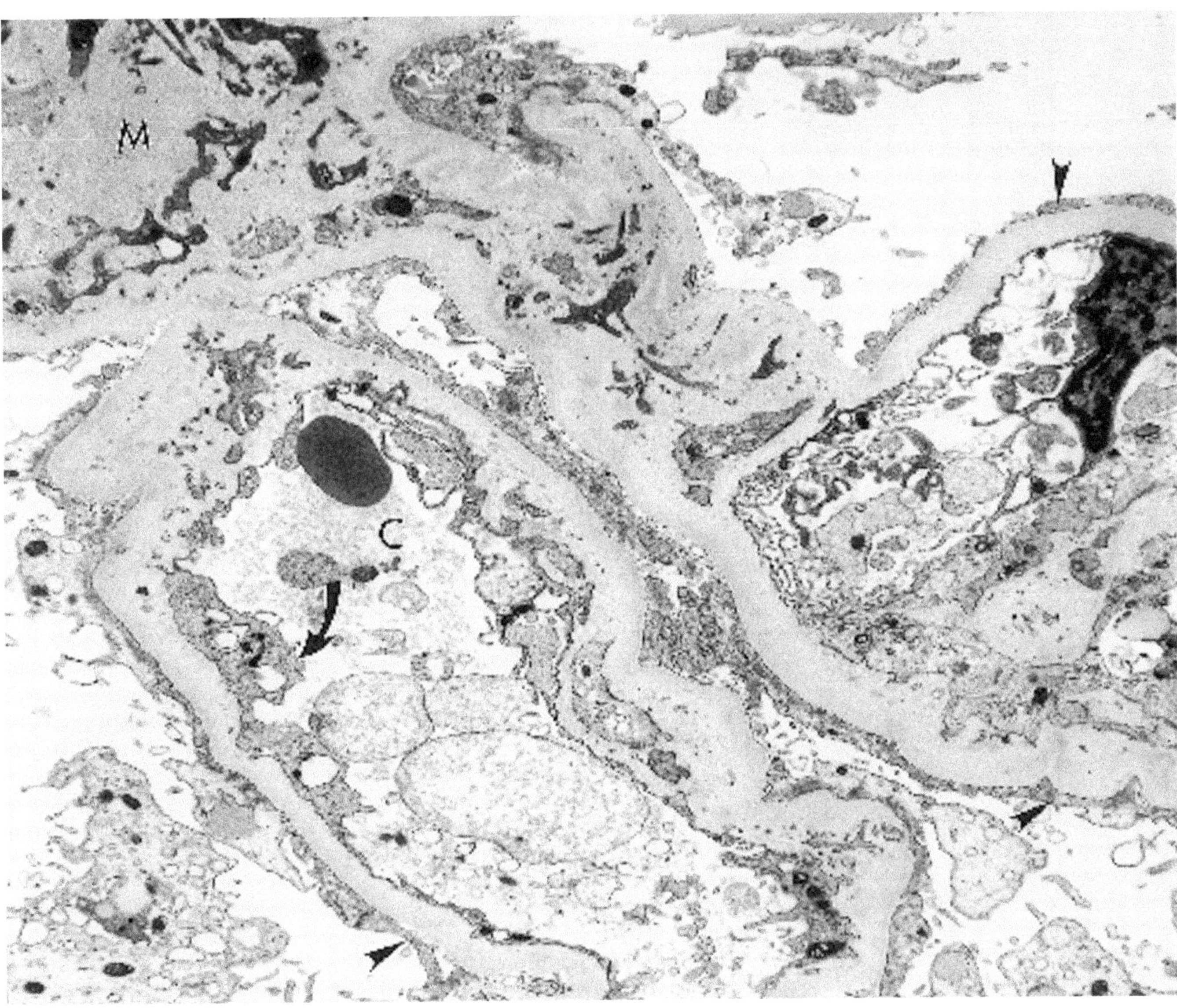

Figure 12.80. Glomerulopathy of preeclampsia (30-year-old female presented at 28 weeks' gestation with proteinuria of 7.3 g/day and rising blood pressures). Elongated glomerular capillary loops (C) that show podocyte foot process effacement (*arrowheads*). The endothelial fenestrae are lost (*arrow*): M = mesangium. (× 7400)

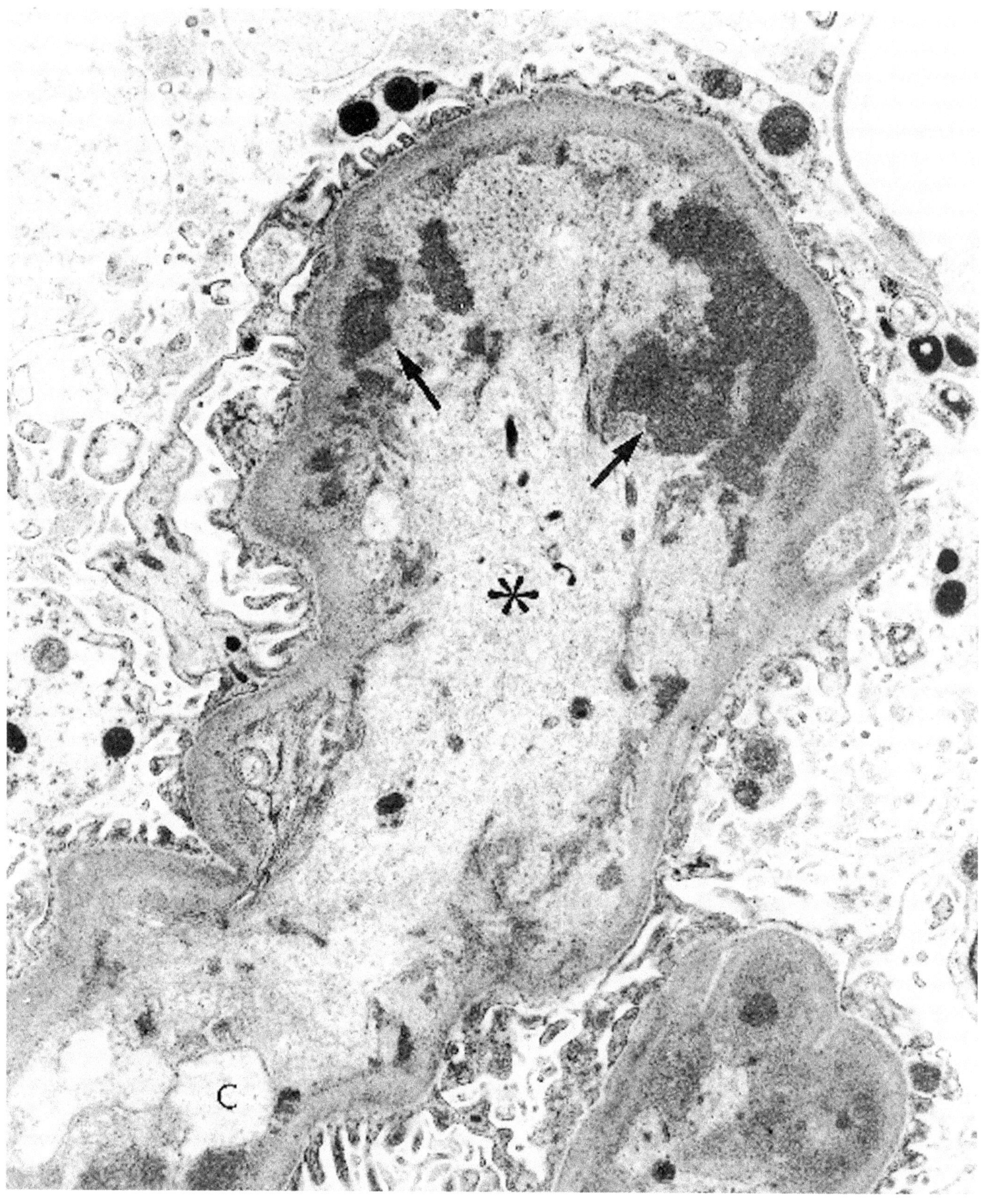

Figure 12.81. Eclampsia/preeclampsia (20-year-old female developed hypertension and proteinuria at 10 weeks' gestation; renal biopsy at 20 weeks' gestation with creatinine = 1.6 mg/dL, 3 g/day proteinuria, and positive antiphospholipid antibody). High magnification of a cigar-shaped glomerular capillary loop with a small lumen (C) and swollen endothelium (*). Amorphous electron-dense deposits (*arrows*), fibrillar and flocculent material, and cell debris accumulate under the endothelium. (× 12,000) (Electron micrograph kindly provided by Dr. A. R. Esparza, Rhode Island Hospital, Providence, Rhode Island.)

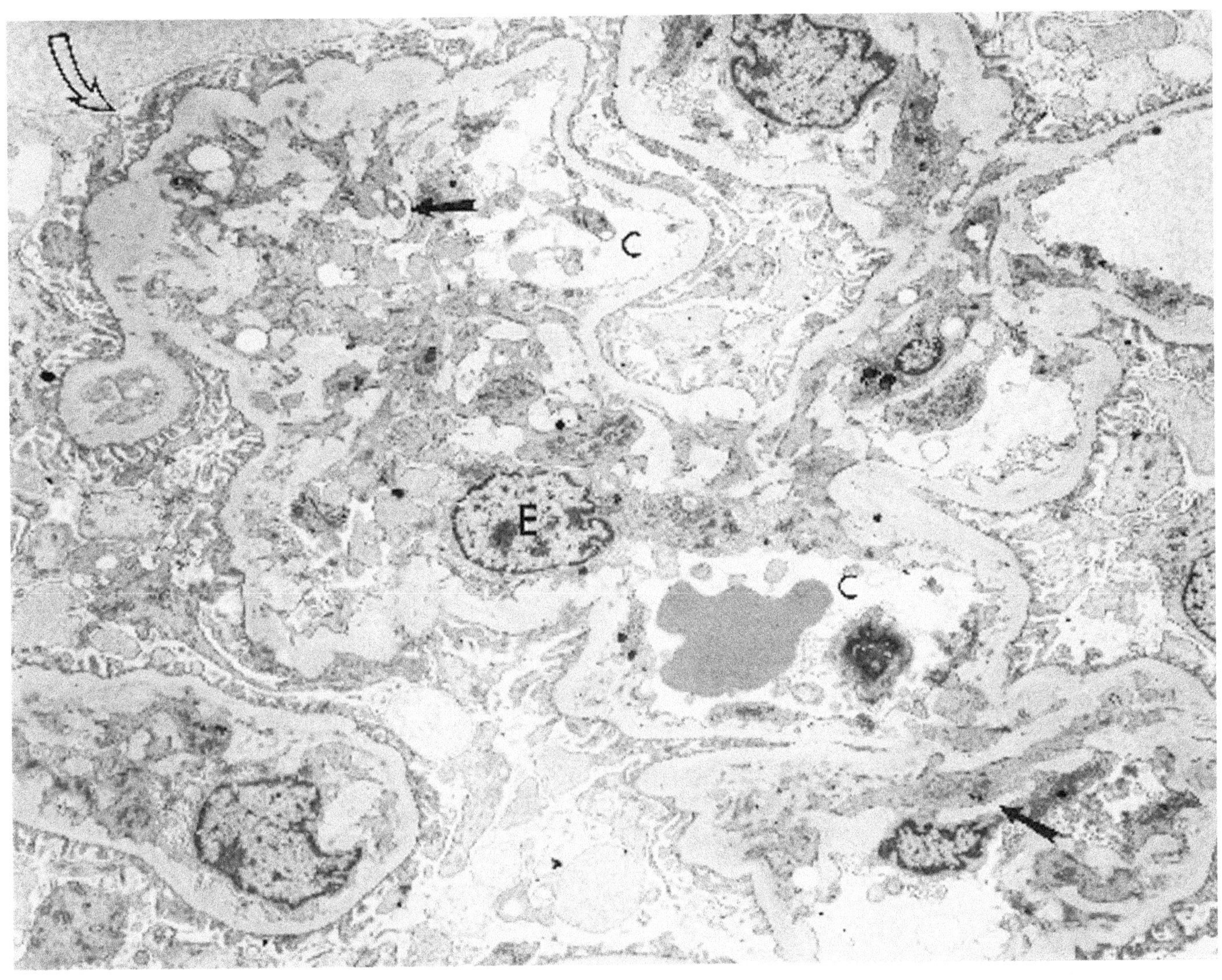

Figure 12.82. Preeclampsia associated focal segmental glomerulosclerosis (21-year-old female with history of preeclampsia at 34 weeks' gestation and persistent high-grade proteinuria postpartum). Segmental collapse of capillary loop (*open arrow*) with GBM duplication (*arrow*). Note also the loss of endothelial fenestrae: C = capillary lumen; E = endothelial cell. (× 4300)

herniation of tufts into the proximal tubule, also called "pouting" (Sheehan and Lynch 1973). Intraglomerular thrombosis is rare in preeclampsia, and when present, usually implies another superimposed process such as antiphospholipid syndrome, endotoxic shock, or other obstetrical complications. Immunofluorescence usually shows the presence of immunoglobulins (IgM), fibrin, and complement in glomeruli. The material in the subendothelial space is believed to be fibrin and related products, cell debris, and possibly immune complexes.

An association between preeclampsia/eclampsia and consequent development of focal segmental glomerulosclerosis (Figure 12.82) with tubular atrophy, interstitial fibrosis, and arteriosclerosis has been described in the literature in patients who were biopsied postpartum (Lee and Kim 1995; Kida et al. 1985).

The Renal Allograft

Acute Allograft Glomerulopathy

(Figure 12.83.)

Diagnostic criteria. (1) Diffusely swollen and reactive glomerular endothelial cells that contain abundant organelles and have enlarged open nuclei with prominent nucleoli (Colvin 1998); the cells almost occlude the capillary loops (Figure 12.83); (2) a thin layer of basement-

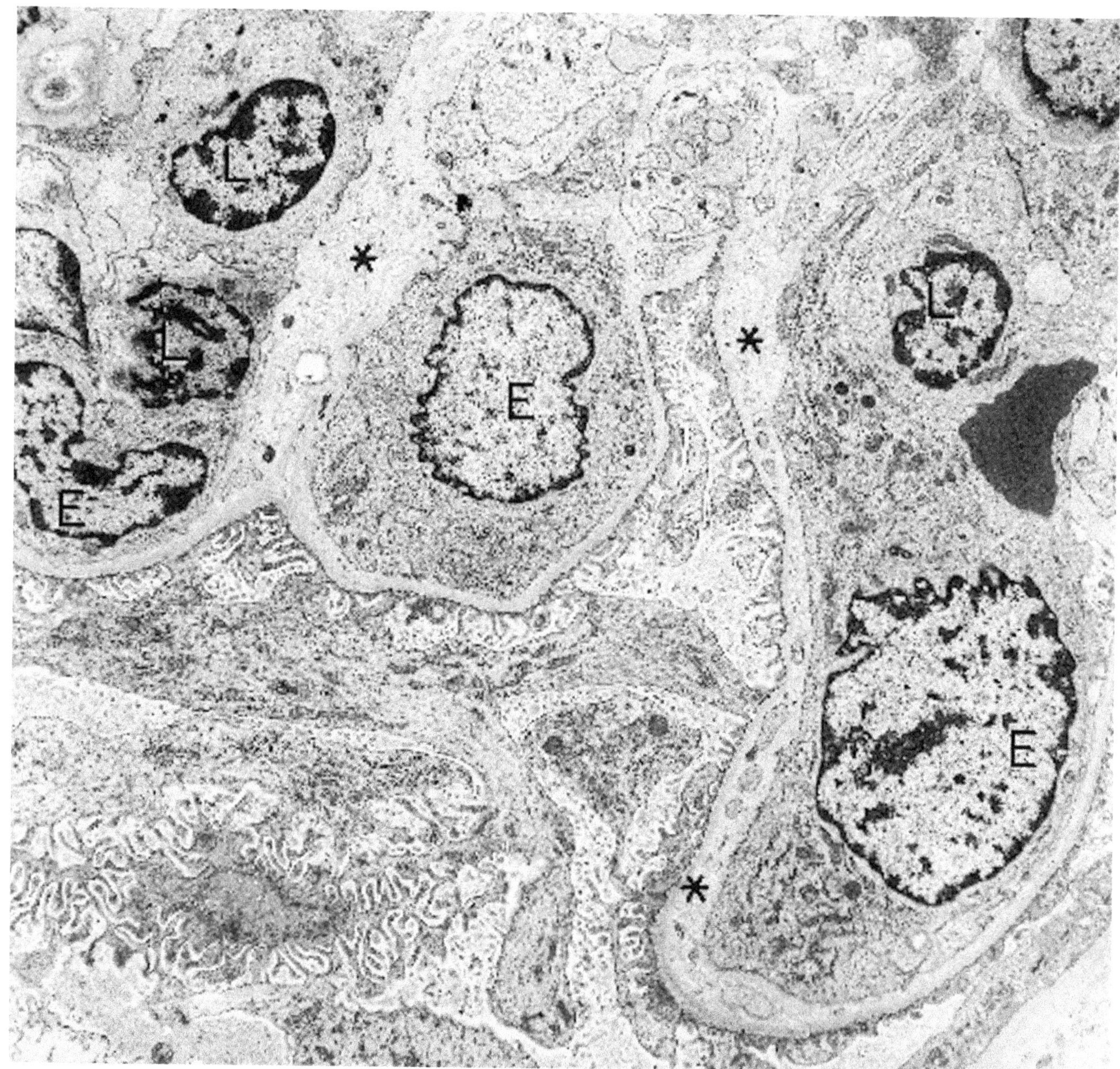

Figure 12.83. Acute allograft glomerulopathy. The endothelial cells (E) are enlarged with abundant organelles and fill the glomerular capillaries. Numerous lymphocytes (L) are in close association with the endothelial cells. Fine laminated material (*) is evident. (× 5088)

membrane-like material close to the surface of the endothelial cell corresponds to the PAS-positive webs seen by light microscopy (Richardson et al. 1981); (3) reactive lymphocytes and monocytes in glomeruli; (4) few or no electron-dense deposits.

Additional points. Capillary loops with fibrin, focal necrosis of endothelial cells, neutrophils, and platelets also may be present. Occasional endothelial cells are separated from the GBM by an accumulation of amorphous material with an electron density similar to that of plasma. The mesangium is somewhat expanded and appears to have the matrix fibrils separated by edema. Monocytes may invade the mesangium. Occasional electron-dense deposits are found. Podocyte foot process fusion is focal. No viral particles are present. In more chronic cases, duplication and cellular interposition may also result, and the features of thrombotic microangiopathy may be seen.

By light microscopy, the glomeruli are hypercellular with endocapillary proliferation and infiltration by mononuclear cells, predominately T-lymphocytes (Tuazon et al. 1987), which are chiefly CD8+. In addition, PAS-positive webs on 2-μ sections are the most definitive. Most or all of the glomeruli are involved. When only a minority are affected, the diagnosis is uncertain. Intimal invasion by mononuclear cells (endarteritis) is commonly associated (>90% of cases) (Colvin et al. 1983). The interstitial infiltrate varies and may be minimal in some cases. Immunofluorescence shows segmental fibrin in glomeruli, with scant IgM and C3 in the mesangium.

Acute allograft glomerulopathy is present in about 4% of renal transplant biopsies and is a form of cell-mediated rejection characterized by the above findings discussed earlier. Glomeruli are typically spared in typical acute cellular rejection. This glomerulopathy has been associated with cytomegalovirus (CMV) viremia in some series [Richardson], but the lesion has been described in patients with no obvious CMV infection and does not occur in autologous kidneys in CMV-infected patients (Colvin 1995). Therefore, the lesions are a form of rejection triggered by CMV and probably other stimuli such as HCV infection (Cosio et al. 1996), perhaps through the enhanced HLA expression induced by interferon.

Thrombotic Microangiopathy and the Renal Allograft

(Figure 12.84.)

Diagnostic criteria for thrombotic microangiopathy in renal allografts are same as those discussed in the previous section on thrombotic microangiopathy. Additionally, in renal allograft rejection and some cases of cyclosporine toxicity, the lesions may be localized to the kidney. Thrombotic microangiopathy in renal allografts has also been found in association with increased anticardiolipin antibodies in hepatitis-C-virus positive patients (Baid et al. 1999).

Acute Humoral Rejection

(Figures 12.85 through 12.88.)

Diagnostic criteria. (1) Neutrophils or mononuclear cells in glomerular capillary loops that may contain fibrin and adherent platelets; (2) fibrinoid necrosis of the arterioles or arteries; (3) neutrophils in dilated peritubular capillaries; (4) severe and extensive endothelial injury and denudation; (5) acute tubular necrosis (Halloran et al. 1990).

Additional points. Reactive endothelial changes, including loss of fenestrae, swelling, and subendothelial lucency, are seen in glomerular, arterial, and peritubular capillaries (Colvin 1998; Trpkov et al. 1996). The absence of significant interstitial mononuclear inflammatory infiltrate is noteworthy. The peritubular capillaries are damaged, and they may disappear with consequent interstitial fibrosis and tubular atrophy.

Acute humoral rejection is a type of delayed-onset antibody-mediated rejection, usually due to circulating anti-HLA class I antibodies (Halloran et al. 1992; Trpkov et al. 1996). Antibodies to HLA class II and endothelial alloantigens have also been implicated. Immunofluorescence for conventional immunoglobulins and complement are not helpful in separating acute humoral rejection (AHR) from acute cellular rejection (ACR). Recently, the authors found (Collins et al. 1999) staining for C4d in peritubular capillaries of renal allografts with AHR but not ACR, supporting classical pathway activation and antibody-mediated injury. Early detection and aggressive management are key to improved survival.

(Text continues on page 885)

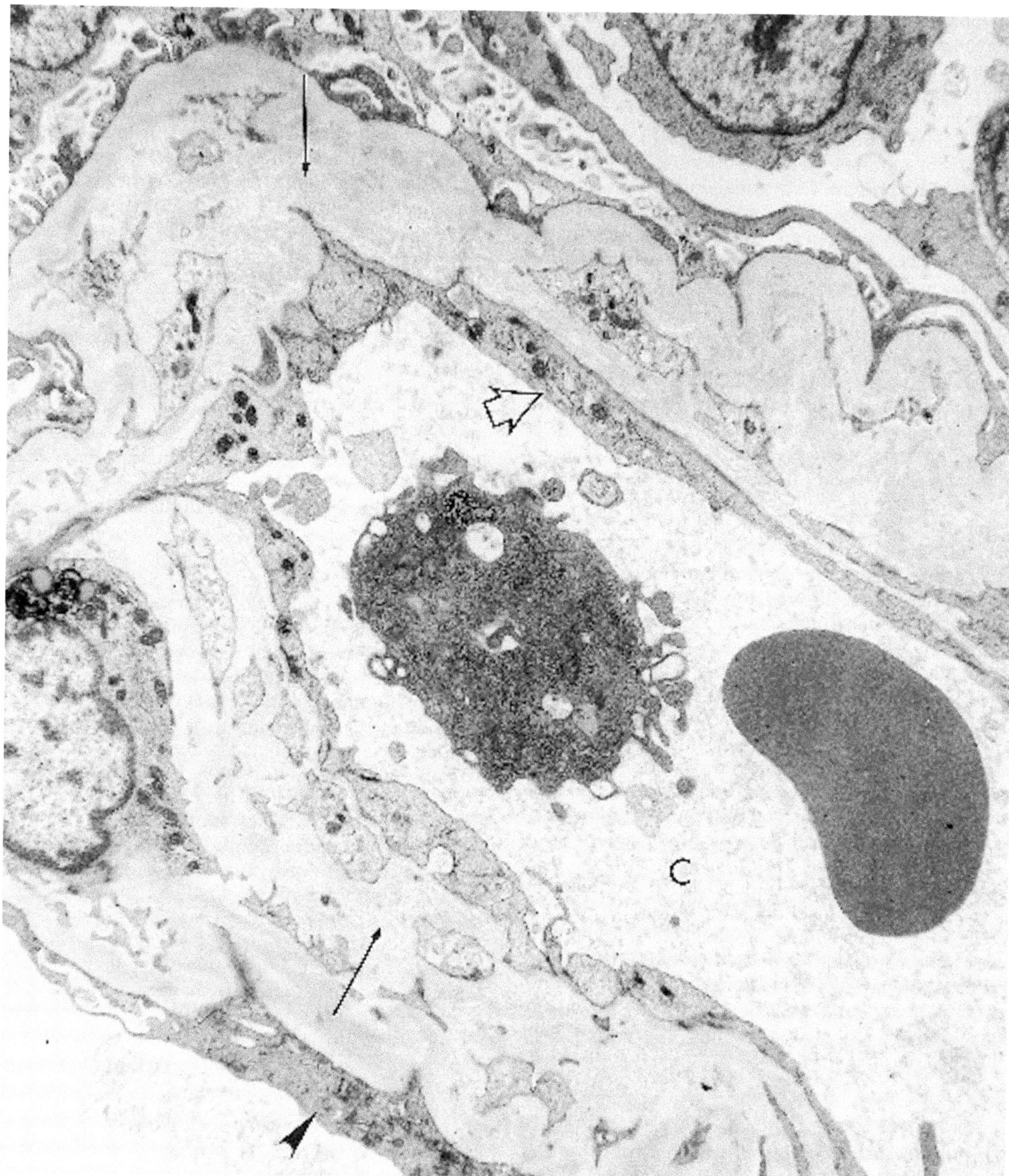

Figure 12.84. Thrombotic microangiopathy associated with acute and chronic rejection (29-year-old female status post cadaveric renal transplantation 6 years ago with creatinine elevation up to 6.5 mg/dL and hepatitis-C virus serology positive). High magnification of a capillary loop showing thickening, wrinkling, and duplication of the GBM. The GBM shows fine laminations (*arrows*). The foot processes are effaced (*arrowhead*). The endothelial fenestrae are lost (*open arrow*): C = capillary lumen. (× 10,100)

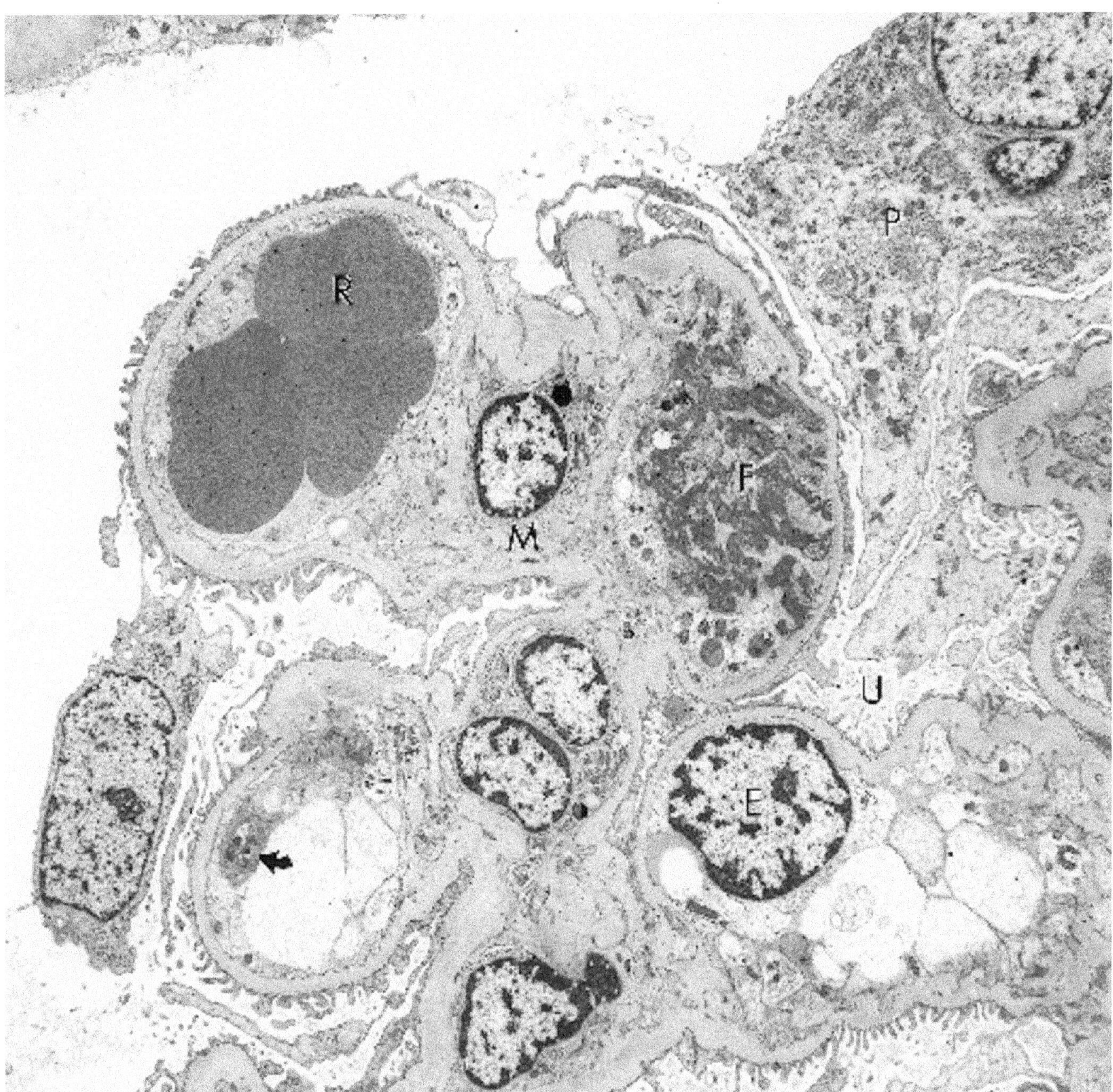

Figure 12.85. Acute humoral rejection (57-year-old female who underwent living related (daughter) renal transplantation and recently developed signs of graft failure with rise in creatinine and cross-match-positive for anti-donor HLA antibodies). Segment of a glomerulus with fibrin thrombus (F) in a capillary loop, another shows platelets (*arrow*) adherent to the endothelium. The endothelial fenestrae are lost. Another capillary loop is congested with red blood cells (R): M = mesangium; P = podocyte; U = urinary space E = endothelial cell. (× 5700)

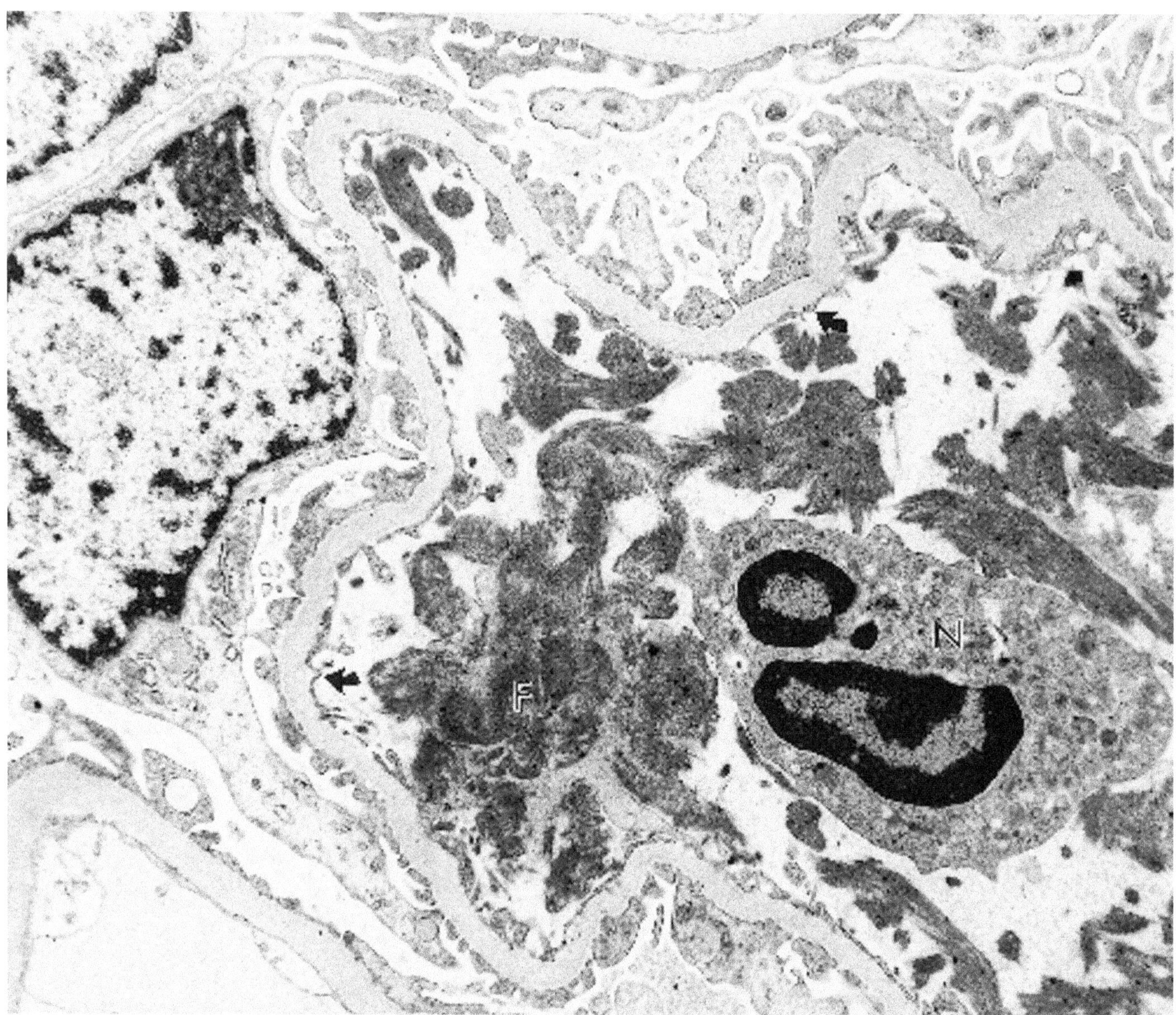

Figure 12.86. Acute humoral rejection (same case as in Figure 12.85). A glomerular capillary loop with luminal fibrin (F) and neutrophil (N). Loss of endothelial fenestrae (*arrows*). (× 11,500)

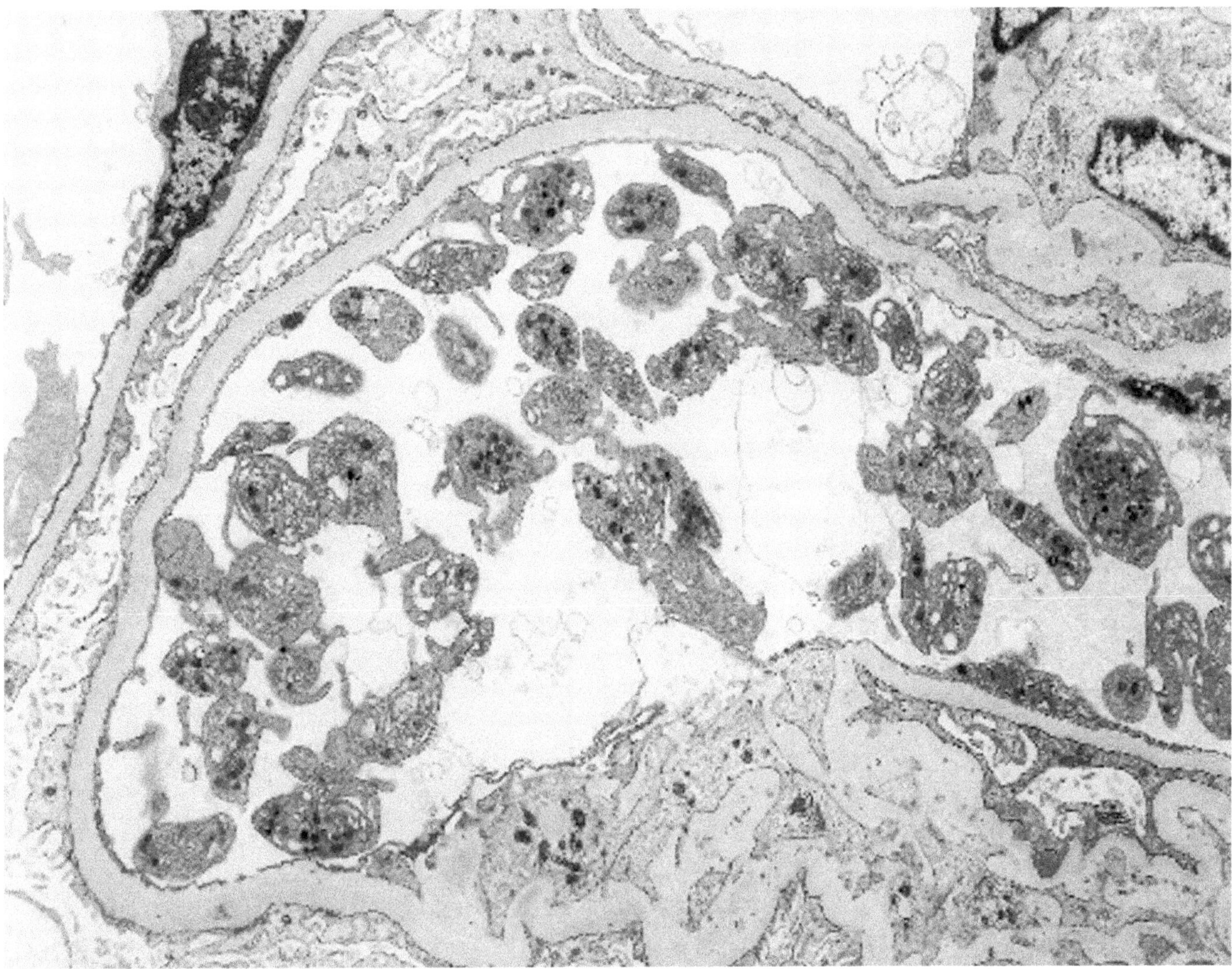

Figure 12.87. Acute humoral rejection (another patient). Numerous platelets are present in the capillary lumen; some are adhering to the endothelium. The endothelial fenestrae are lost. (× 7500)

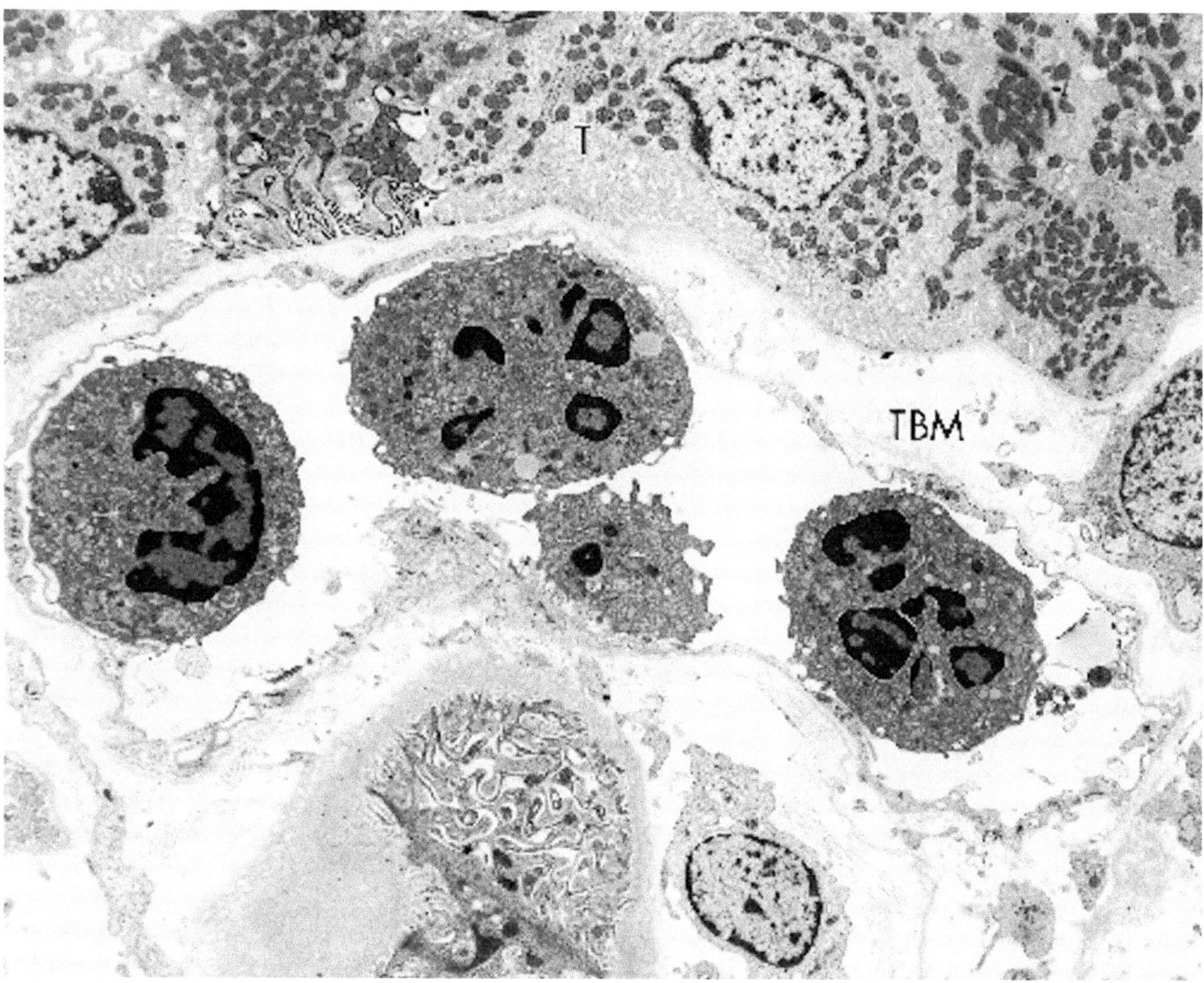

Figure 12.88. Acute humoral rejection (same case as in Figure 12.85). A peritubular capillary contains neutrophils. TBM = tubular basement membrane; T = tubular epithelium. (× 5000)

(Text continued from page 879)

Chronic Allograft Glomerulopathy

(Figures 12.89 through 12.91.)

Diagnostic criteria. (1) Prominent thickening and duplication of the GBM (Colvin 1998; Briner 1987); (2) fine amorphous granular material in the subendothelial regions, eventually involving long segments of the GBM (Figure 12.89); (3) lucent lamina rara interna expansion containing fragments of interposed cells (Figure 12.90), vesicular structures, membranous profiles, fibrin, and microfibrils; (4) laminated appearance of the GBM in advanced cases; (5) increased mesangial cellularity and matrix (Figure 12.91A), sometimes with mesangiolysis; (6) effaced foot processes in patients with heavy proteinuria.

Additional points. The endothelium generally dedifferentiates, as manifested by a loss of fenestrations (Figure 12.89). This alteration should decrease the filtration rate. Electron-dense deposits of all types have been described but are seldom extensive. The GBM is thickened and duplicated from the interpositioned cell processes mentioned in the diagnostic criteria. A distinct pattern, MPGN-type transplant glomerulopathy, is also described (Peng Fei et al. 1988), where lobular architecture, mesangial increase, and cellular interposition closely resemble type I MPGN, but lack immune deposits. The presence of abundant subendothelial deposits in such a case would then suggest a diagnosis of MPGN secondary to HCV, (Cosio et al. 1996; Baid et al. 1999).

The GBM may have abnormal structures or deposits that have been categorized into five ultrastructural types (Olsen et al. 1974): *Type I,* electron-lucent, subendothelial flocculent material is common and also seen in acute rejection. *Type II,* granular electron-dense deposits, similar to immune complexes in other diseases and typically subendothelial or mesangial. The small (*type III*) and large (*type IV*) vesicular particles once thought to be viruses, are probably cell debris. *Type V,* membranous ribbons that may be remnants of the slit diaphragms. Similar basement membrane changes are seen outside the glomerulus, including splitting and multilayering of the peritubular capillary basement membranes (Figure 12.91B) (Monga et al. 1990; Drachenberg et al. 1997) and thickening and lamination of the tubular basement membrane (Nadasdy et al. 1988). Chronic rejection lesions may also take the form of *de novo* membranous glomerulonephritis (see section on other lesions in renal transplants).

By light microscopy, GBM duplication and thickening are discerned. The glomeruli may show global or segmental sclerosis with patchy interstitial fibrosis and tubular atrophy. In cases of chronic rejection, the arteries show intimal proliferation or fibrosis with a mononuclear cell infiltrate. Immunofluorescence shows immunoglobulin and C3 along the GBM in a segmental granular pattern.

The pathogenesis of chronic allograft glomerulopathy is multifactorial. The disease is probably related to histoincompatibility, since it rarely occurs in HLA identical renal allografts. Chronic allograft glomerulopathy is usually seen in association with chronic rejection (Figure 12.91A). T-cells and antibodies may be involved with reactivity to the endothelium; support for the latter comes from animal models. We have recently identified the presence of C4d staining (immunofluorescence microscopy) in 60% of chronic rejection cases with chronic allograft glomerulopathy, also supporting the role of humoral antibody (Mauiyyedi et al. 1999). Chronic allograft glomerulopathy can also occur as a sequelae of acute allograft glomerulopathy. Similar lesions may arise secondary to chronic hepatitis C virus infection or drugs (cyclosporine).

Other Lesions in Renal Transplants

Diabetic Nephropathy

(Figure 12.92.)

Diagnostic criteria. (1) Diffuse thickening of the GBM without lamination or cellular interposition; (2) mesangial matrix increase and hypercellularity.

Additional points. Diabetic nephropathy tends to recur in almost all allografts and typically occurs 6–8 years post transplant, although the hyaline arteriopathy and GBM changes (thickening) can occur in 2 years (Lundgren et al. 1985). *De novo* diabetic glomerulosclerosis, though rare, has been described (Kelly et al. 1992; personal observation) and is usually due to the diabetogenic effects of steroids, cyclosporine, or tacrolimus in a predisposed patient. Differentiating diabetic glomerulosclerosis from chronic allograft glomerulopathy should not be difficult ultrastructurally. The former has diffusely thickened GBM that is otherwise unremarkable, whereas in the latter there is prominent duplication and lamination of the thickened GBM along with cellular interposition. Full-blown nodular diabetic glomerulosclerosis has been described 5–10 years after transplant. Simultaneous pancreas transplant often prevents this complication.

(Text continues on page 891)

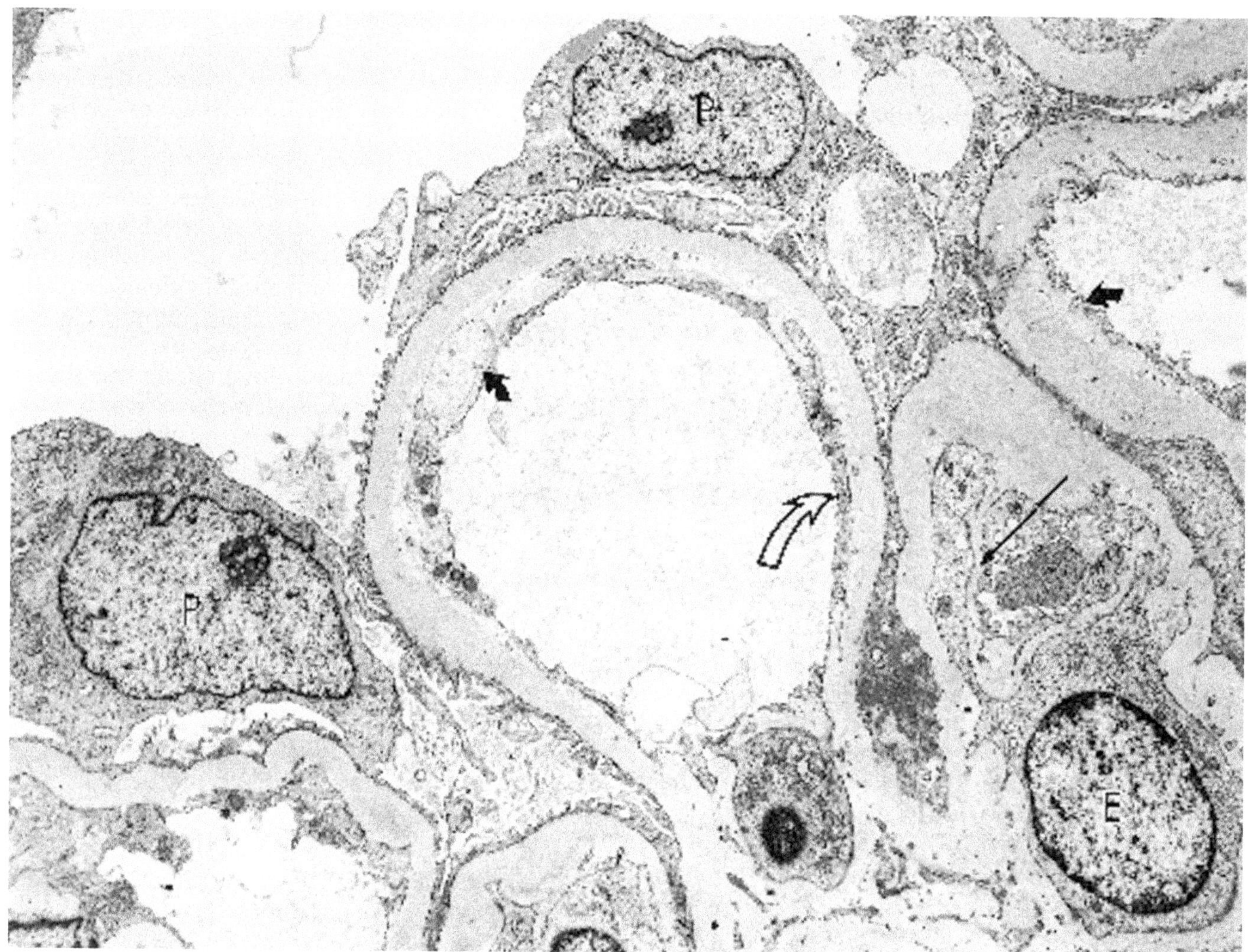

Figure 12.89. Chronic allograft glomerulopathy (second cadaveric renal transplantation 4 years ago, with recent progressive rise in creatinine). The subendothelial space is widened and lucent with lamination and scattered microparticles (*solid arrow*). The thickened GBM is duplicated (*thin arrow*). Endothelial cells (E) show loss of their fenestrations (*open arrow*). Podocytes (P) show hypertrophy and segmental foot process effacement. (× 6000)

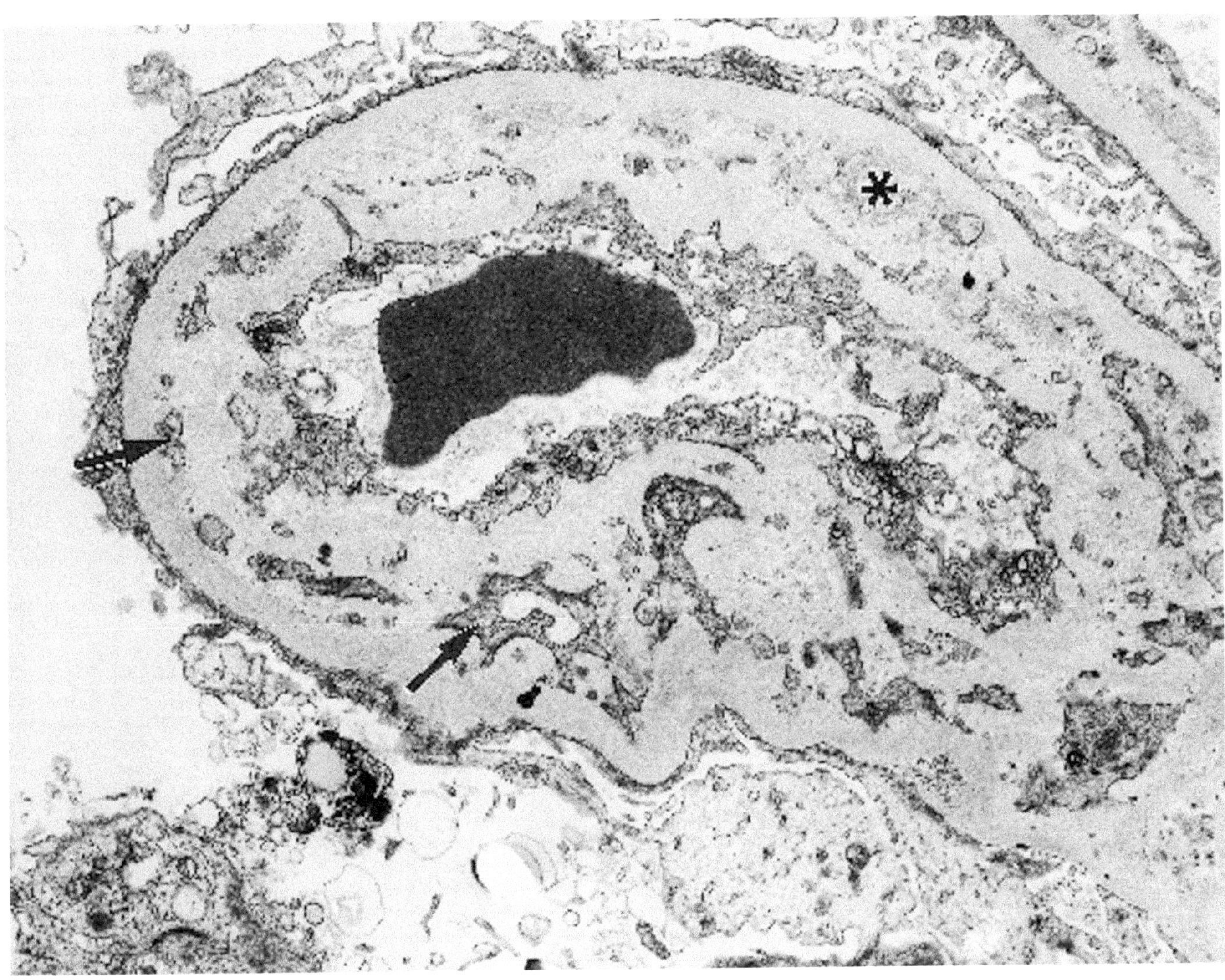

Figure 12.90. Chronic allograft glomerulopathy (same case as in Figure 12.89). A capillary loop with marked thickening and duplication of the GBM with cellular (mesangial or monocyte) interposition (*arrows*) and areas of fine granular material (*). (× 10,400)

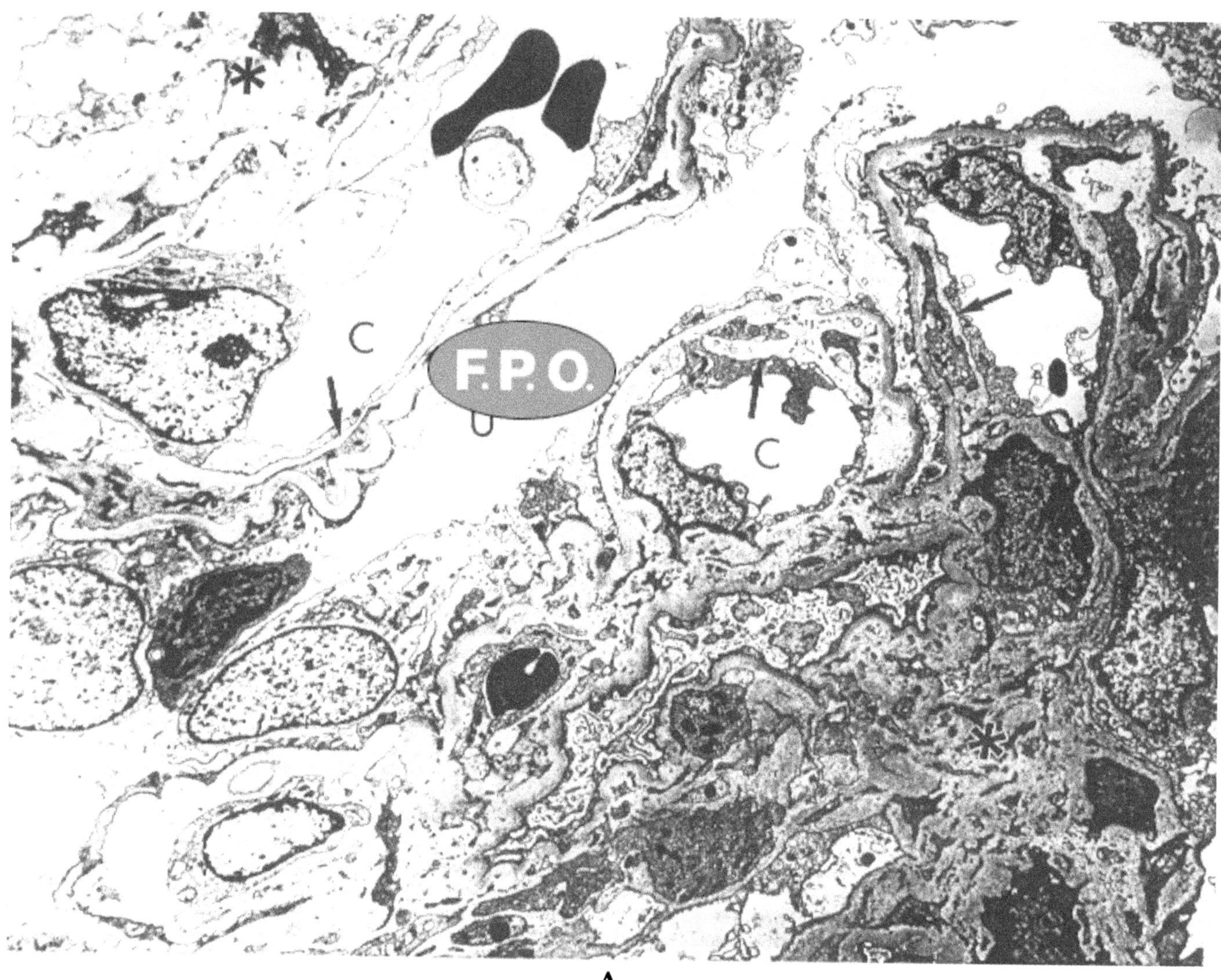

Figure 12.91. Chronic rejection (33-year-old male, who is 9 years post cadaveric renal transplant; has chronic rejection with creatinine of 4.7 mg/dL, hypertension, and 3 g/day proteinuria). **A,** Chronic allograft glomerulopathy with widespread GBM duplication (arrows), cellular interposition, and subendothelial lucency. Increase in mesangium (*) is present. C = glomerular capillary lumen; U = urinary space. (× 3400)

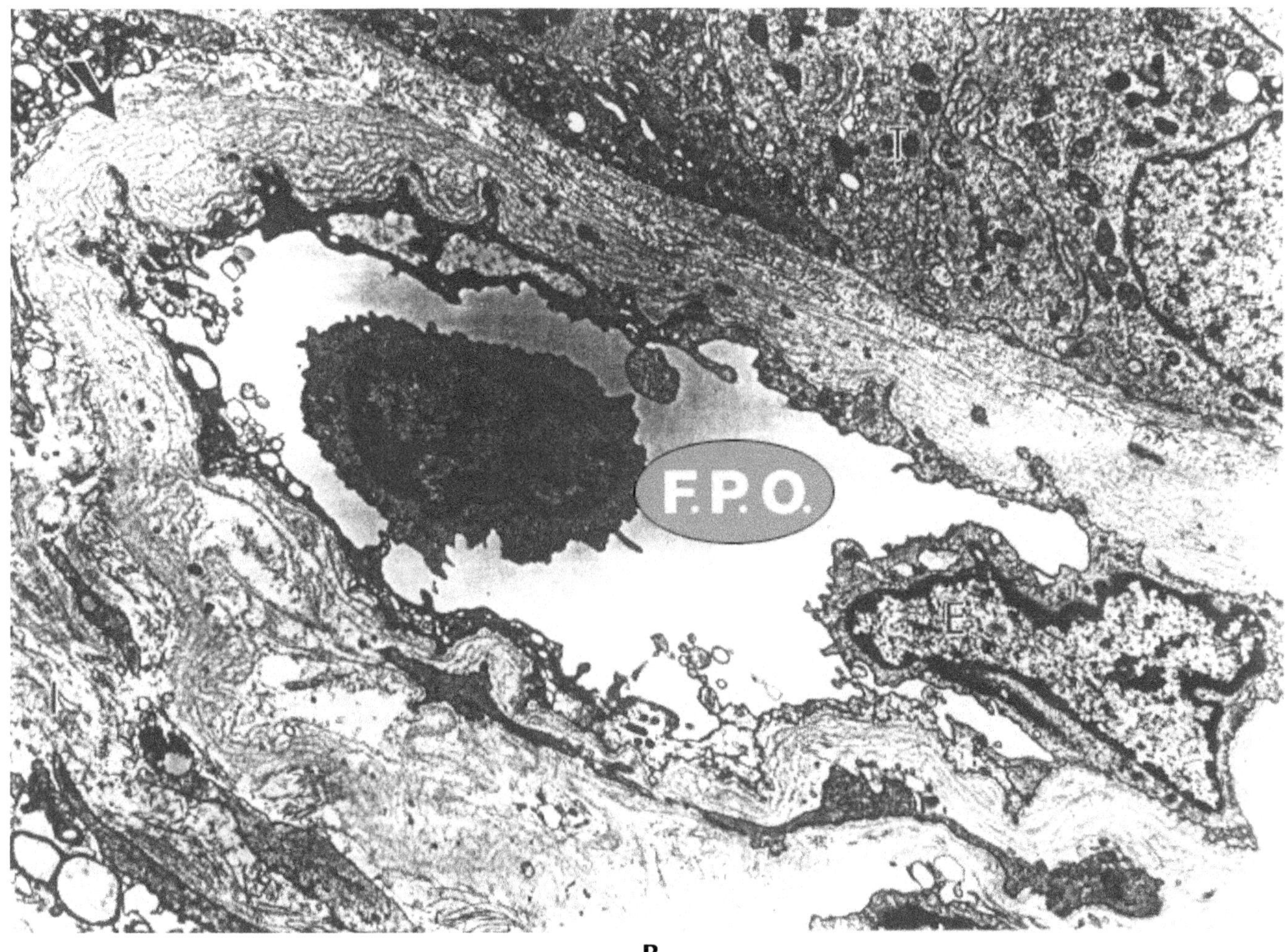

B

Figure 12.91. *(continued)*
B (same case as **A**), Peritubular capillary with circumferential lamination and multilayering of the basement membrane (arrow). E = endothelial cell of peritubular capillary; T = tubule; I = interstitium. (× 7700)

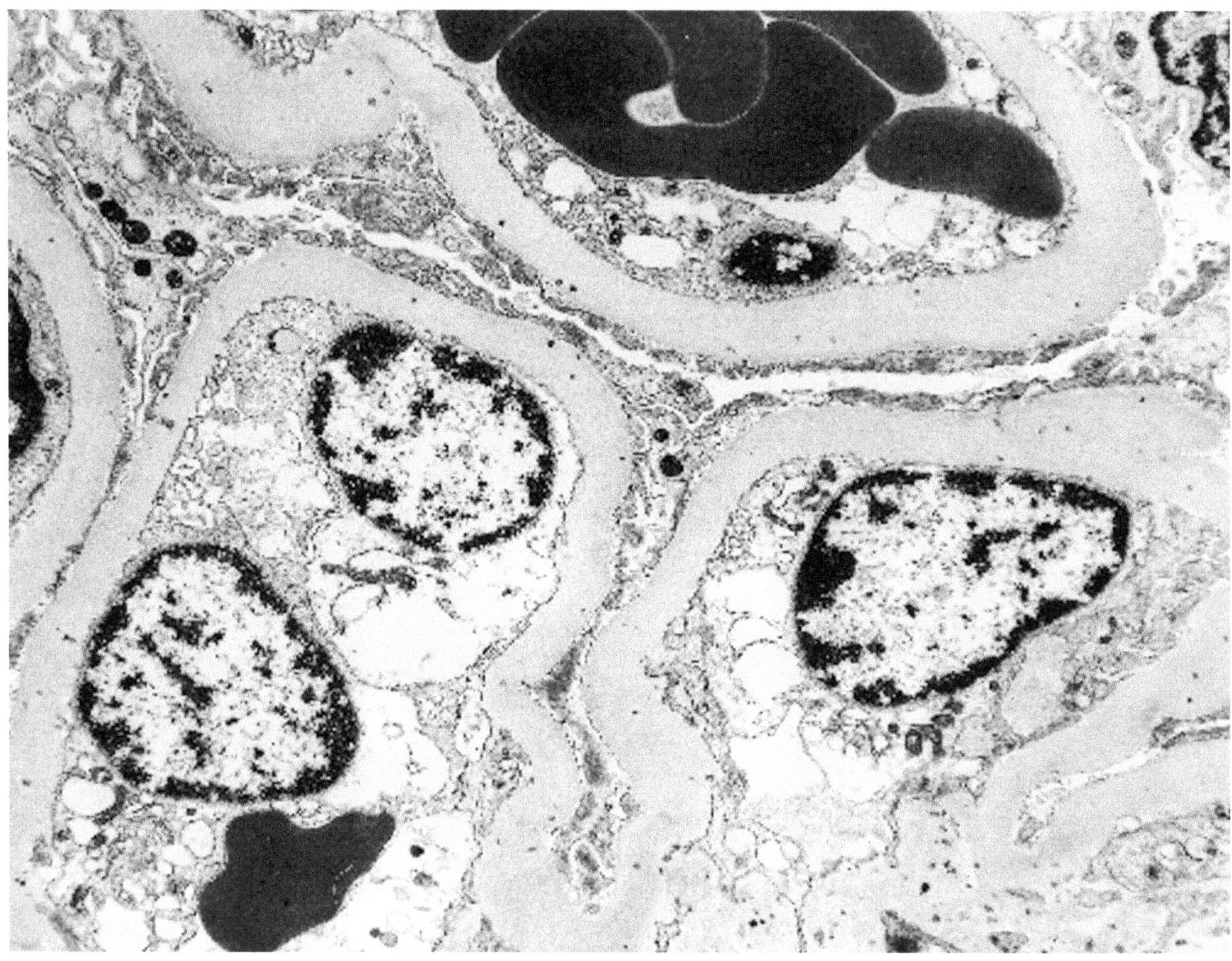

Figure 12.92. Allograft with recurrent diabetes (37-year-old male underwent living related renal transplant 14 years ago; in past year has developed proteinuria, edema, and microscopic hematuria). Segment of a glomerulus with diffusely thickened GBM (823–1176 nm) and podocyte foot process effacement. (× 8500)

(Text continued from page 885)

Membranous Glomerulonephritis

(Figure 12.93.)

Diagnostic criteria. Subepithelial deposits as described in the previous section on membranous glomerulonephritis in native kidneys.

Additional points. Membranous glomerulonephritis has an overall incidence of about 1–2% after transplant, of which *de novo* type accounts for the majority (75%). Both *de novo* and recurrent membranous glomerulonephritis are detected in early stages of the disease. In recurrent disease post transplant, the deposits tend to be more irregularly distributed by electron microscopy. In *de novo* disease, features of chronic rejection are usually seen, and may be a consequence of chronic allograft glomerulopathy (Truong et al. 1989), thus suggesting a special type of rejection. This lesion can be found in the native kidneys of bone marrow transplant recipients, but not in other transplant recipients (heart, lung, liver). The pathogenesis of *de novo* membranous glomerulonephritis post transplant is believed to be an alloantibody response to a non-MHC antigen on the podocyte. The authors have seen one case of donor-derived membranous glomerulonephritis in an allograft lost 3 days post transplant with acute infarction and renal vein thrombosis. Another case of donor-derived membranous glomerulonephritis with good graft survival is reported in the literature (Parker et al. 1995).

BK Virus (Polyomavirus)

(Figure 12.94.)

Diagnostic criteria (Colvin 1998). (1) Intranuclear inclusions in tubular epithelial cells; (2) consisting of spherical viral particles with a diameter of 30–45 nm; (3) arranged in a paracrystalline array.

Additional points. The infected tubular epithelial cell contains a large dark nucleus with loss of nuclear chromatin pattern secondary to the viral inclusions. Usually one or a few cells in a tubule are affected. Associated interstitial mononuclear cell infiltrate, including prominent plasma cells, is found.

BK virus is a polyomavirus, a member of the papovaviridae family of small, nonenveloped viruses, with a covalently closed circular double-stranded DNA genome. Polyomaviruses are ubiquitous in nature and can be isolated from a number of different species, in humans (JC, BK), in monkeys (SV40, CPV), and mice (mouse polyoma virus, K virus) (Demeter 1995). BK and JC virus share 75% homology at the nucleotide sequence level, and each is 70% homologous to SV40 (Khoury et al. 1975). Our illustrated case is a polyoma virus infection (CPV) in a primate kidney (Figure 12.94) that is similar morphologically to the BK virus. BK virus infection most commonly occurs as a consequence of reactivation, rather than primary infection in the recipient or from donor kidney or ureter. Reports of ureteral ulceration and stenosis (Coleman et al. 1978), hemorrhagic cystitis, and interstitial nephritis are described in renal transplant recipients infected with BK virus. BK virus has also caused interstitial nephritis in nontransplant patients who are immunosuppressed.

By light microscopy, infected tubular epithelial cells have large, irregular, dark, glassy, lavender nuclei accompanied by peritubular mononuclear cell infiltrate and plasma cells. Urine cytology is a useful noninvasive method in detecting and screening these patients, with presence of viral inclusion bearing cells also called "decoy" cells. Diagnosis is confirmed by demonstration of polyoma virus antigens (large T-antigen).

Cyclosporine Nephropathy

(Figures 12.95 and 12.96.)

Diagnostic criteria. Three types of cyclosporine toxicity can be seen in the kidney (Colvin 1998). *Acute lesions:* (1) fine isometric vacuolization of tubular epithelial cells; (2) arteriolar smooth muscle cell necrosis/ apoptosis (not shown). *Chronic lesions:* arteriolopathy, with hyaline-like material replacing necrotic smooth muscle cells, forming bead-like, amorphous electron-dense deposits along the media of arterioles and protruding into the adventitia (Figure 12.95). *Glomerulopathy and thrombotic microangiopathy:* (1) reactive glomerular endothelium with vacuolization of the cytoplasm and loss of fenestrae (Figure 12.96); (2) other lesions of thrombotic microangiopathy (Young, et al. 1996) in glomeruli (as discussed in section on thrombotic microangiopathy in renal allografts).

Additional points. It is sometimes difficult to distinguish acute cyclosporine toxicity from ischemia and rejection. Tubular vacuolization may occur in ischemia; however, the arteriolar smooth muscle changes though uncommon, are pathognomonic. Sacchi et al. (1987) observed microdilation and microvacuolization of smooth and rough endoplasmic reticulum in proximal tubular cells with scattered lipid droplets in some tubules. Few of their cases also showed increased numbers and size of mitochondria.

(Text continues on page 896)

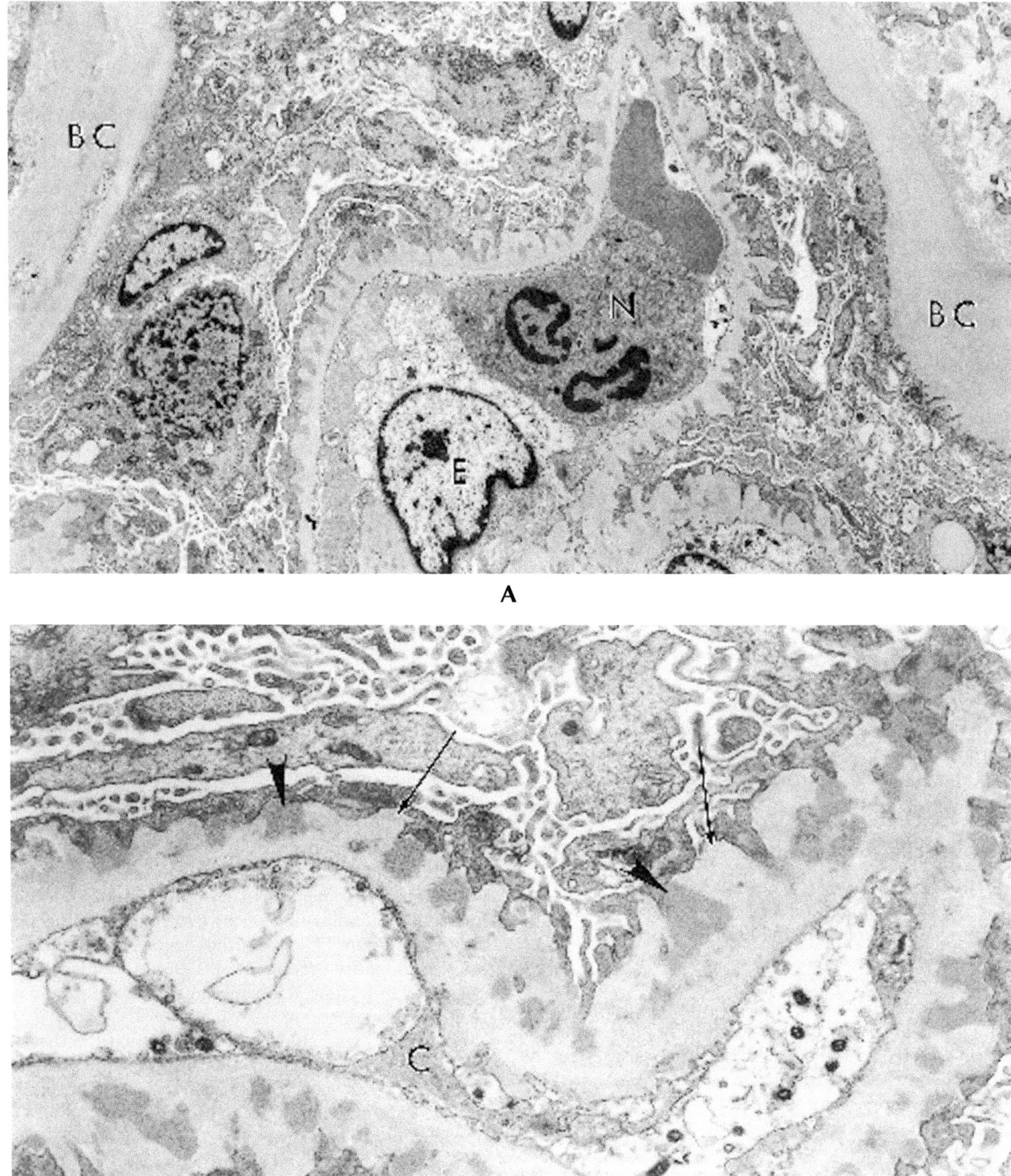

Figure 12.93. Allograft with de novo membranous glomerulonephritis (26-year-old male who developed *de novo* membranous glomerulonephritis 3 years after kidney transplant). **A** and **B,** This biopsy, 2 years later, shows persistence of subepithelial deposits (*arrowheads*), formation of spikes (*thin arrows*), and podocyte foot process effacement. Focal segmental glomerulosclerosis was also present (not shown): E = endothelial cell; N = neutrophil; BC = Bowman's capsule; C = capillary lumen. (**A,** × 5,300; **B,** × 14,700)

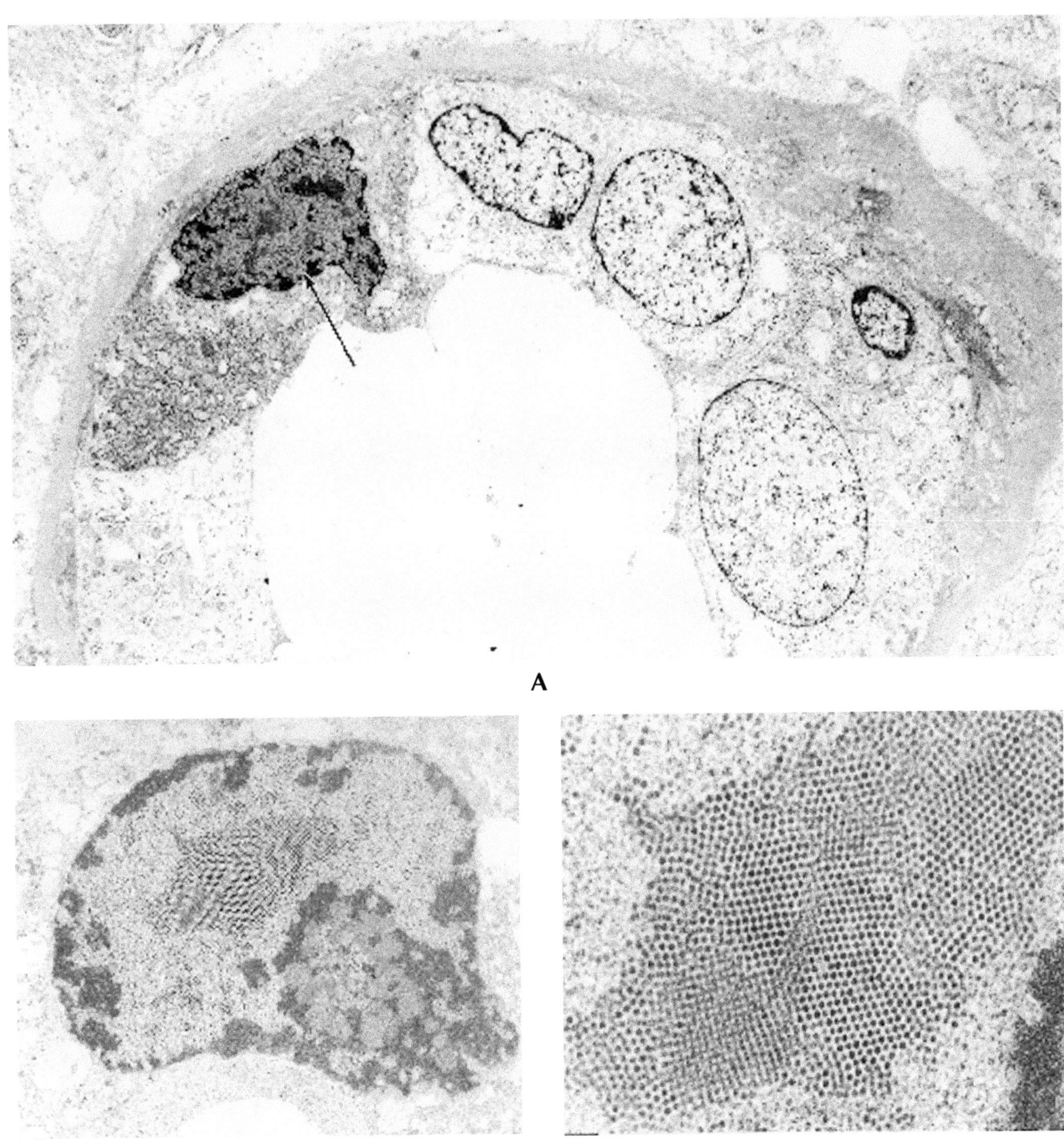

A

B

C

Figure 12.94. Polyomavirus infection (shown is a case of polyomavirus infection in a primate kidney, which is homologous to BK virus; this tissue was retrieved from paraffin). **A,** Low magnification of tubule showing one infected cell containing dark, viral-inclusion-bearing nucleus (*arrow*). (× 5600) **B,** Higher magnification of another nucleus showing the intranuclear spherical viral particles. (12,100) **C,** Higher magnification that shows compact electron-dense virus particles, each with a diameter of about 31–33 nm, that are in a paracrystalline arrangement. (× 42,000)

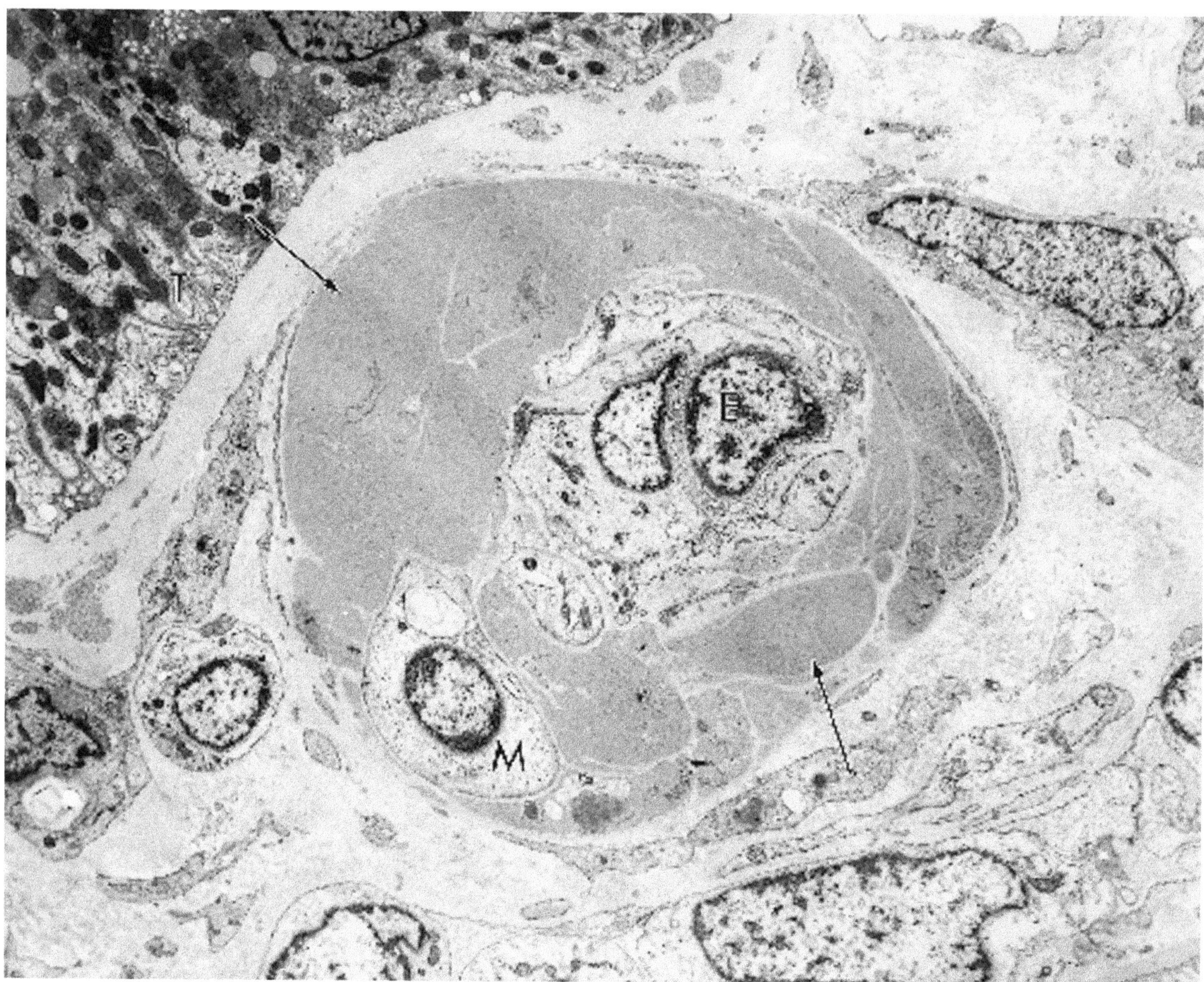

Figure 12.95. Cyclosporine toxicity (52-year-old male status post liver transplant 5 years ago, with rising creatinine). Arteriolopathy; beaded hyaline-like material (*arrows*) is present in the media replacing apoptotic myocytes. M = residual myocyte; E = swollen endothelial cell; T = tubular epithelium. (× 6200)

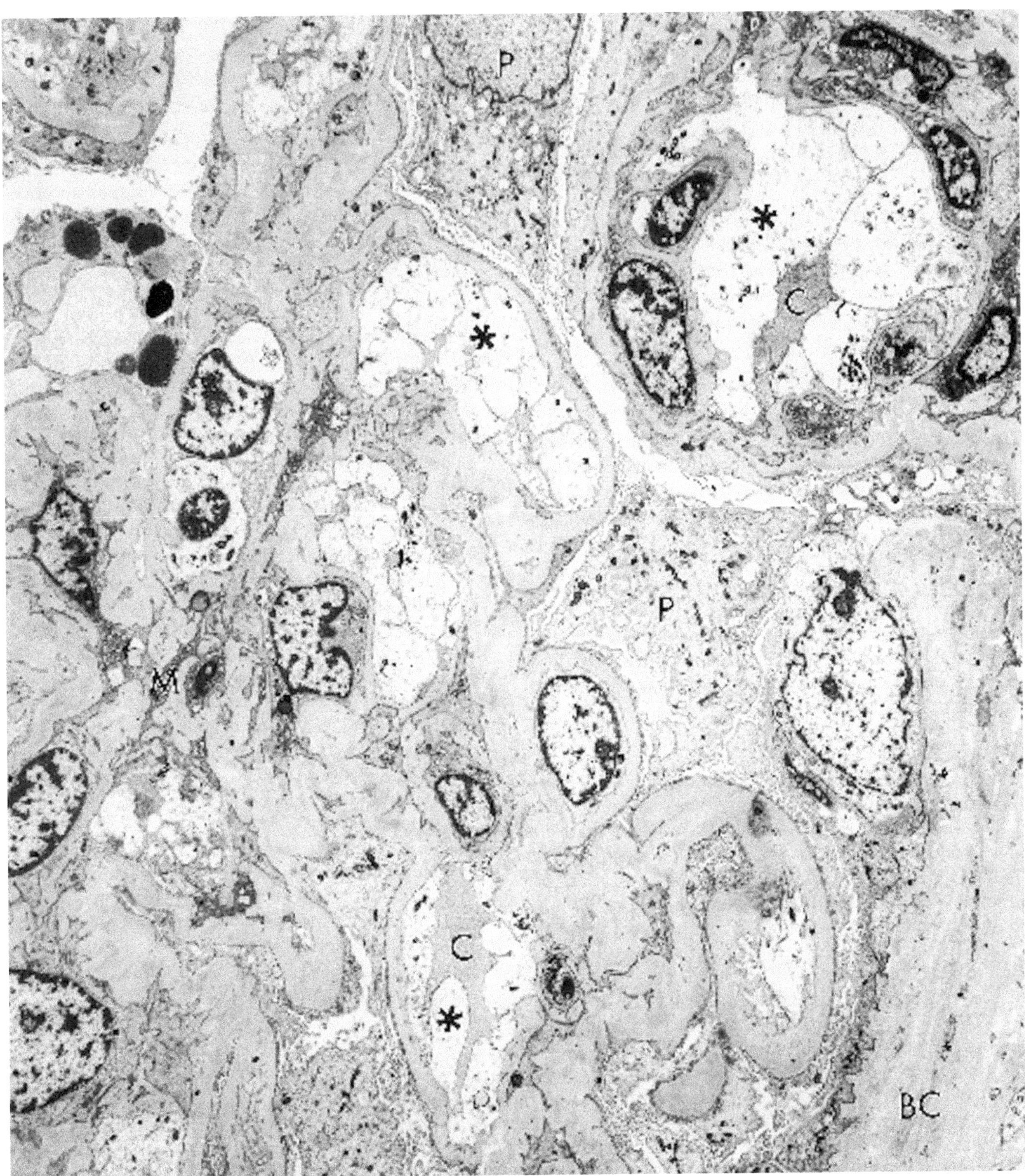

Figure 12.96. Cyclosporine toxicity (56-year-old female who developed cyclosporine toxicity effects in the kidney, status post heart-lung transplantation immunosuppression schedule). Glomerular capillary loops with swollen, reactive endothelial cells (*) and loss of fenestrae. GBM is thickened and focally duplicated. BC = Bowman's capsule; C = capillary loops; P = podocyte; M = mesangium. (× 3700)

(Text continued from page 891)

By light microscopy in the chronic form, tubulointerstitial fibrosis, striped form, with tubular atrophy is present (Myers et al. 1984; Mihatsch et al. 1995), with a variable degree of mononuclear cell infiltrate. This process begins in outer medulla, radiating into the cortical medullary rays that are perpendicular to the renal capsule; it may be a consequence of vascular pathology and therefore nonspecific. The characteristic lesion associated with cyclosporine is the arteriolopathy described above (Mihatsch 1995) predominately affecting afferent arterioles, with narrowing or complete luminal obliteration. The presence of interstitial fibrosis and this type of arteriolopathy together are consistent with cyclosporine nephropathy. The arteriolopathy is usually seen after 4–6 months of drug administration. Focal segmental glomerulosclerosis or global glomerulosclerosis with variable-sized nonsclerosed small and hypertrophied glomeruli are observed in association with the arteriolopathy (Bertani et al. 1991). Collapsing glomerulopathy (Figure 12.10) may also be seen (Mauiyyedi 1999). With prolonged exposure to cyclosporine, progressive increase in arteriolopathy and percentage of globally sclerosed glomeruli has been observed (Falkenhain et al. 1996). GBM wrinkling and thickening may become prominent and difficult to distinguish from chronic allograft glomerulopathy.

The pathophysiology (Bennett et al. 1996) of cyclosporine-associated nephropathy includes preglomerular arteriole vasoconstriction, obliterative arteriolopathy, and chronic ischemia. Various candidates suggested in the literature are thromboxane A2, platelet-derived growth factor, endothelin, angiotensin II, activation of renal sympathomimetic nerves, and reduced nitric oxide and vasodilatory prostaglandins. Others have observed an increase in collagen synthesis (*in vitro*) with small nontoxic cyclosporine doses (Ghiggeri et al. 1994) and production of cytokines that lead to renal interstitial scarring. Early macrophage infiltration and upregulation of macrophage chemoattractant, osteopontin, in proximal tubules (Young et al. 1995); inhibition of calcineurin phosphatase (mechanism of cyclosporine immunosuppression); transforming-growth-factor-beta-mediated mechanisms (Shihab et al. 1996); and inhibition of p-glycoprotein transporter (leading to nephrotoxicity) have also been implicated in the pathophysiology.

Tacrolimus (FK 506) causes a similar spectrum of morphological changes as described for cyclosporine (Randhawa et al. 1993), suggesting that the ultimate mechanism of injury must be closely related.

Miscellaneous Lesions

Microparticles in Deposits

(Figure 12.97.)

Diagnostic criteria. (1) Aggregates of small spherical microparticles present along the GBM in subepithelial, subendothelial, and rarely intramembranous and mesangial locations; (2) usually limited by a membrane around the cluster of microparticles.

Additional points. These microparticles are rounded to oval and appear solid or vesicular (clear center). They may be derived from degenerating cells or represent a nonspecific cellular response to glomerular injury. Some investigators include a viral nature to these particles. The authors have noted their presence in association with electron dense deposits (Figures 12.29, 12.42, and 12.46), especially in membranous glomerulonephritis, lupus nephritis, and chronic allograft glomerulopathy with focal segmental glomerulosclerosis (Figure 12.97). Burkholder et al. (1973) reviewed 476 cases by electron microscopy and found these spherical microparticles in 55 cases. They found the highest incidence in membranous glomerulonephritis, lupus, and focal segmental glomerulonephritis. They believed that the majority of the particles were nonviral (microvesicles from degenerating cells or lipoprotein crystalline bodies) but in some instances could be virus related. Yoshikawa et al. (1982) looked at the ultrastructure of nonsclerotic glomeruli in children with nephrotic syndrome. They found microparticles, striated bodies, and microfilaments more frequently in focal segmental glomerulosclerosis than in minimal change disease or focal global sclerosis.

Various descriptive terms have been used in the literature to describe these microparticles, including virus-like particles, spherical microparticles, extracellular bodies, and lead shots. Using immunogold labeling methods, Nakajima et al. (1991) localized complement components (C3d, C9 and C1s) in these microparticles. Their origins and nature are still unknown.

Gentamicin Bodies

(Figure 12.98.)

Diagnostic criteria. Laminated rounded myeloid bodies in tubular epithelium that may or may not be associated with acute tubular injury. These myeloid bodies are an ultrastructural alteration of lysosomes by phospholipids. They are associated with gentamicin or aminoglycoside use, even without clinical toxicity. Because gentamicin is the most common aminoglycoside used in clinical practice, these structures can also be

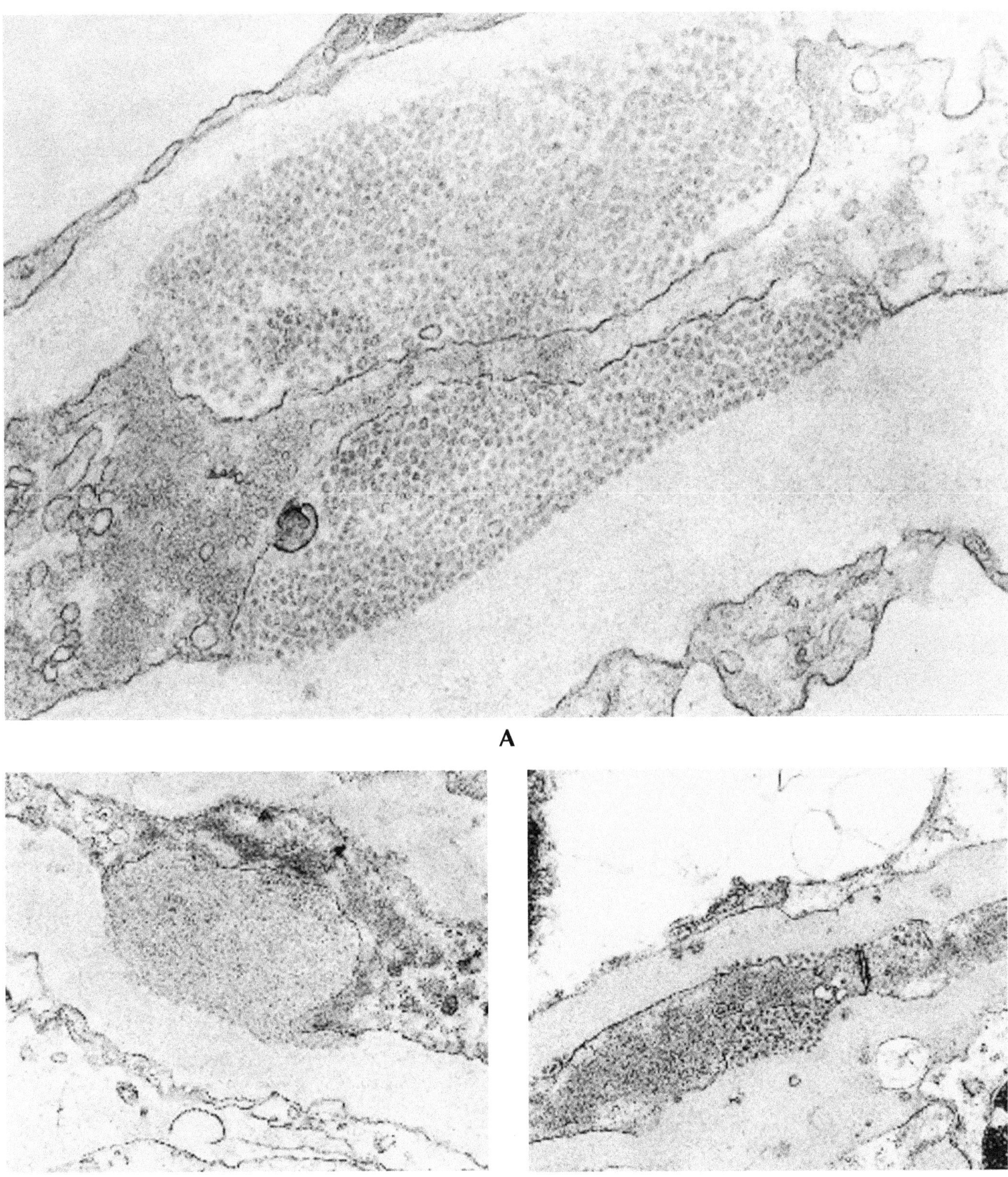

Figure 12.97. Allograft with subepithelial deposits (26-year-old female, second allograft 6 years ago, now has chronic allograft glomerulopathy on biopsy with focal glomerulosclerosis, subepithelial deposits, and arteriolar hyalinosis). **A** and **B,** These subepithelial deposits have a microparticulate substructure, where each rounded microparticle measures about 45 nm and up to 80 nm. The nature of these microparticles is unknown and may represent podocyte cell debris. (**A,** × 44,000; **B,** × 15,000) **C** (first allograft of the same patient), Note similar subepithelial deposits. (× 26,000)

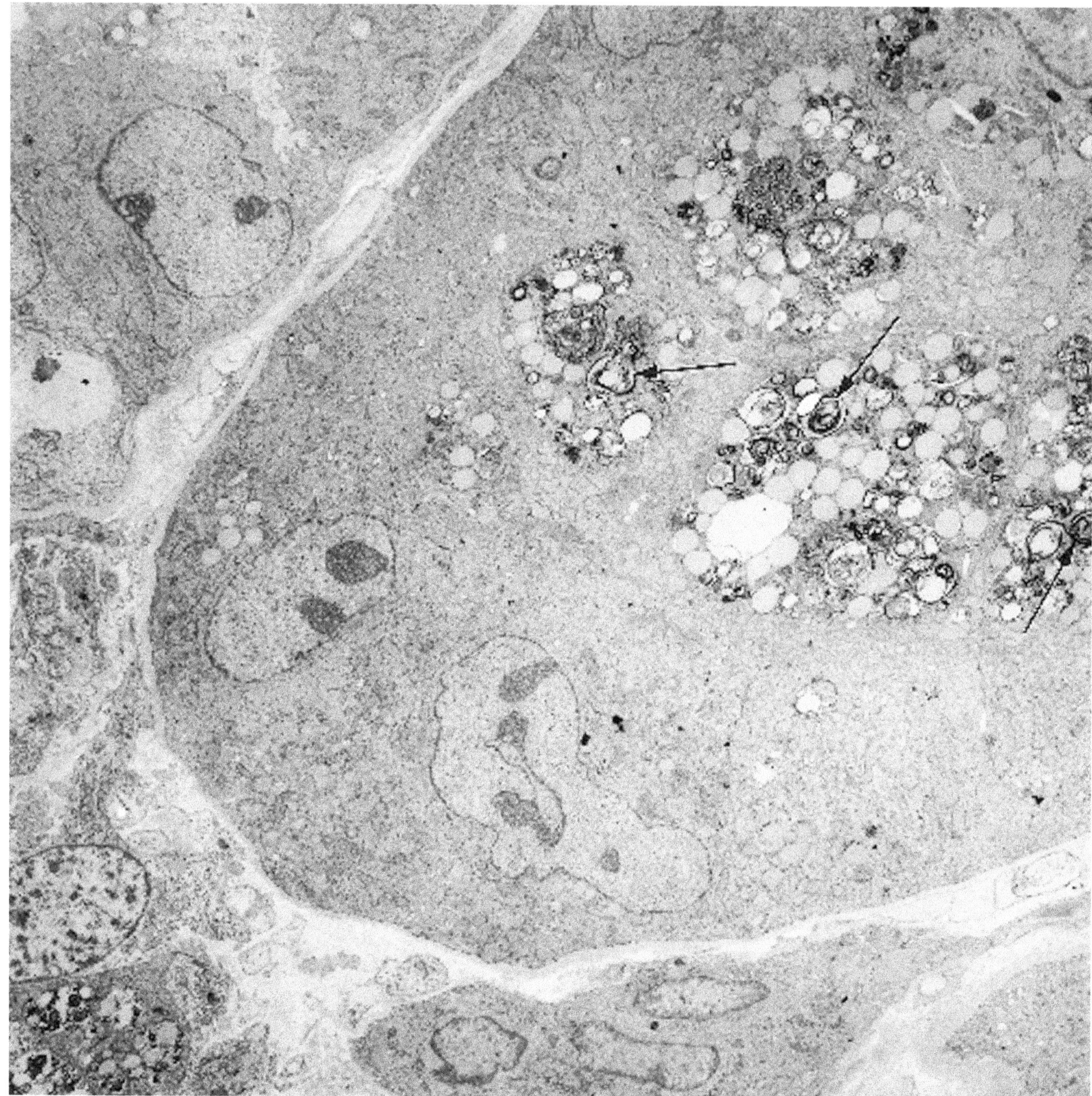

Figure 12.98. Gentamicin bodies. Large whorled crystalline inclusions (*arrows*) in tubular epithelium associated with gentamicin treatment. (× 3800)

called gentamicin bodies. Houghton et al. (1978) found these myeloid bodies by electron microscopy in proximal tubules of 19 cases (N=109), with a history of gentamicin use in 15. Nephrotoxicity was present in only one case. Animal studies have also shown their presence in distal and collecting tubules, but less frequently than in the proximal tubules (Toubeau et al. 1986).

Acute Tubular Injury

(Figure 12.99.)

Diagnostic criteria. (1) Decreased tubular epithelial brush border microvilli; (2) increased cytoplasmic blebs or swelling; (3) increased cytosomes (Jones 1982), which may be osmotic (clear) or autophagic (containing osmiophilic rounded bodies, see Figure 12.99 *inset*); (4) decreased basal and lateral plasma membrane interdigitations; (5) contraction of microfilamentous attachment bodies; (6) apoptotic cells.

Additional points. These lesions may be found in proximal or distal convoluted tubules. Acute tubular injury or acute tubular necrosis may be secondary to a variety of causes, including ischemia, metabolic abnormalities, drugs, radiocontrast material, and other toxins without specific lesions. Associations such as "isometric vacuolization" of tubular epithelium with cyclosporine or

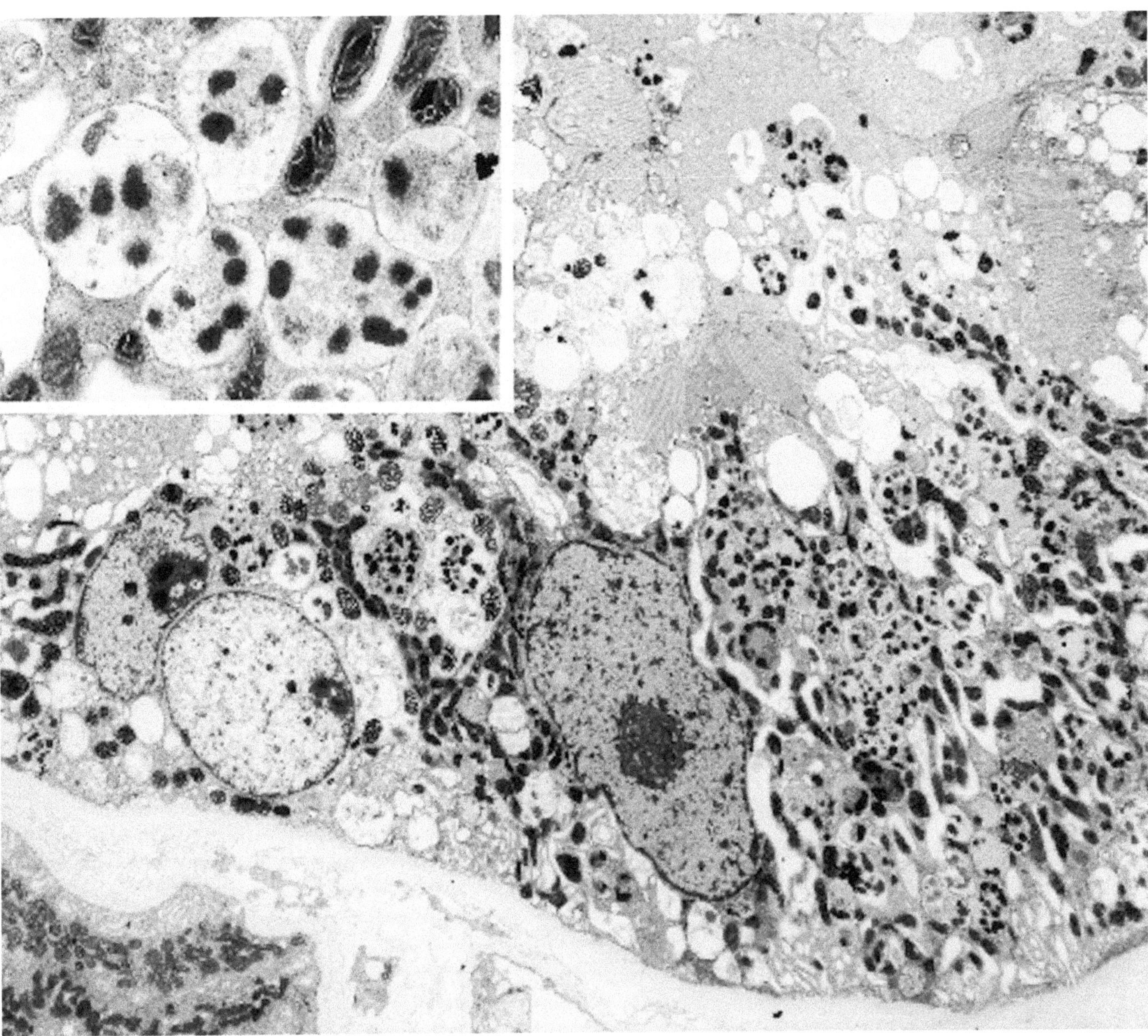

Figure 12.99. Tubular vacuolization associated with magnesium deficiency (61-year-old female has magnesium wasting syndrome and hypokalemia). The tubular epithelium shows numerous vacuoles, many of which contain electron-dense material surrounded by a unit membrane (*inset*). (× 8400) (*inset* × 18,000)

tacrolimus toxicity, though widely described, are not specific. The identification of certain crystals (e.g., uric acid), casts (myeloma), or deposits (light chain deposition disease) allude to a particular etiology. Other lesions are seen in magnesium wasting syndrome and hypokalemia with acute renal failure. Prominent acute tubular injury was noted with numerous scattered autophagic cytosomes in tubular epithelium (Figure 12.99). Schneeberger et al. (1965) in a rat model of experimental magnesium deficiency describe tubular lesions, including destructive epithelial calcification, swollen mitochondria, focally calcified lysosomes or free calcium in cytoplasm (rather than mitochondrial calcification as seen in potassium deficiency), and variable vacuoles. Further discussion of the electron microscopy of tubular lesions is beyond the scope of this chapter. Acute tubular injury cases are usually not biopsied unless of idiopathic origin or associated with glomerular lesions.

REFERENCES

General References

Andres G, Brentjens JR, Caldwell PR, et al: Formation of immune deposits and disease. Lab Invest 55:510–520, 1986.

Churg J: *Renal Disease,* Igaku-Shoin, Tokyo, 1982.

Colvin RB: Renal transplant pathology. In Jennette JC, Olson JL, Schwartz MM, et al, eds: *Heptinstall's Pathology of the Kidney,* 5th ed. Lippincott-Raven, Philadelphia, 1998, pp 1409–1540.

Colvin RB, Bhan AK, McCluskey RT: *Diagnostic Immunopathology,* 2nd ed. Raven Press, New York, 1995.

Ghadially FN: *Ultrastructural Pathology of the Cell and Matrix,* 3rd ed. Butterworths, London, 1988.

Heptinstall RH: *Pathology of the Kidney,* 4th ed. Little, Brown and Co, Boston, 1992.

Meehan S, Colvin RB: Differential diagnosis of renal allograft biopsies. In Hammond EH, ed: *Solid Organ Transplant Pathology,* vol 30. WB Saunders, Philadelphia, 1994, pp 159–185.

Tisher CC, Brenner BM, eds: *Renal Pathology,* 2nd ed. JB Lippincott, Philadelphia, 1994.

Zollinger HU, Mihatsch MJ: *Renal Pathology in Biopsy.* Springer-Verlag, Berlin, 1978.

Introduction

Haas M: A reevaluation of routine electron microscopy in the examination of native renal biopsies. J Am Soc Nephrol 8:70–76, 1997.

Siegel NJ, Spargo BH, Kashgarian M, et al: An evaluation of routine electron microscopy in the examination of renal biopsies. Nephron 10:209–215, 1973.

The Normal Glomerulus

Kriz W, Elger M, Lemley K, et al: Structure of the glomerular mesangium: A biochemical interpretation. Kidney Int 38(suppl 30):S2–S9, 1990.

Mundel P, Kriz W: Structure and function of podocytes: An update: Anat Embryol 192:385–397, 1995.

Osterby R, Gundersen HJG, Kroustrup JP, et al: Enlargement of the glomerular capillary surface and increased glomerular function in early diabetes. In Maunsbach AB, et al: *Functional Ultrastructure of the Kidney.* New York, Academic Press, 1980.

Schneeberger EE, Levey RH, McCluskey RT, et al: The isoporous substructure of the human glomerular slit diaphragm. Kidney Int 8:48–52, 1975.

Steffes MV, Barbosa J, Basken J, et al: Quantitative glomerular morphology of the normal adult kidney. Lab Invest 49:82–86, 1983.

Vogler C, McAdams AJ, Homan SM: Glomerular basement membrane and lamina densa in infants and children: An ultrastructural evaluation. Pediatric Pathol 7:527–534, 1987.

Zhu D, Kim Y, Steffes MV, et al: Application of electron microscopic distribution of type IV and type VI collagen in normal human kidney. J Histochem Cytochem 42: 577–584, 1994.

Minimal Change Disease

Bohman SO, Jaremko G, Bohlin AB, et al: Foot process fusion and glomerular filtration rate in minimal change nephrotic syndrome. Kidney Int 25:696–700, 1984.

Powell HR: Relationship between proteinuria and epithelial cell changes in minimal lesion glomerulopathy. Nephron 16:310–317, 1976.

Seiler MW, Rennke HG, Venkatachalam MA, et al: Pathogenesis of polycation-induced alterations ("fusion") of glomerular epithelium. Lab Invest 36:48–61, 1977.

Yoshikawa N, Cameron AH, White RH: Ultrastructure of the non-sclerotic glomeruli in childhood nephrotic syndrome. J Pathol. 136:133–147, 1982.

IgM Nephropathy

Cohen AM, Border WA, Glassock RJ: Nephrotic syndrome with glomerular mesangial IgM deposits. Lab Invest 38:610–619, 1978.

Trachtman H, Carroll F, Phadke K, et al: Paucity of minimal change lesion in children with early frequently relapsing steroid-responsive nephrotic syndrome. Am J Nephrol 7:13–17, 1987.

Focal and Segmental Glomerulosclerosis (Primary and Secondary Types)

D'Agati V: The many masks of focal segmental glomerulosclerosis. Kidney Int 46:1223–1241, 1994.

Howie AJ, Brewer DB: The glomerular tip lesion: A previously undescribed type of segmental glomerular abnormality. J Pathol 142:205–220, 1984.

Kiprov DD, Colvin RB, McCluskey RT: Focal segmental glomerulosclerosis and proteinuria associated with unilateral renal agenesis. Lab Invest 46:275–281, 1982.

Rennke HG, Klein PS: Pathogenesis and significance of nonprimary focal and segmental glomerulosclerosis. Am J Kidney Dis 13:443–456, 1989.

Schwartz MM, Korbet SM: Primary focal segmental glomerulosclerosis: Pathology, histologic variants, and pathogenesis. Am J Kidney Dis 22:874–883, 1993.

Verani RR: Obesity associated focal segmental glomerulosclerosis: Pathological features of the lesion and relationship with cardiomegaly and hyperlipidemia. Am J Kidney Dis 20:629–634, 1992.

Focal and Segmental Glomerulosclerosis, Collapsing Variant

Chander P, Soni A, Suri A, et al: Renal ultrastructural markers in AIDS-associated nephropathy. Am J Pathol 126:513–526, 1987.

Cohen AH, Nast CC: HIV associated nephropathy: A unique combined glomerular, tubular and interstitial lesion. Mod Pathol 1:87–97, 1988.

D'Agati V, et al: Pathology of HIV associated nephropathy: A detailed morphologic and comparative study. Kidney Int 35:1358–1370, 1989.

Detwiler RK, Falk RJ, Hogan SL, et al: Collapsing glomerulopathy: A clinically and pathologically distinct variant of focal segmental glomerulosclerosis. Kidney Int 45:1416–1424, 1994.

Grishman E, Churg J: Focal glomerular sclerosis in nephrotic patients: An electron microscopic study of glomerular podocytes. Kidney Int 7:111–122, 1975.

Mauiyyedi S, Williams WW, Tolkoff-Rubin N, et al: Glomerulopathy in chronic cyclosporine toxicity in non-renal transplant cases. Lab Invest 79:156, 1999.

Meehan SM, Pascual M, Williams WW, et al: De novo collapsing glomerulopathy in renal allografts. Transplantation 65:1192–1197, 1998.

Congenital Nephrotic Syndrome

Autio-Harmainen H: Renal pathology of fetuses with congenital nephrotic syndrome of the Finnish type. 2. A qualitative and quantitative electron microscopic study. Acta Pathol Microbiol Scand 89A:215–222, 1981.

Autio-Harmainen H, Rapola J: The thickness of the glomerular basement membrane in congenital nephrotic syndrome of the Finnish type. Nephron 34: 48–50, 1983.

Habib R, Bois E: Heterogeneite des syndromes nephrotiques à debut precoce du nourisson (syndrome nephrotique "Infantile"). Etude anatomo-clinique et genetique de 37 observations. Helv Paediatr Acta 28:91–107, 1973.

Diabetic Nephropathy

Kimmelstiel P, Wilson C: Intercapillary lesions in glomeruli of kidney. Am J Pathol 12:83–97, 1936.

Osterby R: Structural changes in the diabetic kidney. Clin Endocrinol Metab 15:733–751, 1986.

Osterby R, Gundersen HJ, Nyberg G, et al: Advanced diabetic glomerulopathy. Quantitative structural characterization of non-occluded glomeruli. Diabetes 36:612–619, 1987.

Schleicher ED, Olgemoller B: Glomerular changes in diabetes mellitus. Eur J Clin Chem Clin Biochem 30: 635–640, 1992.

Thin Basement Membrane Disease (Benign Familial Hematuria)

Basta-Jovanovic G, Venkataseshan VS, Gil J, et al: Morphometric analysis of glomerular basement membranes in thin basement membrane disease. Clin Nephrol 33: 110–114, 1990.

Bodziak KA, Hammond WS, Molitoris BA: Inherited diseases of the glomerular basement membrane. Am J Kidney Dis 23:605–618, 1994.

Hill JS, Jenis EH, Goodloe S Jr: The non-specificity of the ultrastructural alterations in hereditary nephritis with additional observations on benign familial hematuria. Lab Invest 31:516–532, 1974.

Lemmink HH, Schröder CH, Brunner HG, et al: Benign familial hematuria due to mutation of the type IV collagen $\alpha 4$ chain. J Clin Invest 98:1114, 1996.

Milanesi C, Rizzoni G, Braggion F, et al: Electron microscopy for measurement of glomerular basement membrane width in children with benign familial hematuria. Appl Pathol 2:199–204, 1984.

Steffes MV, Barbosa J, Basken J, et al: Quantitative glomerular morphology of the normal adult kidney. Lab Invest 49:82–86, 1983.

Alport's Syndrome (Hereditary Nephritis)

Gregory MC, Terreros DA, Barker DF, et al: Alport syndrome: Clinical phenotypes, incidence and pathology. Contrib Nephrol 117:1–28, 1996.

Grünfeld JP: The clinical spectrum of hereditary nephritis. Kidney Int 27:83–93, 1985.

Gubler M, Levy M, Broyer M, et al: Alport syndrome. A report of 58 cases and review of the literature. Am J Med 70:493–505, 1981.

Hill JS, Jenis EH, Goodloe S Jr; The nonspecificity of the ultrastructural alterations in hereditary nephritis with additional observations on benign familial hematuria. Lab Invest 31:516–532, 1974

Hostikka SL, Eddy RL, Byers MG, et al: Identification of a distinct type IV collagen a chain with restricted kidney distribution and assignment of its gene to the locus of X chromosome-linked Alport syndrome. Proc Nat Acad Sci USA 87:1606–1610, 1990.

Jefferson JA, Lemmink HH, Hughes AE, et al: Autosomal dominant Alport syndrome linked to the type IV collagen alpha 3 and alpha 4 genes (COL4A3 and COL4A4). Nephrol Dial Transplant 12:1595–1599, 1997.

Kashtan CE, Michael AF: Alport syndrome. Kidney Int 50:1445–1463, 1996.

Mochizuki T, Lemmink HH, Mariyama M, et al: Identification of mutations in the $\alpha3$(IV) and $\alpha4$(IV) collagen genes in autosomal recessive Alport syndrome. Nature Genet 8:77, 1994.

Rumpelt HJ, Steinke A, Thoenes W: Alport-type glomerulopathy: Evidence for diminished capillary loop size. Clin Nephrol 37:57–64, 1992.

Spear GS: Pathology of the kidney in Alport's syndrome. Pathol Annu 9:93–138, 1974.

Yoshikawa N, Cameron AH, White RH: The glomerular basal lamina in hereditary nephritis. J Pathol 135: 199–209, 1981.

Glomerular Diseases with Prominent Crescents

Bonsib SM: Glomerular basement necrosis and crescent organization. Kidney Int 33:966, 1988.

Davies DJ, Moran JE, Niall JF, et al: Segmental necrotizing glomerulonephritis with anti-neutrophil antibody: Possible arbovirus etiology. Br Med J 285:606, 1982.

Hall JB, Wadham B, Wood CJ, et al: Vasculitis and glomerulonephritis. A subgroup with an antineutrophil cytoplasmic antibody. Aust NZ J Med 14:277–278, 1984.

Jennette JC, Falk RJ: Antineutrophilic cytoplasmic autoantibodies and associated diseases: A review. Am J Kidney Dis 15:517–529, 1990.

Wegener's Granulomatosis

D'Agati V, Chander P, Nash M, et al: Idiopathic microscopic polyarteritis nodosa: Ultrastructural observations on the renal vascular and glomerular lesions. Am J Kidney Dis 7:95–110, 1986.

Jennette JC, Wilkman AS, Falk RJ: Antineutrophil cytoplasmic autoantibody-associated glomerulonephritis and vasculitis. Am J Pathol 135:921–930, 1989.

Serra A, Cameron JS, Turner DR, et al: Vasculitis affecting the kidney: Presentation, histopathology and long-term outcome. Q J Med 53:181–207, 1984.

van der Woude FJ, van Es LA, Daha MR: The role of the C-ANCA antigen in the pathogenesis of Wegener's granulomatosis: A hypothesis based on both humoral and cellular mechanisms. Neth J Med 36:169–171, 1990.

Weiss MA, Crissman JD: Renal biopsy findings in Wegener's granulomatosis: Segmental necrotizing glomerulonephritis with glomerular thrombosis. Hum Pathol 15:943–956, 1984.

Yoshikawa Y, Watanabe T: Granulomatous glomerulonephritis in Wegener's granulomatosis. Virchows Arch [A] 402:361–372, 1984.

Anti-Glomerular Basement Membrane Nephritis (Goodpasture's Syndrome)

Bazari H, Mauiyyedi S: Case Records of the Massachusetts General Hospital. N Engl J Med, in press.

Bolton KW: Goodpasture's syndrome. Kidney Int 50:1753–1766, 1996.

Bonsib SM: Glomerular basement membrane discontinuities. Am J Pathol 119:357–360, 1985.

Hudson BG, Kalluri R, Gunwar S, et al: Molecular characteristics of the Goodpasture autoantigen. Kidney Int 43:135–139, 1993.

Kalluri R, Wilson CB, Weber M, et al: Identification of the $\alpha3$ chain of type IV collagen as the common autoantigen in anti-basement membrane disease and Goodpasture's syndrome. J Am Soc Nephrol 6(4): 1178–1185, 1995.

McCluskey, Collins AB, Niles JL: Kidney. In Colvin RB, Bhan AK, McCluskey RT, eds: *Diagnostic Immunopathology*, 2nd ed., Raven Press, New York, 1995, pp 117–119.

Rosenblum ND, Colvin RB: Case Records of the Massachusetts General Hospital (Case 16-1993). N Engl J Med 328:1183–1190, 1993.

Senekjian HO, Knight TF, Weinman EJ: The spectrum of renal diseases associated with anti-basement membrane antibodies. Arch Intern Med 140:79–81, 1980.

Sisson S, Dysart NK Jr, Fish AF, et al: Localization of the Good Pasture Antigen by immunoelectron microscopy. Clin Immunol Immunopathol 23:414–429, 1982.

Postinfectious Glomerulonephritis

Churg J, Grishman E: Ultrastructure of immune deposits in renal glomeruli. Ann Intern Med 76:479, 1972.

Sorger K, Gessler U, Hubner FK, et al: Subtypes of acute post-infectious glomerulonephritis, synopsis of clinical and pathological features. Clin Nephrol J 17:114–128, 1982.

Tejani A, Ingulli E: Poststreptococcal glomerulonephritis. Current clinical and pathologic concepts. Nephron 55:1–5, 1990.

Tornroth T: The fate of subepithelial deposits in acute post streptococcal glomerulonephritis. Lab Invest 35: 461–474, 1976.

Membranous Glomerulonephritis

Cavallo T: Membranous nephropathy. Insights from Heyman nephritis. Am J Pathol 144:651–658, 1994.

Collins AB, Bhan AK, Dienstag JL, et al: Hepatitis B immune complex glomerulonephritis: Simultaneous glomerular deposition of hepatitis B surface and e antigens. Clin Immunol Immunopathol 26:137–153, 1983.

Ehrenreich T, Churg J: Pathology of membranous nephropathy. Pathol Annu 3:145–186, 1968.

Graham AR, Nagle RB: Quantitative electron microscopic study of membranous glomerulopathy. Am J Clin Pathol 80:816–821, 1983.

Rosen S: Membranous glomerulonephritis: Current status. Hum Pathol 2:209–231, 1971.

Membranoproliferative Glomerulonephritis, Type I

Burkholder PM, Hyman LR, Kruger RP: Characterization of mixed membranous and proliferative glomerulonephritis: Recognition of three varieties. In Kincaid-Smith P, Mathews TH, Becker EL, eds: *Glomerulonephritis: Morphology, Natural History, and Treatment*. John Wiley and Sons, New York, 1973, pp 557–589.

Churg J, Habib R, White RH: Pathology of the nephrotic syndrome in children: A report for the international study of kidney disease in children. Lancet 760: 1299–1302, 1970.

Jackson EC, McAdams AJ, Strife CF, et al: Differences between membranoproliferative glomerulonephritis Types I and III in clinical presentation, glomerular pathology, and complement perturbation. Am J Kidney Dis 9:115–120, 1987.

Johnson RJ, Gretch DR, Yamabe H, et al: Membranoproliferative glomerulonephritis associated with hepatitis C virus infection. N Engl J Med 328:465–470, 1993.

Kincaid-Smith P: The natural history and treatment of mesangiocapillary glomerulonephritis. In Kincaid-Smith P, Mathew TH, Becker EL, eds: *Glomerulonephritis: Morphology, Natural History, and Treatment*. John Wiley and Sons, New York, 1:591–609, 1973.

Strife CF, McEnery PT, McAdams AJ, et al: Membranoproliferative glomerulonephritis with disruption of the glomerular basement membrane. Clin Nephrol 7:65–72, 1977.

IgA Nephropathy (Berger's Disease)

Beaufils H, Jouanneau C, Katlama C, et al: HIV associated IgA nephropathy—A postmortem study. Nephrol Dial Transplant 10:35–38, 1995.

Berger J: IgA glomerular deposits in renal disease. Transplant Proc 1:939–944, 1969.

Berger J, Hinglais N: Les depots intracapillaires d'IgA. J Urol Nephrol 74:694–695, 1968.

Galla JH: IgA nephropathy. Kidney Int 47:377–387, 1995.

Hara M, Endo Y, Nihei H, et al: IgA nephropathy and subendothelial deposits. Virchows Arch [A] 386: 249–263, 1980.

Hulette C, Carstens PHB: Electron-dense deposits in extraglomerular vascular structures in IgA nephropathy. Nephron 39:179–183, 1985.

Lee HS, Choi Y, Lee SJ, et al: Ultrastructural changes in IgA nephropathy in relation to histologic and clinical data. Kidney Int 35:880–886, 1989.

Lee SM, Rao VM, Franklin WA, et al: IgA nephropathy: Morphologic predictors of progressive renal disease. Hum Pathol 13:314–322, 1982.

Morita M, Sakaguchi H: A quantitative study of GBM changes in IgA nephropathy. J Pathol 154:7–18, 1988.

Ng WL: The ultrastructural morphology of the mesangial cell in IgA nephropathy. Pathology 134:209–217, 1981.

Henoch-Schönlein Purpura

Heaton JM, Turner DR, Cameron JS: Localization of glomerular "deposits" in Henoch-Schönlein nephritis. Histopathology 1:93–104, 1977.

Mihatsch MJ, Imbasciati E, Fogazzi G, et al: Ultrastructural lesions of Henoch-Schönlein syndrome and of IgA nephropathy: Similarities and differences. Contrib Nephrol 40:255–263, 1984.

Yoshikawa N, White RH, Cameron AH: Prognostic significance of the glomerular changes in Henoch-Schönlein nephritis. Clin Nephrol 16:223–229, 1981.

Systemic Lupus Erythematosus

Adler SG, Johnson K, Louie JS, et al: Lupus membranous GN: Different prognostic subgroups obscured by imprecise histologic classifications. Mod Pathol 3: 186–191, 1990.

Alpers CE, Hopper J Jr, Bernstein MJ, et al: Late development of SLE in patients with glomerular fingerprint deposits. Ann Intern Med 100:66, 1984.

Berden JM: Lupus nephritis. Kidney Int 52:538–558, 1997.

Bruijn JA: The glomerular basement membrane in lupus nephritis. Microsc Res Tech 28:178–192, 1994.

Budhai L, Oh K, Davidson A: An in vitro assay for detection of glomerular binding IgG autoantibodies in patients with systemic lupus erythematosus. J Clin Invest 98:1585–1593, 1996.

Churg J, Bernstein J, Glassock RJ: Lupus nephritis. In Churg J, Bernstein J, Glassock RJ, eds: *Renal Diseases: Classification and Atlas of Glomerular Diseases.* Igaku-Shoin, New York, 1995.

Cohen AH, Zamboni L: Ultrastructural appearance and morphogenesis of renal glomerular hematoxylin bodies. Am J Pathol 89:105–118, 1977.

Grishman E, Churg J: Focal segmental lupus nephritis. Clin Nephrol 17:5–13, 1982.

Grishman E, Porush JC, Rosen SM, et al: Lupus nephritis with organized deposits in the kidneys. Lab Invest 16:717, 1967.

Hayslett JP, Carey H: Medical grand rounds: SLE with nephropathy. Yale J Biol Med 68:77–93, 1995.

Hill GS: Systemic lupus erythematosus and mixed connective tissue disease. In Heptinstall RH ed: *Pathology of the Kidney.* Little, Brown and Co, Boston, 1992, pp 871–950.

Jennette JC, Iskander SS, Dalldorf FG: Pathologic differentiation between lupus and non-lupus membranous glomerulopathy. Kidney Int 24:377–385, 1983.

Kim YH, Choi YJ, Reiner L: Ultrastructural "fingerprint" in cryoprecipitate and glomerular deposits: A case report of systemic lupus erythematosus. Hum Pathol 12:86–89, 1981.

Malide D, Londono I, Russo P, et al: Am J Pathol 143: 304–311, 1993.

McCluskey RT: Case Records of the Massachusetts General Hospital. N Engl J Med, in press.

Morita M, Sakaguchi H: Ultrastructure of renal glomerular hematoxylin bodies. Ultrastruct Pathol 7: 13–19, 1984.

Pirani CL: Clinicopathologic correlations in lupus nephritis. Contrib Nephrol 45:185–199, 1985.

Schwartz MM, Fennell JS, Lewis EJ: Pathologic changes in the renal tubule in systemic lupus erythematosus. Hum Pathol 13:534–547, 1982a.

Schwartz MM, Roberts JL, Lewis EJ: Subepithelial electron dense deposits in proliferative glomerulonephritis of SLE. Ultrastruct Pathol 3:105–118, 1982b.

Singh AK, Ucci A, Madias NE: Predominant tubulointerstitial lupus nephritis. Am J Kidney Dis 27:273–278, 1996.

Tojo A, Kimura K, Hirata Y, et al: Silent lupus nephritis with fingerprint deposits. Intern Med 32:323–326, 1993.

van Bruggen MCJ, Kramers C, Walgreen B, et al: Nucleosomes and histones are present in glomerular deposits in human lupus nephritis. Nephrol Dial Transplant 12:57–66, 1997.

Venkataseshan VS, Marquet E, Grishman E: Significance of cytoplasmic inclusions in lupus nephritis. Ultrastruct Pathol, 15:1–14, 1991.

Vlahakos DV, Foster MH, Adams S, et al: Anti-DNA antibodies form immune deposits at distinct glomerular and vascular sites. Kidney Int 41:1690–1700, 1992.

Tubuloreticular Structures/Inclusion Bodies and Tubular Confronting Cisternae

Ghadially FN: Endoplasmic reticulum, tubular confronting cisternae. In Ghadially FN, ed: *Ultrastructural Pathology of Cell and Matrix,* 3rd ed. Butterworths, London, 1988, pp 466–475.

Kippel J, Carette S, Preble OT, et al: Serum alpha interferon and lymphocyte inclusions in SLE. Ann Rheum Dis 44:104–108, 1985.

Luu J, Bockus D, Remington F, et al: Tubuloreticular structures and cylindrical confronting cisternae: A review. Hum Pathol 20:617–627, 1989.

Lyon MG, Bewtra C, Kenik JG, et al: Tubuloreticular inclusions in systemic lupus pneumonitis. Arch Pathol Lab Med 108:599–600, 1984.

Rich SA: De novo synthesis and secretion of a 36-kD protein by cells that form lupus inclusions in response to alpha-interferon. J Clin Invest 95:219–226, 1995.

Rich SA: Human lupus inclusions and interferon. Science 213:772, 1981.

Rich SA, Owens TR, Anzola MC, et al: Induction of lupus inclusions by sera from patients with systemic lupus erythematosus. Arthritis Rheum 29:501–507, 1986.

Rich SA, Owens T, Bartholomew LE, et al: Immune interferon does not stimulate formation of alpha and beta interferon induced human lupus-type inclusions. Lancet 1(8316):127–128, 1983.

Schaff Z, Barry DW, Grimley PM: Cytochemistry of tubuloreticular structures in lymphocytes from patients with SLE and cultured human lymphoid cells: Comparison to paramyxovirus. Lab Invest 29:577, 1973.

Venkataseshan VS, Marquet E, Grishman E: Significance of cytoplasmic inclusions in lupus nephritis. Ultrastruct Pathol 15:1–14, 1991.

Dense Deposit Disease (Membranoproliferative Glomerulonephritis, Type 2)

Berger J, Galle P: Depots denses au sein des membranes basales du rein: Etude en microscopies optique et electronique. La Presse Medicale 71:2351–2354, 1963.

Churg J, Duffy JL, Bernstein J: Identification of dense deposit disease: A report for the international study of kidney diseases in children. Arch Pathol Lab Med 103: 67–72, 1979.

Davis AE, Schneeberger EE, McCluskey RT, et al: Mesangial proliferative glomerulonephritis with irregular intramembranous deposits: Another variant of hypocomplementemic nephritis. Am J Med 63:481–487, 1977.

Duvall-Young J, Short CD, Raines MF, et al: Fundus changes in chronic membranoproliferative glomerulonephritis type II: Clinical and fluorescein angiographic findings. Br J Ophthalmol 73:900–906, 1989.

Galle P: Mise en evidence au microscope electronique d'une lesion singuliere des membranes basales du rein et de la substance hyaline. Thesis Med Paris 1962.

McEnery PT, McAdams AJ: Regression of membranoproliferative glomerulonephritis type II (dense deposit disease): Observation in six children. Am J Kidney Dis 12:138–146, 1988.

Michielsen B, Leys A, Van Damme B, et al: Fundus changes in chronic membranoproliferative glomerulonephritis type II. Doc Opthalmol 76:219–229, 1990–1991.

Muda AO, Barsotti P, Marinozzi V: Ultrastructural histochemical investigations of "dense deposit disease." Pathogenetic approach to a special type of mesangiocapillary glomerulonephritis. Virchows Arch [A] 413: 529–537, 1988.

Sibley RK, Kim Y: Dense intramembranous deposit disease: New pathologic features. Kidney Int 25:660–670, 1984.

Swainson CP, Robson JS, Thomson D, et al: Mesangiocapillary glomerulonephritis: A long term study of 40 cases. J Pathol 141:449–468, 1983.

Thorner P, Baumal R: Extraglomerular dense deposits in dense deposit disease. Arch Pathol Lab Med 106: 628–631, 1982.

Weidner N, Lorentz WB: Three dimensional studies of acellular glomerular basement membranes in dense deposit disease. Virchows Arch [A] 409:595–607, 1986.

Amyloidosis

Dikman SH, Churg J, Kahn I: Morphologic and clinical correlates in renal amyloidosis. Hum Pathol 12:160–169, 1981.

Falk RH, Comenzo RL, Skinner M: Medical progress: The systemic amyloidoses. N Engl J Med 337:898–909, 1997.

Gilmore JD, Hawkins PN, Pepys MB: Amyloidosis: A review of recent diagnostic and therapeutic developments. Br J Haematol 99:245–256, 1997.

Glenner GG: Amyloid deposits and amyloidosis. The beta-fibrilloses. N Engl J Med 302:1283–1292, 1333–1343, 1980.

Jones BA, Shapiro HS, Rosenberg BF, et al: Minimal renal amyloidosis with nephrotic syndrome. Arch Pathol Lab Med 110:889–892, 1986.

Moss J, Shore I, Woodrow D: AA glomerular amyloid. An ultrastructural immunogold study of the colocalization of heparan sulfate proteoglycan and P component with amyloid fibrils together with changes in distribution of type IV collagen and fibronectin. Histopathology 24:427–435, 1994.

Noel LH, Droz D, Ganeval D: Immunohistochemical characterization of renal amyloidosis. Am J Clin Pathol 87:756–761, 1987.

Notling SF, Campbell WG: Subepithelial argyrophylic spicular structures in renal amyloidosis—An aid in diagnosis. Hum Pathol 12:724–734, 1981.

Ozaki S, Abe M, Wolfenbarger D, et al: Preferential expression of human lambda light chain variable region subgroups in multiple myeloma, AL amyloidosis and Waldenström's macroglobulinemia. Clin Immunol Immunopathol 71:183, 1994.

WHO-IUIS Subcommittee: Nomenclature of amyloid and amyloidoses. Bull WHO 71:105, 1993.

Fibrillary Glomerulonephritis

Churg J, Venkataseshan VS: Fibrillary glomerulonephritis without immunoglobulin deposits in the kidney. Kidney Int 44:837–842, 1993.

Coroneos E, Truong L, Olivero J: Fibrillary glomerulonephritis associated with Hepatitis C viral infection. Am J Kidney Dis 29:132–135, 1997.

Duffy J, Khurana E, Susin M, et al: Fibrillary renal deposits and nephritis. Am J Pathol 113:279–290, 1983.

Fogo A, Qureshi N, Horn RG: Morphologic and clinical features of fibrillary glomerulonephritis versus Immunotactoid glomerulopathy. Am J Kidney Dis 22: 367–377, 1993.

Iskandar SS, Falk RJ, Jennette JC: Clinical and pathologic features of fibrillary glomerulonephritis. Kidney Int 42:1401–1407, 1992.

Markovitz GS, Cheng JT, Colvin RB, et al: Hepatitis C viral infection is associated with fibrillary glomerulonephritis and immunotactoid glomerulopathy. J Am Soc Nephrol 9:2244–2252, 1998.

Yang GCH, Nieto R, Stachura I, et al: Ultrastructural immunohistochemical localization of polyclonal IgG, C3, and amyloid P component on the Congo red negative amyloid-like fibrils of fibrillary glomerulopathy. Am J Pathol 141:409–419, 1992.

Immunotactoid Glomerulopathy

Alpers CE: Immunotactoid (microtubular) glomerulopathy: An entity distinct from fibrillary glomerulonephritis? Am J Kidney Dis 19:185–191, 1992.

Alpers CE: Fibrillary glomerulonephritis and immunotactoid glomerulopathy: Two entities, not one. Am J Kidney Dis 22:448–451, 1993.

Duffy J, Khurana E, Susin M, et al: Fibrillary renal deposits and nephritis. Am J Pathol 113:279–290, 1983.

Fogo A, Qureshi N, Horn RG: Morphologic and clinical features of fibrillary glomerulonephritis versus Immunotactoid glomerulopathy. Am J Kidney Dis 22: 367–377, 1993.

Korbet SM, Schwartz MM, Lewis EJ: The fibrillary glomerulopathies. Am J Kidney Dis 23:751–765, 1994.

Korbet SM, Schwartz MM, Rosenberg BF, et al: Immunotactoid glomerulopathy. Medicine 64:228–243, 1985.

Lai FM, Lai KN, Li EK, et al: Immunotactoid glomerulopathy with fingerprint immune deposits. Virchows Arch [A] 415:181–186, 1989.

Schwartz MM: Immunotactoid glomerulopathy: The case for Occam's razor. Am J Kidney Dis 22:446–447, 1993.

Schwartz MM, Lewis EJ: The quarterly case: Nephrotic syndrome in a middle aged man. Ultrastruct Pathol 1:575–582, 1980.

Cryoglobulinemia

Angello V, Churg RT, Kaplan LM: A role for hepatitis C virus infection in type II cryoglobulinemia. N Engl J Med 327:1490–1495, 1992.

Bloch KJ: Cryoglobulinemia and hepatitis C virus. N Engl J Med 327:1521–1522, 1992.

Cordonnier D, Vialtel P, Renversez JC, et al: Renal diseases in 18 patients with mixed type II IgM-IgG cryoglobulinemia: Monoclonal lymphoid infiltration (2 cases) and membranoproliferative glomerulonephritis (14 cases). Adv Nephrol 12:177–204, 1983.

Feiner H, Gallo G: Ultrastructure in glomerulonephritis associated with cryoglobulinemia. Am J Pathol 88: 145–162, 1977.

Ferri C, Zignego AL, Bombardieri S, et al: Etiopathogenetic role of hepatitis C virus in mixed cryoglobulinemia, chronic liver disease and lymphomas. Clin Exp Rheumatol 13(suppl 13):S135–S140, 1995.

Fornasieri A, D'Amico G: Type II mixed cryoglobulinemia, hepatitis C virus infection and glomerulonephritis. Nephrol Dial Transplant 11(suppl 4):25–30, 1996.

Gesualdo L, Grandaliano G, Ranieri E, et al: Monocyte recruitment in cryoglobulinemic membranoproliferative glomerulonephritis: A pathogenetic role for monocyte chemotactic peptide-1. Kidney Int 51:155–163, 1997.

Ishimura E, Nishizawa Y, Shoji S, et al: Heat insoluble cryoglobulin in a patient with essential type I cryoglobulinemia and massive cryoglobulin-occlusive glomerulonephritis. Am J Kidney Dis 26:654–657, 1995.

Miescher PA, Huang YP, Izui S: Type II cryoglobulinemia. Semin Hematol 32:80–85, 1995.

Monga G, Mazzucco G, Casanova S, et al: Ultrastructural glomerular findings in cryoglobulinemic glomerulonephritis. Appl Pathol 5:108–115, 1987.

Stoebner P, Renversez JC, Groulade J, et al: Ultrastructural study of human IgG and IgG-IgM crystalcryoglobulins. Am J Clin Pathol 71:404–410, 1979.

Szymanski IO, Pullman JM, Underwood JM: Electron microscopic and immunochemical studies in a patient with hepatitis C virus infection and mixed cryoglobulinemia type II. Am J Clin Pathol 102:278–283, 1994.

Tomiyoshi Y, Sakemi T, Yoshikawa Y, et al: Fibrillar crystal structure in essential monoclonal IgM kappa cryoglobulinemia. Clin Nephrol 49:325–327, 1998.

Systemic Light Chain Deposition Disease

Bangerter AR, Murphy WM: Kappa light chain nephropathy—A pathologic study. Virchows Arch [A] 410:531–539, 1987.

Bruneval P, Foidart JM, Nochy D, et al: Glomerular matrix proteins in nodular glomerulosclerosis in association with light chain deposition disease and diabetes mellitus. Hum Pathol 16:477–484, 1985.

Buxbaum J: Mechanisms of disease: Monoclonal immunoglobulin deposition disease. Hematol Oncol Clin N Am 6:323, 1992.

Gallo G, Picken M, Fragione B, et al: Nonamyloidotic monoclonal immunoglobulin deposits lack amyloid P component. Mod Pathol 1:453–456, 1988.

Kirkpatrick CJ, Curry A, Galle J, et al: Systemic kappa light chain deposition and amyloidosis in multiple myeloma: Novel morphological observations. Histopathology 10:1065–1076, 1986.

Preud'homme JL, Aucouturier P, Touchard G, et al: Monoclonal immunoglobulin deposition disease (Randall type). Relationship with structural abnormalities of immunoglobulin chains. Kidney Int 46:965–972, 1994.

Randall RE, Williamson WC, Mullinex F, et al: Manifestations of systemic light chain deposition. Am J Med 60:293–299, 1976.

Ronco PM, Aucouturier P, Mougenot B: Monoclonal gammopathies: Multiple myeloma, amyloidosis and related disorders. In Schrier RW, Gottschalk CW, eds: *Diseases of the Kidney*, 6th ed. Little Brown and Co, 1997.

Strom EH, Fogazzi GB, Banfi G, et al: Light chain deposition disease of the kidney. Morphological aspects in 24 patients. Virchows Arch 425:271–280, 1994.

Tubbs RR, Gephardt GN, McMahon JT, et al: Light chain nephropathy. Am J Med 71:263–269, 1981.

Nail Patella Syndrome (Hereditary Osteo-onychodysplasia)

Campeau E, Watkins D, Rouleau GA, et al: Linkage analysis of the nail-patella syndrome. Am J Hum Genet 56:243–247, 1995.

McIntosh I, Clough MV, Schaffer A, et al: Fine mapping of the nail-patella syndrome locus at 9q34. Am J Hum Genet 60:133–142, 1997.

Morita T, Laughlin LO, Kowano K, et al: Nail-patella syndrome. Light and electron microscopic studies of the kidney. Arch Intern Med 131:271–277, 1973.

Taguchi T, Takebayashi S, Nishimura M, et al: Nephropathy of nail-patella syndrome. Ultrastruct Pathol 12: 175–183, 1988.

Collagen Type III Collagenofibrotic Glomerulopathy
Abt AB, Cohen AH: Newer glomerular diseases. Semin Nephrol 16:501–510, 1996.

Arakawa M, Yamanaka N: Collagenofibrotic glomerulopathy: A new type of primary glomerulopathy revealing massive collagen deposits in the renal glomerulus. In Arakawa M, Yamanaka N, eds: *Collagenofibrotic Glomerulopathy* Nishimura Co Ltd, Nigata, Japan, 1991, pp 3–8.

Gubler MC, Dommergues JP, Foulard M, et al: Collagen type III glomerulopathy. Pediatr Nephrol 7:354–360, 1993.

Ikeda K, Yokoyama H, Tomosugi N, et al: Primary glomerular fibrosis: A new nephropathy caused by diffuse intra-glomerular increase in atypical type III collagen fibers. Clin Nephrol 33:155–159, 1990.

Imbasciati E, Gherardi G, Morozumi K, et al: Collagen type III glomerulopathy: A new idiopathic glomerular disease. Am J Nephrol 11:422–429, 1991.

Tamura H, Matsuda A, Kidoguchi N, et al: A family of two sisters with collagenofibrotic glomerulopathy. Am J Kidney Dis 27:588–595, 1996.

Fabry's Disease
Burkholder PM, Updike SJ, Ware RA, et al: Clinicopathologic, enzymatic and genetic features in a case of Fabry's disease. Arch Pathol Lab Med 104:17–25, 1980.

Faraggiana T, Churg J, Grishman E, et al: Light and electron microscopic histochemistry of Fabry's disease. Am J Pathol 103:247–262, 1981.

Gubler MC, Lenoir G, Grunfeld JP, et al: Early renal changes in hemizygous and heterozygous patients with Fabry's disease. Kidney Int 13:223–235, 1978.

Paquin JG, Camirand P, Mandelenakis N, et al: La maladie de Fabry. Etude histologique et ultrastructurale. Union Med Canada 104:1377–1382, 1975.

Tubbs RR, Gephardt GN, McMahon JT, et al: Cytoplasmic inclusions in Fabry's disease. Ultrastructural demonstration of their presence in urine sediment. Arch Pathol Lab Med 105:361–362, 1981.

Cystinosis
Cystinosis Collaborative Research Group: Linkage of the gene for cystinosis to markers on the short arm of chromosome 17. Nature Genet 10:246–248, 1995.

Scotto JM, Stralin HG: Ultrastructure of the liver in a case of childhood cystinosis. Virchows Arch [A] 377: 43–48, 1977.

Spear GS: Pathology of the kidney in cystinosis. Pathol Annu 9:81–92, 1974.

Spear GS, Gubler MC, Habib R, et al: Renal allografts in cystinosis and mesangial demography. Clin Nephrol 32:256–261, 1989.

Spear GS, Slusser RJ, Tousimis AJ, et al: Cystinosis, an ultrastructural and electron probe study of the kidney with unusual findings. Arch Pathol 21:206–221, 1971.

Town M, Jean G, Cherqui S, et al: A novel gene encoding an integral membrane protein is mutated in nephropathic cystinosis. Nature Genet 18:319–324, 1998.

Glomerulopathy of Sickle Cell Disease/Trait
Antonovych TT: Ultrastructural changes in glomeruli of patients with sickle cell disease and nephrotic syndrome. In *Abstracts of the Fifth Annual Meeting of the American Society of Nephrology.* Washington, DC, 1971, p 3.

Bernstein J, Whitten CF: A histologic appraisal of the kidney in sickle cell anemia. Arch Pathol 70:407, 1960.

Elfenbein IB, Patchefsky A, Schwartz W, et al: Pathology of the glomerulus in sickle cell anemia with and without nephrotic syndrome. Am J Pathol 77:357, 1974.

McCoy RC: Ultrastructural alterations in the kidney of patients with sickle cell disease and the nephrotic syndrome. Lab Invest 21:85, 1969.

Pardo V, Strauss J, Kramer H, et al: Nephropathy associated with sickle cell anemia: An autologous immune complex nephritis. II. Clinicopathologic study of seven patients. Am J Med 59:650, 1975.

Thrombotic Microangiopathy (in Hemolytic Uremic Syndrome, Thrombotic Thrombocytopenic Purpura, Acute Scleroderma, Malignant Hypertension, Rejection, and Cyclosporine Toxicity)
Hsu H, Churg J: The ultrastructure of mucoid "onion-skin" intimal lesions in malignant nephrosclerosis. Am J Pathol 99:67–80, 1980.

Sinclair RA, Antonovych TT, Mostofi FK: Renal proliferative arteriopathies and associated glomerular changes: A light and electron microscopic study. Hum Pathol 7:565–588, 1976.

Eclampsia/Preeclampsia
Gaber LW, Spargo BH, Lindheimer MD: Renal pathology in pre-eclampsia. Baillieres Clin Obstet Gynaecol 8:443–468, 1994.

Kida H, Yokoyama H, Tomosugi N, et al: Focal glomerulosclerosis in pre-eclampsia. Clin Nephrol 24: 221–227, 1985.

Lee SH, Kim TS: A morphometric study of pre-eclamptic nephropathy with focal segmental glomerulosclerosis. Clin Nephrol 44:14–21, 1995.

Sheehan HL: Renal morphology in pre-eclampsia. Kidney Int 18:241–252, 1980.

Sheehan HL, Lynch JB: *Pathology of Toxemia of Pregnancy.* Baltimore, Williams and Wilkins, 1973.

Spargo BH, McCartney C, Winemiller R: Glomerular capillary endotheliosis in toxemia of pregnancy. Arch Pathol 13:593–599, 1959.

Tribe CR, Smart GE, Davies DR, et al: A renal biopsy study in toxemia of pregnancy. J Clin Pathol 32:681–692, 1979.

Renal Allograft

Colvin RB: The renal allograft biopsy. Kidney Int 50: 1069–1082, 1996.

Colvin RB: Renal transplant pathology. In Jennette JC, Olson JL, Schwartz MM, et al, eds: *Heptinstall's Pathology of the Kidney,* 5th ed. Lippincott-Raven, Philadelphia, 1998, pp 1409–1540.

Herrera GA, Isaac J, Turbat-Herrera EA: Role of electron microscopy in transplant renal pathology. Ultrastruct Pathol 21:481–498, 1997.

Hsu HC, Suzuki Y, Churg J, et al: Ultrastructure of transplant glomerulopathy. Histopathology 4:351–367, 1980.

Meehan S, Colvin RB: Differential diagnosis of renal allograft biopsies. In Hammond EH, ed: *Solid Organ Transplant Pathology,* vol. 30. WB Saunders, 1994, pp 159–185.

Acute Allograft Glomerulopathy

Colvin RB: Immunopathology of renal allografts. In Colvin RB, Bhan AK, McCluskey RT, eds: *Diagnostic Immunopathology,* 2nd ed. Raven Press, New York, 1995.

Colvin RB, Cosimi AB, Burton RC, et al: Circulating T-cell subsets in human renal allograft recipients: The OKT4+/OKT8+ cell ratio correlates with reversibility of graft injury and glomerulopathy. Transplant Proc 15:1166, 1983.

Cosio FG, Sedmak DD, Henry ML, et al: The high prevalence of severe early post transplant renal allograft pathology in hepatitis C positive recipients. Transplantation 62:1054–1059, 1996.

Richardson WP, Colvin RB, Cheeseman SH, et al: Glomerulopathy associated with cytomegalovirus viremia in renal allografts. N Engl J Med 305:57–63, 1981.

Tuazon TV, Schneeberger EE, Bhan AK, et al: Mononuclear cells in acute allograft glomerulopathy. Am J Pathol 129:119–132, 1987.

Thrombotic Microangiopathy in Renal Allografts

Baid S, Pascual M, Williams WW, Renal thrombotic microangiopathy associated with anticardiolipin antibodies in hepatitis C positive renal allograft recipients. J Am Soc Nephrol 10:146–153, 1999.

Acute Humoral Rejection

Collins AB, Schneeberger EE, Pascual M: Complement activation in acute humoral renal allograft rejection: Diagnostic significance of C4d deposits in peritubular capillaries. JASON 10:2208–2214, 1999.

Halloran PF, Schlaut J, Solez K, et al: The significance of the anti-class I response: II. Clinical and pathologic features of renal transplants with anti-class I-like antibody. Transplantation 53:550–555, 1992.

Halloran PF, Wadgymar A, Ritchie S, et al: The significance of the anti-class I response: I. Clinical and pathologic features of anti-class I mediated rejection. Transplantation 49:85–91, 1990.

Trpkov K, Campbell P, Pazderka F, et al: Pathologic features of acute renal allograft rejection associated with donor-specific antibody: Analysis using the Banff grading schema. Transplantation 61:1586–1592, 1996.

Chronic Allograft Glomerulopathy

Baid S, Pascual M, Mauiyyedi S, et al: Association of HCV infection with chronic transplant glomerulopathy and de novo MPGN lesions in renal allografts. Transplantation 67:27, 1999.

Briner J: Transplant glomerulopathy. Appl Pathol 5: 82–87, 1987.

Cosio FG, Roche Z, Agarwal A, et al: Prevalence of hepatitis C in patients with idiopathic glomerulopathies in native and transplant kidneys. Am J Kidney Dis 28:752–758, 1996.

Drachenberg CB, Hoehn-Saric E, Heffes A, et al: Specificity of intertubular capillary changes: Comparative ultrastructural studies in renal allografts and native kidneys. Ultrastruct Pathol 21:227–233, 1997.

Mauiyyedi S, Della Pelle P, Saidman S, et al: C4d deposits in peritubular capillaries in chronic renal allograft rejection. Lab Invest 79:156, 1999.

Monga G, Mazzucco G, Novara R, et al: Intertubular capillary changes in kidney allografts: An ultrastructural study in patients with transplant glomerulopathy. Ultrastruct Pathol 14:201–209, 1990.

Nadasdy T, Ormos J, Stiller D, et al: Tubular ultrastructure in rejected human renal allografts. Ultrastruct Pathol 12:195–207, 1988.

Olsen S, Bohman SO, Petersen VP: Ultrastructure of the glomerular basement membrane in long-term renal allografts with transplant glomerular disease. Lab Invest 30:176–189, 1974.

Peng-Fei Z, Rao KV, Anderson WR: An ultrastructural study of the membranoproliferative variant of transplant glomerulopathy. Ultrastruct Pathol 12:185–194, 1988.

Other Lesions in Renal Transplants

Diabetic Nephropathy

Kelly JJ, Walker RG, Kincaid-Smith P: De novo diabetic nodular glomerulosclerosis in a renal allograft. Transplantation 53:688–689, 1992.

Lundgren G, Wilczek, Bohman SO, et al: Development of diabetic nephropathy in renal allografts of diabetic patients. Transplant Proc 17:18–20, 1985.

Membranous Glomerulonephritis

Parker SM, Pullman JM, Khauli RB: Successful transplantation of a kidney with early membranous nephropathy. Urology 46:870–872, 1995.

Truong L, Gelfand J, D'Agati V, et al: De novo membranous glomerulopathy in renal allografts: A report of ten cases and review of literature. Am J Kidney Dis 14: 131–144, 1989.

BK Virus Infection

Coleman DV, Mackenzie EFD, Gardner SD, et al: Human polyomavirus (BK) infection and ureteric stenosis in renal allograft recepients. J Clin Pathol 31:338–347, 1978.

Colvin RB: Renal transplant pathology. In Jennette JC, Olson JL, Schwartz MM, et al, eds: *Heptinstall's Pathology of the Kidney,* 5th ed. Lippincott-Raven, Philadelphia, 1998, pp 1409–1540.

Demeter LM: JC, BK, and other polyomaviruses: Progressive multifocal leukoencephalopathy. In Mandell GL, Bennett JE, Dolin R, eds: *Principles and Practice of Infectious Diseases,* 4th ed. Churchill Livingstone, New York, 1995, pp 1400–1406.

Khoury G, Howley PM, Garon C, et al: Homology and relationship between the genomes of papovaviruses, BK virus and simian virus 40. Proc Natl Acad Sci USA 72:2563–2567, 1975.

Cyclosporine Nephropathy

Bennett WM, DeMattos A, Meyer MM, et al: Chronic cyclosporine nephropathy: The Achilles' heel of immunosuppressive therapy. Kidney Int 50:1089–1100, 1996.

Bertani T, Ferrazzi P, Schieppati A, et al: Nature and extent of glomerular injury induced by cyclosporine in heart transplant patients. Kidney Int 40:243–250, 1991.

Falkenhain ME, Cosio G, Sedmak DD: Progressive histologic injury in kidneys from heart transplant and liver transplant recipients receiving cyclosporine. Transplantation 62:364–370, 1996.

Ghiggeri GM, Altieri P, Oleggini R, et al: Selective enhancement by cyclosporin A of collagen expression by mesangial cells "in culture." Eur J Pharmacol 270: 195–201, 1994.

Mauiyyedi S, Williams WW, Tolfkoff-Rubin N, et al: Glomerulopathy in chronic cyclosporine toxicity in non-renal transplant cases. Lab Invest 79:156, 1999.

Mihatsch MJ, Morozumi K, Strom EH, et al: Renal transplant morphology after long-term therapy with cyclosporine. Transplant Proc 27:39–42, 1995.

Mihatsch MJ, Ryffel B, Gudat F: The differential diagnosis between rejection and cyclosporine toxicity. Kidney Int 48(suppl 52):S63–S69, 1995.

Myers BD, Ross J, Newton L, et al: Cyclosporine associated chronic nephropathy. N Engl J Med 311:699–705, 1984.

Randhawa PS, Shapiro R, Jordan ML, et al: The histopathological changes associated with allograft rejection and drug toxicity in renal transplant recipients maintained on FK506. Clinical significance and comparison with cyclosporine. Am J Surg Pathol 17:60–68, 1993.

Sacchi G, Benetti A, Falchetti M, et al: Ultrastructural renal findings in allografted kidneys of patients treated with cyclosporin A. Appl Pathol 5:101–107, 1987.

Shihab FS, Andoh TF, Tanner AM, et al: Role of transforming growth factor beta-1 in experimental chronic cyclosporine nephropathy. Kidney Int 49:1141, 1996.

Young BA, Burdmann EA, Johnson RJ, et al: Cellular proliferation and macrophage influx precede interstitial fibrosis in cyclosporine nephrotoxicity. Kidney Int 48:439–448, 1995.

Young BA, Marsh CL, Alpers CE, Davis CL: Cyclosporine associated thrombotic microangiopathy/hemolytic uremic syndrome following kidney and kidney-pancreas transplantation. Am J Kidney Dis 28:561–571, 1996.

Miscellaneous Lesions

Microparticles

Burkholder PM, Hyman LR, Barber T: Extracellular clusters of spherical microparticles in glomeruli in human renal glomerular diseases. Lab Invest 28:415–425, 1973.

Nakajima M, Hewitson TD, Mathews DC, et al: Localization of complement components in association with glomerular extracellular particles in various renal diseases. Virchows Arch [A] 419:267–272, 1991.

Yoshikawa N, Cameron AH, White RHR: Ultrastructure of the non-sclerotic glomeruli in childhood nephrotic syndrome. J Pathol 136:133–147, 1982.

Gentamicin Bodies

Houghton DC, Campbell-Boswell MV, Bennett WM, et al: Myeloid bodies in renal tubules humans: Relationship to gentamicin therapy. Clin Nephrol 10:140–145, 1978.

Toubeau G, Maldague P, Laurent G, et al: Morphological alterations in distal and collecting tubules of the rat cortex after aminoglycoside administration at low doses. Virchows Arch Cell Pathol 51:475–485, 1986.

Tubular Lesions of Magnesium and Potassium Deficiency

Jones DB: Ultrastructure of human acute renal failure. Lab Invest 46:254, 1982.

Schneeberger EE, Morrison AB: The nephropathy of experimental magnesium deficiency. Lab Invest 14: 674–686, 1965.

13

Diseases of Skeletal Muscle and Peripheral Nerve

Umberto De Girolami
Douglas C. Anthony

This chapter is divided into two sections that discuss the pathology of illnesses that affect *skeletal muscle* and those that involve *peripheral nerve.* Within each section, there is a brief introduction to the general reactions to injury as seen with the electron microscope, followed by an account of important diseases which affect each tissue. The discussion of the most each disease, or category of disease, is divided under three headings: *clinical manifestations, diagnostic criteria, and etiology.* General reference texts and reviews are cited in the introductory comments within each section; in addition, selected recent publications and noteworthy articles dealing with specific aspects of a particular disease entity are referenced in the text.

Skeletal Muscle

(Figure 13.1.)

The normal light microscopic and electron microscopic structure of skeletal muscle is discussed in standard textbooks (Sternberg 1992; Engel and Franzini-Armstrong 1994); several illustrations are given here as a starting point to orient the reader (Figure 13.1). The basic responses of skeletal muscle to injury visible with the electron microscope can be subdivided into the following categories: (1) alterations in sarcolemma (e.g., discontinuities of plasma or basement membrane); (2) alterations in myofilaments (e.g., degeneration and loss of myofilaments, central cores and target formation, ring fibers, sarcoplasmic masses, and contraction bands); (3) Z-band alterations (e.g., streaming and nemaline bodies); (4) nuclear changes (e.g., abnormal location of the nucleus within the muscle fiber and inclusions); (5) mitochondrial changes (e.g., abnormalities in number, size, and structure; intramitochondrial inclusions); (6) abnormalities of sarcoplasmic reticulum and T-system (e.g., tubular aggregates); (7) abnormal accumulations of metabolites (e.g., glycogen and lipid); (8) abnormal cytoplasmic structures (e.g., vacuoles, cytoplasmic bodies, tubular and filamentous inclusions, zebra bodies, concentric laminated bodies, fingerprint bodies, curvilinear bodies).

In general, many of these ultrastructural abnormalities are not specific for a single disease. Electron microscopy can be a valuable adjunct to help the pathologist arrive at a proper interpretation of a muscle biopsy when taken together with all other available clinical, electrophysiologic, and histopathological data. In addition to the pathologic changes that might involve the muscle fibers themselves, many diseases of muscle also simultaneously affect adjoining connective tissue components, blood vessels, and intramuscular nerves. It is, therefore, important to pay particular attention to these structures when examining muscle with the light and electron microscope. For general reference citations that include discussions of the ultrastructural pathology of skeletal muscle, the reader is referred to the references listed at the end of this chapter. These include chapters in textbooks and comprehensive treatises dealing specifically with the pathology of diseases of skeletal muscle (Engel and Franzini-Armstrong, 1994; Carpenter and Karpati 1978; Neville 1979; Dubowitz 1995; De Girolami and Beggs 1997).

A simple classification of diseases of skeletal muscle recognizes two major groups: disorders in which the muscle fiber itself is the primary site of injury—*myopathies*—and diseases in which dysfunction of the muscle cell is secondary to an abnormality of its innervation—*neurogenic atrophy.* The myopathies can be subclassified as follows: (1) hereditary disorders with known or suspected genetic abnormalities, including the muscular dystrophies and the congenital myopathies; (2) hereditary or acquired metabolic and toxic myopathies; and (3) infectious and noninfectious inflammatory myopathies. In the text that follows, the principal light and ultrastructural alterations seen in selected diseases are discussed.

(Text continues on page 920)

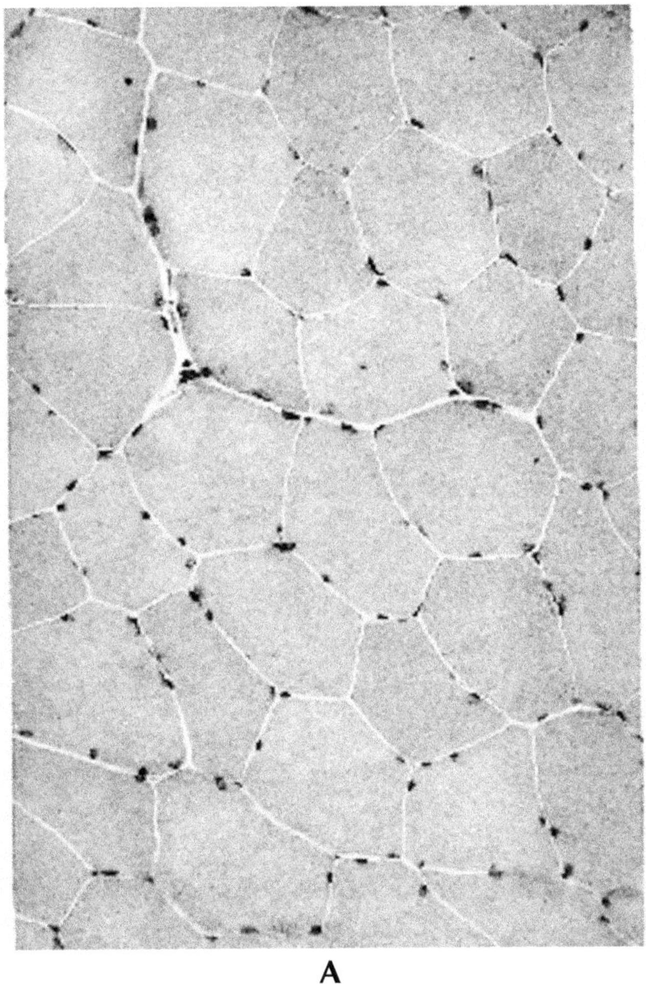

A

Figure 13.1. Normal muscle. **A,** Transverse section of frozen section of skeletal muscle stained with H&E showing normal polygonal contour of adult muscle fibers. (H&E, × 200)

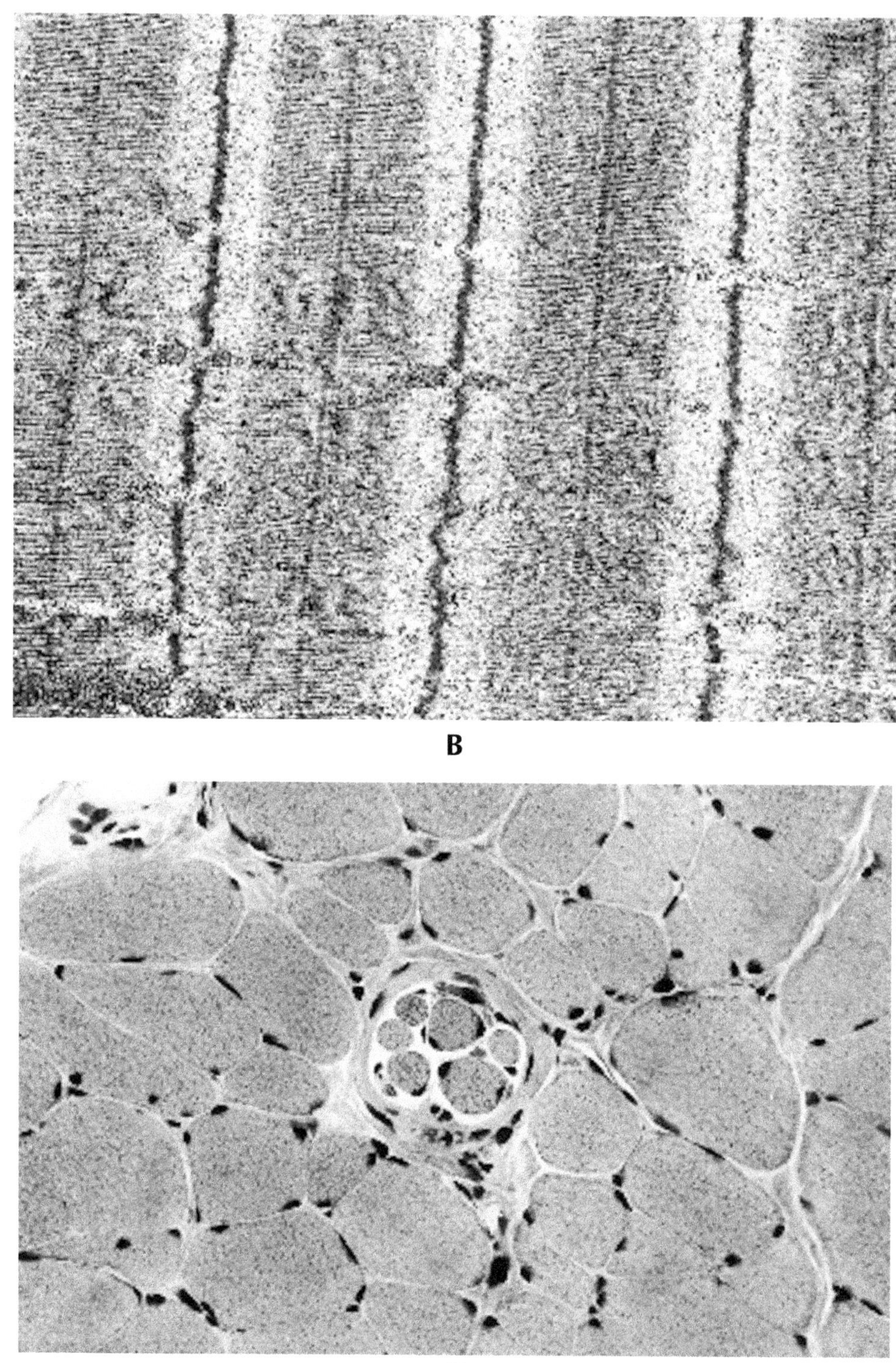

Figure 13.1. *(continued)*
B, Electron micrograph of longitudinal section of normal skeletal muscle showing organization of sarcomere. (× 11,000)
C, Muscle spindle. (H&E, × 200)

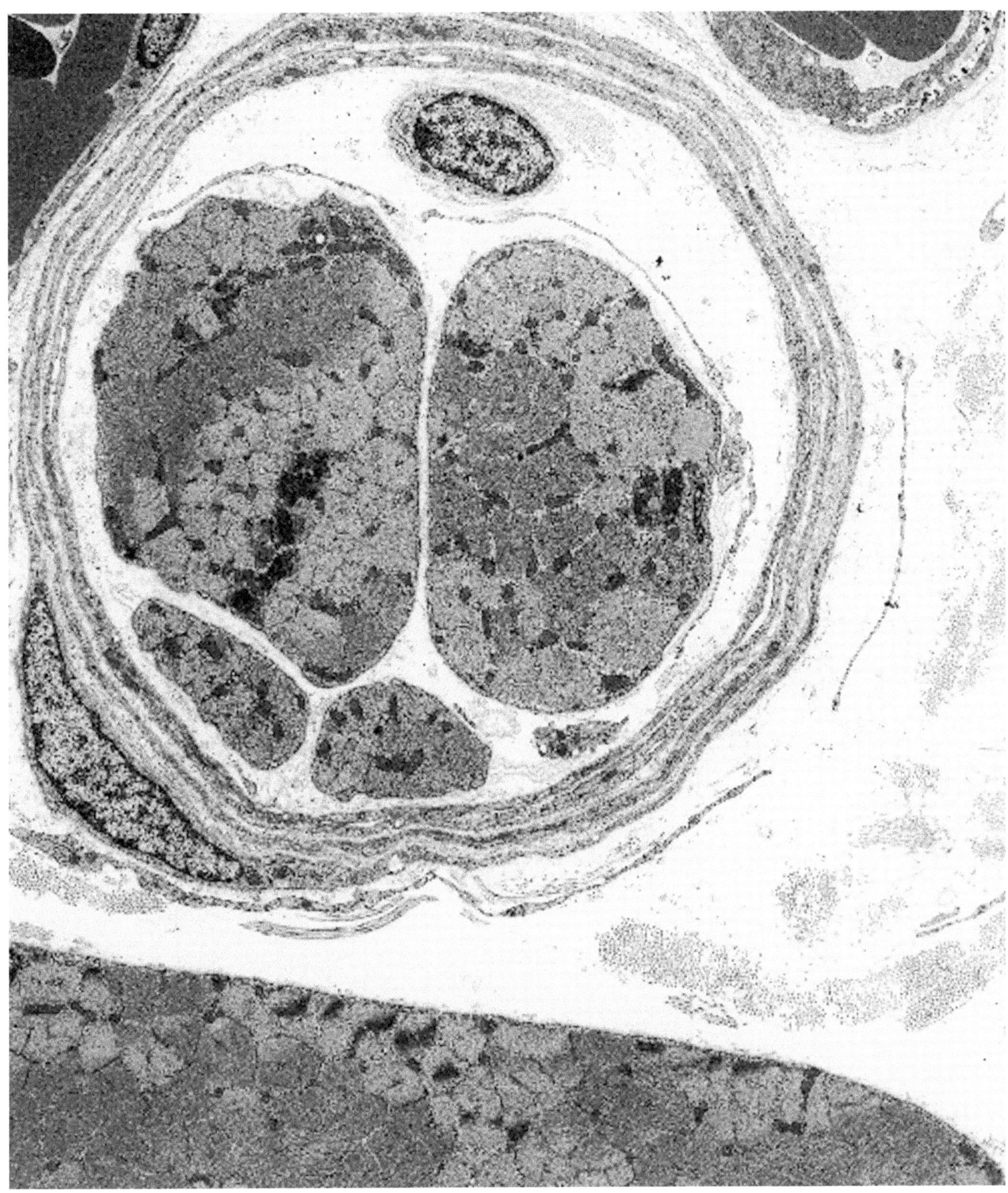

D

Figure 13.1. *(continued)*
D, Electron micrograph of muscle spindle of guinea pig muscle. Note intrafusal fibers surrounded by several layers of capsule cells. (Courtesy of Dr. I. Joris, University of Massachusetts Medical School, Worcester, Mass. (× 5000)

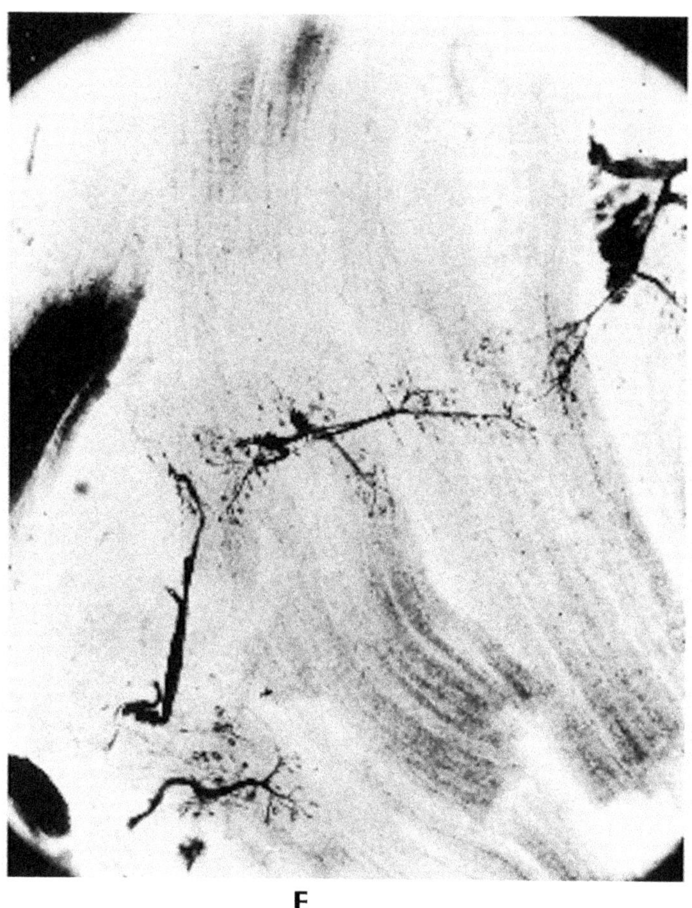

E

Figure 13.1. *(continued)*
E, Neuromuscular junction endings of intercostal muscle of the rat (Ranvier gold chloride method). (Courtesy of Dr. R. D. Adams, Massachusetts General Hospital, Boston, Mass.)

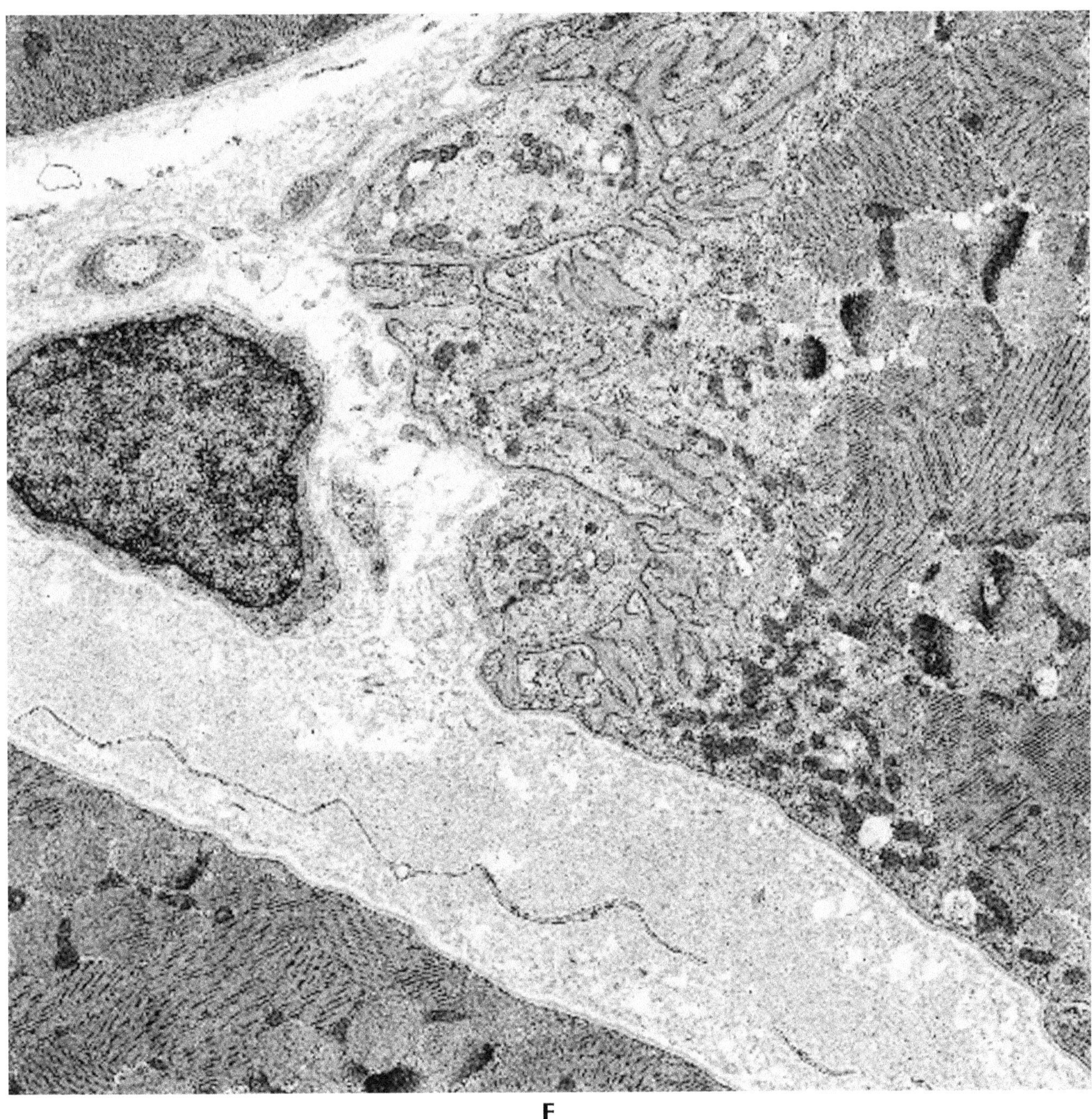

F

Figure 13.1. *(continued)*
F, Electron micrograph of neuromuscular ending. Note folding of muscle cell at ending and overlying axon terminal. (× 10,600)

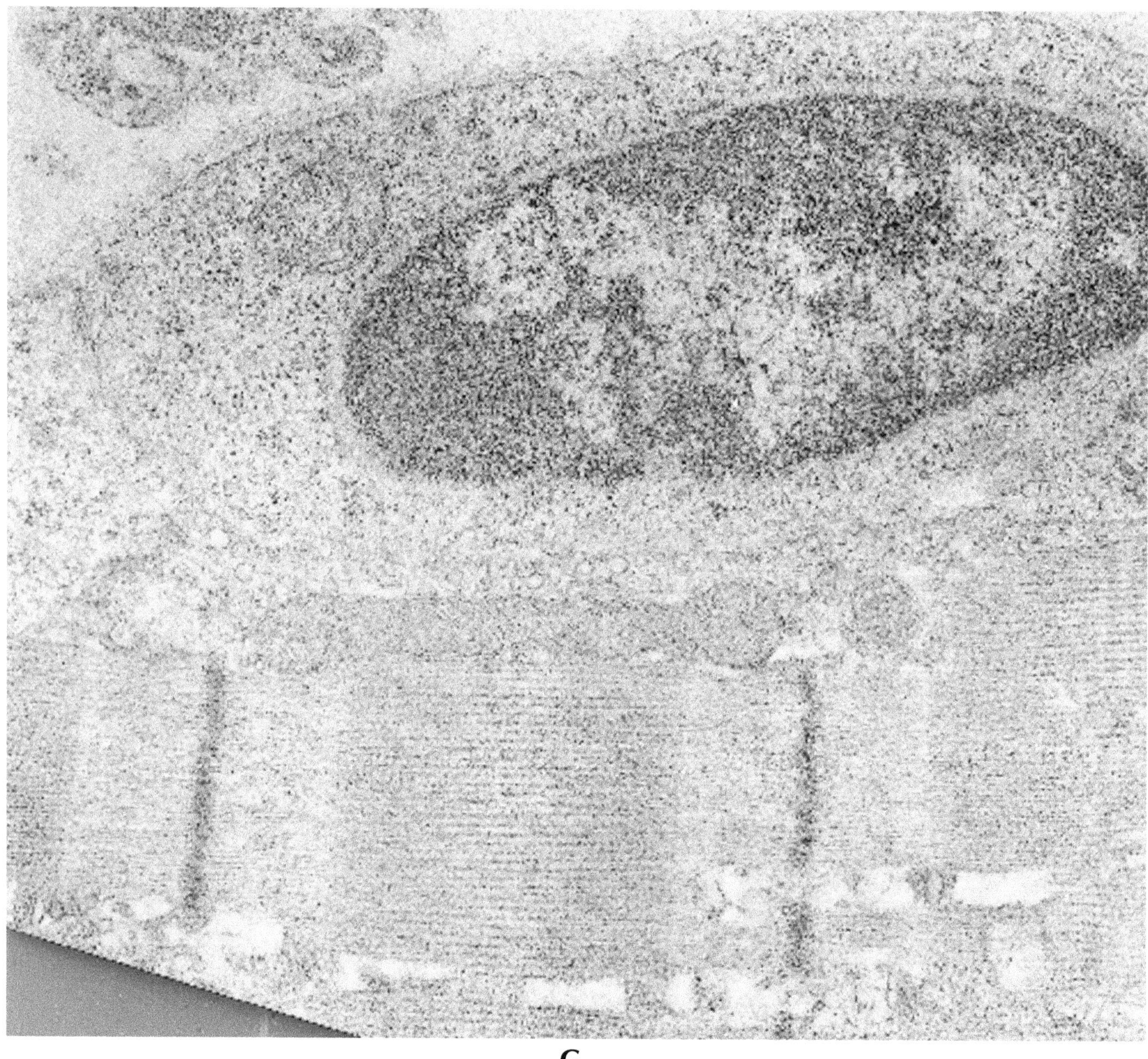

G

Figure 13.1. *(continued)*
G, Electron micrograph of satellite cell. Note nucleated cell enclosed within basement membrane of underlying muscle cell with well-developed sarcomere structure. (× 20,000)

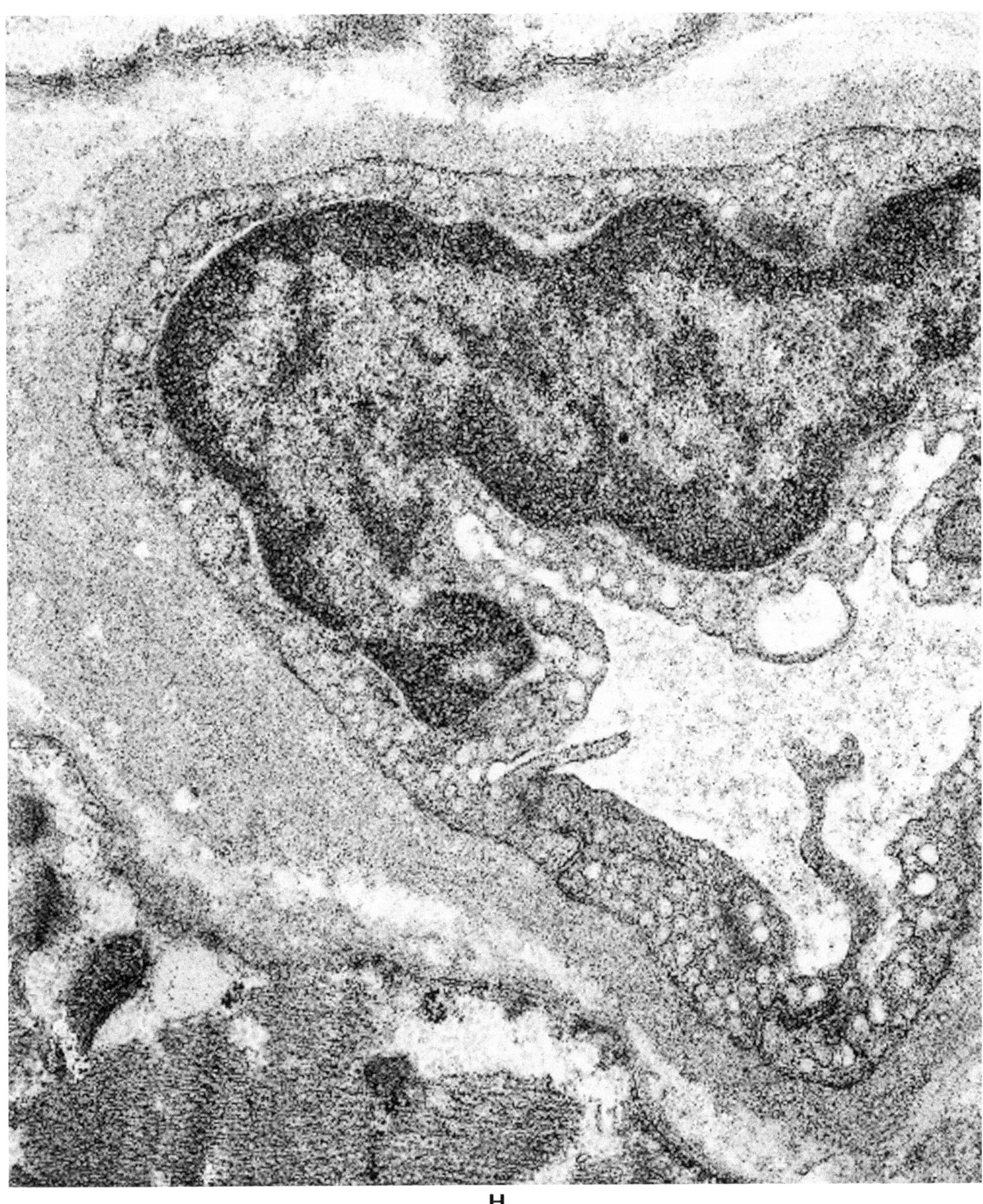

H

Figure 13.1. *(continued)*
H, Electron micrograph of muscle capillary. Note abundant pinocytotic vesicles and junctions between endothelial cells. (× 28,000)

(Text continued from page 913)

Muscular Dystrophy and Congenital Myopathy

Duchenne Muscular Dystrophy

(Figure 13.2.)

Clinical manifestations. The two most common forms of muscular dystrophy are X-linked: Duchenne muscular dystrophy (DMD) and Becker muscular dystrophy (BMD). DMD is the most severe form; the disease becomes clinically manifest by the age of 5 years with weakness that progresses relentlessly until death by the early 20s. Boys with DMD are normal at birth, but walking is delayed. The muscle weakness begins in the pelvic girdle muscles and then also affects the shoulder girdle. An important clinical manifestation of the disease is pseudohypertrophic enlargement of the calf muscles, due, in part, to deposition of connective tissue in and around the damaged muscle fibers. The disease also affects the heart muscle, and patients may develop heart failure or arrhythmias. Death results from respiratory insufficiency, pulmonary infection, and cardiac decompensation.

Although BMD involves the same genetic locus as DMD, it is less common and much less severe than DMD. The onset occurs later in childhood or in adolescence and is accompanied by a slower and more variable rate of progression. Cardiac disease is rare.

Diagnostic criteria. Many of the histopathologic alterations seen in muscle biopsies of patients with DMD are also observed in muscle in the muscular dystrophies as a group. The characteristic light microscopic abnormalities include (1) greater than normal variation in the cross-sectional diameter of muscle fibers due to the intermingling of muscle fibers that are smaller than normal and of greatly enlarged muscle fibers; (2) migration of the sarcolemmal nuclei from their normal subsarcolemmal location to the interior of the fiber; (3) splitting, degeneration, necrosis, and phagocytosis of muscle fibers; (4) regeneration of muscle fibers; and (5) proliferation of endomysial connective tissue. An additional feature, especially characteristic of the X-linked (Duchenne and Becker) muscular dystrophies, is the presence of enlarged fibers that appear rounded and hyaline on transverse sections and have abnormally spaced cross-striations on longitudinal sections. Histochemical stains show no selective fiber type involvement and often poor definition of muscle fiber typing.

Electron microscopic examination of skeletal muscle biopsies in DMD and other muscular dystrophies discloses fibers in various stages of degeneration and regeneration. *Early changes:* The nature of the earliest changes and the exact sequence of events leading up to muscle fiber degeneration and necrosis are not completely understood. Some observers have suggested that the earliest ultrastructural change consists of localized defects in the plasma membrane (Engel and Franzini-Armstrong 1994). Another important abnormality seen in muscle biopsies of patients with DMD is the absence of dystrophin immunohistochemically and by immunoelectronmicroscopy. *Later changes:* As muscle fiber degeneration advances, several different patterns of ultrastructural change become evident. There is severe overcontraction of the muscle such that sarcomeres lose their normal organization, forming dense clumps of packed myofilaments that alternate with zones of discontinuity and rarefaction of the myofilaments. These probably correspond to the round, hyaline fibers seen on light microscopy. There may be preferential loss of Z- and I-bands, or of A-bands, resulting in discontinuity of sarcomeres. There may be complete disorganization of thick and thin myofilaments and Z-bands. These changes may involve the entire muscle fiber or may be confined to a segment of the fiber. The various patterns of degeneration also may coexist at different sites in the same fiber. With further progression of muscle fiber degeneration, various nonspecific changes occur. Dissolution of myofilaments and Z-bands becomes more advanced and there is shrinkage of mitochondria. The sarcoplasmic reticulum becomes swollen, and intracytoplasmic vacuoles form. Eventually there is breakdown of the cell membrane and lysis of nuclei, and the dead cell becomes engulfed by phagocytes. Regenerating muscle fibers are recognized on light microscopy (hematoxylin and eosin [H&E] preparations) because they have slightly basophilic cytoplasm and large vesicular nuclei with prominent nucleoli. Ultrastructurally, the regenerating fiber can be identified by (1) its large nucleus and nucleolus, the nucleus sometimes occupying an internal position in the fiber; (2) large areas of the fiber that may show myofilaments poorly organized into myofibrils; (3) clusters of short thick and thin myofilaments on each side of a thickened Z-band that may be strewn about haphazardly; (4) cytoplasm with an increased number of free ribosomes. In practice, it may be difficult to distinguish ultrastructurally between early degenerating and regenerating muscle fibers. Abnormalities of basement membrane structure (thickening, interruption of continuity) have been demonstrated in some dystrophies.

Etiology. The muscular dystrophies are a heterogeneous group of inherited disorders, often beginning in childhood, and are characterized clinically by progressive muscular weakness and wasting. The gene for DMD and BMD is very large (2.5×10^6 base pairs with more than 80 exons) and is located in the Xp21 region, encoding a 427-kD protein, termed dystrophin (Straub and Campbell 1997). Dystrophin is normally located ad-

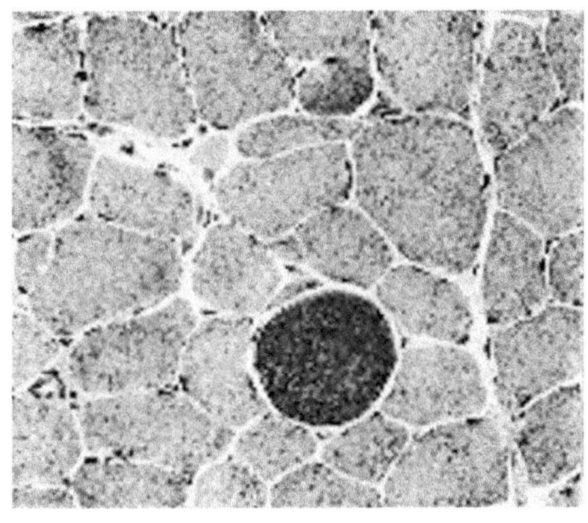

A

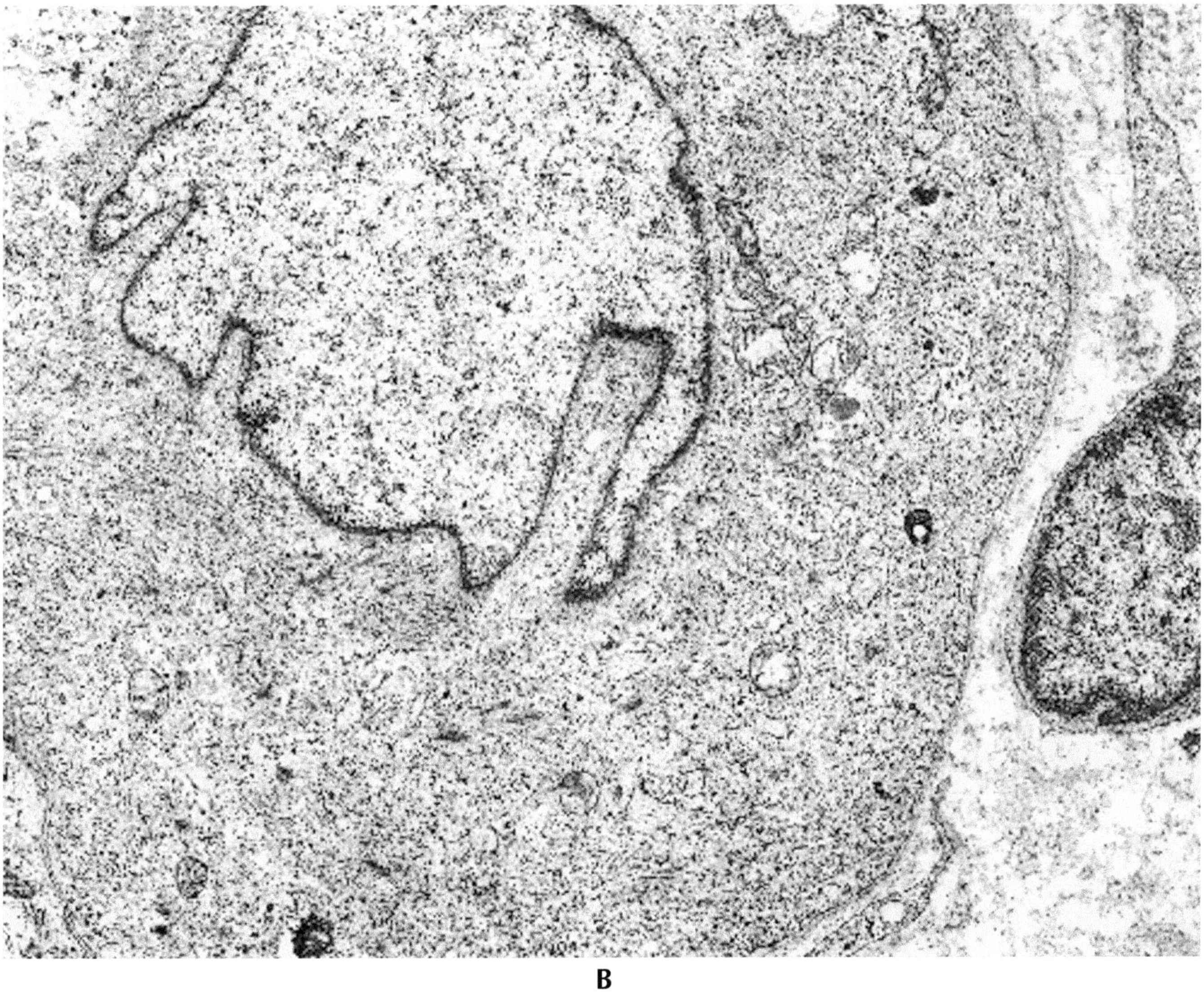

B

Figure 13.2. Duchenne muscular dystrophy. **A,** Variation in fiber size and enlarged round hyaline fiber. (H&E, × 200) **B,** Electron micrograph of dystrophic fiber. There is severe dissolution of normal myofibrillar structure. Note internal nucleus. The basement membrane and sarcolemmal membrane of this fiber are intact. (× 11,000)

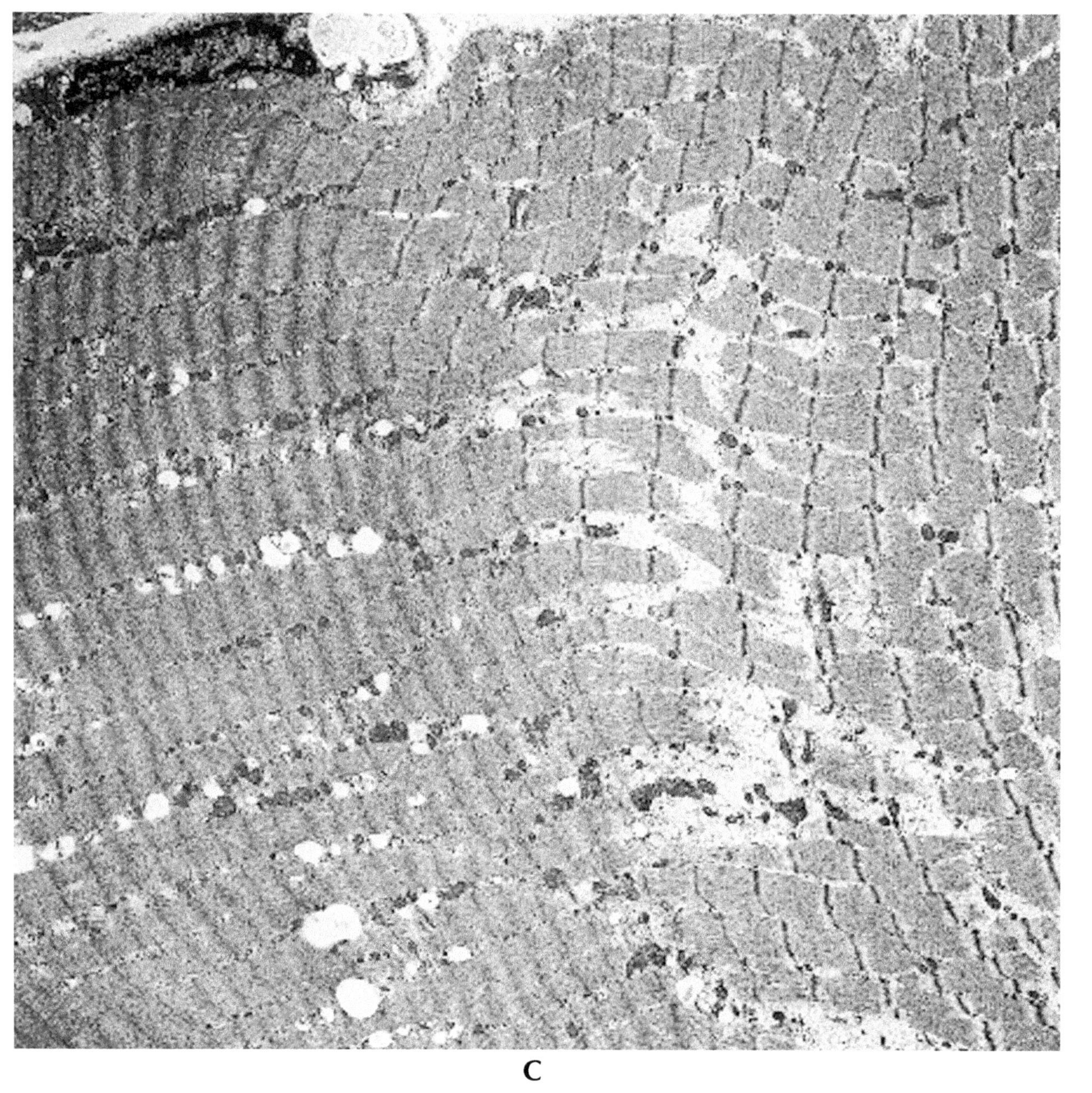

C

Figure 13.12 *(continued)*
C, Hypercontracted fiber. Note abnormally structured and shortened sarcomeres to the left of the fiber. (× 3800)

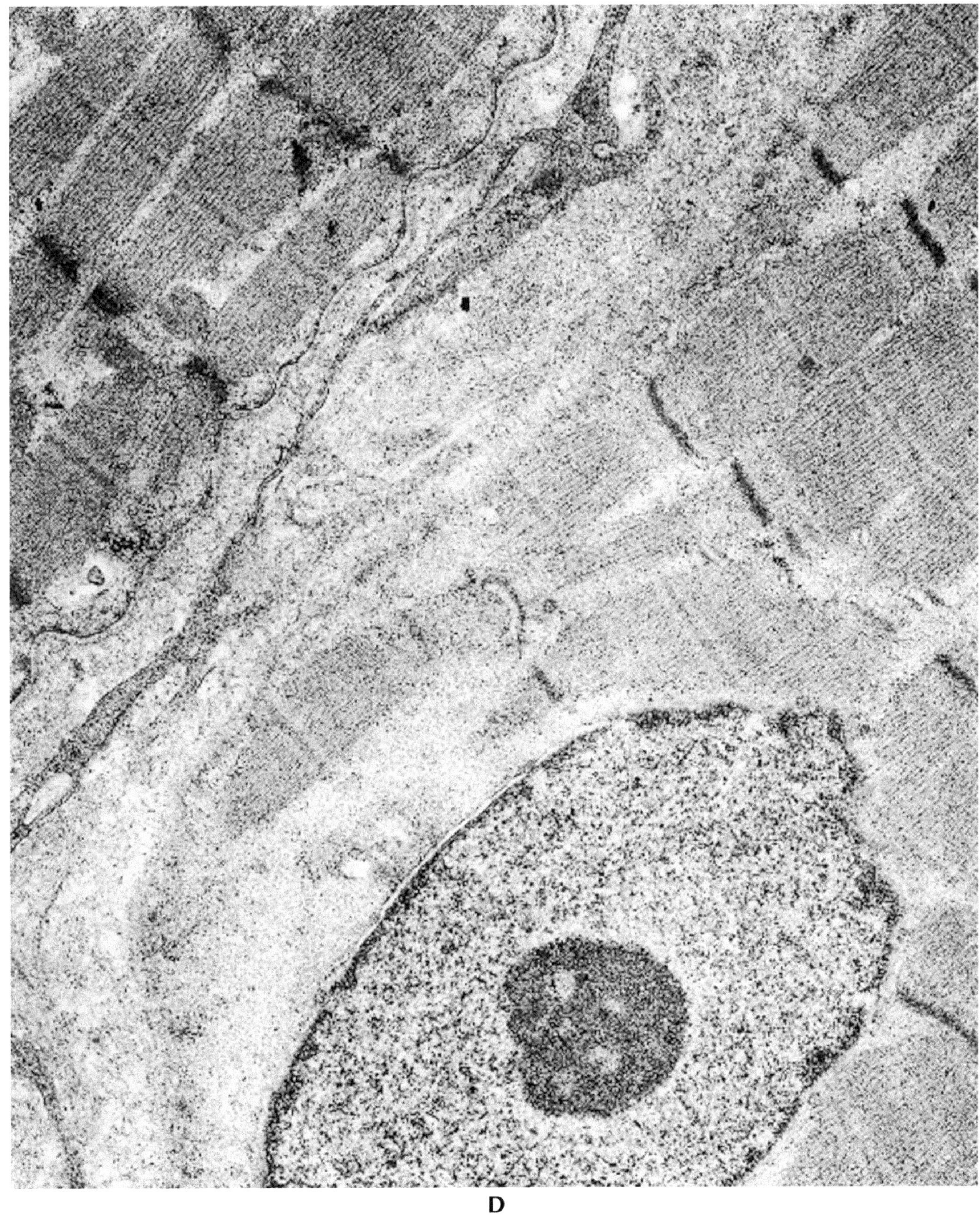

D

Figure 13.2. *(continued)*
D, Muscle fiber regeneration. Note internalized nucleus and prominent nucleolus. (× 30,000)

jacent to the sarcolemmal membrane in myocytes (Worton 1995). Muscle biopsies from patients with DMD show greatly diminished or absent dystrophin by both immunohistochemical staining and biochemical measurements (Western blot).

Closely akin to the dystrophin-related muscular dystrophies are the limb girdle muscular dystrophies (LGMD). In many of the LGMD syndromes, there are mutations involving proteins of the sarcolemma, including a group of proteins, the sarcolgycans, that interact with dystrophin (Bönnemann et al. 1996; Lim and Campbell 1998). The phenotypes of the LGMD range from severe autosomal recessive disorders similar to DMD to more chronic myopathies that present in late adulthood with mild weakness. Also related to this group of diseases are the congenital muscular dystrophies (CMD). In some patients with CMD, there is a malformation of the brain (especially common in Japan and known as Fukuyama CMD) or involvement of the eye (Walker-Warburg CMD and muscle-eye brain disease). In contrast, CMD without clinical involvement of the central nervous system or eye has been identified to have two subtypes: merosin-negative and merosin-positive. In the merosin-negative CMD, there are mutations of the merosin (alpha-laminin) gene, and merosin is not expressed in the basal laminae surrounding individual myocytes. Electron microscopy has demonstrated some discontinuities of the basal lamina in some of these patients (Moretti et al. 1996). These patients with merosin-positivity have no mutations of the merosin gene and express merosin normally. The genetic basis of merosin-positive CMD is not yet understood.

Myotonic Dystrophy

(Figure 13.3.)

Clinical manifestations. Myotonic dystrophy is an autosomal dominant illness characterized clinically by *myotonia,* the sustained involuntary contraction of a group of muscles. The disease begins in late childhood with progressive distal weakness first manifested as gait difficulties, followed by involvement of the intrinsic hand muscles and the muscles of the face. Associated clinical findings include cataracts, frontal balding, and cardiac conduction abnormalities. Gonadal atrophy, smooth muscle involvement, decreased plasma immunoglobulin (Ig) G, and an abnormal glucose tolerance test may also be present.

Diagnostic criteria. Light microscopic examination of skeletal muscle in myotonic dystrophy shows many of the typical features of a dystrophic myopathy, although necrosis, phagocytosis, and regeneration are encountered less often than in DMD. Characteristic features include a marked increase in the number of internalized nuclei, which form conspicuous chains when viewed on longitudinal sections. Histochemical stains show atrophy of type 1 fibers. A distinctive feature of this disorder is the presence of ring fibers (synonyms include Ringbinden, striated annulets). This is an abnormality of the muscle fiber characterized by a circumferential orientation of a portion of the diseased fiber around itself. Both light and electron microscopy show that ring fibers have an outer rim of sarcoplasm where the annulet myofibrils are oriented tangentially relative to the direction of the myofibrils in the principal portion of the fiber. Electron microscopy may be of value in demonstrating the ring fibers when they are not easily visible by light microscopy. Ring fibers are sometimes associated with poorly defined regions of disrupted sarcoplasm extending outward from the ring—sarcoplasmic masses. Ultrastructurally, sarcoplasmic masses consist of disorganized myofilaments, nuclei, cytoplasmic organelles, and dilatation of the T-tubule system and sarcoplasmic reticulum.

Etiology. The gene for myotonic dystrophy, localized to 19q13.2-13.3, encodes a protein kinase termed myotonin-protein kinase. The disease phenotype is correlated with amplification of a trinucleotide repeat located in the 3′ untranslated region of the gene, consisting of (CTG)n. The severity of disease correlates with the magnitude of this expansion. In normal subjects, fewer than 30 trinucleotide repeats are present, whereas in severely affected individuals several thousand may be present. The mutation is not stable within a pedigree; with each generation, more repeats accumulate. This molecular event appears to correspond to the clinical phenomenon of anticipation, whereby the disease tends to increase in severity and come on at a younger age in succeeding generations.

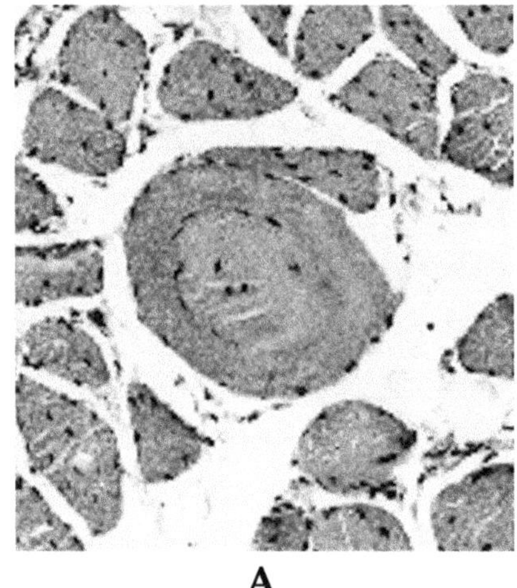

A

Figure 13.3. Myotonic dystrophy. **A,** Ring fiber in center of field. Adjacent fibers contain many internal nuclei. (H&E, × 100)

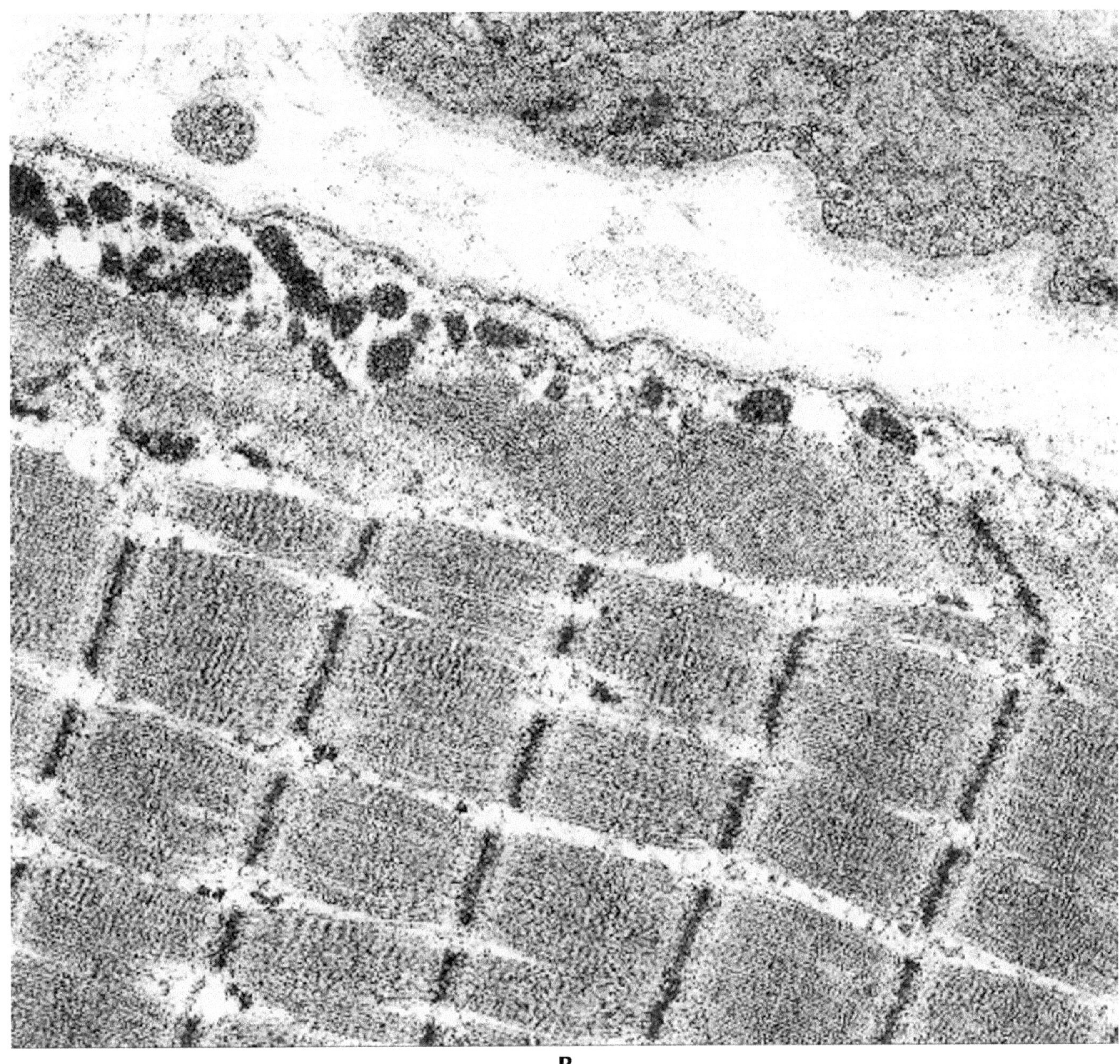

B

Figure 13.3. *(continued)*
B, Ring fiber. Myofibrils oriented perpendicular to the length of the fiber in longitudinal section can be observed beneath the sarcolemma. (× 16,400)

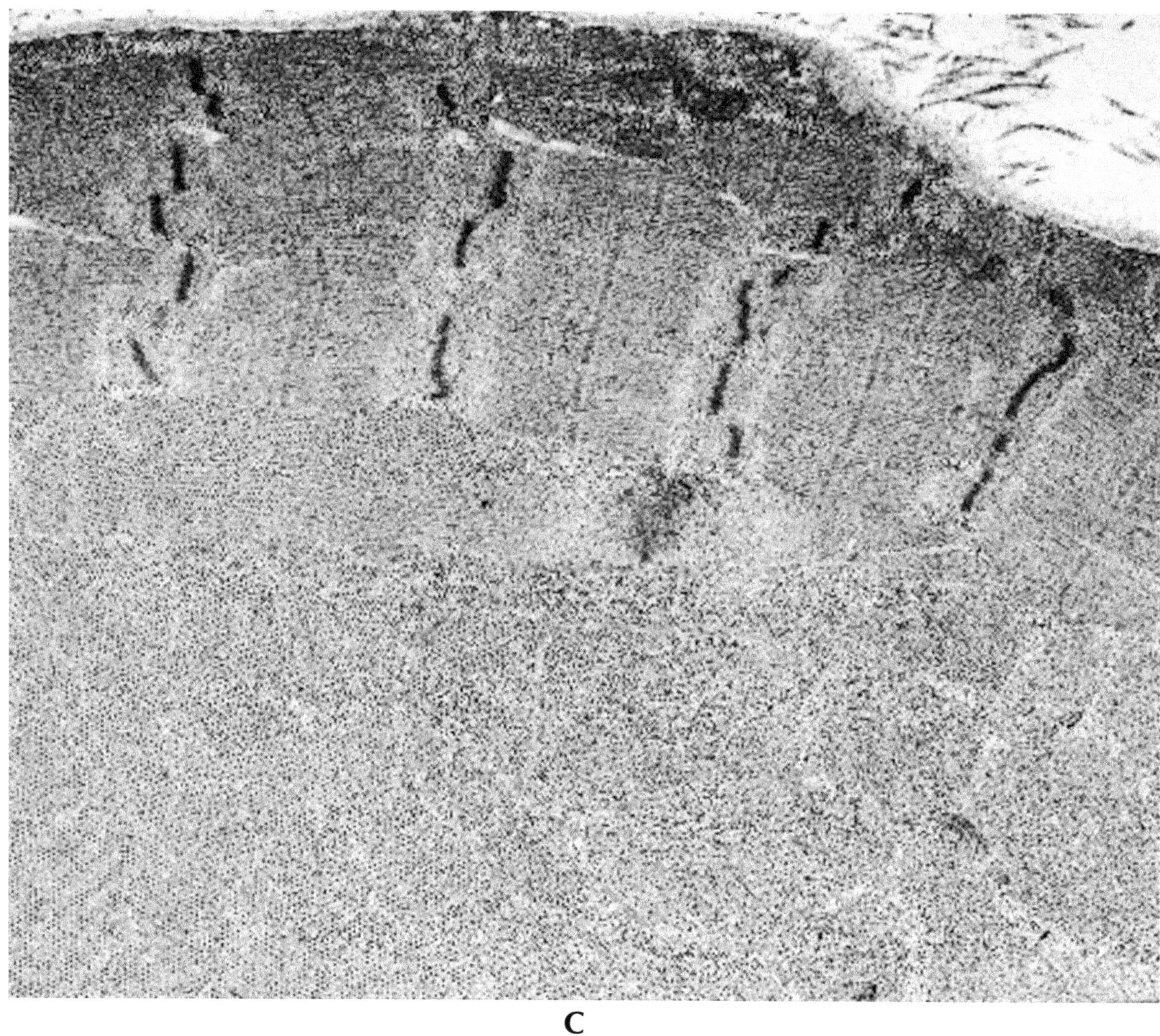

C

Figure 13.3. *(continued)*
C, Ring fiber. Transverse section of fiber showing myofibrils of central portion of fiber cut in cross-section and sarcomeres cut in longitudinal section at the periphery of the fiber. (× 19,000)

Congenital Myopathies

Clinical manifestations. The congenital myopathies comprise a heterogeneous group of disorders, often genetically determined and usually manifest clinically in early life. The clinical course of the illness is ordinarily nonprogressive or slowly progressive. In infancy, patients may present with hypotonia (floppy baby); in childhood the clinical features that distinguish one illness from another are imprecise in many cases. Definitive diagnosis has rested on the demonstration of distinctive structural features identified in the muscle biopsy. Some of these morphologic abnormalities are difficult to visualize by light microscopy; ultrastructural examination is, therefore, an essential part of the evaluation of the skeletal muscle in patients suspected clinically of having one of the congenital myopathies. Molecular characterization of some of these illnesses has advanced rapidly in recent years. The more common forms are described in the next sections.

Central Core Disease

(Figure 13.4.)

Diagnostic criteria. Light microscopy of transverse sections discloses a rounded, centrally placed, amorphous core in many fibers. With histochemical stains, the cores, which are confined to type 1 fibers, lack oxidative and glycolytic enzyme activity. Ultrastructurally, the cores show (1) a disorganized sarcoplasmic pattern; (2) degeneration of myofilaments and Z-bands; and (3) diminution or loss of mitochondria, sarcoplasmic reticulum, glycogen, and T-system. Cores have been subdivided further into structured and unstructured types, depending on the degree of abnormality present. The differential diagnosis generally includes target fibers; however, central cores are homogeneous and involve a much larger length of the myocyte. The clinical context is usually that of congenital myopathy rather than abrupt denervation. Central cores may also be detected in patients susceptible to malignant hyperthermia.

Etiology. The gene for central core disease has been mapped to 19q13.1, and it now appears that most patients have mutations in the ryanodine receptor-1 gene. Not all patients with malignant hyperthermia have central cores or mutations in the ryanodine receptor-1 gene.

Centronuclear (Myotubular) Myopathy

(Figure 13.5.)

Diagnostic criteria. Light microscopy demonstrates a centrally placed nucleus in the majority of muscle fibers. The central nuclei often are surrounded by a clear area that is devoid of adenosine triphosphatase (ATPase) activity. Ultrastructurally, the paranuclear clear area shows: (1) absence of myofilaments; (2) numerous mitochondria; (3) glycogen accumulation. In some cases, especially those in infancy, the central clear zone may be more evident than the internalized nuclei when examining the tissue in cross sections.

Etiology. There are at least two separate genetic loci associated with myotubular myopathy, and these have different clinical phenotypes. The more severe form, X-linked infantile myotubular myopathy, is linked to Xq28, and rapid clinical progression with death before age 2 is common. The gene (MTM1) encodes a protein of at least 620 amino acids known as myotubularin. Autosomal dominant forms of myotubular myopathy, often with later clinical presentation (juvenile) and probably autosomal recessive forms, are also known.

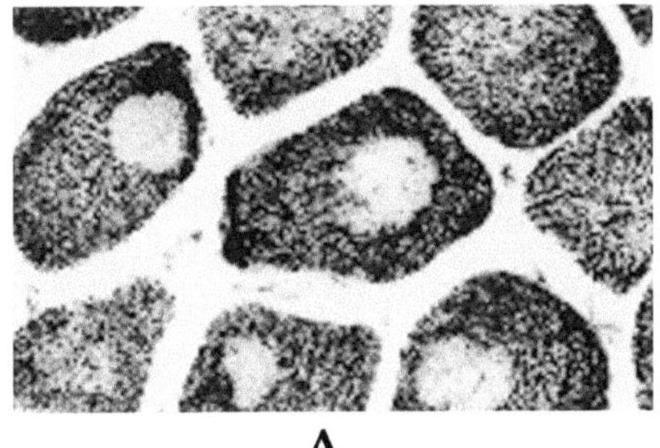

A

Figure 13.4. Central core disease. **A,** Muscle fibers contain centrally placed cores showing absence of oxidative enzyme activity. (NADH-TR, × 300)

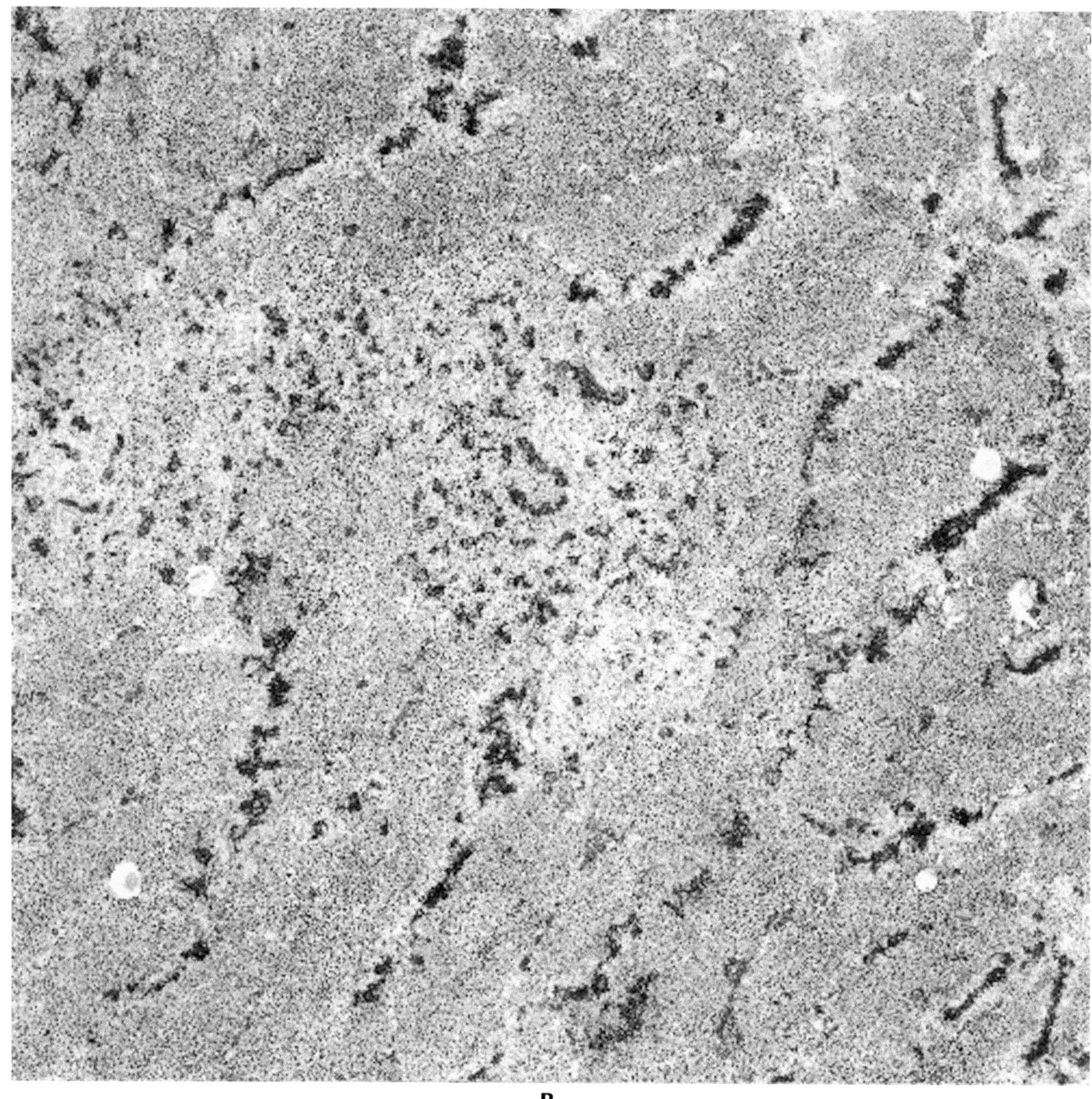

B

Figure 13.4. *(continued)*
B, Muscle fibers with unstructured core. Central area of fiber shows disorganization and loss of myofilaments and abnormal clump of Z-band material. (× 11,000)

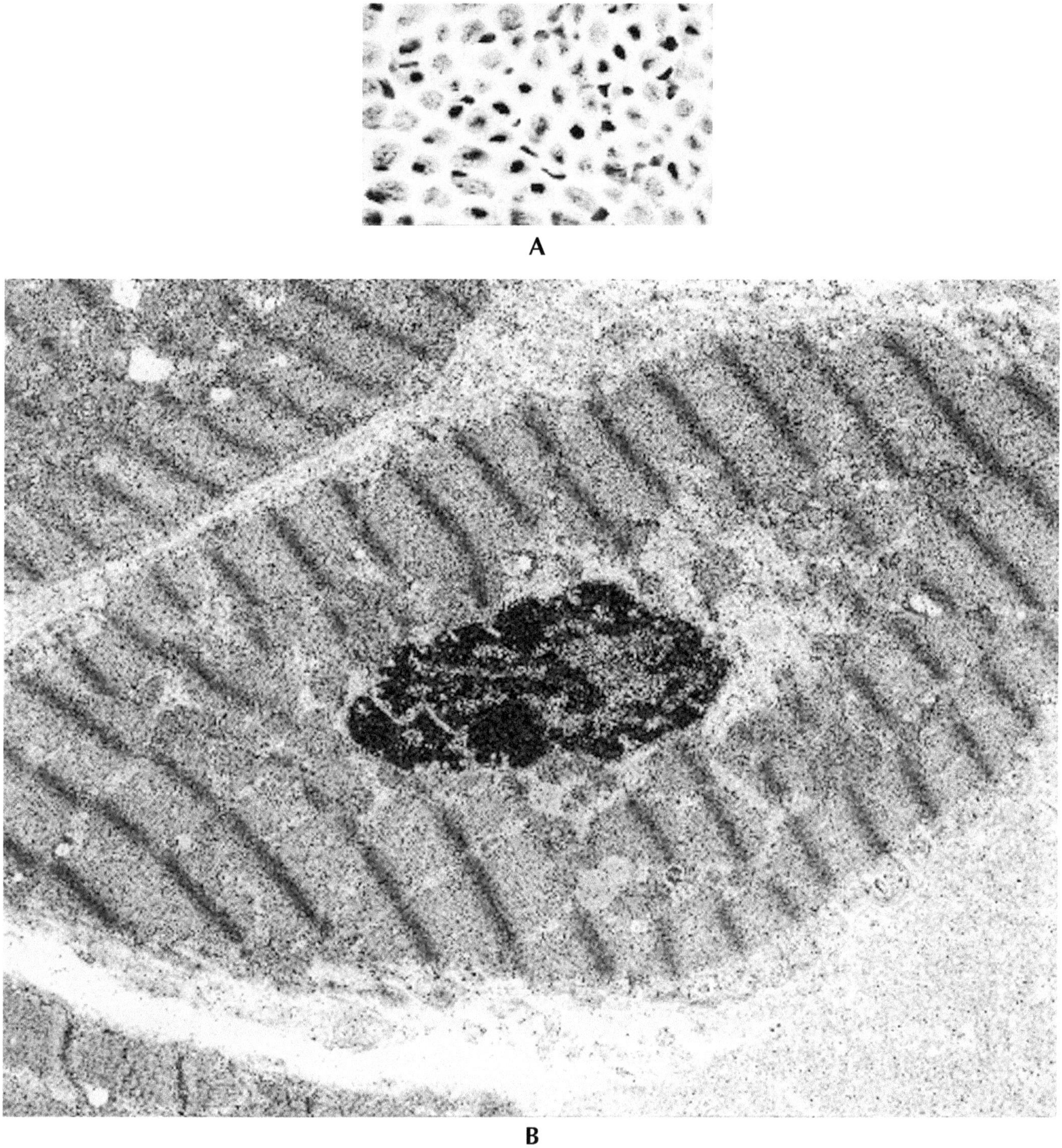

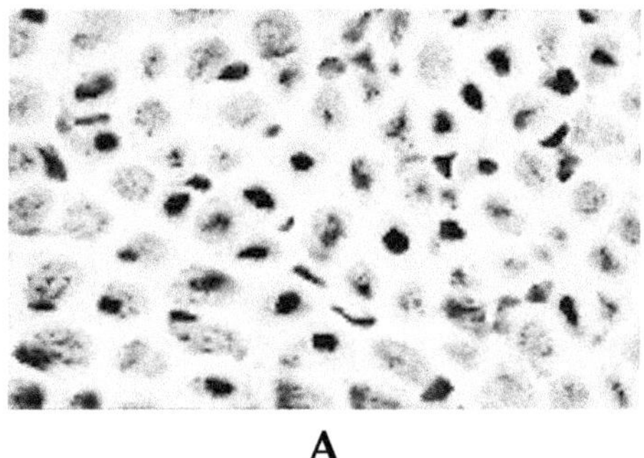

A

B

Figure 13.5. Centronuclear (myotubular) myopathy. **A,** The majority of the small fibers contain centrally placed nuclei. (H&E, × 150) **B,** Muscle fiber with degenerated central nucleus. Note absence of myofilaments and accumulation of glycogen in the paranuclear region to the right. (× 15,000)

Nemaline (Rod Body) Myopathy

(Figure 13.6.)

Diagnostic criteria. The pathologic hallmark of this disorder is the nemaline or rod body. These are the darkly staining, ovoid, or bacilliform structures with a tendency to be located in the subsarcolemmal region of the muscle fiber. They can be difficult to see with H&E and usually are seen more clearly with the modified Gomori trichrome method performed on cryostat sections. The number of fibers containing rod bodies can vary from case to case, irrespective of the severity or duration of the disease. Their definitive identification rests on electron microscopy, which shows them as moderately electron-dense, lattice-like structures with periodic lines oriented parallel and perpendicular to their long axis. They closely resemble normal Z-bands and can be observed in continuity with them. Biochemically, like the Z-disc, nemaline bodies have been shown to consist primarily of alpha-actinin, and antisera are available to demonstrate them on frozen sections by immunohistochemistry. Mutations in tropomyosin-3, alpha-actin, and nebulin have been identified in different families. (Wallgren-Pettersson et al. 1999).

Congenital Fiber-Type Disproportion

(Figure 13.7.)

Diagnostic criteria. The pathologic diagnosis of this condition rests primarily on the enzyme histochemical findings. These show uniformly small type 1 fibers, relatively large type 2 fibers, and type 1 fiber predominance. Ultrastructurally, the small fibers show loss of the normal sarcomeric pattern, with disorganization and loss of the myofilaments, Z-bands, and other organelles; these findings are similar to those seen in neurogenic atrophy. The large fibers are normal ultrastructurally.

Other Congenital Myopathies

In addition to the four types of congenital myopathy defined earlier, several other rare forms have been described that show distinctive structural abnormalities in muscle fibers: multicore or minicore myopathy, fingerprint inclusion myopathy, reducing body myopathy, sarcotubular myopathy, zebra body myopathy, and myopathy with trilaminar fibers (Engel and Franzini-Armstrong 1994).

Multicore (minicore) myopathy is an uncommon, slowly progressive or nonprogressive, congenital myopathy, in which electron microscopic detection of minicores is an important part of the diagnosis. Most families have shown autosomal recessive inheritance, and the condition is usually characterized clinically by proximal muscle weakness in a child, often with external ophthalmoplegia. Light microscopy may reveal small cores, especially on reduced nicotinamide adenine dinucleotide (NADH), which may reach the size of central cores infrequently, but are usually more ill-defined multiple regions with lack of NADH staining within the fiber. By electron microscopy, longitudinal sections demonstrate short (several sarcomere lengths) cores of disorganized filaments and Z-band streaming within the fibers, with exclusion of mitochondria from the minicores.

Metabolic Myopathies

The metabolic myopathies represent clinically heterogeneous diseases affecting skeletal muscle. They are grouped together for purposes of classification based on the notion that there are diseases in which a metabolic abnormality is known to play an important or predominant role in their pathogenesis. Some are associated with a genetically determined, hereditary biochemical defect; in others the metabolic abnormality is acquired, as is the case with endogenous or exogenous toxins. The major diseases included in this category are the ion channel myopathies, glycogen storage diseases, lipid storage myopathies, mitochondrial myopathies, and the endocrine myopathies.

Ion Channel Myopathies

(Figure 13.8.)

Clinical manifestations. The ion channel myopathies (*channelopathies*) are a group of autosomal-dominant familial diseases that include the periodic paralyses, a group of illnesses characterized by short episodes of hypotonic paralysis induced by vigorous exercise, cold, a high carbohydrate meal, or combinations of these. Hypotonia variants of periodic paralysis are recognized that are associated with elevated, depressed, or normal serum potassium levels at the time of the attack—hypokalemic, hyperkalemic, and normokalemic periodic paralysis.

Malignant hyperpyrexia (malignant hyperthermia) is a rare autosomal dominant clinical syndrome characterized by a dramatic hypermetabolic state (tachycardia, tachypnea, muscle spasms, and later hyperpyrexia) triggered by the induction of anesthesia (ordinarily, halogenated inhaled agents and succinylcholine). This clinical syndrome may also occur in predisposed individuals with hereditary muscle diseases, including congenital myopathies, dystrophinopathies, and metabolic myopathies. The only reliable method of diagnosis of the disease using muscle biopsy tissue is the *in vitro* demonstration of muscle contracture upon exposure to anesthetic; ultrastructural study of a muscle biopsy obtained during a hypotonic event shows dilatation of the sarcoplasmic reticulum.

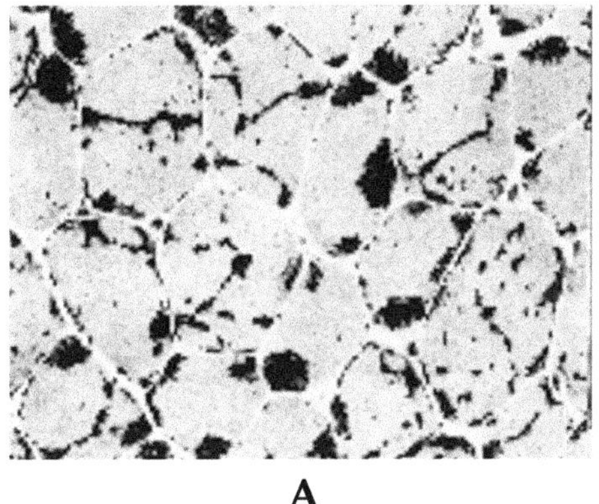

A

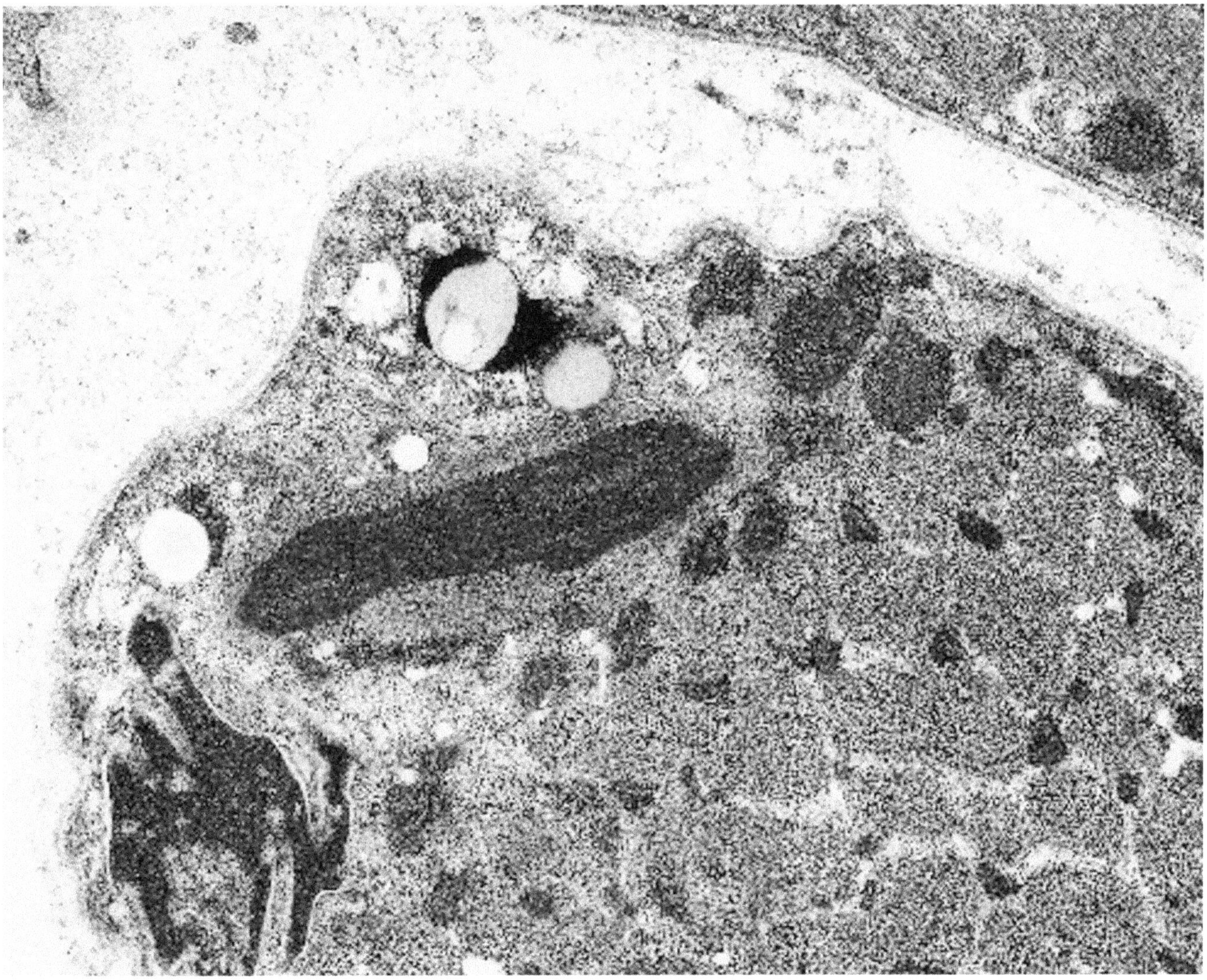

B

Figure 13.6. Nemaline neuropathy. **A,** Clusters of rod-shaped structures can be seen in the subsarcolemmal region of the majority of the muscle fibers. (modified Gomori trichrome, × 200) **B,** Muscle fiber with subsarcolemmal electron-dense nemaline body. (× 19,500)

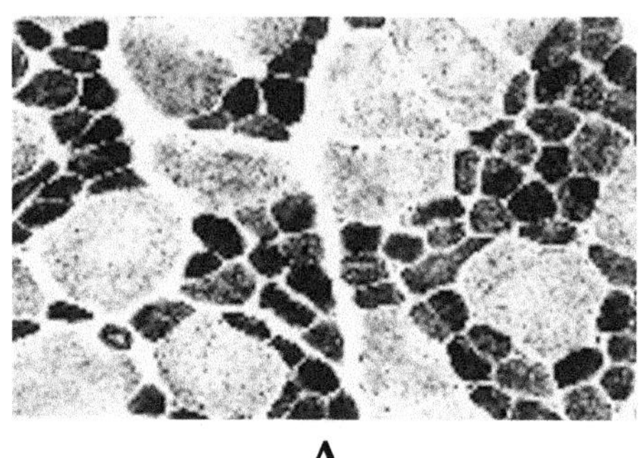

A

B

Figure 13.7. Congenital fiber-type disproportion. **A,** Note two populations of fibers. The small (dark) fibers are exclusively type 1. (NADH-TR, × 100) **B,** Small fiber (*below*) shows disruption of the normal sarcomeric pattern with disorganization and loss of myofilaments, Z-bands, and other organelles. The larger fiber *above* shows normal ultrastructure. (× 12,000)

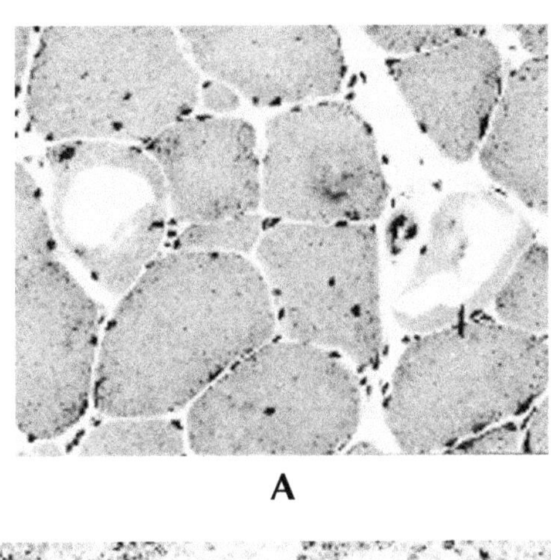

A

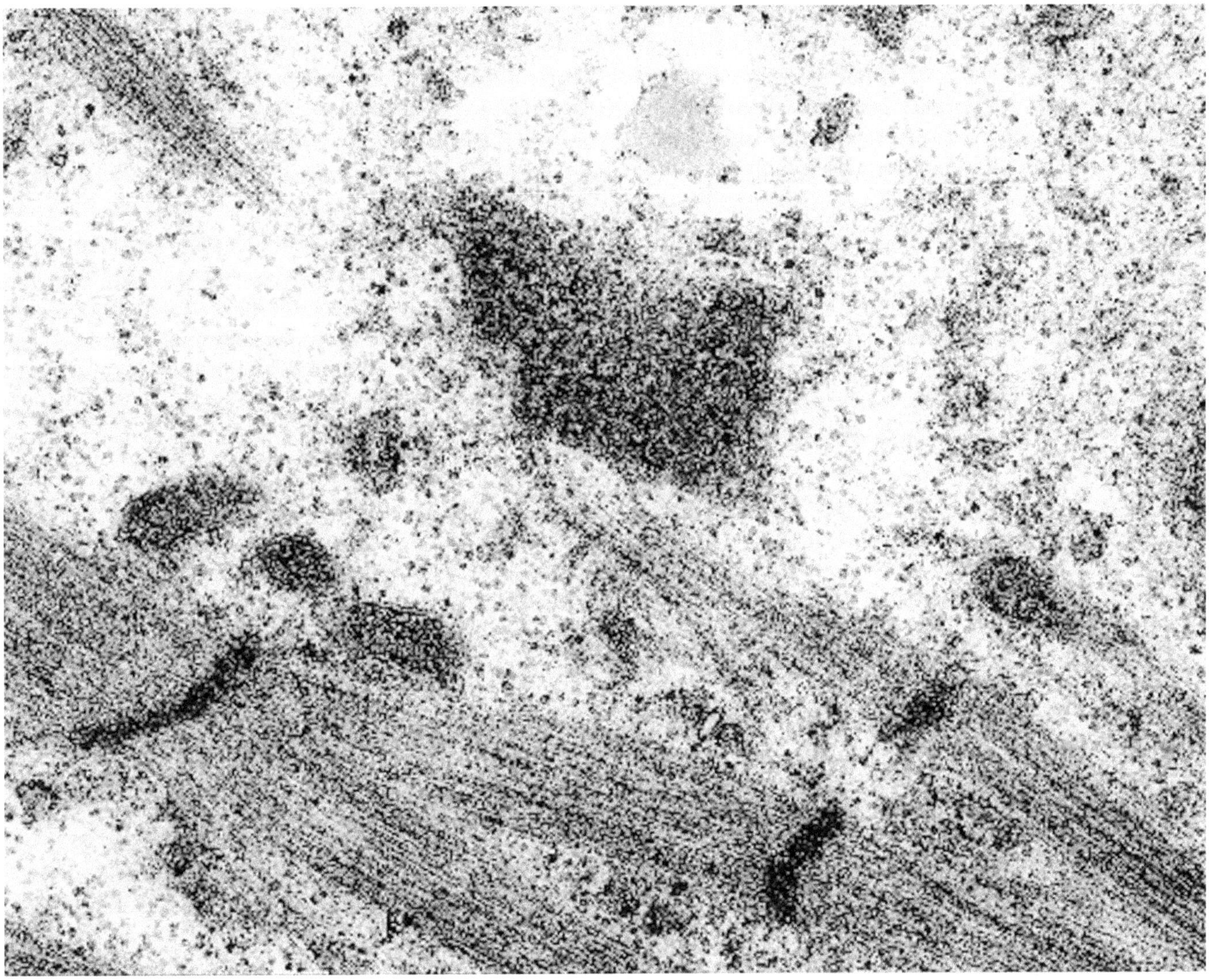

B

Figure 13.8. Hypokalemic periodic paralysis: **A,** Two fibers contain large intracytoplasmic vacuoles. The muscle also shows variation in fiber size and increased numbers of internal nuclei. (H&E, × 300). **B,** Muscle fiber containing small tubular aggregate. Each tubule consists of an electron-dense outer ring and an inner ring that may be less clearly defined. (× 100,000)

Diagnostic criteria. The pathologic findings in all forms of periodic paralysis are qualitatively similar but tend to be most pronounced in the hypokalemic variant. The principal abnormality is the presence of variable amounts of vacuoles within muscle fibers. Ultrastructurally, the vacuoles represent dilatations of the terminal cistern of the sarcoplasmic reticulum. The vacuoles seen in the muscle fibers by light microscopy tend to be most prominent during the attacks of paralysis, but ultrastructural evidence of dilatation of the sarcoplasmic reticulum may be found even between attacks. Another important associated ultrastructural change are tubular aggregates. By light microscopy, tubular aggregates appear as faintly basophilic deposits found both in the interior and periphery of muscle fibers, well demonstrated with nicotinamide adenine dinucleotide dehydrogenase–tetrazolium reductase (NADH-TR) but not with succinic dehydrogenase. Ultrastructurally, they consist of aggregates of parallel, double-walled, straight tubules with a 60–90 nm diameter; on cross-section they form hexagonal profiles. The tubules sometimes can be observed in continuity with the dilated terminal cistern of the sarcoplasmic reticulum, from which they are thought to originate.

Etiology. Different defects on the same gene on chromosome 17 cause hyperkalemic periodic paralysis and paramyotonia congenita. The gene controls the production of a sodium channel protein, which regulates the entry of sodium in muscle during contraction. Hypokalemic periodic paralysis is not related to the sodium channel gene on chromosome 17; it results from mutations, in some families, in the gene encoding a subunit of a specific calcium channel (Ptacek et al. 1994). Both the dominant and recessive forms of myotonia congenita are channelopathies linked to the human skeletal muscle chloride channel on chromosome (Scriver et al. 1995).

Glycogen Storage Diseases

(Figure 13.9.)

Clinical manifestations. The glycogen storage diseases are a group of genetically determined enzymopathies in which there are defects in the availability and integrity of specific enzymes necessary for the metabolism of glycogen, resulting in variable clinical expression depending on the extent of tissue damage caused by glycogen excess in cells. The most commonly encountered glycogen storage diseases that affect skeletal muscle are type II glycogenosis—*acid maltase* (alpha-1,4-glucosidase) *deficiency*—and type V glycogenosis—*myophosphorylase deficiency* (McArdle disease).

The infantile form of acid maltase deficiency (Pompe's disease) is first noted at about 1 month of age, with severe hypotonia, weakness, and heart failure. There is enlargement of the heart, liver, and tongue. Death occurs by 2 years of age from cardiac and respiratory failure. The childhood form of acid maltase deficiency has a later onset. There is a delay in reaching motor milestones, followed by progressive weakness of proximal limb and trunk muscles. The liver and tongue may be enlarged but cardiomegaly is rare. Death occurs before the end of the second decade from respiratory insufficiency. In the adult form of the disease, symptoms begin in the third or fourth decade, with slowly progressive weakness and wasting of proximal limb and trunk muscles with sparing of bulbar musculature. Respiratory muscle weakness may occur and rarely results in death from ventilatory failure. The heart and liver are not affected. The adult variant can present as a limb girdle dystrophy, polymyositis, or spinal muscular atrophy. Serum levels of creatine kinase are elevated in most patients.

Myophosphorylase deficiency (McArdle disease) is caused by a deficiency, restricted to skeletal muscle, of phosphorylase activity. Early in the course of the disease, the clinical manifestations are characterized by easy fatiguability and mild weakness. Later, vigorous activity is accompanied by painful cramps in the exercising muscles. In about one half of patients, muscle necrosis and myoglobinuria occur; renal failure may ensue and is a life-threatening complication of the disease. The ischemic forearm exercise test is very useful in making the diagnosis; it shows that the peak level of venous lactate that normally occurs within 3 to 5 minutes after exercise does not occur because patients with the disease are unable to break down glycogen.

Diagnostic criteria. In Pompe's disease, the muscle fibers appear coarsely vacuolated and contain abundant periodic acid–Schiff (PAS)–positive, diastase-labile material. Characteristically, the vacuoles also are highly

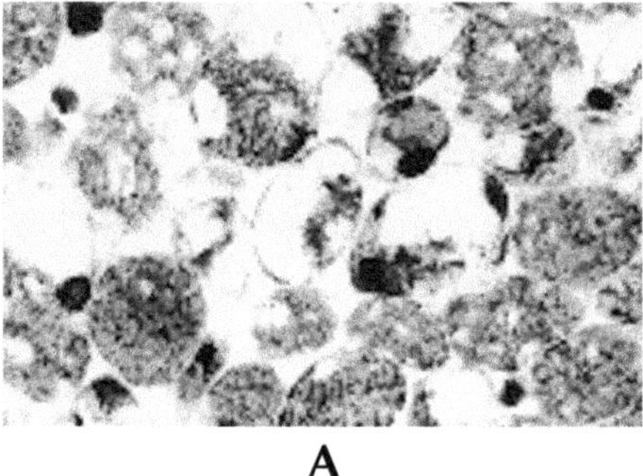

A

Figure 13.9. Pompe's disease (acid maltase deficiency). **A,** The muscle fibers appear coarsely vacuolated. With PAS stain, deposits (glycogen) can be demonstrated in the vacuoles. (H&E, × 500)

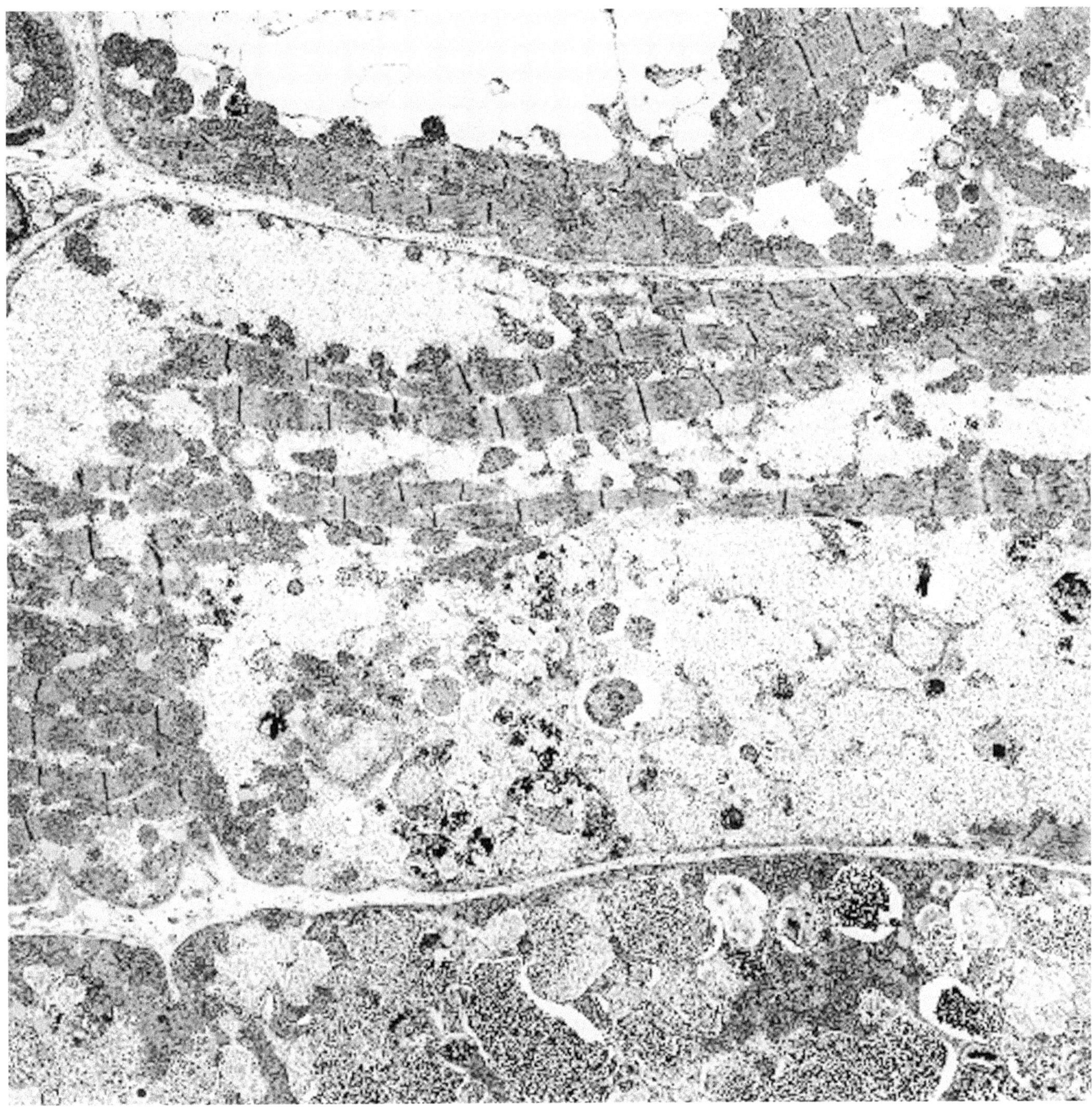

B

Figure 13.9. *(continued)*
B, The muscle fibers contain excessive amounts of glycogen, which is present both free in the cytoplasm and in membrane-bound vacuoles as electron-dense granules. Some of the vacuoles, which may be autophagic, also contain other cytoplasmic degradation products. (× 5000)

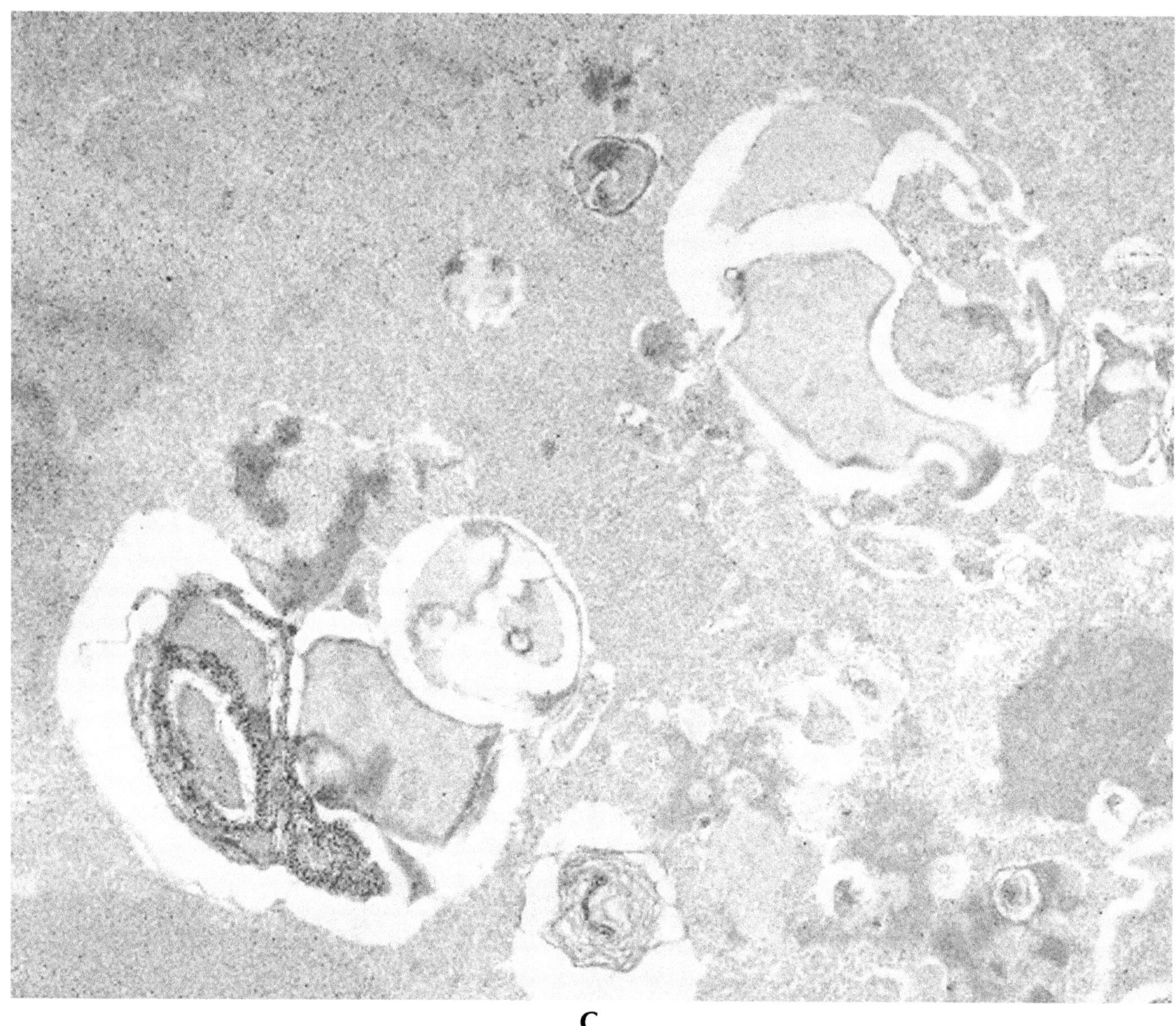

C

Figure 13.9. *(continued)*
C, Electron micrograph showing cytoplasmic and membrane-bound glycogen. (× 70,000)

reactive for the lysosomal enzyme acid phosphatase. By electron microscopy, increased glycogen is present both free in the cytoplasm as well as in distinctive membrane-bound vacuoles, some of which are considered to be lysosomes. There also may be separation and eventual loss of myofibrils. Some fibers contain increased numbers of lipid droplets. In McArdle's disease, free glycogen (i.e., not membrane bound) accumulates predominantly in the subsarcolemmal region of most muscle fibers. These glycogen deposits may be visible on light microscopy as PAS-positive blebs in the periphery of fibers.

Etiology. Both the severe and more benign forms of acid maltase deficiency are inherited as autosomal recessive traits and are due to alterations of the gene that encodes for the enzyme on chromosome 17. There are no molecular genetic data at this time that explain the different phenotypic expressions of the disease.

In most cases of McArdle disease, the illness is inherited as an autosomal recessive trait, although autosomal dominant transmission has been reported. The gene that encodes for myophosphorylase was initially localized to the long arm of chromosome 11; multiple genetic mutations have been demonstrated (Scriver et al. 1995).

Lipid Storage Myopathies

(Figure 13.10.)

Clinical manifestations. Carnitine deficiency may be limited to muscle (myopathic carnitine deficiency) or may be secondary to diminished systemic levels (systemic carnitine deficiency). The cardinal symptom of the myopathic form of disease is weakness; the age at onset is variable. Systemic carnitine deficiency may result from impaired renal reabsorption of carnitine, but more often is secondary to the disorders of beta-oxidation of fatty acids, most commonly medium-chain acyl-CoA dehydrogenase (MCAD) deficiency. *Carnitine palmitoyl transferase deficiency (CPT deficiency)* often presents as recurrent myoglobinuria and is the most common inherited disorder of lipid metabolism affecting muscle. The usual form (CPT II deficiency) presents in teenagers and young adults with episodic acute myonecrosis (rhabdomyolysis) following prolonged exercise and leads to the release of myoglobin into plasma, which when excreted gives the urine an alarmingly dark color (myoglobinuria).

Diagnostic criteria. Staining with oil red O often shows a marked increase in the number of lipid droplets, especially in type 1 fibers. Ultrastructurally, these fibers contain variably sized lipid vacuoles often arranged in parallel rows or large groups. Alterations in mitochondrial structure may sometimes be present.

Etiology. In order to undergo beta-oxidation, cytoplasmic long-chain fatty acyl-CoA esters are conjugated with carnitine through the action of CPT, transported across the outer and inner mitochondrial membranes, re-esterified to acyl-CoA esters, and catabolized to acetyl-CoA units by the acyl-CoA dehydrogenases. Deficiencies affecting the carnitine transport system or deficiencies of the mitochondrial dehydrogenase enzyme systems can lead to the accumulation of lipid droplets within muscle (lipid storage myopathies). CPT II deficiency results from mutation of the CPT II gene locus at chromosome 1p32; the gene product is located on the inner mitochondrial membrane. In contrast, CPT I (locus 11q13) is associated with the outer mitochondrial membrane, and deficiency of CPT I may lead to hypoglycemic hypoketotic coma. The disease is treatable by the administration of medium-chain triglycerides, which do not require the CPT system for crossing the mitochondrial membranes (Scriver et al. 1995).

Mitochondrial Myopathies

(Figure 13.11.)

Clinical manifestations. The mitochondrial encephalomyopathies are a set of heterogeneous disorders with many clinical manifestations and a wide variety of molecular pathogenic mechanisms and patterns of inheritance. Specific mitochondrial disorders for which specific mutations have been identified include MERRF (myoclonic epilepsy and ragged red fibers), MELAS (mitochondrial encephalomyopathy, lactic acidosis, and stroke-like episodes), NARP (developmental delay, retinitis pigmentosa, dementia, seizures, ataxia, proximal neurogenic muscle weakness, and sensory neuropathy), and KSS/CPEO (Kearns-Sayre syndrome/chronic progressive external ophthalmoplegia). The Kearns-Sayre syndrome is a nonfamilial neurologic disorder characterized clinically by onset in the second decade, progressive external ophthalmoplegia, pigmentary retinal degeneration, and heart block and pathologically by the finding of ragged red fibers. Additional clinical manifestations include ataxia, hearing loss, dementia, endocrinologic disturbances, and peripheral neuropathy. Familial cases (maternally inherited via a mitochondrial genetic abnormality or autosomal dominant) of progressive external ophthalmoplegia have been reported.

Diagnostic criteria. These disorders share in common the presence of structurally abnormal mitochondria. Histologically, particularly with the use of frozen sections stained with the modified Gomori trichrome stain, the abnormal mitochondria can be seen as subsarcolemmal aggregates in type 1 muscle cells; with severe involvement they may extend throughout the fiber. They impart a blotchy red appearance to the muscle fiber; furthermore, the muscle fiber contour becomes irregular on cross-section (ragged red fibers). The major ultrastructural changes involving the mitochondria include (1) increased numbers of mitochondria; (2) enlarged and abnormally shaped mitochondria; (3) proliferation and abnormal orientation of cristae, including concentric cristae; and (4) various mitochondrial inclusions, particularly paracrystalline bodies.

Etiology. The underlying mechanism of disease is due to defective oxidative phosphorylation, resulting in deficient production of adenosine triphosphate (ATP). MERRF is a maternally transmitted disease that has been associated with a mutation in a mtDNA gene for a mitochondrial-specific tRNA, resulting in altered function of several of the oxidative complexes. A similar type of mutation of a tRNA gene has been found in MELAS. In Kearns-Sayne syndrome, mtDNA deletions have been found but not in familial chronic progressive external ophthalmoplegia. In the autosomal dominant variant of chronic progressive external ophthalmoplegia, multiple mtDNA deletions are found.

Inflammatory Myopathies

The inflammatory myopathies can be subdivided into two major groups of diseases: infectious myositides (e.g., viral myositis, trichinosis, bacterial myositis) and

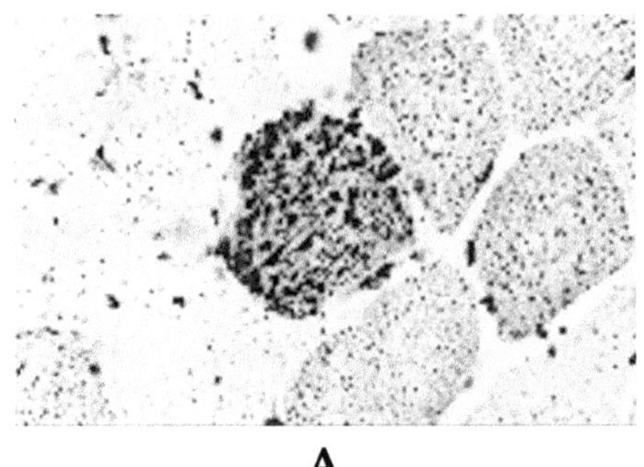

A

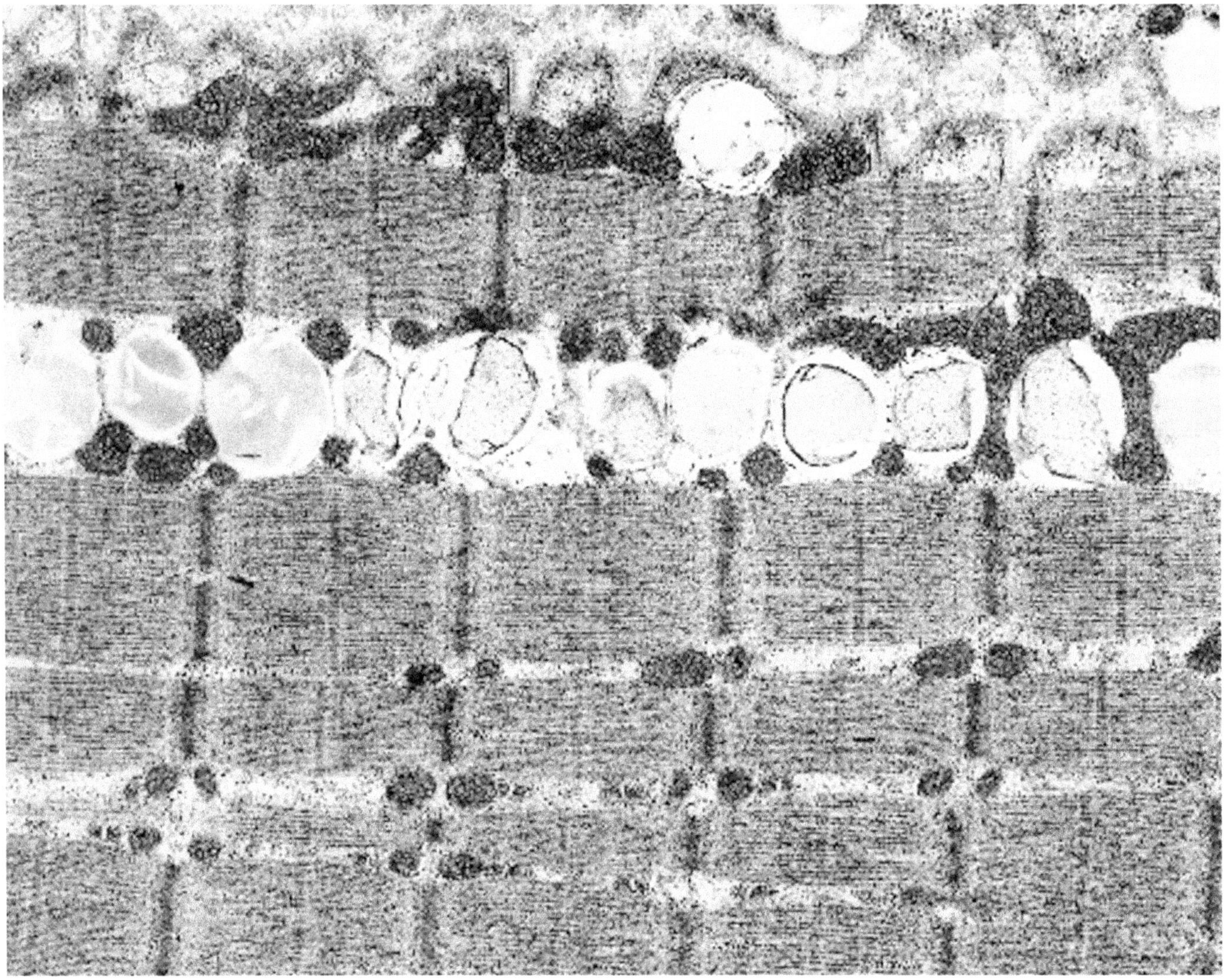

B

Figure 13.10. Lipid storage myopathy. **A,** Muscle fiber in center of field shows excessive number of lipid droplets. (oil red O, × 300) **B,** A large number of irregular electron-lucent lipid vacuoles are present in the intramyofibrillar region of this muscle fiber. (× 23,000)

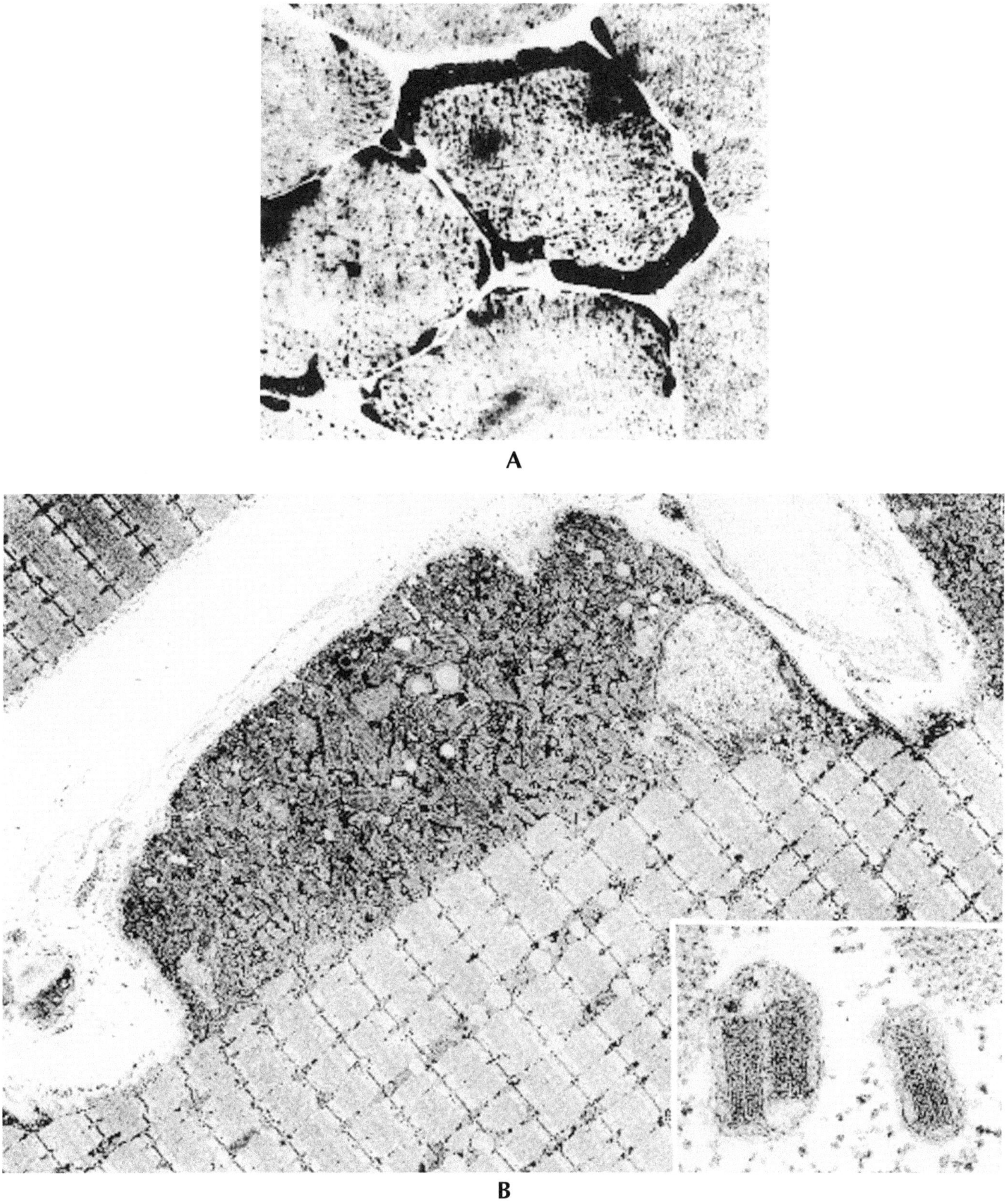

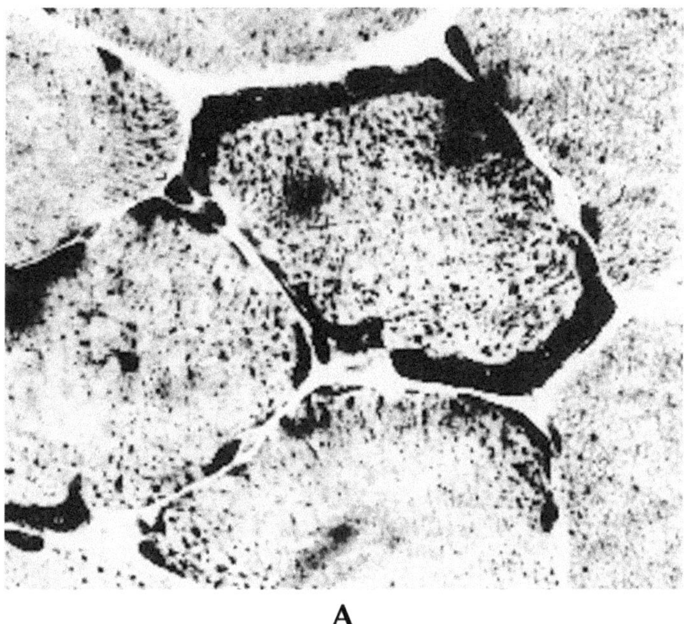

A

B

Figure 13.11. Mitochondrial myopathy. **A,** Ragged red fiber. Several fibers contain prominent subsarcolemmal deposits of granular material that stains red with the modified Gomori trichrome stain. (trichrome, × 600) **B,** Electron micrograph of ragged red fiber showing subsarcolemmal accumulation of abnormal mitochondria. (× 5000) *Inset:* Abnormal mitochondria containing paracrystalline inclusions. (× 50,000)

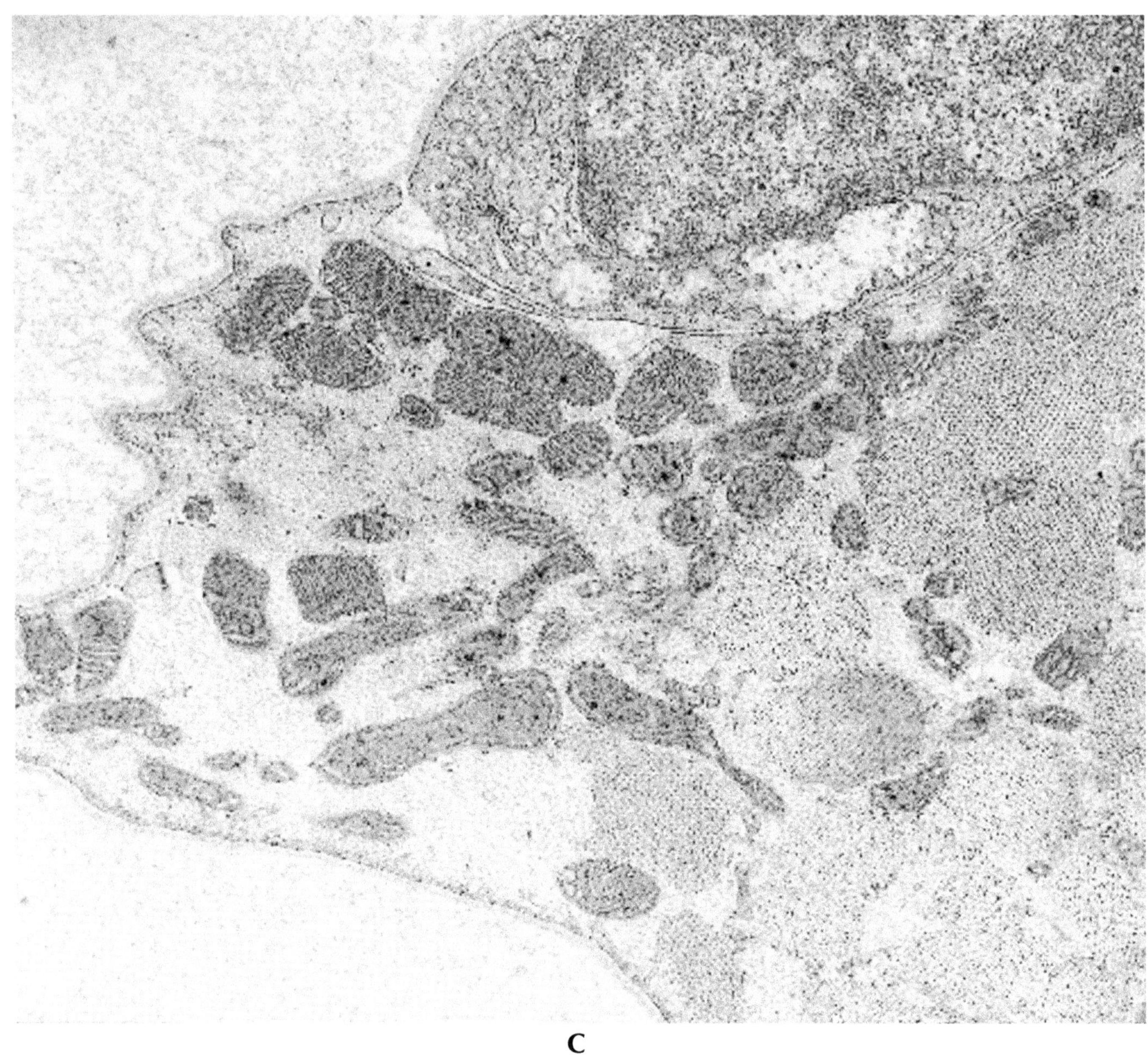

C

Figure 13.11. *(continued)*
C, Electron micrograph showing a range of structurally abnormal mitochondria. (× 70,000)

idiopathic or noninfectious inflammatory myopathies where immune mechanisms play an important part in the pathogenesis. The ultrastructural characteristics of specific infectious diseases and viral and bacterial infectious agents have been described elsewhere in this text. The discussion that follows will address exclusively the noninfectious inflammatory myopathies; these include idiopathic polymyositis/dermatomyositis and inclusion body myositis.

Polymyositis/Dermatomyositis

(Figure 13.12.)

Clinical manifestations. Polymyositis (PM) begins insidiously with asymmetric proximal muscle weakness variably associated with pain in the arms or, less often, leg muscles (difficulty climbing stairs; difficulty raising the arms for combing hair or grooming). Dysphagia occurs in about one third of cases, and weakness of neck

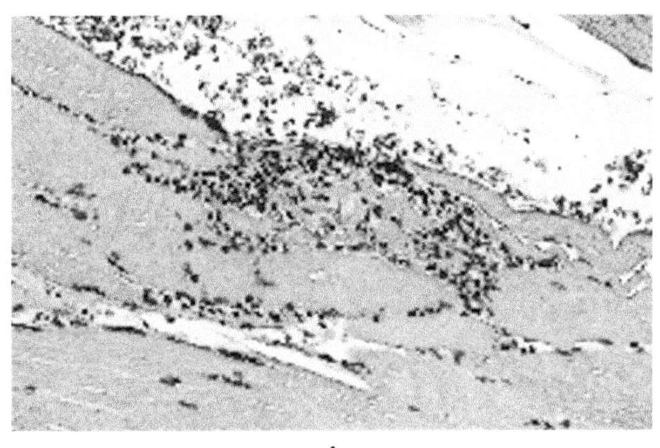

A

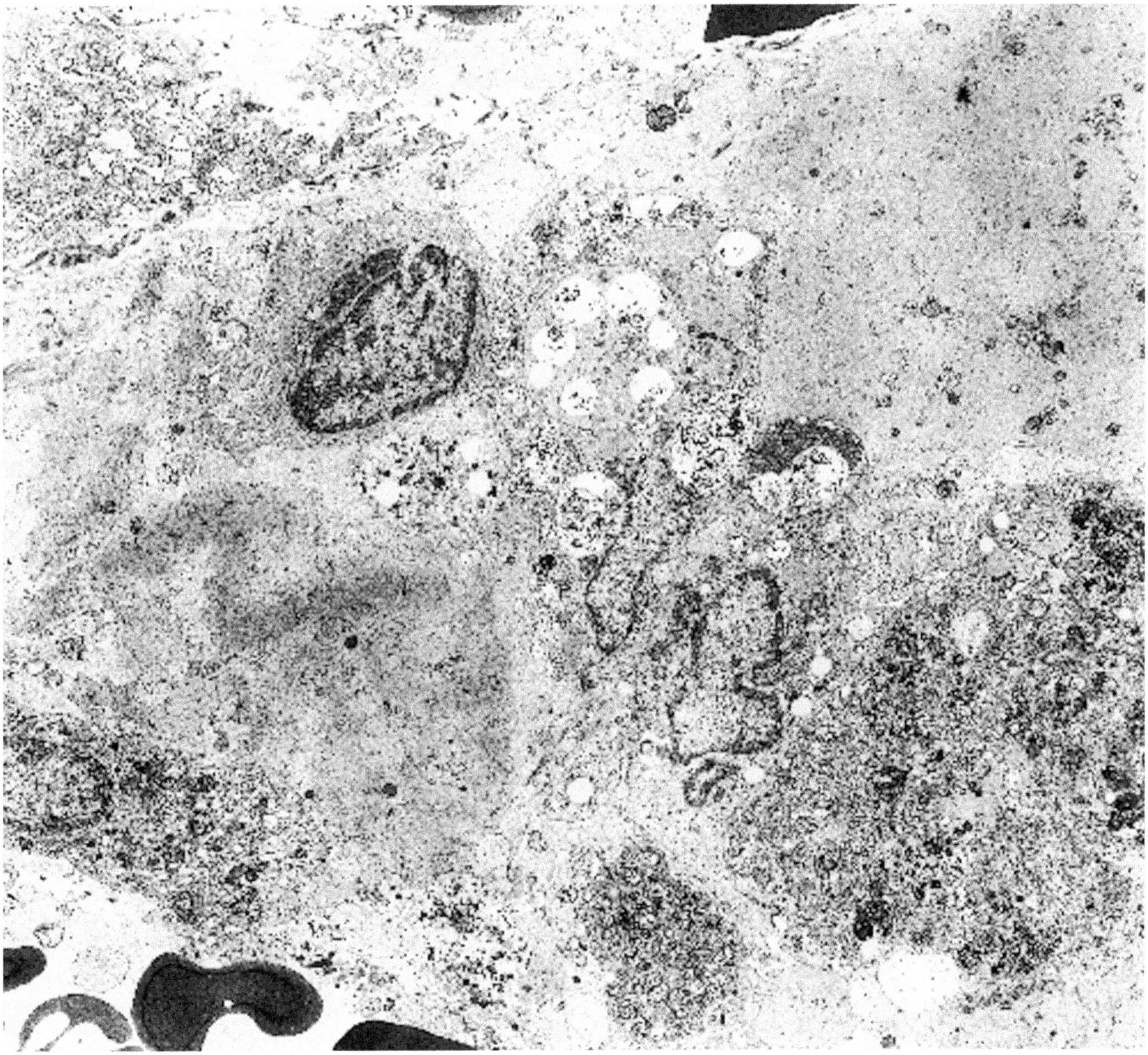

B

Figure 13.12. Inflammatory myopathy (polymyositis). **A,** Necrotic muscle fiber invaded by phagocytic inflammatory cells. (H&E, × 150) **B,** Mononuclear phagocytes can be observed within the cytoplasm of this necrotic muscle fiber. (× 6500)

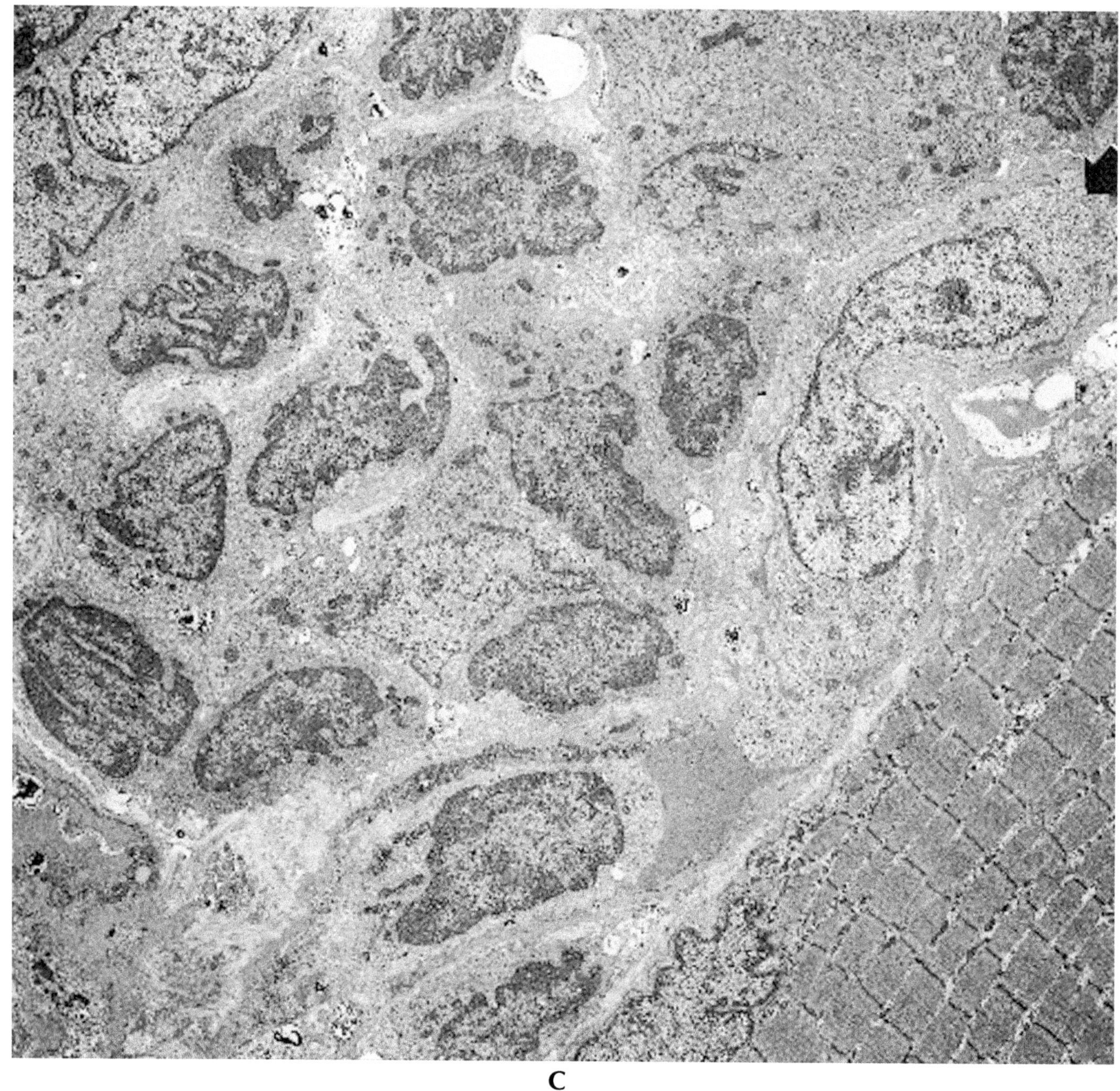

C

Figure 13.12. *(continued)*
C, Inflammatory cells adjacent to regenerating muscle fiber. A relatively normal fiber is seen at the bottom right. (× 3500)

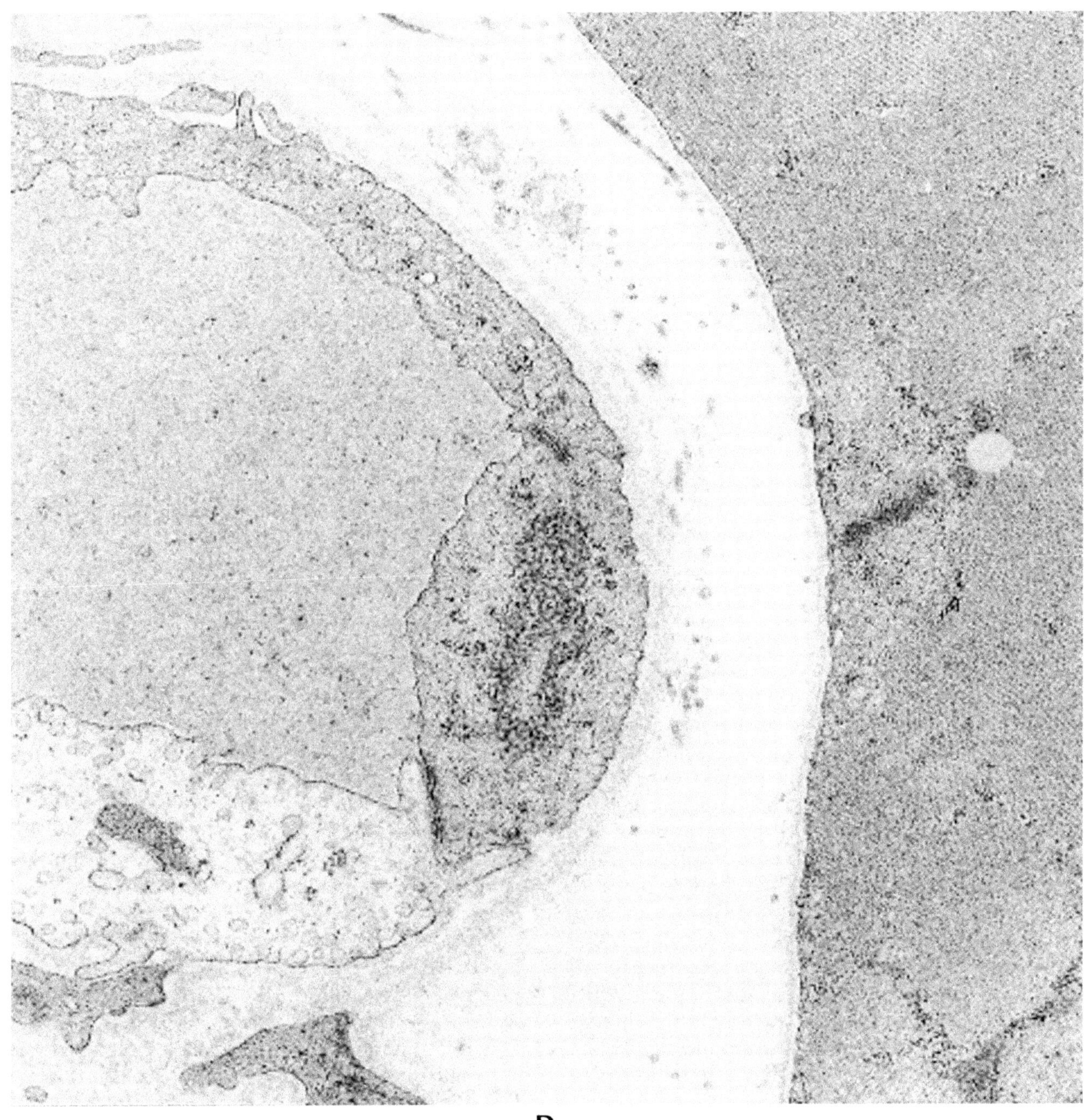

D

Figure 13.12. *(continued)*
D, Endothelial cell with tubuloreticular inclusion in juvenile dermatomyositis. (× 35,900)

flexors is commonly noted. Adult dermatomyositis (DM) is characterized by progressive proximal muscle weakness with erythematous skin lesions that may precede, coincide, or follow the weakness and is accompanied by ungual telangiectases. The facial skin rash begins in the face (periorbital, malar, perioral regions), anterior neck, and chest regions as well as the extensor surfaces of the joints. The skin lesions progress to scaling, brown discoloration, and induration. In both PM and DM, electrocardiographic abnormalities have been described. Case studies from centers in different parts of the world indicate that the incidence of malignancy in middle-aged or elderly patients with PM/DM ranges from 10–40%. Both PM and adult DM may be associated with collagen-vascular diseases such as rheumatoid arthritis, scleroderma, polyarteritis nodosa, Sjögren's syndrome, lupus erythematosus, and mixed connective tissue disease. Laboratory studies show that creatine kinase is elevated, and electromyography (EMG) discloses short, polyphasic potentials with fibrillations, positive waves, and pseudomyotonic bursts. The typical clinical picture of childhood dermatomyositis consists of erythematous malar violaceous; heliotrope rash, also involving the skin over the extensor surfaces of the joints; and muscle weakness of insidious onset relentlessly progressing over a period of weeks or months. In some cases, other associated manifestations include subcutaneous calcifications, gastrointestinal bleeding, and respiratory insufficiency.

Diagnostic criteria. The salient light microscopic features include degenerating and regenerating muscle fibers and mononuclear inflammatory cell infiltrates. On ultrastructural examination, the changes observed in the degenerating and regenerating fibers are similar to those described in the section on muscular dystrophy. The predominant inflammatory cells have the ultrastructural characteristics of activated or transformed lymphoid cells. These are seen mainly in the interstitium between the muscle fibers and around small blood vessels. Similar-appearing cells may also be found internal to the basement membrane, sometimes in contact with muscle fibers. Occasionally, these cells are found within the muscle fiber itself. Other types of inflammatory cells are also present, including histiocytes, plasma cells, and occasionally neutrophils. Nonspecific changes have been described in the small blood vessels in polymyositis: thickening and reduplication of the basement membrane, swelling of endothelial cells, and various inclusions (e.g., autophagic vacuoles and multivesicular bodies). Several reports have described virus-like inclusions in cases of polymyositis. These have usually consisted of 5–7 nm filaments and 8–25 nm filamentous microtubules resembling paramyxovirus nucleocapsids. With few exceptions, viruses have not been isolated from such cases, and the nature of these inclusions remains to be established.

In childhood dermatomyositis there is a characteristic "atrophy" of muscle cells at the periphery of the fascicle. These small fibers often show regenerative changes. The inflammatory infiltrates tend to occur around blood vessels, and myophagocytosis is less common than in polymyosistis. A particular type of inclusion known as a tubuloreticular inclusion characteristically has been seen in endothelial cells.

Etiology. The mechanisms of fiber damage in these diseases have yet to be completely defined, but there is much evidence to implicate immunologic factors (De Bleecker and Engel 1995). In PM, a cell-mediated immunopathogenesis of the muscle fiber destruction is suggested by the available evidence. The predominant inflammatory cells have been demonstrated to be T-cells (about 70%) and macrophages. The T-cells are mostly cytotoxic/suppressor CD8+ and some CD4+ cells. MHC-I expression is demonstrable on and in affected fibers. The focal invasion and destruction of muscle fibers by T-cells and macrophages suggests previous sensitization to muscle-fiber-surface-associated antigens, but no target antigen has yet been identified. Viral infection has long been suspected to be the inciting factor that triggers the immune response, but there is little evidence for this hypothesis in spite of active search. The possible contribution of the humoral immune system in the development of the disease is unresolved at this time. In patients with PM, immunoglobulin deposits have been visualized by immunofluorescence within the walls of small intramuscular blood vessels, but these observations have not been consistent or specific, and therefore such examinations have largely been abandoned as a useful adjunct to diagnosis. In adult DM, humoral immunity appears to play an important part. The inflammatory infiltrate is composed mainly of B-cells and helper T-cells, although macrophages and cytotoxic/suppressor T-cells are also present. In support of the notion that DM is an immune-mediated vasculopathy is the observation that the complement system has been found to be deposited, bound, and activated to completion within the intramuscular microvasculature of patients with childhood DM (to a lesser extent in adult DM). In childhood DM, microcirculatory abnormalities and necrosis and thrombosis of capillaries, small arteries, and venules has been found, especially in those vessels at the periphery of muscle fascicles, presumably accounting for the perifascicular "atrophy." In PM associated with scleroderma, an immunologic response mediated by T-cells appears to be directed against a connective tissue or vascular element and not against the muscle fiber.

Inclusion Body Myositis

(Figure 13.13.)

Clinical manifestations. Inclusion body myositis (IBM) is a distinct inflammatory myopathy that differs clinically from polymyositis in several respects. The disorder comes on in a somewhat older, particularly male, patient population; skin manifestations are not a feature of the illness; the distribution of the weakness tends to be both distal and proximal; the course of the illness is perhaps more protracted. Creatine kinase levels are either in the normal range or mildly elevated, though ordinarily not above levels that are commonly seen in PM and DM. EMG findings are, for the most part, myopathic. The disease appears to be refractory to corticosteroid treatment and other therapeutic measures that are utilized for PM and DM. An association with connective tissue diseases has been noted in a minority of cases, and some familial cases are on record.

Diagnostic criteria. The histopathologic findings on muscle biopsy include sparse or entirely absent interstitial chronic inflammatory cell infiltrates of lymphocytes (largely CD8+ T-cells surround the nonnecrotic fibers) and macrophages, and rarely, myophagocytosis and regenerating fibers. In addition, some muscle fibers contain small vacuoles, known as *rimmed vacuoles,* because they are lined by basophilic granules. Electron microscopy demonstrates aggregates of closely packed 11–18 nm diameter tubular filaments in the cytoplasm and nuclei. Collections of cytoplasmic membranous bodies and abnormal mitochondria with paracrystalline inclusions are also seen within areas of disintegration of the myofibrillar architecture corresponding to the vacuoles seen by light microscopy. Virus-like tubular filaments, 15–18 nm wide, have been noted scattered in the disrupted cytoplasm or within neighboring nuclei. Beta-amyloid has been colocalized to the filaments in both sporadic and familial cases, but it is premature to attribute significance to this finding. The filamentous inclusions do not appear to be specific for IBM. Recent observations have also reported the presence of ubiquitin, but the overall pathogenetic significance of this observation is unknown.

Etiology. Muscle biopsies from patients with IBM, as in PM/DM, have been studied carefully for evidence of paramyxovirus infection, but although studies have demonstrated the presence of virus in some affected individuals, such evidence is lacking in most others (Carpenter 1996). There is a strong possibility that autoimunity plays a role in the disease (Oldfors and Lindberg 1999).

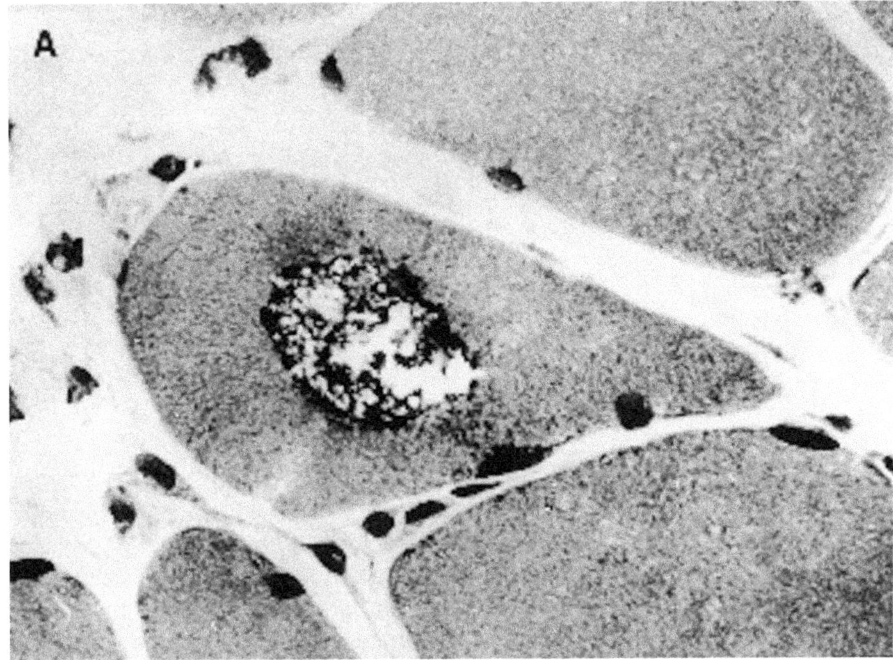

Figure 13.13. Inclusion body myositis. **A,** Muscular fiber containing rimmed vacuole in center of sarcoplasm. (trichrome, × 1600)

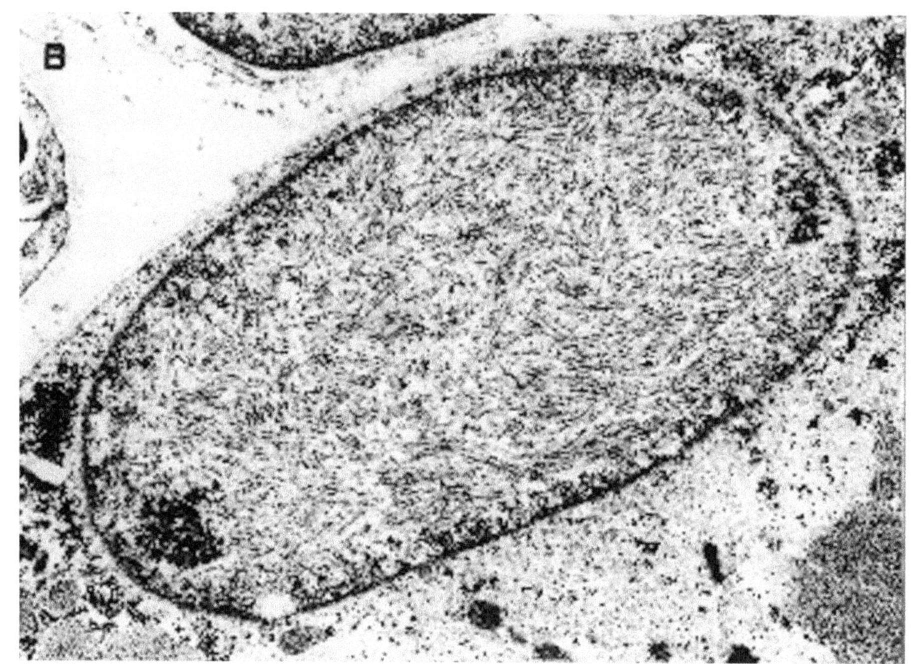

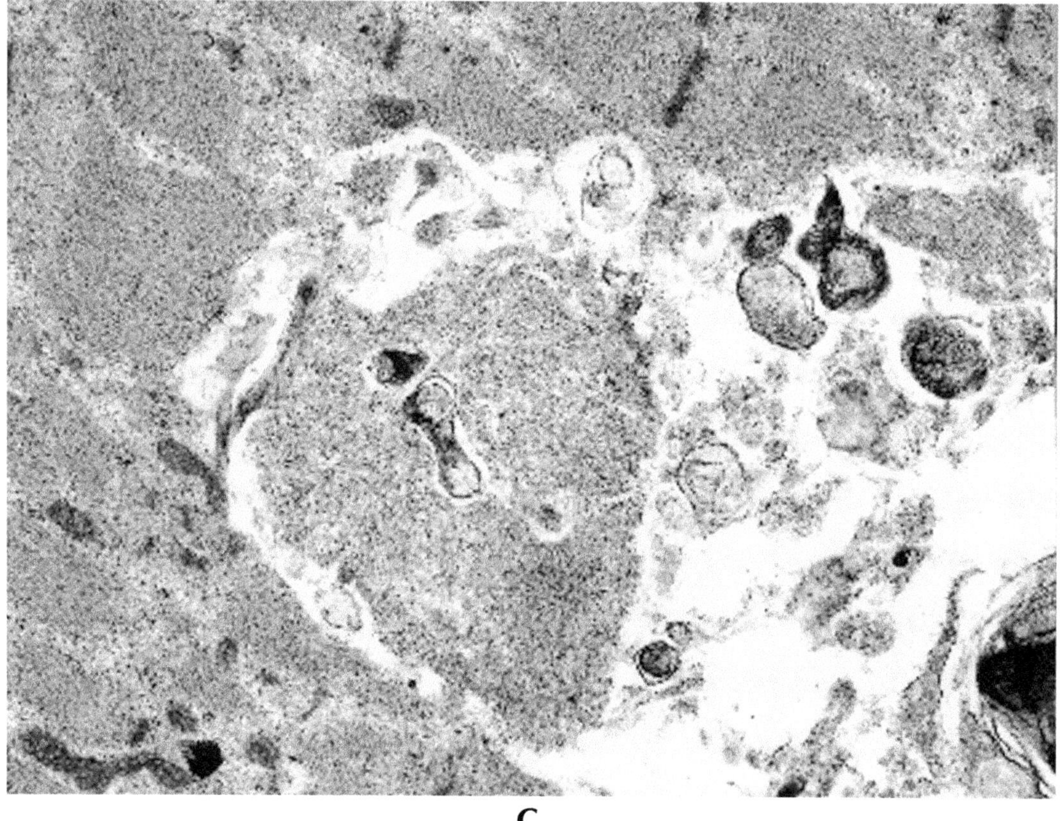

C

Figure 13.13. *(continued)*
B, Nucleus of muscle fiber contains aggregates of 11–18 nm diameter tubular filaments. (× 26,000) **C,** Electron micrograph of content of rimmed vacuole showing broken-down myofibrillar architecture and cytoplasmic membranous bodies. (× 40,000)

Toxic Myopathies

(Figure 13.14.)

Clinical manifestations. Myopathy may be seen in association with the administration of a wide variety of drugs (e.g., colchicine, antimalarials, cholesterol-lowering agents, antiviral agents) (Engel and Franzini-Armstrong 1994). The histopathology of the lesions in these cases are characterized by a monophasic mixed acute and chronic inflammatory reaction with variable

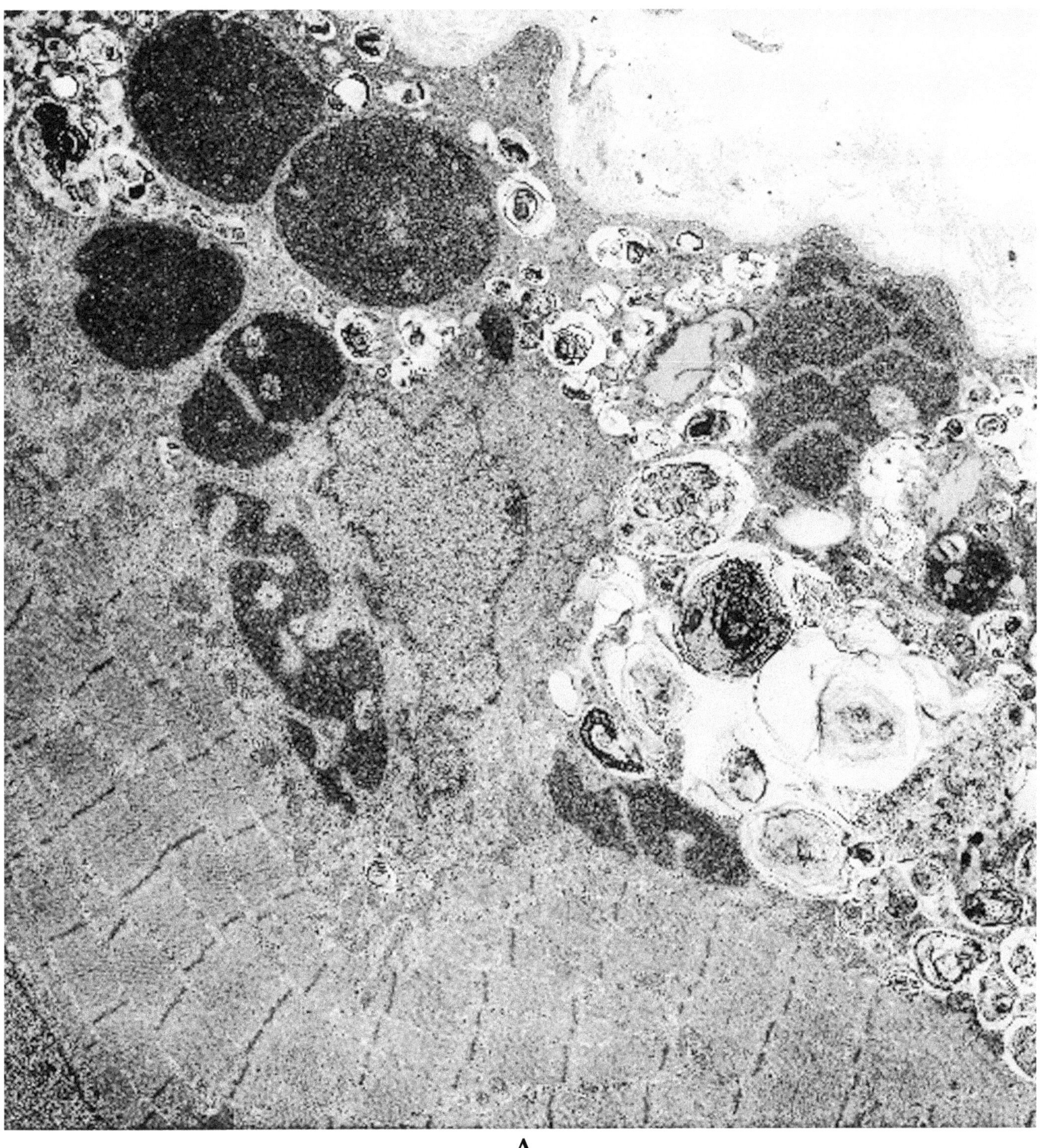

A

Figure 13.14. Toxic myopathy. **A,** Colchicine myopathy. Note focal destruction of the myofibrillar organization and membranous whorls. (× 5800)

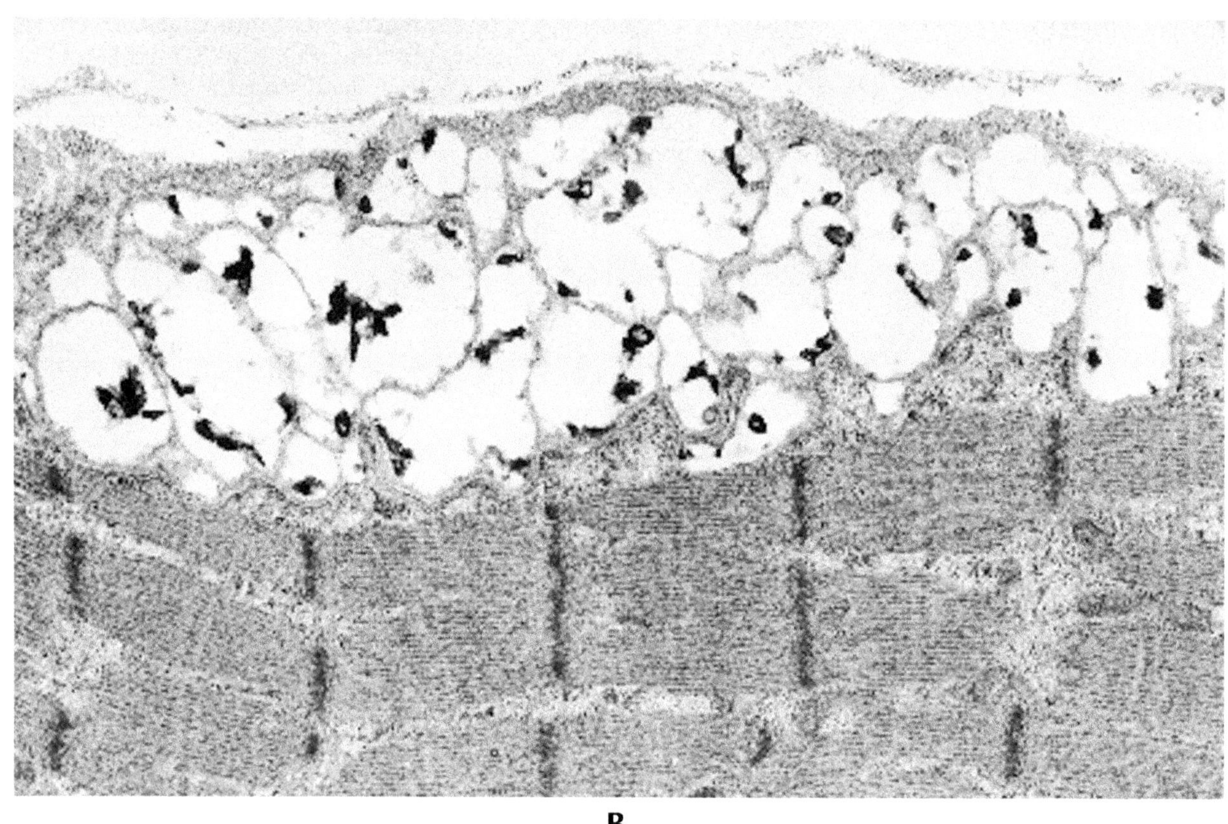

B

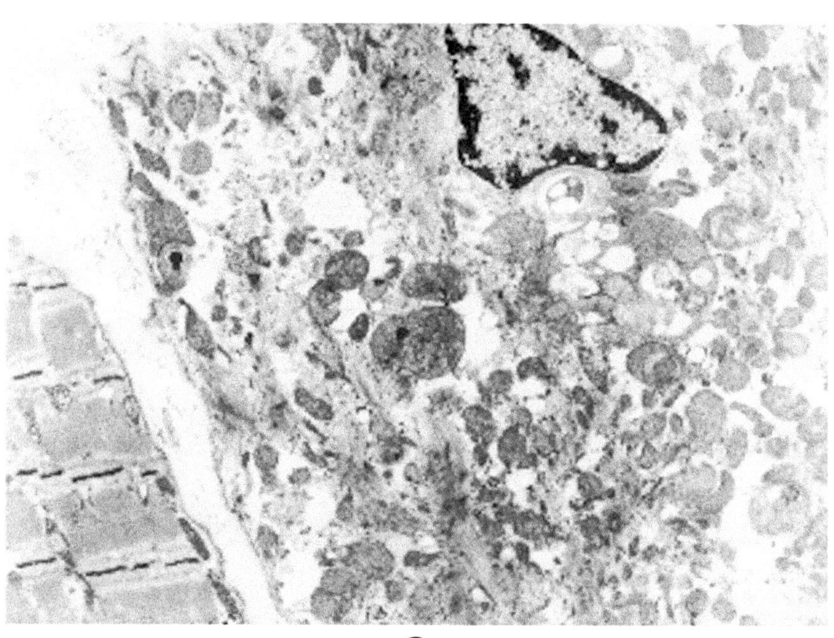

C

Figure 13.14. *(continued)*
B, Chloroquine myopathy. Note dense collections of lysosome-like structures beneath the sarcolemma. (× 14,400)
C, AZT myopathy. Note large collections of abnormally shaped mitochondria. (× 6000)

and multifocal necrosis of muscle fibers. Ultrastructurally, colchicine and chlororoquine myopathies are characterized by extensive deposits of membranous whorls and lysosomes along with focal disintegration of the myofibrillar architecture; in azathioprine myopathy vacuolization of the sarcoplasm and abnormally formed mitochondria predominate.

Neurogenic Atrophy

(Figure 13.15.)

General considerations and etiology. The term *neurogenic atrophy* refers to the pathologic reactions that develop in skeletal muscle secondary to interruption of normal innervation. The principal clinical manifestation of denervation is weakness or paralysis of the affected muscle group. Neurogenic atrophy can ensue after damage to the motor neurons within either the anterior horn of the spinal cord or the cranial nerve motor nuclei in the brain stem, injury to the motor axon, or abnormalities in the neuromuscular ending. The diseases that produce such injuries include all categories of pathologic pro-

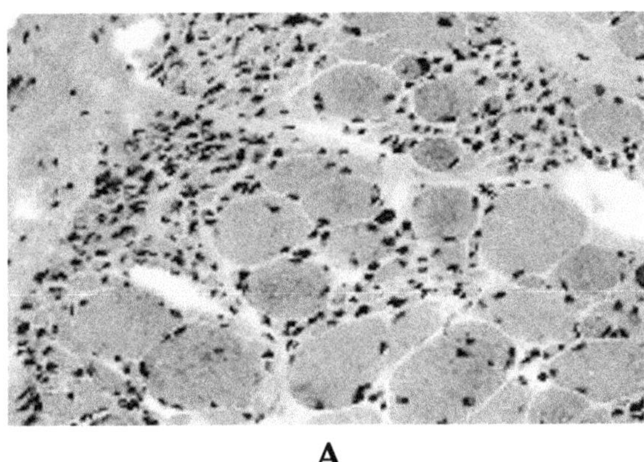

A

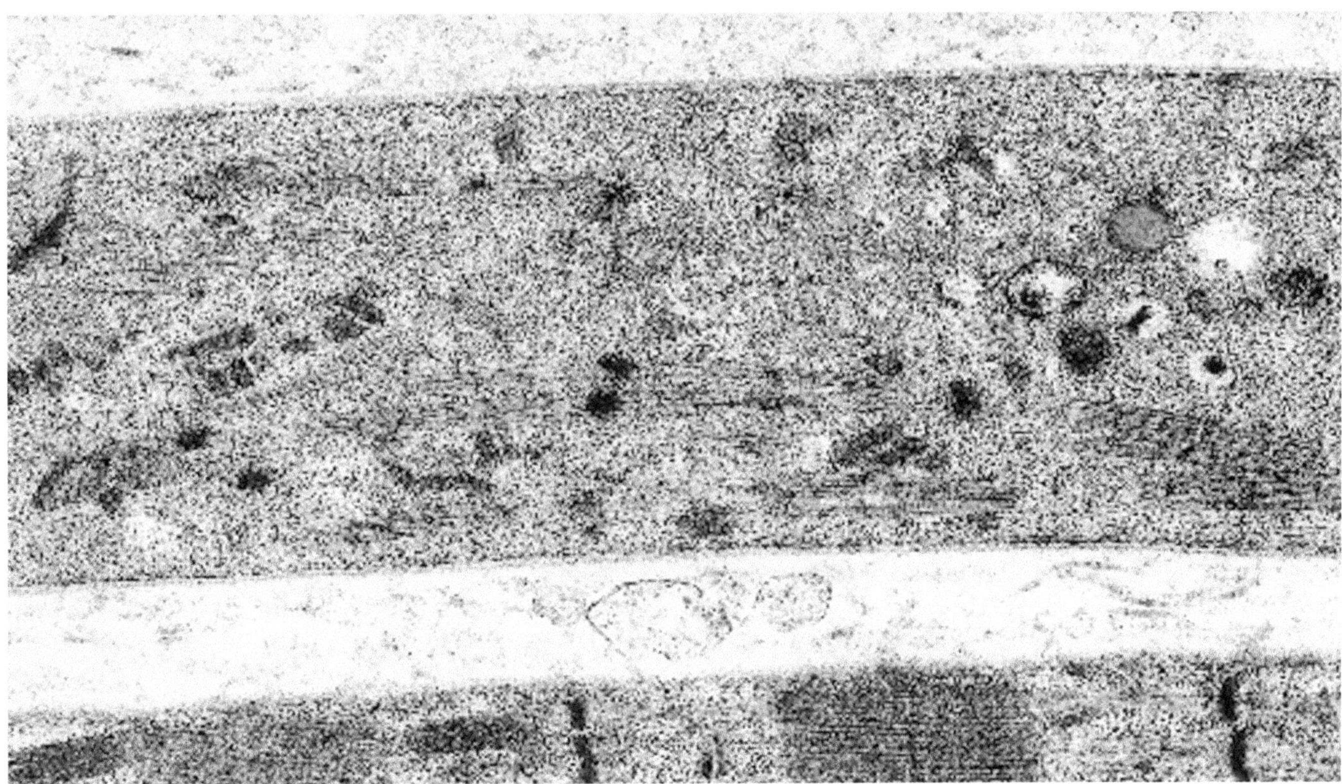

B

Figure 13.15. Neurogenic atrophy. **A,** Group atrophy. Note the clusters of atrophic fibers. (H&E, × 200) **B,** Atrophic fiber showing loss of normal sarcomeric structure. The sarcoplasm contains a few disorganized myofibrils with Z-bands and other organelles. The plasma and basement membranes are intact. (× 21,000)

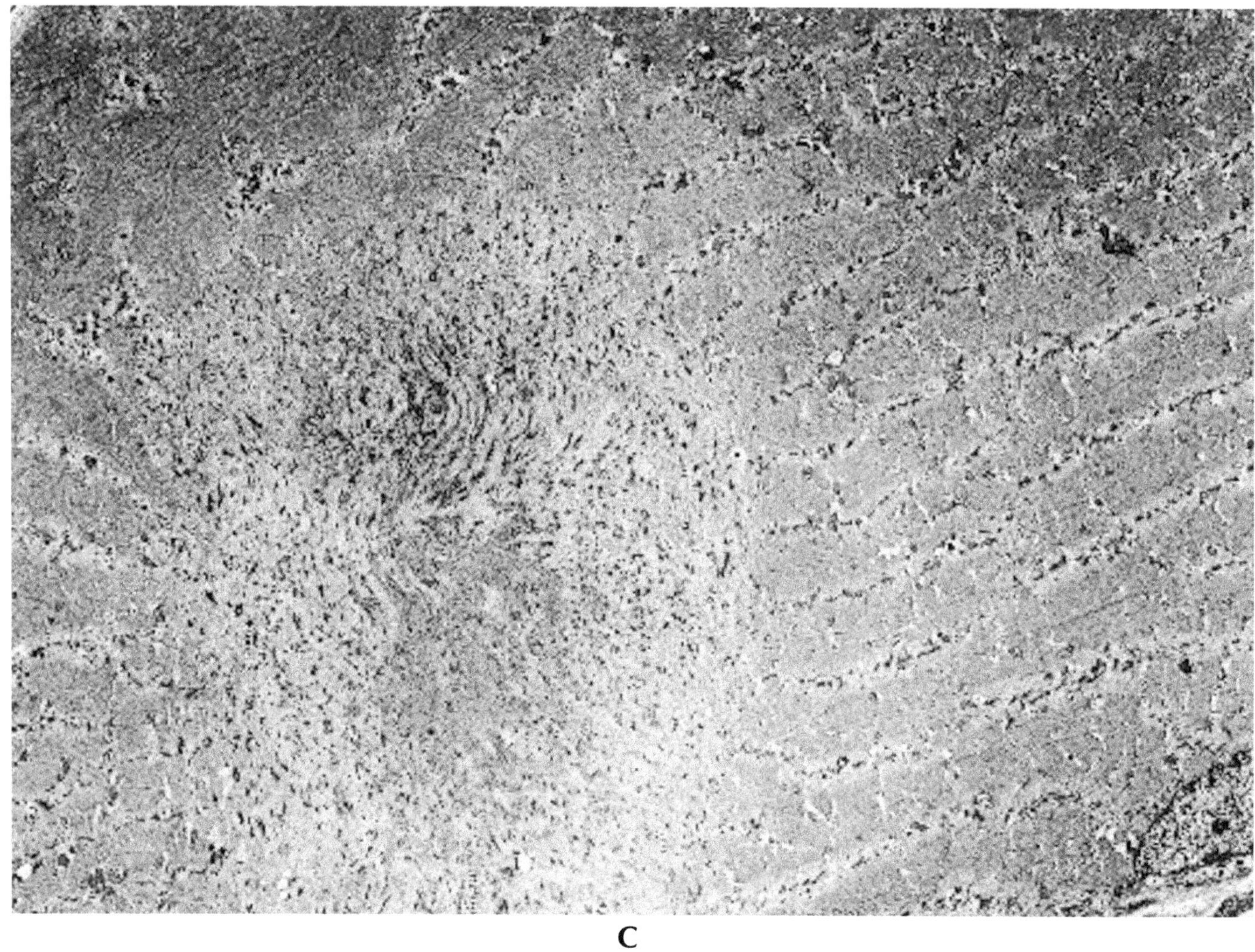

C

Figure 13.15. *(continued)*
C, Target fiber. Note disorganization of central portion of fiber and surrounding accumulation of mitochondria and other organelles. (× 2500)

cesses ranging from hereditary degeneration, to infection, neoplasm, neurotoxicity, and demyelination. Denervation of muscle leads to downregulation of myosin and actin synthesis, with a decrease in cell size and resorption of myofibrils; cells remain viable, however, and can resume normal shape and function upon reinnervation.

Diagnostic criteria. The light microscopic features of neurogenic atrophy are very distinctive. In cross-section, the atrophic fibers are smaller than normal. They lose their normal polygonal contour and develop a triangular (rather than the normal polygonal) shape. In early stages of denervation, the atrophic process is random and affects single fibers. As the disease progresses, and following reinnervation and subsequent denervation of the reinnervated groups of fibers, the atrophic fibers cluster together, a process known as *group atrophy.* Histochemical preparations of muscle in neurogenic disease demonstrate three distinctive phases in the evolution of the disease: (1) in early denervation there is randomly distributed atrophy of both type 1 and type 2 fibers; (2) with reinnervation, there is disruption of the normal checkerboard distribution of intermingled type 1 and type 2 fibers, such that there is now grouping of many normal-sized, contiguous fibers, all with the same histochemical properties (both type 1 and type 2 type grouping occurs); (3) with denervation of the reinnervated groups there is grouping of type 1 or type 2 atrophic fibers.

On ultrastructural examination, the principal change affecting the denervated fiber is loss of the myofibrils. The loss of myofilaments begins at the periphery of the fiber. In the early stages of myofibrillar disintegration, there is an apparently concurrent dissolution of thick and thin myofilaments and of Z-bands. The nucleus is unremarkable, and the number of cytoplasmic organelles is roughly of normal proportion. The sarcolemmal membrane remains intact, although there may be some convolution and reduplication of the basement membrane. In time, as the muscle fiber shrinks, the normal myofibrillar organization is no longer recognizable, and there are aggregates of haphazardly oriented units composed of thickened Z-bands from which streamers of filaments of variable thickness emanate, interspersed with loose myofilaments. These ultrastructural features are not entirely specific for denervated fibers, since they also may be seen in atrophic fibers resulting from disuse of muscle or other causes.

Another pathologic change that may be observed in denervation is the target fiber. This change is best recognized on frozen sections stained for oxidative enzyme such as NADH-TR. These fibers show a central area of decreased or absent oxidative enzyme activity surrounded by a narrow rim of increased activity, which, in turn, is surrounded by the normally staining fiber. These three distinct zones are also evident on electron microscopic examination. The innermost or central zone shows disruption of myofilaments, with loss of the normal banding pattern and smearing of the Z-band. There is a paucity of mitochondria in this region. This ultrastructural appearance is not unlike that seen in the unstructured central core (see the section on central core disease). The central area of myofibrillar disorganization is encircled by cytoplasmic reticulum and mitochondria. The myofilaments are only mildly disrupted. The outermost region consists of normal-appearing muscle.

Peripheral Nerve Disease

The normal light microscopic and electron microscopic structure of peripheral nerve is discussed in standard textbooks (Sternberg 1992; Peters et al. 1991; Dyck et al. 1993). Several illustrations are given here as a starting point to orient the reader (Figure 13.16). The ultrastructural changes detected in peripheral nerve include either the general pathologic responses of peripheral

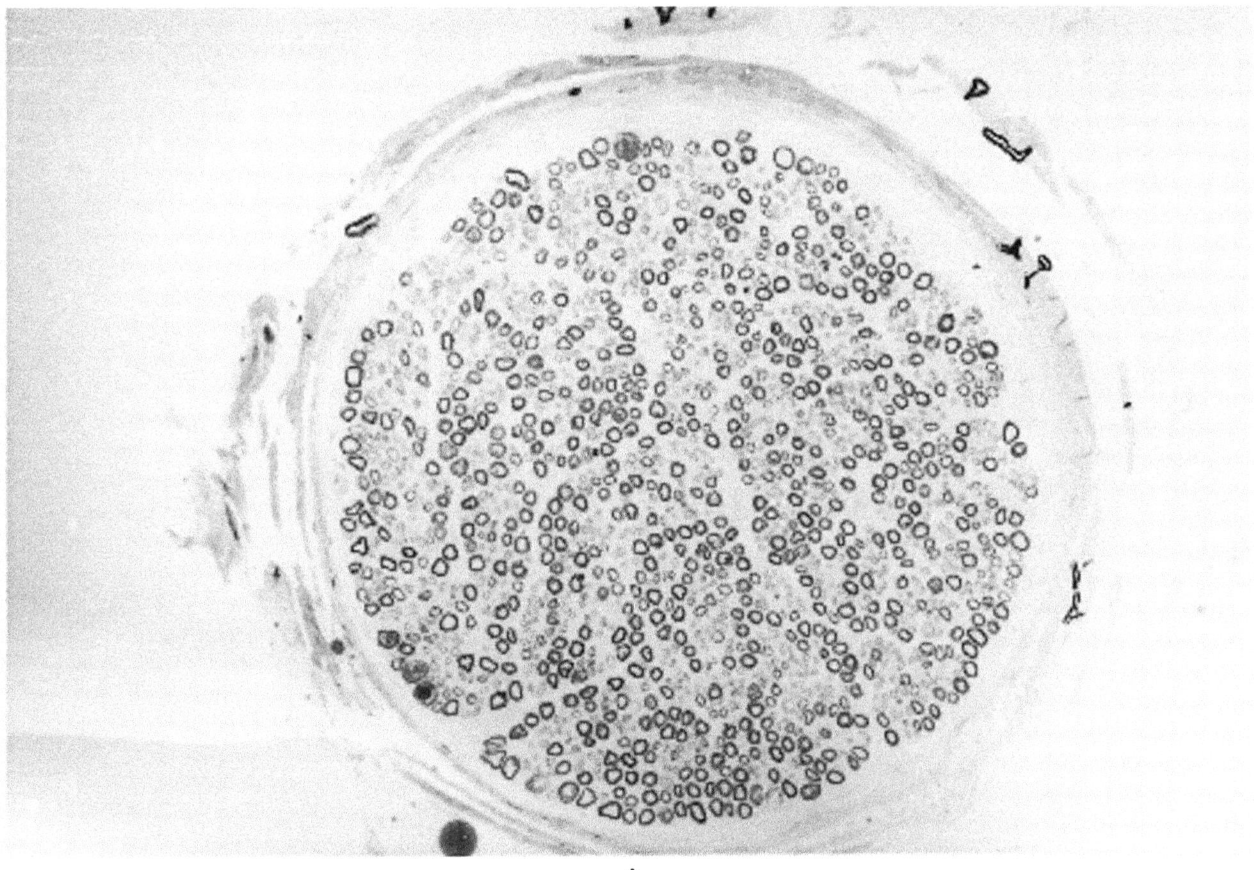

A

Figure 13.16. Normal nerve. **A,** Light micrograph of plastic-embedded sural nerve cut at 1 micron thickness and stained with toluidine blue. Note range of myelinated fiber sizes enclosed within the perineurial connective tissue strands. (× 146)

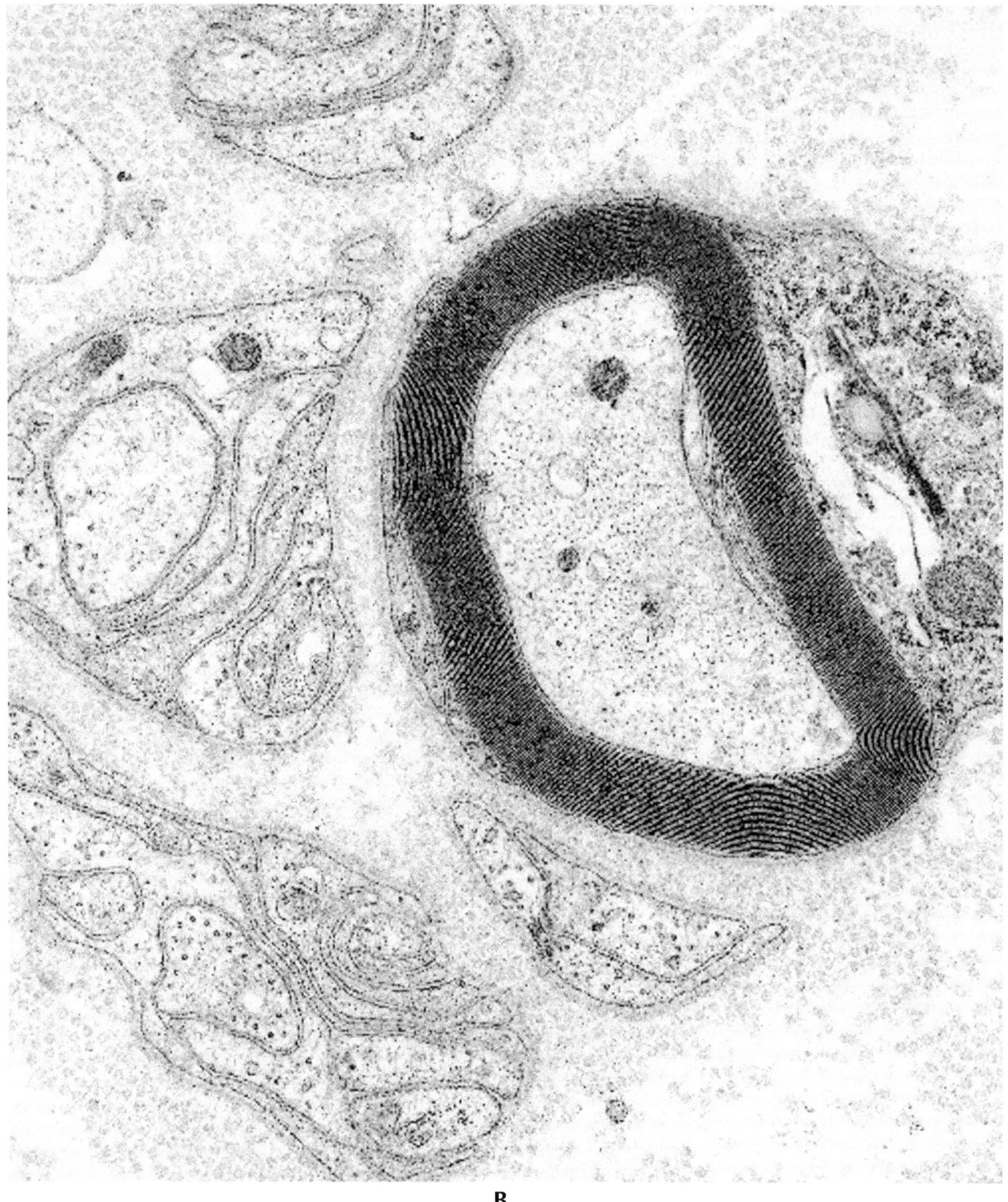

B

Figure 13.16. *(continued)*
B, Electron micrograph of normal sural nerve showing centrally placed myelinated fiber and groups of unmyelinated fibers to the left. (× 75,000)

nerve to injury or the specific pathologic changes associated with a specific disease entity. The general pathologic processes involving peripheral nerve can be subdivided into two broad categories: those that indicate a process primarily affecting the axon—*axonal neuropathy*—and those that indicate a process primarily affecting the Schwann cell or its myelin sheath—*demyelinating neuropathy*. These are diagramatically presented in Figure 13.17. When one of these processes occurs in isolation, it implies that the target of the neuropathy—the primary pathogenetic event—attacks either the axon or the Schwann cell. Not uncommonly, both processes occur concurrently, especially in chronic neuropathies. There are also diseases that cause peripheral neuropathy through damage to the connective tissue elements of peripheral nerves (e.g., inflammatory and metabolic neuropathies). Examination of peripheral nerve biopsies by electron microscopy, therefore, must include evaluation of the axons, myelin and Schwann cells, and the interstitium. Tumors of peripheral nerve (such as Schwannoma and neurofibroma) are discussed in Chapter 6. In this chapter, the ultrastructural findings in the two basic reactions of peripheral nerve (axonal degeneration and demyelination) are discussed with note made of which conditions are most often associated with each type; selected neuropathies characterized by involvement of connective tissue components are also discussed briefly. This is followed by a discussion of specific peripheral neuropathies that have characteristic electron microscopic findings.

For general reference citations that include discussions of the ultrastructural pathology of peripheral nerve, the reader is referred to the references listed at the end of this chapter. These include chapters in textbooks and comprehensive treatises dealing specifically with the pathology of diseases of the peripheral nervous system (Dyck et al. 1993; Bradley 1974; Asbury and Johnson 1978; Ouvrier et al. 1990; Bouche and Vallat 1992; Midroni and Bilbao 1995; Richardson and De Girolami 1995).

Axonal Degeneration and Regeneration

Acute degeneration of the distal portion of an axon was originally described in experimental studies involving nerve transection (Wallerian degeneration), but it is now recognized that this basic pathologic process occurs in a variety of human diseases involving disintegration of an axon. The process involves a single axon at a time and begins at one point of the axon to involve all points distal to it. Both the axon and its myelin sheath break down and undergo phagocytosis. *Chronic*

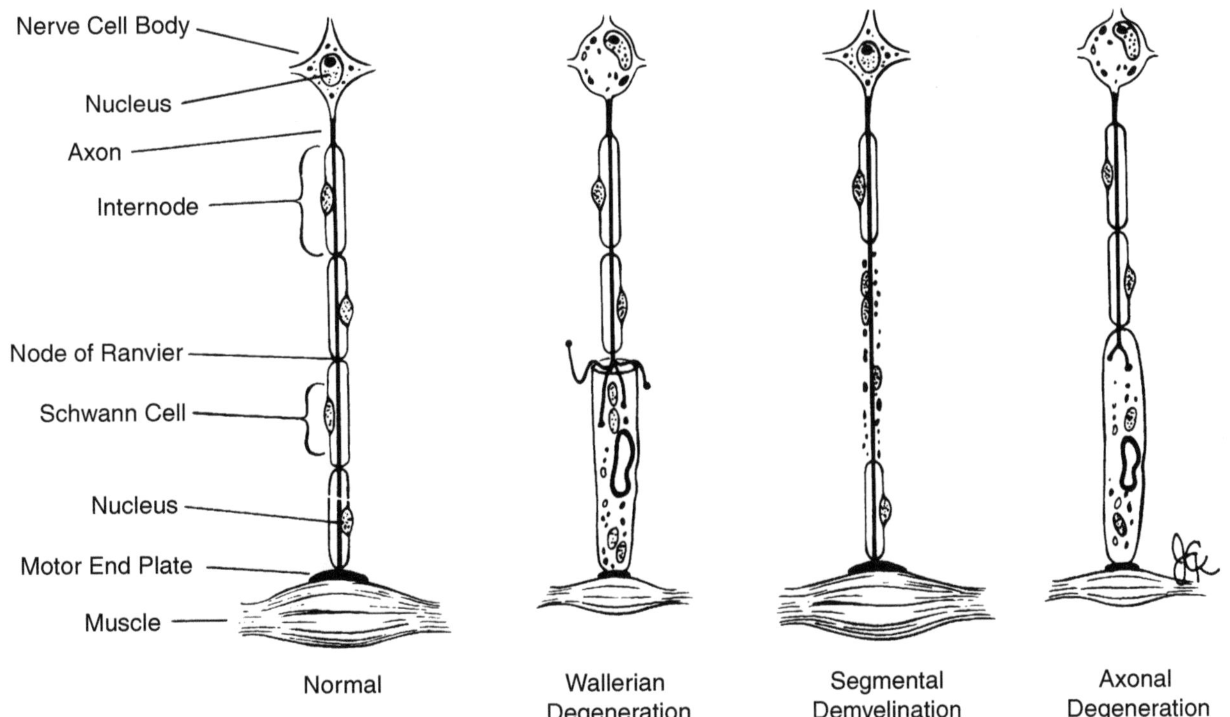

Figure 13.17. Schematic diagram of normal nerve, Wallerian degeneration, segmental demyelination, and axonal degeneration. (Permission for reprinting granted by WB Saunders, Asbury AK, Johnson PC: *Pathology of Peripheral Nerve*, 1978.)

axonal degeneration (or chronic axonal loss) occurs after a prolonged insult directly on the axon or the cell body of that axon, lasting months or years. In both acute and chronic axonal degeneration, because the injury does not necessarily result in the death of the parent neuronal cell body, *regeneration* of the axon may occur, beginning with sprouting of small processes from the tip of the axon (or growth cone) and growing forward from that point toward the target of innervation.

Wallerian Degeneration

(Figure 13.18.)

Clinical manifestations. After the transection of a nerve, transmission of voluntary impulses ceases, and there is a complete loss of motor and sensory modalities in the distribution of that nerve. The distal axon is still intact immediately after transection, and, although disconnected from its cell body, it is capable of conducting externally applied impulses for several days after transection. However, beyond that period, the distal nerve stump loses its ability to conduct impulses as it undergoes the process of Wallerian degeneration.

Diagnostic criteria. In the first days after transection of a peripheral nerve, there is (1) initially swelling of axon with disintegration of tubules and filaments in the distal segments; (2) retraction and disintegration of myelin at nodes of Ranvier; (3) accumulation of vesicles and degenerating mitochondria at the proximal stump; and (4) progressive disintegration of both the axon and myelin at all points distal to the site of transection. Subsequently, but within the first weeks after transection of a peripheral nerve, there is (5) phagocytosis of axon and myelin debris by macrophages; (6) regeneration, consisting of thinly myelinated axonal sprouts; and (7) proliferation of Schwann cells along the course of the disintegrated axon.

Etiology. The sequence of structural changes that follow nerve section, now called Wallerian degeneration, were described by Waller in 1850 in the glossopharyngeal and hypoglossal nerves (Waller 1850). Wallerian degeneration specifically refers to degeneration of the distal segments of a peripheral nerve after severance of the axons from their cell bodies. When the nerve is crushed, the basement membrane of the Schwann cell

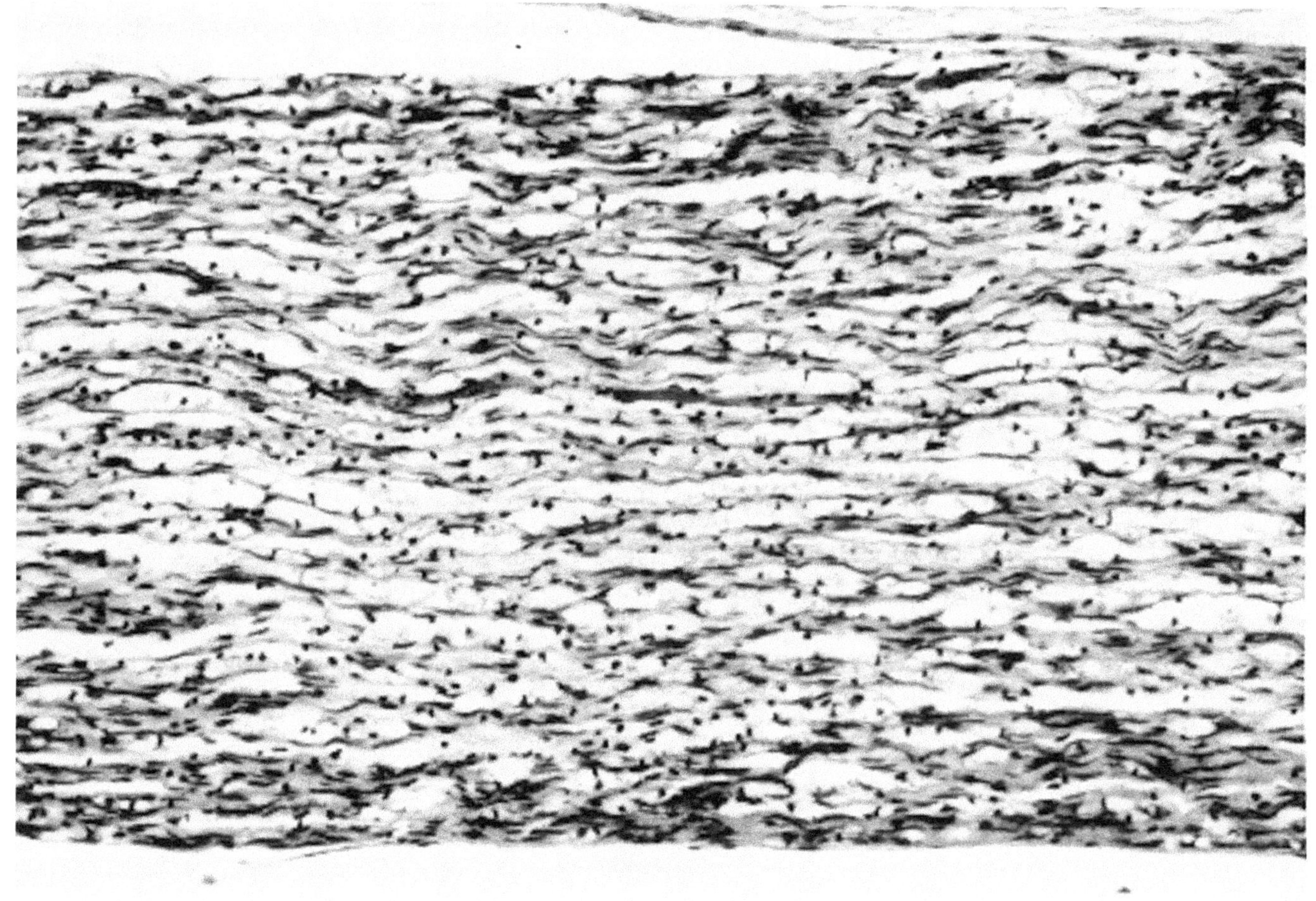

A

Figure 13.18. Axonal degeneration. **A,** Longitudinal section of nerve showing acutely disintegrating myelin. (× 50)

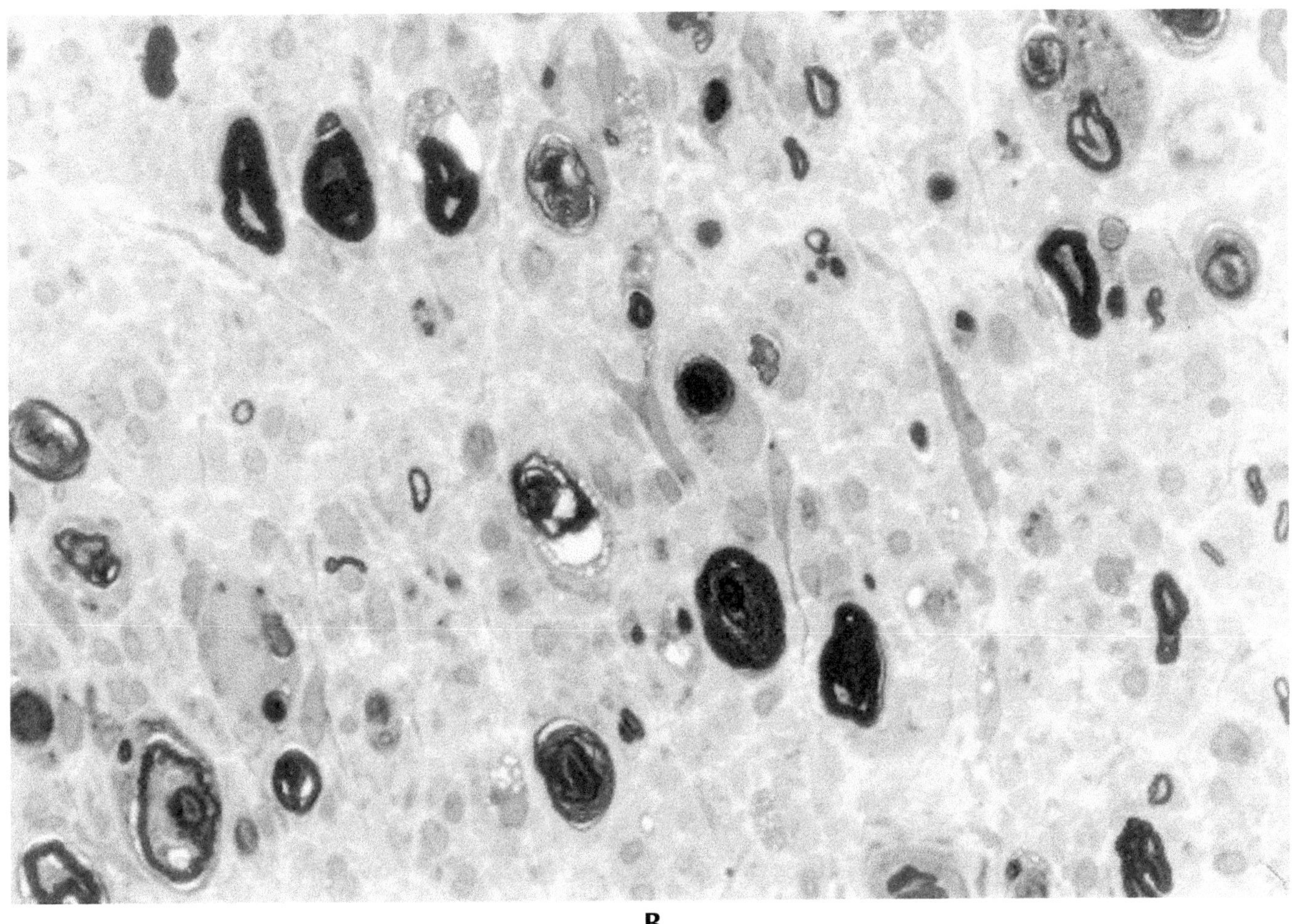

B

Figure 13.18. *(continued)*
B, Transverse section of plastic-embedded nerve showing disintegrated myelinated fibers. (× 500)

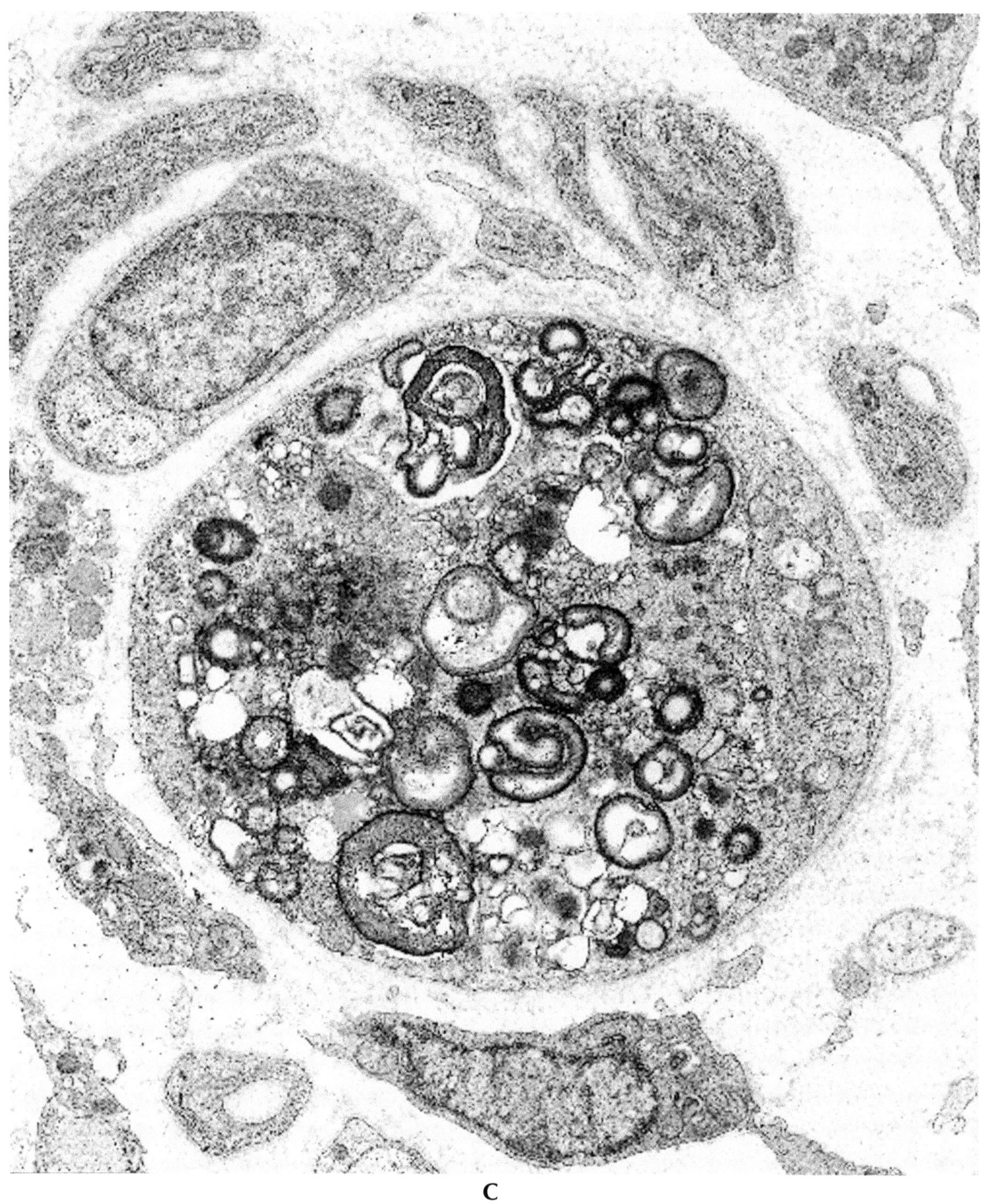

C

Figure 13.18. *(continued)*
C, Electron micrograph showing complete disintegration of the axon and its myelin sheath. Remnants of degenerating fiber consist of vesicular and membranous lamellar structures, including myelin figures. Part of the original Schwann cell still remains (*bottom*). (× 12,000)

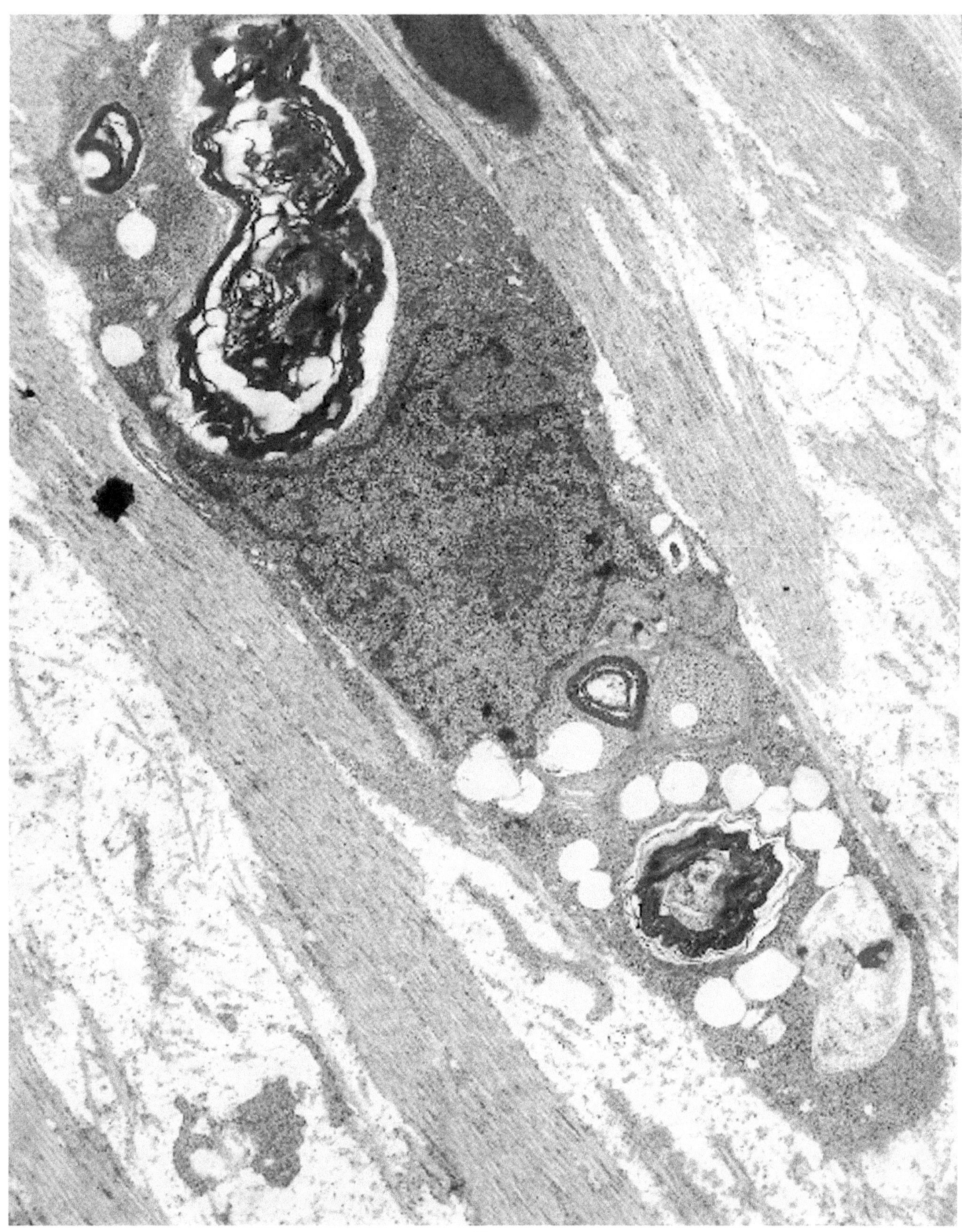

D

Figure 13.18. *(continued)*
D, Electron micrograph showing longitudinal section of degenerating axon with myelin "ovoids," that is, linear arrangement of oval vacuoles of axonal and myelin debris. Note nucleus of Schwann cell between ovoids. (× 10,000)

is preserved, allowing regeneration within the endoneurial tube. In contrast, when the nerve is sectioned, the endoneurial tube (composed of denervated Schwann cells and extracellular matrix) may not be appropriately aligned with the regenerating axons. Axonal regeneration is therefore less efficient after nerve degeneration after sectioning than after crush injuries.

Axonal Degeneration

(Figure 13.19.)

Clinical manifestations. Axonal degeneration may result from any pathologic process in which the primary insult is directed toward the axon itself or the cell body of the axon. Often, especially early in the course of the disease process, damage occurs to only a portion of the axon or affects some axons while sparing others. Furthermore, acquired or hereditary axonopathies may preferentially involve sensory, motor, or autonomic axons. The diminished number of conducting axons results in a diminished amplitude of the conducted impulse when measured electrophysiologically. In general, the extent of the axonal degeneration is greatest in the more distal portions of the peripheral nervous system, and consequently, clinical symptoms are most severe distally, in the hands and feet (known as stocking-glove distribution). With progression of the axonal degeneration to involve more axons and more proximal

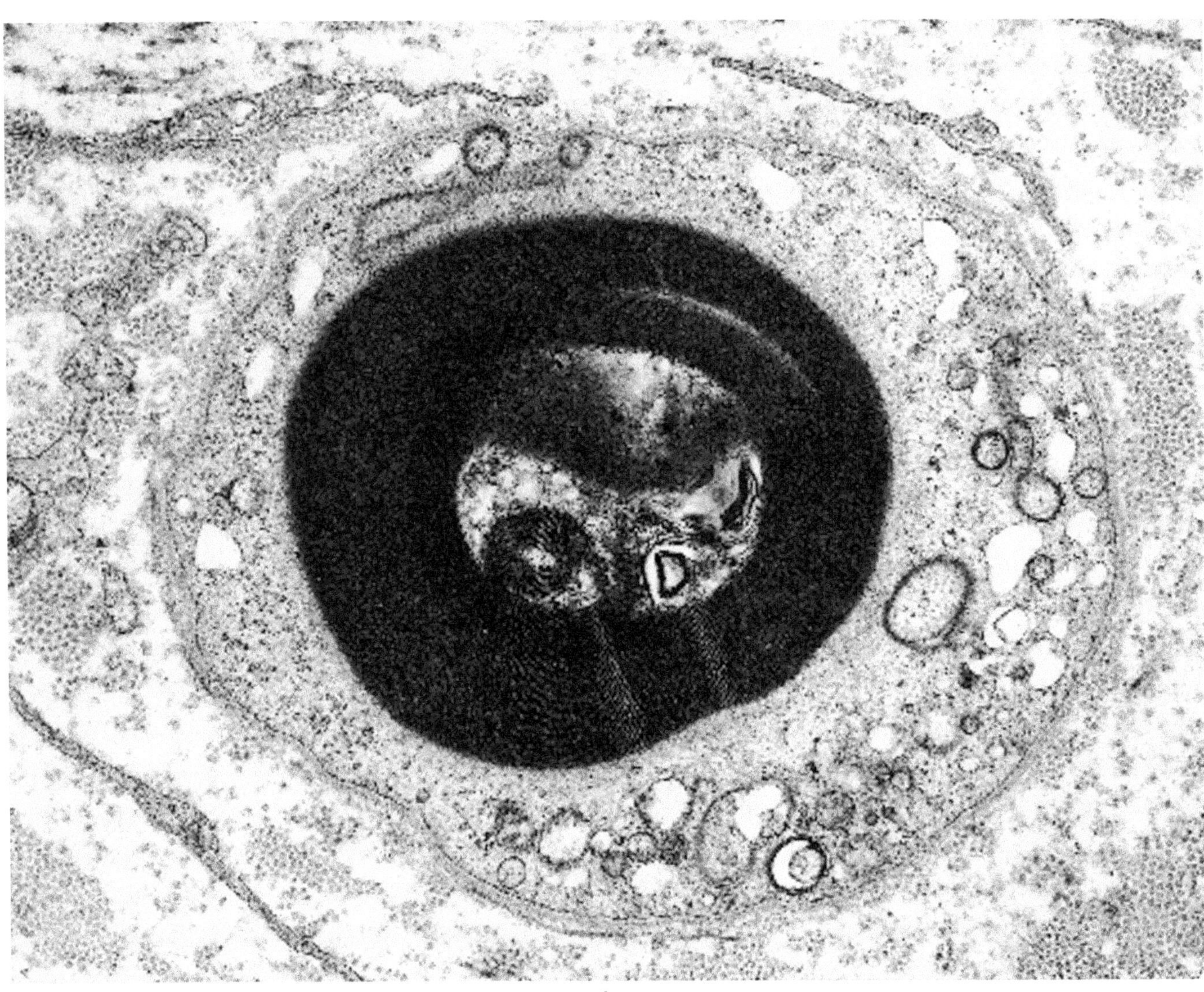

A

Figure 13.19. Axonal degeneration. **A,** Schwann cell containing myelin ovoid. There has been considerable shrinkage and disintegration of the central axoplasm, although the surrounding myelin sheath still retains its normal laminated structure. The Schwann cell cytoplasm contains some vesicular and membranous structures. The ultrastructural features of axonal degeneration are generally indistinguishable from those of Wallerian degeneration. (× 15,000)

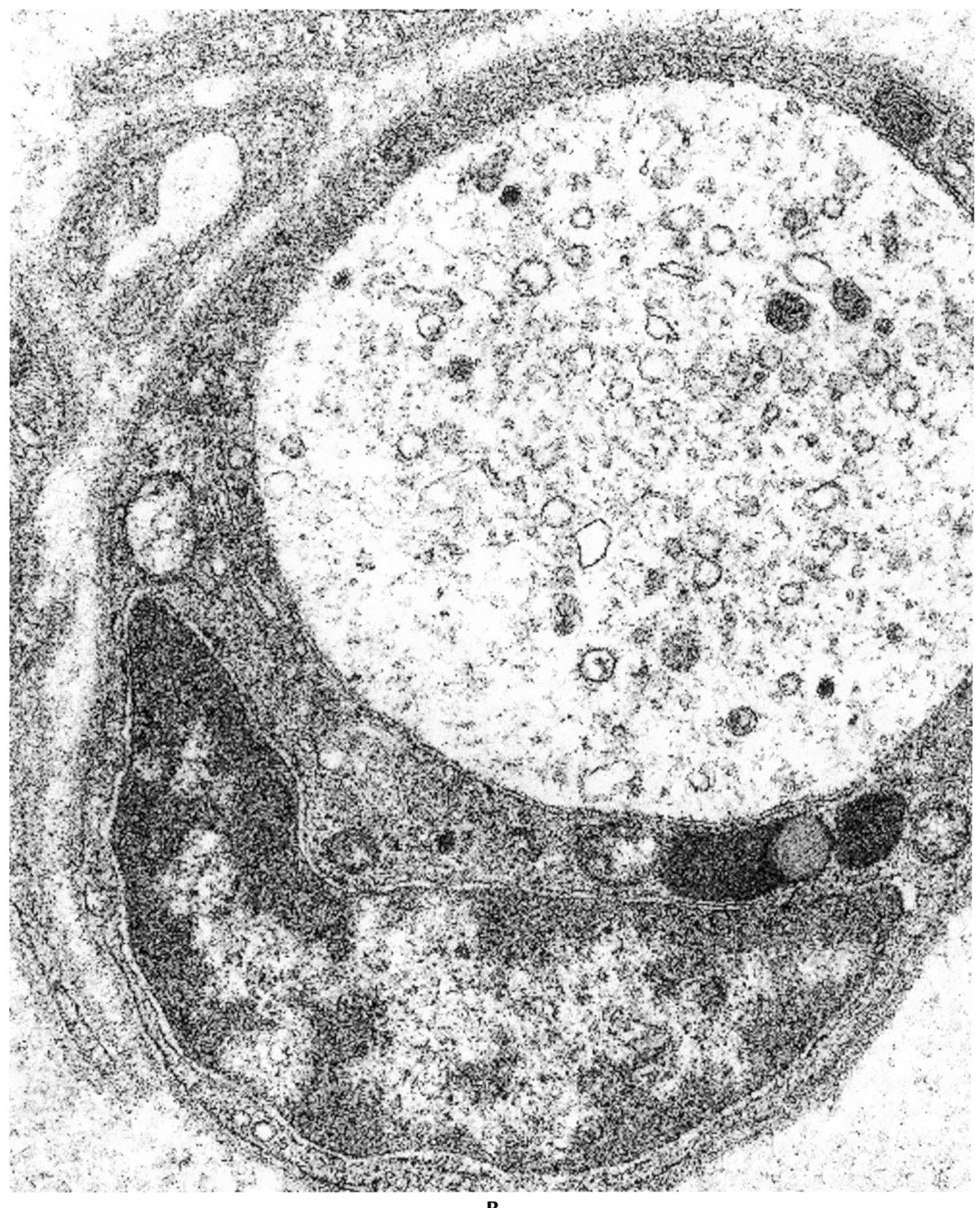

B

Figure 13.19. *(continued)*
B, Nonmyelinated axon with accumulation of organelles and intact Schwann cell. (× 27,000)

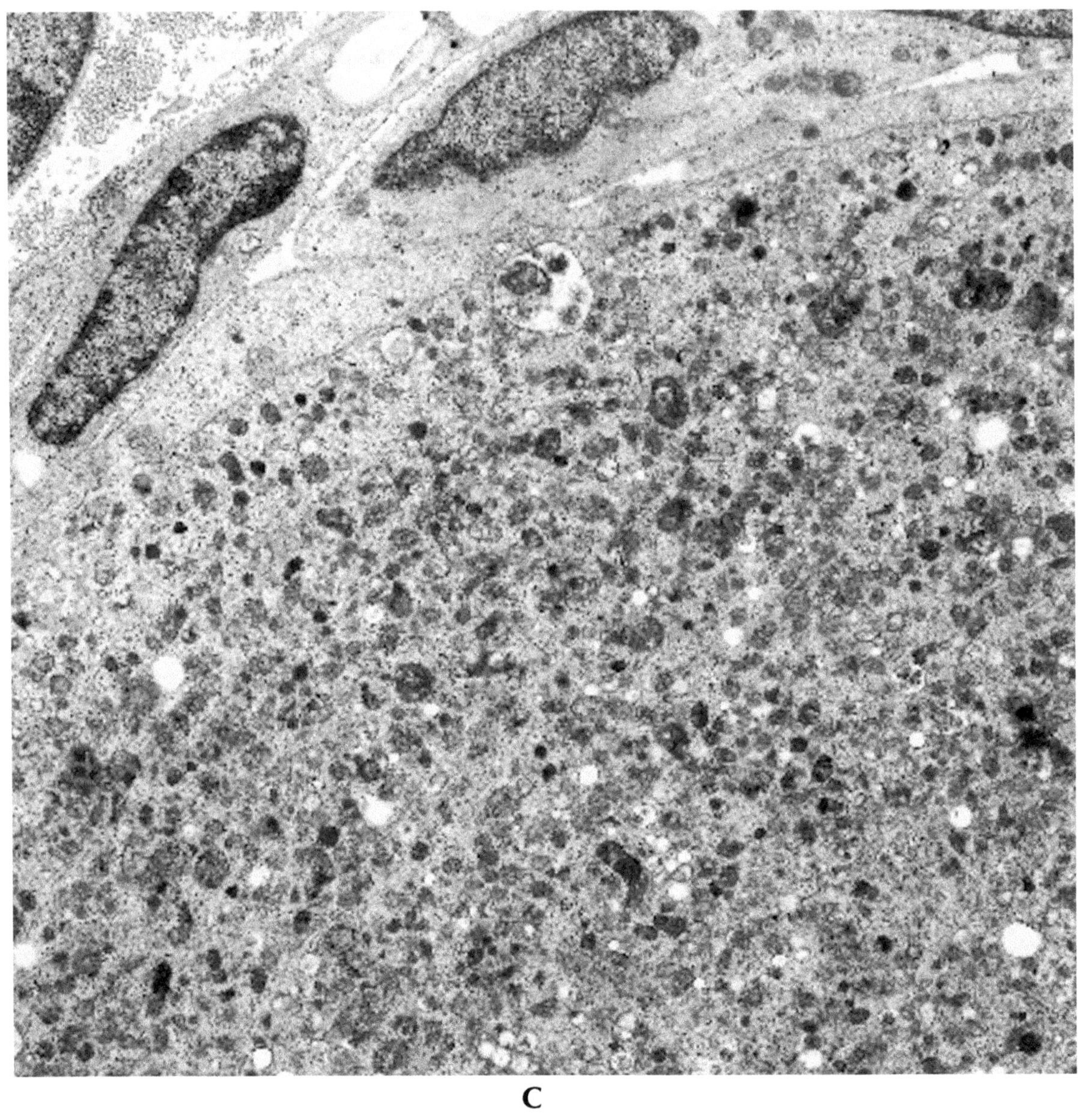

Figure 13.19. *(continued)*
C, Axonal degeneration with massive accumulation of axonal organelles. (× 10,000)

regions of axons, clinical evidence of motor and sensory deficits are manifest as a gradual advance of clinical symptoms to involve more proximal regions of the extremities.

Diagnostic criteria. Features of axonal degeneration include (1) involvement of the most distal portions of the longest nerves; (2) initial increase in neurofilaments followed by (3) decrease of axonal organelles; (4) retraction of the axon from the myelin sheath; (5) disintegration of the axon and myelin sheath; (6) Schwann cell proliferation in distal portions of axons; (7) variable regeneration in the form of axonal sprouting.

Etiology. Sometimes referred to as "dying back polyneuropathy" (Cavanagh 1979) to emphasize the distal-to-proximal progression of the degeneration process, axonal degeneration is described in a wide variety of chronic slowly progressive, symmetric, toxic, and metabolic polyneuropathies. The gradual evolution of the pathologic process to involve proximal portions of the peripheral nervous system appears to be the basis for the clinical manifestations. The pathogenesis is not certain, but dysfunction of the metabolism of the neuron rendering it incapable of supporting the axon or a direct effect on axoplasmic flow within the axon have been proposed as hypotheses. Axonal degeneration is a prominent component in many neuropathies, including those due to toxic exposures, metabolic disease, nutritional deficiencies, inflammatory disorders, and ischemic diseases.

Axonal Regeneration ("Sprouting")

(Figure 13.20.)

Clinical manifestations. Regeneration of the distal axon may occur in the peripheral nervous system, if the inciting cause is identified and removed. Best studied after transection and crush injuries of peripheral nerve, regeneration of the axon occurs by sprouting of the proximal nerve stump and proceeds anterograde at the rate of about 1 mm/day. This rate of axonal regeneration can be used clinically to determine the approximate amount of time that recovery of clinical function will occur after repair of a transected nerve.

Diagnostic criteria. Axonal regeneration typically shows (1) several thinly myelinated axons or axons with undetectable myelin in a cluster; (2) clustered axons with similar diameter sizes and myelin thickness; (3) collagen fibrils interspersed among the myelinated and unmyelinated axons; and (4) a basal lamina tube, which had previously enclosed a healthy axon, surrounding the complex of clustered axons. Aggregates of myelinated fibers that are not completely surrounded by a basal lamina may represent a more advanced stage of axonal regeneration.

Etiology. In chronic axonal neuropathies, it is common for cross-sections of nerve biopsies to show evidence of axon loss (reduced number of myelinated

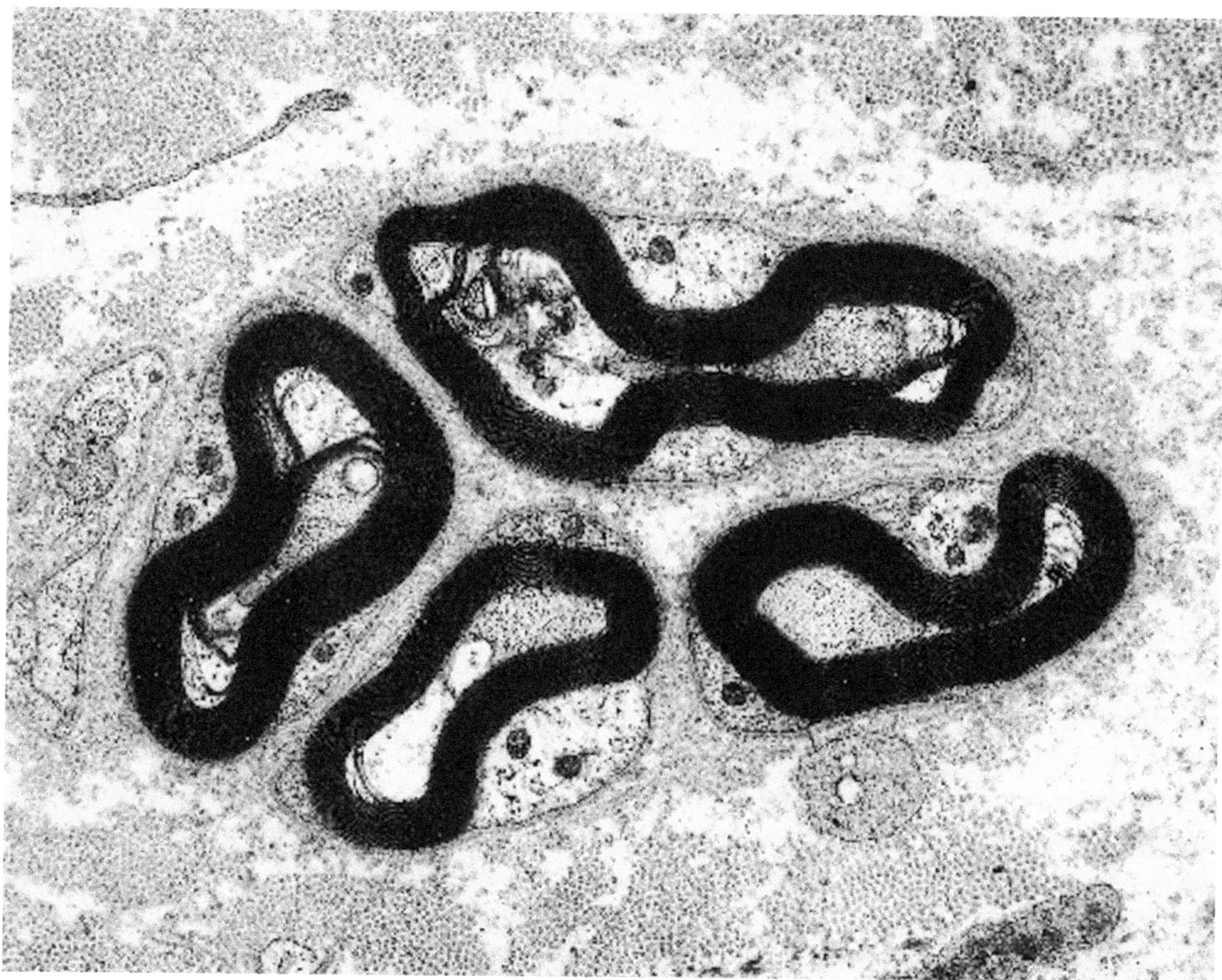

Figure 13.20. Axonal regeneration. Cluster of four small, thinly myelinated regenerating axons. Only focal small remnants of the basement membrane that originally surrounded the cluster complex still remain. (× 13,000)

fibers), axon degeneration (profiles of axonal degeneration), and axonal regeneration (axonal clusters). The groups or clusters of thinly myelinated fibers appear to arise as the proximal axonal stump sprouts multiple small neuritic processes, all of which are surrounded within the Schwann tube of the original axon. Such formations are referred to as regeneration clusters, and are common in cross-sections of nerve biopsies taken from patients with toxic or metabolic neuropathies. Soon after nerve crush or transection (24 hours in experimental animals), regeneration commences in the axons proximal to the site of injury. An axonal growth cone gives rise to multiple axonal sprouts that advance through the site of the injury. Distal to the lesion, Schwann cells begin dividing, filling the endoneurial sheath (which had been previously occupied by an axon) and arranging themselves into cellular tubes known as bands of Büngner. If axonal regeneration does not occur, Schwann cells distal to the injury gradually undergo atrophy, with an increase in endoneurial collagen. In cases of nerve transection, axons that are not appropriately aligned with the distal Schwann cell tube may form sprouts that become misdirected, resulting in a tangled mass of neuritic processes known as a "traumatic neuroma." Axon regeneration also is a prominent feature of many toxic and metabolic neuropathies, such as the neuropathies in drug and chemical intoxication, uremia, diabetes, vitamin deficiency, and malignancy. In these conditions, the major pathologic process is distal axonal degeneration (distal axonopathy). Regeneration is dependent on whether the toxic or metabolic abnormality can be reversed.

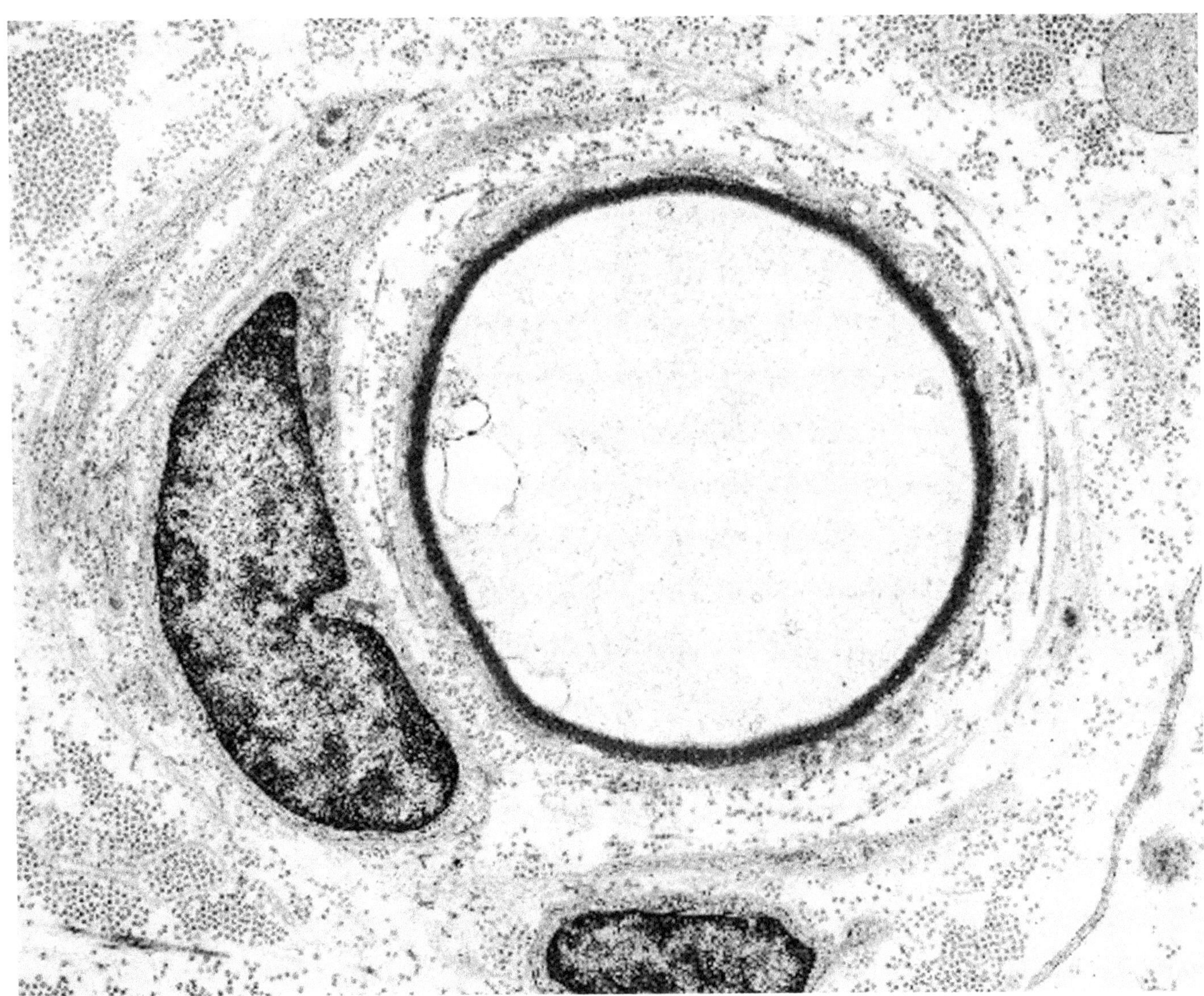

Figure 13.21 Segmental demyelination. Remyelinating axon surrounded by thin myelin sheath. (× 11,000)

Segmental Demyelination of Nerve Fibers

(Figures 13.21 and 13.22.)

Clinical manifestations. Demyelination refers to the primary degeneration of the myelin sheath, with preservation of the axon. This results from an insult that is directed toward the Schwann cell body or its cytoplasmic process, namely the myelin sheath. The reparative or regenerative process involves a proliferation of Schwann cells, which encircle the denuded axon and begin to elaborate myelin, resulting in remyelination of the injured internodes. Because the reparative process (remyelination) begins soon after demyelination, most cases of demyelinating peripheral neuropathies include active remyelination. Although demyelinating neuropathies have a tendency clinically to also involve distal regions of the axon, most likely due to summation of conduction disturbances along the length of the axon, the electrophysiology of demyelinating neuropathies is usually quite different than axonal neuropathies. In demyelinating neuropathies, the amplitude of conduction may be relatively well preserved, but the conduction velocity is usually markedly slowed.

Diagnostic criteria. During segmental demyelination, there is (1) disintegration of myelin with preservation of the axon; (2) absence of central chromatolysis; (3) early changes in myelin sheaths detected at nodes of Ranvier (paranodal retraction); (4) insertion of macrophage cell processes between myelin sheath and axon may occur. Regenerative changes include (1) proliferation of Schwann cells at sites of demyelination; (2) remyelination of demyelinated segments, with shorter internodes and thinner myelin sheaths; (3) Schwann cell hyperplasia (redundant Schwann cell processes around individual axons) as a result of repeated episodes of demyelination and remyelination; (4) progressive onion-bulb formation with multiple concentrically arranged Schwann cell processes surrounding a thinly myelinated axon.

Etiology. Demyelination in the peripheral nervous system is a segmental process, involving a single internode at a time, and compared with Wallerian degeneration, Schwann cell proliferation is minimal and confined to the segment where remyelination occurs. The segmental process is best demonstrated in the longitudinal plane. As a result, it is usually easier to demonstrate segmental demyelination on teased nerve fiber preparations. After the myelin surrounding an internode degenerates, the length of axon originally encircled by one Schwann cell may be remyelinated by two or more newly divided Schwann cells, resulting in a decrease in the internodal length and an increase in the number of internodes. Segmental demyelination is seen typically in the Guillain-Barré syndrome, chronic inflammatory demyelinating polyradiculoneuropathy (CIDP), some hereditary neuropathies, experimental allergic neuritis, and certain toxic neuropathies (such as lead and diphtheria).

Specific Peripheral Neuropathies

Peripheral neuropathy is a very common clinical condition, with an annual incidence of nearly 400 cases per million population (Midroni and Bilbao 1995) and a prevalence in the population that is even higher. The reader is referred to standard texts for a detailed account on the classification and the range of clinical man-

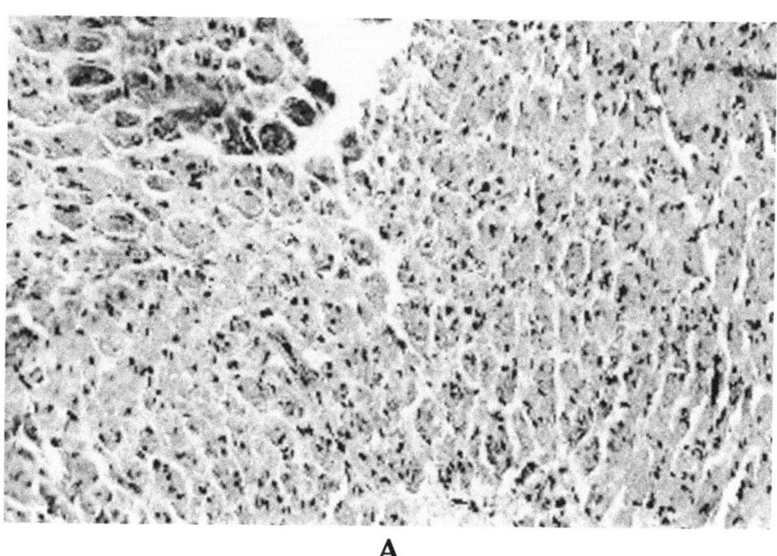

A

Figure 13.22 Hypertrophic neuropathy. **A,** Nerve in transverse section demonstrating numerous onion-bulb formations and increased endoneurial connective tissue. (H&E, × 100)

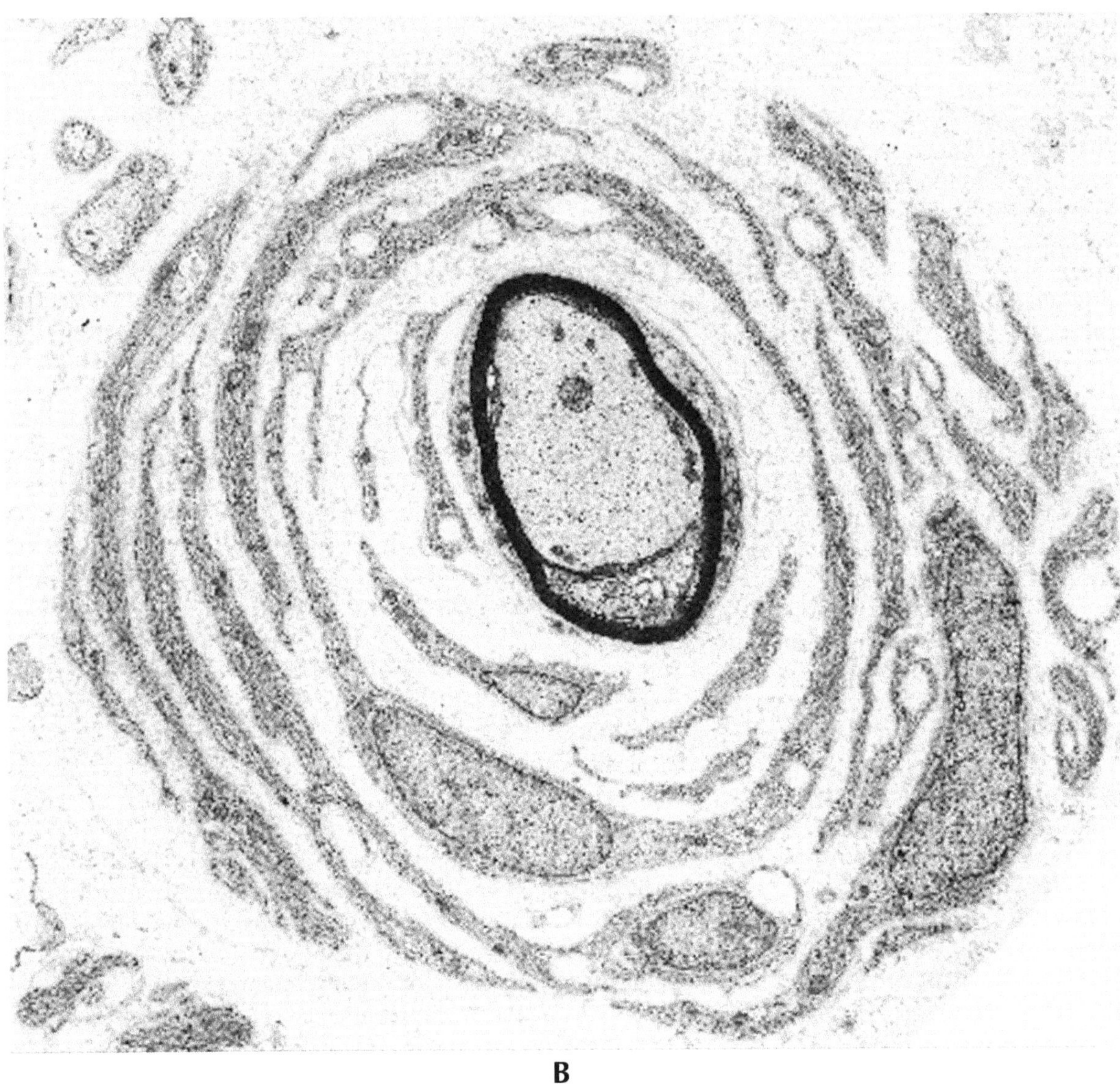

B

Figure 13.22. *(continued)*
B, Onion-bulb formation. Thinly myelinated (remyelinating) central axon is surrounded by concentric Schwann cell processes separated by collagen fibers. (× 6500)

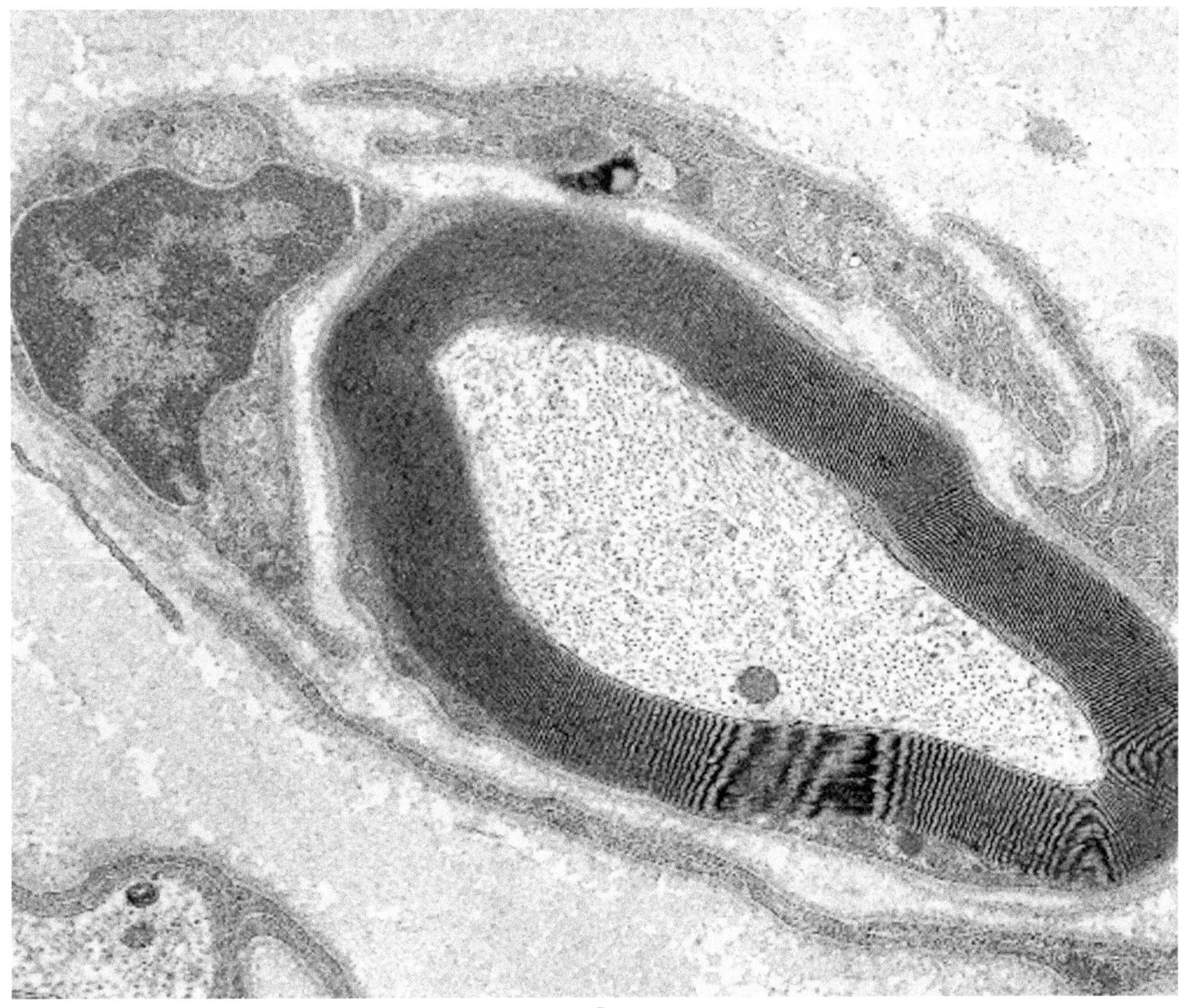

C

Figure 13.22. *(continued)*
C, Schwann cell hyperplasia (early onion bulb formation) with redundant basement membrane and circumferentially arranged Schwann cell processes surrounding myelinated axon. (× 30,000)

ifestations of peripheral neuropathies (Dyck et al. 1993; Vinken and Bruyn 1970). In this chapter, discussion is limited to peripheral neuropathies showing important ultrastructural changes: acquired and hereditary metabolic neuropathies, inflammatory neuropathies, infectious neuropathies, and neuropathies associated with neoplasia and dysproteinemias.

Metabolic Neuropathies Associated with Diabetes Mellitus

(Figures 13.23 and 13.24.)

Clinical manifestations. The prevalence of peripheral neuropathy in patients with diabetes mellitus varies from 10–60% as assessed by clinical criteria, and up to 100% when evaluated by nerve conduction studies. This in turn, is dependent on the duration of the disease; in a recent series, 7% of patients with diabetes mellitus had clinical peripheral neuropathy at the time of diagnosis of the diabetes, and 50% had peripheral neuropathy 25 years after diagnosis. Approximately 17 million people with diabetes in the United States and Europe have peripheral neuropathy.

Although the generic term "diabetic neuropathy" is used widely, it encompasses several distinct clinicopathologic disorders of the peripheral nervous system that occur in diabetes mellitus, often with overlapping features (Dyck and Giannini 1996). The most common type is a chronic progressive symmetric distal sensorimotor peripheral neuropathy. Other patients, however, may develop an asymmetric neuropathy, involvement of multiple single nerves (mononeuritis multiplex), or paralysis of a single nerve (mononeuropathy), especially the third cranial nerve. In each type, vascular changes are usually present, evidenced by thickening of the walls of capillaries. In peripheral nerve biopsies, this vasculopathy appears as a prominent reduplication of the basal lamina that surrounds the endoneurial capillaries. Similar findings are also seen in the microcirculation of skeletal muscle.

Diagnostic criteria. A distal chronic symmetric sensorimotor neuropathy is the most common peripheral neuropathy in diabetic patients, and it is an axonal neuropathy characterized by (1) diminution of myelinated fibers of different calibers; (2) diminution of unmyelinated fibers; (3) absence of specific axonal dystrophic features beyond what is described as characteristic of axonal degeneration (Englestad et al. 1997); (4) an increase in endoneurial connective tissue and Schwann cells later in the course of the neuropathy; and (5) a microangiopathy with thickening of the basal laminae of blood vessels and a similar thickening of basal laminae of perineurial cells.

Etiology. The etiology of the chronic distal neuropathy in diabetes mellitus is unclear; chronic ischemia related to the microangiopathy and direct metabolic effects have been proposed. The more acute or subacute onset of mononeuritis multiplex appears to be related to vascular insufficiency within the vasa nervorum; arteriolar occlusion and nerve infarction have been demonstrated in some cases of mononeuropathy with abrupt onset.

Considerable attention has been devoted to descriptions of the basal laminae of blood vessels and perineurial cells, including well-documented morphometric analysis of the thickness of the basal laminae. The significance of these changes in the basal laminae in the pathogenesis of diabetic neuropathy remains contested. In many cases, Schwann cell destruction leading to segmental demyelination is present in addition to the axonal neuropathy. This pattern of injury is independent of axonal damage and is characterized by (1) disintegration of myelin sheaths; (2) retraction of myelin from the node of Ranvier; and (3) increased quantities of Schwann cell processes. These mixed axonal and demyelinating pathologic changes correlate well with mixed electrophysiologic changes on nerve conduction studies. Both axons and Schwann cells may be affected in diabetes mellitus.

Sensorimotor Neuropathies Associated with Hereditary Metabolic Disease

Clinical manifestations and diagnostic criteria. Several hereditary metabolic diseases have characteristic ultrastructural abnormalities on peripheral nerve biopsy. Although central nervous system (CNS) symptoms are often the dominant clinical feature, nerve biopsies may be performed either because early involvement of the peripheral nervous system is noted or the peripheral nerve represents a more accessible site for biopsy. When a specific storage disorder is suspected clinically, however, biopsy may be unnecessary since biochemical or genetic detection is usually possible.

Krabbe Disease (Globoid Cell Leukodystrophy)

Diagnostic criteria. Characteristic findings in patients with Krabbe disease are (1) demyelinating peripheral neuropathy in infants with leukoencephalopathy; (2) occasional Schwann cell hyperplasia; (3) storage material, usually limited to Schwann cells, appearing as plate-like electron-lucent inclusions. Confirmation is achieved with detection of deficiency of the galactocerebrosidase activity. Globoid cells, which are the hallmark of the disease in the CNS, are not present in the peripheral nervous system (PNS).

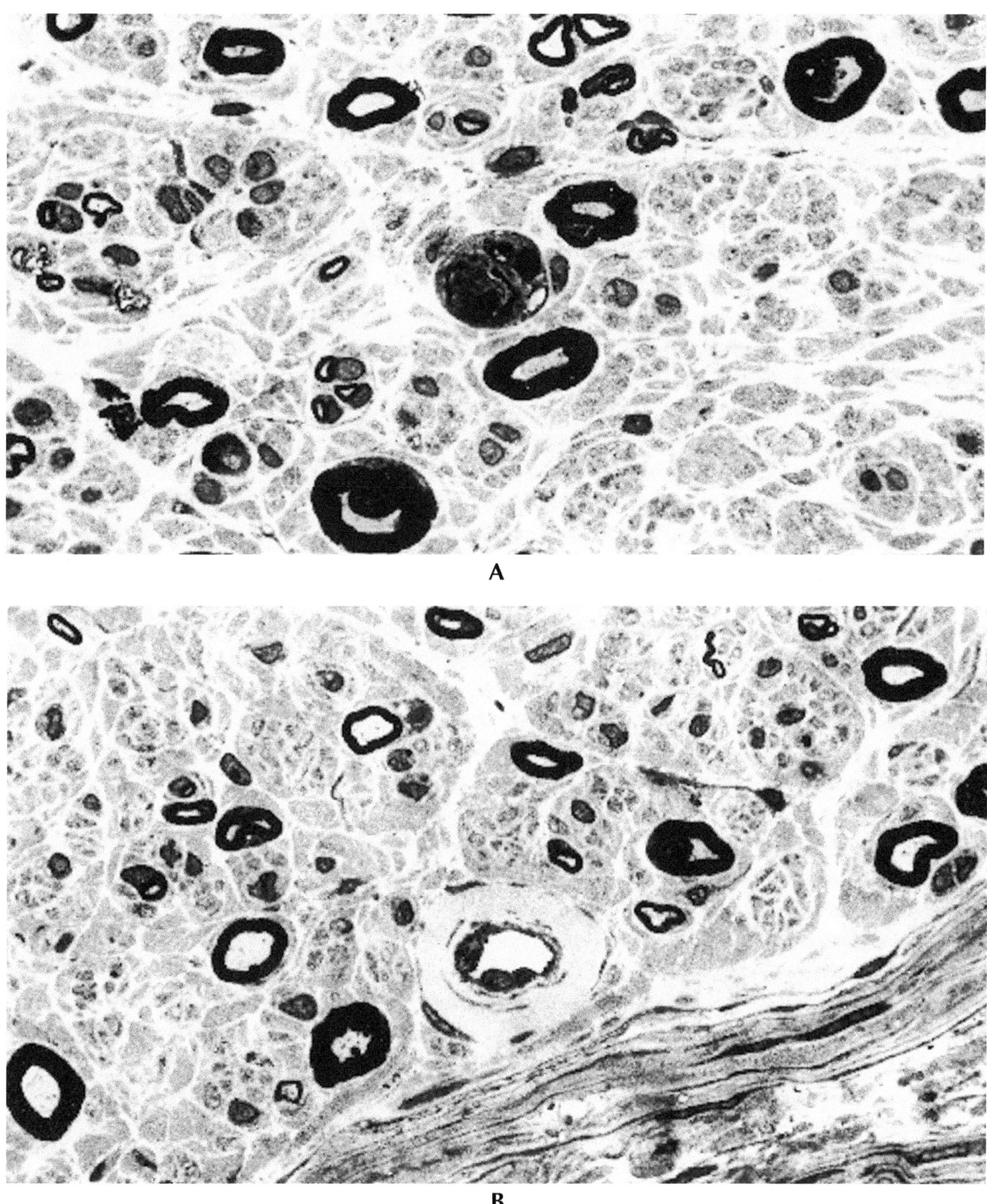

Figure 13.23. Diabetic neuropathy. **A,** One-micron-thick section of nerve showing loss of myelinated fibers. A degenerating axons is present in the center of the field. (toluidine blue, × 1400) **B,** A thickened blood vessel is present beneath the perineurium. (toluidine blue, × 1400)

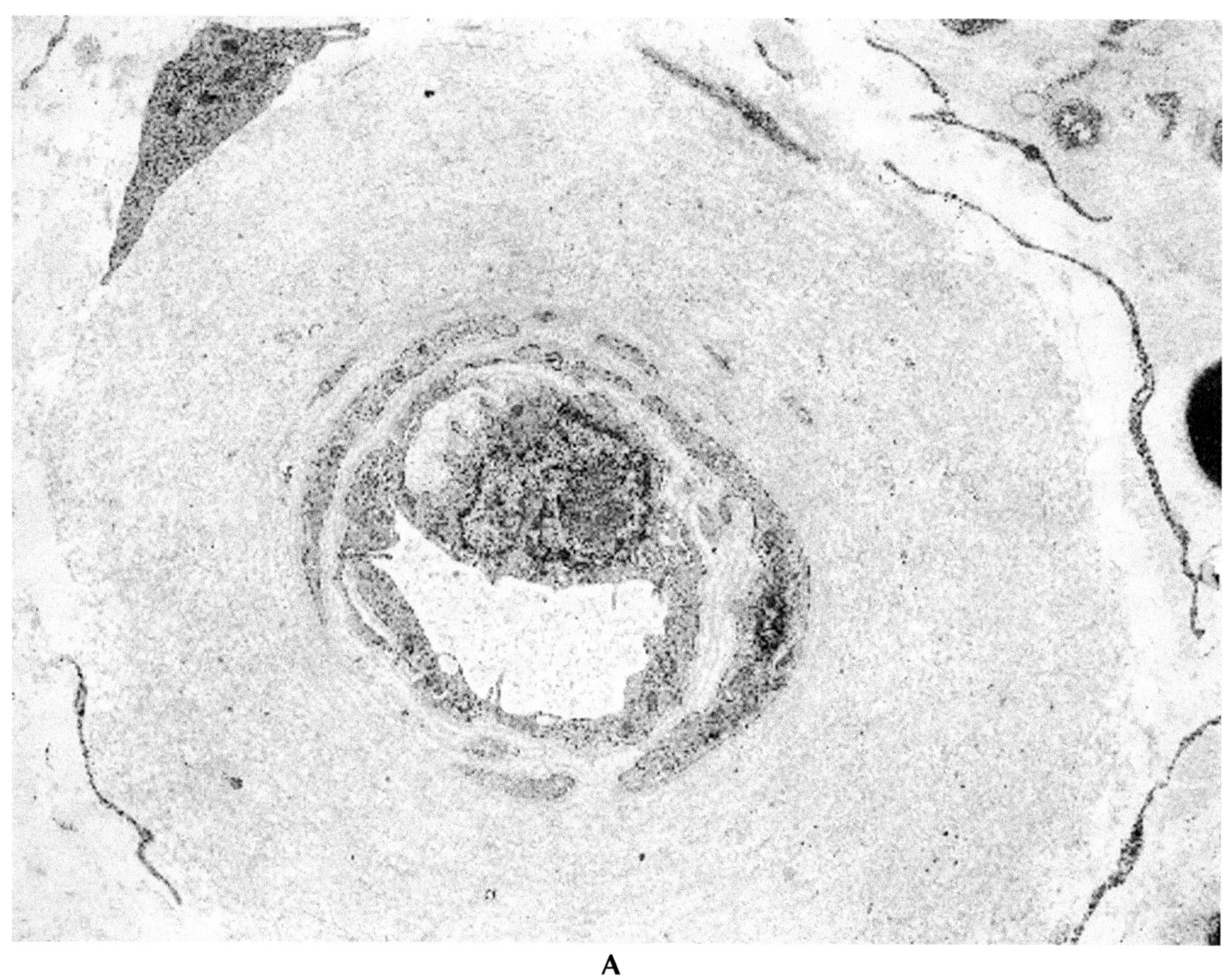

A

Figure 13.24. Diabetic neuropathy. **A,** Small endoneurial vessel surrounded by redundant, thickened basement membranes. (× 6500)

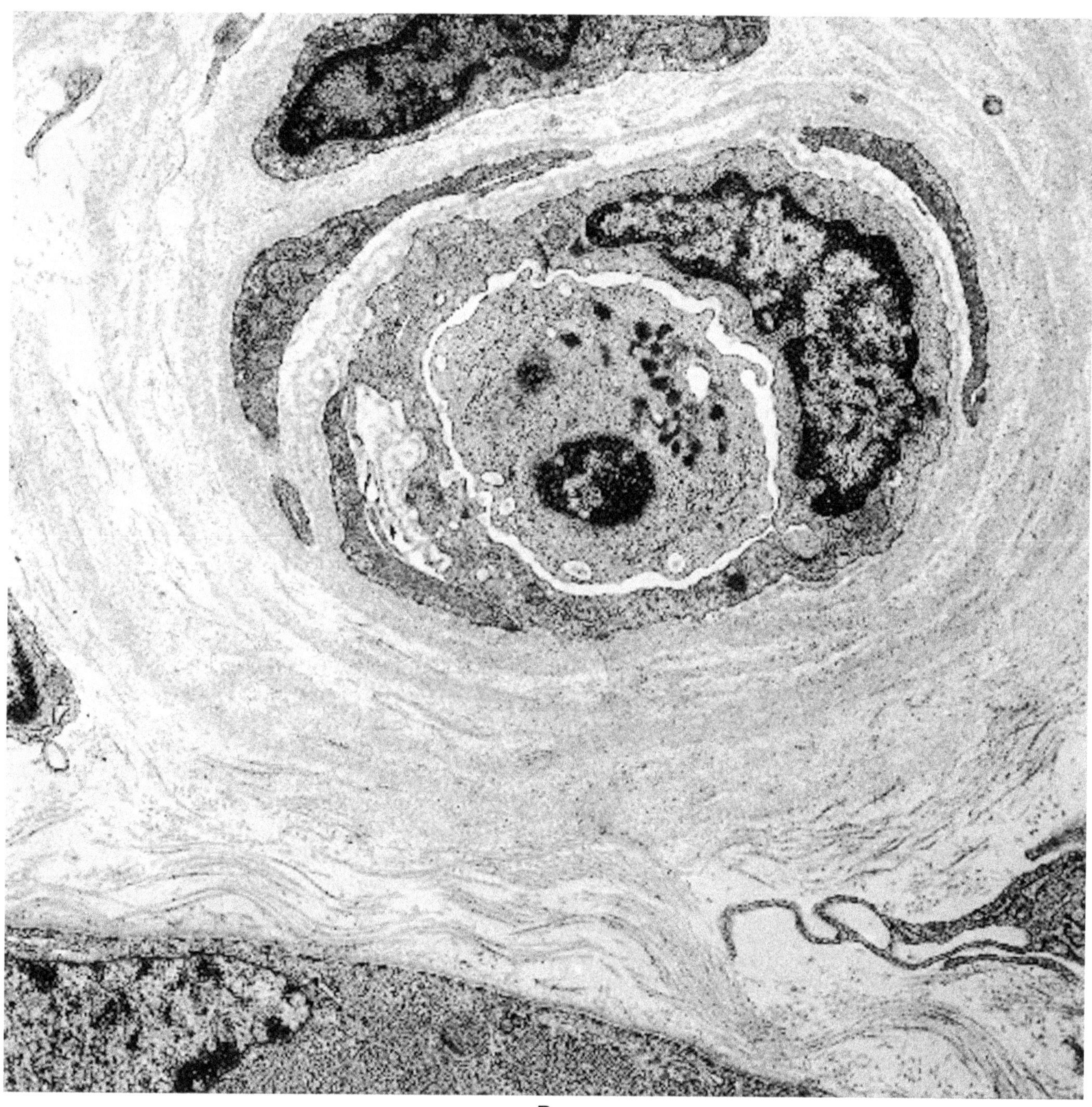

B

Figure 13.24. *(continued)*
B, Endomysial capillary with thickened basement membranes. (× 14,000)

Adrenoleukodystrophy (X-Linked Deficiency of Peroxisomal Catabolism of Very Long Chain (>24 Carbon) Fatty Acids)

(Figure 13.25.)

Diagnostic criteria. Characteristic findings in adrenoleukodystrophy are (1) demyelinating peripheral neuropathy in children with leukoencephalopathy; (2) occasional Schwann cell hyperplasia; (3) needle-shaped electron-lucent inclusions, largely limited to Schwann cells and bounded by leaflets (Mosser et al. 1993; Powers and Moser 1998). Note that the inflammatory component, which is often prominent in the childhood cerebral demyelination, is not usually present in the PNS. Although peripheral nerve involvement may occur in young males with the homozygous (X-linked) disease, peripheral neuropathy is more often clinically symptomatic in heterozygous females and is accompanied by involvement of the long tracts of the spinal cord, adrenal insufficiency, and elevated levels of very long chain fatty acids (adrenomyeloneuropathy).

Metachromatic Leukodystrophy (Autosomal Recessive Deficiency of Arylsulfatase A Activity)

(Figure 13.26.)

Diagnostic criteria. In metachromatic leukodystrophy, there is (1) a demyelinating neuropathy with thin myelin sheaths; (2) proliferation of Schwann cells, at times forming onion bulbs; (3) prismatic, zebra body, and tuffstone storage inclusions. The clinical course is related to the age of onset, with a more rapid course in patients presenting in infancy.

Hereditary Neuropathies

Hereditary neuropathies may involve predominantly motor and sensory functions (hereditary motor and sensory neuropathies [HMSN]) or sensory and autonomic functions (hereditary sensory and autonomic neuropathies [HSAN]). Only the electron microscopic findings in the most common forms of hereditary peripheral neuropathies are discussed (see review by Guzzetta et al. [1995]).

Charcot-Marie-Tooth Disease, Hypertrophic Form (HMSN I, CMT 1)

(Figure 13.27.)

Clinical manifestations. Charcot-Marie-Tooth disease, hypertrophic form (HMSN I, CMT 1) is the most common hereditary neuropathy and is typically an autosomal dominant slowly progressive distal, symmetric neuropathy. Symptoms first appear in early adulthood, and due to the marked atrophy of calf muscles, the term "peroneal muscular atrophy" is often used. In addition to weakness, there are sensory deficits and secondary complications such as pes cavus and neurogenic ulcers.

Diagnostic criteria. The pathologic features of HMSN I include (1) a demyelinating neuropathy; (2) prominent onion bulbs, but they may be less prominent in proximal nerve and in childhood. In young adults, most axons show a diminished amount of myelin surrounding large caliber fibers and onion bulbs surrounding a large proportion of the axons.

Etiology. Nerve conduction velocities have been useful in classifying patients with CMT, since slowing of nerve conduction velocity is typical of the hypertrophic form of disease (HMSN I, CMT 1), but is not typical of the neuronal form (HMSN II). The clinical course is slowly progressive and often impedes ambulation; however, life span is normal.

Identification of the genetic loci now allows molecular subclassification. The majority of HMSN I patients (HMSN IA) have a duplication of a large region of 17p11.2-p12, resulting in "segmental trisomy" of the duplicated region. The duplicated segment includes one of the major proteins of peripheral nerve myelin, peripheral myelin protein-22 (PMP-22). HMSN 1B is the second common form of HMSN 1, and involves a mutation of a Schwann cell protein, P_0 myelin protein, located on chromosome 1. An X-linked form of hypertrophic Charcot-Marie-Tooth disease (CMT 1X) has been identified, localized to Xq13.1 and encoding the protein connexin-32 (Cx32), a gap junction subunit (Bennett 1994).

Hereditary Neuropathy with Liability to Pressure Palsies

(Figure 13.28.)

Clinical manifestations. Hereditary neuropathy with liability to pressure palsies (HNPP) (also known as tomaculous neuropathy) is an autosomal dominant neuropathy with the distinctive clinical feature of susceptibility to numbness and muscular weakness involving a single nerve following minor nerve trauma or compression. The disorder is often detected after several episodes of compression mononeuropathies, and many patients with HNPP are asymptomatic. However, detailed clinical history may reveal a history of compression neuropathy in other members of the family.

Diagnostic criteria. Pathologically, HNPP shows (1) a demyelinating neuropathy; (2) occasional Schwann cell hyperplasia; and (3) focal hypermyelination, which consists of concentric hypermyelination of some fibers or eccentric inner or outer layers of myelin that have a redundant folded pattern. Regardless of whether the

(Text continues on page 976)

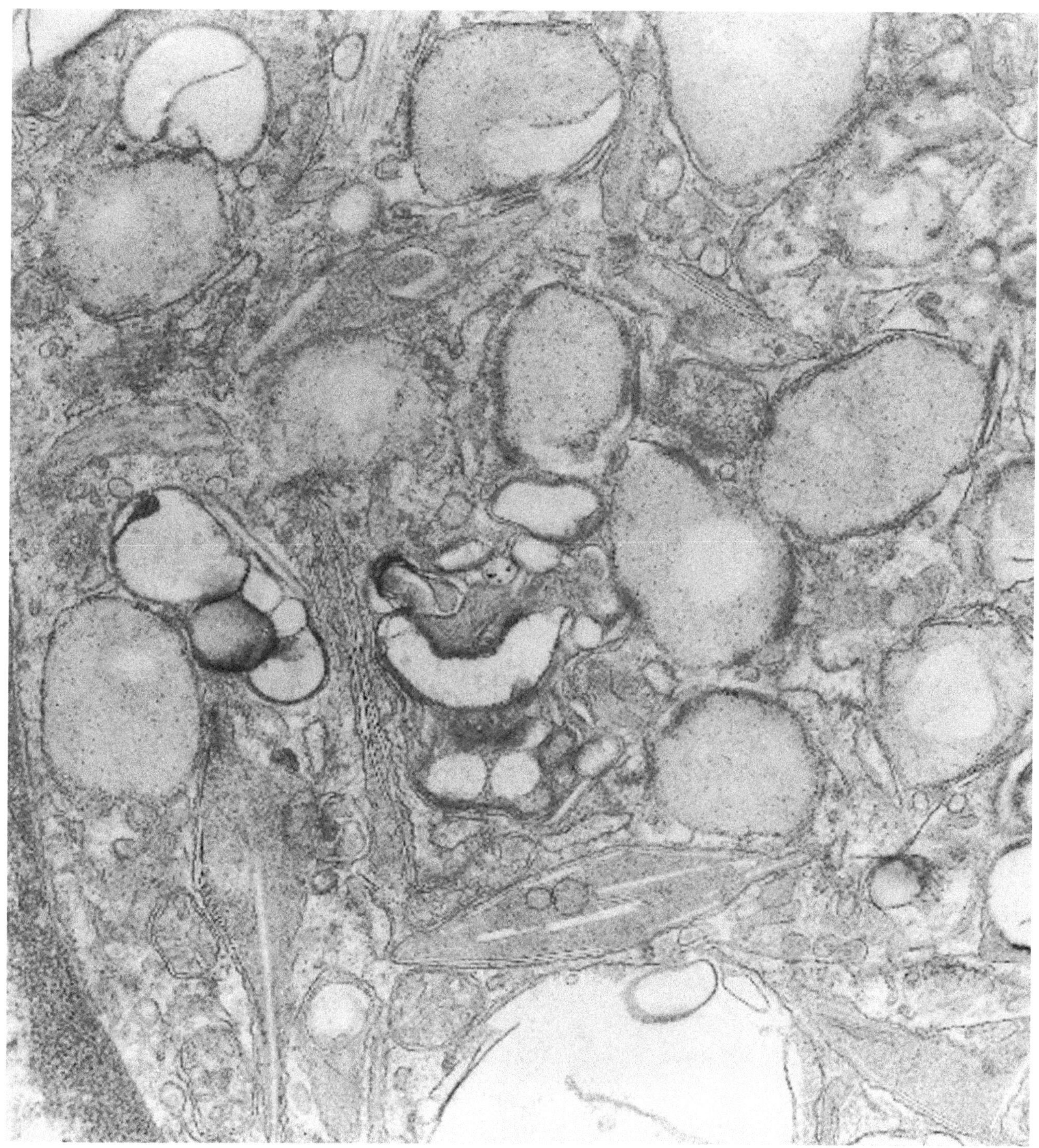

Figure 13.25. Adrenoleukodystrophy. Note abnormal accumulations of lipid and needle-shaped electron-lucent inclusions. (× 40,000)

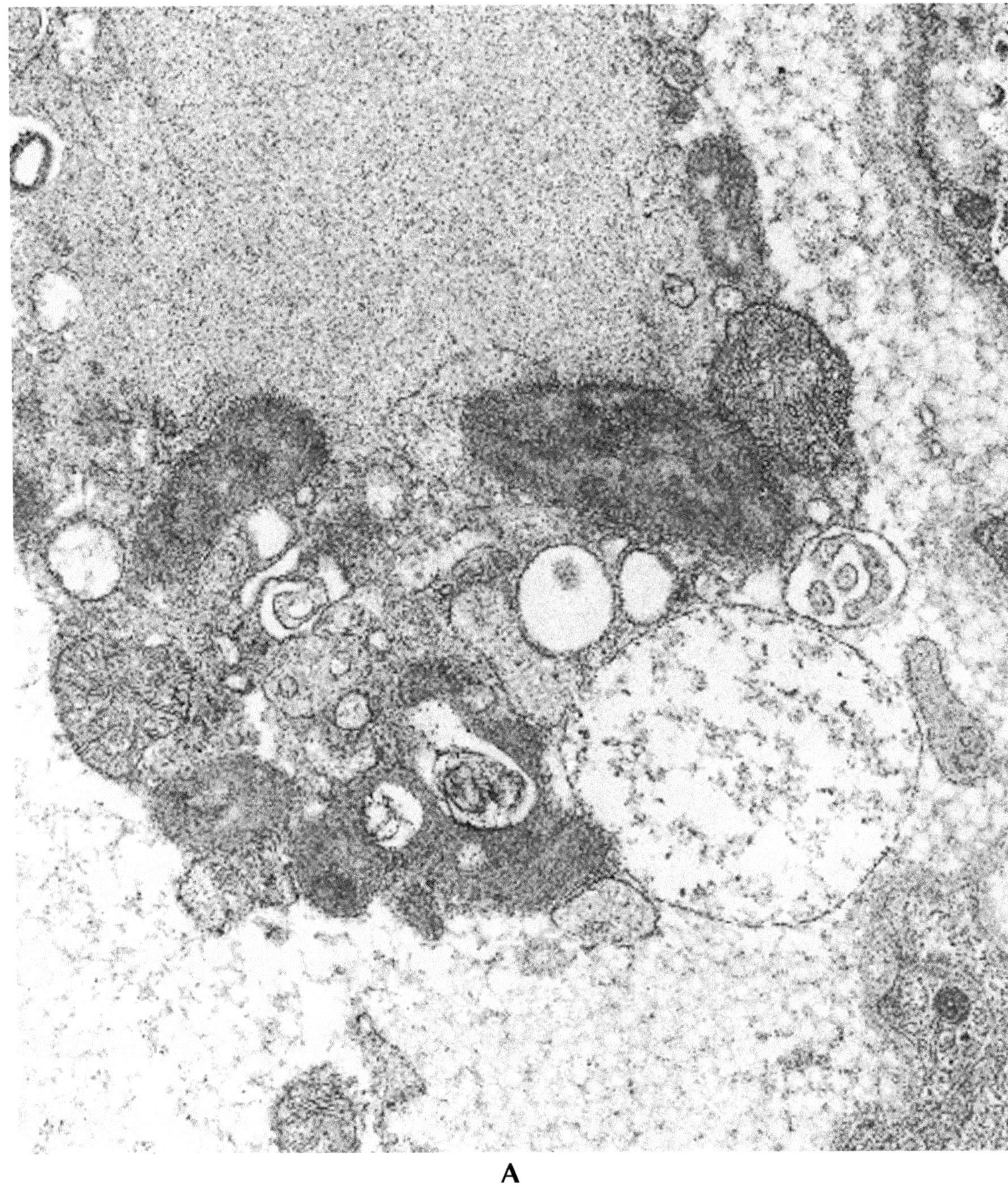

A

Figure 13.26. Metachromatic leukodystrophy. **A,** Tuffstone inclusions within Schwann cell cytoplasm. (× 36,000)

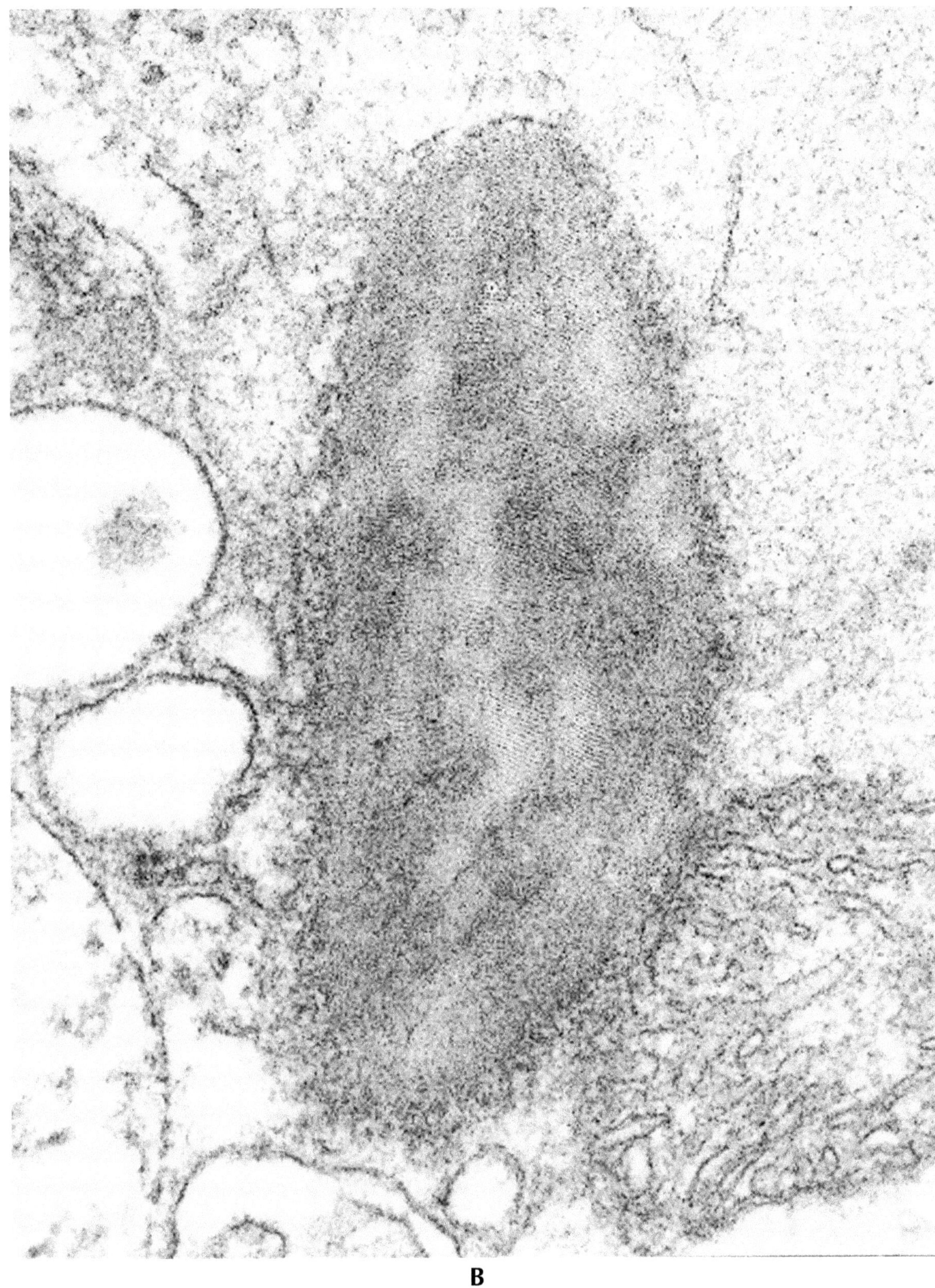

B

Figure 13.26. *(continued)*
B, High magnification of tuffstone inclusions showing the ridges of volcanic-like stone in the inclusion. (× 118,000)

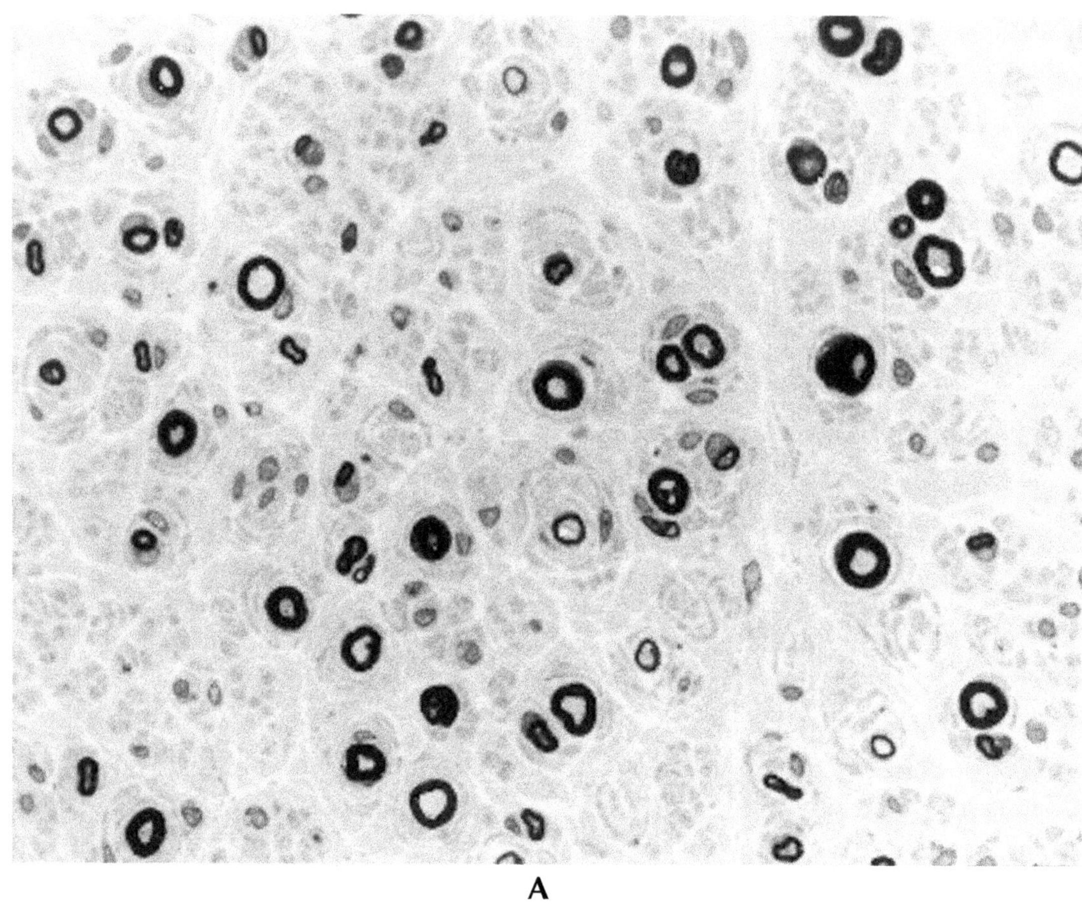

A

Figure 13.27. Charcot-Marie-Tooth Disease, type I (HMSN I). **A,** Light micrograph of plastic-embedded transverse section stained with toluidine blue showing reduction in the number of myelinated fibers and onion bulbs around thinly myelinated fibers (× 200)

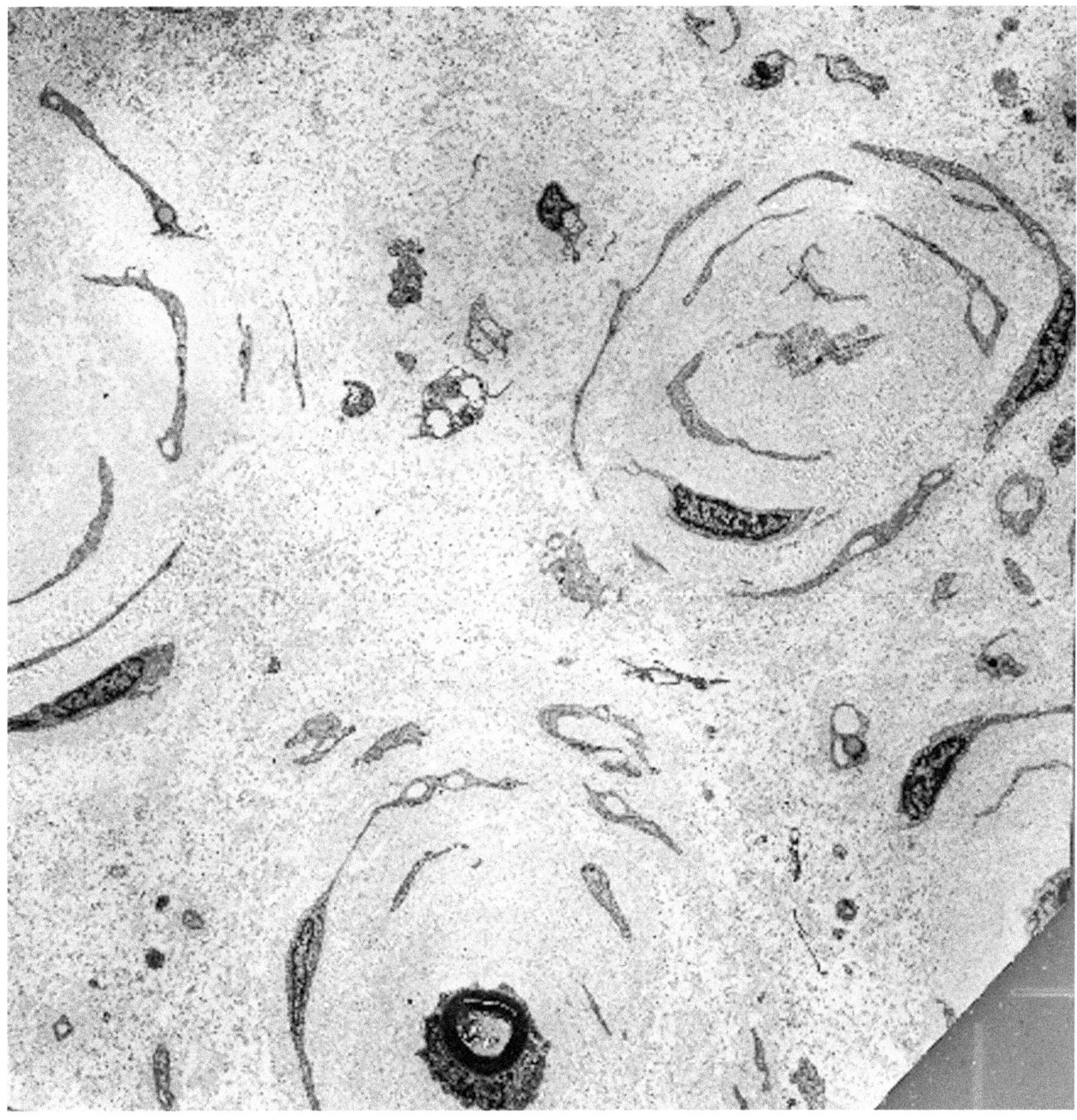

B

Figure 13.27. *(continued)*
B, Electron micrograph showing nerve with marked depletion of axons replaced with abundant collagen and prominent onion bulbs. (× 2500)

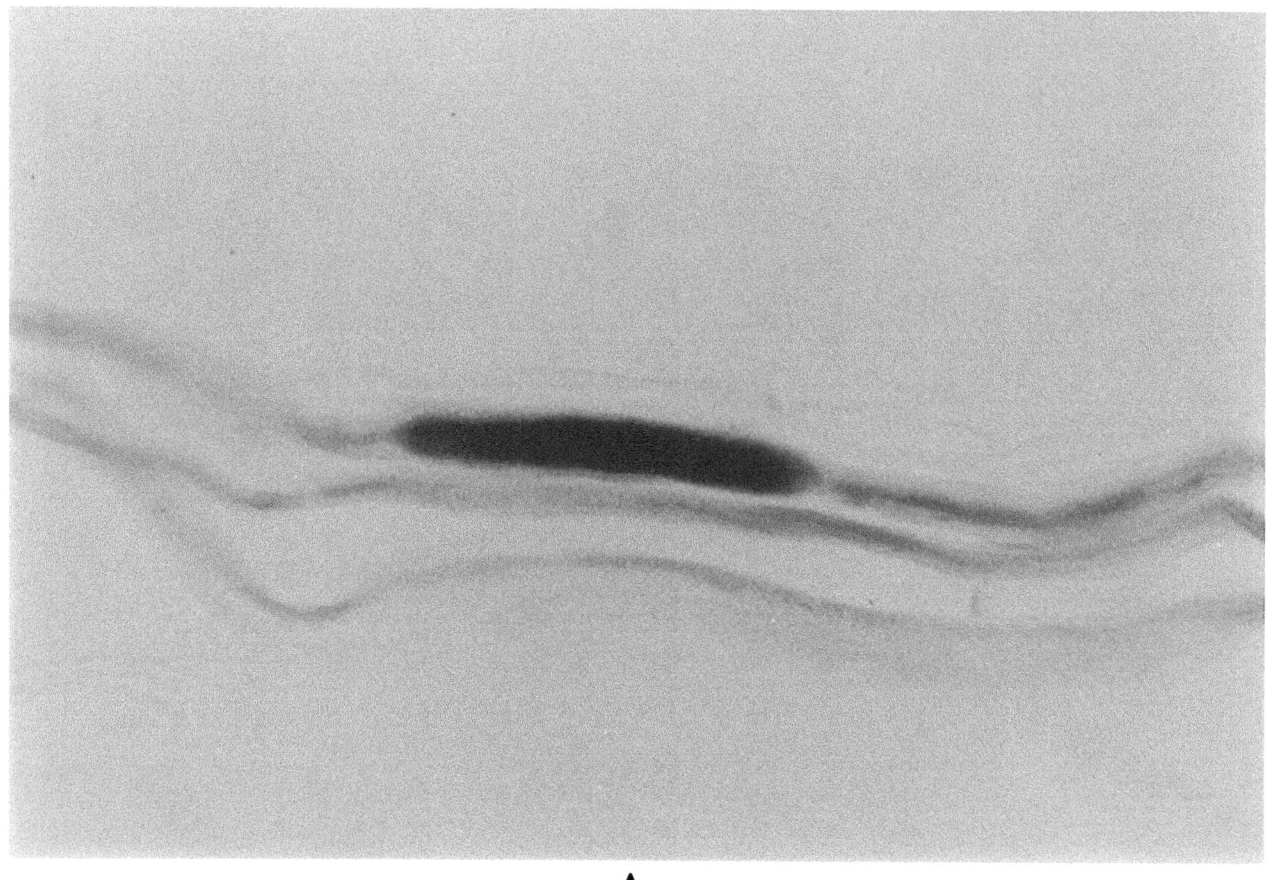

Figure 13.28. Hereditary neuropathy with liability to pressure palsies (HNPP). **A,** Teased nerve fiber preparation showing sausage-like expansion of the myelin. (× 760)

(Text continued from page 970)

hypermyelination is concentric or eccentric, the typical electron microscopic finding is the presence of multiple layers of myelin that loop back, a pattern referred to as "reversal of myelin layers."

Etiology. The locus is for HNPP involves the same site on chromosome 17 as HMSN I; however, instead of duplication of the region, there is a large deletion of 17p11.2-p12. The concordance of the region that is deleted in HNPP and duplicated in HMSN I suggests that these conditions represent complementary genetic alterations resulting from an unequal crossover event during meiosis. It has been argued that the distinct clinical phenotypes represent a gene dosage effect of PMP-22, a major protein component of peripheral nerve myelin, so that elevated gene dosage occurs with the duplication, or "regional trisomy," in HMSN I, and that there is an effect of diminished gene dosage in the "regional monosomy" of HNPP.

Charcot-Marie-Tooth Disease, Neuronal Form

Clinical manifestations. Although less prevalent than HMSN I, the neuronal form of CMT accounts for up to 25% of the total number of cases of Charcot-Marie-Tooth disease. Clinical presentation of type 2 CMT is similar to HMSN I, although the onset is often slightly later and progression is slightly slower.

Diagnostic criteria. Pathologic features of CMT 2 include (1) diminished numbers of large caliber axons; and (2) paucity, or more often, absence of onion bulbs (Ben Othmane et al. 1993).

Etiology. Genetic analysis of CMT 2 has revealed that in some families (CMT 2A), the disease phenotype is linked to a marker on chromosome 1p35-p36 (Engel and Franzini-Armstrong 1994). In other families, the phenotype is not linked to this region of chromosome 1, nor to any of the loci identified for CMT 1, suggesting that there are additional genetic loci, as yet unidentified, in CMT 2.

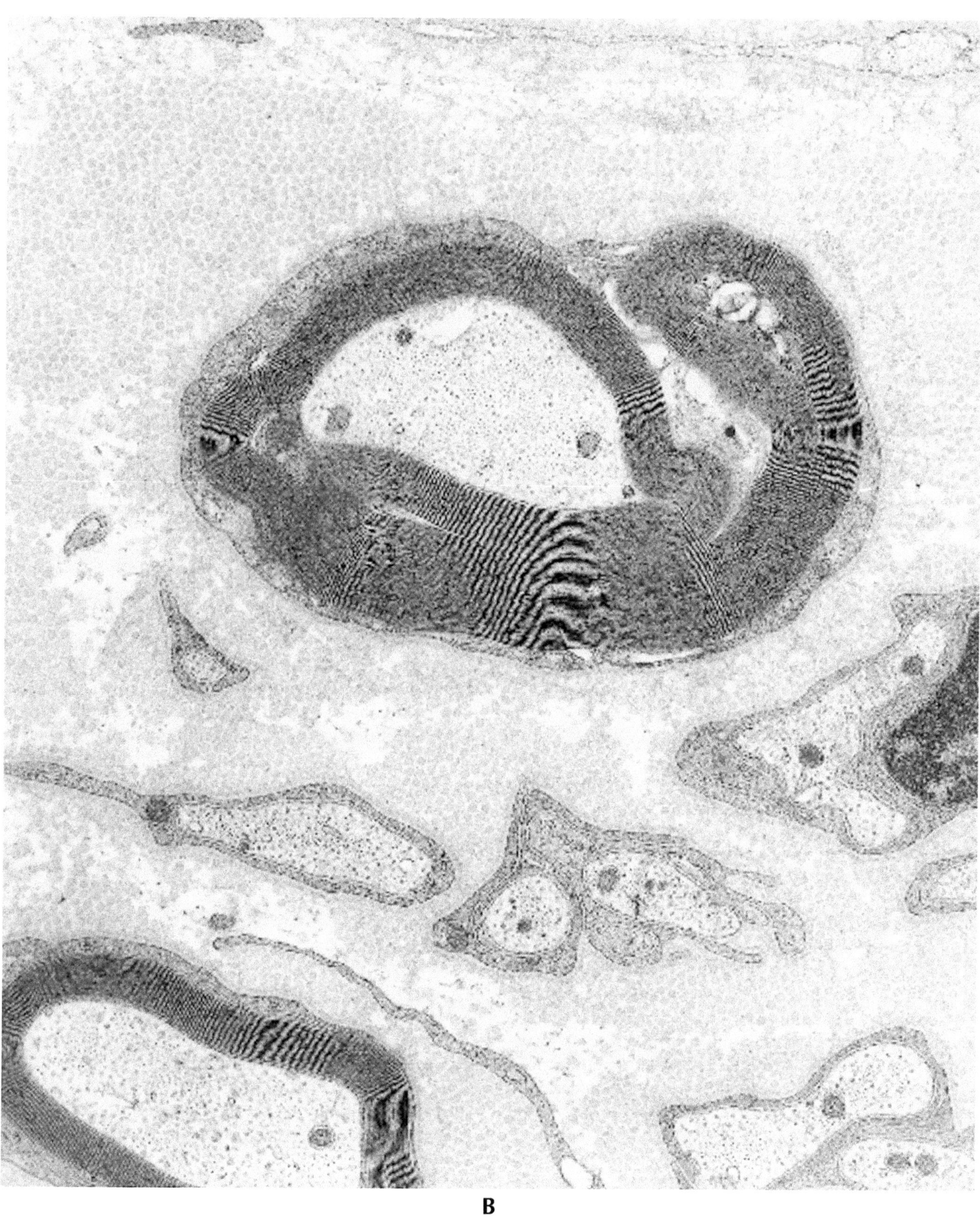

B

Figure 13.28. *(continued)*
B, Expanded "hypermyelinated" region with lack of concentric arrangement of myelin, or "reversal of myelin layers." (× 24,000)

Déjerine-Sottas Disease

(Figure 13.22.)

Clinical manifestations. Déjerine-Sottas disease is a severe hypertrophic neuropathy affecting infants or children, and often presents with delay in the onset of walking or difficulty with running and jumping. This disease is distinguished from CMT by its earlier onset and recessive inheritance pattern. Nerve conduction is slow, and peripheral nerves are so large as to be palpable on physical exam (hypertrophic). Gradual progression of weakness may lead to wheelchair confinement in young adult life.

Diagnostic criteria. The pathologic findings in HMSN III include (1) prominent onion bulb formation, affecting virtually every axon; (2) attenuated, or absent myelin sheaths of large caliber axons; (3) diminished number of axons, especially in later stages of the disease.

Etiology. Just as Déjerine-Sottas disease (HMSN III) and hypertrophic CMT disease (HMSN I) share pathologic and ultrastructural abnormalities, they share molecular features. Mutations in PMP-22 and MPZ have been identified in Déjerine-Sottas disease, the same genes involved in HMSN IA and IB, respectively. For both the PMP-22 and the MPZ genes, the mutations identified in Déjerine-Sottas disease were point mutations that arose *de novo,* suggesting that sporadic occurrence may also be common in HMSN III.

Infantile Neuroaxonal Dystrophy (Schindler Disease, Seitelberger Disease)

(Figure 13.29.)

Clinical manifestations. Neuroaxonal dystrophy (NAD) is characterized by enlarged abnormal axons throughout the nervous system. The most common form is the infantile form (Seitelberger disease), although juvenile and adult forms also occur. In the infantile form, there is usually normal early development, followed by progressive neurologic deterioration, resulting in death during childhood.

Diagnostic criteria. The pathologic findings in nerve biopsies are (1) axons with greatly distended contours; (2) secondary demyelination may occur around the distended axons; (3) intra-axonal tubulovesicular structures, sometimes appearing as prominent membranous structures. Dystrophic axons may also be identified in skin biopsies, and similar dystrophic axons are present throughout the brain.

Etiology. The biochemical basis for one subset of patients with infantile NAD has been identified as a decreased activity of alpha-*N*-acetylgalactosaminidase (alpha-NAGA), a lysosomal enzyme. The gene for this enzyme is located on chromosome 22q and has shown mutations. The subset of infantile NAD with alpha NAGA deficiency has been designated Schindler disease, but other cases of infantile NAD have normal alpha-NAGA activity, and the molecular basis in these cases is not yet determined.

Giant Axonal Neuropathy

Clinical manifestations. Giant axonal neuropathy (GAN) usually arises in infancy or early childhood with an apparent autosomal recessive pattern. Presentation is often from 5–10 years of age with a gradually progressive peripheral neuropathy. Symptoms include weakness, diminished sensation, and absent reflexes. Nerve conduction studies usually show diminished amplitudes, but may also show diminished velocity due to secondary demyelination. Many children with the disorder have unusually curly hair.

Diagnostic criteria. The pathologic findings in GAN are (1) markedly enlarged axons, within both the peripheral and central nervous system, (2) secondary demyelination may occur around the enlarged axons; (3) densely packed neurofilaments completely filling the distended axons.

Etiology. The etiology of GAN of childhood is not yet determined, but appears to involve intermediate filaments in many cell types. It has been linked to chromosome 16q24.1 (Flanigan et al. 1998). In addition to the accumulation of neurofilaments in the nervous system, abnormal aggregates of intermediate filaments occur in other cells, including aggregates of glial fibrillary acidic protein in astrocytes and vimentin in fibroblasts.

Hereditary Sensory and Autonomic Neuropathies

Clinical manifestations. There are three major forms of the hereditary sensory and autonomic neuropathies (HSANs), defined on clinical grounds but having pathologic findings that reflect the predominant fiber type affected.

Ulcerating and Mutilating Acropathy (HSAN 1)

Diagnostic criteria. (1) Degenerating axons and axonal loss in peripheral nerve; (2) predominant involvement of larger caliber axons; (3) relative sparing of unmyelinated axons. Phenotypic variability suggests that multiple genes may be involved; however, no chromosomal or genetic linkage has been identified. Clinically, the principal symptom is numbness, with a slow progression of sensory deficits. Ulcers of the feet may be complicated by osteomyelitis and thrombophlebitis. The disease usually presents in adolescence or young adulthood with symmetric sensory symptoms and minimal autonomic involvement.

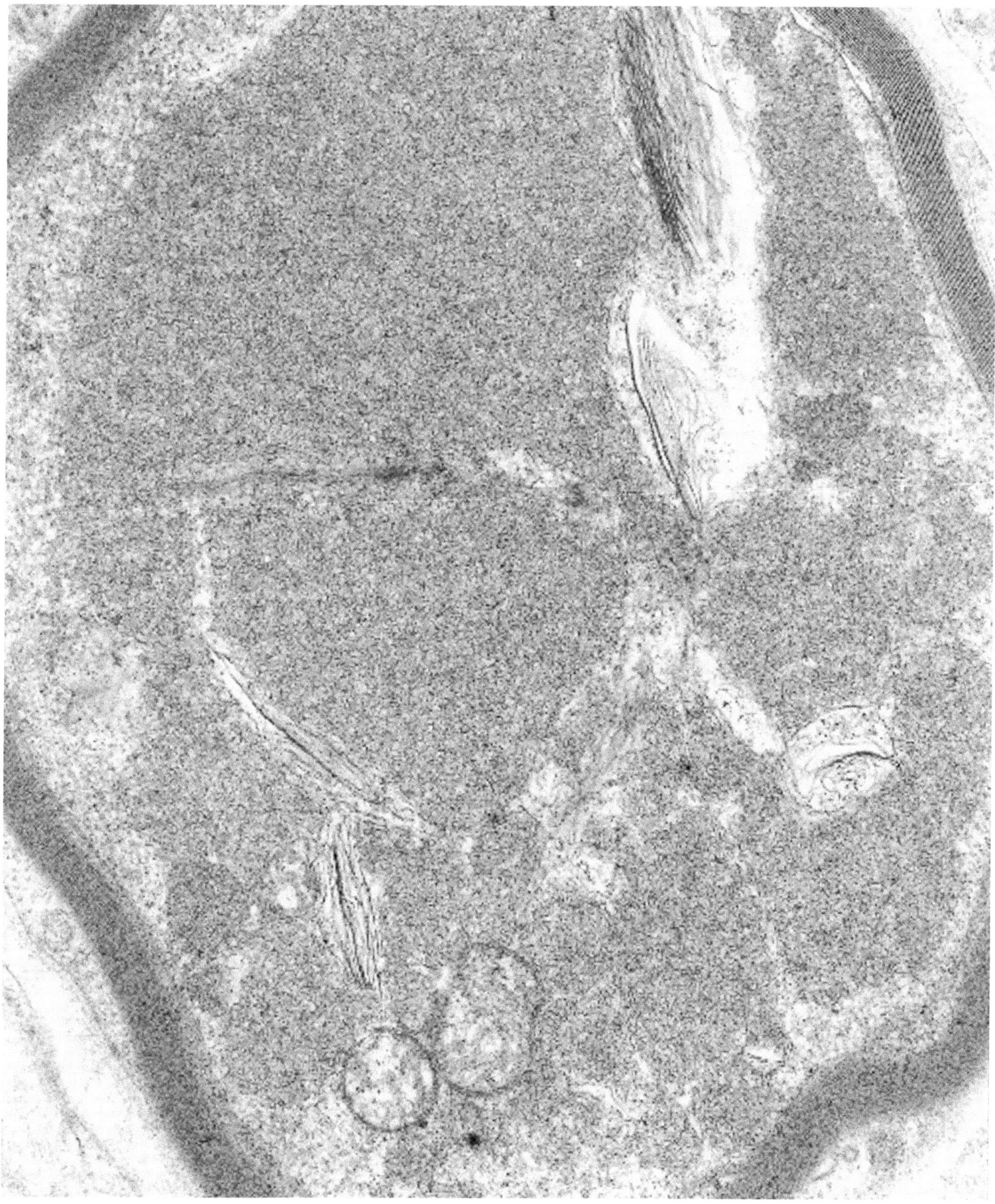

Figure 13.29. Infantile neuroaxonal dystrophy. Note greatly distended axon with intra-axonal tubulovesicular structures. (× 93,000)

Congenital Sensory Neuropathy (HSAN 2)

Diagnostic criteria. (1) Complete absence of large and small myelinated fibers in peripheral nerve; (2) relative preservation of unmyelinated axons. The onset is congenital with a severe sensory deficit, but autonomic features are mild or moderate. Inheritance appears to be autosomal recessive, and the gene is not yet identified. There is progressive involvement of both upper and lower extremities with suppurative lesions developing in the insensitive limbs, but it may be nonprogressive.

Familial Dysautonomia (Riley-Day Syndrome, HSAN 3)

Diagnostic criteria. (1) Loss of neurons and neuronophagia involving the autonomic ganglia; (2) peripheral nerve with a dramatic loss of unmyelinated fibers, (3) relative preservation of myelinated fibers. The disease has been linked to markers on chromosome 9q31-q33. The clinical syndrome typically involves congenital presentation, most commonly in Ashkenazi Jewish families. Autonomic symptoms predominate, including cardiac instability and absence of tears, and the disease is progressive, with death often occurring in infancy.

Amyloid Neuropathy

(Figure 13.30.)

Clinical manifestations. When peripheral neuropathy occurs as a result of amyloidosis, it is usually as a progressive, distal, symmetrical, sensorimotor neuropathy, often with prominent autonomic dysfunction. In the immunocyte-derived form of amyloidosis, there may be damage to other organ systems in addition to peripheral nerve, such as the kidney with nephrotic syndrome, intestines with malabsorption and diarrhea, and heart with congestive cardiac failure. Weakness is usually less of a complaint than sensory symptoms, and on physical examination, it is often pain and temperature sensations that are more affected than touch, vibration, and position. Autonomic signs such as postural hypotension, bladder and bowel hypofunction, impotence, and anhidrosis can be severe.

Diagnostic criteria. Pathologic findings in amyloid neuropathies include (1) a marked reduction in the total number of myelinated fibers; (2) evidence of acute axonal degeneration; (3) a severe reduction in the amount of unmyelinated fibers, leaving many stacks of flattened Schwann cell processes devoid of unmyelinated fibers; (4) amyloid deposits within vessel walls or in the interstitium consisting of rigid aggregates of nonbranching fibrils with hollow cores, 7.5–10 nm wide and of indefinite length.

Etiology. The neuropathy in amyloidosis tends to involve the small myelinated and unmyelinated fibers more than large myelinated ones. This preferential involvement of smaller fibers correlates with greater dysfunction in pain, temperature, and autonomic modalities (all mediated by thinly myelinated and unmyelinated axons) than touch, vibration, position sensation, and motor function (all mediated by large myelinated fibers).

Amyloid is composed of protein with a twisted beta-pleated sheet conformation, assembled into fibrils of 7.5–10 nm diameter and deposited in extracellular location. The fibrils, due to their repeating beta-pleated sheet protein subunit arrangement, bind Congo red dye in a periodic array, producing the characteristic apple green birefringence when examined under polarized light. There are two major types of amyloidosis that involve peripheral nerve: a systemic form of amyloidosis and a localized form. Systemic immunocyte-derived amyloidosis (also known as amyloidosis associated with B-cell dyscrasia, or primary amyloidosis) results from deposition of amyloid fibrils derived predominantly from immunoglobulin light chains, more often lambda than kappa light chains. In immunocyte-derived amyloidosis, common sites of involvement include the heart, gastrointestinal tract, tongue, skin, and peripheral nerves in up to 15% of cases. Peripheral neuropathy is uncommon in reactive systemic amyloidosis (also known as secondary amyloidosis) which results from the deposition of AA (amyloid associated) protein derived from the serum protein, SAA (serum amyloid-associated protein).

The second major type of amyloidosis involving peripheral nerve is localized amyloidosis, in which the amyloid fibrils are derived from the serum protein, transthyretin. Transthyretin-derived amyloid deposition is responsible for most of the familial amyloid polyneuropathies (FAPs), a group of inherited amyloidoses that predominantly affect the peripheral nervous system, though other tissues such as heart or kidney may sometimes be involved. Molecular genetic analysis has revealed over 30 different point mutations in transthyretin (each causing amino acid substitutions), spanning all four exons of the gene, located on chromosome 18q11.2-q12.1. The FAPs were originally classified by geographic designation, recognizing the sites where the affected families were first recognized. However, they may also be classified on the basis of the gene involved and the site of mutation. For example, FAP of theGerman/ Maryland type has a histidine to leucine substitution at position 58 (L58H) of the transthyretin protein.

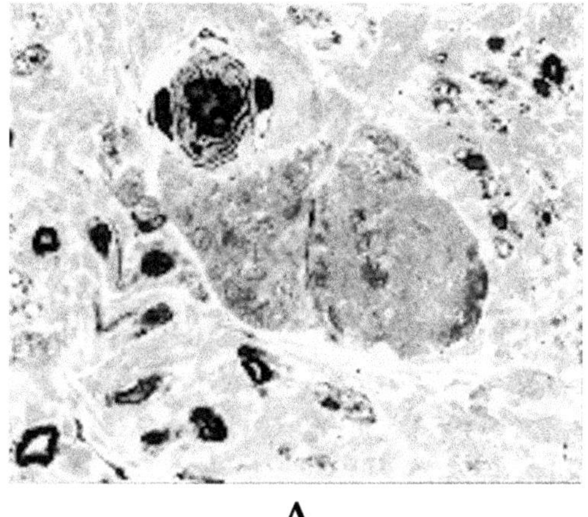

A

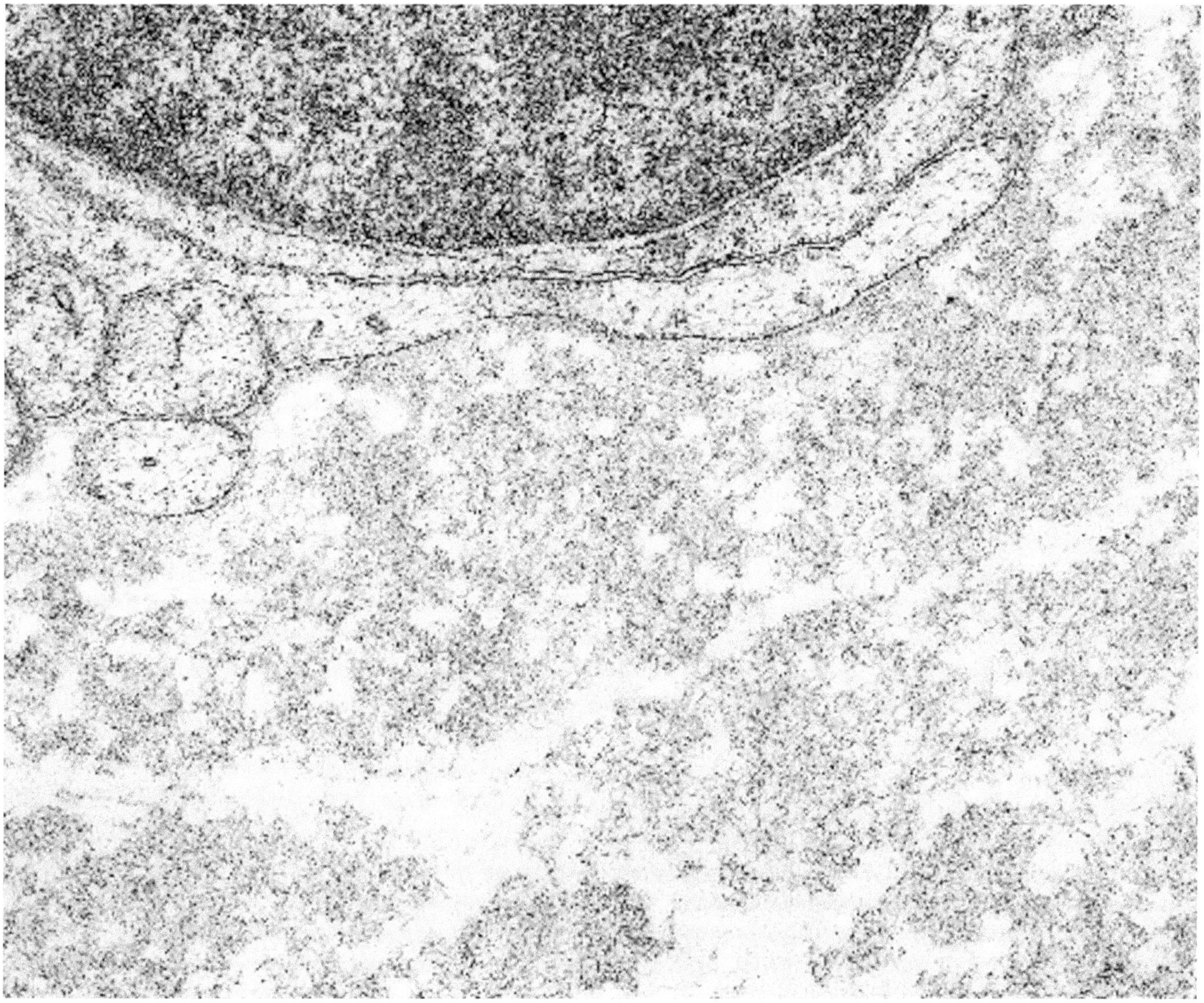

B

Figure 13.30. Amyloid neuropathy. **A,** One-micron-thick section of nerve showing large deposit of amyloid next to endoneurial vessel. (toluidine blue, × 500) **B,** Endoneurium contains numerous amyloid fibrils. (× 38,400)

Immune-Mediated Neuropathies

Guillain-Barré Syndrome

Clinical manifestations. Guillain-Barré syndrome (also known as acute inflammatory demyelinating polyradiculoneuropathy (AIDP) or Landry-Guillain-Barré-Strohl syndrome) is an acute or subacute demyelinating peripheral neuropathy, often following a viral infection (cytomegalovirus, Epstein-Barr virus [mononucleosis], vaccinia, variola, varicella-zoster, measles), *Campylobacter* infections, surgery, or vaccination. It has also been reported at increased frequency in association with acquired immunodeficiency syndrome (AIDS). It is a common form of peripheral neuropathy, with an annual incidence in the United States of 1 to 2 per 100,000, and although the mortality was as high as 25% in the past, the ability to support respiration during the acute phase of the illness has diminished mortality to less than 5% of cases. Usually there is complete recovery; however, up to 15% of patients may have permanent residual neurologic deficits, especially if the onset was particularly rapid or the course was particularly severe.

Diagnostic criteria. Pathologic findings in Guillain-Barré syndrome include (1) acute demyelination affecting a small proportion of fibers on any given cross-section of nerve, but often including both thinly myelinated axons and some axons denuded of myelin; (2) inflammatory infiltrates, often perivascular but variable in intensity and distribution, usually T-cells, with CD4 helper cells outnumbering CD8 cells when markers have been employed); (3) axonal degeneration of variable severity, most prominent in fatal, or other severe cases (Feasby et al. 1993; Griffin et al. 1996); (4) insinuation of macrophage processes beneath the Schwann cell basal lamina, interdigitating between myelin layers. These macrophages may contain myelin debris in the cytoplasm, but the characteristic appearance is the presence of cell processes between the myelin layers or between myelin and the axon.

Etiology. The pathogenesis of Guillain-Barré syndrome involves an immune-mediated response to peripheral nerve myelin (Hartung et al. 1998). This idea is largely supported by the similarities between Guillain-Barré syndrome and experimental allergic neuritis (EAN), an animal model of acute demyelination of peripheral nerve in which T-cell mediated demyelination has been established. In EAN, an acute demyelinating peripheral neuropathy develops approximately ten days following inoculation with P_2 myelin protein. The intensity of inflammatory infiltrates in peripheral nerve, and the degree of demyelination are dependent on the amount of P_2 inoculum, and axonal degeneration may be seen with high doses. An abortive neuropathy may be induced in naive animals by transfer of sensitized T-cells from inoculated animals, supporting that the disorder is mediated by T-cells sensitized to myelin protein.

Although the pathologic changes in Guillain-Barré syndrome are very similar to EAN and support a T-cell mediated demyelination, sera from patients with Guillain-Barré syndrome may initiate the breakdown of myelin in cell cultures. Patients respond clinically to plasmapheresis, suggesting that there may be a humoral factor involved.

Chronic Immune-Mediated Demyelinating Polyneuropathy

Chronic immune-mediated demyelinating polyneuropathy (CIDP) is a chronic peripheral neuropathy with an insidious onset, in contrast to the abrupt onset of Guillain-Barré syndrome. The course is characterized by a symmetric sensorimotor neuropathy with weakness and areflexia in the lower and upper extremities that progresses for at least several months. Nerve conduction velocities are slowed, usually to less than 70% of the lower limit of normal, and CSF protein is usually elevated above 0.45 g/L. Since steroid therapy and plasma exchange have been used with some success in the treatment of CIDP, recognition of this peripheral neuropathy has received a great deal of attention.

Diagnostic criteria. Pathologic findings in CIDP include (1) chronic demyelinating neuropathy, including myelin sheaths of variable thicknesses; (2) often prominent Schwann cell hyperplasia, with the formation of occasional onion bulbs; (3) inflammatory infiltrates in either the endoneurial or epineurial spaces, often in a perivascular distribution, and variably present in individual cases.

Etiology. CIDP is thought to be related to Guillain-Barré syndrome in its immune-mediated pathogenesis and selective demyelination, and it has been referred to as "chronic Guillain-Barré syndrome" by some.

Peripheral Neuropathy and Systemic Immune-Mediated Disorders

Peripheral neuropathy may occur in patients with other immune-mediated disorders, including rheumatoid arthritis, systemic lupus erythematosus (SLE), and Sjögren syndrome. There are four basic forms of peripheral neuropathy in these conditions: (1) vasculitic neuropathy, associated with an abrupt onset of an asymmetric neuropathy, or mononeuritis multiplex; (2) acute ascending motor neuropathy clinically and pathologically indistinguishable from typical Guillain-Barré syndrome; (3) a distal, symmetric, sensorimotor neuropathy of unknown etiology occurring in patients with SLE; (4) a pure sensory neuropathy resulting from inflammation in the dorsal root ganglia associated with Sjögren syndrome.

Infectious Neuropathies

Herpes Infections

Clinical manifestations. The most common viral infections of the peripheral nervous system are those caused by members of the herpes virus family: herpes varicella-zoster, herpes simplex, and cytomegalovirus.

After acute chicken pox infection, varicella-zoster virus (VZV) resides latently in ganglia and may become reactivated at a later time. The predominance of zoster within dermatomes of the face and chest, areas most involved in varicella eruptions, suggests that ganglia with the highest initial infection may be more susceptible to reactivation. The incidence of shingles, which overall is 1/1000 per year, is higher in patients with malignancy, especially lymphomas and leukemias, as well as in patients with AIDS. Although usually limited to a single dermatome, in 2–10% of cases it is disseminated, especially in patients who are immunocompromised.

Herpes simplex is characterized by recurrent vesicular eruptions of the oral or genital mucosa. Latency of type 1 HSV has been documented within trigeminal ganglia in 50% of asymptomatic individuals coming to postmortem examination. In experimental models, reactivation of latent virus is enhanced by nerve injury.

Cytomegalovirus (CMV) is a member of the herpes family of DNA viruses, and it may cause a polyradiculoneuropathy in AIDS patients. The typical clinical syndrome consists of low back pain with radiation into the leg, asymmetric weakness, and sensory deficits of the lower extremities. This syndrome is a reflection of involvement of the cauda equina; disturbances in bladder and bowel functions are common.

Diagnostic criteria. The pathologic findings in peripheral nerve involved by the herpes family of viruses are (1) inflammation of ganglia and nerves; (2) axonal degeneration distally; (3) typical intranuclear inclusions may be difficult to find; (4) encapsulated virus with a core 70 nm in diameter, surrounded by a lucent rim and a 100-nm electron-dense capsid. Immunohistochemistry has been helpful in identifying viral antigen.

Leprosy

(Figure 13.31.)

Clinical manifestations. Although uncommon in the United States, lepromatous neuropathy is the most common infectious peripheral neuropathy worldwide. The infection is caused by *Mycobacterium leprae* and presents with superficial skin lesions and a peripheral neuropathy that is dominated by sensory symptoms with a proclivity to involve multiple superficial nerves, leading to patchy areas of sensory deficit.

Diagnostic criteria. The pathologic findings are different in the different forms of lepromatous involvement of peripheral nerve. In the lepromatous form, the findings include (1) numerous acid-fast bacilli within macrophages, connective tissue, Schwann cells and, rarely, axons; (2) progressive destruction of myelinated and unmyelinated fibers with active phagocytosis; (3) paucity of inflammatory cells; and (4) marked increase in endoneurial collagen. In the tuberculoid form, the findings include (1) well-formed tuberculoid granulomata throughout the nerves and (2) rare acid-fast bacilli.

Etiology. The initial lesion results from invasion of the skin and cutaneous sensory and motor nerves by *M. leprae*. The clinical expression of the disease is dependent on the resistance of the host: it may become aborted, or may progress to tuberculoid leprosy, lepromatous leprosy, or borderline forms (Weddell and Pearson 1975). The distribution of the involvement appears to be related to the temperature dependence of the organism, with a predilection for the cooler locations of the hands, feet, and superficial face.

Lyme Disease

Clinical manifestations. The initial finding in Lyme disease is a red macule surrounding the site of a prior tick bite. The gradual enlargement of this macular region, known as erythema migrans, is characteristic of the initial stage of the infection. There is a subsequent systemic phase, with symptoms of malaise, fever, chills, and headache as the spirochetes disseminate. Meningismus and meningoencephalitis may also occur during this phase and may be accompanied by focal neurologic deficits related to cranial nerves. Up to 50% of patients with Lyme meningoencephalitis have cranial neuropathies, among which the facial nerve is frequently targeted with abrupt symptomatology suggestive of Bell palsy. Other patients develop radicular pain and sensory dysfunctions. The pain usually begins acutely or subacutely; approximately 4 weeks thereafter motor weakness may become evident in a radicular distribution. Still later, a diffuse, distal neuropathy may develop, producing sensory symptoms predominantly in a stocking-glove distribution.

Diagnostic criteria. The typical pathologic alterations are (1) perivascular lymphocytic infiltrate in nerve (Meier et al. 1989); (2) sometimes, extension of inflammatory infiltrates into the vessel walls with thrombosis; (3) axonal degeneration with a predilection for large fibers. The organism has not been demonstrated in peripheral nerve.

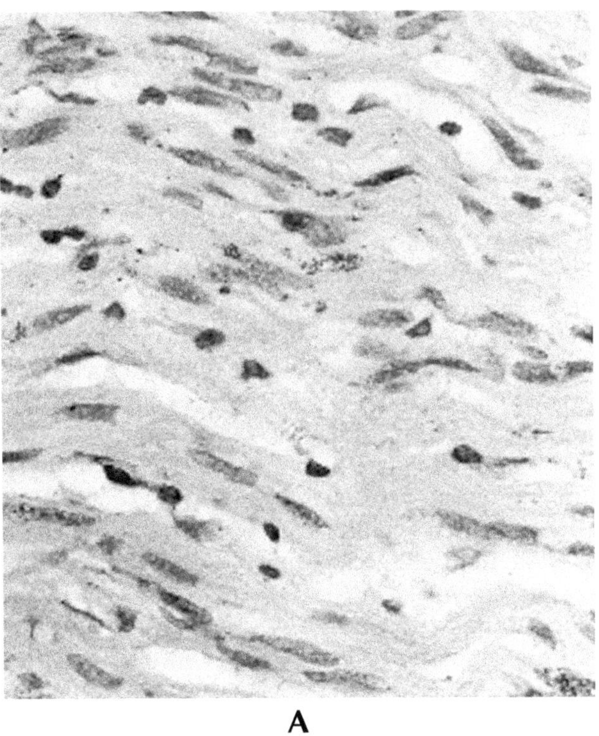

A

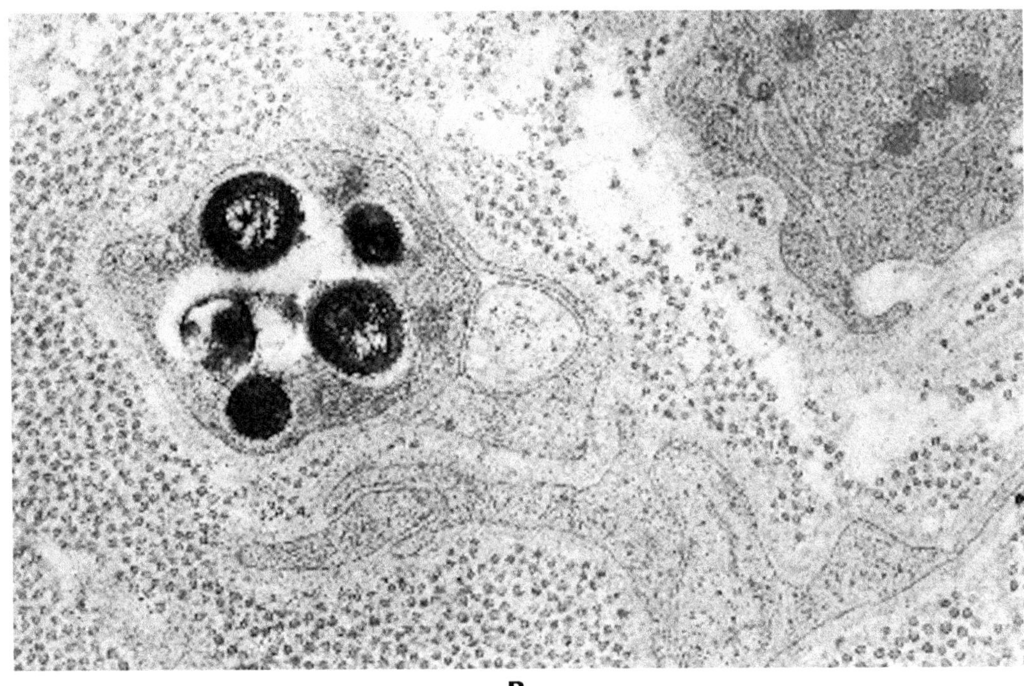

B

Figure 13.31. Lepromatous leprosy. **A,** Light micrograph of paraffin-embedded nerve cut in longitudinal plane and stained for acid-fast bacilli showing intracellular darkly staining structures. (× 150) **B,** An unmyelinated nerve fiber and cross-sections of five *Mycobacterium leprae* are present in the Schwann cell cytoplasm. Note absence of phagosomal membrane around the organism. (× 30,000)

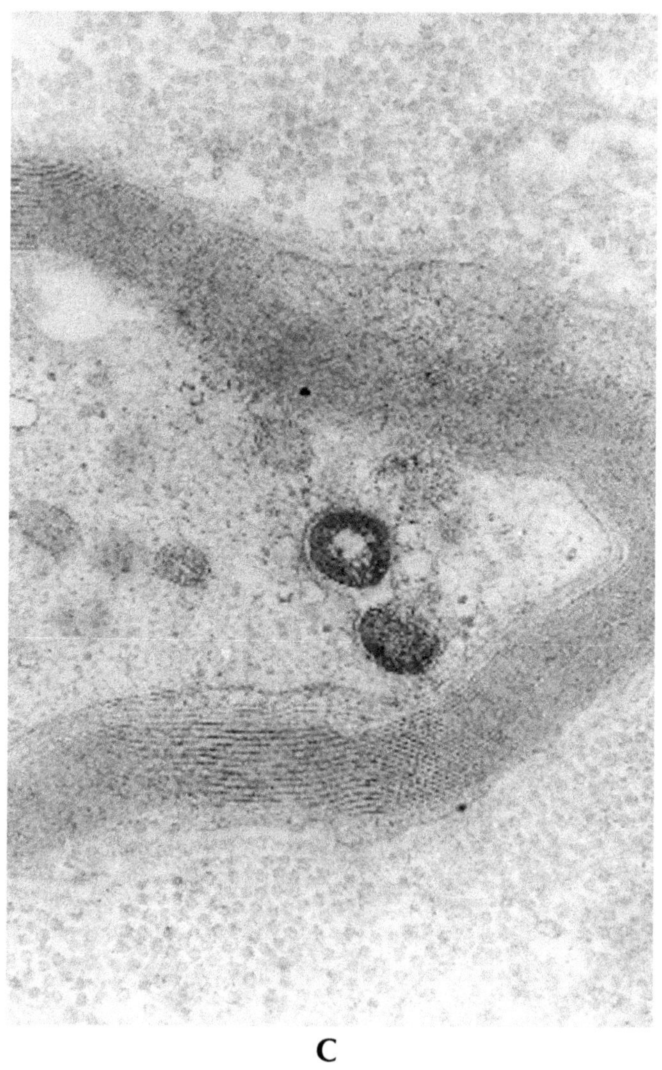

C

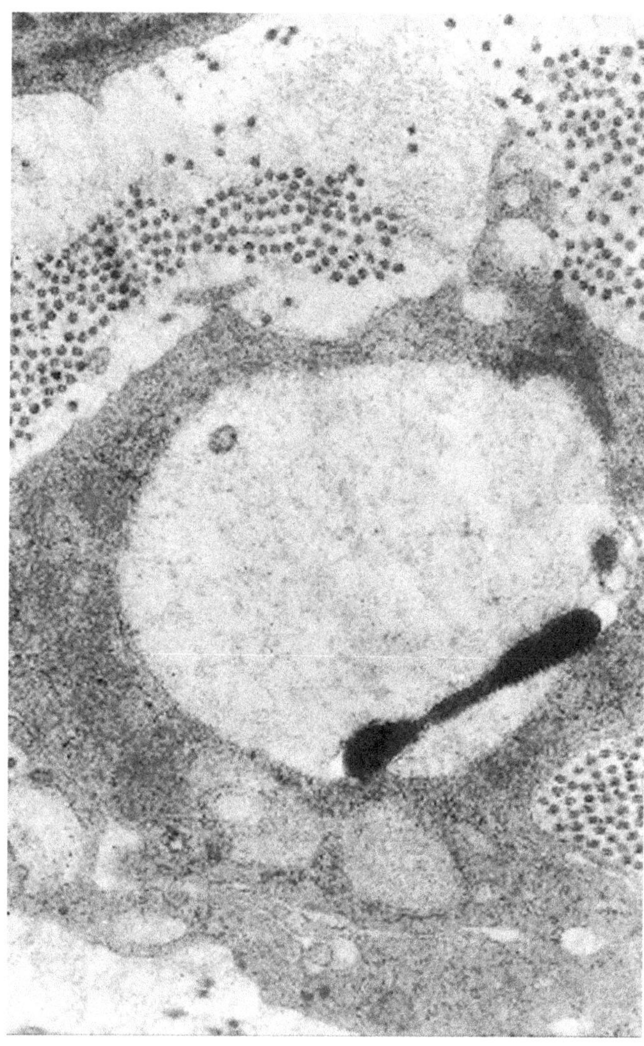

D

Figure 13.29. *(continued)*
C, A myelinated axon shows cross-sections of two *M. leprae* within a phagosome inside the axon. (× 50,000) **D,** Intraneural macrophage containing a large vacuole with a longitudinal section of *M. leprae.* (× 30,000) (Electron micrographs courtesy of Dr. C. K. Job, National Hansen's Disease Center, Carville, La.)

Etiology. Lyme disease is caused by infection with the spirochete *Borrelia burgdorferi,* which is transmitted by the deer tick. It is most common in the northeastern United States, northern Europe, China, and Japan. As many as 80% of the ticks in endemic areas of the United States harbor *B. burgdorferi.* It appears that the cranial nerve and meningoencephalitic involvement is caused by direct invasion by the spirochete, which has been detected in these sites and occurs during the acute dissemination of the spirochete. However, the inability to identify the organism in peripheral nerve, the chronicity of the peripheral neuropathy, and the presence of a shared antigenic determinant by the organism and peripheral nerve have led to the hypothesis that the peripheral neuropathy may be immune mediated.

AIDS Neuropathy

Clinical manifestations. Thirty percent of patients with AIDS have symptoms of peripheral neuropathy; there are at least four distinct neuropathies associated with AIDS (Griffin et al. 1994). A vasculitic neuropathy and Guillain-Barré syndrome are two forms that are indistinguishable from non-AIDS patients with these disorders. The third type of neuropathy in AIDS patients is the cauda equina radiculopathy due to CMV infection (discussed earlier).

The most common neuropathy syndrome in AIDS patients is a distal symmetric axonal neuropathy characterized by sensory symptoms having onset with pain in the soles of the feet, and subsequently spreading to involve the lower leg. The pain may be so severe as to create difficulty in walking, and although clinically limited to the legs, the neuropathy may have a wider distribution by electrophysiological studies. In contrast to the severe sensory symptoms, motor weakness may be minimal although reflexes may be diminished.

Diagnostic criteria. The pathologic findings include (1) loss of large caliber myelinated fibers in the early stages of the disease; (2) axonal degeneration with severe distal axonal loss. In later stages of the disease, at distal sites, there may be few remaining myelinated fibers and the residual often show active axonal degeneration. Neurons of the dorsal root ganglia may show some degeneration, but are generally less severe than the distal axonopathy.

Neuropathy Associated with Paraneoplastic Syndromes and Dysproteinemias

Paraneoplastic Neuropathy

Clinical manifestations. Diffuse paraneoplastic peripheral neuropathy is evident clinically in up to 5% of patients with lung carcinoma (and up to 20–40% by electrophysiologic testing). Paraneoplastic neuropathy may be either purely sensory or may involve both sensory and motor modalities. Sensory paraneoplastic neuropathy, which may be detected up to 1 year before the detection of the carcinoma, usually presents subacutely and includes numbness and pain, but it may also be associated with loss of position and vibration sensations. The amplitude of conduction is reduced in sensory nerves, but not motor nerves electrophysiologically. Most patients with sensory paraneoplastic neuropathy have small cell carcinoma of the lung. The sensorimotor paraneoplastic neuropathies are more common than the purely sensory neuropathy and may occur subacutely or chronically with a progressive or intermittent course. They are most often associated with lung carcinoma, but they may also be seen with carcinomas of the stomach, breast, or gastrointestinal tract.

Diagnostic criteria. The pathologic findings in paraneoplastic neuropathy include (1) severe axonal loss in the peripheral nerve biopsies; (2) active axonal degeneration. The sensory paraneoplastic neuropathy is also characterized by severe neuronal loss in dorsal root ganglia, with an inflammatory ganglionitis composed of lymphocytes and plasma cells and accompanied by a proliferation of satellite cells. The axons of the dorsal root ganglion cells degenerate, resulting in axonal loss not only in the peripheral nervous system, but also in the posterior columns of the spinal cord.

Etiology. Paraneoplastic neuropathy is thought to result from an immune response to tumor antigens that cross-reacts with similar neuronal antigens (Mokri and Engel 1975). Specificity for the dorsal root ganglia may be explained by the entry of immunoglobulins across the incomplete blood nerve barrier of the ganglia. It is clear that these patients have circulating anti-Hu antibodies that recognize a 35-kD protein expressed in both the tumor and in neurons (Hughes et al. 1996).

Neuropathy Associated with Dysproteinemias

(Figure 13.32.)

Clinical manifestations. Patients with monoclonal gammopathy, Waldenstrom's macroglobulinemia, or osteosclerotic form of multiple myeloma may develop a peripheral neuropathy that is predominantly demyelinating. The course is slowly progressive, and the symptoms include distal, symmetrical motor weakness and impairment of sensory modalities of light touch, vibration, and position sense. In these neuropathies, the cerebrospinal fluid protein often is increased, and electrophysiologic studies disclose slowing of nerve conduction velocity, typical of a demyelinating neuropathy. Approximately 10% of patients with idiopathic peripheral neuropathy have a monoclonal gammopathy (MCG), usually in the absence of overt B-cell neoplasia.

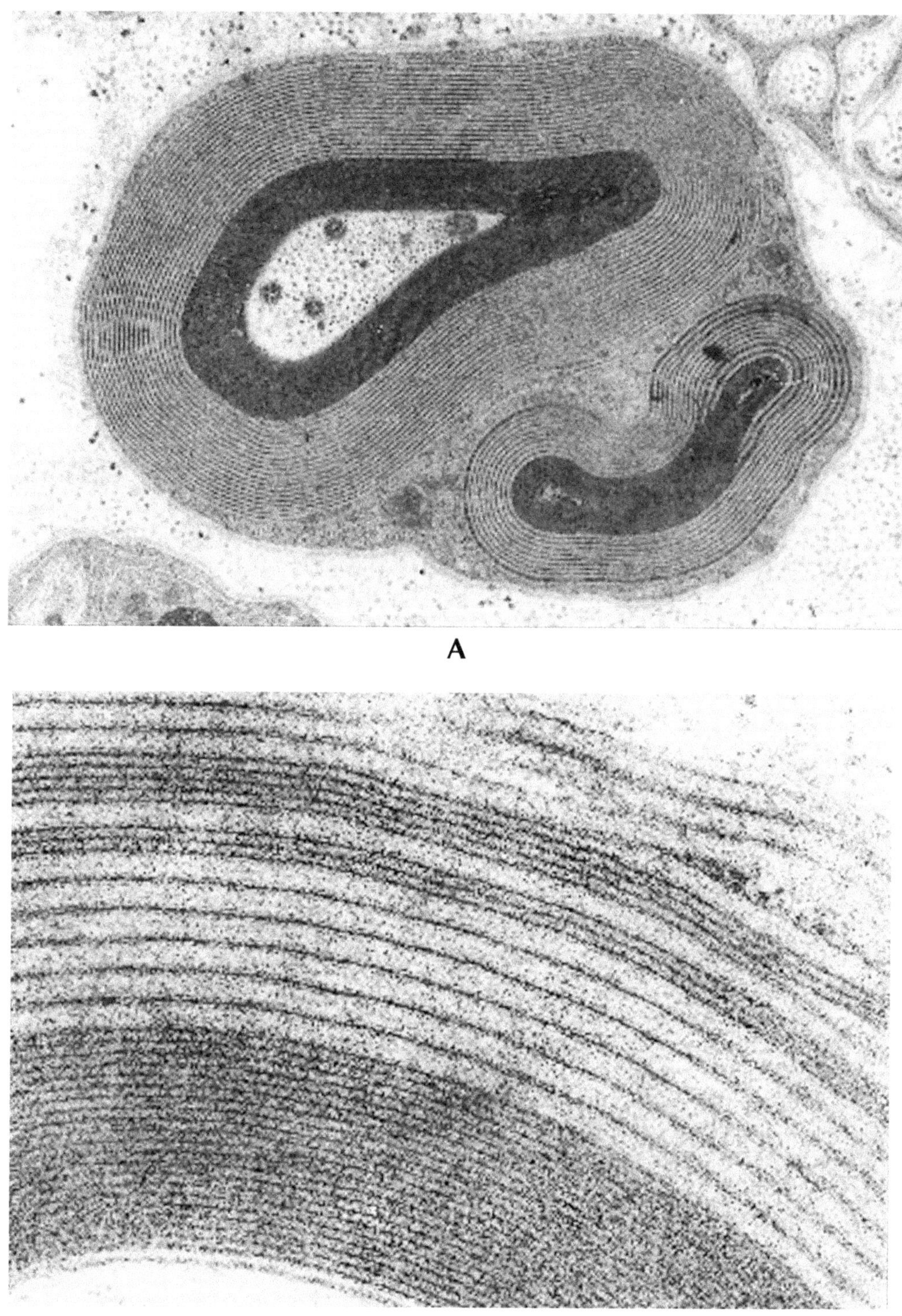

Figure 13.32. Neuropathy associated with paraproteinemia. **A,** Myelinated fibers showing separation of myelin lamellae. (× 47,100) **B,** High magnification of myelin sheath showing opening of the intraperiod lines. (× 122,550) (Permission for reprinting granted by Masson Publishing USA, Vital C, Vallat JM: *Ultrastructural Study of the Human Diseased Peripheral Nerve,* 1980.)

In most of the patients with neuropathies and monoclonal gammopathy, the immunoglobulin is IgM.

Diagnostic criteria. The pathologic findings include (1) mild-to-moderate axonal loss; (2) thinly myelinated large caliber axons and Schwann cell hyperplasia, typical of a chronic demyelinating neuropathy; (3) sparing of unmyelinated axons; (4) noncompacted myelin layers; and (5) sometimes, separation of the myelin lamellae by opening of the intraperiod line, especially in the outer layers. Perivascular mononuclear infiltrates in epineurial and perineurial connective tissues may also be present.

Etiology. The pathogenesis of the dysproteinemic neuropathies is still uncertain, but there is evidence to suggest that the neuropathy results from immunologically mediated deposition of the monoclonal protein within nerve. In 50–90% of patients with IgM paraproteins and neuropathy, the monoclonal protein has specificity for peripheral nerve, and especially the myelin-associated glycoprotein (MAG). A particularly interesting observation is that those patients with widening of myelin lamellae have been demonstrated to have deposition of anti-MAG IgM in the spaces between the myelin layers. The antibody-antigen interaction, therefore, appears to be established in this neuropathy, with remaining questions involving whether the binding is a primary or secondary event in the pathogenesis of this neuropathy (Smith et al. 1983).

Acknowledgments

The authors acknowledge with pleasure the contributions of Ms. Angela Farkas-Ratkowski (EM technician). Dr. T.W. Smith (neuropathologist), and Dr. D. Chad (neurologist) at the University of Massachusetts Medical Center, who helped to provided the foundation of illustrations and text for this chapter in the first edition of this text. We are also grateful to the electron microscopy technicians at the Brigham and Women's Hospital (Ms. Christine Ridolfi and Mr. Anthony Merola) and Children's Hospital (Mr. Howard Mulhern) for their valuable help in the preparation of this edition.

REFERENCES

Asbury AK, Johnson PC: *Pathology of Peripheral Nerve.* WB Saunders, Philadelphia, 1978.

Ben Othmane K, Middleton LT, Loprest LJ, Wilkinson KM, Lennon F, Rosear MP, Stajich JM, Gaskell PC, Roses AD, Pericak-Vance MA, et al. Localization of a gene (CMT2A) for autosomal dominant Charcot-Marie-Tooth disease type 2 to chromosome 1p and evidence of genetic heterogeneity. Genomics 17(2):370–375, 1993.

Bennett MV: Connexins in disease. Nature 368:18–19, 1994.

Bönnemann C, McNally E, Kunkel L: Beyond dystrophin: Current progress in the muscular dystrophies. *Curr Opin Pediatr* 8:569–582, 1996.

Bouche P, Vallat J-M: *Neuropathies Périphériques.* Doin, Paris, 1992.

Bradley WG: *Disorders of Peripheral Nerve.* Blackwell, Oxford, 1974.

Carpenter S: Inclusion body myositis. J Neuropathol Exp Neurol 55:1105–1114, 1996.

Carpenter S, Karpati G: *Pathology of Skeletal Muscle.* Churchill Livingstone, New York, 1978.

Cavanagh JB: The "dying back" process. A common denominator in many naturally occurring and toxic neuropathies. Arch Pathol Lab Med 103:659–64, 1979.

De Bleecker JL, Engel AG: Immunocytochemical study of CD45 T cell isoforms in inflammatory myopathies. *Am J Pathol* 146:1178–1187, 1995.

De Girolami U, Beggs A: Skeletal muscle. In Silverberg SG, Delellis RA, Frable WJ, eds: *Principles and Practice of Surgical Pathology and Cytopathology.* Churchill Livingstone, New York, 1997, pp 943–994.

Dubowitz V: *Muscle Disorders in Childhood.* Philadelphia, Saunders/Baillière Tindall, 1995.

Dyck P, Thomas P, Griffin J, et al, eds: *Peripheral Neuropathy.* WB Saunders, Philadelphia, 1993.

Dyck PJ, Giannini C: Pathologic alterations in the diabetic neuropathies of humans: A review. *Neuropathol Exp Neurol* 55:1181–1193, 1996.

Engel A, Franzini-Armstrong C, eds: *Myology.* McGraw-Hill, New York, 1994.

Engelstad JK, Davies JL, Giannini C, et al: No evidence for axonal atrophy in human diabetic polyneuropathy. J Neuropathol Exp Neurol 56:255–262, 1997.

Feasby T, Hahn A, Brown W, et al: Severe axonal degeneration in acute Guillain-Barré syndrome: Evidence of two different mechanisms? J Neurol Sci 116:185–192, 1993.

Feldman EL, Russell JW, Sullivan KA, et al: New insights into the pathogenesis of diabetic neuropathy. Current Opinion in Neurology 12:553–563, 1999.

Flanigan KM, Crawford TO, Griffin JW, et al: Localization of giant axonal neuropathy gene to chromosome 16q24. Annals of Neurology 43:143–148, 1998.

Griffin J, Wesselingh S, Griffin D, et al: Peripheral nerve disorders in HIV infection. Similarities and contrasts with CNS disorders. Res Publ Assoc Res Nerv Ment Dis 72:159–182, 1994.

Griffin JW, Li CY, Macko C, et al: Early nodal changes in the acute motor axonal neuropathy pattern of the Guillain-Barré syndrome. J Neurocytol 25:33–51, 1996.

Guzzetta F, Rodriguez J, Deodato M, et al: Demyelinating hereditary neuropathies in children: A morphometric and ultrastructural study. Histol Histopathol 10: 91–104. 1995.

Hartung H-P, van der Meché FGA, Pollard JD: Guillain-Barré syndrome, CIDP and other chronic immune-mediated neuropathies. Current Opinion in Neurology 11:497–513, 1998.

Hughes R, Sharrack B, Rubens R: Carcinoma and the peripheral nervous system. J Neurol 243:371–376, 1996.

Meier C, Grahmann F, Engelhardt A, et al: Peripheral nerve disorders in Lyme-Borreliosis. Nerve biopsy studies from eight cases. Acta Neuropathologica 79: 271–278, 1989.

Midroni G, Bilbao JM: *Biopsy Diagnosis of Peripheral Neuropathy.* Butterworth-Heinemann, Boston, 1995, p 477.

Mokri B, Engel A: Duchenne dystrophy: Electron microscopic findings pointing to a basic or early abnormality in the plasma membrane of the muscle fiber. Neurology 25:1111–1120, 1975.

Moretti C, Bado M, Morreale G, et al: Disruption of muscle basal lamina in congenital muscular dystrophy with merosin deficiency. Neurology 46:1354–1358, 1996.

Mosser J, Douar AM, Sarde CO, et al: Putative X-linked adrenoleukodystrophy gene shares unexpected homology with ABC transporters. Nature 361:726–730, 1993.

Neville HE: Ultrastructural changes in diseased human skeletal muscle. In Vinken P, Bruyn G, eds: *Handbook of Clinical Neurology.* Elsevier/North-Holland, Amsterdam, 1979, pp 63–123.

Oldfors A, Lindberg C: Inclusion body myositis. Current Opinion in Neurology 12:527–533, 1999.

Ouvrier R, McLeod J, Pollard J: *Peripheral Neuropathy in Childhood.* Raven Press, New York, 1990.

Peters A, Palay S, Webster H: *The Fine Structure of the Nervous System.* Oxford, New York, 1991.

Powers J, Moser H: Peroxisomal disorders: genotype, phenotype, major neuropathologic lesions, and pathogenesis. Brain Pathol 8:101–120, 1998.

Ptacek LJ, Tawil R, Griggs RC, et al: Dihydropyridine receptor mutations cause hypokalemic periodic paralysis. Cell 77:863–868, 1994.

Richardson E Jr, De Girolami U: *Pathology of the Peripheral Nerve.* WB Saunders, Philadelphia, 1995.

Scriver CR, Beaudet AL, Sly WS, et al: *The Metabolic Basis of Inherited Disease.* McGraw-Hill, New York, 1995.

Smith IS, Kahn SN, Lacey BW, et al: Chronic demyelinating neuropathy associated with benign IgM paraproteinaemia. Brain 106:169–195, 1983.

Sternberg S, ed: *Histology for Pathologists.* Raven Press, New York, 1992.

Straub V, Campbell K: Muscular dystrophies and the dystrophin-glycoprotein complex. Curr Opin Neurol 10:168–175, 1997.

Vinken PJ, Bruyn GW, eds: *Diseases of Nerves, parts I and II. Handbook of Clinical Neurology,* vols 7 and 8. American Elsevier, New York, 1970.

Vital C, Vallat J-M: *Ultrastructural Study of the Human Diseased Peripheral Nerve.* Masson, New York, 1980.

Waller A: Experiments on the section of the glossopharyngeal and hypoglossal nerves of the frog, and observations of the alterations produced thereby in the structure of their primitive fibres. Philos Trans R Soc Lond (Biology) 140:423–429, 1850.

Wallgren-Pettersson C, Pelin K, Hilpela O, et al: Clinical and genetic heterogeneity in autosomal recessive nemaline myopathy. Neuromuscular Disorders 9: 564–572, 1999.

Weddell AGM, Pearson JMH: Leprosy: Histopathologic aspects of nerve involvement. In Hornabrook RW, ed: *Topics on Tropical Neurology.* FA Davis, Philadelphia, 1975, pp 17–28.

Worton R: Muscular dystrophies: Diseases of the dystrophin-glycoprotein complex. *Science* 270:755–756, 1995.

Index

Page numbers in italics refer to figures.

A

D

G

M

T

GPSR Compliance
The European Union's (EU) General Product Safety Regulation (GPSR) is a set of rules that requires consumer products to be safe and our obligations to ensure this.

If you have any concerns about our products, you can contact us on

ProductSafety@springernature.com

In case Publisher is established outside the EU, the EU authorized representative is:

Springer Nature Customer Service Center GmbH
Europaplatz 3
69115 Heidelberg, Germany

www.ingramcontent.com/pod-product-compliance
Ingram Content Group UK Ltd.
Pitfield, Milton Keynes, MK11 3LW, UK
UKHW061831190726
13855UKWH00005B/1752